CLINICALLY ORIENTED anatomy

SECOND EDITION

CLINICALLY ORIENTED
anatomy

SECOND EDITION

Keith L. Moore, Ph.D., F.I.A.C.

Professor of Anatomy and Associate Dean, Basic Sciences,
Faculty of Medicine, University of Toronto, Toronto, Ontario, Canada

Illustrated Primarily by
Dorothy Chubb, Nancy Joy, A.O.C.A., Nina Kilpatrick, B.Sc., AAM., and Angela Cluer, B.Sc., AAM.

Photography by
John Kozie, B.Sc. and Paul Schwartz, B.A.

WILLIAMS & WILKINS
Baltimore • Hong Kong • London • Sydney

Editor: John N. Gardner
Associate Editor: Carol Eckhart
Copy Editor: Shirley Riley Smoot
Design: JoAnne Janowiak
Illustration Planning: Lorraine C. Wrzosek
Production: Anne G. Seitz

Accurate indications, adverse reactions, and dosage schedules for drugs are provided in this book, but it is possible that they may change. The reader is urged to review the package information data of the manufacturers of the medications mentioned.

Printed in the United States of America

First Edition, 1980
Reprinted 1981, 1982, 1983, 1984

Library of Congress Cataloging in Publication Data

Moore, Keith L.
Clinically oriented anatomy.

Includes bibliographies and index.
1. Anatomy, Human. I. Title. [DNLM: 1. Anatomy. QS 4 M822c]
QM23.2.M67 1985 610 84-10438
ISBN 0-683-06132-1

90 91 92 93 94 10 9

Dedicated to the memory of
Professor J. C. Boileau Grant
1886–1973
M.C., M.B., Ch.B., Hon. D.SC. (Man.), F.R.C.S. (Edin.)

Late Professor Emeritus of Anatomy in the University of Toronto. Professor Grant was Professor and Head of Anatomy in the University of Manitoba from 1919–30 before he went to Toronto. He retired as Professor and Chairman of Anatomy in Toronto in 1959 and for several years was Visiting Professor of Anatomy in the University of California, Los Angeles.

The above photograph was taken in 1956 when Professor Grant was awarded an honorary degree by the University of Manitoba.

Grant's Atlas, *Grant's Method*, and *Grant's Dissector* are living memorials to this Great Canadian anatomist.

Preface to the Second Edition

The enthusiastic endorsement of the first edition of *Clinically Oriented Anatomy* by both students and faculty in Canada, the United States, and overseas has been most gratifying. I am particularly pleased to have received so many letters of appreciation from students, many containing very good suggestions for improvement. I hope they will be able to detect the results of their advice. All comments received careful consideration, with the result that this edition should be more useful to students.

The preparation of a book of general suitability is difficult because the teaching of gross anatomy varies according to the objectives of the course. However, the world-wide acceptance of this book indicates that it meets the requirements of most health science students in acquiring a basic knowledge of the structure of the human body, and some appreciation of the practical applications of this anatomical knowledge.

A deliberate attempt has been made to use the clinical sciences to facilitate and encourage understanding of anatomy. The structures described in this book are those which are deemed likely to be of importance to the general practitioner of medicine, dentistry, or other health professions. The references to the clinical significance of various facts were inserted after much consideration and in realization of the fact that there is room for considerable difference of opinion in these areas.

The success this book has had is surely related to the quality of the illustrations, particularly those from Grant's Atlas. These superb illustrations are often much more efficient than text material in conveying complex anatomical information. They also aid in the recognition of structures in dissected specimens and in living persons during surgery. Some illustrations have been omitted, but there are some new illustrations and many others have been improved.

Photographs of living persons are also used extensively because they aid the learning of living anatomy and serve as a bridge to physical diagnosis that uses this aspect of anatomy when examining patients.

It was not surprising to find that students were particularly enthusiastic about the clinical comments in this book because *nothing stimulates students more than the correlation of the anatomy taught with patient oriented problems.* Care has been taken to make the clinical vignettes concise and accurate.

The enthusiasm of anatomy teachers and clinicians for this book is an honor. This undoubtedly results from its clinical orientation that resulted from my 30-year association with many clinical colleagues who assisted me with the *medical considerations of the various anatomical regions.* It will be obvious to specialists that what is presented is only a small portion of the clinical applications that could be made. This material has been limited in order to keep the book a reasonable size. Omission of facts of importance to certain specialists does not imply that they are not of anatomical significance. Write to me if you have strong feelings about the omission of certain basic anatomical facts.

The second edition has fewer pages owing to the careful selection of material based on its use for five years. *A larger page size has also been used to assist in the reduction of the bulk of the book.* The extensive culling of illustrations and revision of the text includes factual changes, as well as those designed to improve clarity and readability.

In accordance with the requests of many students and staff, *tables have been added summarizing the chief attachments, nerve supply, and principle actions of muscles.* Hopefully this organization of material will aid learning of essential facts.

As in the first edition, ***Clinically Oriented Comments are screened for easy recognition***, but the headings have been omitted because many people found these to be annoyingly repetitious. It is hoped that these clinical comments will continue to spark students' interest in learning normal anatomy and understanding disturbed functions of the body when they occur.

The following excerpts from the Preface to the First Edition are worth repeating.

"The title ***Clinically Oriented Anatomy****" was chosen to indicate that the book highlights those features of anatomy which are of clinical importance.*

Surface anatomy is stressed because ignorance of this aspect is a serious handicap when interpreting the results of a physical examination. Furthermore, the performance of "*tests requiring the insertion of needles requires a good knowledge of structures that lie under the skin.*"

The terminology in this book aheres to the internationally accepted Nomina Anatomica (5th ed., Williams & Wilkins, 1983) approved by the Eleventh International Congress of Anatomists held in Mexico City, Mexico in 1980. In accordance with international agreement, the terminology in this book departs from strict Latin in some cases by anglicizing terms, or by using direct English translations. *Eponyms commonly used clinically appear in parentheses*, e.g., **sternal angle** (angle of Louis), to assist students in translating the clinical language used in hospitals and patients' charts. In most cases, the official term is printed in **boldface** as in the example just given.

No book, regardless of its size, can be accomplished without much help. I should like to renew my sincere thanks to several colleagues in my department, especially: *Dr. J. W. A. Duckworth*, Professor Emeritus of Anatomy; *Dr. C. Smith*, Professor Emeritus of Anatomy; *Dr. D. L. McRae*, the late Professor Emeritus of Radiology; *Dr. E. G. Bertram*, Professor of Anatomy and Associate Chairman; *Dr. A. Roberts*, Associate Professor of Anatomy and Subject Supervisor of Gross Anatomy and Embryology; *Dr. W. M. Brown*, Associate Professor of Anatomy; *Dr. K. McCuaig*, Associate Professor of Anatomy; *Dr. I. M. Taylor*, Associate Professor of Anatomy; and *Mrs. E. J. Akesson*, Assistant Professor of Anatomy. Of the many external colleagues who made suggestions for the second edition, I should like to thank Dr. T. V. N. Persaud of Winnipeg and Professor H. E. Peery of Dryden, New York who were especially helpful.

My grateful appreciation is extended again to the medical illustrators who prepared the original illustrations for this book: Mrs. *Dorothy Chubb*; Professor *Nancy Joy*, Professor of Art As Applied to Medicine and Chairman of the Department, University of Toronto; and *Nina Kilpatrick*, Calgary. New illustrations have been created by *Angela Cluer* of Instructional Media Services, University of Toronto. Her professional skill and friendly personality have made the preparation of new illustrations a pleasurable experience. The medical photographers, *Paul Schwartz* and *John Kozie*, have continued to provide superb professional services and help with this edition.

To my secretaries, *Karen McMurray* and my wife Marion, I extend grateful appreciation for their friendly help and patience with the preparation of this edition. *Marion* also spent many hours discussing the manuscript with me and proofreading it. Her support is unique.

Sara Finnegan, President, Toni M. Tracy, Vice President and Editor-in-Chief, Carol Eckhart, Associate Editor, Anne Seitz, Production Sponsor, and Shirley Smoot, Copy Editor, of Williams & Wilkins have spared no efforts in making this second edition better than the first.

University of Toronto Keith L. Moore

Preface to the First Edition

The invitation to write a book on patient oriented anatomy using illustrations from *Grant's Atlas* was one that could not be refused by this medical writer, who believes that learning anatomy can be *especially* exciting when its relevance to medicine is emphasized. This book is written primarily for students studying anatomy for the first time.

The title **Clinically Oriented Anatomy** was chosen to indicate that *the book highlights those features of anatomy which are of clinical importance.* Every good teacher of anatomy recognizes that nothing stimulates the student more than correlating anatomy with a problem presented by a patient. Because there is often difficulty relating what one sees in the cadaver to what is seen in patients during physical examinations and in the operating room, *emphasis is placed on living anatomy.* For example, the pancreas is described as a soft, grayish-pink gland to emphasize that it is not a hard, almost colorless organ that one sees in the cadaver or colored yellow or red as depicted in some books and atlases.

Surface anatomy is stressed because ignorance of this aspect is a serious handicap when interpreting the results of a physical examination. Furthermore, the performance of tests requiring the insertion of needles requires a good knowledge of structures that lie under the skin. Throughout the book, *students are urged to examine their own bodies* and those of others because a good knowledge of surface anatomy makes it unnecessary to learn by rote. Used properly, *the body is a "living textbook."*

This book is not designed primarily as a guide to clinical anatomy which would be of concern to practicing doctors. It is *patient oriented anatomy* that was written to stimulate the interest of beginning students so that they can appreciate what is involved anatomically in nerve injuries, stab wounds, surgical approaches, etc. Care has been taken not to expect students to make diagnoses, and treatments have not been suggested for conditions discussed in the patient oriented problems. Errors in surgery may result from failure to appreciate variations in the body (*e.g.*, in the anatomy of the biliary system); therefore *common variations of form and structure are illustrated.*

The *many radiographs in this book* give a clinical orientation to bones, joints, and organs, enabling students to begin identifying normal structures and how they can change in form and appearance in diseased states. At the University of Toronto, 1st year students are expected to be able to interpret normal radiographs of the body and to recognize obvious fractures and developmental abnormalities such as cervical ribs. With this in mind, radiographs are used to illustrate some of the clinically oriented comments and patient oriented problems.

This book is not intended to be a core textbook of anatomy; several of this kind are available for students who wish to know the minimum amount required for them to pass their 1st year examinations. *Neither is it overly detailed;* several excellent books already fulfill this need. An attempt has been made to cover areas that are most important for students to know and to arouse in them an interest in revising their knowledge of anatomy as they progress in their medical studies. The regional plan has been used because most anatomy courses are based on regional dissection and *the chapters in this book follow the same order as in Grant's Atlas and Grant's Dissector.*

Boldface type and *italics* have been used to highlight important concepts and essential terminology. Explanatory notes and supplementary information appear in intermediate type so they can be read once and passed over during reviews. Clinically oriented comments are screened for special attention and quick referral.

The terminology in this book adheres to the internationally accepted *Nomina Anatomica* (4th ed.) approved by the Tenth International Congress of Anatomists at Tokyo in August, 1975. In accordance with international agreement, the terminology in this book departs from strict Latin in some cases by anglicizing names or by using direct English translations. *Eponyms commonly used clinically appear in parentheses, e.g.,* **sternal angle** (angle of Louis), to assist students in translating the clinical terminology used in hospitals and patients' charts. In all cases, the official term is printed in **boldface** as in the example just given.

This book is freely illustrated because much of the difficulty encountered by students results from their inability to visualize the form and structure of parts of the body. The selected line drawings and photographs based on Grant's dissections are familiar to students, physicians, and surgeons around the world. These illustrations form the nucleus around which this book was written. In addition there are many new illustrations (photographs of models, drawings, radiographs, and clinical photographs). *Studying anatomy at the dissecting table with a good teacher, where the parts may be seen, felt, and dissected, is the best way to learn anatomy.* A well illustrated book with accompanying observations, clinical comments, and discussion of patient oriented problems is probably the next best way. *The legends to the figures from Grant's Atlas have not been significantly changed*; hence students who use this classical atlas will be afforded a review when they re-examine them in the present book. Almost all specimens illustrated in this book may be seen in the anatomy museum of the University of Toronto, and students are encouraged to come and see them. You will observe, as *Professor Grant said in the Preface to his Atlas*, "Little, if any liberty has been taken with the anatomy; that is to say, the illustrations profess a considerable accuracy of detail."

Sir Isaac Newton once said, "*If I have seen further, it is by standing on the shoulders of giants.*" Much of my knowledge of clinically oriented anatomy was taught to me by a "giant in Anatomy," **Professor I. Maclaren Thompson**, former Profesor and Head of Anatomy at the University of Manitoba, who conducted weekly *anatomical clinics* in the Winnipeg General Hospital using patients to illustrate the anatomically related problems. I owe much to this fine gentleman, scholar, and teacher. When I became the Professor of Anatomy and a Consultant at the Health Sciences Centre in Winnipeg, I continued his method of teaching clinically oriented anatomy.

Thanks are due my colleagues in the Department of Anatomy, University of Toronto, especially **Dr. W. M. Brown**, Associate Professor of Anatomy, **Dr. J. W. A. Duckworth,** Professor Emeritus of Anatomy, and **Dr. D. L. McRae,** Professor Emeritus of Radiology. Drs. Duckworth and McRae have commented on the book in the Foreword. Several other members of the Department also gave much help: Mrs. E. J. Akesson, Dr. E. G. Bertram, Dr. B. Liebgott, Dr. R. G. MacKenzie, Dr. A. Roberts, Dr. C. G. Smith, Dr. I. M. Taylor, and Dr. J. S. Thompson. All these colleagues, most of whom have many years of teaching and clinical experience, were generous with their time and thoughts. I thank all of them most sincerely.

I owe much to the following *Williams & Wilkins authors* who kindly consented to let me use illustrations from their books: Drs. J. E. Anderson, T. A. Baramki, M. Bartalos, J. V. Basmajian, R. F. Becker, M. B. Carpenter, W. M. Copenhaver, P. V. Dilts, Jr., J. A. Gehweiler, J. W. Greene, Jr., D. E. Kelly, J. Langman, J. W. Roddick, Jr., R. B. Salter, E. K. Sauerland, J. W. Wilson, and R. L. Wood. I am also grateful to Mr. A. E. Meier, Vice President and Editor-in-Chief, Health Sciences, W. B. Saunders Company, for allowing me to use many illustrations from my book *The Developing Human: Clinically Oriented Embryology*. I should also like to thank all other authors and publishers, acknowledged elsewhere, who have given me permission to use illustrations from their books.

The medical illustrators for this book merit special attention. Mrs. **Dorothy Chubb**, a pupil of Max Brödel, and Professor **Nancy Joy**, Chairman of the Department of Art as Applied to Medicine in the University of Toronto, prepared most of the illustrations in this book. Their expert skill is unsurpassed. Most new illustrations were prepared by Mrs. **Nina Kilpatrick**, a recent graduate of the University of Toronto program in Art as Applied to Medicine. I am grateful to her for her work which was carefully and cheerfully done.

The medical photographers also set high standards. The photographs were taken by Messers. Paul Schwartz, B.A., Mr. John Kozie, B.Sc., Associate Professor of Art as Applied to Medicine and Director of Photographic Services, and Mark Sawyer, B.Sc. Their expertise and friendly help were much appreciated.

My secretaries, especially my wife Marion and Jill Parsons, deserve my most sincere thanks. They have worked hard and cheerfully, often under pressure. *Marion* spent many hours proofreading and discussing the manuscript with me.

Finally I thank the Publishers, Williams & Wilkins—particularly ***Sara A. Finnegan***, the Vice President and Editor-in-Chief—for inviting me to write this book and for her enthusiasm, unfailing courtesy, and consideration in endeavoring to fill my many requests.

Toronto, Canada Keith L. Moore

Acknowledgments

Throughout the text, liberal use has been made of illustrations from the following Williams & Wilkins publications, which the author and publisher acknowledge with sincere thanks:

Anderson JE: *Grant's Atlas of Anatomy*, ed 8, 1983.

Basmajian JV: *Grant's Method of Anatomy*, ed 10, 1980.

Basmajian JV: *Primary Anatomy*, ed 8, 1982.

Basmajian JV: *Surface Anatomy*, ed 2, 1983.

Becker RF, Wilson JW, Gehweiler JA: *The Anatomical Basis of Medical Practice*, 1971.

Carpenter MB: *Core Text of Neuroanatomy*, ed 3, 1985.

Copenhaver WM, Kelly DE, Wood RL: *Bailey's Textbook of Histology*, ed 17, 1978.

Dilts PV, Jr, Greene JW, Jr, Roddick JW, Jr: *Core Studies in Obstetrics and Gynecology*, ed 2, 1977.

Dox I, Melloni BJ, Eisner GM: *Melloni's Illustrated Medical Dictionary*, 1979.

Salter RB: *Textbook of Disorders and Injuries of the Musculoskeletal System*, ed 2, 1983.

Sauerland EK: *Grant's Dissector*, ed 9, 1984.

Stedman's Medical Dictionary, ed 24, 1982.

The author and publisher also gratefully acknowledge the use of illustrations from the following sources:

Connor, T. Women's College Hospital, Toronto (Fig. 6-9).

Eastman Kodak. *Fundamentals of Radiology*, 1980, (Fig. 17).

General Electric Medical Systems, Ltd. (Fig. 21).

Ham AW, Cormack DH: *Histology*, ed 8. Philadelphia, JB Lippincott Co, 1979 (Fig. 7-98).

Haymaker W, Woodhall B: *Peripheral Nerve Injuries*, ed 2. Philadelphia, WB Saunders Co, 1953 (Fig. 6-94).

Healey JE, Jr: *A Synopsis of Clinical Anatomy*. Philadelphia, WB Saunders Co, 1969 (Fig. 3-85, 3-86, 3-90, 4-134).

Laurence KM, Weeks R: Abnormalities of the central nervous system. In Norman AP (ed): *Congenital Abnormalities of Infancy*, ed 2. Oxford, Blackwell Scientific Publications, 1971, (Fig. 7-18).

Laurenson RD: *An Introduction to Clinical Anatomy by Dissection of the Human Body*. Philadelphia, WB Saunders Co, 1968 (Figs. 3-67, 3-72, and 3-76).

Liebgott B: *The Anatomical Basis of Dentistry*, Philadelphia, WB Saunders Co, 1982 (Figs. 7-123, 7-124, and 8-55).

McCredie JA: *Basic Surgery*. New York, Macmillan Publishing Co, 1977 (Fig. 2-49).

Moore KL: *The Developing Human: Clinically Oriented Embryology*, ed 3. Philadelphia, WB Saunders Co, 1982 (Figs. 1-79, 2-17, 2-21, 2-134, 5-26, 5-27, 5-61, 7-82, 7-125, 7-136, 8-43, and 8-79).

Norman D, Korobkin M, Newton TH: *Computed Tomography*. St. Louis, CV. Mosby Co, 1977 (Fig. 7-175).

Smith Kline Corporation. *Essentials of the Neurological Examination*. 1978 (Figs. 7-34, 7-36, and 7-145).

Contents

Overview of Anatomy

ANATOMY IS AN OLD BASIC MEDICAL SCIENCE

As far as is known, *the study of Anatomy originated in Egypt.* Much later (middle of 4th Century B.C.), it was taught in Greece by **Hippocrates**, who is often called the *"Father of Medicine."* In addition to the **Hippocratic Oath** attributed to him, he wrote several Anatomy books and in one of them he stated, *"The nature of the body is the beginning of medical science."*

Another famous Greek physician and scientist, **Aristotle** (384–322 B.C.), was *the founder of comparative anatomy.* He made many new observations, especially concerning developmental anatomy or **embryology**. He is also credited with being the first person to use the word *"anatome,"* a Greek word meaning "a cutting up" or, as we say, *"to dissect."* In those days, to perform an "anatomy" was to do a dissection, a term derived from Latin meaning "to cut asunder." These two words are no longer synonymous.

Dissection is a technique used to study the structure of the body, whereas *anatomy is a discipline* or field of scientific study.

ANATOMY IS THE STUDY OF STRUCTURE AND FUNCTION

Anatomy is the part of biological science that deals primarily with structure and function of the body. In early days the science of anatomy was mainly concerned with the form and function of parts of the body that could be demonstrated by dissection. This technique is still widely used because it provides an orderly display of the parts of the body. It also helps in obtaining a three-dimensional concept of these parts.

You must learn to make and to trust your own observations. During dissection you will observe, feel, move, and dissect the parts of the body. Although this method of studying the body, called macroscopic or **gross anatomy**, is closely associated with surgery, it forms *an essential basis for all branches of medicine and dentistry.*

"Anatomy is to physiology as geography is to history" (Fernel); that is, it provides the setting for the events. At one time **physiology** was part of the discipline of anatomy, but as many new methods of investigating function were developed, it became a separate discipline.

Anatomy is the study of living human beings and cannot be learned completely by studying the bodies of dead persons. *Much can be learned from observing the surface of the body.* The use of **surface anatomy** begins when the doctor, dentist, or other health professional first examines a patient.

To examine a patient without a thorough knowledge of living anatomy is comparable to sailing an unknown sea without charts. Both situations may result in disaster.

Palpation, or examining with the hands and fingers, is a clinical technique you will use when studying living anatomy. *Palpation of arterial pulses is part of every routine evaluation* of the living body.

Before long you will learn how to use various instruments to observe parts of the body (*e.g.*, the eye using an **ophthalmoscope**), or to listen to the functioning parts of the body (*e.g.*, the heart and lungs using a **stethoscope**). You will also learn to use a *reflex hammer* for examining the functional state of nerves and muscles.

Students soon associate living anatomy and descriptive anatomy. Hence their bodies become useful memory aids during examinations.

The aim of surface anatomy is the visualization (in the "mind's eye") of structures which lie beneath the skin and are hidden by it. For example, in patients with stab or *gunshot wounds*, the doctor must visualize the structures beneath the wound that might have been injured.

Surface anatomy is the basis for the physical examination of the body that forms a part of physical diagnosis. The thorough study of a patient's body, with emphasis on the area of complaint, is most helpful in making the correct diagnosis of the anatomical basis for the patient's complaint, *e.g.*, chest pain.

As methods of investigating the structure and function of the body became increasingly complex, and following the development of microscopes and good staining procedures, another branch of anatomy was formed.

The study of the make-up of the tissues and organs of the body under the microscope is called microscopic anatomy or **histology**. Again, microscopy is a *technique* and histology is a *discipline.* **Histology is concerned with the normal structure and function of cells, their growth and differentiation, and their interrelations in the tissues, organs, and systems of the body.**

When x-rays were discovered in the 19th century, some new observations were made about the structure and function of the skeleton of the body. **Bones and joints were readily visualized on radiographs** (x-ray films). Later, by introducing radiopaque or radiolucent substances, the form and function of various organs and cavities could be studied (*e.g.*, during swallowing).

Radiological anatomy *(radiographic anatomy) is the study of the structure and function of the body using radiographic techniques.* Radiological anatomy is an important part of gross anatomy and is **the anatomical basis of radiology**, the branch of medical science dealing with the use of radiant energy in the diagnosis and treatment of disease. The sooner you learn to identify the normal structures of the body on radiographs, the easier it will be for you to recognize and understand the changes visible on radiographs that are caused by disease and injury.

Neuroanatomy *is the study of the structure and function of the nervous system.* Courses in neuroanatomy deal with the gross, microscopic, developmental, and radiological anatomy of the nervous system, with special emphasis on the *central nervous system*, consisting of the brain and spinal cord.

Developmental anatomy *is the study of growth and development.* Much can be learned about the structure and function of the adult body by studying the changes that occur during its development from a single cell (**zygote**) into a multicellular adult person.

Growth and development occur throughout life, but developmental processes are most pronounced during *prenatal life* (before birth), particularly during the **embryonic period** (4 to 8 weeks). The rate of growth and development slows down after birth, but there is active ossification (bone formation) and other important changes during infancy and childhood (*e.g.*, tooth and brain development).

Although the anatomy of the body is studied in various ways and is often taught in three or four different courses, you must **strive to learn anatomy as an integrated subject**, keeping in mind that the various subdisciplines arose as new techniques for studying anatomy evolved and as more knowledge was obtained. Your professors are specialists who will help you learn which facts are of clinical importance and what is **essential knowledge** for you to retain in order to perform adequately as a doctor or a dentist.

Many clinical problems can be understood by using knowledge acquired in anatomy courses. The many references to the clinical significance of anatomy are inserted in this book as **clinically oriented comments and problems** to indicate what is essential knowledge and to add interest to your anatomical studies.

In most anatomy courses the body is examined regionally, *i.e.*, by regions such as the thorax (Chap. 1) and abdomen (Chap. 2). The study of all structures in a region, including their relationships to each other, is known as **regional anatomy**. During regional anatomy, the body is generally divided into the following regions: **thorax, abdomen, perineum and pelvis, lower limb, back, upper limb, head, and neck.**

Although you are familiar with the common or layman's terms for many parts and regions of the body, you should use the internationally adopted nomenclature; *e.g.*, use the word "**axilla**" instead of "armpit" and "**clavicle**" instead of "collar bone." Despite this, you must know what the common terms mean so that you can understand the words your patients use when they describe their complaints to you. In addition, you must be able to explain their problems to them in terms that they can understand.

From the functional standpoint, it is helpful to describe the parts and organs of the body by systems. This is referred to as **systemic anatomy**. The systems of the body are:

1. **The integumentary system**, consisting of the skin and its appendages (*e.g.*, hair and nails).
2. **The skeletal system**, composed of bones and certain cartilaginous parts (*e.g.*, in the nose).
3. **The articular system** consisting of joints or articulations and their associated bones and ligaments.
4. **The muscular system**, comprising the muscles that, with few exceptions, move the joints. Sometimes the muscular and skeletal systems are considered together as the *musculoskeletal system.* The primary function of this system is locomotion.
5. **The nervous system**, consisting of the brain and spinal cord, nerves, and ganglia.
6. **The circulatory system**, comprising the heart and blood vessels and including the **lymphatic system** composed of lymph nodes and lymph vessels. The heart and the blood vessels are often referred to as the *cardiovascular system.*
7. **The digestive system**, composed of a long tube that extends from the mouth to the anus. In addition there are associated glands (*e.g.*, the pancreas and liver). *This system is concerned with the assimilation of food.*
8. **The respiratory system** begins at the nose and ends in the lungs. This system is concerned

with the exchange of oxygen and carbon dioxide.

9. **The urinary system** composed of the kidneys, urinary bladder, and excretory passages. *This system is concerned with the elimination of waste material.*
10. **The reproductive system**, or genital system, is concerned with the *perpetuation of the human species.*

Because of their close association during development and in the adult, especially in the male, the urinary and genital systems are often described together as the **urogenital system** or genitourinary system.

11. **The endocrine system** consists of ductless glands (*e.g., pituitary*) which produce secretions, called *hormones*, that pass into the circulatory system and are carried to all parts of the body.

THE BASIS OF MEDICAL LANGUAGE

Anatomy is the basis of the language of medicine and dentistry. The first book, "*On the Naming of the Parts Of the Body*" was written by Rufus of Ephesus (ca. A.D. 50). **Anatomy students learn a new language** consisting of at least *4500 words.* When you learn these words you will be able to speak the anatomical language fluently. You will feel at ease talking to your clinical colleagues because *the anatomical language constitutes most of the words making up the medical language.* To describe the relationship of one structure to another, **anatomical nomenclature** should be used.

To be understood you must express yourself clearly, using the official terms in the correct way. Many anatomical terms are derived from Latin, Greek, and Arabic. Terms are derived from *Greek* because of the studies of **Hippocrates**, **Aristotle**, and **Galen**, famous Greek physicians. Similarly, many terms come from *Latin* because of the influence of **Galen**, a Greek physician who moved to Rome, and **Vesalius** (1514–1564), a great Flemish anatomist who was the Professor of Anatomy at the University of Padua in Italy for many years.

The study of the derivation of words (**etymology**) can help you remember anatomy and, at the same time, you are likely to find the process enjoyable. The following are good examples.

The term **cecum** is from the Latin word *caecus* meaning "blind." Your cecum is a blind pouch lying inferior to the terminal portion of your **ileum** (from a Latin verb meaning "to roll up or twist"). The ileum is a highly coiled or rolled up part of the small intestines.

Decidua, used to describe the endometrium (G. *endon*, within + *metra*, uterus) or *lining of the uterus during pregnancy*, is from Latin and means "a falling off." This is an appropriate term because this layer "falls off" or is shed after the baby is born, just as the leaves of deciduous trees fall off after the summer.

DESCRIPTIVE TERMS

To describe the body and to indicate the position of its parts and organs relative to each other, *anatomists around the world have agreed to use several terms of position and direction and various planes of the body.* Because **clinicians also use these terms**, it is important for you to take time at the beginning of your professional career to learn them well. Practice using them so that it will be clear what you mean when you describe parts of the body in **patients' histories** or during discussion of patients with your clinical colleagues.

Eponyms should not be used in Anatomy because they give no clue to the structure involved. Who would ever guess that *Wharton's duct* is the **submandibular duct**, the duct of the submandibular salivary gland. In addition, some eponyms are historically inaccurate. Poupart was not the first person to describe the **inguinal ligament** (Poupart's ligament).

Correct terminology will also be required when you write reports for medical and other journals. Persons who examine your reports weeks, months, or years later should be able to understand the description clearly *e.g.*, the exact site of a fracture.

THE ANATOMICAL POSITION

All descriptions of the human body are based on the assumption that the person is in the anatomical position (Figs. 1 and 2). *This position of the body is adopted worldwide* for giving anatomical descriptions and must be understood. By using this position, any part of the body can be related to any other part of it.

A person in the anatomical position is standing erect with the head, eyes, and toes directed forward, the heels and toes together, and the upper limbs hanging by the sides with the palms facing anteriorly. You must always visualize the anatomical position when describing patients (or cadavers) lying on their backs (recumbent or **supine position**), sides, or fronts (**prone position**). Otherwise, confusion as to the meaning of your description will exist and *serious consequences could result.*

Because the anatomical position is not the natural way of standing and is different from the military position of attention, special effort should be made to learn how the body is positioned. In Figures 1 and 2 observe that:

1. *The palms of the hands face anteriorly* (forward) because the forearms and hands are supinated;

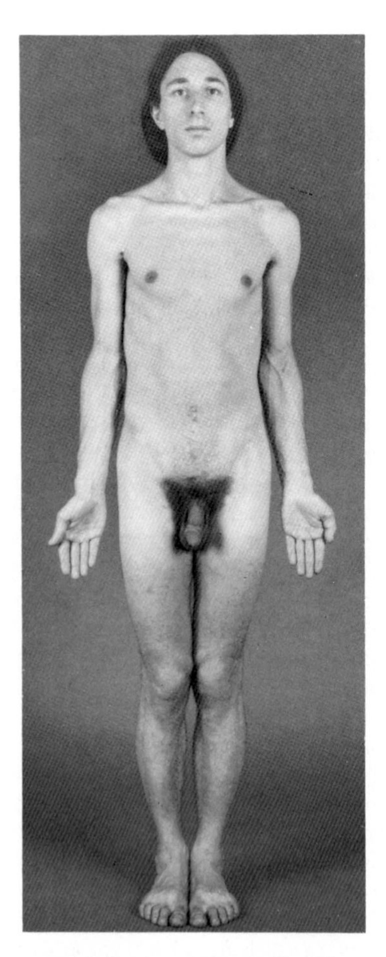
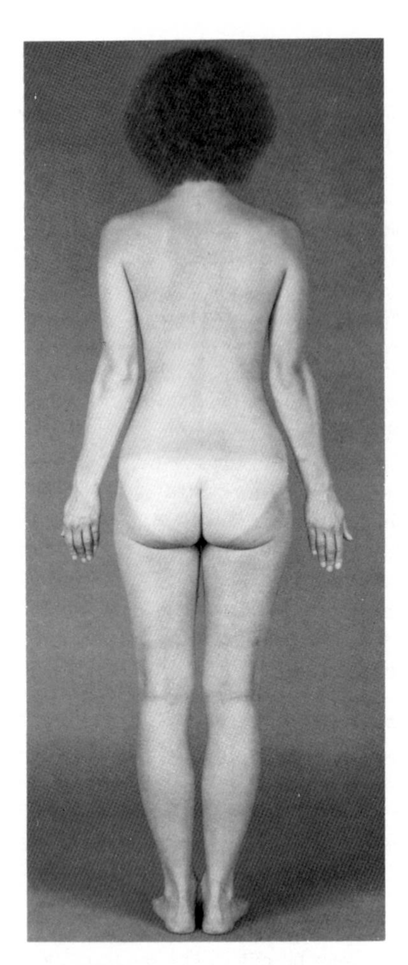

Figure 1. Photographs of a man and woman *standing in the anatomical position. A*, anterior view of a male. Note particularly the bony landmarks formed by his clavicles (collar bones) and the position of his hands and feet. *B*, posterior view of a female. Note the effect ultraviolet has had on the areas of exposed skin in these persons. Sunlight darkens the pigment melanin in the skin and activates the cells (melanocytes) to produce more pigment.

i.e., they have been rotated laterally, away from the median plane of the body. When you stand casually your forearms are partly pronated; *i.e.*, the palms almost face posteriorly (backward).

2. *The great toes touch.* In the usual stance, the heels are together but the great toes are directed anterolaterally (toward the front and sides).

PLANES OF THE BODY

Many descriptions are made using *imaginary planes* passing through the body in the anatomical position. There are median, sagittal, coronal, and horizontal planes.

The Median Plane (Fig. 3). This is the vertical plane passing lengthwise through the midline of the body from front to back, *dividing it into right and left halves*, except for such internal organs as the heart and liver that do not lie in the midline.

The Sagittal Planes (Fig. 3). These are *any vertical planes passing through the body parallel to the median plane.* The sagittal plane that passes through the median plane of the body is called the median sagittal plane or **midsagittal plane**. It is in the same plane as the **sagittal suture**, an arrow-like (L. *saggita*, arrow) fibrous joint between the parietal bones of the skull (Fig. 33*A*).

The sagittal planes that divide the body into right and left portions but do not pass through the median plane of the body are sometimes referred to as paramedian or **parasagittal planes** (G. *para*, beside).

You are certain to hear neurologists, neurosurgeons, and neuroradiologists refer to **parasagittal tumors** (*i.e.*, tumors that are near the midsagittal or median plane). It is always helpful to give a point of reference, *e.g.*, a sagittal plane passing through the midpoint of the clavicle.

The Coronal Planes (Fig. 4). These are *any vertical planes passing through the body at right angles to* *median plane*, dividing it into anterior (front) and posterior (back) portions.

The **coronal suture** of the skull is in a coronal plane (Fig. 33A). The coronal planes of the body pass through the ankles. Hence, the coronal planes of the anterior parts of the feet do not pass through the trunk of the body (Fig. 4).

It is common to hear the coronal plane referred to as the *frontal plane*, probably because several of them pass through the forehead (L. *frons*).

The Horizontal Planes (Fig. 4). These are *any planes passing through the body at right angles to both the median and coronal planes.* A horizontal plane divides the body into superior (upper) and inferior (lower) portions. Again it is always helpful to give a reference point, *e.g.*, a horizontal plane passing through the umbilicus (navel).

It is common to hear the horizontal plane referred to as the *transverse plane*, but it may be erroneous. Note that a transverse section of the hand is in the horizontal plane, but a transverse section of the foot is in the coronal plane (Fig. 4).

SECTIONS OF THE BODY

In order to describe and display many internal structures, sections of the body and its parts are cut in various planes.

Longitudinal Sections (Fig. 5). These sections (L. *longitudo*, length) *run lengthwise in the direction of the long axis of the body or any of its parts* and they are applicable regardless of the position of the body. Longitudinal sections may be cut in the median, sagittal, or coronal planes.

Vertical Sections. These are the same as longitudinal *sagittal sections* except that they denote that the sections are taken through the body, or part of it, in the anatomical position.

Transverse Sections (Figs. 5 and 6*B*). These sections of the body or parts of it are *cut at right angles to the longitudinal axis of the body or its parts*. Transverse sections of the body (*e.g.*, Fig. 2-107) do not necessarily cut vessels or internal organs transversely because they may not be located in the long axis of the body.

A transverse section in a horizontal plane through the chest at the level of the heart does not cut the ventricles horizontally because the heart is not centered in the median plane (Fig. 21). To make a transverse section of an organ or a part (*e.g.*, the upper limb), the section must be cut at right angles to its longitudinal axis (Figs. 5 and 6*B*). Hence a transverse section of the foot is in the coronal plane (Fig. 4).

Oblique Sections (Fig. 6*A*). These are sections of the body or any of its parts that are not cut in one of the main planes of the body; *i.e.*, they slant or deviate from the perpendicular or horizontal.

TERMS OF RELATIONSHIP

Various terms are used to describe the relationship of parts of the body in the anatomical position (Table 1 and Figs. 7 and 8). Because anatomy is a descriptive science, clearly defined and unambiguous terms must be used to indicate the positions of structures to each other and to the body as a whole.

Anterior *(Ventral, Front).* **Nearer to the front** of the body; *e.g.*, the nipples and the umbilicus (navel or "belly button") are on the anterior surface of the body (Figs. 1 and 2). Usually the anterior surface of the hand is called the volar or **palmar surface** (Fig. 2*A*) and the inferior surface of the foot is called the **plantar surface** or sole (Fig. 12*A*).

Ventral is equivalent to anterior and is commonly used in descriptions of embryos because they cannot be placed in the anatomical position. The term ventral is commonly used in neuroanatomy where it has an advantage because it is equally applicable to man and the four-footed animals that are commonly used in research on the nervous system.

Posterior *(Dorsal, Behind).* **Nearer to the back** of the body; *e.g.*, the gluteal region (buttocks) is on the posterior surface (Fig. 1).

Dorsal is interchangeable with posterior and is commonly used in descriptions of embryos and of the nervous system. When describing the posterior or dorsal surface of the hand, the term **dorsum** is commonly used (Fig. 2*B*).

Similarly, when describing the superior surface of the foot, the term dorsum is used (Fig. 2*A*) because this surface faced dorsally in the early embryo. *During the late embryonic period, the upper and lower limbs rotate in different directions.* As a result, following the embryonic period, the dorsal surfaces of the feet come to lie anterosuperiorly.

In the penis the dorsal surface faces anteriorly in the flaccid condition (Fig. 1). This usage is based on comparative anatomy.

Superior *(Cephalic, Cranial, Above).* **Toward the head** or upper part of the body (Fig. 7); *e.g.*, the heart lies superior to the diaphragm, *i.e.*, higher or closer to the head, cranium, or superior end of the body.

Cranial and cephalic are interchangeable adjectives that are commonly used in descriptions of embryos and the nervous system; *e.g.*, the spinothalamic fibers run cranially, *i.e.*, from the spinal cord to the thalamus in the brain. The Greek word for the brain is *enkephalos*. Hence *cephalic* means toward the brain. The suffix "-ad" is sometimes added to positional terms to indicate motion. Hence *cephalad* and *craniad* mean proceeding toward the brain and cranium, respecttively.

Rostral is often used synonymously with anterior, particularly in descriptions of the brain. It is derived from the Latin word *rostrum*, meaning *beak*; thus,

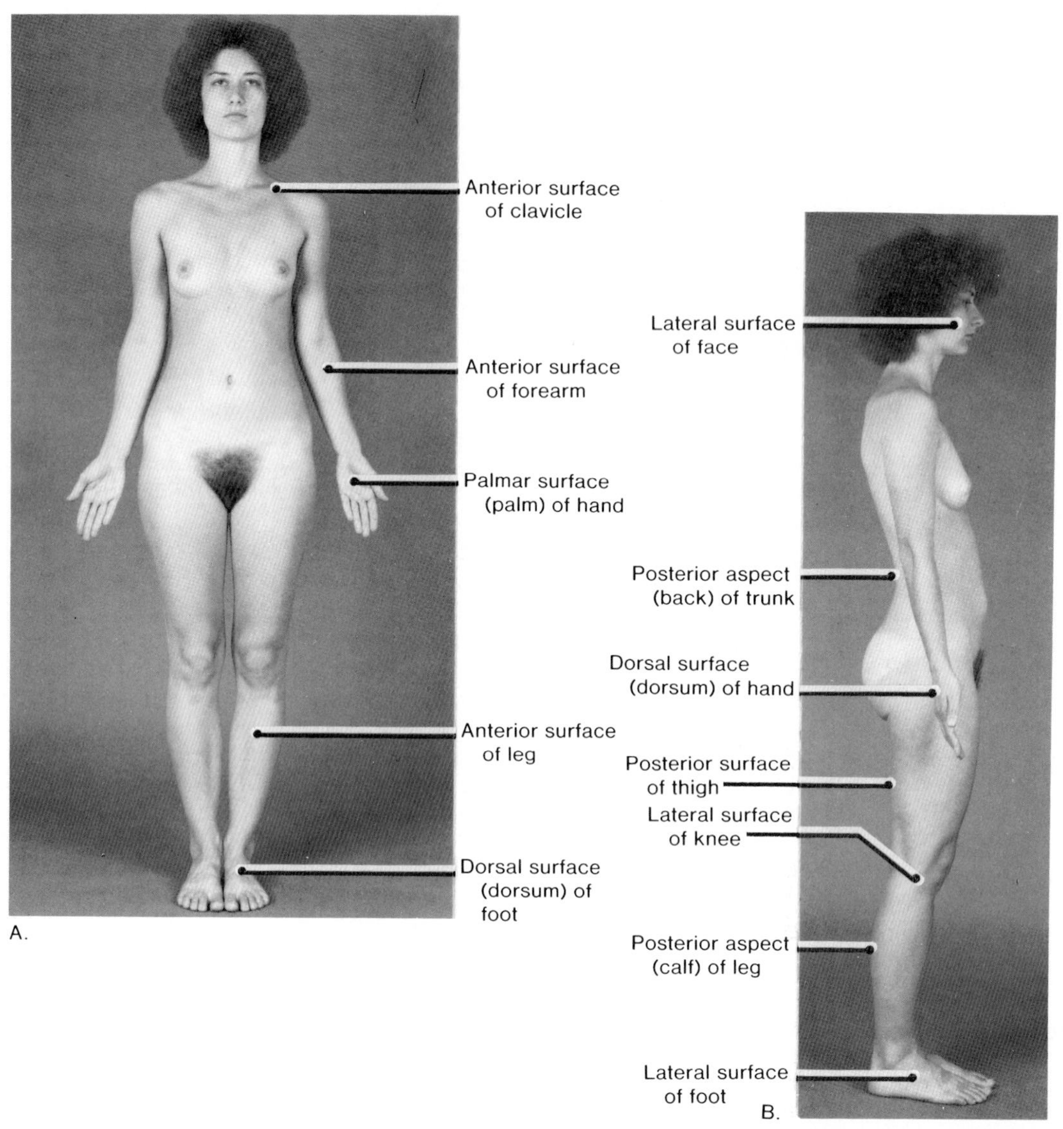

Figure 2. Photographs of a 27-year-old woman demonstrating *the anatomical position* and some anatomical terms. *A*, anterior view. *B*, lateral view. Note: (1) she is standing erect; (2) her face and eyes are directed forward; (3) her hands are by her sides with the palms directed anteriorly; (4) her heels are together; (5) her toes are pointed anteriorly; and (6) her great toes are touching. Owing to rotation of the lower limbs during the embryonic period, the superior (upper) surface of the foot is called the dorsal surface or dorsum.

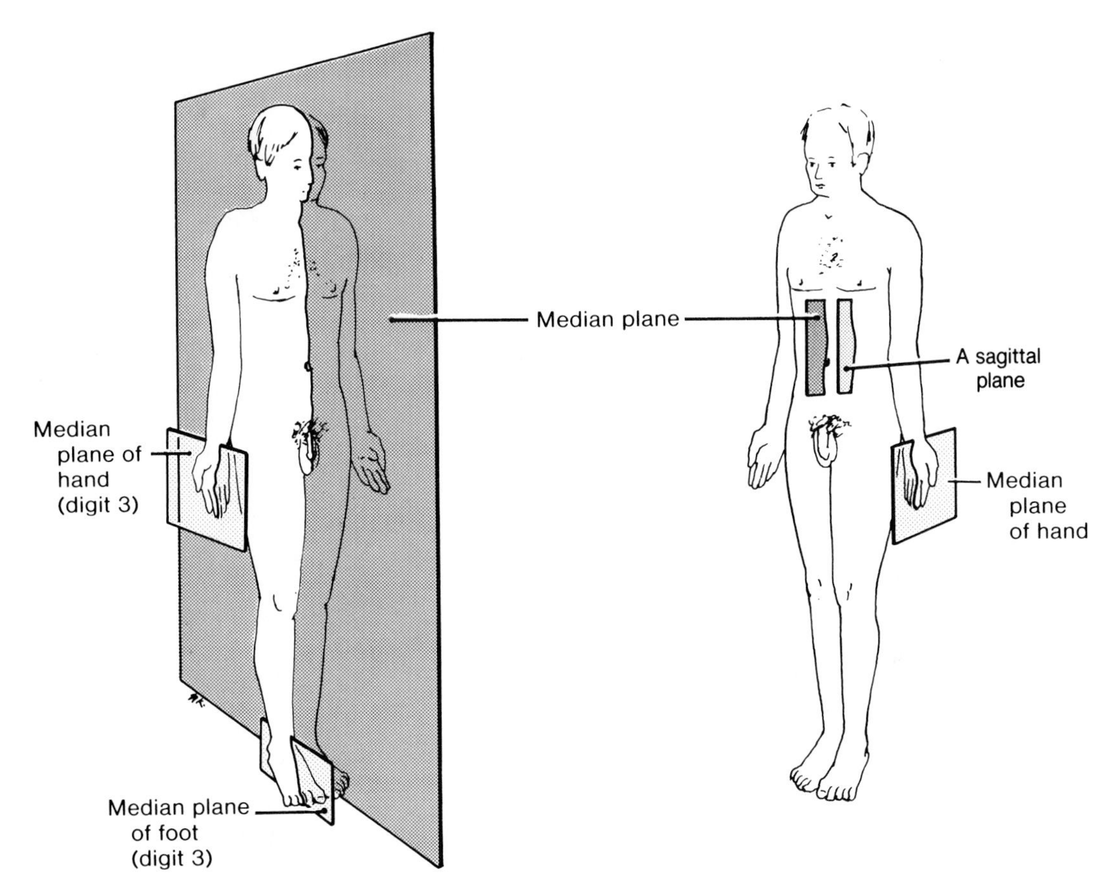

Figure 3. Drawings illustrating *the median and sagittal planes* of the body, two of the four anatomical planes. Observe that the median plane is a vertical plane passing through the body from front to back, dividing it into equal and superficially symmetrical right and left halves. Understand that there are many sagittal planes, because a sagittal plane is *any* vertical plane passing through the body parallel to the median plane. The sagittal plane that passes through the median plane is also called the midsagittal plane. It is a common error to refer to the "midline" of the body when the median plane is meant.

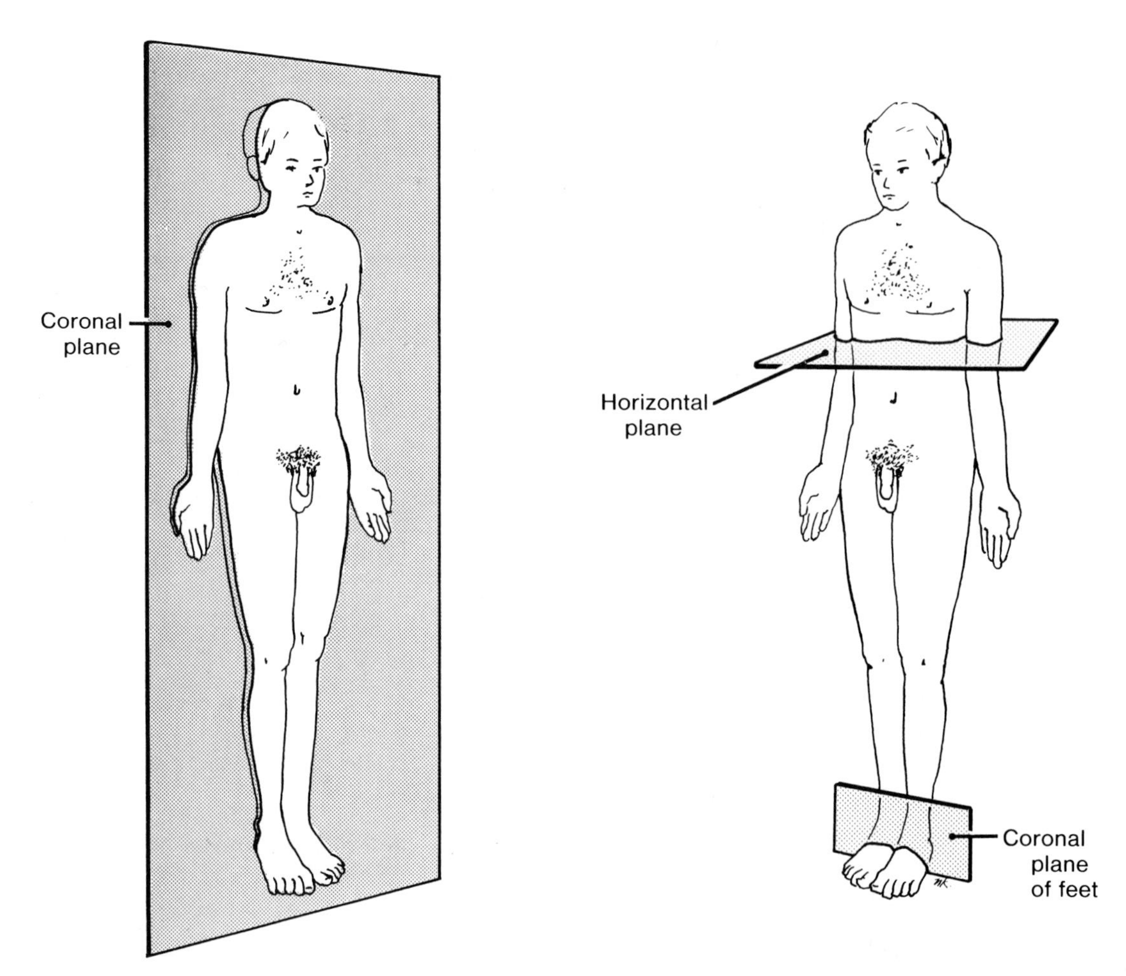

Figure 4. Drawings illustrating *the coronal and horizontal planes. A*, coronal plane is any vertical plane passing through the body at right angles to the median plane. The coronal plane is also referred to as the frontal plane. A horizontal plane is any plane passing through the body at right angles to both the median and coronal planes (*i.e.*, parallel to the surface on which the subject is standing). Note that the coronal plane of the feet *shown here* does not pass through the trunk of the body. Coronal planes of the body pass through the ankles.

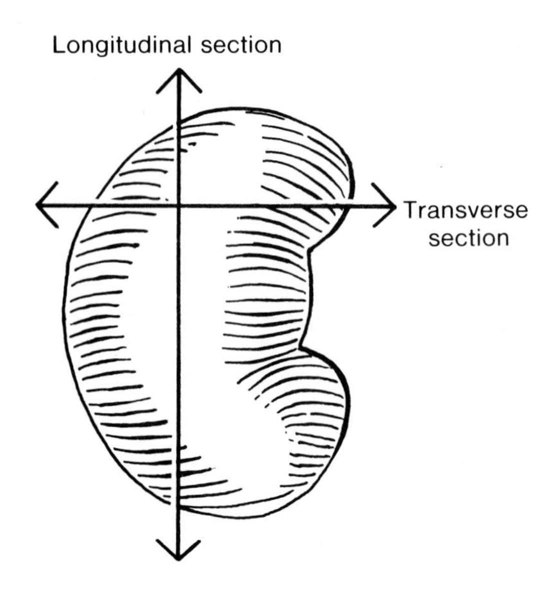

Figure 5. Drawing of a kidney illustrating longitudinal and transverse sections of it. Transverse sections are commonly referred to as "cross sections" because they are cut across the longitudinal axis of a structure.

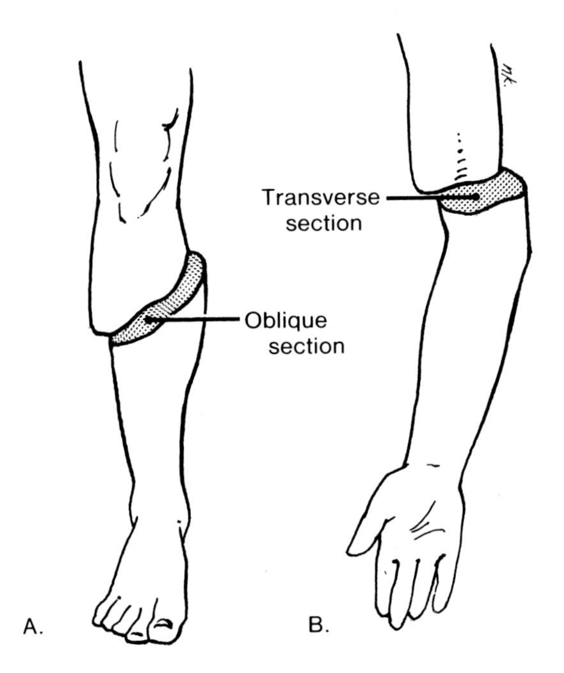

Figure 6. Diagrams illustrating oblique (*A*) and transverse (*B*) sections of the lower and upper limbs, respectively.

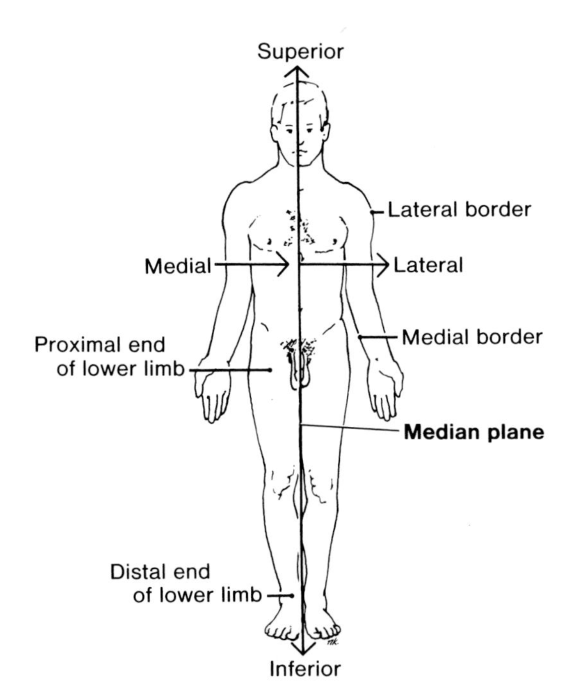

Figure 7. Diagram of an anterior view of a man standing in the anatomical position. The term *medial means toward the median plane* and *lateral means away from the median plane* of the body. The terms median and medial are sometimes confused. Median means "*in* the median plane," whereas medial means "*toward* the median plane."

Table 1
Commonly Used Directional Terms

Term[a]	Definition	Example of usage
Superior	Toward the head	The heart is superior to the stomach.
Inferior	Toward the feet	The stomach is inferior to the heart.
Anterior	Nearer to the front of the body	The sternum is anterior to the heart.
Posterior	Nearer to the back of the body	The heart is posterior to the sternum.
Medial	Nearer the median plane of the body	The ulna is on the medial side of the forearm.
Lateral	Farther away from the median plane of the body	The radius is on the lateral side of the forearm.
Proximal	Nearer the attachment of a limb or a structure	The elbow joint is proximal to the wrist joint.
Distal	Farther from the attachment of a limb or a structure	The wrist joint is distal to the elbow joint
Superficial	Nearer to the surface	The muscles of the arm are superficial to the bone (humerus).
Deep	Farther from the surface	The humerus is deep to the muscles of the arm.
Parietal	Pertaining to the outer wall of a body cavity	The parietal pleura lines the inside of the thoracic wall.
Visceral	Pertaining to the covering of an organ	The visceral pleura covers the external surface of the lungs.

[a] The terms listed above are used to indicate the location of structures in the body with reference to the anatomical position, irrespective of the position of the body of the patient or the cadaver.

structures nearer to the nasal region are rostral to structures that lie posterior to it.

Strictly speaking, the upper limb should be called the superior limb, but few authors use this terminology.

Inferior *(Caudal, Below).* **Toward the feet** or lower part of the body; *e.g.*, the diaphragm is inferior to the heart, *i.e.*, nearer to the feet or inferior end of the body and further from the head.

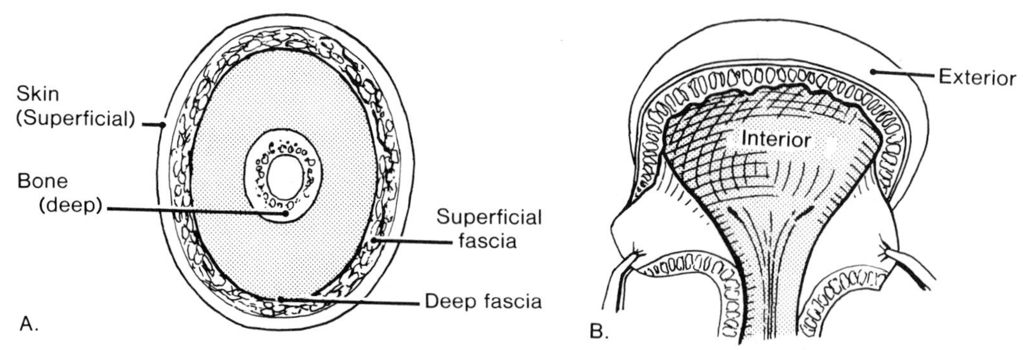

Figure 8. Diagrams illustrating the meaning of the terms superficial and deep and interior and exterior. *A*, when a structure is superficial to another structure, it is nearer to the surface than the other structure, whereas a structure is deep to another structure when it is further away from the surface than the other structure. *B*, the terms interior (internal, inner) and exterior (external, outer) are often used to describe cavities of hollow organs such as the uterus *shown here*. The skull is also described as having an internal surface and an external surface; *e.g.*, a bullet lying on the internal aspect of the skull would lie between the skull and the brain, whereas a bullet lying on the external aspect would lie between the skull and the scalp. The terms inner and outer should not be used when you mean medial (toward the median plane) or lateral (away from the median plane).

The term **caudal** is derived from the Latin word *cauda*, meaning "tail." It is commonly used in descriptions of embryos where the term is applicable because human embryos have a tail until near the end of the embryonic period.

Medial (Figs. 3 and 7). **Toward the median plane** of the body, *e.g.*, the external openings of the nose (**anterior nares** or nostrils) are medial to the eyes. The term mesial (G. *mesos*, middle) is equivalent to medial and is used in dentistry to indicate "*toward the midline of the dental arch.*"

Because medial, as applied to the limbs, is sometimes misinterpreted, depending upon whether one is thinking of the median plane of the body or the midline of the limb, it is common to describe the limbs in terms of the position of their paired bones. In the upper limb, the *radius* is the lateral bone of the forearm (Fig. 16) and the **ulna** is the medial one. Thus, the terms "*ulnar*" and "medial" and "*radial*" and "lateral" are synonymous and refer to the little finger and the thumb sides of the hand, respectively.

Lateral (Figs. 3 and 7). **Farther away from the median plane** of the body. Hence the little toe is lateral to the great toe, but the little finger is medial to the thumb.

Intermediate (Fig. 13). **Between two structures**, one of which is medial and the other lateral; *e.g.*, the ring finger is intermediate between the little and middle fingers. The little finger is medial and the middle finger is lateral to it.

TERMS OF COMPARISON

These terms are used to compare the relative position of two structures to each other.

Proximal (L. *proximus*, next). **Nearest the trunk or point of origin** of a vessel, nerve, limb, or organ. In the limbs, proximal is used to indicate *nearer to the root or attached end* of a limb, *e.g.*, the thigh is at the proximal end of the lower limb (Fig. 2).

Distal (L. *distans*, distant). **Farthest from the trunk or point of origin** of a vessel, nerve, limb, or organ. In the limbs, distal is used to indicate *farther from the root or attached end* of a limb, *e.g.*, the foot is at the distal end of the lower limb.

In dentistry, distal means "farther from the midline of the dental arch." In addition, **labial** means toward the lip (L. *labium*), whereas lingual (L. *lingua*, tongue) means toward the tongue.

Superficial (L. surface). **Nearer to the surface;** *e.g.*, the superficial fascia is closer to the surface of the body than the deep fascia (Fig. 8*A*). Similarly the scalp is superficial to the skull.

Deep (Fig. 8*A*). **Farther from the surface**; *e.g.*, in the arm the bone (humerus) is deep to the muscles and skin.

Interior (*Inside, Inner, Internal*). **Nearer to the center** of an organ or cavity (Fig. 8*B*); *e.g.*, the interior of the urinary bladder. The term is also used to describe structures that pass from the anterior to the posterior surface of the body, or enclose other structures (*e.g.*, as the ribs enclose the thoracic viscera). Hence the internal surface of a rib is the surface farthest from the skin. Also, the **internal carotid artery** passes to the *inside of the skull.*

Exterior *(Outside, Outer, External).* **Farther from the center** of an organ or cavity (Fig. 8*B*); *e.g.*, the external surface of the urinary bladder is farther from the center of the organ than the internal surface. Similarly, the external surface of a rib is the surface closer to the skin and the **external carotid artery** passes to the *outside of the skull.*

Ipsilateral (L. *ipse*, same, + *lateralis*, side). **On the same side** of the body; *e.g.*, the right thumb and the right great (big) toe are ipsilateral.

Contralateral (L. *contra*, opposite + *lateralis*, side). **On the opposite side** of the body; *e.g.*, the right hand and the left hand are contralateral.

Common Expressions. It is not unusual to hear the following common terms: *front* for anterior; *back* for posterior; *upper* for superior; *lower* for inferior; *above* for superior; and *lower* for inferior. These terms are not ambiguous when they are used in reference to the anatomical position, but it is best to employ the commonly used directional terms (Table 1).

Ambiguous Terms. *Do not use the terms on, over, and under* because it may not always be clear what you mean; *e.g.*, when you say a structure passes over another one, do you mean anterior, superior, or superficial to it.

Combined Terms. Often terms are combined to indicate a direction; *e.g.*, **inferomedially** means toward the feet and the median plane. There are many other combinations of terms, *e.g.*, **anteroinferior, anteroposterior**. In a posteroanterior (**PA**) radiograph of the chest, a beam of x-rays passes through the thorax of a patient from posterior to anterior; *i.e.*, the x-ray tube is posterior to the patient and the x-ray film is anterior to him/her.

TERMS OF MOVEMENT

Anatomy is concerned with the living body, hence, there are various terms to describe the different types of movement of the limbs and other parts of the body. Movements take place at certain joints where two or more bones meet one another.

Flexion (Figs. 9 to 12). This movement indicates **bending or making a decreasing angle between the bones or parts of the body.** Usually the movement is an anterior bending in a sagittal plane, *e.g.*, flexion of the forearm at the elbow joint (Fig. 9*B*), but flexion of the leg at the knee joint is a posterior bending of the limb (Fig. 9*C*). To indicate flexion or bending in the dorsal direction, as in the ankle, the term dorsiflexion is used (Fig. 12*A*).

Lateral bending or flexion of the trunk (Fig. 5-9)

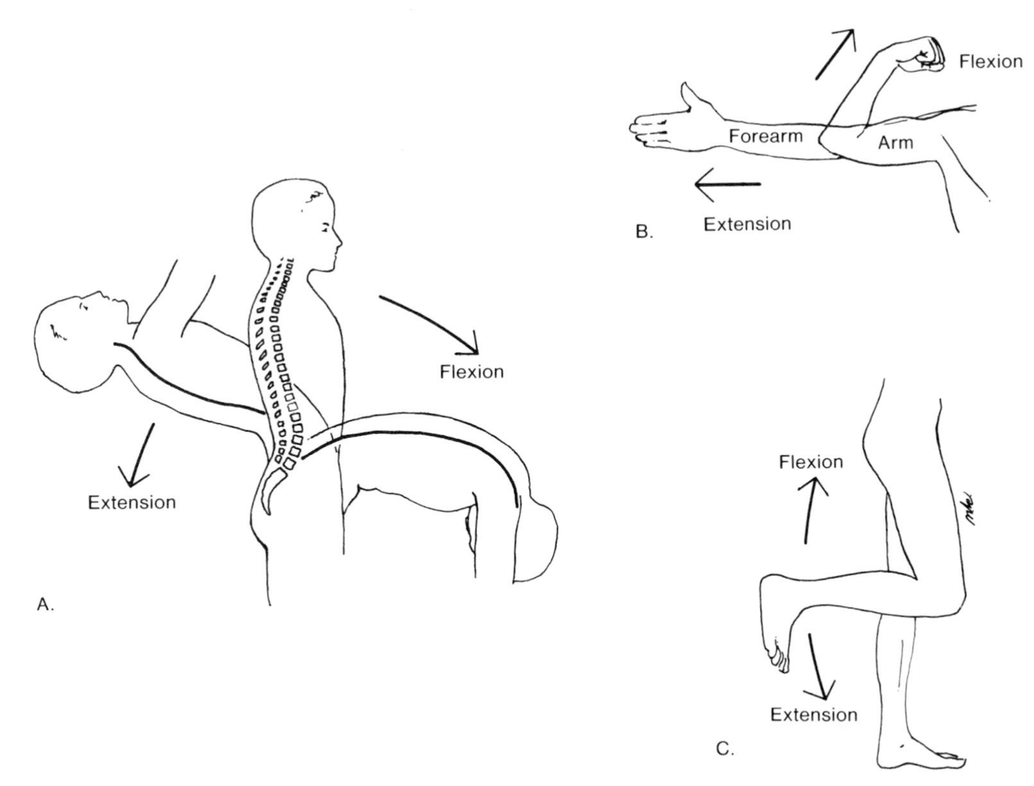

Figure 9. Drawings illustrating *flexion and extension. A*, movements of the trunk or neck in a sagittal plane are known as flexion (forward bending) and extension (backward bending). *B*, movements of the forearm in a sagittal plane are known as flexion and extension. In flexion the angle between the forearm and arm is reduced, whereas in extension the angle is increased. *C*, note that flexion of the leg at the knee joint is a posterior bending of the limb. Hyperextension is extreme or excessive extension of a limb or part. Hyperflexion is forcible overflexion of a limb or part. These movements beyond the normal range may cause injury to the joint or part (*e.g.*, the knee joint or neck). Also see Figure 5-9.

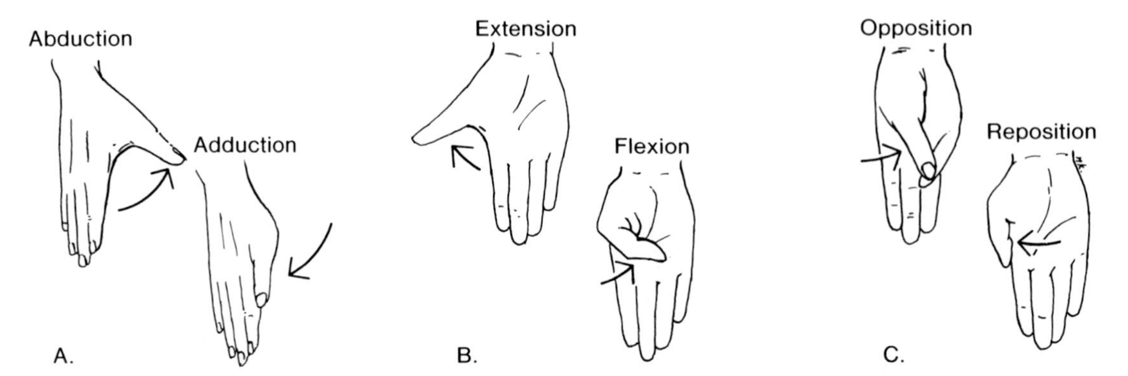

Figure 10. Drawings illustrating movements of the thumb.

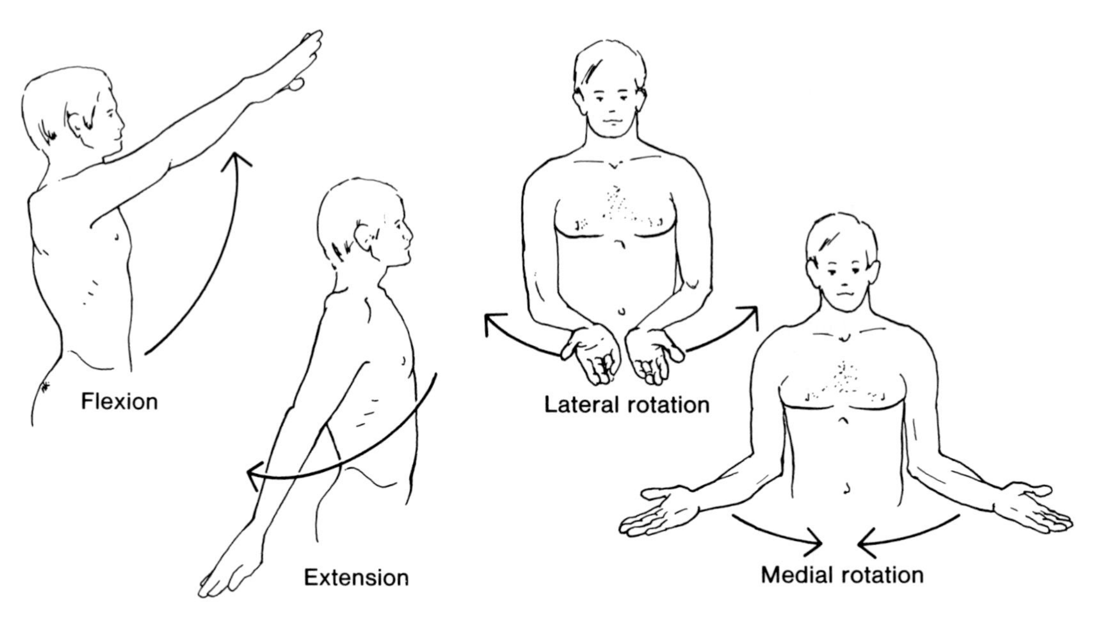

Figure 11. Drawings illustrating some movements at the shoulder joint. Note that the forearms are flexed to 90° during rotation at this joint.

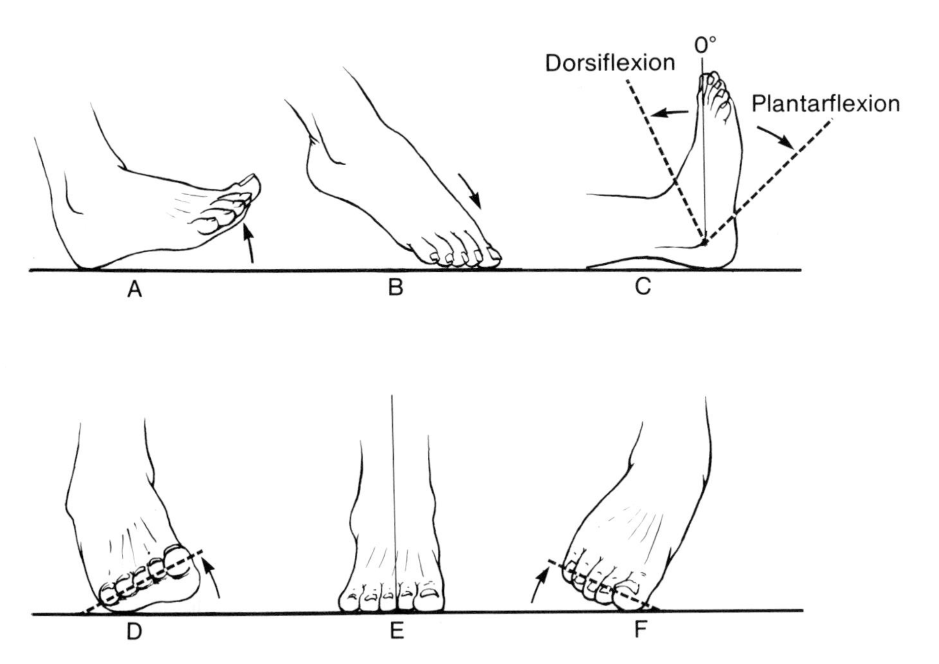

Figure 12. Drawings illustrating movements of the foot. *A*, dorsiflexion; *B*, plantarflexion; *C*, dorsiflexion and plantarflexion; *D*, inversion; *E*, normal; *F*, eversion. For a description of the muscles producing these movements, see Chapter 4. Also see Tables 4-12 and 4-13.

is movement of the trunk laterally (*i.e.*, away from the median plane of the body in the coronal plane).

Extension (Figs. 9 to 12). This is a **straightening of a bent part or making an increasing angle between the bones or parts of the body.** Extension usually occurs in the posterior direction, but extension of the leg at the knee joint is in an anterior direction (Fig. 9*C*). If the movement is continued beyond that which is necessary to straighten the part, it is referred to as **hyperextension** (*e.g.*, as occurs to the neck in rear end collisions. Understand that when the lower limb is completely extended (Fig. 12*B*), *the ankle joint is plantarflexed* (Fig. 12*C*), *e.g.*, when standing on the *tiptoes* as in ballet dancing.

Abduction (Figs. 13*A* and 14*A*). This term means **moving away from the median plane** in the coronal plane, *i.e.*, drawing away laterally, as when moving the upper limb away from the body. The Latin prefix *ab-* means "*away from.*"

In abduction of the fingers or toes, the term means spreading them apart, *i.e.*, moving the fingers away from the middle finger or median plane of the hand (Fig. 13*A*) and the toes away from the second toe or median plane of the foot.

Adduction (Figs. 10*A*, 13*B*, and 14*B*). This is the opposite movement to abduction. It refers to **moving toward the median plane** in a coronal plane, *e.g.*, when moving the upper limb toward the body. The Latin prefix *ad-* means "*to or toward.*" In adduction of the fingers or toes, the term means moving them toward the median plane of the hand or foot, *i.e.*, moving the fingers toward the middle finger and the toes toward the second toe.

Opposition (Fig. 10*C*). This is the movement during which **the thumb pad is brought to a finger pad.** We frequently use this movement to hold a pen, to pinch, or to grasp a cup handle.

Reposition (Fig. 10*C*). This is the term used to describe the movement of the thumb from the position of opposition back to its anatomical position.

Protraction. This is **a movement anteriorly** (forward), as occurs in protruding the mandible (sticking the chin out) or drawing the shoulders forward.

Retraction. This is **a movement posteriorly** (backward) as occurs in drawing the mandible posteriorly (tucking in the chin) or in drawing the shoulders posteriorly, *e.g.*, squaring the shoulders in the military stance.

Elevation. This term means **lifting, raising, or moving a part superiorly** (*e.g.*, elevating the shoulders) as occurs when shrugging them, or in raising the upper limbs superior to the shoulders.

Depression. This is a **letting down, lowering,** or **moving a part inferiorly,** (*e.g.*, depressing or lowering the shoulders as occurs when standing at ease).

Circumduction (Fig. 15). This term is derived from the Latin words *circum*, meaning around, and *duco*, to draw. Hence, it refers to a **circular movement** that means to draw around or to form a circle.

Circumduction is the combination of successive move-

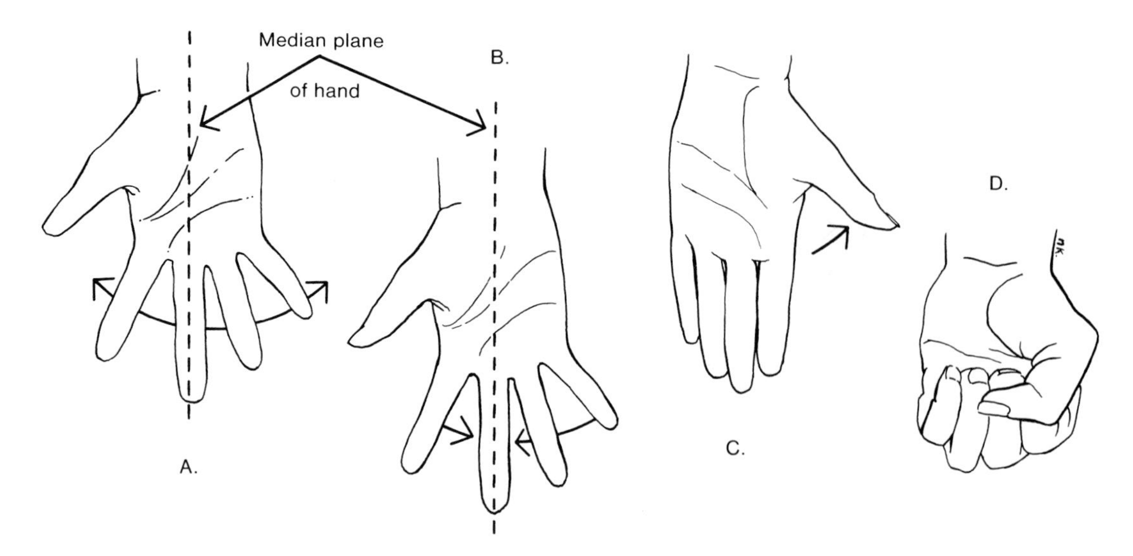

Figure 13. Drawings illustrating various movements of the fingers and thumb. *A*, abduction of the fingers. *B*, adduction of the fingers. *C*, extension of the thumb and fingers. *D*, flexion of the fingers and thumb. Note that the median plane of the hand passes through the midline of the middle finger.

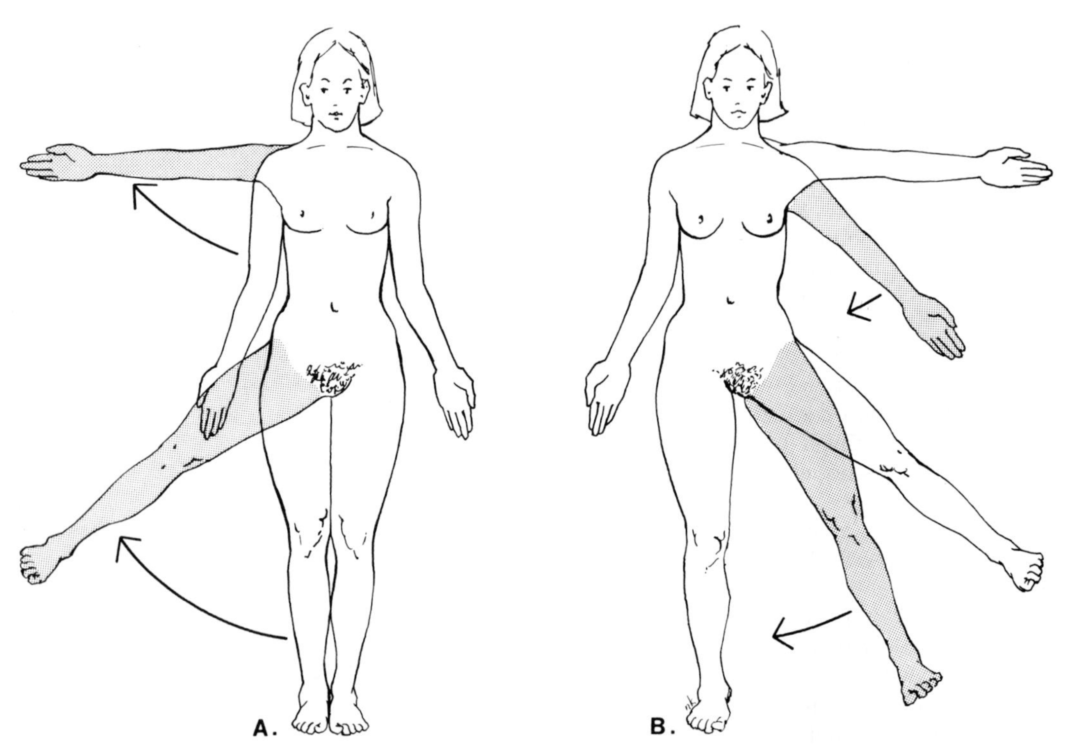

Figure 14. Drawings illustrating *abduction* of the right limbs in *A* and *adduction* of the left limbs in *B*. Observe that abduction of a limb is the *movement away* from the median plane of the body in a coronal plane, whereas *adduction* of a limb is the *movement toward* the median plane of the body in a coronal plane.

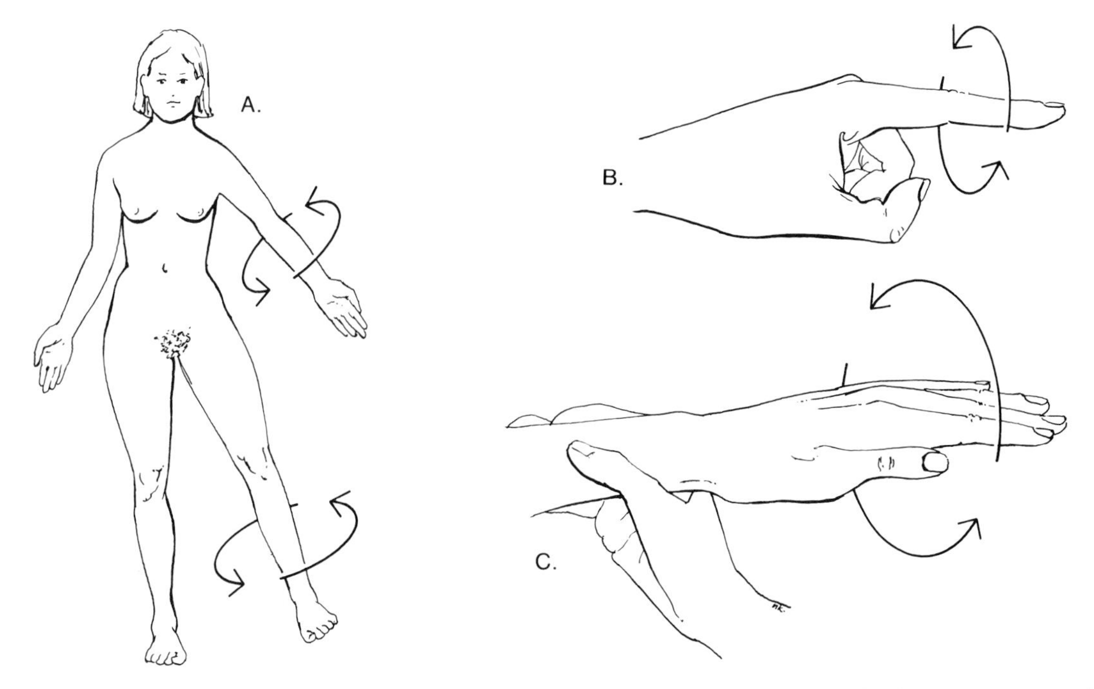

Figure 15. Drawings illustrating the movement called *circumduction*, which is a combination in sequence of the movements of *flexion*, *extension*, *abduction*, and *adduction*. *A*, circumduction of the left upper and lower limbs at the shoulder and hip joints, respectively. *B*, circumduction of the index finger. *C*, circumduction of the hand at the wrist joint.

ments of flexion, abduction, extension, and adduction in such a way that the distal end of the part being moved describes or forms a circle. *This sequence of movements results in a cone of movement.* It can occur at any joint at which the above mentioned four types of movement are possible, *e.g.*, the hip, the wrist, and the metacarpophalangeal joints of the fingers.

You may observe this movement by extending your leg and moving it in such a way as to produce an imaginary circle (Fig. 15*A*).

A person with a paralyzed lower limb swings the stiffly extended limb in a partial arch when walking; this is called a **circumducted gait**.

The thumb and index finger can also be circumducted (Fig. 15*B*), but it is not easy to circumduct the other digits.

Rotation (Fig. 11). This movement involves a turning or **revolving of a part of the body around its long axis**, *e.g.*, rotation of the humerus at the shoulder joint and the femur at the hip joint.

Rotation toward the median plane of the body is **medial rotation**, whereas *rotation away from the median plane is* **lateral rotation**. Medial rotation moves the lateral side of a limb medially, and lateral rotation brings the medial side of a limb laterally.

Eversion of the Foot (Fig. 12*F*). This movement *turns the plantar surface or sole of the foot away from the median plane* of the body (*i.e.*, *the sole faces laterally*).

Inversion of the Foot (Fig. 12*D*). This movement turns the plantar surface or *sole of the foot toward the median plane* of the body (*i.e.*, *the sole faces medially*).

Supination (Fig. 16*A*). This is the movement that rotates the radius of the forearm laterally around its long axis, so that *the dorsum of the hand faces posteriorly and the palm faces anteriorly* when the body is in the anatomical position. When the elbow is flexed 90°, supination moves the forearm so that the palm of the hand is turned superiorly (*i.e.*, faces upward).

Pronation (Fig. 16*B*). This is the movement that rotates the radius of the forearm medially around its long axis so that *the palm of the hand faces posteriorly and its dorsum faces anteriorly* when the upper limb is by the side in the anatomical position. When the elbow is flexed 90°, pronation moves the forearm so that the palm of the hand faces inferiorly (*i.e.*, turned downward) *e.g.*, when the palm of the hand is placed on a table. During pronation the radius of the forearm crosses the ulna diagonally, moving the thumb medially (Fig. 16).

THE MEANING OF TERMS

Anatomy is a descriptive science, hence many anatomical terms are descriptive and indicate the shape,

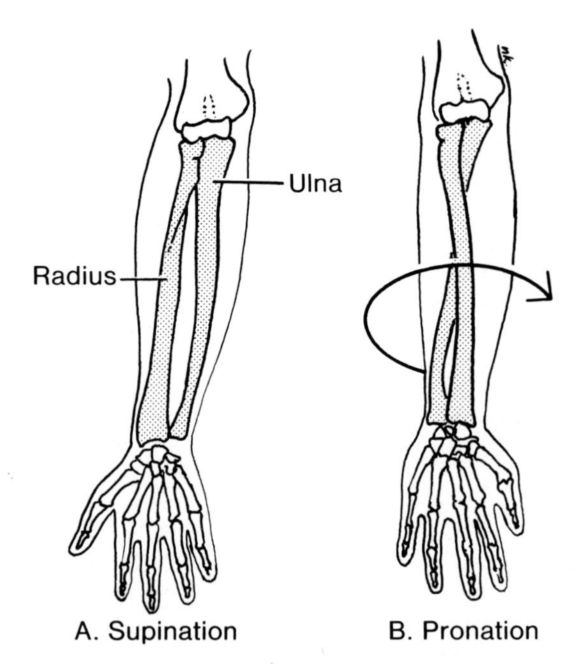

Figure 16. Drawings illustrating *supination* (*A*) and *pronation* (*B*) of the right forearm. Observe that pronation is a medial rotation of the radius from its anatomical position (*A*) so that the dorsum (back) of the hand faces anteriorly. Supination of the pronated forearm returns the hand to its anatomical position shown in *A* and in Figure 2. Note that the positions of the radius and the ulna change during pronation.

size, location, function, or resemblance of a structure to some other structure; *e.g.*, the appendix of the cecum is appropriately referred to as the **vermiform appendix** (L. *vermis*, a worm) because it is shaped like a worm.

You will quickly learn that muscles are given descriptive terms to indicate their characteristics, *e.g.*, the **biceps** (L. *bi*, two, + *caput*, head), the **triceps** (L. *tri*, three) and the **quadriceps** (L. *quattuor*, four) because they have two, three, and four heads, respectively.

Some muscles are named according to their shape, *e.g.*, the **piriformis** (L. pear-like form), the **rectus** (L. straight) abdominis, and the **quadratus** (L. square) femoris. Other muscles are named by location, *e.g.*, the **temporalis** is a muscle in the *temporal region* of the skull (Fig. 7-108).

In other cases actions are used to describe muscles, *e.g.*, the **levator scapulae**, as you have probably guessed, elevates *the scapula.*

Thus, there are good reasons for the names given to the parts of the body and if you learn them and think about them as you dissect, you should have less difficulty remembering their names.

Abbreviations. As abbreviations are commonly used in this and other anatomy books, especially for structures in illustrations, they must be explained. They are easy to understand if you know the word(s) being abbreviated; *e.g.*, "**fl. carpi ulnaris**" refers to the flexor carpi ulnaris muscle in the forearm.

Frequently on illustrations you will also see abbreviations for arteries and ligaments (*e.g.*, "**post. circumflex humeral a.**" refers to the posterior circumflex humeral artery, and "**sup. glenohumeral lig.**" refers to the superior glenohumeral ligament).

As "**a.**" is used for an artery, the abbreviation for arteries is "**aa.**" Similarly, "**v.**" is used for a vein and "**vv.**" for veins, "**m.**" for muscle and "**mm.**" for muscles, "**n.**" for nerve and "**nn.**" for nerves. In most cases the abbreviation "m." is omitted when it is obviously a muscle, external oblique (Fig. 2-9).

When referring to vertebrae and spinal nerves, they are designated both by region and by number, *e.g.*, for the third cervical, it is common to use the abbreviation **C3**. Later you will learn that the ulnar nerve supplying structures in the forearm and hand carries nerve fibers from C7 to T1, meaning that it receives them from the 7th and 8th cervical and first thoracic segments of the spinal cord.

As most medical words stem from Greek (**G.**), Latin (**L.**), Arabic (**A.**), and French (**Fr.**), the derivation of a word is often given because it helps in understanding its meaning; *e.g.*, **cancer** (L. crab) was so named because it gnaws away at the body like a crab. Similarly, **coccyx** (G. cuckoo) indicates that this bone is shaped like the beak of a cuckoo. **Cul-de-sac** (Fr.) means the bottom of a sac and is used to describe a blind pouch or tubular cavity.

Anatomical Variations. There is considerable variation in anatomical material. *No two cadavers are exactly alike.* The different bones of the skeleton vary amongst themselves, not only in their basic shape but also in lesser details of surface structure. Accessory bones are sometimes present.

There is a wide variation in the size, shape, and the form of attachment of muscles. There is also considerable variation in the mode of division of the arteries. *Veins vary greatly* as to where they branch and how they branch.

Typical vertebrae have characteristic features, but the vertebrae of different regions have particular features which make them readily variable. In most descriptions of anatomy you will see the words *usually* or *normally*; this implies that there are variations. In general text books such as this one, the usual or normal anatomy is presented, but variations that are of clinical significance are mentioned in the Clinically Oriented Comments.

RADIOLOGY AND ANATOMY

Anatomy is essential for understanding radiology. During 1st year and when you begin to practice your profession, you will examine *the anatomy of the body in radiographs* nearly every day. You will see anatomical structures this way much more frequently than you will see them displayed at operation or autopsy. Familiarity with normal radiographs allows you to recognize abnormalities, *e.g.*, tumors or fractures.

When faced with an injured patient, *you must be able to visualize in your "mind's eye" the injured part and its surroundings.* When you examine a sick patient, you must be able to visualize the diseased organ and its associated structures. Knowledge of radiological anatomy helps you to do these things.

Each image on normal radiographs should be studied and identified on a skeleton and in your dissection. The essence of the radiological examination is that a highly penetrating beam of x-rays "transilluminates" the patient, showing tissues of differing densities within the body as images of differing densities on x-ray film (Fig. 17).

Density is the ratio of the mass of a homogenous portion of a material to its volume. Its dimension is mass/volume, *e.g.*, the density of water is 1 g/cm^3. It is analogous to specific gravity, the ratio of the weight of a given volume of a homogenous portion of a material to the weight of an equal volume of water. The specific gravity of water is 1.

A tissue or organ that is relatively dense absorbs (stops) more x-rays than a less dense tissue. Consequently, a dense tissue or organ produces a relatively transparent (erroneously called white) area on the x-ray film because relatively fewer x-rays reach the silver salt/gelatin emulsion in the film at that point. Therefore, relatively fewer grains of silver are developed at this area when the film is processed.

A very dense substance is said to be radiopaque, whereas a substance of small density is said to be radiolucent (Table 2). The density of fat is approximately 0.9 g/cm^3, whereas water and most soft tissues of the body have a density of 1.00 to 1.04 g/cm^3. Cancellous bone varies in density, but can be considered to be about 1.3 g/cm^3; compact bone has a density of approximately 1.8 g/cm^3.

As the image is a two-dimensional image of a three dimensional object, images of structures at different depths in the body overlap each other. When "filming" a part of the patient, at least two projections or views must be made, usually at right angles to each other, in order to obviate the overlap of one object by another and to help your mind reconstruct the three-dimensional part (Figs. 18 and 19). **Stereoscopic films** of a part will give a three-dimensional image also.

Body section radiography can also be utilized to give three-dimensional information. There are two types.

Conventional tomography, also called *laminography*, is a method of moving an x-ray tube in the opposite direction to a moving x-ray film during the exposure so that images of a predetermined plane in the body remain stationary on the film, while images in other planes move on the film and are blurred and become invisible (Fig. 20).

Computerized Tomography, using *CT scanners*, shows sections of the body resembling anatomical sections (Fig. 21). In this process a small beam of x-rays is passed through a plane of the body while *the x-ray tube moves in an arc or a circle around the body.* The amount of radiation absorbed by each different volume element of the chosen plane varies with the amounts of fat, water density tissue, and bone in each element. A multitude of linear energy absorptions is measured and passed to a computer. The computer matches the many linear energy absorptions to each point within the section or plane that is scanned and displays the result on a print-out, or on a cathode ray tube.

Nowadays, x-rays for medical diagnosis are produced in a highly evacuated tube containing a hot cathode and a cold anode. *The* ***x-ray tube*** *and its terminals are enclosed in an electrically insulated and x-ray proof housing which is filled with insulating oil.* The terminals of the tube are connected internally to heavily insulated shockproof cables leading to the **x-ray generator**. The rayproof housing of heavy metal has one small area opposite the anode which is made of radiolucent material, often glass, called the port or portal. A slightly divergent beam of x-rays emerges through the **portal**. *No x-rays leave the tube housing elsewhere.*

The **x-ray beam** originates in a heavy metal target, usually tungsten, on the anode. *X-rays are produced when electrons from the hot cathode are pulled to the anode by a momentary large potential difference between the anode and the cathode.* This potential difference varies from 40 to 150 kv, thus the need for a shockproof housing and cables. X-rays leave the anode in all directions, hence a rayproof housing is necessary.

As the x-ray beam emerging through the portal is composed of slightly diverging x-rays, structures far from the x-ray film are magnified and may give you an incorrect idea of their size (Fig. 17). Because some parts of the body are oblique to the x-ray beam when frontal or lateral views are made, their images may be foreshortened, giving you an incorrect idea of their size and shape.

Radiologists sometimes make **oblique projections** of a part of the body to show the image of the part as close as possible to its true shape and size. Oblique projections may also be made to project overlapping images away from the part of greatest interest.

Radiological nomenclature must be understood. An **AP projection** is one in which the x-rays traverse the patient from anterior (**A**) to posterior (**P**). This indicates that the x-ray tube is anterior to the patient and the x-ray film is posterior to the patient.

A **PA projection** is one in which the x-rays traverse the patient from posterior (**P**) to anterior (**A**). Consequently the x-ray tube is posterior to the patient

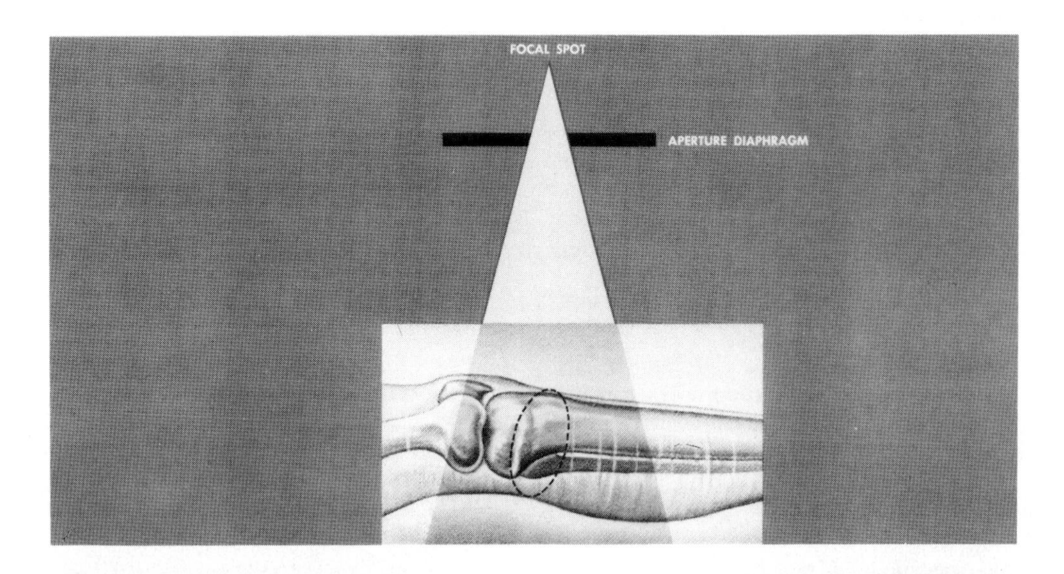

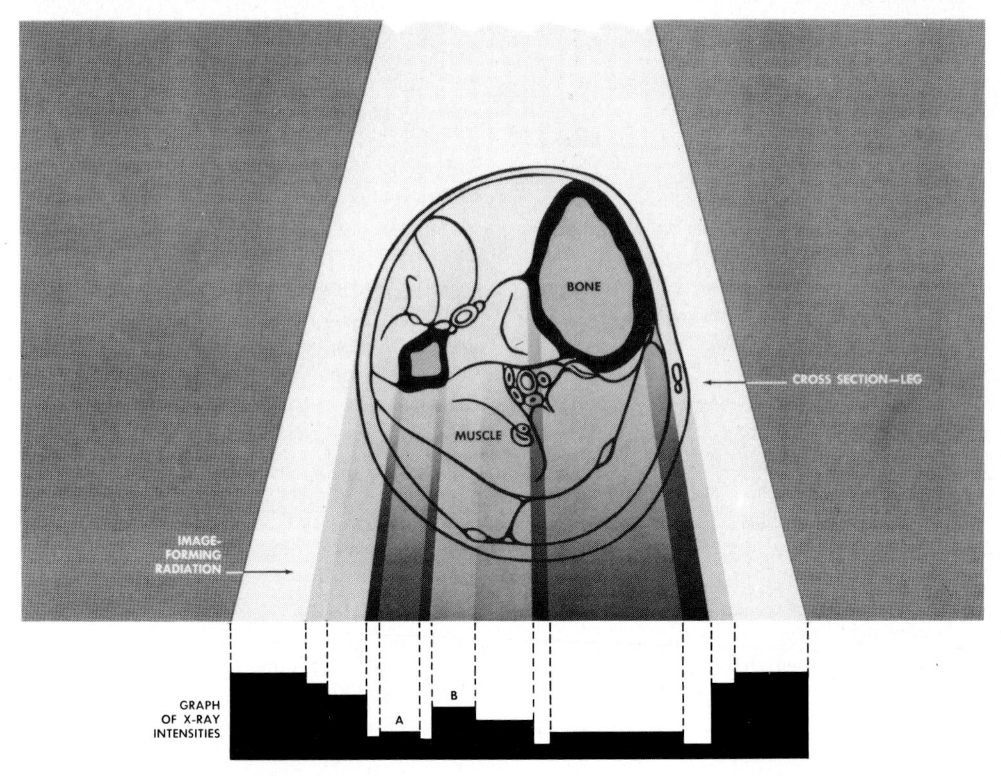

Figure 17. Diagram illustrating how a beam of x-rays passes through the knee and leg, forming images on an x-ray film. On each side of the leg, the x-rays pass through air only, resulting in maximum x-ray intensity in the film and maximum film blackening as indicated on the graph. The greatest thickness of the most dense tissue (compact bone of the cortex of the tibia, the larger bone) stopped the most x-rays, resulting in minimum x-ray intensity in the film and minimum film blackening under the tibia. (Reprinted with permission from Eastman Kodak. *Fundamentals of Radiology*, 1980).

Table 2
The Basic Principles of X-ray Image formation

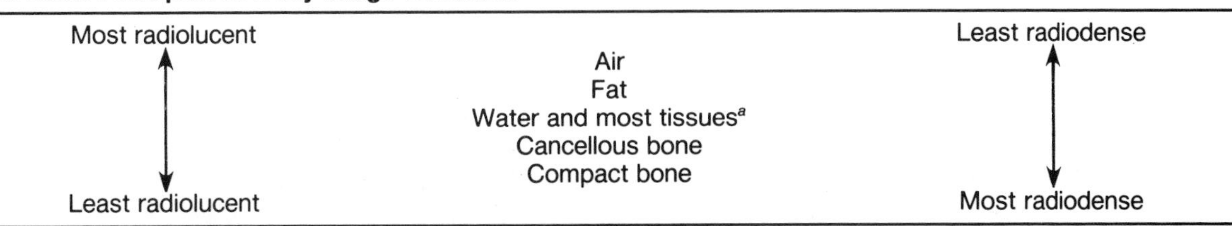

Most radiolucent ↑		Least radiodense ↑
	Air	
	Fat	
	Water and most tissues[a]	
	Cancellous bone	
	Compact bone	
↓ Least radiolucent		↓ Most radiodense

[a] Includes cytoplasm and uncalcified intercellular substances.

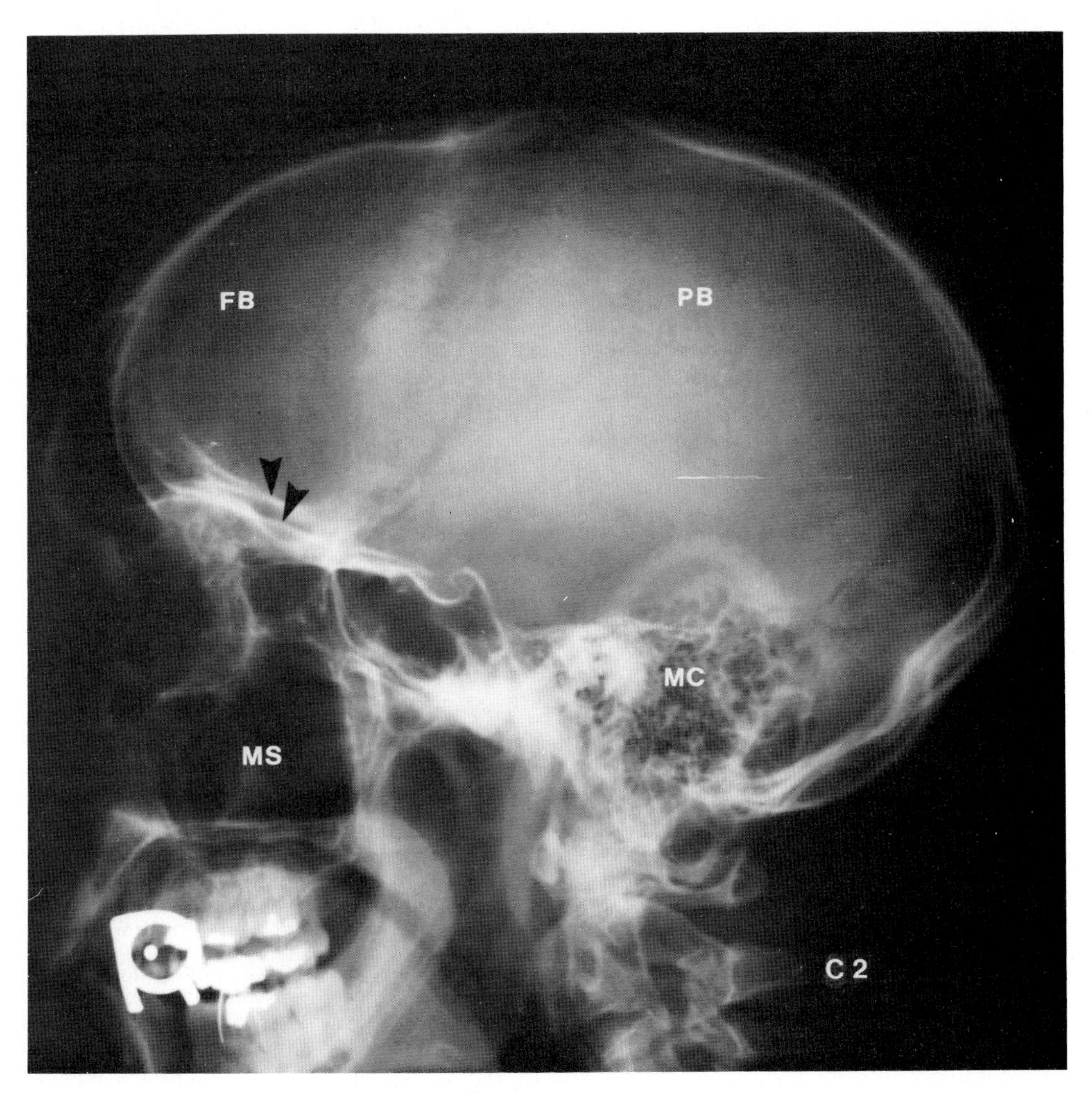

Figure 18. Lateral projection of the skull with arrows pointing to the roof of each orbit (compare with Fig. 19). *FB*, frontal bone; *PB*, parietal bone; *MC*, mastoid cells; *MS*, maxillary sinus. The right and left parietal bones are projected on each other. They can be seen separately if stereo films are made. *C2* is the spinous process of the second cervical vertebra in the neck.

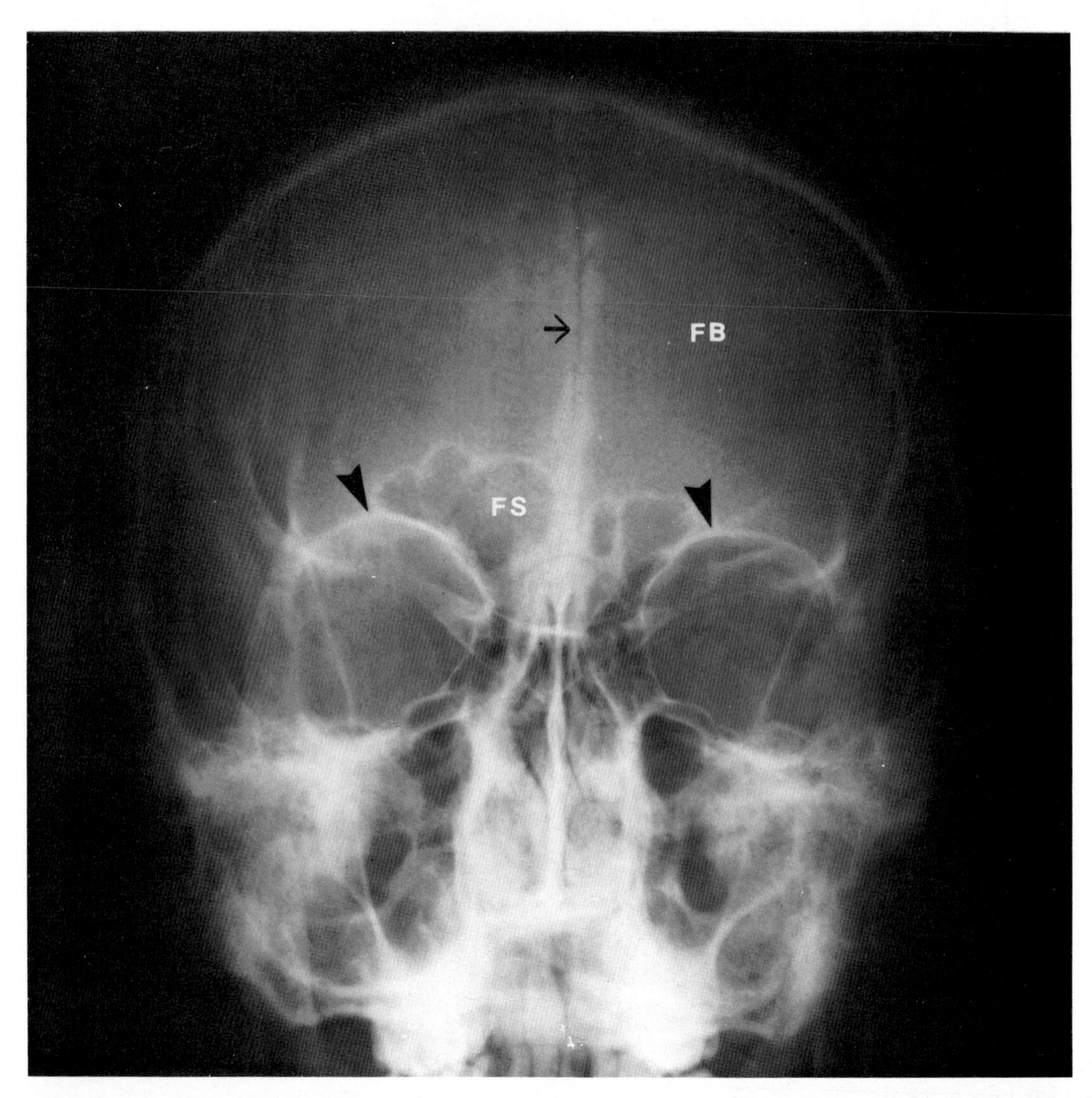

Figure 19. Frontal projection of the skull. The *arrowheads* point to the roofs of the orbits (eye sockets). *FS*, frontal sinus; *FB*, frontal bone. The *arrow* indicates a persistent frontal (metopic) suture between the two halves of the frontal bone. Usually union of the halves of the frontal bone begins during the 2nd year and the suture is usually not visible after the 8th year. The remains of this suture are visible in 1 to 2% of adult people (Fig. 7-1).

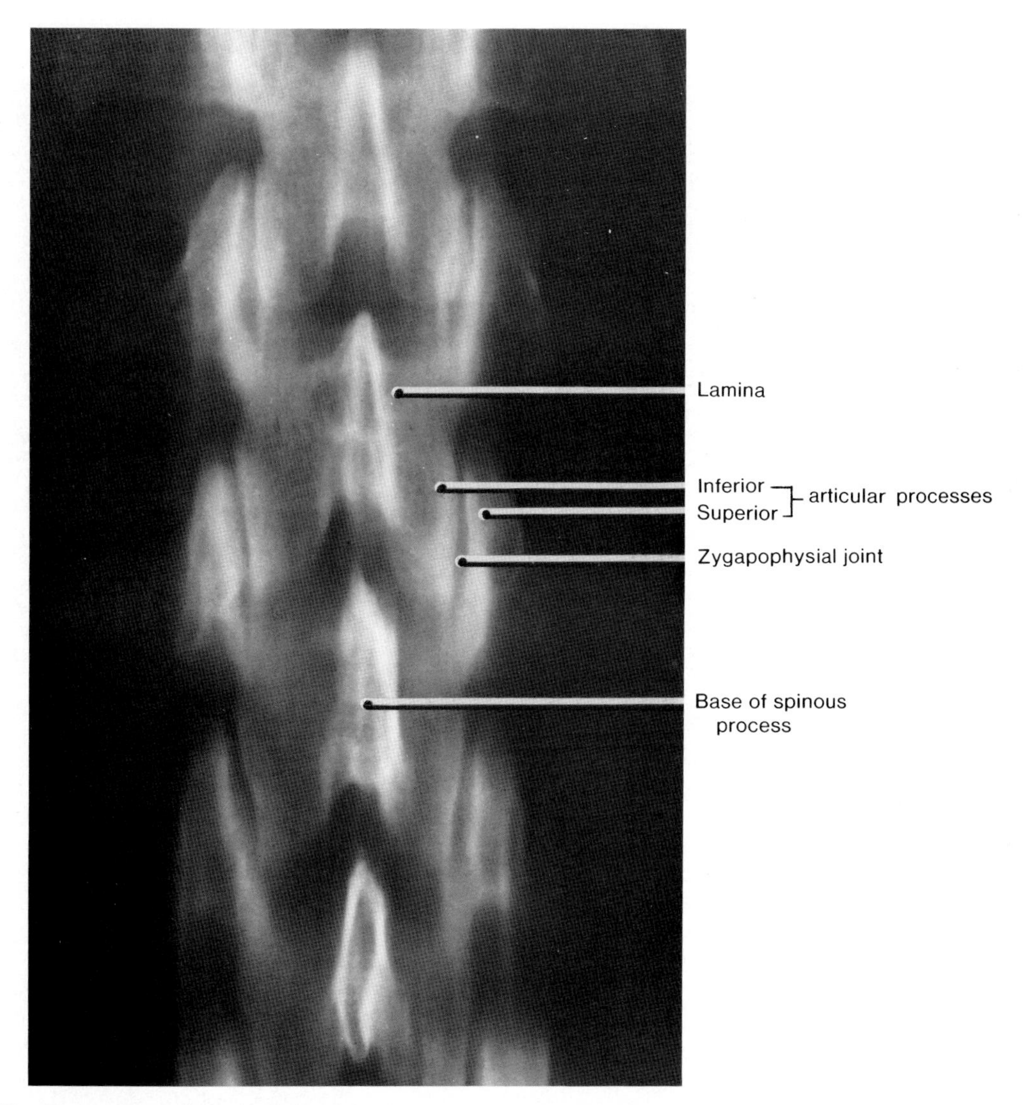

Figure 20. Tomogram of the arches of the lumbar vertebrae. To make the above tomogram, circular movement of the x-ray tube and x-ray film 180° out of phase with each other was made during the exposure. The images of the vertebral bodies are so blurred that they are invisible. Observe that the image of a vertebral arch has a frog-like appearance. The right and left laminae appear as the trunk of the frog, the superior articular processes resembling upper limbs and the inferior articular processes resembling lower limbs. The clarity of the zygapophysial joint "spaces" results from the articular cartilages on the opposing articular facets being of soft tissue density. The "spaces" are spaces in the image on the x-ray film, but in the patient they are cartilage filled areas between bones (the articular processes). See Chapter 5 if you are unfamiliar with the vertebral column (spine).

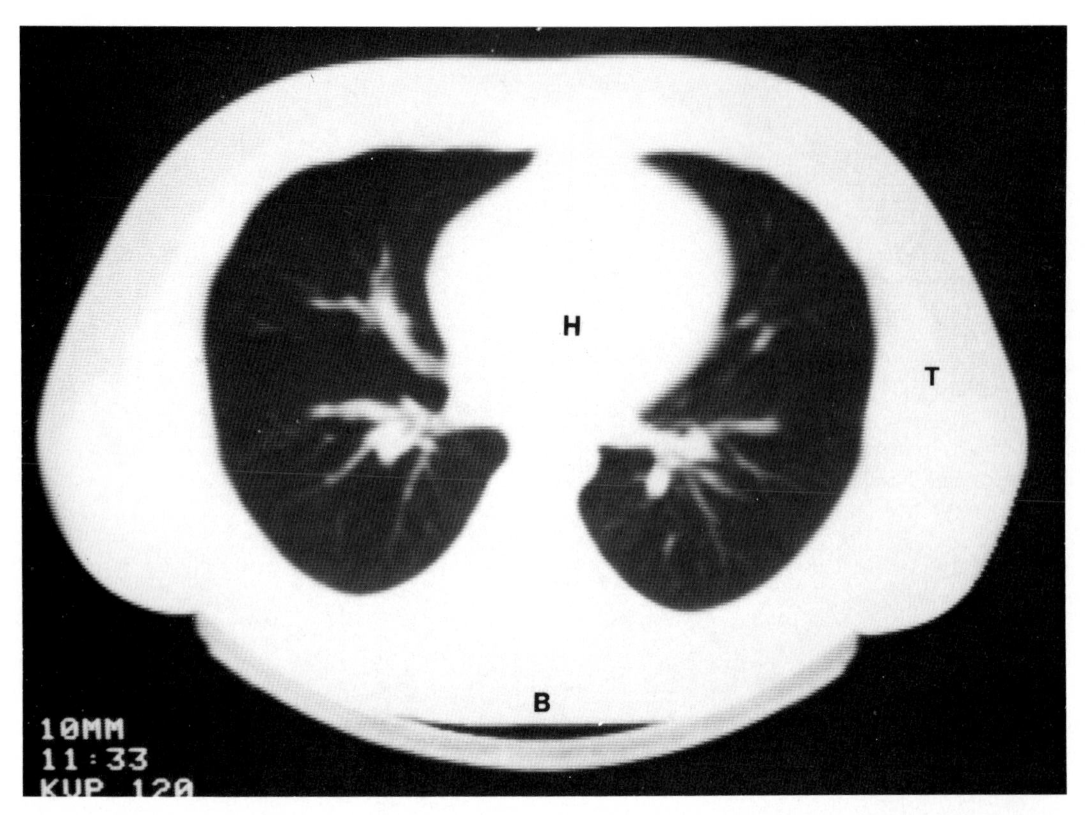

Figure 21. CT scan of the thorax showing the superior part of the heart (*H*) in the midline and bulging to each side. Observe the arteries supplying the lungs which appear dark grey. The thoracic wall (*T*) and the back (*B*) appear featureless. In this CT scan, detail within the chest wall is not visualized because the group of linear absorptions chosen for display was for maximum contrast between air-filled lung and soft tissue densities within it, such as the pulmonary arteries. (From General Electric Medical Systems, Ltd.)

and the x-ray film is anterior to the patient. *Most chest films are PA projections.*

A **left lateral projection** is made with the patient's left side close to the film, and a **right lateral projection** is made with the right side close to the film. A left anterior oblique or **LAO projection** is made with the patient's left anterior surface against the film, whereas a right posterior oblique or **RPO projection** is made with the patient's right posterior surface against the film. The x-rays pass along the same axis of the patient in both LAO and RPO projections, but they pass through it in different directions; therefore structures on different sides are magnified to different degrees.

Historical Note. *X-rays were discovered in 1895 by the Professor of Physics at the University of Würzburg, Germany,* **Dr. Wilhelm Conrad Röentgen**. He discovered them accidentally while investigating cathode rays produced in a Crookes tube, a partly evacuated glass bulb containing two electrodes, a cold cathode and a cold anode.

His paper, entitled "*On a New Kind of Rays*," published on December 28, 1895, about 6 weeks after his first observation, astounded both the scientific and nonscientific world. An x-ray of a hand showing images of the bones appeared on the front pages of newspapers all over the world. Many people began to call the new rays "Röentgen rays" in honor of the discoverer, but **Röentgen** modestly refused to use this designation, preferring the term x-rays. Even today you may hear *radiographs* referred to as *roentgenograms*. For his revolutionary discovery, **Dr. Röentgen was awarded the first Nobel prize in physics in 1901**.

The discovery of x-rays ushered in a new age in medicine. For the first time, **doctors were able to see the images of bones, joints, some organs, and foreign bodies inside the patient**. Soon, as contrast materials were developed (materials denser or lighter than body tissues), nearly every organ of the human body could be demonstrated in the living patient. Contrast materials allowed some physiological processes to be followed (*e.g.*, gastric peristalsis after ingesting a barium meal). *X-ray examinations thus became dynamic as well as static.*

BONES

Bone, or *osseous tissue*, is a rigid form of connective tissue that forms most of the skeleton. The adult

skeletal system or skeleton (G. dried) consists of over 200 bones which *form the supporting framework of the body.*

A few cartilages are also included in the skeletal system (*e.g.*, the **costal cartilages** connecting the anterior ends of the ribs to the *sternum* or breastbone). The junctions between skeletal components are called **joints**; most of them permit movement.

The skeletal system consists of two main parts: (1) **the axial skeleton** consisting of the skull, vertebral column, sternum, and ribs, and (2) **the appendicular skeleton** consisting of the pectoral (shoulder) and pelvic girdles and the limb bones.

The study of bones is called osteology (G. *osteon*, bone, + *logos*, study). Although the bones studied in the laboratory are lifeless and dry, owing to the removal of protein from them, *bones are living organs in the body* that change considerably as one gets older.

Like other organs, bones have blood vessels, lymph vessels, and nerves, and they may become diseased.

Osteomyelitis is an inflammation of the bone marrow and the adjacent bone. When broken or fractured, a bone heals. Unused bones *e.g.*, in a paralyzed limb, **atrophy** (*i.e.*, become thinner and weaker). Bone may be absorbed, as occurs following the loss or extraction of teeth. Bones also **hypertrophy** (*i.e.*, become thicker and stronger) when they have increased weight to support.

Bones from different people exhibit anatomical variations. They vary according to age, sex, physical characteristics, health, diet, race, and with different genetic and endocrinological conditions.

Anatomical variations are useful in identifying skeletal remains, one aspect of **forensic medicine** (the relation and application of medical facts to legal problems).

Living bones are plastic tissues containing organic and inorganic components. They consist essentially of intercellular material impregnated with mineral substances, mainly hydrated calcium phosphate, *i.e.*, $Ca_3(PO_4)_2$.

The collagen fibers in the intercellular material give the bones resilience and toughness, whereas the salt crystals in the form of tubes and rods give them hardness and some rigidity. When a bone is decalcified in the laboratory by submerging it in dilute acid for a few days, its salts are removed, but the organic material remains. The bone retains its shape, but it is so flexible that it can be tied into a knot (Fig 22). A bone that is burned also retains its shape, but its fibrous tissue is destroyed. As a result, the bone becomes brittle, inelastic, and crumbles easily.

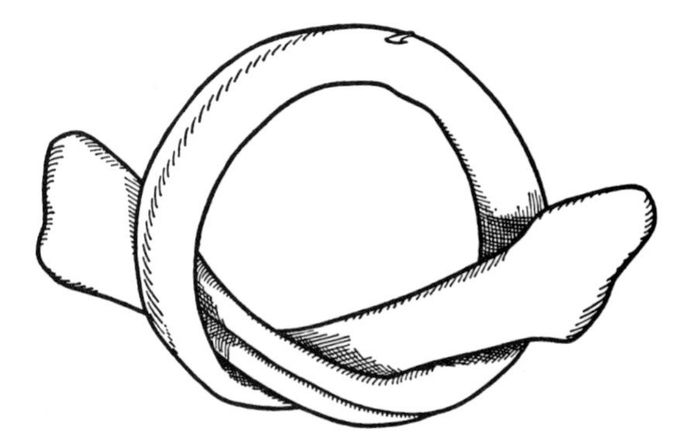

Figure 22. A decalcified fibula, a long slender bone in the leg that has been tied in a knot. The bone was decalcified by putting it in a dilute acid for a few days.

The relative amount of organic to inorganic matter in bones varies with age. Organic matter is greatest in childhood; hence the bones of children will bend somewhat.

In some metabolic disturbances such as **rickets** and **osteomalacia**, there is inadequate calcification of the matrix of bones. As calcium provides the hardness to bones, the uncalcified areas bend, particularly if they are weight-bearing bones. This results in progressive deformities such as knock knees (Fig. 23).

Although the diagnosis of rickets is suggested by clinical enlargement at the sites of the **epiphyseal cartilage plates** (Fig. 26), the diagnosis is confirmed by the typical radiographic changes that occur in the growing ends of the long bones and ribs in these patients.

Fractures are more common in children than in adults owing to the combination of their slender bones and carefree activities. Fortunately many of these breaks are hairline or green-stick fractures that are not so serious. In a green-stick fracture the bone breaks like a willow bough.

Epiphyseal cartilage plate fractures are serious because they may result in premature fusion of the diaphysis and the epiphysis (Fig. 26) with subsequent shortening of the bone; *e.g.*, premature fusion of a radial epiphysis results in progressive radial deviation of the hand as the ulna continues to grow. The existence of unfused epiphyses in young people can be helpful in treating them; *e.g.*, placing staples across the epiphyseal cartilage plate at the knee will arrest growth in the lower limb. It is the normal leg bones that are stapled to permit the bones of the short leg to catch up.

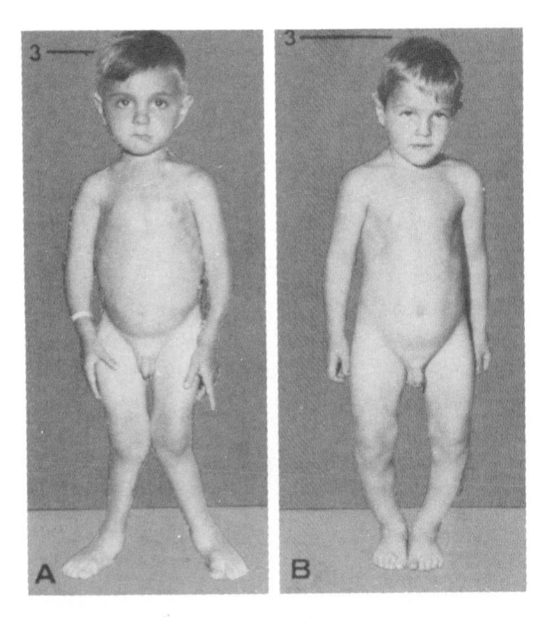

Figure 23. Photographs of boys with rickets. *A*, genu valgum (knock-knee) in a 5-year-old with vitamin D refractory rickets. Note the enlargement at the sites of epiphyseal cartilage plates (*e.g.*, ankles and knees). *B*, genu varum (bowleg) in a 4-year-old with a similar condition.

Fortunately fractures heal more rapidly in children than in adults. A femoral fracture occurring at birth is united in 3 weeks, whereas union takes up to 20 weeks in persons 20 years and older.

During old age both the organic and inorganic components of bone decrease, producing a condition called **osteoporosis** (G. *osteon*, bone, + *poros*, pore). There is a reduction in the quantity of bone (atrophy of skeletal tissue) and as a result, the bones of old people lose their elasticity and fracture easily. For example, elderly people may catch a foot on a slight projection while walking, feel and hear the neck of their femur (thigh bone) snap, and fall to the ground. *Fractures of the neck of the femur are especially common in elderly females* because osteoporosis is more severe in them than in elderly males.

TYPES OF BONE

There are *two main types* of bone, **spongy bone** (cancellous) and **compact bone** (dense), but there are no sharp boundaries between the two types because the differences between them depend upon the relative amount of solid matter and the number and size of the spaces in each of them.

All bones have an outer shell of compact bone around a central mass of cancellous bone, except where the latter is replaced by a marrow cavity or **medullary cavity** (Fig. 24), or an air space, *e.g.*, the **paranasal sinuses** (Fig. 18).

Spongy bone consists of slender, irregular *trabeculae* (bars) of compact bone which branch and unite with one another to form intercommunicating spaces that are filled with **bone marrow** (Fig. 24). The trabeculae of the spongy bone are arranged in lines of pressure and tension (Fig. 31). *In the adult there are two types of bone marrow, red and yellow.*

Red marrow is active in blood formation (hematopoiesis, G. *hemato*, blood, + *poiein*, to form), whereas **yellow marrow is mainly inert and fatty.** In most long bones there is a **medullary cavity** in the body (shaft) of the bone that contains yellow marrow in adult life. In *yellow bone marrow*, most of the hemopoietic tissue has been replaced by fat.

Compact bone appears solid except for microscopic spaces (Fig. 24). Its crystalline structure gives it hardness and rigidity and makes it opaque to x-rays.

Classification of Bones. Bones may be classified regionally as *axial bones* (skull, vertebrae, ribs, and sternum) or as *appendicular bones* (upper and lower limb bones and those associated with them).

Bones are also classified according to their shape.

1. **Long bones** (Figs. 24 and 25) are tubular in shape and have a body (shaft) and two ends (extremities) which are either concave or convex. The length of long bones is greater than their breadth, even though some long bones are short (*e.g.*, in the fingers and toes).

   The ends of long bones articulate with other bones; thus they are enlarged, smooth, and covered with hyaline cartilage. Usually the *body* of a long bone is hollow and typically has three borders separating its three surfaces.
2. **Short bones** are cuboidal in shape and are found only in the foot and wrist, *e.g.*, the carpal or wrist bones. They have six surfaces, four or less of which are articular and two or more are for the attachment of tendons and ligaments and for the entry of blood vessels.
3. **Flat bones** consist of two plates of compact bone with spongy bone and marrow between them, *e.g.*, the bones of the *calvaria* (skullcap), the sternum, and the scapula (except for the thin part of this bone). The marrow space between the tables (external and internal cortices) of flat bones in the skull is known as **diploë** (G. double).

   Most flat bones help to form the walls of cavities (*e.g.*, the cranial cavity); hence, most of them are gently curved rather than flat. In early life a flat bone consists of a thin plate of compact bone, but marrow (*e.g.*, the diploë) appears in it during childhood, resulting in compact plates on each side of the medullary cavity.

The outer table of bone of the **calvaria** in living persons is somewhat resilient. This elasticity, especially in infants, tends to prevent many blows to the head from producing skull fractures. In older persons a **blow on the head** may fracture the outer table of bone.

Fracture of the inner table of bone, involving the brain, is more serious. For more clinically oriented comments on skull fractures, see THE SKULL in Chapter 7.

4. **Irregular bones** have various shapes (*e.g.*, the facial bones and the vertebrae). The bodies of vertebrae have some features of long bones.
5. **Pneumatic bones** contain air cells or sinuses *e.g.*, *mastoid air cells* in the mastoid part of the temporal bone (Fig. 18) and the **paranasal sinuses.** Outgrowths (evaginations) of the mucous membrane of the middle ear and of the nasal cavities invade the medullary cavity, producing the air cells and sinuses, respectively.
6. **Sesamoid bones** are round or oval nodules of bone that develop in certain tendons (*e.g.*, the **patella** or kneecap in the quadriceps femoris tendon, Fig. 4-109, and the **pisiform** in the tendon of the flexor carpi ulnaris muscle, Fig. 6-118).

   They were called *sesamoid bones* because of their resemblance to **sesame seeds**. *Sesamoid bones are commonly found where tendons cross the ends of long bones* in the limbs. *They protect the tendon from excessive wear and they change the angle of the tendon* as it passes to its insertion. This results in a greater mechanical advantage at the joint. The articular surface of a sesamoid bone is covered with articular cartilage, whereas the rest of it is buried in the tendon.
7. **Accessory bones** develop when an additional ossification center appears giving rise to a bone, or one of the usual centers fails to fuse with the main bone. The separated part of the bone gives the appearance of an extra or *supernumerary bone*.

   Accessory bones are common in the foot and it is important to know about them so that they will not be mistaken for bone chips or fractures in radiographs.
8. **Heterotopic bones** are those that do not belong to the main skeleton, but may develop in certain soft tissues and organs as a result of disease. This type of bone may form in scars and chronic inflammation, characteristic of tuberculosis, may produce bony tissue in the lung.

BONE MARKINGS

The surface of bones is not smooth and glossy or even in contour, except over areas covered by cartilage and where tendons, blood vessels, and nerves move in grooves (*e.g.*, the **intertubercular groove** in the head of the humerus and the **groove for the radial nerve** in its body, Fig. 25*B*).

Bones display a variety of bumps, depressions, and holes. Markings appear on dried bones wherever tendons, ligaments, and fascia were attached. The attachment of the fleshy fibers of a muscle make no markings on a bone.

Bone markings start to become prominent during puberty (12 to 16 years) and become progressively more marked as adulthood occurs. The surface features of bones are given names to help distinguish them.

Elevations (Fig. 25). The various kinds of elevation on bones are listed in order of prominence. *Examine each type on a skeleton.*

A **linear elevation** is referred to as a **line** (*e.g.*, the *superior nuchal line* of the occipital bone, Figs. 5-48 and 7-3) or as a **ridge** (*e.g.*, the *medial supracondylar ridge*, Fig. 6-1). Very prominent ridges are called **crests** (*e.g.*, *iliac crest*, Fig. 4-1 and *pubic crest*, Fig. 3-42*B*).

A **rounded elevation** (Fig. 25) is called (1) a **tubercle** (small raised eminence); (2) a **protuberance** (swelling or knob, *e.g.*, *external occipital protuberance*, Fig. 7-3); (3) a **trochanter** (large blunt elevation, *e.g.*, the *greater trochanter* of the femur Fig. 4-1); (4) a **tuberosity or tuber** (large elevation); and (5) a **malleolus** (a hammerhead-like elevation).

A **sharp elevation** or projecting part is called a **spine** (L. *spina*, a thorn), *e.g.*, the *anterior superior iliac spine* (Fig. 4-1), or a **process**, *e.g.*, the *spinous process of a vertebra* (Fig. 5-12).

Facets (F. little faces) are small, smooth, flat areas or surfaces of a bone, especially where it articulates with another bone (Fig. 4-109). **Articular facets** are covered with hyaline cartilage (*e.g.*, the facets of a vertebra, Fig. 5-19).

A **rounded articular area** of a bone is called a **head** (*e.g.*, the *head of the humerus*, Fig. 25) or a **condyle** (G. knuckle), *e.g.*, the *lateral condyle* of the femur (Fig. 4-1). An **epicondyle** (G. *epi*, upon) is a prominent process just proximal to a condyle (Figs. 25 and 4-1).

Depressions (Fig. 25). Small hollows in bones are described as **fossae** (L. pits), whereas long narrow depressions are referred to as **grooves** (L. *sulci*, furrows). An indentation at the edge of a bone is called a notch, *e.g.*, the *acetabular notch* (Fig. 4-4).

Foramina and Canals. When a notch is bridged by a ligament or bone to form a perforation or hole, it is called a **foramen** (*e.g.*, the *foramen magnum*, Fig. 7-7). A foramen that has length is called a **canal** (*e.g.*, the *facial canal*, Fig. 7-168). A canal has an **orifice** (L. an opening) at each end. A **meatus** (L. a passage) is a canal that enters a structure, but does not pass through it, *e.g.*, the *external acoustic meatus* or ear canal (Fig. 7-159).

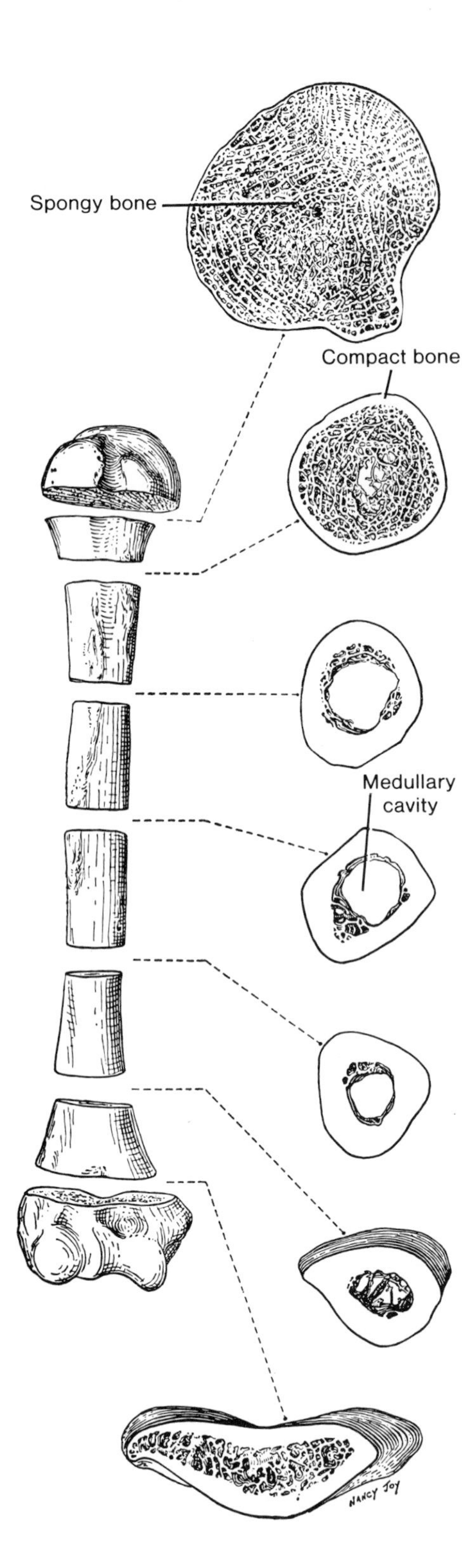

Figure 24. Transverse sections of a dried long bone, the humerus of the arm. Observe that it consists of an outer shell of compact bone, an inner core of spongy bone, and a medullary cavity. These sections demonstrate the variations in thickness of the compact and spongy bone at different levels and at different parts of the same level. The extent of the medullary cavity (marrow cavity) is also demonstrated. In adults the medullary cavity of a long bone contains *yellow marrow* (chiefly fat). Long bones are particularly spongy at their expanded ends, and here the compact bone is very thin and shell-like. The interstices of spongy bone are filled with *red marrow* which produces red blood cells.

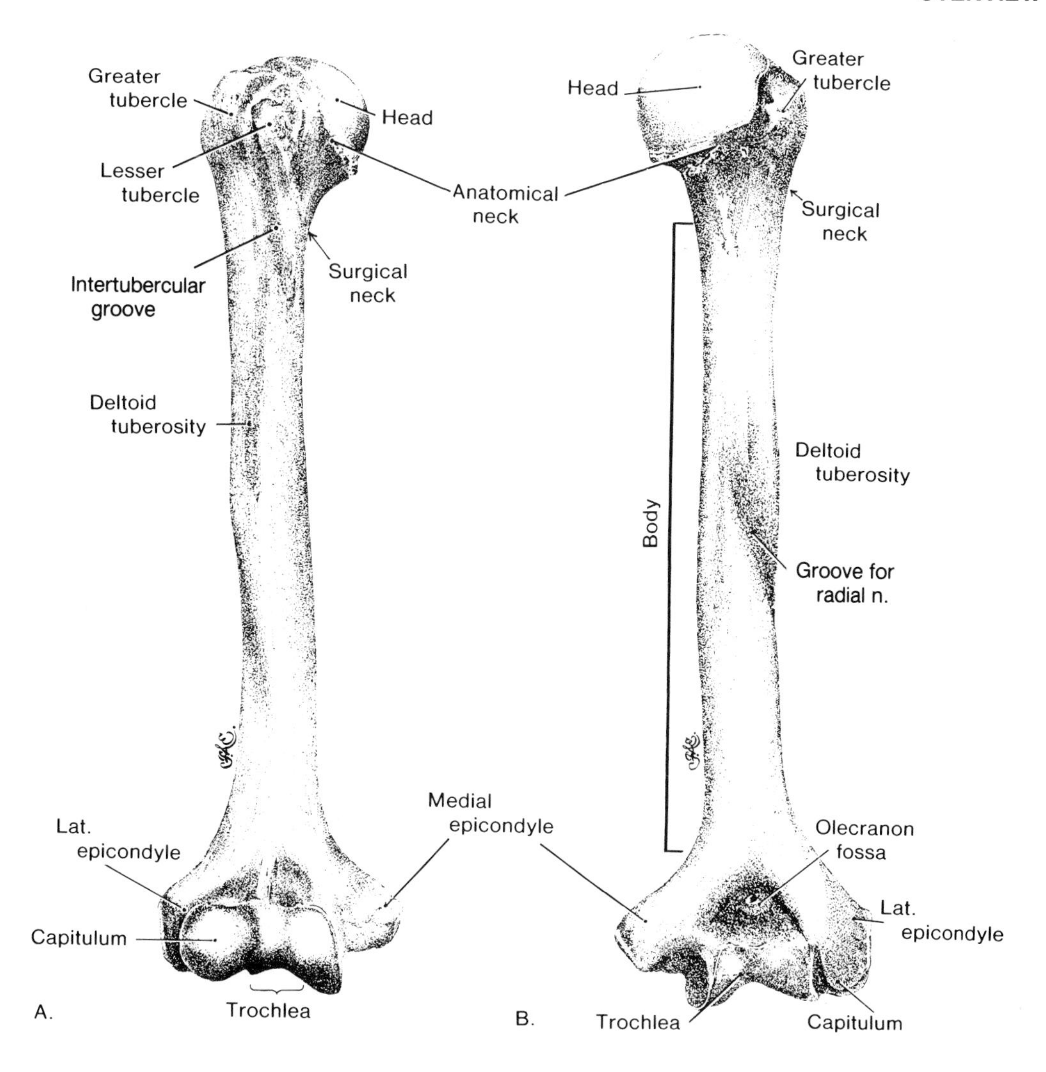

Figure 25. Drawings of the right humerus, *a typical long bone*, showing its features. Note that the articular condyle (G. knuckle) is divided by a ridge into two parts, the capitulum and the trochlea (L. pulley). *A*, anterior view. *B*, posterior view. The bodies (shafts) of long bones are hollow (Fig. 24).

DEVELOPMENT OF BONES

Bones develop from condensations of mesenchyme (embryonic connective tissue). The *mesenchymal model of a bone* (Fig. 26*A*) which forms during the embryonic period may undergo direct ossification, called **intramembranous ossification** (membranous bone formation), or it may be replaced by a *cartilage model* (Fig. 26*B*), that later becomes ossified by **intracartilaginous ossification** (endochondral bone formation).

Simply, bone replaces membrane or cartilage. The process of ossification is similar in each case and the final histological structure of the bone is identical.

Intramembranous ossification occurs rapidly and takes place in bones that are urgently required for protection (*e.g.*, the flat bones of the calvaria or skullcap). *Intracartilaginous ossification*, occurring in most bones of the skeleton, is a much slower process.

Development of Long Bones (Fig. 26). The first indication of ossification in the cartilaginous model of a long bone is visible near the center of the future body or shaft. This is called the **primary center of ossification** (Fig. 26*C*). Primary centers appear at different times in different developing bones, but *most primary centers appear between the 7th and 12th weeks of prenatal life.* Virtually all primary centers are present by birth. By this time, ossification from the pri-

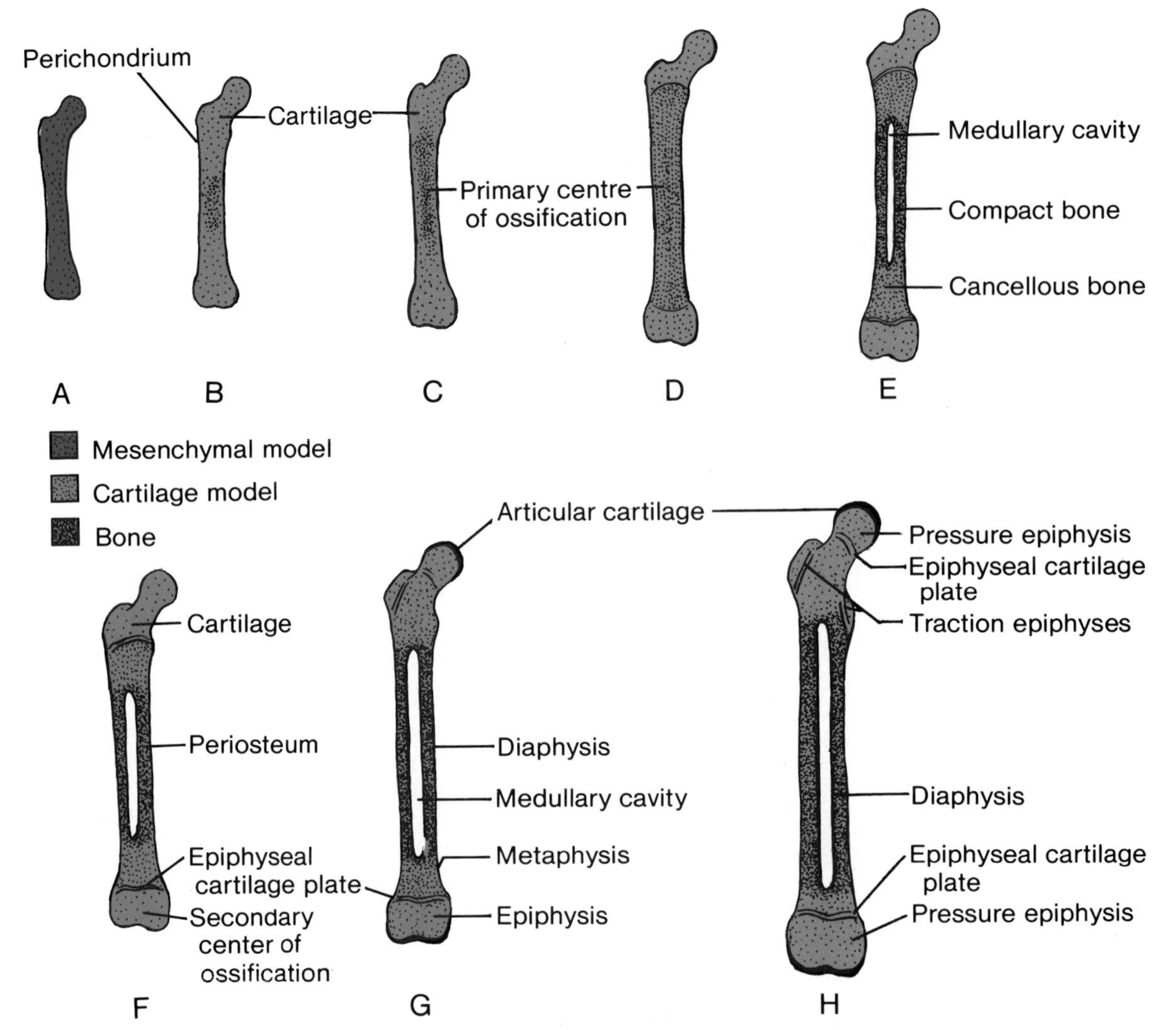

Figure 26. Drawings illustrating the development of the femur, a long bone. *A*, embryo of 5 weeks. *Mesenchymal model* of the bone. *B*, embryo of 6 weeks. Early development of the *cartilaginous model* of the bone. The model grows partly from within (interstitial growth) and partly from cells derived from the perichondrium (appositional growth). *C*, embryo of 8 weeks. Bone first appears in the body (shaft) of the bone (*primary center of ossification*). The process of replacement of cartilage by bone is called *endochondral ossification. D*, fetus of 12 weeks. From the primary center in the body or *diaphysis* of the bone, endochondral ossification advances toward each end of the cartilaginous model. *E*, fetus of 6 months. Resorption of the central part of the bone results in the formation of a *medullary cavity. F*, newborn infant. The diaphysis is ossified, but the periosteum is laying down bone by *intramembranous ossification*. Secondary centers of ossification appear in the cartilaginous epiphyses of some long bones (*e.g.*, the femur). *G*, young child. *Secondary centers of ossification* appear in most epiphyses during childhood. Each epiphysis is separated from the diaphysis by a *epiphyseal cartilage plate. H*, older child. The *articular cartilage* is the growth plate for the epiphysis, whereas the epiphyseal cartilage plate provides growth in the length of the diaphysis. The part of the diaphysis adjacent to the epiphyseal cartilage plate is called the *metaphysis*.

mary center has almost reached the ends of the cartilage model of the long bone (Fig. 26*F*).

The part of a bone ossified from a primary center, the body or shaft, is called the **diaphysis** (G. a growing between).

Around birth additional ossification centers may appear in the cartilaginous ends of a long bone. These are referred to as the epiphyses or **secondary centers of ossification** (Fig. 26*F*).

Most secondary centers of ossification appear after birth. The parts of a bone ossified from secondary centers are called **epiphyses** (G. *epi*, upon, + *physis*, growth). The epiphyses or secondary ossification centers of the bones at the knee (Fig. 4-1) are the first to appear. They may be present at birth.

The cartilaginous epiphyses undergo much the same changes that occur in the diaphysis. As a result, the body of the bone becomes capped at each end by bone, the epiphyses, that develop from the secondary centers of ossification.

The part of the diaphysis nearest the epiphysis is referred to as the **metaphysis** (G. *meta*, beyond, + *physis*, growth).

The diaphysis grows in length by proliferation of cartilage at the metaphysis. To enable growth in length to continue until the adult length of a bone is attained, the bone formed from the primary center in the diaphysis does not fuse with that formed from the secondary centers in the epiphyses until the adult size of the bone is reached.

During the growth of a bone, a plate of cartilage known as the *growth plate* or **epiphyseal cartilage plate,** intervenes between the diaphysis and the epiphysis (Figs. 27 and 28). For brevity, it is often called the epiphyseal plate (disc).

The diaphysis consists of a hollow tube of compact bone surrounding the medullary cavity, whereas the epiphyses and metaphyses consist of spongy bone covered by a thin layer of compact bone (Fig. 24). The compact bone over the articular surfaces of the epiphyses is soon covered with hyaline cartilage called **articular cartilage** (Fig. 26*G* and *H*).

During the first 2 postnatal years, secondary ossification centers appear in the epiphyses that are exposed to pressure (*e.g.*, at the knee and hip, Fig. 26*H*). These centers, often referred to as **pressure epiphyses,** are located at the ends of long bones where they are subjected to pressure from opposing bones at the joint which they form.

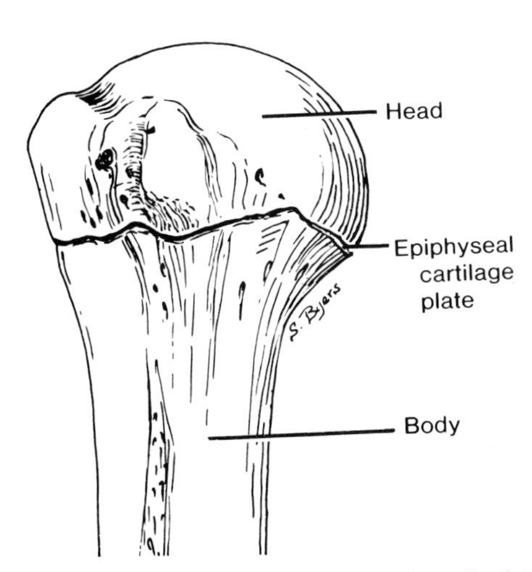

Figure 27. Drawing of the proximal end of the humerus of a 19-year-old man, just before fusion of the epiphysis with the diaphysis. Observe the thin layer of cartilage (epiphyseal cartilage plate) which is responsible for allowing the body or shaft of the bone to lengthen until full growth is obtained. The head unites with the body at 20 to 22 years in males (18 to 20 years in females). In other words, the epiphysis becomes united to the diaphysis by bone when the epiphyseal cartilage plate becomes ossified.

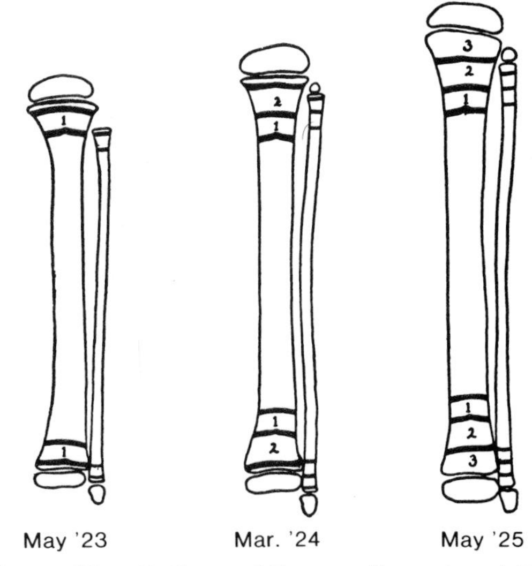

Figure 28. Outlines of three radiographs of the leg bones of a young girl taken over a period of 2 years. Observe that the three lines of arrested growth, denoting three successive illnesses, remain equidistant.

Some secondary ossification centers ossify parts of bone associated with the attachment of muscles and strong tendons. These centers are often referred to as **traction epiphyses** (*e.g.*, the tubercles of the humerus, Fig. 25, and the tuberosities of the femur, Fig. 26*H*). These epiphyses are subjected to traction rather than pressure.

The epiphyseal cartilage plates are eventually replaced by bone development at each of its two sides, diaphyseal and epiphyseal (Fig. 27). When this occurs, growth of the bone ceases and the diaphysis is fused with the epiphyses by bony union or **synostosis** (G. *syn*, with, together + *osteon*, bone, + *osis*, condition).

The bone formed at the site of the epiphyseal cartilage plate is particularly dense and is still recognizable on the radiographs of children and teenagers (Figs. 29 and 4-121). Knowing this prevents confusion with fracture lines.

In general, the epiphysis of a long bone whose center of ossification is the last to appear, is the first one to fuse with the diaphysis. When an epiphysis forms from more than one center (*e.g.*, the proximal end of the humerus), the centers fuse with each other before union of the epiphysis with the diaphysis.

The changes in developing bones are clinically important. Doctors and dentists, especially radiologists, pediatricians, orthodontists, and orthopaedic surgeons, must be knowledgeable about bone growth.

The time of appearance of the various epiphyses varies with chronologic age. As good reference tables are available, it is not useful to memorize the dates of appearance and disappearance of the ossification centers for all bones.

A radiologist determines the bone age of a person by studying the ossification centers. Two criteria are used: (1) *the appearance of calcified material in the diaphysis and/or epiphyses.* The time of its appearance is specific for each epiphysis and diaphysis of each bone for each sex; and (2) *the disappearance of the dark line representing the epiphyseal cartilage plate.* This indicates that the epiphysis has fused to the diaphysis and occurs at specific times for each epiphysis.

Fusion of the epiphyses with the diaphysis occurs 1 to 2 years earlier in females than in males. Determination of bone age is often used in determining the approximate age of human skeletal remains in medicolegal cases.

Some diseases speed up and others slow down ossification times as compared to the chronological age of the individual. The growing skeleton is sensitive to relatively slight and transient illnesses and to periods of **malnutrition.**

Proliferation of cartilage at the metaphysis slows down during starvation and illness, but degeneration of cartilage cells in the columns continues, producing a dense line of provisional calcification which later becomes bone with thicker trabeculae, called **lines of arrested growth** (Fig. 28).

Without a basic knowledge of bone growth and the appearance of bones on radiographs at various ages, *an epiphyseal cartilage plate could be mistaken for a fracture,* and separation of an epiphysis could be interpreted as normal (Fig. 29). If you know the age of a patient and the location of the epiphyses, these errors can be avoided, especially if you note that the edges of the diaphysis and epiphysis are smoothly curved in the region of the epiphyseal cartilage. A fracture leaves a sharp, often uneven edge of bone. An injury that causes a fracture in an adult may cause displacement of an epiphysis in a young person (Fig. 29).

Development of Short Bones. The development of short bones is similar to that of the primary center of long bones and only one bone, the calcaneus (Fig. 4-61), develops a secondary center of ossification.

Blood Supply of Bones (Fig. 30). Bones are richly supplied with blood vessels that pass into them from the **periosteum**, the fibrous connective tissue membrane investing them.

The **periosteal arteries** enter the body at numerous points and are responsible for its nourishment. Hence a *bone from which the periosteum has been removed will die.*

Near the center of the body of a long bone, a **nutrient artery** passes obliquely through the compact bone and reaches the spongy bone and the marrow (Figs. 30 and 4-83).

Some pressure epiphyses are largely covered by hyaline **articular cartilage** (Fig. 30). They receive their blood supply from the region of the epiphyseal cartilage plate. These epiphyses (*e.g.,* head of the femur) are almost completely covered by articular cartilage and receive their blood supply from vessels that penetrate just outside the edge of the articular cartilage.

Loss of blood supply to an epiphysis or to other parts of a bone results in *death of bone tissue,* a condition referred to as **avascular necrosis** (ischemic or aseptic necrosis) of bone. After every fracture, adjacent minute areas of bone undergo avascular necrosis. In a few fractures, there may be avascular necrosis of a large fragment of bone if its blood supply has been cut off. A number of clinical disorders of epiphyses in children result from avascular necrosis of unknown etiology. They are referred to as **osteochondroses** and usually involve a pressure epiphysis at the end of a long bone (Fig. 26*F*).

Nerve Supply of Bones. The periosteum of bones is rich in sensory nerves, called **periosteal nerves.** *This explains why pain from injured bones is usually severe.* The nerves that accompany the arteries into the bones are probably **vasomotor** (*i.e.,* ones causing constriction or dilation of the nutrient vessels).

ARCHITECTURE OF BONES

The structure of bone varies according to its function. In long bones designed for rigidity and providing attachments for muscles and ligaments, compact bone is relatively greatest in amount near the middle of the body (Fig. 24), where it is liable to buckle. The compact bone of the body provides strength architecturally for weight bearing. In addition, as described previously, long bones have elevations (lines, ridges, crests, tubercles, and tuberosities) that serve as buttresses wherever heavy muscles attach (Figs. 25 and 4-19).

Living bones have some elasticity (flexibility) and great rigidity (hardness). Their elasticity results from their organic matter (fibrous tissue) and their rigidity results from their lamellae and tubes of inorganic calcium phosphate. The salts, representing about 60% of the weight of a bone, are deposited in the matrix of collagenic fibers.

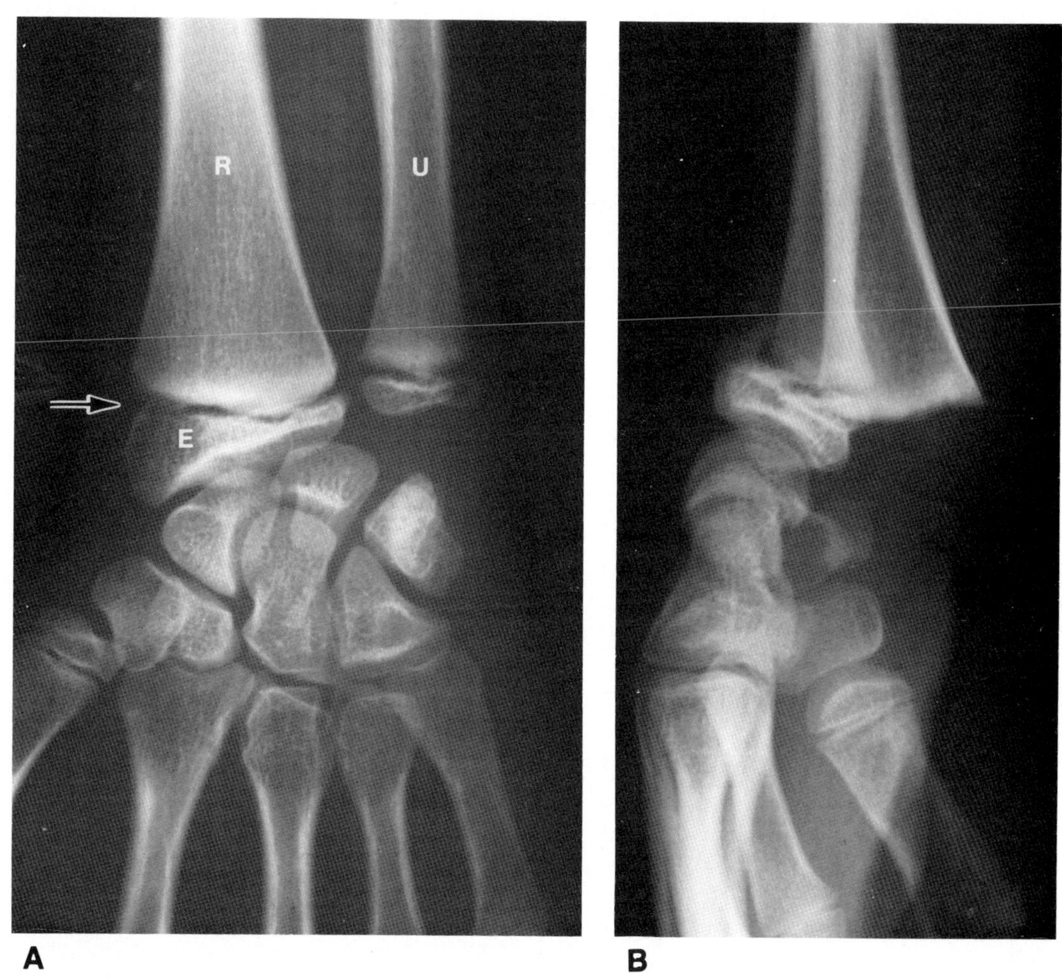

Figure 29. Radiographs of the wrist of a child. *R*, radius; *U*, ulna; *E*, epiphysis at the distal end of the radius. The *arrow* indicates the site of the epiphyseal cartilage plate. *A*, frontal projection of the left wrist showing the apparently normal position of the epiphyses at the distal ends of the radius and ulna. *B*, lateral projection of the right wrist showing dorsal displacement of the epiphysis at the distal end of the radius. As growth of this bone depends on a normally functioning epiphyseal cartilage plate, treatment involves replacing the epiphysis in its normal position.

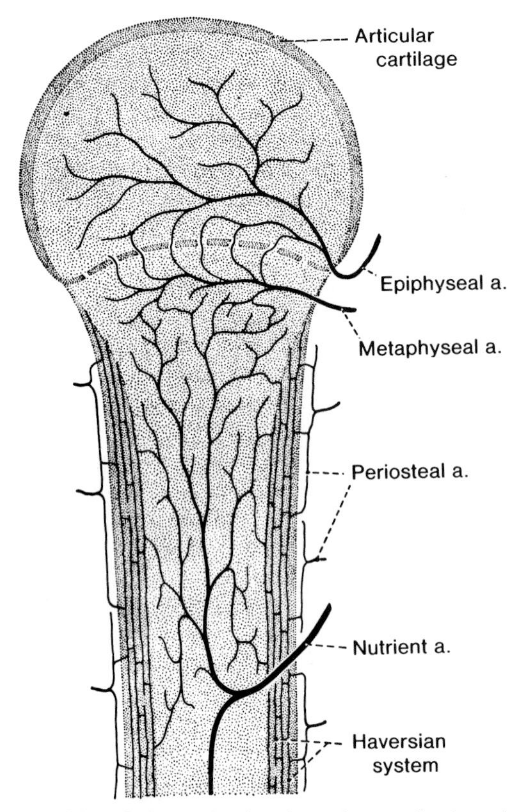

Figure 30. Schematic drawing of part of a long bone illustrating its blood supply. One or two nutrient arteries pierce the body (shaft) of a long bone obliquely through one or two nutrient foramina which lead into nutrient canals.

Bones are like hardwood in resisting tension and like concrete in resisting compression.

Inside the outer shell of **compact bone**, particularly at the ends of long bones, there is **spongy bone** that has a trellis-like appearance (Figs. 24 and 31). Spongy bone is not laid down in a haphazard fashion, but is composed of tubes and lamellae (L. plates) which are arranged like struts along lines of pressure and tension (Figs. 31 and 4-99).

The architecture of the bony trabeculae is distinctive for each person, a fact of value in identifying skeletal remains, an important part of forensic medicine.

Functions of Bones. The main functions of bones are to provide:

1. **Protection** by forming the rigid walls of cavities (*e.g.*, the *cranial cavity*) that contain vital structures (*e.g.*, the brain).
2. **Support** (*e.g.*, the rigid framework for the body).
3. **A mechanical basis for movement** by providing attachments for muscles and serving as levers for ones that produce the movements permitted by joints.
4. **Formation of blood cells.** The red bone marrow in the ends of long bones, in the sternum and ribs, in vertebrae, and in the diploë of the flat bones of the skull are the sites for the development of red blood cells, some lymphocytes, granulocytic white cells, and platelets of the blood.
5. **Storage of salts.** The calcium, phosphorus, and magnesium salts in bones provide a mineral reservoir for the body.

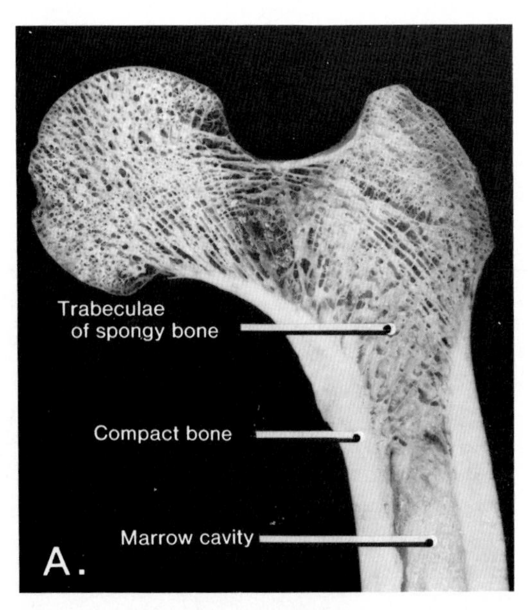

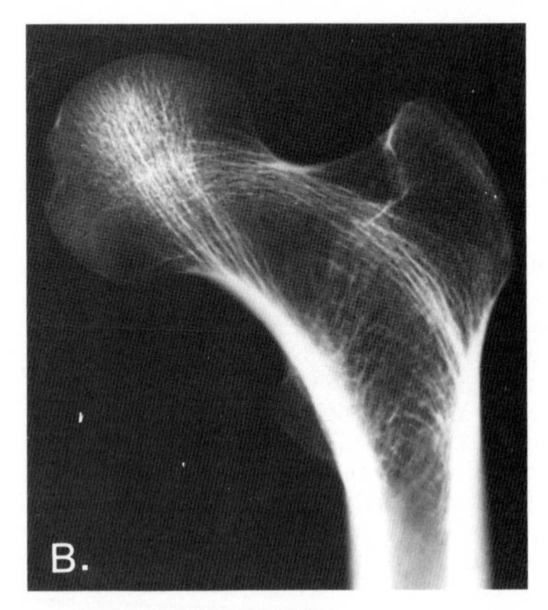

Figure 31. *A*, photograph of a coronal section of the proximal end of the femur showing its bony trabeculae and lines of pressure and tension. *B*, radiograph of this bone showing the same lines. In adults the medullary or marrow cavity contains *yellow marrow* (chiefly fat).

JOINTS

The **articular system** consists of the *articulations or joints* where two or more bones are related to one another at their region of contact. The study of joints is called **arthrology** (G. *arthros*, joints, + *logos*, study).

Joints are classified according to the type of material holding the bones together (*e.g.*, fibrous joints, cartilaginous joints, and synovial joints).

FIBROUS JOINTS

The bones involved in these articulations are *united by fibrous tissue*. The amount of movement permitted at the joint depends on the length of the fibers uniting the bones.

Sutures (Fig. 32*A* and *B*). The bones are separated yet held together by a thin layer of fibrous tissue. The union is extremely tight and there is little or no movement between the bones.

Sutures occur only in the skull; hence they are sometimes referred to as "*skull type*" *joints*. The edges of the bones may overlap (**squamous type**) or interlock in a jigsaw fashion (**serrate type**).

In the skull of a newborn infant, the growing bones of the skullcap do not make full contact with each other (Fig. 33). At places where contact does not occur, the sutures are wide areas of fibrous tissue known as fonticuli or **fontanelles**. The terms *fonticuli* (L.) and *fontanelles* (Fr.) mean "little springs or fountains". Probably they were given this name because in earlier times openings may have been made in the skull at these points in infants with bulging fontanelles resulting from high intracranial pressure. In these cases, the spurting cerebrospinal fluid (CSF) and blood that would well out probably reminded them of a spring of water.

The most prominent fontanelle is the anterior fontanelle, which lay people call the *soft spot*. The separation of the bones at the sutures and fontanelles of the newborn skull allow them to overlap each other during birth, permitting the infant's head to pass through the birth canal. *The anterior fontanelle is not usually present after 18 to 24 months* (*i.e.*, it is the same width as the sutures of the skull). Union of the bones at the **pterion** (Fig. 7-9), located at the site of the anterolateral fontanelle (Fig. 33*B*), has taken place by 6 years in about 50% of children.

Fusion of the bones across the suture lines (**synostosis**) begins on the internal aspect of the calvaria or skullcap during the early 20s and progresses throughout life (Fig. 7-7). Nearly all sutures of the skull are obliterated in very old people (Fig. 7-1).

Syndesmosis (G. *syndesmos*, ligament). In this type of fibrous joint, *the two bones are united by a sheet of fibrous tissue*. The tissue may be a ligament or an interosseous fibrous membrane; *e.g.*, the interosseous border of the radius is attached to the interosseous border of the ulna by an **interosseous membrane** (Fig. 32*B*).

In syndesmoses, slight to considerable movement can be achieved. The degree of movement depends upon the distance between the bones and the degree of flexibility of the uniting fibrous tissue. The interosseous membrane between the radius and ulna in the forearm is broad enough and sufficiently flexible to allow considerable movement, such as occurs during pronation and supination of the forearm (Fig. 16).

CARTILAGINOUS JOINTS

The bones involved in these articulations are *united by cartilage*.

Primary Cartilaginous Joints (Synchondroses). The bones in these joints are united by **hyaline cartilage**, which permits slight bending during early life.

Synchondroses usually represent temporary conditions, e.g., during the period of endochondral development of a long bone. As previously described, an **epiphyseal cartilage plate** separates the ends (epiphyses) and body (diaphysis) of a long bone (Fig. 32*C*).

A synchondrosis type of cartilaginous joint permits growth in the length of the bone. When full growth is achieved, the cartilage is converted into bone and the epiphysis fuses with the diaphysis; *i.e.*, **a synchondrosis is converted into a synostosis.**

Other synchondroses are permanent, *e.g.*, where the costal cartilage of the first rib joins it to the manubrium of the sternum (Fig. 32*C*).

Secondary Cartilaginous Joints (Symphyses). The articular surfaces of the bones in these joints are covered by **hyaline cartilage** and these cartilaginous surfaces are united by fibrous tissue and/or fibrocartilage (Fig. 32*C*).

Symphyses are strong, slightly movable joints. The *anterior intervertebral joints* with their *intervertebral discs* are classified as symphyses (Fig. 5-30). They are designed for strength and shock absorption. The bodies of vertebrae are bound together by longitudinal ligaments and the *anuli fibrosi* of the **intervertebral discs** (Fig. 5-28). Cumulatively, these fibrocartilaginous discs give considerable flexibility to the vertebral column (Fig. 9*A*).

Other examples of symphyses are the **symphysis pubis** between the bodies of the pubic bones (Fig. 3-7), and the **manubriosternal joint** between the manubrium and the body of the sternum (Fig. 1-84).

During pregnancy the symphysis pubis and other joints of the pelvis undergo changes that

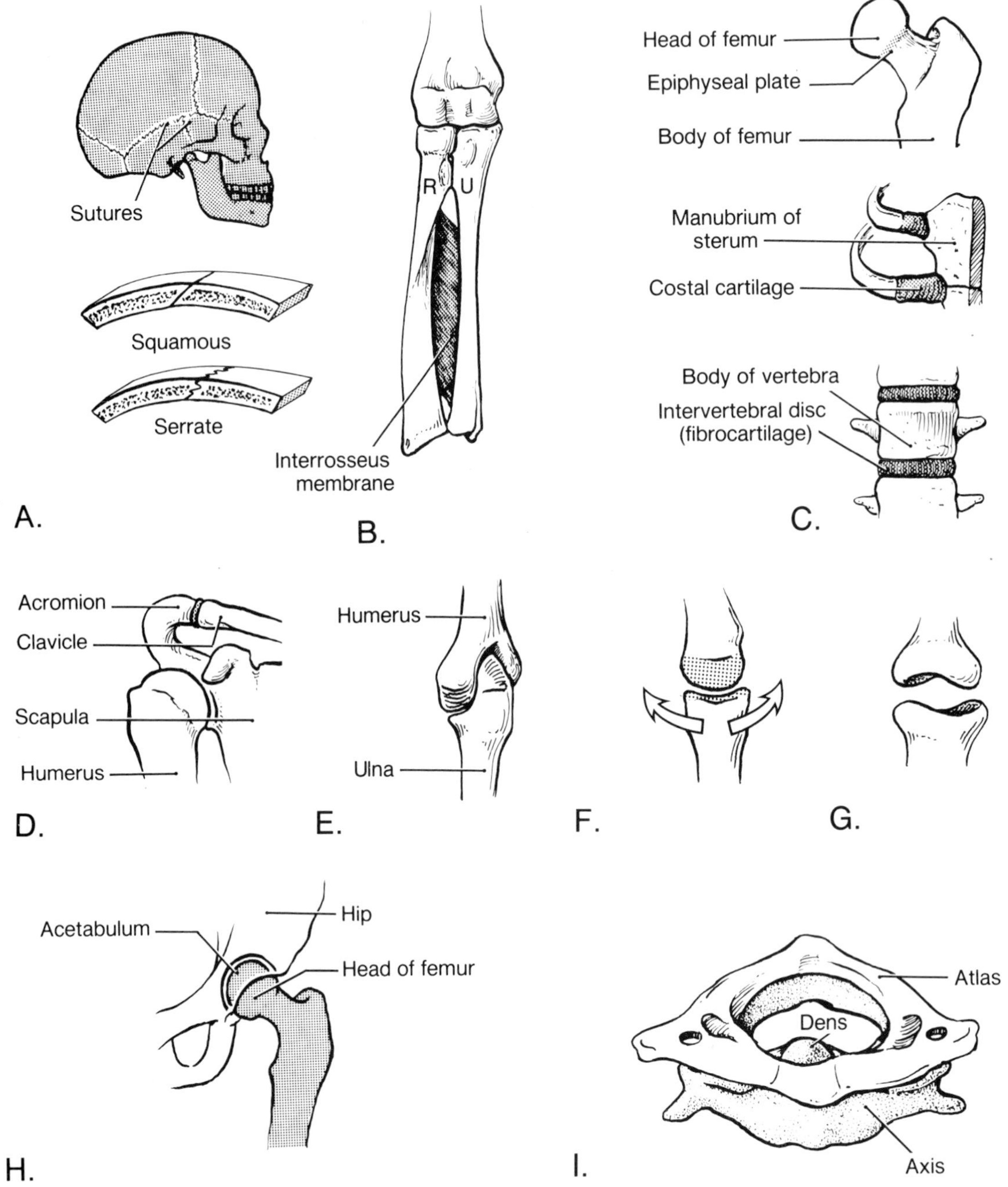

Figure 32. Drawings illustrating the various types of joint. *A* and *B*, *fibrous joints.* Note that in the squamous type of suture the bones overlap, whereas in the serrate type of suture the bones interlock. The interosseous membrane, the fibrous joint between the ulna (*U*) and radius (*R*) of the forearm, is strong but relatively thin. *C*, *cartilaginous joints.* The upper two are primary cartilaginous joints (synchondroses) and the lower one is a secondary cartilaginous joint (symphysis). *D*, *a plane joint.* The acromioclavicular joint is a multiaxial articulation of the scapula. *E*, *a hinge joint*. The humeroulnar articulation, one of the two forming the elbow joint, is a uniaxial one where the rounded trochlea of the humerus articulates with the trochlear notch of the ulna. *F*, *a condyloid joint.* The knuckle-like metacarpophalangeal joint of a finger is a biaxial articulation, permitting movement in two directions at right angles to each other. *G*, *a saddle joint*. The carpometacarpal articulation is a multiaxial joint. Note the saddle-like shape of the articulating bones. *H*, *a ball and socket joint.* The hip joint is a typical multiaxial articulation. The ball-like head of the femur fits into the cup-like acetabulum of the hip bone. *I*, *a pivot joint.* The atlantoaxial articulation is a uniaxial joint of the cervical region of the vertebral column. The finger-like dens of the axis (C2 vertebra) rotates in a collar formed by the anterior arch of the atlas (C1 vertebra) and the transverse ligament (Fig. 5-17).

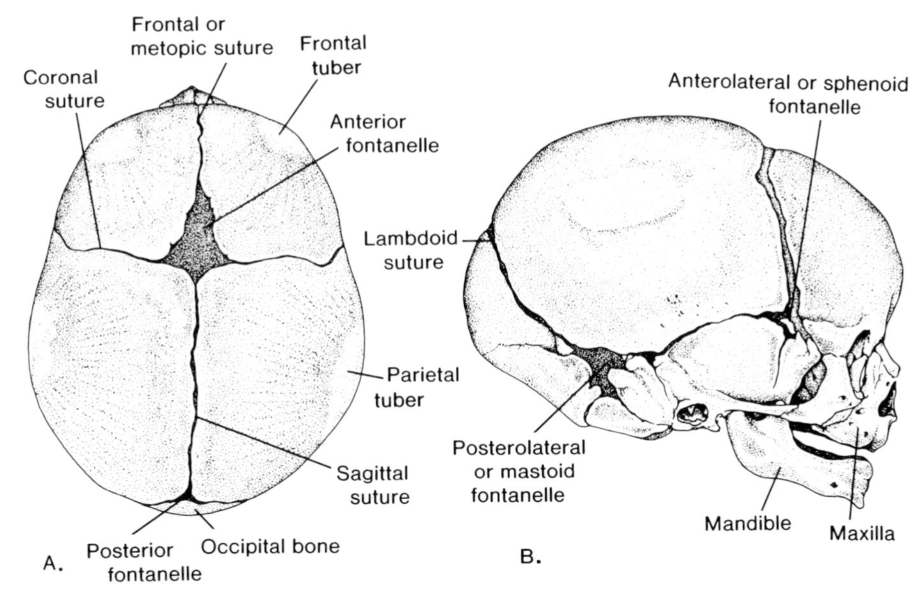

Figure 33. Drawings of the skull of a newborn infant. *A*, superior view. *B*, lateral view. Observe the fontanelles. At these sites the bones are attached to each other by fibrous tissue. The arterial pulsations of the brain may be felt at the anterior fontanelle. The loose construction of the fetal skull and the softness of its bones enable the calvaria (skullcap) to undergo changes of shape during birth. Note that the anterior fontanelle and the sagittal suture together resemble an arrow (L. *sagitta*).

permit freer movement. The ligaments associated with these joints are thought to be "softened" by the hormone **relaxin.** The changes produced in these joints enable the pelvic cavity to enlarge which facilitates the birth process. (See Chapter 3).

SYNOVIAL JOINTS

Synovial joints, *the most common and most important type functionally*, normally provide free movement between the bones they join.

The four distinguishing features of a synovial joint are that they have (1) a **joint cavity,** (2) an **articular cartilage,** (3) a **synovial membrane,** and (4) a **fibrous capsule.**

Friction between the bones in a synovial joint is reduced to a minimum because the articular surfaces are covered with a thin layer of articular cartilage which is lubricated by viscous **synovial fluid** produced by the cells of the *synovial membrane* (Fig. 34).

The articular cartilage is usually hyaline in type, although the matrix contains many collagenous fibers. The articular cartilage has no nerves or blood vessels; it is nourished by the synovial fluid.

The unique feature of synovial joints is the joint cavity (Fig. 34), but it is *normally more potential than real* because it normally contains only a trace of synovial fluid.

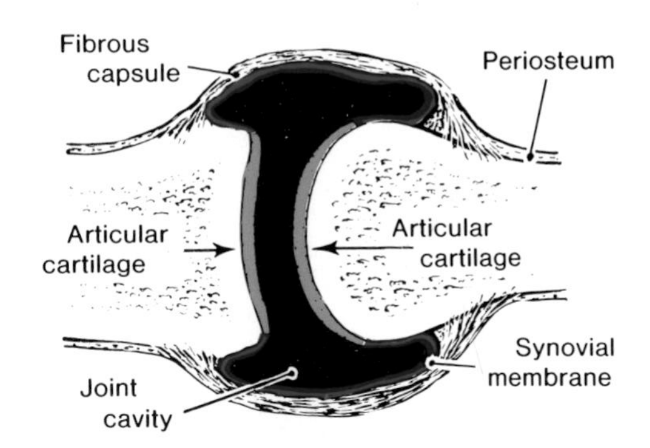

Figure 34. Drawing illustrating the scheme of a synovial joint. For clarity, the two bones have been pulled apart and the capsule has been inflated. Normally the joint cavity is more potential than real. *The articular capsule is composed of two layers.* Its delicate inner layer is the **synovial membrane.** The fibrous connective tissue of its outer layer, called the **fibrous capsule,** is attached to the bones just beyond the limits of the joint cavity.

The fibrous capsules of synovial joints are usually strengthened by thickenings called **accessory ligaments.** These are either part of their fibrous capsules (*intrinsic ligaments*) or separated from them (*extrinsic ligaments*). The ligaments are designed to limit movements of the joint in undesirable directions.

The joint capsule and its associated ligaments are important in maintaining the normal relationship between the articulating bones. Severe trauma to a joint results in **torn ligaments**, a common injury in contact sports such as football (See Fig. 4-134).

Some synovial joints have other features besides the four distinguishing ones listed previously. Three additional features are relatively common.

Articular discs or cartilages are present in some synovial joints (*e.g.*, the articular disc of the wrist joint, Fig. 6-151). They are usually fibrocartilaginous pads that help to hold the bones together, as in the example given. In some cases, however, they are attached to only one of the bones (*e.g.*, the menisci in the knee, Fig. 4-116). The articular cartilages have no nerves except at their attached margins.

Some synovial joints have a fibrocartilaginous ring, called a **labrum** (L. lip), which deepens the articular surface for the bones, *e.g.*, *the glenoidal labrum of the glenoid cavity* of the scapula (Fig. 6-131).

In some synovial joints **a tendon passes within the capsule of the joint**; *e.g.*, the tendon of the long head of the biceps brachii muscle runs within the capsule of the shoulder joint (Fig. 6-133). The intracapsular part of this tendon is covered with synovial membrane.

Types of Synovial Joint. There are six types of synovial joint in the body which are *classified according to the shape of the articulating surfaces and/or the type of movement they permit.* They are described in ascending order of movement as follows:

1. **Plane joints** are numerous and are nearly always small. They *permit gliding or sliding movements*, *e.g.*, the **zygapophysial joints** (*facet joints*) between the articular processes of the vertebrae (Fig. 5-30); the joints between two carpal bones; and the **acromioclavicular joints** (Fig. 32*D*). The opposed surfaces of the bones in these articulations are flat or almost so. Most plane joints may be moved in only one axis; hence they are called **uniaxial joints**. *Movement of plane joints is limited by their tight articular capsules.*
2. **Hinge joints (ginglymus)** permit movement in one axis (*uniaxial joints*) at right angles to the bones involved (*e.g.*, the humeroulnar joint at the elbow, Figure 6-137, and the interphalangeal joints of the fingers, Figure 6-154).

   Hinge joints permit flexion and extension only. The articular capsule of these joints is thin and lax where the movement occurs, but the bones are joined by strong collateral ligaments.
3. **Pivot joints** are *uniaxial joints which allow rotation around a longitudinal axis* through a bone. In pivot joints a rounded process of bone rotates within a sleeve or ring composed of a bony fossa. In the *proximal radioulnar joint* the anular ligament holds the head of the radius in the radial notch of the ulna (Fig. 6-148). In the *atlantoaxial joint* (Fig. 32*I*), the fingerlike dens of axis (C2 vertebra) rotates in a collar formed by the anterior arch of atlas (C1 vertebra) and the transverse ligament (Fig. 5-17).
4. **Condyloid joints** (G. knuckle-like) *are biaxial joints which allow movement in two directions.* Biaxial joints have two axes at right angles to each other.

   Condyloid joints permit flexion and extension, abduction and adduction, and circumduction (Fig. 15).

   Condyloid joints are sometimes referred to as *ellipsoid joints* because the articulating surfaces are ellipsoidal or oval in shape, *e.g.*, the metacarpophalangeal joints or knuckle joints (Fig. 32*F*) between the heads of the metacarpal bones and the bases of the corresponding proximal phalanges.
5. **Saddle joints** are also *biaxial joints* and are appropriately named because the opposing surfaces of the bones are shaped like a saddle, *i.e.*, concave and convex, opposite to each other (Fig. 32*G*). The carpometacarpal joint of the thumb is a good example of a saddle joint (Fig. 6-150).
6. **Ball and socket joints** are *multiaxial articulations* (Fig. 32*H*). In these *highly movable joints* the spheroidal surface of one bone moves within the socket of another (*e.g.*, the hip and shoulder joints). Flexion and extension, abduction and adduction, medial and lateral rotation, and circumduction can occur at ball and socket joints.

NERVE SUPPLY OF JOINTS

There is a rich supply of nerves to joints. *Nerve endings are located in the articular capsule, both in the fibrous capsule and in the synovial membrane.* The nerves supplying a joint (**articular nerves**) are branches of the ones supplying the overlying skin and the muscles that move the joint.

Hilton's law states that *the nerves supplying a joint also supply the muscles moving the joint and the skin covering the insertion of these muscles.*

The main type of sensation from joints is **proprioception** (L. *proprius*, one's own, + *receptor*, receiver), which provides information concerning the movement and position of the parts of the body. Impulses pass from nerve endings in the articular capsule to the spinal cord and the brain, and act in reflex mechanisms concerned with the control of the muscles acting on the joints.

Pain fibers are numerous in the articular capsule and in the associated ligaments. Their sensory endings respond to twisting and stretching such as occurs with distention of the joint with fluid owing to injury.

Excessive stretching and twisting of the articular capsule is very painful. *The fibrous capsule is highly sensitive*, but the synovial membrane is relatively insensitive.

Compared with pain arising in the skin, *joint pain is poorly localized and may be referred to the overlying skin or muscle.* There may be visceral disturbances (*e.g.*, nausea) associated with joint pain.

BLOOD AND LYMPHATIC SUPPLY OF JOINTS

Numerous articular arteries supply the joints. They arise from the vessels around the joint which often anastomose to form a network (*e.g.*, the *anastomoses around the elbow joint*, Fig. 6-65). Diffusion occurs readily between these vessels and the joint cavity.

Veins accompany the arteries and lymphatic networks are present in the articular capsules.

It is generally believed that *changes in temperature, humidity, or pressure make joints more sensitive or painful.* Some people use this phenomenon to predict the weather. It may be that environmental changes reflexly alter the blood flow to the joints.

Most substances in the bloodstream, normal or pathological, easily enter the joint cavity. Similarly traumatic infection of a joint may be followed by **septicemia** ("blood poisoning").

Beginning early in adult life and progressing slowly thereafter, **aging of articular cartilage** occurs on the ends of the bones involved in joints, particularly those of the hip, knee, vertebral column, and hands.

These *degenerative changes* result in the articular cartilage becoming less effective as a shock absorber and a lubricated surface. As a result the joint becomes vulnerable to the repeated friction that occurs during movements and to **subclinical trauma**. In some cases these changes do not produce significant symptoms, but in other cases they cause considerable pain.

Degenerative joint disease (also called *osteoarthritis*, *osteoarthrosis*, and *degenerative arthritis*) is **the most common type of arthritis.**

Degenerative joint disease is most common in weight-bearing joints such as those of the hip, knee, and lumbar region of the vertebral column.

MUSCLES

Muscles enable us to move from place to place (**locomotion**) and to move various parts of our body with respect to other parts. Movement is carried out by specialized cells called **muscle fibers.** *Contractility is highly developed in muscular tissue.*

Two general categories of muscle are recognized: striated and nonstriated. Striated muscle exhibits regular microscopic *transverse bands* along the length of the muscle fibers (Fig. 37), whereas nonstriated or smooth muscle is composed of individual muscle cells without striations.

Striated muscle is commonly subdivided into two types, **skeletal and cardiac.** The contraction of most skeletal muscle fibers is under voluntary control, whereas the rhythmical contraction of cardiac muscle is involuntary. The muscles of the body walls and limbs are composed of **striated skeletal muscle,** whereas the walls of heart are made up of **striated cardiac muscle.**

STRIATED SKELETAL MUSCLE

This type of muscular tissue is what most people refer to as muscle or "flesh". It is commonly called "skeletal" because most of it is attached by at least one end to some part of the skeleton.

Most skeletal muscles move the skeleton. They comprise the red or lean meat of the animals we eat. About 43% of the weight of our bodies is composed of striated skeletal muscle.

Skeletal muscles are often called voluntary muscles because they can usually be controlled voluntarily, *e.g.*, the biceps brachii (Fig. 6-66). However, many of the actions of skeletal muscles are automatic and the actions of some of them are reflex.

At each end of a skeletal muscle the connective tissue blends with the strong connective tissue that anchors it to the structure on which it pulls (*e.g.*, bone, cartilage, or articular capsule). In some cases the muscle tapers into a long **tendon** of collagenous fibers (*e.g.*, the tendo calcaneus or tendon of Achilles, Fig. 4-72).

In certain regions the muscles are attached by sheet-like tendons called **aponeuroses.** They are fibrous membranous expansions of muscle, such as those of the flat muscles of the abdomen (Fig. 2-9). The fleshy part of a muscle is often called its *belly.*

ATTACHMENTS OF SKELETAL MUSCLE

A skeletal muscle has at least two attachments. *Most muscles are attached to bone or cartilage.* Their tendons are anchored by means of **Sharpey's fibers** into the bone or cartilage. These fibers are located where the collagen bundles of the tendon lie buried in bone.

Some muscles (*e.g.*, the facial muscles) are inserted into the dermis of the skin, whereas other muscles (*e.g.*, those of the tongue) are attached to the mucous

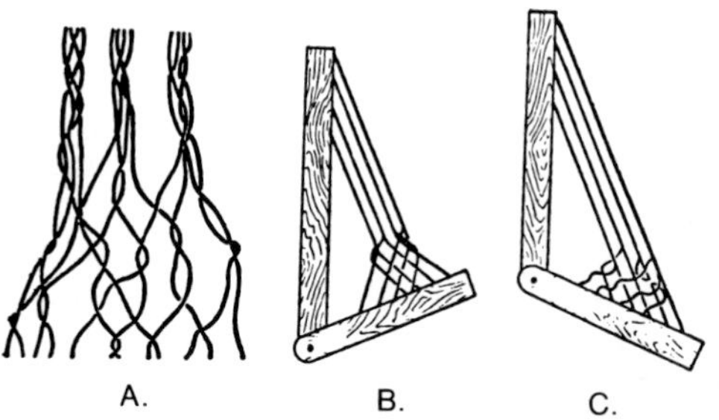

Figure 35. *A*, diagram of the inferior end of a tendon showing that its component fibers are braided or intertwinded. This structure provides added strength. *B* and *C* illustrate that in different positions of a joint, different fibers take the strain. The fan-shaped way most tendons are inserted into a bone ensures that successive parts of it take the strain as the angle of the joint changes.

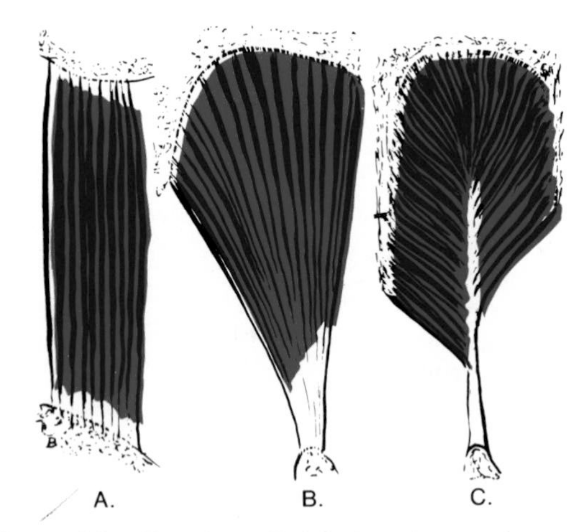

Figure 36. Drawings illustrating the three types of architecture in muscles. *A*, parallel fibers end to end. *B*, nearly parallel fibers, *i.e.*, fan-shaped. *C*, pennate fibers which converge on a central tendon. Note the two chief components of a striated skeletal muscle, the fleshy belly (*red*) and the fibrous tendon (*white*).

membrane. A few muscles are attached to fascia (*e.g.*, the tensor fasciae latae, Fig. 4-21), and other muscles form circular bands called **sphincters** (*e.g.*, the external anal sphincter, Fig. 3-2).

For purposes of description, most muscles are described as having an **origin** and an **insertion**. In all colored illustrations of the bones in this book, *origins of muscles are shown in red* and **insertions of muscles are shown in blue**.

The origin is the attachment that moves the least, whereas the insertion is the attachment that moves the most. Some people *erroneously* refer to the origin as the fixed end and to the insertion as the moving end. This is wrong because one end of a muscle moves in some activities and the other end in others. Because the freedom of movement of a muscle varies, the terms origin and insertion are interchangeable.

Generally the origin is the more proximal attachment and the insertion the more distal.

Muscles are usually inserted near the proximal end of a bone (Fig. 35, *B* and *C*), but some are inserted near the middle of its body or shaft (*e.g.*, the deltoid, Fig. 6-49). Skeletal muscles are frequently part of a lever system; hence, *the power exerted by a muscle depends in part on its angle of pull.* The optimum angle of pull is a right angle and as its angle of pull increases, the power of its pull decreases.

ARCHITECTURE OF SKELETAL MUSCLES (Fig. 36)

Most fibers of some muscles run parallel to the long axis of the muscle, but a few of them run the whole length of a muscle (*i.e.*, they form bundles of overlapping relays).

Muscles with their fibers disposed in parallel can lift a weight through a long distance (Fig. 36*A*).

In some muscles the fibers are oblique to the long axis of the muscle. Because of their resemblance to feathers (Fig. 36*C*), they are called **pennate muscles** (L. *pennatus*, feather). They are known as (1) *unipennate muscles* when their fibers have a linear or narrow origin resembling one-half of a feather; (2) *bipennate muscles* when their fibers arise from a broad surface resembling a whole feather; (3) *multipennate muscles* when septa extend into the origin and insertion of the muscles, dividing them into several feather-like portions *e.g.*, the deltoid, Fig. 6-44); and (4) *circumpennate muscles* when the fibers converge on a tendon extending into their substance.

In other muscles the fibers converge from a wide origin to a fibrous apex (Fig. 36*B*). This type of muscle is referred to as a fan-shaped or **triangular muscle**. On contracting, its fleshy muscle fibers shorten by a third to a half of their resting length. Hence it "swells," as illustrated by your biceps brachii when you flex your forearm (Fig. 9*B*).

MUSCLE ACTION (Fig. 37)

The structural unit of a muscle is a muscle fiber. The **functional unit**, consisting of a *motor neuron and the muscle fibers it controls, is called* a **motor unit**. The number of muscle fibers in a motor unit varies from one to several hundred, but usually there are about 100. This number varies according to the size and the function of the muscle. Large motor units, where one neuron supplies several hundred muscle fibers, are found in the large trunk and thigh muscles, whereas in the small eye and hand muscles, where precision movements are required, the motor units contain only a few muscle fibers.

When a nerve impulse reaches a motor neuron in the spinal cord (Fig. 37), a nerve impulse is initiated which makes *all the muscle fibers of the motor unit*

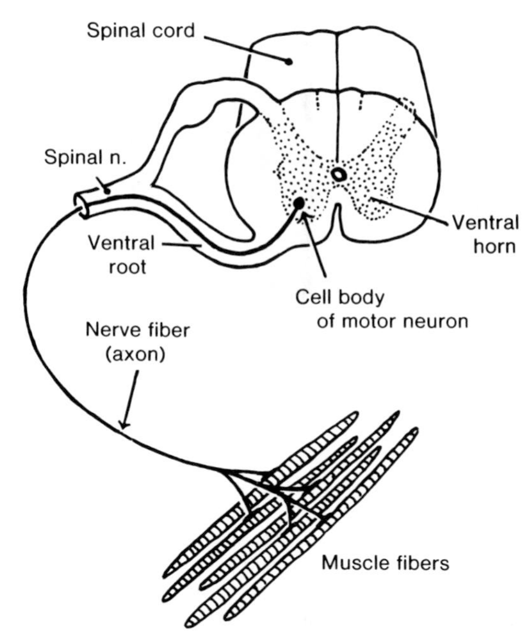

Figure 37. Diagram illustrating one motor unit consisting of a motor neuron and the muscle fibers innervated by it. The fewer muscle fibers per nerve fiber, the more precise is the movement produced by the muscle. Observe that the motor neuron is the ventral horn of the spinal cord

supplied by that motor neuron contract simultaneously. Movements result from an increasing number of motor units being put into action while antagonistic muscles are relaxing.

During maintenance of a given position or posture, the muscles involved are in a state of *reflex contraction* or **tone.** *Tone is the state of excitability of the nervous system controlling or influencing skeletal muscles.* The maintenance of tone depends upon impulses reaching the brain and spinal cord from the sensory endings in the muscles, tendons, and joints.

Tone is abolished by anesthesia; hence dislocated joints or fractured bones are more easily placed in their normal positions (*i.e.*, reduced) when the patient is given an **anesthetic agent**, a compound that reversibly depresses neuronal function.

There are some diseases of the nervous system where tone is increased (*spasticity*), and others where it is decreased (*flaccidity*).

During movements of the body, certain principal muscles are called into action. These muscles, called **prime movers** or agonists (G. *agon*, a contest), contract actively and produce the desired movement.

A muscle that opposes the action of a prime mover is called an **antagonist.** As a prime mover contracts the antagonist progressively relaxes, thereby producing a smooth coordinated movement. The antagonists are called into action at the end of a violent movement to protect the joint involved from injury.

When a prime mover passes over more than one joint, certain muscles prevent movement of the intervening joints. These kinds of muscle are called **synergists** (G. *syn*, together, + *ergon*, work); thus *synergists complement the action of the prime mover.* Other muscles, called **fixators,** steady the proximal parts of a limb (*e.g.*, the arm) while movements are occurring in distal parts (*e.g.*, the hand).

It is important to know that the same muscle may act as a prime mover, antagonist, synergist, or fixator under different conditions.

Muscle actions can be tested in various ways. This is usually done when nerve injuries are suspected. There are *two common ways of determining the status of motor function*: (1) the patient performs certain movements against resistance produced by the examiner, and (2) the examiner performs certain movements against resistance produced by the patient. For example, when testing flexion of the forearm at the elbow joint (Fig. 9*B*), the patient is asked to flex the forearm while the examiner resists the effort. The other way is to ask the patient to keep the forearm flexed while the examiner attempts to extend it. The latter method enables the examiner to gauge the power of the movement.

Electrical stimulation of muscles is used as part of the treatment for restoring the action of muscles. This kind of stimulation is also used for testing muscles.

Electromyography is a good method of testing the action of muscles. Electrodes are inserted in the muscle and then the patient is asked to perform certain movements. The differences in electrical action potentials of the muscles are amplified and recorded. Using this technique, it is possible to analyze the activity of an individual muscle during different movements. *A resting normal muscle shows no activity.*

STRIATED CARDIAC MUSCLE

Heart muscle or myocardium surrounds all the chambers of the heart, but it is much thicker around the ventricles than the atria. The fibers of cardiac muscle fit together so tightly that they give the impression under a light microscope that they form a network or syncytium. However, electron microscopy shows that cardiac muscle fibers are composed of individual cells joined end to end at cell junctions.

Although cardiac muscle is striated, *its rhythmic contractions are not under voluntary control.* The heart rate is regulated by a **pacemaker** composed of special cardiac muscle cells which are innervated by the **autonomic nervous system** (Fig. 1-61). The myocardium will contract spontaneously without any nerve supply.

Heart muscle responds to increased demands by increasing the size of its fibers. This is called **compensatory hypertrophy**. When heart muscle is damaged, fibrous scar tissue is formed. When this muscle becomes **necrotic** (*i.e.*, dies), the lesion is called an infarct. It is usually called a **myocardial infarct (MI)**, even though in most cases there is some involvement of the endocardium and/or epicardium, its inner and outer layers, respectively. A person suffering a heart attack (**MI**) usually describes a crushing substernal pain that does not disappear with rest (see Case 1-1 at the end of Chap. 1).

SMOOTH MUSCLE

Nonstriated or smooth muscle forms the muscular layers of the walls of the digestive tract and of blood vessels. As its name implies, *smooth muscle cells have no microscopic cross striations.*

Smooth muscle is generally arranged in two layers in the walls of the *viscera* (L. soft parts, internal organs). In some parts of the digestive and respiratory tracts and in arteries, the smooth muscle cells are arranged spirally.

Smooth muscle, like cardiac muscle, is innervated by the autonomic nervous system. Hence it is **involuntary muscle** and can contract partly for long periods (*i.e., maintain tonus*). This is important in regulating the size of the lumen of a tubular structure. In the walls of the digestive tract, uterine tubes, and ureters, the smooth muscle cells undergo rhythmic contractions (**peristaltic waves**). This process known as **peristalsis** propels the contents along the tubular structures (*e.g.*, the intestines).

Like skeletal and cardiac muscle, smooth muscle undergoes **compensatory hypertrophy** in response to increased demands. During pregnancy the smooth muscle cells in the wall of the uterus not only increase in size (**hypertrophy**), but they also increase in number (**hyperplasia**).

BLOOD VESSELS

There are three types of blood vessel: *arteries, veins,* and *capillaries.*

ARTERIES

Arteries carry blood away from the heart to the capillary bed. There are three main types of artery: arterioles, muscular arteries, and elastic arteries.

Arterioles (Fig. 40). *These are the smallest type of artery.* They have relatively narrow lumens and thick muscular walls. The degree of pressure within the arterial system (**arterial pressure**) is mainly regulated by the degree of tonus in the smooth muscle in the arterioles. If the tonus of this smooth muscle becomes increased above normal, ***hypertension*** (high blood pressure) results.

Muscular Arteries (Fig. 38). These arteries distribute the blood to various parts of the body. The walls of these arteries consist chiefly of circularly disposed smooth muscle fibers which constrict the lumen of the vessel when they contract.

The smooth muscle in the walls of the muscular arteries is under the control of the ***autonomic nervous system.*** Hence, they regulate the flow of blood to different parts of the body as required (*e.g.*, blood flow to the skeletal muscles in the limbs increases during exercise).

Elastic Arteries. *These are the largest arteries in the body* and their walls consist chiefly of elastin, a yellow elastic fibrous mucoprotein. The maintenance of blood pressure within the arterial system between contractions of the heart results from the elasticity of the elastic arteries. Their elasticity allows them to expand when the heart contracts and to return to normal caliber between cardiac contractions.

VEINS

Veins return blood to the heart from the capillary bed. Contracting skeletal muscles compress the veins, "milking" the blood along. When standing, the venous return from the legs depends largely on the muscular activity of the calf muscles (**"calf pump"**).

The smallest veins are called **venules**. Small veins

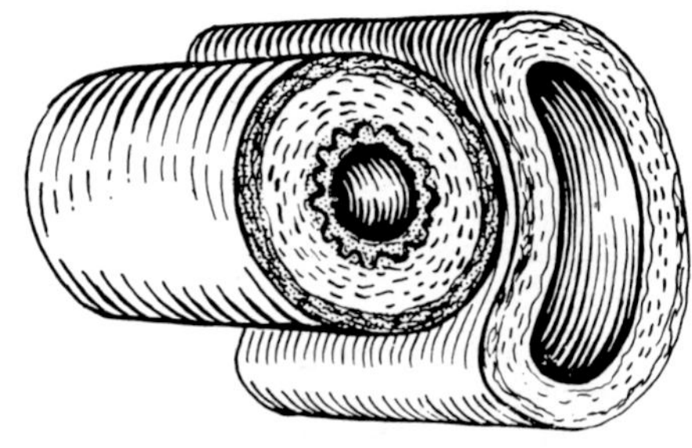

Figure 38. Schematic section of an artery (*left*) and its accompanying vein (vena comitans). Note that the wall of the artery is much thicker than the vein. Observe the three layers in its wall.

or tributaries unite to form larger veins which commonly join together, forming **venous plexuses** or networks (*e.g.*, the dorsal venous network of the hand, Fig. 6-124).

Veins tend to be double or multiple. The two veins that commonly accompany medium-sized arteries, one on each side (Fig. 39*A*), are called **venae comitantes** (L. accompanying veins).

Systemic veins are more variable than the arteries and **anastomoses** (communications) occur more often between them.

Many veins contain valves (Fig. 39*B*) which prevent backflow of blood and encourage the flow of blood toward the heart. The valves support the columns of blood superior to them and divide the columns into smaller amounts of less weight. This assists the return of blood to the heart and prevents the pooling of blood.

Muscular actions in the limbs work with the venous valves to keep the blood moving toward the heart.

> *Varicose veins* have a caliber greater than normal and the cusps of their valves do not meet or they have been destroyed by inflammation. These are said to be **incompetent valves** because they are unable to perform their function competently. *Varicose veins are common in the posterior and medial parts of the lower limbs.*

CAPILLARIES

Capillaries are simple endothelial tubes that connect the arterial and venous sides of the circulation (Fig. 40). Capillaries are generally arranged in communicating networks called **capillary beds**. The blood flowing through a capillary bed is brought to it by arterioles and is carried away by venules.

The volume of blood flowing through a capillary bed is under the control of the autonomic nervous system (discussed subsequently).

As blood is forced through the capillary bed by the hydrostatic pressure in the arterioles, an interchange of materials occurs (*e.g.*, oxygen and carbon dioxide).

In some regions of the body (*e.g.*, in the fingers and the toes), there are direct connections between the arteries and veins; *i.e.*, there are no capillaries between them. The sites of such communications, called **arteriovenous anastomoses** (*AV shunts*), permit blood

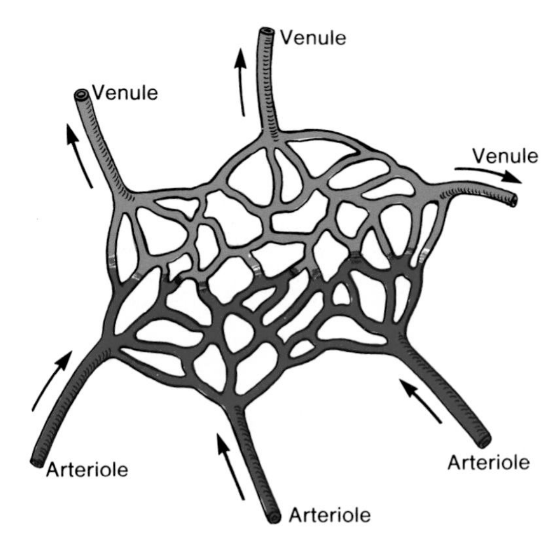

Figure 40. Drawing of a capillary bed or network. Observe that arterioles convey blood into the network and that venules drain blood from it. Venules unite to form veins which return the blood to the heart.

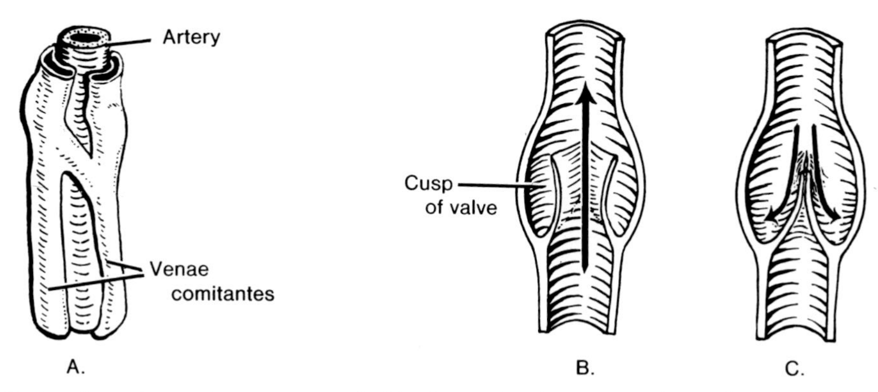

Figure 39. *A*, drawing of an artery accompanied by two veins (venae comitantes), one on each side. Small and medium-sized arteries are generally accompanied by a pair of veins, whereas larger arteries usually have only one accompanying vein (Fig. 38). Some arteries have no companion veins. Observe the communication (anastomosis) between the venae comitantes. Blood flows in one direction only through a vein, *i.e.*, toward the heart. *B*, shows a bicuspid valve (two cusps) in the open position. *C*, shows the valve closed by the back pressure of the blood. The *arrows* indicate the direction of blood flow. Valves are common in limb veins and muscle veins. Valves protect smaller veins from the higher pressure in larger veins and from pressure increases resulting from muscle actions.

to pass directly from the arterial to the venous side of the circulation without having to pass through capillaries. **AV shunts** are fairly numerous in the skin, where they are important in conserving the heat of the body.

LYMPHATICS

As soon as tissue fluid enters a lymph vessel it is called **lymph**. Lymph is carried by lymph capillaries and lymph vessels which are referred to as **lymphatics**. They offer another special *pathway for the return of tissue fluid to the blood stream.*

The Lymphatic System (Fig. 41) consists of: (1) *lymph plexuses* of very small lymph vessels called **lymph capillaries** which begin in the intercellular spaces of most tissues in the body; (2) **lymph nodes** (Fig. 42) composed of small masses of **lymphatic tissue** arranged along the course of the lymph vessels through which lymph passes on its way to the venous system; (3) **aggregations of lymphoid tissue** located in the walls of the alimentary canal and in the spleen and thymus; and (4) **circulating lymphocytes** which are formed in (a) lymphoid tissue located throughout the body (*e.g.*, in lymph nodes, spleen, thymus, and tonsils), and (b) in myeloid tissue located in the bone marrow.

LYMPH VESSELS

Most lymph vessels are not visible in dissections, but they can be demonstrated in vivo (L. in living persons).

Lymph capillaries are small vessels that begin blindly in most tissues of the body. They drain lymph from the tissues and are in turn drained by small lymph vessels. These vessels join to form larger and larger collecting vessels or trunks that pass to nearby or regional lymph nodes.

As a general rule, lymph traverses one or more lymph nodes before it enters the bloodstream (Fig. 41).

The wall of a lymph capillary consists of a single layer of endothelial cells, resembling that of a blood capillary. As the lymph capillaries unite to form larger vessels, connective tissue is added to their walls. The largest lymph vessels, called **lymph trunks and ducts**, also contain smooth muscle in their walls.

The lymphatics carrying lymph to the lymph nodes are called **afferent lymph vessels** (Fig. 42), whereas those leaving a lymph node are called **efferent lymph vessels**. An efferent lymph vessel of one lymph node becomes an afferent lymph vessel of another lymph node in a chain. After traversing one or more lymph nodes, the lymph enters larger lymph vessels called lymph trunks. The trunks unite to form either (1) the **thoracic duct** (Fig. 41) which enters the junction of the left internal jugular and left subclavian veins, or (2) the **right lymphatic duct** (Fig. 41), which enters the junction of the right internal jugular and right subclavian veins.

In general, the thoracic duct drains lymph from the entire body, except for (1) the right side of the head and neck, (2) the right upper limb, and (3) the right half of the thoracic cavity.

Superficial Lymph Vessels. These vessels are located *in the skin* and the subcutaneous tissue (i.e., *deep to the skin*). They also form a network on the deep surface of the epithelium lining cavities. The lymph capillaries run parallel to the superficial vessels of the skin and then join to form slightly larger vessels. The superficial lymph vessels eventually drain into deep lymph vessels.

Deep Lymph Vessels. These vessels run in the *deep fascia* (subcutaneous connective tissue), located between the muscles and the *superficial fascia* (tela subcutanea). The deep vessels also accompany the major blood vessels of the region concerned. These lymph vessels have thick walls containing connective tissue and smooth muscle. They also have valves.

LYMPH NODES

Lymph nodes (once wrongly called lymph "glands") are round, oval, or beanshaped structures that can be easily palpated when they are swollen. Many lymph nodes are located in the **axilla** (armpit) and the **inguinal region** (groin), but there are numerous lymph nodes distributed along the large vessels of the neck (Fig. 41). There are also many lymph nodes in the thorax and abdomen.

Lymph nodes consist of aggregations of lymphatic tissue (Fig. 42), and vary from about the size of a pinhead to a fingerbreadth or more in diameter. Generally there is a slight depression on one side of the lymph node, called the **hilum** (hilus), through which the blood vessels enter and leave the node. The afferent lymph vessels bringing lymph to the node enter different parts of the capsule. The efferent lymph vessel (usually single) carrying lymph from the node, emerges from the hilum.

Each node has a thick outer part called the **cortex** (L. bark) and an inner part called the **medulla** (L. marrow). Connective tissue *trabeculae* (L. little beams) pass inward from the capsule at the hilum, providing support and carrying the blood vessels (Fig. 42). Trabeculae also pass inward from the capsule covering the external surface of the node.

The cortex contains collections of lymphatic cells called **germinal centers** (Fig. 42), whereas the medulla contains cords of cells.

Reticuloendothelial cells are located along the trabeculae and *have the ability to remove foreign material from the lymph* as it passes through the node. The lymph nodes draining the lungs of persons who smoke and/ or inhale dark colored dust have a black-

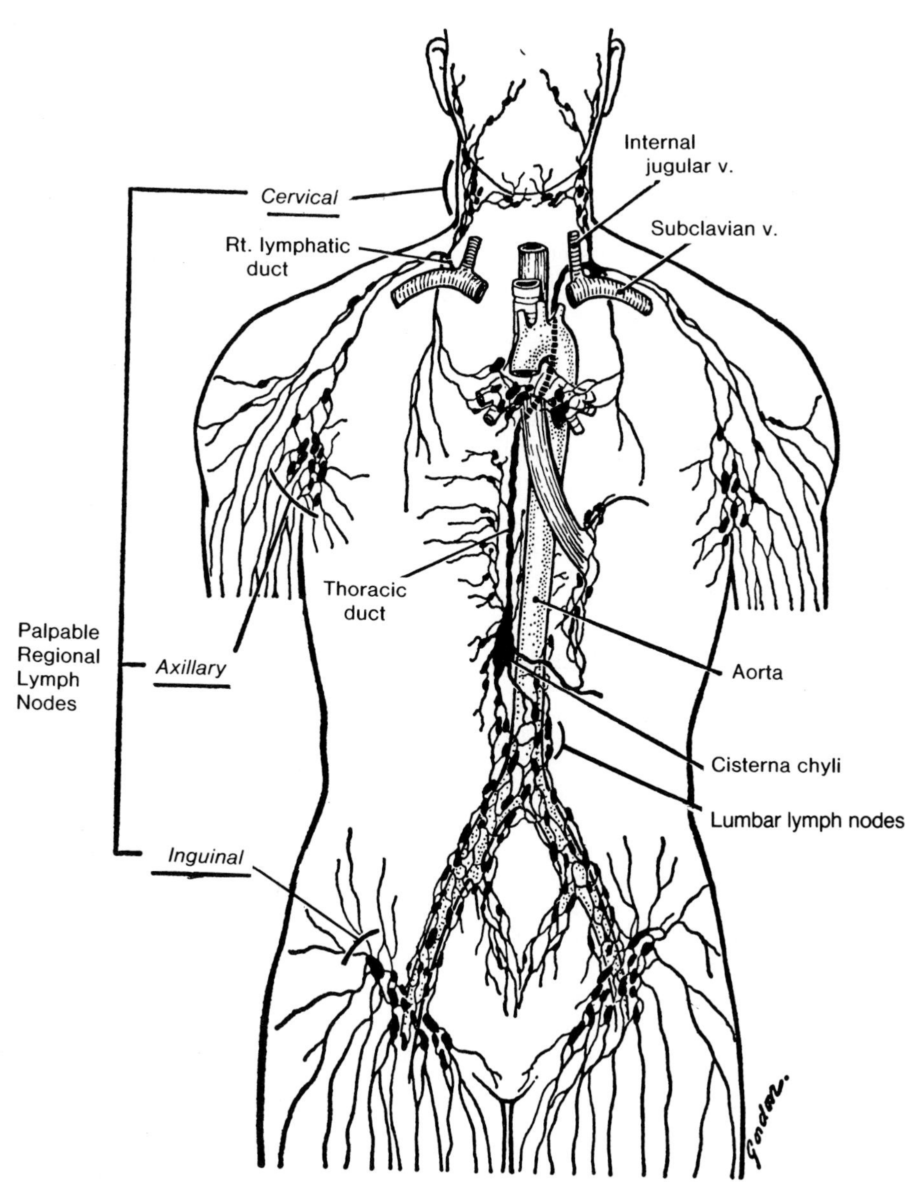

Figure 41. Drawing illustrating the lymphatic system. Note the regional lymph nodes which become palpable when inflamed or invaded by tumor cells. Observe that (1) the lymphatics of the upper limb converge on *axillary lymph nodes* in the axilla (armpit); (2) those of the head and neck drain into *cervical lymph nodes* that are grouped especially around the vessels of the neck (L. *cervix*; hence the adjective cervical). Note that lymph from the body drains into the *thoracic duct* except for the right side of the head and neck, the upper limb, and the right half of the thoracic cavity; (3) most of the lymphatics of the lower limb converge on the *inguinal lymph nodes* in the proximal part of the thigh. The *cisterna chyli* is a dilated sac at the inferior end of the thoracic duct into which the intestinal and two lumbar lymphatic trunks empty. Note that the thoracic duct and the right lymphatic duct open into the venous system.

ened appearance, as do those receiving lymph from a darkly tatooed area of skin).

LYMPH

Lymph (L. *lympha*, clear water) is usually a clear, transparent, watery fluid. Sometimes it is a faintly yellow and slightly opalescent fluid. *Lymph is collected by the lymph capillaries from the intercellular spaces in various tissues of the body.* Usually more tissue fluid is produced at the arterial ends of capillaries than is absorbed at their venous ends. This extra fluid is drained away by the lymph capillaries.

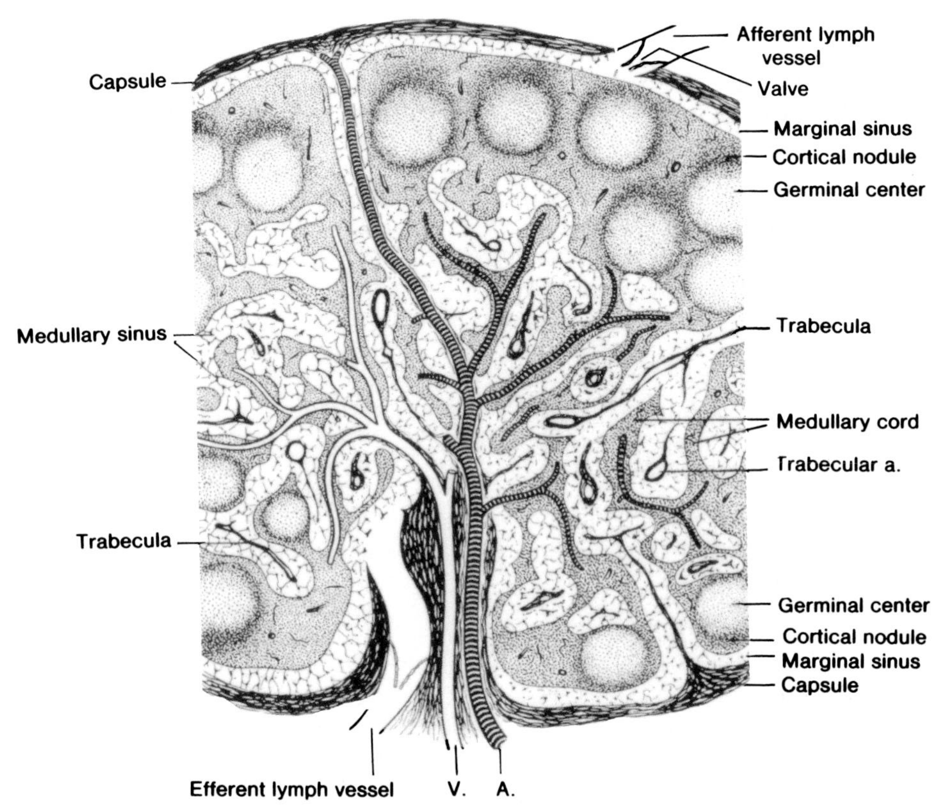

Figure 42. Diagram illustrating the structure of a lymph node. Observe the hilum through which the blood vessels enter and leave the node. The efferent lymph vessel, usually single, also emerges from the node at the hilum. The afferent lymph vessels (usually multiple) bringing lymph to the node enter it at many different parts of the capsule. Both kinds of lymph vessel have flap-type valves which prevent lymph from passing backward toward its point of origin.

Lymph contains the same constituents as blood plasma (*e.g.*, protein and many lymphocytes). Lymph coming from the intestine also contains fat, fatty acids, glycerol, amino acids, glucose, and other substances (*e.g.*, medicine given to a patient).

FUNCTIONS OF LYMPHATICS

Drainage of Tissue Fluid. Lymph capillaries are mainly involved in collecting plasma from the tissue spaces and transporting it back to the venous system. On its way lymph passes through the lymph nodes where particulate matter (*e.g.*, dust inhaled into the lungs) is largely filtered out by the phagocytic (G. *phagein*, to eat) activity of the scavenger cells called **macrophages** in the nodes. In a similar way, *bacteria and other microorganisms drained from an infected area are trapped and ingested by the phagocytic cells*, thereby preventing them from entering the bloodstream.

Absorption and Transport of Fat. The lymphatics draining the intestine contain considerable amounts of emulsified fat after a fatty meal. This creamy lymph, called **chyle** (G. *chylos*, juice), is taken up by the **lacteals** of the intestine during digestion (Fig. 2-91). This lymph and emulsified fat, after passing through various lymph vessels, is conveyed by the **thoracic duct** to the left subclavian vein (Fig. 41), where it enters the blood.

Defense Mechanism of the Body. The lymphatic system provides an *important immune mechanism for the body.* When minute amounts of foreign protein are drained from an infected area by the lymphatic capillaries, **immunologically competent cells** produce a specific antibody to the foreign protein, or lymphocytes are dispatched to the infected area. The antibody is carried to this area via the bloodstream and the tissue fluid. This is referred to as the **humeral mechanism** of the immune rection. When foreign tissue is transplanted, lymphocytes are active in rejecting the graft.

The lymph vessels and lymph nodes draining an infected area often become inflamed. *Inflammation of the lymph vessels* is called **lymphangitis** (G. *angeion*, vessel), whereas *inflammation of a lymph node* is called **lymphadenitis** (G.

adēn, gland). Laypeople usually refer to inflamed lymph nodes as swollen "glands."

Chronic lymphangitis, together with blockage of lymph vessels, can be produced by the escaped ova (L. eggs) of a minute parasitic worm called *Microfilaris nocturna*. The resulting condition, called **elephantiasis**, is common in the tropics and is characterized by enormous enlargement of some parts of the body (*e.g.*, the legs and scrotum in males).

The extensive *spread of cancer* cells via the lymphatic system can also cause blockage of lymph vessels and **lymphedema** (G. *oidēma*, swelling), an accumulation of fluid in the affected region. Similarly, the widespread removal of lymph nodes (e.g., as may occur during a **radical mastectomy**, the surgical removal of a breast) can cause swelling of the upper limbs owing to poor drainage of lymph from them. For a discussion of breast cancer and its spread to the regional lymph nodes, see Chapter 6.

The lymphatic system is involved in the metastasis (spread) of cancer cells. This is referred to as **lymphogenous dissemination of cancer or malignant cells**. Cancer can spread by permeating the lymph vessels as solid cell growths from which minute cellular **emboli** (G. *embolus*, a plug) may break free and pass to regional and distant lymph nodes. This spread of malignant disease from one part of the body to another is called *metastasis*, and the transportation of cancer cells via the lymphatic system is called **lymphogenous metastasis**.

The radiographic study of lymph vessels and lymph nodes, called **lymphography**, is possible after the cannulation of appropriate peripheral lymph vessels and the injection of a radiopaque contrast material.

NERVES

Nerves are strong whitish cords in living persons. They are made up of many **nerve fibers** (axons) arranged in **fasciculi** or *fascicles* (L. bundles) which are held together by a connective tissue sheath (Fig. 43).

Nerves carry impulses to and from the central nervous system (**CNS**), composed of the brain and the spinal cord. The peripheral nervous system (**PNS**) sends impulses to and receives impulses from the CNS. These impulses come and go to all parts of the body. Together the CNS and PNS serve conscious perception, voluntary movement, autonomic functions, and integrate messages received from the body.

The nerves connected with the brain are called **cranial nerves**, whereas those connected with the spinal cord are called **spinal nerves**.

The brain lies in the cranial cavity where it is surrounded by the meninges (G. *membranes*) *and the calvaria* or brain case (Fig. 7-51). The spinal cord is in the vertebral canal where it is *surrounded by the*

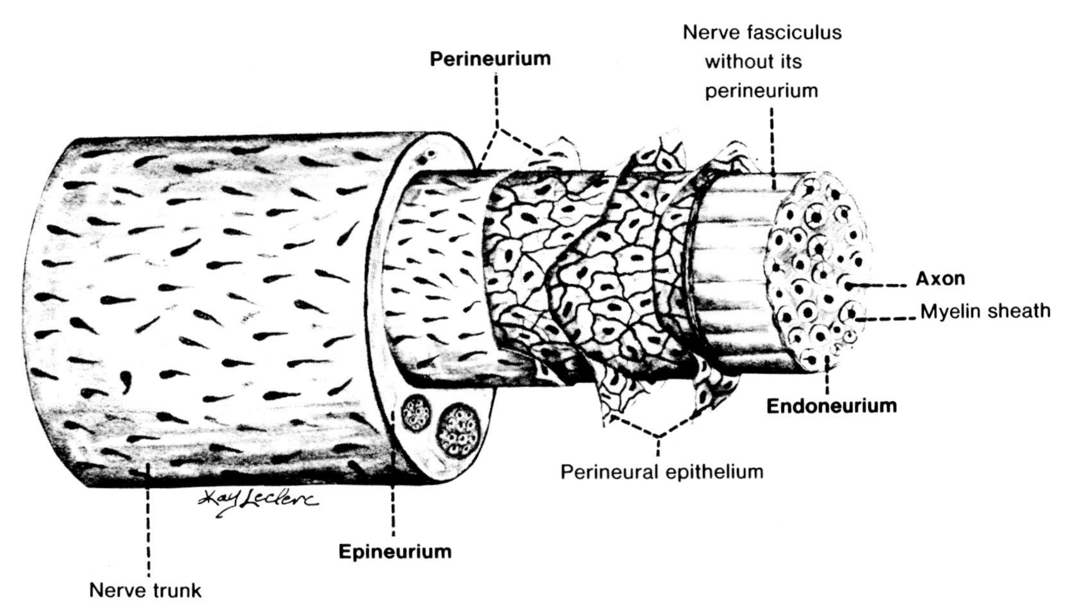

Figure 43. Drawing illustrating the connective tissue sheaths around a peripheral nerve. The cells of the perineurium are disposed as an epithelium which forms a barrier against the penetration of certain materials into the fasciculus (bundle of axons).

meninges, the vertebrae, and their interconnecting ligaments (Figs. 5-51 and 5-54).

The cranial nerves leave the skull through various **cranial foramina** and the spinal nerves leave the vertebral column via the **intervertebral foramina** (Fig. 5-54). The nerves to the limbs form plexuses or networks of nerves (**brachial plexus**, Fig. 6-22, and **sacral plexus**, Fig. 2-125). Hence, the nerves to the limbs contain fibers from several segments of the spinal cord.

The neuron (nerve cell) is the anatomical unit of the nervous system (Fig. 44). *The neuron includes the nerve cell body and all its extensions.* Some neurons are very long, extending from the brain to the inferior end of the spinal cord.

The processes extending from the cell body are specialized for different functions. **Dendrites** (G. *dendron*, tree) *receive impulses and* **axons** *conduct impulses away from the nerve cell body, e.g.*, to the muscles (Fig. 44).

A neuron can be divided into three portions: (1) a *receptive portion* (cell body and dendrites), (2) a *conductile portion* (axon), and (3) an *effector portion* (collateral branch or the end of the axon which produces an effect on other neurons or on a muscle or a gland, Fig. 44).

A **nerve fiber** consists of an *axon*, a *myelin sheath* (in some cases), and a *neurolemmal sheath* (of Schwann). Both the neurolemma and the myelin sheath are components of Schwann cells. The components of all but the smallest peripheral nerves are arranged in nerve **fasciculi** (Fig. 43).

The delicate nerve fibers are strengthened and protected by connective tissue coverings as follows. The entire nerve is surrounded by a thick sheath of loose connective tissue, called the **epineurium**, which contains fatty tissue and blood and lymph vessels (Fig. 43). A more delicate connective tissue sheath enclosing a fasciculus of nerve fibers, called the **perineurium**, provides an effective barrier to the penetration of substances into or out of the nerve fiber. Individual nerve fibers are surrounded by a delicate covering of connective tissue called the **endoneurium**.

The regions of attachment between neurons or between neurons and effector organs are called **synapses** (G. *synapto*, to join). *Synapses are the sites of contact between neurons* at which one neuron is excited or inhibited by another neuron.

A typical spinal nerve arises from the spinal cord by two roots. The *ventral root* contains *motor (efferent)* fibers from motor neurons in the ventral horn of the spinal cord (Fig. 37). The *dorsal root carries sensory (afferent) fibers* of cells in the **spinal ganglion** (Fig. 46). In or close to the *intervertebral foramen* (Fig. 5-51), *the dorsal and ventral roots unite to form a spinal nerve.* As soon as the spinal nerve leaves the intervertebral foramen, it divides into two parts; a ***dorsal primary ramus*** and a ***ventral primary ramus*** (Fig. 46). The dorsal rami supply nerve fibers to the back, whereas the ventral rami supply nerve fibers to the lateral and anterior regions of the trunk and to the limbs.

THE AUTONOMIC NERVOUS SYSTEM

The autonomic nervous system (ANS) is a system of nerves and ganglia *concerned with the distri-*

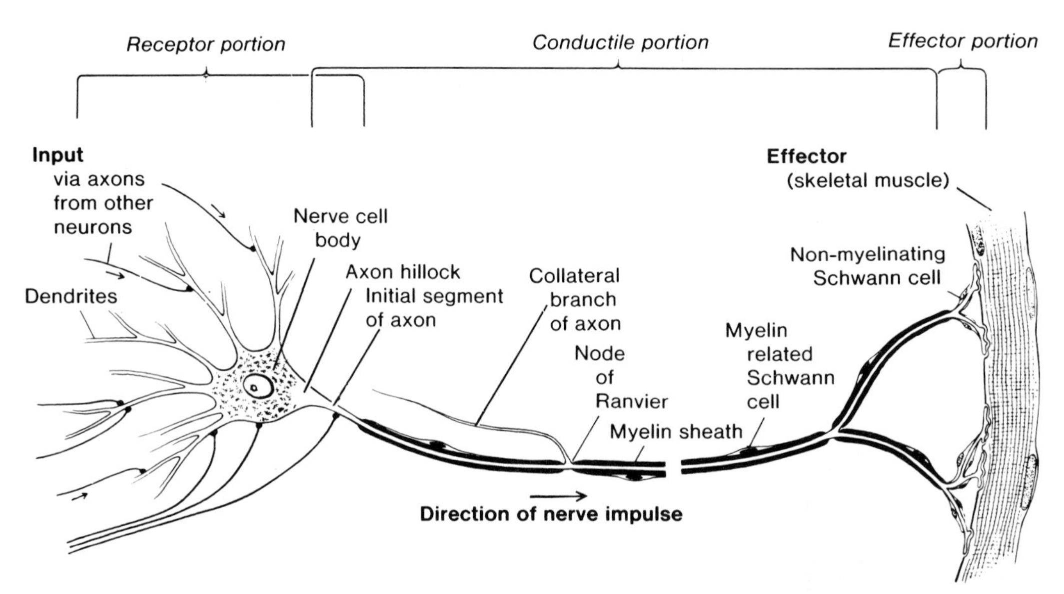

Figure 44. Diagram illustrating the receptor, conductile, and effector portions of a typical large motor neuron. Observe the effector endings on the skeletal muscle. The presence of the myelin sheath on the axon (conductile portion of the neuron) increases conduction velocity.

bution of impulses to (1) the **heart**, (2) **smooth muscle**, and (3) **glands** (Fig. 45). The ANS also receives afferent impulses from these parts of the body.

The autonomic nervous system has two parts: (1) the **sympathetic system** and (2) the **parasympathetic system.** The sympathetic system has connections with the thoracic and lumbar regions of the spinal cord from T1 to L2 or L3 segments. The parasympathetic system has cranial and sacral parts which are connected with the brain through **cranial nerves III, VII, IX,** and **X,** and with the spinal cord through **spinal nerves S2 to S4** (the S2 connection is not always present).

There are neurons in the sympathetic system which correspond to the efferent or motor neurons of the PNS. These efferent neurons are located (1) in **paravertebral ganglia** or ganglia of the sympathetic trunk (Fig. 45); (2) in **prevertebral ganglia** or visceral ganglia (*e.g.*, the celiac ganglion, Fig. 2-99); and (3) in the medulla of the adrenal or **suprarenal gland** (Fig. 2-126).

In Figure 46 observe that the axon of a sympathetic neuron sends a **preganglionic fiber** via the ventral root and the **white ramus communicans,** (connecting branch) to a paravertebral ganglion. Here it may synapse with excitor neurons, but some preganglionic fibers ascend or descend in the **sympathetic trunk** to synapse at other levels (Fig. 45). Other preganglionic fibers pass through the paravertebral ganglia without synapsing to form ***splanchnic nerves*** to the viscera (Fig. 2-114).

The ***postganglionic fibers*** or axons of excitor neurons course through the **gray ramus communicans** (Fig. 46) and pass from the sympathetic ganglion into the spinal nerve. They run to blood vessels, sweat glands, sebaceous glands, and the arrector pili muscles (associated with hairs).

Some postganglionic fibers pass superiorly to the cervical sympathetic ganglia and supply structures in the head. Others pass inferiorly to supply the lower limbs. Hence the sympathetic trunks (Fig. 45) are composed of ascending and descending fibers (preganglionic efferent, postganglionic efferent, and afferent fibers).

The dorsal and ventral primary rami of spinal nerves contain (1) **motor (efferent) fibers** from the ventral

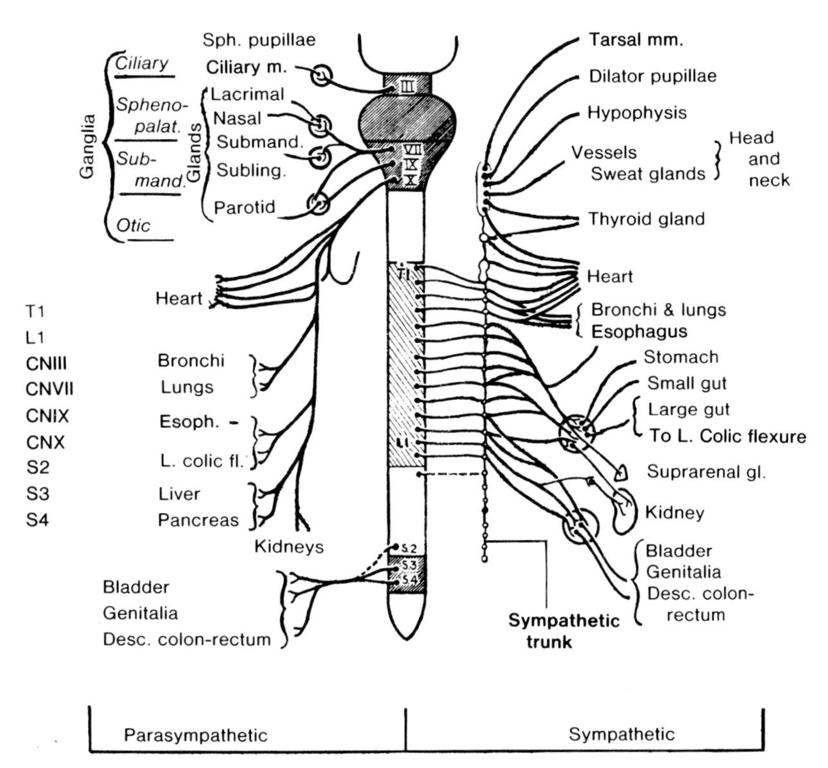

Figure 45. Diagram illustrating the general plan of the autonomic nervous system (ANS). It is so named because it is concerned with involuntary activity. The ANS is a subdivision of the motor portion of the nervous system. Note that it consists of two parts, *sympathetic* and *parasympathetic*. These parts are anatomically separate and usually functionally reciprocal. The ANS carries nerve impulses to the heart, smooth muscle, and glands. The ganglia of the *sympathetic trunk* form a longitudinal series of swellings, one anterior to each spinal nerve (also see Fig. 1-25). The nerve fibers of the parasympathetic division of the ANS leave the CNS in cranial nerves III, VII, IX, and X, and in sacral nerves 2, 3, and 4.

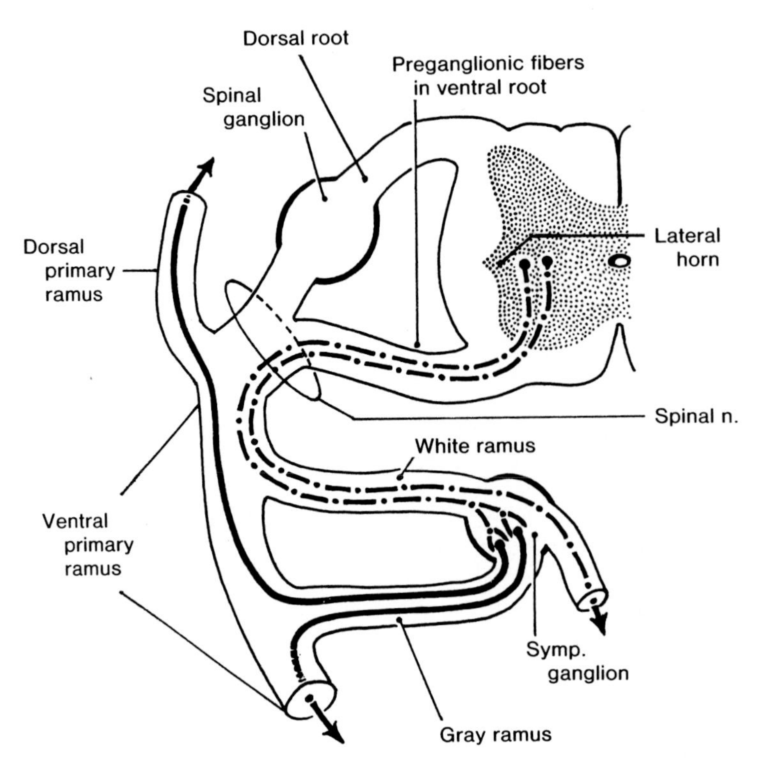

Figure 46. Diagram illustrating the typical sympathetic contribution to a spinal nerve. Most autonomic ganglia resemble spinal ganglia as far as their connective tissue capsule and structure; however, unlike these ganglia, synapses occur in autonomic ganglia. Note that the neurons in the lateral horn of gray matter, extending from segment T1 through L2 or L3, are the source of the preganglionic sympathetic fibers. Observe that these fibers reach the sympathetic trunk by way of the ventral root and the white ramus communicans.

horn cells of the spinal cord, (2) **sensory (afferent) fibers** of spinal ganglion cells, and (3) **autonomic fibers.**

Suggestions for Additional Reading

1. Basmajian JV: *Primary Anatomy*, ed 8. Baltimore, Williams & Wilkins, 1982.

   This book, written as a textbook for elementary courses in anatomy, has been found useful by many medical and dental students as *an introduction to anatomy*. Unlike many anatomy books, including this one, it stresses the systemic approach, which is very helpful to beginning students.

2. Basmajian JV: *Grant's Method of Anatomy*, ed 10. Baltimore, Williams & Wilkins, 1980.

   Grant's method of teaching anatomy made him a world renowned scholar. In addition to this book, his *Atlas* and *Dissector* are still widely used. *His method teaches you to reason anatomically.*

3. Bertram EG, Moore KL: *An Atlas of the Human Brain and Spinal Cord.* Baltimore, Williams & Wilkins, 1982.

   A photographic atlas of the central nervous system illustrating many dissections of the brain and spinal cord.

4. Cormack DH: *Introduction to Histology*. Philadelphia, JB Lippincott Company, 1984.

   The descriptions of bones, joints, muscles, lymphatics, and nerves in the present chapter are summaries. For more details, you are urged to read the descriptions in this concise textbook.

CHAPTER 1

The Thorax

The thorax (G. chest), the superior part of the trunk, is located between the neck and the abdomen. The superficial structures covering the anterior aspect of the bony **thoracic cage** (Fig. 1-1), consisting of the pectoral muscles and the breast, are mainly described with the upper limb (Chap. 6). The pectoral (L. *pectus*, chest) muscles act from the thorax to move the arm.

The bony thoracic cage gives protection to the heart and lungs and provides attachment for many muscles. It also forms a bellows-like chamber that moves during respiration.

Chest pain is a common occurrence, the significance of which varies from negligible to very serious. Most people have had a local sharp, sudden pain in the lateral part of the chest ("stitch in the side"), or a burning pain posterior to the sternum (heartburn owing to gastric upset). However, a person with a **heart attack** (myocardial infarction, **MI**) will probably describe a **crushing substernal pain** that does not disappear with rest. *See Case 1-1* at the end of this chapter (p. 138).

The evaluation of a patient with chest pain is largely concerned with discriminating between serious conditions and the many minor causes of chest pain. To perform a clinical examination, a good knowledge of the anatomy of the thorax is required. Because a patient's life may be "hanging in the balance," there is no opportunity to consult an anatomy book.

The heart and lungs are the most important viscera (L. soft parts) in the thorax (chest). As these organs are constantly moving in living persons, the thorax is one of the most dynamic regions of the body. The thorax also protects the great vessels (aorta and pulmonary trunk), the liver, and the spleen.

THE THORACIC WALL

The thoracic wall provides a good example of the correlation between structure and function in that the anatomical structures provide the mechanism for the functions of breathing, the protection of the underlying viscera, and the venous return from areas inferior to the thorax.

The thoracic wall of infants is thin owing to their underdeveloped muscles. Their osseocartilaginous thoracic cages are also soft and pliable.

BONES OF THE THORACIC WALL

The **thoracic cage** or skeleton of the thorax (Fig. 1-1) is formed by: the ***12 thoracic vertebrae*** and their *intervertebral discs* posteriorly; the ***12 pairs of ribs*** and ***costal cartilages***; and the ***sternum*** anteriorly.

THE THORACIC VERTEBRAE (Figs. 1-1, 1-2, 5-1, 5-2, 5-19, and 5-20)

The vertebrae are described in more detail with other parts of the vertebral column in Chapter 5. Their special features related to the thorax are: (1) **facets on their bodies** for articulation with the heads of ribs; (2) **facets on their transverse processes** for the rib tubercles, except for the inferior two or three; and (3) **long spinous processes** (spines).

The spinous processes vary in their posterior directions. Those of T5 to T8 are nearly vertical and overlap the vertebrae like roof shingles (Figs. 1-2*D* and 5-1). This arrangement covers the small intervals between the laminae of adjacent vertebrae, preventing sharp objects from penetrating the vertebral canal and injuring the spinal cord (Fig. 5-51), when the vertebral column is flexed.

The spinous processes of T1, T2, T11, and T12 are horizontal and those of T3, T4, T9, and T10 are directed obliquely in an inferior direction.

The Costal Facets (Fig. 1-2). Thoracic vertebrae are unique in that they have facets on their bodies and transverse processes for articulation with the ribs (*T11 and T12 are exceptions*).

Two demifacets are located laterally on the bodies of T2 to T9. The **superior demifacet** is for articulation with the head of its own rib (Fig. 1-3), and the **inferior demifacet** is for articulation with the head of the rib immediately inferior to it.

The costal facets of the other vertebrae vary somewhat. T1 has a single costal facet for the head of the first rib and a small demifacet for the cranial part of the second rib (Fig. 1-2*A*). T10 has only one costal

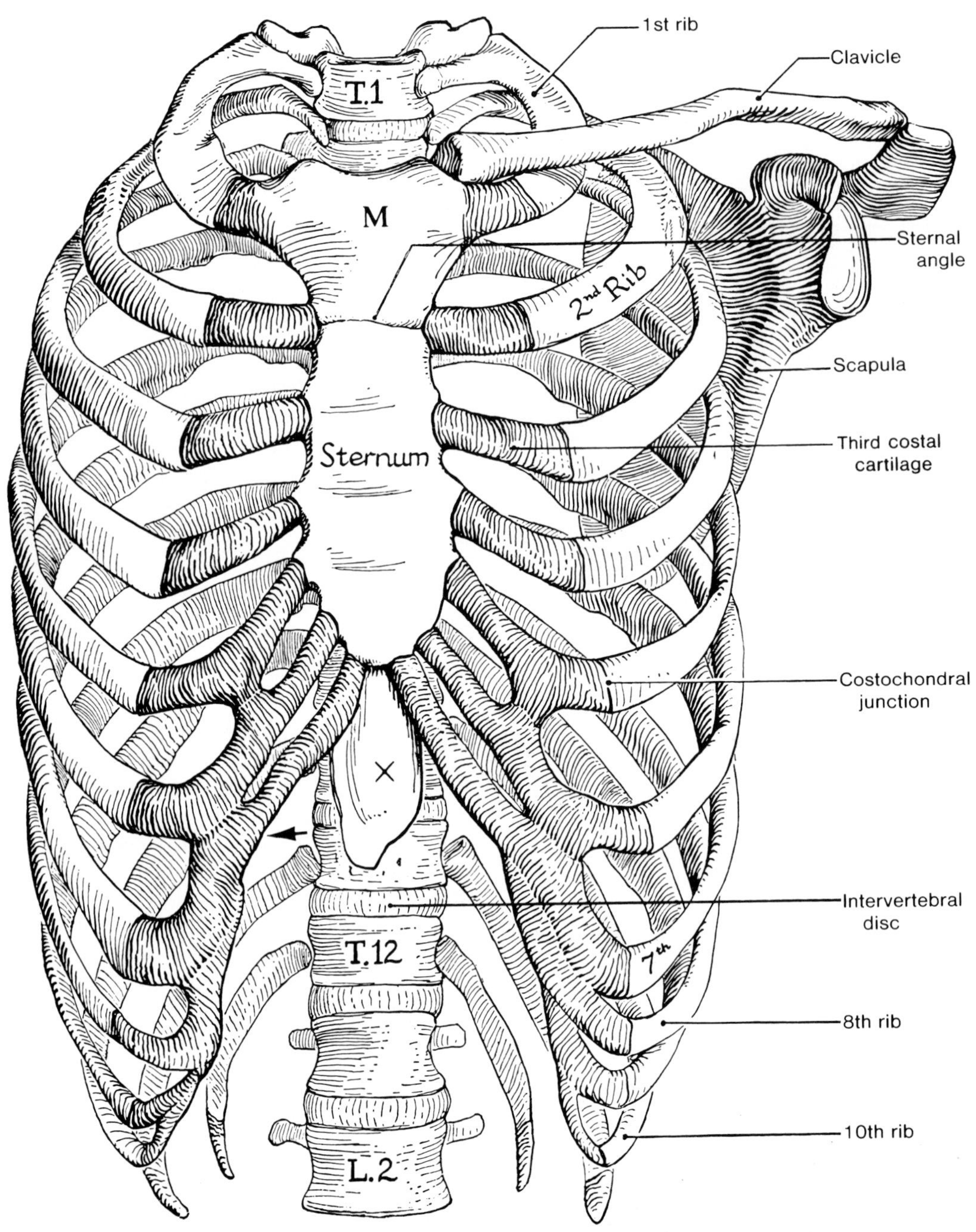

Figure 1-1. Anterior aspect of the bony thorax (thoracic cage). Observe its parts: 12 thoracic vertebrae, 12 pairs of ribs and costal cartilages, and the sternum. Note that each rib articulates posteriorly with the vertebral column and that the costal cartilages of the superior seven pairs articulate directly with the sternum. Note that the 8th, 9th, and 10th costal cartilages articulate with the cartilages superior to them, and that the 11th and 12th cartilages are free anteriorly and have cartilaginous tips. Observe the inferior inclination of all ribs. Note that the costal cartilages of the 3rd to 10th ribs incline superiorly. Observe that the tip of the 12th costal cartilage is at the level of the 2nd lumbar vertebra and that the 10th ribs and costal cartilages are the most inferior ribs and cartilages visible anteriorly. Note that the clavicle lies anterior to the first rib making it difficult to palpate anteriorly in living persons. The second rib is easy to locate because its costal cartilage articulates at the junction of the manubrium (*M*) and the body of the sternum. *The sternal angle, opposite the second costal cartilage, is a key landmark in the chest* that is used when counting the ribs. Observe the costal margin (*arrow*) formed by the upturned costal cartilages of the 7th to 10th ribs. *X* indicates the xiphoid process of the sternum. Also see Figure 1-9. For a drawing of the posterior aspect of the bony thorax, see Figure 5-2.

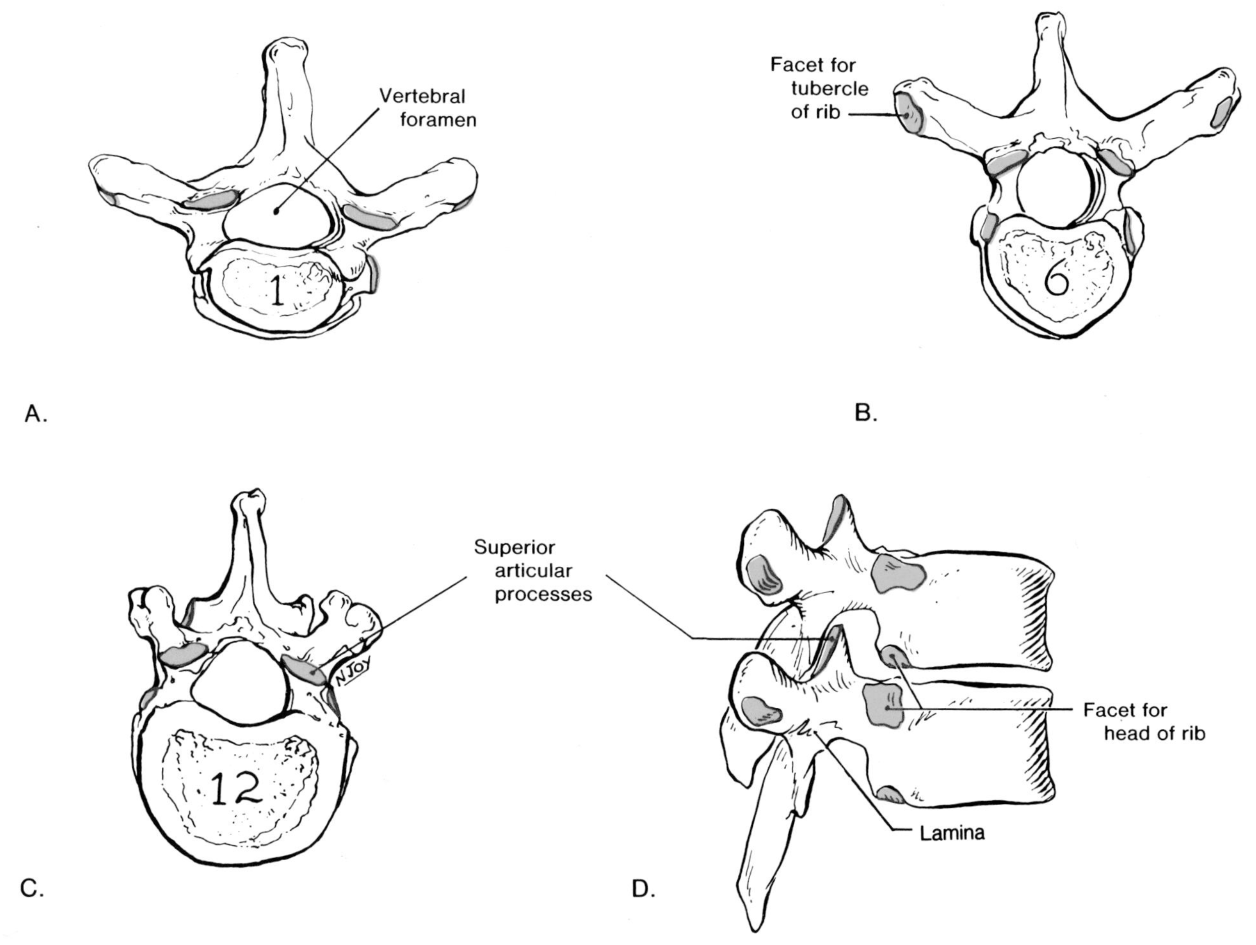

Figure 1-2. *A* to *C*, drawings of superior views of thoracic vertebrae (T1, T6, and T12). *D*, drawing of a lateral view of two typical midthoracic vertebrae. Observe the variation in the size and shape of all these vertebrae and the variation in the shape of the vertebral foramen in T1, T6, and T12. Thoracic vertebrae differ from other vertebrae in that they have costal facets (*yellow*) on their bodies and transverse processes for articulation with ribs, except for T11 and T12, which have only a single costal facet located laterally on their pedicles. Differentiate between costal facets and those for the articular processes, which are also colored yellow.

facet, part of which is on its body and part on its pedicle. T11 and T12 (Fig. 1-2*C*) have only a single costal facet on their pedicles.

Movements between adjacent vertebrae are relatively small in the thoracic region. The limited movements of this part of the vertebral column protects the thoracic viscera.

The bodies of T5 to T8 vertebrae are related to the descending thoracic aorta, which sometimes forms an impression on their left sides (Fig. 1-15). If the aorta develops an **aneurysm** (localized dilation) in this region, the vertebrae may be partly eroded by pressure from this vessel. In some cases these bony changes are visible on chest radiographs. The intervertebral discs are apparently unaffected by pressure from an aneurysm (G., a widening).

Surface Anatomy (Figs. 5-1 and 5-10). The spinous processes of all the thoracic vertebrae can be palpated in the posterior median line. When the vertebral column is flexed, the first spinous process that is visible is usually C7 (**vertebra prominens**), but the spinous process of T1 may be almost as prominent.

THE RIBS (Figs. 1-1 to 1-5)

These elongated flat bones *form the largest part of the* ***thoracic cage***. The sternum, the costal cartilages, and the vertebrae form the other parts of the thoracic

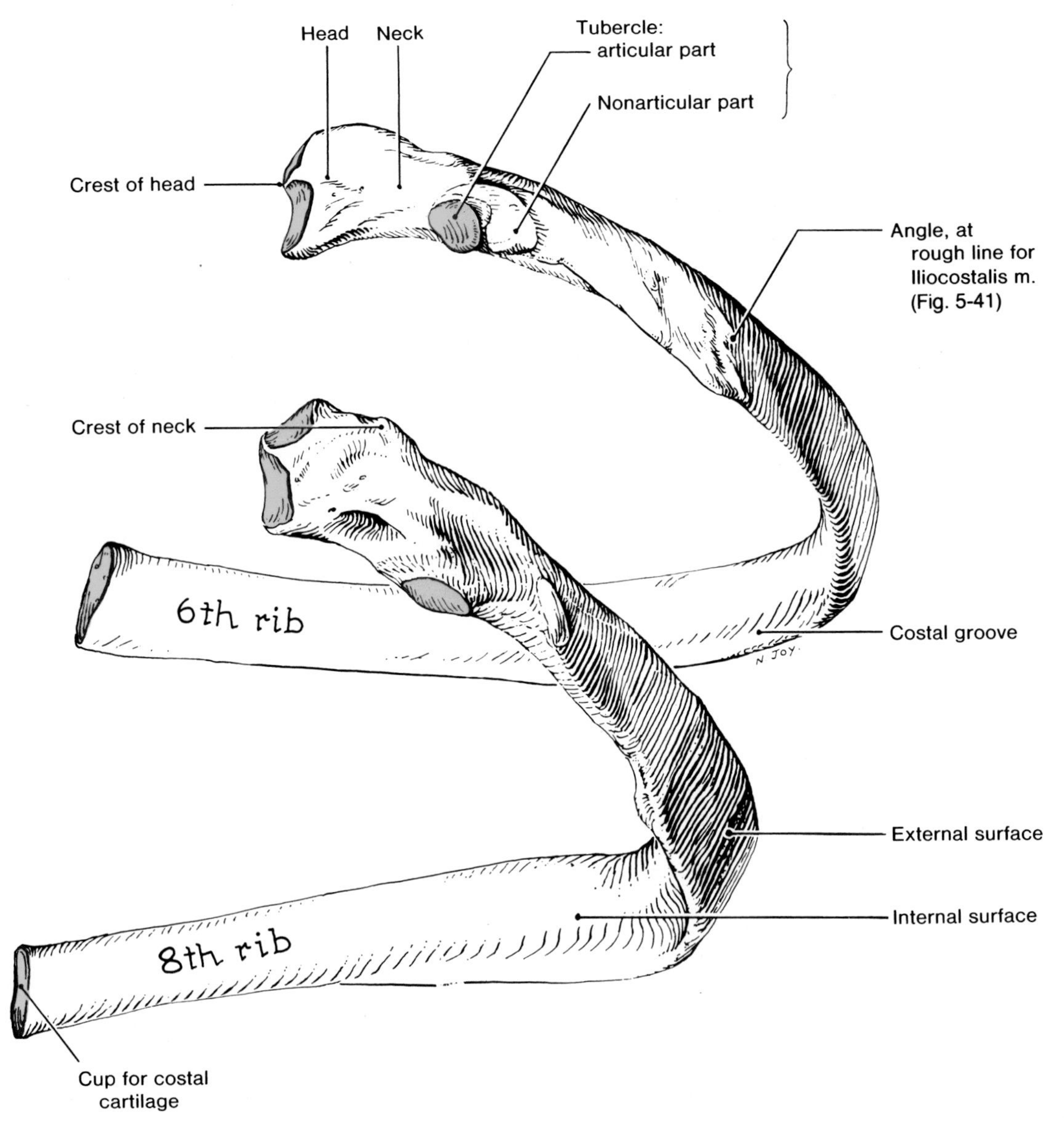

Figure 1-3. Posterior view of two **typical right ribs**. Ribs 3 to 10 are called typical ribs, but the ones shown here are most typical of ribs in general. *They will not lie flat on the table* because they bend relatively sharply at their angles. Observe the wedge-shaped *head*, with a larger inferior facet for its own vertebra and a smaller facet for the vertebra superior to it. Note the crest of the head which is joined to the intervertebral disc by an intra-articular ligament (Fig. 1-82). Examine the posterior surface of the *neck*, noting that it is rough where the costotransverse ligament attaches (Fig. 1-19). Note the sharp *crest of the neck* for the superior costotransverse ligament. Observe the nonarticular part of the tubercle for the lateral costotransverse ligament. Examine the convex articular part of the tubercle (*yellow*) which is on the posterior aspect of the superior seven ribs. Note that the posterior part of the shaft is round on cross-section and that the anterior part is flattened and, being articular, slightly enlarged. Observe the rough line for the attachment of the iliocostalis muscle, where the rib takes not only a bend but also a twist. *Examine the costal groove which lodges the intercostal vein, artery, and nerve* (Fig. 1-22).

cage. The ribs are long, thin, curved, slightly twisted arches of bone. **There are usually 12 pairs of ribs,** but the number may be increased by the development of *cervical ribs* (Fig. 6-35) or lumbar ribs (Fig. 5-44), or it may be decreased by ***agenesis*** (failure of formation) of the 12th pair.

True Ribs (Fig. 1-1). **The first seven pairs of ribs are called true ribs or *vertebrosternal ribs*** because they *articulate with the vertebrae and the sternum.* True ribs increase in length *in a craniocaudal direction.*

False Ribs (Fig. 1-1). **The 8th to 12th pairs of ribs are called false ribs or *vertebrocostal ribs*** because they *articulate with the vertebrae and are* *attached to the sternum through the costal cartilage of another rib, or not at all.* The 8th, 9th, and 10th ribs join the sternum indirectly through articulation of their costal cartilages with each other, and fusion of this combined cartilage with the seventh costal cartilage.

The 11th and 12th ribs are called "floating ribs" because they do not attach to the sternum. Their anterior ends are free and end amongst the muscles of the abdominal wall.

The last two ribs are not really floating because they articulate with the body of their own numbered vertebra (Fig. 1-1).

Variations in the sternocostal junctions are not uncommon. Although the first seven costal cartilages nearly always join the sternum (Fig. 1-1), sometimes only six cartilages articulate with it. In other people, the eighth costal cartilage, especially on the right, joins the sternum (Fig. 1-16). The 10th rib in some people is also normally floating.

Typical Ribs (Fig. 1-3). **Ribs 3 to 10 are typical ribs**; however, the 10th rib is atypical in that it has only one articular facet on its head. *Typical ribs vary slightly in length* and in other characteristics (Fig. 1-1). Each typical rib has a *head*, a *neck*, a *tubercle*, and a *shaft* or a body.

The head of a rib is wedge-shaped and presents *two articular facets* for articulation with the numerically corresponding vertebra and the vertebra superior to it. The facets are separated by the **crest of the head** (Fig. 1-3), which is joined to the intervertebral disc (Fig. 1-1) by an *intra-articular ligament* (Fig. 1-82).

The neck of a rib is the stout, flattened part *located between the head and the tubercle* (Fig. 1-3). The neck lies anterior to the transverse process of the corresponding vertebra. Its superior border, called the **crest of the neck,** is sharp, whereas its inferior border is rounded.

The tubercle of a true rib is on the posterior surface *at the junction of its neck and shaft* and is most prominent on the superior ribs. The tubercles of most ribs have a smooth convex *facet which articulates with the corresponding transverse process* of the vertebra, and a rough nonarticular part for attachment of the lateral costotransverse ligament (Fig. 1-82). The tubercles of the 8th to 10th ribs have *flat facets* for articulation with similar facets on the transverse processes of the vertebrae.

The shaft (body) of a rib is thin, flat, and curved. Forming the greatest part of a rib, the shaft has external and internal surfaces, thick rounded superior borders, and thin, sharp, inferior borders (Fig. 1-3). A short distance beyond the tubercle, the shaft ceases to pass posteriorly and swings rather sharply anteriorly.

The point of greatest change in curvature is called the **angle of the rib.** Here the rib is both curved and twisted. The sharp inferior border of a rib projects inferiorly and is located external to the **costal groove** on the internal surface of the shaft (Fig. 1-3).

The costal groove and the flange formed by the inferior border of the rib are for the accommodation and protection of the intercostal vessels and nerve that accompany the rib (Fig. 1-22). Examine a typical dried rib and note that the costal groove fades out as the anterior end is approached (Fig. 1-3). Also observe that this end is cupped for reception of the costal cartilage.

Atypical Ribs (Fig. 1-4). The 1st, 2nd, 11th, and 12th ribs are atypical.

The first rib (Figs. 1-1 and 1-4) is the *broadest and most curved of all the ribs* and is the shortest of the true ribs.

The first rib is clinically important because so many structures cross or attach to it. It is flat and has a **prominent scalene tubercle** on the inner border of its superior surface *for the insertion of the* ***scalenus anterior muscle*** (Figs. 1-4 and 1-16).

The **subclavian vein** crosses the first rib anterior to the scalene tubercle and the **subclavian artery;** the inferior trunk of the **brachial plexus** (network of nerves to the upper limb, Fig. 1-16) passes posterior to it. Examine the grooves formed by the subclavian vessels on a dried first rib (Fig. 1-4).

The first rib has a single facet which articulates with the body of the first thoracic vertebra and slightly with the intervertebral disc superior to it. The prominent **tubercle of the first rib** articulates with the transverse process of the first thoracic vertebra (Fig. 5-2).

The first rib is difficult to palpate, but the first intercostal space can be felt just inferior to the clavicle. The first rib helps to form the *superior thoracic aperture* (see p. 67).

The second rib (Figs. 1-1 and 1-4) has a curvature similar to that of the first rib, but it *is thinner, much less curved, and about twice as long as the first rib.* It can easily be distinguished from the first rib by the presence of a broad rough eminence, the **tuberosity** for the serratus anterior muscle.

The 11th and 12th ribs are short, especially the 12th pair (Figs. 1-1 and 1-4). They are capped with cartilage and have a single facet on their heads and ***no neck or tubercle.*** The 11th rib has a slight angle and a shallow costal groove. The 12th rib has neither of these features.

The curvature of the ribs is such that a person lying of his/her back is supported by the spinous

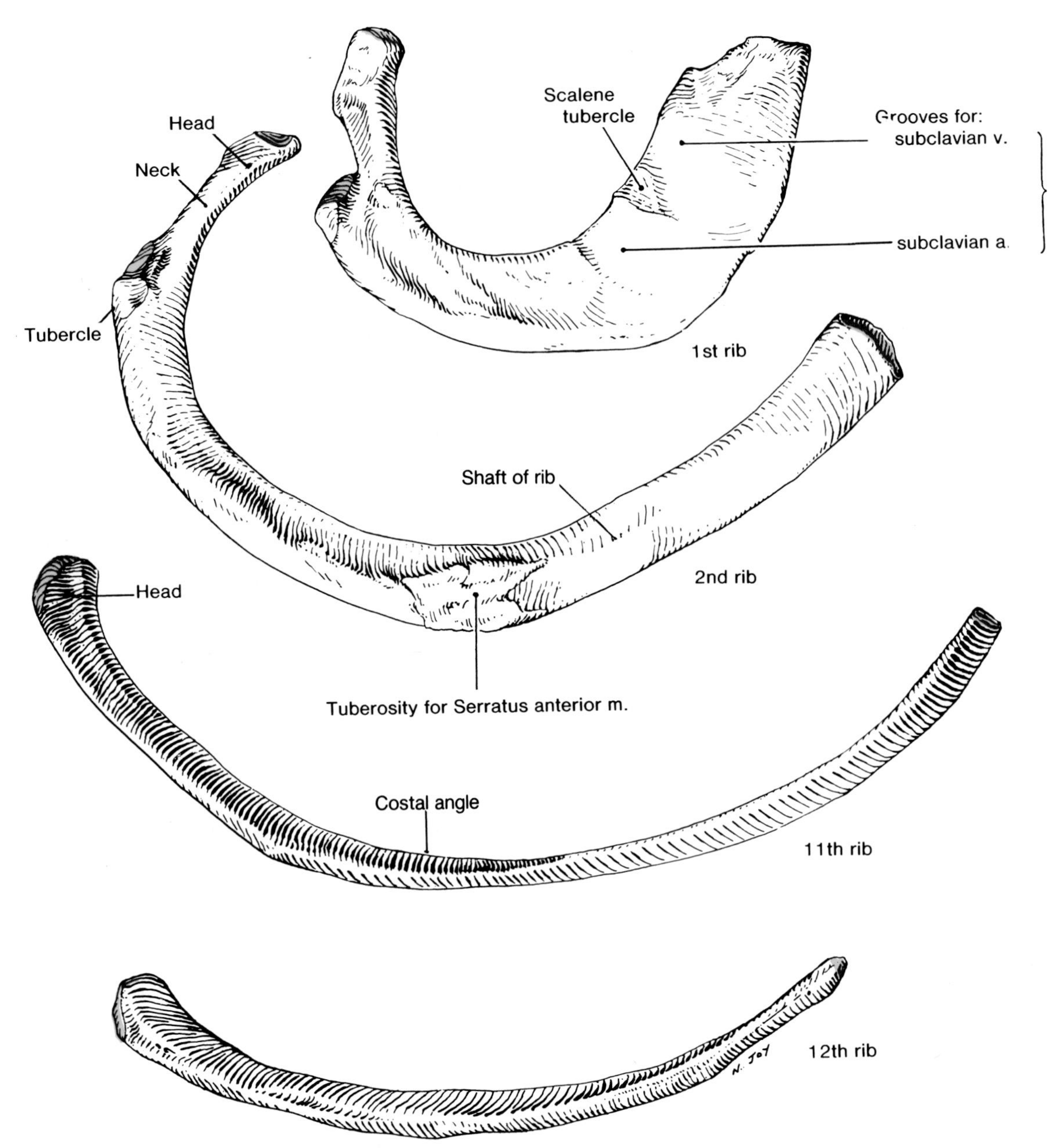

Figure 1-4. Superior view of four **atypical ribs**. These ribs have faint costal angles (*i.e.*, twists); hence *they will lie flat on the table*. Observe that the first rib is the shortest, broadest, strongest, and most curved of these ribs. Note that its head and neck are directed medially and inferiorly and that its head has a single facet. Observe that its tubercle is the most prominent of all, for here the tubercle and angle coincide. The tubercle for the scalenus anterior muscle separates the groove for the subclavian vein anteriorly from the groove for the subclavian artery and the inferior trunk of the brachial plexus posteriorly (Fig. 6-22). Note that the second rib is marked by the tuberosity for the serratus anterior muscle. Observe that the 11th and 12th ribs, called floating ribs, have a single facet on their heads for articulation with the body of their own numbered vertebra. Note that these ribs do not have tubercles, but have tapering ends.

processes of the thoracic vertebrae and the angles of the ribs. *The angles of the ribs are their weakest parts.*

Despite their ability to bend under stress, the ribs may be fractured by direct violence or indirectly by crushing injuries that commonly occur in automobile accidents. Because the ribs in infants and children are elastic, they do not often fracture. On some occasions, ribs have been fractured by muscular men hugging frail women too vigorously.

Fractured ("cracked") ribs are very painful because of their movements during respiration, coughing, and sneezing. *The middle ribs are the ones most commonly fractured.* The first two ribs (which are partially protected by the clavicle) and the last two (which are free and movable) are the least commonly injured.

Crushing injuries tend to break the ribs at their weakest point, *i.e.*, the site of greatest change in curvature, which is just anterior to the costal angle (Fig. 1-3). ***Direct injuries*** *may fracture a rib anywhere and the broken ends may be driven inwardly and injure the thoracic and/or abdominal organs* (*e.g.*, lungs and/or spleen).

Patients with fractured ribs experience considerable pain in the region of the fractures when asked to take a deep breath. Similarly, careful palpation along the broken rib often reveals local tenderness, even when *a fracture may not be visible on a radiograph.*

Flail chest ("stove-in chest") occurs when a sizeable segment of the anterior and/or lateral chest wall is freely movable because of multiple rib fractures. This allows the loose segment of the chest wall to move paradoxically (*i.e.*, inward on inspiration and outward on expiration). This impairs ventilation and thereby affects oxygenation of the blood; if severe, death results. Current treatment is to fix the loose segment of the chest wall by hooks and/or wires so that it cannot move.

Any cutting operation on the thoracic wall is called a **thoracotomy**. To remove a **collection of pus** in the pleural cavity, called **empyema**, sometimes a partial **rib excision** is performed. The piece of rib is removed in a manner similar to that described for excision of a piece of costal cartilage (Fig. 1-16). When removed from its periosteum, access to the pleural cavity (Fig. 1-11) is gained by making an incision in the bed of the rib (*i.e.*, periosteum and fascia).

Sometimes a piece of rib is used for **autogenous bone grafting** (*e.g.*, for reconstruction of the mandible following excision of a tumor). Following the operation, the rib defect regenerates from the remaining periosteum.

Cervical ribs are present in up to 1% of people and articulate with the seventh cervical vertebra (Fig. 6-35). A cervical rib extends into the neck, where its anterior end may be free; be attached to the first rib; its costal cartilage; or even the sternum.

Often cervical ribs are incidental observations on routine radiographs and may not be mentioned unless **neurovascular symptoms** have been reported in the upper limb.

Lumbar ribs are more common than cervical ribs. They articulate with the first lumbar vertebra or are attached to the tips of its transverse processes (Fig. 5-44).

Lumbar ribs have clinical significance in that they may confuse the identification of vertebral levels in radiographs. In addition, a lumbar rib may be erroneously interpreted as a fracture of a lumbar transverse process by an inexperienced observer of radiographs. On rare occasions the 12th ribs fail to develop; hence persons may have only 11 pairs.

Otherwise normal people may have 11 to 14 pairs of ribs. In some persons with congenital abnormalities of the vertebral column, the number may even be less. Accessory cervical or lumbar ribs represent the *costal elements* (Fig. 1-5) that have developed (abnormally) into ribs.

Coarctation of the aorta, or narrowing of the distal part of its arch, results in enlargement of the intercostal arteries and other arteries as a means of providing a **collateral circulation** to inferior parts of the body. These enlarged intercostal arteries erode or notch the inferior borders of the corresponding ribs. Radiographic demonstration of this **notching of the ribs** is very useful in confirming this suspected congenital malformation of the aorta.

Bifid ribs or forked ribs are common (1–2% of people) and this possibility must be considered when a rib count seems to indicate that the cartilages of the superior eight ribs articulate with the sternum. Usually the condition is unilateral; hence, unilaterality of eight apparent true ribs favors the suspicion of a bifid rib. However, in some cases (Fig.1-16) eight normal ribs articulate with the sternum.

Fused ribs are uncommon and are often associated with a vertebral abnormality, *e.g.*, a **hemivertebra** (failure of half of a vertebra to form, Fig. 5-7).

THE COSTAL CARTILAGES (Figs. 1-1 and 1-16)

These *flattened bars of* ***hyaline cartilage*** extend from the anterior ends of the ribs and contribute significantly to the elasticity and mobility of the ribs. *The costal cartilages are unossified parts of the embryonic cartilaginous ribs*, hence their perichondrium blends with the periosteum of the ribs associated with them (Fig. 1-16).

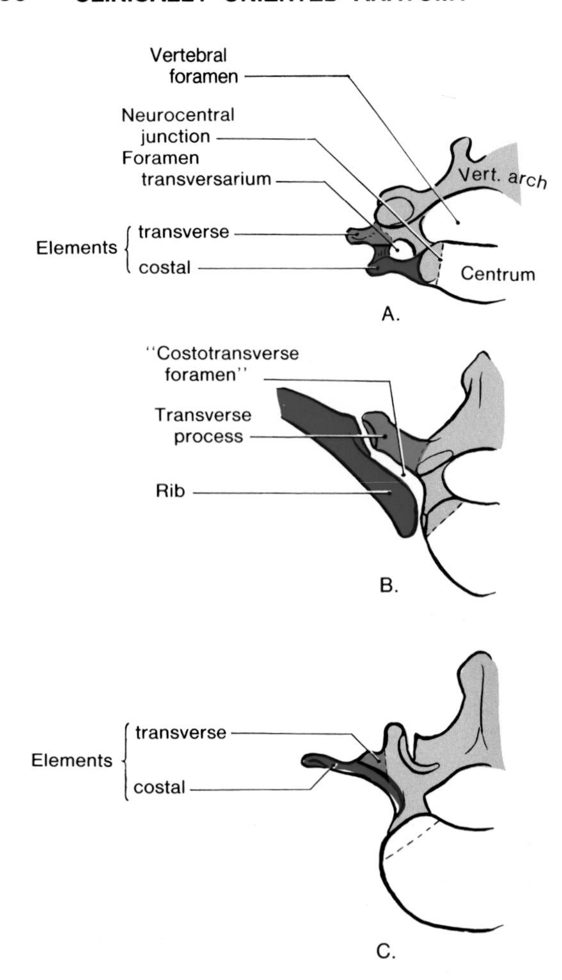

Figure 1-5. Diagrams illustrating the homologous parts of vertebrae. The vertebral arch lies lateral and posterior to the vertebral foramen for the spinal cord and meninges (Fig. 5-51). The centrum is the ossification center of the central part of the body of a vertebra. *Red*, rib or costal element; *blue*, vertebral arch and its processes; *uncolored*, centrum. *A*, in the cervical region the costal element is represented by the anterior part of a transverse process of a cervical vertebra. In up to 1% of people the costal element of C7 develops into a cervical rib (Fig. 6-35). *B*, in the thoracic region the costal element develops into a rib. *C*, in the lumbar region the costal element usually becomes the anterior part of a transverse process, but the costal element of the first lumbar vertebra often develops into a short lumbar rib (Fig. 5-44).

The superior seven costal cartilages are connected with the sternum (Figs. 1-1 and 1-7). Those of the 8th to 10th ribs articulate with the inferior border of the cartilage superior to them. Occasionally the 8th costal cartilage articulates with the xiphoid process of the sternum (Fig. 1-16). The two most inferior costal cartilages have pointed tips which end in the muscular wall of the abdomen (Fig. 1-18). The costal cartilages increase in length from the first to the seventh and then gradually become shorter (Fig. 1-1).

THE COSTAL MARGIN (Figs. 1-1 and 1-9)

The 7th to 10th costal cartilages meet on each side and their inferior edges form the costal margin. The diverging costal margins form an **infrasternal angle** (subcostal angle) inferior to the xiphisternal joint (Fig. 1-46). The apex of the angle is at this joint.

The medial end of the first costal cartilage joins the ***manubrium*** with no joint between them; its perichondrium blends with the periosteum of the first rib and the manubrium of the sternum (Fig. 1-84).

The medial ends of the second to seventh costal cartilages are rounded and *articulate with the body of the sternum* (Figs. 1-1 and 1-7). These sternocostal articulations are *synovial joints*. The costal cartilages of the 8th and 10th ribs do not reach the sternum; they taper to a point and articulate with the cartilages just superior to them. The costal cartilages of the 11th and 12th ribs are unattached and pointed.

The costal cartilages of young people provide resilience to the thoracic cage, preventing many crushing injuries or direct blows from fracturing the ribs and/or the sternum.

In older adults the costal cartilages often undergo superficial calcification. As a result, they lose some of their elasticity and become brittle. This calcification makes them *radiopaque* (*i.e.*, visible on radiographs), which may be confusing to inexperienced observers.

Separation of a rib (separated costal cartilage) refers to a *dislocation of the costochondral junction* (Fig. 1-1) between the rib and its costal cartilage, and not to a lack of continuity in the shaft that may be detected in a fractured rib, especially in a thin person. In a separated rib, one of the costal cartilages of the ribs, usually the 10th, separates from the inferior border of the costal cartilage immediately superior to it. Consequently, the cartilage of the **slipping rib** can move superiorly and override the one superior to it, causing pain. Slipping ribs are most common in young women, but the reason for this is unclear.

THE STERNUM (Figs. 1-1 and 1-6 to 1-10)

The sternum (breastbone) is an elongated flat bone, resembling a short broadsword or dagger, which forms the middle part of the anterior wall of the thorax.

The sternum (G. *sternon*, chest) *consists of three parts*: **manubrium, body, and xiphoid process.** The sternum is covered anteriorly only by skin, superficial fascia, and periosteum, except where the pectoralis major muscle and the sternal head of the sternocleidomastoid muscle arise from it (Fig. 6-18).

The Manubrium of the Sternum (Figs. 1-1, 1-6, and 1-7). The manubrium (L. handle), or superior part of the sternum, is wider and thicker than its inferior two parts (body and xiphoid process). Although generally quadrilateral, the narrow inferior end of the manubrium gives it a somewhat triangular shape.

Broad and thick superiorly, the manubrium slopes inferiorly and anteriorly. The superior surface of the manubrium is indented by a **jugular notch** (suprasternal notch). On each side of this notch (Fig. 1-7*A*) is an oval articular facet, called the **clavicular notch**, for articulation with the medial end of the clavicle. Just inferior to the clavicular notch, the costal cartilage of the first rib is fused with the lateral margin of the manubrium (Figs. 1-1 and 1-7). This is a flexible but strong *primary cartilaginous joint* (***synchondrosis***) where the costal cartilage unites with the manubrium (Fig. 1-84).

The inferior border of the manubrium is oval and rough where it articulates with the body of the sternum at the **manubriosternal joint**. This is a *secondary cartilaginous joint* where fibrocartilage and ligaments join the bones. This articulation is often ossified in older persons (Fig. 1-6*C*).

The manubrium and body of the sternum lie in slightly different planes, hence their line of junction at the manubriosternal joint forms an anteriorly projecting obtuse **sternal angle** (angle of Louis). This very important bony landmark is about 5 cm inferior to the jugular notch (Figs. 1-1 and 1-6 to 1-10).

The sternal angle is an important clinical guide to the accurate numbering of the ribs.

The Body of the Sternum (Figs. 1-1 and 1-6 to 1-9). The body of the sternum is the longest of the three parts of the sternum. It is longer, thinner, and narrower than the manubrium, but its width varies as a result of the scalloping of its lateral borders by the costal notches (Fig. 1-7). It is broadest at about the level of the fifth sternocostal joints and then gradually tapers inferiorly.

The anterior surface of the body is slightly concave from side to side (nearly flat), and is directed slightly anteriorly. It is variably marked by three transverse ridges, representing the lines of fusion of the four **sternebrae**. These sternal ridges are usually absent in older people. The centers of ossification for the sternebrae appear in the fetus (20 to 24 weeks) and begin to fuse at puberty from the inferior end.

By 25 years all the sternebrae have fused to form the body of the sternum. In old age no evidence of the

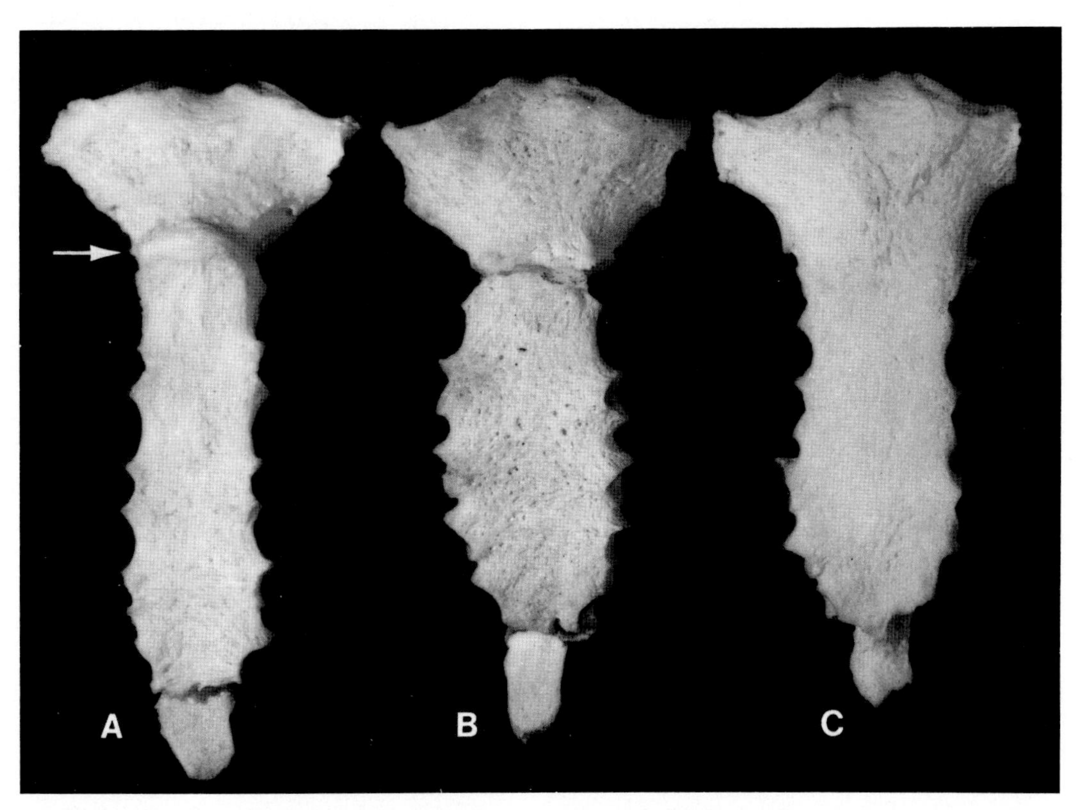

Figure 1-6. Photographs of anterior views of three sterna. *A*, from a male; *B* and *C*, from females. Observe that the bodies of the sterna from the females are shorter and broader than the sternum from the male and less than twice the length of the manubrium. Note that the sternal angle (*arrow*) forms a prominent ridge in *A* and that the union of the manubrium with the body of the sternum has become ossified in *C*. Similarly the union of the xiphoid process with the body of the sternum has become ossified in *B* and *C*. These specimens are from elderly patients.

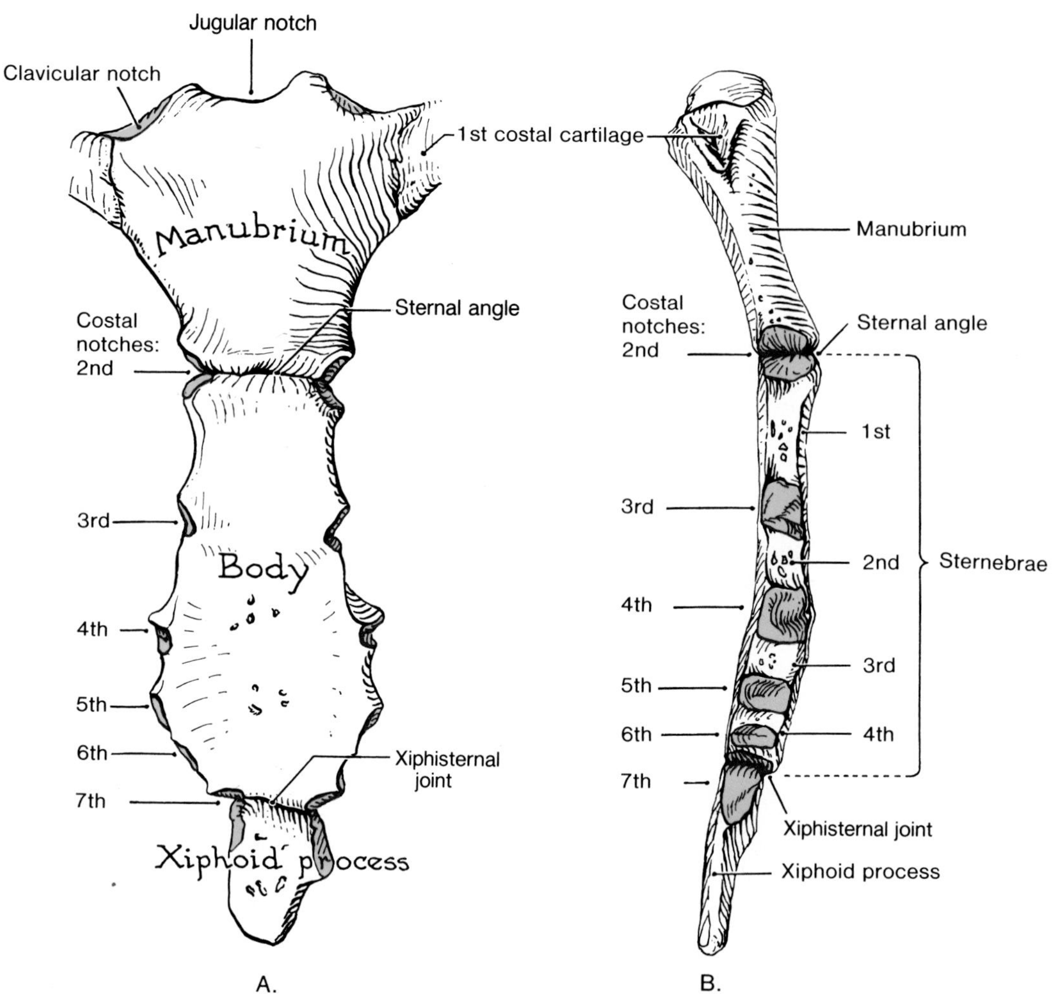

Figure 1-7. Drawings of an adult sternum. Note that it consists of three parts: manubrium, body, and xiphoid process. *A*, anterior view. *B*, lateral view. Observe that the sternum is a flat elongated bone that somewhat resembles a short broadsword. The *manubrium* (L. handle) represents the handle and the *xiphoid* (G. sword-like) process its pointed end. Observe the considerable thickness of the superior third of the manubrium between the clavicular notches. On a dried manubrium verify that its anterior surface is smooth and slightly convex, whereas its posterior surface is concave and featureless. Note the thin and tapering tip and sides of the xiphoid process. Examine two landmarks: (1) the *sternal angle* at the junction of the manubrium and body at the level of the second costal cartilage, and (2) the sharp inferior edge of the body at the *xiphisternal junction*. Observe where the first seven pairs of costal cartilages articulate with the sternum. The other ribs do not articulate with it, except for the eighth which may do so (Fig. 1-16). The sharp inferior edge of the body of the sternum at the xiphisternal junction is palpable. Forceful displacement of the bony xiphoid process in adults (over 40) may lacerate the liver.

former existence of the sternebrae is detectable (Fig. 1-6).

The posterior surface of the body is slightly concave and transverse ridges marking the lines of fusion of the sternebrae may be visible, but they are less distinct than those on the anterior surface.

Frequently there is an opening in the body of the sternum called the **sternal foramen** (Fig. 1-84). This common anomaly resulting from a defect in ossification is of *no clinical significance*, except that its possible presence should be known so that it will not be misinterpreted as a bullet hole.

The body of the sternum is usually significantly shorter in females than in males and is less than twice the length of manubrium (Figs. 1-1 and 1-6). This **sex difference** is useful in identifying the sex of human skeletal remains.

The Xiphoid Process (Figs. 1-1 and 1-6 to 1-9). This thin, sword-shaped (G. *xiphos*, sword) process is the *smallest and most variable part of the sternum.* Although it may be pointed, it is often blunt, bifid, curved, deflected to one side or the other, or directed anteriorly. Although commonly perforated in old persons owing to incomplete ossification, this observation is of no clinical significance.

The xiphoid process is cartilaginous at birth and remains so until early childhood. It may begin to ossify during the third year and then consists of a bony core surrounded by hyaline cartilage. Usually this process does not begin until much later.

The xiphoid process usually ossifies and unites with the body of the sternum around the 40th year, but it may not unite even in old age.

In infants it is not unusual to see the tip of the xiphoid process protruding anteriorly beneath the skin.

Fracture of the sternum is uncommon, except in automobile accidents when the driver's chest is driven into the steering wheel. The body of the sternum is commonly fractured in the region of the sternal angle and it is often a **comminuted fracture** (broken into pieces). Fortunately the ligamentous coverings confine the fragments in most cases so that a compound fracture does not usually occur. A *compound fracture* is one in which there is an open wound leading to the broken bone.

In severe accidents, the body of the sternum separates from the manubrium at the manubriosternal joint and is driven posteriorly, rupturing the aorta and/or injuring the heart and liver. This may result in sufficient loss of blood and/or damage to the heart muscle (**myocardium**) that the patient dies.

Not uncommonly, men in their early forties suddenly detect their ossified xiphoid processes and consult their doctor about the "hard lump in the pit of their stomach." Never having felt it before, they fear they have developed a cancer in the stomach.

The sternum is important to the hematologist because it has a readily accessible medullary cavity (Fig. 1-84). During a procedure called **sternal puncture**, a large bore needle is inserted through the cortex of the sternum into the *red bone marrow* for aspiration of a sample of it for laboratory evaluation.

Access to the anterior mediastinum (Fig. 1-43). To reach the heart and great vessels access may be gained by splitting the sternum in the median plane.

In some people the body of the sternum projects inferiorly and posteriorly rather than inferiorly and anteriorly. This presses the heart posteriorly and widens it, making it appear large on frontal chest radiographs. A lateral chest radiograph reveals the true cause. This rare condition, usually congenital, is called **pectus excavatum** (L. funnel chest) and is characterized by a curving posteriorly of the inferior part of the sternum.

In other people, the chest is flattened on each side and the sternum projects anteriorly. This uncommon keel-like condition, called **pectus carinatum**, was given its name because the chest looks like the keel of a boat (L. *carina*, keel). This malformation is also called "pigeon chest."

SURFACE ANATOMY OF THE RIBS, COSTAL CARTILAGES, AND STERNUM (Figs. 1-8, 1-9, and 1-14)

The smooth superior border of the manubrium is easily felt because it has a shallow concavity which forms the floor of the **jugular notch** at the root of the neck. Palpate this notch with your index finger, verifying that the concavity is deepened by the medial ends of the clavicles. Their knobby ends are superior to their articulation with the manubrium because they are too large for the clavicular notches provided for them (Figs. 1-1 and 1-7). Note that the superior border of the jugular notch is at the level of the junction of the second and third thoracic vertebrae (Fig. 1-10).

Slide your index finger slowly down the midline of your manubrium until you feel a horizontal ridge at the junction of the manubrium and the body of the sternum. As fat does not accumulate anterior to the manubrium, this ridge, indicating the **sternal angle** (Figs. 1-6 to 1-10), is often visible and is usually palpable, especially on inspiration.

The sternal angle is located about 5 cm inferior to the jugular notch and forms where the manubrium joins the body of the sternum at the manubriosternal joint, a symphysis (Figs. 1-6*A* and 1-10).

The sternal angle is situated at the level of the second costal cartilage (Figs. 1-1 and 1-9) and is at the level of the junction of the fourth and fifth thoracic vertebrae (Fig. 1-10). When traced laterally, the horizontal ridge indicating *the sternal angle directs the palpating finger to the second costal cartilage, the starting point from which the ribs are counted.*

The first rib is difficult to palpate because it lies deep to the clavicle and is covered by muscles (Figs. 1-1 and 1–16).

The accurate numbering of ribs posteriorly is more

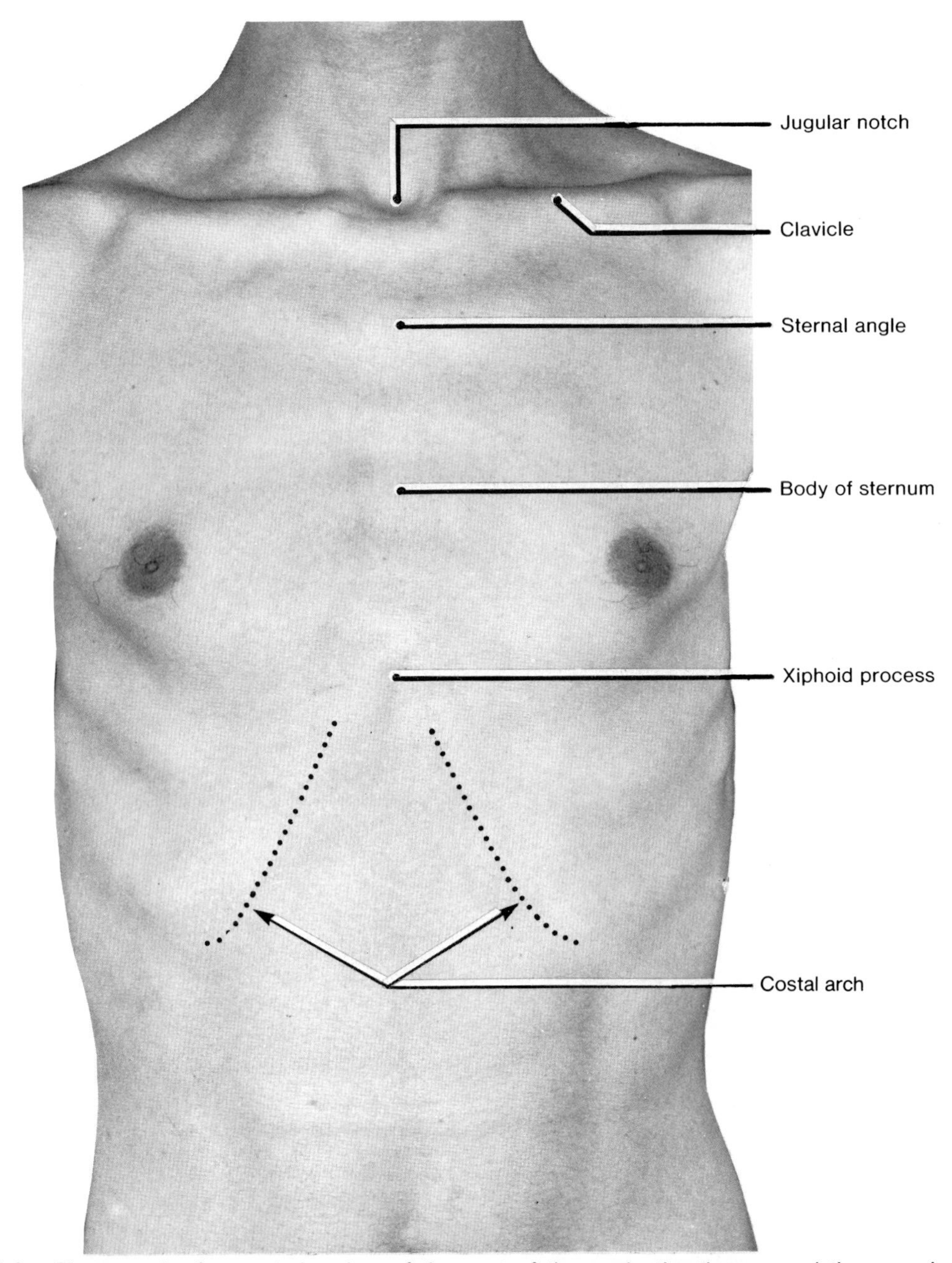

Figure 1-8. Photograph of an anterior view of the root of the neck, the thorax, and the superior part of the abdomen of a 27-year-old man showing the surface features of the clavicles, ribs, and sternum. Observe the sternal angle at the junction of the manubrium and body of the sternum. This ridge forms because the manubrium and the body of the sternum do not articulate in a flat plane. The sternal angle may be visible and is palpable under the skin. It is an important bony landmark clinically in connection with the identification of ribs. Visualize the following two lines of orientation: (1) the *midsternal line* lies in the median plane over the sternum, and (2) the *midclavicular line* runs inferiorly from the midpoint of the clavicle (Fig. 1-9), about 9 cm from the median plane. Together the two costal margins (*broken lines*) form the costal arch.

difficult. The ribs can be counted posteriorly by counting superiorly from the short 12th one, which can be palpated in most people but with difficulty in obese persons. Sometimes the 12th rib is so rudimentary that it does not project beyond the lateral border of the erector spinae muscle (Fig. 5-11), in which case it is not palpable. Also see Figure 2-105.

Count the ribs on the back of a colleague. Note that the spine of the scapula lies over the third rib or third intercostal space (Fig. 5-2). The inferior angle of the

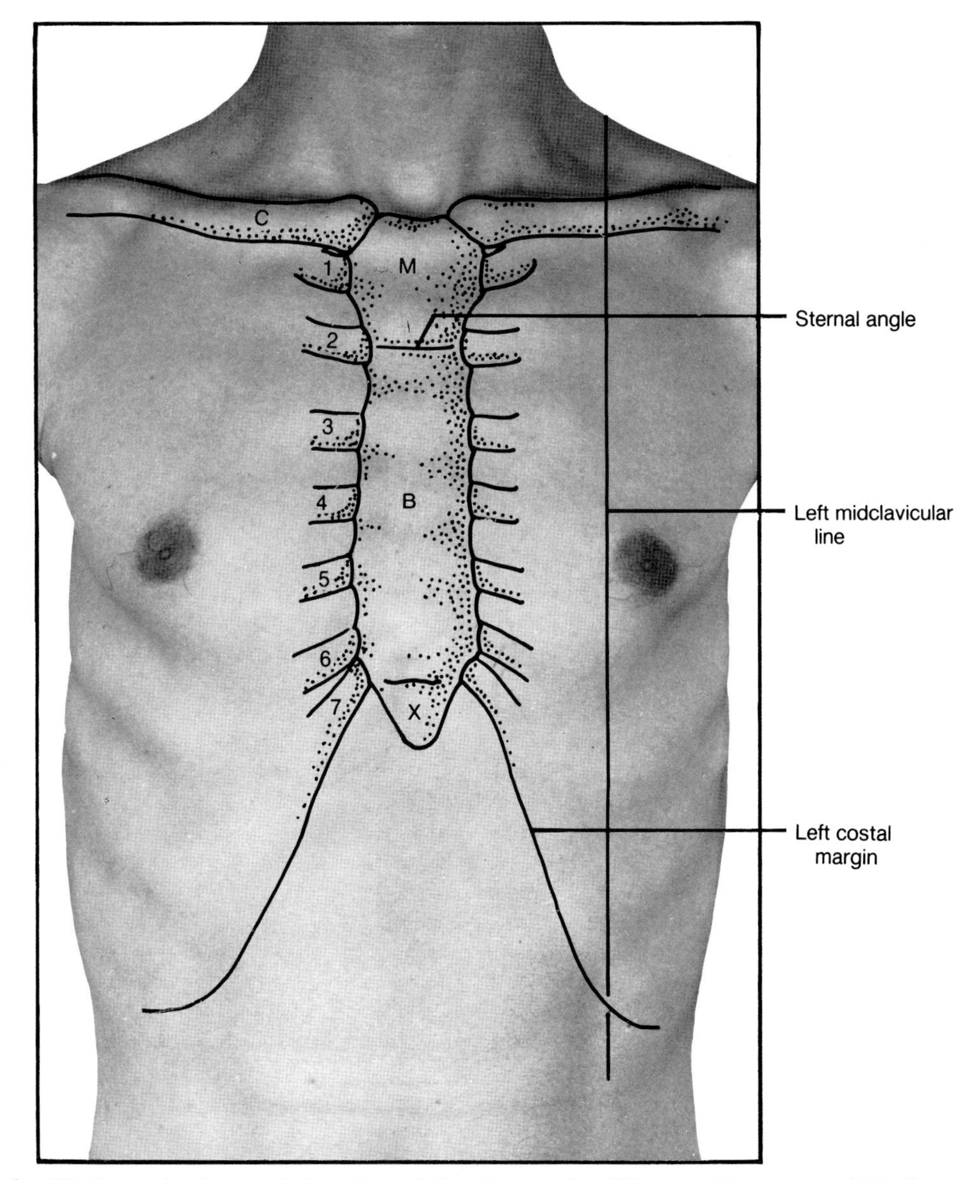

Figure 1-9. Photograph of an anterior view of the thorax of a 27-year-old man on which the outlines of the clavicle, costal cartilages, and sternum have been drawn. *C*, clavicle; *M*, manubrium of sternum; *B*, body of sternum; and *X*, xiphoid process of sternum. The sternal angle is the best guide for numbering the ribs. Note that it lies adjacent to the second costal cartilage.

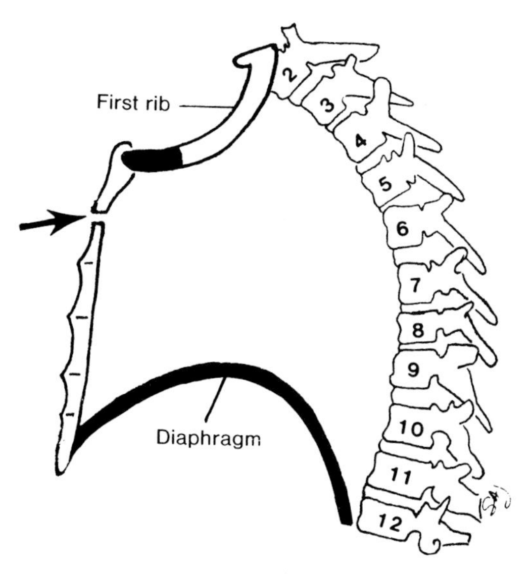

Figure 1-10. Diagrammatic sketch of a median section of the bony thorax, indicating various levels and lengths. The *arrow* indicates the sternal angle where the manubrium articulates with the body of the sternum at the manubriosternal joint, a secondary cartilaginous joint (symphysis). Note the costal cartilage of the first rib (*black tip*) joining the manubrium.

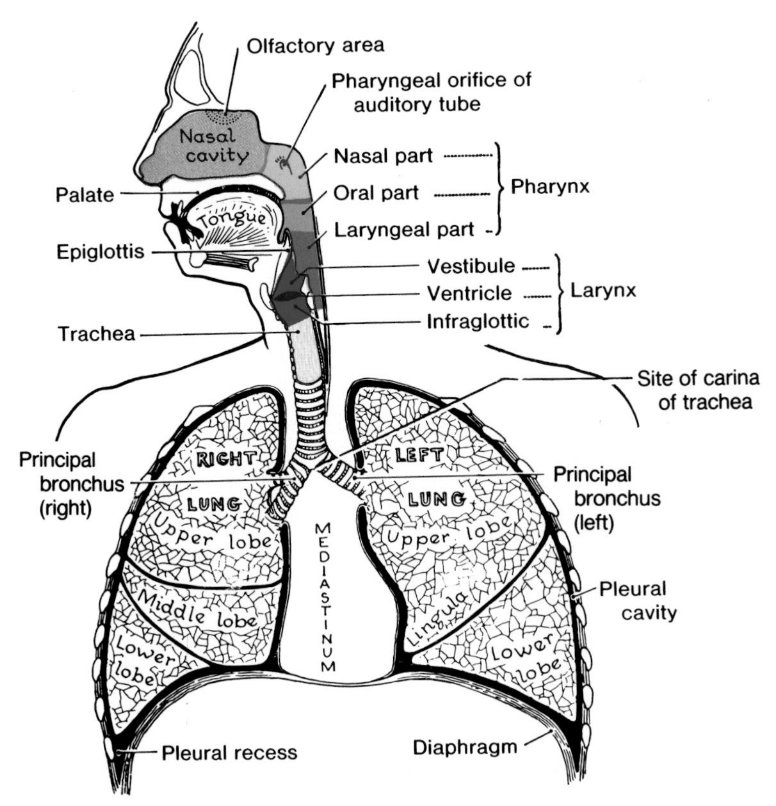

Figure 1-11. Diagram illustrating the relationship of the upper respiratory system (conduction portions) to the lower respiratory system (respiratory portions). The upper respiratory system comprises the air-conducting tubes that connect the exterior of the body with the lungs, where the exchange of gases between the blood and the air takes place. Observe the dome-shaped diaphragm which is attached to the sternum, the ribs, and the vertebral column. This partly muscular partition is the main muscle of respiration. During inspiration the diaphragm contracts and descends, whereas during expiration the relaxed diaphragm moves superiorly. The vertical diameter of the thorax is increased by the descent and flattening of the diaphragm. The carina of the trachea is a ridge on the inside of the trachea, at its bifurcation.

scapula is at the level of the seventh rib and is a good guide to the seventh intercostal space (Fig. 5-2).

The scapular surface markings are not so accurate for identifying the ribs as is the sternal angle.

Often there is a need to count the ribs (*e.g.*, to determine which rib is diseased or injured). First the sternal angle is located and then the palpating finger is passed directly laterally from it on to the second rib. The ribs then are counted inferiorly from this point.

On rare occasions the sternal angle occurs at the level of the third costal cartilage. You should suspect this condition if the manubrium seems longer than usual and the sternal angle is more than 5 cm inferior to the jugular notch.

Access to a lung is gained through the intercostal spaces. In conventional posterolateral **thoracotomy**, the incision is made along the line of the sixth rib. Obviously, accurate identification of this rib is very important (Fig. 5-2).

The costal margins are palpable with ease, extending inferolaterally from the sternum (Figs. 1-1, 1-8, and 1-9). The superior part of the costal margin is usually formed by the seventh costal cartilage and its inferior part is usually formed by the 10th costal cartilage. The degree of prominence of the costal margin varies and should be examined in several persons. It is very obvious in thin persons when they inspire (Figs. 1-8 and 1-14).

The infrasternal angle (xiphisternal angle, Fig. 1-26) also varies in size from person to person and increases during inspiration. This angle (notch) is used in *cardiopulmonary resuscitation* (*CPR*) for locating the proper hand position on the chest. The depressed area between these angles is called the **infrasternal fossa** (epigastric fossa or the "pit of stomach"). The *xiphisternal joint* is felt as a transverse ridge at the superior end of the infrasternal fossa (Figs. 1-7 and 1-8).

The xiphoid process (lying in the infrasternal fossa) can easily be felt extending inferiorly for a varying distance. Observe that the xiphoid process occupies a plane posterior to that of the body of the sternum. The anatomical basis for this is that the xiphoid process is thinner anteroposteriorly than the body of the sternum, and its posterior aspect is at the same level; hence its anterior aspect is posterior to that of the body of the sternum (Fig. 1-6).

MOVEMENTS OF THE THORACIC WALL

Movements of the thoracic wall are primarily concerned with increasing and decreasing the intrathoracic volume. The resulting changes in pressure result in air being alternately drawn into the lungs **(inspiration)** through the nose, mouth, larynx, and trachea and expelled from the lungs **(expiration)** through the same passages (Fig. 1-11).

The chest can increase in diameter in three dimensions to increase the volume of the thorax (vertical, transverse, and anteroposterior diameters). Each of these increases the volume, but most increase occurs when all three dimensions are increased at the same time.

THE VERTICAL DIAMETER (Fig. 1-11)

This dimension of the thorax is **increased primarily when the diaphragm contracts**, lowering it (Fig. 1-51). The superior ribs are raised slightly during very deep breathing, which also increases the vertical diameter of the thorax. During vigorous inspiration the diaphragm descends 5 to 10 cm, forcing the abdominal viscera inferiorly.

During expiration the vertical diameter is returned to normal by the subatmospheric pressure produced in the pleural cavities by the elastic recoil of the lungs. As a result, the diaphragm moves superiorly, diminishing the vertical diameter of the thorax (Fig. 1-52). It is also important to know that the weight of the abdominal viscera will push the diaphragm superiorly in the recumbent position.

THE TRANSVERSE DIAMETER (Fig. 1-12)

This dimension of the thorax is increased by the ribs swinging outward during the so-called "bucket-handle movement," which elevates ribs 2 to 10 approximately and turns their inferior borders outwards. This movement pulls the lateral portions of the ribs away from the midline, thereby increasing the transverse diameter of the thorax.

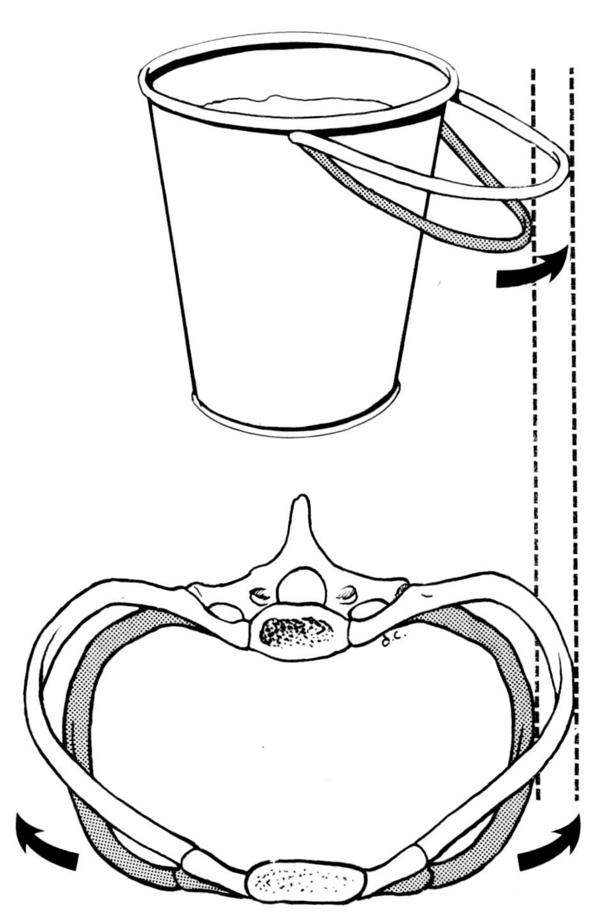

Figure 1-12. Diagrams illustrating the "bucket-handle" inspiratory movement. The inferior ribs move laterally when they are elevated, thereby increasing the transverse diameter of the thorax. Movement at the costovertebral joint leads to elevation of the middle of the rib. This movement takes place mainly at the 7th to 10th costotransverse joints (Figs. 1-82 and 1-83).

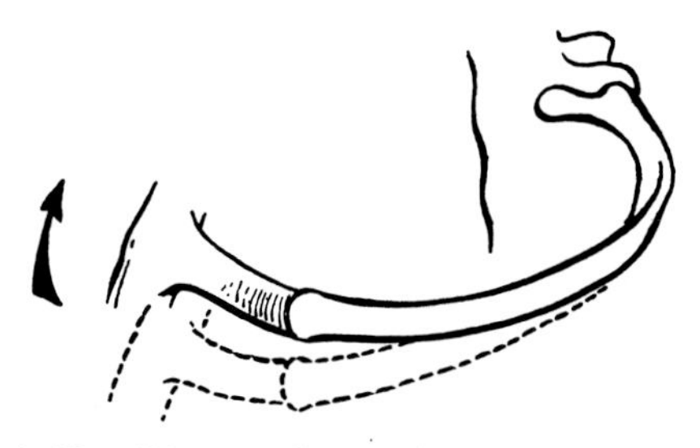

Figure 1-13. Diagram illustrating the "pump-handle" inspiratory movement. When the superior ribs are elevated, the anteroposterior diameter of the thorax is increased. Movement at the costovertebral joint about a side-to-side axis results in raising and lowering of the anterior end of the rib, the so-called pump-handle movement.

THE ANTEROPOSTERIOR DIAMETER (Fig. 1-13).

This dimension of the thorax is also increased by raising the ribs. Movement at the costovertebral joints through the long axes of the necks of the ribs results in raising and lowering their anterior ends, the "pump-handle movement." The second to sixth ribs are the ones mainly involved in this movement.

Because the ribs slope inferiorly, any elevation during inspiration results in a superior movement of the sternum at the manubriosternal joint and an increase

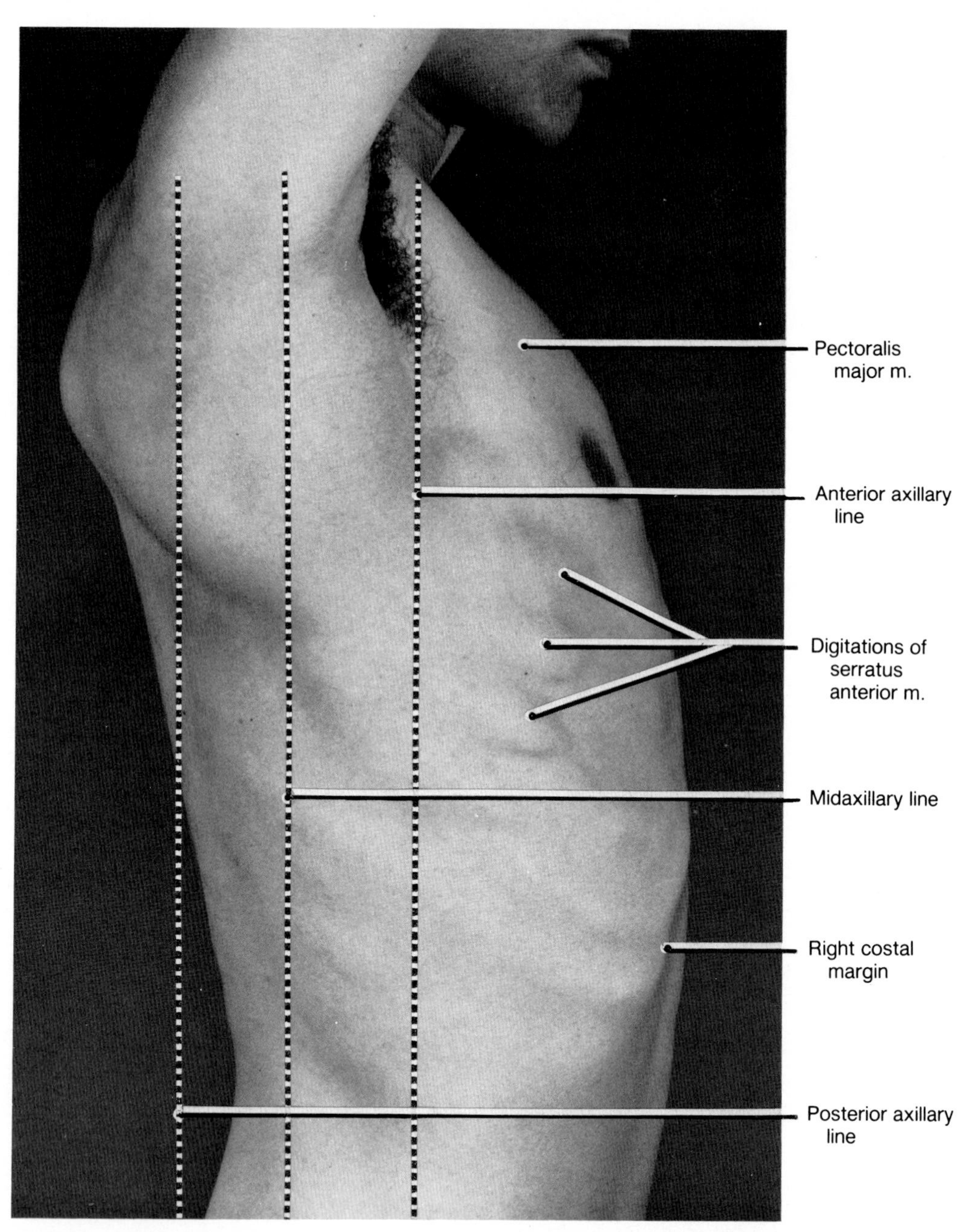

Figure 1-14. Photograph of a lateral view of a 27-year-old man from the right side with his arm raised over his head. Observe the midaxillary line which is an imaginary perpendicular line passing through the center of the axilla (armpit). The superior digitations of the serratus anterior muscle are invisible because they are covered by the pectoral muscles (Fig. 2-6). Most of the contour of the pectoral region in males is formed by the pectoralis major muscle, whereas in females the breast produces varying contours depending upon its development. The nipple in the male usually lies at the level of the fourth intercostal space (*i.e.*, between the fourth and fifth ribs). In the female the position of the nipple is not a reliable guide to this intercostal space because its position varies according to the size and shape of the breast.

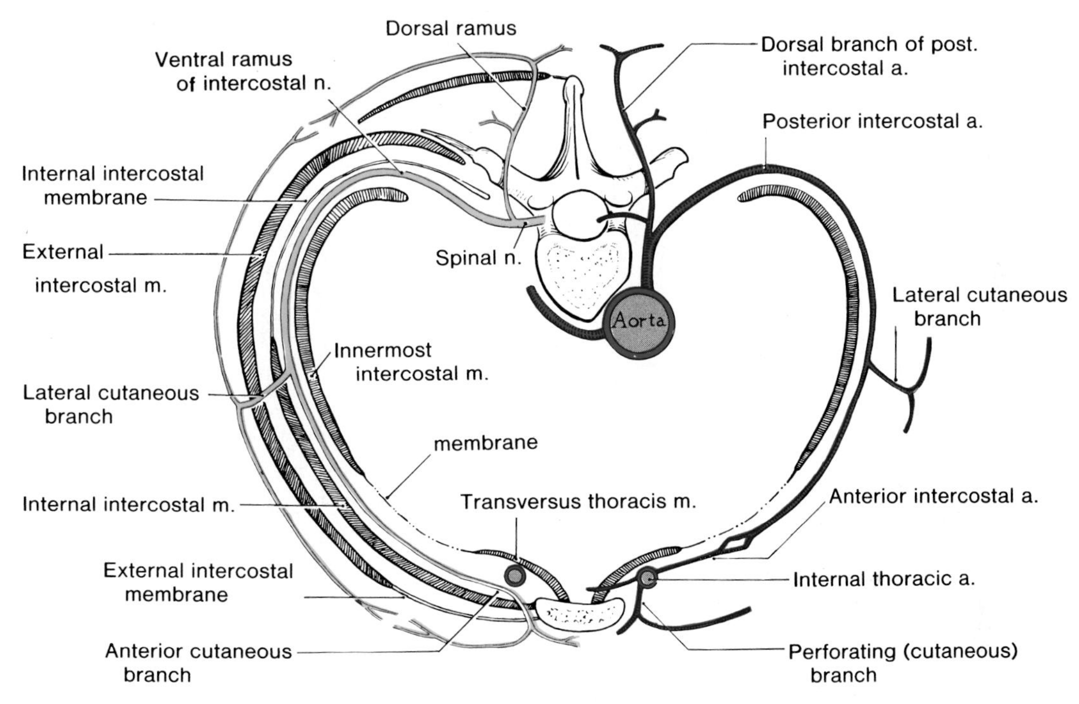

Figure 1-15. Diagram of a horizontal section of the thorax showing the contents of an intercostal space (also see Figs. 1-16 and 1-17). For purposes of illustration, the nerves are shown on the right (*yellow*) and the arteries on the left (*red*). Observe the three muscular layers: (1) external intercostal muscle and external intercostal membrane; (2) internal intercostal muscle and internal intercostal membrane; and (3) innermost intercostal and transversus thoracis muscles and the membrane connecting them. Examine the intercostal vessels and nerves running in the plane between the middle and innermost layers of muscle. The inferior intercostal vessels and nerves occupy the corresponding morphological plane in the abdominal wall. Note that the posterior intercostal arteries are branches of the aorta and that the anterior intercostal arteries are branches of the internal thoracic artery (also see Fig. 1-16).

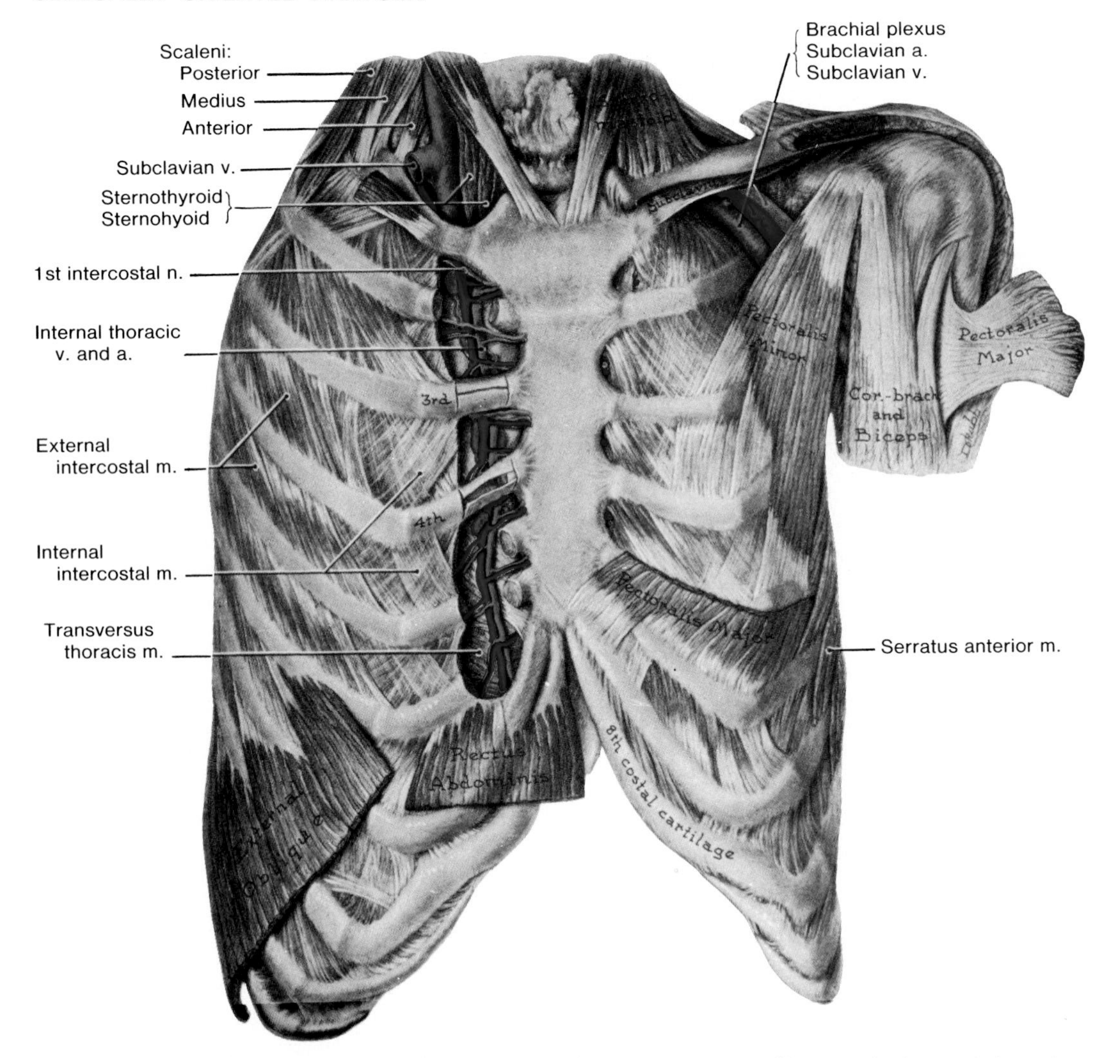

Figure 1-16. Drawing of an anterior view of a dissection of the thoracic wall. Observe the internal thoracic vessels running about 1 cm from the edge of the sternum. Note their intercostal branches. Examine the parasternal lymph nodes (*green*) which receive lymph vessels from the intercostal spaces, the costal pleura, the diaphragm, and the medial part of the breast. It is by this route that a cancer of the breast may spread to the lungs and mediastinum (see Chap. 6). Observe that the subclavian vessels are sandwiched between the first rib and the clavicle (padded by the subclavius muscle). Although the seventh costal cartilage is usually the last one to reach the sternum, it is not uncommon, as in this specimen, for the eighth to do so. The H-shaped cut through the perichondrium of the third and fourth costal cartilages on the right side was used to shell out segments of cartilage as an illustration of a surgical approach to the thoracic cavity.

in the anteroposterior diameter of the thorax. This is most noticeable in young people.

During expiration the elastic recoil of the lungs produces a subatmospheric pressure in the pleural cavities. This and the weight of the thoracic walls cause the lateral and anteroposterior diameters of the thorax to return to normal.

During respiration there is considerable movement of the various joints of the thorax (costovertebral, costochondral, and sternocostal). Hence any disorder that reduces the mobility of these joints interferes with respiration.

The shape of the thorax may be distorted by congenital malformations of the vertebral column and/or ribs. Destructive disease of the vertebral column producing lateral curvature or *scoliosis* (Fig. 5-6*E*) results in a considerable change in the shape of the thorax.

The diaphragm is the most important muscle of respiration, but it is not essential. Paralysis of half of it does not affect the other half because each half has a separate nerve supply.

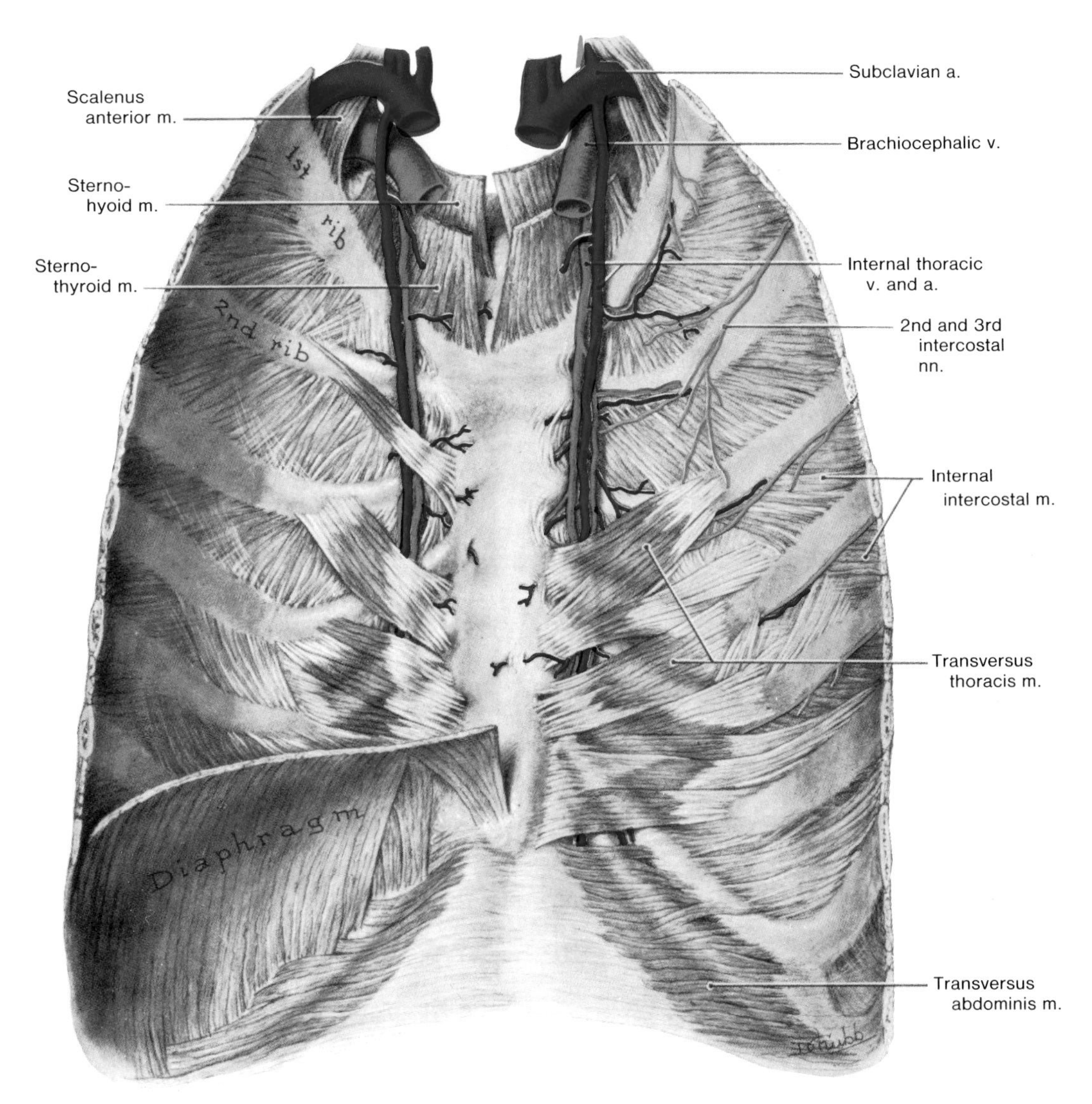

Figure 1-17. Drawing of a posterior view of the anterior thoracic wall. Observe that the internal thoracic artery, arising from the first part of the subclavian artery, is accompanied by two veins (venae comitantes) up to the third or second intercostal space, and superior to this by a single vein (internal thoracic vein) which proceeds to the brachiocephalic vein. Note that the inferior portions of the internal thoracic vessels are covered posteriorly by the transversus thoracis muscle. Observe the continuity of the transversus thoracis muscle with the transversus abdominis muscle, the innermost layer of the three flat muscles of the thoracoabdominal wall (Fig. 2-8).

APERTURES OF THE THORAX

SUPERIOR THORACIC APERTURE

The thoracic cavity communicates with the root of the neck via the superior thoracic aperture, an opening that is commonly called the **thoracic inlet**. Through this relatively small, kidney-shaped opening (about 5 cm anteroposteriorly and 11 cm transversely) pass structures joining the thorax to the upper limbs and neck (*e.g.*, trachea, esophagus, vessels, and nerves). To visualize its size, oppose your thumb and forefinger of one hand with those of the other.

The superior thoracic aperture is bounded by the first thoracic vertebra posteriorly, the medial borders of the ***first ribs*** and their ***costal cartilages*** anterolaterally, and the superior end of the ***manubrium of the sternum*** anteriorly (Fig. 1-1).

The margin of the superior thoracic aperture slopes inferiorly and anteriorly (Fig. 1-1); hence the apex of each lung and its covering of pleura project superiorly through the lateral parts of the superior thoracic inlet (Figs. 1-25 and 1-35).

INFERIOR THORACIC APERTURE

The thoracic cavity communicates with the abdomen via the inferior thoracic opening, a wide aperture commonly called the **thoracic outlet**. In the living state it is closed by the diaphragm which is pierced by the structures passing between the mediastinum (the median partition of the thoracic cavity) and the abdomen. The inferior thoracic aperture is uneven and is much larger than the superior thoracic aperture. It slopes inferiorly and posteriorly.

The inferior thoracic aperture is bounded by the 12th thoracic vertebra posteriorly, the ***12th pair of ribs and costal margins*** anterolaterally, and the ***xiphisternal joint*** anteriorly (Fig. 1-1).

As the inferior trunk of the brachial plexus leaves the neck to enter the axilla, it crosses over the first rib (Fig. 6-22) and may produce a groove in it. The subclavian artery on its way to the upper limb also runs over the first rib, producing a distinct groove (Figs. 1-4 and 1-25).

The subclavian artery may be compressed where it passes over the first rib, producing ***vascular symptoms*** (*e.g.*, pallor, coldness, and *cyanosis* or bluishness of the hands). Less frequently **nerve pressure symptoms** (numbness and tingling in the fingers) *result from pressure on the inferior trunk of the brachial plexus.*

These conditions have been described under several different terms, depending on what was thought to be the cause of the symptoms. Generally the condition is called the **neurovascular compression syndrome.** *The compression generally involves the inferior trunk of the brachial plexus at or close to the level at which it crosses the first rib* (Fig. 6-22). In some cases it is attributed to the shoulder being abnormally low, so that the nerves are pulled over the first rib.

The thoracic inlet syndrome results from the development of a **lymphosarcoma** (malignant lymphatic tumor). There are multiple enlarged lymph nodes and infiltration of tissues in the supraclavicular fossae (Fig. 8-1*B*) and around the thoracic inlet. As a result, blood cannot drain from the head, neck, and upper limbs to the heart so these parts become congested with blood and appear swollen.

MUSCLES OF THE THORACIC WALL

Many muscles are attached to the ribs, including some back muscles (Fig. 5-39) and anterolateral muscles of the abdomen (Fig. 1-16). The *pectoral muscles*, covering the anterior chest wall, act on the upper limb (Fig. 6-19). The *pectoralis major* can be used as an accessory muscle of respiration to expand the thoracic cavity. This large *fan-shaped muscle* (Fig. 6-14) is active when inspiration is deep and forcible.

The *serratus anterior muscle*, important in the protraction of the scapula (Chap. 6), runs around the surface of the thorax from the scapula (Figs. 1-14 and 1-16).

The muscles of the thorax proper are: the serratus posterior, the intercostals, the subcostals, the transversus thoracis, and the levatores costarum. *These intrinsic muscles of the thoracic wall, concerned with the mechanics of breathing, run from rib to rib, sternum to rib, or vertebrae to ribs.*

The Serratus Posterior Muscles (Figs. 2-110 and 5-39). These flat muscles run from the vertebrae to the ribs and are *innervated by intercostal nerves*, the superior ones by the first four and the inferior ones by the last four.

The serratus posterior superior muscle (Fig. 5-39) arises from the inferior part of the ***ligamentum nuchae*** (a median ligamentous band at the back of the neck), and from the spinous processes of the ***seventh cervical and first three thoracic vertebrae.*** The muscle runs inferolaterally to insert into the superior borders of the second to fourth (or fifth) ribs.

The serratus posterior inferior muscle (Figs. 2-110 and 5-39) arises from the spinous processes of the *last two thoracic and first two lumbar vertebrae* (Fig. 5-39). It runs superolaterally to insert into the inferior borders of the inferior three or four ribs near their angles.

Both serratus posterior muscles are inspiratory ones. The serratus posterior superior elevates the superior ribs and the serratus posterior inferior depresses the inferior ribs, preventing them from being pulled superiorly by the diaphgram.

THE INTERCOSTAL MUSCLES (Figs. 1-15 to 1-20 and 1-22)

There are three incomplete layers of muscle in each intercostal space: **external, internal, and innermost intercostal muscles.** In a deeper plane there are two muscles (transversus thoracis and subcostals) that extend over more than one intercostal space. The intercostal muscles are supplied by the corresponding intercostal nerves.

The External Intercostal Muscles (Figs. 1-15, 1-18 to 1-20, and 1-22). Each of the **11 pairs of external intercostal muscles** runs obliquely inferiorly and anteriorly from the **rib above to the rib below.** These muscles arise from the shafts of the ribs, beginning just lateral to their tubercles and extending anteriorly almost to their costal cartilages (Fig. 1-16).

Between the costal cartilages, the external intercostal muscles are replaced by the **external intercostal membranes** (Fig. 1-15). In the inferior intercostal spaces, the external intercostal muscles become continuous with the external oblique muscle of the anterior abdominal wall (Figs. 1-16 and 2-8).

The Internal Intercostal Muscles (Figs. 1-15, 1-19, and 1-22). The **11 pairs of internal intercostal muscles** run deep to and *at right angles to the external intercostal muscles.* The fibers of the internal intercostal muscles run obliquely from the edge of the **costal groove of one rib** inferiorly and posteriorly to the **superior margin of the rib inferior to it.**

The internal intercostal muscles arise from the shafts of the ribs and their costal cartilages, as far anteriorly as the sternum, and as far posteriorly as the angles of the ribs. Between the ribs posteriorly, the internal intercostal muscles are represented by the **internal intercostal membranes** (Figs. 1-15 and 1-19). The inferior internal intercostal muscles are continuous with the internal oblique muscle of the anterior abdominal wall (Fig. 1-18).

The Innermost Intercostal Muscles (Figs. 1-15 to 1-19 and 1-22). The innermost intercostal muscles appear similar to the internal intercostal muscles and in reality are deep portions of them (Figs. 1-15 and

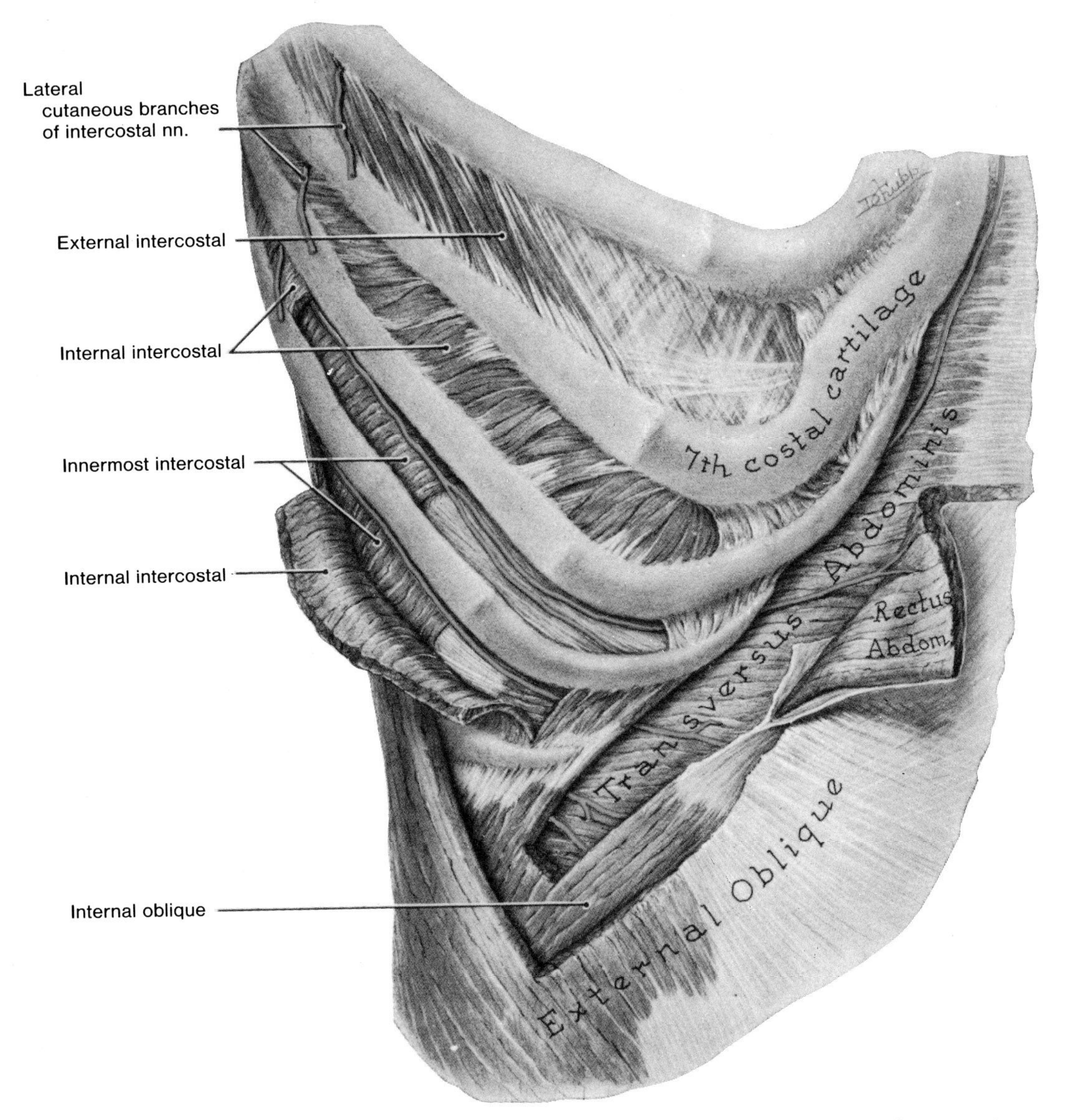

Figure 1-18. Drawing of the anterior ends of the inferior intercostal spaces, right side. Observe the common direction of the fibers of the external intercostal and external oblique muscles, and the continuity of the internal intercostal muscle with the internal oblique muscle at the anterior ends of the 9th to 11th intercostal spaces. Note the morphological plane in which an intercostal nerve lies deep to an internal intercostal muscle, but superficial to an innermost intercostal muscle and either the transversus thoracis or the transversus abdominis, according to the level. Observe the direction in which an intercostal nerve runs: first parallel to the ribs immediately superior and inferior and then parallel fo the costal cartilages. Thus, on gaining the abdominal wall, nerves T7 and T8 continue superiorly, T9 continues nearly horizontally, and T10 continues inferiorly toward the umbilicus (Fig. 2-16*A*).

1-18). The innermost intercostals are sometimes referred to as the *intercostales intimi.*

The innermost intercostal muscles are separated from the internal intercostal muscles by the intercostal nerves and vessels (Figs. 1-18, 1-19, and 1-22). The innermost intercostal muscles pass between the internal surfaces of adjacent ribs and cover approximately the middle two-fourths of the intercostal spaces.

The Subcostal Muscles (Fig. 1-20). The subcostal muscles are thin muscular slips which are variable in size and shape.

The subcostal muscles extend from the inside of the angle of one rib to the internal surface of the rib inferior to it, crossing one or two intercostal spaces. They run in the same direction as the internal intercostal muscles and lie internal to them. These muscles probably depress the ribs.

The Transversus Thoracis Muscle (Figs. 1-15 to 1-17). The transversus thoracis is a thin muscle consisting of four or five slips which arise from the xiphoid process and inferior part of the body of the sternum. These slips of muscle pass superolaterally to the second to sixth costal cartilages.

The transversus thoracis muscle lies on the inner surface of the sternochondral portion of the thorax (Fig. 1-17). **It becomes continuous with the transversus abdominis** (Fig. 1-18). The internal thoracic vessels run anterior to this muscle, between it and the costal cartilages and the internal intercostal muscles (Figs. 1-15 to 1-17). The transversus thoracis muscle probably depresses the second to sixth ribs.

The Levatores Costarum Muscles (Fig. 1-19). The 12 fan-shaped levatores costarum muscles arise from the tansverse processes of C7 and T1 to T11 vertebrae, and pass inferolaterally to insert on the ribs inferiorly, close to their tubercles. As their name levator (L. lifter) indicates, they *elevate the ribs* to which they are attached.

Actions of the Intercostal Muscles. Inspiration and expiration radiographs (Figs. 1-51 and 1-52) show that the intercostal spaces widen on inspiration.

All three layers of intercostal muscles act together in keeping the intercostal spaces rigid, thereby preventing them from bulging out during expiration and from being drawn in during inspiration.

When the first rib is fixed by the scalenus anterior and scalenus medius (neck muscles, Figs. 1-16 and 1-17) during quiet respiration, the intercostal muscles elevate the anterior parts of the other ribs, except for the inferior ones which are held in position by the serratus posterior inferior and abdominal muscles (Figs. 1-18 and 2-110). Elevation of the ribs increases the anteroposterior diameter of the thorax (Figs. 1-13 and 1-51).

When the 12th rib is fixed by the abdominal muscles, the other ribs are lowered by the intercostal muscles during expiration, thereby decreasing the anteroposterior diameter of the thorax (Fig. 1-52).

THE INTERCOSTAL SPACES

The spaces between the ribs (L. *costae*) are called **intercostal spaces** (Figs. 1-15, 1-21, and 1-22). They are deeper anteriorly than posteriorly, and deeper between the superior than the inferior ribs.

The intercostal spaces widen on inspiration (Fig. 1-51). Each space contains three muscles and a neurovascular bundle (Figs. 1-19, 1-20, and 1-22).

It is sometimes necessary to insert a hypodermic needle through an intercostal space to withdraw a *sample of pleural fluid,* or to remove blood or pus from the pleural cavity (Fig. 1-11).

In other cases it may be necessary to anesthetize an intercostal nerve. To avoid damage to the main intercostal vessels and nerve, the needle is inserted superior to the rib (Fig. 1-22).

THE INTERCOSTAL NERVES

As soon as they pass through the intervertebral foramina, the thoracic spinal nerves divide into ventral and dorsal primary divisions or rami (Fig. 1-15).

The dorsal rami pass posteriorly, immediately lateral to the articular processes, to supply the muscles, bones, joints, and skin of the back (Fig. 6-44).

The ventral rami of the first 11 thoracic nerves *are called intercostal nerves because they enter the intercostal spaces.* The large ventral ramus of the 12th nerve, being inferior to the 12th rib, is not intercostal and so is called the **subcostal nerve** (Fig. 2-118).

Each intercostal nerve is connected to a ganglion of the **sympathetic trunk** by small branches, called rami communicantes (Fig. 1-21). The ***white ramus communicans*** carries preganglionic sympathetic fibers to the sympathetic trunk and/or the prevertebral ganglion from the intercostal nerve, whereas the ***gray ramus communicans*** carries postganglionic sympathetic fibers from the sympathetic trunk to the intercostal nerve. These fibers are distributed through all branches of the intercostal nerve to blood vessels, sweat glands, and smooth muscle.

A TYPICAL INTERCOSTAL NERVE (Figs. 1-15 to 1-21 and 1-23)

A typical intercostal nerve (*third to sixth*) enters the intercostal space between the parietal pleura and the internal intercostal membrane (Figs. 1-19 and 1-22). At first it runs more or less in the middle of the intercostal space, across the internal surface of the internal intercostal membrane and muscle (Fig. 1-15). Near the angle of the rib, it passes between the internal intercostal and the innermost intercostal muscles. *Here it enters and is sheltered by the* ***costal groove***

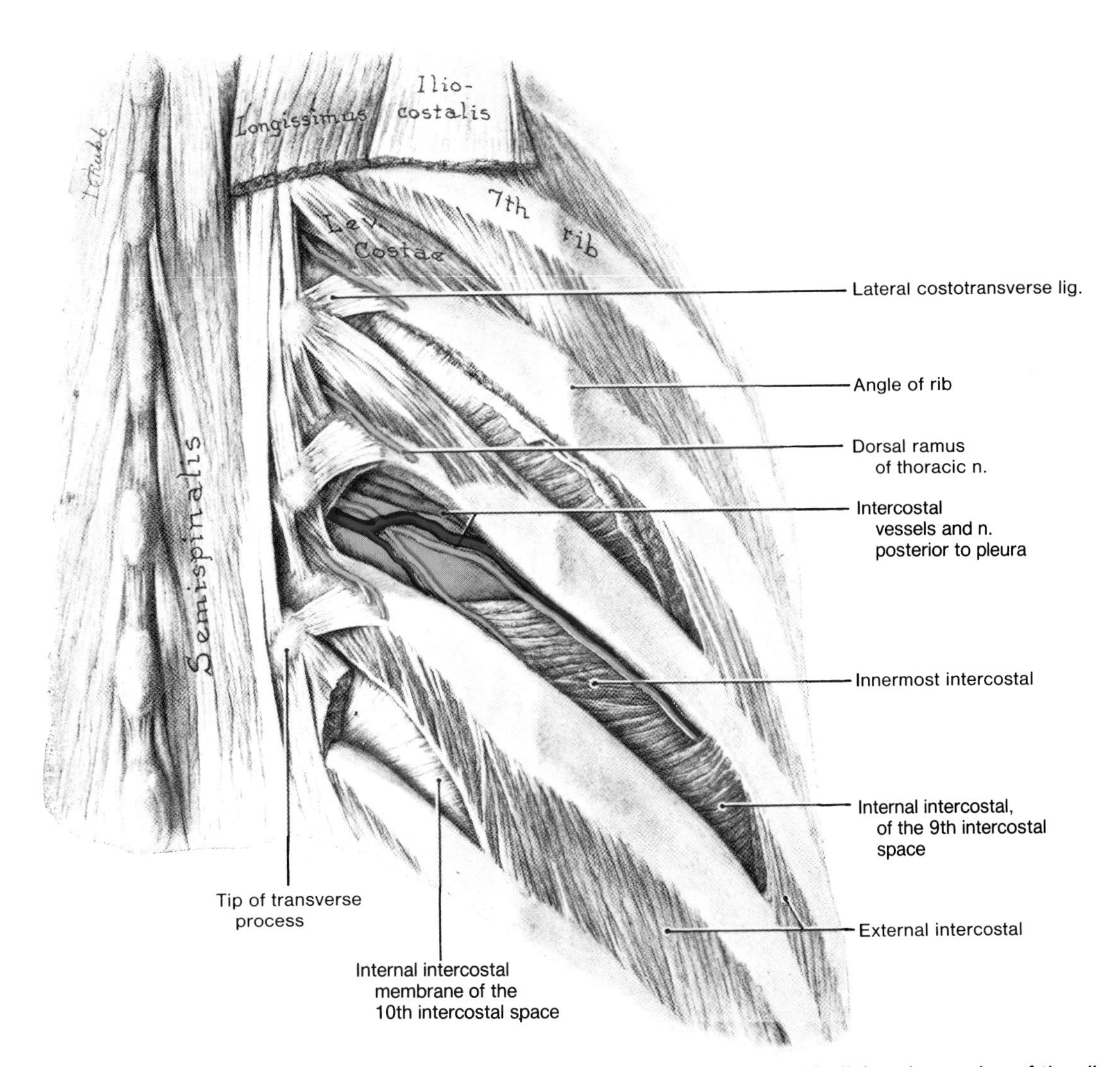

Figure 1-19. Drawing of the posterior ends of the inferior intercostal spaces. Medial to the angles of the ribs, the iliocostalis and longissimus muscles have been removed to expose the levatores costarum muscles. In the five intercostal spaces depicted, observe that (1) the superior two (sixth and seventh) are intact; (2) from the most inferior or 10th intercostal space the levator costae muscle and the underlying part of the external intercostal muscle have been removed to reveal the internal intercostal membrane; (3) from the 8th intercostal space more of the external intercostal muscle has been removed and the internal intercostal membrane is seen extending laterally as an areolar sheet; and (4) in the 9th intercostal space this sheet has been removed in order to show the intercostal vessels and nerve appearing medially between the superior costotransverse ligament and the pleura, and disappearing laterally between the internal intercostal and the innermost intercostal muscles.

(Fig. 1-3) where it lies just inferior to the intercostal artery (Figs. 1-21 and 1-22). It continues anteriorly between the internal and innermost intercostal muscles, giving off *muscular branches* to the intercostal and other muscles, and a *lateral cutaneous branch* (Fig. 1-15). Anteriorly it appears on the inner surface of the internal intercostal muscle (Fig. 1-16), but external to the transversus thoracis muscle and internal thoracic vessels. Near the sternum it turns anteriorly and ends as an anterior cutaneous branch (Fig. 1-15).

Branches of a Typical Intercostal Nerve (Figs. 1-15, 1-18, and 1-21).

1. **Rami communicantes** (Figs. 1-20 and 1-21) connect an intercostal nerve to a sympathetic trunk. The nerve sends a *white ramus communicans* to a ganglion of the *sympathetic trunk* and receives a *gray ramus communicans* from a ganglion of the trunk.
2. **Collateral branches** (Figs. 1-15, 1-18, and 1-22) arise near the angles of the ribs and run along the superior margins of the ribs inferior to them to supply the intercostal muscles.
3. **Lateral cutaneous branches** (Figs. 1-15 and 1-18) arise beyond the angles of the ribs and pierce the internal and external intercostal muscles about halfway around the thorax. The cuta-

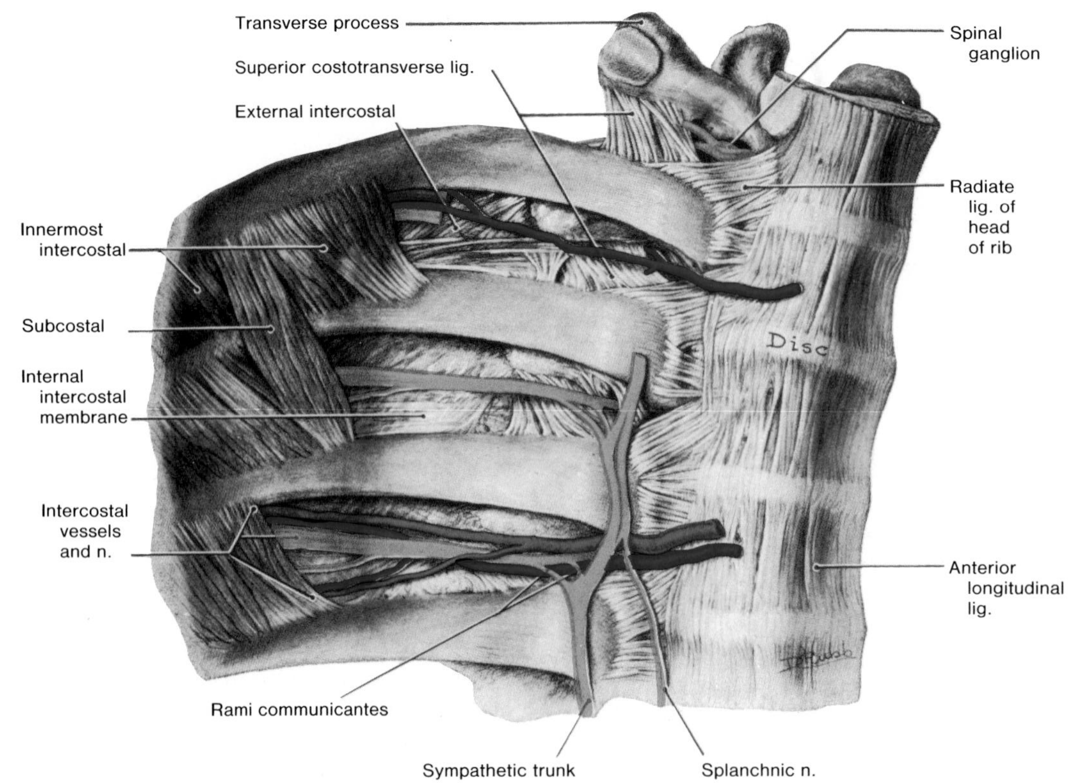

Figure 1-20. Drawing of an anterior view of a dissection of the vertebral end of an intercostal space. Observe that the portions of the innermost intercostal muscle that bridge two intercostal spaces are called subcostal muscles. Note the external intercostal muscle in the superior intercostal space and the internal intercostal membrane in the middle space that is continuous medially with a superior costotransverse ligament. Observe the order of the structures in the lowest space: intercostal vein, artery, and nerve. Note the ventral ramus of a thoracic nerve crossing anterior to a superior costotransverse ligament and a dorsal ramus crossing posterior to it. The rami communicantes are well shown (also seen Fig. 1-21).

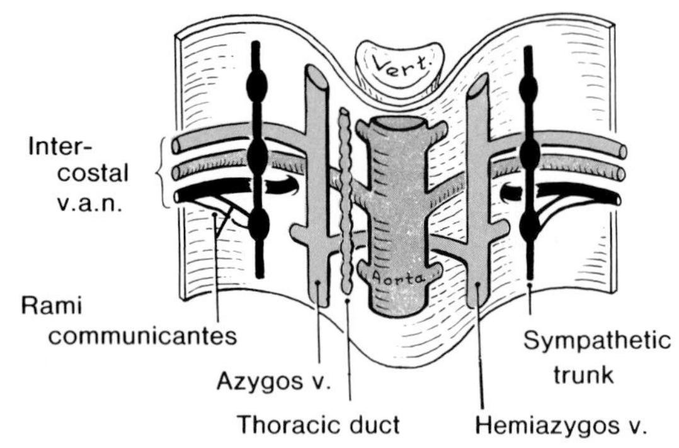

Figure 1-21. Diagram of an intercostal space showing its relationships. From superior to inferior: vein, artery, nerve (VAN). The sympathetic trunk is a bilateral series of sympathetic ganglia that lies a little lateral to the vertebral column.

neous branches divide into anterior and posterior branches which *supply the skin of the thoracic and abdominal walls* (Fig. 2-7).

4. **An anterior cutaneous branch** (Figs. 1-15 and 1-16) *supplies the skin on the anterior aspect of the chest* (Figs. 2-6 and 2-16).
5. **Muscular branches** (Figs. 1-17 and 1-18) supply the subcostal, transversus thoracis, levator costae, and serratus posterior muscles.

ATYPICAL INTERCOSTAL NERVES (Figs. 1-17 to 1-19)

In the first part of their course, the first and second intercostal nerves pass on the internal surfaces of the first and second ribs.

The first intercostal nerve has no anterior cutaneous branch and usually has no lateral cutaneous branch. It divides into a large superior and small inferior part. *The large superior part joins the* ***brachial plexus***, supplying the upper limb (Fig. 6-22). The small inferior part becomes the first intercostal nerve (Fig. 1-16).

The second intercostal nerve may also contribute a small branch to the brachial plexus. *Its lateral cutaneous branch is called the intercostobrachial nerve* (Fig. 6-42) because it supplies the floor of the axilla (Fig. 6-11), and then communicates with the

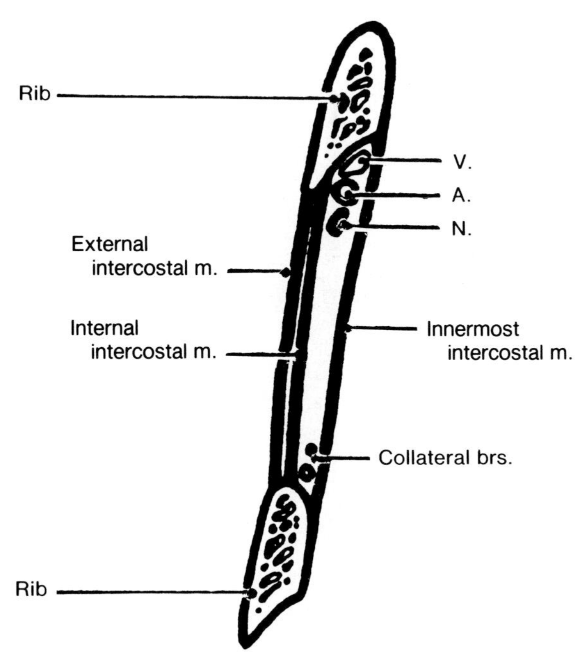

Figure 1-22. Diagrammatic sketch of a coronal section through two ribs illustrating the position of the neurovascular bundle (intercostal nerve and vessels) between the intercostal muscles. The useful key to remembering the relationships of the neurovascular bundle is VAN (*i.e.*, vein, artery, nerve). Observe that a hypodermic needle passed into the chest (*e.g.*, to enter the pleural cavity) superior to the rib will avoid this neurovascular bundle.

medial cutaneous nerve of the arm to supply the medial side of the upper limb as far as the elbow (Figs. 6-92 and 6-93).

The 7th to 11th intercostal nerves (*thoracoabdominal nerves*) supply the abdominal wall as well as the thoracic wall (Fig. 2-16). They pass anteriorly and inferiorly to the anterior ends of the intercostal spaces (Figs. 1-18 and 1-19), where they pass into the anterior abdominal wall deep to the costal cartilages to lie between the internal oblique and transversus abdominus muscles (Fig. 1-18).

The subcostal nerve follows the inferior border of the 12th rib and passes into the abdominal wall (Fig. 2-118). Its lateral cutaneous branch supplies skin in the gluteal region (buttocks).

DERMATOMES (Figs. 1-23 and 5-51)

The area of skin supplied by a dorsal root of a spinal nerve is called a dermatome (G. skin slice). The dermatomes were determined by plotting (1) the areas of vasodilation that resulted from stimulation of individual dorsal nerve roots, and (2) the areas of "remaining sensibility" after cutting three dorsal roots superior to and three inferior to a given nerve root. The areas plotted represent the average finding for each dorsal root based on pain sensation (Fig. 1-23). Areas determined for temperature sensation have similar limits, but those for touch are more extensive.

There is considerable overlapping of contiguous dermatomes; i.e., each segmental nerve overlaps the territories of its neighbors. As a result, no anesthesia results unless two or more consecutive dorsal roots have lost their functions. In Figure 1-23, observe that the first two intercostal nerves (T1 and T2), in addition to supplying the thorax, supply the upper limb. Also note that the inferior six intercostal nerves (T7 to T12) supply the abdominal wall as well as the thoracic wall (Fig. 2-16).

Herpes zoster ("shingles") is a *viral disease of the spinal ganglia.* The term means literally a creeping, girdle-shaped cutaneous eruption. When the *herpes zoster viruses* invade the spinal ganglia of the thoracic spinal nerves, a sharp burning pain is produced in the area of skin supplied by the dorsal roots involved. A few days later, the involved dermatomes become red and vesicular eruptions appear in a segmental distribution of the affected nerves (Fig. 1-23).

Local anesthesia of an intercostal space may be produced by injecting an anesthetic agent around the origin of an intercostal nerve, just lateral to the vertebra concerned (Fig. 1-15). This procedure is known as an **intercostal nerve block.** Because there is considerable overlapping of contiguous dermatomes, complete anesthesia results only when two or more consecutive intercostal nerves are anesthetized.

Pus originating in the region of the vertebral column tends to pass around the thorax along the course of a **neurovascular bundle** (Figs. 1-19 and 1-22), and to pass to the surface where the cutaneous branches of the intercostal nerves pierce the muscles to supply the skin (Fig. 2-16).

THE INTERCOSTAL ARTERIES

*Three arteries, a large **posterior intercostal artery** and small paired **anterior intercostal arteries**, supply each intercostal space.*

THE POSTERIOR INTERCOSTAL ARTERIES (Figs. 1-15, 1-19, 1-25, and 1-78)

The first two posterior intercostal arteries arise from the ***superior intercostal artery***, a branch of the **costocervical trunk** of the subclavian artery (Fig. 1-25).

Nine pairs of large posterior intercostal arteries and one pair of subcostal arteries **arise posteriorly from**

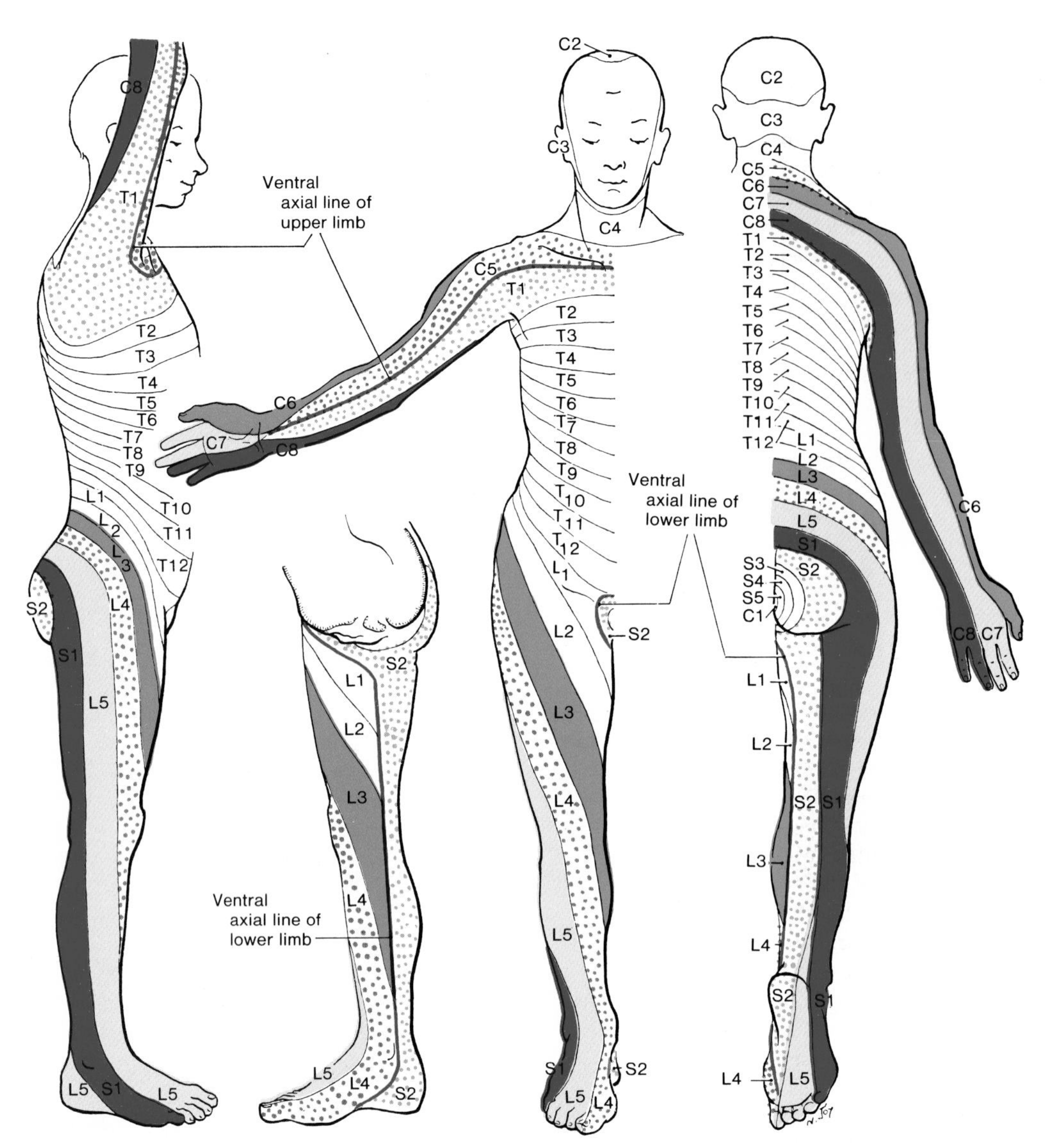

Figure 1-23. Diagrams showing the dermatomes of the body, the strip-like areas of skin supplied by one pair of spinal nerves. When the function of even a single dorsal nerve root is interrupted, faint but definite diminution of sensitivity can be demonstrated in the dermatome. The method for detecting and plotting the area of diminished sensitivity is by using a light pin scratch for pain sensation. An area can be found for temperature and tactile sensation also.

the thoracic aorta (Fig. 1-78). Each posterior intercostal artery gives off a posterior or *dorsal branch* which accompanies the dorsal ramus of the spinal nerve (Fig. 1-15), and *supplies the spinal cord, vertebral column, postvertebral muscles, and skin.*

The anterior or *ventral branch* of the posterior intercostal artery accompanies the intercostal nerve across the intercostal space. Close to the angle of the rib, it enters the **costal groove,** where it lies between the intercostal vein and the intercostal nerve (Figs. 1-21 and 1-22). At first the artery runs between the pleura and the internal intercostal membrane (Figs. 1-15 and 1-19). It then runs between the innermost intercostal and the internal intercostal muscles.

Each posterior intercostal artery also gives off a small *collateral branch* that crosses the intercostal space and runs along the superior border of the rib inferior to the space (Fig. 1-19). The terminal branches of the posterior intercostal artery anastomose anteriorly with the anterior intercostal artery (Fig. 1-15).

THE ANTERIOR INTERCOSTAL ARTERIES (Figs. 1-15 to 1-17 and 1-25)

The anterior intercostal arteries supplying the superior six intercostal spaces are derived *from the* ***internal thoracic arteries*** (Figs. 1-15 and 1-16). *There are two anterior intercostal arteries for each intercostal space.* The arteries supplying the seventh to ninth intercostal spaces are derived from the *musculophrenic arteries*, branches of the internal thoracic arteries. These arteries pass laterally, one near the inferior margin of the superior rib and the other near the superior margin of the inferior rib. At their origins, the first two arteries lie between the pleura and the internal intercostal muscles, whereas the next four arteries are separated from the pleura by the transversus thoracis muscle (Fig. 1-17).

The anterior intercostal arteries supply the intercostal muscles and send branches through them to the pectoral muscles, breasts, and skin (Fig. 6-13).

There are no anterior intercostal arteries in the inferior two intercostal spaces. These spaces are supplied by the posterior intercostal arteries and their collateral branches.

THE INTERNAL THORACIC ARTERY (Figs. 1-15 to 1-17, 1-25, and 1-81)

This artery arises from the inferior surface of the first part of the *subclavian artery* in the root of the neck, at the medial border of the scalenus anterior muscle (Fig. 1-17). It descends into the thorax posterior to the clavicle and the first costal cartilage (Fig. 1-16). It lies on the pleura posteriorly and is *crossed by the phrenic nerve* (Fig. 1-81). It passes inferiorly in the thorax posterior to the superior six costal cartilages and the intervening intercostal muscles, about 1 cm from the margin of the sternum (Figs. 1-17 and 1-18). At the level of the third costal cartilage, it continues inferiorly, anterior to the transverse thoracis muscle, to end in the sixth intercostal space by dividing into superior epigastric and musculophrenic arteries.

The **superior epigastric artery** (Fig. 2-8) supplies the diaphragm and the muscles of the abdominal wall, and the **musculophrenic artery** gives origin to the anterior intercostal arteries which supply the seventh to ninth intercostal spaces (Figs. 1-15 and 1-16).

THE THORACIC CAVITY

The thoracic cavity is divided into three major divisions: **two pleural cavities and the mediastinum.** The pleurae and lungs occupy most of the thoracic cavity (Fig. 1-11), with the heart between them and the esophagus and aorta posterior to the heart.

The mediastinum is the median partition that contains the heart and great vessels (Figs. 1-11 and 1-43) and other important structures (*e.g.,* the trachea, esophagus, thymus, and lymph nodes).

THE PLEURA AND PLEURAL CAVITIES

The pleural cavity on each side is **almost completely filled by a lung.** The two pleural cavities occupy the lateral parts of the thoracic cavity.

Each lung is completely covered with a smooth glistening membrane, the pulmonary or **visceral pleura**, except at the narrow ***root of the lung*** where each lung is attached to the mediastinum. Here, the visceral pleura is continuous with the **parietal pleura** which lines the walls of the pleural cavities (Figs. 1-24 and 1-27).

The *pleural cavities are two separate and closed potential spaces.* Normally there is only a capillary layer of serous lubricating fluid in the pleural cavities.

The pleural fluid lubricates the pleural surfaces and reduces friction between the parietal and visceral layers of pleura. Hence the opposed surfaces of parietal and visceral pleura slide smoothly against each other during respiration. As a result, respiratory movements can occur without interference.

The Parietal Pleura (Figs. 1-24, 1-25, and 1-27 to 1-29). The parietal pleura covers the different parts of the thoracic wall and the thoracic contents. It is given different names according to the parts it covers.

The costal pleura (Fig. 1-24), the strongest part of the parietal pleura, *covers the internal surfaces of the sternum, the* ***costal cartilages****, the ribs, the intercostal muscles, and the sides of the thoracic vertebrae.* It is separated from these structures by a thin layer of loose connective tissue, called the **endothoracic fascia**, which provides a natural cleavage plane for the surgical separation of the costal pleura from the thoracic wall.

The mediastinal pleura (Figs. 1-24 and 1-27) *covers the* ***mediastinum*** (Fig. 1-11). It is continuous with the costal pleura anteriorly and posteriorly, with the diaphragmatic pleura inferiorly, and with the **cupola** or *cervical pleura* superiorly (Fig. 1-24).

Superior to the root of the lung, the mediastinal pleura is a continuous sheet between the sternum and the vertebral column.

At the root of the lung, the mediastinal pleura passes laterally, where it encloses the structures of the lung root and becomes continuous with the visceral pleura (Fig. 1-24).

Inferior to the root of the lung, the mediastinal pleura

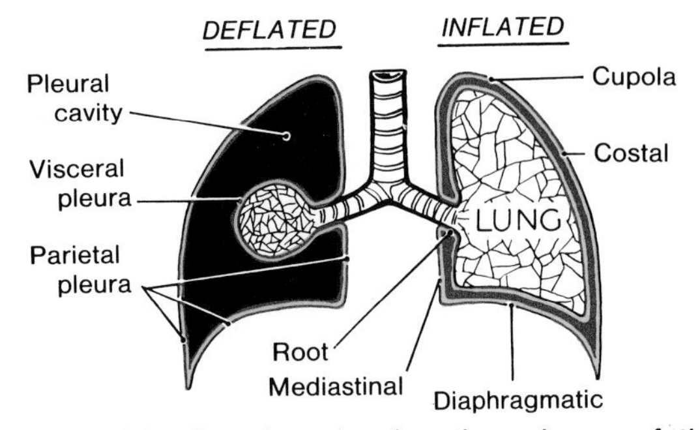

Figure 1-24. Drawing showing the scheme of the pleural cavities and pleurae. In the inflated lung, note that the parietal pleura is applied to the visceral pleura but separated from it by a narrow interval, the pleural cavity, which contains a capillary layer of serous lubricating pleural fluid.

passes laterally as a double layer from the esophagus to the lung, where it is continuous with the visceral pleura. This double layer, called the **pulmonary ligament** (Figs. 1-32 and 1-33), is continuous superiorly with the mediastinal pleura, enclosing the lung root, and ends inferiorly in a free border. The parietal layers of mediastinal pleura come into close apposition between the aorta and esophagus, just superior to the diaphragm (Fig. 1-29).

The diaphragmatic pleura (Figs. 1-24 and 1-29) is thin and *covers the superior surface of the diaphragm lateral to the mediastinum* (Fig. 1-11).

The **pleural cupola** (Fig. 1-35) or *cervical pleura covers the apex of the lung* (Fig. 1-24) and extends through the superior thoracic aperture into the root of the neck (Fig. 1-35). Its summit is 2 to 3 cm superior to the level of the medial third of the clavicle. This *dome-shaped roof* of the pleural cavity is the continuation of the costal and mediastinal parts of the pleura over the apex of the lung.

The cupola extends superiorly but not superior to the neck of the first rib (Figs. 1-25 and 1-35).

Because the first ribs slope inferiorly, the lung and pleura rise 3.5 cm superior to the anterior end of the first rib and 1 to 2 cm superior to the middle third of the clavicle (Figs. 1-33 to 1-35). The cupolae are separated by the trachea and the esophagus and the blood vessels passing to and from the neck (Fig. 1-30*A*).

Each cupola is strengthened by a layer of dense fascia, called the **suprapleural membrane**, which is attached to the inner margin of the first rib and the anterior border of the transverse process of the 7th cervical vertebra. This membrane usually contains some muscle fibers (scalenus minimus) which give it added strength.

Carotid tubercle
Vertebral a.
Costo-cervical trunk
Thyro-cervical trunk
Int. thoracic a.
Common carotid aa.

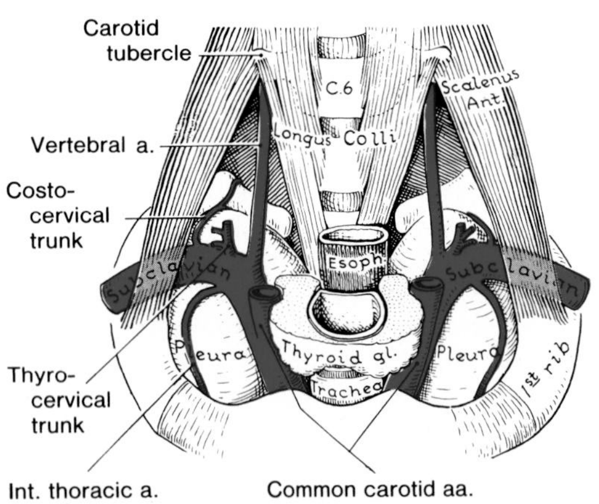

Figure 1-25. Drawing of a dissection showing the subclavian arteries and their main branches. The first two posterior intercostal arteries arise from the superior intercostal artery, a branch of the costocervical trunk of the subclavian artery. The paired anterior intercostal arteries supplying the superior six intercostal spaces are derived from the internal thoracic artery and those supplying the seventh to ninth spaces are derived from its musculophrenic branch.

Because of the inferior slope of the first pair of ribs, the cupolae of the pleurae and the apices of the lungs project superiorly into the neck, posterior to the sternocleidomastoid muscles. (Figs. 1-25 and 1-35). Consequently **the cupolae may be injured in wounds of the neck.**

LINES OF PLEURAL REFLECTION

These are sites at which the costal pleura becomes continuous with the mediastinal pleura anteriorly and posteriorly, and with the diaphragmatic pleura inferiorly (Figs. 1-26, 1-27, and 1-34).

The sternal and costal lines of pleural reflection are important landmarks in clinical medicine.

The sternal pleural reflection is where the costal pleura is continuous with the mediastinal pleura posterior to the sternum. The right and left sternal reflections are indicated by lines that pass inferomedially from the *sternoclavicular joints* to the median line at the level of the ***sternal angle*** (Fig. 1-26). Here the two pleural sacs come in contact and may slightly overlap each other.

On the right side, the sternal reflection continues inferiorly in the midline to the posterior aspect of the xiphoid process. **On the left side**, the sternal reflec-

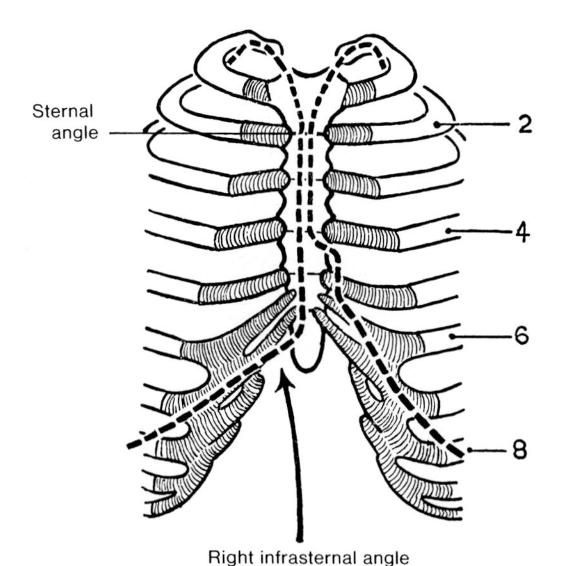

Figure 1-26. Diagram illustrating the sternal and costal lines of pleural reflection. These are very important landmarks in clinical medicine. The numbers indicate the even numbered ribs which are related to the lines of pleural reflection. The truncated apex of the infrasternal angle is at the xiphisternal joint.

tion passes inferiorly in the midline to the level of the *fourth costal cartilage.* Here it passes to the left margin of the sternum and then continues inferiorly to the *sixth costal cartilage* (Fig. 1-26).

The costal pleural reflection is where the costal pleura is continuous with the diphragmatic pleura near the chest margin. It passes obliquely across the *8th rib* in the ***midclavicular line***, the *10th rib* in the ***midaxillary line***, and the *12th rib* at its *neck* or inferior to it.

The pleurae descend inferior to the costal margin in three regions where an abdominal incision might inadvertently enter a pleural sac: (1) the right infrasternal angle (Figs. 1-26 and 1-35); (2) the right costovertebral angle; and (3) the left costovertebral angle (Fig. 1-27). In cases where the 12th rib is very short, the pleura lies inferior to the posterior costal margin after crossing the 11th rib and is therefore in surgical danger.

The parietal pleura is visible radiographically only in certain regions and in special views. The visceral pleura is usually not visible radiographically, except in certain views that demonstrate the fissures of the lung. However, *any disease process that causes thickening of the pleura may make it visible in radiographs.*

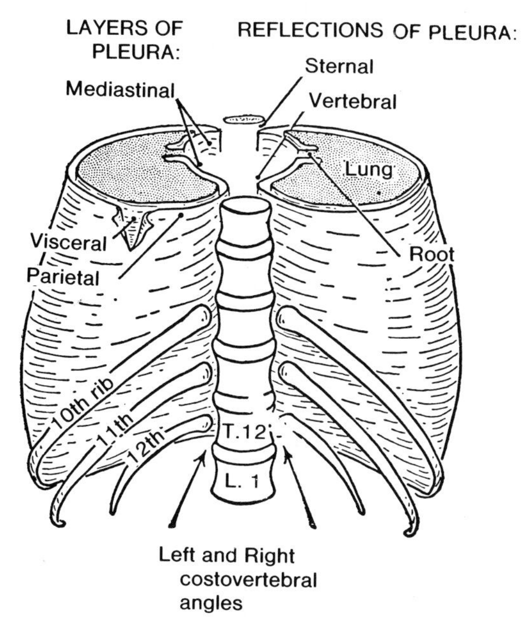

Figure 1-27. Drawing of a posterior view of the thorax, illustrating the layers of pleura and the sternal and vertebral pleural reflections. The bronchi, pulmonary arteries, veins, and nerves, and the bronchial vessels constitute the roots of the lungs. These structures enter and leave at the hilum (Figs. 1-32 and 1-33). The roots of the lungs lie opposite the bodies of the fifth to seventh thoracic vertebrae.

For completeness, two other pleural reflections need to be mentioned. The costal pleura is continuous with the mediastinal pleura along a vertical line which descends along the sides of the bodies of the thoracic vertebrae, just anterior to the heads of the first to twelfth ribs. This is called the ***vertebral pleural reflection*** (Fig. 1-27). The costal pleura is continuous with the diaphragmatic pleura along the line connecting the inferior ends of the sternal and vertebral reflections. This is called the ***mediastinodiaphragmatic pleural reflection.***

The Visceral Pleura (Fig. 1-24). *The visceral pleura* ***covers the lung closely and is adherent to all its surfaces.*** It provides the lungs with a smooth, shiny, slippery surface which enables them to move freely on the parietal pleura.

The visceral pleura dips into the fissures of the lungs so that the lobes are also covered with it. In these fissures the visceral pleura of adjacent lobes is in contact.

The visceral pleura is continuous with the parietal pleura at the ***root of the lung*** (Figs. 1-24 and 1-27), where the bronchi and blood vessels pass from the mediastinum to the lung (Figs. 1-32 and 1-33).

The Pleural Recesses (Figs. 1-11, 1-29, and 1-80). During full inspiration the lungs fill the pleural cavities (Fig. 1-24), but during quiet respiration three parts of them are not occupied by the lungs. At these sites the pleural reflections are so acute that the two portions of parietal pleura are not only in continuous apposition, but are also in contact with one another at their inner or serous surfaces. No lung tissue with its covering of visceral pleura intervenes between the opposed layers of parietal pleura. The three sites of reflection of parietal pleura are known as pleural recesses. In the pleural recesses parietal pleura comes into contact with parietal pleura.

The right and left costodiaphragmatic recesses (Figs. 1-11, 1-29, and 1-80) are *slit-like intervals between the costal and diaphragmatic pleurae on each side.* In these areas the layers of parietal pleura are separated by only a capillary layer of fluid. These recesses become alternately smaller and larger as the lungs move in and out of them during inspiration and expiration (Figs. 1-51 and 1-52).

The costomediastinal recess (Fig. 1-29) *lies along the anterior margin of the pleura.* Here the costal and mediastinal parts of the left pleura come into contact at the **cardiac notch** in the anterior border of the left lung where it overlies the heart (Figs. 1-30*B* and 1-33).

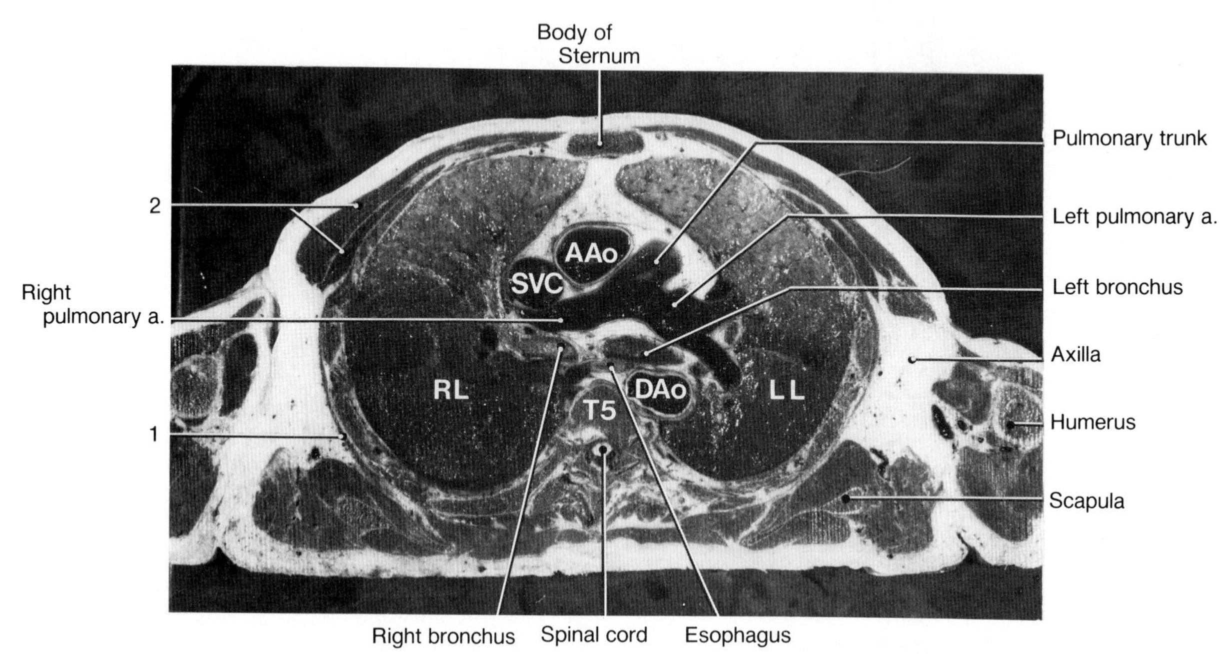

Figure 1-28. Photograph of transverse section of the thorax of an adult male at the level of T5 vertebra. In keeping with radiographic conventions, as in CT scans, the section is viewed from below. Hence the right side of the body appears on the left of the photograph. *AAo*, ascending aorta; *DAo*, descending aorta; *SVC*, superior vena cava; *RL*, right lung; *LL*, left lung; *T5*, fifth thoracic vertebra; 1, serratus anterior muscle; 2, pectoralis major and pectoralis minor muscles.

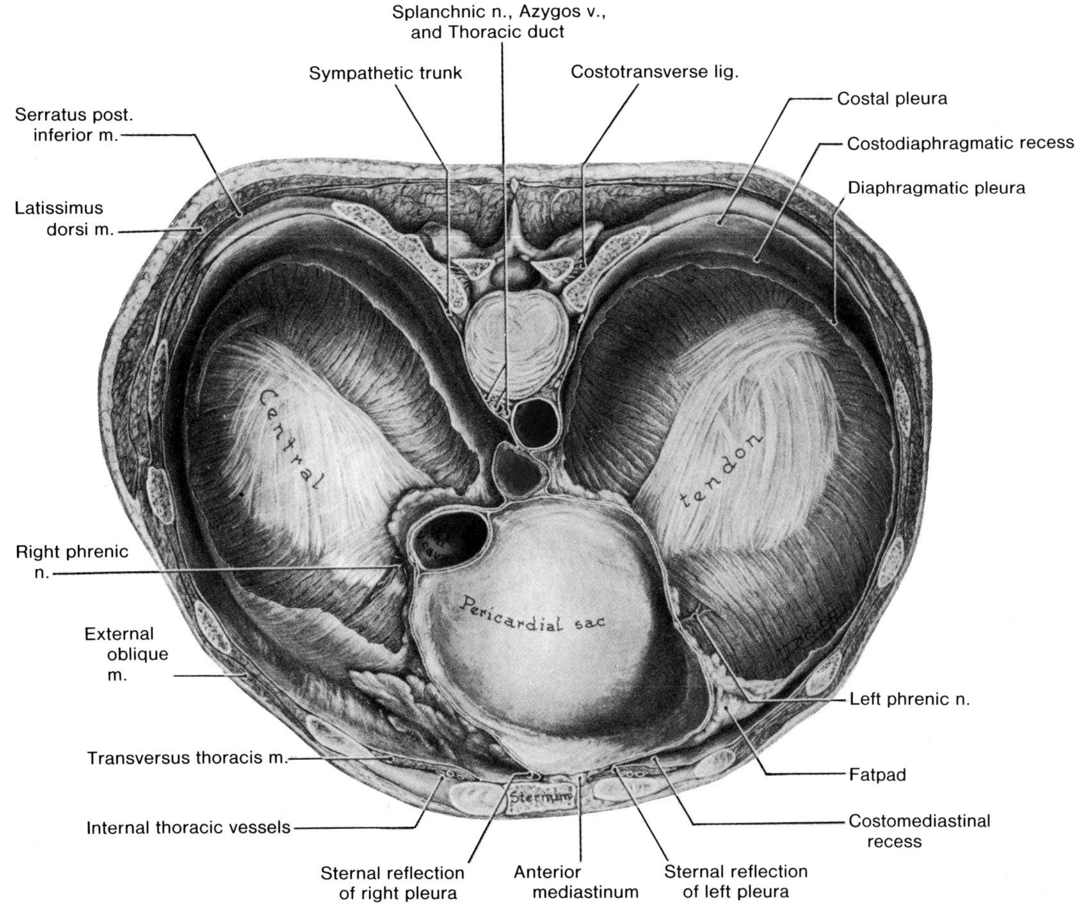

Figure 1-29. A superior view of a dissection of the diaphragm and pericardial sac. Most of the diaphragmatic pleura has been removed. Observe the sternal reflexion of the left pleural sac which fails to meet that of the right sac in the median plane, ventral to the pericardium. Note that the right and left pleural sacs almost meet between the esophagus and the aorta. Examine the costodiaphragmatic recess which is deepest about the midlateral line. Observe that the costal pleura on reaching the vertebral column becomes the mediastinal pleura (see also Fig. 1-27). Note that the central tendon of the diaphragm and the pericardium (pericardial sac) are pierced on the right side by the inferior vena cava posteriorly.

The slit-like costomediastinal recess is to the left of the sternum owing to the bulge of the heart to this side (Fig. 1-29), and lies at the anterior ends of the fourth and fifth interspaces (Fig. 1-26). During inspiration and expiration, a thin edge of the left lung, called the **lingula** (Figs. 1-30*B* and 1-33), slides in and out of the costomediastinal recess.

The pleural cavities are only potential spaces. The pleural spaces, filled with a capillary thin film of pleural fluid, facilitate movement of the lungs, but its obliteration by disease or surgical removal (**pleurectomy**) does not cause appreciable functional consequences.

In other surgical procedures (**pleural poudrage**), adherence of the visceral and parietal layers of pleura is purposely induced by covering the opposing pleural surfaces with a slightly irritating powder. This operation may be performed to prevent recurring spontaneous **pneumothorax** (air in the pleural cavity) resulting from disease of the lung.

During inspiration and expiration, the normal moist, smooth pleurae make no detectable sound during auscultation, however, *inflammation of the pleurae* (**pleuritis** or pleurisy) causes the surfaces to become rough. The resulting friction (**pleural rub** or friction rub) may be heard with a stethoscope.

Irritation of the parietal pleura causes pain that is referred to the thoracoabdominal wall (intercostal nerves) *or to the shoulder* (phrenic nerve). Pleuritis usually leads to the formation of **pleural adhesions** between the parietal and

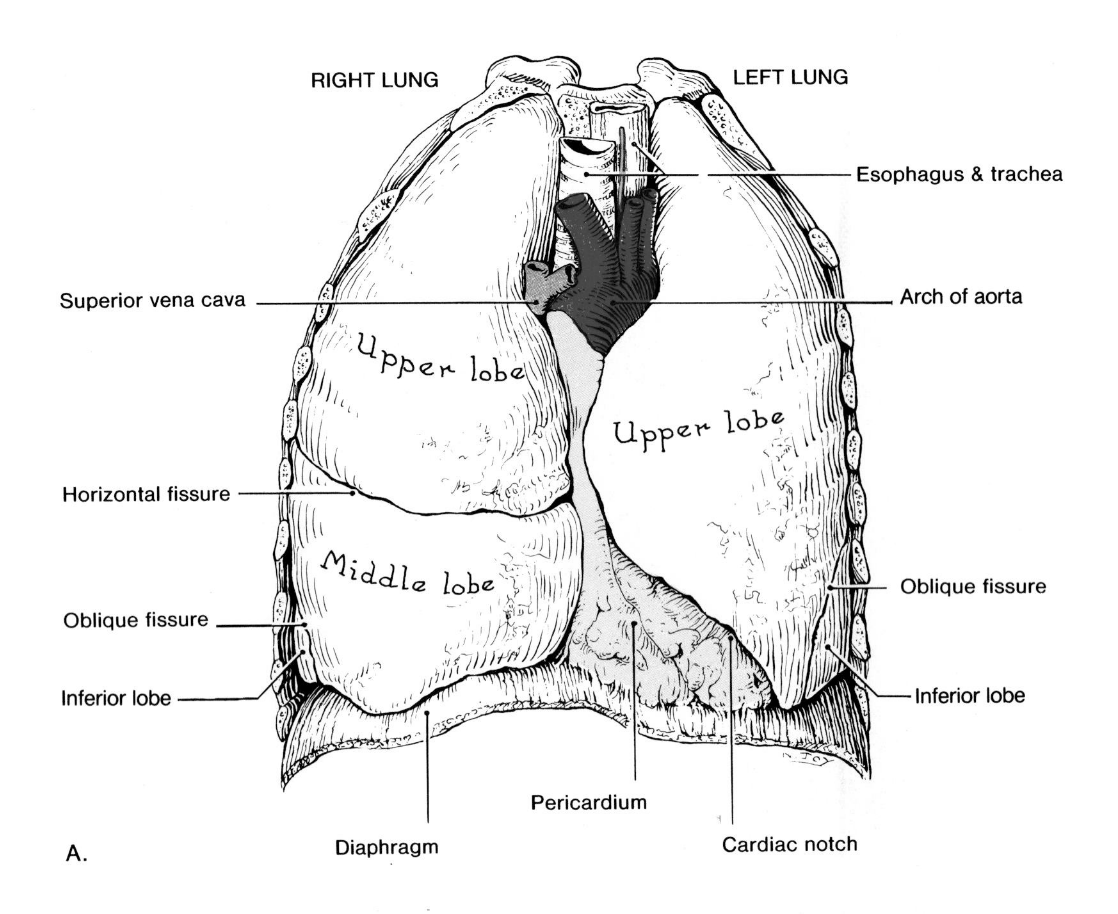

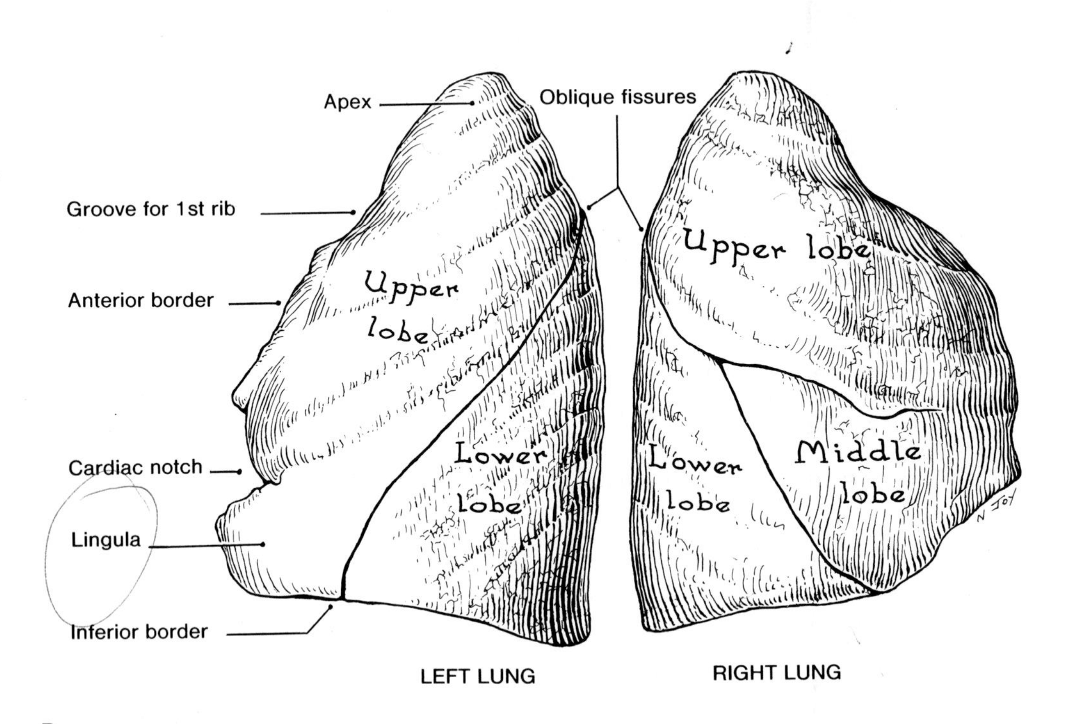

Figure 1-30. *A*, anterior view of the lungs and pericardium. *B*, lateral views of the lungs. Observe the three lobes of the right lung and the two lobes of the left lung. In *A* note that the middle lobe of the right lung lies at the anterior aspect of the thorax. In *B* observe the deficiency of the superior or upper lobe of the left lung, called the cardiac notch. The left lung is deficient here because of the bulge of the heart.

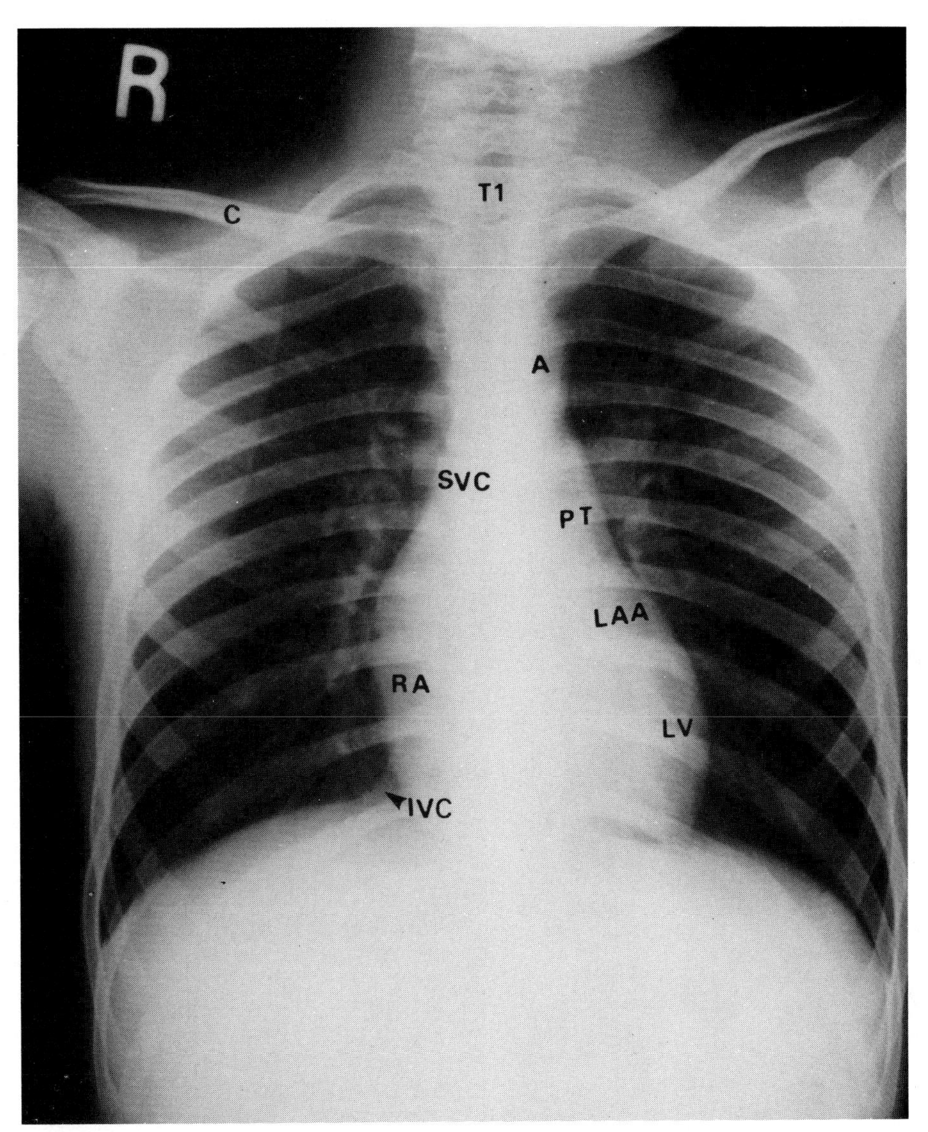

Figure 1-31. Radiograph of the chest, posteroanterior (*PA*) projection. Observe the body of the first thoracic vertebra (*T1*). Follow it laterally to the first rib which curves laterally, then medially, crossing the clavicle (*C*). Note that the dome or cupola of the diaphragm is somewhat higher on the right (see also Fig. 1-42). Observe that the right mediastinal border is formed by the superior vena cava (*SVC*) and the right atrium (*RA*). In the angle between the right atrium and the superior border of the diaphragm, an *arrow* points to the inferior vena cava (*IVC*). Note that the left mediastinal border is formed by the arch of the aorta (*A*), the pulmonary trunk (*PT*), the left auricular appendage (*LAA*), or auricle, and the left ventricle (*LV*). (See Figs. 1-45 and 1-48 for details of the heart's borders).

visceral pleurae. You may encounter pleural adhesions in a cadaver, which must be broken down before the lung can be completely mobilized.

The accumulation of significant amounts of fluid in the pleural cavity (**hydrothorax**) may result from a variety of causes. In advanced cases of pleuritis, serum from the inflamed pleurae may exude or effuse from the blood vessels of the pleurae into the pleural cavity, forming a **pleural exudate**. As fluid accumulates, the subatmospheric pressure is lessened, allowing the lung to retract toward its hilum. After the lung is completely retracted, additional fluid will cause the heart and mediastinum to become displaced toward the opposite side. If the inflamed pleurae become infected, the fluid contains leukocytes and the debris of dead cells (*i.e.*, pus).

Pus in a body cavity is called ***empyema*** and,

when used without qualification, refers to **pyothorax** (pus in a pleural cavity).

A common cause of noninflammatory fluid accumulation in the pleural cavity is **congestive heart failure** (excessive accumulation of fluid resulting from failure of the heart to function normally).

Blood may also appear in the pleural cavity (**hemothorax**) as the result of a chest wound. In rare cases, ***chyle*** (lymph and emulsified fat) may pass into the pleural cavity from a **ruptured thoracic duct** (Fig. 1-42). This condition is called **chylothorax**.

Fluid can be drained from a pleural cavity by inserting a wide-bore needle through an intercostal space (usually posteriorly through the seventh). *Aspiration of serous fluid, blood, or pus from the pleural cavity is called a* **pleural tap**. If the needle is inserted inferior to intercostal space 8 or 9 and pushed too deeply, it is in danger of penetrating the diaphragm. After penetration of the diaphragm, the needle might reach the spleen on the left side or the liver on the right side and injure them.

Entry of air into a pleural cavity (**pneumothorax**), following an external penetrating wound or rupture of a lung, results in partial collapse of the lung.

Fractured ribs often produce a pneumothorax, but the most common type is **spontaneous pneumothorax** that usually results from rupture of bullae (blebs) on the surface of the lungs.

Because the pleural cupolae and the apices of the lungs extend into the neck (Fig. 1-35), they are vulnerable to stab wounds in the root of the neck. These wounds may penetrate the pleural cavity and lung, producing an **open pneumothorax.** In these cases there is communication between the atmosphere and the pleural cavity. Hence you may hear this condition referred to as a **"blowing or sucking" pneumothorax.**

Most cases of uncomplicated pneumothorax are not dangerous, but if the opening in the visceral pleura has a flap over it, air can enter the pleural cavity on inspiration but cannot leave it during expiration. In such cases, the amount of air in the pleural space increases (**positive pressure pneumothorax**), pushing the mediastinum to the other side, compressing the opposite lung, and killing the patient. Thus, *positive pressure pneumothorax is a medical emergency*!

The pleura may also be injured by a needle when a **stellate ganglion block** or a nerve block of the brachial plexus is being performed. The anesthetist's needle tears the pleura as the lungs move during respiration.

Because the pleura crosses the 12th rib (Fig. 1-27), it may be injured during removal of a kidney (**nephrectomy**). When the 12th rib is very short, the 11th rib may be mistaken for it; consequently a posterior incision reaching it would result in opening of the pleural cavity to the atmosphere (**open pneumothorax**).

Vessels and Nerves of the Pleurae (Figs. 1-15 to 1-20). *The arterial supply of the parietal pleura* is from the arteries which supply the thoracic body wall (intercostal, internal thoracic, and musculophrenic arteries).

The nerve supply of the costal pleura and the pleura on the peripheral part of the diaphragmatic pleura is from the **intercostal nerves**, whereas the mediastinal pleura and the central part of the diaphragmatic pleura are supplied by the **phrenic nerves** (Figs. 1-29 and 1-70).

The veins of the parietal pleura join the systemic veins in the neighboring parts of the thoracic wall. Similarly its *lymphatic vessels drain into adjacent lymph nodes* of the thoracic wall (intercostal, parasternal, posterior mediastinal, and diaphragmatic). These nodes (Fig. 1-16) may in turn drain into the *axillary lymph nodes* (Fig. 6-43).

The arterial supply of the visceral pleura is from the bronchial arteries (Fig. 1-72) which are branches of the thoracic aorta (Fig. 1-78).

The veins from the visceral pleura drain into the **pulmonary veins**, and its numerous lymphatic vessels drain into nodes at the hila of the lungs.

The nerve supply of the visceral pleura is derived from the *autonomic nerves* innervating the lung which accompany the bronchial vessels (Fig. 1-74)

The visceral pleura is insensitive to pain or contact because it receives no nerves of general sensation, but *the parietal pleura is very sensitive to pain,* particularly its costal part. Irritation of the costal and peripheral diaphragmatic areas results in local pain and in **referred pain** along the intercostal nerves to the thoracic and abdominal walls (Fig. 2-16). Irritation of mediastinal and central diaphragmatic areas results in referred pain in the root of the neck and over the shoulder. The explanation for this is that the root of the neck and the shoulder are supplied by the same segments of the spinal cord

(C3 to C5) that give origin to the phrenic nerve (Fig. 1-23).

THE LUNGS

The lungs (L. *pulmones*) are the essential **organs of respiration** (Figs. 1-28, 1-130, and 1-31). Within them the inspired air is brought into close relation to the blood in the pulmonary capillaries.

During life the lungs are normally light, soft, and spongy. They are very elastic and will shrink to about one-third when the thoracic cavity is opened (Fig. 1-24).

During early life the lungs are light pink, but often these become dark and mottled during late life owing to the accumulation of inhaled dust particles which become trapped in the fixed phagocytes in the lungs over the years.

Each lung has the shape of a half cone and is contained in its own pleural sac within the thoracic cavity (Figs. 1-24 and 1-28). The lungs are separated from each other by the heart and other structures in the mediastinum (Figs. 1-11, 1-30*A*, and 1-31).

The lungs are attached to the heart and trachea by the roots of the lungs and the ***pulmonary ligaments*** (Figs. 1-30*A* and 1-32); otherwise each lung lies freely within its pleural cavity.

The right lung is larger and heavier than the left lung (Fig. 1-30), but the vertical extent of the right lung is less than that of the left because the right dome of the diaphragm is higher (Figs. 1-11 and 1-43). It is also wider than the left lung because the heart and pericardium bulge more to the left (Figs. 1-30*A* and 1-31).

The hardened lungs in an embalmed cadaver (Figs. 1-32 and 1-33) have impressions formed by adjacent structures (*e.g.*, the ribs and costal cartilages), whereas fresh lungs usually do not. The cadaveric markings are helpful reminders of the relationships of the lungs.

Healthy lungs always contain air, thus excised lungs will float. A diseased lung filled with fluid may not float. The lungs of a stillborn infant are firm and sink when placed in water, whereas those of a liveborn infant will float. *These observations are of medicolegal significance.*

LOBES AND FISSURES OF THE LUNGS

(Figs. 1-11, 1-24, 1-27, 1-28, and 1-30 to 1-40)

The left lung *is divided into superior and inferior lobes by a long deep* ***oblique fissure*** which extends from the costal to the medial surface of the lung.

The superior lobe has a wide **cardiac notch** on its anterior border, where the lung is deficient owing to the bulge of the heart. This leaves part of the anterior aspect of the pericardium uncovered by lung tissue. The anteroinferior part of the superior lobe has a small tongue-like projection called the **lingula** (L. tongue). The inferior lobe of the left lung is larger than the superior lobe and lies inferior and posterior to the oblique fissure (Fig. 1-30*A*).

The right lung *is divided into superior, middle, and inferior lobes by* ***oblique and horizontal fissures.*** The horizontal fissure separates the superior and middle lobes, and the oblique fissure separates the inferior lobe from the middle and superior lobes. The superior lobe is smaller than in the left lung and the middle lobe is wedge-shaped in outline (Fig. 1-30*B*).

Occasionally extra fissures subdivide the lungs. The left lung sometimes has three lobes and the right lung may have only two lobes if a fissure fails to form.

A **lobe of the azygos vein** appears in the right lung in about 1% of people. It develops when the apical bronchus (Fig. 1-37) grows superiorly, medial to the arch of the azygos vein (Fig. 1-42), instead of lateral to it. As a result, the azygos vein comes to lie at the bottom of a deep fissure in the superior lobe. This fissure with the azygos vein at its inferior end produces a linear marking on a radiograph of the chest, which separates the apical part of the lung from the remainder of the superior lobe.

SURFACES AND BORDERS OF THE LUNGS

Each lung presents an *apex*, *three surfaces* (costal, medial, and diaphragmatic), and *three borders* (anterior, inferior, and posterior). *The diaphragmatic surface of the lung is commonly called the base.*

The Apex of the Lung (Figs. 1-30 to 1-33). The rounded superior pole or apex of each lung extends through the superior thoracic aperture into the root of the neck, where it lies in close contact with the pleural **cupola** (Fig. 1-35). Owing to the obliquity of the superior thoracic aperture, the apex of the lung extends superior to the anterior part of the first rib (Fig. 1-25). Its summit lies anterior to the neck of the first rib, about 2.5 cm superior to the medial third of the clavicle (Figs. 1-31 to 1-35).

The apex of the lung is crossed by the subclavian artery and vein (Fig. 1-63). The artery produces a groove in its mediastinal surface (Fig. 1-33). These vessels are separated from the apex of the lung by the

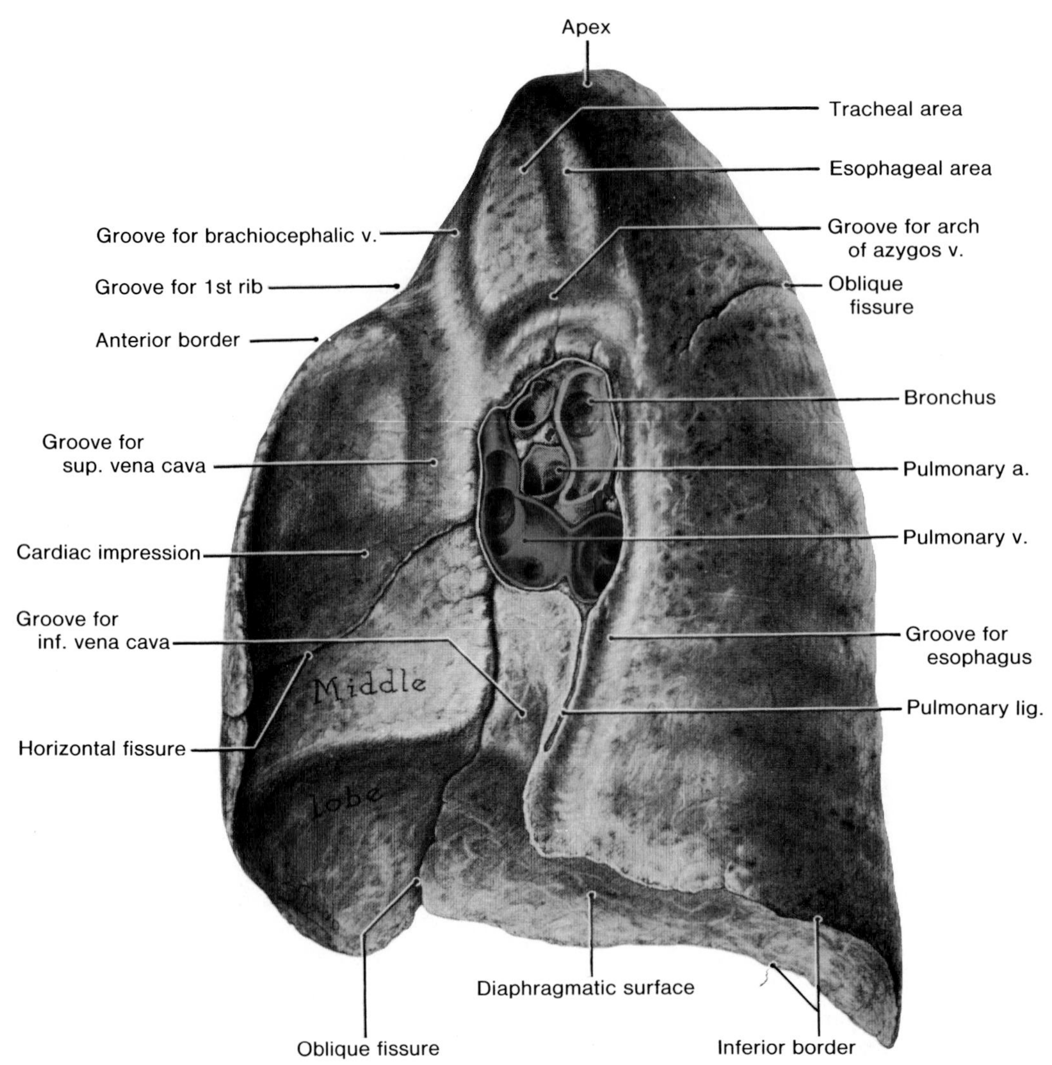

Figure 1-32. Drawing of the *mediastinal surface of the right lung*. Observe that the lung resembles an inflated balloon in that it takes the impressions of the structures with which it comes into contact. Thus the diaphragmatic surface or base of the lung is fashioned by the cupola of the diaphragm (Fig. 1-42), and the costal surface bears the impressions of the ribs. Note that distended vessels leave their marks, whereas empty vessels and nerves do not. Observe the *root of the lung* near the center of the mediastinal surface and the pulmonary ligament descending like a stalk from it. Note the groove for, or line of contact with, the esophagus throughout the length of the lung, except where the arch of the azygos vein intervenes. This groove passes posterior to the root and therefore posterior to the pulmonary ligament, which separates it from the groove for the left vena cava. Observe the oblique fissure, here incomplete. Note the two pulmonary veins, here uniting unusually close to the lung.

cervical pleura and the suprapleural membrane (Fig. 1-25).

Auscultation of the lungs (listening with a stethoscope) must include the root of the neck superior to the medial third of the clavicle, in order that sounds in the apex of the lung may be heard. As mentioned previously, a stab wound superior to this region of the clavicle may pierce both the pleura and the apex of the lung.

The Costal Surface of the Lungs (Figs. 1-30 and 1-32). The costal surface of each lung is large, smooth, and convex. The costal surface includes the bulky posterior part of the lung. It is in contact with the costal pleura and in embalmed cadavers impressions of the ribs are visible because *this surface is related to the curvatures of the ribs* and the intercostal muscles.

The Medial Surface of the Lung (Figs. 1-27, 1-32, and 1-33). This surface of each lung is divided into two parts, a vertebral part and a mediastinal part.

The vertebral part of the medial surface occupies the gutter on each side of the thoracic region

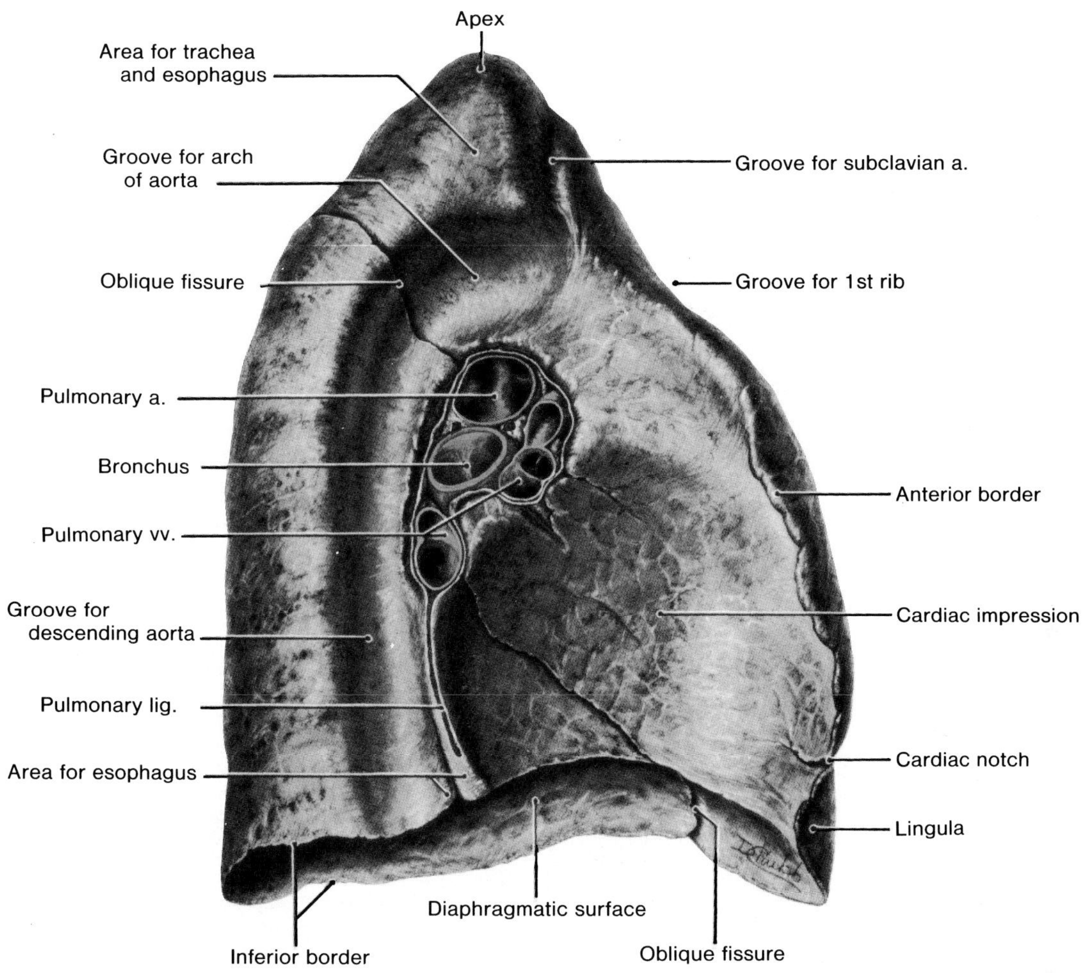

Figure 1-33. Drawing of the *mediastinal surface of the left lung*. Near its center observe the root of the lung and the pulmonary ligament descending from it. Note the lung's site of contact with the esophagus, between the aorta and the inferior end of the pulmonary ligament. Observe the oblique fissure cutting completely through the lung substance. In both the right and left lung roots, the artery is superior, the bronchus is posterior, one vein is anterior, and the other is inferior. In the right root, the bronchus (eparterial) to the superior lobe is the highest structure. The marked groove for the descending aorta was caused by a arteriosclerotic aorta that had deviated from the midline. In living persons the lung lies free within its pleural cavity (Fig. 1-24), attached only by its root and the pulmonary ligament.

of the vertebral column and passes imperceptibly into the costal surface.

The mediastinal part of the medial surface is concave because it *fits against the mediastinum* (Fig. 1-11). **The mediastinal part contains the *root of the lung***, where the visceral and parietal layers of pleura form a "sleeve" around the roots of the lungs containing the vessels and bronchi (Figs. 1-27 and 1-31).

The mediastinal part of the medial surface is indented by the heart and great vessels, especially on the left lung (Figs. 1-32 and 1-33). The deep concavity, called the **cardiac impression**, accommodates the heart and the pericardium (Fig. 1-46).

The cardiac impression is larger and deeper on the left (Fig. 1-33) than on the right (Fig. 1-32), because the heart projects more to the left than to the right (Fig. 1-31).

The **hilum of the lung** (hilus of the lung) is where the bronchi, pulmonary vessels, bronchial vessels, lymph vessels, and nerves enter and leave the lung. It lies posteriorly near the center of the mediastinal part of the lung (Figs. 1-32 and 1-33). The hilum is surrounded by a sleeve of pleura which is reflected off the lung on to the mediastinum (Fig. 1-27).

The pulmonary ligament is the inferior extension of this pleural sleeve (Figs. 1-32 and 1-33).

The structures entering and leaving the lung form the ***root of the lung*** (Fig. 1-27) which is attached at the hilum.

The Diaphragmatic Surface of the Lung (Figs. 1-32 and 1-33). This semilunar concave surface, often spoken of as the **base of the lung**, rests on the cupola or *dome of the diaphragm* (Figs. 1-31 and 1-42). The concavity of the base or diaphragmatic surface is deeper in the right lung than in the left lung because of the slightly higher position of the right dome of the diaphragm (Fig. 1-42). Laterally and posteriorly, the diaphragmatic surface is bounded by a thin sharp margin which projects into the **costodiaphragmatic recess of the pleura** (Figs. 1-11 and 1-29).

The Anterior Border of the Lung (Figs. 1-30 to 1-34). This border of each lung is thin and sharp and *overlaps the pericardium* (Fig. 1-30*A*). It corresponds more or less to the anterior border of the pleura (Fig. 1-34). The anterior border of the left lung has a **cardiac notch** of variable size (Fig. 1-34). Thus, the pericardium is covered only by a double layer of pleura in this area.

The Posterior Border of the Lung (Figs. 1-32 and 1-33). This border of each lung is *thick and rounded.* It lies in a paravertebral gutter and fits against the thoracic region of the vertebral column (Fig. 1-27).

The Inferior Border of the Lung (Figs. 1-30 to 1-34). This border limits the base or diaphragmatic surface of the lung. It is *thin and sharp* where it extends into the **costodiaphragmatic recess** (Figs. 1-11 and 1-29). The inferior border separates the diaphragmatic surface or base of the lung from the costal surface. The inferior border is blunt and rounded medially where it divides the diaphragmatic surface from the mediastinal surface (Figs. 1-32 and 1-33).

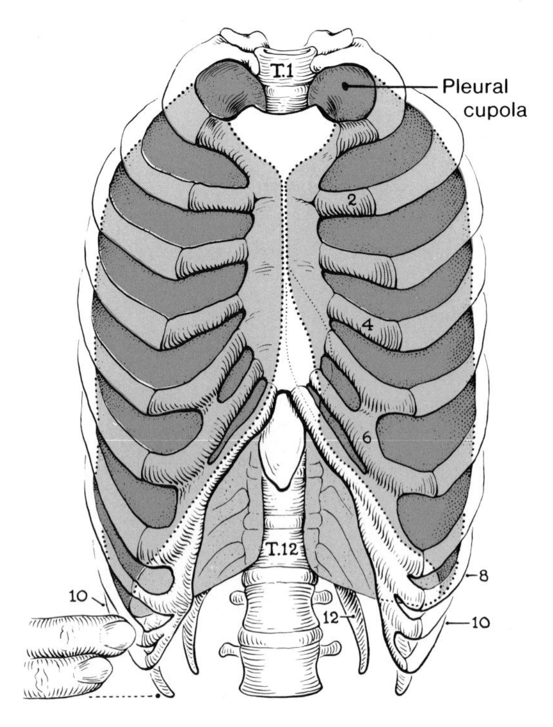

Figure 1-35. Drawing of the thorax illustrating the extent of the pleura in adults. Observe that the pleura rises to but not superior to the neck of the first rib. Note that the right and left sternocostal reflections meet posterior to the sternum, superior to the level of the second ribs, and descend together to the fourth ribs where the left pleura deviates variably to the sixth or seventh rib. Note that the pleura is two fingerbreadths superior to the margin of the bony thorax and that it ascends as the vertebral reflexion on the vertebral bodies of T12 to T1. (Also see Fig. 1-27).

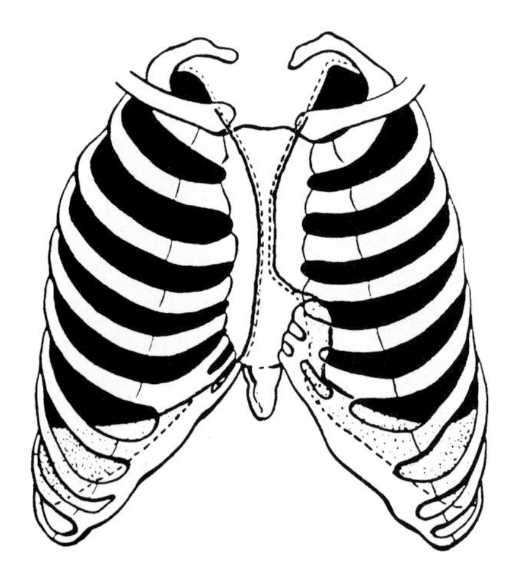

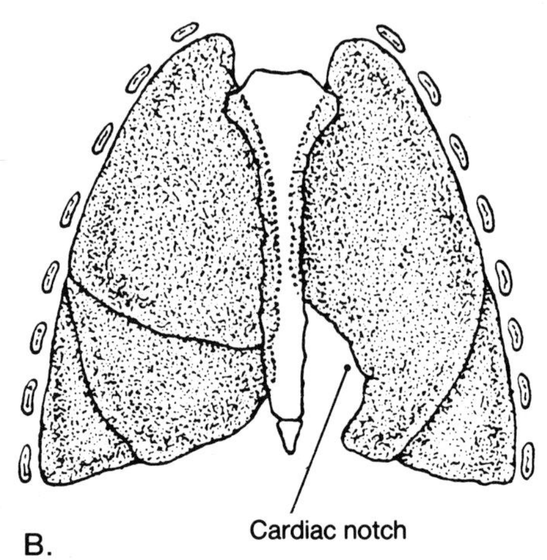

Figure 1-34. Drawings illustrating the lungs and the lines of pleural reflection (*light broken line*). The degree to which the lungs and the pleurae diverge from the sternum varies from one person to another. Note that between the levels of the second and fourth costal cartilages, the anterior margins of the two lungs are near the median line (*dark lines*) and correspond with the anterior borders of the pleurae (*broken lines*). Observe that the anterior margins of the lungs differ inferior to the fourth costal cartilage (see the text for details).

SURFACE MARKINGS OF THE LUNGS (Fig. 1-34)

The anterior borders of the lungs follow the lines of pleural reflection, except at the cardiac notch inferior to the level of the fourth costal cartilage. The location of the lungs can be outlined on the surface of the thorax using the following landmarks. *These markings are clinically important.*

The apex of the lung is represented by a line drawn superolaterally from the sternoclavicular joint to a point 2.5 cm superior to the middle third of the clavicle, and then drawn inferolaterally to the junction of the middle and medial thirds of the clavicle.

The anterior border of the right lung essentially corresponds to the anterior border of the right pleura. Between the levels of the second and fourth cartilage, its anterior margin is near the midline. Inferior to the fourth costal cartilage, the surface of the right lung gradually diverges from the midline and leaves the sternum posterior to the sixth costal cartilage. (Fig. 1-34).

The anterior border of the left lung essentially corresponds to the anterior border of the left pleura as far as the level of the fourth costal cartilage (Fig. 1-34). Here the anterior border of this lung deviates laterally to a point about 2.5 cm lateral to the left border of the sternum to form the **cardiac notch** (Fig. 1-34). It then turns inferiorly and slightly medially to the sixth left costal cartilage.

The inferior border of both lungs is indicated by a line drawn from the inferior end of the line representing the anterior border that crosses the *6th rib in the midclavicular line*, the *8th rib in the midaxillary line*, the *10th rib in the midscapular line*, and ends about 2.5 cm lateral to the spinous process of the *10th thoracic vertebra.*

The inferior borders of the lungs lie two ribs superior to that of the parietal pleura on each of the three vertical lines just mentioned (Fig. 1-34*A*). In children the inferior borders of the lung are about one rib more superior than in adults.

The levels of the inferior borders of the lungs vary according to the phase of respiration (Figs. 1-51 and 1-52).

Surface Markings of the Lung Fissures. *The projection of the oblique fissure* is essentially the same for each lung and is indicated on the surface of the thorax by a line from the spinous process of the second thoracic vertebra around the thorax to the sixth costochondral junction.

The projection of the horizontal fissure in the right lung is indicated by a line on the surface of the thorax that runs from the anterior border of the lung along the fourth costal cartilage to the oblique fissure (Fig. 1-34*B*).

THE BRONCHI (Figs. 1-11, 1-32, 1-33, and 1-37)

The principal bronchi (main bronchi), one to each lung, pass inferolaterally from the termination of the trachea to the hila of the lungs. The principal bronchus enters the hilum and subdivides within the lung to form the "*bronchial tree*" (Fig. 1-37).

Each pulmonary artery passes transversely into the lung, anterior to its bronchus. The two pulmonary veins on each side (superior and inferior) ascend from the hilum of the lung to the left atrium of the heart (Fig. 1-49).

Within the lung the bronchi divide in a constant fashion in constant directions so that each branch supplies a clearly defined sector of the lung (Figs. 1-37 and 1-39). Each principal bronchus divides into secondary bronchi or **lobar bronchi** (two on the left, three on the right), *each of which supplies a lobe* of the lung. Each lobar bronchus then divides into tertiary bronchi or **segmental bronchi** which supply specific sectors of the lung, called **bronchopulmonary segments** (Figs. 1-36 to 1-40).

When the bronchi are examined with a **bronchoscope**, a keel-like ridge called the **carina**

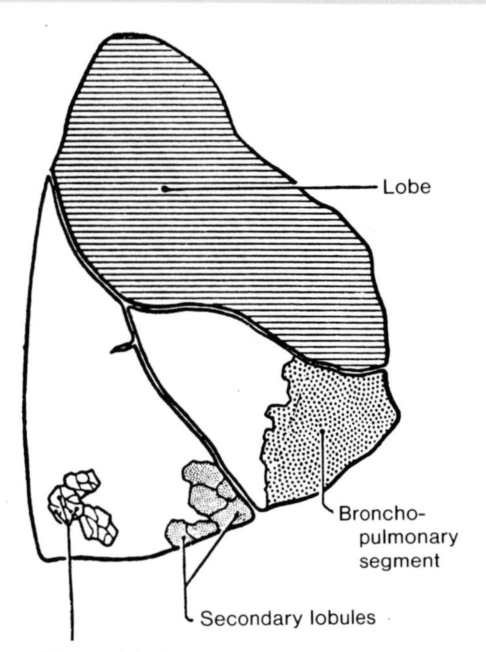

Figure 1-36. Diagram showing the subdivisions of the right lung. It consists of three lobes, each of which is supplied by a lobar bronchus (*e.g.*, the superior lobe bronchus, Fig. 1-39). From the three secondary bronchi, 10 segmental bronchi arise (three for the superior lobe, two for the middle lobe, and five for the inferior lobe). Each segmental bronchus supplies a bronchopulmonary segment (Figs. 1-37 and 1-38).

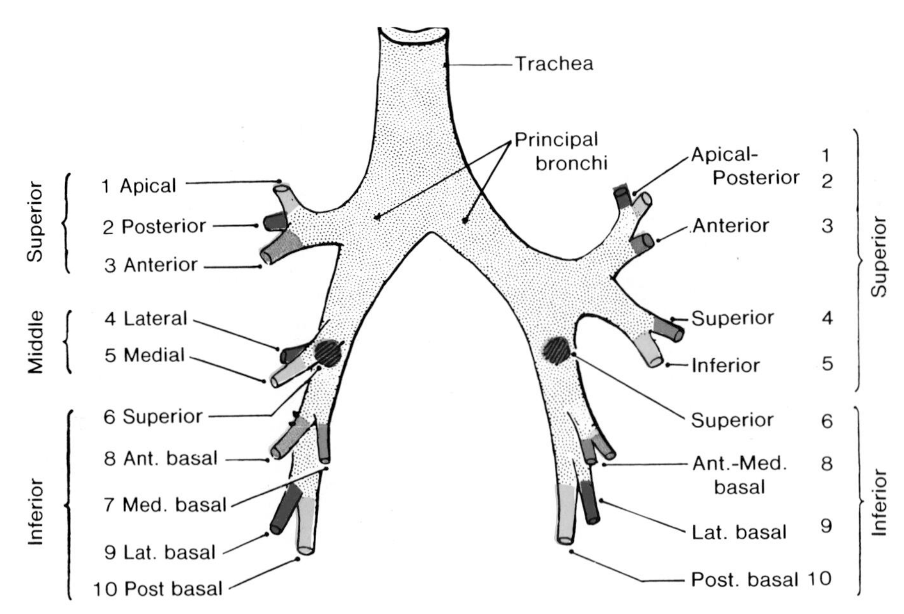

Figure 1-37. Diagram illustrating the tertiary or segmental bronchi for correlation with the bronchopulmonary segments shown in Figure 1-38. Observe that the right lung has three lobes and the left two and that there are 10 segmental bronchi on the right and 8 on the left. Note that on the left the apical and posterior bronchi arise from a single stem, as do the anterior basal and medial basal.

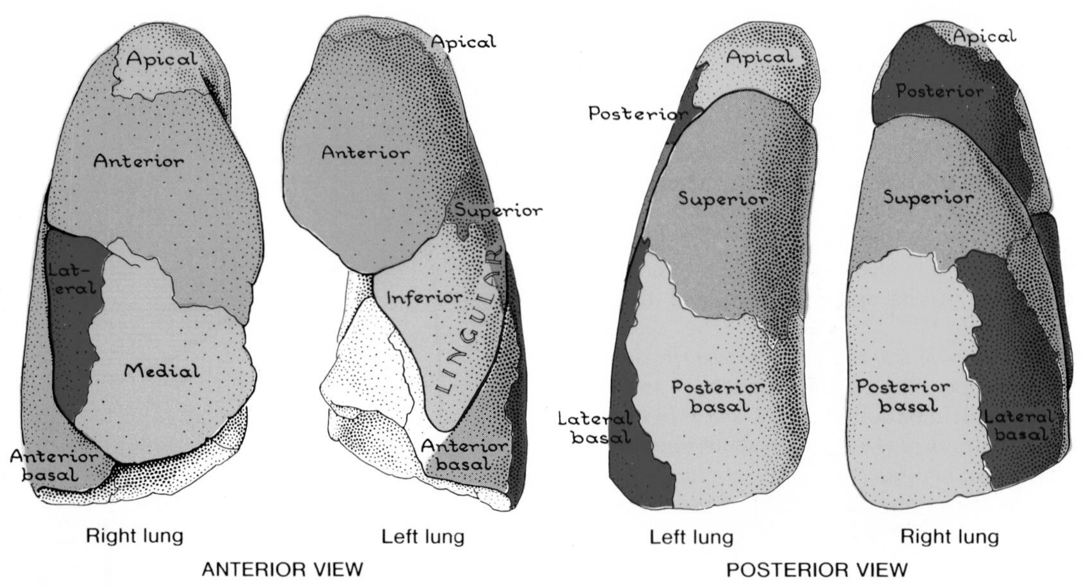

Figure 1-38. Drawings illustrating the bronchopulmonary segments. To prepare these specimens, the tertiary or segmental bronchi (Fig. 1-37) of fresh lungs were isolated within the hilum and injected with latex of various colors. Minor variations in the branching of the bronchi result in variations in the surface patterns in different specimens. A bronchopulmonary segment consists of a tertiary or segmental bronchus, the portion of lung it ventilates, an artery, and a vein (Fig. 1-40). These segments are surgically separable.

(L. keel), is observed at the inferior end of the trachea. The **carina of the trachea** (Fig. 1-11) indicates the origins of the right and left principal bronchi. Normally the carina is in the median plane and has a fairly definite edge. If the **tracheobronchial lymph nodes** (Fig. 1-73) in the angle between the principal bronchi become enlarged (*e.g.*, owing to the lymphoge-

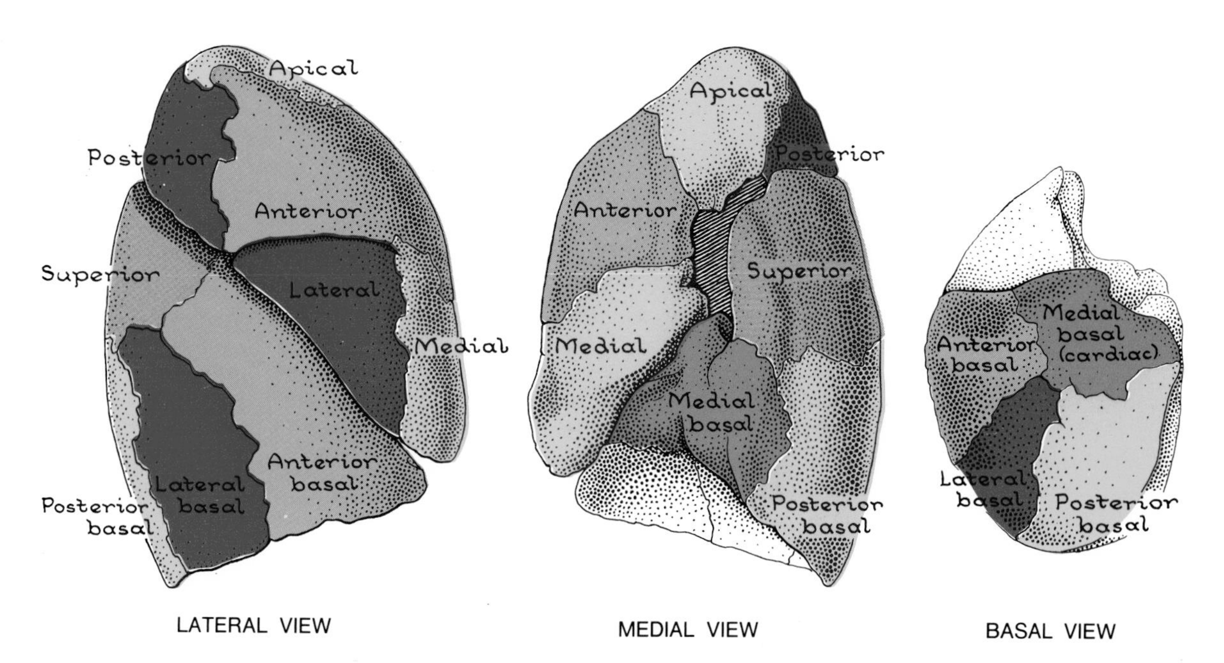

RIGHT BRONCHOPULMONARY SEGMENTS

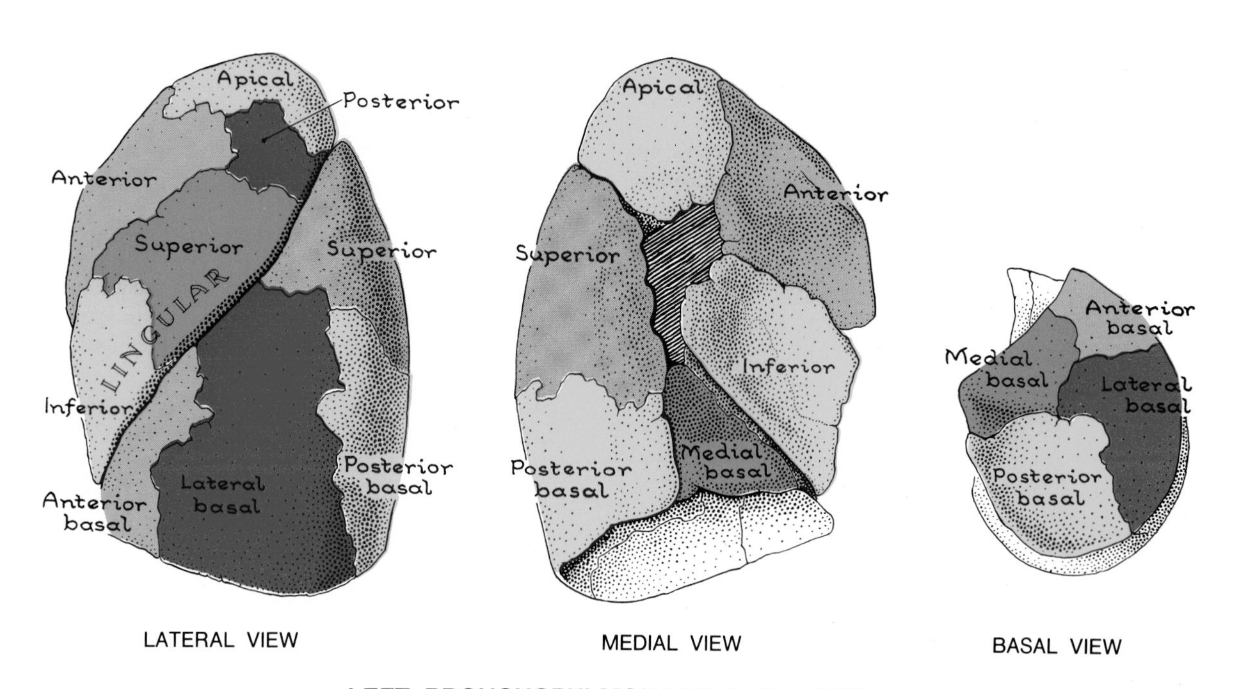

LEFT BRONCHOPULMONARY SEGMENTS

Figure 1-38. Continued.

nous spread of cancer cells from a **bronchogenic carcinoma**), the carina spreads out (Case 1-2). *The morphological changes in the carina are important diagnostic signs* for the **bronchoscopist** in assisting with the differential diagnosis of disease of the respiratory system.

The mucous membrane at the carina is one of the most sensitive areas of the tracheobronchial tree and is associated with the **cough reflex.** For example, when a child aspirates a peanut he/she chokes and coughs. Once the peanut passes the carina coughing usually stops, but the resulting **chemical bronchitis** (inflammation of the bronchus) caused by substances released from the peanut, and the **atelectasis** (collapse) of the lung distal to the foreign body soon causes difficult breathing (**dyspnea**).

The carina is often considered to be the last line of defense and frequently the violent coughing caused by irritation of it results in expulsion of the aspirated foreign body.

At the bifurcation of the trachea, the right principal bronchus is wider and shorter and runs more vertically than does the left principal bronchus (Figs. 1-41 and 1-72). *This is the anatomical reason foreign bodies are more likely to enter the right bronchus than the left* and to lodge in it or in one of its branches.

A common hazard to dentists is an aspirated foreign body (*e.g.*, an extracted tooth or a piece of filling material).

Bronchopulmonary Segments (Figs. 1-36 to 1-40). *The segments of a lung supplied by segmental bronchi are called bronchopulmonary segments.* Within each segment there is further branching of the bronchi (Fig. 1-40). Each segment is pyramidal in shape with its apex pointed toward the root of the lung and its base at the pleural surface.

Each bronchopulmonary segment has its own segmental bronchus, artery, and vein. The names of the various segmental bronchi and bronchopulmonary segments are given in Figures 1-37 and 1-38.

Bronchopulmonary segments are of considerable clinical significance. Bronchial and pulmonary disorders (*e.g.*, a tumor or an abscess) may be localized in one of these segments and the segment may be surgically resected (removed) without seriously disrupting surrounding lung tissue.

Each bronchopulmonary segment is supplied by its own nerve, artery, and vein, but it is important to know that during surgical resection of these segments the planes between them are crossed by branches of pulmonary veins, and sometimes by branches of pulmonary arteries (Fig. 1-40). In addition, the **bronchial arteries** run through the interlobular septa to supply the visceral pleura.

Each bronchopulmonary segment is surrounded by connective tissue that is continuous with the visceral pleura (Fig. 1-40). The connective tissue septa separating the segments prevent air from passing between segments. Therefore, air in a bronchopulmonary segment whose segmental bronchus is obstructed is absorbed by the blood stream. This causes **segmental atelectasis** or collapse of the tissue in the affected segment.

Malignant tumors and certain infections (*e.g.*, ***tuberculosis***) *invade the connective tissue septa separating the bronchopulmonary segments* and involve adjacent segments. In such disorders, surgical resection of several bronchopulmonary segments, a whole lobe (**lobectomy**), or an entire lung (**pneumonectomy**) may be necessary. The Greek word *pneumōn* means lung.

Knowledge of the branching of the bronchial tree is necessary to determine the appropriate postures for draining infected areas of the lung. For example, when a patient with **bronchiectasis** (dilation of bronchi) is positioned in bed on his/her left side, secretions from the right lung and bronchi flow toward the carina of the trachea. As this is a sensitive area, the cough reflex is stimulated and the patient brings up **purulent sputum** (*i.e.*, contains pus), clearing the right bronchial tree.

Alternatively, persons with **bronchiectasis of the lingula** of the left superior lobe (Fig. 1-33) drain it by lying on the right side. The basal bronchi (Fig. 1-37) may be cleared by the patient standing on his/her head for several minutes every morning to promote drainage of the lungs.

Because oropharyngeal and nasopharyngeal contents containing bacteria may be aspirated into the lungs and cause inflammation of the lungs (**pneumonitis** or pneumonia), or a lung abscess in one or more bronchopulmonary segments, the position of very ill and unconscious patients is changed frequently to promote good drainage and aeration of their lungs. In the *prone position* (face downward), the trachea slopes inferiorly; therefore, the usual *supine position* one assumes in bed with the head slightly elevated by a pillow is poor for lung drainage.

A tumor may block a segmental bronchus and cause collapse of the part of the lung distal to it, owing to absorption of the air in the bronchopulmonary segment by the blood which is still circulating through it. The *collapsed portion of lung* can be determined radiographically using a technique known as ***bronchography*** (Fig. 1-39).

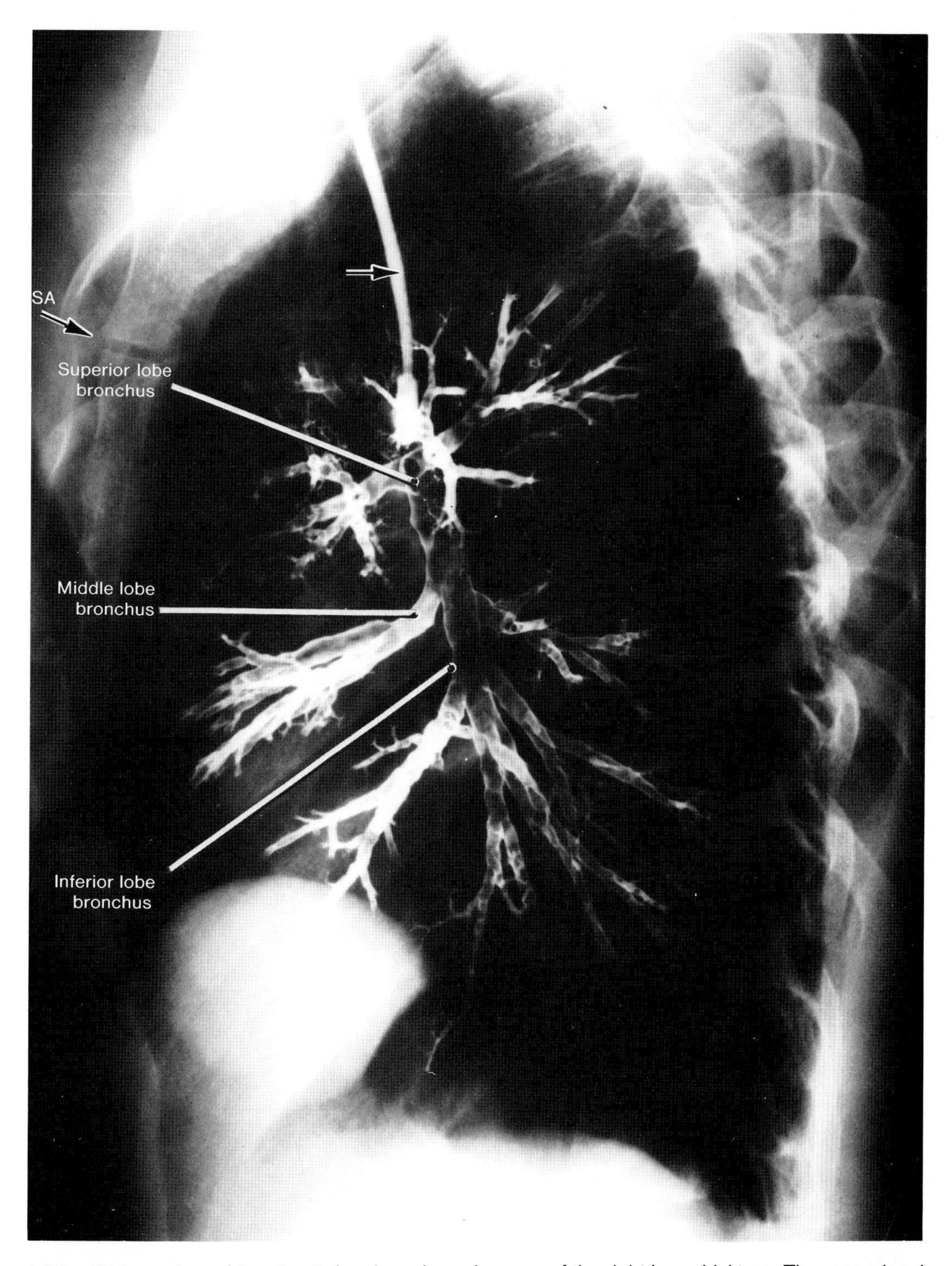

Figure 1-39. Oblique view of the chest showing a bronchogram of the right bronchial tree. The secondary bronchi are labeled. The contrast material was injected via a catheter (*superior arrow*) after topical anesthesia of the nose, pharynx, larynx, and trachea. The injection was performed under fluoroscopic control so that the patient could be postured in various positions to allow the contrast material to flow into all the secondary and segmental (tertiary) branches. *SA* and *arrow* indicate the sternal angle (also see Fig. 1-10).

Bronchography requires *topical anesthesia* (produced by direct application of local anesthetic solutions to the pharynx, larynx, and trachea). The contrast medium may be allowed to flow over the back of the tongue (which has been drawn anteriorly) into the larynx and trachea, or a catheter may be introduced into the trachea after complete anesthetization of the throat, larynx, and trachea. Small quantities of the medium are then introduced and the patient is postured in a suitable position to allow the contrast material to flow (by gravity) into the preselected secondary and segmental bronchi.

Only one side is injected at a time because the contrast medium partly obstructs the flow of air to the segmental bronchi which are being visualized. Impaired oxygenation of the blood may occur if both sides are filled with contrast medium at one time.

The Roots of the Lungs (Figs. 1-27, 1-32, and 1-33). *The root of a lung consists of the structures entering and leaving the hilum.* It connects the medial surface of the lung to the heart and trachea. It is surrounded by pleura which is prolonged inferiorly as the **pulmonary ligament.**

The roots of the lung descend during deep inspiration (Fig. 1-51). The main components of the roots of the lungs are the principal bronchi and the pulmonary vessels (Figs. 1-32 and 1-33). Nerves, bronchial vessels, lymph vessels, and lymph nodes are also in the roots of the lungs.

The root of the right lung (Figs. 1-27 and 1-32) is enclosed by pleura and is composed of *two pulmonary arteries, two pulmonary veins*, a bronchus, *a bronchial artery*, bronchial veins, a pulmonary plexus of nerves, lymph vessels, and lymph nodes enclosed in connective tissue.

The root of the left lung (Figs. 1-27 and 1-33) is similar to the root of the right lung (Fig. 1-32), except that there is only *one pulmonary artery* and there are usually *two bronchial arteries.* Usually the bronchus lies posterior, the artery superior, and the veins inferior.

The ***right superior bronchus*** *has a special location that is more superior than any other bronchus* (Fig. 1-41). It is even more superior than the pulmonary artery and for this reason it is sometimes called the **eparterial bronchus** (G. *epi*, upon).

ARTERIES OF THE LUNGS (Figs. 1-31 to 1-33, 1-40, and 1-41)

The branches of the pulmonary arteries distribute venous blood to the lungs for aeration (changing of venous into arterial blood). These branches follow the bronchi and generally lie on their posterior surfaces.

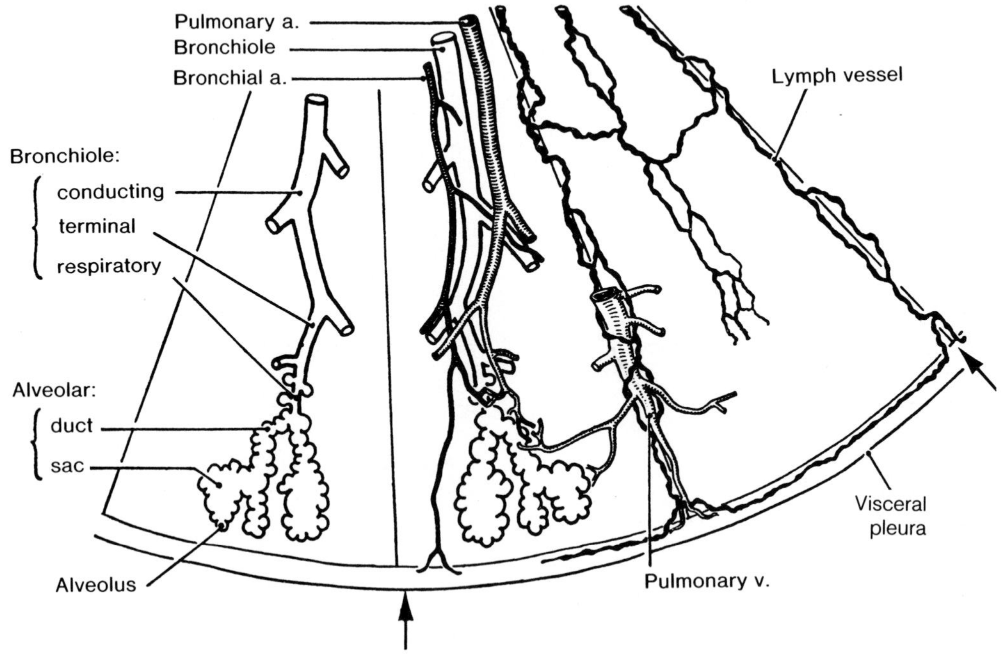

Figure 1-40. Diagrammatic sketch illustrating the structure of a lung. The base of a brochopulmonary segment is between the *arrows*. Within the bronchopulmonary segment, the segmental bronchi divide and decrease in size to form bronchioles.

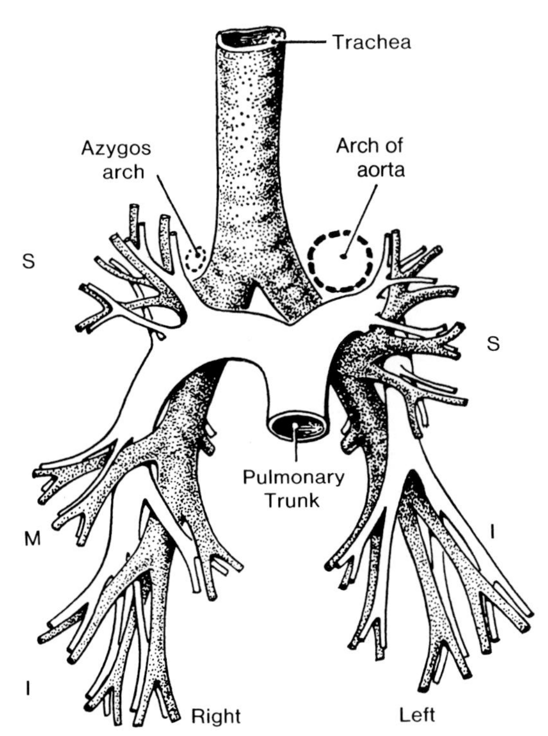

Figure 1-41. Drawing showing the relationship of the pulmonary arteries to the bronchi, anterior view. *S*, *M*, and *L* indicate branches to the superior, middle, and inferior lobes. Compare with Figure 1-39.

Thus there is a branch to each lobe, bronchopulmonary segment, and lobule of the lung (Fig. 1-36).

The terminal branches of the pulmonary arteries divide into capillaries in the walls of the ***alveoli***, the air sacs where gaseous exchange takes place between the blood and the air (Fig. 1-40).

The bronchial arteries supply the substance of the lung (tissue of the bronchial tree and alveoli, Figs. 1-40 and 1-72). They are small and pass along the posterior aspects of the bronchi to supply them as far distally as the **respiratory bronchioles** (Fig. 1-40).

The two left bronchial arteries arise from the thoracic aorta (Fig. 1-78). The single right bronchial artery arises from the first aortic intercostal artery, or from the superior left bronchial artery.

Pulmonary Thromboembolism (PTE) is a common cause of *morbidity* (sickness) and *mortality* (death).

An **embolus** (G. a plug) is produced when a **thrombus** (blood clot), fat globules, or air bubbles are carried from a distant site (*e.g.*, from the leg veins following a fracture or other injury of the lower limb). The thrombus passes through the right side of the heart and passes to a lung via the pulmonary artery. The immediate result of PTE is complete or partial obstruction of the pulmonary arterial blood flow.

Embolic obstruction of a pulmonary artery produces a sector of lung which is ventilated but not perfused (*i.e.*, not functioning).

A large embolus may occlude the pulmonary trunk or one of its main branches (Fig. 1-41). The patient suffers **acute respiratory distress** and may die in a few minutes. A medium-sized embolus may block an artery to a bronchopulmonary segment, producing an **infarct** (area of dead tissue). In healthy people, a ***collateral circulation*** (*i.e.*, an indirect blood supply) often develops so that infarction does not occur. There are abundant anastomoses (communications) with branches of the bronchial arteries in the region of the terminal bronchioles (Fig. 1-40). In sick people in whom circulation in the lung is impaired (*e.g.*, in a person with chronic congestion of the lungs), embolism commonly results in infarction of the lung.

Because an area of pleura is also deprived of blood, it becomes inflamed **(pleuritis)** and rough. The pleuritis results in ***pain*** *in the side of the chest* and the rough pleural surfaces result in a **friction rub** (Case 1-8).

VEINS OF THE LUNGS (Figs. 1-32, 1-33, 1-40, 1-48, and 1-49)

The pulmonary veins drain oxygenated blood from the lungs to the left atrium of the heart. Beginning in the pulmonary capillaries, the veins unite into larger and larger vessels which run mainly in the interlobular septa (Fig. 1-40).

One main vein drains each bronchopulmonary segment, usually on the anterior surface of the corresponding bronchus. The two pulmonary veins on each side, a superior and an inferior one, open into the posterior aspect of the left atrium (Figs. 1-49 and 1-67).

The **superior right pulmonary vein** drains the superior and middle lobes of the right lung, and the **superior left pulmonary vein** drains the superior lobe of the left lung. The right and left **inferior pulmonary veins** drain the respective inferior lobes.

The bronchial veins drain the larger subdivisions of the bronchi. The right bronchial vein drains into the **azygos vein** (Fig. 1-42) and the left bronchial vein drains into the accessory hemiazygos vein or into the left superior intercostal vein (Fig. 1-80).

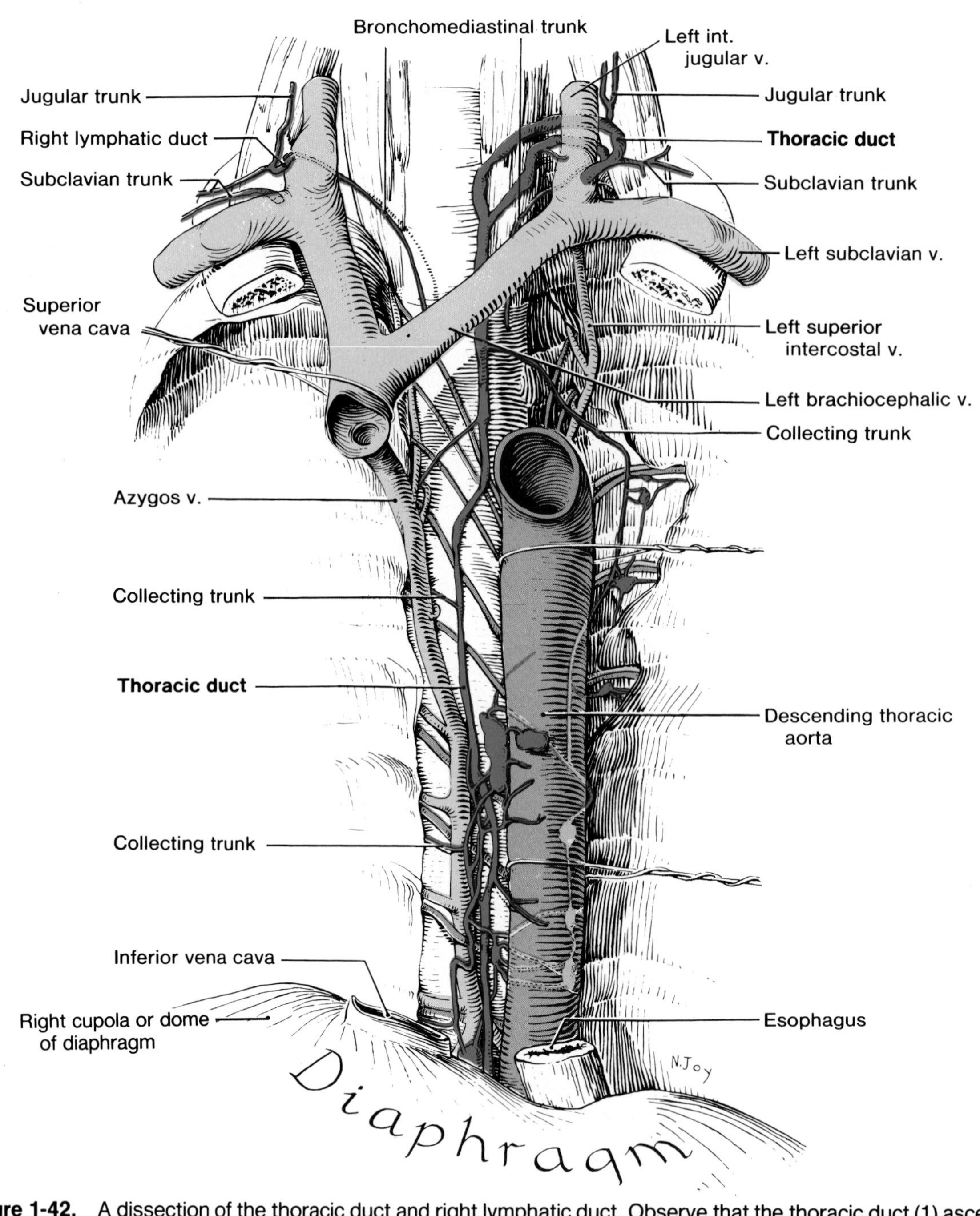

Figure 1-42. A dissection of the thoracic duct and right lymphatic duct. Observe that the thoracic duct (1) ascends on the vertebral column between the azygos vein and the descending aorta, and (2) at the junction of the posterior and superior mediastina, it passes to the left and continues its ascent to the neck, where (3) it arches laterally to open near the angle of union of the internal jugular and subclavian veins. Observe that the thoracic duct is plexiform in the posterior mediastinum and bifurcates in the neck. It receives branches from the intercostal spaces on both sides via several collecting trunks and also from posterior mediastinal structures. Note that the thoracic duct finally receives the jugular, subclavian, and bronchomediastinal trunks. Observe that the right lymphatic duct is very short and is formed by the union of the right jugular, subclavian, and bronchomediastinal trunks. An accessory subclavian lymph trunk is opening directly into the subclavian vein in this specimen.

LYMPHATIC DRAINAGE OF THE LUNGS
(Figs. 1-40 and 1-42)

There are two lymphatic plexuses or networks. ***The superficial plexus*** *lies deep to the visceral pleura* (Fig. 1-40) and the vessels from it drain into the **bronchopulmonary lymph nodes,** located in the hilum of the lung and at the bifurcation of the trachea. These lymph vessels drain the lung tissue and the visceral pleural.

The deep plexus *is located in the submucosa of the bronchi and in the peribronchial connective tissue.* ***No lymph vessels are located in the walls of the pulmonary alveoli.*** (Fig. 1-40). Lymph vessels from the deep plexus follow the bronchi and the pulmonary vessels to the hilum of the lung, where they drain into the **pulmonary lymph nodes** located close to the hilum, and into the **bronchopulmonary lymph nodes** in the hilum.

Lymph vessels then pass to **tracheobronchial lymph nodes** around the trachea and the principal bronchi (Figs. 1-63 and 1-73). The lymph then passes to the right and left **bronchomediastinal lymph trunks** (Fig. 1-42) which are formed by the junction of the efferent lymph vessels from the parasternal, tracheobronchial, and anterior mediastinal lymph nodes. These trunks usually terminate on each side at the junction of the subclavian and internal jugular veins (Fig. 1-42). The left trunk may terminate in the **thoracic duct.**

Lymph from the lungs carries phagocytes containing ingested carbon particles that were deposited in the walls of the alveoli from the inspired air. In older people, especially cigarette smokers and/or city dwellers, the surface of the lung has a mottled gray to black appearance (Fig. 1-32) owing to the presence of these particles. The carbon particles are also carried to the lymph nodes in the hila of the lungs and in the mediastinum, giving them a black appearance.

Bronchogenic carcinoma (cancer of a bronchus) is the *most common cancer in men* and is responsible for about 30% of malignancies.

The major causative factors for bronchogenic carcinoma are cigarette smoking and urban living. Because of the arrangement of the lymphatics, these tumors may metastasize to the pleura, the hila of the lungs, the mediastinum, and from there to distant organs.

Involvement of a closely related phrenic nerve (Figs. 1-29 and 1-46) *by the tumor results in paralysis of half of the diaphragm.*

A tumor may invade the pleura and produce a **pleural effusion.** This *pleural exudate,* which can be sampled by a performing **pleural tap,** may be ***sanguineous*** (bloody) and/or contain exfoliated cancer cells.

Because of the intimate relationship of the **recurrent laryngeal nerve** to the apex of the lung (Fig. 1-63), this nerve may be involved in apical lung cancers resulting in hoarseness owing to paralysis of a vocal fold.

Involvement of the hilar and mediastinal lymph nodes occurs by ***lymphogenous dissemination of cancer cells.*** Lymph from the entire right lung drains into the **tracheobronchial lymph nodes** on the right side, and most lymph from the left lung drains into these nodes on the left side.

Some lymph from the inferior lobe of the left lobe drains into the tracheobronchial lymph nodes on the right side. Thus, tumor cells in the right tracheobronchial lymph nodes can come from the inferior left lobe by lymphogenous dissemination.

Lymph from the lungs drains into the venous system via the right and left bronchomediastinal trunks (Fig. 1-42). Lymph from the lungs may therefore carry cancer cells into the venous system and to the right atrium of the heart. After passing through the pulmonary circulation, the blood returns to the heart for distribution to the body.

Common sites of hematogenous metastasis of cancer cells (spreading via the blood) from a bronchogenic carcinoma are the brain, bones, lungs, and suprarenal glands.

Often the lymph nodes superior to the clavicle (**supraclavicular lymph nodes**) are enlarged and hard when a patient has a carcinoma of a bronchus or the stomach owing to metastases from the primary tumor. The supraclavicular lymph nodes are commonly referred to as **sentinel nodes** because enlargement of them alerts the examiner to the possibility of malignant disease in the thoracic and/or abdominal organs.

The brain is a common site for hematogenous spread of bronchogenic carcinoma. Tumor cells probably enter the systemic circulation by invading the wall of a sinusoid or venule in the lung (Fig. 1-40), and are transported to the brain via the pulmonary veins, the left heart, the aorta, and the cerebral arteries. Once in the brain, the tumor cells probably pass between the endothelial cells lining the capillaries and enter the brain tissue.

THE MEDIASTINUM

The median region between the two pleural sacs is called the mediastinum (L. a middle septum). **This**

thick partition of tissue in and on each side of the median plane contains all structures in the chest, except the lungs and pleurae (Fig. 1-11).

The mediastinum extends from the superior thoracic aperture to the diaphragm inferiorly, and from the sternum and costal cartilages to the anterior surfaces of the 12 thoracic vertebrae posteriorly (Figs. 1-11 and 1-43). *These vertebral bodies are not in the mediastinum.*

The structures in the mediastinum are surrounded by loose connective tissue, lymph nodes, and fat. During life the looseness of the connective tissue and fat, and the elasticity of the lungs and pleurae, enable the mediastinum to accommodate movement and volume changes in the thoracic cavity (*e.g.*, movements of the trachea and bronchi during respiration, pulsations of the great vessels, and volume changes of the esophagus during swallowing).

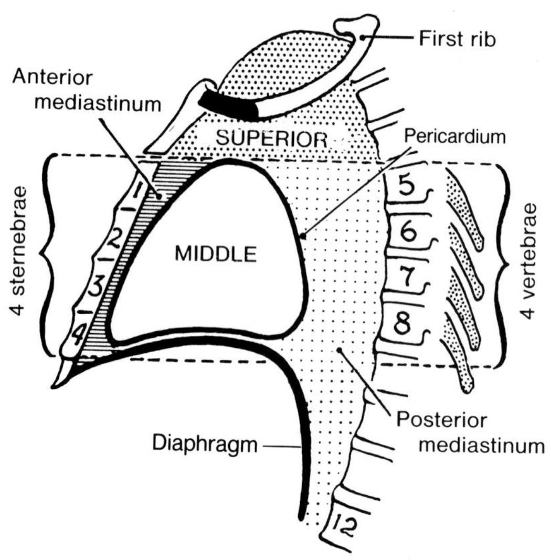

Figure 1-43. Diagram illustrating the four subdivisions of the mediastinum: middle, superior, posterior, and anterior. Observe that a horizontal line at the level of the sternal angle passes through the intervertebral disc between thoracic vertebrae 4 and 5. Superior to this imaginary plane is the superior mediastinum. This plane is especially convenient because it indicates the level of the superior border of the fibrous pericardium and the level of bifurcation of the trachea, *i.e.*, the superior border of the root of the lung. Note the anterior mediastinum, the small portion between the sternum and the pericardium. Observe the middle mediastinum, which contains the pericardium enclosing the heart and the roots of the great vessels. Examine the posterior mediastinum, the portion posterior to the pericardium and anterior to the bodies of the inferior eight thoracic vertebrae. Certain structures traverse the length of the mediastinum (*e.g.*, esophagus, vagus and phrenic nerves, and thoracic duct) and so lie in more than one mediastinal subdivision.

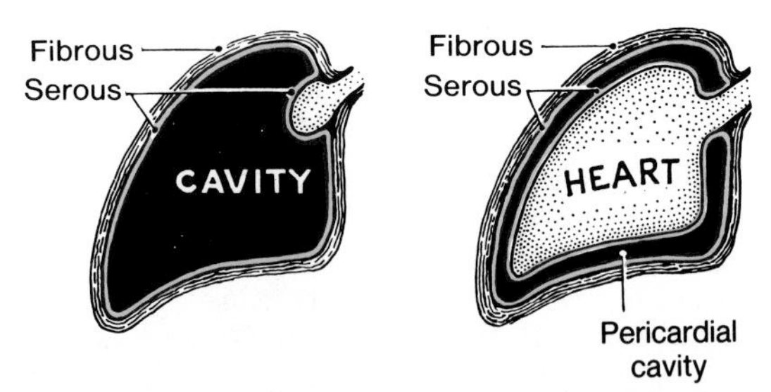

Figure 1-44. Drawings illustrating the scheme of the fibrous and serous pericardia. Observe that the pericardium (pericardial sac) is a double-walled sac enclosing the heart. Its external wall is fibrous and gradually thins out on the surface of the eight vessels that pierce it (Fig. 1-45). The fibrous layer is lined with serous pericardium (*yellow*). At the roots of the great vessels, the serous pericardium is reflected on to the surface of the heart, where it is called the epicardium or visceral pericardium.

SUBDIVISIONS OF THE MEDIASTINUM

For purposes of description, the mediastinum is divided by the pericardium or pericardial sac (Figs. 1-43 and 1-44) into *four subdivisions*: middle, superior, posterior, and anterior.

Certain structures pass through the mediastinum (*e.g.*, esophagus, vagus and phrenic nerves, and thoracic duct) and so lie in more than one subdivision of the mediastinum.

THE MIDDLE MEDIASTINUM

This subdivision of the mediastinum is very important because it contains the pericardium, the heart, and the roots of the great vessels passing to and from the heart (Figs. 1-43 and 1-45).

THE SUPERIOR MEDIASTINUM

As its name indicates, this subdivision is superior to the other three subdivisions of the mediastinum. It is superior to the horizontal line passing from the sternal angle to the inferior border of the fourth thoracic vertebra (Fig. 1-43).

The superior mediastinum contains the thymus, the great vessels related to the heart, the trachea, and the esophagus.

THE POSTERIOR MEDIASTINUM

This subdivision is located posterior to the pericardium and the diaphragm, and anterior to the bodies of the inferior eight thoracic vertebrae (Fig. 1-43). The posterior mediastinum contains the esophagus and the thoracic aorta (Figs. 1-42, 1-72, and 1-75).

THE ANTERIOR MEDIASTINUM

This is the smallest subdivision of the mediastinum. It is located *anterior to the pericardium*, and posterior to the sternum (Fig. 1-43). Although small in the adult,

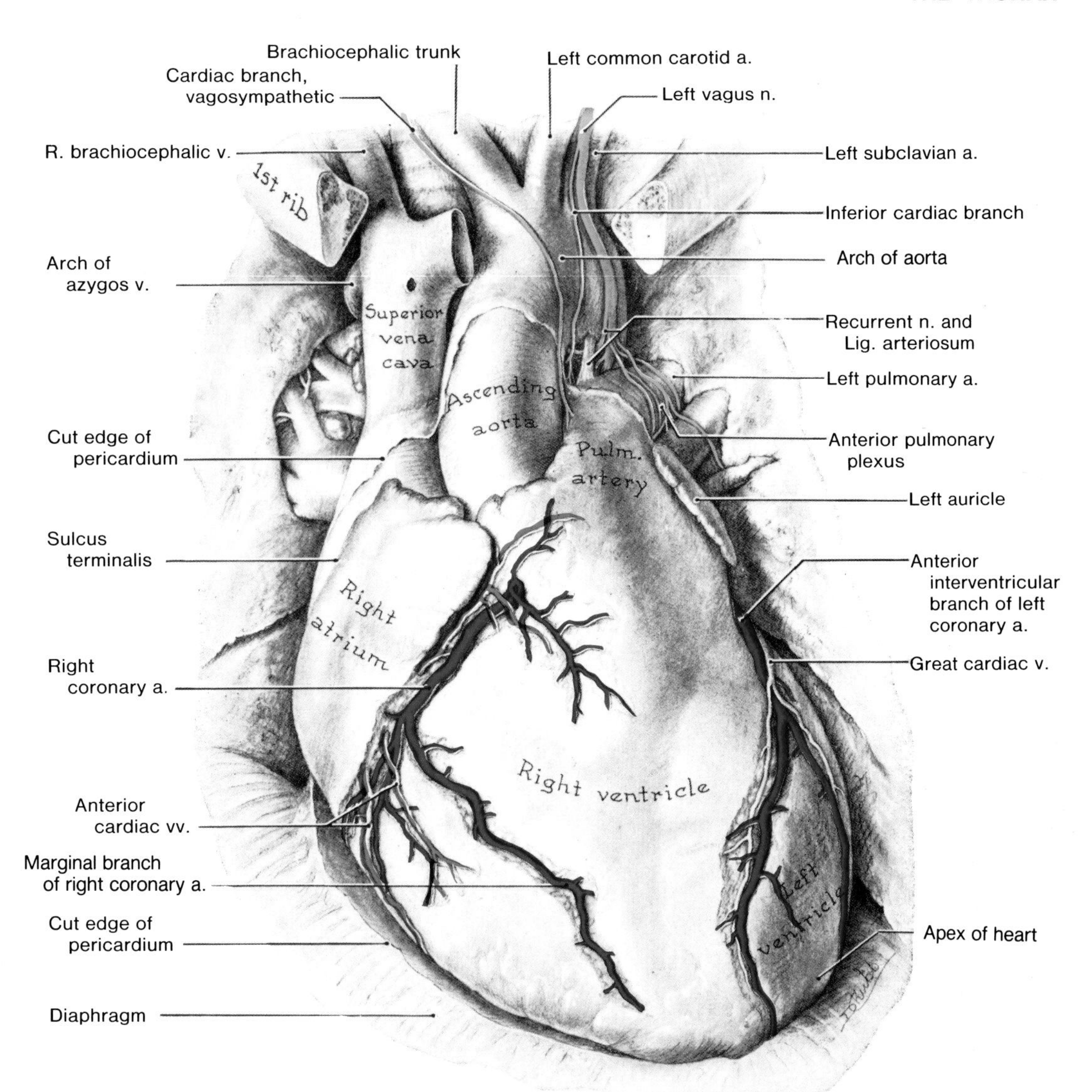

Figure 1-45. Drawing of a dissection of the sternocostal surface of the heart and great vessels, *in situ*. The fibrous pericardium, a cone-shaped sac, is opened. Observe that the apex of the sac is continuous with the external coats of the great vessels and that its base is attached to the central tendon of the diaphragm (also see Fig. 1-29).

it is relatively large during the first few months of life because the thymus extends into it (Fig. 1-69).

In early life the image of the thymus is as wide or wider than that of the heart on chest radiographs.

Much of the mediastinum can be visualized and certain minor surgical procedures can be carried out with a tubular lighted instrument called a ***mediastinoscope***.

Mediastinoscopy is commonly done to obtain tissue from the superior and anterior mediastinal and hilar lymph nodes (*e.g.*, to determine if cancer cells have spread to them from a bronchogenic carcinoma). Usually a midline incision is made at the jugular notch and blunt dissection is performed to reach the area of the tracheal bifurcation (Fig. 1-63).

The mediastinum can also be explored and biopsies can be taken by removing part of a costal cartilage, often the third one (Fig. 1-16).

THE HEART AND PERICARDIUM

The heart and the roots of the great vessels are enclosed in a fibroserous pericardial sac, called the *pericardium*, which is **located in the middle mediastinum** (Figs. 1-43 and 1-46).

The heart is a double, self-adjusting ***muscular pump****, the two parts of which work in unison.* The heart propels the blood through the blood vessels. The right side of the heart receives venous blood and pumps it to the lungs, whereas the left side receives oxygenated blood from the lungs and pumps it into the aorta for distribution to the body.

Each side of the heart consists of an **atrium** (L. antechamber) or receiving area which pumps blood into a **ventricle** (L. little belly) or discharging chamber. The wall of each chamber consists of three layers: an internal layer or *endocardium*; a middle layer or *myocardium* composed of cardiac muscle, and an external layer or *epicardium.* The myocardium or muscular part forms the main mass of the heart.

THE PERICARDIUM

The pericardium (G. around the heart) or **pericardial sac** is a *double walled fibroserous sac* that encloses the heart and the roots of the great vessels (Figs. 1-29 and 1-44 to 1-46). This conical sac is *located in the middle mediastinum, posterior to the body of the sternum and the second to sixth costal cartilages,* and anterior to the fifth to eighth thoracic vertebrae (Fig. 1-43).

The pericardium consists of two layers (Fig.

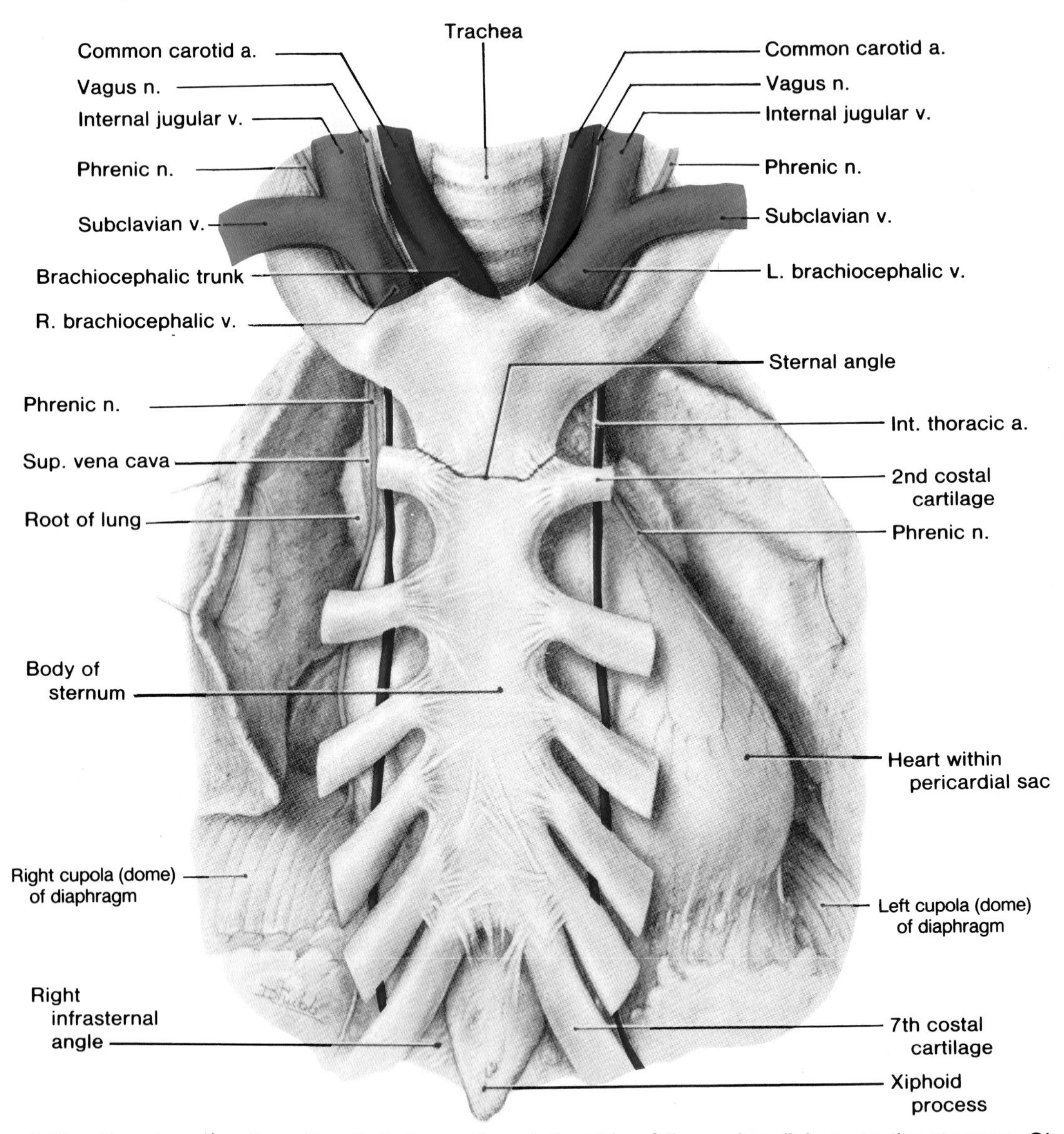

Figure 1-46. Drawing of a dissection that shows the relationship of the pericardial sac to the sternum. Observe that the pericardial sac lies posterior to the body of the sternum from just superior to the manubriosternal joint to the level of the xiphisternal joint. Note that about one-third of the heart lies to the right of the median plane and two-thirds to the left; thus, when the sternum is depressed during cardiopulmonary resuscitation (CPR), blood is forced out of the heart into the great vessels. Observe that the internal thoracic arteries lie a fingerbreadth from the borders of the sternum and that the right and left phrenic nerves are applied to the pericardial sac.

1-44): a strong outer layer composed of tough fibrous tissue called the **fibrous pericardium**, and a closed, *double-layered sac* composed of a transparent membrane called the **serous pericardium.**

THE FIBROUS PERICARDIUM (Figs. 1-29 and 1-43 to 1-47)

The fibrous pericardium is more or less conical; *its truncated* ***apex*** *is pierced by the aorta, the pulmonary trunk, and the superior vena cava.* The ascending aorta carries the pericardium superiorly beyond the heart to the level of the sternal angle (Figs. 1-45 and 1-50).

The ***base*** *of the fibrous pericardium rests on and is fused with the central tendon of the diaphragm* (Fig. 1-29), which separates it from the liver and the fundus of the stomach. Thus, the pericardium is influenced by movements of both the diaphragm and the heart. The central tendon of the diaphragm and the pericardium are *pierced on the right side by the inferior vena cava posteriorly* (Fig. 1-29).

The fibrous pericardium extends 1 to 1.5 cm to the right of the sternum and 5 to 7.5 cm to the left of the median plane at the level of the fifth intercostal space (Figs. 1-46, 1-50, and 1-51). It is separated from the sternum and the costal cartilages of the second to sixth ribs by the lungs and the pleurae (Fig. 1-30*A*), except in the median plane where *it is attached to the posterior surface of the sternum* by the ***sternopericardial ligaments*** and where the bulge of the heart intervenes.

The **cardiac notch** leaves part of the fibrous pericardium uncovered by lung tissue (Fig. 1-30*A*). This area on the left anterior chest is known as the area of **superficial cardiac dullness.** Being completely devoid of overlying lung, it yields a dull note upon **percussion** (tapping with the finger), especially in thin persons who are sitting upright.

Owing to its many connections, *the pericardial sac is firmly anchored within the thoracic cavity.* This keeps the heart in its normal position in the thoracic cavity.

THE SEROUS PERICARDIUM (Fig. 1-44)

The serous pericardium is divided into two layers: parietal pericardium and visceral pericardium in the same way that the pleura is divided into parietal and visceral layers of pleura.

The parietal layer of serous pericardium or *parietal pericardium lines the inner surface of the fibrous pericardium* (Fig. 1-44). The parietal pericardium is so closely adherent to the fibrous pericardium that they are difficult to separate.

The visceral layer of the serous pericardium or *visceral pericardium is reflected on to the heart* where it forms the **epicardium**, the external layer of the wall of the heart. The potential space between the parietal and visceral layers of serous pericardium is called the **pericardial cavity** (Fig. 1-44). It contains a thin film of fluid which enables the heart to move and beat in a frictionless enviroment.

The visceral pericardium is reflected from the heart and great vessels to become continuous with the parietal pericardium where the aorta and pulmonary trunk leave the heart, and the superior and inferior venae cavae and pulmonary veins enter the heart (Figs. 1-45 and 1-47).

Sinuses of the Pericardium (Figs. 1-47, 1-57, and 1-67). The pericardial sinuses develop during the folding of the embryonic heart and result from reflections of the pericardium. The aorta and pulmonary trunk are enclosed by a common sheath of visceral pericardium. When the pericardial sac is opened anteriorly, as in Figure 1-47, a finger can be inserted posterior to the aorta and pulmonary trunk and anterior to the left atrium and superior vena cava. This passage, known as the **transverse pericardial sinus**, connects the two sides of the pericardial cavity.

As the pulmonary veins and the inferior vena cava penetrate the fibrous pericardium to enter the heart, they bulge into the pericardial cavity. Hence, they are partly covered by serous pericardium which forms a somewhat inverted, U-shaped reflection or blind sac called the **oblique pericardial sinus** (Figs. 1-47 and 1-67). It is a wide, slit-like recess posterior to the heart, between the left atrium and the posterior aspect of the pericardium.

The oblique pericardial sinus may be entered inferiorly and will admit several fingers. However, a finger inserted into it cannot pass around any of the vessels because the oblique sinus is a blind pouch (cul-de-sac).

Vessels and Nerves of the Pericardium (Figs. 1-15 to 1-17, 1-42, and 1-46). *The main arterial supply to the pericardium* is derived from the **internal thoracic arteries**, via their pericardiacophrenic and musculophrenic branches. It also receives pericardial branches from the bronchial, esophageal, and superior phrenic arteries (Figs. 1-72 and 1-78).

The visceral layer of the serous pericardium or epicardium is supplied by the coronary arteries (Figs. 1-45 and 1-65).

The veins of the pericardium are tributaries of the *azygos system* (Figs. 1-42 and 1-80). There are also *pericardiacophrenic veins* that enter the internal thoracic veins.

The nerves of the pericardium are derived from the vagus and phrenic nerves and the sympathetic trunks (Figs. 1-70 and 1-74).

Pericardial pain is felt diffusely posterior to the sternum and is referred to as **substernal pain**. The pain may radiate to other areas.

Acute inflammation of the pericardial sac (**pericarditis**) has many causes. It may be associated with metabolic disorders (*e.g.*, uremia), rheu-

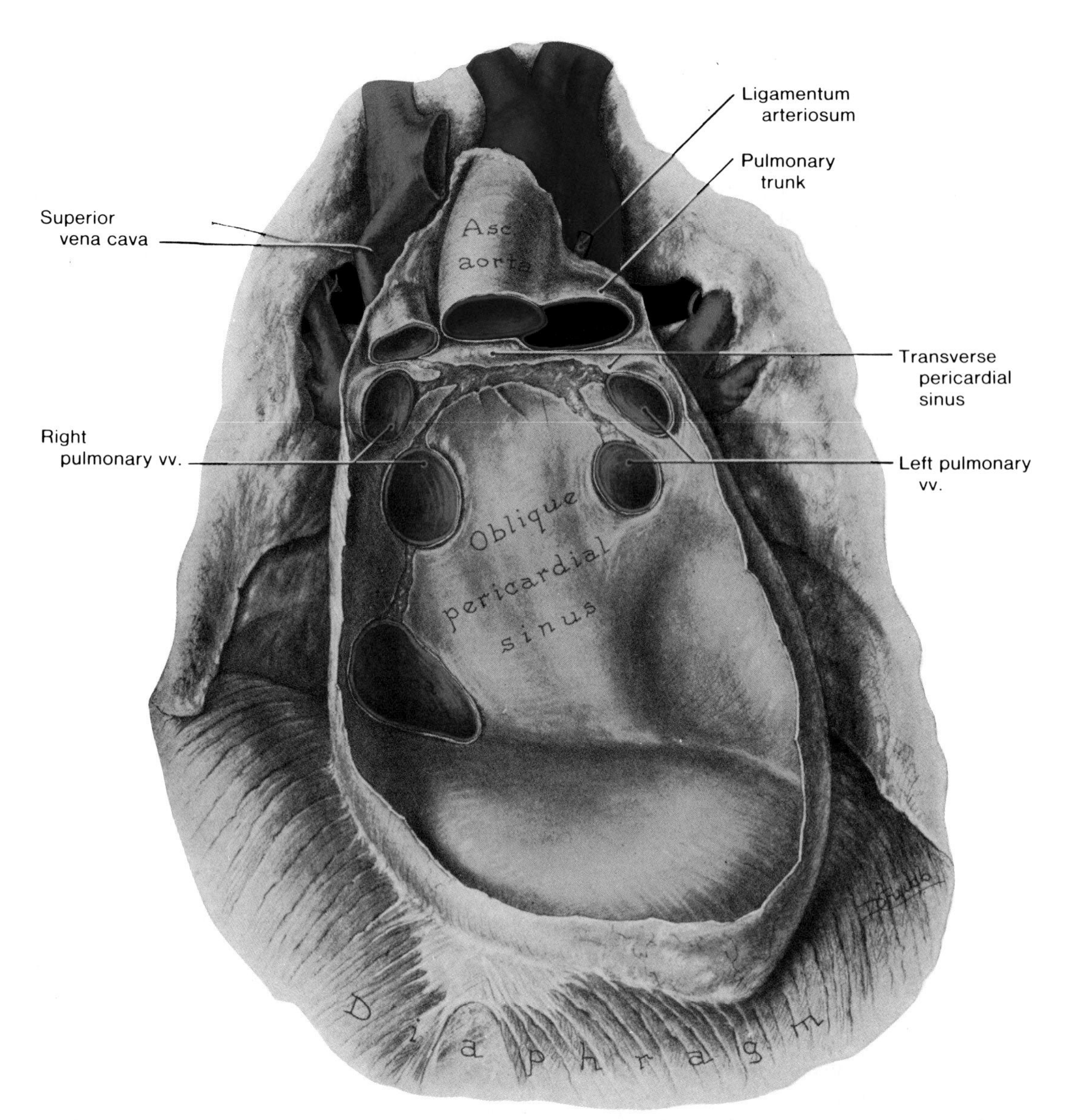

Figure 1-47. Anterior view of the interior of the pericardial sac. When the heart was removed, the eight vessels piercing it were severed (two arteries, two caval veins, and four pulmonary veins). Observe that the oblique pericardial sinus is circumscribed by five veins and is open inferiorly and to the left. Note that the apex of the pericardial sac is near the junction of the ascending aorta and the arch of the aorta. Examine the superior vena cava, which is partly inside and partly outside the pericardium, and the ligamentum arteriosum, which is entirely outside. The ligamentum arteriosum is the remnant of a fetal vessel (ductus arteriosus). Observe the transverse and oblique pericardial sinuses. The heart removed from this pericardial sac is shown in Figure 1-67.

matic fever, and bacterial, viral, or tuberculous infection.

Pericarditis causes ***substernal pain***, frequently severe, and often **pericardial effusion** (passage of fluid from the pericardial capillaries into the pericardial cavity). If the effusion is extensive, the excess fluid in the pericardial cavity may embarass the action of the heart by compressing the pulmonary veins as they cross the pericardial sac and the atria (Fig. 1-49). This condition, occurring because the fibrous pericardium is inelastic, is called **cardiac tamponade**.

A ***pericardial friction rub*** *is an important physical sign of acute pericarditis.* Normally the moist layers of the serous pericardium make no detectable sound during **auscultation** (listening with a stethoscope). However, inflammation of the pericardium causes the surfaces to become rough and the resulting *friction sounds like the rustle of silk*, especially during forced expiration with the patient bending anteriorly.

A penetrating wound of the pericardium (*e.g.*, a knife) commonly pierces the heart and bleeding occurs into the pericardial cavity (Case 1-3, p. 144). As the blood accumulates, the heart is compressed and circulation fails correspondingly. The veins of the face and neck become engorged owing to compression of the superior vena cava as it enters the inelastic pericardium.

Paracentesis of the pericardium (drainage of fluid from the pericardial cavity) is sometimes necessary to relieve the pressure of accumulated fluid on the heart. A wide-bore needle may be inserted through the fifth or sixth intercostal space near the sternum (Fig. 1-46). *Care is taken not to puncture the internal thoracic artery.* The pericardial sac may also be reached by entering the left infrasternal angle and passing the needle superiorly and posteriorly.

THE HEART

The heart is a hollow muscular organ that is somewhat conical in shape and is slightly larger than a clenched fist (Fig. 1-45). The heart is placed obliquely in the thorax and most of its anterior surface consists of the right ventricle.

The heart has four chambers (two atria and two ventricles), **an apex** (the position of the apical impulse or "heart beat"), **a base** formed by the atria, and several surfaces and borders.

The Apex of the Heart (Figs. 1-45, 1-48, 1-51, and 1-52). The apex is formed by the tip of the left ventricle which points inferiorly, anteriorly, and to the left.

The apex is located deep to the left fifth intercostal space, but its location varies with the person's position and with the phase of respiration (Figs. 1-50 to 1-52).

The Sternocostal Surface of the Heart (Figs. 1-45 and 1-48). The sternocostal or anterior surface of the heart is *formed mainly by the right ventricle and right atrium.* The left ventricle and left atrium lie more posteriorly and form only a small strip on the sternocostal surface.

The Diaphragmatic Surface of the Heart (Figs. 1-49 and 1-67). The diaphragmatic or inferior surface of the heart is usually horizontal or slightly concave. *It is formed by both ventricles (mainly the left).* It is separated from the liver and stomach by the diaphragm on which it rests (Fig. 1-45). The posterior interventricular groove (sulcus) divides this surface into a right one-third and a left two-thirds.

The Base of the Heart (Figs. 1-49 and 1-67). The base or posterior surface of the heart is *formed by the atria*, mainly the left one. It lies opposite the middle four thoracic vertebrae (Fig. 1-43).

The base of the heart is *its most superior part*, from which emerge the ascending aorta, the pulmonary trunk, and the superior vena cava (Fig. 1-49). The base of the heart is separated from the diaphragmatic surface of the heart by the posterior part of the **coronary groove** (sulcus).

The heart does not rest on its base; the term "base" derives from the cone shape of the heart, the base being opposite the apex.

Borders of the Heart (Fig. 1-48). The heart has four borders: right, inferior, left, and superior.

The right border is formed by the right atrium. It is slightly convex and is almost in line with the superior and inferior venae cavae.

The inferior border is sharp and thin and is nearly horizontal. It is formed mainly by the right ventricle and slightly by the left ventricle near the apex.

The left border is formed by the left ventricle and very slightly by the left auricle.

The superior border is where the great vessels enter and leave the heart; it is formed by both atria.

SURFACE ANATOMY OF THE HEART (Fig. 1-50)

The apex beat of the heart in adults can be felt or heard in the *fifth left intercostal space*, just medial to the midclavicular line (7 to 9 cm from the midsternal line). In males and immature females, the apex beat is usually slightly inferior and medial to the nipple.

The location of the nipple is not a reliable guide to the position of the heart's apex in most mature females owing to the variation in the size and pedulousness of the breast.

In infants and children, the heart's apex is slightly higher and further laterally than in adults. The position of the apex of the heart almost corresponds to the site of the apex beat. *The apex beat in the newborn* may be palpated in the *fourth left intercostal space* in or just lateral to the midclavicular line. After 2 years of age, the apex beat is usually detected in the fifth intercostal space; slightly inferior and medial to the midclavicular line. In many people the pulsations of the apex are visible.

To perform CPR the surface location of the heart must be known. To assess the heart clini-

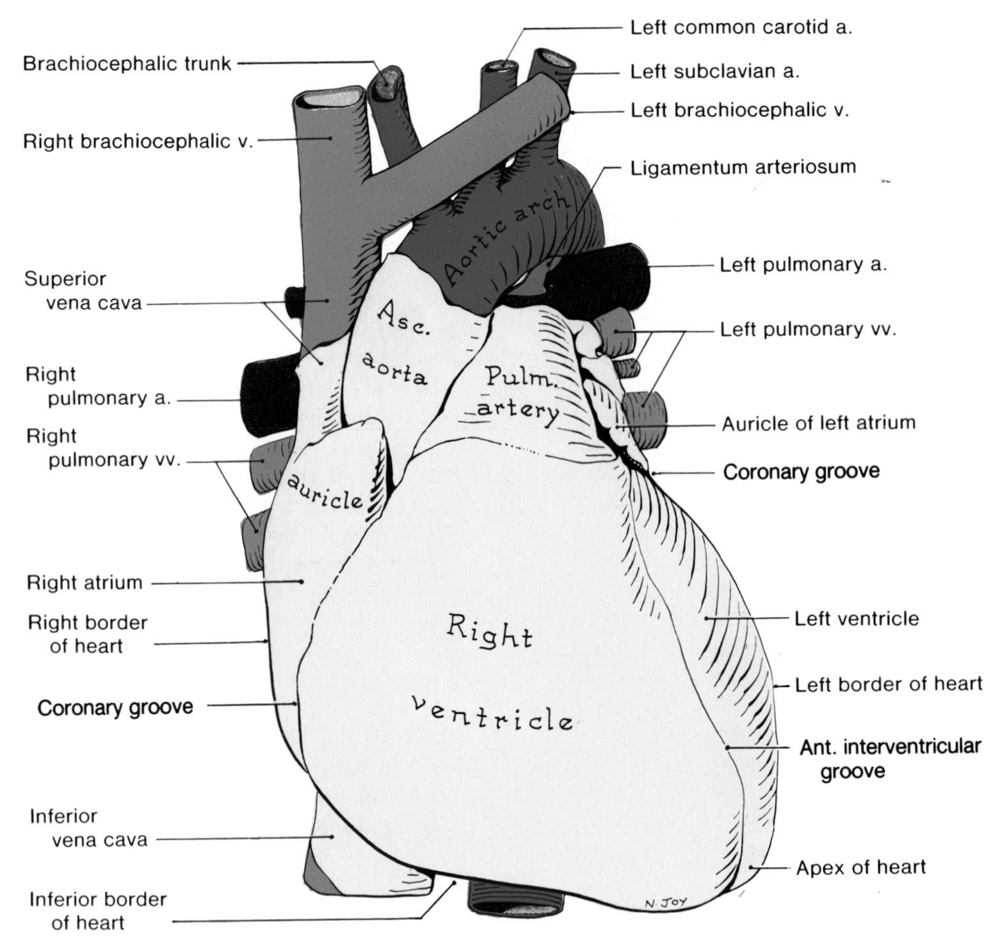

Figure 1-48. Drawing of the sternocostal aspect of the heart and great vessels. The pericardium is colored yellow. Observe that the heart resembles a closed fist in shape and size. Note that each atrium projects anteriorly on each side of the aorta and pulmonary trunk (artery) as an ear-shaped auricle (atrial appendage). The heart lies obliquely in the chest with its apex inferiorly and to the left (Fig. 1-46). Although it is common to use the term *aortic arch* (as on this drawing), it is better to refer to this part of the aorta as the *arch of the aorta* (as in Fig. 1-49) because "aortic arch" is the name given to the artery supplying a branchial arch in the embryo.

cally, detailed surface markings of it have to be known.

Surface Markings of the Sternocostal Surface of the Heart (Fig. 1-50). The outline of the heart is variable and depends partly on the physique of the individual. *The outline of the heart can be traced on the anterior surface of the chest by using the following guidelines.*

The superior border of the heart corresponds to a line connecting the *inferior margin of the second left costal cartilage* (3 cm to the left of the median plane) to the superior border of the *third right costal cartilage* (2 cm from the median plane).

The right border of the heart corresponds to a line drawn from the *third right costal cartilage* (2 cm from the median plane) to the *sixth right costal cartilage* (2 cm from the median plane). This line is slightly convex to the right.

The inferior border of the heart corresponds to a line drawn from the *inferior end of the right border to a point in the fifth intercostal space close to the midclavicular line.* The left end of this line corresponds to the location of the apex beat.

The left border of the heart corresponds to a line connecting the *left ends of the lines representing the superior and inferior borders.*

The preceding surface markings are for the average adult heart. The heart of any individual

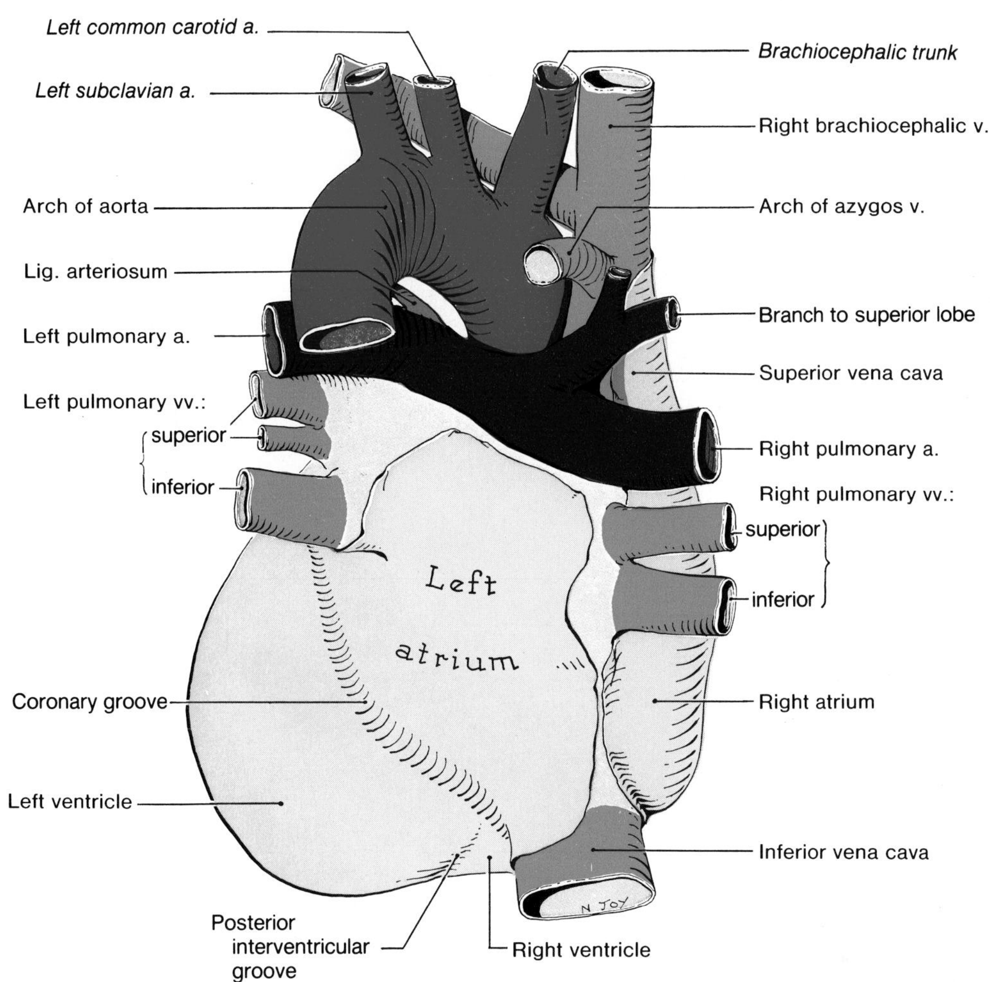

Figure 1-49. Drawing of the posterior aspect or base of the heart and great vessels. Note the branches of the arch of the aorta. The pericardium is colored yellow. Observe the right and left pulmonary veins converging to open into the left atrium. The part of the atrium between them lies to the left of the line of the inferior vena cava and forms the anterior wall of the oblique pericardial sinus (Fig. 1-47). Note that the aorta arches over the left pulmonary vessels and bronchus, and that the azygos vein arches over the right pulmonary vessels and bronchus. The base of the heart is its most superior part. From it emerge the ascending aorta, the pulmonary trunk, and the superior vena cava. the coronary groove (sulcus) separates the base or the posterior surface of the heart from the diaphragmatic or inferior surface of the heart.

may vary slightly from this average and still be *normal for that person.* The surface anatomy of the heart is modified by age, sex, body size and build, respiration, position, and disease of the heart or lungs.

The heart is assessed chiefly by examination through the anterior thoracic wall. When you learn to percuss the heart, you can determine the surface anatomy of a person's heart and compare it with the average learned in anatomy. Later you will learn how to interpret significant departures from the average.

Percussion *is a commonly used diagnostic procedure for determining the density of the heart.* Verify that the character of the sound changes as you tap different areas of the chest. Place the middle finger of your left hand approximately parallel and to the left of the left border of your heart. Now tap it with the middle finger of your other hand. While percussing, move your finger to the right and note how the sound changes as you move from over your lung to over your heart. By noting the areas of cardiac dullness, you can determine the rough outline of your heart. If it were grossly enlarged you could determine this, but a radiological examination

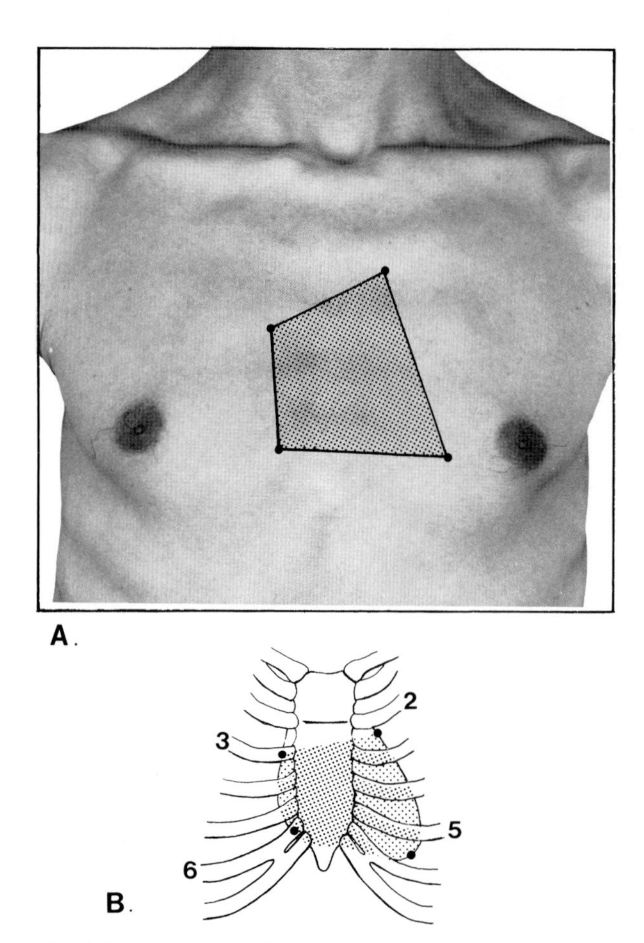

Figure 1-50. *A*, photograph of the thorax of a 27-year-old man showing the surface markings of the sternocostal surface of the heart. The superior border is formed by the superior margins of the left atrium and a small part of the right atrium. The right border is formed by the right atrium. The inferior border is formed by the right ventricle and a small part of the left ventricle. The left border is formed by the left ventricle and the left auricle. *B*, drawing of the thoracic cage showing the landmarks used when drawing the heart on the chest. The apex beat of the heart in adults can be felt or heard in the fifth left intercostal space, just medial to the midclavicular line. In males and immature females, the apex beat is usually slightly inferior and medial to the nipple.

(Fig. 1-51) would be required for a clearer determination of its size and shape.

RADIOLOGICAL ANATOMY OF THE HEART (Figs. 1-31, 1-51, and 1-52)

The heart and great vessels and the blood within them are of the same order of density. Hence, a frontal chest film shows the contour of the heart and great vessels in the middle mediastinum. This is called the **cardiovascular silhouette** or cardiac shadow. As the heart and great vessels are full of blood, the cardiovascular silhouette stands out in contrast to the clearer areas occupied by the air-filled lungs.

Because the fibrous pericardium is attached to the diaphragm, the cardiovascular silhouette becomes longer and narrower during inspiration (Fig. 1-51) and shorter and broader during expiration (Fig. 1-52).

In radiographs of the chest (PA projections), the borders of the cardiovascular silhouette (superior to inferior) are formed by the heart and great vessels (Fig. 1-31). The *right border* is formed by the superior vena cava, right atrium, and inferior vena cava. The *left border* is formed by the arch of the aorta, which produces a prominence known as the *aortic knob*, the pulmonary trunk, the left auricle, and the left ventricle.

The radiographic appearance of the heart varies in different people because the outline of the heart is variable. Its appearance depends partly on the physique of the person.

There are three main types of cardiovascular silhouette: (1) the *oblique type* is present in most people (Fig. 1-31); (2) the *transverse type* is observed in stocky and obese people, pregnant women, and infants; and (3) the *vertical type* is characteristic of thin people with long, narrow chests (Fig. 1-50).

Physicians must know the main structures that form the cardiovascular silhouette so that gross abnormalities can be recognized.

Enlargement of the heart may result from hypertrophy, especially in cases of high blood pressure. The walls grow thicker by increasing the number and size of the cardiac muscle fibers.

Enlargement of the heart can also result from dilation. When blood regurgitates from the aorta into the left ventricle, this chamber dilates to accommodate the extra blood. Similarly, when blood regurgitates through a diseased mitral valve back into the left atrium from the left ventricle, the left atrium dilates to accommodate the additional blood.

As stated previously, a complication of pericarditis may be **hydropericardium**, in which the pericardial sac becomes distended with fluid, interfering with the return of blood to the heart. **Severe pericardial effusion causes an enlargement of the cardiovascular silhouette** and differentiation of it from heart enlargement may be difficult.

In the usual chest radiograph, the separate cavities of the heart are not distinguishable, but **angiocardiography** can be used to study the cavities of the heart and great vessels. When a suitable contrast medium that is miscible with blood is injected via a catheter, the course of the blood can be followed fluoroscopically by either **cineradiography** or serial x-ray films in various projections.

*Most **congenital cardiac defects** can be diagnosed with the aid of cardiac catheterization.* **Right cardiac catheterization** consists of passing a radiopaque catheter under sterile conditions into a peripheral vein (*e.g.*, external iliac) and guiding it with the aid of **fluoroscopy** into the great veins, the right heart chambers, or the pulmonary artery. When a measured dose of contrast material is injected into the right ventricle, it will traverse the pulmonary circulation and enter the left side of the heart.

Congenital cardiovascular abnormalities may cause the contrast material to pass into an abnormal blood vessel or through a defect. For example, if an **ASD** (atrial septal defect) is present, some contrast material will pass through the defect in the interatrial septum into the left atrium.

Left cardiac catheterization consists of inserting a catheter into a peripheral artery (*e.g.*, femoral) and guiding it by fluoroscopy into the ascending aorta and left ventricle. Frequently right and left cardiac catheterizations are performed simultaneously.

CHAMBERS OF THE HEART (Figs. 1-45, 1-48, and 1-49)

There are four chambers, two atria and two ventricles. The **coronary groove** (sulcus), or atrioventricular groove, encircles most of the superior part of the heart, separating the atria from the ventricles (Fig. 1-49). Similarly, the division of the ventricles is indicated by the anterior and posterior interventricular grooves (sulci).

The Right Atrium (Figs. 1-45, 1-48, 1-49, 1-54, and 1-55). This chamber *forms the right border of the heart between the **superior and inferior venae cavae*** (Fig. 1-48). The right atrium receives venous blood from these vessels and the coronary sinus.

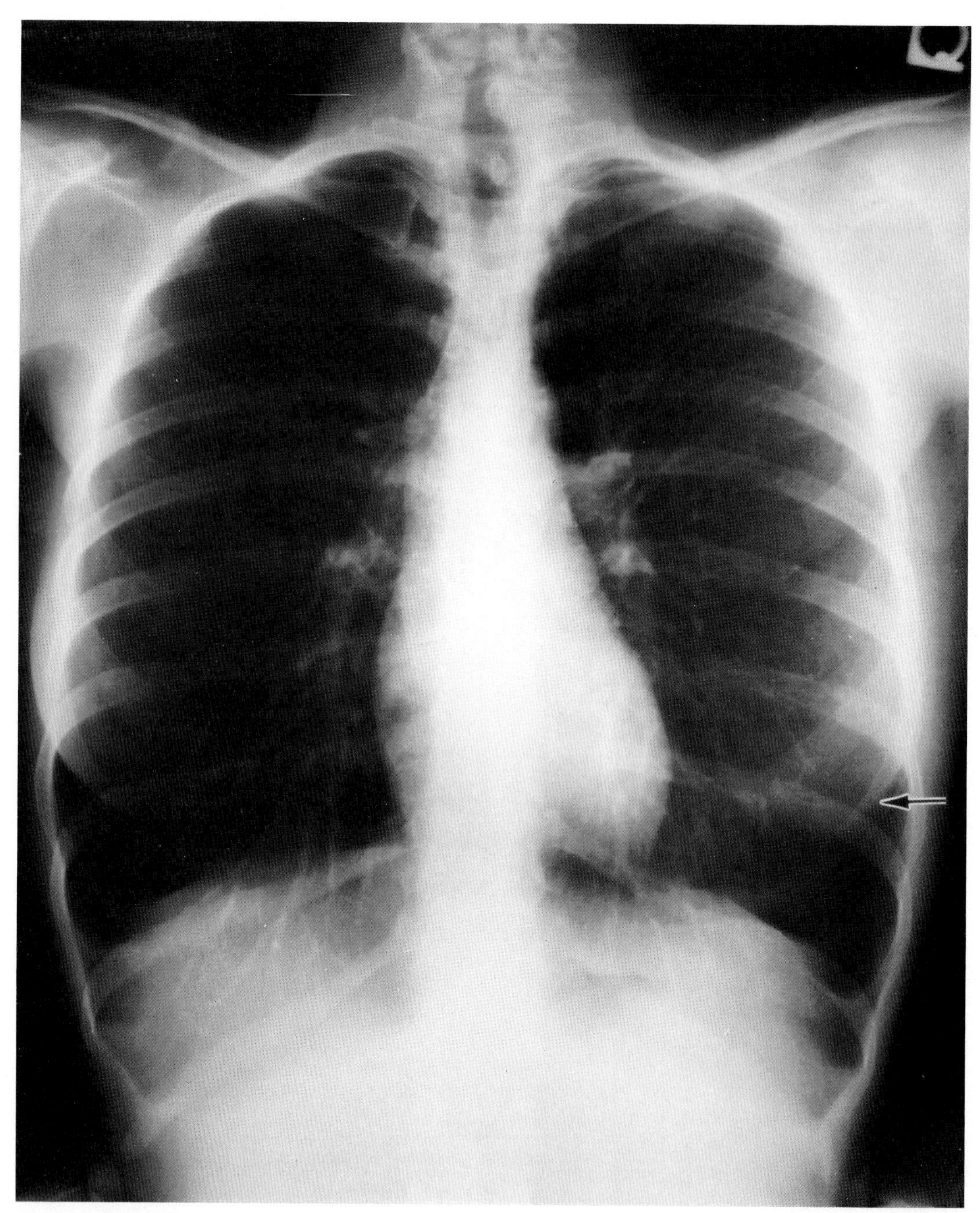

Figure 1-51. Radiograph of the **chest *during inspiration***, posteroanterior (PA) view. This is the *common frontal projection* because in it the heart is close to the x-ray film and is magnified less than in an anteroposterior (AP) projection. It also permits the shoulders to be rolled anteriorly, thereby moving the scapulae away from the lungs. *The lungs are radioucent because of the air they contain.* In inspiration they contain more air than in expiration; therefore, they are more radiolucent during inspiration. Observe that the ribs are splayed out and widely separated. Note the low position of the diaphragm and the recesses of the pleural cavity that are not filled by the lungs. During inspiration the heart appears more vertical because the diaphragm pulls the pericardial sac inferiorly; hence the hila of the lungs are more readily visible. Note the difference between the relation of the left half of the diaphragm and the anterior end of the seventh rib (*arrow*).

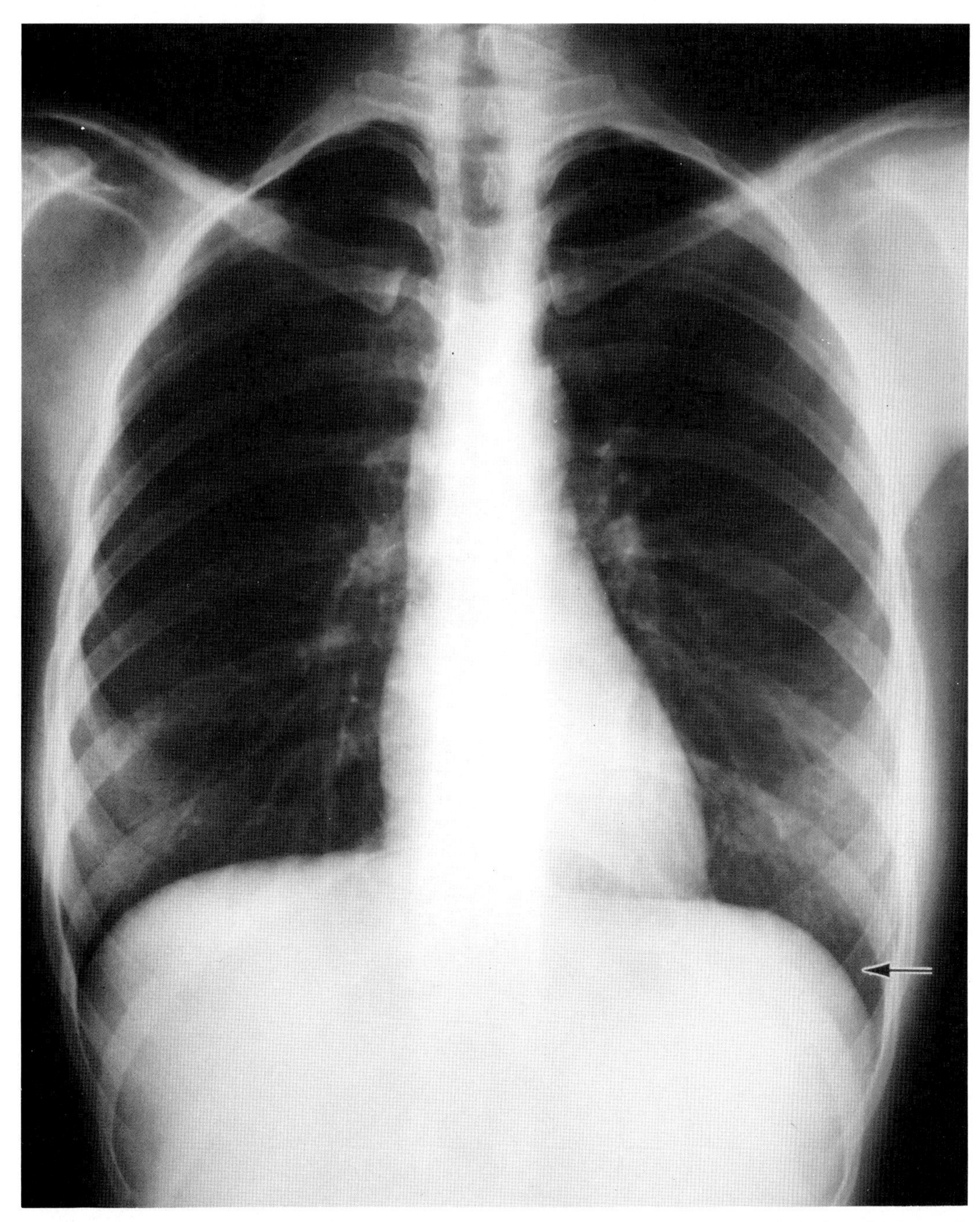

Figure 1-52. Radiograph of the **chest *during expiration***, posteroanterior (PA) view, of the female shown in Figure 1-51. Observe the high position of the diaphragm and lungs, which are not so translucent because they contain less air. Note that the ribs are closer together. Examine the cardiovascular silhouette (shadow). See Figure 1-31 for details. Understand that about one-quarter of the total lung is not visible in these films because it is obscured by the cardiovascular silhouette (shadow) and subdiaphragmatic structures. The *arrow* indicates the seventh rib. Note that the diaphragm casts dome-shaped shadows on each side (slightly higher on the right). Examine the diaphragm in Figure 1-42.

The coronary sinus receives blood from various veins of the heart and opens into the right atrium. The coronary sinus lies in the coronary groove (Figs. 1-49, 1-53, and 1-67).

The right atrium consists of (1) a smooth-walled posterior part, called the **sinus venarum** (sinus of the venae cavae), which receives the venae cavae and the coronary sinus, and (2) a rough-walled anterior part which has internal muscular ridges (***musculi pectinati***) resembling the coarse teeth of a comb. A small conical muscular pouch, the **right auricle** (atrial appendage), projects to the left from the root of the superior vena cava and overlaps the right side of the root of the ascending aorta (Fig. 1-48).

The basis of the two distinctive parts of the right atrium is embryological. The smooth-walled part develops from the absorbed right horn of the embryonic **sinus venosus**, whereas the rough-walled part (including the auricle) develops from the primitive atrium (see Moore, 1982 for details).

The two parts of the right atrium are separated externally by a shallow groove called the **sulcus terminalis** (Fig. 1-45), and internally by a vertical ridge called the **crista terminalis** (Fig. 1-54). The crista terminalis and the valves of the inferior vena cava and coronary sinus represent the remains of the valve of the sinus venosus.

The interatrial septum forms the posteromedial wall of the right atrium. A prominent feature of this wall is the ***fossa ovalis*** (Fig. 1-54), a shallow *translucent depression in the interatrial septum* facing the opening of the inferior vena cava. The thumbprint-sized *fossa ovalis is a remnant of the fetal foramen ovale,* through which oxygenated blood from the placenta passed from the inferior vena cava through the right atrium into the left atrium (see Moore, 1982 for details). The floor of the fossa ovalis is formed by tissue derived from the *valve of the foramen ovale,* a derivative of the septum primum.

The opening of the coronary sinus is located between the atrioventricular orifice and the valve of the inferior vena cava. This opening is partly guarded by a thin semicircular valve, called the *valve of the coronary sinus* (Fig. 1-54).

The superior vena cava *returns blood from the superior half of the body.* It opens into the superior and posterior part of the right atrium, at about the level of the right third costal cartilage (Figs. 1-30, 1-46, and 1-48). Its valveless orifice is directed inferiorly and anteriorly (Fig. 1-54).

The inferior vena cava returns blood from the inferior half of the body. It opens into the inferior part of the right atrium, almost in line with the smaller superior vena cava (Fig. 1-54). The valve of the inferior vena cava is a thin fold of variable size that has no function postnatally. Before birth it directed blood from the inferior vena cava toward the foramen ovale.

Atrial septal defect (ASD) is a common congenital heart malformation. The most common type of ASD is persistent or **patent foramen ovale.** Defects in the area of the fossa

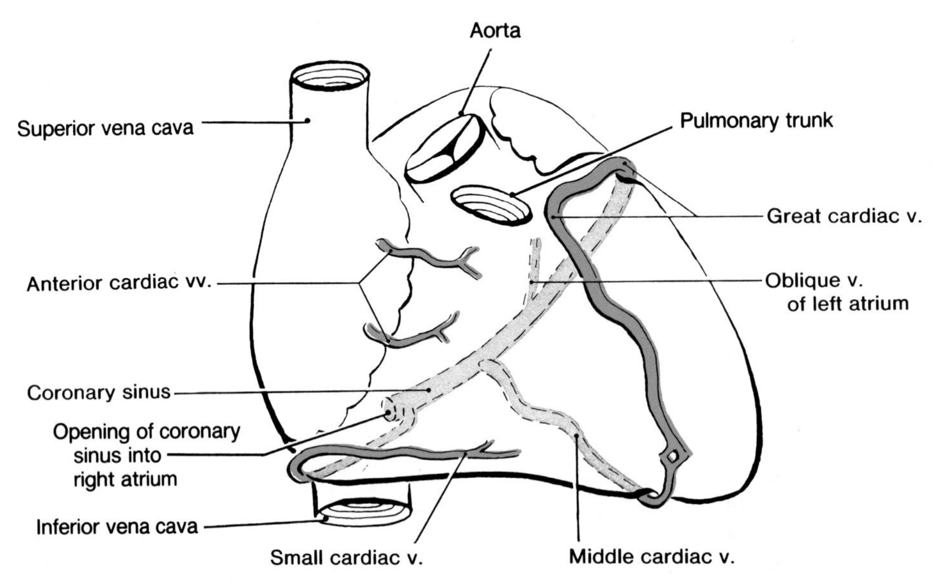

Figure 1-53. Diagram of an anterior view of the heart showing the cardiac veins. Observe the coronary sinus, the main vein of the heart, running from left to right in the posterior part of the coronary groove (sulcus). The heart is drained mainly by veins that empty into the coronary sinus and partly by small veins that empty directly into the chambers of the heart. The coronary sinus enters the right atrium (Fig. 1-54).

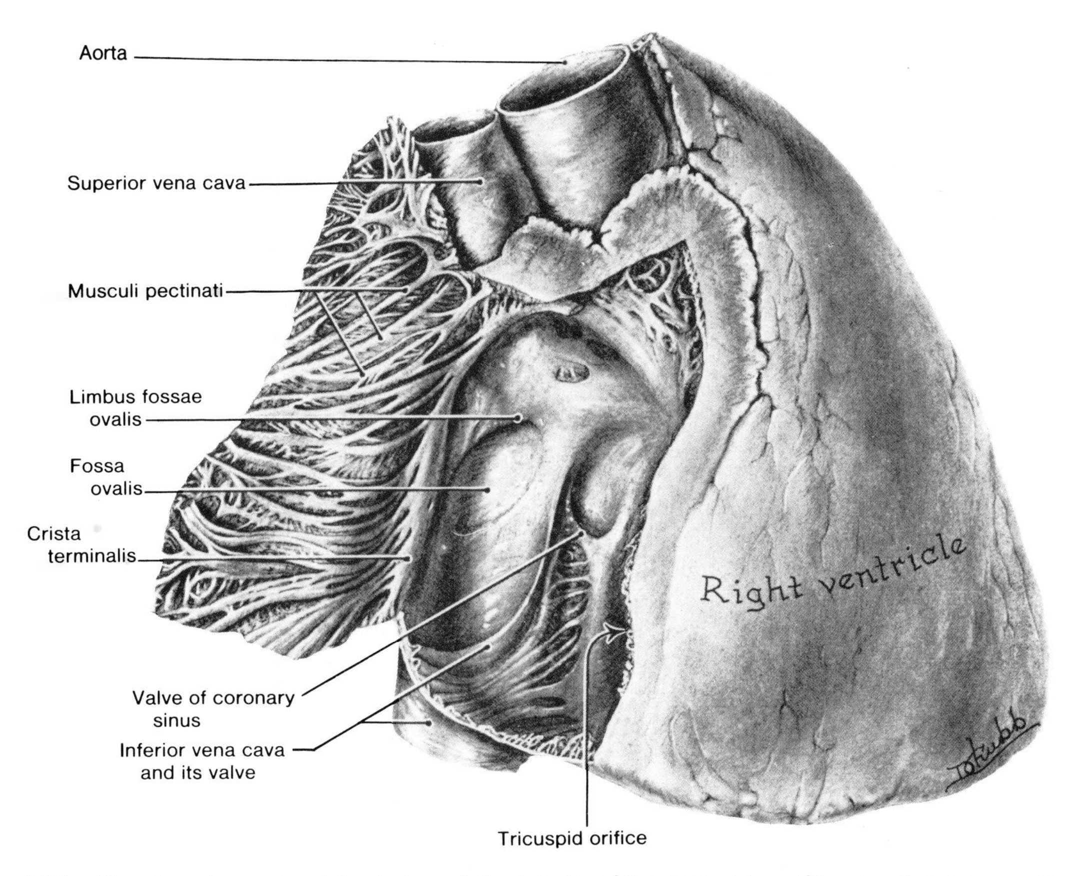

Figure 1-54. Drawing of an anterolateral view of the interior of the right atrium. Observe that the smooth-walled part (derived from the absorbed right horn of the embryonic sinus venosus) is separated from the rough-walled part (derived from the primitive atrium) by a vertical ridge known as the crista terminalis. The crista underlies the sulcus terminalis (Fig. 1-45). Note that the two caval veins and the coronary sinus open into the smooth-walled part of the atrium. Observe the fossa ovalis and the limbus fossae ovalis which are remnants of the prenatal foramen ovale and its valve. Note that the right atrioventricular or tricuspid orifice is situated at the anterior aspect of the atrium.

ovalis are classified as the secundum type of ASD. In clinically significant defects, there is a large opening between the right and left atria.

In up to 25% of adult hearts a probe can be passed obliquely from one atrium to the other through the superior part of the floor of the fossa ovalis. Although this defect is not usually clinically significant, a **probe patent foramen ovale** may be forced open as a result of other cardiac defects and contribute to functional pathology of the heart. Probe patent foramen ovale results from incomplete adhesion between the septum primum and the septum secundum. For illustrations and a complete discussion of the embryological bases of ASD, see Moore (1982).

The Right Ventricle (Figs. 1-45, 1-48, 1-54, and 1-55). This chamber forms the largest part of the sternocostal surface of the heart, a small part of the diaphragmatic surface, and almost all of the inferior border of the heart (Fig. 1-48). Its wall is much thicker than that of the right atrium.

The superior anterior end of the right ventricle tapers into a cone-shaped structure, called the **conus arteriosus** (infundibulum) which *gives origin to the pulmonary trunk* (Fig. 1-55). Internally the conus arteriosus has a funnel-shaped appearance; this explains why it is also called the **infundibulum** (L. a funnel).

The conus arteriosus is the most superior part of the right ventricle. It lies anterior to the root of the aorta and immediately inferior to the root of the pulmonary trunk.

The inner wall of the conus arteriosus is smooth, whereas the rest of the ventricular wall is roughened by a number of irregular muscle bundles (**papillary muscles**) ridges, and bridges (trabeculae carneae). The fleshy **trabeculae carneae** (L. *trabs,* wooden beam + *carneus,* fleshy) project from the ventricular

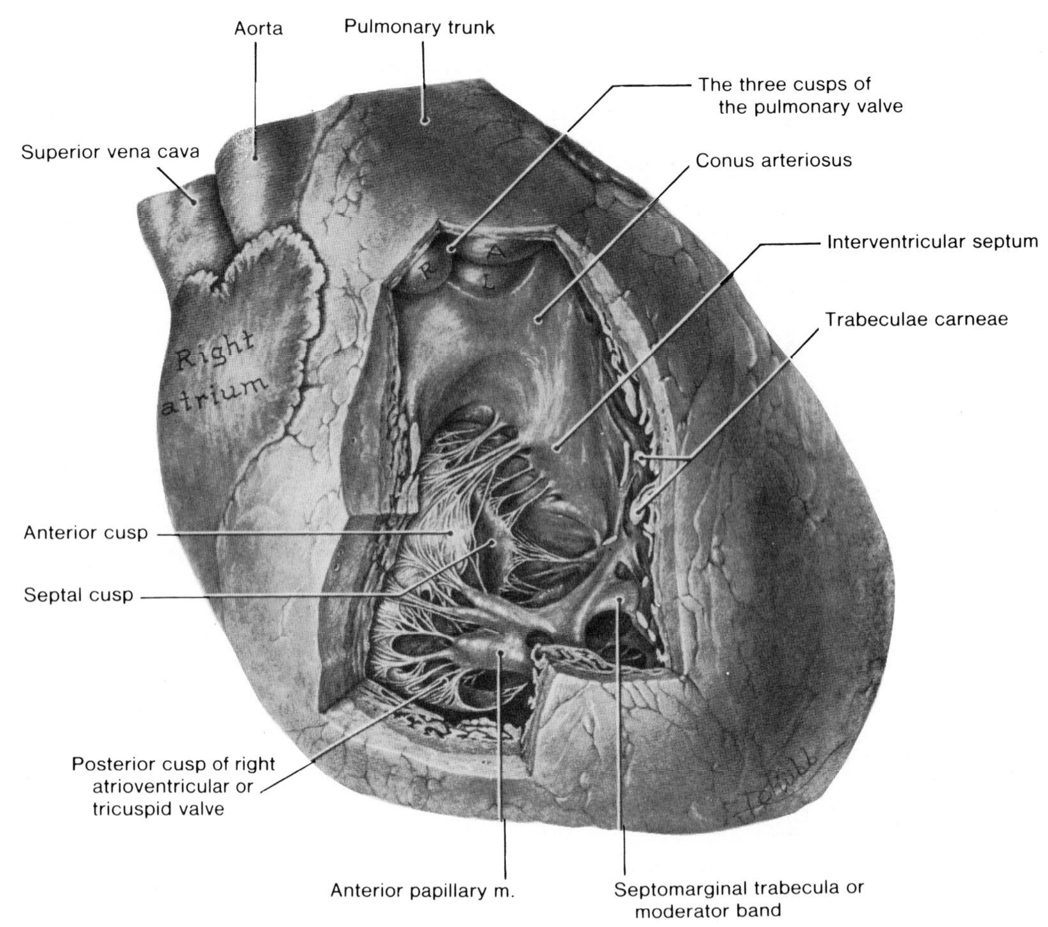

Figure 1-55. Drawing of a dissection of an anterior view of the heart showing the interior of the right ventricle. Observe the entrance to this chamber (right atrioventricular orifice) from the right atrium, situated posteriorly, and the exit (orifice of the pulmonary trunk), situated superiorly. Note the smooth, funnel-shaped wall (conus arteriosus) inferior to the pulmonary orifice and that the remainder of the ventricle has rough fleshy trabeculae carneae. Observe the three types of trabecula: (1) ridges on the ventricular surface, (2) bridges attached at each end, and (3) pillars called papillary muscles. Note the anterior papillary muscle rising from the anterior wall; the posterior (not labeled) rising from the posterior wall; and a series of small septal papillae arising from the septal wall. Observe the septomarginal trabecula, extending from the septum to the base of the anterior papillary muscle, and forming a bridge. It contains the right branch (crus) of the atrioventricular bundle (AV bundle), part of the conducting system of the heart (Fig. 1-61). Note the chordae tendineae (not labeled) passing from the tips of the papillary muscles to the free margins and ventricular surfaces of the three cusps of the tricuspid valve. Each papillary muscle controls the adjacent sides of two cusps.

wall, giving it a coarse sponge-like appearance. One of the trabeculae carneae crosses the cavity of the ventricle from the interventricular septum to the base of the anterior papillary muscle. This **septomarginal trabecula**, or moderator band (Fig. 1-55), *carries the right branch of the* ***atrioventricular bundle*** (Fig. 1-61), part of conducting system of the heart.

The papillary muscles are conical projections which have a number of slender fibrous threads, called **chordae tendineae**, arising from their apices (Figs. 1-55 and 1-56). The chordae tendineae are inserted into the free edges and ventricular surfaces of the cusps of the **right atrioventricular valve** (***tricuspid valve***).

As the chordae tendineae are attached to adjacent sides of two cusps, they prevent their inversion when the papillary muscles and right ventricle contract.

The chordae tendineae prevent the cusps of the right atrioventricular valve from being driven into the right atrium as ventricular pressure rises (*i.e.*, they prevent eversion of the valve).

There are usually three papillary muscles in the right ventricle.

The **anterior papillary muscle** (Fig. 1-55), the largest and most prominent of the three, is attached to the anterior wall of the right ventricle. Its chordae tendineae are attached to the anterior and posterior cusps of the right atrioventricular valve.

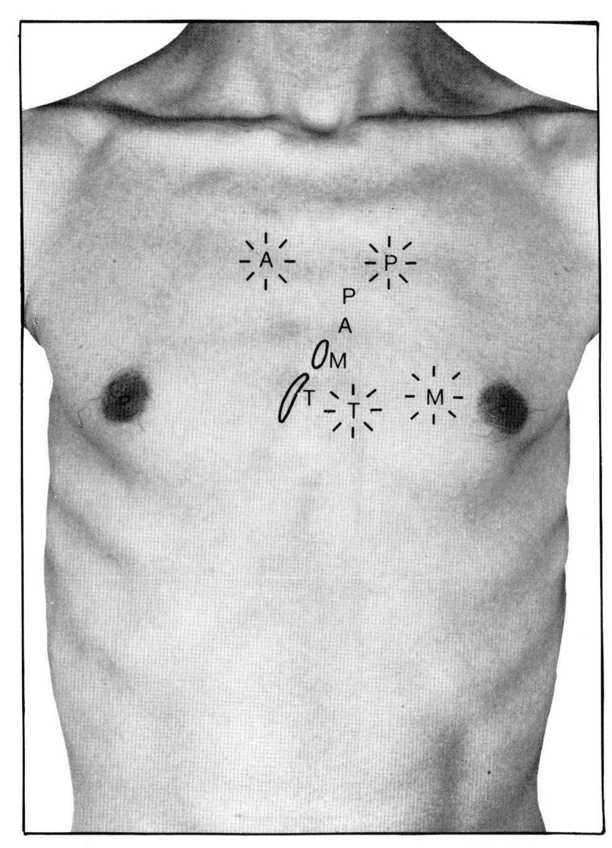

Figure 1-56. Photograph of the thorax of a 27-year-old man showing the surface makings of the orifices of the heart and the areas of auscultation of the valves guarding these orifices. The heart sounds are indicated by a letter and a halo of lines, *e.g.*, one listens to aortic valve (***A***) over the aorta at the second right intercostal space. The heart valves are indicated by an outline of the valve and a letter, *e.g.*, the mitral orifice (***M***) lies posterior to the sternum at the level of the fourth left costal cartilage. ***T***, tricuspid; ***P***, pulmonary.

The **posterior papillary muscle** is smaller than the anterior papillary muscle and may consist of several parts. It is attached to the inferior wall of the right ventricle and its chordae tendineae are attached to the posterior and septal cusps of the right atrioventricular valve (Fig. 1-55).

The **septal papillary muscles**, small and multiple, are attached to the interventricular septum and their chordae tendineae are attached to the anterior and septal cusps of the right atrioventricular valve.

The papillary muscles contract prior to contraction of the ventricle, tightening the chordae tendineae and drawing the cusps together by the time ventricular contraction begins. This prevents ventricular blood from passing back into the right atrium.

The inflow part of the right ventricle receives blood from the right atrium through the right atrioventricular orifice. This opening is surrounded by a ***fibrous ring*** which gives attachment to the cusps of the right atrioventricular or tricuspid valve. This fibrous ring is part of the **fibrous skeleton of the heart** which surrounds both atrioventricular orifices and the pulmonary and aortic orifices.

The right atrioventricular orifice is large enough to admit the tips of three average-sized fingers. It is located posterior to the body of the sternum at the level of the fourth and fifth intercostal spaces.

The right atrioventricular orifice is guarded by **three cusps**, hence its common name **tricuspid valve** (Fig. 1-55). The cusps are more or less triangular in outline and are continuous with one another at their bases, which are attached to the **fibrous ring** that surrounds the atrioventricular orifice. Toward the edges they are disposed as **anterior, septal, and posterior cusps**. The valve cusps project into the right ventricle. Their edges, to which the chordae tendineae are attached, have a serrated appearance (Fig. 1-55).

When the right atrium contracts, the blood in it is forced through the right atrioventricular orifice into the right ventricle, pushing the valve cusps aside like curtains. When the ventricle contracts the papillary muscles contract, pulling on the chordae tendineae, which prevents the cusps of the tricuspid valve from passing into the right atrium.

The pulmonary valve guards the pulmonary orifice which is superior to the aortic orifice and more anterior (Fig. 1-55). It lies at the apex of the conus arteriosus and is about 2.5 cm in diameter.

The pulmonary valve is located at about the level of the third costal cartilage at the left side of the sternum (Fig. 1-56). Blood from the right atrium passes through the right atrioventricular orifice and follows a U-shaped course to pass through the pulmonary orifice.

The pulmonary valve consists of three semilunar valve cusps (*anterior, right,* and *left*), each of which is concave when viewed superiorly (Fig. 1-57). The cusps project into the artery, but lie close to its walls as blood leaves the right ventricle. Following relaxation of the ventricle, the elastic wall of the pulmonary artery forces the blood back toward the heart. However, the valve cusps open up like pockets and completely close the pulmonary orifice. This prevents blood from returning to the right ventricle.

Opposite each valve, the wall of the pulmonary trunk is slightly dilated to form a sinus. The blood in these *pulmonary sinuses* prevents the cusps from sticking to the wall of the pulmonary artery and failing to shut.

In **pulmonary valve stenosis**, the pulmonary valve cusps are fused together, forming a dome with a narrow central opening.

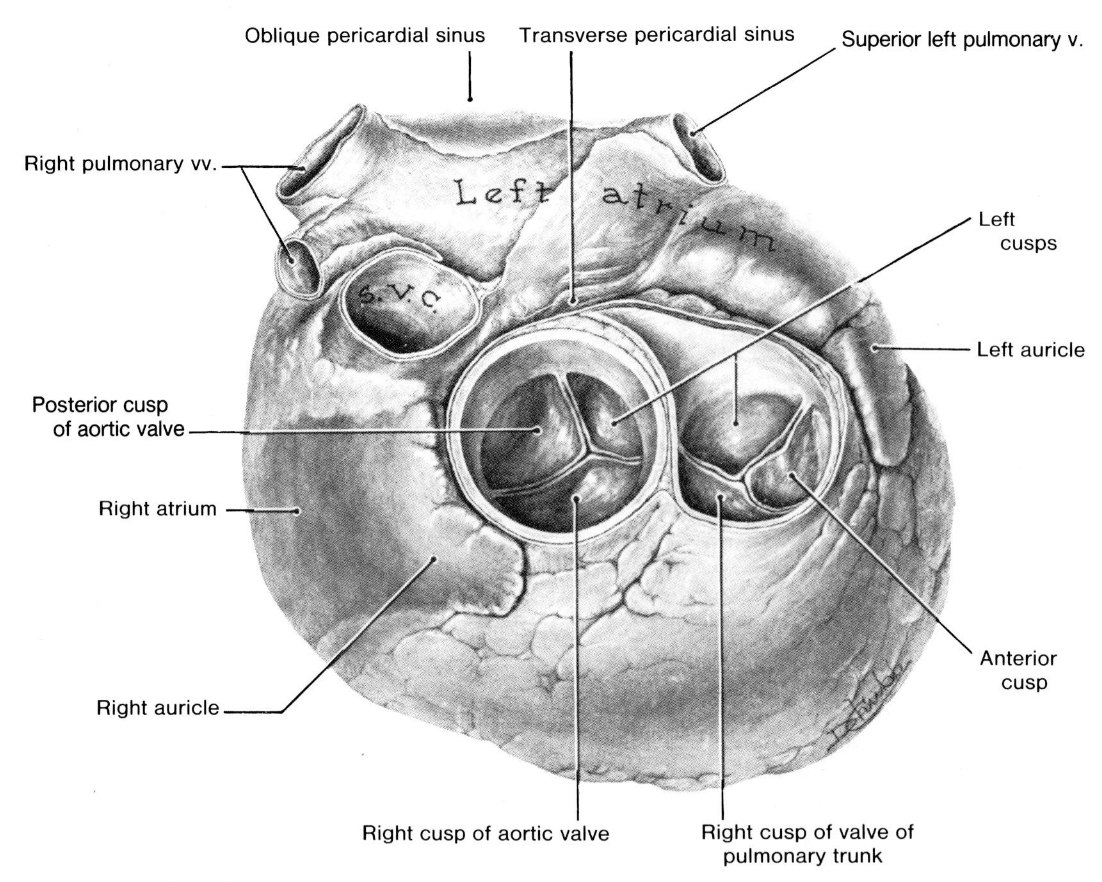

Figure 1-57. Drawing of a superior view of a heart showing the aortic and pulmonary valves. Observe that the cusps of these valves are similar. *The valves are open only while the ventricles are contracting.* Note the anterior position of the ventricles and the posterior position of the atria. Also note that the ascending aorta and pulmonary trunk, which conduct blood from the ventricles, are placed anterior to the atria, superior vena cava and pulmonary veins, which conduct blood to the atria. Observe that the aorta and pulmonary arteries are enclosed within a common tube of serous pericardium and are partly embraced by the auricles of the atria. Examine the transverse pericardial sinus curving posterior to the enclosed stems of the aorta and pulmonary trunk and anterior to the superior vena cava and superior limits of the atria. *Note that each of the semilunar valves (aortic and pulmonary) has three cusps.*

In **infundibular pulmonary stenosis**, the conus arteriosus or infundibulum of the right ventricle is underdeveloped. The two types of pulmonary stenosis may occur together or as separate entities. Depending upon the degree of obstruction to blood flow, there is a variable degree of enlargment or *hypertrophy of the right ventricle.*

The free margins of the pulmonary valve cusps are thin. If these become thickened and inflexible, or damaged owing to disease, the valve will not close completely. As a result, blood can flow back into the right ventricle. This **valvular incompetence** results in a backrush of blood into the right ventricle under high pressure. This *pulmonic regurgitation* may be heard through a stethoscope as a **heart murmur** (cardiac murmur). The murmur results from vibrations set up in the blood in the pulmonary artery, as a result of turbulent blood flow and formation of eddies (small whirlpools).

The Left Atrium (Figs. 1-45, 1-48, 1-49, 1-57, and 1-67). This chamber *forms most of the base or posterior surface of the heart.* The long, tubular **left auricle** (left atrial appendage) forms the superior part of the left border of the heart, which is sometimes visible in radiographs of the chest (Fig. 1-31). The left auricle overlaps the root of the pulmonary trunk (Figs. 1-48 and 1-57).

Four pulmonary veins (two superior and two

inferior) enter the sides of the posterior half of the left atrium (Fig. 1-67).

The wall of the left atrium is slightly thicker than that of the right atrium and its interior is smooth, except for a few **musculi pectinati** in the auricle. The smooth-walled part of the atrium is formed from the absorption of the primitive pulmonary vein during the embryonic period, whereas the rough-walled part, mainly in the auricle, represents part of the primitive atrium (see Moore, 1982 for details).

The interatrial septum slopes posteriorly and to the right; hence much of the left atrium lies posterior to the right atrium.

The left atrioventricular orifice allows oxygenated blood from the left atrium to pass into the left ventricle. It is smaller than the right atrioventricular orifice, but is usually large enough to admit the tips of two medium-sized fingers. The left atrioventricular orifice opens through the inferior half of the anterior wall and is located in the inferior and anterior part of the left atrium (Fig. 1-59).

Thrombi (blood clots) may develop on the walls of the left atrium and in the left auricle in certain types of heart disease. If these thrombi break off, they pass into the systemic circulation and occlude peripheral arteries, large or small, depending on the size of the clot. *A thrombus that obstructs or plugs a vessel is called an **embolus*** (G. *embolos*, a plug).

Occlusion of a small vessel in the brain usually does little damage owing to the numerous anastomoses of arteries in the brain, but if a main artery is occluded, an extensive area of the brain is likely to be involved. **Arterial occlusion results in an embolic stroke or *cerebrovascular accident* (CVA)**, which produces paralysis of the parts of the body previously controlled by the damaged area of the brain.

The Left Ventricle (Figs. 1-31, 1-45, 1-48, 1-49, and 1-59). This chamber *forms the apex of the heart*, nearly all of its left border and surface, and the diaphragmatic surface. The left ventricle forms only a small part of the sternocostal surface of the heart (Fig. 1-48).

The aorta arises from the superior part of the left ventricle, where it is about 2.5 cm in diameter. The aorta is the chief systemic artery of the body (Fig. 2-129). *The arterial supply of the thorax is derived mainly from the branches of the aorta.*

The cavity of the left ventricle is cone-shaped in outline and is longer than that of the right ventricle. Most of its muscular wall is 1 to 1.5 cm thick, but it is much thinner at the apex.

In healthy hearts the wall of the left ventricle is about three times as thick as the wall of the right ventricle (Figs. 1-55, 1-59, and 1-68).

The aortic vestibule is the part of the left ventricular cavity just inferior to the aortic valve. The smooth walls of this region are mainly fibrous (Fig. 1-59). The interior of most of the ventricle is covered with a dense mesh of **trabeculae carneae**, which are finer and more numerous than in the right ventricle. The trabeculae carneae are particularly marked in its inferior half, except at the very apex where the wall is only about 3 mm thick.

There are two large papillary muscles in the left ventricle, anterior and posterior (Fig. 1-59). They are larger than in the right ventricle and their chordae tendineae are thicker but less numerous.

The **anterior papillary muscle** is attached to the anterior part of the left wall, and the **posterior papillary muscle** arises more posteriorly from the inferior wall. The chordae tendineae of each muscle are distributed to the contiguous halves of the two cusps of the left atrioventricular valve.

The left atrioventricular valve or mitral valve (Fig. 1-59) has two obliquely set cusps, anterior and posterior; hence you may hear it called the *bicuspid valve*. **Usually it is called the mitral valve** because its cusps are shaped like a bishop's miter (headdress).

The mitral valve is located posterior to the sternum at the level of the fourth left costal cartilage (Fig. 1-56). It guards the orifice between the left atrium and the left ventricle.

The two cusps of the mitral valve are attached to the fibrous ring which supports the bicuspid orifice. The anterior cusp is the larger one (Fig 1-59). The apices of the cusps project into the left ventricle and as the papillary muscles contract, the chordae tendineae tighten, preventing the cusps from being forced into the left atrium.

The aortic orifice, about 2.5 cm in diameter, lies in the right posterosuperior part of the left ventricle. It is also surrounded by a fibrous ring to which the three cusps of the aortic valve are attached.

The aortic valve (Figs. 1-57 to 1-60) is like the pulmonary valve, except that its cusps are thicker and are placed differently. In addition the **aortic sinuses** (Figs. 1-60 and 1-62), superior to each valve, formed by dilation of the wall of the aorta, are larger than the pulmonary sinuses. The blood in the aortic sinuses prevents the cusps from sticking to the wall of the artery and failing to close. *The aortic valve is located obliquely, posterior to the left side of the sternum at the level of the third intercostal space* (Fig. 1-56).

The mitral valve is the most frequently diseased of the heart valves. **Rheumatic fever** used to

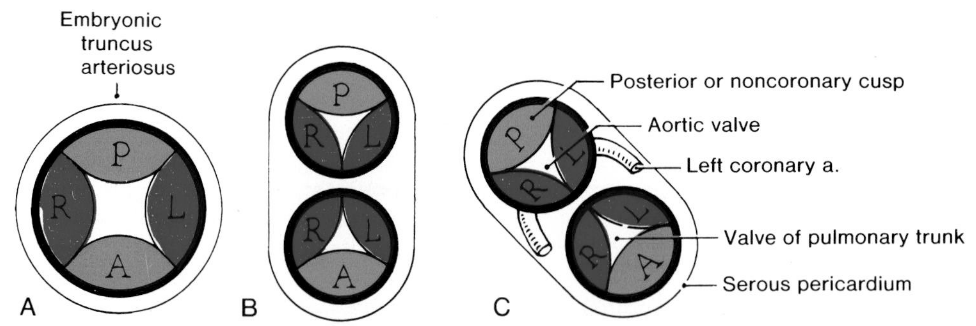

Figure 1-58. Diagrams explaining the embryological basis of the names of the pulmonary and aortic valves. The truncus arteriosus of the embryonic heart with four cusps (*A*) splits to form two valves, each with three cusps (*B*). The heart undergoes partial rotation to the left resulting in the arrangement of cusps shown in *C*.

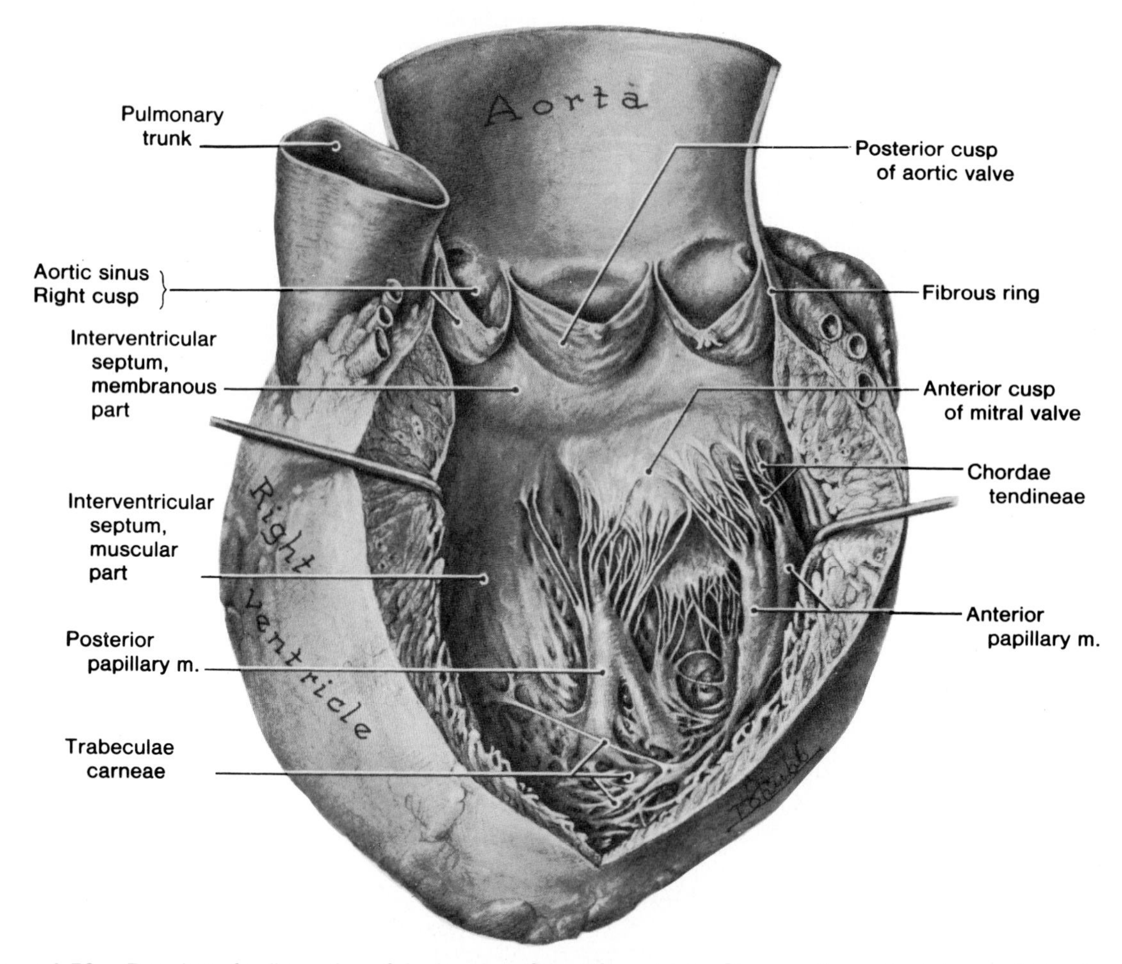

Figure 1-59. Drawing of a dissection of the interior of the left ventricle. Observe the conical shape of this chamber and the entrance (left atrioventricular or mitral orifice) situated posteriorly. Note that the exit (aortic orifice) is situated superiorly. *The ventricular wall is thin and muscular near the apex and thick and muscular superiorly*. Observe that the wall is thin and fibrous at the aortic orifice. Note the trabeculae carneae, as in the right ventricle, forming ridges, bridges, and papillary muscles. Examine the two large papillary muscles, the anterior arising from the anterior wall and the posterior from the posterior wall. Each of these muscles controls (via the chordae tendineae) the adjacent halves of two cusps of the mitral valve. Observe the anterior cusp of the mitral valve intervening between the inlet (mitral orifice) and the outlet (aortic orifice).

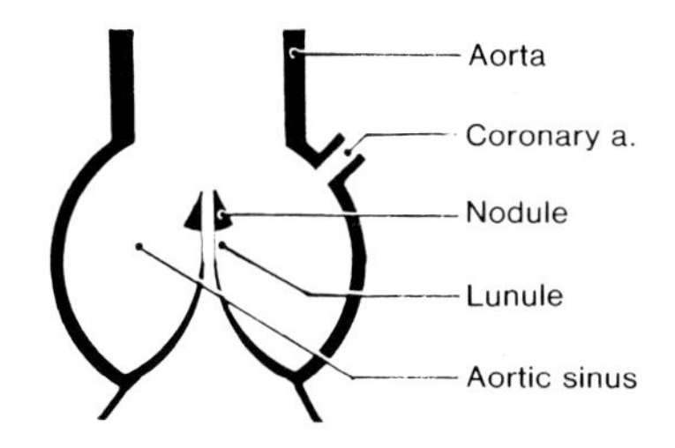

A. Closed, on longitudinal section.

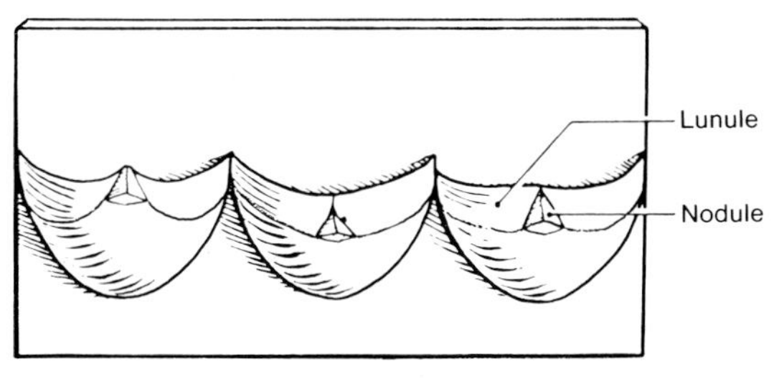

B. Spread out.

Figure 1-60. Diagrams illustrating the aortic valve which, like the valve of the pulmonary trunk (Fig. 1-57), has three semilunar cusps, each with a fibrous nodule at the midpoint of its free edge. When the valve is closed, the nodules meet in the center. Observe that a coronary artery is leaving the aorta from the aortic sinus. The right and left coronary arteries commence at the right and left aortic sinuses. (Also see Fig. 1-62).

be a common cause of this type of **valvular heart disease.** Nodules form on the valve cusps which roughens them and results in irregular blood flow. This produces a *heart murmur* that is audible with a stethoscope. Later the diseased cusps undergo scarring and shortening, resulting in a condition called **valvular incompetence.** In these cases blood in the left ventricle regurgitates into the left atrium, producing a murmur when the ventricles contract.

Further scarring of the cusps of the mitral valve results in progressive narrowing of the orifice (**valvular stenosis**). In these cases blood builds up in the left atrium and lungs, producing **pulmonary congestion** and a strain on the right side of the heart. In addition, when the atria contract, a murmur is produced as the blood is forced through the narrow valvular orifice just before ventricular contraction.

In **aortic valve stenosis** the edges of the valve are usually fused together to form a dome with a narrow opening. This condition may be present at birth (***congenital***) or develop after birth (***acquired***). Valvular stenosis causes extra work for the ventricle and results in *hypertrophy of the left ventricle.* A heart murmur is also produced.

If the aortic valve is damaged by disease, it may not function normally and blood may flow back into the left ventricle. **Aortic valvular incompetence results in *aortic regurgitation*** (a backrush of blood into the left ventricle). This produces a heart murmur and a ***collapsing pulse*** (one with forcible impulse but immediate collapse).

THE INTERVENTRICULAR SEPTUM (Figs. 1-55, 1-59, and 1-61)

The interventricular septum is a strong, obliquely placed partition between the right and left ventricles. It is **composed of muscular and membranous parts** (Fig. 1-61). Its margins correspond with the anterior and posterior interventricular grooves on the surface of the heart (Figs. 1-48 and 1-49). Hence, *the location of the interventricular septum can be mapped out by* the anterior and posterior interventricular *branches of the coronary arteries* which follow these grooves (Figs. 1-45 and 1-65).

The large muscular part of the interventricular septum is thick as well as muscular. It bulges into the right ventricle.

The small, oval-shaped membranous part of the interventricular septum is thin. It is situated just inferior to the attached margins of the right and posterior cusps of the aortic valve (Fig. 1-59).

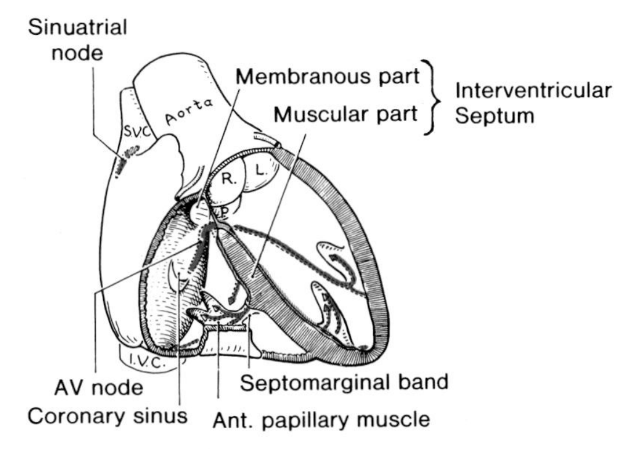

Figure 1-61. Diagram of the conducting system of the heart. The ventricles have been opened and the tricuspid valve has been removed. Note the sinuatrial node (SA node) at the superior end of the crista terminalis (also see Fig. 1-54), and the atrioventricular node (AV node) in the inferior part of the interatrial septum becoming the AV bundle, which divides into right and left limbs (L. *crura*), or bundle branches.

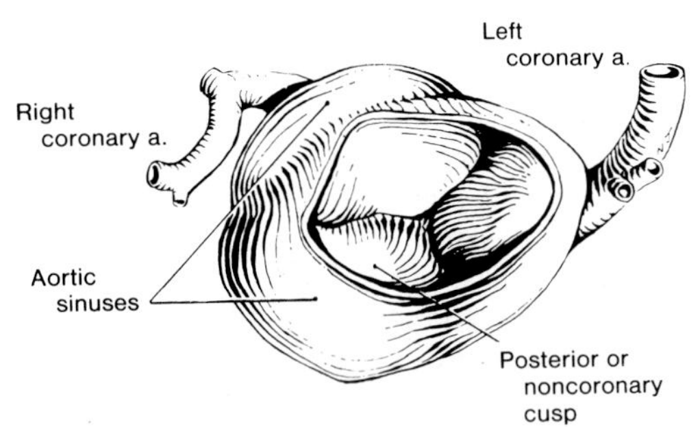

Figure 1-62. Drawing of the ventricular aspect of the closed aortic valve. Observe the coronary arteries arising from the aortic sinuses superior to the valve. Note that the left coronary is larger than the right.

The membranous part of the interventricular septum develops separately from the muscular part of the septum and has a complex embryological origin (see Moore, 1982 for details). Thus, the membranous part is the common site of **VSD (ventricular septal defect).** *VSD, isolated or associated with other cardiac defects, is present in about 50% of all congenital abnormalities of the heart.* Isolated VSD accounts for about 23% of all forms of congenital heart disease. **VSD ranks first on all lists of cardiac defects.** Most cases of VSD result from failure of normal development of the membranous part of the interventricular septum. The size of the defect varies from 1 to 25 mm.

Surface Anatomy of the Heart Valves (Fig. 1-56). The pulmonary, aortic, mitral, and tricuspid valves are located posterior to the sternum on an oblique line joining the third left costal cartilage to the sixth right costal cartilage, but their location is not of too much clinical interest because the sounds produced by these valves are best heard on the chest wall at other sites.

Heart sounds provide valuable clinical knowledge about the heart valves. If the sounds are abnormal they are called **murmurs.** Murmurs do not always indicate a valve problem and some of them have no clinical significance.

Clinicians' interest in the surface anatomy of the cardiac valves arises from their desire to listen to the sounds produced by them. However, the valves are grouped so closely on the oblique line just described that when they are listened to at these sites it is not possible to distinguish clearly the sounds produced at each individual valve. Hence, the valve sounds are listened to at certain **ausculatory areas** (Fig. 1-56) which are as wide apart as possible so that the sounds produced at any given valve may be clearly distinguished from those produced at other valves.

Because the blood tends to carry the sound in the direction of its flow, each area is situated superficial to the chamber or vessel through which the blood has passed, and in a direct line with the valve orifice. For example, the sound produced by the **mitral valve,** located posterior to the middle of the sternum at the level of the fourth costal cartilage, is listened to superficial to the apex beat of the heart (Fig. 1-56). The sound produced by the **aortic valve** is listened for in the second right intercostal space at the edge of the sternum. The **tricuspid valve sound** is heard over the right half of the inferior end of the body of the sternum, and that produced by the pulmonary valve is audible in the second left intercostal space, just to the left of the sternum (Fig. 1-56).

The Cardiac Skeleton. The skeleton of the heart, consisting of fibrous tissue, forms the central support of the heart. Fibrous rings surround the atrioventricular canals and the origins of the aorta and pulmonary trunk (Figs. 1-55 and 1-59).

These fibrous rings give attachment to the valves and prevent the outlets from becoming dilated when the chambers of the heart contract and force blood through them. The cardiac skeleton, together with the membranous part of the interventricular septum, also provide insertion for the cardiac muscle fibers.

The Impulse-Conducting System of the Heart (Fig. 1-61). The conducting system consists of *specialized cardiac muscle fibers* that initiate the normal heart beat and coordinate the contractions of the heart chambers. Both atria contract together, as do both ventricles, but atrial contraction occurs first. *The impulse-conducting system gives the heart its automatic rhythmic beat.*

The sinuatrial node or sinoatrial node (SA node) *initiates the impulse for contraction.* Hence it is the normal **pacemaker of the heart.** Its name, sinuatrial node, is a reminder that it was in the wall of the **sinus venosus** during embryonic development and was absorbed into the right atrium with the sinus venosus (see Moore, 1982).

The SA node consists of a small mass of specialized (i.e., impulse-conducting) cardiac muscle fibers in the wall of the right atrium, located at the superior end of the crista terminalis (Figs. 1-54 and 1-61) to the right of the opening of the superior vena cava.

The SA node is the pacemaker of the heart because it *initiates the impulse* which spreads through the cardiac muscle fibers of both atria causing them to contract. In most people *the SA node gives off an impulse about 70 times/min.* The rate at which the node produces impulses can be altered by nervous stimulation (sympathetic stimulation speeds it up and vagal stimulation slows it down or even stops it).

The atrioventricular node (AV node) is also composed of specialized cardiac muscle fibers. It lies in the posteroinferior part of the interatrial septum, just superior to the opening of the coronary sinus (Fig. 1-61). The impulses from the cardiac muscle fibers of both atria converge on the AV node, which distributes them to the atrioventricular bundle. The AV node conducts the impulses slowly, but sympathetic stimulation speeds up conduction and vagal stimulation slows it down.

The atrioventricular bundle (AV bundle) is a slender strand of *specialized conducting muscle fibers, called Purkinje fibers,* that arise from the AV node and runs anteriorly to the membranous part of the interventricular septum. Here it lies just inferior to the septal cusp of the tricuspid valve. The AV bundle divides into right and left bundle branches or limbs (*L. crura*) which straddle the muscular part of the interventricular septum. Each bundle branch descends deep to the endocardium, one on each side of the interventricular septum (Fig. 1-61).

The **right bundle branch of the AV bundle** innervates the muscle of the interventricular septum, the anterior papillary muscle, and the wall of the right ventricle. The **left bundle branch of the AV bundle** supplies the interventricular septum, the papillary muscles, and the wall of the left ventricle.

Summary of the Impulse-Conducting System of the Heart (Fig. 1-61). The SA node initiates the impulse which is rapidly conducted to the cardiac muscle fibers of the atria, causing them to contract. The impulse enters the AV node and is transmitted through the AV bundle and its branches to the papillary muscles and then throughout the walls of the ventricles. The papillary muscles contract first, tightening the chordae tendineae and drawing the cusps of the atrioventricular valves together. Next, contraction of the ventricular muscle occurs.

The passage of impulses over the heart from the SA node can be amplified and recorded as an **electrocardiogram (ECG)**. The instrument used for recording the potential of the electrical currents that pass through the heart is called an **electrocardiograph**.

Many heart problems involve abnormal functioning of the impulse-conducting system of the heart; hence, *electrocardiograms are of considerable clinical importance in detecting the exact cause of irregularities of the heart beat.*

Patients with a massive myocardial infarction (MI) usually have a crushing substernal chest pain and often an abnormal ECG. The nerve impulses from the heart, which are responsible for producing the pain of myocardial infarction, enter the spinal cord through the superior thoracic ganglia of the sympathetic trunk (Figs. 46, p. 48, and 1-21).

Artificial pacemakers are designed to give off an electrical impulse that will produce ventricular contraction at a predetermined rate. The battery-powered pacemaker, about the size of a pocket watch, is implanted for permanent pacing. An ***electrode catheter*** connected to it is inserted into a vein and followed with a **fluoroscope** (instrument for rendering shadows of x-rays visible when projected on a fluorescent screen). The terminal of the electrode is passed via the vein to the right atrium, through the tricuspid valve to the right ventricle, where it is firmly fixed to the trabeculae carneae on the walls of the ventricle (Fig. 1-55). Here it makes contact with the endocardium.

Fibrillation of the heart refers to multiple, rapid, circuitous contractions or twitchings of cardiac muscular fibers.

In ***atrial fibrillation***, the normal regular rhythmical contractions of the atria are replaced by rapid irregular twitchings of different parts of their walls simultaneously. The ventricles respond at irregular intervals to the dysrhythmic impulses received from the atria, but usually a satisfactory circulation is maintained.

In ***ventricular fibrillation***, the normal ventricular contractions are replaced by rapid, irregular, twitching movements which do not pump (*i.e.*, do not maintain the systemic circulation, including the coronary circulation).

The damaged impulse-conducting system of the heart does not function normally; as a result, an irregular pattern of contractions occurs in all areas of the ventricles simultaneously, except in those that have been infarcted. *Infarction means death of an area of tissue* because of an interrupted blood supply.

Ventricular fibrillation is the most disorganized of all dysrhythmias, and in its presence *no effective cardiac output occurs.* The condition is fatal if allowed to persist. **Brain anoxia** (lack of oxygen to the brain) and *brain* *death* usually occur before the abnormal heart movements cease.

To defibrillate the heart, an electric shock may be given to the heart through the thoracic wall via large electrodes (paddles). This shock causes cessation of all cardiac movements and a few minutes later the heart may begin to beat more normally (regularly). As a result, pumping of the heart is re-established and some degree of systemic (including coronary) circulation results.

The Cardiac Plexus of Nerves (Figs. 1-63 and 1-70). *The impulse-conducting system of the heart is under the control of the* ***cardiac nerves of the autonomic nervous system.*** Many branches of the vagus nerves and sympathetic trunks (Fig. 1-21) form the cardiac plexus. This enables the heart to respond to the changing physiological needs of the body.

Stimulation through the sympathetic nerves increases the heart rate and the force of the heart beat, and causes dilation of the coronary arteries which supply the heart. This results in the supply of more oxygen and nutrients to the myocardium.

Stimulation through the parasympathetic nerves slows the heart rate, reduces the force of the heart beat, and constricts the coronary arteries. The **vagus** (CN X), the parasympathetic cardiac nerve, supplies three branches on each side (Fig. 1-63). The interlacing plexus of sympathetic and parasympathetic nerves lies on the distal part of the trachea, anterior to its bifurcation, and posterior to the arch of the aorta. This plexus contains small ganglia near the SA node which belong chiefly to the parasympathetic system.

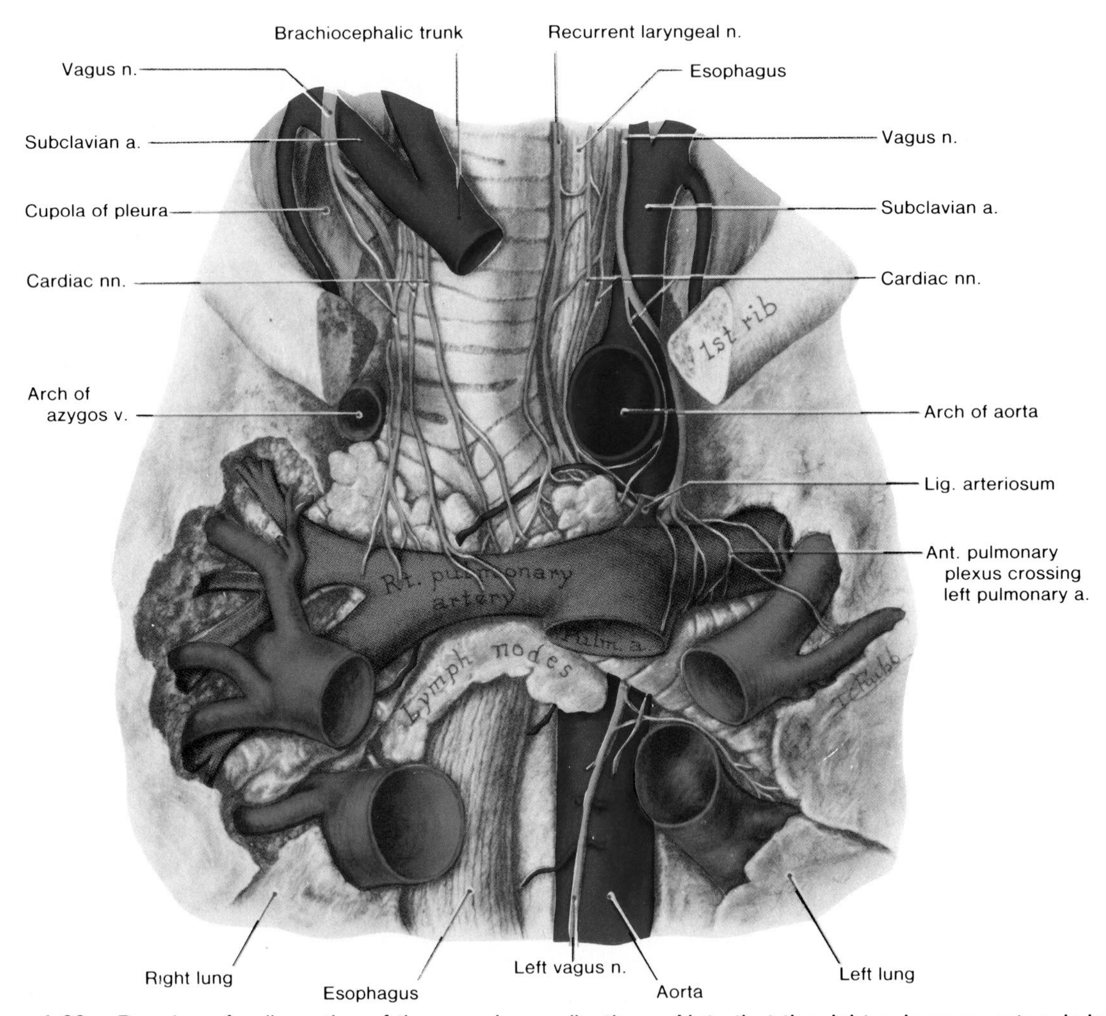

Figure 1-63. Drawing of a dissection of the superior mediastinum. Note that the right pulmonary artery is longer and slightly larger than the left artery and crosses inferior to the bifurcation of the trachea, where it is separated from the esophagus by some lymph nodes. Observe the cardiac branches of the vagus and sympathetic nerves forming the *cardiac plexus of nerves*. Note that the inferior tracheobronchial lymph nodes have fused to form a crescentic mass.

The cardiac plexus receives nerve fibers from: (1) **the sympathetic trunk** (Fig. 1-21) through all its cervical cardiac branches (except the left superior), and from the cardiac branches of the second, third, and fourth thoracic ganglia of both trunks; and (2) **the vagus nerves** through their cervical cardiac branches (except the left inferior), the thoracic cardiac branch of the right vagus, and the cardiac branches of the recurrent laryngeal nerves (Fig. 1-63).

The central nervous system, *via the cardiac plexus, exercises control over the action of the heart and monitors blood pressure and respiration.* This plexus also transmits afferent fibers to the vagus nerves from the great vessels and the lungs. These sensory fibers transmit impulses from *pressure receptors* in the arch of the aorta, superior vena cava, and elsewhere.

The pain of ***angina pectoris*** and **myocardial infarction** commonly radiates from the substernal region and left pectoral region to the left shoulder and the medial aspect of the left arm (Fig. 1-64*A*) This is known as **cardiac referred pain.**

Less commonly the pain radiates to the right shoulder and arm, with or without concomitant pain on the left side (Fig. 1-64*B* and *C*). These cutaneous zones of reference for cardiac referred pain coincide with the segmental distribution of the sensory fibers that enter the same spinal cord segments as the fibers coming from the heart (Fig. 1-23).

The heart is insensitive to touch, cutting, cold, and heat, but ischemia and the resulting accumulation of metabolic products stimulate pain endings in the myocardium. The afferent pain fibers run centrally in the middle and inferior cervical branches and thoracic cardiac branches of the **sympathetic trunk** (Fig. 1-21). The axons of these primary sensory neurons enter spinal cord segments T1 to T4 or T5 on the left side. Cardiac pain is therefore referred to the left side of the chest and the medial aspect of the left arm (Figs. 1-23 and 1-64*A*).

Synaptic contacts may also be made with **commissural neurons** (connector neurons) that conduct impulses to neurons on the right side of comparable areas of the cord. This explains why pain of cardiac origin, although usually referred to the left side, may be referred to the right side or to both sides (Fig. 1-64).

The Coronary Arteries (Figs. 1-45, 1-58, and 1-65 to 1-68). *The heart is supplied by the right and left coronary arteries.* These vessels were called "coronary" (L. *corona,* crown) because they encircle the base of the ventricles like a crown. The coronary arteries, supplying the four chambers of the heart, arise from the right and left **aortic sinuses**, respectively, at the root of the aorta (Figs. 1-60, 1-62, and 1-78).

There is no sharp line of demarcation between the ventricular distribution of the coronary arteries. On leaving the aorta, the coronary arteries pass anteriorly, one on each side of the root of the pulmonary trunk (Figs. 1-45 and 1-65).

Most of the blood in the coronary arteries returns to the chambers of the heart via the **coronary sinus** (Figs. 1-53 and 1-54), but some small venous channels

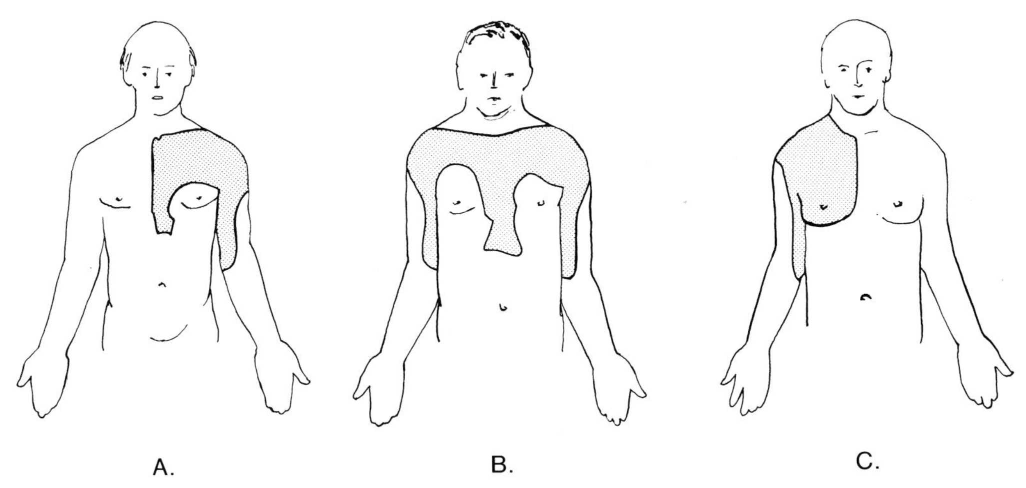

Figure 1-64. Drawings illustrating the common sites of referred pain from the heart. *A*, commonly substernal pain radiates to the left shoulder and the medial aspect of the arm. *B*, less commonly is pain referred to both shoulders and the medial aspects of the arms. *C*, uncommonly, pain is referred to the right shoulder and the medial aspect of the right arm.

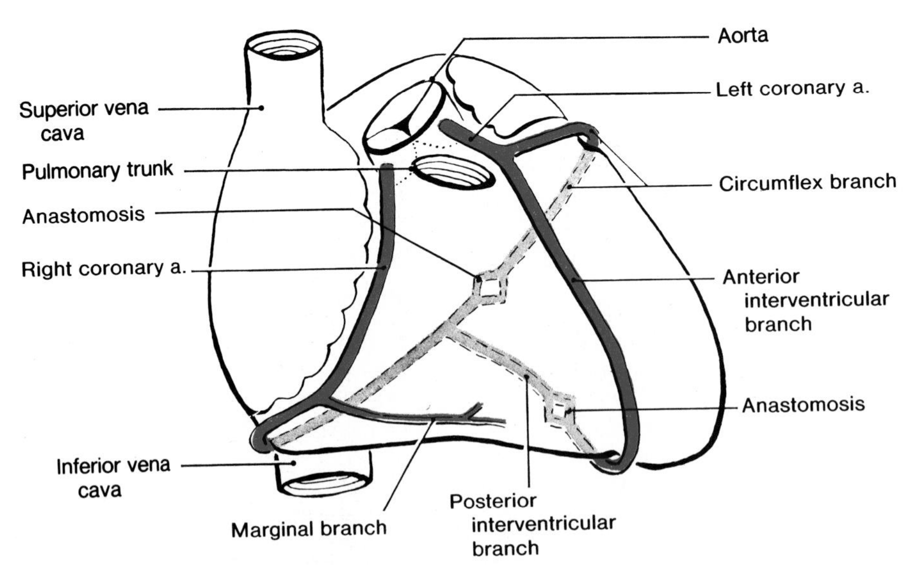

Figure 1-65. Drawing of the coronary arteries. In most cases the right and left coronary arteries share equally in the blood supply to the heart, but preponderance of one vessel may occur. In about 15% of hearts, the left coronary artery is dominant in that the posterior interventricular branch arises from the circumflex branch of the left coronary artery, as in Figure 1-66. Also see Figures 1-45 and 1-67.

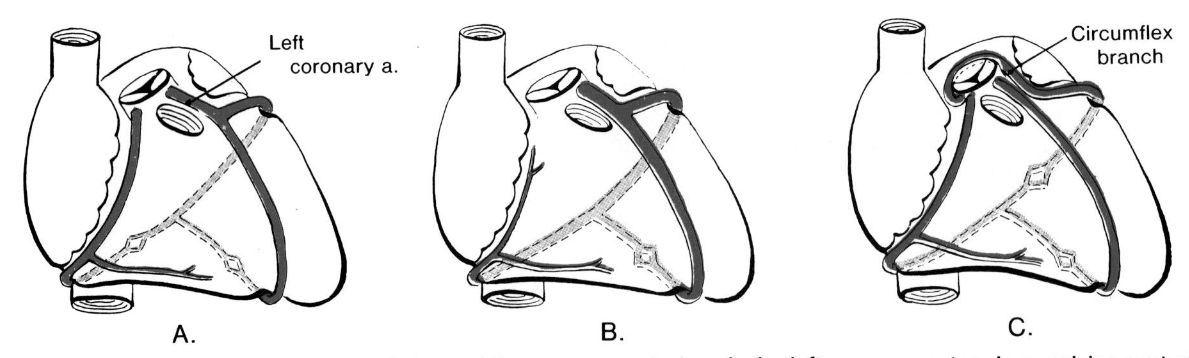

Figure 1-66. Diagrams showing variations of the coronary arteries. *A*, the left coronary artery is supplying part of the usual right coronary artery territory. *B*, there is only one coronary artery. *C*, the circumflex branch arises from the anterior or right aortic sinus. Usually the circumflex branch arises from the left coronary artery (Fig. 1-65).

(*venae cordis minimae and anterior cardiac veins*) empty directly into its chambers (Fig. 1-53).

The right coronary artery arises from the anterior or right aortic sinus (Fig. 1-62) and passes in the **coronary groove** between the right atrium and right ventricle (Figs. 1-45 and 1-65). It then passes to the inferior border of the heart, where it gives off a **marginal branch** that runs toward the apex of the heart (Figs. 1-45 and 1-65). After giving off this branch, the right coronary artery turns to the left in the posterior part of the coronary groove, where it gives off its largest branch, the **posterior interventricular branch** (Figs. 1-65 and 1-67); it descends toward the apex in the ***posterior interventricular groove*** (Fig. 1-49). Near the apex of the heart, the posterior interventricular branch anastomoses with branches of the anterior interventricular branch of the left coronary artery (Fig. 1-65).

Just before giving off its posterior interventricular branch, *the right coronary artery gives off an* ***AV nodal artery*** which enters the posterior part of the atrioventricular groove and *passes* superiorly. **The AV nodal artery supplies *the AV node and the AV bundle.***

The blood supply of the SA node (Fig. 1-61) is often from the right coronary artery, but the **SA nodal artery** may arise from the left coronary artery or its circumflex branch.

The right coronary artery supplies the right atrium, the right ventricle, and a variable amount of the left atrium and left ventricle.

The left coronary artery arises from the left

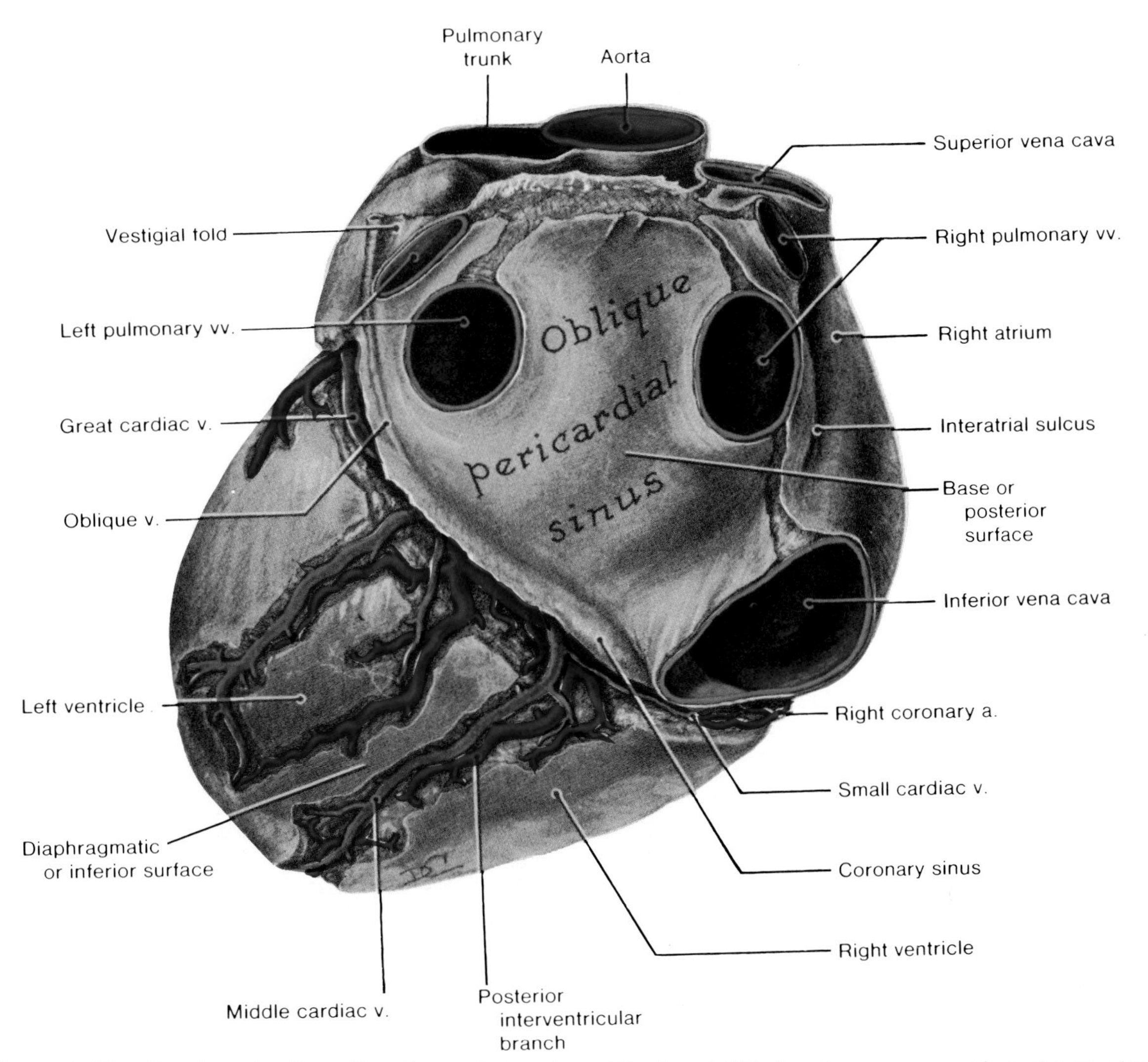

Figure 1-67. Drawing of a dissection of a posterior view of the heart. This heart was removed from the specimen shown in Figure 1-47. *Observe that the oblique pericardial sinus is circumscribed by five veins.* Examine the superior vena cava and the much larger inferior vena cava joining the superior and inferior limits of the right atrium. Note that the left atrium forms the greater part of the base or posterior surface of the heart. Observe the coronary arteries, here irregular in that the left one supplies the posterior interventricular branch.

aortic sinus (Fig. 1-62), and passes between the left auricle and the pulmonary trunk to reach the ***coronary* groove** (Fig. 1-65), where it divides into an anterior interventricular branch and a circumflex branch.

The anterior interventricular branch of the left coronary artery follows the anterior interventricular groove to the apex of the heart, where it anastomoses with the posterior interventricular branch of the right coronary artery (Figs. 1-45 and 1-65). *The anterior interventricular branch supplies both ventricles and the interventricular septum.*

The circumflex branch of the left coronary artery follows the coronary groove around the left border of the heart to the posterior surface (Fig. 1-65). It terminates to the left of the posterior interventricular groove by giving *branches to the left ventricle and left atrium* and anastomosing with the right coronary artery.

The circumflex branch of the left coronary artery gives off a marginal branch which follows the left margin of the heart. *The circumflex branch supplies the left atrium, the left surface of the heart, and the base of the left ventricle inferiorly.*

Variations of the coronary arteries and of their branching patterns are common (Fig. 1-66). In about half the cases, the right

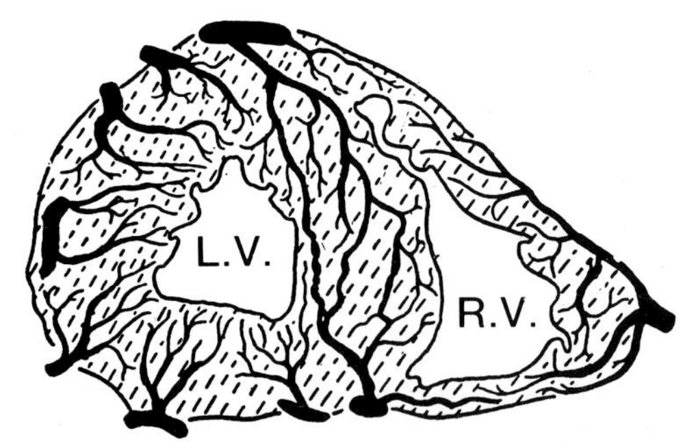

Figure 1-68. Drawing of a transverse section of the ventricles of a heart showing branches of the coronary arteries penetrating the myocardium. *Note that many anastomoses occur between the vessels in the interventricular septum*. When there is myocardial ischemia, these anastomoses enlarge and become functional. Note that the left ventricular wall is about three times as thick as the right one. Because the arterial pressure in the systemic circulation is much higher than in the pulmonary circulation, the left ventricle performs more work, and so is much thicker.

coronary artery is dominant, *i.e.,* it crosses to the left side and supplies the left ventricular wall and the interventricular septum. In about 20% of cases, the left coronary artery is dominant. In addition to supplying the left ventricle and the interventricular septum, it sends branches to the right ventricular wall. In about 30% of cases, the coronary arterial pattern is balanced.

There may be only one coronary artery (Fig. 1-66*B*), and in about 4% of hearts there are *accessory coronary arteries.*

The branches of the coronary arteries are end arteries in the sense that they supply regions of cardiac muscle without overlap from other large branches (Fig. 1-68). Although there is a rich anastomosis between arterioles, this blood supply is inadequate for the requirements of the cardiac muscle when there is a sudden occlusion of a major branch. As a result, *the region supplied by the occluded branch becomes infarcted* (rendered virtually bloodless) and soon undergoes **necrosis.** An area of myocardium that has undergone necrosis is called an **infarct.**

The most common cause of ischemic heart disease is ***coronary insufficiency*** *resulting from atherosclerosis of the coronary arteries. The atherosclerotic process results in lipid accumulations on the inner walls of the coronary arteries.*

Coronary atherosclerosis begins during early adulthood and results in slow narrowing (**stenosis**) of the lumina of these vessels. As coronary atherosclerosis progresses, collateral channels connecting one coronary artery with another expand, permitting adequate perfusion of the heart to continue. Despite this compensatory mechanism, the myocardium may not receive enough oxygen, and when the heart is required to perform increased amounts of work (*e.g.,* during exercise), the inadequate supply of blood to the heart (**myocardial ischemia**) results in substernal discomfort and/or pain.

The coronary arteries are frequent sites of arteriosclerosis (G. *sklērosis,* hardness), with resultant narrowing of their lumina. This reduces blood flow to the parts of the heart supplied by the coronary arteries and their branches. Moderate reduction in blood flow may be asymptomatic until a demand for increased work occurs. The stenotic artery or arteries cannot supply enough blood to meet the increased demand by the heart muscle supplied by these arteries. The result is a characteristic *pain on effort called angina pectoris.*

Angina pectoris, *meaning chest pain,* **is a clinical syndrome characterized by substernal discomfort that results from myocardial ischemia.** Patients commonly describe the discomfort as a tightness or squeezing.

Stress, which produces constriction, is another *common cause of angina pectoris.* Equally common is **strenuous exercise after a heavy meal.** When food enters the stomach, blood flow to the digestive tract is increased. As a result, some blood is diverted away from other organs including the heart. Exercise increases the heart's activity and need for oxygen.

The important feature of angina pectoris is its relation to exertion. It is relieved by 1 or 2 minutes of rest. Sublingual **nitroglycerin** *dilates the coronary arteries,* increases blood flow to the heart, and relieves the pain.

If the supply of oxygen to the myocardium is cut off owing to **coronary occlusion,** the area of muscle concerned undergoes necrosis and infarction. The pain resulting from **myocardial infarction** (MI) is often more severe than with angina pectoris, and it does not disappear after 1 or 2 minutes of rest.

MI may also follow excessive exertion by a person with stenotic coronary arteries. The straining heart muscle demands more oxygen than the stenotic arteries can provide; as a result, the ischemic area of myocardium undergoes infarction and a heart attack occurs.

Coronary occlusion of any but the smallest branches of an artery usually results in death of the cardiac muscle it supplies. The damaged muscle is replaced by fibrous tissue and a scar forms. If parts of the impulse-conducting system are affected by the blockage (*e.g.*, the AV nodal artery), the ventricles may continue to contract independently at their own rate. This is called a **heart block**.

The coronary arteries can be visualized by ***coronary angiography***. Long narrow catheters are passed into the ascending aorta via the femoral or brachial arteries. Under fluoroscopic control, the tip of the catheter is placed just inside the mouth of a coronary artery. A small injection of radiopaque contrast material is made and radiographs or **cineradiographs** are made to show the lumen of the artery and its branches, and any stenotic areas that may be present. The procedure is repeated on the other coronary artery. At some time in the procedure, another catheter is passed into the left ventricle and a large injection of contrast material is given as **cineangiograms** are made to show how well the left ventricular wall is functioning. *If the patient has had a previous* ***cardiac infarct***, *the infarcted area will not contract because it is composed of scar tissue, not muscle.*

In some patients with severe angina pectoris, a **coronary bypass** is carried out. A segment of vein is connected to the aorta or to the proximal part of a coronary artery and then to the coronary artery beyond the stenosis. *A coronary bypass shunts blood from the aorta, or a coronary artery, to a branch of a coronary artery* in order to increase the flow beyond the local obstruction.

Veins of the Heart (Figs. 1-45, 1-53, and 1-67). The heart is drained mainly by veins that empty into the ***coronary sinus***, and partly by small veins (venae cordis minimae and anterior cardiac veins) that open directly into the chambers of the heart, principally those on the right side.

The coronary sinus is the main vein of the heart (Figs. 1-53 and 1-67). This short wide venous channel runs from left to right in the posterior part of the coronary groove. It is the *derivative of the left horn of the embryonic* ***sinus venosus*** (see Moore, 1982 for details). The coronary sinus receives the great cardiac vein at its left end and the middle and small cardiac veins at its right end (Fig. 1-53).

The coronary sinus drains all the venous blood from the heart, except that carried by the anterior cardiac veins and the venae cordis minimae that open directly into the heart.

The coronary sinus opens into the right atrium (Fig. 1-54), immediately to the left of the inferior vena cava and posterior to the right atrioventricular orifice. The one-cusp valve of the coronary sinus is variable in size and form and lies to the right of its opening. This valve, a remnant of the valve of the sinus venosus, *appears to have no function postnatally.*

The great cardiac vein is the main tributary of the coronary sinus (Figs. 1-53 and 1-67). It begins at the apex of the heart and ascends in the *anterior interventricular groove* to enter the left end of the coronary sinus (Fig. 1-67). *The great cardiac vein drains the area of the heart supplied by the left coronary artery.*

The middle cardiac vein also begins at the apex of the heart, but ascends in the *posterior interventricular groove* to enter the right side of the coronary sinus (Figs. 1-53 and 1-67).

The small cardiac vein runs in the coronary groove and enters the coronary sinus to the right of the middle cardiac vein (Figs. 1-53 and 1-67).

The middle and small cardiac veins drain most of the area of the heart supplied by the right coronary artery.

The oblique vein of the left atrium is a small vein which begins over the posterior wall of the left atrium (Figs. 1-53 and 1-67); it descends obliquely to enter the coronary sinus. *The oblique vein is the adult derivative of the embryonic left common cardinal vein* (see Moore, 1982 for a description of the embryonic venous system).

The anterior cardiac veins (two to four) are also small (Fig. 1-53). They begin over the anterior surface of the right ventricle and cross over the coronary groove to end directly in the right atrium.

The venae cordis minimae or smallest cardiac veins (Thebesian veins) are minute vessels that begin in the myocardium and open directly into the chambers of the heart, chiefly the atria. Although called veins, they may also carry blood to the myocardium.

The Lymphatic Drainage of the Heart (Figs. 1–42, 1–72, and 1–73). The lymphatic vessels of the heart form plexuses adjacent to the endocardium and epicardium.

The efferent vessels follow the coronary arteries and empty into the **mediastinal and tracheobronchial lymph nodes.**

THE SUPERIOR MEDIASTINUM

The thymus is located in the superior mediastinum (Figs. 1-43 and 1-69). This mass of lymphoid tissue is a prominent feature of this subdivision of the mediastinum *in early childhood.* It is a flattened, bilobed structure that has a pink, lobulated appearance during early life. It lies immediately posterior to the manubrium in the anterior portion of the superior mediastinum and the adjacent part of the anterior mediastinum (Figs. 1-69 and 1-71).

In newborn infants the thymus may extend superiorly through the superior thoracic aperture into the root of the neck, anterior to the great vessels (Fig. 1-71).

During childhood, particularly as puberty is reached, the thymus begins to diminish in relative size (*i.e.*, undergoes involution). *By adulthood the thymus is often scarcely recognizable.* Usually all that can be found are a few thymic nodules in the loose, irregularly arranged connective tissue in the anterior part of the superior mediastinum.

The blood supply of the thymus is from the *inferior thyroid and internal thoracic arteries* (Figs. 1-16 and 1-46).

The veins of the thymus end in the left brachiocephalic, internal thoracic, and inferior thyroid veins.

The lymph vessels of the thymus end in the brachiocephalic, tracheobronchial, and parasternal lymph nodes (Figs. 1-16 and 1-73).

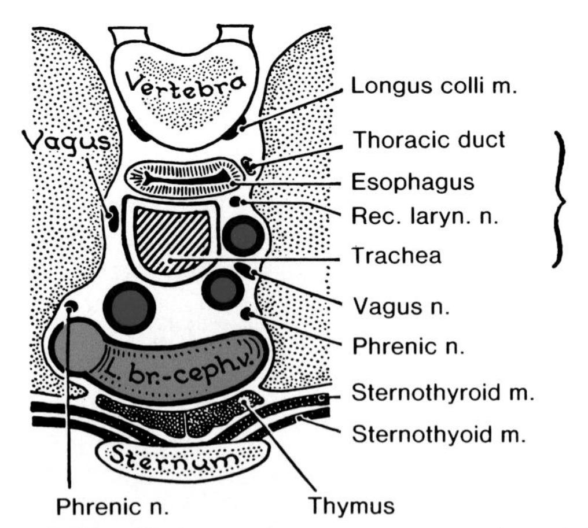

Figure 1-69. Diagram of a transverse section of the superior mediastinum, superior to the level of the arch of the aorta (Fig. 1-75). Observe the posterior relations of the thymus and the structures around the trachea.

The thymus develops from the ventral parts of the embryonic **third pair of pharyngeal pouches,** in common with the *inferior parathyroid glands*, which develop from dorsal parts of these pouches (see Moore, 1982 for a description of the embryonic branchial apparatus). Sometimes the thymus retains a fibrous connection with one or both of the inferior parathyroid glands. Sometimes one of these glands accompanies the thymus into the superior mediastinum.

Thymomas (tumors, usually benign, originating from thymic tissue) are rare, but they may be responsible for vague retrosternal pain, coughing, and **dyspnea** (shortness of breath) owing to pressure on the trachea (Fig. 1–71). They may also compress the superior vena cava, and cause engorgement of the neck veins.

THE BRACHIOCEPHALIC VEINS

The brachiocephalic veins (once called innominate veins) are *located in the superior mediastinum* (Figs. 1-42, 1-46, 1-69, and 1-70). They *arise posterior to the medial ends of the clavicles.* Each brachiocephalic vein is formed by the *union of the internal jugular and subclavian veins.* They have no valves.

At the level of the inferior border of the first right costal cartilage, the two brachiocephalic veins unite to form the **superior vena cava** (Fig. 1-70). The brachiocephalic veins represent the union of the veins from the arm (L. *brachium*), the head (G. *kephale*), and the neck.

Each brachiocephalic vein receives the internal thoracic, vertebral, inferior thyroid, and most superior (highest) intercostal veins.

THE RIGHT BRACHIOCEPHALIC VEIN (Figs. 1-42, 1-46, and 1-70)

This short vein *arises posterior to the right sternoclavicular joint* and descends vertically in the superior mediastinum, posterior to the manubrium and lateral to the arterial brachiocephalic trunk. *The right vagus nerve (CN X) lies between these vessels* and the right phrenic nerve lies posterolateral to the right brachiocephalic vein (Fig. 1-70). The right brachiocephalic vein joins the left brachiocephalic vein to form the superior vena cava at the right border of the sternum and the inferior border of the first costal cartilage (Fig. 1-70). *The right brachiocephalic vein receives the **right lymphatic duct*** (Fig. 1-42).

THE LEFT BRACHIOCEPHALIC VEIN (Figs. 1-42, 1-46, 1-69, and 1-70)

This large vein, about 6 cm long, *arises posterior to the left sternoclavicular joint* and passes to the right and inferiorly, posterior to the manubrium, where it *unites with the right brachiocephalic vein to form the superior vena cava.* The left brachiocephalic vein is over twice as long as the right brachiocephalic vein because it passes obliquely and inferiorly (Figs. 1-42 and 1-70).

During its descent, *the left brachiocephalic vein crosses the left common carotid artery, the brachiocephalic trunk, the left vagus nerve, and the left phrenic nerve* (Fig. 1-70). The left brachiocephalic vein is separated from the manubrium by the thymus, or its remmants, and the origins of the sternohyoid and sternothyroid muscles (Figs. 1-17 and 1-69).

In addition to the tributaries common to both brachiocephalic veins mentioned previously, *the left brachiocephalic vein receives the superior intercostal vein*

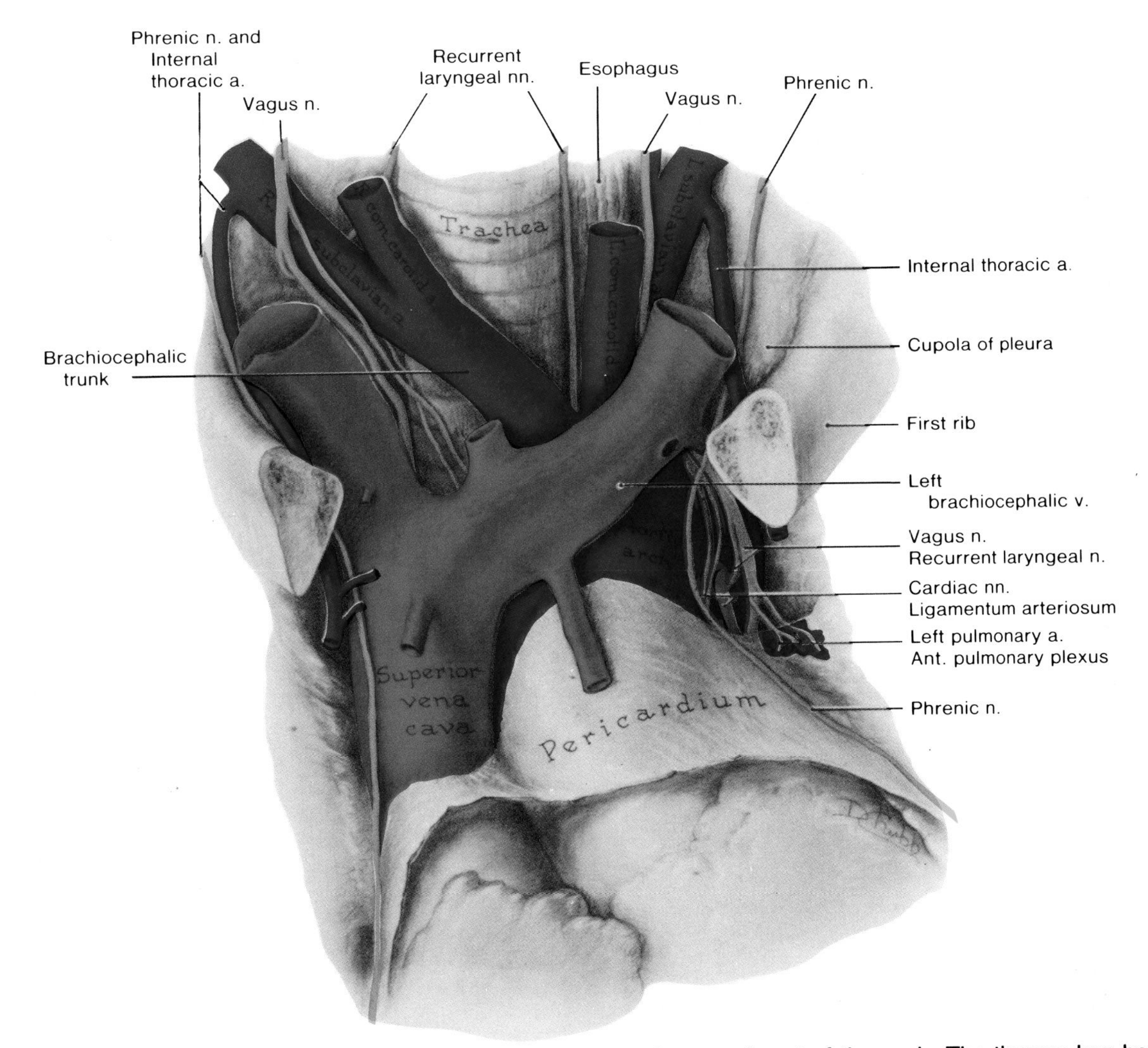

Figure 1-70. Drawing of a dissection of the superior mediastinum and root of the neck. The thymus has been removed. Observe that the great veins are anterior to the great arteries. Note the posterior direction of the arch of the aorta and the nerves crossing its left side. Examine the ligamentum arteriosum, noting that it lies outside the pericardial sac and has the left recurrent nerve on its left side and the vagal and sympathetic branches to the cardiac plexus on its right side. Observe the right vagus nerve crossing the right subclavian artery, giving off its recurrent branch, and then passing medially to reach the trachea and esophagus. Note that the left vagus nerve crosses the arch of the aorta, giving off its recurrent branch, and then passes medially to reach the esophagus. Observe that the left phrenic nerve crosses anterior to the vagus nerve.

and the ***thoracic duct*** (Figs. 1-42 and 1-72), the largest lymph vessel in the body.

THE SUPERIOR VENA CAVA

The superior vena cava is located in the superior mediastinum (Figs. 1-43 and 1-71). This large vein, about 7 cm long, *enters the right atrium of the heart* vertically from its superior aspect (Figs. 1-55 and 1-70).

The superior vena cava returns blood from all structures superior to the diaphragm (*i.e.*, the head, neck, upper limbs, and thoracic wall), ***except the lungs.***

The superior vena cava forms posterior to the first right costal cartilage by the union of the right and left brachiocephalic veins (Figs. 1-46 and 1-70). It passes inferiorly and *ends at the level of the third costal cartilage* by entering the right atrium (Figs. 1-48 and 1-55).

The superior vena cava lies on the right side of the superior mediastinum, anterolateral to the trachea and posterolateral to the ascending aorta (Figs. 1-70 and 1-71). The right phrenic nerve lies between the superior vena cava and the mediastinal parietal pleura, which partly surrounds the right surface of this vessel.

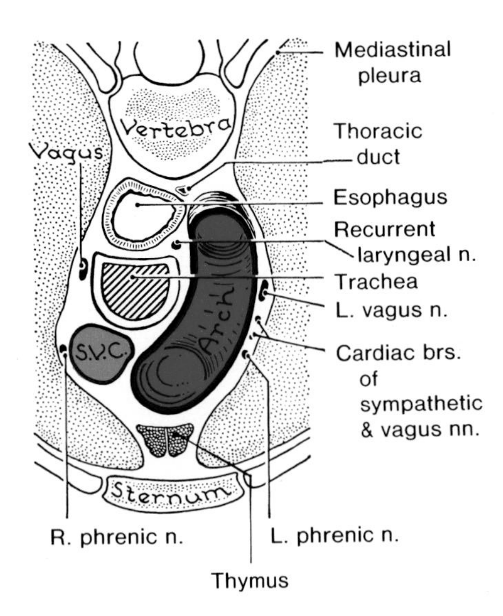

Figure 1-71. Diagram of a transverse section through the superior mediastinum at the level of the arch of the aorta (Fig. 1-75). Observe the four parallel structures: trachea, esophagus, recurrent laryngeal nerve, and thoracic duct (also see Fig. 1-72). SVC, superior vena cava.

The terminal half of the superior vena cava is in the middle mediastinum, where it lies beside the ascending aorta in the pericardium (Fig. 1-49). The shadow cast by the superior vena cava is often seen in radiographs of the chest (Fig. 1-31).

THE ARCH OF THE AORTA

The arch of the aorta (aortic arch) is the curved continuation of the ascending aorta (Figs. 1-48 and 1-73). *The arch of the aorta begins posterior to the right half of the sternal angle* (Fig. 1-63). It arches superiorly and posteriorly with an inclination and convexity to the left.

The arch of the aorta passes to the left of the trachea and esophagus, displacing the trachea to the right, which makes the right principal bronchus almost in line with the trachea (Fig. 1-72).

The arch of the aorta joins the descending aorta on the left of the intervertebral disc between the fourth and fifth thoracic vertebrae, in the same horizontal plane as its origin from the ascending aorta (Figs. 1-73 and 1-75).

The termination of the arch of the aorta corresponds to the sternal end of the second left costal cartilage. The terminal part of the arch of the aorta may be observed in routine radiographs of the chest (Fig. 1-31). The shadow it casts is often called the **aortic knob** (or aortic knuckle).

Anteriorly the arch of the aorta is in contact with the thymus or its remnants (Fig. 1-71). The left brachiocephalic vein crosses just superior to the arch of the aorta (Fig. 1-70), near the origin of its branches. Further posteriorly, the left brachiocephic vein crosses *anterior to the left* ***phrenic nerve, and the cardiac branches*** *of the left vagus and sympathetic nerves* (Figs. 1-63 and 70).

The inferior concave surface of the arch of the aorta curves over the structures passing to the root of the left lung (Fig. 1-74), the bifurcation of the pulmonary trunk, the left pulmonary artery, and the left bronchus (Fig. 1-73).

The ligamentum arteriosum passes from the root of the left pulmonary artery to the inferior concave surface of the arch of the aorta (Figs. 1-45, 1-63 and 1-74). This ligament is the **remnant of the ductus arteriosus**, an embryonic vessel that shunted blood from the left pulmonary artery to the aorta to bypass the lungs (see Moore, 1982 for a description of the fetal circulation and the role of the ductus arteriosus).

The left recurrent laryngeal nerve hooks around the arch of the aorta and the ligamentum arteriosum (Fig. 1-72), and then ascends between the trachea and esophagus (Fig. 1-70).

The most superior part of the arch of the aorta is usually about 2.5 cm inferior to the superior border of the manubrium, but it may be more superior or inferior than this. Sometimes the aorta arches over the root of the right lung and passes inferiorly on the right side. This malformation is called a **right aortic arch**. In some cases this abnormal arch of the aorta, after arching over the root of the right lung, passes posterior to the esophagus to reach its usual position on the left side. For the embryological basis of a right aortic arch, see Moore (1982).

BRANCHES OF THE ARCH OF THE AORTA

These arteries arise from the superior aspect of the arch of the aorta and supply the head and neck, the upper limbs, and part of the body wall. *The arch of the aorta has three branches: the brachiocephalic trunk, the left common carotid artery, and the left subclavian artery.*

The Brachiocephalic Trunk. This is the first and

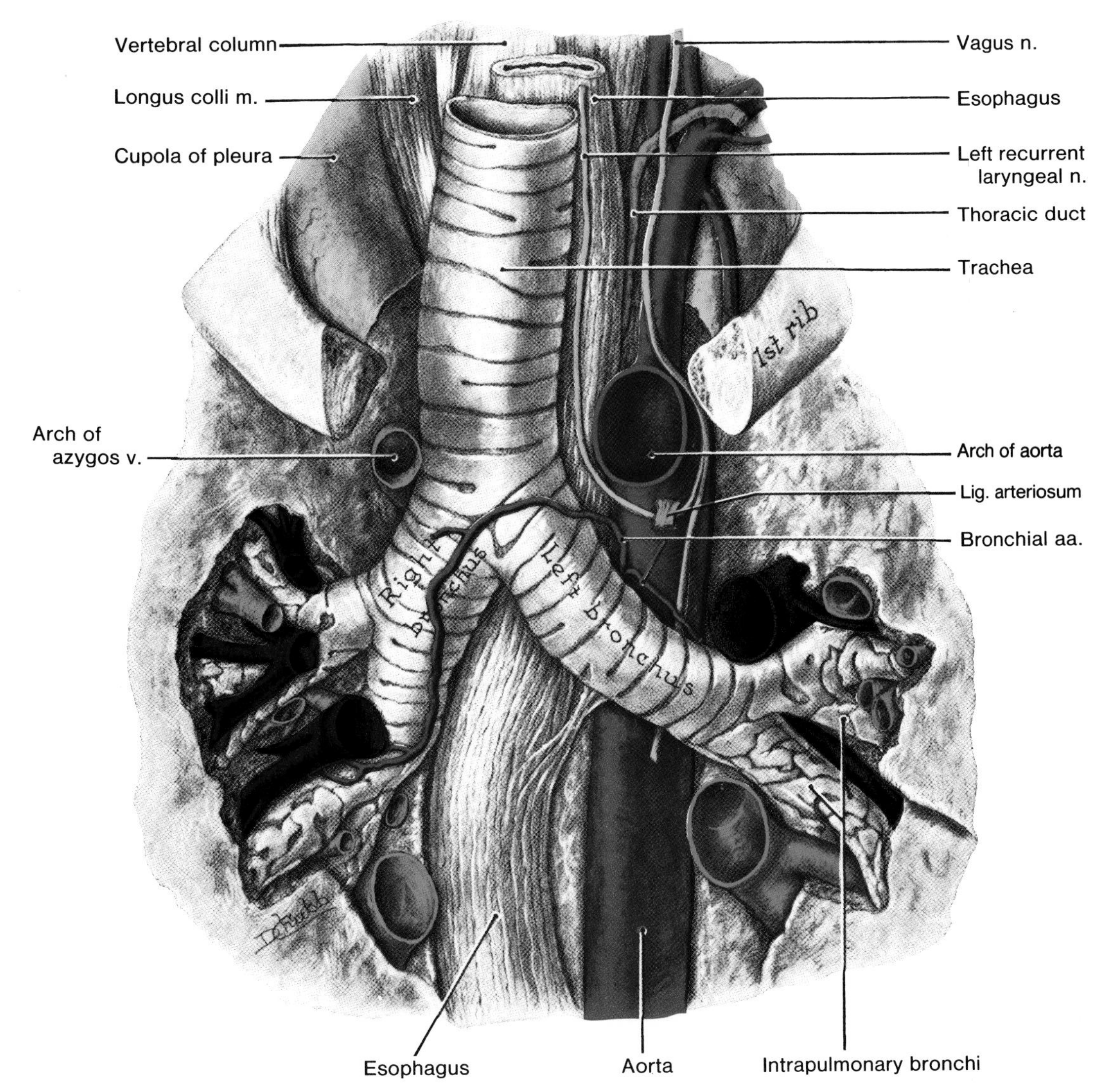

Figure 1-72. Drawing of a deep dissection of the superior mediastinum showing four parallel structures: trachea, esophagus, left recurrent laryngeal nerve, and thoracic duct. Note that the left recurrent nerve lies in the groove between the trachea and esophagus and that the thoracic duct is at the side of the esophagus. Observe that the arch of the aorta runs posteriorly on the left of these four structures and that the arch of the azygos vein passes anteriorly on the right. Examine the trachea inclining to the right and note that the right bronchus is more vertical than the left bronchus, and that its stem is shorter and wider. This is why foreign bodies usually pass into the right bronchus.

largest of the three branches of the arch of the aorta (Figs. 1-70 and 1-73). It arises from the arch of the aorta posterior to the center of the manubrium (Fig. 1-46). Here it is anterior to the trachea and posterior to the left brachiocephalic vein (Fig. 1-70). The brachiocephalic trunk passes superolaterally to reach the right side of the trachea and the right sternoclavicular joint, where it *divides into the right common carotid and right subclavian arteries* (Figs. 1-46, 1-70, and 1-73).

The Left Common Carotid Artery. This artery arises from the arch of the aorta slightly posterior and to the left of the brachiocephalic trunk (Figs. 1-70 and 1-73). It extends superiorly, anterior to the left subclavian artery. The *left common carotid artery* is at first anterior to and then to the left of the trachea (Fig. 1-73). The left common carotid enters the neck by passing posterior to the left sternoclavicular joint (Fig. 1-46).

The Left Subclavian Artery. This artery arises from the posterior part of the arch of the aorta, close to the left common carotid artery (Figs. 1-70 and 1-73). It ascends through the superior mediastinum and lies against the left lung and pleura laterally (Figs. 1-63 and 1-73). In the left lung of a cadaver, the left subclavian artery forms a distinct

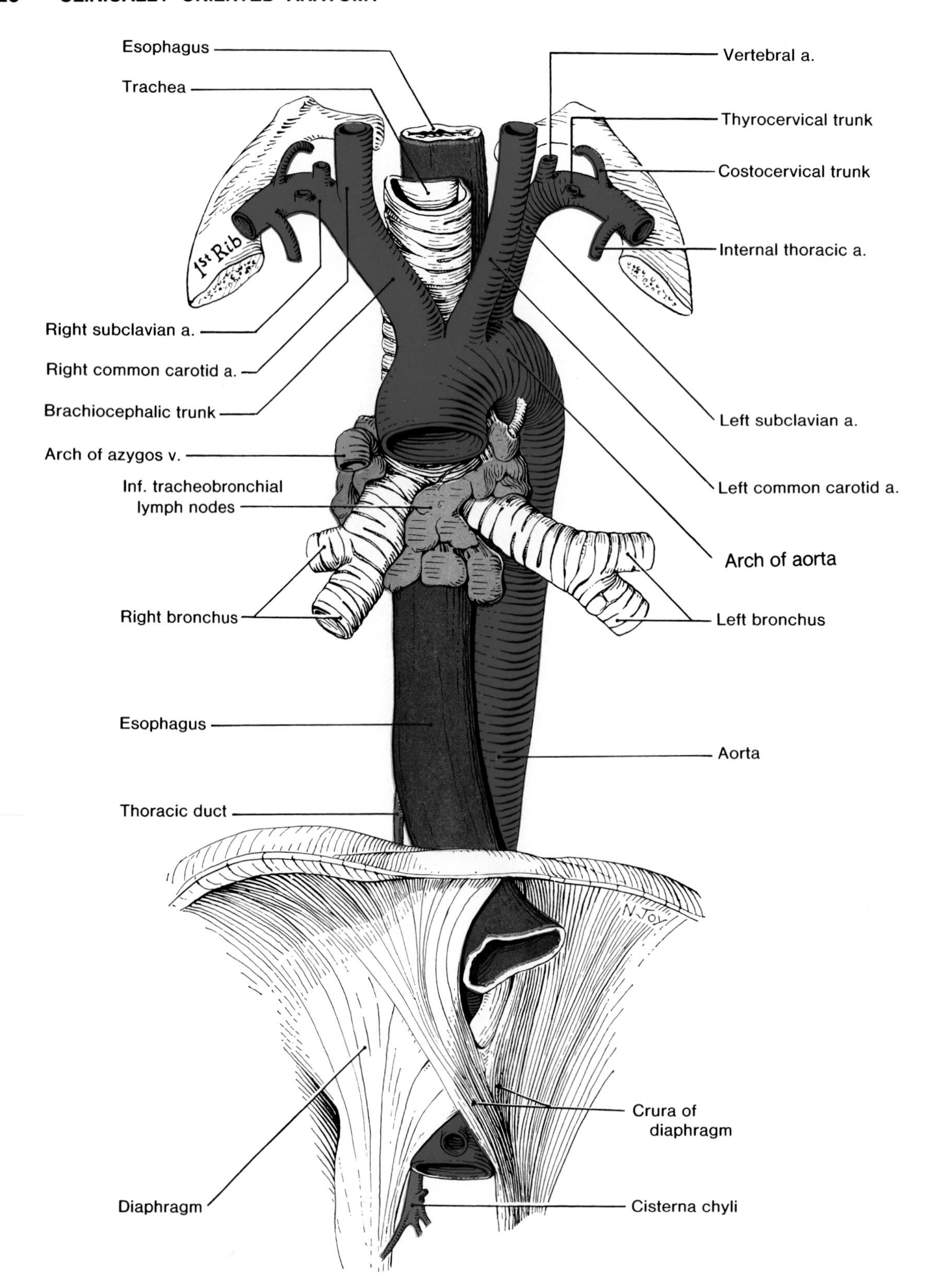

Figure 1-73. Drawing of an anterior view of the thoracic parts of the esophagus, trachea, and aorta. Observe that the arch of the aorta curves posteriorly on the left side of the trachea and esophagus, and that the arch of the azygos vein arches anteriorly on their right sides. Each of these vessels arches superior to the root of a lung. Note that the posterior relation of the trachea is the esophagus, and that the anterior relations of the thoracic part of the esophagus from superior to inferior are: the trachea; the left recurrent laryngeal nerve (Fig. 1-72); the right and left bronchi; the inferior tracheobronchial lymph nodes; the pericardium (removed); and the diaphragm.

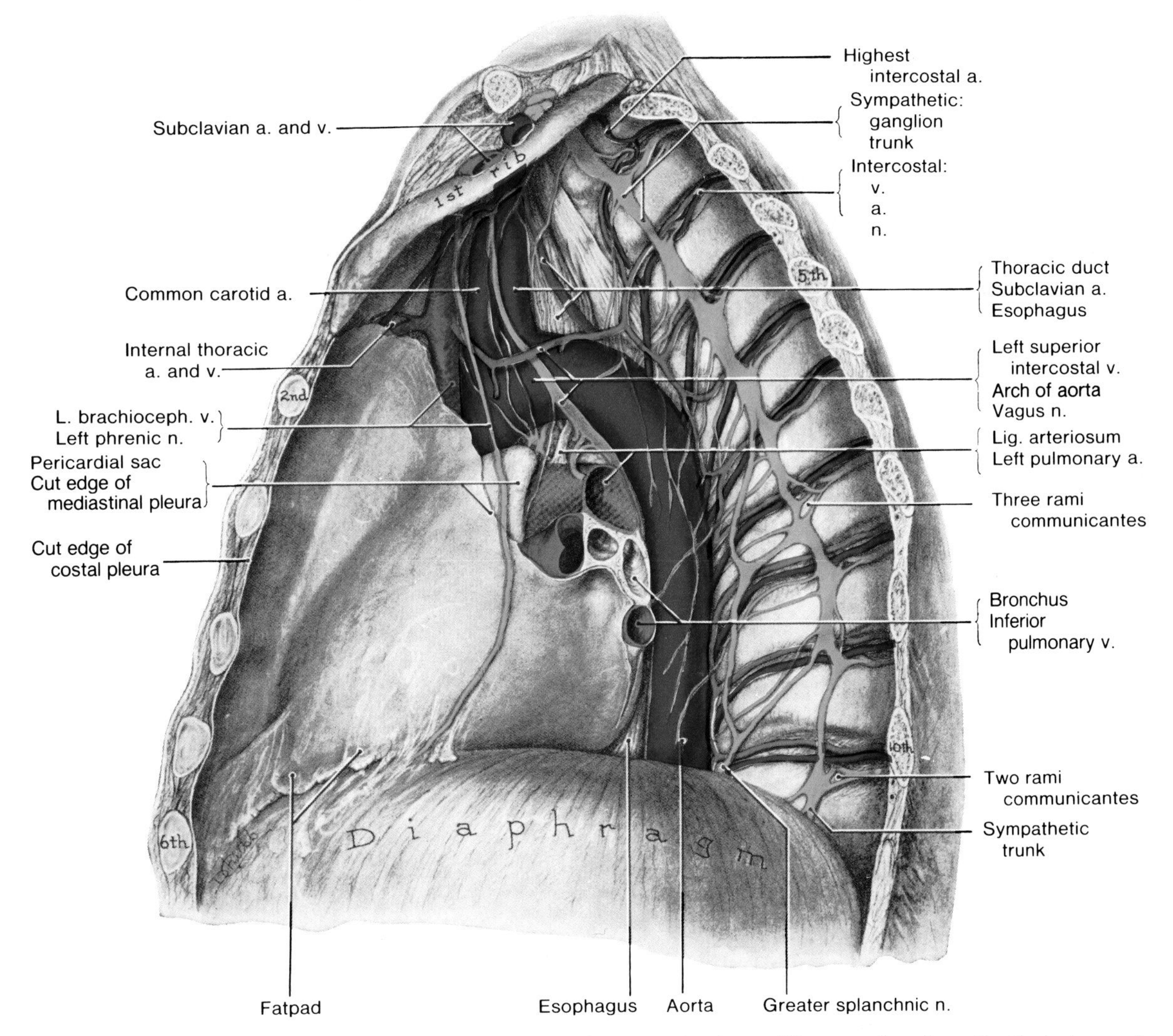

Figure 1-74. Drawing of a dissection of the left side of the mediastinum. The costal and mediastinal pleurae have mostly been removed to expose the underlying structures. Observe that the left side of the mediastinum is the "*red side*" and is dominated by the arch of the aorta, the descending aorta, the left common carotid, and the subclavian arteries. Note that the left phrenic nerve is passing anterior to the root of the lung. Examine the thoracic duct on the side of the esophagus. *Note that the left vagus nerve on the side of the arteries passes posterior to the root of the lung*, sending its recurrent laryngeal branch around the ligamentum arteriosum. Observe that the sympathetic trunk is attached to intercostal nerves by rami communicantes (also see Fig. 1-21).

groove (Fig. 1-33). As it leaves the thorax and enters the root of the neck, the left subclavian artery passes posterior to the left sternoclavicular joint (Fig. 1-46).

Because the arch of the aorta and its branches are derived by transformation of the dorsal aortae and aortic arches of the *embryonic branchial arches*, abnormalities of them are common. For illustrations and a description of the derivatives of the embryonic aortic arches, and of the abnormalities that occur, see Moore (1982).

Abnormalities of the arch of the aorta may develop (*e.g.*, double aortic arch), and several variations in the origins of the branches of the arch of the aorta occur (Fig. 1-76).

A retroesophageal right subclavian artery is not uncommon. As it crosses posterior to the esophagus to reach the right upper limb, the abnormal right subclavian artery may compress the esophagus and cause difficulty in swallowing (*dysphagia*).

A patent ductus arteriosus results from failure of the ductus arteriosus to close

after birth, and become the ligamentum arteriosum. Usually this transformation is completed by the end of the third month after birth.

Coarctation of the aorta (Fig. 1-79) is a congenital abnormality in which the aortic lumen is constricted (stenotic), usually inferior to the origin of the left subclavian artery. When the coarctation is inferior to the ligamentum arteriosum, a collateral circulation develops between the proximal and distal parts of the aorta by way of the intercostal and internal thoracic arteries. The ***postductal type of coarctation*** is compatible with many years of life because the collateral circulation carries blood to the aorta inferior to the stenosis. For illustrations and a description of the embryological basis of coarctation of the aorta, see Moore (1982).

THE VAGUS NERVES

The vagus nerves arise from the medulla of the brain (Chap. 7). The thoracic parts of the vagus nerves descend from the neck posterolateral to the common carotid arteries (Fig. 1-74). *Each nerve enters the superior mediastinum posterior to the sternoclavicular joint* and a brachiocephalic vein (Fig. 1-46).

The right vagus nerve crosses anterior to the origin of the right subclavian artery and posterior to the superior vena cava (Fig. 1-70), to run posteroinferiorly on the right surface of the trachea (Fig. 1-80). The right vagus nerve divides posterior to the trachea into a number of branches that contribute to the ***pulmonary plexuses*** **and the** ***esophageal plexus*** (Figs. 1-63 and 1-81).

Cardiac nerves arise from the right vagus to the right of the trachea and descend to the **cardiac plexus** (Figs. 1-63 and 1-70). After passing anterior to the subclavian artery, the right vagus nerve gives rise to the **right recurrent laryngeal nerve** (Fig. 1-63), which hooks around the right subclavian artery and ascends into the neck between the trachea and the esophagus (Fig. 1-70).

The left vagus nerve descends from the neck posterior to the left common carotid artery, between it and the left subclavian artery (Fig. 1-70). It descends on the left side of the arch of the aorta, between the left common carotid and subclavian arteries. The left vagus nerve is separated laterally from the phrenic nerve by the left superior intercostal vein (Fig. 1-74). The left vagus nerve curves medially at the inferior border of the arch of the aorta and gives off the **left recurrent laryngeal nerve** (Fig. 1-72). This nerve arises from the vagus nerve on the left of the arch of the aorta and ***hooks inferior to the arch of the aorta to the left of the ligamentum arteriosum*** (Figs. 1-63 and 1-72). The *left recurrent laryngeal nerve* ascends through the superior mediastinum in the groove between the trachea and the esophagus (Figs. 1-63 and 1-70).

After giving off the left recurrent laryngeal nerve, the left vagus breaks up into the left *pulmonary plexus,* posterior to the left bronchus (Figs. 1-45 and 1-63). At the inferior border of the root of the lung, it emerges as one or more branches which contribute to the *esophageal plexus* (Figs. 1-63 and 1-80).

THE PHRENIC NERVES

Each phrenic nerve enters the thorax between the subclavian artery and the origin of the brachiocephalic vein (Figs. 1-46 and 1-74).

The phrenic nerves are the sole motor supply to the diaphragm. They arise from the ventral rami of the third, fourth, and fifth cervical nerves (Fig. 6-21).

The right phrenic nerve traverses the thorax posterolateral to the right brachiocephalic vein and superior vena cava (Fig. 1-70), and descends between the parietal pericardium and the mediastinal pleura.

The right phrenic nerve passes anterior to the root of the right lung (Fig. 1-81) and descends on the right of the inferior vena cava to the diaphragm, where it pierces it near the inferior vena caval opening (Fig. 1-29).

The left phrenic nerve descends between the left subclavian and the left common carotid arteries (Fig. 1-74), and crosses the left surface of the arch of the aorta (Fig. 1-70). The left phrenic nerve courses along the pericardium, superficial to the left auricle and ventricle of the heart (Fig. 1-46), and pierces the diaphragm just to the left of the pericardium (Fig. 1-29).

THE TRACHEA

The trachea (windpipe) is a wide tube that *begins in the neck as the continuation of the inferior end of the larynx.* It descends anterior to the esophagus (Fig. 1-70) and enters the superior mediastinum a little to the right of the midline. The posterior surface of the trachea is flat where it is applied to the esophagus (Fig. 1-71). It is kept patent by a series of C-shaped bars of cartilage. The thoracic part of the trachea is 5 to 6 cm long and *ends at the level of the sternal angle by dividing into right and left principal bronchi* (Figs. 1-46 and 1-72).

The arch of the aorta *is at first anterior to the trachea and then on its left side* (Fig. 1-73). Superior to the arch of the aorta, the brachiocephalic trunk and the left common carotid artery are at first anterior and then on the right and left sides of the trachea, respectively. These vessels separate the trachea from the left brachiocephalic vein (Fig. 1-70). The posterior surface of the trachea lies anterior and a little to the right of the esophagus and the left recurrent laryngeal nerve (Figs. 1-63 and 1-70 to 1-74). This nerve sends branches to both the esophagus and trachea.

THE ESOPHAGUS

The esophagus (gullet) extends from the inferior end of the pharynx at the level of the cricoid cartilage (sixth cervical vertebra) to the cardiac orifice of the stomach at the level of the 11th thoracic vertebra (Figs. 1-63, 1-72, and 1-73). Thus *the esophagus has cervical, thoracic, and abdominal parts.*

The esophagus enters the superior mediastinum between the trachea and the vertebral column (Figs. 1-25, 1-69, and 1-71). It passes posterior to the left principal bronchus.

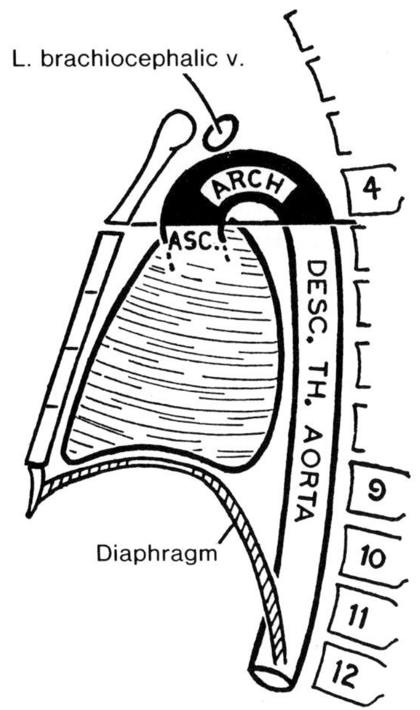

Figure 1-75. Diagram showing the relationships of the ascending thoracic aorta (*ASC*), arch of the aorta, and descending thoracic aorta. The ascending aorta is enclosed within the pericardium. The arch of the aorta is in the superior mediastinum (Fig. 1-43). It becomes the descending thoracic aorta at the level of the intervertebral disc between the fourth and fifth thoracic vertebrae.

The esophagus enters the posterior mediastinum (Figs. 1-43 and 1-72) and descends posteriorly and to the right of the arch of the aorta, and posterior to the pericardium and left atrium (Fig. 1-73). It then deviates to the left and passes through the posterior part of the diaphragm, anterior to the descending thoracic aorta (Figs. 1-29 and 1-74).

The relations of the thoracic part of the esophagus may be summarized as follows.

ANTERIOR RELATIONS OF THE ESOPHAGUS (Fig. 1-69)

In the superior mediastinum the anterior relations of the esophagus are the trachea and the left recurrent laryngeal nerve. In the posterior mediastinum, its anterior relations are the left principal bronchus, the tracheobronchial lymph nodes (Fig. 1-73), the pericardium and left atrium (Fig. 1-30), and the diaphragm inferiorly (Fig. 1-73).

POSTERIOR RELATIONS OF THE ESOPHAGUS (Figs. 1-71 to 1-73)

The posterior relations of the esophagus are the vertebral bodies of T1 to T4, the thoracic duct, the azygos vein (Figs. 1-72 and 1-77), and some right intercostal arteries (Figs. 1-74 and 1-78). At its inferior end, the descending aorta lies between the esophagus and the vertebrae (Fig. 1-77).

The right side of the esophagus is close to the mediastinal pleura and the lung, except where it is crossed by the azygos vein (Figs. 1-72 and 1-77).

The left side of the esophagus is close to the mediastinal pleura superior to the arch of the aorta, except where the thoracic duct and left subclavian artery intervene. The arch of the aorta and descending aorta

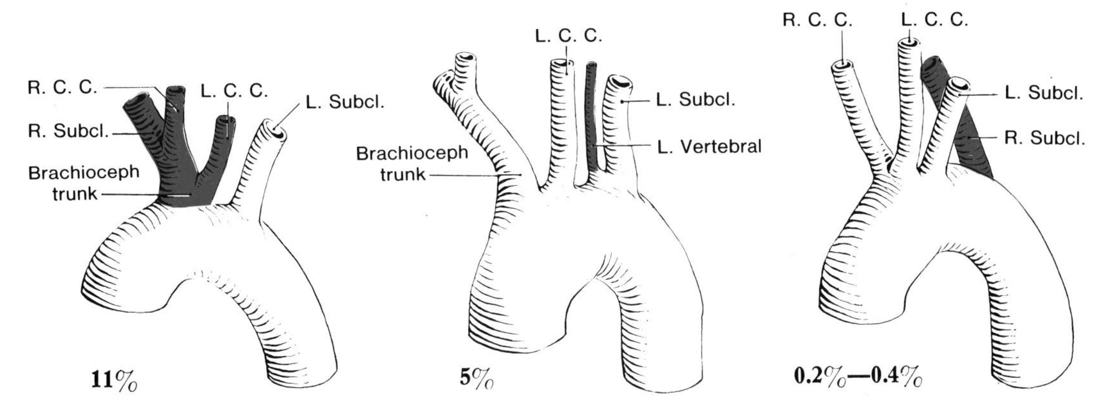

Figure 1-76. Drawings illustrating variations in the origins of the branches of the arch of the aorta. The approximate incidence of these variations is indicated. See Figure 1-73 for the full names of these arteries and for the usual origins of these vessels. The abnormal origin of the right subclavian artery has clinical significance because it passes posterior to the trachea and esophagus to supply the right upper limb. This vascular ring is usually not tight enough to constrict the esophagus and trachea.

lie to the left of the esophagus to the level of the seventh thoracic vertebra (Figs. 1-72 and 1-75).

The esophagus has three "constrictions" in its thoracic part. These may be observed as narrowings of the lumen in oblique radiographs of the chest, taken as barium is being swallowed (Fig. 2-33). The esophagus is compressed by (1) the arch of the aorta, (2) the left principal bronchus, and (3) the diaphragm (Fig. 2-33*A*).

There are no constrictions in the empty collapsed esophagus; as the esophagus expands during filling, the arch of the aorta, left principal bronchus, and diaphragm compress its walls. These "constrictions" disappear as the esophagus empties.

The impressions formed in the esophagus by the structures just mentioned are of clinical interest because of the slower passage of substances through these regions. The "constrictions" indicate where swallowed foreign objects are likely to lodge, and where strictures develop following the accidental drinking of caustic liquids (*e.g.*, lye).

Carcinoma of the esophagus commonly occurs near the inferior end of the esophagus, where it is constricted at the diaphragmatic opening (Figs. 1-73 and 2-33*A*).

The most common congenital abnormality of the esophagus is ***esophageal atresia*** **associated with tracheoesophageal fistula.** This malformation, occurring in about one in 2000 newborn infants, results from incomplete division of the foregut into respiratory and digestive portions (Moore, 1982). In the most common type, the superior portion of the esophagus ends as a blind pouch and a **fistula** (*abnormal canal*) connects the inferior portion of the esophagus to the trachea. As the infant is unable to empty the esophagus into its stomach, saliva and mucous secretions overflow into the larynx and some of them enter the trachea and bronchi in spite of incessant coughing. If attempts are made to feed the infant, the milk fills the esophageal pouch and then overflows into the larynx and trachea, causing more coughing and gagging.

THE POSTERIOR MEDIASTINUM

The posterior mediastinum is the part of the mediastinum located *posterior to the fibrous pericardium* (Fig. 1-43), *inferior to the fourth thoracic vertebra.* Its inferior part lies at a more inferior level than the anteriosuperior part of the diaphragm. The posterior mediastinum ends at the level of the posterior, most inferior part of the diaphragm (Fig. 1-43).

The posterior mediastinum contains: (1) several longitudinal tubular structures (*thoracic aorta, thoracic duct, azygos and hemiazygos veins, esophagus,* and *esophageal plexus*); and (2) several transverse tubular structures (*posterior intercostal arteries, thoracic duct* as it passes from right to left, certain *intercostal veins,* and terminal parts of the hemiazygos veins).

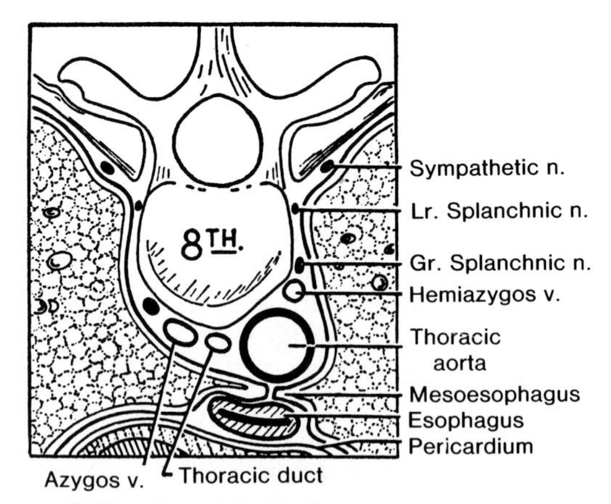

Figure 1-77. Drawing of a transverse section through the posterior mediastinum showing its contents. Its main longitudinal structures are: (1) descending thoracic aorta, (2) esophagus, (3) esophageal plexus of nerves (Fig. 1-80), (4) azygos and hemiazygos veins, and (5) thoracic duct. Note that the posterior mediastinum is located posterior to the pericardium (also see Fig. 1-43).

THE THORACIC AORTA

The ascending aorta arises from the left ventricle in the middle mediastinum (Fig. 1-43). It ends at the level of the sternal angle by becoming the *arch of the aorta* which is located in the superior mediastinum (Fig. 1-75).

The descending thoracic aorta is located in the posterior mediastinum (Figs. 1-43 and 1-75). This large artery is the *continuation of the arch of the aorta.* It begins on the left side of the fourth thoracic intervertebral disc and runs on the left sides of the vertebral bodies (Figs. 1-15, 1-73, and 1-77), commonly producing grooves in them (Fig. 1-15). More inferiorly, the thoracic aorta lies anterior to the vertebral bodies.

The thoracic aorta descends through the posterior mediastinum (Fig. 1-75) against the left pleura, with the thoracic duct and the azygos vein to its right (Figs. 1-74 and 1-81). At first it lies posterior to the root of the left lung and then posterior to the pericardium.

The thoracic aorta passes posterior to the median arcuate ligament of the diaphragm (Fig. 2-118), and enters the abdomen through the most posterior opening in the diaphragm, called the ***aortic hiatus*** (Figs. 1-29 and 1-73).

The **thoracic duct** and the azygos vein lie on the right posterolateral side of the aorta and accompany it through the aortic hiatus.

The thoracic aorta gives off ***bronchial arteries*** *to the lungs* (Figs. 1-72 and 1-78), one or two esophageal branches, and sends twigs to the pericardium and diaphragm. It also gives rise to all the **posterior intercostal arteries** (except the first two pairs) and one pair of subcostal arteries.

THE THORACIC DUCT

This main lymphatic duct conveys most of the lymph of the body to the venous system (Fig. 1-42). It drains the *cisterna chyli,* which lies anterior to the 12th thoracic vertebra, posterior and to the right of the aorta (Fig. 1-73).

The thoracic duct passes superiorly from the cisterna chyli through the aortic hiatus in the diaphragm (Fig. 1-29) on the right side of the aorta (Fig. 1-73).

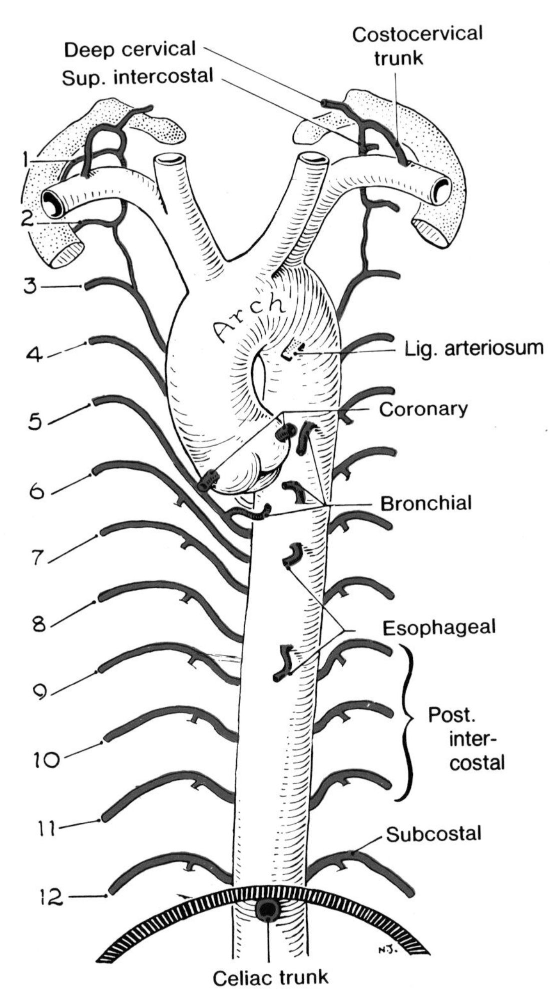

Figure 1-78. Drawing of the thoracic aorta and its branches. The right bronchial artery arises from either the superior left bronchial artery, or the third right posterior intercostal artery (here the fifth), or the aorta directly. The small arteries to the pericardium, tissues in the posterior mediastinum, and the superior surface of the diaphragm are not shown.

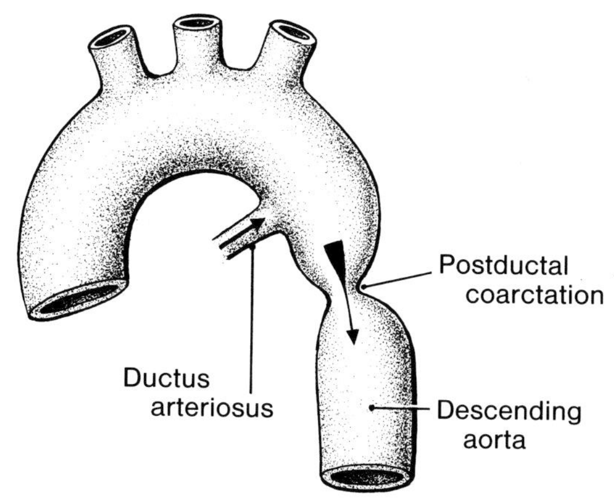

Figure 1-79. Diagram illustrating postductal coarctation (constriction or stenosis) of the aorta. The ductus arteriosus constricts at birth and usually closes in a few days. By the end of the third month, it becomes the ligamentum arteriosum. (Reprinted with permission from Moore KL: *The Developing Human: Clinically Oriented Embryology*, ed 3. Philadelphia, WB Saunders, 1982.)

The thoracic duct reaches the right side of the esophagus and passes superiorly, anterior to the origins of the right posterior intercostal arteries (Fig. 1-74). At about the level of the fifth thoracic vertebra, the thoracic duct deviates to the left, posterior to the esophagus (Figs. 1-69 and 1-71) and ascends through the superior mediastinum into the neck (Fig. 1-42). The thoracic duct passes posterior to the carotid sheath, and anterior to the vertebral artery.

The thoracic duct empties into the venous system at the union of the left internal jugular and subclavian veins (Fig. 1-42).

If the thoracic duct is cut or torn during an accident, **chyle** (L. *chylus,* juice) escapes from it. This milky fluid contains a considerable amount of fine fat droplets.

Leakage of chyle may be prevented by tying off the thoracic duct. The lymph then returns to the venous system by other lymph channels which join the duct superior to the ligature (Fig. 1-42).

During digestion the thoracic duct is distended with chyle, a product of the small intestine (Fig. 2-91).

THE AZYGOS SYSTEM OF VEINS

The azygos system of veins consists of veins on each side of the vertebral column which drain the back and the walls of the thorax and abdomen (Figs. 1-80 and 1-81).

The azygos system of veins exhibits much var-

iation, not only in its origin, but also in its course, tributaries, anastomoses, and termination. The azygos vein and its tributary, the hemiazygos vein, usually arise from the posterior aspect of the inferior vena cava and the renal vein, respectively (Fig. 1-80).

The azygos and hemiazygos veins pass through the diaphragm and present another means of venous drainage from the abdomen and thorax.

The Azygos Vein (Figs. 1-42, 1-73, 1-77, 1-80, and 1-81). This vein ascends on the right side of the bodies of the inferior eight thoracic vertebrae. The azygos vein (G. *a zygon* "unpaired") passes posterior to the root of the right lung and arches over its superior aspect to join the superior vena cava (Figs. 1-73 and 1-80). In addition to the posterior intercostal veins, *the azygos vein receives the vertebral venous plexus* (Fig. 5-58) and the mediastinal, esophageal, and bronchial veins. *The azygos vein drains blood from the thoracic wall.*

The Hemiazygos Vein (Figs. 1-80 and 5-58). This vein arises on the left side in the junction of the left subcostal and ascending lumbar veins. The hemiazygos vein ascends on the left side of the vertebral column, posterior to the thoracic aorta (Fig. 1-77), as far as the ninth thoracic vertebra. Here it crosses to the right, posterior to the aorta, thoracic duct (Fig. 1-42), and esophagus to join the azygos vein (Fig. 1-80).

The hemiazygos vein receives the inferior three posterior intercostal veins, the inferior esophageal veins, and several small mediastinal veins.

The Accessory Hemiazygos Vein (Fig. 1-80). This vein descends on the left side of the vertebral column. It receives tributaries from veins in the fourth to the eighth intercostal spaces, and sometimes from the left bronchial veins. It crosses over the seventh thoracic vertebra to join the azygos vein. Sometimes the accessory hemiazygos joins the hemiazygos vein and opens with it into the azygos vein. The accessory hemiazygos is frequently connected to the left superior intercostal vein, as in Figure 1-80.

THE ESOPHAGUS

The relationships of the esophagus in the posterior mediastinum were described previously. Briefly, it passes from the superior mediastinum and lies on the vertebral bodies down to T10, where it pierces the diaphragm (Fig. 1-73). At its inferior end it lies to the left of the median plane (Fig. 1-43). *Fibers from the vagus nerves* spread out on the esophagus to form an **esophageal plexus** (Fig. 1-81).

THE ANTERIOR MEDIASTINUM

The anterior mediastinum is the part of the mediastinum that lies between the body of the sternum anteriorly and the fibrous pericardium posteriorly (Fig. 1-43). This space is continuous with the superior mediastinum at the sternal angle and is limited inferiorly by the diaphragm.

The anterior mediastinum is very narrow superior to the level of the fourth costal cartilages owing to the closeness of the right and left pleurae (Figs. 1-34 and 1-35).

The anterior mediastinum contains loose areolar tissue, fat, lymph vessels, two or three lymph nodes, sternopericardial ligaments, and a few branches of the internal thoracic artery. In infants and children, the anterior mediastinum may also contain the inferior part of the **thymus gland** (Fig. 1-71). In unusual cases, it may extend as far inferiorly as the fourth costal cartilages. In adults the remains of the thymus gland may be detected.

JOINTS OF THE THORAX

The articulations or joints of the thorax permit movement of the ribs and sternum during respiration. The numerous joints of the thorax are continually moving and any disorder that reduces their movements interferes with respiration.

THE COSTOVERTEBRAL JOINTS

Typically the head of a rib articulates with the sides of the bodies of two thoracic vertebrae, and the tubercle of a rib articulates with the tip of a transverse process (Figs. 1-1, 1-2, and 1-5). Hence there are two articulations of the ribs with the vertebral column. *The costovertebral joints, the plane type of synovial joint*, allow gliding or sliding movements.

JOINTS OF THE HEADS OF THE RIBS (Figs. 1-2, 1-3, and 1-82)

The head of each typical rib articulates with the demifacets of two adjacent vertebrae and the intervertebral disc between them. The crest of the head is attached to the intervertebral disc by an **intra-articular ligament** (Fig. 1-82*A*), which is located within the joint and divides it into two synovial cavities.

An ***articular capsule*** *surrounds each joint and connects the head of the rib with the circumferences of these cavities.* The capsule is strongest anteriorly where the **radiate ligament** (Figs. 1-20 and 1-82*A*) fans out from the anterior margin of the head of the rib to the sides of the bodies of the two vertebrae and the intervertebral disc between them.

There are exceptions to the general arrangement just described. The heads of the first and last three ribs articulate only with their own vertebral bodies.

The heads of the ribs are connected so closely to the bodies of the vertebrae by ligaments that only slight gliding and rotatory movements occur at the joints of the heads of the ribs.

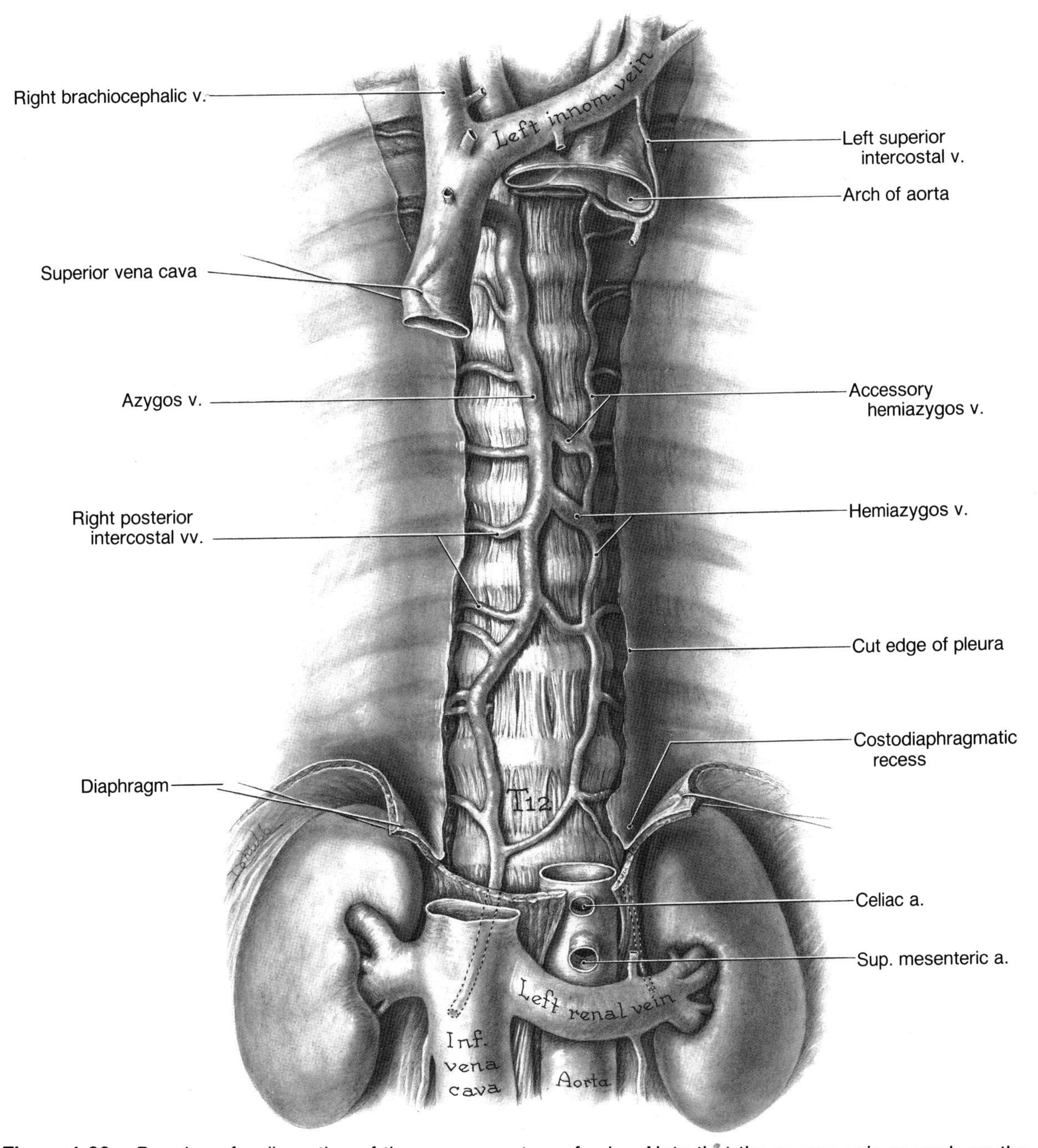

Figure 1-80. Drawing of a dissection of the azygos system of veins. Note that the azygos vein ascends on the right side of the inferior eight thoracic vertebrae, lateral to the thoracic duct and the descending thoracic aorta (shown in Fig. 1-42), and enters the superior vena cava. Observe that on the left side, the veins of the superior intercostal spaces, the accessory hemiazygos vein, and those of the inferior intercostal spaces form the hemiazygos vein. Note that both hemiazygos veins join the azygos vein. The azygos vein forms a direct connection between the inferior vena cava and the superior vena cava. Observe that the left brachiocephalic (left innominate) vein is located anterior to the three branches of the arch of the aorta.

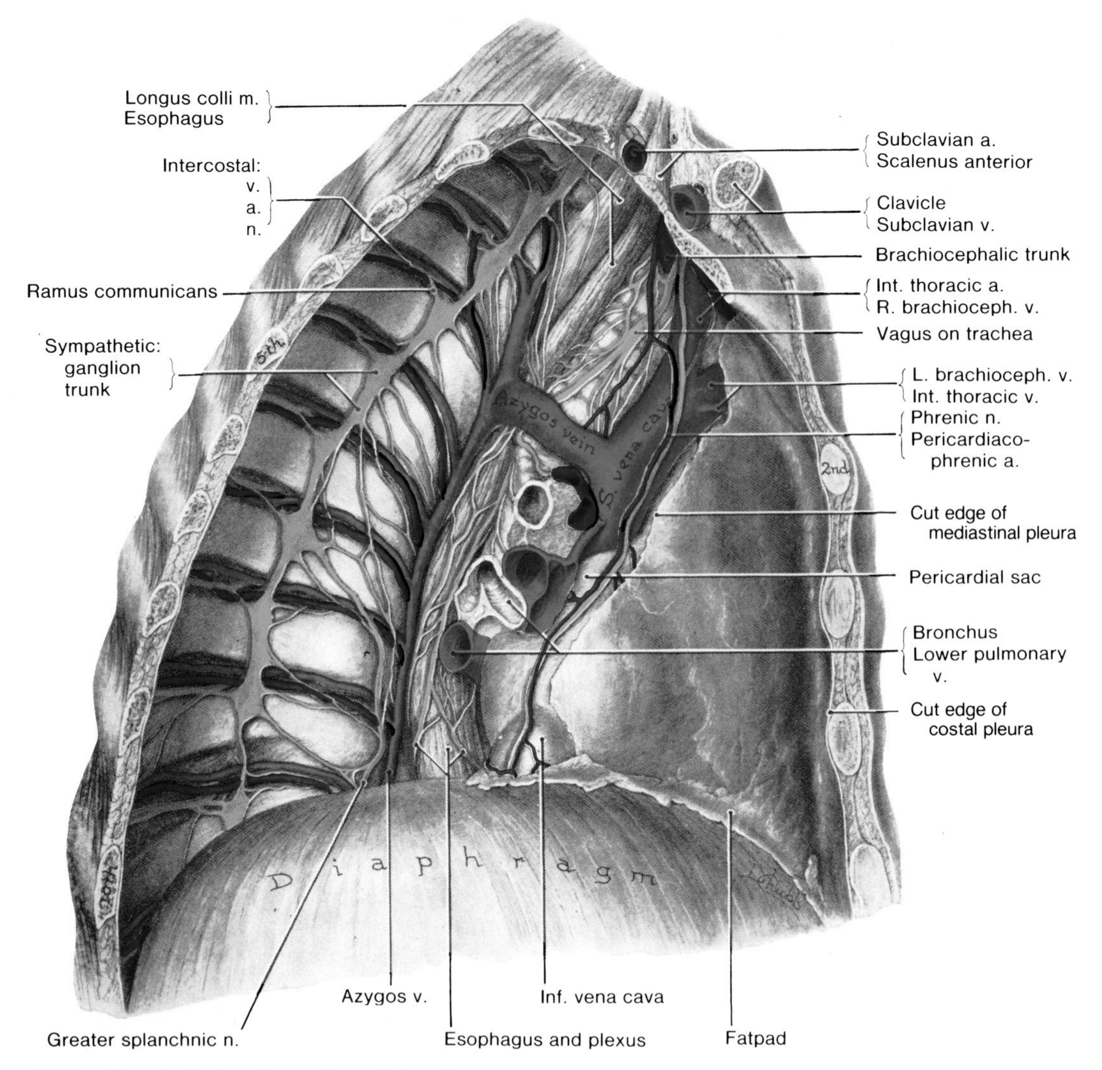

Figure 1-81. Drawing of a dissection of the right side of the mediastinum. Most of the costal and mediastinal pleurae has been removed to expose the underlying structures. Observe that the right side of the mediastinum is the *blue side* and that it is dominated by the arch of the azygos vein, the superior vena cava, and the right atrium. Note than when the mediastinal pleura is removed, the phrenic nerve is free. Observe that the right vagus nerve enters on the trachea, forms a plexus on the esophagus, and is medial to the arch of the azygos vein. Examine the sympathetic trunk and its ganglia.

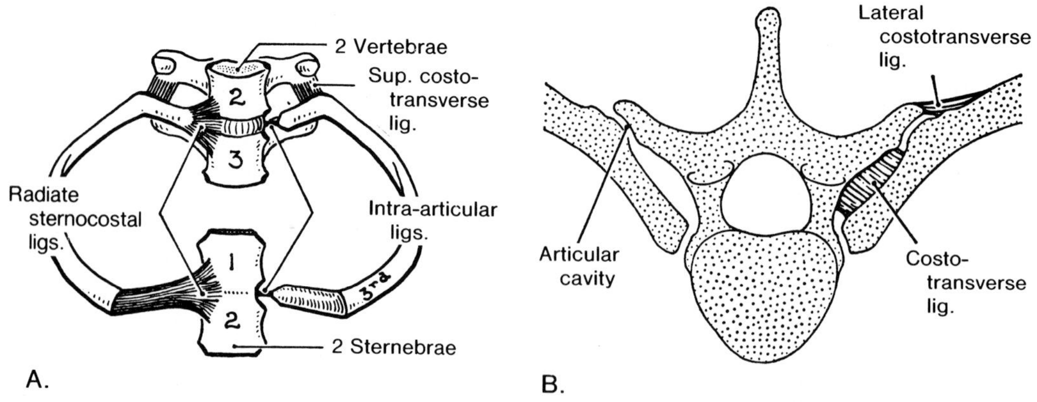

Figure 1-82. Drawings illustrating the joints of the thorax. *A*, compares the articulation at the posterior and anterior ends of a rib and its costal cartilage. *B*, illustrates a costotransverse joint. The articulations of the ribs with the vertebral column are of two types. The joints of the heads of the ribs connect the heads of the ribs with the bodies of the vertebrae, and the costotransverse joints unite the necks and tubercles of the ribs with the transverse processes of the vertebrae.

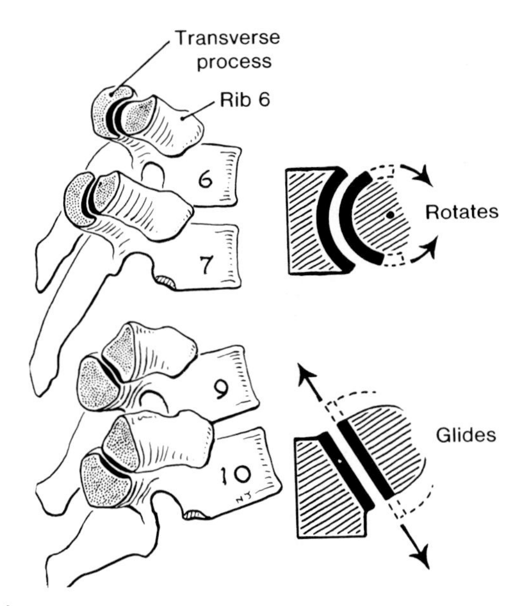

Figure 1-83. Diagrams illustrating the costotransverse joints and showing that the ribs rotate at the superior joints, whereas at the 8th, 9th and 10th joints they glide, increasing the transverse diameter of the superior part of the abdomen.

THE COSTOTRANSVERSE JOINTS (Figs. 1-20, 1-82, and 1-83)

The tubercle of a typical rib articulates with the facet at the tip of the transverse process of its own vertebra (Fig. 1-2*B*) to form a synovial joint. These *small joints* are surrounded by a thin articular capsule which is attached to the edges of the articular facets.

The costotransverse joints are strengthened on each side by a **lateral costotransverse ligament**, passing from the tubercle of the rib to the tip of the transverse process (Figs. 1-19 and 1-82). In addition, a costotransverse ligament unites the posterior surface of the neck of the rib to the anterior surface of the transverse process (Fig. 1-82*B*).

A superior costotransverse ligament joins the crest of the neck of the rib to the transverse process superior to it (Figs. 1-20 and 1-82*A*). The aperture between this ligament and the vertebral column permits passage of the spinal nerve and the dorsal branch of the intercostal artery (Fig. 1-20).

The 11th and 12th ribs do not articulate with transverse processes and have freer movement as a result. The strong ligaments binding the costotransverse joints limit their movements to slight gliding (Fig. 1-82). However the articular surfaces on the tubercles of the superior six ribs are convex and fit into concavities on the transverse processes. Hence some superior and inferior movements of the tubercles are associated with rotation of the ribs.

THE STERNOCOSTAL JOINTS

Usually ribs 1 to 7 articulate via their costal cartilages with the costal notches in the lateral borders of the sternum (Figs. 1-1 and 1-7).

The first pair of costal cartilages *is directly united to the manubrium by a synchondrosis* in which the bones are firmly united by cartilage (Figs. 1-1, 1-82*A*, and 1-84).

The second to seventh pairs of costal cartilages *articulate with the sternum at synovial joints,* but joint cavities are often absent in the inferior ones. The thin articular capsules of these joints are strengthened anteriorly and posteriorly by the **radiate sternocostal ligaments** (Figs. 1-82 and 1-84). These thin, broad membranous bands pass from the costal cartilages to the anterior and posterior surfaces of the sternum.

THE INTERCHONDRAL JOINTS

The articulations between the adjacent borders of the sixth and seventh, the seventh and eighth, and the eighth and ninth costal cartilages are *plane synovial joints.* Each of these articulations is enclosed within a fibrous capsule that is lined with a synovial membrane. The joints are strengthened by interchondral ligaments (Fig. 1-84).

The articulation between the ninth and tenth costal cartilages is a fibrous joint in which the cartilages are joined by fibrous tissue.

THE COSTOCHONDRAL JOINTS

Each rib has a cup-shaped depression at its anterior end into which its costal cartilage fits (Figs. 1-1 and 1-3). *The rib and its costal cartilage are firmly bound together by the continuity of the periosteum of the rib with the perichondrium of the costal cartilage.* No movement normally occurs at these joints.

THE STERNAL JOINTS

THE MANUBRIOSTERNAL JOINT (Figs. 1-1, 1-7, 1-10, and 1-84)

This articulation is between the manubrium and the body of the sternum. The bony ridge indicating the manubriosternal joint, called the **sternal angle**, is very important clinically because it is the *best guide to the numbering of ribs.*

The manubriosternal joint is classified as a symphysis or a **secondary cartilaginous joint** in which the connecting material is a disc of fibrocartilage. The articulating surfaces of the bones are covered by hyaline cartilage. In about 30% of people, the central part of the disc undergoes absorption forming a cavity, but this is not a synovial joint cavity. The manubriosternal joint is strengthened by anterior and posterior fibers from the periosteum.

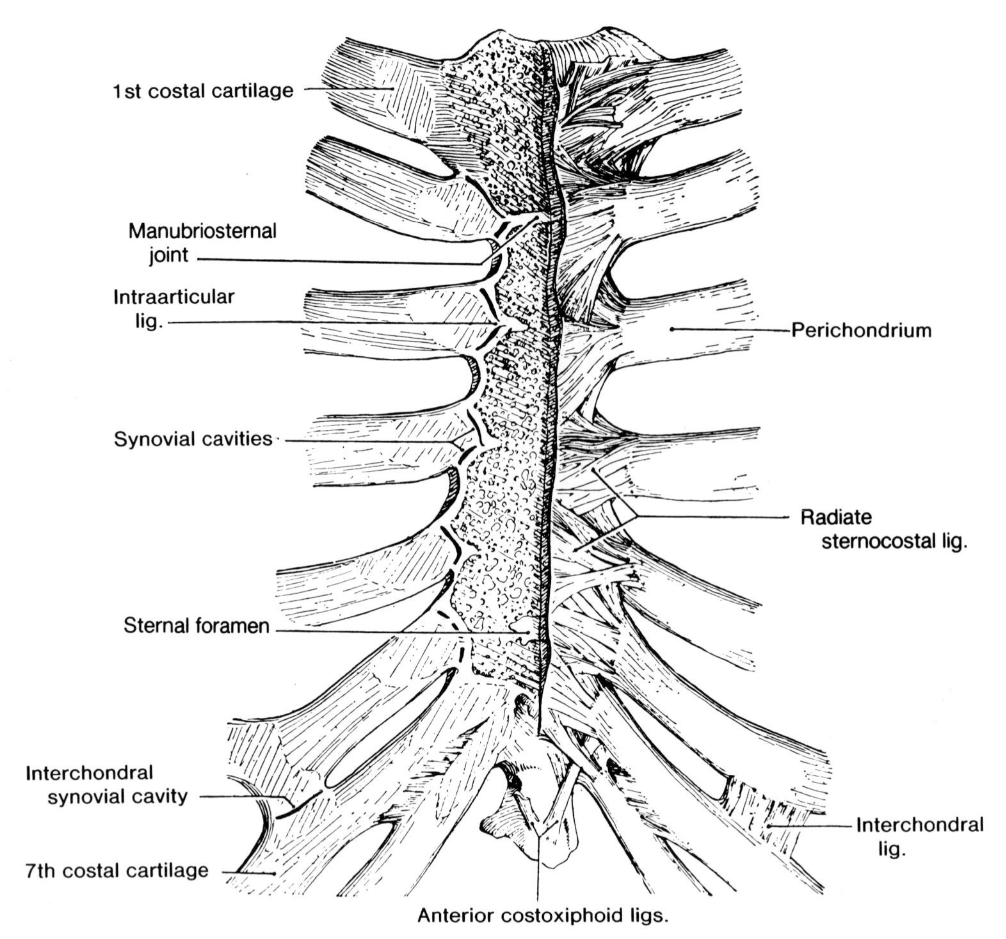

Figure 1-84. Drawing illustrating an anterior view of the sternocostal and interchondral joints. The cortex of the right half of the sternum and costal cartilages has been shaved away. To obtain a specimen of bone marrow, a *sternal puncture* is made through the thin cortical bone into the spongy bone. On the left side, the dissection shows that the fibers of the perichondrium terminate as radiate sternocostal ligaments. Three types of joint are demonstrated here: (1) *primary cartilaginous joints* or synchondroses between the first costal cartilages and the manubrium, and between the seventh costal cartilage and the sternum; (2) a *secondary cartilaginous joint* or symphysis at the manubriosternal joint; and (3) *synovial joints* at the other sternocostal joints and the interchondral joints.

In most people the manubriosternal joint moves slightly during respiration, but in some adults (especially old persons) the joint becomes ossified. In these cases the manubrium is united to the body of the sternum by bone and no movement occurs (Fig. 1-6*C*).

THE XIPHISTERNAL JOINT (Figs. 1-7 and 1-84)

This articulation between the xiphoid process and the body of the sternum is *also a symphysis* or secondary cartilaginous joint. The body of the sternum and the xiphoid process are jointed by hyaline cartilage.

The xiphisternal joint usually ossifies after the 40th year and fuses with the body of the sternum, *i.e.*, it becomes a **synostosis** (Fig. 1-6*C*), but the xiphoid process may remain separate from the body of the sternum even in very old people (Fig. 1-6*A*).

PRESENTATION OF PATIENT ORIENTED PROBLEMS

Case 1-1

While having a heated argument with a client, a 48-year-old male lawyer experienced a sudden *crushing pain* in his chest (**substernal pain**) and epigastrium that radiated along the medial aspect of his left arm. The client helped him to the chesterfield where the lawyer attempted to relieve the pain by squirming, stretching, and belching. When his secretary noted he was pale, perspiring, and writhing in pain, she called his doctor and an ambulance.

The ambulance attendants administered oxygen and rushed him to the hospital, where he was admitted to the **intensive care unit (ICU)**. He was placed under observation with *ECG monitoring* for detection of potential fatal **arrhythmias** (irregularities of the heart rhythm). The patient's blood pressure was low (a sign of shock).

On questioning, the resident learned that the patient had had previous attacks of substernal discomfort (heaviness and pressure) during activity, which he was reluctant to describe as pain. As this discomfort always passed when he rested, he had not complained of it. The resident taking the history recognized these symptoms as a clinical syndrome known as **angina pectoris,** *a symptom of ischemic heart disease* (inadequate perfusion of a portion of the myocardium).

When the resident asked the patient to describe his present chest pain, he said that it was *the worst pain he had ever felt* and clenched his fist to demonstrate the vise-like nature of the squeezing pain. He said that when the pain struck, he had a feeling of weakness and nausea. On auscultation the resident detected an occasional arrhythmia. The ECG was also abnormal.

A diagnosis of acute myocardial infarction owing to coronary insufficiency caused by coronary atherosclerosis was made.

Problems. Define acute myocardial infarction and coronary atherosclerosis. Explain the referral of pain from the heart to the left side of the chest, left shoulder, and medial aspect of the arm. *These problems are discussed on page 142.*

Case 1-2

A 58-year-old man who had lived in an industrial area all his life consulted his doctor because he was coughing up blood (**hemoptysis**), and was experiencing shortness of breath recently during exertion (**dyspnea on exertion**).

On questioning it was learned that he had been a heavy cigarette smoker for over 40 years, and had had a ***smoker's cough*** for several years. He stated that his cough had been getting worse for the last few months and so had his shortness of breath. He first noticed that his ***sputum was blood-streaked*** about 3 weeks ago, and stated that he experienced vague chest pain on the left side at that time.

Physical examination revealed that his left medial **supraclavicular lymph nodes were slightly enlarged** and more firm than usual. His breath sounds and resonance were diminished on the left side compared with the right side.

The doctor requested chest films. The radiologist reported an **obscuration of the hilum of the left lung by a mass.** The normal left mediastinal contours superior to the hilum could not be recognized, and there was slight radiolucency of the remainder of the left lung. The mediastinum was shifted slightly to the left. The radiologist suggested that this was most likely caused by *a tumor in the left superior lobe bronchus with* ***metastases to the left hilar lymph nodes.***

On examination of the interior of the principal bronchi under local anesthesia with a bronchoscope, the **otolaryngologist** (a physician who specializes in diseases of the ear, nose, and throat) observed a growth obstructing the origin of the left superior lobe bronchus (Fig. 1-39). Through the **bronchoscope** he obtained a biopsy of the tumor. *The enlarged supraclavicular lymph nodes were also biopsied for microscopic examination.*

The pathologist reported **bronchogenic carcinoma** in the bronchial biopsy, but the supraclavicular lymph nodes did not show definite tumor involvement.

Examination of the mediastinum through a suprasternal incision under anesthesia (**mediastinoscopy**) revealed some enlarged lymph nodes. Through the **mediastinoscope**, the surgeon removed pieces of tissue (biopsies) from the nodes. The pathologist reported that these nodes showed the presence of many tumor cells, a sign that **metastasis of the tumor** beyond the primary growth had occurred.

In view of the clear evidence of metastases to the mediastinal lymph nodes, it was decided that the tumor was inoperable; surgical removal of the lung (*pneumonectomy*) was not done.

Problems. Bronchogenic carcinoma metastasizes through the lymph (***lymphogenous metastasis***) and the blood (***hematogenous metastasis***). Using your knowledge of the anatomical relations of the lungs, state which structures are likely to be involved by direct extension of a malignant tumor of the bronchus. Where would you expect tumor cells to spread via the lymph and blood?

What is unusual about the lymph drainage of the left inferior lobe of the lung? Explain the probable anatomical basis for metastasis of tumor cells from a bronchogenic carcinoma to the brain. *These problems are discussed on page 143.*

Case 1-3

During an argument with his wife, a 44-year-old inebriated man was *stabbed with a paring knife*, the blade of which was 9 cm long. The knife, penetrating the fourth intercostal space along the left sternal border, produced little external bleeding. By the time he was taken to the emergency room of the hospital, the patient was semiconscious, in shock, and gasping for breath. In a few moments he became unconscious and died.

Problems. Using your knowledge of surface anatomy, what organ(s) would you expect to be punctured by the knife? Where would the blood accumulate? Speculate on the cause of death. *These problems are discussed on page 144.*

Case 1-4

A short thin man with spindly limbs (***gracile habitus***), 42 years of age, complained about recent difficulties in breathing during exercise (**exertional dyspnea**) and about fatigue. He stated that other than being physically underdeveloped, he had been well most of his life (***asymptomatic***), until the last year or so when he had had several respiratory infections.

Physical examination revealed a prominent right ventricular cardiac impulse. A moderately loud **midsystolic murmur** was heard over the second and third intercostal spaces along the inferior left sternal border.

Radiographs revealed enlargement of the right side of the heart, especially of the outflow tract of the right ventricle, a small aortic knob, **dilation of the pulmonary artery** and its major branches, and increased pulmonary vascular markings. *The ECG showed changes suggestive of right ventricular hypertrophy.*

During **right cardiac catheterization**, the catheter easily passed from the right atrium into the left atrium at about the center of the interatrial septum. Serial samples of blood for determination of oxygen saturation were taken as the catheter was withdrawn from the left atrium into the right atrium and then into the inferior vena cava.

These studies revealed increased oxygen saturation of the right atrial blood compared with blood in the inferior vena cava. Serial determinations of pressures showed unequal pressures in the atria (slightly higher in the left atrium).

Atrial septal defect (ASD) was diagnosed. It was classified as the secundum type with *a left atrium to right atrium shunt of blood.*

Problems. Was this man likely born with this defect in his interatrial septum; *i.e.*, ***is this a congenital defect?*** Where else may defects occur in the interatrial septum? What additional complications do you think might occur in this patient in view of the left to right shunt of blood? *These problems are discussed on page 144.*

Case 1-5

During the physical examination of a 15-year-old girl for summer camp, a "***machinery-like***" ***murmur*** was heard during auscultation at the second intercostal space near the left sternal edge. On palpation, a continuous thrill (vibration) was felt at the same location. Other physical findings were normal.

Radiographs of the chest revealed slight left ventricular enlargement and slight prominence of the pulmonary artery and aortic knob. An ECG indicated a moderate degree of left ventricular hypertrophy.

On questioning, the girl said she had always been well, although she feels that she gets "out of breath" faster than other girls. Following consultation with her parents and a cardiologist, the family physician decided to conduct further investigations.

Angiocardiography was performed. The radiologist passed a heart catheter via the femoral vein and inferior vena cava into her right atrium, right ventricle, and pulmonary artery. As he continued, the catheter passed to the left, superiorly, and posteriorly. A small injection of contrast showed the tip of the catheter to be in the descending thoracic aorta.

The catheter was drawn back to the right atrium and a **right angiocardiogram** was performed which showed an essentially normal right heart. Another catheter was passed via the femoral artery into the ascending aorta and contrast medium injected into it (**aortography**). The ascending aorta and the arch of the aorta appeared normal, but the left and right pulmonary arteries were opacified, as well as the descending thoracic aorta.

The radiologist concluded that there was *left to right shunting of blood through a patent ductus arteriosus.*

Problems. Discuss the location of the ductus arteriosus and its embryological origin, prenatal function, and postnatal closure. What caused the characteristic "machinery-like" murmur and left ventricular enlargement? How do you think this left to right shunting of blood could be stopped surgically? What clinical condition do you think might cause **right to left shunting of blood** through the ductus arteriosus? *These problems are discussed on page 145.*

Case 1-6

A 16-month-old boy was "helping" his mother clean up the morning after a cocktail party when he *suddenly started to choke and cough.* Thinking he must have something in his throat, she straddled him over her forearm and gave him several back blows. Although he seemed to be somewhat better after this, it was not long before he began coughing again. When she observed that he was having difficulty breathing (**dyspnea**), she called her pediatrician who arranged to meet her at the hospital.

When asked what the child had been eating when he began to choke, the mother replied, "Nothing! But he could have picked up something that had been dropped onto the floor such as an hors d'oeuvre or a peanut."

On examination it was obvious that the child was in ***respiratory distress*** which was characterized by coughing and **dyspnea**. On subsequent examination, the pediatrician noted limited movement of the right side of the chest.

Auscultation disclosed reduced breath sounds over the right lung anteriorly and posteriorly. On ***percussion*** he thought there was slight hyperresonance over the right lung. He requested a **fluoroscopic examination of the thorax** and inspiration and expiration chest films.

The radiologist reported that there was *overinflation (hyperinflation) of the middle and inferior lobes of the right lung, with a shift of the heart and mediastinal structures to the left which decreased on inspiration.*

In view of these observations, the radiologist suggested that there was likely a **foreign body** lodged in the right middle lobe bronchus, inferior to the origin of the superior lobe bronchus (Fig. 1-39).

Under general anesthesia, the interior of the tracheobronchial tree was examined with a **bronchoscope**. A peanut was observed in the right principal bronchus at the site suggested by the radiologist. The ***bronchoscopist*** removed the peanut with some difficulty, using forceps passed through the bronchoscope.

Speaking to the mother later, the pediatrician urged her to keep small objects out of the reach of her child. He explained that inhalation of nuts, beans, and dry watermelon seeds was harmful because they rapidly swell to several times their usual size, causing **bronchial obstruction.**

When asked for details about this condition, the pediatrician made simple sketches (Fig. 1-85) showing how some air can usually go into the lung because the bronchus expands during inspiration. However, because the caliber of the bronchus becomes less during expiration, the bronchial wall closes on the peanut, trapping the air distal to it. He further explained that aspiration of certain materials such as peanuts may cause both a local and a systemic reaction.

The pediatrician later explained to an intern that *overinflation of a sizeable portion of a lung may be progressive and compress the rest of the lung,* and even the other lung (by displacement of the mediastinum), to the point of **respiratory failure** and death. In some cases, the peanut swells so much that it completely obstructs the bronchus, producing **atelectasis** (collapse) of the lung distal to the peanut. In addition, peanuts release substances that induce a *chemical bronchitis and pneumonitis.*

Problems. When foreign bodies are aspirated, the right lung is involved more often than the left (*3:1*).

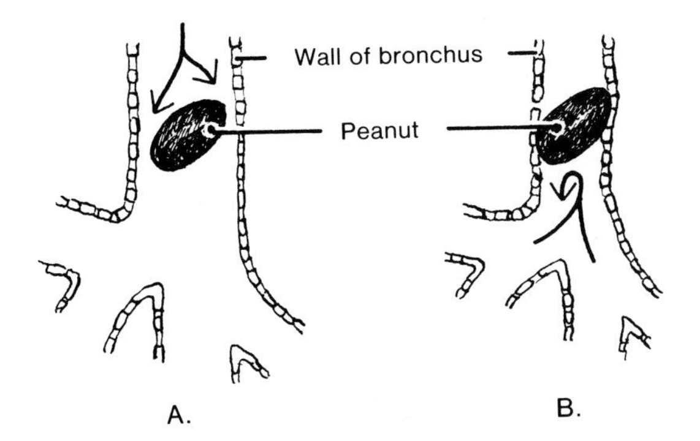

Figure 1-85. Diagrams illustrating bronchial obstruction and the check valve action of the peanut. *A*, the bronchus expands during inspiration allowing some air to pass around the peanut into the lung. *B*, the bronchus contracts during expiration preventing little if any air to leave the lung.

What is the anatomical basis for foreign bodies usually entering the right principal bronchus and involving the middle and inferior lobes of the right lung? If the peanut had not been removed, the right middle and inferior lobes of the infant's lung would later have collapsed (**obstructive atelectasis**).

Explain why the lung collapses. What would be the appearance of the atelectatic lobes in a radiograph? What effect would atelectasis have on the position of the heart, other mediastinal structures, and the diaphragm? *These problems are discussed on page 145.*

Case 1-7

A 10-year-old girl, wrapped in a blanket, was carried into the outpatient department. The nurse immediately took her into an examining room and called a doctor. As the nurse prepared the child for physical examination, she observed that the child was shivering (***chills***) and was holding the right side of her chest. She noted that her respirations were rapid (**tachypnea**) but shallow. The girl had a hacking cough and brought up some sputum containing some blood-tinged mucous material. Her temperature was 41.5°C and her pulse rate was 115.

During questioning the doctor learned that the child had had a really bad cold (**upper respiratory tract infection**) for about a week. The mother said that she did not become worried about her child until she developed a chill and a fever with a hacking cough and **chest pain.**

On percussion of the thorax, the doctor noted dullness over the right inferoposterior region of the child's chest. ***On auscultation,*** he noted suppression of breath sounds on the right side and a **pleural rub** (friction caused by rubbing of the inflamed pleurae together).

When asked to describe the pain, the patient said it was a ***sharp, stabbing pain*** that became worse when she breathed in deeply, coughed, or sneezed. When asked where she first felt pain, she placed her hand over the inferior part of her chest. When asked where else she experienced pain, she pointed to her umbilical area and to her right shoulder.

The doctor requested a complete blood count, a sputum culture, and chest films in both prone and upright positions. He told the nurse that he wanted these investigations STAT (L. *statum*, at once).

The patient's white cell count was elevated (**leukocytosis**) and many *pneumococci* were seen in the sputum. The radiographs revealed an area of consolidation (**airless lung**) in the posterior part of the base of the right lung. There was also a slight shift of the heart and mediastinal structures to the right side.

A diagnosis of **pleurisy** or pleuritis (inflammation of the pleurae) caused by pneumococcal pneumonia or *pneumonitis* (inflammation of the lungs) was made.

Problems. What is the function of the pleurae?

Using your knowledge of the nerve supply to the pleurae, explain the referral of pain to the right side of the chest, periumbilical area, and right shoulder. Explain why a slight shift of the heart and mediastinal structures occurs with pneumonitis. *These problems are discussed on page 146.*

Case 1-8

During a lengthy trip in a car, a 38-year-old woman experienced a pressing **substernal discomfort**, pain in her right chest, and breathlessness (***dyspnea***). She said that she felt sick to her stomach (***nausea***) and that she was going to faint (**syncope**). Believing she may have been having a heart attack, her husband drove her to the closest hospital.

On physical examination, the doctor observed evidence of shock and rapid breathing (**tachypnea**). He also noted swollen, tender veins, particularly in her right thigh and calf in the area drained by the great saphenous vein (Fig. 4-11), which are signs and symptoms of **thrombophlebitis**.

On questioning he learned that she had painful **varicose veins** in her legs for some time, and that they became very painful during her recent long car ride. He also learned that she had been taking birth control pills for about 9 years.

Examination of her lungs revealed a few small, moist **atelectatic rales** (transitory, light crackling sounds) in the right side of her chest. ***Auscultation*** *also revealed a pleural rub on the right side.*

On cardiac examination the doctor detected **tachycardia** (rapid beating of the heart) and *arrhythmia* (irregularity of the heart beat). Her **ECG** suggested some right heart strain. Chest radiographs showed some increase in radiolucency of the right lung.

Fluoroscopy of the lungs *revealed poor or absent pulsations in the descending branch of the right pulmonary artery,* and relative anemia of the right lung which was consistent with the clinical impression of **pulmonary thromboembolism (PTE).**

Photoscans (scintigrams) were obtained after intravenous injection of radioactive iodinated (^{131}**I**) human albumin microparticles. The image produced on the *gamma camera* showed practically **no pulmonary blood flow in the right lung**.

To determine the exact site and size of the thrombus, ***pulmonary angiography*** was ordered. Radiopaque contrast material, injected into the right ventricle of the heart, revealed a **filling defect in the right main pulmonary artery** at its bifurcation in the right hilum. The left pulmonary artery and its branches were grossly normal.

As the patient was in critical condition and considered **in danger of an embolus to the remaining lung,** it was felt that immediate prophylactic, lifesaving surgery was absolutely necessary. Consequently, surgical *venous interruption* was performed to prevent recurrence of PTE by **plication of the inferior vena cava** (narrowing of the lumen by clips or sutures inferior to the renal veins.

Problems. Thinking anatomically, how do you think the radiologist injected the contrast material into the right ventricle of the heart? *What probably caused the patient's severe substernal discomfort and shoulder pain?* Following plication of the patient's inferior vena cava, how would adequate venous drainage of the patient's lower limbs be accomplished? *These problems are discussed on page 147.*

DISCUSSION OF PATIENT ORIENTED PROBLEMS

Case 1-1

Acute myocardial infarction is a disease of the myocardium, characterized by *necrosis of ventricular muscle* that results from sudden occlusion of a part of the coronary circulation (Fig. 1-65).

Blockage of the coronary circulation results in dysfunction of the heart as a pump. If a large branch of a coronary artery is involved, the infarcted area may be so extensive that cardiac function is severely disrupted and death occurs.

Myocardial infarction may also result from **excessive exertion** (*e.g.*, running to catch a train) by a person with stenotic coronary arteries. The straining heart muscle is using more oxygen than the stenotic arteries can supply. As a result, the tissue becomes **anoxic** and soon undergoes necrosis.

The sudden occlusion of a coronary artery by an ***embolus*** composed of a detached clot, or its more gradual obstruction by arterial disease or ***thrombosis***, is a common cause of death in persons 45 years and over. If obstruction of a coronary artery is incomplete, the patient often suffers from **angina pectoris,** a typical substernal pain that is initiated by exertion.

Coronary atherosclerosis (lipid deposits in the intima of the first 3 to 5 cm of a coronary artery) usually begins early in adult life. Gross evidence of this condition is almost always present in persons over 45 years of age.

An atheroma is a lipid deposit *that produces a swelling on the endothelial surface of the blood vessel.* Ulceration of the atheroma results in the release of **atheromatous debris** that is carried along the coronary artery until it reaches the stenotic part, usually where the vessel bifurcates, and then it stops. Because it blocks the vessel (*one type of coronary embolus*), no blood can pass to the myocardium and myocardial infarction occurs unless a good collateral circulation has developed previously.

Arteriolar anastomoses exist between the terminations of the right and left coronary arteries in the coronary *groove* (Fig. 1-68), and between the interventricular branches around the apex in about 10% of apparently normal hearts. Thus, an important factor in determining whether or not **ischemic heart disease** develops during coronary atherosclerosis is the presence or absence of these anastomoses. The potential for the development of this collateral circulation probably exists in most if not all hearts.

In very slow occlusion of a coronary artery, *the collateral circulation has time to increase* so that there will be adequate perfusion of the myocardium and infarction usually does not result. However, when there is sudden blockage of a large coronary branch (Fig. 1-65), some **myocardial infarction** results, but the extent of the area damaged depends on the degree of development of collateral channels that has occurred previously.

If large branches of both coronary arteries are partially obstructed, there is an **extracardiac collateral circulation** that may be utilized to supply blood to the heart. These collaterals connect the coronary arteries with the **vasa vasorum** in the tunica adventitia of the aorta and pulmonary arteries, and with branches of the internal thoracic, bronchial, and phrenic arteries. However, unless these collaterals have dilated in response to **pre-existent ischemic heart disease**, they are unlikely to be able to supply sufficient blood to the heart to prevent myocardial infarction.

*The dominant symptom of myocardial infarction is **deep visceral pain**.* Afferent pain fibers from the heart run centrally to the middle and inferior cervical branches and the thoracic branches of the sympathetic trunks of the thorax and neck (Fig. 1-81). Axons of these primary sensory neurons enter spinal cord segments T1 to T4 or T5 on the left side.

Pain of cardiac origin is referred to the left side of the chest and along the medial aspect of the arm and upper forearm (Fig. 1-64*A*), as these are the areas of the body which send sensory impulses to the same segments of the spinal cord that receive cardiac sensation (Fig. 1-23). Radiation of visceral pain to cutaneous areas is called "**referred pain.**"

Case 1-2

In view of the anatomical relations of the lung, some cancers of this organ extend directly into the thoracic wall, the diaphragm, or the mediastinum and its contents.

*Involvement of a phrenic nerve in the mediastinum results in **paralysis of half of the diaphragm**.* Direct infiltration of the pleura produces **pleural effusion** (escape of fluid from the pleural blood and lymphatic vessels) into the pleural cavity. This *pleural exudate* may be bloody (**sanguineous**) and *may contain exfoliated malignant cells.*

Because of the close relationship of the **recurrent laryngeal nerves** to the apices of the lungs, they may be involved in cancer involving this region of a lung. *This produces hoarseness by paralyzing the vocal folds.*

The left recurrent laryngeal nerve passes around the arch of the aorta to the left of the ***ligamentum arteriosum***, the adult derivative of the ductus arteriosus, and then passes superiorly. Although more superior, the right recurrent laryngeal nerve passes around the right subclavian artery and is closely related to the apex of the right lung and the cervical pleura (Fig. 1-63).

If tumors of the apices of the lungs invade locally, there may be involvement of the superior thoracic nerves, the thoracic sympathetic chain, and the stellate ganglion. If this occurs, there is likely to be pain in the shoulder and axilla and signs of the **Horner syndrome** (a drooping eyelid or ***ptosis miosis***, or pupillary constriction; ***anhidrosis*** or absence of sweating; and slight ***enophthalmos*** or recession of the eyeball).

Involvement of the hilar and mediastinal lymph nodes occurs by **lymphogenous dissemination**. The lymph vessels of the lungs originate in superficial and deep plexuses, accompanying small blood vessels. Lymph then drains into **bronchopulmonary lymph nodes** in the hilum (Fig. 1-40), which are often referred to clinically as *hilar nodes*. As these nodes enlarge, they increase the size of the hilum of the lung, giving it a lumpy appearance.

The bronchopulmonary lymph nodes drain into inferior and superior groups of tracheobronchial nodes that lie in the angles between the trachea and bronchi (Fig. 1-63). They form part of the mediastinal group of lymph nodes that are scattered throughout the mediastinum.

Clinically, the inferior group of **tracheobronchial nodes** are commonly referred to as *carinal nodes* because of their relationship to the **carina**, the ridge separating the right and left principal bronchi at their junction with the trachea (Fig. 1-11). Splaying and fixation of the carina of the trachea may be associated with **cancer of a bronchus** when it has metastasized to the carinal nodes. These abnormalities can be seen bronchoscopically and radiologically.

Enlarged mediastinal lymph nodes may indent the esophagus which can be observed radiologically as the patient swallows a barium sulfate emulsion.

As lymph from vessels in the costal parietal pleura reaches the **parasternal lymph nodes** (Fig. 1-16) via intercostal lymph vessels, lymphogenous metastatic spread of cancer may involve these nodes also.

Lymph from the entire right lung drains into tracheobronchial nodes on the right side, and most lymph from the left lung drains into nodes on the left side, but some lymph from the inferior lobe of the left lung also drains into nodes on the right side. Thus, tumor cells in tracheobronchial lymph nodes on the right side may spread by lymphogenous dissemination from the inferior lobe of the left lung.

The right and left bronchomediastinal trunks, drain lymph from the thoracic viscera and lymph nodes. The right bronchomediastinal trunk may join the **right lymphatic duct** and the left trunk may join the thoracic duct (Fig. 1-42), but more commonly they open independently into the *junction of the internal jugular and subclavian veins* of their own side.

Thus, lymph from the lungs and pleurae containing tumor cells soon enters the venous system and heart. After passing through the pulmonary circulation, the blood returns to the heart for distribution to the body.

Common sites of hematogenous metastasis from bronchogenic carcinoma are the brain, bones, lungs, and suprarenal glands.

Often the **medial supraclavicular lymph nodes,** particularly on the left side, are enlarged and hard because they are tumorous when there is carcinoma of the bronchus, stomach, or occasionally other abdominal organs. For this reason, these lymph nodes are often referred to as **sentinel nodes** because enlargement of them alerts the examiner to the possibility of malignant disease from the thoracic and/or abdominal organs.

The anatomical basis for involvement of *the sentinel lymph nodes* is that lymph passes cranially from the thoracic and abdominal viscera via the bronchomediastinal trunks and the **thoracic duct** to reach the venous system (Fig. 1-42).

Backflow of lymph from the thoracic duct can pass into the deep supraclavicular nodes, posterior to the sternocleidomastoid muscles. This is probably the reason why nodes on the left side are most commonly involved.

The brain is a common site for hematogenous spread of bronchogenic carcinoma. Tumor cells probably enter the blood through the wall of a capillary or venule in the lung and are transported to the brain via the internal carotid artery and vertebral artery systems (see Chap. 7). Once in the brain, the tumor cells probably pass between the endothelial cells of the capillaries and enter the brain.

Although most cancer cells from the lung are likely transported to the brain via the arterial system, others may be carried by the venous system. It has been suggested that constant coughing and **enlarged mediastinal lymph nodes compress the superior and inferior venae cavae,** causing the blood draining the bronchi to reverse its flow and pass via the bronchial veins into the azygos venous system (Fig. 1-80).

The azygos system of veins drains primarily the thoracic wall. From here, blood and tumor cells pass to the **extradural vertebral plexus of veins** around the spinal dura mater (Fig. 5-58). As this plexus communicates with the **cranial venous sinuses,** tumor cells can be transported to the brain when the patient is lying down. In this position, the normal negative pressure in the cranial dural venous sinuses becomes equal to the pressure in the vertebral plexus of veins. From the dural venous sinuses tumor cells pass into the cerebral veins and through their walls into the brain where they establish secondary tumors (**metastases**).

The passage of tumor cells to the veins of the vertebral column also explains the frequency of metastases of tumor cells to vertebrae.

Case 1-3

The knife, entering the fourth intercostal space at the left sternal border (Fig. 1-34*A*), did not penetrate the left lung because of the **cardiac notch** in its anterior border which begins at the fourth costal cartilage (Fig. 1-30*A*). The knife nicked the parietal layer of pleura of the left lung (Fig. 1-35), and then passed through the conus arteriosus of the right ventricle and the **aortic vestibule** of the left ventricle, immediately inferior to the aortic orifice (Figs. 1-55, 1-59, and 1-86).

Blood passed from the wounds in both ventricles into the ***pericardial sac.*** As blood accumulated in the pericardial cavity, severe compression of the heart (**cardiac tamponade**) and great veins occurred. This pressure increased until it exceeded the pressure in these large veins, preventing normal venous return to the heart and outflow of blood from the heart to the lungs and systemic circulation. This explains the patient's shock and gasping for breath prior to his death.

Case 1-4

ASD is a congenital abnormality because it is an imperfection in the interatrial septum that developed during embryonic life; thus it was present at birth (L. *congenitus*, born with).

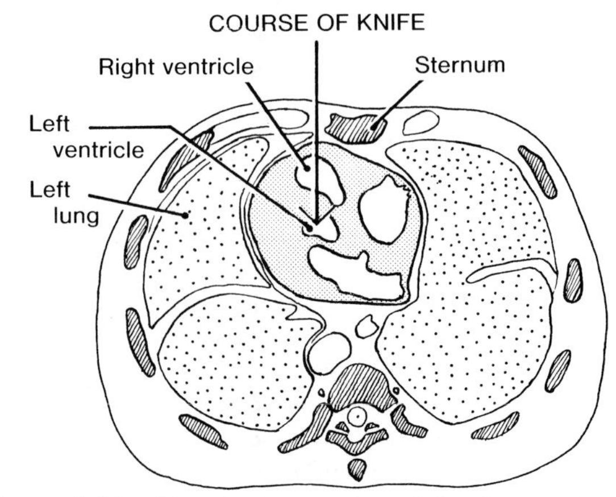

Figure 1-86. Transverse section of the thorax showing the course taken by the knife (*arrow*), passing to the left of the sternum in the fourth intercostal space, and entering the superior part of the right ventricle and then the left ventricle.

The common form of ASD is the secundum type, so classified because it results from abnormal development of the foramen ovale and septum secundum. The septum primum and the septum secundum normally fuse in such a way that no opening remains between the right and left atrium. The site of the prenatal opening, and where most defects occur, is represented by the **fossa ovalis** of the adult heart (Fig. 1-54).

Secundum type ASD is more common in females (2:1); the reason for this is not known. The ASD probably results from an abnormally short valve of the foramen ovale and/or an unusually large foramen ovale.

The **left to right shunt of blood** occurs because left atrial pressure exceeds that in the right atrium during the major part of the cardiac cycle. *Some of the patient's blood therefore makes two circuits through the lungs.* As a result of this shunting, the workload of the right ventricle is increased and its muscular wall hypertrophies.

The cavities of the right atrium, right ventricle, and pulmonary artery are dilated to accommodate the excess amount of blood in them.

Despite the increase in pulmonary blood flow, most ASD patients are asymptomatic in early life, although there may be some physical underdevelopment, as in the present case. After 40 years, many patients develop **pulmonary arterial hypertension** and usually complain of fatigue and dyspnea on exertion.

During the first 40 years, a majority of ASD patients have a moderate to good exercise tolerance, despite the fact that the opening in the interatrial septum is 2 to 4 cm in diameter. *Usually patients with ASDs do not exhibit* ***cyanosis*** (dark bluish coloration of the skin owing to deficient oxygenation of the blood).

Pulmonary vascular disease (arteriosclerosis) is likely to develop with increased pulmonary artery pressure, particularly if recurrent respiratory infections occur. Severe pulmonary hypertension may eventually result in higher pressure in the right atrium than in the left, reversing the shunt and causing cyanosis, severe disability, and heart failure.

Case 1-5

The ***ductus arteriosus*** *is a fetal vessel that connects the left pulmonary artery to the aortic arch,* just distal to the origin of the left subclavian artery. At birth the ductus arteriosus may be equal or larger in diameter than either the pulmonary artery or the aorta.

The antenatal function of the ductus arteriosus *is to allow most of the blood in the left pulmonary artery to bypass the uninflated lungs.* Because of the relatively high pulmonary vascular resistance to blood flow through the uninflated lungs, and the relatively low resistance in the embryonic thoracic and abdominal aorta and umbilical arteries, blood easily flows from the pulmonary artery into the arch of the aorta and the descending thoracic aorta. This shunting of blood in this way provides a more direct route for oxygenation of the fetal blood via the umbilical arteries to the placenta.

Patency of the ductus arteriosus after the perinatal period is a relatively common congenital abnormality, occurring about once in every 3,000 births, and more often in females than in males.

Patent ductus arteriosus is the most common malformation associated with ***maternal rubella*** infection during early pregnancy. Although this malformation occurs more frequently as an isolated abnormality, it may coexist with other malformations.

The typical continuous loud **"machinery-like"** murmur results from turbulent flow of blood from a high pressure vessel (aorta) to a low pressure vessel (pulmonary artery) via the ductus arteriosus. As the pressure gradient exists during both systole and diastole, the murmur is continuous. The left to right shunt increases the workload of the left ventricle; as a result, it enlarges and its walls thicken. The left atrium may also enlarge owing to the increased volume of blood returning from the lungs.

Patent ductus arteriosus may result in cardiac failure and pulmonary edema in the premature infant, but its presence is compatible with survival until adult life in most cases. However, as the leading cause of death in adults with this malformation is cardiac failure and/or **bacterial endocarditis** (inflammation of the endocardium of the heart, its valves and great vessels), ligation or division of the ductus arteriosus is commonly performed.

Pulmonary vascular disease (arteriosclerosis) may develop in a patient with a patent ductus arteriosus, in which case the high pulmonary vascular resistance results in an increase in pressure in the right ventricle and pulmonary artery. This causes a reversal of blood flow through the ductus (*i.e.*, right to left). Consequently, unoxygenated blood is shunted from the left pulmonary artery into the arch of the aorta and descending thoracic aorta.

As the ductus arteriosus enters the arch of the aorta distal to the origin of the left subclavian artery, the toes (but not the fingers) become **cyanotic** (bluish owing to oxygen deficiency) and **clubbed** (broadened and thickened). The finding of cyanosis in the toes, but not in the fingers, is referred to as **differential cyanosis**.

Case 1-6

As the right principal bronchus is wider, shorter, and more vertical than the left principal bronchus (because the trachea is displaced a little to the right by the arch of the aorta, Fig. 1-72), *foreign bodies more frequently pass into the right than into the left principal bronchus.* Foreign bodies that are commonly found are nuts, hardware, pins, crayons, and dental material (*e.g.*, part of a tooth).

The right middle and inferior lobes of the right lung are usually involved because (1) the right inferior lobe bronchus is in line with the right principal bronchus (Figs. 1-39 and 1-41), which is almost in line with the trachea, and (2) the foreign body often lodges in the inferior lobe bronchus, superior to the origin of the middle lobe bronchus.

When there is complete obstruction of a principal bronchus, the entire lung eventually collapses, becoming a nonaerated or *atelectatic lung.*

Airlessness of the lung owing to absorption of air from the alveoli, as in the present case, is called **atelectasis.** Collapse of the lung occurs when the gas (air, *i.e.*, oxygen and nitrogen) in the trapped part of the lung is absorbed over the next few hours into the blood perfusing the lung. Depending on the site of obstruction, collapse may involve an entire lung, a lobe, or a **bronchopulmonary segment.**

Because a collapsed lung is of soft tissue density, *atelectatic lungs, lobes, or segments appear as homogeneous dense shadows on radiographs,* in contrast to normal air-filled lung which is relatively lucent and appears dark on films.

When atelectasis of a sizeable segment of lung occurs, the heart and mediastinum are drawn toward the obstructed side and remain there during inspiration and expiration. The diaphragm on the normal side moves normally, whereas on the opposite side it moves much less.

Case 1-7

The pleurae are continuous with each other around and inferior to the root of the lung (Fig. 1-27). Normally, these layers are in contact during all phases of respiration. The potential space between them (**pleural cavity**) contains a capillary film of fluid. The visceral pleura normally slides smoothly on the parietal pleura during respiration, facilitating movement of the lung. When the pleurae are inflamed (**pleuritis**), the pleural surfaces become rough and the rubbing of their surfaces produces friction which is audible as a **pleural rub** during auscultation. Occasionally vibrations produced by rubbing of these roughened pleurae (**fremitus**) can be felt with the hand.

The pleural cavity is not usually visible on radiographs, but when air, fluid, pus, or blood collect between the visceral and parietal layers of pleura, the pleural cavity becomes apparent. If the inflammatory process in the present case had not been treated, an effusion or exudation of serum would have occurred from the blood vessels supplying the pleura.

Pleural exudate collects in the pleural cavity and is visible on radiographs as a more or less homogeneous density that obscures the normal markings of the lung. *Large* ***pleural effusions*** *are associated with a shift of the heart and mediastinal structures to the opposite side.*

If the inflamed pleurae become infected, pus accumulates in the pleural cavity (**empyema**). A small empyema may be drained by ***thoracentesis*** during which a wide bore needle is inserted posteriorly through the seventh intercostal space, along the superior border of the eighth rib.

The insertion of a needle close to the superior border of the rib avoids injury to the intercostal nerves and vessels (Fig. 1-22).

The parietal pleura, particularly its costal part, is very sensitive to pain, whereas the pulmonary pleura is insensitive. Afferent fibers from pain endings in the costal pleura and the pleura on the peripheral part of the diaphragm are conveyed through the thoracic wall as fine twigs of the intercostal nerves.

Irritation of nerve endings in the pleura resulting from rubbing of the inflamed pleurae together, particularly during inspiration, produces stabbing pain. Similarly, laughing or coughing may produce paroxysms (sharp spasms) of pain.

Referred pain from the pleura is felt in the thoracic and abdominal walls, the areas of skin innervated by the intercostal nerves (Figs. 1-15 and 2-16). Pain around the umbilicus is explained by the fact that the 10th intercostal nerve supplies the band of skin which includes the umbilicus (Fig. 1-23).

The mediastinal pleura and the pleura on the central part of the diaphragm are supplied by sensory fibers from the phrenic nerve (C3, **C4**, and C5). Irritation of these areas of pleura, as in the present case, stimulates the nerve endings in the pleura, resulting in pain being referred to the root of the neck and over the shoulder. *These areas of skin are supplied by the* ***supraclavicular nerves*** (C3, C4), derived from two of the segments of the spinal cord that give origin to the phrenic nerve.

Disease of the liver or gallbladder may also irritate the peritoneum covering the diaphragm. The resulting pain is felt in the inferior part of the chest if it originates in the periphery of the diaphragm because this area of peritoneum, like the costal pleura and the pleura on the peripheral part of the diaphragm, is supplied by sensory fibers from the inferior intercostal nerves (Fig. 1-87). However, if the central area of the diaphragm is affected, the pain is referred to the shoulder and root of the neck because the peritoneum and pleura related to the central part of the diaphragm are supplied by sensory branches of the phrenic nerves (Figs. 1-46 and 1-87).

The heart and mediastinal structures shift to the affected side in **pneumonitis** and occupy the space created by the slight loss of volume of the consolidated lung tissue resulting from the loss of air from alveoli. If **pleural effusion** or ***empyema*** occurs, the heart

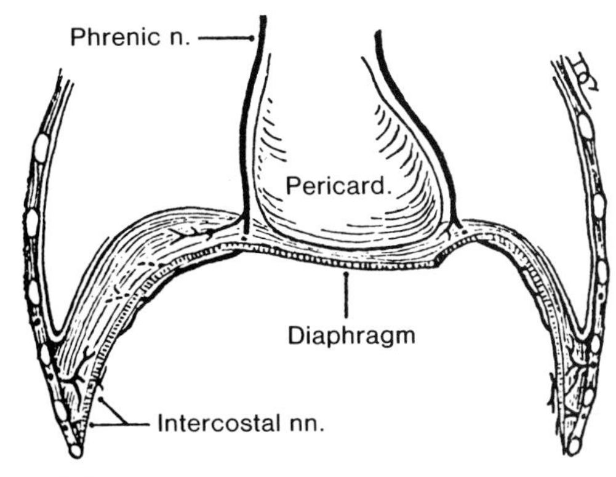

Figure 1-87. Diagram illustrating the nerve supply of the diaphragm. Parts of the pleura and peritoneum related to the diaphragm are supplied by sensory fibers of these nerves.

and mediastinal structures are pushed toward the opposite side by the accumulated serum or pus, respectively.

Case 1-8

Pulmonary thromboembolism (PTE) is an important cause of *morbidity* (sickness, disease) and *mortality* (death) in patients confined to bed, in pregnant patients, and in women who have been taking birth control pills for a long time.

In the present case, it is probable that *the thrombus was released from the great saphenous vein and transported to the pulmonary circulation,* where it caused complete or almost complete obstruction of the right pulmonary artery. This lead to various respiratory and hemodynamic disturbances (*e.g.*, dyspnea, tachypnea, arrhythmias, and tachycardia). In the opinion of many, *varicose veins rarely produce emboli.*

Three factors are involved in thrombus formation (thrombogenesis): stasis, abnormalities of the wall of the vessel, and alterations in the blood coagulation system. Various conditions are associated with thromboembolism (*e.g.*, pregnancy, fractures of the pelvis or lower limbs, abdominal operations, and the use of oral contraceptives.

Possibly the lengthy car trip with a safety belt around her abdomen and the birth control pills were contributory factors to thrombogenesis in the present case.

Likely the radiologist visualized the right ventricle by **right cardiac catheterization**. A radiopaque catheter was likely inserted into the left femoral vein just inferior to the inguinal ligament (Fig. 4-15). It would then have been guided with the aid of **fluoroscopy** into the inferior vena cava, right atrium, and right ventricle.

The patient's chest discomfort probably resulted from right heart strain and distention of the left pulmonary artery and its branches. As the right main pulmonary artery was obstructed, most of the blood was going through the left pulmonary artery. When the right lung failed to receive an adequate blood supply, changes very likely resulted in the pleura and pulmonary tissue. In view of the ***pleural rub***, there was likely some **pleuritic chest pain** caused by irritation of nerve endings of pain fibers in the costal pleura. The **referred pain** was felt in the thoracic wall, the area of skin innervated by the intercostal nerves (Figs. 1-15, 1-18, and 1-23).

When the inferior vena cava is narrowed by **plication** (L. *plica*, a fold), inferior to the renal veins, adequate venous drainage of the lower limbs is maintained through the collateral lumbar veins which connect with the **azygos system of veins** (Fig. 1-80).

Although sufficient to permit adequate venous return from inferior regions of the body, the plicated inferior vena cava and the collateral venous circulation does not prevent the passage of small emboli to the lungs. However, most of these pulmonary emboli are dissolved by the **fibrinolytic system** in about 2 weeks. In some cases, the thrombus is transformed into a linear or plate-like fibrous scar in the lungs.

Suggestions for Additional Reading

1. Maden RE: Cardiopulmonary surgery. In McCredie JA (ed): *Basic Surgery.* New York, Macmillan Publishing Co, Inc, 1977.

   This chapter presents a overview of intrathoracic surgery, describing how pulmonary operations are performed, and discusses infectious diseases of the lung (*e.g.*, bronchiectasis). There is also a concise account of tumors of the lung and diseases of the mediastinum. The most important congenital malformations of the heart are also discussed, as is acquired heart disease.

2. Moore KL: *The Developing Human: Clinically Oriented Embryology*, ed 3. Philadelphia, WB Saunders Company, 1982.

   The embryological bases of congenital malformations of the lower respiratory tract and of the heart and great vessels are fully discussed. Review of these accounts will amplify many of the points briefly referred to in the present text (*e.g.*, ASD and VSD).

3. Swartz MA, Moore ME: *Medical Emergency Manual Differential Diagnosis and Treatment*, ed. 3. Baltimore, Williams & Wilkins, 1983.

   This book presents clearly and succinctly the management of patients with chest pain. It explains how doctors evaluate a patient with chest pain and how they discriminate between minor causes of pain and those that are potentially life-threatening.

4. Ross RS, Lesch M, Braunwald E: Acute myocardial infarction. In Thorn GW, Adams RD, Braunwald E, Isselbacher KJ, Petersdorf G (eds): *Harrison's Principles of Internal Medicine*, ed 10. New York, McGraw-Hill Book Company, 1983.

   Myocardial infarction (MI) is one of several

dangerous causes of chest pain. If you read this brief account, you will know the classic characteristics of MI. You will also discover that chest pain is not always present with MI.

5. Squire LF, Colaiace WM, Strutynsky N: *Exercises in Diagnostic Radiology. 1. The Chest,* Philadelphia, WB Saunders Company, 1970.

   This book presents typical problems faced daily by radiologists. Although the exercises are designed for persons late in their medical training, they will give you some understanding of how radiological diagnoses are made.

CHAPTER 2

The Abdomen

Abdominal pain *is the most common presenting symptom associated with intraabdominal disease.* For this reason it is desirable to develop the habit of visualizing each abdominal organ in your "mind's eye."

Appendicitis (inflammation of the appendix) ranks high on the list of diseases leading to hospitalization and is a major cause of abdominal pain. However, any abdominal organ is subject to disease or injury.

Rupture of a hollow gastrointestinal (GI) organ *permits its irritating contents to escape into the peritoneal cavity,* producing **peritonitis** (inflammation of the peritoneum, the serous sac which lines the abdominal cavity and covers most of the viscera).

GI rupture may result in free air inferior to the diaphragm which can be observed in an upright abdominal radiograph.

Often a doctor's reputation depends on his/her ability to diagnose and treat the common causes of abdominal pain. Some patients requiring an abdominal operation have preoperative chest and/or abdominal radiographs taken because chest films may disclose an unsuspected cause for an abdominal pain (*e.g.*, **pneumonia**). Abdominal radiographs may indicate that rupture of an organ has occurred, or disclose a **calculus** (gallstone, urinary stone), or **fecalith** in the appendix (concretion formed from feces).

The abdomen is the region between the ***diaphragm*** *(thoracoabdominal diaphragm) and the pelvis.* The diaphragm forms a roof for the abdominal cavity; it has no floor. However, *the abdominal cavity is continuous with the pelvic cavity,* which has a floor (pelvic diaphragm).

The abdominal cavity extends superiorly, under the thoracic cage, to about the fifth anterior intercostal space when the person is supine. Hence, *a considerable part of the abdominal cavity lies under cover of the bony thoracic cage.* Parts of the liver, stomach, and spleen are in this location (Fig. 2–4).

The abdominal cavity is separated from the thoracic cavity by the diaphragm, and joins the pelvic cavity inferiorly at the plane of the ***superior pelvic aperture*** (Fig. 3–42*A*).

ANTEROLATERAL ABDOMINAL WALL

When a patient's abdomen is examined, the anterior abdominal wall is inspected and palpated. When the abdomen is operated on, the anterior abdominal wall is incised. Obviously a clear understanding of the structure of this wall is essential knowledge for every physician and surgeon.

The anterior abdominal wall consists of skin, subcutaneous tissue, muscles and fascia, extraperitoneal tissue, and peritoneum. *The layers of the abdominal wall have surgical importance.*

The abdominal wall expands slowly during pregnancy and obesity and when there is an abdominal tumor or an accumulation of fluid in the peritoneal cavity (**ascites**).

If you examine your anterior abdominal wall, you will observe that there are *cleavage lines* (Langer's lines) in the skin which run almost horizontally around the abdomen. Surgeons use these lines because an incision along or between them will heal better than one that crosses them. The reason for this is that the bundles of collagen in the dermis of the skin are arranged in rows and a surgical incision that crosses them disrupts them, resulting in the formation of fresh collagen that induces the formation of scar tissue. In some people a broad ugly scar results from the formation of a **keloid** (a mass of hyperplastic scar tissue).

BOUNDARIES OF ANTERIOR ABDOMINAL WALL

The anterior abdominal wall is bounded ***superiorly*** by the ***xiphoid process*** of the sternum and the ***costal cartilages of the 7th to 10th ribs.*** The anterior abdominal wall is bounded ***inferiorly*** (on each side) by the ***iliac crest,*** the ***anterior superior***

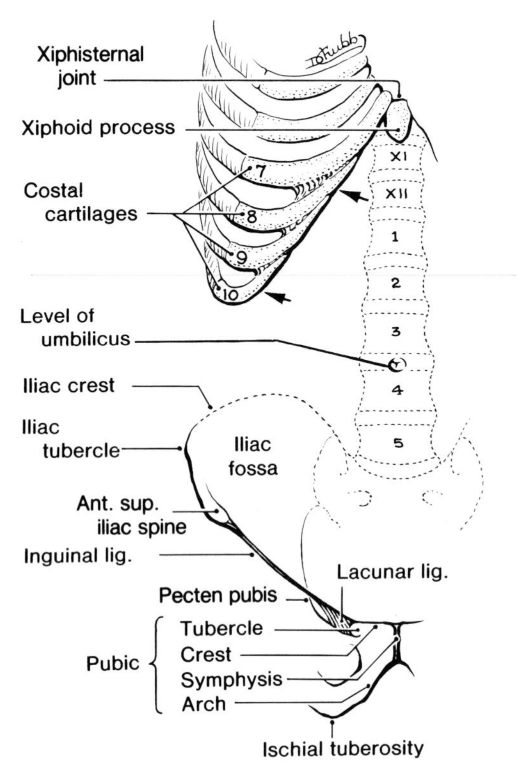

Figure 2-1. Drawing of the skeleton of the abdomen illustrating the boundaries and the bony landmarks of the anterior abdominal wall. Observe that it is bounded superiorly by the xiphoid process and the costal cartilages of the 7th to 10th ribs, and inferiorly on each side by the iliac crest, the anterior superior iliac spine, the inguinal ligament, the pubic tubercle, the pubic crest, and the symphysis pubis (pubic symphysis). The costal margin (*arrows*) is formed by the 7th to 10th costal cartilages. Together the two costal margins form the costal arch. Note that the most superior point of the iliac crest is at the level of the fourth lumbar vertebra.

iliac spine, the ***inguinal ligament,*** the ***pubic tubercle,*** the ***pubic crest,*** and the ***pubic symphysis*** (Figs. 2-1 and 2-2).

The inguinal ligament (Figs. 2-1, 2-2, and 2-9) is the inferior edge of the *aponeurosis of the external oblique muscle,* the most superficial muscle of the anterior abdominal wall. *The inguinal ligament extends from the pubic tubercle to the anterior superior iliac spine.*

The inguinal fold of skin (Fig. 2-2), overlying the inguinal ligament, indicates the separation between the anterior abdominal wall and the anterior aspect of the thigh. In obese people, the distended abdomen overhangs and conceals the inguinal fold.

SURFACE ANATOMY OF ANTERIOR ABDOMINAL WALL

The ***umbilicus*** (navel, belly-button) is the most obvious surface marking on the anterior abdominal wall of most people. This puckered scar represents the *former site of attachment of the* ***umbilical cord*** *in the fetus.*

The position of the umbilicus varies somewhat, depending on such factors as the degree of obesity, the tone of the anterior abdominal muscles, and the distention of the abdomen. **The position of the umbilicus varies considerably in obese people,** depending on whether they are in the supine or erect position. Furthermore, the umbilicus tends to be more inferior in children and old persons. *In most people the umbilicus lies at the level of the intervertebral disc between the third and fourth lumbar vertebrae* (Figs. 2-1 and 2-4).

The position of the linea alba (L. white line) is indicated by a slight groove or furrow in the anterior median line, which is particularly obvious in thin muscular persons (Fig. 2-3). This groove in the skin extends from the *xiphoid process* of the sternum to the *symphysis pubis.* The linea alba can not be seen as a white line until the skin is reflected in dissection (Figs. 2-8 to 2-10).

The lineae semilunares are slight surface depressions which sweep in gentle curves, convex laterally, from the ninth costal cartilages to the pubic tubercles (Figs. 2-2 and 2-3).

The lineae semilunares indicate the lateral borders of the rectus abdominis muscles (Fig. 2-10*A*). To identify your lineae semilunares, lie on your back and then sit up without using your upper limbs. This contracts your rectus abdominis muscles, making their lateral borders stand out. These lines are easily visible in persons with good abdominal muscular development (Fig. 2-3).

There are usually three transverse grooves in the skin overlying the ***tendinous intersections*** *of the rectus abdominis muscles* (Figs. 2-3 and 2-7). They extend laterally from the linea alba.

The inguinal ligament, indicated by the *inguinal fold* (Fig. 2-2), extends from the anterior superior iliac spine to the pubic tubercle (Figs. 2-1, 2-2, and 2-9). The inguinal ligament may be felt along its length.

The pubic tubercle may be felt about 2.5 cm lateral to the symphysis pubis (Figs. 2-1 and 2-2).

To palpate your inguinal ligament, lie on a table in the supine position and let one of your lower limbs drop to the floor. You can easily feel your inguinal ligament at its medial end, where it attaches to the pubic tubercle, and you may be able to see it separating your anterior abdominal wall from your thigh.

The symphysis pubis (pubic symphysis) is felt as a firm resistance in the anterior midline at the inferior extremity of the anterior abdominal wall (Figs. 2-1, 2-2, and 2-9). In many people the fat covering this cartilaginous joint makes it difficult to palpate.

The pubic crest is felt extending laterally from the symphysis pubis for about 2.5 cm (Fig. 2-1). The crest

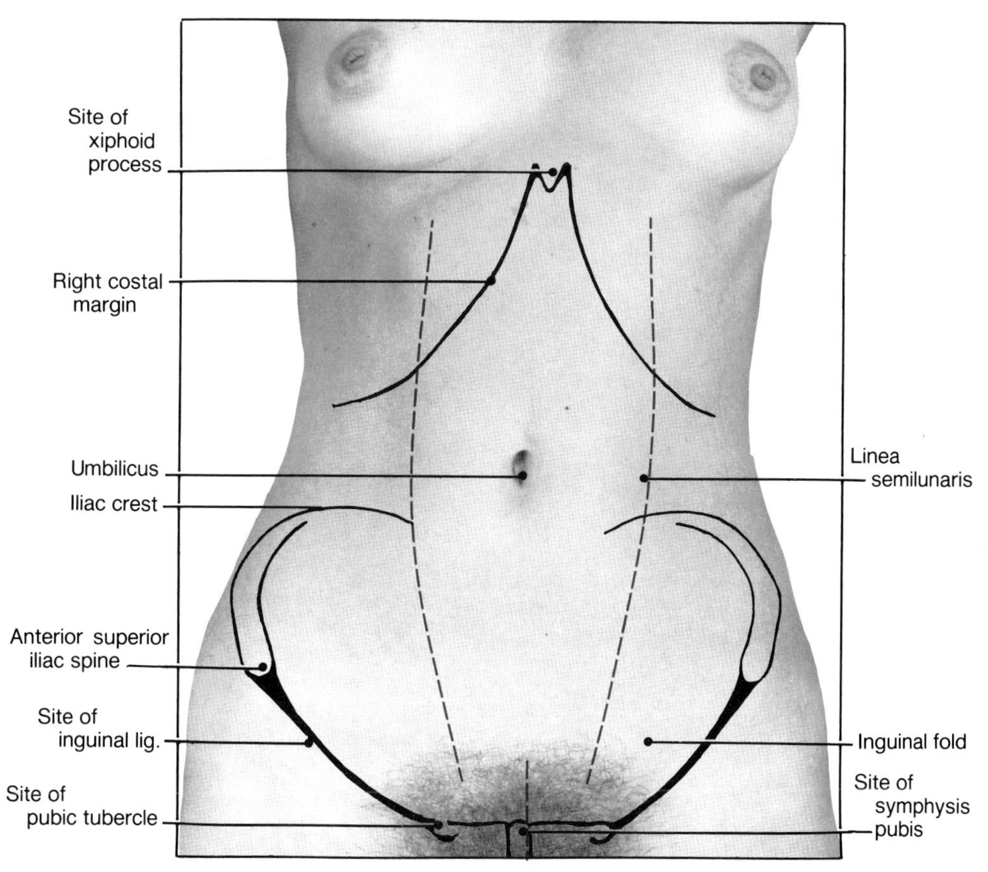

Figure 2-2. Photograph of a 27-year-old woman showing the surface features of her anterior abdominal wall. The costal margin is formed by the upturned costal cartilages of ribs 7 to 10. The symphysis pubis marks the inferior limit of the anterior abdominal wall in the median plane. The inguinal ligament stretches from the pubic tubercle to the anterior superior iliac spine. The shape of the abdomen varies considerably in persons of both sexes and in the same person between the erect and supine positions. The surface features recognizable in the above woman may be difficult or impossible to observe or palpate in obese persons. Observe the lateral borders of her rectus abdominis muscle (lineae semilunares; also see, Fig. 2-10*A*).

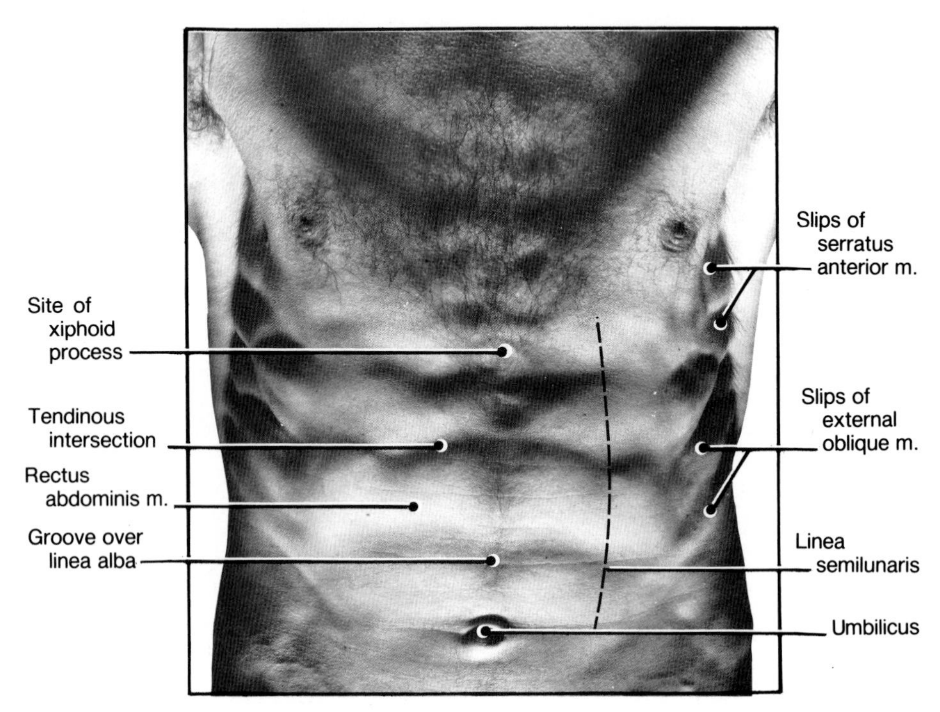

Figure 2-3. Photograph of the anterior abdominal wall of a 46-year-old man showing the surface features. One of the three tendinous intersections of the rectus abdominis muscle is indicated. Three or more of these intersections are usually present (Fig. 2–7). The lateral borders of the rectus abdominis muscles form slight surface depressions, called the linea semilunares (Fig. 2–10*A*).

terminates at the pubic tubercle to which the inguinal ligament attaches.

The iliac crest (Figs. 2–1 and 2–2) can be easily felt throughout its length, extending posteriorly from the anterior superior iliac spine. Laterally it forms the superior end of the hip region (Fig. 4–7).

The iliac tubercle (Figs. 2–1 and 4–1) can be palpated about 6 cm posterior to the anterior superior iliac spine.

The epigastric fossa ("pit of stomach") is a slight depression just inferior to the xiphoid process of the sternum. It is particularly noticeable when a person is in the supine position because the viscera move posteriorly and laterally, drawing the anterior abdominal wall inward in this region.

PLANES AND POINTS OF REFERENCE FOR THE ABDOMEN

The abdomen is divided into **nine regions** *by two vertical and two horizontal planes* (Figs. 2–4*A* and 2–5*A*). These regions are helpful for the localization of a pain or a swelling and for indicating the location of deep structures (*e.g.*, the duodenum).

You are urged to use the vertebral column as a scale and to refer structures to their vertebral levels. Certain horizontal planes are reliable guides to vertebral levels.

The Horizontal Planes (Fig. 2–4). Of the various horizontal planes to be described, *the transpyloric, transtubercular, and supracristal planes are the ones that are commonly used clinically as landmarks* when examining the abdomen. The names of these planes are printed in boldface type in the following descriptions.

The transpyloric plane (Fig. 2–4*B*) is a horizontal plane situated midway between the **jugular notch** of the sternum (Fig. 1–8) and the **symphysis pubis.** In most people this plane is *approximately* midway between the xiphisternal joint and the umbilicus.

The transpyloric plane runs through the level of the first lumbar vertebra (L1). It usually passes through the **pylorus** of the stomach (Fig. 2–4*B*). The transpyloric plane also passes through the tips of the ninth costal cartilages, the duodenojejunal junction, the neck of the pancreas, and the hila of the kidneys (Fig. 2–61).

The subcostal plane (Fig. 2–4*A*) joins the most inferior point of the costal margin on each side. Hence, *the subcostal plane indicates the inferior margins of the 10th costal cartilages* (Fig. 1–1). The subcostal plane also indicates the level of L3 vertebrae (Figs. 2–1 and 2–4*A*).

The transumbilical plane (Fig. 2–4*B*) passes through the umbilicus. *In persons with a reasonably firm anterior abdominal wall, the transumbilical plane indicates the level of the intervertebral disc between L3 and L4 vertebrae.*

The transtubercular plane (Fig. 2–4*A*) *passes through the iliac tubercles and lies at the level of the body of L5 vertebra.*

The interspinous plane passes through the anterior superior iliac spines and the promontory of the sacrum (Fig. 5–24*A*).

The supracristal plane (Fig. 2–4*B*) passes between the most superior points of the iliac crests and *passes through L4 vertebra.* This plane is generally used as a landmark on the posterior surface of the body for identifying the vertebral spinous processes.

The plane of the fifth intercostal space (anteriorly) gives an approximate indication of the level of the superior limit of the abdomen and the dome of the diaphragm (Fig. 2–35) when a person is supine. *The fifth space lies at the level of T10 or T11 vertebrae.*

The Vertical Planes (Fig. 2–4*A*). The two vertical planes used to divide the abdomen into regions are the **midclavicular lines.** *They extend inferiorly from the midpoints of the clavicles to the midinguinal points* (the midpoints of the lines joining the anterior superior iliac spines and the symphysis pubis).

QUADRANTS OF THE ABDOMEN

A simple and commonly used clinical method of dividing the abdomen for descriptive purposes is to divide it into **four quadrants** (Fig. 2–5*B*), using the median and the transumbilical planes. These quadrants are useful to clinicians when describing the location of a pain or a tumor (*e.g.*, the pain of acute **appendicitis** usually starts in the periumbilical region and later *localizes in the right lower quadrant*).

REGIONS OF THE ABDOMEN

For more detailed localization of pains, swellings, or the positions of organs, the abdomen is divided into regions, as shown in Figure 2–5*A*.

Nine abdominal regions are mapped out using two horizontal planes (subcostal and transtubercular planes) and two vertical planes, right and left midclavicular lines. The commonly used regions are: **the epigastric region** (G. *epi*, upon + *gastēr*, stomach); right and left **hypochondriac regions** (G. *hypo*, under + *chondros*, cartilage); **the umbilical region**; right and left **lumbar regions**; **the hypogastric region** (G. *hypo*, under + *gastēr*, stomach) or suprapubic region; and right and left iliac or **inguinal regions**.

The various organs in the abdomen can be related to the abdominal regions. For example, *most of the liver lies in the right hypochondriac and epigastric regions* (Figs. 2–4 and 2–5), and

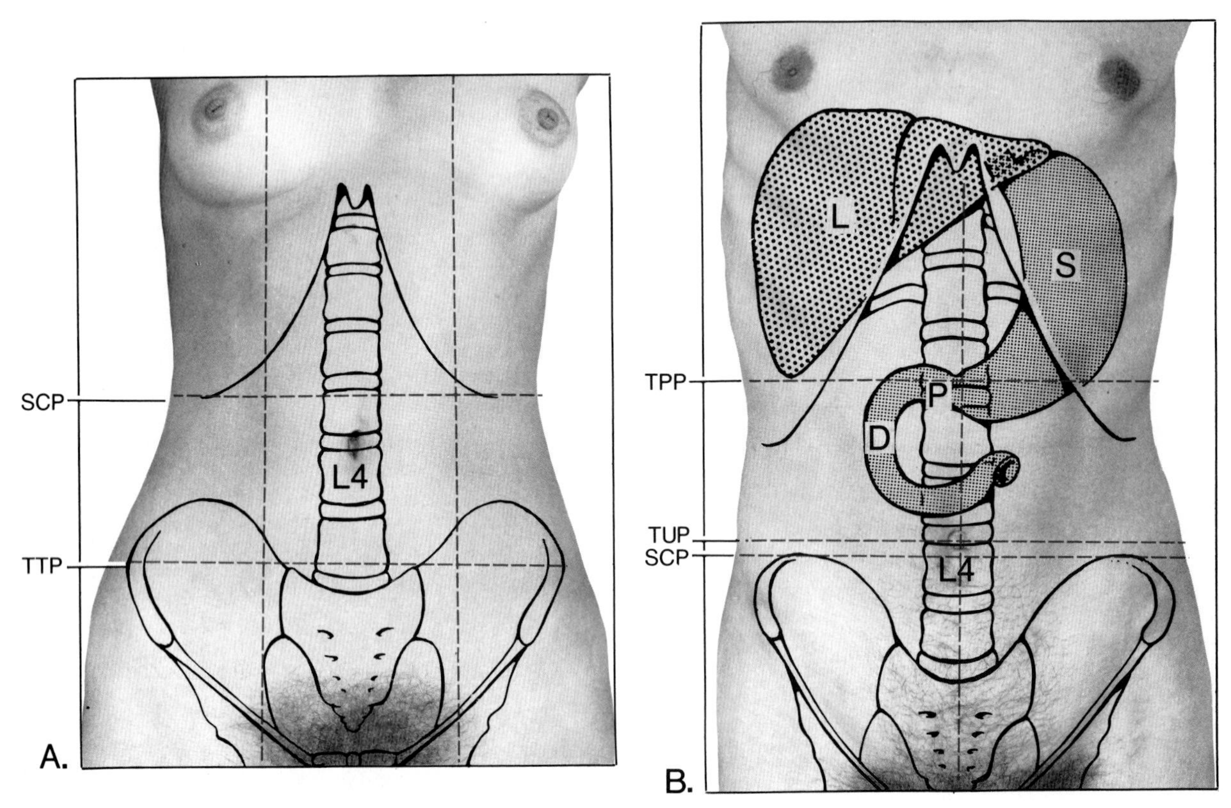

Figure 2-4. Photographs illustrating the common planes (*red*) and points of reference for the abdomen. *A*, shows the lines and planes used for dividing the abdomen into the nine regions shown in Figure 2-5*A*. The vertical planes are the midclavicular lines which pass through the midpoints of the clavicles. The horizontal lines are the subcostal plane (*SCP*) joining the most inferior points of the costal margins and passing through L3 vertebra, and the transtubercular plane (*TTP*) joining the tubercles of the iliac crests (Fig. 2-1) and passing through L5 vertebra. *B*, shows three other horizontal planes and their relationship to the vertebral column. Liver (*L*), stomach (*S*), and duodenum (*D*). The transpyloric plane (*TPP*) usually passes through the pylorus (*P*) of the stomach and L1 vertebra. The transumbilical plane (*TUP*) passes through the umbilicus and indicates the level of the intervertebral disc between L3 and L4 vertebrae. The supracristal plane (SCP) passes through the most superior points of the iliac crests and L4 vertebra.

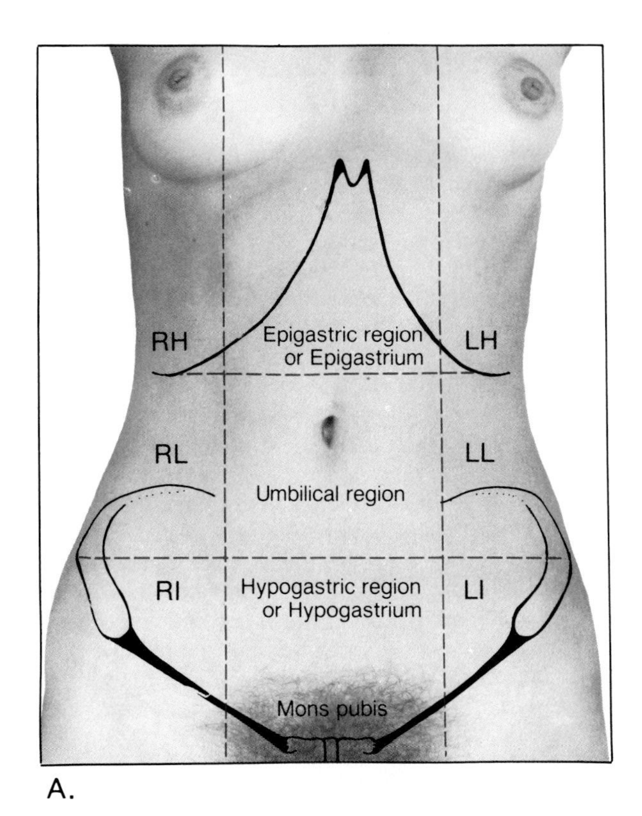

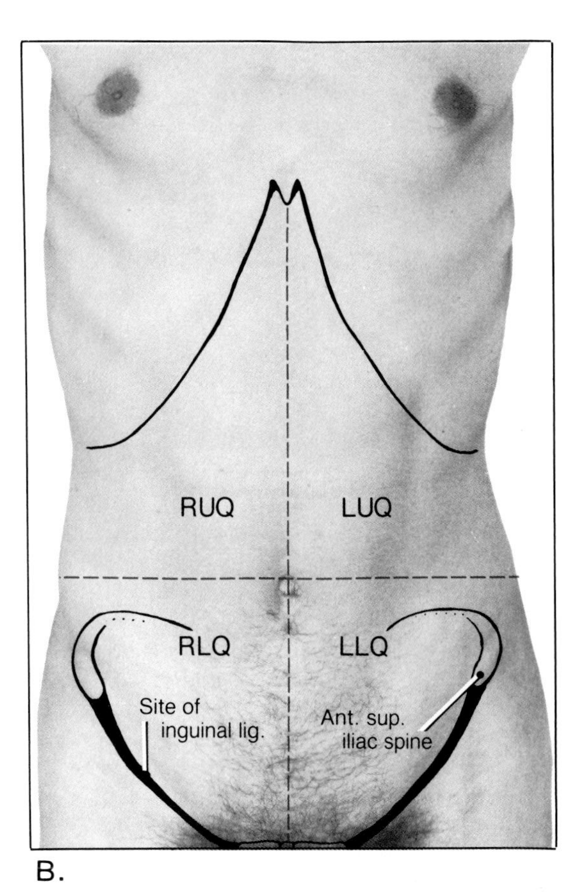

Figure 2-5. Photographs of a 27-year-old female and a 27-year-old male, illustrating the common clinical methods of dividing the abdomen for descriptive purposes. *A*, for accurate localization of a finding, the abdomen is divided into nine regions. The subcostal and transtubercular planes are the two horizontal planes and the midclavicular planes are the two vertical planes. Terms for three of these regions are commonly used; epigastric, umbilical, and hypogastric. The inferior part of the hypogastric region, just superior to the mons pubis, is often referred to as the suprapubic region. The other regions are: *LH*, left hypochondriac, *RH*, right hypochondriac, *LL*, left lumbar, *RL*, right lumbar, *RI*, right inguinal, or iliac, *LI*, left inguinal or iliac. Some of these regions are referred to in two ways: epigastric region or epigastrium; hypochondriac region or hypochondrium; and hypogastric region or hypogastrium. *B*, for general localization of a finding, the abdomen is divided into four quadrants (right upper, right lower, left upper, and left lower) by imaginary lines passing through the umbilicus. The vertical line is in the median plane and the horizontal line is in the transumbilical plane.

the body and fundus of the stomach are situated in the left hypochondriac region.

Similarly if one speaks of right inguinal or iliac pain, other physicians regardless of their country of origin will know where the pain is located.

Every physician knows that pain in the umbilical region followed by vomiting are early symptoms of appendicitis. Later the patient describes **right lower quadrant pain.** Hence one must know the abdominal regions to describe the symptoms.

CONTOUR OF THE ABDOMEN (Figs. 2–2, 2–3, and 2–5)

In persons with good abdominal muscular development, the contour of the abdomen is flat from its superior to inferior ends and is evenly rounded from side to side. The contour varies in other people.

Protuberance of the abdomen is normal in infants and young children because their GI tracts contain considerable amounts of air. Also their abdominal cavities are still enlarging and their abdominal muscles are gaining strength. The child's liver is also relatively large which accounts for some of the abdominal protrusion.

Protrusion of the abdomen in women occurs during pregnancy owing to the ***fetus***, and expands in both sexes following the deposition of fat or the accumulation of feces, fluid, or flatus (gas in the GI tract). Note that five common causes of abdominal protrusion all begin with the letter "**f**." During old age, abdominal muscle laxity also contributes to the abdominal protuberance.

Abdominal or pelvic tumors and accumulation of fluid in the peritoneal cavity (**ascites**) also result in abdominal enlargement. As this happens the anterior muscles thin out, but the skin grows and the nerves and blood vessels lengthen. Usually the umbilicus is depressed, hence one that is protruding may indicate the presence of ascites.

Reddish elongate lines called **striae gravidarum** (L. *gravidus*, heavy) may appear in the anterolateral abdominal skin of pregnant women. These striae gradually change into thin, white, scar-like lines called **lineae albicantes** ("stretch marks"). They also appear in the skin of the abdomens and thighs of obese men and women.

When examining the abdomen, the hands should be warm so that the patient's muscles will not become tense. To relax the anterior abdominal wall, patients are examined while lying in the supine position with their thighs semiflexed by a pillow under their knees. If this is not done, the deep fascia of the thighs pulls on the deep layer of superficial fascia of the abdomen and tenses the anterior abdominal wall.

FASCIA OF ABDOMINAL WALL

The fascia of the anterior abdominal wall consists of superficial and deep layers, but the deep layer is unremarkable.

The Superficial Fascia. Over the greater part of the anterior abdominal wall, the superficial fascia consists of one layer containing a variable amount of fat. *In some persons the fat in the superficial fascia is several inches thick* and can be separated from the thin deep fascia covering the external oblique muscle.

The superficial fascia of the abdominal wall just superior to the inguinal ligament may be divided into two layers: (1) **a fatty superficial layer** (Camper's fascia) containing a variable amount of fat, and (2) **a membranous deep layer** (Scarpa's fascia) containing fibrous tissue and very little fat. The superficial layer of fascia is adherent with the superficial fascia of the thigh, and the deep layer is continuous with the deep fascia of the thigh, called the **fascia lata** (Fig. 4–16). The deep layer of fascia is also continuous with the superficial perineal fascia (Colle's fascia), and with that investing the scrotum and penis (Fig. 3–8). The deep layer of superficial fascia fuses with the deep fascia of the abdomen.

Surgeons often use the membranous deep layer of the superficial fascia for holding sutures during closure of an abdominal skin incision.

Between the deep layer of the superficial fascia and the deep fascia of the abdomen, there is a potential space in which fluid may accumulate. For example, urine may pass through a rupture in the spongy urethra (Fig. 2–102), called **extravasation of urine,** into this space and extend superiorly into the anterior abdominal wall between the deep layer of the superficial fascia and the deep fascia.

Falling astride a picket fence or a steel beam and car accidents are common causes of a ruptured urethra. (See Case 3–2 at the end of Chap. 3, p. 391).

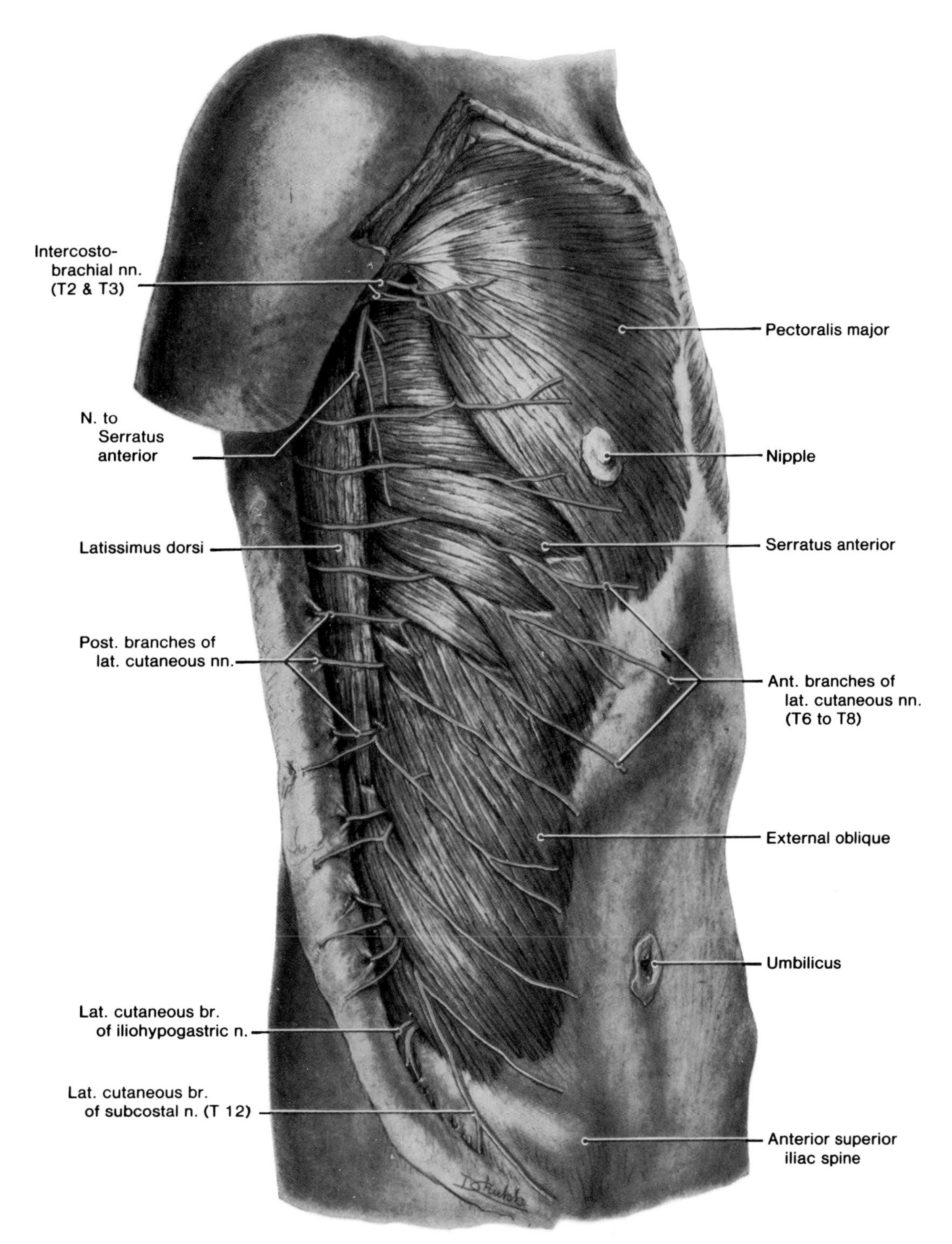

Figure 2-6. Drawing of a lateral view of a superficial dissection of the trunk showing the serratus anterior and external oblique muscles and the lateral cutaneous nerves. Observe that the fleshy fibers of the external oblique originate superior to the costal margin (Fig. 2-1). Note that the superior four digitations of the external oblique interdigitate with the serratus anterior, and that the inferior four digitations interdigitate with the latissimus dorsi muscle. Between the digitations of the external oblique, observe the lateral cutaneous nerves (T7 to T12). The posterior branches of these nerves turn posteriorly over the latissimus dorsi, and the anterior branches descend in line with the fibers of the external oblique. See Figure 1-14 for an illustration of the surface anatomy of the region that is dissected in this drawing.

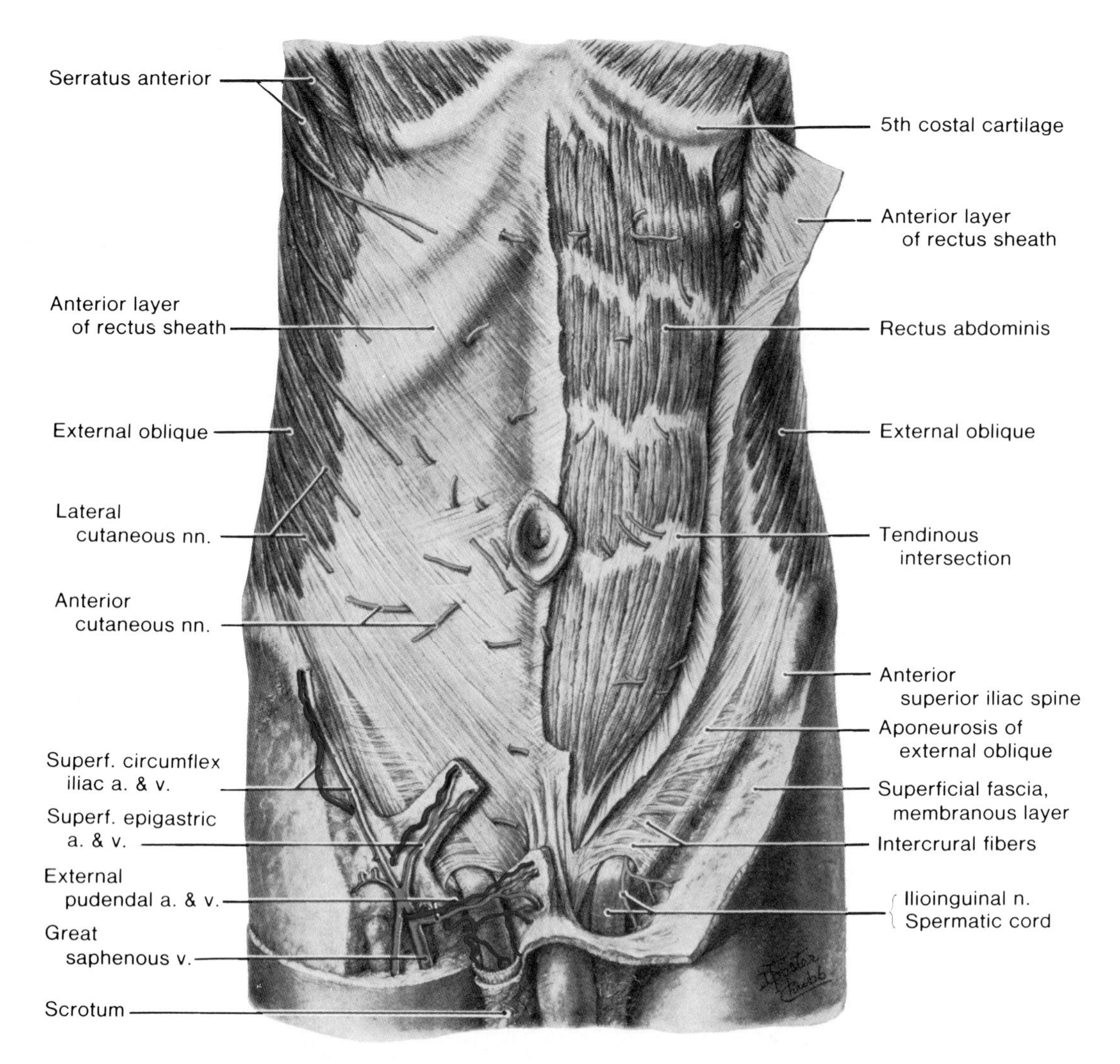

Figure 2-7. Drawing of a dissection of the anterolateral abdominal wall. The anterior layer of the rectus sheath is reflected on the left side. Observe that the external oblique muscle is aponeurotic medial to a line that curves superiorly from a point 2.5 cm lateral to the anterior superior iliac spine to the fifth rib. Observe the anterior cutaneous nerves (T7 to T12) piercing the rectus abdominis muscle and the anterior layer of the rectus sheath. T10 supplies the region of the umbilicus (Fig. 2–16). Note that the external pudendal artery and vein cross the spermatic cord. Observe that the membranous deep layer of the superficial fascia blends with the fascia lata of the thigh, a fingerbreadth inferior to the inguinal ligament.

The Deep Fascia. Little can be said about the deep fascia of the anterior abdominal wall, except that it forms a thin layer over the external oblique muscle.

MUSCLES OF ANTEROLATERAL ABDOMINAL WALL

Four muscles form parts of the anterolateral abdominal wall (Figs. 2–3 and 2–6 to 2–15): **three flat muscles** (external oblique, internal oblique, and transversus abdominis), and **one straplike muscle** (rectus abdominis).

The combination of muscles and aponeuroses (sheet-like tendons) in the anterolateral abdominal wall affords considerable protection to the abdominal viscera, especially when the muscles are in good physical condition. The flat muscles cross each other in such a way (somewhat like a three-plied corset) that strengthens the anterolateral abdominal wall and diminishes the risk of hernial protrusion between separated muscle bundles.

The three flat abdominal muscles are: the *external oblique, internal oblique, and transversus abdominis.* Anteriorly the rectus abdominis muscle and its sheath contribute to the anterior abdominal wall (Figs. 2–7 and 2–8).

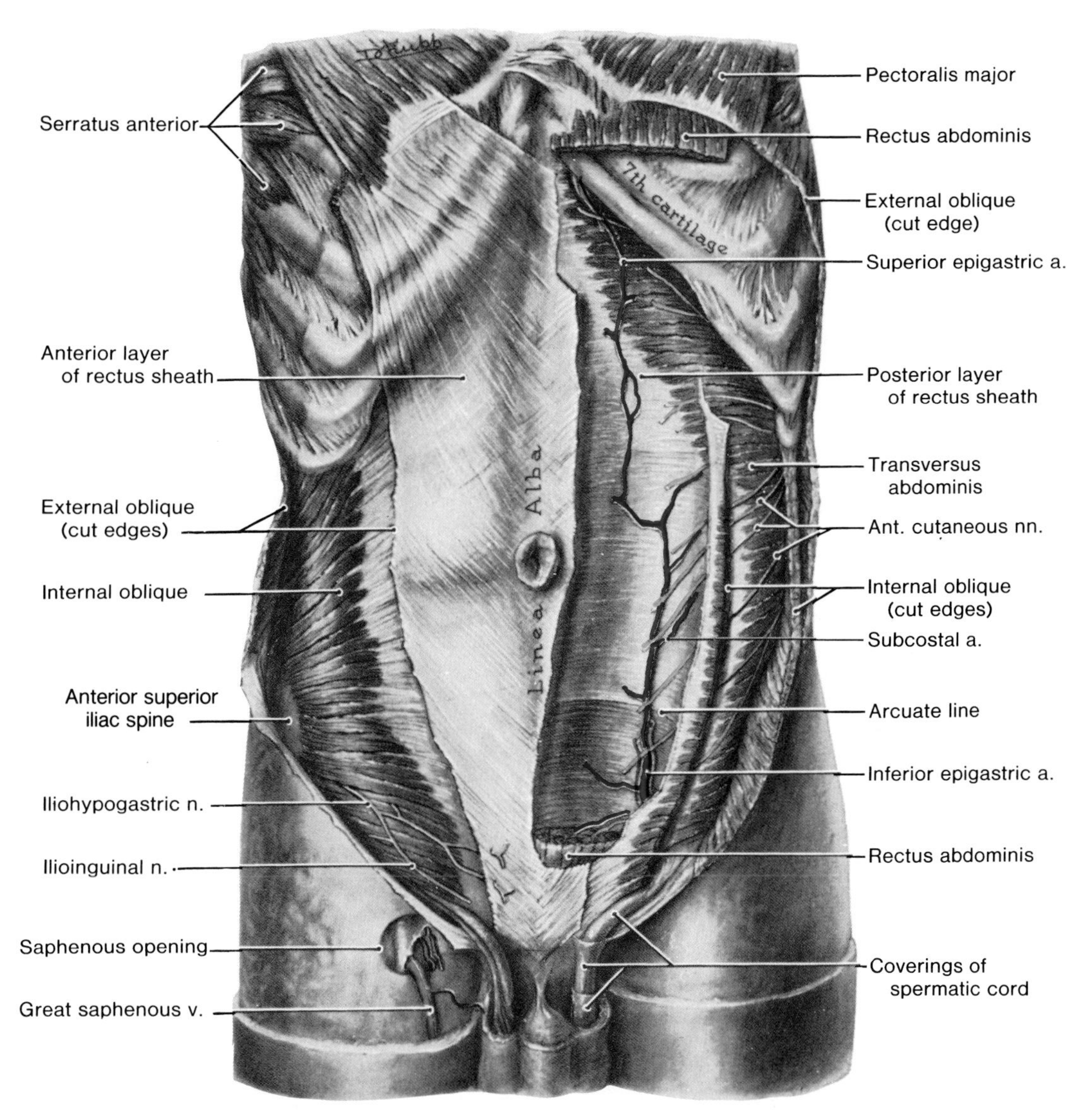

Figure 2-8. Drawing of a deeper dissection than that in Figure 2-7. Most of the external oblique muscle is excised on the right side. The rectus abdominis is excised on the left side and the internal oblique is divided. *Note the anastomosis between the superior and inferior epigastric arteries* which indirectly unites the arteries of the upper limb to those of the lower limb. Observe that nerves T7 to T12, but not L1, enter the rectus sheath.

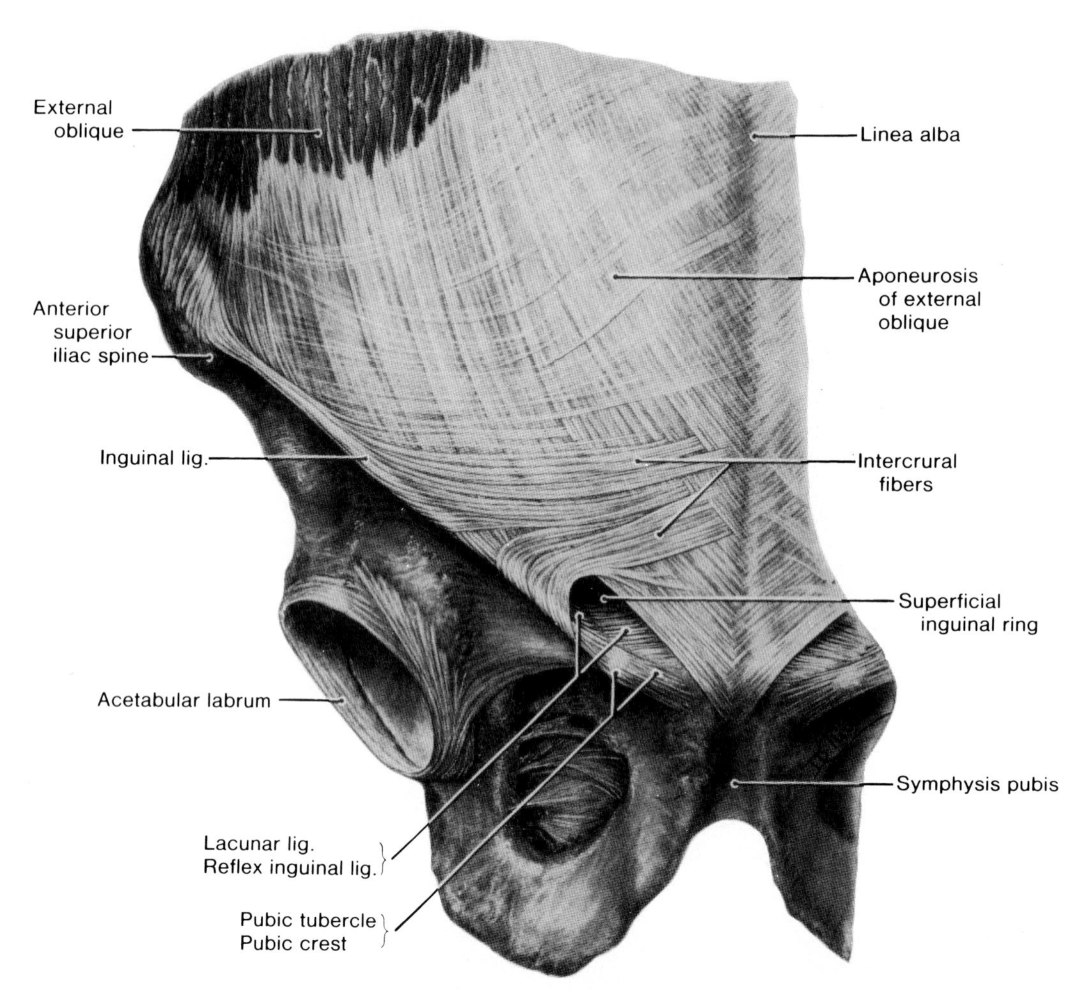

Figure 2-9. Drawing of a superficial dissection of the inguinal region showing the aponeurosis of the external oblique and the superficial inguinal ring. Observe the linea alba, a ligamentous raphe uniting the xiphoid process to the symphysis pubis. Note that the intercrural fibers which prevent the crura of the superficial inguinal ring from spreading. Observe that the superficial inguinal ring is triangular in shape and that: (1) its central point is superior to the pubic tubercle, (2) its base is the lateral half of the pubic crest, (3) its lateral crus is the inguinal ligament, and (4) its medial crus is formed by fibers of the external oblique aponeurosis that cross the pubic crest at its midpoint.

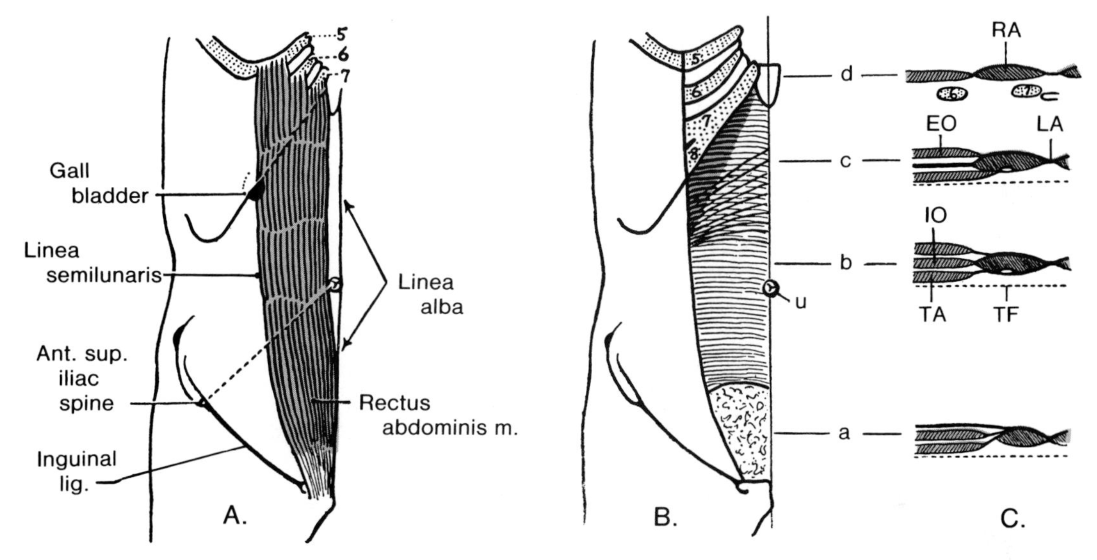

Figure 2-10. Drawings illustrating the rectus abdominis muscle and the rectus sheath. *A*, anterior view of the right part of the rectus abdominis muscle. *B*, the posterior wall of the rectus sheath. *C*, transverse sections of the rectus sheath at four levels. *EO*, external oblique; *IO*, internal oblique; *TA*, transversus abdominis; *RA*, rectus abdominis; *TF*, transversalis fascia. *LA*, linea alba; *U*, umbilicus. The linea alba is a fibrous raphe stretching from the xiphoid process of the sternum to the symphysis pubis (also see Figs. 2-8 and 2-9). It forms the central anterior attachment for the muscles of the abdomen. In *C*, note that the linea alba (*LA*) is formed by the interlacing fibers of the aponeuroses of the right and left oblique and transversus abdominis muscles. Also, observe that the rectus sheath is formed by the aponeuroses of the three flat abdominal muscles which surround the rectus abdominis.

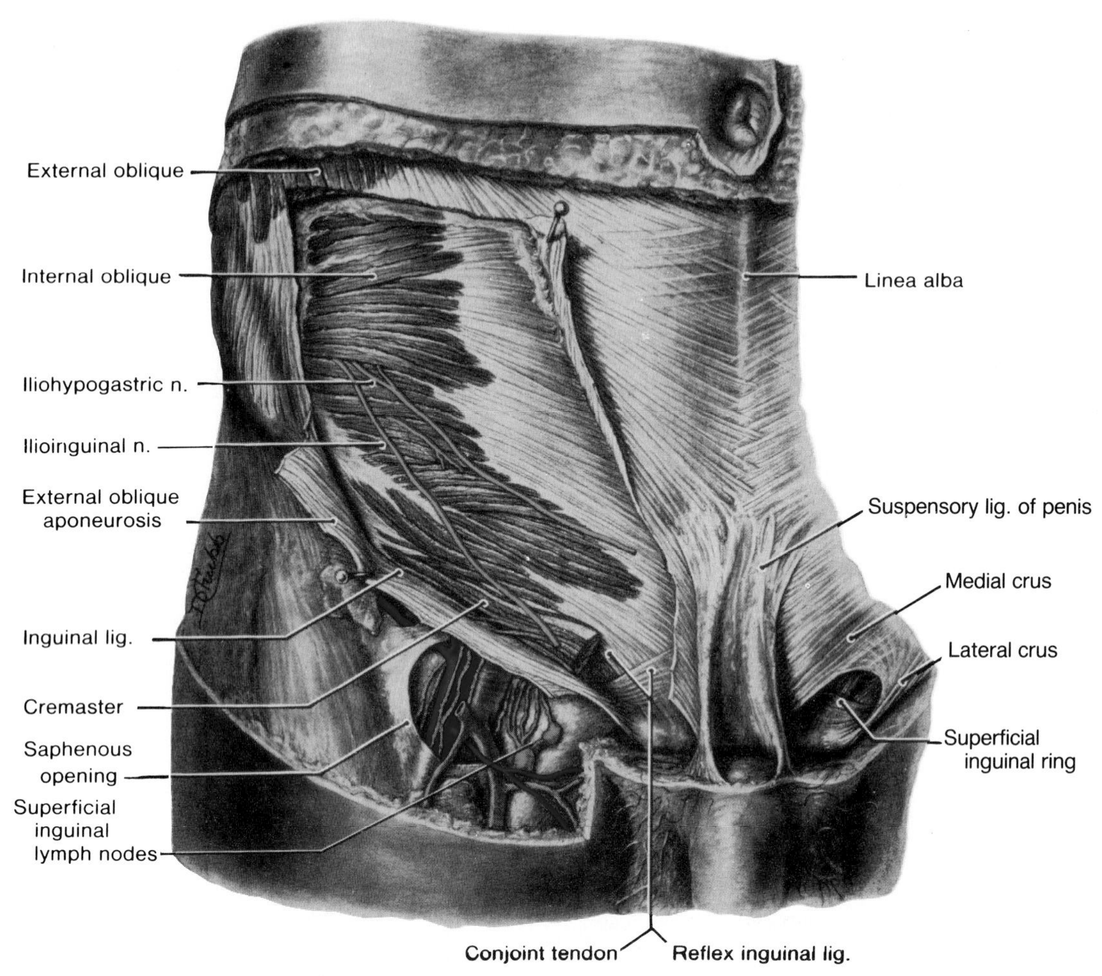

Figure 2-11. Drawing of a dissection of the inguinal region. The external oblique aponeurosis is partly cut away and the spermatic cord is cut short. Observe the laminated, fundiform ligament of the penis descending to the junction of the fixed and mobile parts of the organ. Examine the reflex inguinal ligament which represents the external oblique, lying anterior to the conjoint tendon which represents the internal oblique and transversus abdominis muscles. Note that only two structures course between the external and internal oblique muscles, namely the iliohypogastric and ilioinguinal branches of the first lumbar nerve. They are sensory from this point to their terminations. Observe that the fleshy fibers of the internal oblique at the level of the anterior superior iliac spine run horizontally; those from the iliac crest pass mediocranially, and those from the inguinal ligament arch mediocaudally. Note the cremaster muscle covering the spermatic cord (also see Fig. 2–12), and filling the arched space between the conjoint tendon and the inguinal ligament.

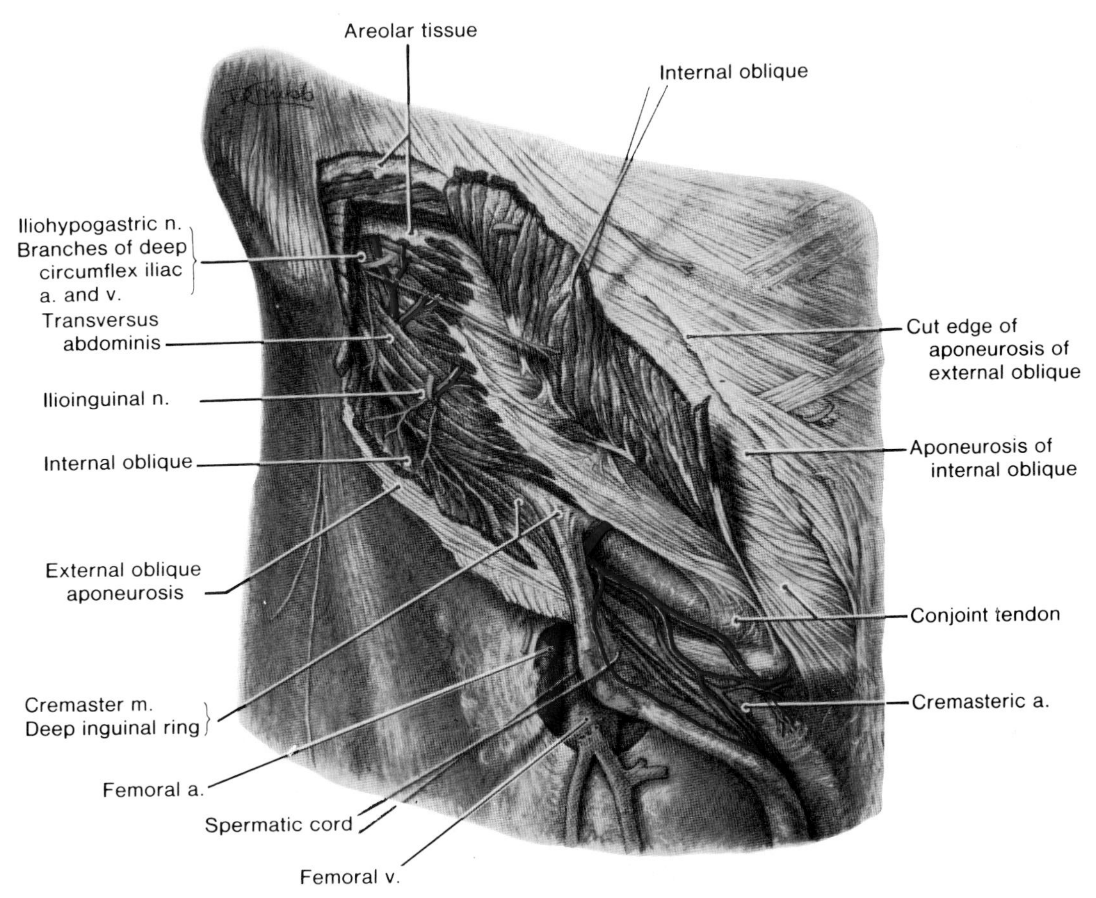

Figure 2-12. Drawing of a dissection of the inguinal region with the internal oblique muscle reflected and the spermatic cord retracted. Observe the transversus abdominis muscle fibers taking, in this region, the same common mediocaudal direction as the fibers of the external oblique aponeurosis and the internal oblique. Note that the transversus abdominis has a less extensive origin from the inguinal ligament than the internal oblique muscle. Observe that the internal oblique portion of the conjoint tendon is attached to the pubic crest and that the transversus abdominis portion extends laterally along the pecten pubis (Fig. 2-1). Note that the conjoint tendon is not sharply defined from the fascia transversalis but blends with it. Observe that lumbar segment 1, via the iliohypogastric and ilioinguinal nerves, supplies the fibers of the internal oblique and transversus abdominis, and therefore controls the conjoint tendon. Observe that the fascia transversalis is evaginated to form the tubular internal spermatic fascia; the mouth of the tube, called the deep inguinal ring, is situated lateral to the inferior epigastric vessels. Note that the cremasteric artery, a branch of the inferior epigastric, anastomoses with the testicular artery and the artery to the ductus deferens.

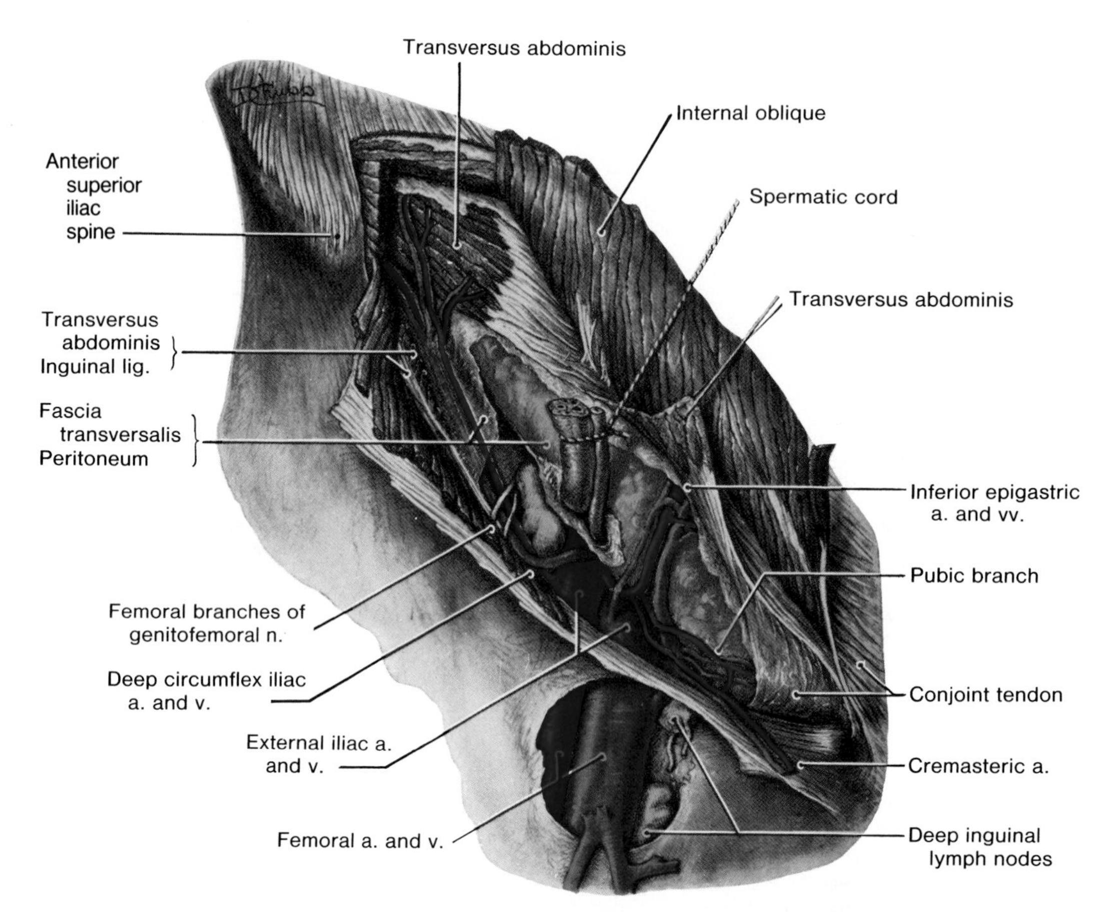

Figure 2-13. Drawing of a deep dissection of the inguinal region. The inguinal part of the transversus abdominis muscle and the fascia transversalis are partly cut away and the spermatic cord is excised. Observe that the inferior limit of the peritoneal sac lies some distance superior to the inguinal ligament laterally but close to it medially. Note the location of the deep inguinal ring, a fingerbreadth superior to the inguinal ligament at the midpoint between the anterior superior iliac spine and the pubic tubercle. Observe the proximity of the external iliac artery and vein to the inguinal canal. The fascia transversalis or transversalis fascia is the internal investing layer which lines the entire abdominal wall (Fig. 2–16, *B* and *C*).

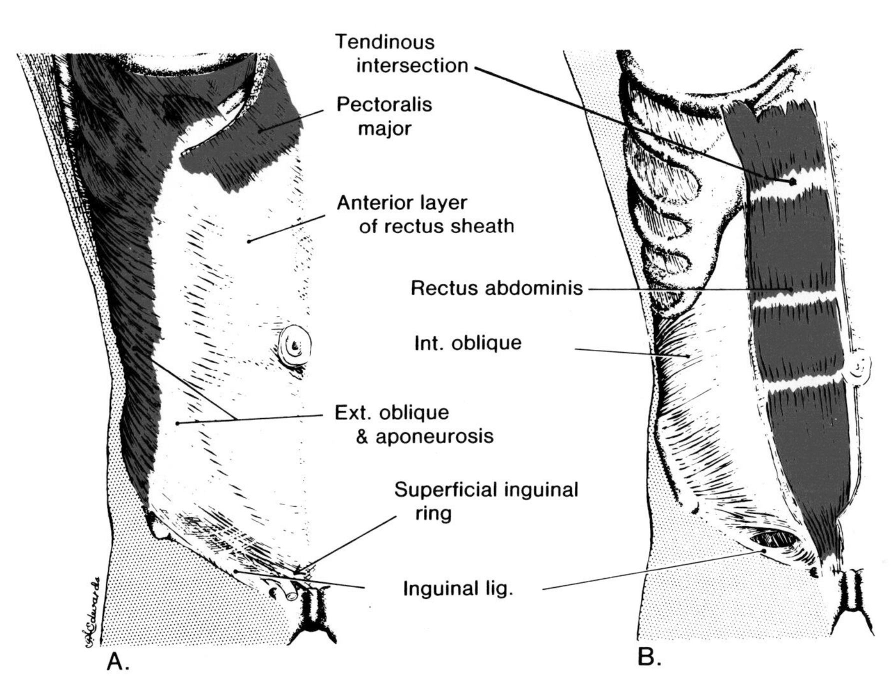

Figure 2-14. Drawing of a dissection of the anterior abdominal wall. *A*, shows the external oblique muscle and its aponeurosis contributing to the anterior layer of the rectus sheath. *B*, shows the internal oblique and rectus abdominis muscles. The right side of the rectus abdominis was exposed by removing the anterior layer of the rectus sheath formed by the aponeuroses of the two oblique muscles (Fig. 2-10*C*). The anterior layer of the rectus sheath is firmly attached to the rectus muscle at the tendinous intersections. Observe that the external oblique muscle arises from the *external* surfaces of the inferior eight ribs (*A*), and that the internal oblique inserts into the *inferior* surface of the inferior three ribs (*B*).

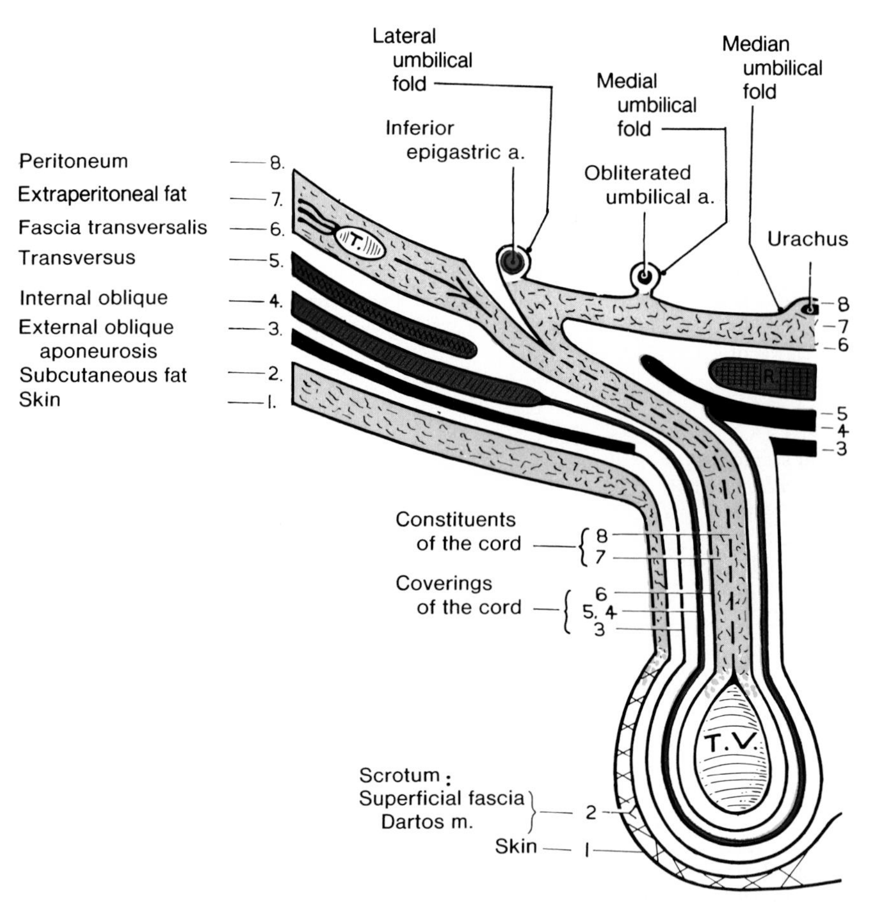

Figure 2-15. Drawing illustrating the scheme of the inguinal canal and the origin of the coverings of the spermatic cord. In this schematic horizontal section, the scrotum and testis are assumed to have been raised to the level of the superficial inguinal ring. The drawing shows the eight layers of the abdominal wall and their three evaginations; the scrotum, the coverings of the spermatic cord, and the constituents of the spermatic cord. *T*, testis. *R*, rectus abdominis, and *T.V.*, tunica vaginalis. With regard to the coverings of the cord, note that *3.* External Spermatic Fascia is derived from the external oblique aponeurosis; *4, 5.* Cremaster muscle and cremasteric fascia are derived from the internal oblique and its fascia; and *6.* Internal Spermatic Fascia is derived from the fascia transversalis. Also see Table 2-2.

The anterior abdominal wall may be the site of certain congenital malformations, such as ventral hernias. Most hernias occur in the umbilical or inguinal regions. Inguinal hernias are discussed subsequently.

Umbilical hernias are very common and are usually congenital. They result from incomplete closure of the anterior abdominal wall after the umbilical cord is ligated at birth. The hernia occurs through the defect created by the degenerating umbilical vessels.

Hernias may also occur through defects in the linea alba (Fig. 2–9), where they are called **median hernias.** If the hernia occurs in the epigastric region (Fig. 2–5), it is called an **epigastric hernia.** These hernias tend to occur after 40 years of age and are often associated with obesity. Extraperitoneal fat and/or omentum protrude through the defect in the linea alba.

The External Oblique Muscle (Figs. 2–3, 2–6 to 2–9, and Table 2–1). This is the largest and most superficial of the three flat abdominal muscles. This large muscle is located in the anterior and lateral parts of the abdominal wall. Its fleshy part forms the lateral portion.

Origin (Figs. 2–6 and 2–7). **External surfaces of ribs 5 to 12** by fleshy digitations or slips. The superior four slips of muscle interdigitate with the serratus anterior muscle, and the inferior four slips of muscle interdigitate with the latissimus dorsi muscle (Figs. 2–

Table 2-1
Muscles of the Anterolateral Abdominal Wall

Muscle	Origin	Insertion	Nerve Supply	Action(s)
External oblique (Figs. 2-6 and 2-7)	External surfaces of ribs 5 to 12 (Fig. 2-6)	Linea alba, pubic tubercle, and anterior half of iliac crest (Figs. 2-8 and 2-9)	Inferior five intercostal nn., subcostal n., and iliohypogastric n. (Figs. 2-6 and 2-8)	Compress and support abdominal viscera Flex and rotate vertebral column
Internal oblique (Figs. 2-6 to 2-14)	Thoracolumbar fascia (Fig. 2-111), iliac crest, and lateral two-thirds of inguinal ligament (Fig. 2-8)	Inferior borders of ribs 10 to 12, linea alba, and pubis via conjoint tendon (Figs. 2-8 and 2-11)	Inferior five intercostal nn., subcostal, iliohypogastric, ilioinguinal, and subcostal nn. (Figs. 2-6, 2-8 and 2-110)	
Transversus abdominis (Figs. 2-8 and 2-10)	Internal surfaces of inferior six costal cartilages, thoracolumbar fascia, iliac crest, and lateral third of inguinal ligament (Fig. 2-8).	Linea alba with aponeurosis of internal oblique, pubic crest, and pectin pubis via conjoint tendon (Figs. 2-1 and 2-13)		Compresses and supports abdominal viscera
Rectus abdominis (Figs. 2-3, 2-10, and 2-14)	Symphysis pubis and pubic crest (Figs. 2-1 and 2-10)	Xiphoid process and 5th to 7th costal cartilages (Figs. 2-10 and 2-14)	Inferior five intercostal nn. and subcostal n. (Figs. 2-6 and 2-110)	Flexes trunk and tenses anterior abdominal wall
Pyramidalis[1]	Body of pubis	Linea alba about halfway to umbilicus	Subcostal n.[2] (Fig. 2-110)	Tenses linea alba

[1] The pyramidalis is a small unimportant muscle which lies anterior to the inferior part of the rectus abdominis muscle. It is absent in about 18% of people.
[2] The subcostal nerve is the ventral ramus of the twelfth thoracic spinal nerve.

3 and 2–6). All these digitations unite to form a broad, somewhat fan-shaped muscle.

Insertion (Figs. 2–1 and 2–9). Most fibers of the external oblique radiate inferiorly, anteriorly, and medially to **insert into the linea alba, the pubic tubercle, and the anterior half of the iliac crest.** This is the direction taken by your outstretched fingers when put into the pockets of your pants.

Most muscle fibers end in the broad aponeurosis of the external oblique (Fig. 2–9). By means of this aponeurosis, the external oblique reaches the midline, where it fuses with the aponeuroses of the other two anterior muscles on its same side and with those on the opposite side to form the **linea alba** (Figs. 2–8 to 2–10).

The linea alba is a tendinous raphe stretching from the xiphoid process to the symphysis pubis. The aponeurosis of the external oblique is also attached to the superior border of the symphysis pubis and to the pubic crest as far as the pubic tubercle (Figs. 2–1 and 2–9).

The inferior turned under part of the aponeurosis of the external oblique forms the inguinal ligament. The **inguinal ligament** is attached to the anterior superior iliac spine and the pubic tubercle (Figs. 2–1 and 2–9).

The **superficial inguinal ring** (Fig. 2–9) lies at the end of a triangular cleft in the external oblique aponeurosis, immediately superior to the pubic tubercle. This ring represents a weakness in the aponeurosis of the external oblique and the anterior abdominal wall.

Nerve Supply (Figs. 2–6 to 2–8). Ventral rami of *inferior five thoracic spinal nerves, the subcostal nerve, and the iliohypogastric nerve.*

The Internal Oblique Muscle (Figs. 2–6 to 2–14 and Table 2–1). This is the middle one of the three flat abdominal muscles.

Origin (Fig. 2–111). Posterior layer of **thoracolumbar fascia,** anterior two-thirds of **iliac crest,** and lateral two-thirds of **inguinal ligament.** Its fibers run at right angles to those of the external oblique muscle, fanning out from their origins and radiating superiorly and anteriorly.

Insertion. Posterior fibers insert into the **inferior borders of inferior three or four ribs** (usually 10 to 12), where they are continuous with the inferior three internal intercostal muscles (Fig. 1–18). The remaining fibers end in a broad aponeurosis that inserts into the **linea alba** and the *pubis* via the conjoint tendon. The superior fibers of the aponeurosis of the internal oblique split to enclose the rectus abdominis muscle (Fig. 2–10), and come together again at the linea alba. The inferior fibers of the aponeurosis arch over the spermatic cord as it lies in the **inguinal canal** (Fig. 2–8), and then descend posterior to the superficial inguinal ring to insert into the **pubic crest** and the **pecten pubis** (Fig. 2–1).

The most inferior tendinous fibers of the internal oblique muscle join with aponeurotic fibers of the transversus abdominis muscle to form the **conjoint tendon** (Figs. 2–11 to 2–13), which turns inferiorly to insert into the pubic crest and the pecten pubis.

Nerve Supply (Figs. 2–6 and 2–8). Ventral rami of *inferior five thoracic spinal nerves, the subcostal nerve, the iliohypogastric nerve, and the ilioinguinal nerve.*

The Transversus Abdominis Muscle (Figs. 2–8, 2–10, and Table 2–1). This is the third and innermost of the three flat abdominal muscles.

Origin (Fig. 2–8). **Internal surfaces of inferior six costal cartilages** where they interdigitate with the fibers of origin of the diaphragm (Fig. 1–17). *Fibers also arise from the* ***thoracolumbar fascia,*** the **iliac crest,** and the **lateral third of inguinal ligament.** The fibers run more or less horizontally, except for the most inferior ones, which pass inferiorly and run parallel to those of the internal oblique muscle. Muscle fibers of the transversus abdominis end in an aponeurosis which contributes to the formation of the rectus sheath (Fig. 2–10).

Insertion (Fig. 2–13). **Linea alba** with aponeurosis of internal oblique, the **pubic crest** and **pecten pubis** via the conjoint tendon.

Nerve Supply (Figs. 2–6 and 2–8). Ventral rami of *inferior five thoracic spinal nerves, the subcostal nerve, the iliohypogastric nerve, and the ilioinguinal nerve.*

Actions of the Flat Abdominal Muscles. The anterior abdominal wall is unsupported and unprotected by bone; however *the three flat muscles and their extensive aponeuroses form a strong but expansible* ***support*** *and considerable* ***protection*** *for the viscera,* especially when the muscles are in good condition. The strong expansible, anterolateral abdominal wall stretches to accommodate a full stomach and/or a pregnant uterus.

The abdominal muscles also play an *important role in movements of the trunk* (flexion, extension, twisting, and lateral bending), *and in the maintenance of posture.*

Normally there are quiet rhythmic movements of the anterior abdominal wall accompanying respirations. When the diaphragm contracts during inspiration, its dome flattens and descends, increasing the vertical dimension of the thorax (Fig. 1–51). To make room for the abdominal viscera, the anterior abdominal wall expands as its muscles relax. When the thoracic cage and the diaphragm relax during expiration (Fig. 1–52), the anterior abdominal wall sinks in passively; however, in the forced expiration that occurs during coughing, sneezing, vomiting, and straining, all the anterior abdominal muscles act strongly.

Acting together, *the flat abdominal muscles increase the intra-abdominal pressure.* When the ribs and the diaphragm are fixed, compression of the viscera by the anterior abdominal muscles occurs which raises the intra-abdominal pressure. This action produces the force required for **defecation** (bowel movement), **micturition** (urination), and **parturition** (childbirth).

Acting separately, *the flat abdominal muscles move the vertebral column; e.g.,* contraction of one internal oblique muscle produces a combination of flexion and rotation of the trunk to its side.

The Rectus Abdominis Muscle (Figs. 2–3, 2–7, 2–8, 2–10, 2–11, 2–14, and Table 2–1). This long strap muscle is the *principal vertical muscle of the anterior abdominal wall.* The two parts of the muscle lie edge to edge inferiorly, but broaden out superiorly where they are separated by the linea alba. One part of the muscle lies on each side of the linea alba. The rectus is three times as wide superiorly as inferiorly.

The lateral borders of the rectus and its sheath are convex and form clinically important surface markings known as the **lineae semilunares** (Figs. 2–2, 2–3, and 2–10*A*).

The rectus abdominis muscle is largely enclosed in the rectus sheath (Figs. 2–8 and 2–10), formed by the aponeuroses of the three flat abdominal muscles.

Origin (Figs. 2–10, 2–14, and Table 2–1). **Symphysis pubis** and **pubic crest.**

Insertion. Anterior surfaces of **xiphoid process** and the **5th to 7th costal cartilages.**

The anterior layer of the rectus sheath is firmly attached to the rectus muscle at three **tendinous intersections**[1] (Fig. 2–7). When the rectus muscle is tensed in muscular persons, each stretch of muscle between the tendinous intersections bulges out (Fig. 2–3). *The location of the tendinous intersections is indicated by the grooves between the muscle bulges.* They are usually located at the level of (1) the xiphoid process, (2) the umbilicus, and (3) about halfway between these structures (Figs. 2–3 and 2–7).

Nerve Supply. Ventral rami of **inferior five intercostal nerves and the subcostal nerve.**

Actions. Flexes trunk and tenses anterior abdominal wall. The rectus is the powerful flexor of the trunk, but the oblique muscles assist it during this

[1] Some people have four or more tendinous intersections and a few have only two, but this is unimportant.

movement. When you lie on your back and flex your trunk, you can easily feel your rectus abdominis.

The rectus abdominis also ***depresses the ribs and stabilizes the pelvis*** *during walking.* This fixation of the pelvis enables the thigh muscles to act effectively. Similarly, during leg lifts from the supine position, the rectus abdominis muscles contract to prevent tilting of the pelvis by the weight of the lower limbs.

The Linea Alba and Rectus Sheath (Figs. 2–3, 2–8 to 2–11, and 2–14). These two structures have been mentioned several times and briefly described; however, owing to their clinical importance, a more detailed description of the rectus sheath and its relationship to the linea alba will be given.

The rectus sheath *is the strong, incomplete fibrous compartment of the rectus abdominis muscle.* It forms by the fusion and separation of the aponeuroses of the flat abdominal muscles (Fig. 2–10). At its lateral margin, the internal oblique aponeurosis splits into two layers, one passing anterior to the rectus muscle and one passing posterior to it (Fig. 2–10*C*). The anterior layer joins with the aponeurosis of the external oblique to form the ***anterior wall of the rectus sheath,*** and the posterior layer joins with the aponeurosis of the transversus abdominis muscle to form the ***posterior wall of the rectus sheath.*** The anterior and posterior walls of the sheath fuse in the anterior median line to form a tendinous raphe, called the **linea alba** (Figs. 2–8 to 2–11). It is narrow inferior to the umbilicus, but is wide superior to it. A groove in the skin is visible anterior to it in thin, muscular persons (Fig. 2–3). *The linea alba lies between the two parts of the rectus abdominis muscle.* The **umbilicus** is located just inferior to the midpoint of the linea alba (Figs. 2–7 and 2–8).

Superior to the costal margin, the posterior wall of the rectus sheath is deficient because the transversus abdominis muscle passes internal to the costal cartilages, and the internal oblique muscle is attached to the costal margin. Hence, *superior to the costal margin, the rectus abdominis muscle lies directly on the thoracic wall* (Fig. 2–8).

The inferior one-fourth of the rectus sheath is also deficient because the internal oblique aponeurosis does not split here to enclose the rectus muscle (Fig. 2–10*C*). The inferior limit of the posterior wall of the rectus sheath is marked by a crescentic border called the **arcuate line** (Fig. 2–8). *The position of the arcuate line is usually midway between the umbilicus and the pubic crest* (Fig. 2–1).

Inferior to the arcuate line, the aponeuroses of the three flat muscles pass anterior to the rectus muscle to form the anterior layer of the rectus sheath (Figs. 2–8 and 2–10*C*).

Within the rectus sheath there is usually a small triangular muscle, called the *pyramidalis,* which lies on the anterior surface of the inferior part of the rectus abdominis muscle. It arises from the body of the pubis and inserts into the linea alba (Table 2–1). Although it tenses the linea alba, the reason for this is unknown. *The pyramidalis muscle is unimportant* and is absent in about 18% of people.

Important Structures In The Rectus Sheath. In addition to the **rectus abdominis muscle,** the rectus sheath contains the **superior and inferior epigastric vessels** (Fig. 2–8), and the terminal parts of the *inferior five intercostal and subcostal vessels and nerves* (Fig. 2–118).

The Transversalis Fascia (Figs. 2–10*C*, 2–13, 2–15, and 2–16). This somewhat transparent fascia is the **internal investing layer which lines the entire abdominal wall.** It covers the deep surface of the transversus abdominis muscle and its aponeurosis and is *continuous from side to side deep to the linea alba.*

Each part of the transversalis fascia is named according to the structures on which it lies; hence it is called the **diaphragmatic fascia** on the diaphragm; the **iliac fascia** on the iliacus; the **psoas fascia** on the psoas; and the **pelvic fascia** in the pelvis. It is also prolonged into the thigh with the iliac fascia to form the **femoral sheath** (Fig. 4–16). It also passes through the inguinal canal to form the **internal spermatic fascia,** part of the covering of the spermatic cord (Figs. 2–8, 2–15, and 2–18).

Internal to the transversalis fascia is the **peritoneum** (Fig. 2–15), the extensive serous membrane which lines the abdominal and pelvic cavities. The transversalis fascia is separated from the peritoneum by a variable amount of subperitoneal fat, referred to as **extraperitoneal fat** (Fig. 2–15).

NERVES OF ANTEROLATERAL ABDOMINAL WALL

The skin and muscles of the anterolateral abdominal wall are supplied almost entirely by the continuation of the **inferior intercostal nerves** (T7 to T11) and the **subcostal nerves** (T12). The inferior part of the anterolateral abdominal wall is supplied by the first lumbar nerve via the **iliohypogastric and ilioinguinal nerves** (Figs. 2–8, 2–11, and 2–16).

The main trunks of the intercostal nerves pass anteriorly from the intercostal spaces and then run between the internal oblique and transversus abdominis muscles (Figs. 2–8 and 2–16). The common nerve supply of the skin and muscles of the anterolateral abdominal wall explains why palpating the abdomen with cold hands is sufficient to cause contraction of the abdominal muscles.

All nerves of the anterolateral abdominal wall pass between or through muscles to reach the rectus sheath. They supply the three flat abdominal muscles, as well as the rectus abdominis.

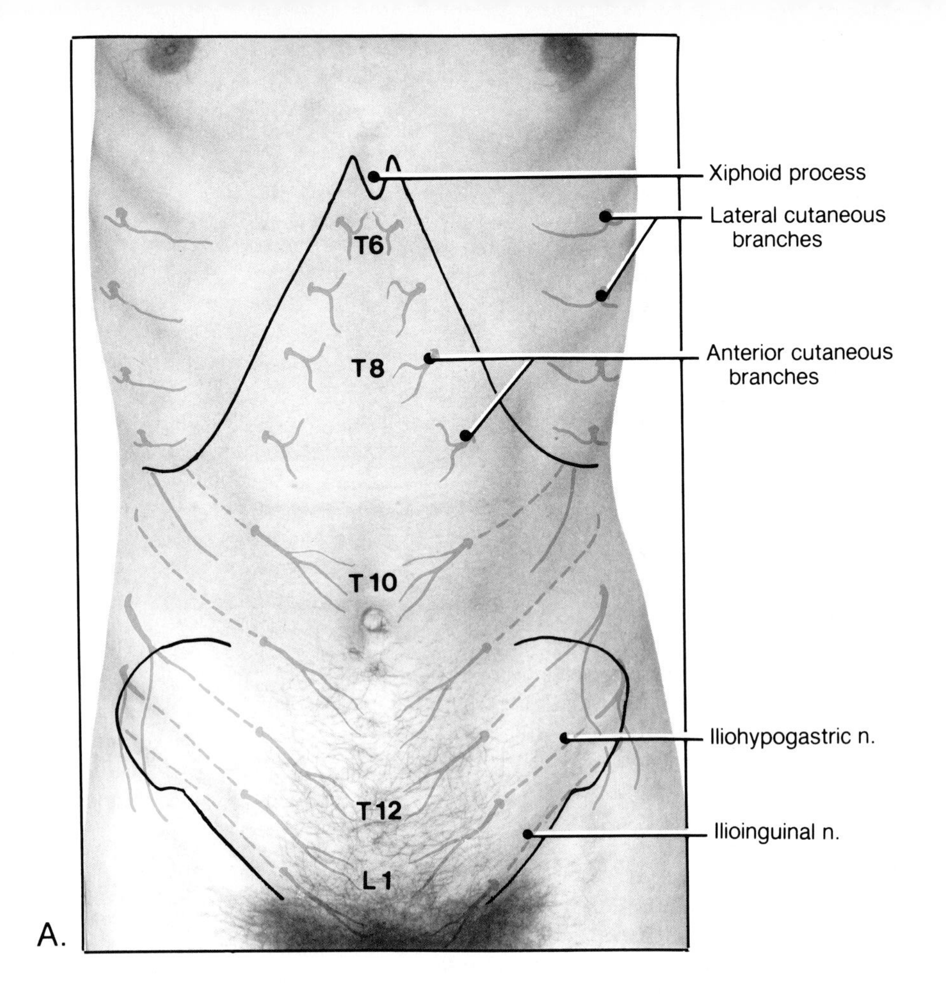

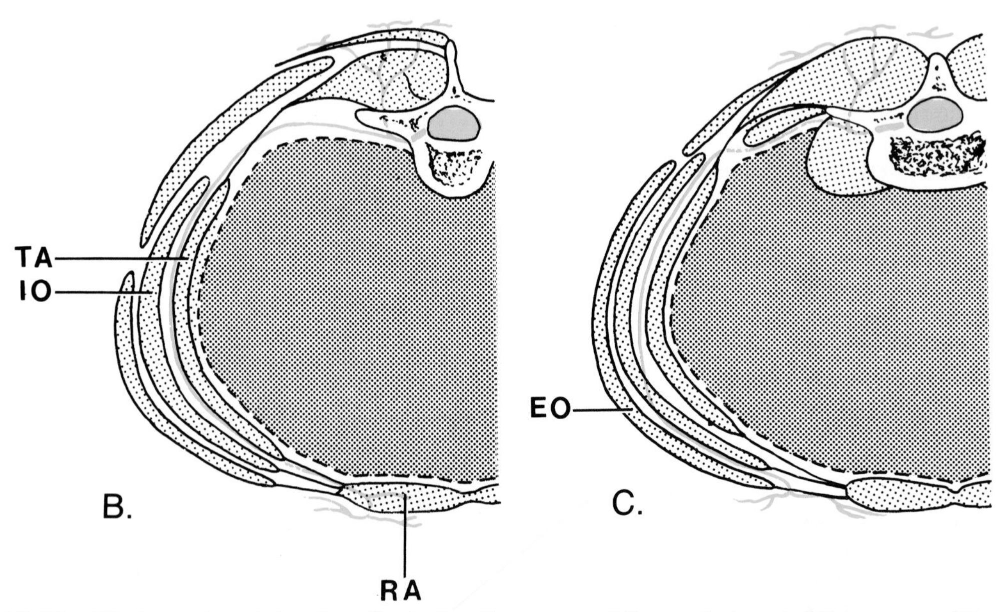

Figure 2-16. Photograph and drawings illustrating the course of the ventral rami of the nerves of the anterolateral abdominal wall (also see Figs. 2–6 to 2–8). *A*, photograph of a 27-year-old man illustrating the distribution of the thoracoabdominal nerves. Note that the cutaneous nerve supply to the anterior abdominal wall is derived from the ventral rami of the inferior six thoracic and the first lumbar nerves. *B*, diagram of an oblique section of the trunk illustrating the course of the inferior six thoracic nerves. Note that the ventral rami of these nerves run between the internal oblique (*IO*) and transversus abdominis (*TA*) muscles to the lateral margin of the rectus abdominis (*RA*) muscle. *C*, schematic diagram of a section of the lower abdomen showing the course of the first lumbar nerve. This lumbar nerve is represented by the iliohypogastric and ilioinguinal nerves. Note that the ventral ramus of the first lumbar nerve pierces the internal oblique and then runs between this muscle and the external oblique (*EO*) to pierce the external oblique aponeurosis. The *broken lines* (— — — —) in *B* and *C* indicate the transversalis fascia.

The anterior cutaneous nerves pierce the rectus sheath a short distance from the median plane (Figs. 2–7 and 2–16). The anterior cutaneous branches of *T7 to T9* supply the skin *superior to the umbilicus* (Fig. 1–23); *T10* innervates the skin at the *level of the umbilicus;* and *T11, T12, and L1* supply the skin *inferior to the umbilicus* (Figs. 2–8 and 2–16*A*). *The first lumbar nerve appears superior to the inguinal ligament, where it is called the* **iliohypogastric nerve** (Figs. 2–8 and 2–16*A*).

When incising the anterior abdominal wall during surgery, the incision is made through the part of the wall which gives the freest access to the organ concerned, with the least disturbance of the nerve supply to the abdominal muscles. Because there are communications between the intercostal nerves in the intercostal spaces and in the anterior abdominal wall (Fig. 2–6), one or two branches of the cutaneous nerves may be cut without noticeable loss of sensation.

Little if any communication occurs between nerves from the lateral border of the rectus abdominis muscle to the midline. For this reason, **a transverse incision through the rectus abdominis causes the least possible damage to its nerve supply.** A vertical incision through the lateral portion of the rectus abdominis muscle (**pararectus incision**) denervates the portion of the muscle medial to the incision (Fig. 2–7). **A pararectus incision is therefore inadvisable anatomically.**

The rectus abdominis muscle may be divided transversely without serious damage because when rejoined, a new transverse band forms similar to the tendinous intersections, provided the nerve supply is intact.

The common incisions used by surgeons are **median incisions** and right or left paramedian incisions. These incisions can be made rapidly because no major blood vessels are cut.

A **paramedian incision** passes through the anterior layer of the rectus sheath (Fig. 2–7) and the muscle is freed from its sheath and retracted laterally. *This avoids injury to the nerve supply of the rectus muscle* (Fig. 2–8). The posterior layer of the rectus sheath is then incised to enter the abdominal and peritoneal cavities. In this way, the nerve supply is not interfered with and a strong repair normally results.

Another common incision is the small *muscle-and-aponeurosis-splitting right lower quadrant incision* used for **appendectomy** (Fig. 2–43). The external oblique aponeurosis is split obliquely in the direction of its fibers and retracted. The internal oblique and transversus abdominis musculoaponeurotic fibers, lying at right angles to those of the external oblique (Fig. 2–13), can then be separated parallel to their course. Carefully made, *the entire exposure cuts no musculoaponeuronitic fibers,* so that when the incision is closed the abdominal wall is as strong after the operation as it was before.

An **incisional hernia** is a protrusion of an organ or tissue through an incision (surgical wound). *The surgeon who has a thorough knowledge of the anatomy of the anterior abdominal wall and makes incisions accordingly will only occasionally have to operate on this kind of hernia.* (For more information on the pathological anatomy of incisional hernias, see Anson and McVay, 1971).

Inflammation of the peritoneum lining the abdominopelvic cavity (**peritonitis**) causes pain in the overlying skin, and a *reflex increase in the tone of the anterior abdominal muscles.* Normally, rhythmic movements of the anterior abdominal wall accompany respirations (*i.e.*, the abdomen moves out with the chest). If the abdomen goes in as the chest goes out (**paradoxical abdominothoracic rhythm**) and muscle rigidity is present, it is probable that peritonitis or pneumonitis is present.

A special feature on examination of an **acute abdomen** (patient with very intense abdominal pain) is the finding of spasm of the anterolateral abdominal muscles, sometimes amounting to a boardlike **rigidity.**

VESSELS OF ANTEROLATERAL ABDOMINAL WALL

Small arteries arise from anterior and collateral branches of the **posterior intercostal arteries** of the 10th and 11th intercostal spaces, and from anterior branches of the **subcostal arteries,** to supply the muscles of the anterolateral abdominal wall (Figs. 1–15 and 2–8). They anastomose with the **superior epigastric arteries,** with the ***superior lumbar arteries,*** and with each other (Figs. 2–7, 2–8, and 2–12).

The main arteries of the anterolateral abdominal wall are the inferior epigastric and deep circumflex iliac arteries (Fig. 2–8), *branches of the* ***external iliac artery*** (Fig. 2–127), *and the superior epigastric artery, a terminal branch of the* ***internal thoracic artery*** (Figs. 1–16 and 1–17).

The inferior epigastric artery runs superiorly in the transversalis fascia to reach the arcuate line; there it enters the rectus sheath (Figs. 2–8 and 2–13).

The deep circumflex iliac artery runs on the deep aspect of the anterior abdominal wall, parallel to the inguinal ligament (Fig. 2–13), and along the iliac crest between the transversus abdominis and internal oblique muscles.

The superior epigastric artery enters the rectus sheath superiorly (Fig. 2–8), just inferior to the seventh costal cartilage.

The superficial epigastric and lateral thoracic veins (Fig. 2–7) anastomose, thereby uniting the veins of the superior and inferior halves of the body.

The three superficial inguinal veins end in the **great saphenous vein** of the lower limb (Figs. 2–7 and 2–8).

The superficial lymph vessels of the anterolateral abdominal wall, *superior to the umbilicus,* pass to the **axillary lymph nodes** (Figs. 6–8 and 6–43), whereas those *inferior to the umbilicus* drain into the **inguinal lymph nodes** (Figs. 2–11 and 2–13). Lymph from the tissues of the abdominal wall, in general, drains to the **lumbar lymph nodes** or to the common and external *iliac lymph nodes* (Figs. 2–121 and 3–73).

INTERNAL SURFACE OF ANTEROLATERAL ABDOMINAL WALL (Figs. 2–15 and 2–16)

The internal surface of the anterolateral abdominal wall exhibits several folds, some of which contain obliterated fetal vessels that carried blood to and from the placenta before birth.

The term fold is usually used to describe an elevation of peritoneum with a free edge. Most folds are raised (formed) by underlying blood vessels and do not provide much strength.

Superior to the umbilicus there is a median fold or elevation of peritoneum, called the **falciform ligament** (Fig. 2–28). It passes from the umbilical region to the liver and holds the liver against the anterior abdominal wall. **The falciform ligament contains the ligamentum teres, the obliterated umbilical vein.** The umbilical vein carried oxygenated blood from the placenta to fetus. For an illustration and description of the fetal circulation, see Moore, 1982. The umbilical vein is patent for some time after birth and may be used for **exchange transfusions** during early infancy (*e.g.*, in infants with ***erythroblastosis fetalis*** or hemolytic disease of the fetus).

*Inferior to the umbilicus there are **five umbilical folds*** (two on each side and one in the median plane) which pass superiorly toward the umbilicus. The **lateral umbilical folds** are formed by elevations of peritoneum covering the **inferior epigastric vessels** (Figs. 2–13 and 2–15).

The **medial umbilical folds** are formed by elevations of peritoneum covering the **medial umbilical ligaments,** the obliterated umbilical arteries. These vessels carried blood from the fetus to the placenta for oxygenation before birth.

The **median umbilical fold** is formed by an elevation of peritoneum covering the **median umbilical ligament,** the remnant of the **urachus** (Figs. 2–15 and 3–15), which developed from the intra-abdominal part of the embryonic ***allantois.*** The *urachus* and the median umbilical ligament extend from the internal aspect of the umbilicus to the apex of the urinary bladder (Figs. 3–62 and 3–66).

THE INGUINAL REGION

The inguinal region is very important surgically because it is where **inguinal hernias** (ruptures) occur, usually in males.

Nine out of ten hernias are inguinal hernias. This results from the fact that the **inguinal region is an area of weakness in the anterior abdominal wall,** owing to the perforation of the wall in males by the ductus deferens and its accompanying vessels and nerves (Fig. 2–13), and in females by the round ligament of the uterus (Fig. 2–22).

All the abdominal muscles, chiefly their aponeuroses, contribute inferiorly to the inguinal ligament and the inguinal canal. The inferior free edge of the aponeurosis of the external oblique muscle forms a thick band known as the **inguinal ligament** (Fig. 2–9), *which extends from the anterior superior iliac spine to the pubic tubercle* (Figs. 2–1, 2–2, 2–9, 2–10, and 2–14). Fibers are reflected from the medial end of the inguinal ligament to the pecten pubis to form the ***lacunar ligament*** (Figs. 2–1 and 2–9).

THE SUPERFICIAL INGUINAL RING (Figs. 2–7, 2–9, 2–11, and 2–19A)

Although called a ring, the superficial inguinal ring (external inguinal ring) is a more or less triangular aperture or **deficiency in the aponeurosis of the external oblique muscle.**

The superficial inguinal ring is triangular in shape. The base of the triangle is formed by the **pubic crest** and its apex is directed superolaterally (Fig. 2–9). The sides of the superficial inguinal ring are formed by the *medial and lateral crura* (L. legs) of the superficial inguinal ring (Fig. 2–11).

Emerging from the superficial inguinal ring is the **spermatic cord** in the male (Fig. 2–7) and the **round ligament of the uterus** in the female (Fig. 2–22). The central point of the superficial inguinal ring is superior to the pubic tubercle (Figs. 2–9 and 2–22).

The lateral crus of the superficial inguinal ring is formed by the part of the external oblique aponeurosis that attaches to the pubic tubercle via the inguinal ligament (Fig. 2–9). The spermatic cord rests on the inferior part of the lateral crus (Fig. 2–7).

The medial crus of the superficial inguinal ring is formed by the part of the aponeurosis that diverges to attach to the pubic bone and the pubic crest, medial to the pubic tubercle (Fig. 2–9). **Intercrural fibers** from the inguinal ligament arch superiorly and medially across the superficial inguinal ring. *The intercrural fibers prevent the crura from spreading apart* (Figs. 2–7 and 2–9).

The superficial inguinal ring is palpable just superior and lateral to the pubic tubercle. *The superficial inguinal ring in adult males can be examined by invaginating the skin of the scrotum with the tip of the index finger* (Fig. 2–20*C*), and by probing gently superiorly with the finger along the **spermatic cord.** If the ring is somewhat enlarged, it may admit the index finger without causing pain.

In women and children, the dimensions of the superficial inguinal ring are much less than in adult males and palpation of it is difficult. In male infants, the superficial inguinal ring does not normally admit the tip of the index finger.

DESCENT OF THE TESTES

To understand the inguinal canal, an understanding of the migration and descent of the testes is essential.

The testes develop in the lumbar region inside the abdominal cavity, deep to the transversalis fascia, between it and the peritoneum. They normally migrate through the **inguinal canals** (Figs. 2–15 and 2–17) into the scrotum just before birth.

The site of the inguinal canal is first indicated by a ligament, the **gubernaculum,** which extends from the testis through the anterior abdominal wall and inserts into the internal surface of the scrotum. Later, a finger-like *outpouching or diverticulum of peritoneum,* called the **processus vaginalis,** follows the gubernaculum and evaginates (protrudes through) the anterior abdominal wall forming the inguinal canal. The processus vaginalis carries extensions of the layers of the anterior abdominal wall before it (Fig. 2–17).

In males these prolongations of the layers of the abdominal wall become the coverings of the spermatic cord (Figs. 2–15 and 2–17). *In both sexes,* the opening produced by the processus vaginalis in the external oblique aponeurosis forms the **superficial inguinal ring** (Fig. 2–9).

Before birth, the testes normally enter the inguinal canals and pass inferomedially through them to enter the scrotum (Fig. 2–17*C*). Normally the stalk of the processus vaginalis obliterates shortly after birth, leaving only the part surrounding the testis, which becomes the **tunica vaginalis** (Figs. 2–17*D* and 2–18*B*).

Maldescent of the testis (undescended testis or ***cryptorchidism***) is a common abnormality. The testes are undescended at birth in about 3% of full-term and 30% of premature infants.

Undescended testes are usually located in the pelvic cavity or somewhere in the inguinal canal.

Most undescended testes descend during the first few weeks or months after birth. A few more descend at puberty owing to stimulation by testicular androgens.

The seminiferous tubules (Fig. 2–18*B*) do not develop fully in testes that remain undescended after puberty, and infertility results when the condition is bilateral. *Androgen secretion by undescended testes is usually unimpaired.*

Uncommonly, the gubernaculum does not attach to the scrotum. Instead it enters the perineum or the thigh. The testis later follows it through the inguinal canal to an abnormal site. Hence, an **ectopic testis** may be located in the perineum, in the pubopenile area, or in the femoral region.

PRENATAL MIGRATION OF THE OVARIES

The ovaries migrate from their place of origin in the abdominal cavity to a point just inferior to the pelvic brim (Fig. 3–50); however they do enter the inguinal canals. The processus vaginalis normally obliterates completely and the **gubernaculum** becomes attached to the uterus. It is divided into the *ligament of the ovary* and the *round ligament of the uterus* (Figs. 2–22 and 3–77). **The round ligament of the uterus passes through the inguinal canal** and attaches to the internal surface of the labium majus (homologous to half of the scrotum).

Persistence of the processus vaginalis in a female, often called the **canal of Nuck,** may result in a *congenital inguinal hernia* as in a male (Fig. 2–21*B*). **Cysts in the inguinal canal** and the labium majus may develop from remnants of the processus vaginalis.

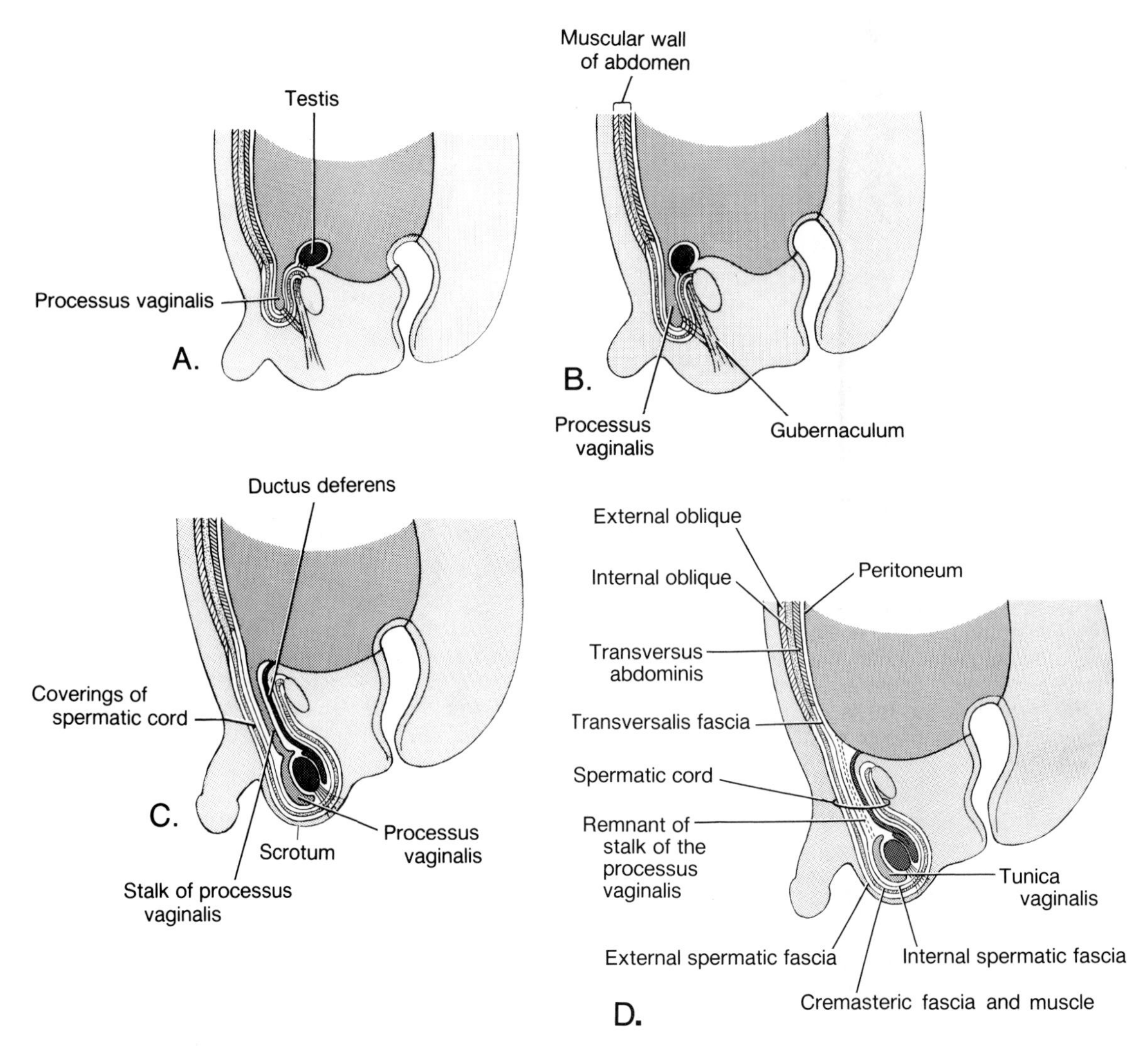

Figure 2-17. Schematic drawings of sagittal sections illustrating formation of the inguinal canals and descent of the testes. The testes develop in the abdominal cavity and migrate into the scrotum just before birth. An evagination of the peritoneum, the processus vaginalis, evaginates the abdominal wall and carries fascial layers of the wall before it. Normally the stalk of the processus obliterates shortly after birth. The tunica vaginalis is the remains of the inferior end of the processus vaginalis. It is therefore a peritoneal sac surrounding the testis. (Reprinted with permission from Moore KL: *The Developing Human: Clinically Oriented Embryology*, ed 3. Philadelphia, WB Saunders Co, 1982.)

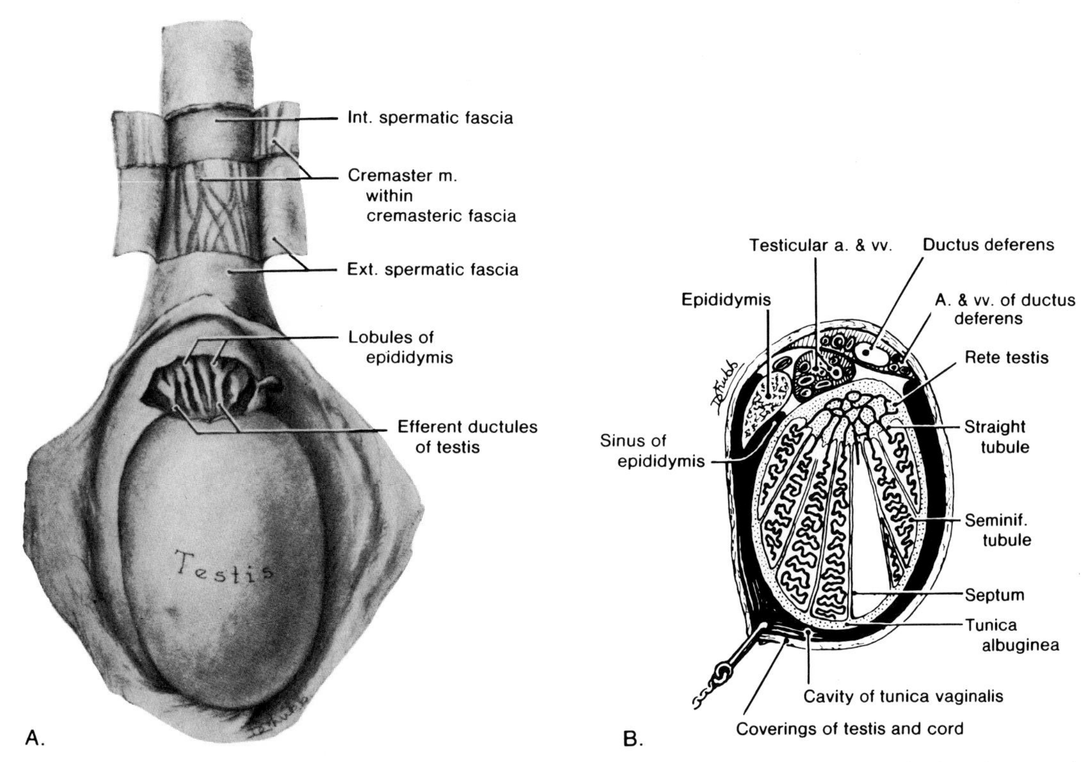

Figure 2-18. *A*, drawing of a dissection of the testis and spermatic cord illustrating, in particular, the coverings of the cord. *B*, drawing of a cross-section of the right testis. Observe the cavity of the tunica vaginalis surrounding the testis anteriorly and at the sides, and extending between the testis and the epididymis as the sinus of the epididymis. Examine the epididymis lying posterolateral to the testis. The cavity of the tunica vaginalis is artificially distended to show its visceral and parietal layers. Observe the ductus deferens with its small lumen and thick wall lying posteromedial to the testis. Note the pyramidal compartments for the seminiferous tubules. Each of the 250 compartments contains two or three hair-like seminiferous tubules which join in the mediastinum testis to form the rete testis.

THE DEEP INGUINAL RING (Figs. 2–12, 2–13, and 2–19)

The deep inguinal ring is a slit-like opening in the transversalis fascia, located just lateral to the **inferior epigastric artery** (Fig. 2–13). The deep inguinal ring is immediately superior to the midpoint of the inguinal ligament and medial to the origin of the transversus abdominis muscle from the inguinal ligament.

The deep inguinal ring (internal inguinal ring) is the opening or mouth of a finger-like diverticulum of the transversalis fascia (Fig. 2–19*C*). The deep ring formed when the processus vaginalis evaginated the transversalis fascia (Fig. 2–17*B*).

The margins of the deep inguinal ring are not sharply defined, as are those of the superficial inguinal ring. When the external oblique is reflected and the epigastric vessels are displaced, it ceases to exist as a ring; however from the inner aspect, a dimple in the peritoneum often marks the site of the ring (Fig. 2–15).

COVERINGS OF THE SPERMATIC CORD (Figs. 2–8, 2–10, 2–15, 2–17, and 2–18*A*)

The bundle of structures passing to and from the testis (ductus deferens and the associated nerves and vessels), called the **spermatic cord,** *is covered by three concentric layers of fascia derived from the anterior abdominal wall.*

The coverings of the spermatic cord formed as the processus vaginalis evaginated the abdominal wall and carried extensions of its layers into the scrotum (Fig. 2–17). The coverings are not easily separable from one another, either in a cadaver or a living man.

The Internal Spermatic Fascia (Figs. 2–15, 2–17, and 2–18*A*). As the processus vaginalis evaginates the **transversalis fascia** at the deep inguinal ring, it carries a thin layer of fascia before it that becomes the **internal spermatic fascia.** It constitutes the filmy *innermost covering of the spermatic cord* (Fig. 2–18*A*).

The Cremaster Muscle and Cremasteric Fas-

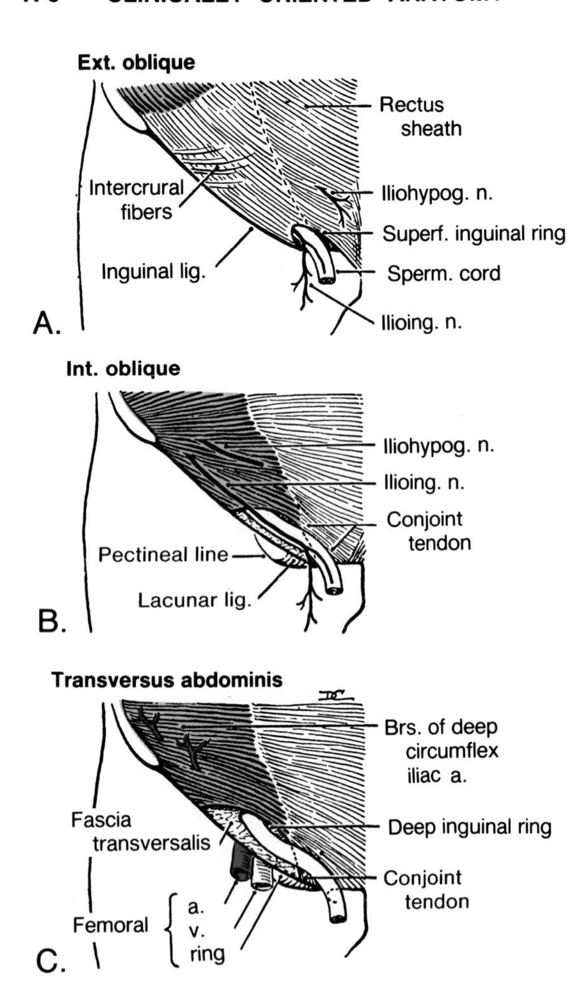

Figure 2-19. Drawings of progressive dissections of the three flat muscles of the anterior abdominal wall, inferior to the level of the anterior superior iliac spine. Note that the anterior wall of inguinal canal is formed throughout by the external oblique aponeurosis, and that its posterior wall is formed by the fascia transversalis and conjoint tendon.

cia (Figs. 2–11, 2–12, and 2–18*A*). As the processus vaginalis, with its covering of fascia transversalis (future internal spermatic fascia), evaginates under the edge of the **internal oblique muscle,** it acquires a few of this muscle's fibers and some of its investing fascia. These muscle fibers and fascia form the **cremaster muscle** and **cremasteric fascia,** respectively.

The cremasteric fascia forms the middle or second covering of the spermatic cord, which contains loops of cremaster muscle (Figs. 2–12 and 2–18*A*). The cremaster muscle, which is continuous with the internal oblique, *reflexly draws the testis to a more superior position in the scrotum,* particularly in the cold. Contraction of the cremaster muscle can be produced by lightly scratching the skin on the medial aspect of the superior part of the thigh, in the area supplied by the ilioinguinal nerve (Figs. 2–16, 4–13, and 4–14). This results in reflex contraction of the cremaster muscle supplied by the genital branch of the genitofemoral nerve (L1 and L2). *The reflex raising of the testis is called* **the cremasteric reflex.**

The testis, located outside the pelvis in the scrotum, is sensitive to cold; hence the cremaster muscle draws it superiorly in the scrotum for warmth and protection against injury.

Testing of the cremasteric reflex is part of every routine physical examination in male patients because abdominal and cremasteric reflexes may be absent in both upper and lower *motor neuron disorders.*

The cremasteric reflex is much more active in children. During childhood, **hyperactive cremasteric reflexes** may cause apparent undescended testes. The cremasteric reflex can be abolished by having the child sit in a crosslegged squatting position. If the testis is normal, it can be palpated in the scrotum.

The cremasteric reflex is often sluggish or absent in older men. Unilateral absence of this reflex is not *in itself* indicative of neurological disease.

The External Spermatic Fascia (Figs. 2–15, 2–17, and 2–18*A*). As the processus vaginalis evaginates the external oblique aponeurosis and forms the superficial inguinal ring, it carries an extension of this aponeurosis before it which forms *the external spermatic fascia, the thin outermost covering of the spermatic cord.* The external spermatic fascia is attached superiorly to the crura of the superficial inguinal ring (Figs. 2–7 to 2–9), and is continuous with the deep fascia covering the external oblique muscle.

THE INGUINAL CANAL (Figs. 2-9, 2–11 to 2–13, 2–15, 2–19, and 2–20)

The inguinal canal is an oblique intermuscular passage through the inferior part of the anterior abdominal wall between the deep and superficial inguinal rings. It is located immediately superior to the medial half of the inguinal ligament, with which it is parallel.

The inguinal canal forms during the fetal period as the processus vaginalis evaginated the layers of the anterior abdominal wall (Fig. 2–17). The testis normally descends through it just before birth, invaginating the posterior wall of the processus vaginalis. The testis takes with it the ductus deferens, blood and lymph vessels, and nerves which become the constituents of the spermatic cord (Figs. 2–17 and 2–18).

The inguinal canal begins at the deep inguinal ring and ends at the superficial inguinal ring. It is 4 to 5 cm long in adults and runs inferomedially to the superficial inguinal ring.

The anterior wall of the inguinal canal is formed mainly by the ***aponeurosis of the external oblique*** *muscle* (Figs. 2–11, 2–15, and 2–19). The anterior wall is reinforced laterally by fibers of the internal oblique muscle.

The posterior wall of the inguinal canal is formed by the ***fascia transversalis,*** *which is reinforced medially by the* ***conjoint tendon*** (Figs. 2–11 to 13 and 2–19), *the common tendon of insertion of the internal oblique and transversus abdominis muscles.*

The floor or inferior wall of the inguinal canal is formed by the superior surface of the ***inguinal ligament*** and the ***lacunar ligament*** (Fig. 2–9).

The roof or superior wall of the inguinal canal is formed by the inferior arching fibers of the ***internal oblique and transversus abdominis muscles*** (Figs. 2–12, 2–13, and 2–19).

The inferior epigastric artery (Figs. 2–8 and 2–13) and the obliterated umbilical artery run posterior to the posterior wall of the inguinal canal (Figs. 2–13 and 2–15). ***The inferior epigastric artery lies at the medial boundary of the deep inguinal ring*** (Fig. 2–13). Hence, pulsations of the inferior epigastric artery form a useful landmark during surgery for determining the location of the deep inguinal ring.

The presence of the inguinal canal produces a potential weakness in the inferior part of the anterior abdominal wall, however the anatomical structure of the canal compensates somewhat for this weakness.

Owing to the obliquity of the inguinal canal, the deep and superficial inguinal rings do not coincide; thus, increases in intra-abdominal pressure act on the deep inguinal ring, forcing the posterior wall of the canal against the anterior wall. This strengthens this potentially weak part of the anterior abdominal wall.

The inguinal canal may be likened to an arcade of three arches formed by the three flat abdominal muscles. Contraction of the external oblique muscle approximates the anterior wall (formed by the aponeurosis of the external oblique) to the posterior wall (formed by the transversalis fascia and conjoint tendon). Contraction of the internal oblique and transversus abdominis muscles makes them taut; as a result, the roof of the canal is lower and the passage is constricted.

During standing there is continuous contraction of the internal oblique and transversus abdominis muscles in the inguinal region.

During coughing and straining, the raised intra-abdominal pressure threatens to force some of the abdominal contents through the canal producing a hernia. However, vigorous contraction of the arched fleshy fibers of the internal oblique and transversus abdominis muscles "clamp down", without damaging the spermatic cord. The action is that of a half-sphincter (Fig. 2–19) and tends to prevent herniation.

The superficial inguinal ring has the conjoint tendon immediately posterior to it (Fig. 2–12), and the rectus abdominis muscle is posterior to the conjoint tendon. When intraabdominal pressure rises, the flat muscles of the abdomen all contract, forcing the external oblique aponeurosis against the conjoint tendon, which then pushes against the rectus abdominis. Hence, the conjoint tendon and rectus abdominis reinforce the posterior surface of the superficial inguinal ring, tending to prevent herniation of the abdominal contents through it.

In spite of the factors that strengthen the inguinal region, **inguinal hernia** is the most common defect of the anterior abdominal wall.

A hernia is a protrusion of a structure, viscus, or organ from the cavity in which it belongs. The term is derived from the Greek word meaning offshoot. Some laypeople refer to a hernia as a *rupture,* indicating that it is like a blowout caused by force or pressure.

Because the scrotum and the layers within it represent outpouchings of the anterior abdominal wall (Figs. 2–15 and 2–17), hernias into the scrotum or through the abdominal wall in the inguinal region are particularly common in males.

The labia majora in females are homologous with the scrotum, but they consist mostly of fat (Fig. 2–22). Hence, the potential weak part of the abdominal wall is small and *inguinal hernia is much less common in females than in males.*

An inguinal hernia typically contains part of a viscus, most commonly part of the small intestine (Fig. 2–21*B*).

There are two types of inguinal hernia, indirect and direct. **Indirect inguinal hernia is the more common type of hernia** at all ages and in both sexes. *It is most common in male children and makes up about 75% of inguinal hernias.* As its name indicates, **an indirect hernia takes an indirect course through the anterior abdominal wall.** It follows the route normally taken by the processus vaginalis before birth (Fig. 2–17).

An indirect inguinal hernia leaves the abdominal cavity lateral to the inferior epigastric vessels. It traverses the deep inguinal ring, the inguinal canal, and the superficial inguinal ring *within the spermatic cord* (Figs. 2–20*B* and 2–21*B*). Hence an indirect inguinal

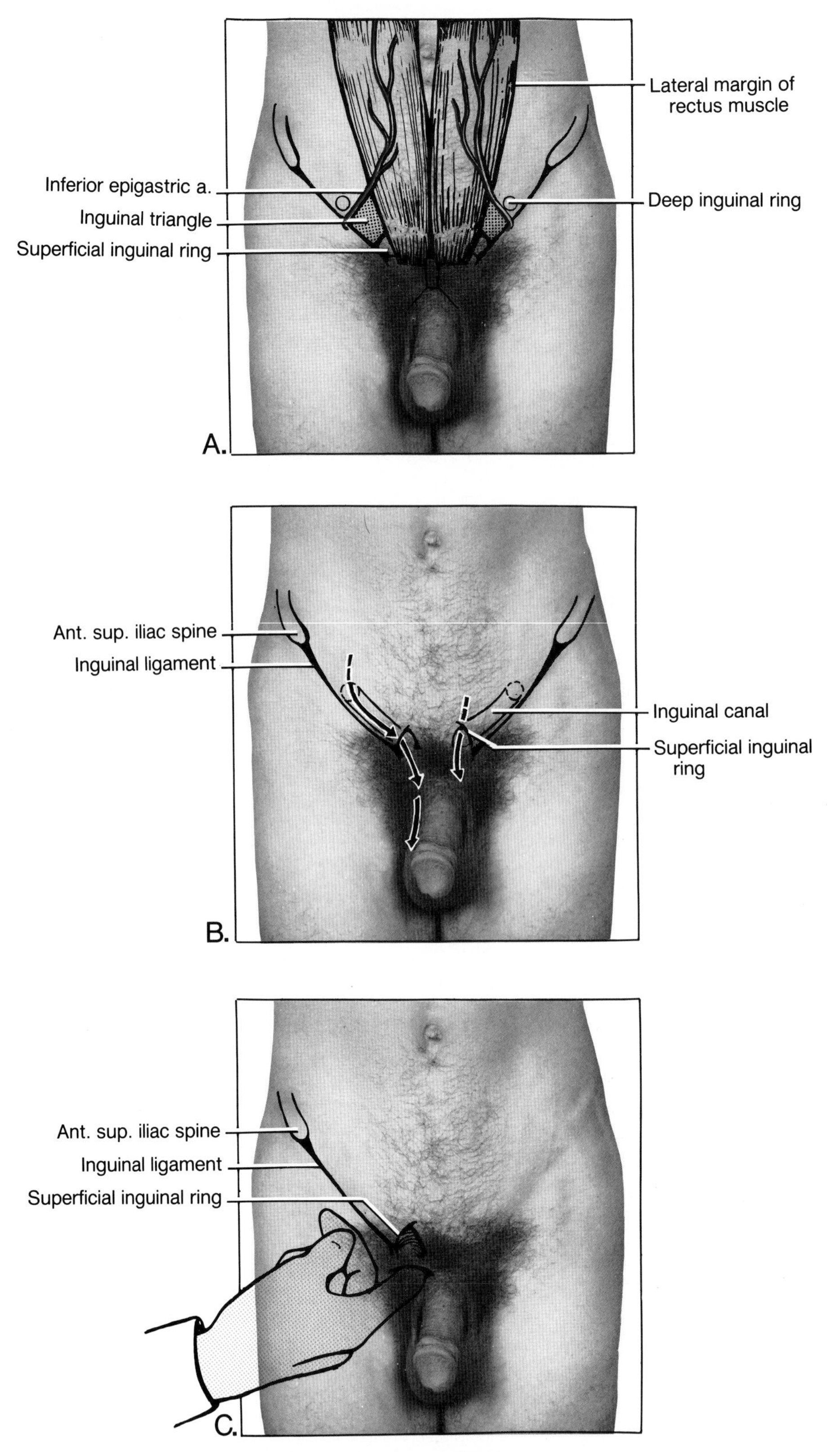

Figure 2-20. Drawings illustrating the inguinal region, inguinal canals, and types of inguinal hernia. *A*, illustrates the boundaries of the inguinal triangle (screened): inguinal ligament inferiorly; inferior epigastric artery laterally, and rectus abdominis muscle medially. *B*, illustrates the course and presentation of inguinal hernias. The *indirect inguinal hernia* (right side) comes down the inguinal canal, often into the scrotum. The *direct inguinal hernia* (left side) bulges anteriorly through the inferior part of the inguinal triangle. *C*, illustrates invagination of the scrotal skin during palpation of the superficial inguinal ring. If an indirect inguinal hernia is present it comes down when the patient coughs and touches the examiner's fingertip.

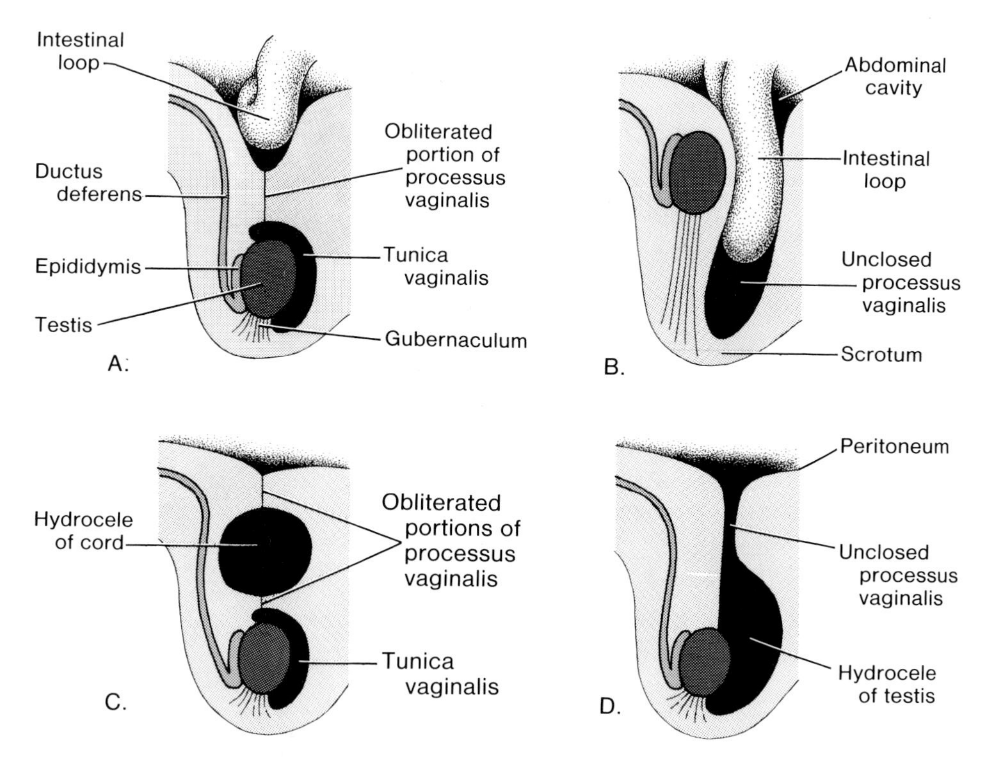

Figure 2-21. Drawings of sagittal sections of the inguinal region of fetuses. *A*, incomplete indirect inguinal hernia resulting from persistence of the proximal part of the processus vaginalis. *B*, indirect inguinal hernia into the scrotum resulting from persistence of the entire processus vaginalis. Cryptorchidism (undescended testis), a commonly associated malformation, is also illustrated. *C*, large cyst derived from an unobliterated portion of the processus vaginalis. This condition is called a hydrocele of the spermatic cord. *D*, hydrocele of the testis and spermatic cord resulting from peritoneal fluid passing into the unclosed processus vaginalis. (Reprinted with permission from Moore KL: *The Developing Human: Clinically Oriented Embryology*, ed 3. Philadelphia, WB Saunders Co, 1982.)

hernia is covered by all three layers of the spermatic cord.

Indirect inguinal hernia has an embryological basis. An evagination of the peritoneum, called the **processus vaginalis,** forms the inguinal canal (Fig. 2–17) and creates the potential passageway for an indirect inguinal hernia.

The hernial sac represents the remains of the processus vaginalis, which normally obliterates, except for the part that forms the tunica vaginalis (Figs. 2–17*D*, 2–18*B*, and 2–21*A*). If the processus vaginalis does not obliterate, the hernia is complete and extends into the scrotum (Fig. 2–21*B*).

Indirect inguinal hernia is about 20 times more common in males than in females. The main reason for this is that the ovary and its vessels do not descend through the inguinal canal; hence, the potential weak part of the abdominal wall is not so great as in the male. However, *if the processus vaginalis persists in a female, a hernia may enter it and follow it through the inguinal canal into the labium majus.*

An indirect inguinal hernia can also occur after normal obliteration of the processus vaginalis because the passage of the spermatic cord (or the round ligament of the uterus) through the inguinal canal produces a weakness in the anterior abdominal wall that predisposes it to inguinal hernia.

Inguinal hernias are less common in females because the inguinal region has a much firmer construction, which results from the passage of fewer and smaller structures through the inguinal canal (Fig. 2–22).

A direct inguinal hernia protrudes anteriorly through the posterior wall of the inguinal canal, and leaves the abdominal cavity medial to the inferior epigastric vessels (Figs. 2–13, 2–20, and 2–133). The protru-

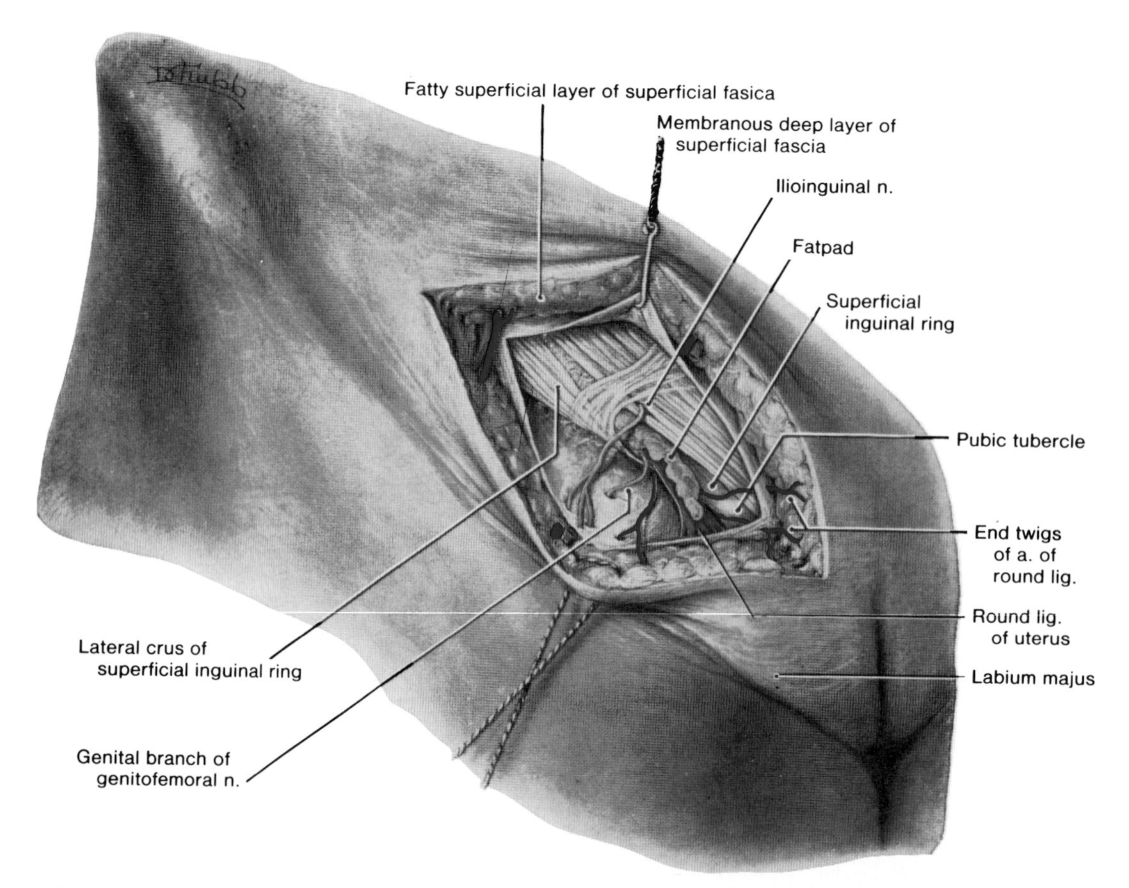

Figure 2-22. Drawing of a dissection of the female inguinal canal. Observe that the superficial inguinal ring is smaller than in the male (Figs. 2–9 and 2–11). Note the following structures issuing from the superficial inguinal ring: (1) the round ligament of the uterus, (2) a closely applied pad of fat; (3) the genital branch of the genitofemoral nerve, and (4) the artery of the round ligament of the uterus. Also see Figure 3–77.

sion passes through some part of the **inguinal triangle** (Hesselbach's triangle), usually the inferior part.

The inguinal triangle is bounded by the inguinal ligament inferiorly, the inferior epigastric artery laterally, and the rectus abdominis muscle medially (Fig. 2–20*B*). The inguinal triangle lies just posterior to the superficial inguinal ring; it is the area of the posterior wall of the inguinal canal that is formed only by transversalis fascia. Obviously, *the inguinal triangle is a potential weak area of the anterior abdominal wall* and the site of a direct inguinal hernia.

In a direct inguinal hernia, the hernial sac does not pass through the deep inguinal ring but passes through or around the conjoint tendon and directly to the superficial inguinal ring. If the hernia passes lateral to the conjoint tendon, it pushes the peritoneum and transveralis fascia before it, and emerges through the superficial inguinal ring, either superior or inferior to the spermatic cord.

If the hernia passes through the fibers of the conjoint tendon, it is covered by peritoneum, transversalis fascia, and fibers of the conjoint tendon.

Direct inguinal hernia is much less common than indirect inguinal hernia and it usually occurs in men over forty. This type of hernia is rare in women.

Direct inguinal hernia usually results from weakening of the conjoint tendon, a condition that is most common in older men.

Some people have considerable difficulty remembering *the difference between direct and indirect inguinal hernias* because they think "direct" means the hernia goes directly through the inguinal canal which is a potentially weak part of the anterior abdominal wall. This is wrong.

"Direct" means that the hernia passes

directly through the anterior abdominal wall. *The hernia passes through the posterior wall of the inguinal canal, medial to the internal ring and the inferior epigastric, to enter the inguinal canal.* Hence when entering the inguinal canal, **a direct inguinal hernia bypasses the deep inguinal ring** by passing through the canal's posterior wall. A direct hernia may produce a general bulging of the abdominal wall, or the hernial sac may emerge from the superficial inguinal ring and turn superiorly. Some direct inguinal hernias enter the scrotum, but this is rare.

"Indirect" means that the hernia passes indirectly through the anterior abdominal wall by entering the deep inguinal ring, passing through the inguinal canal, and emerging through the superficial inguinal ring. *An indirect inguinal hernia follows the indirect course taken by the testis* and ductus deferens in the perinatal period (Figs. 2–17 and 2–21*B*).

There are two types of indirect inguinal hernia, *congenital and acquired.* Although the term "congenital" means "present at birth," the herniation of a loop of bowel into a patent processus vaginalis (Fig. 2–21*B*) may not be present at birth; the herniation may occur during infancy, childhood, or even adulthood. In other words, *the potential for inguinal herniation is present at birth because of the patent processus vaginalis.*

In *acquired indirect inguinal hernia,* the loop of bowel herniates through the deep ring, the inguinal canal, and the superficial ring, pushing before it the peritoneum of the anterior abdominal wall (Fig. 2–15). In these people the processus vaginalis closed after birth, but its stalk did not obliterate completely as shown in Figure 2–21*A*. The hernia therefore resects the stalk of the processus vaginalis and enters the scrotum, but does not enter the tunica vaginalis. In females the hernia resects the stalk of the processus vaginalis and enters the labium majus.

SCROTUM AND TESTES

The scrotum and testes are considered in this chapter because their development is related to the abdomen. The scrotum and the layers within it represent outpouchings of the anterior abdominal wall in the inguinal region (Fig. 2-17).

THE SCROTUM

The scrotum develops from a cutaneous outpouching of the skin of the abdomen (Fig. 2-17); hence *it is a pendulous pouch or sac of skin* that contains the testes (Figs. 2-15, 2-20, and 3-27). The testes are mobile organs (sex glands) within the scrotum. *The scrotum consists of two layers,* **skin** and **superficial fascia** (Fig. 2-15 and Table 2-2). The thin skin is dark colored and rugose (wrinkled) in young males. The scrotum is divided on its surface into right and left halves by a **scrotal raphé**. This cutaneous ridge indicates the bilateral origin of the scrotum from the *labioscrotal swellings* (Fig. 3-26). The superficial fascia is devoid of fat, but it contains a sheet of smooth muscle called the **dartos muscle**. Its fibers are united to the skin and contraction of them causes the scrotal skin to wrinkle when cold. This helps to regulate the loss of heat through the skin of the scrotum. *Normal spermatogenesis requires a controlled temperature.*

The superficial fascia of the scrotum is continuous anteriorly with the membranous layer of superficial fascia of the anterior abdominal wall, and posteriorly

Table 2-2
Corresponding Layers of the Anterior Abdominal Wall, Spermatic Cord, and Scrotum[1]

Layers of Anterior Abdominal Wall	Scrotum and Coverings of Testis	Coverings of Spermatic Cord
Skin	Skin } SCROTUM[2]	
Superficial fascia	Superficial fascia and dartos muscle } SCROTUM[2]	
External oblique aponeurosis	External spermatic fascia	External spermatic fascia
Internal oblique muscle	Cremaster muscle	Cremaster muscle
Fascia of internal oblique muscle	Cremasteric fascia	Cremasteric fascia
Transversus abdominis muscle		
Transversalis fascia	Internal spermatic fascia	Internal spermatic fascia
Extraperitoneal tissue		
Peritoneum	Tunica vaginalis	

[1] Also see Figures 2-15, 2-17, 2-18, and 2-21.
[2] The scrotum consists of skin and superficial fascia. The smooth involuntary dartos muscle is located in the superficial fascia.

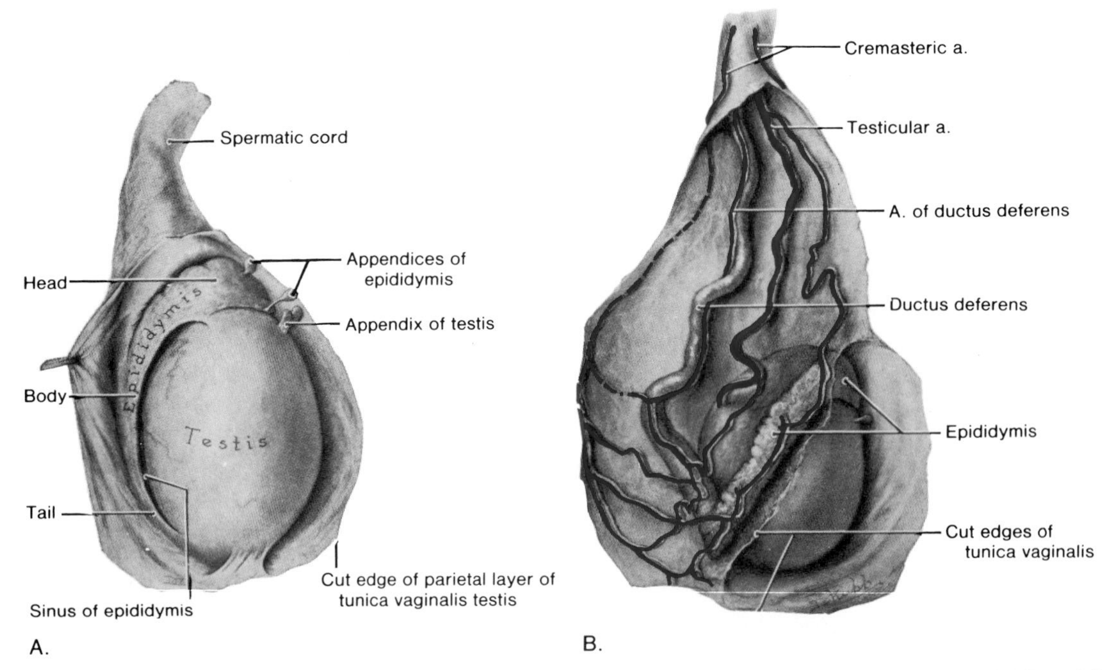

Figure 2-23. *A*, drawing of a lateral view of a dissection of the testis. The tunica vaginalis (Figs. 2-15 and 2-18*B*) has been incised longitudinally. *B*, dissection of the testis and spermatic cord showing the blood supply of the testis. The epididymis is displaced slightly to the lateral side. Note the free anastomosis between the three arteries.

with the superficial fascia of the perineum. The superficial fascia, including the dartos muscle, forms an incomplete scrotal septum that divides the scrotum into right and left halves, one for each testis.

The coverings of the testis are continuous with the coverings of the spermatic cord (Figs. 2-15 and 2-18*A*). The outermost covering of the testis, the **external spermatic fascia**, is continuous with this covering of the spermatic cord (Table 2-2), which is continuous with the external oblique aponeurosis at the superficial ring (Figs. 2-15 and 2-17*D*). Internal to this layer is the **cremaster muscle** and **cremasteric fascia**; inside this layer is the **internal spermatic fascia**, which is continuous with this covering of the spermatic cord and the transversalis fascia (Fig. 2-15).

Inside the internal spermatic fascia is the **tunica vaginalis testis** (Fig. 2-18*B*). This closed serous sac of peritoneum is a derivative of the processus vaginalis (Fig. 2-17). The tunica vaginalis is a peritoneal sac surrounding the testis (Fig. 2-18*B*). It consists of two layers: the **parietal layer** which is superficial and is *adjacent to the internal spermatic fascia*, and the **visceral layer** which is *adherent to the testis and the epididymis*, a coiled tubular structure that stores the sperms (Figs. 2-23 and 2-24). Laterally the visceral layer of the tunica vaginalis passes between the testis and the epididymis to form the **sinus of the epididymis** (Figs. 2-18*B* and 2-23*A*). A capillary layer of fluid normally separates the visceral and parietal layers of the tunica vaginalis.

Arteries and Veins of the Scrotum. The skin and dartos muscle of the scrotum are supplied partly by the *perineal branch of the internal pudendal artery* (Figs. 3-57 and 3-84), and partly by the *external pudendal branches of the femoral artery* (Fig. 2-7). The scrotum is also supplied by the cremasteric branch of the inferior epigastric artery (Fig. 2-8).

The veins accompany the arteries. The *external pudendal veins* enter the great saphenous vein (Fig. 2-7).

Nerves of the Scrotum. Several nerves supply the scrotum. *The genital branch of the genitofemoral nerve* sends sensory branches to the anterior and lateral surfaces of the scrotum (Figs. 4-14 and 4-16), and also supplies the cremaster muscle. The anterior surface of the scrotum is also supplied by the anterior scrotal branches of the *ilioinguinal nerve* (Fig. 2-16*A*).

The perineal branch of the pudendal nerve supplies the posterior surface of the scrotum and the *perineal branches of the posterior femoral cutaneous nerve* supply the inferior surface of the scrotum (Fig. 4-42).

Because the anterior third of the scrotum is supplied mainly by the L1 segment of the spinal cord via the ilioinguinal and genitofemoral nerves, and the posterior two-thirds of the scrotum are supplied mainly by the L3 segment through the perineal and posterior femoral cu-

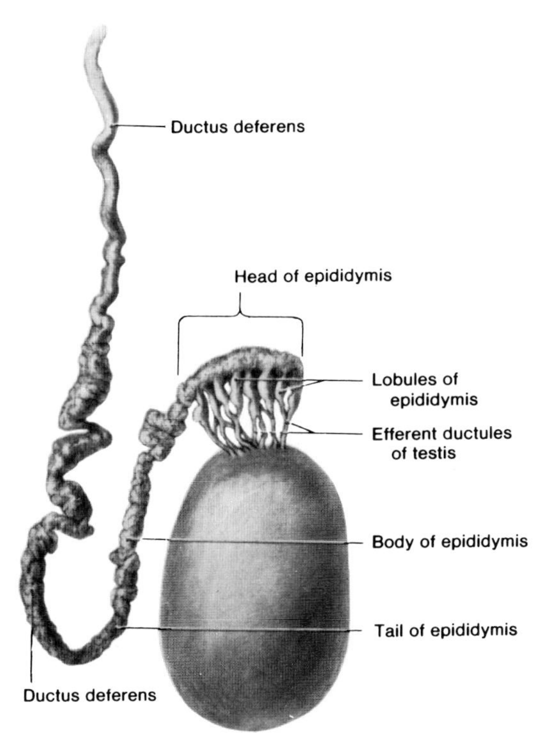

Figure 2-24. Drawing of a testis, epididymis, and ductus deferens after removal of their coverings (shown in Fig. 2-23). Observe the efferent ductules passing from the rete testis (Fig. 2-18*B*) to the head of the epididymis. The epididymis is attached to the posterolateral aspect of the testis (Fig. 2-23). The body of the epididymis lies against the body of the testis, partially separated from it by the sinus of the epididymis (Fig. 2-23*A*). The inferior portion of the epididymis, called the tail, is attached to the testis. Here the epididymis becomes the ductus deferens.

taneous nerves, a spinal anesthetic must be injected more superiorly to anesthetize the anterior surface of the scrotum.

Lymph Vessels of the Scrotum (Figs. 2-11 and 4-13). The lymph vessels ascend in the superficial fascia of the scrotum and empty into the *superficial inguinal lymph nodes* (Fig. 4-35).

The presence of excess fluid anywhere in the processus vaginalis after birth is called a hydrocele (Fig. 2-21). Infants with an obliterated processus vaginalis may have residual peritoneal fluid in the cavity of their tunica vaginalis testis (**noncommunicating hydrocele**), but this fluid usually absorbs during the first year.

If the processus vaginalis remains patent (Fig. 2-21*B* and *D*), peritoneal fluid may be forced into it, forming a **communicating hydrocele.** An indirect inguinal hernia is often associated with this congenital condition (see Case 2-3 on pages 289 and 292). The extent of the hydrocele depends upon how much of the processus vaginalis remains patent.

The proximal and distal ends of the stalk of the processus vaginalis may become obliterated, leaving an intermediate cytic area called a hydrocele of the cord (Fig. 2-21*C*).

Certain pathological conditions (*e.g.*, injury and/or inflammation of the epididymis) may result in an increase in the fluid in the tunica vaginalis, producing a **hydrocele of the testis** and scrotal enlargement (Fig. 2-21*D*). Surgical treatment of this condition may be required. To remove excess fluid from the cavity of the tunica vaginalis, an instrument called a trocar and cannula are inserted through the scrotum.

THE TESTES

The testes (testicles), *the main reproductive organs*, are paired ovoid glands that are *suspended in the scrotum by the spermatic cords* (Fig. 2-23). The surface of each testis is covered by the visceral layer of the tunica vaginalis, except where it is attached to the epididymis and the spermatic cord (Fig. 2-18*B*). Internal to this layer of the tunica vaginalis is the *tunica albuginea*, a connective tissue coat.

The testes produce male germ cells called **sperms** (spermatozoa) and male sex hormones called **androgens.** The sperms are formed in the **seminiferous tubules** which join to form a network of canals known as the **rete testis** (Fig. 2-18*B*). Small **efferent ductules** (15 to 20) connect the rete testis to the head of the epididymis.

The Epididymis (Figs. 2-18 and 2-23 to 2-25). This *comma-shaped structure* is applied to the superior and posterolateral surfaces of the testis. The ***body*** of the epididymis consists of the highly convoluted ***duct of the epididymis.*** The sperms are stored in this duct where they undergo the final stages of maturation as they pass slowly through it. The ***tail of the epididymis*** is continuous with the **ductus deferens** (vas deferens), the duct which transports the sperms to the ***ejaculatory duct*** for expulsion into the urethra. The *sinus of the epididymis* is a small recess of the tunica vaginalis (Figs. 2-18*B* and 2-23*A*).

THE SPERMATIC CORD

The spermatic cord consists of the structures running to and from the testis. They are surrounded by

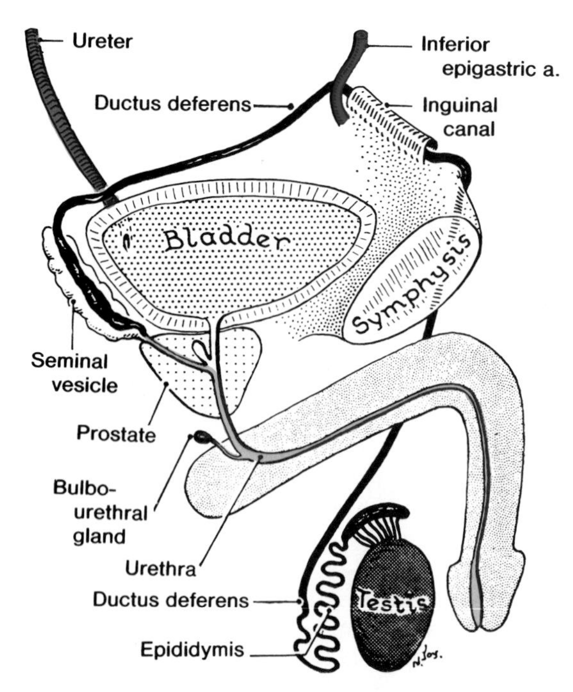

Figure 2-25. Diagram of the male genital system. Note that the ductus deferens passes superiorly, enters the superficial inguinal ring, and passes through the inguinal canal. The sperms are produced in the testis and enter the epididymis where they undergo maturation and are stored until ejaculation. They are then propelled through the ductus deferens and ejaculatory duct into the urethra, which traverses the prostate gland and the penis.

the coverings derived from the layers of the anterior abdominal wall (Figs. 2-12, 2-18, 2-23, and Table 2-2).

The spermatic cord begins at the deep inguinal ring, lateral to the inferior epigastric artery, where its constituents assemble, *and ends at the posterior border of the testis.* The spermatic cord suspends the testis in the scrotum.

The spermatic cord passes through the inguinal canal and emerges at the superficial inguinal ring to descend in the scrotum to the testis. As the spermatic cord leaves the inguinal canal, it acquires its covering of external spermatic fascia. The spermatic cord can be felt as a firm cord along its course from the epididymis to the superficial inguinal ring (Figs. 2-7 and 2-8).

Constituents of the Spermatic Cord (Figs. 2-12, 2-13, 2-18, and 2-23). Within the coverings of the spermatic cord are:

1. **Ductus deferens** (Figs. 2-7, 2-8, and 2-24). This large *duct of the testis* lies in the posterior part of the spermatic cord. It is easily palpable because of its *thick wall of smooth muscle.*
2. **Arteries** (Figs. 2-23*B* and 2-101). The ***testicular artery*** is a long slender vessel that arises from the anterior aspect of the aorta at the level of the second lumbar vertebra (Fig. 2-51), where the testis started to develop in the embryo. *The testicular artery is the main vessel supplying the testis and epididymis.*

   The artery of the ductus deferens *is a slender vessel that arises from the inferior vesical artery* (Figs. 2-23*B* and 3-57). It accompanies the ductus deferens throughout its course and anastomoses with the testicular artery near the testis.

   The cremasteric artery *is a small vessel that arises from the inferior epigastric artery* (Figs. 2-8 and 2-23*B*). It accompanies the spermatic cord and supplies the cremaster muscle and other coverings of this cord. It also anastomoses with the testicular artery near the testis.
3. **Veins** (Figs. 2-101 and 2-116). *Up to 12 veins* from the posterior surface of the testis *anastomose to form a* **pampiniform plexus.** This unusual name is derived from the Latin word "*pampinus*" meaning tendril, the spirally coiling organ of a climbing plant. This large vine-like venous plexus, forming much of the bulk of the spermatic cord, surrounds the ductus deferens and the arteries in the spermatic cord. It is located within the internal spermatic fascia and *ends in the testicular vein* (Fig. 2-116).
4. **Nerves.** There are sympathetic nerve fibers on the arteries and sympathetic and parasympathetic fibers on the ductus deferens. These autonomic sensory nerves carry the impulses that normally produce deep visceral pain when the testis is squeezed, and produce excruciating pain and a sickening sensation when the testis is hit.

   The genital branch of the genitofemoral nerve passes into the spermatic cord and supplies the cremasteric muscle (Fig. 2-118).
5. **Lymph vessels** (Figs. 2-26 and 2-121). The lymph vessels draining the testis and immediately associated structures pass superiorly in the spermatic cord. These vessels end in the *lateral aortic and preaortic lymph nodes,* situated between the common iliac and renal veins.
6. **Remnants of the processus vaginalis.** Normally the stalk of the processus vaginalis in the spermatic cord obliterates during the perinatal period. If if persists the tunica vaginalis is in direct communication with the peritoneal cavity (Fig. 2-21*D*).

Occasionally the processus vaginalis persists and forms a hernial sac for an **indirect inguinal hernia** (this was discussed previously on p. 179). Sometimes isolated remnants of the stalk of the processus vaginalis persist and one or more of these may become swollen with fluid

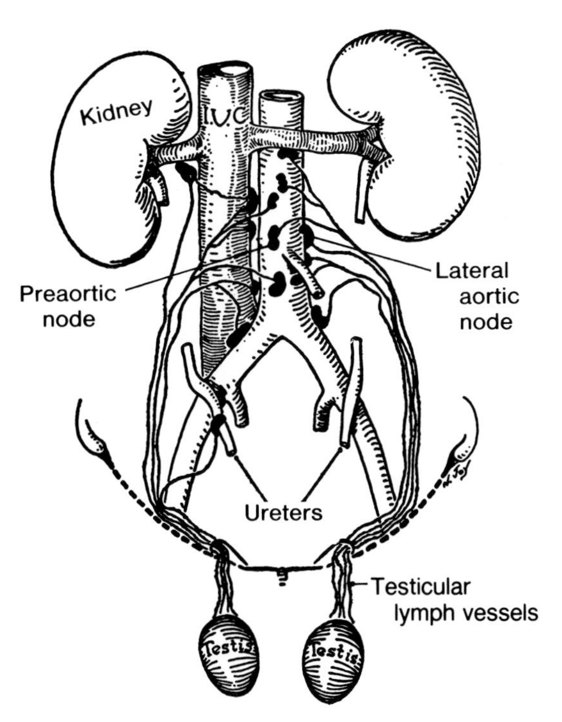

Figure 2-26. Drawing illustrating the lymphatic drainage of the testes. The testicular lymph vessels ascend with the testicular vessels in the spermatic cord, and end in the lateral aortic and preaortic lymph nodes which are scattered along the abdominal aorta. The lymph vessels of the skin of the scrotum drain into the superficial inguinal lymph nodes (Fig. 2-11).

and form **hydroceles of the spermatic cord** (Fig. 2-21*C*).

The pampiniform plexus of veins sometimes becomes varicose (dilated and tortuous), producing a condition known as **varicocele** that reminds one of a "bag of worms." Variocele (varicose veins of the spermatic cord) is more common on the left side and often results from defective valves in the testicular vein. The **worm-like swelling** disappears when the person lies down and the scrotum is elevated. Persons claiming to have two testis on one side usually have a varicocele or a testicular tumor. Rarely, a varicocele may result from blockage of the renal vein owing to a tumor of the left kidney. This blockage interferes with drainage of the left testicular vein (Fig. 2-116).

A hematocele of the testis is a collection of blood in the cavity of the tunica vaginalis, often resulting from trauma to the testis. This damages the vessels around the testis (Fig. 2-23*B*) and blood enters the tunica vaginalis.

The difference in the lymphatic drainage of the testis and the scrotum is clinically important. Cells from a **testicular tumor** may spread by lymphogenous dissemination to the ***lumbar lymph nodes*** (Fig. 2-121), whereas a **cancer of the scrotum** metastasizes to the ***superficial inguinal lymph nodes*** (Figs. 2-11 and 4-35).

Rudimentary structures may be observed around the testis and epididymis (Fig. 2-23*A*). The ***appendix of the testis*** and the ***appendices of the epididymis*** are visible when the tunica vaginalis is opened. These vestigial remnants of the genital ducts in the embryo are rarely observed, unless pathological changes occur in them, because normally they are tiny structures.

The **appendix of the testis** is a vesicular remnant of the cranial end of the paramesonephric duct (embryonic female genital duct) which is attached to the superior pole of the testis (Fig. 2-23*A*).

The **appendix of the epididymis** is the remnant of the cranial end of the mesonephric duct (embryonic male genital duct) which is attached to the head of the epididymis (Fig. 2-23*A*).

Another remnant, the **paradidymis,** may be detected when the ductus epididymis is unraveled. If present, it is located between the efferent ductules and the body of the epididymis (Fig. 2-24). The paradidymis forms from embryonic mesonephric tubules that do not become efferent ductules. In most people these unused embryonic tubules degenerate (see Moore, 1982).

The ductus deferens is ligated bilaterally when **sterilization of a male** is desired. In performing this operation, called a **vasectomy,** the ductus deferens (vas deferens) is isolated through the superior wall of the scrotum. Following the operation, sperms can no longer pass to the urethra; they degenerate in the epididymis and the ductus deferens. But the secretions of the auxillary genital glands (*e.g.,* seminal vesicles and prostate) can still be ejaculated.

THE PERITONEUM

The peritoneum is a thin, translucent, serous membrane. The peritoneum lining the abdominal wall is called **parietal peritoneum,** whereas the peritoneum covering a viscus or an organ is called **visceral peritoneum.** Both types consist of a single layer of simple squamous epithelium, called **mesothelium.**

The parietal and visceral layers of peritoneum are separated from each other by a *capillary film of serous fluid.* This fluid lubricates the peritoneal surfaces, enabling the intraabdominal organs to move upon each other without friction.

If the mesothelium is damaged or removed in any area (*e.g.*, during an operation), the two layers of peritoneum may adhere to each other forming an **adhesion** (L. *adhaesio* from *adhaerere*, to stick to). Such an adhesion may interfere with the normal movements of the viscera. The cutting or division of adhesions at an operation is called an **adhesiotomy**.

Commonly during dissection, the opposing surfaces of the peritoneal layers are found to be adherent (Fig. 2-48). These adhesions (strands of fibrous tissue) probably resulted from inflammatory processes (*e.g.*, peritonitis). You can usually break down these adhesions with your fingers.

THE PERITONEAL CAVITY

The *potential space* between the visceral and parietal layers of peritoneum, normally containing only a small amount of *peritoneal fluid*, is called the **peritoneal cavity.** In males this is a closed cavity, but in females there is a communication with the exterior through the uterine tubes, uterus, and vagina (Fig. 3-78).

For descriptive purposes, the peritoneal cavity is divided into two parts, the *greater sac* of the peritoneal cavity and the *lesser sac* of the peritoneal cavity. The lesser sac or **omental bursa**, the smaller of the two cavities, lies posterior to the stomach (Fig. 2-30). It is continuous with the greater sac through the **epiploic foramen** (Fig. 2-35).

During embryonic development the organs invaginated the peritoneal sac, reducing the embryonic peritoneal cavity to the merest interval between the visceral and parietal layers of peritoneum. It was during this invagination that the viscera received their covering of visceral peritoneum.

To visualize this process, push your fist into a partially inflated balloon which contains a few drops of water. The inner wall of the balloon surrounding your fist is comparable to the visceral peritoneum. The outer wall of the balloon is comparable to the parietal peritoneum, and the cavity of the balloon represents the peritoneal cavity. Observe that as you push your fist further, the two walls come into contact and the cavity practically disappears. Note also that the two layers form a double fold at your wrist, and that they are continuous with each other.

Understand from this demonstration that the peritoneal cavity normally has a very small volume; however, it may be distended with air or fluid as far as the abdominal walls will permit, just as you could inject air or fluid into the balloon.

Under certain pathological conditions, the potential space of the peritoneal cavity may be distended to form an actual space containing several liters of fluid.

Ascites *is an accumulation of serous fluid in the peritoneal cavity* **(hydroperitoneum).** Widespread metastases of cancer cells cause exudation of ascitic fluid that contains cancer cells and is often blood stained. Fluid in the peritoneal cavity represents an imbalance between the rate of fluid formation and its absorption by the peritoneum. *Patients with liver disease (e.g., cirrhosis) usually have ascites.* If the intestine (bowel) ruptures owing to penetrating wounds or a closed abdominal injury results from automobile accidents, gas and intestinal material enter the peritoneal cavity.

Generalized peritonitis and ileus of the bowel results. *Ileus is an obstruction of the bowel* associated with severe **colicky pain**, vomiting, and often fever and dehydration. ***Distention of the bowel*** occurs with ileus.

Several terms are used to describe the different parts of the peritoneum. These are: mesentery, omentum, ligaments, and folds (L. *plicae*).

Mesentery (Figs. 2-27 and 2-39). *A mesentery is a* ***double layer of peritoneum*** *which encloses an organ and connects it with the abdominal wall.* Mesenteries are covered on both sides by mesothelium and have a core of loose connective tissue containing a variable number of fat cells and lymph nodes, together with blood vessels, lymphatic vessels, and nerves passing to and from the viscera (L. soft parts, internal organs).

The most mobile parts of the intestine have a mesentery (*e.g.*, the small intestines). Some viscera have

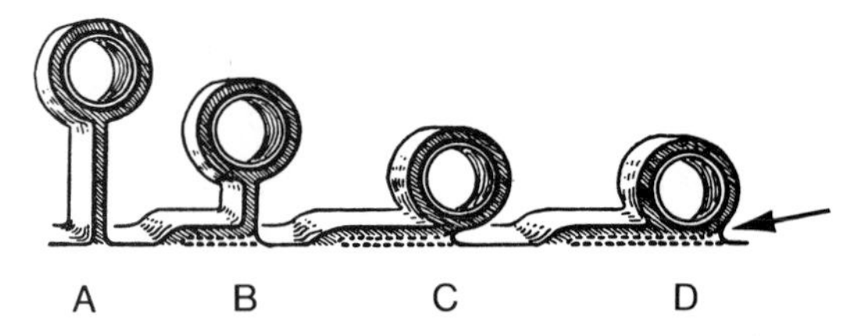

Figure 2-27. Diagrams illustrating the primitive mesentery of the large intestine in various stages of absorption. The extent to which the ascending and descending colons lose their primitive mesenteries varies. The *arrow* indicates the paracolic gutter where the visceral peritoneum is attached to the parietal peritoneum, and where an incision is made during mobilization of the intestine prior to surgery on the ascending colon or descending colon.

no mesentery and are extraperitoneal or retroperitoneal (*e.g.*, the ascending colon and kidneys). These organs lie on the posterior abdominal wall and are covered by peritoneum anteriorly (Fig. 2-27*D*).

Omentum (Figs. 2-28 and 2-36 to 2-38). *This is a mesentery or double layer of peritoneum that extends from the stomach to adjacent organs.*

The lesser omentum (Figs. 2-36 and 2-37) joins the lesser curvature of the stomach and the proximal part of the duodenum to the liver.

The greater omentum (Figs. 2-28, 2-35, and 2-37) is attached to the greater curvature of the stomach and hangs like a *vascular apron* from the transverse colon; hence its name (L. *omentum*, caul or veil). The greater omentum is translucent or filled with fat depending on the nutritional state of the person.

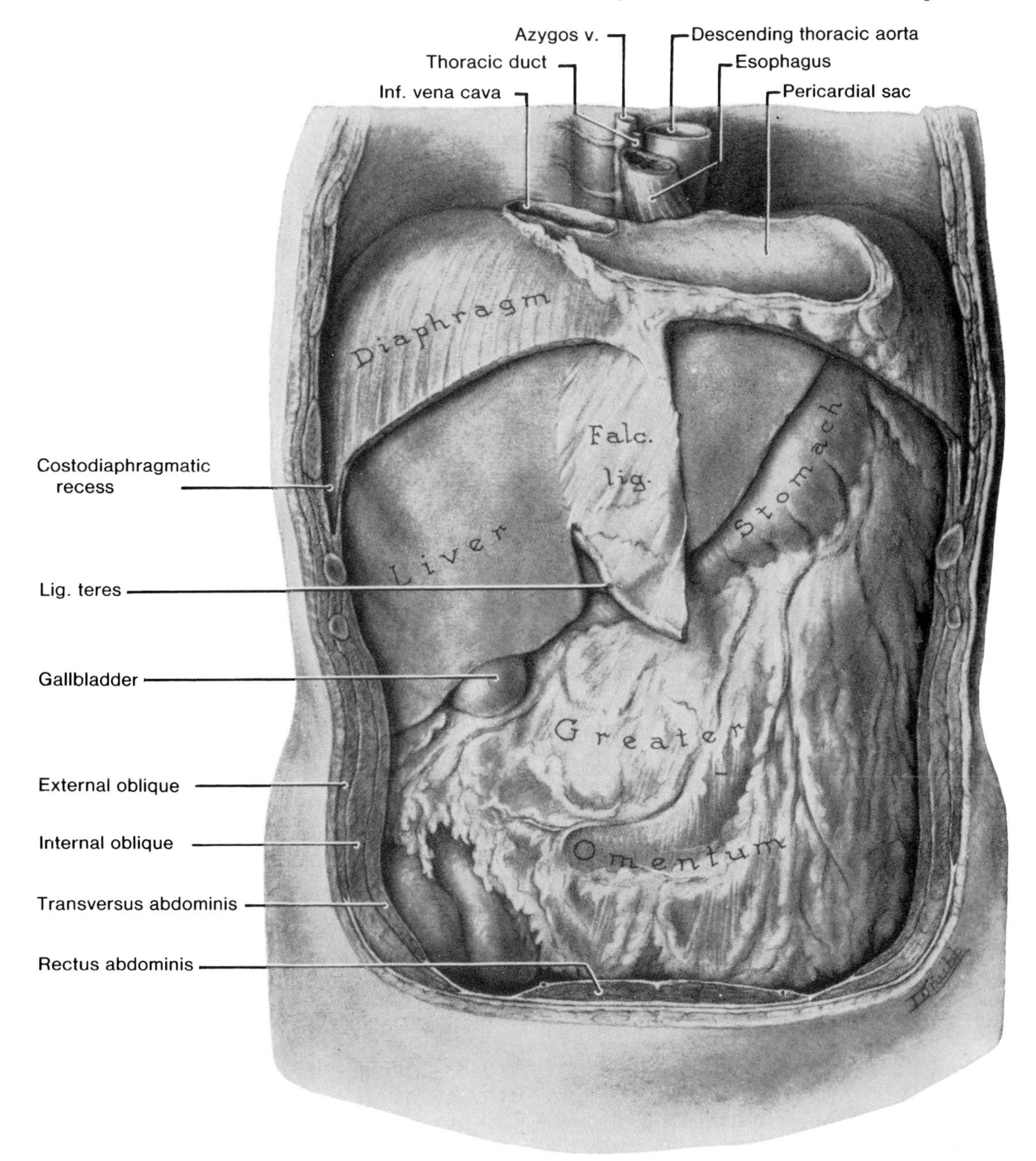

Figure 2-28. Drawing of a dissection illustrating the abdominal contents. The anterior and thoracic walls are cut away. Observe the falciform ligament with the ligamentum teres in its free edge, which was severed at its attachment to the anterior abdominal wall. Note the gallbladder projecting inferior to the sharp, inferior border of the liver. Observe that the internal oblique is the thickest of the three flat anterior abdominal muscles.

Ligaments (Fig. 2-38). All double layers or folds of peritoneum that are not called mesenteries or omenta are referred to as **peritoneal ligaments.**

Peritoneal ligaments are double layers of peritoneum that run from one viscus to another (*e.g.*, the **gastrolienal ligament** passes from the stomach to the spleen, Fig. 2-38), or from a viscus to the body wall (*e.g.*, the **falciform ligament** passes from the liver to the anterior abdominal wall, Fig. 2-28. Ligaments may contain blood vessels or remnants of vessels (*e.g.*, the ligamentum teres, shown in Fig. 2-28 is the remnant of the fetal umbilical vein).

Folds (Figs. 2-15, 2-39, 2-41, and 2-63). Folds are reflections of peritoneum with more or less sharp borders. Often they are formed by peritoneum covering blood vessels and ducts, some of which represent obliterated fetal vessels (Figs. 2-15 and 2-39). In these places the peritoneum is lifted off the body wall.

Retroperitoneal Organs (L. *retro*, back or behind). These organs are posterior to *the peritoneal sac*; thus they are covered anteriorly with peritoneum. The pancreas is retroperitoneal (Fig. 2-37).

Viscera without free mesenteries are retroperitoneal; *e.g.*, most of the duodenum, the ascending colon, and the descending colon (Fig. 2-87).

Peritoneal Recesses (Figs. 2-35 to 2-37 and 2-63). In certain places the peritoneum folds to form blind pouches (**culs-de-sacs**), or tubular cavities that are closed at one end and yet have an opening into the main part of the peritoneal cavity.

The largest peritoneal recess is the **omental bursa** (lesser sac) which lies posterior to the lesser omentum and the stomach (Fig. 2-30). It communicates with the general peritoneal cavity (greater sac) via the **epiploic foramen** (Fig. 2-35).

The duodenojejunal area often has two or three peritoneal recesses which are produced by accessory peritoneal folds (Fig. 2-63). In the ileocecal area there is usually a **retrocecal recess** where the peritoneal cavity extends superiorly, posterior to the cecum. *Frequently the vermiform appendix lies in the retrocecal recess* (Fig. 2-42). Usually there are one or two **ileocecal recesses** (Fig. 2-41). The superior ileocecal recess opens inferiorly, just superior to the terminal part of the ileum. The inferior ileocecal recess also opens inferiorly and is produced by the **ileocecal fold** which extends from the anterior and inferior parts of the ileum to the **mesentery of the appendix** (Fig. 2-41). At the inferior aspect of the apex of the sigmoid mesocolon (Fig. 2-39), there is often a pocket-like extension of the peritoneal cavity, called the **intersigmoid recess**, that passes superiorly, posterior to the root of the sigmoid mesocolon.

Awareness of the various peritoneal recesses is clinically important because of the possible occurrence of **internal hernias.** A loop of gut, usually the small intestine, may pass into one of these recesses and become strangulated. The intestine becomes twisted around the herniated loop, resulting in an obstruction of the bowel. The patient usually presents with severe colicky pain, owing to the **acute obstruction of the bowel.** Operative relief of the obstruction and closure of the hernial defect is required.

The peritoneal recesses are of clinical importance in connection with the spread of pathological fluids. Normally these recesses are in communication with each other, but they may become separated from each other by adhesions between the adjacent peritoneum and viscera.

The fluid accumulated in a peritoneal recess, if excessive, may have to be aspirated (*e.g.*, **culdocentesis** or transvaginal aspiration of fluid from the recess between the uterus and the rectum, called the rectouterine pouch (Fig. 2-31).

THE ABDOMINAL CAVITY

The abdominal cavity is the larger part of the abdominopelvic cavity. *The abdominal cavity is located superior to the superior aperture of the pelvis* (Figs. 2-29 and 3-42). It is limited superiorly by the diaphragm and is continuous inferiorly with the pelvic cavity, the smaller part of the abdominopelvic cavity, at the pelvic brim (Figs. 2-29 and 2-31).

Note that a large part of the abdominal cavity is under cover of the thoracic cage (Figs. 2-28 and 2-31).

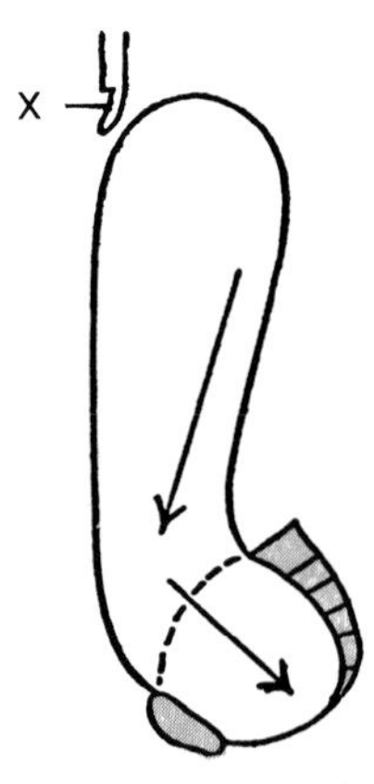

Figure 2-29. Diagram illustrating the abdominal and pelvic cavities as viewed in a median section of the body. The *broken line* (---) indicates the plane of the superior aperture of the pelvis which divides the abdominopelvic cavity into the abdominal cavity (*top arrow*) and the pelvic cavity (*bottom arrow*). *X*, xiphoid process.

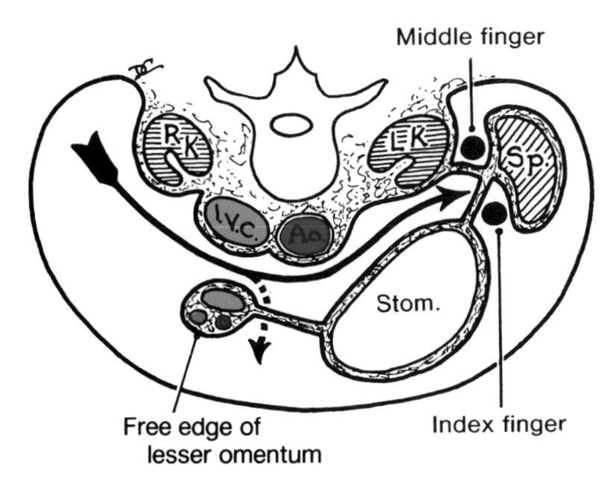

Figure 2-30. Diagram of a transverse section of the abdomen at the level of the epiploic foramen (Fig. 2-35). Note that the abdominal cavity is kidney-shaped because the vertebral column and the vessels anterior to it protrude into the cavity in the midline posteriorly. This diagram shows how one can palpate the hilum of the spleen while its pedicle is held between the two fingers of the right hand. The *arrow* indicates the path taken by a finger passed through the epiploic foramen (Fig. 2-35) into the omental bursa to reach the hilum of the spleen. The *broken arrow* indicates the passage into the superior recess of the omental bursa (Fig. 2-37). *RK, LK,* right kidney, left kidney. *Sp.*, spleen. Note that the bile duct (*green*), hepatic artery (*red*), and portal vein (*blue*), are located in the free edge of the lesser omentum.

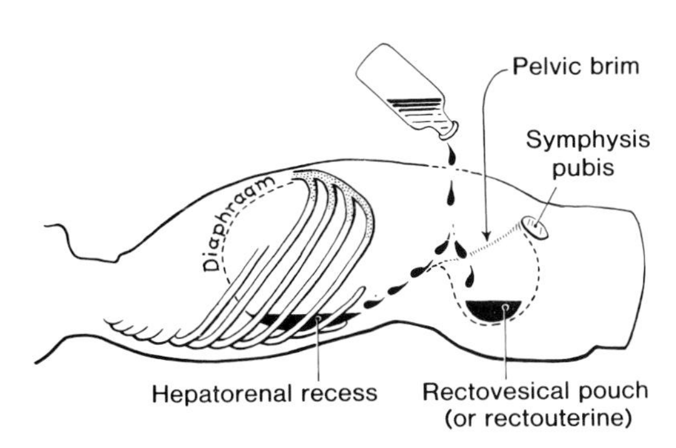

Figure 2-31. Diagram illustrating the two most posterior parts of the peritoneal cavity when a person is in the supine position. Serous fluid or blood from ruptured abdominal organs gravitates to the superior part of the abdomen in this position because the paravertebral grooves (gutters) slope posterosuperiorly. The hepatorenal recess is in the greater peritoneal sac just to the right of the epiploic foramen (Fig. 2-35). Its medial margin is the right kidney and its superior boundary is the liver. Fluid drains from this recess into the pelvis via the right paracolic gutter that lies to the right of the ascending colon. In a male the fluid accumulates in the rectovesical pouch, whereas in a female it collects in the rectouterine pouch. Observe that a large part of the abdominal cavity is under cover of the thoracic cage.

The abdominal cavity, as opposed to the peritoneal cavity, is *the part of the body cavity within the abdominal walls.*

Peritoneum lines the walls of the abdominal cavity; hence ***the peritoneal sac and peritoneal cavity are within the abdominal cavity.*** This concept is fundamental to the understanding of the abdomen.

Neither the abdominal cavity nor the peritoneal cavity is actually a cavity. The peritoneal cavity contains a capillary layer of fluid and the abdominal cavity is occupied by the closely packed viscera.

The abdominal cavity is kidney-shaped in cross-section because the vertebral column protrudes into it (Fig. 2-30). On each side of the vertebral column is a **paravertebral groove** (gutter) containing a kidney, a ureter, and part of the colon. These grooves slope posterosuperiorly so that when fluid accumulates in the peritoneal cavity it follows these grooves to the superior part of the abdomen when a patient lies in the supine position (Fig. 2-31).

THE ABDOMINAL VISCERA

Before considering the viscera in detail, a survey of them will be given for orientation.

THE LIVER

The liver is the largest gland in the body and is *larger than any other abdominal organ.* Most of the liver lies in the right upper quadrant of the abdomen, occupying almost all of the right hypochondriac region (Figs. 2-4*B* and 2-44). It extends inferiorly as far as the right costal margin (Figs. 2-28 and 2-32). Its smooth surfaces are in contact with the diaphragm and with the anterior abdominal wall. The **falciform ligament** attaches the liver to both of these structures (Figs. 2-28 and 2-35). The main attachment of the liver to the diaphragm is through the **coronary ligaments** (Fig. 2-71).

The stomach and the abdominal part of the esophagus are in contact with the left lobe of the liver, whereas the right lobe is in contact with the right colic flexure and the duodenum, close to the gallbladder (Fig. 2-32). The right kidney and suprarenal gland are also in contact with the right lobe of the liver.

Most of the liver is covered with peritoneum (Figs. 2-28 and 2-37).

In living persons the liver is reddish brown and is easily lacerated (e.g., by stab wounds or fractured ribs). Owing to its great vascularity, *liver lacerations often cause considerable hemorrhage* resulting in right upper quadrant pain. Because the liver is friable, wounds of the liver are not sutured too tightly.

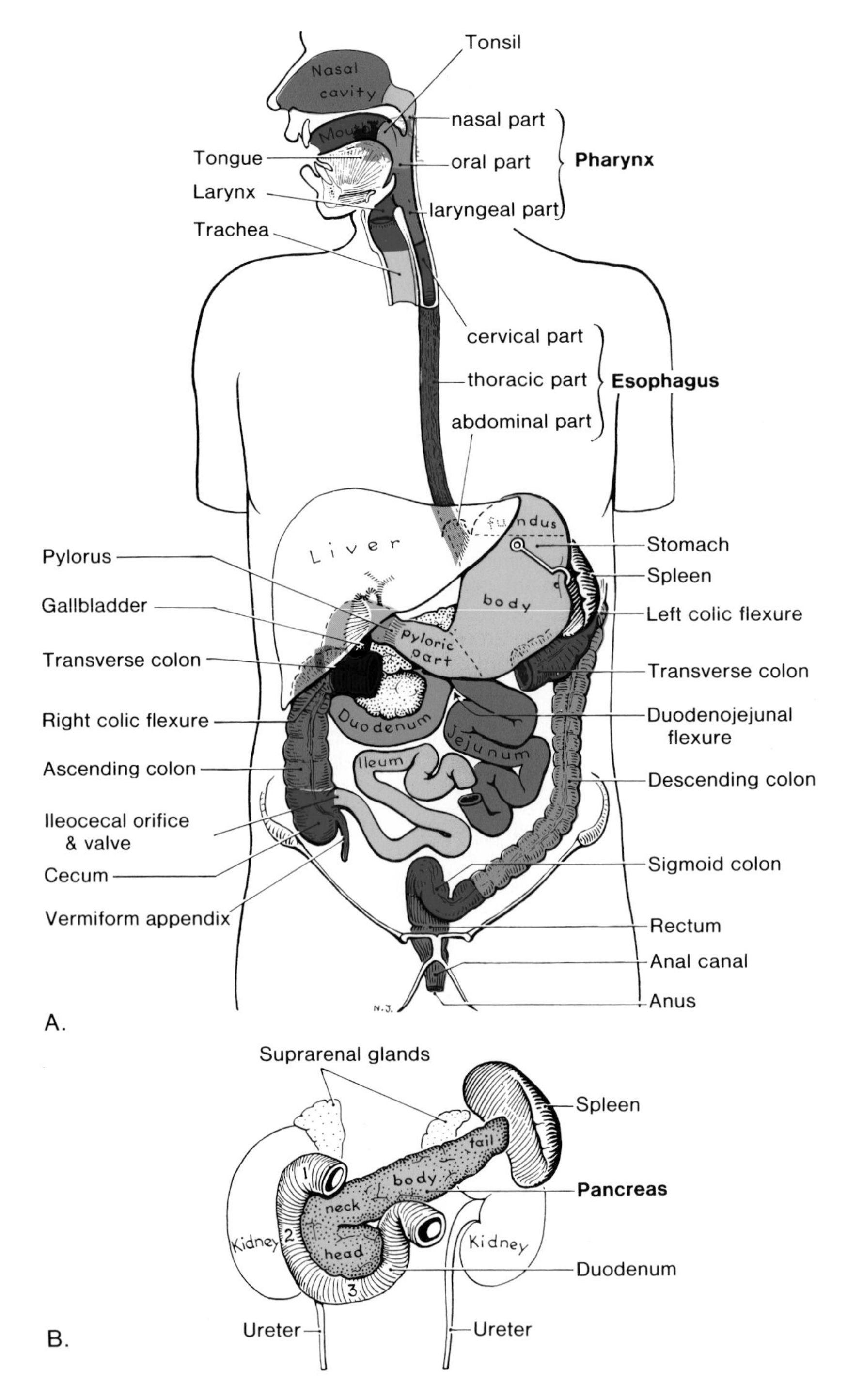

Figure 2-32. Diagram of the digestive system. Its relationship to the upper respiratory system (nasal cavity, nasal part of pharynx, larynx, and trachea) is also illustrated. *A*, shows that the digestive tract extends from the lips to the anus. In *B*, the spleen, pancreas, and duodenum are drawn after removal of the stomach and transverse colon. The abdominal viscera are: the stomach, intestines, liver, pancreas, spleen, kidneys, and suprarenal glands. Observe that the kidneys, ureters, and suprarenal glands lie on the posterior abdominal wall, where they are retroperitoneal and are enclosed in the fascial lining of the abdominal cavity. Note that other viscera lie anterior to these structures, where they are surrounded to a greater or lesser extent by the peritoneal cavity. The size of the peritoneal cavity in *A* appears exaggerated because most of the small intestine is not illustrated.

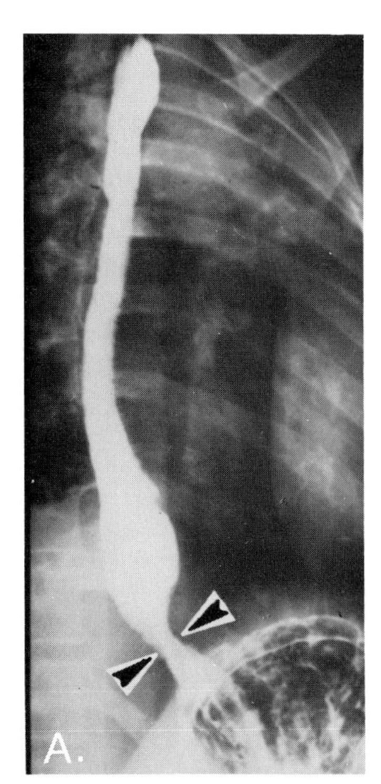

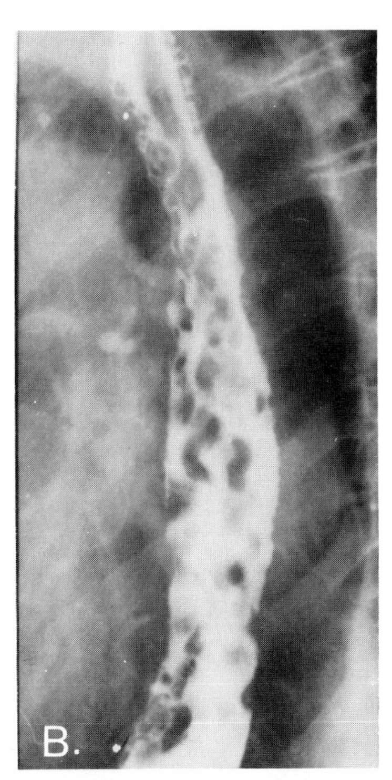

Figure 2-33. Radiographs of the esophagus. *A*, a normal esophagus after swallowing barium. Note the constriction where it passes through the esophageal hiatus in the diaphragm (*arrows*), and its short course (1.5 to 2.5 cm) in the abdomen before entering the stomach at the cardiac orifice. Note the bubble of gas in the fundus of the stomach. *B*, abnormal esophagus showing the barium outlining distended veins (esophageal varices) which are encroaching on the lumen. See Figures 2-32 and 2-101 for orientation.

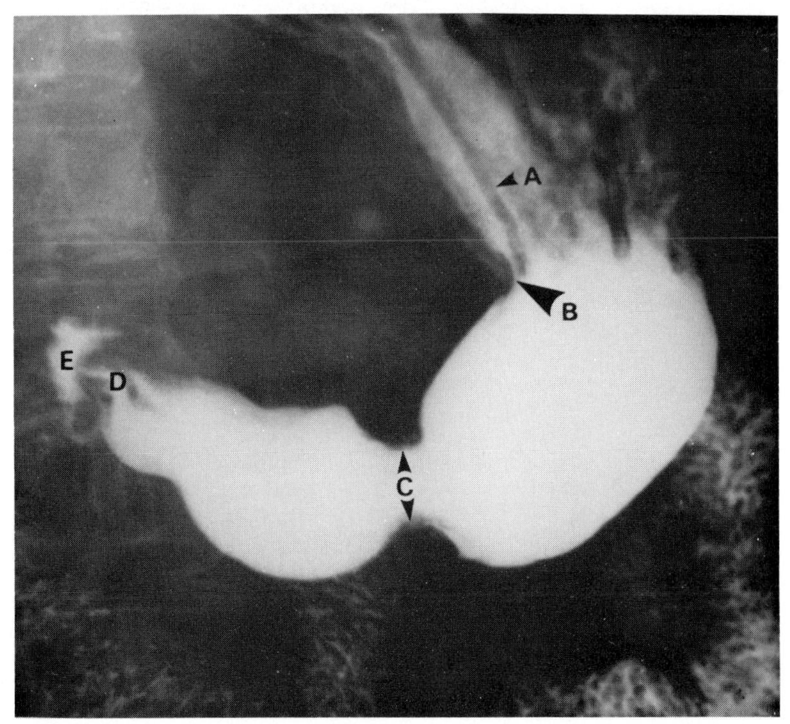

Figure 2-34. Radiograph of the stomach following a barium meal. The barium (*white*) is in a low anterior position in the stomach and the air (*gray* to *black*) is in a high posterior position. Observe *A*, longitudinal ridges (rugae) of mucous membrane (also see Fig. 2-46); *B*, the angular notch (incisura angularis); *C*, a peristaltic wave traveling toward the pylorus; *D*, the pylorus; *E*, the duodenal "cap" (the first portion of the superior part of the duodenum).

THE ESOPHAGUS

The esophagus is a fairly straight *muscular tube that extends from the pharynx to the stomach* (Figs. 2-28, 2-32, 2-33*A*, and 2-35). As the food is previously mixed with saliva, it passes rapidly down the esophagus. Its wall contains only a few mucous glands to supply additional lubrication.

The esophagus pierces the diaphragm just to the left of the midline (Figs. 1-29 and 2-35). The short abdominal part of the esophagus grooves the left lobe of the liver and enters the stomach at its cardiac orifice (Fig. 2-46). Here the esophagus is covered anteriorly and laterally by peritoneum and is encircled by the **esophageal plexus of nerves** (Fig. 1-81).

Heartburn is an esophageal symptom commonly experienced by most people. The term describes *a feeling of warmth or burning in the substernal region.* Heartburn is often accompanied by regurgitation of small amounts of food or fluid into the pharynx.

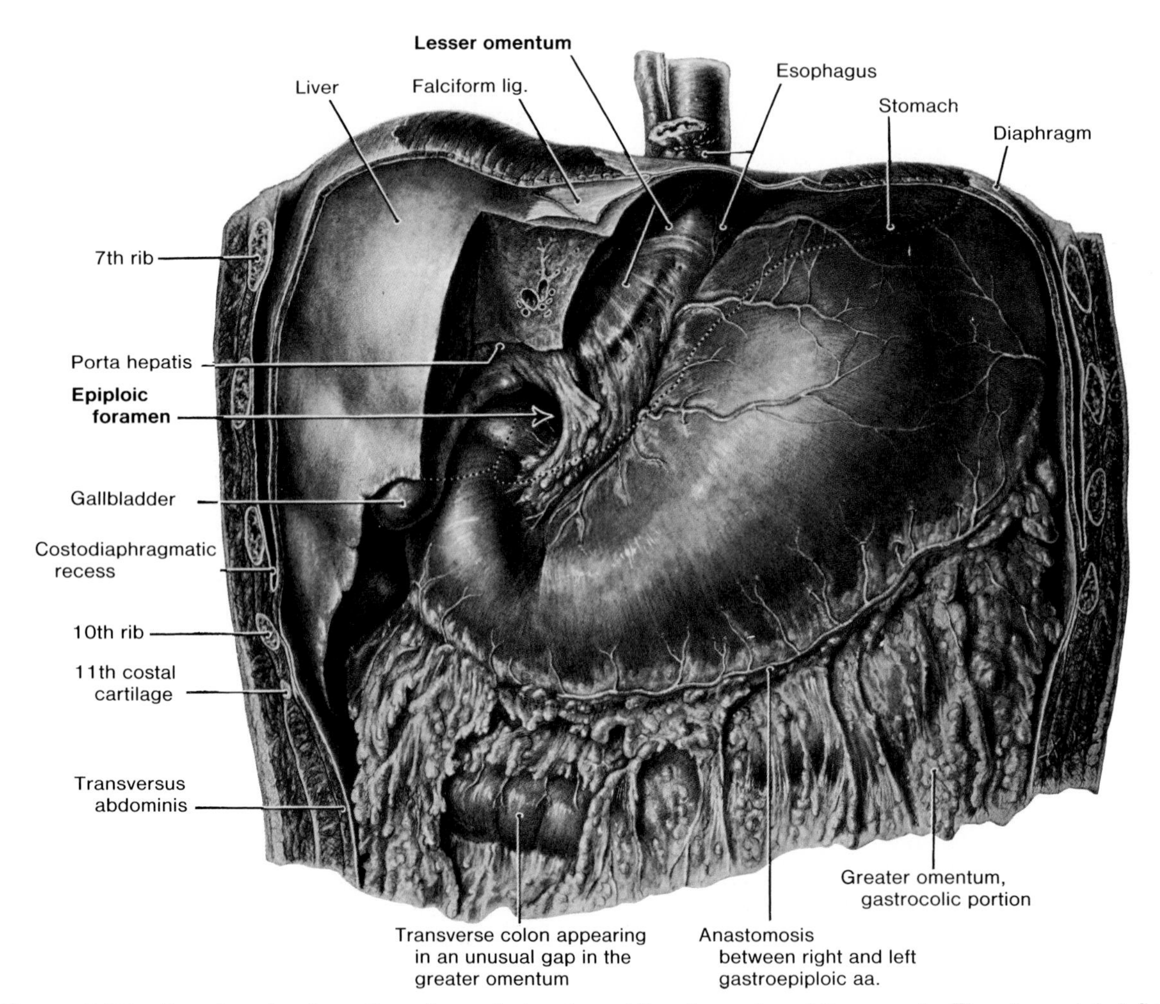

Figure 2-35. Drawing of a dissection of an anterior view of the stomach and the omenta. The stomach is inflated with air and the left part of the liver is cut away. Observe that the pyloric end of the stomach lies inferoposterior to the gallbladder. Note that the first or superior part of the duodenum almost occludes the epiploic foramen. The *arrowhead* indicates the epiploic foramen or mouth of omental bursa. Observe that the gallbladder, followed superiorly, leads to the free margin of the lesser omentum and, hence, acts as a guide to the epiploic foramen lying posterior to that free margin. Examine the lesser omentum passing from the lesser curvature of the stomach and the first 2 cm of the duodenum to the fissure for the ligamentum venosum and the porta hepatis. This omentum is thickened at its free margin where it forms the anterior lip of the epiploic foramen. Elsewhere the omentum is perforated so that the caudate lobe of the liver is visible through it. Observe the greater omentum attached to the greater curvature of the stomach and hanging from the transverse colon over the small intestine (see Fig. 2-37).

THE STOMACH

The stomach is the expanded portion of the alimentary canal between the esophagus and the small intestine (Figs. 2-28, 2-30, 2-32, 2-34 to 2-38, and 2-48).

The stomach lies in the left upper quadrant in the epigastric, umbilical, and left hypochondriac regions (Figs. 2-4 and 2-5). Its capacity commonly reaches about 1500 ml in adults. The **pylorus** (G. *pylorōs*, gatekeeper) is the distal portion of the stomach that opens into the duodenum (Figs. 2-36, 2-46, and 2-47).

In most people the stomach is J-shaped and its pyloric part lies horizontally or ascends to the proximal part of the duodenum (Fig. 2-32). The most inferior part of the greater curvature of the stomach may extend into the pelvis major in the erect position. Hence, *the position and shape of the stomach vary in different people* and in the same person, depending on its contents and the position of the body.

The stomach has a complete covering of peritoneum and is connected to other abdominal organs by peritoneal ligaments and omenta (Figs. 2-36 to 2-38).

The Lesser Omentum (Figs. 2-35 to 2-38). This *double layer of peritoneum* embraces the abdominal part of the esophagus and connects the lesser curvature of the stomach and the first 2 cm of the duodenum to the liver. *The bile duct, hepatic artery, and portal vein are in its free edge* (Figs. 2-30 and 2-58).

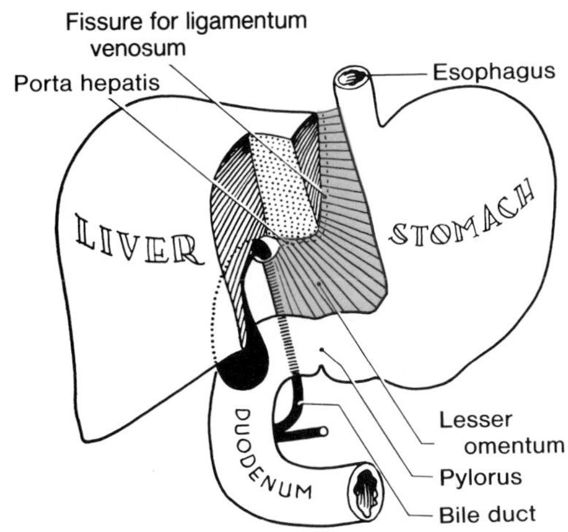

Figure 2-36. Diagram illustrating the attachments of the lesser omentum. Two sagittal cuts have been made through the liver: one at the fissure for the ligamentum venosum, the other at the right limit of the porta hepatis. These two cuts have been joined by a coronal cut. Note that the bile duct occupies the free edge of the lesser omentum. Observe that the lesser omentum extends from the lesser curvature of the stomach and the first 2 cm of the duodenum. The part of the lesser omentum attached to the body of the stomach passes to the fissure for the ligamentum venosum, and that attached to the pyloric part of the stomach and duodenum passes to the porta hepatis.

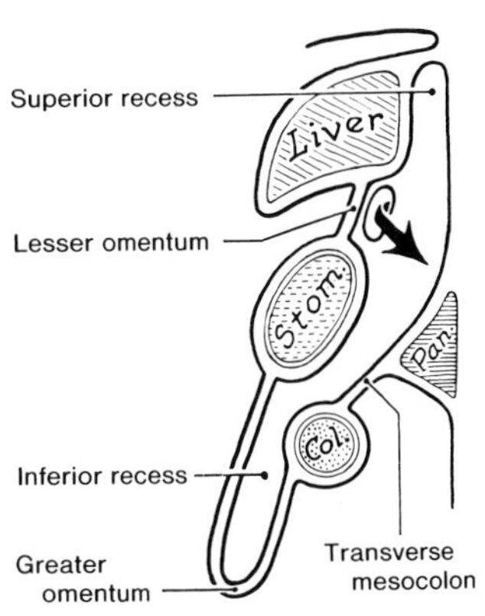

Figure 2-37. Diagram of a median section of the abdomen showing the vertical extent of the omental bursa (lesser sac). The *arrow* passes through the epiploic foramen (opening into omental bursa from the greater sac of the peritoneal cavity). The transverse colon (*Col.*), its mesentery, and the greater omentum are also illustrated. The two double layers of peritoneum of the greater omentum normally fuse during the fetal period, thereby obliterating the inferior recess of the omental bursa. As a result, the greater omentum is composed of four layers of peritoneum. They can be separated only partially in adults, and with difficulty, but this offers a surgical entrance to the omental bursa and the posterior aspect of the stomach.

The lesser omentum lies posterior to the left lobe of the liver (Figs. 2-35 and 2-36). It is attached to the liver in the fissure for the **ligamentum venosum** (remnant of fetal ductus venosus) and to the **porta hepatis** (Fig. 2-72). *The lesser omentum is a double-layered sheet of peritoneum which may be divided into two parts* (Fig. 2-38): the **hepatogastric ligament** and the **hepatoduodenal ligament.** The lesser omentum ends in a free edge between the porta hepatis and the duodenum (Figs. 2-35 and 2-36).

The portal vein, the hepatic artery, and the bile duct *run between the layers of the lesser omentum near its free edge* (Figs. 2-30 and 2-81).

The Greater Omentum (Figs. 2-28, 2-35, and 2-37 to 2-39). When the anterior abdominal wall is removed, the intestines are more or less hidden by this ***fatty, double-layered, vascular apron*** that hangs over them from the greater curvature of the stomach (Fig. 2-38). Hanging inferiorly over the transverse colon, the two layers of peritoneum are attached to the posterior abdominal wall (Fig. 2-37). The two layers of the greater omentum descend and then turn superiorly (Fig. 2-37), where they come to lie against each other (Fig. 2-38). Hence *the greater omentum is a four-layered omental apron* (Fig. 2-28) within which a potential space separates the two sets of layers (Fig. 2-38).

The greater omentum often contains a considerable amount of extraperitoneal tissue and fat. It may be

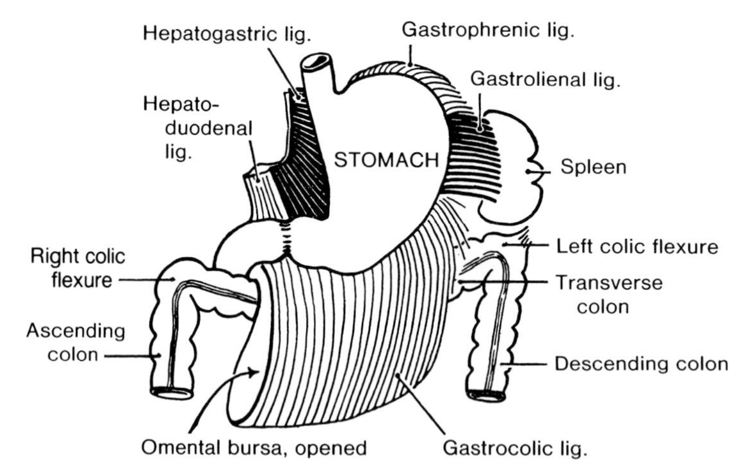

Figure 2-38. Diagram illustrating the two parts of the lesser omentum (hepatogastric and hepatoduodenal ligaments), and the three parts of the greater omentum (gastrophrenic, gastrolienal, and gastrocolic ligaments). *Gaster* is the Greek word for belly (stomach) and is the basis of many medical terms; *e.g.*, gastritis means inflammation, especially mucosal, of the stomach. From the transverse colon the two folds of the double layered gastrocolic ligament hang down over the coils of small intestine. All four leaves of peritoneum become adherent to each other and can be separated only partially and with difficulty in the adult to open the inferior recess of the omental bursa as shown in this illustration and in Figure 2-37.

short or long enough to reach the pelvic brim. In emaciated persons, the greater omentum may be as thin as a piece of paper, where *in obese persons the greater omentum is of considerable thickness and weight.*

After passing inferiorly, the greater omentum loops back on itself (Figs. 2-37 and 2-38), overlying and attaching to the transverse part of the large intestine (**transverse colon**), which runs across the abdomen just inferior to the stomach (Figs. 2-35 and 2-38).

The greater omentum, extending from the greater curvature of the stomach, may be divided into three parts (Fig. 2-38): (1) *the apron-like part,* called the **gastrocolic ligament**, is attached to the transverse colon. This is the part usually referred to when the term greater omentum is used; (2) *the left part,* called the **gastrolienal ligament** (gastrosplenic ligament), is attached to the spleen (L. *lien*). This part connects the spleen to the greater curvature of the stomach (Fig. 2-38) and (3) *the superior part,* called the **gastrophrenic ligament**, is attached to the diaphragm (G. *phrēn*).

Some people believe that the greater omentum prevents the intestines from adhering to the parietal peritoneum on the anterior abdominal wall. The greater omentum has considerable mobility and tends to migrate to any inflamed area and wrap itself around an inflamed organ such as the appendix. Because it "walls off" an infection from other organs, it has been called the "policeman of the abdomen".

THE SMALL INTESTINE

The pylorus of the stomach empties into the intestinal tract. *The convoluted tube, extending from the pylorus to the ileocecal valve, is called the small intestine* (small bowel). It consists of three parts: **duodenum, jejunum, and ileum.**

Most digestion occurs in the small intestine. The length of the small intestine varies in different persons, but the average is 6 to 7 meters. Excision of up to one-third of it is compatable with a normal life.

The Duodenum (Figs. 2-5*B*, 2-28, 2-32, 2-34 to 2-36, 2-38, and 2-39). This first part of the small intestine pursues a *C-shaped course from the pylorus around the head of the* ***pancreas*** *to become continuous with the jejunum.* The duodenum (L. *duodeni*, twelve) was given its name because it is about 12 fingerbreadths long (25 cm).

The position of the duodenum is variable, but it **begins on the right side** (2 to 3 cm from the median plane) and **ends on the left side** at the duodenojejunal junction (2 to 3 cm from the median plane. *Although about 25 cm long, the two ends of the duodenum are only about 5 cm apart* (Figs. 2-59 to 2-61).

The duodenum is fixed and retroperitoneal, *i.e., most of it does not have a mesentery.* It lost its mesentery during fetal development and came to lie retroperitoneally against the posterior abdominal wall.

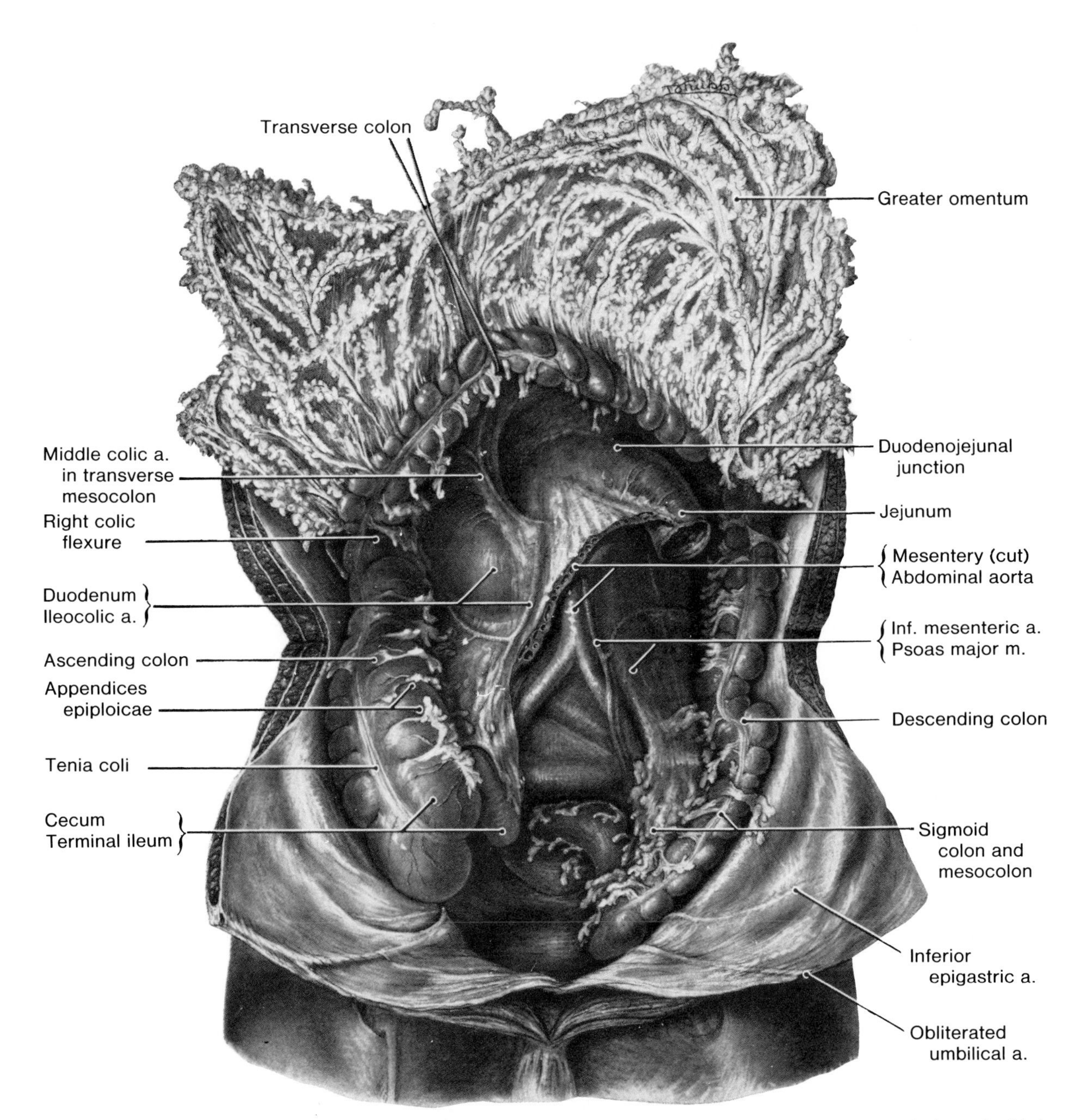

Figure 2-39. Drawing of a dissection of the intestines. The greater omentum is reflected superiorly and with it the transverse colon and the transverse mesocolon (mesentery of the transverse colon). The jejunum and ileum are cut away, except their end pieces and the mesentery is cut short. Observe the duodenojejunal junction, situated to the left of the median plane and immediately inferior to the root of the transverse mesocolon. Note that the first few centimeters of the jejunum descend inferiorly and to the left, anterior to the left kidney, and that the last few centimeters of the ileum ascend to the right from the pelvic cavity. Observe that the large intestine forms 3½ sides of a square picture frame around the jejunum and ileum (removed); the missing half-side being between the cecum and the sigmoid colon. Note the following distinguishing features of the large intestine: (1) its position around the small intestine; (2) the teniae coli or longitudinal muscle bands; (3) the sacculations or haustra; and (4) the appendices epiploicae (fat-filled pouches). The vermiform appendix was removed at operation.

The beginning of the duodenum, called the **duodenal cap** (Fig. 2-34*E*), is attached to the liver by the *hepatoduodenal ligament* (Fig. 2-38). This ligament is part of the lesser omentum; thus this part of the duodenum is mobile.

The Jejunum (Figs. 2-32 and 2-39). The jejunum (L. *jejunus*, empty) begins at the duodenojejunal flexure. It constitutes *about two-fifths of the small intestine.* Most of the jejunum occupies the umbilical region (Fig. 2-5*A*).

The Ileum (Figs. 2-32, 2-39, and 2-40). The ileum (L. rolled up, twisted) *comprises the distal three-fifths of the small intestine.* Most of the ileum lies in the hypogastric region (Figs. 2-5*A* and 2-32), but its distal part is nearly always in the pelvis, from which it ascends to end in the medial aspect of the cecum (Figs. 2-32, 2-39, and 2-41).

The jejunum and ileum are attached to the posterior abdominal wall by a fan-shaped mesentery, the root of which runs obliquely and to the right from the upper left quadrant to the lower right quadrant of the abdomen (Figs. 2-5*B* and 2-39).

THE LARGE INTESTINE

The large intestine (large bowel) consists of the cecum, the colon (ascending, transverse, descending, and sigmoid), the rectum, and the anal canal (Figs. 2-32 and 2-39 to 2-41). The large intestine forms an arch or almost complete frame for the coils of small intestine (Fig. 2-32).

The large intestine can easily be distinguished from the small intestine by (1) its *three thickened bands of longitudinal muscle*, called **teniae coli** or taeniae coli (Figs. 2-39 and 2-41); (2) the sacculations of its wall between the teniae coli, called **haustra** (Figs. 2-40 and 2-41); and (3) the small pouches of peritoneum filled with fat, called *appendices epiploicae* (Fig. 2-41).

The Cecum (Figs. 2-32 and 2-39 to 2-41). The cecum (L. *caecus*, blind) is *the sac-like extension of the large intestine*, 5 to 7 cm in length, that lies in the right lower quadrant in the right iliac fossa (Figs. 2-41, 2-43, and 2-69). The ileum joins the cecum at the **ileocecal valve**. The cecum is the commencement of the large intestine.

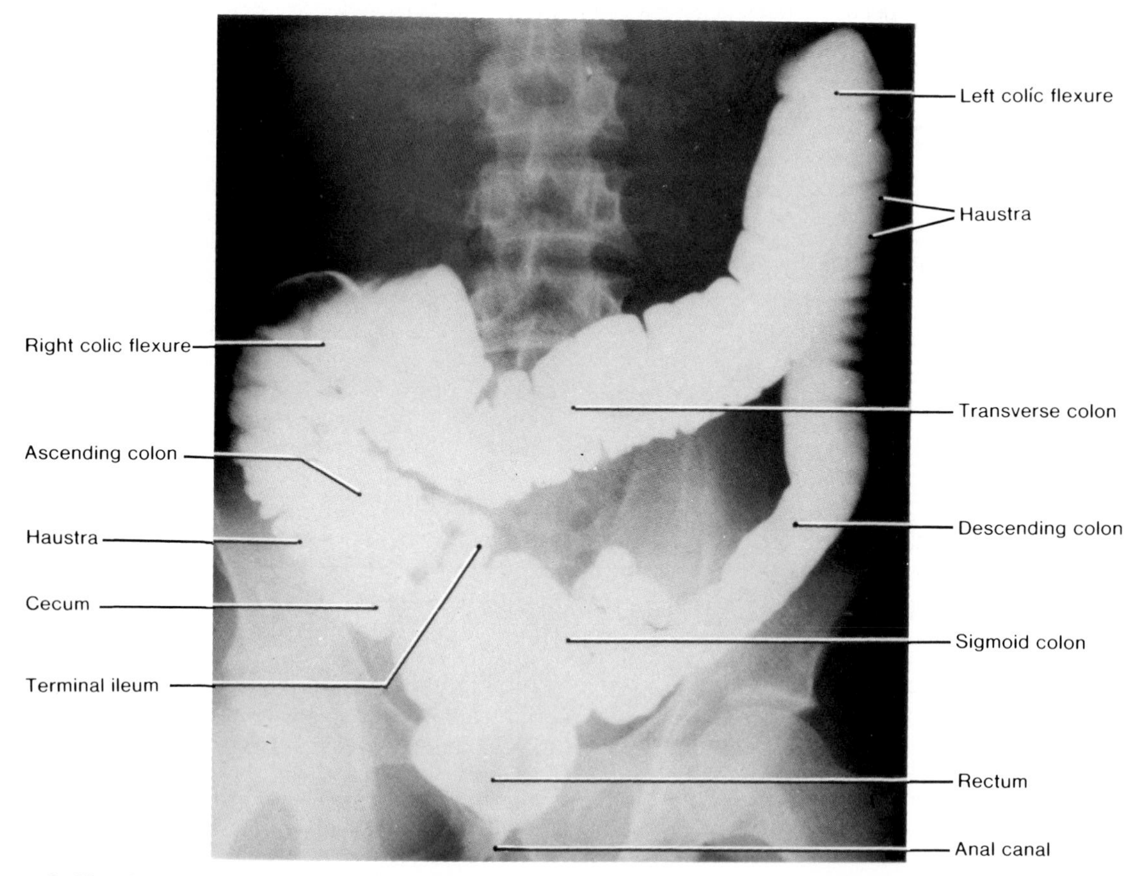

Figure 2-40. Anteroposterior radiograph of the abdomen following a barium enema with the patient in the supine position. Observe the haustra (sacculations) in the wall of the colon. The right colic flexure lies inferior to the liver (Fig. 2-32) and lies more inferiorly than the left colic flexure which lies inferior to the spleen (Figs. 2-32 and 2-38).

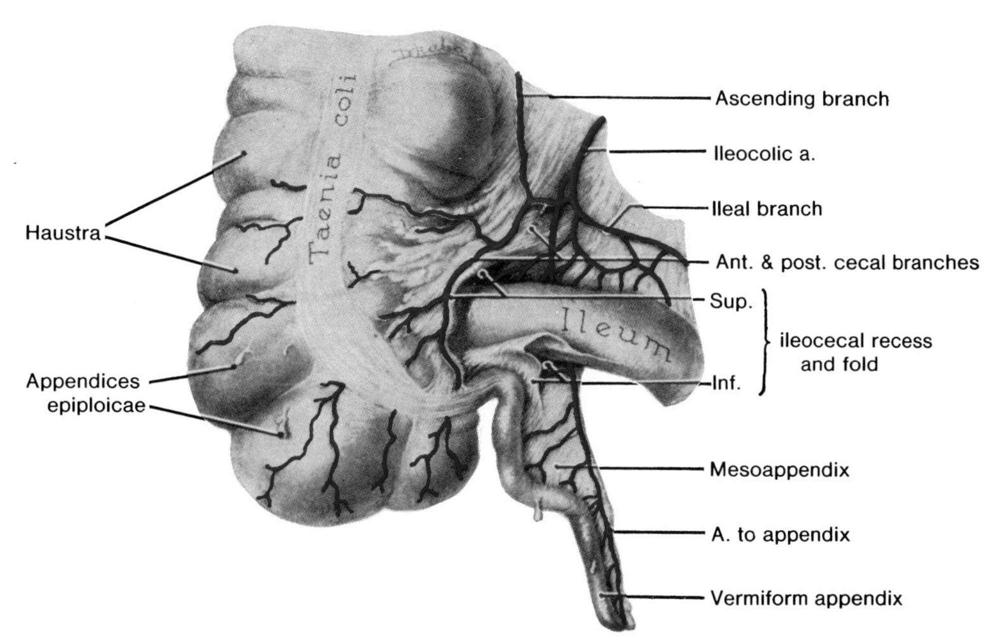

Figure 2-41. Drawing of a dissection of the ileocecal region. Observe the vermiform appendix is in one free border of the mesoappendix and its artery is in the other. Note that the anterior tenia coli or taenia coli leads to the appendix; this is a guide to the appendix during appendectomy. Observe the inferior ileocecal fold extending from the ileum to the mesoappendix. Although this is often called the "bloodless fold," one should be alert to the possibility of the presence of a small artery in it. Note that the superior ileocecal fold is vascular and that the artery to the appendix is a branch of the ileocolic artery. It represents the entire vascular supply to the appendix.

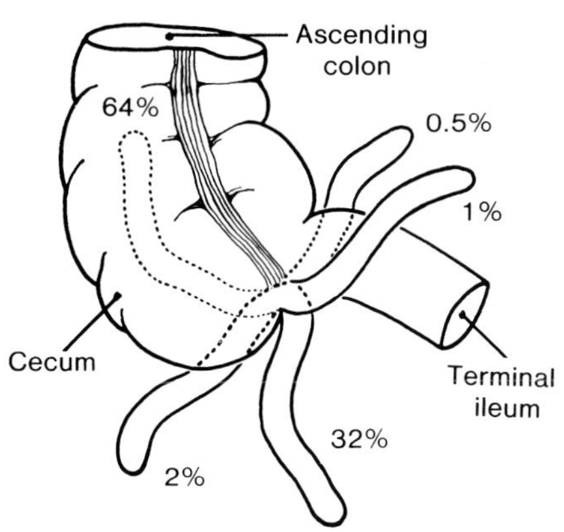

Figure 2-42. Diagram showing the approximate incidence of the various locations of the vermiform appendix. Like the hands of a clock, it may point in any direction. It may be long (20 cm) or short (5 cm), but the appendix averages 8 cm in length. Note that in most people (64%) the appendix is found posterior to the cecum (*i.e.*, retrocecal) in the retrocecal recess. When long enough it may lie posterior to the inferior part of the ascending colon (*i.e.*, retrocolic). When it extends into the pelvis (32%), it often lies close to the ovary and the uterine tube in females and to the ureter in both sexes.

Usually the cecum is almost entirely enveloped by peritoneum, but *it does not possess a mesentery.* However, the cecum is frequently attached by peritoneum to the iliac fossa laterally and medially, producing a small cue-de-sac of the peritoneal cavity, called the **retrocecal recess**. It lies posterior to the cecum and may extend superiorly, posterior to the inferior end of the ascending colon. Often the retrocecal recess is deep enough to admit a finger and in most people the vermiform appendix lies in it (Fig. 2-42).

The Vermiform Appendix (Figs. 2-32 and 2-41 to 2-43). This narrow, worm-shaped (L. *vermis*, worm + *forma*, form), blind tube of variable length (averages 8 cm) joins the cecum 2 to 3 cm inferior to the ileocecal junction. It is longer in infants and children than in adults.

The appendix has its own short mesentery called the **mesoappendix** (Fig. 2-41) which connects it to the inferior part of the mesentery of the ileum.

The position of the appendix is variable (Fig. 2-42). **Usually the appendix is retrocecal or pelvic** (*i.e.*, hangs over the pelvic brim into the pelvis minor).

The base (root) of the appendix usually lies deep to **McBurney's point** (Fig. 2-43), which is at the junction of the lateral and middle thirds of the line joining the anterior superior iliac spine and the umbilicus.

The vermiform appendix is probably not a vestigial organ. In infants and children it has the appearance of a well-developed lymphoid organ and may have important immunological functions.

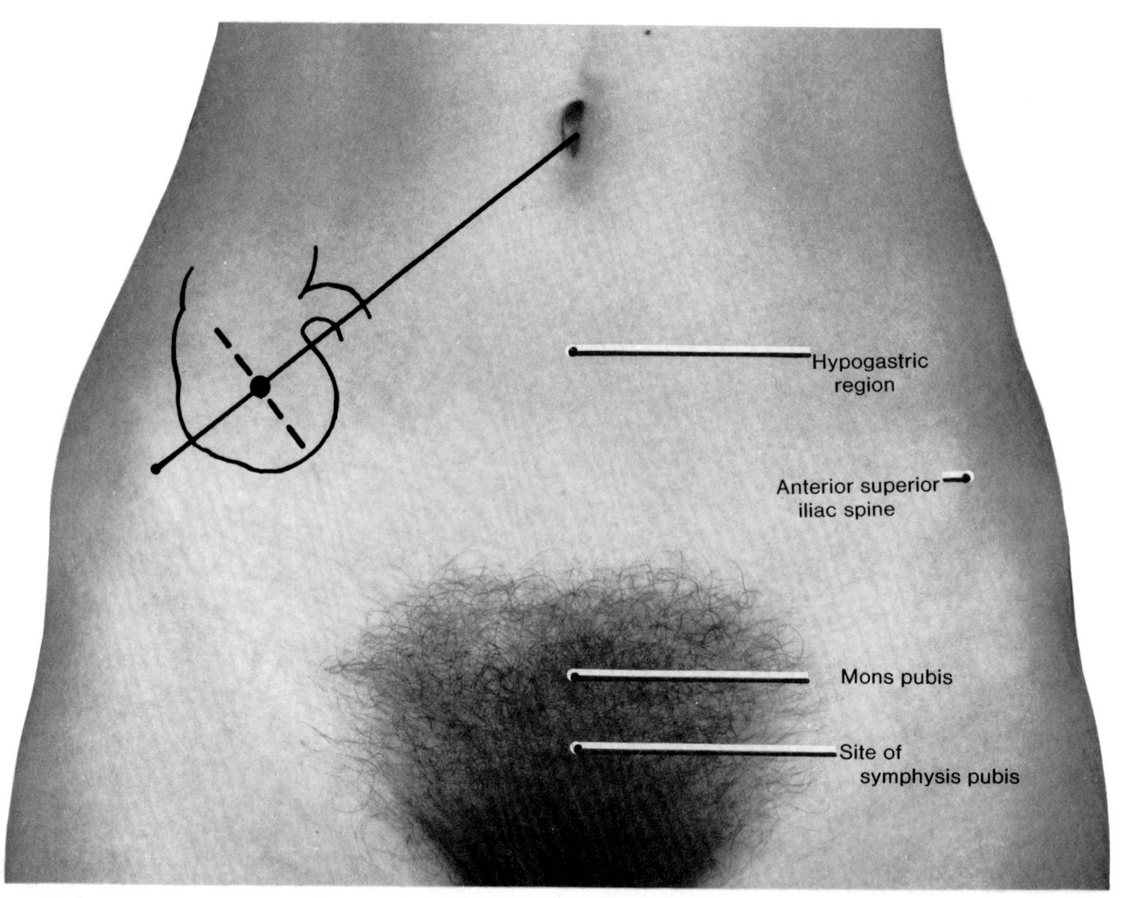

Figure 2-43. Photograph of the abdomen of a 27-year-old woman showing a projection of the cecum and terminal ileum, and the location of the base of the appendix as indicated by McBurney's point (*black dot*). Note that this clinically important landmark is at the junction of the lateral and middle thirds of a line joining the anterior superior iliac spine and the umbilicus. McBurney's point indicates where the vermiform appendix usually opens into the cecum (see Fig. 2-41), and where pressure of the finger elicits tenderness in acute appendicitis. The broken line (----) indicates the classical McBurney oblique skin incision centered at McBurney's point. The symphysis pubis is the cartilaginous joint (Fig. 2-1) that is located in the midline between the bodies of the pubic bone. The mons (L. mountain) pubis is a prominence caused by a pad of fatty tissue over the symphysis pubis in the female (Fig. 3-5).

The structure of the appendix varies with age. During old age the lymphoid tissue atrophies and is replaced largely by connective tissue.

Inflammation of the vermiform appendix (**appendicitis**) is one of the most common causes of an "**acute abdomen.**" *The pain of acute appendicitis usually commences in the periumbilical region* (**periumbilical pain**) and then localizes in the right lower quadrant.

In typical cases, fingertip pressure over **McBurney's point** (Fig. 2-43) registers the maximum abdominal tenderness, but in cases of retrocecal appendix (Fig. 2-42), the maximum tenderness may be in the *flank* (the region between the ribs and the iliac crest, Fig. 2-1). The pain is felt low in the right side when the appendix hangs in the pelvis (Fig. 2-42).

You must keep in mind that in unusual cases of malrotation or **incomplete rotation of the cecum**, the base of the appendix is not located at McBurney's point. *When the cecum is high* (***subhepatic cecum***), *the appendix is located in the right hypochondriac region*, and in these cases, the pain is localized there, not in the right lower quadrant (Fig. 2-5).

Acute infection of the appendix may result in **thrombosis of the appendicular artery** and development of **gangrene** (necrosis owing to obstruction of blood supply).

Rupture of an inflamed appendix results in infection of part or all of the peritoneum (*i.e.*, a local or *general peritonitis*), increased abdominal pain, and *abdominal rigidity*.

Appendectomy is performed through a muscle splitting technique (p. 171) in the right lower quadrant which is centered at McBurney's point (Fig. 2-43). The cecum is delivered into the wound and the mesentery of the appendix containing the appendicular vessels (Fig. 2-41) is firmly ligated and divided. The base of the appendix is tied, the appendix is excised, and its stump is usually cauterized and invaginated into the cecum.

The Ascending Colon (Figs. 2-32, 2-38 to 2-40, 2-52, and 2-69). This part of the colon (G. *kolos*, large intestine, hollow), varies from 12 to 20 cm in length. It extends superiorly on the right side of the abdominal cavity from the cecum to the **right colic flexure**. Its anterior and lateral surfaces are covered with peritoneum (Fig. 2-27*D*).

It has no mesentery and lies retroperitoneally along the right side of the posterior abdominal wall. Its original fetal mesentery became adherent to the peritoneum of the posterior abdominal wall (Fig. 2-27*D*).

Prior to **resection** (surgical excision) of all or part of the ascending colon (*e.g.*, owing to *colonic cancer*), it has to be mobilized. The basic principle of **mobilization of the colon** is reconstruction of its primitive mesentery to the stage shown in Figure 2-27*A*. An incision is made along the attachment of the visceral peritoneum to the parietal peritoneum in the right paracolic gutter (Fig. 2-27*D*), and the colon is reflected medially by cleavage of the fused layers of fascia.

During mobilization of the ascending colon, neither the vessels supplying it nor the ureter and vessels supplying the kidney are disturbed, because they lie posterior to the separated layers of fascia (Figs. 2-27*D* and 2-45).

The Transverse Colon (Figs. 2-32, 2-35, 2-37 to 2-40, and 2-69). The transverse colon, about 45 cm in length, is the largest and most movable part of the large intestine. *The transverse colon crosses the abdomen from the* ***right colic flexure*** *to the* ***left colic flexure***, where it bends inferiorly to become the descending colon. The left colic flexure is attached to the diaphragm by the **phrenicocolic ligament** (Fig. 2-48), which also forms a supporting shelf for the spleen.

The left colic flexure is at a more superior level and in a more posterior plane than the right colic flexure (Figs. 2-32, 2-40, and 2-69). Between these two flexures, the transverse colon is freely movable and forms a loop that is directed inferiorly and anteriorly.

The transverse colon has a mesentery, called the **transverse mesocolon**, which is connected to the inferior border of the pancreas and to the greater omentum that covers it anteriorly (Figs. 2-35, 2-37, and 2-39).

Because it is freely movable, the transverse colon is very variable in position. It may be at the level of the transpyloric plane (Fig. 2-4*B*), or it may hang down as far as the pelvic brim (Figs. 2-31, 2-40, and 2-69).

The Descending Colon (Figs. 2-32, 2-38 to 2-40, and 2-69). The descending colon, varying from 22-30 cm in length, descends from the sharply curved left colic flexure into the left iliac fossa (Fig. 2-5*A*), where it is continuous with the sigmoid colon. The caliber of the descending colon is considerably smaller than that of the ascending colon (Figs. 2-39 and 2-40).

The descending colon has no mesentery and lies retroperitoneally along the left side of the posterior abdominal wall. Its posterior surface is attached to the posterior abdominal wall, but the descending colon can be mobilized surgically as just discussed for the ascending colon (Fig. 2-27).

The Sigmoid Colon (Figs. 2-32, 2-39, and 2-40). The sigmoid colon (pelvic colon) forms a loop of variable length (15 to 80 cm) that reminded early anatomists of the Greek letter **sigma**. *The sigmoid colon is the portion of the large intestine between the descending colon and the rectum* (Figs. 2-32, 2-40, and 2-69). The sigmoid colon has a mesentery (**sigmoid mesocolon**) and therefore considerable freedom of movement. The sigmoid mesentery has a Λ-shaped attachment, superiorly along the external iliac vessels and inferiorly from the bifurcation of the common iliac vessels to the anterior aspect of the sacrum.

The **appendices epiploicae are very long in the sigmoid colon** (Fig. 2-39). These appendices (L. appendages) are little processes or sacs of peritoneum which are generally distended with fat.

The Rectum (Figs. 2-32, 2-40, and 2-69). The rectum (L. *rectus*, straight) is only partially covered with peritoneum and *has no mesentery.* The inferior part of the rectum passes through the pelvic floor to become the anal canal. As the rectum is a pelvic organ, it is described with the other contents of the pelvic cavity in Chapter 3 (p. 381).

The Anal Canal (Figs. 2-32 and 2-40). The anal canal is *the terminal part of the digestive tract.* The anal canal terminates at the anus in the perineum (Figs. 3-2 and 3-5). The anal canal is described with the perineum in Chapter 3 (p. 384).

THE KIDNEYS AND SUPRARENAL GLANDS

(Figs. 2-30, 2-32*B*, 2-61, and 2-105)

The kidneys (L. *renes*, kidneys), one on each side of the vertebral column, lie in the **paravertebral grooves** at the level of T12 to L3 vertebrae. Their long axes are almost parallel with the long axis of the body, but *their superior poles are slightly more medial* *than their inferior poles.* Owing to the bulk of the liver, the right kidney usually lies at a slightly lower level than the left (Fig. 2-105).

The kidneys lie in a mass of perirenal fat, *external or posterior to the peritoneum; i.e.,* **the kidneys are retroperitoneal.** A ureter runs inferiorly from each kidney and passes over the pelvic brim at the bifurcation of the common iliac artery (Fig. 3-57). It runs along the sidewall of the pelvis and enters the urinary bladder (Fig. 2-25). *The ureter is covered with peritoneum on its anterior surface throughout its entire length in the abdomen.*

The Suprarenal Glands (Fig. 2-32*B*). Each of the triangular suprarenal glands (adrenal glands) lies against the superomedial surface of the corresponding kidney, forming a cap over its superior pole (Figs. 2-61 and 2-62*B*). *Like the kidneys, these endocrine glands are retroperitoneal.*

THE OMENTAL BURSA (Figs. 2-30, 2-37, 2-38, and 2-45)

The omental bursa (lesser sac) is a compartment or recess of the peritoneal cavity located between the stomach and the posterior abdominal wall. It develops as the stomach rotates in the embryo. It is an extension of the peritoneal cavity into the right side of the dorsal mesentery of the stomach and soon expands to the left and inferiorly.

The inferior extension of the omental bursa, called the **inferior recess** (Fig. 2-37), is between the duplicated layers of the gastrocolic ligament of the greater omentum.

In adults the inferior recess is *a potential space.* It is usually shut off from the main part of the bursa by adhesion of the layers of the gastrocolic ligament.

The omental bursa also has a **superior recess** (Fig. 2-37), which is limited superiorly by the diaphragm and the posterior layers of the **coronary ligament** (Fig. 2-71).

The omental bursa or lesser sac is in communication with the main peritoneal cavity or greater sac through the **epiploic foramen,** *located posterior to the free edge of the lesser omentum* (Figs. 2-35, 2-37, and 2-47).

The boundaries of the epiploic foramen are: ***anteriorly,*** the portal vein, hepatic artery, and bile duct (all in the free edge of the lesser omentum); ***posteriorly,*** the inferior vena cava and the right crus of the diaphragm; ***superiorly,*** the caudate lobe of the liver (Fig. 2-72); and ***inferiorly,*** the superior part of the duodenum, portal vein, hepatic artery, and bile duct on their way to and from the lesser omentum.

The omental bursa is located posterior to the lesser omentum and the stomach (Figs. 2-35 and 2-37). As the anterior and posterior walls of the bursa slide smoothly on each other, *the omental bursa gives considerable movement to the stomach,* permitting it to slide freely on it during contraction and distention.

A loop of small intestine occasionally passes through the epiploic foramen into the omental bursa and becomes strangulated by the edges of the foramen. As none of the boundaries of epi-

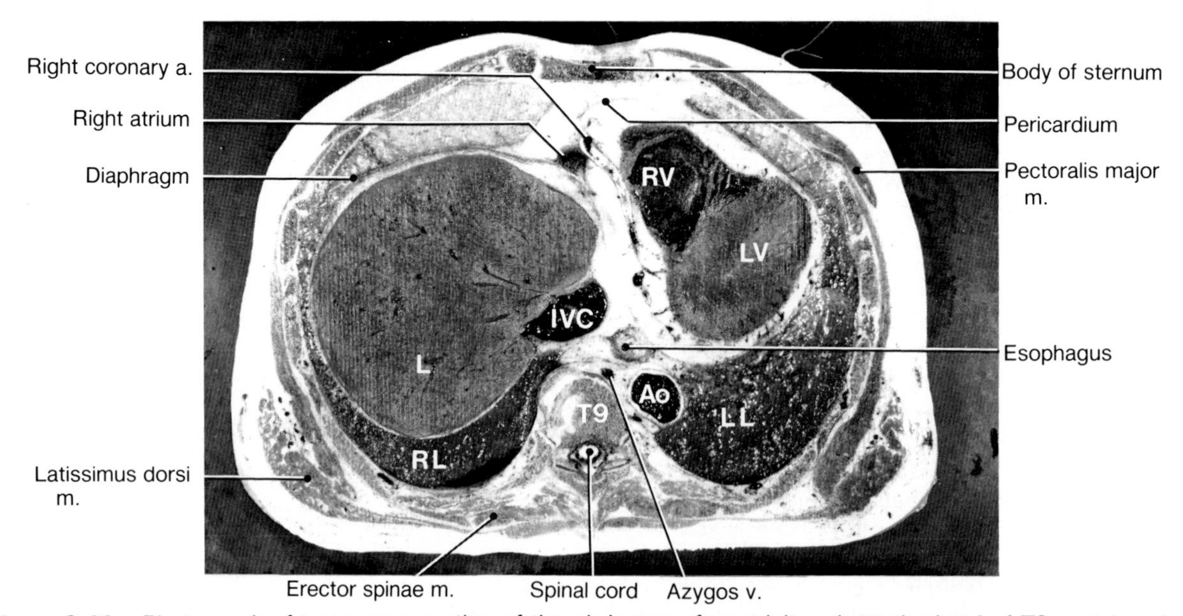

Figure 2-44. Photograph of transverse section of the abdomen of an adult male at the level of T9 vertebra. In keeping with radiographic convention, as in CT scans, the section is viewed from below. Hence the right side of the body appears on the left side of the photograph. *IVC*, inferior vena cava; *Ao*, descending aorta; *RV*, right ventricle; *LV*, left ventricle; *L*, liver; *RL*, right lung; *LL*, left lung; *T9*, ninth thoracic vertebra.

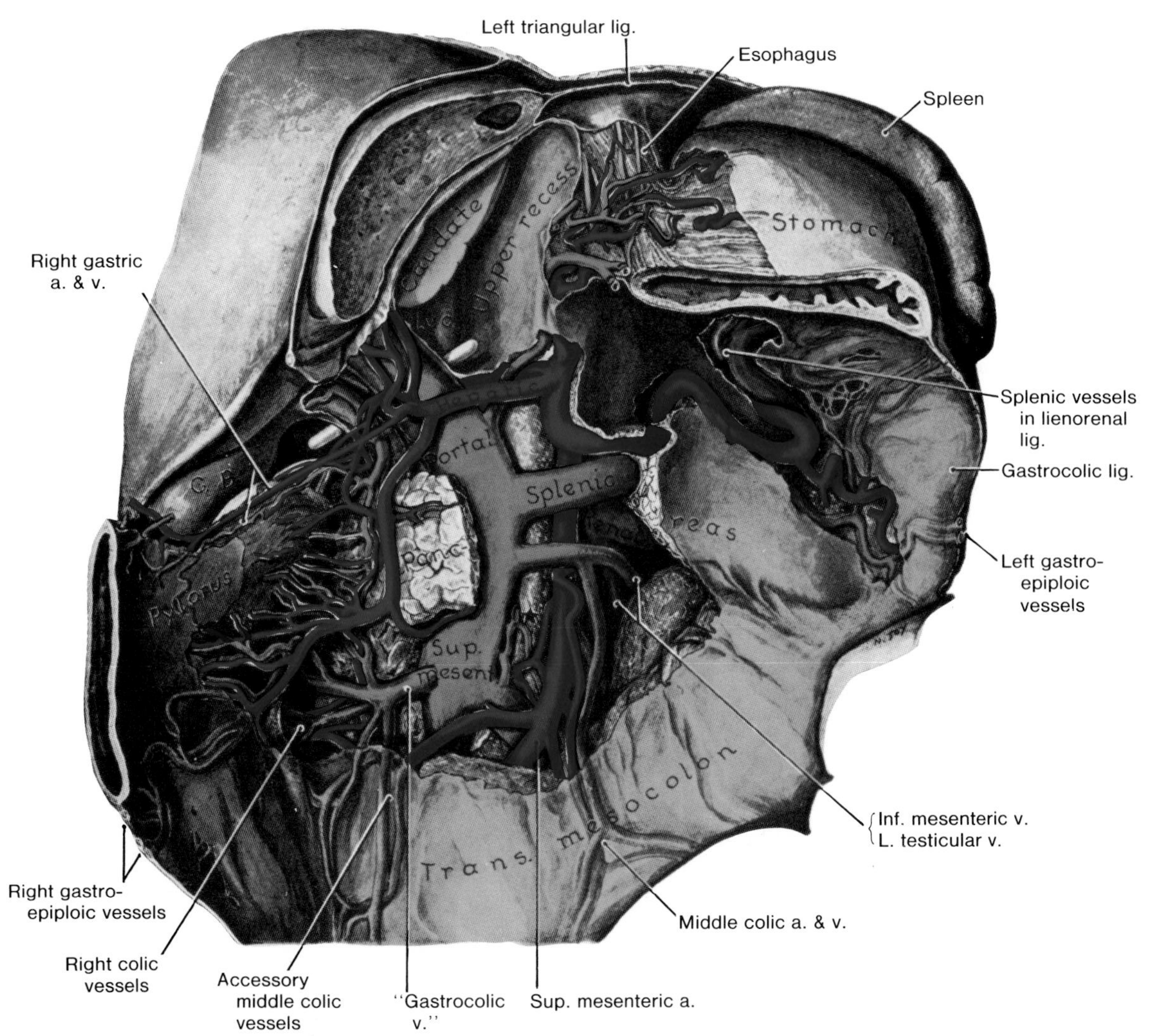

Figure 2-45. Drawing of a dissection of the posterior wall of the omental bursa. The peritoneum of the posterior wall has largely been removed and a section of the stomach and pancreas has been excised. The pyloric end of the stomach has been turned to the right. Observe that a *white rod* had been passed through the epiploic foramen. Note the esophageal branches of the left gastric vessels and the vagal trunks applied to the esophagus. Observe that the portal vein is formed posterior to the neck of the pancreas by the union of the superior mesenteric and splenic veins, with the inferior mesenteric vein joining at or near the angle of union. Here, the left gastric vein joins it. Note that the neck of the gallbladder (*GB*) serves as a guide to the epiploic foramen (also see Figs. 2-35 and 2-52).

ploic foramen can be incised because of the presence of blood vessels in them, the swollen intestine is usually decompressed by a needle so it may be returned to the main part of the peritoneal cavity through the epiploic foramen.

When the cystic artery is accidently severed during **cholecystectomy** (removal of gallbladder), hemorrhage from it can be controlled by compressing the hepatic artery between the index finger in the epiploic foramen and the thumb on the anterior wall of the foramen.

Following is a more detailed description of the abdominal viscera.

THE ESOPHAGUS

This *relatively straight muscular tube*, 23 to 25 cm long, is continuous with the inferior end of the pharynx (Fig. 2-32). It follows the curve of the vertebral column as it descends through the neck and posterior mediastinum to perforate the diaphragm (Figs. 2-28 and 2-35). It ends a little further inferiorly by opening into the **cardiac orifice** of the stomach (Fig. 2-46), posterior to the seventh left costal cartilage about 2.5 cm from the median plane (Fig. 2-44).

When the esophagus is full, it is constricted in four places: (1) at its beginning in the neck, (2) where it is crossed by the arch of the aorta (Fig. 1-73), (3) where it is crossed by the left principal bronchus (Fig. 1-72), and (4) where it pierces the diaphragm (Figs. 2-28 and 2-33*A*).

Awareness of the sites of these indentations in the esophagus is important clinically when passing instruments along the esophagus (*e.g.*, an **esophagoscope** or a ***gastroscope***). Otherwise the wall of the esophagus may be damaged.

These narrowings of the lumen of the esophagus are not anatomical constrictions; they are visible only when the esophagus is full. They can be observed in radiographs taken as barium is being swallowed (Fig. 2-33*A*), and they disappear as the esophagus empties.

The abdominal part of the esophagus, 1.5 to 2.5 cm long (Figs. 2-32, 2-33*A*, and 2-35), is conical in shape. The right border of the esophagus is continuous with the lesser curvature of the stomach (Fig. 2-46), but its left border is separated from the fundus of the stomach by the ***cardiac notch***.

At the inferior end of the esophagus, the *esophagogastric junction*, there is a *physiological* mechanism known as the **esophageal sphincter** which is capable of contraction and relaxation. Radiological studies show that swallowed food stops momentarily at this sphincter and that it is quite efficient in preventing reflux of gastric contents into the esophagus.

Arterial Supply of Esophagus (Figs. 2-50, 2-51, and 2-58). The arteries supplying the abdominal part of the esophagus are derived from the left gastric branch of the *celiac artery* and from the left *inferior phrenic* branch of the abdominal aorta.

Venous Drainage of Esophagus (Figs. 1-80 and 2-58). The veins from the abdominal part of the esophagus drain partly into the *azygos vein* and partly into the *left gastric vein.*

Lymphatic Drainage of Esophagus (Fig. 2-53). The lymph vessels from the abdominal part of the esophagus drain into the *left gastric lymph nodes.* Some vessels pass directly into the thoracic duct (Figs. 1-42 and 1-72).

Nerves of Esophagus (Figs. 1-74, 1-81, 2-54, and 7-135). The nerves supplying the esophagus are derived from the vagus (CN X) and the sympathetic trunks. The abdominal part of the esophagus is supplied by the *vagal trunks* (anterior and posterior gastric nerves), the thoracic *sympathetic trunks*, the greater and lesser *splanchnic nerves*, and the plexus of nerves around the left gastric and inferior phrenic arteries.

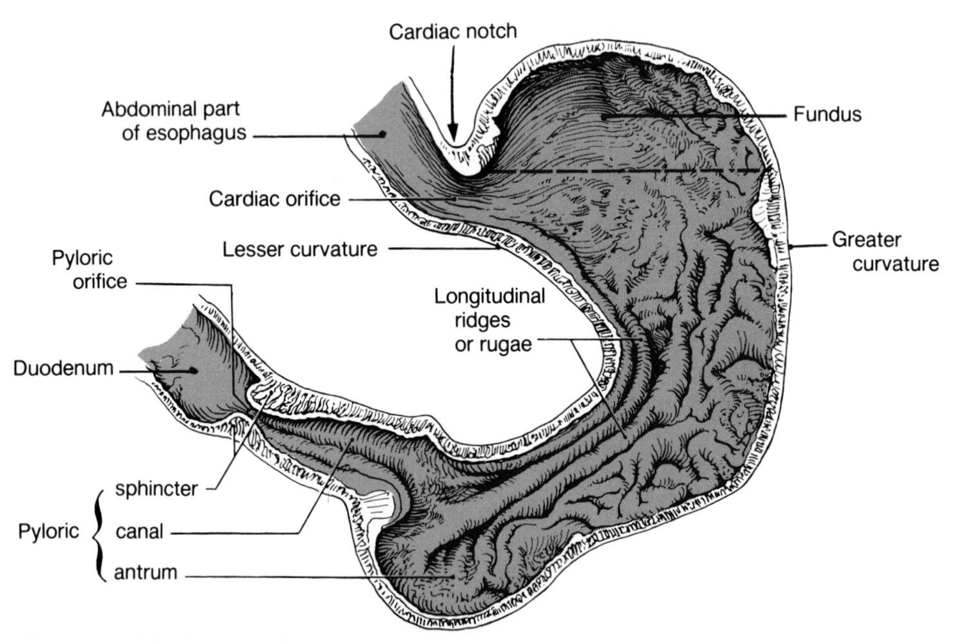

Figure 2-46. Drawing of the interior of the abdominal part of the esophagus, the stomach, and the first portion of the superior part of the duodenum ("duodenal cap"). Observe the longitudinal ridges extending from the esophagus to the pylorus. Elsewhere note that the mucous membrane is rugose when the stomach is empty. The numerous prominent ridges and folds of the lining mucosa are called rugae (L. wrinkles); they tend to disappear or become smaller when the stomach is distended. The rugae are also visible in radiographs of the stomach taken after a barium meal (Fig. 2-34).

The incidence of **cancer of the esophagus** is low in North America, but high in certain countries (*e.g.*, Japan). The onset of difficulty in swallowing (**dysphagia**) in anyone over 45, especially in males, raises suspicion of esophageal cancer. A *barium swallow* (Fig. 2-33) shows a persistent filling defect owing to narrowing of the lumen. During **esophagoscopy** a tumor may be observed and biopsied.

Cancer cells from a tumor of the abdominal part of the esophagus metastasize to the *left gastric lymph nodes* (Fig. 2-53) and to hepatic lymph nodes.

Heartburn or *pyrosis is the most common type of esophageal pain.* This is a burning sensation posterior to the inferior part of the sternum. Esophageal pain often accompanies dysphagia and may be severe if food is unable to pass a constricted segment of the esophagus.

THE STOMACH

The stomach acts as a food blender and a reservoir where the digestive juices act on the food.

The stomach is a very distensible organ. The empty stomach is only of slightly larger caliber than the large intestine, but it is capable of considerable expansion and can hold 2 to 3 liters of material.

The newborn infant's stomach is only the size of a lemon, but it can hold about 30 ml of fluid.

The stomach has two curvatures (Fig. 2-46). The **lesser curvature of the stomach** is continuous with the right border of the esophagus and *forms the right or concave border of the stomach.* The **greater curvature of the stomach** is continuous with the left border of the esophagus and *forms the left or convex border of the stomach.* The greater curvature is four to five times longer than the lesser curvature.

The cardiac part of the stomach or cardia,[2] is a rather indefinite region around the **cardiac orifice** (Fig. 2-46).

The fundus of the stomach is the dilated portion to the left of and superior to the cardiac orifice (Fig. 2-46). The fundus is the most superior portion of the stomach which rests against the left dome of the diaphragm (Fig. 2-35). It usually contains a bubble of gas (Figs. 2-33 and 2-34).

The body of the stomach (Fig. 2-32*A*), the major portion, lies between the fundus and the pyloric antrum (Fig. 2-46).

The pyloric part of the stomach (Fig. 2-32*A*) consists of a wider portion, the ***pyloric antrum***, and a narrow portion, the ***pyloric canal*** (Fig. 2-46). The pyloric canal (1 to 2 cm long) is continuous with the ***pylorus***, the sphincteric region of the stomach which separates the stomach from the duodenum (Figs. 2-32*A*, 2-46, and 2-47). The **angular notch** is a sharp angulation of the lesser curvature (Fig. 2-34*B*) which indicates the junction of the body and the pyloric part of the stomach.

The pylorus (G. gatekeeper), the *distal sphincteric region*, is thick because it contains an increased amount of smooth circular muscle. The circular middle layer in the muscularis externa is greatly thickened at the pylorus to form the **pyloric sphincter**, which *controls the rate of discharge of stomach contents* into the duodenum. **The pylorus is normally in tonic contraction** (*i.e.*, it is closed except when emitting the semifluid contents of the stomach).

Pylorospasm (spasmodic contraction of the pylorus) sometimes occurs in infants, usually between 2 to 12 weeks. It is characterized by failure of the smooth muscle fibers encircling the pyloric canal to relax normally. As a result food does not pass easily from the stomach into the duodenum and the stomach becomes overly full. This usually results in vomiting. Often drugs are given to relax the smooth muscle of the pylorus.

A semisolid ***bolus*** (L. lump) of food from the esoph-

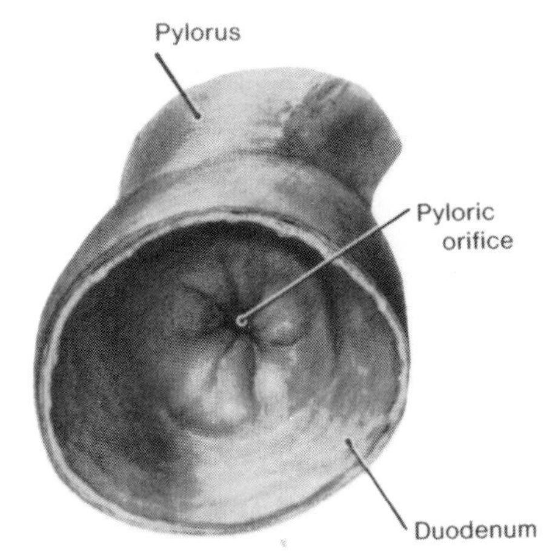

Figure 2-47. Drawing of the pylorus, the distal sphincteric region of the stomach, and the first or superior part of the duodenum. Observe that the pyloric orifice is narrow and is surrounded by a thick ring of circular muscle which forms the pyloric sphincter that controls the rate of discharge of stomach contents into the duodenum.

---

[2] This part of the stomach was so named because it lies near the diaphragm where the heart rests (Fig. 1-46). The Greek word *kardia* means heart.

agus enters the stomach, where it is mixed with and partly digested by the **gastric juices** until it has the consistency of gruel (thin porridge). At irregular intervals, gastric peristalsis passes this semifluid mass of partly digested food, called **chyme** (G. juice), through the pyloric canal into the small intestine for further mixing, digestion, and absorption.

The shape of the stomach varies in position and shape in different persons and in the same individual, depending upon its contents and whether the person is in the supine or erect position. In the supine position, the stomach commonly lies in the left upper quadrant. In the erect position, the J-shaped stomach moves inferiorly 1 to 16 cm. In asthenic (weak) persons, the body of the stomach may extend into the pelvis major.

The mucosa of the empty contracted stomach is thrown into numerous ridges and folds called **rugae** (Fig. 2-46). The rugae (L. wrinkles) can usually be observed in radiographs after a barium meal (Fig. 2-34*A*).

Surface Anatomy of the Stomach (Figs. 2-4*B* and 2-69). The surface features of the stomach vary greatly because its size and position change under a variety of circumstances.

The cardiac orifice *of the stomach is located posterior to the seventh left costal cartilage,* 2 to 4 cm from the midline, at the level of the 10th or 11th thoracic vertebra (Figs. 2-4*A* and 2-69).

The most superior part of the fundus of the stomach is located *posterior to the fifth left rib* in the midclavicular line (Figs. 2-35 and 2-69).

The pyloris of the stomach usually lies at the level of the ***transpyloric plane*** (Fig. 2-4*B*).

The normal stomach is not palpable because its walls are flat and rather flabby. **Fluorscopic studies of the stomach** show that the cardiac orifice (Fig. 2-46) remains nearly stationary, but the level of the pylorus varies from L1 to L3 vertebrae in the supine position. In the erect position, *the location of the pylorus varies from about the level of L2 to L4.* It is usually on the right side, but occasionally it is in the midline.

When the body or the pyloric part of the stomach contains a tumor, the mass may be palpable. In infants with **hypertrophic pyloric stenosis,** the overgrown pylorus may also be felt as a small hard mass about the size of an olive, usually as it descends during inspiration.

The incidence of carcinoma of the stomach is higher in certain countries (*e.g.,* Japan and Scandinavian countries) than in others (*e.g.,* North America). The disease is more common in men than women. The cause (etiology) is not known, but appears to be related to diet. Smoked

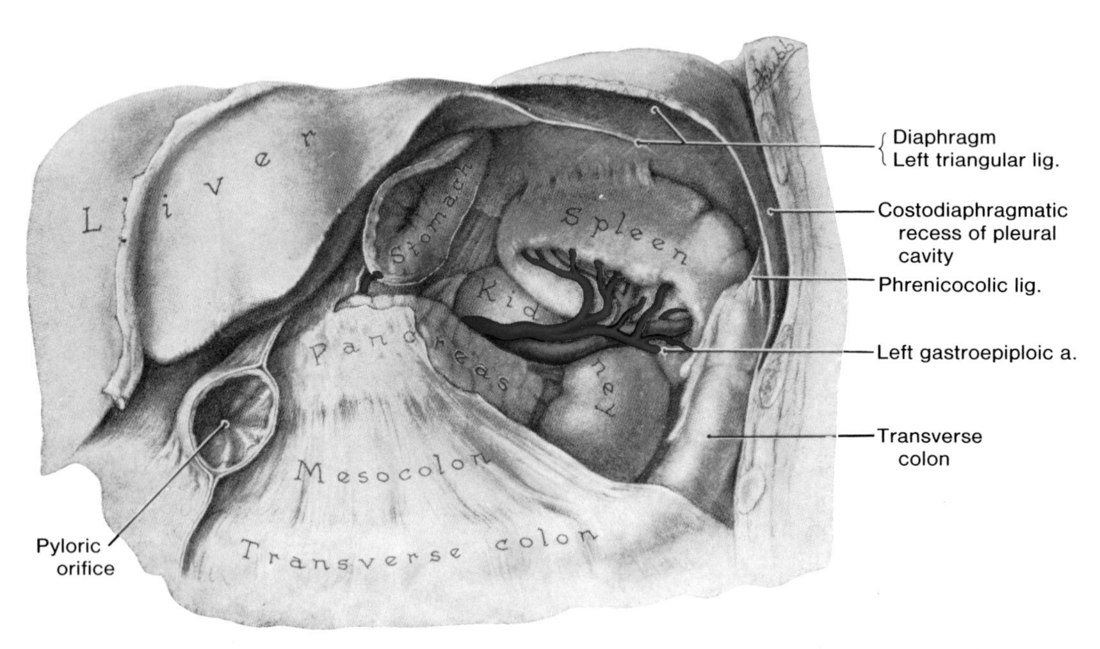

Figure 2-48. Drawing of a dissection of the stomach bed. The stomach has been excised and the peritoneum of the omental bursa covering the stomach bed has been largely removed; as has the peritoneum covering the inferior part of the kidney and pancreas. In this specimen the pancreas is unusually short and there are adhesions binding the spleen to the diaphragm; these are pathological but not unusual. Observe the lesser omentum attached to the superior border of the pylorus and the greater omentum to its inferior border. Note the costodiaphragmatic recess of the pleural cavity separating the spleen and the diaphragm from the thoracic wall. Observe the notched superior border of the spleen. (Also see Fig. 2-55).

fish and spices have been implicated in areas that show a high incidence of gastric cancer.

Since the development of flexible *fiberendoscopes*, **gastroscopy** has become common. It enables physicians to observe gastric lesions and to take biopsies.

Relations of the Stomach (Figs. 2-28, 2-30, 2-32, 2-35 to 2-38, 2-45, 2-48, and 2-77). The stomach is covered entirely by peritoneum, except where the blood vessels run along its curvatures and at a small bare area posterior to the cardiac orifice.

The two layers of the lesser omentum surround the stomach and leave the greater curvature as the greater omentum (Fig. 2-37).

The stomach bed (Fig. 2-48) is formed by the posterior wall of the omental bursa and the retroperitoneal structures between it and the posterior abdominal wall (*e.g.*, the pancreas, Fig. 2-37).

Superiorly the stomach bed consists of the **diaphragm, spleen, superior pole of the left kidney, and the left suprarenal gland.**

Inferiorly the stomach bed consists of the **body and tail of the pancreas and the transverse mesocolon.**

The fundus of the stomach and a portion of the cardiac part are in contact with the diaphragm (Fig. 2-35), posterior to the inferior left costal cartilages. In radiographs of the chest, there is often a gas bubble inferior to the left dome of the diaphragm (Fig. 1-51). This bubble outlines the fundus of the stomach and is also visible in radiographs of the esophagus (Fig. 2-33*A*) and the stomach (Fig. 2-34). The longitudinal **rugae** (Fig. 2-46) are also outlined by gas in radiographs of the stomach (Fig. 2-34*A*).

The anterior surface of the stomach is in contact with (1) the **diaphragm** in the fundic region, (2) the **left lobe of the liver**, and (3) the **anterior abdominal wall** (Figs. 2-28, 2-32, and 2-35).

Malformations of the stomach are very rare, except for **congenital hypertrophic pyloric stenosis.** This marked thickening of the pylorus affects approximately 1 in every 150 male infants, and 1 in every 750 female infants. The elongated, overgrown pylorus is hard and *there is severe stenosis (narrowing) of the pyloric canal* (Fig. 2-46) owing to hypertrophy of the circular musclar layer. The stomach is usually secondarily dilated owing to the obstruction.

Although the cause of congenital hypertrophic pyloric stenosis in unknown, genetic factors appear to be involved because of its high incidence in both infants of monozygotic twins.

Part of the stomach may be herniated through the diaphragm at birth owing to a congenitally large esophageal hiatus. Sometimes the stomach may enter the thorax through a large posterolateral defect in the diaphragm (Fig. 2-134). This type of **congenital diaphragmatic hernia** occurs about once in every 2000 newborn infants.

Acquired hiatal hernias are quite common, and occur most often in middle-aged people owing to *weakening and widening of the esophageal hiatus in the diaphragm* (Fig. 2-117). A portion of the fundus of the stomach may herniate through the esophageal hiatus into the chest. There are two main types of hernia.

In **sliding hiatal hernia** (Fig. 2-49*A*), *the gastroesophageal region slides superiorly into the chest* through the lax esophageal hiatus when the person lies down or bends over. There is often regurgitation of acid from the stomach into the esophagus because the clamping action of the crura of the diaphragm on the inferior end of the esophagus is lost (Fig. 1-73).

In **paraesophageal hiatal hernia** (Fig. 2-49*B*), which is far less common, *the gastroesophageal region remains in its normal position.* But a pouch of peritoneum, often containing the fundus of the stomach, extends through the esophageal hiatus, anterior to the esophagus. In these cases there is no regurgitation because the cardiac "sphincter" is in its normal position. There may be pain, nausea, vomiting, and **dysphagia.**

Arteries of the Stomach (Figs. 2-50, 2-51, 2-56, and 2-57). *The stomach has a rich blood supply from all three branches of the celiac trunk*: (1) **the left gastric artery,** the smallest branch of the celiac trunk. As it passes along the lesser curvature of the stomach, it supplies five branches to the stomach (Fig. 2-50); (2) the **right gastric and right gastroepiploic arteries** (branches of the common hepatic artery); and (3) the **left gastroepiploic** and **short gastric arteries** (branches of the splenic artery).

The left gastric artery, a small branch of the celiac trunk (Fig. 2-50), passes superiorly and to the left, across the posterior wall of the omental bursa. It lies in the floor of this bursa, posterior to the parietal peritoneum (Fig. 2-54). The left gastric artery passes from the posterior abdominal wall to the cardiac part of the stomach. *It then runs inferiorly between the layers of the lesser omentum (**hepatogastric ligament**) along the lesser curvature*, frequently as two branches, to the pylorus. It supplies both surfaces of

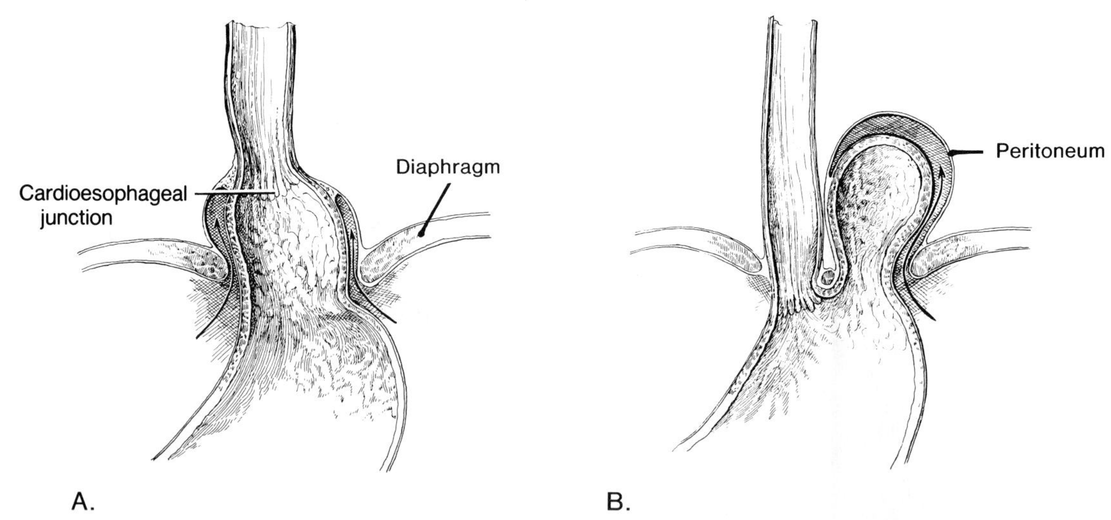

Figure 2-49. Drawings illustrating the two types of hiatal hernia. *A*, sliding hiatal hernia showing the cardioesophageal junction situated superior to the esophageal hiatus in the diaphragm. *B*, paraesophageal hiatal hernia with the cardioesophageal junction in its normal position. Note that the pouch of peritoneum extending through the esophageal hiatus in the diaphragm into the chest contains a portion of the fundus of the stomach. Sliding hiatal hernias (*A*) are more common than paraesophageal hernias (*B*), but a combination of both types is frequent. (Modified and reprinted with permission from McCredie JA: *Basic Surgery.* New York, Macmillan Publishing Company, 1977.)

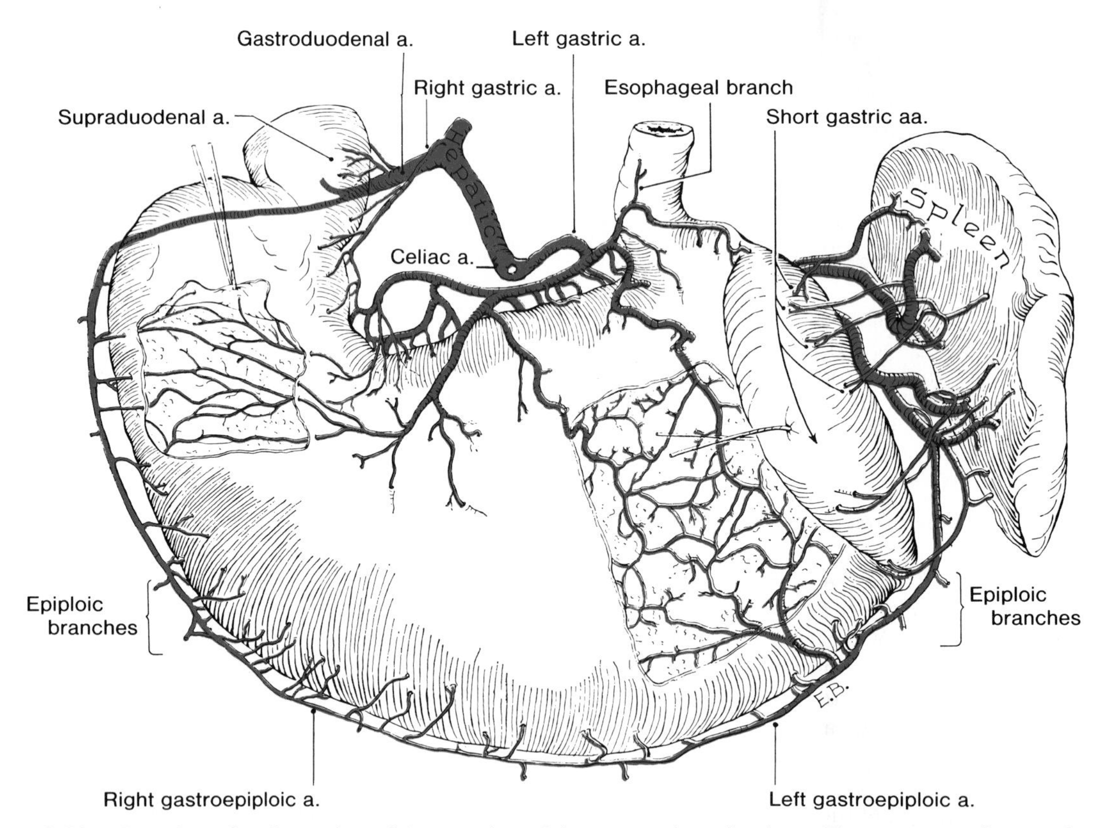

Figure 2-50. Drawing of a dissection of the arteries of the stomach and spleen. The serous and muscular coats have been removed from two areas of the stomach, to reveal the anastomotic networks in the submucous coat. Observe the arterial arch on the lesser curvature formed by the left gastric artery and the much smaller right gastric artery. Note that the arterial arch on the greater curvature is formed equally by the right and the left gastroepiploic arteries. The anastomosis between their two trunks is attenuated; commonly it is absent. Observe the anastomoses between the branches of the two foregoing arterial arches taking place in the submucous coat, two-thirds of the distance from the lesser to the greater curvature of the stomach. Note the four or five tenuous short gastric arteries leaving the terminal branches of the splenic artery close to the spleen. Also observe the left gastroepiploic artery arising within 2.5 cm of the hilum of the spleen.

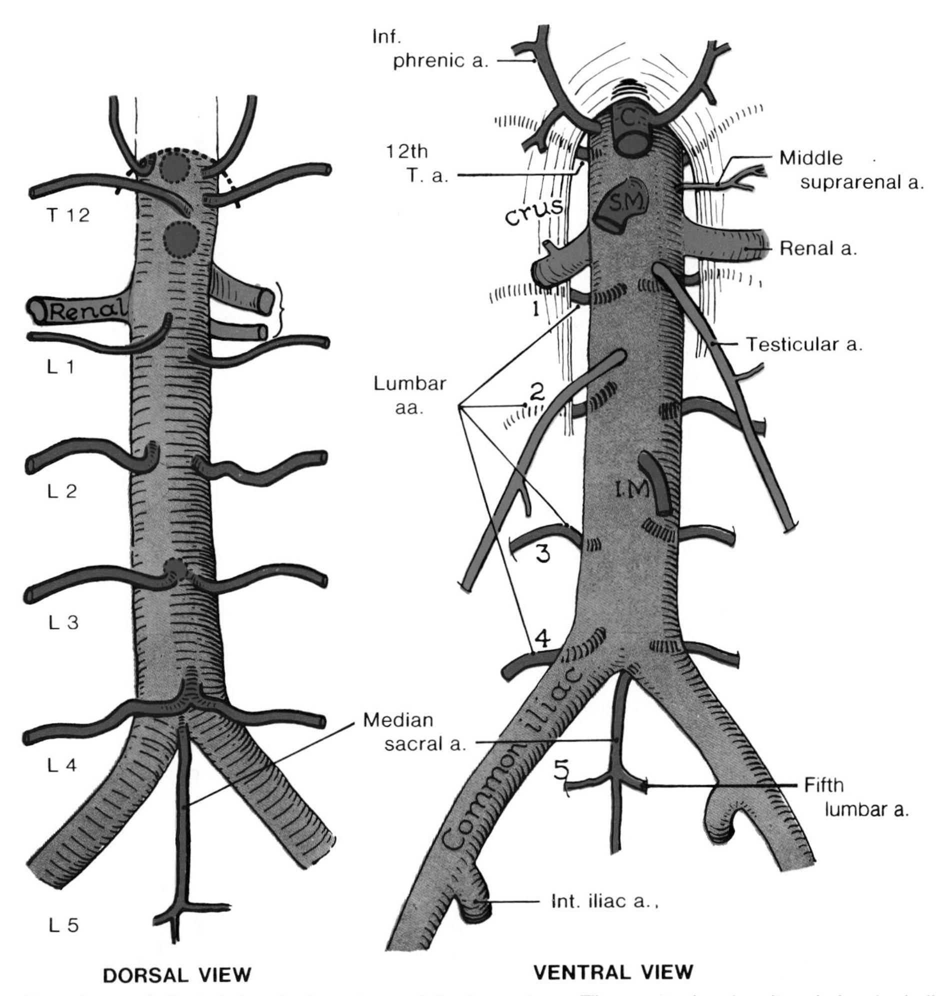

Figure 2-51. Drawings of the abdominal aorta and its branches. The aorta begins its abdominal distribution of blood after passing through the aortic hiatus in the diaphragm (Fig. 2-128). The relatively short abdominal aorta ends by dividing into right and left common iliac arteries, usually anterior to the body of the fourth lumbar vertebra (Fig. 2-127). Besides its several paired branches, it has three unpaired branches for the supply of the GI system. These are the celiac trunk (*C*), the superior (*SM*), and inferior mesenteric (*IM*) arteries.

the stomach and anastomoses with the right gastric artery (Fig. 2-50).

The right and left gastroepiploic arteries *run along the greater curvature of the stomach*, supplying both its surfaces (Fig. 2-50). These branches run between the layers of the greater omentum, a short distance from its attachment to the stomach.

The right gastroepiploic artery, a branch of the gastroduodenal, runs to the left and anastomoses with the left gastroepiploic artery (Fig. 2-50). It sends branches to the right part of the stomach, the superior part of the duodenum, and the greater omentum.

The left gastroepiploic artery (Fig. 2-50), a branch of the splenic, runs between the layers of the gastrolienal ligament (Fig. 2-38) to the greater curvature of the stomach. It runs to the right within the greater omentum, supplying the stomach and the omentum, and ends by anastomosing with the right gastroepiploic artery (Fig. 2-50).

The short gastric arteries (4 to 5) are also branches of the splenic artery (Fig. 2-50). They pass between the layers of the gastrolienal ligament to the fundus of the stomach, where they anastomose with branches of the left gastric and left gastroepiploic arteries.

Because the anastomoses of the various arteries of the stomach provide a good **collateral circulation**, one or more of the major arteries may be ligated without seriously affecting its blood supply. During a **partial gastrectomy** (*e.g.*, excision of the pyloric antrum), the greater omentum is incised inferior to the right gastroepiploic artery (Figs. 2-35 and 2-54). Even though all the omental branches of this artery are ligated, the greater omentum does not degenerate because the omental branches of the left gastroepiploic artery are still intact.

Veins of the Stomach (Fig. 2-45). The veins of the stomach parallel the arteries in position and in course. The gastric veins drain into the **portal system of veins** (Fig. 2-52).

Venous blood from the stomach and other parts of the GI tract enters the liver via the portal vein and passes through its capillary network before continuing on to the heart in the inferior vena cava (Fig. 2-52).

There is considerable variation in the way the gastric veins drain into the portal system of veins. Usually the right and left gastric veins drain directly into the portal vein. The right gastroepiploic vein usually drains into the superior mesenteric vein, but the right gastroepiploic vein may enter the portal vein directly or join the splenic vein. The left gastroepiploic vein and the short gastric vein drain into the splenic vein or its tributaries.

Lymphatic Drainage of the Stomach (Fig. 2-53). The lymph vessels of the stomach accompany the arteries of the stomach along its greater and lesser curvatures (Fig. 2-50).

There are four major areas of lymphatic drainage,

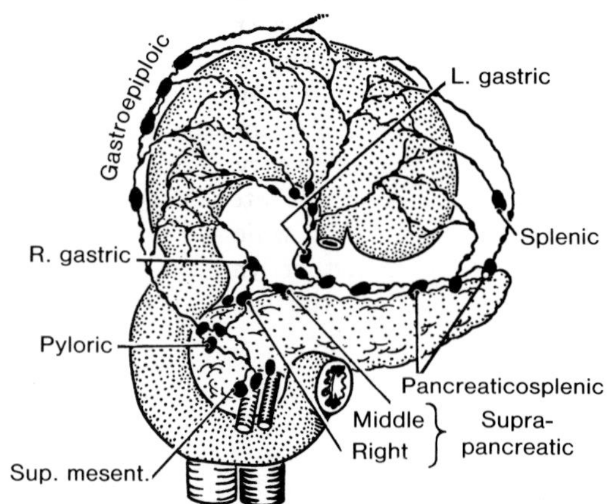

Figure 2-53. Diagram illustrating the lymphatic drainage of the stomach, pancreas, and spleen. The stomach has been reflected superiorly. The gastric nodes consist of right and left gastric, right gastroepiploic, and pyloric groups. The pyloric lymph nodes are especially important because they receive lymph from the pyloric region, where carcinoma of the stomach is most frequent. The collecting vessels from the capsule of the spleen end in the pancreaticosplenic lymph nodes.

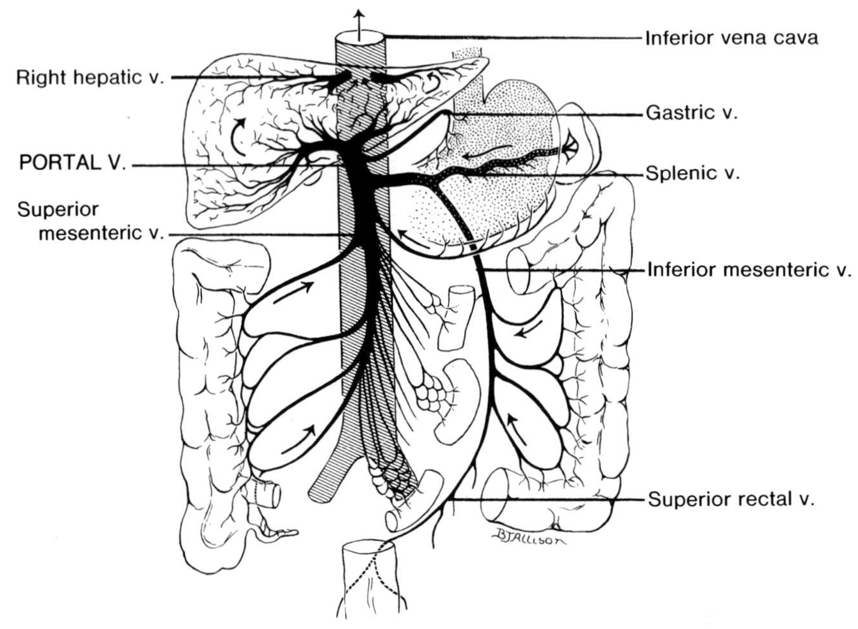

Figure 2-52. Scheme of the portal circulation. The portal vein drains only the gastrointestinal tract and its unpaired glands, the liver of course excepted. It returns the blood to the liver delivered by the celiac, superior mesenteric, and inferior mesenteric arteries to these organs. The portal vein is formed posterior to the neck of the pancreas by the union of the splenic and superior mesenteric veins. It ascends to the right end of the porta hepatis (Fig. 2–45), where it divides into the right and left portal veins. The right vein enters the right lobe of the liver and the left vein passes transversely to the left end of the porta hepatis to supply the caudate, quadrate, and left lobes. There are no functioning valves in the portal system.

each of which has its own regional lymph nodes. The lymph vessels of the stomach drain lymph from its anterior and posterior surfaces toward its curvatures, where many of the lymph nodes are located.

The largest area of lymphatic drainage is from the lesser curvature and a large part of the body of the stomach into the **left gastric lymph nodes** (Fig. 2-53), which lie along the left gastric artery (Fig. 2-54).

The *next largest area of lymphatic drainage* is from the right part of the greater curvature and most of the pyloric part of the stomach into the **gastroepiploic lymph nodes** (Fig. 2-53), which lie along the right gastroepiploic vessels. Other lymph vessels from this area pass directly into the *very important* **pyloric lymph nodes**, located on the anterior surface of the head of the pancreas, close to the pylorus (Fig. 2-53).

The third area of lymphatic drainage is smaller than the previous two. Lymph vessels from the left part of the greater curvature pass to the **gastroepiploic lymph nodes**, which lie along the left gastroepiploic vessels, and to the **pancreaticolienal lymph nodes** which lie along the splenic vessels.

The fourth and smallest area of drainage is from the lesser curvature related to the pyloric part of the stomach. Lymph vessels from this area pass to the **right gastric lymph nodes** which lie along the right gastric artery (Figs. 2-53 and 2-54).

Lymph from all four major groups of lymph

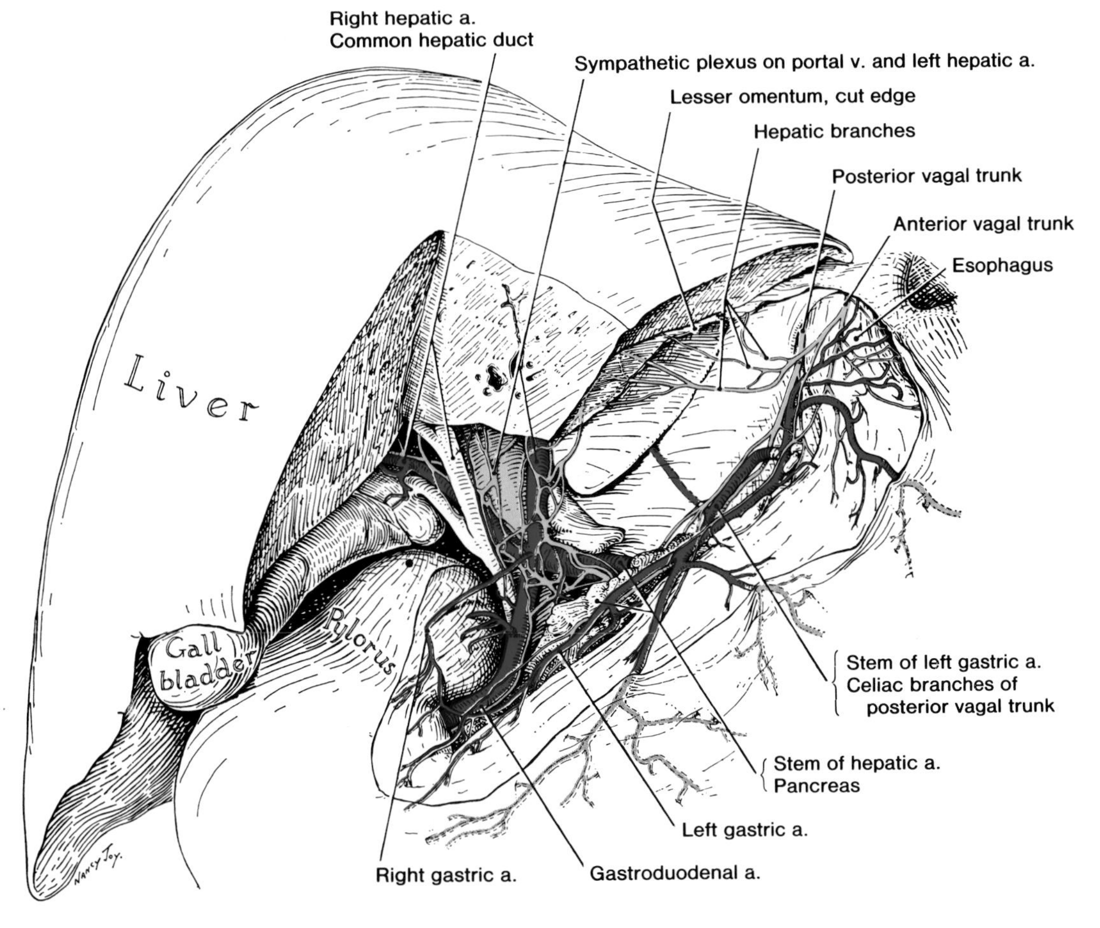

Figure 2-54. Drawing of a dissection illustrating the vagus nerves within the abdomen. The nerves to the stomach are branches of the anterior and posterior vagal trunks. The parasympathetic fibers arrive via the right and left vagus nerves and the sympathetic fibers come via preganglionic fibers from the right and left sympathetic trunks (splanchnic nerves), which synapse in preaortic ganglia. Both kinds of nerve mingle in the rich tangle of nerve plexuses on the anterior aspect of the aorta, especially around the celiac trunk, where they form the *celiac plexus*. Laymen refer to this as the "solar plexus." Right and left *celiac ganglia* connect with the celiac plexus medially and send large plexuses into the suprarenal glands. Both kinds of fiber are distributed by following the walls of branches of the abdominal aorta to their destinations. Observe that the posterior and anterior vagal trunks enter on the esophagus and supply gastric branches, and that the celiac branch of the posterior vagal trunk leaves to contribute to the preaortic plexuses. Note that the hepatic branches of the anterior vagal trunk are joined by sympathetic fibers from the celiac plexus. Also see the scheme of distribution of the vagus nerve (Fig. 7-135).

Table 2-3
Three Types of Splanchnic Nerves

Names	Type	Origin
1. Thoracic (greater, lesser and lowest) splanchnics	Sympathetic	Branches of thoracic sympathetic ganglia 5 to 12.
2. Lumbar splanchnics	Sympathetic	Branches of the four lumbar sympathetic ganglia.
3. Pelvic splanchnics (Fig. 2-114)	Parasympathetic	Branches of ventral rami of sacral spinal nerves, S2, S3, (S4).

nodes drains into the celiac lymph nodes (Fig. 2-58) around the origin of the celiac trunk. Lymph from these nodes passes with that from other parts of the GI tract to the **cisterna chyli** and **thoracic duct** (Fig. 1-42).

Resection of the stomach for carcinoma involves the *removal of all involved regional lymph nodes.* The **pyloric nodes** are especially important because they receive lymph from the pyloric region of the stomach where carcinoma is most frequent. As lymph from this region also drains into the **right gastroepiploic nodes,** they are very frequently involved by carcinoma of the stomach. In more advanced cases, cancer cells spread by lymphogenous dissemination to the **celiac lymph nodes** grouped around the origin of the celiac trunk (Fig. 2-58) because all groups of lymph nodes around the stomach drain into them.

Nerves to the Stomach (Figs. 2-54 and 7-135). *The **parasympathetic nerve supply*** is derived from the anterior and posterior **vagal trunks** and their branches. Both vagal trunks can often be found close to where the left gastric artery reaches the stomach.

The ***sympathetic nerve supply*** is mainly from the **celiac plexus** through the plexuses around the gastric and gastroepiploic arteries (Fig. 2-99). The efferent sympathetic fibers to the stomach arise from the *sixth to the ninth thoracic segments of the spinal cord* (Table 2-3).

The anterior vagal trunk (Fig. 7-135), derived mainly from the left vagus nerve, usually enters the abdomen as a single branch that lies on the anterior surface of the esophagus (Fig. 2-54). It runs toward the lesser curvature of the stomach where it gives off hepatic and duodenal branches that leave the stomach in the hepatoduodenal ligament. The rest of the anterior vagal trunk continues along the lesser curvature of the stomach, giving rise to anterior gastric branches.

The posterior vagal trunk, derived mainly from the right vagus nerve (Fig. 7-135), enters the abdomen on the posterior surface of the esophagus and passes toward the lesser curvature of the stomach (Fig. 2-54). The posterior vagal trunk gives off a celiac branch that runs to the *celiac plexus* (Fig. 2-99). It then continues along the lesser curvature, giving rise to posterior gastric branches.

The secretion of acid by the parietal cells of the stomach is largely controlled by the vagus nerves; hence section of the vagal trunks **(vagotomy)** as they enter the abdomen is sometimes performed to reduce the production of acid in persons with **peptic ulcers** in the stomach or duodenum (Fig. 2-131).

Mucus in the stomach covers the mucosa, forming a barrier between the acid and the cells. Sometimes this protection is inadequate and the gastric juices erode the stomach mucosa forming a **gastric ulcer.** Excess acid secretion is associated with these lesions of the gastric mucosa.

Vagotomy is often done in conjunction with resection of the ulcerated area. Often a ***selective vagotomy*** is done during which only the gastric branches of the vagus are sectioned. This has the desired effect on the acid-producing cells of the stomach without affecting other abdominal structures supplied by the vagus (Fig. 7-135).

Pain impulses from the stomach seem to be carried in the sympathetic nerves because the pain of a recurrent peptic ulcer may persist after complete vagotomy, whereas patients who have had a **bilateral sympathectomy** may have a perforated peptic ulcer with no pain.

THE SPLEEN

The spleen (G. *splēn* and L. *lien*) is a large soft **vascular lymphatic organ** in the left upper quadrant. *It is the largest single mass of lymph tissue in the body* (Figs. 2-48, 2-55, and 2-77).

The spleen is located between the layers of the dorsal mesogastrium (Fig. 2-45) which suspends the

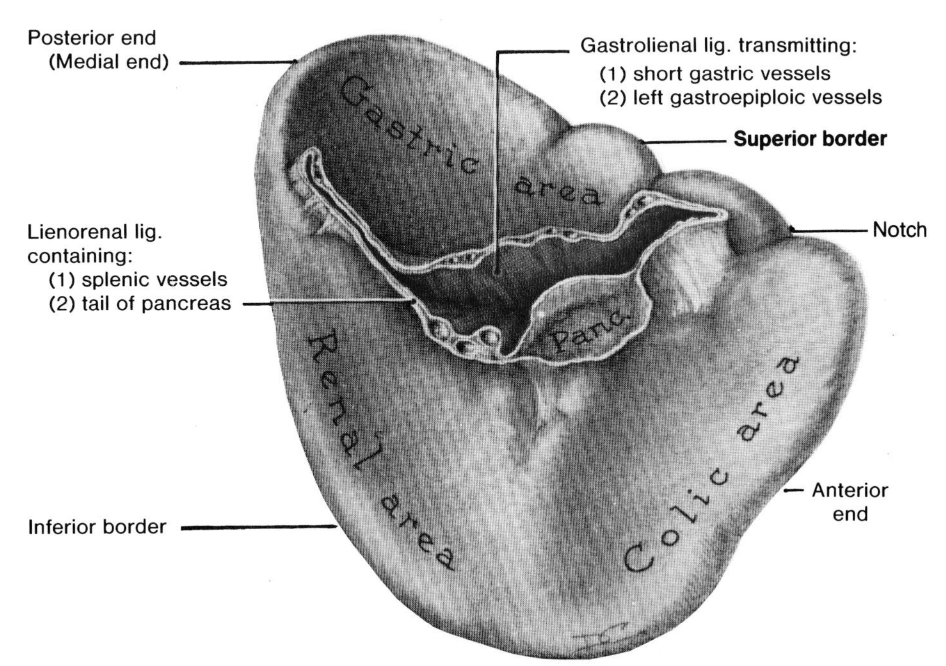

Figure 2-55. Drawing of the visceral surface of the spleen (also see Fig. 2-48). Observe its inferior, superior, and anterior borders, separating its visceral surface (shown here) from its diaphragmatic surface. Note the notches which are characteristic of the superior border. Examine the left limit of the omental bursa at the hilum of the spleen between the lienorenal and gastrolienal ligaments. Observe that this cadaveric spleen takes the impressions of the structures in contact with it. The large colic area presumably resulted from the colon being full of gas.

stomach from the posterior abdominal wall (Fig. 2-30). The spleen is in contact with the posterior wall of the stomach and is connected to its greater curvature by the gastrolienal ligament (Figs. 2-38 and 2-55). It is attached to the left kidney by the lienorenal ligament (Figs. 2-38 and 2-55).

During life the spleen is soft, purplish, freely movable, and considerably larger than in most cadavers. It is located in the left hypochondrium, posterior to the stomach and anterior to the superior part of the left kidney (Figs. 2-48, 2-61, and 2-77).

The spleen lies against the diaphragm laterally which separates it from the pleural cavity. It is related to ribs 9 to 11 on the left side (Figs. 2-32, 2-48, 2-61, and 2-105).

The diaphragmatic surface of the spleen is convexly curved to fit the concavity of the diaphragm. *The anterior and superior borders of the spleen are sharp and are often notched* (Figs. 2-48 and 2-55). These notches represent the remains of the lobulated fetal spleen. The posterior and inferior borders of the spleen are rounded (Fig. 2-55).

The gastrolienal and lienorenal ligaments are attached to the ***hilum*** *on the medial aspect of the spleen* (Fig. 2-55), where the branches of the splenic artery enter and the tributaries of the splenic vein leave the spleen (Figs. 2-57 and 2-58). Except at the hilum (hilus), where these vessels enter and leave, the spleen is completely surrounded by peritoneum. *The hilum of the spleen is usually intimately related to the tail of the pancreas* (Figs. 2-32*B* and 2-64).

The spleen varies in size and shape, but it is usually about 12 cm long and 7 cm wide and fits into one's cupped hand. *It normally contains a large amount of blood.* Its capsule and trabeculae contain some smooth muscle which enables it to expel some of its blood into the circulation.

The shape of the spleen is affected by the fullness of the stomach and the transverse colon at the left colic flexure (Fig. 2-32*A*). The distended stomach gives the spleen the shape of a segment of orange, whereas when the colon is full the spleen has a tetrahedral or four-sided appearance (Fig. 2-55).

Surface Anatomy of the Spleen (Figs. 2-61 and 2-105). The spleen is normally under the shelter of the ribs. It lies deep to the 9th, 10th, and 11th ribs and its external surface is convex to fit these ribs (Figs. 2-45 and 2-77). *The long axis of the spleen lies in the line of the 10th rib,* where it rests on the left colic flexure (Figs. 2-32, 2-40, and 2-61).

Normally the spleen does not extend inferior to the left costal margin, and so a normal spleen is seldom palpable through the abdominal wall. The anterior tip of the spleen usually does not extend farther medially than the midclavicular line (Fig. 2-61).

Arterial Supply of the Spleen (Figs. 2-48, 2-50,

and 2-55 to 2-58). The ***splenic artery*** *is the largest branch of the celiac trunk.* It follows a tortuous course posterior to the omental bursa (Fig. 2-45), anterior to the left kidney, and along the superior border of the pancreas (Fig. 2-48). Between the layers of the lienorenal ligament (Fig. 2-55), *the splenic artery divides into five or more branches* which enter the hilum of the spleen (Figs. 2-48 and 2-57). The branches of the splenic artery supply the individual elements of the spleen as **end arteries**.

There is no anastomoses between the small branches of the splenic arteries. Consequently obstruction of them results in death of splenic tissue (*splenic infarction*).

Venous Drainage of the Spleen (Figs. 2-45, 2-48, and 2-52). The **splenic vein** is formed by several tributaries that emerge from the hilum of the spleen. It is joined by the inferior mesenteric vein and runs posterior to the body and tail of the pancreas throughout most of its course.

The splenic vein unites with the superior mesenteric vein posterior to the neck of the pancreas to form the *portal vein* (Figs. 2-52 and 2-59).

Lymphatic Drainage of the Spleen (Fig. 2-53). The lymph vessels arise from the capsule and trabeculae of the spleen and pass along the splenic vessels to drain into the ***pancreaticosplenic lymph nodes.*** These nodes are related to the posterior surface and superior border of the pancreas (Figs. 2-53 and 2-59).

Nerve Supply of the Spleen. The nerves to the spleen come from the **celiac plexus** (Fig. 2-99). They are distributed mainly to the branches of the splenic artery and are vasomotor in function.

Removal of part of the spleen is followed by rapid regeneration. Even total removal of the spleen **(splenectomy)** does not produce serious effects, because its functions are assumed by other reticuloendothelial organs.

When the spleen is diseased it may be 10 or more times its normal size (**splenomegaly**). In some cases of splenomegaly, the spleen may fill

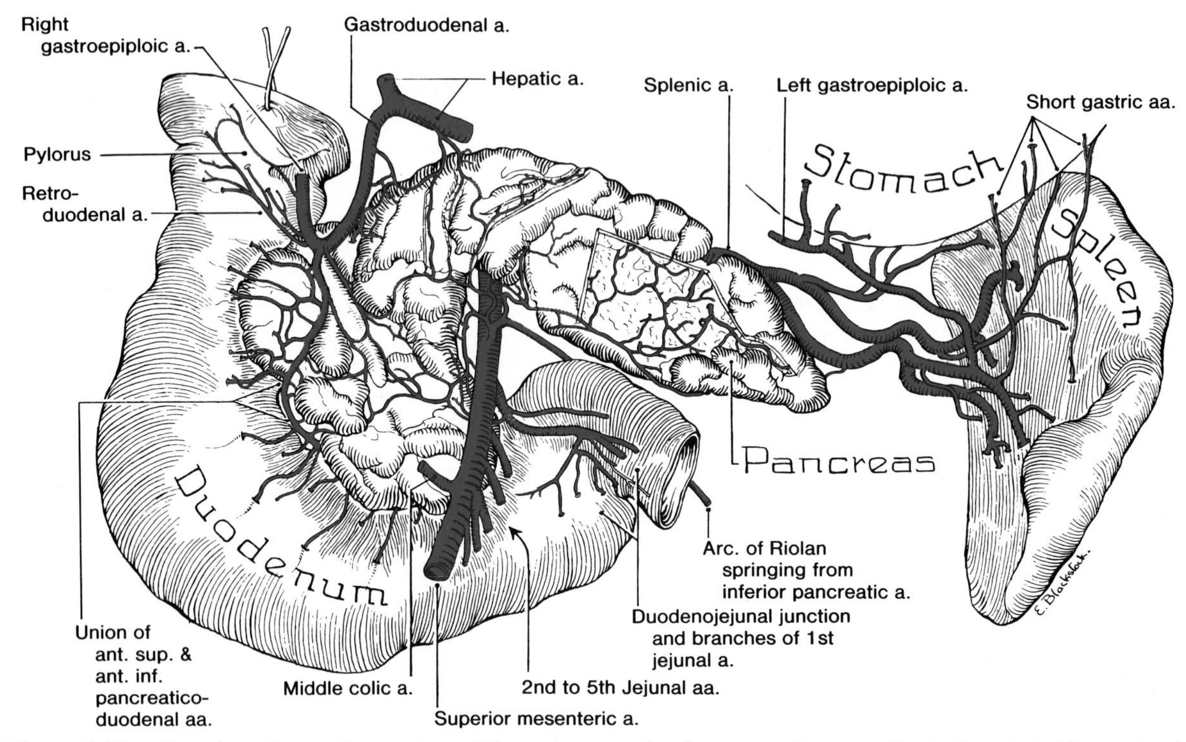

Figure 2-56. Drawing of an *anterior view* of the pancreas, duodenum, and spleen illustrating their blood supply. A slice has been removed from the tail of the pancreas to show the internal arrangement of the vessels. Observe the territories supplied by the hepatic, splenic, and superior mesenteric arteries. Note that several retroduodenal branches spring from the right gastroepiploic artery. Observe that the anterior superior pancreaticoduodenal branch of the gastroduodenal artery and the anterior inferior pancreaticoduodenal branch of the superior mesenteric artery form an arch anterior to the head of the pancreas. Observe the many arteries entering the hilum of the spleen; these are end arteries which do not have significant anastomoses in the substance of the spleen.

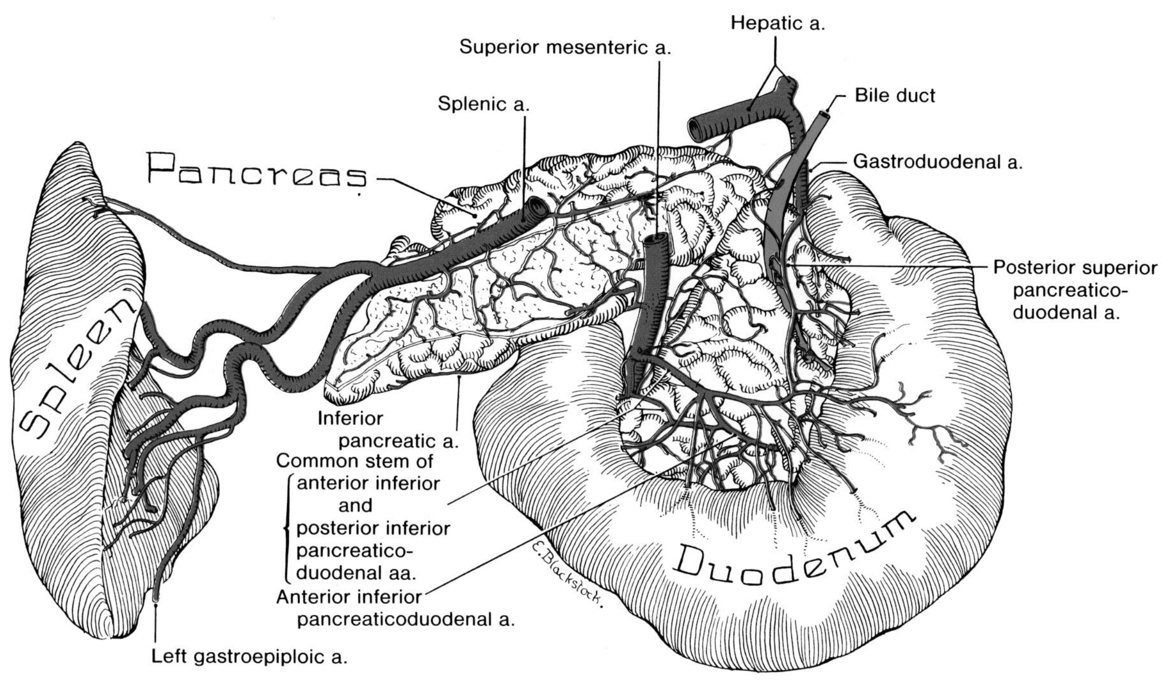

Figure 2-57. Drawing of a *posterior view* of the pancreas, duodenum, and spleen illustrating their blood supply. A slice has been removed from the pancreas to show the internal arrangement of the vessels. Observe that the posterior superior and posterior inferior branches of the gastroduodenal artery and the superior mesenteric artery form an arch posterior to the pancreas. Note the two inferior arteries here, as is usual, spring from a common stem. From the arch thus formed, straight vessels called vasa recta duodeni pass to the posterior surface of the second, third, and fourth parts of the duodenum. Observe that the duodenojejunal junction is supplied by the large branching vessel depicted. Examine the fine network of arteries that pervades the pancreas. It is derived from: (1) the common stem of the hepatic artery, (2) the gastroduodenal artery, (3) the pancreaticoduodenal arches, (4) the splenic artery, and (5) the superior mesenteric artery.

the left half of the abdomen. *When a spleen is grossly enlarged it projects inferior to the left costal margin* and its notched superior border (Fig. 2-55) faces inferomedially.

The notched border is very helpful when palpating an enlarged spleen because when the patient takes a deep breath, these notches can be palpated as it moves inferiorly and anteriorly.

Certain conditions (trauma, tumors, and certain hematological diseases) require *removal of the spleen* (**splenectomy**). During this operation, the surgeon has to be aware of the intimate relationship of the tail of the pancreas to the hilum of the spleen (Fig. 2-64) to avoid injury to this important digestive gland.

Although well protected by the ribs (Fig. 2-105), *the spleen is the most frequently injured organ in the abdomen* when severe blows are received to the left hypochondrium. Sometimes football players have their spleens ruptured when they are tackled from the left side.

Rupture of the spleen causes severe intraperitoneal hemorrhage and shock. Splenectomy is often performed to prevent the patient from bleeding to death.

The spleen ruptures spontaneously in some patients with **infectious mononucleosis**, malaria, and septicemia ("blood poisoning") because it is large and friable under these conditions.

Accessory spleens (one or more) occur most commonly near the hilum of the spleen or they may be embedded partly or wholly in the tail of the pancreas. They may also be found between the layers of the gastrolienal ligament.

Accessory spleens occur in about 10% of people and usually are about 1 cm in diameter. Awareness of their possible presence is important because if not removed during splenectomy, they may result in persistence of the symptoms which indicated removal of the spleen (*e.g.*, **splenic anemia**).

The relationship of the costodiaphragmatic recess of the pleural cavity to the spleen is clinically important (Figs. 2-35 and 2-48). This *potential cleft* occurs at the level of the 10th rib in the midaxillary line. It must be kept in mind when doing a **splenic needle biopsy**, and when injecting radiopaque material into the spleen for visualization of the portal vein (**splenoportography**). If care is not taken, the material may enter the pleural cavity.

THE DUODENUM

The duodenum, the first part of the small intestine, joins the pylorus of the stomach to the jejunum (Fig. 2-32*A*). *The duodenum is the shortest, widest, and most fixed part of the small intestine.*

The duodenum forms a **U-shaped loop** (about 25 cm long) which is molded around the head of the pancreas. Its concavity faces superiorly and to the left (Figs. 2-61 and 2-64).

The duodenum is particularly important because it receives the openings of the bile and pancreatic ducts (Fig. 2-66). All but the first 2.5 cm of the duodenum is posterior to the peritoneum (*i.e.*, **retroperitoneal**) and the omental bursa (Fig. 2-62*A*).

For purposes of description, the duodenum is divided into four parts which are related to the vertebral column as follows (Figs. 2-61 and 2-64): superior or first part anterolateral to the body of L1; descending or second part to the right of the bodies of L1, L2, and L3; horizontal or third part anterior to L3; and ascending or fourth part to the left of the body of L3 rising as high as L2. Hence the superior and ascending parts of the duodenum are only about 5 cm apart.

The Superior or First Part of the Duodenum. The superior part of the duodenum is only 2.5 to 3 cm long. *It is the most movable of the four parts of the duodenum.* It begins at the pylorus and passes superiorly, posteriorly, and to the right toward the neck of the gallbladder (Figs. 2-60 and 2-66). In right anterior oblique radiographs, the superior part of the duodenum appears much shorter because of its oblique direction (Fig. 2-34).

Radiologists refer to the beginning or ampulla of the superior part of the duodenum as the **duodenal cap** *or bulb* (Fig. 2-34*E*).

The proximal half of the superior part of the duo-

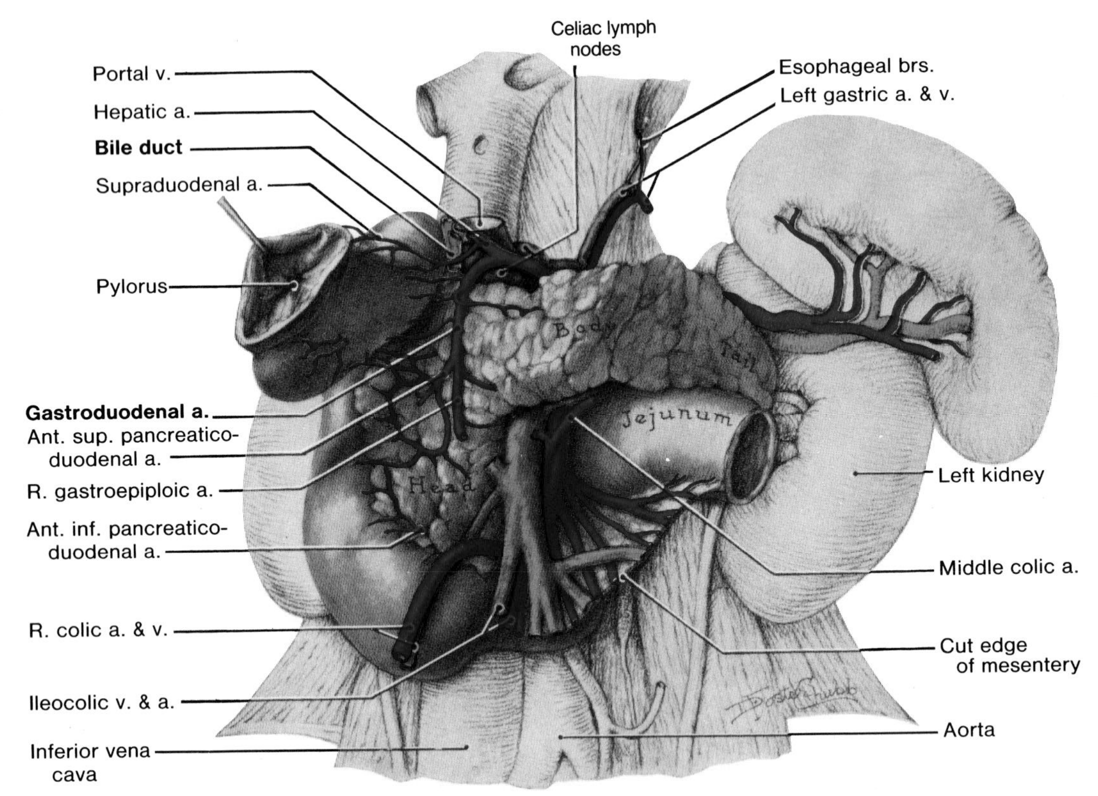

Figure 2-58. Drawing of an *anterior view* of a dissection of the duodenum and pancreas in situ. Observe that the duodenum is molded around the head of the pancreas. Its first or superior part (retracted) is overlapping the pancreas and passing posteriorly, superiorly, and to the right. Its remaining parts (second to fourth) are overlapped by the pancreas. Note that near the junction of its third and fourth parts, the duodenum is crossed by the superior mesenteric vessels. Observe that the pancreas is very short; usually it touches the spleen (Fig. 2-64). Note that the pancreas is arched anteriorly where it crosses the vertebral column and the aorta.

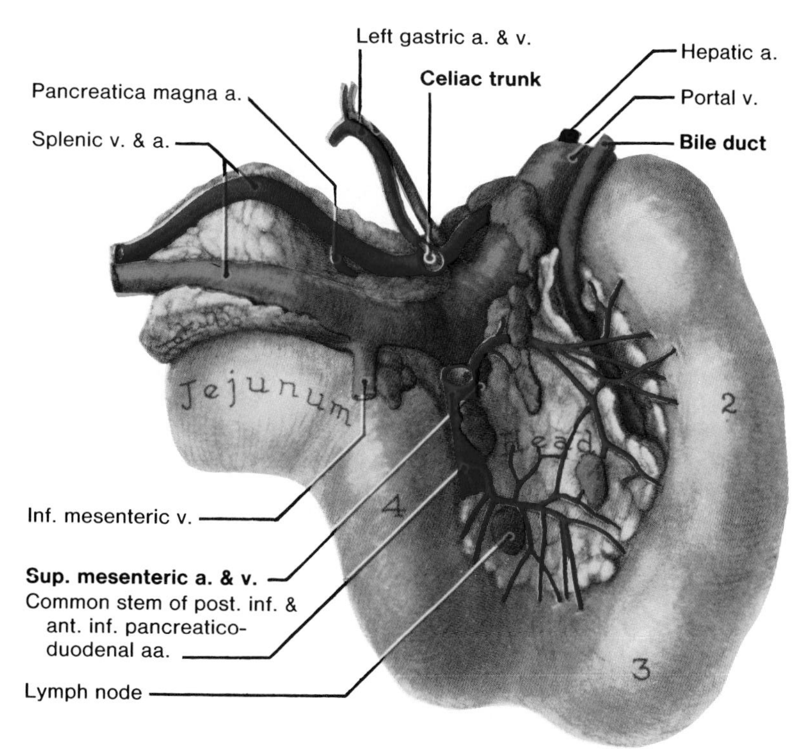

Figure 2-59. Drawing of a *posterior view* of the duodenum, pancreas, and bile duct. Observe that only the end of the superior or first part of the duodenum is in view and that the bile duct is descending in a long fissure (opened up here) in the posterior part of the head of the pancreas.

denum has a mesentery. The greater omentum and the hepatoduodenal ligament are attached to this part (Fig. 2-38); hence it is free to move with the stomach. For this reason, **the beginning of the first part of the duodenum is often called the free part.**

The distal half of the superior part of the duodenum has no mesentery and so is not freely movable. It is attached to the posterior abdominal wall (Fig. 2-62*A*).

The principal relations of the superior part of the duodenum (Figs. 2-35, 2-36, 2-57 to 2-60, and 2-62*A*). *Anteriorly,* peritoneum, gallbladder, and quadrate lobe of liver; *Posteriorly,* bile duct, portal vein, inferior vena cava, and gastroduodenal artery; *Superiorly,* neck of gallbladder; *Inferiorly,* pancreas.

Because of its close relationship to the gallbladder (Fig. 2-60), the anterior surface of the superior part of the duodenum in cadavers is commonly stained with bile.

The Descending or Second Part of the Duodenum. The descending part is 8 to 10 cm long. **It has no mesentery.** It *descends retroperitoneally* along the right side of the first, second, and third lumbar vertebrae (Fig. 2-64). During its descent it passes parallel to the inferior vena cava and to the right of it. It lies directly on the medial part of the right kidney (Figs. 2-58, 2-61, 2-62, and 2-64).

The principal relations of the descending part of the duodenum. *Anteriorly,* transverse colon (Figs. 2-32 and 2-38), transverse mesocolon, and some coils

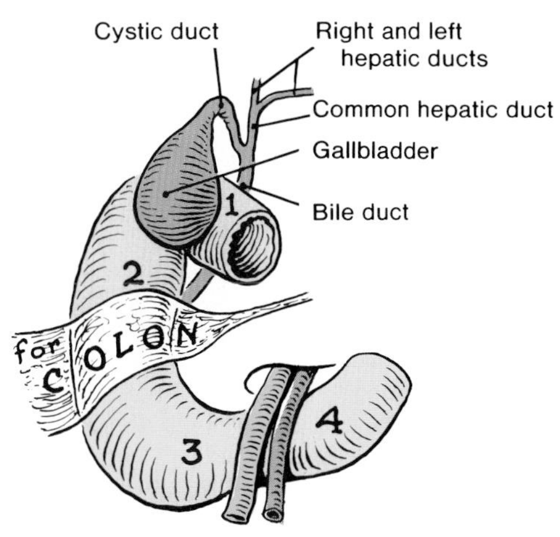

Figure 2-60. Drawing showing the three notable structures that are related anteriorly to the four parts of the duodenum. The pear-shaped gallbladder is suspended by the cystic duct so that its inferior end or fundus is its most inferior part (also see Fig. 2-35).

of small intestine (Fig. 2-39); *posteriorly,* hilum of right kidney, renal vessels, ureter, and psoas major muscle (Fig. 2-58 and 2-62); and *medially,* head of pancreas, pancreatic duct, and bile duct (Fig. 2-58).

The bile duct and the main pancreatic duct enter the posteromedial wall of the duodenum

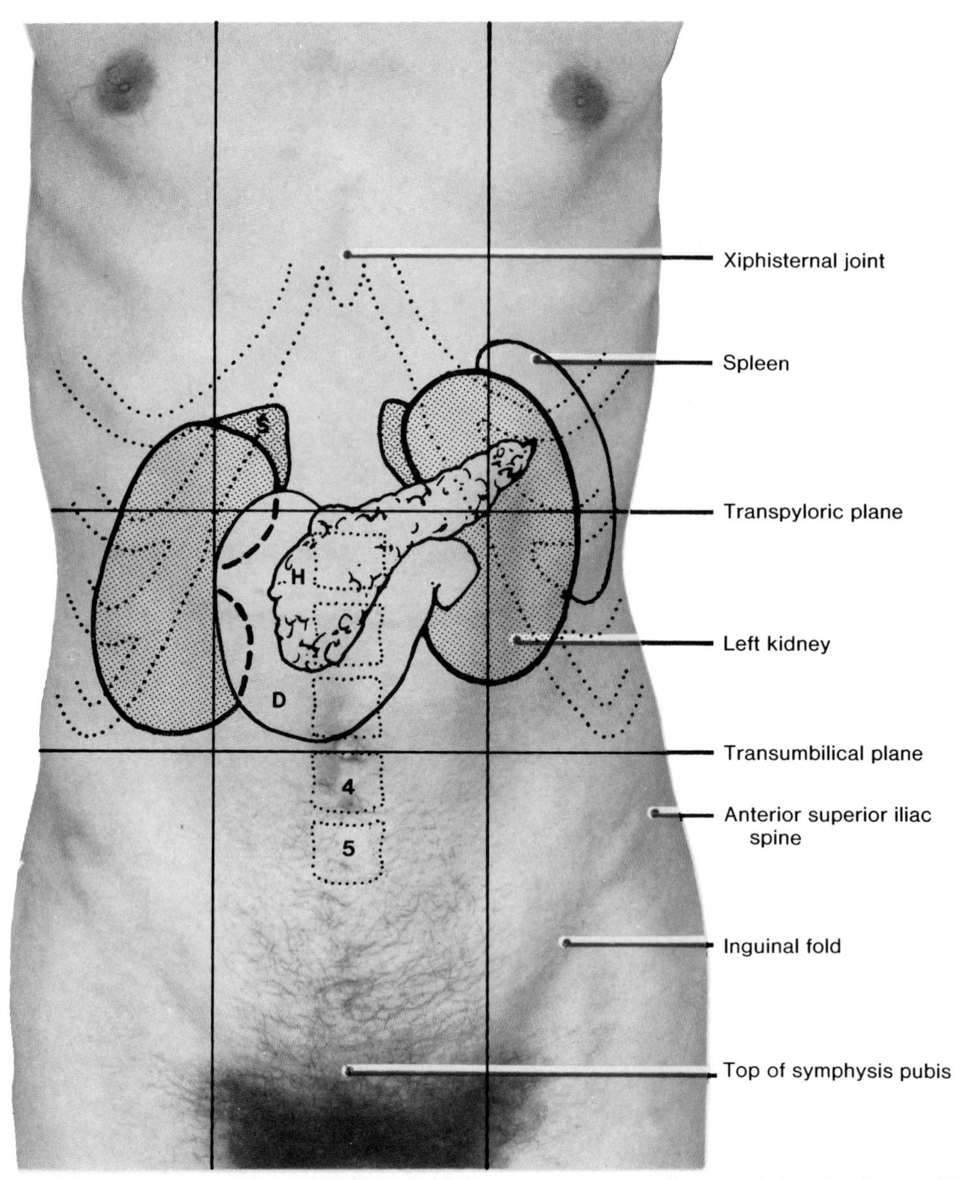

Figure 2-61. Photograph of a 27-year-old man showing the surface projection of the duodenum (*D*), pancreas, kidneys, spleen, and suprarenal glands (*S*). The inferior ribs and their costal cartilages and the lumbar vertebrae are also shown. The right and left vertical planes are the midclavicular lines. The transpyloric plane is approximately midway between the xiphisternal joint and the umbilicus (also see Fig. 2-4*B*). Note that the U-shaped duodenum (*D*) enclosing the head (*H*) of the pancreas, lies entirely superior to the umbilicus and that its two ends are not very far apart (usually about 5 cm).

about two-thirds of the way along its length (Fig. 2-66). These ducts enter the wall obliquely and usually unite to form a short dilated tube, known as the **hepatopancreatic ampulla** (Figs. 2-66 and 2-79). This ampulla opens on the summit of the ***major duodenal papilla*** (Fig. 2-67), located 8 to 10 cm distal to the pylorus.

The opening of the major duodenal papilla is guarded by the ***sphincter of the hepatopancreatic ampulla****.* It is capable of constricting the ampulla and thereby controlling the discharge of bile and pancreatic secretions into the duodenum. In some cases the bile and pancreatic ducts do not join but open separately on the major duodenal papilla.

The Horizontal or Third Part of the Duodenum. The horizontal part of the duodenum, about 10 cm long, runs horizontally from right to left across the third lumbar vertebra (Figs. 2-61 and 2-64). *The horizontal part is retroperitoneal and is adherent to the posterior abdominal wall.*

called the ***pancr***
other around the
hepatopancrea
The accessor
able (Fig. 2-65, (
the pancreas. Usı
connected to the
9% of people it is
65*G*) which open
of the **minor du**
Vessels and N
ies of the pancre
pancreaticodu
branches from th
tail of the pancre
The **superior**
the *gastroduode*
aticoduodenal a
artery supply the
57, and 2-68). Th
tomose with each
The veins of
splenic, and supe
2-59), but *most o*
the splenic vein.
The lymph v
blood vessels to
nodes along the
the **pyloric nod**
and celiac nodes
celiac arteries (F
Most lymph ve

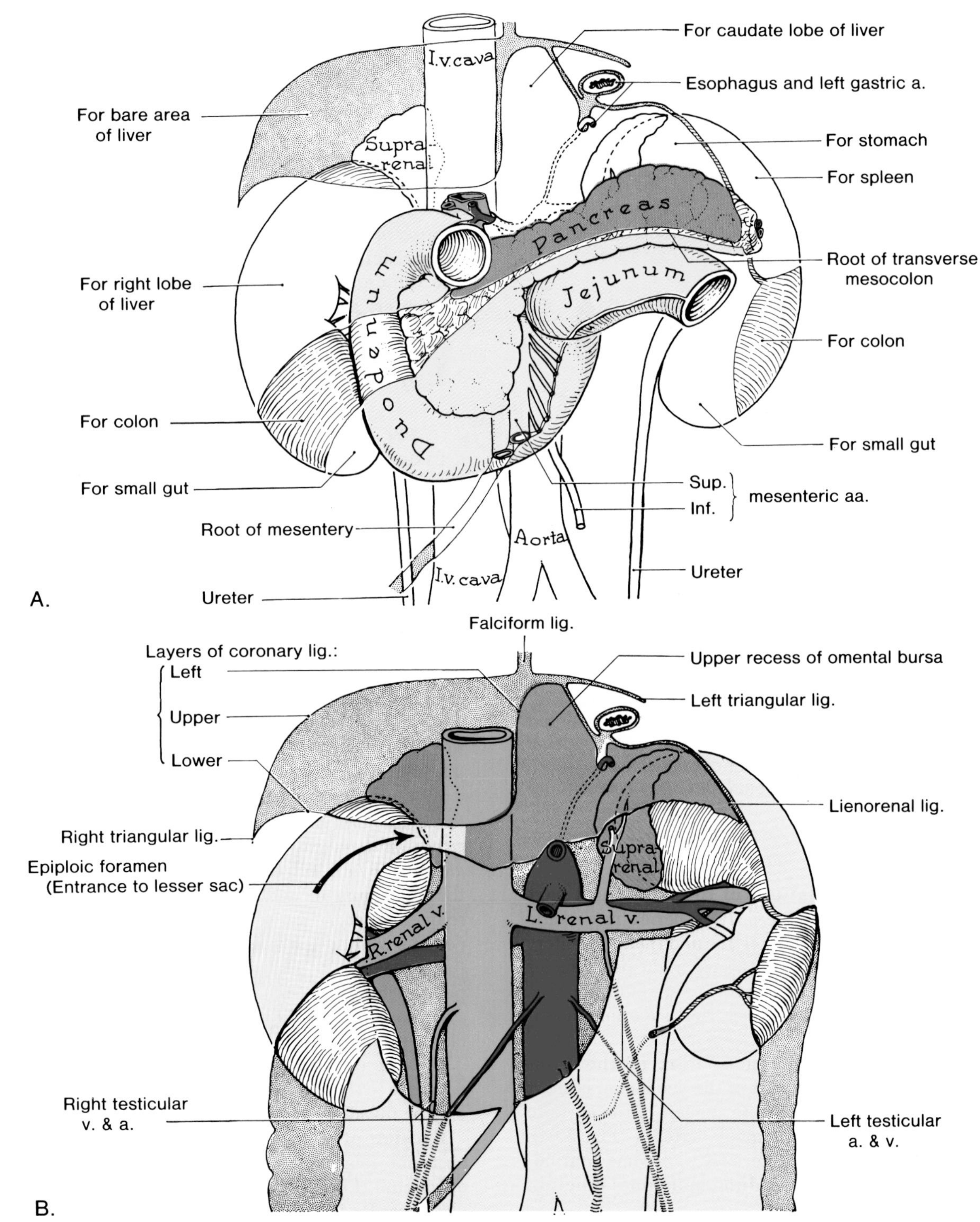

Figure 2-62. Drawings of dissections of the posterior abdominal viscera and their relations. *A,* duodenum and pancreas in situ. *B,* duodenum and pancreas removed. Observe the peritoneal covering (*yellow*) of the pancreas and duodenum. Examine the colic area of the right kidney, the descending or second part of the duodenum, and the head of the pancreas. Note the line of attachment of the transverse mesocolon to the body and tail of the pancreas and to the colic area of the left kidney. Observe the anterior relations of the kidneys and suprarenal glands. Observe that the three parts of the coronary ligaments are attached to the diaphragm, except where the inferior vena cava, the suprarenal gland, and the kidney intervene.

Figure 2-68. D
are supplied by
posterior superio
inferior pancreati

the pancreas occludes the main pancreatic duct. This blockage soon results in pancreatitis involving the body and tail of the pancreas.

If the accessory pancreatic duct connects with the main pancreatic duct (Fig. 2-65*E*) and opens into the duodenum, it may compensate for an obstructed main pancreatic duct or spasm of the hepatopancreatic sphincter.

Pancreatic injury may occur when there is sudden forceful compression of the abdomen (Fig. 2-61), as occurs in an auto accident when a person is thrown against the steering wheel. Because the pancreas lies transversely across the posterior abdominal wall, the vertebral column acts like an anvil and a traumatic force may rupture the pancreas.

Rupture of the pancreas frequently tears its duct system, allowing pancreatic juice to enter the substance of the gland and invade adjacent tissues. *Digestion of tissues by pancreatic juice is very serious and painful.*

Owing to the posterior relations of the head of the pancreas (Fig. 2-62), cysts or tumors of it may cause symptoms by pressing on the portal vein, bile duct, or inferior vena cava. Pressure on the portal vein may cause **ascites**, an accumulation of serous fluid in the peritoneal cavity.

Cancer of the pancreas usually involves its head and accounts for most cases of extrahepatic obstruction of the biliary system.

Cancer of the head of the pancreas frequently results in obstruction of the bile duct and/or the hepatopancreatic ampulla. This results in the retention of bile pigments and the yellow staining of most tissues of the body. The gallbladder enlarges and is often palpable.

Jaundice (F. *jaune*, yellow) is the name given to the *yellow or bronze color of the skin*, the mucous membranes, and the conjunctiva that results from retention of bile.

Cancer of the body of the pancreas is often not diagnosed until the tumor is large and has infiltrated the somatic nerves of the posterior abdominal wall, producing **midback pain**. An extensive tumor in the body of the pancreas may cause **obstruction of the inferior vena cava** because the body of the pancreas rests against this large vein (Figs. 2-62 and 2-64).

The neck of the pancreas is inferior and posterior to the pylorus (Fig. 2-58); hence tumorous enlargement of this region may cause obstruction of the pylorus of the stomach.

In rare cases, the two primordia of the pancreas (Fig. 2-65*A*) may completely surround the descending part of the duodenum, forming an **anular pancreas** (L. *anus*, ring). This may produce duodenal obstruction during the perinatal period or later if inflammation or malignant disease develops in this ring-like pancreas.

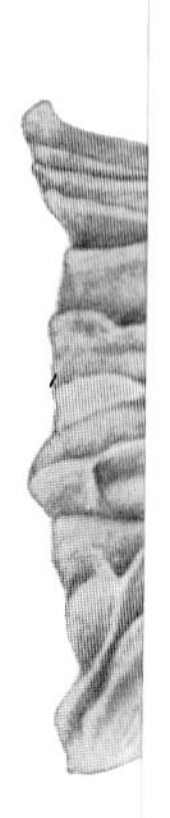

THE LIVER

The liver (G. *hepar*) is a huge glandular organ belonging to the GI system which occupies most of the right upper quadrant of the abdomen (Figs. 2-69 and 2-77). In most living persons the liver is a soft reddish organ.

The liver is the largest gland in the body, accounting for about 2% of the body weight in an adult and 5% in infants. As it is relatively large in infants, it produces most of the prominence of their abdomens.

The liver receives venous blood returning from the GI tract through the ***portal vein*** (Fig. 2-52) which is laden with the products of digestion. In addition to its many metabolic activities, ***the liver is a storehouse for glycogen and it secretes bile.*** Bile is an important agent in digestion, especially of fats. ***Liver bile*** passes via the hepatic ducts and the cystic duct to the gallbladder (Fig. 2-75), where it is concentrated by absorption of water and inorganic salts and stored.

When fat-containing chyme enters the duodenum from the stomach, a hormone (**cholecystokinin**) is released by the upper GI tract. It stimulates contraction of the gallbladder, forcing concentrated ***gallbladder bile*** into the duodenum along with some liver bile already in the bile duct. *It is clinically important to distinguish between liver bile and the more concentrated gallbladder bile.*

Surface Anatomy of the Liver (Fig. 2-69). *In living persons the liver is a soft pliable organ that is molded by the structures related to it* (Fig. 2-72). It conforms to the cavity of the right dome of the diaphragm and *rises to its highest point posterior to the right fifth rib* (Fig. 2-69), just inferior to the nipple.

The liver occupies almost all of the right hypochondrium and much of the epigastrium: it also extends into the left hypochondrium. The liver is somewhat triangular in shape with its base at the right and its apex toward the left. The liver is hidden and protected by the thoracic cage (Figs. 2-69 and 2-105).

The liver moves with respiration because it is connected to the diaphragm; *it also shifts its position with any postural change that affects the diaphragm. The major part of the liver lies on the right side under cover of ribs 5 to 10 and the diaphragm* (Figs. 2-28 and 2-69).

The liver is the easiest abdominal organ to palpate. The patient is asked to inspire deeply while you feel for the inferior edge of the liver.

The liver is enlarged in many diseases and

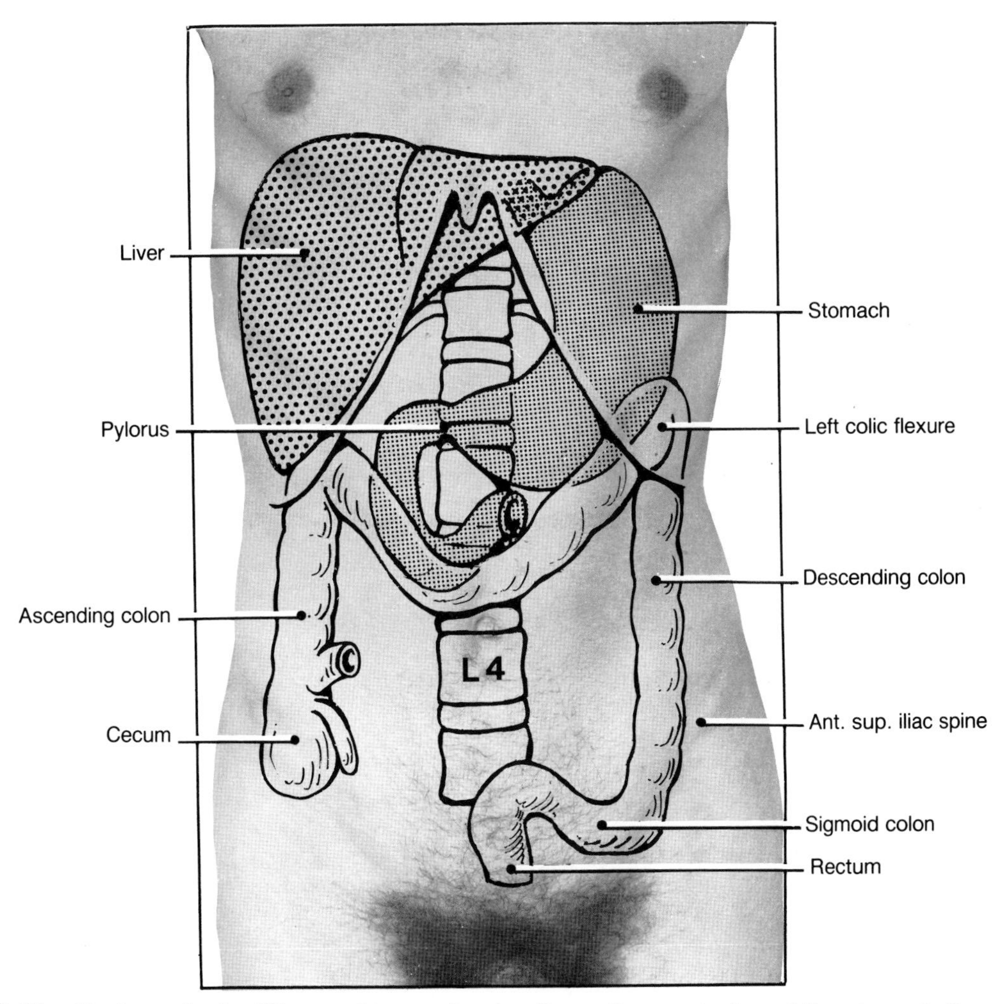

Figure 2-69. Photograph of a 27-year-old man showing the surface projection of the stomach, liver, and large intestines. The outlines of the vertebrae are also shown. The transverse colon forms a loop (as in Fig. 2-40).

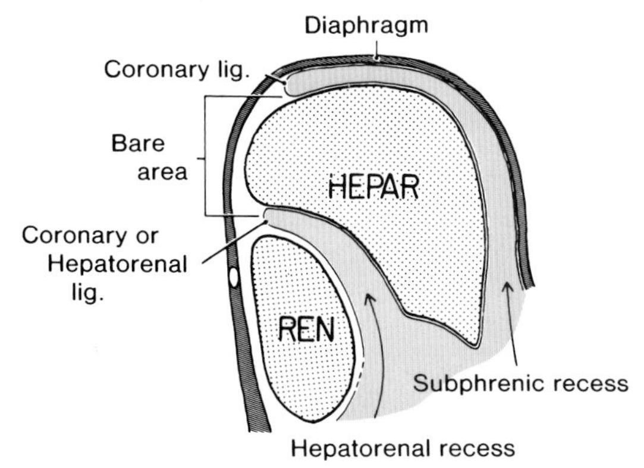

Figure 2-70. Diagram of a sagittal section through the diaphragm, liver, and right kidney (L. *ren*) showing that the bare area of liver (L. *hepar*) is situated between the posterior ends of two peritoneal recesses. Note that the diaphragmatic surface of the liver is dome-shaped and conforms to the concavity of the dome of the diaphragm.

liver enlargement is frequently associated with heart failure. In massive liver enlargement the inferior edge may reach the right lower quadrant.

Owing to the location of the liver (Fig. 2-69), injury to the chest wall inferior to the fifth intercostal space on the right side may involve the liver.

Surfaces and Borders of the Liver (Figs. 2-70 and 2-72). *The liver has two surfaces, **diaphragmatic and visceral**.* The diaphragmatic and visceral surfaces of the liver are separated from each other by the *sharp inferior border*, except posteriorly.

The diaphragmatic surface of the liver is smooth and dome-shaped because *it conforms to the concavity of the inferior surface of the dia-*

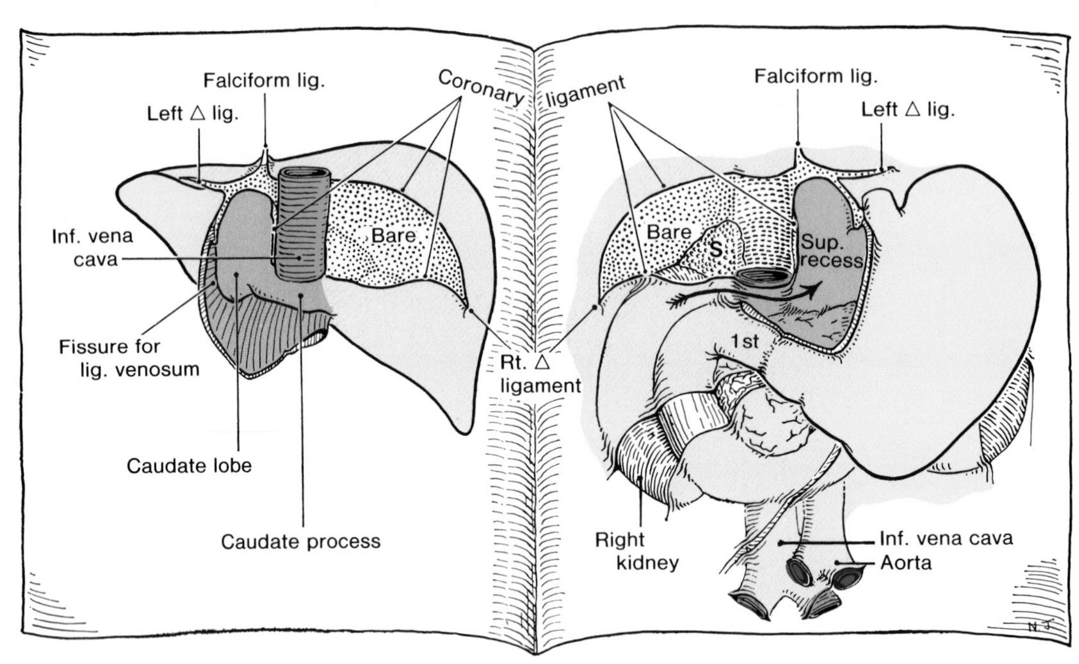

Figure 2-71. Diagram illustrating the peritoneal ligaments of the liver. The attachments of the liver are cut through and the liver is turned to the right side, as you would turn the page of a book. Hence, the posterior aspect of the liver is shown on the left and its posterior relations on the right. On the right diagram, observe the *arrow* passing from the greater sac of peritoneum (*yellow*) into the superior recess of the lesser sac of peritoneum or omental **bursa (***orange*). Note that most of the superior recess lies to the left of the median plane of the body. Observe that the aorta lies posterior to the left lobe of the liver and the superior recess. Note that the inferior vena cava occupies the left or medial limit of the bare area of the liver. Observe that the bare area is triangular; hence the so-called coronary ligament surrounding it is not a crown (L. *corona*) but is three-sided. Its left side or base is between the inferior vena cava and the caudate lobe of the liver and can be palpated when the right index finger is inserted into the superior recess. Its apex is at the right triangular ligament, where the cranial and caudal layers of the coronary ligament meet. The inferior or caudal layer of the coronary ligament is reflected from the liver onto the diaphragm, the right kidney, and the right suprarenal gland (*S*). It is often called the hepatorenal ligament. Followed medially, this layer crosses the inferior vena cava at the epiploic foramen and turning cranially becomes the left or basal layer of the ligament.

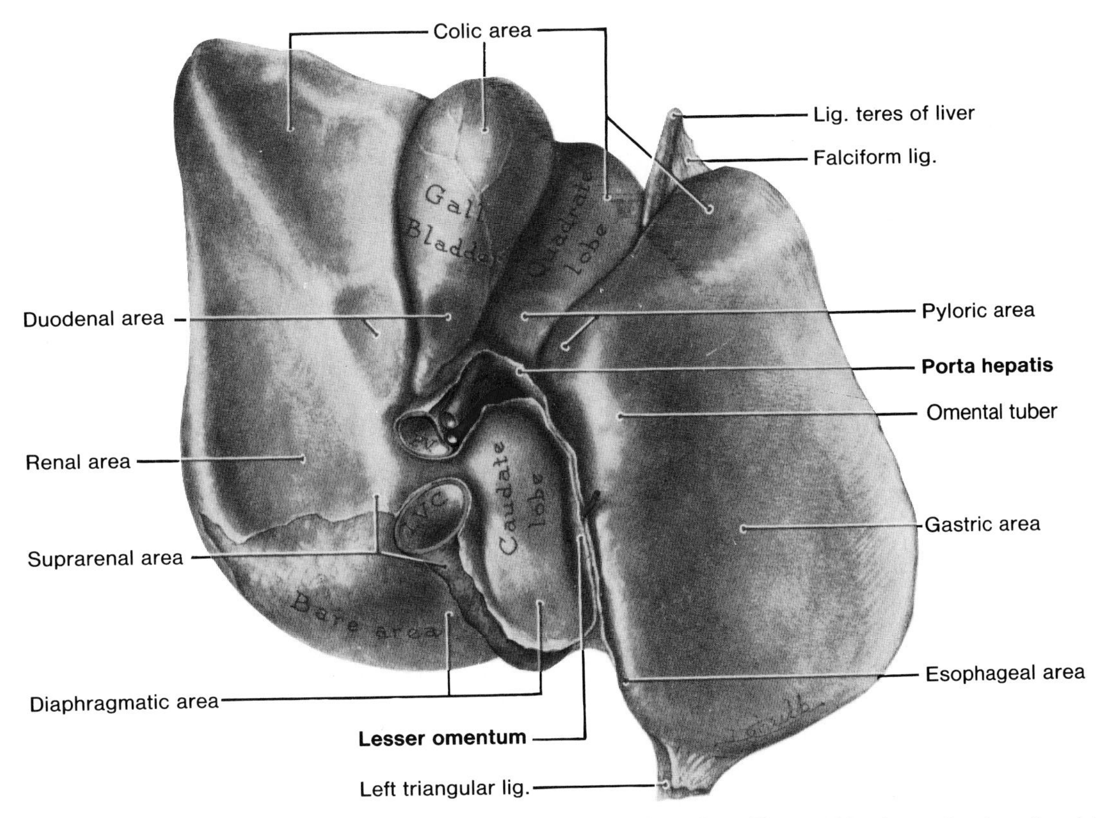

Figure 2-72. Drawing of a dissection of the visceral surface of the liver. To get this view, stand on the right side of a cadaver and face the head; then divide the attachments of the liver and raise its sharp inferior border. Observe the visceral areas: (1) for the esophagus, stomach, pylorus, and duodenum; (2) for the transverse colon; and (3) for the right kidney and the right suprarenal gland. The gallbladder rests on the transverse colon and the duodenum (Fig. 2-35). Examine the posterior surface comprising: (1) the bare area occupied on its left by the inferior vena cava, (2) the caudate lobe, and (3) the groove for the esophagus. Note that the caudate lobe is separated from the quadrate lobe by the porta hepatis and is joined to the right lobe by the caudate process (Fig. 2-74), which is squeezed between the inferior vena cava and the portal vein. At the right end of the bare area, the right triangular ligament (not labeled) bifurcates into the superior and inferior layers of the coronary ligament. The inferior layer crosses the renal and suprarenal areas and, after passing anterior to the inferior vena cava, turns cranially as the left layer or base of the coronary liagment. Followed to the left, this layer of peritoneum forms the superior limit of the superior recess and then turns caudally as the posterior layer of the lesser omentum. This omentum is attached to the fissure for the ligamentum venosum and to the porta hepatis.

phragm. The diaphragm separates the superior part of the liver from the structures in the thorax.

The superior part of the liver is covered with peritoneum, except posteriorly at the edge of the bare area (Figs. 2-70 and 2-71).

The posterior part of the liver includes most of the bare area that is between the reflections of the **coronary ligament** (Figs. 2-70 and 2-71).

In the bare area, the liver is in direct contact with the diaphragm and the inferior vena cava occupies a fossa in the left part of the bare area (Fig. 2-71), just to the right of the median plane (Fig. 2-80).

The visceral surface of the liver is directed inferiorly, posteriorly, and to the left. It is separated from the diaphragmatic surface of the liver by the inferior border. *Under cover of the visceral surface of the liver are:* (1) the superior right portion of the anterior surface of the ***stomach***; (2) the superior part of the ***duodenum***; (3) the ***lesser omentum***; (4) the ***gallbladder***; (5) the ***right colic flexure***; and (6) many associated vessels and nerves. The visceral surface of the liver is covered with peritoneum, except at the gallbladder and the porta hepatis (Fig. 2-72).

There are many irregularities on the visceral surface of the liver. **It has an H-shaped group of deep fissures and wide sulci** which define the four lobes of the liver. The crossbar of the H is the **porta hepatis** (Figs. 2-72 and 2-73*B*).

The porta hepatis is a deep transverse fissure, about 5 cm long, that contains the ***portal vein,*** *the* ***hepatic artery proper,*** *the* ***hepatic nerve plexus,*** *the* ***hepatic ducts,*** *and* ***lymphatic vessels.*** The left sagit-

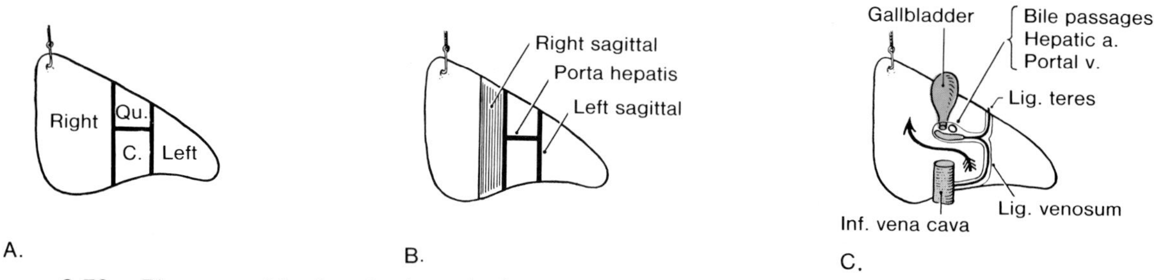

Figure 2-73. Diagrams of the liver to show the features of its inferior surface. *A*, illustrates its four lobes: right, quadrate, caudate, and left. *B*, demonstrates the H-shaped deep fissures and wide sulci defining the lobes. *C*, shows the structures in the fissures and sulci. The *arrow* traverses the epiploic foramen and passes into the omental bursa (see Fig. 2-71).

tal limbs of the H are deep fissures containing the **ligamentum teres** and the **ligamentum venosum.** The right sagittal limbs of the H are fossae for the gallbladder and the inferior vena cava (Fig. 2-73*C*).

Lobes of the Liver (Figs. 2-72 to 2-75). For descriptive purposes the liver is divided into a large right lobe and a smaller left lobe by the falciform ligament. The right lobe is subdivided into a **quadrate lobe,** lying between the gallbladder and the falciform ligament, and a **caudate lobe** lying between the inferior vena cava and the fissure for the ligamentum venosum. *The quadrate lobe is separated from the caudate lobe by the porta hepatis* (Figs. 2-72 and 2-73).

Although the quadrate and caudate lobes are considered to be part of the right lobe by many anatomists, it must be understood that according to internal morphology, based mainly on the distribution of blood vessels, they belong to the left lobe of the liver.

The right and left halves of the liver are functionally separate. Each half receives its own arterial and portal venous supply and has its own venous drainage. There is little (if any) overlap between the two halves of the liver. Similarly, the right hepatic duct drains bile from the right half of the liver and the left hepatic duct drains it from the left half of the liver. The pattern of blood vessel distribution also forms a basis for dividing the liver into ***hepatic segments*** which are of surgical significance.

The caudate lobe is the lobe with a tail (L. *cauda*), called the ***caudate process*** (Fig. 2-74). This narrow isthmus of liver bounds the epiploic foramen superiorly and connects the caudate lobe to the visceral surface (Fig. 2-72).

Peritoneal Relations of the Liver. The liver is almost entirely covered with peritoneum and so it presents a smooth glistening appearance in the fresh state. Its peritoneal relationships are not difficult to understand when development of the liver is understood.

The developing liver grows between the two layers of the ***ventral mesentery*** and much of this double-layered sheet becomes the peritoneal covering (visceral peritoneum) of the liver (Fig. 2-37). The remainder of the ventral mesentery forms the **lesser omentum** and the **falciform ligament** (Figs. 2-28, 2-36, and 2-72).

The various reflections of peritoneum from the liver constitute its ligaments. The passage of the ***embryonic umbilical vein*** from the umbilicus to the liver produces a sickle-shaped fold, the **falciform ligament,** that connects the liver to the anterior abdominal wall and to the diaphragm (Fig. 2-28).

The line of attachment of the double-layered falciform ligament to the liver is the line of division of the organ into *anatomical* right and left lobes. Along this line the two layers of the falciform ligament separate to enclose the liver, forming the visceral peritoneum of the liver (Fig. 2-36).

The falciform ligament encloses a few small *paraumbilical veins* (Fig. 2-100) and the umbilical vein in its free border during prenatal life. Several weeks after birth the umbilical vein obliterates close to the umbilicus, but it usually remains patent in the free edge of the falciform ligament as far as the left branch of the portal vein.

The obliterated portion of the umbilical vein is known as the ligamentum teres of the liver. On the visceral surface (Figs. 2-72 and 2-73*C*), the layers of the falciform ligament reflect onto the liver along the line of the fissure for the ligamentum teres, as far as the porta hepatis. At the superior end of the falciform ligament, its two layers separate from each other, exposing a hand-sized triangular area on the superior surface of the liver called the **bare area** (Fig. 2-71), because it is *devoid of peritoneum.* Here, the two peritoneal layers, diverge laterally and are reflected onto the diaphragm to form the **coronary ligament** (Fig. 2-71).

Because the liver is applied directly to the diaphragm in the bare area, *the right and left layers of the coronary ligament are spread apart and surround the bare area* (Fig. 2-71).

The right reflection of the falciform ligament constitutes the anterior layer of the coronary ligament. It

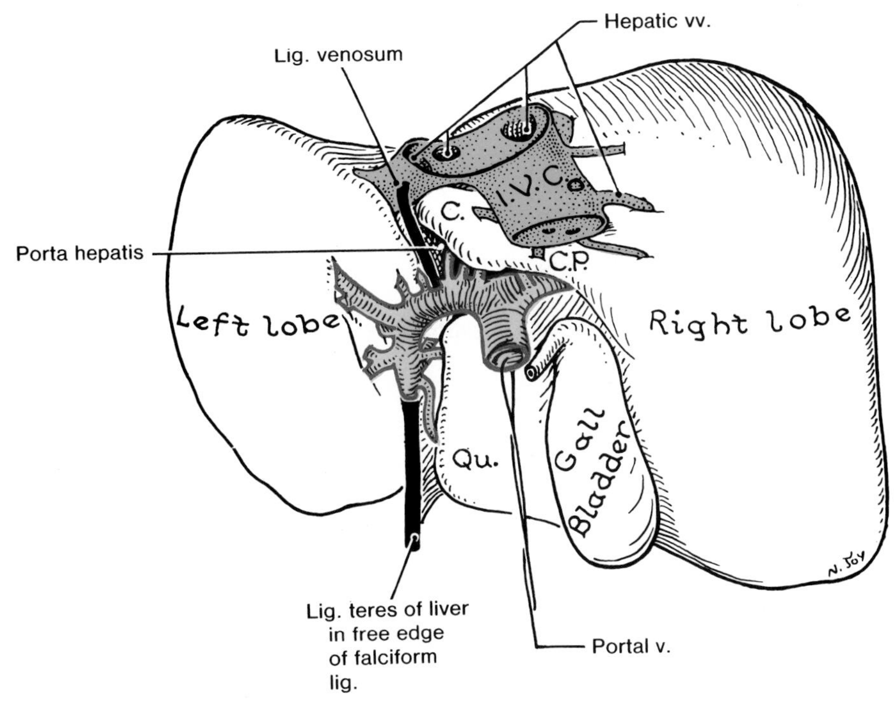

Figure 2-74. Drawing of a posteroinferior view of the liver showing its veins. Observe the branches of the portal vein (*yellow*) entering the liver and the three large and several small hepatic veins (*blue*) leaving the liver to join the inferior vena cava. *C*, caudate lobe; *CP*, caudate process; *Qu*, quadrate lobe. Also see Figure 2-52.

passes to the right and then bends sharply at the right triangular ligament to become the posterior layer of the coronary ligament.

The left reflection of the falciform ligament forms the left triangular ligament, which becomes continuous with the posterior layer of the coronary ligament.

From the porta hepatis the peritoneal reflections pass to the lesser curvature of the stomach and the superior part of the duodenum as the lesser omentum (Fig. 2-36). *The left triangular ligament (posterior layer) is continuous with the lesser omentum* (Fig. 2-71). The portion of the lesser omentum, extending between the liver and the stomach is called the **hepatogastric ligament**, and the portion between the liver and the duodenum is called the **hepatoduodenal ligament** (Fig. 2-38).

In *the free edge of the lesser omentum*, the two layers enclose the hepatic artery, the portal vein, the bile duct (Figs. 2-30 and 2-78), a few lymph nodes and vessels, and the hepatic plexus of nerves.

BLOOD VESSELS OF THE LIVER

The liver has a double blood supply from the **hepatic artery** (30%) and the **portal vein** (70%). The hepatic artery carries oxygenated blood to the liver (Fig. 2-78), and the portal vein carries venous blood containing the products of digestion absorbed from the GI tract (Fig. 2-52). The arterial blood is conducted to the **central vein** of each liver lobule. The central veins drain into the hepatic veins which open into the inferior vena cava (Figs. 2-52 and 2-72).

The common hepatic artery arises from the celiac trunk and passes anteriorly to the right in the posterior wall of the omental bursa (Fig. 2-45). *It runs inferior to the epiploic foramen* to reach the superior part of the duodenum. After giving off the gastroduodenal artery, it passes between the layers of the lesser omentum as the **hepatic artery proper** (Fig. 2-78). This artery ascends in the free edge of the lesser omentum anterior to the portal vein and to the left of the bile duct.

Near the porta hepatis the hepatic artery proper divides into right and left terminal branches, called the **right and left hepatic arteries** (Fig. 2-75).

In about 11% of cases the left hepatic artery arises from the left gastric artery in the vicinity of the gastroesophageal junction (Fig. 2-76*A*), and passes between the layers of the superior part of the lesser omentum to the left lobe of the liver.

The portal vein, supplying most of the blood to the liver, is formed posterior to the neck of the pancreas by the union of the superior mesenteric and splenic veins (Fig. 2-52). The portal vein runs in the free right edge of the lesser omentum

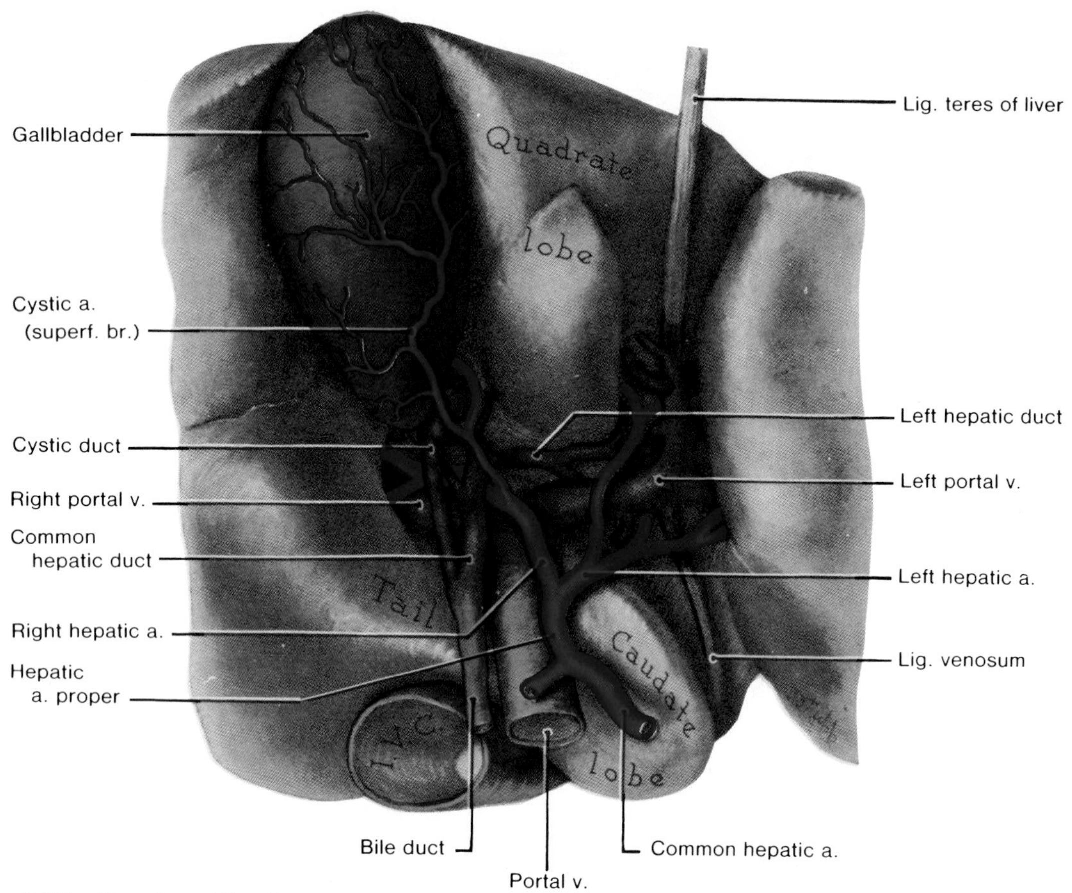

Figure 2-75. Drawing of the visceral surface of the liver showing the porta hepatis and the cystic artery. Observe that the tail of the caudate lobe, called the caudate process (Fig. 2-74), forms the superior boundary of the epiploic foramen and lies between the superior end of the portal vein and the inferior vena cava. The structures entering the liver at the porta hepatis are sometimes referred to clinically as the hepatic pedicle. Note the relation of the structures as they ascend to the porta hepatis: duct to the right, artery to the left, and vein posterior. Observe the order of the structures at the porta hepatis: bile duct, artery, vein from anterior to posterior. Note that the left portal vein and the left hepatic artery supply the quadrate and caudate lobes en route to the left lobe, and that they are accompanied by the tributaries of the left hepatic duct. Examine the ligamentum teres passing to the left portal vein and the ligamentum venosum arising opposite it and ascending to the inferior vena cava. Observe the cystic artery springing from the right hepatic artery and dividing into superficial and deep branches on the respective surfaces of the gallbladder. Note that the cystic duct is sinuous at its origin.

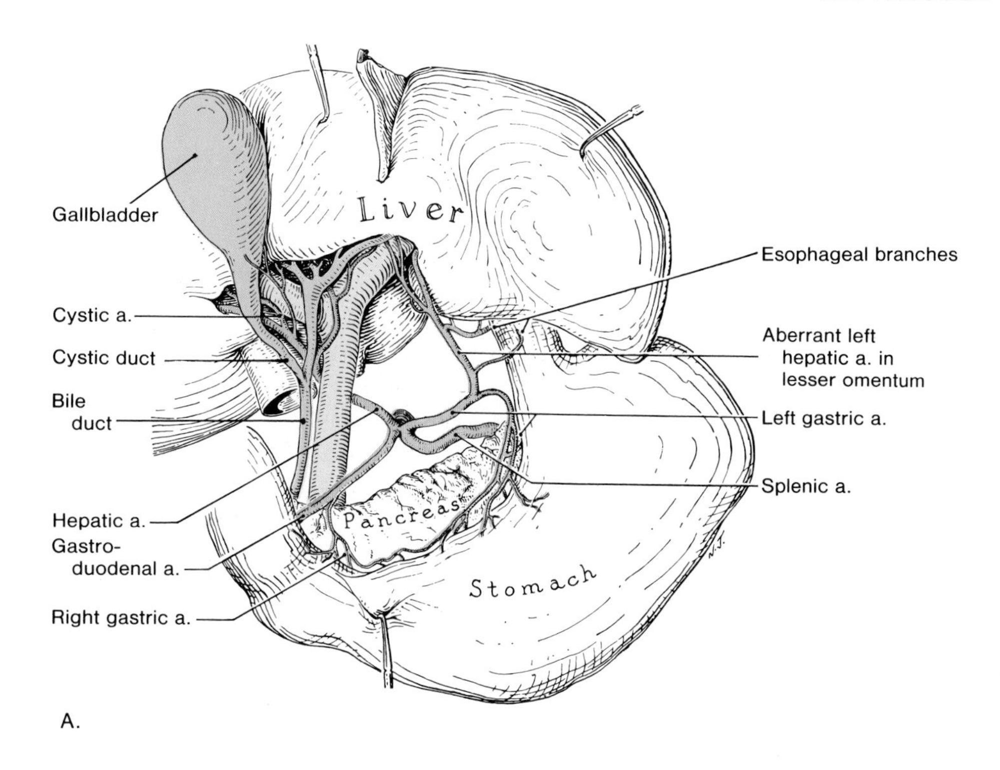

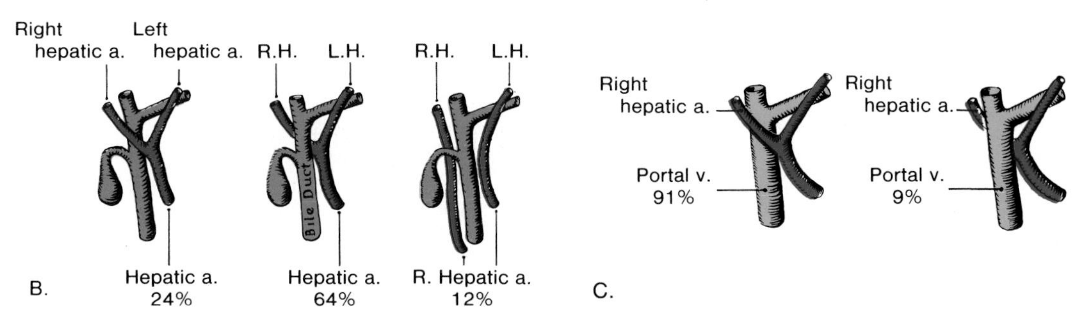

Figure 2-76. Drawings of variations in the hepatic arteries. *A*, aberrant left hepatic artery. The left hepatic artery was entirely replaced by a branch of the left gastric artery, as in this specimen, in 11.5% of 200 cadavers, and in another 11.5% it was partially replaced. *B*, right hepatic artery variants. In a study of 165 cadavers, three patterns were observed: in 24%, the right hepatic artery crossed ventral to the bile passages; in 64%, the right hepatic artery crossed dorsal to the bile passages and in 12%, the aberrant artery arose from the superior mesenteric artery. *C*, the artery crossed ventral to the portal vein in 91% of 165 specimens and dorsal to it in 9% of cases.

(Fig. 2-30), posterior to the bile duct and the hepatic artery, and anterior to the epiploic foramen. At the right end of the porta hepatis, *the portal vein terminates by dividing into two branches (right and left), each of which supplies about half of the liver* (Fig. 2-52).

The hepatic veins draining blood from the liver are formed by the union of the central veins of the lobules of the liver. *The hepatic veins open into the inferior vena cava just inferior to the diaphragm* (Figs. 2-52 and 2-74). There are superior and inferior groups of hepatic veins. The superior group may consist of only right and left veins, but usually there is a middle vein from the caudate lobe. The inferior group consists of 6 to 18 small veins which drain blood from the right and caudate lobes of the liver.

LYMPH VESSELS OF THE LIVER

Lymph vessels from the liver pass in several directions: (1) those from the superior and anterior parts of the *diaphragmatic surface of the liver* pass through the falciform ligament to the inferior **parasternal lymph nodes** (Fig. 1-16); (2) those from the *visceral surface and deep parts of the liver* drain into the **hepatic lymph nodes** in the porta hepatis, into the lymph nodes around the hepatic artery, and into the **gastric lymph nodes** (Fig. 2-53) via the lesser omentum; and (3) those from the *bare area of the liver* drain into the **middle diaphragmatic lymph nodes** on the thoracic surface of the diaphragm, and also into the **celiac lymph nodes** (Fig. 2-58).

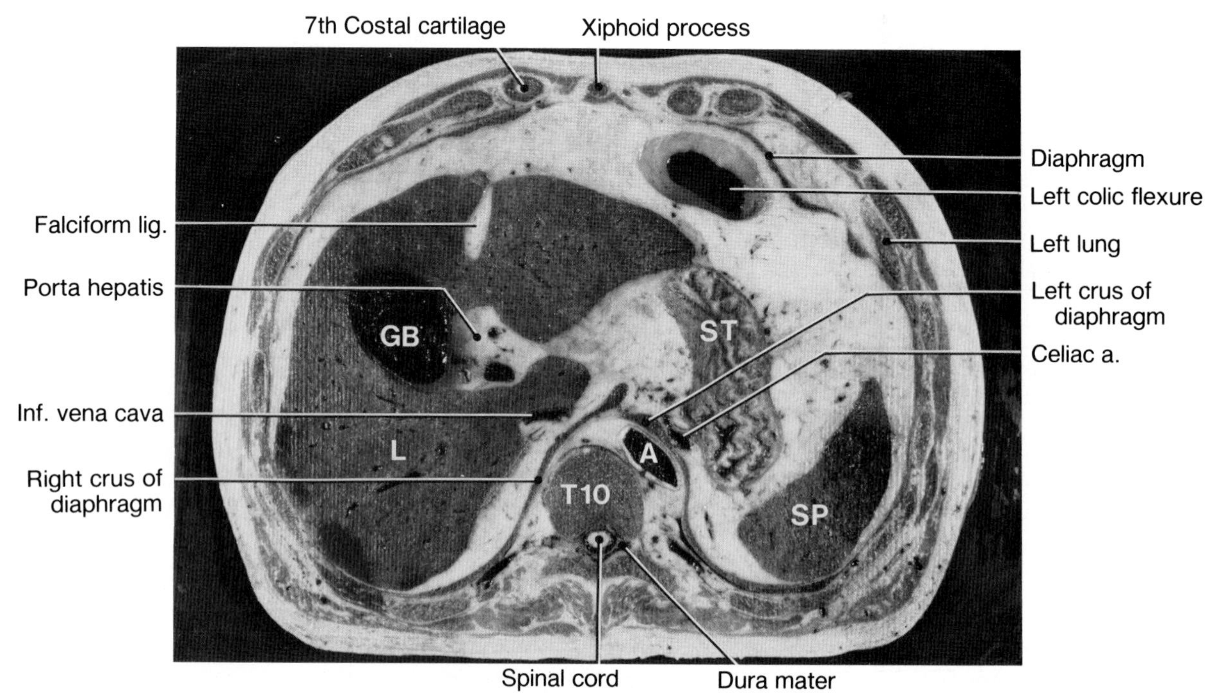

Figure 2-77. Photograph of a transverse section of the abdomen of an adult male at the level of T10 vertebra. In keeping with radiographic convention, as in CT scans, the section is viewed from below. Hence the right side of the body appears on the left side of the photograph. *GB*, gallbladder; *ST*, stomach; *A*, descending aorta; *L*, liver; *SP*, spleen; *T10*, tenth thoracic vertebra.

Most of the lymph from the liver eventually drains into the thoracic duct (Fig. 1-42). It has been estimated that up to one-half of the lymph entering the thoracic duct is derived from the liver.

NERVES OF THE LIVER

The nerves to the liver contain both *sympathetic and parasympathetic* (vagal) *fibers.* These nerves reach the liver via the **hepatic plexus**, the largest derivative of the celiac plexus (Figs. 2-99 and 2-114), which also receives filaments from the left and right vagus and right phrenic nerves (Figs. 2-120 and 7-135).

The hepatic plexus of nerves accompanies the hepatic artery and portal vein and their branches, and enters the liver at the porta hepatis (Figs. 2-54 and 2-82).

Variations in the origin and the course of the hepatic arteries are common (Fig. 2-76) and knowledge of them is important to the surgeon.

Liver tissue for diagnostic purposes may be obtained by **liver biopsy**. The needle puncture is commonly made through the right seventh, eighth, or ninth intercostal space in the midaxillary line (Figs. 2-35 and 2-105). The biopsy is taken while the patient is holding his/her breath in full expiration to reduce the costodiaphragmatic recess (Figs. 2-28, 2-35, and 2-101), and to lessen the possibility of damaging the lung and contaminating the pleural cavity.

The liver is a common site of metastatic carcinoma from structures drained by the portal system of veins (Fig. 2-52). Cancer cells may also travel to the liver from the thorax or the breast, owing to communications between thoracic lymph nodes and the lymph vessels draining the bare area of the liver.

Normally the liver is a soft mass with a jelly-like consistency, almost all of which is protected by the rib cage (Figs. 2-69 and 2-105). In some normal people the inferior edge of the liver may be palpable, 1 to 2 cm inferior to the right costal margin; hence, *a palpable inferior border of the liver does not by itself indicate liver enlargement* (**hepatomegaly**). Large livers are associated with carcinoma of the liver, congestive heart failure, fatty infiltration, and Hodgkin's disease.

Injury to the chest wall inferior to the fifth intercostal space on the right side may involve the liver (Figs. 2-69 and 2-105), which is separated from the thoracic cage only by the diaphragm (Fig. 2-48).

The liver is easily ruptured because it is large, fixed in position, and friable. The bleeding may

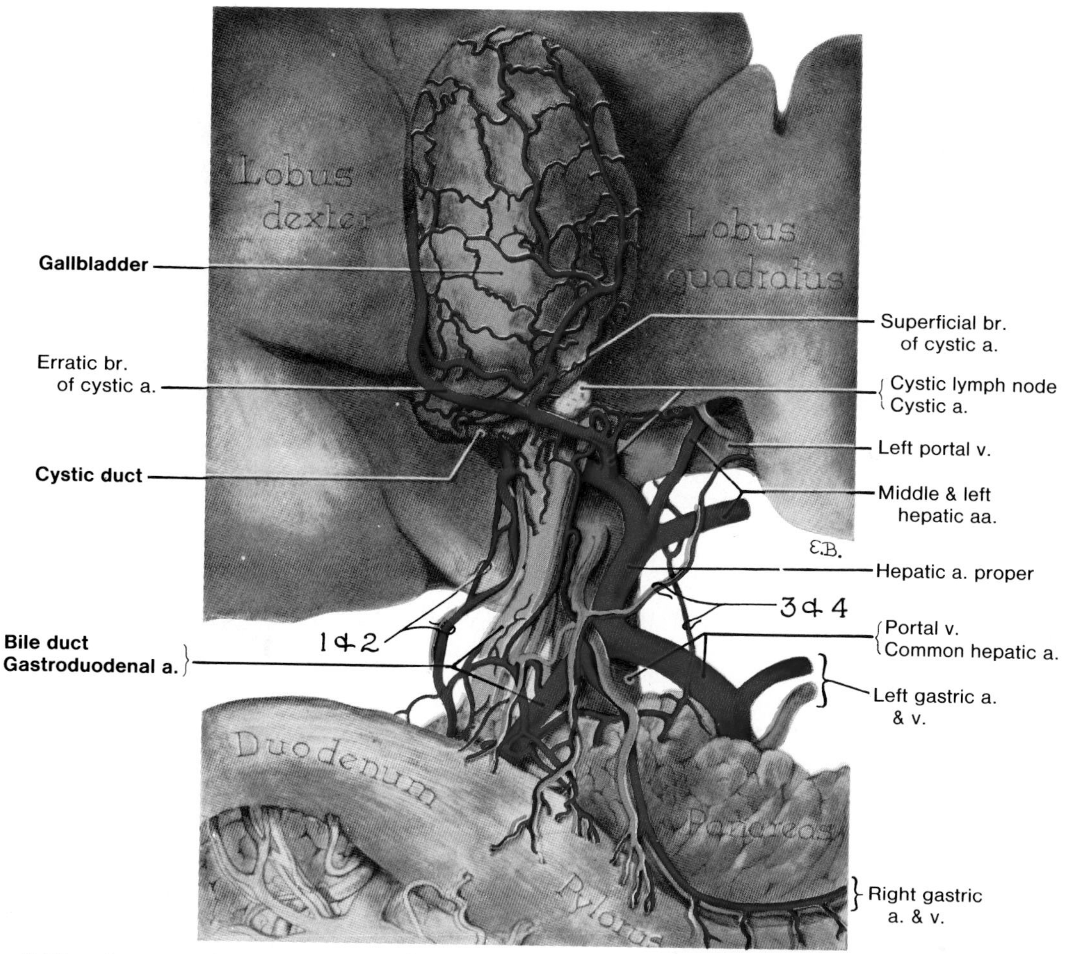

Figure 2-78. Drawing of a dissection of the gallbladder, the biliary ducts, and the related blood vessels. The liver is reflected superiorly with the gallbladder and the duodenum is pulled inferiorly. Note the large erratic deep branch of the cystic artery crossing superficial to the neck of the gallbladder. Examine the many fine sinuous arterial twigs supplying the biliary ducts. Observe that the right gastric artery arises from the gastroduodenal artery.

be severe because the hepatic veins are located in rigid canals and are unable to contract. *Often the liver is torn by the end of a broken rib* which perforates the diaphragm.

In cirrhosis of the liver there is progressive *destruction of hepatocytes* and replacement of them by fibrous tissue. This tissue surrounds the intrahepatic blood vessels and biliary radicles, impeding circulation of blood through the liver. The consistency of *the liver is very firm owing to the large amounts of fibrous tissue.* The surface of the liver has a nodular appearance, accounting for the term "*hob-nail liver.*" There is often a reddish yellow or tawny color to the liver, hence the name **cirrhosis** (G. *kirrhos*, tawny, orange colored + *osis*, condition).

Peritonitis may result in the formation of localized **abscesses** (collections of pus) in various parts of the peritoneal cavity. *A common site for an abscess is in the* ***subphrenic recess*** (Fig. 2-70), located between the diaphragmatic surface of the liver and the diaphragm.

Subphrenic abscesses occur much more frequently on the right side because of the frequency of ruptured appendices, duodenal ulcers, and gallbladders. As the right and left subphrenic recesses are continuous with the ***hepatorenal recesses*** (Figs. 2-70 and 2-81), pus from a subphrenic abscess may drain into one of these recesses when the patient is supine, because in this position the hepatorenal recesses are the most posterior parts of the peritoneal cavity (Fig. 2-31). *Pus in a hepatorenal recess may be drained through a posterior incision* superior to the 12th rib (Fig. 2-105).

THE BILIARY DUCTS AND GALLBLADDER

Bile *is secreted by the hepatic cells into the bile canaliculi,* the smallest branches of the intrahepatic duct system. Most of the canaliculi drain into small **interlobular bile ducts** which join with others to form progressively larger ducts. Eventually, **right and left hepatic ducts** are formed which emerge from the porta hepatis (Figs. 2-79 to 2-81). The right hepatic duct drains approximately the right half of the liver, and the left hepatic duct drains approximately the left half of the liver. The areas of the liver drained by the hepatic ducts are the same as those supplied by the right and left branches of the hepatic artery and the portal vein.

Shortly after leaving the porta hepatis, the right and left hepatic ducts unite to form the **common hepatic duct** (Figs. 2-79 and 2-80). About 4 cm in length, it passes inferiorly and to the right, between the layers of the lesser omentum, where it is joined on the right side at an acute angle by the **cystic duct** from the gallbladder to form the large **bile duct** (common bile duct). The bile duct is 8 to 10 cm long and 5 to 6 mm in diameter (Figs. 2-79 to 2-82).

The bile duct passes in the free edge of the lesser omentum with the hepatic artery and the portal vein (Figs. 2-30 and 2-81). The bile duct passes inferiorly, *anterior to the epiploic foramen* (Figs. 2-35 and 2-37), where it is anterior to the right edge of the portal vein and on the right side of the hepatic artery (Fig. 2-78).

The bile duct passes posterior to the superior part of the duodenum and the head of the pancreas (Fig. 2-64), occupying a groove in the posterior part of the head or embedded in it (Fig. 2-59). As it runs posterior to the duodenum, *the bile duct lies to the right of the gastroduodenal artery* (Figs. 2-58 and 2-78). On the left side of the descending part of the duodenum, the bile duct comes into contact with the pancreatic duct and the two of them run obliquely through the wall of

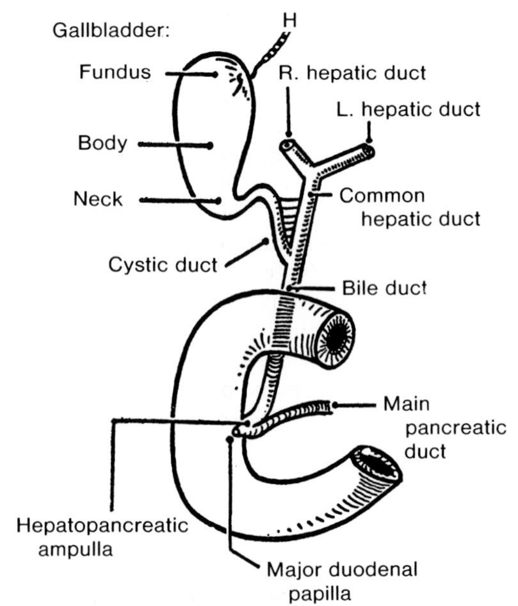

Figure 2-79. Diagram showing the extrahepatic parts of the biliary system and the relationship of the bile and pancreatic ducts. The gallbladder has been pulled superiorly with a hook (*H*). Observe that the cystic duct joins the common hepatic duct at an acute angle to form the bile duct. Note that the bile duct runs posterior to the superior part of the duodenum and passes obliquely through the duodenal wall, where it is joined by the main pancreatic duct. Usually the two ducts empty into a common channel, the hepatopancreatic ampulla, which opens into the duodenum at the apex of the major duodenal papilla (Fig. 2-67).

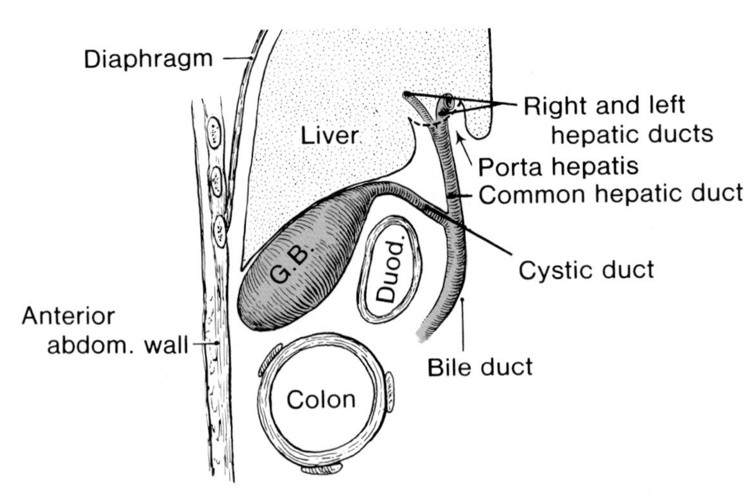

Figure 2-80. Drawing illustrating the relations of the gallbladder (*GB*): **anteriorly**, anterior abdominal wall and visceral surface of the liver; **posteriorly**, transverse colon and superior and descending parts of the duodenum (also see Fig. 2-60). You may hear the bile duct called the choledochal duct.

the duodenum, where they usually unite (Figs. 2-65*D* and 2-66) to form the **hepatopancreatic ampulla** (Fig. 2-79). The distal constricted end of the ampulla opens into the descending part of the duodenum at the summit of the **major duodenal papilla,** 8 to 10 cm from the pylorus (Figs. 2-66 and 2-67).

The circular muscle around the distal end of the bile duct is thickened to form the **choledochal sphincter** (G. *cholēdochus*, containing bile), a muscular sheath that surrounds the bile duct, just before and after it penetrates the duodenal wall.

There is also a sphincter around the hepatopancreatic ampulla called the **sphincter of the hepatopancreatic ampulla** (sphincter of Oddi). This sphincter controls the discharge of bile and pancreatic juice into the duodenum. When the choledochal sphincter contracts, bile cannot enter the ampulla and/or the duodenum; hence it backs up and passes along the **cystic duct** into the gallbladder for concentration and storage of the bile.

As the distal constricted end of the hepatopancreatic ampulla is the narrowest part of the biliary passages, it is a common site for impaction of a gallstone. Stones in the ampulla can usually be removed through an incision in the supraduodenal part of the bile duct (Figs. 2-60 and 2-78). Sometimes the duodenum and the head of the pancreas have to be mobilized, as shown in Figures 2-81 and 2-82, to expose the bile duct.

Blockage of the bile duct may cause severe pain in the right upper quadrant of the abdomen; however, in 10 to 15% of patients the pain ***(biliary colic)*** is not severe. *Understand that it is not the stone that causes the pain, but the distention of the bile duct* between the wave of contraction and the point of obstruction.

Jaundice may occur if the blockage has been present for some time. **Extrahepatic bile duct obstruction** may also result from extramural or intramural causes (*e.g.*, cancer).

Accessory hepatic ducts are common and awareness of their possible presence is of surgical importance (Fig. 2-85*D* to *F*). In some cases the cystic duct opens into one of these accessory ducts rather than joining the common hepatic duct to form the bile duct.

THE GALLBLADDER

The gallbladder is a pear-shaped sac that lies along the right edge of the quadrate lobe of the liver in a shallow fossa on its visceral surface.

The gallbladder concentrates the bile secreted by the liver by absorption of fluid by its epithelium. **It stores the bile** in the intervals between active phases of digestion. The gallbladder has a capacity of 30 to 60 ml. For descriptive purposes the gallbladder is divided into a *fundus, body, and neck* (Fig. 2-66).

The fundus of the gallbladder is its *wide end* which projects from the inferior border of the liver (Fig. 2-28). *The gallbladder is usually located at the tip of the ninth costal cartilage in the midclavicular line, where the linea semilunaris meets the costal margin* (Fig. 2-35). It is directed inferiorly, anteriorly, and to the right where it comes into relationship with the posterior surface of the anterior abdominal wall and the descending part of the duodenum (Fig. 2-80).

The body of the gallbladder is main part which is directed superiorly, posteriorly, and to the left from the fundus. *The body of the gallbladder lies in contact with the visceral surface of the liver* to which it is attached by loose connective tissue. It also contacts the right part of the transverse colon (Fig. 2-72) and the superior part of the duodenum (Fig. 2-80).

The neck of the gallbladder is narrow and tapered (Fig. 2-79), and is directed toward the porta hepatis (Fig. 2-80). It makes an S-shaped bend and is somewhat constricted as it *becomes continuous with the **cystic duct.*** The neck of the gallbladder is twisted in such a way that its mucosa is thrown into a **spiral fold** (Fig. 2-66).

The neck of the gallbladder serves as a guide to the epiploic foramen which lies immediately to its left (Figs. 2-35 and 2-45), posterior to the free margin of the lesser omentum.

The gallbladder is normally covered on its posterior and inferior surfaces by peritoneum. Occasionally the gallbladder is completely invested with peritoneum, and may even be connected to the liver by a short mesentery.

Although the fundus of the gallbladder is usually located at the level of the tip of the ninth costal cartilage (Figs. 2-28, 2-32*A*, and 2-35), it may lie as far inferiorly as the iliac crest.

If the gallbladder has a mesentery it is easy to remove surgically, but if it is deeply embedded in the liver it may be very difficult to remove.

The Cystic Duct (Figs. 2-60, 2-66, 2-81, and 2-83). The cystic duct, 2 to 4 cm long, *first* runs superiorly and to the left from the gallbladder, *then* it turns posteriorly and *finally* inferiorly to join the ***common hepatic duct*** (Fig. 2-80), forming the ***bile duct.*** Occasionally the cystic duct joins the right hepatic duct (Fig. 2-85*B*).

The cystic duct runs between the layers of the lesser omentum, usually parallel to the common hepatic duct,

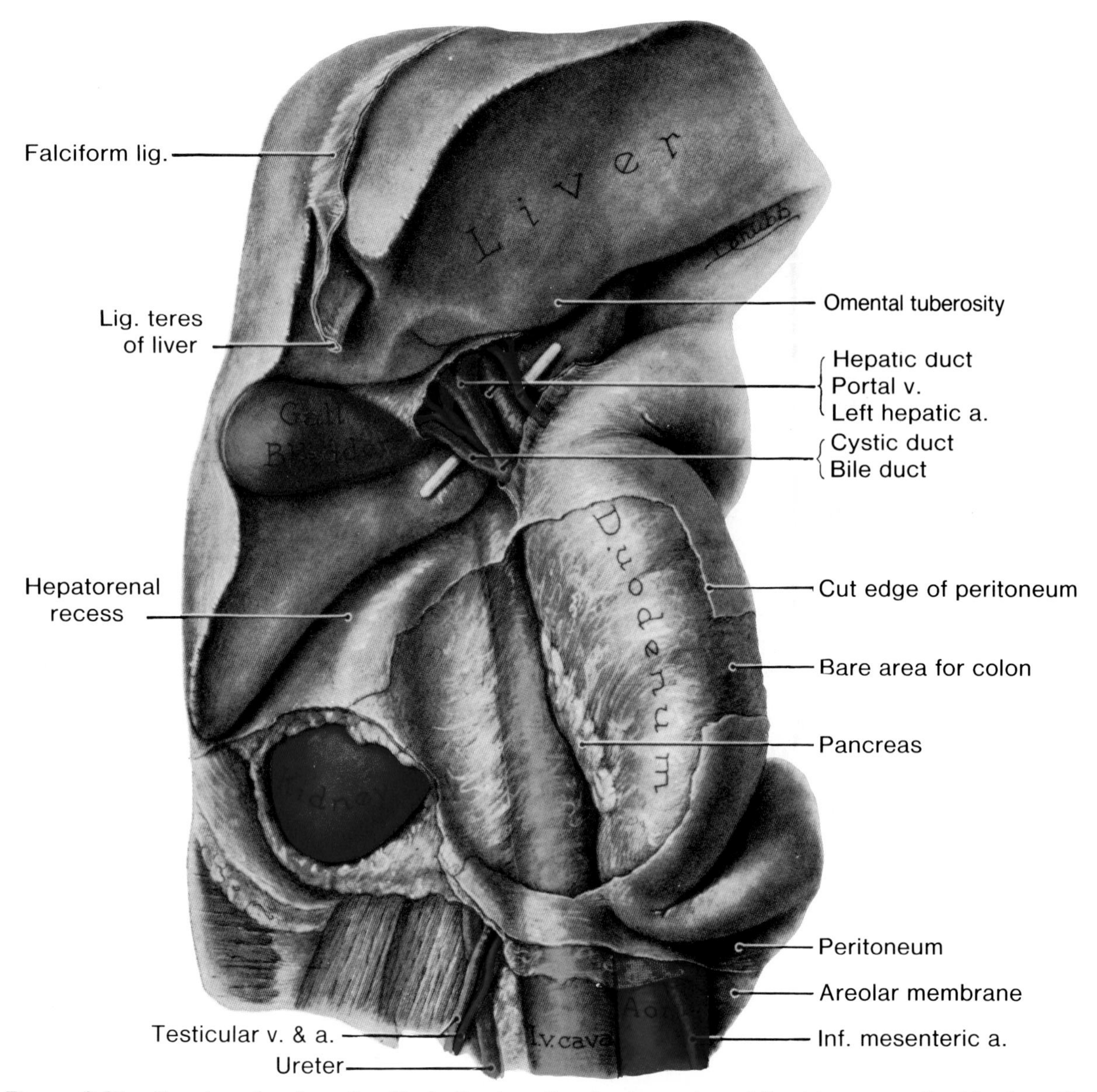

Figure 2-81. Drawing of a dissection illustrating how the duodenum is mobilized to expose the bile duct. The lesser omentum has been removed and the transverse colon has been separated from the descending part of the duodenum. The peritoneum has been cut along the right convex border of the descending part of the duodenum and this part of the duodenum has been swung anteriorly like a door on a hinge. Observe the three main structures that are exposed when the anterior wall of the lesser omentum is removed: the portal vein is posterior, the hepatic artery ascends from the left, and the biliary ducts descend to the right.

before joining it just inferior to the porta hepatis (Figs. 2-72 and 2-81). The mucous membrane of the cystic duct is thrown into a **spiral fold** with a core of smooth muscle (Fig. 2-66). This fold is continuous with a similar one in the neck of the gallbladder and it coils along the cystic duct, giving it the tortuous *appearance of a spiral valve.*

The spiral fold keeps the cystic duct constantly open so that: (1) bile can easily pass into the gallbladder when the bile duct is closed by the choledochal sphincter and/or the hepatopancreatic sphincter; and (2) bile can pass in the opposite direction into the duodenum when the gallbladder contracts.

A hormonal mechanism is involved in gallbladder contraction. Eating fat is particularly effective in producing gallbladder contraction. While digesting fatty food, a hormone called **cholecystokinin** is produced by the intestinal mucosa. It passes to the gallbladder,

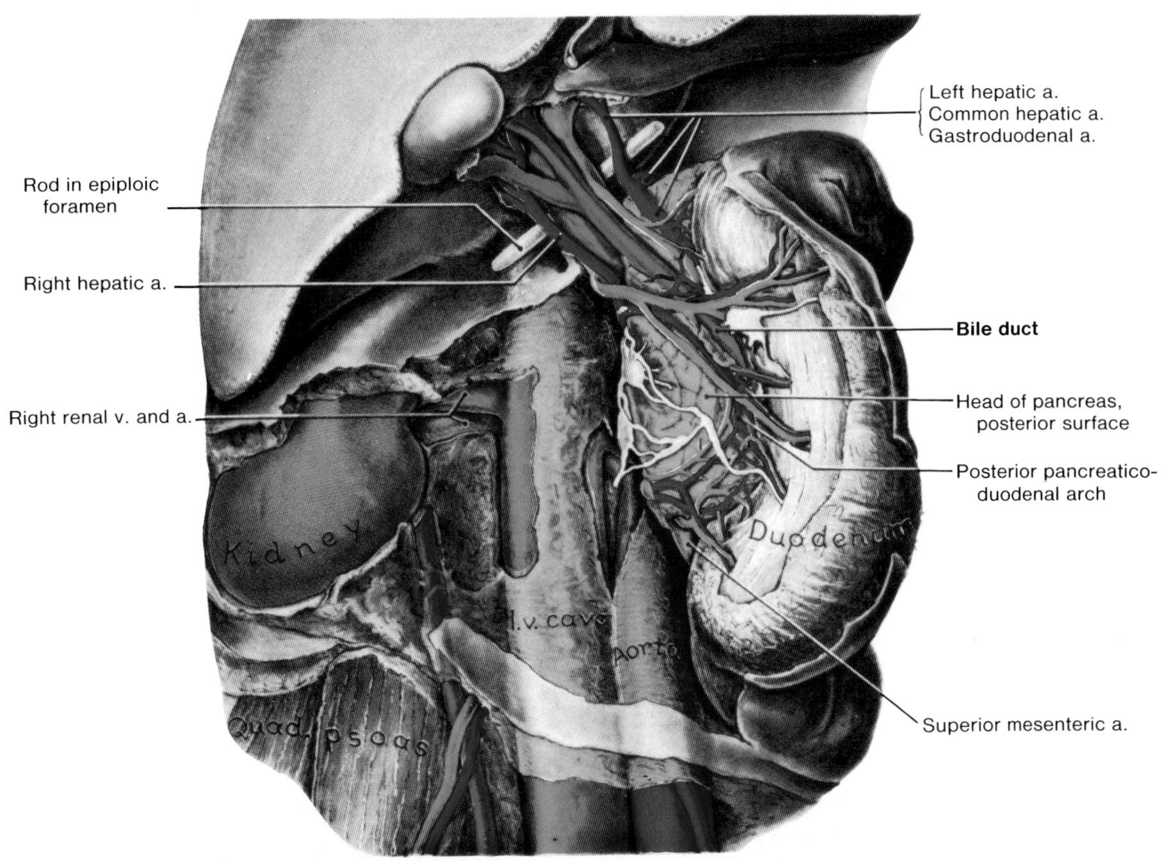

Figure 2-82. Drawing of a dissection exposing the posterior aspect of the bile duct (see Fig. 2-81 for an earlier stage of this dissection). The duodenum has been pulled anteriorly and to the left, taking the head of the pancreas with it. The areolar membrane covering these two organs is largely removed and that covering the great vessels is partly removed. Examine the bile duct descending in a groove on the head of the pancreas. Observe the very close posterior relationship of the inferior vena cava to the portal vein and the bile duct.

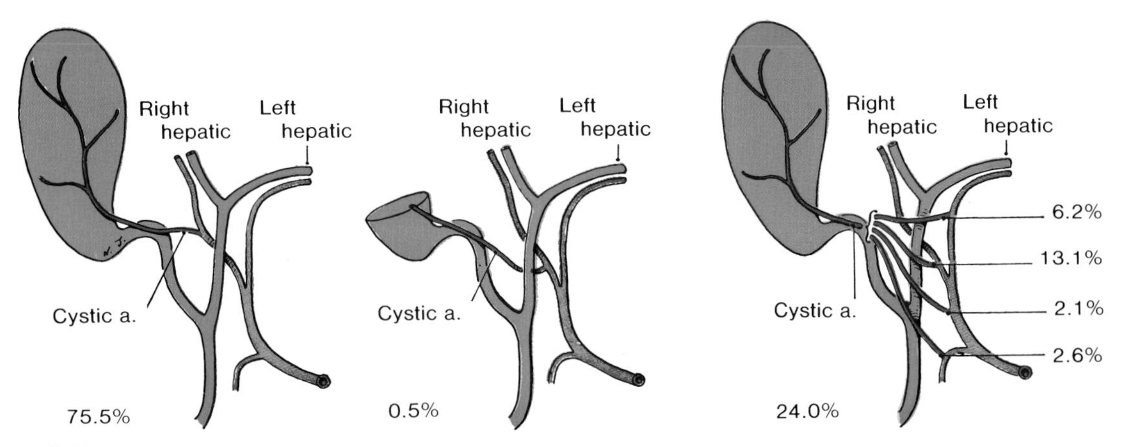

Figure 2-83. Drawings illustrating variations in the origin and course of the cystic artery in 580 cadavers. The cystic artery usually arises from the right hepatic artery in the angle between the common hepatic duct and the cystic duct. However, when it arises on the left of the biliary ducts, it usually crosses anterior to these ducts. The right-hand diagram is a *composite* of the variations of the cystic artery passing anterior to the common hepatic duct. Because of the above variations, the arterial supply to the gallbladder must be clearly visualized before removal. Otherwise, the right hepatic artery may be inadvertently ligated. If this is done, half of the liver will probably degenerate.

causing it to contract and release bile into the cystic and bile ducts.

Blood Vessels of the Gallbladder (Figs. 2-54, 2-75, 2-78, 2-83, and 2-84). *The gallbladder is supplied by the **cystic artery,*** which commonly arises from the right hepatic artery in the angle between the common hepatic duct and the cystic duct. *Variations in the origin and course of the cystic artery are common* (Fig. 2-83); one should be aware of them.

The veins draining the biliary ducts and the neck of the gallbladder join veins which connect the gastric, duodenal, and pancreatic veins partly to the liver directly and partly via a portal vein. The veins of the fundus and body of the gallbladder pass directly into the liver (Fig. 2-84).

Lymph Vessels of the Gallbladder. Lymph passes to the ***cystic lymph node*** at the neck of the gallbladder (Fig. 2-78) and to the ***node of the epiploic foramen.*** From these nodes lymph passes to the hepatic nodes and then to the celiac lymph nodes (Fig. 2-58).

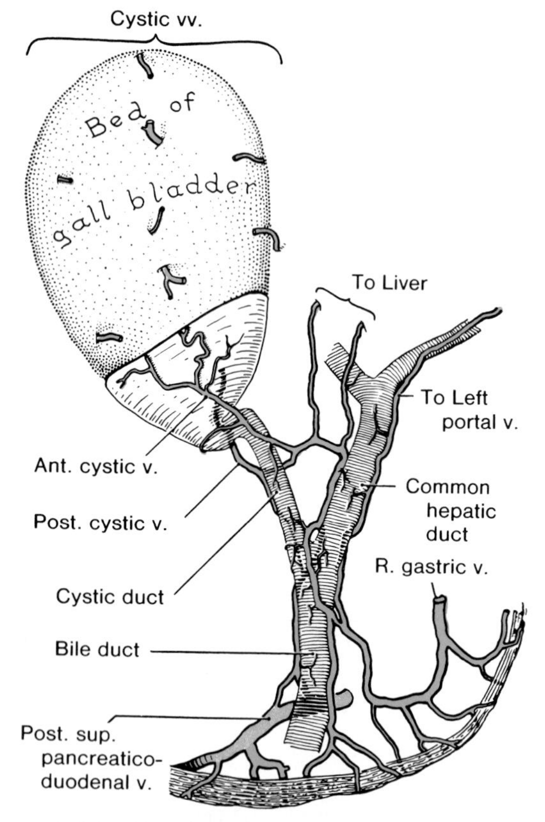

Figure 2-84. Drawing of the veins of the biliary ducts and gallbladder. Observe that the venous twigs draining the ducts and the neck of the gallbladder join veins which connect the gastric, duodenal, and pancreatic veins partly to the liver directly and partly via a portal vein. The veins of the fundus and body of the gallbladder pass directly into the liver.

Nerves of the Gallbladder (Figs. 2-99 and 7-135). The nerves pass along the cystic artery from the **celiac plexus** (sympathetic), the **vagus** (parasympathetic), and the **right phrenic nerve** (sensory).

Blood Vessels of the Bile Duct (Figs. 2-78 and 2-82). The bile duct is supplied by several arteries: (1) the distal or *retroduodenal part* is supplied by the **posterior superior pancreaticoduodenal artery** (Fig. 2-57); (2) the *middle part* is supplied by the **right hepatic artery;** and (3) the *proximal part* is supplied by the **cystic artery** (Fig. 2-78).

The supraduodenal branches of the *gastroduodenal artery* also give branches to the bile duct as it passes posterior to the superior part of the duodenum (Figs. 2-57 and 2-78). *There is considerable variation in the arrangement of these arteries.*

The veins from the proximal part of the bile duct and the hepatic ducts generally enter the liver directly (Fig. 2-84). They are connected inferiorly with the *posterior superior pancreaticoduodenal vein.* Veins from the distal part of the bile duct drain into the portal vein (Fig. 2-52).

Lymph Vessels of the Bile Duct. Lymph passes to the *cystic lymph node* (Fig. 2-78), the node of the epiploic foramen, the hepatic lymph nodes, and the **celiac lymph nodes** (Fig. 2-58).

After the gallbladder is removed at operation **(cholecystectomy)**, the hepatic ducts and the bile duct often dilate in order to store bile.

Variations in the arterial supply of the liver, gallbladder, and extrahepatic bile passages are common and clinically important. (Figs. 2-76 and 2-83). Most errors in gallbladder surgery result from failure to appreciate the common variations in the gross anatomy of the biliary system (Figs. 2-76, 2-83 and 2-85).

Before dividing any structures and removing the gallbladder, surgeons clearly identify all three biliary ducts, including the cystic and hepatic arteries.

Hemorrhage during cholecystectomy may be controlled by compressing the hepatic artery in the anterior wall of the epiploic foramen (Fig. 2-82). **Practice this in a cadaver!** Put your index finger in the epiploic foramen and your thumb on its anterior edge and pinch. This procedure stops the bleeding so the torn artery can be ligated.

The gallbladder and cystic duct have the radiolucency of water and *only a few gallstones (about 30%) produce radiodense shadows on plain radiographs of the abdomen.* To visualize the gallbladder and its contents radiographically, a technique called **oral cholecystography** is

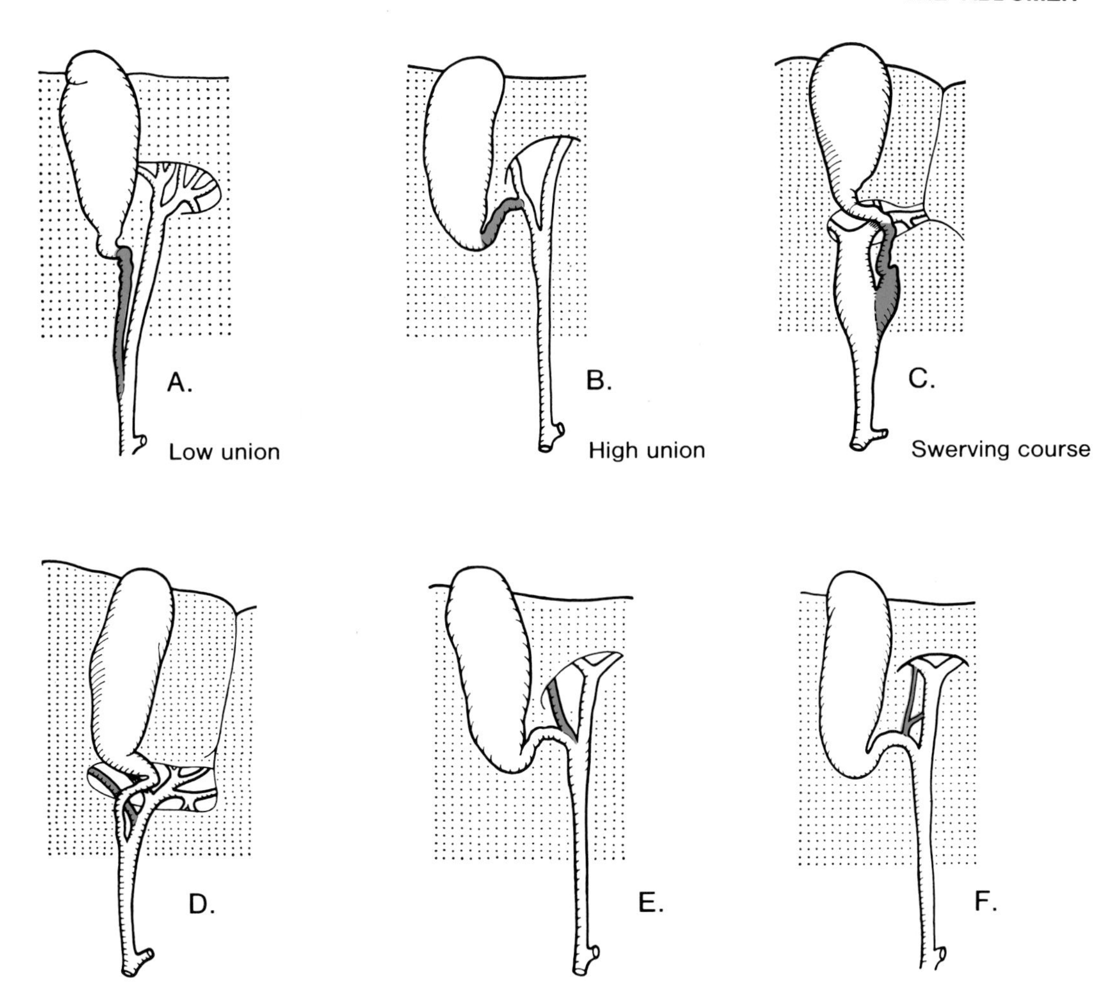

Figure 2-85. *A* to *C*, drawings illustrating variations in the length and course of the cystic duct. The cystic duct (*green*) usually lies on the right of the common hepatic duct and joins it just superior to the duodenum, but this varies as shown here. *D* to *E*, drawings showing some common varieties of accessory hepatic ducts (*green*). Of 95 gallbladders and biliary ducts injected in situ with melted paraffin wax and then dissected, 7 had accessory hepatic ducts in positions of surgical danger: *D*, four joined the common hepatic duct near the cystic duct; *E*, two joined the cystic duct; and *F*, one was an anastomosing duct.

commonly used. Patients are given a contrast medium by mouth after a fat-free evening meal.

The contrast material is absorbed by the small intestine, carried to the liver by the portal venous system, and secreted from the blood into the dilute **liver bile.** The bile passes to the gallbladder where water is absorbed from it (if it is not diseased). The resulting **gallbladder bile** is sufficiently concentrated (10 to 12 times) to be radiopaque. *Radiolucent gallstones produce negative images in the opacified gallbladder bile.*

Following these studies, the patient *may* be given a fatty meal to cause contraction of the gallbladder, forcing the radiopaque bile into the cystic and bile ducts. Usually these ducts can be visualized radiographically 15 to 20 min after ingestion of the fatty meal.

*Using **cholecystography**, it has been shown that the form and position of the gallbladder vary in normal individuals.* In short persons with good muscular development **(hypersthenic physical type),** the gallbladder tends to be broad and to lie superolaterally, opposite the first lumbar vertebra. In tall slender persons with poor muscular development **(asthenic physical type),** the gallbladder tends to be narrow and to lie closer to the vertebral column, as far inferiorly as the fourth lumbar vertebra.

If the gallbladder cannot be visualized or is poorly visualized by ***oral cholecystography,*** the contrast material may be given intravenously **(intravenous cholangiography).** *If the biliary duct system is obstructed,* it is usually dilated and contrast material may be injected directly into a bile duct via a slender needle passed through the skin toward the porta hepatis. This technique is called **transhepatic cholangiography.**

Radiographic studies of the gallbladder may also reveal congenital abnormalities such as double gallbladder (very rare), **folded gallbladder** (present in about 15% of people), and abnormal position of the gallbladder.

The junction of the cystic and common hepatic ducts varies. When the junction is low (Fig. 2-85*A*), the two ducts may be closely connected by fibrous tissue, making it difficult to clamp the cystic duct without injuring the hepatic or bile ducts. Failure to recognize a high union of the cystic duct (Fig. 2-85*B*) would result in no drainage of bile from the left half of the liver, if the left hepatic duct was ligated.

Occasionally there are some narrow channels **(accessory hepatic ducts)** running from the right lobe of the liver into the anterior surface of the body of the gallbladder. As they carry bile directly from the liver to the gallbladder, they are a cause of **bile leakage** after cholecystectomy. This is usually controlled by cauterizing the gallbladder bed and then sewing a patch over it, which is taken from the greater omentum.

From the right wall of the neck of the gallbladder, a dilation may be present called **Hartmann's pouch.** It projects inferiorly and posteriorly toward the duodenum.

Gallstones frequently lodge in Hartmann's pouch and if the gallbladder is inflamed **(cholecystitis),** this pouch may adhere to the cystic duct. As Hartmann's pouch is almost always present when the gallbladder is dilated, it is used as a point of traction for making identification of the cystic duct easier.

THE JEJUNUM AND ILEUM

The jejunum and ileum are *the greatly coiled parts of the small intestine.* The jejunum is continuous with the ascending part of the duodenum at the duodenojejunal flexure (Fig. 2-63), and the ileum joins the cecum at the ileocecal valve (Fig. 2-92).

There is no clear line of demarcation between the jejunum and the ileum, but the character of the small intestine does change gradually. The jejunum and ileum are 6 to 7 m long, the superior two-fifths of which is considered to be jejunum.

As *intestinal localization is of surgical importance,* the gross characteristics of the jejunum and the ileum will be described. The jejunum is often empty, hence its name (L. *jejunus,* empty). *The jejunum is thicker, more vascular, and redder in living persons than is the ileum.* Most of the jejunum usually lies in the umbilical region of the abdomen, whereas the ileum occupies much of the hypogastric and pelvic regions.

The terminal part of the ileum usually lies in the pelvis and ascends over the right psoas major muscle and right iliac vessels to enter the cecum (Figs. 2-40 and 2-92).

The **circular folds (plicae circulares)** of the mucous membrane are large and well developed in the superior part of the jejunum (Fig. 2-90), whereas they are small in the superior part of the ileum and absent in the terminal ileum.

When the jejunum of a living person is grasped between the forefinger and thumb, *the plicae circulares can be felt distinctly through the wall of the superior part of the jejunum.* As they are low and sparse in the superior part of the ileum and absent in the inferior part of the ileum, it is possible to distinguish the superior part from the inferior part of the jejunum and ileum.

The Mesentery of the Jejunum and Ileum. The jejunum and ileum are suspended from the posterior abdominal wall by a fan-shaped mesentery (Fig. 2-87). The **root of the mesentery,** attached to the posterior abdominal wall, is about 15 cm long.

The root of the mesentery is directed obliquely inferiorly and to the right from the left side of the second lumbar vertebra to the right sacroiliac joint. Between these two points, **the root of the mesentery crosses** (1) the *horizontal part of the duodenum,* (2) *aorta,* (3) *inferior vena cava,* (4) *psoas major muscle,* (5) *right ureter,* and (6) *right testicular or ovarian vessels.*

The mesentery of the jejunum and ileum consists of two layers of peritoneum, between which are the jejunal and ileal blood vessels, lymph nodes and vessels, nerves, and extraperitoneal fatty tissue.

The jejunal mesentery contains less fat than that of the ileum; thus **the arterial arcades of the jejunum are easier to observe than in the ileum** (Fig. 2-88). This anatomical characteristic helps surgeons to differentiate the jejunum from the ileum.

Arteries of the Jejunum and Ileum. The arteries to the jejunum and ileum arise from the **superior mesenteric artery,** the second of the unpaired branches of the abdominal aorta (Fig. 2-86).

The superior mesenteric artery usually arises from the aorta at the level of the first lumbar vertebra, about 1 cm inferior to the celiac trunk and posterior to the body of the pancreas and the splenic vein (Fig. 2-51). It descends across the left renal vein, the uncinate process of the pancreas, and the horizontal part of the duodenum to enter the mesentery (Figs. 2-64 and 2-

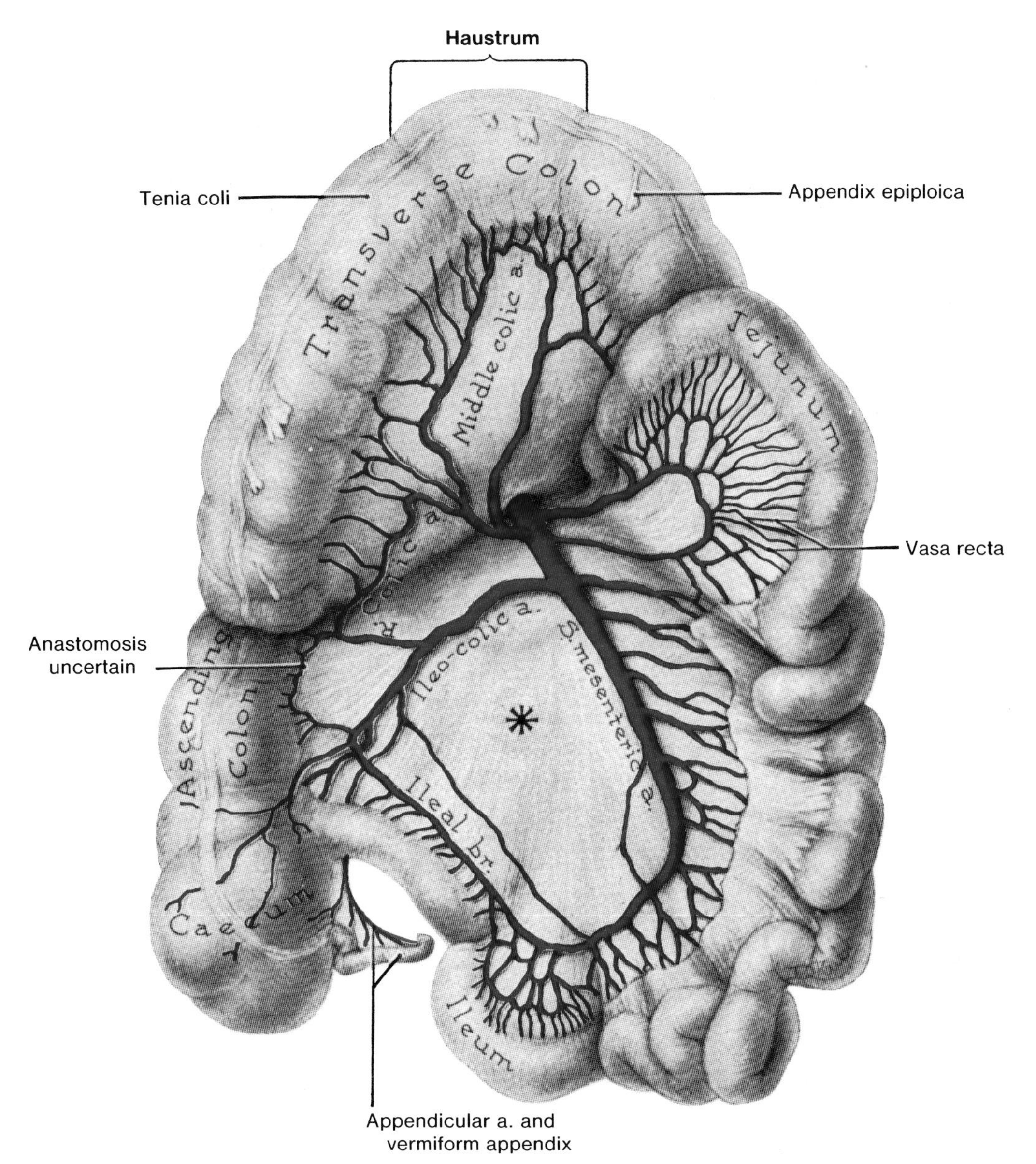

Figure 2-86. Drawing of a dissection of the *superior mesenteric artery.* The peritoneum is partly stripped off. Observe the extensive field of supply of the superior mesenteric artery. Note that it ends by anastomosing with one of its own branches, the ileal branch of the ileocolic artery. Examine the branches of the superior mesenteric artery: (1) *from its left side*, 12 to 20 jejunal and ileal branches which anastomose to form arcades from which vasa recta pass to the small intestine; (2) *from its right side*, the middle colic, the ileocolic, and commonly, but not here, an independent right colic artery; and (3) the two inferior pancreaticoduodenal arteries (see Fig. 2-59) arise from the main artery either directly or in conjunction with the jejunal branch. Note the teniae coli, haustra, and appendices epiploicae, which distinguish the large intestine from the small intestine. The *asterisk* indicates the area of mesentery between the superior mesenteric and ileocolic arteries which is almost avascular and liable to break down and produce a hole through which the coils of small intestine may herniate. This is one type of *internal hernia.*

86). *The superior mesenteric artery runs obliquely in the root of the mesentery to the right iliac fossa, sending branches to the intestines* (Figs. 2-86 and 2-87). Its last ileal branch anastomoses with a branch of the ileocolic artery (Fig. 2-86).

The superior mesenteric artery and its branches are surrounded by a plexus of sympathetic and parasympathetic nerve fibers (Figs. 2-99 and 2-114).

The 15 to 18 jejunal and ileal branches arise from the left side of the superior mesenteric artery (Figs. 2-86 and 2-87), and pass between the two layers of the mesentery. The arteries unite to form

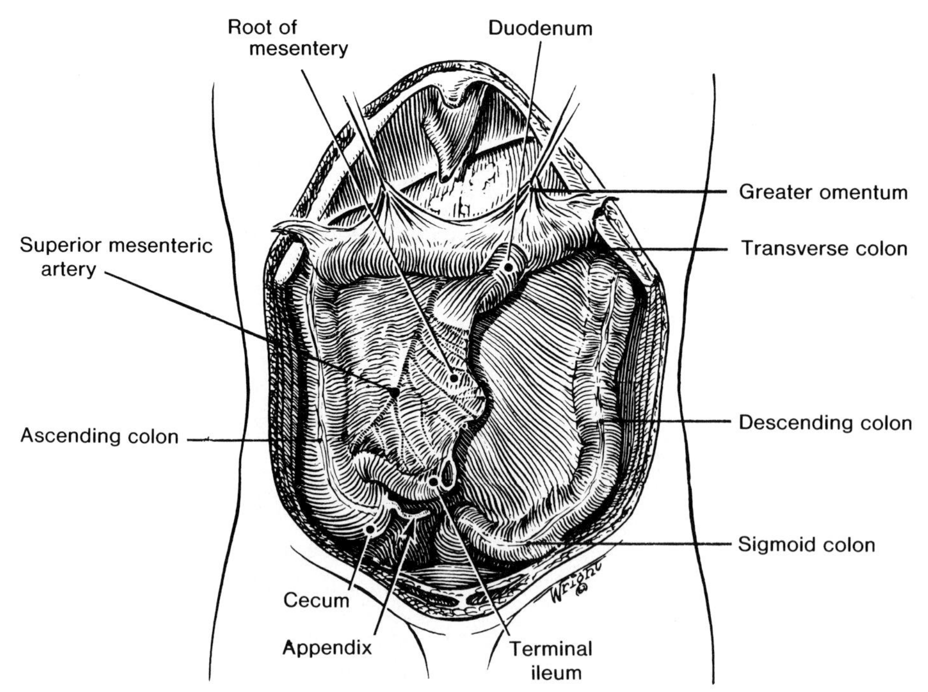

Figure 2-87. Drawing illustrating the root of the mesentery of the small intestine. Most of the jejunum and ileum are removed. Note that the root of the mesentery extends in an oblique manner from the left side, at the level of the second lumbar vertebra to the right sacroiliac joint, a distance of about 15 cm.

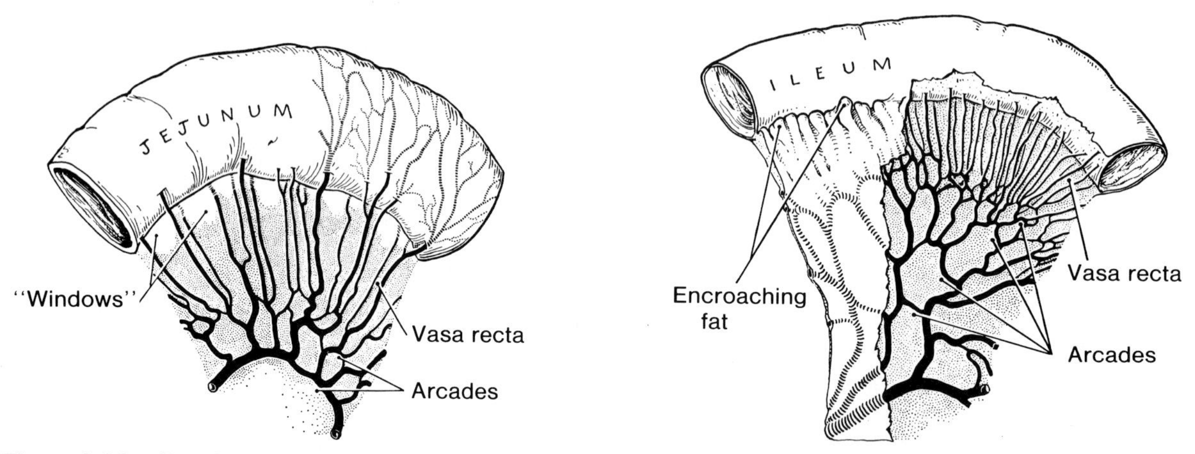

Figure 2-88. Drawings contrasting the arteries of the jejunum and ileum. Compare the diameter, thickness of the wall, number of arterial arcades, long or short vasa recta, presence of translucent (fat-free) areas at the mesenteric border, and fat encroaching on the wall of the gut. Note that the arterial arcades in the ileum are more complex and that the vasa recta are shorter than in the jejunum.

loops or arches called **arterial arcades** (Figs. 2-86, 2-88, and 2-89), from which vasa recta (L. straight vessels) arise.

The vasta recta do not anastomose within the mesentery; hence "windows" appear between the vessels (Fig. 2-88). The **vasa recta** pass from the arcades to the mesenteric border of the intestine, where they pass more or less alternately to opposite sides.

There are many anastomoses of the blood vessels in the wall of the intestine (Figs. 2-88 and 2-89). The vascularity of the wall is greater in the jejunum than in the ileum, but *the arterial arcades are shorter and more complex in the ileum.*

Veins of the Jejunum and Ileum (Figs. 2-52 and 2-94). *The superior mesenteric vein drains the jejunum and ileum.* It accompanies the superior mesenteric

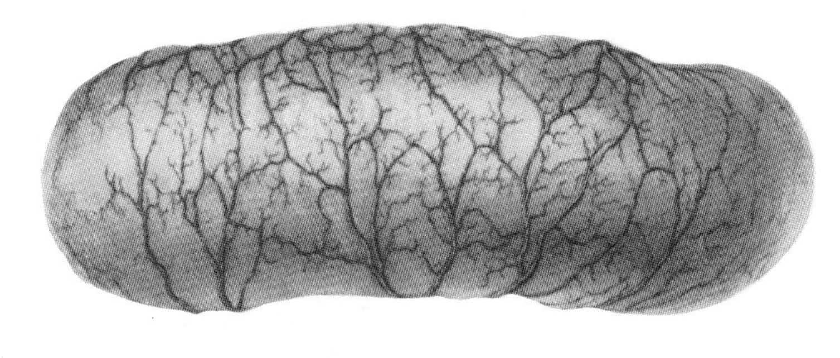

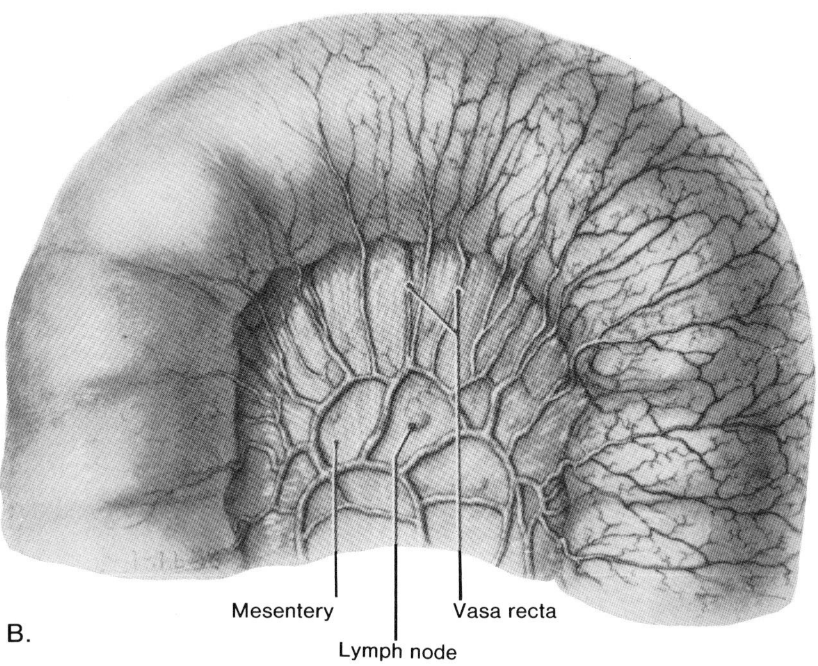

Figure 2-89. Drawings illustrating the blood supply of the small intestine. *A*, antimesenteric border. *B*, segment of jejunum with its mesentery and arteries. Observe the series of anastomotic arterial arches, called arcades in the mesentery. Note that the vasa recta proceed from the arcades to the mesenteric border of the intestine and then pass more or less alternately to opposite sides of the intestine. Examine the arborizations of the vasa recta in the wall of the intestine and the fine anastomoses effected between adjacent branches. Observe the efficient anastomoses across the antimesenteric border. Note the mesenteric lymph nodes (also see Fig. 2-7); enlargement of these nodes occurs in many diseases of the intestine.

artery, lying anterior and to its right in the root of the mesentery (Fig. 2-94). **The superior mesenteric vein** crosses the horizontal part of the duodenum and the uncinate process of the pancreas (Fig. 2-64). It terminates posterior to the neck of the pancreas by uniting with the splenic vein to form the portal vein (Fig. 2-52). The tributaries of the superior mesenteric vein have an arrangement similar to the branches of the superior mesenteric artery, and they drain the same area supplied by the arteries.

Lymph Vessels of the Jejunum and Ileum. The lymphatics in the intestinal villi, called **lacteals** (L. *lactis*, milk), empty their milk-like fluid into a plexus of lymph vessels in the wall of the jejunum and ileum (Fig. 2-91). The lymph vessels then pass between the two layers of the mesentery to the **mesenteric lymph nodes** (Figs. 2-89 and 2-97).

The mesenteric lymph nodes are in three locations: (1) close to the wall of the intestine, (2) amongst the arterial arcades, and (3) along the superior part of the superior mesenteric artery.

Lymph vessels from the terminal ileum follow the ileal branch of the ileocolic artery to the ileocolic lymph nodes (Fig. 2-97).

Nerves of the Jejunum and Ileum (Figs. 2-99, 2-114, and 7-135). The nerves of the jejunum and ileum are derived from the **vagus** and the **splanchnic nerves** through the *celiac ganglion* and the plexuses around the superior mesenteric artery.

The superior mesenteric nerve plexus receives its parasympathetic fibers from the celiac division of the ***posterior vagal trunk*** and its sympathetic fibers from the *superior mesenteric ganglion.*

Congenital malformations resulting from abnormal rotation of the midgut during the fetal period are not uncommon. Sometimes the intestinal loops fail to return to the abdominal cavity from the umbilical cord during the 10th fetal week. This hernia, called an **omphalocele,** may consist of a single loop of intestine or may contain most of them. *The wall of the hernial sac is formed by the* ***amnion*** *covering the umbilical cord.* (See Moore, 1982 for details).

Sometimes the intestines return to the abdominal cavity during the fetal period, but there is incomplete closure of the anterior abdominal wall in the umbilical region. As a result, an intestinal loop herniates after birth, forming a **congenital umbilical hernia,** often no larger than a cherry. This type of hernia is covered by skin and is easily detected when the infant cries.

Abnormal rotation and lack of fixation of the intestines sometimes occurs when they return to the abdominal cavity during the 10th fetal week. *Incomplete rotation usually results in nonfixation of the mesentery,* and the entire midgut loop may hang from a narrow pedicle. This often results in twisting of the intestine and the superior mesenteric vessels. This twisting, called a **volvulus** (L. *volvo,* to roll), may render the small intestine **ischemic** (deficient in blood) and even necrotic.

Rupture of the intestinal wall may occur if the volvulus is not corrected promptly after symptoms of intestinal obstruction begin.

Meckel's diverticulum (Fig. 2-93) is one of the most common malformations of the digestive tract. *A Meckel's diverticulum, representing the remnant of the proximal part of the embryonic yolk stalk,* is of clinical significance because it sometimes becomes inflamed and it may *cause symptoms mimicking appendicitis.* The wall of a Meckel's diverticulum contains all layers of the ileum and it may contain patches of gastric-type epithelium and pancreatic tissue. The gastric mucosa may secrete acid, producing ulceration and bleeding from the diverticulum.

Typically, a Meckel's diverticulum is a finger-like pouch, 3 to 6 cm long, projecting from the antimesenteric border of the ileum, within 50 cm of the ileocecal junction (Fig. 2-93). A Meckel's diverticulum may be connected to the umbilicus by a fibrous cord or a fistula (persistent yolk stalk). See Moore (1982) for details.

Occlusion of a series of vasa recta (arteries or veins) results in poor nutrition or drainage of the part of the intestine concerned. An embolus in an artery or a thrombosis of a vein *may* lead to necrosis of the segment of bowel concerned and **ileus** of the paralytic type (G. *eileos,* intestinal colic). If the condition can be diagnosed early enough (*e.g.*, using a **superior mesenteric arteriogram**), the obstructed portion of the vessel may be cleared surgically.

Owing to the many anastomoses of blood vessels (Fig. 2-98), blockage of a single vessel or a small group of vessels is not usually followed by **gangrene** of the intestine.

THE LARGE INTESTINE

The large intestine, about 1.5 meters long, *extends from the cecum in the right iliac fossa to the anus in the perineum* (Fig. 2-32A).

The three characteristic distinguishing features of the large intestine are: **teniae coli, haustra, and appendices epiploicae** (Figs. 2-39 to 2-41, 2-94, and 2-95). These terms are explained on page 196.

THE CECUM AND VERMIFORM APPENDIX

The cecum is a blind sac, continuous with the ascending colon, into which the terminal ileum and the vermiform appendix enter posteromedially (Figs. 2-41, 2-42, 2-86, 2-87, and 2-92). The ileum enters the cecum obliquely and is partly invaginated into it, forming lips superior and inferior to the ***ileocecal orifice.*** These lips, which form the **ileocecal valve,** meet medially and laterally to form ridges called the **frenula of the ileocecal valve** (Fig. 2-92).

When the cecum is distended the frenula tighten, drawing the lips of the valve together. However, as the circular muscle is poorly developed in these lips, *the ileocecal valve has very little sphincteric action.* Following a barium enema, barium commonly enters the terminal ileum (Fig. 2-40); consequently, contraction of the circular muscle of the terminal ileum is probably more important in preventing excessive reflux of cecal material into the ileum.

Vessels of the Cecum and Vermiform Appendix (Figs. 2-41, 2-86, and 2-94). The cecum and appendix are supplied by the **ileocolic artery,** a branch of the superior mesenteric artery (Fig. 2-86).

A tributary of the superior mesenteric vein, the

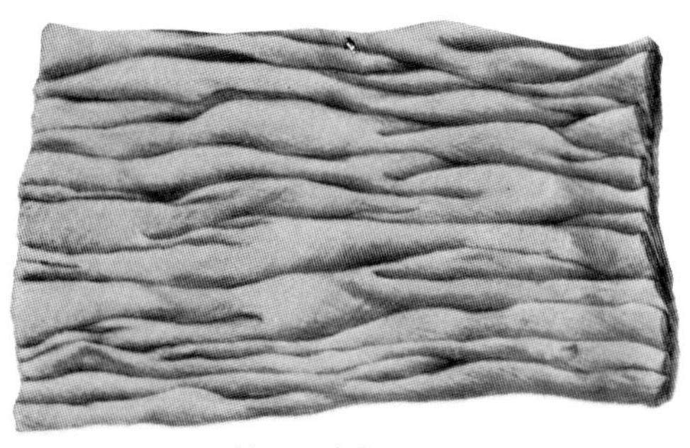

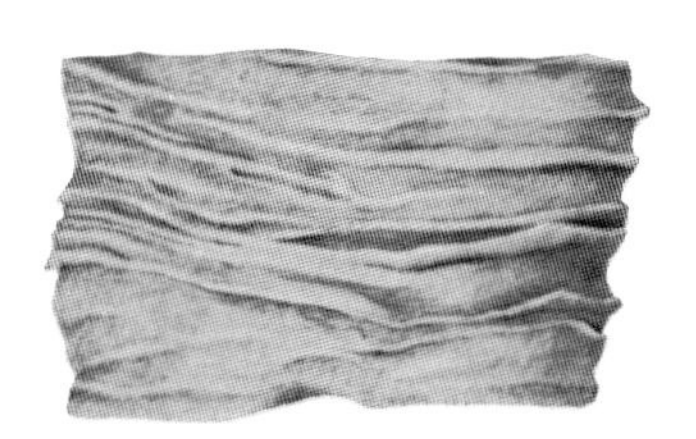

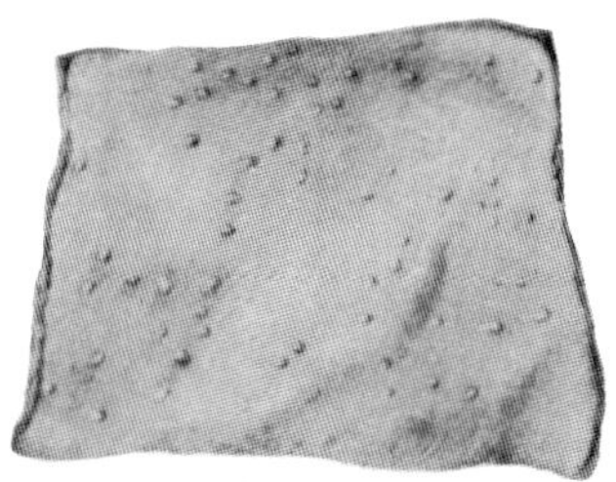

Figure 2-90. Drawings of the interior of the jejunum and ileum. Observe that the permanent circular folds or plicae circulares are: large, tall, and closely packed in the upper jejunum; low and sparse in the upper ileum; and absent in the lower ileum. Note that the caliber of the ileum is reduced compared with the jejunum. Observe the solitary lymph nodules which stud the wall of the lower ileum.

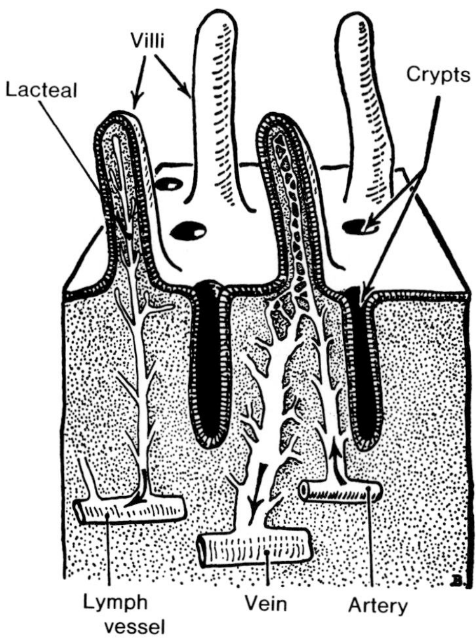

Figure 2-91. Three-dimensional illustration of a microscopic piece of small intestine. The villi are drained by veins and lymphatics. Small fat particles pass between the endothelial cells into the lymphatic capillaries called lacteals. The milk-like lymph then passes to a lymph vessel in the wall of the intestine.

ileocolic vein drains blood from the cecum and appendix (Fig. 2-52).

Lymph vessels from the cecum and appendix pass to lymph nodes in the mesentery of the appendix and to those along the ileocolic artery (Fig. 2-97).

The nerves to the cecum and appendix are derived from the *celiac and superior mesenteric ganglia* (Figs. 2-99 and 7-135).

THE ASCENDING COLON

The ascending colon, 12 to 20 cm long, and narrower than the cecum, *extends from the ileocecal valve to the right colic flexure* (Fig. 2-32*A*). The ascending colon ascends retroperitoneally on the posterior abdominal wall in the right paravertebral groove (Fig. 2-87). The ascending colon is separated from the muscles posteriorly by the kidney and the nerves of the posterior abdominal wall (Figs. 2-107 and 2-121). It is usually separated from the anterior abdominal wall by coils of small intestine and the greater omentum (Figs. 2-28 and 2-107).

The ascending colon is covered by peritoneum anteriorly and on its sides, which attaches it to the posterior abdominal wall. On the lateral side of the ascending colon, the peritoneum forms a trench called the **right paracolic gutter** (Fig. 2-27). The depth of the paracolic gutter depends on how much gas the ascending colon contains.

Fluid in the right ***hepatorenal recess*** (Figs. 2-31 and 2-70) passes along the right paracolic gutter to the **rectouterine or rectovesical pouch** when the person is in an inclined position.

Vessels of the Ascending Colon. The ascending colon and right colic flexure are supplied by the **ileocolic and right colic arteries,** branches of the superior mesenteric artery (Fig. 2-86).

The **ileocolic and right colic veins,** tributaries of the superior mesenteric vein, drain blood from the ascending colon (Figs. 2-52 and 2-94).

The lymph vessels of the ascending colon pass to the *paracolic and epicolic lymph nodes* (Figs. 2-94 and 2-97) and from them to the *superior mesenteric lymph nodes* (Fig. 2-97).

The nerves of the ascending colon are the same as just described for the caecum and appendix (Fig. 2-99).

THE TRANSVERSE COLON

The transverse colon, 40 to 50 cm long, is the *largest and most mobile part of the large intestine.* **It extends between the right and left colic flexures,** forming a loop that hangs inferiorly and anteriorly (Figs. 2-40

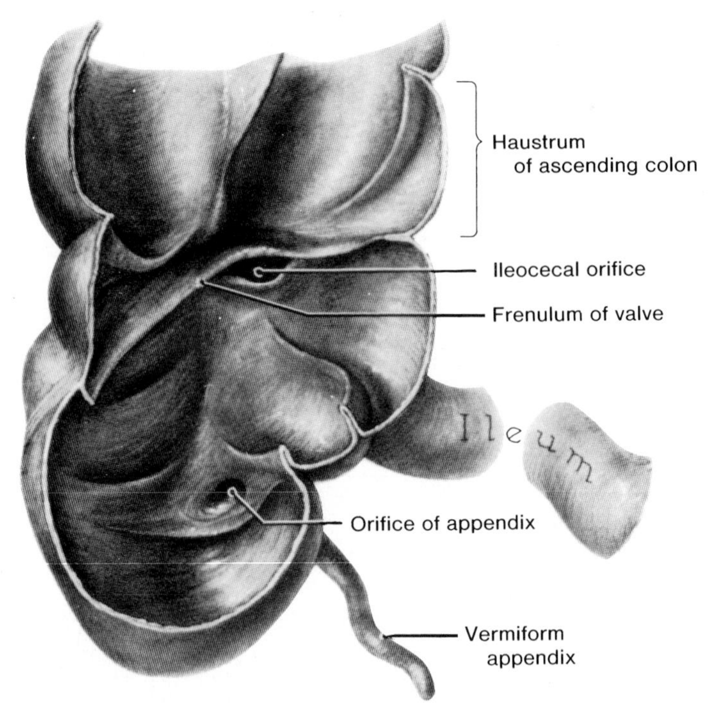

Figure 2-92. Drawing of the interior of a dried cecum. Observe the ileocecal valve guarding the iliocecal orifice.

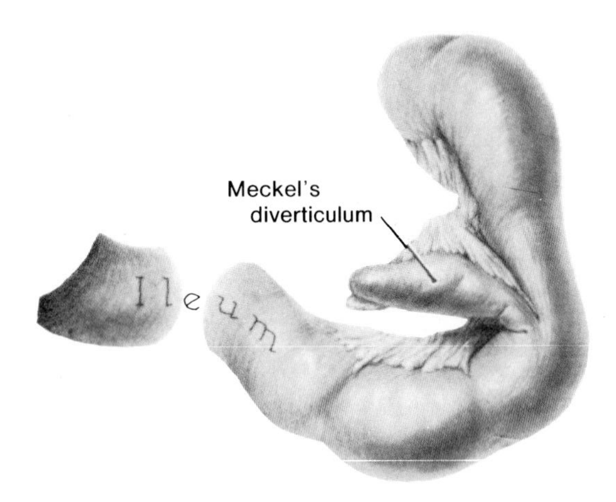

Figure 2-93. Drawing of the terminal part of the ileum illustrating a Meckel's diverticulum. This outpouching of the ileum, found in 2 to 4% of people, is the remains of the embryonic yolk stalk. It projects from the antimesenteric border of the ileum within 50 cm of the ileocecal junction.

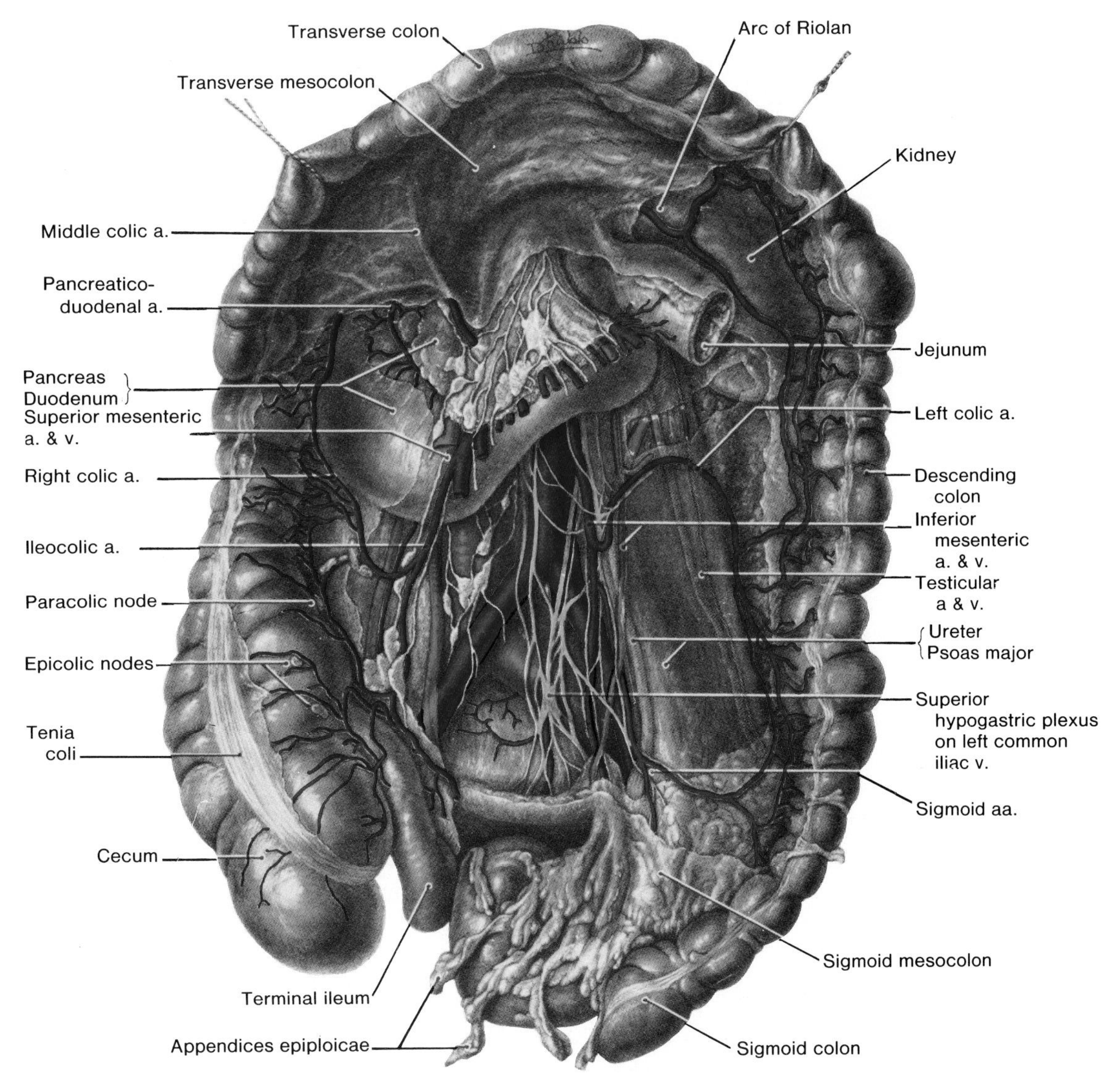

Figure 2-94. Drawing of a dissection of structures on the posterior abdominal wall. Observe the jejunal and ileal branches (cut) passing from the left side of the superior mesenteric artery and the right colic artery, here, as commonly, a branch of the ileocolic artery. Note the accessory artery, called the arc of Riolan, which connects the superior mesenteric artery to the left colic artery. On the right side observe the small epicolic lymph nodes on the colon, the small paracolic nodes beside the colon, and the lymph nodes along the ileocolic artery which drain into the main nodes ventral to the pancreas. Note that the intestines and the intestinal vessels lie on a plane anterior to that of the testicular vessels, and that these in turn lie anterior to the plane of the kidney, its vessels, and the ureter. Observe that the right and left ureters are asymmetrically placed in this specimen. Examine the superior hypogastric plexus (presacral nerve) lying within the fork of the aorta and ventral to the left common iliac vein, the body of the fifth lumbar vertebra, and the fifth intervertebral disc.

and 2-95). It may drop as low as the pelvis in some people (Fig. 2-40).

The transverse colon is suspended posteriorly by the ***transverse mesocolon,*** a broad fold of peritoneum that passes forward from the pancreas (Figs. 2-37, 2-39, and 2-48). The transverse mesocolon is fused to the posterior surface of the greater omentum (Figs. 2-35, 2-37, and 2-48).

Vessels of the Transverse Colon. The transverse colon is supplied mainly by the **middle colic artery,** a branch of the ***superior mesenteric artery,*** but it also receives blood from the **right and left colic arteries** (Fig. 2-86). *The left colic artery is a branch of the inferior mesenteric artery* (Fig. 2-96).

Blood is drained from the transverse colon via the **superior mesenteric vein** (Fig. 2-52).

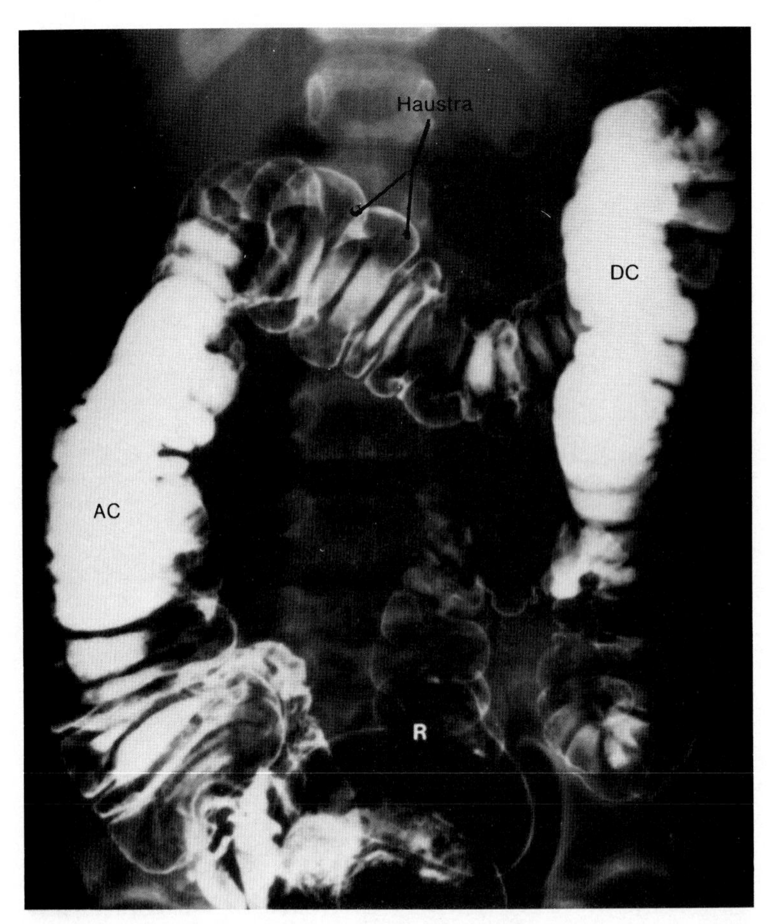

Figure 2-95. Radiograph of the large intestine. After evacuating the barium introduced by a barium enema, the colon was inflated with air and a radiograph was made in the supine position. Observe that the barium clinging to the walls of the air-filled parts of the colon gives a good view of the hausta. As the colon seldom can be completely evacuated, as in this patient, parts of the right and left colon are still filled with the barium sulphate emulsion (*white areas*). *AC*, ascending colon, *DC*, descending colon. *R*, rectum (also see Fig. 2-40).

The lymph vessels of the transverse colon *end in the superior mesenteric lymph nodes* after traversing nodes located along the middle colic artery (Fig. 2-97).

The nerves of the transverse colon, which follow the right and middle colic arteries, are derived from the ***superior mesenteric plexus*** (Fig. 2-99) and transmit sympathetic and vagal nerve fibers (Fig. 7-135). The nerves which follow the left colic artery are derived from the ***inferior mesenteric plexus*** and carry sympathetic and pelvic parasympathetic fibers.

THE DESCENDING COLON

The descending colon passes from the left colic flexure to the brim of the pelvis, where it *joins the sigmoid colon* (Figs. 2-32, 2-39, 2-40, 2-69, 2-96, 2-98, and 2-101). It is usually narrower than other parts of the large intestine and, like the ascending colon, it **lies retroperitoneally.**

The descending colon passes inferiorly, anterior to the lateral border of the left kidney and the transversus abdominis and quadratus lumborum muscles to the left iliac fossa.

Vessels of the Descending Colon (Figs. 2-94 and 2-96). The descending colon is supplied by the **left colic and superior sigmoid arteries,** branches of the inferior mesenteric artery.

The **inferior mesenteric vein** drains the descending colon (Fig. 2-52).

The lymph vessels from the descending colon pass to the lymph nodes along the left colic artery and then to the **inferior mesenteric lymph nodes** around the inferior mesenteric artery (Fig. 2-97). However, those from the left colic flexure also drain to the **superior mesenteric lymph nodes** by vessels that accompany the superior mesenteric vein.

THE SIGMOID COLON

This *S-shaped part of the large intestine* varies in length from 15 to 80 cm, but averages about 30 cm.

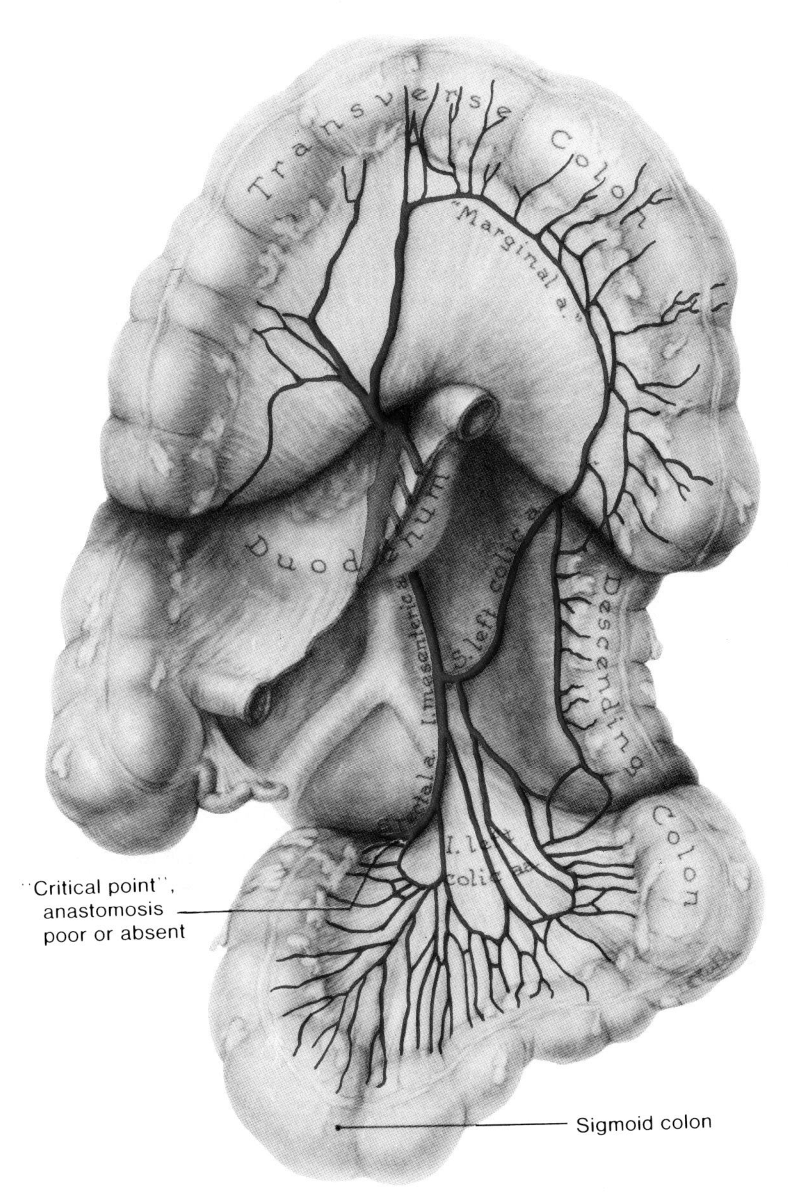

Figure 2-96. Drawing of a dissection illustrating the inferior mesenteric artery. The mesentery has been cut at its root and discarded with the jejunum and ileum. Observe the inferior mesenteric artery arising posterior to the duodenum, superior to the bifurcation of the aorta. On crossing the left common iliac artery, it becomes the superior rectal artery. Note the branches of the inferior mesenteric artery: (1) a superior left colic artery and (2) several sigmoid arteries (inferior left colic arteries) arising from its left side.

The sigmoid (pelvic colon) extends from the descending colon to the third segment of the sacrum, where it *joins the rectum* (Figs. 2-32 2-94 to 2-97, and 2-101). It is suspended from the pelvic wall by a Λ-shaped mesentery, the **sigmoid mesocolon** (Fig. 2-94).

The apex of the sigmoid mesentery lies anterior to the left ureter (Figs. 2-94 and 2-101) and the division of the left common iliac artery.

The sigmoid colon forms an S-shaped loop, the shape and position of which depend upon how full it is. Feces are usually stored in the sigmoid colon until just before defecation. Usually the sigmoid colon lies relatively free within the pelvis minor, inferior to the small intestine (Fig. 2-32*A*). Posterior to the sigmoid colon are the left external iliac vessels, the left sacral plexus, and the left piriformis muscle (Fig. 2-101).

Vessels of the Sigmoid Colon (Figs. 2-94, 2-96, and 2-97). Two to three **sigmoid arteries** supply the sigmoid colon. The **inferior mesenteric vein** returns blood from the sigmoid colon (Fig. 2-52).

Lymph vessels from the sigmoid colon pass to lymph nodes on the branches of the left colic arteries

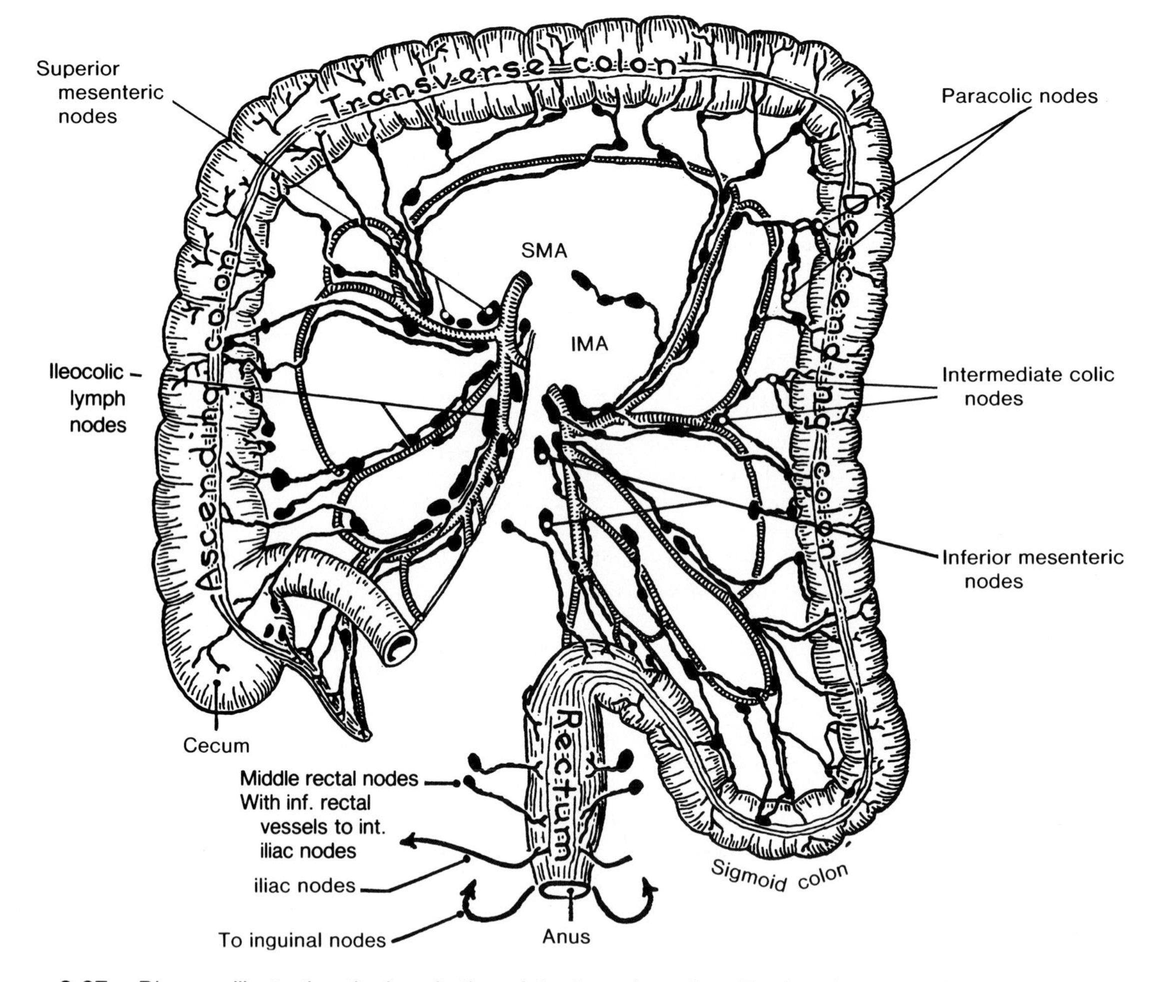

Figure 2-97. Diagram illustrating the lymphatics of the large intestine. The lymph nodes of the colon are in four groups: (1) *epicolic nodes* on the wall of the gut; (2) *paracolic nodes* along the medial borders of the ascending and descending colon; (3) *intermediate nodes* along the right, middle, and left colic arteries; and (4) *terminal colic nodes* near the main trunks of the superior (*SMA*) and inferior (*IMA*) mesenteric arteries.

and **end in the inferior mesenteric lymph nodes** around the inferior mesenteric artery (Fig. 2-97).

THE RECTUM AND ANAL CANAL

Brief descriptions of these regions of the large intestine have been given on page 199. Detailed descriptions of these pelvic viscera are given in Chapter 3.

A chronic disease of the colon, called **ulcerative colitis**, is characterized by severe *inflammation and ulceration of the colon and rectum.* In some of these patients, a **total colectomy** is performed during which the terminal ileum, ascending, transverse, descending, and sigmoid colon, as well as the rectum and anal canal, are removed.

An **ileostomy** is then constructed to establish an opening between the ileum and the skin of the anterior abdominal wall. In **subtotal colectomy**, the rectum and anal canal are preserved and the ileum is joined to the rectum by an *end-to-end anastomosis.*

Colostomies establish an opening between the colon and the skin of the anterior abdominal wall, creating a temporary *fecal fistula* or "**artificial anus.**"

The most mobile parts of the colon are used; thus the usual colostomies are: *cecostomy, transverse colostomy,* and *sigmoidostomy*). The type of colostomy constructed depends upon which part of the colon has been resected. When parts of the colon are resected, it must be ensured that an adequate blood supply to the remaining segments is preserved. Failure to do so results in necrosis of the segment of intestine concerned.

Herniations of the mucosa through the muscularis mucosae of the sigmoid colon are common. Known as **diverticulosis**, it can be observed in about 10% of persons over 40 years of age who have barium enemas for radiological studies. When one or more of the diverticula become inflamed, the condition is known as **diverticulitis.**

THE PORTAL VEIN AND PORTAL-SYSTEMIC ANASTOMOSES

The portal vein drains the abdominal and pelvic parts of the GI system (Fig. 2-52), except for the distal part of the anal canal. It also drains blood from the spleen, pancreas, and gallbladder.

The portal vein is formed posterior to the neck of the pancreas by the *union of the splenic and superior mesenteric veins* (Fig. 2-58). The inferior mesenteric vein opens into (1) the splenic vein (Fig. 2-52), (2) the junction of the splenic and superior mesenteric veins, *or* (3) the superior mesenteric vein (Fig. 2-45); ***The portal vein carries blood from three major veins*** (splenic, superior mesenteric, and inferior mesenteric).

The portal vein ascends to the liver in the free margin of the less omentum, posterior to the bile duct and the hepatic artery (Fig. 2-81). At the **porta hepatis** (Fig. 2-75), *the portal vein divides into right and left branches* which empty their blood into the **hepatic sinusoids.** This blood contains the products of digestion of carbohydrates, fats, and protein from the intestine, and the products of red cell destruction from the spleen.

In several locations the portal venous system communicates with the systemic venous system (Fig. 2-100). *These anastomoses are very important clinically.* When the portal circulation is obstructed (*e.g.*, owing to liver disease), blood from the GI tract can still reach the right side of the heart through the inferior vena cava via a number of **collateral routes.**

The portal vein and its tributaries have no valves; hence, blood can flow from the obstructed liver to the inferior vena cava via these alternate routes. In *portal hypertension* venous pressure in the portal venous system is increased; consequently some blood in the portal venous system may reverse its direction and pass through the portal-systemic anastomoses into the systemic venous system. This causes the veins in the portal-systemic anastomotic areas to dilate and become varicose.

The following are the portal-systemic anastomotic areas (Fig. 2-100).

1. In the gastroesophageal region, *the esophageal tributaries of the* ***left gastric vein*** *anastomose with the* ***esophageal veins***, which empty into the **azygos vein** (Figs. 1-42 and 1-80).
2. In the anorectal region, *the* ***superior rectal vein*** *anastomoses with the* ***middle and inferior rectal veins*** (Fig. 2-100) which are tributaries of the internal iliac and internal pudendal veins, respectively.
3. In the paraumbilical region, *the* ***paraumbilical veins*** *in the falciform ligament anastomose with subcutaneous veins in the anterior abdominal wall.*
4. In the retroperitoneal region, *tributaries of the* ***splenic and pancreatic veins*** *anastomose with the* ***left renal vein***. Short veins also connect the splenic and colic veins to the lumbar veins of the posterior abdominal wall. The veins of the bare area of the liver also communicate with the veins of the diaphragm and the right internal thoracic vein (Fig. 2-100).

In patients with **portal hypertension**, the abnormal increase in pressure in the portal vein and its tributaries is often caused by **cirrhosis of the liver**, a decrease characterized by progressive destruction of hepatic parenchymal cells and replacement of them by fibrous tissue.

Portal hypertension and the diseases causing it are serious conditions, but the symptoms are modified by the portal-systemic anastomoses providing alternative pathways for the blood to flow. Because *there are no functionally competent valves in the portal venous system*, the increase in portal pressure is reflected throughout the system. Blood tends to be diverted into the systemic venous system in the regions where **portal-systemic anastomoses** occur (Fig. 2-100). The veins in these areas tend to become dilated and tortuous and are called varicose veins.

Varicose veins in the region of the anal canal are called **hemorrhoids** (piles). Varicose veins in the gastroesophageal region are called **esophageal varices.** The veins in both locations may become so dilated that their walls rupture, resulting in hemorrhage. *Bleeding from esophageal varices is often severe and may be fatal.*

In severe cases of portal obstruction, even the *paraumbilical veins may become varicose* (Fig. 2-132), which look somewhat like snakes under the skin. This condition is referred to as **caput medusae** because of its resemblance to the serpents on the head of Medusa, a character in Greek mythology.

A common way of reducing portal pressure is to divert blood from the portal venous system to the systemic venous system by creating a communication between the portal vein and the inferior vena cava. This **portacaval anastomosis** may be done where these vessels lie close to each other posterior to the liver (Figs. 2-75 and 2-82).

Another good way of reducing portal pressure is to join the splenic vein to the left renal vein (**splenorenal anastomosis**), following removal of the spleen (*splenectomy*). Other anastomoses between the portal and systemic venous systems are sometimes created.

THE KIDNEYS AND URETERS

The kidneys remove excess water, salts, and products of protein metabolism from the blood and maintain its pH. The waste products removed from the blood are conveyed in the urine to the urinary bladder by the ureters (Fig. 2-102).

Position, Form, and Size of the Kidneys. Each kidney lies posterior to the peritoneum (**retroperitoneally**) on the posterior abdominal wall. *The bean-shaped kidneys lie alongside the vertebral column, against the psoas major muscles* (Figs. 2-98, 2-101, 2-105, and 2-107). They occupy the superior parts of the **paravertebral grooves** (gutters), anterior to the diaphragm. The long axes of the kidneys are almost parallel to the long axis of the body and to the lateral border of the psoas muscle (Figs. 2-101 and 2-108*A*).

The superior parts of the kidneys are protected by the bony thorax and are tilted so that their superior poles are nearer the midline than their inferior poles (Fig. 2-105). Owing to the large size of the right lobe of the liver, the right kidney lies at a slightly lower level than the left kidney.

Fresh adult kidneys are reddish-brown in color and measure about 10 cm in length, 5 cm in width, and 2.5 cm in thickness. The left kidney is often slighty longer than the right kidney.

Each kidney is ovoid in outline, but its indented medial margin gives it a somewhat bean-shaped appearance. At this concave middle part of each kidney there is a vertical cleft, the **renal hilum** (renal hilus), through which the renal artery enters and the renal vein and the renal pelvis leave the kidney.

The hilum of the kidney leads into a space within the kidney called the **renal sinus**, which is about 2.5 cm deep (Figs. 2-103 and 2-104). *The renal sinus is occupied by the renal pelvis, the renal calyces, the renal vessels and nerves, and varying amounts of* fat (Fig. 2-104*B*).

The lobes of the kidney are demarcated on the surface of fetal and infantile kidneys. This external evidence of the lobes usually disappears by the end of the first year. In rare instances evidence of the fetal lobes may be visible in adult kidneys (Fig. 2-115*B*). Several indentations along the lateral margin of the kidney correspond to the renal pyramids of the medulla.

Occasionally the left kidney is somewhat triangular in shape, probably as the result of molding by the spleen. The superior pole of the kidney is narrow and the inferior pole is wide, giving the kidney a *humpback appearance*. This abnormally-shaped kidney, called a **dromedary kidney**, has no clinical significance.

Surface Anatomy of the Kidneys (Figs. 2-61 and 2-105). In very muscular and/or obese people, the kidneys may be impalpable. In thin adults with poorly developed abdominal musculature, ***the inferior pole of the right kidney is usually palpable by bimanual examination in the right lumbar region*** as a firm, smooth, somewhat rounded mass that descends during inspiration. The normal left kidney is usually not palpable.

When the kidney is enlarged (*e.g.*, owing to a tumor, Fig. 2-108*B*) or is abnormally mobile, it can usually be felt as it slides between the opposed fingers of the examiner's hands.

The levels of the kidneys change during respiration and with changes in posture. Each kidney moves about 3 cm in a vertical direction during the movement of the diaphragm that occurs with deep breathing.

The hila of the kidneys lie in the neighborhood of the transpyloric plane (Fig. 2-61), about 5 cm from the median plane. As the posterior approach to the kidney is the usual one, it is helpful to know that the inferior pole of the right kidney is about a fingerbreadth superior to the crest of the ilium, and that its superior pole is superior to the 12th rib (Figs. 2-105 and 2-111).

Surfaces and Margins of the Kidneys (Fig. 2-103). Each kidney has *anterior and posterior surfaces, medial* and *lateral margins* (borders), and *superior* and *inferior poles.* The lateral margin is convex and the medial margin is indented or concave where the renal sinus and renal pelvis are located.

Structure of the Kidneys (Figs. 2-103, 2-104, and 2-107). The kidneys are closely invested by a **strong fibrous capsule** which gives the fresh kidney a glistening appearance. The fibrous capsule strips easily from a normal kidney. It passes over the **lips of the hilum** (Fig. 2-103*B*) to line the renal sinus and become continuous with the walls of the calyces.

The kidney and its capsule are surrounded by fat, but it is sparce on the anterior surface (Fig. 2-107). This **perirenal fat** is less dense (lower specific gravity than the kidney); thus *an outline of the kidney is usually visible in radiographs* (Fig. 2-106).

The renal pelvis is the superior expanded end of the ureter. It is surrounded by the fat, vessels, and nerves in the renal sinus (Fig. 2-104*B*). The word *pelvis* is derived from the Greek word *pyelos*, meaning a tub or basin. Hence a **pyelogram** (Fig. 2-106) is a radiograph of the renal pelvis and ureter, and **pyelonephritis** indicates inflammation of the renal pelvis and kidney.

Within the renal sinus, the renal pelvis divides into two wide, funnel-like tubes called **major calyces** (G. cups of flowers). Occasionally there is a third major calyx. Each major calyx is subdivided into 7 to 14 **minor calyces.** The urine empties into a minor calyx from collecting tubules which pierce the tip of a **renal papilla** obliquely. It then passes through the major calyx, renal pelvis, and ureter to enter the urinary bladder.

THE URETERS

The ureters are muscular ducts that carry urine from the kidneys to the urinary bladder (Figs. 2-101 to 2-106). As the urine passes along the ureters, peristaltic waves occur in their walls. Each ureter begins in the renal pelvis, its funnel-shaped superior end.

The abdominal part of the ureter is about 12.5 cm long and 5 mm wide. The inferior half or *pelvic part of the ureter* is described in Chapter 3 (p. 357). The ureter is an expansile, thick-walled muscular tube with a narrow lumen (Fig. 2-106).

The ureter is retroperitoneal throughout its entire length. It adheres closely to the peritoneum and is usually lifted with it. As it is pale colored and somewhat resembles a blood vessel, **the ureter is in danger during surgery in the abdomen and pelvis.**

The ureter descends almost vertically along the psoas major muscle (Figs. 2-101 and 2-109), anterior to the tips of the transverse processes of L2 to L5 vertebrae (Fig. 2-105).

In Figure 2-121, note that **as the right ureter descends, it lies in close relationship to the inferior vena cava, the lumbar lymph nodes, and the sympathetic trunk.**

In Figure 2-101, observe that the ureter crosses the brim of the pelvis and the external iliac artery, just beyond the bifurcation of the common iliac artery.

Rapid excessive distention of the ureter causes severe rhythmic pain (**ureteric colic**). Frequently the colic results from a kidney stone or **ureteric calculus** (L. pebble) that is usually composed of calcium oxalate, calcium phosphate, and/or uric acid.

Urinary calculi (kidney stones) may be located in the calyces, renal pelvis, ureter, or urinary bladder. Calcium-containing stones are radiopaque, whereas those composed of uric acid are radiolucent.

Ureteric stones may cause complete or in-

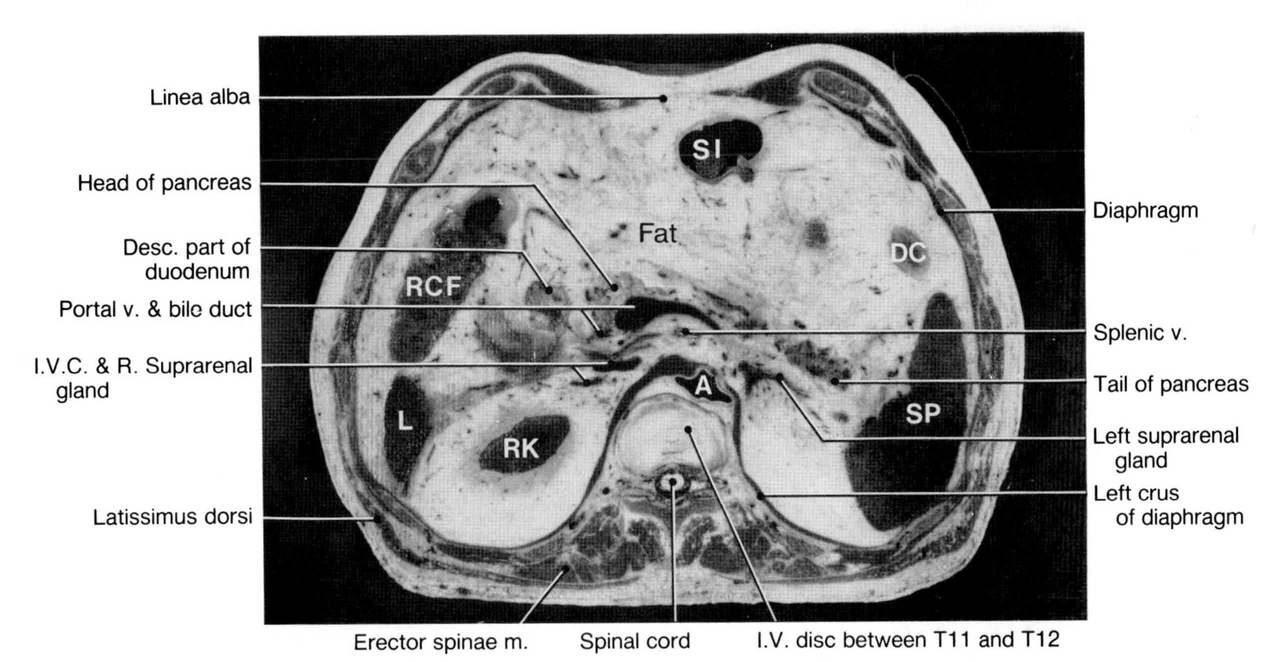

Figure 2-98. Photograph of a transverse section of the abdomen of an adult male at the level of the intervertebral disc between T11 and T12 vertebrae. In keeping with radiographic convention, as in CT scans, the section is viewed from below. Hence the right side of the body appears on the left side of the photograph. *SI*, small intestine; *DC*, descending colon; *SP*, spleen; *A* descending aorta; *RK*, right kidney; *L*, liver; *RCF*, right colic flexure.

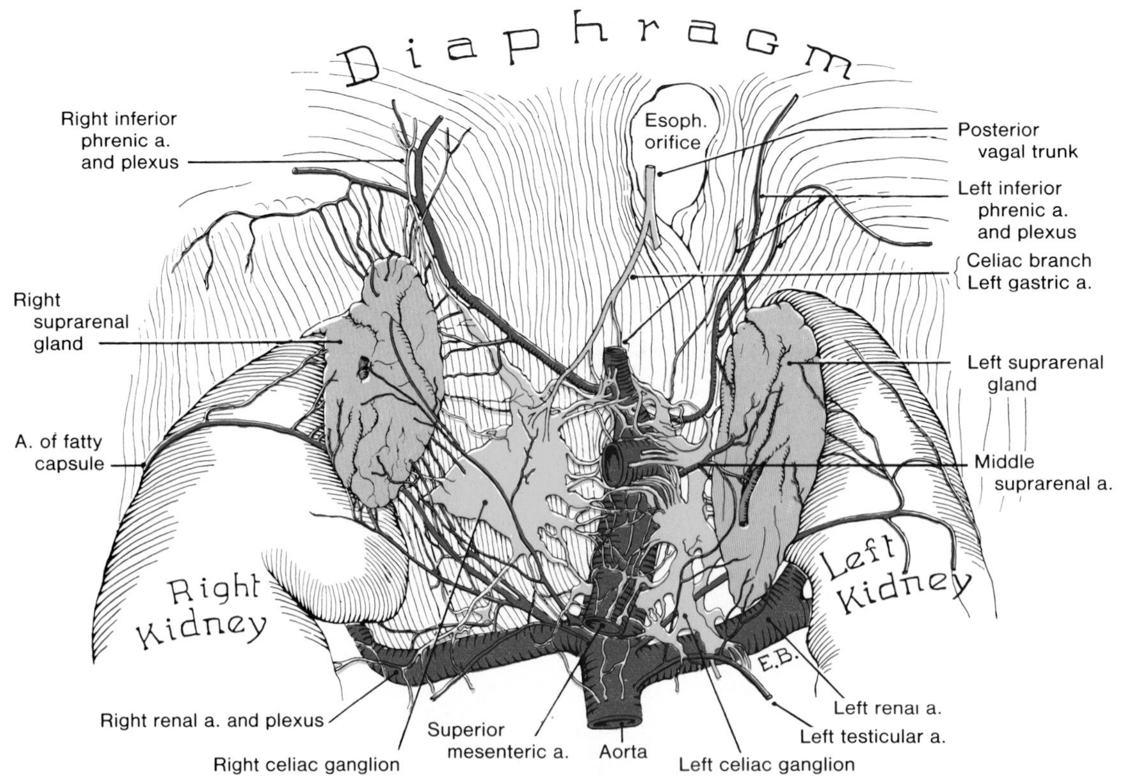

Figure 2-99. Drawing of a dissection showing the celiac trunk, celiac plexus, celiac ganglia, and the suprarenal glands. Observe the celiac plexus of nerves surrounding the celiac trunk (not labelled) and connecting the right and left celiac ganglia. Note a stout branch from the posterior vagal trunk descending along the stem of the left gastric artery and conveying vagal (parasympathetic) fibers to the celiac ganglia. Examine the nerves extending along the arteries to the viscera and down the aorta. The nerves to the suprarenal glands are mostly preglanglionic.

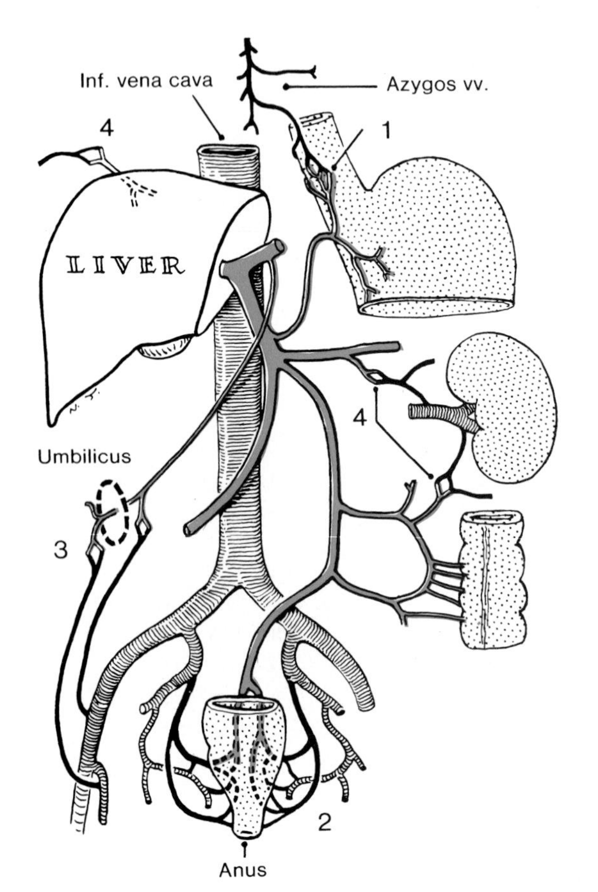

Figure 2-100. Diagram illustrating the portal-systemic anastomoses which provide a collateral portal circulation or bypass in cases of obstruction in the liver or the portal vein. In this diagram, portal tributaries are *blue*, systemic tributaries are *striped*, and communicating veins are *black*. In portal hypertension (as in hepatic cirrhosis), the anastomotic veins become varicose and may rupture. The sites of anastomosis shown are: (*1*) *between the esophageal veins* (when dilated these veins are called esophageal varices); (*2*) *between the rectal veins* (when dilated these veins are called hemorrhoids); (*3*) *paraumbilical veins* (when dilated these veins may produce the "caput medusae" (Fig. 2-132); and (*4*) *twigs of colic* and *splenic veins with renal veins* and veins of the bare area of the liver with the right internal thoracic vein.

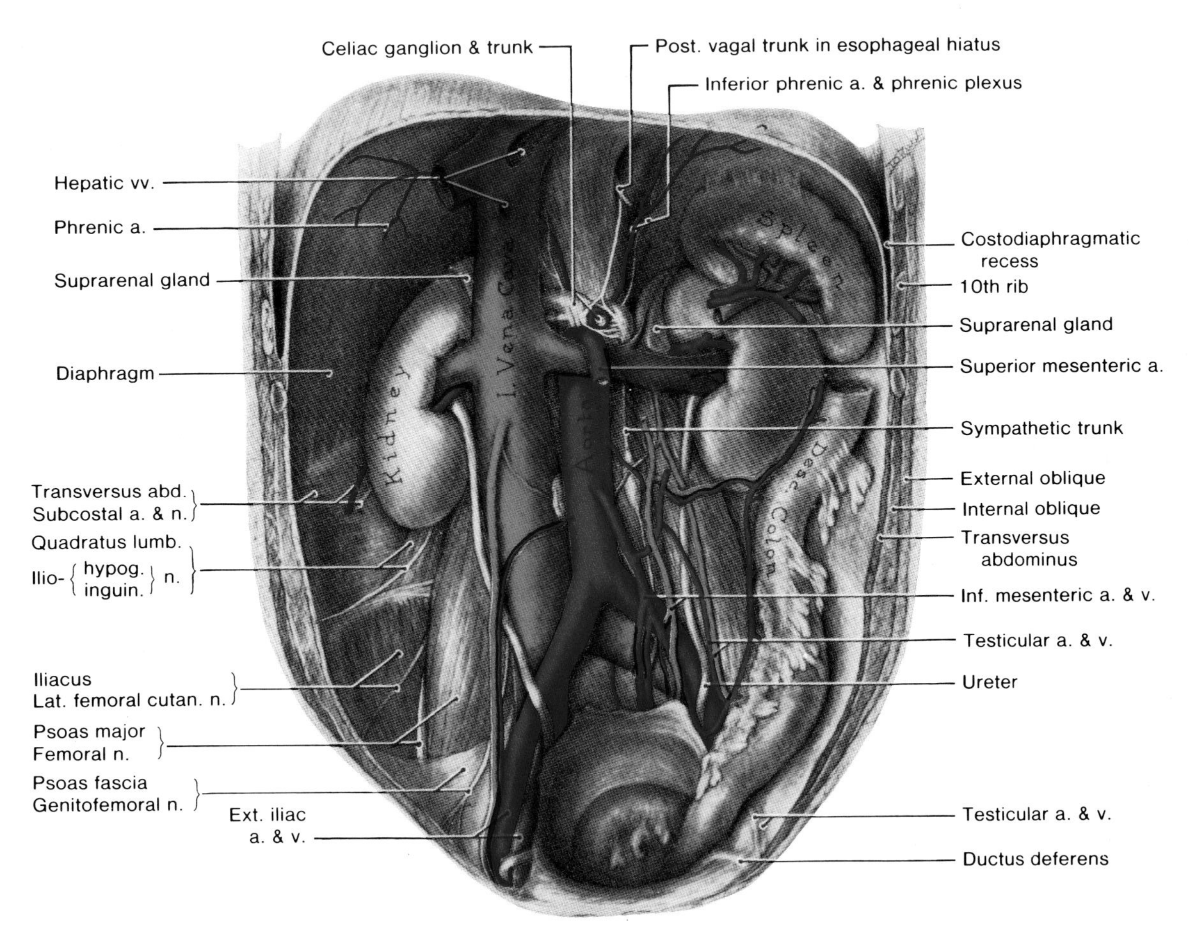

Figure 2-101. Drawing of a dissection of the abdomen of a male showing the great vessels, the kidneys, the ureters, and the suprarenal glands. Most of the fascia has been removed from the posterior abdominal wall. Note that the superior mesenteric artery arises just inferior to the celiac trunk. Observe that the inferior mesenteric artery arises about 4 cm superior to the aortic bifurcation and crosses the left common iliac vessels to become the superior rectal artery. Note that the kidneys lie anterior to the diaphragm, the transversus aponeurosis, the quadratus lumborum, and the psoas major. Observe that the ureter crosses the external iliac artery just beyond the common iliac bifurcation, and that the testicular vessels cross anterior to the ureter, and at the deep inguinal ring they pass with the ductus deferens into the inguinal canal. The ureter in living persons can be identified by its thick muscular wall which undergoes worm-like movements when it is gently stroked or squeezed. The ureter adheres to the parietal peritoneum during its abdominal and upper pelvic course and remains with it when the peritoneum is dissected and pulled superiorly.

termittent obstruction of urinary flow. The obstruction may occur at the **ureteropelvic junction** or anywhere along the ureter, but most often it occurs (1) where the ureter crosses the iliac vessels and the brim of the pelvis (Fig. 2-101), and (2) where it passes obliquely through the wall of the urinary bladder (Fig. 3-61).

***Ureteric colic is usually a sharp, stabbing pain** which follows along the course of the ureter i.e., from the loin (between the ribs and pelvis) to the groin* (inguinal region) as the stone is gradually forced down the ureter.

In males, the pain is frequently also referred to the scrotum and in females, it may radiate to the labia majora (Fig. 2-22).

The ureter is supplied with pain afferents that are included in the **lowest splanchnic nerve** (Fig. 2-114). Impulses enter via T12 and L1 segments which explains why the spasmodic and agonizing pain is referred to the loin and the groin, the cutaneous areas supplied by T12 and L1 (Fig. 2-16*A*).

To study the kidneys, ureters, and urinary bladder, a radiopaque contrast medium is injected intravenously (**intravenous urography**), and serial radiographs (*urograms*) are

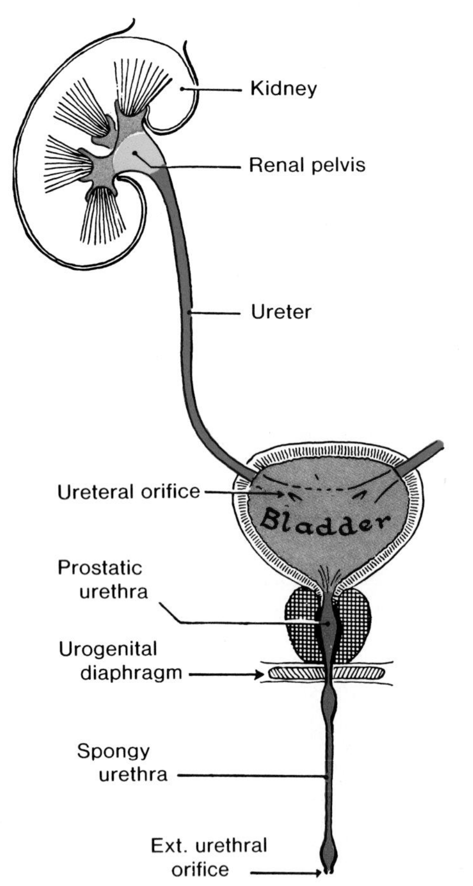

Figure 2-102. Diagram of the male urinary system. The urinary organs are: the *kidneys* where the urine is formed; the *ureters* which convey the urine to the urinary bladder; the *urinary bladder* where the urine is temporarily stored; and the *urethra* through which the urine passes to the exterior. The membranous urethra is the part surrounded by the urogenital diaphragm.

taken at intervals (Fig. 2-106), usually with the patient in different positions.

Various abnormalities of the kidney and ureter may be detected radiologically, *e.g.*, **duplications of the upper urinary tract** (Fig. 2-115, *C* and *E*). Bulges on the renal surface representing tumors or cysts, depression of the renal surface representing scars, and displacements by retroperitoneal masses can also be detected radiographically.

Acute obstruction of the ureter results in an abnormally dense and persistent kidney shadow on that side during the first few minutes (**nephrogram phase**) after the injection of contrast material.

Opacification of the renal calyces and pelvis is delayed and is poor when compared to the normal side (Fig. 2-108*A*). Obstruction of the lower urinary tract, *e.g.*, resulting from hypertrophy and/or cancer of the prostate gland, is usually chronic. In such cases **urography** usually shows dilation and poor opacification of the upper urinary tract.

In **retrograde urography** (pyelography), the contrast medium is injected through a cath-

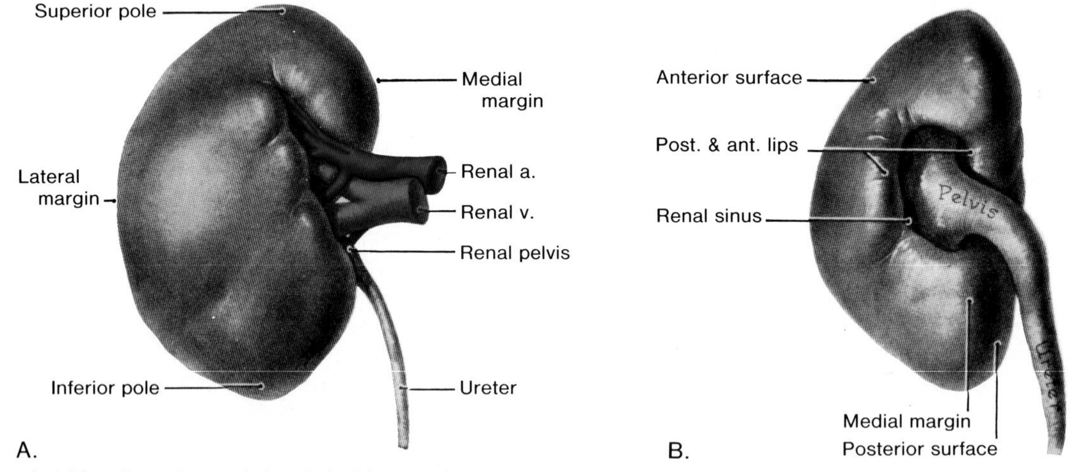

Figure 2-103. Drawings of the right kidney. *A*, anterior view showing the order of the structures at the hilum (entrance to the renal sinus): vein, artery, and pelvis of ureter. Note that a branch of the artery crosses posterior to the renal pelvis. *B*, anteromedial view showing the renal sinus which is a considerable space making up a large part of the interior of the kidney. Note that it contains the greater part of the renal pelvis, the calyces (Fig. 2-104), the blood and lymph vessels, and the nerves of the kidney. There are also variable amounts of fat in the renal sinus.

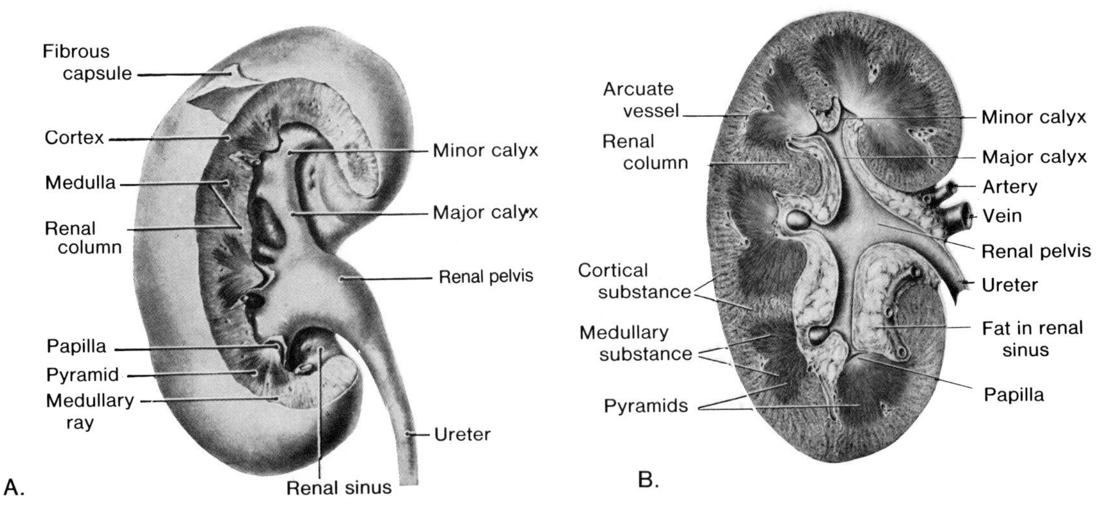

Figure 2-104. *A*, drawing of a right kidney from which the anterior lip of the renal sinus is cut away to show the structure of the kidney. *B*, longitudinal section of the kidney. In these drawings observe that the outer one-third of the renal substance is cortex and the inner two-thirds is medulla. Note that the cortical tissue (composed of glomeruli and convoluted tubules) is granular on section and extends as renal columns through the medulla to the renal sinus. The medulla contains 7 to 14 pyramids which are striated because of the collecting tubules they contain. Each pyramid ends as a papilla on which a dozen or more of the largest collecting tubules open. One to four papillae project into each minor calyx. Usually there are from 8 to 18 renal papillae and 7 to 15 minor calyces. Several minor calyces unite to form a major calyx. There are usually two major calyces which are directed toward the superior and inferior parts of the kidney. Occasionally there is a third major calyx.

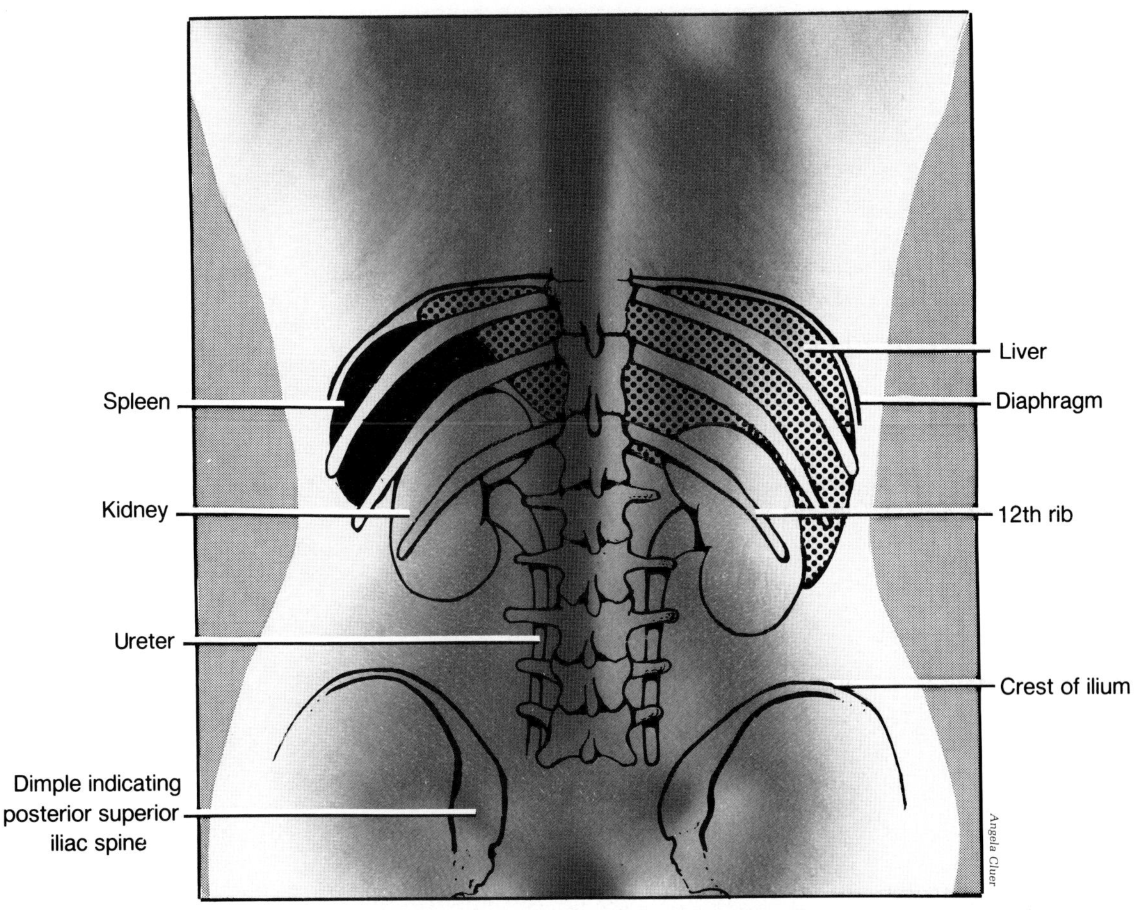

Figure 2-105. Photograph of the back of a 21-year-old woman illustrating the surface anatomy. Observe that the kidneys lie on each side of the vertebral column from T12 to L3 vertebrae and that the right kidney is slightly inferior to the left one owing to the bulk of the right lobe of the liver. Note that the superior poles of the kidneys are protected by the 11th and 12th ribs, and that the interior poles are about a fingerbreadth superior to the crest of the ilium. Observe that the ureters arise from the renal pelves in the hila which lie near the transpyloric plane and descend anterior to the tips of the transverse processes of L2 to L5 vertebrae.

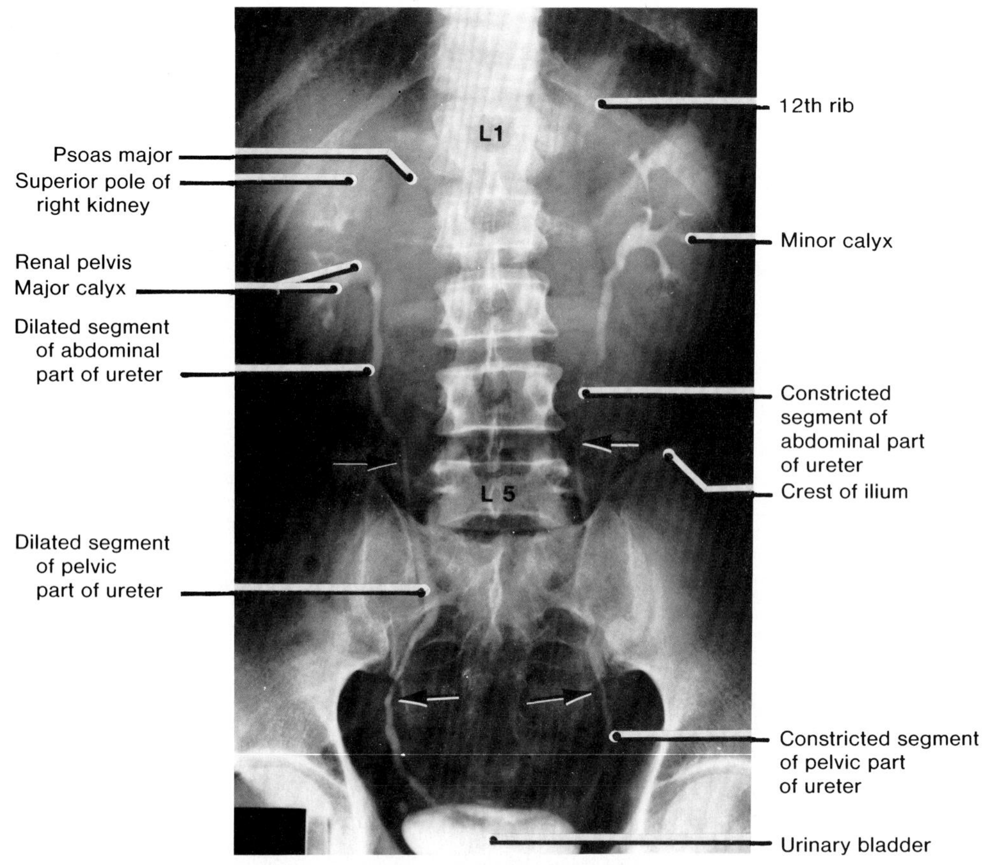

Figure 2-106. Intravenous urogram (pyelogram). The contrast medium was injected intravenously and was concentrated and excreted by the kidneys. This anteroposterior projection shows the calyces, renal pelves, and ureters outlined by the contrast medium filling their lumina. Note the difference in shape and level of the renal pelves and the constrictions and dilations in the ureter resulting from peristaltic contractions of their smooth muscle walls. Observe the relation of the ureters to the transverse processes of L3 to L5 vertebrae. The *arrows* indicate narrowings of the lumen resulting from peristalitic contractions (courtesy of Dr. John Campbell, Sunnybrook Medical Centre, Toronto).

eter that is inserted via a **cystoscope** into the urinary bladder and the ureter. The contrast medium is usually injected when the tip of the catheter enters the renal pelvis. Retrograde urographic studies, although not commonly used, are helpful when visualization of the upper urinary tract is poor during **intravenous urography** (Fig. 2-106). The passage of a catheter into the ureter is also useful for locating the position of a stone in the ureter during surgery.

Lithotripsy is now available that uses sound waves to pulverize ureteric stones painlessly, making surgery unnecessary in all cases. The sound waves break down the stone to sediment which is passed in the urine.

Renal Fascia and Renal Fat (Figs. 2-104 and 2-107). The kidney, invested by a ***fibrous renal capsule***, is embedded in a substantial mass of perirenal fat that constitutes a ***fatty renal capsule***. Very little fatty tissue lies anterior to the kidney (Fig. 2-107). The fatty renal capsule is in turn covered by fibroareolar tissue called **renal fascia**.

The renal fascia encloses the kidney, its surrounding fibrous and fatty capsules, and the suprarenal gland. These coverings help to maintain these organs in position. The renal fascia is adherent to the parietal peritoneum and for this reason the kidney is easily lifted from the posterior abdominal wall in a retroperitoneal approach (Fig. 2-121).

Superiorly the renal fascia is continuous with the fascia on the inferior surface of the diaphragm (**diaphragmatic fascia**). Medially the anterior layers of fascia on the right and left sides blend with each other anterior to the abdominal aorta and inferior vena cava.

The posterior layer of renal fascia fuses medially

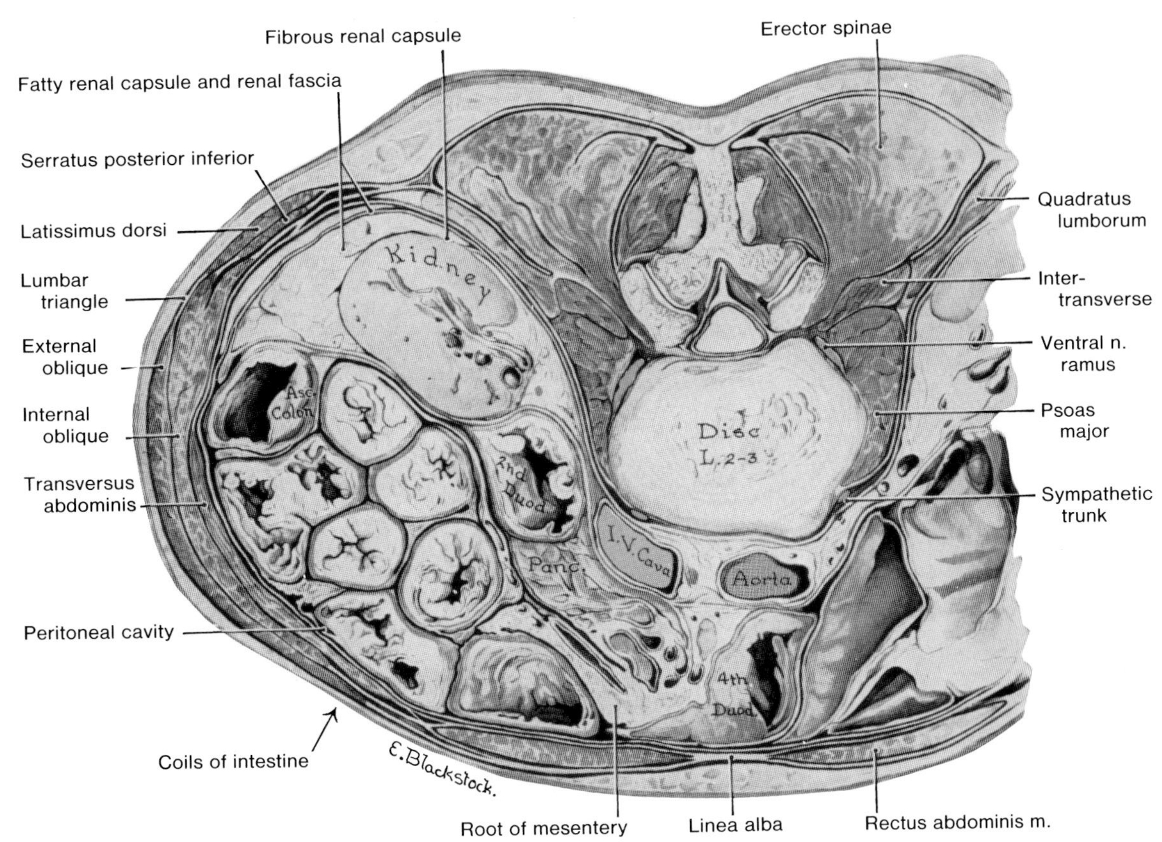

Figure 2-107. Drawing of a transverse section through the abdomen at the level of the disc between L2 and L3 vertebrae. Observe that the anterior aspect of the verebral column is nearer to the anterior surface of the body than to the posterior surface in this thin recumbent cadaver. Observe that the anterior and posterior surfaces of the kidney do not face anteriorly and posteriorly but ventrolaterally and dorsomedially. Examine the fatty renal capsule (perirenal fat) massed along the borders of the kidney and leaving the anterior surface close to the peritoneum. Note that little fatty tissue lies anterior to the kidney. Observe that the descending part of the duodenum overlaps this surface of the kidney. Note the sympathetic trunk lying along the anterior border of the psoas muscle and on the right side posterior to the inferior vena cava. Also see the photograph of a transverse section of the abdomen of a fatter male (Fig. 2-98), at the level of the intervertebral disc between T11 and T12 vertebrae.

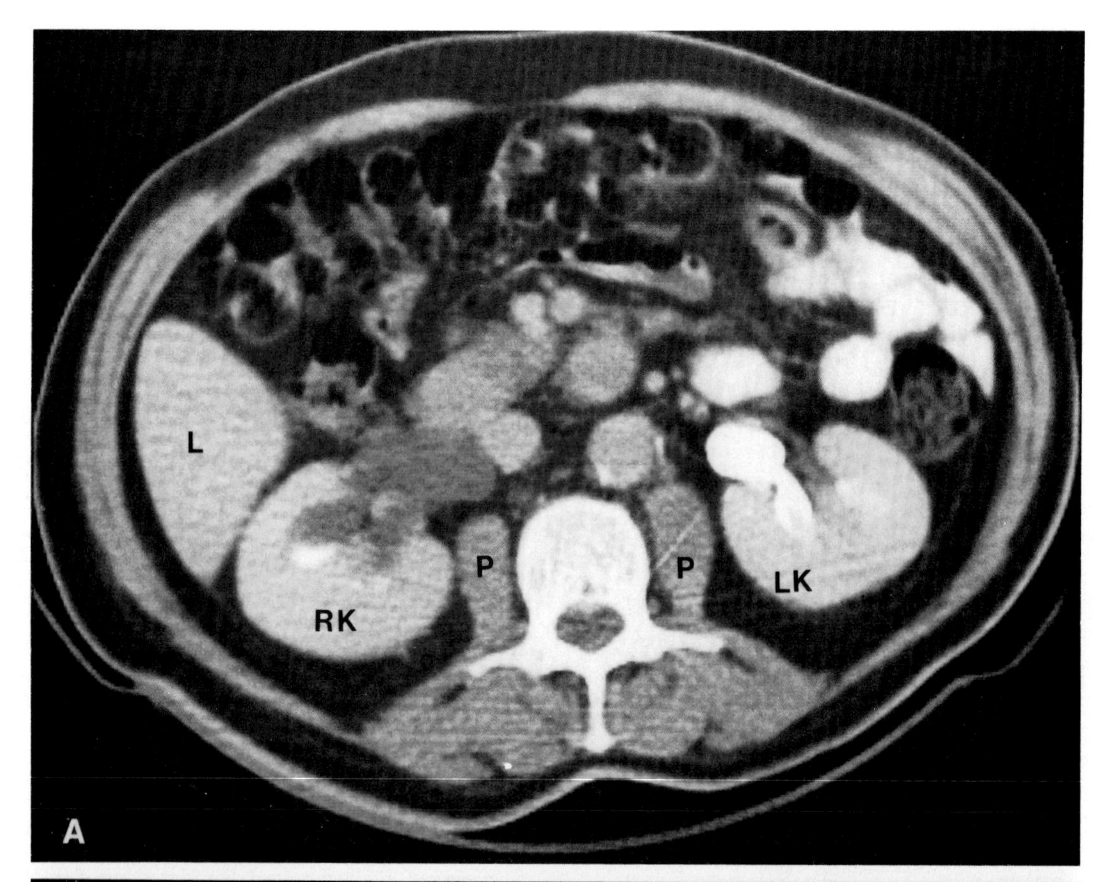
L
P
P
LK
RK
A

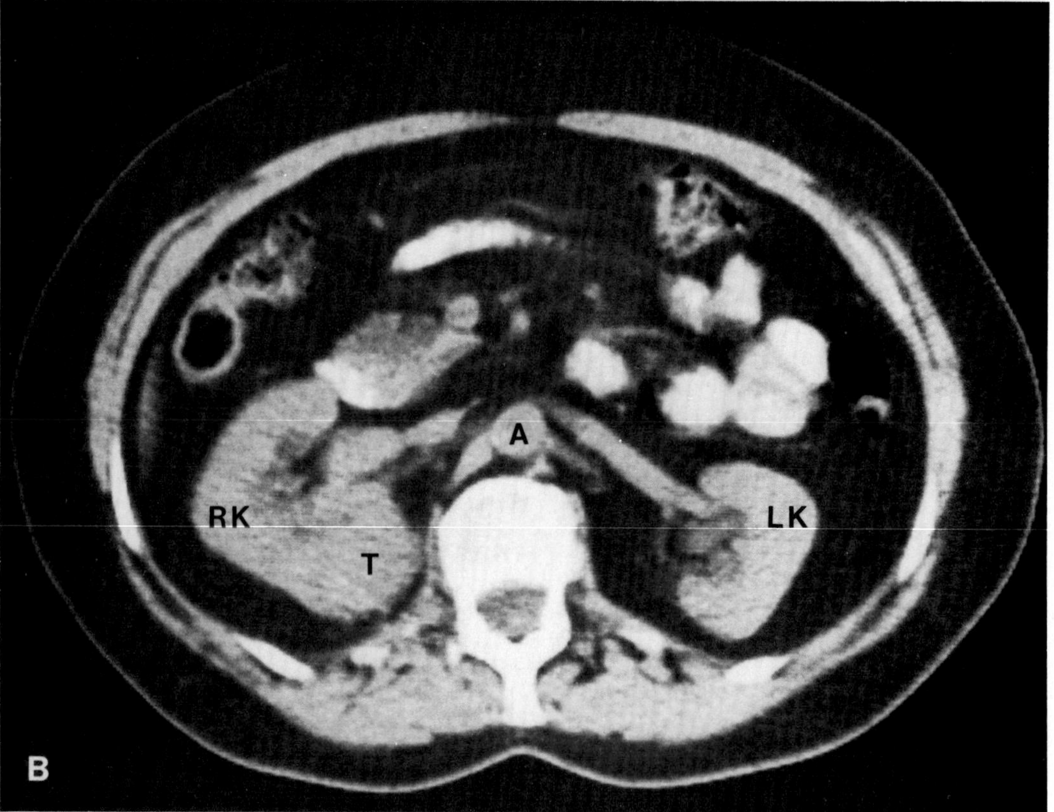
A
RK
LK
T
B

with the fascia over the psoas major muscle (Fig. 2-124). The layers of renal fascia are loosely united inferiorly and may be easily separated inferior to the kidney.

At body temperature the perirenal fat is liquid; hence, *the kidneys move superiorly and inferiorly slightly during respiration* because they lie in contact with the diaphragm (Figs. 2-109 and 2-121). It has been determined radiographically that *the normal renal mobility is about 3 cm* (roughly the height of one vertebral body).

The encasement of the kidney in fat is an important factor in anchoring it in position. The amount of fat in the fatty capsule varies with the individual. The fat outside the renal fascia (**pararenal fat**) is located between the peritoneum of the posterior abdominal wall and the renal fascia. In emaciated persons there may be very little fat around the kidneys.

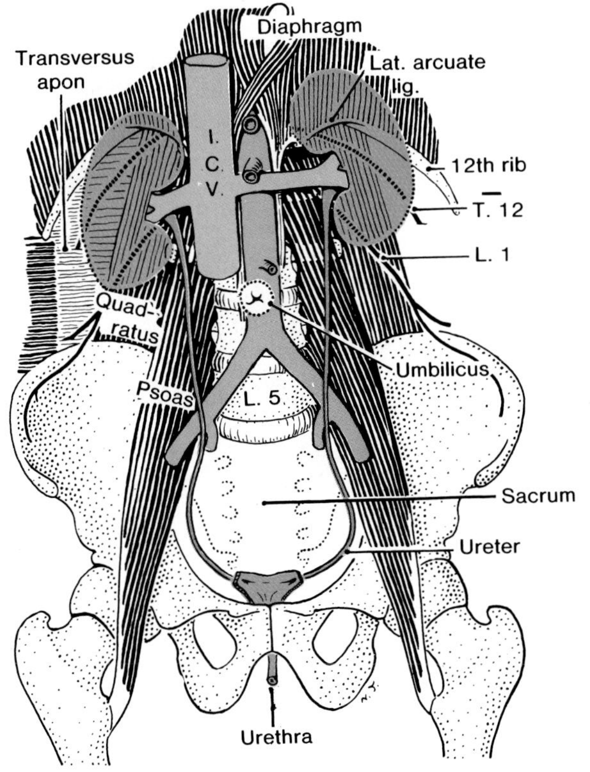

Figure 2-109. Drawing illustrating the posterior relations of the kidneys and the course of the ureters.

Gases injected into the extraperitoneal tissue anterior to the sacrum will rise to the level of the kidneys, and diffuse amongst the extraperitoneal pararenal fat in 12 to 24 hr. Radiologists utilize this knowledge to outline the kidney and suprarenal gland with gas for visualization of these structures. For **insufflation of gases** around the renal and suprarenal areas, the needle is inserted through the perianal skin ((Fig. 3-3) into the extraperitoneal tissue. *As gas embolism may occur during this technique, it is performed only by specialists.*

If the fatty renal capsule is thin or absent, as occurs in emaciated patients, the kidneys are difficult to see radiologically, because the perirenal fat does not produce an outline of the kidney as in Figure 2-106.

When the fatty renal capsule is absent, *the kidney may descend to an abnormally low level*, where it is supported primarily by the renal vessels. This downward displacement of the kidney is called *nephroptosis* (*G. ptōsis*, a falling).

In **hypermobility of the kidney** ("floating kidney"), the organ moves up and down within the renal fascia in a vertical plane more than is normal. It does not move from side to side. In excessively low positions, the ureter may be kinked, *possibly* resulting in some obstruction to the flow of urine into the urinary bladder.

Blood from an injured kidney or pus from a **perinephric abscess** distends the renal fascia and may force its way inferiorly into the pelvis, between the anterior and posterior layers of

Figure 2-108. CT scans of the upper abdomen showing the cross-sectional radiographic anatomy of living patients. These sections are visualized as if you were looking at cross-sections of the abdomen from below, with the right side to your left. The images of the vertebrae indicate the anterior and posterior surfaces. *A*, fairly normal patient who has had a contrast agent injected intravenously a few minutes previously. There is slight *hydronephrosis* (dilation of the renal pelvis and calyces) of the right kidney (*RK*). In a right calyx, observe the layer of heavier opacified urine posterior to the lighter, less dense urine in the dilated calyx. On the left (*LK*), note a normal calyx and renal pelvis filled with opacified urine (*white*). The renal pelvis seems to lie more outside the kidney than you might expect, but this is normal. Observe the psoas muscles (*P*) medial to the kidneys on the sides of the vertebra. Note the tip of the right lobe of the liver (*L*). *B*, patient with a malignant tumor (*T*) of the right kidney (*RK*). This scan was made at a slightly higher level than in *A*. Note that the psoas muscle is not visible. The soft tissues at the sides of the vertebra are the crura of the diaphragm. Observe the renal artery to the left kidney (*LK*) passing to the aorta (*A*). Note the much larger renal vein lying anterior to the artery. The liver is not visible because the tumor has pushed it superiorly and the patient's habitus is different from the one shown in *A*.

pelvic fascia. The attachment of the renal fascia in the midline prevents extravasation or spread of blood and/or pus to the opposite side.

Relations of the Kidneys (Figs. 2-98, 2-101, 2-105, and 2-107 to 2-112). *Posteriorly, each kidney lies on muscle.* The posterior surface of the superior pole is related to the **diaphragm**, which separates it from the pleural cavity and the 12th rib. More inferiorly, the kidney is related posteriorly to the **quadratus lumborum** muscle, sometimes encroaching slightly on the **psoas major** muscle medially and the **transversus abdominis** muscle laterally.

The subcostal nerve and vessels and the iliohypogastric and ilioinguinal nerves descend diagonally across the posterior surface of the kidney (Figs. 2-101 and 2-109).

Anteriorly, *the relations of the kidneys differ on the two sides*, except that the anterior and medial aspects of *the superior pole of each kidney is covered by the corresponding suprarenal gland* (Figs. 2-99 and 2-101).

The superior pole of the right kidney is related to the inferior surface of the liver (Figs. 2-62 and 2-72). Except for the superior pole, **the right kidney is separated from the liver by the hepatorenal recess** (Figs. 2-70 and 2-81). More inferiorly the descending part of the duodenum passes across the hilum of the right kidney (Fig. 2-107).

The **right colic flexure** lies anterior to the lateral border and inferior pole of the right kidney. In Figure 2-62 observe that *the suprarenal, duodenal, and colic areas of the kidney are not covered by peritoneum.*

Part of the small intestine lies across the inferior pole of the right kidney; it is separated from it by a film of peritoneal fluid and peritoneum (Fig. 2-107).

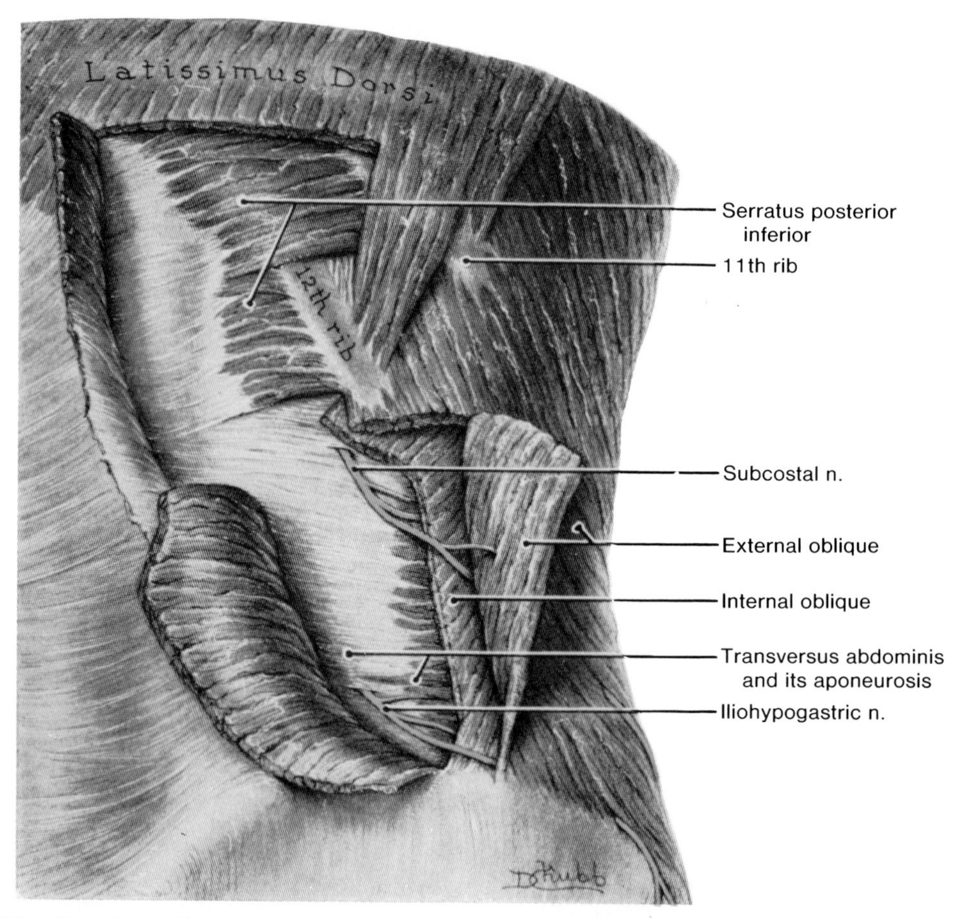

Figure 2-110. Drawing of the right posterior abdominal wall. The external oblique muscle has been incised and turned laterally and the internal oblique muscle incised and turned medially, exposing the transversus abdominis muscle and its posterior aponeurosis. Observe that the subcostal (T12) and iliohypogastric (L1) nerves give off motor twigs and lateral cutaneous branches, before continuing anteriorly between the internal oblique and the transversus abdominis muscles.

The anterior relations of the left kidney are (Fig. 2-62): the left ***suprarenal gland, stomach, spleen, pancreas, jejunum,*** **and** ***descending colon.*** The gastric, splenic, and jejunal areas are covered by peritoneum. The left kidney, along with the pancreas and spleen, is in the **stomach bed** (Fig. 2-48), where it is covered by the posterior wall of the omental bursa.

The close relationship of the kidneys to the psoas major muscles explains why extension of the thigh may increase pain resulting from inflammation in the pararenal regions.

The common surgical approach to kidney is the lumbar renal or retroperitoneal approach (Figs. 2-110 and 2-111). **Lumbar nephrectomy** (removal of a kidney via the lumbar route) is indicated when contamination of the peritoneal cavity is likely (*e.g.*, if there is **inflammatory renal disease** and/or renal calculi). During this surgery the subcostal, iliohypogastric, and ilioinguinal nerves are vulnerable to injury. The transabdominal approach to the kidney is used for surgery of the renal vessels or the ureter (Fig. 2-121).

Renal transplantation is now an established operation for the treatment of selected cases of chronic renal failure. Rejection is the main problem in transplants, but this can usually be controlled by drugs which the recipient has to take for the rest of his/her life. A long survival of a **kidney transplant** is most likely when the kidney is obtained from a monozygotic twin (identical twin).

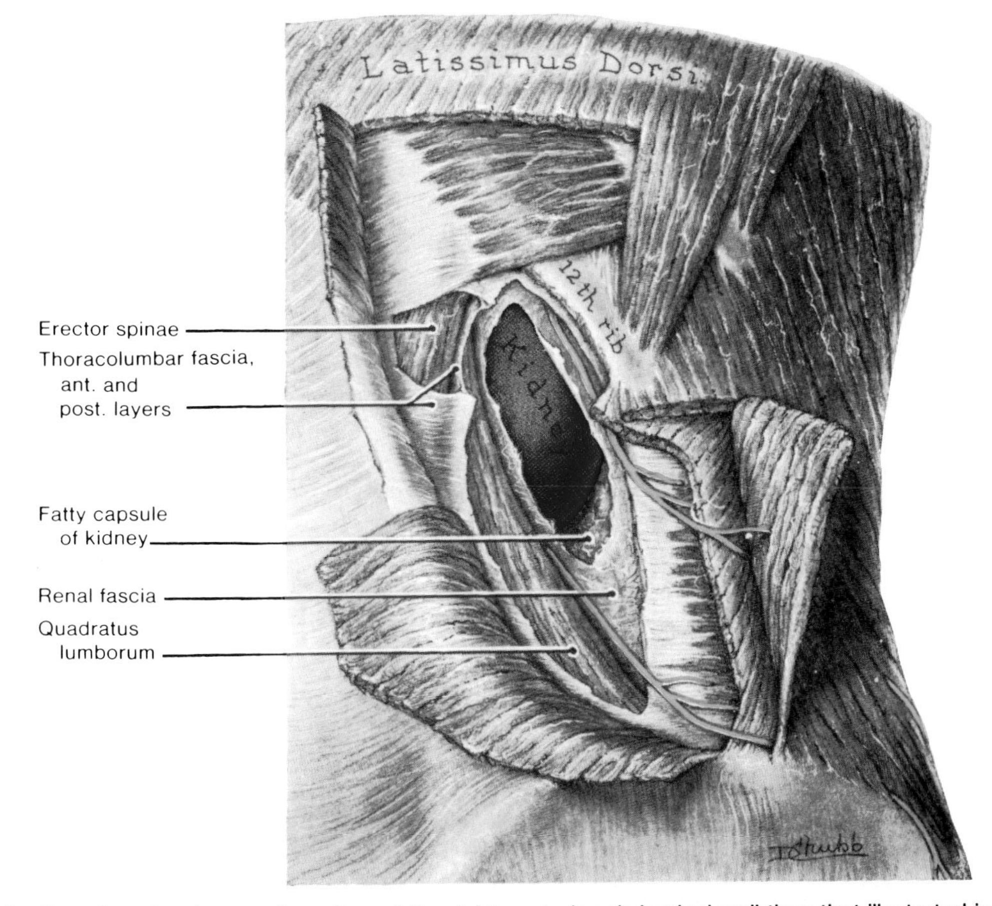

Figure 2-111. Drawing of a deeper dissection of the right posterior abdominal wall than that illustrated in Figure 2-110. Observe that on dividing the posterior aponeurosis of the transversus abdominis muscle between the subcostal and iliohypogastric nerves, and lateral to the oblique lateral border of the quadratus lumborum muscle, the retroperitoneal fat surrounding the kidney is exposed. The renal fascia is within this fat (Fig. 2-107). The portion of fat inside the renal fascia is termed the *fatty renal capsule* (perirenal fat); the fat outside the renal fascia is *pararenal fat* (also see Figs. 2-98 and 2-107).

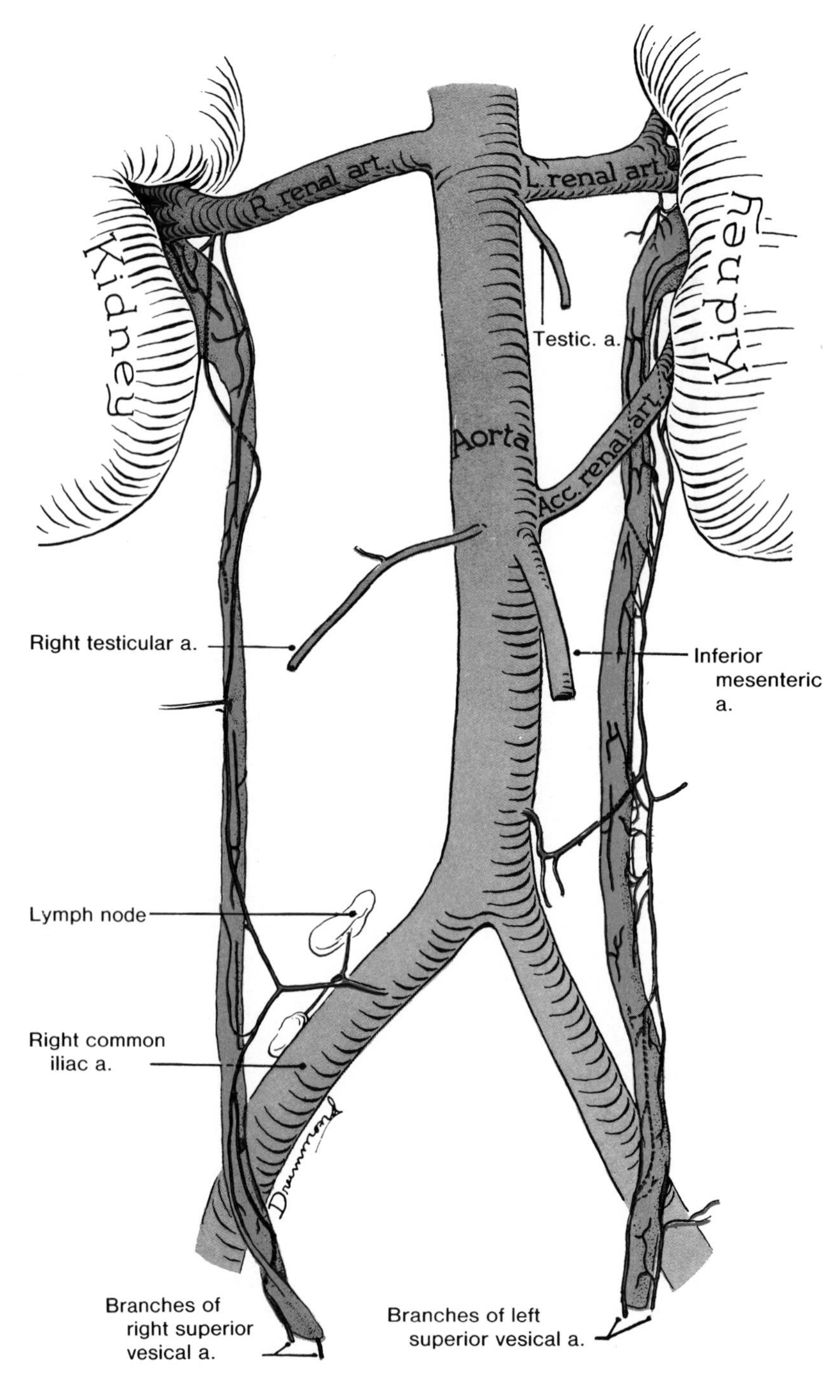

Figure 2-112. Drawing illustrating the blood supply of the kidneys and ureters. The arterial system was injected with latex via the femoral artery. Observe that the blood supply to the ureter comes from three main sources: (1) the *renal artery* superiorly, (2) the *superior vesical artery* inferiorly, and (3) near its middle from either the *common iliac artery* or the *aorta*. Note that these branches approach the ureter from the medial site. In this specimen an accessory renal artery also supplies blood to the left kidney. In some men the testicular artery also contributes a branch. Examine the excellent *anastomotic chain* made by these long tenuous branches.

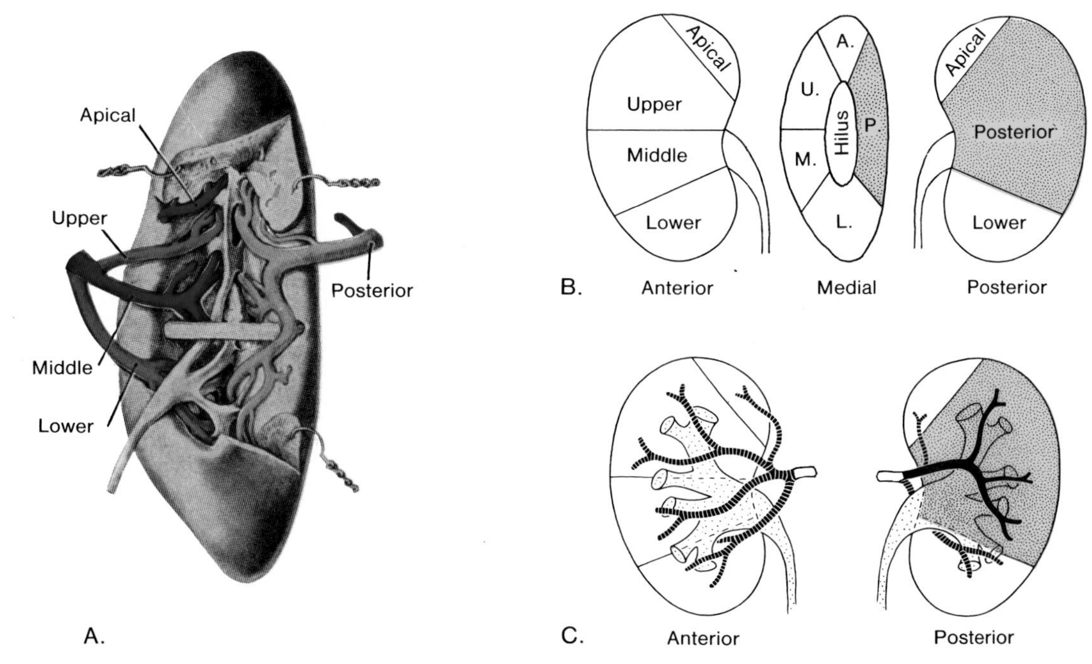

Figure 2-113. Drawings of the kidneys illustrating their arterial supply. *A*, the branches of the renal artery within the renal sinus. Typically, there are five segmental arteries. The posterior lip of the renal sinus has been incised, superiorly and inferiorly, near the limits of the territory of the posterior segmental artery. *B*, diagram of the segments of the kidney. According to its arterial supply the kidney has five segments: (1) apical (superior); (2) superior (anterosuperior); (3) middle (anteroinferior); (4) lower or inferior; and (5) posterior. *C*, diagram of the segmental arteries supplying the segments of the kidney shown in *B*. Only the apical and inferior arteries supply the whole thickness of the kidney. Note that the posterior artery crosses superior to the renal pelvis to reach its segment.

Vessels of the Kidneys and Ureters (Figs. 2-103 and 2-113). *The renal arteries are large vessels that arise from the aorta at right angles, at the level of the intervertebral disc between L1 and L2 vertebrae* (Figs. 2-112 and 2-127). Typically each renal artery divides close to the hilum into *five segmental arteries.* Most of these pass anterior to the pelvis of the kidney (Fig. 2-113), but one or two pass posterior to it.

Based on the arterial distribution, segments of the kidney are described (Fig. 2-113*B*), each of which is supplied by a **segmental artery**. The initial branches of the segmental arteries are called **lobar arteries.** Each of these divides into *interlobar arteries* which enter the kidney and ascend at the sides of the pyramids. The interlobar arteries become **arcuate arteries** which pass between the cortex and medulla of the kidney.

The arcuate arteries give rise to **interlobular arteries** from which, side branches called *intralobular arteries* arise. Each of these branches into one or more afferent *glomerular arteries.*

The blood supply to the ureter comes from several arteries; usually there are three main sources as shown in Figure 2-112, but branches to the ureter may arise from any of the following arteries (*main sources appear in bold face*): **renal**, testicular or ovarian, **aorta**, internal and **common iliac**, and **vesical** or **uterine** arteries.

Usually these long arteries form such an excellent anastomotic chain that some of the branches may be ligated without interfering with the blood supply to the ureter. In some people the longitudinal anastomoses along the ureter are poor owing to the wide spacing of the vessels supplying it.

The renal veins lie anterior to the renal arteries, and *the left renal vein passes anterior to the aorta,* just inferior to the origin of the superior mesenteric artery (Fig. 2-101). Each renal vein joins the inferior vena cava.

The lymph vessels of the kidney follow the renal vein and drain into the lateral **aortic lymph nodes** (Figs. 2-26 and 2-121). *Lymph vessels from the superior part of the ureter may join those of the kidney* or pass directly to the lateral aortic lymph nodes.

Lymph vessels from the middle part of the ureter usually drain into the **common iliac lymph nodes,** whereas *lymph vessels from the inferior part of the ureter* drain into the common, external, or internal iliac lymph nodes.

The Nerve Supply of the Kidneys and Ureters. Nerves to the kidneys and ureters come from the **renal plexus** (Figs. 2-101 and 2-114) and consist of

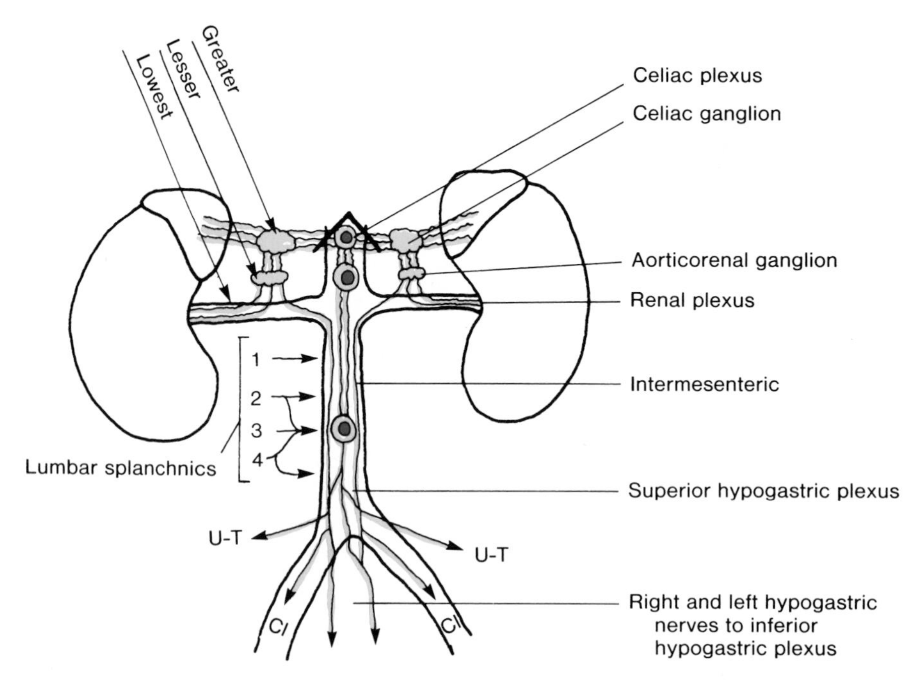

Figure 2-114. Diagram illustrating the autonomic supply to the abdomen and pelvis. Both sympathetic and parasympathetic fibers are delivered via a complex tangle of nerves around the abdominal and pelvic arteries. This network is variable, difficult to dissect, peculiarly named, and described differently by different authors. The above simplified plan is for orientation. Observe the interconnected plexuses on the abdominal arteries. Note that the stems of the celiac, superior mesenteric, and inferior mesenteric arteries are surrounded by nerve fibers. Observe that the sympathetic fibers here synapse *outside* the sympathetic trunk in ganglia, some of which are small and scattered, but two of which are large and named: celiac and aorticorenal. The *arrows* show the source of sympathetic input: greater, lesser, lowest splanchnic nerves, and four lumbar splanchnic nerves. Although paired, they are shown here only on one side. Note that the superior hypogastric plexus supplies the ureteric and testicular plexuses (*U-T*) and a plexus on each of the common iliac arteries (*C*). The parasympathetic nerve supply is not shown. Branches of the vagus nerve are distributed to the foregut and midgut. Pelvic splanchnic nerves join the lower part of the nerve network shown and supply the hindgut and pelvic viscera. The term "splanchnic" means "viscera."

sympathetic and parasympathetic fibers. The renal plexus is supplied by fibers from the lesser and lowest **splanchnic nerves** which pass along the renal artery to supply the kidney.

The renal pelvis and calyces accommodate about 8 ml of fluid. If more than this amount of radioopaque contrast material is injected into them during **retrograde urography** (see previous comments on this radiographic technique), the excessive pressure may tear the epithelial junction of the minor calyces with the papillae (Fig. 2-104*B*), allowing the material to enter the adjacent renal veins. This is undesirable but not a serious occurrence.

The embryonic kidneys develop in the pelvis and normally ascend to their final position in the abdomen by the beginning of the fetal period (9th week). During this ascent the kidneys receive their blood supply from successively more superior vessels. Usually the inferior vessels degenerate as the superior ones develop.

Failure of degeneration of some of these vessels results in **multiple renal arteries** (Fig. 2-115*B*). Variations in the number of renal arteries and in their position with respect to the renal veins are common. Supernumerary arteries, usually two or three, are about twice as common as supernumerary veins, and they usually arise at the level of the kidney.

When first formed in the pelvis the kidneys are close together. In 1 in about 600 fetuses, the kidney poles fuse across the midline to form a **horseshoe kidney** (Fig. 2-115*A*). The large U-shaped kidney usually lies at the level of the lower lumbar vertebrae because normal ascent was prevented by the root of the inferior mesenteric artery.

Horseshoe kidney usually produces no symptoms, but there may be associated abnormalities

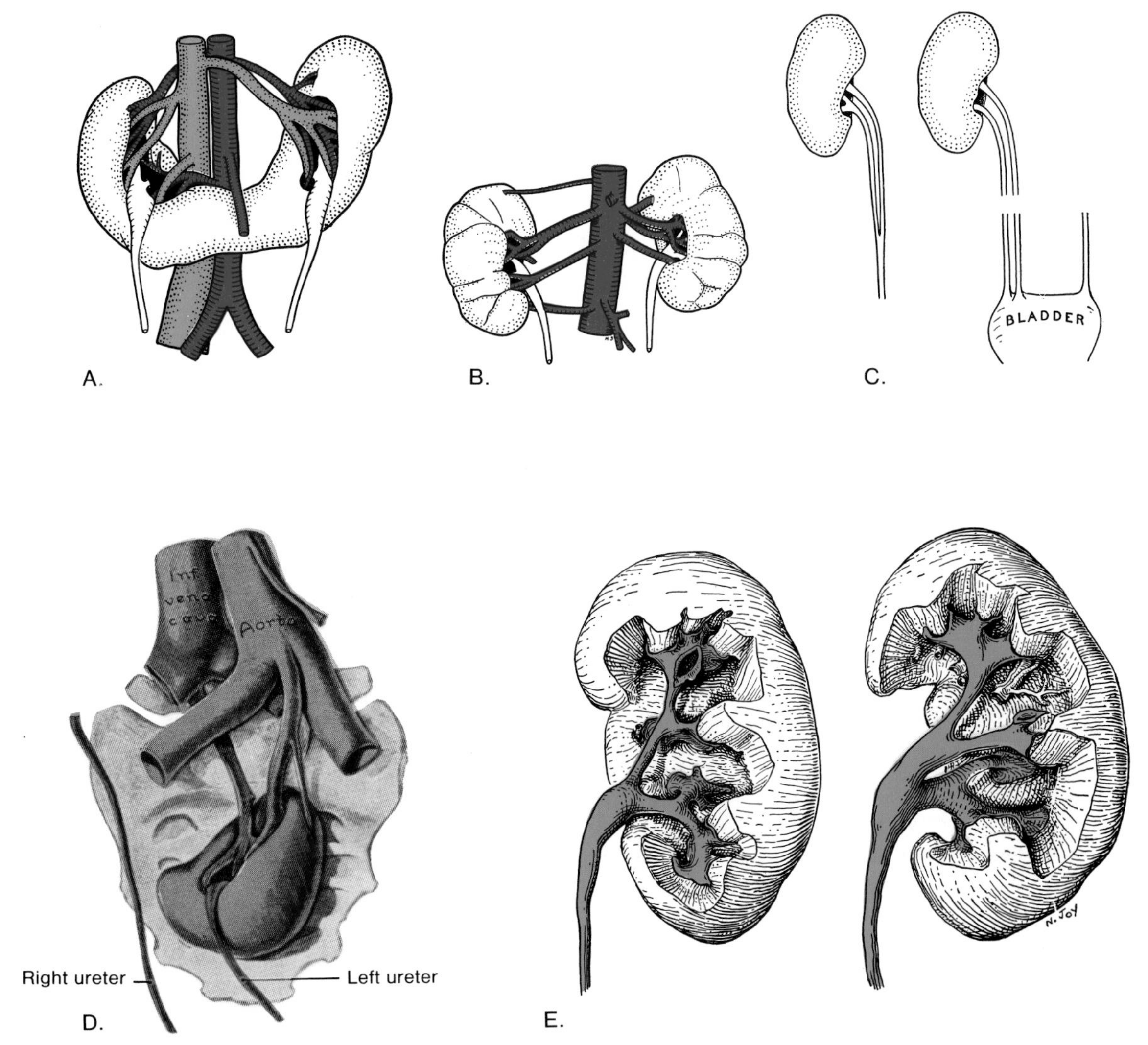

Figure 2-115. Drawings illustrating congenital abnormalities of the kidney and ureter. *A*, horseshoe kidney resulting from fusion of the inferior poles of the kidneys when they were in the embryonic pelvis. *B*, multiple renal arteries and persistence of fetal lobulation. About 25% of kidneys receive two to four branches from the aorta which enter either through the renal sinus or the superior or inferior pole (also see Fig. 2-112). *C*, bifid and duplicated ureters. *D*, ectopic pelvic kidney. This *rare* condition results from failure of the embryonic kidney to ascend from the pelvis. *E*, bifid renal pelves. The pelves are almost replaced by two long major calyces which lie entirely within the renal sinus (*left*) and partly within and partly outside it (*right*).

of the renal pelvis and kidney which may favor development of obstruction and/or infection.

Sometimes the embryonic kidney on one or both sides fails to ascend into the abdomen (Fig. 2-115*D*), and lies in the hollow of the sacrum. This ectopic **pelvic kidney** must not be mistaken for a pelvic tumor. In addition, a pelvic kidney in a woman may be injured or cause obstruction to the passage of the infant's head during childbirth.

Pelvic kidneys do not receive their blood supply from the usual source. The arteries arise from one of the following: the inferior end of the aorta, the iliac artery, or the median sacral artery (Fig. 2-127).

Duplication of the abdominal part of the ureter and renal pelvis is common (Fig. 2-115*C* and *E*), but a *supernumerary kidney* is rare. These abnormalities result from division of the **metanephric diverticulum** (ureteric bud), the primordium of the ureter and renal pelvis. The extent of ureteral duplication depends on how complete the division of the ureteric bud is. Incomplete division of the bud results in the formation of a **bifid ureter** and/or renal pelvis.

THE SUPRARENAL GLANDS

The paired suprarenal (adrenal) glands, 3 to 5 cm long, ***lie on each side of the vertebral column against the superomedial surface of the corresponding kidney*** (Figs. 2-32*B*, 2-62, 2-64, and 2-101). *In vivo*, the suprarenal glands are yellowish in color owing to the presence of lipoid substances in them.

Each suprarenal gland is enclosed with the kidney within the renal fascia. A little fatty connective tissue separates it from the superior pole of the kidney; hence it can be easily separated from this organ. The shape and relations of the suprarenal glands differ on the two sides.

The Right Suprarenal Gland. This gland is *pyramidal in shape* with its apex superiorly and its base on the kidney. It lies between the diaphragm posteromedially and the inferior vena cava anteromedially. The **bare area of the liver** is anterior to the right suprarenal gland (Fig. 2-71) and the kidney is inferior to it. Superiorly the gland lies on the bare area of the liver (Fig. 2-62*A*). Its inferior end is covered by peritoneum reflected onto it from the liver. Its hilum is on its anterior surface and from it the right suprarenal vein leaves to drain into the inferior vena cava (Fig. 2-116).

The Left Suprarenal Gland. This gland is *semilunar in shape* and extends further inferiorly on the medial margin of the kidney than does the right gland (Fig. 2-62*B*). *The left suprarenal gland is related anteriorly to the stomach and pancreas and posteriorly to the diaphragm.* Its hilum is also anterior and from it the left suprarenal vein leaves to drain into the left renal vein (Fig. 2-116).

Structure of the Suprarenal Glands. A sectioned, fresh suprarenal gland shows two distinct regions, an outer **cortex** and an inner **medulla.** These regions are distinct embryologically, structurally, and functionally.

Each gland is surrounded by a tough connective tissue capsule. The cortex, the major part of the gland, secretes several steroid hormones and is essential to life. The medulla, *derived from neural crest cells* in the embryo, secretes *adrenaline and noradrenaline.*

Vessels and Nerves of the Suprarenal Glands (Fig. 2-99). The suprarenal glands have an **abundant arterial supply** from *three sources.* Each gland has direct branches from the **aorta** (up to 10) and also

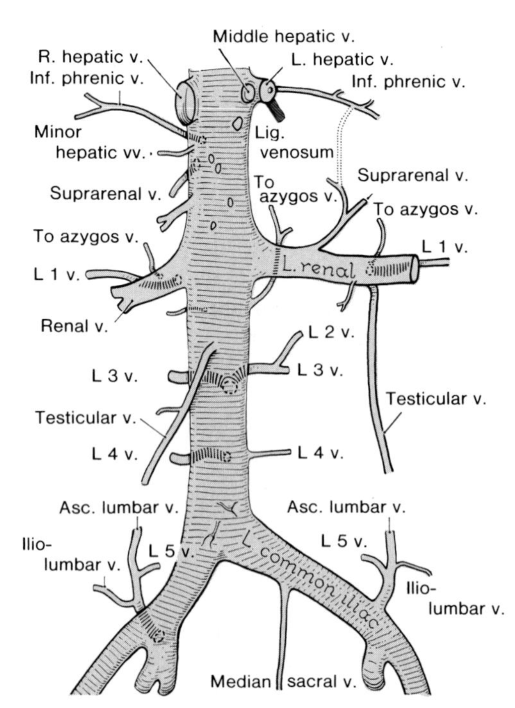

Figure 2-116. Drawing of the inferior vena cava and its tributaries. This is the widest vein in the body. It drains blood from the lower limbs, most of the abdominal wall, the urogenital system, and the suprarenal glands.

receives branches from the **inferior phrenic artery** (up to 27) and the **renal artery** (up to 30).

Although each suprarenal gland may receive over 60 arteries, *it is drained by a single, large central vein.* The right suprarenal vein drains into the superior vena cava, whereas the left suprarenal vein joins the left renal vein (Fig. 2-116).

Each suprarenal gland also has a rich nerve supply, mainly preganglionic sympathetic fibers from the splanchnic nerves and the celiac plexus (Figs. 2-99 and 2-114). These nerve fibers pass through the hilum into the medulla of the gland. Apparently the cortex receives no nerve supply.

Many **lymph vessels** leave the suprarenal glands. Most of them end in the ***aortic lymph nodes*** (Figs. 2-26 and 2-121).

THE DIAPHRAGM

The diaphragm is a dome-shaped, musculotendinous partition separating the thoracic and abdominal cavities (Figs. 2-28 and 2-117).

The diaphragm is the principal muscle of respiration. *It forms the floor of the thorax and the roof of the abdomen.* The diaphragm is pierced by the structures passing between the thorax and the abdomen (Figs. 1-42, 2-35, and 2-119).

The diaphragm rises and falls during respiration, *alternately decreasing and increasing the vertical dimension of the thoracic cavity.* The heart lies on the central part of the diaphragm and slightly depresses it (Fig. 2-120).

STRUCTURE OF THE DIAPHRAGM

The diaphragm is composed of a **muscular portion,** consisting of a sheet of radiating muscle fibers extending from the inferior border of the thorax and the superior lumbar vertebrae. The muscular portion converges on a trefoil-shaped **aponeurotic portion** called the central tendon.

The Central Tendon. The central tendon is *a strong aponeurosis with interlacing tendinous fibers* (Figs. 1-29 and 2-117). All muscle fibers of the diaphragm converge on the central tendon which is incompletely divided into three leaves (*i.e.,* it is trefoiled). This gives the central tendon a C-shaped appearance, somewhat like a boomerang. The middle leaf is anterior and intermediate in size, whereas the right lateral one is the largest and the left lateral leaf is the smallest. The right and left leaves curve posteriorly in the corresponding halves of the diaphragm.

The middle leaf of the central tendon lies just inferior to the heart (Figs. 1-45 and 2-120). The heart is held closely to the diaphragm because the fibrous pericardium is fused with the central tendon.

The central tendon has no bony attachments. The *foramen for the inferior vena cava* is in the right side of the middle leaf (Fig. 2-117).

The Muscular Portion. The muscular portion of the diaphragm **inserts into the central tendon.** It is *divided into three parts* according to the origin of its fibers.

1. **The sternal part of the diaphragm** consists of *two small muscular slips* that arise from the posterior aspect of the **xiphoid process** of the sternum, and pass posteriorly to insert into the central tendon (Fig. 2-117). In the cadaver the sternal part often appears to ascend from its origin because of the postmortem relaxation and ascent of the diaphragm. On each side of these small muscular slips, there is a small anterolateral gap or hiatus known as the **sternocostal hiatus** (Fig. 2-117).

   The superior epigastric branch of the ***internal thoracic artery*** (Fig. 1-17), its ***venae comitantes,*** and some **lymph vessels** from the abdominal wall and the superior surface of the liver pass through the sternocostal hiatus.
2. **The costal part of the diaphragm** consists of *wide muscular slips* that arise from the internal surface of the **inferior six ribs** at the costal margin. *These muscular slips interdigitate with the slips of the transversus abdominis muscles* at their costal attachments (Figs. 1-17 and 2-118). The costal part of the muscular portion of the diaphragm forms the right and left hemidiaphragms or domes (Figs. 1-29 and 1-42) that are visible on radiographs of the chest (Figs. 1-31, 1-51, and 1-52).
3. **The lumbar part of the diaphragm** *arises from the lumbar vertebrae by* ***two crura*** *and the* ***arcuate ligaments*** (Fig. 2-117). The musculotendinous crura (L. legs) are attached on each side of the aorta to the anterolateral surfaces of the superior two (left) or three (right) lumbar vertebrae and the intervening intervertebral discs (Fig. 2-117).

The crura of the diaphragm blend with the anterior longitudinal ligament of the vertebral column (Fig. 2-118). The right crus is broader and longer than the left crus. The fibers of the right crus surround the esophageal hiatus (Fig. 2-117).

The Arcuate Ligaments (Fig. 2-118). Three arcuate ligaments give rise to fibers of the diaphragm.

The median arcuate ligament is a tendinous band (Fig. 2-117) that unites the medial sides of the two crura. It passes over the anterior surface of the aorta and gives origin to some fibers of the right crus of the diaphragm (Figs. 2-117 and 2-118).

The medial arcuate ligament (medial lumbosacral arch) on each side is *a thickening of the anterior layer of the thoracolumbar fascia over the superior part of the psoas major muscle* (Fig. 2-118). It forms a fibrous arch that runs from the crus of the diaphragm, superficial to the psoas major muscle, and attaches to the transverse process of the first lumbar vertebra.

The lateral arcuate ligament (lateral lumbosacral arch) on each side is *a thickening of the anterior layer of the thoracolumbar fascia over the superior part of the quadratus lumborum muscle* (Fig. 2-118). It forms a fibrous arch running from the transverse process of the first lumbar vertebra to the 12th rib.

Superior to the lateral arcuate ligament, the muscular portion of the diaphragm is often thin, especially superior to the left kidney (Fig. 2-109), because the lateral arcuate ligament does not reach the tip of the 12th rib. This triangular deficiency, known as the **vertebrocostal triangle,** consists only of an areolar membrane separating the left kidney from the pleura. *This triangular area is the usual site for a diaphragmatic hernia* and eventration of the diaphragm.

Diaphragmatic rupture or tearing can result from a sudden, dramatic increase in either the intrathoracic or intraabdominal pressure. The most common cause of this injury is blunt trauma to the thorax or abdomen during a motor

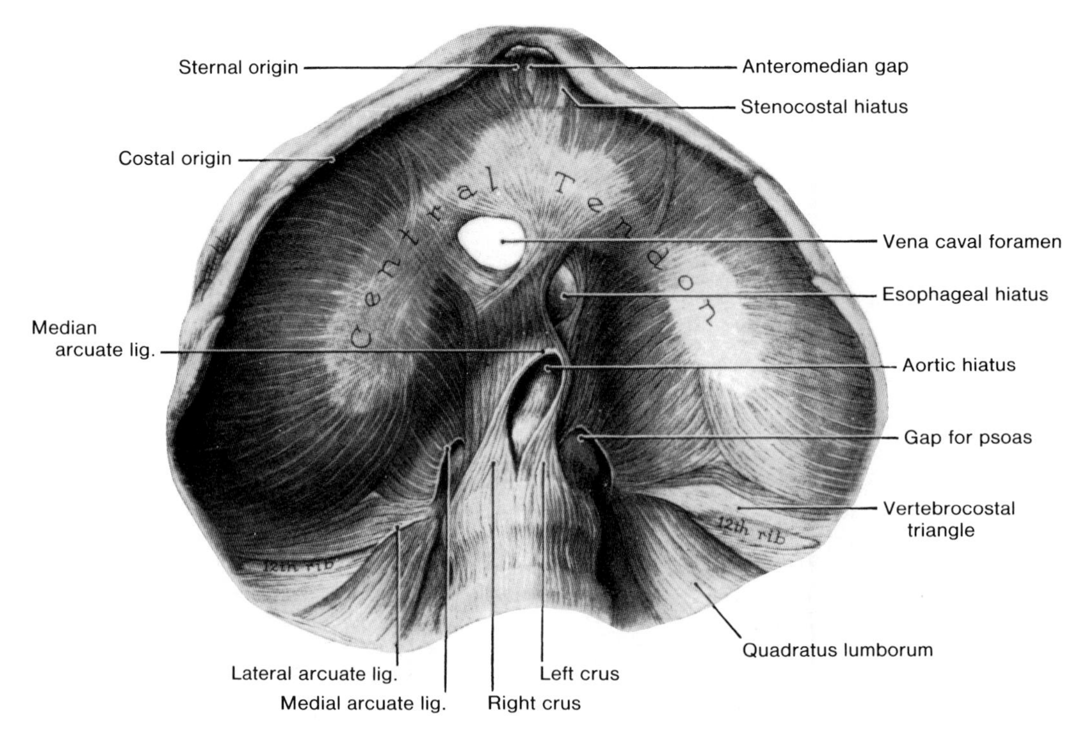

Figure 2-117. Drawing of the inferior surface of the diaphragm. Observe the trefoil-shaped, aponeurotic portion, called the *central tendon*, and fleshly fibers of the sternal, costal, and lumbar origins that converge on this tendon. Examine the three large openings in the diaphragm: (1) the vena caval foramen in the central tendon; (2) the esophageal hiatus surrounded by fibers of one or both crura; and (3) the aortic hiatus lying posterior to the diaphragm in the median plane. Observe the right and left crura on the sides of the aortic hiatus which are united superiorly by a fibrous arch, the median arcuate ligament. Note the thickenings of the psoas and quadratus lumborum fasciae, called the medial and lateral arcuate ligaments.

vehicle accident. About 95% of diaphragmatic ruptures are left sided, usually at the apex or posterior portion of the diaphragm. After diaphragmatic rupture, the following structures (in decreasing order of frequency) herniate into the thorax: stomach, colon, small intestine, mesentery, and spleen.

APERTURES IN THE DIAPHRAGM

There are three major apertures or foramina in the diaphragm (Figs. 1-29 and 2-117).

The Vena Caval Foramen. The opening for the inferior vena cava is in the right side of the middle leaf of the central tendon of the diaphragm. *It is located approximately at the level of the intervertebral disc between the eighth and ninth thoracic vertebrae,* 2 to 3 cm to the right of the median plane (Figs. 2-117 and 2-119). It is the most superior of the three large openings in the diaphragm.

The inferior vena cava is adherent to the margin of the vena caval foramen; consequently, when the diaphragm contracts during inspiration, it pulls the vena caval foramen open and stretches and dilates the inferior vena cava. These changes facilitate the flow of blood through the inferior vena cava.

Some branches of the right phrenic nerve and some lymph vessels from the liver pass through the vena caval foramen. Sometimes the right hepatic vein passes through this foramen before it drains into the inferior vena cava (Fig. 2-52).

The Esophageal Hiatus (Figs. 1-73 and 2-117). The esophagus passes obliquely through this oval aperture in the muscular portion of the diaphragm, just posterior to the central tendon.

The esophageal hiatus is located in the right crus of the diaphragm, 2 to 3 cm to the left of the median plane, approximately *at the level of the 10th thoracic vertebra* (Fig. 2-119). The esophageal hiatus also transmits the **anterior and posterior vagal trunks** (Figs. 2-99 and 7-135), and the esophageal branches of the **left gastric vessels** (Fig. 2-50).

The esophagus is circled by the fleshy fibers of the right crus as they swing across the midline. These fibers constrict the distal end of the esophagus during inspiration, helping to prevent the reflux of gastric contents into the esophagus. This constriction can be

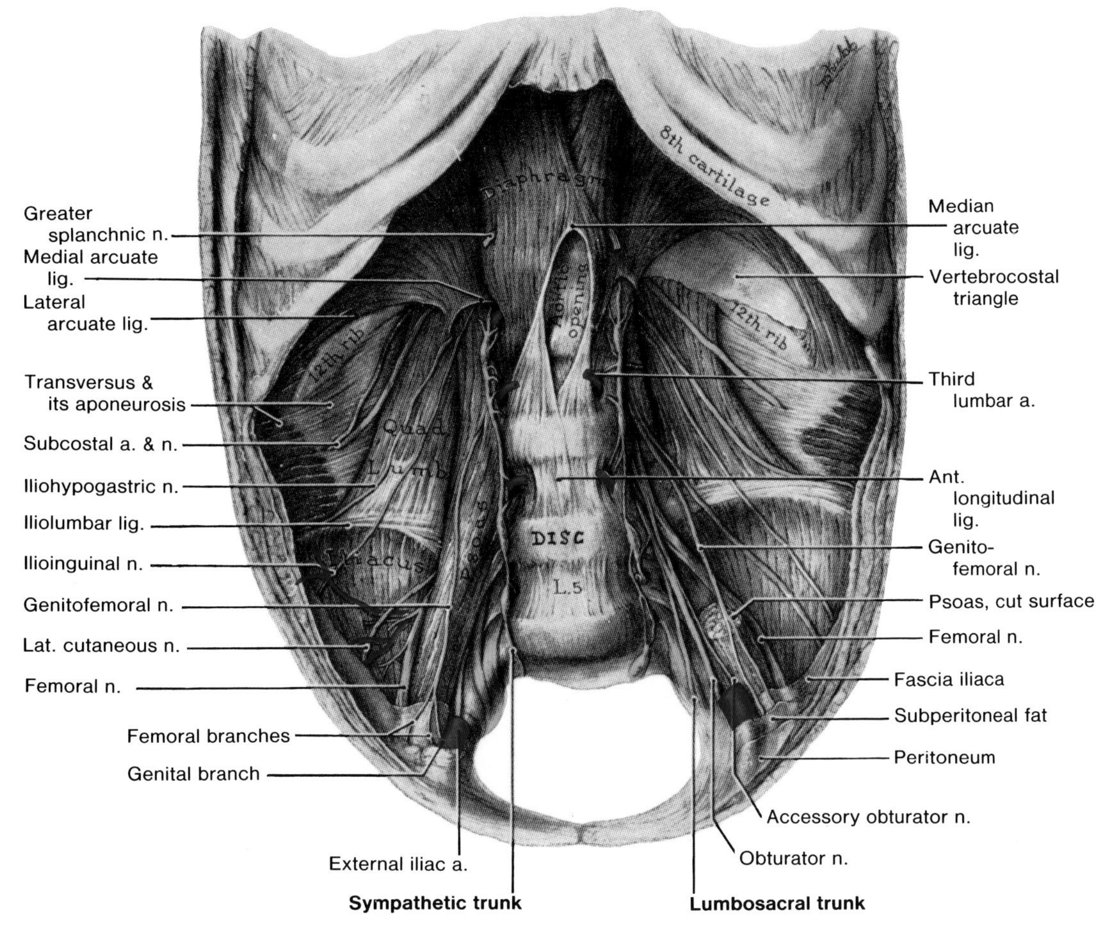

Figure 2-118. Drawing of a dissection of the posterior abdominal wall showing the lumbar plexus. Most of the fascia has been removed. Note that most of the left psoas major muscle has been removed to show the lumbar plexus and that the transversus abdominis muscle becomes aponeurotic on a line dropped from the tip of the 12th rib. Observe that the quadratus lumborum muscle has an oblique lateral border and that its fascia is thickened to form the lateral arcuate ligament superiorly and the iliolumbar ligament inferiorly. Note that the iliacus muscle lies inferior to the iliac crest and that the psoas major rises superior to the crest and extends superior to the medial arcuate ligament, which is thickened psoas fascia. Observe that the subcostal nerve (T12) passes posterior to the lateral arcuate ligament and runs at some distance inferior to the 12th rib with its artery. Note that the next four nerves appear at the lateral border of the psoas major muscle and that of these, the iliohypogastric (T12, L1) takes the characteristic course shown here. The ilioinguinal (L1) and the lateral femoral cutaneous nerve (L2, L3) are variable; the femoral (L2, L3, L4) descends in the angle between the iliacus and the psoas major muscles. Observe that the genitofemoral nerve (L1, L2) pierces the psoas major and its fascia anteriorly. Note that the obturator nerve (L2, L3, L4) and a branch of L4 that joins with L5 to form the lumbosacral trunk, appear at the medial border of the psoas major and cross the ala of the sacrum to enter the pelvis. Now examine the *sympathetic trunk*. Observe that it enters the abdomen with the psoas major muscle posterior to the medial arcuate ligament and descends on the vertebral bodies and the intervertebral discs, following closely the attached border of the psoas major to enter the pelvis. Note that its rami communicantes run dorsally with or near the lumbar arteries to join the lumbar nerves.

easily observed during barium studies of the esophagus (Fig. 2-33).

The Aortic Hiatus (Figs. 1-73 and 2-117). The aorta does not pierce the diaphragm but *passes posterior to the median arcuate ligament and anterior to the 12th thoracic vertebra* (Fig. 2-119), just to the left of the midline. The aorta is unaffected by contraction of the diaphragm because it lies posterior to it. The aortic hiatus also transmits the **thoracic duct**, the ***azygos vein***, and ***lymph vessels*** descending from the thorax to the *cysterna chyli* (Fig. 1-73).

Other Structures Passing Through or Around the Diaphragm. The **right phrenic nerve** passes through the *central tendon* of the diaphragm, either

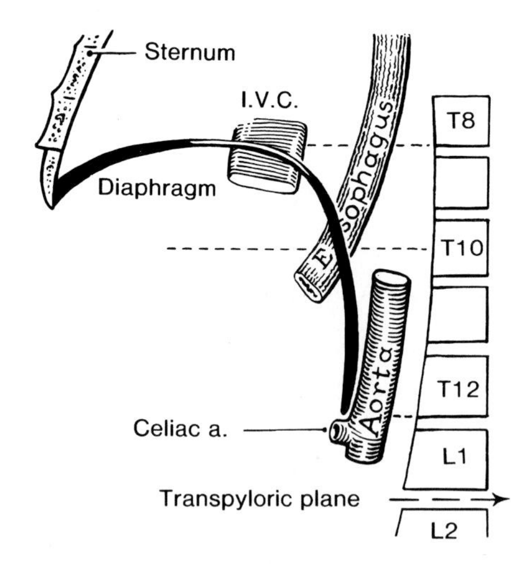

Figure 2-119. Diagram showing the three openings in the diaphragm and their vertebral levels. Observe that the more superior the vertebral level, the more anterior is the opening in the diaphragm. The vertebral levels are **T8, T10, and T12** for the inferior vena cava, esophagus, and aorta, respectively.

through the vena caval foramen or just lateral to it. The **left phrenic nerve** passes through the *muscular portion of the diaphragm,* anterior to the central tendon, just lateral to the pericardium of the heart (Fig. 2-120).

The **superior epigastric vessels** (Fig. 2-8) pass through the *sternocostal hiatus*, the interval between the sternal and costal origins of the muscular part of the diaphragm (Fig. 2-120). The **musculophrenic vessels** perforate the diaphragm near the ninth costal cartilage.

The inferior five **intercostal nerves** pass between the muscle slips of the diaphragm arising deep to the costal cartilages of the inferior six ribs.

The **subcostal nerves and vessels** pass through the diaphragm *posterior to the lateral arcuate ligament* (Fig. 2-118).

The **sympathetic trunks** pass through the diaphragm *posterior to the medial arcuate ligaments* (Fig. 2-118).

The **splanchnic nerves** pierce the crura and the **hemiazygos vein** passes through the left crus.

VESSELS AND NERVES OF THE DIAPHRAGM

The main arteries supplying the diaphragm are the **phrenic arteries** which arise anteriorly from the aorta (Fig. 2-127), and two branches of the internal thoracic artery, the **musculophrenic and pericardiophrenic arteries.** Each pericardiophrenic artery accompanies a phrenic nerve between the pleura and the pericardium to the diaphragm.

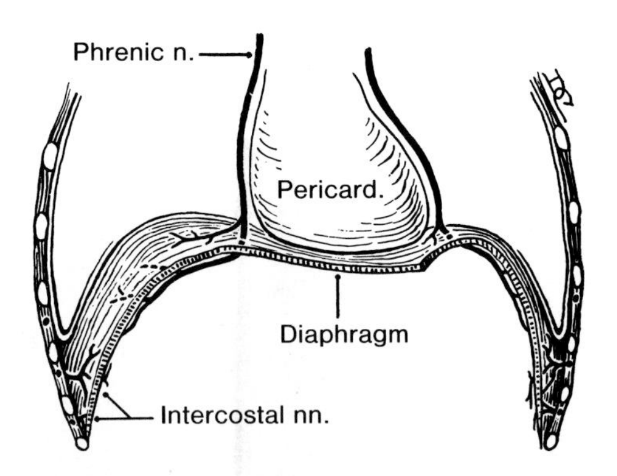

Figure 2-120. Diagram illustrating the nerve supply to the diaphragm. Note that each phrenic nerve (C3, C4, and C5) is the sole motor nerve to its own half of the diaphragm. It is also sensory to the greater part of its own half of the diaphragm, including the pleura on the thoracic surface and the peritoneum on the abdominal surface. Observe that the inferior intercostal nerves are sensory to the peripheral part of the diaphragm.

There is an extensive network of lymph vessels on the thoracic and abdominal surfaces of the diaphragm. These vessels drain into the phrenic or **diaphragmatic lymph nodes,** located on the thoracic surface of the diaphragm.

Lymph vessels from the bare area of the liver, which is in direct contact with the diaphragm, also drain into *middle diaphragmatic lymph nodes* on the thoracic surface of the diaphragm. Lymph from these nodes drains into the *parasternal, middle phrenic, and posterior mediastinal lymph nodes* (Figs. 1-16 and 1-42). A few lymph vessels from the abdominal surface of the diaphragm drain into the *lumbar lymph nodes* (Fig. 2-121).

The motor supply to the diaphragm is from the phrenic nerves (Fig. 2-120), which arise from the ventral rami of C3, **C4**, and C5. *The phrenic nerve is the sole motor supply to the diaphragm.* The contribution to the phrenic nerve from the ventral ramus of C5 may be derived as a branch from the nerve to the subclavius muscle. This is called the **accessory phrenic nerve** (Fig. 8-13).

The superior level of origin of the phrenic nerves from the cervical segments results from the caudal migration of the developing diaphragm relative to the vertebral column. In addition to the sensory nerve supply from the phrenic nerves, the diaphragm receives sensory fibers from the **intercostal nerves** (Fig. 2-120). They supply the peripheral fringes of the diaphragm that develop from the lateral body walls.

ACTIONS OF THE DIAPHRAGM

The diaphragm is the chief muscle of inspiration. During inspiration its muscular portion contracts, drawing its central tendon inferiorly and ante-

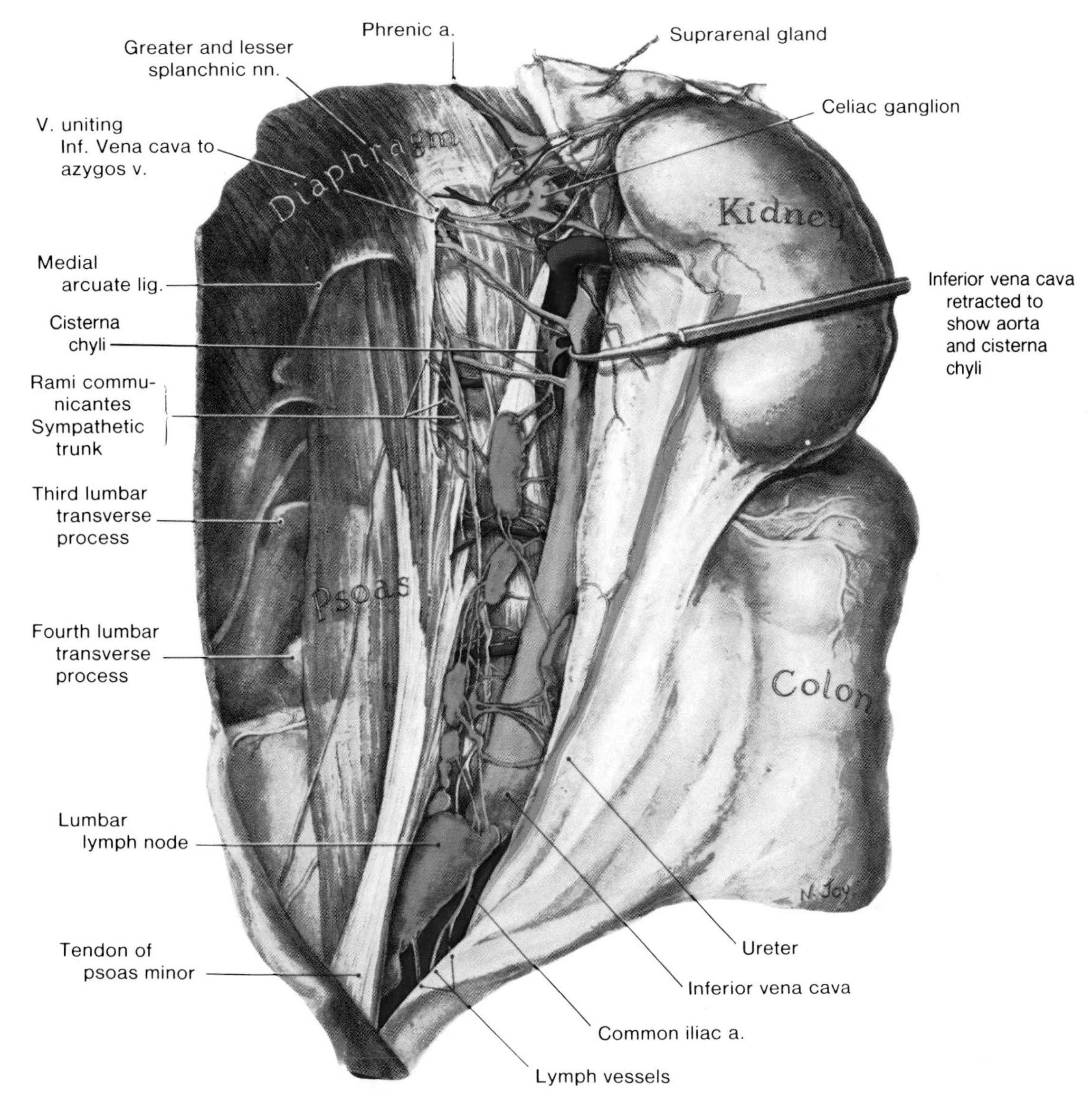

Figure 2-121. Drawing of a dissection of the posterior abdominal wall showing the right celiac ganglion, the splanchnic nerves, and the sympathetic trunk. The right suprarenal gland, kidney, ureter, and colon are turned to the left like the page of a book so that the posterior surface of the right kidney is facing you. The inferior vena cava is pulled medially and the third and fourth lumbar veins are removed. Observe the greater splanchnic nerve ending in the celiac ganglion. Usually the greater splanchnic nerve pierces the crus of the diaphragm at the level of the celiac trunk. The lesser splanchnic nerve passes inferolateral to this and the sympathetic trunk enters with the psoas major muscle. Examine the sympathetic trunk lying on the bodies of the vertebrae, the lumbar vessels alone intervening, and descending along the anterior border of the psoas major muscle. Note that the sympathetic trunk is slender where it enters the abdomen and that its ganglia are ill defined.

riorly. As the dome of the diaphragm moves inferiorly, it pushes the abdominal viscera before it. This ***increases*** *the volume of the thoracic cavity* and ***decreases*** *the intrathoracic pressure*, resulting in air being taken into the lungs. In addition, the volume of the abdominal cavity is somewhat decreased and the intraabdominal pressure is somewhat raised because **the anterior abdominal wall moves reciprocally with the diaphragm.**

Diaphragmatic movements are also important in the circulation of blood because the increased abdominal pressure and decreased thoracic pressure accompanying contraction of the diaphragm are helpful in returning blood to the heart. When the diaphragm contracts compressing the abdominal viscera, blood in the inferior vena cava is forced superiorly into the heart. This movement is facilitated by the enlargement of the vena caval foramen and the dilation of the inferior vena cava that occurs when the diaphragm contracts.

The diaphragm is an important muscle in abdominal straining. It assists the anterior abdominal muscles in raising the intraabdominal pressure during ***micturition*** (urination), ***defecation***, and ***parturition*** (childbirth). During these processes, a person often inspires deeply which closes the glottis (opening in the **larynx**). This traps air in the respiratory tract and prevents the diaphragm from rising. A grunt is produced when some air escapes from the respiratory tract.

The diaphragm is used during weight lifting. A person about to lift a heavy object takes a deep breath to raise intraabdominal pressure. *Increased intraabdominal pressure gives additional support to the vertebral column and helps to prevent its flexion.*

POSITION OF THE DIAPHRAGM

It is clinically important to acquire a three-dimensional concept of the diaphragm, particularly for viewing radiographs of the thorax and abdomen. The posterior attachment of the dome-shaped diaphragm is considerably more inferior than its anterior attachment; hence there is much of it that cannot be seen in posteroanterior radiographs (Figs. 1-51 and 1-52). Normally the right hemidiaphragm bulges more superiorly into the thorax than does the left hemidiaphragm (Fig. 2-120).

The level of the diaphragm on both side varies in relation to the ribs and the vertebrae according to: (1) *the phase of respiration*, (2) the posture assumed, and (3) the size and degree of distention of the abdominal viscera. The diaphragm is most superior when a person is supine because the abdominal viscera push the diaphragm superiorly into the thorax. When a person lies on one side, the hemidiaphragm next to the bed or table rises to a higher level owing to the superior push of the viscera on that side. The diaphragm assumes an inferior level when the person is sitting or is in an erect position. This explains why patients with **dyspnea** (difficult breathing) prefer to sit up rather than lie down.

The diaphragm develops from four sources. Failure of fusion of these parts occurs about once in every 2000 fetuses. This results in a **congenital diaphragmatic hernia** (Fig. 2-134), usually on the left side.

Defective formation and/or fusion of the left pleuroperitoneal membrane with other parts of the diaphragm is the usual cause of developmental defects in the diaphragm (see Moore, 1982).

A *posterolateral defect of the diaphragm occurs five times more often on the left side than on the right.* It is located at the periphery of the diaphragm in the region of the **vertebrocostal triangle** (Fig. 2-117). The intestines and occasionally other abdominal viscera pass through this **posterolateral defect of the diaphragm** (foramen of Bochdalek) into the thoracic cavity.

Radiographs of the chest reveal gas-filled loops of bowel in the left pleural cavity and displacement of the thoracic viscera. The herniated bowel prevents inflation of the lung when the baby is born. After a few minutes or hours, the infant swallows air which distends the gut somewhat, and the underinflated lung is then compressed. These conditions frequently result in the infant suffering very **severe respiratory distress.**

When the hernia is detected, the herniated viscera are replaced into the abdominal cavity and the defect in the diaphragm is closed surgically. *Expansion of the hypoplastic lung may take several days.*

When muscle fails to grow into the pleuroperitoneal membrane in the region of the **vertebrocostal triangle**, only an areolar membrane separates the kidney and other abdominal viscera from the thoracic cavity. *This creates a potential site for eventration.* If this occurs, the abdominal viscera herniate into the thoracic cavity, but they are covered by the thin areolar membrane that closed the vertebrocostal triangle. This condition, also more common on the left, is called **eventration of the diaphragm.** This abnormality may be present at birth or develop later in life. Eventrations may also occur through a very thin part of the central tendon.

A *rare type of hernia* is through the **sternocostal hiatus** for the superior epigastric vessels (Fig. 2-117). This uncommon type of hernia usually occurs on the right side because of the attachment of the pericardium to the diaphragm on the left (Fig. 1-46).

Acquired hiatal hernias are common. This condition develops in people around middle age in whom *weakening* and *widening of the esophageal hiatus* has occurred (Fig. 2-49).

Section of the phrenic nerve in the neck results in complete paralysis and atrophy of the muscle of the corresponding half of the diaphragm, except in persons who have an **accessory phrenic nerve** (Fig. 8-13).

Paralysis of a hemidiaphragm can be recog-

nized radiographically by its permanent elevation and paradoxical movement. Instead of descending on inspiration, it is forced superiorly by the increased intraabdominal pressure secondary to descent of the unparalyzed opposite hemidiaphragm.

Because there are many communications between the lymphatics on the abdominal surface of the diaphragm with those on its thoracic surface, and with the lymphatics in the thorax, a **subphrenic abscess** or collection of pus in the **subphrenic recess** (Fig. 2-70) may lead to **pleuritis**. Similarly, **empyema** or pus in the pleural cavity, called *pyothorax*, may lead to development of a subphrenic abscess.

POSTERIOR ABDOMINAL WALL

The posterior abdominal wall is composed principally of muscles and fascia attached to the vertebrae, the hip bones, and the ribs. There are also important ***nerves***, ***vessels***, and ***lymph nodes*** on the posterior abdominal wall.

MUSCLES OF THE POSTERIOR ABDOMINAL WALL

There are three paired muscles in the posterior wall of the abdomen that are clinically important: psoas major, iliacus, and quadratus lumborum.

The Psoas Major Muscle (Figs. 2-101, 2-106, to 2-109, 2-118, 2-121, and 2-124). This long, thick, fusiform muscle *lies lateral to the lumbar region* of the vertebral column. Psoas is a Greek work meaning "*muscle of the loin.*" Butchers refer to the psoas muscle in animals as the tenderloin.

Origin. Transverse processes and borders of the **bodies of L1 to L5 vertebrae** and sides and associated **intervertebral discs of T12 to L5** vertebrae. It descends along the brim of the pelvis and enters the thigh by passing posterior to the inguinal ligament.

Insertion (Fig. 4-28). **Lesser trochanter of femur.**

Nerve Supply (Fig. 2-118). **Ventral rami of lumbar nerves L2 to L4.**

Actions. Flexes the thigh at the hip joint. *Acts with the iliacus muscle as part of the iliopsoas* muscle. **Bends vertebral column anteriorly** (*e.g.*, when sitting up from supine position). **Bends lumbar region laterally.** This action is used to maintain the balance of the trunk when sitting.

The Iliacus Muscle (Figs. 2-101 and 2-118). This large triangular sheet of muscle *lies along the lateral side of the psoas* major muscle.

Origin. Superior part of iliac fossa. Its fibers pass inferomedially posterior to the inguinal ligament. Most of them attach to the side of the psoas tendon and the two muscles are then called the iliopsoas muscle.

Insertion (Fig. 4-30). **Lesser trochanter of femur** and just inferior to trochanter.

Nerve Supply (Fig. 2-118). **Femoral nerve** (L2, L3).

Actions. Acting with the psoas major, **flexes the thigh at the hip joint.** *The iliopsoas muscle is the most powerful flexor of the thigh.*

The Psoas Minor Muscle (Fig. 2-121). This small *weak muscle* with a short belly and a long tendon is present in about 60% of people. It lies anterior to the psoas major in the abdomen.

Origin. Sides of **bodies of T12 and L1 vertebrae** and the intervertebral disc between them.

Insertion (Fig. 4-1). **Pecten pubis and the iliopubic eminence.**

Nerve Supply. Ventral ramus of **first lumbar nerve.**

Action. Weak flexor of the pelvis and the lumbar region of vertebral column.

The iliopsoas muscle has extensive and clinically important relations (Figs. 2-101 and 2-121). If the kidneys, ureters, cecum, appendix, sigmoid colon, pancreas, lumbar lymph nodes, or nerves of the posterior abdominal wall are diseased, *movements of the iliopsoas muscle may be accompanied by pain.* As it lies along the vertebral column and crosses the sacroiliac joint, disease of the intervertebral and sacroiliac joints may cause **spasm of the iliopsoas muscle**, a *protective reflex.*

Although there has been a great fall in the prevalance of *tuberculosis* in North America in recent years, this infection still occurs and may spread via the blood (**hematogenous spread**) to the vertebrae, particularly during childhood. An **abscess** caused by tuberculosis in the lumbar region of the vertebral column tends to spread from the vertebrae into the psoas major muscle (**psoas abscess**). As a consequence the psoas fascia (Fig. 2-124) thickens to form a strong stocking-like tube. The pus then passes inferiorly along the psoas major muscle within this fascial tube over the pelvic brim and deep to the inguinal ligament. The pus *points* (surfaces) in the **femoral triangle** (Fig. 4-15) in the superior thigh region.

The inferior part of the **iliac fascia** (Fig. 4-16) is often tense and raises a fold that passes to the inner aspect of the iliac crest (Fig. 2-101). The superior part of the iliac fascia is loose and may form a pocket, sometimes called the ***fossa iliacosubfascialis***, posterior to the above mentioned fold into which a portion of bowel may become trapped (cecum and/or appendix on the right and sigmoid colon on the left).

The Quadratus Lumborum Muscle. This is a thick quadrilateral muscular sheet (hence its name). *It lies adjacent to the transverse processes of the lumbar vertebrae* and is broader inferiorly.

Origin (Figs. 2-118 and 2-122). **Iliolumbar ligament, adjacent part of iliac crest,** and inferior two to four **lumbar transverse processes.**

Insertion (Figs. 2-122 and 2-124). Medial part of **anterior surface of 12th rib** and tips of **transverse processes of L1 to L4 vertebrae.** It narrows as it ascends posterior to the inferior margin of the diaphragm, *i.e.*, posterior to the lateral arcuate ligament.

Nerve Supply (Fig. 2-118). Ventral rami of T12 (**subcostal nerve**) and **superior lumbar nerves.**

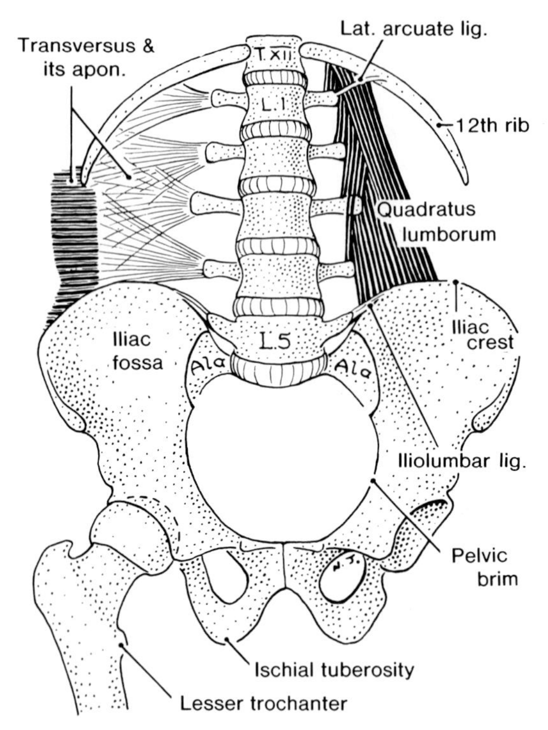

Figure 2-122. Diagram illustrating some muscles of the posterior abdominal wall. Note that the quadratus lumborum, a four-sided muscle, lies adjacent to the transverse processes of the lumbar vertebrae and that it is broader inferiorly than superiorly. It bends the lumbar region of the vertebral column laterally and fixes the 12th rib, preventing it from rising with the other ribs. In this way it helps to elongate the thorax during deep inspiration.

Actions. Fixes the 12th rib in relation to the pelvis, holding it down against the traction exerted by the diaphragm when it contracts during inspiration. *The quadratus lumborum is a* ***muscle of inspiration*** because it increases the vertical diameter of the thorax. ***Acting alone***, it bends the trunk toward the same side (*i.e.*, laterally bends it). ***Acting together***, the two muscles help to extend the lumbar region of the vertebral column and to give it lateral stability.

The Transversus Abdominis Muscle (Figs. 2-101, 2-110, and 2-122). *This is the third and innermost of the three flat muscles of the anterior abdominal wall* (Table 2-1). In Figures 2-107 and 2-121 note that the aponeurosis of the transversus abdominis muscle runs horizontally, posterior to the oblique borders of the quadratus lumborum muscle.

FASCIA OF THE POSTERIOR ABDOMINAL WALL

Each of the muscles forming the posterior abdominal wall is enclosed in fascia.

The Fascia Iliaca (Fig. 2-118). The fascia iliaca (iliac fascia) ***covers the psoas and iliacus muscles.*** Although thin superiorly, it thickens inferiorly as it approaches the inguinal ligament.

The part of the iliac fascia covering the psoas major muscle is one sheet that is attached medially to the lumbar vertebrae and the pelvic brim (Figs. 2-122 and 2-124). This fascial sheet is fused laterally with the anterior layer of the thoracolumbar fascia and, inferior to the iliac crest, it is continuous with the part of the fascia iliaca covering the iliacus muscle. The part of (Fig. 4-16) the fascia covering the psoas major muscle also blends with the fascia covering the quadratus lumborum muscle. Superiorly this part of the fascia iliaca is thickened to form the **medial arcuate ligament of the diaphragm** (Figs. 2-117, 2-118, and 2-121).

The fascia iliaca continues inferiorly into the thigh. The dense part of the fascia iliaca covering the iliacus muscle is attached to the iliac crest and the pelvic brim (Fig. 2-121) and is continuous with the transversalis fascia (Figs. 2-13 and 2-15). The posterior margin of the transversalis fascia is attached to the inguinal ligament and is continuous there with the fascia iliaca as it passes into the thigh.

The Quadratus Lumborum Fascia (Figs. 2-111, 2-118, and 2-124). The fascia covering the quadratus lumborum muscle is a *dense membranous layer* that is continuous laterally with the anterior layer of the **thoracolumbar fascia.** *The quadratus lumborum fascia is attached to the anterior surfaces of the transverse processes of the lumbar vertebrae, the iliac crest, the 12th rib, and the transversalis fascia.* The quadratus lumborum fascia is thickened to form the **lateral arcuate ligament** superiorly and is adherent to the *iliolumbar ligament* inferiorly (Figs. 2-117 and 2-118).

The Thoracolumbar Fascia (Figs. 2-111, 2-123, and 2-124). The thoracolumbar fascia (lumbar fascia) is an extensive sheet of fascia covering the deep muscles of the back (Fig. 6-47).

The lumbar part of the thoracolumbar fascia extends between the 12th rib and the iliac crest. Laterally it is attached to the internal oblique and transversus abdominis muscles.

The thoracolumbar fascia splits into three layers medially (Fig. 2-124). The quadratus lumborum muscle lies between the anterior and middle layers. *The deep back muscles are enclosed between the middle and posterior layers of the thoracolumbar fascia.* The thin anterior layer of thoracolumbar fascia (forming the quadratus lumborum fascia) is attached, along with the psoas fascia, to the anterior surfaces of the lumbar transverse processes. The thick **middle layer** of thoracolumbar fascia is attached to the tips of the transverse processes. The dense **posterior layer** is attached to the spinous processes of the lumbar and sacral vertebrae, and to the supraspinous ligament.

NERVES OF THE POSTERIOR ABDOMINAL WALL

There are two types of nerves in the posterior abdominal wall; **somatic nerves of the lumbar plexus** and its branches (Figs. 2-118 and 2-125), and visceral or **splanchnic nerves of the autonomic nervous system** (Figs. 2-114 and 2-118).

The *five lumbar nerves* pass from the spinal cord through the intervertebral foramina, inferior to the corresponding vertebrae, and divide into dorsal and ventral primary rami. Each ramus contains sensory and motor fibers.

The dorsal primary rami pass posteriorly to supply the muscles and skin of the back (Figs. 2-123 and 6-44), whereas **the ventral primary rami** extend into the posterior part of the psoas major muscle. Here *they are connected to the sympathetic trunk by rami communicantes* (Figs. 2-118 and 2-121). The ventral rami give branches to the psoas major, quadratus lumborum, and intertransverse muscles.

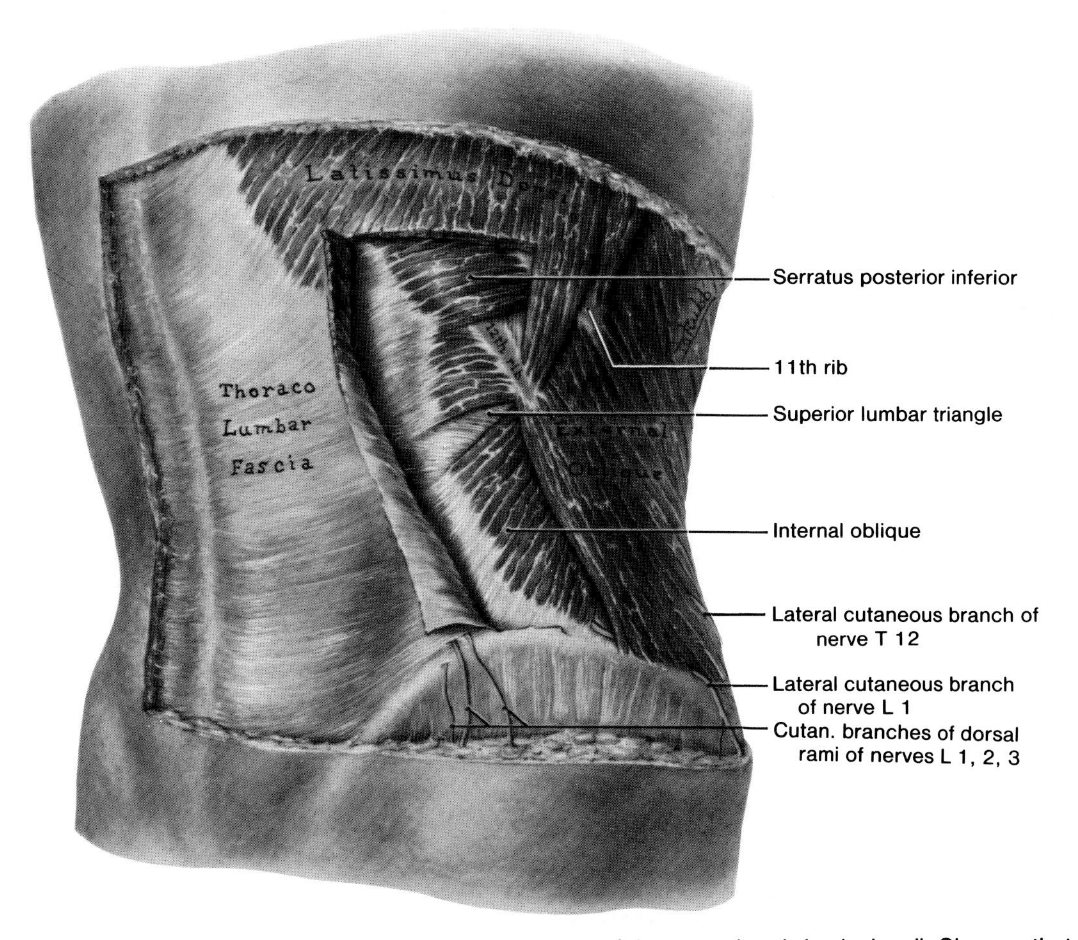

Figure 2-123. Drawing of a dissection of a posterolateral view of the posterior abdominal wall. Observe that the external oblique muscle has an oblique, free posterior border which extends from the tip of the 12th rib to the midpoint of the iliac crest. Note that the internal oblique muscle extends posterior to the external oblique muscle.

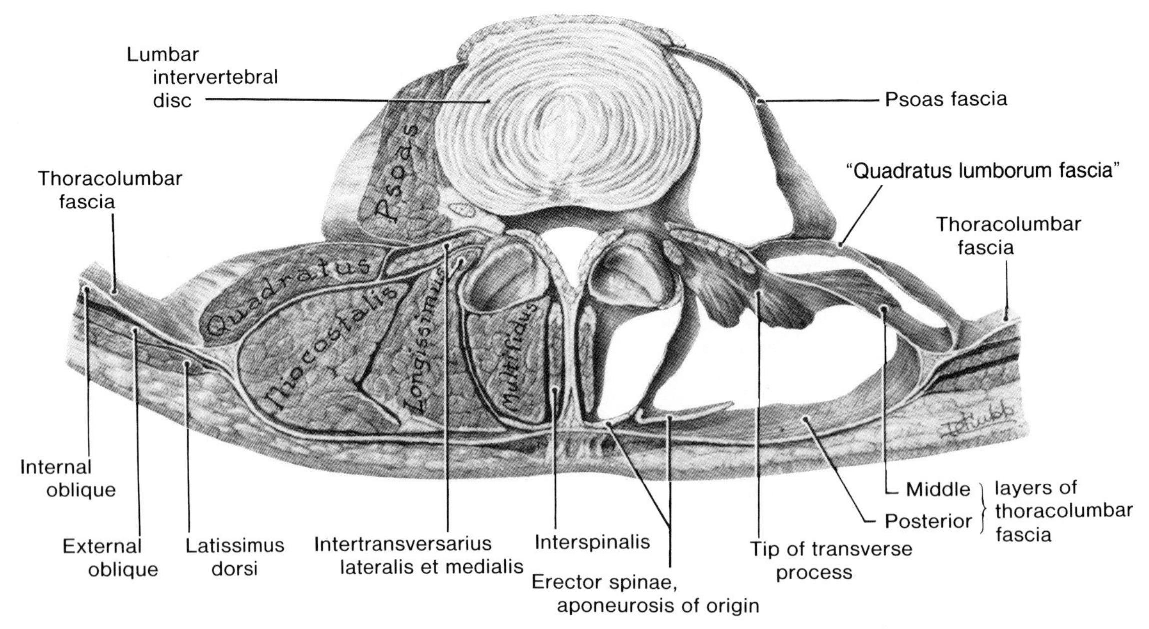

Figure 2-124. Drawing of a dissection of the muscles of the back in a cross-section of the superior lumbar region. On the *right side* the empty sheaths are seen. Observe the anterior, middle, and posterior layers of the thoracolumbar fascia which enclose the deep muscles of the back. Note that the posterior layer is reinforced by the latissimus dorsi muscle and at a superior level by the serratus posterior inferior muscle. Observe the fascial layers covering the quadratus lumborum and psoas muscles. Note that the ends of the intertransversarius, longissimus, and quadratus lumborum muscles are attached to the transverse process.

The ventral rami of L1 to L3 nerves and the superior branch of L4 form the **lumbar plexus** (Figs. 2-118 and 2-125). The inferior branch of L4 and all of L5 form the **lumbosacral trunk** (Fig. 2-118), which descends to the **sacral plexus** (Fig. 2-125).

The subcostal nerve is the ventral ramus of T12. *It passes posterior to the lateral arcuate ligament of the diaphragm,* about 1 cm caudal to the 12th rib (Fig. 2-118). Usually it sends a branch to the ventral ramus of the first lumbar nerve, and then runs inferolaterally across the anterior surface of the quadratus lumborum muscle.

The subcostal nerve pierces the transversus abdominis muscle and runs in the anterior abdominal wall between this muscle and the internal oblique. *The subcostal nerve supplies the abdominal wall inferior to the umbilicus and superior to the pubic symphysis* (Figs. 2-6, 2-16*A*, and 2-43).

THE LUMBAR PLEXUS OF NERVES

The lumbar plexus is formed within the psoas major muscle, anterior to the transverse processes of the lumbar vertebrae (Figs. 2-118 and 2-125). Hence the origin of the nerves contributing to it can be studied only when the psoas muscle is carefully removed. *The lumbar plexus of nerves passes through the psoas major muscle at different levels.*

The lumbar plexus is formed by the ventral rami of the first three lumbar nerves and the superior part of the fourth lumbar nerve. In about 50% of people there is a *contribution from the subcostal nerve* (the large ventral ramus of T12). All five of these ventral rami receive gray rami communicantes from the sympathetic trunk, and the superior two send white rami communicantes to the sympathetic trunk.

The largest and most important branches of the lumbar plexus are the obturator and femoral nerves, which are derived from the same spinal cord segments.

The obturator nerve (L2, L3, and L4) descends through the psoas major muscle, leaving the medial border of this muscle at the brim of the pelvis (Fig. 2-118). It pierces the psoas fascia, crosses the sacroiliac joint, passes lateral to the internal iliac vessels and the ureter, and enters the pelvis minor.

In the pelvis minor the obturator nerve lies in the extraperitoneal fat where it is liable to be injured in surgical operations designed to remove the pelvic lymphatics. *The obturator nerve leaves the pelvis by passing through the obturator foramen into the thigh* (Fig. 3-25).

The femoral nerve (L2, L3, and L4) pierces the psoas major muscle, runs inferolaterally within it to emerge between the psoas major and the iliacus (Fig. 2-118), just superior to the inguinal ligament. It enters the thigh lateral to the femoral artery and the femoral sheath. In the abdomen *the femoral nerve supplies the*

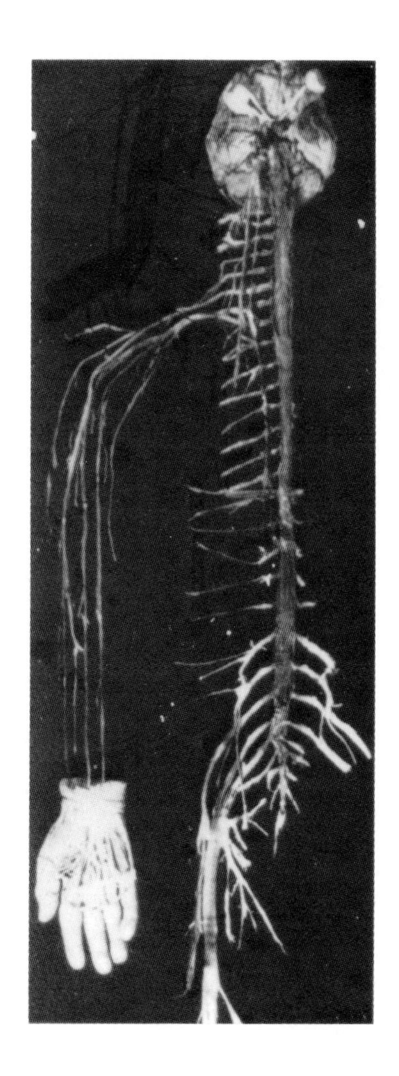

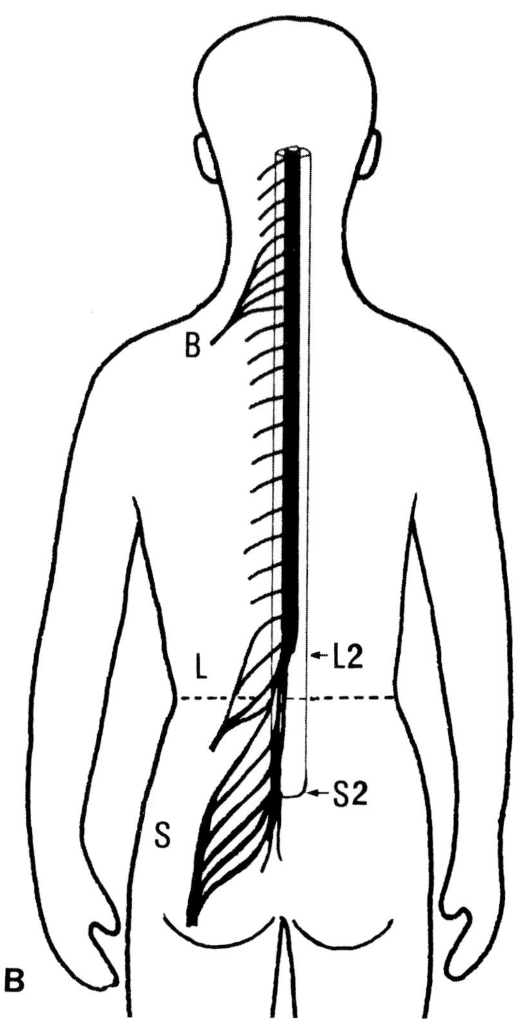

Figure 2-125. *A*, photograph of the brain, spinal cord, and the major nerves which were dissected and then removed from a cadaver. (Prepared by B. S. Jadon, Department of Anatomy, McMaster University, Hamilton, Ontario.) *B*, schematic outline of the spinal cord superimposed on a human figure. Observe the brachial (*B*), lumbar (*L*), and sacral (*S*) plexuses. The *dotted line* joining the superior parts of the iliac crests is a guide to the space between L3 and L4 vertebrae.

psoas and iliacus muscles. (See p. 435 for its distribution in the thigh).

The ilioinguinal and iliohypogastric nerves are both derived from L1, often by a common stem. *They enter the abdomen posterior to the medial arcuate ligament* of the diaphragm (Fig. 2-118), ***and pass inferolaterally, anterior to the quadratus lumborum muscle.*** Often the two nerves do not separate until they are under cover of the transversus abdominis muscle. They pierce this muscle near the anterior superior iliac spine and then pass through the internal and external oblique muscles to *supply the skin of the suprapubic and inguinal regions* (Fig. 2-16). Both nerves also supply branches to the abdominal musculature.

The iliohypogastric nerve (L1) sends a *lateral branch* to supply the *skin of the gluteal region* and an *anterior branch* to the *skin of the hypogastric region* (Figs. 2-6, 2-8, and 2-16).

The ilioinguinal nerve (L1) passes through the superficial inguinal ring and supplies the *skin of the groin and scrotum or labium majus* (Figs. 2-7 and 2-16*A*).

The genitofemoral nerve (L1 and L2) pierces the fascia iliaca and the anterior surface of the psoas major muscle (Fig. 2-118). It runs inferiorly in the muscle and divides lateral to the common and external iliac arteries into *two branches*, **femoral** and **genital** (Fig. 2-118).

The lateral femoral cutaneous nerve (L2 and L3) passes through the psoas major muscle, emerging superior to the **iliac crest** (Figs. 2-101 and 2-118). It runs inferolaterally on the iliacus muscle and enters the thigh posterior to or through the inguinal ligament, just medial to the anterior superior iliac spine (Fig. 2-13). It *supplies skin over the anterior and lateral parts of the thigh* (Fig. 4-14).

The Lumbosacral Trunk (L4 and L5). This large flat nerve trunk is *formed by the inferior part of the ventral ramus of the fourth lumbar nerve and the ventral primary ramus of L5* (Fig. 2-118). The L4 component descends through the psoas major muscle on the medial part of the transverse process of L5 vertebra, and then passes over the **ala of the sacrum**, to which it is closely applied (Fig. 2-118), to join the first sacral nerve.

The iliohypogastric nerve is in danger during an appendectomy. It may be injured during the **McBurney incision** (Figs. 2-16 and 2-43), as it passes between the external and internal oblique muscles (Fig. 2-11) in the anterior abdominal wall. If the iliohypogastric nerve is severed, the resulting weakness of the muscles in the region of the inguinal canal may result in the development of a **direct inguinal hernia** (Fig. 2-20*B*).

In persons in whom the lateral cutaneous nerve of the thigh passes through the inguinal ligament, the nerve may be compressed and irritated at this site, particularly in obese persons owing to the pressure of the bulging abdomen. This condition, called **meralgia paraesthetica,** is characterized by numbness and tingling on the lateral side of the inferior part of the thigh. These symptoms can usually be relieved by flexing the thigh.

THE AUTONOMIC NERVOUS SYSTEM IN THE POSTERIOR ABDOMINAL WALL

The autonomic nervous system in the posterior abdominal wall consists of sympathetic and parasympathetic portions. The efferent nerves of the viscera are part of the autonomic nervous system. These nerves emerge from the spinal cord and the brain stem as fibers of certain spinal nerves (Figs. 2-99 and 2-118).

The sympathetic and parasympathetic nerves are distributed to the abdominal viscera by a rich tangle of nerve plexuses and ganglia along the anterior surface of the abdominal aorta (Figs. 2-94 and 2-114).

The principal part of this system is the celiac plexus (solar plexus) and its ganglia located on each side of the celiac trunk, at the level of the superior part of the first lumbar vertebra (Figs. 2-99 and 2-126).

The Sympathetic Nerves (Figs. 2-114, 2-118, and Table 2-3). The **thoracic splanchnic nerves** are the

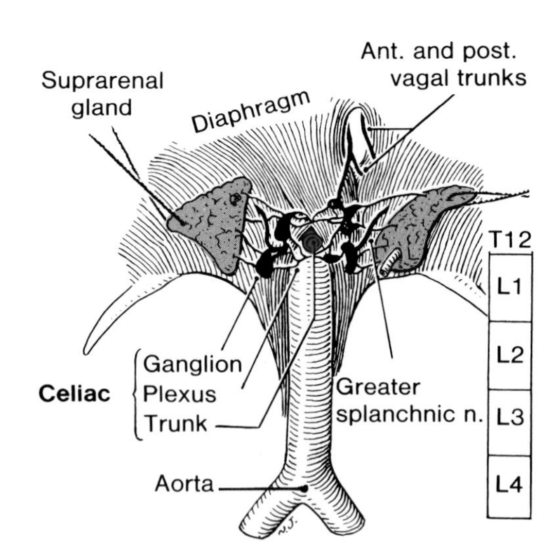

Figure 2-126. Drawing of the celiac plexus and its ganglia located around and on each side of the celiac trunk, a short artery that arises from the abdominal aorta at the level of the 12th thoracic vertebra (Figs. 2-127 and 2-128). Note that the celiac plexus and its ganglia are located at the level of the superior part of the first lumbar vertebra. (Also see Fig. 2-99).

main source of sympathetic nerves in the abdomen. *The greater, lesser, and lowest splanchnic nerves are branches of thoracic sympathetic ganglia 5 to 12.* The splanchnic nerves are **preganglionic fibers** which come from the spinal cord via ***white rami communicantes*** *and pass through the sympathetic ganglia without stopping.* They end in the celiac and aorticorenal ganglia, from which they are relayed as unmyelinated postganglionic fibers.

The greater splanchnic nerve (Figs. 2-118 and 2-121) is usually formed by branches that run anteroinferiorly on the bodies of the vertebra **from the fifth to the ninth sympathetic ganglia.** It runs inferiorly *just lateral to the* azygos *vein, pierces the crus of the diaphragm (Fig. 2-118), and ends in the celiac ganglion* (Figs. 2-118 and 2-121).

The lesser splanchnic nerve usually arises from the 9th and 10th sympathetic ganglia, and runs inferiorly, lateral to the greater splanchnic nerve (Fig. 2-121). *It pierces the crus of the diaphragm* and ends in the inferior part of the celiac ganglion, which is called the *aorticorenal ganglion* (Fig. 2-114).

The lowest splanchnic nerve is formed by branches from the 11th and/or 12th sympathetic ganglion. It pierces the crus of the diaphragm near or with the lesser splanchnic nerve and *ends in the renal plexus* (Figs. 2-99 and 2-114).

The Abdominal Sympathetic Trunks. *The paired sympathetic trunks receive their preganglionic fibers from white rami communicantes* arising in the spinal cord. Each trunk consists of a series of ganglia

(11 or 12 in the thorax and 5 in the abdomen), connected by nerve fibers, and lies on the anterolateral surfaces of the vertebrae.

The sympathetic trunks enter the abdomen by passing posterior to the medial arcuate ligament (Fig. 2-118), or through the crura of the diaphragm (Fig. 2-121). Each trunk descends anterior to the psoas major muscle or in a groove between it and the vertebral bodies.

The right sympathetic trunk lies posterior to the inferior vena cava, the lumbar lymph nodes, and *the right ureter* (Fig. 2-121). **This relationship is surgically important.** Both trunks pass anterior to the small lumbar vessels supplying the posterior abdominal wall (Fig. 2-118), and then *run posterior to the common iliac vessels* to enter the pelvis (Fig. 2-121). The two trunks unite in the median ***ganglion impar*** on the coccyx.

The medial branches passing from the lumbar sympathetic ganglia are called **lumbar splanchnic nerves** (Fig. 2-114). Usually their synapses are in the superior or inferior mesenteric ganglia. They pass to the ***intermesenteric or hypogastric plexuses*** (Fig. 2-114). Branches from these plexus pass along the arteries to the viscera (Figs. 3-79 and 3-82).

The Abdominal Autonomic Plexuses. These plexuses surround the abdominal aorta and its major branches. *They receive parasympathetic fibers from the vagus nerve* (Fig. 7-135) *and the sacral parasympathetic outflow. The sympathetic fibers are derived from the greater, lesser, and lowest splanchnic nerves and from the lumbar splanchnic nerves* (Fig. 2-114). The plexuses are named according to the arteries they surround or accompany (*e.g.*, celiac and aorticorenal).

Collections of nerve cells (sympathetic ganglia) are scattered amongst the celiac and intermesenteric plexuses (Figs. 2-99 and 2-114). *The parasympathetic ganglia are in the walls of the viscera*, e.g., the myenteric plexus (of Auerbach) in the muscular coat of the stomach and intestines.

The intermesenteric plexus (aortic plexus) consists of 4 to 12 nerves on the anterior and anterolateral aspects of the aorta (Fig. 2-114), *between the superior and inferior mesenteric arteries.* **The intermesenteric plexus receives contributions from the first two lumbar splanchnic nerves** and gives rise to renal, testicular (or ovarian), and ureteric branches. Occasionally, it gives off branches to the duodenum, pancreas, aorta, and inferior vena cava.

The superior hypogastric plexus (presacral plexus) is continuous with the intermesenteric plexus (Figs. 2-114 and 3-82). It lies anterior to the inferior part of the aorta, its bifurcation, and the median sacral vessels. It *receives the inferior two lumbar splanchnic nerves* ***and divides into right and left hypogastric nerves*** which pass to the inferior hypogastric plexus. The superior hypogastric plexus supplies ***ureteric and testicular plexuses*** and a plexus on each common iliac artery (Figs. 2-114 and 3-82).

The inferior hypogastric plexuses are *formed by the right and left hypogastric nerves* (Fig. 3-80). There are sympathetic ganglia within these plexuses which surround the corresponding internal iliac artery. *Each plexus receives small branches from the superior sacral sympathetic ganglia and the* ***sacral parasympathetic outflow*** *from S2 to S4* (pelvic splanchnic nerves). Extensions of the inferior hypogastric plexus send autonomic fibers along the blood vessels which form visceral plexuses on the walls of the pelvic viscera (*e.g.*, the rectal plexus and the vesical plexus).

Afferent Fibers in Sympathetic Nerves. Although the sympathetic nerves are motor, they also carry some sensory fibers from sense organs in the viscera. These fibers pass toward the spinal cord via the splanchnic nerves (Fig. 2-114) as far as the sympathetic trunk (Fig. 2-118). They then pass superiorly or inferiorly to reach the level of the spinal cord which is to convey the impulses conducted by them to the central nervous system. They then *leave the sympathetic trunk in a white ramus communicans* (Fig. 2-121) and enter a spinal nerve and then the spinal cord via its dorsal root. *The cell bodies of the visceral sensory fibers are in the spinal ganglia.*

Afferent Fibers in Parasympathetic Nerves. Although the parasympathetic nerves are visceral efferent (*i.e.*, motor to smooth muscle and glands), the viscera supplied by them contain sense organs and the afferent fibers from them pass back to the central nervous system in the parasympathetic nerves.

All the sensory fibers in parasympathetic nerves have their cell bodies in the sensory ganglion of the nerve supplying the viscus with parasympathetic fibers (*e.g.*, in the sensory ganglion of the vagus nerve, Fig. 7-135, and in the spinal ganglia of S2 to S4 nerves).

Pain arising from an abdominal viscus varies from dull to very severe and is poorly localized. It radiates to the part of the body served by somatic sensory fibers associated with the same segment of the spinal cord which receives visceral sensory fibers from the viscus concerned (Fig. 1-23). This is called **visceral referred pain.** Hence, a knowledge of the segmental origin of the sensory nerve fibers of each viscus is helpful in interpreting pain referred to the abdominal wall by a diseased organ.

Pain from the stomach (*e.g.*, owing to a gastric ulcer) is referred to the **epigastric region** because the stomach is supplied by pain afferents which reach the seventh and eighth thoracic segments of the spinal cord via the ***greater splanchnic nerve***. The pain is interpreted by the brain as if the irritation occurred in the area of skin supplied by the dorsal roots of the seventh to ninth thoracic nerves (Figs. 1-23 and 2-16*A*).

Pain from an inflamed vermiform appendix pass centrally in the lesser splanchnic nerve on the right side and *is referred* ***initially*** *to the umbilical region*, which lies in the T10 dermatome (Figs. 1-23 and 2-16*A*). *Pain is* ***later*** *referred to the lower right quadrant* when the parietal peritoneum in contact with the appendix becomes inflamed. Pain arising from the parietal peritoneum is of the somatic type and is usually severe. It can be precisely located to the site of origin. The anatomical basis for this is that *the parietal peritoneum is supplied by somatic sensory fibers through the thoracic nerves*, whereas the appendix is supplied by visceral sensory fibers in the lesser splanchnic nerve.

Inflamed parietal peritoneum is extremely sensitive to stretching. Hence, when pressure is applied to the anterior abdominal wall over the site of inflammation (*e.g.*, **McBurney's point**, Fig. 2-43) and suddenly removed, extreme local pain is usually felt. As the inflamed parietal peritoneum is stretched by pressure and then rebounds, pain is produced. This pain is called **rebound tenderness.**

The treatment of some patients with arterial disease in the lower limbs occasionally includes *the surgical removal of two or more lumbar sympathetic ganglia* with division of their rami communicantes. This operation is called a *lumbar sympathectomy.*

Surgical access to the sympathetic trunks is commonly through a lateral **extraperitoneal approach** because the sympathetic trunks lie retroperitoneally in the extraperitoneal fatty tissue. The muscles of the anterior abdominal wall are split and the peritoneum is swept medially and anteriorly to expose the medial edge of the psoas major muscle, along which the sympathetic trunk lies (Figs. 2-107 and 2-118). The left trunk is slightly overlapped by the aorta (Fig. 2-101) and sometimes by a **persisting left inferior vena cava** (Fig. 2-130). The right sympathetic trunk is covered by the inferior vena cava (Figs. 2-101, 2-107, and 2-121). Hence the surgeon has to retract these structures medially to expose the sympathetic trunks. They usually lie in the groove between the psoas major muscle laterally and the lumbar vertebral bodies medially, but they are often obscured by fat and lymphatic tissue (Fig. 2-121).

Identification of the sympathetic trunks in the abdomen is not easy. Great care must be taken not to remove inadvertently pieces of the *genitofemoral nerve* (Fig. 2-118), the lumbar lymphatics, or the *ureter* (Fig. 2-121).

The intimate relationship of the sympathetic trunks to the aorta and the inferior vena cava (Figs. 2-107 and 2-121) also make these large vessels vulnerable to injury during a lumbar sympathectomy.

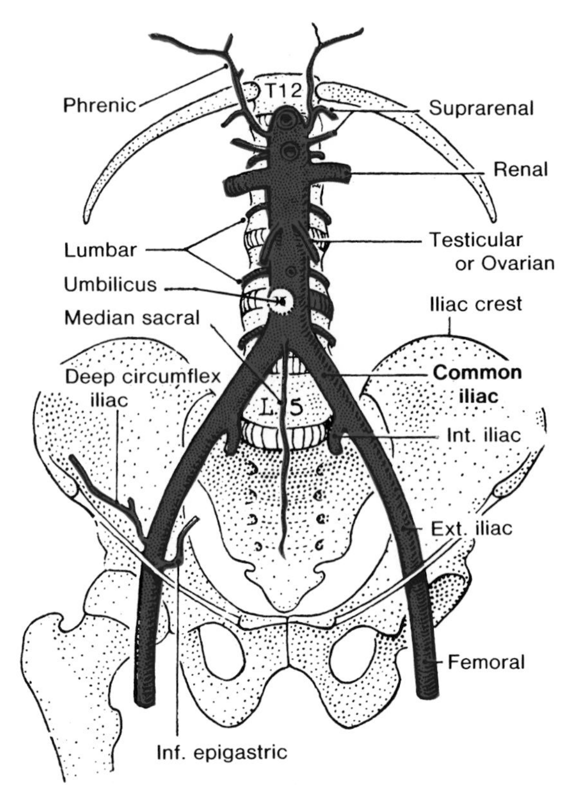

Figure 2-127. Drawing of the abdominal aorta showing its branches and relationship to the vertebral column. Note that it is only about 13 cm in length. Observe the stems of the celiac, superior mesenteric, and inferior mesenteric arteries. Note that the abdominal aorta begins in the median plane anterior to the inferior border of the body of *T12* vertebra and ends at the level of the body of *L4* vertebra by dividing into two common iliac arteries. Note that this division occurs slightly to the left and inferior to the umbilicus (*U*). It is very unusual for the median sacral artery to arise from the anterior surface of the aorta as here. (See Fig. 2-51 for its usual origin.)

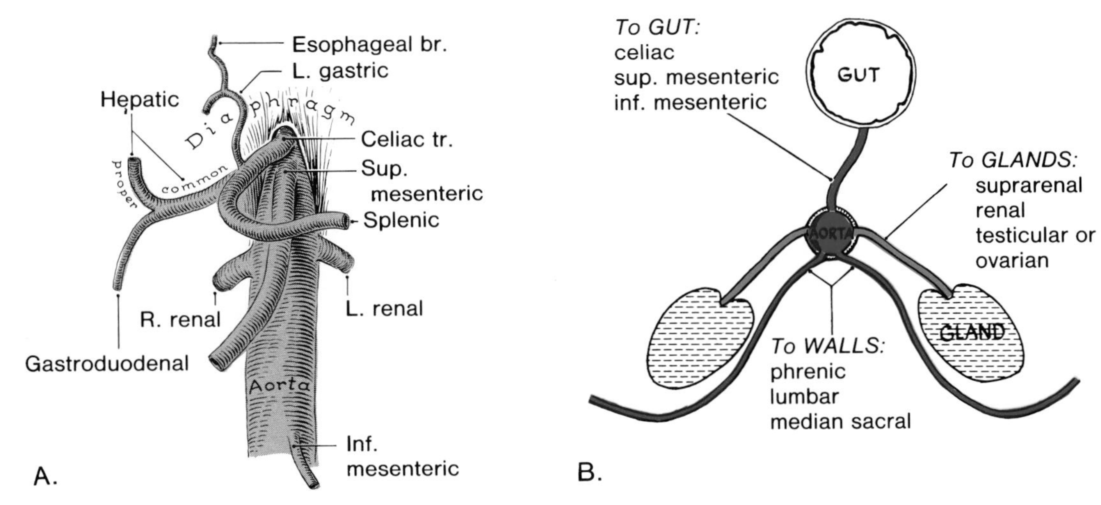

Figure 2-128. Drawing of the abdominal aorta illustrating its branches. *A*, note that the abdominal aorta begins where the diaphragm rests on the celiac trunk at the level of the intervertebral disc between T12 and L1 vertebrae. *B*, shows three of the four types of branches of the aorta: (1) unpaired visceral branches to the intestines; (2) paired visceral branches to the glands; and (3) paired parietal branches to the walls. The fourth type (unpaired parietal) is represented by the median sacral artery (Fig. 2-127).

VESSELS OF THE POSTERIOR ABDOMINAL WALL

The arteries of the posterior abdominal wall arise from the abdominal aorta (Fig. 2-127), except for the subcostal arteries. The veins are tributaries of the **inferior vena cava** (Figs. 2-116 and 2-129), except for the left testicular (or ovarian vein) which enters the renal vein.

The Subcostal Arteries (Figs. 1-78, 2-8, 2-101, and 2-118). These arteries are the *last branches of the descending thoracic aorta.* They were given their name because they are *situated inferior to the intercostal arteries, the 12th rib,* and the costal cartilages. Each artery runs laterally over the body of the 12th thoracic vertebra and posterior to the splanchnic nerves, the sympathetic trunk, the pleura, and the diaphragm.

Each subcostal artery enters the abdomen posterior to the lateral arcuate ligament of the diaphragm, with the subcostal nerve (T12). They then run anterior to the quadratus lumborum muscle and posterior to the kidney (Figs. 2-101 and 2-109), before piercing the aponeurosis of origin of the transversus abdominis muscle to run between that muscle and the internal oblique. The subcostal arteries anastomose anteriorly with the inferior epigastric and inferior intercostal arteries (Fig. 2-8), and posteriorly with the lumbar arteries (Fig. 2-51).

THE ABDOMINAL AORTA

The abdominal aorta is the direct continuation of the descending thoracic aorta (Fig. 2-129).

The abdominal aorta ***begins at the aortic hiatus*** *in the diaphragm at the level of the intervertebral disc between T12 and L1 vertebrae,* and **ends at about the level of L4 vertebra** *by dividing into the two common iliac arteries.*

The relations of the abdominal aorta are important, particularly to surgeons, *e.g.*, when excising part of the aorta (called an ***aortectomy***).

Anteriorly, the abdominal aorta is related to the **celiac trunk** and its branches (Fig. 2-128*A*), the **celiac plexus** (Fig. 2-99), the **omental bursa** (Fig. 2-45), the **pancreas**, the **left renal vein** (Fig. 2-62*B*), the ascending part of the **duodenum** (Fig. 2-64), the **root of the mesentery** (Figs. 2-62*A* and 2-87), and the **intermesenteric plexus** of nerves (Fig. 2-114).

Posteriorly, the abdominal aorta descends anterior to the bodies of **L1 to L4 vertebrae,** the ***intervertebral discs*** between them (Fig. 2-127), and the corresponding part of the anterior longitudinal ligament (Fig. 2-118).

On the right, the abdominal aorta is related superiorly to the **cisterna chyli** (Fig. 1-73), the **thoracic duct,** and the **right crus of the diaphragm** (Figs. 2-109, 2-117, and 2-128*A*). Inferiorly, it is related posteriorly to the ***inferior vena cava*** (Fig. 2-129).

On the left, the abdominal aorta is related superiorly to the **left crus of the diaphragm** (Fig. 2-109) and the **left celiac ganglion** (Fig. 2-99). The **duodenojejunal flexure** is on its left opposite the second lumbar vertebra (Figs. 2-39 and 2-64), and the **sympathetic trunk** course along its left side (Figs. 2-101 and 2-107).

The surface anatomy of the abdominal aorta may be represented by a broad band, about 2 cm wide, extending from a midline point about 2.5 cm superior to the **transpyloric plane** to a point slightly inferior

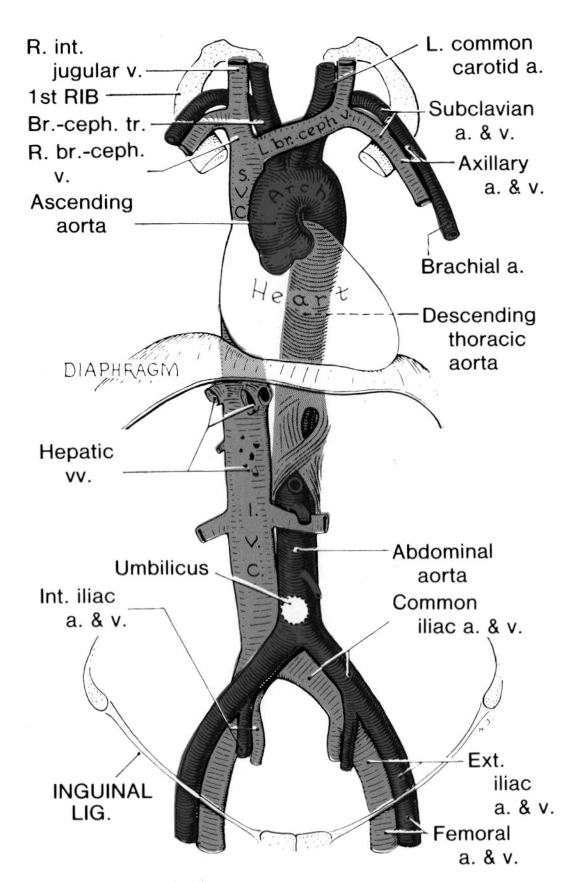

Figure 2-129. Diagram illustrating the great arteries and veins. The abdominal aorta, a continuation of the thoracic aorta, *begins* in the median plane at the aortic hiatus of the diaphragm (Fig. 2-117) and *ends* by dividing into two common iliac arteries. Observe that the inferior vena cava begins slightly inferior and to the right of the bifurcation of the aorta by the union of the two common iliac veins.

and to the left of the **umbilicus** (Fig. 2-127). This point indicates the level of the bifurcation of the aorta into the common iliac arteries. *The aortic bifurcation is just to the left of the midpoint of the line joining the highest points of the iliac crests.* This line is very helpful when examining obese persons in whom the umbilicus is usually not a reliable landmark.

When the anterior abdominal wall is relaxed, particularly in children and thin adults, *the most inferior part of the abdominal aorta may be readily compressed against the body of the fourth lumbar vertebra* by firm pressure on the anterior abdominal wall, just superior to the umbilicus. The pulsations of the abdominal aorta can be felt, however, too much pressure should not be applied or the pulsations will be obliterated.

The branches of the abdominal aorta may be grouped into four types: (1) *unpaired visceral branches*; (2) *paired visceral branches*; (3) *paired parietal branches*; and (4) an *unpaired parietal branch.*

The unpaired visceral branches arise from the anterior surface of the aorta. (Fig. 2-128) They are the *celiac trunk*, the *superior mesenteric artery*, and the *inferior mesenteric artery.* They arise at the following vertebral levels: the celiac trunk (**T12**); the superior mesenteric artery (**L1**); and the inferior mesenteric artery (**L3**).

The paired visceral branches arise from the lateral surfaces of the aorta at the following vertebral levels (Figs. 2-51 and 2-127): (1) **the middle suprarenal arteries** (**L1**), one or more on each side, arise close to the origin of the superior mesenteric artery (Fig. 2-99). They run laterally on the crura of the diaphragm to the suprarenal glands; (2) **the renal arteries** (**L1**) arise just inferior to the superior mesenteric artery (Figs. 2-51*B*, 2-99, and 2-128*A*). Occasionally there is an *accessory renal artery*, particularly on the left side (Fig. 2-112); (3) **the testicular or ovarian arteries** (**L2**) are long slender vessels that arise from the aorta, usually a short distance inferior to the renal arteries (Figs. 2-51 and 2-112). They pass inferiorly either anterior (Fig. 2-101) or posterior (Fig. 2-94) to the inferior vena cava on the right side across the psoas major muscle, but adherent to the parietal peritoneum (Fig. 2-101).

The testicular artery (Fig. 2-112) passes through the deep inguinal ring to *enter the inguinal canal* and become part of the **spermatic cord** (Fig. 2-23*B*).

The ovarian artery (Fig. 3-59) follows a similar course through the abdomen, but crosses the proximal ends of the external iliac vessels to enter the pelvis minor, where it runs in the *infundibulopelvic ligament* to reach (Fig. 3-62) and supply the ovary.

The paired parietal branches arise from the posterolateral surfaces of the aorta (Fig. 2-51). The **inferior phrenic arteries** (Figs. 2-99 and 2-101) arise just inferior to the diaphragm and pass superolaterally over the crura of the diaphragm. Each inferior phrenic artery gives rise to several **superior suprarenal arteries** (Fig. 2-99), and then spreads out on the inferior surface of the diaphragm.

The four pairs of lumbar arteries (Fig. 2-51) arise from the posterolateral surfaces of the abdominal aorta. Each pair passes around the sides of the superior four lumbar vertebrae (Fig. 2-118). *The lumbar arteries pass posteromedial to the sympathetic trunks and,* ***on the right,*** *run posterior to the inferior vena cava* (Fig. 2-121). The lumbar arteries divide between the transverse processes of the lumbar vertebrae into anterior and posterior branches.

Each anterior branch of a lumbar artery *passes deep to the quadratus lumborum muscle and then around the abdominal wall* between the internal oblique and transversus abdominis muscles, where it anastomoses in the posterior part of the rectus abdominus muscle with the inferior epigastric arteries (Fig.

2-8). The anterior branches supply the anterolateral walls of the inferior half of the abdomen.

Each posterior branch of a lumbar artery *passes posteriorly, lateral to the articular processes, and supplies the spinal cord, the cauda equina, the spinal meninges, the erector spinae muscles, and the overlying skin* (Figs. 5-56 and 5-57). **The spinal arteries** arise **from the posterior branches of the lumbar arteries** and pass through the intervertebral foramina to supply the vertebrae. Important branches of the spinal arteries, called **radicular arteries**, join either the anterior or posterior spinal arteries (Fig. 5-56). The largest of the radicular arteries, called the great radicular artery or **arteria radicularis magna** (spinal artery of Adamkiewicz), joins the inferior half of the anterior spinal artery and *supplies most of the blood to the inferior part of the spinal cord*, including the lumbosacral enlargement (for details see p. 618).

The unpaired parietal artery **is the median sacral artery** (Figs. 2-51 and 2-127). This tiny vessel, originally the part of the dorsal aorta in the sacral region of the embryo, typically arises from the posterior surface of the aorta just proximal to its bifurcation (Fig. 2-51). It descends in the midline anterior to L4 and L5 vertebrae, usually giving off a small lumbar artery on each side called the **fifth lumbar artery** or arteria lumbalis ima (L. *ima*, lowest). Their distribution is similar to that of the lumbar arteries.

The abdominal aorta and its branches can be studied radiologically by injecting radiopaque contrast medium into them. When the distal part of the abdominal aorta and/or the external iliac artery are not occluded, a catheter is passed into a femoral artery just inferior to the inguinal ligament using the *Seldinger technique.*

A percutaneous puncture of the femoral artery is made with a size 16 needle. A spring wire guide is passed through the needle and up the external iliac artery for several cm. The needle is then slipped off, leaving the guide in the artery. Next, a size 16 plastic catheter with a curved tip is threaded on the guide up to the artery wall. The catheter and guide are advanced into the lumen of the artery. The guide is then withdrawn, leaving the catheter in the artery.

Seldinger was a medical student when he devised this method of angiography. Under fluoroscopic control the catheter is then passed to the desired level of the aorta for injection of the contrast medium, or it can be manipulated until it is in an aortic branch for **selective angiography** of one branch only.

When there is **atherosclerosis** (narrowing of the aorta/or the iliac arteries owing to lipid deposits in their walls), the Seldinger technique may not be feasible. In *some* of these cases, the abdominal aorta is injected with contrast material by *direct needle puncture* (**translumbar aortography**). The level of puncture is from L1 to L3 vertebra, but it is sometimes performed at T12/L1 level because of the constancy of the position of the aorta at this level in the **aortic hiatus** of the diaphragm. The needle puncture is commonly made 1 cm below the 12th rib, 6 to 8 cm to the left of the midline.

Atheromatous occlusion (blockage owing to lipid deposits in the intima of arteries) of one of the common iliac arteries, just distal to the bifurcation of the aorta, is common. Patients with this **arteriosclerotic occlusive disease** complain of calf, thigh, or hip pain on exertion (**claudication**). *This pain disappears when they rest.* Sometimes this condition is treated by inserting a **prosthetic graft** into the artery which bypasses the obstructed area.

Aneurysm of the abdominal aorta (localized dilatation) distal to the renal arteries (Fig. 5-65) may also be resected and replaced by a prosthetic graft.

During certain surgical procedures in the abdomen, it may be necessary to ligate, or to retract for several minutes, one or more lumbar arteries. When doing this, it must be kept in mind that the **arteria radicularis magna** supplying the inferior two-thirds of the spinal cord may arise from one of the lumbar arteries. The origin of this **large artery** varies, but it *commonly arises from one of the inferior intercostal or superior lumbar arteries.* In most cases it arises on the left side and joins or forms the inferior half of the anterior spinal artery (Fig. 5-56). It usually gives off a small branch that anastomoses with the posterior spinal artery.

Prolonged pressure or ligation of the lumbar artery giving rise to the arteria radicularis magna leads to ***circulatory impairment*** *of the inferior part of the spinal cord, which may result in an area of necrosis* (**infarction**). This could result in paralysis of the lower limbs (**paraplegia**), and loss of all sensation inferior to the infarcted area.

THE INFERIOR VENA CAVA

The inferior vena cava is the largest vein in the body. It returns blood from the lower limbs, most of the abdominal walls, and the abdominopelvic viscera. Blood from the viscera passes through the **portal system** and the liver (Fig. 2-52) before entering the inferior vena cava via the *hepatic veins* (Fig. 2-74).

The inferior vena cava *begins anterior to the fifth lumbar vertebra* by the union of the common iliac veins, inferior to the bifurcation of the aorta and the proximal part of the right common iliac artery (Figs. 2-116 and 2-129).

The inferior vena cava ascends to the right of the median plane, pierces the central tendon of the diaphragm (**vena caval foramen**) at the level of the eighth thoracic vertebra (Figs. 2-117 and 2-119), and enters the right atrium of the heart (Fig. 2-129).

RELATIONS OF INFERIOR VENA CAVA

Posteriorly, inferior vena cava lies on the bodies of ***L3 to L5 vertebrae*** (Fig. 2-64), the ***right psoas major muscle*** (Fig. 2-101), the ***right sympathetic trunk*** (Figs. 2-107 and 2-121), the ***right renal artery***, the ***right suprarenal gland***, the ***right celiac ganglion*** (Fig. 2-101), and the ***right crus of the diaphragm*** as it passes to the vena caval foramen in the central tendon of the diaphragm.

Anteriorly, the relations of the inferior vena cava are the ***peritoneum*** (Fig. 2-62*B*), the ***superior mesenteric vessels*** in the root of the mesentery (Fig. 2-64), and the horizontal ***part of the duodenum*** and the head of the **pancreas,** with the portal vein and the bile duct intervening (Fig. 2-58).

Superior to the first part of the duodenum, ***the inferior vena cava lies in the posterior boundary of the epiploic foramen*** (Figs. 2-81 and 2-82). It then enters a groove on the inferior surface of the liver, between the right and caudate lobes (Figs. 2-71 to 2-75).

To the left of the inferior vena cava is the aorta (Fig. 2-129); *to its right* are the right **ureter** and **kidney**, and the descending part of the **duodenum** (Figs. 2-107 and 2-121).

Tributaries of the Inferior Vena Cava are: (1) the *common iliac veins*; (2) the third and fourth *lumbar veins*; (3) the right *testicular* or *ovarian vein;* (4) the *renal veins*; (5) the *azygos vein*; (6) the right *suprarenal vein*; (7) the *inferior phrenic veins*; and (8) the *hepatic veins* (Figs. 2-116 and 2-129).

The right and left common iliac veins are formed by the *union of the external and internal iliac veins* (Figs. 2-116 and 2-129). **The right testicular or ovarian vein and the right suprarenal vein** usually drain into the inferior vena cava, whereas *these veins on the left side usually drain into the left renal vein* (Fig. 2-116).

The renal veins drain into the inferior vena cava at the *level of L2 vertebra.* They lie anterior to the corresponding renal artery (Fig. 2-101). The right renal vein receives few if any tributaries other than those from the kidney, whereas *the left renal vein also drains blood from the left suprarenal gland and the testis or ovary.*

The azygos and hemiazygos veins (Figs. 1-42, 1-80, and 2-116) usually commence in the abdomen. The *azygos vein commonly arises from the inferior vena cava at the level of the renal vein,* but it may begin as the continuation of the right subcostal vein or from the junction of that vein and the right ascending lumbar vein.

The azygos vein enters the thorax through the ***aortic hiatus*** or the right crus of the diaphragm. *The inferior hemiazygos vein arises from the posterior surface of the left renal vein,* but it may also arise from the union of the left subcostal and the left ascending lumbar vein.

The right suprarenal vein (Fig. 2-116) is short and drains into the posterior aspect of the inferior vena cava, whereas the **left suprarenal vein** is long and *usually drains into the left renal vein,* but it may also drain into the inferior vena cava.

The inferior phrenic veins drain blood from the abdominal surface of the diaphragm. The *right inferior phrenic vein* generally empties into the inferior vena cava, whereas the *left inferior phrenic vein usually joins the left suprarenal vein.*

The hepatic veins (Figs. 2-52, 2-116, and 2-129) are short trunks which open into the inferior vena cava, just as it passes through the diaphragm. The right hepatic vein sometimes passes through the vena caval foramen before entering the inferior vena cava.

The lumbar veins consist of four or five segmental pairs (Fig. 2-116). Their dorsal branches drain the back and communicate with the **vertebral venous plexuses** (Fig. 5-58). The mode of termination of the lumbar veins varies. They may drain separately into the inferior vena cava or the common iliac vein (Fig. 2-116), but they are generally united on each side by a vertical connecting vein, the **ascending lumbar vein**, which lies posterior to the psoas major muscle. Each ascending lumbar vein passes posterior to the medial arcuate ligament of the diaphragm to enter the thorax. *The right ascending lumbar vein joins the right subcostal vein to form the azygos vein* (Fig. 1-80), whereas the left ascending lumbar vein unites with the left subcostal vein to form the **hemiazygos vein.**

There are three collateral routes available for venous blood to pass to the right side of the heart if the inferior vena cava is obstructed, or if ligation of the inferior vena cava is necessary. In such cases an extensive collateral venous circulation is soon established owing to enlargement of superficial and/or deep veins.

The first route bypassing the inferior vena cava is via various anastomoses in the pelvis and the abdomen which enable blood to reach the ***superficial and inferior epigastric veins*** (Fig. 2-7), and to ascend in them to the thoracoepigastric and superior epigastric veins and the superior vena cava.

The second route bypassing the inferior vena cava is via tributaries of the inferior vena cava that anastomose with the **vertebral system of veins** (Fig. 5-58). These veins, passing *within the vertebral canal and the vertebral bodies,* can also provide a route for metastasis of cancer cells to the vertebral bodies, or to the brain from an abdominal or pelvic tumor.

The third route bypassing the inferior vena cava is via the *lateral thoracic vein* which connects the circumflex iliac veins with the axillary vein (Fig. 2-7).

Sometimes the inferior vena cava is ligated or plicated (folded to reduce its size) in order to prevent **pulmonary emboli** following thrombosis of the veins in the pelvis or lower limbs from reaching the lungs and producing **pulmonary infarcts.** For details see page 93.

The inferior vena cava and the common iliac veins are vulnerable to injury during **repair of a herniated nucleus pulposis** of an intervertebral disc (Fig. 5-68). As these vessels lie anterior to the fifth intervertebral disc (Fig. 2-127), they could be injured by a **rongeur** that is unintentionally pushed through the L4/L5 intervertebral disc during removal of its herniated nucleus pulposis. A rongeur is a strong biting forceps used for gouging away bone and for removing herniated intervertebral discs.

LYMPHATIC DRAINAGE OF THE POSTERIOR ABDOMINAL WALL

The lymph nodes of the posterior abdominal wall lie along the iliac vessels, the aorta, and the inferior vena cava (Figs. 2-97, 2-112 and 2-121).

The external iliac lymph nodes are scattered along the external iliac vessels. Inferiorly, *the medial lymph nodes* receive lymph from the lower limbs and the pelvic viscera, whereas *the lateral lymph nodes* receive lymph from the areas supplied by the inferior epigastric and deep circumflex iliac vessels.

The common iliac lymph nodes are scattered along the common iliac vessels. *They receive lymph from the external and internal iliac lymph nodes.* The medial group of common iliac lymph nodes also drains lymph directly from the pelvis. Lymph from the common iliac lymph nodes passes to the lumbar lymph nodes.

The lumbar lymph nodes (Fig. 2-121) lie along the abdominal aorta and the inferior vena cava. They receive lymph directly from the posterior abdominal wall, the kidneys and ureters, the testes or ovaries, the uterus, and the uterine tubes. They also receive lymph from the descending colon, the pelvis, and the lower limbs through the inferior mesenteric and common iliac lymph nodes (Fig. 2-97). *Efferent lymph vessels from these large lymph nodes form the right and left lumbar lymph trunks.*

The Cisterna Chyli (Figs. 1-73 and 2-121). The sac-like cisterna chyli, about 5 cm long and 6 mm wide, is *located between the origin of the abdominal aorta and the azygos vein.* It lies on the right sides of the bodies of the first two lumbar vertebrae and is *usually located posterior to the right crus of the diaphragm.* **The thoracic duct** (Figs. 1-42 and 1-73) begins in the cisterna chyli, ascends through the **aortic hiatus** in the diaphragm into the thorax to open near or at the angle of union of the internal jugular and subclavian veins.

The cisterna chyli receives lymph from the right and left lumbar lymph trunks, the intestinal trunk, and a pair of vessels that descend from the inferior intercostal lymph nodes. Lymph from the digestive tract first passes to the lymph nodes close to the viscera (Figs. 2-53, 2-65 and 2-97). It then passes along lymph vessels that follow the major blood vessels to the **mesenteric lymph nodes** and from them into the preaortic lymph nodes (Fig. 2-26).

PRESENTATION OF PATIENT ORIENTED PROBLEMS

Case 2-1

A 32-year-old accountant complained to his doctor about a steady, gnawing, ***burning pain*** *of about 2 weeks duration in the "pit of this stomach"* (**epigastric region**). On careful questioning it was revealed that the pain usually began about 2 hr after the patient eats and then disappears when he eats again or drinks a glass of milk. He said that he did not become overly concerned about this pain until it started to awake him in the middle of the night.

Except for *mild tenderness in the right upper quadrant,* just lateral to his xiphoid process, the physical examination was normal. Suspecting a **peptic ulcer** (Fig. 2-131), the doctor ordered plain radiographs of the patient's abdomen and upper GI studies (gastrointestinal series of x-ray examinations with **fluoroscopy** and films during and after the swallowing of a barium sulphate emulsion, often called a **barium meal**).

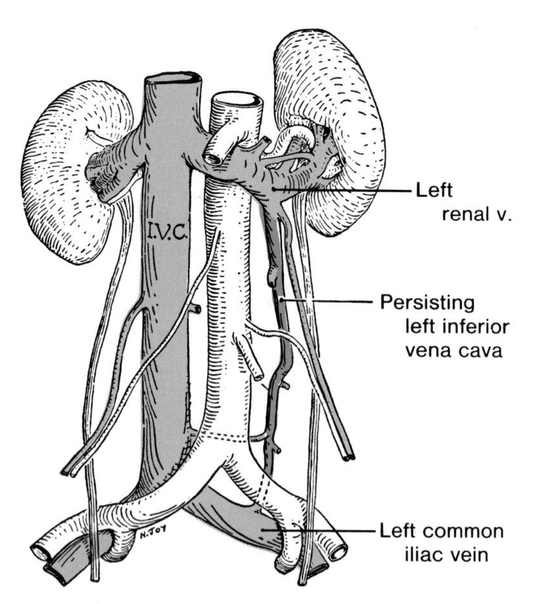

Figure 2-130. Drawing of a dissection of a persisting left inferior vena cava which joins the left common iliac vein to the left renal vein. This abnormal vein may be small, as in this specimen, or it may be large. A persistent left inferior vena cava results from persistence of the left embryonic supracardinal vein which normally disappears. The possible presence of this vessel is of surgical significance during left lumbar sympathectomy because it may conceal the sympathetic trunk.

The plain radiographs were normal, but the GI studies showed the presence of an **active ulcer** (lesion of the mucosa caused by loss of tissue) in a moderately deformed duodenal cap (Fig. 2-34). *A diagnosis of active **duodenal ulcer** was made.*

The patient responded fairly well to medical treatment (*e.g.*, antacids, frequent bland feedings, and abstinence from smoking and drinking alcohol). Although the patient curtailed his responsibilities for 2 months, he again began to work long hours, smoke heavily, and consume excessive amounts of coffee and alcohol. His symptoms became worse again and vomiting sometimes occurred when the pain was severe.

One evening the patient developed a sudden upper abdominal pain, vomited, and fainted. He was rushed to the emergency room of the hospital.

Examination revealed extreme pain and ***rigidity** of the abdomen and **rebound tenderness*** (sharp pain elicited when the depressed palpating hand is quickly removed). On questioning, the patient revealed that his ulcer had been "acting up" lately and that he had noticed blood in his vomitus (**hematemesis**).

Emergency surgery was performed and a perforated duodenal ulcer was found. There was a generalized **chemical peritonitis** resulting from the escape of bile and the contents of the GI tract into the peritoneal cavity.

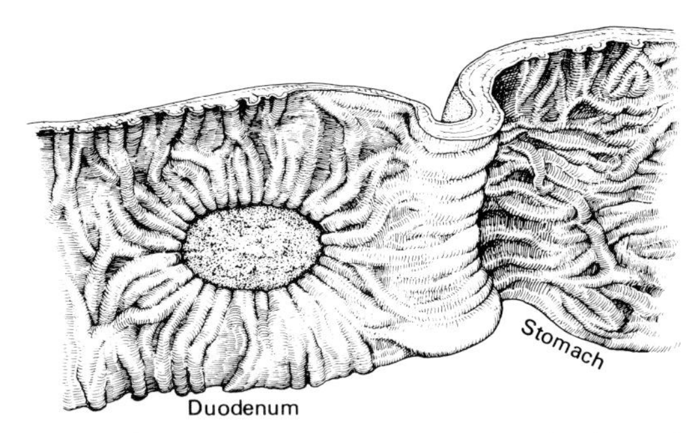

Figure 2-131. Drawing illustrating a peptic ulcer of the duodenum. About 80% of peptic ulcers occur in the duodenum, usually within 3 cm of the pylorus. Peptic ulcers also occur in the stomach.

Problems. Define **peptic ulcer**, with special emphasis on duodenal ulcer. What structures, closely related to the superior part of the duodenum, might be eroded by a ***perforated duodeal ulcer***? Name the congenital malformation of the ileum in which peptic ulcers commonly develop. Explain the anatomical basis for abdominal pain in the right upper and lower quadrants. What nerves might be cut during surgical procedures to reduce the secretion of acid by the parietal cells of the stomach. *These problems are discussed on page* 291.

Case 2-2

On the way home from work, a 42-year-old office worker suddenly experienced a sharp, **lancinating pain in his right loin.** The pain was so excruciating that he doubled up and moaned in agony. A fellow worker helped him to the emergency room of the hospital.

When the doctor asked him to describe the onset of pain, the patient said that he first felt a slight pain, and then it gradually increased (**exacerbation of pain**) until it was so severe that it brought tears to his eyes. He said that this unbearable pain lasted for several minutes and then suddenly eased. He explained that the pain comes and goes, but it seemed to be moving toward his groin.

During the physical examination, the doctor noted that the patient's abdomen moved with his respirations, and that there was some **tenderness and guarding** (abdominal muscle spasm) in the right lower quadrant, but there was no rigidity. While palpating the tender area as deeply as possible, the doctor suddenly removed his hand. Instead of wincing, the patient seemed relieved that the probing had stopped (**absence of rebound tenderness**). By this time the patient reported that he felt the pain in his right groin and testis and along the medial side of his thigh. The doctor noted that the right testis was unusually tender and was retracted.

When asked to produce a urine sample, the patient stated that it was difficult and painful for him to urinate (**dysuria**). The nurse reported that his urine sample contained blood (**hematuria**). Although the doctor was quite certain that the diagnosis was **ureteral colic**, he ordered plain radiographs of the abdomen. It was reported that a small calcified object, compatible with a **ureteral calculus** was visible in the region of the inferior end of the right ureter.

Problems. What probably caused the patient's attack of excruciating pain? At what other sites would a ureteral calculus apt to become lodged? Explain the intermittent exacerbation of pain and the course taken by the pain from his loin to his groin. Briefly discuss **referred pain from the ureter.** Explain the anatomical basis of the absence of rebound tenderness in the abdomen in cases of ureteric colic. *These problems are discussed on page* 292.

Case 2-3

A 14-year-old boy suffered **pain in his right groin** while attempting to lift a heavy weight. As soon as he noticed a lump in the region where he felt pain, he decided to lie down. The bulge soon disappeared and he went home. On the way he blew his nose very hard and again experienced pain and felt the swelling in his right groin. Fearing that he may have developed a "rupture," his father called the family doctor.

During the physical examination, the doctor inserted the tip of his little finger through the ***superficial inguinal ring*** and along the patient's inguinal canal toward the ***deep inguinal ring.*** Nothing was felt until he asked the patient to cough; he then felt an impulse with the tip of his finger.

When the patient was in a horizontal position the bulge disappeared, but on straining, a plum-sized bulge appeared in the right inguinal region. It was palpable inferior to the superficial inguinal ring and medial to the pubic tubercle. A diagnosis of complete *indirect inguinal hernia* was made.

Problems. Define the condition called indirect inguinal hernia. *Explain the embryological basis of this kind of hernia.* What layers of the spermatic cord cover the hernial sac? What structures are endangered during an operation for repair of an indirect inguinal hernia? *These problems are discussed on page* 292.

Case 2-4

A 22-year-old married female medical student woke up one morning not feeling as well as usual; she was **anorexic** and had crampy abdominal pains. As this coincided with the time of her menses (L. *mensis*, month), she thought at first that these cramps were the beginning of her usual painful menstruation (**dysmenorrhea**). Because she had missed her last menstrual period, she also thought that she might be having early symptoms of a ruptured **ectopic pregnancy.** She had a slight fever and felt dizzy, and uncomfortable owing to the cramps. Consequently, she decided to stay in bed.

The pain soon localized around her umbilicus. By evening the site of pain shifted to the right lower quadrant of her abdomen and she suspected **acute appendicitis.** As she was in considerable pain, her husband decided to take her to the hospital.

On examination it was found that there was a slight elevation of her temperature, an increased pulse rate, and an abnormally high white blood cell count (**leukocytosis**).

When asked to indicate where the pain began, she circled her umbilical area. When asked where she now felt pain, she put her finger on **McBurney's point** (Fig. 2-43).

During gentle palpation of her abdomen, the doctor detected ***localized rigidity*** (muscle spasm) and tenderness in her right lower quadrant. When the doctor suddenly removed her palpating hand from the area of McBurney's point, the patient winced in pain (**rebound tenderness**). On rectal examination, there was slight right-sided tenderness in her ***rectouterine pouch.*** As all these findings were highly suggestive of acute appendicitis, an immediate operation for excision of her vermiform appendix was performed (*appendectomy*).

Problems. What type of incision would the surgeon most likely have made to expose the appendix? Why is this a good incision anatomically? *How would you locate McBurney's point.* What part of the appendix is usually deep to this point? *Based on your knowledge of dissection, how do you think the appendix would be exposed?* Where is the appendix most likely to be located? What position of an inflamed appendix might give rise to pelvic or rectal pain? *Discuss **referred pain** from the appendix.* If this patient had had her appendix removed previously, what other appendage of the intestine could become inflamed and produce signs and symptoms similar to appendicitis? *These problems are discussed on page* 293.

Case 2-5

A 58-year-old obese man with a history of heartburn, indigestion (**dyspepsia**), and belching after heavy meals complained to his doctor about recent ***epigastric and retrosternal pain.*** He stated that the pain behind his breastbone (posterior to his sternum) developed recently, and that it was most severe after dinner when he is lying or stooping down. Fearing these might be heart pains (**angina pectoris**), his wife insisted that he consult a doctor.

When asked if he had noticed any other abnormalities, he stated that he often brought up (regurgitates) small amounts of sour or bitter-tasting substances (*gastroesophageal reflux*), particularly when he stooped to tie his shoes. He also reported that he recently had been having bouts of hiccups (hiccoughs) and that he occasionally had difficulty swallowing (***dysphagia***). He said that these symptoms were similar to those of his brother who had a peptic ulcer.

An ***ECG*** (electrocardiogram) showed no evidence of heart disease. Plain radiographs of the abdomen were negative, but **fluoroscopic examination** with the patient standing erect behind a fluoroscope showed a round space filled with gas and fluid in the inferior part of the patient's posterior mediastinum. On swallowing a barium sulphate emulsion, the barium was seen to enter this space, which he now identified as *the gastroesophageal region of the stomach*. There was no radiological evidence of gastric or duodenal ulcers. A diagnosis of *sliding hiatal hernia* was made.

Problems. Define diaphragmatic hernia. Discuss the embryological basis of congenital diaphragmatic hernia. Is a **hiatal hernia** usually present at birth? What caused the patient's epigastric and retrosternal pain and hiccups? Based on your anatomical knowledge, what structures do you think would be endangered in the surgical repair of a hiatal hernia? *These problems are discussed on page* 294.

Case 2-6

A fair, fat, flatulent, 40-year-old woman with five children was rushed to hospital with ***severe colicky pain*** in the right upper quadrant of her abdomen. When asked where she first felt pain, she pointed to the superior middle part of her abdomen (***epigastrium***). When asked where the pain went (was referred), she ran her fingers under her right ribs (***hypochondrium***) and around her right side to her back, stating that the pain was felt near the lower end of the her shoulder blade (inferior angle of her scapula).

On questioning, she said the sharp midline pain followed a heavy meal containing several fatty foods, after which she felt nauseated and vomited. Gradually there was an increase (**exacerbation**) in pain; when it became excruciating, her husband rushed her to the hospital.

During gentle palpation of her abdomen, the doctor noted ***rigidity and tenderness*** in the right upper quadrant, especially during inspiration. The radiologist reported that plain radiographs of her abdomen showed that there was *probably a small calculus in her cystic duct*.

In view of this observation and her symptoms, a diagnosis of **biliary colic** (intense pain from impaction of a gallstone in the cystic duct) was made. A **cholecystectomy** (removal of gallbladder) was performed and a sound (metal probe) was passed down her bile duct and up her hepatic ducts to remove any stones that may have been lodged in them.

Problems. Why was the radiologist not able to state clearly that the stone was in the cystic duct? What is a gallstone? Explain the anatomical basis of the patient's pain in: (1) the epigastric region, (2) the right hypochondrium, and (3) the infrascapular region. Does peritoneum separate the gallbladder from the liver? What structures are endangered during cholecystectomy? *These problems are discussed on page* 295.

Case 2–7

A 54-year-old mechanic was admitted to hospital because of severe epigastric pain and vomiting of blood (***hematemesis***). It was obvious that he had been drinking heavily.

On examination, it was noted that *the blood in his vomitus was bright red*. On questioning, it was learned that the patient had exhibited upper gastrointestinal bleeding on previous occasions, but never so profusely. *His blood pressure was low and his pulse rate was high.*

The patient's skin and conjunctivae were slightly yellow (**jaundiced**). His eyes appeared to be slightly sunken. Spider nevi or ***angiomas*** (branching arterioles) were present on his cheeks, neck, shoulders, and arms. His abdomen was large and there was protrusion and considerable downward displacement of his umbilicus (Fig. 2-132). Palpation of the patient's abdomen revealed some enlargement of the liver (***hepatomegaly***) and the spleen (***splenomegaly***).

Several bluish, dilated varicose veins radiated from his umbilicus, forming a **caput medusae** (Fig. 2-132). During a proctoscopic examination (inspection of the rectum and anal canal with a proctoscope), **internal hemorrhoids** were observed. On questioning, the patient said that he sometimes saw *blood in his bowel movements* (feces, stools). At other times his stools, he said were black and shiny. A diagnosis of alcoholic *cirrhosis of the liver* was made.

Problems. Briefly define hepatic cirrhosis. *Discuss anatomically* the basis of the patient's hematemesis, hemorrhoids, bloody stools, and caput medusae. What is the likely cause of the ***ascites*** and **splenomegaly**? Thinking anatomically, how would you suggest that blood pressure in the portal system could be reduced? *These problems are discussed on page* 296.

Case 2–8

The presenting complaint of a 54-year-old man was *an oval swelling in his left groin* (Fig. 2-133). He stated that this painless swelling enlarged when he coughed and disappeared when he lay down.

During examination of the patient in the standing position, the doctor put his little finger through the patient's left superficial inguinal ring. He got the sensation that his finger was going directly back into the abdomen, rather than along the inguinal canal.

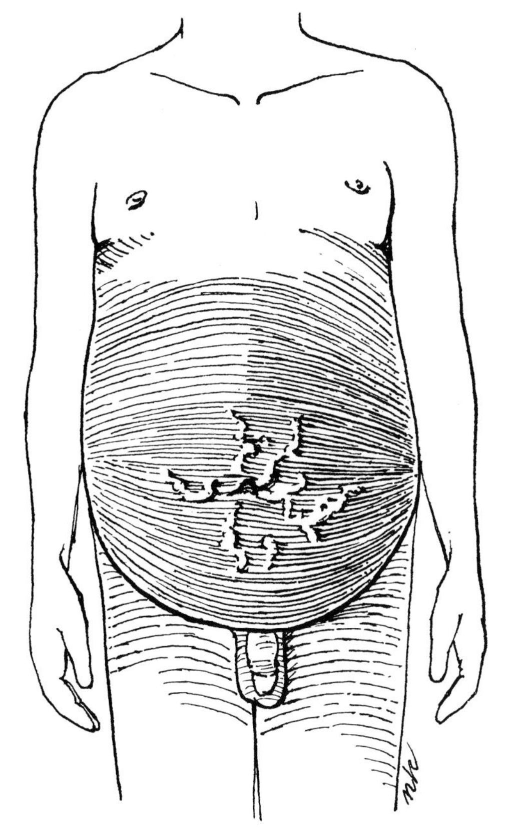

Figure 2-132. Drawing of patient's enlarged abdomen owing to ascites. Note the snake-like varicose veins radiating from his umbilicus (caput medusae).

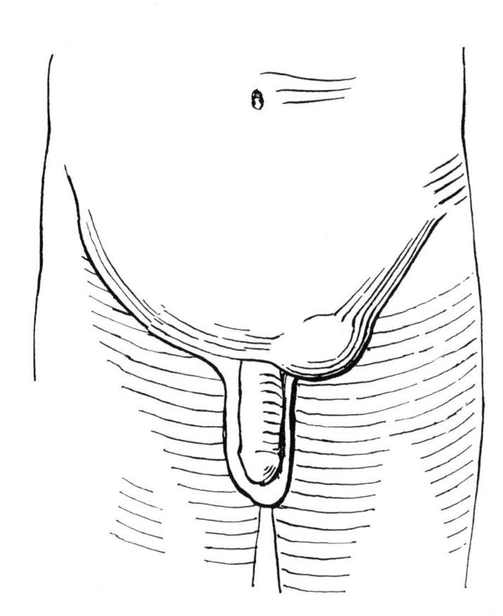

Figure 2-133. Illustration of a direct inguinal hernia protruding through the inguinal triangle in the anterior abdominal wall. It does not pass through the deep inguinal ring, but passes through the posterior wall of the inguinal canal either lateral to the conjoint tendon or through it.

When the patient coughed, the doctor felt a mass strike the pad of his finger, which was against the posterior wall of the inguinal canal. When the patient was asked to lie down, the mass reduced itself immediately. The doctor then placed his fingers over the **inguinal triangle** (Fig. 2-20*A*) and instructed the patient to hold his nose and blow it. The doctor felt a mass protruding from the inferior portion of this triangle. A diagnosis of *direct inguinal hernia* was made.

Problems. Explain what is meant by the term direct inguinal hernia. *How does it differ from an indirect inguinal hernia*? Does a direct inguinal hernia have an embroylogical basis? **What is the relationship of a direct inguinal hernia to the inferior epigastric artery**? Is this relationship different from that of an indirect inguinal hernia? Inadvertent injury to which nerves of the abdominal wall during surgery may predispose to the development of inguinal hernia? *These problems are discussed on page* 296.

DISCUSSION OF PATIENT ORIENTED PROBLEMS

Case 2-1

*A **peptic ulcer** is an ulceration of the mucous membrane of the stomach or duodenum* and occurs only in tissues in contact with gastric juice. Ulcers are common in the stomach (***gastric ulcers***) and duodenum. But about 80% of peptic ulcers occur in the duodenum (***duodenal ulcers***). Ulcers are usually found within 3 cm of the pylorus and occur more commonly in males. The depth of the ulcer varies from a shallow erosion of the mucosa to complete penetration of the wall of the duodenum (**perforated duodenal ulcer**). Both gastric and duodenal ulcers tend to bleed. Sometimes organs and vessels adjacent to the duodenum, usually the pancreas, become adherent to an ulcer and are eroded; *e.g.*, a posterior penetrating ulcer may erode the gastroduodenal artery or one of its branches. This causes sudden massive hemorrhage, which may be fatal.

*Peptic ulcers occur in a **Meckel's diverticulum***, a remnant of the yolk stalk attached to the ileum, which is present in about 2% of persons (Fig. 2-93).

Gastric tissue may be present in the wall of a Meckel's diverticulum, which may secrete acid that causes ulcer formation. Hemorrhage is a common complication of a Meckel's diverticulum, particularly in males 10 years and under. A Meckel's diverticulum is not often visualized in a GI series, but those that are bleeding and contain gastric tissue can be detected by radioisotope studies.

Pain is the distressing symptom of peptic ulceration, varying from a slight discomfort to a boring, gnawing ache. Some authorities attribute the pain to the contact of hydrochloric acid with the ulcerated surface;

others believe the pain results from contractions of the stomach. It seems certain that pain occurs when there is marked inflammatory reaction or penetration of surrounding organs (*e.g.*, the pancreas).

Pain resulting from a gastric ulcer is referred to the epigastric region because the stomach is supplied with pain afferents that reach the seventh and eighth thoracic segments through the greater splanchnic branch of the sympathetic trunk. *Pain resulting from a peptic ulcer is referred to the anterior abdominal wall superior to the umbilicus* because both the duodenum and this area of skin are supplied by the ninth and tenth thoracic nerves (Fig. 1-23).

When a duodenal ulcer perforates there may be pain all over the abdomen. Sometimes the peritoneal gutter associated with the ascending colon (Fig. 2-27*D*) may act as a watershed and direct the escaping inflammatory material into the right iliac fossa. The leakage of duodenal contents through such a perforation leads to **acute chemical peritonitis**. This explains why pain from an anterior perforation of a duodenal ulcer may cause right upper and lower quadrant pain. In such cases, the differential diagnosis between a perforated duodenal ulcer and a perforated appendix may be difficult.

As the ***vagus nerves*** *largely control the secretion of acid by the parietal cells of the stomach, and as excess acid secretion is associated with peptic ulcers, section of the vagus nerves* (**vagotomy**) as they enter the abdomen is sometimes performed to reduce acid production. Vagotomy is often performed in conjunction with ***resection*** (excision) of the ulcerated area and the acid-producing part of the stomach. Often only the gastric branches of the vagus nerves are cut (**selective vagotomy**), thereby avoiding adverse effects on other organs. (*e.g.*, dilation of the gallbladder).

Case 2-2

The patient's initial attack of excruciating pain was almost certainly caused by *passage of the calculus* from his renal pelvis into the superior end of his right upper ureter, a narrow cylindrical tube (Fig. 2-112). Calculi that are larger than the lumen of the ureter (3 mm) cause severe pain when they attempt to pass through it. *The pain moves from the loin to the groin* as the calculus passes along the ureter. The pain ceases when it passes into the urinary bladder, although tenderness along the course of the ureter often persists for some time. During a subsequent micturition, the stone passes through the urethra.

The patient very likely would experience severe pain when the calculus was temporarily impeded owing to angulation of the ureter as it crosses the brim of the pelvis minor and when it became wedged in the ureter, where it passed through the wall of the urinary bladder. At the inferior end of the ureter, there is a definite narrowing of the lumen; this it is a common site of obstuction (Figs. 3-61 and 3-64).

The ureteral pain results from passage of the calculus through the ureter. As the ureter is a muscular tube in which peristaltic contractions normally convey urine from the kidney to the urinary bladder, *pain results from distention of the ureter by the calculus and the urine that is unable to pass by it.* The smooth muscular coat of the ureter normally undergoes peristaltic contractions from its superior to its inferior end. As the peristaltic wave approaches the obstruction, forceful smooth contraction causes excessive dilation of the ureter between the wave and the stone. **It is the ureteral distention that produces the lancinating pain; exacerbation of pain occurs as distention increases.** The peristaltic contractions of the ureter explain the intermittent exacerbation of pain.

The afferent pain fibers supplying the ureter are included in the **lesser splanchnic nerve** (Figs. 2-114 and 2-121). Impulses also enter the first and second lumbar segments of the spinal cord and the pain is felt in the cutaneous areas innervated by the inferior intercostal nerves (T11, T12), the ***iliohypogastric*** *and* ***ilioninguinal nerves*** (L1), and the genitofemoral nerve (L1, L2). These are the same regions of the spinal cord that supply the ureter (T11 to L2); hence the pain commences in the loin and radiates inferiorly and anteriorly to the groin and the scrotum (Figs. 1-123 and 2-16*A*). The retraction of the testis by the cremaster muscle and the pain along the medial part of the front of the thigh in the present case indicates that the genital and femoral branches of the **genitofemoral nerve** (L1, L2) were involved.

Ureteral colic is caused by ureteral distention which stimulates pain afferents in its wall. As there was no peritonitis, there was no rigidity and no rebound tenderness. When peritonitis is present, pressing the hand into the abdominal wall and rapidly releasing it causes pain when the abdominal musculature springs back into place, carrying the inflamed peritoneum with it. Hence, the **abdominal rebound test** is useful in the differentiation of ureteral colic from appendicitis and intestinal colic.

Case 2-3

A complete indirect inguinal hernia is an outpouching of the peritoneal sac that enters the deep inguinal ring, traverses the inguinal canal, exits through the superficial inguinal ring, and enters the scrotum. It is referred to as an indirect hernia because it pursues an oblique course through the anterior abdominal wall and the inguinal canal (Fig. 2-20*B*).

A direct inguinal hernia does not pass through the deep inguinal ring; it protrudes through the anterior abdominal wall, medial to the deep ring and the inferior epigastric artery (Figs. 2-20, *A* and *B*).

The embryological basis of an indirect inguinal hernia is persistence of the processus vaginalis, a diverticulum of the peritoneum which pushes through the abdominal wall and forms the inguinal canal (Fig. 2-17) in preparation for later descent of the testis through it. The processus vaginalis evaginates all layers of the abdominal wall before it, and in males they become the coverings of the spermatic cord (Fig. 2-18).

The opening produced in the transversalis fascia by the processus vaginalis becomes the deep inguinal ring, and the opening it forms in the aponeurosis of the external oblique becomes the superficial inguinal ring. *In females, the entire processus vaginalis normally disappears*, whereas in males the inferior portion persists and becomes the **tunica vaginalis** (Fig. 2-18*B*).

If the stalk of the processus vaginalis does not obliterate after birth, a loop of intestine may herniate into it and enter the scrotum (Fig. 2-21*B*), or the labium majus.

The **hernial sac** (former processus vaginalis) may vary from a short one not extending beyond the superficial ring to one that extends into the scrotum (or the labium majus), where it is continuous with the tunica vaginalis.

A persistent processus vaginalis predisposes to indirect inguinal hernia by creating a weakness in the anterior abdominal wall, and a hernial sac into which abdominal contents may herniate if the intraabdominal pressure becomes very high, as occurs during straining while lifting a heavy object.

The obliquity of the inguinal canal and the contraction of the abdominal muscles usually prevent herniation of abdominal contents during rises of intraabdominal pressure that occur during coughing, straining, or nose blowing. When infants become more active at 2 to 3 months, increases in intraabdominal pressure, as occur during crying and coughing, may force part of the greater omentum and/or a loop of bowel, into the **patent processus vaginalis**. The hernial sac then appears as a bulge in the inguinal region extending into the scrotum or labium majus. The herniated loop of intestine in the sac may become constricted, resulting in interference with its blood supply and formation of a **strangulated hernia**.

Once the deep inguinal ring has been enlarged by a herniation of abdominal contents, coughing may cause herniation to occur again. This is the basis of the test done during physical diagnosis, where the examiner's finger is inserted through the superficial ring into the inguinal canal, and the patient is asked to cough (Fig. 2-20*C*).

In males, the testis and spermatic cord normally pass through the inguinal canal before birth. The spermatic cord and testis are therefore covered by extensions of the abdominal wall.

Passage of the spermatic cord through the inguinal canal enlarges the canal and weakens the entrance to it (deep inguinal ring). This explains why indirect inguinal hernia is more common in males (about 20:1). The hernial sac lies within the coverings of the spermatic cord; thus, it is covered by internal spermatic fascia and the cremaster muscle and cremasteric fascia. As the testis passes posterior to the processus vaginalis, descending through the inguinal canal before or shortly after birth, *the ductus deferens lies immediately posterior to the hernial sac*.

During the surgical repair of an indirect inguinal hernia, the genital branch of the **genitofemoral nerve** is endangered because it traverses the inguinal canal in both sexes, and exits through the **superficial inguinal ring** (Figs. 2-13 and 2-22).

The ilioinguinal nerve may also be injured. This nerve supplies the skin of the superomedial area of the thigh, the skin over the root of the penis, and the part of the scrotum with sensory fibers (Figs. 2-16*A* and 4-14). In the female, the ilioinguinal nerve supplies the same area of the thigh, the skin of the mons pubis, and the adjoining part of the labium majus (Fig. 2-22). If this nerve is injured, anesthesia of these areas of skin will likely result. If the nerve is constricted by a suture, **postoperative neuritic pain** may also occur in these areas.

Injury to the external iliac artery or to the inferior epigastric artery, one of its two main branches, is uncommon during the repair of the inguinal hernia, but their relationship to the deep inguinal ring makes them liable to injury, *e.g.*, by a suture that is placed too deeply. Serious extraperitoneal bleeding can result from an unrecognized tear of these vessels.

As the ***ductus deferens*** *lies immediately posterior to the hernial sac, it may be damaged when the sac is freed, ligated, and excised.* As the hernial sac is within the spermatic cord, the **pampiniform plexus** of veins and the testicular artery may also be injured (Fig. 2-23), resulting in impairment of circulation to the testis. Injury to the vessels of the spermatic cord may result in ***atrophy of the testis*** on that side.

Case 2-4

The type of skin incision used for an appendectomy depends on the type of patient and the certainty of the diagnosis. Usually the McBurney or **gridiron incision** is made (Fig. 2-43), which is an oblique or almost transverse one that follows *Langer's lines* (cleavage lines of the skin). The center of the incision is at **McBurney's point**, which is at the junction of the lateral and middle thirds of the line joining the anterior superior iliac spine and the umbilicus (Fig. 2-

43). In most cases, this point overlies the base of the appendix.

Following incision of the skin and the superficial fascia, *the aponeurosis of the external oblique muscle is incised in the direction of the fibers of this muscle* (Fig. 2-9). The other two muscles of the anterior abdominal wall (internal oblique and transversus abdominis) are then split (not cut) in the direction of their fibers. This lessens the chances of injuring the nerves supplying them. Next the transversalis fascia and the parietal peritoneum are incised to expose the cecum. *The base of the appendix is indicated by the point of convergence of the three teniae coli* (Fig. 2-41).

The vermiform appendix varies in length and in position (Fig. 2-42). It usually lies posterior to the cecum and, if long enough, posterior to the inferior part of the ascending colon (*i.e.*, ***retrocecal appendix*** or ***retrocolic appendix***), but it may descend over the brim of the pelvis minor (pelvic appendix). In the female, a ***pelvic appendix*** lies close to the right uterine tube and ovary.

The variations in length and position of the appendix may give rise to varying signs and symptoms in appendicitis. For example, the site of maximum tenderness in cases of **retrocecal appendix** may be just superomedial to the anterior superior iliac spine, even as far superior as the level of the umbilicus.

Subhepatic cecum and appendix are uncommon. This unusual condition results from incomplete rotation of the gut during the fetal period. *In these cases the site of pain is likely to be in the right upper quadrant* of the abdomen.

If the appendix is long (10 to 15 cm) and extends into the pelvis minor, the site of pain in a female might suggest peritoneal irritation resulting from a ruptured **ectopic pregnancy**. As the appendix crosses the psoas major muscle, the patient often flexes the right thigh to relieve the pain. Thus, *hyperextension of the thigh (psoas test) causes pain because it stretches the muscle and its inflamed fascia.*

Tenderness on the right side during a rectal examination may indicate an inflamed pelvic appendix.

Initially, the pain of typical acute appendicitis is referred to the periumbilical region of the abdomen; later the site of pain usually shifts to the right lower quadrant. Afferent nerve fibers from the appendix are carried in the lesser splanchnic nerve and impulses enter the 10th thoracic segment of the spinal cord. As impulses from the skin in the periumbilical region are also sent to this region of the spinal cord (Fig. 1-23), the pain is interpreted as somatic rather than visceral, apparently because impulses of cutaneous origin are more often received by the thalamus.

The shift of pain to the right lower quadrant is caused by irritation of the parietal peritoneum, usually on the posterior abdominal wall. Afferent fibers from this region of peritoneum and skin are carried in the inferior intercostal, subcostal, and first lumbar nerves. The pain during palpation results from stimulation of pain receptors in the skin and the peritoneum, whereas the increased tenderness detected in the right side of the **rectouterine pouch** (rectovesical pouch in the male, Fig. 2-31) is caused by irritation of the parietal peritoneum in this pouch. When the abdominal wall is depressed and allowed to rebound, the patient usually winces because, as the abdominal muscles spring back, the inflamed peritoneum is carried with it.

If the patient had previously had her appendix removed, an **inflamed Meckel's diverticulum** could give rise to signs and symptoms similar to appendicitis. This ileal diverticulum is one of the most common abnormalities of the intestinal tract, occurring in 2 to 4% of people (Fig. 2-93).

A Meckel's diverticulum represents the remnant of the proximal portion of the yolk stalk and appears as a finger-like projection (usually 3 to 6 cm long) from the antimesenteric border of the ileum, 40 to 50 cm from the ileocecal junction.

Case 2-5

Diaphragmatic hernia is a herniation of abdominal viscera into the thorax through an opening in the diaphragm. A **congenital diaphragmatic hernia** is present in about one of 2000 infants. It results from *defective formation and/or fusion of a pleuroperitoneal membrane* with the dorsal mesentery of the esophagus and the septum transversum during development of the diaphragm. As a result there is a **posterolateral defect in the diaphragm**, usually on the left (Fig. 2-134), through which abdominal organs, usually the intestines, herniate into the *thorax*.

Hiatal hernia is common, particularly in older people. Usually the gastroesophageal region of the stomach herniates through the esophageal hiatus into the inferior part of the thorax (Fig. 2-49).

Hiatal hernia is usually acquired, but a congenitally enlarged esophageal hiatus may be a predisposing factor. The esophageal hiatus is in the muscular part of the diaphragm (Fig. 2-117). Understand that the right crus passes to the left of the midline; hence the esophageal hiatus and hernia are to the left of the midline even though they are within the right crus. In most cases, herniation appears after 50 years of age and results from **widening of the esophageal hiatus** owing to weakening of muscle fibers of the right crus of the diaphragm.

There are two main types of hiatal hernia (1) *sliding hiatal hernia* and (2) *paraesophageal hernia*, but some hernias present characteristics of both types and are referred to as *mixed hiatal hernias*.

Sliding hiatal hernia (Fig. 2-49*A*) is the most common type (up to 10 times more common than the paraesophageal and mixed types combined). In sliding hiatal hernia, as in the present case, **the gastro-**

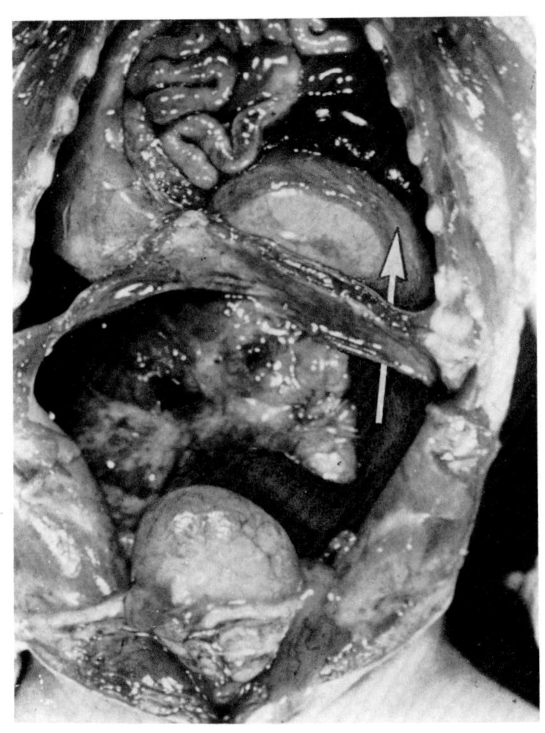

Figure 2-134. Photograph of an infant's thorax and abdomen taken at autopsy. The *arrow* passes through a large posterolateral defect in the diaphragm. The liver has been removed. Note the intestines in the thorax. (Reprinted with permission from Moore KL: *The Developing Human: Clinically Oriented Embryology*, ed. 3. Philadelphia, WB Saunders Co, 1982.)

esophageal junction herniates into the thorax. The hernial sac consists of parietal peritoneum anteriorly and the anterior wall of the stomach posteriorly.

Paraesophageal hiatal hernia is much less common than sliding hiatal hernia, but it is more common in women than in men (10 to 1). In this type, **the gastroesophageal junction remains in its normal position**, but a pouch of peritoneum herniates through the esophageal hiatus into the inferior part of the thorax. *Usually the hernial sac contains a portion of the fundus of the stomach* (Fig. 2-49*B*).

The thoracic region of the vertebral column becomes shorter with age owing to **dessication of the intervertebral discs**, and the abdominal fat generally increases during middle age. Both of these occurrences favor development of hiatal hernias.

Most of the present patient's complaints (heartburn, belching, regurgitation, and epigastric pain) resulted from irritation of the esophageal mucosa by the **reflux of gastric juice**. The irritant effect of the gastric juice produces **esophageal spasm**, resulting in **dysphagia** and retrosternal pain. Pain endings in the esophagus are stimulated by the forcible contractions of the smooth muscle in esophageal wall occurring during spasm.

Pain of gastroesophageal origin is referred to the epigastric and retrosternal regions, the cutaneous zones of reference for these regions of the viscera. The patient's **hiccups** are caused by ***spasmodic contractions of the diaphragm***, which result from pressure created by the hernia. Probably enlargement of the esophageal hiatus stimulates fibers of the phrenic nerves supplying the diaphragm (Fig. 2-120).

An incidence of hiatal hernia as high as 70% has been reported in routine radiographic studies, but in most cases these hernias are relatively asymptomatic (*i.e.*, produce only heartburn and mild indigestion). Most hiatal hernias either require no treatment or can be managed medically by decreasing gastroesophageal reflux (*e.g.*, through weight reduction).

As the esophageal hiatus also transmits the vagus nerves and the esophageal branches of the left gastric vessels, these structures as well as the esophagus must be protected from injury during surgical repair of hiatal hernias.

Case 2-6

The radiologist was unable to identify the calculus in the cystic duct because it was small and not faceted or laminated. This is typical of the **nidus** (L. nest) around which other substances are deposited to produce a larger, more typical gallstone.

Obese middle-aged women who have had several children (**multipara**), are most prone to gallbladder disease, but tall thin men, virgin women, and children may also develop the disease. Thus, the common aphorism "*forty, flatulent, female, fertile,* and *fat*" describes most patients, but it does not characterize all patients with gallstones. Biliary calculi are more common in females over 20, but this is not necessarily so after 50 years of age.

In about 50% of persons, gallstones are "silent" (asymptomatic). A gallstone is a concretion in the gallbladder, cystic duct, or bile duct, composed chiefly of cholesterol crystals.

The pain is severe when a biliary calculus is lodged in the cystic or bile duct. The patient's sudden severe pain in the epigastric region (**biliary colic**) was caused by a gallstone wedged in the cystic duct.

The pain referred to the right upper quadrant and scapular region results from inflammation of the gallbladder and distention of the cystic duct. The nerve impulses pass centrally in the **greater splanchnic nerve** on the right side (Fig. 2-121), and enter the spinal cord through the dorsal roots of the seventh and eighth thoracic nerves. This **visceral referred pain** is felt in the right upper quadrant of the abdomen and in the right infrascapular region because the source of the stimuli entering this region of the cord is wrongly interpreted as cutaneous (Fig. 1-23).

Often the inflamed gallbladder irritates the peritoneum covering the peripheral part of the diaphragam, resulting in a **parietal referred pain** in the inferior part of the thoracic wall. This part of the peritoneum is supplied by the inferior **intercostal nerves**. In other cases, the peritoneum covering the diaphragm is irritated and the pain is referred to the shoulder region because this area of peritoneum is supplied by the **phrenic nerve**. The skin of the shoulder region is supplied by the **supraclavicular nerves** (C3 and C4), the same segments of the cord that receive pain afferents from the central portion of the diaphragm.

Although a calculus in the cystic duct is radiopaque, *only one in five gallstones contain enough calcium to be visible on plain radiographs* of the abdomen.

When fat enters the duodenum, **cholecystokinin** causes contraction of the gallbladder. In the present case, it is very likely that the patient's gallbladder contracted vigorously after her fatty meal, squeezing a stone into her cystic duct.

Acute cholecystitis is associated with a gallstone impacted in the cystic duct in a high percentage of cases. *The impacted calculus causes sudden distention of the gallbladder which compromises its blood and lymphatic supply.*

Usually peritoneum does not separate the gallbladder from the liver. The gallbladder lies in a fossa on the inferior surface of the right hepatic lobe (Fig. 2-75), and the peritoneum on this surface of the liver passes over the inferior surface of the gallbladder, leaving the superior surface attached to the liver by connective tissue.

The abdominal rigidity detected in the present case resulted from involuntary contractions of the muscles of the anterior abdominal wall, particularly the rectus abdominis. This muscle spasm was a reflex response to stimulation of nerve endings in the peritoneum associated with the dilated gallbladder.

Anatomical variations in the gallbladder and cystic duct and in the arteries supplying them are very common (Figs. 2-83 and 2-85). Because of this, surgeons must determine the existing anatomical pattern and identify the cystic, bile, and hepatic ducts and the cystic and hepatic arteries before dividing the cystic and duct and its artery.

Important variations occur in the length and course of the cystic duct (Fig. 2-85*A* to *C*). If these abnormalities are not recognized, the bile duct may be mistaken for the cystic duct and be ligated and divided. This results in severe **jaundice** (F. *jaune*, yellow) and death if the bile duct is not reopened.

Variations in the origin and course of the cystic artery occur in up to 25% of persons (Fig. 2-83). As there may be accessory cystic branches from the hepatic arteries, unexpected hemorrhage may occur during cholecystectomy. A more serious complication is postoperative hemorrhage following unrecognized injury during surgery to an abdominal blood vessel.

Case 2-7

Hepatic cirrhosis is a disease characterized by progressive destruction of hepatic parenchymal cells. These hepatic cells are replaced by fibrous tissue (**fibrosis**) which contracts and hardens. The fibrous tissue surrounds the intrahepatic blood vessels and biliary radicles (roots). As this process advances, circulation of blood through the branches of the portal vein and of bile through the **biliary radicles** in the liver (Fig. 2-77) is impeded. As pressure in the portal vein rises (**portal hypertension**), the liver becomes more dependent on the hepatic artery for its blood supply and blood pressure in the portal vein rises, reversing blood flow in the normal **portacaval anastomoses**. This results in portal blood entering the systemic circulation (Fig. 2-100). As these anastomotic veins seldom possess valves, they can conduct blood in either direction. This causes enlargement of the veins (**varicose veins**) forming these anastomoses at the inferior end of the esophagus (**esophageal varices**), the inferior end of the rectum and anal canal (**hemorrhoids**), and around the umbilicus (**caput medusae**).

Because of pressure during swallowing and defecation, the esophageal varices and hemorrhoids, respectively, may rupture, resulting in **bloody vomitus** and/or bleeding from the anus and **bloody stools** (feces).

Internal hemorrhoids are varicosities of the tributaries of the **superior rectal vein** (Fig. 3-86). Blood may also pass in a retrograde direction in the **paraumbilical veins** (small tributaries of the portal vein) via the vein in the ligamentum teres.

In **portal hypertension** the paraumbilical veins may become varicose, forming a radiating venous pattern at the umbilicus, called a **caput medusae** (Fig. 2-132), owing to its resemblance to the snakes adorning the head of Medusa, a mythological character.

In **cirrhosis of the liver**, the ramifications of the portal vein are compressed by the contraction of the fibrous tissue in the portal canals. As a result, *there is increased pressure in the splenic and superior and inferior mesenteric veins* (Fig. 2-52). Fluid is forced out of the capillary beds drained by these veins into the peritoneal cavity. *Accumulation of fluid in the peritoneal cavity is called* **ascites.**

The spleen usually enlarges (**splenomegaly**) when there is hepatic cirrhosis because of increased pressure in the splenic vein. As there are *no valves in the portal system*, pressure in the splenic vein is equal to that in the portal vein. **A common method of reducing portal pressure** is by diverting blood from the portal vein to the inferior vena cava through a surgically-created anastomosis (***portacaval anastomosis***). Similarly, the splenic vein may be anastomosed to the left renal vein (***splenorenal anastomosis***).

Case 2-8

A **direct inguinal hernia** enters the inguinal canal through its posterior wall, whereas an **indirect in-**

guinal hernia enters the inguinal canal through the deep inguinal ring (Fig. 2-20*B*).

Direct inguinal hernia is much less common than indirect inguinal hernia; both types occur more often in men than in women. Usually a direct inguinal hernia is acquired and usually occurs in men over 40 years of age.

The sac of a direct inguinal hernia is formed by the peritoneum lining the anterior abdominal wall. The sac protrudes through the **inguinal triangle.** *This triangle is bounded medially by the* ***lateral border of the rectus abdominus*** *muscle, inferiorly by the* ***inguinal ligament****, and laterally by the* ***inferior epigastric artery*** (Fig. 2-20*A*). The hernia may protrude through the anterior abdominal wall and escape from the abdomen on the lateral side of the conjoint tendon to pass through the posterior wall of the inguinal canal. In this case, the hernial sac is covered by transversalis fascia, cremaster muscle, cremasteric fascia, and external spermatic fascia. Occasionally the hernial sac is forced through the fibers of the conjoint tendon to enter the end of the inguinal canal. In this case, it is covered by transversalis fascia, conjoint tendon, and external spermatic fascia.

Direct inguinal hernias usually protrude anteriorly through the inferior part of the **inguinal triangle** and extend toward the superficial inguinal ring, but they may pass through this ring and enter the scrotum or the labium majus.

A direct hernia is acquired and results from ***weakness of the anterior abdominal wall***, *e.g.*, of the transversalis fascia with atrophy of the conjoint tendon. There is no known embryological basis for this type of hernia.

The type of inguinal hernia (direct or indirect) can often be determined by the relationship of the hernial sac to the inferior epigastric artery. The pulsations of this artery can usually be felt by the tip of the examiner's finger in the inguinal canal. *In direct inguinal hernia, the neck of the hernial sac is in the inguinal triangle and lies* **medial** to the inferior epigastric artery, whereas *in indirect inguinal hernia the neck of the hernial sac is in the deep inguinal ring and lies* **lateral** to the inferior epigastric artery.

As the inferior intercostal nerves and the iliohypogastric and ilioinguinal nerves from the first lumbar nerve supply the abdominal musculature (Fig. 2-8), injury to any of them during surgery or an accident could result in weakening of muscles in the inguinal region, predisposing to development of direct inguinal hernia.

The ilioinguinal nerve also gives motor branches to the fibers of the internal oblique muscle which are inserted into the lateral border of the conjoint tendon. Division of this nerve paralyzes these fibers and relaxes the conjoint tendon; this may result in a direct inguinal hernia.

Suggestions for Additional Reading

1. Anson BJ, McVay CB: *Surgical Anatomy*, ed 5. Philadelphia, WB Saunders Co, 1971, vol 1.

   This classical textbook gives good accounts of the various *types of abdominal incision* and discusses the anatomical basis for them. There is a *well-illustrated account of hernias* and how they are repaired. Various methods of anastomosing the stomach to the intestine after partial gastric section are also described. *Kidney transplantation is discussed* with emphasis on the importance of understanding the renal vascular variations and abnormalities of the upper urinary tract.
2. Healey JE: *A Synopsis of Clinical Anatomy*. Philadelphia, WB Saunders Co, 1969.

   This book gives a good account of the *clinical applications of gross anatomy*. It is particularly well illustrated and has a good account of the *embryology of the GI tract*. Areas are pointed out where surgical injury can occur, *e.g.*, he states "Surgical injury to the ductal system is far too frequent and often tragic."
3. Lowenfels AB: *Comparison Guide to Surgical Diagnosis*. Baltimore, Williams & Wilkins, 1975.

   This small monograph *reviews common diagnostic problems encountered in surgical patients.* Emphasis is placed on a systematic approach to data gathering. There are good accounts of the types of *abdominal hernia*, *abdominal incisions*, *wound drainage*, *abscesses*, *fistulas*, abdominal injuries, the acute abdomen, intestinal obstruction, and abdominal masses.
4. Moore KL: *The Developing Human: Clinically Oriented Embryology*, ed 3. Philadelphia, WB Saunders Co, 1982.

   In order to appreciate the normal relationships of the abdominal viscera and the many anatomical variations and congenital abnormalities which may occur, a good knowledge of the development and rotation of the embryonic gut is imperative.
5. Sparberg M: Examination of the abdomen. In *A Primer of Clinical Diagnosis*. Buckingham WB, Sparberg M, Brandfonbrenner M (eds): New York, Harper & Row Publishers, Inc, 1971.

   The technique of examining a patient's abdomen is clearly described and illustrated. Hepatomegaly, splenomegaly, gallbladder enlargement, kidney enlargement, aortic enlargement, and enlargement of the stomach, colon, pancreas, and umbilicus are discussed. Bowel sounds, *rebound tenderness*, *guarding*, *paralytic ileus*, and other clinical terms are defined.
6. Squire LF, Colaiace WM, Strutynsky N: *Exercises in Diagnostic Radiology. 2. The Abdomen*, Philadelphia, WB Saunders Co, 1971.

   This short book of exercises includes radiographs of groups of three patients presenting with the same symptom or symptom complex.

CHAPTER 3

The Perineum and Pelvis

THE PERINEUM

The perineum is the region overlying the inferior pelvic aperture (pelvic outlet). Its posterior part, called the **anal region** (anal triangle), contains the termination of the *anal canal*; its anterior part, called the **urogenital region** (urogenital triangle), contains the *external urogenital organs.*

In the anatomical position the perineum is a narrow area between the thighs, but when the thighs are abducted (Fig. 3-1) *the perineum is a diamond-shaped region extending from the symphysis pubis to the tip of the coccyx* (Figs. 3-2 to 3-6). The perineum includes the openings of the *anal canal* (**anus**) and of the *urogenital system* (**vagina** and/or **urethra**). The perineum is best examined and dissected with the person in the **lithotomy position** (Fig. 3-1).

The male perineum is of special importance to certain surgical specialists, e.g., **proctologists** (surgeons who deal with the rectum and anus and their diseases), and **urologists** (surgeons who are concerned with the urogenital organs).

The relationship of the male perineum to the urethra, prostate, seminal vesicles, urinary bladder, rectum, and anal canal is essential knowledge for all doctors.

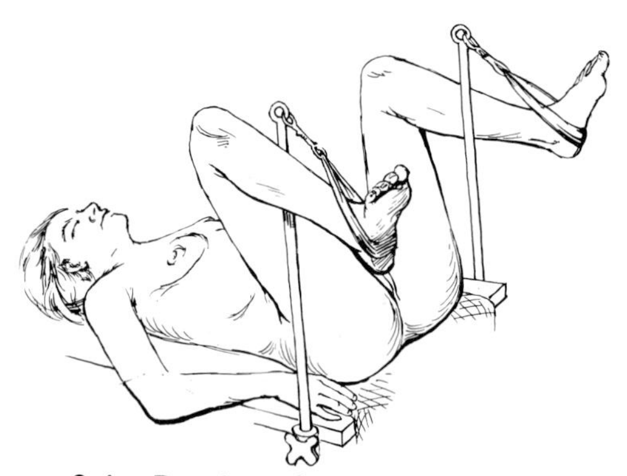

Figure 3-1. Drawing of a woman in the lithotomy position which is used for pelvic examinations and delivering babies. The feet are held by straps (as here) or by metal stirrups. The essentials of this position are: (1) the patient is on her back, (2) her lower limbs are flexed at the knees and hips, and (3) her thighs are abducted.

The female perineum is of special importance to ***obstetricians***, specialists who care for women during pregnancy and parturition (childbirth); **gynecologists** who deal with diseases of the genital tract in women; and urologists and proctologists.

Knowledge of clinically oriented anatomy, including the bony landmarks of the pelvis, is a prime requisite for the clinical management of parturition and many diseases involving the bladder, urethra, vagina, rectum, and anal canal.

BOUNDARIES OF THE PERINEUM

The boundaries of the perineum, illustrated in Figures 3-2 and 3-4, are: (1) the *symphysis pubis*; (2) the

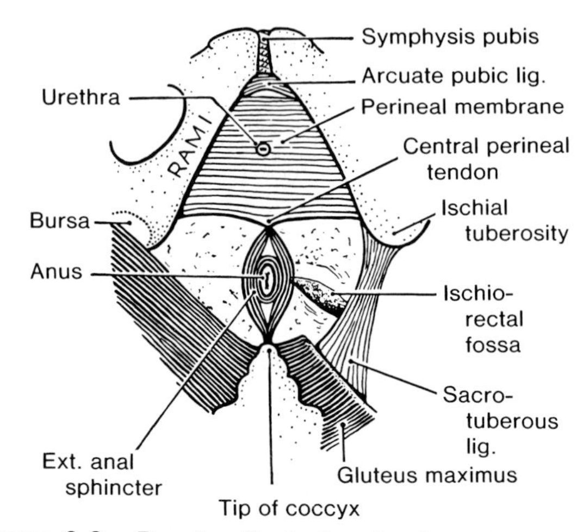

Figure 3-2. Drawing illustrating the boundaries and subdivisions of the perineum. The perineum is the diamond-shaped region at whose angles are the arcuate pubic ligament, the tip of coccyx, and the ischial tuberosities. The anterior half of the perineum is the urogenital region or triangle and the posterior half is the anal region or triangle. The deep fascia on the inferior or superficial surface of the urogenital diaphragm is thickened to form the dense perineal membrane. It is continuous with the deep layer of fascia anteriorly and posteriorly. Observe that the perineal membrane is pierced by the urethra. In the female the vagina also pierces this fascia (also see Fig. 3-14*D*).

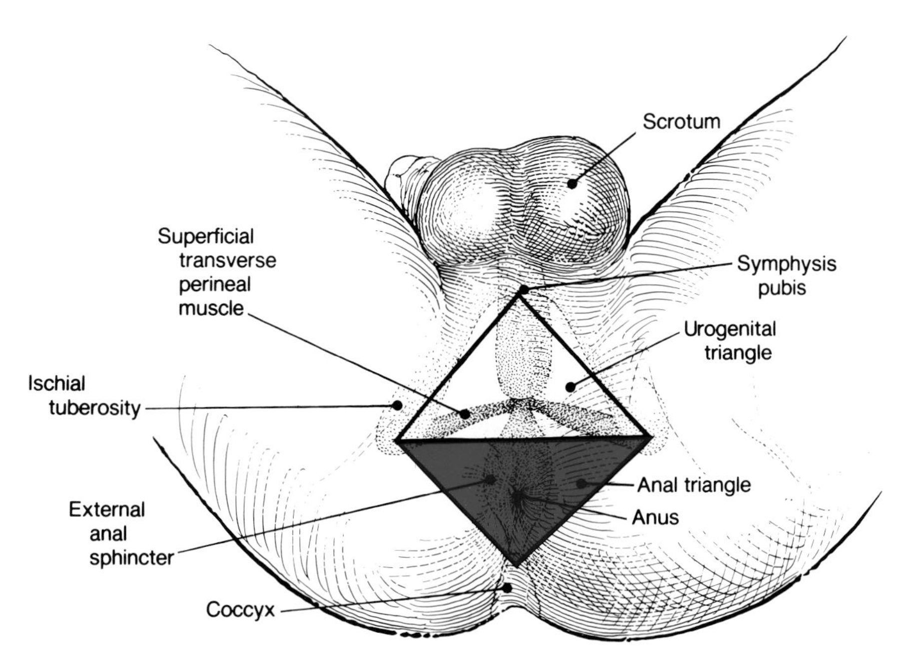

Figure 3-3. Diagram illustrating the diamond-shaped perineum, extending from the symphysis pubis to the coccyx (also see Figs. 3-1 and 3-2). Note that a transverse line between the right and left ischial tuberosities divides the perineum into two triangular areas, the urogenital region (triangle) anteriorly and the anal region (triangle) posteriorly.

inferior pubic rami; (3) the *ischial rami*; (4) the ischial tuberosities; (5) the *sacrotuberous ligaments*, and (6) the *coccyx*.

For descriptive purposes the perineum is divided into two triangular parts by a transverse line joining the ischial tuberosities (Figs. 3-2 and 3-3). The *anal region*, containing the anus, is posterior to this line and the *urogenital region*, containing the root of the scrotum and the penis (or the vulva in the female), is anterior to this line.

The inferior pelvic aperture (Fig. 3-7) is closed, except where it transmits the urethra and the anal canal (Fig. 3-6); *in the female* it also transmits the vagina (Fig. 3-5). The anterior half of the inferior pelvic aperture is closed by the **urogenital diaphragm** (Figs. 3-8, 3-11, and 3-14), and the posterior half is closed by the levatores ani muscles.

The levatores ani muscles (Figs. 3-15 and 3-17) run posteroinferiorly from the anterolateral walls of the pelvis minor and meet one another in the median plane from the posterior margin of the urogenital diaphragm to the coccyx.

THE PELVIC DIAPHRAGM

The two levatores ani muscles and the two coccygeus muscles form the pelvic diaphragm (Figs. 3-4 and 3-13), which closes the inferior pelvic outlet somewhat like a funnel would if it were placed in the pelvic cavity. The levator ani and coccygeus muscles are described on pages 342 and 347, respectively.

The pelvic diaphragm divides the pelvic cavity into two parts: (1) a superior part containing the pelvic viscera, and (2) an inferior part containing mainly fat, called the ***ischiorectal fossae*** (Figs. 3-2 and 3-5).

The pelvic diaphragm forms the V-shaped floor of the pelvic cavity and the Λ-shaped roof of each ischiorectal fossa (Fig. 3-11).

THE UROGENITAL DIAPHRAGM

The urogenital diaphragm is a thin sheet of striated muscle stretching between the two sides of the pubic arch (Fig. 3-14*C*). The urogenital diaphragm covers the anterior part of the inferior pelvic aperture (Figs. 3-7 and 3-8). The most anterior and posterior fibers of the urogenital diaphragm (*deep transversus perinei muscle*) run transversely (Fig. 3-14), whereas its middle fibers (*sphincter urethrae muscle*) encircle the urethra.

THE SPHINCTER URETHRAE MUSCLE (Figs. 3-12 to 3-14*C*)

The sphincter urethrae muscle arises from the medial surface of **inferior pubic ramus**. Its fibers pass medially toward the urethra, where they meet the fibers from the opposite side.

The sphincter urethrae encircles the membranous urethra in the male. The inferior half of the sphincter urethrae in the female blends with the anterolateral walls of the vagina (Fig. 3-13).

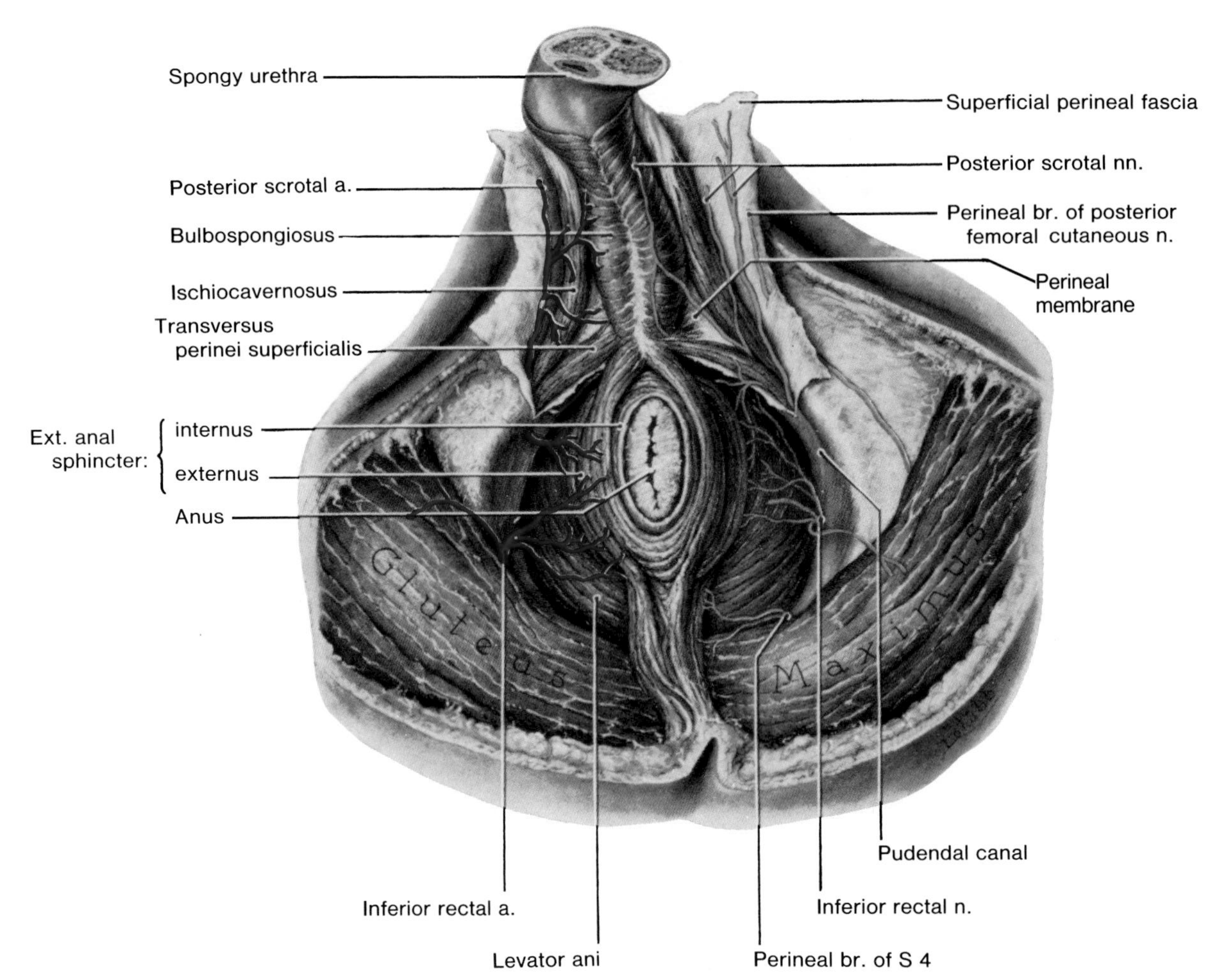

Figure 3-4. Drawing of a dissection of the male perineum. In the anal region observe the anal orifice (anus) at the center, surrounded by the external anal sphincter and an ischiorectal fossa on each side. Note that the superficial fibers of the external anal sphincter anchor the anus anteriorly to the perineal body (central tendon of the perineum) and to the coccyx, here to the skin. Examine the ischiorectal fossa filled with fat, which is bounded (1) *medially* by the levator ani and external anal sphincter muscles, (2) *laterally* by the obturator internus fascia, (3) posteriorly by the gluteus maximus muscle overlying the sacrotuberous ligament, and (4) anteriorly by the base of the perineal membrane. The apex or roof of the anal region is where the medial and lateral walls meet; its base or floor, formed by tough skin and deep fascia, is removed. In the urogenital region observe that the superficial perineal fascia is (1) incised in the midline, (2) freed from its attachment to the base of the perineal membrane, and (3) reflected.

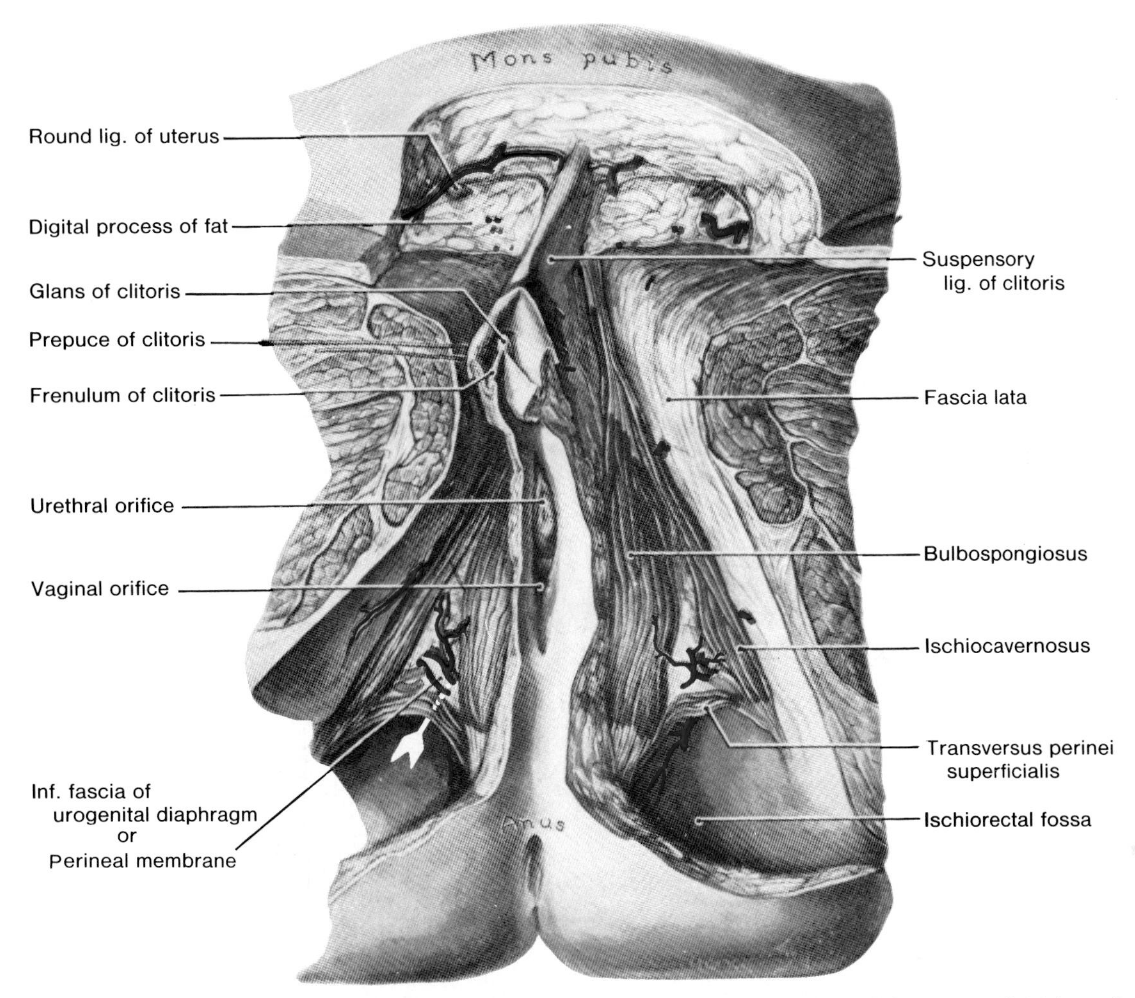

Figure 3-5. Drawing of a dissection of the female perineum. Observe the thickness of the superficial fatty tissue at the mons pubis and the encapsulated digital process of fat deep to this. Note that the suspensory ligament of the clitoris descends from the linea alba and the symphysis pubis. Examine the prepuce of the clitoris which forms a hood over the clitoris. Note that the anterior ends of the labia minora unite to form the frenulum of the clitoris. Observe the three muscles on each side: bulbospongiosus, ischiocavernosus, and transversus perinei superficialis which, when slightly separated, reveal the inferior fascia of the urogenital diaphragm or perineal membrane. The bulbospongiosus muscle overlies the bulb of the vestibule. Note that the vestibule and the orifice of the vagina separate the muscles of the two sides. Examine the ischiorectal fossa lateral to the anus. The anterior recess of this wedge-shaped fascial space is indicated on the left by the tail of the *white arrow*. The arrowhead (invisible) is lying in the anterior recess of the ischiorectal fossa, deep to the perineal membrane.

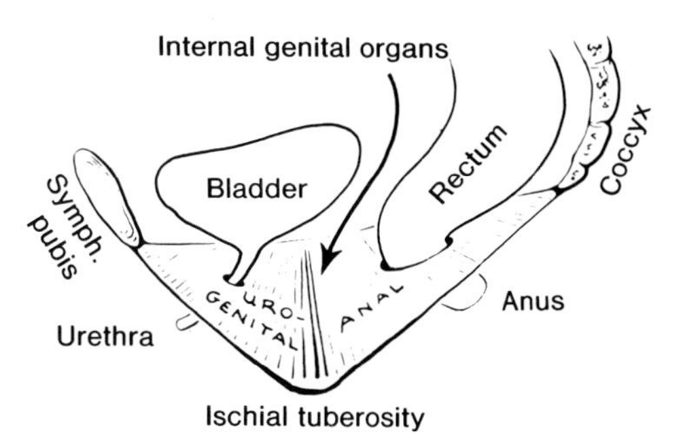

Figure 3-6. Schematic median section of the male pelvis showing the urethra passing through the urogenital region and the terminal part of the digestive system traversing the anal region of the diamond-shaped perineum (Figs. 3-2 and 3-3). Note that the urogenital region faces anteroinferiorly, whereas the anal region faces posteroinferiorly.

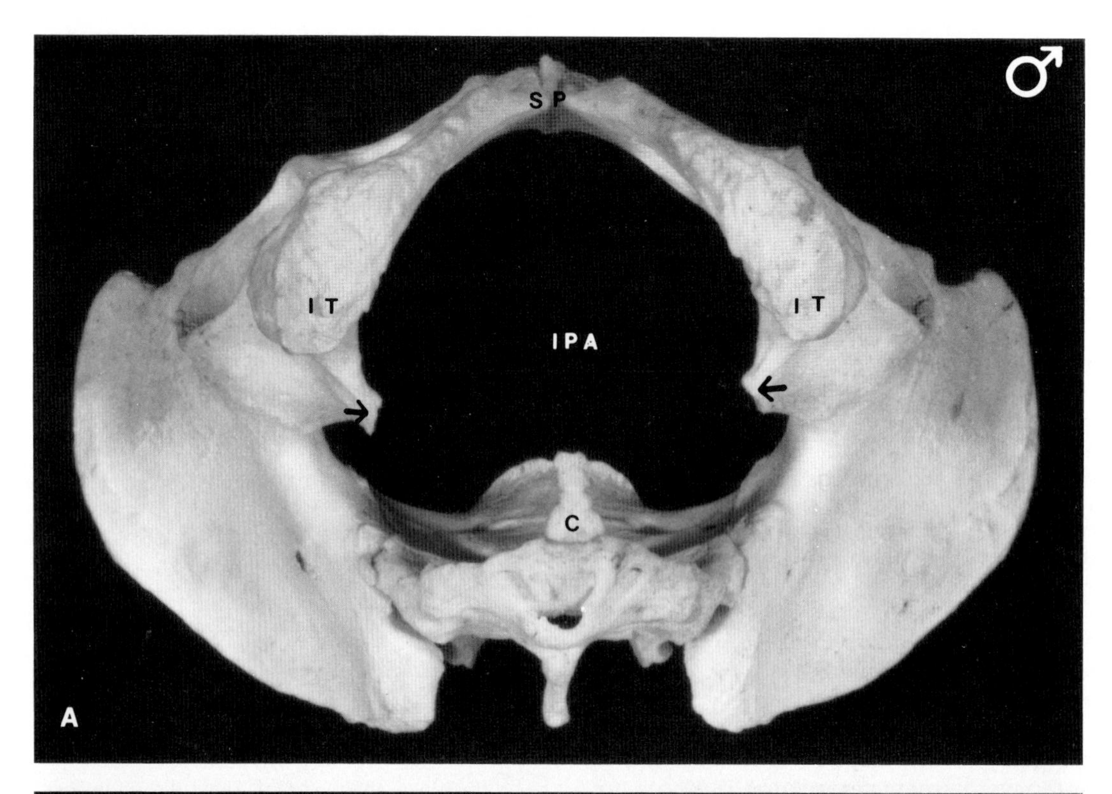

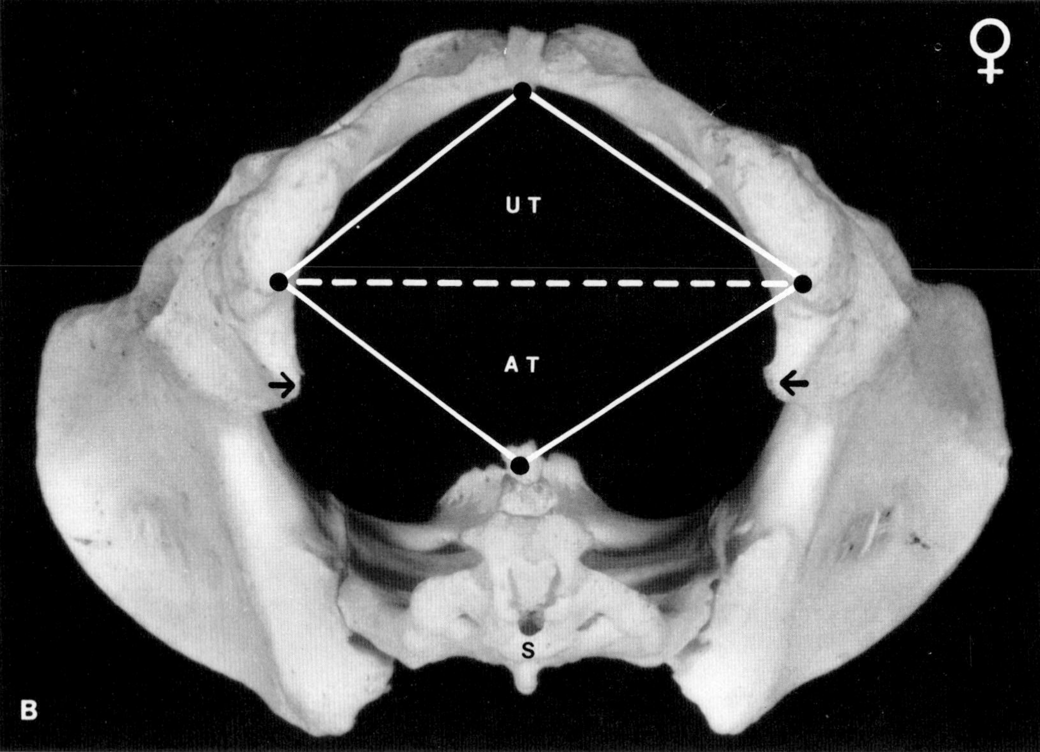

Figure 3-7. Photographs of male (♂) and female (♀) pelves showing their inferior apertures (pelvic outlets). These are closed in living persons by the soft tissues of the perineum. Observe the difference in the size of the inferior pelvic aperture (*IPA*) in the male and female. The view of the pelvis shown in *B* is the one that the obstetrician visualizes in his/her "mind's eye" when the patient is in the lithotomy position (Fig. 3-1). Note that at the angles of the inferior pelvic aperture are the symphysis pubis (*SP*), the coccyx (*C*), and the ischial tuberosities (*IT*). Note that the *broken transverse white line* between the right and left ischial tuberosities (*IT*) divides the diamond-shaped perineum into two triangles or regions, the urogenital triangle (*UT*) and the anal triangle (*AT*). The *arrows* indicate the ischial spines to which the sacrosphinous ligaments are attached (Fig. 3-16). Note that the sacrum (*S*) is wedged between the iliac bones.

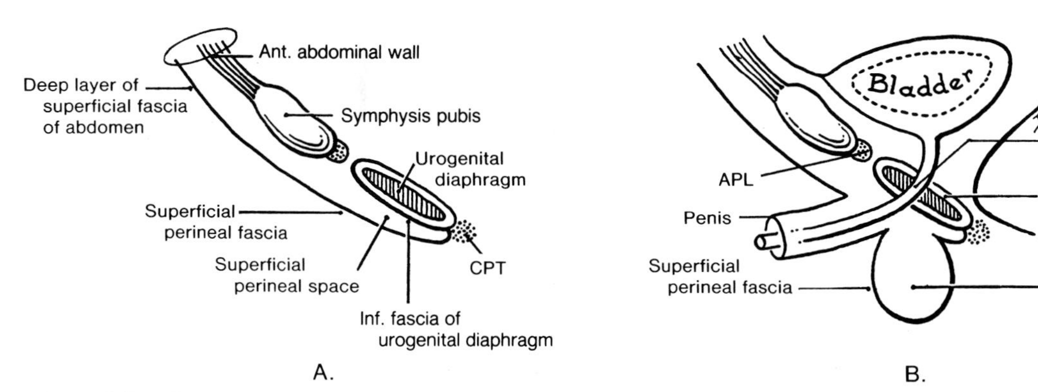

Figure 3-8. Schematic drawings of midline sections of the pelvis showing the urogenital diaphragm and perineal spaces (pouches). Note that the superficial perineal fascia (Colles' fascia) is a continuation of the deep layer of the superficial fascia (Scarpa's fascia) of the abdomen. *CPT*, central perineal tendon or perineal body (see Figs. 3-10 and 3-35); *APL*, arcuate pubic ligament (see Fig. 3-2).

Nerve Supply (Fig. 3-24). Perineal branch of the **pudendal nerve** (S2, S3, and S4).

Action. Constricts membranous urethra. This is the *voluntary sphincter of the urethra* which is thought to expel the last drops of urine or semen.

In the female the sphincter urethrae does not surround the urethra and so does not constrict it.

THE DEEP TRANSVERSUS PERINEI MUSCLE (Figs. 3-9, 3-10, and 3-14*C*)

The transverse *muscle fibers posterior to the urethra*, called the deep transversus perinei, ***arise from the medial surface of the ramus of the ischium, run transversely, and insert into the central perineal tendon***. In the female some fibers are also inserted into the vaginal wall.

Nerve Supply (Fig. 3-24). Perineal branch of the **pudendal nerve** (S2, S3, and S4).

Actions (Fig. 3-10). **Steadies central perineal tendon**, thereby contributing to the general supportive role of the perineum.

THE CENTRAL PERINEAL TENDON (Figs. 3-2 and 3-10)

This fibromuscular node, often called the perineal body, is a small wedge-shaped mass of fibrous tissue located at the center of the perineum.

The central perineal tendon is the landmark of the perineum where several muscles converge (transverse perineal muscles, bulbospongiosus, some fibers of the external anal sphincter, and the levator ani muscles of both sides).

The central perineal tendon is a particularly important structure in the female. Tearing or stretching of it during childbirth (parturition) removes support from the inferior part of the posterior wall of the vagina. As a result, **prolapse of the vagina** through the vaginal orifice may occur (Fig. 3-56).

When a tear of the perineum including the central perineal tendon appears inevitable during childbirth, an incision is often made in the perineum because *a clean surgical incision is preferable to a jagged tear.* This relaxing or appeasing incision is called an **episiotomy** (Fig. 3-41).

In a *median episiotomy*, the incision passes posteriorly from the frenulum of the labia minora (Figs. 3-10 and 3-41*A*) through the vaginal mucosa and the central perineal tendon. The incision does not reach the external anal sphincter and rectum. In cases where there is a possibility of the incision tearing posteriorly and involving the anal sphincter, a *mediolateral episiotomy* is done (Fig. 3-41*B*).

THE PERINEAL FASCIA (Figs. 3-2 to 3-5, 3-8, and 3-11)

The urogenital diaphragm, like other muscles, is surrounded by deep fascia. The perineal fascia consists of two sheets, the *inferior and superior fasciae of the urogenital diaphragm.*

The inferior fascia of the urogenital diaphragm (perineal membrane) is continuous with the deep layer of fascia anteriorly and posteriorly. It is also attached laterally to the pubic arch (Fig. 3-9).

The superficial perineal fascia (Figs. 3-4 and 3-8), or the membranous layer of the subcutaneous connective tissue of the perineum, *formerly called Colles' fascia*, is continuous with the membranous layer of the subcutaneous connective tissue of the inferior

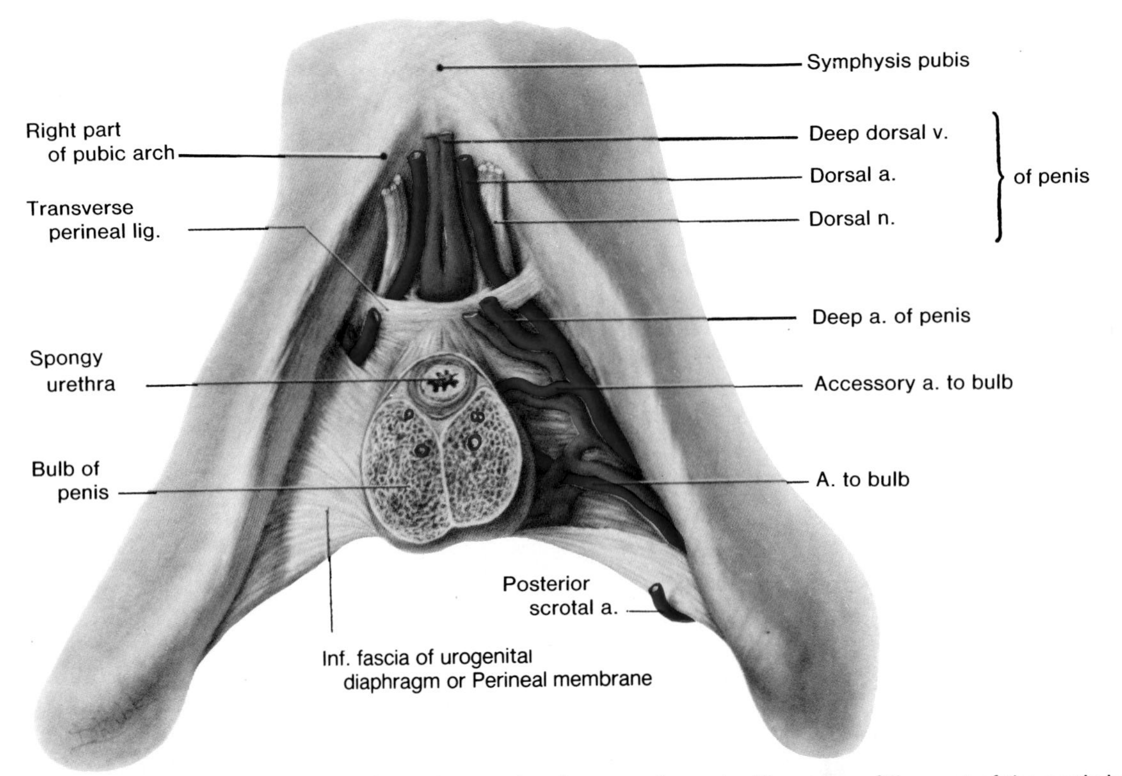

Figure 3-9. Drawing of a dissection of the deep perineal space of a male. The crura of the root of the penis have been removed (For orientation of this section, see Fig. 3-30). On the *right* the perineal membrane is partly removed and the deep perineal space is thereby opened. Observe the fibers of the perineal membrane converging on the bulb of the penis and mooring it to the pubic arch. Note that the urethra is bound to the dorsum of the bulb of the penis. Observe the artery to the bulb (here double); the artery to the crus, called the deep artery; and the dorsal artery which ends in the glans penis. Note the deep dorsal vein which ends in the prostatic plexus.

anterior abdominal wall, *formerly called Scarpa's fascia.*

The superficial perineal fascia is attached to: (1) *the **fascia lata*** enveloping the muscles of the thigh (Fig. 4-16); (2) *the **pubic arch***; and (3) the base of *the **perineal membrane***. Anteriorly the superficial perineal fascia is prolonged over the penis and scrotum (Figs. 3-4 and 3-8*B*), thereby forming a covering for the testes and the spermatic cords.

If the urethra ruptures into the **superficial perineal space,** *the attachments of the perineal fascia determine the direction of flow of the extravasated urine.* Hence it may pass into the areolar tissue in the scrotum, around the penis, and superiorly into the anterior abdominal wall (Fig. 3-8*B*). The urine cannot pass into the thighs because the deep layer of the superficial fascia of the anterior abdominal wall blends with the fascia lata enveloping the thigh muscles, just distal to the inguinal ligament (Fig. 4-16). In addition, the urine cannot pass posteriorly into the anal triangle because the two layers of fascia are continuous with each other around the superficial perineal muscles.

THE SUPERFICIAL PERINEAL SPACE (Figs. 3-8 and 3-14)

This is the *fascial space between the superficial perineal fascia and the perineal membrane.*

In the male the superficial perineal space contains the root of the penis and the muscles associated with it, the proximal part of the spongy uretha, branches of the internal pudendal vessels, and the pudendal nerves.

In the female the superficial perineal space contains the superficial transversus perinei, ischiocavernosus, and the bulbospongiosus muscles, and the *greater vestibular glands* (Fig. 3-39).

THE DEEP PERINEAL SPACE (Figs. 3-8*B*, 3-9, 3-14, and 3-15)

This is the fascial *space enclosed by the superior and inferior fasciae of the urogenital diaphragm.*

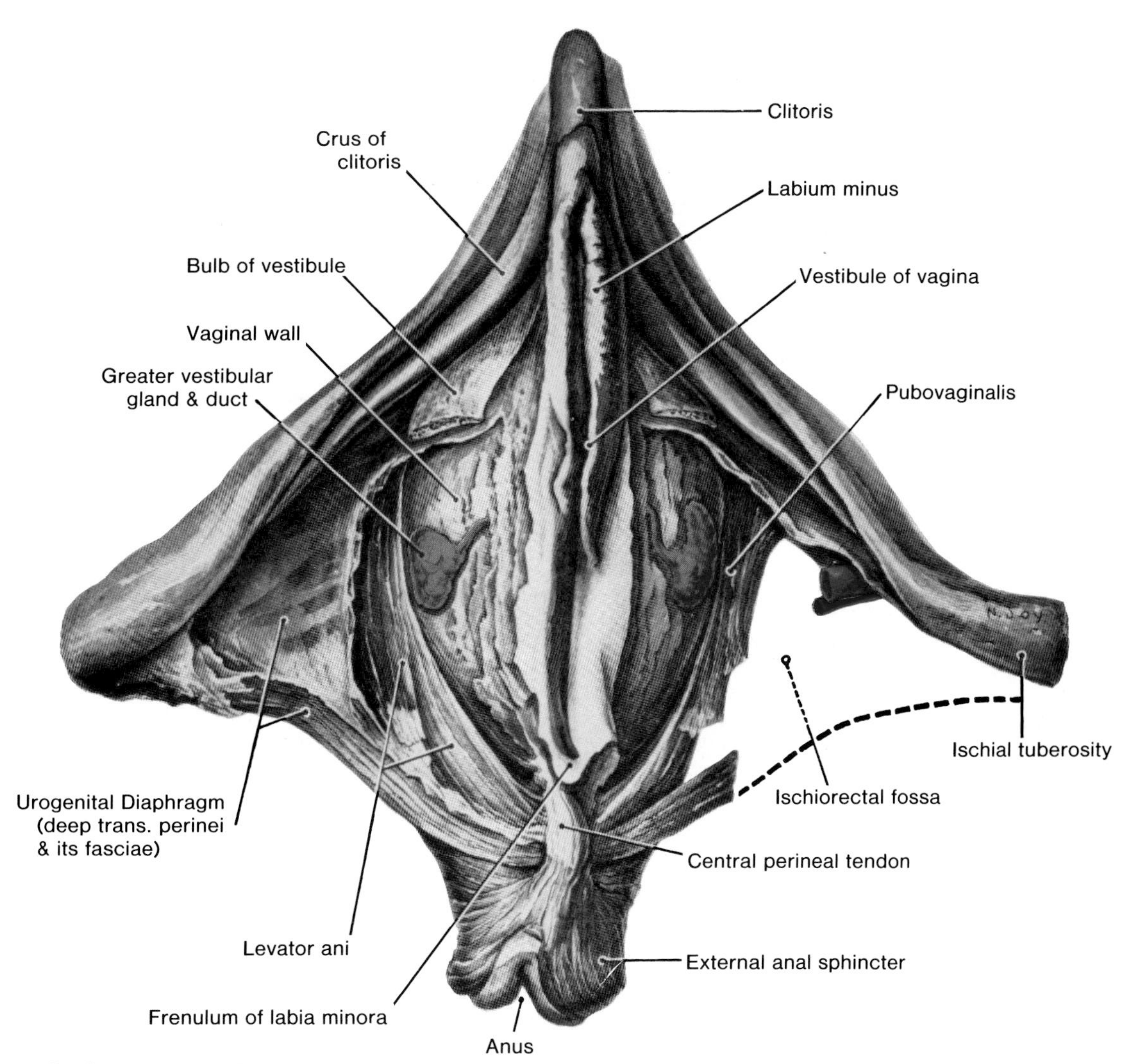

Figure 3-10. Drawing of the female perineum showing the urogenital diaphragm. This diaphragm and its fasciae are partly cut away on the right side and are extensively cut away on the left side. Examine the urogenital diaphragm, a sheet of striated muscle, mainly deep transversus perinei, placed between a superior and an inferior sheet of fascia and having two parts: (1) a posterior part which is a strong fleshy band that meets its fellow in the central tendon of perineum (perineal body), and (2) an anterior part which is more areolar than fleshy. Medially, the urogenital diaphragm and its fasciae have been raised from the sloping inferior surface of the levator ani muscle and detached from the sloping external wall of the vagina with which it fuses. Observe the anterior parts of the levatores ani muscles (pubovaginales) meeting posterior to the vaginal orifice. Also see Figure 3-14.

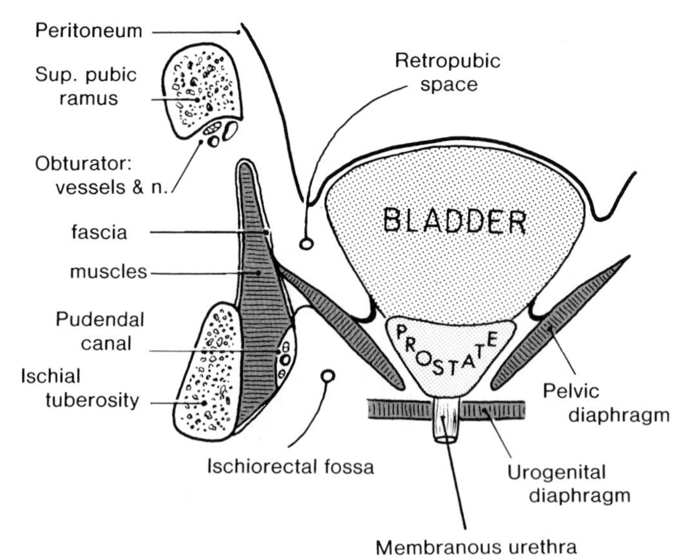

Figure 3-11. Diagrammatic coronal section of the pelvis illustrating the main part of the funnel-shaped pelvic diaphragm. The pelvic diaphragm forms the floor of the abdominal and pelvic cavities and consists of the paired levatores ani and coccygeus muscles together with their superior and inferior fasciae. Observe that only the pelvic diaphragm (levator ani portion) intervenes between the ischiorectal fossa and the retropubic space. The urogenital diaphragm is formed basically by the sphincter urethrae muscle (Fig. 3-12), which is attached to the rami of the pubis and ischium.

In the male the deep perineal space is occupied by the *membranous urethra,* the *sphincter urethrae,* the *bulbourethral glands,* and the deep transversus perinei muscles (Fig. 3-14*C* and 3-15).

In the female the deep perineal space is occupied by *part of the urethra,* the *sphincter urethrae muscle,* the inferior *part of the vagina,* and the *deep transversus perinei muscles.*

The deep perineal space in both sexes also contains the blood vessels and nerves associated with the structures within it.

THE ANAL REGION

The anal region of the perineum (Figs. 3-2 to 3-6) is often called *the anal triangle.*

The boundaries of the anal region are: *posteriorly, the tip of the* **coccyx** and *anteriorly,* the line joining the **ischial tuberosities.**

The anal region is related anteriorly to the posterior border of the **urogenital diaphragm** (Fig. 3-14) and posterolaterally to the **sacrotuberous ligaments** (Figs. 3-2 and 3-16). Overlying the anal region is the gluteus maximus muscle (Figs. 3-1 and 3-21).

The anal region contains the anal orifice (*anus*), *the external anal sphincter, and the ischiorectal fossae* (Figs. 3-2, 3-11, and 3-17). The anal canal passes through the floor of the pelvis and opens on the surface of the perineum as the anus (Figs. 3-2 to 3-6). The Latin word *anus* means a ring.

The perianal skin contains large sebaceous and sweat glands and is pigmented. This dark skin is thrown into radiating folds, giving it a characteristic puckered appearance. The folds are produced by the pull of the underlying **fibroelastic septa** (Fig. 3-17).

THE SPHINCTER ANI EXTERNUS MUSCLE
(Figs. 3-2, 3-4, 3-10, 3-12, and 3-17 to 3-20)

The large external anal sphincter is under voluntary control. It surrounds the inferior end of the anal canal and lies in the perineum. It forms a broad band (2 to 3 cm wide) on each side of the anal canal which *consists of three parts*: *subcutaneous, superficial,* and *deep,* but they are not distinctly separated from each other. Many branches of the inferior rectal nerve and vessels pass between the superficial and deep parts of this muscle (Fig. 3-4).

In general the fibers of the external anal sphincter run from the central perineal tendon to the anococcygeal ligament (Fig. 3-20).

The subcutaneous part of the external anal sphincter (Figs. 3-4, 3-12, and 3-17*A*) is slender and surrounds the anus. These subcutaneous fibers, which cross anterior and posterior to the anus, have no bony attachments.

The superficial part of the external anal sphincter is elliptical or oval in shape. *Its fibers extend anteriorly from the tip of the coccyx and the anococcygeal ligament around the anus to the central perineal tendon* (Figs. 3-10 and 3-20), mooring the anus to the median plane.

The deep part of the external anal sphincter surrounds the anal canal like a collar. Some of

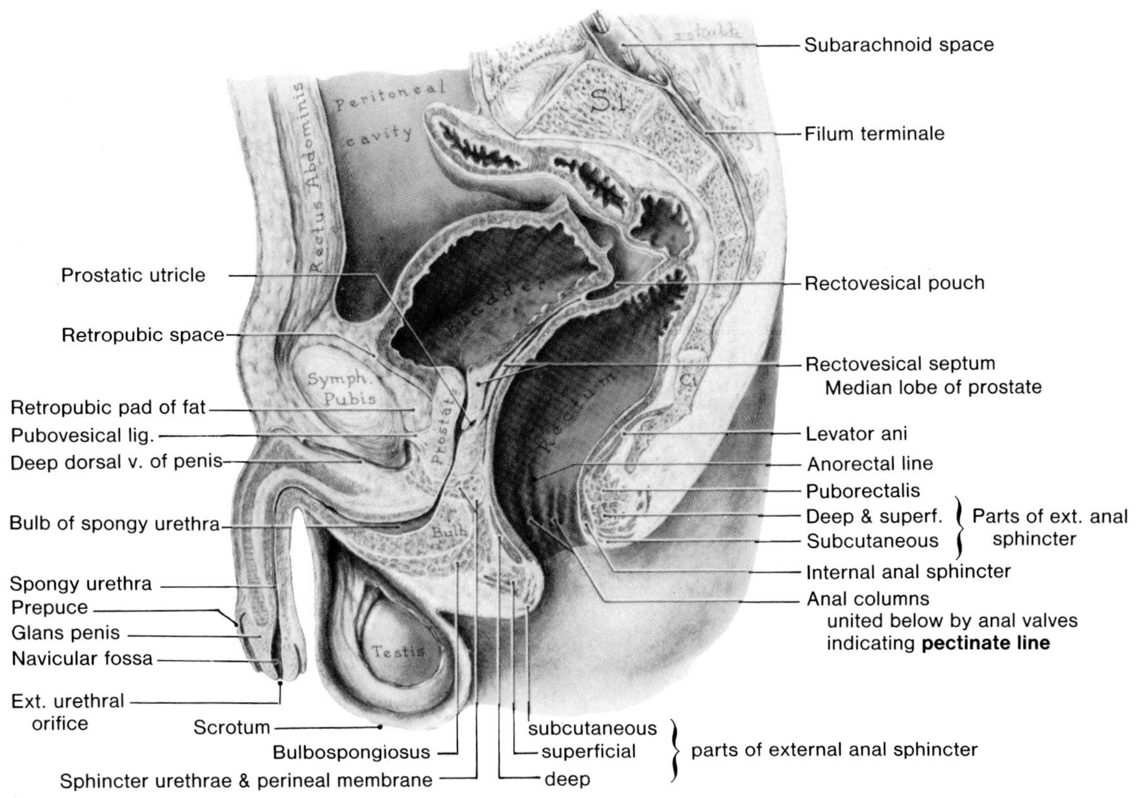

Figure 3-12. Drawing of a median section of the *male pelvis*. Note that the involuntary internal anal sphincter muscle does not descend so far as the voluntary external anal sphincter muscle, and that it is separated from it by an areolar layer. Examine the two layers of rectovesical septum (fascia) in the median plane between the bladder and the rectum. Note that on each side it contains the ductus deferens, the seminal vesicle, and the vesical vessels. Observe the peritoneum passing from the abdominal wall superior to the symphysis pubis to the distended bladder, over the bladder to the bottom of the rectovesical pouch, and up the anterior aspect of the rectum.

these fibers cross to join the opposite superficial transverse perineal muscle. *The deep part of the external anal sphincter arises from the central perineal tendon* (Fig. 3-10) and fuses with the puborectalis part of the levator ani (Fig. 3-17*A*). Superiorly it blends with the levator ani and is not always sharply defined from it.

Nerve Supply. Perineal branch of **fourth sacral nerve and inferior rectal nerves.**

Actions. Closes anus and draws anal canal anteriorly, thereby increasing *the anorectal angle* (the angle between the anal canal and rectum). The deep part of the external anal sphincter is assisted in this action by the ***puborectalis muscle***, part of the levator ani, which forms the sling occupying the anorectal angle (Fig. 3-55).

THE ISCHIORECTAL FOSSA (Figs. 3-2, 3-10, 3-11, 3-15, 3-18, 3-21, and 3-22)

On each side of the anal canal and rectum there is a large *wedge-shaped, fascia-lined space* called the ischiorectal fossa.

The ischiorectal fossa is located between the skin of the anal region and the pelvic diaphragm (Figs. 3-11 and 3-21).

The apex of the ischiorectal fossa lies superiorly, where the levator ani muscle arises from the obturator fascia (Figs. 3-11, 3-21, and 3-22).

Because the levator ani is shaped like a funnel, the ischiorectal fossa on each side is wide inferiorly and narrow superiorly. The apex is located about 6 cm superior to the ischial tuberosity (Fig. 3-11).

The base of the ischiorectal fossa is formed by the skin and deep fascia of the perineum. *Anteriorly* the ischiorectal fossa continues superior to the urogenital diaphragm as the **anterior recess of the ischiorectal fossa** (Fig. 3-5). This space is filled with loose areolar tissue. There is also a **posterior recess** lateroposteriorly where the gluteus maximus muscle overhangs the ischiorectal fossa (Fig. 3-23).

The ischiorectal fossae of the two sides communicate with each other over the anococcygeal ligament (Fig. 3-20). *Posteriorly* each fossa is continuous with the **lesser sciatic foramen**, superior to the sacrotuberous ligament (Fig. 3-16).

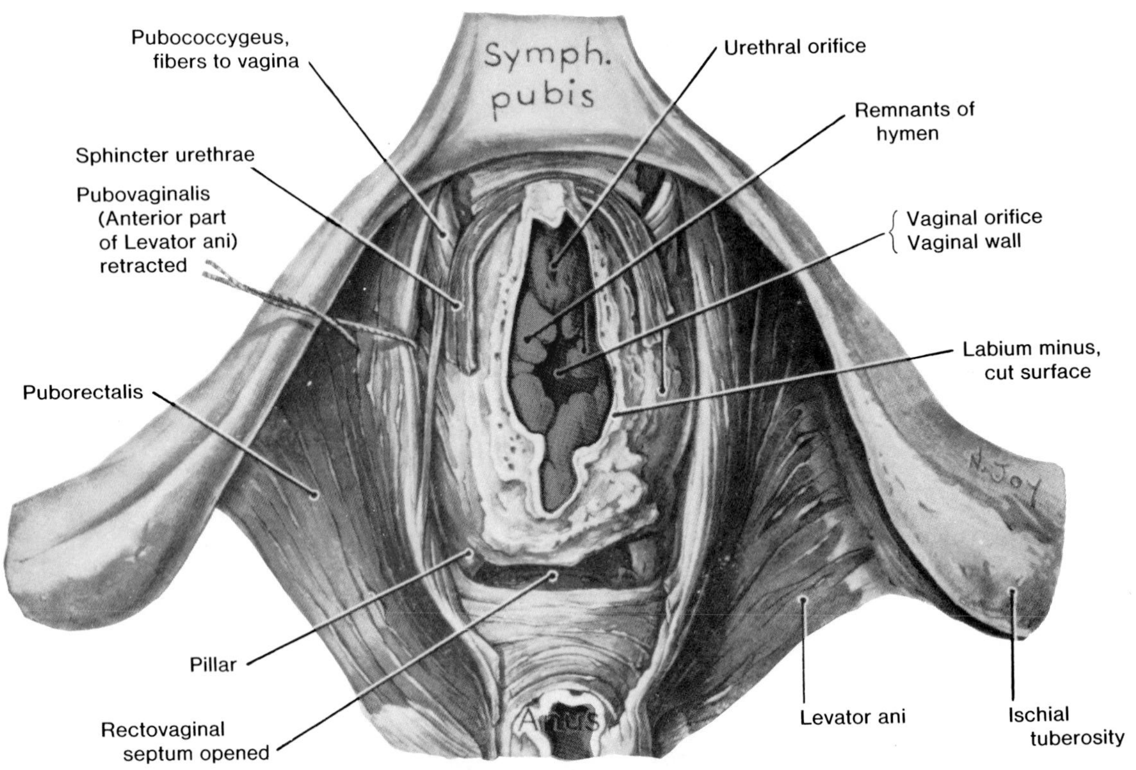

Figure 3-13. Drawing of a dissection of the female perineum. Examine the sphincter urethrae muscle arising from the inferior pubic ramus. Its fibers run medially and some of them cross each other between the urethra and the vaginal orifice. Some of these fibers are attached to the wall of the vagina. Observe the fragments of the torn hymen around the margins of the vaginal orifice. You may hear the vaginal orifice referred to as the *introitus*, the Latin word for entrance. *Vagina* is the Latin term for sheath. In Rome it was used to describe a scabbard or sheath for a sword (L. *gladius*), the common term used by Romans for the penis. *Hymen* was the Greek god of marriages; this explains the relationship of the hymen to virginity and marriage.

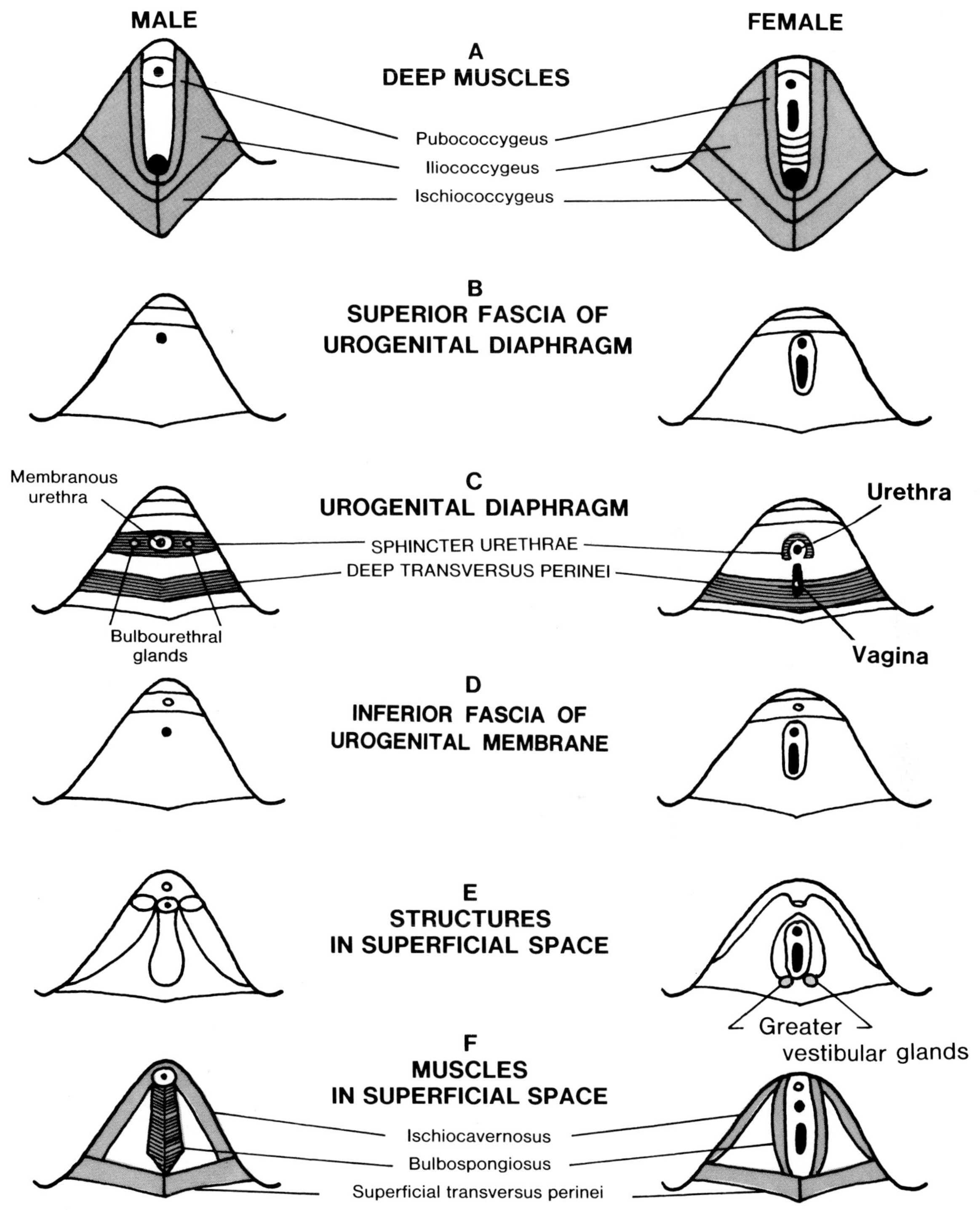

Figure 3-14. Schematic diagrams illustrating the layers of the perineum in the male and female and showing the layers of the perineum built up from deep to superficial. In *A*, the angle between the two ischiopubic rami is almost filled by the three coccygeus muscles. The urethra (and vagina in the female) peers through anteriorly, the rectum posteriorly. A superior layer of fascia *B* and an inferior layer of fascia *D* enclose a deep perineal space *C* containing two muscles and, *in the male*, the bulbourethral glands. The "sandwich" formed by the two layers of fascia and the contents of the deep space comprise the urogenital diaphragm. Observe that the superficial and deep layers of perineal fascia are attached to the ischiopubic ramus and to the posterior margin of the urogenital diaphragm; hence they enclose the superficial perineal space which contains the structures shown in *E* and the muscles shown in *F*. *In the female* the greater vestibular glands lie posterior to the bulb of the vestibule (Fig. 3-39).

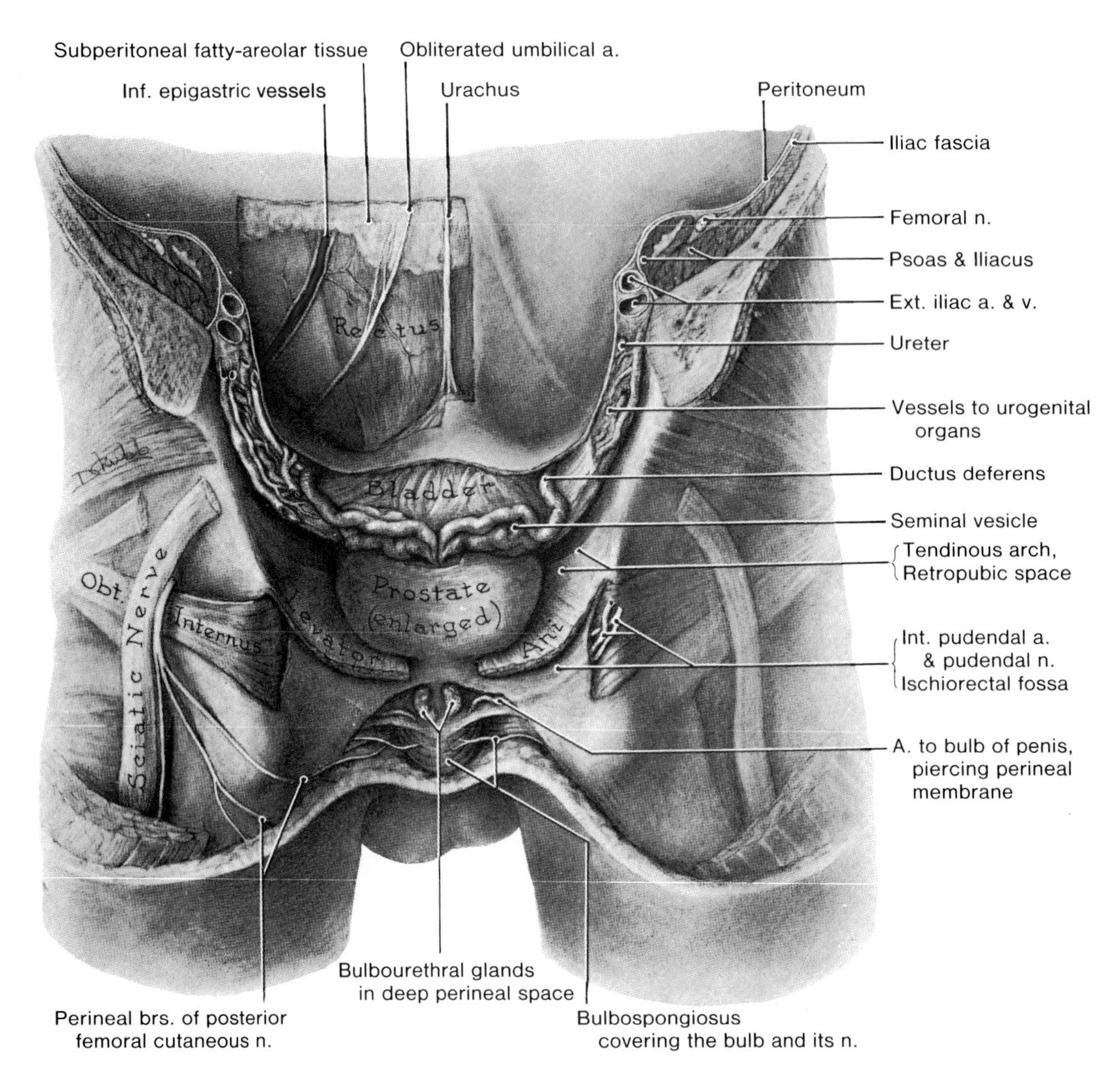

Figure 3-15. Drawing of a posterior view of a coronal section of the male pelvis just anterior to the rectum. Observe the inferior epigastric artery and its venae comitantes entering the rectus sheath. Note that the obliterated umbilical artery and the urachus, like the urinary bladder, are in the subperitoneal fatty-areolar tissue. Examine the femoral nerve lying between the psoas and iliac muscles, outside the psoas fascia, which is attached to the pelvic brim. Note that the external iliac artery and vein lie inside this fascia. Observe that the ductus deferens and the ureter are both subperitoneal. Near the bladder note that the ureter is accompanied by several vesical vessels enclosed in rectovesical fascia. Examine the levator ani and its fascial coverings separating the retropubic space from the ischiorectal fossa. Observe the free anterior borders of the levatores ani which are the width of a scalpel handle apart. Examine the bulbourethral glands and the artery to the bulb lying superior to the perineal membrane (inferior fascia of the urogenital diaphragm), *i.e.*, in the deep perineal space.

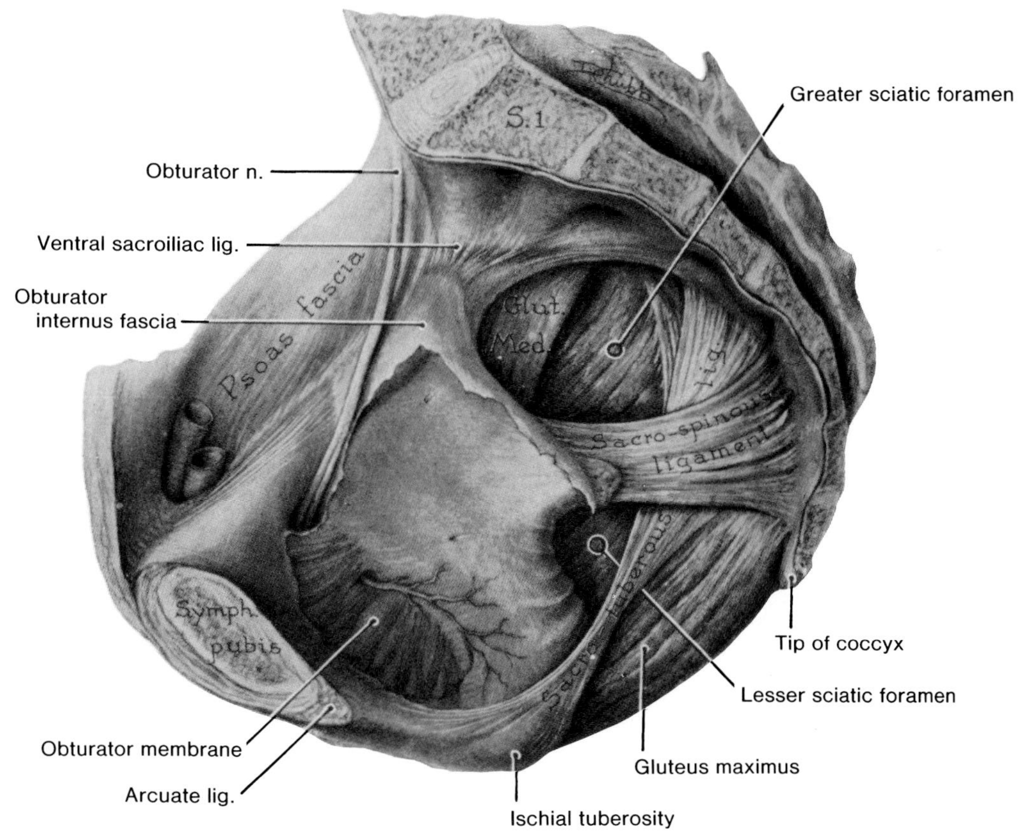

Figure 3-16. Drawing of a dissection showing the bony and ligamentous walls of the pelvis minor of a female. Observe that posterolaterally the coccyx and inferior part of the sacrum are connected to the ischial tuberosity by the sacrotuberous ligament and to the ischial spine by the sacrospinous ligament. Note that part of the sacrum is joined to the ilium by the ventral sacroiliac ligament. Anterior to the sacrotuberous ligament are the greater and lesser sciatic foramina, the one being superior and the other inferior to the sacrospinous ligament. Note that anterolaterally the fascia covering the obturator internus muscle is snipped away and the obturator internus is removed from its osseofascial pocket, thereby exposing the ischium and the obturator membrane. The mouth of this pocket in the lesser sciatic foramen through which the obturator internus escapes from the pelvis. The grooves made by its tendon are conspicuous. Observe that the obturator internus fascia is attached along the line of the obturator nerve superiorly, to the sacrotuberous ligament inferiorly, and to the posterior border of the body of the ischium posteriorly.

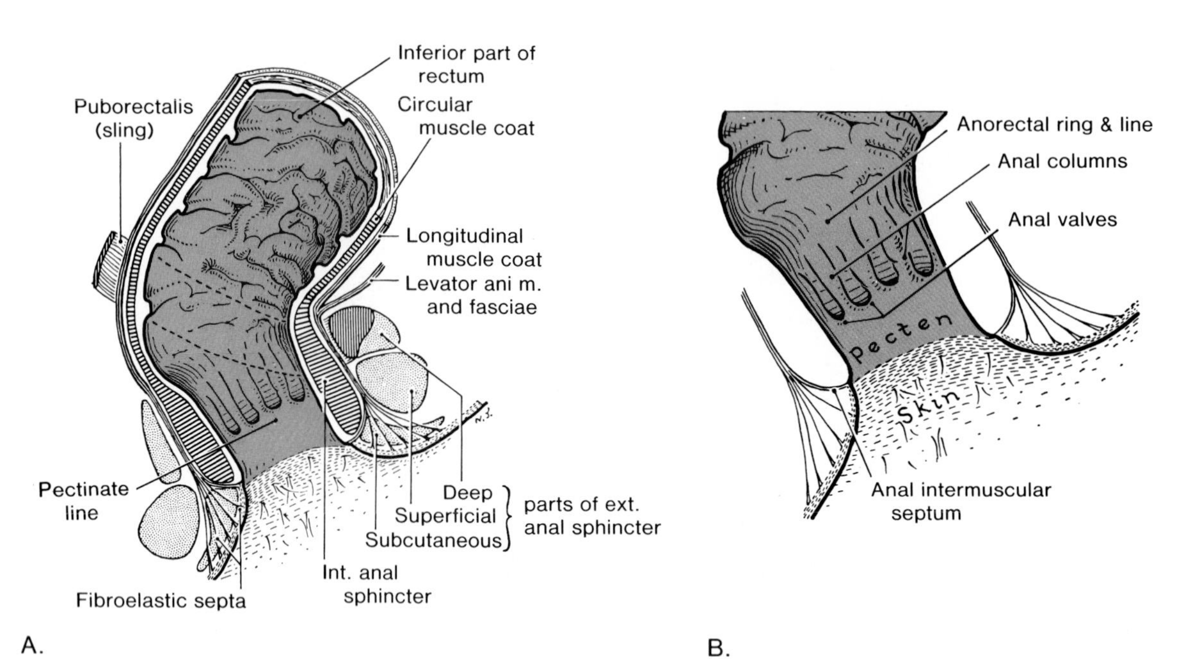

Figure 3-17. Drawings of median sections of the inferior part of the rectum and the anal canal. In *A*, observe that the anal canal, about 3 cm long, extends posteroinferiorly from the distal end of the rectum. Note that the thickened inferior part of the inner circular muscular layer forms an *involuntary* internal anal sphincter. Examine the large *voluntary* external sphincter forming a broad band on each side of the anal canal. Observe that it has three parts and that its deep part is associated with the puborectalis muscle posteriorly (also see Fig. 3-55). The puborectalis is part of the levator ani muscle. In *B*, examine the anal columns, noting that they are vertical folds of mucosa containing twigs of the superior rectal artery and vein. Varicosity of these veins produces *internal hemorrhoids.* Note that the inferior ends of the anal columns are joined by crescentic folds called anal valves. The concavities of the valves (*black*) are called anal sinuses (crypts). The comb-shaped inferior limit of the anal valves forms the *pectinate line* (L. *pecten*, comb). This line indicates the approximate site of the *anal membrane* which normally ruptures toward the end of the embryonic period. The pecten is the 15 mm or so of the anal mucosa inferior to the pectinate line. It has a shiny bluish appearance in living persons. At the anus the moist, hairless mucosa of the anal canal changes to dry, hairy skin.

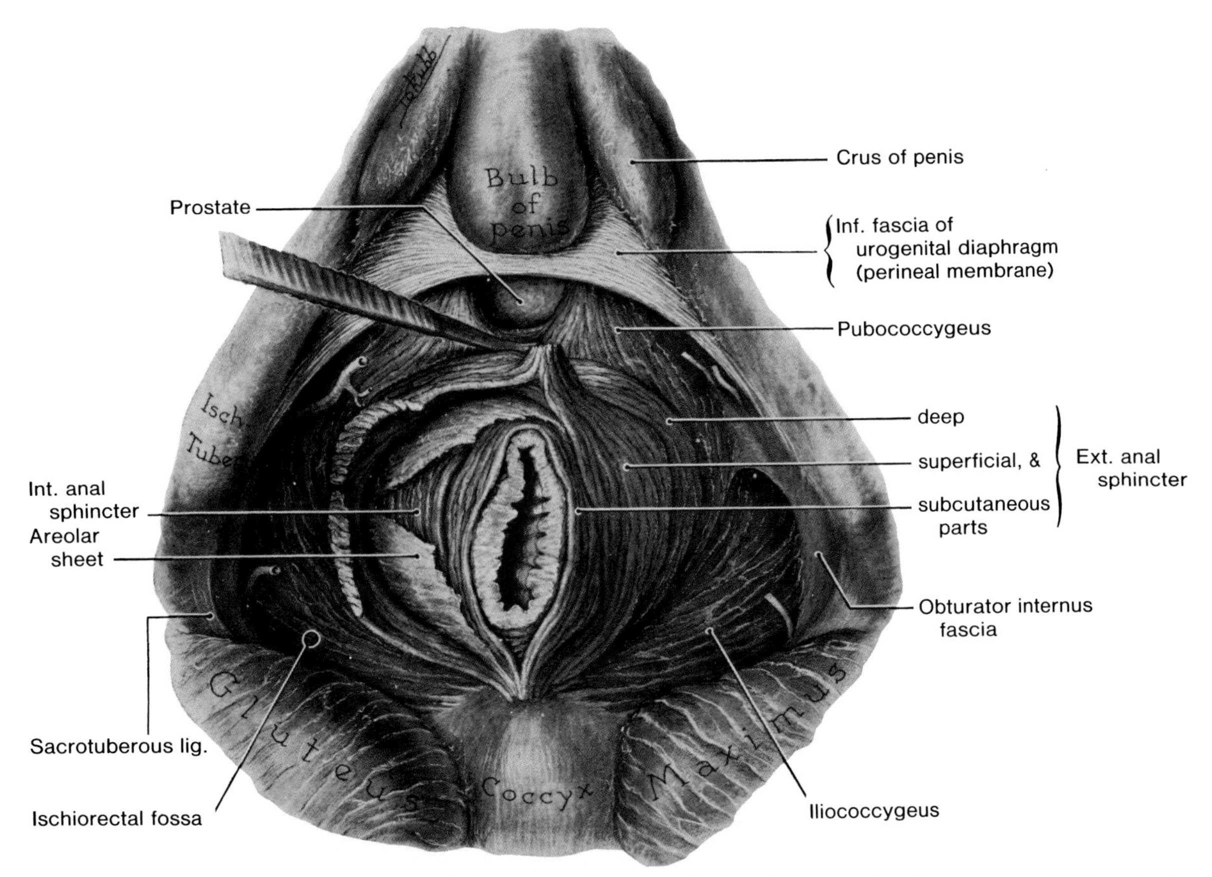

Figure 3-18. Drawing of a dissection of the external anal sphincter. Observe the three parts of this *voluntary sphincter*: (1) subcutaneous, encircling the anal orifice; (2) superficial, anchoring the anus in the median plane to the central perineal tendon anteriorly and to the coccyx posteriorly; and (3) deep, forming a wide encircling band. *On the left*, the superificial and deep parts of the sphincter are reflected and the underlying sheet, consisting of areolar tissue, levator ani fibers, and an outer longitudinal muscular coat of the gut, is cut in order to reveal the inner circular muscular coat of the gut which is thickened to form the *involuntary internal anal sphincter*. Examine the anterior free borders of the levatores ani muscles meeting anterior to the anal canal and pressed posteriorly in order to expose the prostate. *On the right*, observe the remains of the "false roof" of the ischiorectal fossa, *ie*., a layer of fascia that stretches from the obturator internus fascia to the thin fascia covering the levator ani muscle.

Boundaries of the Ischiorectal Fossa. The ischiorectal fossa is bounded ***laterally by the ischium*** and the inferior part of the *obturator internus* (Figs. 3-11, 3-21, and 3-22); ***medially by the rectum and anal canal*** to which the levator ani and sphincter ani externus muscles are applied (Figs. 3-4, 3-21, and 3-22); ***posteriorly by the sacrotuberous ligament*** (Figs. 3-16 and 3-19) and the overlying ***gluteus maximus muscle*** (Figs. 3-2, 3-16, and 3-21); and ***anteriorly by the base of the urogenital diaphragm*** and its fasciae (Figs. 3-4 and 3-11).

Contents of the Ischiorectal Fossa. This wedge-shaped fascial space is *filled with soft fat* (Fig. 3-21), called the **ischiorectal pad of fat**, which is traversed by many tough, fibrous bands and septa (Fig. 3-17).

The ischiorectal pad of fat supports the anal canal, but it is readily displaced to allow feces to pass.

The ischiorectal fossa also contains the internal pudendal vessels and the pudendal nerve. These structures run on the lateral wall of the fossa in a fibrous canal called the **pudendal canal** (Figs. 3-4, 3-11, 3-21, and 3-24). Posteriorly these vessels and the nerve give off the **inferior rectal vessels and nerve**, which pass through the ischiorectal fossa on their way to the anal region (Fig. 3-4). These structures become more and more superficial as they pass toward the surface to supply the external anal sphincter and the perianal skin.

Two other cutaneous nerves, the perforating branch of the second and third sacral nerves and the perineal branch of the fourth sacral nerve, also emerge through the ischiorectal fossa (Fig. 3-4).

The ischiorectal fossa is occasionally the site of infection which may result in the formation of an **ischiorectal abscess** (collection of pus in

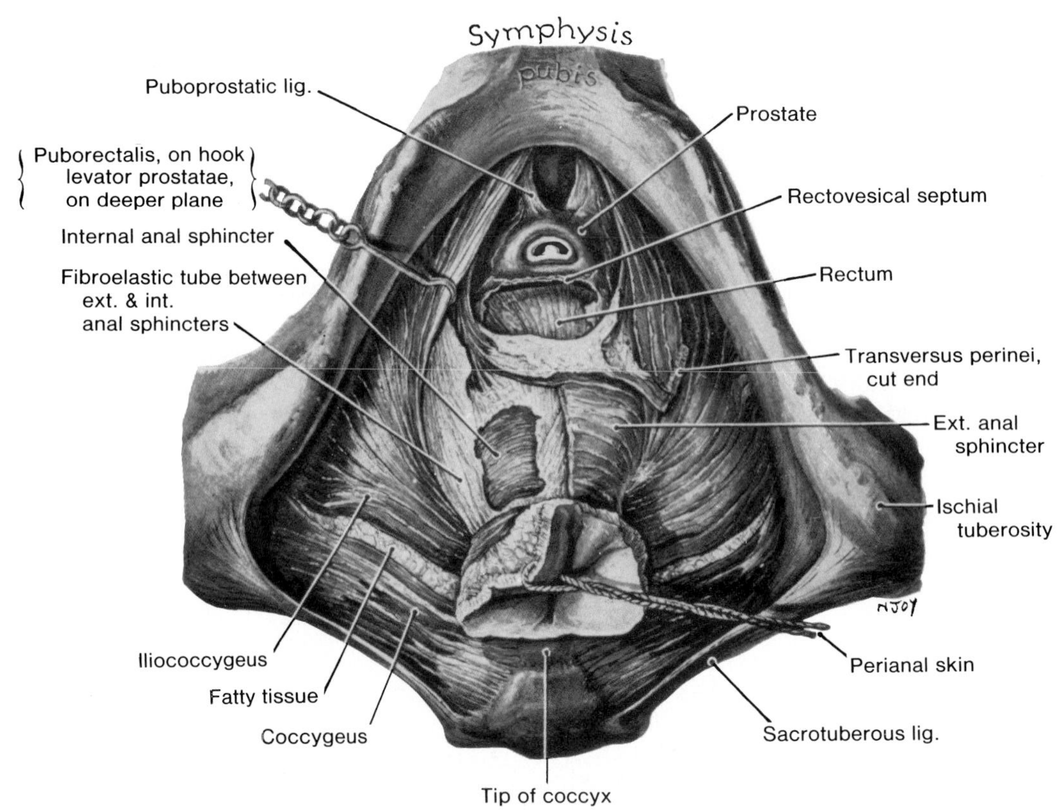

Figure 3-19. Drawing of a dissection of the levatores ani and coccygeus muscles of a male. The urogenital diaphragm and its fasciae have been removed. *Observe that the anal canal is guarded by two sphincters*: (1) the internal anal sphincter and (2) the external anal sphincter extending inferior to the internal anal sphincter. Examine the puborectalis, the sling-like muscle uniting with the external anal sphincter posteriorly and at the sides. The rectovesical septum is a fascial septum which passes from the central perineal tendon to the peritoneum on the floor of the rectovesical pouch. It lies between the inferior portion of the rectum and anal canal posteriorly and the prostate and seminal vesicles anteriorly.

the ischiorectal fossa). Ischiorectal abscesses are annoying and painful (Fig. 3-85).

The fat in the ischiorectal fossa, being near the rectum, is liable to infection. The infection may reach the ischiorectal fossa: (1) following **cryptitis** or inflammation of the anal sinuses (Fig. 3-85); (2) from extension of a **pelvirectal abscess**; (3) following a tear in the anal mucous membrane; or (4) from a penetrating wound in the anal region.

Diagnostic signs of an ischiorectal abscess are fullness and tenderness between the anus and the ischial tuberosity. An ischiorectal abscess may spontaneously open into: (1) the anal canal, (2) the rectum, (3) the skin in the perineum near the anus, or (4) into all these places (Fig. 3-85). Usually these abscesses are opened for free drainage of pus.

Because the ischiorectal fossae communicate, *an abscess in one ischiorectal fossa may spread to the other one* and involve a semicircular area around the posterior aspect of the anus.

THE PUDENDAL CANAL (Figs. 3-4, 3-11, 3-21, and 3-24)

The pudendal canal is a fibrous tunnel on the lateral wall of the ischiorectal fossa through which the pudendal vessels and the pudendal nerve pass. They run anteromedially in the lateral wall of the ischiorectal fossa to the anal canal (Fig. 3-21).

The pudendal canal begins at the posterior border of the ischiorectal fossa and runs from the lesser sciatic notch, adjacent to the ischial spine, to the posterior edge of the urogenital diaphragm (Figs. 3-4 and 3-11).

The pudendal canal is *formed by a splitting of the thickened inferior portion of the obturator fascia* which forms the lateral wall of the ischiorectal fossa and covers the obturator internus muscle (Fig. 3-21).

There are three structures at the posterior end of the pudendal canal: *the internal pudendal artery, the internal pudendal vein, and the pudendal nerve* (Figs. 3-22 and 3-24). These structures pass along the ischiopubic ramus toward the urogenital diaphragm.

The pudendal nerve supplies most of the innervation to the perineum. Toward the distal end of the pudendal canal, **the pudendal nerve splits to form the dorsal nerve of the penis** (or of the clitoris) and the **perineal nerve** (Figs. 3-9, 3-24, and 3-39). These nerves run anteriorly on each side of the internal pudendal artery. The perineal nerve gives off scrotal (or labial) branches and continues to supply the muscles of the urogenital diaphragm (Fig. 3-25).

The *dorsal nerve of the penis (or of the clitoris)*, a sensory nerve, runs through the deep perineal space (Figs. 3-8*B* and 3-24) to reach its area of supply.

The inferior rectal vein passes from the inferior end of the anal canal to empty into the internal pudendal vein in the pudendal canal. This vein forms anastomoses with the superior rectal veins of the *portal venous system* (Fig. 2-100).

THE MALE PERINEUM

The male perineum is easier to understand than the female; it will therefore be considered first. Many structures in the male have comparable parts in the female. The male perineum contains the anus and the root of the scrotum and penis.

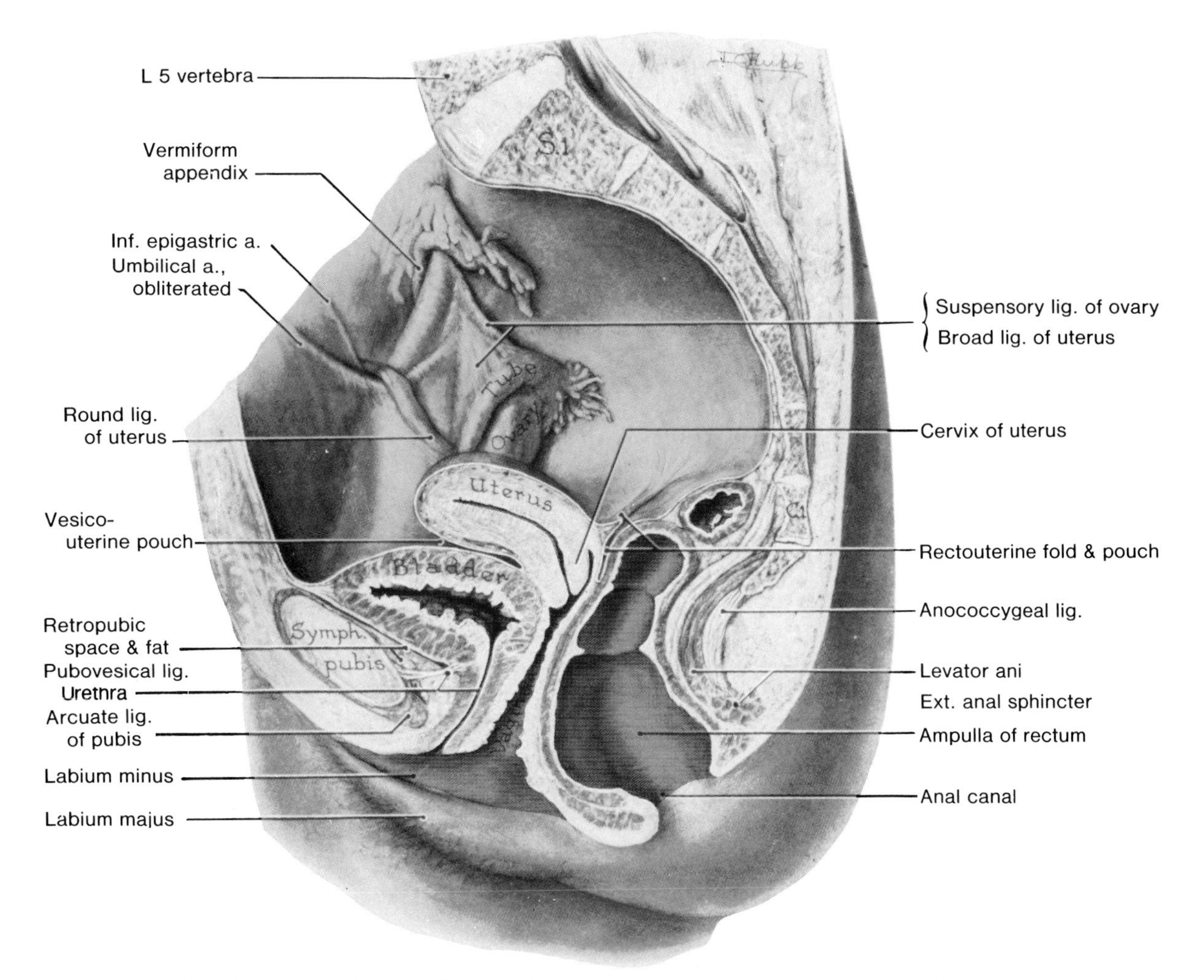

Figure 3-20. Drawing of a median section of a *female pelvis*. The uterus was sectioned in its own median plane and depicted as though this coincided with the median plane of the body, which is seldom the case. Observe the uterine tube and the ovary on the side wall of the pelvis, *i.e.*, in the angle between the ureter and the umbilical artery and medial to the obturator nerve and vessels. Examine the uterus which is bent on itself at the junction of its body and cervix. Note the cervix opening on the anterior wall of the vagina. Observe the ostium (external os) of the uterus at the level of the superior end of the symphysis pubis. Examine the posterior fornix covered with 1.25 cm or more of the rectouterine pouch. Observe that the urethra, vagina, and rectum are parallel to one another. Note that the uterus is nearly at right angles to them when the bladder is empty.

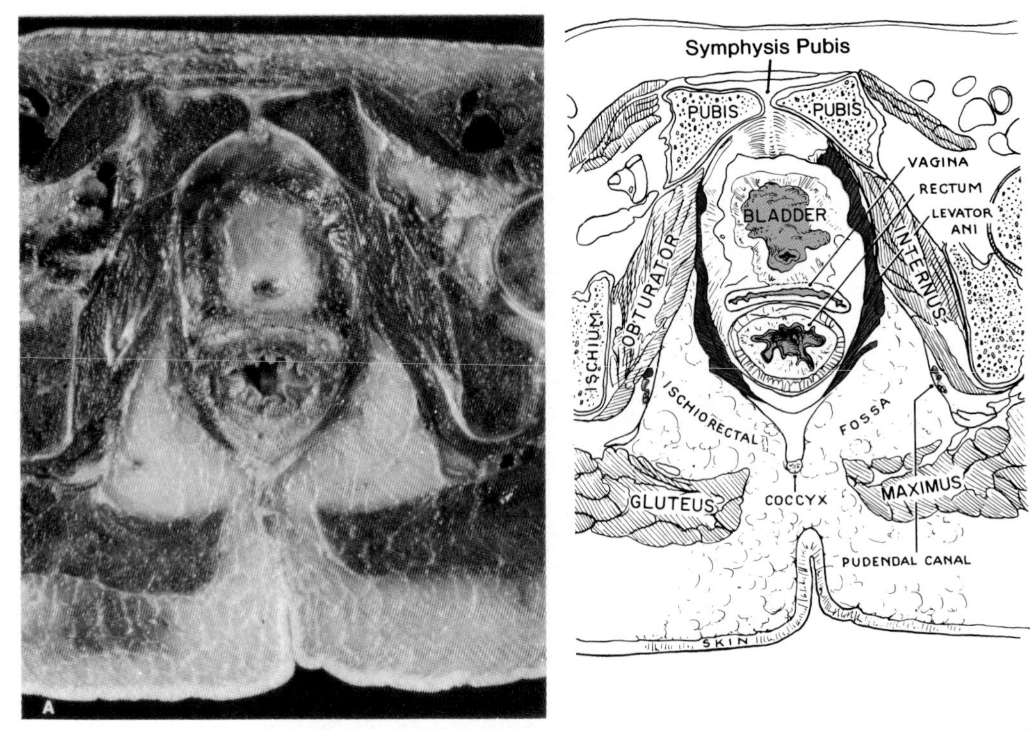

Figure 3-21. Horizontal section of a female pelvis. *A*, photograph. *B*, explanatory drawing. Examine the ischiorectal fossae noting that they are wedge-shaped fascial spaces, one on each side of the rectum and the anal canal. Observe that they are filled with fat which allows the rectum to become distended and to empty. Note that each fossa is bounded *laterally* by the ischium (from which the obturator internus muscle arises); *medially* by the rectum (anal canal inferiorly) to which the levator ani and external anal sphincter are applied; *posteriorly* by the sacrotuberous ligament and the overlying gluteus maximus muscle; and *anteriorly* by the base of the urogenital diaphragm and its fasciae.

STRUCTURES IN THE SUPERFICIAL PERINEAL SPACE (Figs. 3-8 and 3-14)

The superficial perineal space is the fascial space between the superficial perineal fascia and the inferior fascia of the urogenital diaphragm (perineal membrane). It contains the *root of the penis* and the *muscles of the penis* (Fig. 3-4), the proximal part of the *spongy urethra*, and branches of the *internal pudendal vessels* and *pudendal nerves.*

STRUCTURES IN THE DEEP PERINEAL SPACE (Figs. 3-8, 3-14, and 3-15)

The deep perineal space is the fascial space enclosed by the superior and inferior fasciae of the urogenital diaphragm. It contains the sphincter urethrae and the deep transversus perinei muscles, the bulbourethral glands (Fig. 3-15), and the **membranous urethra** (Figs. 3-8, 3-29, and 3-30). It also contains the *internal pudendal artery, branches of the perineal nerve* supplying the sphincter urethrae and deep transversus perinei muscles, and the *dorsal nerve of the penis* (Fig. 3-24).

THE SCROTUM (Figs. 3-12, 3-24, and 3-27)

The contents of the scrotum, the ***testes*** and their coverings, were described in Chapter 2 because the scrotum is a pouch that develops from an outpouching of the skin of the anterior abdominal wall (Figs. 2-17 and 2-18).

The scrotum is a cutaneous and fibromuscular sac which is situated posteroinferior to the penis and inferior to the symphysis pubis. The bilateral formation of the scrotum is indicated by the midline **scrotal raphe** (Fig. 3-26), which continues anteriorly on the ventral surface of the penis as the **penile raphe**, and posteriorly along the median line of the perineum to the anus.

The scrotum is composed of skin and the dartos muscle. As the result of cold or exercise, the wall of the scrotum becomes contracted and firm, and its skin becomes rugose.

The dartos muscle (Fig. 2-15) is firmly attached to the skin. *The dartos consists largely of smooth muscle fibers which contract under the influence of cold,* exercise, and sexual stimulation. In old men the dartos

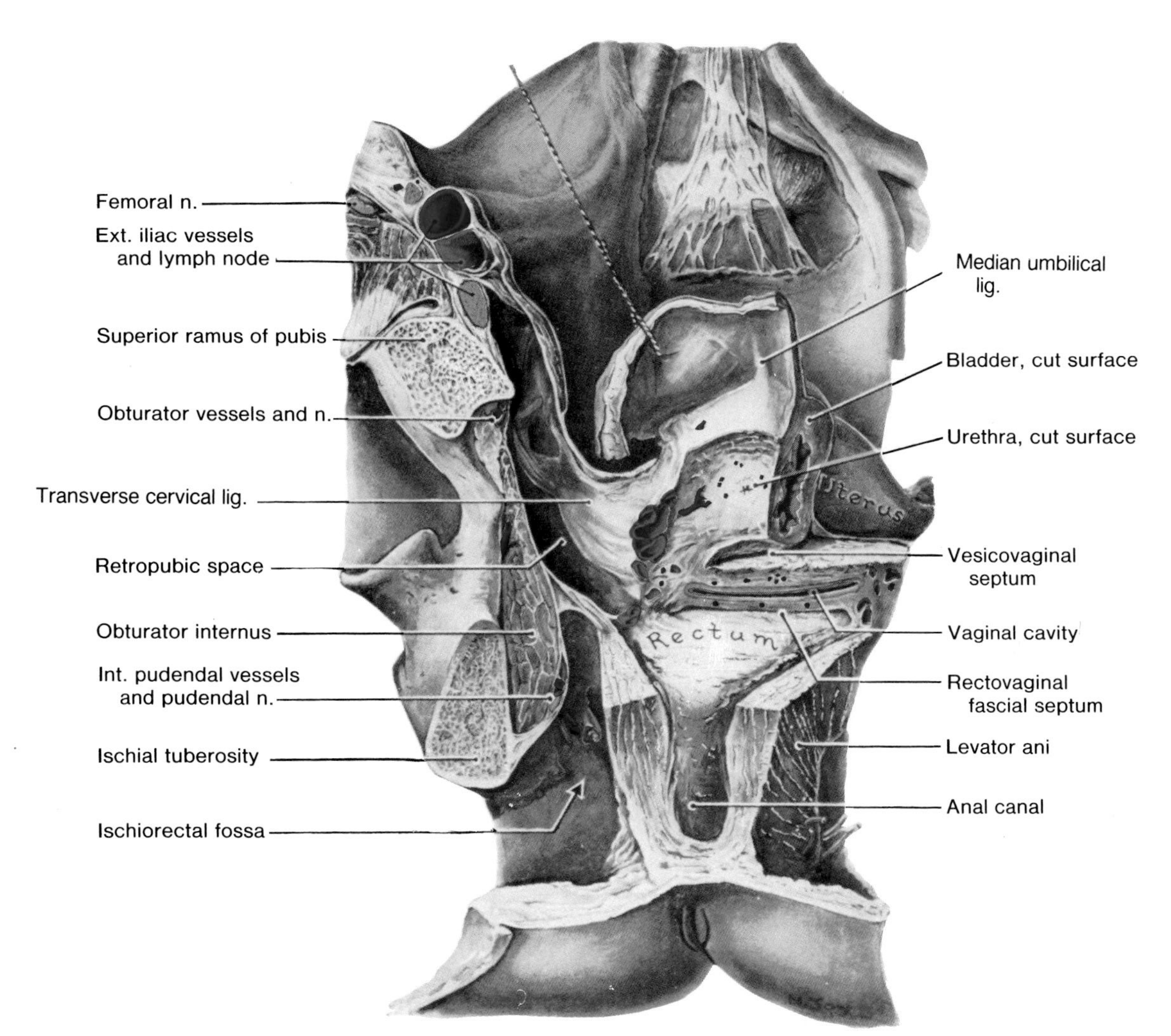

Figure 3-22. Drawing of a dissection of approximately a coronal section of the pelvis showing the suspensory and supporting mechanisms of the vagina. The neck of the bladder and the vagina are cut transversely; the bladder is divided sagittally and is rotated posteriorly. Observe that the partition separating the retropubic space from the ischiorectal fossa is formed by the thin origin of the levator ani muscle from the obturator internus fascia and its areolar coverings. Note: (1) the rectum supporting the posterior wall of the vagina; (2) the posterior wall supporting the anterior wall; and (3) the anterior wall supporting the bladder. Observe the dense areolar tissue within which the vesicovaginal plexus of veins passes posterosuperiorly to join the internal iliac veins. This acts as a suspensory ligament for the cervix and vagina and is called the cardinal ligament. This ligament blends with the fascia that encapsules the vagina and with adjacent fasciae. Note that anteriorly the fascia encapsuling the vagina blends with the vesical fascia and adheres intimately to the urethra; posteriorly it is attached loosely to the rectal fascia.

muscle loses its tone and the scrotum tends to be smooth and hangs down further.

The Blood Vessels of the Scrotum (Figs. 3-4, 3-9, and 3-57). The *external pudendal arteries* supply the anterior aspect of the scrotum and the *internal pudendal arteries* supply the posterior aspect of the scrotum. Branches of the testicular and cremasteric arteries also supply the scrotum (Fig. 2-23*B*).

The scrotal veins accompany the arteries and join the external pudendal veins.

The Nerve Supply of the Scrotum (Figs. 3-4 and 3-24). The anterior part of the scrotum is supplied by the **ilioinguinal nerve** (Fig. 2-7). Its posterior part is supplied by the medial and lateral scrotal branches of the **perineal nerve** (Fig. 3-24), and by the perineal branch of the *posterior femoral cutaneous nerve* (Fig. 3-15).

The Lymphatics of the Scrotum (Figs. 3-73 and 4-35). The lymph vessels from the scrotum drain into the superficial inguinal lymph nodes.

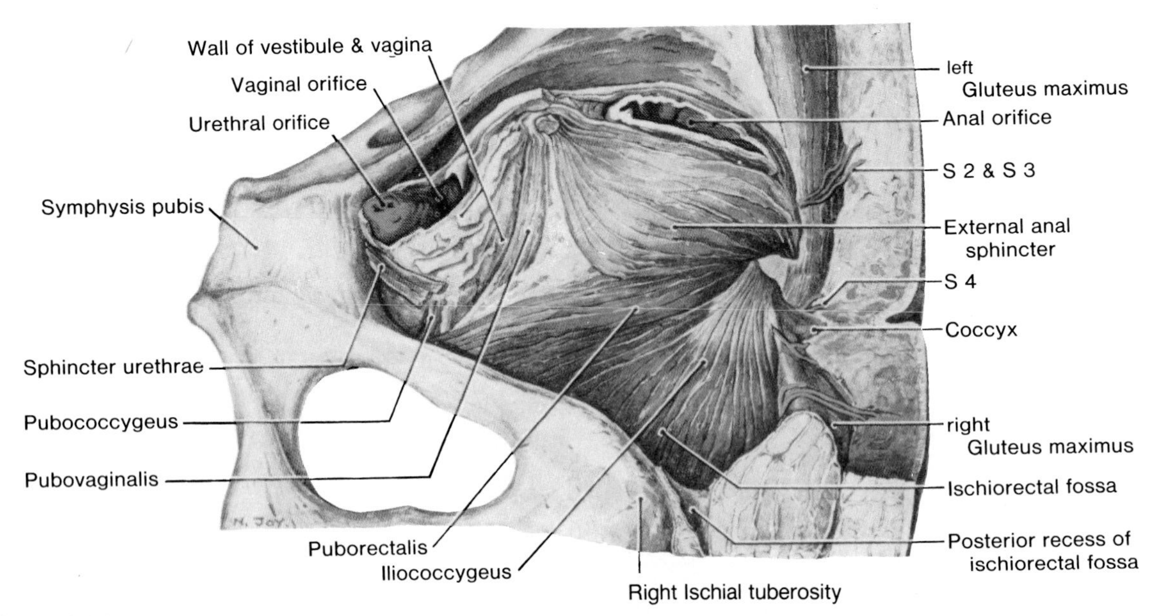

Figure 3-23. Drawing of a dissection of an obliquely tilted lateral view of the pelvis of a female, showing one levator ani muscle. Observe the sphincter urethrae muscle resting like a saddle on the urethra and straddling the vagina. Examine the parts of the levator ani muscle: pubococcygeus, puborectalis, and iliococcygeous muscles. The pubovaginalis muscle is formed by anterior fibers of the levator ani that meet posterior to the vaginal orifice. The puborectal fibers of opposite sides meet in the anorectal junction to form the puborectal sling (Fig. 3-55). The iliococcygeus meets its fellow in an aponeurosis between the rectum and the coccyx (Fig. 3-52).

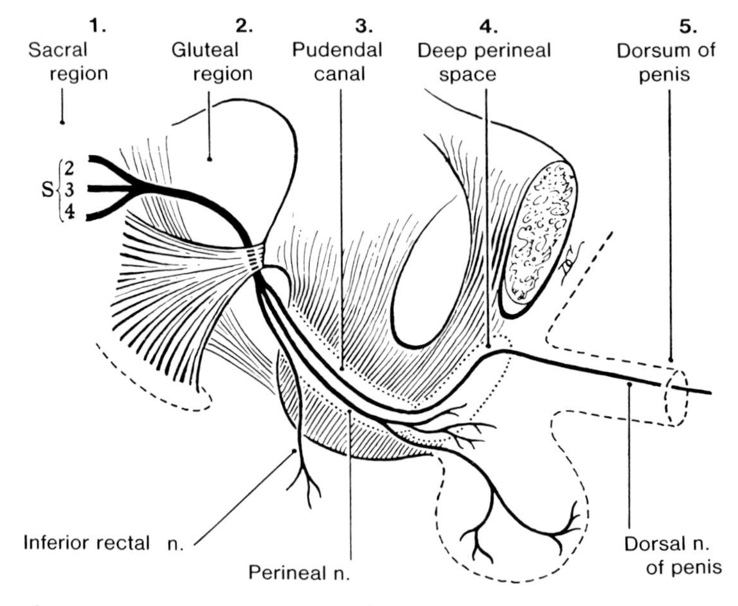

Figure 3-24. Diagram of the *pudendal nerve.* Note the five regions in which it runs and the three divisions into which it divides. Observe the pudendal canal for the pudendal nerve and the internal pudendal vessels (not illustrated here). Observe that the thickened inferior portion of the fascia of the obturator internus muscle in the ischiorectal fossa splits to form the fibrous pudendal canal. Note that the dorsal nerve of the penis runs through the deep perineal space to reach the penis. Also see Figures 3-21 and 3-22.

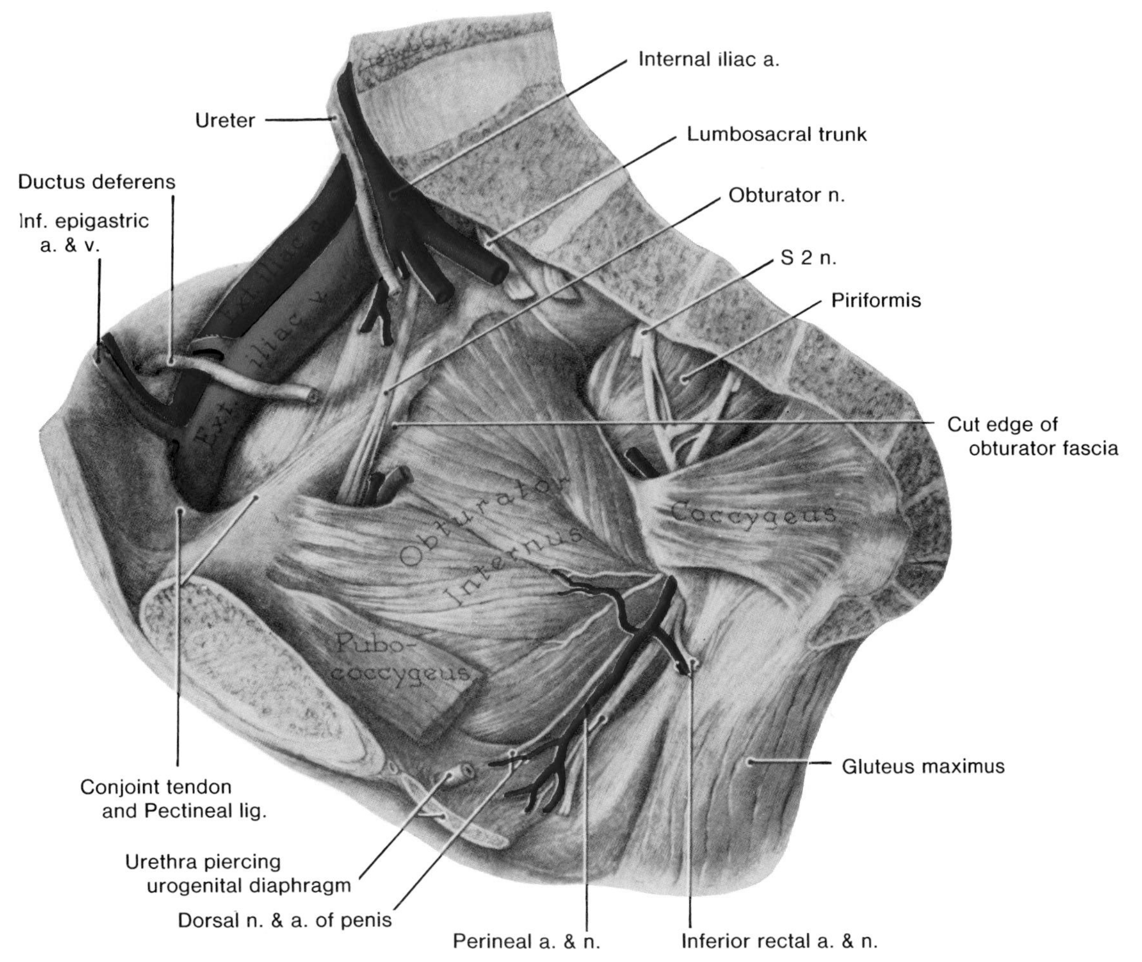

Figure 3-25. Drawing of a dissection of the muscles of the pelvis minor of a male. Observe the obturator internus muscle padding the side wall of the pelvis and escaping through the lesser sciatic foramen; its nerve is also visible. Examine the cut edge of the obturator fascia which is attached to the pelvic brim. The obturator nerve is running in the extraperitoneal fat, medial to the obturator fascia. Note the piriformis muscle padding the posterior wall of the pelvis and escaping through the greater sciatic foramen. Observe the coccygeus muscle concealing the sacrospinous ligament and the pubococcygeus muscle. It is the chief and strongest part of the levator ani muscle springing from the body of the pubis. Examine the obturator nerve, artery, and vein escaping through the obturator foramen. Note the internal pudendal artery and the pudendal nerve exiting through the greater sciatic foramen, re-entering through the lesser sciatic foramen, and taking an anterior course (in the pudendal canal) within the obturator internus fascia to the urogenital diaphragm. In the pelvis major, observe the ductus deferens and the ureter descending across the external iliac artery and vein, the psoas fascia, and the pelvic brim to enter the pelvis minor (true pelvis).

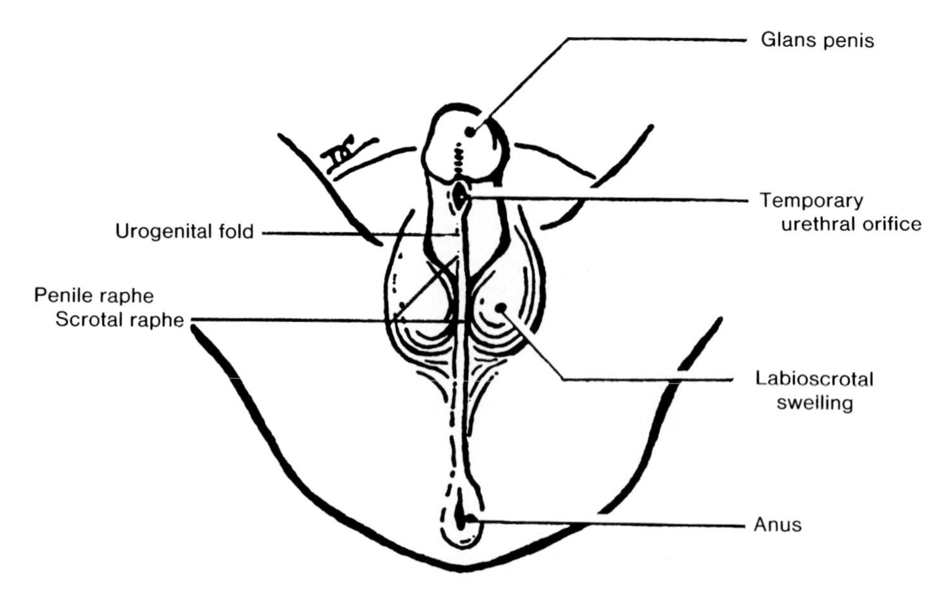

Figure 3-26. Drawing of the developing male external genitalia. The labioscrotal swellings have fused in the midline to form the scrotum. The scrotal raphe marks the site of fusion. The urogenital folds are fusing and enclosing the urethra in the penis. The site of fusion of these folds is indicated by a dark line called the penile raphe. The temporary urethral orifice normally closes during the early fetal period (12th week). If it is present at birth, a congenital malformation called *hypospadias* exists (Fig. 3-68).

THE PENIS (Figs. 3-4, 3-9, 3-12, and 3-26 to 3-35)

The penis (L. tail) is the *organ of copulation* in the male and serves as the common outlet for urine and **semen** (seminal fluid).

The penis is composed of three cylindrical bodies (L. *corpora*) *of erectile, cavernous tissue* that are enclosed in a dense white fibrous capsule, the tunica albuginea (Fig. 3-34). Superficial to the tunica albuginea is the **deep fascia of the penis** which forms a common covering for the corpora cavernosa and the corpus spongiosum (Figs. 3-32 and 3-34).

The skin of the adult penis is very thin, dark in color, and loose (Fig. 3-27). Two of the three erectile bodies, the **corpora cavernosa penis**, are arranged side by side in the dorsal part of the organ (Fig. 3-34). The **corpus spongiosum penis** (corpus cavernosum urethrae) lies ventrally in the median plane.

The corpora cavernosa are fused with each other in the median plane, except posteriorly where they separate to form two **crura** (L. legs). The crura of the penis are attached on each side to the conjoint rami of the pubis and ischium (Figs. 3-28 to 3-31). *The crura of the penis support the corpus spongiosum penis* lying between and inferior to the conjoint rami.

The penis consists of a root and a body. The surface of the penis, which faces posterosuperiorly when the penis is erect, and anteriorly when the penis is flaccid, is called the *dorsum of the penis* (Fig. 3-27). The other aspect is referred to as the ventral surface (*urethral surface*).

The root of the penis, the attached portion (Figs. 3-27 to 3-31), is *located in the superficial perineal space* (Fig. 3-8*B*), between the inferior fascia of the urogenital diaphragm superiorly and the superficial perineal fascia inferiorly (Fig. 3-8).

The root of the penis consists of the two crura of the penis, the bulb of the penis, and the muscles associated with them (Figs. 3-29 and 3-30).

The bulb of the penis is located between the two crura in the superficial perineal space (Figs. 3-15 and 3-29). The enlarged posterior part of the bulb is penetrated superiorly by the urethra (Figs. 3-9 and 3-30).

The body of the penis is the *free part* which is pendulous in the usual flaccid condition (Figs. 3-12 and 3-27). It consists of the corpora cavernosa and the corpus spongiosum (Fig. 3-34).

The dorsum of the penis is continuous with the anterior abdominal wall and faces anteriorly when the penis is flaccid. The ventral surface or urethral aspect faces posteriorly in the flaccid condition and anterosuperiorly when the penis is erect.

The median **penile raphe** on its ventral surface is continuous with the scrotal raphe (Fig. 3-26). *The penile raphe indicates where the urogenital folds fused during the fetal period.*

The spongy urethra runs within the corpus spongiosum penis; hence it is the longest part of the urethra (Figs. 2-102, 3-12, and 3-34). Distally the corpus spongiosum penis expands to form the conical **glans penis,** the concavity of which covers the free blunt ends of

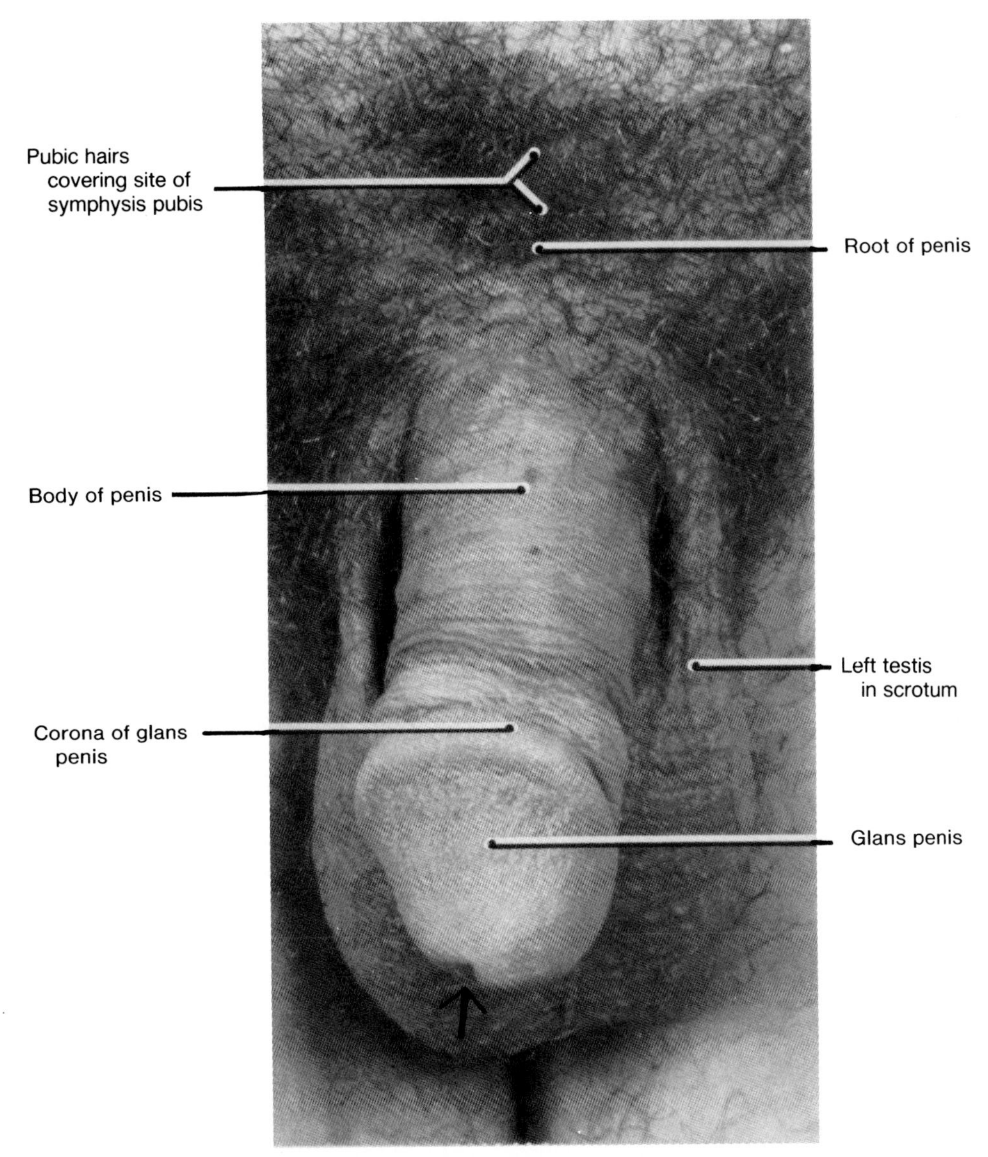

Figure 3-27. Photograph of the penis in the flaccid state and the scrotum of a 27-year-old man (actual size). The glans (L. acorn) penis is bare because the prepuce was removed by circumcision during infancy. The *arrow* indicates the external urethral orifice. The dorsum of the body of the penis (shown here) faces anteriorly when the organ is flaccid as here. The surface closest to the spongy urethra (Fig. 3-32) is called the ventral surface or the urethral surface. Observe that the penile skin is thin and has no hairs except near its root. The length of the erect penis varies, but it is usually 15 to 16 cm long and about 3.75 cm in diameter.

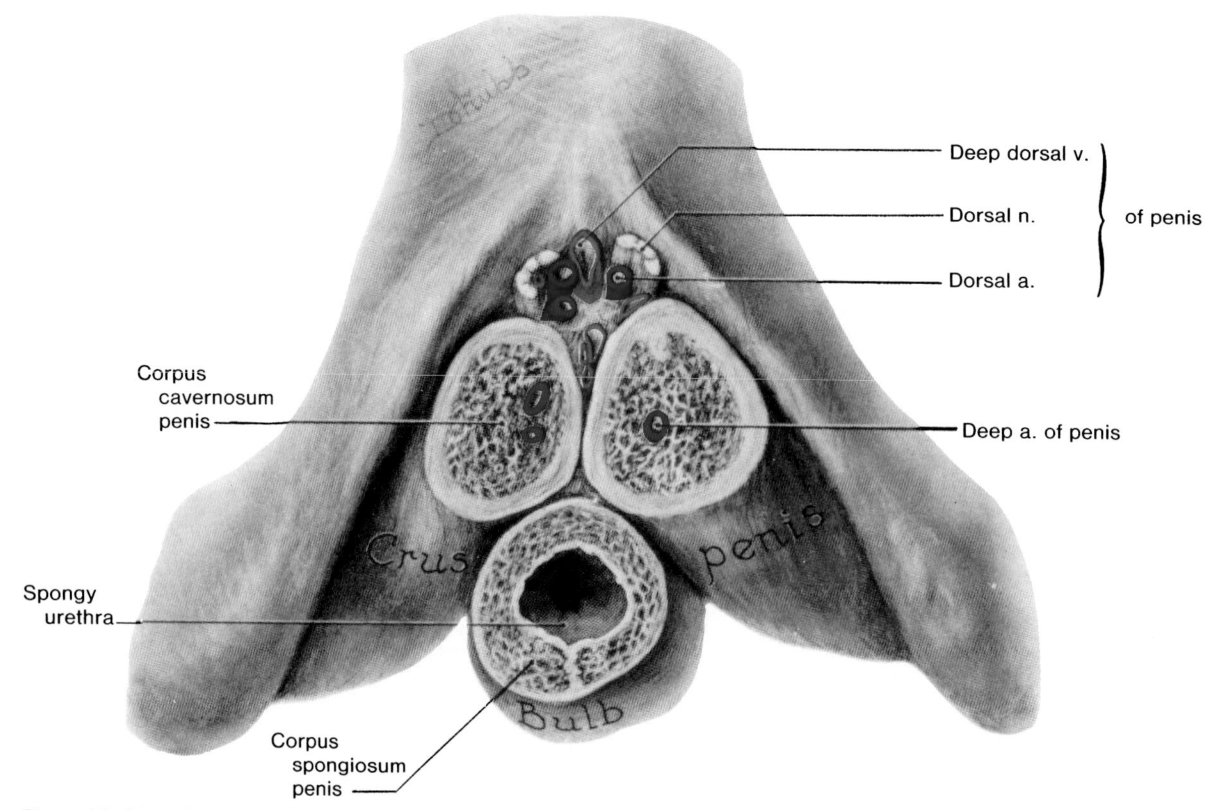

Figure 3-28. Drawing of a section through the root of the penis. Observe that the urethra is dilated within the bulb of the penis, the expanded proximal part of the corpus spongiosum of the penis (Fig. 3-29). Note that the substance of the penis consists essentially of three cylindrical bodies of erectile tissue, two corpora cavernosa and one corpus spongiosum. The fibrous tissue that binds these bodies together and the penile skin have been removed.

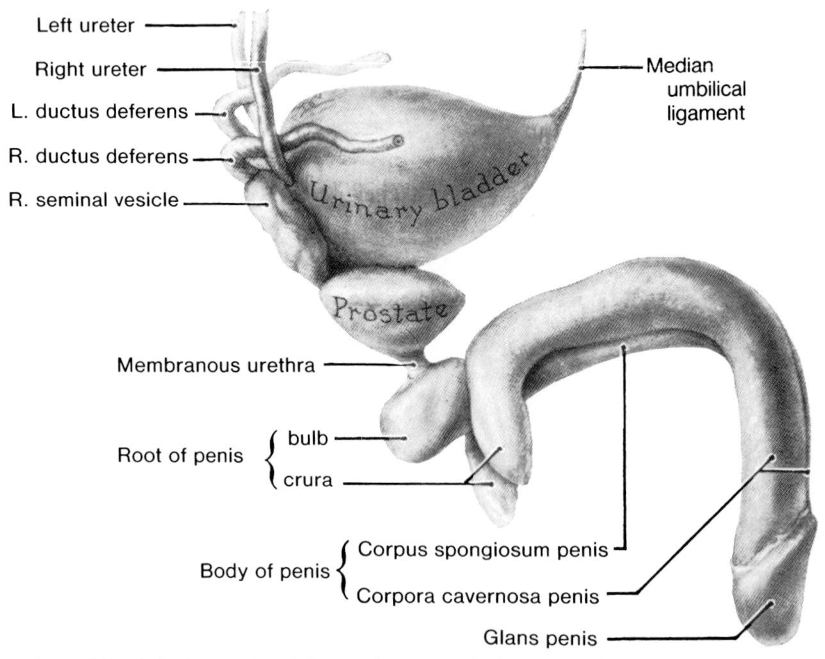

Figure 3-29. Drawing of the inferior parts of the male urogenital organs. Observe that the two corpora cavernosa and the corpus spongiosum form the body of the penis which has an expanded terminal part called the glans penis. The corpus spongiosum penis contains the spongy urethra (Figs. 3-28 and 3-34).

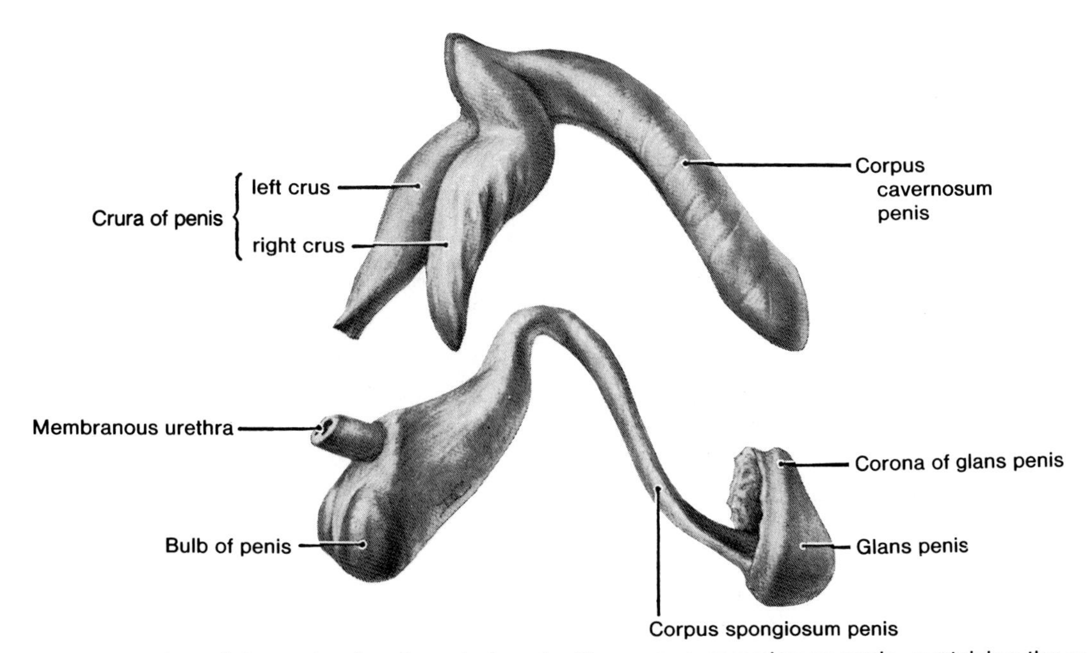

Figure 3-30. Drawing of the parts of a dissected penis. The corpus spongiosum penis, containing the spongy urethra (Fig. 3-34), is separated from the corpora cavernosa penis. Observe that the corpora cavernosa penis are bent where the penis is slung by the suspensory ligament of the penis (Fig. 3-31), and that they are grooved by the encircling vessels (Fig. 3-32). The corpus spongiosum penis is enlarged to form the bulb of the penis and the glans penis. The glans fits like a cap on the blunt ends of the corpora cavernosa penis (Fig. 3-29).

the corpora cavernosa (Figs. 3-29 and 3-30). The prominent margin of the glans penis, called the **corona of the glans**, projects beyond the ends of the corpora cavernosa penis. The corona of the glans overhangs an obliquely grooved constriction called the ***neck of the penis.***

The slit-like opening of the spongy urethra, called the **external urethral orifice**, is near the tip of the glans penis (Figs. 3-27, 3-32 and 3-33).

The skin and fasciae of the penis are prolonged as a free fold or *double layer of skin,* called the **prepuce** (L. *praeputium,* foreskin), which covers the glans penis for a variable extent (Figs. 3-12, 3-32, 3-33, and 3-35). A median fold, called the **frenulum of the prepuce** (Fig. 3-32), passes from the deep layer of the prepuce to a point just inferior to the external urethra orifice.

The weight of the body of the penis is supported by two ligaments which are continuous with the fascia of the penis (Figs. 2-11 and 3-31).

The fundiform ligament (L. sling-like) of the penis arises from the inferior part of the *linea alba* (Fig. 2-11) and splits into two parts which pass on each side of the penis.

The suspensory ligament of the penis (Fig. 3-31) is a condensation of superficial fascia in the form of a thick, triangular fibroelastic band. *The suspensory ligament arises from the anterior surface of the symphysis pubis* and passes inferiorly, splitting to form a sling which attaches to the deep fascia of the penis at the junction of its fixed and mobile parts (*i.e.,* where it bends in the flaccid state, Figs. 3-27 and 3-31).

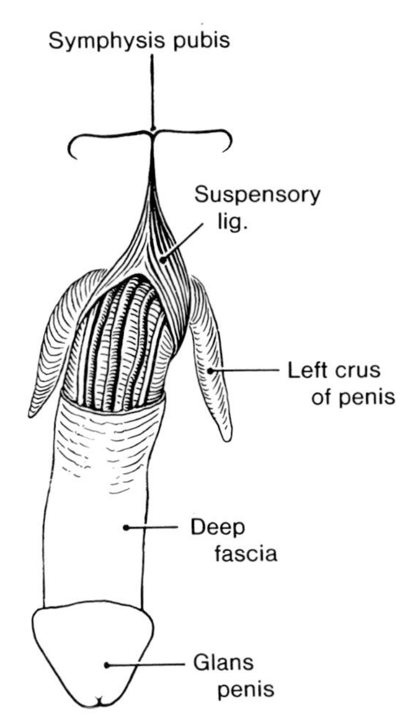

Figure 3-31. Drawing of the penis illustrating its suspensory ligament. The skin and superficial dorsal vein have been removed. The suspensory ligament is a fibroelastic structure which spreads out from the anterior surface of the symphysis pubis and fuses with the deep fascia on the dorsum and sides of the penis. Note that the deep dorsal vessels and nerves lie deep to the suspensory ligament. This ligament is also visible, but not labelled, in the dissection of the anterior abdominal wall shown in Figure 2-8.

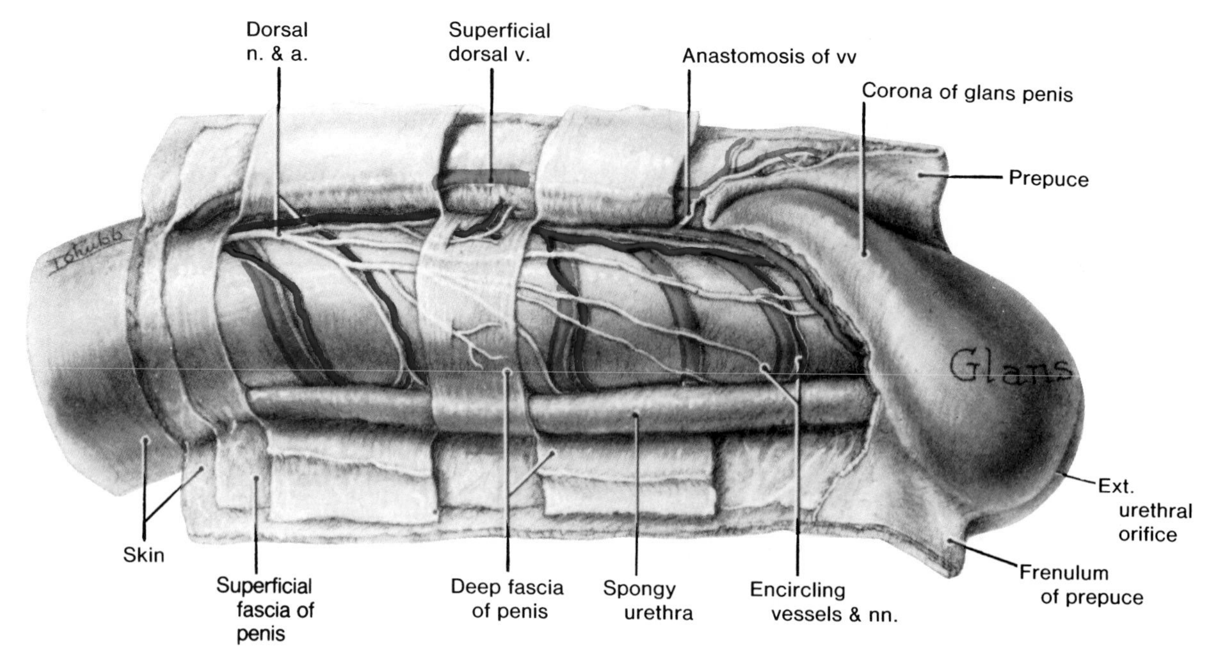

Figure 3-32. Drawing of a lateral view of a dissection of the body of the penis. The three tubular coverings of the penis are reflected. Observe that the penile skin is folded upon itself to form the prepuce. Note the loose, laminated, subcutaneous areolar tissue (dartos muscle and fascia), called the superficial fascia of the penis, which is carried into the prepuce. Observe that the superficial dorsal vein begins in the prepuce and anastomoses with the deep dorsal vein. Examine the deep fascia of the penis which ends at the glans penis. Observe the large encircling tributaries of the deep dorsal vein, the thread-like companion arteries, and the numerous oblique nerves.

In *Peyronie's disease* plaques or strands of dense fibrous tissue form in the corpora cavernosa. This causes a curvature of the penis (chordee) and often pain on erection resulting from the lack of distensibility of the corpora cavernosa. The curvature is most apparent on erection.

The prepuce of the penis is usually sufficiently elastic to permit it to be retracted over the glans penis; however in some cases it is not and fits tightly over the glans and cannot be retracted. This condition is called **phimosis** (F. *phimos*, a muzzle). As there are modified sebaceous glands in the internal surface of the prepuce, the secretions from them usually accumulate in persons with phimosis and may cause irritation.

In some persons there is a **narrow preputial opening** and retraction of the prepuce over the glans penis constricts the neck of the penis so much that interference with the drainage of blood and tissue fluid from the glans penis occurs. In patients with this condition, called **paraphimosis**, the glans enlarges so much that the prepuce cannot be retracted over it. Circumcision is commonly performed in such cases.

Balanitis (G. *balanos*, acorn + *itis*, inflammation), the term for inflammation of the glans penis, may result from irritation owing to phimosis and or uncleanliness.

Circumcision (L. *circumcido*, to cut around) is the operation of removing the prepuce (Fig. 3-27). This is the most commonly performed operation on male infants in North America. In adults circumcision is performed when there is an obvious phimosis.

Amputation of the penis may be required when there is *carcinoma of the penis.*

The penis feels spongy because it is composed of cavernous **erectile tissue** (Fig. 3-34). It consists of interlacing and intercommunicating **cavernous spaces** that are lined by an endothelium which is continuous with that of the veins draining the cavernous spaces.

The cavernous spaces in the three corpora of the penis are separated by fibrous trabeculae. The cavernous spaces are filled with blood in the erect penis, but many of them are empty when the penis is flaccid.

The arteries to the penis are: (1) the ***dorsal arteries*** (Figs. 3-28, 3-32, and 3-34) which run in the interval between the corpora cavernosa penis *on each side of the deep dorsal vein;* and (2) the ***deep arteries,*** which run within each of the corpora cavernosa penis

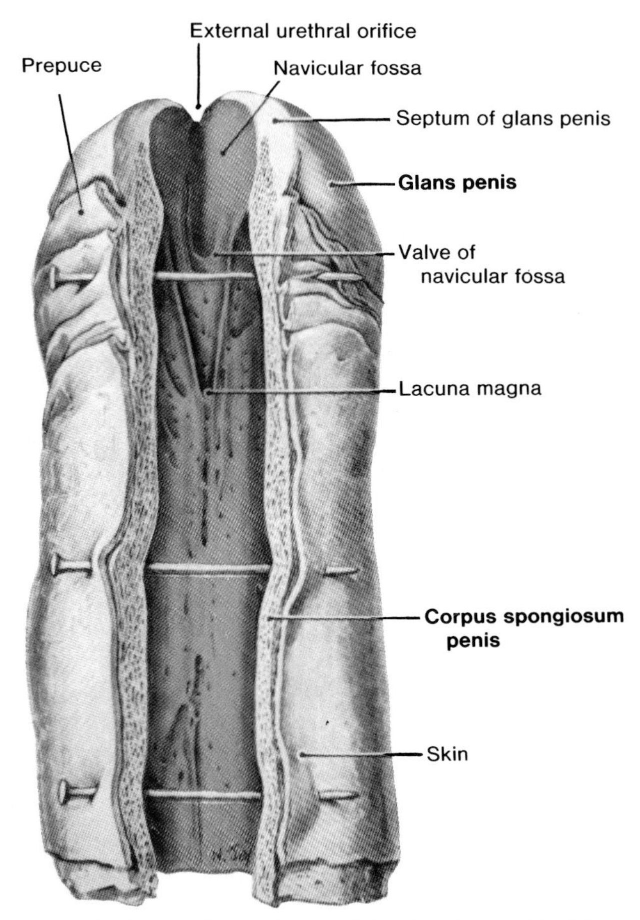

Figure 3-33. Drawing of part of the body of a penis in which a longitudinal incision has been made on its ventral surface and carried through the floor of the urethra. Hence, the view is of the dorsal surface of the interior of the spongy urethra. Observe that the urethra is expanded within the glans penis to form the navicular fossa. Note the large recess (lacuna magna) which may impede the passage of a catheter along the urethra.

(Figs. 3-25 and 3-34). *The dorsal and deep arteries of the penis are* ***branches of the internal pudendal arteries***, which arise in the pelvis from the internal iliac arteries (Figs. 3-25 and 3-57).

The deep arteries of the penis are the principal vessels that supply the cavernous spaces and are involved in an erection. They give off numerous branches that open directly into these spaces. When the penis is flaccid most of these arterial branches have a spiral course; hence they are called **helicine arteries** (G. *helix*, a coil). The smooth muscle of the helicine arteries and trabeculae is supplied both by sympathetic and parasympathetic fibers.

Blood from the cavernous spaces is drained by a ***venous plexus*** that joins the ***deep dorsal vein*** located in the *tunica albuginea* (Figs. 3-28 and 3-34).

The anatomical basis of an erection is as follows. When a male is stimulated erotically, *the smooth muscle of the trabeculae and helicine arteries* relaxes owing to **parasympathetic stimulation**. As a result, the arteries straighten and their lumina enlarge, *allowing blood to flow freely into the cavernous spaces.*

The blood engorges and dilates the cavernous spaces, compressing the venous plexuses at the periphery of the corpora cavernosa and preventing the return of blood. As a result, *the three corpora become rigid and enlarged and the penis erects.*

Following **orgasm** (climax of sexual act) and ejaculation, or passage of erotic thoughts, the penis gradually returns to its flaccid state, a subsiding process called **detumescence**. This results from sympathetic stimulation which causes constriction of the smooth muscle in the helicine arteries and allows blood to flow into the veins. Blood is slowly drained from the cavernous spaces into the deep dorsal vein.

The dorsal nerve of the penis is one of the two terminal branches of the pudendal nerve; the perineal nerve is the other (Figs. 3-24, 3-28, and 3-34).

The dorsal nerve of the penis arises in the pudendal canal and passes anteriorly into the deep perineal space.

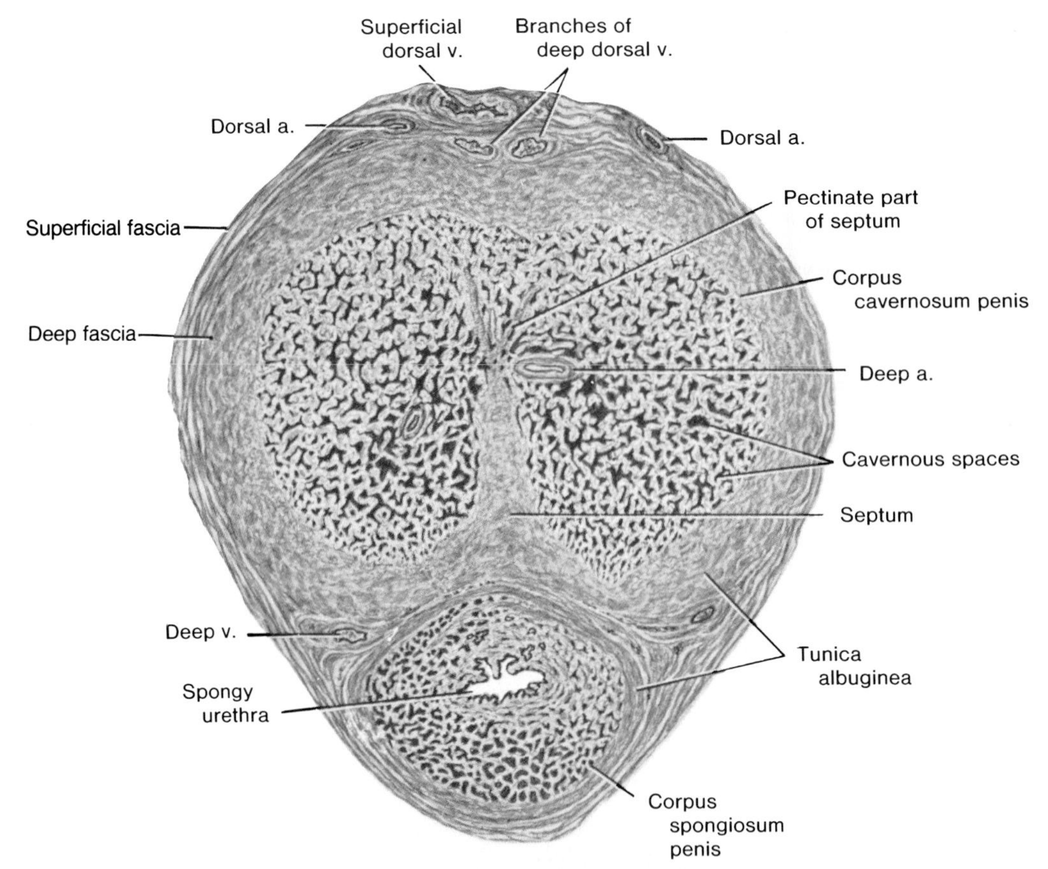

Figure 3-34. Drawing of a cross-section of the body of the penis of an adult male from which the skin has been removed. Observe that the corpora cavernosa consist of masses of cavernous erectile tissue, enclosed in a dense fibrous capsule called the tunica albuginea. Note that the tunicae albugineae fuse medially to form an incomplete median septum; hence the cavernous erectile tissue of the two corpora is continuous (also see Fig. 3-32). × 4.5.

The dorsal nerve of the penis then passes to the dorsum of the penis where it runs lateral to the arteries. It supplies both the skin and the glans penis (Fig. 3-32). The penis is richly provided with a great variety of sensory nerve endings; thus it is very sensitive.

THE SUPERFICIAL PERINEAL MUSCLES (Figs. 3-4, 3-14, 3-15, and 3-18)

There are three superficial perineal muscles: (1) the *superficial transversus perinei*; (2) the *bulbospongiosus*; and (3) the *ischiocavernosus*. **The last two are the muscles of the penis.**

The superficial perineal muscles lie in the superficial perineal space which is bounded inferiorly by the superficial perineal fascia and superiorly by the inferior fascia of the urogenital diaphragm (perineal membrane).

The perineal nerve supplies all three superficial perineal muscles (Fig. 3-24).

The superficial transversus perinei are slender, narrow, muscular strips which pass transversely, anterior to the anus (Figs. 3-4 and 3-5). *Each muscle extends from the ischial tuberosity to the central perineal tendon*, and probably helps to fix this wedge-shaped mass of fibrous tissue (Figs. 3-10 and 3-14*F*).

The bulbospongiosus muscle (Figs. 3-4, 3-14*F*, 3-15, and 3-35), *the muscle of the bulb of the penis*, lies in the median plane of the perineum, anterior to the anus. It consists of two symmetrical parts that are united by a **median tendinous raphe** inferior to the bulb of the penis (Fig. 3-4). *The bulbospongiosus arises from this median raphe and the central perineal tendon.*

The paired bulbospongiosus muscles form a sphincter which compresses the bulb and the corpus spongiosum of the penis, thereby emptying the spongy urethra of residual urine and/or semen.

The anterior fibers of the bulbospongiosus also assist with erection of the penis by increasing the pressure on the erectile tissue and compressing the deep dorsal vein of the penis. This impedes the venous drainage of the cavernous spaces and helps to promote the enlargement and turgidity of the penis.

The ischiocavernosus muscles (Figs. 3-4 and 3-14*F*) surround the crura of the penis. *Each muscle*

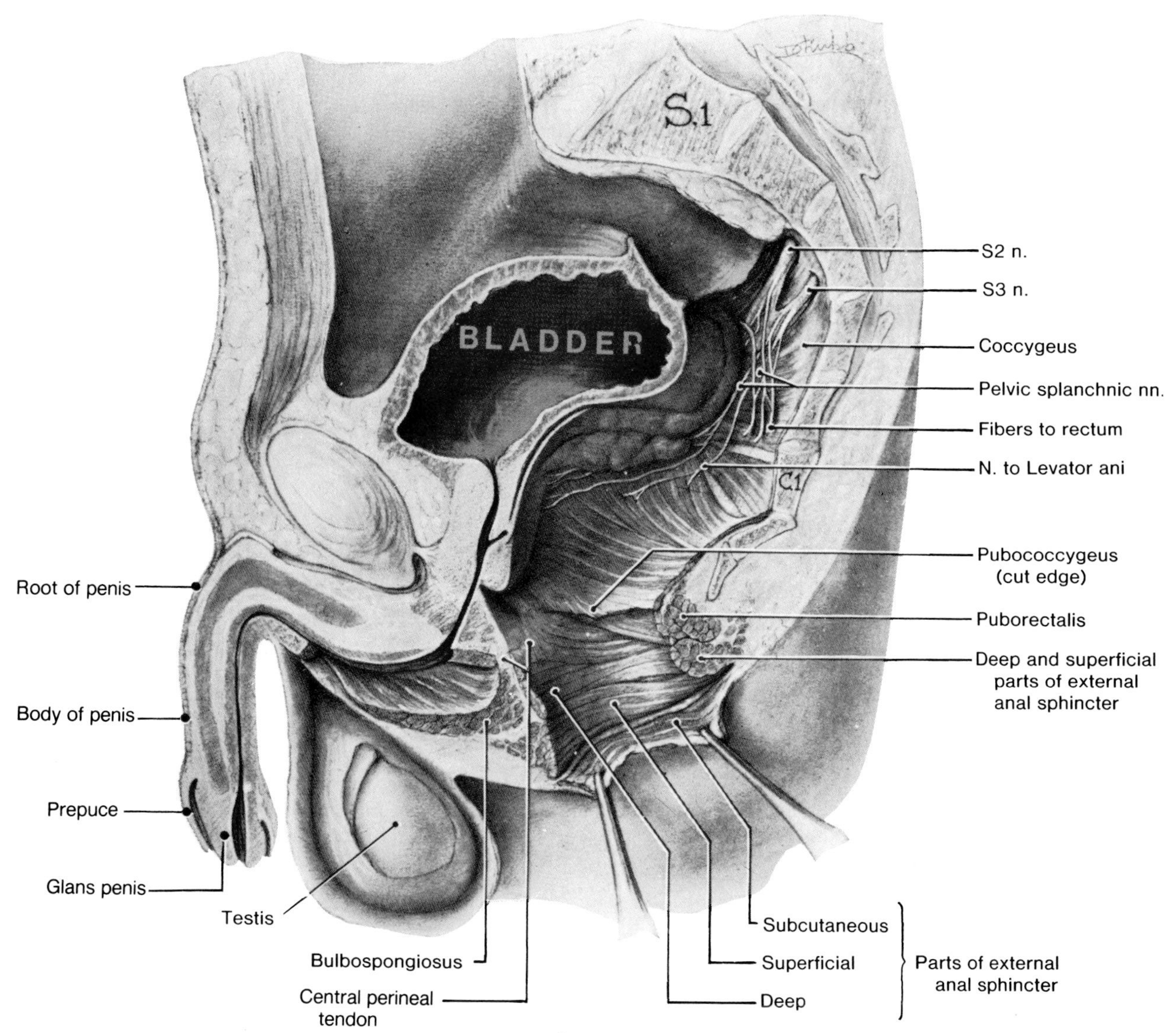

Figure 3-35. Drawing of a median section of a male pelvis from which the rectum, anal canal, and bulb of the penis have been removed. Examine the external anal sphincter and the levatores ani muscles. Observe the subcutaneous fibers of the external anal sphincter which are reflected with forceps; the superficial fibers mingling posteriorly with deep fibers; and the deep fibers mingling with the puborectalis muscle which forms a sling that occupies the anorectal angle between the anal canal and the rectum (also see Fig. 3-55). Note that the pubococcygeus is divided to allow for the removal of the anal canal to which it is in part attached. The penis has a fixed root consisting of the bulb of the penis, the crura (Fig. 3-29), and the muscles associated with them. The body of the penis, which hangs free, consists of two corpora cavernosa and the corpus spongiosum.

arises from the internal surface of the ischial tuberosity and the ischial ramus and passes anteriorly on the crus of the penis to be *inserted into the sides and ventral surface of the crus* (Fig. 3-4).

The ischiocavernosus muscles force blood from the cavernous spaces in the crura of the penis into the distal parts of the corpora cavernosa penis, thereby increasing the turgidity of the penis. Contraction of the ischiocavernosus muscles also compresses the deep dorsal vein of the penis as it leaves the crus of the penis (Figs. 3-28 and 3-34), thereby cutting off the venous return from the penis and helping to maintain the erection.

THE FEMALE PERINEUM

The muscles, nerves, and vessels of the female perineum are almost identical with those of the male. The genital organs also have similarities to those of males because they develop from similar primordia. The main differences are: (1) *the vagina pierces the urogenital diaphragm* (Fig. 3-14C); (2) **the urethra is in the anterior wall of the vagina** (Fig. 3-13); and (3) *the clitoris does not contain the urethra.*

The female external genital organs *are known collectively as the vulva (pudendum).* **The vulva consists of:** the *mons pubis, labia majora, labia minora, vestibule of the vagina, clitoris, bulb of the vestibule,* and *greater vestibular glands* (Figs. 3-10 and 3-36).

STRUCTURES IN THE SUPERFICIAL PERINEAL SPACE (Figs. 3-8 and 3-14*E*)

As in the male, the superficial perineal space is the fascial space between the superficial perineal fascia and the inferior fascia of the urogenital diaphragm (perineal membrane).

The superficial perineal space contains the ***superficial transversus perinei, ischiocavernosus, and bulbospongiosus muscles, and the greater vestibular glands.***

In the female, the bulbospongiosus is separated from its contralateral part by the vagina (Figs. 3-5, 3-14, and 3-39). *The bulbospongiosus muscle arises from the central perineal tendon, passes around the vagina, and inserts into the clitoris* (Fig. 3-5). The ischiocavernosus muscle inserts into the crus of the clitoris (Fig. 3-39).

STRUCTURES IN THE DEEP PERINEAL SPACE (Figs. 3-8 and 3-14)

The deep perineal space is the fascial space enclosed by the superior and inferior fasciae of the urogenital diaphragm.

The deep perineal space contains the ***urethra,*** *the inferior part of the* ***vagina,*** *and the* ***deep transversus perinei muscles.*** The deep perineal space also contains the *internal pudendal vessels,* the *dorsal nerve of the clitoris,* and branches of the *perineal nerve* supplying the sphincter urethrae and deep transversus perinei muscles.

THE FEMALE EXTERNAL GENITAL ORGANS (Fig. 3-36)

The Mons Pubis (Figs. 3-5 and 3-36). The mons (L. mountain) pubis is ***a rounded, fatty eminence lying anterior to the symphysis pubis.*** It is formed mainly by *a pad of fatty tissue* deep to the skin (Fig. 3-5). The mons pubis becomes *covered with coarse pubic hairs* during puberty (Fig. 2-43).

The Labia Majora (Figs. 3-20, 3-36, 3-38, and 3-61). The labia (L. lips) majora are two large folds of skin filled largely with subcutaneous fat, which pass posteriorly from the mons pubis. They are joined anteriorly by the *anterior labial commissure* (Fig. 3-36). The labia majora do not join posteriorly but a transverse fold of skin, called the *posterior labial commissure,* passes between them (Fig. 3-36).

Embryologically, the labia majora are homologous to the scrotum of the male. They develop from the unfused labioscrotal folds (Fig. 3-26).

The **pudendal cleft** is the slit or *opening between the labia majora* into which the vestibule of the vagina opens (Figs. 3-36 and 3-38).

The round ligaments of the uterus pass through the inguinal canals and enter the labia majora, where they end as branching bands of fascia that are attached to the skin (Figs. 2-22, 3-5, and 3-38).

The Labia Minora (Figs. 3-10, 3-20, 3-36, 3-38, and 3-61). The labia minora are two thin delicate folds of fat free, hairless skin. Usually the labia minora are completely hidden in the pudendal cleft by the labia majora. Sebaceous and sweat glands open on both surfaces of the labia minora.

The labia minora lie between the labia majora and their external surfaces are in contact with the smooth moist internal surfaces of the labia majora. Although the internal surface of each labium majus consists of thin skin, it has the typical pink color of a mucous membrane.

The labia minora enclose the vestibule of the vagina and lie on each side of the orifices of the urethra and vagina. In young females the labia minora are usually concealed by the labia majora, but in **parous women** (ones who have borne children), the labia minora *may* protrude through the pudendal cleft.

Posteriorly the labia minora are united by a small fold of the skin, the **frenulum of the labia minora** (Figs. 3-36 and 3-41*B*). Some obstetricians and gynecologists refer to the frenulum (L. small bridle) of the labia minora as the *fourchette* (F. fork). The frenulum of the labia minora is frequently torn or incised during childbirth (Fig. 3-41).

The Vestibule of the Vagina (Figs. 3-10, 3-36, and 3-38). The vestibule (L. *vestibulum,* antechamber) of the vagina is *the space between the labia minora.* **The urethra, vagina, and ducts of the greater vestibular glands open into the vestibule** *of the vagina.*

The external urethral orifice, usually a median aperture, is located 2 to 3 cm posterior to the clitoris and immediately *anterior to the vaginal orifice* (Figs. 3-5 and 3-36). On each side of the urethral orifice are the openings of the ducts of the **paraurethral glands** (Skene's glands). These glands are *homologous to the prostate gland* in the male.

The vaginal orifice (Figs. 3-5, 3-20, and 3-36) is much larger than the external urethral orifice, and is located inferior and posterior to it. The size and appearance of the vaginal orifice depend upon the woman's age and condition of the **hymen,** a thin incomplete fold of mucous membrane surrounding the vaginal orifice (Figs. 3-13 and 3-36).

The **greater vestibular glands** (Bartholin's glands) are located on each side of the vestibule of the vagina (Figs. 3-10 and 3-39), posterolateral to the vaginal orifice. There are several small **lesser vestibular glands** on each side which open into the vesti-

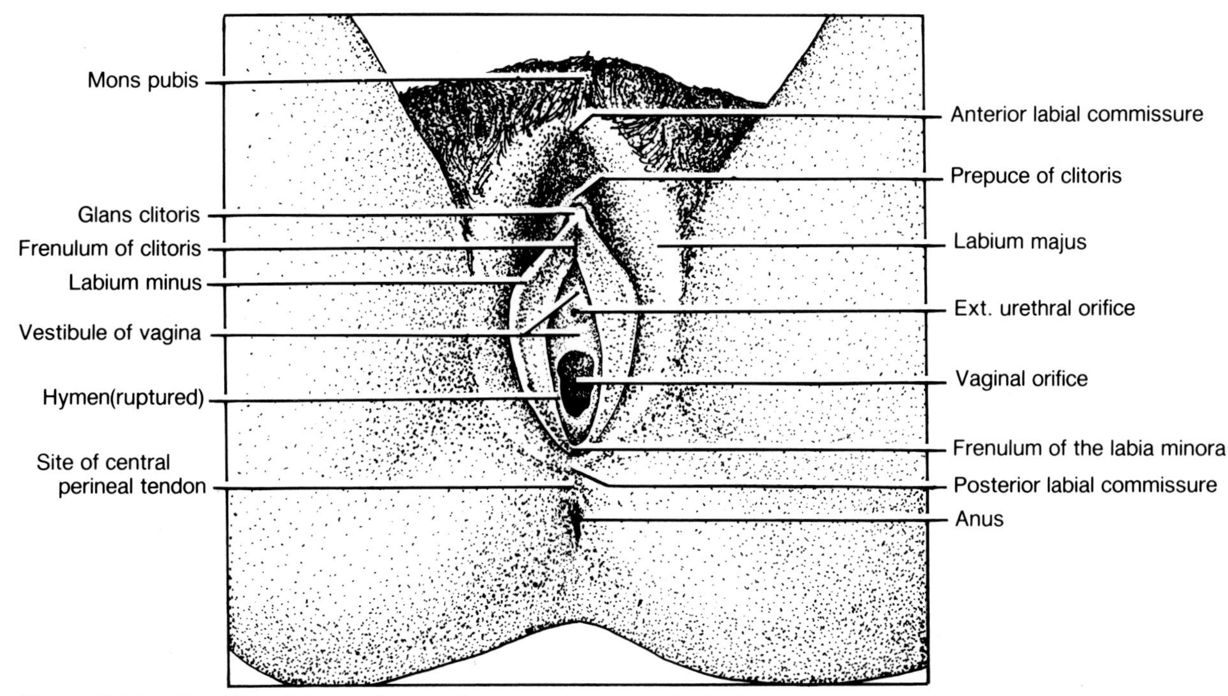

Figure 3-36. Drawing of the perineum of a *woman* as seen in the lithotomy position (Fig. 3-1). The labia majora and minora are spread apart to show the vestibule of the vagina which leads to both the urethral and vaginal orifices. Examine the different parts of the external genitalia, known collectively as the vulva or pudendum.

bule of the vagina between the urethral and vaginal orifices.

In virgins the vaginal orifice is partially closed by the hymen, a thin crescentic or anular fold (G. *hymen,* membrane). The hymen is variable in size and shape. Sports activities, pelvic examinations, sexual intercourse, and trauma may tear the hymen, resulting in varying amounts of bleeding in prepubertal girls. In women who have borne children via the vagina, the vaginal orifice is enlarged and the hymen is usually represented by only a few tags or rounded elevations of mucous membrane, called **hymeneal caruncles** (Fig. 3-13). The Latin word *carunculae* means **"small fleshy masses."**

In virgins the aperture in the hymen varies from a pinpoint in size to one admitting one or two fingers. During initial sexual intercourse the hymen usually lacerates in several places and a variable amount of bleeding may occur. In some cases the hymen may fail to rupture, resulting in difficult and/or painful intercourse known as **dyspareunia** (G. *dyspareunos,* badly mated).

In some female infants the hymen has no aperture, a persistent fetal condition known as **imperforate hymen.** In these cases there is no communication between the vagina and its vestibule. With the **onset of menses** (menstrual bleeding), there is an accumulation of menstrual blood in the vagina (**hematocolpos**).

The Clitoris (Figs. 3-5, 3-10, and 3-36 to 3-40). The clitoris, 2 to 3 cm in length, is homologous with the penis in the male. Unlike the penis, *the clitoris is not traversed by the urethra.* It is located posteroinferior to the anterior labial commissure (Fig. 3-36).

This small sexual organ is composed of erectile tissue and, like the penis, is capable of enlargement upon tactile stimulation. It is highly sensitive and **important in the sexual excitement of the female.**

The clitoris consists of two crura, two corpora cavernosa, and a glans and, like the penis (Fig. 3-31), is suspended by a suspensory ligament (Fig. 3-5). *The clitoris has no corpus spongiosum and is not associated with the urethra.*

The clitoris is located between the anterior ends of the labia minora (Figs. 3-36 and 3-38). The parts of the labia minora passing anterior to the clitoris form the **prepuce of the clitoris** (homologous to the prepuce of the penis), whereas the parts passing posterior to it form the ***frenulum of the clitoris*** (Fig. 3-36).

The Bulbs of the Vestibule (Figs. 3-10 and 3-39). The bulbs of the vestibule *consist of two elongated*

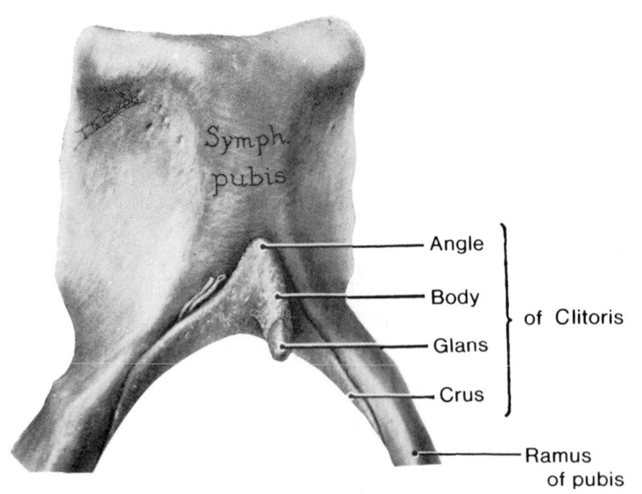

Figure 3-37. Drawing of a dissection of the clitoris, the homologue of the penis. Compare its parts with those of the penis (Fig. 3-29). The clitoris is composed of two corpora cavernosa; it contains no corpus spongiosum and *the urethra does not enter it.* During sexual arousal the clitoris becomes erect by a mechanism similar to that described for erection of the penis.

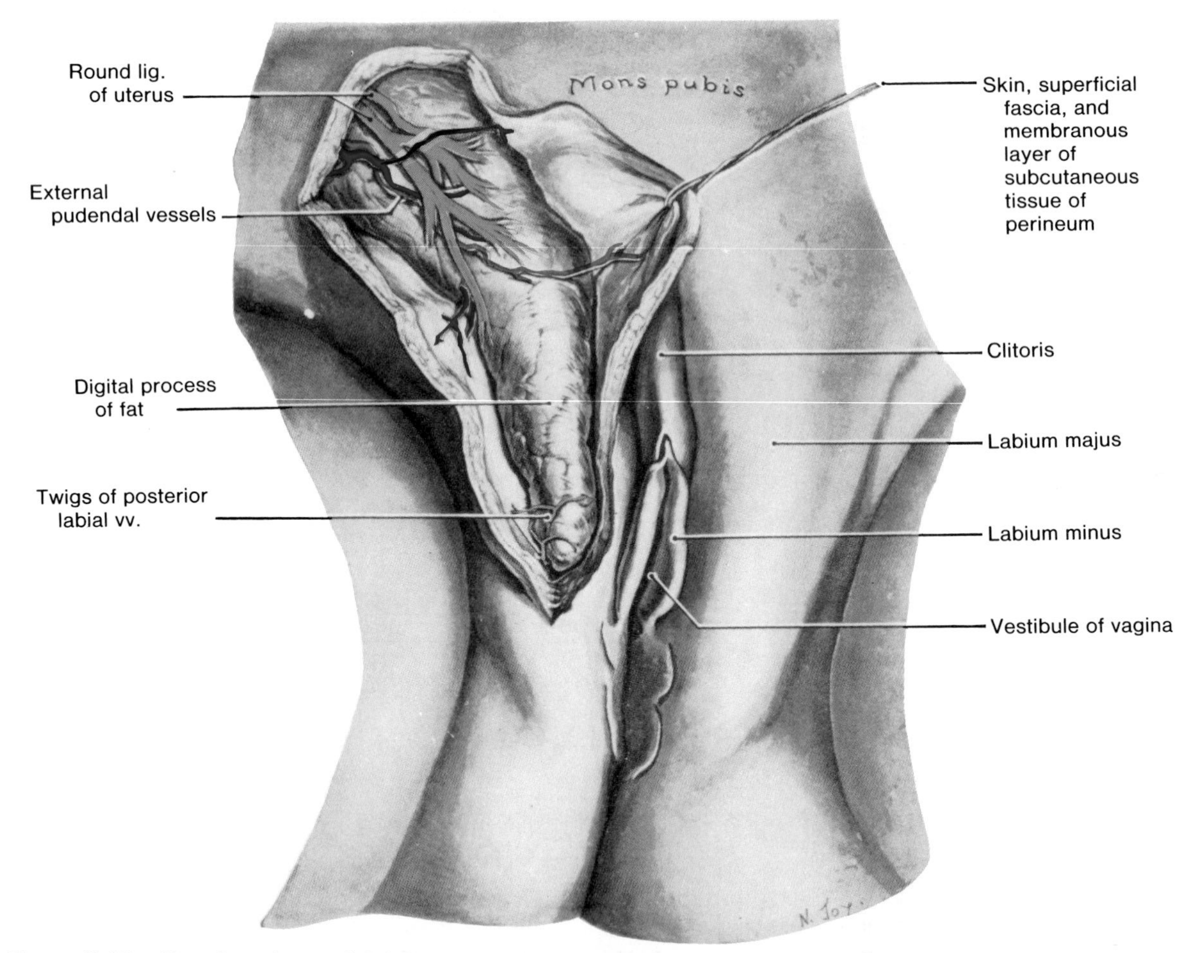

Figure 3-38. Drawing of superficial dissection of the right labium majus. Observe the long digital process of fat lying deep to the subcutaneous fatty tissue and descending far into the labium majus. Observe the round ligament of the uterus, noting that it ends as a branching band of fascia. Note the external pudendal vessels crossing the process of fat. Examine the vestibule of the vagina, the region between the labia minora and external to the hymen (Fig. 3-13). Also see Figures 2-22 and 3-77.

masses of erectile tissue, about 3 cm in length, lying along the sides of the vaginal orifice, deep to the bulbospongiosus muscle (Fig. 3-5).

The bulbs of the vestibule are homologous to the bulb and corpus spongiosum of the penis (Fig. 3-30). *The posterior ends of the bulbs of the vestibule are in contact with the greater vestibular glands* (Fig. 3-39).

The Greater Vestibular Glands (Figs. 3-10, 3-14*E*, and 3-39). The two greater vestibular glands (Bartholin's glands) are *located in the superficial per-*

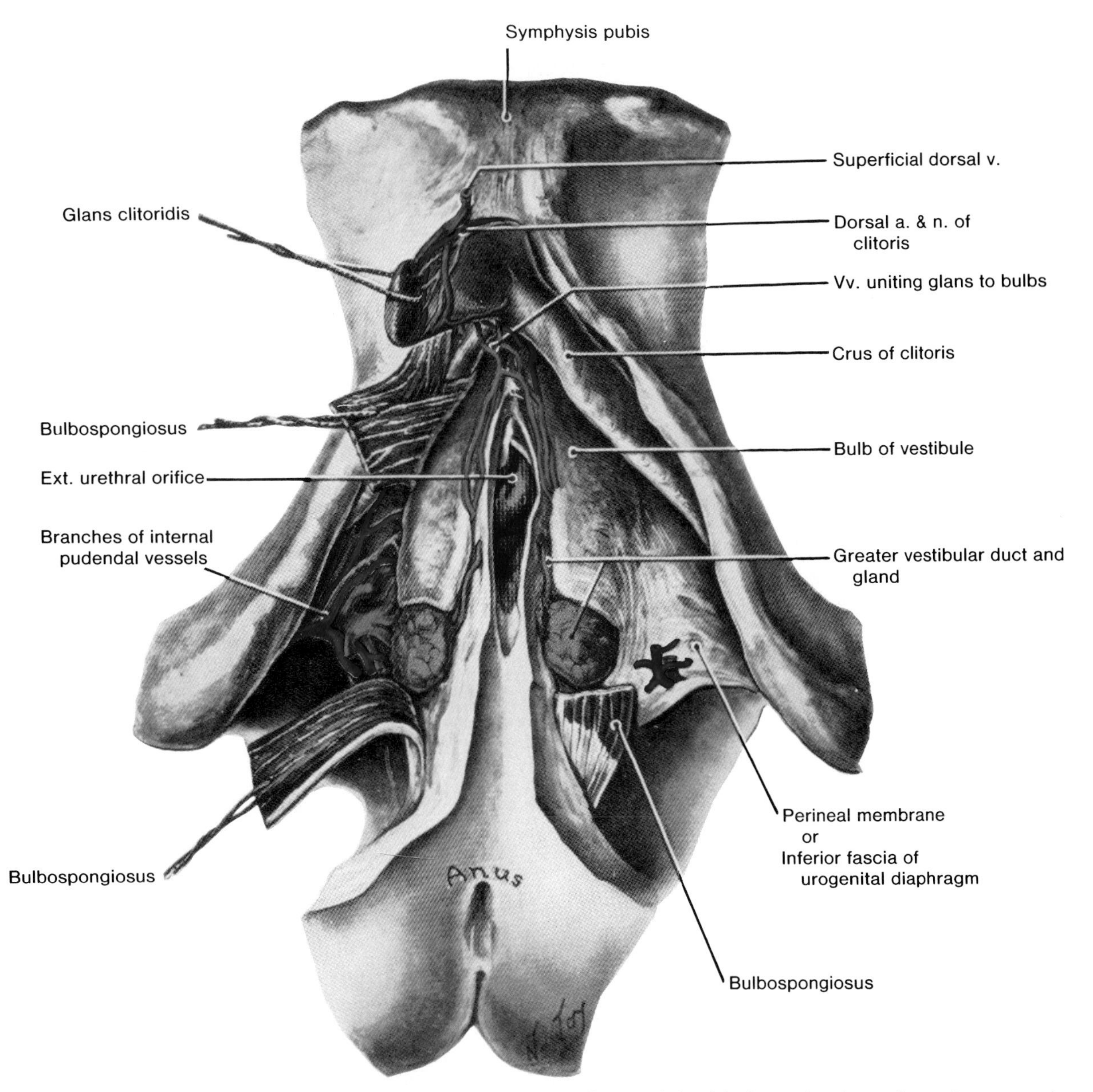

Figure 3-39. Drawing of a dissection of the female perineum. On the *right side* the perineal membrane is removed. Observe the paired bulbospongiosus muscles which are divided and reflected on the right side and largely excised on the left side. Note that the glans clitoris is pulled over to the right side to show the dorsal vessels and nerve of the clitoris running to it. Observe the bulbs of the vestibule, one on each side of the vestibule of the vagina. Note the veins connecting the bulbs of the vestibule to the glans of the clitoris. Observe the greater vestibular gland situated at the posterior blunt end of the bulb and like it, covered by the bulbospongiosus muscle. Note that this gland has a long duct which opens into the vestibule of the vagina. On the *right side* the perineal membrane is cut away to illustrate the vessels of the bulb and the dorsal nerve and vessels of the clitoris within the deep perineal space.

ineal space, one on each side, just posterior to the bulb of the vestibule and partly under the cover of its posterior part. These rounded or ovoid *tubuloalveolar glands*, secrete lubricating mucus into the vestibule of the vagina.

The greater vestibular glands are homologous with the bulbourethral glands of the male (Fig. 3-15).

The greater vestibular glands are not usually palpable, but become readily so when infected.

Infection of the greater vestibular glands (***Bartholinitis***) may result from a number of pathogenic organisms. Infected glands may enlarge to a diameter of 4 to 5 cm and impinge the wall of the rectum.

Vessels and Nerves of the Female External Genitalia (Figs. 3-39 and 3-40). *The rich arterial supply to the vulva is from two external pudendal arteries and one internal pudendal artery on each side.* The ***labial arteries*** are branches of the internal pudendal artery.

The labial veins are tributaries of the internal pudendal veins and venae comitantes of the internal pudendal artery (Fig. 3-40).

The nerves to the vulva are branches of the

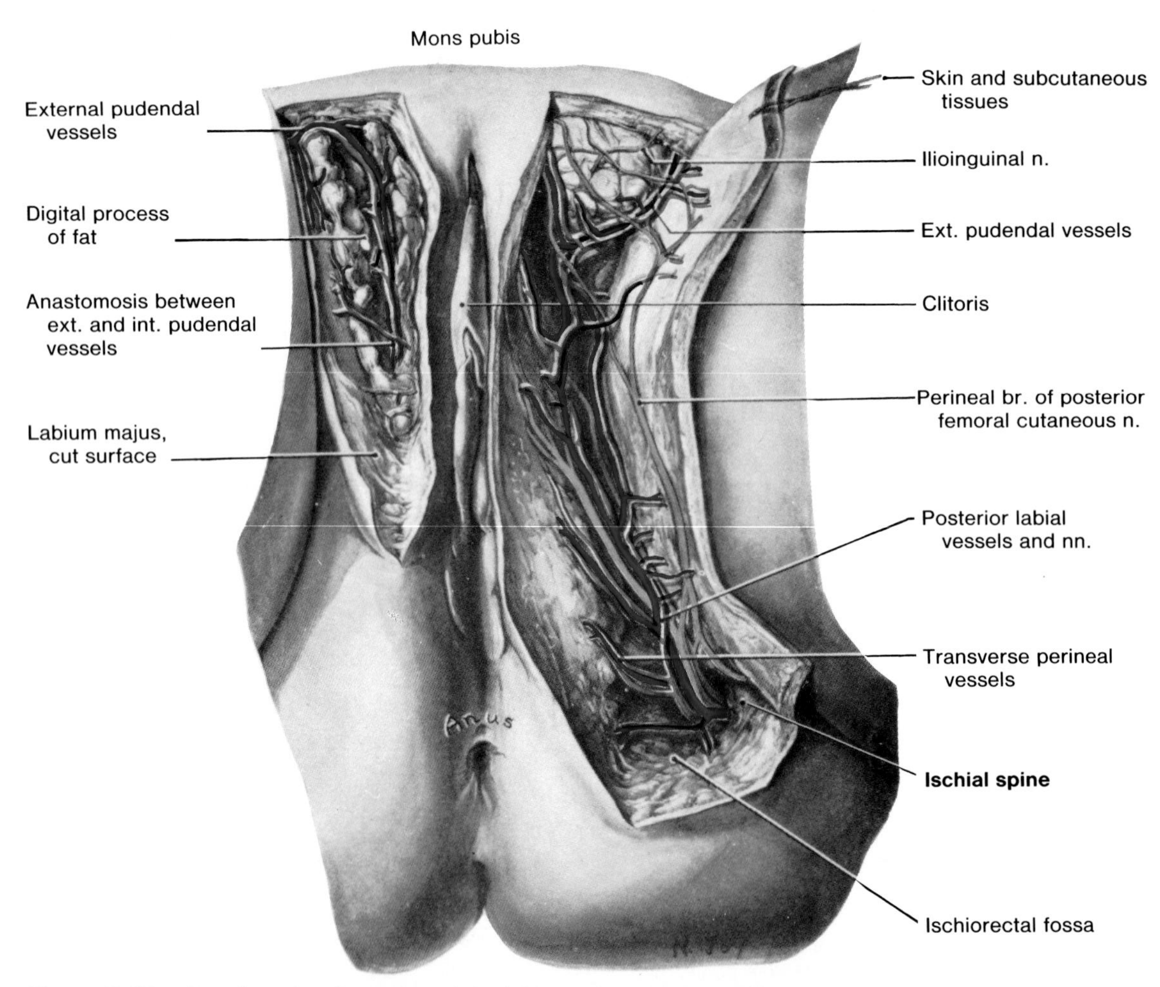

Figure 3-40. Drawing of a dissection of the labia majora demonstrating their vessels and nerves. *On the left* the lobulated digital process of fat has been opened to show the anastomotic vessels which unite the external and internal pudendal vessels. *On the right* the digital process of fat is largely removed. Observe that the posterior labial vessels and nerves (S2 and S3) are joined by the perineal branch of the posterior cutaneous nerve of the thigh (S1, S2, and S3), and run almost to the mons pubis. Here the vessels anastomose with the external pudendal vessels and the nerves meet the ilioinguinal nerve.

ilioinguinal nerve, the genital branch of the *genitofemoral nerve*, the perineal branch of the *femoral cutaneous nerve of thigh*, and the ***perineal nerve***. As the labia majora are homologous to the scrotum, their nerves are similar to those supplying the scrotum (Fig. 3-24).

Because there is a rich arterial supply to the labia majora and minora, hemorrhage from injuries to them may be severe. During parturition painful labor often occurs and the most anguish is often felt when the fetal head passes through the vulva owing to stretching of its parts.

To relieve the pain associated with childbirth, **pudendal block anesthesia** may be performed by injecting a local anesthetic agent into the tissues surrounding the pudendal nerve (Fig. 3-90). The injection is usually made where the pudendal nerve crosses the lateral aspect of the sacrospinous ligament (Fig. 3-16), near its attachment to the ischial spine (Fig. 3-40).

To ease delivery of a fetus and to avoid laceration of the perineum, an **episiotomy** or relaxing incision is frequently made in the perineum (Fig. 3-41). This enlarges the distal end of the birth canal and prevents serious damage to perineal structures, especially the rectum and external anal sphincter.

Two types of episiotomy are commonly performed, median and mediolateral (Fig. 3-41). In a ***median episiotomy*** the incision is in the median plane of the perineum (Fig. 3-41*A*), beginning at the frenulum of the labia minora and passing posteriorly through the skin, vaginal mucosa, and central perineal tendon. *The incision stops well short of the external anal sphincter* and rectum. In a **mediolateral episiotomy** (Fig. 3-41*B*), the incision begins at the midpoint of the frenulum of the labia minora (Fig. 3-10) and extends toward the ischiorectal fossa and the ischial tuberosity. *The incision passes through*: (1) the posterior vaginal wall, (2) the perineal skin, (3) the bulbospongiosus muscle, (4) the superficial transversus perinei muscle, (5) the perineal membrane, and (6) the deep transversus perinei muscle. *The incision stops well short of the levator ani* because of the important role of this muscle in supporting the pelvic floor. After delivering the baby, the incision is carefully sutured in layers.

THE PELVIS

The pelvis (L. basin) is the inferior basin-shaped division of the abdominopelvic cavity (Figs. 2-31, 3-42, and 3-43). *The skeleton of the pelvis is referred to as the bony pelvis.*

BONES OF THE PELVIS

The bony pelvis is formed by the two ossa coxae or hip bones anteriorly and laterally, and by the *sacrum and coccyx posteriorly (Figs. 3-7, 3-42, and 3-43). The sacrum and coccyx, parts of the vertebral column*, are interposed dorsally between the hip bones (Fig. 3-42).

The os coxae or hip bone (Fig. 3-44) is a large,

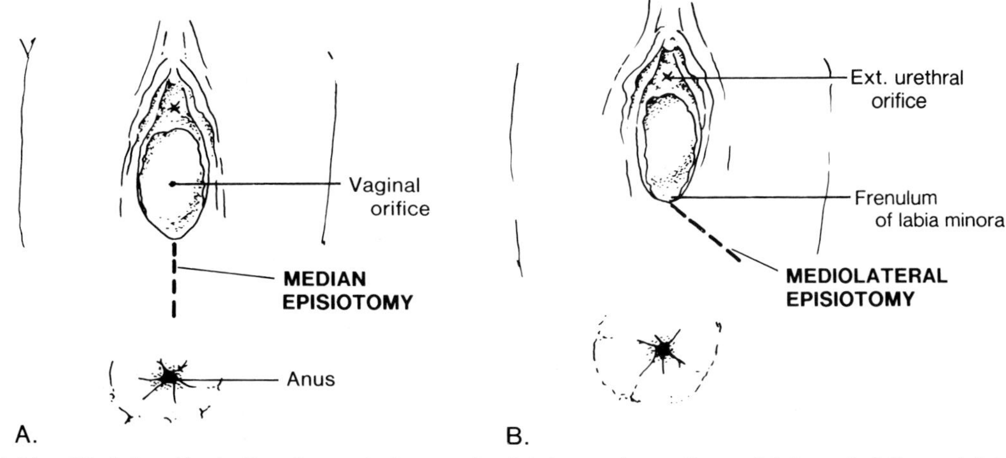

Figure 3-41. Sketches illustrating the main types of episiotomy. *A*, *median episiotomy* (midline episiotomy). The incision passes posteriorly from the frenulum of the labia minora (fourchette) toward the anus, dividing the central perineal tendon in the midline. *B*, *mediolateral episiotomy*. The incision is at a 45° angle from the midline of the frenulum of the labia minora and passes toward the ischial tuberosity.

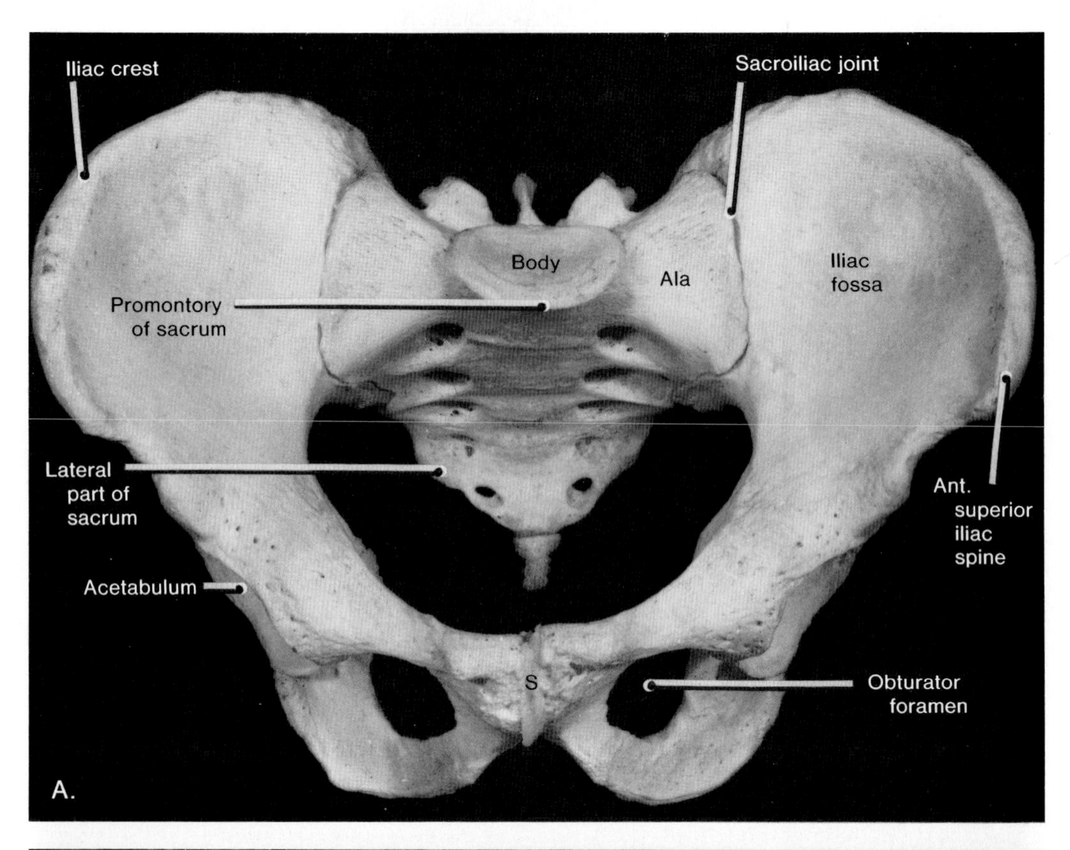

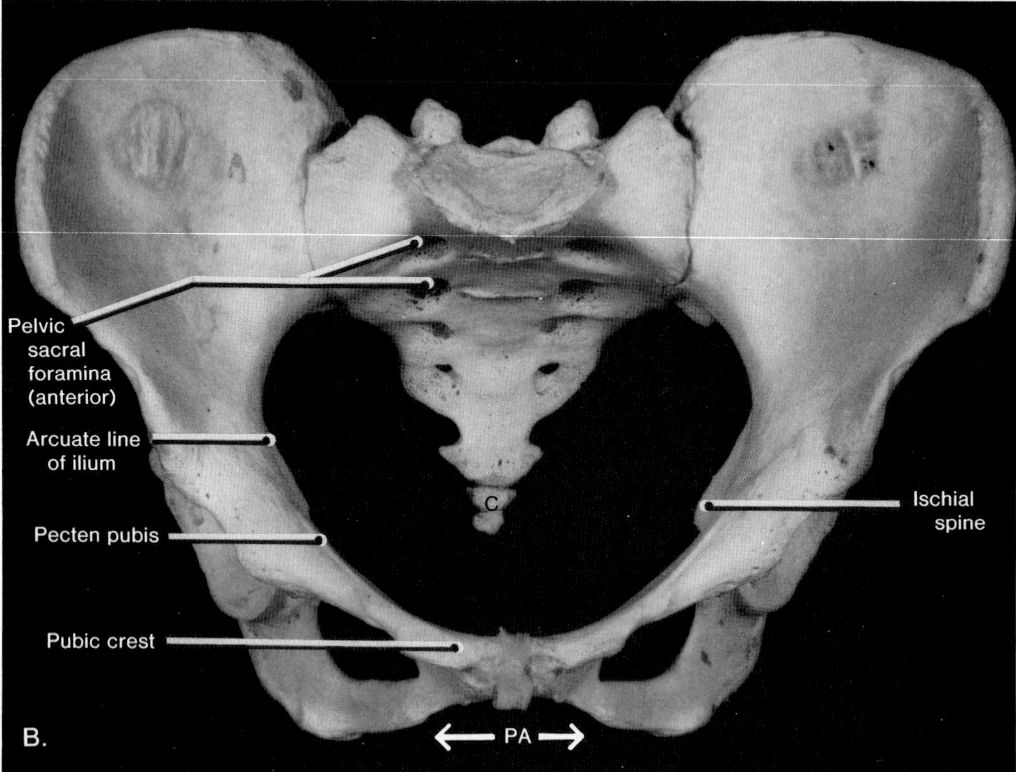

Figure 3-42. Photographs of anterior views of male (*A*) and female (*B*) bony pelves in the anatomical position (*i.e.*, the anterior superior iliac spines and the superior end of the symphysis pubis (*S*) lie in the same coronal plane. Observe that the superior pelvic aperture (pelvic inlet) is larger in the female. This aperture (outlined by the pelvic brim, Fig. 3-50) is the boundary between the pelvis major and pelvis minor. Note that the pubic arch (*PA*) is very wide in the female pelvis; consequently, the subpubic angle in the female is much wider than in the male.

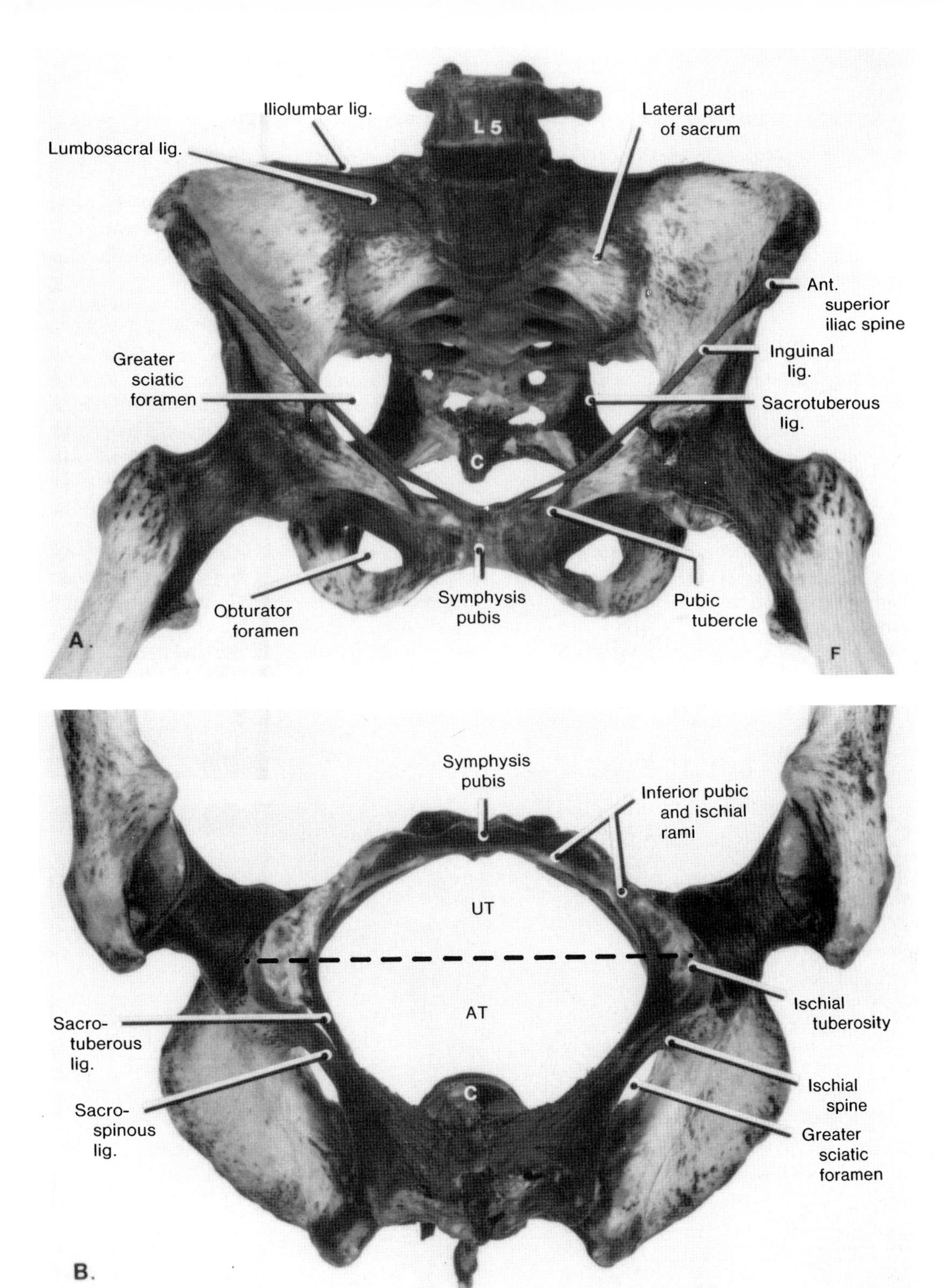

Figure 3-43. Photographs of a bony female pelvis with the femora (*F*) of the lower limbs and the ligaments of the pelvis and hip joint attached. *A*, *anterior view*. Observe the inguinal ligament extending from the pubic tubercle to the anterior superior iliac spine. Note the strong iliolumbar ligament which unites the transverse process of L5 vertebra to the internal lip of the iliac crest. Its inferior fibers attach to the lateral part of the sacrum as the lateral lumbosacral ligament. *B*, *inferior view* showing the inferior pelvic aperture as visualized by the obstetrician when the patient is in the lithotomy position (Fig. 3-1). The perineum fills this large aperture (Figs. 3-2 and 3-5), which lies between the superior parts of the thighs and the inferior parts of the buttocks. Here, the diamond-shaped perineal region is divided by a transverse *broken line* (- - - - -) passing through the ischial tuberosities, immediately anterior to the anus, into (1) an anterior urogenital region or triangle (*UT*) and (2) a posterior anal region or triangle (*AT*). Examine the sacrotuberous ligaments which have a wide origin from the dorsal surfaces of the sacrum and coccyx (*C*) and from both posterior superior iliac spines (Fig. 3-42). Note that they pass to the impressions on the ischial tuberosities and extend along their medial margins.

irregularly shaped bone which *consists of three parts*: *ilium*, *ischium*, and *pubis*. The three parts meet at the **acetabulum**, the cup-shaped cavity for the head of the femur (Figs. 3-42*A*, 3-45*B*, and 3-46).

In infants and children the three parts of the hip bone are not fused with each other (Fig. 4-3). Fusion occurs at 15 to 17 years; the bones are firmly joined in the adult (Fig. 3-44).

The three parts of the bony pelvis are bound together by dense ligaments (Fig. 3-43). *The bones are united by four articulations*: **two synovial joints** (the *sacroiliac joints*, Fig. 3-43*A*) and **two symphyses** (the *symphysis pubis*, Figs. 3-7, 3-42, and 3-43, and the *sacrococcygeal joint*). The bones involved in the last two joints are connected by a fibrocartilaginous disc, as well as by ligaments.

The pelvis is divided into a **pelvis major** (false pelvis) and a **pelvis minor** (*true pelvis*). Some obstetricians and other doctors refer to the female pelvis minor as the "obstetric pelvis."

The *pelvis major* is the portion of the pelvis located *superior to the pelvic brim* (Fig. 3-50), and the *pelvis minor* is the portion of the pelvis that is located *inferior to the pelvic brim*.

The dividing line between the pelvis major and pelvis minor is an oblique plane that passes through the sacral promontory posteriorly and the lineae terminales laterally and anteriorly (Fig. 3-48).

The pelvic brim (Fig. 3-50), surrounding the superior pelvic aperture, extends from the ***sacral promontory*** posteriorly to the top of the ***symphysis pubis*** anteriorly (Fig. 3-42).

The three parts of the pelvic brim are (Fig. 3-42): (1) the anterior border of the ala of the sacrum (*sacral part*); (2) the arcuate line of the ilium (*iliac part*); and (3) the pecten pubis and pubic crest (*pubic part)*.

SEX DIFFERENCES BETWEEN MALE AND FEMALE PELVES (Figs. 3-7, 3-42, 3-43, and 3-46)

The pelves of males and females differ in several respects that are related mainly to the heavier build and stronger muscles in the male, and to the adaptation of the female pelvis for childbearing.

The general structure of the male pelvis is heavy and

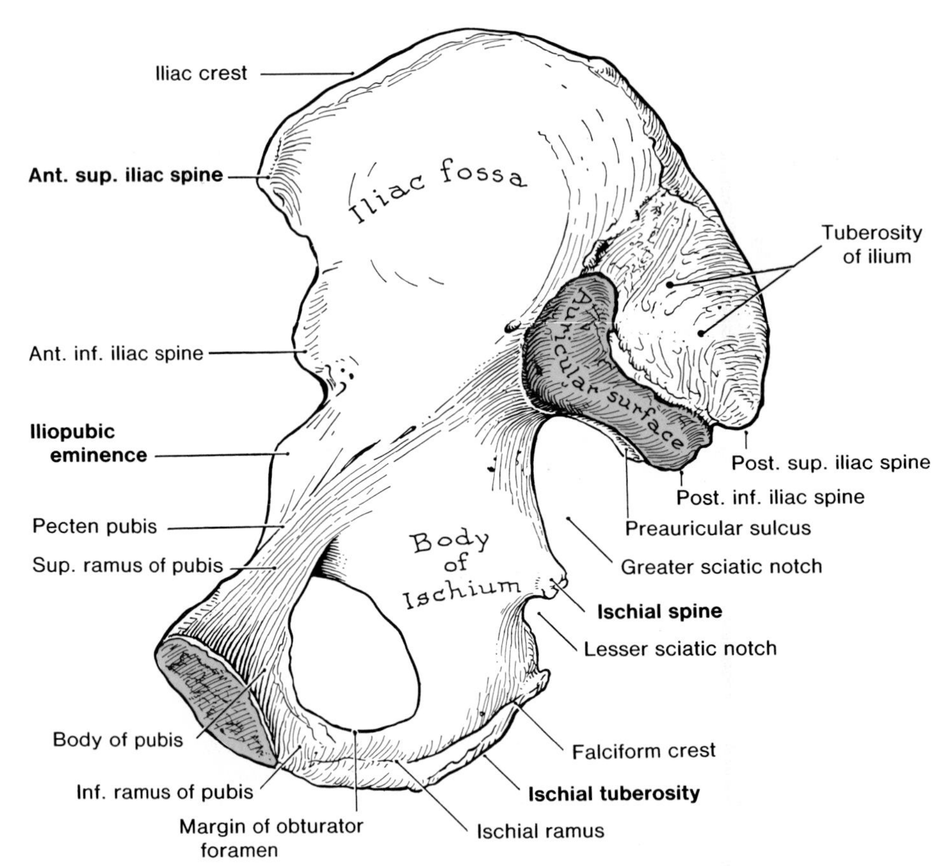

Figure 3-44. Drawing of the medial aspect of the right os coxae or hip bone. This bone consists of three parts: the *ilium*, *ischium*, and *pubis* (Fig. 4-3). These bones are fused together in adults. Fusion occurs around the 16th year. Note that the ilium is fan-shaped. The spread of the fan is called the *ala* (L. wing) and its broad handle is called the body. The iliac crest is the rim of the fan and the iliac fossa is the concavity of the ala.

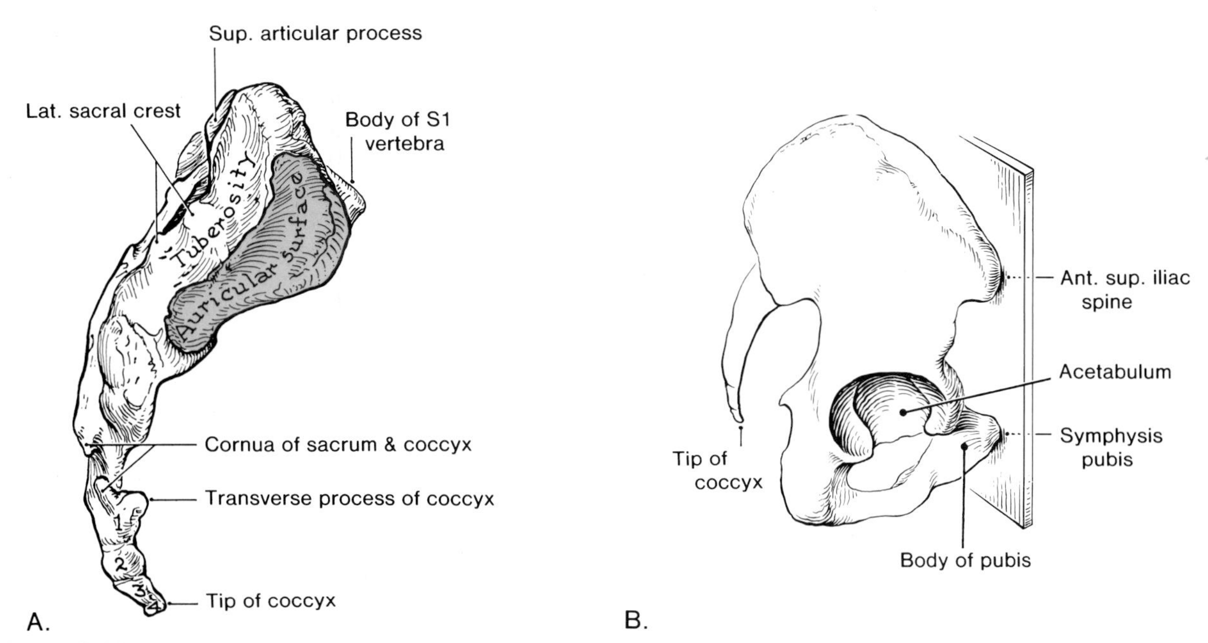

Figure 3-45. *A*, drawing of the right lateral aspect of the sacrum and coccyx. *B*, drawing of the lateral aspect of the right os coxae or hip bone and coccyx, demonstrating that in the anatomical position the anterior superior iliac spine and the superior margin of the symphysis pubis lie in the same coronal plane, as though against a wall as illustrated. Note that the tip of coccyx is on a level with the superior half of the body of the pubis.

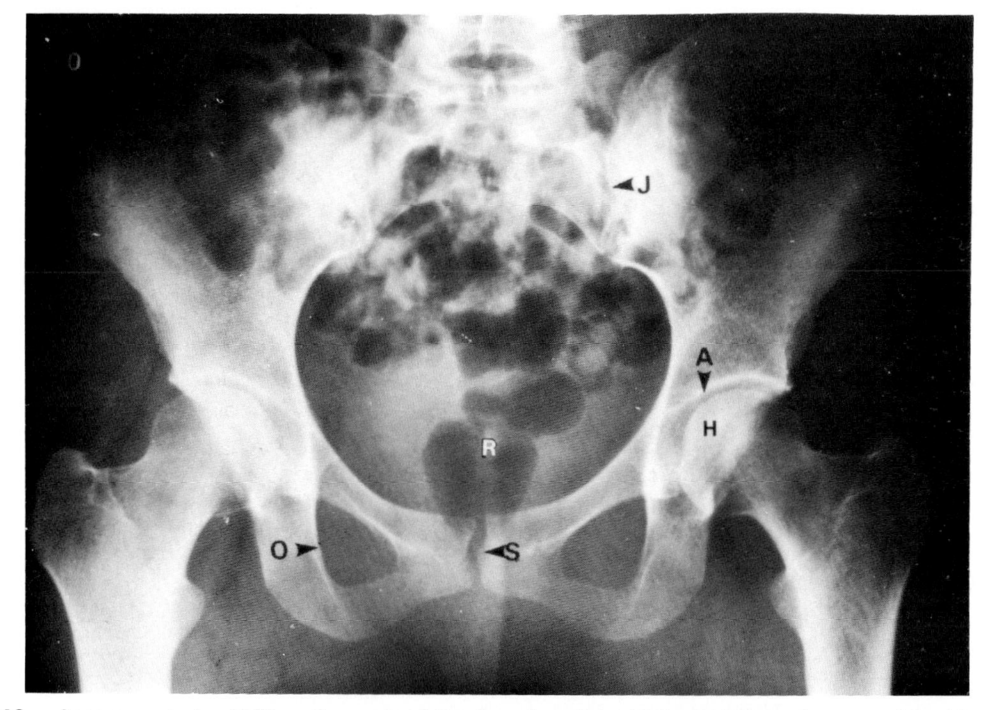

Figure 3-46. Anteroposterior (*AP*) radiograph of the female pelvis. Note that there is a considerable amount of air in the rectum (*R*). The *arrows* point to the symphysis pubis (*S*), the sacroiliac joint (*J*), the roof or superior surface of the cup-shaped acetabulum (*A*), and the margin of the obturator foramen (*O*). *H* indicates the part of the head of the femur which is overlapped by the anterior and posterior parts of the acetabulum.

thick and it has more prominent bone markings than the female pelvis.

The female pelvis is wider, shallower, and has larger superior and inferior pelvic apertures than the male pelvis, because it surrounds and limits the size of the birth canal.

Using typical male and female pelves (Fig. 3-42), observe that **in females:** (1) *the hip bones are farther apart* owing to the broader sacrum, which explains the relatively wider hips of females; (2) *the ischial tuberosities are farther apart* because of the greater subpubic angle of the pubic arch (Fig. 3-42*B*); and (3) *the sacrum is less curved* which increases the size of the inferior pelvic aperture.

Although there are usually clear-cut anatomical differences between male and female pelves, the pelvis of any person may have anatomical features distinctive of the opposite sex. The diagnosis of pelvic type (Fig. 3-47) is made on overall architecture and not on the appearance of the superior pelvic aperture alone. The presence of certain male characteristics in a female pelvis may offer hazards to successful pelvic delivery of a fetus (discussed on p. 340).

The Pelvis Major (Figs. 3-42 and 3-46). The pelvis major *lies superior to the superior pelvic aperture and pelvic brim* (Fig. 3-50). It is formed on each side by the ala (L. wing) and the base of the sacrum.

The cavity of the pelvis major is part of the abdominal cavity; hence it contains abdominal viscera (*e.g.*, the terminal ileum and sigmoid colon, Fig. 3-46).

The pelvis major is bounded anteriorly by the abdominal wall, laterally by the iliac fossae, and posteriorly by L5 and S1 vertebrae. Hence its anterior wall is markedly longer than its posterior wall.

The Pelvis Minor (Figs. 3-42 and 3-46). The pelvis minor *lies inferior to the superior pelvic aperture and pelvic brim* (Fig. 3-50). It is limited inferiorly by the **inferior pelvic aperture** (pelvic outlet).

The pelvis minor contains the pelvic viscera (urinary bladder, rectum, and parts of the urogenital organs), blood vessels, lymphatics, and nerves.

The cavity of the pelvis minor (pelvic cavity proper) is short and curved and forms a basin-like

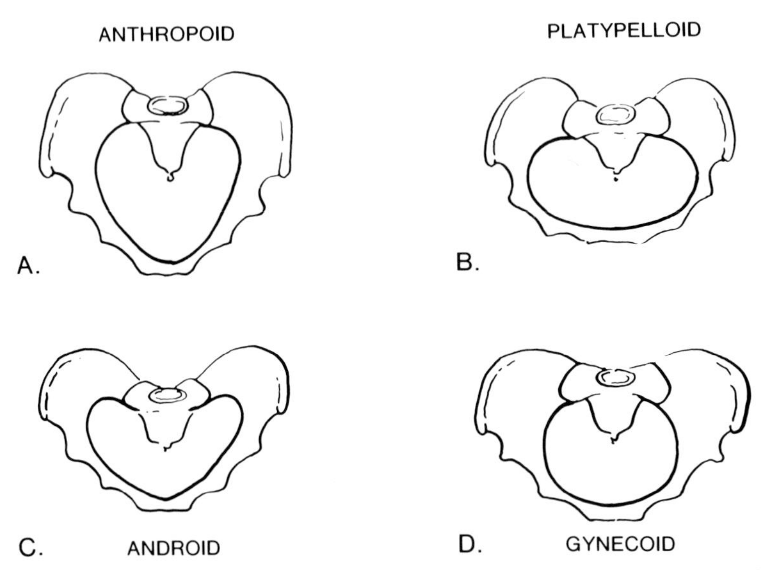

Figure 3-47. Drawings illustrating the four types of pelvis. The types shown in *A* and *C* are most common in males, whereas those illustrated in *C* and *D* are common in females. *B* type is uncommon in both sexes. *A*, **anthropoid pelvis** (present in some males and in about 23% of females). Note that the AP diameter of the superior pelvic aperture is greater than the transverse diameter. *B*, **platypelloid pelvis** (rare in males; present in about 2% of females). Observe that the pelvis is flat because its transverse diameter is greater than its anteroposterior diameter. *C*, **android pelvis** (present in most males and in about 32% of females). Note that the superior pelvic aperture has a wide transverse diameter, but the posterior part of the aperture is narrow with an almost triangular anterior segment. Observe that the superior pelvic aperture has the shape of a "heart" or valentine. *D*, **gynecoid pelvis** (present in about 43% of females). This is the most common type of pelvis in females and is the most roomy obstetrically. Usually a woman with a gynecoid pelvis (G. *gyne*, woman + *eidos*, resemblance) has an uneventful delivery of a fetus.

cavity that is tilted anteroinferiorly in the anatomical position (Fig. 3-42).

The posterior wall of the cavity of the pelvis minor is notably longer than its anterior wall.

The posterior wall of the pelvis minor is formed by the concave anterior or *pelvic surface of the sacrum and coccyx* (Fig. 3-42*B*).

The anterior wall of the pelvis minor is formed by the *symphysis pubis*, the *body of the pubis*, and the *pubic rami* (Figs. 3-42, 3-44, and 3-46).

The sides of the pelvis minor are formed by the pelvic aspects of the ilium and ischium.

The ischial spines project inferiorly and slightly medially (Figs. 3-7 and 3-42 to 3-44). The **sacrospinous ligaments** are attached to the ischial spines and to the inferolateral border of the sacrum (Fig. 3-16).

The Superior Pelvic Aperture (pelvic inlet) is *variable in contour*. It is heart-shaped in males and in some females (Figs. 3-42*A* and 3-47*C*), but *in most females the superior pelvic aperature is larger than in males and is rounded or oval in contour* (Fig. 3-47D). The superior pelvic aperture is encroached upon by the promontory of the sacrum (Fig. 3-42*A*).

The periphery of the superior pelvic aperture (formed by the pelvic brim) is indicated by the lineae terminales (Fig. 3-48). Each linea terminalis is formed by: (1) the **pubic crest**, (2) the **pecten pubis**, and (3) the **arcuate line** of the ilium (Fig. 3-42).

The superior pelvic aperture is routinely measured for obstetric reasons during a pelvic examination (Fig. 3-48). Do not try to memorize these measurements at this time.

The anteroposterior (AP) diameter of the superior pelvic aperture is the measurement from the *midpoint of the superior border of the symphysis pubis to the midpoint of the sacral promontory* (Fig. 3-48*A*).

The AP diameter is determined by inserting the index and middle fingers into the vagina until the middle finger touches the sacral promontory and the upper edge of the index finger is against the inferoposterior border of the symphysis pubis. The AP diameter of the pelvis can

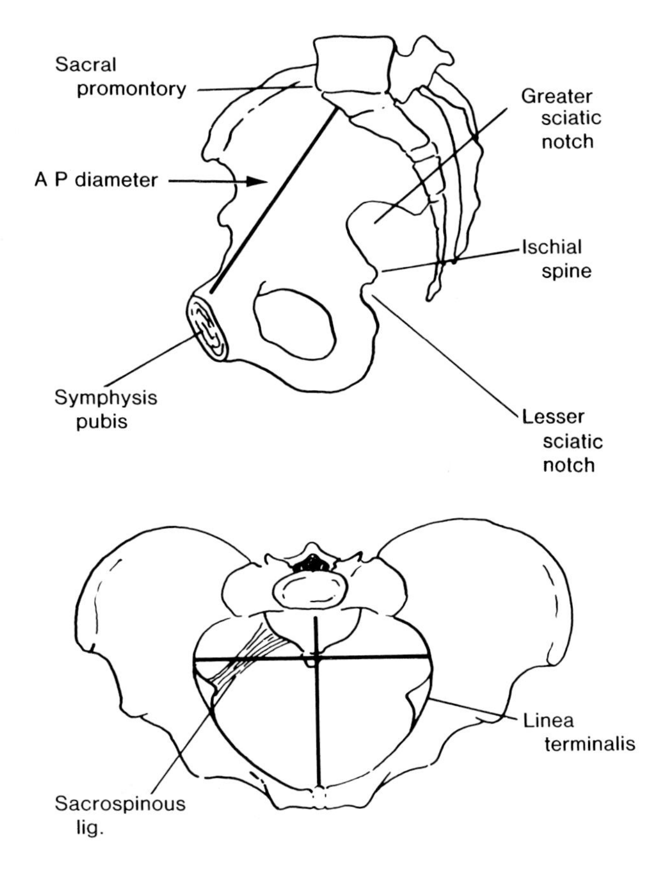

Figure 3-48. Drawings of a female pelvis illustrating the anteroposterior (AP) and transverse diameters of the superior pelvic aperture. These pelvic planes are useful as obstetrical measurements. Observe that the transverse diameter of the superior pelvic aperture is greater than the AP diameter. A pelvis in which any major plane is reduced below normal is called a *contracted pelvis*.

be estimated by measuring the distance between the fingers and subtracting 1.5 cm.

The transverse diameter of the superior pelvic aperture is its greatest width, measured transversely from the linea terminalis on one side to this line on the opposite side (Fig. 3-48*B*).

The oblique diameter of the superior pelvic aperture is measured from one **iliopubic eminence** (Fig. 3-44) to the opposite **sacroiliac joint** (Fig. 3-42*A*). In some cases pelvic measurements are made using radiographs (Fig. 3-46).

A general idea of the shape and position of the sacrum can be developed by palpating the sacral concavity during a **pelvic examination.** The *ischial spines* can be palpated and a general concept of their prominence can be obtained (Figs. 3-42 to 3-44).

The midplane diameter (interspinous diameter) of the pelvis (plane of least pelvic dimension) cannot be measured, but *the midplane diameter may be estimated by palpating the sacrospinous ligament during a vaginal examination* (Fig. 3-43*B*). Its length is equal to about half the midplane diameter. *The ischial spines may be a barrier to the passage of the fetal head* during childbirth if they are closer than 9.5 cm.

The Inferior Pelvic Aperture (pelvic outlet) does not have a smooth contour (Figs. 3-7 and 3-43*B*) because it is bounded posteriorly by the *coccyx* and *sacrum*, anteriorly by the *symphysis pubis*, and laterally by the *ischial tuberosities*.

The plane of the inferior pelvic aperture makes an angle of 10 to 15° with the horizontal when the pelvis is in the anatomical position (Fig. 3-42*B*).

Observe the two large **sciatic notches** between the sacrum and coccyx and the ischial tuberosities (Fig. 3-44). Note that they are divided into **greater and lesser sciatic foramina** by the sacrotuberous and sacrospinous ligaments (Figs. 3-16 and 3-43). These ligaments give the inferior pelvic aperture a diamond shape.

The anteroposterior (AP) diameter of the inferior pelvic aperture is usually about 11.5 cm. This is measured from the symphysis pubis to the anterior surface of the sacrum (Fig. 3-48).

In Figures 3-7, 3-42, and 3-46 examine the **pubic arch.** *Note that the subpubic angle is narrow in the male and wide in the female.* Also observe that this angle is rounded in the female. Verify that the subpubic angle of a male pelvis is about one fingerbreadth (Fig. 3-42*A*), whereas this angle in most females accommodates three average fingers. (Fig. 3-42*B*).

In forsenic medicine the bony pelvis is a reliable indicator of sex; even parts of a pelvis may give good clues as to the sex of the person from whom it came.

To pass through the birth canal (cervix and vagina) from the superior pelvic aperture to the inferior pelvic aperture, the fetal head must make almost a 90° turn.

The gynecoid pelvis (Fig. 3-47*D*) has a wide, circular superior pelvic aperture with a wide subpubic arch, and widely-spaced ischial spines (Figs. 3-7 and 3-43).

The gynecoid pelvis is the most common type in females and is the most roomy obstetrically; hence *a woman with a gynecoid pelvis has a reasonably uneventful delivery.*

The android pelvis (Fig. 3-47*C*) has a heart-shaped superior pelvic aperture and somewhat resembles the male pelvis (Fig. 3-42*A*). The ischial spines are usually quite prominent and the subpubic angle is narrow. In females with this type of pelvis, the fetal head has difficulty entering the superior pelvic aperture. As a result, labor is likely to be difficult.

The anthropoid pelvis, or ape-like pelvis (Fig. 3-47*A*), is fairly common. Its sides are long and narrow and the AP diameter of the superior pelvic aperture is greater than the transverse diameter. The sacrum is also long, consequently the pelvic cavity is deep. The subpubic angle is narrow and the ischial spines are prominent.

Females with anthropoid pelves often have difficulty in delivery owing to the narrowness of the transverse diameter of the superior pelvic aperture.

The platypelloid pelvis (Fig. 3-47*B*) is a flattened type of pelvis present in about 2.5% of females. The adjective *platypelloid* is derived from the Greek words *platys*, meaning broad or flat, and *pellis*, meaning bowl. Hence a platypelloid pelvis resembles a shallow, flat bowl.

The AP diameter of the superior pelvic aperture of the platypelloid is short and the transverse diameter is long. In patients with this type of pelvis, the fetal head may have difficulty entering the superior pelvic aperture and delivery of the fetus by **cesarean section** may be necessary. This incision is through the anterior abdominal wall and the anterior wall of the uterus through which the fetus is removed.

The bony pelvis is able to resist considerable trauma. Violent injuries are required to fracture the adult pelvis (*e.g.*, as occur in automobile accidents, Fig. 3-49). Lateral parts of the hip bone are the strongest.

Pelvic weak areas are (1) the sacroiliac region, (2) the ala of the ilium, and (3) the pubic rami. *Fractures of the pelvis in the pubo-obturator area are relatively common* and are often complicated because of their relationship to the urinary bladder (Fig. 3-89).

Anteroposterior compression of the pelvis occurs during "squeezing accidents" and commonly fractures the pubic rami. When the pelvis is compressed from the side, the acetabula and ilia are squeezed toward each other and may be broken. Some fractures of the pelvis result from a tearing away of bone by the posterior ligaments associated with the **sacroiliac joints** (Fig. 3-43).

In falls on the feet or the ischial tuberosities, (1) the pubic rami may be fractured, (2) the acetabula may be injured, and (3) the femora may be driven through the acetabula (Figs. 3-46 and 3-49) into the pelvis, injuring the pelvic organs.

In persons under 17 years, the acetabula may fracture into their three developmental parts (Fig. 4-3), or the acetabular margins may be torn away.

Direct violence may also fracture the sacrum, iliac crest, or any other part of the bony pelvis. *Pelvic fractures are often complicated by damage to the pelvic viscera* (*e.g.*, rupture of the bladder and/or urethra, Fig. 3-81), or to the large pelvic vessels; this results in extensive internal hemorrhage.

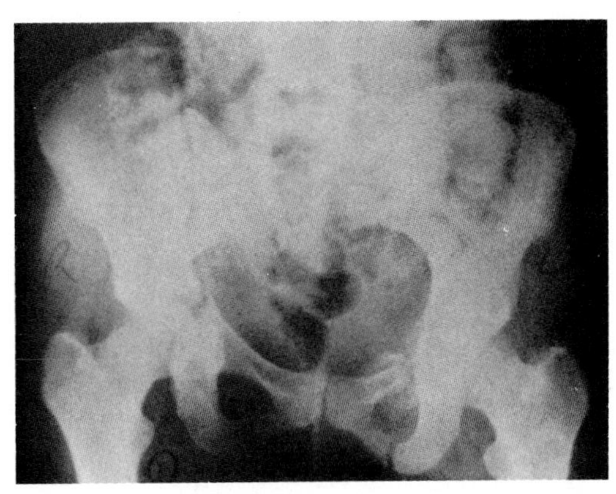

Figure 3-49. Radiograph of the pelvis of a 44-year-old woman who was involved in a serious automobile accident. Observe the vertical fracture of her ilium just lateral to the right sacroiliac joint. Also note that the superior and inferior pubic rami are fractured on the left side and that there is a fracture of the right acetabulum.

THE VERTEBROPELVIC LIGAMENTS (Figs. 3-16 and 3-43)

The parts of the bony pelvis are bound together by dense ligaments. The ilium is united to L5 vertebra by the *iliolumbar ligament* and the sacrum is joined to the ischium by the *sacrotuberous and sacrospinous ligaments.*

The Iliolumbar Ligament (Fig. 3-43). This strong triangular ligament connects the tip and the inferior and anterior parts of each transverse process of L5 vertebra to the internal lip of each iliac crest posteriorly. Occasionally this ligament is also attached to L4 vertebra. The inferior fibers of the iliolumbar ligament are attached to the lateral part of the sacrum; this band is called the **lateral lumbosacral ligament**.

The iliolumbar ligaments are important because they (1) limit axial rotation of L5 vertebra on the sacrum, and (2) *assist the vertebral articular processes in preventing anterior gliding of L5* vertebra on the sacrum.

The Sacrotuberous Ligament (Figs. 3-6, 3-7, 3-19, and 3-43). This ligament *passes from the sacrum to the ischial tuberosity*. It has a wide origin from the dorsal surfaces of the sacrum and coccyx and the posterior superior iliac spines. Its fibers run inferolaterally to the superior medial impression on the ischial tuberosity and then extend along its medial margin.

The Sacrospinous Ligament (Figs. 3-16 and 3-43). This thin triangular ligament *extends from the lateral margin of the sacrum and coccyx to the ischial spine*. It is related anteriorly to the coccygeus muscle.

The sacrum is wedged between the iliac bones and is held in position by the powerful **interosseous and dorsal sacroiliac ligaments** (Fig. 3-88).

The sacrotuberous and sacrospinous ligaments bind the sacrum to the ischium and resist posterior rotation of the inferior end of the sacrum. They also hold the posterior part of the sacrum inferiorly, thereby preventing the body weight from depressing its anterior part at the sacroiliac joints. The sacrotuberous and sacrospinous ligaments permit some movement of the sacrum, giving resilience to this region when sudden weight increases are applied to the vertebral column (*e.g.*, when landing on the feet during a fall).

The vertebropelvic ligaments relax progressively during pregnancy and movements between the vertebral column and the pelvis become freer. Furthermore, the symphysis pubis relaxes owing to a hormone called **relaxin** and the distance between the pubic bones increases considerably. These changes facilitate passage of the fetus through the birth canal during parturition.

MUSCLES OF THE LATERAL PELVIC WALLS

The walls of the pelvic cavity are lined in part with muscles (Figs. 3-50 to 3-53). The muscles of the lateral pelvic walls (piriformis and obturator internus) pass from the pelvis into the gluteal region, where they form part of the muscle group which rotates the thigh at the hip joint.

THE PIRIFORMIS MUSCLE (Figs. 3-25, 3-52, and 3-53)

This small pear-shaped muscle (L. *pirum*, pear + *forma*, form) *occupies a key position in the gluteal region* (see p. 446). The piriformis muscle is located partly within the pelvis minor (on its posterior wall) and partly posterior to the hip joint.

Origin (Figs. 3-25, 3-52, and 3-53). Anterior aspect of the second, third, and fourth **lateral masses of the sacrum and sacrotuberous ligament.**

Insertion (Fig. 4-28). Superior border of **greater trochanter of femur**. *The piriformis leaves the pelvis via the greater sciatic foramen* (Fig. 3-16) and passes posterior to the head of the femur to reach its insertion. Within the pelvis *the piriformis forms a muscular bed for the sacral plexus* of nerves (Fig. 3-54).

Nerve Supply (Fig. 3-54). Branches of ventral rami of **first and second sacral nerves.**

Actions. Laterally rotates the thigh when the hip joint is extended and **abducts the thigh** when the hip joint is flexed. *Assists in holding the head of the femur in the acetabulum.*

THE OBTURATOR INTERNUS MUSCLE (Figs. 3-25, 3-50, and 3-51)

This *thick, fan-shaped muscle* is situated partly within the pelvis minor and partly posterior to the hip joint. It covers most of the side wall of the pelvis minor.

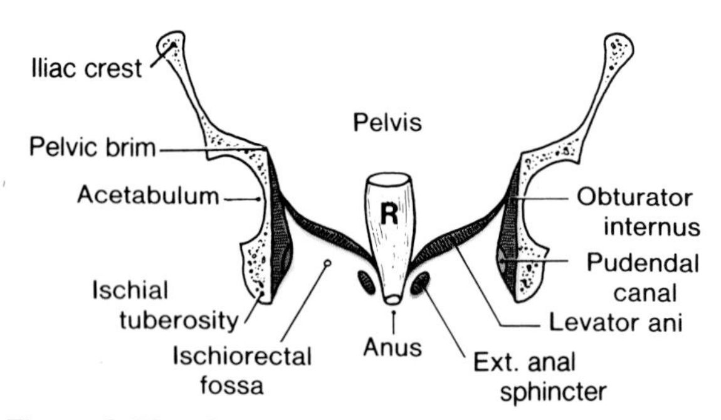

Figure 3-50. Schematic drawing of a coronal section of the pelvis showing the ischiorectal fossae and the funnel-shaped pelvic diaphragm formed by the two levatores ani and the two coccygeus muscles. The pelvis minor is situated inferior to the superior pelvic aperture (pelvic brim). *R* indicates the ampulla of the rectum which rests on and is anchored to the pelvic diaphragm.

Origin (Fig. 3-50). Almost entire internal surface of the **anterolateral wall of pelvis minor**, including the margins of the obturator foramen and the **obturator membrane.**

Insertion (Fig. 4-28). **Medial surface of the greater trochanter of the femur.** From its wide origin, the fibers of the obturator internus converge on a strong tendon which *passes through the lesser sciatic foramen.* It makes a right angle turn around the lesser sciatic notch to enter the gluteal region and pass to its insertion.

Nerve Supply (Fig. 3-54). **Nerve to obturator internus** (L5 and S1).

Actions. Laterally rotates the thigh when the hip joint is flexed. ***Assists in holding the head of the femur in the acetabulum.***

MUSCLES OF THE PELVIC FLOOR

The pelvic diaphragm forms the fibromuscular pelvic floor (Figs. 3-11 and 3-50) which supports the pelvic contents (Figs. 3-12, 3-20, and 3-46).

The two levatores ani muscles and the two coccygeus muscles, *with their superior and inferior investing fasciae,* ***form the funnel-shaped pelvic diaphragm*** (Figs. 3-11, 3-15, and 3-50 to 3-52).

THE LEVATOR ANI MUSCLE (Figs. 3-4, 3-11 to 3-13, 3-15, 3-21, 3-22, and 3-50 to 3-53)

This broad, thin curved sheet of muscle unites with its partner to form the largest and most important part of the pelvic diaphragm. The levatores ani muscles stretch between the pubis anteriorly and the coccyx posteriorly, and from one side wall of the pelvis to the other.

The levatores ani muscles form most of the floor of the pelvic cavity which separates it from the ischiorectal fossae. (Figs. 3-15 and 3-50).

The funnel-shaped pelvic diaphragm is perforated by the ***urethra*** *and the* ***anal canal*** *in the male* (Fig. 3-6), *and by the* ***urethra, vagina, and anal canal*** *in the female* (Fig. 3-13).

Origin (Figs. 3-50 to 3-53). **Pelvic surface of the body of the pubis to the ischial spine.** Between these bony attachments, *it arises from a* ***tendinous arch*** *formed by a thickening of the parietal pelvic fascia* covering the obturator internus muscle.

Insertion (Figs. 3-10 and 3-51 to 3-53). The two muscles converge and are inserted into (1) the **central perineal tendon**, (2) the **wall of the anal canal,** (3) the **anococcygeal ligament**, and (4) the **coccyx.** The anterior fibers, passing inferior to the prostate and inserting into the central perineal tendon, constitute the **levator prostatae muscle** (Figs. 3-19 and 3-52).

In the female these fibers, crossing the sides of the vagina before ending in the central perineal tendon, constitute the **pubovaginalis muscle** (Fig. 3-13).

For descriptive purposes three parts of the levator

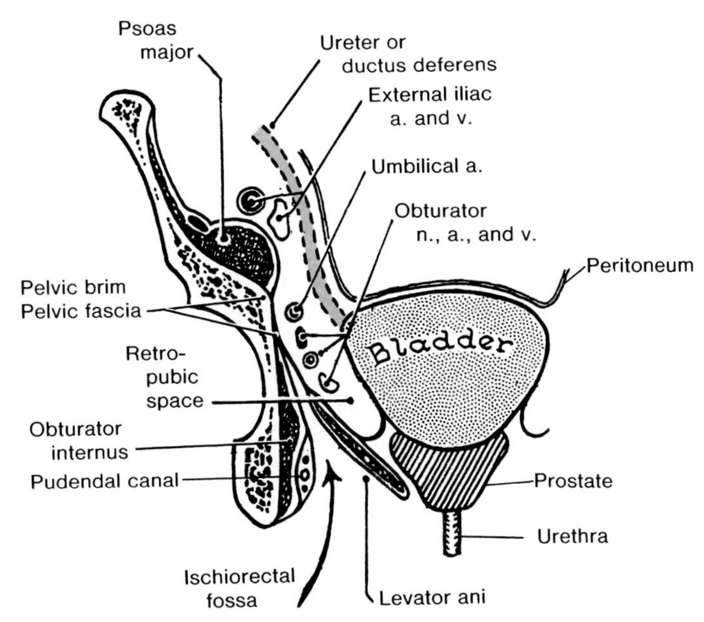

Figure 3-51. Diagrammatic drawing of the side wall of the male pelvis in coronal section. Observe that the intrapelvic surfaces of the muscles lining the walls of the pelvic cavity are covered with pelvic fascia and that this fascia is firmly attached to the pelvic brim.

ani muscle are named: *pubococcygeus, puborectalis, and iliococcygeus.*

The pubococcygeus muscle (Figs. 3-13, 3-14*A*, 3-18, and 3-53) is the main part of the levator ani. It ***arises from the pubis*** and runs posteromedially to ***insert into the coccyx and the anococcygeal ligament.***

The anococcygeal ligament (Fig. 3-20) *is the median fibrous intersection of the pubococcygeus muscles from the two sides*; it is *located between the anal canal and the tip of the coccyx.* As it courses inferiorly and medially in the female (Fig. 3-52), the pubococcygeus encircles the urethra, vagina, and anus and merges into the central perineal tendon.

The puborectalis muscle (Figs. 3-12 to 3-14, 3-17*A*, 3-19, 3-35, and 3-52) *is the part of the levator ani that lies medial to and at a more inferior level than the pubococcygeus muscle.* Like the pubococcygeus, it ***arises from the pubis*** and passes posteriorly; however, instead of inserting into the coccyx, *the muscles from the two sides loop around the posterior surface of the anorectal junction, forming a U-shaped rectal sling* (Fig. 3-55).

The iliococcygeus muscle (Figs. 3-14*A*, 3-19, 3-23, and 3-52) is *the thin part of the levator ani that* ***arises from the tendinous arch*** *of the parietal pelvic fascia and the* ***ischial spine*** (Fig. 3-52). The muscle on each side passes medially and posteriorly and **inserts into the coccyx and the anococcygeal ligament.**

Nerve Supply (Fig. 3-52). Fibers from the **third and fourth sacral nerves** which enter its pelvic surface and the **inferior rectal nerve** which enters its perineal surface.

Actions. With the coccygeus muscle, all parts of the levatores ani muscles **form the pelvic diaphragm**, which constitutes the pelvic floor.

The muscular pelvic diaphragm supports the pelvic viscera and resists the inferior thrust accompanying increases in the intraabdominal pressure (*e.g.*, as occurs in forced expiration and coughing).

Acting together the levatores ani muscles, forming the most important part of the pelvic diaphragm, **raise the pelvic floor**, thereby assisting the anterior abdominal muscles in compressing the abdominal and pelvic contents. This action is an important part of forced expiration, coughing, vomiting, urinating, and fixation of the trunk during strong movements of the upper limbs (*e.g.*, when lifting a heavy object).

The part of the levator ani that inserts into the central perineal tendon **supports the prostate** (*levator prostatae* (Fig. 3-19) **and the posterior wall of the vagina** (*pubovaginalis*, Fig. 3-13).

When the part of the levator ani that inserts into the wall of the anal canal and the central perineal tendon (puborectalis) contracts, it *raises the anal canal over a descending mass of feces*, thereby aiding defecation.

The puborectalis part of the levator ani holds the anorectal junction anteriorly (Fig. 3-55), thereby increasing the angle between the rectum and anal canal. This prevents passage of feces from the rectum into the anal canal when defecation is not desired or is inconvenient.

The anorectal angle supports most of the weight of the fecal mass, thereby relieving much pressure from the **external anal sphincter** (Figs. 3-12 and 3-17*A*).

During parturition the levatores ani muscles support

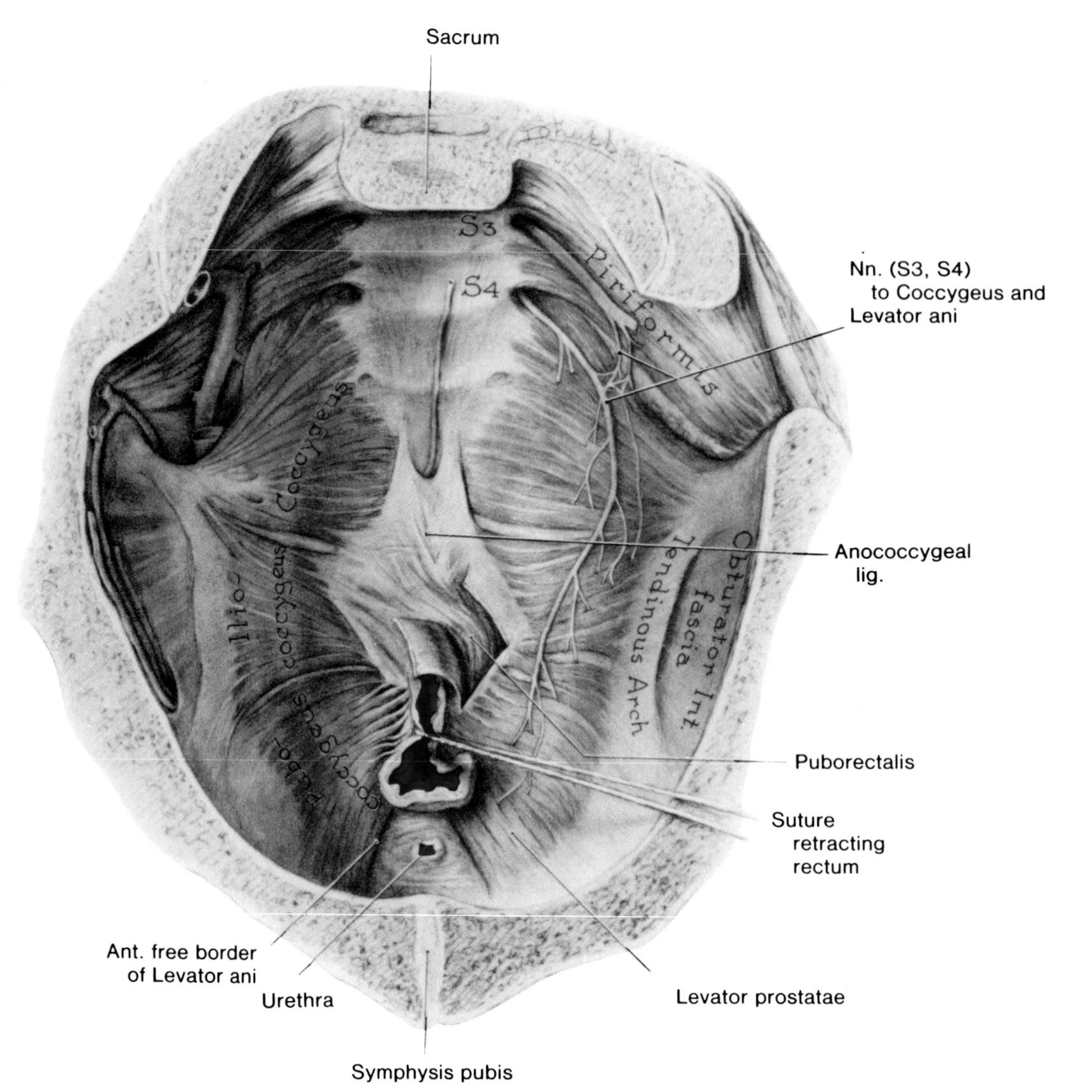

Figure 3-52. Drawing of a dissection of *a superior view the floor of the male pelvis*. The pelvic viscera are removed and the bony pelvis is sawn through transversely. Observe the pubococcygeus, the part of the levator ani muscle arising mainly from the pubic bone, the coccygeus arising from the ischial spine and the iliococcygeus arising from the tendinous arch in between. The pubococcygeus is strong, the iliococcygeus is weak. The coccygeus is largely transformed into the sacrospinous ligament (Fig. 3-16). Note that the urethra passes between the anterior borders of the pubococcygei and that the rectum perforates the pubococcygei: thus (1) the anterior fibers of the muscles of opposite sides meet and unite in the central perineal tendon anterior to the rectum; (2) the posterior fibers unite posterior to the rectum in an aponeurosis that extends posteriorly to the anterior sacrococcygeal ligament; (3) the middle fibers blend with the external wall of the anal canal and pass between the internal and external sphincters of the anus; and (4) on the left side this aponeurosis is reflected to show the puborectalis. Examine the branches of S3 and S4 nerves supplying the levator ani and coccygeus muscles. (The pudendal nerve, via its perineal branch, also supplies the levator ani).

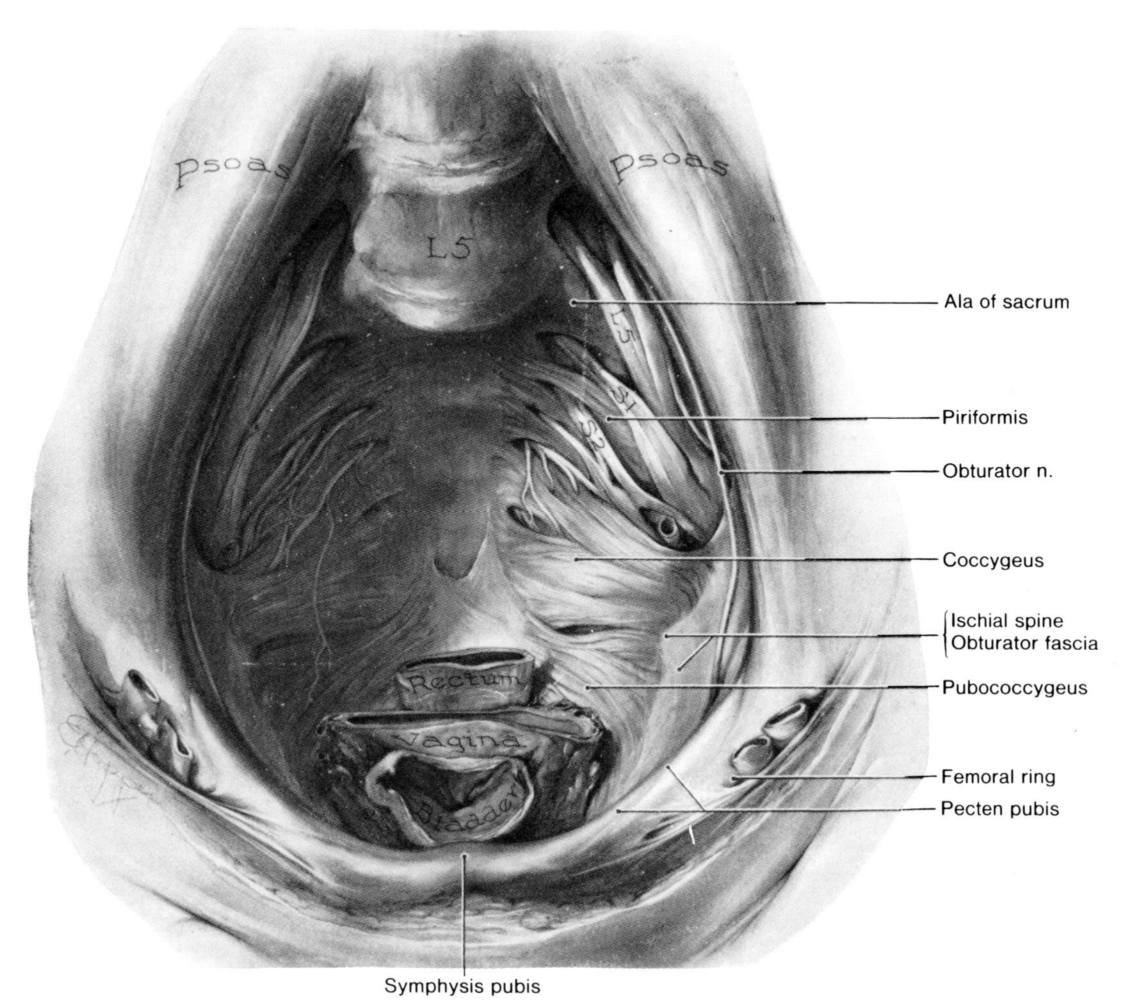

Figure 3-53. Drawing of a dissection of *the floor of the female pelvis*. Note the relative positions of the bladder, vagina, and rectum. Observe the muscles of the pelvic floor (levatores ani and coccygeus). Examine the obturator nerve, derived from nerves L2, L3, and L4, running along the side wall of the pelvis to enter the thigh through the obturator foramen.

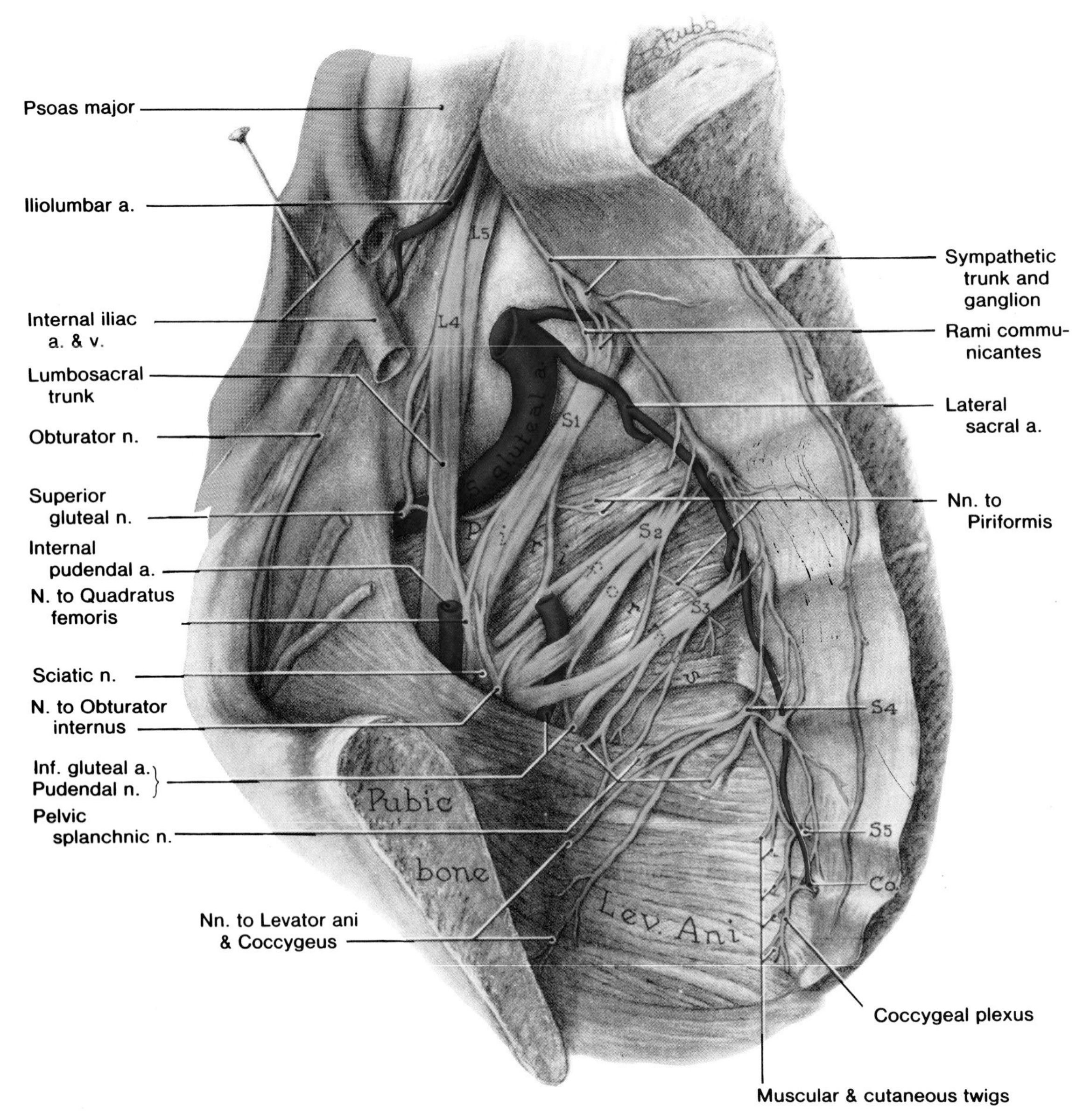

Figure 3-54. Drawing of a dissection of the sacral and coccygeal nerve plexuses. Observe that either the sympathetic trunk or its ganglia send gray rami communicantes to each sacral nerve and to the coccygeal nerve. Note the branch from L4 joining L5 to form the lumbosacral trunk. Observe that the roots of S1 and S2 supply the piriformis muscle and that S3 and S4 supply the coccygeus and levatores ani muscles. Note that S2, S3, and S4 each contribute a branch to the formation of the pelvic sphanchnic nerve. Observe the sciatic nerve arising from segments L4, L5, S1, S2, and S3 and the pudendal nerve arising from S2, S3, and S4. Examine the coccygeal plexus (*CO*) formed from S4 and S5 segments.

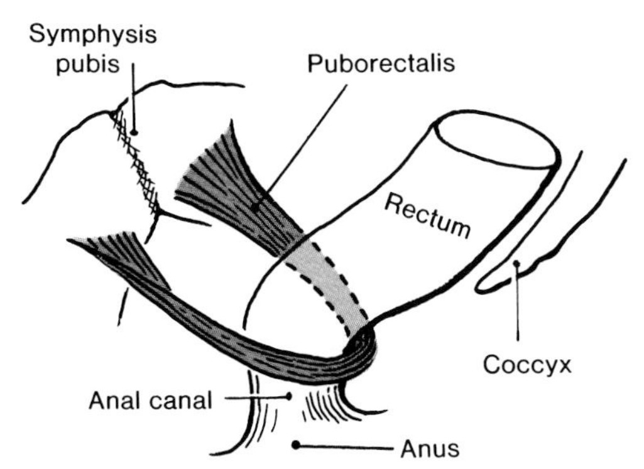

Figure 3-55. Diagram illustrating how the puborectalis muscle, part of the levator ani, forms a *U-shaped sling* around the anorectal junction which helps to hold the inferior part of the rectum anteriorly. The puborectal sling keeps the anorectal angle closed, except during defecation. When it relaxes it allows the anorectal junction to straighten, while other fibers draw the anal canal over the feces that is being expelled.

the fetal head while the cervix of the uterus is dilating to permit delivery of the baby.

THE COCCYGEUS MUSCLE (Figs. 3-19, 3-25, 3-52, and 3-53)

This *triangular sheet of muscle* lies against the posterior part of the iliococcygeus muscle and is continuous with it (Fig. 3-52).

The coccygeus muscle forms the posterior and smaller part of the pelvic diaphragm. On its external surface it blends with the **sacrospinous ligament** (Figs. 3-16 and 3-25).

Origin (Figs. 3-52 and 3-53). Pelvic surface of the **ischial spine** and the **sacrospinous ligament.**

Insertion (Figs. 3-19 and 3-25). Lateral margins of **the fifth sacral vertebra and the coccyx.**

Nerve Supply (Fig. 3-52). Branches from ventral rami of **fourth and fifth sacral nerves.**

Actions. Forms the most inferior part of posterior wall of pelvis minor and the posterior part of pelvic diaphragm. **The coccygeus muscle supports the coccyx and pulls it anteriorly** after it has been pressed posteriorly during childbirth.

Labor and childbirth may injure the supporting structures of the bladder, urethra, vagina, and rectum.

The pubococcygeus muscle, the main part of the levator ani, is obstetrically important because it encircles and supports the urethra, vagina, and anus. *The pubococcygeus muscle may be damaged during parturition.* **A median episiotomy** (Fig. 3-41*A*) involves the superficial transversus perinei muscle, the central perineal tendon, and often the pubococcygeus muscle (Fig. 3-53).

A mediolateral episiotomy (Fig. 3-41*B*) involves the same structures, as well as the bulbocavernosus muscle and *possibly* the deep transversus perinei muscle.

Injuries to the pelvic fascia and the pubococcygeus muscle may result in **cystocele** (herniation of urinary bladder, Fig. 3-56). When the urethra is also involved, the condition is called **cystourethrocele** (urethrocystocele). Herniation of the rectum (**rectocele**) results from damage to the middle third of the vagina and the wall of the pelvic diaphragm.

Episiotomy is performed to enlarge the external opening of the birth canal (vestibule of the vagina and pudendal cleft), and to prevent serious damage to the structures supporting the bladder, urethra, and rectum which may result later in cystocele and rectocele (Fig. 3-56).

Urinary stress incontinence may accompany weakening of the pelvic diaphragm. Stress incontinence is a disease *characterized by dribbling of urine* whenever the intraabdominal pressure is raised (*e.g.*, during coughing, sneezing, and lifting). This condition often results from *weakening of the supporting structures of the bladder and urethra* which are stretched and occasionally lacerated during parturition.

Weakening of the vesicourethral junction results in these patients being unable to prevent dribbling of urine when the intraabdominal pressure rises.

NERVES OF THE PELVIS

The pelvis is innervated mainly by the sacral and coccygeal nerves and by the pelvic part of the autonomic nervous system.

The piriformis muscle, described previously, *pads the posterior wall of the pelvis and forms a bed for the sacral and coccygeal nerve plexuses* (Figs. 3-25 and 3-54). The ventral rami of S2 and S3 nerves emerge between digitations of the piriformis muscle.

THE LUMBOSACRAL TRUNK (Figs. 2-118 and 3-54)

The descending part of the fourth lumbar nerve unites with the ventral ramus of the fifth lumbar nerve to form this thick, cord-like trunk. It passes inferiorly, anterior to the ala of the sacrum, to join the sacral plexus (Fig. 3-54). It descends obliquely over the *sacroiliac joint* and passes into the pelvis posterior to the pelvic fascia (Fig. 3-54). It then crosses the superior gluteal vessels to join the first sacral nerve which lies on a bed formed by the piriformis muscle.

THE SACRAL PLEXUS (Figs. 2-125 and 3-54)

This large plexus of nerves is located in the pelvis minor, where it is closely related to the anterior surface of the piriformis muscle.

The sacral plexus is formed by the lumbosacral trunk, the ventral rami of the first three sacral nerves, and the descending part of the fourth sacral nerve. The main nerves of the sacral plexus lie *external to the parietal pelvic fascia. All branches of the sacral plexus leave the pelvis through the greater sciatic foramen, except for the nerve to the piriformis muscle (S2), the perforating cutaneous nerves (S2, S3), and those to the pelvic diaphragm (Fig. 3-54).*

The sciatic nerve is formed by the ventral rami of L4 through S3, which converge on the anterior surface of the piriformis muscle. *The largest nerve in the body, the sciatic passes through the greater sciatic foramen inferior to the piriformis muscle* along with the **inferior gluteal nerve** (see p. 448 in Chap. 4 for details on these nerves).

The pudendal nerve, described previously, arises from the sacral plexus by separate branches from *ventral rami S2, S3, and S4*. It accompanies the internal pudendal artery (Fig. 3-54) and leaves the pelvis between the piriformis and coccygeus muscles, *hooking around the sacrospinous ligament to enter the perineum through the lesser sciatic foramen* (Fig. 3-16). Here it *supplies the muscles of the perineum*, including the external anal sphincter, and *ends as the dorsal nerve of the penis or clitoris* (Fig. 3-24).

The pudendal nerve is important because it supplies most of the perineum, including the external anal sphincter; it is also sensory to the external genitalia.

The superior gluteal nerve (L4, L5, and S1) leaves the pelvis via the greater sciatic foramen, superior to the piriformis muscle (Fig. 3–54). It supplies the gluteus medius and gluteus minimus muscles, and the tensor fasciae latae muscle (see Tables 4-1 and 4-3).

Other components of the sacral plexus (Fig. 3–54) include: (1) twigs to the piriformis muscle (S1 and S2); (2) twigs to the pelvic diaphragm (S3 and S4); (3) the nerve to the quadratus femoris muscle (L4, L5, and S1); and (4) the nerve to the obturator internus muscle (L5, S1, and S2).

Injuries to the sacral plexus are uncommon; however the sacral plexus and the lumbosacral trunk may be compressed by **pelvic tumors.** This compression usually causes pain in the lower limb and when associated with malignant pelvic tumors, the pain may be excruciating.

The head of a fetus may compress the nerves of the sacral plexus, producing aching pains in the mother's lower limbs.

THE OBTURATOR NERVE (Figs. 2-118 and 3-54)

This nerve is not derived from the sacral plexus; it *arises from the lumbar plexus in the abdomen* (L2, L3, and L4) and enters the pelvis minor. It *runs along the side wall of the pelvis in the extraperitoneal* fat, which divides it into anterior and posterior parts (Fig. 3-54). They leave the pelvis via the obturator foramen (Fig. 3-46) to supply the thigh (see Chap. 4).

The obturator nerve supplies the obturator externus and the adductor muscles of the thigh. It lies on the side wall of the pelvis (Figs. 3-54 and 3-58). Here it is vulnerable to injury during removal of **cancerous lymph nodes** from the side wall of the pelvis (Fig. 3-57) in patients with **malignant pelvis disease**. Inadvertent removal of part of this nerve results in deficient adduction of the thigh on the affected side.

Because the obturator nerve lies posterolateral to the ovary, it may be involved in pathological changes in this reproductive gland.

The Coccygeal Plexus (Fig. 3-54). This small plexus, which is of little importance, is formed by the ventral rami of S4 and S5 and the coccygeal nerve. It lies on the pelvic surface of the coccygeus muscle and supplies this muscle, part of the levator ani, and the sacrococcygeal joint. It then pierces the coccygeus muscle to supply the skin in the region of the coccyx.

THE PELVIC FASCIA

The pelvic fascia lines the pelvic cavity (Figs. 3-15, 3-22, and 3-51) as far inferiorly as the ischiopubic rami. It is attached to the periosteum just inferior to the pelvic brim (Figs. 3-11 and 3-51).

The pelvic fascia extends onto the superior surface of the pelvic diaphragm and is attached to the pelvic viscera.

The fascia lining the abdominal and pelvic cavities is continuous. It is anchored at the pelvic brim (Fig. 3-51) and is separated from the parietal peritoneum by extraperitoneal fat. In addition to enclosing these cavities, the pelvic fascia encloses the pelvic viscera and great vessels.

Superiorly the pelvic fascia is continuous with the transversalis fascia (Fig. 2-13) and is anchored to the periosteum on the posterior aspect of the body of the pubis.

The attachment of the pelvic fascia to the body of the pubis prevents the spread of infection from the anterior abdominal wall into the pelvis.

The pelvic fascia is divided into two layers: (1) *the parietal pelvic fascia* forming the fascial sheaths of the pelvic muscles, and (2) *the visceral pelvic fascia* forming the fascial coverings of the pelvic viscera.

The parietal pelvic fascia covers the pelvic surfaces of the obturator internus, piriformis, coccygeus, sphincter urethrae, deep transversus perinei, and levatores ani muscles (Figs. 3-11, 3-15, 3-22, and 3-51). The fascia covering the obturator internus muscle, (the **obturator fascia**) is thicker than other parts of the parietal pelvic fascia. The obturator fascia is separated superiorly from the psoas fascia by its attachment to the periosteum, just inferior to the pelvic brim (Fig. 3-51).

As described previously, *the levator ani muscles arise from a thickening of the obturator fascia, known as the tendinous arch,* stretching between the body of the pubis and the ischial spine (Fig. 3-52).

Superior to the tendinous arch the obturator fascia is thick and tough, whereas inferior to it the fascia is thin and lines the lateral wall of the **ischiorectal fossa**; it forms the medial wall of the *pudendal canal* (Figs. 3-11, 3-21, 3-24, and 3-51).

The fascia of the pelvic diaphragm covers both surfaces of the levator ani (Figs. 3-11 and 3-51). The fascia lining the superior surface of the levator ani is called the *superior fascia of the pelvic diaphragm.*

In the female this fascia is attached to the posterior aspect of the body of the pubis, the neck of the urinary bladder, the vagina, and the rectum.

In the male the superior fascia of the pelvic diaphragm is attached to the prostate and the rectum.

At the neck of the bladder the superior fascia of the pelvic diaphragm is thickened to form two cord-like bands, called the ***pubovesical ligaments*** in the female and the ***puboprostatic ligaments*** in the male (Figs. 3-12, 3-19, and 3-20). *These ligaments, one on each side of the median plane, anchor the neck of the urinary bladder to the pubis.* The pubovesical ligaments also attach to the wall of the vagina.

The inferior fascia of the pelvic diaphragm covers the inferior surface of the levator ani muscle, and forms the medial wall of the ischiorectal fossa (Figs. 3-2, 3-15, 3-18, 3-21, and 3-22). The inferior fascia of the pelvic diaphragm is continuous with the fascia on the medial surface of the inferior half of the obturator internus muscle (Figs. 3-11 and 3-51), and on the inferior surface of the external anal sphincter (Fig. 3-4).

The space between the pelvic fascia and the anterior surface of the bladder is called the retropubic space (Figs. 3-12, 3-15, 3-20, and 3-22).

The retropubic space contains extraperitoneal fat, loose areolar tissue, blood vessels, and nerves. This fat and areolar tissue accommodates the expansion of the urinary bladder.

ARTERIES OF THE PELVIS

Four arteries enter the pelvis minor: (1) *internal iliac* (paired); (2) *median sacral;* (3) *superior rectal;* and (4) *ovarian* (paired).

THE INTERNAL ILIAC ARTERY (Figs. 2-127 and 3-57 to 3-63)

The internal iliac artery supplies most of the blood to the pelvic viscera. **It is a terminal branch of the common iliac artery** which arises from the bifurcation of this vessel, medial to the psoas major muscle and anterior to the sacroiliac joint (Fig. 2-109).

The internal iliac artery begins at the level of the intervertebral disc between L5 and S1 (Fig. 2-127), where it is crossed by the ureter (Fig. 3-58). It is separated from the sacroiliac joint by the internal iliac vein and lumbosacral trunk (Figs. 3-54 and 3-58).

The internal iliac artery passes posteromedially into the pelvis minor, medial to the external iliac vein and obturator nerve (Figs. 3-57 and 3-58), and *lateral to the peritoneum.*

The branches of the internal iliac artery include both visceral ones and those supplying the body wall and the lower limb.

The arrangement of the visceral branches of the internal iliac artery is variable. The following branches of the internal iliac artery are listed in the order in which they commonly arise.

The Umbilical Artery (Figs. 3-57, 3-62, and 3-66*B*). This vessel runs anteroinferiorly between the urinary bladder and the lateral wall of the pelvis. It gives off the **superior vesical artery** which supplies numerous branches to the superior part of the urinary bladder (Fig. 3-57).

Prenatally the umbilical arteries carry blood to the placenta for reoxygenation. Postnatally their distal parts atrophy and become fibrous cords, called the **medial umbilical ligaments (obliterated umbilical arteries),** which run on the deep surface of the anterior abdominal wall (Figs. 2-15, 3-15, 3-57, and 3-62).

The Obturator Artery (Figs. 3-57 and 5-31A). The origin of this vessel is variable; usually it arises close to the umbilical artery and runs along the lateral wall of the pelvis, where **it is crossed by the ureter near its origin.**

The obturator artery passes anteroinferiorly on the obturator fascia, between the obturator nerve and vein (Fig. 3-51), and passes through the obturator foramen to supply muscles of the thigh and the ligament of the head of the femur (see p. 439 and 542).

Within the pelvis the obturator artery gives off some muscular branches, a nutrient artery to the ilium, and a pubic branch. This artery ascends on the pelvic surface of the ilium to anastomose with the pubic branch of the inferior epigastric artery of the external iliac. This anastomosis may be quite large and may replace part or all of the obturator artery (Fig. 4-39B).

The Inferior Vesical Artery (Fig. 3-57). This vessel occurs, as a named artery, only in the male. *It corresponds to the vaginal artery in the female* (Fig. 3-61).

The inferior vesical artery passes to the base of the urinary bladder and ***supplies the seminal vesicle, the prostate, and the posteroinferior part of the bladder.*** It also gives rise to the artery of the ductus deferens (Figs. 2-23 and 3-57).

The Vaginal Artery (Fig. 3-61). This vessel passes anteriorly and then along the side of the vagina, where it divides into numerous branches which supply the anterior and posterior surface of the vagina, postero-inferior parts of the urinary bladder, and the pelvic part of the urethra.

The Uterine Artery (Figs. 3-60 and 3-61). The uterine artery usually arises separately from the internal iliac, but it may arise from the umbilical artery.

The uterine artery descends on the lateral wall of the

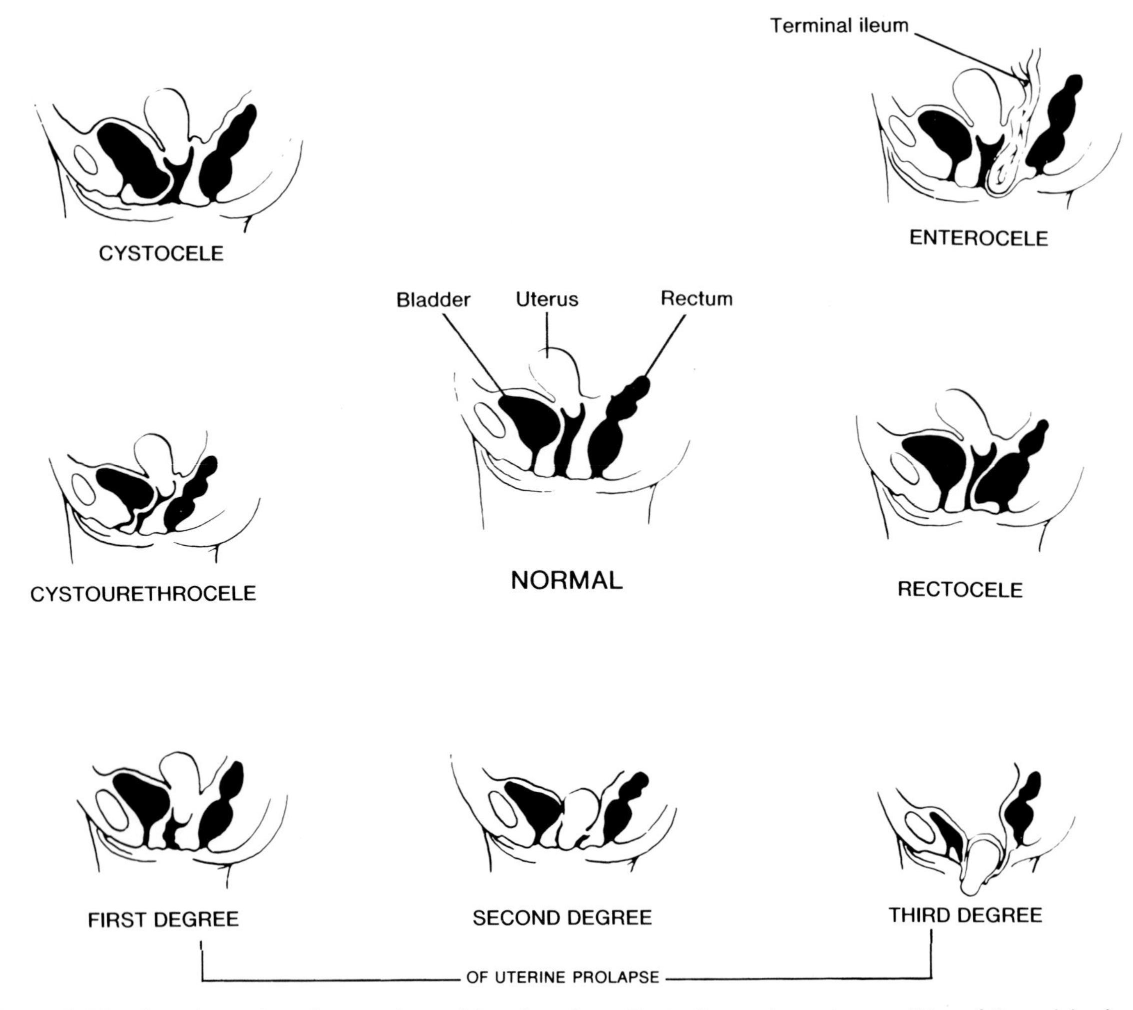

Figure 3-56. Drawings of median sections of female pelves, illustrating various abnormalities of the pelvic viscera resulting from weakening of the pelvic floor following injuries that occur during delivery of a fetus. The most commonly encountered abnormalities are cystocele (herniation of bladder), cystourethrocele (herniation of bladder and urethra), and rectocele (herniation of rectum). In third degree uterine prolapse, the cervix protrudes through the vaginal orifice and the pudendal cleft.

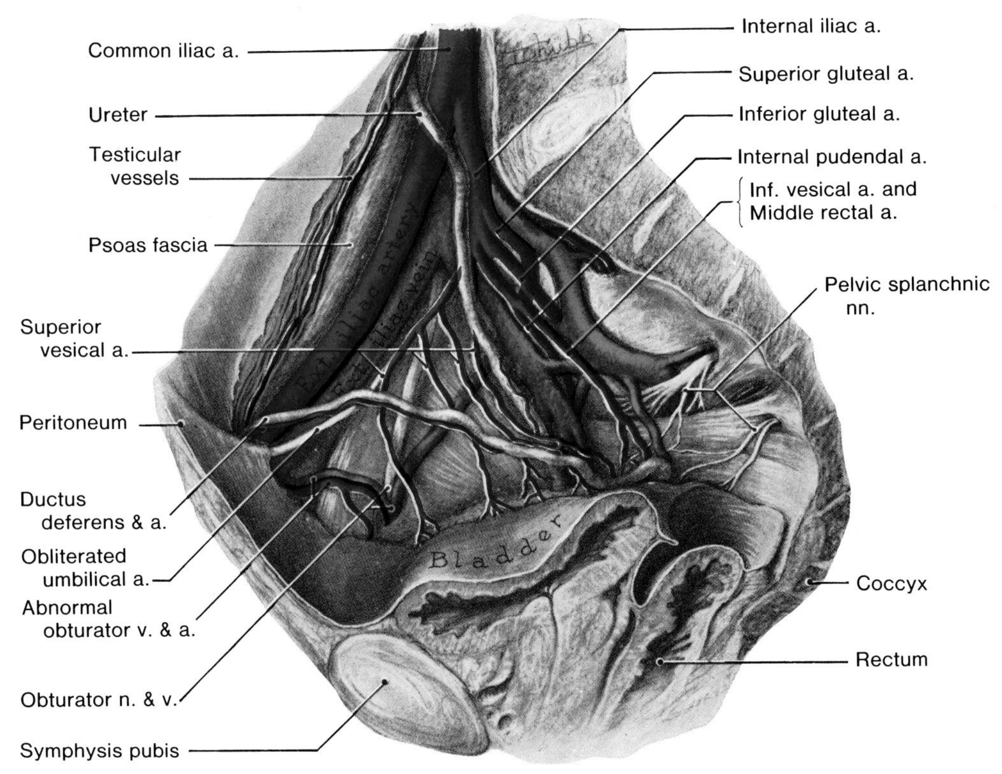

Figure 3-57. Drawing of a dissection of the lateral wall of the *male pelvis*. Observe the ureter and the ductus deferens running a subperitoneal course across the external iliac vessels, umbilical artery, obturator nerve and vessels and each receiving a branch from a vesical artery. Note that the ureter crosses the external iliac artery at its origin and the ductus deferens crosses it at its termination. Note the obturator artery, here springing from the inferior epigastric artery; i.e., the artery is "abnormal." Distally the umbilical artery is obliterated.

pelvis, anterior to the internal iliac artery (Fig. 3-58), and enters the root of the broad ligament where it passes superior to the lateral fornix of the vagina to reach the lateral margin of the uterus (Figs. 3-60 and 3-61).

It is clinically important to observe that the *uterine artery passes anterior to and superior to the ureter near the lateral fornix* of the vagina (Fig. 3-59). **This point of crossing lies about 2 cm superior to the ischial spine.** On reaching the side of the cervix, the uterine artery divides into (1) a large *superior branch supplying the body and fundus of the uterus,* and (2) a smaller *vaginal branch supplying the cervix and vagina.* The uterine artery pursues a tortuous course along the lateral margin of the uterus (Fig. 3-60), and ends when *its ovarian branch anastomoses with the ovarian artery* between the layers of the broad ligament (Figs. 3-60 and 3-62).

The fact that the uterine artery crosses anterior to and superior to the ureter near the lateral fornix of the vagina (Fig. 3-59) is clinically important because ***the ureter is in danger of being inadvertently clamped* or severed during a hysterectomy** (*G. hystera,* uterus + *ektomē,* excision) when the uterine artery is tied off. The left ureter is particularly vulnerable because it is very close to the lateral aspect of the cervix (Fig. 3-59).

The ureter is also vulnerable to injury when the ovarian vessels are being tied off during surgery (e.g., during an **ovariectomy**) because these structures lie very close to each other where they cross the pelvic brim (Fig. 3-62).

The Middle Rectal Artery (Figs. 3-57 and 3-84). This small vessel runs medially to the rectum. It also sends branches to the prostate and seminal vesicle in males, and to the vagina in females.

The Internal Pudendal Artery (Figs. 3-15 and 3-57). This vessel is larger in the male than the female. It passes inferolaterally, anterior to the piriformis muscle and sacral plexus (Fig. 3-54), and *leaves the pelvis between the piriformis and coccygeus muscles* by passing through the inferior part of the *greater sciatic foramen.*

The internal pudendal artery passes around the

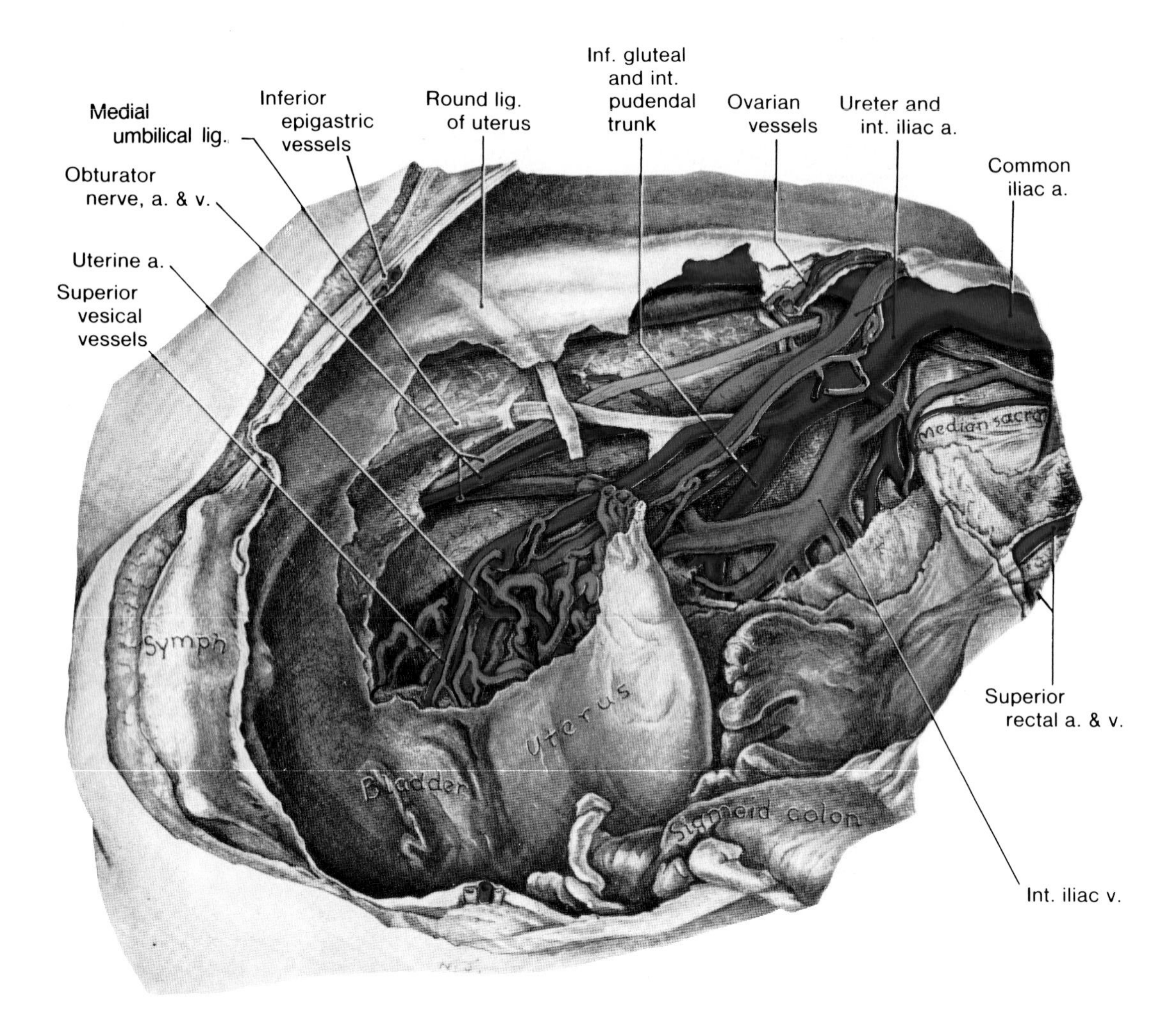

Figure 3-58. Drawing of a dissection of the blood vessels on the side wall of the *female pelvis,* viewed from the left side. In this old person the uterus is retroverted (inclined posteriorly). Note that the inferior gluteal and internal pudendal arteries arise from a common trunk. The *uterine plexus of veins* communicates with the superior rectal veins, thereby providing an additional area of portacaval anastomosis in the female (see Fig. 2-100).

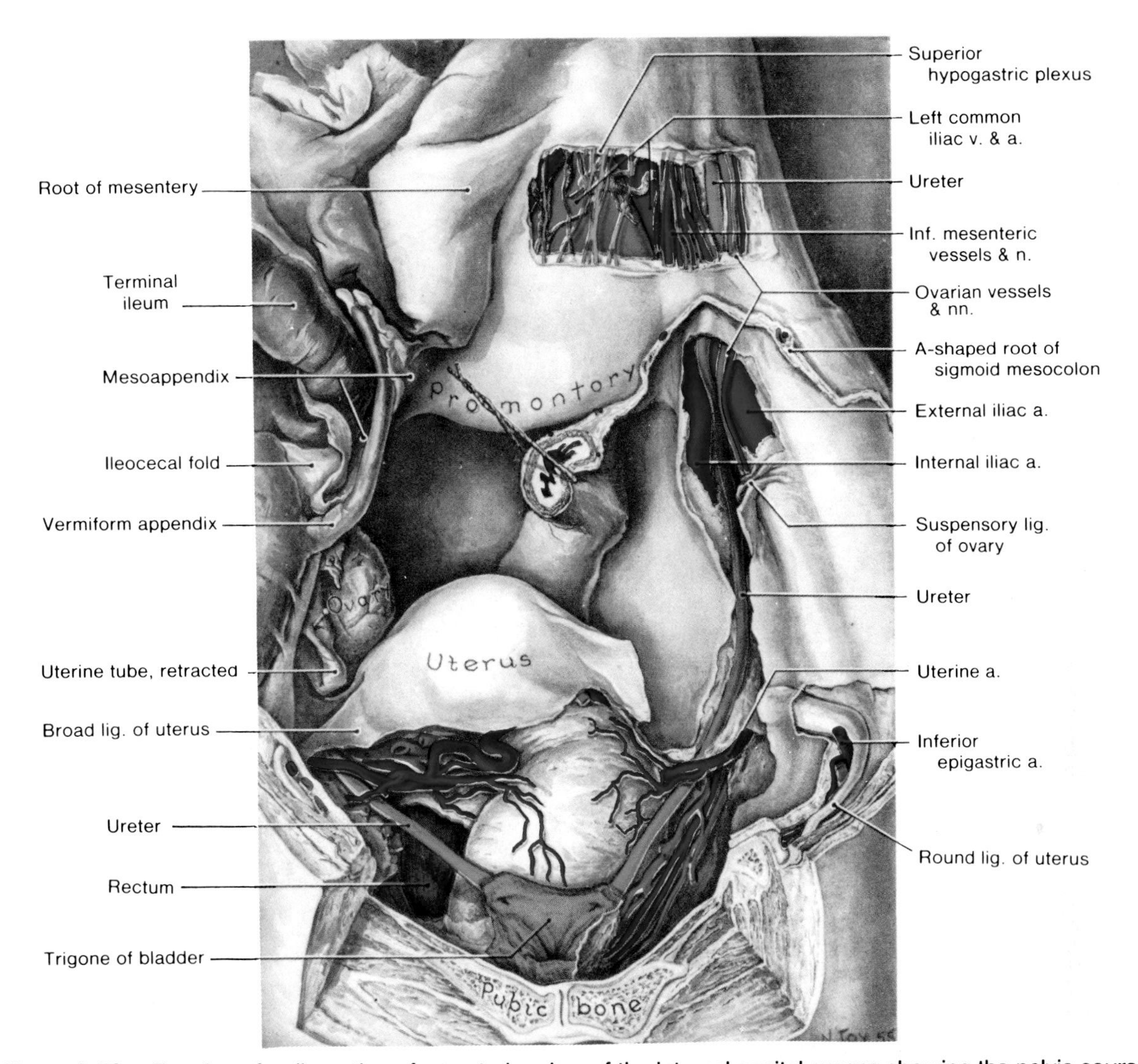

Figure 3-59. Drawing of a dissection of an anterior view of the internal genital organs showing the pelvic course of the ureter in the female. Observe the superior hypogastric plexus and some lymph vessels anterior to the left common iliac vein. Note that the left ureter is crossed by sigmoid branches of the inferior mesenteric artery, the ovarian vessels, and the uterine artery. Examine the apex of the Λ-shaped root of the sigmoid mesocolon, which is situated anterior to the left ureter and acts as a guide to it. Observe the ureter crossing the external iliac artery at the bifurcation of the common iliac artery and the ovarian vessels descending anterior to the internal iliac artery.

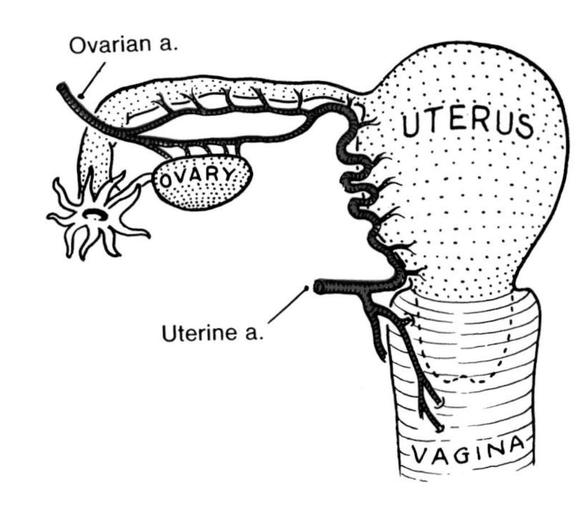

Figure 3-60. Drawing of an anterior view of the female internal genital organs illustrating the anastomosis of the ovarian and uterine arteries. This occurs in the broad ligament of the uterus (Figs. 3-61 and 3-62). Each ovarian artery arises from the abdominal aorta inferior to the renal artery, whereas the uterine artery is a branch of the internal iliac artery.

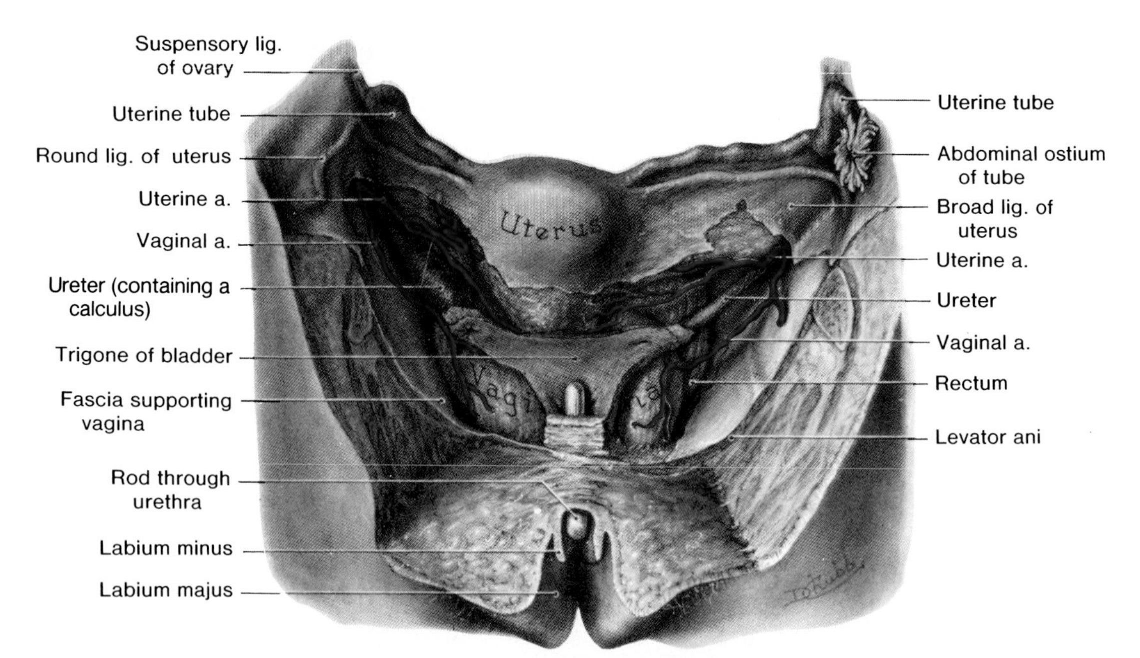

Figure 3-61. Drawing of an anterior view of the uterus and uterine tubes. The pubic bones and urinary bladder, except for the trigone, are removed. Examine the abdominal ostium of the left uterine tube which here happens to face anteriorly. The ostium is located at the bottom of the funnel-shaped infundibulum, the margins of which have irregular processes called fimbriae. Note that the right ureter contains a calculus or stone.

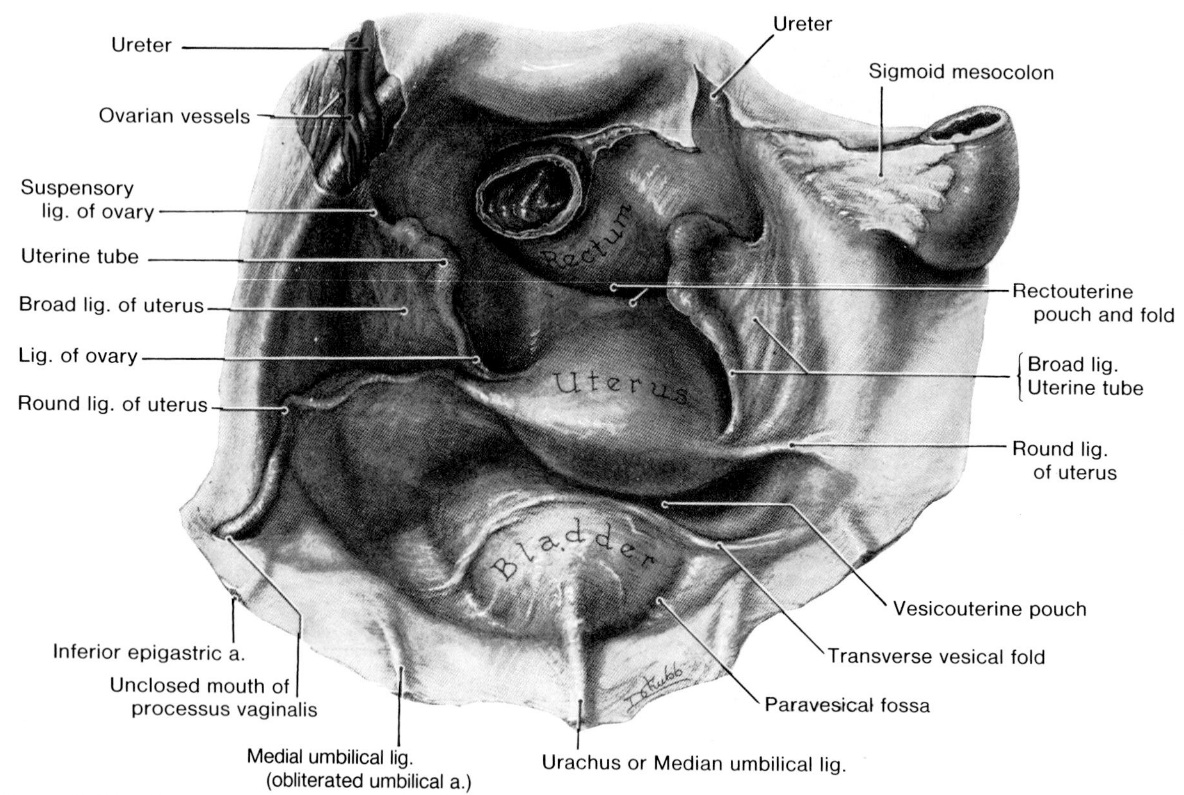

Figure 3-62. Drawing of a dissection of the female pelvis minor, superior aspect. Observe the pear-shaped uterus which, as usual, is asymmetrically placed, here leaning to the left. Note the right round ligament of the uterus which, in this specimen, is longer than the left and has an acquired "mesentery." The round ligament of the uterus takes the same subperitoneal course as the ductus deferens in the male. Examine the ovarian vessels in the suspensory ligament of the ovary. Observe the ovarian vessels crossing the external iliac vessels very close to the ureter. Note that the left ureter crosses at the apex of the Λ-shaped root of the sigmoid mesocolon.

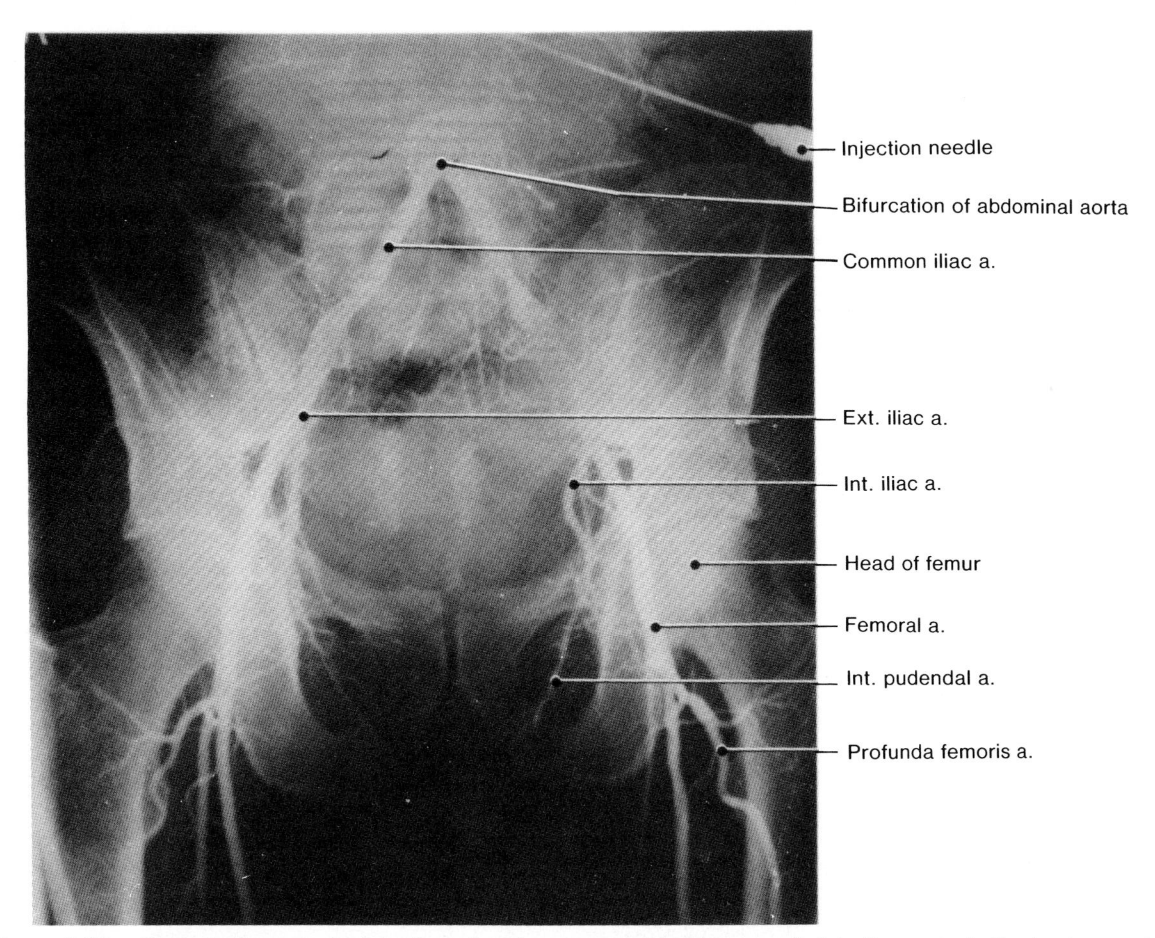

Figure 3-63. Iliac arteriogram. An injection of radiopaque material was made into the aorta in the lumbar region (note needle). Observe the bifurcation of the aorta into right and left common iliac arteries (anterior to L4 vertebra), and the common iliacs into internal and external iliac arteries opposite the sacroiliac joint at the level of the lumbosacral intervertebral disc. Also see Figures 2-109, 2-127, and 2-129.

posterior aspect of the ischial spine or sacrospinous ligament (Figs. 3-15 and 3-16) *to enter the ischiorectal fossa through the lesser sciatic foramen* (Fig. 3-40).

The internal pudendal artery, along with the internal pudendal veins and branches of the pudendal nerve, **passes through the pudendal canal in the lateral wall of the *ischiorectal fossa*** (Figs. 3-21 and 3-51). Just before it reaches the symphysis pubis, *the internal pudendal artery divides into its terminal branches: the* ***deep** and **dorsal arteries of the penis*** (Figs. 3-9 and 3-28), or *of the* ***clitoris*** (Fig. 3-39).

The Inferior Gluteal Artery (Figs. 3-54 and 3-57). This vessel passes posteriorly between the sacral nerves (usually S2 and S3), and *leaves the pelvis through the inferior part of the greater sciatic foramen, inferior to the piriformis muscle* (Fig. 3-54). It supplies the muscles and skin of the buttock and the posterior surface of the thigh (see Fig. 4-41).

The Superior Gluteal Artery (Figs. 3-54 and 3-57). This large artery passes posteriorly and ***runs between the lumbosacral trunk and the ventral ramus of the first sacral nerve.*** It leaves the pelvis through the superior part of the greater sciatic foramen, superior to the piriformis muscle to supply muscles in the buttock (see Fig. 4-41).

The Iliolumbar Artery (Fig. 3-54). This vessel runs superolaterally to the iliac fossa, passing anterior to the sacroiliac joint and posterior to the psoas major muscle. Here **it separates the obturator nerve from the lumbosacral trunk.**

Within the iliac fossa the iliolumbar artery divides into (1) an *iliac branch* supplying the iliacus muscle and the ilium, and (2) a *lumbar branch* supplying the psoas major and quadratus lumborum muscles.

The Lateral Sacral Arteries (Fig. 3-54). These vessels, usually a superior one and an inferior one on each side, may arise from a common trunk. They pass medially and **descend anterior to the sacral ventral rami**, giving off spinal branches that pass through the pelvic sacral foramina and *supply the spinal meninges and the roots of the sacral nerves.* Some branches of the lateral sacral arteries pass from the sacral canal through the dorsal sacral foramina to supply the muscles and skin overlying the sacrum.

The Median Sacral Artery (Figs. 2-51 and 3-58). This small unpaired artery arises from the posterior surface of the abdominal aorta, just superior to its bifurcation, and runs anterior to the body of the sacrum to end in a series of anastomoses that form the *coccygeal body*, the function of which is unclear at this time.

The median sacral artery represents the caudal end of the dorsal aorta in the embryo. It ends in the *coccygeal body*, a small cellular and vascular mass located anterior to the tip of the coccyx. Before the median sacral artery enters the pelvis, it sometimes *gives rise to the fifth lumbar arteries* (Fig. 2-51) and sends small branches to the posterior part of the rectum. The median sacral artery anastomoses with the lateral sacral arteries.

THE SUPERIOR RECTAL ARTERY (Figs. 3-58 and 3-84)

This artery is the *direct continuation of the inferior mesenteric artery.* It crosses the left common iliac vessels and descends into the pelvis minor in the sigmoid mesocolon.

At the level of S3 vertebra, the superior rectal artery divides into two branches which descend on each side of the rectum, supplying it as far inferiorly as the internal anal sphincter. The superior rectal artery anastomoses with branches of the middle rectal artery, a branch of the internal iliac artery, and with the inferior rectal artery, a branch of the internal pudendal artery (Fig. 3-84).

THE OVARIAN ARTERY (Figs. 2-127, 3-58, 3-59, 3-60, and 3-62)

The ovarian artery arises from the abdominal aorta inferior to the renal artery. It passes inferiorly *adherent to the parietal peritoneum and anterior to the ureter* on the posterior abdominal wall. It crosses the proximal ends of the external iliac vessels and enters the pelvis minor. It then *runs medially in the suspensory ligament of the ovary* and enters the superolateral part of the broad ligament and *supplies the ovary and the uterine tube.* **The ovarian artery anastomoses with the uterine artery** (Fig. 3-60).

VEINS OF THE PELVIS

The pelvis is mainly drained by the internal iliac veins and their tributaries (Figs. 2-166 and 2-129), but there is some drainage through the *superior rectal, median sacral, and ovarian veins.* Some blood from the pelvis also passes to the *internal vertebral venous plexus* (Fig. 5-58).

THE INTERNAL ILIAC VEIN (Figs. 2-116, 2-129 and 3-58)

The internal iliac vein joins the external iliac vein to form the common iliac vein which unites with its partner to form the **inferior vena cava** at the level of L5 vertebra (Figs. 2-101 and 2-116). *The internal iliac vein lies posteroinferior to the internal iliac artery* (Fig. 3-58) and its tributaries are similar to the branches of this artery, except for the fetal **umbilical vein**, which drains into the left branch of the portal vein (Fig. 2-100), and the iliolumbar vein which usually drains into the common iliac vein (Fig. 5-58).

The superior gluteal veins, the vena comitantes of the superior gluteal arteries, are the largest tributaries of the internal iliac veins, except during pregnancy when the uterine veins become larger.

Pelvic venous plexuses are formed by the veins in the pelvis (Figs. 3-22 and 3-64). *These intercommunicating networks of veins are clinically important.* The various plexuses (vesical, prostatic or uterine, vaginal, and rectal) unite and drain mainly into the internal iliac vein, but some plexuses drain via the ***superior rectal vein*** into the inferior mesenteric vein (Fig. 2-100).

THE PELVIC LYMPHATICS

There are many lymph nodes and lymph vessels in the pelvis, but they are not easy to demonstrate in dissections of old cadavers.

In general the pelvic organs drain through the external and internal iliac lymph nodes and the sacral lymph nodes (Fig. 3-73). In addition there are small lymph nodes between the layers of the broad ligament and in the fascial sheaths of the bladder and rectum. *From all these nodes, lymph drains to the common iliac and lumbar lymph nodes* (Fig. 2-121).

The external iliac lymph nodes (8 to 10) lie on the corresponding external iliac vessels (Fig. 3-73), and *drain lymph from the lower limb, abdominal wall, bladder, and prostate* or the *uterus* and *vagina.*

The internal iliac lymph nodes surround the internal iliac vessels and their branches (Fig. 3-73). *The internal iliac lymph nodes receive lymph from all the pelvic viscera*, deep parts of the perineum, and the gluteal and thigh regions.

The sacral lymph nodes lie on the median and lateral sacral arteries. *They receive lymph from the posterior pelvic wall, rectum, neck of the bladder, and the prostate (or cervix).*

The common iliac lymph nodes form two groups: (1) a *lateral group* lies along the common iliac vessels, and (2) a *median group* is located in the angle between these vessels (Fig. 3-73).

The lateral group of common iliac lymph nodes receives lymph from the lower limb and pelvis via the external and internal iliac lymph nodes, whereas the medial group of common iliac lymph nodes receives lymph directly from the pelvic viscera and indirectly through the internal iliac and sacral lymph nodes.

The lumbar lymph nodes (Fig. 2-121) lie along

the abdominal aorta and the inferior vena cava and *receive lymph from the common iliac lymph nodes.* The efferent vessels from the lumber lymph nodes form right and left lumbar trunks which drain into the ***cisterna chyli*** (Fig. 1-73), a lymph sac lying on the first two lumbar vertebrae.

THE PELVIC VISCERA

The pelvic viscera are the organs that cannot be removed from the pelvis without cutting. Hence the intestines which descend into the pelvis are not pelvic viscera because they can be easily removed from the pelvis without cutting their mesenteries.

THE URINARY ORGANS

The urinary system (Fig. 2-102) consists of: (1) the two excretory organs or ***kidneys*** (described on p. 252), (2) the ***ureters*** which convey the urine to (3) the ***urinary bladder***, a reservoir for temporarily holding urine which passes through (4) the ***urethra*** to reach the exterior.

THE URETERS (Figs. 2-101 to 2-106, 2-112, 2-121, 3-59, 3-61, and 3-62)

The abdominal parts of the ureters, about half of these 25-cm long muscular tubes, are described on page 253. When leaving the abdomen to enter the pelvis minor, *the ureters pass over the pelvic brim, anterior to the origins of external iliac arteries* (Figs. 3-58 and 3-62).

The pelvic part of the ureter courses postero-inferiorly, external to the parietal peritoneum on the lateral wall of the pelvis and anterior to the internal iliac artery (Figs. 3-59 and 3-61). It continues this course to a point about 1.5 cm superior to the ischial spine (Fig. 3-57), and then curves anteromedially, superior to the levator ani muscle.

The ureter is closely adherent to the peritoneum. The only structure that passes between the ureter and the peritoneum is the ductus deferens (Fig. 3-57).

In the male the ureter lies lateral to the ductus deferens and enters the posterosuperior angle of the bladder, just superior to the seminal vesicle (Figs. 3-57, 3-64, and 3-65).

In the female the ureter descends on the lateral wall of the pelvis minor (Fig. 3-58), where it forms the posterior boundary of the **ovarian fossa** (Fig. 3-74*B*). As it descends, *the ureter passes medial to the origin of the uterine artery* (Fig. 3-58) and continues to a point at the level of the ischial spine, where it *is crossed superiorly by the uterine artery* (Fig. 3-59). It then passes close to the lateral fornix of the vagina, especially on the left side, and enters the posterosuperior angle of the bladder (Fig. 3-59).

Vessels of the Ureters (Figs. 2-112 and 3-57). The blood supply to the ureter comes from ***three main sources***: (1) the *renal artery* (superior end), (2) the *common iliac artery* or the *aorta,* and (3) the *vesical arteries.* **In the pelvis the arteries supplying the ureter approach it from the lateral side** (Fig. 2-112).

In the female the most constant arteries supplying the pelvic part of the ureter are branches of the **uterine artery** in the floor of the pelvis minor (Fig. 3-58).

In the male similar branches are derived from the **inferior vesical artery** (Fig. 3-57).

Veins accompany all the above arteries and have corresponding names.

Nerves of the Ureters (Figs. 2-114 and 3-80). The nerve supply of the ureter is derived from adjacent *autonomic plexuses* (renal, testicular or ovarian, and inferior hypogastric, which contain pain fibers. The afferent fibers reach the spinal cord through the dorsal roots of T11, T12, and L1 nerves.

The ureters are expansile muscular tubes which become dilated if obstructed. An acute obstruction results from a **ureteric calculus** or kidney stone (Fig. 3-61).

Although the passage of some calculi (**stones**) causes little or no pain, most cause people to rush to a hospital owing to the severe pain.

The symptoms and severity depend on the location, type, and size of the stones and on whether they are smooth or spiky. The pain caused by a stone is referred to as colicky pain or colic.

The colic results from hyperperistalsis in the ureter, superior to the point of the obstruction. Usually **colicky pain** is accompaned or followed by a dull, more constant pain owing to distention of the ureter and renal pelvis.

Ureteric calculi may cause complete or intermittent obstruction of urinary flow. The obstruction may occur anywhere along the ureter, but it occurs most often (1) where the ureter crosses the external iliac artery and the brim of the pelvis (Fig. 3-57), and (2) where it passes obliquely through the wall of the urinary bladder (Figs. 3-61 and 3-62).

The presence of kidney stones can often be confirmed by abdominal radiographs or by an *intravenous urogram* (Fig. 2-106).

Surgical removal of ureteric stones, involving a general anesthetic and abdominal incision, is required for some stones. A recent innovation enables stones to be removed from the kidney without major surgery using a *nephroscope.* For stones too large to extract, an *ultrasound probe*

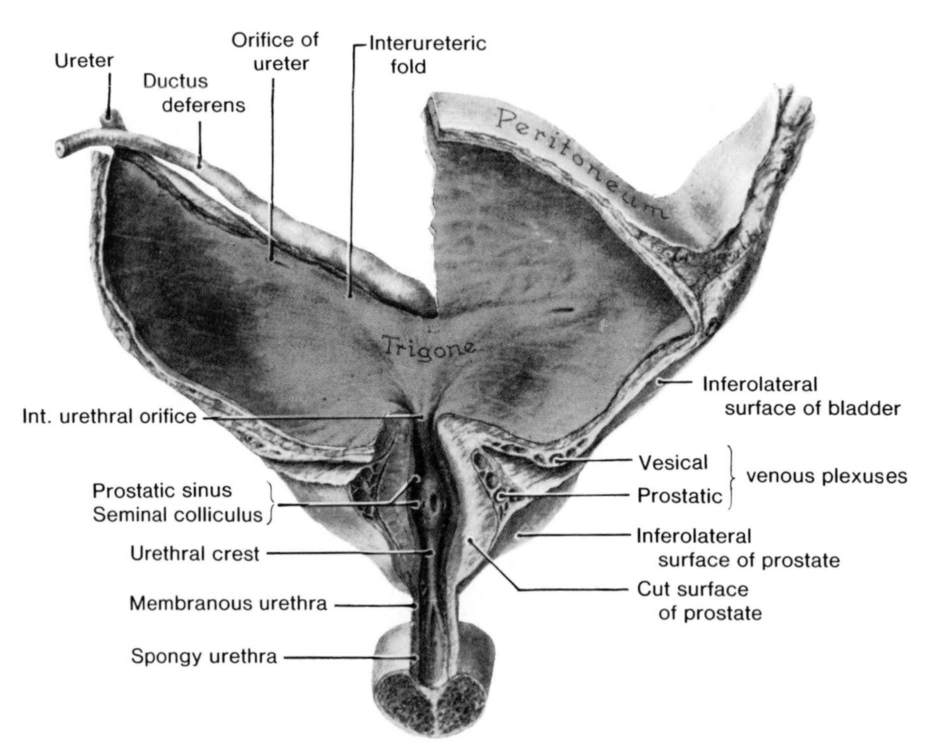

Figure 3-64. Drawing of the interior of the male urinary bladder and prostatic urethra. The anterior parts of the bladder, prostate, and urethra are cut away. The knife was then carried through the posterior wall of the bladder at the superior border of the right ureter and interureteric fold. This fold unites the ureters along the superior limit of the trigone. Observe that the right ureter does not join the bladder wall but traverses it obliquely, as far as its slitlike orifice, situated 2.5 to 3 cm from the left orifice. Note that the mucous membrane is smooth over the trigone and rugose elsewhere. Observe the slight fullness posterior to the internal urethral orifice which, when exaggerated, becomes the uvula vesicae. This small projection is caused by the median lobe of the prostate. Examine the orifice of the prostatic utricle at the summit of the seminal colliculus on the urethral crest. This orifice is homologous with the vaginal orifice in the female. Observe the tiny orifices of the ejaculatory ducts, one on each side of the prostatic utricle (also see Fig. 3-12).

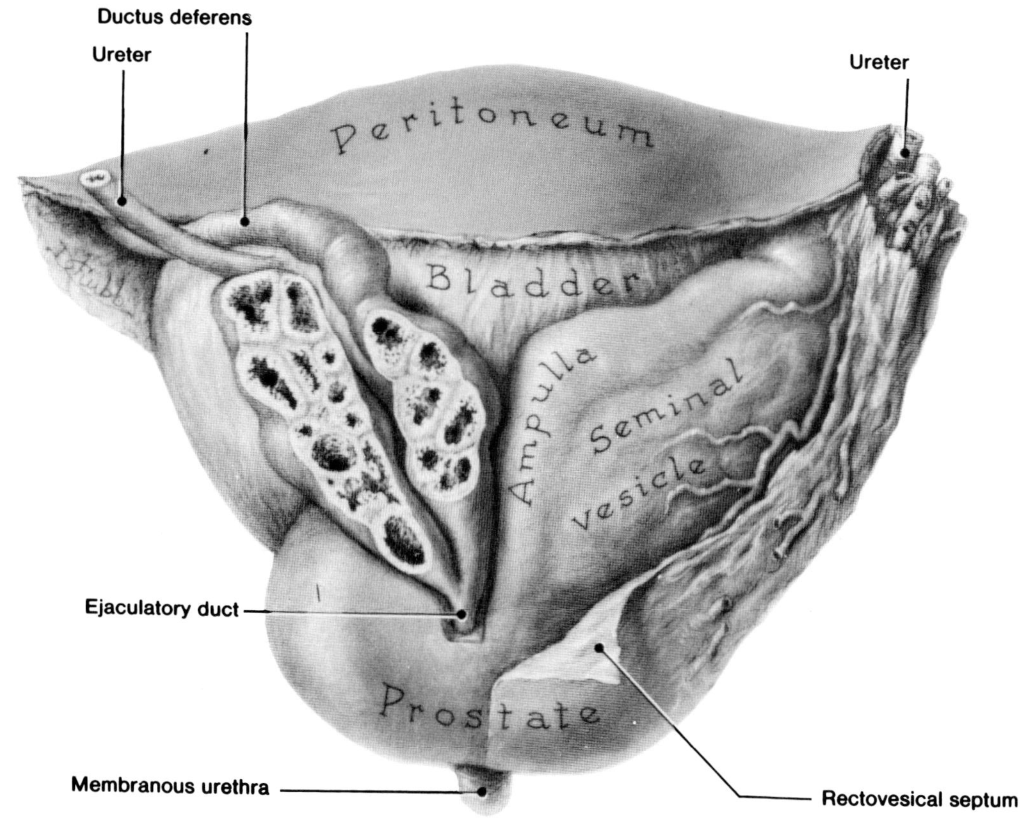

Figure 3-65. Drawing of a posterior view of a dissection of the urinary bladder, ductus deferens, seminal vesicles, and prostate gland. The left seminal vesicle and the ampulla of the ductus deferens are dissected free and sliced open. The ureters pass obliquely through the bladder wall in an anteromedial direction to open into the corresponding superolateral angle of the trigone of the bladder (Figs. 3-61 and 3-64).

may be used through the nephroscope to break the stones up before removal.

THE URINARY BLADDER (Figs. 3-12, 3-20, 3-29, 3-57, 3-61, 3-62, and 3-64 to 3-67)

The urinary bladder (L. *vesica*) is **a muscular sac or vesicle for urine storage.** *In the adult* the empty bladder lies in the pelvis minor, posterior to the pubic bones, from which it is separated by the **retropubic space** (Figs. 3-20 and 3-22).

In infants and children the urinary bladder is in the abdomen even when empty (Fig. 3-66*B*). The bladder begins to enter the pelvis major at about 6 years of age, but it is not entirely within the pelvis minor until after puberty.

The urinary bladder is a hollow viscus with strong muscular walls that is characterized by its distensibility. *Its shape, size, position, and relations vary with the amount of urine it contains and with the age of the person.*

The mucous membrane lining the bladder is loosely connected to its muscular wall, except in a triangular area at its base called the **trigone of the bladder** (Figs. 3-61 and 3-64). The mucous membrane in the empty contracted bladder is thrown into numerous folds or rugae, except in the trigone where the mucous membrane is always smooth, because it is firmly attached here to the muscular wall.

An empty bladder lies almost entirely in the pelvis. It is located in the anteroinferior part of the pelvis minor, *inferior to the peritoneum* (Figs. 3-57 and 3-65). It rests on the pelvic floor posterior to the symphysis pubis. As it fills, it ascends into the abdomen. *A full bladder may reach the level of the umbilicus.*

In the female, the peritoneum is reflected from the superior surface of the bladder near its posterior border on to the anterior wall of the uterus, at the junction of the body and cervix (Fig. 3-20).

The vesicouterine pouch of peritoneum extends between the bladder and the uterus (Fig. 3-20). This pouch is empty except when the uterus is retroverted (Fig. 3-58). In this case a loop of bowel may lie in it.

In the male, the peritoneum is reflected from the bladder over the superior surfaces of the ductus defer-

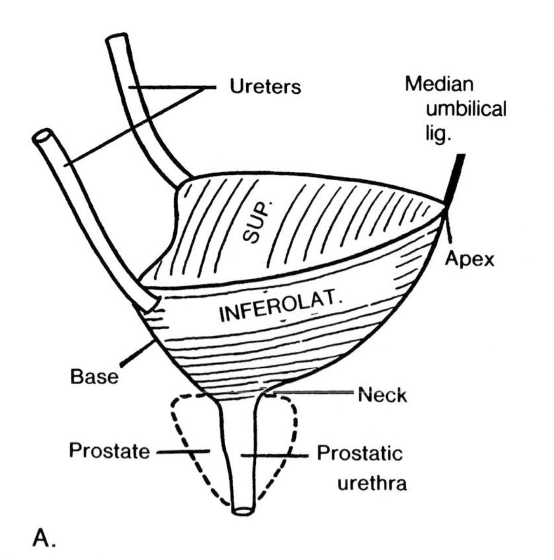

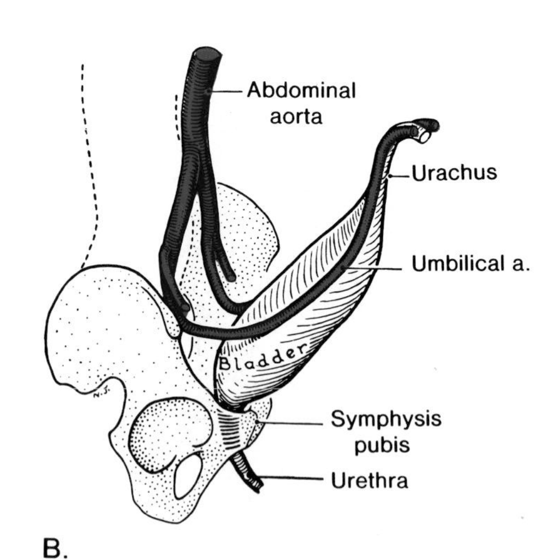

Figure 3-66. *A*, drawing of an empty, contracted adult bladder. Observe that it has the form of a triangular pyramid with its base posteriorly. It reminds one of the forecastle of a ship. Note that it has four surfaces, four angles, and three ducts (ureters and urethra). *B*, drawing of the empty bladder of an infant. Note that it is fusiform in shape and is located in the abdomen. The urinary bladder begins to enter the enlarging pelvis minor at 6 years, but does not become a pelvic organ until after puberty. Most of the intra-abdominal parts of the umbilical arteries form the medial umbilical ligaments. The proximal parts of these vessels persist as the superior vesical arteries (Fig. 3-57). The urachus is the remnant of the embryonic allantois. It usually loses its lumen and is represented in the adult (A) by a fibrous cord, called the median umbilical ligament (also see Fig. 3-62).

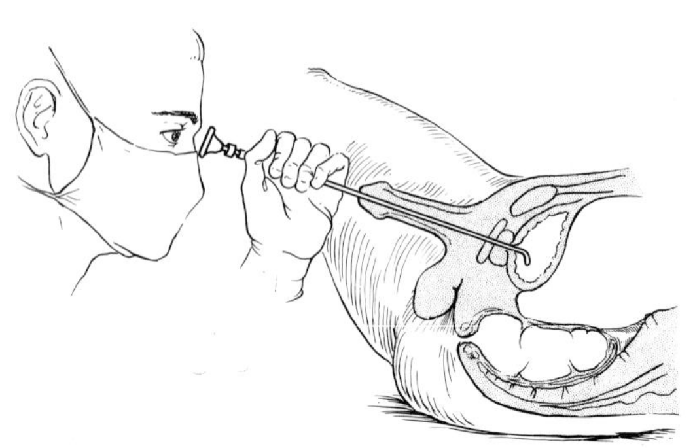

Figure 3-67. Drawing illustrating the use of a cytoscope to examine the interior of the urinary bladder. Various surgical procedures may also be performed through a cytoscope with the aid of carefully designed instruments that are inserted along the barrel of the instrument. A catheter may also be inserted into the ureter through the cytoscope to obtain a sample of urine from the pelvis of the kidney, and to inject radiopaque contrast material for retrograde pyelography. (Illustrated by Mrs. D. M. Hutchinson. Reprinted with permission from Laurenson RD: *An Introduction to Clinical Anatomy by Dissection of the Human Body.* Philadelphia, WB Saunders, 1968.)

entes and the seminal vesicles (Figs. 3-15, 3-57, and 3-65). The bladder is relatively free within the loose extraperitoneal fatty tissue (Fig. 3-51), except for its neck, which is held firmly by the **puboprostatic ligaments** in the male (Fig. 3-19) and the **pubovesical ligaments** in the female (Fig. 3-20). Hence, as the bladder fills it can expand superiorly into the extraperitoneal fatty tissue of the anterior abdominal wall (Figs. 3-12, 3-15, and 3-20). This lifts the peritoneum from the transversalis fascia of the anterior abdominal wall (Fig. 2-15).

In fixed cadaveric specimens, the empty bladder has the form of a triangular pyramid (Fig. 3-66*A*). *In living persons* the bladder always contains some urine; hence it is usually more or less rounded (Fig. 3-62).

The empty bladder has four surfaces (Fig. 3-66*A*): a **superior surface** facing superiorly; **two inferolateral surfaces** facing inferiorly, laterally, and anteriorly; and a **posterior surface** facing posteriorly and slightly inferiorly. The inferolateral surfaces of the bladder are in contact with the fascia covering the levatores ani muscles (Figs. 3-21 and 3-51).

The posterior surface of the bladder is referred to as the **base of the bladder** and its anterior end is known as the **apex of the bladder** (Fig. 3-66*A*). The inferior part of the bladder, where the base and the inferolateral surfaces converge is called the **neck of the bladder.** This is where the lumen of the bladder opens into the urethra bladder (Fig. 3-64). *In the male* the neck of the bladder rests on the prostate gland (Figs. 3-12 and 3-66).

The Bladder Bed (Figs. 3-15 and 3-51). The shape of the bladder is largely determined by the structures closely related to it. The entire organ is enveloped by areolar tissue, called **vesical fascia,** in which is located the *vesical venous plexus* (Fig. 3-64). The bladder bed is formed on each side by the pubic bones and the **obturator internus and levator ani muscles**, and posteriorly by the rectum (Figs. 3-12, 3-14, and 3-21).

In the female the base of the bladder is separated from the rectum by the cervix and the superior part of the vagina (Fig. 3-20).

In the male the base of the bladder is separated from the rectum by the ampullae of the ductus deferentes and the seminal vesicles (Figs. 3-15 and 3-65).

Structure of the Urinary Bladder. The wall of the bladder is composed chiefly of smooth muscle, called the **detrusor muscle** (L. *detrudere*, to thrust out). It consists of three layers running in many directions. There are external and internal layers of longitudinal fibers and a middle layer of circular fibers. Toward the neck of the bladder, these muscle fibers form the involuntary **internal sphincter of the urinary bladder.** Some of these fibers run radially and assist in the opening of the **internal urethral orifice** (Fig. 3-64).

In the male the muscle fibers in the neck region of the urinary bladder are continuous with the connective tissue stroma of the prostate, whereas *in the female* they are continuous with the muscle fibers in the wall of the urethra.

The mucous membrane of the bladder, lined with transitional epithelium, can undergo considerable stretching. The openings of the ureters (*ureteric orifices*) and urethra (*internal urethral orifice*) are located at the base of the bladder, where they form the **angles of the trigone** (Fig. 3-61).

The ureters pass obliquely through the bladder wall in an inferomedial direction, which helps to prevent urine from backing up into the ureters. An increase in bladder pressure presses the walls of the ureters together, preventing the pressure in the bladder from forcing urine up the ureters and damaging the kidneys.

The orifices of the ureters are connected by a narrow **interureteric fold** (ridge) (Fig. 3-64) which forms the superior margin of the trigone.

Vessels of the Urinary Bladder (Figs. 3-57 to 3-59). *The main arteries supplying the bladder are branches of the internal iliac arteries.* The ***superior vesical arteries***, branches of the umbilical arteries, supply anterosuperior parts of the bladder, and the ***inferior vesical arteries***, branches of the internal iliac arteries, supply the base of the bladder.

The *obturator and inferior gluteal arteries* also supply small branches to the bladder. In the female the *uterine and vaginal arteries* send small branches to the bladder (Fig. 3-59).

The veins of the urinary bladder correspond to the arteries and are *tributaries of the internal iliac veins.* They form the ***vesical venous plexuses.***

The vesical venous plexus in the male (Fig. 3-64) *envelops the base of the bladder and prostate, the seminal vesicles, ductus deferentes, and the inferior ends of the ureters.* The vesicle venous plexus is connected with the prostatic venous plexus. It mainly drains through the inferior vesical veins into the internal iliac veins, but it may drain via the sacral veins into the vertebral venous plexus (Fig. 5-58).

The vesical venous plexus in the female (Fig. 3-22) *envelops the pelvic part of the urethra and the neck of the bladder.* It receives blood from the dorsal vein of the clitoris and communicates with the vaginal plexus.

The lymph vessels from superior parts of the urinary bladder drain to the *external iliac lymph nodes* (Fig. 3-73), whereas *the lymph vessels from inferior parts of the bladder drain to the internal iliac lymph nodes.* Some lymph vessels from the neck region of the bladder drain into the *sacral or common iliac lymph nodes.*

Nerves of the Urinary Bladder (Figs. 3-35, 3-70, and 3-80). The nerve supply of the bladder is from the *pelvic splanchnic nerves (parasympathetic fibers).* They are *motor to the detrusor muscle* and *inhibitory to the internal sphincter of the bladder.* Hence when these fibers are stimulated by stretching, the bladder contracts, the internal sphincter relaxes, and urine flows from the bladder into the urethra.

The sympathetic fibers to the bladder are derived from T11, T12, L1, and L2 nerves. These fibers are probably inhibitory to the bladder.

The sensory fibers from the bladder are visceral and transmit pain sensations (*e.g.*, from overdistention of the bladder). The nerves supplying the bladder form the **vesical nerve plexus**, consist of both sympathetic and parasympathetic fibers. This plexus is continuous with the *inferior hypogastric plexus* (Figs. 2-114 and 3-80).

As it expands, the bladder rises from the pelvis within the extraperitoneal fat. When excessively distended, it rises to the level of the umbilicus or even higher in some cases. In so doing, it lifts several centimeters of parietal peritoneum from the suprapubic part of the anterior abdominal wall. The bladder then lies adjacent to this wall without the intervention of peritoneum. Thus the distended bladder may be punctured (**suprapubic cystostomy**), or approached superior to the symphysis pubis surgically for the introduction of instruments into it without traversing the peritoneum and involving the peritoneal cavity.

Urinary calculi, foreign bodies, and small tumors may also be removed from the bladder by this suprapubic route.

Because of the high position of the distended bladder, it may be ruptured by injuries to the inferior part of the anterior abdominal wall, or by fractured bones of the pelvis. The rupture may result in the escape of urine extraperitoneally or intraperitoneally (Fig. 3-89). The majority of bladder tears are extraperitoneal.

Rupture of the superior part of the bladder frequently tears the peritoneum, resulting in extravasation of urine into the peritoneal cavity (Fig. 3-89). **Posterior rupture of the bladder** usually results in escape of urine extraperitoneally.

The interior of the bladder can be examined with a **cytoscope** (Fig. 3-67) inserted through the urethra. The bladder can also be examined radiographically after it has been filled with radiopaque material via a catheter (Fig. 3-89). **A cystogram** will reveal extravasation of dye in the peritoneal cavity, or at the base of the bladder with spreading into the retroperitoneal space.

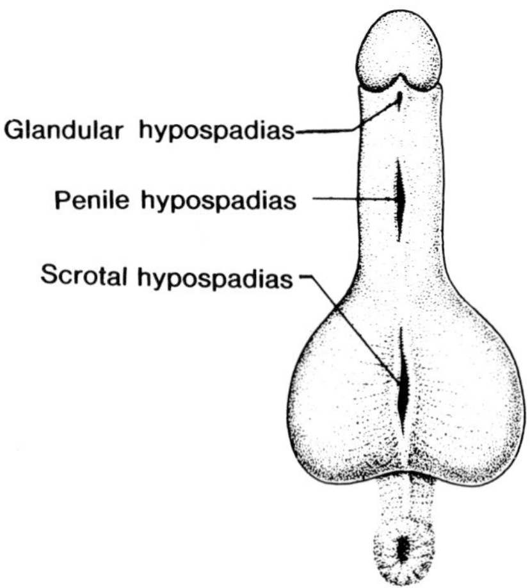

Figure 3-68. Drawing of the ventral surface (urethral surface) of the penis and the scrotum, illustrating various abnormal locations of the urethral opening. The glandular and penile types of hypospadias result from failure of fusion of the urogenital folds; the scrotal type results from failure of fusion of the labioscrotal swellings (see Fig. 3-26).

THE MALE URETHRA (Figs. 2-25, 2-102, 3-12, and 3-32)

The urethra in the male, 15 to 20 cm long, *conveys urine from the urinary bladder to the external urethral orifice* located at the tip of the glans penis. The male urethra also *provides an exit for semen* (seminal fluid). For descriptive purposes the male urethra is divided into three parts.

The Prostatic Part of the Urethra (Figs. 2-102, 3-12, and 3-64). *This first part of the urethra,* about 3 cm long, *begins at the internal urethral orifice at the apex of the trigone* of the bladder and **descends through the prostate gland**, describing a gentle curve that is concave anteriorly.

The prostatic part of the urethra ends at the superior layer of deep fascia of the sphincter urethrae muscle (Fig. 3-12). Its lumen is narrower superiorly and inferiorly than in the middle, but it is contracted except when fluid is passing through it. The prostatic part is the most dilatable part of the urethra.

The posterior wall of the prostatic part of the urethra has notable features. There is a median, longitudinal ridge called the **urethral crest** (Fig. 3-64) with a groove on each side of it, called a **prostatic sinus.** Most of the ducts of the prostate gland open into the prostatic sinuses (Fig. 3-69); some **prostatic ducts** open along the sides of the urethral crest.

In the central part of the urethral crest, there is a rounded eminence, called the **seminal colliculus**, on which there is a small slit-like opening (Fig. 3-64). This opening leads into a small *cul-de-sac* called the **prostatic utricle** (Fig. 3-12).

Explanatory Note. The seminal colliculus used to be called the *verumontanum.* The term **utricle** is derived from a Latin word meaning a *small leather bag.* **The prostatic utricle is a vestigial structure** that is a remnant of the *uterovaginal canal* in the embryo. As it is homologous to the uterus and vagina in the female, you may hear it called by its old names, "vagina masculina" or "uterus masculinus."

On each side of the orifice of the prostatic utricle are the minute ***openings of the ejaculatory ducts.*** The openings of these ducts are so small they are difficult to see in a cadaver; however, they can be catheterized via the urethra in living persons.

Owing to the close relationship of the prostate gland to the prostatic part of the urethra (Fig. 3-64), enlargement of the prostate (**hypertrophy of the prostate**) may obstruct the urethra by compressing it (Fig. 3-15). Often an obstruction can be relieved by an instrument called a *resectoscope* that is inserted into the urethra. This operation is called a **transurethral resection of the prostate gland.**

The Membranous Part of the Urethra (Figs. 3-8*B*, 3-11, 3-29, 3-30, 3-64, 3-65, and 3-69). *This second part of the urethra* is the shortest portion of the urethra (about 1 cm long) and is the **least dilatable part.** Except for the *external urethral orifice*, the membranous part is the narrowest portion of the urethra. It descends from the apex of the prostate to the bulb of the penis (Fig. 3-29) and *traverses the sphincter ure-*

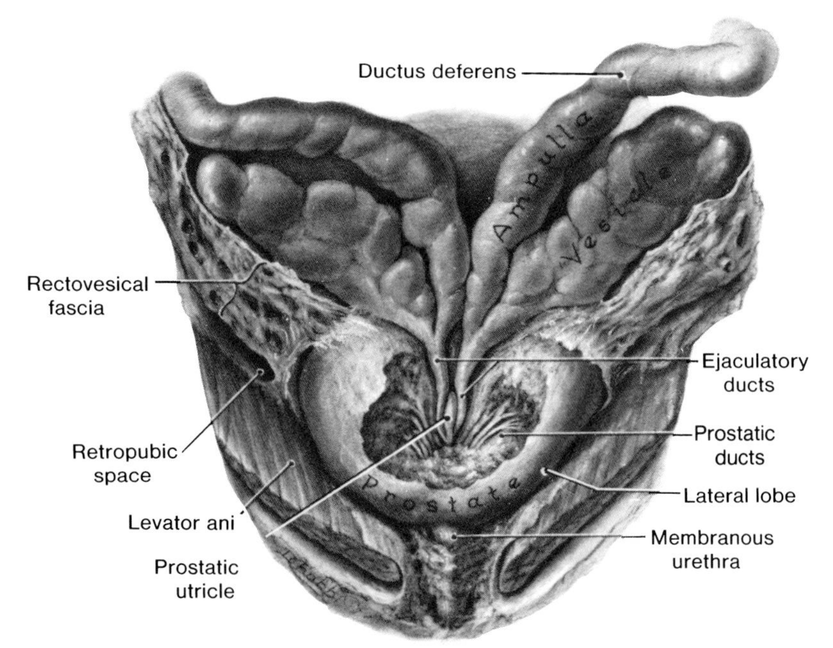

Figure 3-69. Drawing of a dissection of the posterior surface of the prostate gland. Observe the right and left ejaculatory ducts, each formed where the duct of a seminal vesicle joins the ductus deferens. Note the vestigial prostatic utricle lying between the ends of the two ejaculatory ducts. All three structures open into the prostatic urethra. The prostatic ducts mostly open into the prostatic sinuses on each side of the urethral crest (Fig. 3-64).

thrae muscle and inferior fascia of the urogenital diaphragm (Figs. 3-8, 3-9, and 3-14). On each side of the membranous urethra is a small **bulbourethral gland** (Figs. 3-14*C* and 3-15).

The membranous urethra is the narrowest part of the urethra. Its narrowness results from contraction of the sphincter urethrae (Figs. 3-12 and 3-14*C*). This circular investment of muscle also makes the membranous part the least distensible part of the urethra. The most inferior part of the membranous urethra is vulnerable to rupture or penetration by a urethral catheter.

The Spongy Part of the Urethra (Figs. 3-9, 3-12, 3-28, and 3-32 to 3-35). This *third part of the urethra* is the longest portion of the urethra. It passes through the bulb of the penis and the full length of the corpus spongiosum of the penis, and ends at the external urethral orifice.

The lumen of the spongy urethra is about 5 mm in most places, but is expanded in the bulb of the penis to form the **bulb of the urethra** and in the glans penis to form the navicular fossa (Figs. 3-12 and 3-33).

The ducts of the bulbourethral glands open into the ventral wall of the proximal part of the spongy urethra (Fig. 3-15). The orifices of these ducts are very small. There are also minute openings of the ducts of the mucus-secreting **urethral glands**. These are most numerous on the dorsal surface of the spongy urethra (Fig. 3-33).

Blood Vessels Supplying the Male Urethra. The arteries are derived from the structures the urethra traverses. Hence, branches of prostatic vessels supply it as it passes through the prostate gland and the artery of the bulb and the urethral artery, branches of the internal pudendal, supply its other two parts (Fig. 3-15).

The nerves of the male urethra are branches of the pudendal nerve (Figs. 3-24 and 3-70). Most afferent fibers from the urethra run in the *pelvic splanchnic nerves* (Fig. 3-70). Pain fibers course in the pelvic splanchnic and pudendal nerves.

The normal male urethra, about 5 mm in diameter, will expand enough to permit the passage of an instrument about 8 mm in diameter. The external urethral orifice is the narrowest and least distensible part of the urethra; hence an instrument that passes through this opening should pass through all other parts of the urethra.

Urethral stricture may occur as the result of external trauma or infection; instruments called *sounds* are used to dilate the urethra. The

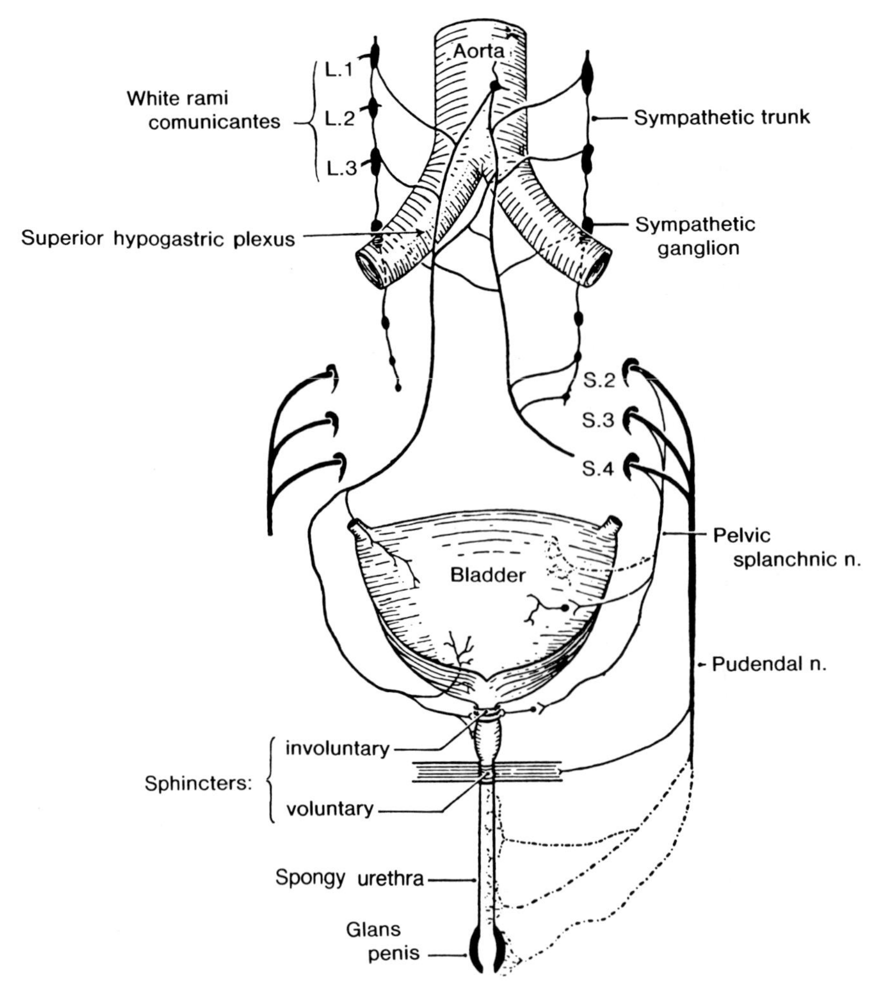

Figure 3-70. Diagram of the nerve supply to the urinary bladder and male urethra. The *broken lines* indicate afferent fibers. The *parasympathetic fibers* are in the pelvic splanchnic nerves (S2, S3, and S4) and are the motor nerves to the bladder. When they are stimulated, the bladder empties, the blood vessels dilate, and the penis erects. They are also the sensory nerves of the bladder. The *sympathetic fibers* through the superior hypogastric plexus (presacral nerve; lower thoracic; L1, L2, and L3) are motor to a continuous muscle sheet comprising the ureteric musculature, the trigonal muscle, and the muscle of the urethral crest. They also supply the muscle of the epididymis, ductus deferens, seminal vesicle, and prostate gland. When the superior hypogastric plexus is stimulated, seminal fluid is ejaculated into the urethra, but is hindered from entering the bladder, perhaps by the muscle sheet which is drawn toward the internal urethral orifice. The sympathetic fibers are also vasoconstrictor and to some slight extent are sensory to the trigone region. The pudendal nerve is motor to the sphincter urethrae and sensory to the glans penis and the urethra. It seems that the sympathetic supply to the bladder has a vasoconstrictor and sexual effect, and with regard to micturition, it is not antagonistic to the parasympathetic supply.

interior of the urethra may be observed by an electrically lit instrument called an *urethroscope*.

Urethral catheterization is frequently done to remove urine from a patient who is unable to micturate. It is also performed to irrigate the bladder and to obtain an uncontaminated sample of urine. *When inserting catheters and sounds, the curves in the urethra must be considered.* The membranous part runs inferiorly and anteriorly as it passes through the urogenital diaphragm (Figs. 3-8*B*, 3-11, and 3-29), and the prostatic part takes a slight curve which is concave anteriorly as it traverses the prostate (Fig. 3-12). Just inferior to the perineal membrane (Fig. 3-9), the spongy urethra is well covered inferiorly and posteriorly by the erectile tissue of the bulb of the penis, but a short segment of it is unprotected superiorly. Because the urethral wall is thin and distensible here, it is vulnerable to injury during instrumentation.

Rupture of the spongy urethra in the bulb of the penis (Figs. 3-28 and 3-30) is fairly common in **straddle injuries.** The urethra is torn when it is caught between a hard object (*e.g.*, a steel beam or the crossbar of a bicycle) and the person's equally hard pubic arch. Urine escapes into the superficial perineal space (Fig. 3-8) and from there inferiorly into the scrotum and superiorly into the anterior abdominal wall, deep to the superficial fascia.

The most *common congenital abnormality* of the urethra is **hypospadias** (Fig. 3-68). In these males there is a defect in the skin and ventral wall of the spongy urethra. Hence the urethral opening is in a more proximal position on the ventral surface of the penis, or in the scrotum.

In the *glandular type of hypospadias*, the prepuce is deformed and the frenulum may be absent. The external urethral orifice is also absent.

Hypospadias results from failure of normal fusion of the urogenital folds and/or labioscrotal swellings (Fig. 3-26). For photographs of the different types of hypospadias and illustrations explaining their embryological bases, see Moore (1982).

THE FEMALE URETHRA (Figs. 3-5, 3-20, 3-39, and 3-61)

The female urethra is a short muscular tube (about 4 cm long) lined by mucous membrane. It corresponds to the prostatic and membraneous parts of the male urethra.

The female urethra passes anteroinferiorly from the urinary bladder, posterior and then inferior to the symphysis pubis.

The external urethral orifice is located between the labia minora, just anterior to the vaginal orifice (Figs. 3-20 and 3-36) *and inferoposterior to the clitoris* (Fig. 3-39). The urethra, 5 to 6 mm in diameter, is closed except during micturition.

The urethra lies anterior to the vagina and is separated from it superiorly by a ***vesicovaginal space*** (Fig. 3-22). Inferiorly it is so intimately associated with the vagina that it appears to be embedded in it (Fig. 3-13).

The urethra passes with the vagina through the pelvic and urogenital diaphragms and the perineal membrane (Figs. 3-14, 3-23, and 3-53). The inferior end of the urethra is surrounded by the **sphincter urethrae muscle** (Fig. 3-13) and some of its fibers enclose both the urethra and vagina (Fig. 3-23). **Urethral glands** are present, particularly in its superior part. One group of these glands on each side, called the *paraurethral glands*, are homologous to the prostate gland in the male. These glands have a common *paraurethral duct* which opens (one on each side) near the external urethral orifice.

The blood supply to the female urethra is from the *inferior vesical, internal pudendal*, and *vaginal arteries* (Fig. 3-61).

Lymph vessels from the female urethra pass to the *sacral and internal iliac lymph nodes* (Fig. 3-73); a few lymph vessels also pass to the *inguinal lymph nodes.*

The female urethra is easily exposed to infection. **Urethritis** (inflammation of the urethra) may result from pathological organisms, *e.g.*, owing to a *gonococcal infection.* The gonococci may enter the ducts of the many urethral glands and, although the patient may be asymptomatic, can enter and infect the male urethra during sexual intercourse.

If the ducts of the paraurethral glands become infected and their openings obstructed, a **periurethral abscess** may develop which can cause *urinary retention.*

The short female urethra is very distensible because it contains much elastic tissue as well as smooth muscle. Thus, it can easily be dilated to 1 cm without injuring it.

The passage of catheters or cystoscopes is much easier in females than in males. Hence urine may be readily removed from a distended female bladder by passing a catheter through the urethra into the bladder.

Catheterization of the urinary bladder is often performed before pelvic operations to decompress the bladder.

THE MALE GENITAL ORGANS

The male genital organs comprise the testes, ductus deferentes (plural of ductus deferens), seminal vesicles, ejaculatory ducts, and penis.

The testis and scrotum are described in Chapter 2 (p. 181). The penis was described earlier in the present chapter (p. 320).

THE DUCTUS DEFERENS (Figs. 2-24, 2-25, 3-17, 3-25, 3-57, 3-64, and 3-65)

The ductus deferens (vas deferens) is a *thick-walled muscular tube* which is the continuation of the duct of the epididymis, an irregularly twisted tube that forms the epididymis (Figs. 2-23 and 2-24).

The ductus deferens begins in the tail of the epididymis and ends by joining the duct of the seminal vesicle to form in the ejaculatory duct.

The ductus deferens, about 45 cm long, ascends in the spermatic cord, passes through the inguinal canal (Figs. 2-23 to 2-25), and crosses over the external iliac vessels to enter the pelvis minor (Fig. 3-57).

The ductus deferens passes along the lateral wall of the pelvis, where it lies external but adherent to the parietal peritoneum and medial to the vessels and nerves.

The ductus deferens crosses the ureter near the posterolateral angle of the bladder, *running between the ureter and the peritoneum* (Figs. 3-57 and 3-65) to reach the base of the urinary bladder (Figs. 3-57, 3-64, and 3-65). At first it lies superior to the seminal vesicle and then it descends medial to the ureter and this vesicle.

The ductus deferens enlarges to form the **ampulla of the ductus deferens** as it passes posterior to the bladder (Fig. 3-65). The ductus deferens then narrows and **joins the duct of the seminal vesicle to form the ejaculatory duct** (Figs. 3-65 and 3-69).

The tiny *artery to the ductus deferens* is closely applied to its surface (Fig. 2-23). It arises from the *umbilical artery* (Fig. 3-57) and terminates by anastomosing with the testicular artery posterior to the testis (Fig. 2-23*B*).

The ductus deferens is richly innervated by autonomic nerve fibers, thereby facilitating its rapid contraction for expulsion of sperms during ejaculation.

A common method of sterilizing males is deferentectomy or **vasectomy** during which part of the ductus deferens (vas deferens) is excised. In this operation an incision is made in the scrotum, the ductus deferentes are located and each is tied in two places. The portion between the sutures is excised. Although the production of sperms continues, they cannot pass into the ejaculatory ducts and the urethra and be ejaculated. Hence the ejaculated fluid (from the seminal vesicles, prostate gland, and bulbourethral glands) contains no sperms. The unexpelled sperms degenerate in the epididymis and in the remaining part of the ductus deferens.

THE SEMINAL VESICLES (Figs. 3-15, 3-29, 3-65, and 3-69)

Each seminal vesicle consists of a long tube (15 cm) which is coiled to form a vesicle-like mass between the posterior surface of the bladder and the rectum.

The seminal vesicles do not store sperms. They secrete a thick secretion which mixes with the sperms as they pass along the ejaculatory ducts. This secretion is expelled when the seminal vesicles contract during **orgasm.** *The secretions of the seminal vesicles form a large part of the semen.*

The superior ends of the seminal vesicles are covered with peritoneum and lie posterior to the ureters, where they are separated from the rectum by peritoneum of the **rectovesical pouch** (Fig. 3-12). The inferior ends of the seminal vesicles are closely related to the rectum, and are separated from it by only a layer called the **rectovesical septum** (Figs. 3-12, 3-19 and 3-65).

The duct of each seminal vesicle joins the ductus deferens to form the ejaculatory duct, which opens into the posterior wall of the prostatic part of the urethra near the opening of the prostatic utricle (Figs. 3-12, 3-64, and 3-69).

The artery to the ductus deferens also supplies the seminal vesicle (Fig. 3-57). The muscular walls of the seminal vesicles contain a plexus of nerve fibers and some sympathetic ganglia. The preganglionic sympathetic fibers emerge from the *superior lumbar nerves* and the parasympathetic fibers emerge from the *pelvic splanchnic nerves* (Fig. 3-70).

THE EJACULATORY DUCTS (Figs. 3-65 and 3-69)

Each of these ducts is a slender tube formed by the union of the duct of the seminal vesicle and the ductus deferens. The ejaculatory ducts, 2 to 2.5 cm long, are formed near the neck of the bladder. The ducts run close together as they pass anteroinferiorly through the prostate and along the sides of the prostatic utricle (Fig. 3-69).

The ejaculatory ducts open by slit-like apertures into the prostatic urethra, one on each side of the orifice of the prostatic utricle (Fig. 3-64).

THE PROSTATE GLAND (Figs. 3-12, 3-15, 3-18, 3-29, 3-64 to 3-66, and 3-69)

The prostate is the largest accessory gland of the male reproductive tract. It is a partly glandular and partly fibromuscular organ, about the size of a walnut. **The prostate gland surrounds the prostatic part of the urethra.** It is enclosed by a dense *prostatic sheath or fascial sheath of the prostate.* This fascial sheath is continuous inferiorly with the superior fascia of the urogenital diaphragm (Fig. 3-11). Posteriorly the prostatic sheath is part of the *rectovesical septum* (Fig. 3-65) which separates the bladder, seminal vesicles, and prostate gland from the rectum (Fig. 3-19).

The prostatic venous plexus (Fig. 3-64) lies between the capsule of the prostate and its fascial sheath. *The prostate gland is somewhat conical in shape* and has a base, an apex, and four surfaces (posterior, anterior, and two inferolateral surfaces).

The base of the prostate or its vesicular surface *is related to the neck of the bladder* (Fig. 3-66*A*). The prostatic urethra enters the middle of the base of the prostate near its anterior surface (Fig. 3-12).

The apex of the prostate is inferior and *is related to the superior fascia of the urogenital diaphragm.* The apex rests on the sphincter urethrae muscle (Figs. 3-11 and 3-12). The prostate is embraced by the medial margins of the levatores ani muscles (Figs. 3-11 and 3-51).

The posterior surface of the prostate is triangular and flattened transversely, facing posteriorly and slightly inferiorly toward the urogenital diaphragm. (Figs. 3-11, 3-29, and 3-64).

The anterior surface of the prostate is transversely narrow and convex, and extends from the apex to the base.

The inferolateral surfaces of the prostate meet anteriorly with the convex anterior surface and rest on the fascia covering the levatores ani (Fig. 3-51).

The prostatic part of the urethra and the ejaculatory ducts pass through the substance of the prostate gland, dividing it into median and lateral lobes.

Explanatory Note. Definitions of the divisions of the prostate are subject to much controversy. All divisions are arbitrary and not structurally distinct. In addition to the usual median and lateral lobes, some clinicians refer to a posterior lobe which is the part of the lateral lobes that can be palpated through the rectum (Fig. 3-72).

The median lobe of the prostate lies anterosuperior to the prostatic utricle and ejaculatory ducts (Figs. 3-64 and 3-69). In old men *the median lobe commonly produces a projection into the cavity of the bladder*, just posterior to the internal urethral orifice (Fig. 3-64), called the uvula vesicae or **uvula of the bladder.** Here the median lobe is in contact superiorly with the inferior part of the trigone of the bladder. The **prostatic utricle,** the vestigial remains of the embryonic uterovaginal canal, is located in the substance of the median lobe (Figs. 3-64 and 3-69).

The lateral lobes of the prostate gland form the main mass of the gland. These lobes are continuous posteriorly (Fig. 3-65) and are separated by the prostatic portion of the urethra (Fig. 3-64).

The 20 to 30 prostatic ducts *open chiefly into the prostatic sinuses on each side of the urethral crest on the posterior wall of the prostatic urethra* (Fig. 3-64). This occurs because most of the glandular tissue is located posterior and lateral to the prostatic urethra. *The prostatic secretion is a thin, milky fluid* which is discharged into the urethra by contraction of the smooth muscle in the prostate (Fig. 3-71). Prostatic fluid constitutes up to one-third of the semen.

The arteries of the prostate gland are derived from the *internal pudendal, inferior vesical,* and *middle rectal* arteries (Fig. 3-57).

The veins of the prostate form the *prostatic venous plexus* around the sides and base of the prostate (Fig. 3-64). *The prostatic venous plexus drains into the internal iliac veins, but it also communicates with the vesical venous plexus and the vertebral venous plexuses* (Figs. 3-64 and 5-58).

The lymph vessels of the prostate gland terminate chiefly in the *internal iliac and sacral lymph nodes* (Fig. 3-73). Some vessels from the posterior surface pass with lymph vessels of the bladder to the *external iliac lymph nodes.*

The nerves of the prostate gland are derived from the *inferior hypogastric plexuses* (Fig. 3-80).

The prostate gland is of great medical interest because *benign nodular hyperplasia of the prostate is a common condition in older men.* This disease results in varying degrees of *obstruction of the neck of the urinary bladder.*

The prostate is small at birth but rapidly enlarges at puberty. In most males the prostate progressively enlarges (undergoes **hypertrophy**) after the mid 40s, but in some males it becomes more fibrous and smaller (undergoes **atrophy**). The cause of these changes is not known, but they are probably related to endocrine changes, primarily in the sex hormones.

Benign prostatic hypertrophy (Fig. 3-15) affects a high proportion of older men and is a common cause of urethral obstruction leading to ***nocturia*** (need to void during the night), ***dysuria*** (difficulty and/or pain during urination), and ***urgency*** (sudden desire to void).

The enlarged prostate gland projects into the urinary bladder impeding the urinary flow by elevating the internal urethral orifice and lengthening and distorting the prostatic urethra. Sometimes the enlarged prostate gland mainly involves the median lobe and forms a valve-like mechanism at the internal urethral orifice. Consequently as the patient strains the *obstruction of the internal urethral orifice* increases.

Partial or complete surgical removal of the prostate gland is called a *prostatectomy.*

Cancer of the prostate gland is one of the most common tumors of men, being found *microscopically* at autopsy in about 60% of men over 80 years of age.

*Cancer of the prostate gland spreads (metastasizes) via both blood **(hematogenous spread)** and lymph vessels **(lymphogenous spread).*** An anatomical basis for metastases to the vertebral column and pelvis is via the valveless *venous communications between the prostatic venous plexus and the vertebral venous plexuses*

(Fig. 5-58), especially the internal vertebral plexus. The main connections are via the pelvic and common iliac veins to the ascending lumbar vein. It is thought that straining to urinate may cause the blood draining the **prostatic venous plexus** (Fig. 3-64) to reverse its flow and pass via the lumbar veins into the **vertebral venous plexuses** (Fig. 5-58).

Cancer cells can spread via the pelvic lymphatics (**lymphogenous metastases**) to the lymph nodes around the internal iliac and common iliac arteries and the aorta (Fig. 3-73).

Tumor cells from the prostate may also pass via the pelvic veins to the inferior vena cava, the right heart, the lungs, the left heart, and throughout the body (**hematogenous metastases**). Generally prostatic cancer cells are more apt to produce ***bone metastases*** than organ metastases, and are more likely to cause an *increase in bone density* than bone destruction.

Because the posterior surface of the prostate is in contact with the rectum, it is *palpable through the rectum* (Fig. 3-72). Only the anterior wall of the rectum and the rectovesical septum separate the examiner's gloved finger from the prostate gland. Palpating the prostate rectally provides information about its size and consistency.

A normal prostate is an elastic swelling, whereas a malignant prostate feels hard and nodular.

Prostatitis (inflammation of the prostate) results in an enlarged, tender "*hot prostate.*" In some patients (*e.g.*, those suspected of having *gonorrhea*), **prostatic massage** is performed to obtain prostatic fluid for microscopical and bacteriological examination.

The position of the prostate gland depends on the fullness of the bladder. *A full bladder displaces the gland inferiorly so that it is more readily palpable.* An unwary examiner may assume the gland is hypertrophied because it is so easily palpated when it is actually a normal, displaced gland.

In histological sections of the prostate it is common to see spherical or ellipsoid lamellated bodies, called **prostatic concretions** or corpora amylacea (Fig. 3-71). Their number increases with age. Small prostatic concretions pass out of the gland with its secretions and appear in the semen. Large concretions are unable to pass through the prostatic ducts (Fig. 3-69); they remain in the prostate. If the prostatic concretions become calcified, they are known as **prostatic calculi** (stones). Some of these become very large and may be palpated during prostatic massage if they are near the posterior surface. Because they are firmly embedded in the fibrous stroma of the prostate, *large prostatic calculi simulate the irregular hardness of a carcinoma.* In some cases the calculi are relatively free and give the palpating finger the impression of a bean bag.

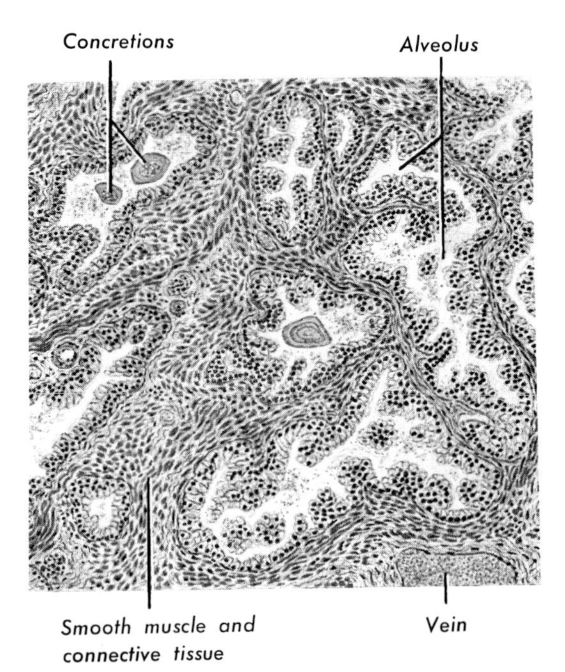

Figure 3-71. Drawing of a section of the prostate gland removed from a 54-year-old person killed in an accident. Observe the prostatic concentrations in the alveoli of the gland. Hematoxylin-eosin, ×65.

THE BULBOURETHRAL GLANDS (Figs. 2-25, 3-14*C*, and 3-15)

The two pea-sized, yellowish, bulbourethral glands (Cowper's glands), *lie posterolateral to the membranous urethra.* They are superior to the bulb of the penis, within the fibers of the sphincter urethrae muscle (Fig. 3-14*C*). Their relatively long ducts (2.5 to 3 cm) pass through the inferior fascia of the urogenital diaphragm (perineal membrane) with the urethra, and through the bulb of the penis *to open by minute apertures into the proximal part of the spongy urethra* (Figs. 2-25 and 3-15).

The blood supply of the bulbourethral glands is from the *arteries to the bulb of the penis* (Figs. 3-9 and 3-15).

THE FEMALE GENITAL ORGANS

The female genital organs consist of internal and external structures (Fig. 3-20).

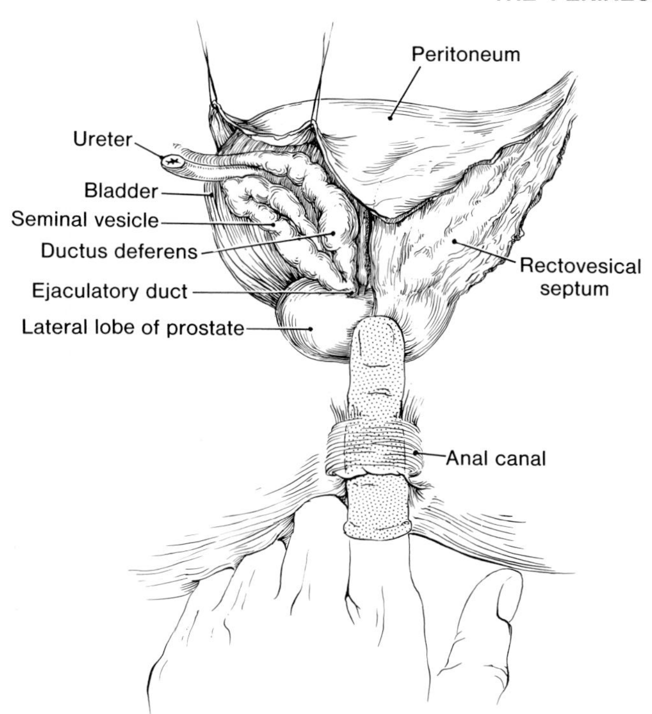

Figure 3-72. Drawing illustrating how the prostate gland is palpated during a rectal examination. Only the anterior wall of the rectum and rectovesical septum separate the gloved finger from the posterior surface of the gland. During prostatic massage, the prostate is massaged with slow, firm strokes. This releases its glandular secretions which can be milked from the urethra and collected for microscopical and bacteriological examination. If the examiner has a long finger, the seminal vesicles can also be palpated through the rectum, particularly if they are enlarged, and a specimen of their secretions can be obtained for study. (Illustrated by Mrs. D. M. Hutchinson. Reprinted with permission from Laurenson RD: *An Introduction to Clinical Anatomy by Dissection of the Human Body.* Philadelphia, WB Saunders, 1968).

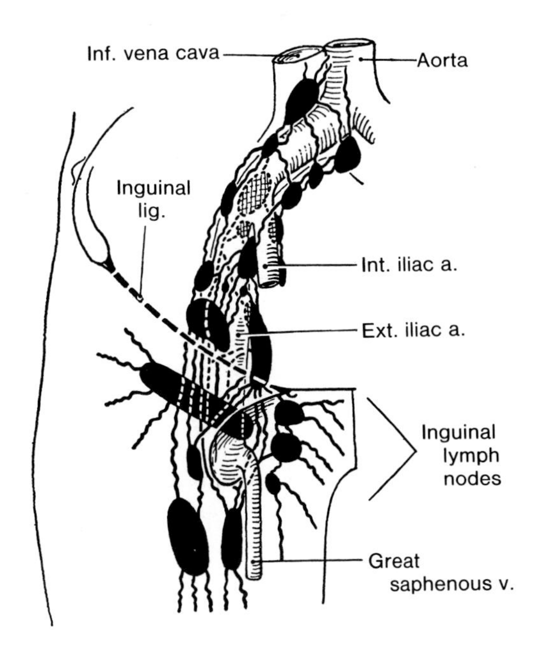

Figure 3-73. Drawing of a dissection of the inguinal and pelvic lymphatics. Observe that there are *two groups of pelvic lymph nodes*: (1) near the pelvic brim and (2) within the pelvic cavity. The lymph nodes near the pelvic brim (12 or more) are the external and common iliac lymph nodes which are arranged around the blood vessels and the nodes superior to the sacral promontory. The nodes within the pelvic cavity, not shown (internal iliac, lateral sacral, and median sacral) are arranged: (1) on the respective blood vessels; (2) in the vesical fascia; (3) in the rectal fascia mainly posterior to the rectum (pararectal nodes); (4) on the course of the superior rectal artery in the broad ligament near the cervix; and (5) between the prostate gland and the rectum. The lymph nodes arranged along the external, internal, and common iliac vessels form a complex that is often referred to clinically as the *iliopelvic plexus of lymph nodes.* Also see Figure 41 on page 43.

The external female genitalia (**vulva**) have been described previously (p. 328), and are illustrated in Figures 3-20 and 3-36 to 3-40.

The internal female genitalia consist of the vagina, uterus, uterine tubes, and *ovaries.*

THE VAGINA (Figs. 3-20 to 3-23 and 3-61)

The vagina, the **female organ of copulation,** is a musculomembranous tube (usually 7-9 cm in length) which forms the inferior portion of the female genital tract and ***birth canal.***

In the anatomical position the vagina descends anteroinferiorly from the **rectouterine pouch** (Fig. 3-20). Its anterior and posterior walls are normally in apposition (Fig. 3-21*B*), except at its superior end, where the cervix of the uterus enters its cavity (Figs. 3-20 and 3-75).

The vagina communicates superiorly with the cavity of the uterus (Figs. 3-20 and 3-75) and opens inferiorly into the **vestibule of the vagina** between the labia minora (Figs. 3-10, 3-20, and 3-36).

A thin *fold of mucous membrane*, the **hymen** (G. membrane), surrounds the *vaginal orifice* (Fig. 3-13). After childbirth the hymen usually consists only of tabs (*hymenal caruncles*).

The vagina lies posterior to the urinary bladder and anterior to the rectum (Figs. 3-20 and 3-53), ***and passes between the medial margins of the levatores ani muscles*** (Fig. 3-21). The vagina pierces the urogenital diaphragm with the sphincter urethrae muscle, the posterior fibers of which are attached to the vaginal wall (Figs. 3-13, 3-14*C*, and 3-23).

The cervix of the uterus projects into the superior part of the anterior wall of the vagina, slightly separating its walls (Fig. 3-20). As a result, *the uterus lies almost at a right angle to the axis of the vagina in its normal anteverted position.* This uterine angle increases as the urinary bladder fills and raises the fundus of the uterus (Fig. 3-75).

The vaginal recess anterior to the cervix is called the **anterior fornix** (L. *fornix*, arch); the recess posterior to it is known as the **posterior fornix** (Fig. 3-74*A*); and the recesses at its sides are referred to as the **lateral fornices** (Fig. 3-75). *The posterior fornix is the deepest and is related to the rectouterine pouch* (Fig. 3-20). Understand that the four fornices are parts of a continuous vaginal recess surrounding the cervix.

The relations of the vagina are clinically important. *Its anterior wall is in contact with the cervix, the base of the bladder, the terminal parts of the ureters, and the urethra* (Figs. 3-20 and 3-59).

The superior limit of the vagina (the 1 to 2 cm of its posterior wall covering the posterior fornix) *is usually covered by peritoneum* (Fig. 3-20). Thus, injuries to this part of the vagina may involve the peritoneal cavity.

Inferior to the posterior fornix, only the loose connective tissue of **the rectovaginal septum separates the posterior wall of the vagina from the rectum** (Fig. 3-20). Thus it can be palpated via the rectum. The vagina is related inferiorly to the central perineal tendon (Fig. 3-10).

The narrow lateral walls of the vagina in the region of the lateral fornices *are attached to the broad ligament of the uterus*, where it contains the ureters and the uterine vessels (Fig. 3-59). Inferiorly, the lateral walls of the vagina *are in contact with the levator ani muscles* (Fig. 3-53), *the greater vestibular glands, and the bulbs of the vestibule* (Figs. 3-10 and 3-39).

Contraction of the levator ani muscles (pubococcygeus part, Fig. 3-53) decreases the size of the vaginal lumen by drawing the walls of the vagina together.

The Arteries of the Vagina (Figs. 3-60 and 3-61). The *vaginal artery*, the vaginal branch of the *uterine artery*, the *internal pudendal artery*, and the vaginal branches of the *middle rectal artery* supply the vagina. All these vessels are branches of the internal iliac arteries.

The Veins of the Vagina (Fig. 3-59). The vaginal veins form *vaginal venous plexuses* along the sides of the vagina (Fig. 3-22), which drain into the internal iliac veins.

The vaginal venous plexuses lie along the sides of the vagina and within its mucosa (Fig. 3-59). They communicate with the vesical, uterine, and rectal venous plexuses.

The Lymph Vessels of the Vagina. These vessels are in three groups: (1) those from the *superior part of the vagina* accompany the uterine artery and drain into the **internal and external iliac lymph nodes** (Fig. 3-73); (2) those from the *middle part of the vagina* accompany the vaginal artery and drain into the **internal iliac lymph nodes**; and (3) those from the *vestibule of the vagina* drain mainly into the **superficial inguinal lymph nodes** (Fig. 3-73). Some lymph vessels from the vestibule of the vagina drain into the sacral and common iliac lymph nodes.

The Nerves of the Vagina. The vaginal nerves are derived from the *uterovaginal plexus* which lies in the base of the broad ligament *on each side of the supravaginal part of the cervix* (Figs. 3-79 and 3-80). Sympathetic, parasympathetic, and afferent fibers pass through this plexus. The inferior nerve fibers from this plexus supply the cervix and the superior part of the vagina.

The vagina is the longest part of the birth canal and can be markedly distended by the fetal head, particularly in an anteroposterior direction. *Distention of the vagina laterally is limited by the presence of the ischial spines and*

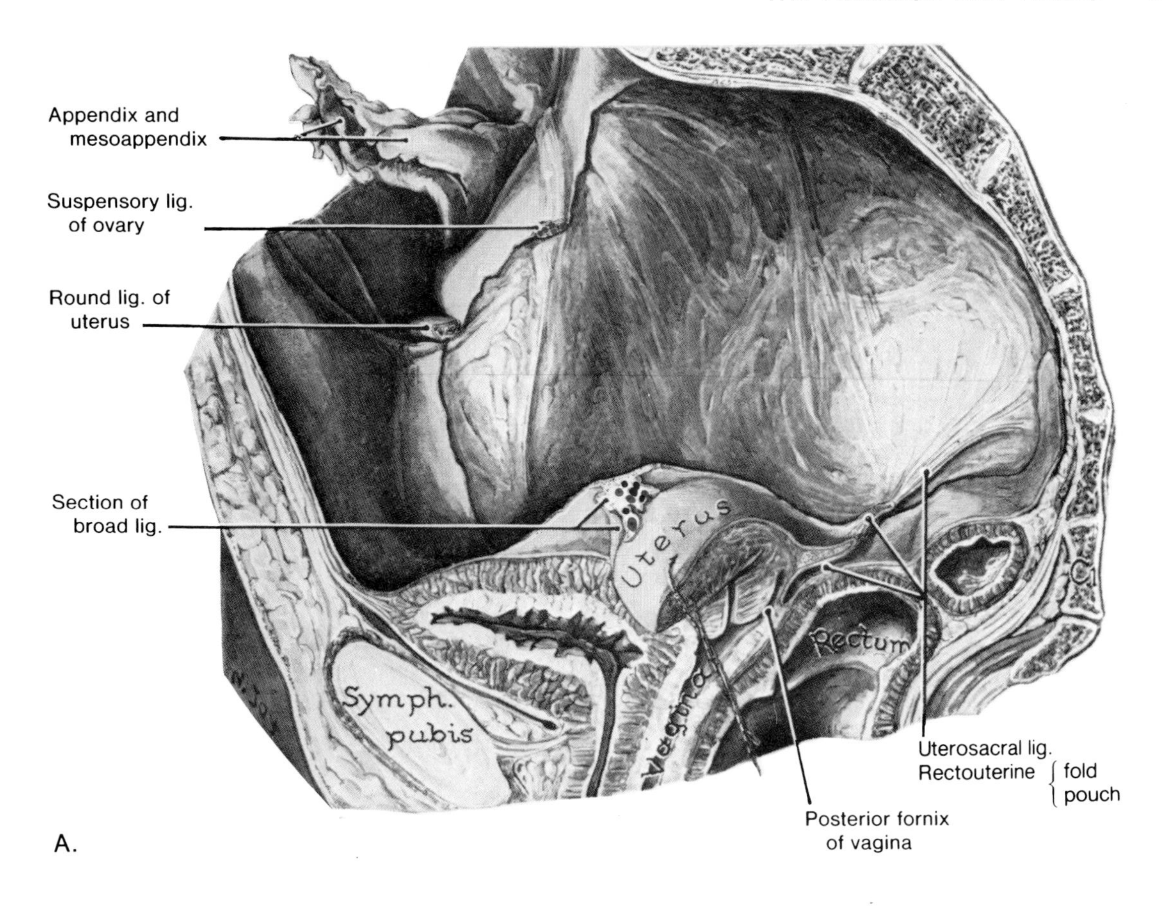

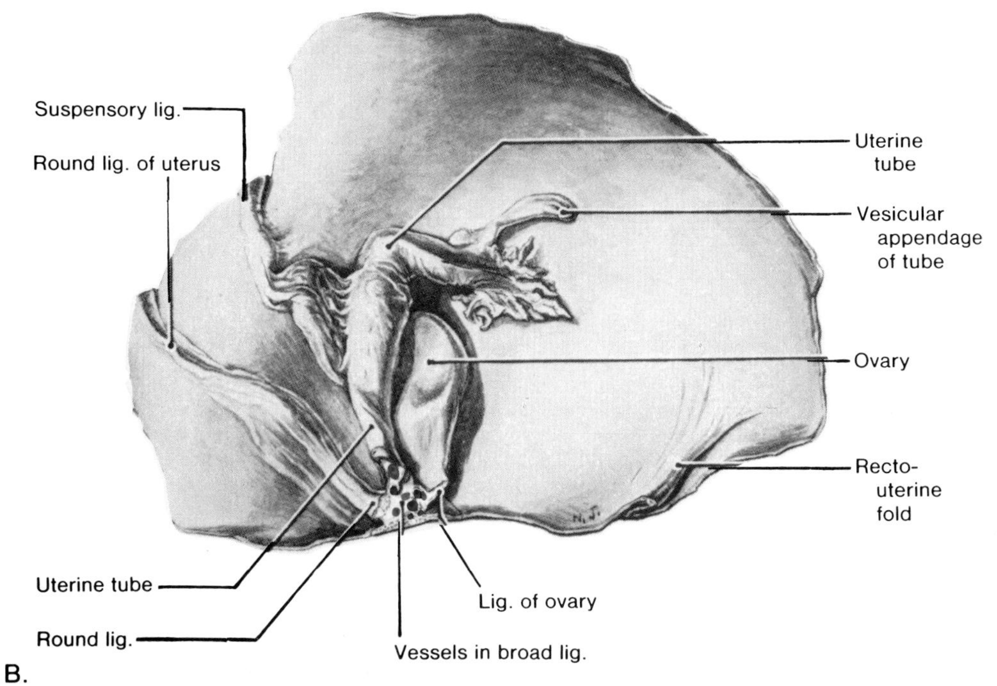

Figure 3-74. *A*, drawing of a median section of the female pelvis. Note that the following structures have been divided at the pelvic brim: the round ligament of the uterus and the suspensory ligament of the ovary with the contained ovarian vessels and nerves. The following structures have been divided at the side of the uterus: the broad ligament with the uterine tube in its free margin; the round ligament of the uterus anteriorly; the suspensory ligament of the ovary posteriorly; and several branches of the uterine vessels. *B*, drawing of the broad ligament and related structures which were peeled off the side wall of the pelvis of the specimen shown in *A*.

the sacrospinous ligaments (Figs. 3-7, 3-42, and 3-43).

The interior of the vagina and the vaginal part of the cervix of the uterus can be examined through a **vaginal speculum** (Fig. 3-76), or by palpation with the fingers in the vagina or the rectum. *Pulsations of the uterine arteries* may be felt via the lateral fornices (Fig. 3-60).

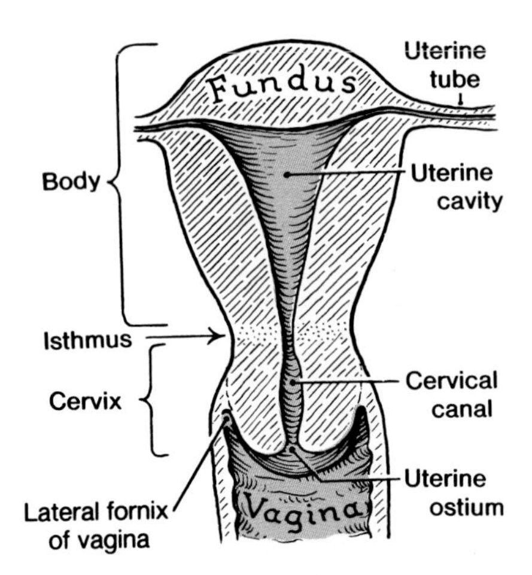

Figure 3-75. Diagram of a section of the uterus and vagina illustrating the parts of the uterus and the relationship of its cervix (L. neck) to the superior end of the vagina (L. sheath). Observe that the superior end of the vagina surrounds the vaginal portion of the cervix. For brevity clinicians often call the uterine ostium (L. entrance) the *external os*, and the opening of the cervical canal into the uterine cavity, the *internal os*.

Owing to the anatomical relationships of the vagina, a instrument directed posteriorly into the vagina may be pushed through the posterior wall of the vagina into the rectouterine pouch and the peritoneal cavity (Figs. 3-20 and 3-76). This leads to **peritonitis.** This has occurred when attempts have been made to induce an **abortion** with a hatpin, coathanger, or some other sharp object.

Abscesses in the rectouterine pouch (Figs. 3-74*A* and 3-76) may be drained by incising the posterior vaginal wall in the posterior fornix. Via this same route, a peritoneoscope or **culdoscope** may be inserted to examine the ovaries (*e.g.*, for cysts or tumors), or the uterine tubes (*e.g.*, for a tubal pregnancy). This procedure, called **posterior colpotomy**, is used in the diagnosis of a number of pelvic diseases.

Surgical operations on the vagina are relatively common and they are usually performed via the *perineal approach.* These operations are commonly done to correct abnormal relaxation of the anterior and posterior vaginal walls when childbearing has resulted in weakening of the pelvic diaphragm. This may result in a bulging of the bladder into the anterior wall of the vagina (**cystocele**), or a bulging of the anterior wall of the rectum into the posterior wall of the vagina, a **rectocele** (Fig. 3-56).

In rare instances the hymen completely covers the vaginal orifice, a congenital condition known as **imperforate hymen.** Surgery is required to open the orifice to permit the discharge of menstrual fluid and to make sexual intercourse pos-

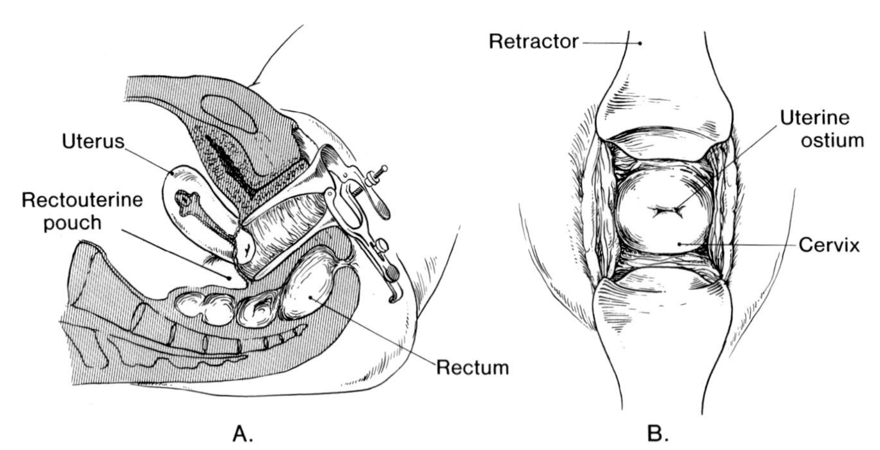

Figure 3-76. Drawings illustrating how the vagina and the vaginal aspect of the cervix of the uterus are examined. *A*, the patient in the lithotomy position (Fig. 3-1) with a vaginal speculum separating the anterior and posterior walls of the vagina. The uterine ostium is usually at the level of the ischial spines which can be palpated through the wall of the vagina. *B*, a view of the cervix from the vagina with the retractors in place. (Illustrated by Mrs. D. M. Hutchinson. Reprinted with permission from Laurenson RD: *An Introduction to Clinical Anatomy by Dissection of the Human Body.* Philadelphia, WB Saunders, 1968).

sible. In **virgins** the vaginal orifice may be only a few millimeters in diameter. More often the opening in the hymen will admit the tip of one finger. Dilation of the vaginal orifice may be necessary to prevent painful tearing of the hymen during intercourse.

THE UTERUS (Figs. 3-20, 3-59 to 3-62, and 3-74 to 3-80)

The uterus (L. womb) of a nonpregnant woman is a hollow, thick-walled, **pear-shaped muscular organ.** It is 7 to 8 cm long, 5 to 7 cm wide, and 2 to 3 cm thick. The uterus normally projects superiorly and anteriorly, superior to the urinary bladder from the superior end of the vagina (Fig. 3-20).

During pregnancy the uterus enlarges greatly to accommodate the embryo and later the fetus.

The uterus consists of two major parts (Fig. 3-75): (1) the expanded superior two-thirds known as the **body**, and (2) the cylindrical inferior third called the **cervix** (L. neck). Because the cervix projects into the vagina, it is divided into *vaginal and supravaginal parts* for descriptive purposes. The vaginal part of the cervix communicates with the vagina via the **uterine ostium** (L. mouth) or *external os* (Fig. 3-75).

The fundus (L. bottom) of the uterus is the rounded superior *part of the body* located superior to the line joining the points of *entrance of the uterine tubes* (Figs. 3-75 and 3-77). The region of the body of the uterus on each side, where the uterine tube enters, is called the **cornu** (L. horn) of the uterus.

The isthmus of the uterus is the narrow zone of transition between the body and cervix of the uterus (Fig. 3-75). This slight constriction is most obvious prior to the first pregnancy.

The uterus is normally bent anteriorly (**anteflexed**) between the cervix and the body (Fig. 3-20), and the entire uterus is normally bent or inclined anteriorly (**anteverted**). The uterus is frequently retroverted in older women (Fig. 3-58). Retroversion is a turning posteriorly of the uterus without flexing the organ.

The wall of the uterus consists of three layers: (1) the outer serous coat called the **perimetrium**, consisting of peritoneum supported by a thin layer of connective tissue; (2) the middle muscular coat called the **myometrium**, consisting of 12 to 15 mm of smooth muscle; and (3) the inner mucous coat called **endometrium** which is firmly adherent to the underlying myometrium.

The myometrium increases greatly during pregnancy. The main branches of the blood vessels and nerves of the uterus are located in this layer.

The uterine cervix is lined by simple columnar, *mucus-secreting epithelium.* The endometrium lines only the body of the uterus. The middle coat of the cervix consists mostly of fibrous tissue. There are only small amounts of smooth muscle in the cervix. Owing to this structure, the cervix is more firm and rigid than the body of the uterus.

The endometrium is partly sloughed off each month during menstruation. *The blastocyst implants in the endometrium of the uterus*, where it develops into an embryo and its membranes. For details of implantation of the blastocyst, see Moore (1982).

The uterus has an anteroinferior or **vesical surface** related to the urinary bladder, and a posterosuperior or **intestinal surface** related to the intestines. These convex surfaces are separated by *right and left borders.*

The uterine tubes extend laterally from the sides of the body of the uterus (Fig. 3-77). Each tube opens at one end into the peritoneal cavity near the ovary and at the other end into the anterolateral part of the uterine cavity (Fig. 3-75).

The **ligaments of the ovaries** are attached to the uterus, posteroinferior to the *uterotubal junction.* The **round ligaments of the uterus** are attached anteroinferiorly to this junction (Fig. 3-77). *These ligaments are continuous within the wall of the uterus and are both derived from the gubernaculum* in the embryo (Fig. 2-17*B*). Each round ligament runs between the layers of the broad ligament and across the pelvic wall to the ***deep inguinal ring.*** After traversing the inguinal canal, the round ligament of the uterus merges with subcutaneous tissues of the labium majus (Figs. 2-22 and 3-38).

The body of the uterus is enclosed between the layers of the broad ligament (Figs. 3-61 and 3-77). It is freely movable; hence, as the bladder fills the uterus rises, and when the bladder is fully distended, the uterus is inclined posteriorly (*retroverted*) and lies in line with the vagina. As the bladder empties it moves to its normal anteverted position (Fig. 3-20).

The cervix of the uterus is not very mobile because it is held in position by several ligaments.

The transverse cervical ligaments (lateral cervical ligaments) extend from the cervix and lateral fornices of the vagina to the lateral walls of the pelvis.

The uterosacral ligaments (rectouterine ligaments) pass from the sides of the cervix toward the sacrum. They are deep to the peritoneum and superior to the levator ani muscles. These ligaments can be palpated through the rectum as they pass posteriorly at the sides of the rectum (Fig. 3-74*A*). *The uterosacral ligaments tend to hold the cervix in its normal relationship to the sacrum.*

The principal supports of the uterus are the *pelvic floor and the pelvic viscera surrounding the uterus* (Figs. 3-20 and 3-53). ***The levator ani muscle, the coccygeus muscle, and the muscles of the urogenital diaphragm are particularly important in supporting the uterus*** (Figs. 3-13 and 3-20).

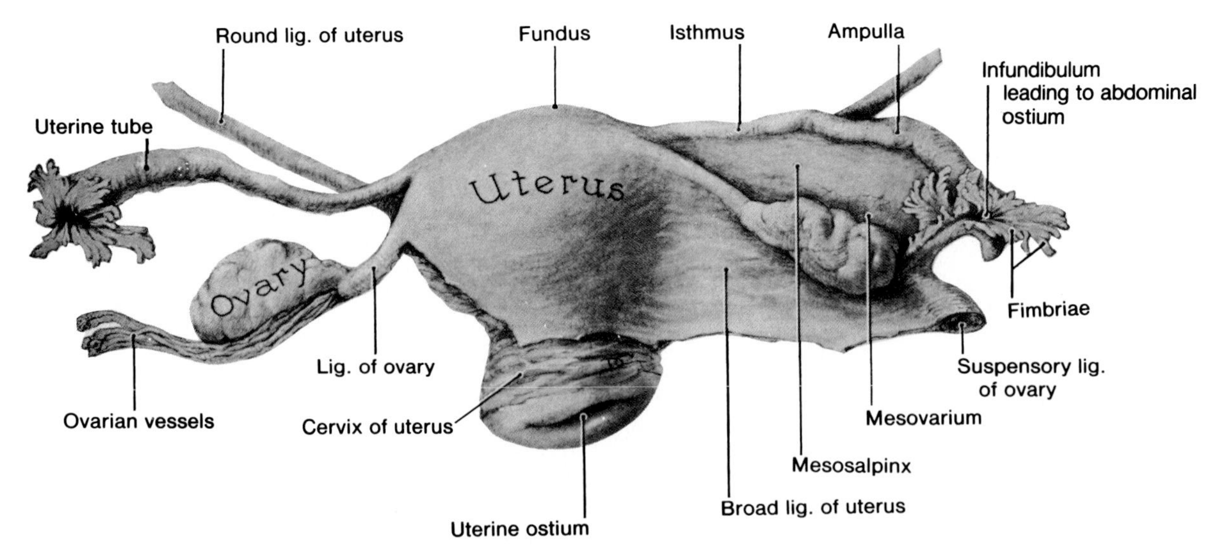

Figure 3-77. Drawing of a posterior view of the uterus, ovaries, uterine tubes, and related structures. *On the left side* the broad ligament of the uterus is removed, thereby "setting free" the uterine tube, the round ligament of the uterus, and the ligament of the ovary. These three structures are attached to the side of the uterus close together, at the junction of its fundus and body. *On the right side* observe the "mesentery" of the uterus and uterine tube, called the broad ligament. Note that the ovary is attached (1) to the broad ligament by a mesentery of its own called the mesovarium; (2) to the uterus by the ligament of the ovary; and (3) near the pelvic brim by the suspensory ligament of the ovary which transmits the ovarian vessels. The part of the broad ligament superior to the level of the mesovarium is called the mesosalpinx.

Peritoneum covers the uterus anteriorly and superiorly, except for the vaginal part of the cervix (Figs. 3-20, 3-58, and 3-62). The peritoneum is reflected anteriorly from the uterus onto the bladder and posteriorly over the posterior fornix of the vagina onto the rectum (Fig. 3-20).

The broad ligaments of the uterus (Figs. 3-59, 3-61, 3-62, and 3-77) *are folds of peritoneum with mesothelium on their anterior and posterior surfaces.* These **double-layered sheets of peritoneum** extend from the sides of the uterus to the lateral walls and the floor of the pelvis.

The broad ligaments hold the uterus in its relatively normal position. The two layers of the broad ligament are continuous with each other at a free edge which is directed anteriorly and superiorly to surround the uterine tube.

Enclosed in the free edge of each broad ligament is a uterine tube. The ovarian ligament or the *ligament of the ovary* lies posterosuperiorly and the *round ligament of the uterus* lies anteroinferiorly within the broad ligament (Figs. 3-61, 3-62, and 3-77). The broad ligaments contain extraperitoneal tissue (connective tissue and smooth muscle) called the **parametrium.**

The broad ligament gives attachment to the ovary through the **mesovarium** (Fig. 3-77); this short peritoneal fold connects the anterior border of the ovary with the posterior layer of the broad ligament. The **mesosalpinx** is the part of the broad ligament between the ligament of the ovary, the ovary, and the uterine tube (Fig. 3-77).

The relationships of the uterus are important clinically. *Anteriorly the body of the uterus is separated from the urinary bladder by the* **vesicouterine pouch** (Figs. 3-20 and 3-62). Here the peritoneum is reflected from the uterus onto the posterior margin of the superior surface of the bladder.

The vesicouterine pouch is empty when the uterus is in its normal position (Fig. 3-20), but it usually contains a loop of intestine when the uterus is retroverted (Fig. 3-58).

Posteriorly the body of the uterus and the supravaginal part of the cervix are separated from the *sigmoid colon* by a layer of peritoneum and the peritoneal cavity (Fig. 3-59). The uterus is separated from the rectum by the **rectouterine pouch** (pouch of Douglas). The inferior part of the rectouterine pouch is on the posterior fornix of the vagina (Figs. 3-20 and 3-62).

The close relationship of the ureter to the uterine artery is very important. *The ureter is crossed superiorly by the uterine artery at the side of the cervix* (Fig. 3-59).

The blood supply of the uterus is derived mainly from the uterine arteries, *which are branches of the internal iliac arteries* (Figs. 3-59 to 3-61). They enter the broad ligaments beside the lateral

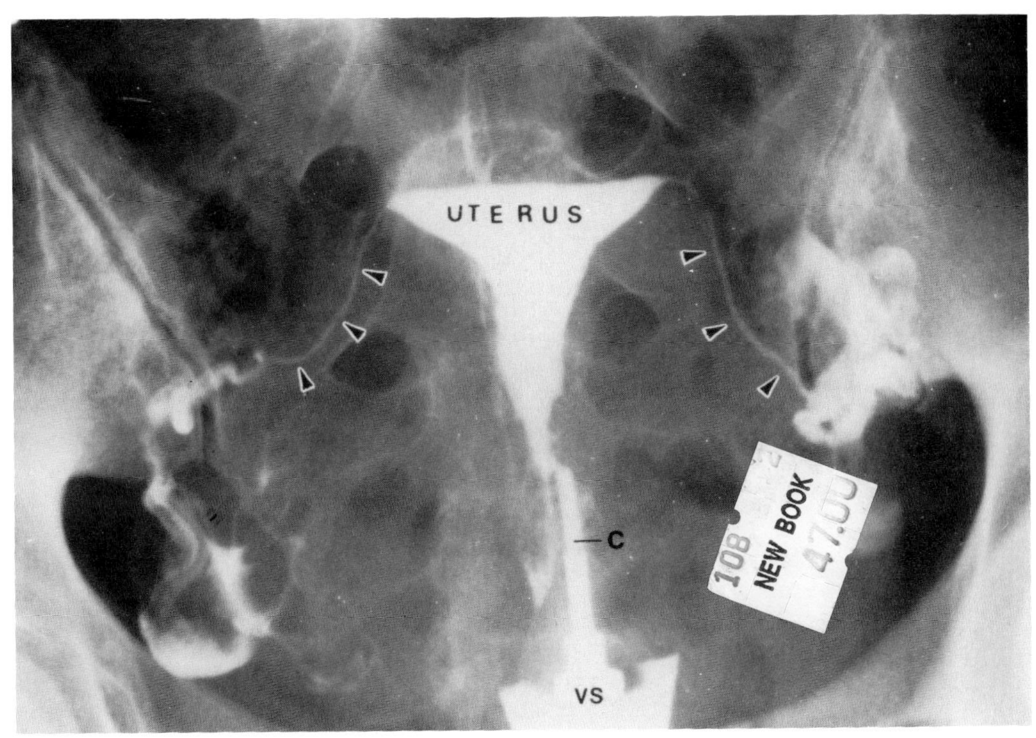

Figure 3-78. *Hysterosalpingogram* (radiograph of the uterus and uterine tubes) taken after the injection of radiopaque material into the uterus through the uterine ostium. The triangular uterine cavity is clearly outlined. The contrast medium has travelled through the uterine tubes, indicated by the *arrowheads*, to the infundibulum (Fig. 3-77) and passed into the peritoneal cavity on both sides (lateral to the arrow). In comparison with the size of the uterus, the uterine cavity is small; this is related to the great thickness of the uterine wall (Fig. 3-20). The cavity of the uterus (when viewed from the side) is a mere slit, whereas when viewed posteriorly, the cavity is triangular with its base in the fundus and its apex continuous with the cervical canal. The lateral angles of the triangle indicate the entrance to the uterine tubes. *C*, catheter in cervical canal; *VS*, vaginal speculum in vagina (also see Fig. 76*A*). Understand that the female genital tract is a direct communication into the peritoneal cavity from the exterior, and is therefore a potential pathway for infection and the development of *pelvic inflammatory disease.*

fornices of the vagina, *superior to the ureters* (Fig. 3-61).

At the isthmus of the uterus (Fig. 3-75), the uterine artery divides into a larger **ascending branch** that *supplies the body of the uterus* and a smaller **descending branch** that *supplies the cervix and vagina* (Fig. 3-60). The uterus is also supplied by the **ovarian arteries** which are branches of the aorta (Fig. 2-127).

The uterine arteries pass along the sides of the uterus within the broad ligament (Fig. 3-61), and then turn laterally at the entrance to the uterine tubes, where they *anastomose with ovarian arteries* (Fig. 3-60).

The veins of the uterus enter the broad ligaments with the uterine arteries (Figs. 3-58, 3-59, and 3-61). ***The uterine veins form a uterine venous plexus on each side of the cervix*** and their tributaries drain into the *internal iliac veins.* The uterine venous plexus is connected with the superior rectal vein, forming a *portal-systemic anastomosis* (Fig. 2-100).

The lymph vessels of the uterus follow three main routes. (1) Most lymph vessels from the **fundus** of the uterus pass with the ovarian vessels to the aortic or **lumbar lymph nodes** (Fig. 2-121), but some lymph vessels pass to the *external iliac lymph nodes* or run along the round ligament of the uterus to the *superficial inguinal lymph nodes* (Fig. 3-73). (2) Lymph vessels from the **body** of the uterus pass through the broad ligament to the **external iliac lymph nodes.** On the way to these nodes, the lymph may pass through *parauterine lymph nodes.* (3) Lymph vessels from the **cervix** pass to the *internal iliac and sacral lymph nodes* (Fig. 3-73).

The nerves of the uterus arise from the *inferior hypogastric plexus* (Fig. 3-80), largely from the **uterovaginal plexus** which lies in the broad ligament on each side of the cervix. Sympathetic, parasympathetic, and afferent fibers pass through this plexus.

The nerves to the cervix form a plexus in which are located small *paracervical ganglia*, one of which is often large and is called the ***uterine cervical ganglion.*** The autonomic fibers of the uterovaginal plexus

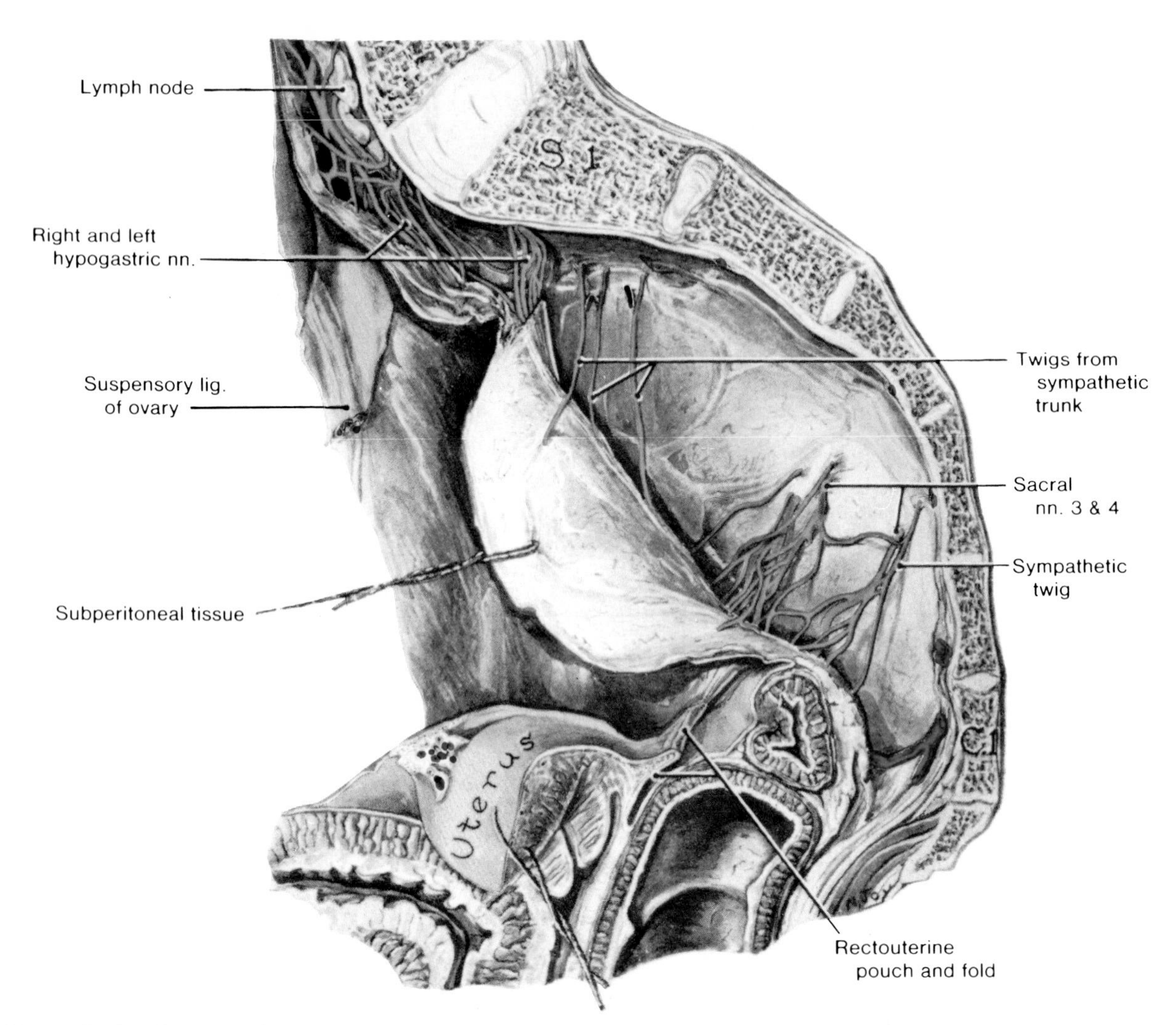

Figure 3-79. Drawing of a dissection of the autonomic nerves in the female pelvis. Observe that the rectum and subperitoneal fatty-areolar tissue have been pulled anteriorly, thereby making taut the pelvic splanchnic nerve (from S3 and S4), sympathetic twigs, and right hypogastric nerve.

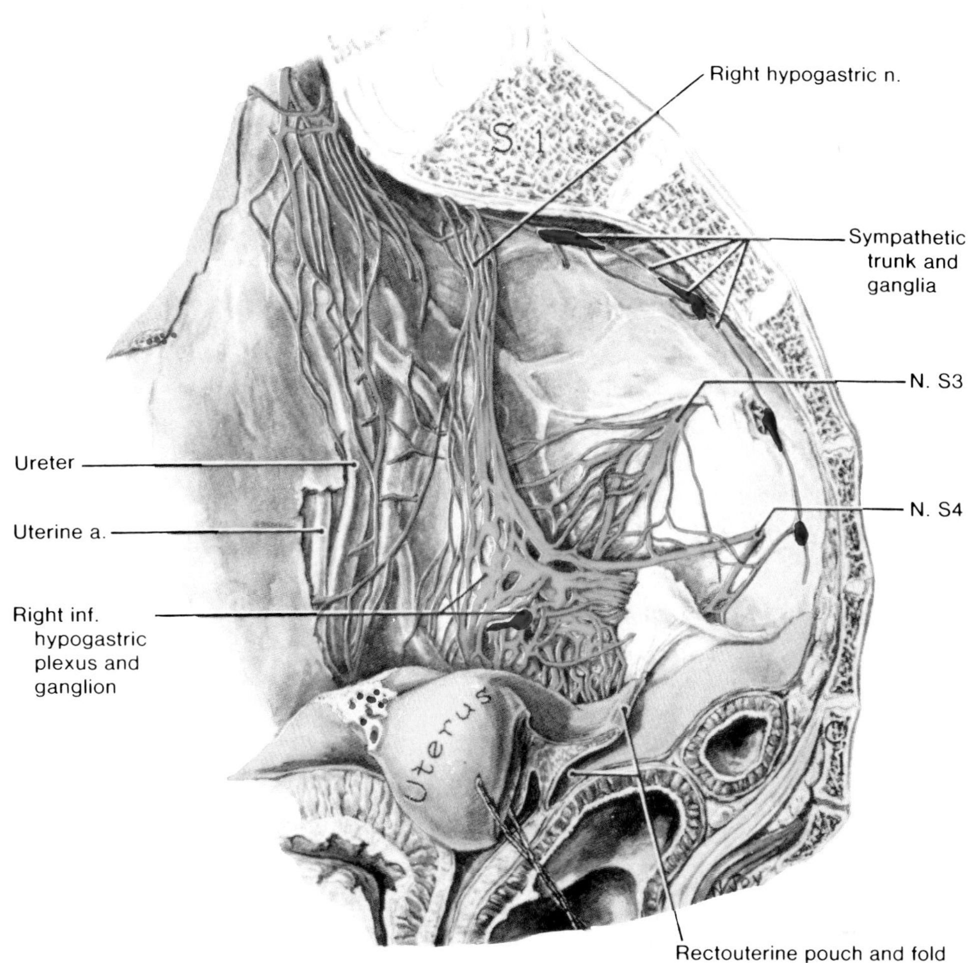

Figure 3-80. Drawing of a later stage of a dissection of the autonomic nerves in the female pelvis. The inferior hypogastric plexus lies in the extraperitoneal connective tissue. Each plexus is situated on the side of the rectum, cervix, fornix of the vagina, posterior part of the bladder, and extends into the base of the broad ligament. *In the male*, each plexus is located on the side of the rectum, seminal vesicle, prostate, and posterior part of the bladder.

are mainly vasomotor. Most of the afferent fibers ascend through the hypogastric plexus and enter the spinal cord via T10 to T12 and L1 spinal nerves (Figs. 3-70, 3-79, and 3-80).

The uterus is mainly an abdominal organ during infancy and the cervix is relatively large. **During puberty the uterus grows rapidly.** During *menopause* (46 to 52 years), the uterus becomes inactive and decreases in size (undergoes *atrophy*). Most female cadavers have atrophic uteri; some have none because they were removed surgically during an operation called **hysterectomy** (G. *hystera*, uterus + *ektomē*, excision).

In rare cases the ***paramesonephric ducts*** degenerate in the female embryo, as they normally do in the male; this results in ***absence of the uterus and vagina***. More commonly, fusion of these embryonic ducts is incomplete and a variety of congenital malformations results (Fig. 3-81).

During pregnancy the uterus increases rapidly in size and weight and rises high in the abdomen. During the last weeks of pregnancy, the uterus is about 20 cm in length and may weigh as much as 1 kg. After childbirth the uterus contracts but it is still large for several weeks. By 8 weeks after childbirth, the uterus is close to its normal size and weight.

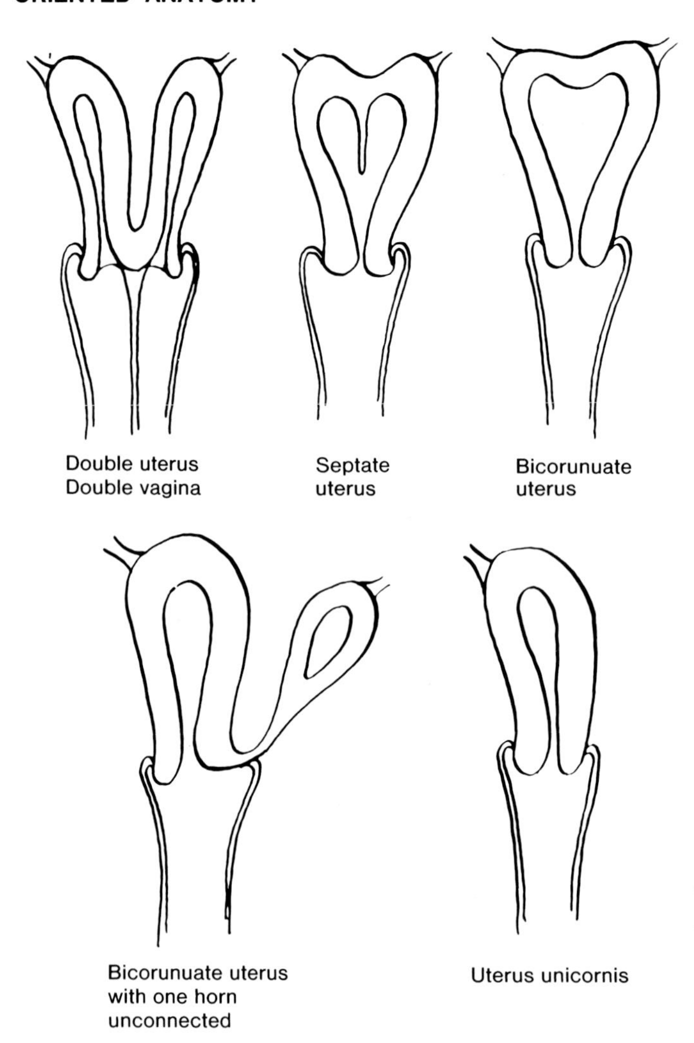

Figure 3-81. Sketches illustrating the common types of congenital malformation of the uterus and vagina which result from failure of fusion of the paramesonephric ducts during the early fetal period. The uterus with a single horn and uterine tube (uterus unicornis) results when only one paramesonephric duct develops.

Prolapse of the uterus (Fig. 3-56) is rare. The uterus descends to an abnormally inferior level in the pelvis, and in advanced cases the cervix protrudes through the vagina and pudendal cleft. Prolapse of the uterus usually results from severe stretching or tearing of the pelvic floor during childbirth.

The cervix and body of the uterus may be examined by **bimanual palpation**. Two fingers of the right hand are passed high in the vagina while the other hand is pressed inferiorly and posteriorly on the hypogastric region of the anterior abdominal wall, just superior to the symphysis pubis. The size and other characteristics of the uterus can be determined in this way (*e.g.*, whether the uterus is in its normal anteverted position as in Fig. 3-20, or retroverted as in Fig. 3-58).

Owing to *softening of the isthmus of the uterus* (**Hegar's sign**) in early pregnancy, the cervix feels as though it were separate from the body of the uterus during bimanual examination. Softening of the uterine isthmus is an **early sign of pregnancy**.

The part of the mesonephric duct which forms the ductus deferens and ejaculatory duct in the male may persist in the female as a **duct of Gartner**. It lies between the layers of the broad ligament along the lateral wall of the uterus, or in the wall of the vagina. Vestigial remnants of the mesonephric ducts may also give rise to **Gartner's duct cysts** (for details see Moore, 1982).

The degree of dilation of the cervix during labor can be determined by rectal examination. As the cervical canal dilates, a fingertip may be inserted into it. The finger in the rectum can be passed across the fetal head as it presents in the uterine ostium, and the degree of dilation can be estimated accurately without entering the vagina. If there is doubt as to the findings, a *sterile vaginal examination* of the cervix and fetal head may be necessary.

THE UTERINE TUBES (Figs. 3-20, 3-60 to 3-63, and 3-77)

The uterine tubes (Fallopian tubes) are ducts, 10 to 12 cm long and 1 cm in diameter, which **extend laterally from the cornua or horns of the uterus.** The uterine tubes conduct oocytes from the ovaries and sperms in the uterus to the *fertilization site* in the ampulla of the uterine tube (Fig. 3-77).

The uterine tube also conveys the dividing zygote to the uterine cavity. Each tube opens at its proximal end into the horn (L. cornu) of the uterus and at its distal end into the **peritoneal cavity** near the ovary. Consequently, *the uterine tubes allow communication between the peritoneal cavity and the exterior of the body* (Figs. 3-20 and 3-78).

The uterine tube is divided into four parts, from lateral to medial (Fig. 3-77).

The infundibulum (L. funnel) is the funnel-shaped distal end of the uterine tube which is *closely related to the ovary.* Its opening into the peritoneal cavity is called the **abdominal ostium** (Fig. 3-61). About 2 mm in diameter, the ostium lies at the bottom of the infundibulum. The margins of the infundibulum have fringed edges or folds called **fimbriae** (L. fringes). These finger-like processes spread over most of the surface of the ovary and a large one, **the ovarian fimbria,** is attached to the superior pole of the ovary. *During ovulation the fimbriae trap the oocyte* and sweep it through the abdominal ostium into the tube.

The ampulla of the uterine tube receives the oocyte from the infundibulum. It is in this part that *fertilization of the oocyte* by a sperm usually occurs. ***The ampulla is the widest and longest part of the uterine tube,*** making up about two-thirds of its length.

The isthmus (G. narrow passage) of the uterine tube is the short (2 to 5 cm), narrow, thick-walled part of the uterine tube which joins the *horn of the uterus.*

The uterine part (*intramural portion*) of the uterine tube is the short segment of the uterine tube which passes through the wall of the uterus. The **uterine ostium** of the tube is smaller than the abdominal ostium. The lumen of the uterine tube increases gradually in width from the uterus toward the ovary (Figs. 3-77 and 3-78).

The uterine tubes lie in the free edges of the broad ligaments of the uterus (Figs. 3-61, 3-74, and 3-77). The part of the broad ligament attached to the uterine tube is called the **mesosalpinx** (G. *salpinx,* tube or trumpet), or mesentery of the uterine tube. The uterine tubes extend posterolaterally to the lateral walls of the pelvis and then ascend and arch over the ovary. Except for their uterine parts, the uterine tubes are clothed in peritoneum (Figs. 3-61 and 3-77).

The arteries of the uterine tube are derived from the *uterine and ovarian arteries* which anastomose (Fig. 3-60). The tubal branches pass along the tube between the layers of the mesosalpinx (Fig. 3-61).

The veins of the uterine tube are arranged similarly to the arteries and drain into the *uterine and ovarian veins* (Figs. 3-58 and 3-59).

The lymph vessels of the uterine tubes run with those of the fundus of the uterus and the ovary to the *lumbar lymph nodes* (Fig. 2-121).

Usually the oocyte is fertilized in the ampulla of the uterine tube and the dividing zygote passes slowly along it into the uterus. Fertilization of an oocyte cannot occur when both tubes are blocked because the sperms cannot reach the oocyte.

One of the major causes of infertility in women is blockage of the uterine tubes resulting from infection. The patency of a uterine tube may be determined by a procedure in which CO_2 gas is introduced into the uterus to see if it enters the peritoneal cavity. Tubal patency can also be determined by injecting a radiopaque material into the uterus (**hysterosalpingography**, Fig. 3-78).

Ligation of the uterine tubes is one method of birth control. Oocytes discharged from the ovarian follicles in these patients die in the uterine tube and soon disappear.

Because the female genital tract is in direct communication with the peritoneal cavity via the abdominal ostia of the uterine tubes, infections of the vagina, uterus, and tubes may result in **peritonitis**. Conversely, inflammation of the uterine tube (**salpingitis**) may result from infections that spread from the peritoneal cavity. In some cases collections of pus may develop in the uterine tube (**pyosalpinx**) and the tube may be partly occluded by *adhesions.* In these cases the zygote may not pass to the uterus and the blastocyst may implant in the mucosa of the uterine tube, producing an **ectopic tubal pregnancy**. Although implantation may occur in any part of the tube, the common site is in the ampulla.

Tubal pregnancy is the most common type of ectopic gestation and occurs about once in every 250 pregnancies in North America. Ectopic tubal pregnancies usually result in **rupture of the uterine tube** and *hemorrhage into the abdominopelvic cavity* during the first 8 weeks of gestation. Tubal rupture and the associated severe hemorrhage constitute a threat to the mother's life, and results in death of the embryo.

The **epoophoron** lies in the mesosalpinx between the uterine tube and the ovary. *This vestigial structure* consists of a number of small rudimentary tubules which are *remnants of cranial mesonephric tubules.* These tubules were associated with the mesonephric kidney in the embryo. They are homologous with the efferent ductules of the testis (Fig. 2-18*A*) and they may give rise to *paraovarian cysts.*

A vesicular appendage is sometimes attached to the infundibulum of the uterine tube (Fig. 3-74*B*). It represents the *remains of the cranial end of the mesonephric duct* which forms the ductus epididymis in the male, but normally degenerates in females.

THE OVARIES (Figs. 3-20, 3-59 to 3-61, 3-74*B*, and 3-77)

The ovaries are oval, almond-shaped, pinkish-white glands about 3 cm long, 1.5 cm wide, and 1 cm thick. They are located one on each side, close to the lateral wall of the pelvis minor, in a recess called the **ovarian fossa** (Fig. 3-74*B*).

The ovarian fossa is bounded anteriorly by the medial umbilical ligament (obliterated umbilical artery), and posteriorly by the ureter and internal iliac artery (Fig. 3-58). The ampulla of the uterine tube curves over the lateral end of the ovary so that the infundibulum curls around the ovary (Fig. 3-77). The **ovarian fimbria** of the infundibulum attaches the tube to the ovary.

Each ovary is attached to the posterosuperior aspect of the broad ligament and is suspended from the posterior layer of this ligament by a fold of peritoneum, called the **mesovarium.** The ovarian vessels pass to and from the ovary via the mesovarium (Figs. 3-74*B* and 3-77).

Each ovary is also attached to the uterus by a band of fibrous tissue, the **ligament of the ovary** which runs in the broad ligament. This ligament connects the uterine pole of the ovary to the lateral wall of the uterus.

Near the pelvic brim the ovary is attached by the **suspensory ligament of the ovary** (Figs. 3-62 and 3-77), a thickening of fibrous tissue which passes over the iliac vessels and psoas major muscle.

The suspensory ligament of the ovary contains the ovarian vessels and nerves which pass to the lateral end of the ovary and into the mesovarium and the *hilum of the ovary* (Fig. 3-59, 3-62, and 3-77).

The surface of the ovary is not covered by peritoneum; hence during ovulation oocytes are expelled into the peritoneal cavity. The surface of the ovary in young women is covered by cuboidal epithelium which is continuous with the flattened mesothelium of the peritoneum forming the mesovarium.

The surface epithelium of the ovary does not give rise to oocytes. Oogonia develop during the fetal period from **primordial germ cells** that migrate into the embryonic ovaries (for details see Moore, 1982).

Before puberty the surface of the ovary is smooth, whereas after puberty the ovary becomes progressively scarred and distorted as successive **corpora lutea** degenerate (Fig. 3-77). *The corpora lutea are endocrine structures* that develop from ovarian follicles that have expelled their oocytes.

The ovarian arteries arise from the abdominal aorta at about the level of L2 vertebra (Fig. 2-127), and descend on the posterior abdominal wall. On reaching the pelvic brim, *the ovarian arteries cross over the external iliac vessels and enter the **suspensory ligaments*** (Figs. 3-59 and 3-62). At the level of the ovary, the ovarian artery sends branches through the mesovarium to the ovary and continues medially in the broad ligament to supply the uterine tube and to anastomose with the uterine artery (Fig. 3-60).

The ovarian veins leave the hila of the ovaries and form a leash of vessels, called the ***pampiniform plexus***, in the broad ligament near the ovary and the uterine tube. This plexus of veins (Fig. 3-59) communicates with the uterine plexus of veins. Each ovarian vein arises from the pampiniform plexus and leaves the pelvis minor with the ovarian artery (Fig. 3-62).

The right ovarian vein ascends to the inferior vena cava, whereas the left ovarian vein drains into the left renal vein, as do the testicular veins (Fig. 2-116).

The lymph vessels of the ovary follow the blood vessels and join those from the uterine tubes and the fundus of the uterus. They ascend with the ovarian veins to the ***lumbar lymph nodes*** near the aorta and inferior vena cava (Figs. 2-121 and 3-79).

The nerves of the ovary descend along the ovarian vessels from the *abdominal sympathetic plexus* (Fig. 2-114). The **ovarian plexus of nerves** lies on, and follows, the ovarian artery. It supplies the ovary, broad ligament, and uterine tube (Fig. 3-80). The parasympathetic fibers of the ovarian plexus are derived from the vagus (Fig. 7-135).

At ovulation some women experience paraumbilical pain, called **mittelschmerz** (Ger. *mittel*, middle and *schmerz*, pain), owing to stretching of the ovarian wall. As afferent impulses from the ovary reach the CNS through the dorsal root of the tenth thoracic nerve, the pain is referred to the dermatome of T10 (Figs. 1-23 and 2-116*A*).

The position of the ovaries varies considerably in women who have borne children. During pregnancy the broad ligaments and the ovaries are carried superiorly with the enlarging uterus. After childbirth the ovaries descend as the uterus contracts, but they may not return to their original locations. The ovaries are also very mobile and may be displaced by the intestine.

On the right side the vermiform appendix may lie very close to the ovary (Figs. 3-59 and 3-74*A*). After **menopause** the formation of ovarian follicles, corpora lutea, and corpora albicantia ceases and the ovaries gradually atrophy. The ovaries become small and shriveled, as they appear in most female cadavers (Fig. 3-59).

THE RECTUM

The rectum, continuous superiorly with the sigmoid colon, begins at about the level of the third part of the sacrum (S3 vertebral level). About 12 cm long, the rectum follows the curve of the sacrum and coccyx to about 3 cm beyond the tip of the coccyx (Fig. 3-12).

The rectum ends by turning posteroinferiorly to become the anal canal (Figs. 3-12 and 3-20).

The ***puborectalis muscle*** forms a sling at the junction of the rectum and the anal canal, producing the anorectal angle (Figs. 3-17*A* and 3-55).

Inferiorly the rectum lies immediately posterior to the prostate gland in the male (Fig. 3-18) *and the vagina in the female* (Fig. 3-53). The termination of the rectum lies posterior to the **central perineal tendon** in both sexes, and the **apex of the prostate** in the male (Figs. 3-2, 3-12, and 3-35).

PERITONEAL COVERING OF THE RECTUM (Figs. 3-12, 3-20, 3-80, and 3-82)

Peritoneum covers the superior third of the rectum on its anterior and lateral surfaces. The middle third has peritoneum on its anterior surface only. ***The inferior third of the rectum has no peritoneal covering.***

In the male the peritoneum is reflected from the anterior surface of the rectum to the posterior wall of the bladder, where it forms the floor of the **rectovesical pouch** (Figs. 3-12 and 3-82). In male children, in whom the bladder is in the abdomen (Fig. 3-66*B*), the peritoneum extends inferiorly as far as the base of the prostate. As the bladder moves into the pelvis minor during puberty, the adult peritoneal relationship is attained (Fig. 3-12).

In the female the peritoneum is reflected from the rectum to the posterior fornix of the vagina, where it forms the floor of the **rectouterine pouch** (Figs. 3-20, 3-62, 3-76*A*, and 3-79, and 3-80).

In both sexes the lateral reflections of peritoneum from the rectum form **pararectal fossae** on each side of the rectum in its superior one-third (Fig. 3-82). The pararectal fossae permit the rectum to distend.

SHAPE AND FLEXURES OF THE RECTUM (Figs. 3-82 to 3-84)

Despite the origin of its name (L. *rectus*, straight) the rectum is curved to follow the sacrococcygeal curve. *Its terminal part bends sharply in a posterior direction, where it joins the anal canal* (Fig. 3-55). Although generally smaller in caliber than the sigmoid colon (Fig. 3-82), the rectum increases in diameter as it passes inferiorly. Its inferior part, the **rectal ampulla**, is very distensible (Fig. 3-20).

The rectum is S-shaped in the coronal plane (Figs. 3-82 and 3-83). At the three concavities in the rectum indicated by the flexures, there are internal folds (valves) of the mucous membrane that are called **transverse rectal folds** (plicae transversales), which partly close the lumen of the rectum (Fig. 3-83). The rectal folds consist of mucous membrane covering circular smooth muscle. Their form is maintained by prolongations of the **teniae coli** (muscular bands) in the anterior and posterior walls of the rectum (Fig. 3-82).

RELATIONS OF THE RECTUM

Posteriorly the rectum rests on the inferior three *sacral vertebrae, the coccyx, the anococcygeal ligament* (Figs. 3-20 and 3-52), *the median sacral vessels*, branches of the *superior rectal artery* (Fig. 3-84), and the inferior ends of the *sympathetic trunks* (Figs. 3-80, and 3-82). The rectum is surrounded by a fascial sheath and is loosely attached to the anterior surface of the sacrum.

Anteriorly, in the male, the rectum is related to the base of the *urinary bladder* (Figs. 3-12 and 3-82), terminal parts of the *ureters*, ductus deferentes, *seminal vesicles*, and the *prostate gland* (Fig. 3-72).

The two layers of the **rectovesical septum** (Figs. 3-12 and 3-72) lie in the median plane between the bladder and rectum and are closely associated with the seminal vesicles and prostate gland.

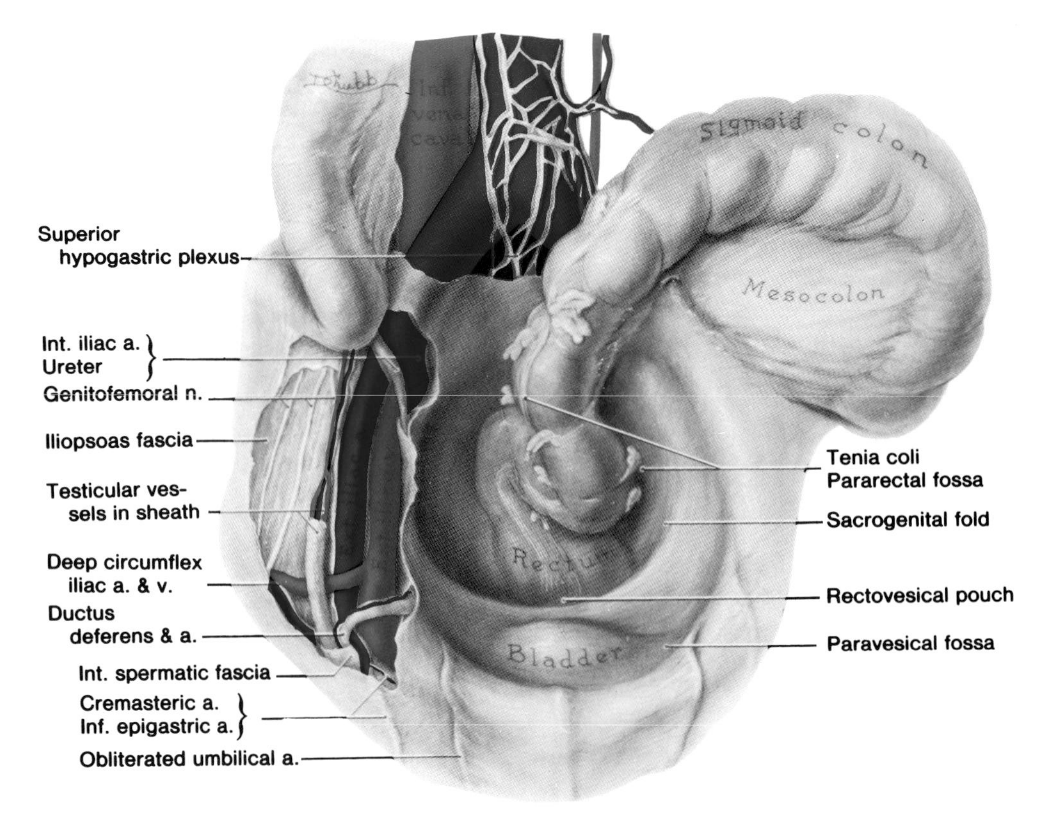

Figure 3-82. Drawing of an anterosuperior view of a dissection of the male pelvis. Observe one limb of the Λ-shaped root of the sigmoid mesocolon ascending near the external iliac vessels, and the other descending to the third piece of the sacrum. Note the crescentic fold of peritoneum called the sacrogenital fold. Observe the superior hypogastric plexus lying in the fork of the aorta and anterior to the left common iliac vein (also see Fig. 2-114). Note the ureter adhering to the peritoneum, crossing the external iliac vessels, and descending anterior to the internal iliac artery. Observe that the ductus deferens and its artery also adhere to the peritoneum, cross the external iliac vessels, and then hook around the inferior epigastric artery to join the other constituents of the spermatic cord. The rectum begins where the sigmoid colon loses its mesentery; this occurs at about the S3 vertebral level (Fig. 3-46).

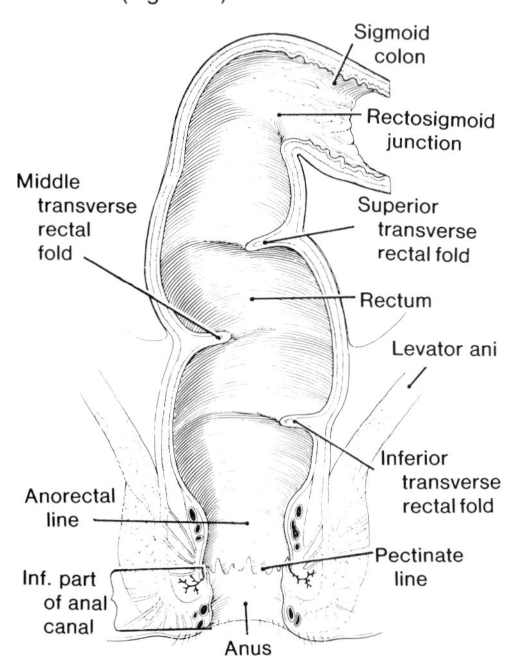

Figure 3-83. Drawing of a coronal section of the rectum and anal canal. Observe the curvature of the rectum and the acute flexion at the rectosigmoid junction. Note the three transverse rectal folds, two on the right and one on the left, which aid in the support of the feces. The transverse rectal folds may also aid in the separation of the feces from gas (L. flatus, *a blowing*). The part of the anal canal superior to the pectinate line develops from the hindgut, and the part of it inferior to this line develops from the proctodeum. Because of this difference in embryological origin, the pectinate line is a clinically important vascular, lymphatic, and nerve boundary (see text for details).

The rectovesical septum represents a potential cleavage plane between the rectum and the prostate gland.

In the female the anterior relation of the rectum is the vagina (Figs. 3-20 and 3-53). The rectum is separated from the posterior fornix of the vagina and the cervix of the uterus by the **rectouterine pouch** (Figs. 3-20, 3-74*A*, and 3-76). Inferior to this pouch, the **rectovaginal septum** separates the vagina and the rectum (Fig. 3-22).

THE ARTERIAL SUPPLY OF THE RECTUM (Figs. 3-57 and 3-84)

There are *five rectal arteries*, illustrated below, which anastomose freely with each other.

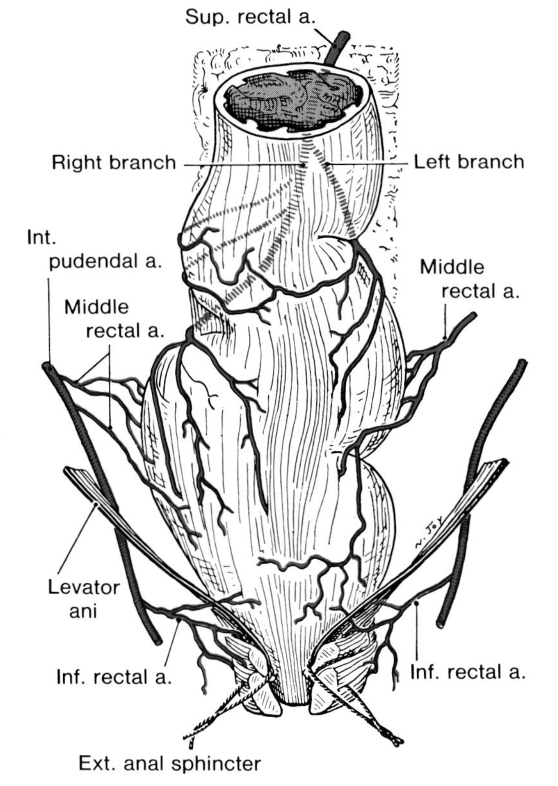

Figure 3-84. Drawing of anterior view of the arteries of the rectum and anal canal. Observe the branches of the right and left divisions of the superior rectal artery obliquely encircling the rectum. The middle rectal arteries, usually branches of the internal iliac arteries, are small. In this specimen the right artery is represented by two small branches of the internal pudendal artery. Note that the inferior rectal arteries, branches of the internal pudendal arteries, mainly supply the inferior part of the anal canal. Observe the lateral flexures of the rectum. These flexures and the transverse rectal folds (Fig. 3-83) help to support the weight of the feces. The anastomosis of arteries in the wall of the rectum is so extensive that the middle and inferior rectal arteries can supply the entire rectum if the inferior mesenteric artery (and thereby the superior rectal artery) is clamped.

The superior rectal artery, *the continuation of the inferior mesenteric artery*, supplies the terminal part of the sigmoid colon and the superior part of the rectum (Fig. 3-84). It crosses the left common iliac vessels and descends into the pelvis minor within the sigmoid mesocolon (Figs. 2-96 and 3-82). At about the level of S3 vertebra, the superior rectal artery divides into two branches which descend each side of the rectum (Fig. 3-84).

The two middle rectal arteries, *branches of the internal iliacs* (Figs. 3-57 and 3-84), supply the middle and inferior parts of the rectum.

The two inferior rectal arteries, *branches of the internal pudendal arteries* (Fig. 3-84), originate in the ischiorectal fossae (Fig. 3-25). They supply the inferior part of the rectum and anal canal.

THE VENOUS DRAINAGE OF THE RECTUM (Figs. 3-58 and 3-86)

The rectum is drained via superior, middle, and inferior rectal veins. There are many anastomoses between these vessels.

The rectal venous plexus (Fig. 3-86) surrounds the rectum and is joined to the *vesical venous plexus* in the male (Fig. 3-64) and the *uterovaginal venous plexus* in the female (Fig. 3-58).

The rectal venous plexus consists of two parts: (1) an **internal rectal venous plexus** that is deep to the epithelium of the rectum, and (2) an **external rectal venous plexus** that is external to the muscular coats of the rectum.

The internal rectal plexus drains into the superior rectal vein, but communicates freely with the external rectal venous plexus. The superior part of the external rectal venous plexus drains into the superior rectal vein, the beginning of the superior mesenteric vein. *The inferior part of the external rectal venous plexus drains into the internal pudendal vein*, and the middle part of the external rectal venous plexus drains into the middle rectal vein and then into the internal iliac vein.

Because *the superior rectal vein drains into the portal system* (Fig. 2-100), whereas the middle and inferior veins drain into the systemic system, this is an important area of portacaval anastomosis (see p. 251 in Chap. 2).

LYMPHATIC DRAINAGE OF THE RECTUM (Figs. 2-97 and 3-73)

Most lymph vessels from the *superior half* or more of the rectum *ascend along the superior rectal vessels* to the **pararectal lymph nodes**, and then pass to the lymph nodes in the inferior part of the mesentery of the sigmoid colon and to the **inferior mesenteric lymph nodes** (Fig. 2-97).

The lymph vessels from the *inferior half of the rectum* pass superiorly with the middle rectal arteries and drain into the **internal iliac lymph nodes** (Fig. 3-73). Lymph from these nodes passes to the *common iliac and aortic lymph nodes* (Fig. 2-121).

THE NERVES OF THE RECTUM (Figs. 3-35, 3-54, 3-57, 3-79, and 3-80)

The nerve supply to the rectum is derived from the sympathetic and parasympathetic systems. The middle rectus plexus is an offshoot from the **inferior hygogastric plexus** (Fig. 3-80). Four to eight nerves pass directly from this plexus to the rectum.

The *parasympathetic nerves* are derived from the second, third, and fourth sacral nerves and run in the **pelvic splanchnic nerves** to join the *inferior hypogastric plexus*. The sensory nerves follow the path of the parasympathetic nerves and their fibers are stimulated by distention of the wall of the rectum.

After rectal surgery some men are unable to ejaculate owing to damage to the pelvic splanchnic nerves (Fig. 3-57) which also supply the glans penis and initiate ejaculation (Fig. 3-70).

Many of the structures related to the inferior part of the rectum may be palpated through its walls, e.g., the prostate and seminal vesicles in males and the cervix in females (Fig. 3-72).

In both sexes, the pelvic surfaces of the sacrum and coccyx may be felt posteriorly. The **ischial spines and ischial tuberosities** (Fig. 3-42) may be palpated. *Enlarged internal iliac lymph nodes*, pathological thickening of the ureters, swellings in the ischiorectal fossae (*e.g.*, **ischiorectal abscesses**, and abnormal contents in the **rectovesical pouch** in the male (Fig. 3-12) or the **rectouterine pouch** in the female (Fig. 3-76*A*) may be detected. Tenderness of an *inflamed vermiform appendix* can be detected rectally if this organ lies in the pelvis (Fig. 2-42).

The rectum can also be examined with a **proctoscope** (G. *prōktos*, anus + *skopēo*, to view), and biopsies of lesions may be taken through it. During insertion of a *sigmoidoscope* for inspection of the rectum and sigmoid colon, the curvatures of the rectum (Fig. 3-82) and the acute flexion at the **rectosigmoid junction** have to be kept in mind (Fig. 3-83) so that the patient will not undergo unnecessary discomfort. One must also know that the **transverse rectal folds** may temporarily impede passage of an instrument.

Herniation or prolapse of the rectum in females, called a **rectocele** (Fig. 3-56), occurs when there is a weakness of the fibromuscular layer of the posterior wall of the vagina. The vagina tends to bulge through the vaginal orifice with the attached wall of the rectum. In some cases defecation is difficult unless the patient presses on the rectocele with her fingers in her vagina.

THE ANAL CANAL

The anal canal, 2.5 to 4 cm long in the adult, is *the terminal part of the large intestine* (Fig. 2-32*A*). The anal canal begins where the rectal ampulla narrows abruptly at the level of the U-shaped sling formed by the **puborectalis muscle** (Figs. 3-20 and 3-55). It extends from this muscle to the anus. The anal canal and anus are contracted and form an anteroposterior slit (Fig. 3-4), except during the passage of feces.

The anal canal is surrounded by internal and external anal sphincters (Fig. 3-17*A*) and descends posteroinferiorly between the anococcygeal ligament (Fig. 3-20) and the central perineal tendon (Fig. 3-10).

The anal canal is surrounded by the levatores ani muscles which form the main part of the pelvic diaphragm (Figs. 3-50 and 3-83).

The internal anal sphincter (Figs. 3-4 and 3-17 to 3-19) cannot be contracted voluntarily. It consists of a thickening of the circular smooth muscle of the intestine. This *involuntary sphincter of the anal canal* (about 2.5 cm long) surrounds the superior two-thirds of the anal canal. The internal anal sphincter relaxes when it is stimulated by the parasympathetic nerves supplying it.

The external anal sphincter (Figs. 3-4, 3-12, 3-17 to 3-19, and 3-35) was described with the anal region (p. 306). *The external anal sphincter can be contracted voluntarily.* Together with the puborectalis muscle, the external anal sphincter forms a muscular **anorectal ring**. It surrounds the inferior two-thirds of the anal canal, forming a broad band on each side that *overlaps the internal anal sphincter and the fibers of the levator ani* (Fig. 3-17*A*).

The external anal sphincter has three parts: subcutaneous, superficial, and deep. The **superficial part** is attached posteriorly to the coccyx and anteriorly to the central perineal tendon. Hence *the anorectal junction has an anorectal ring composed of voluntary and involuntary muscle fibers* which are responsible for maintaining rectal continence.

The innervation of the external anal sphincter is primarily by S4 via the **inferior rectal nerve** (Fig. 3-4). *This nerve leaves the pudendal canal and runs anteromedially and superficially across the ischiorectal*

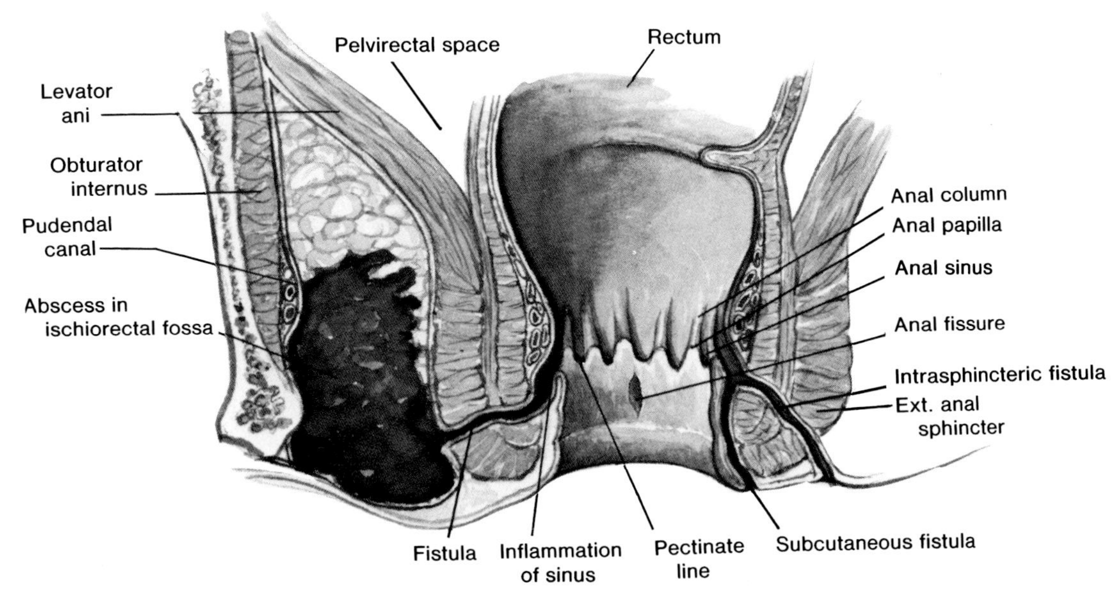

Figure 3-85. Drawing of a coronal section of the rectum and anal canal, illustrating an ischiorectal abscess in the right ischiorectal fossa. Infection and inflammation of anal sinuses or crypts (*cryptitis*) spread into the ischiorectal fossa. As pus develops it aggregates to form an abscess (collection of pus). Anal fistulae are pathological channels that may develop following infection and carry the infection from the anal canal to one or more of the following sites: ischiorectal fossa, pelvirectal space, perineum, and buttock. Observe the tear in the anal mucosa called an anal fissure; over 90% of them occur posteriorly in the midline. The anal mucosa appears to be vulnerable to injury here because the superficial part of the external anal sphincter inserts posteriorly into the coccyx supporting the anal mucosa. As a result, it is more easily torn by hard fecal material than elsewhere. The small epithelial projections on the anal valves, called anal papillae, may be remnants of the *embryonic anal membrane*. Sometimes they become hypertrophied as illustrated in Figure 3-86. (Reprinted with permission from Healey JE, Jr: *A Synopsis of Clinical Anatomy*. Philadelphia, WB Saunders, 1969).

fossa to supply the external anal sphincter. Hence the inferior rectal nerve is vulnerable during surgical treatment of an ischiorectal abscess (Fig. 3-85).

THE INTERIOR OF THE ANAL CANAL (Figs. 3-12, 3-17, 3-20, 3-83, and 3-85)

The superior half of the mucous membrane of the anal canal is characterized by a series of longitudinal ridges or folds called **anal columns**. These columns contain the terminal branches of the superior rectal artery and vein. *It is here that the superior rectal veins of the portal system anastomose with the middle and inferior rectal veins of the caval system* (Fig. 2-100).

The superior ends of the anal columns indicate the **anorectal line** (Figs. 3-17*B* and 3-83), where the rectum joins the anal canal. The rectum and superior part of the anal canal are derived from the **hindgut** in the embryo.

The inferior ends of the anal columns are joined to each other by semilunar folds of epithelium called **anal valves** (Figs. 3-12 and 3-17*B*). Superior to the valves are small recesses called **anal sinuses** (Fig. 3-85). When compressed by feces, the mucous-containing anal sinuses exude mucous which aids in evacuation of the anal canal.

The anal valves are in the area formerly occupied by the **anal membrane** in the embryo (Fig. 3-17*B*). The inferior comb-shaped limit of the anal valves is known as the **pectinate line** (Fig. 3-83).

The pectinate line indicates the site of junction of the superior part of the anal canal (derived from the embryonic hindgut) and the inferior part of the anal canal (derived from the proctodeum or anal pit). The pectinate line also approximates the line of junction of the columnar epithelium of the superior part of the anal canal and the stratified squamous epithelium of the inferior part. At the anus the moist, hairless mucosa of the anal canal becomes dry, hairy skin (Fig. 3-17*B*).

As described subsequently, *the part of the anal canal superior to the pectinate line differs from the part inferior to the pectinate line in its arterial supply, innervation, and in its venous and lymphatic drainage.* This results from their different embryological origins.

ARTERIAL SUPPLY OF THE ANAL CANAL (Figs. 3-25, 3-57, 3-84, and 3-85)

The superior rectal artery supplies the superior part of the anal canal. Its terminal branches run in the anal columns as far as the anal valves, where they form anastomotic loops.

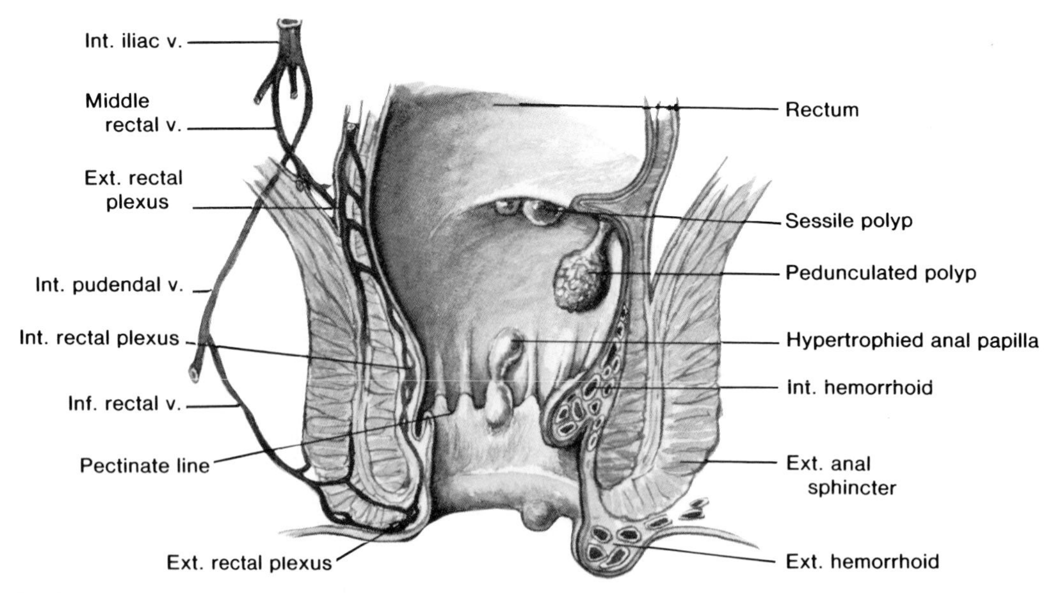

Figure 3-86. Drawing of a coronal section of the rectum and anal canal, showing its venous drainage and illustrating various anorectal problems. Polyps are protruding growths from the mucous membrane. Internal hemorrhoids are covered with mucosa, whereas external hemorrhoids are covered with modified anal skin. *Internal hemorrhoids* are varicosities of tributaries of the superior rectal vein and *external hemorrhoids* are varicosities of the inferior rectal vein. (Reprinted with permission from Healey, JE, Jr: *A Synopsis of Clinical Anatomy*. Philadelphia, WB Saunders, 1969).

The two inferior rectal arteries supply the inferior part of the anal canal (distal to the pectinate line), as well as the surrounding muscles and perianal skin.

The **middle rectal arteries** assist with the blood supply to the anal canal by forming anastomoses with the superior and inferior rectal arteries.

THE VENOUS DRAINAGE OF THE ANAL CANAL (Fig. 3-86)

The internal rectal plexus drains in both directions from the level of the pectinate line (Fig. 3-83).

Superior to the pectinate line the internal rectal plexus of veins drains chiefly into the **superior rectal vein**, *and thereby into the portal system* (Fig. 2-100).

Inferior to the pectinate line the internal rectal plexus of veins drains into **inferior rectal veins** *around the margin of the external anal sphincter.*

The *middle rectal veins*, tributaries of the internal iliac veins, drain chiefly the muscular walls of the ampulla of the rectum (Figs. 3-20 and 3-86). They form anastomoses with the superior and inferior rectal veins.

LYMPHATIC DRAINAGE OF THE ANAL CANAL (Figs. 3-73 and 4-35)

Superior to the pectinate line the lymph vessels drain into the **internal iliac lymph nodes** and through them into the common iliac and aortic lymph nodes, whereas the lymph vessels *inferior to the pectinate line* drain into the **inguinal lymph nodes** (Fig. 4-35).

NERVES OF THE ANAL CANAL (Figs. 3-57, 3-79, and 3-80)

The nerve supply of the anal canal superior to the pectinate line is the same as that for the rectum. The **sympathetic nerves** pass mainly along the inferior mesenteric and superior rectal arteries. The **parasympathetic nerves** are from S2 to S4 and run in the pelvic splanchnic nerves to join the inferior hypogastric plexus. The superior part of the anal canal is sensitive only to stretching.

The nerve supply of the anal canal inferior to the pectinate line is derived from the inferior rectal branches of the **pudendal nerve** (Fig. 3-24). This part of the anal canal is sensitive to pain, touch, and temperature.

As has been mentioned, *the anal canal superior to the pectinate line develops from the hindgut* (endoderm), as does the rectum, whereas *the anal canal inferior to the pectinate line develops from the proctodeum* (ectoderm). The pectinate line, indicated by the anal valves, indicates the approximate site of the anal membrane in the embryo.

Because of its hindgut origin the superior part of the anal canal is supplied by the superior rectal artery, the terminal branch of the **inferior mesenteric artery** (hindgut artery), whereas the inferior part of the canal derived from the proctodeum is supplied by the inferior rectal arteries, branches of the **internal pudendal artery**. As described previously, the venous and lymphatic drainage and the nerve supply of these regions also differ because of the different embryological origins of the superior and inferior parts of the anal canal.

In the embryo the rectum is separated from the exterior by the **anal membrane**, but this normally breaks down at the end of the 8th week. If it persists, a condition known as **imperforate anus** exists. Some form of imperforate anus occurs once in about 5000 births and is more common in males.

Most anorectal malformations result from abnormal development of the urorectal septum, resulting in incomplete separation of the cloaca into urogenital and anorectal portions (see Moore, 1982 for illustrations and details).

Internal hemorrhoids (piles) are *varicosities of the tributaries of the superior rectal veins* and are covered by mucous membrane (Fig. 3-86).

External hemorrhoids *are varicosities of the tributaries of the inferior rectal veins* and are covered by skin. As there are multiple anastomoses between the venous plexuses of the rectal veins, these communicating veins may also be dilated.

Hemorrhoids prolapsing through the external anal sphincter are often compressed, impeding the blood supply. As a result they tend to ulcerate and strangulate.

Thrombus formation is more common in external hemorrhoids than in internal hemorrhoids.

The anastomoses between the superior, middle, and inferior rectal veins form a clinically important communication between the portal and systemic systems (Fig. 2-100). The superior rectal vein drains into the inferior mesenteric vein, whereas the middle and inferior rectal veins drain through the systemic system to the inferior vena cava. Any abnormal increase in pressure in the valveless portal system may cause enlargement of the superior rectal veins, resulting in internal hemorrhoids. In **portal hypertension**, as in *hepatic cirrhosis*, the tiny anastomotic veins in the anal canal and elsewhere (Fig. 2-100) become varicose and may rupture.

In chronically constipated persons, the anal valves may be torn by hard fecal material and the anal mucosa may also be torn. The slit-like lesion, known as an **anal fissure** (Fig. 3-85), is usually inferior to the anal valves and is very painful because this region is supplied by the inferior rectal nerve (Fig. 3-24).

Perianal abscesses (collections of pus) may follow infection of anal fissures and the infection may spread to the **ischiorectal fossae** or into the pelvis, forming ischiorectal and pelvirectal abscesses, respectively (Fig. 3-85).

An **anal fistula** may develop as a result of the spread of an infection. One end of the abnormal canal opens into the anal canal and the other end opens into an abscess in the ischiorectal fossa, or into the perianal skin (Fig. 3-85).

As the anal canal superior to the pectinate line is supplied through *autonomic nerve plexuses*, an incision or a needle insertion in this region is painless.

The anal canal inferior to the pectinate line is very sensitive (*e.g.*, to the prick of a hypodermic needle) because it is supplied by the *inferior rectal nerve* which contains sensory fibers, including those carrying pain.

THE PELVIC AUTONOMIC NERVES

THE SYMPATHETIC TRUNKS

The sacral sympathetic trunks are directly continuous with the lumbar sympathetic trunks posterior to the common iliac vessels (Figs. 2-101 and 2-118). The sacral trunks are smaller than the lumbar trunks and each one has four ganglia. The trunks descend on the pelvic surface of the sacrum, just medial to the pelvic sacral foramina, and converge to form the small median **ganglion impar** (L. unequal, *i.e.*, unpaired) on the coccyx. The sympathetic trunks run in the presacral fascia, external to the peritoneum (Fig. 3-82).

BRANCHES OF THE SACRAL SYMPATHETIC TRUNKS (Figs. 3-54, 3-70, 3-79, and 3-80)

The sympathetic trunks send ***gray rami communicantes*** *to each of the ventral rami of the sacral and coccygeal nerves.* They also send small branches to the median sacral artery and to the *inferior hypogastric plexuses* (Fig. 3-80). A few branches from the ganglion impar pass to the *coccygeal body* which lies anterior to the apex of the coccyx.

THE HYPOGASTRIC PLEXUSES (Figs. 3-59, 3-70, 3-79, 3-80, and 3-82)

The ***superior hypogastric plexus*** (presacral plexus) descends into the pelvis and lies just inferior to the bifurcation of the aorta (Fig. 3-82). It is the inferior prolongation of the *intermesenteric plexus* (Fig. 2-114) which is joined by L3 and L4 splanchnic nerves (Fig. 3-70).

Branches from the superior hypogastric plexus enter the pelvis and descend anterior to the sacrum as the **right and left hypogastric nerves** (Figs. 3-79 and 3-80). These nerves descend on the lateral walls of the pelvis where they mingle with the **pelvic splanchnic nerves** (Figs. 3-35 and 3-57) to form the right and left **inferior hypogastric plexuses** (Fig. 3-80).

The pelvic splanchnic nerves are parasympathetic and are derived from S2, S3, and S4 (Figs. 3-57 and 3-70). Hence the inferior hypogastric plexuses contain both sympathetic and parasympathetic fibers.

Each inferior hypogastric plexus surrounds the corresponding internal iliac artery. There are small ganglia within these plexuses (Fig. 3-80), and each plexus receives small branches from the superior sacral ganglia of the sympathetic trunks. Branches from the hypogastric plexuses, containing sympathetic and parasympathetic fibers, are distributed to the pelvic viscera along the branches of the internal iliac artery (Fig. 3-57).

The visceral plexuses are extensions of the inferior hypogastric plexuses in the walls of the pelvic viscera.

The *middle rectal plexus* arises from the superior part of the inferior hypogastric plexus and extends inferiorly as far as the internal anal sphincter. Branches from the middle rectal plexus pass directly to the rectum or along the middle rectal artery. Parasympathetic fibers also pass from this plexus to the sigmoid and descending parts of the colon.

The **vesical plexus** arises from the anterior part of the inferior hypogastric plexus and branches from it pass to the *urinary bladder* along the vesical arteries (Fig. 3-57). Branches from the vesical plexus also pass to the seminal vesicles, the ductus deferentes, and the prostate gland (Fig. 3-70).

The **prostatic plexus** arises from the inferior part of the hypogastric plexus and is composed of rather large nerves which enter the base and sides of the prostate gland. These nerves are also distributed to the seminal vesicles, ejaculatory ducts, urethra, bulbourethral glands, and penis (Fig. 3-70). The nerves supplying the corpora cavernosa of the penis, called the *cavernous nerves*, arise from the anterior part of the prostatic plexus and join with branches from the **pudendal nerve** (Figs. 3-24 and 3-70). These nerves pass along the membranous urethra to the penis.

The **uterovaginal plexus** arises mainly from the part of the inferior hypogastric plexus that lies in the base of the broad ligament (Fig. 3-79). Some nerves from the uterovaginal plexus pass inferiorly to the vagina and cervix with the vaginal arteries; other nerves pass directly to the cervix or superiorly with the uterine arteries to the body of the uterus. Some of these nerves also supply medial parts of the uterine tube.

The uterovaginal plexus is homologous with the prostatic plexus. The **vaginal nerves** from the uterovaginal plexus also supply the urethra, bulbs of the vestibule, greater vestibular glands, and the clitoris.

JOINTS OF THE PELVIS

The articulations of the pelvis are the lumbosacral, sacrococcygeal, and sacroiliac joints, and the symphysis pubis.

THE LUMBROSACRAL JOINTS

The fifth lumbar vertebra (L5) and the first sacral vertebra (S1) articulate with one another by an *anterior fibrocartilaginous joint* (intervertebral disc) between their bodies, and by two *posterior synovial joints* between their articular processes. **The intervertebral disc between L5 and S1 vertebrae is wedge-shaped** because it is thicker anteriorly (Fig. 3-20).

The **zygapophysial joints** (Fig. 5-30) are synovial joints between the inferior articular processes of L5 and the superior articular processes of the S1 part of the sacrum (Fig. 5-24). The S1 facets face posteriorly and medially, thereby preventing L5 vertebra from sliding anteriorly (Fig. 5-1).

The fifth lumbar vertebra is attached to the ilium and sacrum by the **iliolumbar ligaments** (Fig. 3-43*A*). *The strong iliolumbar ligaments unite each thick transverse process of L5 vertebra to the internal lip of the iliac crest posteriorly.* The fibers of each iliolumbar ligament, often called the **lumbosacral ligament**, descend to the anterior part of the ala of the sacrum (Fig. 3-43*A*).

The iliolumbar ligaments help to stabilize the lumbosacral joint and are important because they limit axial rotation of L5 vertebra on the sacrum.

Large transverse processes on L5 are not uncommon and are more likely to strengthen the lumbosacral joint than to weaken it.

A condition called **spondylolysis** is found in the inferior lumbar region in about 5% of white North American adults. It occurs more frequently in certain races, *e.g.*, Eskimos, Australian aborigines, and South African bushmen.

In spondylolysis a defect is found in the vertebral arch between the superior and inferior facets, the area called the **pars interarticularis.** When bilateral the defects result in the vertebra being in two pieces (Fig. 5-37). The posterior piece consists of the laminae, inferior articular processes, and the spinous process, and the anterior piece represents the remainder of the vertebra.

If the two pieces of the lumbar vertebra separate, the condition is called **spondylolisthesis** (Fig. 3-87). The resulting anteroinferior displacement of the anterior piece of L5 reduces the AP (anterior posterior) diameter of the superior pelvic aperture and *may interfere with parturition.* Obstetricians test for spondylolisthesis by running their fingers down the lumbar spinous processes of the pregnant patient's back. If the spinous process of L5 is prominent, it indicates that the anterior part of L5 vertebra and the vertebral column superior to it have moved anteriorly (Fig. 3-87). Radiographs are then made to confirm the diagnosis and to measure the AP diameter of the superior pelvic aperture

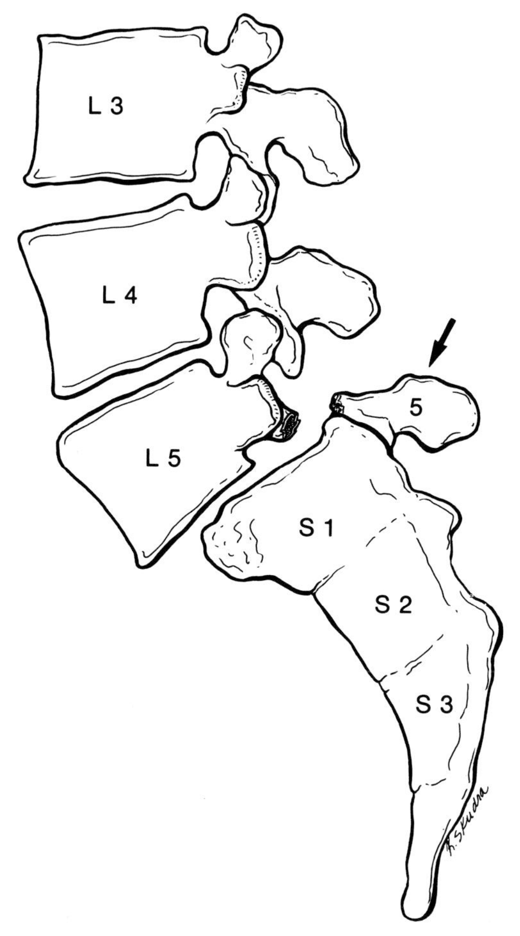

Figure 3-87. Drawing made using a lateral radiograph of a patient with spondylolisthesis of L5 vertebra. Observe that the anterior piece of L5 (body, pedicles, transverse processes, and superior articular processes) has moved anteriorly taking the vertebrae superior to this level with it. Note that the posterior piece of L5 (laminae, inferior articular processes, and spinous process) has stayed with the sacrum. The *arrow* points to the spinous process of L5 which is on a vertical plane posterior to the other lumbar spinous processes; thus it appears unduly prominent.

THE SACROCOCCYGEAL JOINT

The sacrococcygeal joint is usually a symphysis, a type of cartilaginous joint in which fibrocartilage and ligaments join the bones. The apex of the sacrum and the base of the coccyx are united by a thin fibrocartilaginous intervertebral disc (Figs. 3-20 and 5-25) that is slightly thicker anteriorly.

The **sacrococcygeal ligaments** correspond to the anterior and posterior longitudinal ligaments of the other intervertebral joints (Fig. 5-28).

The *sacral and coccygeal cornua* are also united by **intercornual ligaments.** In persons up to middle age, there is slight movement of the coccyx posteriorly on defecation, and during childbirth there is considerable posterior movement of the coccyx.

In elderly persons the first coccygeal vertebra is frequently fused to the apex of the sacrum, thereby eliminating the sacrococcygeal joint (Fig. 5-1).

THE SACROILIAC JOINTS

The sacroiliac articulations are very strong synovial joints between the articular surfaces of the sacrum and the ilium (Figs. 3-44, 3-45*A*, and 3-46). These articular surfaces have irregular elevations with depressions in the opposing surfaces.

The **articular capsule** is attached close to the articulating surfaces of the sacrum and ilium. The sacrum is suspended between the iliac bones (Fig. 3-43) and the bones are firmly held together by very strong **interosseous and dorsal sacroiliac ligaments.**

THE INTEROSSEOUS SACROILIAC LIGAMENTS (Fig. 3-88)

These massive ligaments uniting the iliac and sacral tuberosities are ***very strong.*** They consist of short, strong bundles of fibers which blend with the dorsal sacroiliac ligament posteriorly. The bundles of fibers radiating from the iliac tuberosity to the lateral aspect of the ala of the sacrum, posterior to the auricular facets of the sacrum, suspend the sacrum between the two ilia.

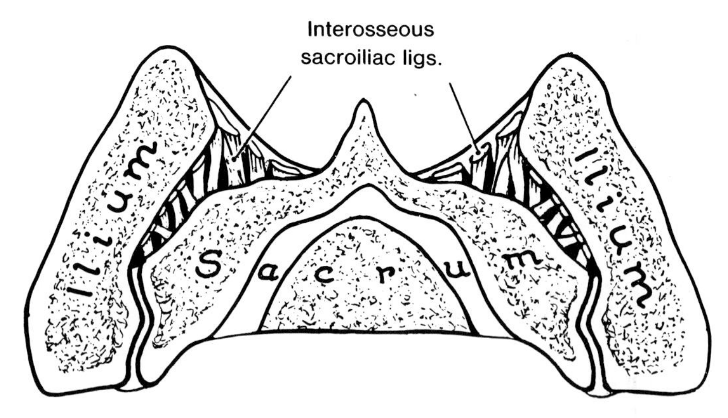

Figure 3-88. Drawing of a transverse section of the sacroiliac joints illustrating the powerful interosseous ligaments. Each ligament consists of short strong fibers passing between the tuberosities of the sacrum and ilium. The sacrum is suspended between the iliac bones by the very strong interosseous and dorsal sacroiliac ligaments.

THE DORSAL SACROILIAC LIGAMENTS

These ligaments are composed of: (1) strong, short, transverse fibers joining the ilium and the first and second tubercles of the lateral crest of the sacrum; and (2) long vertical fibers uniting the third and fourth transverse tubercles of the sacrum to the posterior iliac spine (Fig. 4-27). The dorsal sacroiliac ligaments blend with the ***sacrotuberous ligament*** (Figs. 3-16, 3-18, and 3-43*B*).

THE VENTRAL SACROILIAC LIGAMENTS (Fig. 3-16)

This wide, thin sheet of transverse fibers is located on the anterior and inferior aspects of the sacroiliac joint. It covers the abdominopelvic surface of this articulation.

The sacrotuberous and sacrospinous ligaments (Figs. 3-16, 3-18, and 3-43*B*) are accessory ligaments of the sacroiliac joints. The sacroiliac joints are covered posteriorly by the erector spinae and gluteus maximus muscles (Figs. 5-40 and 5-44). The **skin dimples** indicating the **posterior superior iliac spines** (Figs. 2-105 and 5-11), lie at the level of the middle of the sacroiliac joints.

The sacroiliac joints are strong weightbearing joints which are responsible for transmitting the weight of the body to the hip bones.

MOVEMENTS OF THE SACROILIAC JOINTS

Little movement occurs at these joints because they are designed primarily for weight bearing.

Movement of the sacroiliac joints is limited to a slight gliding and rotary movement, except when a considerable force is applied as occurs during a jump from a height. In this case the force is transmitted via the vertebral column to the superior end of the sacrum, which tends to rotate anteriorly. This rotation is counterbalanced by the strong sacrotuberous and sacrospinous ligaments, allowing the force to be transmitted to each ilium and lower limb.

ARTERIAL SUPPLY OF THE SACROILIAC JOINTS (Fig. 3-57)

The articular branches to the sacroiliac joints are derived from the superior gluteal, iliolumbar, and lateral sacral arteries.

NERVE SUPPLY OF THE SACROILIAC JOINTS (Fig. 3-54)

The articular branches are derived from the *superior gluteal nerves*, the *sacral plexus*, and the dorsal rami of *S1* and *S2 nerves.*

When one falls from a height and lands on one's feet, the sacroiliac joints transfer most of the body weight to the hip bones. Flexion of the lower limbs also helps to prevent injury to the vertebral column. The resilience of the sacrotuberous and sacrospinous ligaments (Fig. 3-43*B*) also cushions the shock to the vertebral column.

The sacroiliac ligaments are thought to become softer and more yielding during the late stages of pregnancy, increasing the range of movement of the sacroiliac joints. Combined with similar changes in the pubic symphysis and the associated ligaments (Fig. 3-43), the passsage of the fetus through the birth canal is facilitated.

The sacroiliac joints often become partially ossified during old age, especially in men. Calcification may occur in the ventral sacroiliac ligament making the joint space less visible on radiographs, even though it is still present.

THE SYMPHYSIS PUBIS

The symphysis pubis is a **cartilaginous joint** between the bodies of the pubic bones (Figs. 3-12, 3-20, 3-43*B*, 3-46, 3-52, and 3-53). Each articular surface is covered by a thin layer of hyaline cartilage which is connected to the cartilage of the other side by a thick fibrocartilaginous **interpubic disc** (Fig. 3-52). The disc is generally thicker in females than in males.

The **superior pubic ligament** connects the pubic bones superiorly and extends as far as the pubic tubercles (Fig. 3-43*A*).

The **arcuate pubic ligament** (Figs. 3-2, 3-16, and 3-20) is a thick arch of fibers: (1) connecting the inferior borders of the joint, (2) rounding off the subpubic angle (Fig. 3-43*B*), and (3) forming the superior border of the pubic arch.

PRESENTATION OF PATIENT ORIENTED PROBLEMS

Case 3-1

A 23-year-old **primigravida** (L. *primus*, first + *gravida*, pregnancy) had been in labor for nearly 24 hr. The crown of the fetal head was visible through the vaginal orifice. The obstetrician, fearing that her perineum might be torn, decided to perform a **mediolateral episiotomy** (Fig. 3-41*B*) to enlarge the inferior opening of the birth canal.

Problems. What structures would probably be cut during this surgical procedure? What perineal structures might have been injured if the perineum had been allowed to tear in an uncontrolled fashion? In severe **perineal lacerations**, what muscles may be torn? *These problems are discussed on page* 392.

Case 3-2

A 31-year-old construction worker was walking along a steel beam when he fell, straddling it. He was in severe pain owing to *trauma to his testes and perineum*. Later he observed swelling and discoloration of his scrotum and when he attempted to urinate, only a few drops of bloody urine appeared.

Urethrography, performed by gently injecting a radiopaque contrast solution into his urethra with a syringe, revealed a **rupture of the spongy urethra** just inferior to the inferior fascia of the urogenital diaphragm (Fig. 3-9). The **urethrograms** showed passage of contrast material out of the urethra into the surrounding tissues of the perineum, a process known as *extravasation of urine*.

Problems. When the patient tried to urinate, practically no urine came from his external urethral orifice. Where did it go? Explain why extravasated urine cannot pass posteriorly, laterally, or into the pelvis minor. *These problems are discussed on page* 392.

Case 3-3

A 49-year-old family physician noted *tenderness and pain to the right of his anus*. The pain was aggravated by defecation and sitting. As he had a history of **hemorrhoids** and ***pruritus ani*** (L. *pruritus*, an itching), he suspected that he might be developing an *abscess in his ischiorectal fossa*.

After he had explained his symptoms and history, his physician examined the rectum and anal canal. When he asked the patient to strain as if to defecate, **prolapsing internal hemorrhoids** (Fig. 3-86) came into view.

During digital examination of the rectum, the doctor detected some *swelling in the patient's right ischiorectal fossa*. The swelling produced severe pain when it was compressed.

A diagnosis of **ischiorectal abscess** was made (Fig. 3-85). The abscess was drained through an incision in the skin between the anus and the ischial tuberosity.

Problems. *Differentiate between internal and external hemorrhoids.* What is an ischiorectal abscess? What nerve is vulnerable to injury during surgical treatment of an ischiorectal abscess? If this nerve were severed, what structure(s) would be partly denervated? *These problems are discussed on page* 393.

Case 3-4

During the examination of a male child, a congenital malformation of the penis known as **hypospadias** was detected. The urethra opened just proximal to the site where the frenulum usually attaches the **prepuce** to the ventral or urethral surface of the penis (Fig. 3-32). There was a slight indentation at the site where the **external urethral orifice** is normally located. In addition, there was a slight ventral curvature of the penis (*chordee*).

Micturition *was essentially normal* except that he dribbled if he urinated while standing up, wetting his clothing and shoes.

Problems. What is the *embryological basis of hypospadias*? What type of hypospadias is present? Discuss its etiology and other types of hypospadias. Do you think this condition would subsequently interfere with retroductive function? *These problems are discussed on page* 393.

Case 3-5

A 40-year-old unconscious woman was rushed to the hospital because she had sustained *multiple injuries during an automobile accident*. Priority was given to securing a patent airway by inserting an endotracheal tube. Next, emergency care was directed toward controlling bleeding and treating the shock.

When the patient's general condition had stabilized, radiographs were taken of the injured regions of her body. Because she had not urinated since being admitted, she was catheterized. The presence of blood in her urine (**hematuria**) suggested rupture of the urinary bladder. Therefore, 100 ml of sterile dilute contrast solution were injected through a catheter and radiographs of the pelvis and abdomen were taken.

The radiologist reported that there were **fractures of the pubic rami** on both sides and that the *cystogram showed extravasation of contrast material* from the superior surface of the bladder (Fig. 3-89).

Problems. Where would the extravasated urine go? What covers the superior surface of the urinary bladder? Thinking anatomically, what route do you think the surgeon would take in repairing the ruptured bladder? *These problems are discusssed on page* 394.

Case 3-6

A 28-year-old woman was experiencing pregnancy for the first time (*primigravida*). Toward the end of the gestational period, she suffered painful uterine contractions at night which subsided toward morning (**false pains**). When she called her doctor, he told her that her *labor was imminent.*

In a few days she observed a discharge of mucus and some blood, referred to as a *show*. This vaginal discharge results from release of the **cervical plug** which had filled the cervical canal during pregnancy, forming a barrier between the uterus and vagina.

When she reported that her "pains" (**uterine contractions**) were occuring every 10 minutes, her obstetrician asked her to go to the hospital. Following admission, the doctor palpated her cervix rectally and informed the intern that the **uterine ostium** was open about one fingertip, and that she was still in the **first stage of labor** (period of dilation of the uterine ostium). Later a large volume of fluid (*rupture of the fetal membranes*) was expelled.

When the patient entered the **second stage of labor** (period of expulsive effort beginning with complete dilation of the cervix and ending with delivery of the baby), she began to experience considerable pain. Although she had wanted to have a *natural birth* without the use of anesthetics, she was unable to bear the pain and asked for relief. Medication for pain relief was administered as ordered by her physician.

When it was determined that her contractions were 2 minutes apart and lasting 40 to 60 sec, she was moved to the case room and placed on a delivery table. As the fetal head dilated the **birth canal**, it was obvious that the woman was suffering intense pain. The obstetrician decided to do a **mediolateral episiotomy** (Fig. 3-41*B*) when it appeared possible that a tear might occur in her perineum. He gave her an intradermal injection of an anesthetic agent into the perineum. Although the local anesthetic enabled the incision to be made without pain, it did not alleviate the severe pain of her labor.

Although **extradural anesthesia** is often used in obstetrics because it relieves the pelvic pains without interfering with uterine contractions, the obstetrician in this case decided to do **bilateral pudendal nerve blocks**. Thereafter the patient completed the second stage and proceeded through the **third stage of labor** (beginning after delivery of the child and ending with expulsion of the placenta and fetal membranes).

Problems. What membranes rupture during the first stage of labor or at the end of it? What fluid escapes? *Name the structures supplied by the pudendal nerve.* Based on your knowledge of the anatomy of this nerve, where would you inject the anesthetic agent to make a *pudendal nerve block*? What is the principal landmark in the perineal route.

When complete perineal anesthesia is required, branches of what other nerves would have to be blocked? *These problems are discussed on page* 395.

DISCUSSION OF PATIENT ORIENTED PROBLEMS

Case 3-1

During a mediolateral episiotomy the following structures are usually cut: (1) the perineal skin, (2) the posterior wall of the vagina, (3) the bulbospongiosus muscle, (4) the superficial transversus perinei muscle, and (5) the inferior fascia of the urogenital membrane (Figs. 3-39 and 3-41). Part of the deep transversus perinei muscle may be cut in some cases (Fig. 3-14).

An episiotomy is performed to ease delivery of a fetus and/or when a perineal laceration seems inevitable. If a tear is allowed to occur spontaneously in whatever direction it may, ***the central perineal tendon, the external anal sphincter, and the wall of the rectum may be torn***. Episiotomy thus makes a clean cut away from important structures.

Because the levator ani and coccygeus muscles form the **pelvic diaphragm**, the chief function of which is to form the floor of the pelvis, it is important during suturing of a lacerated perineum (**perineorrhaphy**) to repair the medial part of the levator ani muscle if it has been torn. Failure to do this results in poor perineal support for the pelvic organs, which may cause sagging of the pelvic floor in later life. This could lead to difficulty in bladder control, occasionally in control of the rectum, and could be the basis of a *prolapse of the uterus* in rare cases (Fig. 3-56).

Case 3-2

The traumatic rupture of the man's spongy urethra within the bulb of his penis (Figs. 3-28 to 3-30) resulted in superficial or subcutaneous **extravasation of urine** when he attempted to urinate.

Urine from the torn urethra passed into the perineum, *superficial to the inferior fascia of the urogenital diaphragm*, but *deep to the superficial perineal fascia* (Fig. 3-8). The urine in the superficial perineal space passes *inferiorly* into the areolar tissue of the scrotum, *anteriorly* into the penis, and *superiorly* into the anterior wall of the abdomen.

The inferior fascia of the urogenital diaphragm (Figs. 3-8 and 3-9) and the superficial fascia of the perineum are firmly attached to the ischiopubic rami. The urine cannot pass posteriorly because *the two layers are continuous with each other around the superficial transversus perinei muscles.* The urine does not extend laterally because these two layers are connected to the rami of the pubis and ischium, and with the deep fascia of the thigh (**fascia lata**), where it is continuous with the membranous layer of the superficial fascia of the abdomen. It cannot extend into the pelvis minor because the opening into this cavity is closed by the inferior fascia of the urogenital diaphragm (Fig. 3-8).

Urine cannot pass into the thighs because the membranous layer of the superficial fascia of the anterior abdominal wall blends with the fascia lata, just distal to the inguinal ligament. The fascia lata is the strong fascia enveloping the muscles of the thigh (Fig. 4-16).

Case 3-3

Hemorrhoids are varicosities of one or more of the veins draining the anal canal. **Internal hemorrhoids are varicosities of the tributaries of the *superior rectal vein*.** This vein becomes the inferior mesenteric vein and belongs to the portal system of veins (Fig. 2-52). The tributaries of the superior rectal vein arise in the *internal rectal plexus*, which lies in the anal columns (Fig. 3-17*B*). Here they frequently become varicose (dilated and tortuous).

Internal hemorrhoids are covered by mucous membrane. At first they are contained in the anal canal (Fig. 3-86), but as they enlarge they may protrude through the anal canal on straining during defecation. Bleeding from these hemorrhoids is common.

Chronic constipation with prolonged straining is a common predisposing factor in persons with internal hemorrhoids, but there are several other causes (*e.g.*, pregnancy and portal obstruction related to cirrhosis of the liver).

Most hemorrhoids do not result from portal obstruction. They frequently occur in members of the same family, suggesting that there is an increased susceptibility to the development of hemorrhoids.

External hemorrhoids are varicosities of the tributaries of the *inferior rectal vein* arising in the *external* rectal plexus (Fig. 3-86), which drains the inferior part of the anal canal.

External hemorrhoids are covered by anal skin and are usually not painful unless they undergo thrombosis (G. clotting). These are referred to as **thrombosed hemorrhoids**.

Perianal abscesses often result from injury to the anal mucosa by hardened fecal material. Inflammation of the anal sinuses may result, producing a condition called **cryptitis**. The infection may spread through a small crack or lesion in the anal mucosa and pass through the anal wall into the *ischiorectal fossa*, producing an **ischiorectal abscess** (Fig. 3-85).

Infections in the fat in the ischiorectal fossa are not uncommon. They can result from tears of the anal mucosa, disease of the perineal skin, and rarely from infection brought via the blood stream. The abscess may be connected medially, forming an **anorectal fistula** which joins the ischiorectal abscess to the anal canal and/or to the skin of the perineum. Untreated ischiorectal abscesses may extend superiorly into the **pelvirectal space** (Fig. 3-85), producing a supralevator or **pelvirectal abscess**.

The ischiorectal fossa is a wedge-shaped space lateral to the anus and levator ani muscle (Figs. 3-18 and 3-21). *The main component of the ischiorectal fossae is fat.* The branches of the nerves and vessels (pudendal nerve, internal pudendal vessels, and the nerve to the *obturator internus muscle*) enter the ischiorectal fossa through the **lesser sciatic foramen** (Fig. 3-16). The pudendal nerve and internal pudendal vessels pass in the *pudendal canal* lying in the lateral wall of the ischiorectal fossa (Fig. 3-51).

The **inferior rectal nerve** leaves the pudendal canal and runs anteromedially and superficially across the ischiorectal fossa. It passes to the **external anal sphincter** and supplies it (Fig. 3-4). *Damage to the inferior rectal nerve results in impaired action of this voluntary anal sphincter.*

Case 3-4

In 1 in 300 male infants, the external urethral orifice is on the ventral surface of the penis. Most often the defect is of the glandular type, as in the present case (Fig. 3-68). In other patients the opening is on the body of the penis (**penile hypospadias**), or in the perineum (**penoscrotal hypospadias**). In *scrotal hypospadias* the halves of the scrotum are not completely fused (Fig. 3-68).

The embryological basis of glandular and penile hypospadias is failure of the ***urogenital folds*** *to fuse on the ventral surface of the developing penis* and form the spongy urethra (Fig. 3-26). Urine is not discharged from the tip of the penis, but from an opening on the ventral surface of the penis (Fig. 3-68).

The embryological basis of scrotal hypospadias is failure of the labioscrotal folds to fuse and form the scrotum (Fig. 3-26).

The developmental cause of hypospadias is not clearly understood, but it appears to have a **multifac-**

torial etiology (*i.e.*, genetic and environmental factors are involved). Close relatives of patients with hypospadias are more likely than the general population to have the abnormality. It is generally believed that *hypospadias is associated with an inadequate production of androgens by the fetal testes.* Differences in the timing and degree of hormonal insufficiency probably account for the different types of hypospadias.

Because the urethral orifice is not located at the tip of the glans and there is ventral bowing of the penis (**chordee**), which is more marked when the penis is erect, reproduction by persons with this malformation is difficult. In some cases the degree of curvature is so severe during erection that **intromission** (insertion of the penis into the vagina) and natural **insemination** are impossible.

Surgical correction of the chordee to produce a straight penis and repair of the urethra (**urethroplasty**) were recommended in the present case before the boy started school, so that it would be possible for him to urinate normally in the standing position and later be able to reproduce.

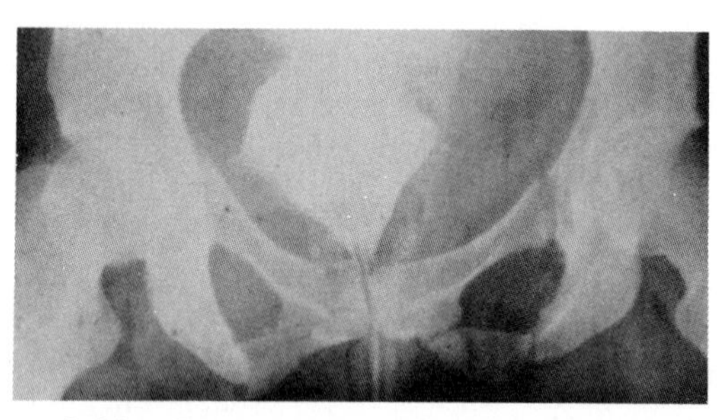

Figure 3-89. Radiograph of the pelvis showing fractures of the pubic rami on both sides. Contrast medium injected through the catheter in the urinary bladder has passed out of the superior portion of the bladder into the peritoneal cavity.

Case 3-5

The urine which escaped from the superior surface of the patient's ruptured urinary bladder would pass into the peritoneal cavity (Fig 3-89). Although *pelvic fractures are sometimes complicated by **bladder rupture***, the radiographs indicated that the rupture was not likely caused by a sharp bone fragment. Probably the bladder was ruptured by the same compressive blow to the region of the symphysis pubis that fractured the pelvis.

A full bladder is especially liable to rupture at its superior surface following a nonpenetrating blow.

The superior surface of the bladder is almost completely covered with peritoneum. In the female the peritoneum is reflected onto the uterus at the junction of its body and cervix, forming the vesicouterine pouch (Fig. 3-20). In the male, the peritoneum is reflected from the bladder over the superior surfaces of the ductus deferentes and seminal vesicles.

In a patient with an *intraperitoneal bladder rupture*, signs and symptoms of peritoneal irritation are likely to develop. **Septic peritonitis** (G. *sepsis*, putrifaction) may develop if there are pathogenic organisms in the urine. As the urine accumulates in the peritoneal cavity, dullness will be detected over the paracolic gutters during percussion of the abdomen. This dullness will disappear from the left side when the patient is rolled onto the right side and vice versa, indicating

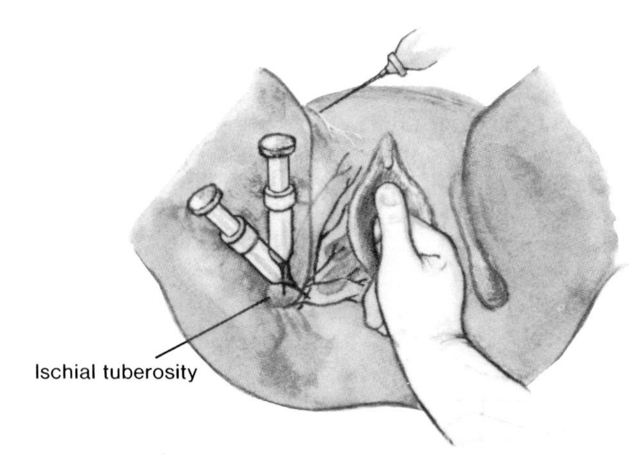

Figure 3-90. Drawing illustrating how a pudendal nerve block may be performed. The chief bony landmark is the ischial tuberosity. The needle is inserted toward and just medial to the tuberosity, where the pudendal nerve emerges from the pudendal canal (Fig. 3-40). Some obstetricians guide the needle by the fingers placed in the vagina until its tip is posterior and inferior to the ischial spine where the pudendal nerve lies. The injection at the top of the drawing blocks or anesthetizes the ilioinguinal nerve and the branches of the genitofemoral nerve supplying the vulva. (Reprinted with permission from Healey JE, Jr: *A Synopsis of Clinical Anatomy*. Philadelphia, WB Saunders, 1969).

free fluid in the peritoneal cavity from a ruptured viscus (the bladder in this case).

Access to the urinary bladder for surgical repair of its ruptured superior wall would most likely be via the **suprapubic route**. The bladder is separated from the pubic bones by a thin layer of areolar tissue which may contain fat (Fig. 3-20). When the bladder is full, its anteroinferior surface is in contact with the anterior abdominal wall, without the interposition of peritoneum.

Case 3-6

When the "bag of waters" breaks, the amniotic and chorionic sacs rupture, permitting the amniotic fluid to escape (up to 1000 ml in most cases).

The amniotic and chorionic sacs protrude into the cervical canal during the first stage of labor and help to dilate it.

The pudendal nerve, arising from the sacral plexus (S2, S3, and S4), is ***the main nerve of the perineum***. It is both motor and sensory to this region and also carries some postganglionic sympathetic fibers to the perineum.

In the female the pudendal nerve divides into the perineal nerve and the dorsal nerve of the clitoris. The **perineal nerve** gives off two *posterior labial nerves* (Fig. 3-40), and divides into small terminal muscular branches which enter the superficial and deep perineal spaces to supply the muscles in them and the *bulb of the vestibule* (Fig. 3-39).

The **dorsal nerve of the clitoris** supplies the prepuce and glans of the clitoris (Fig. 3-36) and the associated skin.

When the pudendal nerve is blocked via the perineal route, *the chief bony landmark is the ischial tuberosity* (Fig. 3-90). With the patient in the lithotomy position (Fig. 3-1), the **ischial tuberosity** is palpated and a skin wheal (lesion produced by an intradermal injection) is raised just medial to the tuberosity. Here the *pudendal nerve* emerges from the *pudendal canal* to distribute itself over the perineum (Fig. 3-40). The needle is inserted superomedially for about 2.5 cm before the injection is made.

When complete perineal anesthesia is required, *genital branches of the genitofemoral nerve*, the *ilioinguinal nerve* (Fig. 2-8), and the *perineal branch of the posterior cutaneous nerve of the thigh* (Fig. 3-40) must also be anesthetized by making an injection along the lateral margin of the labia majora.

Suggestions for Additional Reading

1. Dilts PV, Jr, Greene JW, Jr, Roddick JW, Jr: *Core Studies in Obstetrics and Gynecology*, ed 3. Baltimore, Williams & Wilkins, 1981.

This relatively small book (242 pp.) presents the basic information about obstetrics and gynecology. It serves as an introduction to these specialities and clearly describes important basic material.

2. Lowenfels AB: *Companion Guide to Surgical Diagnosis*. Baltimore, Williams & Wilkins, 1975.

It is usually more difficult to make a differential diagnosis of abdominopelvic pain in women than in men because there are more genital organs in the female pelvis. Gynecological diseases such as acute salpingitis, hemorrhage into or torsion of an ovarian cyst, and ectopic pregnancy on the right side may produce symptoms similar to appendicitis.

Enlargement of the uterus during pregnancy displaces the cecum and appendix into the superior part of the abdomen. If these clinically oriented problems interest you, refer to this relatively small text (276 pp.). The approach is simple and informal.

3. Moore KL: *The Developing Human: Clinically Oriented Anatomy*, ed. 3. Philadelphia, WB Saunders Co, 1982.

Malformations of the genital organs are relatively common and frequently occur in association with abnormalities of the urinary system. To understand these conditions, including intersexuality, a good understanding of normal sex development is needed. If you are not clear about the role of the testes in sexual differentiation and hypospadias, or if you are unsure why congenital adrenal hyperplasia may cause masculization of a female fetus, you are urged to do additional reading.

CHAPTER 4

The Lower Limb

The lower limb is the organ of locomotion and is specialized for bearing the weight of the body and for maintaining equilibrium.

The lower limb consists of four major parts (Figs. 4-1 and 4-2): (1) hip, containing the *hip bone* (innominate bone or *os coxae*), and connecting the skeleton of the lower limb to the vertebral column; (2) **the thigh**, containing the *femur* and connecting the hip and knee; (3) **the leg** (L. *crus*), containing the *tibia* ("shin bone") and *fibula* ("splint bone"), and connecting the knee and ankle; and (4) **the foot** (L. *pes*), containing the *tarsal bones, metatarsal bones*, and *phalanges* (bones of the toes), which is distal to the ankle.

The Latin word for the foot (*pes*) is the basis of the words *peduncle* and *pedicle*. The Greek word for the foot (*podos*) appears in the terms *podiatry* (treatment of the foot) and *podalgia* (pain in the foot).

As the lower limb is involved in weight bearing, some movement has been sacrificed to acquire stability (*e.g.*, compare the mobility of your toes and fingers). Although people use the term leg to refer to their lower limb, understand that the leg is only the part of the lower limb between the knee and the ankle.

Because injuries and degenerative disorders of the lower limbs are so common, you must understand clearly the normal and abnormal functioning of these organs. To understand why a patient cannot walk or has an **abnormal gait**, you must know what structures are involved in normal walking.

The parts of the lower limb are comparable to those of the upper limb (*e.g.*, foot and hand; knee and elbow). This is understandable only after you learn that the limbs rotated in opposite directions during embryonic development. The upper limbs rotate laterally through 90° on their longitudinal axes bringing the thumb to the lateral side, whereas the lower limbs rotate medially through almost 90° bringing the great toe to the medial side. Hence the knee faces anteriorly when one stands in the anatomical position and the extensor muscles lie on the anterior aspect of the lower limb. Therefore, *extension occurs in opposite directions in the upper and lower limbs.*

As it is concerned with movement of the whole body and bearing its weight, the lower limb is stronger and heavier than the upper limb (*e.g.*, compare the size of the muscles of the thigh and the arm and the weight of their bones (femur and humerus).

The bones of the lower limb form the inferior part of the ***appendicular skeleton*** (Figs. 4-1 and 4-2).

The hip bones articulate posteriorly with the **sacrum** (Fig. 3-7) and meet inferiorly and anteriorly at the **symphysis pubis** (Fig. 4-1). The hip bones, the sacrum, and the coccyx form the skeleton of the **bony pelvis** (Figs. 3-42 and 3-43) which *encloses the pelvis,* the inferior basin-shaped division of the abdominopelvic cavity (Fig. 2-31), and *contains the pelvic viscera* (Chap. 3).

On the lateral aspect of the hip bone is the socket of the **hip joint**, where the head of the ***femur*** articulates with the cup-shaped **acetabulum** (Figs. 4-1 and 4-4).

THE HIP AND THIGH

The hip and thigh include *the area from the iliac crest to the knee* (Figs. 4-1 and 4-7). The hip is the region between the iliac crest and the greater trochanter of the femur, and the thigh is the region between the greater trochanter and the knee.

BONES AND SURFACE ANATOMY OF THE HIP

The hip bone (os coxae), large and irregularly shaped, consists of three bones during childhood (Fig. 4-3): *the* ***ilium*** (os ilium), *the* ***ischium*** (os ischium), and *the* ***pubis*** (os pubis). These bones fuse at 15 to 17 years and are indistinguishably joined in the adult (Fig. 4-4).

THE ILIUM (Figs. 4-1 to 4-4)

The ilium is fan-shaped; its *ala* (L. wing) resembling the spread of a fan and its *body* representing the handle. The ilium forms the superior two-thirds of the hip bone and the superior two-fifths of the cup-shaped ***acetabulum.***

When you put your hand on your hip, it rests on the superior margin of the ilium, called the **iliac crest** (Figs. 4-1 to 4-4, 4-7, and 4-9).

The margin of the ilium has internal and external lips and a curved superior margin or **iliac crest** between them (Fig. 4-4). This crest is easily palpated as it extends through the inferior margin of the flank or side of the trunk (Fig. 4-7). Its highest point, as palpated posteriorly, is at the level of the fourth lumbar vertebra (Fig. 4-9). *Clinically, this level is commonly used as a surface marking for lumbar punctures* (Fig. 5-60).

The iliac crest ends anteriorly in a rounded ***anterior superior iliac spine*** (Figs. 4-1 and 4-4), which is easily felt and may be visible (Figs. 4-5 and 4-7).

The iliac crest ends posteriorly in a sharp **posterior superior iliac spine** (Fig. 4-2), which is difficult to palpate in most people, but its position can always be determined because it lies at the bottom of a *skin dimple*, about 4 cm lateral to the median plane (Figs. 2-105 and 4-6). These dimples form because the skin and underlying fascia are attached to the posterior superior iliac spine. *A line connecting the right and left skin dimples is at the level of the second sacral vertebra and the middle of the sacroiliac joints* (Fig. 3-42*A*).

The skin dimples (Fig. 4-6) are useful landmarks for a doctor wishing to obtain bone marrow from the ilium. The needle is inserted 1 cm inferolateral to the dimple into the bone and marrow is aspirated for examination.

Another palpable bony landmark, the **tubercle of the crest** (Figs. 4-1 and 4-2), is located about 5 cm posterior to the anterior superior iliac spine. The ***anterior inferior iliac spines*** (Fig. 4-1) and the ***posterior inferior iliac spines*** (Fig. 4-2) are usually difficult to identify by palpation.

The posterior part of the internal surface of the ilium articulates with the side of the sacrum at the **sacroiliac joint** (Fig. 3-42*A*). Just inferior to this joint is the large **greater sciatic notch** (Fig. 4-2) through which pass the sciatic nerve and other important structures.

THE ISCHIUM (Figs. 4-1 to 4-4)

This bone forms the posteroinferior third of the hip bone and the posterior two-fifths of the acetabulum. The ischium (G. hip) is the roughly L-shaped part of the hip bone which passes inferiorly from the acetabulum, and then turns anteriorly to join the pubis.

The ischium consists of two parts, a body and a ramus. *The body of the ischium* is fused with the ilium and the pubis (Fig. 4-3). The inferior end of the body of the ischium has a rough, blunt projection called the ischial tuberosity.

The ischial tuberosity (Figs. 4-2 and 4-4) is covered by the gluteus maximus muscle when the thigh is extended (Fig. 4-9), but it is uncovered when the thigh is flexed. It bears the weight of the body when one sits (Fig. 5-3) and can be felt through the distal part of the gluteus maximus, just superior to the medial portion of the **gluteal fold,** a prominent skin fold delimiting the buttock inferiorly (Figs. 4-7 and 4-9). *The gluteal fold coincides with the inferior border of the gluteus maximus muscle.* Inferior to the gluteal fold is the **gluteal sulcus**, a groove or crease which separates the buttock from the posterior aspect of the thigh.

The ischial spine separates the ***greater sciatic notch*** superiorly from the ***lesser sciatic notch*** inferiorly (Figs. 4-2 and 4-4). The lesser sciatic notch is located between the ischial spine and the ischial tuberosity.

The sacrospinous ligament spans the **greater sciatic notch** (Fig. 4-2) converting it into the **greater sciatic foramen** (Fig. 3-16) through which pass the *piriformis muscle* and the vessels and nerves to the gluteal region and the thigh.

The sacrotuberous and sacrospinous ligaments convert the **lesser sciatic notch** (Fig. 4-2) into the **lesser sciatic foramen** (Fig. 3-16) through which pass the tendon and nerve of the *obturator internus muscle*, the pudendal nerve, and the internal pudendal vessels (Figs. 3-15 and 3-16).

The ramus of the ischium extends medially from the body and joins the inferior ramus of the pubis to form the **ischiopubic ramus,** which completes the *obturator foramen* (Figs. 4-1 and 4-4).

Alteration in the degree of prominence of the gluteal fold occurs in certain abnormal conditions; for example, wasting of the gluteus maximus resulting from **spinal poliomyelitis.** The muscular atrophy results from denervation of the muscle plus the *atrophy of disuse* owing to the paralysis.

THE PUBIS (Figs. 4-1, 4-3, and 4-4)

This bone forms the anterior part of the hip bone and the anteromedial one-fifth of the acetabulum.

The pubis consists of three parts: a body and two rami. The flattened body lies medially.

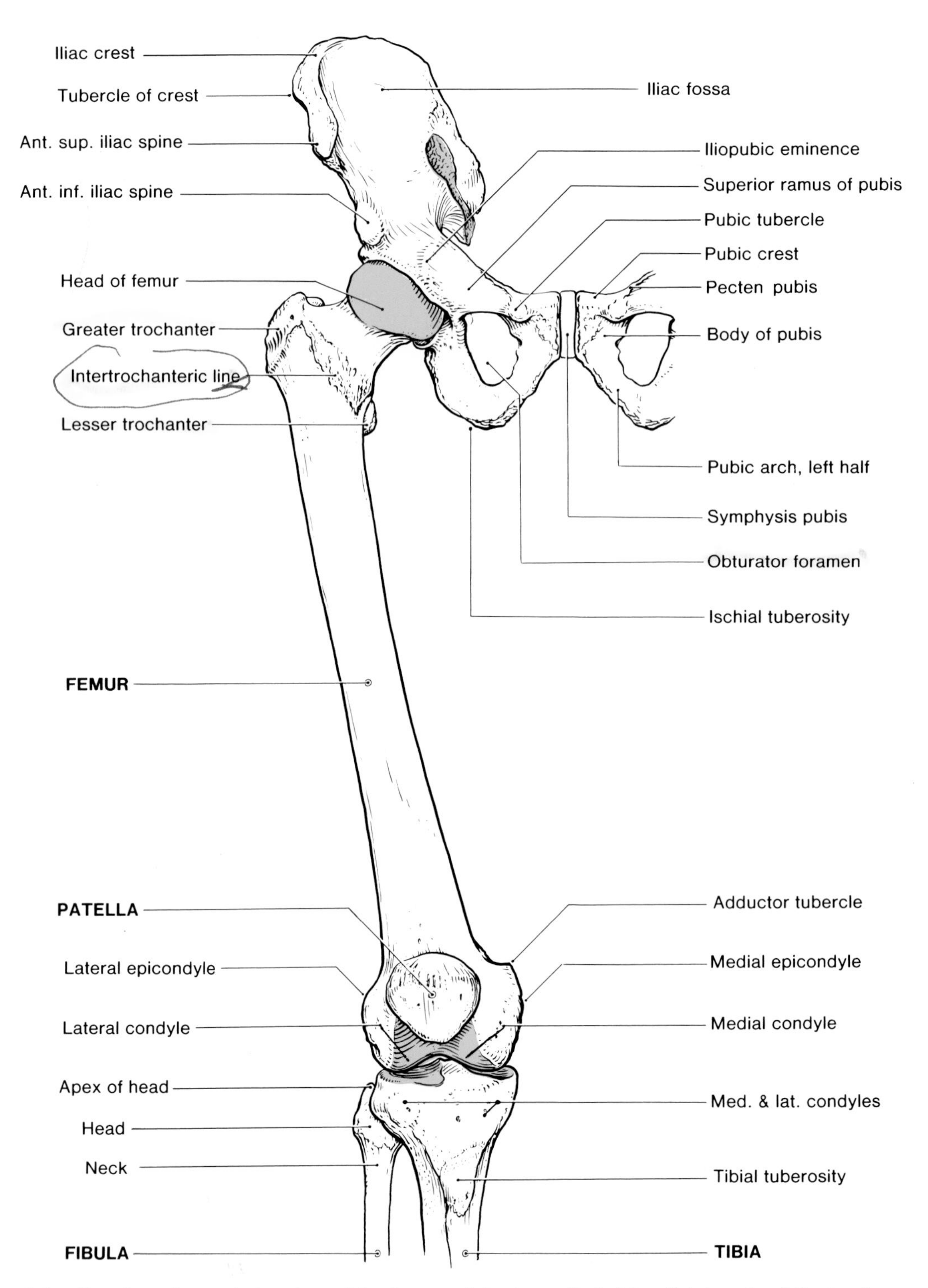

Figure 4-1. Drawing of an anterior view of the bones of the lower limb. The distal parts of the leg bones are not illustrated. The skeleton of the limb is connected to the vertebral column by the pelvic girdle which is formed by the hip bones. Note that the hip bones meet at the symphysis pubis, a secondary cartilaginous joint. There is a fibrocartilaginous disc between the bodies of the two pubic bones and strong anterior and inferior ligaments. The posterior and superior ligaments are relatively weak. See Figure 3-43 and page 391.

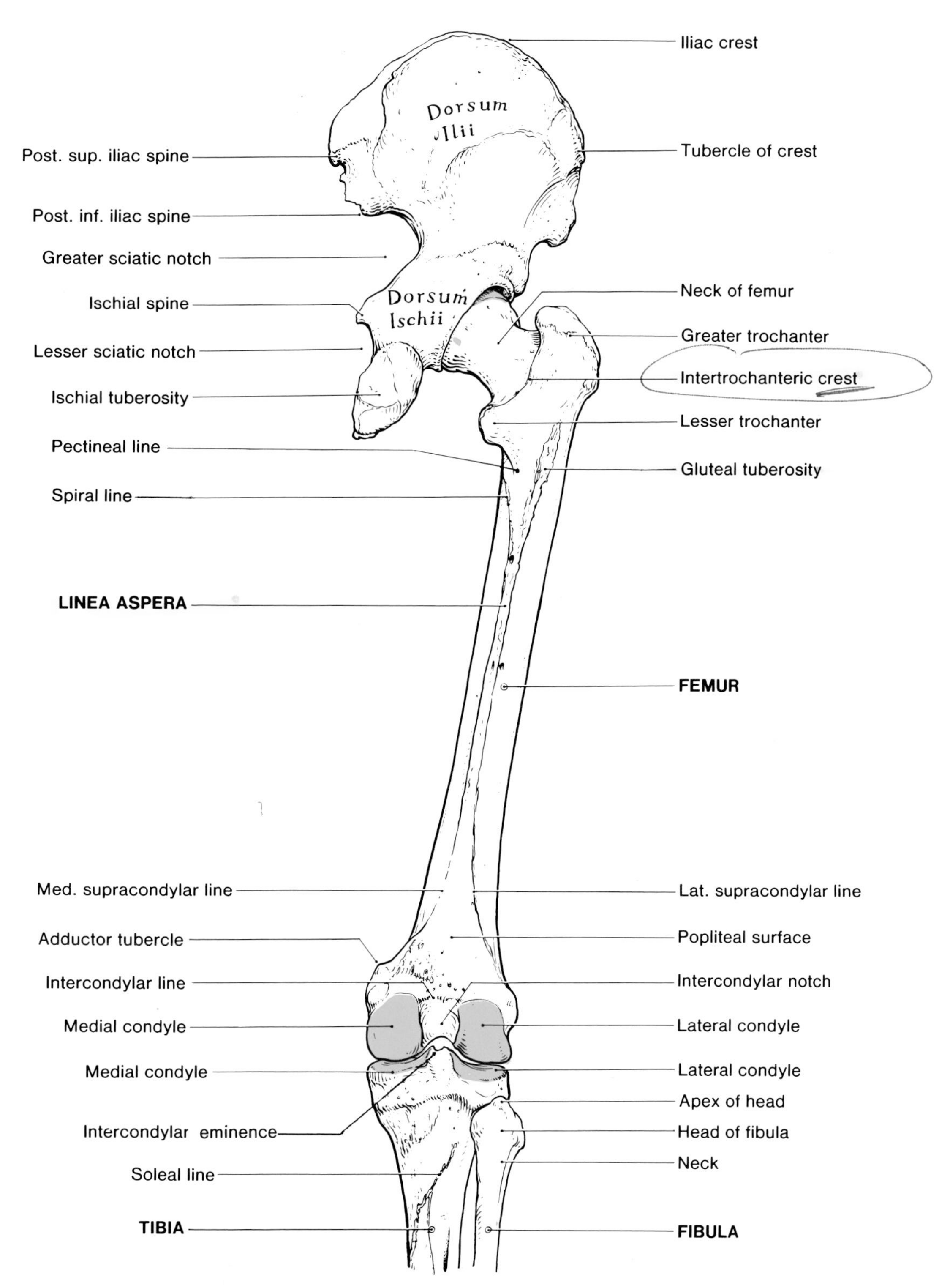

Figure 4-2. Drawing of a posterior view of the bones of the lower limb. The distal parts of the tibia and fibula are illustrated in Figure 4-55. With the sacrum and coccyx (not shown), the two hip bones form the skeleton of the bony pelvis (see Fig. 3-42). The articular cartilages of the condyles of the femur and tibia are colored *yellow*, as they are in Figure 4-1. The posterior superior iliac spine lies in the floor of a skin dimple (Fig. 4-6).

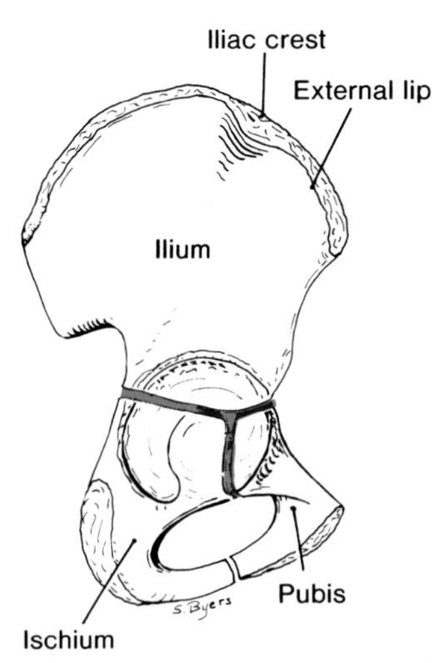

Figure 4-3. Drawing of the right hip bone of a child. Observe that it is composed of three bones (ilium, ischium, and pubis) which meet at the cup-shaped acetabulum (Fig. 4-4), the socket for the head of the femur. Note that the bones have not fused at this stage of development and are united by cartilage (*blue*) along a Y-shaped line in the acetabulum. Fusion of these bones to form the hip bone occurs around the 16th year. The lines of fusion may be visible in the adult bone. Note that the ilium is roughly fan-shaped. The ala (L. wing) represents the spread of the fan and its body is the broad "handle" that forms the superior part of the acetabulum.

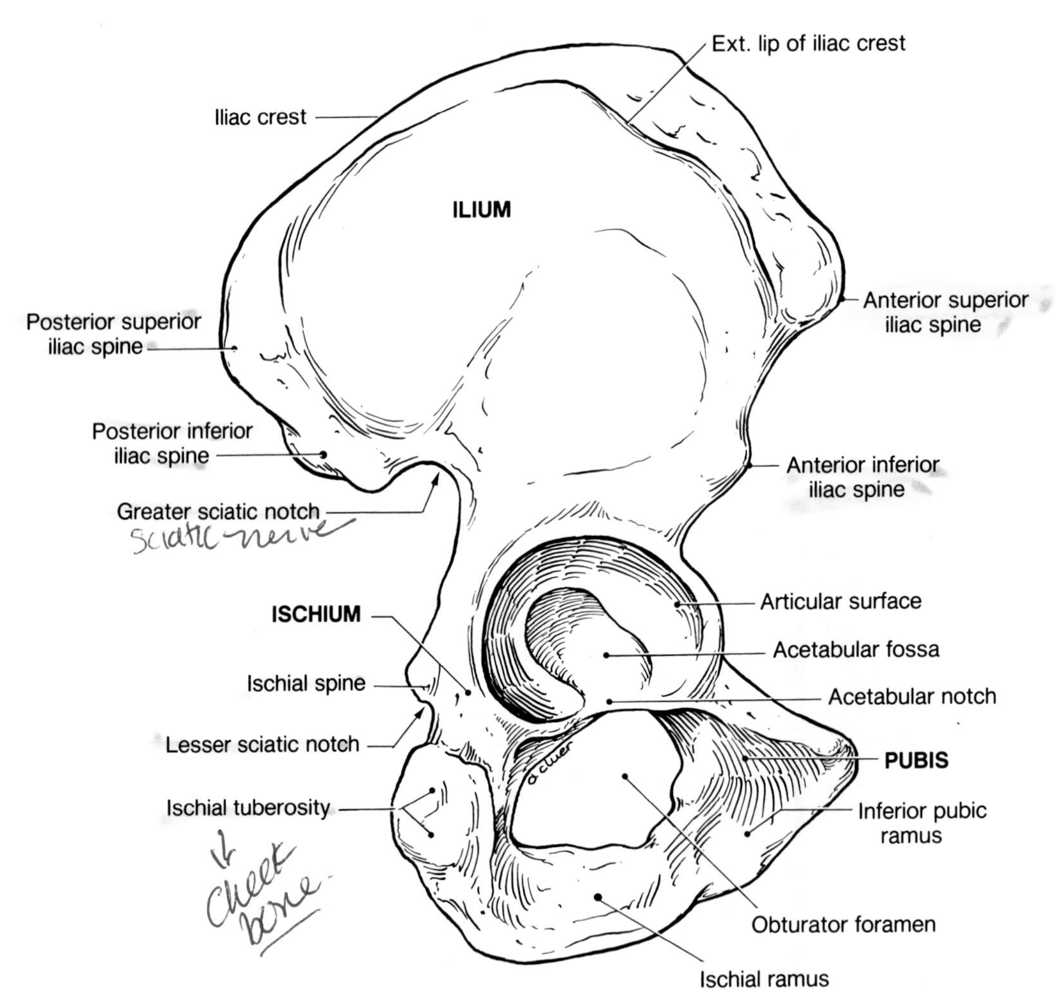

Figure 4-4. Drawing of the lateral aspect of a right hip bone (os coxae) from an adult in the anatomical position. The fan-shaped part of the ilium is called the wing (L. ala). Note that the acetabulum, the socket for the head of the femur, faces laterally, inferiorly, and slightly anteriorly. The hip bone looks somewhat like a boat propeller owing to its expanded and oppositely bent ends. The hip is the region of the lower limb between the iliac crest and the level of the greater trochanter of the femur, and the thigh is the region between the level of the greater trochanter and the knee (Fig. 4-7).

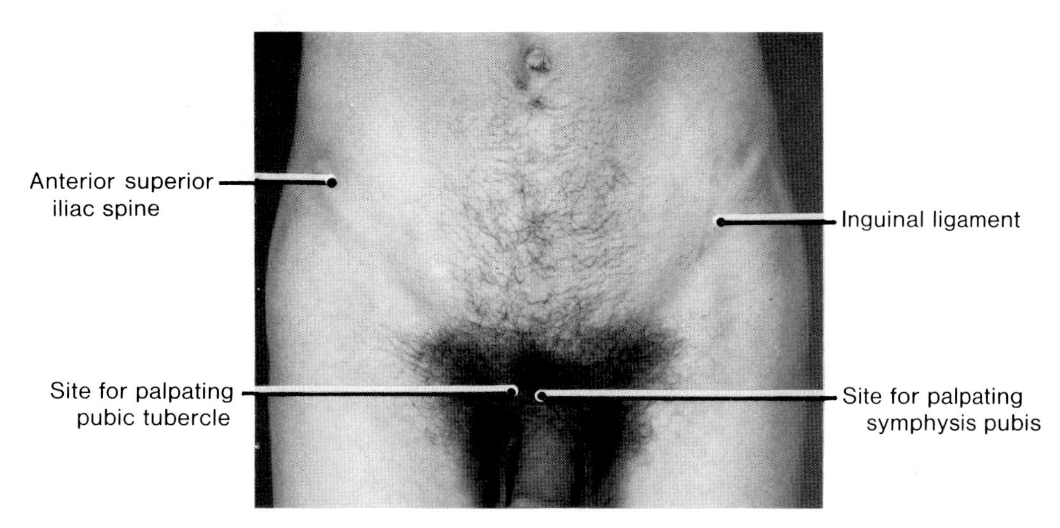

Figure 4-5. Photograph of the pelvic, hip, and thigh regions of a 27-year-old man showing the principal surface landmarks. Observe the anterior end of the iliac crest, called the anterior superior iliac spine (Fig. 4-1). This important bony landmark indicates where the inguinal ligament attaches; its other end is attached to the pubic tubercle (Fig. 4-15). The symphysis pubis, (pubic symphysis) readily felt in the midline, is a secondary cartilaginous union. The pubic tubercle can be palpated as a small protuberance on the superior border of the body of the pubis, about 2.5 cm lateral to the symphysis pubis (Fig. 4-1).

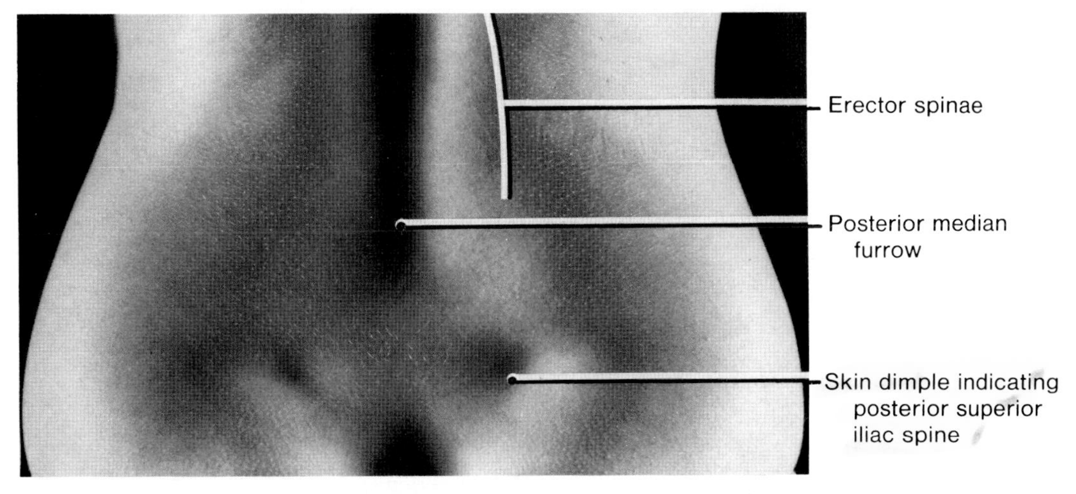

Figure 4-6. Photograph of the inferior back region of a 21-year-old woman showing the principal surface features. The posterior superior iliac spines (Fig. 4-2) lie in the floor of skin dimples which are located about 4 cm from the posteromedian line (center of posterior median furrow). These dimples indicate where the deep fascia is attached to the posterior superior iliac spines. The line joining these dimples is at the level of the spinous process of the second sacral vertebra. Also see Figure 2-105.

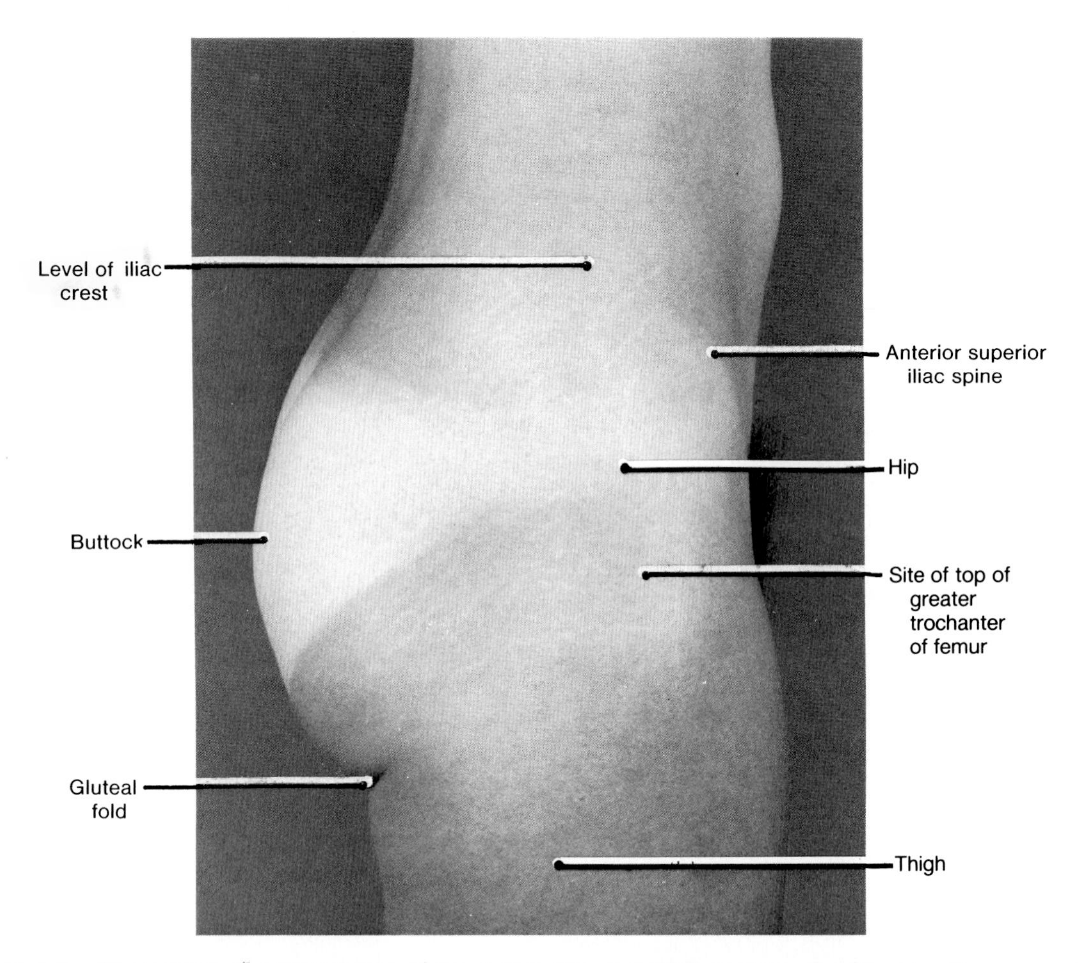

Figure 4-7. Photograph of a lateral view of the guteal, hip, and thigh regions of a 27-year-old woman showing the principal surface features. The iliac crest can be easily palpated by tracing posterosuperiorly from the anterior superior iliac spine. To determine approximately where *the ilium, ischium, and pubis united to form the hip bone around the 16th year*, see Figure 4-3. Note the prominence (buttock) formed mainly by the gluteus maximus muscle. When a person is standing, the gluteus maximus covers the ischial tuberosity (Fig. 4-2), but it can be felt by deep palpation through this large muscle just superior to the medial portion of the gluteal fold. This prominent fold coincides with the inferior border of the gluteus maximus. The groove or crease inferior to the gluteal fold is the gluteal sulcus; it demarcates the buttock from the thigh.

The superior ramus of the pubis passes superolaterally to the acetabulum, where it is fused with the ilium and the ischium. ***The inferior ramus*** passes posteriorly, inferiorly, and laterally to join the ramus of the ischium and form half of the pubic arch (Fig. 4-1). The body of the pubis joins the body of the opposite pubis in the median plane at a cartilaginous joint called the **symphysis pubis** (Figs. 4-1, 4-3, and 4-5).

The anterior border of the body of the pubis is thickened to form a **pubic crest** (Fig. 4-1). At its lateral end there is a projection, known as the **pubic tubercle**, which provides *the main pubic attachment for the inguinal ligament* (Fig. 3-43*A*). The pubic tubercle can be palpated about 2.5 cm from the median plane (Fig. 4-5).

The pubic tubercle is a very important bony landmark in cases of ***inguinal hernia*** (Fig. 2-20).

ORIENTATION OF THE HIP BONE (Figs. 3-45 and 4-4)

To place the hip bone in the anatomical position, move it until the anterior superior iliac spine and the symphysis pubis are in the same coronal plane. In this position the medial aspect of the body of the pubis faces almost directly superiorly (Fig. 4-1).

THE OBTURATOR FORAMEN (Figs. 4-1, 4-3, and 4-4)

This large oval aperture is surrounded by the bodies and rami of the pubis and ischium. It lies inferomedial to the acetabulum and *is nearly closed by the fibrous* ***obturator membrane***, which is attached to its margins (Figs. 3-16 and 4-100).

THE ACETABULUM (Figs. 4-3 and 4-4)

This cup-shaped cavity in the hip bone articulates with the head of the femur (Fig. 4-1). It was given its name because it *resembles a Roman vinegar cup called an acetabulum.* Until puberty the ilium, ischium, and pubis are united by a Y-shaped hyaline cartilage (Fig. 4-3). At 15 to 17 years the three bones fuse to form the hip bone and the cartilage is replaced by bone.

Fractures of the hip bone are common in serious automobile or truck accidents (Fig. 3-49). Anteroposterior compression of the hip bones commonly fractures the pubic rami. Lateral compression of the pelvis may fracture the acetabulum, as do *"falls on the feet"* (*e.g.*, from a roof). See pages 341 and 392 for a fuller description of **pelvic fractures.**

BONE AND SURFACE ANATOMY OF THE THIGH

THE FEMUR (Figs. 4-1, 4-2, and 4-8)

The femur (thigh bone) is the longest, strongest, and heaviest bone in the body. *A person's height is roughly four times the length of his/her femur.* It extends from the hip joint, where its rounded head articulates with the acetabulum, to the knee joint where its condyles articulate with the tibia.

The femur consists of a **body** or shaft and **two ends** (extremities). The proximal end of the femur

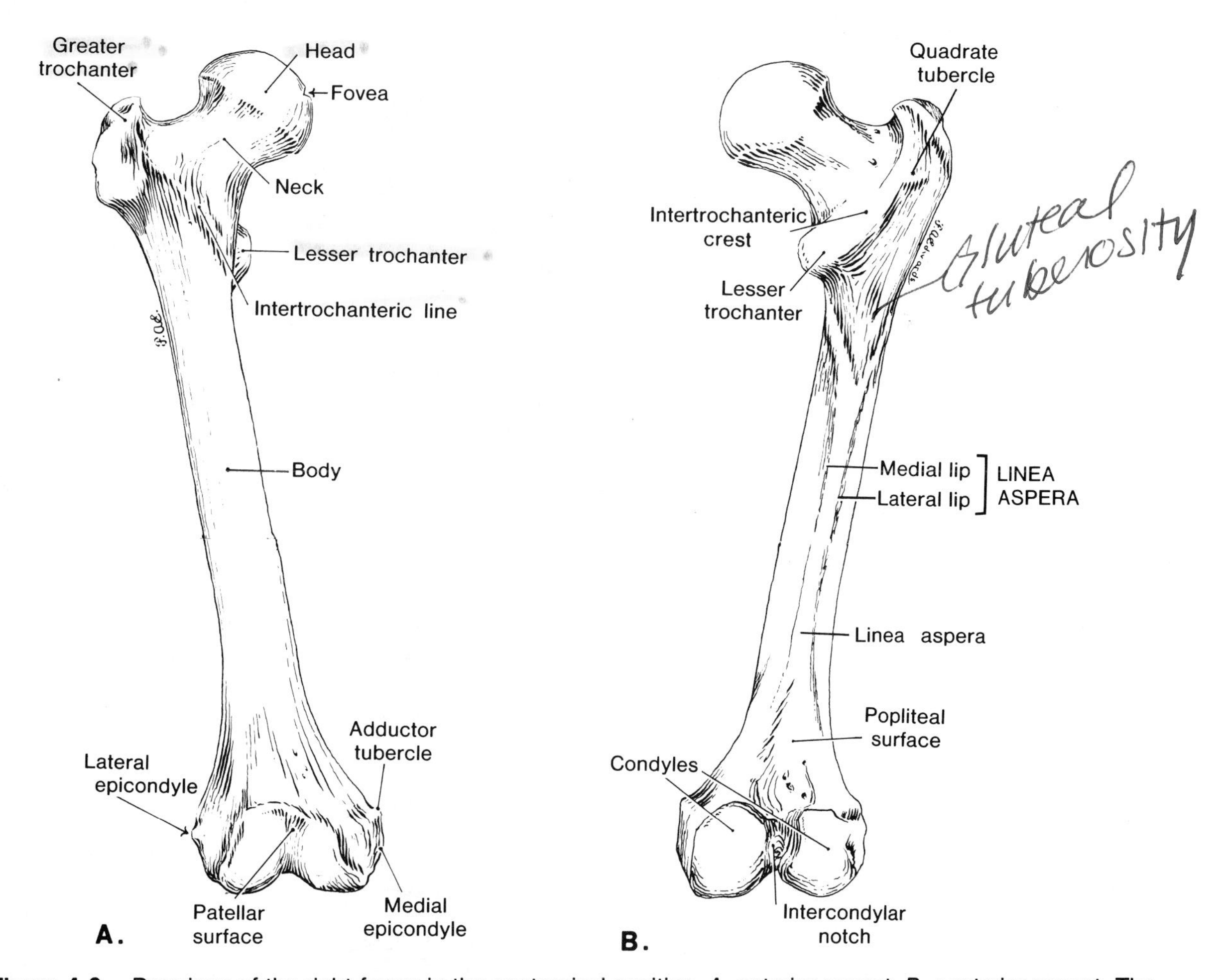

Figure 4-8. Drawings of the right femur in the anatomical position. *A*, anterior aspect, *B*, posterior aspect. The linea aspera is a broad rough ridge that usually forms a crest-like projection with distinct medial and lateral lips (also see Fig. 4-2). It gives attachment to several muscles (*e.g.*, adductor magnus, Fig. 4-20).

consists of a head, neck, and greater and lesser trochanters. The distal end of the femur is broadened where it articulates with the tibia and the patella to form the *knee joint* (Fig. 4-1).

The head of the femur is smooth and forms about two-thirds of a sphere. It is directed medially, superiorly, and slightly anteriorly to fit deeply into the acetabulum of the hip bone (Figs. 4-1 and 4-2). A little inferior and posterior to its center is a **fovea** (L. a pit) where the ligament of the head is attached (Figs. 4-8*A*, 4-98, and 4-99). The head of the femur can sometimes be palpated, particularly in thin males, when the thigh is rotated laterally.

The neck of the femur connects the head to the body (Figs. 4-2 and 4-8*A*). It runs inferolaterally to meet the body at an angle of about 125°. The neck is limited laterally by the **greater trochanter** and is narrowest in diameter at its middle.

The neck of the femur is frequently fractured in older persons. The neck has many pits, especially posterosuperiorly, for the entrance of blood vessels (Fig. 4-31). These are vulnerable to injury during fractures of the neck of the femur.

A broad, rough **intertrochanteric line** runs inferomedially from the greater trochanter toward the lesser trochanter (Figs. 4-1 and 4-8*A*). This line passes inferior to the lesser trochanter and becomes *continuous with the spiral line* on the posterior aspect of the femur (Fig. 4-2). It is produced by the attachment of the massive **iliofemoral ligament.**

The intertrochanteric line separates the anterior surface of the neck from the body of the femur (Fig. 4-8*A*). A prominent ridge, the **intertrochanteric crest**, unites the two trochanters posteriorly (Figs. 4-2 and 4-8*B*).

The greater trochanter of the femur is a large, somewhat rectangular projection from the junction of the neck and the body. It provides an insertion for several muscles of the gluteal region (Figs. 4-19 and 4-20).

The greater trochanter lies laterally, close to the skin, and can be easily palpated on the lateral side of the thigh (Figs. 4-7 and 4-9). Because it is the most lateral point of the hip region, the greater trochanter makes you uncomfortable when you lie on your side on a hard surface.

In the anatomical position, a line joining the tips of the greater trochanters normally passes through the center of the heads of the femora and the pubic tubercles. Verify this using Figure 4-1. The degree of prominence of the greater trochanter is increased when the

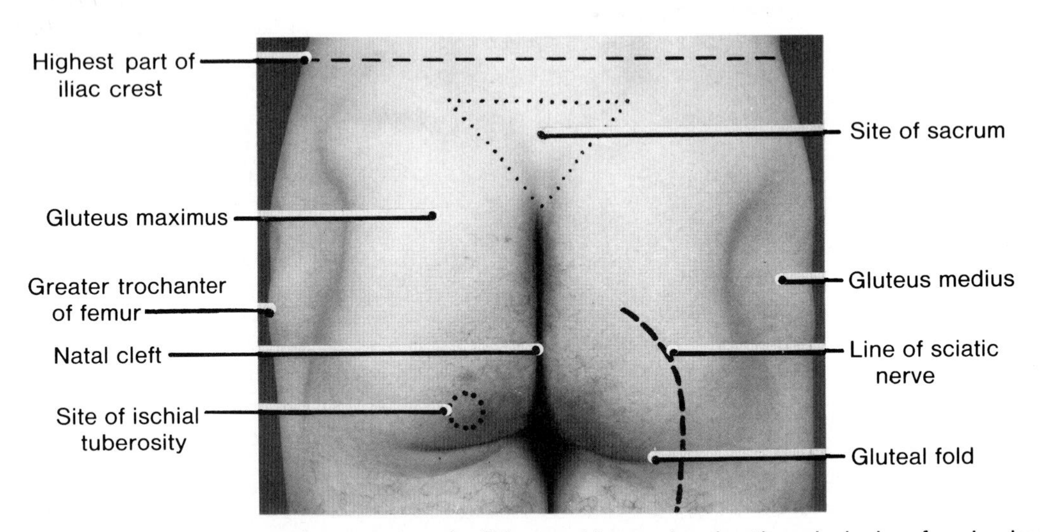

Figure 4-9. Photograph of the gluteal region of a 27-year-old man showing the principal surface landmarks. To produce these features the subject was asked to stand in the anatomical position and to press his heels together (*i.e.*, he was doing an isometric contraction of his lateral femoral rotators). Observe that the greater trochanter is the most lateral point of the hip region. The term buttock (L. *natis*) refers to the prominence produced by the underlying gluteal muscles. A line (– – –) joining the most superior parts of the iliac crests usually crosses the fourth lumbar vertebra, generally at the interval between the spinous processes of the third and fourth lumbar vertebrae. Consequently, it serves as a useful guide for determining the site for inserting a lumbar puncture needle to obtain a sample of cerebrospinal fluid (Fig. 5-60). Observe the prominent gluteal fold that delimits the buttock inferiorly. The groove or crease inferior to the gluteal fold is called the gluteal sulcus; it separates the buttock from the posterior aspect of the thigh.

gluteal muscles atrophy (*e.g.*, waste away owing to injury to the gluteal nerves).

The lesser trochanter of the femur projects from the posteromedial surface of the femur at the inferior end of the intertrochanteric crest (Figs. 4-1, 4-2, and 4-8). It is located in the angle between the neck and body of the femur.

The body (shaft) of the femur is slightly bowed anteriorly and is narrowest at its midpoint. Its middle two-quarters are approximately circular in cross-section. Inferior to the neck, the body is smooth and featureless except for a *rough ridge of bone,* called the **linea aspera** (L. rough line), in the middle of its posterior surface (Figs. 4-2 and 4-8*B*). The linea aspera has medial and lateral lips which diverge inferiorly to form the **supracondylar lines** of the femur (Fig. 4-2). The body of the femur is not usually palpable because it is so well covered with large muscles.

The pectineal line of the femur runs from the lesser trochanter to the medial lip of the linea aspera (Fig. 4-2). The tendon of the pectineus muscle inserts into it (Fig. 4-20).

The distal end of the femur is broadened for articulation with the tibia (Figs. 4-1, 4-2, and 4-8). Two large, oblong **condyles** (G. knuckles) project posteriorly and are separated by a deep U-shaped **intercondylar notch** (Figs. 4-2 and 4-8*B*). The medial and lateral condyles blend with each other anteriorly, and with the body of the femur superiorly. Although the articular surfaces are confluent anteriorly (Figs. 4-1 and 4-8*A*), each condyle is separated from the **patellar surface** by a slight groove. The patellar surface is where the patella (kneecap) slides during flexion and extension of the leg at the knee joint. The lateral and medial margins of the patellar surface can be palpated when the leg is flexed.

The adductor tubercle, a small prominence of bone (Figs. 4-1, 4-2, and 4-8*A*), may be felt at the superior part of the medial femoral condyle. Stand beside a skeleton and observe its adductor tubercle as you feel yours.

The medial and lateral ***condyles*** *of the femur are subcutaneous and easily palpable.* Palpate them as you flex and extend your knee joint. At the center of each condyle is a prominent **epicondyle,** to which the tibial and fibular *collateral ligaments* of the knee joint are attached. The medial and lateral epicondyles are easily palpable.

The femur is large and strong, particularly its body, but a **violent direct injury may fracture the femur** and it may take up to 20 weeks for firm union of the fragments to occur.

Fractures of the neck of the femur, or between the greater and lesser trochanters (***intertrochanteric fractures***), or through the trochanters (***pertrochanteric fractures***), are common in persons over 60 years of age. Fractures of the neck of the femur usually result from indirect violence and often occur through tripping over something.

Fractures of the neck of the femur are more common in older women than in men because their bones become markedly weakened owing to senile and **postmenopausal osteoporosis.** In this condition, bone resorption is greater than bone formation. When one hears that an old person has a "broken hip", the usual injury is a fracture of the femoral neck (see Case 4-1 on page 560).

The distal end of the femur constantly undergoes ossification just before birth. The visibility of this center of ossification in radiographs is commonly used as medicolegal evidence that a newborn infant found dead was viable.

FASCIA OF THE THIGH

THE SUPERFICIAL FASCIA OF THE THIGH

The subcutaneous connective tissue of the thigh, including that over certain bones and prominences, constitutes the superficial fascia of the thigh.

The superficial fascia lies deep to the dermis of the skin and consists of loose connective tissue, containing a considerable amount of fat. Over the ischial tuberosities (Figs. 4-2 and 4-9) the fat is within much fibrous tissue.

In certain regions the superficial fascia splits into two layers, between which run the superficial vessels and nerves (Fig. 4-10). These layers are thick in the inguinal region where the superficial layer is continuous with the superficial fascia of the abdomen (Figs. 2-7 and 2-22). Also between the layers of superficial fascia are the **superficial inguinal lymph nodes** and the great saphenous vein.

The superficial veins of the lower limb terminate in the two **saphenous veins** (Fig. 4-10), which drain most of the blood from the superficial fascia. *There is only one saphenous vein in the thigh because the small saphenous vein passes from the foot to the posterior aspect of the knee, where it ends in the popliteal vein* (Fig. 4-10*A*).

The Great Saphenous Vein (Figs. 4-10 to 4-16). This large vein, *the longest in the body,* is also called the long saphenous vein. It ascends from the foot to the groin in the subcutaneous connective tissue.

The great saphenous vein begins at the medial end of the dorsal venous arch of the foot and passes anterior to the medial malleolus of the

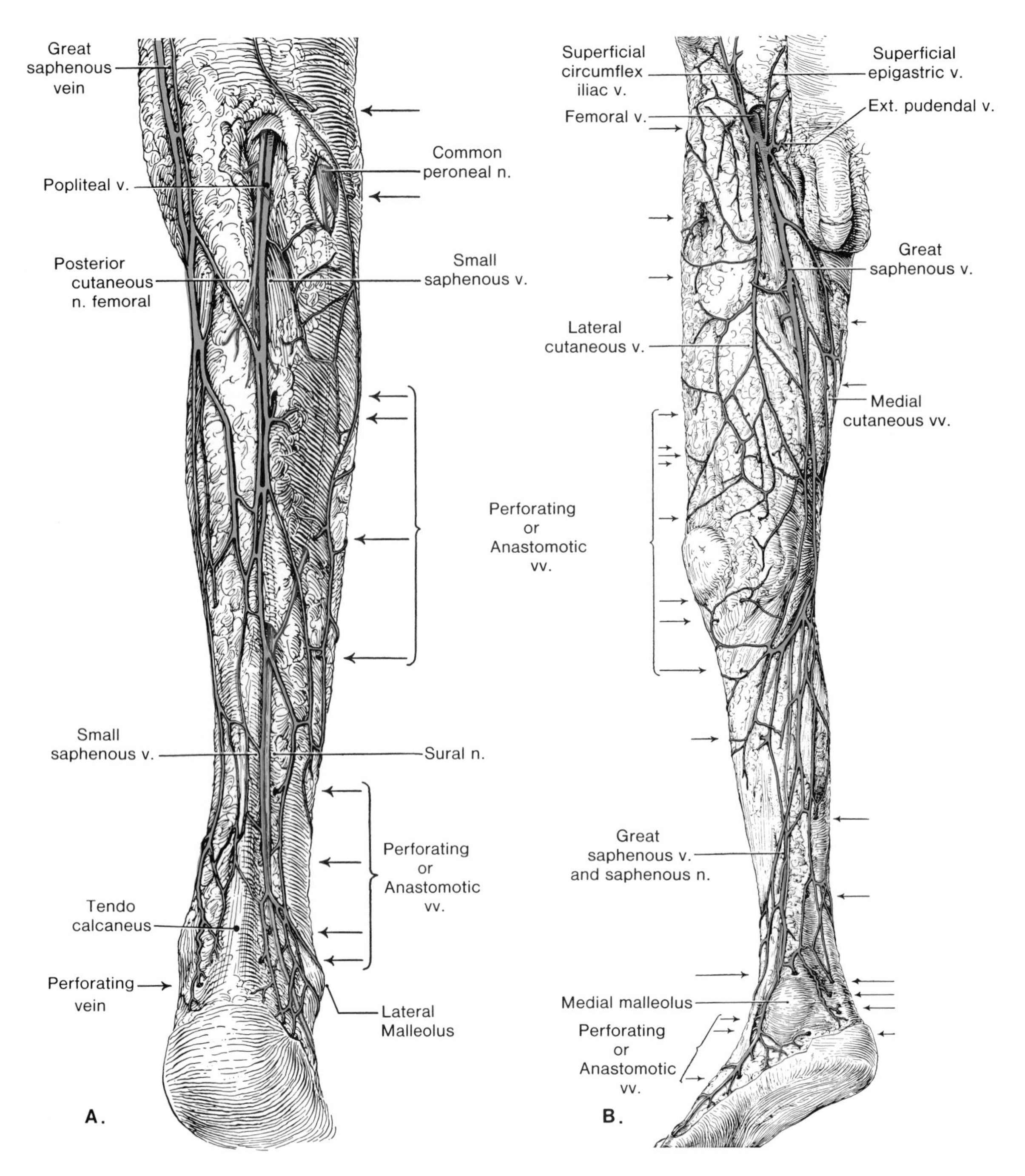

Figure 4-10. Drawings of the superficial veins of the lower limb; some nerves are also shown. *A*, posterior view. *B*, anteromedial view. Note the clinically important relationship of the great saphenous vein to the medial malleolus (also see Fig. 4-11). The *arrows* indicate where perforating veins pierce the deep fascia, bringing the superficial and deep veins into communication with each other. Note that the short saphenous vein accompanies the sural nerve and ends posterior to the knee in the popliteal vein in the popliteal fossa. Often the saphenous veins become dilated and tortuous and their valves become incompetent (*i.e.*, their cusps do not meet and close the vein). Veins that become dilated and tortuous are referred to as varicose veins. Note that the saphenous nerve (*yellow*) descends with the great saphenous vein on the medial side of the leg (also see Fig. 4-14).

tibia (Figs. 4-10 and 4-11*B*). It then ascends obliquely across the inferior third of the tibia to the medial aspect of the knee. Here it lies superficial to the medial epicondyle, about 10 cm posterior to the medial border of the patella (Fig. 4-11*A*). From here it runs superolaterally to the anterior midline of the thigh and the **saphenous opening** (fossa ovalis) in the deep fascia (Figs. 4-11*A*, 4-13, and 4-14).

The great saphenous vein perforates the sieve-like **cribriform fascia** (L. *cribum*, a sieve) and the femoral sheath (Fig. 4-16), parts of the deep fascia of the thigh, and **ends in the femoral vein** (Fig. 4-11). The cribriform fascia is a thin portion of the deep fascia that covers the saphenous opening (Figs. 4-10, 4-11, and 4-15). It is perforated by the great saphenous vein, one or more superficial arteries, and some lymph vessels (Fig. 4-13).

In Figure 4-10*A* observe that the great saphenous vein anastomoses freely with the small saphenous vein. Examine the clinically important **perforating veins** (anastomotic veins) which *connect the superficial veins with the deep veins* (Fig. 4-10). The main perforating veins from the great saphenous vein are arranged in three sets: one related to the adductor canal (Fig. 4-17), one related to the calf muscles, and one just proximal to the ankle joint.

The great saphenous vein is often duplicated, especially distal to the knee, and has 10 to 20 valves which

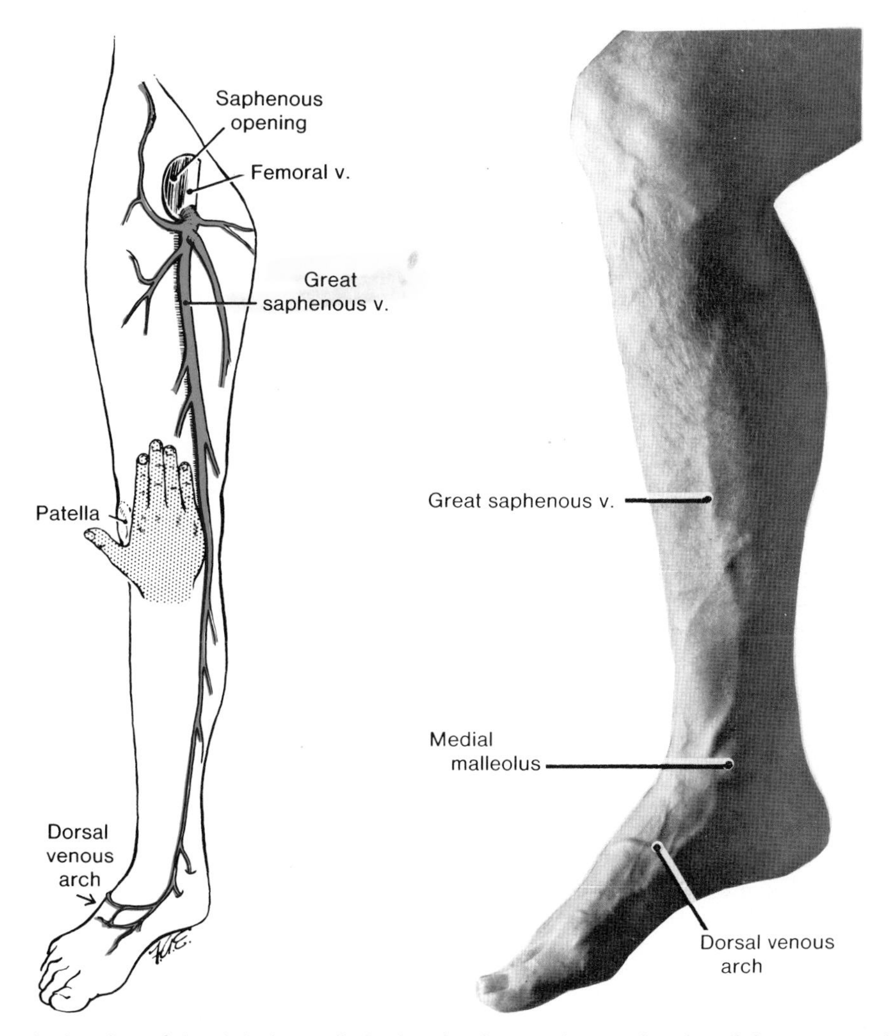

Figure 4-11. *A*, drawing of the right lower limb showing how to locate the site of the great saphenous vein at the knee. Observe that it is located about 10 cm (*i.e.*, about a handbreadth) posterior to the medial border of the patella (kneecap). *B*, photograph of the right leg of a 75-year-old man. Note that the great saphenous vein begins at the medial end of the dorsal venous arch of the foot and passes anterior to the medial malleolus. It passes posterior to the knee, spirals around the medial convexity of the thigh, and ends by passing through the cribriform fascia to open into the medial side of the femoral vein. The cribriform fascia is a thin layer of deep fascia that covers the saphenous opening (also see Figs. 4-12 and 4-13).

are more numerous in the leg than in the thigh. *The perforating veins also have valves.*

When the valves of the perforating veins in the lower limb become "**incompetent**" (*i.e.*, dilated so that their cusps do not meet and close the veins), contractions of the calf muscles which normally propel the blood superiorly cause a **reverse flow through the perforating veins** (*i.e.*, deep to superficial). This makes the perforating and superficial veins become tortuous and dilated (*i.e.*, produce **varicose veins**). For a typical history of a patient with varicose veins, and a description of the signs and symptons of **thrombophlebitis**, see Case 1-8 on page 142.

Prior to therapy for varicose veins, radiographic visualization of the veins after injection of a radiopaque substance, a technique called **venography**, is often done to locate all the perforating veins which are to be treated surgically (*e.g.*, ligated), or medically (injected with corrosive fluids), to prevent the passage of blood from the deep to the superficial veins.

Vein grafts using the great saphenous vein have been used to bypass obstructions in blood vessels (*e.g.*, atheromatous occlusion of the femoral and coronary arteries).

Coronary bypass surgery is becoming relatively common in some countries. When a portion of the great saphenous vein is removed and inserted as a bypass, *the vein is reversed so that its valve cusps do not obstruct blood flow.* Following removal of the great saphenous vein for vein grafting, blood from superficial parts of the lower limb reach the deep veins via the perforating veins (Fig. 4-10).

It is clinically very important to know that *the great saphenous vein lies immediately anterior to the medial malleolus* (Figs. 4-10 and 4-11). Even when it may not be visible in infants and obese persons, or in patients in shock whose veins are collapsed, the great saphenous vein can always be located and punctured by making a skin incision anterior to the medial malleolus. This common clinical procedure, often referred to as a "**saphenous cutdown**," is used to insert a cannula for prolonged administration of blood, plasma expanders, electrolytes, or drugs.

In Figures 4-10*B* and 4-14, observe that **the saphenous nerve accompanies the great saphenous vein anterior to the medial malleolus.** Should this nerve be caught by a ligature during a saphenous cutdown, the patient is likely to complain of pain along the medial border of the foot.

The Small Saphenous Vein (Fig. 4-10*A*). This vein, also referred to as the short saphenous vein, *begins posterior to the lateral malleolus.* It is formed by the union of veins arising from the lateral part of the **dorsal venous arch** (Fig. 4-11*A*).

The small saphenous vein passes along the lateral side of the foot with the sural nerve, posterior to the **lateral malleolus** (Fig. 4-10*A*), and ascends along the lateral side of the **tendo calcaneus** (Achilles tendon).[1] The small saphenous vein passes to the **popliteal fossa**, the diamond-shaped area posterior to the knee, where it perforates the deep popliteal fascia and **ends in the popliteal vein** (Fig. 4-10*A*).

The small saphenous vein has several communications with the great saphenous vein and the deep veins (Fig. 4-10*A*). Just before piercing the popliteal fascia, the small saphenous vein frequently gives off a branch which unites with another vein to form the **accessory saphenous vein**. When present, this vein becomes the main communication between the great and small saphenous veins.

In the standing position, the venous return from the lower limbs depends almost completely on muscular activity, especially of the calf muscles. The action of this "*calf pump*" is assisted by the tight sleeve of deep fascia surrounding these muscles.

Varicose veins are common in the posterior and medial parts of the lower limb, particularly in older persons. They cause considerable discomfort. Varicose saphenous veins have a caliber greater than normal and their cusps are incompetent; hence they allow blood to run from the deep veins into the superficial veins.

A localized dilation of the terminal part of the great saphenous vein, known as **a saphenous varix**, causes a swelling in the femoral triangle, located just inferior to the inguinal ligament (Fig. 4-15). A saphenous varix may be confused with other groin swellings (*e.g.*, a femoral hernia). Varicose veins elsewhere in the lower limb usually indicate the cause of the swelling.

Varicose veins have various causes, e.g., where the normal venous return is impeded as in con-

[1] This common tendon for the gastrocnemius and soleus muscles is often referred to as Achilles tendon after the mythical Greek warrior who was vulnerable only in the heel. You will also hear about the Achilles reflex (ankle jerk), a contraction of the calf muscles that occurs when the tendo calcaneus is struck sharply.

stipated persons, pregnant females, and persons with large abdominal tumors. Varicose veins often develop in members of the same family, which may indicate a genetic weakness in the walls of the veins. *Varicose veins and hemorrhoids (varicose rectal veins) are often associated.*

Thrombophlebitis (inflammation of a vein with secondary thrombus formation) of the deep veins of the leg destroys their valves, resulting in much of the blood from the leg being returned to the femoral vein via the superficial veins. This causes them to become varicose. Should such a **thrombus** break loose, it will be carried through the right side of the heart to the lung. Here it will be stopped as the branches of the pulmonary artery become progressively smaller. If the **pulmonary embolus** is small it may produce few or no symptoms, but if it is very large it may result in sudden death.

The Cutaneous Nerves of the Thigh (Figs. 4-13 to 4-16). Several cutaneous nerves in the superficial fascia supply the skin on the anterior, medial, and lateral aspects of the thigh.

The ilioinguinal nerve is distributed to the skin of the superomedial area of the thigh. Femoral branches of the **genitofemoral nerve** supply the skin just inferior to the middle part of the inguinal ligament (Fig. 4-16).

The lateral femoral cutaneous nerve enters the thigh region deep to the lateral end of the inguinal ligament, near the anterior superior iliac spine (Figs. 4-14 to 4-17), to supply the skin on the anterior and lateral aspects of the thigh.

The femoral nerve, through its anterior cutaneous branches, supplies the skin on the anterior and medial aspects of the thigh (Fig. 4-14).

Knowledge of the anatomy of the lateral femoral cutaneous nerve is clinically important in a condition called **meralgia paresthetica**, a tingling and painful itching in the anterolateral region of the thigh, the area of distribution of this nerve (Figs. 4-14 to 4-16). This condition often occurs as people get older and fatter, especially when the abdomen bulges over the inguinal ligament and compresses this nerve as it enters the thigh deep to the lateral end of the inguinal ligament (Fig. 4-16).

THE DEEP FASCIA OF THE THIGH (Figs. 4-13 to 4-18)

The deep fascia of the thigh, known as the **fascia lata** (L. broad band), is a *strong, dense, broad layer* which invests the muscles of the thigh like a stocking. The fascia lata is extremely strong laterally where it runs from the tubercle of the iliac crest to the tibia (Figs. 4-1 and 4-21*A*). This part of the fascia lata, known as the **iliotibial tract** (Fig. 4-15), receives tendinous reinforcements from the tensor fasciae latae and gluteus maximus muscles (Figs. 4-21*A* and 4-40). The distal end of the straplike iliotibial tract is attached to the lateral condyle of the tibia (Figs. 4-19 and 4-21*A*). To palpate your iliotibial tract, raise your heel from the floor with your knee flexed (Fig. 4-25).

Just posterior and inferior to the anterior superior iliac spine, the fascia lata encases the **tensor fasciae latae muscle** (Figs. 4-17 and 4-21*A*). *The tensor fasciae latae muscle pulls on the iliotibial tract, thereby steadying the trunk on the thigh and preventing posterior displacement of the iliotibial tract by the gluteus maximus muscle*, three-quarters of which inserts into the iliotibial tract (Fig. 4-40).

The Saphenous Opening in the Fascia Lata (Figs. 4-11 to 4-13). Just inferior to the inguinal ligament, there is a *gap or deficiency in the fascia lata* or deep fascia of the thigh, known as the **saphenous opening**, through which the great saphenous vein passes to join the femoral vein. The center of the saphenous opening is located about 6 cm inferolateral to the pubic tubercle (Figs. 4-1 and 4-5). The saphenous opening is about 4 cm long and 1 to 2 cm wide (Figs. 4-11 and 4-13). Its medial margin is smooth but its superior, lateral, and inferior margins form a sharp crescentic edge, called the **falciform margin** (L. *falx*, sickle + *forma*, form). This sickle-shaped margin of the saphenous opening is joined to its medial margin by fibrous and fatty tissue known as the **cribriform fascia** (Figs. 4-12 and 4-13). This thin part of the deep fascia spreads over the saphenous opening.

The large powerful thigh muscles can be organized into three main groups (*anterior, medial, and posterior*) on the basis of their location, action, and nerve supply (Tables 4-1, 4-2, and 4-4). These muscle groups are separated by three intermuscular septa (Fig. 4-18). The lateral intermuscular septum is strong; the other two are relatively weak.

THE ANTERIOR THIGH MUSCLES

The anterior group of muscles in the thigh consists of the **tensor fasciae latae**, the **sartorius**, and the **quadriceps femoris** (Figs. 4-15 and 4-17).

The anterior part of the thigh also contains the terminations of the two muscles of the posterior abdominal wall (*iliacus and psoas major*) which together are called the **iliopsoas muscle** (Table 4-1).

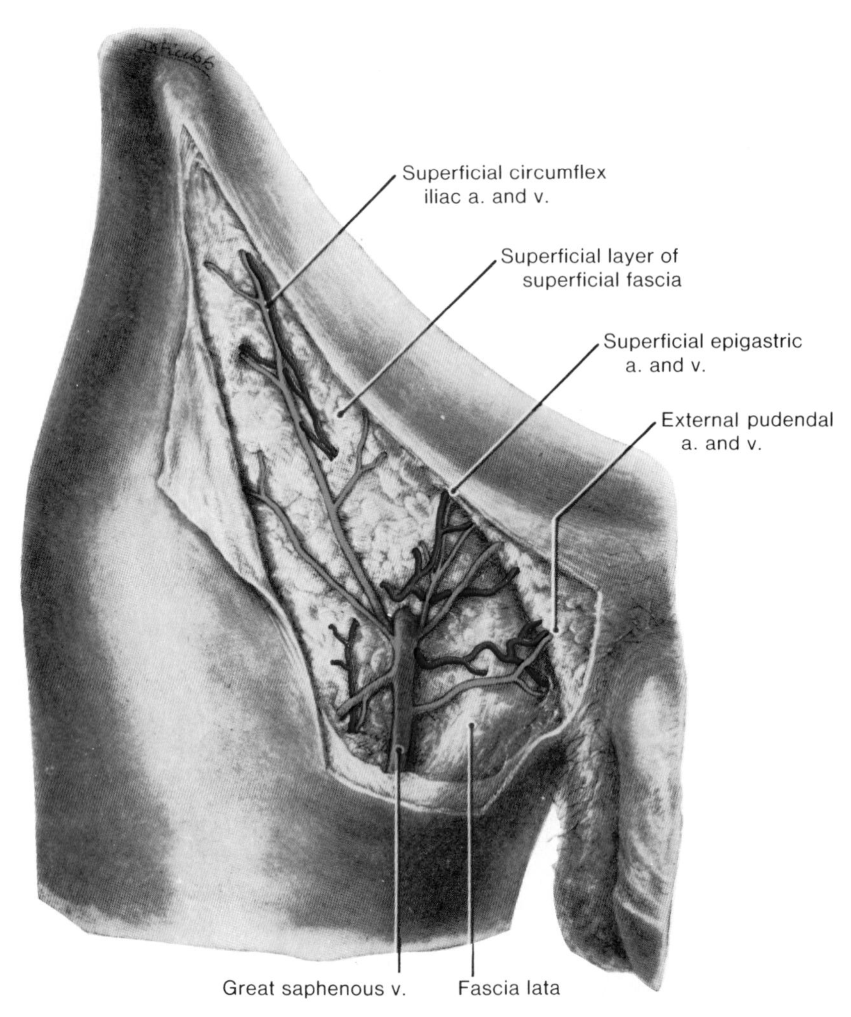

Figure 4-12. Drawing of a dissection of the superficial inguinal arteries and veins. The arteries are branches of the femoral artery and the veins are tributaries of the great saphenous vein. Observe that this vein passes through the cribriform fascia to enter the femoral vein (also see Fig. 4-10*B*). The cribriform fascia is a thin, sieve-like layer of deep fascia that fills the saphenous opening, an aperture through which pass the great saphenous vein and other vessels (Fig. 4-13).

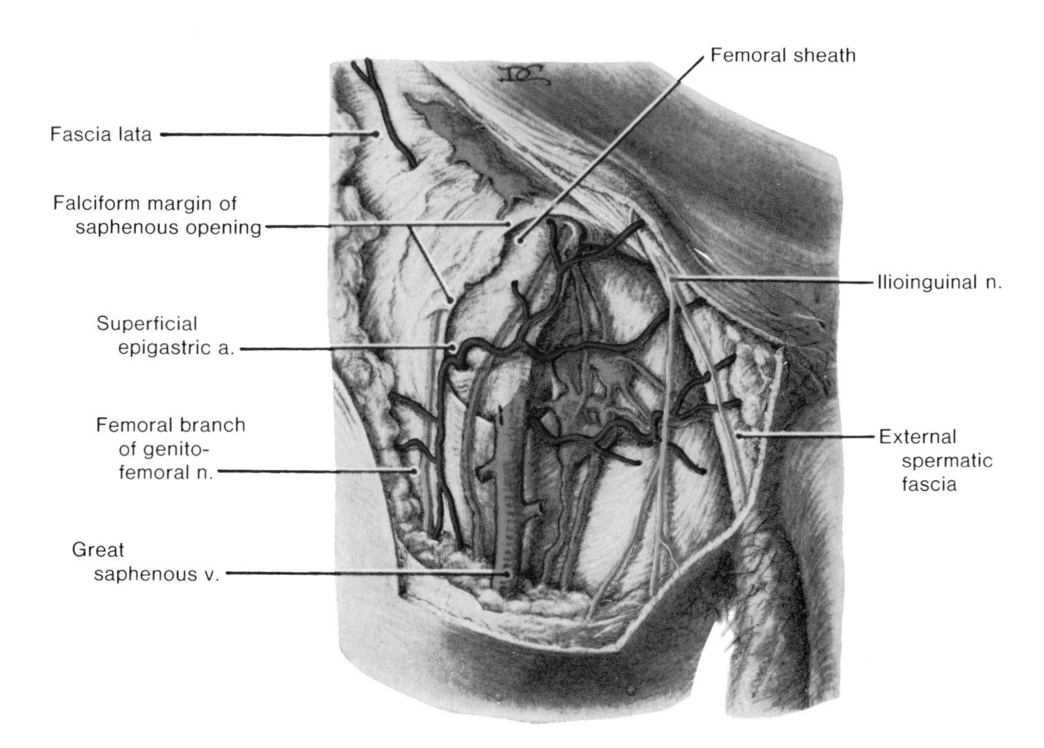

Figure 4-13. Drawing of a deeper dissection of the superficial inguinal arteries, veins, and nerves. Observe the saphenous opening in the fascia lata through which the great saphenous vein and other vessels pass through the cribriform fascia to enter the femoral vein (Fig. 4-10*B*). Note that the superior, inferior, and lateral margins of the saphenous opening form a sharp crescentic edge of deep fascia, known as the falciform margin (also see Fig. 4-16). Observe the arrangement of the inguinal lymph nodes (*green*). See Figure 4-35 for a description of these important lymph nodes.

Table 4-1
Anterior Thigh Muscles

Muscle	Origin	Insertion	Nerve Supply	Actions
Iliopsoas				
Psoas major (Figs. 4-17 and 4-21*A*)	T12 to L5 vertebrae and intervertebral discs (Fig. 4-26)	Lesser trochanter of femur (Fig. 4-28)	Ventral rami of lumbar nerves (**L1**, **L2**, and L3)[1]	These two muscles act conjointly in flexing the thighs
Iliacus (Figs. 4-15, 4-17, and 4-21)	Iliac crest, iliac fossa, and ala of sacrum (Fig. 4-19)	Tendon of psoas major and femur, inferior and anterior to lesser trochanter (Fig. 4-28)	Femoral nerve (**L2** and L3)	
Quadriceps femoris				
Rectus femoris (Figs. 4-17 and 4-22)	Anterior inferior iliac spine and groove superior to acetabulum (Fig. 4-20)	Base of patella and via patellar ligament into tibial tuberosity (Figs. 4-19 and 4-45)	Femoral nerve (L2, **L3**, and **L4**)	Extend leg; the rectus femoris also flexes the thigh
Vastus lateralis (Figs. 4-17, 4-21, and 4-22)	Greater trochanter and lateral lip of linea aspera of femur (Fig. 4-20)			
Vastus medialis (Figs. 4-21 to 4-23)	Intertrochanteric line (Fig. 4-1) and medial lip of linea aspera of femur (Fig. 4-20)			
Vastus intermedius (Fig. 4-21)	Anterior and lateral surfaces of body of femur (Fig. 4-20)			
Tensor fasciae latae (Figs. 4-12 and 4-21)	Anterior superior iliac spine and external lip of iliac crest (Fig. 4-20)	Iliotibial tract which inserts into lateral condyle of tibia (Figs. 4-19, 4-21, and 4-45)	Superior gluteal nerve (L4 and L5)	Abducts and flexes thigh; helps to keep knee extended; steadies trunk on the thigh
Sartorius (Figs. 4-15, 4-21, and 4-22)	Anterior superior iliac spine (Fig. 4-20)	Superior part of medial surface of tibia (Fig. 4-24)	Femoral nerve (L2 and L3)	Flexes thigh and leg; aids in abducting and rotating thigh

[1] In this and subsequent tables, the numbers indicate the spinal cord segmental innervation of the nerves. For example, **L1**, **L2**, and L3 indicate that nerves supplying the psoas major muscle are derived from the first three lumbar segments of the spinal cord. The **bold face type** indicates the main segmental innervation. Note that the muscles are innervated from more than one segment of the spinal cord. Damage to these segments or to the motor nerve roots arising from them result in paralysis of the muscles concerned.

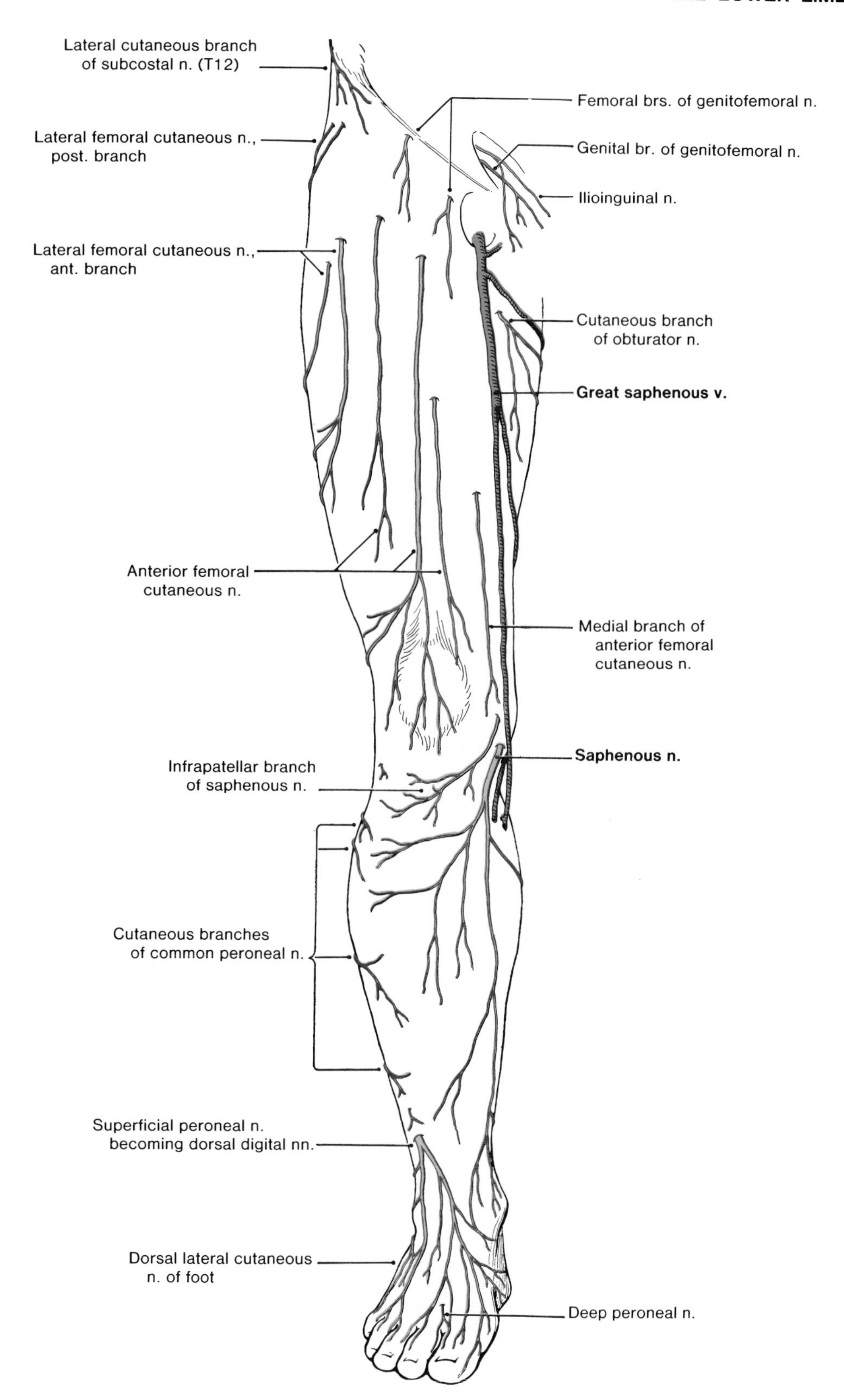

Figure 4-14. Drawing of an anterior view of the cutaneous nerves of the lower limb (see Fig. 4-42 for a posterior view). The superior half of the great saphenous vein is also shown. The saphenous opening is indicated (also see Figs. 4-11 and 4-13).

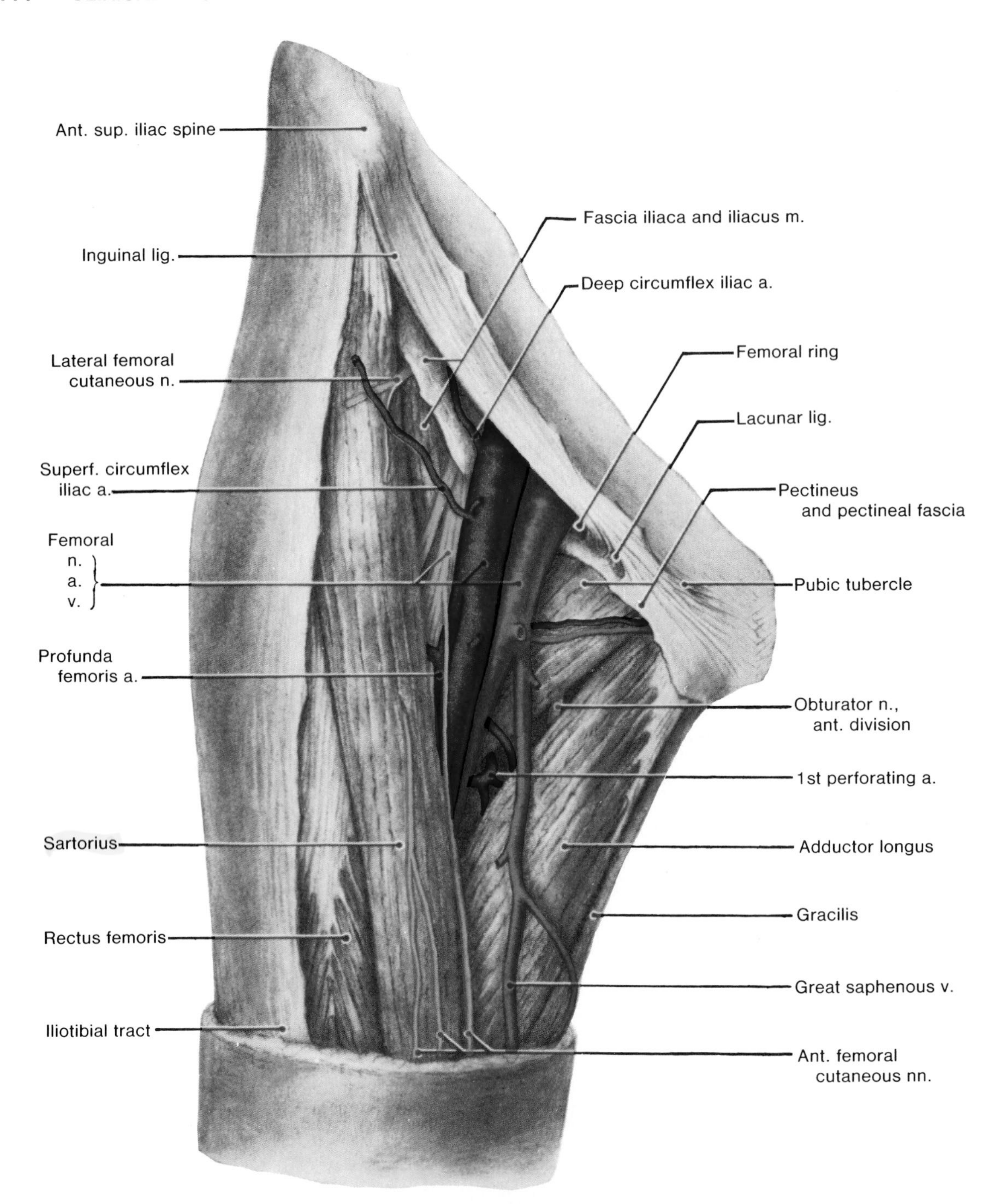

Figure 4-15. Drawing of a dissection of the femoral triangle. Observe its boundaries: the inguinal ligament superiorly; the medial border of the adductor longus medially; and the medial border of the sartorius, laterally. The head of the femur lies posterior to the point where the inguinal ligament crosses the femoral artery (Fig. 4-30). Note that the femoral artery and vein lie anterior to the fascia covering the iliopsoas and pectineus muscles, respectively, and that the femoral nerve lies posterior to it. *Observe that the femoral artery is about midway between the anterior superior iliac spine and the pubic tubercle* and that it disappears into the adductor canal where the sartorius muscle crosses the adductor longus muscle.

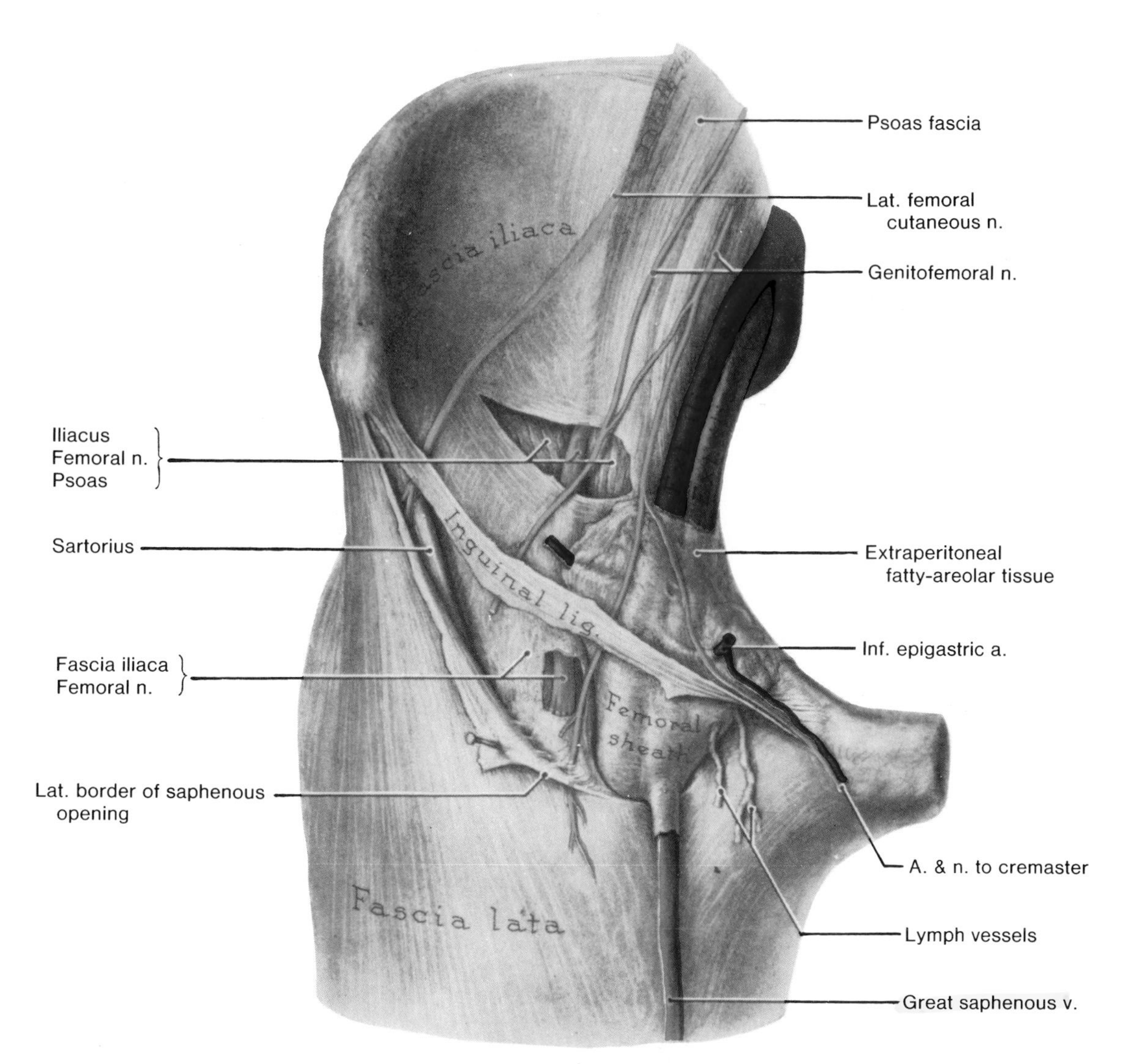

Figure 4-16. Drawing of a dissection of the femoral sheath. The falciform margin of the saphenous opening in the fascia lata is cut and reflected. Observe that the fascia iliaca or iliac fascia is continuous medially with the psoas fascia and that it is carried inferiorly, anterior to the iliacus muscle into the thigh. Observe the delicate funnel-shaped sac, called the femoral sheath, which is loosely adherent to the inguinal ligament anteriorly and to the pecten pubis (Fig. 4-1) posteriorly. Note that the femoral nerve, is external to the femoral sheath. The *contents of the femoral sheath* are illustrated in Figures 4-33 and 4-36.

THE PSOAS MAJOR MUSCLE (Figs. 4-16, 4-17, 4-21, 4-26, and Table 4-1)

This long, thick, powerful muscle passes from the abdomen into the thigh *deep to the inguinal ligament.* Acting conjointly with the iliacus, the psoas major is the *chief flexor of the thigh at the hip joint.*

Origin (Figs. 2-118 and 4-26). Transverse processes, sides of **vertebral bodies, and intervertebral discs of T12 to L5** vertebrae.

Insertion (Figs. 4-19 and 4-28). **Lesser trochanter of femur** via iliopsoas tendon.

Nerve Supply. Ventral rami of **lumbar nerves (L1, L2,** and L3).

Actions. Flexes the thigh and stabilizes the trunk on the thigh. The psoas major acts conjointly with the iliacus muscle. The muscle formed by these two muscles is called the *iliopsoas.*

THE ILIACUS MUSCLE (Figs. 4-15 to 4-17, 4-21, and Table 4-1)

This large triangular or fan-shaped muscle lies along the lateral side of the psoas major in the pelvis.

Origin (Figs. 2-118 and 4-19). **Iliac crest, iliac fossa, ala of sacrum,** and iliolumbar and ventral sacroiliac ligaments.

Insertion (Figs. 4-19 and 4-28). Most fibers insert into the lateral side of the **tendon of the psoas major.** Some fibers insert into the **femur, inferior and anterior to the lesser trochanter.**

Nerve Supply. Femoral nerve (L2 and L3).

Action. Flexes thigh (assisted by psoas major). When the iliacus and psoas major muscles of both sides act inferiorly, they **flex the vertebral column.** When these muscles act on one side, they bend the vertebral column laterally.

THE TENSOR FASCIAE LATAE MUSCLE (Figs. 4-17, 4-21 to 4-23, and Table 4-1)

This fusiform muscle lies on the lateral side of the anterior part of the thigh, enclosed between two layers of the fascia lata.

Origin (Fig. 4-20). **Anterior part of external lip of iliac crest,** lateral surface of **anterior superior iliac spine,** and the notch inferior to it.

Insertion (Fig. 4-19). **Iliotibial tract** which is attached to the lateral condyle of the tibia.

Nerve Supply. Superior gluteal nerve (L4 and L5).

Actions. Abducts and flexes thigh and *helps to keep the knee extended in erect posture by making the iliotibial tract taut.*

The tensor fasciae latae **also steadies the trunk on the thigh** and counteracts the posterior pull of the gluteus maximus on the iliotibial tract.

THE SARTORIUS MUSCLE (Figs. 4-17 to 4-22, 4-26, and Table 4-1)

This narrow, strap-like muscle is the *longest in the body* and is the most superficial muscle in the anterior part of the thigh. It was given its name because it is used to cross the legs in the tailor's cross-legged sitting position (L. *sartor,* a tailor). Throughout much of its course, **the sartorius covers the femoral artery** as it runs in the adductor (subsartorial) canal (Fig. 4-17).

Origin (Figs. 4-17 to 4-22). **Anterior superior iliac spine** and the superior part of the notch inferior to it.

Insertion (Fig. 4-24). **Superior part of the medial surface of tibia,** anterior to the insertion of the gracilis and semitendinosus muscles.

Nerve Supply. Femoral nerve (L2 and L3).

Actions. Flexes thigh and leg. It also aids in abducting the thigh and rotating it laterally. You use your sartorius muscles when you sit in a cross-legged position.

THE QUADRICEPS FEMORIS MUSCLE (Figs. 4-15, 4-17 to 4-23, 4-26, and Table 4-1)

The quadriceps femoris consists of four muscles **(rectus femoris, vastus lateralis, vastus medialis, and vastus intermedius).** *This great extensor muscle of the leg covers the anterior and lateral parts of the femur* (Fig. 4-18).

The names of the parts of the quadriceps muscle indicate their location: the **rectus femoris** (L. *rectus,* straight) has deep fibers that run straight down the thigh (Fig. 4-17); the **vastus lateralis** lies on the lateral side of the thigh; the **vastus medialis** covers the medial aspect of the thigh; and the **vastus intermedius** is located between the vastus medialis and the vastus lateralis (Fig. 4-21*B*). *Vastus* is a Latin term meaning great or large. *Quadriceps* means four (L. *quadri*) heads (L. *cipital* or *-ceps* from L. *caput,* head).

Origin (Figs. 4-19, 4-20, and 4-28). **Rectus femoris:** *straight head,* **anterior inferior iliac spine;** *reflected head,* **groove superior to acetabulum.**

The three vasti muscles (*lateralis, medialis, and intermedius*) **arise from the body of the femur** (see Table 4-1 and Figs. 4-19 and 4-20 for details).

A small muscle, the **articularis genus** (Fig. 4-21*C*), is sometimes blended with the vastus intermedius. It arises from the inferior part of the femur and inserts into the synovial capsule of the knee joint and the walls of the **suprapatellar bursa** (Fig. 4-108). This small muscle pulls the synovial capsule superiorly during extension of the leg at the knee joint.

Insertion (Figs. 4-19, 4-21, and 4-24). **Base of the patella and the tibial tuberosity.** The tendons of all four parts of the muscle unite.

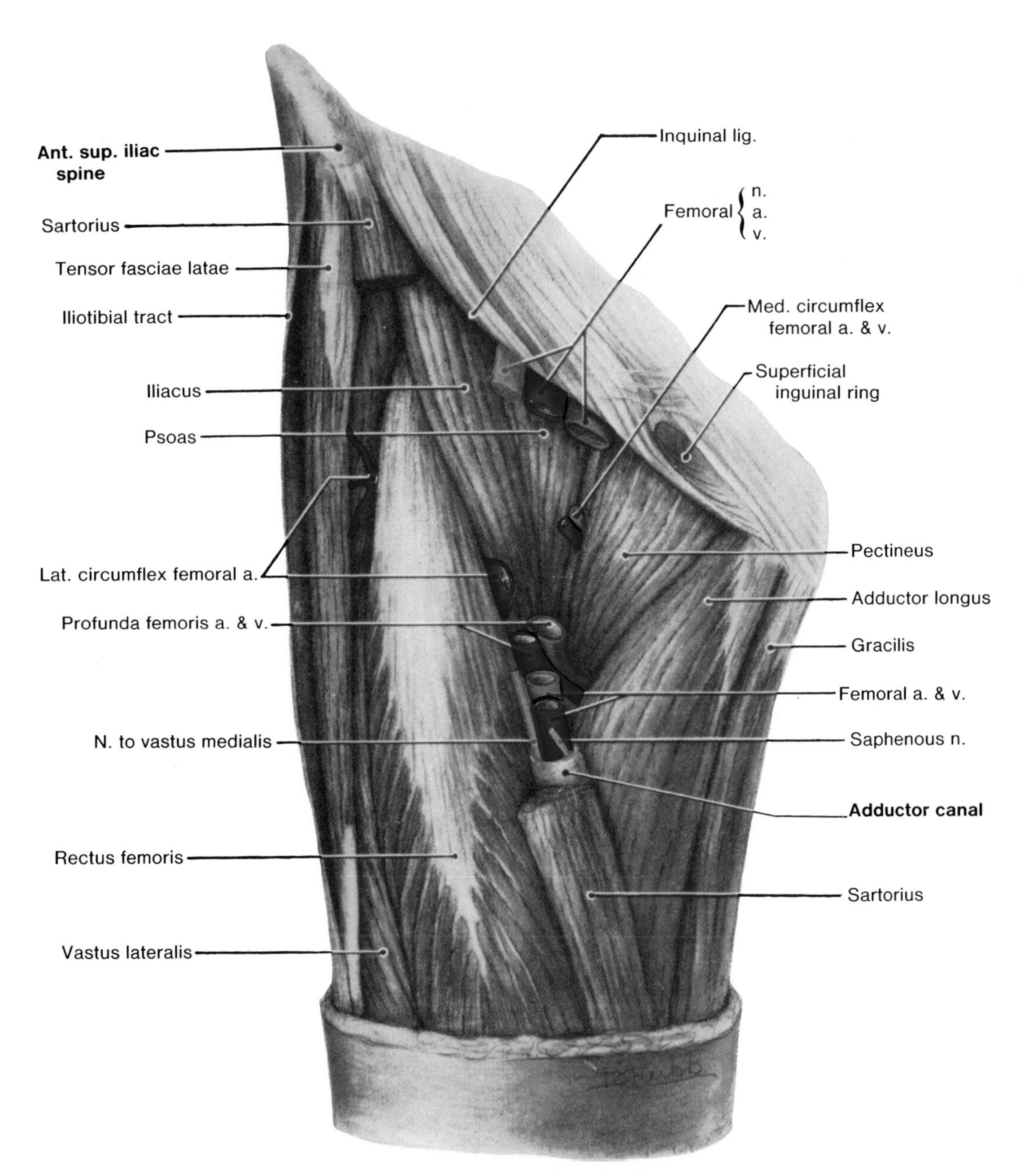

Figure 4-17. Drawing of a dissection of the femoral triangle. Sections are removed from the sartorius and the femoral vessels and nerve. The lateral boundary of the femoral triangle is the medial border of the sartorius and its medial boundary is the medial border of the adductor longus. Observe that the floor of the femoral triangle is composed of muscles (adductor longus, pectineus, psoas major, and iliacus). Note that the triangle is shallow at its base and deep at its apex (also see Fig. 4-30). Observe that the femoral artery is entering the adductor canal. It is accompanied through this canal by the femoral vein and saphenous nerve.

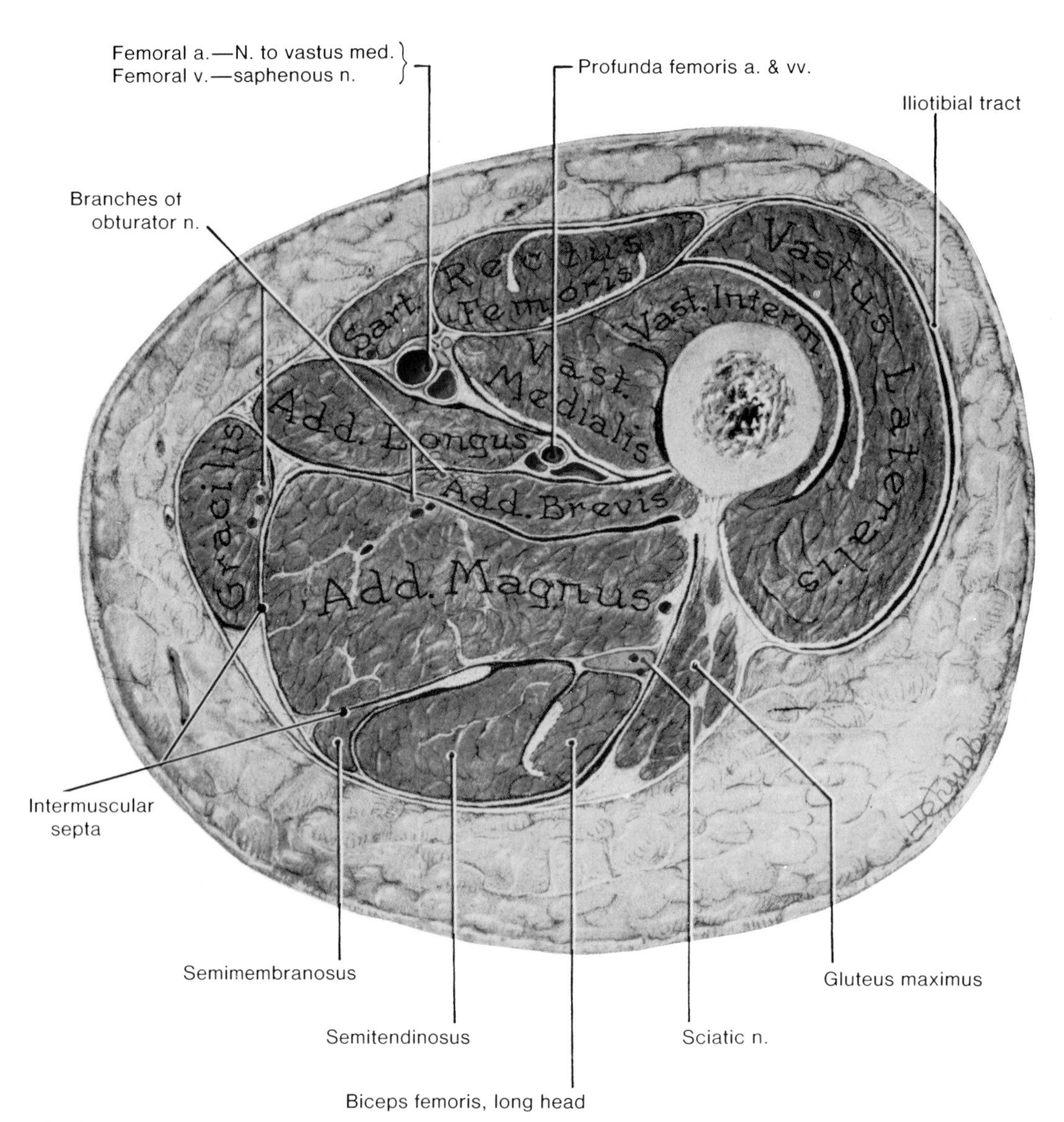

Figure 4-18. Drawing of a cross-section through the thigh of a woman, 10 to 15 cm inferior to the inguinal ligament. Observe the intermuscular fascial septa. Note the adductor longus muscle intervening between the femoral and profunda femoris vessels, and the adductor brevis muscle intervening between the anterior and posterior divisions of the obturator nerve. Note that the aponeurosis of the semimembranosus muscle is similar to the sciatic nerve and could be mistaken for it. Observe the vastus intermedius muscle arising from the anterior and lateral surfaces of the body of the femur. Note that the vastus medialis covers the medial surface of the body of the femur, but does not arise from it. Observe the femoral vessels and the saphenous nerve in the adductor canal (also see Fig. 4-17).

The quadriceps tendon is inserted into the patella. This *common tendon* continues inferiorly as the **patellar ligament** or ligamentum patellae which *inserts into the tibial tuberosity* (Fig. 4-21*B*). In addition, expansions of the aponeuroses of the vasti muscles, called the medial and lateral **retinacula of the patella**, insert into the condyles of the tibia.

Nerve Supply (Fig. 4-15). **All four components of the quadriceps femoris muscle are supplied by the femoral nerve** (L2, **L3**, and **L4**).

Actions. Extends leg at knee joint. All four parts of the quadriceps contribute to this action by pulling on the patella and through it the patellar ligament extends the leg at the knee joint. Only one part of the quadriceps (rectus femoris) crosses the hip joint; thus, **the rectus femoris also flexes the thigh.**

All parts of the quadriceps are used during climbing, running, jumping, and rising from a chair. Put your hands on your quadriceps as you rise from a chair and feel these large muscles contract.

If the quadriceps femoris is paralyzed the leg cannot be extended, but the patient can stand erect because the body weight tends to overextend the knee joint. The patient can also walk with short steps if the pelvis is rotated to prevent extension of the hip so far as to flex the knee.

Patients with paralysis of the quadriceps femoris muscle often press on the distal end of their thigh during walking to prevent flexion of their knee joint.

In football broadcasts, you often hear about an injury known as a "**hip pointer**." This is *a contusion or bruise over the iliac crest, particularly its anterior superior iliac spine,* from which the inguinal ligament and the sartorius muscle arise.

Another common term used by sports broadcasters is "**Charley horse.**" This is a contusion and tearing of muscle fibers that result in the formation of a **hematoma** (a local mass of blood that escapes into the muscle from damaged vessels). *The most common site of "Charley horse" involves the quadriceps muscle.* It is associated with localized pain and/or muscle stiffness, and commonly follows direct trauma (*e.g.*, a tackle in football).

The Patella (Figs. 4-21*A* and 4-23). The patella is a triangular sesamoid bone with its apex pointing inferiorly. It is **embedded in the quadriceps femoris tendon**. The *patellar ligament* (ligamentum patellae), which attaches the patella (L. little plate) to the **tibial tuberosity** (Fig. 4-24), is really just a continuation of the tendon of the quadriceps muscle.

The apex of the patella indicates the level of the knee joint when the patellar ligament is taut.

The patella is subcutaneous and can be easily palpated. It lies anterior to the distal end of the femur, hence it articulates posteriorly with the condyles of the femur (Fig. 4-1). The patella is thought to increase the power of the already strong quadriceps femoris muscle by increasing its leverage.

Palpate your patella as you flex your leg, noting that it is pulled down. Stand in a relaxed position and note that you can move your patella from side to side because the quadriceps femorus muscle is relaxed. Kneel on the floor and verify that it is the tibial tuberosity and the patellar ligament that bear most of the weight; not the patella.

The patella is cartilaginous at birth and becomes ossified during the 3rd to 6th years, frequently from more than one center. Although these centers usually coalesce forming a single bone, they may remain separate on one or both sides, giving rise to a bipartite or tripartite patella. An unwary observer might interpret this condition on a radiograph as a fracture of the patella.

A direct blow on the patella may fracture it in two or more fragments. The patella may also be fractured transversely by sudden contraction of the quadriceps (*e.g.*, when one slips and attempts to prevent a backward fall). In these cases the proximal fragment is pulled superiorly with the quadriceps tendon, and the distal fragment remains with the patellar ligament. This condition would have more tendency to occur in persons with unfused or poorly fused ossification centers in this bone.

Fracture of the patella with separation of the fragments leads to a decrease in the power of the quadriceps femoris muscle. Treatment consists of wiring the patellar fragments together or excising one or all fragments, depending on the severity of the injury.

Excision of the patella and repair of the quadriceps femoris tendon and patellar ligament result in practically no functional deficiency of the quadriceps femoris muscle.

Because of the tendency of the developmental parts of a fractured patella to separate, the position of extension (in which the quadriceps femoris is relaxed) is used to reduce a fractured patella.

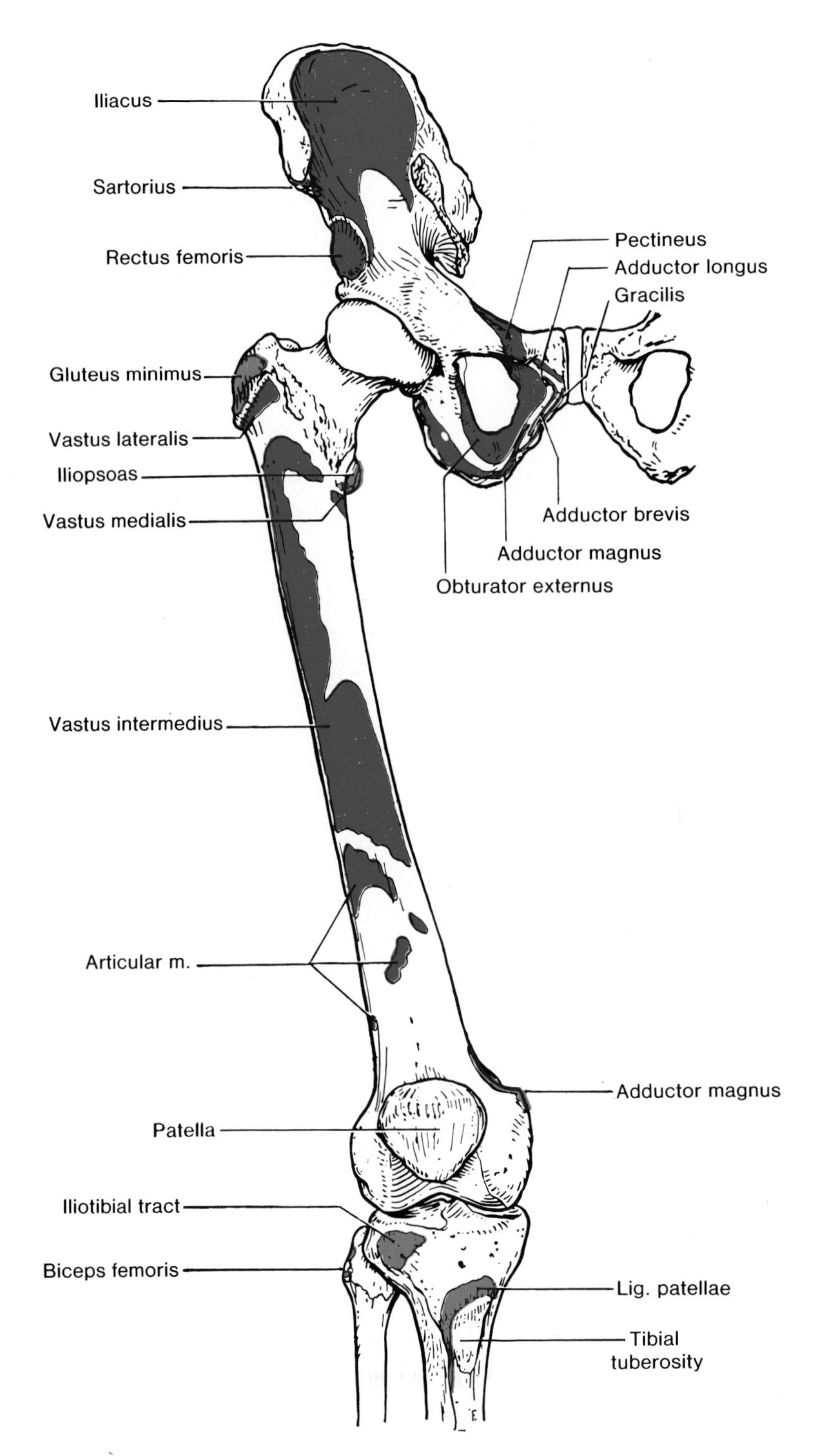

Figure 4-19. Drawing of an anterior view of the bones of the lower limb showing the attachments of muscles. Note that the rectus femoris arises from the ilium, whereas the vastus lateralis, medialis, and intermedius muscles arise from the femur (see Fig. 4-20 also). *Red* indicates origins of muscles and *blue*, insertions.

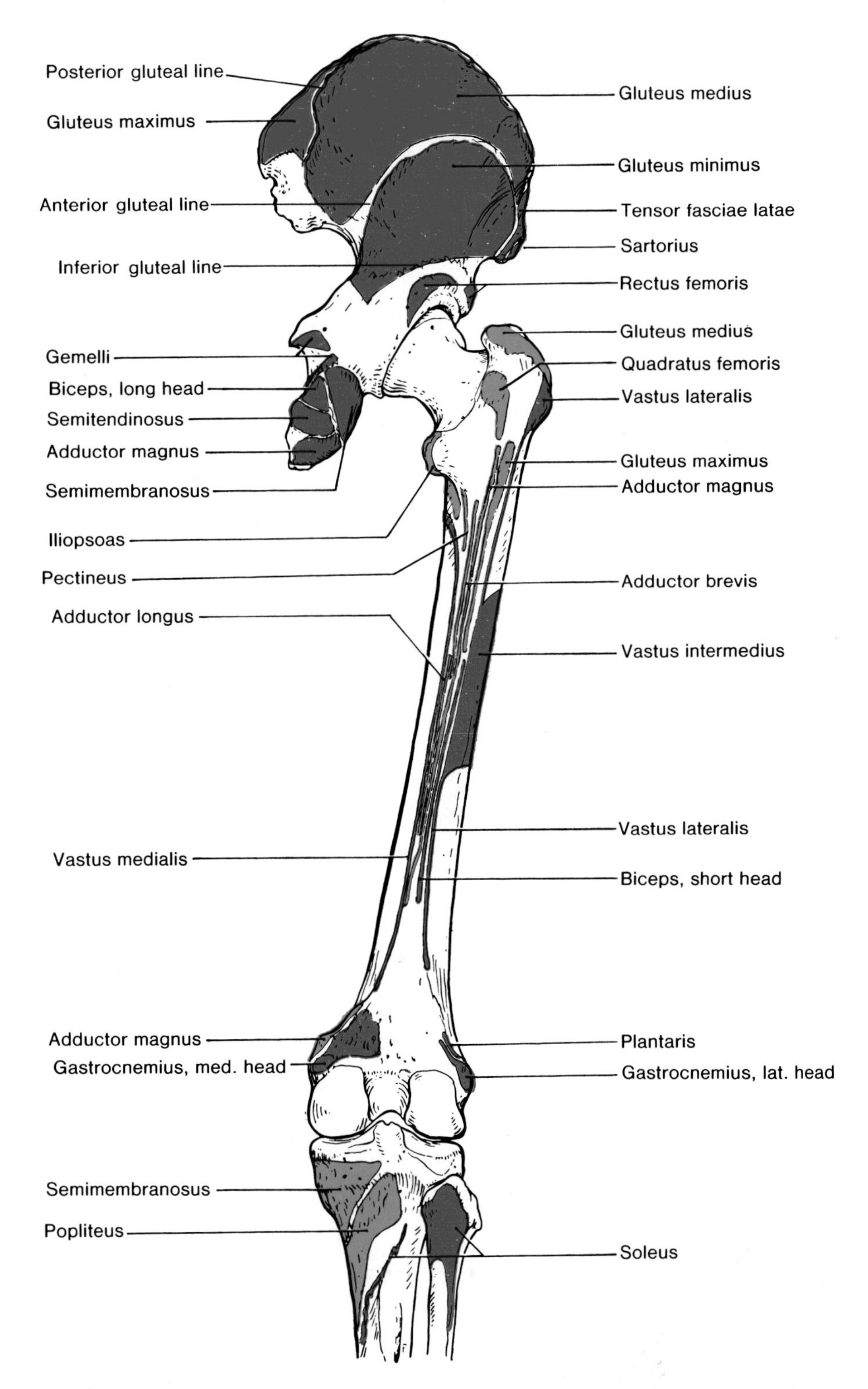

Figure 4-20. Drawing of a posterior view of the bones of the lower limb showing the attachments of muscles. Observe that several muscles are attached to the linea aspera (Figs. 4-2 and 4-8*B*). Note the extensive origin of the vastus lateralis, the largest of the four muscles making up the quadriceps femoris (also see Fig. 4-18).

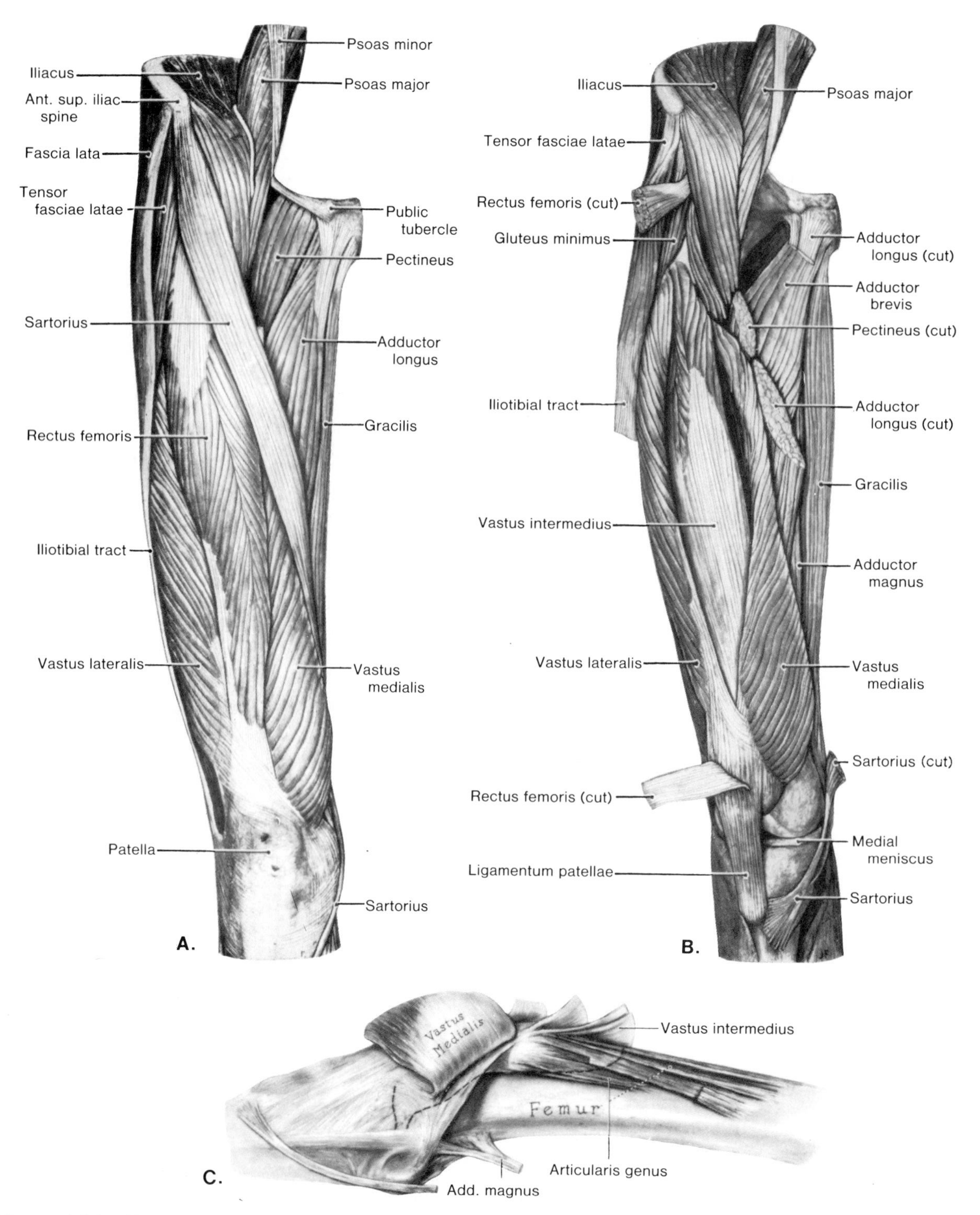

Figure 4-21. Drawings of the muscles of the anterior aspect of the thigh. *A*, superficial dissection. *B*, deeper dissection with sections of the sartorius, rectus femoris, pectineus, and adductor longus muscles excised. The pectineus, adductor longus, and gracilis muscles originate from a curved line on the pubic bone (Fig. 4-19). *C*, deep dissection of a medial view of the knee region showing the articularis genus muscle which sometimes blends with the vastus intermedius muscle. The articularis genus is inserted into the synovial capsule and retracts it during extension of the leg at the knee joint.

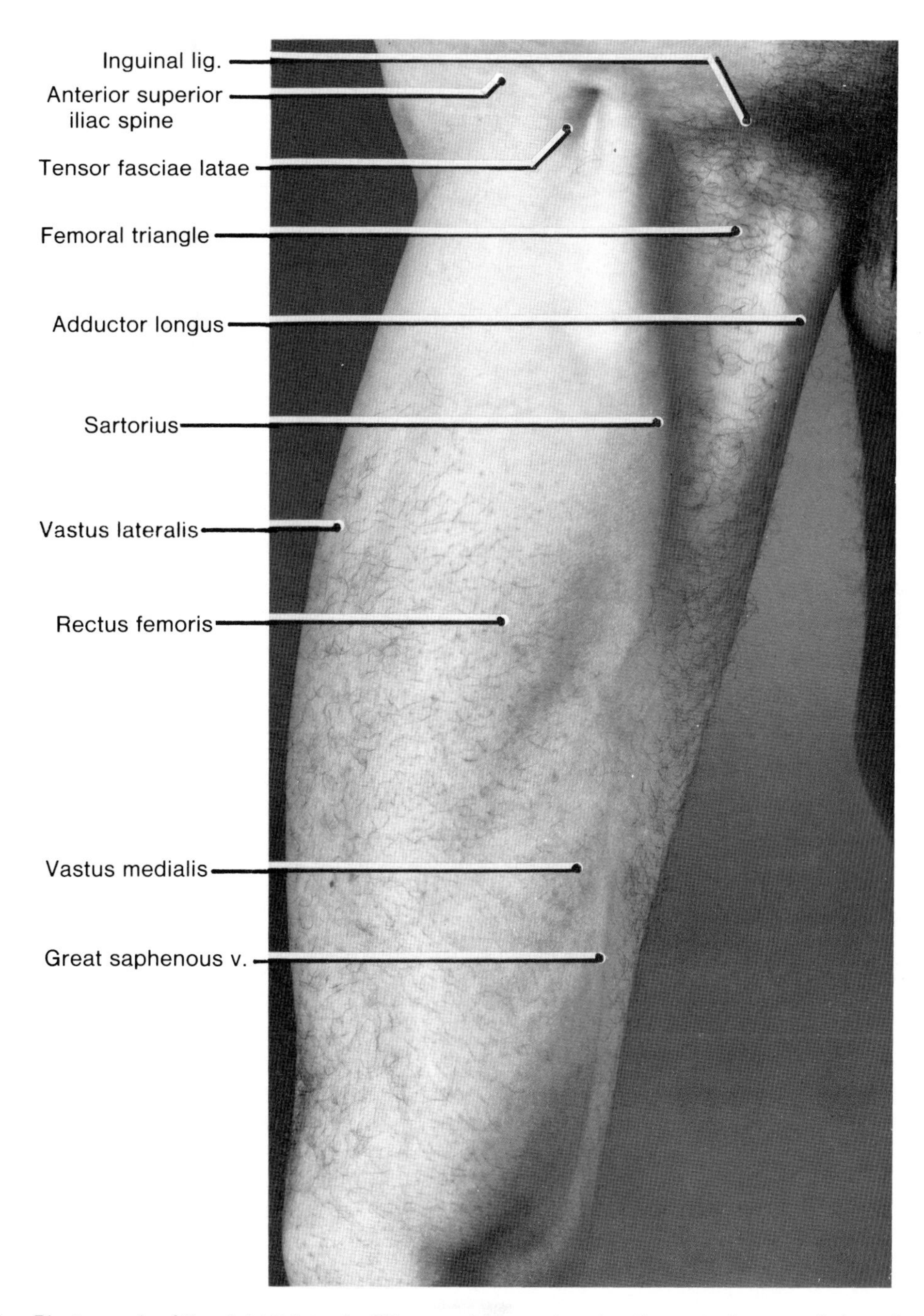

Figure 4-22. Photograph of the right thigh of a 27-year-old man showing the sartorius muscle in action. To display this muscle, he was asked to flex, abduct, and laterally rotate his thigh. This photograph was taken with his pelvis somewhat rotated to the left, bringing the anterior superior iliac spine *apparently* near the midline of the thigh. The femoral triangle can be seen as a depression inferior to the inguinal ligament which forms its base. The lateral boundary of this triangle is the sartorius and its medial boundary is the adductor longus (also see Fig. 4-15). The appearance of a lump in the femoral triangle may be the first indication of tuberculosis of the lumbar region of the vertebral column. An abscess caused by infection tends to spread from the lumbar vertebra to the psoas muscle. It then tracks along this muscle within the psoas fascia to the head of the femur, which lies posterior to the femoral triangle (Fig. 4-30).

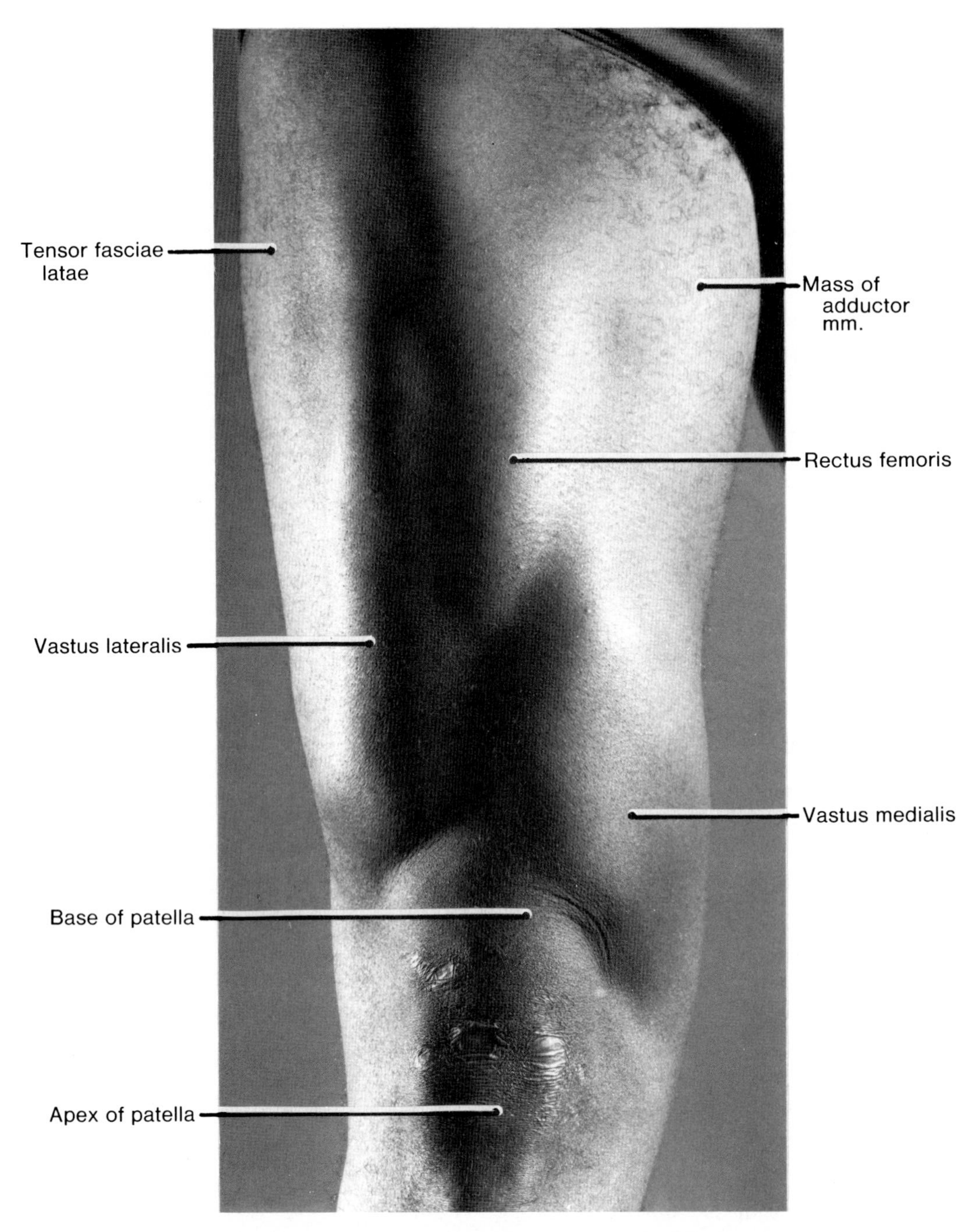

Figure 4-23. Photograph of the anterior aspect of the thigh and knee of a 35-year-old man showing the principal surface features. Compare with Figure 4-21*A*. The parts of the quadriceps femoris became visible when he extended his knee against resistance on his leg. Observe that the fleshy bulk of the vastus medialis extends more distally than does the vastus lateralis. This helps to counteract lateral displacement of the patella by the pull of the rectus femoris and vastus lateralis. When this muscle is relaxed, the patella is freely movable from side to side. All four parts of the quadriceps converge on the base of the patella, a sesamoid bone in the quadriceps tendon (Fig. 4-109). The deepest portion of the quadriceps (vastus intermedius) is not visible from the surface (view it in Fig. 4-21*B*).

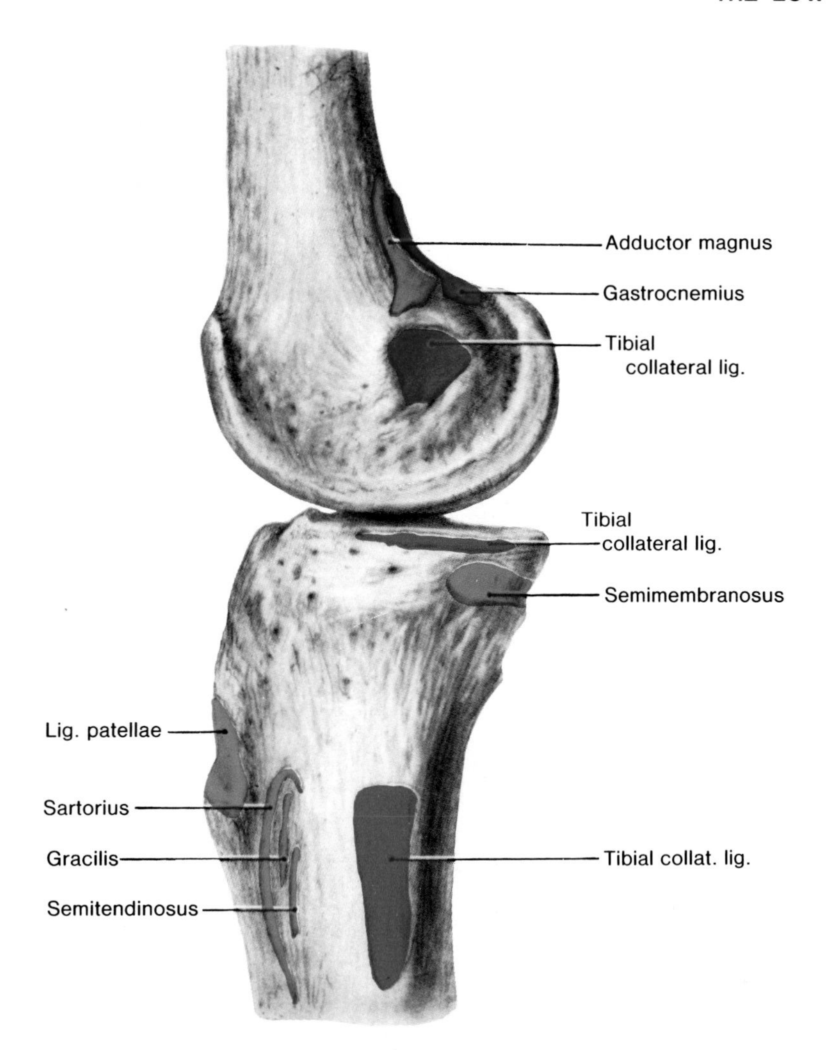

Figure 4-24. Drawing of a medial view of two of the bones of the right knee joint showing the sites of attachment of muscles and ligaments to them. The other bone involved in the knee joint is the patella (see Fig. 4-19).

Tapping the patellar ligament with a percussion hammer normally elicits the **quadriceps reflex** (patellar reflex, ***knee jerk***, knee reflex). This reflex is routinely tested during a physical examination as follows. The patient is seated on the edge of a table or bed with the legs hanging loosely, or the patient's knee is flexed over the supporting forearm of the examiner with the heel resting lightly on the bed. The patellar ligament is tapped briskly until contraction of the quadriceps femoris is elicited. This results in extension of the leg at the knee joint.

Tapping the patellar ligament activates muscle spindles in the quadriceps femoris muscle. Afferent impulses from these spindles travel in the femoral nerve to the spinal cord (L2, **L3**, and **L4** segments). From here, efferent impulses are transmitted via motor fibers in the femoral nerve to the quadriceps femoris, resulting in a jerk-like contraction of the muscle and extension of the leg at the knee joint. Diminution or **absence of the quadriceps reflex** may result from any lesion which interupts the reflex arc just described, *e.g., peripheral nerve disease.*

THE MEDIAL THIGH MUSCLES

The main action of the medial group of muscles is **adduction of the thigh** (*e.g.*, holding oneself on a horse). Hence they constitute the **adductor group of thigh muscles** (Table 4-2) which includes the pectineus, gracilis and adductors magnus, brevis, and longus muscles.

Table 4-2
Medial Thigh Muscles[1]

Muscle	Origin	Insertion	Nerve Supply	Actions
Pectineus (Figs. 4-15, 4-17, and 4-21)	Pecten pubis (Figs. 4-1 and 4-19)	Pectineal line of femur (Figs. 4-2 and 4-20)	Femoral nerve (**L2** and L3); may receive a branch from the obturator nerve	Adducts and flexes thigh
Gracilis (Figs. 4-17, 4-21, and 4-26)	Body and inferior ramus of pubis (Fig. 4-19)	Superior part of medial surface of tibia (Fig. 4-24)	Obturator nerve (**L2** and L3)	Adducts thigh and flexes leg
Adductor longus (Figs. 4-15, 4-17, and 4-21)	Body of pubis, inferior to pubic crest (Fig. 4-19)	Middle third of linea aspera of femur (Fig. 4-20)	Obturator nerve (L2, **L3**, and L4)	Adducts and flexes thigh
Adductor brevis (Fig. 4-21*B*)	Body and inferior ramus of pubis (Fig. 4-19)	Pectineal line and proximal part of linea aspera of femur (Figs. 4-20 and 4-28)	Obturator nerve (L2, **L3**, and L4)	Adducts thigh and to some extent flexes it
Adductor magnus (Fig. 4-26)	Inferior ramus of pubis, ramus of ischium, and ischial tuberosity (Figs. 4-19 and 4-20)	Gluteal tuberosity, linea aspera, medial supracondylar line, and adductor tubercle of femur (Figs. 4-19 and 4-28)	Obturator nerve (L2 and **L3**) and sciatic nerve (L2, **L3**, and **L4**)	Adducts and extends thigh
Obturator externus (Figs. 4-27 and 4-38)	Margins of obturator foramen and obturator membrane (Figs. 4-19 and 4-38)	Trochanteric fossa of femur (Fig. 4-28)	Obturator nerve (L3 and **L4**)	Laterally rotates thigh

[1] The first five muscles listed are known collectively as *the adductors of the thigh*, but their actions are more complex than this, *e.g.*, they act as *fixators of the hip* during flexion of the knee joint and are active during walking. See the text for other actions of these muscles.

All adductors of the thigh, except the pectineus, are supplied by the obturator nerve (L2, L3, and L4). *The pectineus is supplied by the femoral nerve* (L2 and L3). The "hamstring" part of the adductor magnus is supplied by the sciatic nerve.

THE PECTINEUS MUSCLE (Figs. 4-15, 4-17, 4-21, and Table 4-2)

This short, flat quadrangular muscle is **in the floor of the femoral triangle** (Figs. 4-21 and 4-22).

Origin (Figs. 4-1 and 4-19). **Pecten pubis** (pectineal line of pubis). The pecten pubis is a sharp ridge on the pubic bone (L. *pecten*, comb).

Insertion (Figs. 4-2, 4-20, and 4-28). **Pectineal line of femur.**

Nerve Supply (Fig. 4-15). **Femoral nerve (L2** and L3); sometimes it receives a branch from the obturator nerve (L2 and L3).

Actions. Adducts and flexes thigh.

THE GRACILIS MUSCLE (Figs. 4-15, 4-17, 4-18, 4-21, 4-26, and Table 4-2)

The gracilis (L. slender) is a *long strap-like muscle* that lies along the medial side of the thigh and knee. *The gracilis is the most superficial of the adductor group of muscles.* It is the weakest member and *is the only one of the adductor group to cross the knee joint.*

Origin (Fig. 4-19). **Body and inferior ramus of pubis.**

Insertion (Fig. 4-24). Superior part of **medial surface of tibia,** posterior to the sartorius.

Nerve Supply (Figs. 4-15 and 4-34). Anterior division of **obturator nerve (L2** and L3).

Actions. Adducts thigh and flexes leg.

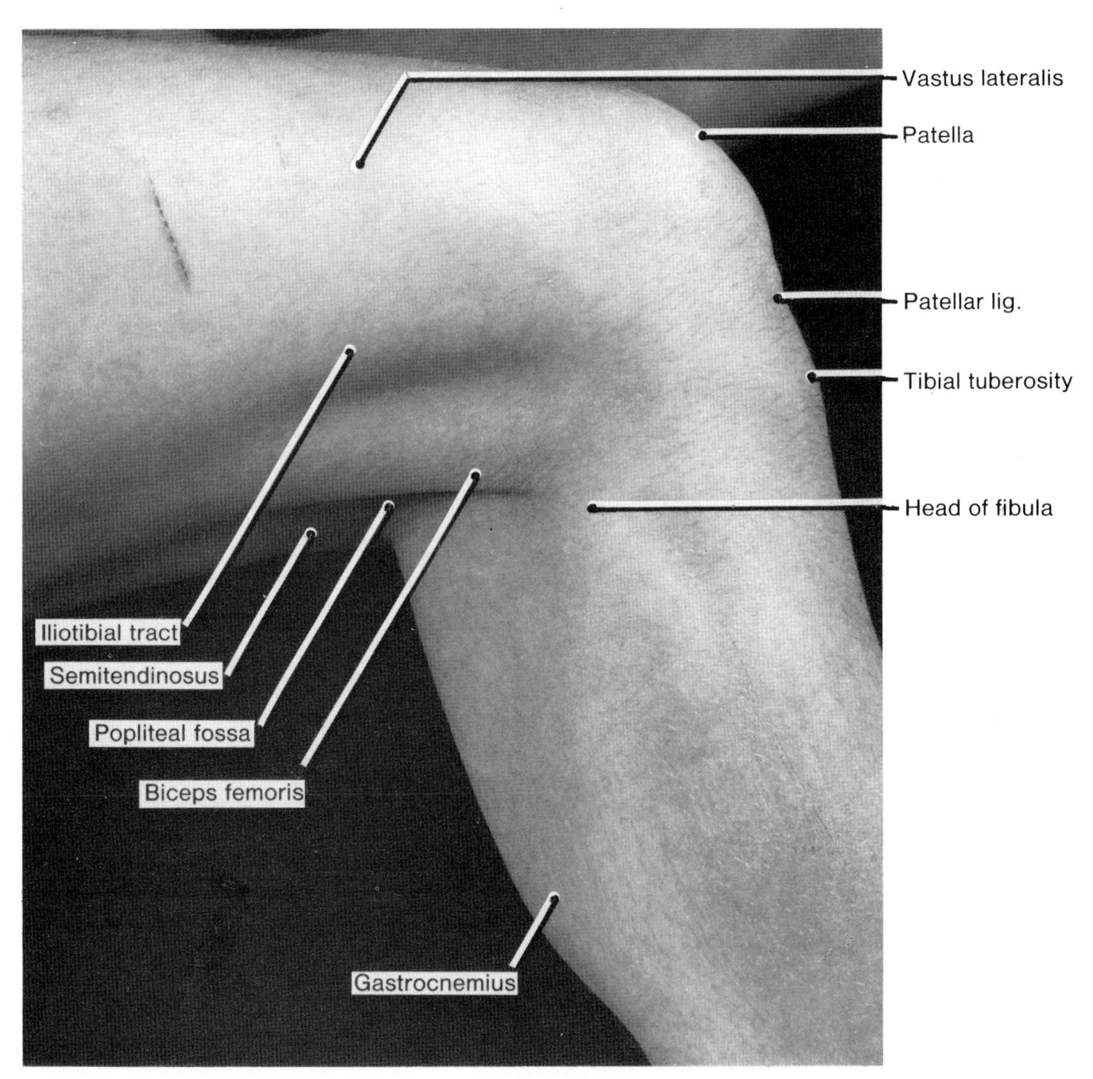

Figure 4-25. Photograph of the lateral aspect of the distal part of the thigh and the proximal part of the right leg of a 12-year-old girl showing the principal surface features. The quadriceps femoris (rectus femoris and vastus lateralis parts are shown) constitutes the major anterolateral mass of muscle that acts on the knee joint to extend the leg. All four of its parts insert into the base of the patella (Fig. 4-23), which in turn attaches to the tibial tuberosity via the patellar ligament (Figs. 4-21 and 4-24). When this ligament is tapped with a reflex hammer, the quadriceps reflex (knee jerk) is elicited. The biceps femoris tendon may be traced distally by palpation to its insertion into the head of the fibula.

Because the gracilis is a relatively weak member of the adductor group, it can be removed without noticeable loss of function. Hence, *surgeons often transplant the gracilis* with its nerves and blood vessels to replace a damaged muscle (*e.g., in the forearm*).

THE ADDUCTOR LONGUS MUSCLE (Figs. 4-15, 4-17, 4-18 to 4-22, and Table 4-2)

This triangular muscle is *the most anterior of the adductor group.* Its medial border forms the medial boundary of the femoral triangle (Fig. 4-30).

Origin (Figs. 4-1 and 4-19). Anterior surface of **body of pubis,** just inferior to the pubic crest.

Insertion (Figs. 4-2 and 4-20). **Middle third of linea aspera of femur.**

Nerve supply (Fig. 4-15). **Obturator nerve** (L2, **L3,** and L4).

Actions. Adducts and flexes the thigh. *It is also a fixator of the hip* during flexion of the knee joint, and it can laterally rotate the flexed thigh.

THE ADDUCTOR BREVIS MUSCLE (Figs. 4-18, 4-21*B*, and Table 4-2)

This short adductor *lies deep to the pectineus and adductor longus* and anterior to the adductor magnus.

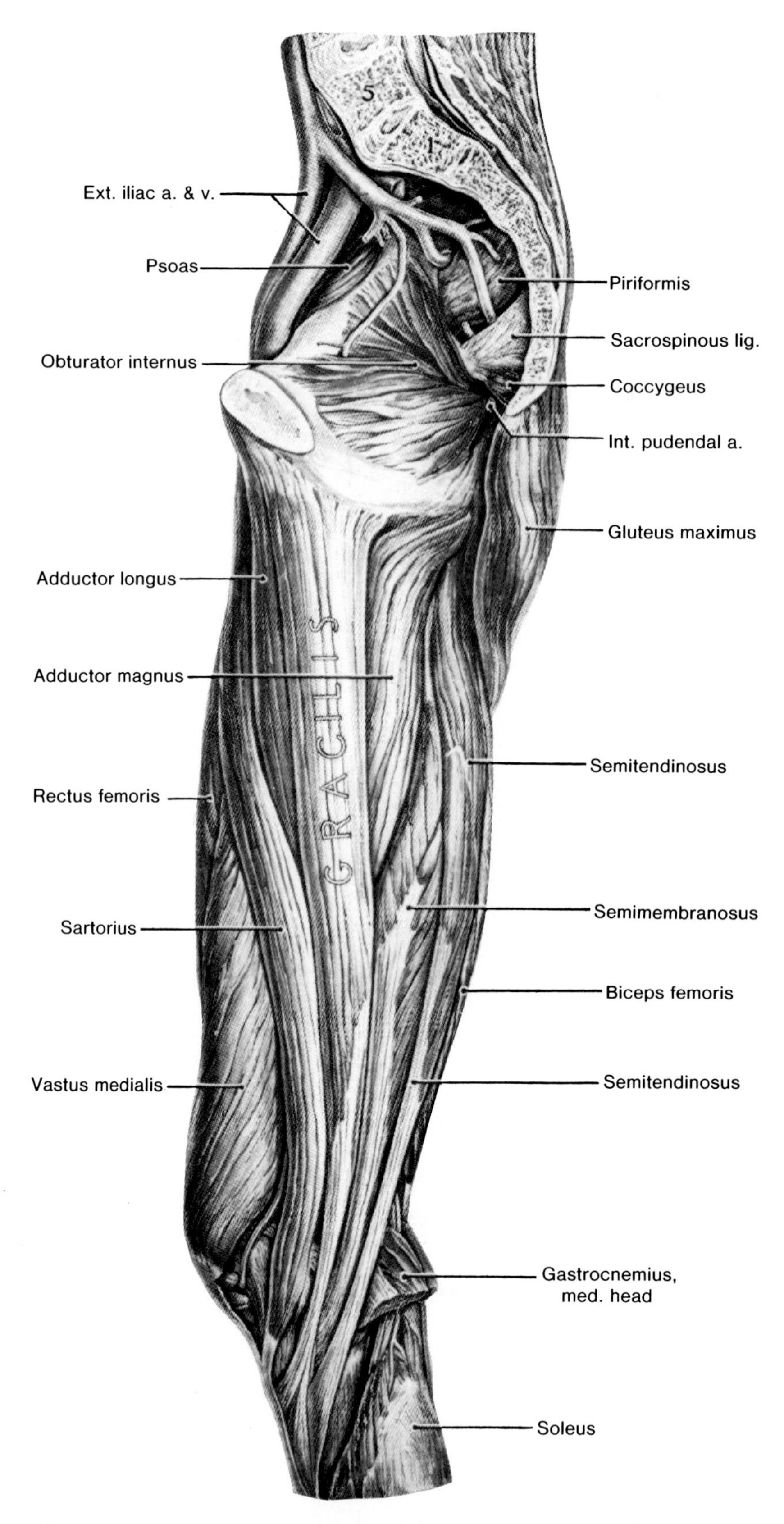

Figure 4-26. Drawing of a dissection of the muscles on the medial side of the right thigh. Observe that the adductor group of muscles forms a large mass that adducts the thigh at the hip joint (also see Fig. 4-22 and Table 4-2).

Origin (Figs. 4-19 and 4-21*B*). **Body and inferior ramus of pubis.**

Insertion (Figs. 4-20 and 4-28). **Pectineal line and proximal part of linea aspera of femur.**

Nerve Supply (Fig. 4-15). **Obturator nerve** (L2, **L3,** and L4).

Actions. Adducts thigh at hip joint and flexes it to some extent.

THE ADDUCTOR MAGNUS MUSCLE (Figs. 4-18, 4-21*B*, 4-26, and Table 4-2)

As its name indicates, this massive muscle is *the largest of the adductors.* Actually it is a composite muscle (part adductor and part hamstring).

Origin (Figs. 4-19 and 4-20). **Inferior ramus of pubis, ramus of ischium, and ischial tuberosity.**

Insertion (Figs. 4-2, 4-19, 4-20, 4-24, and 4-28). **Gluteal tuberosity, linea aspera, medial supracondylar line, and adductor tubercle of femur.**

There is a hiatus (L. aperture) in the aponeurotic insertion of the adductor magnus to the supracondylar line, called the **adductor hiatus** (Fig. 4-29). *This opening enables the femoral vessels to pass into the popliteal fossa.*

Nerve Supply (Figs. 4-38 and 4-40). Posterior division of **obturator nerve** (L2, and **L3**), except for the vertical "hamstring" part passing from the ischial tuberosity to the adductor tubercle, which is supplied by the tibial division of the **sciatic nerve** (L2, **L3,** and **L4**).

Actions. *Adductor part* **adducts thigh** at the hip joint and *hamstring part* **extends thigh** at the hip joint.

THE OBTURATOR EXTERNUS MUSCLE (Fig. 4-27 and Table 4-2)

This *fan-shaped, relatively small muscle,* is deeply placed in the superomedial part of the thigh.

Origin (Fig. 4-19). **Margins of obturator foramen and obturator membrane.**

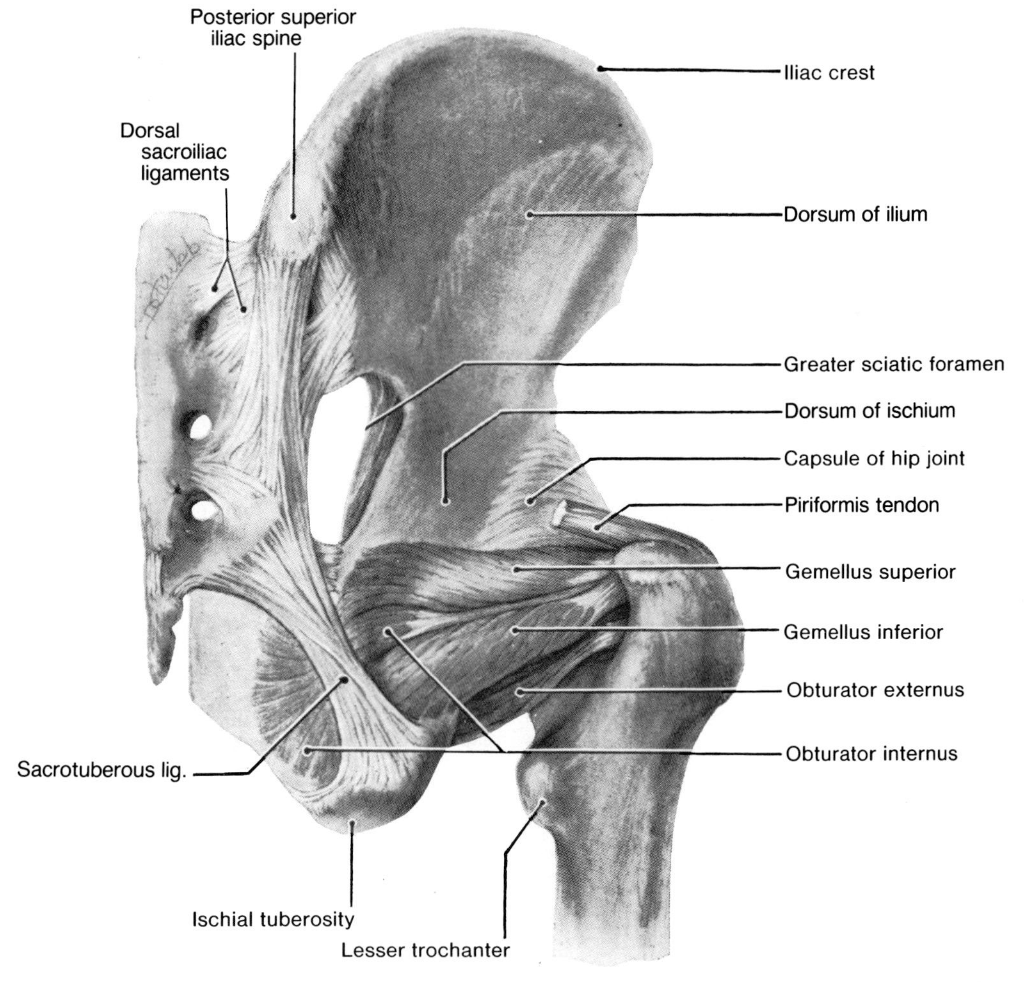

Figure 4-27. Drawing of a posterior view of a dissection of the obturator muscles. Note that the inferior end of the ischial tuberosity is at the same level as the lesser trochanter of the femur.

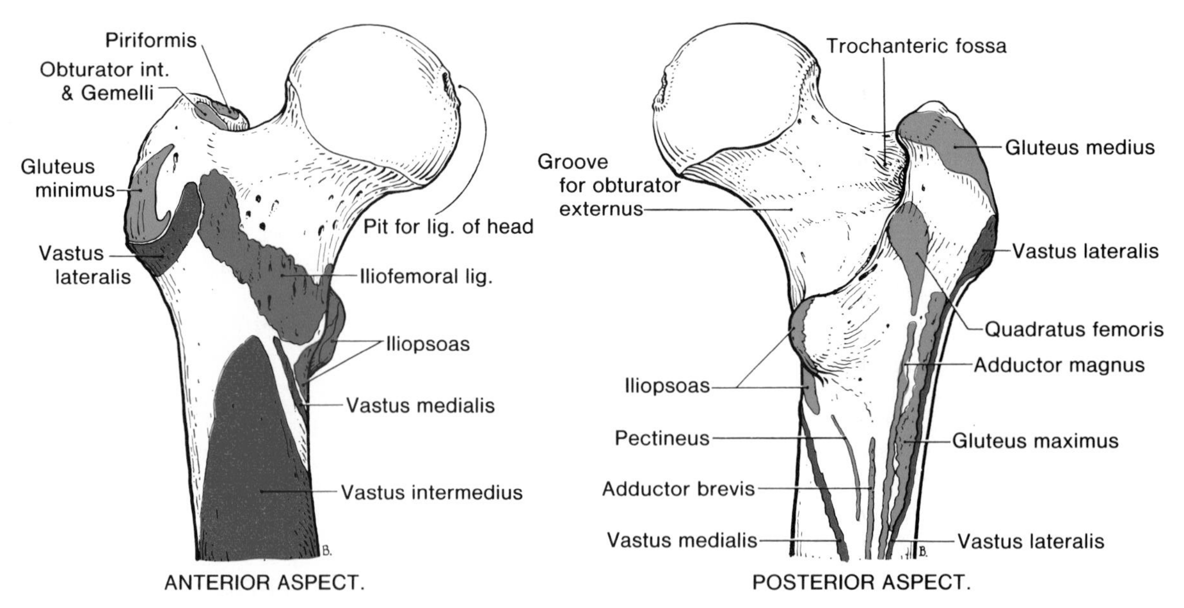

Figure 4-28. Drawings of the proximal end of the right femur showing the sites of attachment of muscles to it. *Red* indicates origins, and *blue* insertions of muscles.

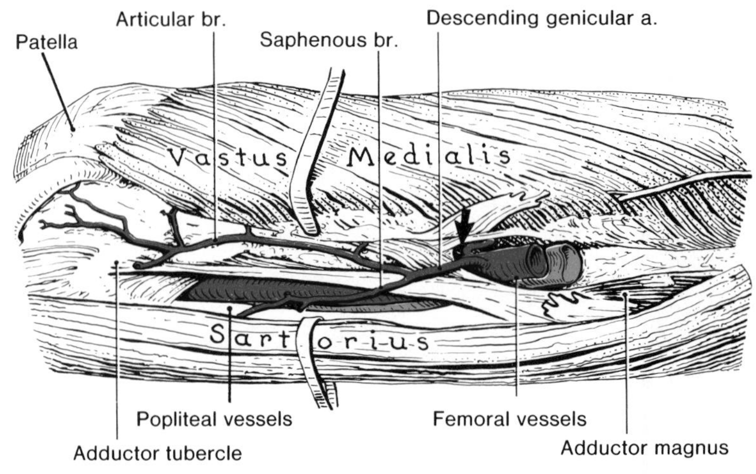

Figure 4-29. Drawing of an anterior view of a dissection of the distal end of the right thigh showing the adductor hiatus (*arrow*) between the insertion of the adductor magnus muscle and the femur. This opening allows the femoral vessels to pass into the popliteal fossa, the diamond-shaped space posterior to the knee joint (Figs. 4-47 and 4-49). Once in this fossa the femoral vessels are known as popliteal vessels.

Insertion (Fig. 4-28). **Trochanteric fossa** on the posterior aspect of the femur.

Nerve Supply (Fig. 4-32). **Obturator nerve** (L3 and **L4**).

Action. Laterally rotates thigh at the hip joint.

Occasionally you will hear about an injury called a "**pulled groin**." Usually this means that there has been a strain, stretching, and probably some tearing away of the tendinous origins of the adductor group of muscles. *They arise in the groin,* the junction of the thigh and abdomen (Fig. 4-19).

Ossification sometimes occurs in the tendons of the adductor longus muscles. You may hear these ossified tendons referred to as "*rider's bones*" because the condition is common in persons who ride horses and adduct their thighs.

THE FEMORAL TRIANGLE

The femoral triangle is a *clinically important region* in the superomedial part of the thigh. It appears as a depression inferior to the inguinal ligament (Figs. 4-17 and 4-22). *The femoral triangle contains the femoral vessels, the femoral nerve, and some inguinal lymph nodes* (Fig. 4-35).

BOUNDARIES OF THE FEMORAL TRIANGLE (Figs. 4-15, 4-17, 4-22, and 4-30)

The femoral triangle is bounded *superiorly* by the **inguinal ligament**, *medially* by the medial border of the **adductor longus muscle**, and ***laterally*** by the medial border of the **sartorius muscle.**

The base of the femoral triangle is formed by the **inguinal ligament**; *its apex is located where the medial borders of the sartorius and the adductor longus muscles meet* (Fig. 4-30). Note that the apex is distal to the point where the femoral vessels enter the adductor canal (Fig. 4-34).

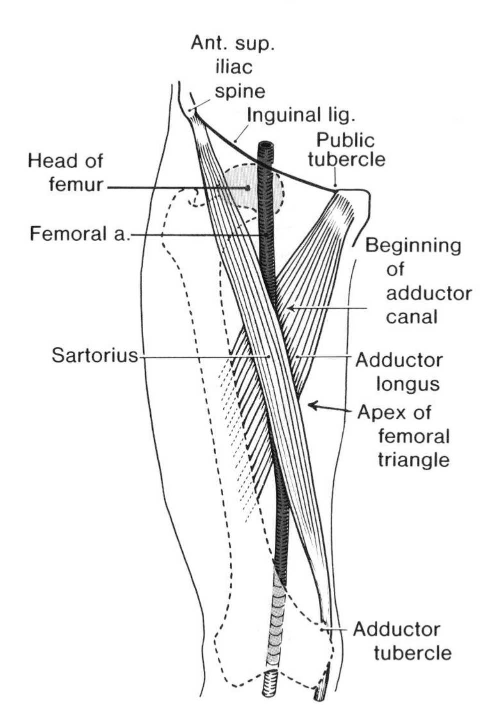

Figure 4-30. Drawing of the right thigh showing *the boundaries of the femoral triangle*. Its base is formed by the inguinal ligament and its sides are formed by the medial border of the sartorius and the medial border of the adductor longus (some authors regard the lateral border of the adductor longus as the medial side of the femoral triangle). Observe the beginning of the adductor canal through which the femoral vessels pass to enter the popliteal fossa (Fig. 4-29). Note that the head of the femur (*yellow*) lies posterior to the femoral artery, just inferior to the midpoint of the inguinal ligament. Compression of the femoral artery is easily effected at this site.

The floor of the femoral triangle is gutter-shaped and is *formed medially by* ***the adductor longus*** **and** ***pectineus muscles*** *and laterally by the iliopsoas muscle*, the collective name for the iliacus and psoas major muscles (Table 4-1).

The roof of the femoral triangle is formed by skin and fascia (superfical and deep).

SURFACE ANATOMY OF THE FEMORAL TRIANGLE (Fig. 4-22)

When a person stands with the thigh somewhat flexed, abducted, and laterally rotated, the femoral triangle appears as a depression in the proximal third of the thigh. You can easily palpate and often observe its base, the inguinal ligament (Fig. 4-22). Its lateral boundary, the medial border of the sartorius, is obvious in most people, but its medial boundary (medial border of adductor longus) is not usually so easy to identify.

You can easily palpate the femoral pulse in the femoral triangle (Fig. 4-30), 2 to 3 cm inferior to the midpoint of the inguinal ligament. The head of the femur lies posterior to the **femoral artery** at this site, making compression of the vessel easy.

CONTENTS OF THE FEMORAL TRIANGLE (Figs. 4-30 and 4-34)

This triangular area on the anterior aspect of the thigh contains the **femoral vessels** and the **femoral nerve**. The femoral triangle also contains the **profunda femoris vessels**, inguinal lymph nodes (Fig. 4-35), and some cutaneous nerves.

The Femoral Artery (Figs. 4-30 to 4-34, 4-36, and 4-37). This large vessel **provides the chief arterial supply to the lower limb.** *The femoral artery is the continuation of the* ***external iliac artery*** (Fig. 4-31). It enters the femoral triangle deep to the midpoint of the inguinal ligament and lateral to the femoral vein (Fig. 4-30).

The femoral artery is posterior to the deep fascia (Fig. 4-37), whereas the great saphenous vein is in the superficial fascia. These relationships are clinically important.

The femoral artery bisects the femoral triangle as it passes through it and runs deep to the sartorius muscle within the adductor canal (Fig. 4-30). Toward the inferior end of the femoral triangle, the femoral artery crosses the femoral vein so that at the apex of the femoral triangle, it lies anterior to it.

The profunda femoris artery, the largest branch of the femoral artery, is *the chief artery to the thigh.* It arises from the lateral side of the femoral artery within the femoral triangle, about 4 cm inferior to the

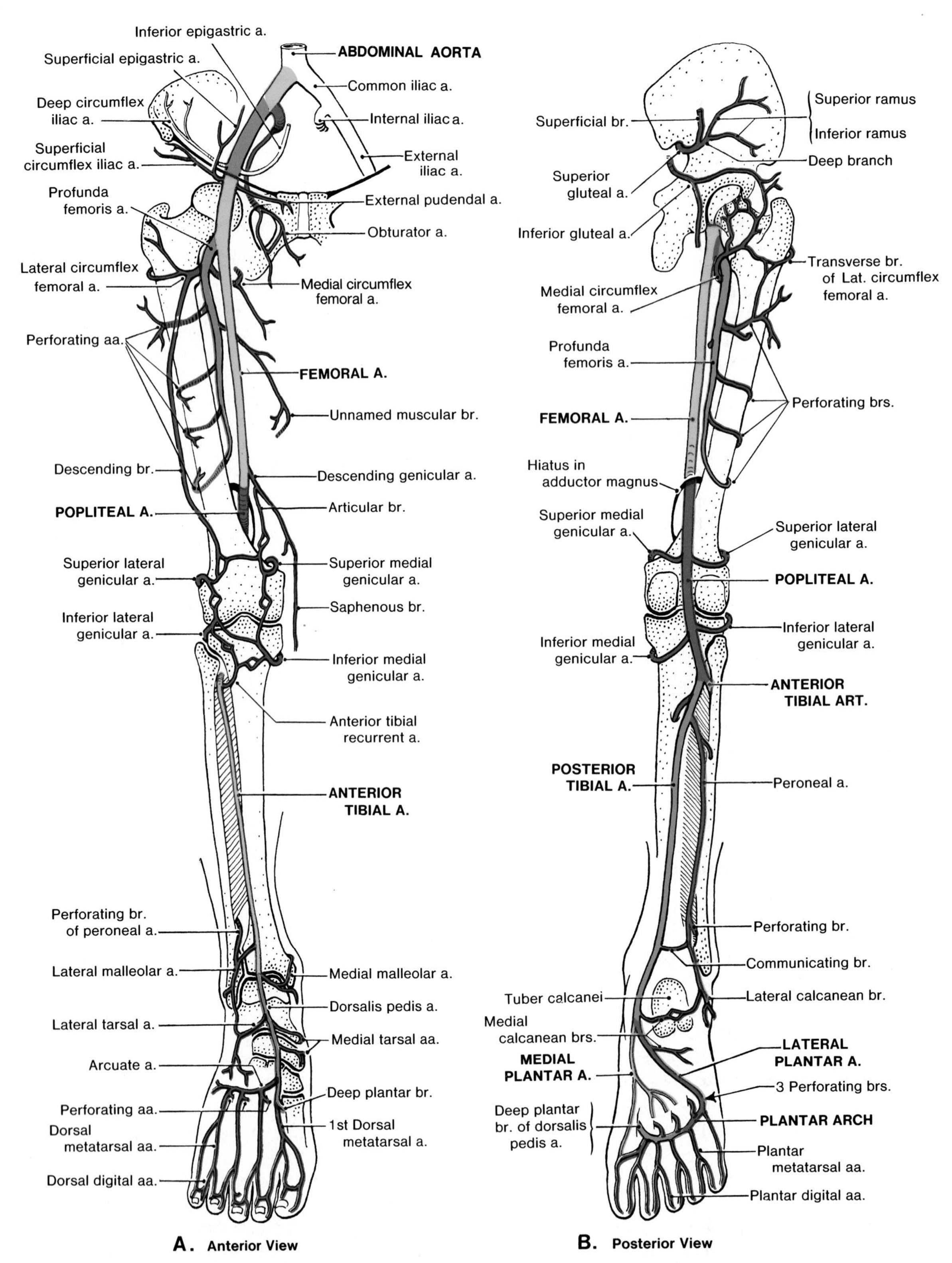

Figure 4-31. Drawings of the arteries of the lower limb. Note the following three arteries: (1) profunda femoris artery; (2) lateral circumflex femoral artery; and (3) medial circumflex femoral artery. In *B*, observe the anastomoses between the internal iliac and femoral arteries via the superior and inferior gluteal branches of the internal iliac and the medial and lateral circumflex branches of the profunda femoris. These anastomoses are important in the blood supply to the head and neck of the femur.

inguinal ligament (Figs. 4-32 and 4-34). It runs lateral to the femoral artery, and then passes posterior to it and the femoral vein. The profunda femoris artery leaves the femoral triangle between the pectineus and adductor longus muscles (Fig. 4-17), and descends posterior to the latter muscle, giving off *perforating arteries* which supply the adductor magnus and hamstring muscles (Figs. 4-31 and 4-34).

The medial and lateral circumflex femoral arteries, branches of the profunda femoris, supply the thigh muscles and *the proximal end of the femur* (Figs. 4-31 and 4-34). ***The medial circumflex femoral artery*** passes deeply between the iliopsoas and pectineus muscles to reach the posterior part of the thigh.

The medial circumflex artery is clinically important because it *supplies most of the blood to the head and neck of the femur* (Fig. 4-31*B*).

The lateral circumflex femoral artery passes laterally, deep to the sartorius and rectus femoris muscles, and between the branches of the femoral nerve. Here it divides into branches that supply the muscles on the lateral side of the thigh and the head of the femur (Figs. 4-31 and 4-34).

For **left cardiac angiography** and renal angiography, a long slender catheter is inserted into the femoral artery as it passes through the femoral triangle. Also see page 285.

The superficial position of the femoral artery in the femoral triangle renders it vulnerable to lacerations and puncture by gunshot wounds (Fig. 4-15). Commonly, both the femoral artery and vein are torn owing to their closeness in the femoral sheath (Figs. 4-16 and 4-33). In some of these cases, an **arteriovenous shunt** occurs as the result of communication between the injured vessels.

If it is necessary to ligate the femoral artery, blood is supplied to the lower limb through the **cruciate anastomosis**, the union of the medial and lateral circumflex femoral arteries with the inferior gluteal artery superiorly and the first perforating artery inferiorly (Fig. 4-31*B*). Nowadays, injured or obstructed segments of the femoral artery are commonly replaced by grafts rather than by ligating the proximal part of the vessel.

The Femoral Vein (Figs. 4-32 to 4-34). This large vessel *ends posterior to the inguinal ligament*, where it **becomes the external iliac vein** (Fig. 2-129). It leaves the femoral triangle a little medial to the midinguinal point and the femoral artery. In the inferior part of the femoral triangle, the femoral vein lies deep to the femoral artery (Fig. 4-17). While within the femoral triangle, the femoral vein receives the profunda femoris and great saphenous veins and other tributaries (Figs. 4-10*B*, 4-15, and 4-35).

To secure blood samples and take pressure recordings from the right chambers of the heart and/or pulmonary artery, or for right cardiac angiography, a long slender catheter is inserted into the femoral vein as it passes through the femoral triangle. The catheter is passed under fluoroscopic control through the external and common iliac veins and the inferior vena cava into the right atrium of the heart (see p. 105).

For left cardiac angiography, a series of radiographs is made as a contrast medium is injected into the right atrium. These radiographs record the passage of the opacified blood from the right atrium to the right ventricle, and to the pulmonary arteries and their branches. Pressure records are made as the catheter is slowly withdrawn from a small pulmonary artery to a main pulmonary artery, to the right ventricle, and finally into the right atrium.

The Inguinal Lymph Nodes (Figs. 4-35 and 4-37). The inguinal lymph nodes drain the lower limb, the perineum, the anterior abdominal wall as far superiorly as the umbilicus, the gluteal region, and parts of the anal canal. The lymph drainage of the skin covering all these areas is into the superficial inguinal lymph nodes (Fig. 41 on p. 43).

The superficial inguinal lymph nodes lie about 2 cm inferior to the inguinal ligament (proximal group), and along each side of the great saphenous vein (distal group). The efferent lymph vessels from the superficial inguinal lymph nodes drain into the deep inguinal lymph nodes.

The deep inguinal lymph nodes, one to three in number, lie on the medial side of the femoral vein, *within as well as inferior to the **femoral canal***. There are many anastomoses between the lymph vessels in the inguinal region.

About 24 efferent vessels leave the superficial and deep inguinal nodes. They pass deep to the inguinal ligament, and enter the **external iliac lymph nodes** (Fig. 3-73). Less than half these vessels pass through the femoral canal (Fig. 4-33); most of them ascend alongside the femoral artery and vein, some inside and some outside the femoral sheath.

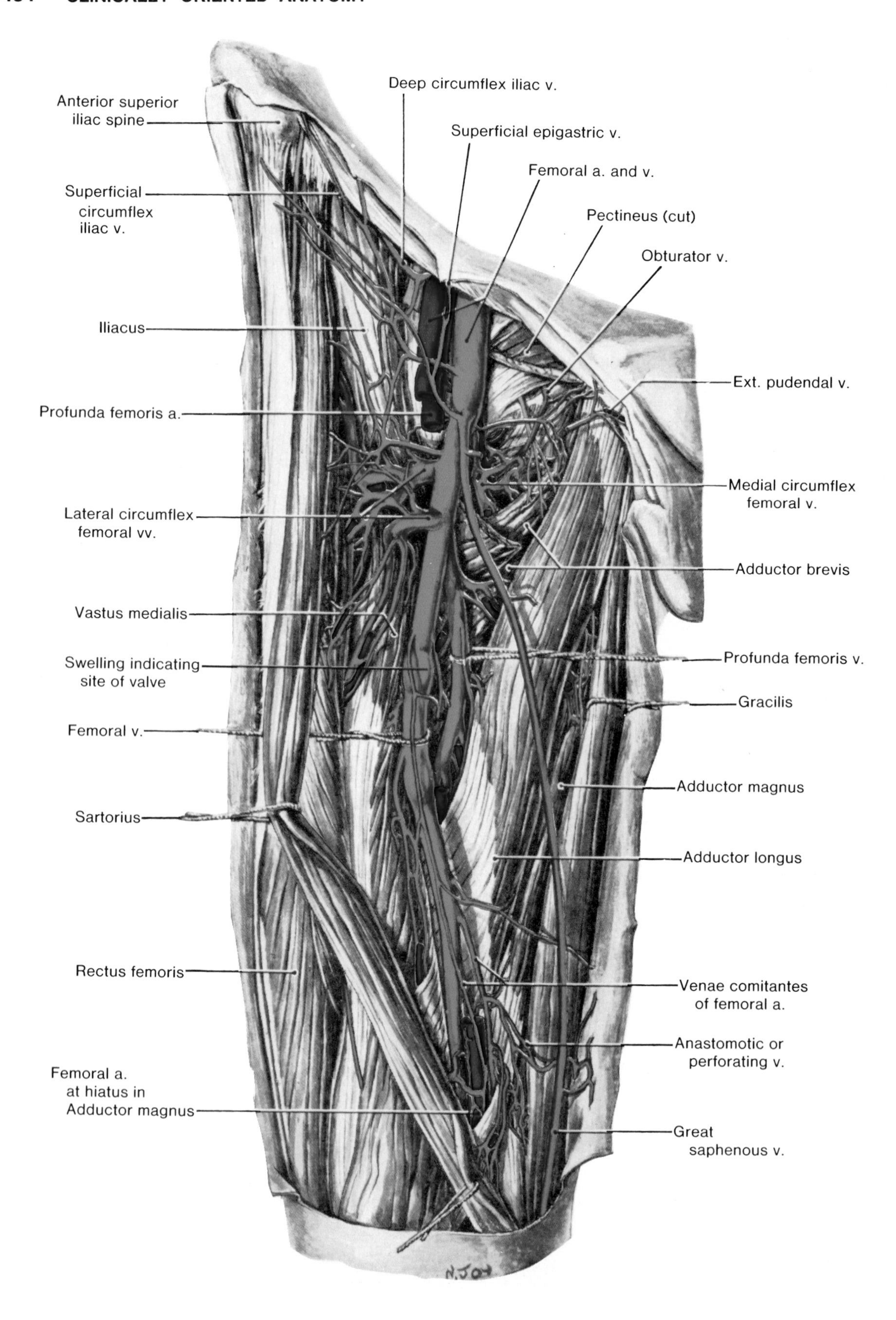
Deep circumflex iliac v.
Superficial epigastric v.
Femoral a. and v.
Pectineus (cut)
Obturator v.
Anterior superior iliac spine
Superficial circumflex iliac v.
Iliacus
Profunda femoris a.
Ext. pudendal v.
Medial circumflex femoral v.
Lateral circumflex femoral vv.
Adductor brevis
Vastus medialis
Swelling indicating site of valve
Profunda femoris v.
Gracilis
Femoral v.
Adductor magnus
Sartorius
Adductor longus
Rectus femoris
Venae comitantes of femoral a.
Anastomotic or perforating v.
Femoral a. at hiatus in Adductor magnus
Great saphenous v.

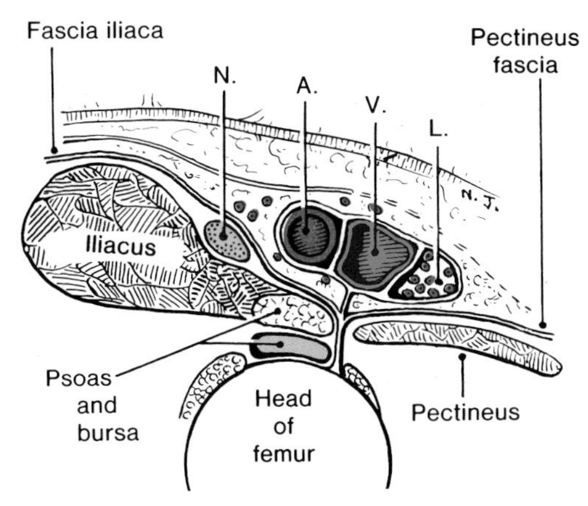

Figure 4-33. Diagram of a horizontal section of the superior part of the thigh showing the femoral sheath and the relationship of the femoral vessels and femoral nerve to each other and the head of the femur. Note that the psoas tendon and psoas bursa separate the femoral artery from the head of the femur. From medial to lateral the *contents of the femoral sheath* are: (1) the *femoral canal* containing lymph vessels (*L*); (2) the *femoral vein* (*V*); and (3) the *femoral artery* (*A*). Note that the femoral nerve (*N*) lies deep to the iliac fascia, the deep fascia surrounding the iliacus muscle (fascia iliaca), *lateral to the femoral sheath*.

The inguinal lymph nodes become enlarged in diseases of the areas which they drain. **Minor sepsis** (presence in the blood or other tissues of pathogenic microorganisms or their toxins), and abrasions of the lower limb produce slight enlargement of these nodes in otherwise healthy people.

Malignancies of the external genitalia and perineal abscesses result in enlargement of the superficial inguinal lymph nodes.

The Femoral Nerve (Figs. 4-32 to 4-34 and 4-36). The femoral nerve (L2, L3, and L4), ***the largest branch of the lumbar plexus***, forms in the abdomen within the substance of the psoas major muscle (Fig. 2-101). The femoral nerve descends posterolaterally through the pelvis to the midpoint of the inguinal ligament. It then passes lateral to the femoral vessels and **outside the femoral sheath** (Fig. 4-33).

After passing distally in the femoral triangle, the femoral nerve "breaks up" into several terminal branches. *The femoral nerve supplies the anterior femoral muscles* (Fig. 4-34 and Table 4-1). It also sends articular branches to the hip and knee joints and gives several branches to the skin on the anteromedial side of the lower limb.

The saphenous nerve (Fig. 4-34), a cutaneous branch of the femoral nerve, descends through the femoral triangle, lateral to the femoral sheath containing the femoral vessels. *The saphenous nerve accompanies the femoral artery in the adductor canal* (Fig. 4-17). It becomes superficial by passing between the sartorius and gracilis muscles (Fig. 4-26). *The saphenous nerve passes anteroinferiorly to supply the skin and fascia of the anterior and medial aspects of the knee, leg, and foot.*

THE FEMORAL SHEATH

The femoral sheath is an oval, funnel-shaped, **fascial tube that encloses the femoral vessels and femoral canal.** *Note that the femoral sheath does not enclose the femoral nerve* (Figs. 4-33 and 4-36). The femoral sheath is a diverticulum or inferior prolongation of the fasciae lining the abdomen (transversalis fascia anteriorly and iliac fascia posteriorly). *The femoral sheath is covered by the* ***fascia lata*** (Fig. 4-36).

The femoral sheath ends about 4 cm inferior to the inguinal ligament by becoming continuous with the adventitia of the femoral vessels. The medial wall of the femoral sheath is pierced by the great saphenous vein and lymphatic vessels.

COMPARTMENTS OF THE FEMORAL SHEATH (Figs. 4-33, 4-36, and 4-37)

The femoral sheath is subdivided by two vertical septa into **three compartments:** (1) **a lateral compartment** for the *femoral artery*; (2) **an intermediate compartment** for the *femoral vein*; and (3) **a medial compartment** called the *femoral canal.*

The Femoral Canal (Figs. 4-33, 4-36, and 4-37). This short, conical, **medial compartment of the femoral sheath allows the femoral vein to expand.** It contains a few lymph vessels, a lymph node, loose connective tissue, and fat.

Figure 4-32. Drawing of an anterior view of a dissection of the thigh primarily to illustrate the femoral vein. The thigh is rotated laterally. Only stumps of the arteries remain. Note the profunda femoris vein joining the femoral vein inferior to the inguinal ligament. Observe the large size of the lateral circumflex femoral vein, here double and ending in the femoral vein. Note the many long, slender, paired venae comitantes that accompany the various arteries. Observe the superficial circumflex iliac vein; in this specimen it receives the superficial epigastric vein, communicates with the veins proximal and distal to it, and ends both in the great saphenous and femoral veins. Note the great saphenous vein communicating with a vena comitans of the femoral artery, and the medial circumflex femoral vein communicating with the obturator vein. Observe the perforating vein connecting the great saphenous and femoral veins via the venae comitantes.

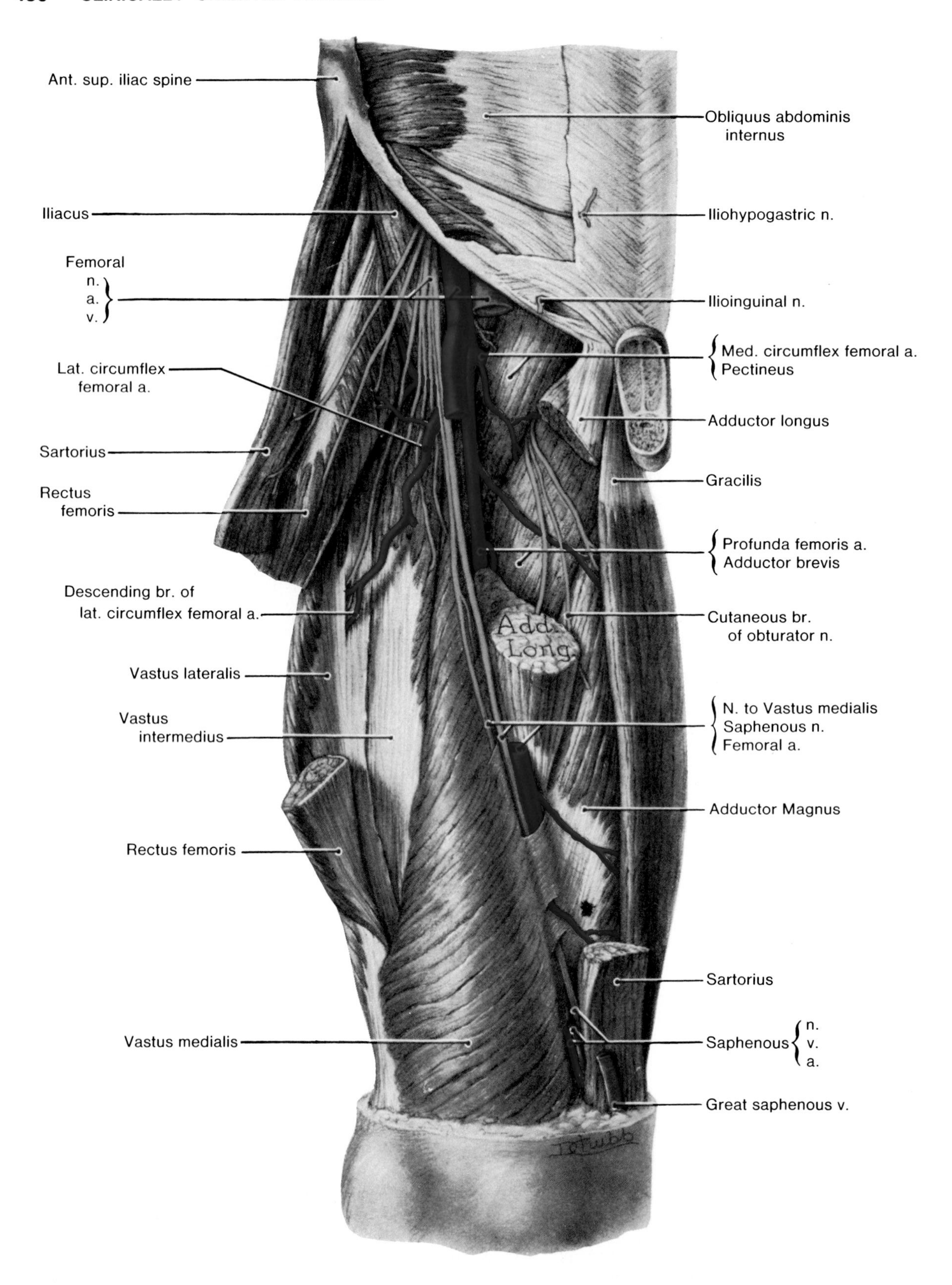
Ant. sup. iliac spine
Obliquus abdominis internus
Iliacus
Iliohypogastric n.
Femoral n. a. v.
Ilioinguinal n.
Med. circumflex femoral a.
Pectineus
Lat. circumflex femoral a.
Adductor longus
Sartorius
Gracilis
Rectus femoris
Profunda femoris a.
Adductor brevis
Descending br. of lat. circumflex femoral a.
Cutaneous br. of obturator n.
Add. Long.
Vastus lateralis
N. to Vastus medialis
Saphenous n.
Femoral a.
Vastus intermedius
Adductor Magnus
Rectus femoris
Sartorius
Vastus medialis
Saphenous n. v. a.
Great saphenous v.

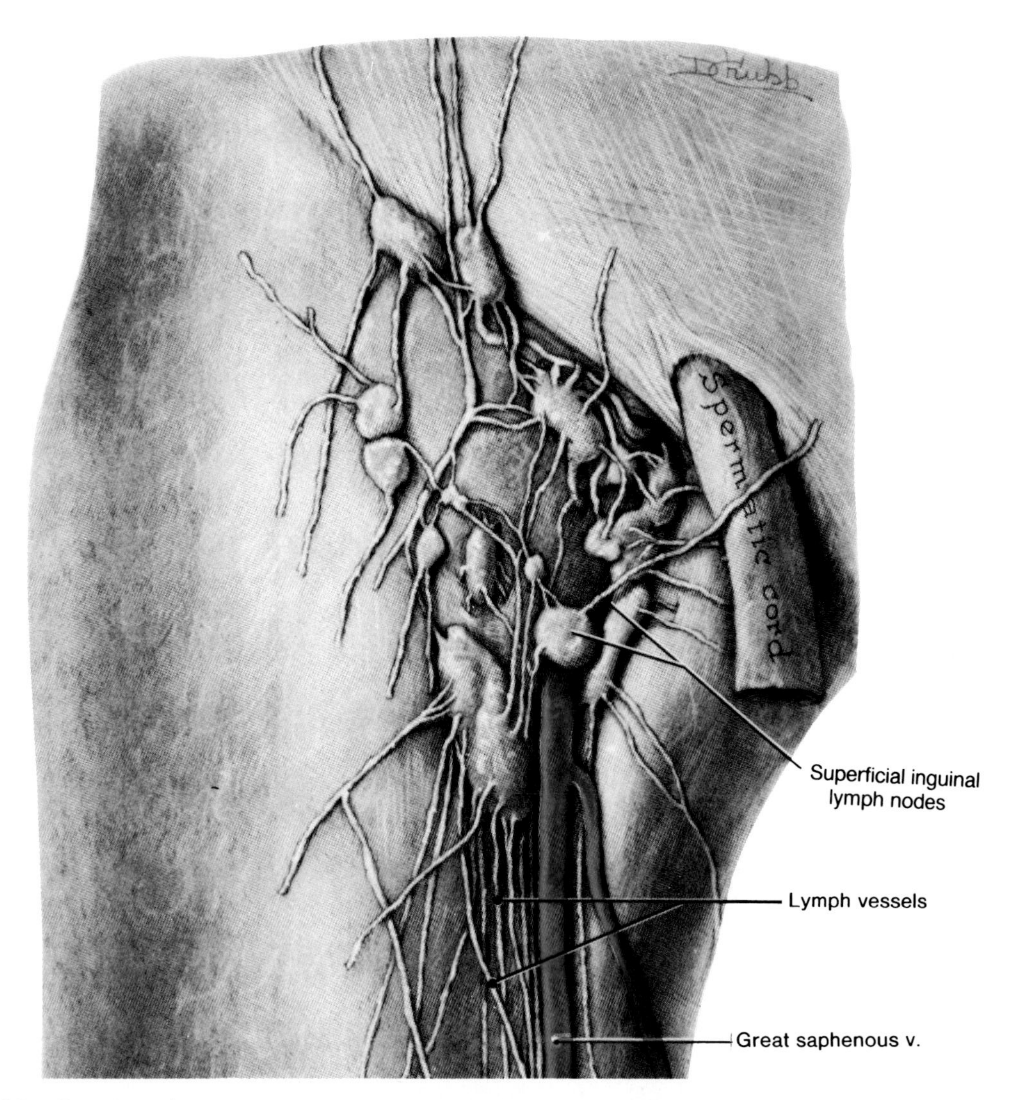

Figure 4-35. Drawing of a dissection of the inguinal lymph nodes on the right side. Observe: (1) a proximal chain parallel to the inguinal ligament (*superficial inguinal nodes*); (2) a distal chain along the sides of the great saphenous vein (*superficial subinguinal nodes*); and (3) more deeply, a chain of two or three nodes on the medial side of the femoral vein (*deep inguinal nodes*), one inferior to the femoral canal and one or two within it (Fig. 4-37). Minor sepsis (presence of pus-forming organisms) and abrasions of the lower limb are so common that it is not unusual to find enlarged inguinal lymph nodes in healthy people (also see Fig. 2-13).

Figure 4-34. Drawing of an anterior view of a dissection of the right thigh and adductor region. The limb is rotated laterally. Observe the femoral nerve breaking up into several nerves on entering the thigh. Note the femoral artery lying between two motor territories, that of the obturator nerve which is medial and that of the femoral nerve which is lateral. Observe that no motor nerve crosses anterior to the femoral artery, but the twig to the pectineus muscle crosses posterior to it. Note that the nerve to vastus medialis and the saphenous nerve accompany the femoral artery into the adductor canal. Observe the saphenous nerve and artery and their companion anastomotic vein emerging from the inferior end of the adductor canal and becoming superficial between the sartorius and gracilis muscles. Note the profunda femoris artery arising about 4 cm inferior to the inguinal ligament, lying posterior to the femoral artery, and disappearing posterior to the adductor longus muscle.

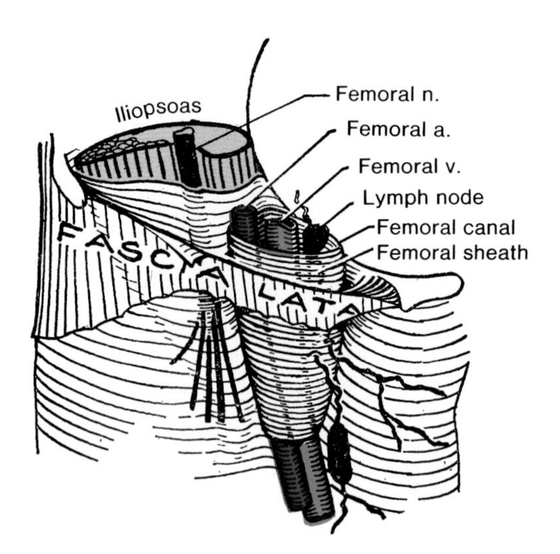

Figure 4-36. Drawing of the superior part of the right thigh showing the femoral sheath and its contents (also see Fig. 4-16). Note that this fascial wrapping for the femoral vessels and the femoral canal is funnel-shaped and has three compartments: a lateral one for the femoral artery, an intermediate one for the femoral vein, and a medial one (the femoral canal) for lymph vessels and a lymph node. Note that the femoral sheath does not surround the femoral nerve.

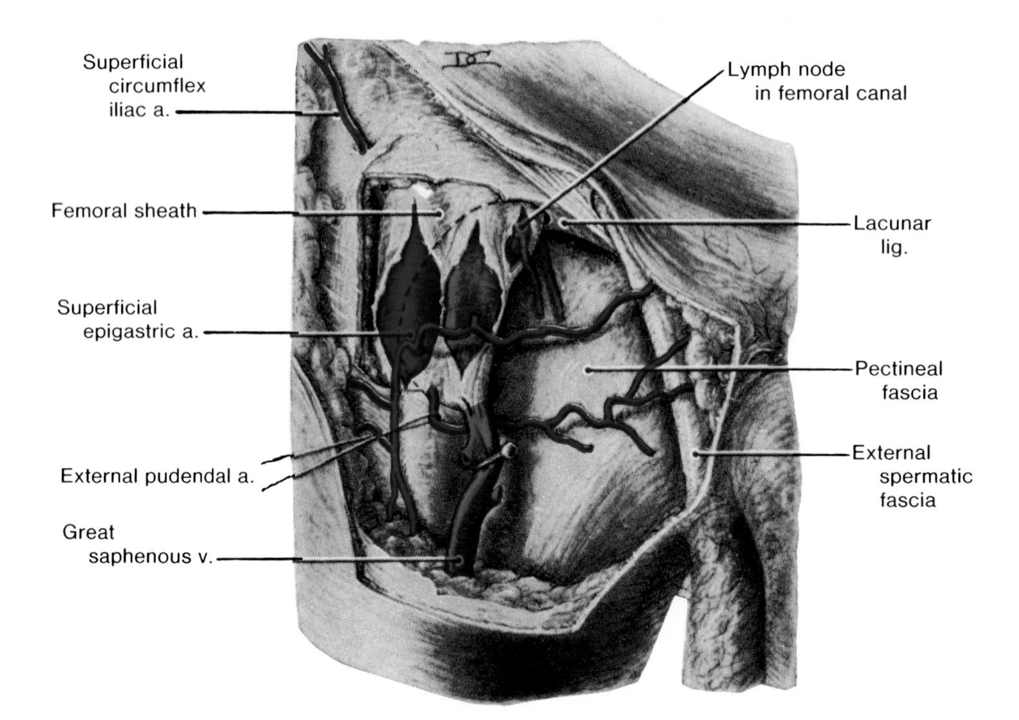

Figure 4-37. Drawing of a dissection of the superior part of the right thigh showing the femoral sheath, femoral canal, and femoral ring. The falciform edge of the saphenous opening is cut away. Note the superior horn of the opening passing toward the pubic tubercle and blending with the inguinal and lacunar ligaments. Observe the medial border of the opening formed by the fascia covering the pectineus muscle and as such passing laterally posterior to the femoral sheath. Note the three compartments of the sheath: (1) the lateral one for the femoral artery; (2) the intermediate one for the femoral vein; and (3) the medial one, called the femoral canal, for lymph vessels and a lymph node. Observe the proximal end (opening) of the femoral canal, called the femoral ring, bounded medially by the lacunar ligament, anteriorly by the inguinal ligament, posteriorly by the pectineal fascia, and laterally by the femoral vein. It is into the femoral canal that a femoral hernia may protrude (Fig. 4-133).

The femoral canal is widest at its abdominal end and extends distally to the level of the proximal end of the saphenous opening (Fig. 4-13).

The femoral ring (Fig. 4-37) is the small *abdominal opening or mouth of the femoral canal.* It is closed by extraperitoneal tissue called the **femoral septum,** which is pierced by the lymph vessels connecting the inguinal and external iliac lymph nodes (Figs. 4-33 and 4-37).

The boundaries of the femoral ring are: *laterally,* the **femoral vein** (Fig. 4-37); *posteriorly,* the **superior ramus of the pubis** covered by the pectineus muscle and its fascia; *medially,* the **lacunar ligament and conjoint tendon** (Fig. 4-37); and *anteriorly,* the **inguinal ligament and spermatic cord**.

The femoral ring is a weak area in the anterior abdominal wall that normally admits the tip of the little finger. This is important in understanding the mechanism of **femoral hernia, a protrusion of abdominal viscera (often small intestine) through the femoral ring into the femoral canal.**

A femoral hernia passes through the femoral ring into the femoral canal, pushing a covering of peritoneum before it. The hernial sac compresses the contents of the femoral canal (lymph vessels, connective tissue, and fat) and distends its wall.

*A **femoral hernia** can usually be palpated just inferior to the inguinal ligament.* Initially it is relatively small because it is contained within the femoral canal, but it can enlarge by passing inferiorly through the saphenous opening into the loose connective tissue of the thigh.

Strangulation of a femoral hernia may occur owing to the sharp, rigid boundaries of the femoral ring, particularly the concave margin of the lacunar ligament (Fig. 4-15).

Strangulation of a femoral hernia interferes with the blood supply to the herniated bowel, and this vascular impairment may result in death of the tissues concerned. Because the distal end of the femoral canal reaches the proximal part of the saphenous opening in the fascia lata (Fig. 4-37), a femoral hernia may herniate through the cribriform fascia and bulge anteriorly under the skin over the saphenous opening.

A femoral hernia presents as a mass inferolateral to the pubic tubercle and medial to the femoral vein. This type of hernia is more common in women than men, largely because the femoral ring is larger owing to the greater breadth of the female pelvis.

The **obturator artery** (Fig. 4-31*A*), a branch of the internal iliac, passes through the obturator foramen to supply adjacent muscles. In 20 to 30% of persons, an enlarged pubic branch of the inferior epigastric artery takes the place of the obturator artery or forms an **abnormal or accessory obturator artery** (Fig. 4-39*B*). This artery runs close to or across the femoral ring to reach the obturator foramen. Here it is closely related to the free margin of the lacunar ligament and the neck of a femoral hernia. Consequently this artery could be involved in a strangulated femoral hernia.

A more important clinical concern is the vulnerability of an abnormal obturator artery to damage during the surgical repair of a femoral hernia.

THE ADDUCTOR CANAL

The adductor canal (subsartorial canal), about 15 cm in length, is a **narrow fascial tunnel** (Fig. 4-17), deep to the sartorius muscle in the middle third of the medial part of the thigh.

*The adductor canal is an intermuscular cleft through which the **femoral vessels** pass to reach the popliteal fossa,* where they become the popliteal vessels (Figs. 4-29, 4-30, and 4-49).

The adductor canal begins about 15 cm inferior to the inguinal ligament, where the sartorius muscle crosses over the adductor longus muscle (Fig. 4-30). The adductor canal *ends at the **adductor hiatus*** in the tendon of the adductor magnus muscle (Fig. 4-29).

BOUNDARIES OF THE ADDUCTOR CANAL (Fig. 4-34)

The adductor canal is bounded *laterally* by the **vastus medialis,** *posteromedially* by the **adductor longus and adductor magnus** muscles, and *anteriorly* by the **sartorius** muscle. The sartorius and subsartorial fascia form the roof of the adductor canal (Fig. 4-18); hence *the adductor canal is often called the subsartorial canal.*

CONTENTS OF THE ADDUCTOR CANAL (Figs. 4-29 and 4-34)

The ***femoral vessels*** enter the adductor canal where the sartorius muscle crosses over the adductor longus muscle (Fig. 4-30), the vein lying posterior to the artery. The femoral artery and vein leave the

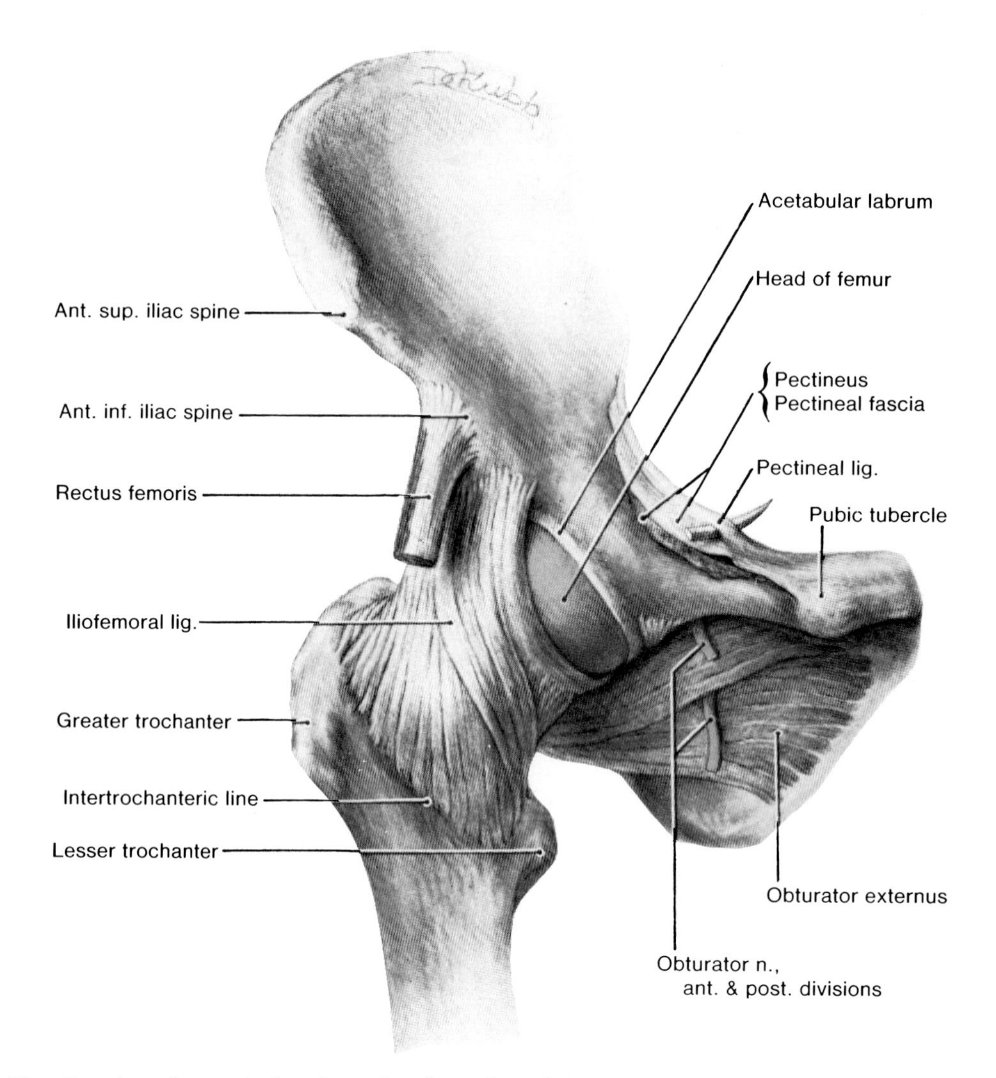

Figure 4-38. Drawing of an anterior view of a dissection of the right hip joint. Observe the head of the femur, exposed just medial to the iliofemoral ligament, and facing not only superiorly and medially, but also anteriorly. Note the obturator externus muscle crossing obliquely, inferior to the neck of the femur. Observe the thinness of the pectineus muscle and the blending of its fascia with the pectineal ligament along the pecten pubis (Fig. 4-1).

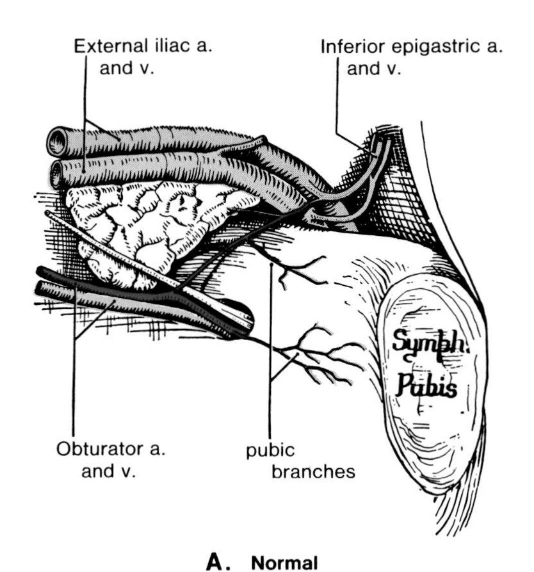

A. Normal

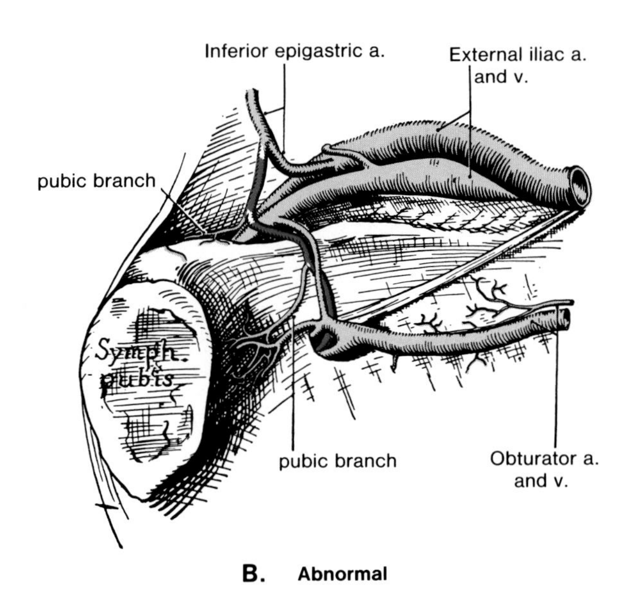

B. Abnormal

Figure 4-39. Drawing of dissections showing normal and abnormal obturator arteries. In *B*, the obturator artery arises from the inferior epigastric artery via the pubic anastomoses.

adductor canal through the **adductor hiatus** (Fig. 4-29), the tendinous opening in the adductor magnus muscle.

As soon as the femoral artery enters the ***popliteal fossa,*** *it is called the* ***popliteal artery***; similarly the femoral vein becomes the popliteal vein.

In Figure 4-17, observe that *the profunda femoris artery and vein do not enter the adductor canal.* The perforating branches of these deep vessels pierce the fibers of the adductor muscles to reach the posterior aspect of the thigh (Figs. 4-15 and 4-31).

The saphenous nerve, a cutaneous branch of the femoral nerve, accompanies the femoral vessels through the adductor canal (Fig. 4-34). It enters the adductor canal lateral to the vessels, crosses them anteriorly, and lies medial to them at the distal end of the canal (Fig. 4-34).

The saphenous nerve does not leave the adductor canal via the adductor hiatus. It passes between the sartorius and gracilis muscles (Fig. 4-34), pierces the deep fascia on the medial aspect of the knee, and passes down the medial side of the leg with the saphenous vein.

The ***nerve to the vastus medialis*** *accompanies the femoral artery through the proximal part of the adductor canal* (Fig. 4-34), and divides into branches that supply this muscle and the knee joint.

John Hunter, a renowned Scottish anatomist and surgeon (1728 to 1793), was the first person to describe the adductor canal and to ligate the femoral artery within it to reduce pressure in the vessel when there was an **aneurysm** in the popliteal artery (Fig. 4-31*B*). For years the adductor canal was referred to as Hunter's canal.

MUSCLES OF THE GLUTEAL REGION

The gluteal region or buttock is the prominence on each side formed by the gluteal muscles (Fig. 4-40). It is *bounded superiorly* by the **iliac crest** and *inferiorly* by the inferior border of the **gluteus maximus muscle.** The gluteal groove or crease, located inferior to the gluteal fold, indicates the inferior border of the gluteus maximus.

The gluteal region or buttock is largely formed by the gluteus maximus muscle. The gluteus medius forms the superolateral part of the gluteal region (Fig. 4-40). The terms *buttock* ("butt"), *nates*, *clunis*, and "*rump*" all refer to the gluteal region. It is very important clinically to know the boundaries of the buttock (Fig. 4-44). Some people wrongly consider the buttock to be only the prominence formed by the gluteus maximus. The buttock is a much larger area.

BONY LANDMARKS AND SURFACE MARKINGS OF THE GLUTEAL REGION (Fig. 4-27)

Revise your knowledge of the following: iliac crest, anterior superior iliac spine, greater sciatic notch, lesser sciatic notch, ischial spine, and posterior superior iliac spine (Figs. 3-42 to 3-46).

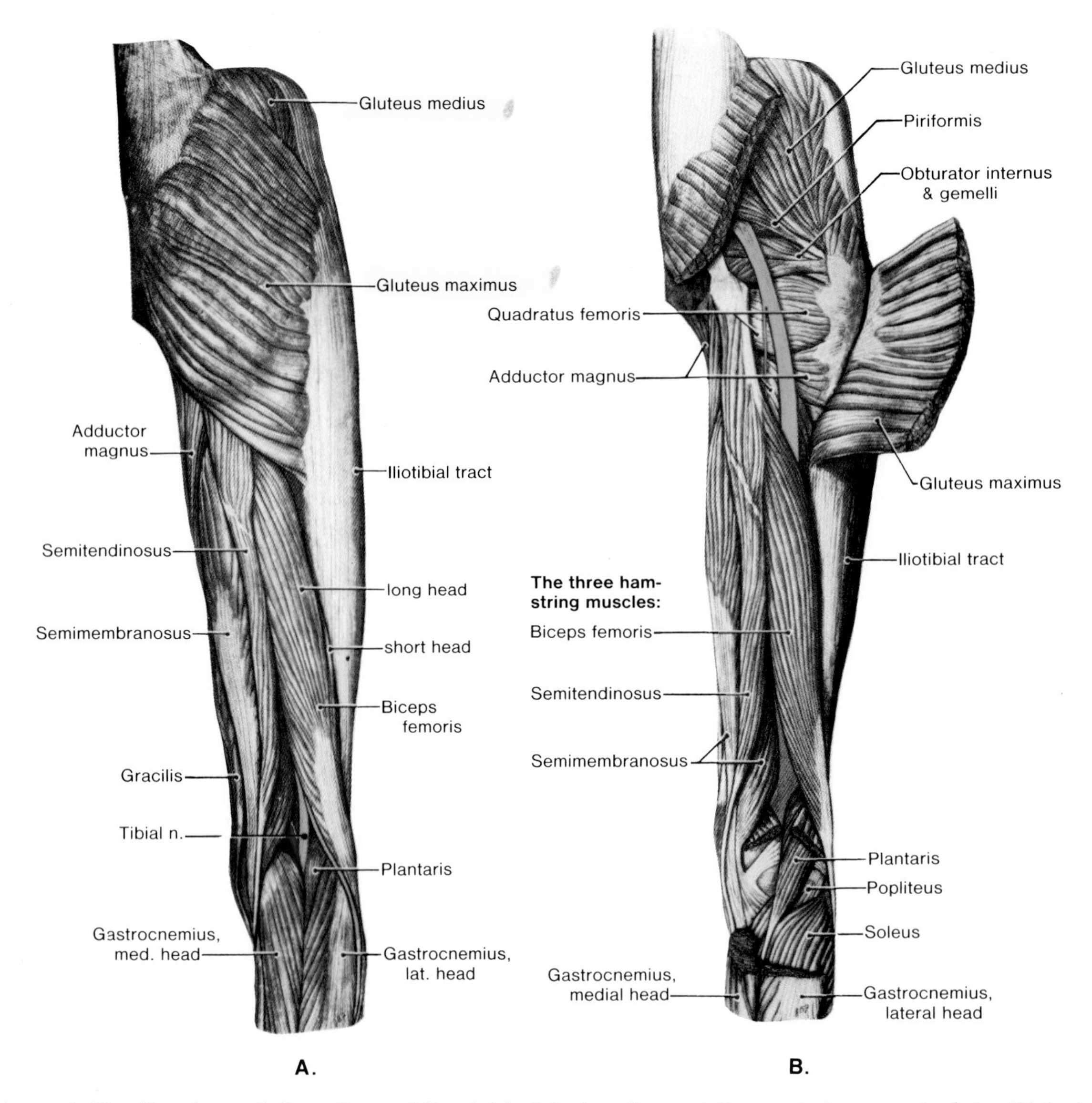

Figure 4-40. Drawings of dissections of the right gluteal region and the posterior aspect of the thigh. In *B*, observe that the large gluteus maximus muscle overlies the sciatic nerve (*yellow*) and several other muscles. The common term "hamstrings" is a collective term for the three separate muscles listed.

Palpate your anterior and posterior **superior iliac spines** and iliac crests. The iliac crest extends from the anterior superior iliac spine to the posterior superior iliac spine, indicated by the skin dimple (Figs. 2-105 and 4-6). Between the two posterior superior iliac spines is the posterior aspect of the sacrum (Fig. 4-9). Recall that the spinous process of the second sacral vertebra is at the level of the posterior superior iliac spines, and that the summit of the iliac crest corresponds to the level of the spinous process of the fourth lumbar vertebra (Fig. 4-44). Inferomedially, the **ischial tuberosity** can be palpated deep to the gluteus maximus muscles (Fig. 4-40).

LIGAMENTS OF THE GLUTEAL REGION (Figs. 3-16, 3-43, and 4-27)

The parts of the bony pelvis are bound together by dense ligaments. For a description of these ligaments, see page 341.

The region between the sacrum and the bony pelvis is bridged by two important accessory ligaments of the sacroiliac joint called the **sacrotuberous and sacrospinous ligaments**. They close the sciatic notches of the hip bones, producing the greater and lesser sciatic foramina (Fig. 4-27).

The greater sciatic foramen is a *passageway for*

structures entering or leaving the pelvis (*e.g.*, the sciatic nerve, Fig. 4-40), whereas **the lesser sciatic foramen** is a doorway for structures entering or leaving the perineum (*e.g.*, the pudendal nerve, Fig. 3-24).

The greater and lesser sciatic notches are separated by a beak-like process, the **ischial spine** (Figs. 4-2 and 4-4), which gives attachment to the sacrospinous ligament.

The Sacrotuberous Ligament (Figs. 3-16, 4-27, and 4-41). This long band extends from the posterior superior and posterior inferior iliac spines, and from the dorsum and sides of the sacrum and coccyx to the ischial tuberosity.

The two sacrotuberous ligaments resist posterior rotation of the inferior end of the sacrum. They also provide an origin for the gluteus maximus muscle.

The Sacrospinous Ligament (Figs. 3-16 and 3-19). This triangular ligament is "sandwiched" between the sacrotuberous ligament and the coccygeus muscle. Coextensive with the coccygeus, it extends from the ischial spine to the inferolateral border of the sacrum and coccyx and *lies on the pelvic surface of the sacrotuberous ligament.*

The muscles of the gluteal region are listed in Table 4-3 and are illustrated in Figures 4-40 and 4-41.

THE GLUTEUS MAXIMUS MUSCLE (Fig. 4-40, 4-41, 4-54, and Table 4-3)

The gluteus maximus is the *largest, heaviest, and most coarsely fibered muscle in the body.* It forms a **thick, quadrilateral pad over the ischial tuberosity** when the thigh is extended (Fig. 4-40*A*).

The ischial tuberosity can be felt on deep palpation through the inferior portion of the gluteus maximus, just superior to the medial portion of the gluteal fold (Fig. 4-9). When the thigh is flexed, the distal border of the gluteus maximus moves superiorly, leaving the ischial tuberosity subcutaneous.

Origin (Figs. 4-20 and 4-41). **External surface of ilium**, posterior to the posterior gluteal line, including **iliac crest**; dorsal surfaces of **sacrum and coccyx**; and **sacrotuberous ligament**.

Insertion (Figs. 4-20, 4-28, and 4-40). **Iliotibial tract** (most fibers) which inserts into the lateral condyle of the tibia (Figs. 4-45 and 4-46) and the **gluteal tuberosity of the femur.**

Nerve Supply (Fig. 4-41). **Inferior gluteal nerve** (L5, **S1**, and **S2**).

Actions. Extends the thigh and steadies it. *The gluteus maximus is the chief extensor of the thigh.* When acting with its insertion fixed, it is also a strong extensor of the pelvis (*e.g.*, when rising from the seated or stooped position). *The gluteus maximus also assists with lateral rotation of the thigh at the hip joint.*

Bursae Associated with the Gluteus Maximus. Usually there are three bursae separating this large muscle from underlying structures: (1) **the trochanteric bursa** separates it from the lateral side of the greater trochanter of the femur; (2) **the gluteofemoral bursa** separates it from the superior part of the origin of the vastus lateralis muscle; and (3) **the ischial bursa** separates it from the ischial tuberosity. The purpose of these bursae (small sacs with synovial linings and containing a small amount of fluid) is to reduce friction where the muscle passes over bones or tendons.

The ***ischial bursa,*** *superficial to the ischial tuberosity, may become inflamed as the result of excessive friction.* This produces a friction bursitis known as **ischial bursitis** (weaver's bottom). Weavers extend first one lower limb and then the other. This repeated friction on the ischial bursae may lead to inflammation of their walls. The **inflammatory reaction** results in painful swelling of the bursae.

As the ischial tuberosities bear the weight during sitting (Fig. 5-3), these pressure points may lead to **pressure sores** in debilitated patients, particularly paraplegic persons, if good nursing care is not exercised. Sitting on an air cushion is often helpful.

Trochanteric bursitis causes diffuse deep pain in the gluteal and lateral thigh regions. This type of **pelvic girdle pain** is characterized by tenderness over the greater trochanter of the femur (Fig. 4-9).

THE GLUTEUS MEDIUS MUSCLE (Figs. 4-40, 4-41, 4-54, and Table 4-3)

Most of this thick, fan-shaped muscle lies deep to the gluteus maximus on the external surface of the ilium (Fig. 4-40).

Origin (Figs. 4-20 and 4-40). **External surface of ilium** between the anterior and posterior gluteal lines.

Insertion (Figs. 4-20 and 4-28). **Lateral surface of greater trochanter of femur**. A bursa separates its tendon from the trochanter.

Nerve Supply (Fig. 4-41). **Superior gluteal nerve (L5,** and S1).

Actions. Abducts and medially rotates thigh. It also ***steadies the pelvis*** so that it does not sag when the foot on the opposite side is raised (*e.g.*, during walking).

THE GLUTEUS MINIMUS MUSCLE (Figs. 4-41, 4-54, and Table 4-3)

This fan-shaped muscle, *the smallest of the gluteal muscles*, lies deep to the gluteus medius.

Origin (Figs. 4-20 and 4-54). **External surface of ilium** between anterior and inferior gluteal lines.

Table 4-3
Muscles of the Gluteal Region

Muscle	Origin	Insertion	Nerve Supply	Actions
Gluteus maximus (Fig. 4-40)	External surface of ilium, including iliac crest, dorsal surface of sacrum and coccyx, and sacrotuberous ligament (Fig. 4-20)	Iliotibial tract (most fibers) which inserts into lateral condyle of tibia (Fig. 4-45); some fibers insert on gluteal tuberosity of femur (Figs. 4-20 and 4-28)	Inferior gluteal nerve (L5, **S1**, and **S2**)	Extends thigh and steadies it
Gluteus medius (Figs. 4-40 and 4-41)	External surface of ilium between anterior and posterior gluteal lines (Fig. 4-20)	Lateral surface of greater trochanter of femur (Figs. 4-20 and 4-28)	Superior gluteal nerve (**L5** and S1)	Abduct and medially rotate thigh; steady the pelvis
Gluteus minimus (Figs. 4-41 and 4-54)	External surface of ilium between anterior and inferior glutal lines (Fig. 4-20)	Anterior surface of greater trochanter of femur (Figs. 4-19 and 4-28)	Superior gluteal nerve (**L5** and S1)	
Piriformis (Figs. 4-26, 4-40, and 4-41)	Anterior surface of sacrum and sacrotuberous ligament (Figs. 3-25 and 3-52)	Superior border of greater trochanter of femur (Fig. 4-28)	Branches from ventral rami of **S1** and S2	Laterally rotate extended thigh and abduct flexed thigh
Obturator internus (Figs. 4-27, 4-40*B*, and 4-41)	Pelvic surface of obturator membrane and surrounding bones (Fig. 4-40*B*)	Medial surface of greater trochanter of femur (Fig. 4-28)	Nerve to obturator internus (L5 and **S1**)	
Quadratus femoris (Figs. 4-40*B*, 4-41, and 4-54)	Lateral border of ischial tuberosity (Fig. 4-40*B*)	Quadrate tubercle of femur (Fig. 4-28)	Nerve to quadratus femoris (L5 and S1)	Laterally rotates thigh

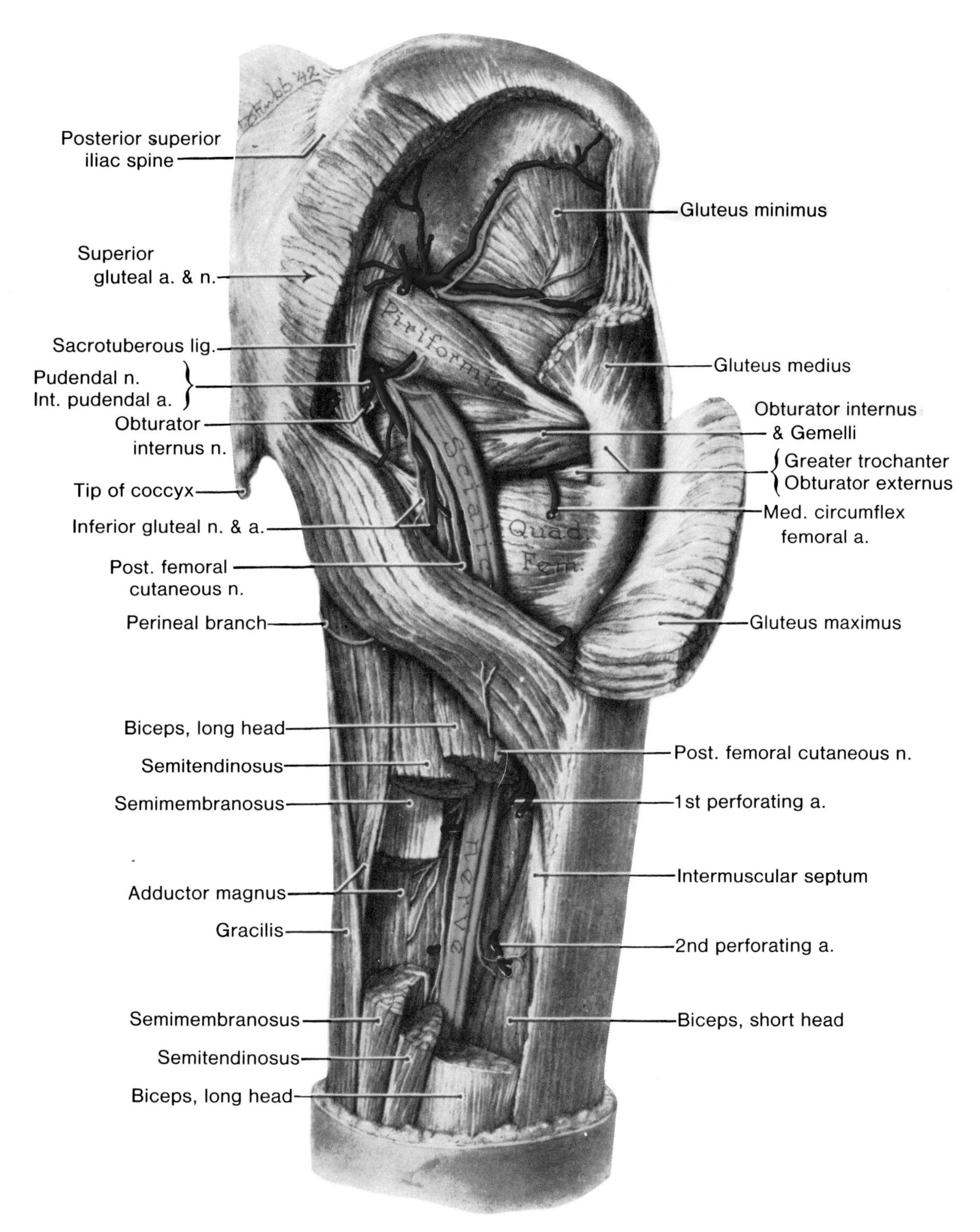

Figure 4-41. Drawing of a dissection of the right gluteal region and the posterior aspect of the thigh. Most of the gluteus maximus is reflected and parts of the gluteus medius and hamstring muscles are excised. Observe that *the superior gluteal vessels and nerve appear superior to the piriformis* muscle and that all other vessels and nerves appear inferior to it. Note that there are no nerves or vessels of importance lateral to the sciatic nerve.

Insertion (Figs. 4-28). **Anterior surface of greater trochanter of femur.** A bursa separates its tendon from the trochanter.

Nerve Supply (Fig. 4-41). **Superior gluteal nerve** (**L5** and S1).

Actions. Abducts and medially rotates thigh and steadies the pelvis. Note that its actions are the same as the gluteus medius; it acts with this muscle during walking.

If the gluteus medius and minimus muscles are paralyzed (*e.g.*, owing to injury of the superior gluteal nerve, or by a disease such as **poliomyelitis**), the supportive and steadying effect of these muscles on the pelvis is lost. Hence when the foot is raised on the normal side, the pelvis falls on that side. Similarly, when the person walks there is a waddling gait known as a **gluteus medius limp** (gluteal gait), characterized by falling of the pelvis toward the unaffected side at each step.

A similar gait occurs in persons with unilateral **posterior dislocation of the hip joint** (Fig. 4-104). This condition prevents normal functioning of the gluteus medius and minimus muscles.

THE PIRIFORMIS MUSCLE (Figs. 4-40*B*, 4-41, 4-54, and Table 4-3)

Because of its position, *this small, pear-shaped* (L. *piriform*) *muscle is used as the* ***landmark of the gluteal region***. It is the key to understanding relationships in this area because it determines the names of the blood vessels and nerves; *e.g.*, the superior gluteal vessels and nerve emerge superior to it and the inferior gluteal vessels and nerve emerge inferior to it.

The surface marking of the superior border of the piriformis muscle is indicated by a line joining the skin dimple, formed by the posterior superior iliac spine, to the top of the greater trochanter of the femur (Fig. 4-41).

Because the piriformis occupies a key position in the gluteal region, it is clinically important to know its location in the buttock (*i.e.*, be able to visualize it in your "mind's eye").

Origin (Figs. 4-41 and 4-54). **Anterior surface of sacrum** (opposite greater sciatic notch) **and sacrotuberous ligament**.

Insertion (Fig. 4-28). **Superior border and medial surface of greater trochanter of femur.** It leaves the pelvis via the greater sciatic notch and passes posterior to the head of the femur to reach the greater trochanter.

Nerve Supply. Branches of **ventral rami of S1** and S2.

Actions. Laterally rotates thigh when it is extended, and **abducts thigh** when it is flexed. **Helps to hold the head of the femur in the acetabulum**; *i.e.*, it stabilizes the hip joint.

THE OBTURATOR INTERNUS MUSCLE (Figs. 4-40*B*, 4-41, and 4-54, and Table 4-3).

This muscle, like the piriformis, is one of the muscles of the pelvic wall. The two *gemelli* (L. twins) muscles are extrapelvic parts of the obturator internus.

Origin (Figs. 4-40*B* and 4-41). **Pelvic surface of obturator membrane and surrounding bones.** As it leaves the pelvis through the lesser sciatic foramen, it becomes tendinous and makes a sharp turn around the ischium, just inferior to its spine. It is separated from this bone by a bursa.

The superior and inferior gemelli (Fig. 4-27) arise from the **ischial spine** and the **ischial tuberosity**, respectively.

Insertion (Fig. 4-28). **Medial surface of greater trochanter of femur.** The gemelli insert into the tendon of the obturator internus.

Nerve Supply. Nerve to obturator internus (L5 and **S1**). The superior gemellus is also supplied by this nerve, but the inferior gemellus is supplied by the **nerve to the quadratus femoris** (L5 and **S1**).

Actions. Laterally rotates the extended thigh and abducts thigh when it is flexed. **Helps to hold the head of the femur in acetabulum**; *i.e.*, it stabilizes hip joint. The gemelli assist the obturator internus in the performance of their actions.

THE QUADRATUS FEMORIS MUSCLE (Figs. 4-40*B*, 4-41, 4-54, and Table 4-3)

This short, flat, *quadrilateral muscle* is located inferior to the obturator internus and gemelli muscles.

Origin (Figs. 4-40*B* and 4-54). **Lateral border of ischial tuberosity.**

Insertion (Figs. 4-20 and 4-28). **Quadrate tubercle of femur** (Fig. 4-8*B*).

Nerve Supply. Nerve to quadratus femoris (L5 and S1).

Action. Laterally rotates thigh.

NERVES OF GLUTEAL REGION

There are many nerves in the gluteal region; the deep ones are the most important clinically.

THE SUPERFICIAL GLUTEAL NERVES (Figs. 4-41 and 4-42)

The skin of the gluteal region is richly innervated. It receives cutaneous branches from several lumbar and sacral segments. These cutaneous nerves are sometimes called clunial nerves (L. *clunis*, buttock).

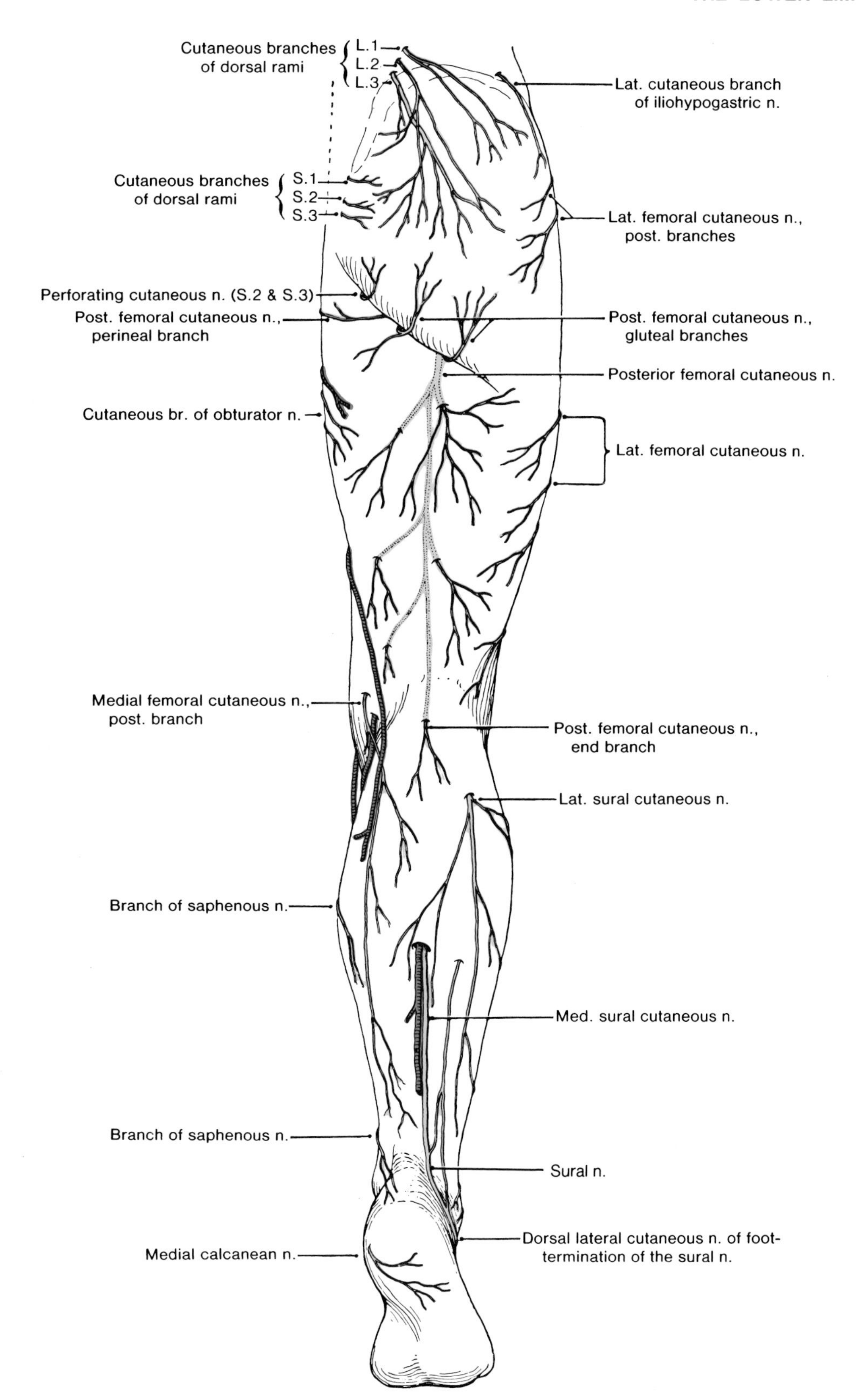

Figure 4-42. Drawing of a posterior view of the cutaneous nerves of the right lower limb (see Fig. 4-14 for an anterior view). Observe that the chief supply to the buttock is from the dorsal primary rami of lumbar and sacral nerves. The cutaneous nerves to buttock are sometimes called clunial nerves. (L. *clunis*, buttock).

The superior cutaneous nerves are lateral branches of the *dorsal rami of the first three lumbar nerves.* They emerge from the deep fascia just superior to the iliac crest and supply the skin on the superior two-thirds of the buttock (Fig. 4-42).

The middle cutaneous nerves are also lateral branches of the *dorsal rami of the first three sacral nerves.* They become cutaneous along a line connecting the posterior superior iliac spine and the tip of the coccyx and supply the skin over the sacrum and adjacent areas of the buttock (Fig. 4-41).

The inferior cutaneous nerves are gluteal branches of the *posterior femoral cutaneous nerves,* and are larger than the other cutaneous nerves. They are branches of the *ventral rami of the first three sacral nerves.* These nerves become cutaneous and curve around the inferior border of the gluteus maximus muscle, just superior to the *gluteal fold* (Fig. 4-7), to supply the inferior one-third of the buttock (Fig. 4-42).

The iliohypogastric nerve arises from the first lumbar ventral ramus, with a small contribution from the 12th thoracic. It supplies cutaneous branches to the lateral area of the hip over the crest of the ilium and the greater trochanter of the femur (Figs. 4-34 and 4-42).

THE DEEP GLUTEAL NERVES (Figs. 4-40 and 4-41)

There are seven deep gluteal nerves which are branches of the ***sacral plexus*** and leave the pelvis via the greater sciatic foramen.

Except for the superior gluteal nerve, **they emerge inferior to the piriformis muscle.** The site of exit of these nerves can be detected by deep pressure (usually painful), just superior to the midpoint of a line joining the posterior superior iliac spine and the ischial tuberosity.

The Superior Gluteal Nerve (L4, L5, and S1). This nerve leaves the pelvis through the superior part of the greater sciatic foramen, ***superior to the piriformis muscle*** (Fig. 4-41). It runs laterally between the gluteus medius and minimus muscles with the deep branch of the superior gluteal artery. The superior gluteal nerve divides into a ***superior branch*** that supplies the gluteus medius, and an ***inferior branch*** which supplies the gluteus minimus and tensor fasciae latae muscles.

The Inferior Gluteal Nerve (L5, S1, and S2). This nerve leaves the pelvis through the inferior part of the greater sciatic foramen, ***inferior to the piriformis muscle*** and superficial to the sciatic nerve (Fig. 4-41). It immediately breaks up into several branches that supply the overlying gluteus maximus muscle.

The Sciatic Nerve (L4 and L5; S1, S2, and S3). ***This very important nerve***, about 2 cm in diameter at its commencement, is *the largest nerve in the body* (Figs. 4-40*B*, 4-41, and 4-44).

The sciatic nerve leaves the pelvis through the inferior part of the greater sciatic foramen and *enters the gluteal region inferior to the piriformis muscle.* It runs inferolaterally deep to the gluteus maximus muscle, *midway between the greater trochanter of the femur and the ischial tuberosity.* The sciatic nerve rests on the ischium and then passes posterior to the obturator internus, quadratus femoris, and adductor magnus muscles. *The sciatic nerve usually supplies no structures in the gluteal region.*

The sciatic nerve is really two nerves, the tibial and common peroneal, which are bound together in the same connective tissue sheath (*epineurium*). The two nerves usually separate from each other in the inferior third of the thigh (Fig. 4-52), but occasionally they are separate when they leave the pelvis (Fig. 4-43, *B* and *C*). In these cases the tibial nerve passes inferior to the piriformis and the common peroneal pierces this muscle or passes superior to it.

The Posterior Femoral Cutaneous Nerve (S1, S2, and S3). This nerve leaves the pelvis with the inferior gluteal nerve and vessels and the sciatic nerve (Fig. 4-41). It gives branches to the skin of the inferior part of the buttock (Fig. 4-42) and continues inferiorly where *it supplies the skin of the posterior thigh and popliteal regions.*

The Nerve to the Quadratus Femoris Muscle (L4, L5, and S1). This nerve passes deep to the sciatic nerve and the obturator internus muscle, and over the posterior surface of the hip joint. It supplies an articular branch to this joint and innervates the quadratus femoris and inferior gemellus muscles.

The Nerve to the Obturator Internus (L5, S1, and S2). This nerve passes through the greater sciatic foramen inferior to the piriformis muscle and across the base of the ischial spine (Fig. 4-41). It supplies the superior gemellus muscle and then passes posterior to the ischial spine, re-entering the pelvis via the lesser sciatic foramen to supply the obturator internus muscle.

The Pudendal Nerve (S2, S3, and S4). This nerve is the most medial structure to pass through the greater sciatic foramen inferior to the piriformis muscle (Fig. 4-41). It passes lateral to the sacrospinous ligament, re-entering the pelvis via the lesser sciatic foramen, to supply structures in the perineum (Fig. 3-24).

Injury to the deep gluteal nerves and to the sciatic nerve, which passes through the gluteal region, may occur in wounds of the but-

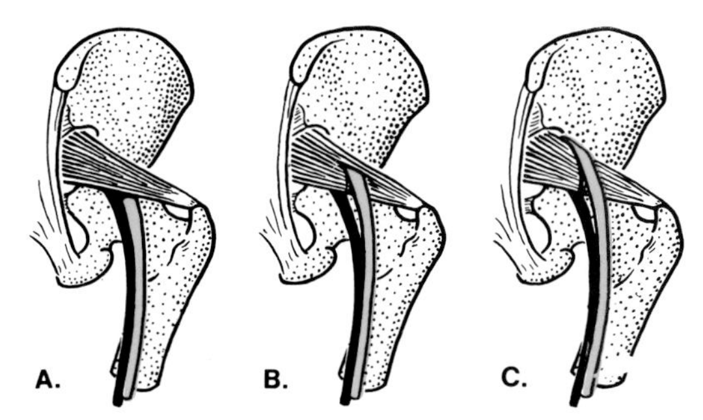

Figure 4-43. Diagrams illustrating the relationship of the sciatic nerve to the piriformis muscle. In most cases it passes inferior to the muscle (*A*). In 10 to 12% of cases the sciatic divides before entering the gluteal region and the common peroneal division (*yellow*) passes through the piriformis muscle (*B*). In 0.5% of cases the common peroneal division (*yellow*) passes superior to the muscle, where it is vulnerable to injury during intragluteal injections. The safe area for giving injections is illustrated in Figure 4-44, where there are no important nerves or vessels (as shown in Fig. 4-41).

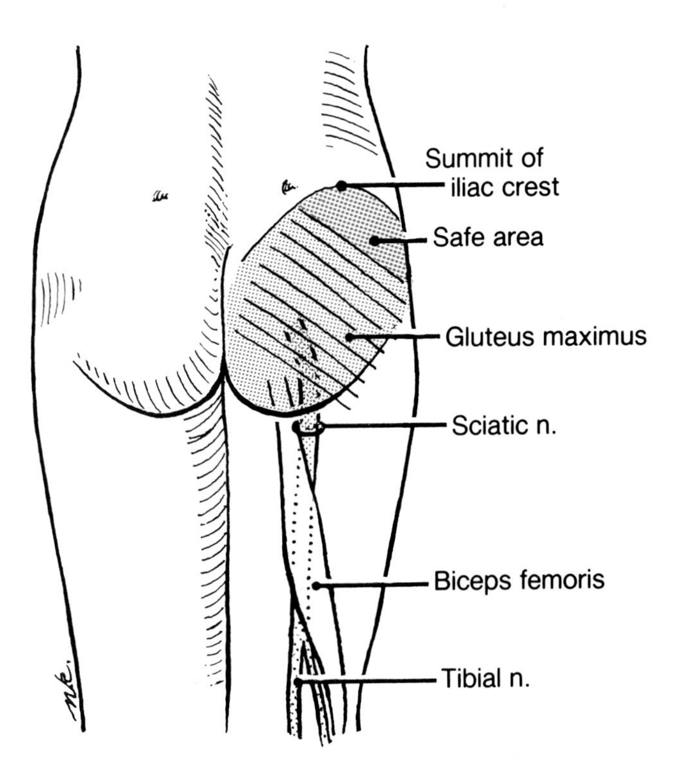

Figure 4-44. Diagram showing the extent of the gluteal region or buttock and the location of the sciatic nerve deep to the gluteus maximus muscle. The safe area for giving intramuscular injections is shown in *green.* The area to be avoided is shown in *red.* If injections are given into the rounded inferior part (summit or prominence) of the buttock, the sciatic or other nerves and vessels may be injured (Fig. 4-41).

tock (*e.g.*, gunshot or stab wounds). Consequently, knowledge of their course through the gluteal region is clinically important (Fig. 4-41).

With respect to the sciatic nerve, *the buttock has a* ***side of safety*** *(its lateral side) and a* ***side of danger*** *(its medial side).* Wounds or surgery on the medial side are liable to injure the sciatic nerve and its branches to the hamstring muscles on the posterior aspect of the thigh (Fig. 4-40). Paralysis of these muscles results in impairment of extension of the thigh and flexion of the leg (Table 4-4).

Sciatica *is a term used to describe pain in the area of distribution of the sciatic nerve.* The pain is generally in: the gluteal region, particularly in the area of the greater sciatic notch; the posterior aspect of the thigh; the posterior and lateral aspects of the leg; and the lateral part of the foot, particularly around the lateral malleolus (Fig. 5-64).

Although sciatica may result from irritation of the sciatic nerve owing to inflammation, **sciata is usually caused by pressure on a dorsal root and/or a ventral root of** one of the nerves which form it (Fig. 5-68).

As the sciatic nerve is derived from the ventral rami of L4, L5, S1, S2, and S3 nerves, the location of the pain felt by a patient varies according to the nerve root(s) involved (Fig. 1-23). For example, if the intervertebral disc between L5 and S1 herniates posteriorly, pressure would likely be exerted on the S1 roots of the sciatic nerve. As a result the patient would probably experience pain over the posterior region of the thigh and the posterolateral region of the leg (Fig. 5-64).

Sciatica can also result from pressure on the sciatic nerve in the pelvis, in the gluteal region, or in the thigh.

The gluteal region is a common site for the intramuscular injection of drugs. Intramuscular injections penetrate the skin, subcutaneous tissue, and muscles. Muscles in the gluteal region are thick and large (Figs. 4-40 and 4-41); consequently they provide a large surface area for absorption of drugs.

It is essential to know the extent of the gluteal region and the safe region for giving injections. Too many people restrict the area of the buttock to the "cheek" or most prominent part; this is a dangerous concept. The full extent of the buttock is shown in Figure 4-44.

As there are a number of important nerves and blood vessels in the gluteal region, **injections can only be made safely into the superior portion of the gluteus medius muscle**, the part which is not covered by the gluteus maximus muscle (Figs. 4-40 and 4-44).

Injections into other areas could endanger and possibly injure the sciatic or other nerves and vessels (Fig. 4-41). It used to be recommended that injections be given anywhere in the superolateral quadrant of the buttock. This is not recommended now because the sciatic nerve sometimes divides within the pelvis into the tibial and common peroneal nerves, in which case the common peroneal nerve passes through the piriformis muscle or superior to it. Hence injection into the inferomedial part of the superolateral quadrant could injure the common peroneal nerve (Fig. 4-43*C*) and produce **footdrop.**

Improper intragluteal injections may also cause injury to the gluteal branches of the **posterior femoral cutaneous nerve** resulting in ***pain and dyesthesia*** (loss of sensation) in the area of skin supplied by its gluteal branches (Fig. 4-42).

The hazards of injecting drugs into the gluteal region of small infants are well recognized. Because of the danger of injuring the nerves, drugs are commonly injected into the muscles of the anterolateral region of the thigh. To avoid injury to the femoral nerve, injections are given inferolateral to the anterior superior iliac spine into the tensor fasciae latae and vastus lateralis muscles (Fig. 4-21*A*). Some doctors also use this site for giving intramuscular injections to adults.

ARTERIES OF THE GLUTEAL REGION

The arteries supplying the gluteal region directly are *branches of the* ***internal iliac artery***.

THE SUPERIOR GLUTEAL ARTERY (Figs. 4-31*B* and 4-41)

This short artery is ***the largest branch of the internal iliac artery*** (Fig. 3-57). It passes posteriorly between the lumbosacral trunk and the first sacral ventral ramus.

The superior gluteal artery leaves the pelvis through the greater sciatic foramen, superior to the piriformis muscle. The superior gluteal artery divides immediately into superficial and deep branches (Fig. 4-31*B*). The superficial branch supplies the gluteus maximus muscle and the skin over this muscle's origin (Fig. 4-40); the deep branch supplies the gluteus medius, gluteus minimus, and tensor fasciae latae muscles (Fig. 4-41). The superior gluteal artery anastomoses with the inferior gluteal and medial circumflex femoral arteries (Fig. 4-31*B*).

THE INFERIOR GLUTEAL ARTERY (Figs. 4-31*B* and 4-41)

This artery arises from the internal iliac artery and passes posteriorly between the first and second (or second and third) sacral ventral rami (Fig. 3-54).

The inferior gluteal artery leaves the pelvis through the greater sciatic foramen, inferior to the piriformis muscle. It supplies the gluteus maximus, obturator internus, quadratus femoris, and superior parts of the hamstring muscles. The inferior gluteal artery anastomoses with the superior gluteal artery and participates in the **cruciate anastomosis of the thigh,** involving the first perforating arteries of the profunda femoris and the medial and lateral circumflex femoral arteries.

THE INTERNAL PUDENDAL ARTERY (Figs. 3-54 and 4-41)

This vessel arises from the internal iliac artery, passes lateral to the pudendal nerve, and *leaves the pelvis via the greater sciatic foramen inferior to the piriformis muscle.* The internal pudendal artery then *descends posterior to the ischial spine, re-enters the pelvis via the lesser sciatic foramen,* and then enters the perineum with the pudendal nerve. It supplies the external genitalia (Figs. 3-15 and 3-22) and muscles in the pelvic and gluteal regions (Fig. 4-41).

VEINS OF THE GLUTEAL REGION

The veins supplying the gluteal region are tributaries of the internal iliac veins (Figs. 2-129 and 3-58).

THE GLUTEAL VEINS

The superior and inferior gluteal veins accompany the corresponding arteries through the greater sciatic foramen. Usually each of these veins is double, *i.e.,* **venae comitantes** (L. accompanying veins). They communicate with the tributaries of the femoral vein and thus can provide an alternate route for blood return from the lower limb if the femoral vein is occluded or has to be ligated.

THE PUDENDAL VEINS

The internal pudendal veins accompany the corresponding arteries and join to form a single vein that enters the internal iliac vein. They drain blood from the external genitalia and perineal region (Fig. 3-22).

THE POSTERIOR THIGH MUSCLES

Three large femoral muscles make up the **hamstring muscles,** *semitendinosus, semimembranosus,* and *biceps femoris* (Figs. 4-47, 4-50, and Table 4-4).

The hamstrings (the common term for these muscles) can be made to stand out by flexing the leg against resistance.

The hamstrings have a *common site of origin* from the **ischial tuberosity** (Fig. 4-20) deep to the gluteus maximus, but one of them, the biceps femoris, has an additional origin from the femur (Fig. 4-40). They also have a *common nerve supply* from the **sciatic nerve** (Figs. 4-40*B* and 4-82).

The hamstring muscles span two joints, the hip joint and knee joint; hence they are **extensors of the thigh and flexors of the leg.** You cannot perform both actions fully at the same time. The hamstring muscles descend in the posterior aspect of the thigh and their tendons are visible posterior to the knee (Figs. 4-40, 4-47, and 4-49).

The posterior thigh muscles became known as the hamstrings because their tendons posterior to the knee were used to hang up hams (hip and thigh regions) of animals such as pigs. Furthermore, in ancient times it was common for soldiers to slash their opponent's horse posterior to the knees in order to cut the tendons of their hamstrings. This would bring the horses and their riders down. Similarly, they cut the same tendons of the soldiers so they could not run; this was called "hamstringing" the enemy.

THE SEMITENDINOSUS MUSCLE (Figs. 4-40, 4-41, 4-46 to 4-52, and Table 4-4)

As its name indicates, *this muscle is half tendinous.* Its slender, cord-like tendon begins about two-thirds of the way down the thigh.

Origin (Fig. 4-40). **Ischial tuberosity** by a common tendon with the long head of the biceps femoris muscle.

Insertion (Fig. 4-26). **Medial surface of superior part of tibia**, posterior to the insertions of the sartorius and gracilis muscles.

Nerve Supply (Fig. 4-50). **Tibial division of sciatic nerve (L5, S1**, and S2).

Actions. Extends thigh at hip joint, **flexes leg** at knee joint, and with the semimembranosus, *it can **medially rotate the tibia on the femur*** (particularly when the knee is semiflexed).

THE SEMIMEMBRANOSUS MUSCLE (Figs. 4-4-40, 4-46 to 4-52, and Table 4-4)

As its name indicates, *this muscle is half membranous.*

Origin (Fig. 4-20). **Ischial tuberosity.**

Insertion (Fig. 4-54). **Posterior part of medial condyle of tibia.**

Nerve Supply (Fig. 4-50). **Tibial division of sciatic nerve (L5, S1**, and S2).

Actions. Extends thigh at hip joint, **flexes leg** at knee joint, and with the semitendinosus *can **medially rotate the tibia*** on the femur (particularly when the knee is semiflexed).

THE BICEPS FEMORIS MUSCLE (Figs. 4-40, 4-46 to 4-52, and Table 4-4)

As its name indicates, *this muscle has two heads of origin*, long and short.

Origin (Fig. 4-40). *Long head*, **ischial tuberosity** by common tendon with semitendinosus muscle. *Short*

Table 4-4
Posterior Thigh Muscles

Muscle	Origin	Insertion	Nerve Supply	Actions
Semitendinosus (Figs. 4-25, 4-26, 4-40, and 4-52)	Ischial tuberosity (Fig. 4-20)	Medial surface of superior part of tibia (Fig. 4-24)	Tibial division of sciatic nerve (**L5**, **S1**, and S2)	Extend thigh and flex leg[1]
Semimembranosus (Figs. 4-26, 4-40, and 4–51)	Ischial tuberosity (Fig. 4-20)	Posterior part of medial condyle of tibia (Fig. 4-24)	Tibial division of sciatic nerve (**L5**, **S1**, and S2)	Extend thigh and flex leg[1]
Biceps femoris (Figs. 4-40, 4-51, and 4-52)	*Long head*: Ischial tuberosity *Short head*: Lateral lip of linea aspera and lateral supracondylar line (Figs. 4-2 and 4-20)	Head of fibula (Figs. 4-19 and 4-45)	*Long head*, tibial division of sciatic nerve (L5, **S1** and S2) *Short head*, common peroneal division of sciatic nerve (L5, **S1**, and S2)	Extend thigh and flex leg[1]

[1] See the text for other actions of these muscles.

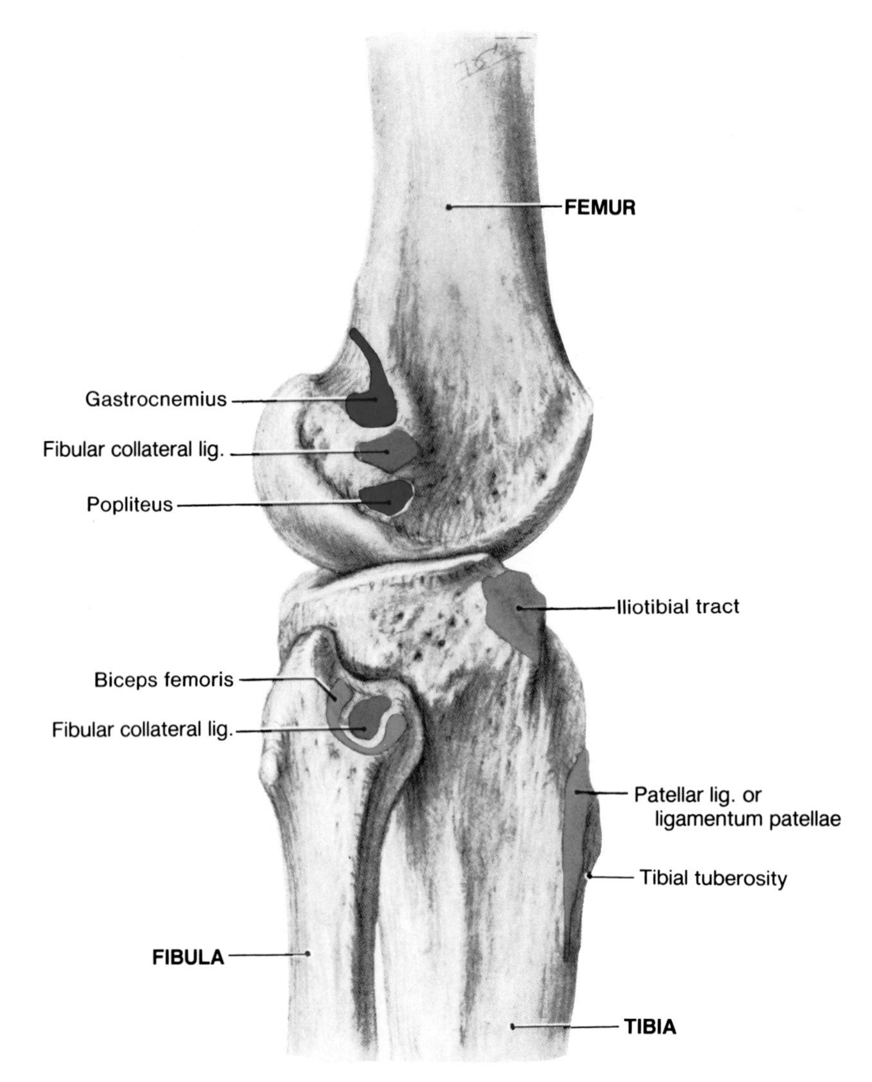

Figure 4-45. Drawing of a lateral view of the bones of the right knee joint showing the attachments of muscles and ligaments. Origins of muscles are shown in *red* and insertions in *blue*. Attachments of the fibular collateral ligament are shown in *green*.

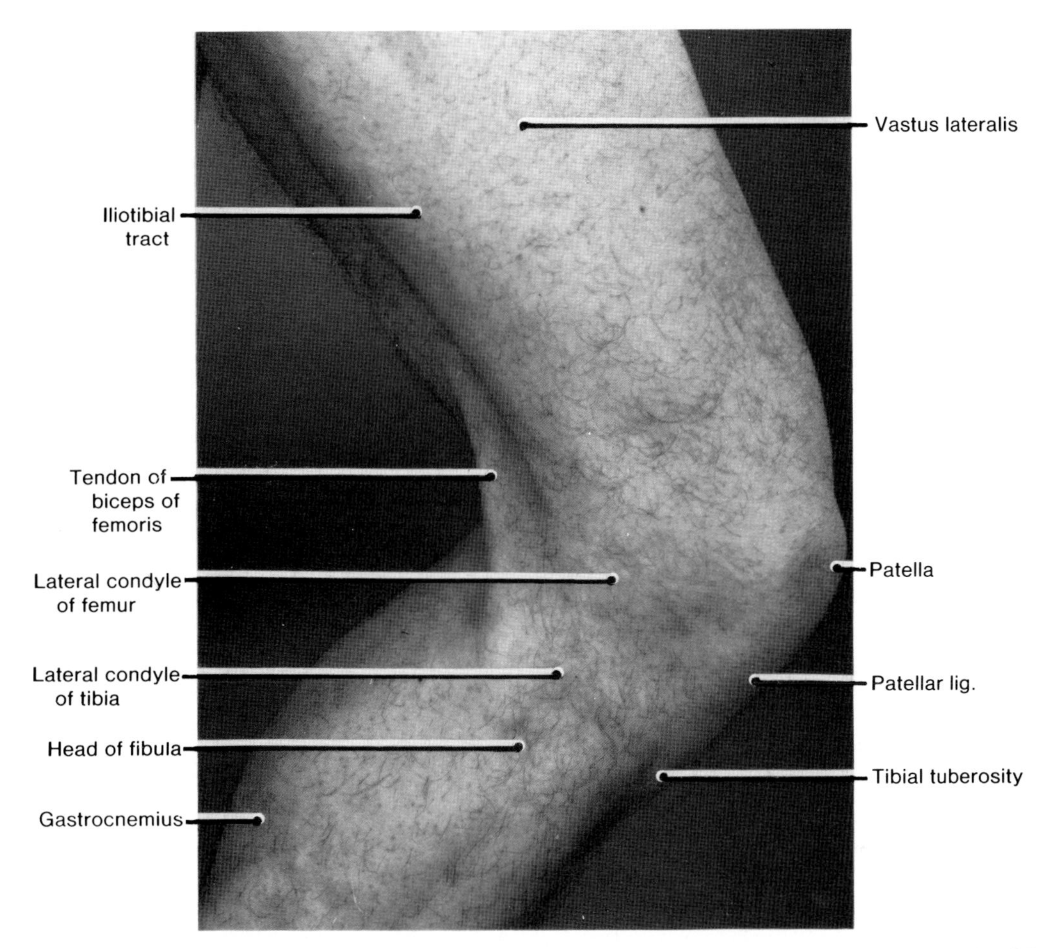

Figure 4-46. Photograph of the lateral surface of the right thigh and proximal part of the leg of a 27-year-old man, showing the principal surface features of the muscles and bones. The gluteus maximus and tensor fasciae latae muscles insert into the iliotibial tract posteriorly and anteriorly, respectively (Fig. 4-40). The iliotibial tract (band of deep fascia) inserts into the lateral condyle of the tibia (Fig. 4-45). The tendon of the biceps femoris may be traced by palpation from the posterior aspect of the distal part of the thigh, inferolaterally to its insertion into the head of the fibula (Fig. 4-45). Verify by palpation that the lateral condyles of the femur and the tibia, as well as the head of the fibula, are subcutaneous. The neck of the fibula can be felt just distal to its head. The tendon of the biceps femoris and the proximal end of the fibula are important guides to the common peroneal nerve. It is indicated by a line drawn along the tendon of the biceps femoris, posterior to the head of the fibula, and then around the lateral aspect of the neck of the fibula to its anterior aspect (also see Figs. 4-57*A* and 4-67). You should be able to feel this commonly injured nerve lateral to the neck of the fibula. It is particularly exposed to injury here because it is subcutaneous and is related deeply to bone (fibula). See Figure 4-57*A* for the relationship of the common peroneal nerve to the head and neck of the fibula.

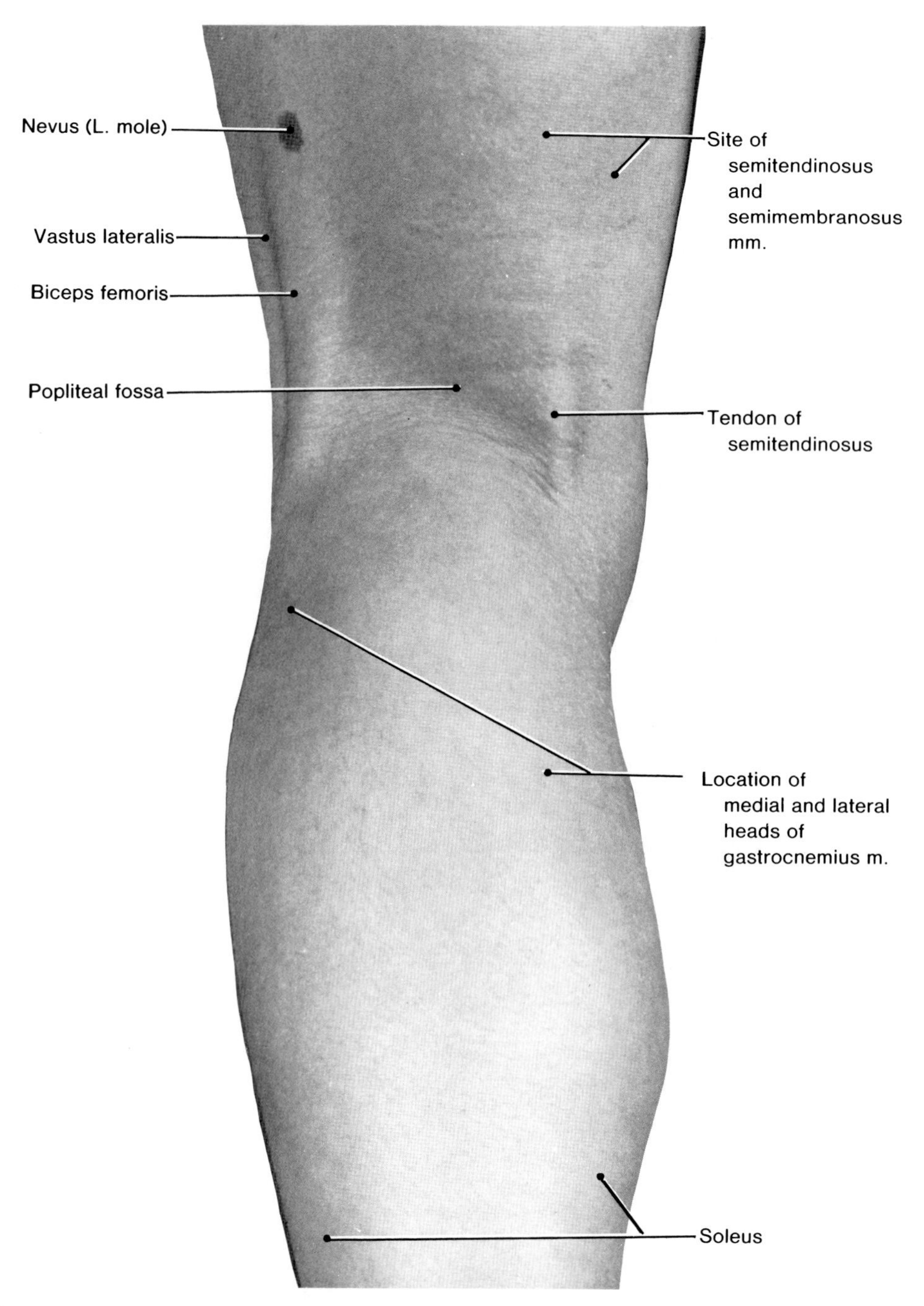

Figure 4-47. Photograph of the posterior aspect of the distal part of the left thigh and proximal part of the leg of a 12-year-old girl, showing the principal surface features of the muscles and the popliteal fossa. This fossa is visible only when the knee is flexed (as here). Note that the popliteal fossa is bounded superiorly and laterally by the biceps femoris muscle and superiorly and medially by the semitendinosus and semimembranosus muscles. To feel the tendons of your hamstrings as displayed here, sit on a chair and press your heel firmly against the chair leg. The common peroneal nerve passes on the medial side of the biceps femoris tendon and then passes posterior to the head of the fibula (Figs. 4-49 and 4-57*A*). The tendons of the semitendinosus and semimembranosus can be traced medially toward their insertion into the proximal part of the tibia (Fig. 4-24). Near its insertion, the tendon of the semitendinosus is lateral to that of the semimembranosus and, as it is on a more superficial plane, is clearly visible. The gastrocnemius, forming the inferior boundary of the popliteal fossa along with the soleus muscle, makes up the triceps surae which forms the prominence of the calf. The popliteal artery (Fig. 4-52) lies deeply in the popliteal fossa and the popliteal pulse may be palpated here; however, its pulsations are difficult to feel.

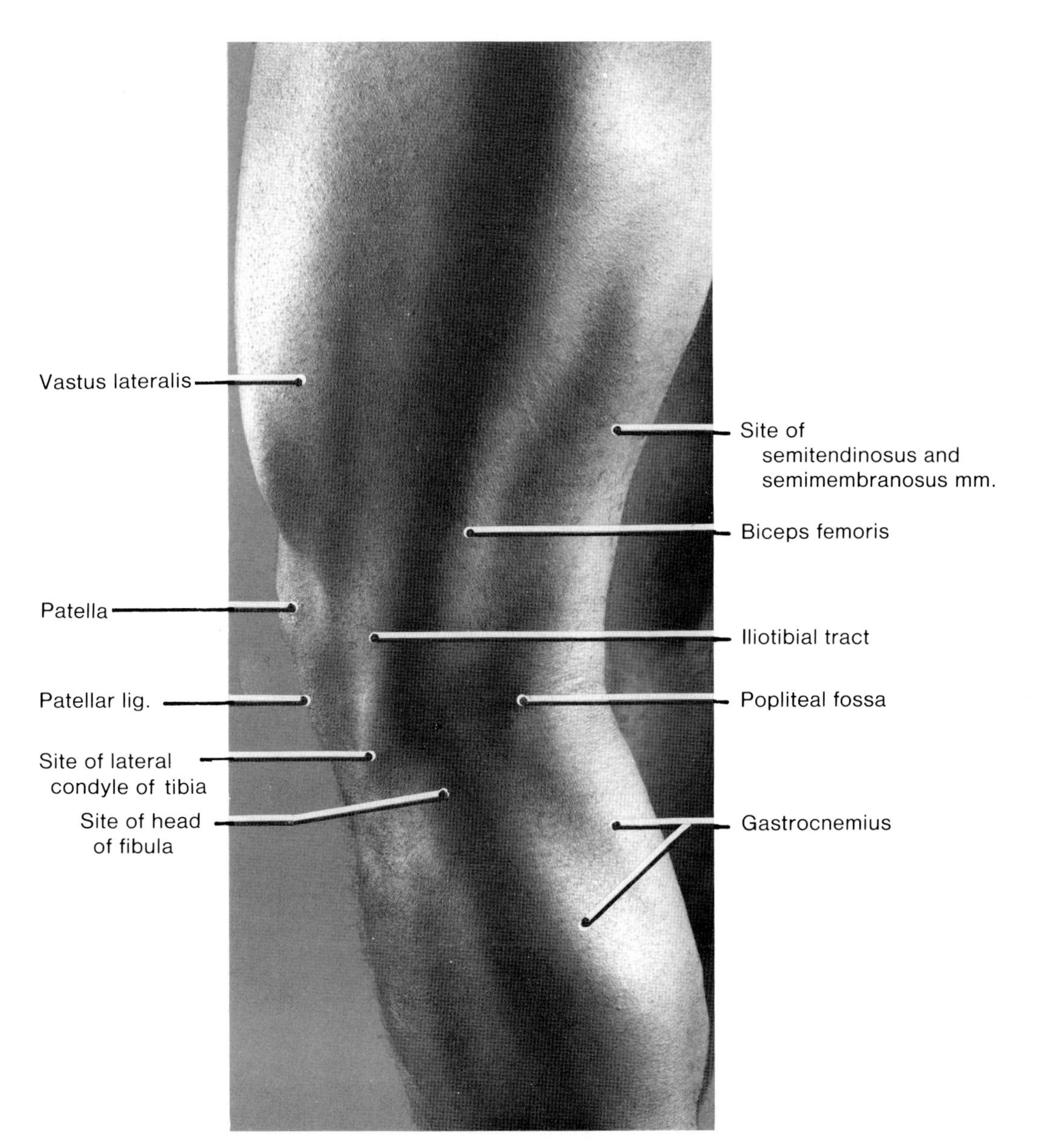

Figure 4-48. Photograph of the lateral aspect of the left thigh and proximal half of the leg of a 35-year-old man showing the principal surface features of the muscles and bones when the knee is slightly flexed. Observe the prominent posterior border of the iliotibial tract which inserts into the lateral condyle of the tibia (Fig. 4-45), the fibrous capsule of the knee joint, and the patella. About one fingerbreadth posterior to the iliotibial tract, observe the biceps femoris tendon that inserts on the head of the fibula (Fig. 4-45). The popliteal fossa is the diamond-shaped region posterior to the knee (see the diagram in Fig. 4-51). Its contents are shown in the following two illustrations.

head, **lateral lip of linea aspera and lateral supracondylar line** of femur.

Insertion (Fig. 4-45). **Head of fibula.** *The tendon of the biceps femoris muscle is split by the fibular collateral ligament of the knee.*

The rounded tendon of the biceps femoris can easily be seen and felt where it passes the knee to insert into the head of the fibula, especially when the knee is flexed against resistance (Figs. 4-46 and 4-47).

Nerve Supply (Fig. 4-50). *Long head*, **tibial division of sciatic nerve** (L5, **S1**, and S2) like the other hamstrings. *Short head*, **common peroneal division of the sciatic nerve** (L5, **S1**, and S2).

Actions. *Long head*, like the other hamstrings, **extends the thigh** at hip joint. *Both heads*, **flex the leg** at the knee joint, and ***laterally rotate the leg*** when the leg is flexed.

Because the *heads of the biceps femoris have a different nerve supply* (*i.e.*, from different divisions of the sciatic nerve), a wound in the thigh may sever a nerve, paralyzing one head and not the other.

The length of the hamstring muscles varies considerably. In some people they will not stretch enough to allow them to touch their toes when they flex their vertebral column and keep their knees straight. In other people the hamstrings are long and they can easily touch the floor with their palms, or do a high kick with comparative ease. Some athletes are unable to excel in some sports (*e.g.*, gymnastics) despite rigorous exercises because of their short hamstrings.

"Pulled hamstrings" are common sports injuries in persons who run very hard or kick (*e.g.*, footballs). Often the violent muscular exertion required to excel in sports tears or **avulses part of the tendinous origin of the hamstrings from the ischial tuberosity.** Usually there is also contusion and tearing of muscle fibers, resulting in rupture of some blood vessels supplying the muscles. The resultant **hematoma** (collection of blood) is contained by the dense fascia lata (Figs. 4-16 and 4-18). *The tearing of fibers of a hamstring muscle is so painful, the runner will often fall and writhe in pain.*

Contracture of the tendons of the hamstring muscles is a common complication of diseases of the knee joint. This causes flexion of the leg at the knee joint and a partial posterior dislocation of the tibia with some lateral rotation. The rotation is probably caused by contraction of the biceps femoris muscle.

THE POPLITEAL FOSSA

The popliteal fossa is **the diamond-shaped region posterior to the knee** (Figs. 4-47 to 4-52). The popliteal fossa lies posterior to the distal third of the femur, the knee joint, and the proximal part of the tibia. The fossa appears as a hollow space when the knee joint is flexed.

Flex your leg and palpate the hollow (popliteal fossa) posterior to it, between the borders of the semitendinosus and biceps femoris tendons (Fig. 4-47). Palpate deeply in the fossa for pulsations of the popliteal artery. You may find your "**popliteal pulse**" difficult to feel because the artery lies so deeply in it (Figs. 4-49 and 4-52).

Slowly extend your leg and feel the prominence formed partly by the fleshy semimembranosus muscle as it pushes laterally and bulges posteriorly.

Roof of the Popliteal Fossa (Figs. 4-47 and 4-49). The roof or posterior wall of the popliteal fossa is *formed by skin and fascia* which are stretched during extension of the leg.

The **superficial popliteal fascia** contains fat, the small saphenous vein, and three cutaneous nerves (Figs. 4-10*A* and 4-49).

The small saphenous vein perforates the deep popliteal fascia and ends in the popliteal vein.

The **deep popliteal fascia** forms a strong, dense sheet that affords a *protective covering for the neurovascular structures passing from the thigh to the leg* (Fig. 4-49). When the leg is extended, the semimembranosus muscle moves laterally offering further protection to these structures. The *deep fascia of the thigh* is strengthened posterior to the knee by transverse fibers. Here it forms one of the layers of the roof of the popliteal fossa and is known as the *deep popliteal fascia.* Most of this fascia was removed during the superficial dissection of the popliteal fossa illustrated in Figure 4-49.

Floor of the Popliteal Fossa (Figs. 4-52 and 4-54). The floor or anterior wall of the popliteal fossa is formed by the *popliteal surface of the femur*, the *oblique popliteal ligament*, an expansion of the semimembranosus tendon, and the *popliteus fascia.*

Because the deep popliteal fascia is strong, and does not permit expansion, pain from an abscess or tumor in the popliteal fossa is usually severe. In addition, **popliteal abscesses** tend to spread superiorly and inferiorly because of the tough deep popliteal fascia enclosing them.

Because the floor of the popliteal fossa is related to the knee joint, fluid may escape from the synovial cavity and accumulate in the pop-

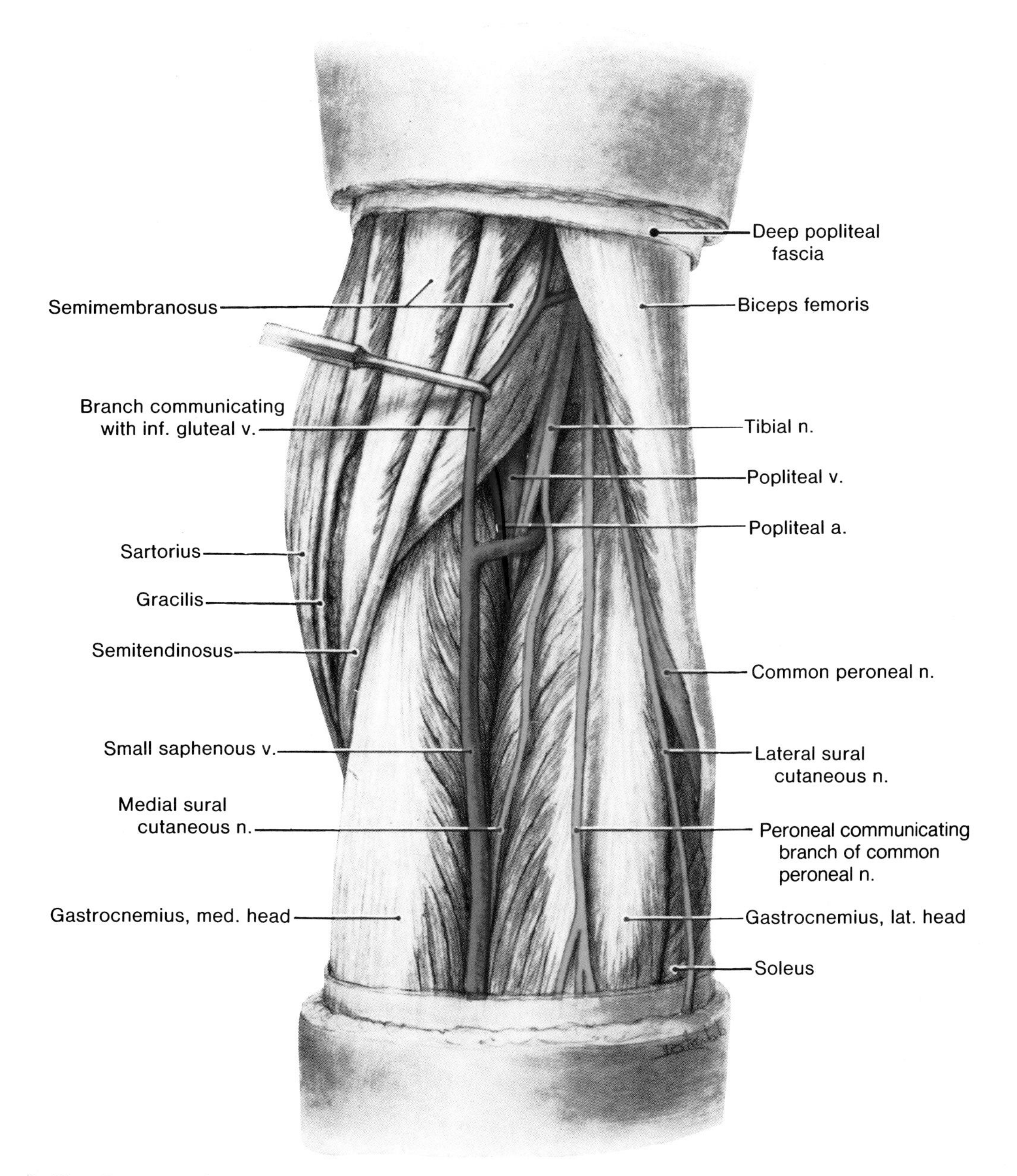

Figure 4-49. Drawing of a superficial dissection of the right popliteal fossa. Note that only a small part of the popliteal artery is not covered by muscles. Observe the two heads of the gastrocnemius, the semimembranosus, the semitendinosus, and the biceps femoris. These muscles form the boundaries of the diamond or lozenge-shaped popliteal fossa (Fig. 4-51). To expose the contents of the fossa, the pad of fat which surrounds the nerves and vessels was carefully removed. Note the small saphenous vein running between the heads of the gastrocnemius. Deep to this vein is the medial sural cutaneous nerve which leads proximally to the tibial nerve. Observe that the tibial nerve is superficial to the popliteal vein, which in turn is superficial to the popliteal artery. Note the common peroneal nerve following the posterior border of the biceps femoris and giving off two cutaneous branches: the lateral sural cutaneous nerve and the peroneal communicating branch. Usually the peroneal communicating branch unites with the medial sural cutaneous nerve (branch of tibial) to form the sural nerve. The level of union varies: (1) in the popliteal fossa (Fig. 4-50); (2) in the middle of the calf (Fig. 4-10*A*); or (3) posterosuperior to the lateral malleolus (Fig. 4-42).

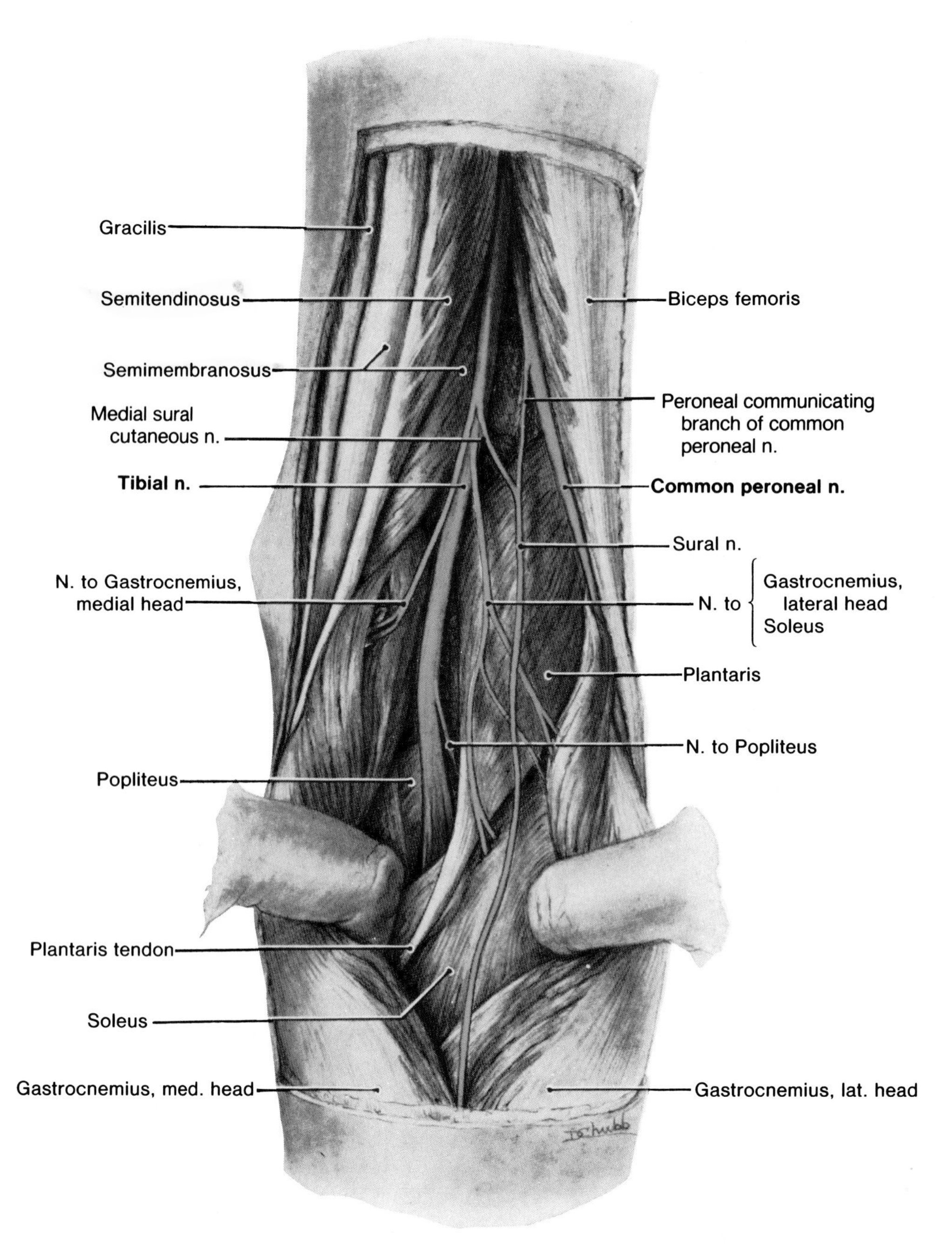

Figure 4-50. Drawing of a deeper dissection of the popliteal fossa than that shown in Figure 4-49. The two most important nerves are the tibial and the common peroneal (terminal branches of sciatic nerve). To expose them the medial and lateral heads of the gastrocnemius muscle were pulled forcibly apart. Observe the medial sural cutaneous nerve, a branch of the tibial nerve joining the peroneal communicating branch of the common peroneal nerve to form the sural nerve. This junction is high; usually it is 5 to 8 cm superior to the ankle (Fig. 4-10*A*). *Sura is Latin for calf.* Note that all motor branches in this region arise from the tibial nerve, one from its medial side and the others from its lateral side.

liteal fossa, forming a fluid-filled swelling called a **popliteal cyst.** This condition is discussed with the knee joint (p. 529).

BOUNDARIES OF THE POPLITEAL FOSSA (Figs. 4-47 to 4-52)

The muscles surrounding the popliteal fossa delineate a diamond-shaped space that is bounded *superolaterally* by the **biceps femoris** muscle, *superomedially* by the **semimembranosus and semitendinosus** muscles, and *inferolaterally and inferomedially* by the lateral and medial **heads of the gastrocnemius** muscle, respectively. Palpate the superior boundaries formed by the diverging hamstring tendons (Fig. 4-47), and the inferior boundaries formed by the converging heads of the gastrocnemius (Fig. 4-49).

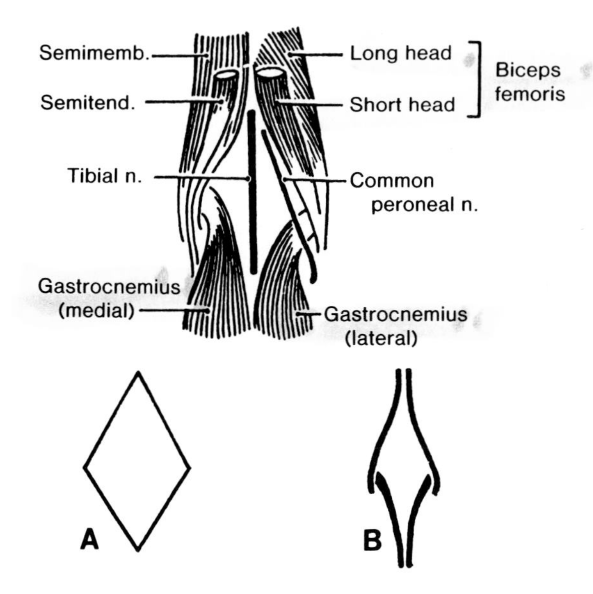

Figure 4-51. Drawings illustrating the boundaries of the right popliteal fossa and its two important nerves (tibial and common peroneal). The muscular boundaries have been pulled apart (as in Fig. 4-50) to illustrate its diamond-shape. Normally these muscles are closely packed together (Fig. 4-49) and the fossa is only a small interval between the diverging hamstring muscles and the converging gastrocnemius muscles. *A* and *B* illustrate the diamond shape of the fossa. Superolaterally the fossa is bounded by the biceps femoris muscle; superomedially it is bounded by the semitendinosus and semimembranosus muscles; and inferolaterally and inferomedially it is bounded by the lateral and medial heads of the gastrocnemius muscle. Note that the tibial nerve, the larger division of the sciatic nerve, passes through the middle of the popliteal fossa (also see Fig. 4-50).

CONTENTS OF THE POPLITEAL FOSSA (Figs. 4-49 to 4-52)

When the muscles forming the boundaries of the fossa are pulled apart, especially the heads of the gastrocnemius, the popliteal fossa and its contents can be observed (Fig. 4-50). Although the fossa appears large when this is done, normally the muscles are packed closely together and the fossa is relatively small (Fig. 4-49).

The contents of the popliteal fossa are: **fat**, the **popliteal vessels** (artery, vein, and lymph), the **tibial and common peroneal nerves**, the **small saphenous vein**, the end branch of the **posterior femoral cutaneous nerve** (Fig. 4-42), an articular branch of the **obturator nerve**, four to six *popliteal lymph nodes*, and the *popliteus bursa* (Fig. 4-110).

The Popliteal Artery (Figs. 4-31*B*, 4-40*A*, 4-49, and 4-52). The popliteal artery begins as soon as the femoral artery passes through the adductor hiatus in the tendon of the adductor magnus muscle (Fig. 4-29). ***The popliteal artery is the direct continuation of the femoral artery.*** From its *origin at the adductor hiatus*, the popliteal artery passes inferolaterally through the fat of the popliteal fossa.

The popliteal artery ends by dividing into the anterior and posterior tibial arteries at the inferior border of the popliteus muscle (Figs. 4-31*B* and 4-52).

The popliteal artery is located deeply throughout its course (Figs. 4-40*A*, 4-49, and 4-52). Anteriorly, from proximal to distal, it lies against the fat on the posterior surface of the femur, the fibrous capsule of the knee joint, and the popliteus fascia (Fig. 4-52). Posteriorly, from proximal to distal, it lies deep to the semimembranosus muscle, popliteal vein, tibial nerve, and gastrocnemius muscle.

Branches of the popliteal artery are numerous (Fig. 4-53). They supply the skin on the posterior aspect of the leg and the muscles of the thigh and leg.

Genicular branches of the popliteal artery (L. *genu*, knee) supply the articular capsule and the ligaments of the knee joint. *The genicular arteries are named as follows:* ***lateral superior and inferior, medial superior and inferior, and middle genicular arteries*** (Figs. 4-52 and 4-53).

Muscular branches of the popliteal artery supply the hamstring, gastrocnemius, soleus, and plantaris muscles. The arteries supplying the calf muscles are called **sural arteries** (L. *sura*, the calf). The superior muscular branches of the popliteal artery have clinically important anastomoses with the terminal part of the profunda femoris and gluteal arteries (Fig. 4-31).

A *cutaneous branch* of the popliteal artery, the **superficial sural artery**, accompanies the small saphenous vein.

The Genicular Anastomoses (Figs. 4-31 and 4-

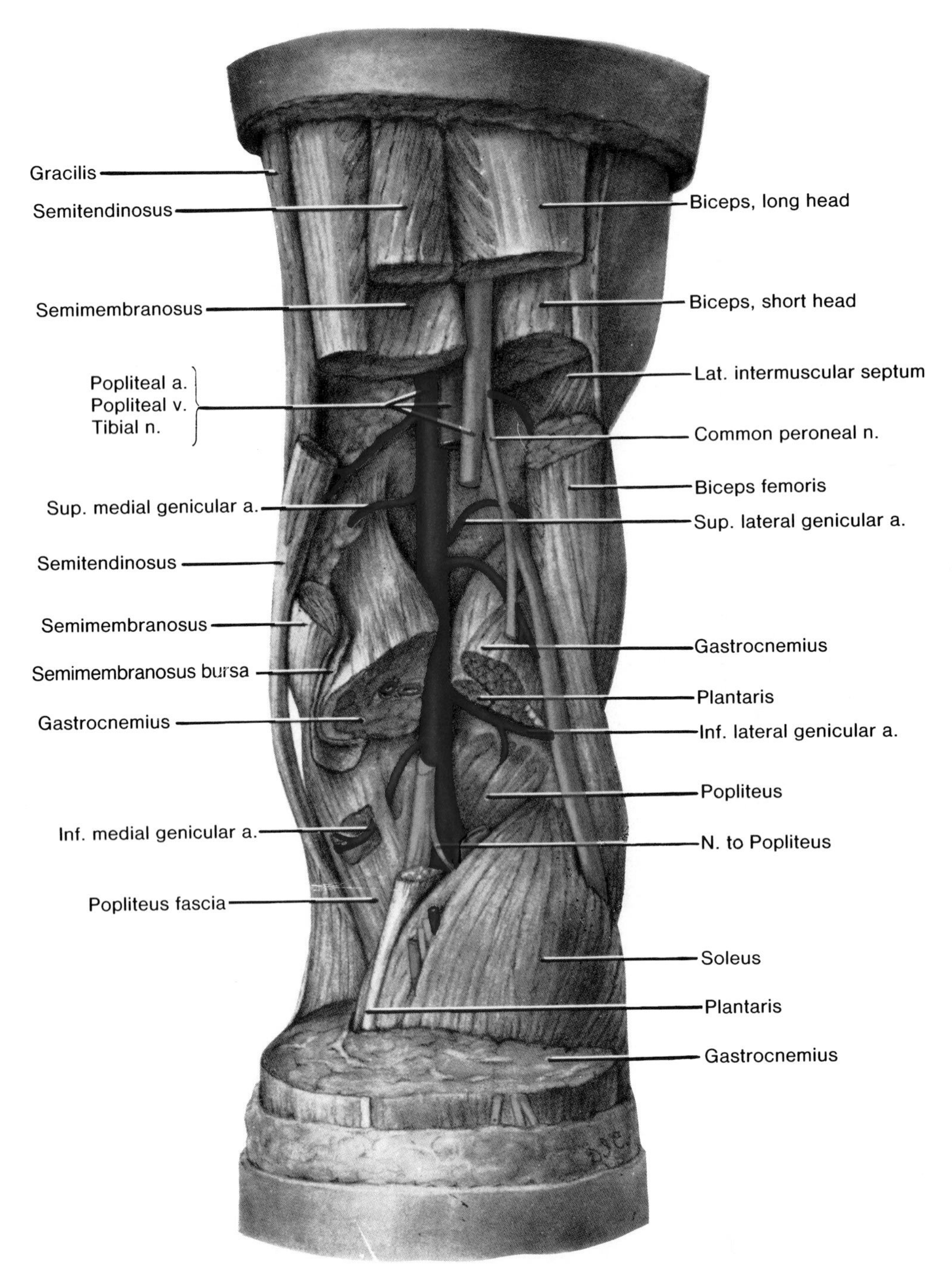

Figure 4-52. Drawing of a deep dissection of the right popliteal fossa. Observe the popliteal artery and its genicular branches lying deep on the floor of the fossa, formed by the popliteal surface of the femur, the capsule of the knee joint, and the popliteal fascia.

53). *This is an important network of arterial vessels around the knee involving 10 vessels.* The genicular anastomoses are located around the patella and the proximal ends of the tibia and fibula. There is a **superficial network of arterial vessels** between the fascia and the skin, superior and inferior to the patella, and in the fat posterior to the patella. There is also a **deep network of arterial vessels** lying on the articular capsule of the knee joint, and on the condyles of the femur and tibia. In addition to supplying the capsule of the knee joint, these arteries supply the adjacent bones.

Occasionally there is little or no blood flow through the popliteal artery owing to obstructive disease resulting from **atherosclerosis.** A common site for atheromatous occlusion in the femoral artery is at the adductor hiatus (Fig. 4-29), near the origin of the popliteal artery (Fig. 4-31*A*).

When **obstructive disease of the popliteal artery** is suspected, an obvious sign would be *loss of the popliteal pulse.* Normally the pulsations of this artery can be felt on deep palpation in the popliteal fossa when the leg is flexed. Palpation of this pulse is commonly done by placing the patient in the prone position on a table with the leg partly flexed to relax the popliteal fascia and the hamstring muscles.

Popliteal aneurysm (localized dilation of the popliteal artery) can cause swelling and pain in the popliteal fossa. To prevent rupture of the aneurysm, **the femoral artery may be ligated in the adductor canal.**

When the femoral artery has been ligated, blood can bypass the occlusion via the genicular anastomoses (Fig. 4-53) and reach the popliteal artery beyond the blockage. This is known as an indirect or **collateral circulation.**

When an artery is blocked by an embolus (G.

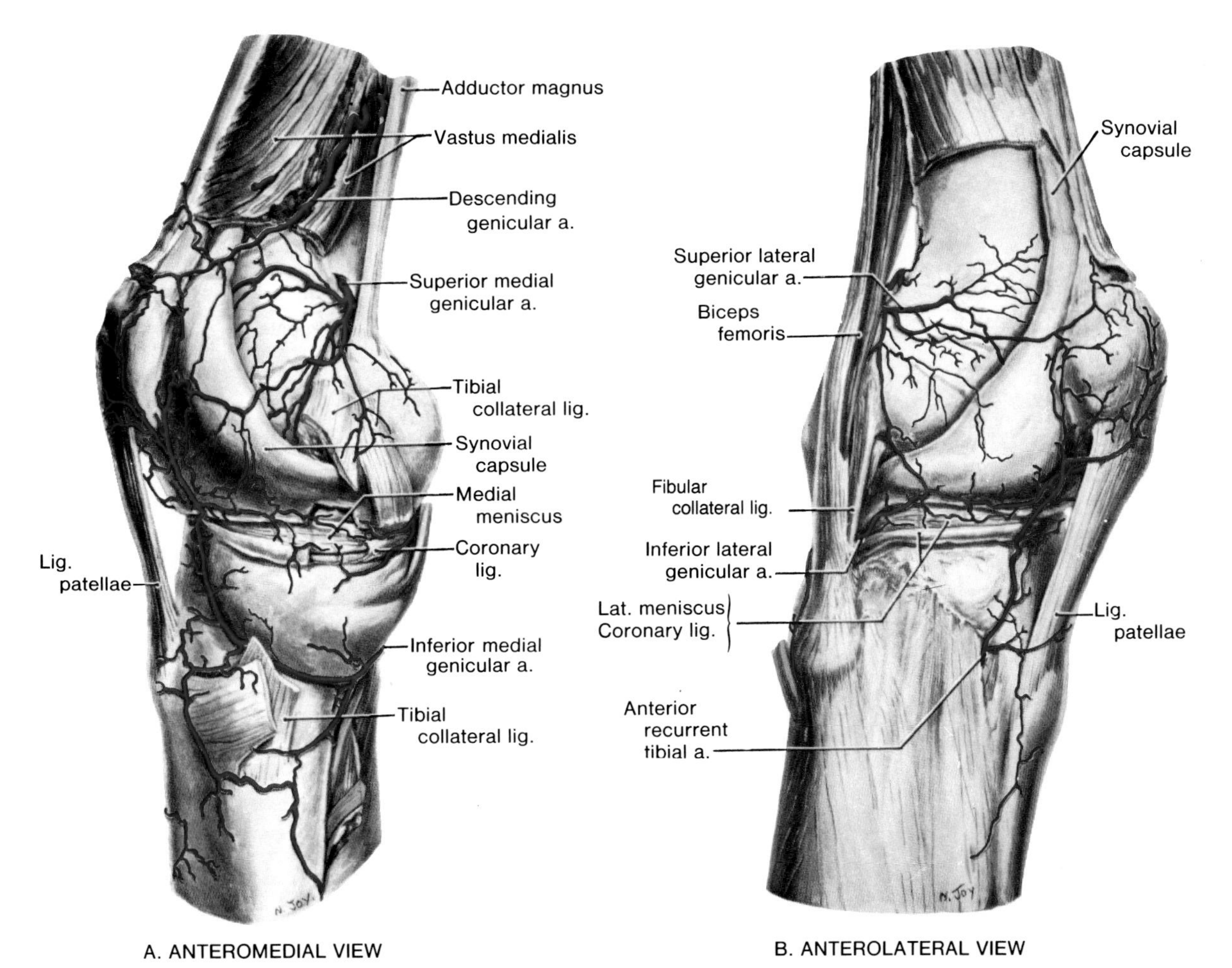

Figure 4-53. Illustrations of the clinically important genicular anastomoses (L. *genu*, knee). Observe the two named genicular branches of the popliteal artery on each side, superior and inferior. Note also the descending genicular artery, a branch of the femoral artery superomedially, and the anterior recurrent branch of the anterior tibial artery, inferolaterally. Observe the inferior lateral genicular artery running along the lateral meniscus and an unnamed artery running similarly along the medial meniscus.

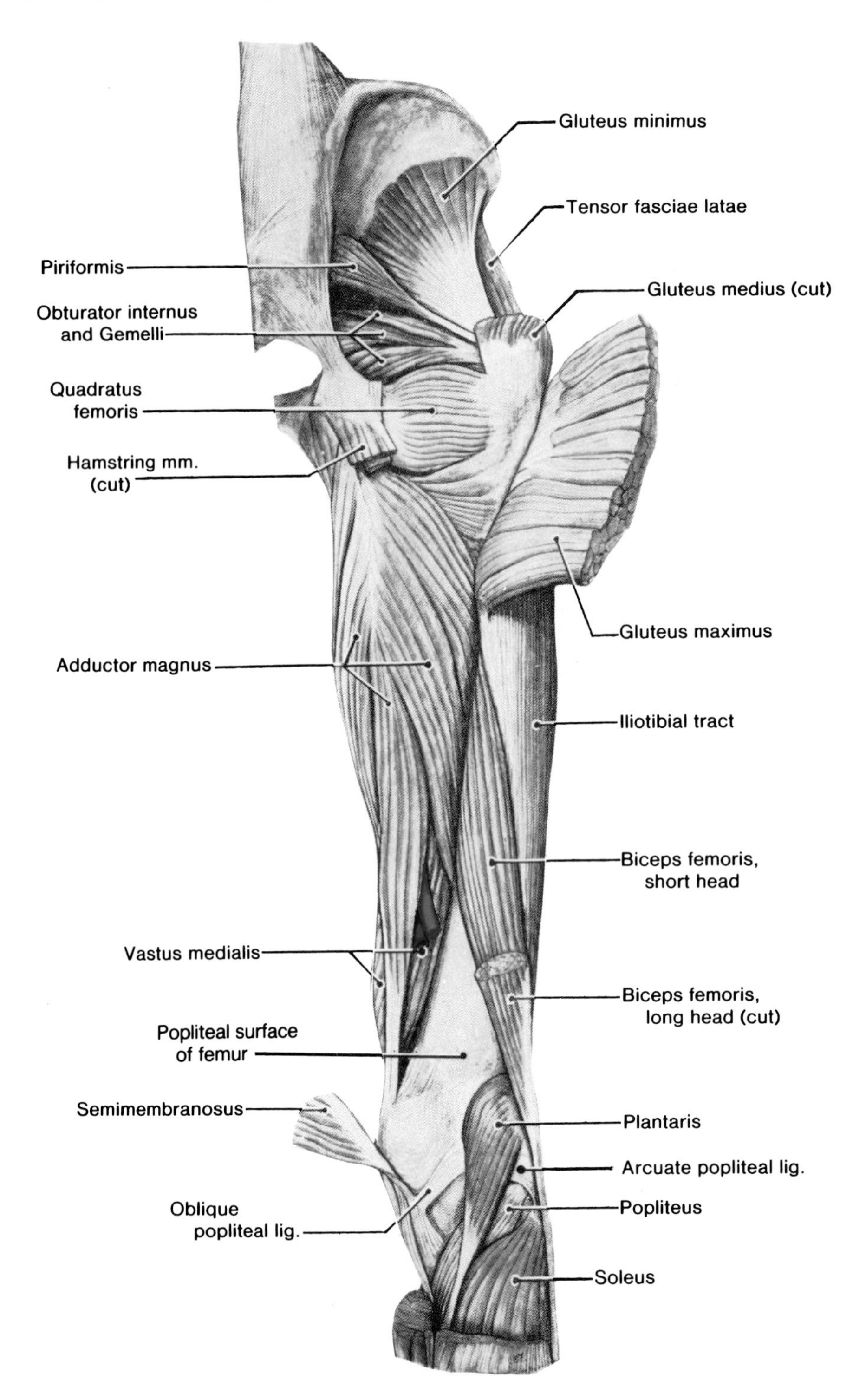

Figure 4-54. Drawing of a deep dissection of the muscles of the right gluteal and posterior thigh regions. Most of the hamstring muscles have been cut and reflected to give a posterior view of the adductor magnus muscle. The floor of the popliteal fossa is also shown. It is formed by the popliteal surface of the femur, the oblique popliteal ligament, and the popliteus fascia (Fig. 4-52). The proximal parts of the popliteal vessels are also shown.

a plug) or is ligated, the collateral anastomosing vessels become wider. However sudden blockage gives much less time for enlargement of the collateral vessels and the development of new ones than does gradual narrowing owing to obstructive arterial disease.

A collateral circulation often provides an inadequate blood supply to the leg which, if very poor, can result in death of tissue (**necrosis**). Sometimes an *atherosclerotic aneurysm of the popliteal artery* is excised and the artery is reconstructed with a **prosthetic graft**, or with a piece of the patient's great saphenous vein.

The Popliteal Vein (Figs. 4-49, 4-52, and 4-54). The popliteal vein is formed at the distal border of the popliteus muscle by the union of the venae comitantes of the anterior and posterior tibial veins. As it ascends through the popliteal fossa, *the popliteal vein crosses from the medial to the lateral side of the popliteal artery* (Fig. 4-54). Throughout its course, it lies superficial to and in the same fibrous sheath as the popliteal artery. *The popliteal vein ends at the adductor hiatus where it becomes the femoral vein* (Fig. 4-29).

The small saphenous vein (Figs. 4-10*A* and 4-49) pierces the roof of the popliteal fossa and *drains into the **popliteal vein.*** The other tributaries of the popliteal vein correspond with the branches of the popliteal artery.

A varicosity of the popliteal vein, or more likely a **varicose short saphenous vein**, may cause a swelling in the popliteal fossa. For a discussion about the causes of varicose veins, see page 408.

The Popliteal Nerves (Figs. 4-49 to 4-52). The sciatic nerve usually ends at the superior angle of the popliteal fossa by dividing into the tibial and common peroneal nerves.

The tibial nerve (medial popliteal nerve), the *larger of the two terminal branches of the sciatic nerve*, descends through the center of the popliteal fossa (Figs. 4-50 and 4-51).

The tibial nerve (L4, L5, S1, S2, and S3) arises from the ventral surface of the sacral plexus. *The tibial nerve is the most superficial of the three main central components of the popliteal fossa* (*i.e.*, nerve, vein, and artery). It lies immediately deep to the popliteal fascia.

At first the tibial nerve is covered by the semimembranosus muscle. It then passes obliquely, superficial to the popliteal vessels, and comes to lie medial to them, where it is covered by the converging heads of the gastrocnemius muscle (Fig. 4-50).

The tibial nerve gives three articular branches to the knee joint. The **genicular branches of the tibial nerve** accompany the superior and inferior medial and middle genicular vessels to the knee joint.

The muscular branches of the tibial nerve supply the gastrocnemius, plantaris, popliteus, and soleus muscles.

The medial sural cutaneous nerve arises from the tibial nerve in the popliteal fossa. *It descends in the groove between the two heads of the gastrocnemius muscle* to the middle of the leg, where it usually joins the peroneal communicating branch of the common peroneal nerve to form the **sural nerve** (Fig. 4-42). This nerve supplies the lateral aspect of the ankle and foot. The level of junction of the medial sural cutaneous with the peroneal communicating branch of the common peroneal nerve to form the sural nerve is variable. Most commonly the junction is low (Fig. 4-42), but it may be high (Fig. 4-50) or in the middle of the calf.

Pieces of the sural nerve are often used for **nerve grafts** (*e.g.*, in the repair of defects in nerves resulting from wounds). For this reason, *variations in the level of formation of the sural nerve must be understood* (Fig. 4-49). Sometimes the medial sural cutaneous does not unite with the peroneal communicating branch of the common peroneal nerve; as a result no sural nerve is present. The skin it normally innervates is supplied by the branches that usually join to form the sural nerve (Fig. 4-50).

The common peroneal nerve (lateral popliteal nerve), the *smaller of the two terminal branches of the sciatic* nerve, arises from the dorsal divisions of the sacral plexus (L4, L5, S1, and S2). The common peroneal nerve begins at the superior angle of the popliteal fossa and follows the medial border of the biceps femoris muscle and its tendon along the superolateral boundary of the popliteal fossa (Figs. 4-49, 4-50, and 4-51). It leaves the fossa by passing superficial to the lateral head of the gastrocnemius muscle (Fig. 4-49). The common peroneal nerve then passes over the posterior aspect of the head of the fibula before winding around the lateral surface of the neck of this bone (Fig. 4-57*A*). It then runs deep to the superior part of the peroneus longus muscle (Fig. 4-67).

The common peroneal nerve is palpable where it winds around the neck of the fibula, and where it is vulnerable to injury (Figs. 4-57*A* and 4-67). The common peroneal nerve can be rolled against the bone posterolateral to the neck of the fibula. In this region **the common peroneal nerve ends by dividing into the superficial and deep peroneal nerves** (Figs. 4-57*A* and 4-70).

Within the popliteal fossa (*i.e.*, before dividing), the common peroneal nerve gives off articular branches to the knee and proximal tibiofibular joints. These **genicular branches of the common peroneal nerve** accompany the superior and inferior lateral genicular vessels to the knee joint.

Within the popliteal fossa the common peroneal nerve also gives off the ***lateral sural cutaneous nerve*** to the skin of the calf and the *peroneal communicating branch.*

The common peroneal nerve *is the most commonly injured nerve in the lower limb*, mainly because it winds superficially around the neck of the fibula (Fig. 4-57). Even when there is injury to the sciatic nerve, its common peroneal division is usually more severely affected than its tibial division.

The common peroneal nerve may be lacerated at the side of the knee, **severed during fracture of the neck of the fibula**, or *severely stretched subsequent to rupture of the fibular collateral ligament.* In addition, it is susceptible to pressure exerted in the region of the head of the fibula by a tightly applied **plaster cast**, or by the buckle of a restraining strap on the operating table.

When surgically manipulating the biceps femoris tendon, *it is important to keep in mind the close relationship of the common peroneal nerve to the medial border of the biceps femoris tendon* (Fig. 4-50).

Severance of the common peroneal nerve results in paralysis of all the dorsiflexor and evertor muscles of the foot (Table 4-5, p. 483).

The loss of eversion and dorsiflexion of the foot causes the foot to hang down, a condition known as "**foot-drop**." The patient has a *high steppage gait* in which the foot is raised higher than is necessary so the toes do not hit the ground, and the foot is brought down suddenly, producing a distinctive "clop."

There is also a variable loss of sensation on the anterolateral aspect of the leg and the dorsum of the foot, areas supplied by the superficial peroneal and deep peroneal nerves (Figs. 4-70

The tibial nerve is not commonly injured because of its protected position in the popliteal fossa (Fig. 4-49). It may however be injured by deep lacerations.

Severance of the tibial nerve results in paralysis of the flexor muscles in the leg and the intrinsic muscles of the sole of the foot (Tables 4-7 and 4-8). Persons with tibial nerve injury are unable to plantarflex their foot or toes (Fig. 4-65*B*). There is also a loss of sensation on the sole of the foot.

THE LEG

The leg (L. *crus*) is the part of the lower limb between the knee and ankle joints. **The bones of the leg are the tibia and fibula** (Fig. 4-55).

The tibia ("shin bone") supports most of the weight. It articulates with the condyles of the femur superiorly and the talus inferiorly.

The fibula ("calf bone") is mainly for the attachment of muscles (Figs. 4-56 and 4-57), but it also provides stability to the ankle joint.

The bodies of the tibia and fibula are connected by an interosseous membrane composed of strong oblique fibers (Fig. 4-64).

THE TIBIA

The tibia is located on the anteromedial side of the leg. It is the second largest bone of the skeleton. *The proximal end of the tibia is large* and has an almost flat superior surface (Figs. 4-1 and 4-55). The medial and lateral **condyles of the tibia** articulate with the condyles of the femur.

The superior surface of the tibia is flat (Fig. 4-55) and consists of medial and lateral **tibial plateaus.** In Figures 4-2 and 4-55, note that the *intercondylar eminence* of the tibia fits into the *intercondylar notch* between the femoral condyles.

In Figure 4-57*B*, observe that the lateral condyle of the tibia has a facet inferiorly for the head of the fibula. Note the prominent **tibial tuberosity** anteriorly into which the *patellar ligament* or ligamentum patellae inserts (Fig. 4-46).

The distal end of the tibia is small and has facets for the fibula and the talus (Fig. 4-57*B*). The distal end projects medially and inferiorly as the **medial malleolus**, which has a facet on its lateral surface for articulation with the talus (Fig. 4-55).

The body (shaft) of the tibia is approximately triangular in cross-section (Fig. 4-64) and has medial, lateral, and posterior surfaces. In Figures 4-57*B* and 4-64, observe that muscles attach to the lateral surface of the tibia. The lateral border of the tibia is sharp where it gives attachment to the **interosseous membrane**, uniting the two leg bones (Fig. 4-64). This border is commonly referred to as the **interosseous border**.

On the posterior surface of the proximal part of the body of the tibia, observe a rough diagonal ridge known as the **soleal line** (Fig. 4-55). It runs inferomedially to the medial border, about a third of the way down the body.

The nutrient foramen of the tibia is the largest in the skeleton (Fig. 4-55). It is located on the posterior surface of the superior third of the bone. The nutrient canal runs a long inferior course in the compact bone before opening into the medullary cavity.

Fracture of the tibia through the nutrient canal predisposes to nonunion of the fragments owing to damage to the nutrient artery. The body of the tibia is narrowest at the junction of its middle and inferior thirds. This is the most frequent site of fracture and the region where **rickets** (a disease of growing bone) has its effect during infancy and childhood) (Fig. 23, p. 24).

March fractures of the inferior third of the tibia are common in persons who take long walks when they are not used to this activity. The strain of this strenuous activity may fracture the anterior cortex of the tibia.

Indirect violence may be applied to the tibia when the body turns during a fall with the foot fixed (*e.g.*, when tackled in a football game). In addition, severe torsion during skiing may produce a **spiral fracture of the tibia** at the junction of the middle and inferior thirds, and a fracture of the neck of the fibula. An anterior or posterior fall may produce a **"boot-top fracture"** owing to the rigidity of most ski boots.

Fractures of the tibia *may also result from a direct blow, e.g.*, when the bumper of a car strikes the leg. Because of this common cause, they are often called "**bumper fractures.**" Because the tibia lies subcutaneously (Fig. 4-64), the blow often tears the skin, permitting the bone fragments to protrude (**compound fracture**).

As the body of the tibia is unprotected anteromedially throughout its course and is relatively slender at the junction of its inferior and middle thirds, it is not surprising that **the tibia is the most common long bone to be fractured and to suffer compound injury**. Because of its extensive subcutaneous surface, **the tibia is accessible for obtaining pieces of bone for grafting**.

In fracture-dislocations of the ankle (*e.g.*. **Pott's fracture**), the medial malleolus of the tibia may be avulsed (pulled off) by the strong deltoid ligament (Fig. 4-137).

THE FIBULA

This long, pin-like bone (L. *fibula*, pin or skewer) lies posterolateral to the tibia (Fig. 4-64). Because of its deep location, sportswriters call it the "calf bone."

The fibula is the lateral bone of the leg (Figs. 4-55 and 4-64). Its slender body (shaft) has little or no function in weight bearing, but its malleolus helps to hold the talus in its socket (Fig. 4-55).

The fibula is mainly for muscle attachments. It also acts as a brace in providing support for the tibia. The fibula enables the tibia to withstand some bending and twisting. *Without fibular support, tibial fractures would occur more frequently.*

The slightly constricted part of the body near the head of the fibula is the **neck** (Figs. 4-55 and 4-57*A*). The sharp **interosseous border** of the fibula is for the attachment of the **interosseous membrane** (Figs. 4-57*B* and 4-64). A small *nutrient foramen* is usually present in the middle third of the fibula, entering on the posterior surface (Fig. 4-55).

The proximal end or **head of the fibula** is irregular in shape and knob-like. The head has a facet on its superior surface for articulation with the inferior surface of the lateral tibial condyle (Fig. 4-57*B*).

The distal end of the fibula or **lateral malleolus** (Figs. 4-55 and 4-57) forms a knob-like subcutaneous prominence on the lateral surface of the ankle.

The medial surface of the fibula articulates with the lateral side of the tibia and the talus (Fig. 4-57). Posteroinferior to the facet for the talus is a depression, called the **malleolar fossa**, for the posterior talofibular ligament (Fig. 4-57*B*). In Figure 4-55, *observe that the lateral malleolus lies more inferior and posterior than does the medial malleolus.*

Fractures of the fibula commonly occur 2 to 5 cm proximal to the distal end of the lateral malleolus, and are often associated with fracture-dislocations of the ankle joint, *e.g.*, **Pott's fracture**. The fibula breaks when the talus is forcibly tilted against the lateral malleolus, pushing it laterally. The ***posterior tibiofibular ligament*** acts as a fulcrum, translating the lateral force on the malleolus to a medial force on the body of the fibula, just proximal to the ankle joint, where it often breaks (Fig. 4-137).

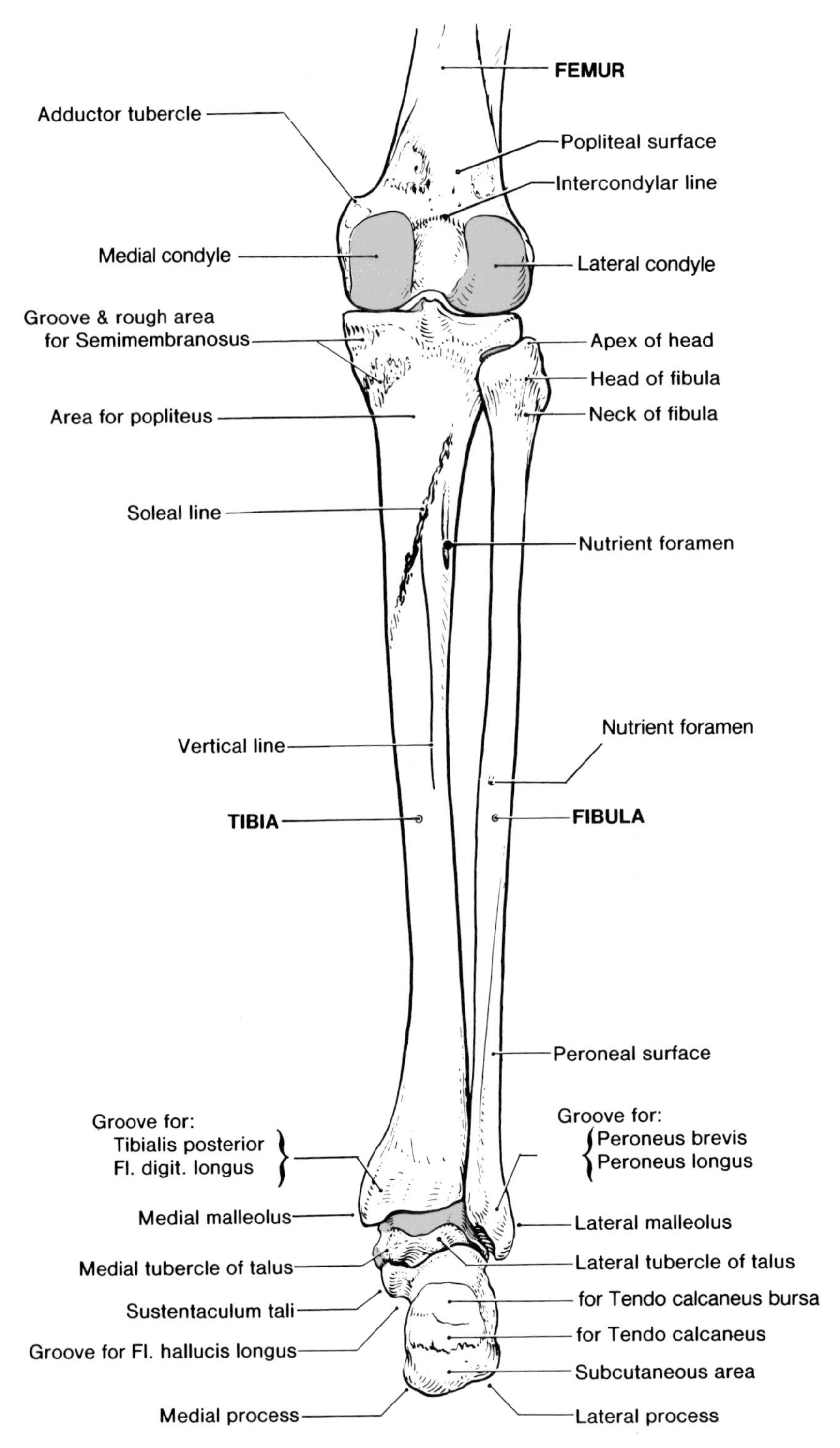

Figure 4-55. Drawing of a posterior view of the bones of the right lower limb. The larger tibia lies medially and its medial surface is subcutaneous. The smaller fibula is deeply placed. In the body the tibia and the fibula are connected by an interosseous membrane (Fig. 4-64). Observe that the lateral malleolus lies more inferior (1 to 2 cm) and posterior than does the medial malleolus. Note that the talus is held between the two sides of the mortise formed by the inferior surface of the tibia and the medial and lateral malleoli of the tibia and fibula, respectively.

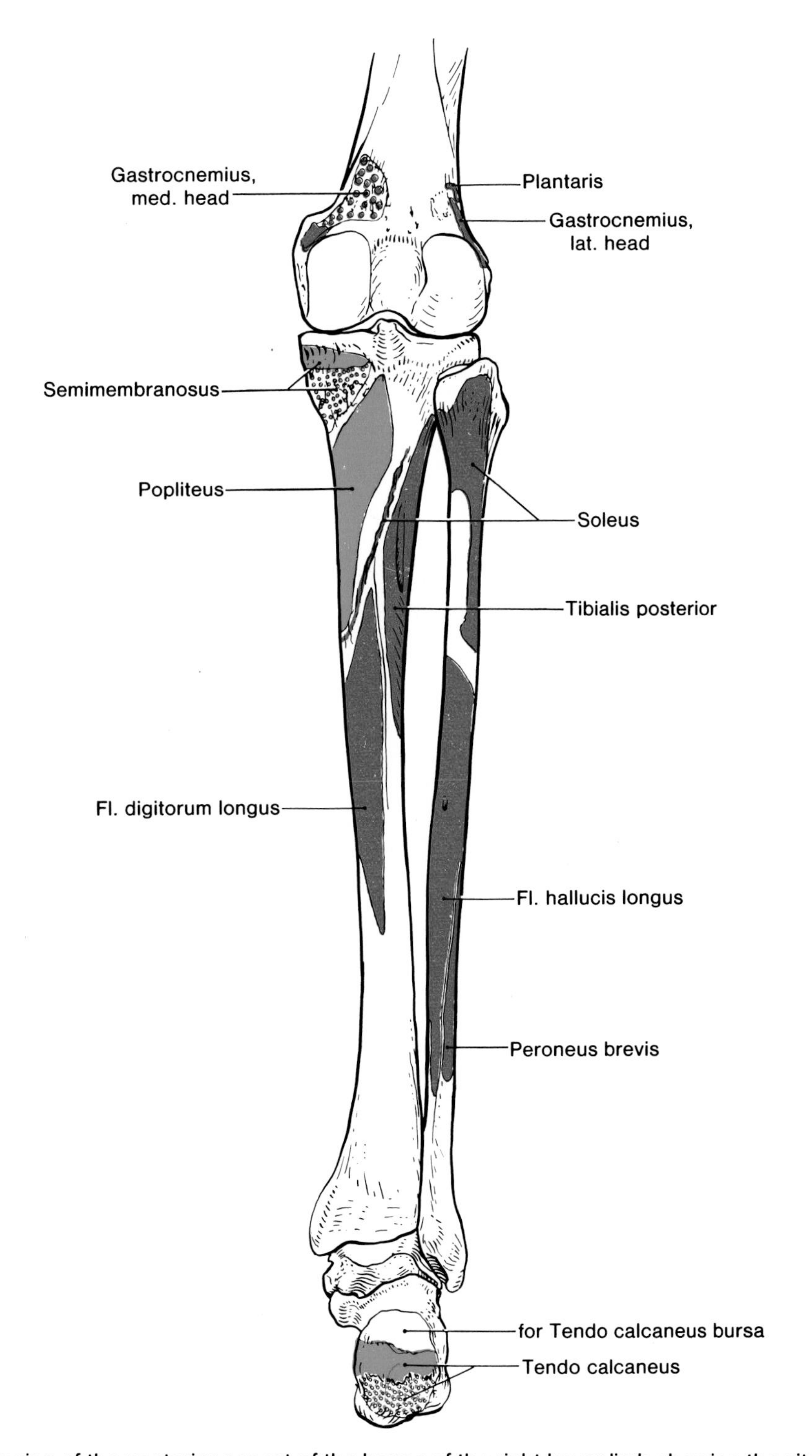

Figure 4-56. Drawing of the posterior aspect of the bones of the right lower limb showing the sites of attachment of muscles to it. Only the distal end of the bone of the thigh (femur) is shown. The main attachment of the tendo calcaneus into the middle of the posterior surface of the calcaneus is shown in *blue.* The origins of muscles are shown in *red* and insertions in *blue.*

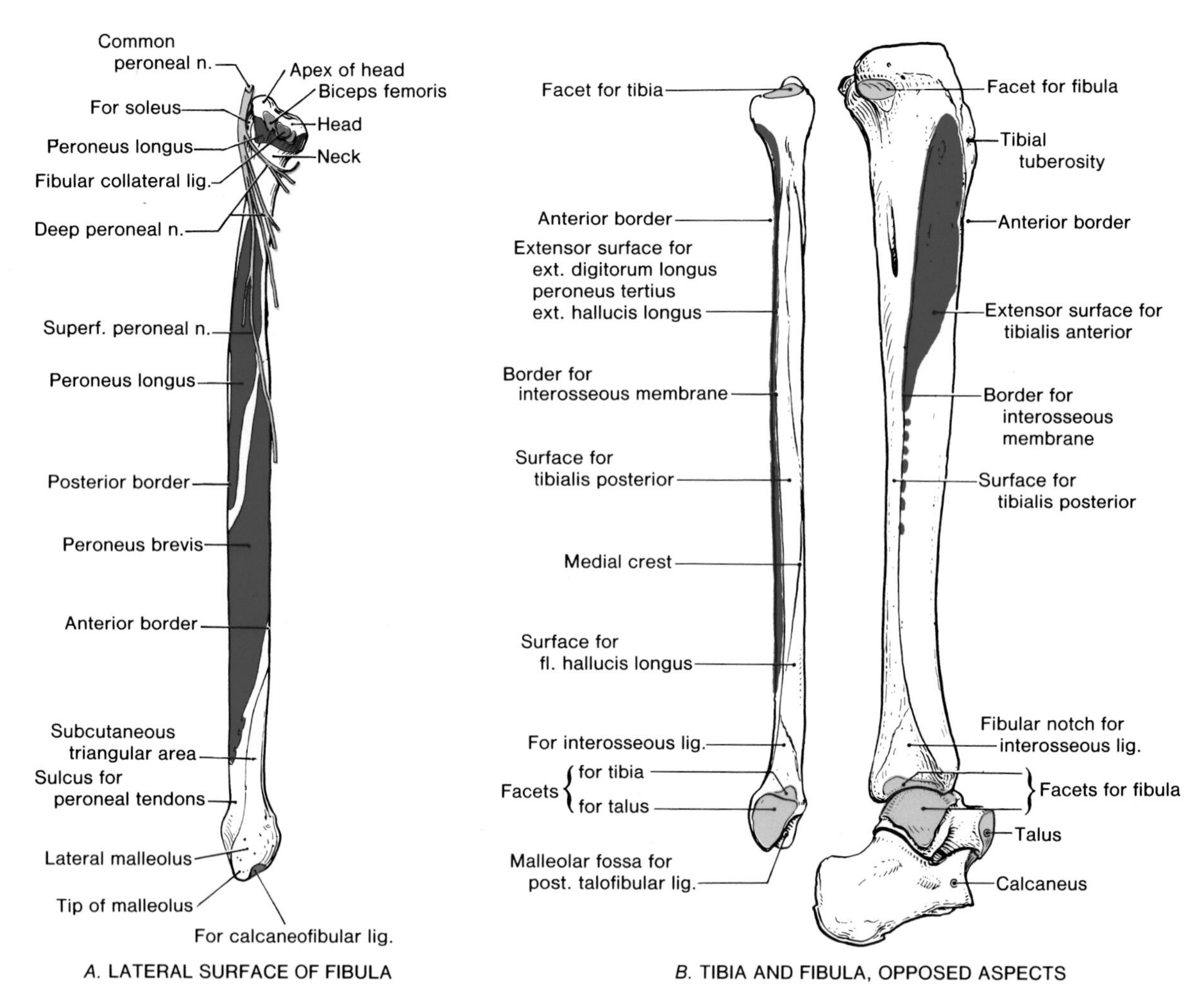

Figure 4-57. Drawings of the leg bones showing the attachments of muscles and ligaments. Some bones of the foot are shown in *B.* The tibia, the weight bearer of the leg, has a proximal expanded end which articulates with the femoral condyles and the head of the fibula (Fig. 4-56). The fibula, the lateral bone of the leg, bears little or no weight and plays no part in the knee joint, but its malleolus forms part of the mortise for the talus. In *A*, note the *clinically important relationship of the common peroneal nerve* to the proximal end of the fibula. Observe that this nerve passes posterior to the head of the fibula and around the lateral aspect of its neck. The nerve is particularly exposed to injury here, being subcutaneous and related so closely to the fibula.

When a person slips and the foot is forced into an excessively inverted position, the ankle ligaments tear and the talus is forcibly tilted against the lateral malleolus, shearing it off. **Fracture of the lateral malleolus** is relatively common in certain athletes (*e.g.*, soccer and basketball players).

The fibula is a common source of bone for grafting. Free vascularized fibulae have been used to restore skeletal integrity to upper and lower limbs in which congenital bone defects exist, or to replace segments of bone following trauma or tumor excision. Removal of portions of the fibula does not affect leg or foot function. The missing piece of bone usually does not regenerate because the periosteum and nutrient artery are generally removed so that the graft will remain alive when transplanted.

Awareness of the location of the *nutrient foramen in the fibula* (Fig. 4-55) is important when performing **free vascularized fibular transfers**. As the nutrient foramen is located in the middle third of the fibula in most cases, this segment of the bone should be used for transplanting when it is desirable for the graft to include an endosteal as well as a periosteal blood supply.

SURFACE ANATOMY OF THE LEG

A thorough physical examination of the leg requires a good knowledge of its surface features.

SURFACE ANATOMY OF THE TIBIA

The surface features of the tibia are more or less familiar to most people. Palpate the anteromedial surface of your tibia (Fig. 4-64), noting that it is subcutaneous, smooth and flat. The skin covering it is freely movable. As you run your hand distally, feel the prominence at the ankle known as the **medial malleolus.** Note that it is also subcutaneous and that its inferior end is blunt.

You may be able to observe the *great saphenous vein* crossing the inferior third of the medial surface of your tibia obliquely (Fig. 4-81). The ankle joint is about 1.5 cm proximal to the tip of the medial malleolus.

Run your hand proximally along the medial surface of your tibia until you reach its medial condyle. Verify that it is also subcutaneous. Feel the anterior border of your tibia (Fig. 4-64), noting that it is sharp and subcutaneous. The skin here is very close to the bone. No wonder it hurts so much and bruises so easily when you hit the anteromedial surface of your leg (shin) on something hard.

Run your hand proximally along the anterior aspect of the tibia until you feel the rounded elevation called the **tibial tuberosity** (Fig. 4-46). It is about 5 cm distal to the apex of the patella.

Palpate your ligamentum patellae or **patellar ligament** which extends from the apex and margins of the patella to the tuberosity of the tibia. It is most easily felt when your leg is extended. Flex your leg (*i.e.*, **genuflex**) and feel the depression on each side of the patellar ligament. Usually some indentation is visible at these sites when the leg is extended. The capsule of the knee joint is very superficial in these depressions (Fig. 4-53).

*The **tibial tuberosity** is a useful bony landmark because it roughly indicates the level of the division of the popliteal artery into its terminal branches, the anterior and posterior tibial arteries*, at the distal border of the popliteus muscle.

The smooth superior part of the tibial tuberosity is at the level of the head of the fibula. The subcutaneous, rough inferior part of the tibial tuberosity, which bears the weight during kneeling, is at the level of the neck of the fibula.

SURFACE ANATOMY OF THE FIBULA

You can easily palpate the head of the fibula at the level of the superior part of the tibial tuberosity, because this knob of bone is subcutaneous at the posterolateral aspect of the knee (Figs. 4-46 and 4-48). *A good guide to the location of the head of the fibula is the distal end of the tendon of the biceps femoris* (Figs. 4-46 and 4-67).

The neck of the fibula (Figs. 4-46, 4-57*A*, and 4-67) can be palpated just distal to the head. The **common peroneal nerve** may be rolled under your finger here. Usually this causes a tingling sensation on the anterolateral aspect of the leg and the dorsal surfaces of the toes, the areas of skin supplied by branches of this nerve.

Only the distal portion of the body of the fibula is subcutaneous. Hence **the part of the fibula just proximal to the lateral maleolus is commonly fractured.** Palpate your lateral malleolus noting that it is subcutaneous and that its inferior end is sharp (Fig. 4-55).

The tip of the lateral malleolus extends further distally (1 to 2 cm) and posteriorly than does the tip of the medial malleolus (Fig. 4-55). This relationship is important in the diagnosis and treatment of injuries in the ankle region (*e.g.*, **Pott's fracture-dislocation of the ankle.**

BONES OF THE FOOT

Because many leg muscles insert in the foot, the bones of the ankle and foot are described here so the insertions of these muscles will be understood better.

The bones of the foot comprise the **tarsus, metatarsus, and phalanges.** Using an articulated foot (Fig. 4-58), observe that its medial border is almost straight. Note that the line joining the midpoints of the medial and lateral borders of the foot is oblique and that the metatarsal bones and phalanges are located anterior to this line and the tarsal bones are posterior to it.

The foot is relatively stable because there is little free movement between the bones of the foot owing to the way they fit together and are held in position by ligaments. The foot must absorb the shock of the body weight everytime one takes a step. Consequently *the foot is adapted for weight bearing, locomotion, and maintaining equilibrium.*

The tarsus (G. *tarsos*, flat) **consists of seven tarsal bones** (Figs. 4-58 to 4-62): talus, calcaneus, cuboid, navicular, and three cuneiforms. Only one of them, the talus, articulates with the leg bones.

THE TALUS (Figs. 4-55 to 4-62)

The talus (L. ankle bone) has a ***body***, a ***neck***, and a ***head***. It looks somewhat like a saddle when viewed from its dorsal aspect (Fig. 4-58). The talus rests on the anterior two-thirds of the calcaneus (Figs. 4-61 and 4-62.) It also articulates with the tibia, fibula, and navicular bone. The saddle-shaped superior surface of the talus bears the weight of the body transmitted via the tibia (Figs. 4-57*B* and 4-62).

The body of the talus is cuboidal in shape. Its pulley-shaped superior or trochlear surface, often called the **trochlea** (L. pulley), articulates with the inferior surface of the tibia as part of the ankle joint. The body has three continuous facets for articulation (Fig. 4-55): one for the facet on the inferior surface of the tibia, one for the facet on the lateral surface of the medial malleolus, and one for the facet on the medial surface of the lateral malleolus.

The inferior surface of the body of the talus has an oval, deeply concave area for articulation with the calcaneus. The body also has a **posterior process** that has *medial and lateral tubercles* (Fig. 4-58). There is a groove between these tubercles for the tendon of the flexor hallucis longus muscle (Figs. 4-55 and 4-60).

The head of the talus is its rounded anterior end, which is directed anteromedially (Figs. 4-58 to 4-61). It has a large facet for articulation with the navicular bone and one for articulation with the shelf-like projection of the calcaneus, known as the **sustentaculum tali** (Figs. 4-55 and 4-59), as well as a small facet for articulation with the plantar calcaneonavicular ligament.

The neck of the talus is the slightly constricted part between the head and body (Fig. 4-61). Inferiorly there is a deep groove called the **sulcus tali** for the interosseous ligaments between the talus and calcaneus.

Occasionally during ossification, the lateral tubercle of the posterior process (Fig. 4-61) fails to unite with the body of the talus. This results in an extra bone, known as the *os trigonum*, which could be misinterpreted as a fracture by an inexperienced viewer of radiographs.

Fractures of the neck of the talus occur during severe dorsiflexion of the ankle (*e.g.*, when a person is pressing hard on the brake pedal of a car during a head-on collision). In some injuries the body of the talus is dislocated posteriorly.

THE CALCANEUS (Figs. 4-55 to 4-62)

The calcaneus (calcaneum) is the largest and strongest bone of the foot. The calcaneus (L. heel) lies inferior to the talus; thus its superior surface has articular facets for it. The posterior facet is demarcated anteriorly by a groove, the **sulcus calcanei**. Anterior to this sulcus is the **sustenaculum tali** (Figs. 4-55, 4-59, and 4-60), which helps to support the talus.

The calcaneus projects posteriorly and forms the prominence of the heel (Figs. 4-74 and 4-80). At the posterior end of the inferior surface, there is a **tuber calcanei** projecting inferiorly (Figs. 4-58 and 4-60). It is deep to the fibrous tissue and fat of the heel pad and its inferior part transmits the weight of the body to the floor or ground. The tuber calcanei (tuberosity of the calcaneus) has medial and lateral processes for the attachment of muscles.

On the medial surface of the calcaneus there is a groove on the inferior surface of the **sustentaculum tali** for the flexor hallucis longus tendon (Fig. 4-60), and on its lateral surface there is a tubercle, the **peroneal trochlea** (Fig. 4-61).

Persons who fall on their heels may fracture their calcanei. The calcaneus often breaks into several fragments. A *calcaneal* **or calcanean** *fracture* is usually very disabling because of disruption of the subtalar joint (Fig. 4-123).

THE NAVICULAR (Figs. 4-58 to 4-62)

The navicular (L. little ship) is a flattened, oval bone that is shaped somewhat like a boat. Located between the head of the talus and the three cuneiform bones (Fig. 4-58), it has facets for articulation with each of them. The navicular also has an occasional facet for articulation with the cuboid bone. Medially and inferiorly, there is a rough **navicular tuberosity** to which the tendon of the tibialis posterior muscle attaches (Figs. 4-59 and 4-60).

THE CUBOID (Figs. 4-58, 4-59, 4-61, and 4-62)

This bone, approximately cubical in shape, is *the most lateral bone in the distal row of the tarsus.* Posteriorly it presents an articular surface for the calcaneus and anteriorly two facets for the fourth and fifth metatarsals. On its medial surface there are facets for the lateral cuneiform and navicular bones. Anterior to the **tuberosity of the cuboid**, on the lateral and inferior surfaces of the bone (Fig. 4-61), there is a **groove for the tendon of the peroneus longus muscle** (Fig. 4-61).

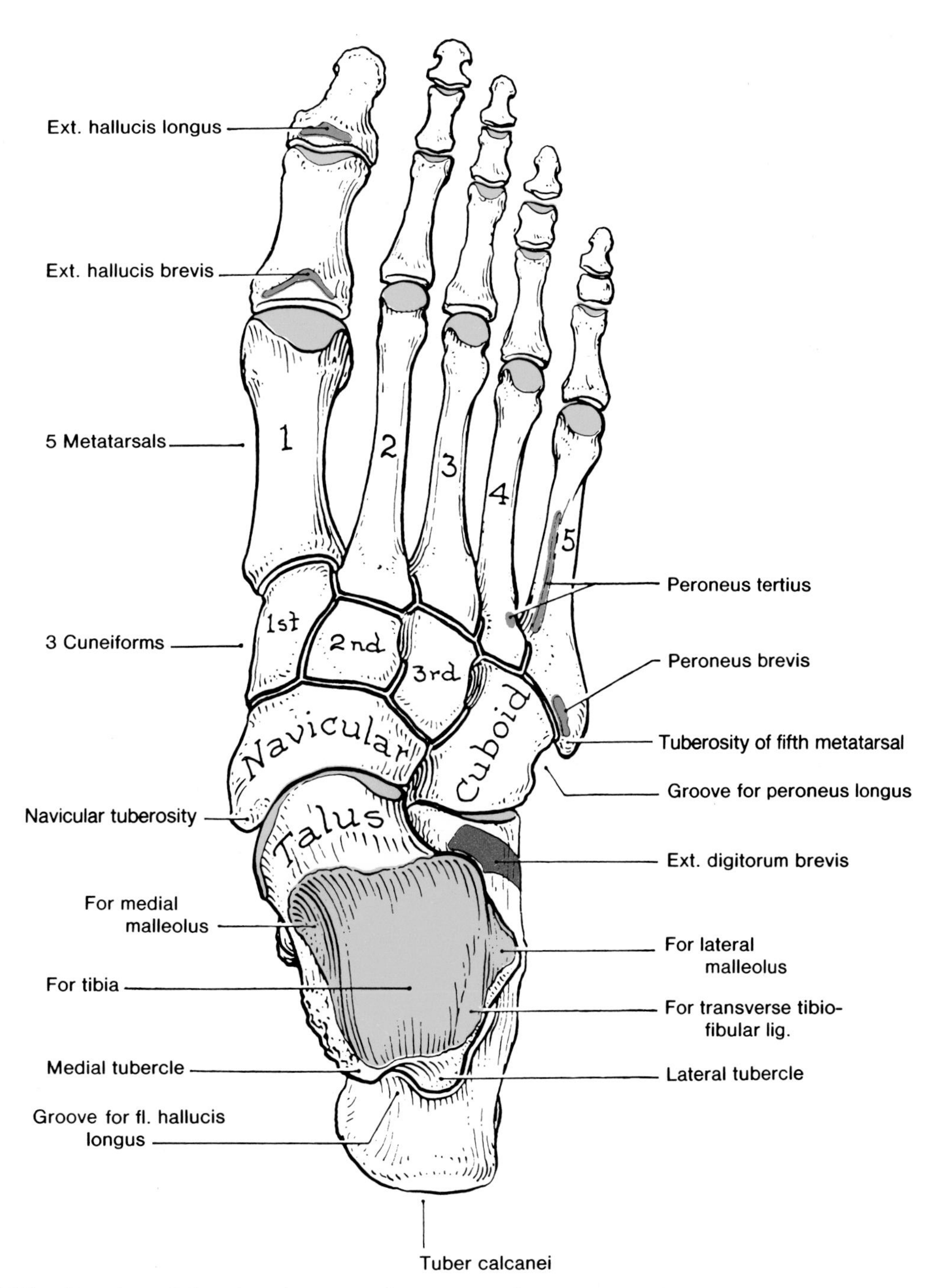

Figure 4-58. Drawing of the dorsal or superior aspect of the bones of the right foot, showing the muscle attachments and articular cartilages (*yellow*). The origin of the extensor digitorum brevis is shown in *red*; insertions of muscles are shown in *blue*.

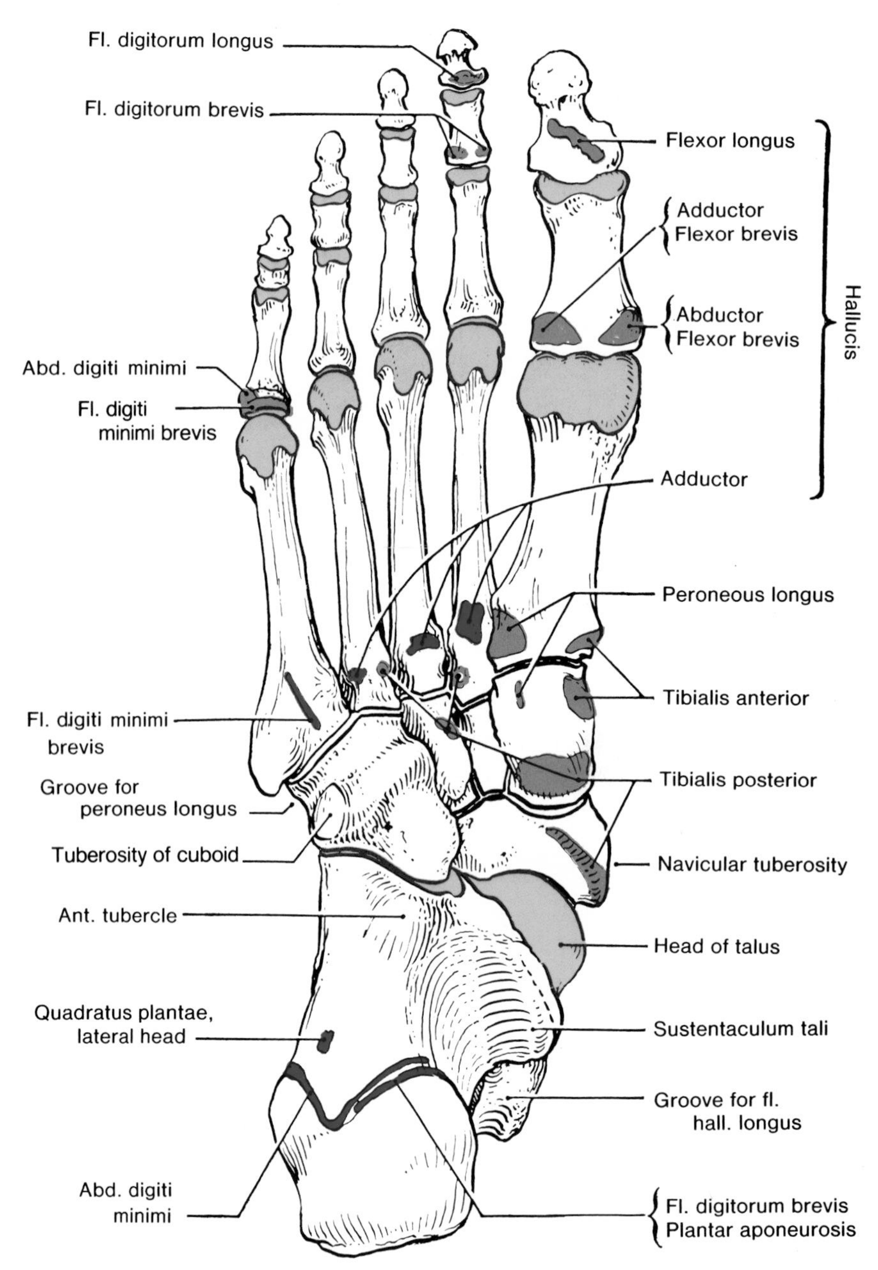

Figure 4-59. Drawing of the plantar or inferior aspect of the bones of the right foot showing the attachments of muscles and bony features.

THE CUNEIFORM BONES (Figs. 4-58 to 4-62)

The name of these three bones is derived from a Latin word meaning "*wedge-shaped.*" They are referred to as the **medial** (first), **intermediate** (second), and **lateral** (third) cuneiforms. The *medial cuneiform is the largest bone*, and the intermediate cuneiform the smallest. Each cuneiform articulates with the navicular bone posteriorly and with the base of its appropriate metatarsal anteriorly. In addition the lateral cuneiform articulates with the cuboid bone.

The metatarsus consists of five metatarsal bones (Figs. 4-58 to 4-62). In Figure 4-58, note that these *miniature long bones* are numbered from the medial side of the foot, and that each bone consists of a base proximally, a body (shaft), and a head distally (Fig. 4-61). The bases of the metatarsals articulate with the cuneiform and cuboid bones, and their heads articulate with the proximal phalanges.

In Figure 4-58, observe that the second metatarsal bone is wedged between the medial and lateral cuneiforms, and between the first and third metatarsals. Also note that the second metatarsal bone is the longest metatarsal bone.

On the plantar surface of the head of the first metatarsal bone, there are prominent medial and lateral **sesamoid bones** (Fig. 4-60). The heads of the metatarsal bones bear some of the weight of the body.

Observe that the base of the fifth metatarsal has a large tuberosity which projects over the lateral margin of the cuboid bone (Figs. 4-58 and 4-61). The **tuberosity of the fifth metatarsal bone** provides attachment on its dorsal surface for the peroneus brevis tendon (Figs. 4-58 and 4-70).

The sesamoid bones of the great (big) toe take the weight of the body, especially during the latter part of the stance phase of walking (Fig. 4-62).

Occasionally a supernumerary or accessory bone, called the *os vesalianum pedis* (Vesalius' bone), appears near the base of the fifth metatarsal. When examining radiographs, it is important to know of its possible presence so it will not be diagnosed as a fracture of this tuberosity. When this accessory bone is large, the tuberosity is small.

Fractures of the metatarsals usually occur when a heavy object falls on the foot or the foot is run over by a metal wheel. When the foot is suddenly and violently inverted, the tuberosity of the fifth metatarsal may be avulsed (pulled off) by the tendon of the peroneus brevis muscle.

March fractures of the second metatarsal bone are common in persons who go on long marches when they are unaccustomed to prolonged strenuous activity.

THE PHALANGES (Figs. 4-58 to 4-62)

There are 14 phalanges: the big or great toe (L. *hallux*), has two strong phalanges (proximal and distal), and the other four digits have three each (proximal, middle, and distal). Each phalanx consists of a base proximally, a body (shaft), and a head distally.

SURFACE ANATOMY OF THE FOOT

Because pains, injuries, and deformities of the feet are so common (*e.g.*, fractures, clubfoot, and flatfoot), familiarity with the surface anatomy of the bones of the feet is essential knowledge, especially for orthopaedists and podiatrists.

THE TALUS

Its head is often visible and is palpable in two places: anteromedial to the proximal part of the lateral malleolus on inversion of the foot (Fig. 4-65*D*), and anterior to the medial malleolus on eversion of the foot (Fig. 4-65*C*).

The head of the talus occupies the space between the sustentaculum tali and the navicular tuberosity (Figs. 4-60 and 4-80). When the foot is plantarflexed (Fig. 4-65*B*), the superior surface of the **body of the talus** can be palpated with difficulty on the anterior aspect of the ankle, anterior to the inferior end of the tibia.

THE CALCANEUS

The posterior, medial, and lateral surfaces of this large bone can be easily palpated, but the inferior surface is not easily felt owing to the overlying plantar aponeurosis and plantar muscles.

The sustentaculum tali (Figs. 4-55 and 4-60) can be felt as a small prominence distal to the tip of the medial malleolus.

The peroneal trochlea may be detectable as a small tubercle on the lateral aspect of the calcaneus (Fig. 4-61). It lies anteroinferior to the tip of the lateral malleolus. Evert your foot (Fig. 4-65*C*) and palpate the **tendon of the peroneus longus** (Fig. 4-69) and the tendon of the peroneus brevis, which are separated by this small tubercle.

THE NAVICULAR

The ***tuberosity of the navicular bone*** is easily seen and palpated on the medial aspect of the foot, inferior and anterior to the tip of the medial malleolus.

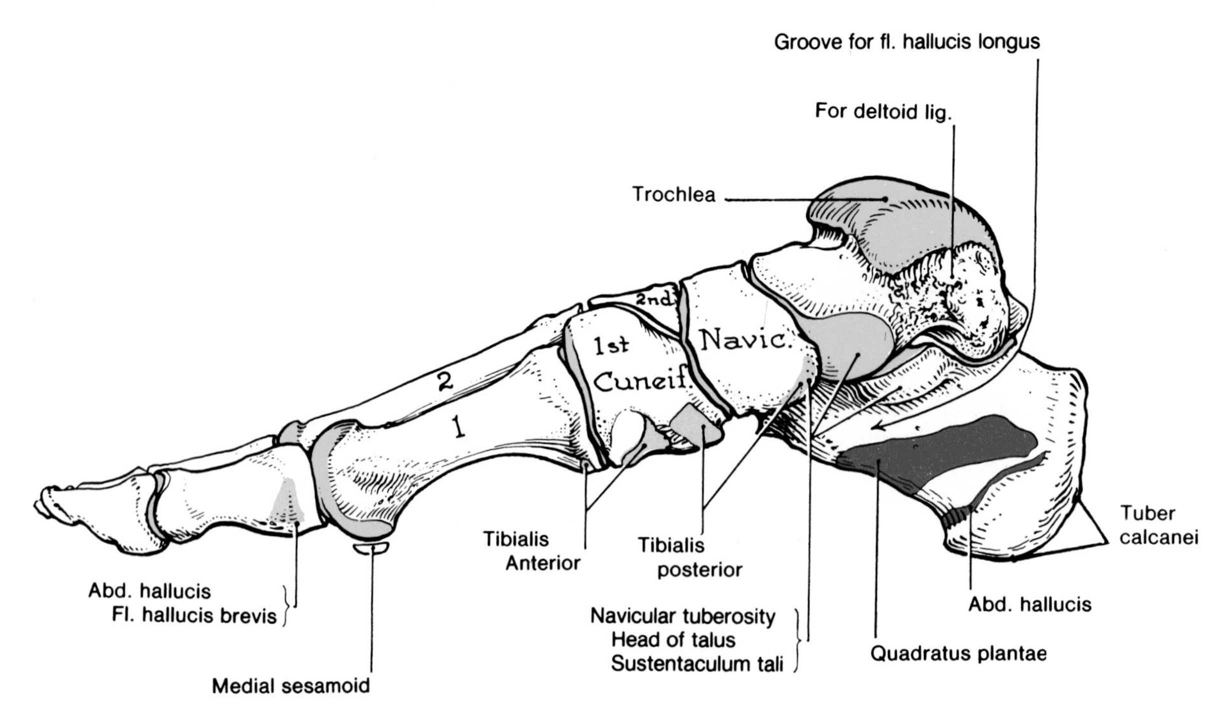

Figure 4-60. Drawing of the medial aspect of the bones of the right foot. The trochlea is the superior part of the body of the talus that articulates with the ankle mortise. The trochlea of the talus has a superior part, a medial malleolar part, and a lateral malleolar part. In this and other drawings the colors indicate the articular cartilages (*yellow*), the origin of muscles (*red*), and the insertions of muscles (*blue*).

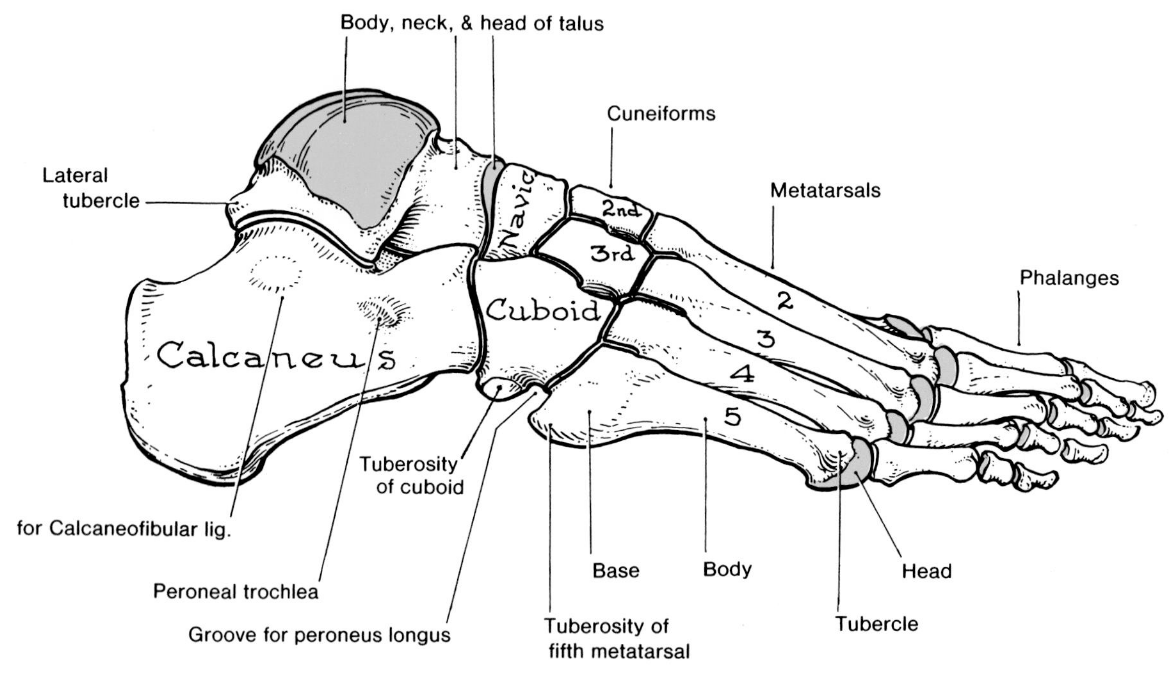

Figure 4-61. Drawing of the lateral aspect of the bones of the right foot. Observe that the calcaneus is the largest bone of the foot. There are two large bones (talus and calcaneus) and five small bones in the tarsus (instep). The part of the body of the talus indicated here is the trochlea which articulates with the tibia and malleoli.

It is a prominent and *important bony landmark in the foot.* Actively invert your foot and palpate the tendon of the tibialis posterior muscle passing to and inserting into this tuberosity.

THE CUBOID AND CUNEIFORMS (Figs. 4-58 and 4-62)

These bones are difficult to identify individually by palpation. The cuboid can be felt somewhat indistinctly on the lateral aspect of the foot, posterior to the base of the fifth metatarsal. The medial cuneiform can be indistinctly palpated between the tuberosity of the navicular and the base of the first metatarsal bone.

THE METATARSUS (Figs. 4-80 and 4-84)

The *head of the* ***first metatarsal bone*** *forms a prominence on the medial aspect of the foot.* The medial and lateral sesamoids inferior to the head of this metatarsal (Fig. 4-60) can be felt to slide when the great toe is moved passively with the fingers.

The base of the fifth metatarsal forms a prominent landmark on the lateral aspect of the foot (Figs. 4-61 and 4-69), and the large tuberosity of the fifth metatarsal can easily be palpated at the midpoint of the lateral border of the foot. In some people it even produces a prominence in their shoe.

The bodies of the metatarsals (Fig. 4-61) can be felt indistinctly on the dorsum of the foot between the extensor tendons (Figs. 4-69 and 4-70).

THE PHALANGES (Fig. 4-61)

The dorsal surfaces of these bones can be felt indistinctly through the extensor tendons.

In **flatfoot**, weakening of the supporting muscles and the plantar calcaneonavicular ligament (Fig. 4-125) removes the support for the head of the talus. As a result it descends, stretching the ligaments and flattening the longitudinal arch of the foot. The depressed head of the talus can then be palpated just anterior to the medial malleolus. For more on the anatomical basis of flatfoot, see page 556).

Clubfoot (talipes equinovarus), in which the sole of the foot is turned medially and inverted (Fig. 4-63), is quite common (1 in 1500). The foot is bent so that the tuberosity of the navicular bone comes close to the sustenaculum tali (Fig. 4-59).

Fractures of the phalanges most often result from heavy objects falling on them or from stubbing the bare toes (*e.g.*, on a doorstep).

THE CRURAL FASCIA

The deep fascia of the leg or crural fascia (Fig. 4-67) is continuous with the deep fascia of the thigh or **fascia lata**, but it does not completely invest the leg. The crural fascia (L. *crura*, legs) is attached to the anterior and medial borders of the tibia (Fig. 4-64), where it is continuous with its periosteum. Deep fascia is absent over the subcutaneous part of the medial surface of the tibia, and over the triangular subcutaneous surface of the inferior quarter of the fibula (Fig. 4-71). Here it is attached to the borders of the fibula.

The crural fascia is very thick in the proximal part of the anterior aspect of the leg (Fig. 4-64), where it forms part of the origin of the underlying muscles (*e.g.*, tibialis anterior, Fig. 4-67). Although thin in the distal part of the leg, it is thickened where it forms the superior and inferior retinacula (Fig. 4-66).

THE SUPERIOR EXTENSOR RETINACULUM (Figs. 4-66 to 4-68)

This strong *broad band of deep fascia passes from the fibula to the tibia,* proximal to the malleoli. It binds down the tendons of muscles in the anterior crural compartment (Table 4-5), preventing them from bowstringing anteriorly during dorsiflexion of the ankle joint (Fig. 4-65*A*).

THE INFERIOR EXTENSOR RETINACULUM (Figs. 4-66 to 4-68)

This *Y-shaped band of deep fascia is attached laterally to the anterosuperior surface of the calcaneus.* It forms a strong loop around the tendons of the peroneus tertius and the extensor digitorum longus (Figs. 4-67 and 4-71).

The proximal limb of the inferior extensor retinaculum is attached to the medial malleolus. During its superomedial course it passes over the tendons of the tibialis anterior and extensor hallucis longus muscles, the dorsalis pedis vessels, and the deep peroneal nerve (Fig. 4-66).

The **distal limb of the inferior extensor retinaculum is attached medially to the plantar aponeurosis** (Fig. 4-85). During its inferomedial course, it also passes over the tendons of the tibialis anterior and extensor hallucis longus muscles, the dorsalis pedis vessels, and the deep peroneal nerve (Fig. 4-66).

The tibia and fibula, the interosseous membrane, and the crural intermuscular septa divide the leg (L. *crus*) into **three crural compartments**: anterior, lateral, and posterior (Fig. 4-64).

The fascial septa, called crural intermuscular septa or simply crural septa, are attached superficially to the ensheathing deep fascia and the fibula.

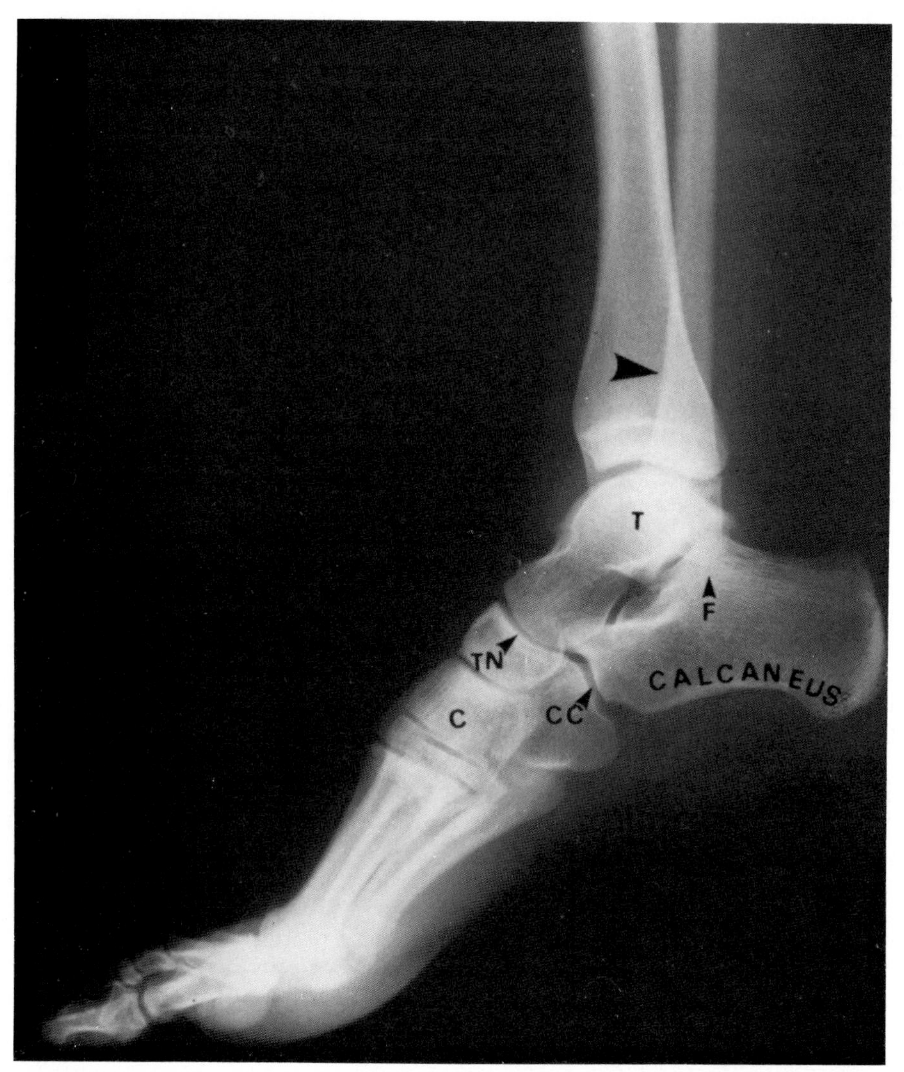

Figure 4-62. Lateral radiograph of the bones of the right leg and foot. This radiograph was taken with the foot raised as in walking. The *large arrow* points to the edge of the triangular area where the tibia and fibula are superimposed on each other. The *small arrow* (*F*) indicates how far the fibula extends distally. The talus (*T*) participates in the talonavicular joint (*TN*) and the calcaneus in the calcaneocuboid (*CC*) joint. The cuneiforms (*C*) and the proximal ends of the metatarsals are superimposed upon each other.

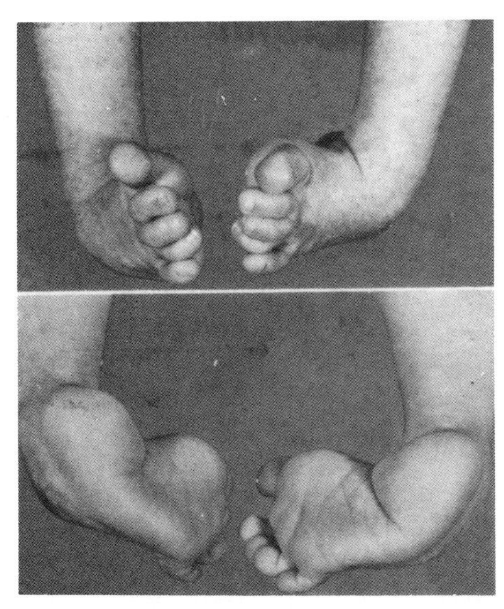

Figure 4-63. Photograph of the legs and feet of a newborn infant with the most common type of congenital clubfeet (*i.e.*, talipes equinovarus). Note that the feet are inverted and the ankles plantarflexed.

The anterolateral part of the leg contains the anterior and lateral crural compartments, which are separated by the **anterior crural intermuscular septum** (Fig. 4-64). The peroneal muscles in the lateral crural compartment are separated from muscles in the posterior crural compartment by the **posterior crural intermuscular septum.** The much larger posterior compartment of the leg, often called the "*calf,*" is subdivided by a broad **transverse crural intermuscular septum** (*deep transverse fascia of leg*) into superficial and deep posterior crural compartments containing the superficial and deep crural muscles, respectively (Fig. 4-64 and Tables 4-7 and 4-8).

THE ANTERIOR CRURAL COMPARTMENT

The anterior crural compartment, *located anterior to the interosseous membrane*, is between the lateral surface of the tibia and the anterior crural intermuscular septum (Fig. 4-64).

The anterior crural compartment contains the tibialis anterior, extensor hallucis longus, extensor digitorum longus, and peroneus tertius muscles (Fig. 4-64). These muscles are mainly concerned with dorsiflexion of the ankle joint and extension of the toes (Table 4-5). They are supplied by the **deep peroneal nerve**, a branch of the common peroneal, and by the anterior tibial vessels (Figs. 4-64 and 4-70).

THE TIBIALIS ANTERIOR MUSCLE (Figs. 4-64, 4-66 to 4-68 and Table 4-5)

This long, thick muscle lies against the lateral surface of the tibia, where it is easy to palpate.

Origin (Fig. 4-57*B*). **Lateral condyle and superior half of lateral surface of tibia.** Some fibers also arise from the deep fascia of the leg and from the interosseous membrane.

Insertion (Figs. 4-59, 4-60, and 4-68). **Medial and inferior surfaces of medial cuneiform and base of first metatarsal bone.** The tendon of the tibialis anterior passes deep to the extensor retinacula.

Nerve Supply (Figs. 4-64 and 4-70). **Deep peroneal nerve** (L4 and L5).

Actions (Fig. 4-65, *A* and *D*). **Dorsiflexes and inverts the foot.**

When the tibialis anterior is paralyzed owing to **injury of the common peroneal nerve** or to its branch, the deep peroneal nerve, the foot drops (*i.e.*, it falls into plantarflexion when it is raised from the ground). This condition, known as **foot-drop**, was discussed previously (p. 464).

"**Shin splints**" is a lay term for a painful condition of the anterior compartment of the leg that follows vigorous and/or lengthy exercise. Often persons who lead sedentary lives develop pains in the anterior part of their legs when they undertake long walks. Their anterior tibial muscles swell from sudden overuse and the swollen muscles in the anterior crural compartment reduce the blood flow to the muscles. Cramps may develop if use of the muscles is continued. Even at rest the swollen muscles are painful and tender to pressure.

"Shin splints" may also occur in trained athletes who do not warm up adequately or warm down sufficiently.

People living in Alaska and Northern Canada commonly get "shin splints" after the first heavy snow fall that requires them to use snowshoes. Their anterior tibial muscles, and probably their peroneal muscles, are used in **snowshoeing** to raise the toes of the snowshoes and to keep them straight. Because these muscles have not been used much during the summer, the severe demand on these muscles causes "shin splints;" this is often referred to as "**snowshoe leg.**" Trappers sometimes treat their swollen and painful muscles by heating the overlying skin with a hot iron. This is an example of the use of **counterirritation** (irritation of the skin aimed at relieving a deep inflammatory process).

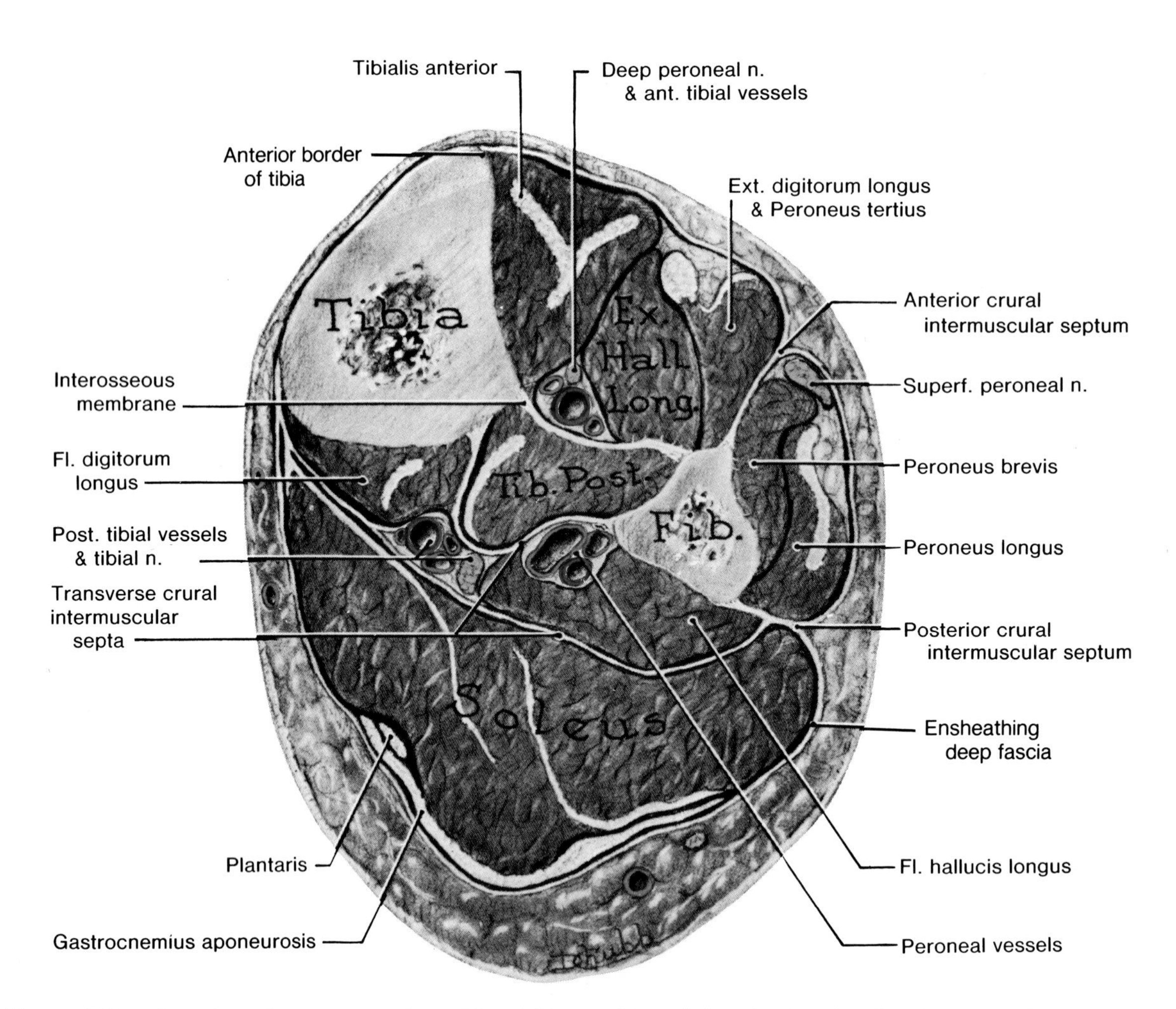

Figure 4-64. Drawing of a cross-section of the right leg of an adult male showing the three crural compartments and their contents. Observe the *anterior crural compartment* bounded by the tibia, interosseous membrane, fibula, anterior crural intermuscular septum, and deep fascia. It contains the tibialis anterior, extensor hallucis longus extensor digitorum longus, and peroneus tertius muscles. It also contains the anterior tibial vessels and the deep peroneal nerve. Note the *lateral crural compartment* bounded by the fibula, anterior and posterior crural intermuscular septa, and deep fascia. It contains the peroneal muscles and the superficial peroneal nerve. Observe the *posterior crural compartment* bounded by the tibia, interosseous membrane, fibula, posterior crural intermuscular septum, and deep fascia. This compartment is subdivided by two coronal intermuscular septa into three subcompartments: (1) the deepest compartment contains the tibialis posterior muscle; (2) the intermediate compartment contains the flexor hallucis longus and flexor digitorum longus muscles, the posterior tibial vessels, and the tibial nerve; and (3) the superficial compartment contains the soleus, gastrocnemius, and plantaris muscles. Only the aponeurosis of the gastrocnemius muscle is visible because the section is inferior to the fleshy belly of the muscle (see Fig. 4-72).

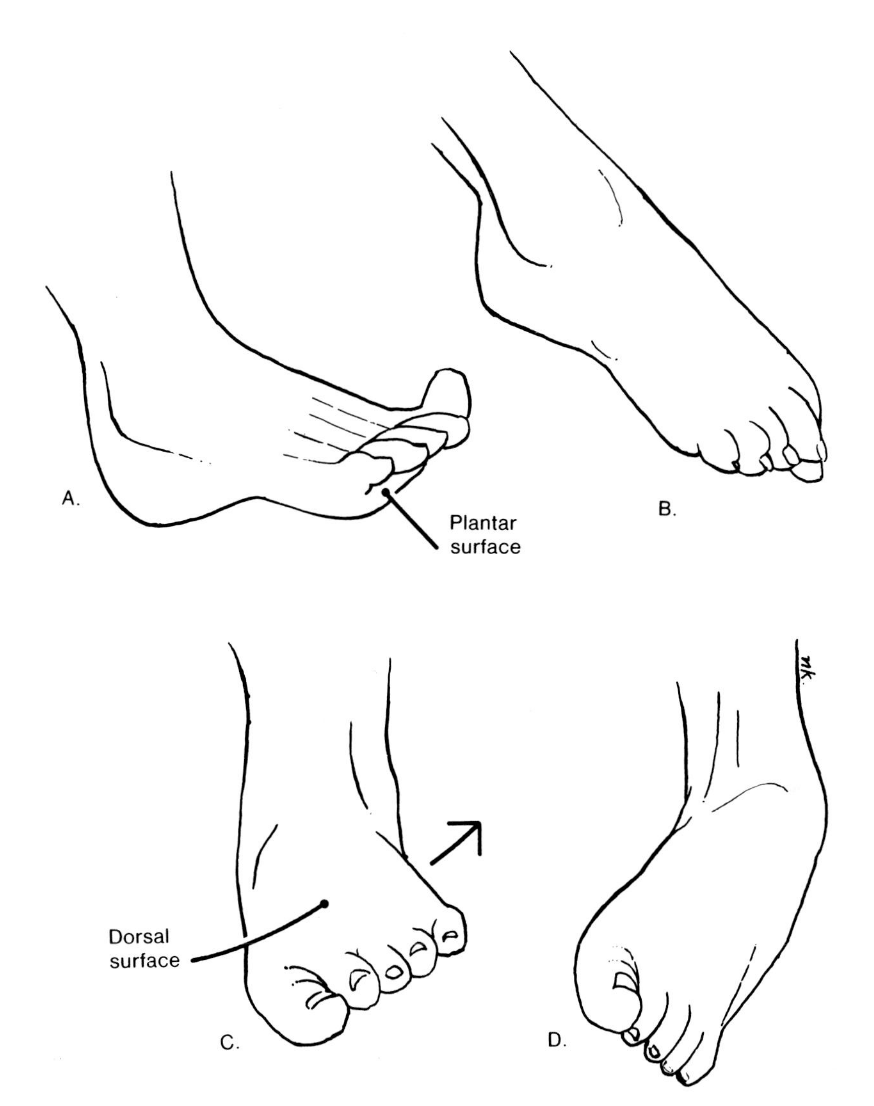

Figure 4-65. Drawings illustrating movements of the foot. *A*, dorsiflexion. *B*, plantarflexion. *C*, eversion. *D*, inversion. Also see Figure 12 on page 13.

THE EXTENSOR HALLUCIS LONGUS MUSCLE (Figs. 4-64, 4-66 to 4-68, and Table 4-5)

This thin muscle lies between and partly deep to the tibialis anterior and extensor digitorum longus muscles.

Origin (Fig. 4-57*B*). **Middle half of anterior surface of fibula and interosseous membrane.**

Insertion (Figs. 4-58 and 4-66). **Dorsal aspect of base of distal phalanx of great toe.** Its tendon passes deep to the superior and inferior extensor retinacula, where it is easy to observe and palpate when the great toe is dorsiflexed.

Nerve Supply (Fig. 4-66). **Deep peroneal nerve** (L5 and S1).

Actions (Fig. 4-65*A*). **Extends great toe and dorsiflexes the foot.**

THE EXTENSOR DIGITORUM LONGUS MUSCLE (Figs. 4-64, 4-66 to 4-69, and Table 4-5)

This muscle lies lateral to the tibialis anterior and can be easily palpated. Its tendons may be seen and felt when the toes are dorsiflexed (Fig. 4-69).

Origin (Fig. 4-57*B*). **Lateral condyle of tibia, superior three-fourths of anterior surface of fibula, and interosseous membrane.**

Insertion (Figs. 4-66 to 4-68). **Middle and distal phalanges of lateral four toes.** Its tendon passes deep to the superior and the inferior extensor retinacula and divides into four tendons or slips which run to the lateral four toes.

A **common synovial sheath** surrounds the four tendons (Fig. 4-68) which diverge on the dorsum of the foot as they pass to their insertions. Each tendon forms a membranous **extensor expansion** over the dorsum of the proximal phalanx (Fig. 4-66) which divides into two lateral slips and one central slip. The central slip inserts into the base of the middle phalanx and the lateral slips converge to insert into the base of the distal phalanx.

Nerve Supply (Figs. 4-66 and 4-70). **Deep peroneal nerve** (L5 and S1).

Actions (Figs. 4-65*A* and 4-69). **Extends lateral four toes** at the metatarsophalangeal joints **and dorsiflexes foot.**

THE PERONEUS TERTIUS MUSCLE (Figs. 4-64, 4-66 to 4-68, and Table 4-5)

This small muscle (sometimes missing) is a partially separated part of the extensor digitorum longus muscle.

Origin (Fig. 4-57*B*). **Inferior third of anterior surface of fibula and interosseous membrane.**

Insertion (Fig. 4-58). **Dorsum of base of fifth metatarsal bone** and often along the body of the bone. Its tendon runs with those of the extensor digitorum longus through the inferior extensor retinaculum.

Nerve Supply (Figs. 4-66 and 4-70). **Deep peroneal nerve** (L5 and S1).

Actions (Fig. 4-65, *A* and *C*). **Dorsiflexes foot and aids in eversion of the foot.**

THE DEEP PERONEAL NERVE (Figs. 4-57*A*, 4-64, 4-66, and 4-70)

This is the nerve of the anterior crural compartment. It is one of the two terminal branches of the common peroneal nerve. **The deep peroneal nerve begins between the fibula and the superior part of the peroneus longus muscle** (Figs. 4-57*A* and 4-70). It then runs inferomedially on the fibula, deep to the extensor digitorum longus. After piercing the anterior crural intermuscular septum and the extensor digitorum longus, the deep peroneal nerve descends anterior to the interosseous membrane in the anterior crural compartment (Figs. 4-64 and 4-70), where it accompanies the anterior tibial artery between the extensor hallucis longus and tibialis anterior muscles. The deep peroneal nerve passes deep to the extensor retinacula with the anterior tibial artery (Fig. 4-70), where it ends by dividing into medial and lateral branches.

In addition to supplying the muscles in the anterior crural compartment, the deep peroneal nerve gives branches to the posterior tibial and peroneal arteries. It also sends articular branches to the ankle joint and other joints it crosses and supplies the skin between the great and second toes.

THE ANTERIOR TIBIAL ARTERY (Figs. 4-64, 4-66, and 4-70)

Structures in the anterior crural department are supplied by the anterior tibial artery and its branches.

The smaller of the terminal branches of the popliteal artery, *the anterior tibial, begins opposite the inferior border of the popliteus muscle. It ends at the ankle joint,* midway between the malleoli where it **becomes the dorsalis pedis artery** (Figs. 4-66 and 4-70).

From its origin in the posterior part of the leg (Fig. 4-77), the anterior tibial artery passes anteriorly through the interosseous membrane. It then descends on the anterior surface of this membrane between the extensor hallucis longus and tibialis anterior muscles with the deep peroneal nerve (Fig. 4-70). In the distal part of the leg, the anterior tibial artery lies on the tibia.

In addition to supplying muscles in the anterior crural compartment, the anterior tibial artery has several other named branches. The anterior and posterior **tibial recurrent arteries** join the anastomoses around the knee (Fig. 4-53), and the medial and lateral anterior **malleolar arteries** ramify over the medial

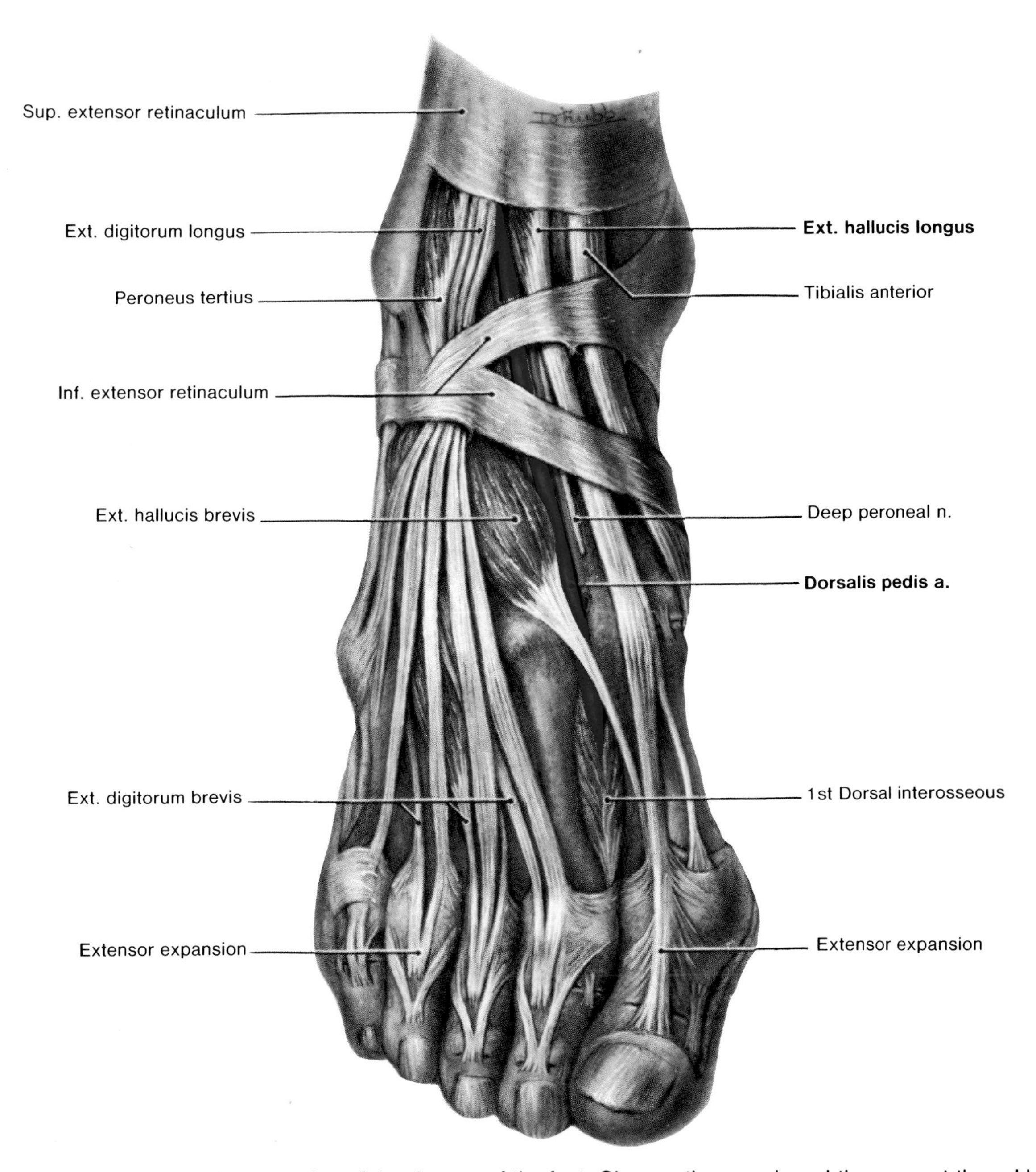

Figure 4-66. Drawing of a dissection of the dorsum of the foot. Observe the vessels and the nerve at the ankle lying midway between the malleoli with two tendons on each side of them. Observe that the dorsalis pedis artery, the continuation of the anterior tibial, is crossed by the tendon of the extensor hallucis brevis muscle and then disappears between the two heads of the first dorsal interosseous muscle to end in the sole of the foot. The pulsations of this artery can easily be felt in most people just lateral to the tendon of the extensor hallucis longus muscle, where it passes over the navicular and cuneiform bones. A knowledge of how to feel the dorsalis pedis pulse is clinically important in cases of suspected arterial disease of the lower limb, and of threatened or established gangrene of a toe or toes. Observe the inferior extensor retinaculum restraining the tendons from bowstringing anteriorly and/or medially. *The extensor hallucis brevis is the medial part of the extensor digitorum brevis muscle, the only muscle on the dorsum of the foot.*

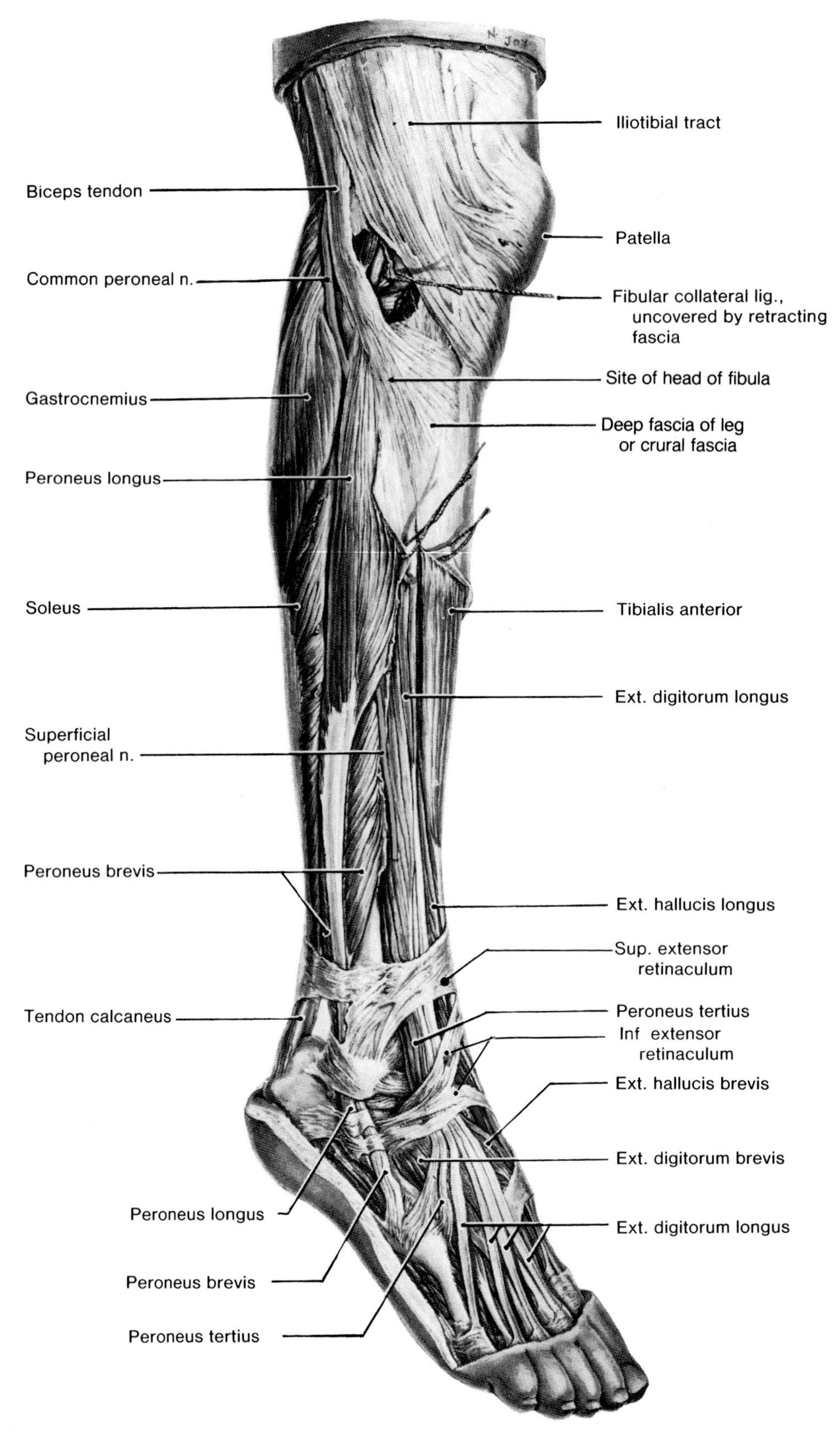

Figure 4-67. Drawing of an anterolateral view of a dissection of the muscles of the right leg and foot. Observe the superficial position of the common peroneal nerve, a terminal branch of the sciatic nerve (also see Figs. 4-57*A* and 4-77). Because the common peroneal nerve is subcutaneous in the knee region, it is commonly injured when the neck of the fibula is fractured or when there is a deep laceration on the lateral side of the knee.

Table 4-5
Muscles of Anterior Crural Compartment[1]

Muscle	Origin	Insertion	Nerve Supply	Actions
Tibialis anterior (Figs. 4-64 and 4-66 to 4-68)	Lateral condyle and superior half of lateral surface of tibia (Fig. 4-57*B*)	Medial and inferior surfaces of medial cuneiform and base of first metatarsal bones (Figs. 4-59 and 4-60)	Deep peroneal nerve (**L4** and L5)	Dorsiflexes and inverts foot (Fig. 4-65*A* and *D*)
Extensor hallucis longus (Figs. 4-64 and 4-66 to 4-68)	Middle half of anterior surface of fibula and interosseous membrane (Fig. 4-57*B*)	Dorsal aspect of base of distal phalanx of great toe (Fig. 4-58)	Deep peroneal nerve (L5 and S1)	Extends great toe and dorsiflexes foot (Fig. 4-65*A*)
Extensor digitorum longus (Figs. 4-64 and 4-66 to 4-69)	Lateral condyle of tibia, superior three-fourths of anterior surface of fibula, and interosseous membrane (Fig. 4-57*B*)	Middle and distal phalanges of lateral four toes (Figs. 4-59 and 4-66)	Deep peroneal nerve (L5 and S1)	Extends lateral four toes and dorsiflexes foot (Fig. 4-65*A*)
Peroneus tertius (Figs. 4-64 and 4-66 to 4-68)	Inferior third of anterior surface of fibula and interosseous membrane (Fig. 4-57*B*)	Dorsum of base of fifth metatarsal bone (Fig. 4-58)	Deep peroneal nerve (L5 and S1)	Dorsiflexes foot and aids in eversion of foot (Fig. 4-65, *A* and *D*)

[1] Extensions of the crural fascia or fasica of the legs (L. *crura*, legs) form anterior and posterior intermuscular septa (Fig. 4-64), thereby giving rise to **three crural compartments**: (1) the extensor or *anterior crural compartment* containing muscles supplied by the *deep peroneal nerve*; (2) the peroneal or *lateral crural compartment* containing muscles supplied by the *superficial peroneal nerve*; and (3) the flexor or *posterior crural compartment* containing muscles supplied by the *tibial nerve.* You are likely to hear these compartments referred to simply as the anterior, lateral, and posterior compartments of the leg.

and lateral malleoli, respectively, (Fig. 4-70), contributing to the arterial networks around the ankle.

THE LATERAL CRURAL COMPARTMENT

The lateral crural compartment is bounded by the lateral surface of the fibula, the anterior and posterior crural intermuscular septa, and the crural fascia (Fig. 4-64).

The lateral crural compartment contains the peroneus longus and peroneus brevis muscles which plantarflex and evert the foot (Fig. 4-65, *B* and *C*). They are supplied by the **superficial peroneal nerve,** a branch of the common peroneal (Fig. 4-64 and Table 4-6).

THE PERONEUS LONGUS MUSCLE (Figs. 4-64, 4-67 to 4-72, and Table 4-6)

This is the more superficial of the two peroneal muscles and it arises more superiorly on the fibula (Figs. 4-57*A* and 4-64).

The peroneus longus is a long, narrow muscle, extending from the fibula to the sole of the foot. Its tendon can easily be palpated and observed proximal and posterior to the lateral malleolus (Fig. 4-69).

Origin (Fig. 4-57*A*). **Head and superior two-thirds of lateral surface of fibula.** It also arises from the anterior and posterior crural intermuscular septa.

Insertion (Fig. 4-59). **Base of first metatarsal and medial cuneiform bone.** Its long tendon runs deep to the superior peroneal retinaculum (Fig. 4-71) as it curves posterior to the lateral malleolus, which it uses as a pulley.

The peroneus longus is *enclosed in a common synovial sheath with the peroneus brevis muscle* (Fig. 4-68). The peroneus longus **passes inferior to the peroneal trochlea** on the calcaneous (Fig. 4-61) to enter a groove on the anteroinferior aspect of the cuboid bone (Fig. 4-59). The peroneus longus then crosses the sole of the foot, running obliquely and distally to reach its insertion on the first metatarsal and medial cuneiform bones.

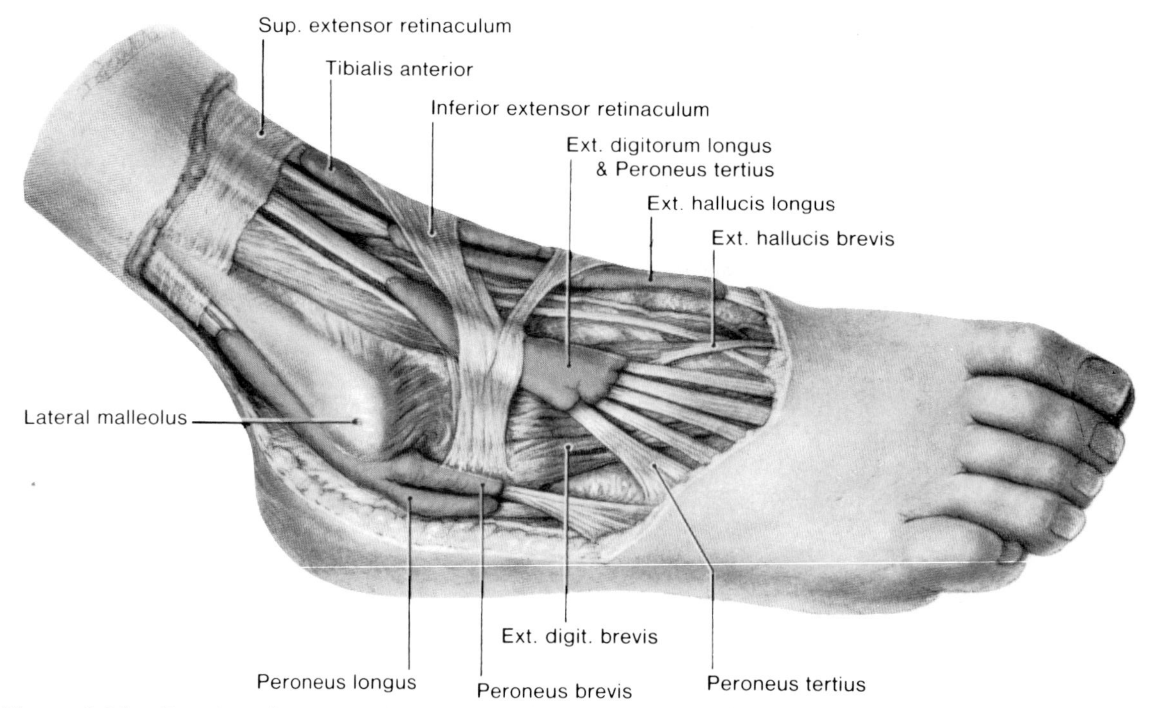

Figure 4-68. Drawing of an anterolateral view of a dissection of the right foot showing the synovial sheaths of the tendons. Observe that the tendons of the peroneus longus and peroneus brevis muscles are enclosed in a common synovial sheath, posterior to the lateral malleolus. Distal to this the synovial sheath splits into two, one for each tendon. Note that the tendon of the small peroneus tertius muscle runs with the tendons of the extensor digitorum longus within a common synovial sheath. The peroneus tertius muscle is a partially separated portion of the extensor digitorum longus.

Nerve Supply (Figs. 4-64, 4-67, and 4-70). **Superficial peroneal nerve (L5, S1**, and S2).

Actions (Fig. 4-65, *B* and *C*). **Plantarflexes and everts foot.** Its tendon crosses obliquely in the sole of the foot, thereby helping to maintain the transverse and longitudinal arches of the foot.

THE PERONEUS BREVIS MUSCLE (Figs. 4-64, 4-67, 4-68, 4-70 to 4-72, and Table 4-6)

This muscle lies deep to the peroneus longus and, as its name indicates, is shorter and smaller than its partner in the lateral crural compartment.

Origin (Figs. 4-56 and 4-57*A*). **Inferior two-thirds of lateral surface of fibula.** It also arises from the anterior and posterior crural intermuscular septa.

Insertion (Figs. 4-58, 4-67, and 4-71). **Dorsal surface of tuberosity on lateral surface of base of fifth metatarsal bone.** Its tendon grooves the posterior aspect of the lateral malleolus and can be felt inferior to the lateral malleolus, where it *lies in a common tendon sheath with the peroneus longus* (Figs. 4-68 and 4-71). The tendon of the peroneus brevis can be easily traced to its insertion into the base of the fifth metatarsal bone. A slip from the muscle often inserts into the long extensor tendon of the small toe and is known as the *peroneus digiti minimi* (Fig. 4-70).

Nerve Supply (Figs. 4-64, 4-67, and 4-70). **Superficial peroneal nerve (L5, S1**, and S2).

Actions (Fig. 4-65, *B* and *C*). **Plantarflexes and everts foot.** Its actions are the same as the peroneus longus, except that it appears to be less important in supporting the arches of the foot because its tendon does not pass through the sole of the foot.

The tuberosity of the fifth metatarsal bone may be avulsed by the peroneus brevis tendon during violent eversion of the foot. This kind of fracture is associated with a severely sprained ankle.

Injury to the superficial peroneal nerve (Fig. 4-67) results in an inverted foot owing to paralysis of the peroneal muscles (the evertors of the

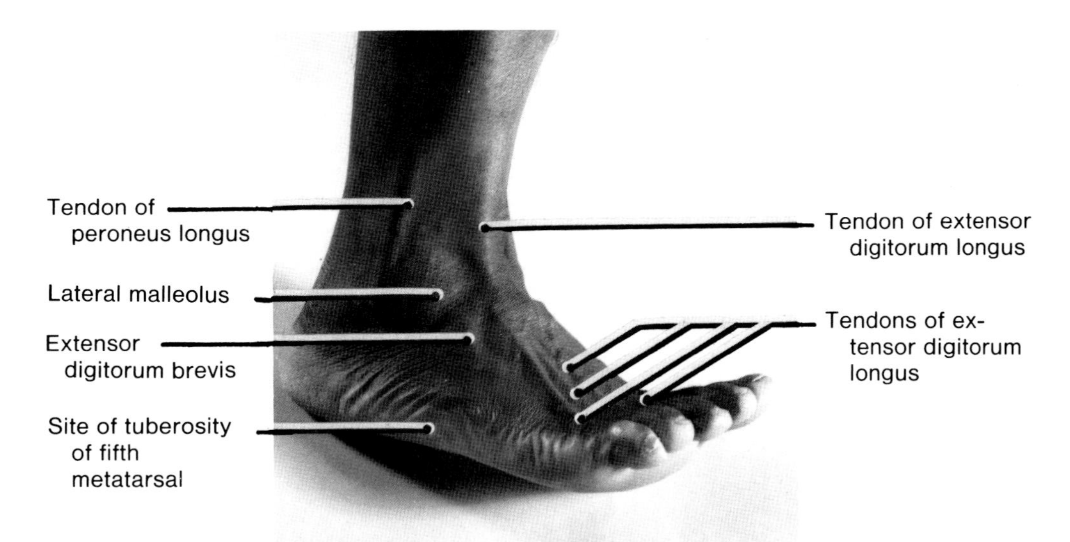

Figure 4-69. Photograph of the right lower leg and foot of a 35-year-old man with his toes dorsiflexed and foot everted. Observe the tendons of the extensor digitorum longus which run to the lateral four toes. A common synovial sheath surrounds the four tendons (Fig. 4-68). The tuberosity of the fifth metatarsal bone (Fig. 4-61) can be felt half-way along the lateral border of the foot. This tuberosity may be avulsed in violent inversion of the foot.

foot) in the lateral crural compartment (Fig. 4-64 and Table 4-6).

In children and adolescents there is often a secondary ossification center for the lateral surface of the tuberosity of the fifth metatarsal bone. This results in the formation of a ***chip-like piece of bone*** which should not be mistaken in a radiograph for a **flake fracture of the tuberosity** of the fifth metatarsal bone. The presence of similar secondary centers in both feet usually indicates that a fracture is not present. These centers are not usually observed in adults because they have fused with the tuberosity by this stage.

THE SUPERFICIAL PERONEAL NERVE (Figs. 4-64, 4-67, and 4-70)

This is the nerve of the lateral crural compartment. It is one of the two terminal branches of the common peroneal nerve.

The superficial peroneal nerve begins between the peroneus longus muscle and the fibula (Fig. 4-70) and descends posterolateral to the anterior crural intermuscular septum. It lies anterolateral to the fibula between the peroneal muscles and the extensor digitorum longus (Fig. 4-64).

The superficial peroneal nerve supplies the peroneal muscles and then pierces the deep fascia to become superficial in the distal third of the leg (Fig. 4-67), where it is vulnerable to injury. It passes in the superficial fascia to supply the skin on the distal part of the anterior surface of the leg, nearly all the dorsum of the foot, and most of the toes.

There are no arteries in the lateral crural compartment, except for muscular branches to the peroneal muscles from the peroneal artery (Fig. 4-76), a branch of the posterior tibial artery.

THE POSTERIOR CRURAL COMPARTMENT

From medial to lateral, the posterior crural compartment lies posterior to the tibia, the interosseous membrane, the fibula, and the posterior crural intermuscular septum (Fig. 4-64).

The muscles in the posterior crural compartment are divided into superficial and deep groups by the transverse crural intermuscular septum (Fig. 4-64), which is also known as the *deep transverse fascia of the leg.*

The superficial group of muscles in the posterior crural compartment consists of the gastrocnemius, plantaris, and soleus, and the deep group of muscles consists of the tibialis posterior, flexor digitorum longus, and flexor hallucis longus (Tables 4-7 and 4-8).

The superficial group of muscles forms a powerful mass in the calf of the leg which plantarflexes the foot. (Fig. 4-72 and Table 4-7). The large size of the superficial group of muscles is a human characteristic and is directly related to our upright stance. They are strong and heavy because they must support and move the weight of the body.

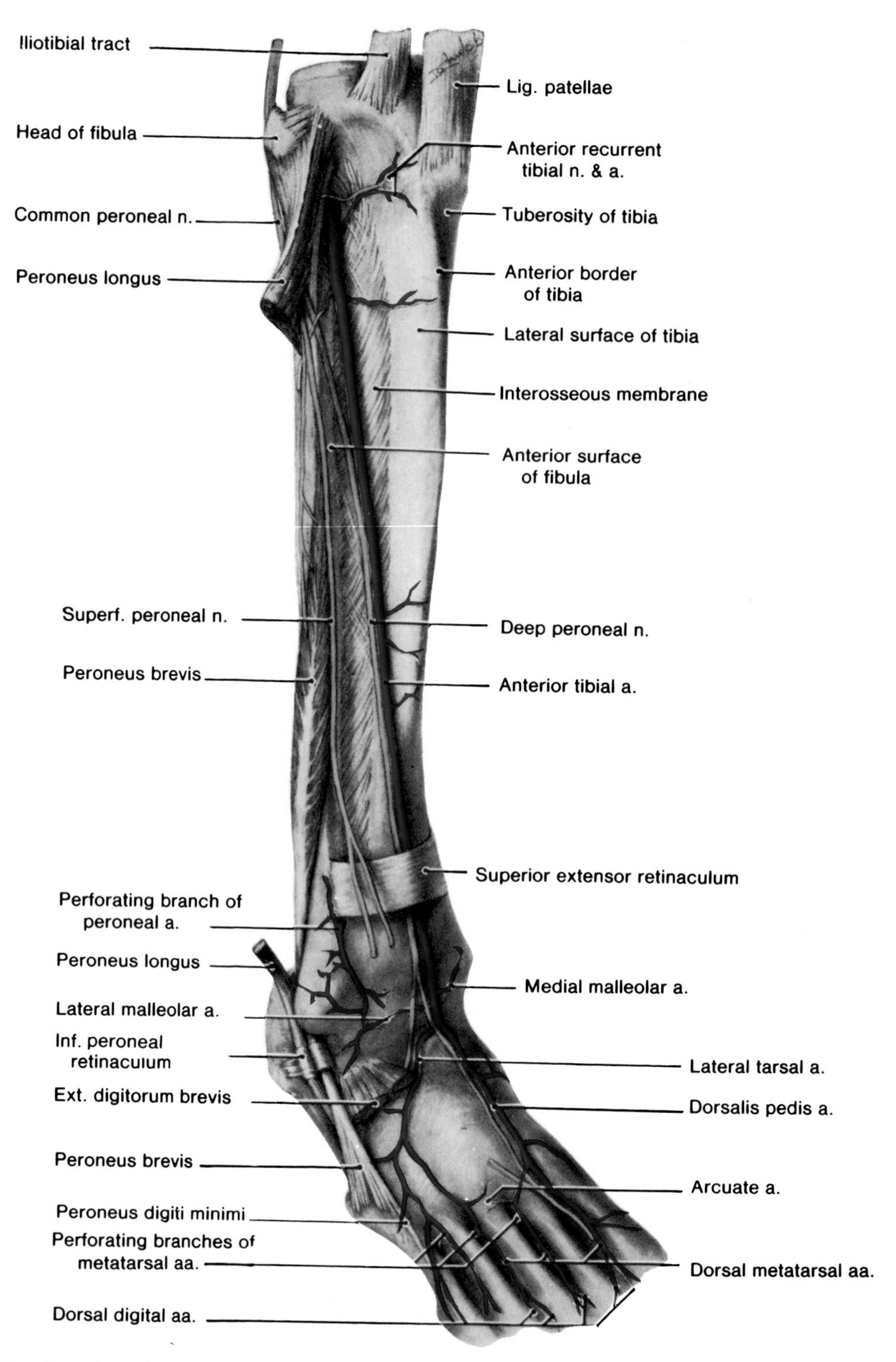

Figure 4-70. Drawing of a dissection of the right leg and dorsum of the foot. The anterior crural muscles are removed and the peroneus longus is excised. Observe the anterior tibial artery entering the region in contact with the medial side of the neck of the fibula and the common peroneal nerve in contact with its lateral side. Note that the artery and nerve and their named branches lie on the skeletal plane and are undisturbed by removal of the muscles. Observe the superficial peroneal nerve following the anterior border of the peroneus brevis muscle which guides it to the surface, a variable distance superior to the triangular subcutaneous area of the fibula. Note that the fibers of the interosseous membrane are so directed to allow the fibula to be forced superiorly but not pulled inferiorly.

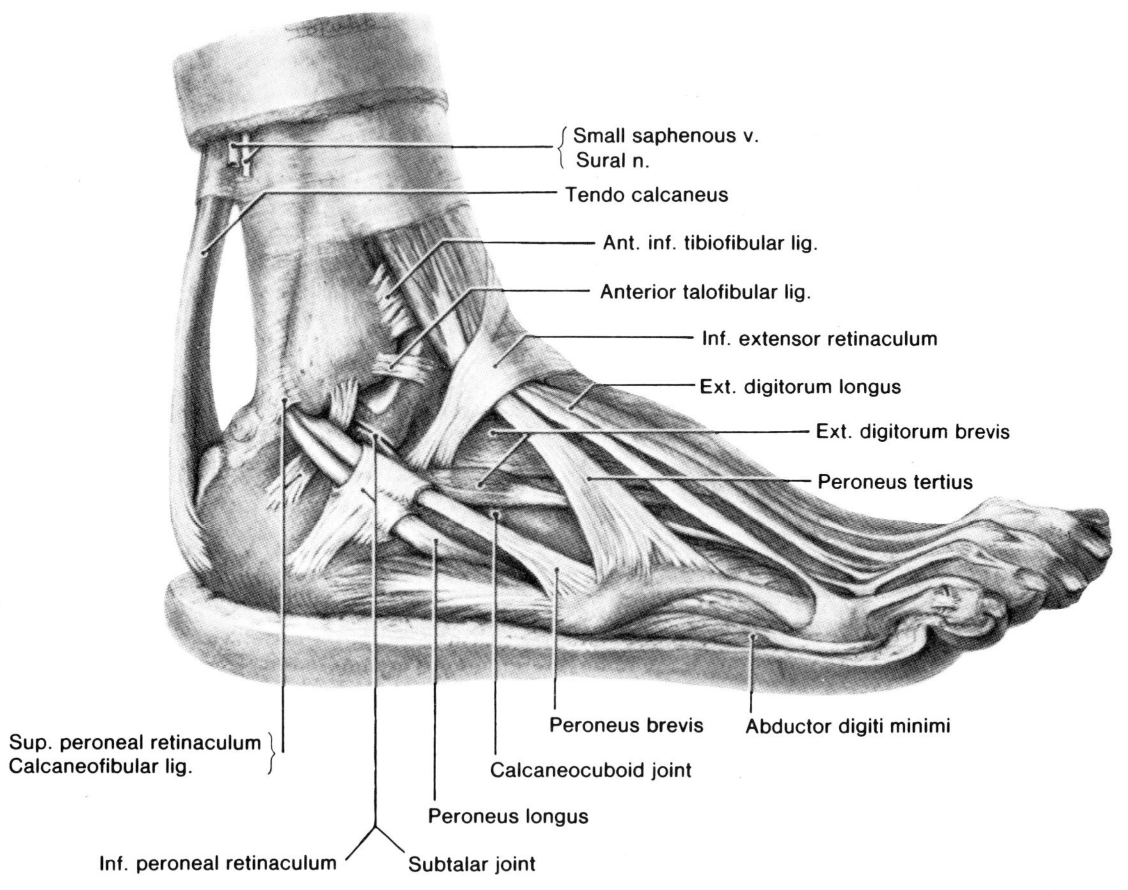

Figure 4-71. Drawing of a lateral view of a dissection of the dorsum of the right foot showing the ankle, subtalar, and calcaneocuboid joints. Observe the round calcaneofibular ligament attached anteriorly to the lateral malleolus, and that its tip overlaps the peroneal tendons, preventing them from slipping anteriorly. Note that the inferior peroneal retinaculum is attached to the lateral surface of the calcaneus and is in line with the inferior extensor retinaculum which is attached to its superolateral surface.

Table 4-6
Muscles of Lateral Crural Compartment

Muscle	Origin	Insertion	Nerve Supply	Actions
Peroneus longus (Figs. 4-64 and 4-67 to 4-69)	Head and superior two-thirds of lateral surface of fibula (Fig. 4-57*A*)	Base of first metatarsal bone and medial cuneiform bone (Figs. 4-59 and 4-78)	Superificial peroneal nerve (**L5**, **S1**, and S2)	Plantarflex and evert foot
Peroneus brevis (Figs. 4-64, 4-67, and 4-71)	Inferior two-thirds of lateral surface of fibula (Fig. 4-57*A*)	Dorsal surface of tuberosity on lateral side of base of the fifth metatarsal bone (Figs. 4-58 and 4-71)		

In Figures 4-64 and 4-77, observe that the tibial nerve and posterior tibial vessels supply both divisions of the posterior crural compartment and run between the superficial and deep groups of muscle. Observe that the tibial nerve and posterior tibial vessels are deep to the **transverse crural intermuscular septum,** and that the superficial muscles are much larger than the deep ones.

Three muscles comprise the superficial group: the gastrocnemius, soleus, and plantaris. The gastrocnemius and soleus form a tripartite muscle which is sometimes referred to as the **triceps surae muscle**. Note that it forms the prominence of the calf (Figs. 4-72 and 4-73).

The muscles of the superficial crural compartment act together in plantarflexing the foot at the ankle joint (Table 4-7). They raise the heel against the weight of the body, *e.g.*, in walking, dancing, and standing on the toes.

THE GASTROCNEMIUS MUSCLE (Figs. 4-72 to 4-75 and Table 4-7)

The gastrocnemius, *the most superficial of the muscles in the posterior crural compartment*, forms most of the prominence of the calf.

The gastrocnemius has *two heads of origin*; its medial head is slightly larger and extends a little more distally than does its lateral head. The two heads of the muscle come together at the inferior margin of the **popliteal fossa** (Figs. 4-49 to 4-51), where the converging heads form the inferolateral and inferomedial boundaries of the popliteal fossa.

Origin (Fig. 4-56). The ***lateral head* arises from the *lateral surface* of the lateral condyle** of the femur and the ***medial head* arises from the *popliteal surface of the femur,*** superior to the medial condyle.

The lateral head often contains a *sesamoid bone*, called a **fabella** (L. bean), close to its origin (Fig. 4-106*C*) which is usually visible on lateral radiographs of the knee.

Insertion (Figs. 4-71 to 4-78). **Posterior surface of calcaneus** via tendo calcaneus (Achilles tendon). It shares this very strong tendon with the soleus muscle. The tendo calcaneus is the thickest and strongest tendon in the body.

The inferior expanded end of the tendo calcaneus attaches to the middle of the posterior surface of the calcaneus (Fig. 4-56). A **tendo calcaneus bursa** (Figs. 4-55) separates the tendo calcaneus from the superior part of this surface.

Nerve Supply (Figs. 4-75 and 4-76). **Tibial nerve** (S1 and **S2**).

Actions (Figs. 4-65*B*, 4-73, and 4-74). **Plantarflexes the foot, raises the heel during walking, and flexes the knee joint.** It acts with the soleus muscle in plantarflexing the ankle (*e.g.*, in walking) and contracts to produce this movement when this action is resisted (*e.g.*, when standing on the toes). Although the gastrocnemius acts on the knee and ankle joints, it is unable to exert its full power on both joints at the same time.

THE SOLEUS MUSCLE (Figs. 4-64, 4-72 to 4-77, and Table 4-7)

This broad, flat, fleshy muscle was named because of its resemblance to the sole, a flat fish. The soleus lies deep to the gastrocnemius and can be palpated on each side of this muscle and inferior to the midcalf, when a person is standing on his/her tiptoes (Figs. 4-73 and 4-74).

Origin (Figs. 4-55 and 4-56). **Posterior aspect of head and superior fourth of fibula, soleal line, and middle third of medial border of tibia.** It also arises from the tendinous arch between the tibia and fibula which arches over the tibial vessels.

The soleus has a horseshoe-shaped origin from the tibia and fibula, just inferior to the knee.

Insertion (Figs. 4-72 to 4-78). **Posterior surface of calcaneus** via tendo calcaneus. It shares this strong tendon with the gastrocnemius muscle.

Nerve Supply (Figs. 4-75 and 4-76). **Tibial nerve** (S1 and **S2**).

Actions. Plantarflexes the foot at the ankle joint and steadies the leg on the foot during standing. It acts with the gastrocnemius in plantarflexing the ankle (*e.g.*, in walking and dancing). *The soleus muscle does not act on the knee joint.*

The soleus appears to be concerned with the maintenance of posture, *e.g.*, preventing the body from falling anteriorly when standing.

Rupture of the tendo calcaneus is not uncommon. Although it often occurs during games such as squash, it may happen during a stumble or when a person is startled, causing him/her to jump or suddenly start to run (*e.g.*, when crossing a street).

Complete rupture of the tendo calcaneus usually results in *abrupt pain in the posterior aspect of the leg,* an inability to walk, and a lump or increase in the prominence of the calf owing to shortening of the triceps surae muscle (gastrocnemius and soleus).

The tendon may also rupture in young persons, frequently at the start of a 100-m dash when the tendon is severely stressed during take-off. Ruptures of the tendon also occur when

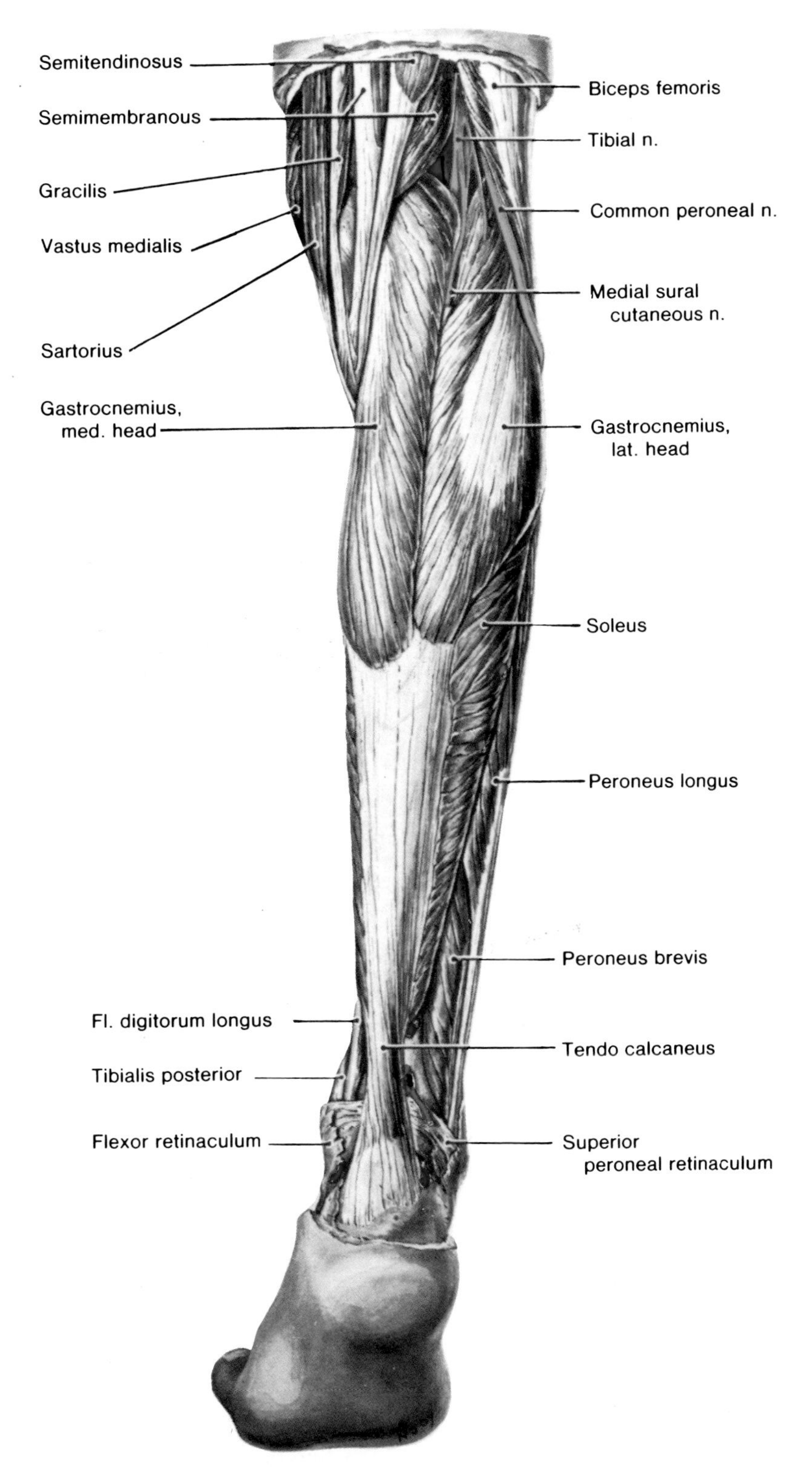

Figure 4-72. Drawing of a superficial dissection of the right leg showing the muscles of the superficial posterior crural compartment. The gastrocnemius and soleus muscles (triceps surae) form the prominence of the calf. Note that the tendon of the gastrocnemius joins the tendon of the soleus to form the tendo calcaneus.

it becomes weaker than normal owing to **ischemia** (lack of good blood supply).

The rupture of the tendo calcaneus usually occurs about 3 cm proximal to its insertion into the calcaneus. A distinct gap can easily be felt in it.

Following rupture of the tendo calcaneus, the foot can be dorsiflexed to a greater extent than normal and the patient is unable to plantarflex his/her foot against resistance.

The ankle reflex (ankle jerk) is the *twitch of the triceps surae* muscle and is induced by striking the tendo calcaneus with a reflex hammer. *The reflex center for the ankle reflex is the S1 and S2 segments of the spinal cord.*

Persons who continually wear high heels may develop shortening of the triceps surae because the origin of this muscle is continually brought closer to its insertion. If this occurs, **transitory calf pain** may be experienced when walking without shoes or in flat shoes owing to tightness of the triceps surae muscle.

Tennis leg is a painful calf injury resulting from partial *tearing of the medial belly of the gastrocnemius* at or near the musculotendinous junction. It is caused by overstretching the muscle by concomitant full extension of the knee and dorsiflexion of the ankle joint. Contributory factors appear to be muscle fatigue and degenerative changes in the muscle. It usually occurs when a middle-aged tennis player is serving the ball or stretches for a difficult shot.

The gastrocnemius is one of a few muscles with only one source of blood supply, (*i.e.*, the sural arteries, branches of the popliteal). They are virtually end arteries, *i.e.*, no anastomoses except by capillaries. If one branch is blocked, the part supplied by it dies. The soleus muscle is also supplied by the sural arteries, but in addition it receives branches from the peroneal artery (Figs. 4-76 and 4-77).

Swelling of the tendo calcaneus bursa, called **calcaneal bursitis,** is fairly common in long distance runners and Scottish dancers, owing to excessive friction on the bursa as the tendon slides over it.

When standing the venous return of the leg depends largely on muscular activity, especially of the triceps surae muscles. The efficiency of this "**calf pump**" is improved by the tight sleeve of deep fascia covering these muscles (Fig. 4-64). When the calf muscles contract, blood is pumped superiorly in the **deep veins.** Normally blood is prevented from flowing into the **superficial veins** by the valves in the communicating veins (Fig. 4-10). If these valves become incompetent, blood is forced into the superficial veins during contraction of the triceps surae muscles and by hydrostatic pressure when straining or standing. The distented perforating and superficial veins are called **varicose veins.**

THE PLANTARIS MUSCLE (Fig. 4-64 and Table 4-7)

This small muscle is variable in size and extent. It may be absent. When present it has a fleshy belly (about 10 cm long) and a long, slender tendon which runs obliquely between the gastrocnemius and soleus muscles. *This feeble muscle is of no practical importance.*

Origin (Figs. 4-54 and 4-56). **Inferior end of the lateral supracondylar line** and the oblique popliteal ligament.

Insertion (Fig. 4-56). **Posterior surface of calcaneus** as part of tendo calcaneus, or separately on the medial side of it.

Nerve Supply (Figs. 4-75 and 4-76). **Tibial nerve** (S1 and **S2**).

Actions. Assists with flexion of the knee joint and plantarflexion of the foot. It acts with the gastrocnemius and soleus muscles, but its role is minor.

The clinical importance of the plantaris lies in the possibility of its rupture during violent ankle movements. Sudden dorsiflexion of the ankle joint may rupture the slender tendon of this feeble muscle. In most cases of apparent **rupture of the plantaris tendon,** muscle fibers of the triceps surae are also torn. This injury is common in basketball players, sprinters, and ballet dancers. Surprisingly the pain following rupture may be so severe that the person is unable to bear weight on the foot. Usually some fibers of the gastrocnemius are also torn, resulting in internal bleeding and ***pain in the calf.*** This is called a "**Charley horse**" by athletes, as is pain in the anterior aspect of the thigh, following injury to the quadriceps femoris muscle. The pain of a Charley horse is more severe and prolonged than a **cramp** (painful muscle spasm).

The long tendon of the plantaris muscle is commonly used in reconstructive surgery of the tendons of the hand. It can be removed completely without causing any disability of knee or ankle movements.

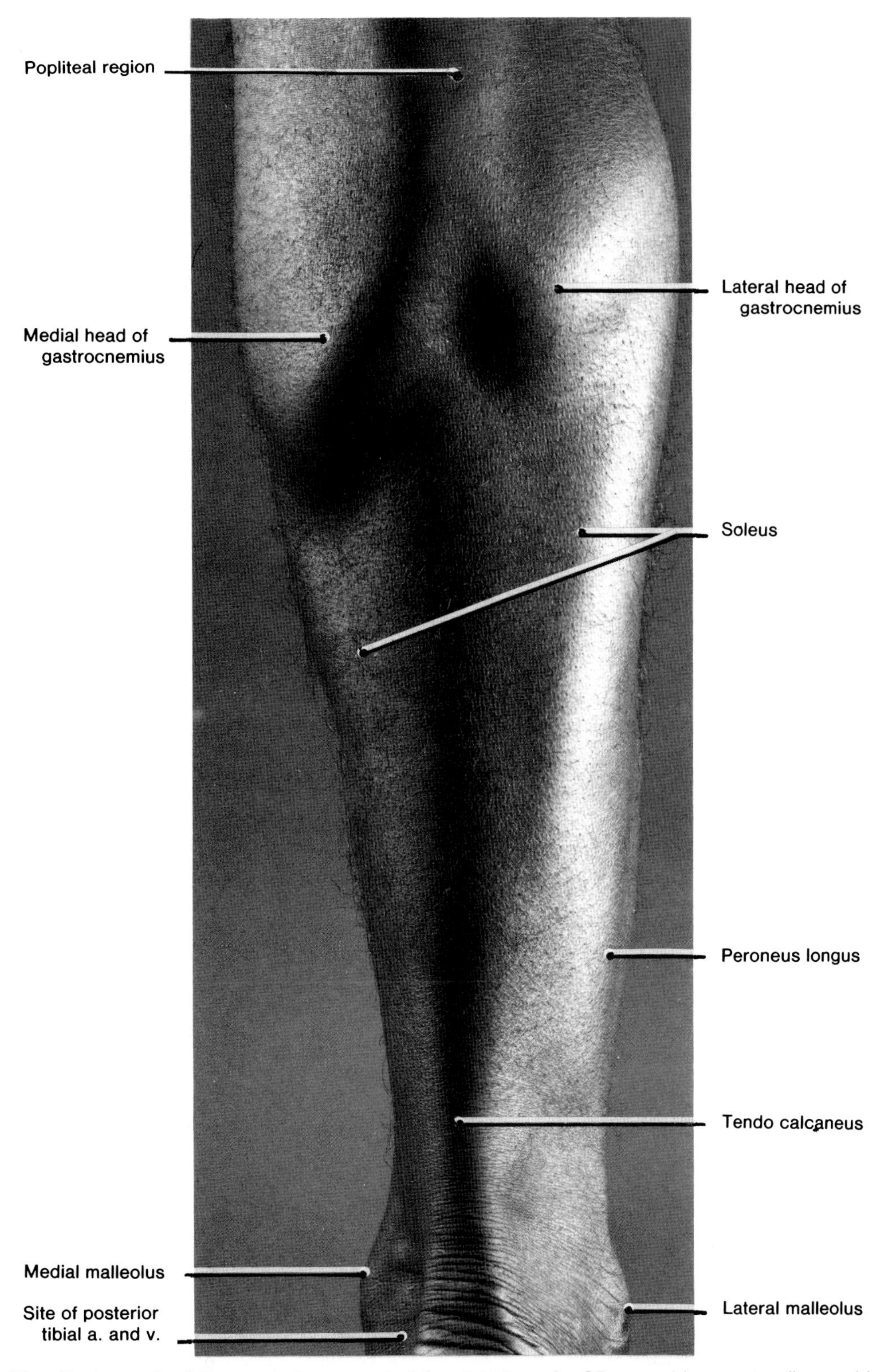

Figure 4-73. Photograph of the posterior aspect of the right leg of a 35-year-old man standing on his tiptoes to show the principal surface features of his muscles. Observe that the medial and lateral malleoli are subcutaneous and prominent. Note that the lateral malleolus is more distal than the medial malleolus. Observe that the two heads of the gastrocnemius come together to form a single muscle (also see Fig. 4-72), the tendon of which joins the tendon of the soleus muscle to form the tendo calcaneus that inserts into the posterior surface of the calcaneus (Figs. 4-56 and 4-71). The two-headed gastrocnemius, together with the soleus, form the triceps surae muscle, which produces the prominence of the calf.

Table 4-7
Superficial Muscles of Posterior Crural Compartment

Muscle	Origin	Insertion	Nerve Supply	Actions
Gastrocnemius (Figs. 4-49 to 4-52 and 4-72 to 4-75)	*Lateral head*: Lateral aspect of lateral condyle of femur *Medial head*: Popliteal surface of femur superior to medial condyle (Figs. 4-20 and 4-56)	Posterior surface of calcaneus via tendo calcaneus (Figs. 4-56 and 4-71)	Tibial nerve (S1 and **S2**)	Plantarflexes foot, raises heel during walking, and flexes knee joint
Soleus (Figs. 4-50, 4-52, and 4-72 to 4-75)	Posterior aspect of head of fibula, superior fourth of posterior surface of fibula, soleal line, and middle third of medial border of tibia (Figs. 4-20 and 4-56)			Plantarflexes foot and steadies the leg on the foot
Plantaris (Figs. 4-40, 4-50, 4-52, and 4-64)	Inferior end of lateral supracondylar line and oblique popliteal ligament (Figs. 4-20, 4-54, and 4-56)			Assists gastrocnemius in plantarflexing foot and flexing knee joint

Four muscles comprise the deep group in the posterior crural compartment (Fig. 4-64 and Table 4-8): the **popliteus, flexor digitorum longus, flexor hallucis longus**, and **tibialis posterior**. The popliteus acts on the knee, whereas the other muscles act on the ankle and foot joints.

THE POPLITEUS MUSCLE (Fig. 4-77 and Table 4-8)

This thin, triangular muscle *forms the floor of the inferior part of the* ***popliteal fossa.***

Origin (Figs. 4-45 and 4-56). **Lateral surface of lateral condyle of femur,** just inferior to the attachment of the fibular collateral ligament of the knee. Some fibers arise from the posterior surface of the **lateral meniscus of the knee joint.** Its stout cord-like tendon arises from a rough pit on the condyle known as the **popliteal groove of the femur.**

The origin of the popliteus is inside the fibrous capsule of the knee joint, deep to the fibular collateral ligament; thus the deep surface of its tendon is covered by synovial membrane (Fig. 4-108).

Insertion (Fig. 4-56). **Posterior surface of tibia, superior to soleal line**; hence, the inferior border of the popliteus is adjacent to the superior border of the soleus (Fig. 4-77).

Nerve Supply. (Fig. 4-77). **Tibial nerve** (L4, **L5**, and S1).

Actions. Flexes and rotates the knee joint. The popliteus flexes the knee and **unlocks the locked knee by rotating the tibia medially on the femur.** It is a lateral rotator of the femur on the tibia when the foot is fixed (*e.g.*, on the ground).

Because of the attachment of the popliteus to the lateral meniscus, it is thought to pull this cartilage posteriorly during flexion of the knee and lateral rotation of the tibia so it will not be crushed.

The following three muscles (*flexor hallucis longus, flexor digitorum longus, and tibialis posterior) all pass deep to the flexor retinaculum* (Figs. 4-76 and 4-77). This thickening of the deep crural fascia passes from the medial side of the calcaneus to the medial malleolus (Fig. 4-76).

THE FLEXOR HALLUCIS LONGUS MUSCLE (Figs. 4-76, 4-77, and Table 4-8)

This long powerful muscle, the largest of the three deep muscles, *lies laterally and is closely attached to the fibula.*

Origin (Figs. 4-56 and 4-57*B*). **Inferior two-thirds of the posterior surface of the fibula, inferior part of the interosseous membrane,** and the intermuscular septa. Its tendon passes posterior to the distal end of the tibia and deep to the flexor retinaculum (Fig. 4-76). The tendon occupies a shallow groove on the posterior surface of the talus which is continuous with the groove on the plantar surface of the sustenaculum tali.

The tendon of the flexor hallucis longus then crosses deep to the tendon of the flexor digitorum longus in the sole of the foot, giving a tendinous slip to its tendon (Fig. 4-78). As it passes to the great toe, the tendon runs between two sesamoid bones in the tendons of the flexor hallucis brevis. These small bones protect the tendon from the pressure of the head of the first metatarsal bone.

Table 4-8
Deep Muscles of Posterior Crural Compartment

Muscle	Origin	Insertion	Nerve Supply	Actions
Popliteus (Figs. 4-50, 4-54, and 4-77)	Lateral surface of lateral condyle of femur and lateral meniscus (Figs. 4-45 and 4-56)	Posterior surface of tibia, superior to soleal line (Fig. 4-56)	Tibial nerve (L4, **L5**, and S1)	Flexes knee and unlocks the locked knee by rotating tibia medially on the femur
Flexor hallucis longus[1] (Figs. 4-64, 4-76, and 4-77)	Inferior two-thirds of posterior surface of fibula and inferior part of interosseous membrane (Figs. 4-56 and 4-57*B*)	Base of distal phalanx of great toe (Figs. 4-59 and 4-78)	Tibial nerve (**S2** and S3)	Flexes great toe and plantarflexes foot (Fig. 4-65*B*)
Flexor digitorum longus[1] (Figs. 4-75 to 4-78)	Medial part of posterior surface of tibia, inferior to soleal line (Fig. 4-56)	Bases of distal phalanges of lateral four digits (Fig. 4-59)	Tibial nerve (**S2** and S3)	Flexes lateral four toes and plantarflexes foot (Fig. 4-65*B*)
Tibialis posterior[1] (Figs. 4-76 to 4-79)	Interosseous membrane, lateral part of posterior surface of tibia, and superior two-thirds of medial surface of fibula (Figs. 4-56 and 4-79)	Tuberosity of navicular, cuneiforms, cuboid, and bases of second, third, and fourth metatarsals (Figs. 4-60 and 4-79)	Tibial Nerve (L4 and L5)	Plantarflexes and inverts foot (Fig. 4-65, *B* and *D*)

[1] All three of these muscles pass deep to the flexor retinaculum (Figs. 4-76 and 4-77) and all three serve to steady the leg on the foot when standing.

Insertion (Fig. 4-78). **Base of the distal phalanx of the great toe.**

Nerve Supply (Figs. 4-76 and 4-77). **Tibial nerve** (**S2** and S3).

Actions. Flexes great toe at all joints, acting primarily on the distal phalanx, **and plantarflexes the foot**. It also helps to maintain the medial longitudinal arch of foot.

The flexor hallucis longus is the "push-off" muscle during walking and running and provides much of the spring to the step. It is also important in holding the leg in the normal position on the foot.

THE FLEXOR DIGITORUM LONGUS MUSCLE (Figs. 4-75 to 4-78 and Table 4-8)

This long flexor of the toes *lies medially and is closely attached to the tibia.* It is smaller than the flexor hallucis longus, even though it moves four toes.

Origin (Fig. 4-56). **Medial part of the posterior surface of the tibia**, inferior to the soleal line.

Insertion (Fig. 4-78). **Bases of distal phalanges of lateral four digits.** Its tendon runs inferiorly, *passing posterior to the tibialis posterior tendon and the medial malleolus.* It then passes diagonally in the sole of the foot, superficial to the tendon of the flexor hallucis longus (Fig. 4-78). As the tendon reaches the middle of the sole, it divides into four tendons which pass to the distal phalanges of the lateral four digits.

Nerve Supply (Figs. 4-76 and 4-77). **Tibial nerve** (**S1** and S2).

Actions. Flexes lateral four toes, acting primarily on the distal phalanges, **plantarflexes the foot** at the ankle joint, and *helps to maintain the medial longitudinal arch of foot.* Flexion of the toes is important in walking and running (especially with bare feet) because they give the foot a grip on the ground.

THE TIBIALIS POSTERIOR MUSCLE (Figs. 4-76 to 4-79 and Table 4-8)

This large muscle, *the deepest one in the posterior crural compartment,* lies between the flexor digitorum longus and the flexor hallucis longus in the same plane as the tibia and fibula.

Origin (Figs. 4-56, 4-57*B*, and 4-79). **Interosseous membrane, lateral part of the posterior surface**

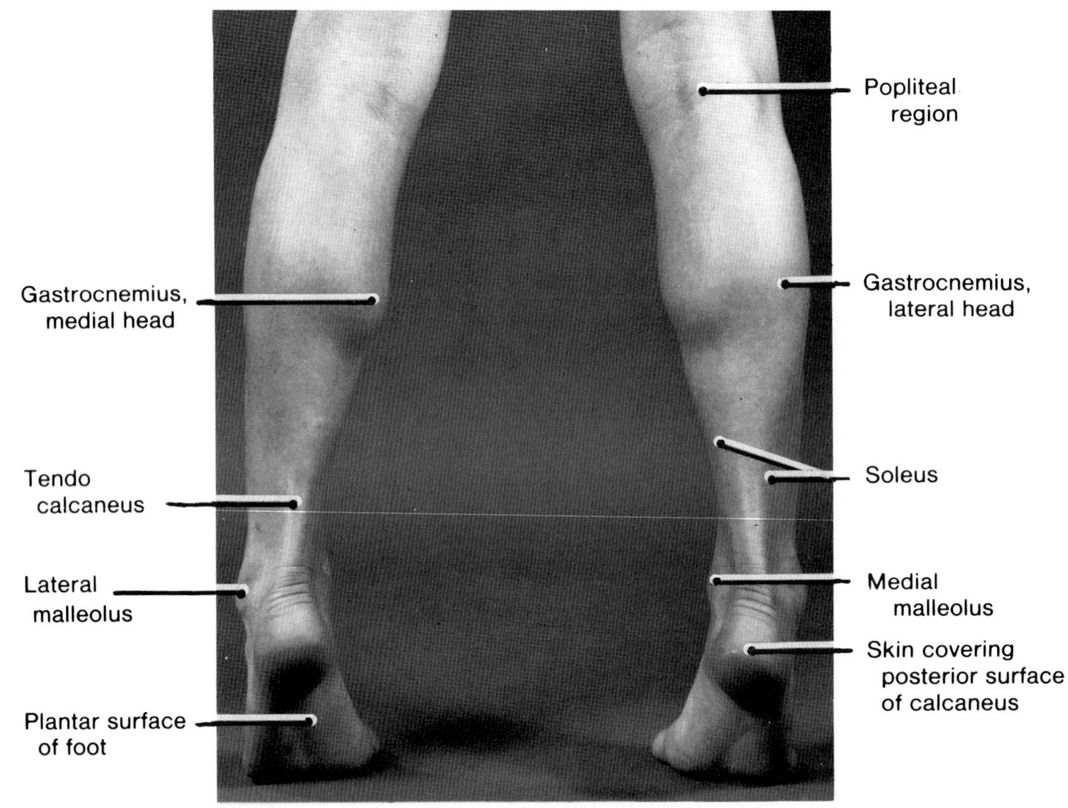

Figure 4-74. Photograph of the posterior aspect of the legs of a 12-year-old girl who is standing on her tiptoes to show the principal surface features of her muscles. The fleshy posterior part of the leg, formed by the gastrocnemius and soleus muscles (collectively known as the triceps surae), is often referred to as the "calf of the leg." The triceps surae is the principal plantarflexor of the foot (Fig. 4-65*B*); hence when it is paralyzed, the patient cannot stand on the tiptoes.

of the tibia (inferior to the soleal line), and **superior two-thirds of the medial surface of the fibula.**

Insertion (Figs. 4-60 and 4-79). **Tuberosity of navicular, cuneiforms, cuboid, and the bases of second, third, and fourth metatarsals.**

Nerve Supply (Figs. 4-76 and 4-77). **Tibial nerve** (L4 and L5).

Actions (Fig. 4-65, *B* and *D*). **Plantarflexes and inverts foot.** *It also helps to maintain the medial longitudinal arch of the foot.*

THE TIBIAL NERVE (Figs. 4-76 and 4-77)

The tibial nerve supplies all muscles in the posterior crural compartment (Fig. 4-64 and Tables 4-7 and 4-8). The *larger terminal branch of the sciatic nerve,* the tibial arises from the ventral branches of the ventral rami of L4, L5, S1, S2, and S3 (Fig. 4-82).

The tibial nerve descends through the middle of the popliteal fossa, posterior to the popliteal vein and artery (Figs. 4-49 to 4-51 and 4-77). At the distal border of the popliteus muscle, the tibial nerve passes with the posterior tibial vessels deep to the tendinous arch of the soleus muscle (Figs. 4-75 and 4-76). It then descends deep to the soleus muscle and runs inferiorly on the tibialis posterior muscle in company with the posterior tibial vessels.

The tibial nerve leaves the posterior crural compartment by passing deep to the flexor retinaculum in the interval between the medial malleolus and calcaneus (Figs. 4-76 and 4-77). The tibial nerve lies between the posterior tibial vessels and the tendon of the flexor hallucis longus muscle (Fig. 4-81).

Posteroinferior to the medial malleolus, the tibial nerve divides into the ***medial and lateral plantar nerves*** (Figs. 4-81 and 4-82).

The tibial nerve gives branches to all muscles in the posterior crural compartment (Fig. 4-82). A cutaneous branch of the tibial, the *medial sural cutaneous nerve,* unites with the peroneal communicating branch of the common peroneal nerve to form the **sural nerve** (Fig. 4-50). The sural nerve supplies the skin of the lateral

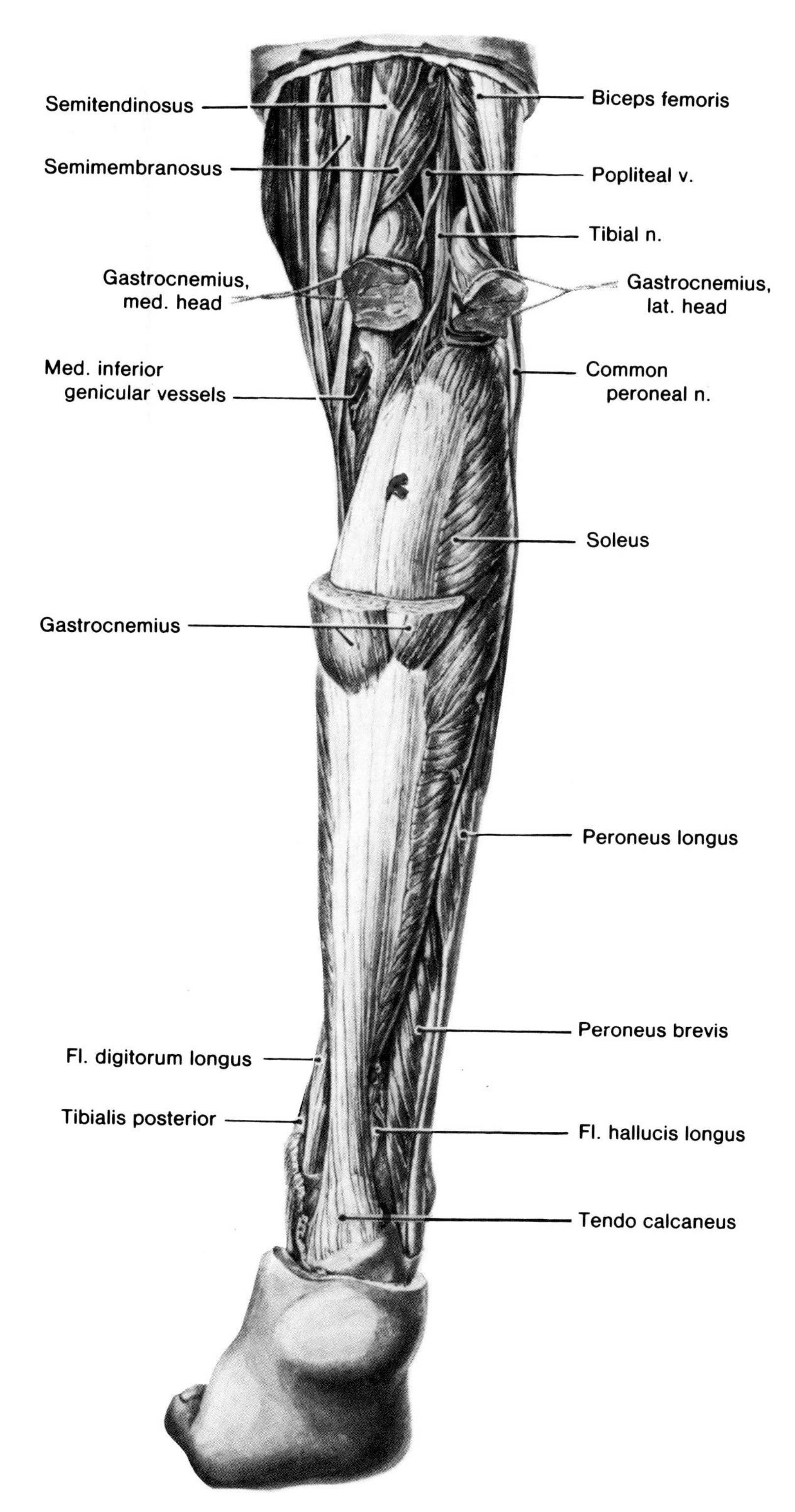

Figure 4-75. Drawing of a dissection of the muscles of the superficial posterior crural compartment. The fleshy bellies of the gastrocnemius muscle are largely excised, exposing the origin of the soleus muscle.

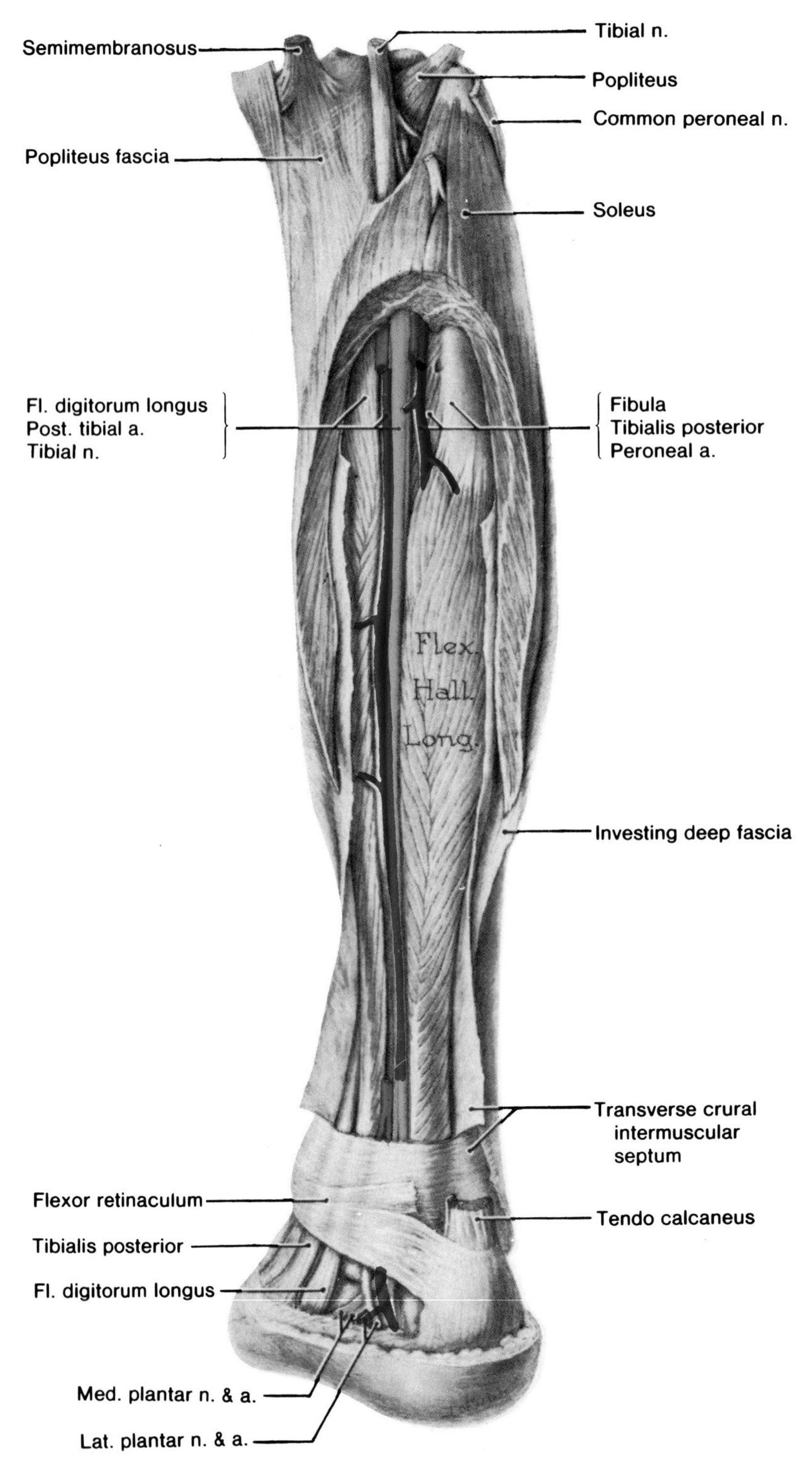

Figure 4-76. Drawing of a deep dissection of the posterior compartment of the right leg. The tendo calcaneus is divided and the gastrocnemius muscle and a horseshoe-shaped section of the soleus muscle are removed. Observe the posterior tibial artery and tibial nerve descending between the large flexor hallucis longus muscle and the smaller flexor digitorum longus muscle. Note the transverse crural intermuscular septum blends medially with the weaker investing deep fascia to form the flexor retinaculum.

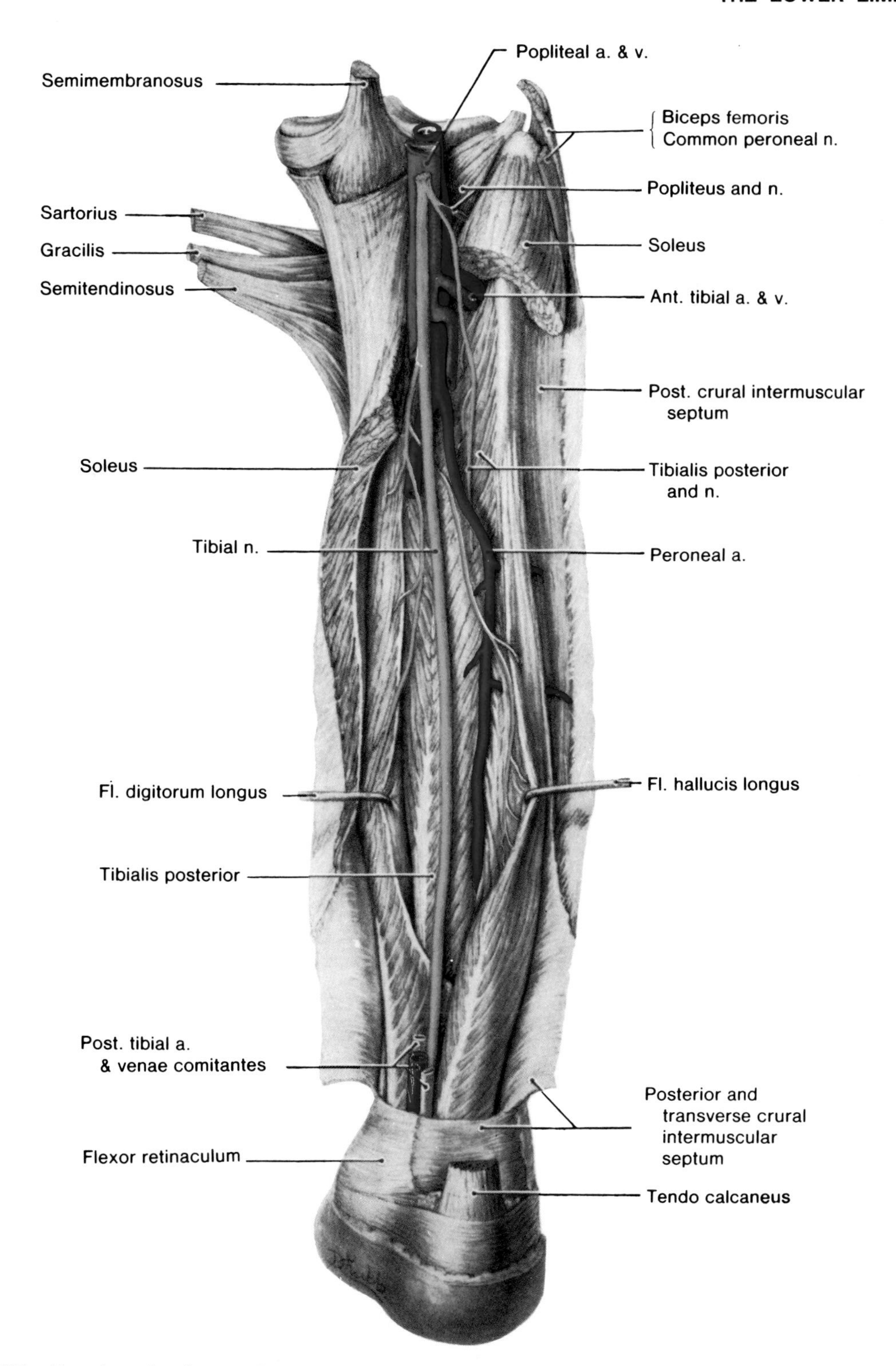

Figure 4-77. Drawing of a deeper dissection of the posterior compartment of the right leg. The soleus muscle is largely cut away, the two long digital flexors are pulled apart, and most of the posterior tibial artery is excised. Observe the tibialis posterior muscle lying deep to the long digital flexors and that the peroneal artery is overlapped by the flexor hallucis longus muscle. Note that the nerve to the tibialis posterior here arises in conjunction with the nerve to the popliteus, and that the nerve to the flexor digitorum longus arises in conjunction with the nerve to the flexor hallucis longus. Observe that the tibial nerve is superficial to the popliteal artery in the popliteal fossa, whereas at the ankle the posterior tibial artery is superficial to the tibial nerve.

and posterior part of the inferior third of the leg and the lateral side of the foot (Figs. 4-10 and 4-94).

Articular branches of the tibial nerve supply the knee joint and medial calcaneal branches supply the skin of the heel.

Because the tibial nerve is deep and well protected, it is not commonly injured. However, lacerations in the popliteal fossa or **posterior dislocations of the knee joint** may damage this nerve, producing paralysis of all muscles in the posterior compartment of the leg and the intrinsic muscles in the sole of the foot (Tables 4-8 and 4-9). When the plantarflexors of the foot are paralyzed, the patient is unable to curl the toes or stand on them. In addition, there is loss of sensation in the sole of the foot making it vulnerable to the development of pressure sores.

THE POSTERIOR TIBIAL ARTERY (Figs. 4-76, 4-77, 4-81, 4-83, and 4-96)

The posterior tibial artery begins at the distal border of the popliteus muscle, between the tibia and fibula, as the **larger terminal branch of the popliteal artery**.

As the direct continuation of the popliteal, *the posterior tibial artery passes deep to the origin of the soleus* muscle and, after giving off the **peroneal artery,** its largest branch, it passes inferomedially on the posterior surface of the tibialis posterior muscle (Fig. 4-83).

During its descent, the posterior tibial artery is accompanied by the tibial nerve and two venae comitantes, deep to the transverse crural intermuscular septum (Fig. 4-64). At the ankle *the posterior tibial artery runs posterior to the medial malleolus,* from which it is separated by the tendons of the tibialis posterior and flexor digitorum longus muscles (Fig. 4-81). *Inferior to the medial malleolus, it runs between the tendons of the flexor hallucis longus and flexor digitorum longus* muscles.

Deep to flexor retinaculum and the origin of the abductor hallucis muscle, **the posterior tibial artery divides into medial and lateral plantar arteries** (Figs. 4-76 and 4-81).

Branches of Posterior Tibial Artery (Figs. 4-76, 4-77, and 4-83). Most of the muscular branches of the posterior tibial artery are unnamed, but its most important one is called the peroneal artery.

The Peroneal Artery (Figs. 4-77 and 4-83). The peroneal artery is the largest and ***most important branch of the posterior tibial artery*** (Figs. 4-77 and 4-83). It begins inferior to the distal border of the popliteus muscle and the tendinous arch of the soleus (Fig. 4-77). It descends obliquely toward the fibula and passes along its medial side within the flexor hallucis longus muscle, or between it and the intermuscular septum and tibialis posterior muscle.

The peroneal artery gives off muscular branches to the popliteus and to other muscles in the posterior and lateral compartments of the leg. **It also supplies a nutrient artery to the fibula** and a communicating branch which joins that of the posterior tibial artery (Fig. 4-83). The peroneal artery usually pierces the interosseous membrane and passes to the dorsum of the foot, where it anastomoses with the ***arcuate artery*** (Fig. 4-70).

The Circumflex Fibular Artery. The circumflex fibular branch of the posterior tibial artery arises at the knee and passes laterally over the neck of the fibula to the anastomoses around the knee (Fig. 4-53).

The Nutrient Artery of the Tibia (Figs. 4-31*B* and 4-83). The nutrient artery of the tibia arises from the posterior tibial artery near its origin. It is the *largest nutrient artery in the body.* The nutrient foramen through which it passes is just distal to the soleal line on the posterior surface of the tibia (Fig. 4-55).

Other branches of the posterior tibial artery are the **calcanean arteries** which supply the tissues of the heel (Figs. 4-81 and 4-83). They pass medial and posterior to the tendo calcaneus and anastomose with branches of the peroneal artery. A **malleolar branch** joins the network of vessels on the medial malleolus (Fig. 4-31).

Absence of the posterior tibial artery with compensatory enlargement of the peroneal artery occurs in about 5% of people (see *Grant's Atlas of Anatomy, ed 8,* Fig. 4-129).

The pulse of the posterior tibial artery can usually be palpated about half way between the posterior surface of the medial malleolus and the medial border of the tendo calcaneus (Figs. 4-81 and 4-84). Curve your fingers posterior and slightly inferior to the medial malleolus. This pulse is usually easy to feel in children and its pulsations may be visible. In older persons, in whom the pulse may be difficult to palpate with the foot relaxed, it is usually easier to feel with the patient's foot dorsiflexed and inverted.

Knowing how to palpate the **posterior tibial pulse** is essential for examining patients with a condition called **intermittent claudication**. It is caused by ischemia to the leg muscles owing to **arteriosclerotic stenosis** or occlusion of the leg arteries. *Adminished or absent pulse suggests arterial insufficiency;* however this pulse may be congenitally absent.

Intermittent claudication is characterized by

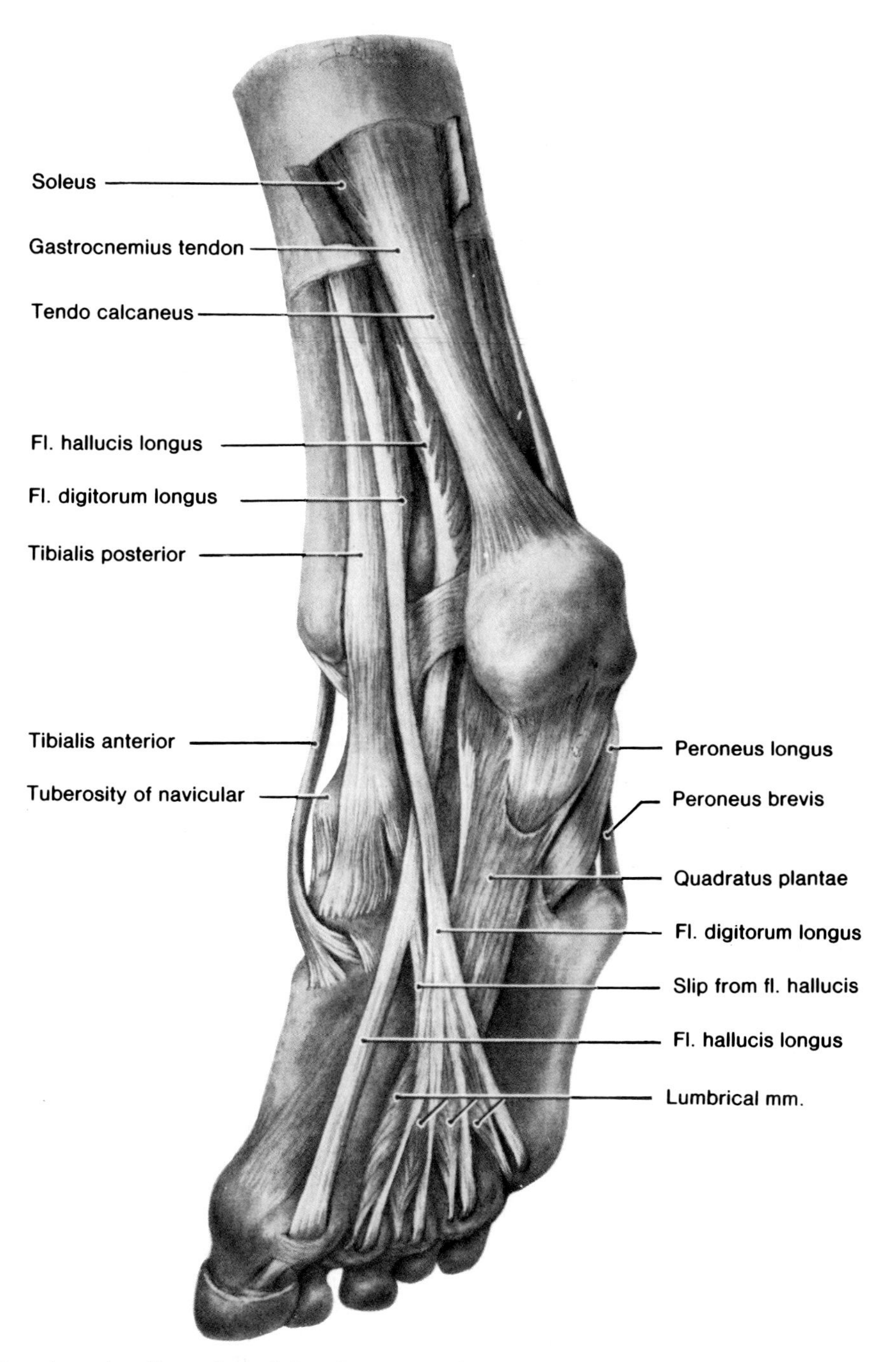

Figure 4-78. Drawing of a dissection of the distal part of the right leg and foot, displaying the second layer of plantar muscles, consisting of the flexor hallucis longus, the flexor digitorum longus, the four lumbricals, and the quadratus plantae (L. *quadratus*, square). The quadratus plantae is called the flexor digitorum accessorius by some authors. Observe that the flexor digitorum longus crosses superficial to the tibialis posterior muscle, posterior to the medial malleolus, and superficial to the flexor hallucis longus. Note the four lumbrical muscles passing to the medial sides of the toes. Observe that the flexor hallucis longus sends a strong tendinous slip to the flexor digitorum longus muscle.

leg cramps which develop during walking and disappear soon after rest. When a few more steps are taken, the cramps reappear. When a patient complains of **leg cramps**, often the peripheral *pulses in the leg are absent or markedly diminished.* These cramps develop during exercise because the narrowed arteries and collateral circulation to the leg muscles are unable to supply the extra blood that is required by the actively contracting leg muscles. When examining a patient with leg cramps, it is important to determine whether the circulatory failure in the leg results from narrowing (**arteriosclerosis**) or spasm of the muscular arteries.

THE FOOT

The foot is the part of the lower limb distal to the ankle joint. It is concerned mainly with support and locomotion of the body.

The bones of the foot are described and illustrated on pages 469 to 475. The clinical importance of the foot is indicated by the estimate that the average orthopaedic surgeon devotes about 20% of his/her practice to foot problems, and the entire practice of podiatry is concerned with the diagnosis and treatment of diseases, injuries, and abnormalities of the foot.

SKIN OF THE FOOT

The skin on the dorsal surface or dorsum of the foot is thin and mobile; hair is sparse and there is relatively little subcutaneous fat. Because of the thinness of the skin and superficial fascia on the dorsum of the foot, the tendons are usually visible, especially during dorsiflexion (Fig. 4-80).

The skin on the plantar surface or sole of the foot is thin on the toes and instep, but thick over the heel and ball of the foot (base of great toe). The plantar skin contains many sweat glands and much fat in the subcutaneous tissue, especially over the heel (Fig. 4-85), which is firmly bound down to underlying structures by fibrous connective tissue. The sole of the foot is designed for weight bearing and for protection of the underlying nerves and vessels.

FASCIA OF THE FOOT

The deep fascia of the foot is continuous with that of the ankle. It is thin on the dorsum of the foot (Fig. 4-81), where it is continuous with the inferior extensor retinaculum (Fig. 4-71). Over the lateral and posterior aspects of the foot, the deep fascia is continuous with the **plantar fascia** or deep fascia of the sole (Fig. 4-85).

THE PLANTAR APONEUROSIS (Fig. 4-85)

The central part of the plantar fascia is greatly thickened to form the plantar aponeurosis. It consists of a strong, thick central part and weaker and thinner medial and lateral portions.

The plantar aponeurosis consists of longitudinally arranged bands of dense fibrous connective tissue. *The plantar aponeurosis helps to support the longitudinal arches* of the foot *and to hold the parts of the foot together.* It extends anteriorly from the tuber calcanei (Fig. 4-60), becoming broader and somewhat thinner.

The plantar aponeurosis divides into five bands for the digits (Fig. 4-85). These bands split to enclose the digital tendons and are attached to the margins of the fibrous digital sheaths and to the sesamoids of the great toe (Figs. 4-90 and 4-92).

From the margins of the central part of the plantar aponeurosis, vertical septa extend deeply to form **three compartments of the sole of the foot**: the medial compartment, lateral compartment, and central compartment. The muscles, nerves, and vessels in the sole may be described according to these compartments, but the muscles of the sole are usually dissected and described by layers (Fig. 4-86).

MUSCLE OF THE DORSUM OF THE FOOT

There is only one small muscle on the dorsum of the foot, the extensor digitorum brevis, and it is relatively unimportant. The medial part of this muscle is usually more or less distinct and is sometimes called the *extensor hallucis brevis muscle* (Figs. 4-66 and 4-68).

THE EXTENSOR DIGITORUM BREVIS MUSCLE (Figs. 4-66 to 4-71)

This broad thin muscle forms a fleshy mass on the lateral part of the dorsum of the foot.

Origin (Figs. 4-58 and 4-68). **Anterior part of dorsal surface of calcaneus**, anteromedial to the lateral malleolus. It also arises from the **inferior extensor retinacutum.**

Insertion (Fig. 4-66). **Base of the proximal phalanx of the great toe and the tendons of the extensor digitorum longus to toes 2 to 4.**

Nerve Supply (Figs. 4-66 and 4-70). **Deep peroneal nerve** (S1 and S2).

Action. Aids in extending toes 1 to 4, ***i.e., it*** assists the extensor hallucis longus and extensor digitorum longus muscles.

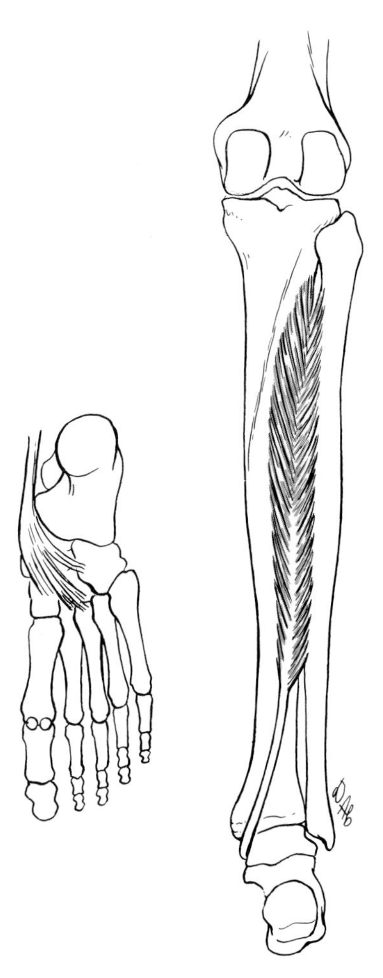

Figure 4-79. Drawings illustrating the origin and insertion of the right tibialis posterior muscle, the deepest muscle in the posterior crural compartment (Fig. 4-64). It arises from the interosseous membrane and the adjoining tibia and fibula and inserts into the tuberosity of the navicular bone (Fig. 4-78). Note that the tendon sends slips to adjacent bones and to the bases of the second, third, and fourth metatarsal bones.

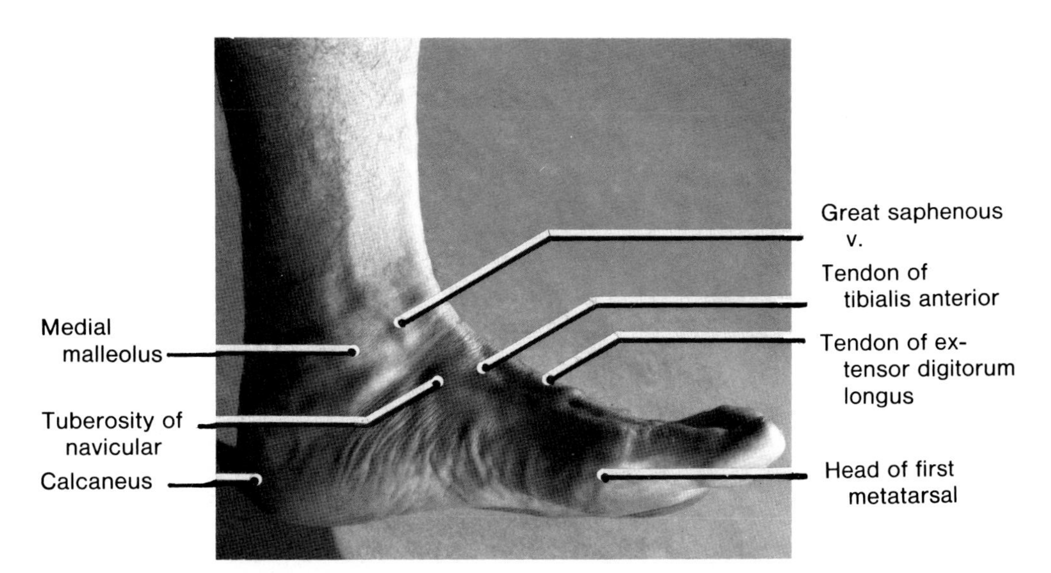

Figure 4-80. Photograph of the medial aspect of the left ankle of a 45-year-old man showing the principal surface features visible when the foot is dorsiflexed (Fig. 4-65*A*). The great saphenous vein is very constant in position, as here, anterior to the medial malleolus. During a "cutdown" (venostomy), this vein is dissected in order to insert a cannula for prolonged administration of intravenous fluids or medication. Observe this vein in Figure 4-81.

The extensor digitorum brevis muscle is functionally unimportant. Probably the only clinical reason for knowing about its presence is that contusion and tearing of its fibers result in a **hematoma**, often producing a swelling anteromedial to the lateral malleolus. Persons who have injured their ankles commonly think the swollen muscles, which have not been seen before, indicate sprained ankles.

MUSCLES OF THE SOLE OF THE FOOT

There are four layers of muscles in the sole of the foot (Table 4-9). They are specialized to help maintain the arches of the foot and to enable one to stand on uneven ground. Consequently the muscles in the sole of the foot have gross functions rather than delicate individual functions like those in the hand.

The specialization of the plantar muscles has resulted in the loss of many of their functions; as a result, several muscles in the foot have names implying functions that they rarely perform, or are unable to perform. The muscles of the sole of the foot are of little importance individually because the fine control of the individual toes is not important to most people.

The first layer of plantar muscles contains three muscles, all of which extend from the posterior part of the calcaneus to the phalanges (Figs. 4-86, 4-88, and Table 4-9). This layer consists of the abductors of the great and small toes and the short flexor of the toes. *These muscles comprise a functional group that acts as an elastic spring for supporting the arches of the foot and maintaining the concavity of the foot* (Fig. 4-87).

THE ABDUCTOR HALLUCIS MUSCLE (Figs. 4-86 to 4-88 and Table 4-9)

This *abductor of the great toe* (L. *hallux*) lies superficially along the medial border of the foot.

Origin (Figs. 4-60 and 4-87). **Medial process of tuber calcanei, flexor retinaculum, and plantar aponeurosis.**

Insertion (Figs. 4-59 and 4-60). **Medial side of the base of the proximal phalanx of the great toe.**

Nerve Supply (Figs. 4-88 and 4-90). **Medial plantar nerve** (S2 and **S3**).

Actions. Abducts and flexes great toe. When the foot is bearing weight, it supports the medial longitudinal arch of the foot (Fig. 4-87).

THE FLEXOR DIGITORUM BREVIS MUSCLE (Figs. 4-86, 4-88, 4-91, and Table 4-9)

This *short flexor of the toes* lies between the abductor hallucis and abductor digiti minimi muscles.

Origin (Fig. 4-59). **Medial process of tuber calcanei, plantar aponeurosis, and intermuscular septa.**

Insertion (Fig. 4-59). **Both sides of the middle phalanges of lateral four toes.** Each of the tendons splits to allow a tendon of the flexor digitorum longus to pass to the distal phalanx (Fig. 4-92).

Nerve Supply (Figs. 4-88 and 4-91). **Medial plantar nerve** (S2 and **S3**).

Action. Flexes lateral four toes at the proximal interphalangeal joints. When the foot is bearing weight, it supports the medial and lateral longitudinal arches of the foot.

THE ABDUCTOR DIGITI MINIMI MUSCLE (Figs. 4-88 to 4-91 and Table 4-9)

This *abductor of the small toe* is the most lateral of the three muscles in the first layer.

Origin (Fig. 4-59). **Medial and lateral processes of tuber calcanei, plantar aponeurosis, and intramuscular septum.**

Insertion (Fig. 4-59). **Lateral side of the base of the proximal phalanx of small toe.**

Nerve Supply (Figs. 4-88 and 4-91). **Lateral plantar nerve** (S2 and **S3**).

Actions. Abducts and flexes small toe. When the foot is bearing weight, it supports the lateral longitudinal arch of the foot.

The second layer of plantar muscles located deep to the first layer, consists of the quadratus plantae and the lumbricales or lumbrical muscles (Figs. 4-86, 4-89, and Table 4-9). The tendons of two leg muscles (flexor hallucis longus and flexor digitorum longus) are also in this layer (Fig. 4-78). The tendon of the flexor hallucis longus crosses deep to the tendon of the flexor digitorum longus as it passes to the great toe.

THE QUADRATUS PLANTAE MUSCLE (Figs. 4-78, 4-89, 4-91, and Table 4-9)

This small, flat muscle joins the tendon of the flexor digitorum longus to the calcaneus (Fig. 4-89), and forms a fleshy sheet of muscle in the posterior half of the foot. ***Its two heads of origin embrace the calcaneus*** (Fig. 4-78).

Origin (Figs. 4-59 and 4-60). *Medial head,* **medial surface of calcaneus;** *Lateral head,* **lateral margin of plantar surface of calcaneus** and long plantar ligament (Fig. 4-92).

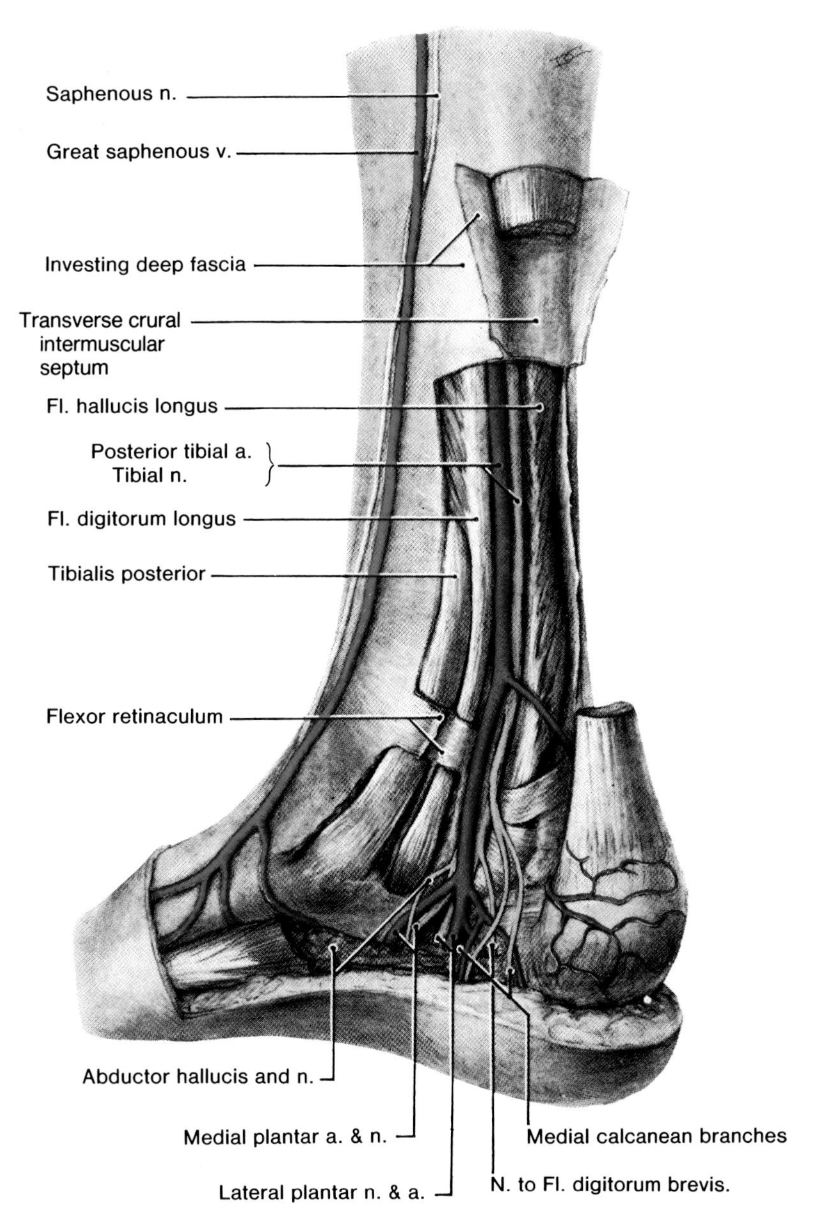

Figure 4-81. Drawing of a medial view of a dissection of the right leg, ankle, and heel. The posterior part of the abductor hallucis muscle is excised. Observe the posterior tibial artery and tibial nerve lying between the flexor digitorum longus and flexor hallucis longus muscles, and dividing into medial and lateral plantar branches. Note that the tibialis posterior and flexor digitorum longus occupy separate osseofibrous tunnels posterior to the medial malleolus, which acts as their pulley. Observe the medial and lateral plantar nerves lying in the fork formed by the medial and lateral plantar arteries. Note the deep veins of the foot emerging to join the great saphenous vein.

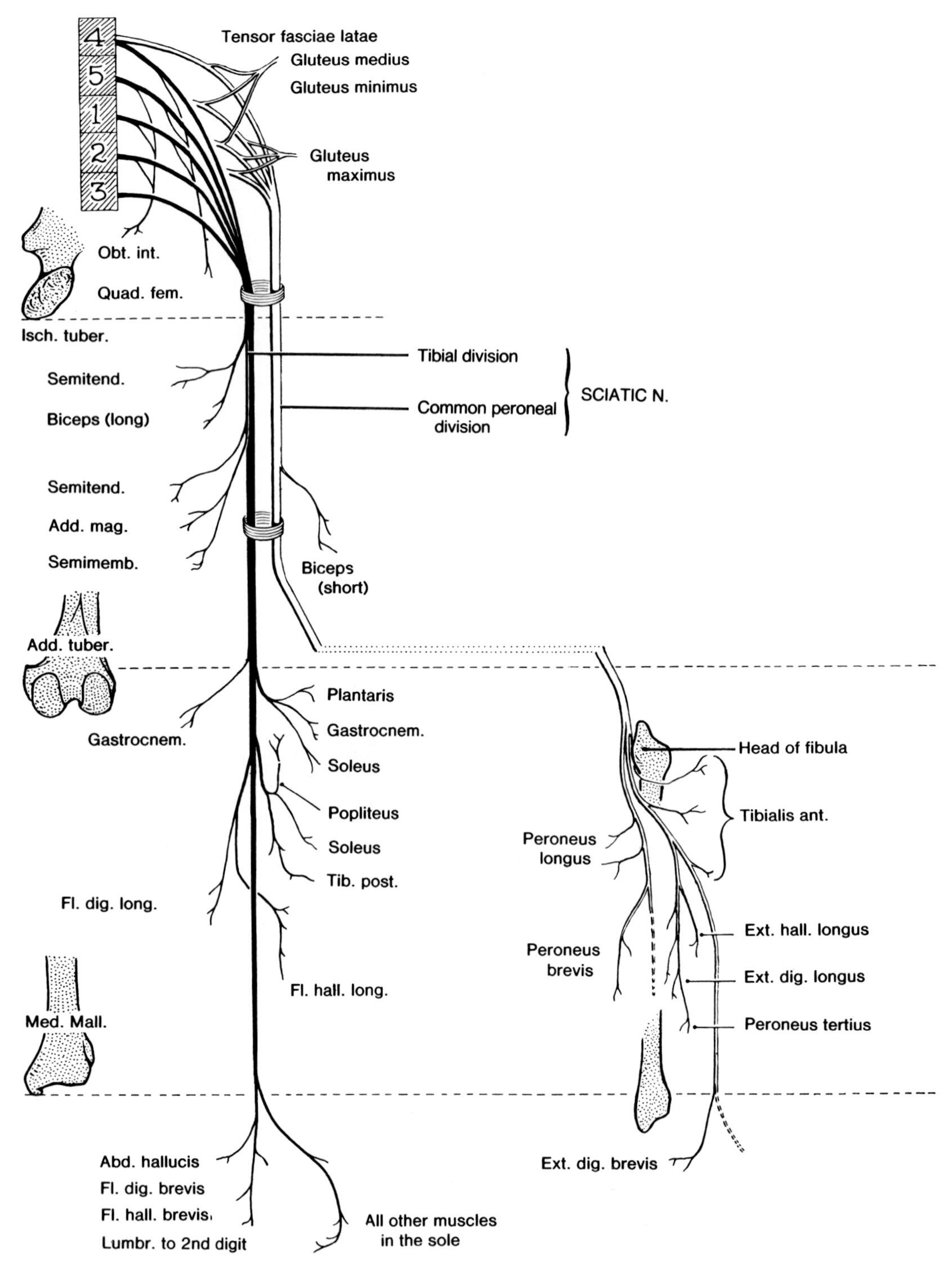

Figure 4-82. Scheme of the motor distribution of the sciatic nerve and its branches. The tibial nerve, the larger of the two terminal branches of the sciatic nerve, arises from the ventral branches of the ventral rami of L4, L5, S1, S2, and S3. The common peroneal nerve, about half the size of the tibial nerve, is derived from the dorsal branches of the ventral rami of L4, L5, S1, and S2. Observe that the sciatic is really two nerves enclosed in a sheath of fascia. Note that the tibial nerve divides posterior to the medial malleolus into the medial and lateral plantar nerves.

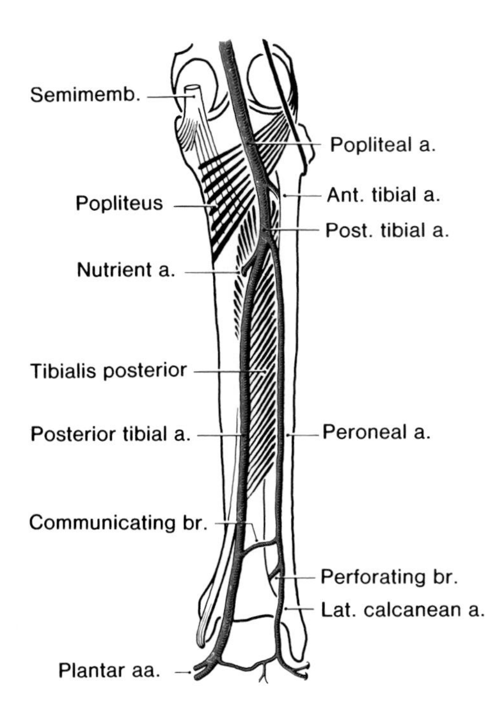

Figure 4-83. Drawing of the arteries in the posterior crural compartment. Note that the posterior tibial artery, the larger of the two terminal branches of the popliteal artery, begins at the inferior border of the popliteus muscle. Observe the peroneal artery, its largest branch, and the nutrient artery to the tibia. In some people the peroneal artery may be larger than the continuation of the posterior tibial artery. Note the important anastomoses posterior to the heel between the posterior tibial and peroneal arteries.

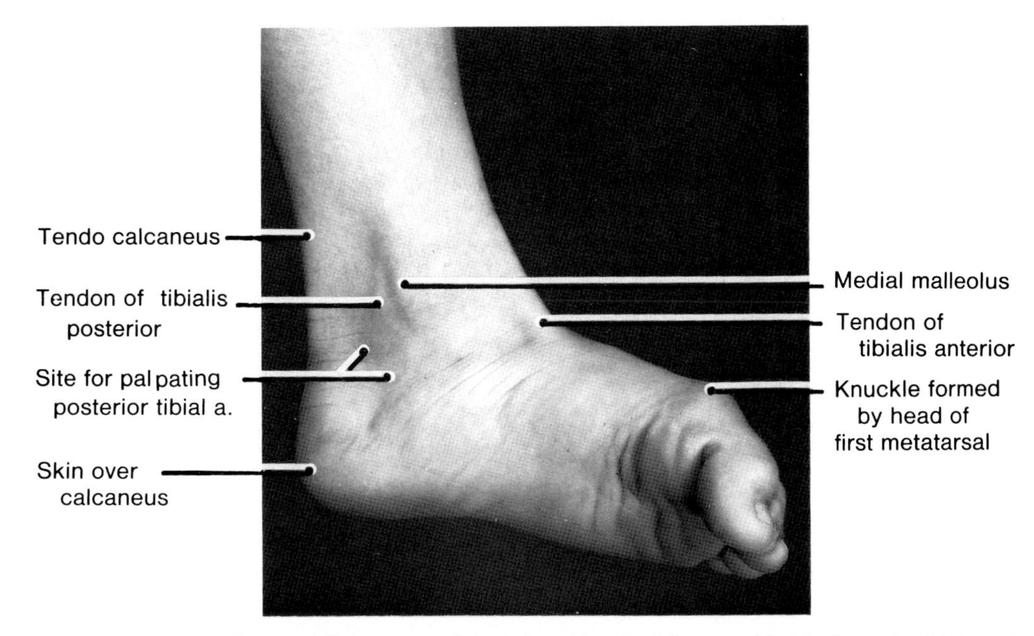

Figure 4-84. Photograph of the medial aspect of the left ankle of a 12-year-old girl whose foot is inverted. During this action the tendon of the tibialis posterior (an invertor of the foot) may be observed and palpated as it passes posterior to the medial malleolus. The tendon can be easily traced to its insertion into the navicular bone (Figs. 4-78 and 4-79). The pulse of the posterior tibial artery may be palpable where it passes posteroinferior to the medial malleolus (Fig. 4-81). It can be felt about a fingerbreadth posterior to the medial malleolus. An examination of the posterior tibial pulse may yield valuable information in cases of suspected arteriosclerotic stenosis of the posterior tibial artery. A diminished or absent pulse suggests arterial insufficiency, but one must keep in mind that this pulse is often difficult to palpate in normal people.

Insertion (Figs. 4-78 and 4-89). **Posterolateral margin of the tendon of the flexor digitorum longus.**

Nerve Supply (Fig. 4-91). **Lateral plantar nerve** (S2 and **S3**).

Action. Assists the flexor digitorum longus muscle in flexing the lateral four toes by adjusting the pull of the flexor digitorum longus (*i.e.*, it brings its tendon more directly in line with the long axes of the digits).

THE LUMBRICAL MUSCLES (Figs. 4-78, 4-88, and Table 4-9)

There are four of these small, worm-like muscles called lumbricales or **lumbricals** (L. *lumbricus*, earthworm).

Origin (Fig. 4-78). **Tendons of the flexor digitorum longus muscle.**

Insertion (Figs. 4-66 and 4-78). **Medial sides of bases of proximal phalanges of lateral four toes and extensor expansions** of the tendons of the extensor digitorum longus muscle.

Nerve Supply (Figs. 4-88 and 4-90). *Medial one,* **medial plantar nerve** (S2 and **S3**); *Lateral three,* lateral plantar nerve (S2 and **S3**).

Actions. Flex proximal phalanges at metatarsophalangeal joints and **extend middle and distal phalanges of lateral four toes** at interphalangeal joints. The lumbricals probably assist the interosseous muscles in flexing the proximal phalanges.

The third layer of plantar muscles consists of the short muscles of the great and small toes which lie in the anterior half of the sole of the foot (Figs. 4-91 and 4-92).

THE FLEXOR HALLUCIS BREVIS MUSCLE (Figs. 4-91, 4-92, and Table 4-9)

This fleshy muscle has two heads which cover the plantar surface of the first metatarsal bone.

Origin (Figs. 4-90 and 4-92). **Cuboid and lateral cuneiform bones and the tendon of the tibialis posterior muscle.**

Insertion (Figs. 4-59 and 4-92). **Both sides of base of proximal phalanx of great toe** by two tendons, the medial one inserting with the abductor hallucis and the lateral one with the adductor hallucis. A ***sesamoid bone*** adheres to each of these tendons (Figs. 4-89, 4-90, and 4-92). The sesamoids protect the tendons from pressure from the head of the first metatarsal bone during standing and walking.

Nerve Supply (Figs. 4-88 and 4-91). **Medial plantar nerve** (S1 and **S2**).

Action. Flexes great toe at the metatarsophalangeal joint. This muscle could also prevent excessive extension at the metatarsophalangeal of the great toe.

THE ADDUCTOR HALLUCIS MUSCLE (Figs. 4-90, 4-91, and Table 4-9)

This adductor of the great toe has **two heads of origin.**

Origin (Figs. 4-59 and 4-90). *Oblique head:* **bases of second, third, and fourth metatarsal bones,** and the fibrous sheath of the peroneus longus tendon. *Transverse head:* **plantar ligaments of the lateral four metatarsophalangeal joints.**

Insertion (Figs. 4-59 and 4-90). **Lateral side of base of proximal phalanx of great toe.** The tendons of both heads insert with the lateral tendon of the flexor hallucis brevis.

Nerve Supply (Fig. 4-91). Deep branch of **lateral plantar nerve** (S2 and **S3**).

Actions. Adducts great toe (*i.e.*, toward second toe), **flexes metatarsophalangeal joint,** and ***helps to maintain the transverse arch of the foot.***

THE FLEXOR DIGITI MINIMI BREVIS MUSCLE (Figs. 4-90 to 4-92 and Table 4-9)

This relatively insignificant muscle is only a fleshy slip.

Origin (Fig. 4-59). **Base of fifth metatarsal bone** and sheath of peroneus longus tendon.

Insertion (Fig. 4-59). **Base of proximal phalanx of the small toe.**

Nerve Supply (Fig. 4-91). Superficial branch of **lateral plantar nerve** (S2 and **S3**).

Action. Flexes small toe at metatarsophalangeal joint.

The fourth layer of plantar muscles consists of the interosseous muscles (interossei) and the tendons of the peroneus longus and tibialis posterior muscles which cross the sole of the foot to reach their insertions (Figs. 4-78, 4-92, and 4-93).

THE INTEROSSEOUS MUSCLES (Figs. 4-66, 4-92, and 4-93)

There are three plantar and four dorsal interossei and, as their name indicates, they occupy the spaces between bones (metatarsals). The dorsal interossei are larger than the plantar interossei and arise by two heads.

Origin (Fig. 4-93). *Plantar interossei:* **bases and medial sides of the third, fourth, and fifth metatarsal bones.** *Dorsal interossei:* **adjacent sides of the metatarsal bones.**

Insertion (Figs. 4-66, 4-92, and 4-93). *Plantar interossei:* **medial sides of bases of proximal phalanges of the third, fourth,** and **fifth toes.**

Dorsal interossei: first, **medial side** and *second to*

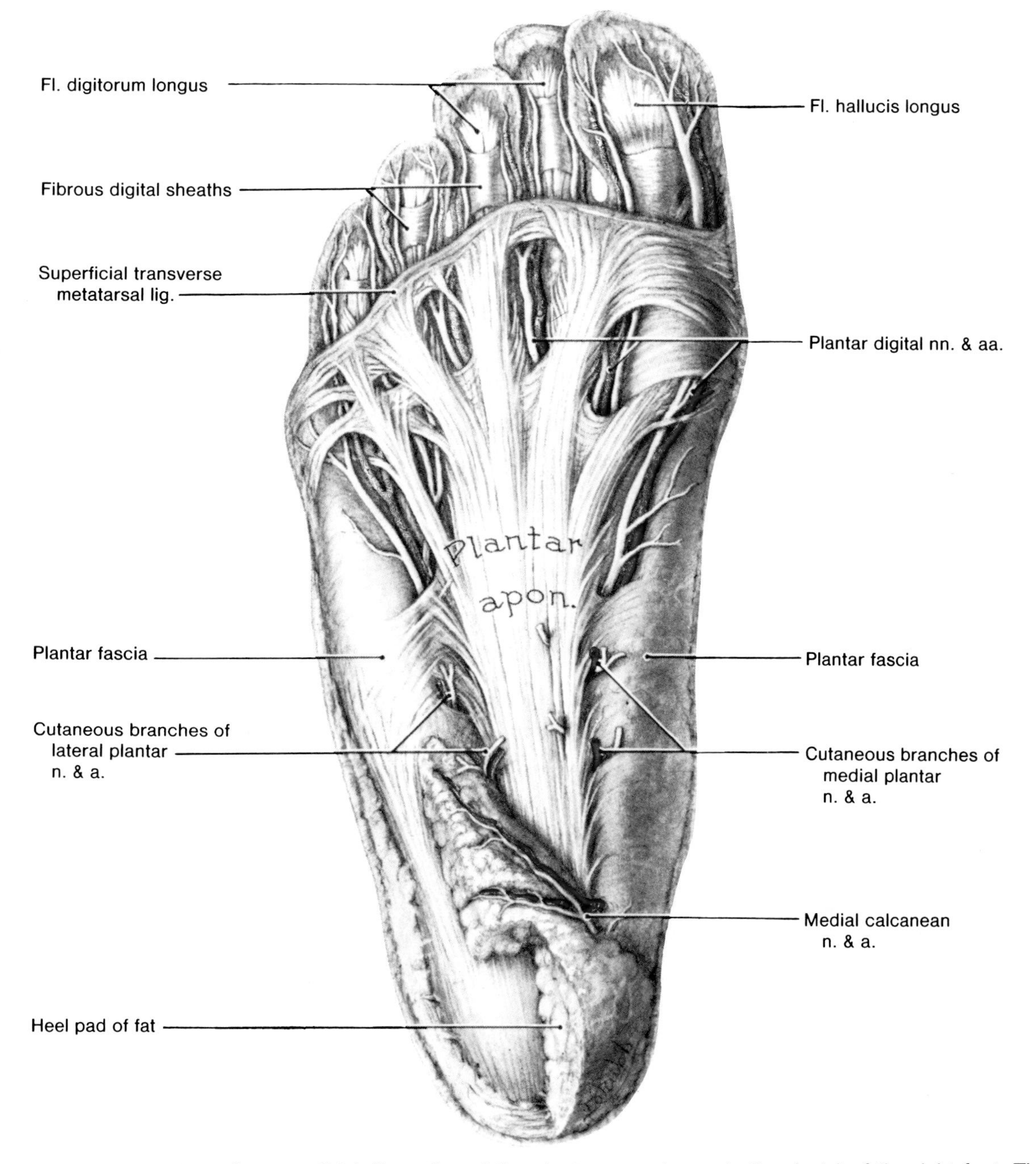

Figure 4-85. Drawing of a superficial dissection of the plantar aspect or sole (L. *planta*) of the right foot. The plantar aponeurosis acts as a strong tie for the maintenance of the longitudinal arches of the foot. To fill this role it is extremely sturdy. It stretches from the calcaneus posteriorly to the five digits anteriorly.

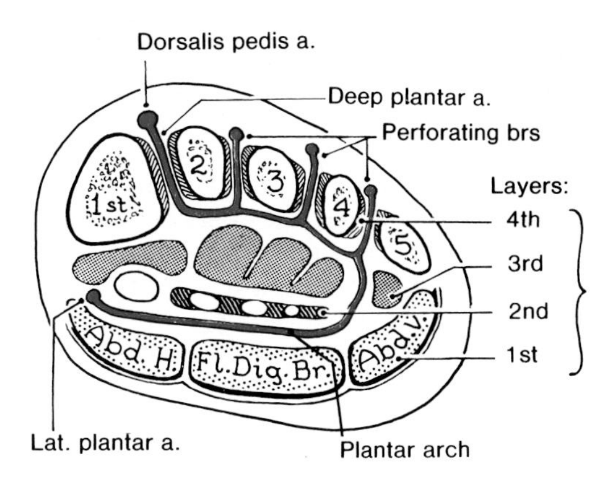

Figure 4-86. Drawing of a cross-section of the right foot near the bases of the metatarsal bones (Fig. 4-95), showing the four layers of muscles in the sole of the foot and the plantar arteries.

fourth, **lateral sides of the proximal phalanx of second *to fourth toes*.**

All interossei are also attached to the *extensor expansions of the corresponding toes* (Fig. 4-66).

Nerve Supply (Figs. 4-88 and 4-91). **Lateral plantar nerve** (S2 and **S3**).

Actions. The plantar interossei adduct the toes (**PAD** is the key, **P**lantar **AD**duct) and the dorsal interossei abduct the toes (**DAB** is the key, **D**orsal **AB**duct). *Adduction* is moving the toes toward the second toe and *abduction* is moving the toes away from the second toe. These actions are not important to most people, however **the** interossei ***maintain the integrity of the forefoot*** by approximating the bones during weight bearing. *The interossei also flex the toes at the metatarsophalangeal joints.*

NERVES OF THE FOOT

The tibial nerve divides posterior to the medial maleolus into medial and lateral plantar nerves (Figs. 4-76, 4-81, 4-88, and 4-91). They supply the intrinsic muscles of the foot, except for the extensor digitorum brevis, which is supplied by the deep peroneal nerve. These nerves also supply the skin of the foot (Fig. 4-94).

THE MEDIAL PLANTAR NERVE (Figs. 4-76, 4-81, 4-82, 4-85, 4-88, 4-91, and 4-94)

*This is the larger of the two terminal branches of the **tibial nerve**.* It passes deep to the abductor hallucis muscle and runs anteriorly between this muscle and the flexor digitorum brevis on the lateral side of the medial plantar artery (Figs. 4-81 and 4-88).

*The medial plantar nerve terminates near the bases of the metatarsal bones by dividing into **three digital nerves*** (Fig. 4-88), which supply cutaneous branches to the medial three and a half digits (Fig. 4-94), and motor branches to the abductor hallucis, flexor digitorum brevis, flexor hallucis brevis muscles, and the most medial lumbrical muscle.

THE LATERAL PLANTAR NERVE (Figs. 4-76, 4-81, 4-82, 4-85, 4-88, 4-91, and 4-94)

*This is the smaller of the two terminal branches of the **tibial nerve**.* It begins deep to the flexor retinaculum and abductor hallucis muscle and runs anterolaterally, medial to the lateral plantar artery and between the first and second layers of plantar muscles (Figs. 4-88 and 4-91).

The lateral plantar nerve terminates by dividing into a superficial and a deep branch. The superficial branch divides into two digital nerves which send cutaneous branches to the lateral one and a half toes. The superficial and deep branches of the lateral plantar nerve supply motor branches to muscles of the sole that are not supplied by the medial plantar nerve (Fig. 4-82 and Table 4-9).

THE SURAL NERVE (Fig. 4-94)

The sural nerve usually forms in the popliteal fossa and *descends between the two heads of the gastrocnemius* muscle. It pierces the deep fascia around the middle of the posterior aspect of the leg, where it is joined by the *peroneal communicating branch of the common peroneal nerve* (Figs. 4-49 and 4-50). The sural nerve supplies skin on the lateral and posterior part of the inferior one third of the leg (Fig. 4-94).

The sural nerve enters the foot posterior to the lateral malleolus and *supplies the skin along the lateral margin of the foot and the lateral side of the small toe* (Fig. 4-94).

THE SAPHENOUS NERVE (Fig. 4-94)

This is *the largest cutaneous branch of the femoral* nerve. In addition to supplying skin and fascia on the anterior and medial sides of the leg, the saphenous nerve passes to the dorsum of the foot, **anterior to the medial malleolus.**

The saphenous nerve supplies the skin along the medial side of the foot as far anteriorly as the head of the first metatarsal bone (Figs. 4-84 and 4-94).

ARTERIES OF THE FOOT

The arteries of the foot are the terminal branches of the anterior and posterior tibial arteries.

THE DORSALIS PEDIS ARTERY (Figs. 4-70 and 4-95)

This vessel is *the direct continuation of the **anterior tibial artery*** distal to the ankle joint. **The dorsalis pedis artery begins midway between the malleoli** (Fig. 4-95) and runs anteromedially, deep to the inferior extensor retinaculum, to the posterior end of the first interosseous space. Here it divides into a ***deep plantar artery***, which passes to the sole of the foot, and an ***arcuate artery***.

The arcuate artery (Fig. 4-95) runs laterally across the bases of the metatarsal bones, deep to the extensor tendons, where it gives off the second, third, and fourth **dorsal metatarsal arteries** (Fig. 4-95). These vessels run to the clefts of the toes where each of them divides into two **dorsal digital arteries** for the sides of adjoining toes.

The deep plantar artery (Fig. 4-95) passes deeply through the first interosseous space to join the lateral plantar artery and form the **deep plantar arch** (Figs. 4-95 and 4-96).

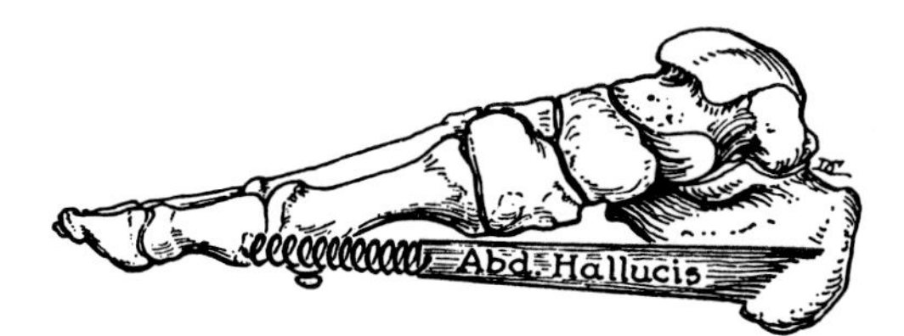

Figure 4-87. Drawing of the bones of the right foot illustrating how the abductor hallucis, and other muscles in the first layer of plantar muscles, act as an elastic spring in helping to maintain the arches of the foot.

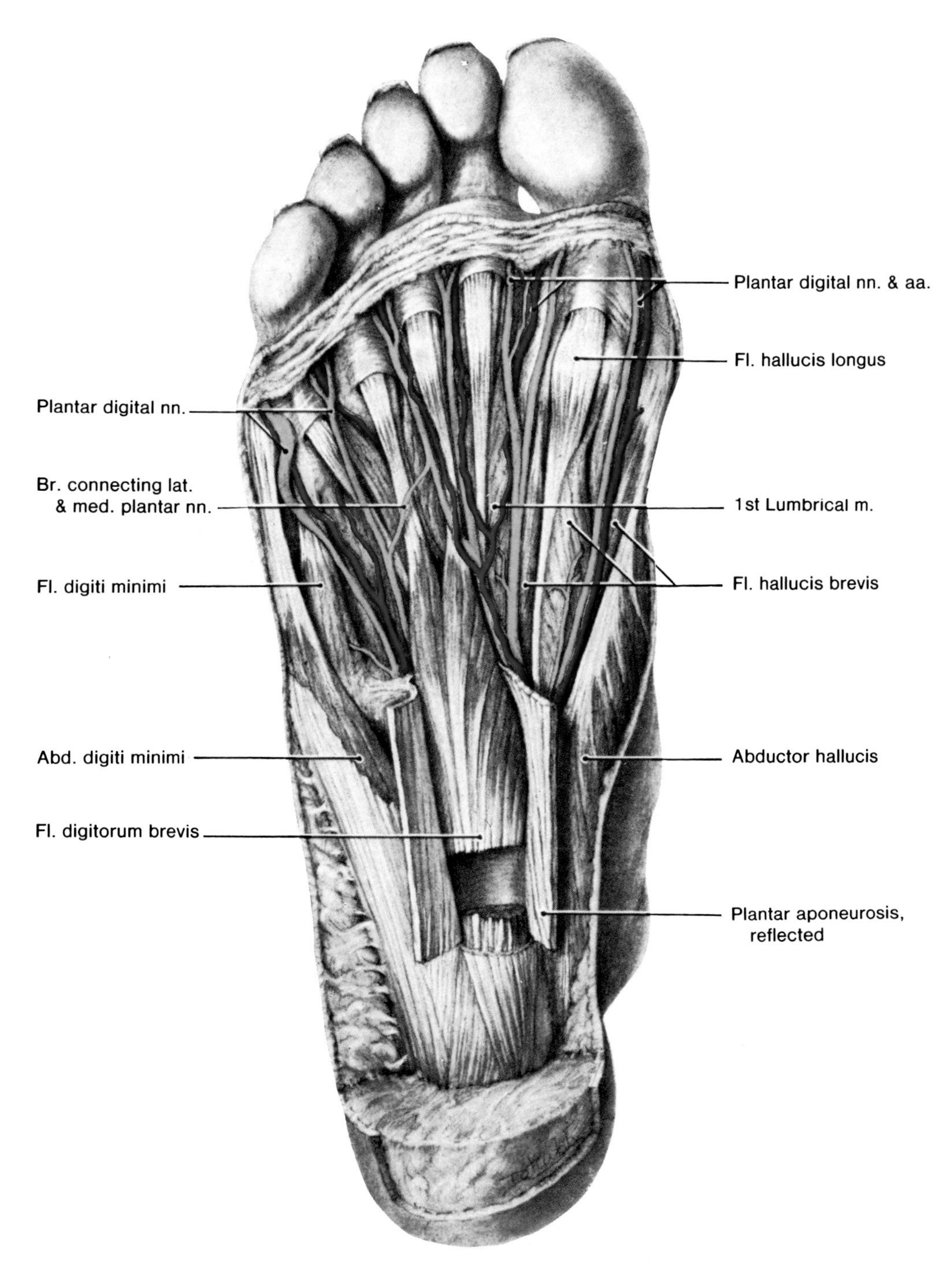

Figure 4-88. Drawing of a dissection of the plantar aspect of the right foot demonstrating the *first layer of plantar muscles*, and the digital nerves and arteries. The plantar aponeurosis and fascia are reflected or removed and a section is removed from the flexor digitorum brevis (Fig. 4-90). Note that the lateral and medial plantar digital nerves supply one and a half and three and a half digits respectively and are united by a connecting branch. In this specimen the flexor digitorum brevis does not send a tendon to the small toe.

Table 4-9
Muscles of the Sole of the Foot

Muscle	Origin	Insertion	Nerve Supply	Actions
First Layer				
Abductor hallucis (Figs. 4-86 to 4-88)	Medial process of tuber calcanei, flexor retinaculum, and plantar aponeurosis (Figs. 4-60 and 4-87)	Medial side of base of proximal phalanx of great toe (Figs. 4-59 and 4-60)	Medial plantar nerve (S2 and **S3**)	Abducts and flexes great toe
Flexor digitorum brevis (Figs. 4-88 and 4-91)	Medial process of tuber calcanei, plantar aponeurosis, and intermuscular septa (Fig. 4-59)	Both sides of middle phalanges of lateral four toes (Fig. 4-59)	Medial plantar nerve (S2 and **S3**)	Flexes lateral four toes
Abductor digiti minimi (Figs. 4-88 and 4-91)	Medial and lateral processes of tuber calcanei, plantar aponeurosis, and intermuscular septum (Fig. 4-59)	Lateral side of base of proximal phalanx of small toe (Fig. 4-59)	Lateral plantar nerve (S2 and **S3**)	Abducts and flexes small toe
Second Layer				
Quadratus plantae (Figs. 4-78, 4-89, and 4-91)	Medial surface and lateral margin of plantar surface of calcaneus (Figs. 4-59 and 4-60)	Posterolateral margin of tendon flexor digitorum longus (Fig. 4-89)	Lateral plantar nerve (S2 and **S3**)	Assists flexor digitorum longus in flexing lateral four toes
Lumbricales (Lumbrical muscles) (Figs. 4-78 and 4-88)	Tendons of flexor digitorum longus (Fig. 4-78)	Medial sides of bases of proximal phalanges of lateral four toes and extensor expansions of tendons and extensor digitorum longus (Figs. 4-66 and 4-78)	*Medial one*: medial plantar nerve (S2 and **S3**) *Lateral three*: lateral plantar nerve (S2 and **S3**)	Flex proximal phalanges and extend middle and distal phalanges of lateral four toes
Third Layer				
Flexor hallucis brevis (Figs. 4-90 and 4-92)	Cuboid and lateral cuneiform bones and tendon of tibialis posterior (Fig. 4-92)	Both sides of base of proximal phalanx of great toe (Figs. 4-59 and 4-92)	Medial plantar nerve (S1 and **S2**)	Flexes great toe
Adductor hallucis (Figs. 4-90 and 4-91)	*Oblique head*: Bases of metatarsals 2–4 *Transverse head*: Plantar ligaments of metatarsophalangeal joints 2–4 (Fig. 4-59)	Both heads: lateral side of base of proximal phalanx of great toe (Fig. 4-59)	Deep branch of lateral plantar nerve (S2 and **S3**)	Adducts and flexes great toe
Flexor digiti minimi brevis (Figs. 4-90 and 4-91)	Base of fifth metatarsal (Fig. 4-59)	Base of proximal phalanx of small toe (Fig. 4-59)	Superficial branch of lateral plantar nerve (S2 and **S3**)	Flexes small toe

Table 4-9—*Continued*

Fourth Layer				
Plantar interossei (3 muscles) (Figs. 4-92 and 4-93)	Bases and medial sides of metatarsals 3–5	Medial sides of bases of proximal phalanges of toes 3 to 5	Lateral plantar nerve (S2 and **S3**)	Adduct toes and flex metatarsophalangeal joints
Dorsal interossei (4 muscles) (Figs. 4-92 and 4-93)	Adjacent sides of metatarsals 1–5	*1st*, medial side of proximal phalanx of second toe *2nd–4th*, lateral sides of second to fourth toes		Abduct toes and flex metatarsophalangeal joints

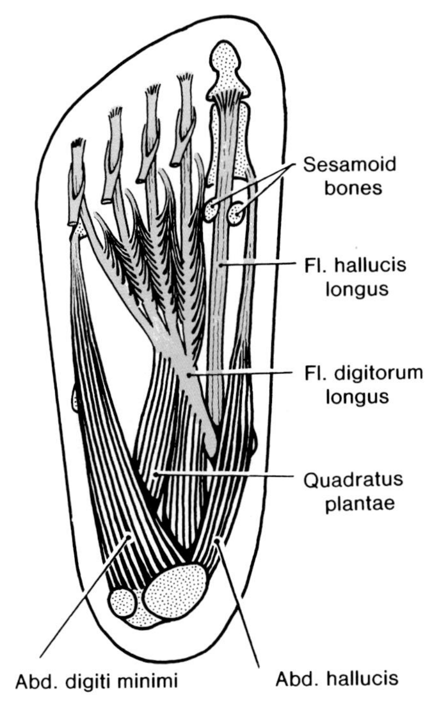

Figure 4-89. Drawing of some of the muscles of the sole of the right foot. The *second layer of plantar muscles*, displayed by removal of the flexor digitorum brevis (Fig. 4-88), is framed by the abductor muscles of the great and small toes.

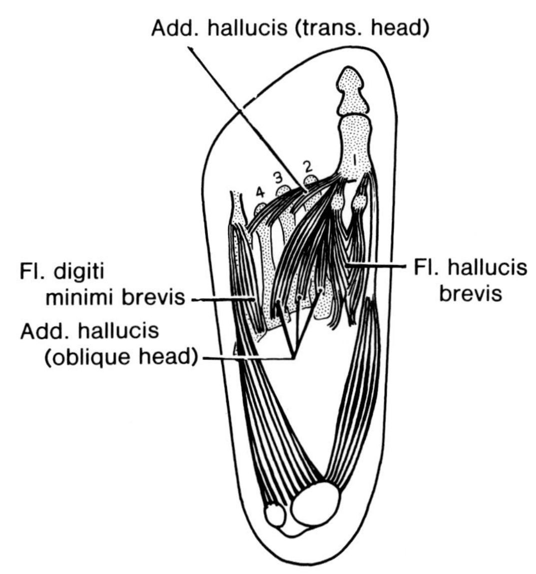

Figure 4-90. Diagramatic illustration of the *third layer of plantar muscles* in the right foot. Note that these muscles form three sides of a square in the anterior half of the sole which is largely filled by the oblique head of the abductor hallucis muscle.

*The ability to palpate the **pulse of the dorsalis pedis** artery is essential for clinical practice*, particularly in cases of **intermittent claudication** (cramps in the calf brought on by exercise and relieved by rest).

The dorsalis pedis pulse can usually be felt on the dorsum of the foot, where it passes over the navicular and cuneiform bones just lateral to the extensor hallucis longus tendon (Fig. 4-66). It may also be felt distal to this at the proximal end of the first interosseous space (Fig. 4-95). A diminished or absent dorsalis pedis pulse suggests arterial insufficiency. In some people the dorsalis pedis artery may be too small to palpate, or it may not be in its usual position. Consequently, *failure to detect a dorsalis pedis pulse* does not always indicate the presence of **arteriosclerotic disease.**

The arteries of the sole of the foot are derived from the posterior tibial artery. It divides deep to the abductor hallucis muscle to form the medial and

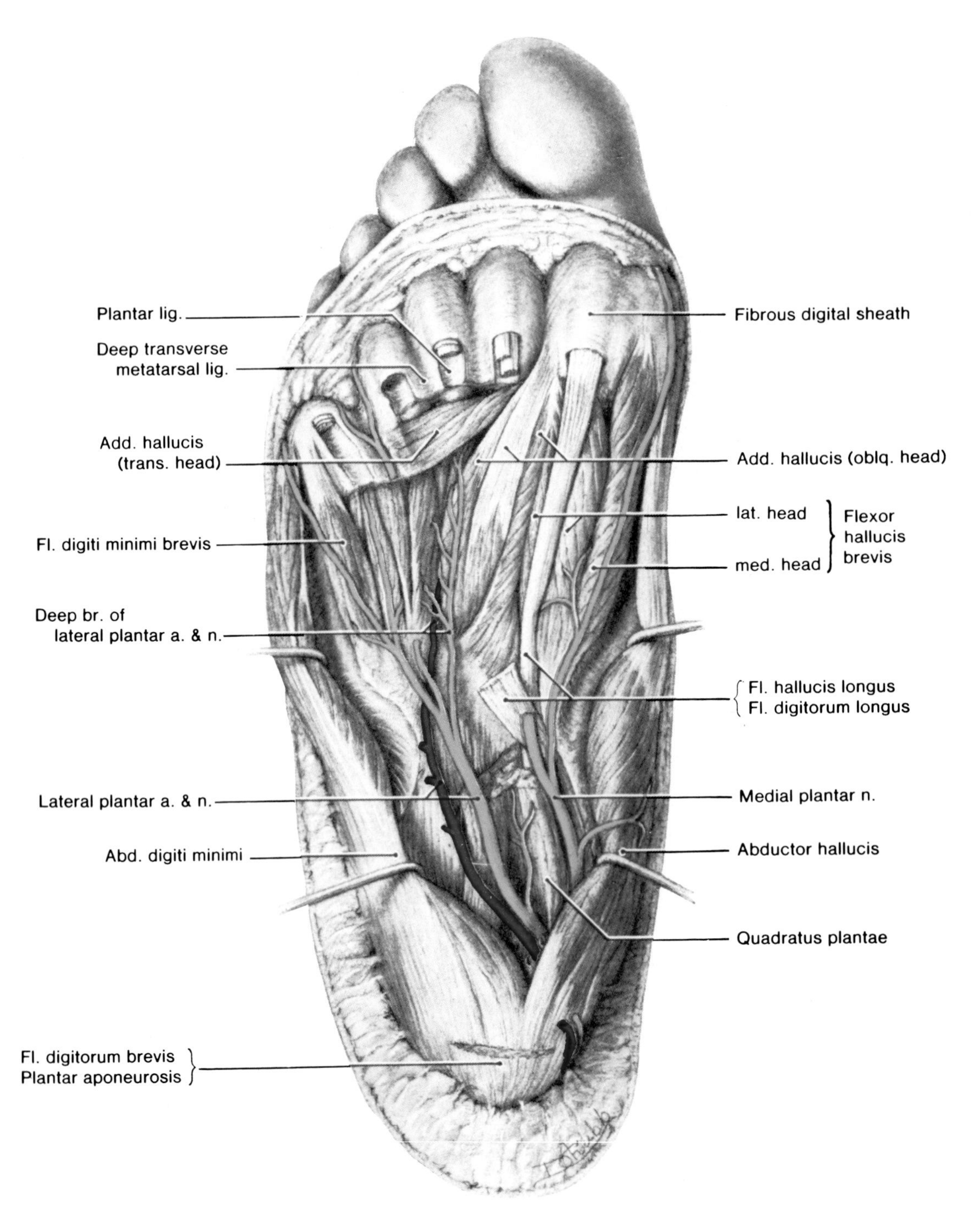

Figure 4-91. Drawing of a dissection of the *third layer of plantar muscles* in the right foot. The abductor digiti minimi and the abductor hallucis muscles of the first layer are pulled aside and the flexor digitorum brevis is cut short. The flexor digitorum longus and lumbricals of the second layer are excised and the quadratus plantae is cut. Note that the lateral plantar nerve and artery course laterally between the muscles of the first and second layers.

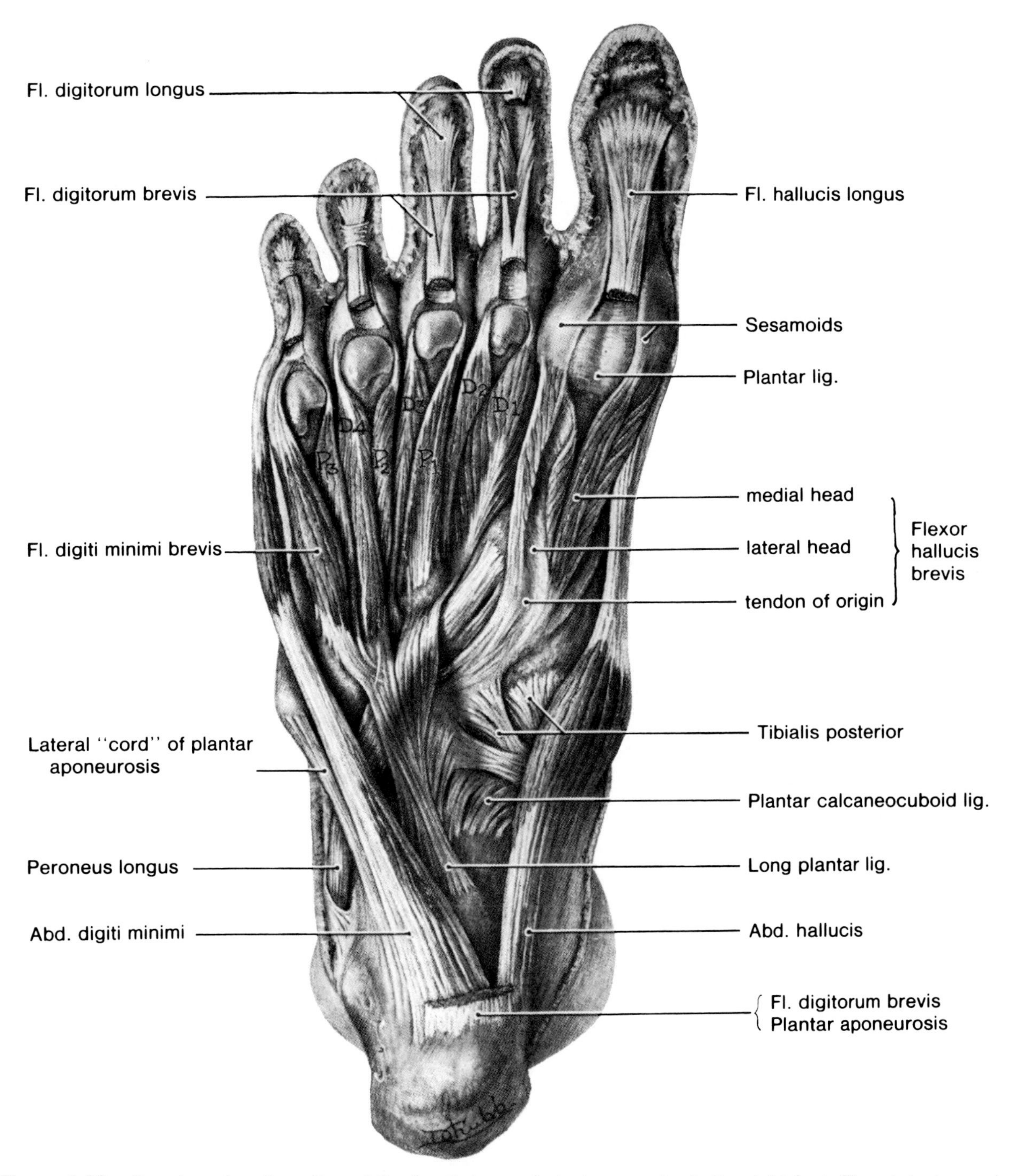

Figure 4-92. Drawing of a dissection of the *fourth layer of plantar muscles* in the right foot. The abductor and flexor brevis of the small toe and the abductor and flexor brevis of the great toe of the first and third layers of muscles remain for purposes of orientation. Observe the muscles of the fourth layer: three plantar and four dorsal interossei in the anterior half of the foot, and the tendons of peroneus longus and tibialis posterior in the posterior half. The plantar interossei adduct the three lateral toes and the dorsal interossei abduct them.

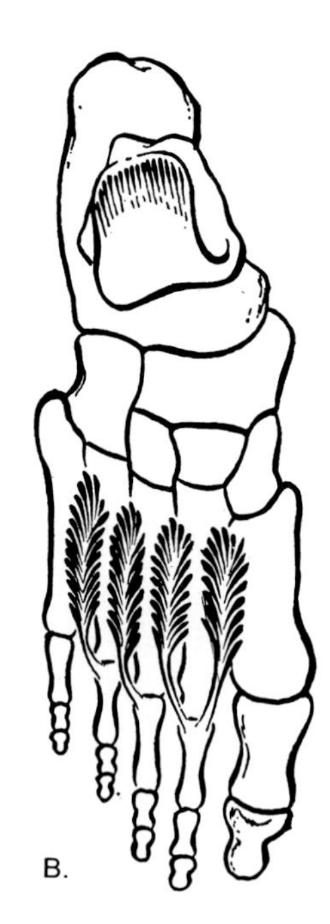

Figure 4-93. Drawings illustrating the interossei of the fourth layer of muscles in the sole of the right foot. *A*, plantar interossei. *B*, dorsal interossei. The axis of abduction and adduction in the toes is the second digit rather than the third as in the hand. All interosseous muscles are supplied by the lateral plantar nerve (Table 4-9).

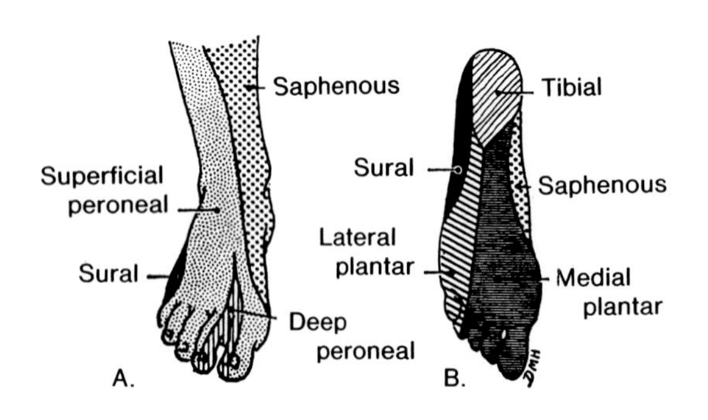

Figure 4-94. Diagrams illustrating the cutaneous distribution of the nerves of the distal part of the leg and foot (also see Fig. 4-14). The superficial and deep peroneal nerves are branches of the common peroneal nerve, the most commonly injured nerve in the lower limb. *A*, dorsum of right leg and foot. *B*, plantar surface of left foot.

Figure 4-95. Drawing of the arteries of the dorsum of the right foot. Observe that the anterior tibial artery takes a straight course anterior to the ankle joint and ends midway between the malleoli as the *dorsalis pedis artery*. This clinically important artery can be palpated where it passes over the navicular and cuneiform bones. As the artery can be easily compressed against bone here, this is the common site for taking the pedal pulse.

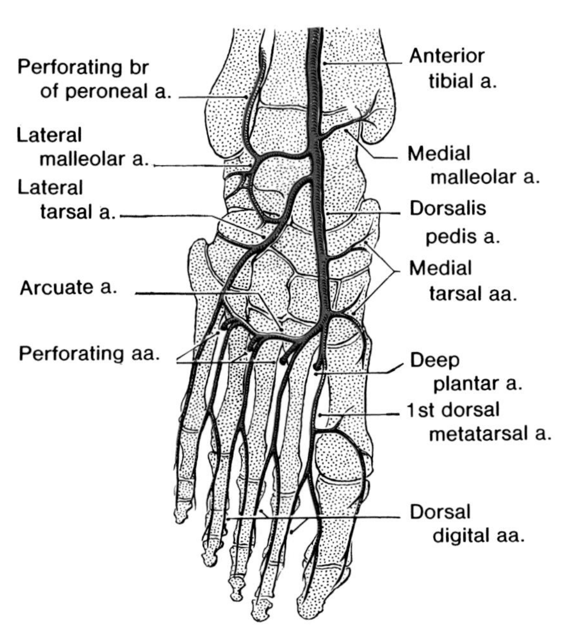

lateral plantar arteries which run parallel to the similar-named nerves (Figs. 4-81, 4-88, and 4-96). These arteries supply the plantar muscles and anastomose with arteries on the dorsum of the foot (Figs. 4-86 and 4-96).

THE MEDIAL PLANTAR ARTERY (Figs. 4-31*B*, 4-76, 4-81, and 4-96)

This vessel, *the smaller of the two terminal branches of the* ***posterior tibial artery***, arises deep to the flexor retinaculum, midway between the medial malleolus and the prominence of the heel. It passes distally on the medial side of the foot between the abductor hallucis and flexor digitorum brevis muscles (Fig. 4-88). The medial plantar artery supplies branches to the medial side of the great toe and gives off muscular, cutaneous, and articular branches during its course.

THE LATERAL PLANTAR ARTERY (Figs. 4-91 and 4-96)

This vessel, *the larger of the two terminal branches of the* ***posterior tibial artery***, arises deep to the flexor retinaculum and runs obliquely across the sole of the foot on the lateral side of the lateral plantar nerve (Fig. 4-91), between the flexor digitorum brevis and quadratus plantae muscles. The lateral plantar artery gives off calcaneal, cutaneous, muscular, and articular branches. When it reaches the base of the fifth metatarsal bone, it curves medially between the third and fourth muscular layers. It terminates at the base of the first metatarsal bone, where it joins the deep plantar branch of the dorsalis pedis artery to form the plantar arterial arch (Figs. 4-86 and 4-96).

THE PLANTAR ARTERIAL ARCH (Figs. 4-86 and 4-96)

The plantar arterial arch *begins opposite the base of the fifth metatarsal bone as the continuation of the lateral plantar artery.* It is completed medially by its union with the deep plantar artery, a branch of the dorsalis pedis artery (Figs. 4-86 and 4-95).

As is crosses the foot, *the plantar arch gives off four* ***plantar metatarsal arteries*** (Fig. 4-96), *three perforating arteries, and branches to the tarsal joints and the muscles in the sole of the foot.* These arteries join with the superficial branches of the medial and lateral plantar arteries to form **digital arteries**.

Wounds of the foot involving the plantar arterial arch result in serious bleeding. Ligature of the arch is difficult owing to the depth of the vessel and the structures that are related to it (Fig. 4-86).

VEINS OF THE FOOT

The **dorsal digital veins** run along the dorsal margins of each toe and unite in their webs to form **common dorsal digital veins** (Fig. 4-97). These join to form a **dorsal venous arch** on the dorsum of the foot from Veins leave the dorsal venous arch and converge medially to form the **great saphenous vein** and laterally to form the ***small saphenous vein*** (Figs. 4-81 and 4-97).

The superficial veins of the sole unite to form a **plantar venous arch**, from which efferents pass to medial and lateral marginal veins that join the great and small saphenous veins (Fig. 4-97).

The deep veins of the sole of the foot begin as **plantar digital veins** on the plantar aspects of the toes. They communicate with the dorsal digital veins via **perforating veins** (Fig. 4-97). Most blood returns from the foot via the deep veins which are connected with the superficial veins by these perforating veins.

JOINTS OF THE LOWER LIMB

As the **pelvic girdle** or lower limb girdle, consisting of the two hip bones, connects the lower limb to the trunk, its joints are included with those of the lower limb. The pelvic joints are described in Chapter 3 with the pelvis.

THE HIP JOINT

The hip joint is a multiaxial ball and socket type of synovial joint between the head of the femur and the acetabulum of the hip bone.

ARTICULAR SURFACES OF THE HIP BONE (Figs. 4-98 to 4-100)

The globular head of the femur articulates with the cup-like acetabulum of the hip bone. The head of the femur forms about two-thirds of a sphere and is completely covered with hyaline cartilage, except over the roughened **fovea** or pit (Fig. 4-98), to which the ***ligament of the head of the femur*** is attached (Fig. 4-99). More than half of the femoral head is contained within the **acetabulum**, which is deepened by the fibrocartilaginous **acetabular labrum** (Figs. 4-99 and 4-100).

The articular or ***lunate surface of the acetabulum*** *is horseshoe-shaped* (Fig. 4-100).The acetabulum has a centrally located nonarticular **acetabular fossa** which is occupied by a fatpad that is covered with synovial membrane. The acetabular fossa is closed inferiorly by the **transverse acetabular ligament**.

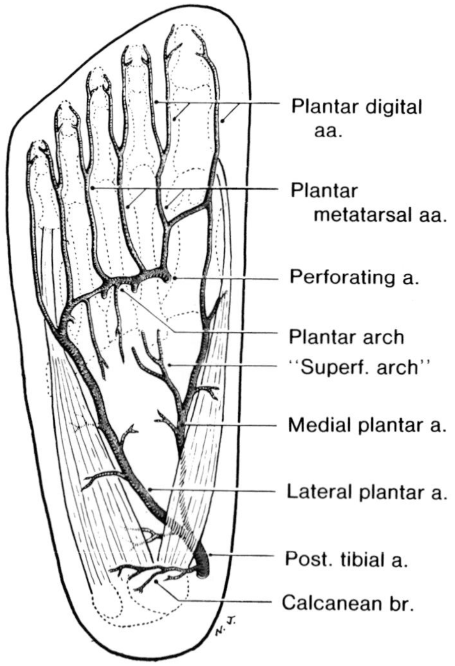

Plantar digital aa.
Plantar metatarsal aa.
Perforating a.
Plantar arch
"Superf. arch"
Medial plantar a.
Lateral plantar a.
Post. tibial a.
Calcanean br.

Figure 4-96. Diagram illustrating the distribution of the plantar arteries in the right foot. Observe that the lateral plantar artery runs across the sole of the foot to the base of the fifth metatarsal bone. It then turns medially and forms the plantar arterial arch.

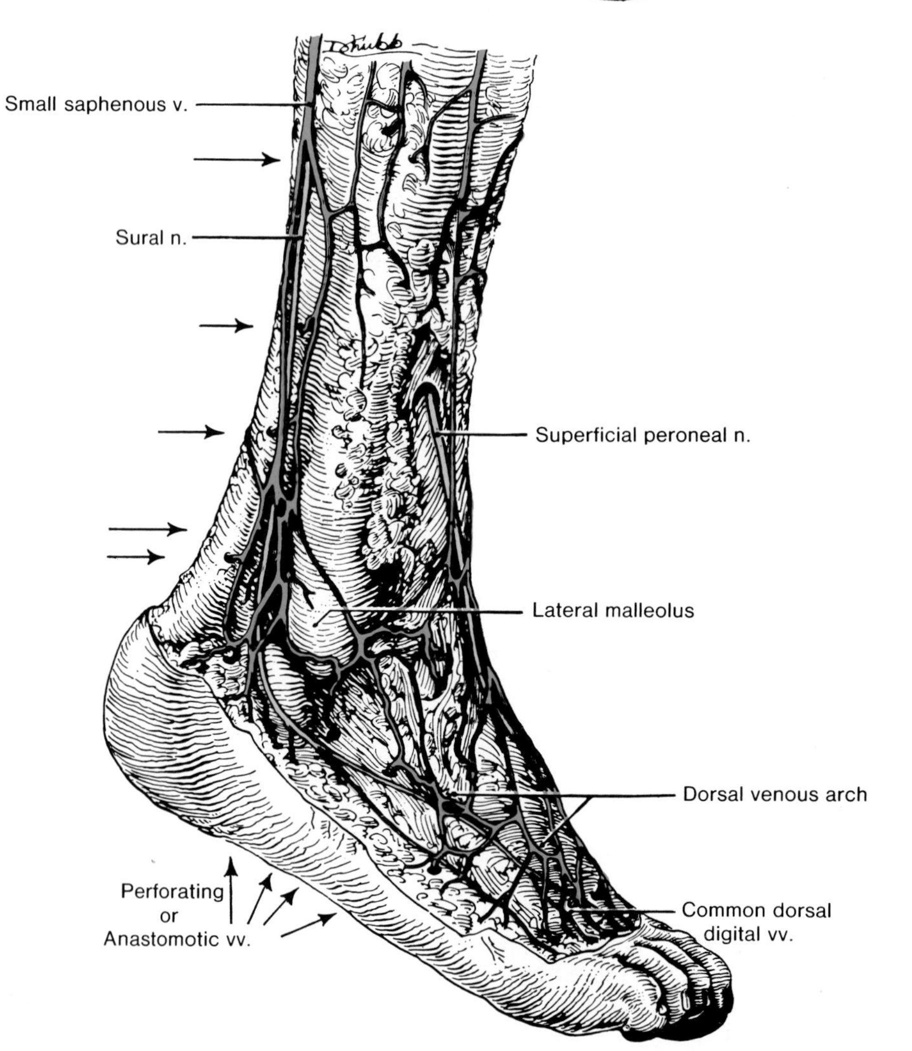

Figure 4-97. Drawing of an anterolateral view of the superficial veins of the ankle and dorsum of the right foot. Observe that the short saphenous vein arises posterior to the lateral malleolus from the lateral part of the dorsal venous arch.

Damaged or diseased hip joints may be surgically replaced either partly or completely, *e.g.*, to relieve the pain and deformity of **osteoarthritis** (a degenerative joint disease).

The artificial parts (*e.g.*, head and neck of femur) are inserted after removal of the broken and/or diseased parts of the articulating bones. The new metal or plastic hip joint functions almost as well as a normal joint.

The Acetabular Labrum (Figs. 4-99 and 4-100). The depth of the acetabulum is increased by the *fibrocartilaginous* ***acetabular labrum*** (L. lip). It is attached to the bony rim of the acetabulum and to the transverse acetabular ligament (Fig. 4-100); thus the labrum completes and deepens the socket for the femoral head. Its free thin edge clasps around the head and helps to hold it firmly in the acetabulum.

MOVEMENTS OF THE HIP JOINT (Figs. 4-101, 4-103, and Table 4-10)

The range of movement of the hip joint is decreased somewhat compared with the shoulder joint, in order to provide greater stability and strength.

The movements of the thigh at the hip joint are flexion-extension, abduction-adduction, medial and lateral rotation, and circumduction (Fig. 4-103).

The iliopsoas muscle (iliacus and psoas major muscles) is *the strongest flexor of the thigh at the hip joint* (Tables 4-1 and 4-10).

The chief extensors of the thigh at the hip joint are the *gluteus maximus and hamstrings* (Fig. 4-40). The gluteus maximus is a powerful extensor of the thigh, but it is relatively inactive unless forceful extension is required (*e.g.*, straightening up after bending over).

The chief abductors of the thigh at the hip joint are the *gluteus medius and the gluteus minimus.*

The chief adductors of the thigh at the hip joint are the *adductor muscles* (longus, brevis, and magnus). The widest range of adduction is obtained when the thigh is flexed.

The chief medial rotators of the thigh at the hip joint are the *tensor fasciae latae, gluteus medius, and gluteus minimus.* The role of other muscles in this movement is controversial.

The chief lateral rotators of the thigh at the hip joint are the *obturator muscles, gemelli, and quadratus femoris.* This powerful movement is limited by the tension of the medial rotators of the thigh at the hip joint and the iliofemoral ligament.

THE ARTICULAR CAPSULE OF THE HIP JOINT (Figs. 4-99 and 4-102)

The fibrous capsule is strong and dense. Proximally the fibrous capsule is attached to the edge of the acetabulum, just distal to the acetabular labrum, and to the transverse acetabular ligament. *Distally the fibrous capsule is attached to the neck of the femur* as follows: *anteriorly* to the intertrochanteric line and the root of the greater trochanter; *posteriorly* to the neck proximal to the intertrochanteric crest.

The fibrous capsule forms a cylindrical sleeve that encloses the hip joint **and most of the neck of the femur** (Figs. 4-99 and 4-102). Most fibers of the capsule take a spiral course from the hip bone to the lateral portion of the intertrochanteric line of the femur, but some deep fibers forming an **orbicular zone**, pass circularly around the neck of the femur (Figs. 4-99 and 4-102). These fibers form a collar around the neck of the femur which constricts the capsule and helps to hold the femoral head in the acetabulum.

Some deep longitudinal fibers of the capsule (**retinacula**) are reflected superiorly along the neck of the femur as longitudinal bands. ***The retinacula of the neck of the femur contain blood vessels*** *that supply the head and neck of the femur.*

Four main groups of longitudinal capsular fibers (**intrinsic ligaments**) are given names according to the region of the hip bone that is attached to the femur. ***The intrinsic ligaments are thickened parts of the fibrous capsule which strengthen the hip joint.***

THE ILIOFEMORAL LIGAMENT (Fig. 4-102)

The iliofemoral ligament is a very strong band that covers the anterior aspect of the hip joint. It is shaped like a Y and is attached proximally to the anterior inferior iliac spine and the acetabular rim. The iliofemoral ligament is attached distally to the intertrochanteric line of the femur.

The capsule of the hip joint is taut and the iliofemoral ligament is tense in full extension of the hip joint. **The iliofemoral ligament is very strong and has an important role in preventing overextension of the hip joint** during standing (*i.e.*, it helps to maintain the erect posture).

When standing at ease, one tends to thrust the pelvis anteriorly, thereby slightly hyperextending the thighs. In this position, much of the weight of the body is borne by the exceedingly strong iliofemoral ligaments. Because no mus-

Table 4-10
Muscles Producing Movements of the Hip Joint[1]

Flexion	Extension	Abduction	Adduction	Medial Rotation	Lateral Rotation
Iliopsoas **Iliacus** **Psoas major** **Tensor fasciae latae** **Rectus femoris** Sartorius Adductor longus Adductor brevis Pectineus	**Gluteus maximus** **Semitendinosus** **Semimembranosus** **Biceps femoris** (long head) Adductor magnus (ischial fibers)	**Gluteus medius** **Gluteus minimus** Tensor fasciae latae Sartorius Piriformis (in flexion) Obturator externus (in flexion)	**Adductor magnus** **Adductor longus** **Adductor brevis** Pectineus Gracilis	**Tensor fasciae latae** **Gluteus medius** **Gluteus minimus** (anterior fibers)	**Obturator internus and gemelli** **Obturator externus** **Quadratus femoris** Piriformis Gluteus maximus Sartorius

[1] **Bold face** indicates muscles chiefly responsible for the movement; the other muscles assist with the movement.

cles need to contract to maintain the stand easy position, it is a relaxing way to stand. This fact is the anatomical basis of the traditional stand easy position used by the armed forces.

THE PUBOFEMORAL LIGAMENT

This ligament arises from the pubic part of the acetabular rim and the iliopubic eminence (Fig. 4-1) and blends with the medial part of the iliofemoral ligaments (Fig. 4-38).

The pubofemoral ligament strengthens the inferior and anterior parts of the fibrous capsule of the hip joint. The pubofemoral ligament tightens during extension of the hip joint and becomes tense during abduction. Although it is a relatively weak ligament, it tends to prevent overabduction of the thigh at the hip joint.

THE ISCHIOFEMORAL LIGAMENT (Fig. 4-102)

The ischiofemoral ligament reinforces the fibrous capsule of the hip joint posteriorly. It arises from the ischial portion of the acetabular rim and spirals superolaterally to the neck of the femur, medial to the base of the greater trochanter. This anatomical construction tends to screw the femoral head medially into the acetabulum during extension of the thigh at the hip joint, thereby resisting hyperextension of it.

THE LIGAMENT OF THE HEAD OF FEMUR (Figs. 4-99 and 4-100)

The ligament of the head of the femur (ligamentum teres femoris) is a weak ligament which appears to be of little importance in strengthening the hip joint. Its wide end is attached to the margins of the acetabular notch and to the transverse acetabular ligament (Fig. 4-100), and its narrow end is attached to the fovea of the femur (Fig. 4-98). Usually it contains a small **artery to the head of the femur**, which is a branch of the obturator artery (*Grant's Atlas of Anatomy*, Fig. 4-47). The ligament of the head is stretched when the flexed thigh is adducted or laterally rotated.

The ligament of the head of the femur is inside the fibrous capsule of the hip joint (intracapsular) and is surrounded by synovial membrane (Fig. 4-99).

The ligament of the head of the femur varies in size and strength in different people; sometimes it is absent. Its function, other than as a pathway for the artery to the femoral head, is unclear. It appears to be of limited value in strengthening the hip joint because no appreciable disability occurs when it ruptures or does not develop.

Fractures of the femoral neck close to the head often disrupt the blood supply to the head of the femur (Fig. 4-31). *In some cases the blood supplied via the artery in the ligament of the head may be the only blood received by the proximal fragment of the femoral head.* If the ligament is ruptured, the fragment of bone may receive no blood and undergo **aseptic necrosis** (death in the absence of infection).

The synovial capsule of the hip joint (Figs. 4-99, 4-100, and 4-102) lines the internal surface of the fibrous capsule and is reflected from it on the neck of the femur. The synovial capsule forms a sleeve for the ligament of the head of the femur which attaches to the margins of the fovea (Figs. 4-98 and 4-99).

The synovial capsule lines the acetabular fossa and covers the fatty pad in the acetabular notch (Figs. 4-99 and 4-100). It is attached to the edges of the acetabular fossa and to the transverse acetabular ligament.

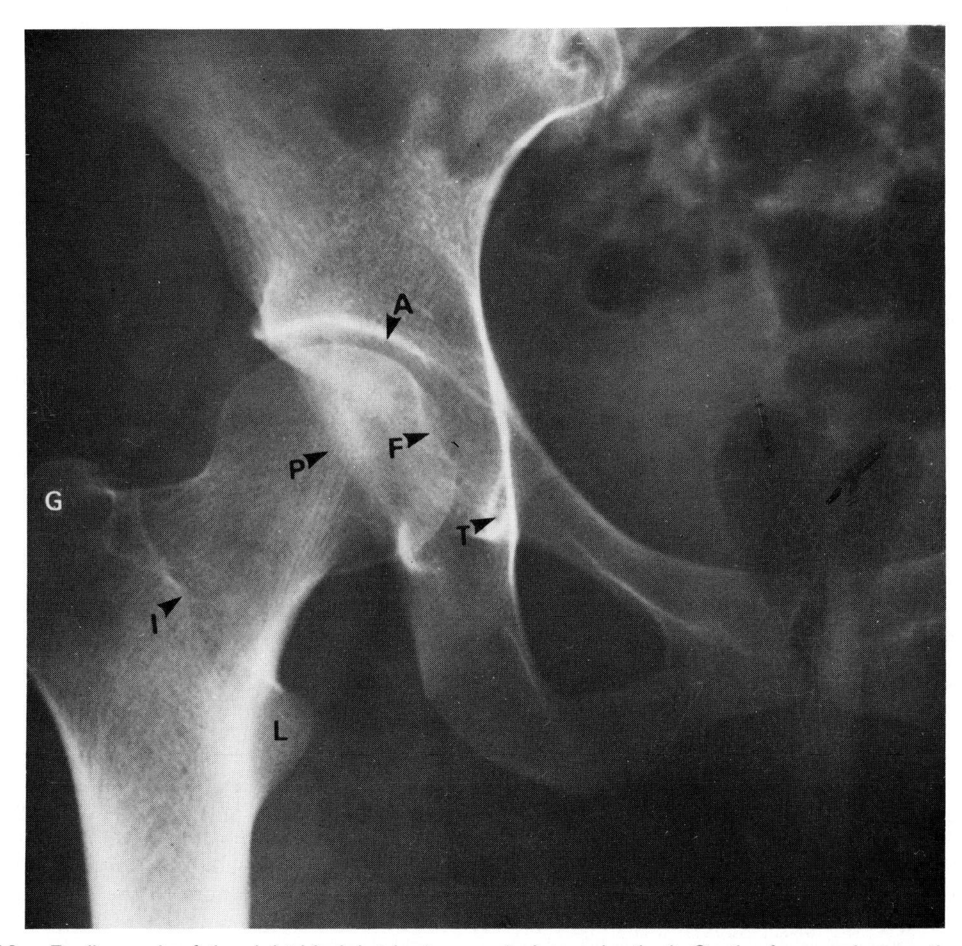

Figure 4-98. Radiograph of the right hip joint (anteroposterior projection). *On the femur,* observe the greater (*G*) and lesser (*L*) trochanters, the intertrochanteric crest (*I*), and the fovea (*F*) for the ligament of the head. *On the pelvis*, observe the roof (*A*) and posterior rim (*P*) of the acetabulum and the "teardrop" appearance (*T*) caused by the superimposition of structures at the inferior margin of the acetabulum. The hip joint is the best example in the body of a ball-and-socket joint. Its strength depends largely upon the depth of the acetabulum and the strength of the surrounding ligaments and muscles (also see Figs. 4-1 and 4-99).

The synovial capsule protrudes inferior to the fibrous capsule posteriorly (Fig. 4-102), where it forms a bursa which protects the tendon of the obturator externus muscle.

STABILITY OF THE HIP JOINT (Figs. 4-38, 4-100, and 4-102).

The hip joint is a very strong and stable articulation. It is surrounded by powerful muscles and the articulating bones are united by a dense fibrous capsule that is strengthened by strong intrinsic ligaments, particularly the iliofemoral ligament.

BLOOD SUPPLY OF THE HIP JOINT

The articular arteries are branches of the medial and lateral circumflex femoral arteries, the deep division of the superior gluteal artery, and the inferior gluteal artery. The artery to the head of the femur is a branch of the posterior division of the **obturator artery.**

NERVE SUPPLY OF THE HIP JOINT (Figs. 4-34 and 4-41)

The articular nerves are derived from the **femoral nerve** via the nerve to the rectus femoris muscle; the **obturator nerve** via its anterior division; the **sciatic nerve** via the nerve to the quadratus femoris muscle; and the **superior gluteal nerve.**

Because the femoral, sciatic, and obturator nerves also supply the knee joint, *hip disease may cause referred pain to the knee.*

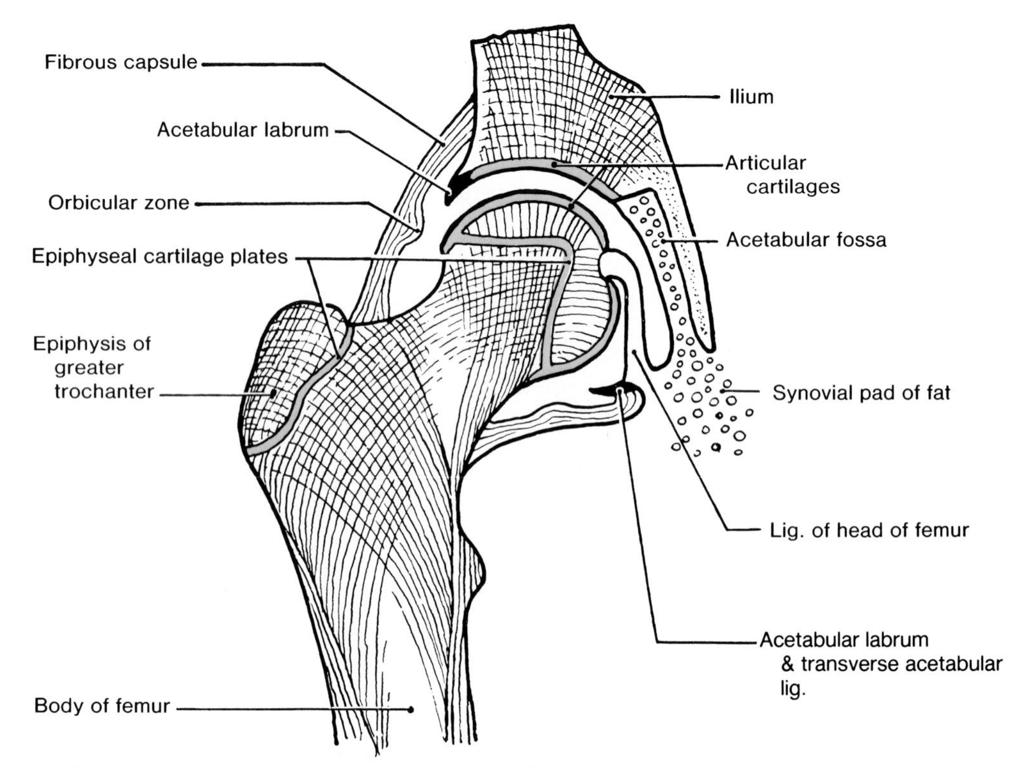

Figure 4-99. Drawing of a coronal section of the right hip joint. Observe that the bony trabeculae of the ilium are projected into the head of the femur as lines of pressure and that the trabeculae of the femur cross these as lines of tension. Note that the epiphysis of the head of the femur is entirely within the fibrous capsule of the joint. Observe that the ligament of the head of the femur is a synovial tube that is fixed at the fovea in the head of the femur (Fig. 4-98). Note that it is open inferiorly where it is continuous with the synovial membrane covering the fat in the acetabular fossa and with the synovial membrane covering the transverse acetabular ligament. The ligament of the head becomes taut during adduction of the hip joint (*e.g.*, when crossing the legs).

Congenital dislocation of the hip joint is common, occurring in about 1.5 per 1000 live births. This abnormality is bilateral in about half the cases.

Despite its name, congenital dislocation of the hip is not usually obvious at birth and may not be for several months. *Females are affected much more often than males* (8:1). Some studies have shown that the articular capsule of the hip joint is loose at birth and that there is *hypoplasia of the acetabulum and the femoral head.*

A characteristic clinical sign of congenital dislocation of the hip joint is inability to abduct the thigh. In addition, the affected leg *seems* to be shorter because the dislocated femoral head is more superior than on the normal side.

Acquired dislocation of the hip joint is uncommon because this articulation is very strong and stable. Nevertheless *dislocation of the hip joint may occur during an* **automobile accident** when the hip joint is flexed, adducted, and medially rotated (Fig. 4-104). When the person's *knee strikes the dashboard* with the thigh in this position, the force transmitted superiorly along the femur drives the femoral head out of the acetabulum. In this position the femoral head is covered posteriorly by capsule rather than bone. As a result, the capsule ruptures inferiorly and posteriorly allowing the femoral head to pass through the tear in the capsule and over the posterior margin of the acetabulum.

Often the acetabular margin fractures, producing a **fracture-dislocation of the hip joint.** When the femoral head dislocates, it usually carries the acetabular bone fragment and the acetabular labrum with it.

Owing to **the close relationship of the**

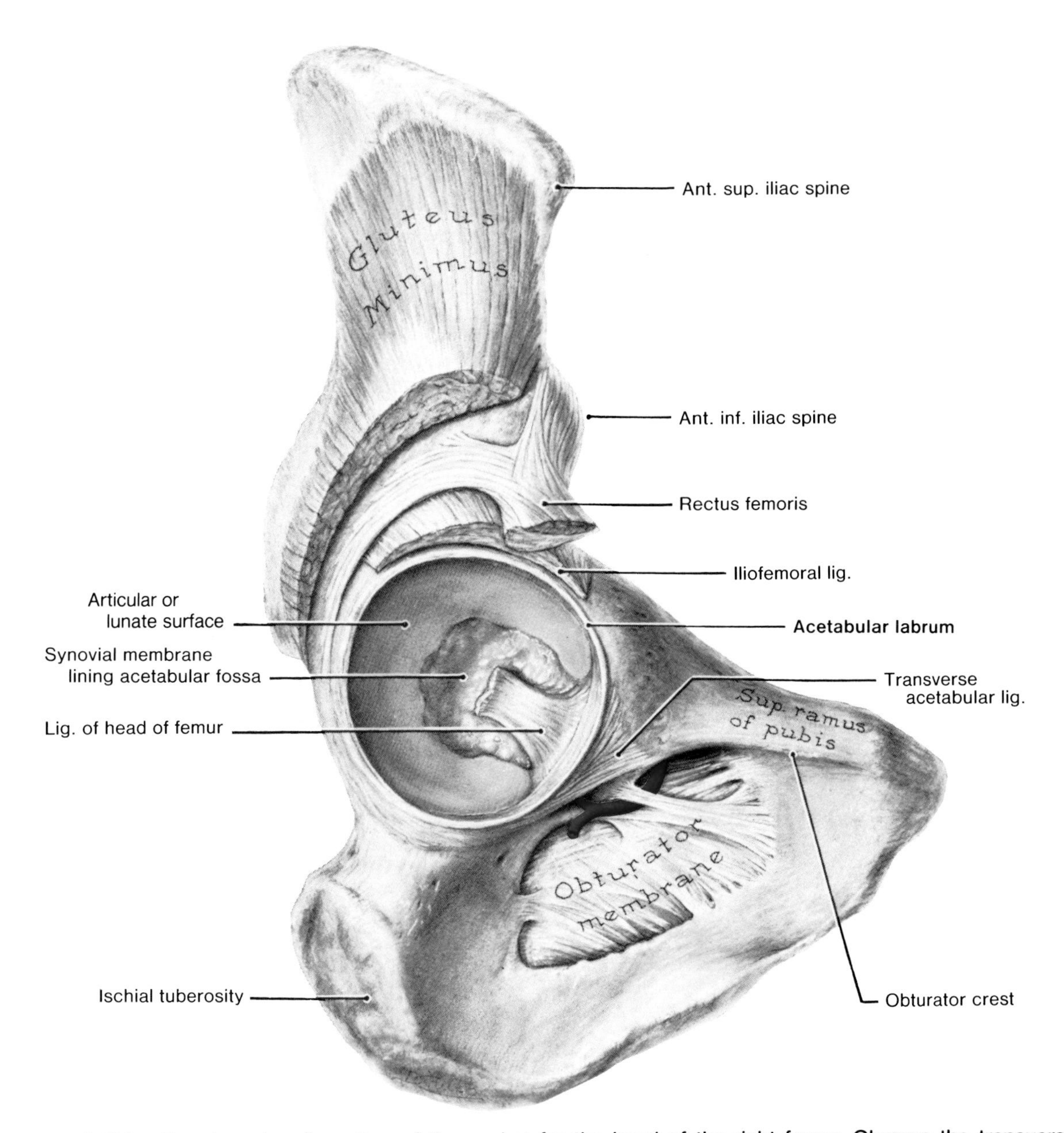

Figure 4-100. Drawing of a dissection of the socket for the head of the right femur. Observe the transverse acetabular ligament, the fibers of which convert the acetabular notch (Fig. 4-4) into the acetabular foramen. Note the acetabular labrum attached to the acetabular rim and to the transverse acetabular ligament. The acetabular labrum forms a complete ring around the head of the femur (Fig. 4-98). Observe the articular or lunate surface of the acetabulum and the synovial membrane attached to the margin of the articular cartilage and covering the pad of fat and vessels in the acetabular fossa (also see Fig. 4-99). Note the ligament of the head of the femur, which is a hollow cone of synovial membrane compressed between the head of the femur and its socket. Through it passes the artery to the head of the femur.

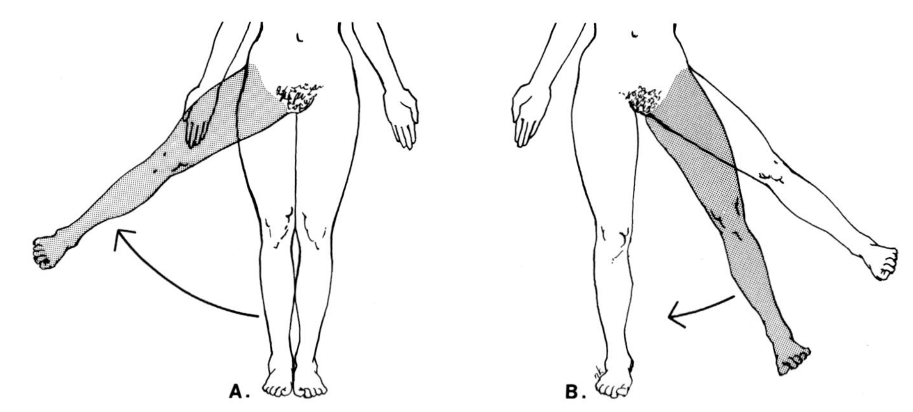

Figure 4-101. Drawings illustrating abduction (*A*) and adduction (*B*) of the thigh at the hip joint. Note that abduction takes the limb away from the median plane of the body and adduction brings it back toward the median plane. See Table 4-10 for the muscles which produce these movements.

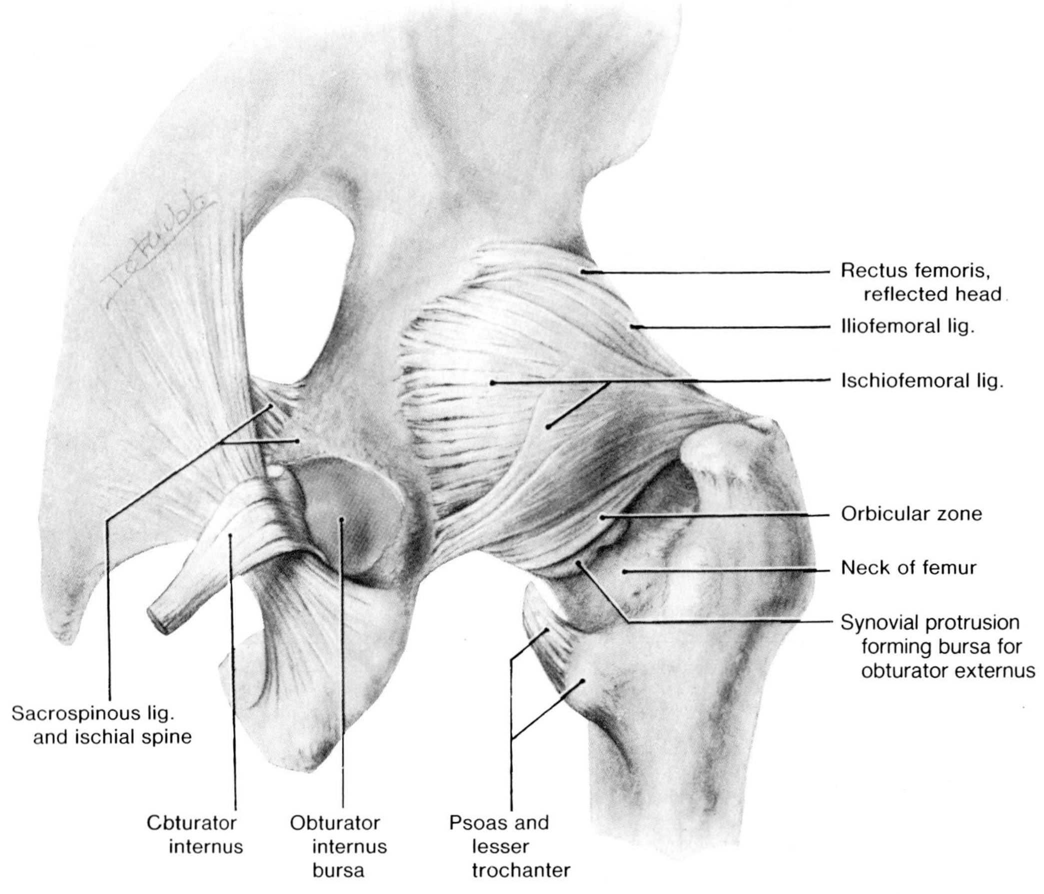

Figure 4-102. Drawing of a posterior view of a dissection of the right hip joint. Observe that the fibrous capsule is directed spirally so it becomes taut during extension and medial rotation of the femur. Note that the fibers cross the neck posteriorly, just superior to the intertrochanteric crest. Observe that the synovial membrane (*blue*) protrudes inferior to the fibrous capsule forming a bursa for the tendon of the obturator externus muscle. The intrinsic ligaments of the articular capsule, particularly the very strong iliofemoral ligament, reinforce the capsule of the hip joint.

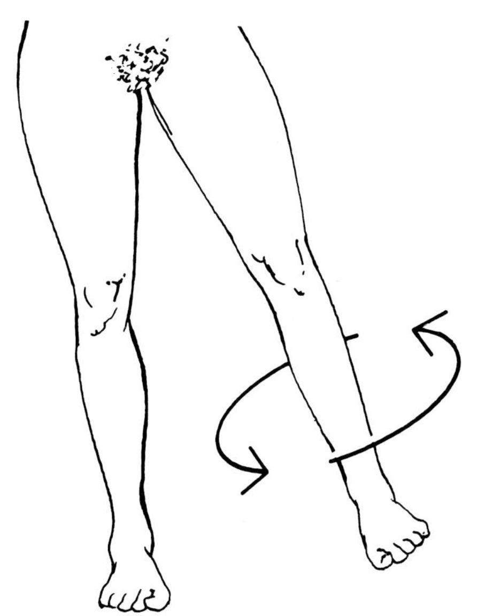

Figure 4-103. Drawing illustrating circumduction of the thigh at the hip joint. The cone of movement results from a combination of flexion-extension and abduction-adduction.

sciatic nerve to the hip joint (Fig. 4-41), it may be injured (stretched and/or compressed) during posterior dislocations or fracture-dislocations of the hip joint. This may result in **paralysis of the hamstring muscles** (Fig. 4-40*B*) and of the muscles distal to the knee supplied by the sciatic nerve (Fig. 4-82). There may also be sensory changes in the skin over the posterior and lateral aspects of the leg and over much of the foot (Fig. 4-94).

THE KNEE JOINT

The knee joint is a hinge type of synovial joint that permits some rotation. The structure is complicated because it consists of three joints, an intermediate one between the patella and femur, and lateral and medial ones between the femoral and tibial condyles.

ARTICULAR SURFACES OF THE KNEE JOINT (Figs. 4-24, 4-45, and 4-105 to 4-107)

The bones involved in the knee joint are the femur, tibia, and patella. The articular surfaces are the large curved condyles of the femur, the flattened condyles of the tibia, and the facets of the patella.

When you stand in the anatomical position your knees are in contact, but your femora are set obliquely because their heads are separated by the width of the pelvis (Fig. 4-1). This produces an open angle at the lateral side of the knee, toward which the patella tends to be displaced when the quadriceps femoris muscle contracts.

The knee joint is mechanically relatively weak because of the configurations of its articular surfaces. It relies on the ligaments that bind the femur to the tibia for strength (Figs. 4-108 and 4-109). As the knee joint is subjected to many stresses and strains, especially during sports, it is not surprising that **knee injury is a common sports injury**.

The angle between the vertical axes of the femur and tibia is exaggerated in **genu valgum** or knock-knee (Fig. 23*A*, p. 24), whereas the knees are widely separated in *genu varum* or bow-leg (Fig. 23*B*).

The long axis of the body of the femur is at an angle of 80° to the horizontal in males and 76° in females. This angular difference in females results from their wider pelves which are designed for child bearing. *This sex difference in the femora has medicolegal significance for identifying skeletal remains.*

Dislocation of the patella is uncommon, but it tends to occur more often in females because of the previously mentioned angle of the femur. The vastus medialis tends to prevent lateral dislocation of the patella because its muscle fibers are attached to the medial border of the patella and blend with the quadriceps femoris tendon (Figs. 4-21, 4-23, and 4-111). Consequently, **weakness or paralysis of the vastus medialis** predisposes to patellar dislocation.

On the superior surface of each tibial condyle there is an articular area for the corresponding femoral condyle (Figs. 4-105 and 4-114). These articular areas, commonly referred to as the medial and lateral **tibial plateaus**, are separated from each other by a narrow nonarticular area which widens anteriorly and posteriorly into anterior and posterior **intercondylar areas,** respectively (Fig. 4-114*A*).

SURFACE ANATOMY AND RADIOGRAPHIC ANATOMY OF THE KNEE JOINT (Figs. 4-23, 4-46 to 4-48, and 4-106)

The knee joint may be felt as a slight gap on each side between the corresponding femoral and tibial

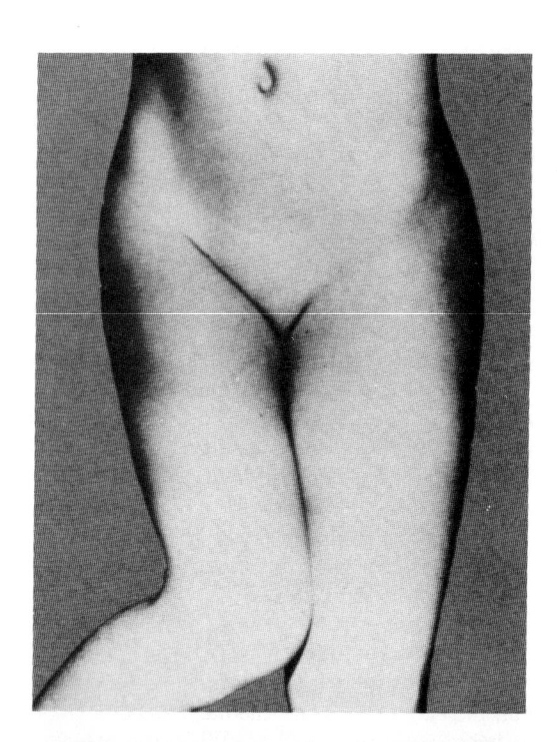

A.

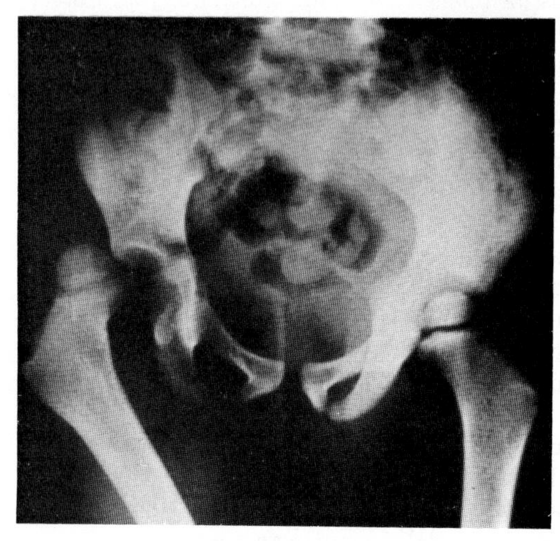

B.

Figure 4-104. *A*, photograph of a female child showing the typical deformity associated with a traumatic *posterior dislocation of the right hip joint.* Observe that her hip is medially rotated, adducted, and slightly flexed. *B*, radiograph of the pelvis of this child (anteroposterior projection) showing the right femoral head out of the acetabulum and displaced posteriorly. On the *left side*, the epiphysis of the femoral head is clearly visible (*black line*) and the head is in its socket (acetabulum).

condyles. When the leg is flexed or extended, a depression appears on each side of the patellar ligament (Fig. 4-48). The articular capsule of the knee joint is very superficial in these depressions.

The knee joint lies deep to the apex of the patella (Figs. 4-21, 4-23, and 4-48).

MOVEMENTS OF THE KNEE JOINT (Table 4-11)

The principal movements occurring at the knee joint are **flexion and extension of the leg,** but some medial and lateral **rotation** also occurs.

Flexion and extension of the knee joint are very free movements. Flexion of the leg normally stops when the calf contacts the thigh. Extension of the leg is stopped by the ligaments of the knee (Fig. 4-109).

The knee is one of the least secure joints in the body. Because it is formed between the two longest bones in the body, considerable leverage can occur. This can result in tearing of the ligaments. Hyperflexion and hyperextension and/or trauma can strain or tear the menisci and ligaments of the knee (Figs. 4-109 and 4-115). These injuries are discussed subsequently.

Sit on a table with your legs dangling over the edge and rotate one leg as far medially as you can and then as far laterally as it will go. While doing this, put the fingers of both your hands on the sides of your knee joint and feel the tibia rotate medially and laterally on the femur. Thus some rotation is permitted while the knee is flexed or semiflexed. A slight degree of medial rotation of the femur is also necessary to achieve complete extension of the knee joint.

When the knee is fully extended, as when sitting on a chair with the heel resting on another chair, the skin anterior to the patella is loose and can easily be picked up between the fingers and the thumb. This laxity of the skin helps flexion to occur. Note that the slackness disappears as the leg is flexed.

When some people fully extend their knee, it "locks" owing to medial rotation of the femur on the tibia. This makes the lower limb a solid column and more adapted for weight bearing. To "unlock" the knee the popliteus muscle contracts, thereby rotating the femur laterally so that flexion of the knee can occur.

THE ARTICULAR CAPSULE OF THE KNEE JOINT (Figs. 4-107 and 4-108)

***The fibrous capsule* of the knee joint is fairly strong**, especially where local thickenings form ligaments. *Ligaments continuous with the fibrous capsule are called intrinsic ligaments of the knee joint.*

Superiorly the fibrous capsule is attached to the femur, just proximal to the articular margins of the condyles and to the intercondylar line posteriorly. It is deficient on the lateral condyle to allow the tendon of the popliteus muscle to pass out of the joint and insert into the tibia (Fig. 4-108).

Inferiorly the fibrous capsule is attached to the articular margin of the tibia, except where the tendon of the popliteus muscle crosses the bone (Fig. 4-109). Here the capsule is prolonged inferolaterally over the popliteus to the head of the fibula, forming the *arcuate popliteal ligament* (Fig. 4-54).

The fibrous capsule is supplemented and strengthened by five ligaments: ligamentum patellae, fibular collateral ligament, tibial collateral ligament, oblique popliteal ligament, and arcuate popliteal ligament. They are often called external ligaments to differentiate them from the internal ligaments (*e.g.*, the cruciate ligaments) which are inside the fibrous capsule (Fig. 4-109).

The Ligamentum Patellae or Patellar Ligament (Figs. 4-108 and 4-109). The ligamentum patellae or patellar ligament is a very strong, thick band which is really the *continuation of the tendon of the quadriceps femoris* muscle. The patella is a sesamoid bone in this tendon. The patellar ligament is continuous with the fibrous capsule of the knee joint (Fig. 4-107).

Verify by palpation that the patellar ligament extends from the inferior border or apex of the patella to the tuberosity of the tibia.

The patellar ligament is most easily felt when the leg is extended. The superior part of its deep surface is separated from the synovial membrane of the knee joint by a mass of loose fatty tissue called the **infrapatellar fatpad** (Fig. 4-107). The inferior part of the patellar ligament is separated from the anterior surface of the tibia by the **deep infrapatellar bursa** (Fig. 4-110).

The patellar ligament is struck, when the leg is flexed, to elicit a knee jerk (quadriceps, patellar, or **knee reflex**). This reflex results in extension of the leg.

The knee reflex is blocked by damage to the femoral nerve which supplies the quadriceps muscle (Fig. 4-15 and Table 4-1). Similarly, damage to the reflex centers in the spinal cord (L2, L3, L4) will affect the knee reflex.

The Fibular Collateral Ligament (Figs. 4-108, and 4-109). The fibular collateral ligament (lateral ligament) is a round, pencil-like cord. *It extends inferiorly from the* ***lateral epicondyle of the femur*** *to the lateral surface of the* **head of the fibula**.

In Figure 4-108, observe that the tendon of the popliteus muscle passes deep to the fibular collateral ligament, separating it from the lateral meniscus. Also observe that the tendon of the biceps femoris muscle is split into two parts by this ligament. As the rounded tendon of the biceps femoris can be easily observed and felt, it serves as a guide to the attachment of the inferior end of the fibular collateral ligament (Fig. 4-110).

Superiorly the fibular collateral ligament is fused with the fibrous capsule of the knee joint; hence this part is an intrinsic ligament. Inferiorly the fibular collateral ligament is separated from the fibrous capsule by fatty tissue; hence this part is an extrinsic ligament. In Figure 4–109, observe that **the fibular collateral ligament is not attached to the lateral meniscus.**

The fibular collateral ligament is not commonly torn because it is very strong. Furthermore, severe blows to the medial side of the knee that might force it laterally and tear it are uncommon. However, lesions (*e.g.*, sprains or tears) of the fibular collateral ligament can have serious consequences. Usually it is the distal end of the ligament that tears, and sometimes the head of the fibula is pulled off (avulsed) because the ligament is stronger than bone.

Complete tears of the fibular collateral ligament are often associated with stretching of the common peroneal nerve (Fig. 4-67). This affects the muscles of the anterior and lateral compartments of the leg (Tables 4-5 and 4-6), and may produce **footdrop** owing to paralysis of the dorsiflexor and evertor muscles of the foot.

The Tibial Collateral Ligament (Figs. 4-107, 4-109, and 4-111). The tibial collateral ligament (medial ligament) is a strong, flat band which extends from the medial epicondyle of the femur to the medial condyle and superior part of the medial surface of the tibia.

The tibial collateral ligament is a thickening of the fibrous capsule of the knee joint and is partly continuous with the tendon of the adductor magnus muscle (Fig. 4-111). The inferior end of the tibial collateral ligament is separated from the tibia by the medial inferior genicular vessels and nerve (Fig. 4-111).

The deep fibers of the tibial collateral ligament are firmly attached to the medial meniscus and the fibrous capsule of the knee joint (Figs. 4-107 and 4-109).

The tibial and fibular collateral ligaments normally prevent disruption of the sides of the knee joint. They are tightly stretched when the leg is extended and their directions (tibial collateral ligament passing inferoanteriorly and fibular collateral ligament passing inferoposteriorly) prevents rotation of the tibia laterally or the femur medially.

As the collateral ligaments are slack during flexion of the leg, they permit some rotation of the tibia on the femur in this position.

The firm attachment of the tibial collateral ligament to the medial meniscus is of considerable clinical significance because injury to the ligament frequently results in concomitant injury to the medial meniscus.

Rupture of the tibial collateral ligament, often associated with tearing of the medial meniscus and the anterior cruciate ligament, is the ***most common type of football injury.*** The damage is frequently caused by a blow to the lateral side of the knee (Fig. 4-134).

When considering soft tissue injuries of the knee, **always think of the three C's** which indicate structures that may be damaged: Collateral ligaments, Cruciate ligaments, and Cartilages or menisci (Fig. 4-114*B*).

Sprains of the tibial collateral ligament result in tenderness over the femoral or tibial attachments of this ligament (Figs. 4-24 and 4-111), owing to tearing of parts of it.

The Oblique Popliteal Ligament (Fig. 4-54). This broad band is an expansion of the tendon of the semimembranosus muscle. *The oblique popliteal ligament strengthens the fibrous capsule of the knee joint posteriorly.* The oblique popliteal ligament arises posterior to the medial condyle of the tibia and passes superolaterally to attach to the central part of the posterior aspect of the fibrous capsule of the knee joint.

The Arcuate Popliteal Ligament (Fig. 4-54). This Y-shaped band of fibers *also strengthens the fibrous capsule posteriorly.* The stem of the ligament arises from the posterior aspect of the head of the fibula. As it passes superomedially over the tendon of the popliteus muscle, the arcuate popliteal ligament spreads out over the posterior surface of the knee joint. It inserts into the intercondylar area of the tibia

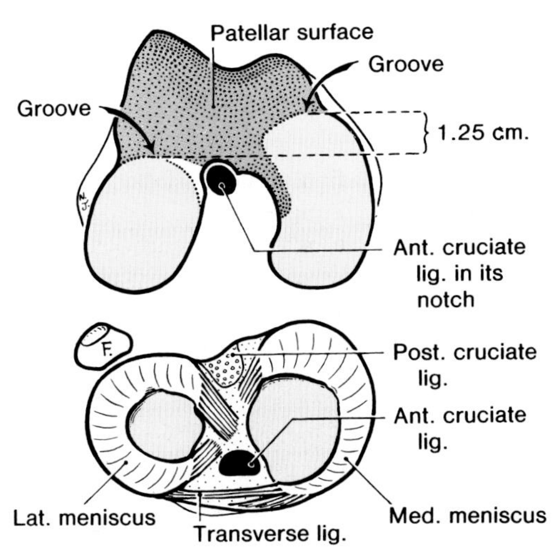

Figure 4-105. Diagrams illustrating the menisci and articular surfaces of the right knee joint (*yellow*).

and the posterior aspect of the lateral epicondyle of the femur.

The synovial capsule of the knee joint is extensive (Figs. 4-107 and 4-108). It lines the inner aspect of the fibrous capsule and reflects onto the articulating bones as far as the edges of their articular cartilages (Fig. 4-105).

The synovial capsule of the knee joint is more extensive than that of any other joint; thus the synovial cavity of the knee joint is the largest joint space in the body.

The synovial capsule is also attached to the periphery of the patella. (Figs. 4-107 and 4-108). The synovial capsule is separated from the patellar ligament by the *infrapatellar fatpad* (Fig. 4-107).

With the knee flexed to a right angle, you can palpate and often observe a depression on each side of the ligamentum patellae. The joint cavity is very superficial on each side of this ligament.

Injections are made into the synovial cavity of the knee joint for diagnostic or therapeutic purposes. When a joint is inflamed (*e.g.*, owing to infection or arthritis), the amount of synovial fluid may increase. Aspiration of this fluid may be necessary to relieve pressure in the joint, or to obtain a sample of it for diagnostic studies. In other cases it may be necessary to evacuate blood that has entered the joint, *e.g.*, after a tibial fracture in which the fracture line extends into the joint cavity (Fig. 4-131).

In **pneumoarthrography** (air contrast study of a joint) air is injected into the joint cavity of the knee to facilitate study of the structures in it (Fig. 4-106*A*). Being less opaque than these structures (*e.g.*, the menisci), it appears black on radiographs and outlines the soft tissues which appear as gray images.

Sometimes the cavity of the knee joint is examined for injury or disease using an **arthroscope.** The knee joint cavity can also be irrigated or washed out using saline.

Steroids are sometimes injected into the joint cavity for treatment of **noninfectious joint diseases.** Injections are generally given into the lateral side of the knee joint, with the patient sitting on the side of a table with the knee flexed and the leg hanging. *To determine the site of injection*, three bony points are located: the apex of the patella, the lateral tibial plateau, and the anterior prominence of the lateral femoral condyle. Joining these three points forms a triangle, the center of which indicates the usual injection site.

BURSAE AROUND THE KNEE (Figs. 4-107 to 4-110)

There are several bursae around the knee joint because most tendons around it run parallel to the bones and pull lengthwise across the joint. *Four bursae communicate with the synovial cavity of the knee joint*; these lie deep to the tendons of the quadriceps femoris, popliteus, and the medial head of the gastrocnemius muscle.

The Suprapatellar Bursa (Figs. 4-106*A*, 4-108, and 4-110). This large saccular extension of the synovial capsule passes superiorly between the femur and the tendon of the quadriceps femoris muscle. It extends about 8 cm (four fingerbreadths) superior to the base of the patella (Fig. 4-108).

***The suprapatellar bursa* permits free movement of the quadriceps tendon over the distal end of the femur** (Fig. 4-108), and facilitates full extension and flexion of the knee joint. The bursa is held in position by the part of the vastus intermedius muscle, called the *articularis genus muscle* (Fig. 4-21*C*).

Because the suprapatellar bursa communicates freely with the synovial cavity of the knee

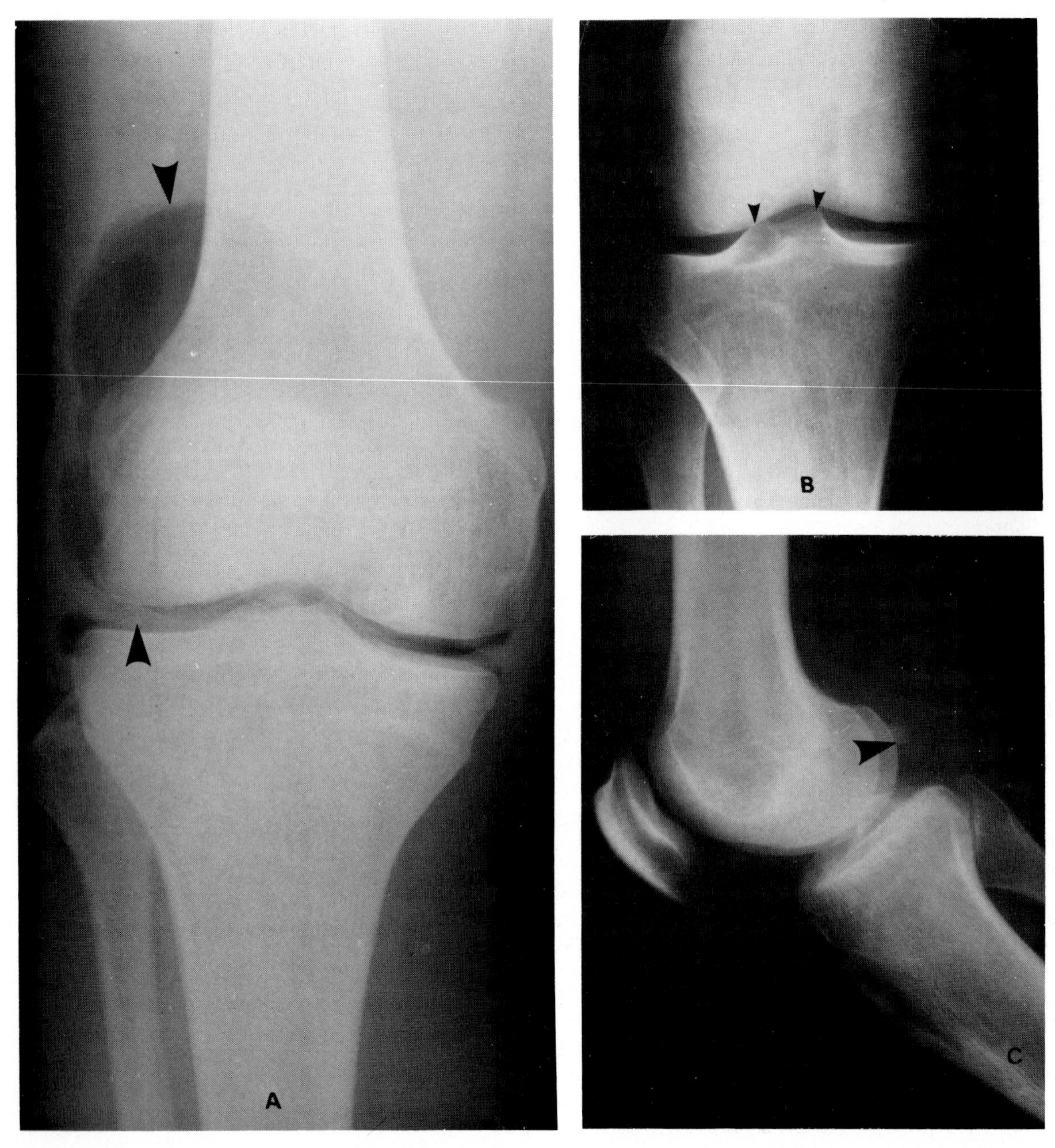

Figure 4-106. Radiographs of the knee. Before taking film *A*, air was injected into the joint cavity. Being less opaque than bone and other tissues, it appears black. The superior *arrow* points to the superior margin of the suprapatellar bursa (see Fig. 4-108). The inferior *arrow* indicates the lateral meniscus that is outlined with air, *B*, *arrows* point to the lateral and medial intercondylar tubercles in this AP view. These tubercles are illustrated in Figure 4-114*A*. *C*, lateral view of the flexed knee joint. The *arrow* points to a fabella which is a sesamoid bone in the tendon of the lateral head of the gastrocnemius muscle.

Table 4-11
Muscles Producing Movements of the Knee Joint[1]

Flexion	Extension	Medial Rotation of Flexed Leg	Lateral Rotation of Flexed Leg
Hamstrings	**Quadricep Femoris**	**Popliteus**	**Biceps femoris**
Semimembranosus	**Rectus femoris**	**Semimembranosus**	
Semitendinosus	**Vastus lateralis**	**Semitendinosus**	
Biceps femoris	**Vastus intermedius**	Sartorius	
Gracilis	**Vastus medialis**	Gracilis	
Sartorius	Tensor fasciae latae		
Popliteus			

[1] **Bold face** indicates muscles chiefly responsible for movement; the other muscles assist with the movement.

joint, it is regarded as a part of it. Hence *stab or puncture wounds proximal to the patella in the anterior part of the thigh may infect the knee joint via the suprapatellar bursa.* It may also be involved in fractures of the distal end of the femur, resulting in **hemarthrosis** (blood in the joint).

The Popliteus Bursa (Fig. 4-110). This subpopliteal extension of the synovial cavity of the knee joint *lies between the tendon of the popliteus muscle and the lateral condyle of the tibia.* The popliteus bursa opens into the lateral part of the synovial cavity of the knee joint, inferior to the lateral meniscus. Sometimes this bursa is also continuous with the synovial cavity of the proximal tibiofibular joint owing to perforation of the partition between the popliteus bursa and its joint cavity.

The Gastrocnemius Bursa. This extension of the synovial cavity of the knee joint lies deep to the origin of the tendon of the medial head of the gastrocnemius muscle. As it separates the tendon from the femur, it is often called the ***subtendinous bursa*** of the gastrocnemius.

The Semimembranous Bursa (Fig. 4-52). This bursa is related to the insertion of the semimembranosus muscle and is located between the medial head of the gastrocnemius and the semimembranosus tendon. Frequently it is a prolongation of the medial gastrocnemius bursa and communicates with the knee joint cavity.

Occasionally synovial fluid escapes from the knee joint and accumulates in the popliteal fossa. It becomes enclosed in a membranous sac, known as a **popliteal cyst** (Baker's cyst). Popliteal cysts often develop in relation to the semimembranosus bursa or the popliteus bursa (Fig. 4-110). Such popliteal cysts are common in children and, although often quite large, they seldom produce symptoms.

In adults, popliteal cysts usually communicate with the synovial cavity of the knee joint by a narrow stalk which passes through the fibrous capsule of the joint. This type, in a sense, is a "*synovial hernia.*" In cases where there is an escape of synovial fluid (**synovial effusion**), owing to either rheumatoid or degenerative joint disease, the popliteal cyst becomes distended by the effused fluid and may extend inferiorly as far as the middle of the calf. In such cases the large popliteal cyst usually interferes with functioning of the knee and ankle joints; operative excision of the cyst is often performed.

The following bursae do not normally communicate with the synovial cavity of the knee joint. They are all related to the patella.

The Subcutaneous Prepatellar Bursa (Figs. 4-107, 4-108, and 4-110). This bursa lies between the skin and the anterior surface of the patella. It allows free movement of the skin over the patella during flexion and extension of the leg.

Because of its superficial and exposed position, the subcutaneous prepatellar bursa may become inflamed after prolonged working on the hands and knees (*e.g.*, when scrubbing a hard floor).

Prepatellar bursitis is a *friction bursitis* caused by friction between the skin and the patella. If the inflammation is chronic, the bursa becomes distended with fluid and forms a soft

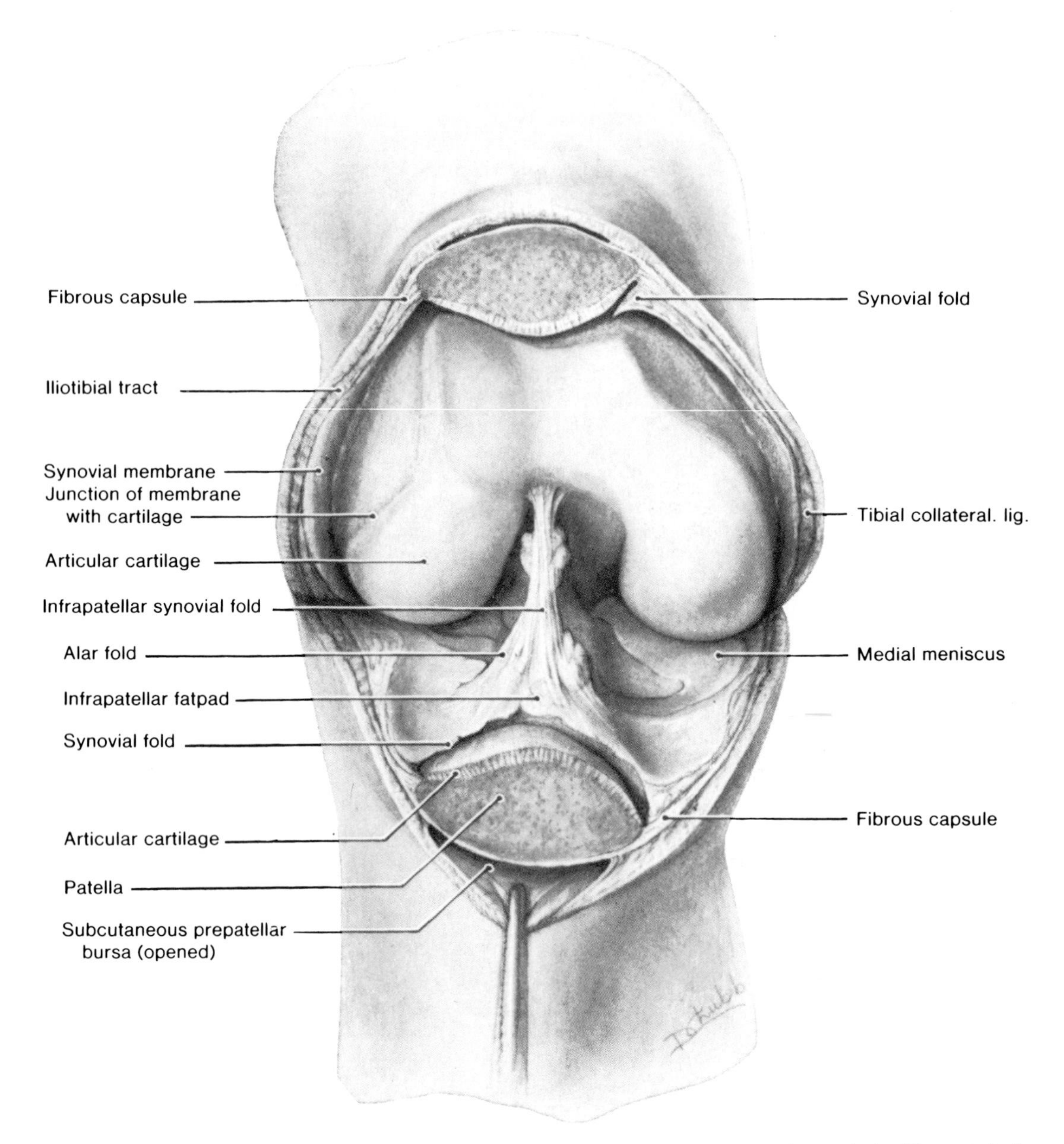

Figure 4-107. Drawing of a dissection of a right knee joint which has been opened anteriorly. The patella is sawn through, the skin and fibrous capsule are cut through, and the joint is flexed. Note the infrapatellar synovial fold resembling a partially collapsed bell-tent whose apex is attached to the intercondylar notch and whose base is inferior to the patella. Observe that the infrapatellar fatpad is continued into the "tent." Understand that a fracture of the patella would bring the prepatellar bursa into communication with the knee joint cavity. Note that the articular cartilage and synovial membrane are continuous with each other on the side of the condyle, as in other joints. Note that the tibial collateral ligament is simply a thickening of the fibrous capsule of the knee joint, but is unique in that it is attached to the medial meniscus. The fibular collateral ligament is not visible in this dissection because it is located more posteriorly than the tibial collateral ligament (Fig. 4-109).

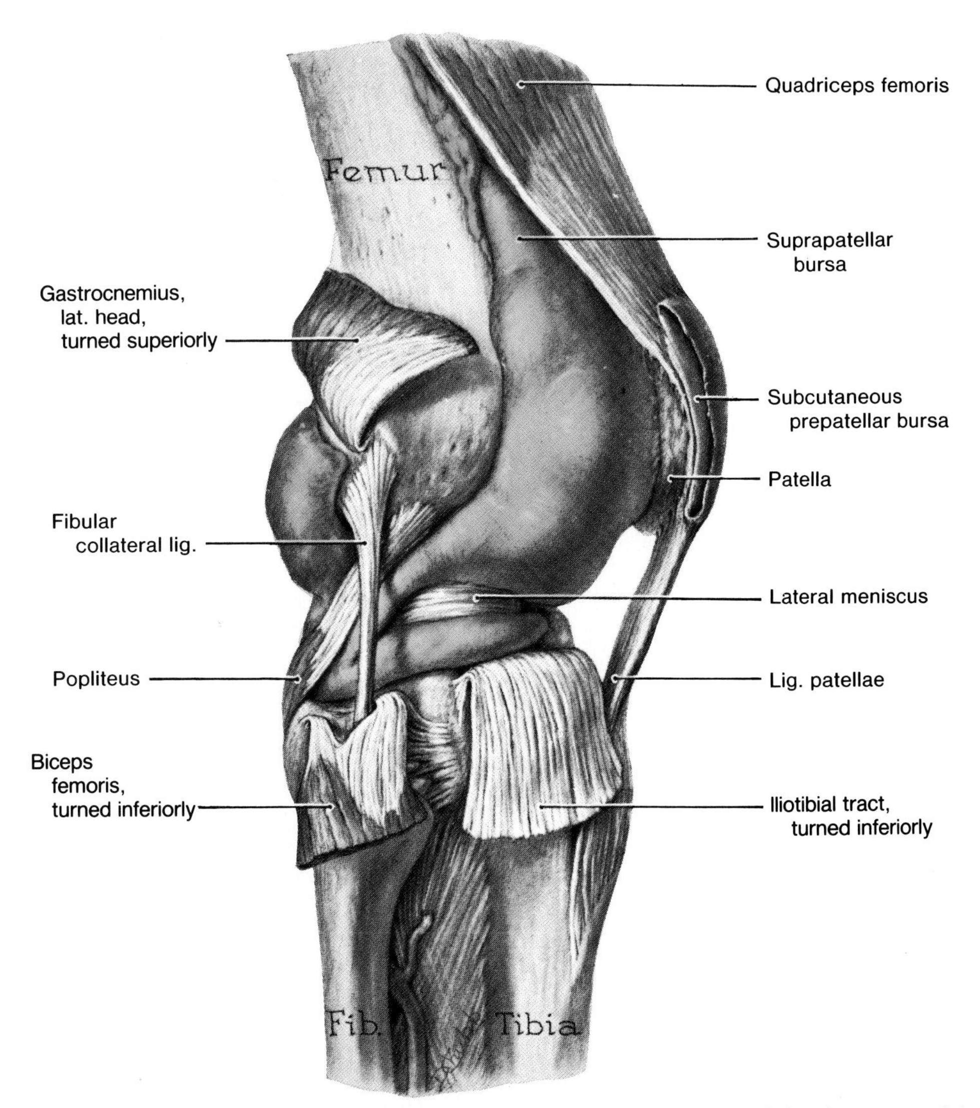

Figure 4-108. Drawing of a lateral view of a dissection of a distended right knee joint. Latex was injected into the joint cavity and fixed with acetic acid. The gastrocnemius is turned superiorly and the biceps and iliotibial tract are turned inferiorly. The latex has flowed into the superior tibiofibular joint cavity. Observe the extent of the synovial capsule: (1) *superiorly* it arises about two fingerbreadths superior to the base of the patella, where it rests on a layer of fat (*yellow*) which allows it to glide freely in movements of the joint. This superior part is called the suprapatellar bursa; (2) *posteriorly* it rises as far superiorly as the origin of the gastrocnemius; (3) *laterally* it curves inferior to the lateral femoral epicondyle where the popliteus tendon and fibular collateral ligament are attached; and (4) *inferiorly* it bulges inferior to the lateral meniscus.

fluctuant swelling anterior to the knee (Fig. 4-112). This condition is commonly called "housemaid's knee," but miners and other people who work on "all fours" also develop prepatellar bursitis. They call it a worn-out or *"beat knee."*

The Subcutaneous Infrapatellar Bursa (Fig. 4-110). This bursa lies between the skin and the tibial tuberosity. This bursa allows the skin to glide over the tibial tuberosity and to withstand pressure when kneeling with the trunk upright (*e.g.*, if you genuflect during praying).

Subcutaneous infrapatellar bursitis results from excessive friction between the skin and the tibial tuberosity. The swelling occurs over the proximal end of the tibia. This condition has been called *"clergyman's knee,"* but it occurs more often in roofers, floor tilers, and carpet layers who do not wear kneepads.

The Deep Infrapatellar Bursa (Fig. 4-110). This small bursa lies between the ligamentum patellae and the anterior surface of the tibia, superior to the tibial tuberosity. It is separated from the knee joint by the infrapatellar fatpad (Figs. 4-107 and 4-110).

Deep infrapatellar bursitis results in a swelling between the patellar ligament and the tibia, superior to the tibial tuberosity. The swelling is usually less pronounced than that associated with superficial prepatellar bursitis. Enlargement of this bursa obliterates the dimples that are present on each side of the patellar ligament when the leg is extended. Obliteration of these dimples also results from **effusion of the knee joint** (*i.e.*, escape of fluid from adjacent blood vessels around the joint cavity.

THE CRUCIATE LIGAMENTS OF THE KNEE JOINT (Figs. 4-105, 4-109, 4-113, 4-114, and 4-116)

Within the capsule of the knee joint, there are two cruciate ligaments that attach the femur to the tibia.

It is important that you understand that *the* ***cruciate ligaments*** *are within the capsule, but outside the synovial cavity of the knee joint.* They are located between the medial and lateral condyles and are separated from the joint cavity by synovial membrane. The synovial capsule lines the fibrous capsule, except posteriorly where it is reflected anteriorly around the cruciate ligaments.

The cruciate (L. resembling a cross) ligaments are strong, rounded bands that cross each other obliquely like the limbs of St. Andrew's cross or an X (Fig. 4-109). They are named anterior and posterior according to their site of attachment to the tibia, *i.e. the* ***anterior cruciate ligament*** *attaches to the tibia anteriorly, and the* ***posterior cruciate ligament*** *attaches to the tibia posteriorly.*

The Anterior Cruciate Ligament (Figs. 4-105, 4-109, 4-113, and 4-114). *This is the weaker of the two cruciate ligaments.* The anterior cruciate ligament arises from the anterior part of the intercondylar area of the tibia (Fig. 4-114), just posterior to the attachment of the medial meniscus. It extends superiorly, posteriorly, and laterally to attach to the posterior part of the medial side of the lateral condyle of the femur (Fig. 4-113). The anterior cruciate ligament is slack when the knee is flexed and taut when the knee is fully extended.

The anterior cruciate ligament prevents posterior displacement of the femur on the tibia and hyperextension of the knee joint. When the knee joint is flexed to a right angle, the tibia cannot be pulled anteriorly because it is held by the anterior cruciate ligament.

The relatively weak anterior cruciate ligament is *sometimes torn when the tibial collateral ligament* ruptures after the knee is hit hard from the lateral side while the foot is on the ground (Fig. 4–134).

First the tibial collateral ligament ruptures, opening the knee joint on the medial side. This may tear the medial meniscus and the anterior cruciate ligament.

The anterior cruciate ligament may also be torn when (1) the tibia is driven anteriorly on the femur, (2) the femur is driven posteriorly on the tibia, and (3) the knee joint is severely hyperextended. Use Figure 4-113*A* to visualize how these tears of the anterior cruciate ligament could occur.

The knee joint becomes very unstable when the anterior cruciate ligament is torn (Fig. 4-115). To test its stability, the tibia is pulled in an anterior direction. If there is anterior movement (*"anterior drawer sign"*), a tear of the anterior cruciate ligament is suggested.

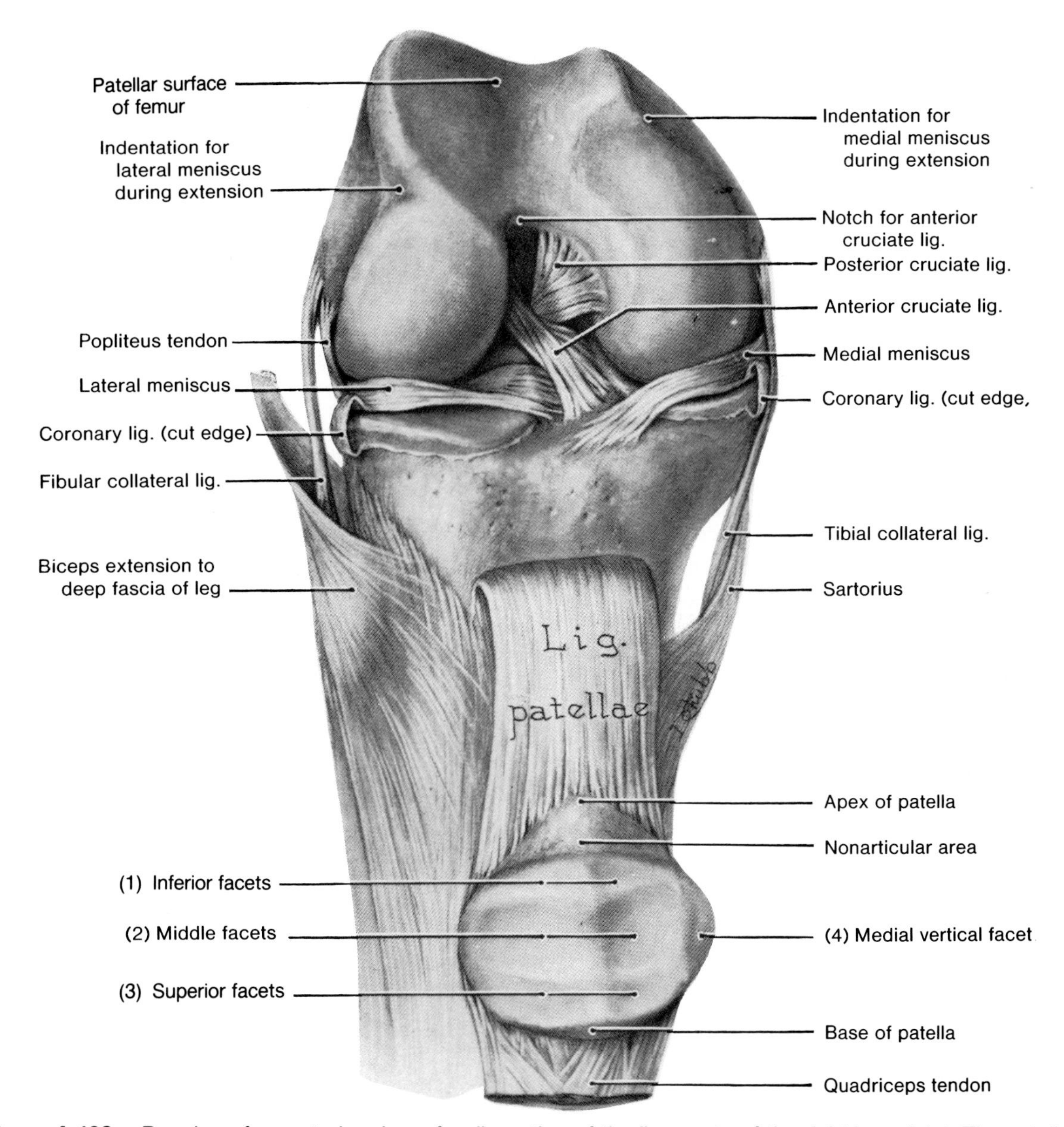

Figure 4-109. Drawing of an anterior view of a dissection of the ligaments of the right knee joint. The patella is turned inferiorly and the joint is flexed. Observe the subsidiary notch at the anterolateral part of the intercondylar notch for the reception of the anterior cruciate ligament on full extension. Note the three paired facets on the posterior surface of the patella for articulation with the patellar surface of the femur.

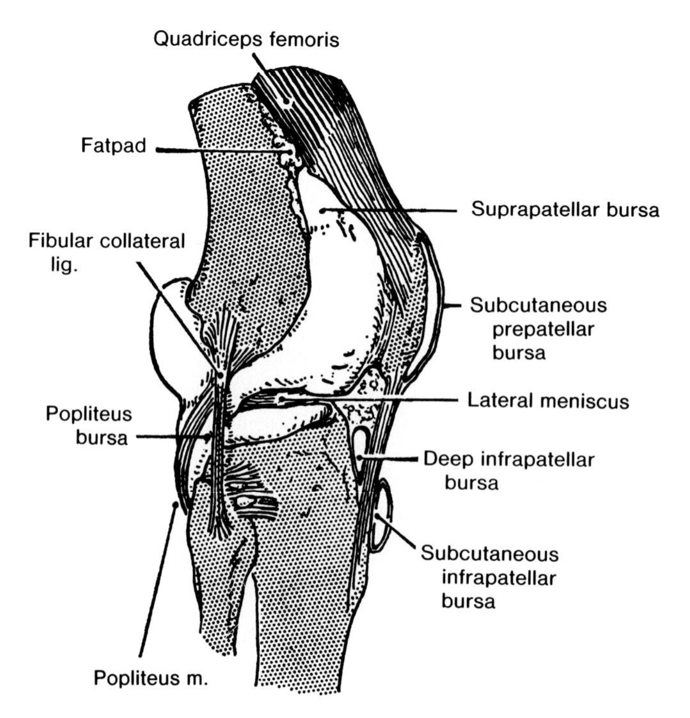

Figure 4-110. Drawing of a lateral view of a right knee joint filled with latex to show the extent of the synovial capsule. The fibular collateral ligament prevents disruption of the joint at the side. It is not commonly torn because it is very strong. Observe that the fibular collateral ligament extends from the lateral epicondyle of the femur to the head of the fibula.

The Posterior Cruciate Ligament (Figs. 4-105, 4-113, 4-114, and 4-116). *This is the stronger of the two cruciate ligaments.* The posterior cruciate ligament arises from the posterior part of the intercondylar area of the tibia and passes superiorly and anteriorly on the medial side of the anterior cruciate ligament to attach to the anterior part of the lateral surface of the medial condyle of the femur (Fig. 4-113*A*). The posterior cruciate ligament is the first structure observed when the knee joint is opened posteriorly (Figs. 4-113*A* and 4-114*B*).

The posterior cruciate ligament tightens during flexion of the knee joint, preventing anterior displacement of the femur or posterior displacement of the tibia. *The posterior cruciate ligament prevents anterior displacement of the femur on the tibia and hyperflexion of the knee joint.*

The posterior cruciate ligament may be injured when the superior part of the tibia is struck with the knee flexed. This kind of injury may occur when a passenger's leg is driven against the dashboard, but dislocation of the hip is more likely to occur (Fig. 4-104).

If the tibia is driven posteriorly on the femur, or the femur is driven anteriorly on the tibia, or the knee joint is severely hyperflexed, the posterior cruciate ligament may be torn.

The flexed knee is unstable when the posterior cruciate ligament is torn. To test its stability, the tibia is forced in a posterior direction. If there is posterior movement ("*posterior drawer sign*"), a tear of the posterior cruciate ligament is indicated.

THE MENISCI OF THE KNEE JOINT

The medial and lateral menisci (G. crescents) *are crescentic plates of fibrocartilage that lie on the articular surface of the tibia* (Figs. 4-105, 4-107, and 4-114). They act like shock absorbers.

Because of their shape, they are sometimes called the *semilunar cartilages.* Wedge-shaped in cross-section, **the menisci are firmly attached at their ends to the intercondylar area of the tibia** (Fig. 4-114).

The menisci deepen the articular surfaces of the tibia where they articulate with the femoral condyles. Their superior surfaces are slightly concave for reception of

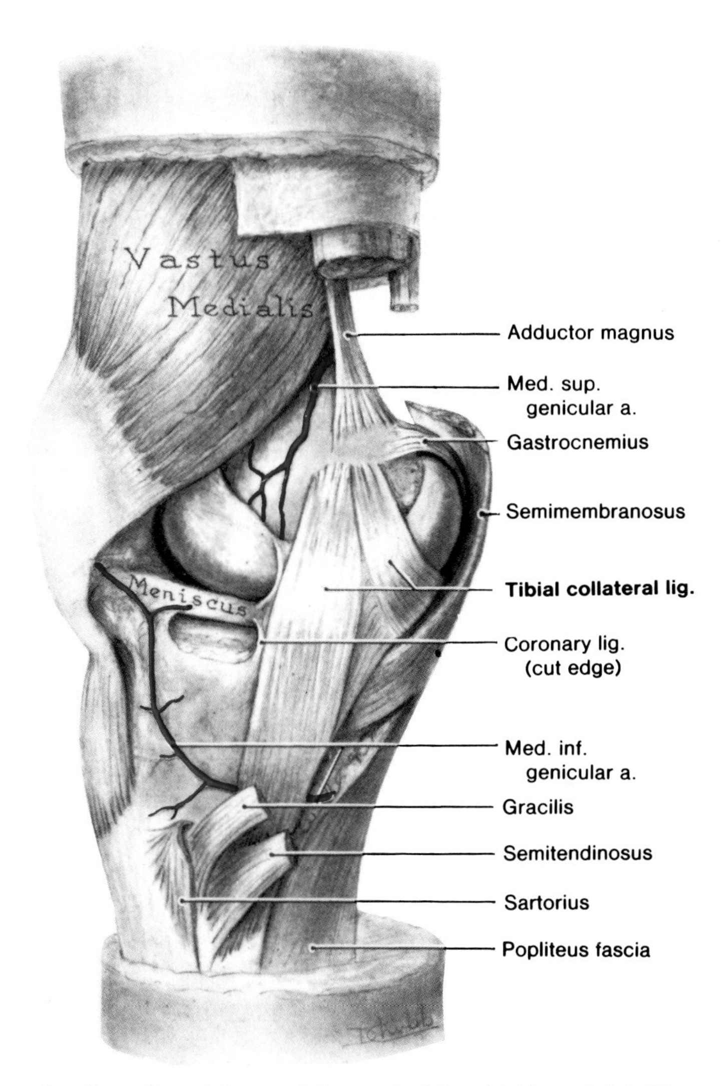

Figure 4-111. Drawing of a dissection of the medial aspect of the right knee joint. Observe the band-like part of the tibial collateral ligament which is attached to the medial epicondyle, almost in line with the adductor magnus tendon and crossing the insertion of the semimembranosus. Note that tibial collateral ligament crosses the medial inferior genicular artery and is crossed by the tendons of the three medial rotators (sartorius, gracilis, and semitendinosus), each of which is supplied by a different nerve (femoral, obturator, and sciatic). *Observe that the medial meniscus is firmly attached to the tibial collateral ligament;* this relationship is more obvious in Figure 4-116. Because of this attachment, rupture of this ligament commonly results in a concomitant tearing of the medial meniscus.

the femoral condyles, whereas their inferior surfaces that rest on the tibial condyles are flatter.

The menisci are thick at their peripheral attached margins and thin at their internal unattached edges. Being smooth and slightly movable, *the menisci fill the gaps between the femur and tibia that would otherwise be present during movements of the knee joint.* Their external margins are attached to the fibrous capsule of the knee joint and through it to the edges of the articular surfaces of the tibia. The capsular fibers which attach the menisci to the tibial condyles are called medial and lateral **coronary ligaments** (Figs. 4-109 and 4-111).

The transverse ligament of the knee joins the anterior parts of the two menisci (Fig. 4-105). This connection allows them to move together during movements of the femur on the tibia. The thickness of the transverse ligament varies in different people; sometimes it is absent.

The thick peripheral margins of the menisci are vascularized by genicular branches of the popliteal artery (Figs. 4-53 and 4-111), but their thin unattached edges in the interior of the joint are avascular.

The Medial Meniscus (Figs. 4-105, 4-107, 4-109, 4-111, 4-114, and 4-116). The medial meniscus is a *C-shaped cartilage* that is broader posteriorly than anteriorly. Its anterior end or horn (L. *cornu*) is attached to the *anterior intercondylar area of the tibia*, anterior to the attachment of the anterior cruciate ligament (Fig. 4-114). Its posterior end or horn is attached to the *posterior intercondylar area*, anterior to the attachment of the posterior cruciate ligament, and between the attachments of the lateral meniscus and the posterior cruciate ligament (Figs. 4-105 and 4-114).

The medial meniscus is firmly adherent to the deep surface of the tibial collateral ligament (Figs. 4-109, 4-111, 4-114*B*, and 4-116).

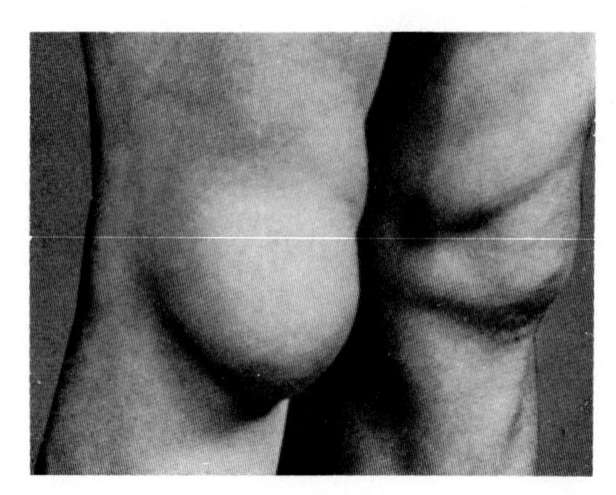

Figure 4-112. Photograph of the knees of a patient with prepatellar bursitis ("housemaid's knee"), resulting from inflammation of the subcutaneous prepatellar bursae (Figs. 4-108 and 4-110). The amount of fluid in the bursa over her right patella is greater than is usually seen in this type of friction bursitis.

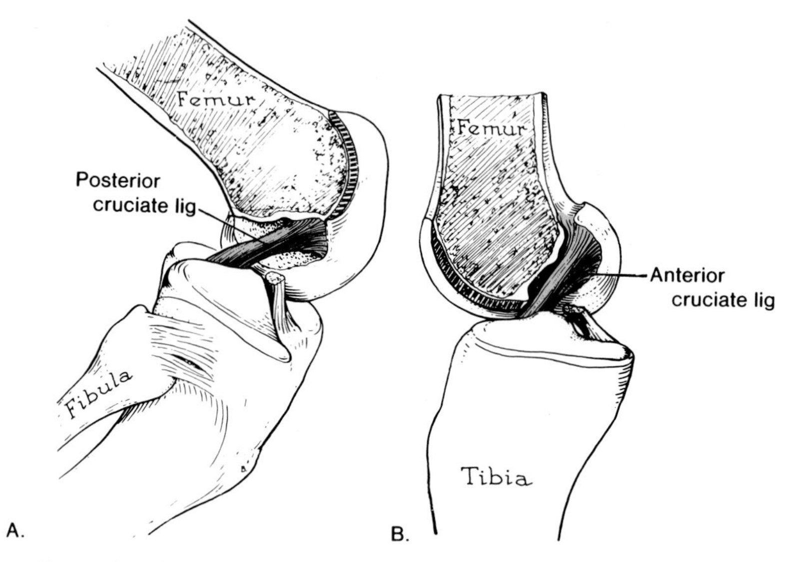

Figure 4-113. Drawings illustrating the cruciate ligaments. In each illustration, half of the right femur is removed with the proximal part of the corresponding cruciate ligament. *A*, lateral view showing that the posterior cruciate attaches the femur to the tibia posteriorly and prevents anterior sliding of the femur, particularly when the knee is flexed. *B*, medial view showing that the anterior cruciate ligament attaches the femur to the tibia anteriorly and prevents posterior sliding of the femur and hyperextension of the knee. It also limits medial rotation of the extended knee when the foot is on the ground (*i.e.*, when the leg is fixed).

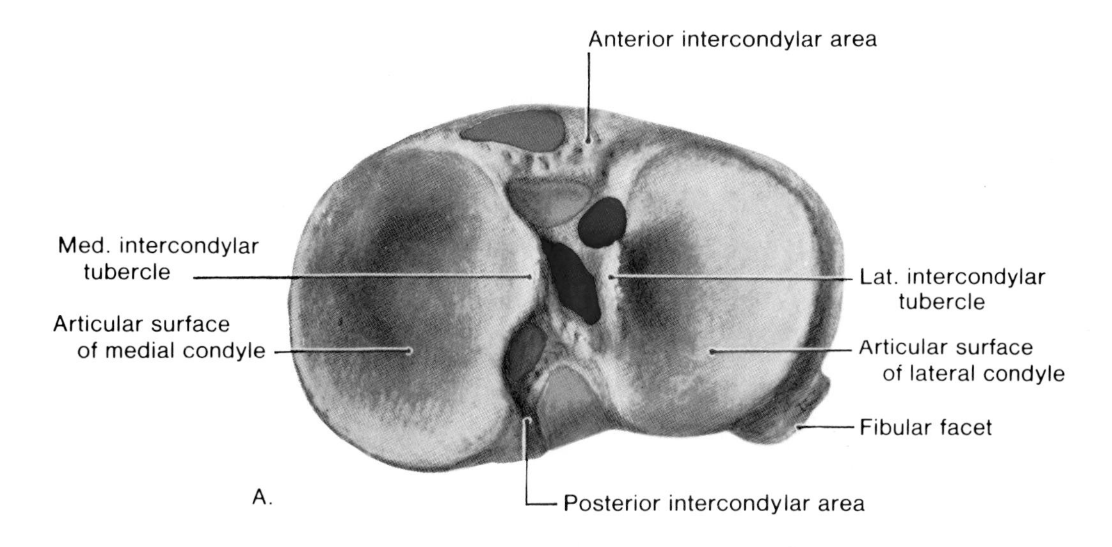

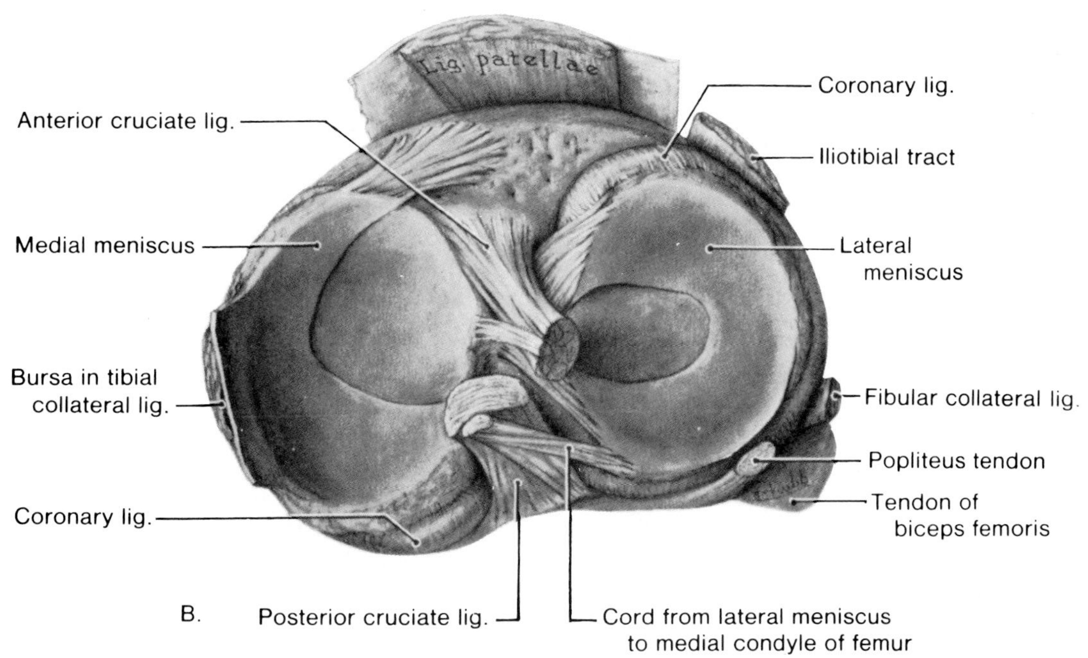

Figure 4-114. Drawings of dissections of the right knee joint demonstrating the cruciate ligaments and menisci. *A*, superior aspect of the proximal end of the tibia showing the medial and lateral plateaus (articular surfaces). Sites of attachment of the cruciate ligaments are colored *yellow*; those of the medial meniscus, *blue*; and those of the lateral meniscus, *red*. *B*, this view shows the menisci and their attachments to the intercondylar area of the tibia. The tibial attachments of the cruciate ligaments are also shown. The menisci are cartilaginous and tough where they are compressed between the femur and tibia, but they are ligamentous and pliable at their attachments. The menisci conform to the shapes of the articular surfaces on which they rest. As the horns of the lateral meniscus are attached close together and its coronary ligament is slack, this meniscus can slide anteriorly and posteriorly on the condyle. Because the horns of the medial meniscus are attached far apart, its movements on the condyle are restricted. Movement of the medial meniscus is also restricted because of its attachment to the tibial collateral ligament; hence it is commonly torn.

Injuries to the knee joint are common because it is a major weight-bearing joint and its stability depends almost entirely upon its associated ligaments and muscles.

Ligamentous injuries of the knee joint may result from any blow that forces it to move in an abnormal plane. A blow on the lateral side of the knee when a person is bearing weight on the leg stresses the tibial collateral ligament. If the blow is relatively minor, the fibers are stretched and some of them may be torn; this is called a **sprained tibial collateral ligament**. When the blow is severe, all the fibers may be torn partially or completely. This is called a **torn tibial collateral ligament**; the tear usually occurs near its attachment to the medial epicondyle of the femur (Fig. 4-116).

Localized tenderness and pain in the flexed knee on the medial side of the patellar ligament, just proximal to the medial tibial plateau, may indicate **injury to the medial meniscus**. Injury to this cartilage is a common occurrence (about 20 times more common than injury to the lateral meniscus).

*Injury to the medial meniscus, results from a **twisting strain*** that is applied to the knee joint when it is flexed. **Because the medial meniscus is firmly adherent to the tibial collateral ligament, twisting strains of this ligament may tear and/or detach the medial meniscus from the fibrous capsule.** Part of the torn cartilage may become displaced toward the center of the joint and become lodged between the tibial and femoral condyles. This "locks the knee" in the flexed position, preventing the patient from fully extending the knee. People who have a minor form of this condition say they have a "*clicking knee.*"

Because the internal edges of the menisci are poorly supplied with blood, tears in them heal poorly. Tears near the peripheral margin, which are vascularized by genicular branches of the popliteal artery (Figs. 4-53 and 4-111), usually heal well.

When weight is borne by the flexed knee joint, a sudden twist of the knee may cause rupture of the medial meniscus, usually splitting it longitudinally. This injury is common in athletes who twist their flexed knees while running. It also occurs in coal miners and other persons who topple over when they are working in a crouched or squatting position.

Usually a torn medial meniscus is surgically excised (**medial meniscectomy**) because repeated displacement of the torn part of the meniscus is temporarily disabling, and may lead to degenerative disease of the joint.

The Lateral Meniscus (Figs. 4-105, 4-108 to 4-110, 4-114, and 4-116). The lateral meniscus is *nearly circular in shape,* conforming to the more circular lateral tibial condyle (Figs. 4-105 and 4-114*B*). Note that the lateral meniscus covers a larger area of articular surface than does the medial meniscus.

The tendon of the popliteus muscle separates the lateral meniscus from the fibular collateral ligament (Figs. 4-108 to 4-110).

The anterior and posterior ends or horns of the lateral meniscus are attached close together in the anterior and posterior intercondylar areas (Fig. 4-114). A strong tendinous slip, called the *posterior meniscofemoral ligament,* joins the lateral meniscus to the posterior cruciate ligament and the medial femoral condyle (Fig. 4-116).

The lateral meniscus, smaller and more freely movable than the medial meniscus, is less likely to be torn, mainly because it is not attached to the fibular collateral ligament. The popliteus tendon and popliteus bursa intervene between the lateral meniscus and this ligament (Figs. 4-108 to 4-110).

When air and/or dense contrast material are injected into the synovial cavity of the knee joint before radiographs are taken, the menisci can be observed (Figs. 4-106 and 4-115).

Pneumoarthrograms (radiographs taken after air injection) or ***double contrast arthrograms*** are helpful in demonstrating soft tissue lesions of the knee joint. Because air is less opaque than the menisci, it appears black in the radiograph and outlines the soft tissues (*e.g.*, the menisci). When dense contrast materials are used, the articular cartilages and menisci appear as radiolucent images within the dense contrast medium.

STABILITY OF THE KNEE JOINT

Although the knee joint is well constructed and one of the strongest joints in the body, particularly when extended, its function is commonly deranged (*e.g.*, in body contact sports such as hockey and football).

The stability of the knee joint depends upon the strength of the surrounding muscles and

ligaments. Of these supports, the muscles are most important. Thus, many sports injuries are preventable through appropriate conditioning and training.

The most important muscle in stabilizing the knee joint is the quadriceps femoris, particularly the inferior fibers of the vastus medialis and vastus lateralis (Figs. 4-21 and 4-23). Evidence for this is that the knee joint will function surprisingly well following a strain of the ligaments if the quadriceps femoris is well developed.

The quadriceps femoris muscle undergoes considerable atrophy during periods of disuse (e.g., while the lower limb is in a cast). Thus it must be exercised to prevent disuse atrophy and instability of the knee. Without good muscular support, repaired knee ligaments may be more easily sprained or retorn, either partially or completely. ***Residual instability of the knee joint*** *is one of the most troublesome complications of ligamentous injuries of the knee.*

Traumatic dislocation of the knee joint is not common; however this injury may occur, *e.g.*, during an automobile accident (Fig. 4-117). In addition to disruption of the ligaments of the knee, the popliteal artery and tibial nerve may be injured, resulting in partial or complete paralysis of the muscles supplied by this nerve.

BLOOD SUPPLY OF THE KNEE JOINT (Figs. 4-53 and 4-111)

The articular arteries to the knee joint are branches of the vessels that form the ***genicular anastomoses*** *around the knee.* The middle genicular artery, a branch of the popliteal artery, penetrates the fibrous capsule and supplies the cruciate ligaments, synovial capsule, and the margins of the menisci.

NERVE SUPPLY OF THE KNEE JOINT (Fig. 4-50)

The articular nerves are branches of the obturator, femoral, tibial, and common peroneal nerves.

THE SUPERIOR TIBIOFIBULAR JOINT

This articulation is a *plane type of synovial joint* between the head of the fibula and the lateral condyle of the tibia.

ARTICULAR SURFACES OF THE SUPERIOR TIBIOFIBULAR JOINT (Figs. 4-110 and 4-116)

The flat, oval-to-circular facet on the head of the fibula articulates with a similar facet located posterolaterally on the inferior aspect of the lateral condyle of the tibia.

MOVEMENTS OF THE SUPERIOR TIBIOFIBULAR JOINT

Slight movement occurs at the superior tibiofibular joint during dorsiflexion of the foot at the ankle joint, which presses the lateral malleolus laterally and causes movement of the body and head of the fibula. Some movement of the joint also occurs during plantarflexion of the foot.

THE ARTICULAR CAPSULE OF THE SUPERIOR TIBIOFIBULAR JOINT (Figs. 4-108, 4-110, and 4-116)

The *fibrous capsule* of the superior tibiofibular joint surrounds the joint and is attached to the margins of the articular facets on the fibula and tibia. The fibrous capsule is strengthened by the anterior and posterior **ligaments of the head of the fibula.** The fibers of these ligaments run superomedially from the fibula to the tibia (Fig. 4-116). The tendon of the popliteus muscle is intimately related to the posterosuperior aspect of the proximal tibiofibular joint (Fig. 4-110).

The synovial capsule of the superior tibiofibular joint lines the fibrous capsule (Fig. 4-108). The pouch of synovial membrane prolonged under the tendon of the popliteus muscle, known as the **popliteus bursa** (Fig. 4-110), sometimes communicates with the synovial cavity of the proximal tibiofibular joint through an opening in the superior part of the capsule. Consequently, the superior tibiofibular joint may be indirectly in communication with the synovial cavity of the knee joint.

BLOOD SUPPLY OF THE SUPERIOR TIBIOFIBULAR JOINT (Fig. 4-53)

The articular arteries to this joint are derived from the inferior **lateral genicular and anterior tibial recurrent arteries.**

NERVE SUPPLY OF THE SUPERIOR TIBIOFIBULAR JOINT (Figs. 4-50 and 4-70)

The articular nerves are derived from the ***common peroneal nerve and the nerve to the popliteus.***

THE INFERIOR TIBIOFIBULAR JOINT

This articulation is a fibrous joint of the syndesmosis type (*i.e.*, the bony surfaces are held together by fibrous tissue). It is located between the inferior ends of the tibia and fibula (Figs. 4-118, 4-119, and 4-121).

ARTICULAR SURFACES OF THE INFERIOR TIBIOFIBULAR JOINT

The rough, convex, triangular articular area on the medial surface of the inferior end of the fibula articulates with a facet at the inferior end of the tibia.

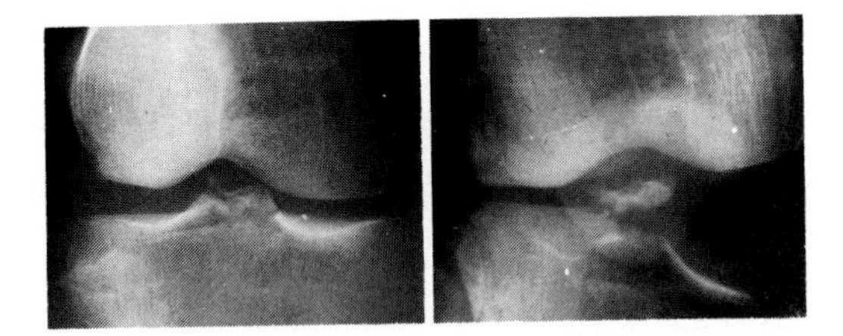

Figure 4-115. Radiographs of the right knee (frontal projections) of a 21-year-old football player who was hit hard from the lateral side. *A*, shows a normal relationship between the femur and tibia. *B*, shows the same knee radiographed while it was being stressed in abduction with the patient under anesthesia. Note that the knee joint has opened up on the medial side, indicating a complete tear of the tibial collateral ligament. Note the displacement of the lateral tubercle of the intercondylar eminence which is fractured (see Fig. 4-114*A*). This displacement indicates that the anterior cruciate ligament and the bone which attaches it to the tibia have been torn away (Fig. 4-113*B*).

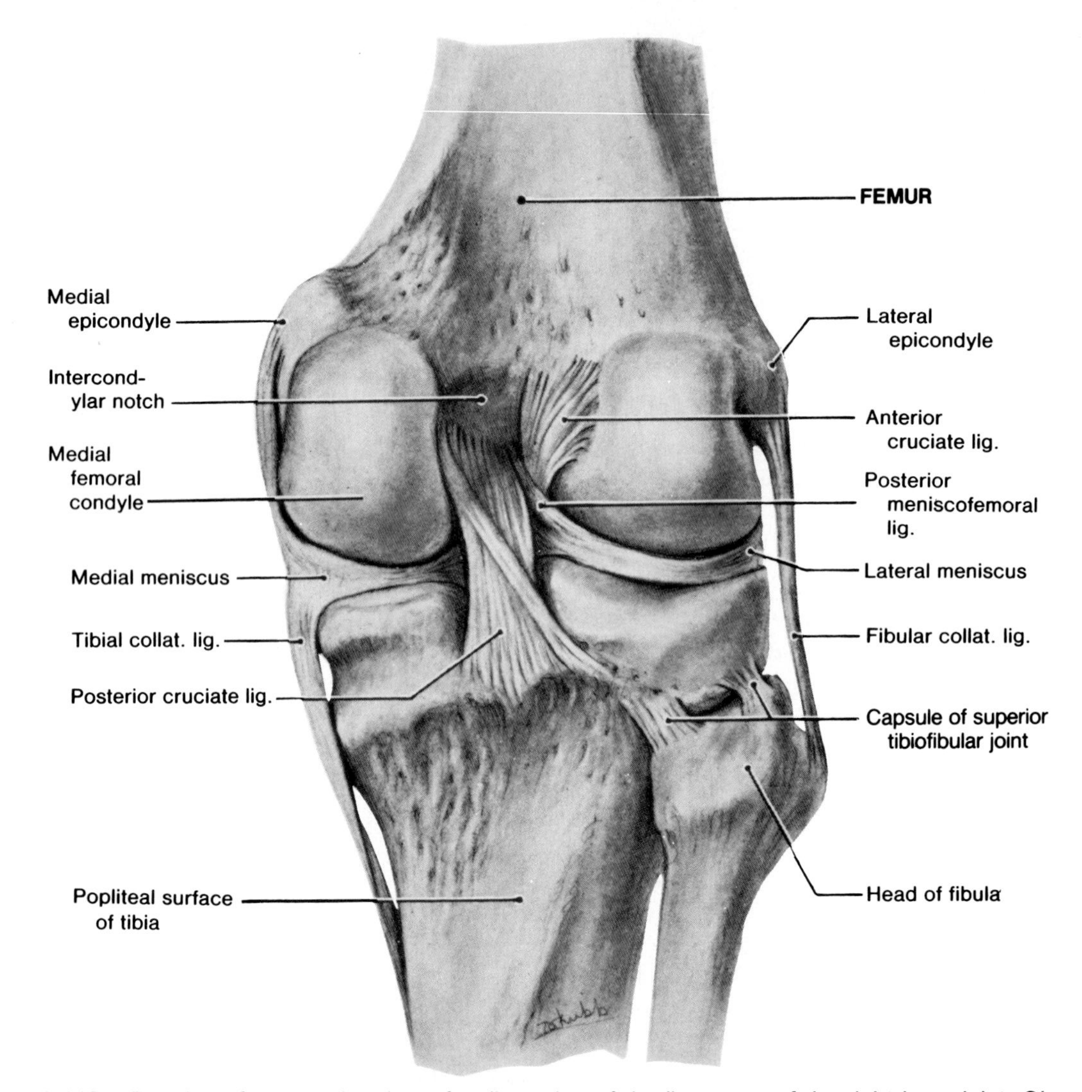

Figure 4-116. Drawing of a posterior view of a dissection of the ligaments of the right knee joint. *Observe the band-like tibial collateral ligament attached to the medial meniscus.* Note that the cord-like fibular collateral ligament is separated from the lateral meniscus. Thus, *undue stress on the tibial collateral ligament may tear the medial meniscus*, whereas the lateral meniscus is usually not torn by undue stress on the fibular collateral ligament. Note that the posterior cruciate ligament is joined by the posterior meniscofemoral ligament that runs from the lateral meniscus to the lateral surface of the medial femoral condyle. Observe the attachment of the anterior cruciate ligament to the lateral femoral condyle. Athletic injuries are the most frequent cause of trauma to the collateral and cruciate ligaments.

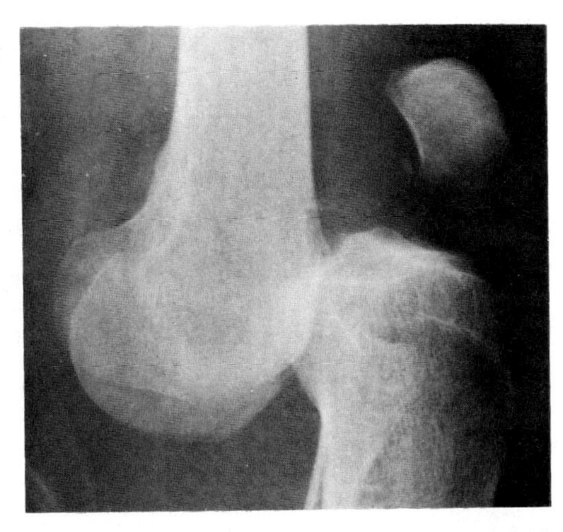

Figure 4-117. Radiograph showing an anterior dislocation of the knee joint in a 28-year-old man who sustained multiple injuries in an automobile accident. The degree of dislocation indicates that the collateral and cruciate ligaments were torn. In addition the menisci were torn and displaced. There was also severe injury to the popliteal artery, which required surgical reconstruction with an arterial prosthesis.

There is a small superior projection of the synovial capsule of the ankle joint into the inferior part of the inferior tibiofibular joint (Fig. 4-119).

A strong **interosseous ligament** (Figs. 4-118 and 4-119), continuous superiorly with the interosseous membrane, *forms the principal connection between the tibia and fibula* at this joint. It consists of strong bands which extend from the fibular notch of the tibia to the medial surface of the distal end of the fibula.

The inferior tibiofibular joint is also strengthened anteriorly and posteriorly by the strong **anterior and posterior tibiofibular ligaments** (Figs. 4-118 and 4-119). They extend from the borders of the fibular notch of the tibia (Fig. 4-57*B*)to the anterior and posterior surfaces of the lateral malleolus, respectively.

The inferior, deep part of the posterior tibiofibular ligament is called the **transverse tibiofibular ligament** (see *Grant's Atlas of Anatomy*, Fig. 4-89). This strong band closes the posterior angle between the tibia and the fibula.

The posterior tibiofibular ligament is much stronger than the anterior tibiofibular ligament. In severe ankle injuries, the posterior tibiofibular ligament may avulse the posteroinferior part of the tibia. In these cases the fracture enters the ankle joint. If, in addition, the medial and lateral malleoli are fractured, the injury is sometimes referred to as a **trimalleolar fracture** (*i.e.*, a fracture of both the malleoli and the posterior part of the inferior border of the tibia).

STABILITY OF THE INFERIOR TIBIOFIBULAR JOINT

This articulation forms a ***strong union*** between the distal ends of the tibia and fibula; much of the strength of the ankle joint is dependent on this union.

MOVEMENT OF THE INFERIOR TIBIOFIBULAR JOINT

Slight movement of the distal tibiofibular joint occurs to accommodate the talus during dorsiflexion of the foot at the ankle joint.

BLOOD SUPPLY OF THE INFERIOR TIBIOFIBULAR JOINT (Figs. 4-70 and 4-83)

The articular arteries are derived from the *perforating branch of the* ***peroneal artery*** *and the medial malleolar branches of the anterior and posterior* ***tibial arteries***.

NERVE SUPPLY OF THE INFERIOR TIBIOFIBULAR JOINT (Figs. 4-70 and 4-81)

The articular nerves are derived from the ***deep peroneal, tibial, and saphenous nerves***.

THE ANKLE JOINT

The ankle joint or talocrural joint is a hinge type of synovial joint. It is located between the inferior ends of the tibia and fibula and the superior part of the talus. The ankle joint (talocrural joint) can be felt between the tendons on the anterior surface of the ankle as a slight depression, about 1 cm proximal to the tip of the medial malleolus (Figs. 4-80 and 4-121).

ARTICULAR SURFACES OF THE ANKLE JOINT (Figs. 4-118 to 4-122)

The inferior ends of the tibia and fibula form a deep socket or *box-like* ***mortise*** into which the superior pulley-shaped **trochlea** (L. pulley) of the talus fits (Fig. 4-121).

The ***fibula*** *has an articular facet on its lateral malleolus* which faces medially and articulates with the facet on the lateral surface of the talus (Fig. 4-120).

The ***tibia*** *articulates with the talus in two places* (Figs. 4-120 and 4-121). Its inferior surface forms the roof of the mortise which is wider anteriorly than posteriorly and slightly concave from anterior to posterior. The lateral surface of its medial malleolus also articulates with the talus (Fig. 4-120).

*The **talus** has three articular facets which articulate with the inferior surface of the tibia and the malleoli* (Fig. 4-120). The superior articular or trochlear surface of the talus is often called the *trochlea* (L. pulley) because of its pulley-like shape. It is wider anteriorly than posteriorly, convex from anterior to posterior, and slightly concave from side to side (Fig. 4-120).

MOVEMENTS OF THE ANKLE JOINT (Fig. 4-65, *A* and *B* and Table 4-12)

The ankle joint is uniaxial; its movements are **dorsiflexion and plantarflexion**. During dorsiflexion the trochlea rocks posteriorly in its mortise (Fig. 4-121), and the malleoli tend to be forced apart because the superior articular surface of the talus is wider anteriorly than posteriorly. The lateral malleolus moves laterally during dorsiflexion as the tibia and fibula are forced slightly apart. This separation of the malleoli requires some movement of the superior tibiofibular joint (Fig. 4-116).

When the foot is plantarflexed, some rotation, abduction, and adduction of the ankle joint are possible.

The range of plantarflexion is greater than that of dorsiflexion, but there is considerable variation in these movements. Some people cannot dorsiflex their feet beyond a right angle, especially women who wear high heels much of the time. These persons may feel *calf pain* when they switch from high-heeled to low-heeled shoes owing to straining of the triceps surae muscle (Fig. 4-72).

Table 4-12
Muscles Producing Movements of the Ankle Joint[1]

Dorsiflexion	Plantarflexion
Tibialis anterior	**Triceps surae**
Extension digitorum longus	**Gastrocnemius**
	Soleus
Extensor hallucis longus	Plantaris
	Tibialis posterior
Peroneus tertius	Flexor hallucis longus
	Flexor digitorum longus

[1] **Bold face** indicates muscles chiefly responsible for the movement; the other muscles assist with the movement.

THE ARTICULAR CAPSULE OF THE ANKLE JOINT (Figs. 4-118 and 4-119)

The fibrous capsule of the ankle joint is thin anteriorly and posteriorly, but *it is supported on each side by strong collateral ligaments.*

The fibrous capsule is attached superiorly to the borders of the articular surfaces of the tibia and malleoli, and inferiorly to the talus close to the superior articular surface (Fig. 4-120), except anteroinferiorly, where it is attached to the dorsum of the neck of the talus (Fig. 4-122).

*The fibrous capsule is strengthened medially and laterally by two strong **collateral ligaments*** (the deltoid and lateral ligaments).

The Deltoid Ligament (Fig. 4-122). The strong deltoid ligament ***attaches the medial malleolus to the tarsus*** (tarsal bones of the instep). The apex of the deltoid ligament (medial ligament) is attached to the margins and tip of the *medial malleolus,* and its broad base fans out to attach to *three tarsal bones* (talus, navicular, and calcaneus).

The deltoid ligament consists of three parts which are named according to their bony attachments (Fig. 4-122): **tibionavicular, posterior tibiotalar and tibiocalcanean ligaments.**

The deltoid ligament strengthens the ankle joint and holds the calcaneus and navicular bones against the talus. In addition, it helps to maintain the medial side of the foot and the medial longitudinal arch (Fig. 4-130).

A bony outgrowth commonly develops on the anterior aspect of the distal end of the tibia and on the superior surface of the neck of the talus in persons who repeatedly kick a football or soccer ball. The plantarflexion associated with kicking a ball pulls on the attachments of the tibiotalar ligament (Fig. 4-119), inducing a characteristic bony outgrowth. The condition is called a **foot-baller's ankle.**

Medial dislocations of the ankle joint are not common owing to the strength of the deltoid ligament. In **fracture-dislocations of the ankle joint**, the distal end of the tibia and/or the fibula is usually fractured. The common **Pott's fracture** occurs when the foot is forcibly everted (Fig. 4-137). This pulls on the extremely strong deltoid ligament, often tearing off the medial malleolus. The talus then moves laterally, shearing off the lateral malleolus or, more commonly, breaking the fibula superior to the inferior tibiofibular joint. If the tibia is carried anteriorly, the posterior margin of the distal end of the tibia is also sheared off by the talus.

The Lateral Ligament (Figs. 4-118 and 4-119). The tripartite lateral ligament ***attaches the lateral malleolus to the talus and calcaneus***. It is not nearly so strong as the deltoid ligament. The lateral ligament consists of three distinct parts (*anterior and posterior talofibular ligaments and the calcaneofibular ligaments*). You will hear these parts of the lateral ligament referred to as the *lateral ligaments.*

The anterior talofibular ligament (Figs. 4-118 and 4-119) extends anteromedially from the lateral malleolus to the neck of the talus. It is not very strong.

The posterior talofibular ligament (Fig. 4-119) is thick and fairly strong. It runs horizontally medially and slightly posteriorly from the malleolar fossa to the lateral tubercle of the posterior process of the talus.

The calcaneofibular ligament (Figs. 4-118 and 4-119) is a round cord that passes posteroinferiorly from the tip of the lateral malleolus to the lateral surface of the calcaneus. The calcaneofibular ligament is crossed superficially by the tendons of the peroneus longus and brevis muscles.

The ankle is the most frequently injured major joint in the body. The lateral ligament of the ankle joint is the one most frequently injured. One or more of its three parts may be stretched and/or torn (Figs. 4-136 and 4-137).

A sprained ankle results from twisting of the weight-bearing foot and is nearly always an **inversion injury** (*i.e.*, the foot is forcefully inverted). A typical history is as follows: (1) The person steps on an uneven surface and falls. (2) This stretches most of the fibers of the lateral ligaments and tears some of them. (3) The ankle soon becomes painful and localized swelling and tenderness appear anteroinferior to the tip of the lateral malleolus.

In severe ankle sprains some fibers of the lateral ligament are torn, either partially or completely, resulting in instability of the ankle joint. The foot has to be placed in the everted position until the torn ligaments heal. If not, a fracture-dislocation of the ankle may occur if the ankle is severely inverted again.

The synovial capsule of the ankle joint (Fig. 4-119) lines the fibrous capsule and projects superiorly between the tibia and fibula for a short distance.

The synovial cavity of the ankle joint is somewhat superficial on each side of the tendo calcaneus (Fig. 4-119). Hence, when the ankle joint is inflamed (*e.g.*, owing to **arthritis**), the synovial fluid may increase, causing swelling in these locations.

STABILITY OF THE ANKLE JOINT (Fig. 4-118)

The ankle joint is very strong during dorsiflexion because it is supported by powerful ligaments and is crossed by several tendons that are tightly bound down by thickenings of the deep fascia called retinacula. Its stability is also greatest in dorsiflexion because in this position the trochlea of the talus fills the **mortise formed by the malleoli** (Fig. 4-121). The malleoli grip the talus tightly as it rocks anteriorly and posteriorly during movements of the ankle joint.

The grip of the malleoli on the trochlea of the talus is strongest during dorsiflexion of the foot because this movement forces the anterior part of the trochlea posteriorly, spreading the tibia and fibula slightly apart. This spreading is limited by the strong interosseous ligament and by the anterior and posterior tibiofibular ligaments that unite the leg bones (Figs. 4-57*B*, 4-118, and 4-119).

The ankle joint is relatively unstable during plantarflexion. During this movement the trochlea of the talus moves anteriorly in the mortise, causing the malleoli to come together. However, the grip of the malleoli on the trochlea is not so strong as during dorsiflexion. Some side movement can be demonstrated in full plantarflexion of the foot. The wedge-shaped form of the trochlea (Figs. 4-58 and 4-60) assists the ligaments in preventing posterior displacement of the foot when jumping and stopping suddenly.

Although the ankle joint is unstable in plantarflexion, appropriate training and conditioning strengthens the joint in this position (*e.g.*, ballet dancers and persons who wear shoes with high heels).

As inversion of the foot tends to occur with active plantarflexion, **inversion injuries of the ankle tend to occur more often than eversion injuries when the foot is in plan-**

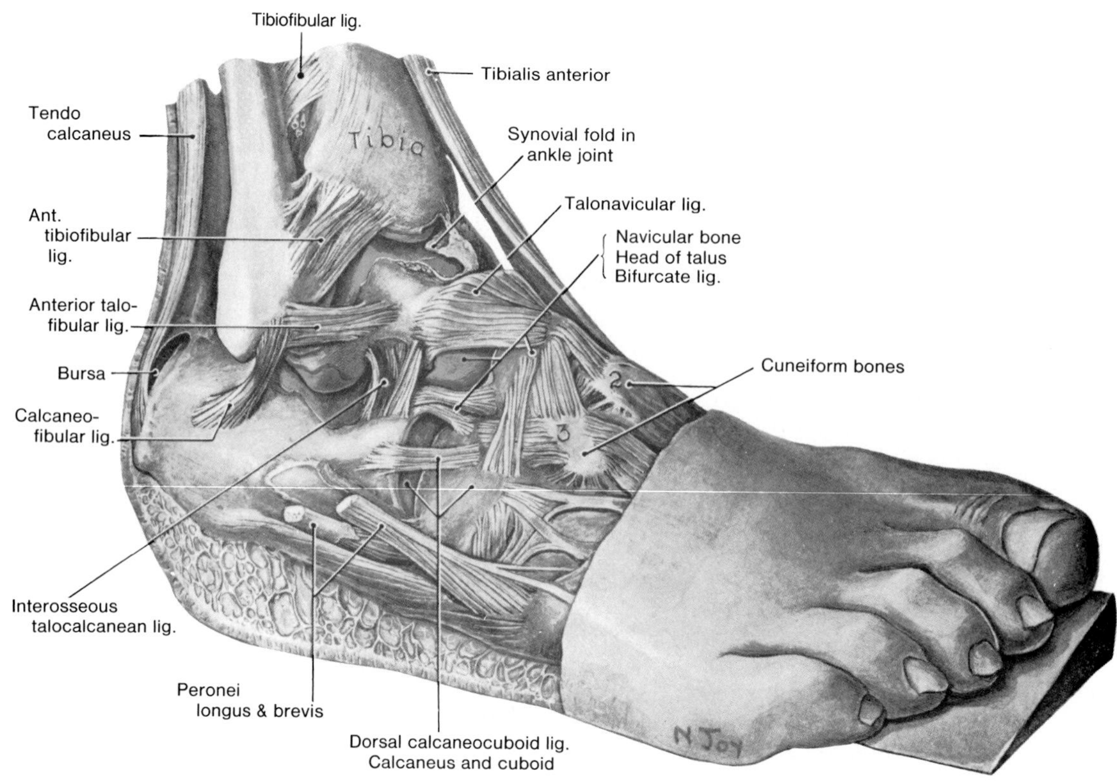

Figure 4-118. Drawing of a dissection of a lateral view of the right ankle joint and the joints of inversion and eversion. The foot has been inverted in order to demonstrate the articular areas (*yellow*) and the ligaments that become taut during inversion of the foot. The joints of inversion and eversion are (1) the subtalar (posterior talocalcanean) joint, (2) the talocalcaneonavicular (combined anterior talocalcanean and talonavicular) joint, and (3) the transverse tarsal (combined calcaneocuboid and talonavicular) joint. Note that the talonavicular joint is involved twice. The articular areas colored *yellow* are: (1) the posterior talar facet of the calcaneus, (2) the anterior surface of the calcaneus, (3) the head of the talus, and (4) the superior and lateral parts of the trochlea of the talus. The anterior talofibular and dorsal calcaneocuboid ligaments are weak and easily torn, giving rise to a sprained ankle.

tarflexion. When one or more of the three components of the lateral ligament are stretched or torn, the ankle becomes very unstable.

Fracture-dislocation of the ankle joint may occur in severe injuries in which the tip of the lateral malleolus is avulsed (Fig. 4-136). This injury often occurs when the foot is fixed against some object (*e.g.*, a large stone) and is thrown into an inverted position. The body weight is then violently transferred to the lateral ligament of the ankle. *Usually the* ***calcaneofibular ligament*** *tears, partially or completely, often along with the anterior talofibular ligament.* As the latter ligament is fused with the fibrous capsule of the ankle joint, this part of the capsule may also be torn.

The peroneus brevis muscle tends to prevent overinversion of the foot (Figs. 4-67 and 4-71), thereby aiding the lateral ligament in preventing inversion injuries of the ankle joint.

Violent inversion of the foot may result in avulsion of the tuberosity of the fifth metatarsal bone, into which the tendon of peroneus brevis muscle inserts. For this reason, radiologists routinely ensure that this bone is visible in radiographs of persons with ankle injuries so they may look for avulsion of the tuberosity of the fifth metatarsal.

BLOOD SUPPLY OF THE ANKLE JOINT (Figs. 4-70 and 4-95)

The articular arteries are derived from the *malleolar branches of the* ***peroneal and anterior and posterior tibial arteries.***

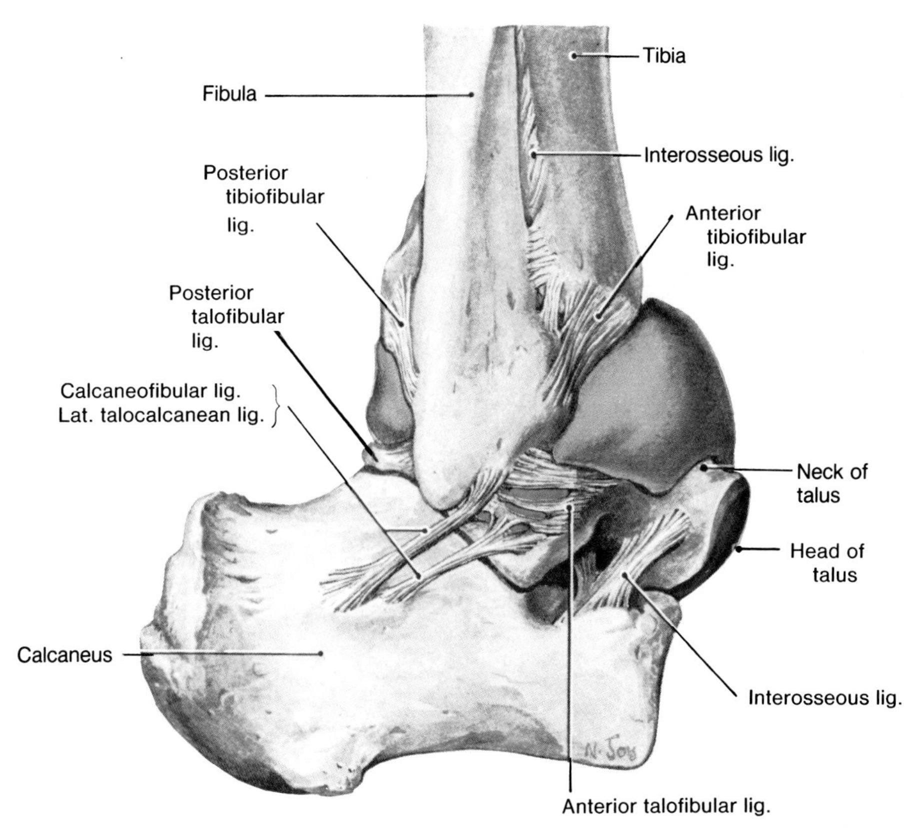

Figure 4-119. Drawing of a dissection of a lateral view of a distended right ankle joint. Observe the anterior extension of the synovial cavity over the neck of the talus. Posteriorly and laterally, note the closeness of the synovial cavity of the ankle joint to the subtalar (posterior talocalcanean) joint. Note that the lateral ligament consists of three separate ligaments: anterior talofibular ligament, calcaneofibular ligament, and posterior talofibular ligament. For this reason, you may hear the lateral ligament referred to as the lateral ligaments.

NERVE SUPPLY OF THE ANKLE JOINT (Fig. 4-81)

The articular nerves are derived from the ***tibial nerve*** and the ***deep peroneal nerve***, a division of the common peroneal nerve.

JOINTS OF THE FOOT

There are many joints in the foot involving the tarsal and metatarsal bones and the phalanges.

The important intertarsal joints are the subtalar, talocalcaneonavicular, and calcaneocuboid joints. The last two constitute the *transverse tarsal joint* (midtarsal joint) shown in Figure 4-123.

The other intertarsal joints are relatively small and are so tightly joined by ligaments that only slight movement occurs between them. All the foot bones are united by **dorsal and plantar ligaments** (Figs. 4-118 and 4-125).

THE SUBTALAR JOINT (Figs. 4-118, 4-123, and 4-124)

The subtalar joint (talocalcanean joint) is distal to the ankle joint, where the talus rests on and articulates with the calcaneus.

The subtalar joint is a gliding type of synovial joint between the inferior surface of the body of the talus and the superior surface of the calcaneus. It is surrounded by an articular capsule which is attached near the margins of the articular facets.

The fibrous capsule of the subtalar joint is supported by medial, lateral, and posterior **talocalcaneal ligaments** (Fig. 4-118). In addition, the subtalar joint is supported anteriorly by the *interosseous talocalcanean ligament* (Fig. 4-118).

Movements of the Subtalar Joint (Fig. 4-65). *Inversion* and *eversion* of the foot occur at this joint. The muscles producing these movements are given in Table 4-13. Movements of the subtalar joint are closely

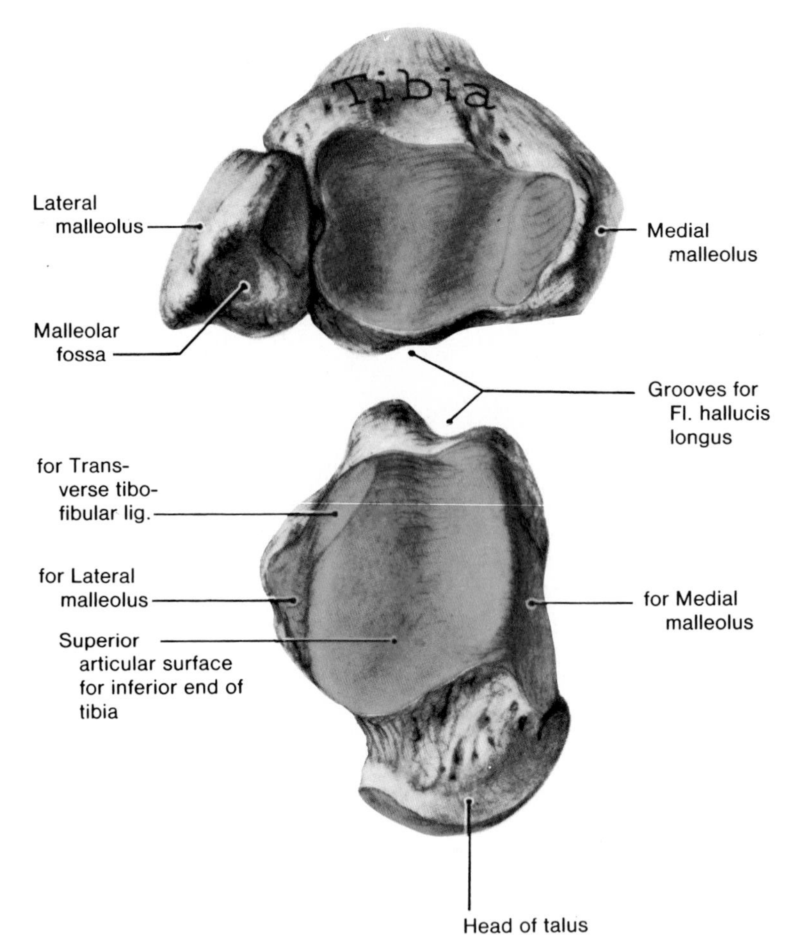

Figure 4-120. Drawings illustrating the articular surfaces of the right ankle joint. Note that the superior articular surface of the trochlea of the talus is broader anteriorly than posteriorly. Hence the tibial and fibular malleoli which form a mortise for the talus (Fig. 4-121) are forced apart in dorsiflexion. The inferior tibiofibular joint, on which the brunt of the strain then falls, gives resilience to the ankle joint.

associated with those at the talocalcaneonavicular and calcaneocuboid joints.

Table 4-13
Muscles Producing Movements of the Intertarsal Joints[1]

Inversion[2]	Eversion[2]	Plantarflexion
Tibialis anterior	**Peroneus longus**	**Peroneus longus**
Tibialis posterior	**Peroneus brevis**	**Tibialis posterior**
	Peroneus tertius	**Abductor hallucis**
		Abductor digiti minimi
		Flexor digitorum brevis
		Peroneus brevis

[1] **Bold face** indicates muscles chiefly responsible for the movement; the other muscles assist with the movement.

[2] Inversion and eversion occur mainly at the subtalar and transverse tarsal joints (Fig. 4-123).

THE TALOCALCANEONAVICULAR JOINT (Figs. 4-118 and 4-122 to 4-124)

This joint is located where the head of the talus articulates with the head of the posterior surface of the navicular bone, the superior surface of the plantar calcaneonavicular ligament, the sustentaculum tali, and the articular surface of the calcaneus.

The head of the talus has three facets, one for the navicular and two for the calcaneus. All these articular surfaces are surrounded by a single articular capsule which blends with the interosseous talocalcaneal ligament posteriorly (Fig. 4-118).

The talocalcaneonavicular joint is reinforced dorsally by the dorsal talonavicular ligament, a broad band connecting the neck of the talus and the dorsal surface of the navicular bone (Fig. 4-118).

The plantar calcaneonavicular ligament, or spring ligament, is a triangular sheet that extends from the *sustenaculum tali* to the posteroinferior surface of the *navicular bone* (Fig. 4-122). The spring

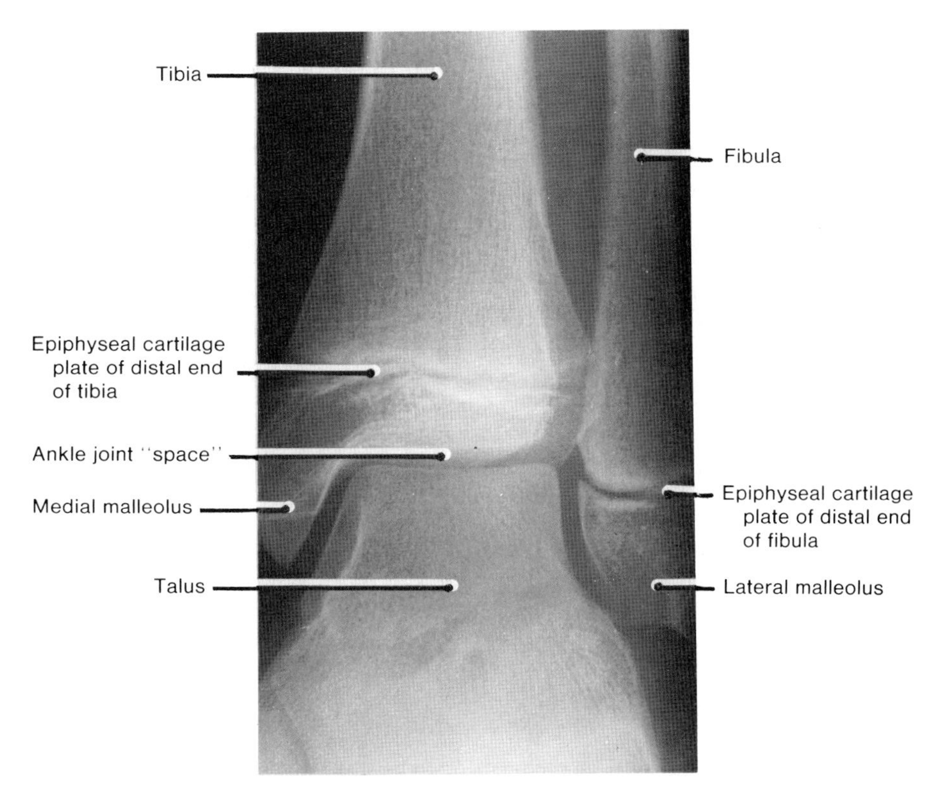

Figure 4-121. A slightly oblique frontal radiograph of the ankle joint of a 14-year-old boy showing how the trochlea of the body of the talus (Figs. 4-60 and 4-61) fits into the mortise formed by the medial and lateral malleoli. After fracture and/or dislocation, it is essential to reduce the displaced fragments so that the normal mortise is regained. The radiolucent lines crossing the tibia and fibula near their inferior ends are the epiphyseal cartilage plates separating the primary ossification centers (diaphyses) from the secondary ossification centers (epiphyses). The apparent "spaces" between the talus and the malleoli represent the radiolucent articular cartilages.

ligament blends with the deltoid ligament medially and forms part of the socket for the head of the talus.

The plantar calcaneonavicular or spring ligament plays an important role in maintaining the longitudinal arch of the foot (Fig. 4-130).

THE CALCANEOCUBOID JOINT (Figs. 4-118, 4-123, and 4-124)

This joint is between the anterior surface of the **calcaneus** and the posterior surface of the **cuboid.** The capsule of the calcaneocuboid joint is strengthened by the dorsal calcaneocuboid ligament (Fig. 4-118) and plantar calcaneocuboid ligaments (Figs. 4-122 and 4-124). The calcaneocuboid joint is also supported by the long plantar ligament (Fig. 4-126).

The long plantar ligament passes from the plantar surface of the calcaneus, including the anterior tubercle, to both lips of the groove on the cuboid bone (Fig. 4-126). Some of its fibers extend to the bases of the second, third, and fourth metatarsal bones, thereby forming a tunnel for the tendon of the peroneus longus muscle. This tendon passes through the groove in the cuboid bone to insert into the base of the first metatarsal and the adjoining part of the medial cuneiform bone (Fig. 4-126). *The long plantar ligament is important in maintaining the arches of the foot* (Fig. 4-130).

The plantar calcaneocuboid ligament (short plantar ligament) is deep to the long plantar ligament (Figs. 4-125 and 4-126). It extends from the anterior aspect of the inferior surface of the calcaneus to the inferior surface of the cuboid.

THE TRANSVERSE TARSAL JOINT (Figs. 4-60, 4-118, and 4-123)

The talonavicular and calcaneocuboid joints are separated from each other, but together they **constitute the transverse tarsal joint** (midtarsal joint). They extend right across the tarsus and lie in almost the same transverse plane.

Movements of the Transverse Tarsal Joint (Fig. 4-65, *C* and *D* and Table 4-13). Movements occurring at the transverse tarsal joint produce inversion and eversion of the foot.

The joints of inversion and eversion are (1) the *subtalar joint;* (2) the *talocalcaneonavicular joint;* and (3) the *transverse tarsal joint.*

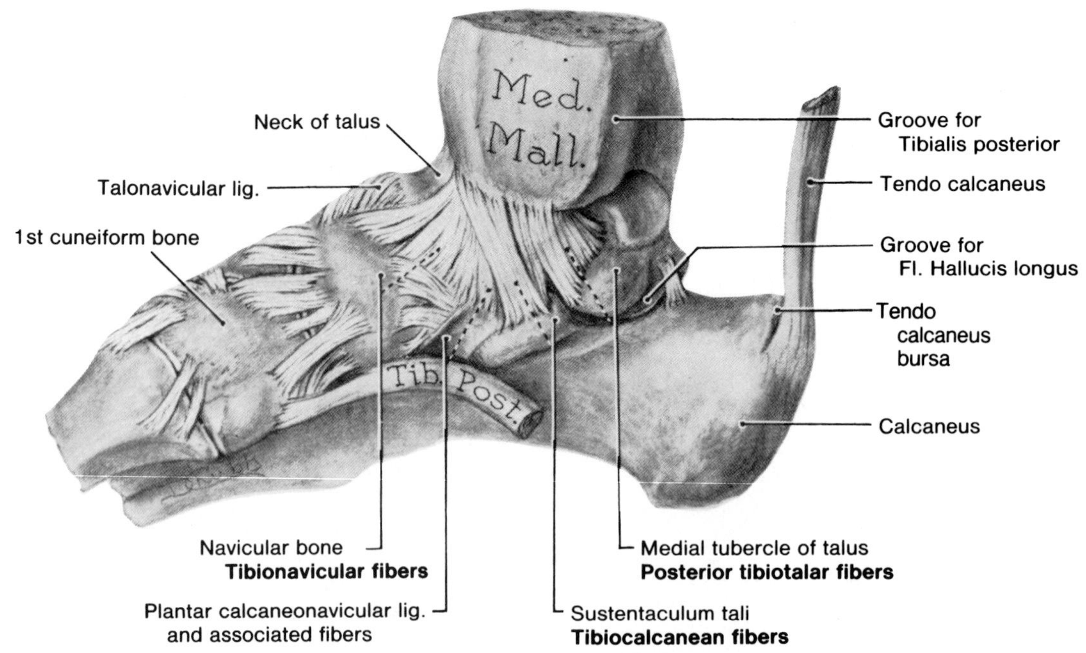

Figure 4-122. Drawing of a dissection of the right ankle joint and tarsal joints from the medial side. Observe the parts of the extremely strong deltoid ligament (*bold face type*); another part (the anterior tibiotalar ligament) is not visible here. Note that the deltoid ligament radiates from the inferior border of the medial malleolus and inserts into three foot bones, and that the deltoid ligament is attached to the medial side of the plantar calcaneonavicular ligament. The deltoid ligament is so strong that instead of rupturing when the foot is strongly everted, it often avulses the tip of the medial malleolus (Fig. 4-137).

Inversion and eversion of the foot occur mainly at the subtalar and transverse tarsal joints. The muscles producing these movements are listed in Table 4-13.

In inversion the foot is adducted and directed so that its medial border is raised and its lateral border is depressed. In other words, inversion directs the sole of the foot toward the median plane of the body (Fig. 4-84).

In eversion the foot is abducted and directed so that the lateral border is raised and the medial border lowered. In other words, eversion directs the sole of the foot away from the median plane of the body.

Inversion and eversion of the feet commonly occur during walking on cobblestones or rough ground in adjusting the feet to the stones or depressions.

The strong ***deltoid ligament*** *tends to prevent overeversion of the foot; the weaker* ***lateral ligament*** *(with the assistance of the peroneus longus and brevis muscles) tends to prevent overinversion of the foot.* Consequently, inversion injuries of the foot are much more common than eversion injuries.

THE TARSOMETATARSAL JOINTS (Figs. 4-118 and 4-123 to 4-127)

These articulations are the plane type of synovial joint which permit only gliding or sliding movements. In Figures 4-123 and 4-127, note that the four anterior tarsal bones articulate with the bases of the metatarsal bones. The metatarsal bones are firmly attached to the tarsal bones by dorsal, plantar, and interosseous ligaments (Figs. 4-125 and 4-126). *There are three separate tarsometatarsal joint cavities.*

The medial tarsometatarsal joint occurs between the medial cuneiform bone and the base of the first metatarsal bone (Figs. 4-123 and 4-125 to 4-127). The medial tarsometatarsal joint has more range of movement than the other two joints.

The intermediate tarsometatarsal joint is between all three cuneiform bones and the second and third metatarsal bones. In Figure 4-123, observe that the base of the second metatarsal bone fits into a socket formed by the cuneiform bones; thus *the intermediate tarsometatarsal joint is the strongest of the three tarsometatarsal joints.* As the second metartarsal bone is firmly attached to the cuneiform bones, it has little independent movement.

The cavity of the intermediate tarsometatarsal joint is continuous with that between the two medial cuneiform bones and through it with the synovial cavity of the cuneonavicular joint (Fig. 4-123).

The lateral tarsometatarsal joint occurs between the cuboid and the fourth and fifth metatarsal bones (Figs. 4-123 and 4-125 to 4-127). There is more movement permitted at this joint than at the inter-

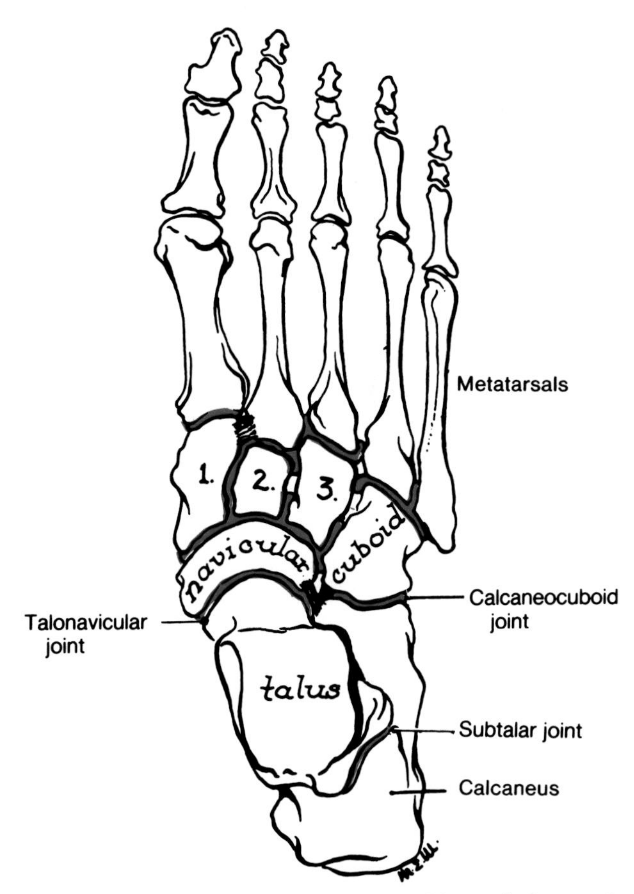

Figure 4-123. Drawing of the dorsal (superior) aspect of the bones of the right foot, showing the six separate joint cavities (*red*). Also see Figure 4-58. The **transverse tarsal joint** is the articular plane that extends from side to side across the foot; it is composed of the *talonavicular joint* medially and the *calcaneocuboid joint* laterally. Although anatomically separate, these joints act together during movements of the foot.

mediate tarsometatarsal joint, but not so much as at the medial tarsometatarsal joint.

Because the second metatarsal bone has little movement at the second tarsometatarsal joint, it is particularly liable to fracture when sudden, unaccustomed stresses are applied to the distal part of the foot. For example, when a person who is "out of condition" begins to participate in strenuous exercise programs, long walks, track and field activities, or ballet dancing, a **stress fracture** may occur in one of the weight-bearing bones (Fig. 4-62). Stress fractures of the metatarsals are sometimes referred to as "**march fractures**" or "fatigue fractures."

Resections of part of the foot (**amputations of the foot**) may be necessary following severe trauma (*e.g.*, crush injuries) or in cases of cancer, arteriosclerosis, or peripheral vascular diseases of the foot. In one type, *Lisfranc's amputation,* the foot is resected through the tarsometatarsal joints (Fig. 4-127). In another, *Syme's amputation,* the foot is resected through the transverse tarsal joint (Fig. 4-123).

THE INTERMETATARSAL JOINTS (Figs. 4-123, 4-125, and 4-126)

These articulations between the bases of the metatarsal bones are the **plane type of synovial joint** which permit a slight gliding movement. Their joint cavities are extensions of the tarsometatarsal joints (Fig. 4-123).

The bases of the second to fifth metatarsal bones are very firmly bound together by dorsal, plantar, and interosseous ligaments (Figs. 4-125 and 4-126).

The **deep transverse metatarsal ligament** connects the heads of the metatarsal bones. The deep transverse metatarsal ligament, along with the interosseous ligaments, helps to maintain the transverse arch of the foot. Owing to the tight binding of the bases of the metatarsal bones together (Figs. 4-125 and 4-126), little individual movement of the metatarsal bones is possible.

THE METATARSOPHALANGEAL JOINTS (Fig. 4-123)

These articulations between the heads of the metatarsal bones and the bases of the proximal phalanges are the knuckle-like **condyloid type of synovial joint.** These joints permit flexion, extension, and some abduction, adduction, and circumduction (Table 4-14).

The first metatarsophalangeal joint is by far the largest articulation owing to the size of the head of the first metatarsal bone, and to the presence of the sesamoid bones in the two tendons of the flexor hallucis brevis muscle (Fig. 4-92).

An ***articular capsule*** *surrounds each joint* which is attached near the margins of the articular surfaces of the involved bones. In Figures 4-58 and 4-59, note that the articular surfaces pass well onto the dorsal and plantar surfaces of the metatarsal bones. The articular surfaces of the first metatarsophalangeal joint is particularly large and is related to the dorsiflexion of the great toe during walking.

The fibrous capsules *of the metatarsophalangeal joints are strengthened on each side by thick collateral ligaments. The plantar part of the capsule is greatly thickened to form the* **plantar ligament** (Figs. 4-90 and 4-92). This fibrocartilaginous plate is firmly attached to the proximal border of the phalanx and forms part of the socket for the head of the first

Table 4-14
Muscles Producing Movements of the Metatarsophalangeal Joints[1]

Flexion	Extension	Abduction	Adduction
Flexor digitorum longus	**Extensor hallucis longus**	**Abductor hallucis**	**Adductor hallucis**
Lumbricales	**Extensor digitorum**	**Abductor digiti minimi**	**Plantar interossei**
Interossei	**Extensor digitorum brevis**	**Dorsal interossei**	
Flexor hallucis brevis			
Flexor digitorum brevis			
Flexor hallucis longus			
Flexor digiti minimi brevis			
Flexor digitorum longus			
Quadratus plantae			

[1] **Bold face** indicates muscles chiefly responsible for the movement; the other muscles assist with the movement.

metatarsal. The margins of the plantar ligament give attachment to the fibrous flexor sheath, to slips of the plantar aponeurosis, and to the deep transverse metatarsal ligaments (Figs. 4-85 and 4-90).

The first metatarsophalangeal joint may become enlarged and deformed with permanent lateral displacement of the great toe (L. *hallux*). This condition, known as **hallux valgus** (Fig. 4-128), is common in persons who wear pointed shoes. They are unable to move their great toe away from their second toe because the sesamoids under the head of the first metatarsal bone (Fig. 4-92) are usually displaced and lie in the space between the heads of the first and second metatarsal bones. When hallux valgus develops during adolescence, it is usually progessive. As the prognosis is poor, the deformity is usually corrected surgically.

Gout, a metabolic disorder, is *characterized by urate deposits in connective tissue*, including cartilage and bone. Gout commonly affects the first metatarsophalangeal joint which becomes swollen and painful, a condition referred to as *gouty arthritis.*

Degenerative joint disease (osteoarthritis) is also common in the metatarsophalangeal joint of the great toe. Often it is the first joint to be affected. When this painful condition is present without deformity, it is known as **hallux rigidus.**

THE INTERPHALANGEAL JOINTS (Fig. 4-123 and Table 4-15).

The interphalangeal joints are between the head of one phalanx and the base of the one distal to it. They are the **hinge type of synovial joint** permitting only flexion and extension. In most people the lateral four toes are, to a varying degree, partially flexed at the interphalangeal joints at all times.

Table 4-15
Muscles Producing Movements of the Interphalangeal Joints[1]

Flexion	Extension
Flexor hallucis longus	**Extensor hallucis longus**
Flexor digitorum longus	**Extensor digitorum longus**
Flexor digitorum brevis	**Extensor digitorum brevis**
Quadratus plantae	

[1] **Bold face** indicates muscles chiefly responsible for the movement; the other muscle assists with the movement.

Hammer toe is a common deformity in which the proximal phalanx is permanently dorsiflexed at the metatarsophalangeal joint, and the middle phalanx is plantarflexed at the interphalangeal joint. The distal phalanx is also flexed or extended, giving the toe (usually the second) a **hammer-like appearance**. This deformity may result from weakness of the lumbrical and interosseus muscles, which flex the metatarsophalangeal joints and extend the interphalangeal joints (Table 4–9).

THE ARCHES OF THE FOOT

The bones of the foot are arranged in longitudinal and transverse arches that are designed as **shock absorbers** for supporting the weight of the body and for propelling it during movement.

The design of the foot makes it adaptable to surface and weight changes. The resilient arches of the foot

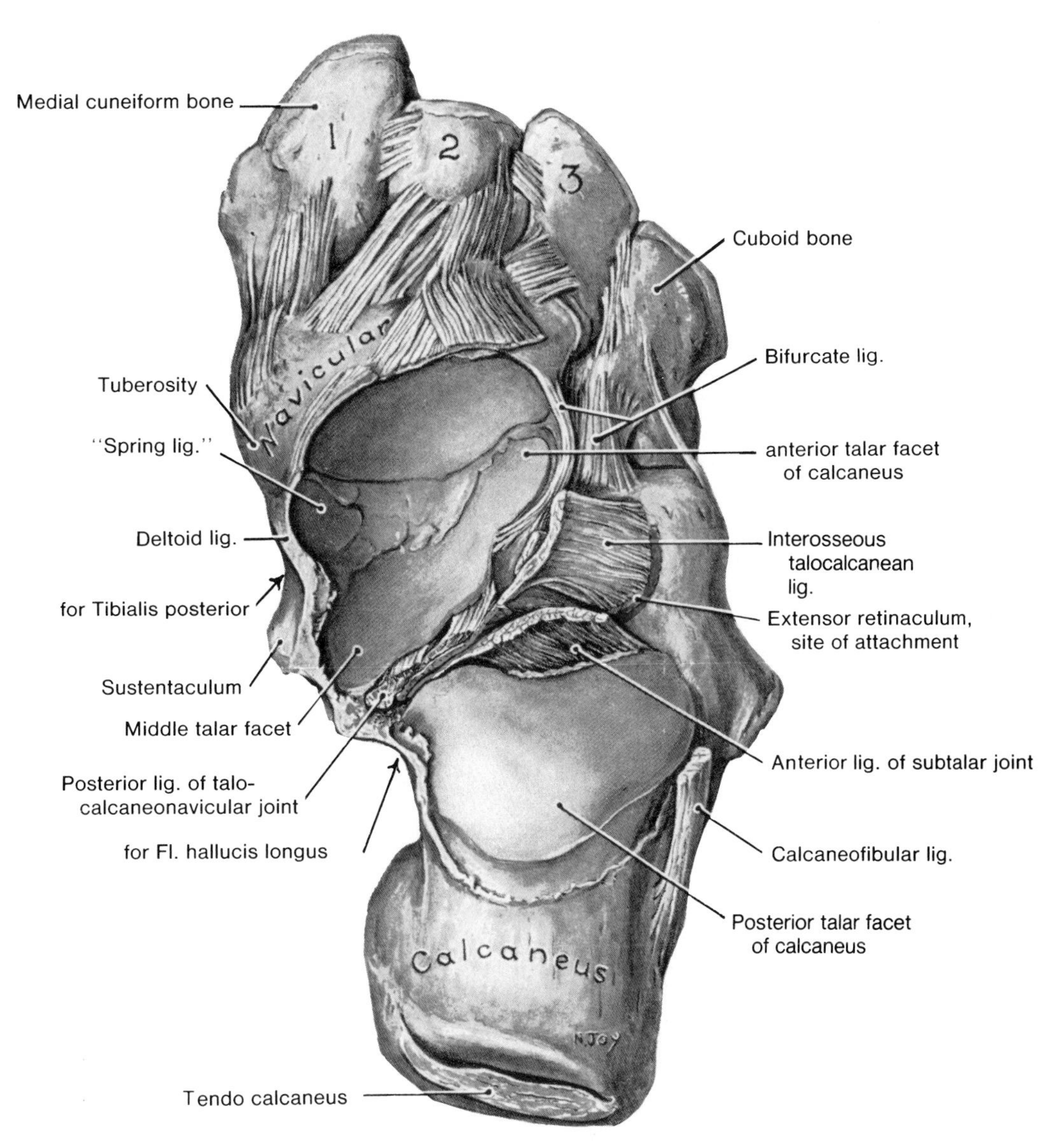

Figure 4-124. Drawing of a dissection of the right foot showing the joints of inversion and eversion. This specimen was prepared by sawing through the body of the talus and, after discarding it, nibbling away the neck and head of the talus. At the wide lateral end of the tarsal sinus, note the strong interosseous talocalcanean ligament and the site of attachment of the extensor retinaculum (*blue*) which extends medially between the posterior ligament of the anterior talocalcanean joint and the anterior ligament of the subtalar joint. Observe that the subtalar joint has a synovial cavity to itself, whereas the talonavicular and anterior talocalcanean joints share a common synovial cavity; hence the collective name, talocalcaneonavicular joint. Note that the angular space between the navicular bone and the middle talar facet on the sustenaculum tali is bridged by the plantar calcaneonavicular or spring ligament.

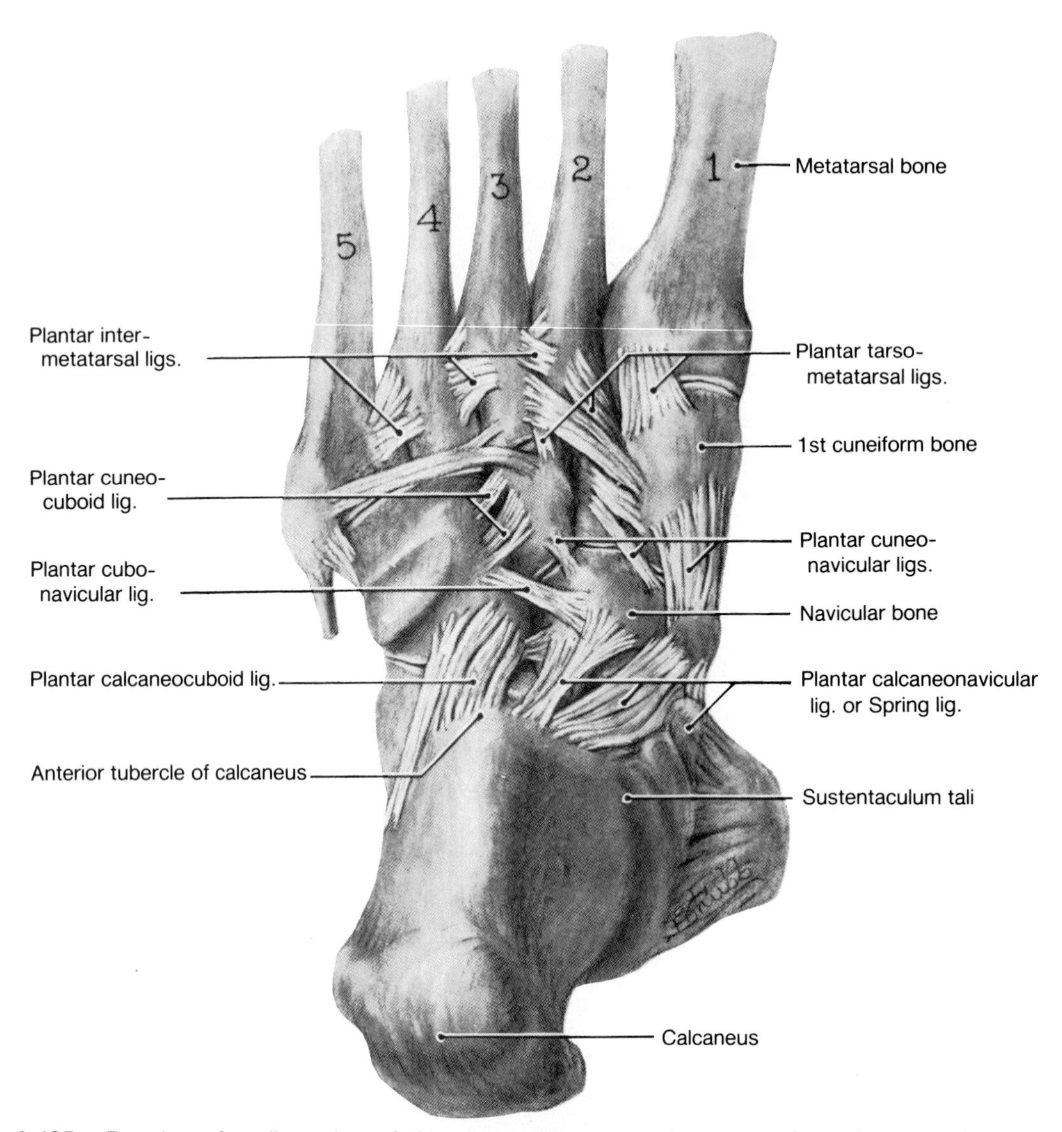

Figure 4-125. Drawing of a dissection of the plantar ligaments of the right foot. Observe that the plantar calcaneocuboid (short plantar) ligament and the plantar calcaneonavicular (spring) ligament are the inferior ligaments of the transverse tarsal joint (Fig. 4-123). Having a common purpose, they have a common direction. Note that the ligaments in the anterior part of the foot diverge posteriorly from each side of the long axis of the third metatarsal and third cuneiform bones. Hence, a posterior thrust to the first metatarsal (*e.g.*, when rising on the great toe in walking) is transmitted directly to the navicular and talus by the first cuneiform, and indirectly by the second metatarsal and second cuneiform and also by the third metatarsal and third cuneiform. A posterior thrust to the fourth and fifth metatarsals is transmitted directly to the cuboid and calcaneus. These four bones of the lateral longitudinal arch of the foot (Fig. 4-130*B*) are not displaced posteriorly because of the adjoining ligaments.

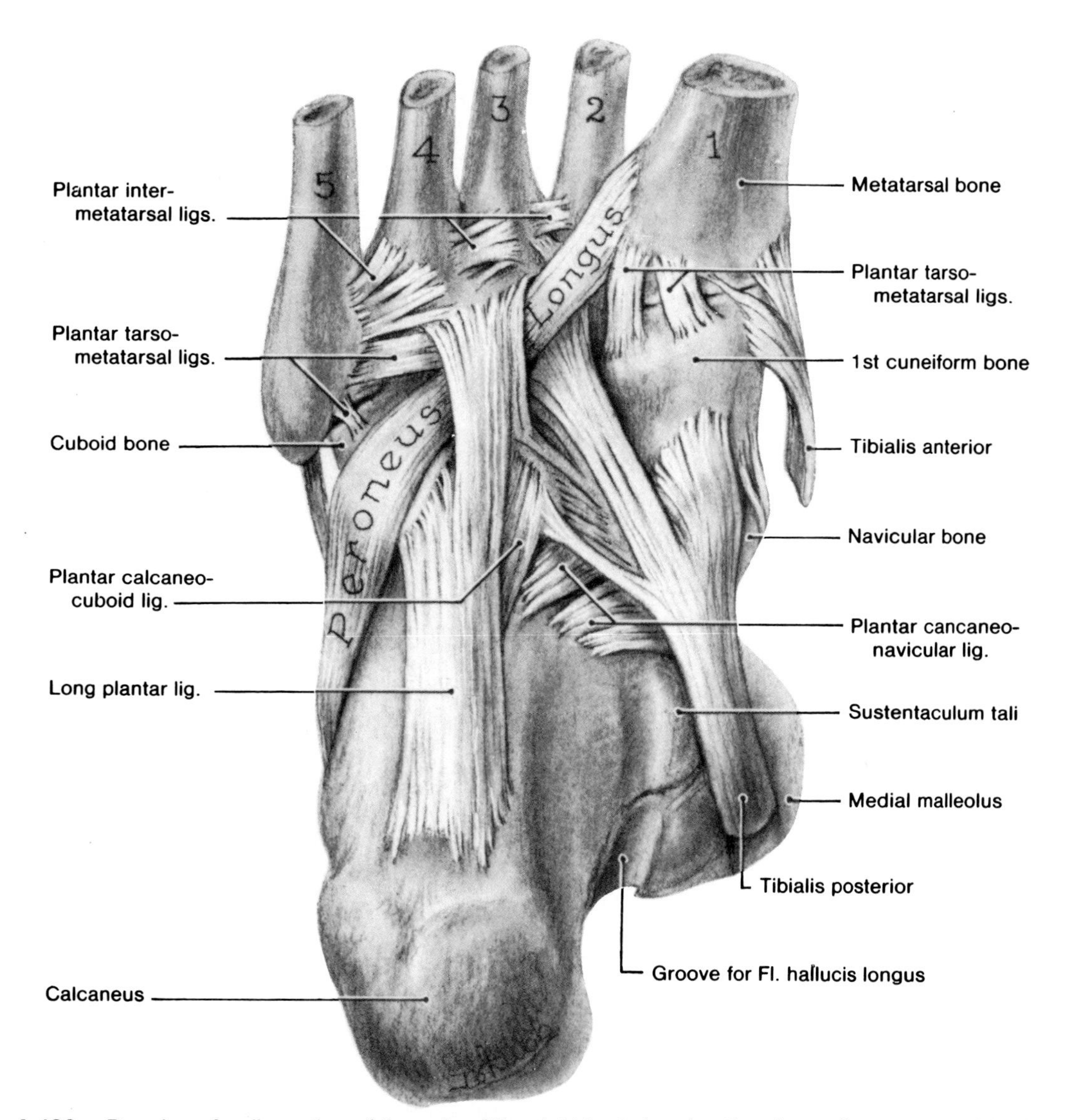

Figure 4-126. Drawing of a dissection of the sole of the right foot showing the plantar ligaments and the insertion of three long tendons (peroneus longus, tibialis anterior, and tibialis posterior). Observe the tendon of the peroneus longus crossing the sole of the foot in a groove in the cuboid bone (Fig. 4-59). Note that it is bridged by some fibers of the long plantar ligament. Note that the tendon of the peroneus longus is inserted into the base of the first metatarsal bone. Recall that it also inserts into the adjoining part of the medial cuneiform bone (Fig. 4-59). Observe that slips of the tendon of the tibialis posterior muscle extend like fingers to grasp the bones anterior to the transverse tarsal joint (also see Fig. 4-79).

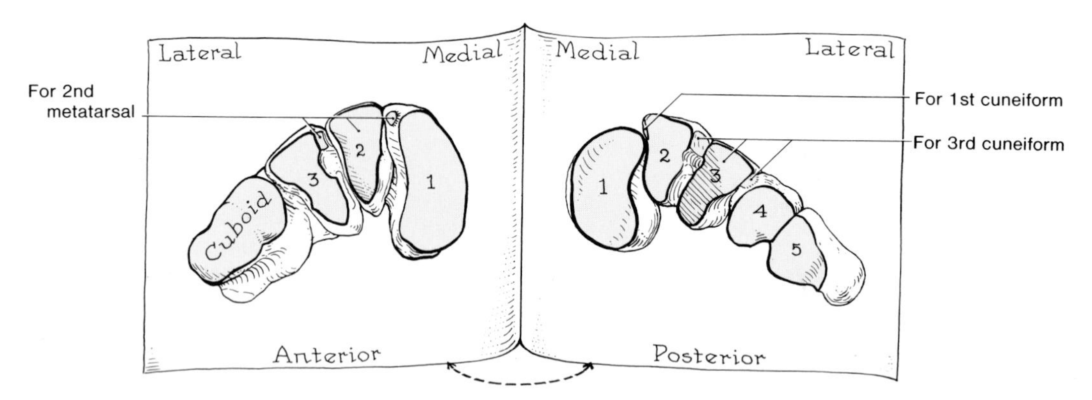

Figure 4-127. Drawings of the bony surfaces of the tarsometatarsal joints of the right foot. The anterior surfaces of the cuboid, the three cuneiform bones and the posterior surfaces of the bases of the five metatarsal bones are displayed like the pages of a book. Note that the bases of the first three metatarsal bones articulate with the three cuneiform bones and that the bases of the fourth and fifth metacarpals articulate with the cuboid bone. Observe that the bones are wedge-shaped; this bony shape helps to maintain the transverse arch of the foot.

provide it with this adaptability. The weight of the body is transmitted to the talus from the tibia and fibula (Fig. 4-121). Then the weight is transmitted posteroinferiorly to the calcaneus or anteroinferiorly to the heads of the metatarsal bones (Fig. 4-129).

Body weight is divided about equally between the calcaneus and the heads of the metatarsal bones (Fig. 4-130). Between these weight-bearing points are the arches of the foot, formed by the tarsal and metatarsal bones. The relatively elastic arches, convex superiorly and disposed both longitudinally and transversely, become slightly flattened by the body weight during standing, but they normally resume their curvature when the body weight is removed (*e.g.*, during sitting). There are two arches, longitudinal and transverse.

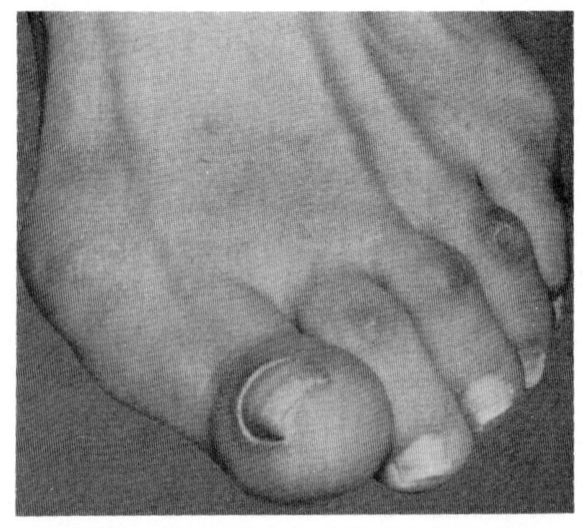

Figure 4-128. Photograph of the left foot of a woman with hallux valgus and a localized swelling (bunion) on the medial aspect of her first metatarsophalangeal joint. A bunion is a type of friction bursitis. Note also the corns caused by pressure on the skin from the patient's shoe over the underlying proximal interphalangeal joints of toes 2 to 4. Observe the tendons of the extensor hallucis longus and extensor digitorum longus (also see Figs. 4-66 to 4-69).

THE LONGITUDINAL ARCH OF THE FOOT (Fig. 4-130)

For purposes of description, the longitudinal arch is regarded as being composed of medial and lateral parts. The medial longitudinal arch is higher and more important. Functionally, both parts of the longitudinal arch act as a unit with the transverse arch, spreading the weight in all directions.

The medial longitudinal arch is composed of the calcaneus, talus, navicular, three cuneiforms, and three metatarsal bones (Figs. 4-60, 4-62, and 4-130*A*). *The head of the talus*, located at the summit of the medial longitudinal arch, is the "*keystone*" and receives the weight of the body. At the articular surfaces between the talus and navicular, and also between the navicular and the three cuneiforms, the medial longitudinal arch yields slightly when weight is put on it and recoils when it is removed.

The lateral longitudinal arch rests on the ground during standing, producing the usual footprint (Fig. 4-129). It is composed of the calcaneus, cuboid, and lateral two metatarsal bones (Figs. 4-61 and 4-130*B*).

THE TRANSVERSE ARCH OF THE FOOT (Fig. 4-127)

This arch runs from side to side. It is formed by the cuboid, the three cuneiforms, and the bases of the metatarsal bones. The medial and lateral parts of the longitudinal arch serve as pillars for the transverse arch. The peroneus longus muscle (Fig. 4-126) helps to maintain the curvature of the transverse arch.

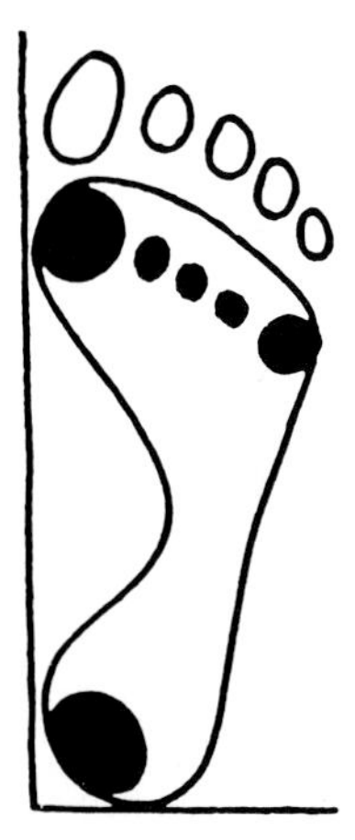

Figure 4-129. Diagram of a footprint illustrating the weight-bearing points of the right foot. Body weight is divided about equally between the calcaneus and the heads of the metatarsal bones . The anterior part of the foot has six points of contact with the ground: two with the sesamoid bones associated with the head of the first metatarsal bone (Figs. 4-60 and 4-92), and four with the heads of the lateral four metatarsals. Hence the first metatarsal supports a double load.

MAINTENANCE OF THE ARCHES OF THE FOOT (Figs. 4-125 to 4-127)

The integrity of the bony arches of the foot is maintained by: (1) ***the shape of the interlocking bones***; (2) ***the strength of the plantar ligaments and the plantar aponeurosis*** (Fig. 4-85); and (3) ***the action of muscles*** through the bracing action of their tendons.

Of these three factors, the plantar ligaments and the plantar aponeurosis bear the greatest stress and are most important in maintaining the arches of the foot while standing (Fig. 4-126).

Electromyographic studies indicate that the muscles are relatively inactive until walking begins. The invertor and evertor muscles of the foot (Table 4-13) appear to control weight distribution in the foot (*e.g.*, when walking on rough ground).

The maintenance of the arches of the foot is also dependent on the intertarsal, tarsometatarsal, and intermetatarsal joints, because at these articulations the bones are bound together as parts of the arches of the foot. The plantar ligaments of these joints are the strongest and they are supported by robust interosseous ligaments.

The following fibrous structures, listed in order of importance, are essential for maintaining the arches of the foot.

The plantar calcaneonavicular ligament is the most important ligament in the foot because it is the main supporter of the medial longitudinal arch of the foot (Figs. 4-122, 4-125, and 4-126). Its principal attachments are the sustentaculum tali of the calcaneus and the tuberosity of the navicular bone (Fig. 4-60).

The plantar calcaneonavicular ligament is a strong fibrocartilaginous band, that acts as a tie between the calcaneus and navicular bones, preventing collapse of the medial longitudinal arch. Because of the resilience the plantar calcaneonavicular ligament gives to the medial longitudinal arch when it is stressed, it is commonly called **"the spring ligament."** It is supported by the tendon of the tibialis posterior muscle and the deltoid ligament (Fig. 4-126).

The long plantar ligament (Fig. 4-126) is the next most important ligament for supporting the arches of the foot. Its principal attachments are the tubercle of the calcaneus and the plantar surface of the cuboid bone. The long plantar ligament is longer and more superficial than the plantar calcaneonavicular ligament (spring ligament). It stretches like a tie beam under nearly the whole length of the lateral longitudinal arch, thus *the long plantar ligament provides the main support for the lateral longitudinal arch of the foot.*

The plantar aponeurosis (Fig. 4-85) *acts as a strong tie beam for the maintenance of the longitudinal arch.* One part of it, a dense fibrous band known as the **calcaneometatarsal ligament**, extends from the lateral process of the tuberosity of the calcaneus to the tuberosity of the fifth metatarsal bone. It is particularly important in helping the long plantar liga-

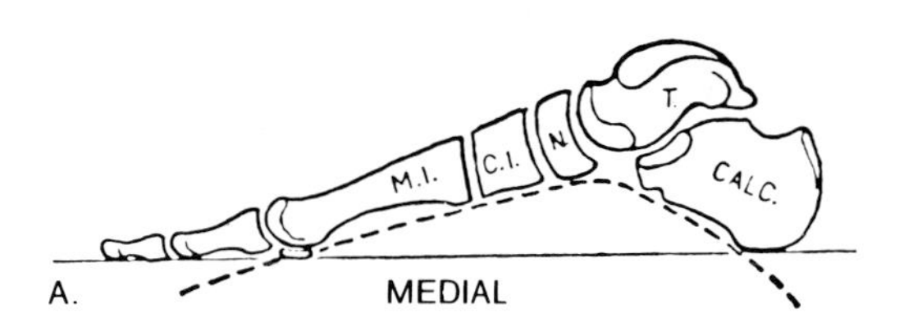

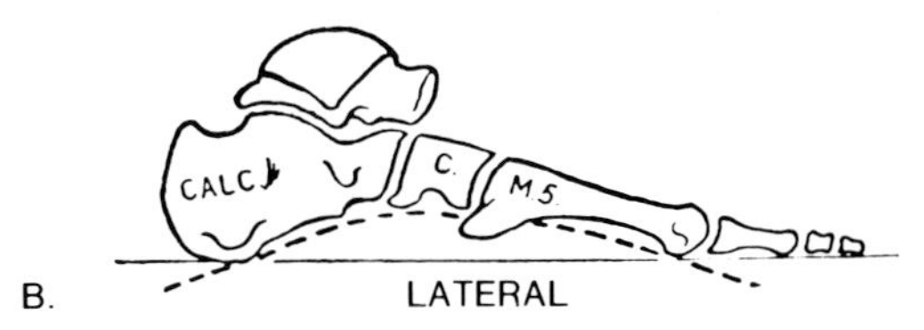

Figure 4-130. Drawings illustrating the medial and lateral longitudinal arches of the right foot. Observe that the foot is arched longitudinally and that the posterior pillar of both arches is the calcaneus. Note that the medial side of the longitudinal arch is high, whereas the lateral side is low. This explains the appearance of footprints of people with normal feet. The arches of the foot act somewhat like springs (*i.e.*, normally they give a little but not completely, unless the person has flatfeet).

ment maintain the lateral longitudinal arch. The medial part of the plantar aponeurosis, through its attachment to the sesamoid bones of the flexor hallucis brevis, is important in strengthening the medial longitudinal arch when standing on the toes.

The plantar calcaneocuboid ligament (short plantar ligament) is broader than the long plantar ligament (Figs. 4-125 and 4-126). Its main attachments are the anterior end of the calcaneus and the proximal edge of the cuboid bone. This ligament aids the plantar calcaneonavicular ligament (spring ligament) and the long plantar ligament in supporting the longitudinal arch.

The peroneus longus tendon forms a sling beneath the lateral part of the longitudinal arch and acts as a tie beam for the transverse arch. (Fig. 4-126).

In infants the flat appearance of the feet is normal and results from the subcutaneous fatpads in the soles of their feet. The arches of the foot are present at birth, but do not become visible until after the infant has walked for a few months.

Flatfoot (pes planus) is common and is not a painful condition in some people. Flatfeet in adolescents and adults are caused by **"fallen arches,"** usually the medial longitudinal arches. During standing the plantar ligaments and plantar aponeurosis, important in maintaining the arches of the foot, stretch somewhat under the body weight. If these *ligaments become abnormally stretched during long periods of standing,* the plantar calcaneonavicular ligament (spring ligament) can no longer adequately support the head of the talus. As a result, some flattening of the medial longitudinal arch occurs and there is concomitant lateral deviation of the forefoot.

In the ***common type of flatfoot,*** *the foot resumes its arched form when the weight is removed from it.* Flatfeet are common in older persons, particularly if they undertake much unaccustomed standing or gain weight rapidly. This results from the added stress on the muscles and the increased strain on the ligaments supporting the arches. Fallen arches cause pain owing to stretching of the plantar muscles and straining of the ligaments.

Flattening of the transverse arch of the foot may also occur. Frequently callus formation occurs on the plantar surfaces of the heads of the lateral four metatarsal bones. This thickening of the skin develops as a protective measure where abnormal pressure is exerted.

PRESENTATION OF PATIENT ORIENTED PROBLEMS

Case 4-1

While visiting your home, your elderly grandmother slipped on the polished floor in the front hall. As you approached her, she was lying on her back in severe pain. Her right lower limb immediately attracted your attention because it was laterally rotated and noticeably shorter than her left lower limb. She was unable to get up or lift her limb off the floor, and when she attempted to do so, she experienced considerable pain.

Fortunately you had spent a summer in the emergency department of the local hospital and realized the probable seriousness of your grandmother's injury. You asked your mother to call the doctor, indicating that your grandmother may have a **"fractured hip."** In the meantime you made her comfortable on the floor, resisting your sister's demands that she be moved to her bed.

Problems. What bone was probably fractured? Name the common fracture site of this bone in elderly people. Why is this bone so fragile in older persons? Explain anatomically why her injured limb was shorter than the other one. What are the anatomical reasons for the complications commonly associated with these fractures? *These problems are discussed on page 560.*

Case 4-2

While playing in an old-timer's hockey game, a 45-year-old man was accidentally kicked with a skate on the lateral surface of his right leg just inferior to the knee.

The ***superficial wound*** was treated by the trainer, but the man was unable to continue playing because of pain in the region of the laceration and loss of power in his leg and foot. He also had numbness and tingling on the lateral surface of his leg and the dorsum of his foot. When he removed his skates, he found that he was unable to move his right foot or his toes superiorly (*i.e.*, dorsiflex them). He was advised to see his doctor forthwith.

As he walked into the examining room, the doctor observed that *he had an* ***abnormal gait,*** in that he raised his right foot higher than usual and brought it down suddenly, making a flapping noise.

During the physical examination, the doctor detected tenderness over the neck region of the patient's fibula and a sensory deficit on the lateral side of the distal part of his leg, including the dorsum of his foot.

Radiographs revealed a fracture of the neck of the fibula.

Problems. What is the anatomical basis of the loss

of sensation and impaired function in the patient's foot? What nerve appears to have been injured? What is its relationship to the neck of the fibula? If the skate blade had not severed the nerve, what probably would have injured it if he had continued to play? What is the name given to the foot condition exhibited by the patient when he walked? *These problems are discussed on page 561.*

Case 4-3

While a 26-year-old worker was loading a heavy crate, it fell on his knee. He suffered severe pain and was unable to get up. The first aid team carried him to the doctor's office on a stretcher.

Following a physical examination, the doctor requested x-rays of the man's knee. The radiographs showed **comminuted fractures of the proximal end of the tibia** and a fracture of the neck of the fibula (Fig. 4-131).

Problems. What artery or arteries might have been torn by the bone fragments? Using your anatomical knowledge, where would you check the patient's pulse to determine whether there has been damage to these arteries? What nerve may have been injured by the fracture of the fibula? *These problems are discussed on page 561.*

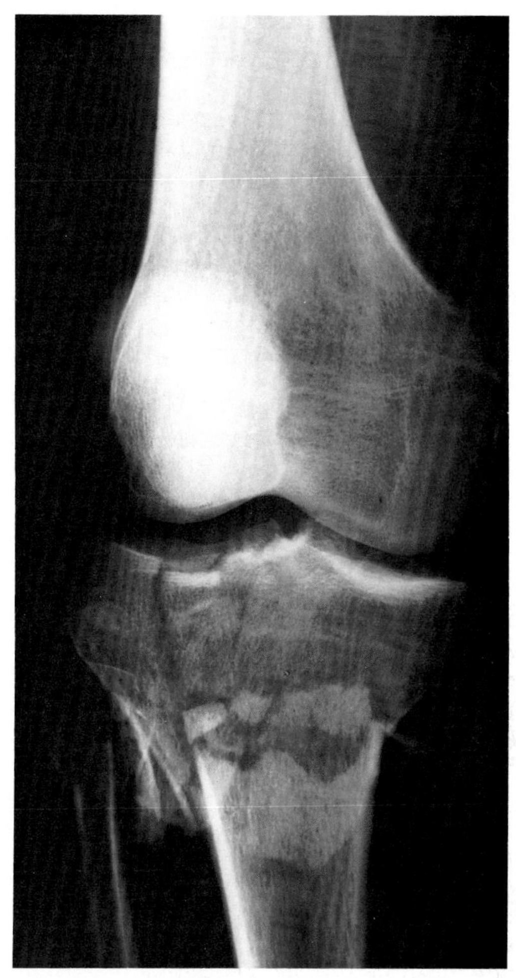

Figure 4-131. Radiograph of the right knee region showing comminuted fractures of the proximal ends of the tibia and fibula. Note the increased width of the superior end of the tibia caused by the comminution (breaking into fragments) and spreading of the pieces of broken bone.

Case 4-4

A 32-year-old man slipped on a patch of ice and fell. After he was helped up, he was unable to bear weight on his right foot. When he noticed that his ankle was beginning to swell, he hailed a cab and went to the hospital for treatment of what he thought was a badly **"sprained ankle."**

On examination it was found that the patient could barely move his ankle because of pain. Maximum tenderness was located over the lateral malleolus, about 2 cm proximal to its tip.

Radiographs of the ankle revealed a transverse **fracture of the lateral malleolus** at the level of the superior articular surface of the talus.

Problems. What excessive movement usually results in a sprained ankle? Discuss what is meant by the term "sprain." Explain anatomically how this fracture probably occurred. What structures were probably torn or ruptured? Does the patient have what is usually referred to as a Pott's fracture? *These problems are discussed on page 561.*

Case 4-5

A 22-year-old woman was a front seat passenger in a car that was involved in a head-on collision. Although she sustained head injuries, **her chief complaint was a sore right hip** which prevented her from standing up. Believing that she might have broken her hip, she was rushed to the nearest hospital.

A physical examination revealed that her lower limb was slightly flexed, adducted, medially rotated, and appeared shorter than the other limb (Fig. 4-132).

Radiographs showed that there was a **posterior dislocation of her right hip joint** with a fracture of the posterior margin of her acetabulum.

Problems. Explain anatomically how this injury probably occurred. What nerve may have been injured? When paralysis of this nerve is complete, which is rare, what muscles would be paralyzed. Where may cutaneous sensation be lost? *These problems are discussed on page 562.*

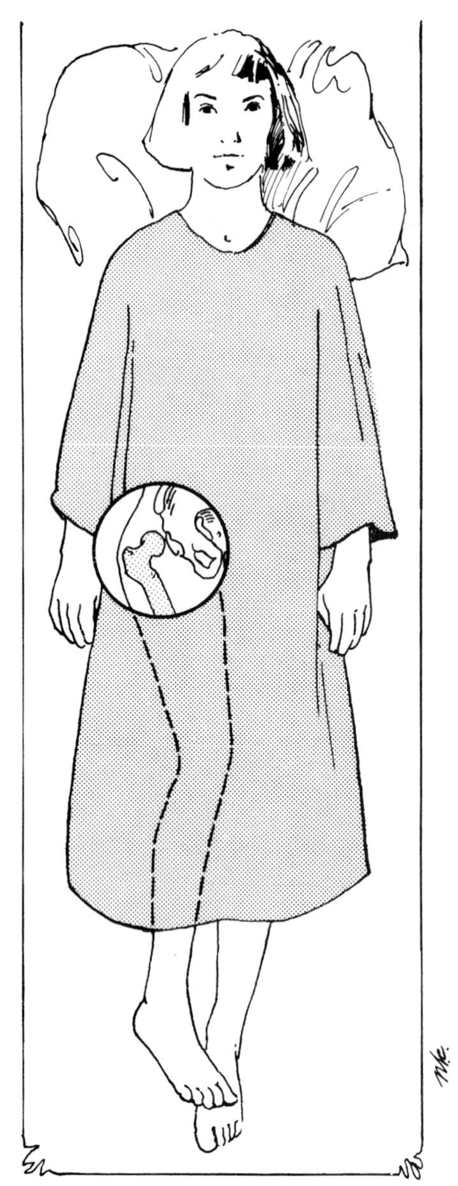

Figure 4-132. Drawing illustrating a posterior dislocation of the right hip. Observe that the women's lower limb is flexed, adducted, medially rotated, and *appears* shorter than her left limb. Also see Figure 4-104.

Case 4-6

A 62-year-old man presented with an **aching pain in his left buttock** which extended along the posterior aspect of his left thigh.

During the examination the patient pointed to the area where he felt most pain, which was in the region of the greater sciatic notch. Tenderness was also elicited by pressure along a line beginning from a point midway between the top of the greater trochanter of his femur and the ischial tuberosity to a point in the midline of the thigh about half-way to the knee. When seated, the patient was unable to extend his left leg fully because of severe pain.

With the patient in the supine position, the doctor grasped the patient's left ankle and placed his other hand on the anterior aspect of the knee in order to keep the leg straight. He then slowly raised the left lower limb; when it reached about 75°, the man grimaced with pain. Even more pain was elicited when the patient's foot was dorsiflexed.

Problems. What nerve is involved in this case? From which segments of the spinal cord does it arise? Why does the *straight leg-raising test* elicit pain? Why did the pain increase when his foot was dorsiflexed? What **back lesion** probably produced the pain in the buttock and posterior thigh region? Thinking anatomically, what other lesions (*e.g.*, resulting from disease or injury) do you think might have caused the patient's symptoms? *These problems are discussed on page 562.*

Case 4-7

A 55-year-old woman complained of a globular **swelling in her right groin** (Fig. 4-133*A*). She stated that the swelling became smaller when she lay down, but that it never completely disappeared. She also said that the mass occasionally got quite large and bulged under the skin on the anterior aspect of her thigh. When this occurred, she stated that she got a pain down the medial side of her thigh.

On examination the doctor noted that the swelling was inferior to the medial third of the inguinal ligament, and **lateral to the pubic tubercle**. When he inserted his index finger into superficial ring of her inguinal canal and asked her to cough, he felt no mass or protruding gut, but observed a slight increase in the size of the swelling. When the patient was asked to point to the site where the swelling first appeared, she placed her finger over the site of her femoral ring (Fig. 4-37). When asked in which direction the swelling came down when she felt pain, she ran her finger along her thigh to the region of the **saphenous opening** (Fig. 4-11).

The doctor applied extremely gentle manual pressure to the swelling with the thigh flexed and medially rotated, but was unable to reduce the protrusion. A diagnosis of irreducible, *complete femoral hernia* was made.

Problems. Define the terms femoral ring, femoral canal, and femoral hernia. What are the usual contents of the femoral canal? Use your anatomical knowledge to explain why a femoral hernia curves superiorly (Fig. 4-133*B*). Can you think of any anatomical reasons why femoral hernias are more common in females than in males? Explain anatomically why strangulation of this type of hernia is common. Enlargement of what structure in the femoral canal might be mistaken for a femoral hernia? *These problems are discussed on page 563.*

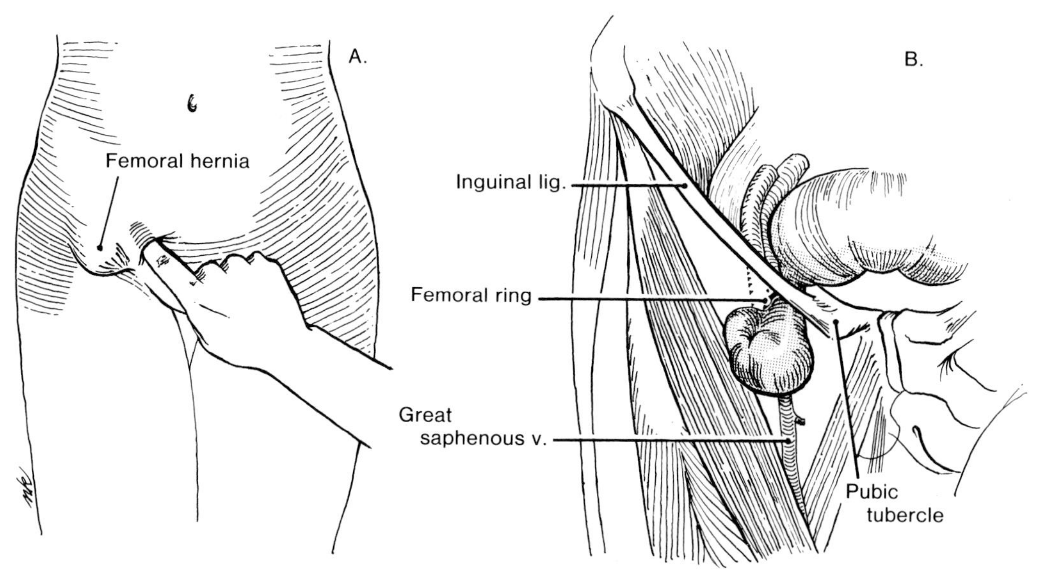

Figure 4-133. *A*, drawing of the external appearance of the femoral hernia. Note that it lies inferolateral to the pubic tubercle and the external ring of the inguinal canal. No mass was palpable in this canal during the invagination test illustrated. *B,* dissection of the right thigh and pelvic regions showing the loop of bowel passing deep to the inguinal ligament into the femoral canal. For review of the femoral triangle, femoral sheath, and femoral canal, see Figures 4-15 to 4-18, 4-36, and 4-37.

Figure 4-134. Drawing illustrating the way knee injuries occur. This infraction of football rules is called "clipping." Note that the runner's right knee was hit from the side. The severe strain on the tibial collateral ligament may rupture it and tear the intra-articular ligaments and cartilages. (Reprinted with permission from Healey JE, Jr: *A Synopsis of Clinical Anatomy*. Philadelphia, WB Saunders Co, 1969.)

Case 4-8

A football player was clipped (blocked from the rear) as he was about to tackle the ball carrier (Fig. 4-134). The lineman's hip hit the runner's knee from the side. It was obvious on the slow motion videotape replay that the tackler's knee was slightly flexed and his foot firmly implanted in the turf when he was hit. As he lay on the ground clutching his knee, it was obvious from his face that he was in severe pain.

While he was being helped to the sidelines, you said to your friend, "I'm afraid he has **torn knee ligaments.**" Not knowing much about the functioning of the knee joint, your friend said, "Which knee ligaments are probably torn?"

Problems. How would you explain this injury to your friend, assuming that he has little knowledge of the anatomy of the knee joint? What ligament was probably ruptured? What ligament may have been torn? Would the menisci be injured? *These problems are discussed on page 563.*

DISCUSSION OF PATIENT ORIENTED PROBLEMS

Case 4-1

Very likely your grandmother **fractured the neck of her femur**, one of the common fractures in elderly women (Fig. 4-135). This injury is often wrongly referred to as a "*fractured hip*," implying that the hip bone is fractured.

As your grandmother stumbled and tried to "catch herself," she exerted a torsional force on one hip, producing a fracture of the femoral neck, the most fragile part of the femur in elderly persons. She fell when the bone fractured; hence the fracture was probably the cause of her fall rather than the result of it.

The lateral rotation and shortening of her limb are characteristic clinical features following fractures of the neck of the femur. The rotation results from the change in the axis of the limb owing to the separation of the body and head of the femur.

The shortening of the lower limb results from the superior pull of the muscles connecting the femur to the hip bone (Table 4-2). **Spasm of the muscles** (sudden involuntary muscular contraction) causes the pull.

Bones lose mineral with advancing age, a disorder called **osteoporosis;** as a result the neck of the femur becomes weaker. Consequently, fractures of the proximal part of the femur can result from little or no trauma. In this bone disorder of postmenopausal women and elderly men, *absorption of bone is greater than bone formation.*

The blood vessels to the proximal part of the femur are derived mostly from the medial and lateral circumflex femoral arteries (Fig. 4-31). Branches of these arteries run in the retinacula (reflections) of the fibrous capsule of the hip joint. A variable amount of blood may reach the femoral head through a branch of the obturator artery that runs in the ligament of the head of the femur, called the **artery of the ligament of the head.** This ligament may be ruptured during fractures of the femoral neck or head. Furthermore, this vessel is often not patent in elderly patients because they commonly have **arteriosclerosis.**

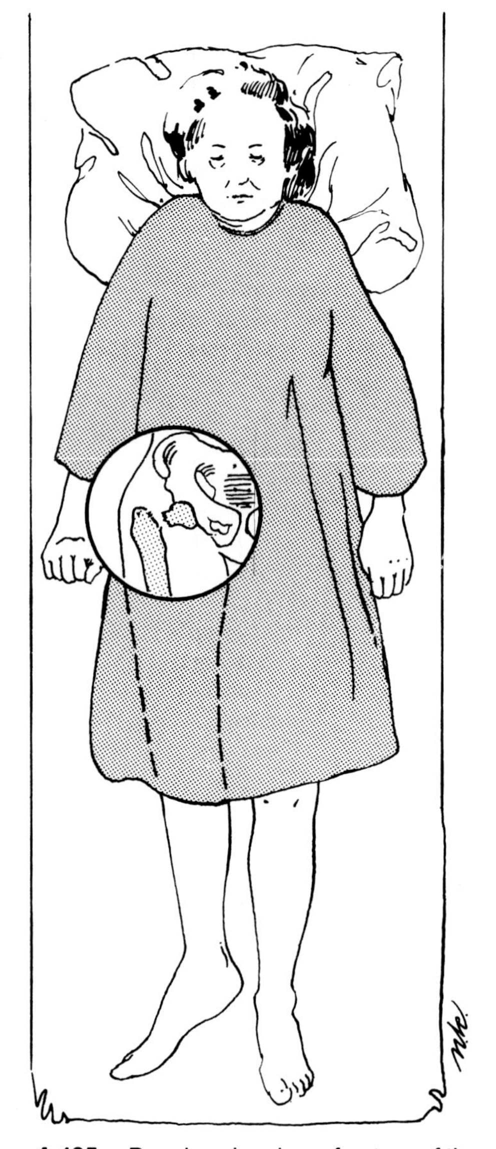

Figure 4-135. Drawing showing a fracture of the neck of the right femur. Observe that there is also shortening and lateral rotation of her injured lower limb.

Sometimes other blood vessels supplying the femoral head are torn when the femoral neck fractures. Generally, the more proximal the fracture, the greater are the chances of interrupting the vascular supply. A poor blood supply may result in nonunion and **avascular necrosis of the femoral head** (death and collapse of the proximal bone fragment owing to poor blood supply).

Intracapsular fractures (high in the neck) almost always present healing problems because they usually interfere with the blood supply to the proximal bone fragment. The importance of preserving the blood supply to the proximal part of the femur is one reason why patients with this type of injury are handled with extreme care; another is that this injury is very painful.

Case 4-2

The close relationship of the **common peroneal nerve** to the neck of the fibula (Fig. 4-57*A*) makes it vulnerable to injury when this region of the bone is fractured. As the nerve lies on the lateral aspect of the neck of the fibula (Fig. 4-67), it can be easily injured by superficial lacerations.

This patient's signs and symptoms make it obvious that the common peroneal nerve was injured. Superficial wounds, prolonged pressure by hard objects (*e.g.*, the sharp edge of a bed during sleep), or compression by a tight plaster cast may present similar clinical features.

Injury to the common peroneal nerve affects the muscles in the lateral crural compartment (peroneus longus and brevis supplied by the superficial peroneal nerve), **and in the anterior crural compartment** (muscles in the anterior part of the leg and the extensor digitorum brevis supplied by the deep peroneal nerve). Consequently, eversion and dorsiflexion of the foot, and extension of the toes are impaired (Tables 4-5 and 4-6).

This patient showed a characteristic **foot-drop** (plantarflexion and slight inversion) and **steppage gait**. As the patient walked, his toes dragged and his foot slapped the floor. In attempting to prevent this from happening, he raised his foot higher than usual.

The **dysesthesia** (impairment of sensation) on the patient's leg and foot resulted from injury to sensory fibers in the cutaneous branches of the common peroneal nerve (Fig. 4-94). The injury to the nerve resulted from the skate grazing the nerve or the nerve being compressed or torn by the bone fragments.

Although the fibula is not a weight-bearing bone, fractures of its proximal end cause pain on walking because the pull of muscles attached to it causes the fragments to move, which is painful.

Case 4-3

As the ***popliteal artery*** lies deep in the popliteal fossa against the fibrous capsule of the knee joint, it could have been torn by fragments from the comminuted fractures of the proximal ends of the tibia and fibula (Fig. 4-131). As the popliteal artery divides into its terminal branches (anterior and posterior tibial arteries) at the inferior end of the **popliteal fossa**, they may also have been torn when the bones fractured. Undoubtedly one or more of the five **genicular arteries,** branches of the popliteal artery, would have also been torn. They supply the capsule and ligaments of the knee joint (Figs. 4-53 and 4-111).

Pulsations of the posterior tibial artery may be felt halfway between the medial malleolus and the heel (Figs. 4-81 and 4-84). The **dorsalis pedis artery**, the continuation of the anterior tibial artery, can also be palpated where it passes over the navicular and cuneiform bones in the foot (Fig. 4-66). These are good places to take the pulse of the arteries because they are superficial here and can be compressed against the bones.

Loss of a pulse in these arteries in the present case would suggest a torn popliteal and/or tibial arteries.

There is also a chance that the **tibial nerve** would be injured in this patient, as it is the most superficial of the three main central structures in the popliteal fossa.

Severance of *the tibial nerve would result in* ***paralysis of the popliteus and muscles of the calf*** (gastrocnemius, soleus, flexor hallucis longus, and tibialis posterior), together with those in the sole of the foot (Fig. 4-82). Probably some of the genicular branches (articular nerves) to the knee joint would also be served.

The close relationship of the **common peroneal nerve** to the neck of the fibula (Fig. 4-57*A*) makes it vulnerable to injury when this region of the bone is fractured. For the signs and symptoms resulting from severance of this nerve, see Case 4-2.

Case 4-4

The usual sprained ankle results from **excessive inversion of the weight-bearing foot**, which causes rupture of the anterolateral portion of the fibrous capsule of the ankle joint and the calcaneofibular and talofibular ligaments (Fig. 4-136).

The term "*sprain*" is used to indicate some degree of tearing of the ligaments without fracture or dislocation. In severe sprains many fibers of the ligaments are completely torn and often considerable instability of the ankle joint results.

In the present case, the sprain and fracture occurred when the patient slipped in such a way that his foot was forced into an excessively inverted position. His body weight then caused a forceful inversion of the ankle joint.

The calcaneofibular and anterior talofibular ligaments were torn partly or completely. Normally the

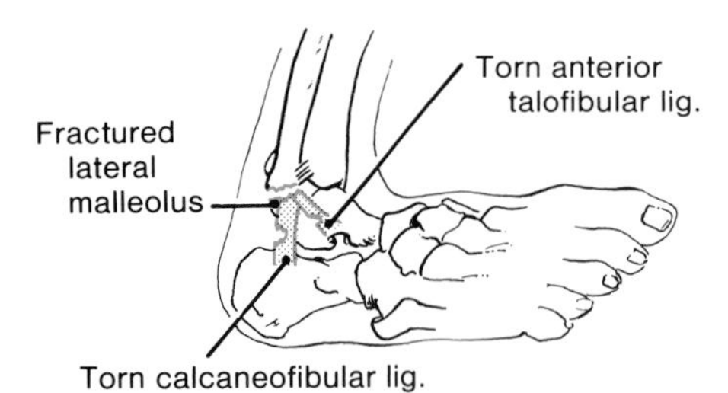

Figure 4-136. Drawing illustrating torn calcaneofibular and talofibular ligaments and a fracture of the lateral malleolus, resulting from forced inversion of the right foot. For a better view of these ligaments, see Figure 4-118.

deep mortise formed by the distal end of the tibia and malleoli holds the talus firmly in position (Fig. 4-121). When the ankle ligaments tear, the talus is forcibly tilted against the lateral malleolus shearing it off (Fig. 4-136).

Had the man's ankle been forced in the opposite direction (*i.e.*, in an extremely everted position), the strong deltoid ligament may have avulsed the medial malleolus (Fig. 4-137). As the force continued, it would have tilted the talus, moving it and the lateral malleolus laterally. Because the interosseous tibiofibular ligament acts as a pivot, the fibula breaks proximal to the inferior tibiofibular joint.

This injury is really a **fracture-dislocation of the ankle joint** caused by forceful eversion of the foot, and is what is usually called a **Pott's fracture** because it is the kind of fracture-dislocation sustained and described by Dr. Percival Pott, an English surgeon. However, the term Pott's fracture is often used loosely to describe most fractures and fracture-dislocations of the malleoli.

Case 4-5

Dislocation of the hip joint is uncommon owing to the stability of this articulation. The head of the femur is deeply seated in the acetabulum and is held there by an exceedingly strong fibrous capsule.

Traumatic dislocations of the hip joint commonly occur during automobile accidents when the hip joint is flexed, and the thigh is adducted and *medially rotated.* Very likely, the patient's knee struck the dashboard when her right lower limb was in the position just described. Consequently the force was transmitted up the femur, driving its head and the posterior margin of the acetabulum posteriorly. As the head in this position is covered posteriorly by capsule rather than bone, the articular capsule probably ruptured inferiorly and posteriorly. This permitted the head to dislocate posteriorly and carry the fractured posterior margin of the acetabulum and acetabular labrum with it. As a result, the head of the femur came to lie on the gluteal surface of the ilium.

The close relationship of the **sciatic nerve** (L4 to S3) to the posterior surface of the hip joint makes it vulnerable to injury in posterior dislocations. If the paralysis is complete, which is rarely the case, the hamstring muscles and those distal to the knee would all be paralyzed (Fig. 4-82). In addition there would probably be anesthesia in the lower leg and foot, except for the skin on the medial side which is supplied by the saphenous nerve (L3 and L4), a terminal branch of the femoral nerve (Fig. 4-94).

Case 4-6

The site of the patient's pain and its course down the posterior aspect of the thigh clearly implicates the **sciatic nerve**, the largest branch of the sacral plexus. It arises from spinal cord segments L4 to S3. The sciatic nerve leaves the pelvis through the inferior part of the greater sciatic notch and extends from the inferior border of the piriformis muscle to the distal third of the thigh, along the course clearly indicated by the patient's pain.

The straight leg-raising test elicits pain because the sciatic nerve is stretched when the limb is raised. Dorsiflexion of the foot further increases the pull on the sciatic nerve and its roots.

Sciatica is the name given to *pain in the area of distribution of the sciatic nerve.* Variation occurs in the location of the pain in different patients owing to involvement of different nerve roots.

A posterior protrusion of an intervertebral disc is a common cause of sciatica, most often affecting the first sacral nerve roots. A protruding L5/S1 disc exerts pressure on the dorsal and ventral roots, producing sciatica, which may be accompanied by "*lumbago*" (low back pain).

Sciatic pain can also result from pressure (*e.g.*, a tumor) on the sciatic nerve or its components in the

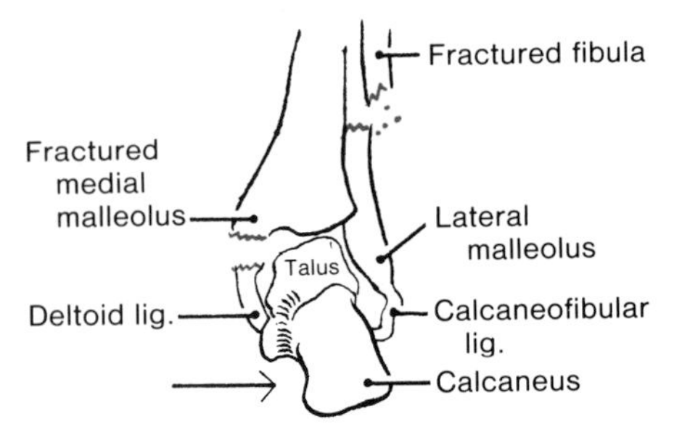

Figure 4-137. Drawing illustrating a fracture-dislocation (Pott's fracture) of the ankle caused by forced eversion of the right foot. Note that the strong deltoid ligament has not ruptured but has avulsed the medial malleolus. Observe the associated fracture of the fibula.

pelvis, in the gluteal region, or in the thigh. Pain could also be produced by irritation of the sciatic nerve resulting from inflammation of the nerve **(neuritis)** or its sheath.

Case 4-7

A femoral hernia is a protrusion of fat, peritoneum, omentum, and usually a loop of intestine through the femoral ring into the femoral canal (Fig. 4-133*B*). **Bowel sounds** may be heard with a stethoscope when there is intestine in the hernial sac.

The ***femoral canal*** is a short, blind, potential space in the medial compartment of the ***femoral sheath*** (Fig. 4-36), which is a prolongation of the fascial lining of the interior of the abdomen (fascia transversalis anteriorly and fascia iliaca posteriorly). Normally this space contains lymph vessels and at least one lymph node embedded in connective tissue.

The femoral canal is a source of weakness in the abdominal wall; thus, when intraabdominal pressure rises very high (as may occur when a chronically constipated person attempts to defecate), abdominal contents may be forced through the femoral ring into the femoral canal. The intestine carries a pouch of peritoneum before it as it descends along the femoral canal and through the saphenous opening. Being prevented from extending further down by the deep fascia of the thigh, the hernial sac is directed anteriorly and then superiorly, forming a swelling inferior to the inguinal ligament.

While the hernia is in the femoral canal (**incomplete femoral hernia**) it is usually small, but after it passes anteriorly through the saphenous opening into the loose areolar tissue of the thigh, it becomes much larger (**complete femoral hernia**). The differential diagnosis between indirect inguinal hernia and complete femoral hernia is at times difficult, because advanced types of femoral hernia sometimes produce a swelling superior to the inguinal ligament. The swelling produced by femoral hernia is more lateral, however, than one caused by indirect inguinal hernia.

The **pubic tubercle** is the important bony landmark in differentiating an inguinal from a femoral hernia. The neck of an **inguinal hernial sac is** *superomedial to the pubic tubercle* at the superficial inguinal ring, whereas the neck of a **femoral hernial sac** is *inferolateral to the pubic tubercle* at this site. In addition, if a hernia does not present in the inguinal canal during the **invagination test** (Figs. 2-20*C* and 4-133*A*), as in the present case, the hernia cannot be an indirect inguinal hernia.

If the hand is placed gently over the hernia and eased inferiorly, the fold of the groin produced by the inguinal ligament will be seen passing superior to a femoral hernia, whereas if the hand is eased superiorly, the fold of the groin will be seen passing inferior to an inguinal hernia.

Femoral hernia is more common in females than in males (about 3:1) because the femoral ring is larger in women than in men owing to the greater breadth of the female pelvis, the smaller size of their femoral vessels, and the changes that occur in the associated tissues when pregnancy occurs.

Strangulation of a complete femoral hernia is common. The tendency for this type of hernia to strangulate (*i.e.*, compress the vessels of the hernia) results from the sharp boundaries of the femoral ring (*e.g.*, the inguinal ligament anteriorly and the lacunar ligament medially). Stricture of the hernia may also be produced by the sharp edges of the saphenous opening (Fig. 4-13).

Because of the relatively small size of the femoral ring and the saphenous opening, and the rigidity of the surrounding structures, blockage of the venous return from the protruded loop of intestine often occurs. As arterial blood continues to pass into the loop, it becomes engorged with blood and circulation through it soon stops. Early surgical intervention is required to prevent **necrosis of the strangulated loop** of intestine.

A soft enlarged lymph node in the femoral canal (Fig. 4-37) could be mistaken for a femoral hernia, although the node would likely be more firm. Cancer (Fig. 4-37) could be mistaken for a femoral hernia, although the node would likely be more firm. Cancer or infections in the areas drained by these nodes could cause one or more of them to enlarge.

Case 4-8

The knee joint is one of the most secure joints in the body, particularly when it is extended. Although these bones are bounded together by strong ligaments (Fig. 4-111), the knee is subject to a wide range of injuries because of the severe strain that is placed on its attachments, especially in contact sports like hockey and football.

The blow on the lateral side of the tackler's knee by the lineman's hip occurred while he was running and his foot was fixed in the ground (Fig. 4-134). As he was bearing weight on his leg, the hard blow bent the runner's knee medially relative to the fixed tibia. This severely stressed the ligament on the inside of the knee, known as the **tibial collateral ligament.** Some fibers of this ligament may have ruptured and he may only have a sprain; however because of the severity of the blow, probably the entire ligament ruptured near its attachment to the medial femoral epicondyle (Fig. 4-138). Because the **medial meniscus**, a fibrocartilaginous articular disc, is attached to this ligament, the meniscus may tear or become detached from it.

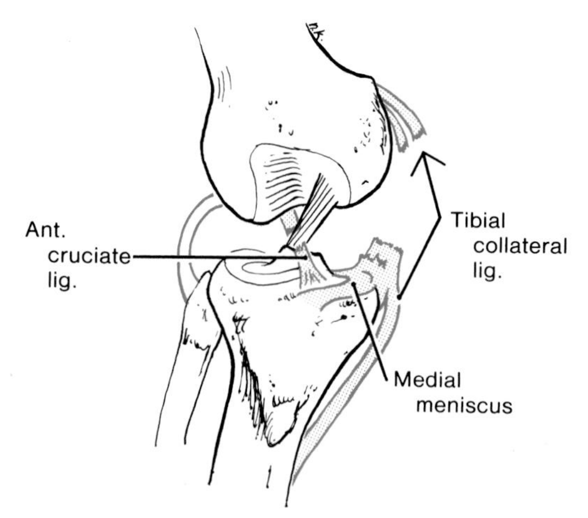

Figure 4-138. Diagrammatic illustration showing how the force from a hard blow on the lateral side of the right knee (Fig. 4-134) probably ruptured the tibial collateral ligament, medial meniscus, and anterior cruciate ligament.

This may have been the full extent of the athelete's injury, but he may also have ruptured his **anterior cruciate ligament** (Fig. 4-138). This ligament, which prevents posterior displacement of the femur and hyperextension of the knee joint, is sometimes torn when the knee is hit hard from the lateral side.

In summary, the forced abduction and lateral rotation of the runner's leg by the block probably resulted in the simultaneous rupture of three structures: the tibial collateral ligament, the medial meniscus, and the anterior cruciate ligament.

Suggestions for Additional Reading

1. Anderson JE: *Grant's Atlas of Anatomy*, ed 8. Baltimore, Williams & Wilkins, 1983.

   This world famous atlas should be referred to for more illustrations of dissections of the lower limb.

2. Haymaker W, Woodhall B: *Peripheral Nerve Injuries. Principles of Diagnosis*, ed 2. Philadelphia, WB Saunders Co, 1953.

   This is a very good source of information concerning injuries to nerves.

3. Hollinshead, WH, Jenkins, DB: The lower limb. In *Functional Anatomy of the Limbs and Back*, ed 5. Philadelphia, WB Saunders Co, 1981.

   The chief emphasis is on the muscles of the lower limb and their actions, including discussions of the skeletal, nervous, and vascular systems.

4. Salter, RB: *Textbook of Disorders and Injuries of the Musculoskeletal System*, ed 2. Baltimore, Williams & Wilkins, 1983.

   You will find this an excellent introduction to orthopaedics, rheumatology, metabolic bone disease, rehabilitation, and fractures. Although written for medical students, it will be of interest to all who are involved in the care of patients with disabilities of the musculoskeletal system. The author is recognized internationally for fundamental scientific investigations of disorders and injuries of the musculoskeletal system.

5. Williams, JGP: *Sports Medicine*, ed 2. Baltimore, Williams & Wilkins, 1976.

   This book discusses the etiology, diagnosis, treatment, and prevention of disorders and injuries of athletes and describes methods of athletic training. It also gives an insight into the speciality of sports medicine.

CHAPTER 5

The Back

The term *back* is used for referring to the posterior aspect of the *trunk*, the main part of the body to which the head and limbs are attached. The back consists of skin, muscles, vertebrae, ribs (in thoracic region), intervertebral discs, vessels, and nerves (Fig. 2–107).

Low back pain is a commonly encountered complaint in medical practice. To understand the anatomical basis of back problems that cause disabling pain, a good understanding of the structure and function of the back is required.

The cervical and lumbar regions of the vertebral column are the most mobile and the common sites of aches and pains.

THE VERTEBRAL COLUMN

The vertebral column, **commonly called the spine** (backbone), forms the skeleton of the back and is *part of the axial skeleton* (Figs. 5–1 and 5–2). It consists of a number of bones called **vertebrae** which are united by a series of intervertebral joints.

The vertebral column forms a strong but flexible shaft that supports the trunk and the limbs. It extends from the base of the skull through the whole length of the neck and trunk (Figs. 5–9 and 5–10).

The vertebral column is stabilized by ligaments which limit somewhat the movements produced by the back muscles. The ***spinal cord***, spinal nerve roots, and their coverings or *meninges* are located within the **vertebral canal**, the canal formed by the foramina in the successive vertebrae (Figs. 5–50 and 5–51).

The ***spinal nerves*** and their branches are located outside the vertebral canal, except for the meningeal nerves, which return through the intervertebral foramina to innervate the ***spinal meninges*** (G. membranes.

The vertebral column provides a partly rigid and partly flexible axis for the body and a pivot for the head. Thus, *the vertebral column has an important role in posture, in support of body weight, in locomotion, and in protection of the spinal cord and spinal nerve roots.*

During sitting the vertebral column transmits the weight of the body across the **sacroiliac joints** to the ilia, and then to the ischial tuberosities (Fig. 5–3). When standing, body weight is transferred from the sacroiliac joints to the acetabula and then to the femora (Figs. 3–42*A* and 4–1).

The adult vertebral column usually consists of 33 vertebrae, but only 24 of them (7 cervical, 12 thoracic, and 5 lumbar) are movable. The five sacral vertebrae are fused to form the **sacrum** (Fig. 5–24), which can be palpated in the low back region (Fig. 5–11).

Usually the four coccygeal vertebrae are fused to form a slender tapering bone called the **coccyx** (Figs. 5–1, 5–3, and 5–24), but the first coccygeal vertebra may be separate from the rest.

The 24 presacral vertebrae are movable and hence give the vertebral column flexibility (Figs. 5–9 and 5–10). Stability of the vertebral column is provided by intervertebral discs, ligaments, muscles, and the shape of the vertebrae.

The presacral vertebrae are connected by resilient intervertebral discs (Figs. 5–4 and 5–8), which play an important role in movements between the vertebrae, and in absorbing shocks transmitted up or down the vertebral column. The presacral vertebrae are also connected to each other by paired, posterior **zygapophysial joints** between the articular processes (Figs. 5–4 and 5–30), and by anterior and posterior *longitudinal ligaments* (Figs. 5–28 and 5–29). These ligaments extend the length of the vertebral column and are attached to the intervertebral joints and to the vertebral bodies.

The intervertebral ligaments and joints generally prevent excessive flexion and extension of the vertebral column (*i.e.*, hyperflexion and hyperextension). Movements beyond the normal limits for an individual may cause damage to the joints and ligaments connecting the vertebrae, and to the associated muscles, nerves, and vessels, *e.g.*, **hyperextension of the neck** (Fig. 5–62).

The bodies of the vertebrae contribute about three-fourths of the length of the presacral part of the vertebral column, and the intervertebral discs contribute the other one-fourth.

When counting vertebrae in vivo, it is important to

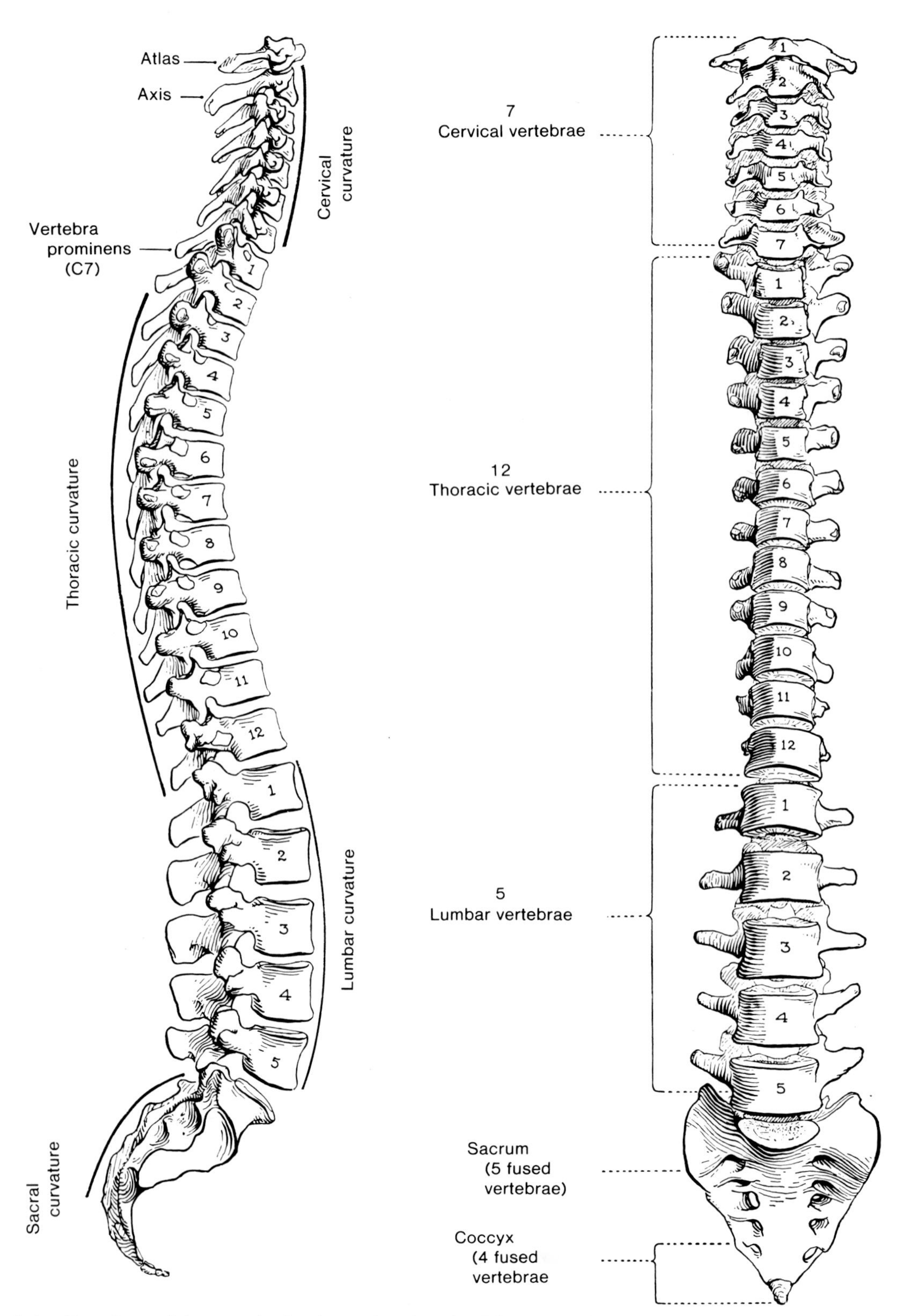

Figure 5-1. Drawings of the vertebral column showing the 24 presacral vertebrae, the sacrum, the coccyx, and the curvatures of the adult vertebral column. Note that the first coccygeal vertebra has fused with the sacrum. Most vertebral columns range between 72 and 75 cm in length, of which about one-fourth is contributed by the fibrocartilaginous intervertebral discs (Fig. 5-4). The vertebral column supports the skull and transmits the weight of the body through the pelvis to the lower limbs.

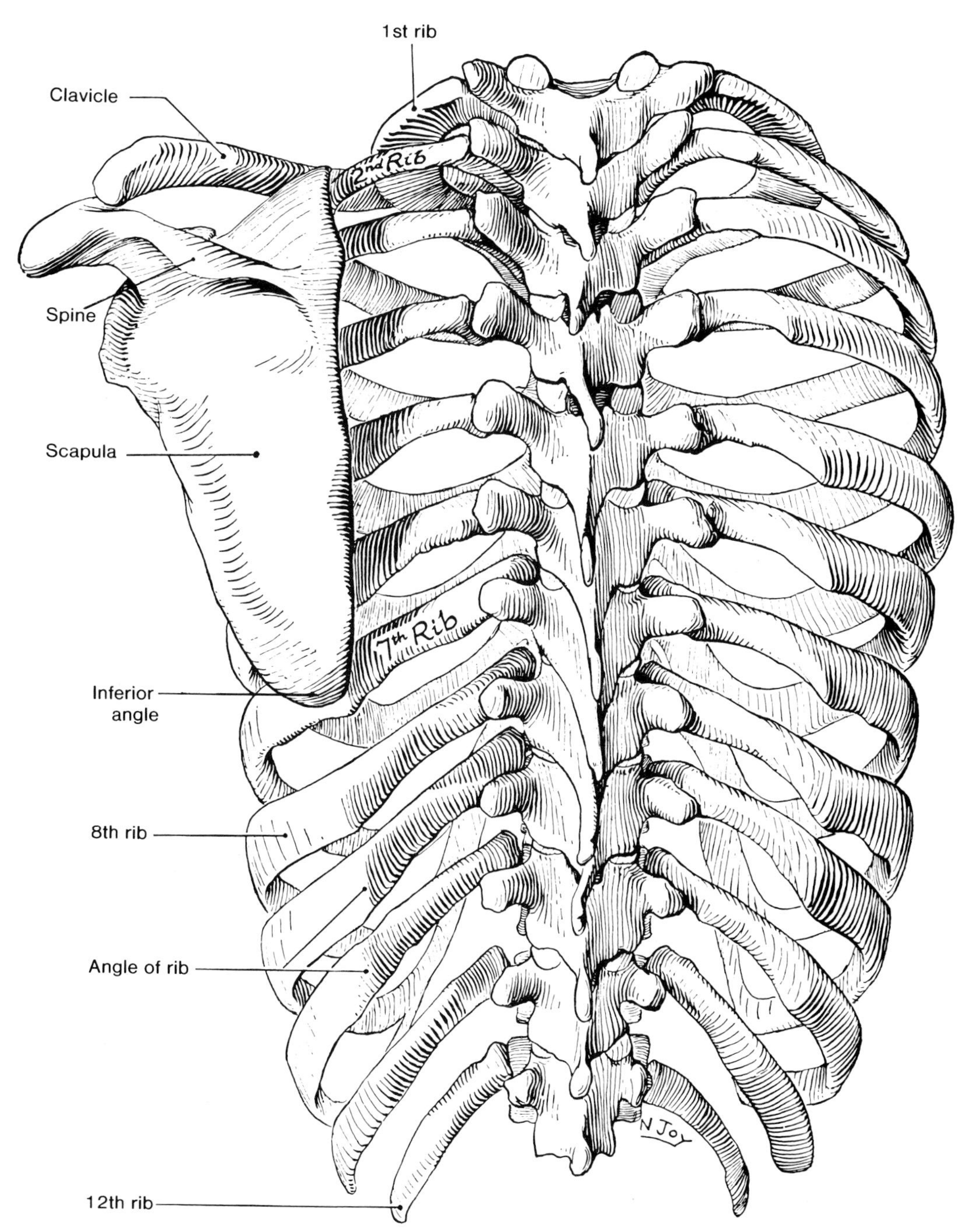

Figure 5-2. Drawing of the posterior aspect of the bony thorax. The left clavicle and scapula are also illustrated. Observe the progressive increase in length of the first seven ribs and costal cartilages. Although the seventh rib is the longest rib, the eighth and ninth ones are more oblique. Note that the scapula crosses the second to seventh ribs and that the seventh intercostal space is located just inferior to the tip of the inferior angle of the scapula.

begin at the base of the neck (Fig. 5–10), because what may appear to be an extra lumbar vertebra in a radiograph may be an extra throracic or sacral vertebra. The spinous processes (spines) of the vertebrae can be palpated as far inferiorly as the sacrum.

In Figure 5–1 observe that the vertebral bodies gradually become larger as the sacrum is approached, and then become progressively smaller toward the coccyx. These structural differences are related to the fact that *the lumbosacral region of the vertebral column carries more weight than the cervical and thoracic regions.*

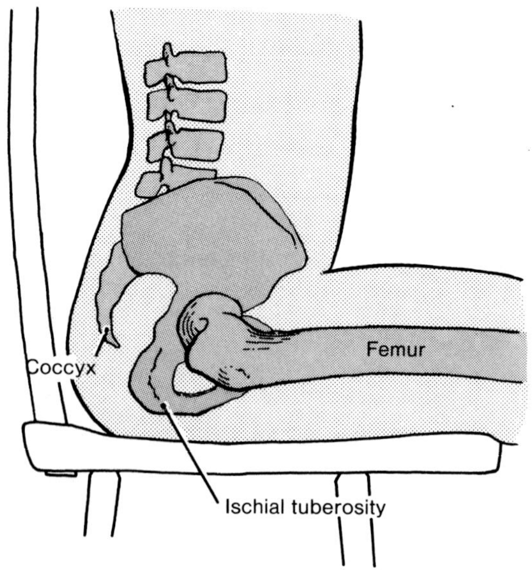

Figure 5-3. Diagram illustrating the relationship of the vertebral column to the pelvis during sitting. The coccyx was given its name because of its resemblance to a cuckoo's bill (G. *kokkyx*, a cuckoo). The vertebral column transmits the weight of the body through the pelvis to the lower limbs. When sitting, the weight is borne by the ischial tuberosities. Also see Figure 4–1.

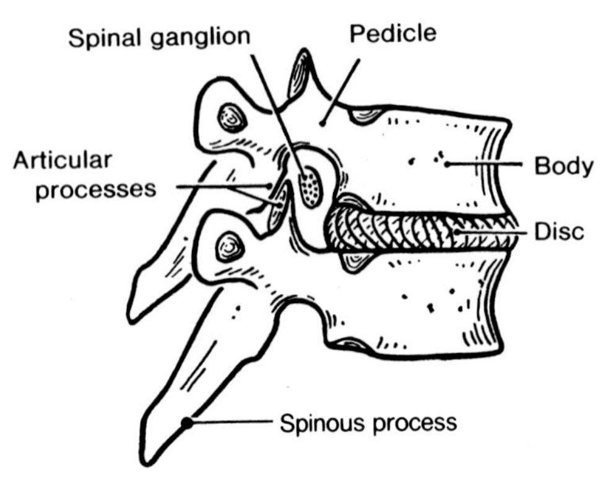

Figure 5-4. Lateral view of two thoracic vertebrae illustrating the intervertebral disc between them and the intervertebral foramen containing a spinal ganglion. The intervertebral discs are resilient pads that form fibrocartilaginous joints between the bodies of adjacent vertebrae. The vertebral arches are connected at articulations called zygapophysial joints (Figs. 5-30 and 5-31), between the articular processes (zygapophyses).

Not everyone has 33 vertebrae, but the number of cervical vertebrae is very constant in mammals, including man. Even the giraffe has only 7 cervical vertebrae, but they are long! However, variations occur in the number of thoracic, lumbar, and sacral vertebrae in about 5% of otherwise normal people. Differences in number can be either a change in one region (+ or −) without change in other regions, or a change in one region at the expense of another.

Although numerical variations may be clinically important, many have been detected at autopsy or in radiographs of persons with no history of back problems. Caution, is therefore required in ascribing symptoms (backache) to numerical variations of vertebrae.

In some people, the fifth lumbar vertebra is partly or completely incorporated into the sacrum (**sacralization of fifth lumbar vertebra,** Fig. 5–22). In other people, the first sacral vertebra is separated from the sacrum (**lumbarization of first sacral vertebra,** Fig. 5–23). The relationship between sacralization or lumbarization and back symptoms is unclear. Caution is therefore also required in ascribing symptoms to these conditions.

CURVATURES OF THE VERTEBRAL COLUMN

In the articulated vertebral column and in lateral radiographs, four anteroposterior curvatures are visible in the adult. The thoracic and sacral curvatures are concave anteriorly, whereas the cervical and lumbar curvatures are concave posteriorly (Fig. 5–1).

The thoracic and sacral curvatures develop during the embryonic period (Fig. 5–5); thus they are **primary curvatures.** The cervical and lumbar curvatures, called ***secondary curvatures***, begin to appear in the cervical and lumbar regions during the fetal period, but they are not very obvious until infancy.

The *cervical curvature* is accentuated as the infant

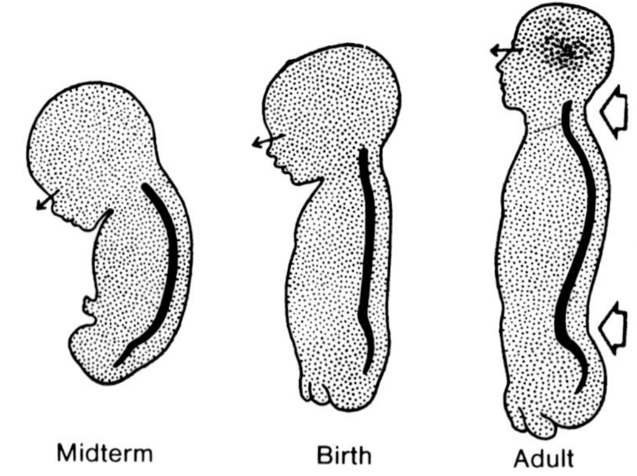

Figure 5-5. Drawings illustrating the development of the curvatures of the vertebral column. In the drawing of the adult, the cervical and lumbar curvatures are indicated by *open arrows.* All the adult curvatures are illustrated in Figure 5-1.

begins to hold its head erect, and the *lumbar curvature* becomes obvious when the child begins to walk.

The vertebral column in some people does not have normal curvatures because abnormal development and pathological processes, *e.g.*, **osteoporosis** (L. *os*, bone + *porosis*, porous), affect the bodies of one or more vertebrae in such a way that abnormal curvatures develop.

Kyphosis (G. humpback) is characterized by an abnormal curvature that is convex posteriorly (*i.e.*, **a posterior curvature of the vertebral column**). It usually occurs in the thoracic region (Fig. 5–6*D*). Thoracic kyphosis often develops in elderly people and is generally more marked in women (*e.g.*, the "*dowager's hump*").

Scoliosis (G. crookedness) indicates that the curvature is convex to the side (*i.e.*, **a lateral curvature of the vertebral column** (Fig. 5–6*E*). This is the most common type of abnormal curvature. Scoliosis may result from an asymmetric weakness of the vertebral muscles (*myopathic scoliosis*), or from failure of one half of the body and arch of a vertebra to develop (**hemivertebra**, Fig. 5–7). Many cases of scoliosis are of unknown origin and are referred to as *idiopathic scoliosis.*

Lordosis (G. backward bending) is characterized by an increased curvature of the vertebral column that is convex anteriorly (*i.e.*, **an anterior curvature of the vertebral column** (Fig. 5–6*C*). Excessive "hollowing" of the back or lordosis, often referred to as swayback or saddleback, generally occurs in the lumbar region.

Pregnant women often develop a temporary lumbar lordosis during the later stages of preg-

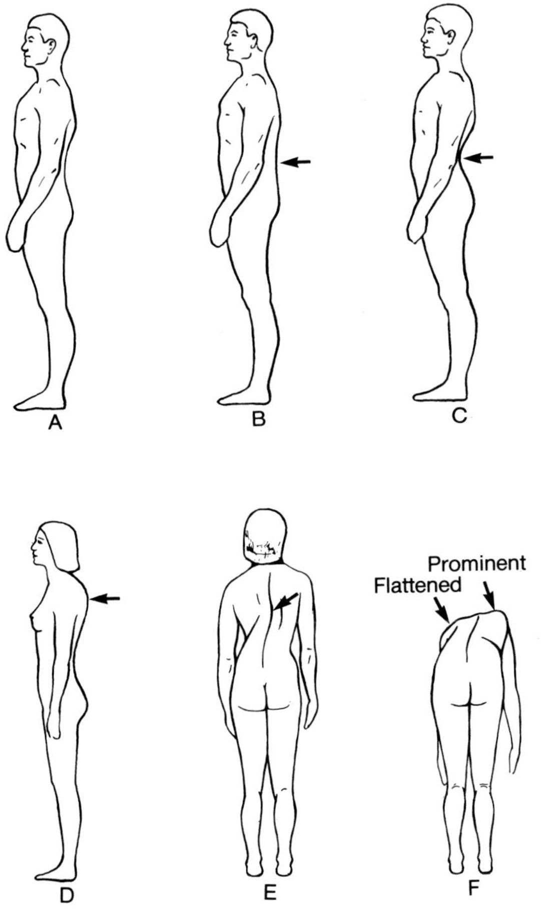

Figure 5-6. Drawings illustrating normal and abnormal curvatures of the vertebral column. *A*, normal curvatures of the back. *B*, flattening of the lumbar curvature. *C*, lumbar lordosis (accentuation of the normal curvature). *D*, thoracic kyphosis (rounded thoracic convexity). *E*, scoliosis with convexity to the right. *F*, accentuation of structural scoliosus occurs during flexion of the vertebral column. A scoliosis secondary to unequal length of the lower limbs would disappear (Modified from Bates B: *A Guide To Physical Examination*, ed 2. Philadelphia, JB Lippincott Co, 1979).

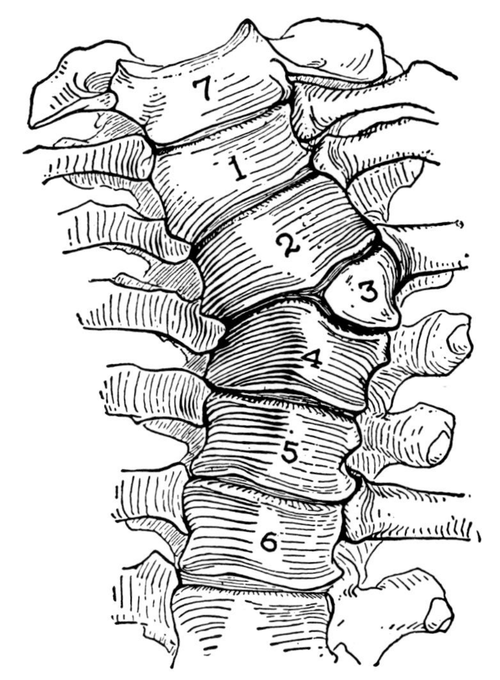

Figure 5-7. Drawing of a half vertebra (hemivertebra) that has produced a scoliosis (lateral curvature of the vertebral column) in the thoracic region. Half of the third thoracic vertebra and the corresponding rib are absent. In some cases, hemivertebrae are in addition to the normal vertebrae (also see Fig. 5-6*E*). Developmental failure of half a vertebra is one cause of scoliosus.

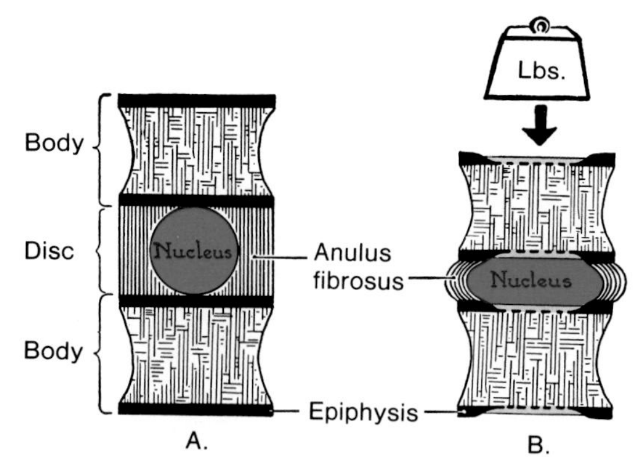

Figure 5-8. Diagrams of a fibrocartilaginous intervertebral disc, composed of a ring of fibrous tissue (anulus fibrosus) surrounding an internal semifluid mass (nucleus pulposus). Drawing *B* illustrates the cushioning value of the nucleus pulposus during weight bearing.

nancy in attempting to restore their line of gravity to the normal position. This abnormal curvature of the vertebral column disappears after childbirth.

Extreme obesity can also result in lumbar lordosis and low back pain owing to the increased weight of the abdominal contents. Loss of weight usually corrects the condition.

MOVEMENTS OF THE VERTEBRAL COLUMN

Movements between adjacent vertebrae take place on the resilient **nuclei pulposi** of the intervertebral discs (Figs. 5–4 and 5–8), and at the zygapophysial joints (Figs. 5–30 and 5–31). Although movements between adjacent vertebrae are relatively small, especially in the thoracic region, the result of all the small movements is a considerable range of movement of the vertebral column as a whole (Figs. 5–9 and 5–10). Movements of the vertebral column are freer in the cervical and lumbar regions than elsewhere; as stated previously, *these regions are the most frequent sites of aches.*

The main movements of the vertebral column are ***flexion*** (forward bending), ***extension*** (backward bending), **lateral bending** (lateral flexion), and ***rotation*** (twisting of the vertebrae relative to each other). Some ***circumduction***, a combination of flexion-extension and lateral bending, also occurs.

The thoracic region is relatively stable owing to its connection to the sternum via the ribs and costal cartilages. In addition, the intervertebral discs in the thoracic region are slightly thinner and the spinous processes of the thoracic vertebrae overlap (Figs. 5–1 and 5–2).

PALPATION OF THE VERTEBRAL COLUMN

The spinous processes of most of the presacral vertebrae can be palpated (Fig. 5–10). In the adult there is often a slight hollow posterior to the first cervical vertebra (**C1** or **atlas**), just inferior to the external occipital protuberance (Fig. 5–47), but the atlas is not palpable posteriorly because it has no spinous process (Fig. 5–16). The transverse processes of the atlas can usually be felt on deep palpation about 1.5 cm inferior to the tips of the mastoid processes (Figs. 5–16 and 5–47).

The spinous process of the second cervical vertebra (**C2** or **axis**) can be palpated about 10 cm inferior to the external occipital protuberance (Fig. 5–47). The spinous process of the axis is the first bony point that can be felt in the neck inferior to this protuberance.

The spinous processes of the third to fifth cervical vertebrae are difficult to palpate because they are short and lie at a considerable depth from the surface, from which they are separated by the *ligamentum nuchae* (Fig. 6–44). The spinous process of the sixth cervical vertebra is easily palpable in some people.

In most people the distinct posterior bony projection at the base of the neck is the spinous process of the seventh cervical vertebra (Fig. 5–10). **Because of the prominence of its spinous process, C7 vertebra is called the vertebra prominens** (Figs. 5–10, 5–16, and 5–18). Its spinous process can be palpated and observed most easily when the neck is flexed as far as possible.

Thoracic and lumbar spinous processes can be palpated, particularly in thin persons. In some people the spinous process of T1 may be as prominent as that of C7 (Fig. 5–10). Hence, when the spinous processes of two vertebrae appear equally prominent, they are C7 and T1. **To count the vertebrae,** *begin just inferior to the vertebra prominens* (*i.e.*, at T1) because the number of cervical vertebrae is constant, except when there is a congenital malformation of the cervical region, *e.g.*, the Klippe-Feil syndrome or **brevicollis** (see Moore, 1982 for details).

Because the spinous processes of T4 through T12 angle inferiorly, each overlies not its own vertebra but the body of the vertebra inferior to it. For example, the spinous process of T6 overlies the body of T7 and is adjacent to the seventh rib (Fig. 5–2).

The median crest of the sacrum (Fig. 5–24*B*), formed by the reduced spinous processes of three or four of its vertebrae, can be palpated but it is difficult to count the sacral vertebrae with certainty. *A line joining the skin dimples formed by the posterior superior iliac spines crosses the spinous process of the second sacral vertebra* (Figs. 5–11 and 5–42). The superior end of the natal cleft between the buttocks usually lies over the proximal end of the coccyx and the **sacral hiatus,** the aperture to the inferior end of the sacral canal (Fig. 5–24).

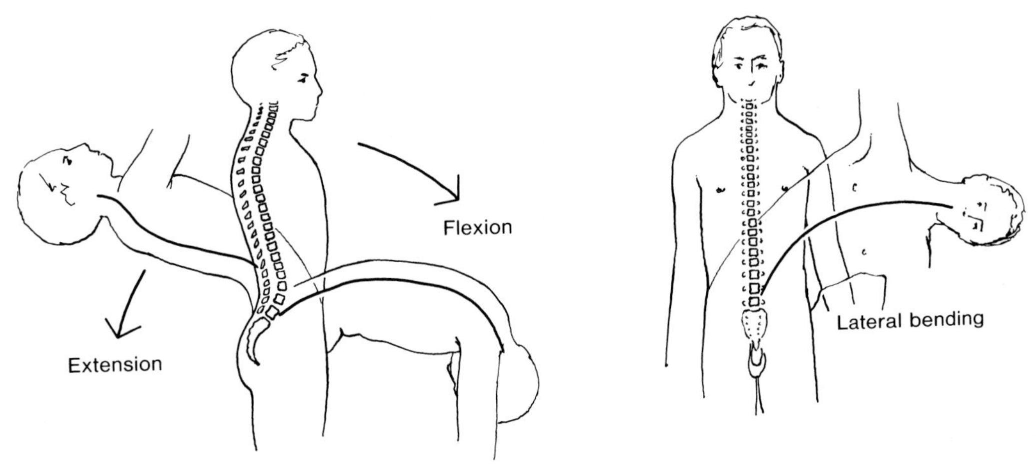

Figure 5-9. Drawings illustrating the main movements of the vertebral column: flexion (forward bending), extension (backward bending), and lateral bending (lateral flexion). Rotation and circumduction of the vertebral column are also possible. The range of movement varies according to the individual, being extraordinary in some (*e.g.*, acrobats who train from early in life). The range is limited by the thickness and compressibility of the intervertebral discs (Fig. 5-8), and by muscles and ligaments. *Extension* can be carried out to a greater degree than flexion, especially in the lumbar region. Extreme or excessive extension, causing damage to the intervertebral discs, back muscles, and ligaments is called *hyperextension* of the vertebral column. Some people use the term hyperextension to refer to posterior bending of the vertebral column beyond the position of maximum erectness, but *clinically hyperextension refers to extension beyond the normal limit* (*i.e.*, forcible overextension that strains or tears the ligaments). Hyperextension injuries occur most commonly in the cervical region (Fig. 5–62). These injuries rarely occur in the thoracic region owing to the extra support provided by the ribs.

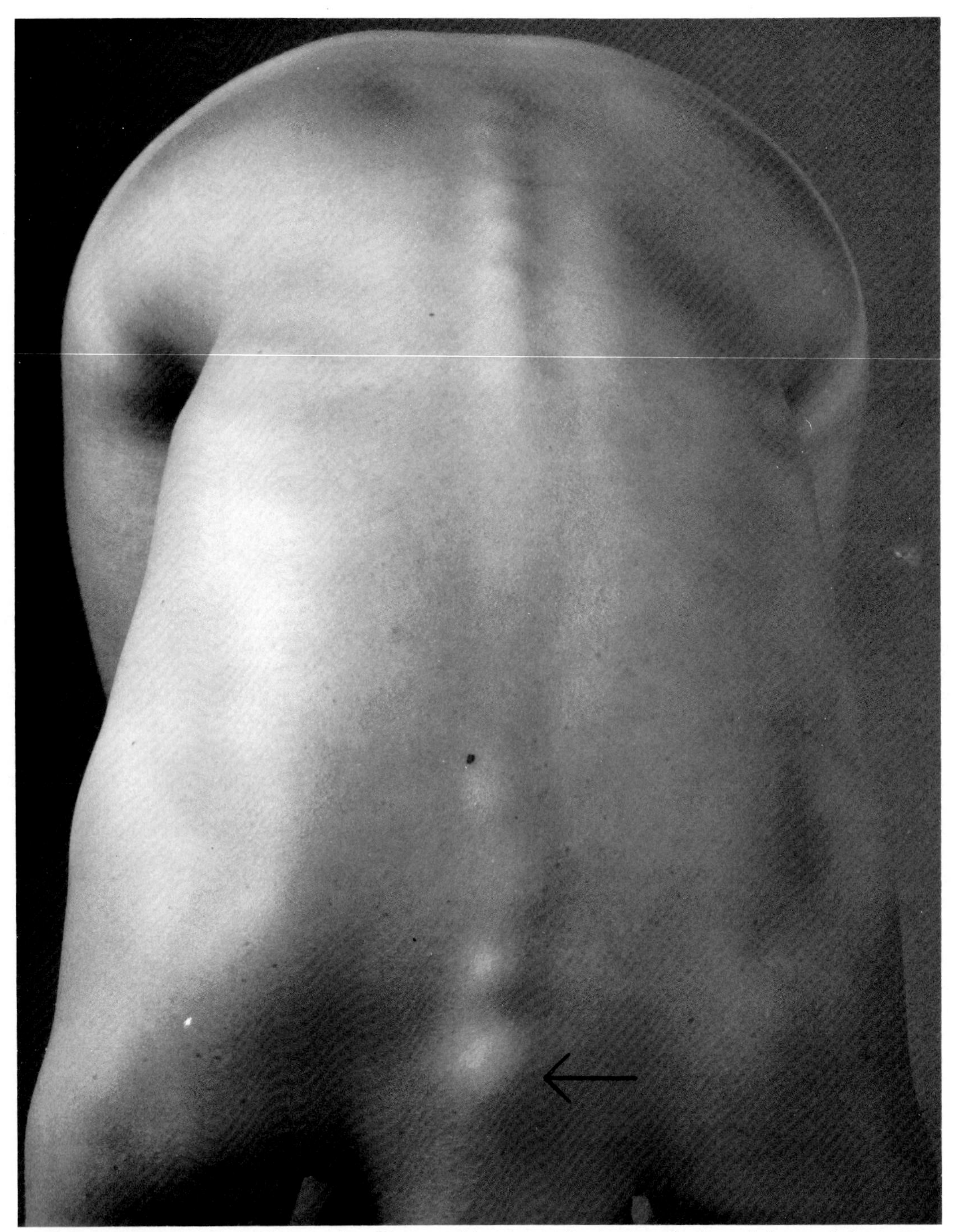

Figure 5-10. Photograph of the back of a 27-year-old woman illustrating flexion of the vertebral column and showing the vertebra prominens (*arrow*). Note that the spinous process of T1 is also prominent and that those of the lumbar vertebrae are visible in the "small of her back." Note that the spinous process of T2 vertebra deviates slightly from the median plane. A variation of this sort may be normal, but it could also be the result of a previous fracture/dislocation of the vertebra.

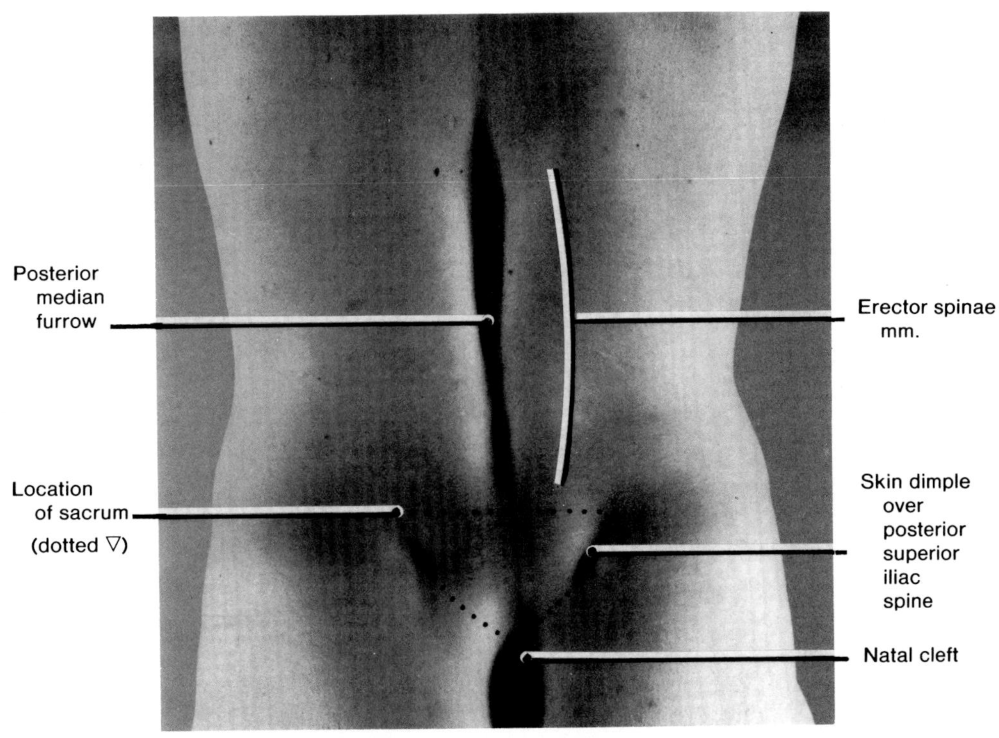

Figure 5-11. Photograph of the back of a 28-year-old man showing the principal surface landmarks. The dimples indicating the posterior superior iliac spines are very pronounced in this person. A line joining the skin dimples formed by the posterior superior iliac spines crosses the spinous process of the second sacral vertebra. Consequently these dimples are very useful landmarks clinically. Compare with the back of the woman shown in Figures 2–105 and 5-42. For a dissection showing the erector spinal muscles, see Figure 5-40.

During life the sacral hiatus is filled with fibrous tissue through which a local anesthetic can be injected in the sacral canal (**spinal anesthesia** or extradural anesthesia, Fig. 5–60*B*).

The curvatures of the vertebral column can be observed by studying the surface anatomy of the spinous processes of the vertebrae. Because abnormalities of these curvatures are relatively common and of great importance (Fig. 5–6), a thorough acquaintance with the normal curvatures is essential.

THE VERTEBRAE

A clear understanding of the general characteristics of vertebrae in the different regions of the vertebral column is essential knowledge. Should you specialize later (*e.g.*, in radiology, orthopaedics, or neurosurgery), you will need to learn the distinctive features of all vertebrae.

A TYPICAL VERTEBRA (Fig. 5–12)

The "typical" vertebra (*e.g.*, a midlumbar vertebra) is composed of two parts, a **body** and a **vertebral arch** which has several processes. Each part has a specific function.

Typical vertebrae vary in size and other characteristics from one region to another, and to a lesser degree within each region. Examine the thoracic vertebrae in Figure 5–1, noting that their bodies gradually become larger, particularly from T4 inferiorly.

Parts of a Typical Vertebra (Fig. 5–12). A typical vertebra consists of a body anteriorly and a vertebral arch posteriorly. Seven processes arise from the vertebral arch.

The body (Figs. 5–12 to 5–15) is the heavy anterior part of a vertebra which resembles a *short, long bone*. Its function, like long bones, is to support weight. The bodies of the vertebrae from C3 to S1 become progressively larger in order to bear progressively greater weight (Fig. 5–1).

The vertebral arch (Figs. 5–12 to 5–14) is the posterior part of a vertebra. The vertebral arch (neural arch) is attached to the body. *The vertebral arch protects the neural tissues* (spinal cord and spinal nerve roots) from injury.

The vertebral arch is formed by two **pedicles.** (L. little feet) and two **laminae** (L. thin plates). Four articular processes, two transverse processes, and one

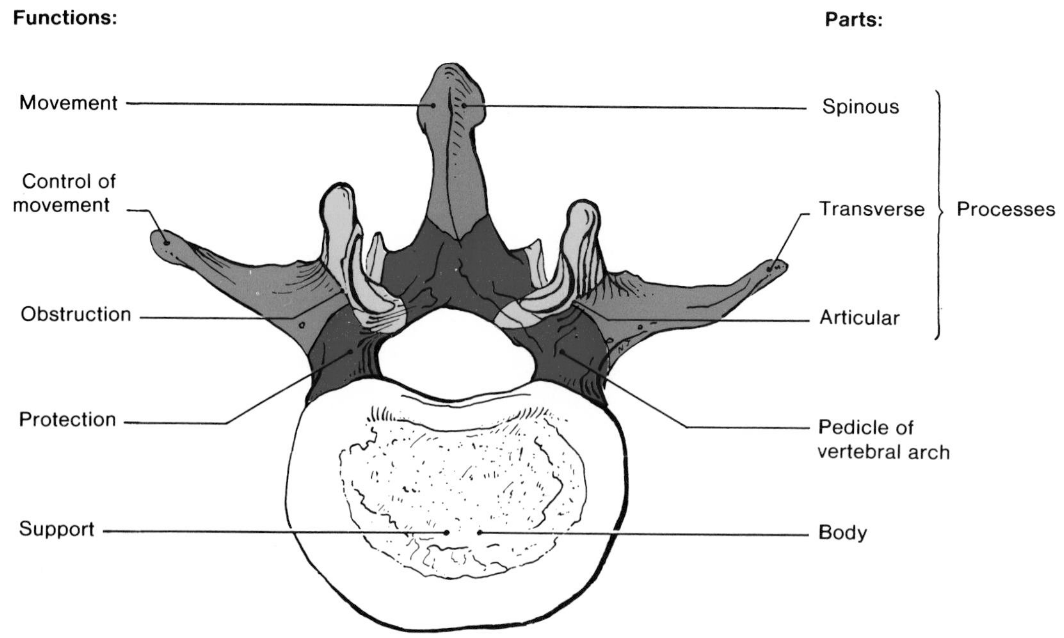

Figure 5-12. Drawing of a typical midlumbar vertebra illustrating the functions of its parts. Hyaline cartilage covers the rough superior and inferior surfaces of the body, whereas the smooth rounded rims are covered with fibrocartilage.

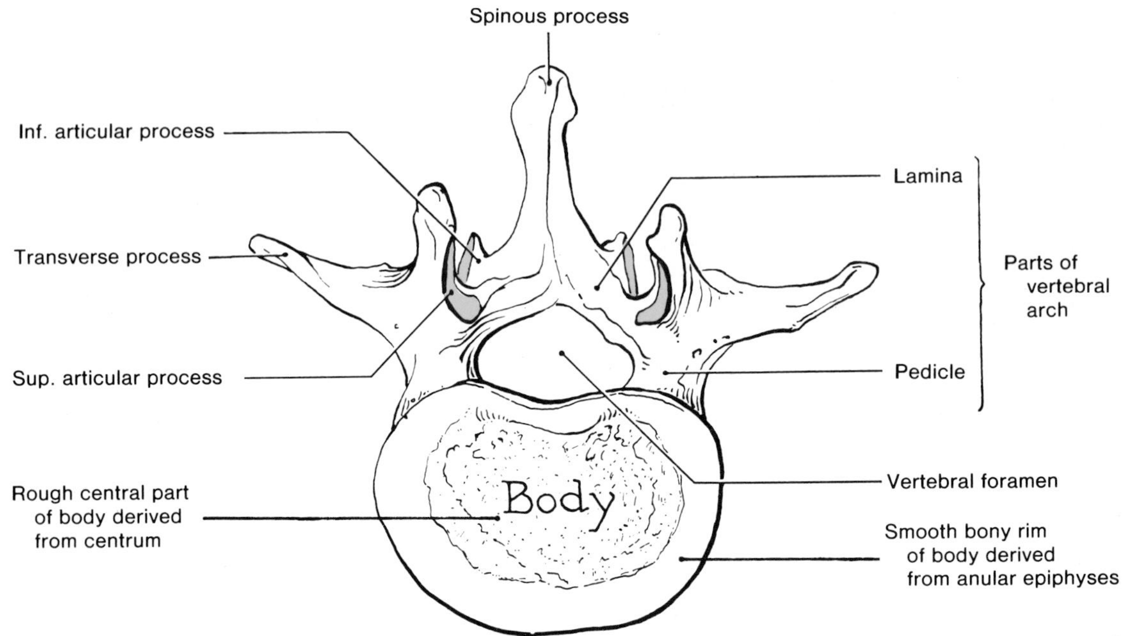

Figure 5-13. Drawing showing the parts of the second lumbar vertebra, superior view. For an explanation of the terms centrum and anular epiphyses, see Figure 5-26.

spinous process (spine) arise from the vertebral arch (Fig. 5–12).

The vertebral arch encloses an aperture known as the **vertebral foramen** (Figs. 5–13 and 5–19). Successive vertebral foramina form the **vertebral canal** (spinal canal), which contains the spinal cord and its meninges, nerve roots, and blood vessels (Figs. 5–50 to 5–54).

The pedicles of the vertebral arch are attached anteriorly to the body and are continuous posteriorly with the flat laminae. There is a small notch superior to each pedicle, the **superior vertebral notch** (Fig.

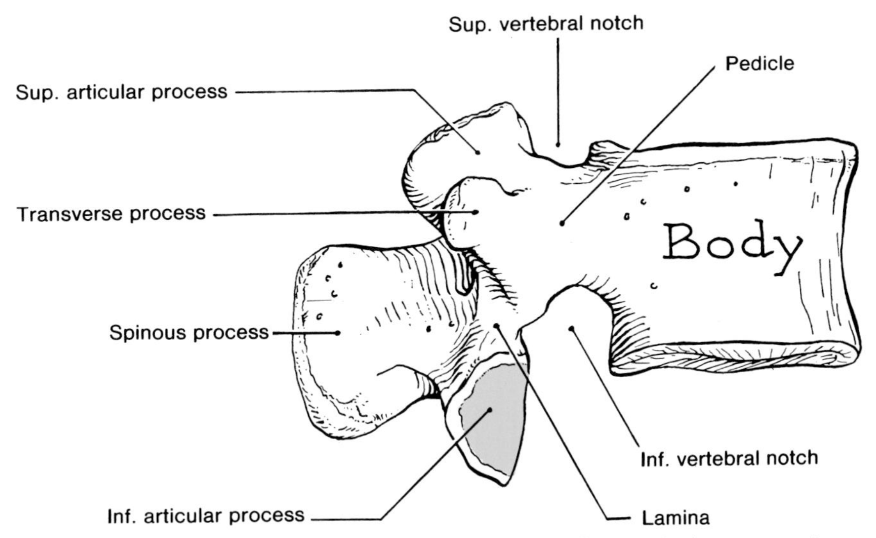

Figure 5-14. Drawing of a lateral view of a second lumbar vertebra. Each articular process has an articular facet. The facet (*yellow*) of the inferior articular process is illustrated. The articular processes (zygapophyses) articulate at the zygapophysial joints (Figs. 5-30 and 5-31); for brevity, clinicians often call them facet joints.

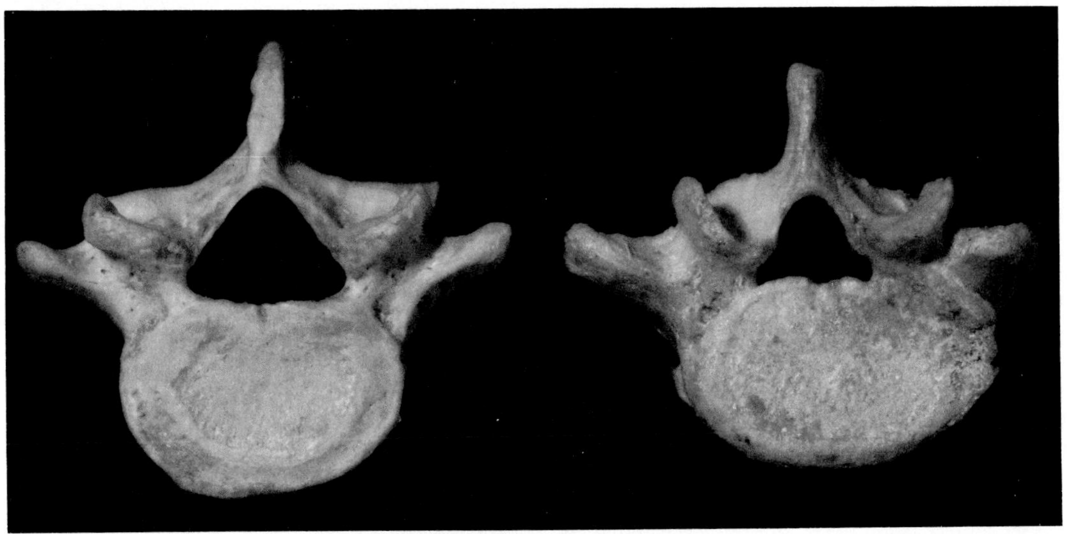

Figure 5-15. Photographs of fifth lumbar vertebrae illustrating the varying sizes and shapes of the vertebral canal in different persons. The common oval-triangular form is shown on the left, whereas the uncommon trifoliate (trefoil) pattern is illustrated on the right. A small trefoil-shaped canal is a contributory factor in cases of sciatica and/or cauda equina claudication (lower limb pain resulting from pressure on spinal nerve roots).

5–14), which varies in size at different levels of the vertebral column. The notch inferior to each pedicle, the **inferior vertebral notch**, is usually larger. When two vertebrae are in articulation (Fig. 5–4), the vertebral notches are adjacent to each other and form an almost complete bony ring, the **intervertebral foramen** (Figs. 5–4 and 5–30).

The dorsal and ventral nerve roots are in the vertebral canal and the **spinal ganglia** (dorsal root ganglia) are in the intervertebral foramina (Figs. 5–4 and 5–51). The nerve roots join each other at the external edge of the intervertebral foramen to form a **spinal nerve** (Figs. 5–33 and 5–51).

The vertebral processes (Figs. 5–12 to 5–14) arise from the vertebral arch. Typical vertebrae have seven processes: three are lever-like (spinous process and two transverse processes), and four are articular. Muscles and ligaments attach to the lever-like processes which act as levers during movements of the vertebrae.

The spinous processes project posteriorly or posteroinferiorly (Fig. 5–1), usually in the median plane, from the place of union of the laminae (Figs. 5–12 to

5–14). The spinous processes serve as attachments for the interspinous and supraspinous ligaments, and many muscles (Figs. 5-29 and 5-60). These ligaments strengthen the vertebral column posteriorly.

The transverse processes project laterally from the junctions of the pedicles and laminae (Figs. 5-12 to 5-14). The transverse processes act as attachments for muscles, helping to increase their leverage over the vertebral column.

The articular processes (zygapophyses) arise near the junction of the pedicles and laminae (Figs. 5–12 to 5–14). Each articular process bears an articular facet. The contact between the superior and inferior **articular facets** of an articulated vertebral column helps to prevent anterior movement of an superior vertebra on an inferior one, especially in the thoracic and lumbar regions. The articular facets allow some flexion and extension as well as varying degrees of lateral bending (lateral flexion) and rotation.

REGIONAL CHARACTERISTICS OF THE VERTEBRAE FORMING THE VERTEBRAL COLUMN

The vertebrae in the various regions of the vertebral column have distinctive characteristics which enable them to be identified with relative ease. In addition, there are obvious differences in the size of the vertebral foramen in the same regions of different persons (Fig. 5–15). There are also differences in the size of the vertebral foramina in the various regions. Verify this by inserting your finger into the vertebral foramen of a midthoracic vertebra, and then into the foramen of a first and a twelfth thoracic vertebra. Note that the vertebral foramina are larger in these regions because the spinal cord is enlarged in these locations for innervation of the limbs (Fig. 5–56).

Although there are characteristic regional differences in the vertebrae, only an expert can differentiate a fifth from a sixth thoracic vertebra, or a second from a third lumbar vertebra. However, *everyone* who takes an anatomy course should be able to distinguish a cervical vertebra from a thoracic vertebra and a thoracic vertebra from a lumbar vertebra.

The Cervical Vertebrae (Figs. 5–16 to 5–18). *Cervical vertebrae form the bony axis of the neck.* **Their distinctive feature is the foramen transversarium** (transverse foramen) in each transverse process. The foramina transversaria in C7 are smaller than those of the other cervical vertebrae and occasionally they are absent.

The vertebral arteries pass through the foramina transversaria (Figs. 5–51 and 7–37), except in C7, which transmit only small accessory vertebral veins.

The spinous processes of the third to sixth cervical vertebrae are short and bifid (L. *bifidus*, cleft in two parts). The spinous process of the seventh cervical vertebra is long and non-bifid (Fig. 5–16). It produces the superior prominence in the posterior aspect of the neck (Figs. 5–10 and 5–18). Because of this feature, it is called the **vertebra prominens.** Another characteristic of cervical vertebrae is the almost equal sizes of their superior and inferior vertebral notches.

The first, second, and seventh cervical vertebrae are atypical.

The first cervical vertebra, a ring-shaped bone, is called the **atlas** (Figs. 5-16 to 5-18). Because it supports the skull, it was named after a Titan called Atlas who, according to Greek mythology, supported the earth on his shoulders. The kidney-shaped, concave, superior articular facets of the atlas receive the **occipital condyles** (Figs. 5-17, 5-47, 7-8, and 7-12).

The atlas has no spinous process or body; it consists of anterior and posterior arches, each of which has a tubercle and a lateral mass (Figs. 5-16 and 5-17).

The second cervical vertebra, the strongest of the cervical vertebrae, is known as the **axis** (Fig. 5-16). It has two flat bearing surfaces, the superior articular facets, upon which the atlas rotates. Its distinguishing feature is the blunt tooth-like **dens** (odontoid process). The strong dens (G. tooth) is held in position by the transverse ligament of the atlas (Fig. 5-17), which prevents horizontal displacement of the atlas. The second cervical vertebra or axis has a large bifid spinous process (Fig. 5-16), which is the first one that can be felt in the posterior groove of the neck (*nuchal furrow*). The Latin term **nucha** refers to the *nape* (the posterior aspect or "scruff" of the neck).

The seventh cervical vertebra is usually called the vertebra prominens because its long spinous process is visible through the skin (Fig. 5-10). C7 also has large transverse processes (Fig. 5-16). The vertebra prominens is easily recognized in a lateral radiograph of the neck (Fig. 5-18).

The Thoracic Vertebrae (Figs. 5-19 and 5-20). *All 12 thoracic vertebrae* ***articulate with ribs***; thus, they are characterized by articular facets for them (Figs. 5-2 and 5-19). There is one or more facet on each side of the body for articulation with the head of a rib, and one on each transverse process of the superior 10 vertebrae for the tubercle of a rib. The spinous processes tend to be long and slender, and those of the middle ones are directed inferiorly over the vertebral arches of the vertebrae inferior to them (Figs. 5-2 and 5-19*B*).

The middle four thoracic vertebra are typical. The outline of their bodies, viewed from the superior aspect, is heart-shaped and their vertebral foramina are circular (Fig. 5-19*A*). Sometimes an impression is visible on the left sides of the bodies of the middle thoracic vertebrae, which is produced by the descending thoracic aorta (Fig. 1-15).

The first four thoracic vertebra have some features of cervical vertebrae. The first thoracic vertebra differs from typical thoracic vertebrae, in that it has an

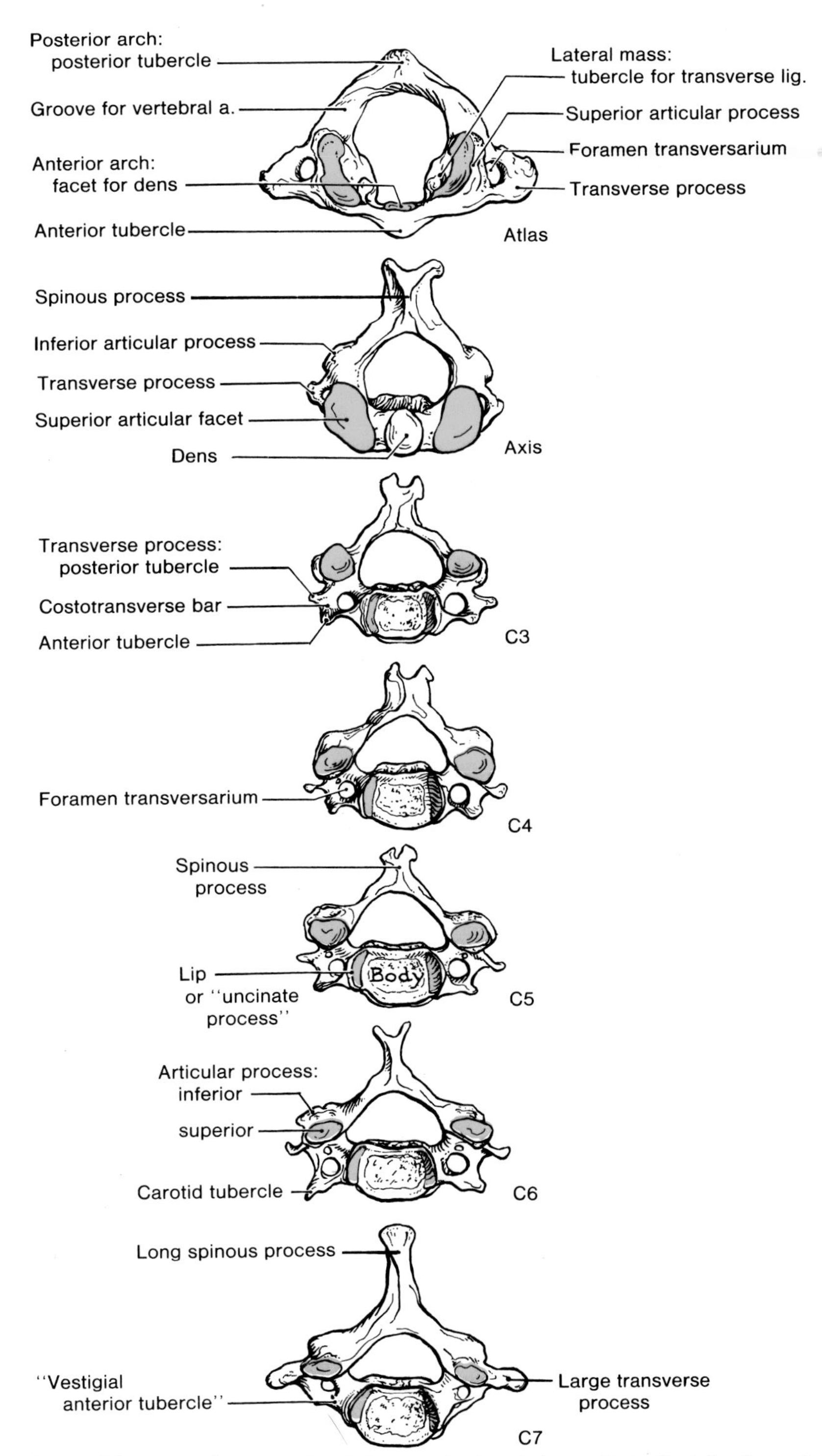

Figure 5-16. Drawings of the superior aspects of the cervical vertebrae. Note that the foramina transversaria in C7 are smaller than those of the other cervical vertebrae; occasionally they are absent. Observe that the seventh cervical vertebra has a long spinous process and large transverse processes. It is often called the vertebra prominens (Figs. 5-10 and 5-18).

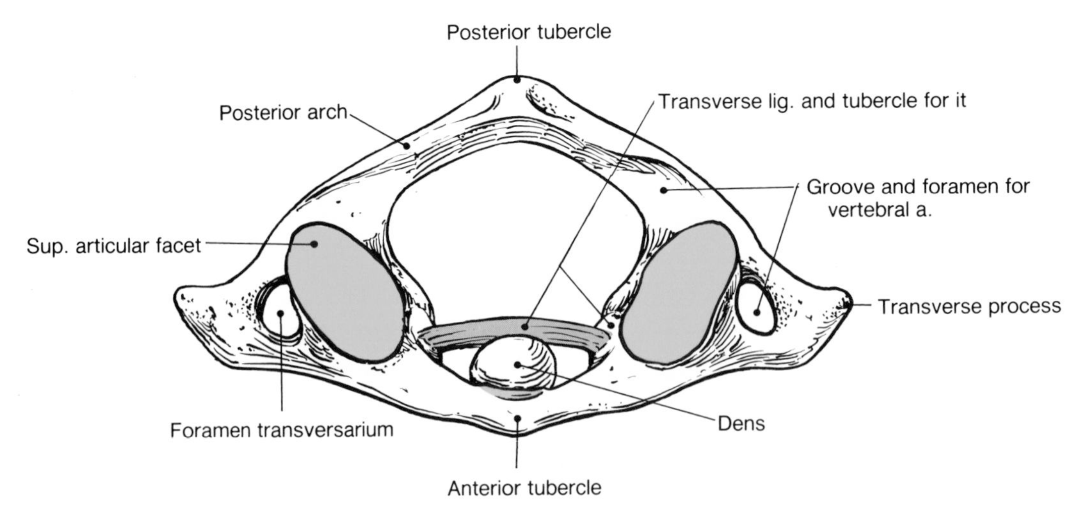

Figure 5-17. Drawing of a superior view of the first cervical vertebra or atlas. It has no body or spinous process. Note that the atlas is a ring of bone consisting of slender anterior and posterior arches united by a lateral mass. The dens of the atlas (C2) is held in place by the transverse ligament. Also see Figure 5–35.

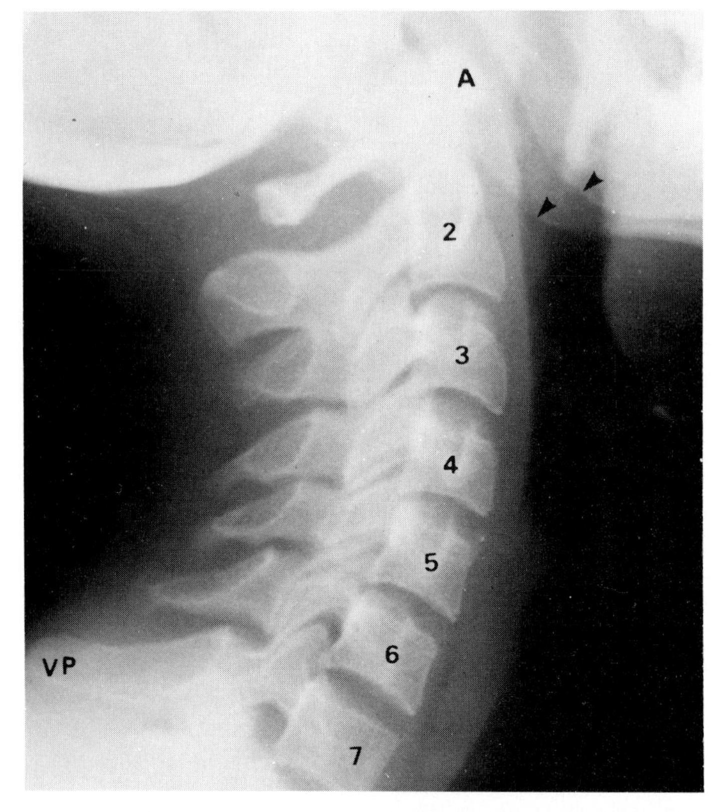

Figure 5-18. Lateral radiograph of the cervical region of the vertebral column in which the bodies of cervical vertebrae 2 to 7 have been numbered. Note that the anterior arch of the atlas (*A*) is in a plane anterior to the curved line joining the anterior aspects of the bodies of the other cervical vertebrae. Note also the vertebra prominens (C7), characterized by its long spinous process, labeled *VP*. This process is visible through the skin at the inferior end of the nuchal furrow (Fig. 5-10). The *arrows* point to the angles of the mandible which are not perfectly superimposed upon each other.

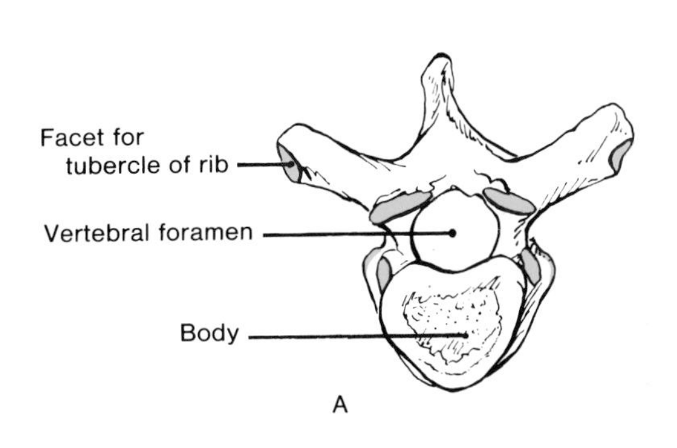

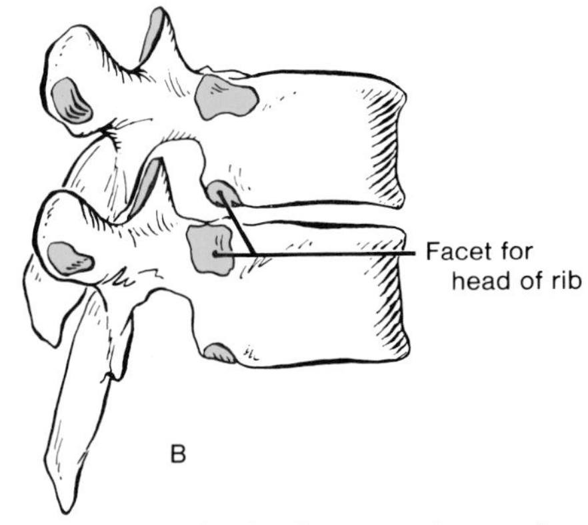

Figure 5-19. Drawings of middle or typical thoracic vertebrae showing their distinctive features. *A*, superior view. *B*, lateral view. In *A*, note the heart-shaped body and the circular vertebral foramen.

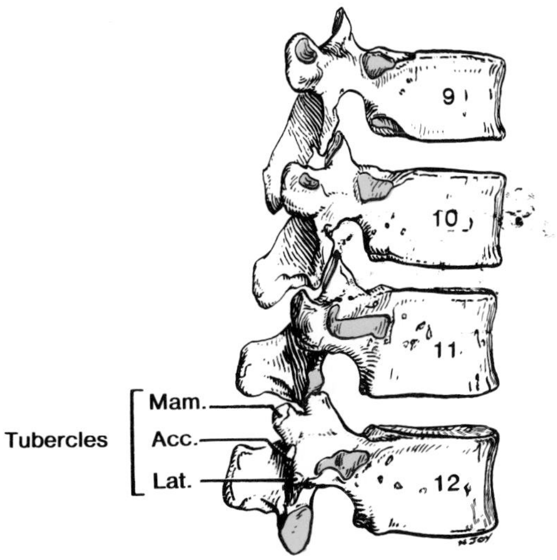

Figure 5-20. Drawings of the inferior four thoracic vertebrae. Observe that they are atypical; compare them with the typical thoracic vertebrae shown in Figure 5-19.

almost horizontal spinous process and long transverse processes. The spinous process of the first thoracic vertebra may be as long as that of the vertebra prominens (Fig. 5-10). The first thoracic vertebra has a complete costal facet on the superior edge of the body for the first rib, and a demifacet on the inferior edge which contributes to the articular surface for the second rib.

The inferior four thoracic vertebrae are atypical (Fig. 5-20). They often have features of lumbar vertebrae and possess mammillary, accessory, and lateral tubercles.

The spinous processes of all the thoracic vertebrae can be palpated and observed through the skin when the trunk is flexed (Fig. 5-10).

The Lumbar Vertebrae (Figs. 5-12 to 5-15 and 5-21). These vertebrae are in the "small of the back," and their spinous processes are prominent when the vertebral column is flexed (Fig. 5-10).

Lumbar vertebrae may be distinguished by their relatively large bodies, as compared with cervical and thoracic vertebrae, *and by the absence of costal facets.* Their vertebral bodies, viewed from their superior aspects, are kidney-shaped and their vertebral foramina are oval to triangular (Fig. 5-21).

The fifth lumbar vertebra, the largest of all movable vertebrae, is *characterized by stout transverse processes* which arise from the lateral and posterolateral aspects of its body (Figs. 5-15 and 5-21). The fifth lumbar vertebra is largely responsible for the **lumbosacral angle** between the lumbar region of the vertebral column and the sacrum (Figs. 5-1 and 5-25*B*).

In about 5% of otherwise normal persons, L5 vertebra is partly or completely incorporated into the sacrum, conditions known as hemisacralization or **sacralization of the fifth lumbar vertebra** (Fig. 5-22).

The Sacrum (Fig. 5-24). This large, triangular, *wedge-shaped bone* is usually composed of five fused sacral vertebrae in the adult. The sacrum (L. lit. sacred bone) *provides strength and stability to the pelvis* and transmits the weight of the body to the pelvic girdles through the sacroiliac joints (Figs. 3-42*A* and 4-1).

On the pelvic (Fig. 5-24*A*) and dorsal surfaces (Fig. 5-24*B*), there are typically four pairs of foramina for the exit of the anterior and posterior primary divisions of the sacral nerves (Fig. 3-54). Observe that the pelvic foramina are larger than the dorsal ones. Also notice that there are four transverse lines on the pelvic surface which indicate where fusion of the five sacral vertebrae occurred after the 20th year.

The base of the sacrum is formed by the superior

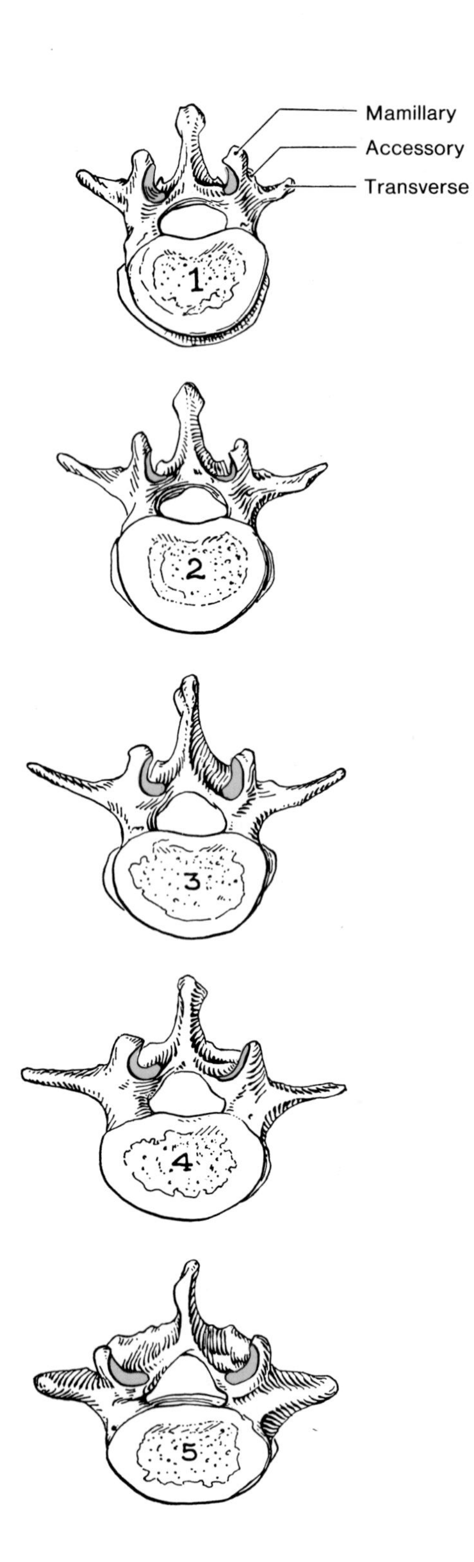

Figure 5-21. Drawings of superior views of the lumbar vertebrae. Observe the stout transverse processes of L5 that are connected to the whole of the lateral surface of the pedicle of the vertebral arch, and to the posterolateral part of the body. Lumbar vertebrae may be distinguished by their relatively *large bodies*, as compared with cervical and thoracic vertebrae, and by the *absence of costal facets*.

surface of the first sacral vertebra. The superior articular processes that articulate with the inferior articular processes of the fifth lumbar vertebra (L5) project superiorly from the base (Fig. 5-24*A*). The projecting anterior edge of the body of the first sacral vertebra is called the **sacral promontory** (L. *promontorium*, mountain ridge). See Figures 5-24*A* and 5-25.

The sacrum supports the vertebral column and forms the posterior part of the bony pelvis (Figs. 3-7, 5-1, 5-25, and 5-42). It is tilted so that it articulates with L5 vertebra at an angle, the **lumbosacral angle** (Fig. 5-1). The sacrum is often wider in proportion to length in the female than in the male, but the body of the first sacral vertebra is usually larger in males.

The dorsal surface of the sacrum is rough, convex, and marked by five prominent longitudinal ridges (Fig. 5-24*B*). The central one, **the median sacral crest,** represents the fused spinous processes of the sacral vertebra. *The median sacral crest can be palpated in the median plane*, superior to the natal cleft (Fig. 5-11).

The *intermediate sacral crest* represents the fused articular processes, and the *lateral sacral crest* represents portions of the transverse processes of the sacral vertebrae. The tips of these processes appear as a row of tubercles (Fig. 5-24*B*).

Because the articular surface of the lateral aspect of the sacrum looks somewhat like an auricle (L. external ear), it is called the **auricular surface** (Figs. 3-45*A* and 5-24*B*). This is the site of the *sacroiliac joint* (Figs. 3-42 and 3-46), located between the sacrum and the ilium.

Observe the Λ-shaped **sacral hiatus** (L. an aperture) on the dorsal surface of the sacrum, and the **sacral cornua** (L. horns). These horns, representing the inferior articular pocesses of S5 vertebra, project inferiorly on each side of the hiatus (Fig. 5-23*B*). Palpate the sacral hiatus and cornua in the superior end of your natal cleft (Fig. 5-11).

Anesthetic agents are sometimes injected through the sacral hiatus, a procedure called **extradural anesthesia** or epidural anesthesia or caudal anesthesia (Fig. 5-60*B*). The anesthetic acts on the sacral and coccygeal nerves (Fig. 5-54). As the sacral hiatus is between the sacral cornua, the horns are important bony landmarks for locating the hiatus. Anesthetic agents can also be injected through the dorsal sacral foramina (Fig. 5-24*B*).

In some people the first sacral vertebra is more or less separated from the sacrum **(lumbarization of the first sacral vertebra).** In some cases of lumbarization of S1 (Fig. 5-23), or of sacralization of L5 (Fig. 5-22), it is the first "normal articulation" (intervertebral disc plus two zygapophysial joints) that takes the strain. Consequently this articulation may degenerate prematurely or extensively. For example, when L5 is sacralized (Fig. 5-22) the L5/S1 level is strong and the L4/L5 level degenerates, often producing symptoms (*e.g.*, backache).

You are encouraged to become familar with the main radiographic features of the lumbosacral region of the vertebral column because low back pain is such a common complaint (Fig. 5-25). The planes of the lumbar **zygapophysial joints** are variable, but this variation is of debatable clinical significance. In perfect lateral views of the lumbosacral region, the images of the right and left articular processes and the intervertebral foramina are projected on each other so they appear as single rather than double images (Fig. 5-25*B*).

The Coccyx (Figs. 5-1, 5-24, and 5-25*B*). The coccyx (tailbone) is the remnant of the tail which human embryos have until the beginning of the 8th week. Usually four rudimentary vertebrae are present, but there may be one less or one more. The vertebrae consist of bodies only, except for the first coccygeal vertebra, which has **cornua** that represent remnants of the pedicles and transverse processes of the vertebra (Fig. 5-24*B*).

The three inferior coccygeal vertebrae often fuse during middle life, forming a beak-like bone; this accounts for the name coccyx, the Greek word for a cuckoo. During old age, the first coccygeal vertebra often fuses with the sacrum (Fig. 5-1).

The coccyx gives no support to the vertebral column, but it provides origin for part of the gluteus maximus and coccygeus muscles and the anococcygeal ligament (Figs. 3-20 and 3-52). The inferior surface and tip of the coccyx can be palpated in the natal cleft, 2–5 cm posterior to the anus. The anterior surface of the coccyx can be palpated with the gloved forefinger in the rectum and the thumb in the natal cleft.

The coccyx gives origin to part of the sacrotuberous ligament and provides an attachment for muscles of the pelvic floor (Figs. 3-52 and 3-54).

The coccyx is joined to the sacrum by cartilage (cartilaginous joint); therefore it can bend to some extent during childbirth. In unusual cases the coccyx may separate from the sacrum, producing **coccydynia** (pain in the coccygeal re-

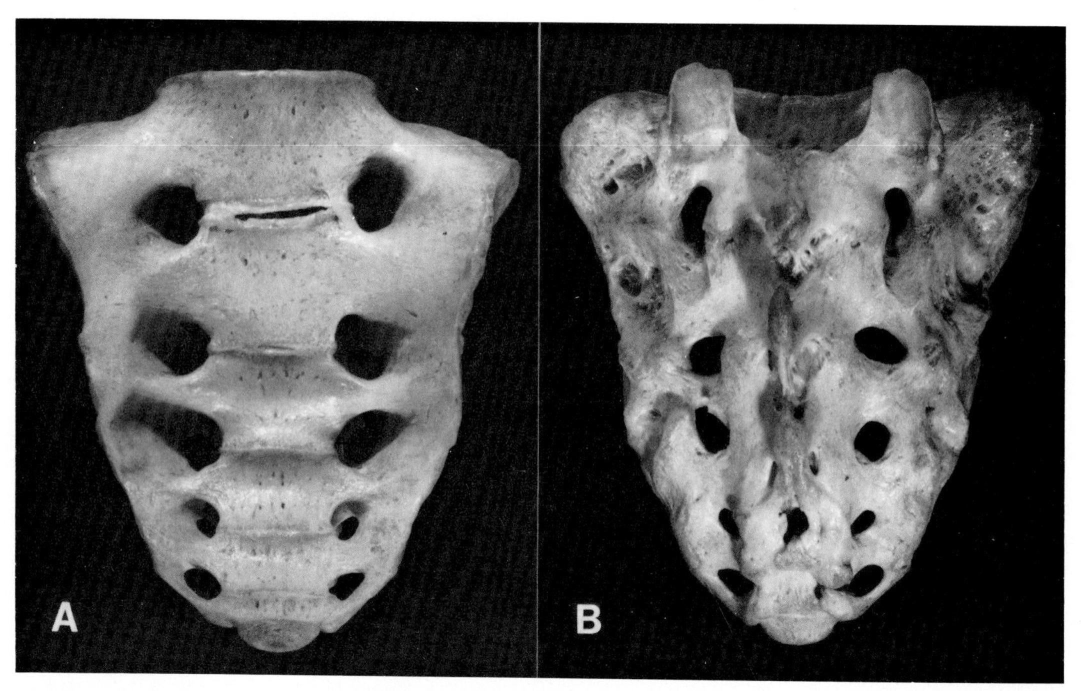

Figure 5-22. Photographs of a sacrum showing sacralization of the fifth lumbar vertebra. *A*, pelvic (ventral) surface. *B*, dorsal surface. As fusion of the fifth lumbar vertebra with the sacrum is almost complete in this case, only four lumbar-type vertebrae would probably be recognized in a radiograph.

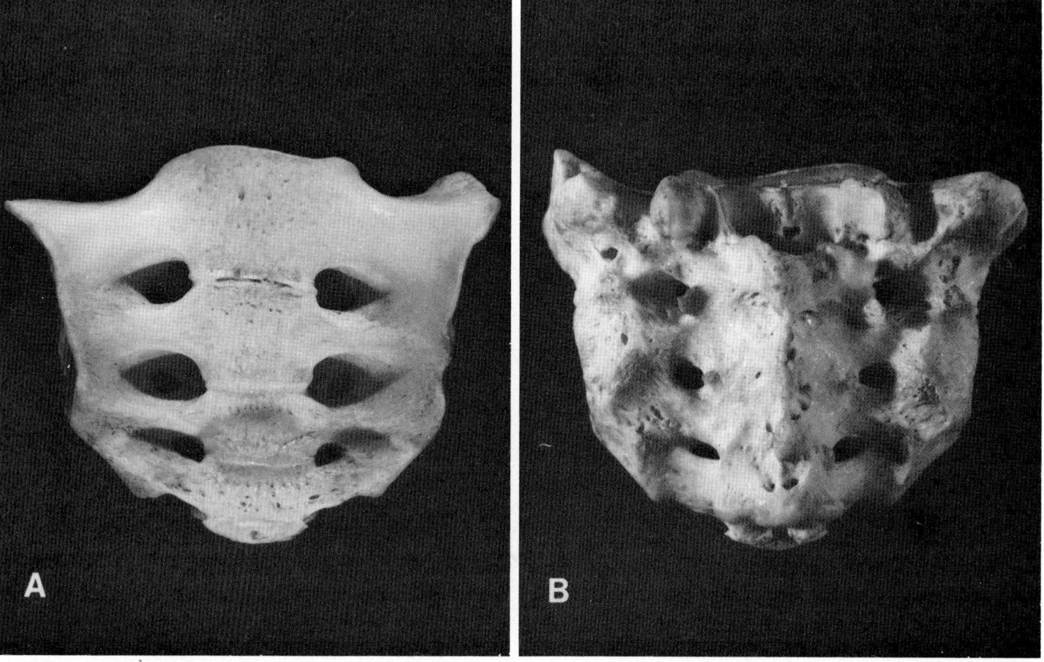

Figure 5-23. Photographs of a sacrum showing only four fused vertebrae, probably as a result of lumbarization of the first sacral vertebra. *A*, pelvic surface; *B*, dorsal surface. Compare with a normal sacrum (Fig. 5-24).

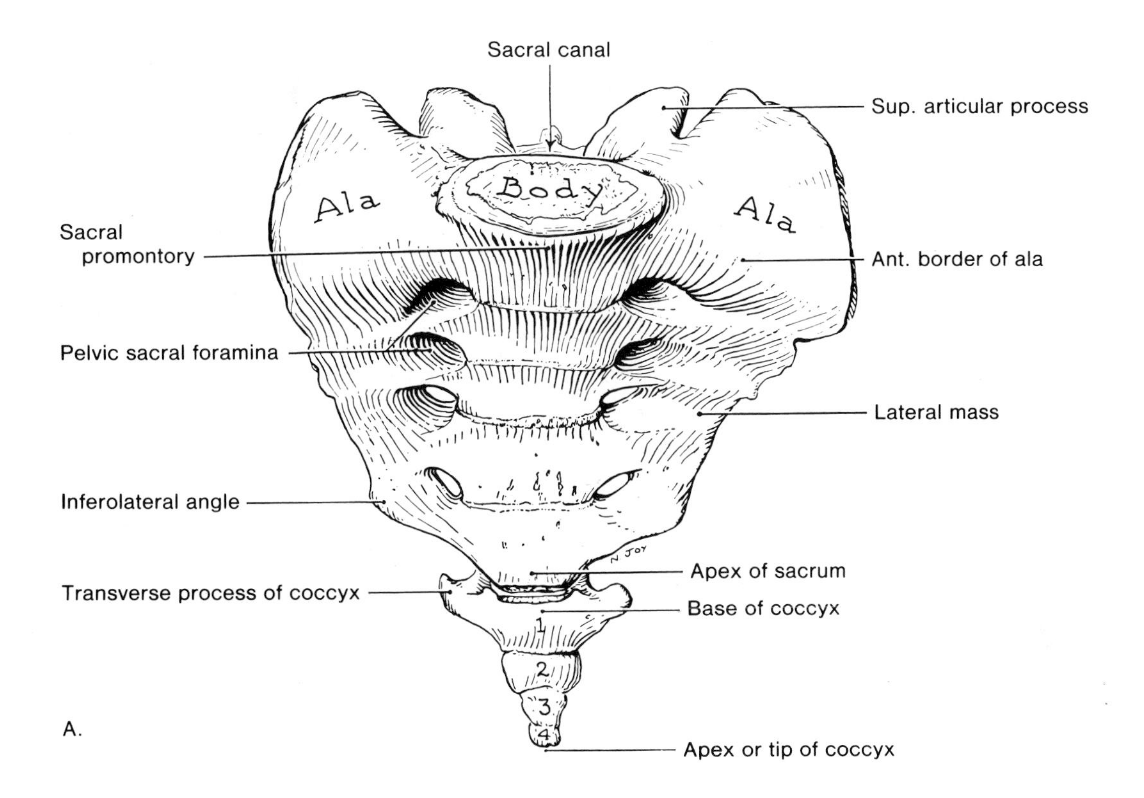

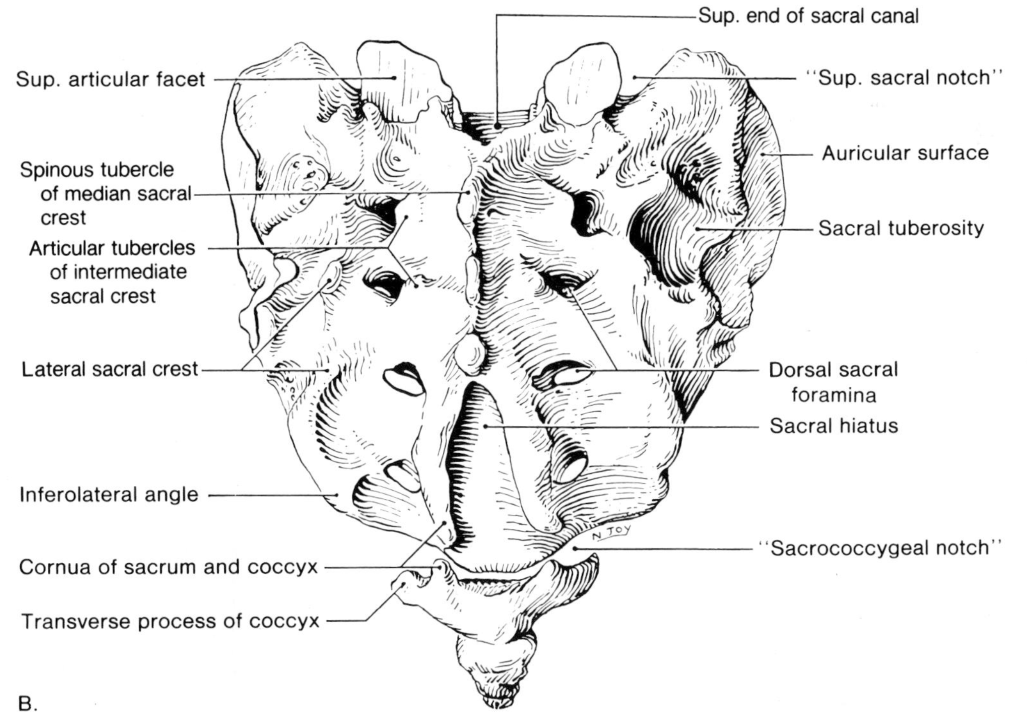

Figure 5-24. Drawings of the sacrum and coccyx. *A*, the pelvic (ventral) surface faces inferiorly and anteriorly and is concave from superior to inferior and from side to side (Fig. 5-1). *B*, the dorsal surface is rough and has a series of longitudinal ridges formed by the fused posterior parts of the five fused vertebrae.

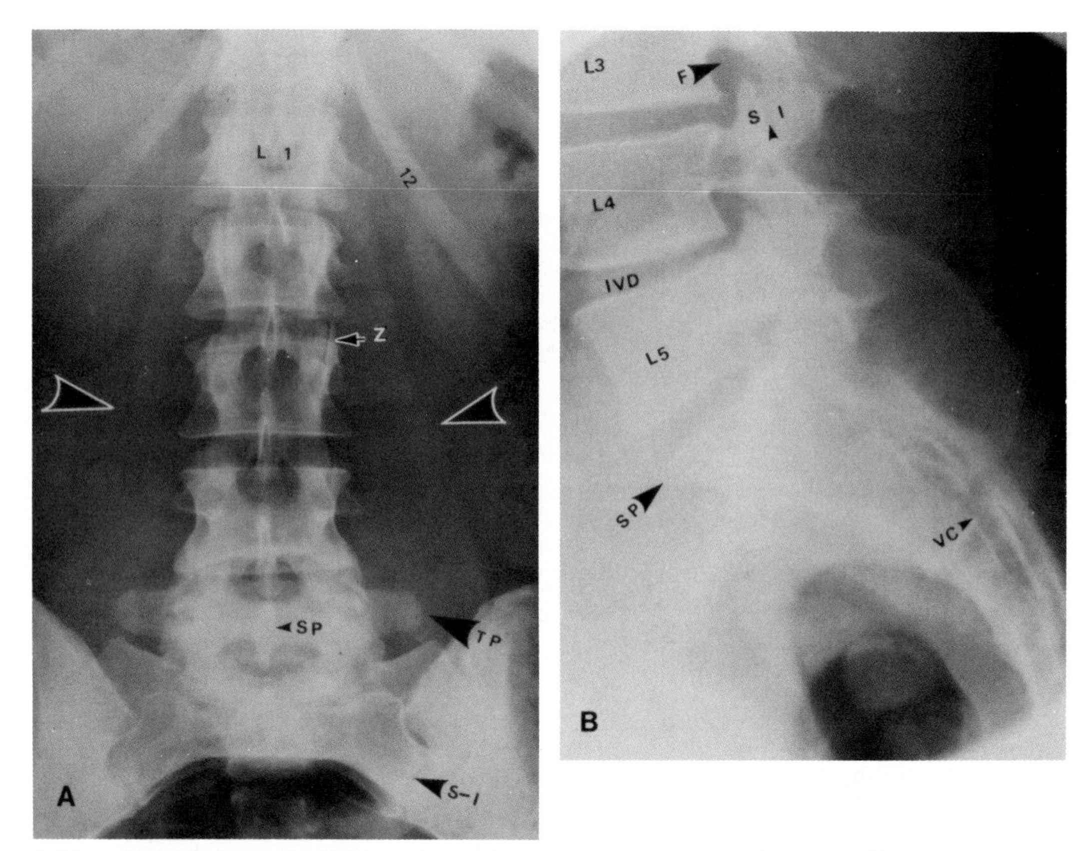

Figure 5-25. Radiographs of the lumbosacral region. *A*, anteroposterior (*AP*) view. Observe the articulation of the last (12th) "floating" rib with the last thoracic vertebra and the bodies and processes of the five lumbar vertebrae. The spinous process (*SP*) and transverse process (*TP*) of L5 vertebra are labeled. The left zygapophysial joint (*Z*) between L2/L3 vertebrae is indicated. The *large arrows* indicate the lateral margins of the psoas muscles. *B*, lateral view. Observe the last three lumbar vertebrae and the spaces representing the intervertebral discs (IVD). The IVD between L4/L5 is marked. Note the angulation at the lumbosacral junction, producing the sacral promontory (*SP*). An *arrow* points to the joint between the superior articular process of L4 (*S*) and the inferior articular process of L3 (*I*). A *small arrow* points to the anterior margin of the vertebral canal (*VC*) and a large arrow points to the intervertebral foramen (*F*).

gion, usually marked during sitting. Coccydynia more commonly results from a fall directly on the coccyx.

DEVELOPMENT OF VERTEBRAE

The vertebrae begin to develop during the embryonic period as condensations of mesenchyme around the notochord (Fig. 5-26*A*). Chondrification centers soon appear and form a cartilaginous vertebra (Fig. 5-26*B*).

Typical vertebrae begin to ossify toward the end of the embryonic period (7 to 8 weeks). **Three primary ossification centers** develop in each cartilaginous vertebra, *one in the centrum and one in each half of the vertebral arch* (Fig. 5-26*C*). At birth the inferior sacral vertebrae and all the coccygeal vertebrae are cartilaginous. They begin to ossify during infancy.

At birth each typical vertebra consists of three bony parts, united by hyaline cartilage (Fig. 5-26*D*). The halves of the vertebral arch begin to fuse in the cervical region during the 1st year, and fusion is usually complete in the lumbar region by the 6th year. In children the vertebral arch articulates with the centrum at **neurocentral joints** which are *synchondroses* (joints where the articulating surfaces are connected by plates of cartilage). The vertebral arch fuses with the centrum during childhood (5 to 8 years).

Five secondary ossification centers develop during puberty (12 to 16 years) in each typical vertebra—*one at the tip of the spinous process, one at the tip of each transverse process*, and *two* ***anular epiph-***

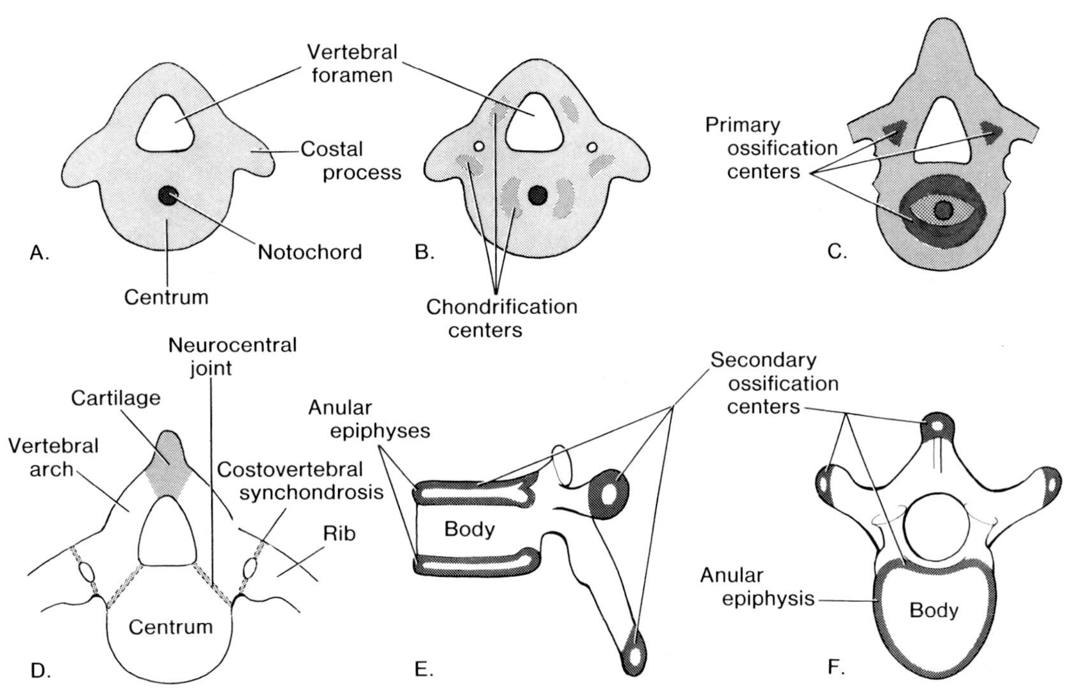

Figure 5-26. Drawings illustrating the stages of vertebral development. *A*, precartilaginous stage (7 weeks). The vertebra is composed of mesenchyme (embryonic connective tissue). *B*, chondrification centers appear. A cartilaginous vertebra is developing. *C*, primary ossification centers appear at 7 weeks. A bony vertebra is forming. *D*, a midthoracic vertebra at birth consisting of three bony parts. Note the cartilage between the halves of the vertebral arch and between the centrum and vertebral arch. *E* and *F*, lateral and superior views of a midthoracic vertebra showing the location of the secondary centers of ossification. The anular epiphyses form the smooth rounded rims on the circumferences of the superior and inferior surfaces of the body. (From Moore KL: *The Developing Human: Clinically Oriented Embryology*, ed 3. Philadelphia, WB Saunders Co, 1982).

yses (ring epiphyses), one on the superior and one on the inferior edge of the centrum (Fig. 5-26*E* and *F*).

The body of the vertebra forms mainly from growth of the centrum, but the anular epiphyses form the smooth rounded rims on the circumference of its superior and inferior surfaces (Figs. 5-13 and 5-15).

All secondary ossification centers are usually united with the vertebra by the 25th year, but the times of their union are variable. Caution must be exercised so that a persistent epiphysis is not mistaken for a fracture in a radiograph.

Exceptions to the typical ossification of vertebrae occur in the atlas, the axis, the seventh cervical vertebra, the lumbar vertebrae, the sacrum, and the coccyx. For details of their ossification, consult Williams and Warwick (1980).

The common developmental abnormality of vertebrae is **spina bifida occulta** (Fig. 5-27). The defect in the vertebral arch of L5 and/or S1 vertebrae occurs in about 10% *of people.* This defect results from failure of the halves of the vertebral arch to grow enough to meet each other and fuse.

The defect in the vertebral arch is concealed by skin, but it is often marked by a tuft of hair. The vertebral defect varies from a slight deficiency to almost complete absence of the vertebral arch. Although most persons with spina bifida occulta have no back problems, some people with this vertebral defect present with low back pain. For descriptions and illustrations of the types of spina bifida associated with neurological symptoms, see Moore (1982).

JOINTS OF THE VERTEBRAL COLUMN

The vertebrae from C2 to S1 articulate with each other at ***three joints***: one anterior intervertebral joint, the **intervertebral disc**; and two posterior intervertebral joints, the **zygapophysial joints** (apophysial or facet joints).

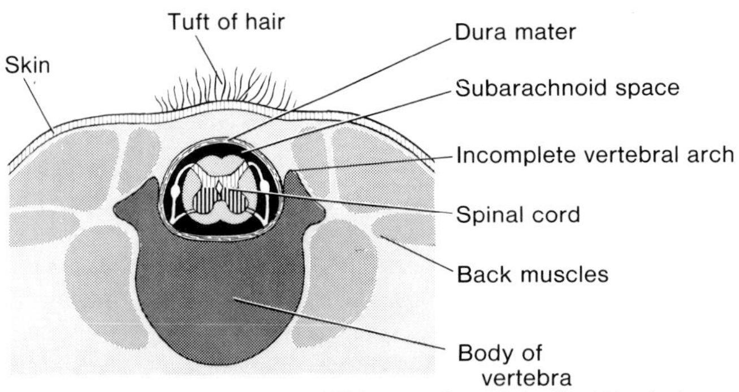

Figure 5-27. Diagrammatic sketch illustrating spina bifida occulta. Note that the halves of the vertebral arch have failed to develop fully and fuse with each other. The spinous process of the vertebral arch is also absent. (From Moore KL: *The Developing Human: Clinically Oriented Embryology*, ed 3. Philadelphia, WB Saunders Co, 1982).

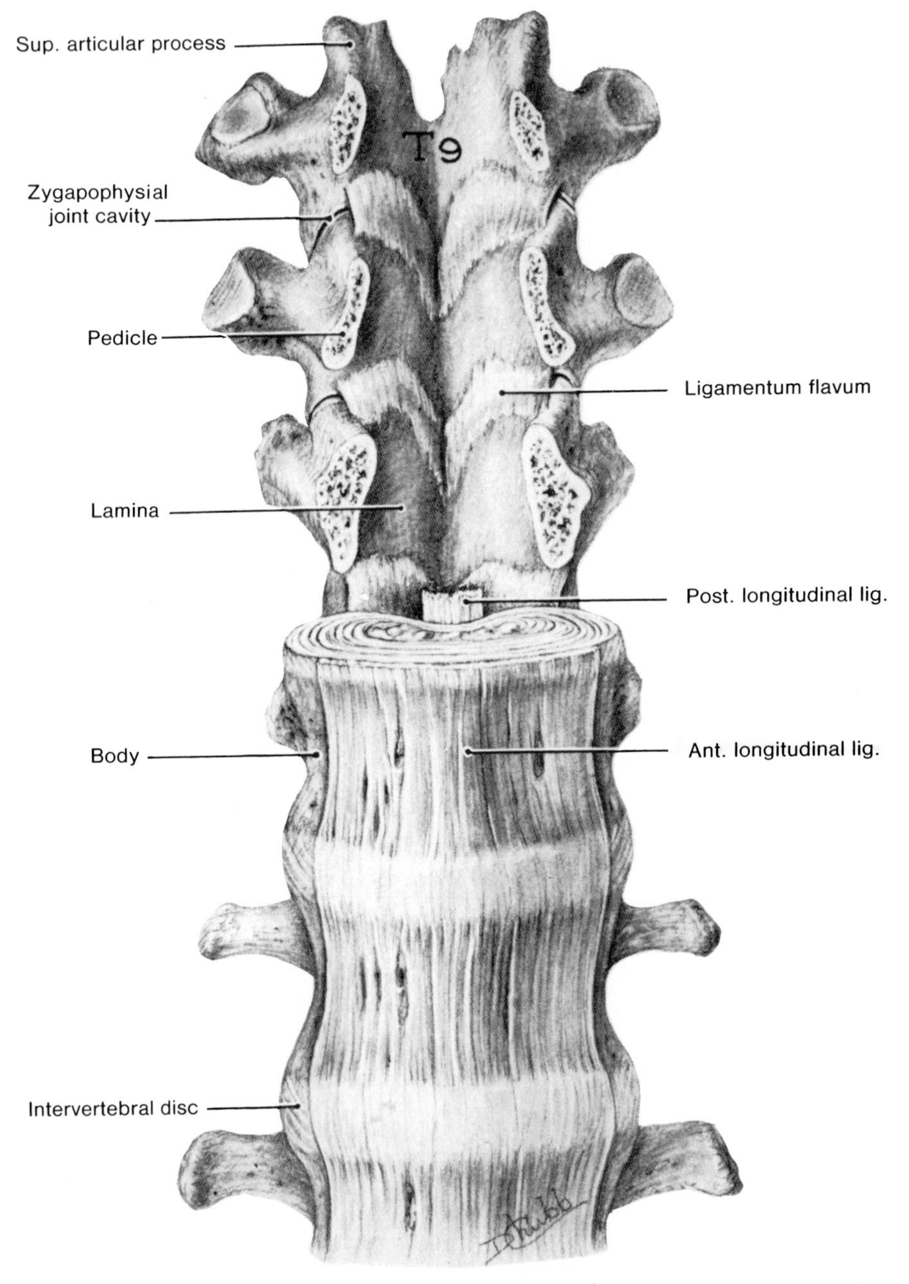

Figure 5-28. Drawing of the thoracic and lumbar regions of the vertebral column showing its joints and ligaments. The pedicles of T9 to T11 vertebrae have been sawn through and their bodies have been discarded. Note that the anterior longitudinal ligament is broad, whereas the posterior longitudinal ligament is narrow.

JOINTS OF THE VERTEBRAL BODIES (Figs. 5-28 to 5-32)

The anterior intervertebral joints or *intervertebral discs* are classified as **symphyses** which are *fibrocartilaginous articulations* designed for strength.

The intervertebral discs (disks), interposed between adjacent surfaces of the vertebral bodies, provide the strongest attachment between the vertebrae. In addition to the intervertebral discs, the bodies of the vertebrae are united by anterior and posterior longitudinal ligaments.

The anterior longitudinal ligament (Figs. 5-28 and 5-29) is a strong, broad, fibrous band that runs longitudinally along the anterior surfaces of the intervertebral discs and bodies of the vertebrae. This ligament extends from the pelvic surface of the sacrum to the anterior tubercle of the atlas and the base of the skull (Fig. 5-34).

The fibers of the anterior longitudinal ligament are firmly fixed to the intervertebral discs and to the periosteum of the vertebral bodies. **The anterior longitudinal ligament tends to prevent hyperextension of the vertebral column** (Figs. 5-9 and 5-62).

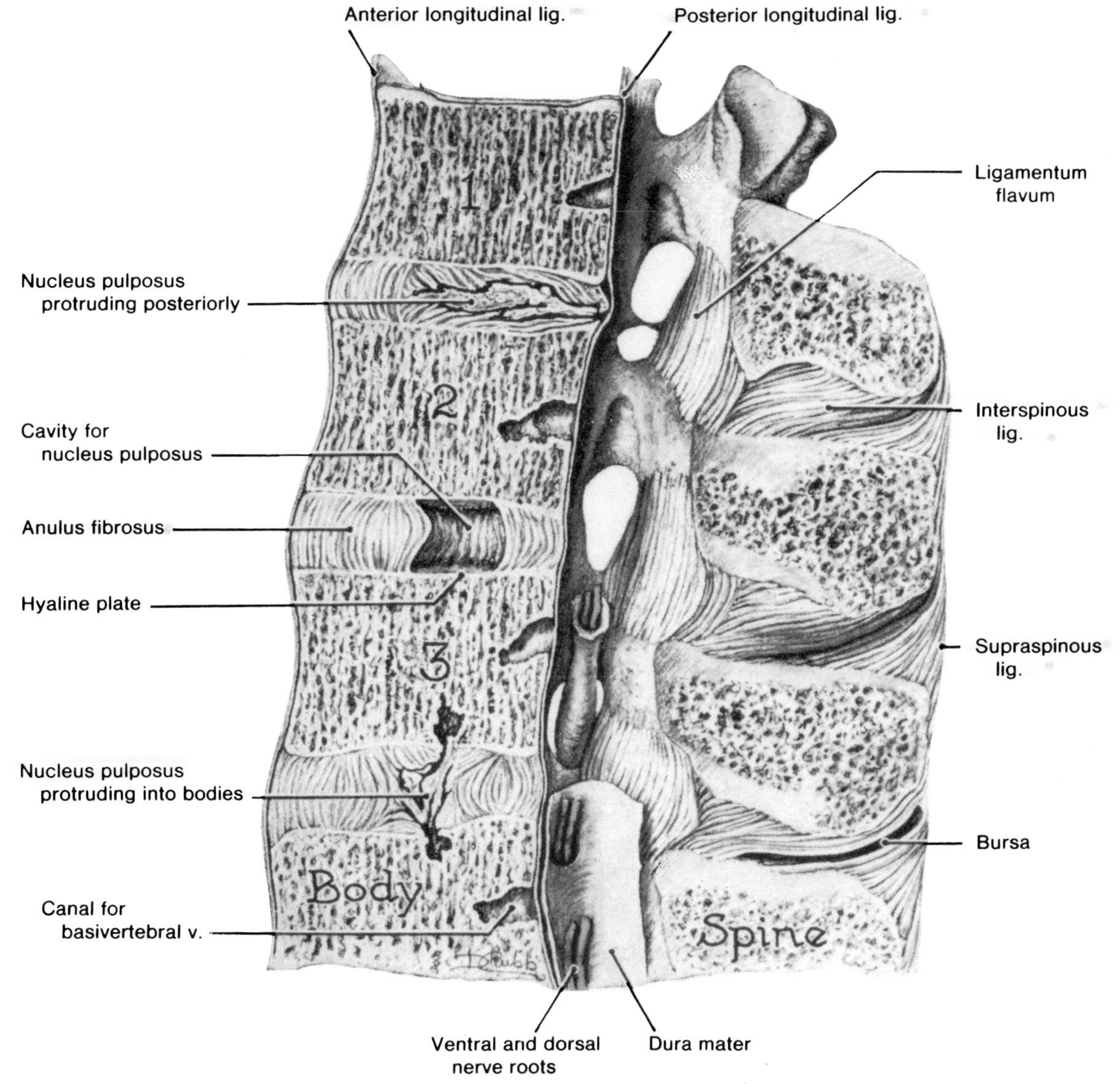

Figure 5-29. Drawing of a median section of the lumbar region of the vertebral column showing the ligaments and anuli fibrosi of the intervertebral discs that connect the vertebrae. Note the protrusions of the nuclei pulposi (called "slipped discs" by laymen) between L1/L2 and L3/L4 resulting from degenerative changes in the anuli fibrosi. The intervertebral discs represent about one-fourth of the length of the vertebral column. Observe the bursa, superior to the spinous process (spine), between the interspinous and supraspinous ligaments. This bursa reduces friction between these ligaments during movements of the vertebral column.

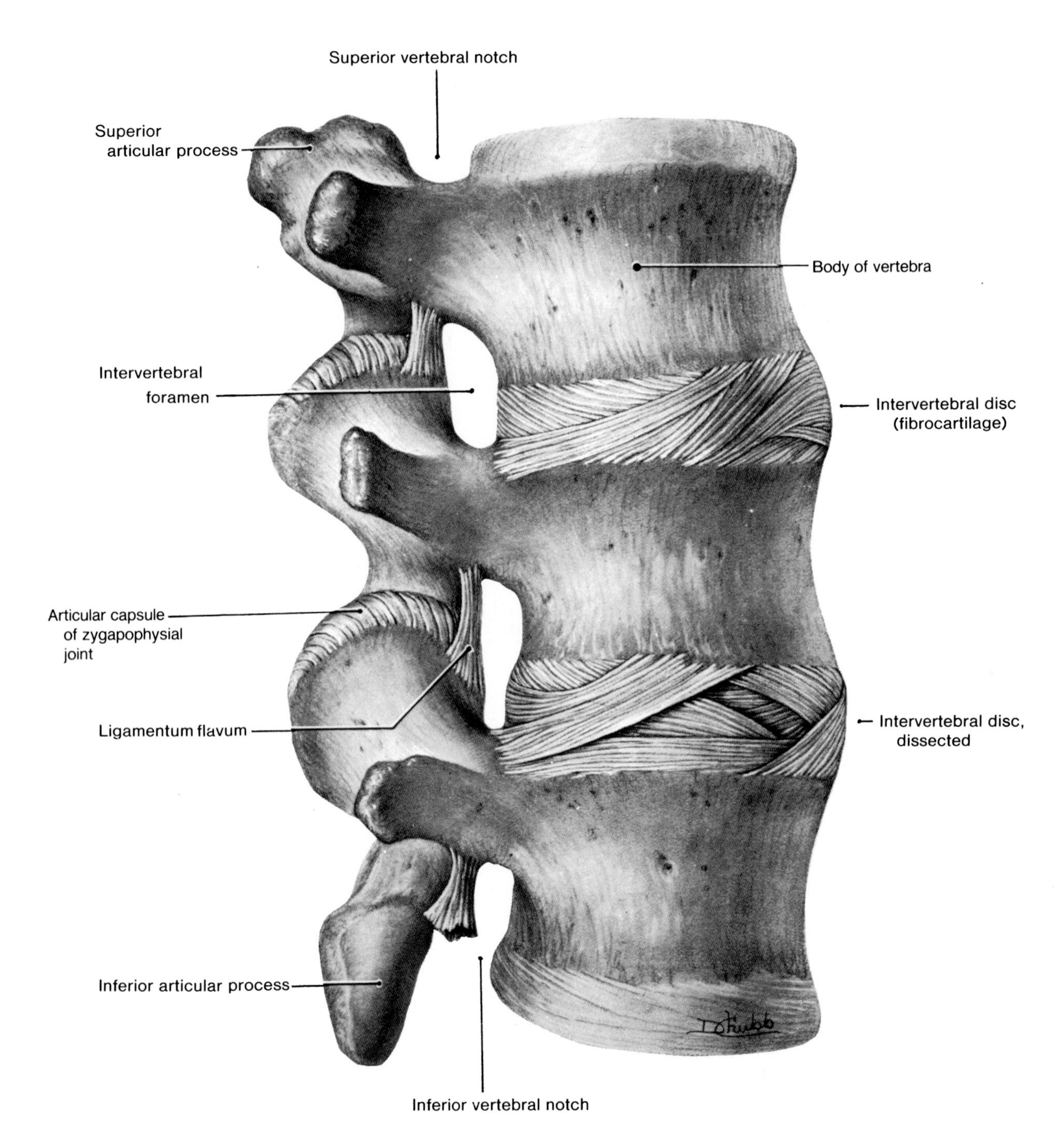

Figure 5-30. Drawing of a portion of the superior lumbar region of the vertebral column, lateral view, primarily to show the structure of the anuli fibrosi of the intervertebral discs. The inferior intervertebral disc has been dissected to show the directions taken by the fibers of the anulus fibrosus. The intervertebral discs vary in size and in thickness in the various regions of the vertebral column. The intervertebral discs, the anterior and posterior longitudinal ligaments, and the other ligaments (Figs. 5-28 and 5-29) hold the vertebrae in proper alignment during movements of the vertebral column (Fig. 5-9).

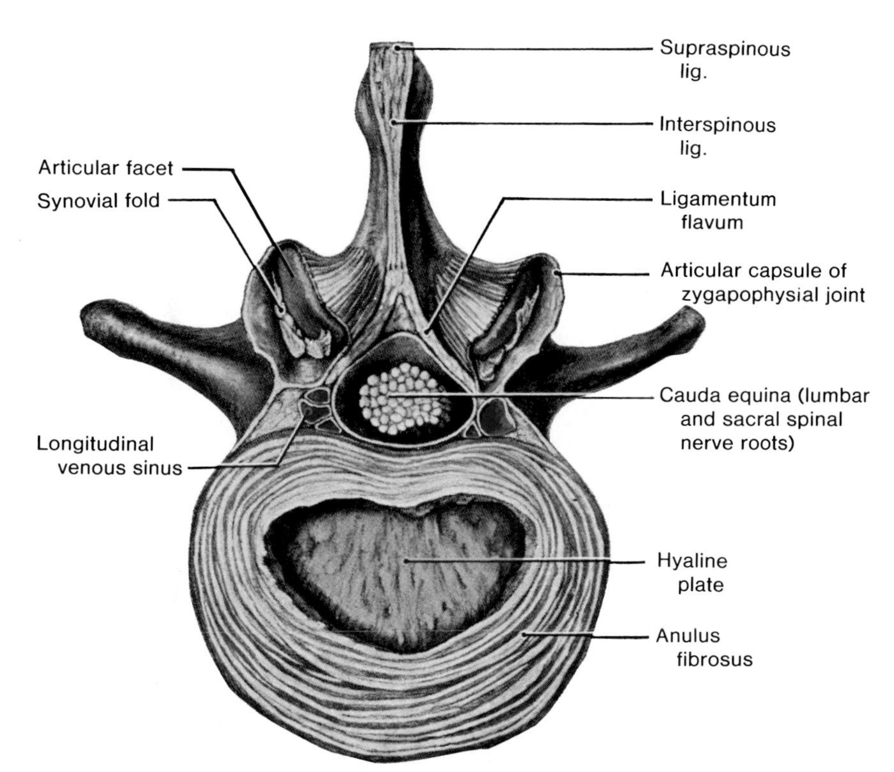

Figure 5-31. Drawing of a cross-section of an intervertebral disc and the associated intervertebral ligaments. The nucleus pulposus has been scooped out to show the hyaline cartilage plate of the superior surface of the vertebral body. The nucleus pulposus (Fig. 5-8) contacts the hyaline plate of the articular cartilage, which is attached to the rough part of the vertebral body (Fig. 5-13). The anulus fibrosus inserts into the smooth bony rim of the body which is derived from the anular epiphysis.

The anterior longitudinal ligament is severely stretched and is sometimes torn during severe **hyperextension of the neck** (Fig. 5-62). The association of rear end automobile collision and this injury is well known, especially to litigation lawyers. For a discussion of the so-called "whiplash" injury, see page 622.

The posterior longitudinal ligament (Figs. 5-28, 5-29, and 5-32), is a narrow, somewhat weaker band than the anterior longitudinal ligament. *The posterior longitudinal ligament runs inside the vertebral canal* from the atlas to the sacrum (Fig. 5-34), and is attached to the intervertebral discs and the posterior edges of the vertebral bodies. **The posterior longitudinal ligament tends to prevent hyperflexion of the vertebral column** (Fig. 5-9).

The fibrocartilaginous **intervertebral discs** are composed of **anuli fibrosi**, enclosing gelatinous **nuclei pulposi** (Fig. 5-46).

The anuli fibrosi insert into the smooth, rounded rims on the articular surfaces of the vertebral bodies. The nuclei pulposi contact the hyaline plates of articular cartilage (Fig. 5-31) which are attached to the rough parts of the vertebral bodies (*i.e.*, inside the smooth bony rims shown in Fig. 5-13). There are poorly developed intervertebral discs between the bodies of the sacral vertebrae in young persons, but they usually ossify with advancing age (Fig. 5-24).

The anulus fibrosus of the intervertebral disc (Figs. 5-30, 5-31, and 5-46) is composed of concentric lamellae of fibrocartilage which run obliquely from one vertebra to another. Some fibers in one lamella are at right angles to those in adjacent ones. This arrangement, while allowing some movement between adjacent vertebrae, provides a strong bond between them. The lamellae of the anulus fibrosus are thinner and less numerous posteriorly than they are anteriorly or laterally. Peripheral parts of the anulus fibrosus are supplied by adjacent blood vessels (Fig. 5-32).

The nucleus pulposus of the intervertebral disc acts like a shock absorber for axial forces and like a semifluid ball bearing during flexion, extension, and lateral bending of the vertebral column (Fig. 5-9). It becomes broader under compression (Fig. 5-8*B*).

The nucleus pulposus is a derivative of the notochord in the embryo. Up to the 10th year, a few notochordal cells can be observed in the mucoid material of the

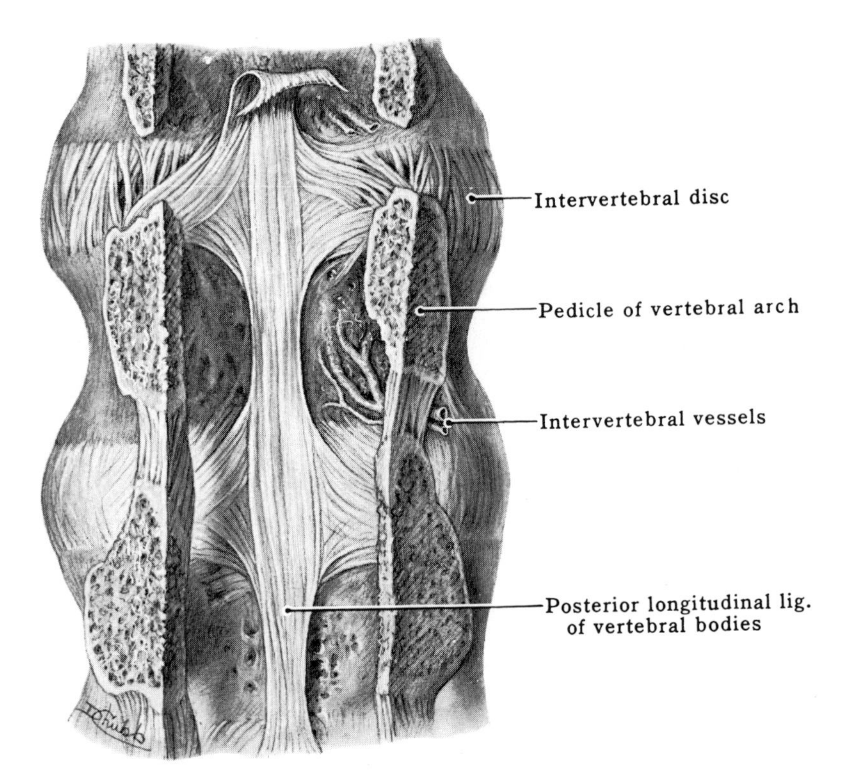

Figure 5-32. Drawing of a posterior view of a dissection of the posterior longitudinal ligament. The vertebral arches have been sawn through, leaving parts of the pedicles. Observe the diamond shape taken by the ligament posterior to each disc, where it both sends out fibers and receives them.

nucleus pulposus (L. *pulpa*, fleshy). After this, the mucoid material is gradually replaced by fibrocartilage from the anulus fibrosus.

The nucleus pulposus is avascular. It receives its nourishment by diffusion from blood vessels at the periphery of the anulus fibrosus and adjacent surfaces of the vertebral bodies.

In young persons the intervertebral discs are very strong and the water content of their nuclei pulposi is about 88%. This gives them great turgor (fullness). In healthy young adults, the intervertebral discs are so strong that the vertebrae will break before the discs will rupture during a fall. However, during **violent hyperflexion of the vertebral column**, rupture of an intervertebral disc may occur without fracturing the adjacent vertebral bodies.

As people get older, their nuclei pulposi lose their turgor and become thinner owing to dehydration and degeneration. These age changes account in part for the slight loss in height that occurs during old age.

After the 20th year, the intervertebral discs are also subject to pathological changes that may result in protrusion of the nucleus pulposus into or through the anulus fibrosus (Fig. 5-68). **A herniated or protruding intervertebral disc** is often incorrectly called a "slipped disc." Protrusions of the nucleus pulposus usually occur posterolaterally, where the anulus fibrosus is weak and poorly supported by the posterior longitudinal ligament (Figs. 5-28, 5-29, and 5-68). The protruding part of the nucleus pulposus may compress an adjacent spinal nerve root, causing leg pain (*sciatica*) and/or low back pain (Figs. 5-64 and 5-68).

Acute low back pain (*lumbago*) and **sciatica** (pain in the *area of the sciatic foramen*, Fig. 3-16, that radiates down the posterior aspect of the thigh) are often caused by a posterolateral protrusion of a lumbar intervertebral disc (Fig. 5-68). The clinical picture varies considerably, but pain of acute onset in the lower back is a common presenting symptom.

The terms *lumbar* and *lumbago* are derived from the Latin word *lumbus*, meaning the loin (the part of the back between the ribs and the iliac crest).

Because of the **muscle spasm** associated with low back pain, the lumbar region of the vertebral column becomes rigid and movement is painful. With treatment, this type of pain usually begins to fade after a few days, but it may be gradually replaced by sciatica.

Sciatica, as stated, is an ache or pain in the area of the sciatic foramen (lateral to the ischial tuberosity) and in the posterior aspect of the thigh, often spreading distal to the knee, calf, ankle, and foot (Fig. 5-64).

Sciatica often results from a posterolateral herniation of the nucleus pulposus through a rupture in the anulus fibrosus of the intervertebral disc (Fig. 5-68). A lumbar intervertebral disc that herniates posterolaterally exerts pressure on the spinal nerve root.

About 95% of lumbar disc protrusions occur at the L4/L5 or L5/S1 levels. The remaining protrusions are usually at the L3/L4 level. Any maneuver that stretches the sciatic nerve, such as flexing the thigh with the leg extended, will reproduce or exacerbate the pain of a disc protrusion.

Intervertebral discs are often damaged by violent twisting or flexing of the vertebral column. Posterior herniations of the nucleus pulposus may also exert pressure on the spinal cord at cervical or thoracic levels, and on the cauda equina at inferior levels (Figs. 5-29 and 5-31).

Symptom-producing disc protrusions occur in the cervical region almost as often as in the lumbar region. A forcible flexion in the cervical region may rupture the disc posteriorly without fracturing the vertebral body. The cervical discs most commonly ruptured are those between C5/C6 and C6/C7, compressing spinal nerve roots C6 and C7, respectively. These protrusions result in pain in the neck, shoulder, arm, and hand.

If a surgical operation is required to alleviate the symptoms of disc disease, the laminae of the involved vertebra(e) may be removed. The operation consists of excision of the protruding disc and curretting part of the core of the disc to remove the nucleus pulposus.

In selected cases an enzyme, *chymopapain*, is often injected into the protruding part of the intervertebral disc. This enzyme is thought to shrink the disc and relieve the pressure on the nerve root(s). This technique is being evaluated for the treatment of disc disease.

JOINTS OF THE VERTEBRAL ARCHES (Figs. 5-30, 5-31, and 5-33)

Synovial joints of the plane variety, known as **zygapophysial joints**, are formed by the opposing articular processes (zygapophyses) of adjacent vertebral arches. Because the contact surfaces of the articular processes have articular facets (Fig. 5-14), clinicians often refer to zygapophysial joints as "*facet joints*" for brevity. The articular surfaces consist of smooth, shiny compact bone covered with hyaline cartilage.

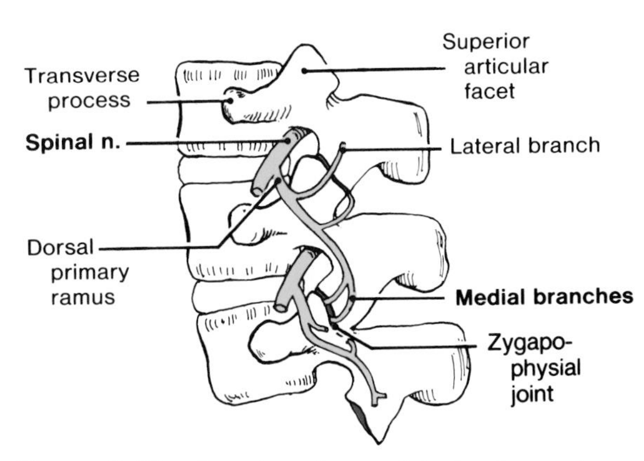

Figure 5-33. Drawing of part of the lumbar region of the vertebral column showing the innervation of the zygapophysial joints. Observe that the dorsal primary ramus arises from the spinal nerve outside the intervertebral foramen, and then divides into medial and lateral branches (Fig. 5-51). The medial branch descends in a groove posterior to the transverse process beside the superior articular process. It sends small articular branches to the capsule of the zygapophysial joint beside it and to the subjacent joint as well.

Each zygapophysial joint is surrounded by a thin, loose **articular capsule** which is attached to the articular margins of the articular processes (Figs. 5-30 and 5-31). These fibrous capsules are longer and looser in the cervical region than in the thoracic and lumbar regions; consequently *flexion is most extensive in the cervical region.* The fibrous capsule of the zygapophysial joint is lined by a synovial membrane (Fig. 5-31).

In the cervical and lumbar regions, the zygapophysial joints bear some weight, sharing this function with the intervertebral discs. The zygapophysial joints help to control flexion, extension, and rotation of adjacent cervical and lumbar vertebrae. Most of the movement of the vertebral column takes place in these regions.

The laminae of adjacent vertebral arches are joined by broad, elastic bands called **ligamenta flava** (Figs. 5-28 and 5-31). Their fibers extend to the articular capsules of the zygapophysial joints and contribute to the posterior boundaries of the intervertebral foramina (Fig. 5-30).

The adjacent edges of the spinous processes are joined by weak **interspinous ligaments**, and the tips of the spinous processes are joined by a strong **supraspinous ligament** (Fig. 5-29). The interspinous and supraspinous ligaments are represented superiorly by the **ligamentum nuchae**, a triangular median septum between the muscles on each side of the posterior aspect of the neck (Fig. 6-44). The ligamentum nuchae extends from the spinous process of C7 to the posterior border of the foramen magnum, the external occipital crest, and the external occipital protuberance (Fig. 5-48).

The **intertransverse ligaments** connect adjacent transverse processes. They consist of a few scattered fibers, except in the lumbar region, where they are thin and membranous.

Innervation of the Zygapophysial Joints (Fig. 5-33). These joints are innervated by nerves that arise from the ***medial branches of the dorsal primary rami of the spinal nerves*** (Fig. 5-51). As these nerves pass posteroinferiorly, they lie in grooves on the posterior surfaces of the medial parts of the transverse processes. Each articular branch supplies the zygapophysial joint nearby, and it may send twigs to the subjacent joint as well.

Low back pain may also result from disease of the zygapophysial joints in the lumbar region (*e.g.*, owing to osteoarthritis).

Denervation of lumbar zygapophysial joints is a procedure currently being evaluated for treatment of low back pain thought to be caused by disease of the zygapophysial joints. In some cases the nerves are sectioned near these joints; in other cases the nerves are destroyed by radiofrequency percutaneous **rhizolysis** (G. *rhiza*, root + *lysis*, dissolution). In each procedure the destructive process is directed at the medial branches of the dorsal primary rami of the spinal nerves which supply these joints (Figs. 5-33 and 5-51), usually at the L1 to L4 level.

JOINTS OF THE SUBOCCIPITAL REGION (Figs. 5-34 to 5-36 and Table 5-1)

These suboccipital joints are between the skull and the atlas, and between the atlas and the axis.

The Atlantooccipital Joints (Figs. 5-34 to 5-36 and Table 5-1). These articulations between the atlas and occipital condyles permit nodding of the head (*i.e.*, flexion and extension of the neck that occurs when indicating approval). The atlantooccipital joints, one on each side, are between the superior articular facets

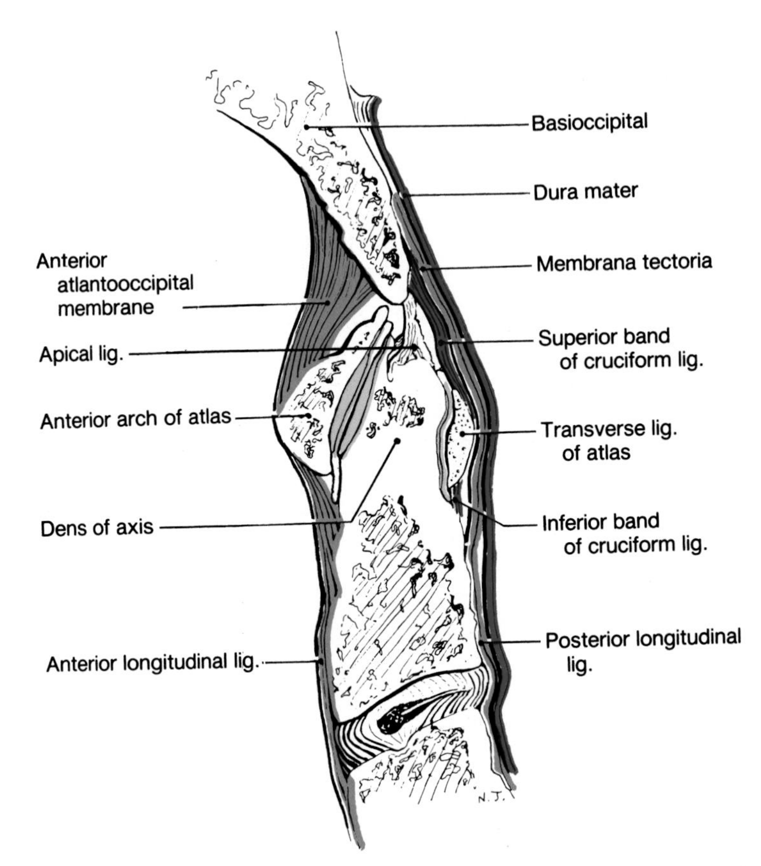

Figure 5-34. Drawing of a median section of the ligaments and joints of the suboccipital region. There are two joints: the *atlantooccipital joint* between the atlas and occipital bone, and the *atlantoaxial joint* between the atlas and axis. Note that the anterior longitudinal ligament is attached to the anterior tubercle of the atlas, and that the posterior longitudinal ligament is continuous with the membrana tectoria.

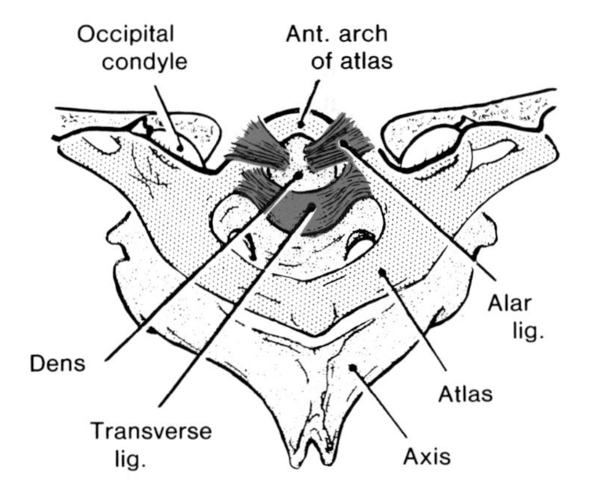

Figure 5-35. Drawing of the craniovertebral joints showing the transverse ligament of the atlas and the alar ligament. Note that the transverse ligament embraces the dens of the axis. This is a uniaxial pivot joint. Also see Figure 5-17.

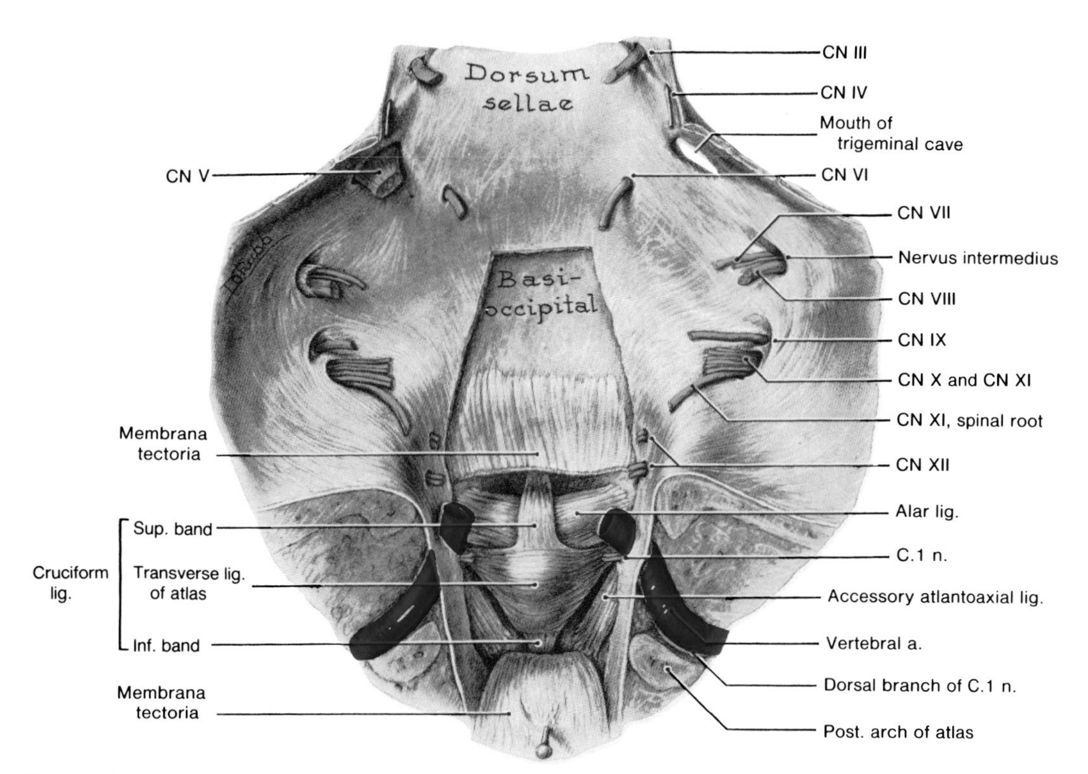

Figure 5-36. Drawing of a superior view of the craniovertebral region showing the ligaments of the atlantoaxial and atlantooccipital joints. Observe the cross-shaped cruciform ligament. See Figure 5-34 for orientation.

on the lateral masses of the atlas and the ***occipital condyles*** (Figs. 5-17 and 5-35). They are synovial joints with thin loose articular capsules.

The skull and atlas are also connected by anterior and posterior **atlantooccipital membranes**, which extend from the anterior and posterior arches of the atlas to the anterior and posterior margins of the foramen magnum (Fig. 5-34).

The transverse ligament of the atlas (Figs. 5-16, 5-17, 5-34, and 5-35) is a strong band extending between the tubercles on the lateral masses of the atlas. *The transverse ligament of the atlas holds the dens of the axis against the anterior arch of the atlas.* There is a synovial joint between them. Vertically oriented superior and inferior bands pass from the transverse ligament to the occipital bone superiorly and to the body of the axis inferiorly. They form the **cruciform ligament** (L. *crux*, cross) which was given this name because of its resemblance to a cross (Fig. 5-36).

The alar ligaments extend from the dens to the lateral margins of the foramen magnum (Figs. 5-35 and 5-36). These strong, short, rounded cords *check lateral rotation and side-to-side movements of the head* and attach the skull to the axis.

The membrana tectoria is the continuation of the

Table 5-1
Muscles Producing Movements of the Atlantooccipital and Atlantoaxial Joints[1]

Flexion	Extension	Rotation and Lateral Bending (Lateral Flexion)
Longus capitis	**Semispinalis capitis**	**Sternocleidomastoid**
Rectus capitis anterior	**Splenus capitis**	**Obliquus capitis inferior**
Sternocleidomastoid (anterior fibers)	**Rectus capitis posterior major**	**Obliquus capitis superior**
	Rectus capitis posterior minor	**Rectus capitis lateralis**
	Obliquus capitis superior	**Longissimus capitis**
	Longissimus capitis	**Splenius capitis**
	Trapezius (superior part)	Semispinalis capitis
	Sternocleidomastoid (posterior fibers)	Trapezius (superior part)

[1] The principal muscles producing these movements appear in **bold face**. *Atlanto* is a combining form relating to the atlas.

posterior longitudinal ligament (Figs. 5-34 and 5-36). It runs from the body of the axis to the internal surface of the occipital bone, and covers the alar and transverse ligaments (Fig. 5-36).

The Atlantoaxial Joints (Figs. 5-34 to 5-36 and Table 5-1). There are two lateral joints and one medial joint. *The atlantoaxial joints permit the head to be turned from side to side* (*e.g.*, when rotating the head to indicate disapproval). During this movement, the skull and atlas rotate as a unit on the axis.

The apical ligament of the dens extends from the tip of the dens to the internal surface of the occipital bone (Fig. 5-34).

Movements between the skull and atlas and between the atlas and axis, occurring at the craniovertebral joints, are augmented by the flexibility of the neck owing to movements of the vertebral joints in the middle and inferior cervical regions.

During rotation of the head, the dens of the axis is held in a collar formed by the anterior arch of the atlas and the transverse ligament of the atlas. The articulation of the dens with the atlas is a pivot joint (Fig. 5-35).

If the transverse ligament of the atlas ruptures, the dens may be driven into the cervical region of the spinal cord, causing **quadriplegia** (paralysis of the upper and lower limbs), or into the inferior end of the medulla of the brain, causing sudden death. The latter injury often occurs when persons hang themselves. There is controversy about the so-called "*hangmans's fracture*," but it is generally agreed that it is a complex fracture of the axis.

All back and neck injuries are potentially serious because of the possibility of fracturing vertebrae and injuring the spinal cord and/or the cauda equina (Figs. 5-31, 5-51, and 5-53).

Some regions of the vertebral column tend to have greater numbers of certain types of fracture owing to their mobility and curvatures.

The cervical region of the vertebral column is especially vulnerable to injury. Injuries to the spinal cord in the cervical region, associated with **compression or transection of the spinal cord**, may result in loss of all sensation and voluntary movement inferior to the lesion, or in sudden death, depending on the level of injury.

Compression of any part of the central nervous system, rendering it ischemic for 3 to 5 minutes results in death of nervous tissue, particularly nerve cells.

Fractures, dislocations, and fracture-dislocations of the vertebral column often result from sudden forceful flexion, as may occur in a car accident, or from a violent blow on the back of the head. The common fracture is a *compression fracture* of the body of one or more vertebrae.

In **severe flexion injuries** the posterior longitudinal and interspinous ligaments are often torn, and the vertebral arches may be dislocated and/or fractured along with fractures of the vertebral bodies. *With most severe flexion injuries, there are injuries to the spinal cord.*

When a person falls from a height and lands on the crown of his/her head, the violence is transmitted along the axis of the vertebral column. Falling on the feet or the buttocks from a height, or hitting the head while diving, produces a similar axial force.

A fall on the head may force the occipital condyles into the atlas, splitting it into two or more fragments. In other cases, the thin bone around the occipital condyles fractures (Fig. 5-47), and the condyles pass superiorly with the atlas and axis into the posterior cranial cavity containing the cerebellum, pons, and medulla (Fig. 7-86).

Injuries can also be caused by extension forces to the vertebral column. **Extension fractures and/or dislocations** vary from one vertebral region to another, but posterior portions of the vertebral column are most likely to be injured.

Severe hyperextension of the neck may pinch the posterior arch of the atlas between the occiput and the axis. The atlas usually breaks at one or both grooves for the vertebral arteries (Figs. 5-16 and 5-17). If the force is continued, the arch of the axis also may be pinched off and break at the isthmus between the lateral mass and the inferior articular processes, unilaterally or bilaterally. If the force is greater, the anterior longitudinal ligament and adjacent anulus fibrosus of the C2/C3 intervertebral disc may rupture. At the moment of impact and *severe* hyperextension of the neck, the skull, atlas, and axis are separated from the rest of the axial skeleton and the spinal cord is usually severed. Such patients seldom survive more than 5 minutes because the injury to the spinal cord is superior to the **phrenic outflow**, the origin of the phrenic nerves (C3, C4, and C5). As these nerves innervate the diaphragm (Fig. 1-74), respiration is severely affected.

Sometimes when the neck is hyperextended while the head is turned to one side, *the ganglion of the second cervical nerve on the opposite side is compressed between the posterior arches of the atlas and the axis* (Fig. 5-48). This may be followed by prolonged headaches in the occipital region so severe that they may result in suicidal tendencies.

Hyperextension injuries of the vertebral column rarely occur in the thoracic region owing to the extra support provided by the ribs. However, the last two thoracic vertebrae, being freer to move than the others, may sustain injuries similar to lumbar vertebrae.

Extension injuries of the lumbar vertebral column tend to injure the posterior parts of the vertebrae. In addition to fractures of the laminae and articular processes, there are sometimes small compression fractures of posterior parts of the vertebral bodies. In severe hyperextension, the forces on the anterior longitudinal ligament and adjacent anulus may tear off the anterosuperior or anteroinferior corner of a lumbar vertebral body. In these cases, one can deduce that there was a great displacement of the vertebrae and their fragments at the moment of impact, and that the spinal cord and/or cauda equina was probably damaged.

Dislocation of vertebrae without fracture is rare, except in the cervical region, because of the interlocking of the thoracic and lumbar articular processes. *The vertebral canal in the cervical region is usually somewhat larger than the spinal cord* (Fig. 5-51). Thus there can be some displacement of the vertebrae without causing serious damage to the spinal cord (Fig. 5-66).

Displacement of the Vertebral Column (Fig. 5-37). In a few people, particularly Australian aborigines, South African bushmen and Eskimos, the fifth lumbar vertebra consists of two parts. The posterior fragment, consisting of the spinous process, remains in normal relation to the arch of the sacrum. The anterior fragment and the superimposed vertebra may move anteriorly (Fig. 3-87). This anterior displacement of most of the vertebral column is called **spondylolisthesis** (G. *spondylos*, vertebra + *olisthesis*, a slipping and falling). If the anterior part of the bone does not move anteriorly, the condition is called **spondylolysis**.

Spondylolisthesis at L5 may result in pressure

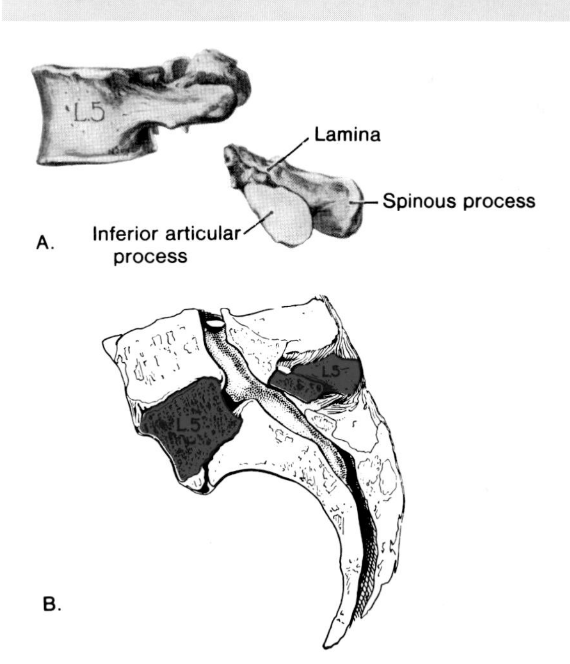

Figure 5-37. *A*, bipartite fifth lumbar vertebra. Note that one piece, consisting of the spinous process, laminae, and inferior articular processes, is separated from the body, pedicles, and superior articular processes. *B*, the body and superimposed vertebrae tend to slide anteriorly, a condition known as spondylolisthesis (also see Fig. 3-87).

on the spinal nerves as they pass into the superior part of the sacrum. The cause of this condition, which develops postnatally and may cause acute low back pain, is unknown. Most people believe it is a strain or "**march fracture**" of L5 vertebra (Fig. 3-87).

MUSCLES OF THE BACK

For descriptive purposes, the muscles of the back are divided into three groups: *superficial, intermediate, and deep.* The superficial and intermediate groups are **extrinsic back muscles** that are concerned with movements of the limbs and with respiration. The deep group constitutes the **intrinsic back muscles** that are concerned with movements of the vertebral column (Figs. 5-9 and 5-38).

The superficial and intermediate groups of muscles (extrinsic back muscles) are superficial to the intrinsic muscles of the back (Fig. 5-38).

EXTRINSIC BACK MUSCLES

SUPERFICIAL EXTRINSIC BACK MUSCLES (Figs. 5-38, 6-44, 6-45, and 6-47)

The trapezius, latissimus dorsi, levator scapulae, and rhomboid muscles connect the upper limb to the axial skeleton. As they are concerned with movements of the upper limbs, they are described in Chapter 6.

INTERMEDIATE EXTRINSIC BACK MUSCLES (Figs. 5-38, 5-39, and 6-47)

As the serratus posterior superior and inferior are respiratory muscles, they are described with the thorax in Chapter 1.

INTRINSIC BACK MUSCLES

The intrinsic muscles are the true back muscles *because they are concerned with the maintenance of posture and movements of the vertebral column.* The intrinsic back muscles are covered posteriorly by a tough sheet of fascia which fuses with the aponeuroses of several extrinsic muscles to form the **thoracolumbar fascia** (Figs. 2-123, 2-124, 5-40, 5-44, and 5-46). When this fascia is removed, paired muscular columns are exposed which lie in longitudinal bands on each side of the spinous processes (Figs. 5-38 to 5-40). They are supplied by branches of the dorsal primary rami of the spinal nerves that pass through them to supply the skin (Figs. 5-45, 5-51, and 6-44).

The three layers of intrinsic back muscles *are named according to their relationship to the surface and can be identified by the direction of their fibers* (Fig. 5-40): (1) a ***superficial layer,*** with fibers passing superolaterally; (2) *an* ***intermediate layer*** with fibers running longitudinally, parallel to the long axis

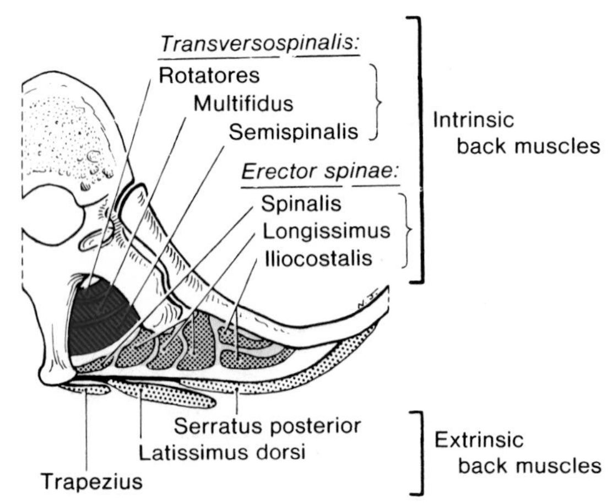

Figure 5-38. Schematic cross-section of the back, illustrating the disposition of the intrinsic and extrinsic muscles.

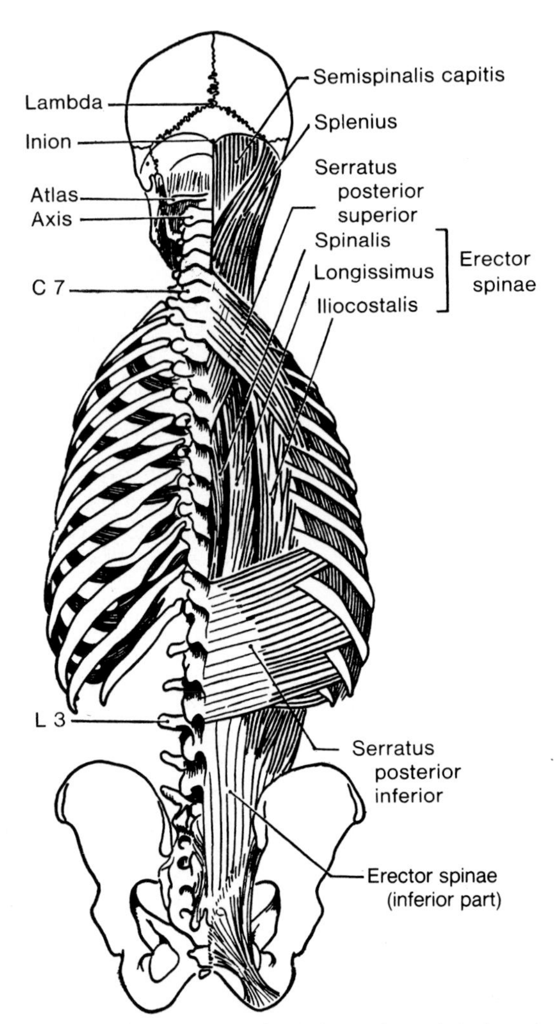

Figure 5-39. Drawing of a dissection showing the muscles of the back. The serratus posterior superior and inferior are muscles of the thorax. The deep or intrinsic muscles are covered by the serratus muscles in the thoracic region and by the splenius muscles in the neck region.

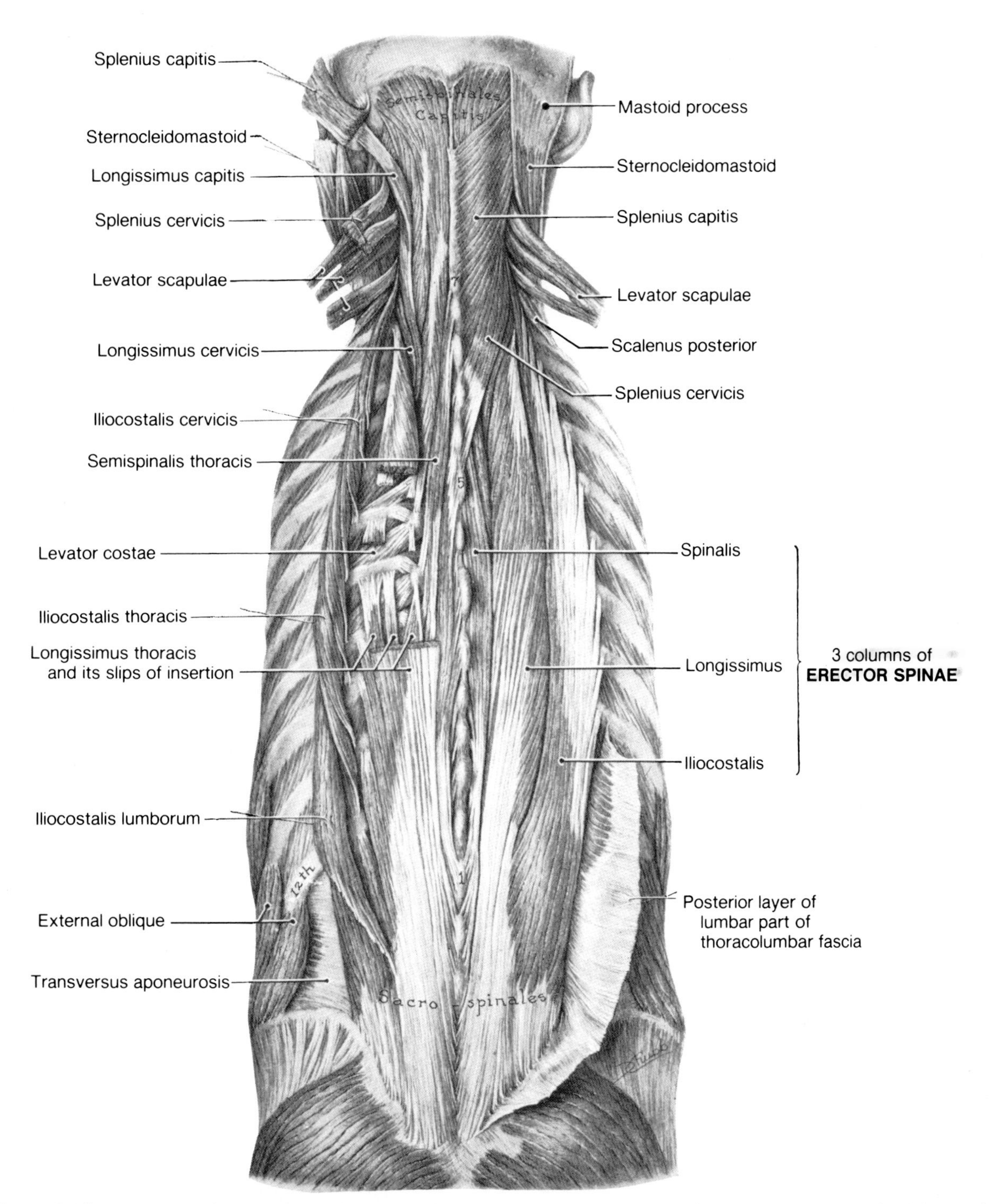

Figure 5-40. Drawing of a dissection illustrating the deep or intrinsic muscles of the back. The superficial layer of deep muscles (splenius capitis and cervicis) is reflected on the left side. The intermediate layer of deep muscles (erector spinae) is intact on the right side, lying between the spinous processes of the vertebrae medially and the angles of the ribs laterally. Note that the erector spinae (sacrospinalis) consists of three columns of muscles, and that its fibers run parallel to the spinous processes of the vertebrae. The muscles of the back are responsible for the maintenance of posture and for movements of the vertebral column. Also see Tables 5-2 and 5-3.

of the vertebral column; and (3) *a* ***deep layer***, with fibers passing superomedially.

SUPERFICIAL LAYER OF INTRINSIC BACK MUSCLES (Figs. 5-39 and 5-40)

The splenii (G. *splenion*, bandage) consist of two large muscles in the back of the neck. They somewhat resemble bandages as they ascend from the midline of the neck to the base of the skull and the transverse processes of the superior cervical vertebrae.

The Splenius Capitis and Splenius Cervicis Muscle (Fig. 5-40 and Table 5-2). This broad thin muscle, composed of two parts (capitis and cervicis), passes superolaterally over the deeper muscles of the neck.

Origin. (Figs. 5-40 and 6-47). **Inferior half of ligamentum nuchae and spinous processes of C7 to T6 vertebrae.**

Insertion (Fig. 5-40). The splenius capitis (L. *caput*, head) inserts into the **mastoid process** of the temporal bone (Fig. 5-47) and into the adjacent **occipital bone**, whereas the splenius cervicis (L. *cervix*, neck) inserts into the **transverse processes or the superior two to four cervical vertebrae.**

Actions. *Acting alone,* **the splenii laterally bend and rotate the neck**, turning the face to the same side. *Acting together,* **the splenii pull the head posteriorly** (*i.e.*, extend it).

Nerve Supply (Fig. 6-44). **Cervical spinal nerves.** Lateral branches of dorsal rami of the middle (splenius capitis) and inferior (splenius cervicis) cervical nerves.

INTERMEDIATE LAYER OF INTRINSIC BACK MUSCLES (Figs. 5-38 to 5-42)

This muscle layer is formed by the large erector spinae muscle (sacrospinalis) lying in the groove on each side of the vertebral column. It is characterized by long muscle bundles running vertically, parallel to the spinous processes of the vertebrae.

The erector spinae is a massive muscle, extending from the pelvis to the skull, which can be readily palpated. Its lateral border is obvious in most people (Figs. 5-11 and 5-42).

The large erector spinae muscle divides into three columns in the superior lumbar region (Fig. 5-41): *iliocostalis, longissimus, and spinalis.*

The Iliocostalis Muscle (Figs. 5-38 to 5-41 and Table 5-3). As its name indicates, this lateral column of the erector spinae muscle arises from the iliac crest and inserts into the ribs (L. *costae*).

The iliocostalis consists of three parts: (1) the **iliocostalis lumborum,** arising from the iliac crest and inserting into the angles of the inferior six or seven ribs; (2) the **iliocostalis thoracis**, arising from the inferior six ribs and inserting into the angles of the superior six ribs and the transverse process of C7; and (3) the **iliocostalis cervicis**, arising from the third to sixth ribs and inserting into the transverse processes of C6 to C4 vertebrae.

The Longissimus Muscle (Figs. 5-38 to 5-41 and Table 5-2). The longissimus (L. longest), the intermediate column of the erector spinae muscle, runs mainly between the transverse processes of the vertebrae. The longissimus muscle may also be divided into *three parts*: **longissimus thoracis, longissimus cervicis, and longissimus capitis,** which indicate the sites of insertion of its fibers.

The longissimus thoracis inserts into the tips of the transverse processes of all the thoracic vertebrae, and into the inferior 9 to 10 ribs between their tubercles and angles.

The longissimus cervicis inserts into the posterior tubercles of the transverse processes of C2 to C6 vertebrae.

The longissimus capitis inserts into the mastoid process of the temporal bone (Figs. 5-41 and 5-47).

The longissimus thoracis and cervicis ***extend the vertebral column*** *and, acting on one side,* ***bend it laterally,*** *whereas the longissimus capitis* ***extends***

Table 5-2
Muscles Producing Movements of the Cervical Intervertebral Joints[1]

Flexion	Extension	Rotation and Lateral Bending (Lateral Flexion)
Sternocleidomastoid	**Splenius capitis**	**Sternocleidomastoid**
Longus colli	**Splenius cervicis**	**Scalene muscles**
Longus capitis	**Semispinalis capitis**	**Splenius capitis**
	Semispinalis cervicis	**Splenius cervicis**
	Iliocostalis cervicis	**Longissimus capitis**
	Longissimus capitus	**Longissimus cervicis**
	Longissimus cervicis	**Iliocostalis cervicis**
	Trapezius	**Multifidus**
	Interspinales	Levator scapulae
		Longus colli
		Intertransversarii

[1] The principal muscles producing these movements appear in **bold face**.

the head and turns the face *toward the same side.*

The Spinalis Muscle (Figs. 5-38 to 5-42). This medial column of the erector spinae muscle is relatively insignificant. It arises from spinous processes in the superior lumbar and inferior thoracic regions and inserts into spinous processes in the superior thoracic region (Fig. 5-41). It may also be divided into three parts (spinalis thoracis, cervicis, and capitis).

All three columns of the erector spinae extend the vertebral column and, acting on one side, bend the vertebral column laterally (Fig. 5-9 and Tables 5-2 and 5-3). The erector spinae is the chief extensor of the back. It also "pays out" during flexion of the vertebral column, thereby permitting slow, controlled flexion to occur.

DEEP LAYER OF INTRINSIC BACK MUSCLES (Figs. 5-38 and 5-43 to 5-46)

When the massive erector spinae muscles are removed, several short muscles (semispinalis, multifidus, and rotatores) are visible in the groove between the processes of the vertebrae (Fig. 5-38).

Collectively this obliquely disposed group of muscles is known as the **transversospinal muscles** because *their fibers run from the transverse processes to the spinous processes* of most vertebrae.

The Semispinalis Muscle (Figs. 5-38, 5-43 to 5-48, and Table 5-2). As its name indicates, this muscle ***originates from about half the vertebral column*** **or spine** (*i.e.*, T10 superiorly).

The semispinalis muscle can be divided into *three parts* according to its insertions: **semispinalis thoracis** into the cervical (C7) and thoracic (T1 to T4) spinous processes; **semispinalis cervicis** into the spinous processes of C2 to C5; and **semispinalis capitis** into the occipital bone.

The semispinalis thoracis and cervicis extend the thoracic and cervical regions of the vertebral column and rotate them toward the opposite side (Table 5-2).

The semispinalis capitis forms the largest muscle mass in the posterior aspect of the neck (Fig. 5-48).

The semispinalis capitis extends the head and turns the face toward the opposite side.

The Multifidus Muscle (Figs. 5-38, 5-43, 5-44, and Tables 5-2 and 5-3). The name multifidus (L. *multus*, many + *findo*, to cleave) indicates that this muscle is divided into several bundles. The multifidus muscle extends the entire length of the vertebral column, but it is heaviest in the lumbar region. It occupies the groove on each side of the spinous processes of the vertebrae from the sacrum to the axis (Figs. 5-38 and 5-44).

The multifidus muscle arises mainly from the sacrum and the mammillary processes of L5 to T12 vertebrae (Figs. 5-43 and 5-44). Some muscle fibers arise from the transverse processes of the thoracic vertebrae and from the articular processes of the cervical vertebrae. The bundles of multifidus muscle fibers pass superiorly over two to five vertebrae and then insert into the spinous processes of the vertebrae.

The multifidus muscle rotates the vertebral column slightly toward the opposite side. The multifidus muscles also stabilize the vertebral column.

The Rotatores Muscles (Figs. 5-38, 5-45, and Table 5-3). These short muscles, the deepest group in the groove between the spinous and transverse processes, run the entire length of the vertebral column. They can be demonstrated best in the thoracic region. The rotatores are the shortest muscles in the transversospinal group.

The rotatores muscles arise from the transverse process of one vertebra and *insert into the base of the spinous process* of the vertebra superior to it (Fig. 5-45).

The rotatores muscles extend the vertebral column and rotate it. They also stabilize the vertebral column.

The Interspinales and Intertransversarii Muscles (Figs. 5-44 and 5-46). These small insignificant muscles unite the spinous and transverse processes of the vertebrae, respectively. They are well developed only in the cervical region. The interspinales can extend the vertebral column and the intertransversarii can produce lateral bending of the vertebral column.

THE LEVATORES COSTARUM MUSCLES (Figs. 1-19 and 5-45)

These muscles in the thoracic region, represent the posterior intertransversarius muscles of the neck. They are described in Chapter 1.

GENERAL COMMENTS ABOUT THE BACK MUSCLES

Try to gain an overview of the intrinsic back muscles and their actions because acute backache caused by muscle strain is such a common complaint.

It may help you understand the arrangement of these muscles if you *visualize the erector spinae muscles as fanning out from the inferior end of the vertebral column* (Fig. 5-41). They span the vertebral column and the posterior aspect of the thorax as they extend superolaterally.

The deeper muscles (transversospinalis) span shorter distances and are directed superomedially (Fig. 5-43). The deepest muscles (rotatores) span the shortest distance and run from a transverse process inferolaterally to a spinous process superomedially (Fig. 5-45).

The vertebral column is controlled by the intrinsic muscles of the back (i.e., they regulate its posture). When you are standing at attention,

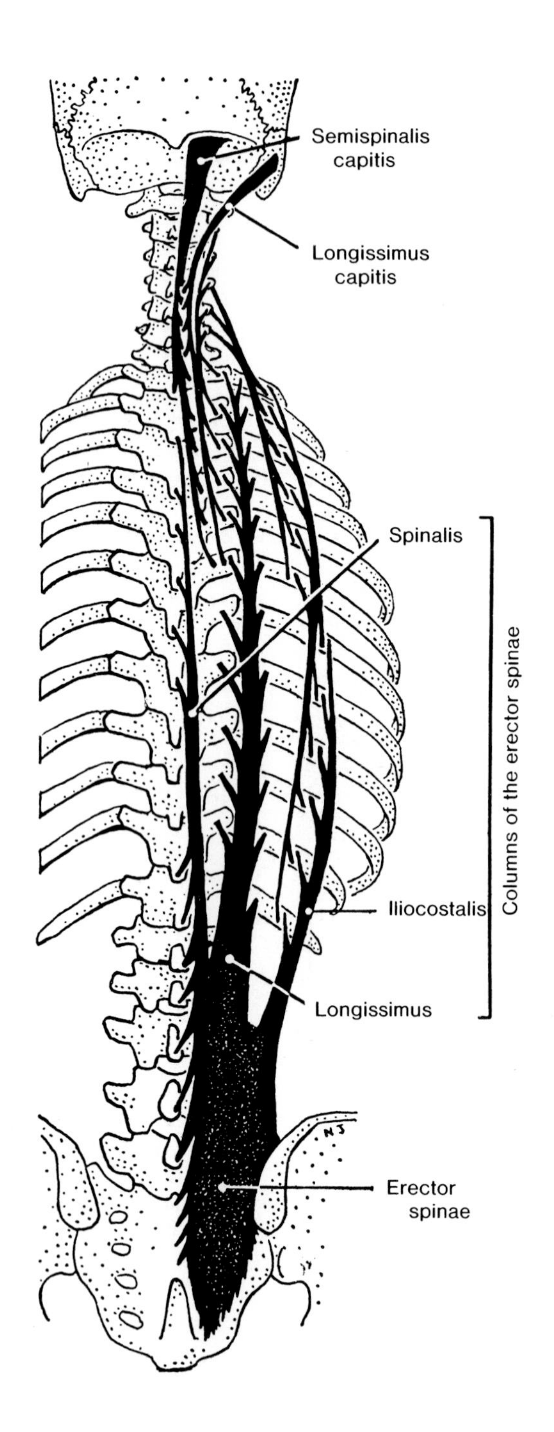

Figure 5-41. Drawing of the erector spinae to show the massive and complex form of this muscle, which lies directly deep to the posterior layer of the thoracolumbar fascia (Fig. 5-40). The erector spinae ascends throughout the length of the back, but its three columns are composed of rope-like fascicles of shorter length. The erector spinae extends the vertebral column and, acting on one side, bends the vertebral column to that side (Table 5-3). The erector spinae is the largest muscular mass of the back. Also see Figure 5-40.

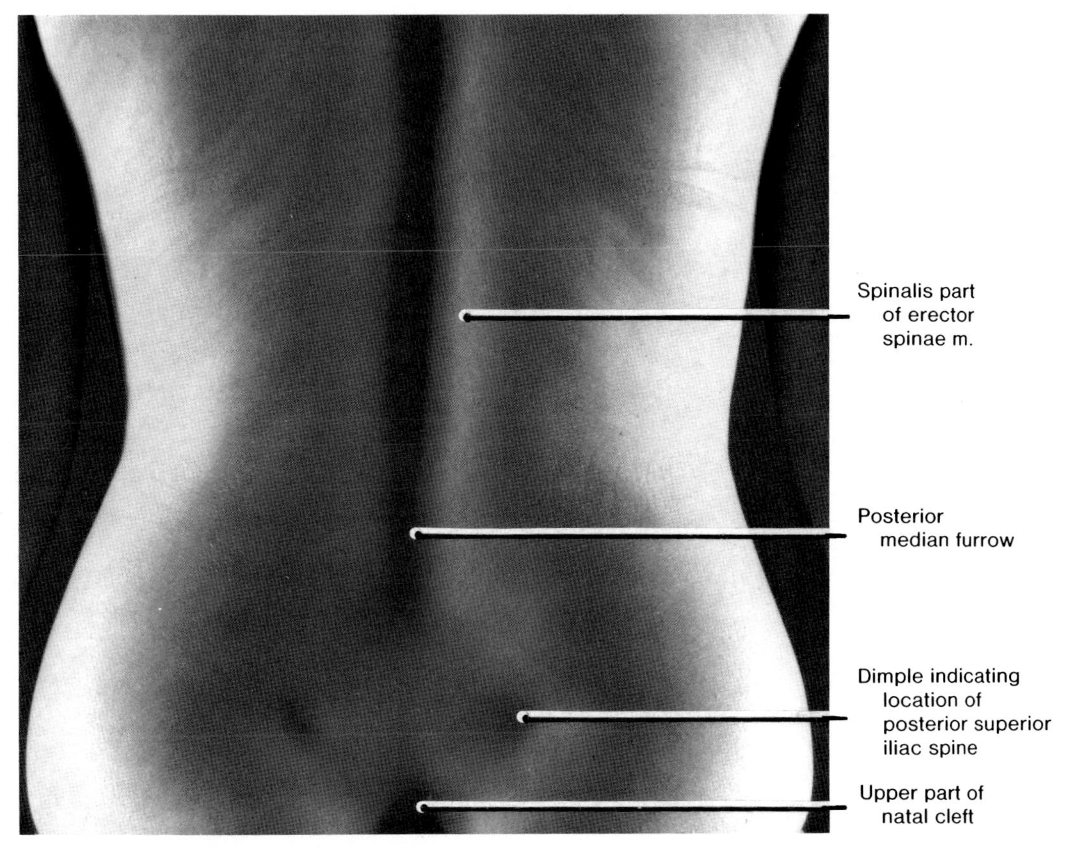

Figure 5-42. Photograph of the back of a 21-year old woman showing the principal surface landmarks and the external appearance of the erector spinae muscles. The erector spinae muscle is the chief extensor of the back. The dimples indicating the posterior superior iliac spines are obvious because the skin and underlying fascia are attached to bone in this area (Figs. 2-105 and 5-44). A line joining these dimples crosses the spinous process of the second sacral vertebra. The spinalis part of the erector spinae muscle is about 2 cm wide and is inserted into the spinous processes (Fig. 5-41).

Table 5-3
Muscles Producing Movements of the Thoracic and Lumbar Intervertebral Joints[1]

Flexion	Extension	Rotation and Lateral Bending[2]
Rectus abdominis The other anterior abdominal muscles may aid in flexion, but like the rectus they are seldom called upon. Flexion occurs mainly by gravity under the control of the back muscles.	**Erector spinae** **Quadratus lumborum** Trapezius	**Psoas major** **Quadratus lumborum** **External oblique and internal oblique** of anterior abdominal wall **Multifidus** **Iliocostalis lumborum** **Iliocostalis thoracis** **Rotatores** Intertransversarii

[1] The principal muscles producing these movements appear in **bold face**.
[2] Lateral bending is also called lateral flexion (Fig. 5-9).

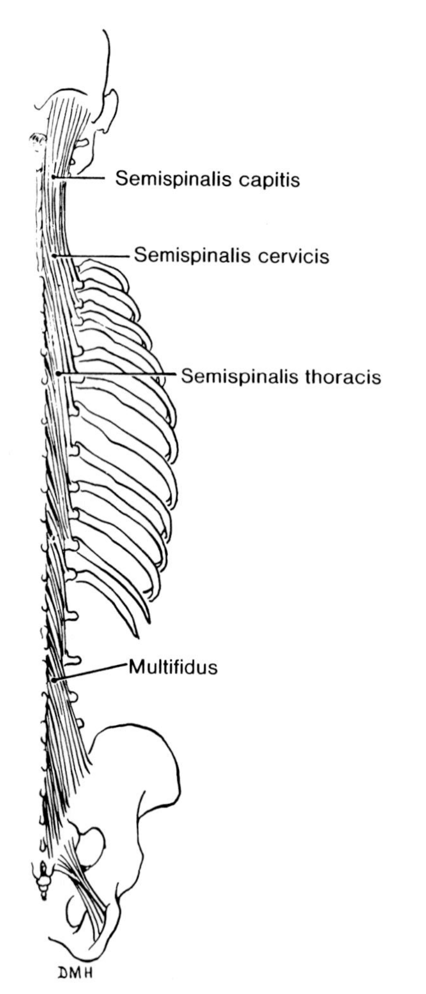

Figure 5-43. Drawing of the back muscles which lie deep to the erector spinae muscles. As most of them originate from transverse processes and insert into the spinous processes of vertebrae, they are known collectively as the transversospinal muscles.

many of your back muscles are active, whereas when you stand at ease, many of them are inactive. They relax completely when you flex your vertebral column as far as possible (Figs. 5-9 and 5-10), because in this position the ligaments support the vertebral column (Figs. 5-28 and 5-45).

THE SUBOCCIPITAL REGION

The suboccipital region (Fig. 5-48) is the triangular area around the articulation between the skull and the superior end of the vertebral column. It is located between the occipital portion of the skull and the posterior aspects of the atlas and axis, deep to the trapezius and semispinalis capitis muscles (Fig. 5-40).

THE BONES OF THE SUBOCCIPITAL REGION (Figs. 5-16, 5-17, 5-34 to 5-36, and 5-47 to 5-49)

The bones of the suboccipital region are the *occipital condyles* of the **occipital bone** of the skull superiorly, and the **atlas and axis** inferiorly.

THE MUSCLES OF THE SUBOCCIPITAL REGION (Figs. 5-40, 5-48, and 5-49)

Four small muscles, lying deep to the semispinalis capitis, *extend and rotate the head.*

The rectus capitis posterior major (*rectus major*) originates from the spinous process of the axis, whereas the **rectus capitis posterior minor** (*rectus minor*) arises from the posterior tubercle of the atlas (Figs. 5-16, 5-17, and 5-49). These muscles insert, side by side, into the occipital bone inferior to the inferior nuchal line (Figs. 5-47 to 5-49).

The obliquus capitis inferior (*inferior oblique*) is a thick muscle that arises from the spinous process of the axis and runs obliquely superiorly and anteriorly to insert into the tip of the transverse process of the atlas (Figs. 5-48 and 5-49). Although not attached to the skull, *the obliquus capitus inferior muscle rotates the head by pulling on the atlas. It also turns the face to the same side.*

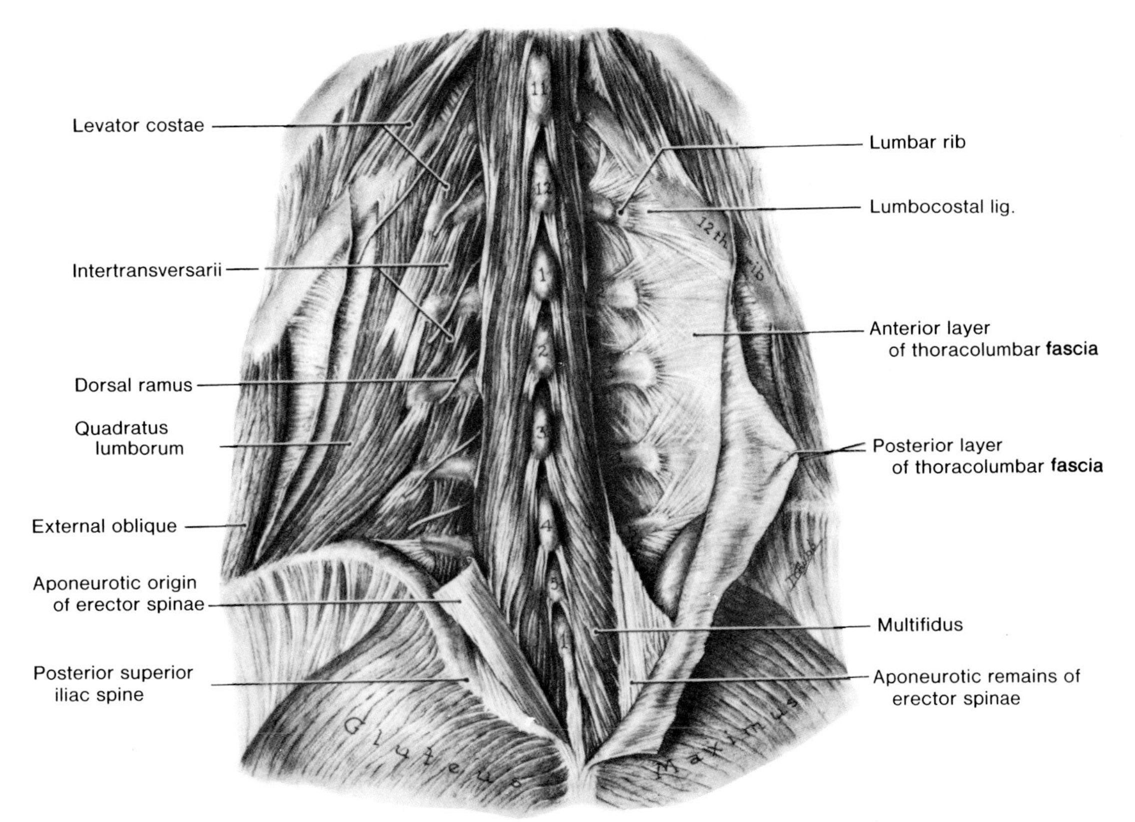

Figure 5-44. Drawing of a deep dissection of the back showing the multifidus and other intrinsic back muscles. Note that a short lumbar rib is present. The multifidus muscle extends throughout the length of the vertebral column, but it is heaviest in the lumbar region.

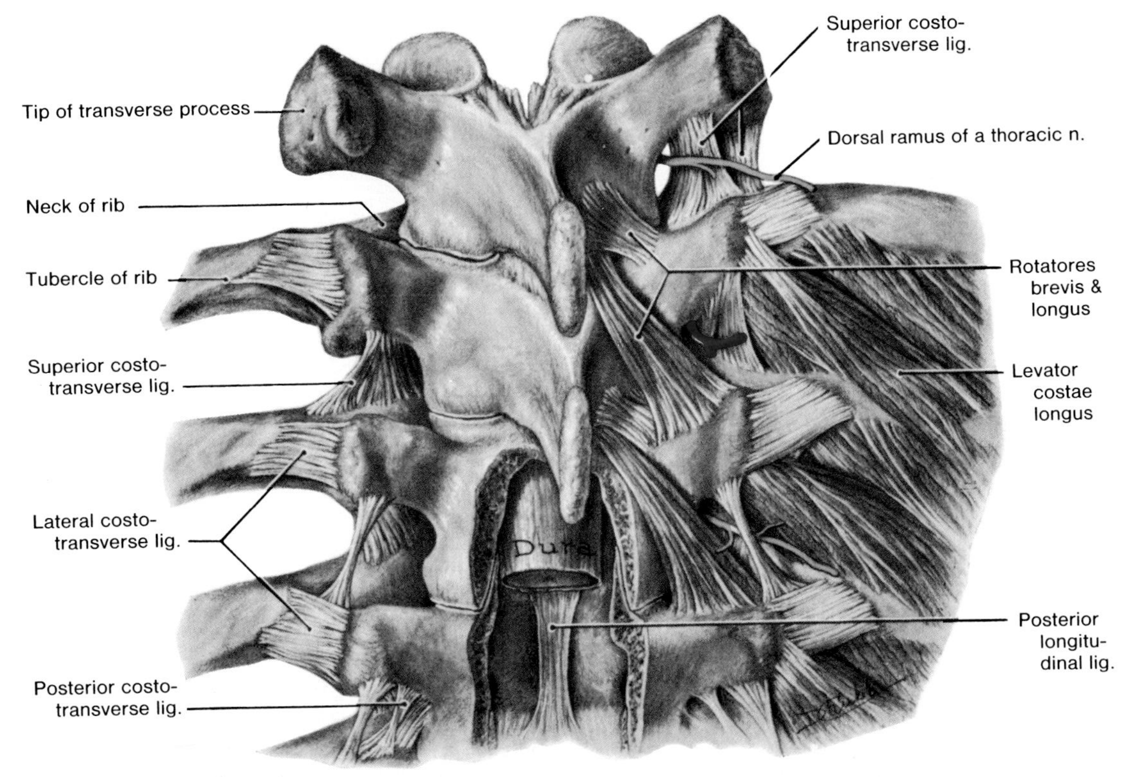

Figure 5-45. Drawing of a dissection of the back showing the rotatores muscles and the costotransverse ligaments. Of the three layers of transversospinalis muscles (semispinales, multifidus, and rotatores), the rotatores are the deepest and shortest. They pass from the root of the transverse process of one vertebra to the junction of the transverse process and the lamina of the vertebra superior to it. The rotatores muscles rotate the vertebral column and bend it laterally.

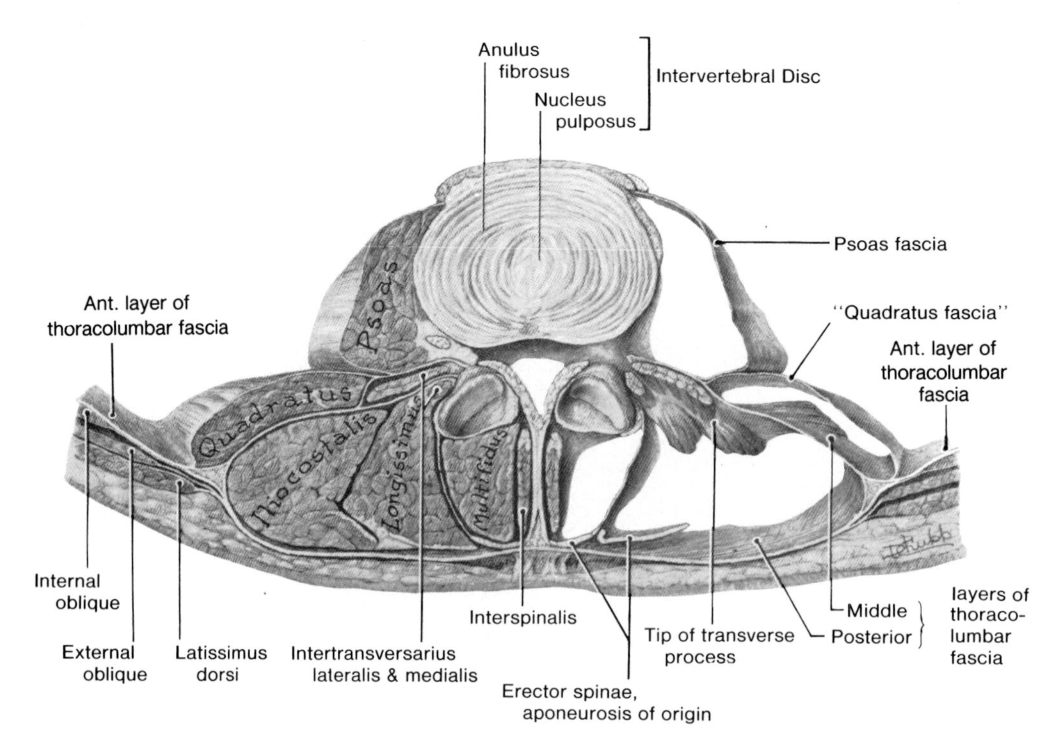

Figure 5-46. Drawing of a transverse section of a dissection of the deep muscles of the back. The muscles have been removed from their fascial sheaths on the right side. Observe the laminated form of the anulus fibrosus of the intervertebral disc. No organization is visible in the nucleus pulposus which consists mainly of fibrocartilage in older persons. Note that the anulus fibrosus is strongest anteriorly and laterally. Protrusion of the intervertebral disc commonly occurs posterolaterally, where the annulus fibrosus is thin (Fig. 5-68).

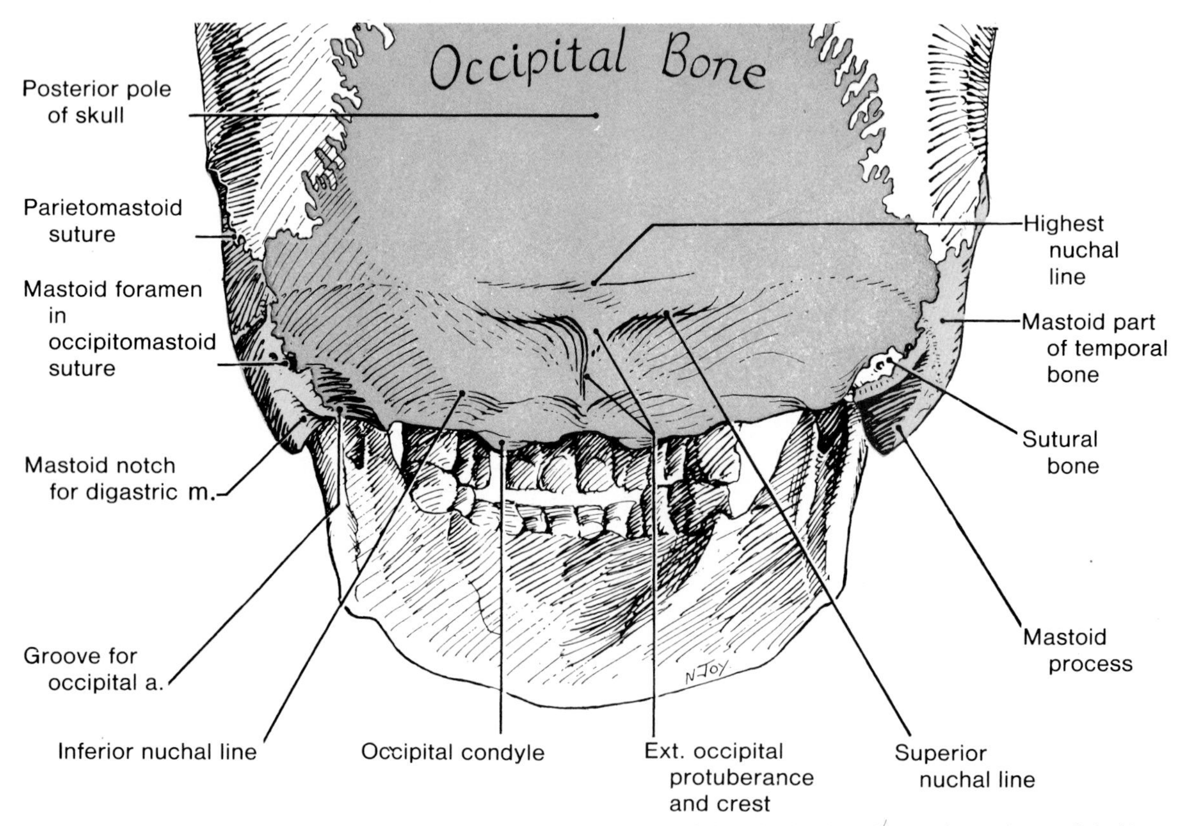

Figure 5-47. Drawing of the inferior half of the posterior aspect of the skull, primarily to show the occipital bone and the mastoid processes of the temporal bones.

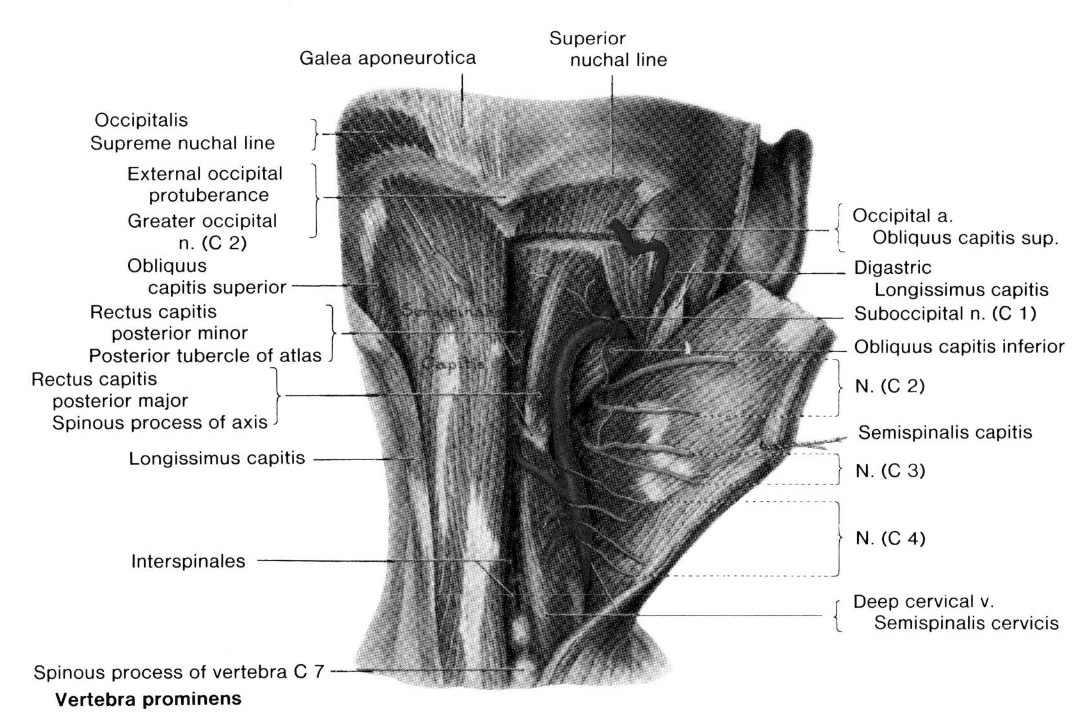

Figure 5-48. Drawing of a dissection of the suboccipital region. The trapezius, sternocleidomastoid, and splenius muscles have been removed. (For illustrations of them, see Figs. 6-44 and 6-47). The semispinalis capitis is the largest muscle mass in the posterior aspect of the neck, and is largely responsible for the longitudinal bulge on each side of the median plane. The semispinalis capitis is a very powerful extensor of the head (Table 5-1).

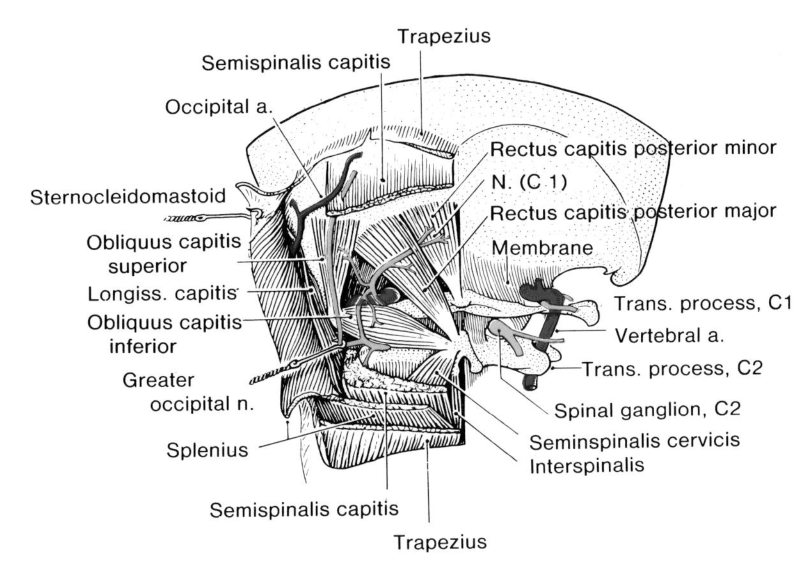

Figure 5-49. Diagram of the suboccipital region. The suboccipital triangle is bounded by three muscles: obliquus capitis inferior, obliquus capitis superior, and rectus capitis posterior major. The suboccipital triangle contains the vertebral artery and the suboccipital nerve (Fig. 5-48), both lying in a groove on the superior surface of the posterior arch of the atlas (Figs. 5-16 and 5-17). Observe the vertebral artery winding posterior to the superior articular process of the atlas to enter the foramen magnum of the skull.

The obliquus capitis superior (*superior oblique*) is a flat triangular muscle that arises from the tip of the transverse process of the atlas and runs obliquely superiorly and posteriorly to insert into the occipital bone (Figs. 5-48 and 5-49). *The obliquus capitus superior bends the head posteriorly and to the same side.*

THE SUBOCCIPITAL TRIANGLE (Figs. 5-48 and 5-49)

The boundaries of the suboccipital triangle are formed by three muscles: **rectus capitis posterior major,** superiorly and medially; **obliquus capitus superior,** superiorly and laterally; and **obliquus capitis inferior,** inferiorly and laterally.

The floor of the suboccipital triangle is formed by the posterior atlantooccipital membrane and the posterior arch of the atlas (Figs. 5-36 and 5-49).

The roof of the suboccipital triangle is formed by the semispinalis capitis muscle.

The suboccipital triangle is important clinically because it contains the ***vertebral artery*** *and the* ***suboccipital nerve*** (dorsal ramus of the first cervical nerve). These structures lie in a groove on the superior surface of the posterior arch of the atlas (Figs. 5-16, 5-48, and 5-49).

The curves in the vertebral arteries as they wind their way from the vertebral column, posterior to the superior articular process of the atlas, to enter the foramen magnum of the skull (Fig. 5-48), may be clinically significant when blood flow through them is reduced (*e.g.*, owing to **arteriosclerosis.** Under these conditions, prolonged turning of the head as may occur when backing up a car, may cause dizziness and other symptoms because of interference with the blood supply to the brain stem.

A needle can be inserted in the midline, superior to the arch of the atlas, into the subarachnoid space for the collection of cerebrospinal fluid (CSF). This method of obtaining CSF is called a **cisternal puncture** because the cerebellomedullary cistern (Figs. 7-70 and 7-86) is punctured with the needle.

CSF is collected for determination of alterations and variations in cells or in the concentration of chemicals in it. Cisternal puncture and *injection of a contrast material into the cerebellomedullary cistern* are also performed by radiologists to outline the superior end of a tumor in the vertebral column (*intraspinal tumor*). This procedure is performed after a myelogram has shown that the inferior end of the tumor completely blocks the vertebral canal.

A myelogram is a radiograph of the spinal cord taken after the injection of a contrast medium into the subarachnoid space (Figs. 5-53 and 5-55*D*).

CSF is usually obtained by lumbar puncture (Fig. 5-60), a procedure during which a needle is inserted between the spinous processes of L3 and L4 (or L4 and L5) into the subarachnoid space inferior to the spinal cord.

THE SPINAL CORD AND MENINGES

The spinal cord is the part of the central nervous system that lies in the vertebral canal formed by the superimposed vertebrae (Figs. 5-50 and 5-51). The spinal cord is a cylindrical structure that is slightly flattened anteriorly and posteriorly (Fig. 5-52). It is protected by the vertebrae, their ligaments, the spinal meninges (membranes), and CSF.

The spinal cord begins as a continuation of the medulla, the inferior part of the brain stem, and extends from the foramen magnum of the occipital bone to the level of the second lumbar vertebra (Fig. 5-50). The spinal cord ranges from 42–45 cm in length.

The spinal cord in adults usually ends opposite the intervertebral disc between L1 and L2 vertebrae, but it may terminate at T12 or at L3 (Fig. 5-60*B*).

Note that the spinal cord occupies only the superior two-thirds of the vertebral canal (Fig. 5-50), and that it is enlarged in two regions for innervation of the limbs (Fig. 5-56).

The cervical enlargement extends from C4 to T1 segments of the spinal cord, and most of the corresponding spinal nerves form the brachial plexus for innervation of the upper limb (Fig. 6-22).

The lumbosacral enlargement extends from L2 to S3 segments of the spinal cord, and the corresponding nerves make up the lumbar and sacral plexuses for innervation of the lower limb (Figs. 2-118 and 2-125).

It is clinically important to understand that the spinal cord segments do not correspond with the vertebral levels, e.g., the lumbosacral enlargement (L2 to S3 segments of the spinal cord) extends from about the body of T11 vertebra to the level of the body of L1. In Figure 5-50, note that the thoracic region of the spinal cord is the longest part, and that *the sacral region of the spinal cord is considerably superior to the sacrum.* Also note that the sacral region is the shortest part of the spinal cord (Figs. 5-50 and 5-54).

THE STRUCTURE OF THE SPINAL NERVES (Figs. 5-50 to 5-54)

There are 31 pairs of spinal nerves attached to the spinal cord by dorsal and ventral roots. The **ventral roots** leaving the cord contain efferent (motor) fibers,

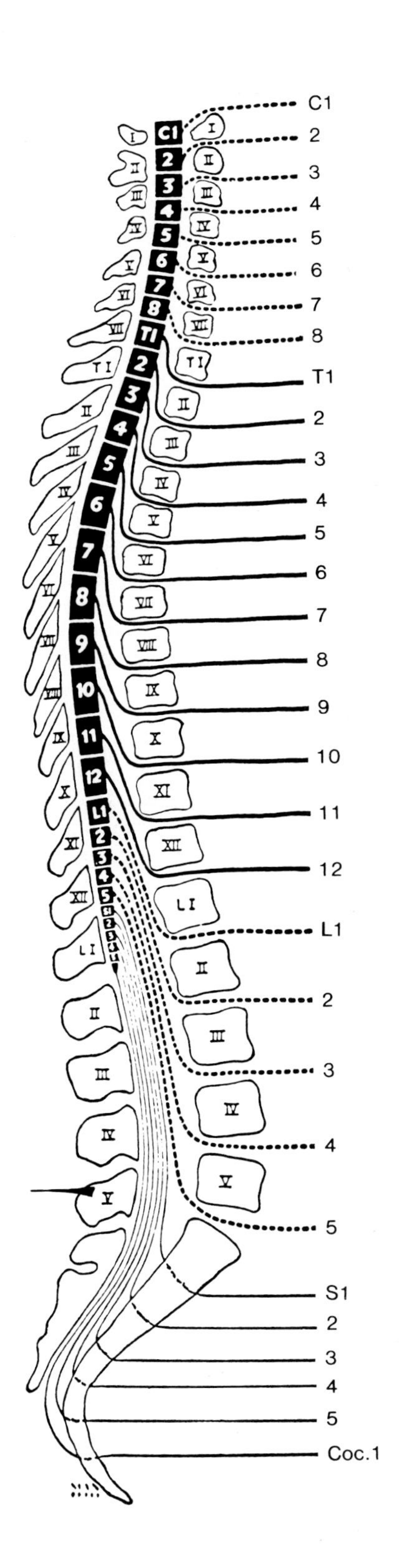

Figure 5-50. Diagram illustrating the relation of segments of the spinal cord and spinal nerves to the vertebral column. Note that there are 31 pairs of spinal nerves.

whereas the **dorsal roots** entering the cord contain afferent (sensory) fibers.

The cell bodies of axons making up the ventral roots are in the ventral gray horn of the spinal cord (Figs. 5-51 and 5-52), whereas the cell bodies of axons making up the dorsal roots are outside the spinal cord in the **spinal ganglia** (dorsal root ganglia).

The dorsal root of each spinal nerve has a spinal ganglion which is located in the intervertebral foramen, where it rests on the pedicle of the vertebral arch (Figs. 5-51, 5-53, and 5-54). Distal to the spinal ganglion and just outside the intervertebral foramen, the dorsal and ventral nerve roots unite to form a **spinal nerve.** The spinal nerve divides almost immediately into a **ventral ramus** (L. branch) and a **dorsal ramus** (Fig. 5-51).

POSITIONAL CHANGES OF THE DEVELOPING SPINAL CORD (Fig. 5-55)

In the embryo the spinal cord extends the entire length of the vertebral canal, and the spinal nerves form just outside the intervertebral foramina at their levels of origin (Fig. 5-55*A*). Because the vertebral column grows more rapidly than the spinal cord in the embryo and fetus, this relationship does not persist. The inferior end of the spinal cord comes to lie at relatively higher levels. At 24 weeks it ends at the level of S1 (Fig. 5-55*B*).

In the newborn infant the spinal cord terminates at L2 or L3 (Fig. 5-55*C*), and *in the adult it usually ends at the inferior border of L1 vertebra* (Figs. 5-50 and 5-55*D*).

The position of the spinal cord during development determines the direction of the spinal nerve roots in the subarachnoid space. The dorsal and ventral roots of the spinal nerves from C1 to C7 segments of the spinal cord leave the vertebral canal through the intervertebral foramina, superior to the corresponding pedicles of the vertebrae (Figs. 5-50 and 5-51). The spinal nerves from C1 and C2 segments lie on the vertebral arches of the atlas and the axis, respectively. The dorsal and ventral roots of the eighth cervical nerve pass through the intervertebral foramen between C7 and T1 vertebrae, because there are eight cervical nerves and only seven cervical vertebrae.

Owing to the inequality in the length of the adult spinal cord and vertebral column, there is a progressive obliquity of the dorsal and ventral nerve roots within the subarachnoid space (Figs. 5-50 and 5-53). Consequently, *the length and obliquity of the nerve roots increase progressively as the inferior end of the vertebral column is approached,* because of the increasing distance between the spinal cord segments and the corresponding vertebrae. Thus, the lumbar and sacral spinal nerve roots are the longest; they form a bundle of nerve roots in the subarachnoid space, caudal to the termination of the spinal cord, called the **cauda equina** (L. horse's tail). *This bundle of spinal nerve roots* extends into the part of the adult vertebral canal distal to L2 vertebra (Figs. 5-31, 5-50, 5-53, and 5-54).

The inferior end of the spinal cord tapers rather abruptly into the **conus medullaris** (Figs. 5-53 and 5-55). From its inferior end, a slender fibrous strand, called the **filum terminale** (L. *filum,* a thread), descends within the cauda equina (Figs. 5-53 to 5-56). Its subarachnoid part ends at S2, where it is attached to the inferior end of the **dural sac,** and its extradural prolongation inserts into the dorsum of the coccyx (Figs. 5-54 and 5-55). The filum terminale, which has no functional significance, consists of connective tissue, pia mater, and neuroglial elements.

THE BLOOD SUPPLY OF THE SPINAL CORD (Figs. 5-56 and 5-57)

The spinal cord is supplied by three longitudinal vessels, an anterior spinal artery and two posterior spinal arteries. These vessels are reinforced by blood from segmental vessels called radicular arteries.

The anterior spinal artery is formed by the union of two small branches of the vertebral arteries (Fig. 5-56*A*). It runs the length of the spinal cord in the ventral median fissure (Figs. 5-52 and 5-57).

The anterior spinal artery supplies the anterior two-thirds of the spinal cord. The caliber of the anterior spinal artery varies according to its proximity to a major radicular artery. It is usually smallest in the T4 to T8 regions of the spinal cord.

The two posterior spinal arteries arise as small branches of either the vertebral or the posterior inferior cerebellar arteries (Fig. 5-56*B*). The posterior spinal arteries anastomose frequently with each other and with the anterior spinal artery (Fig. 5-57).

The posterior spinal arteries supply the posterior one-third of the spinal cord.

The blood supplied by the anterior and posterior spinal arteries is sufficient only for the superior cervical segments of the spinal cord. The remaining segments receive most of their blood from the many radicular arteries.

The anterior radicular arteries supply the anterior spinal artery and the posterior radicular arteries contribute blood to the posterior spinal arteries (Figs. 5-56 and 5-57).

The radicular arteries arise from the spinal branches of the vertebral, deep cervical, ascending cervical, posterior intercostal, lumbar, and lateral sacral arteries (Fig. 5-56). The radicular arteries enter the vertebral canal through the intervertebral foramina, and divide into *anterior and posterior radicular arteries* (Fig. 5-57).

The radicular arteries supply the vertebrae, meninges, and spinal arteries (Figs. 5-56 and 5-57). The radicular arteries pass along the dorsal and

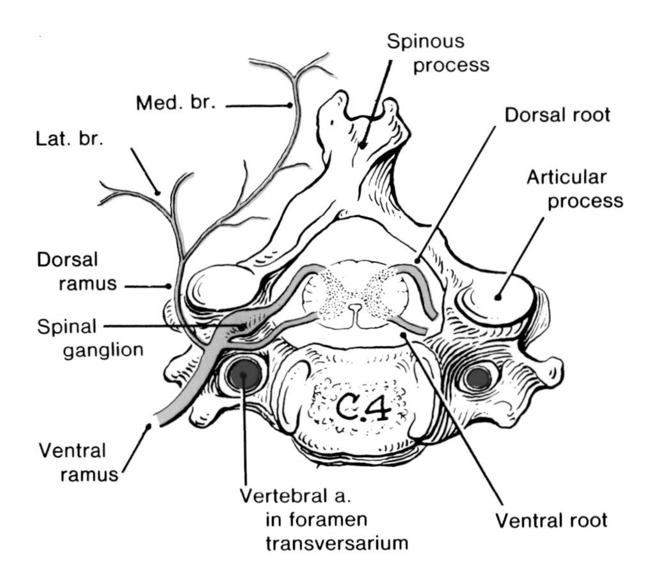

Figure 5-51. Drawing of the superior surface of the fourth cervical vertebra showing the spinal cord in its vertebral foramen. The meninges are not illustrated here (see Fig. 5-52). Note that a typical spinal nerve arises from the spinal cord by dorsal and ventral roots. The spinal ganglion is located on the dorsal root and is within the intervertebral foramen. Note also that the dorsal and ventral roots join at about the level of the lateral edge of the intervertebral foramen to form a spinal nerve. Observe that the fourth cervical nerve leaves the intervertebral foramen superior to the pedicle of C4. Just outside the foramen, the nerve divides into dorsal and ventral rami.

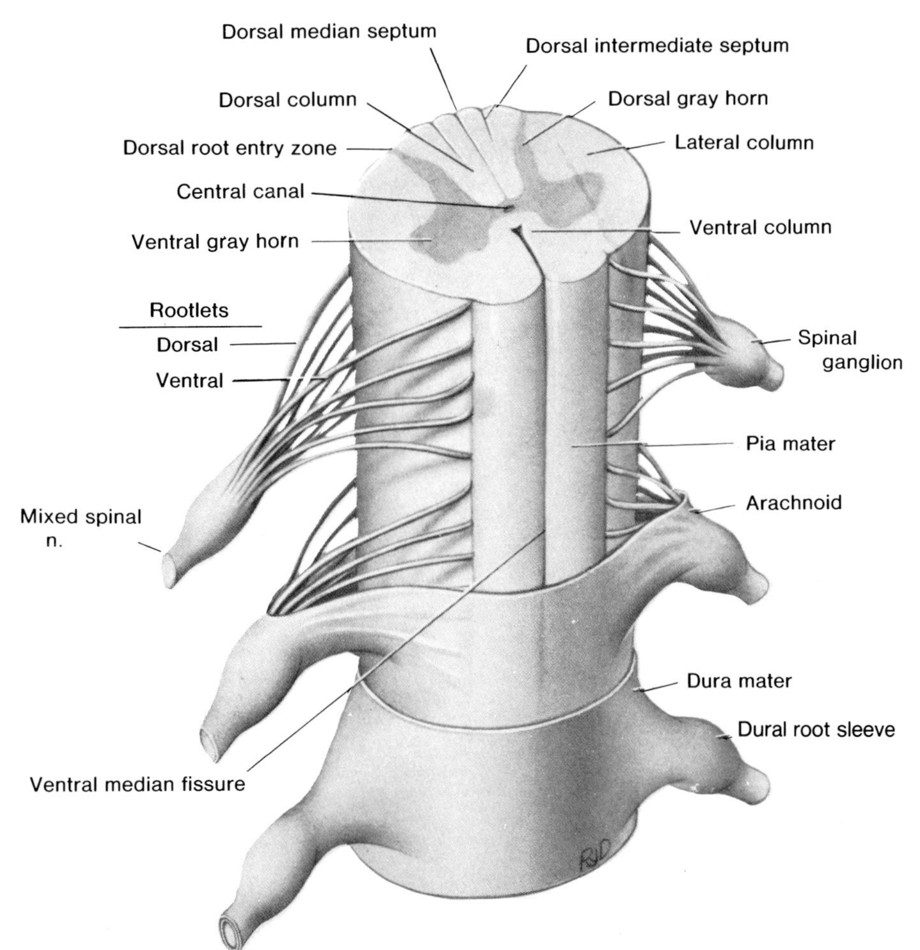

Figure 5-52. Drawing of the spinal cord, nerve roots, and meninges illustrating their general structure. Observe that each nerve root emerges from the spinal cord as a series of rootlets and that each spinal nerve is formed by the union of dorsal and ventral spinal nerve rootlets. In the section of the spinal cord, observe that the gray matter has an H-shaped or butterfly outline and is surrounded by white matter. Note that the dorsal and ventral gray horns divide the white matter into three columns: dorsal, lateral, and ventral. The spinal pia mater is connective tissue that closely invests the spinal cord, enmeshing its blood vessels (Fig. 5-57).

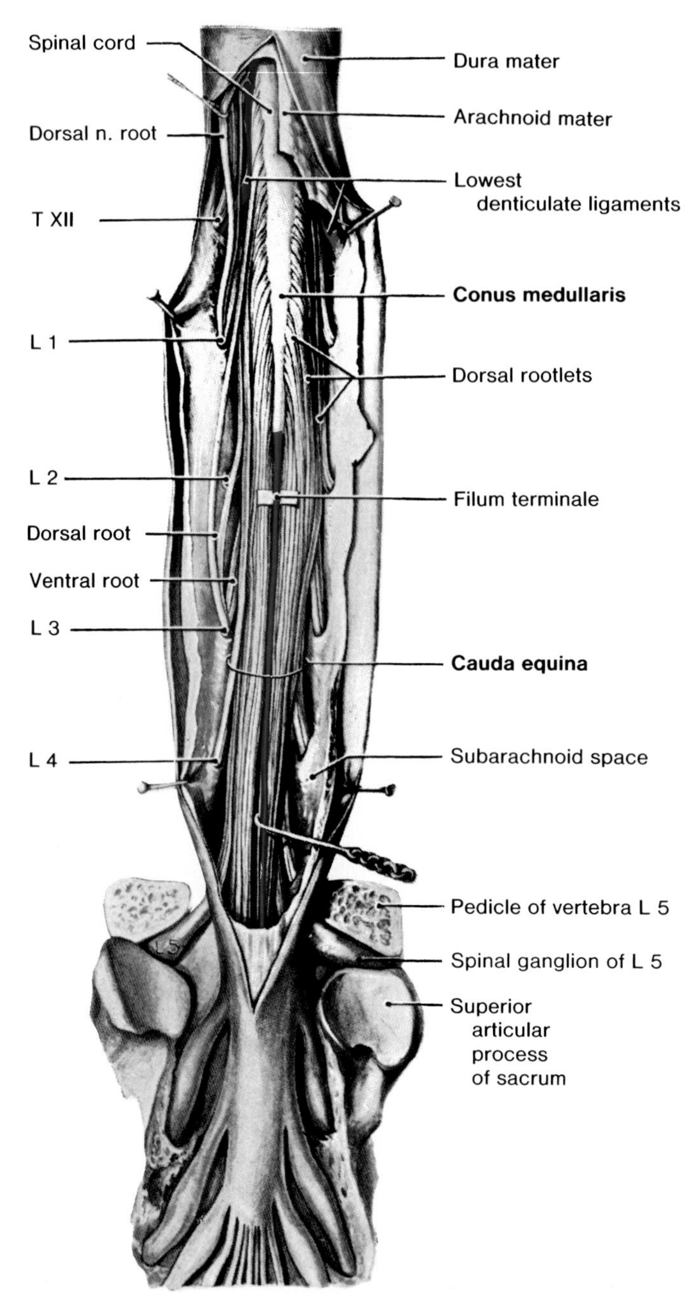

Figure 5-53. Drawing of a dissection of the inferior end of the spinal cord and cauda equina (L. horse's tail). The dura mater and arachnoid have been incised to expose the spinal cord and the cauda equina. Observe that the compact array of spinal nerve roots and the filum terminale resemble a horse's tail which accounts for its name.

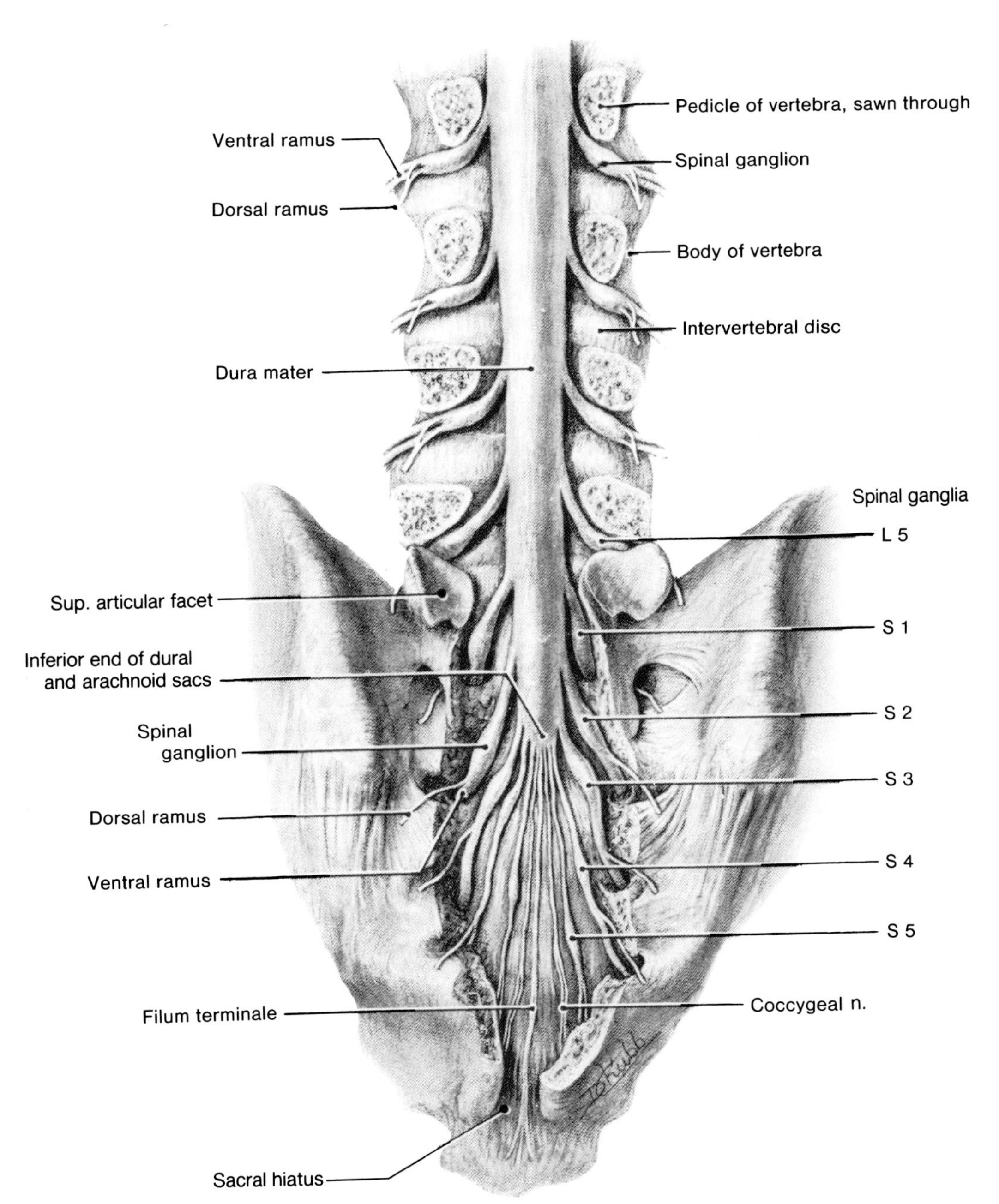

Figure 5-54. Drawing of a dissection of back to show the inferior end of the spinal cord in the dural and arachnoid sacs, the spinal ganglia, and the spinal nerves. In the adult the spinal cord usually ends at the intervertebral disc between L1 and L2 vertebrae (Figs. 5-50 and 5-53). The posterior parts of the lumbar and sacral vertebrae have been removed. The lumbar spinal ganglia are in the intervertebral foramina; the sacral spinal ganglia are in the sacral canal. The sacral canal is formed by the vertebral foramina of the sacral vertebrae (Fig. 5-24). Observe the inferior median opening of the sacral canal, called the sacral hiatus.

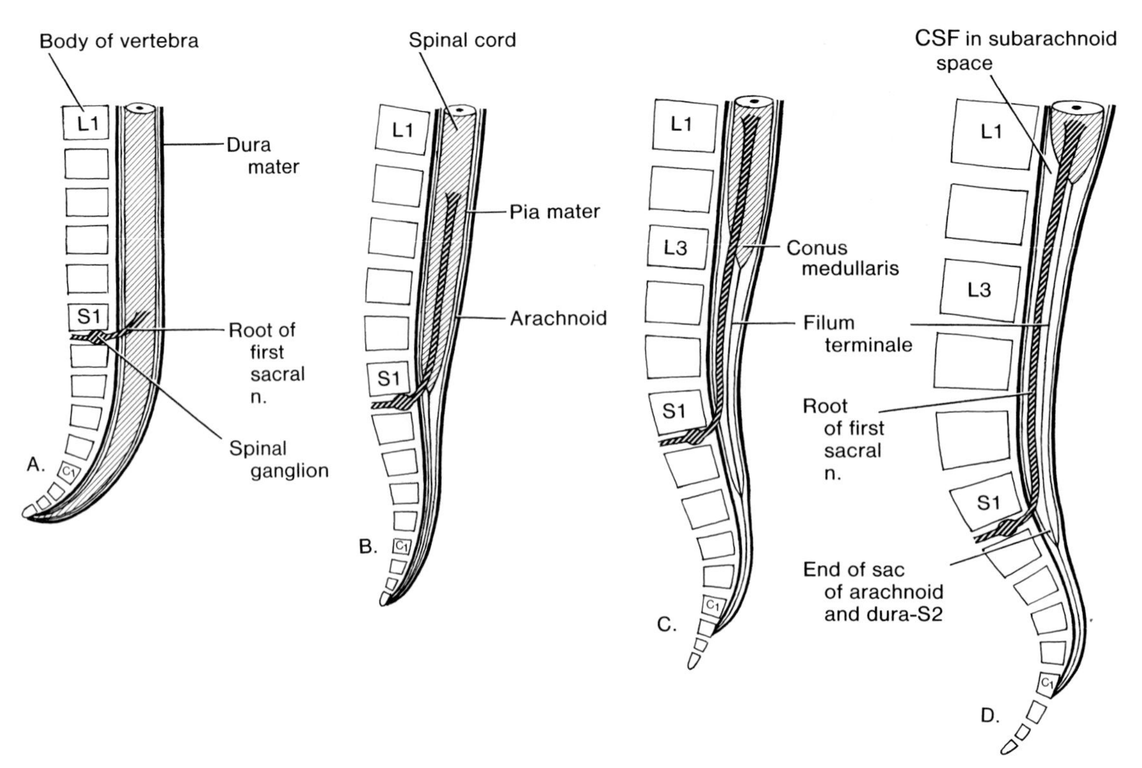

Figure 5-55. Drawings showing the position of the inferior end of the spinal cord in relation to the vertebral column and meninges at various stages of development. The increasing length and obliquity of the nerve roots is also shown. *A,* 8 weeks; *B,* 24 weeks; *C,* newborn; *D,* adult. (From Moore KL: *The Developing Human: Clinically Oriented Embryology*, ed 3. Philadelphia, WB Saunders Co, 1982).

ventral roots of the spinal nerves to reach the spinal cord (Fig. 5-57).

Usually one of the anterior radicular arteries supplying the lumbosacral enlargement of the cord is much larger than the others. This *great radicular artery or* **arteria radicularis magna** (artery of Adamkiewicz) arises more frequently on the left side from an inferior intercostal or a superior lumbar artery.

The great radicular artery or arteria radicularis magna is clinically important because it usually makes a major contribution to the anterior spinal artery, and provides the main blood supply to the inferior two-thirds of the spinal cord.

The small number of radicular arteries supplying the midthoracic region of the spinal cord means that **the midthoracic region is the watershed area of the spinal cord**. It is a junctional zone between the segments superior and inferior to it that are well supplied with blood.

The segmental reinforcements of blood supply from the radicular arteries are very important in feeding the anterior and posterior spinal arteries. Fractures, dislocations, and fracture-dislocations may interfere with the blood supply from the anterior and posterior spinal arteries to the spinal cord. Ischemia affects its function and can lead to paralysis of muscles.

The spinal cord may also suffer circulatory impairment if the radicular arteries, particularly the **arteria radicularis magna,** are compromised by obstructive arterial disease, or by ligation during surgery of the posterior intercostal or lumbar arteries which give rise to them (Fig. 5-56). These patients may lose all sensation and voluntary movement inferior to the level of impaired blood supply to the spinal cord.

***When there is a severe drop in systemic blood pressure* for 3 to 5 minutes, blood flow through the small radicular arteries supplying the watershed area of the spinal cord may be reduced or stopped.** This may also result in necrosis of neurons in the midthoracic region of the cord, and these patients may also lose sensation and voluntary move-

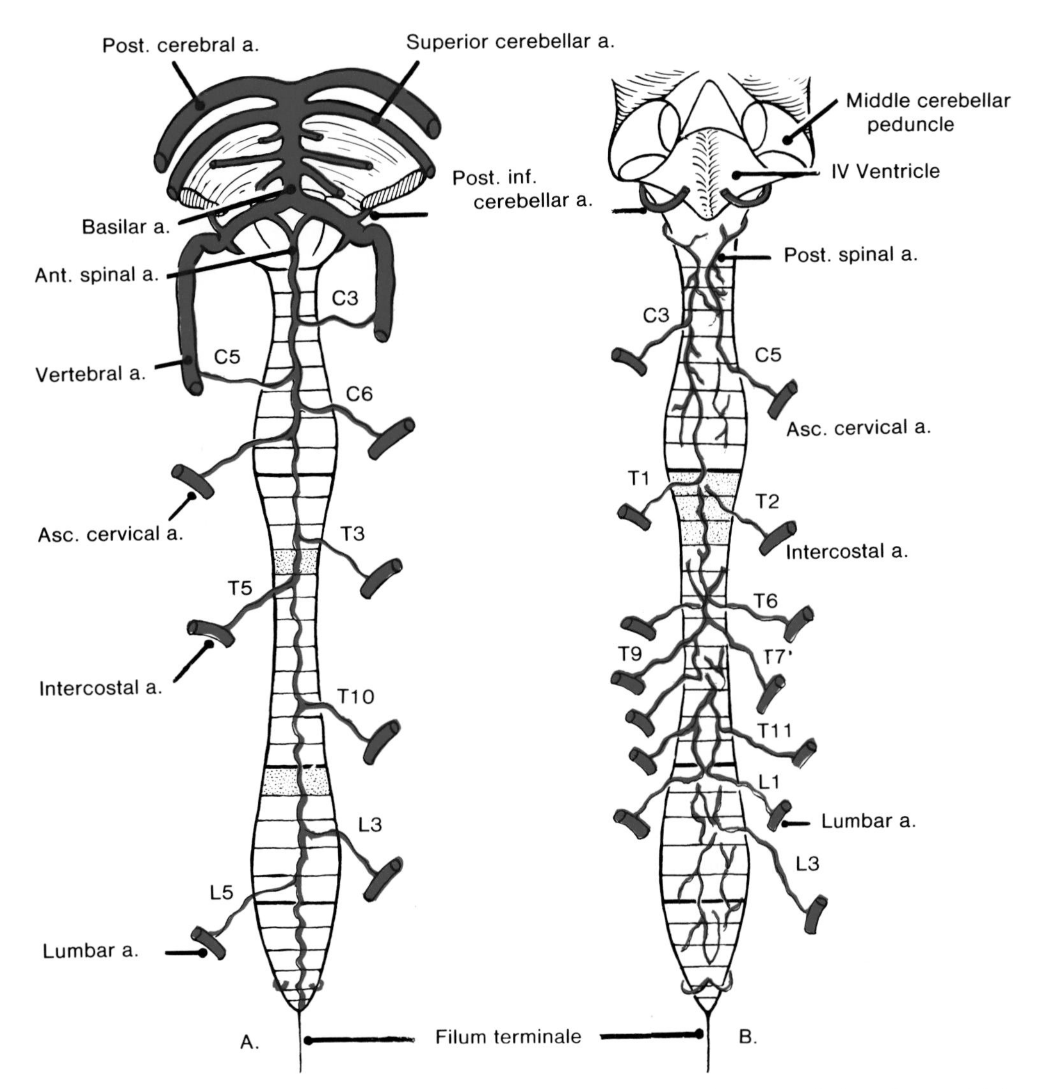

Figure 5-56. Drawings illustrating the arteries of the spinal cord. *A,* ventral aspect; *B,* dorsal aspect. The stippled areas of the cord indicate the regions most vulnerable to vascular deprivation when contributing arteries are injured. The levels of entry of the common radicular branches are shown (*e.g.*, C5 and T5). Note that the spinal cord is enlarged in two regions for innervation of the limbs. The cervical enlargement extends from C4 to T1 and the lumbosacral enlargement extends from L2 to S3 segments of the spinal cord.

ment in the areas supplied by the affected level of the spinal cord.

VENOUS DRAINAGE OF THE SPINAL CORD (Figs. 5-57 and 5-58)

The spinal veins have a distribution somewhat similar to that of the spinal arteries. There are usually *three anterior and three posterior spinal veins.* They are arranged longitudinally, communicate freely with each other, and are drained by numerous radicular veins.

The vertebral canal contains a profuse plexus of thin-walled, valveless veins which surround the spinal dura mater (Figs. 5-57 and 5-58). These veins communicate via the anterior and posterior longitudinal sinuses with the venous sinuses of the dura mater (Fig. 7-39).

The anterior and posterior spinal veins and vertebral venous plexuses drain into intervertebral veins, and via them into the **vertebral veins, ascending lumbar veins, and the azygos venous system** (Figs. 1-80 and 5-58).

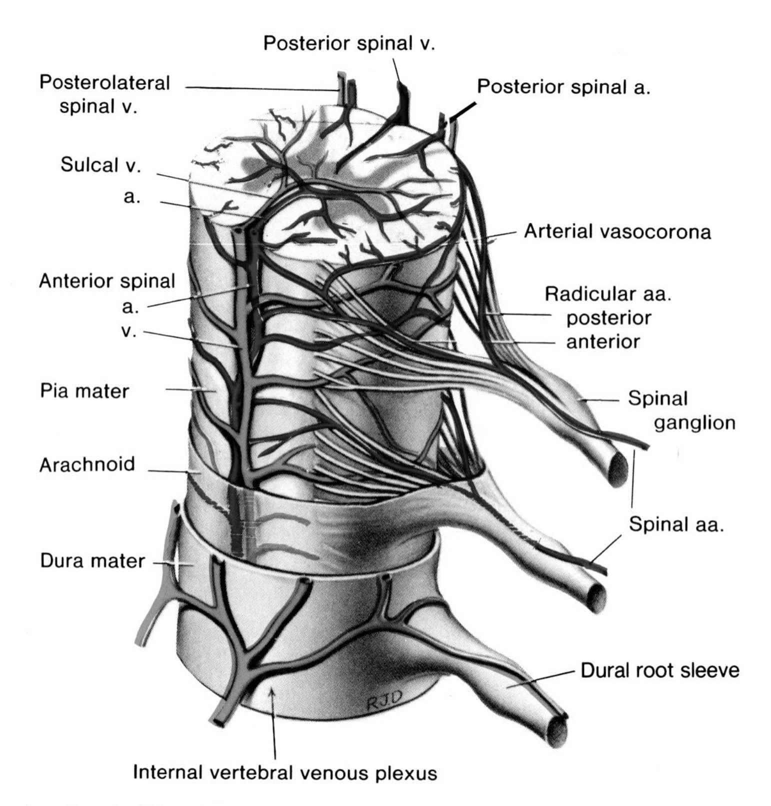

Figure 5-57. Drawing showing the blood supply and venous drainage of the spinal cord. The anterior spinal artery and the paired posterior spinal arteries run longitudinally throughout the length of the spinal cord. Observe that the anterior and posterior radicular arteries run along the dorsal and ventral roots of the spinal nerves to reach the spinal cord. Note that the dura mater evaginates along the dorsal and ventral nerve roots of the spinal nerves and the spinal ganglia to form dural root sleeves (dorsal root sleeves).

The vertebral venous plexuses are important clinically because blood may return from the pelvis or abdomen through them and reach the heart through the superior venal cava (Figs. 1-80 and 1-81). When the prostate gland is cancerous, blood may pass from it to the vertebral venous plexuses and the superior vena cava, instead of by its usual route via the inferior vena cava. Tumor cells from a prostatic cancer may be deposited via these veins in the vertebrae and develop secondary cancers **(metastases).**

The vertebral canal varies considerably in size and shape from level to level, particularly in the cervical and lumbar regions (Figs. 5-15, 5-16, and 5-21). A small vertebral canal in the cervical region, into which the spinal cord fits as tightly as a finger in a glove, is potentially dangerous because a minor fracture and/or dislocation of the cervical vertebrae may damage the spinal cord (Fig. 5-66).

The transitory protrusion of a cervical intervertebral disc following a neck injury may cause "*spinal cord shock,*" associated with paralysis inferior to the site of the lesion. In these cases no fracture or dislocation of cervical vertebrae can be found. If the patient dies, a **softening of the spinal cord** may be found at the site of the disc protrusion.

A small, trefoil-shaped vertebral foramen is sometimes found in the fifth lumbar vertebra (Fig. 5-15*B*), but rarely in L4. Such a small foramen may be a contributory factor in cases of **sciatica** (pain along the distribution of the sciatic nerve) and/or **cauda equina claudication** (L. *claudicatio*, to limp) owing to pressure on some of the spinal nerve roots.

Encroachment of the vertebral canal by a protruding intervertebral disc, swollen ligamenta flava, or as a result of **osteoarthritis** of the zygapophysial joints, may exert pressure on one or more spinal nerve roots of the cauda equina. Pressure may produce sensory and motor symptoms in the areas concerned. This group

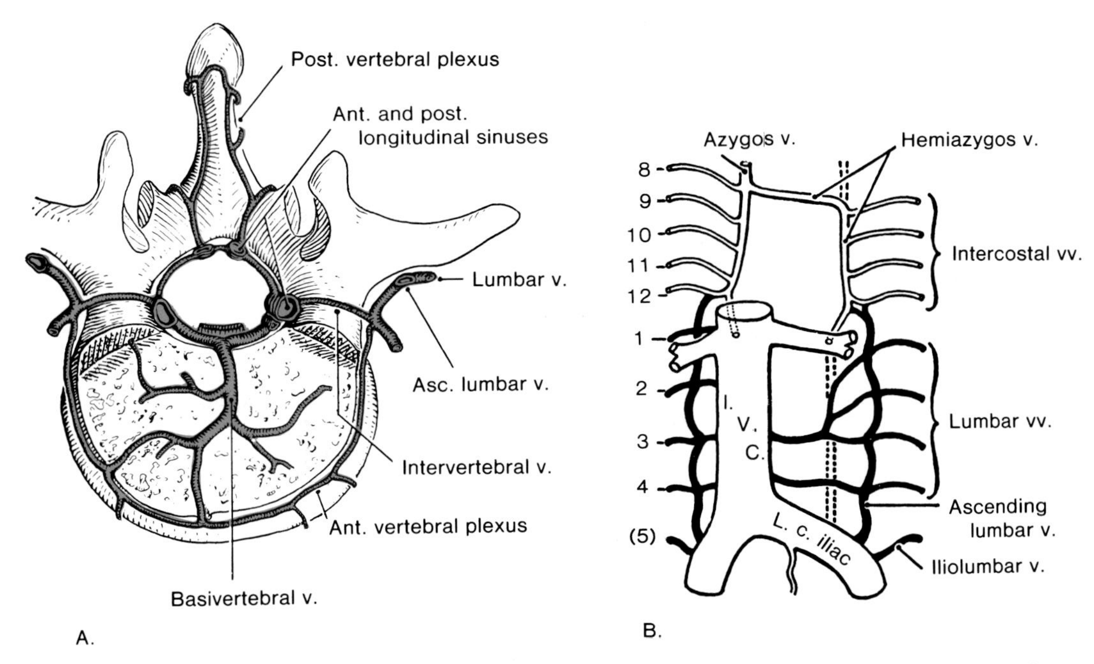

Figure 5-58. Drawings illustrating the vertebral venous plexuses and the azygos venous system (also see Figs. 1-80 and 1-81). *A,* the vertebral venous plexuses. Note that the vertebral canal contains an internal vertebral venous plexus; it surrounds the spinal dura mater (also see Fig. 5-57). This plexus communicates with the occipital and basilar sinuses (Fig. 7-39). Observe also the communications of the external plexus of veins. *B,* the ascending lumbar vein and the azygos venous system. Note how the vertebral venous plexuses connect with this system.

of bone and joint abnormalities is often called **lumbar spondylosis** (degenerative joint disease affecting the lumbar vertebrae and intervertebral discs that causes pain and stiffness).

In some old people, the nuclei pulposi of the intervertebral discs degenerate, the vertebrae come together, and the anuli fibrosi bulge anteriorly, posteriorly, and laterally (Fig. 5-29). This leads to the formation of bony outgrowths called ***osteophytes*** that may produce pressure on the spinal nerve roots and cause sensory and motor symptoms.

Cervical spondylosis is often accompanied by swollen ligamenta flava and osteoarthritis of the zygapophysial joints. In these conditions, there is often encroachment on the intervertebral foramina and/or vertebral canal. This may cause pressure on the cervical spinal nerve roots and/or spinal cord, resulting in various neurological symptoms and signs (For more information, see Brain and Wilkinson, 1967).

Complete transection of the spinal cord results in loss of all sensation and voluntary movement inferior to the lesion. The patient is *quadriplegic* (upper and lower limbs paralyzed) if the cervical cord superior to C3 is transected, and the patient may die owing to respiratory failure. The patient is *paraplegic* (lower limbs paralyzed) if the transection is between the cervical and lumbosacral enlargements (Fig. 5-56). The abdominal and back muscles are also affected, causing additional problems for the patient.

THE SPINAL MENINGES AND CEREBROSPINAL FLUID

The dura mater, arachnoid (mater), and pia mater are known collectively as the meninges (G. membranes).

The spinal meninges surround and support the spinal cord (Figs. 5-52 to 5-55, 5-57, 5-59, and 5-60). Between the dura mater and arachnoid there is a *potential space,* called the **subdural space,** containing only a capillary layer of fluid. Between the arachnoid and pia mater there is an *actual space,* called the **subarachnoid space,** containing cerebrospinal fluid (CSF) and the vessels of the spinal cord (Figs. 5-55, 5-57, and 5-60).

THE DURA MATER (L. *dura,* hard + *mater,* mother)

The dura mater, the outermost covering of the spinal cord, is a tough fibrous membrane composed of white fibrous and elastic tissue. **The spinal dura mater** forms a long tubular sheath or dural sac that is free within the vertebral canal (Figs. 5-45, 5-53 to 5-55, and 5-60). The dura mater is adherent to the margin of the foramen magnum of the skull, where it is **continuous with the cranial dura mater.**

The spinal dura mater hangs down from the skull like a tube, with a closed inferior end that usually terminates at the level of the inferior border of S2 vertebra in adults (Figs. 5-55*D* and 5-60).

The spinal cord is suspended in this dural sac by a saw-toothed **denticulate ligament** (L. *dentatus*, toothed) on each side (Fig. 5-59). This ribbon-like ligament, composed of pia mater, is attached along the lateral surface of the spinal cord, midway between the dorsal and ventral nerve roots. The lateral edge of the denticulate ligament is notched or serrated (L. *serratus*, a saw). The 21 tooth-like processes of the denticulate ligament are attached to the dura mater between the foramen magnum and the level at which the dura is pierced by the nerve roots of the S1 segment of the spinal cord.

The spinal dura mater evaginates along the dorsal and ventral nerve roots of the spinal nerves and spinal ganglia to form **dural root sleeves** (dorsal root sleeves), which continue into the intervertebral foramina (Figs. 5-52 and 5-57). The dural root sleeves are adherent to the periosteum lining the intervertebral foramina and end by blending with the epineurium of the spinal nerves (Fig. 43, p. 45).

Sometimes it is necessary to expose a patient's spinal dura, spinal cord, and/or spinal nerve roots. **Laminectomy,** the surgical procedure used to make this exposure, was given this name because the spinous processes and laminae of the vertebral arch are removed. Using this procedure, pressure on neural structures from bony fragments, protruding nuclei pulposi of intervertebral discs, tumors, or hematomas may be relieved.

Sometimes operations are performed on the spinal cord to relieve **intractable pain** (*e.g.*, in the late stages of malignant disease of a pelvic viscus). In surgical section of the spinal cord **(open chordotomy),** the cut is made in the ventrolateral portion of the cord in order to interrupt the pain pathway ***(lateral spinothalamic tract).*** When the operation is completed, the edges of the dura mater are carefully joined and the muscles are replaced before the skin is sutured.

Percutaneous chordotomy. The use of high frequency electricity, has almost completely replaced open chordotomy. Because this procedure can be performed on a conscious patient, the position and size of the lesion in the spinal cord may be controlled by asking the patient what she/he feels while the electrode is in place. The electrode is usually inserted between C1 and C2 vertebrae (atlas and axis) to destroy all ascending pain fibers in the spinal cord on one side.

THE ARACHNOID MATER

The spinal arachnoid (G. spider-like) is the delicate, filamentous, avascular covering of the spinal cord (Figs. 5-57 and 5-59). It is composed of white fibrous and elastic tissue and is coextensive in length with the dura mater. However, it is separated from this layer by a potential **subdural space** (Fig. 5-59).

The arachnoid is separated from the pia mater by an actual space, the **subarachnoid space** (Figs. 5-53 and 5-55), but the two layers are connected by delicate strands of connective tissue called **arachnoid trabeculae,** as they are in the brain (Fig. 7-67). The arachnoid covers the spinal nerve roots and the spinal ganglia, and blends with the sheaths of the spinal nerves. The arachnoid also ensheaths the cauda equina (Fig. 5-53).

The arachnoid and pia mater develop as one layer in the embryo and then separate, but numerous connections (arachnoid trabeculae) remain. Together the pia mater and the arachnoid are called the **leptomeninges** (G. slender membranes) or pia-archnoid.

THE PIA MATER

The spinal pia mater (L. *pius*, tender + *mater*, mother) is composed of two fused layers of loose connective tissue. It encloses a fine network of blood vessels and adheres to the surface of the spinal cord (Figs. 5-52 and 5-57). It also covers the roots of the spinal nerves and spinal blood vessels. The pia mater ensheaths the anterior spinal artery occupying the ventral median fissure of the spinal cord (Figs. 5-52 and 5-57). This investment of pia mater is called the *linea splendens.*

The denticulate ligament (discussed previously) is continuous with the pia mater on each side of the spinal cord (Fig. 5-59), and its lateral border is fixed at intervals to the spinal dura mater.

The subarachnoid space (Figs. 5-53 and 5-55) is between the arachnoid and pia mater. It contains CSF, a clear slightly alkaline fluid. The spinal subarachnoid space extending from L2 to S2 vertebrae is known as the **lumbar cistern** (L. *cisterna*, a reservoir). In addition to CSF, the lumbar cistern contains the cauda equina and the filum terminale (Figs. 5-53 to 5-55).

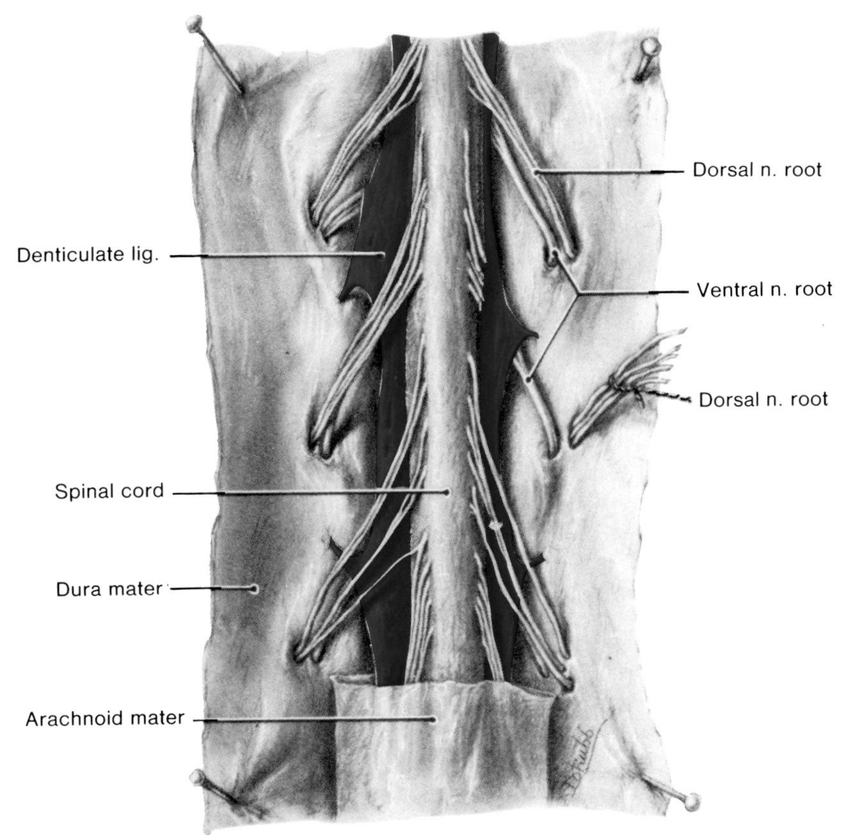

Figure 5-59. Drawing of a dissection of the spinal cord within the spinal meninges, posterior view. The dura mater and arachnoid (mater) have been split and pinned to expose the spinal cord and nerve roots. Observe the denticulate ligament (*red*), running like a band along each side of the spinal cord. Note that it anchors the spinal cord to the dura mater by means of strong tooth-like processes between successive nerve roots. Observe that the ventral nerve roots lie anterior to the denticulate ligament and that the dorsal nerve roots lie posterior to it.

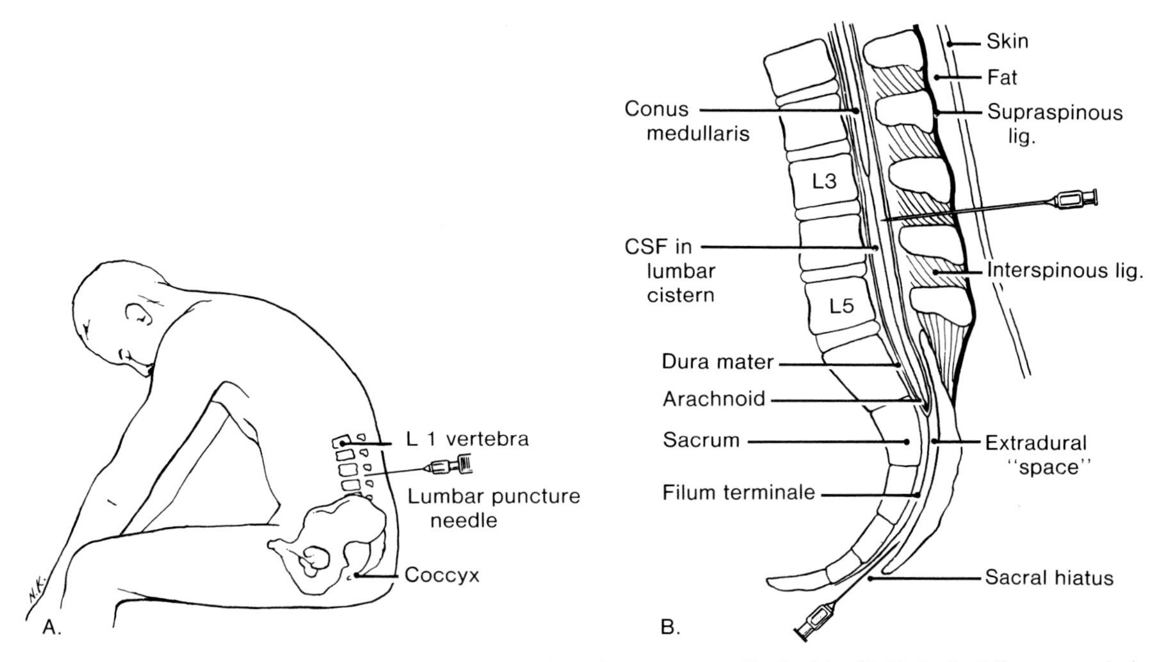

Figure 5-60. *A,* drawing illustrating the technique of lumbar puncture ("spinal tap"). Note that the person's back is flexed to open the spaces between the spinous processes and laminae of the vertebrae. *B,* median section of the inferior end of the vertebral column containing the spinal cord and its membranes. A lumbar puncture needle has been inserted between L3/L4 vertebrae for withdrawal of CSF. A needle is also shown in the sacral hiatus, the site sometimes used for extradural anesthesia. Note that the spinal cord in this person ends at the body of L3 vertebra, which is more inferior than usual. Insertion of the needle between L2/L3 vertebra would damage the spinal cord in this patient.

CSF can be obtained from the lumbar cistern located in the inferior part of the vertebral canal (Fig. 5-60*B*). As the spinal cord usually ends between L1 and L2 vertebrae, there is no danger of injuring the spinal cord when a **lumbar puncture needle** is inserted between the spinous processes of L3/L4 or L4/L5 vertebrae into the lumbar cistern to obtain a sample of CSF (Fig. 5-60). In lumbar puncture there is seldom any damage to the spinal nerve roots because, being suspended in CSF, they tend to move away from its point. The needle is more likely to touch a spinal nerve root if it is not inserted exactly in the midline. Touching a nerve root with the point of the needle may result in sharp pain in the **dermatome** or area of skin supplied by the nerve root concerned (Fig. 1-23).

Lumbar puncture is performed to obtain CSF for diagnostic procedures. During **pneumoencephalography** (G. *pneuma*, air + *enkephalos*, brain + *graphē*, a drawing), some CSF is removed and replaced by air. Skull radiographs are then taken in various positions. These special radiographs, called **pneumoencephalograms,** permit visualization of the ventricles of the brain and the subarachnoid space (Fig. 7-83). Filling defects in the subarachnoid space may locate tumors or other masses. **Spinal anesthetics** can also be injected by lumbar puncture.

During **myelography,** a radiopaque substance is injected into the lumbar cistern and then a *fluoroscopic examination* is performed on a tilting table. For these examinations, an apparatus known as **fluoroscope** is used which makes the shadows of organs visible on a fluorescent screen. These special radiographs, called **myelograms,** permit visualization of the spinal cord and subarachnoid space. The spinal cord used to be called the spinal medulla (G. *myelos*, medulla), which explains the use of the prefix "myelo" in the term myelogram.

Spina bifida (L. *bifidus*, cleft in two parts) is used to describe a wide range of developmental defects of the vertebral column. In its most simple form, **spina bifida occulta,** the halves of the vertebral arch fail to develop fully (Fig. 5-27).

Spina bifida cystica is a more serious abnormality in which there is herniation of the meninges and/or spinal cord through the defect in the vertebral arch (Fig. 5-61). When the meninges alone are herniated, the condition is known as **spina bifida with meningocele** (Fig. 5-61*A*), whereas when the meninges and spinal cord are herniated, the condition is known as **spina bifida with meningomyelocele** (Fig. 5-61*B*). Patients with this severe malformation often exhibit spinal cord or spinal nerve root malfunction, *e.g.*, paralysis of the limbs and incontinence of urine and feces.

Spina bifida cystica (meningocele or meningomyelocele), involving the meninges and/or the spinal cord, occurs about once in every 1000 births. Most clinically significant cases of spina

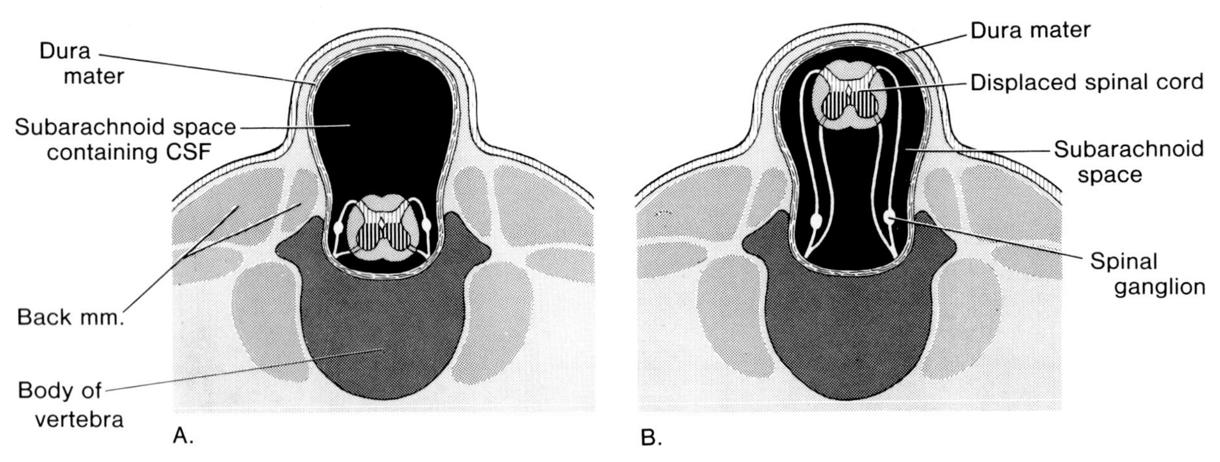

Figure 5-61. Diagrammatic sketches illustrating two types of spina bifida. Severe types of spina bifida, involving protrusion of the meninges and/or spinal cord through the defect in the vertebral arches, are referred to as spina bifida cystica because of the cyst-like protrusion or sac. *A,* spina bifida with meningocele. *B,* spina bifida with meningomyelocele. (From Moore KL: *The Developing Human: Clinically Oriented Embryology*, ed 3. Philadelphia, WB Saunders Co, 1982).

bifida probably result from failure of neural tube closure **(neural tube defect),** owing to a local overgrowth of the developing neural tube. The excessive development of the neural tube results in underdevelopment and nonfusion of the halves of the vertebral arch (Fig. 5-61). Usually several vertebrae in the thoracic, lumbar, or sacral regions are involved when there is a meningocele or a meningomyelocele. For photographs of infants with these malformations, see Moore (1982).

PRESENTATION OF PATIENT ORIENTED PROBLEMS

Case 5-1

During a fight a 16-year-old boy was stabbed in the back of the neck with a knife. As he ducked attempting to avoid his attacker, he flexed his neck. Much to the surprise of his assailant, the boy fell to the ground and was completely immobilized from the neck down.

Problems. How did this serious injury probably occur? Using your anatomical knowledge of the vertebral column and its contents, explain the basis of the injury. How might this knowledge be utilized in the diagnosis and treatment of diseases of the nervous system, and in the administration of anesthetic agents? *These problems are discussed on page 621.*

Case 5-2

A 51-year-old man was relaxing while waiting for a traffic light to turn green, when his car was "rear ended." His body was pushed forward and his head was thrown violently backward (Fig. 5-62). He suffered a slight **concussion** and felt shaky. When he talked to the man who had hit his car from behind and to the traffic officer, he informed them that he was not badly hurt. The officer noted that the headrest in his car was below the level of his head (Fig. 5-62).

The next morning his neck was stiff and painful, and there was pain in the region of his left trapezius muscle and in his left arm. The **neck pain** was aggravated by movement of his head. Gradually he developed a "crick in his neck" (Fig. 5-63). On examination the doctor noted that he held his head rigidly and tilted to the right. He also observed that his chin was pointed to the left and that his neck was slightly flexed.

Palpation of the posterior aspect of his neck revealed some tenderness over the spinous processes of his inferior cervical vertebrae. **The biceps reflex was weak on his left side.**

Radiographs showed thin intervertebral discs at C5/C6 and C6/C7, with small fringes of bone on the opposing edges of the bodies of C5, C6, and C7.

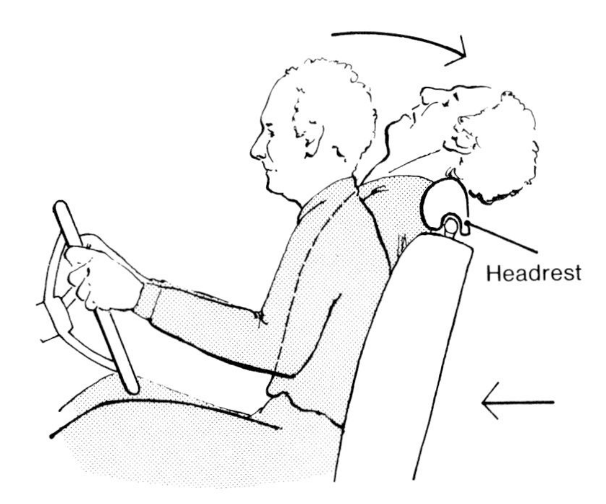

Figure 5-62. Illustration showing how the head is thrown violently backward during a rear end collision, producing a hyperextension injury of the neck. Note that the headrest was not raised to a position where it would have prevented hyperextension of the neck.

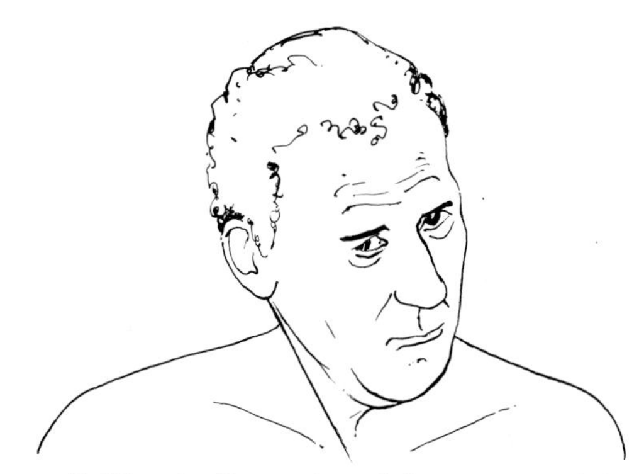

Figure 5-63. An illustration of the posture of the patient's head and neck, indicating that his neck muscles were probably injured during the accident (Case 5-2).

A diagnosis of hyperextension injury of the neck was made.

Problems. What is the anatomical basis of the patient's concussion, stiff neck, and pain in the neck and arm? What spinal nerve root was probably compressed? What muscles were likely injured? What probably caused the thinning of the patient's intervertebral discs and the formation of the bony fringes on the edges of his cervical vertebral bodies. *These problems are discussed on page 622.*

Case 5-3

While helping you carry a heavy box of books, your father suddenly experienced a severe pain in his lower back. Later he developed a dull ache in the posterior and lateral aspects of his left thigh and leg (Fig. 5-64).

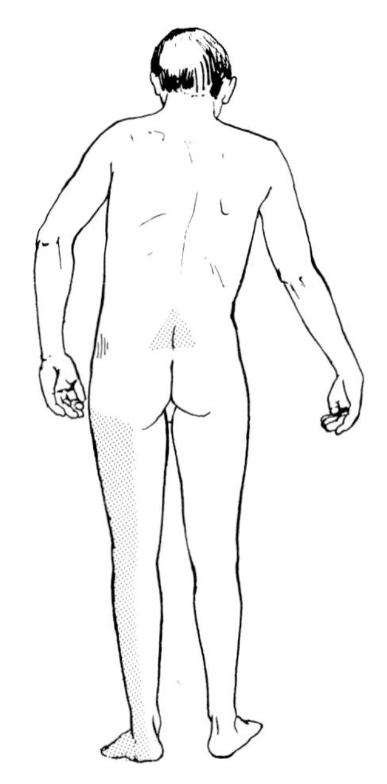

Figure 5-64. Drawing showing the patient's lumbar deviation and the regions where he felt pain (*stippled*).

A lateral deviation or tilt of the lumbar region of his vertebral column was also observed. He limped when he walked because he did not fully extend his thigh.

During the examination the orthopaedist told you that your father's back muscles were in spasm. When asked to indicate the site of most severe pain, your father pointed to his lower lumbar region (Fig. 5-64).

During the examination you noted that your father had **no ankle reflex** on the left side, and that he experienced increased pain when the doctor raised his lower limb on that side.

The radiographs showed slight narrowing of the space between the vertebral bodies of L5 and S1. The orthopaedist explained that the nucleus pulposus of one of your father's intervertebral discs was protruding, and that he would be confined to bed for several weeks.

The doctor urged you not to refer to your father's **protruding disc** as a "*slipped disc*" because this expression gives an erroneous concept of the condition. He emphasized that part of your father's intervertebral disc had protruded, not slipped.

Problems. What is the anatomical basis of protrusion of an intervertebral disc and the resulting low back pain? What produced the lumbar deviation? Why did the patient experience pain in his thigh and leg? Why did the pain increase when the doctor raised the patient's lower limb? *These problems are discussed on page 622.*

Case 5-4

An 18-year-old man was thrown from a horse and sustained a spinal cord injury as the result of severe hyperextension of his neck. He died in about 5 minutes.

Problems. What vertebrae were most likely fractured and dislocated? What associated structures of the vertebral column were probably also ruptured? Although one would expect the patient to be quadriplegic following a superior **cervical spinal cord transection,** what probably caused his death? *These problems are discussed on page 623.*

Case 5-5

During surgery for resection of an abdominal **aortic aneurysm** (Fig. 5-65), there was extensive mobilization of the aorta and several arteries were ligated. Although the aneurysm was successfully removed and replaced by a prosthesis, the patient was **paraplegic** and his bladder and bowel functions were no longer under voluntary control.

Problems. What is the anatomical basis of the patient's paraplegia? What arteries were probably ligated? Name the important artery supplying the spinal cord that was likely deprived of blood. Why is its supply to the spinal cord so important? *These problems are discussed on page 624.*

Case 5-6

A 28-year-old man was involved in a head-on collision. When removed from the car, he complained of a sore neck and loss of sensation and voluntary movements in his lower limbs. There was also impaired

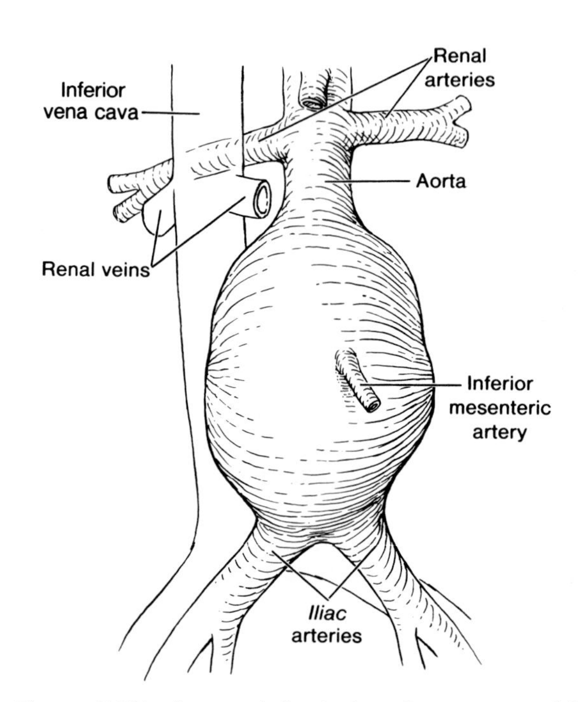

Figure 5-65. Large abdominal aortic aneurysm (circumscribed dilation of abdominal aorta).

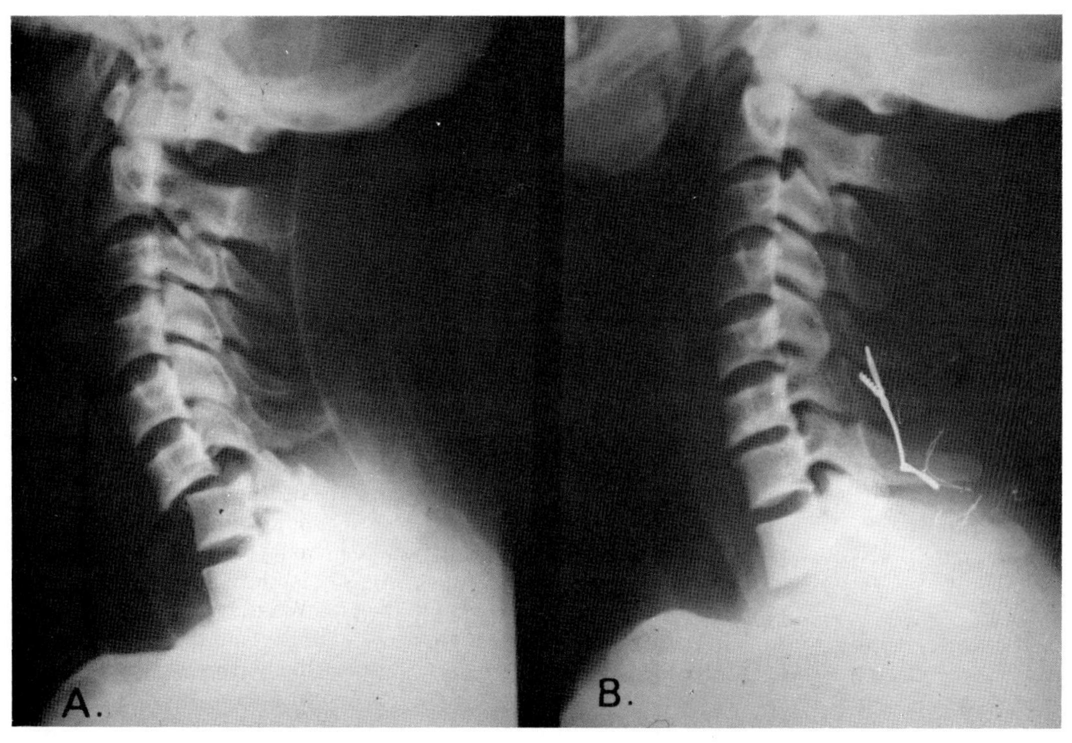

Figure 5-66. Radiographs of the cervical region of the vertebral column, lateral views. *A,* dislocation of C6 on C7. Note the small fragment of bone that has broken off the anterosuperior corner of the body of C7 vertebra. *B,* the spinous processes of C6 and C7 have been wired to hold the vertebrae in their normal position.

ability of upper limb movements, particularly in his hands.

Radiographs showed dislocation of C6 vertebra on C7, and a chip fracture of the anterosuperior corner of the body of C7 (Fig. 5-66*A*). Open reduction was carried out and the spinous processes of C6 and C7 were wired together to hold the vertebrae in normal relation to each other (Fig. 5-66*B*). The reduction was maintained by immobilization of the neck in a plastic collar, thereby allowing the patient to exercise his upper limbs and to sit up within a day or so after the injury.

Problems. What name is applied to the condition of patients with paralysis of both lower limbs? What joints of the vertebral column were dislocated? What ligaments binding the vertebrae together were probably torn? What was the most likely cause of the patient's paralysis? What other functions would no longer be under voluntary control? *These problems are discussed on page 624.*

DISCUSSION OF PATIENT ORIENTED PROBLEMS

Case 5-1

The spinous processes and laminae of the vertebral arches usually protect the spinal cord from injury to the posterior aspect of the neck. However when the neck is flexed, the spaces between the vertebral arches increase. This would permit a knife to pass between two adjacent vertebral arches and enter the vertebral canal. A knife entering the cervical region of the canal would partly or completely sever the spinal cord.

You can verify this movement of the cervical vertebrae by placing your hand on the back of your neck and then flexing it. Note that the space between the external occipital protuberance and the spinous process of the axis widens, as do the spaces between the spinous processes of C2 to C7 vertebrae. Had the knife severed the spinal cord superior to C3, the lesion would have stopped the patient's breathing because it would have interfered with the **phrenic outflow** (C3, C4, and C5), the nerve supply to the diaphragm. As a result, the patient would have died in a few minutes.

Complete transection of the spinal cord *results in loss of all sensation and voluntary movement inferior to the lesion.* The patient is **quadriplegic** (upper and lower limbs paralyzed) when the lesion is superior to C5 segment, because the **brachial plexus of nerves** supplying the upper limb is derived from C5 to T1 segments of the spinal cord (Fig. 6-22).

Had the boy been stabbed in the same place when his head was erect, he might not have been severely injured. Very likely the knife would have struck the spinous processes or the laminae of the cervical vertebrae, and perhaps glanced off without damaging the spinal cord.

Similar gaps exist between the lumbar spinous processes when the back is flexed. They are clinically important because they enable clinicians to insert a **lumbar puncture needle.** In adults the needle is usually inserted between the spinous processes of L3 and L4 vertebrae into the **subarachnoid space** (Fig. 5-60), inferior to the termination of the spinal cord. This procedure, known as **lumbar puncture,** is performed to obtain a sample of CSF. Lumbar punctures are performed during the investigation of some diseases of the nervous system (*e.g.*, **meningitis**).

Local anesthetic solutions may be injected into the **extradural space** (epidural space). The extradural space is not a real space because it is filled with areolar tissue, fat, and veins.

Extradural anesthesia *is often used in obstetrics because it relieves pelvic pains without interfering with uterine contractions.* The needle may also be inserted through the **sacral hiatus** at the inferior end of the sacrum (Fig. 5-54). This is inferior to the level of the dural sac in most patients (Fig. 5-60*B*). In some people the dura mater and the subarachnoid space extend into the inferior part of the sacral canal. Consequently if the needle is inserted too deeply into the sacral hiatus in such a person, the anesthetic solution could be injected into the subarachnoid space and mix with the CSF. If the anesthetic should reach the cervical part of the spinal cord, death would probably occur owing to interruption of the motor innervation of the diaphragm (*i.e.*, the phrenic outflow).

Case 5-2

The association of rear end automobile collisions and hyperextension injuries of the soft tissues of the cervical region of the vertebral column is well-known. Headrests and bucket seats have been designed to minimize these injuries; however a headrest is useless if it is not raised so that the occipital region of the head will hit it if a rear end collision occurs.

The mechanism of injury is primarily one of rapid hyperextension of the neck. Because the headrest was not in the correct position, there was nothing to restrict the posterior movement of the head. In addition the muscles of the neck, the chief stabilizers of the cervical region of the vertebral column, were relatively relaxed because the patient was caught off guard when his car was hit from behind.

As a result, his anterior longitudinal ligament and neck muscles were severely stretched and some fibers were probably torn, leading to small hemorrhages in these muscles. The resulting muscle spasms would give him a stiff and painful neck.

The concussion experienced by the patient probably resulted from the sudden impact of the frontal and sphenoid bones against the frontal and temporal poles of his brain (Fig. 5-67).

The pain in his left shoulder and the weakness of the biceps reflex on the left very likely resulted from *compression of the left sixth cervical nerve root*, probably by a posterolateral protrusion of the intervertebral disc between the fifth and sixth cervical vertebrae.

The musculocutaneous nerve (C5 and C6) supplies the biceps brachii muscle and the **biceps reflex** is also mediated through C5 and C6 (Figs. 6-20 and 6-29).

A hyperextension injury of the neck is popularly called a *"whiplash injury,"* especially by litigation lawyers. Many doctors consider this term an unacceptable medical designation because there is no well defined clinical syndrome or fixed pathology associated with the injury.

The thinning of the intervertebral discs in the cervical region probably resulted from desiccation of the nuclei pulposi of the intervertebral discs. This often occurs with advancing age (Fig. 5-29) and produces bulging of the anuli fibrosi of the intervertebral discs. The formation of fringes of subperiosteal new bone on the edges of the vertebral bodies also occurs.

Case 5-3

Low back pain, commonly called "*lumbago*," was probably caused by rupture of the posterolateral part of the anulus fibrosus and protrusion of the nucleus pulposus of the intervertebral disc between L5 and S1 vertebrae (Fig. 5-68).

The lumbar deviation of the vertebral column was produced by spasm of the intrinsic back muscles (Fig.

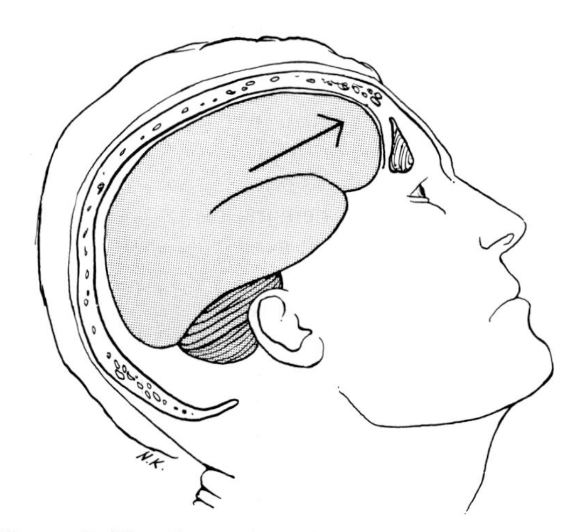

Figure 5-67. Illustration of the way contusion of the brain occurs as the result of sudden pressure on the frontal and temporal lobes when they are compressed against the anterior aspect of the skull during hyperextension of the neck (Fig. 5-62).

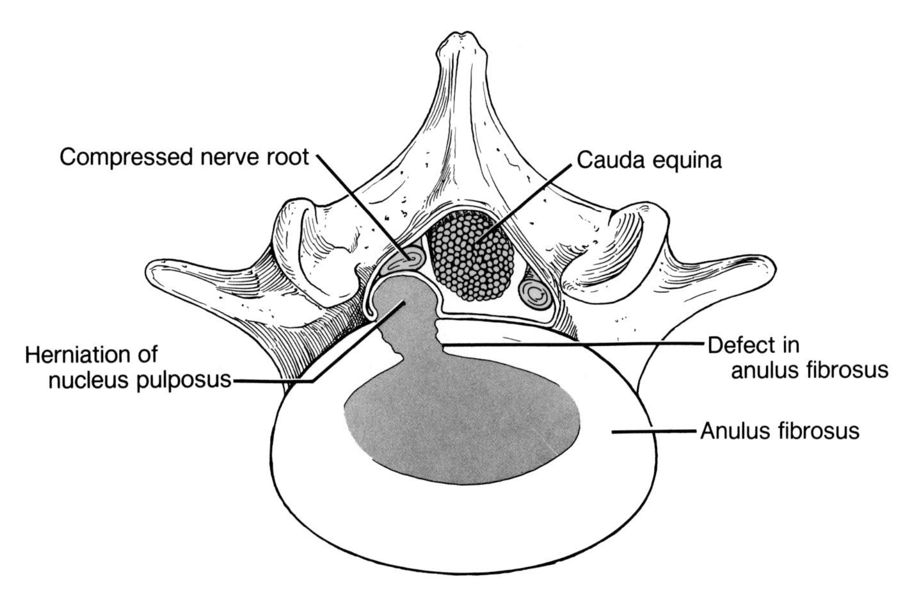

Figure 5-68. Drawing illustrating how a posterolateral herniation of the nucleus pulposus through a defect in the anulus fibrosus of an intervertebral disc compresses a nerve root. Also see the anulus fibrosus and cauda equina in Figure 5-31.

5-40). **Muscle spasm has a protective splinting effect on the vertebral column.** As your father lifted the heavy box of books, the strain on his intervertebral disc was so severe that the anulus fibrosus tore, resulting in **herniation of the nucleus pulposus** (Fig. 5-68).

The disc protrusion exerted pressure on the S1 component of the sciatic nerve. Disc protrusions most commonly occur posterolaterally where the anulus fibrosus is thin. The posterior longitudinal ligament strengthens the median portion of the posterior part of the anulus fibrosus (Figs. 5-28 and 5-32). Consequently median posterior protrusions of lumbar discs are unusual.

As the dorsal and ventral nerve roots cross the posterolateral region (Fig. 5-51), the protruding nucleus pulposus often affects one or more spinal nerve roots (Fig. 5-68). Some hemorrhage, muscle spasm, and edema would be present at the site of the rupture. This probably caused some initial back pain.

In the present case, pressure appears to have affected the S1 component of the sciatic nerve as it passes inferiorly, posterior to the L5/S1 intervertebral disc (Fig. 5-54). As a result, the patient experienced pain over the posterolateral region of his thigh and leg (Figs. 1-23 and 5-64). When the doctor raised the patient's lower limb, the sciatic nerve was stretched. As its S1 component is compressed by the protruding disc (Fig. 5-54), the lower limb pain increased because of stretching of the compressed fibers in that root.

Sciatica is the name given to pain in the area of distribution of the sciatic nerve (L4 to S3, Fig. 4-82). Pain is felt in one or more of the following areas: the buttock, especially the region of the greater sciatic notch, the posterior aspect of the thigh, the posterior and lateral aspects of the leg, and usually parts of the lateral aspect of the ankle and foot. The variation in the location of the pain is caused by the fact that a posterolateral protrusion of a single lumbar disc presses on only one nerve root. However, the sciatic nerve is composed of several inferior lumbar and superior sacral roots.

The paravertebral muscle spasm and pain is caused by the muscles being in continuous tonic contraction to prevent the vertebrae from moving and causing severe pain.

The narrowing or thinning of the space between the vertebral bodies noted in the radiographs is caused by the reduction of disc material between the adjacent vertebral bodies which normally occurs with advancing age.

Case 5-4

Severe ***hyperextension of the neck*** *usually causes a fracture of the atlas* at one or both grooves for the vertebral arteries (Fig. 5-16). The vertebral arch of the axis may break at the isthmus between the lateral mass and the inferior articular process. Probably the patient's anterior longitudinal ligament and the anterior part of the C2/C3 intervertebral disc were also ruptured.

As the patient hit the ground, hyperextending his neck, his skull, atlas, and axis were probably separated from the rest of his vertebral column. As a result, *his spinal cord was probably torn* in the *superior cervical region.* Patients with this severe injury rarely survive more than a few minutes because the injury to the spinal cord is superior to the **phrenic outflow** (origin of the phrenic nerves). As these nerves are the sole motor supply to the diaphragm, respiration is severely affected; in addition, the action of the intercostal muscles is lost.

Case 5-5

During certain surgical procedures in the abdomen, it is necessary to ligate aortic segmental branches (*e.g.*, the lumbar arteries). If the **arteria radicularis magna** arises from one of the intercostal or lumbar arteries that has been ligated, the blood supply to the lumbosacral enlargement of the spinal cord may be severely impaired. As a result, **spinal cord infarction** (necrosis of nervous tissue), ***paraplegia,*** and *loss of sensation inferior to the lesion may follow.*

Arising more frequently on the left from an inferior intercostal (T6 to T12) or lumbar (L1 to L3) artery, the arteria radicularis magna enters the vertebral canal through an intervertebral foramen. It supplies blood mainly to the inferior two-thirds of the spinal cord (Figs. 5-56 and 5-57), therefore it is understandable why function is lost in the lower limbs, bladder, and bowels when this artery and part of the spinal cord are deprived of blood.

Case 5-6

The patient is paraplegic and the condition is known as **paraplegia.** Both the intervertebral disc and the ***zygapophysial joints*** between the bodies and vertebral arches of C6 and C7, respectively, were dislocated in this case. Probably the posterior longitudinal and interspinous ligaments, as well as the anulus fibrosus, ligamenta flava, and articular capsules of the zygapophysial joints were torn.

The cervical region of the vertebral column, being the most mobile part, is the most vulnerable to injuries such as dislocations and fracture-dislocations. Most of these injuries occur when a person's head moves forward suddenly and violently, as in the present case, or when the head is struck by a hard blow.

In hyperflexion injuries of the neck, the anterior longitudinal ligament is usually not torn, and when the patient's neck is placed in a position of extension, this ligament tightens and tends to hold the vertebrae together.

Open operation was carried out in order to visualize the contents of the vertebral canal and to reduce the dislocation of the vertebrae. The spinous processes of C6 and C7 were wired together to help stabilize the vertebral column during the initial part of the rehabilitation program (Fig. 5-66*B*), and to promote healing of the torn ligaments and intervertebral disc.

The vertebral bodies are bound together by the longitudinal ligaments and the anuli fibrosi of the intervertebral discs (Fig. 5-28). The **posterior longitudinal ligament,** a narrower and weaker band than the anterior longitudinal ligament, is attached to the intervertebral discs and to the edges of the vertebral bodies (Fig. 5-32). It lies inside the vertebral canal and tends to prevent excessive flexion of the vertebral column. As dislocation occurred in this case, the posterior longitudinal ligament and the ligamenta flava were severely stretched and probably torn.

As the anulus fibrosus of the intervertebral disc attaches to the compact bony rims on the articular surfaces of the vertebral bodies, its posterior part would also have been stretched and probably torn at the C6/C7 level. It is possible that **protrusion of the nucleus pulposus of the intervertebral disc** between these vertebrae also occurred, because these nuclei are semifluid in young adults.

Because the vertebral canal in the cervical region is usually larger than the spinal cord, there can be some displacement of the vertebrae without causing damage to the spinal cord. In view of the patient's **paraplegia,** it is likely that the spinal cord was severely stretched and/or torn. At the moment of impact, the displacement of C6 on C7 was undoubtedly greater than shown in the radiograph (Fig. 5-66*A*).

There is an initial period of **spinal shock** in these cases, lasting from a few days to several weeks, during which all somatic and visceral activity is abolished. On return of reflex activity, there is spasticity of muscles and exaggerated tendon reflexes inferior to the level of the lesion. In addition bladder and bowel functions are no longer under voluntary control.

Suggestions for Additional Reading

1. Armstrong JR: *Lumbar Disc Lesions. Pathogenesis and Treatment of Low Back Pain and Sciatica,* ed. 3. Edinburgh, E & S Livingstone, Ltd, 1965.
   An authoritative and comprehensive account of lesions of the lumbar intervertebral discs.
2. Brain WR, Wilkinson M: *Cervical Spondylosis.* London, William Heinemann Medical Books, Ltd, 1967.
   A classical book on the neck dealing with disorders of the cervical region of the vertebral column.
3. Moore KL: *The Developing Human: Clinically Oriented Embryology,* ed 3. Philadelphia, WB Saunders Co, 1982.
   A standard textbook of embryology explaining the embryological bases of developmental defects of the vertebral column.
4. Rothman RH, Simeone FA (Eds): *The Spine.*

Philadelphia, WB Saunders Co, vol. 1, 1975.

A comprehensive textbook on the vertebral column, with chapters on its development, applied anatomy, and on the diagnosis and treatment of spinal disease.

5. Williams PL, Warwick R (Eds): *Gray's Anatomy,* ed 36 (British). Philadelphia, WB Saunders Co, 1980.

This is an ideal reference book if you wish details that are not included in the present book. In the chapter on osteology there is a good illustrated account of the bones of the vertebral column, including details of their ossification.

6. Weinstein PR, Ehi G, Wilson LB: *Lumbar Spondylosis.* Chicago, Year Book Medical Publishers, 1977.

A good reference for the pathology of the lumbar region of the vertebral column. There is also a good discussion of the surgical treatment of lumbar spinal stenosis and spondylosis.

CHAPTER 6

The Upper Limb

The upper limb is the organ of manual activity. It is freely movable, especially the hand, which is adapted for grasping and manipulating. For purposes of description, the upper limb is divided into the **shoulder** (junction of arm and trunk), **arm** (brachium), **forearm** (antebrachium), and **hand** (manus). Although many people refer to the upper limb as the arm, it should be noted that *the arm is only the superior part of the upper limb.*

As the upper limb is not usually involved in weight bearing, its stability has been sacrificed to gain mobility. The digits (thumb and fingers) are the most mobile, but other parts are still more mobile than comparable parts of the lower limb.

Because the disabling effect of injury to the upper limb, particularly the hand, is far out of proportion to the extent of the injury, it is important to obtain a sound understanding of the structure and function of this limb. Knowledge of its structure without an understanding of its functions is almost useless clinically, because the aim of treating injured limbs is to preserve or restore their functions.

BONES OF THE UPPER LIMB

The bones of the upper limb form the *superior part of the appendicular skeleton* (Fig. 6-1). They are the **clavicle** (collar bone) and **scapula** (shoulder blade) in the pectoral girdle (shoulder girdle); the **humerus** in the arm; the **radius and ulna** in the forearm; the **carpal bones** in the carpus (wrist); the **metacarpal bones** in the hand; and the **phalanges** in the digits. Descriptions of the bones are given with the regions of the upper limb.

THE PECTORAL REGION

The pectoral region (L. *pectus,* chest) is located anteriorly on the thoracic wall (chest wall), extending from the root of the neck to the **axilla** (armpit) laterally and to the costal margin (L. *costa,* rib) inferiorly.

The pectoral region includes the breast and the pectoral muscles (pectoralis major, pectoralis minor, subclavius, and serratus anterior). The muscles act on the upper limb and connect it to the thoracic skeleton (thoracic vertebrae, ribs, and sternum). All these muscles insert into the pectoral girdle, except the pectoralis major, which inserts into the humerus (Fig. 6-18 and Table 6-1).

BONES OF THE PECTORAL GIRDLE

THE CLAVICLE (Figs. 1-1, 6-1, and 6-3)

The clavicle (collar bone) extends from the manubrium of the sternum to the acromion of the scapula. *The clavicle connects the upper limb to the trunk and props the shoulder away from the chest* (Fig. 6-2).

The rounded medial end (sternal end) of the clavicle articulates with the sternum at the **sternoclavicular joint.** The medial two-thirds of the body (shaft) of the clavicle is convex anteriorly, whereas the lateral one-third of the body is flattened and concave anteriorly. The lateral end (acromial end) of the clavicle articulates with the acromion of the scapula at the acromioclavicular joint.

The clavicle is the first bone in the body to ossify. Intramembranous ossification begins during the seventh embryonic week.

Fractures of the clavicle are relatively common (Fig. 6-5). The weakest part of the clavicle is the junction of the middle and lateral thirds. One function of the clavicle is to transmit forces from the upper limb to the axial skeleton; hence in falls on the shoulder or hand, if the force is greater than the strength of the clavicle, fracture results.

Fractures of the clavicle medial to the attachment of the coracoclavicular ligament are common, especially in children and young adults (Fig. 6-5). In children the fracture is often incomplete, *i.e.,* a **green-stick fracture** in which

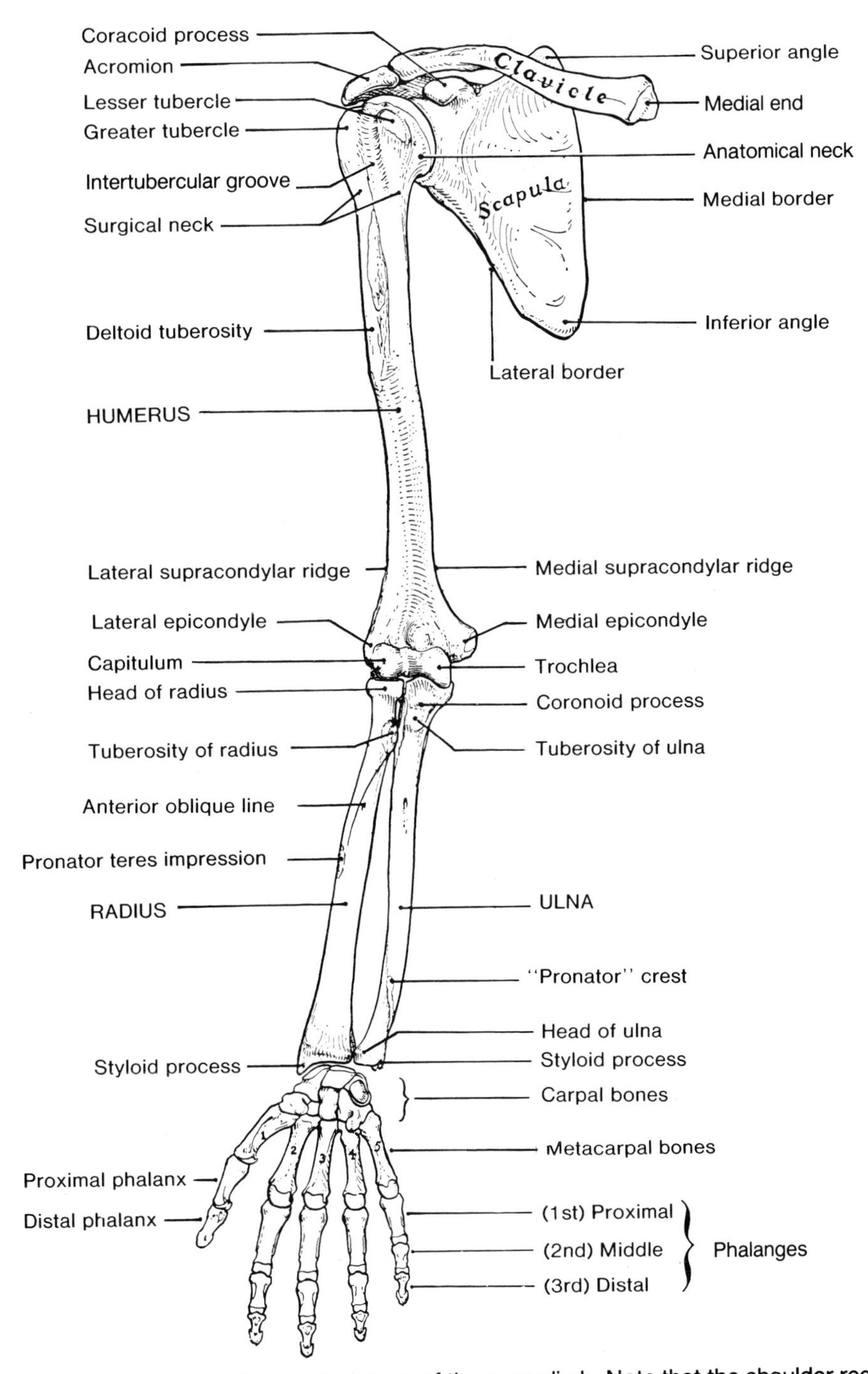

Figure 6-1. Drawing of an anterior view of the bones of the upper limb. Note that the shoulder region is supported by the bones of the pectoral girdle (scapula and clavicle). Observe the beak-like coracoid process of the scapula; coracoid is derived from a Greek word meaning "a crow." Clinically the arm is measured from the tip of the acromion to the lateral epicondyle of the humerus.

one cortex of the bone breaks and the opposite one bends.

After fracture of the clavicle, the clavicular head of the sternocleidomastoid muscle elevates the medial fragment of bone. As the trapezius muscle is unable to hold up the lateral fragment owing to the weight of the upper limb, it drops (Fig. 6-5). Patients with a fractured clavicle frequently present with their upper limb in a sling, or supporting the sagging limb with the other hand.

In addition to being depressed, the lateral fragment of the clavicle is pulled medially by the adductors of the arm, principally the latissimus dorsi and pectoralis major muscles. This overriding of the bone fragments shortens the clavicle (Fig. 6-5).

Occasionally a communicating vein from the cephalic vein in the deltopectoral triangle (Fig. 6-14) passes anterior to the clavicle to join the external jugular vein (Fig. 6-40). This communicating vein may be torn when the clavicle fractures and give rise to a subcutaneous collection of blood (**hematoma**).

In rare instances the clavicle is incomplete or absent. This congenital abnormality is often associated with delayed ossification of the skull. The combined condition, known as **cleidocranial dysostosis**, is characterized by drooping and excessive mobility of the shoulders (Fig. 6-126). Sometimes only the middle of the clavicle is absent and the two ends are joined by a fibrous band.

THE SCAPULA (Figs. 5-2, 6-1, 6-3, 6-4, and 6-46)

The scapula (shoulder blade) is a flattened, triangular bone that lies on the posterolateral aspect of the thorax (Fig. 6-3), covering parts of the 2nd to 7th ribs. *The scapula connects the clavicle to the humerus.*

The body of the scapula ("blade" of the scapula) is thin and translucent. The scapula has a concave costal surface (**subscapular fossa**) and a dorsal surface from which the **spine of the scapula** projects. The smaller dorsal part superior to the spine is called the **supraspinous fossa,** and the larger dorsal part inferior to the spine is called the **infraspinous fossa**. The spine of the scapula continues laterally into the **acromion** (acromion process) which projects anteriorly and articulates with the clavicle (Fig. 6-1).

Superolaterally the scapula has a shallow **glenoid cavity** (glenoid fossa) for articulation with the head of the humerus (Fig. 6-131). This part of the scapula, called the *head*, is connected to the blade-like *body* of the scapula by a short *neck*. **The coracoid process,** resembling a bird's beak or a bent finger, arises from the superior border of the head of the scapula and projects anteriorly and slightly laterally (Fig. 6-1).

SURFACE ANATOMY OF THE PECTORAL GIRDLE

THE CLAVICLE

The clavicle (L. little key) is located at the thoracocervical junction (root of the neck) and can be palpated throughout its entire length (Fig. 6-2). The medial end of the clavicle can be easily palpated because it projects superior to the **manubrium sterni.** Between the two medial elevations is the deep **jugular notch** (suprasternal notch).

As the clavicle passes laterally in the horizontal plane, its medial part can be felt to be convex anteriorly. The large vessels and nerves to the upper limb pass posterior to this convexity (Fig. 6-20).

The lateral end of the clavicle does not reach the "*point of the shoulder*" which is formed by the **acromion** of the scapula (Figs. 6-1 to 6-3). The acromion is subcutaneous and usually obvious (Fig. 6-2). The small ***acromioclavicular joint***, where the clavicle and acromion articulate, can be palpated 2 to 3 cm medial to the lateral border of the acromion, particularly when the upper limb is swung slowly anteriorly and posteriorly. Either or both ends of the clavicle may be prominent (Fig. 6-2).

The anterior axillary fold (Figs. 6-2 and 6-13) is formed by the lateral border of the pectoralis major muscle (Figs. 6-14 and 6-17). The anterior axillary fold can easily be felt between your index finger and thumb. Make the fold more prominent by putting your hand on your hip and pressing hard against it. You can feel your pectoralis major muscle contracting within the anterior axillary fold.

The posterior axillary fold (Figs. 6-4 and 6-13 to 6-15) contains the latissimus dorsi and teres major muscles. It can also be easily palpated.

THE SCAPULA

The scapula is a *highly mobile bone.* The posterior border or *crest of the spine* of the scapula is subcutaneous throughout and is easily felt. The medial end of the spine, often called the *root* (Fig. 6-3), is opposite the spinous process of the third thoracic vertebra when the arm is by the side (Fig. 6-3).

The **acromion**, projecting anteriorly from the lateral end of the spine of the scapula (Fig. 6-3), is clearly visible in some people (Fig. 6-2). Because of this feature, **the acromion is often referred to as the "point of the shoulder."**

The acromion is important because it is the proximal point from which clinicians measure the length of the upper limb. Inferior to the acromion is the smooth

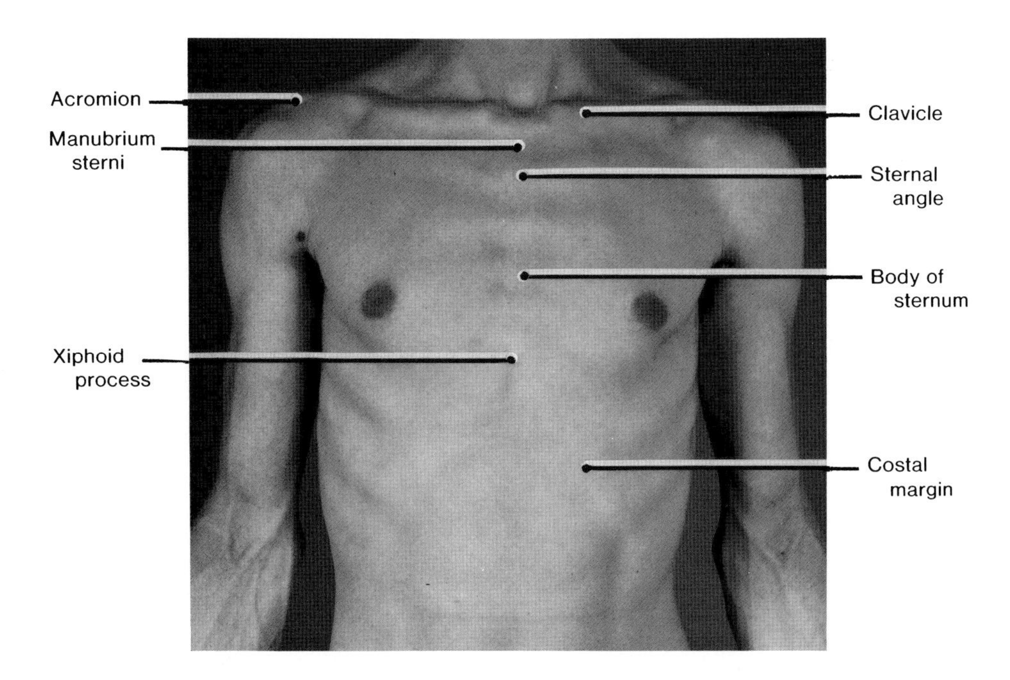

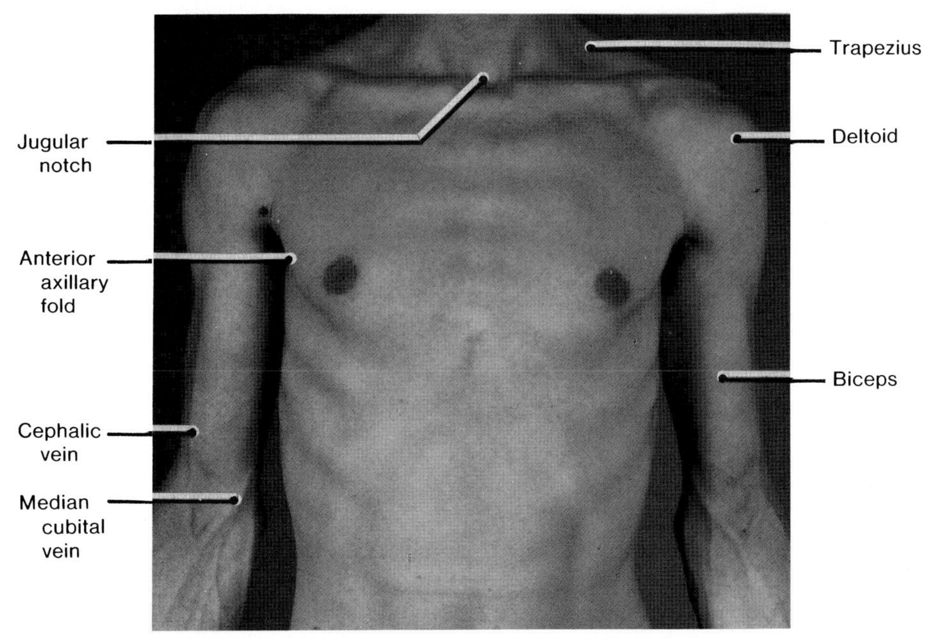

Figure 6-2. Photograph of a 27-year-old man illustrating the surface anatomy of the pectoral region and arm. The term acromion is derived from Greek words meaning "tip of the shoulder." Lay people refer to it as the "point of the shoulder" because of its prominence. Understand that it is part of the scapula, not the clavicle (see Figs. 6-1 and 6-3). The costal margin (L. *costa*, rib) extending inferolaterally from the sternum, is palpable with ease (also see Fig. 1-9). The clavicle, which can be easily palpated, is one of the most commonly fractured bones in the body (see Fig. 6-5).

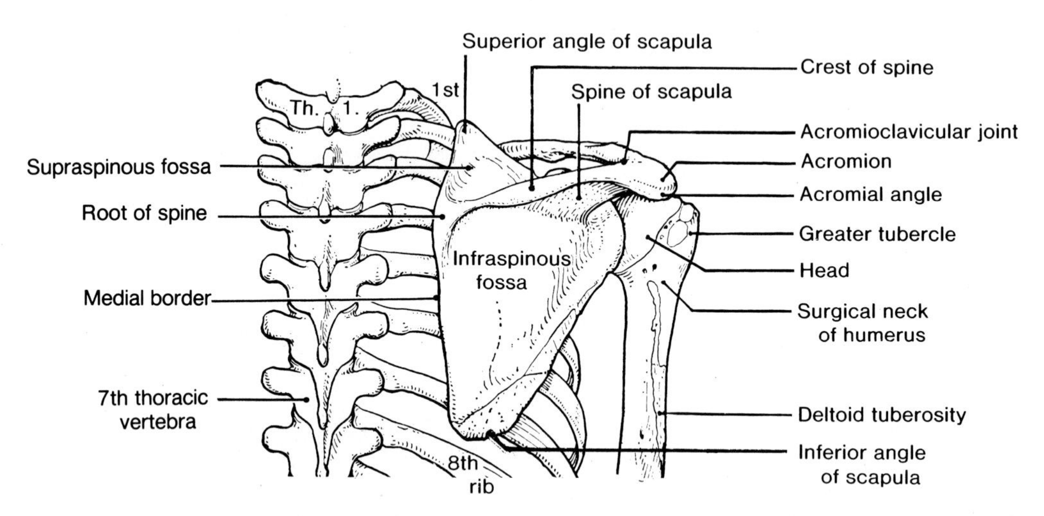

Figure 6-3. Drawing of a posterior view of the bones of the thorax and the proximal part of the upper limb. The posterior border or crest of the spine is subcutaneous and easily palpated. Note that the greater tubercle of the humerus is the most lateral bony point of the shoulder. The scapula is a triangular bone which lies over the posterior aspects of the second to seventh ribs. Along with the subscapularis muscle, the scapula forms a main part of the posterior wall of the axilla (Fig. 6-10 and p. 644). Because the inferior angle of the scapula is visible (Fig. 6-4) and is easily palpated, it is a useful bony landmark. Observe that the spine of the scapula is continuous laterally with the acromion which overhangs the shoulder joint. The acromion is another obvious bony landmark (Fig. 6-2).

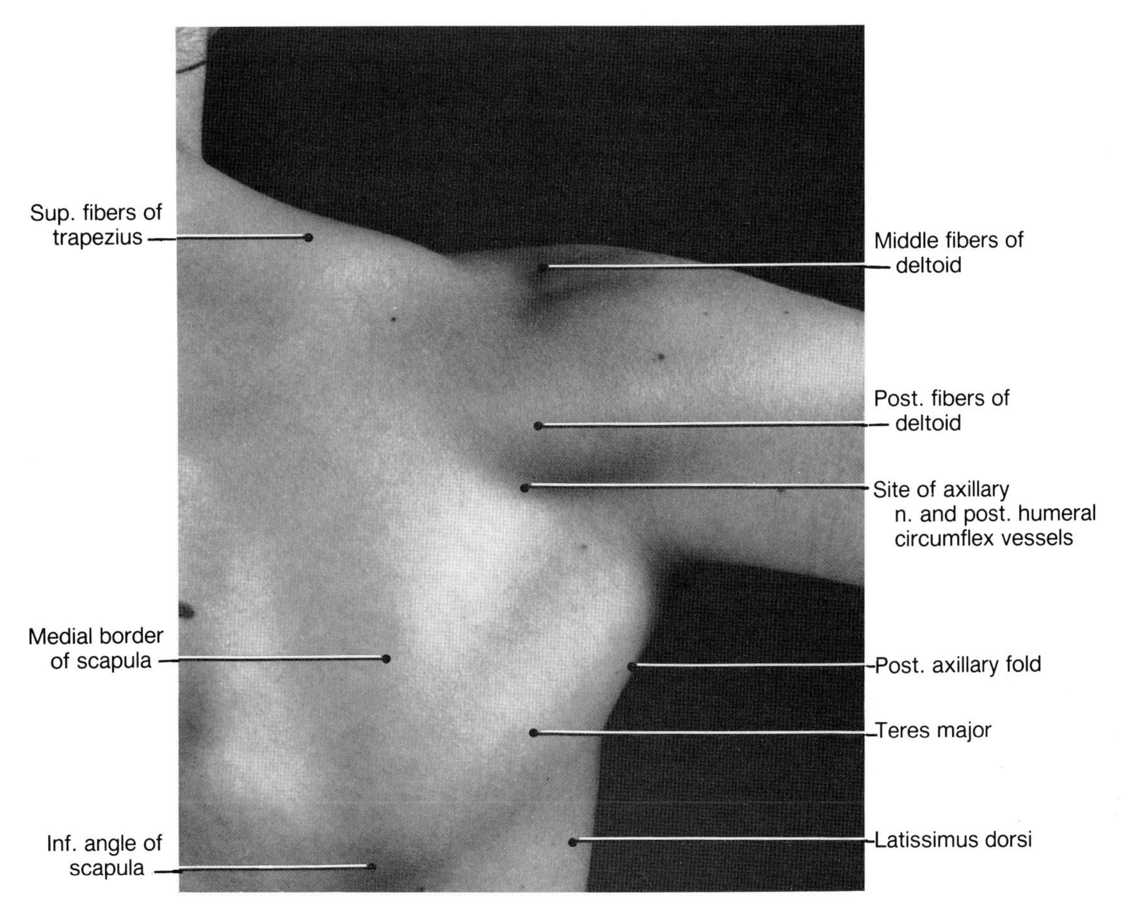

Figure 6-4. Photograph of the shoulder and scapular regions of a 12-year-old girl. To make her deltoid muscle stand out, she abducted her arm against resistance. Her muscles are well developed because she was a competitive gymnast when this photograph was taken. Note that when the middle fibers of the deltoid muscle contract they produce longitudinal skin furrows. This occurs because the middle fibers of the deltoid are short and multipennate in architecture. They take origin from small bony eminences on the acromion of the scapula, as well as from tendinous intersections within the muscle (Figs. 6-18 and 6-19).

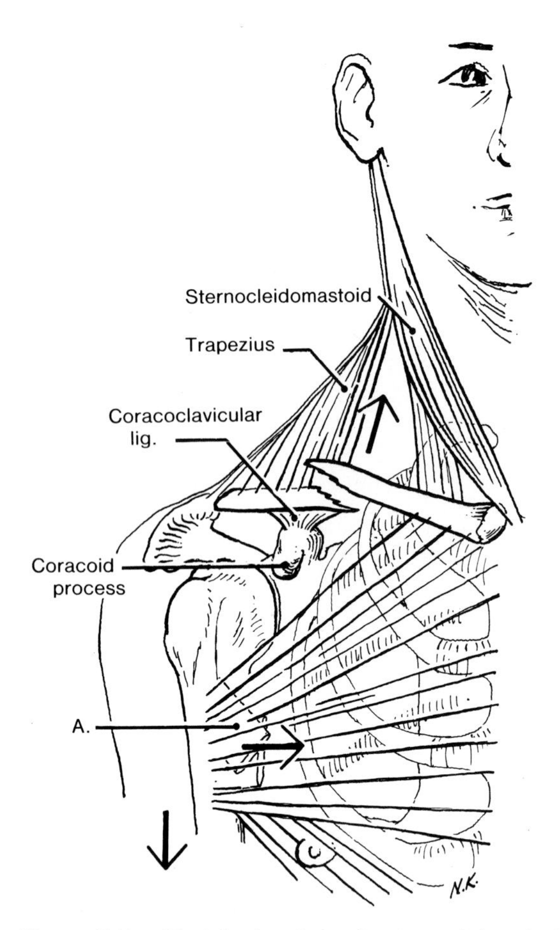

Figure 6-5. Sketch showing a fracture of the clavicle near the junction of the middle and lateral thirds, its weakest part. Note that the patient's shoulder has sagged slightly owing to the weight of the limb. *A*, pectoralis major, an adductor and medial rotator of the upper limb. (Fig. 6-19 shows a dissection of this large muscle).

rounded curve of the shoulder formed by the deltoid muscle (Figs. 6-2 and 6-4).

The **superior angle of the scapula** lies at about the level of T2 vertebra (Fig. 6-3). The **inferior angle** lies at the spinous process of T7 vertebra.

The inferior angle of the scapula is usually a good guide posteriorly to the seventh intercostal space when the arm is by the side (Figs. 6-3 and 6-4). Also see Figure 5-2.

The superior half of the **medial border of the scapula** (vertebral border) is covered by the trapezius muscle (Fig. 6-4), but it can be easily palpated and observed from the superior to the inferior angles of the scapula. Note that the medial border of the scapula crosses ribs 2 to 7 (Fig. 6-3). The **lateral border** (axillary border) of the scapula is not easily palpated, except for its inferior part.

The tip of the **coracoid process of the scapula** can be palpated by pressing deeply just under the lateral border of the **deltopectoral triangle** (Figs. 6-14 and 6-17).

THE BREASTS AND MAMMARY GLANDS

The breasts are situated on the anterior surface of the thorax, overlying the pectoral muscles (Figs. 6-6 to 6-8). The **mammary glands** are accessory organs of the female reproductive system, located within the **breasts** (L. mammae). The amount of fat (adipose tissue) surrounding the glands determines the size of the breast. At the end of pregnancy the mammary glands produce milk. Although the breasts lie anterior to the thorax, they are usually described with the upper limb because they must be removed during dissection to study the pectoral muscles.

Both males and females have breasts; normally only females have well-developed mammary glands. These glands in males are normally rudimentary throughout life and consist of only a few small ducts.

In many patients with the Klinefelter syndrome, the breasts begin to enlarge at or soon after the onset of puberty. The breasts usually slowly increase in size over a period of several years and then remain stationary. Enlargement of the breasts (**gynecomastia**) is often the presenting symptom in these males who usually have an XXY sex chromosome complex. (See Moore, 1982 for more details).

THE FEMALE BREASTS OR MAMMAE (Figs. 6-6 to 6-9)

At puberty (12 to 15 years), the breasts normally grow and *the circular areas of skin around the nipples* called **areolae**, enlarge and become more pigmented. The lactiferous ducts give rise to buds which form 15 to 20 lobules of glandular tissue (mammary glands). Each lobule is drained by a **lactiferous duct** (Fig. 6-7), each of which opens on the nipple.

The lactiferous ducts extend from the nipple like the spokes of a wheel. Under the areola each duct has a dilated portion, called the **lactiferous sinus**, in which milk accumulates during lactation (milk production).

The areola (L. dim of *area*), composed of pigmented skin, contain numerous **areolar glands**. These sebaceous glands enlarge during pregnancy and secrete an oily substance that provides a protective lubricant for the areola and nipple. The areolae, variable in size, are pink in white **nulliparous women** (those who have not borne children). During pregnancy the areolae enlarge and become deep brown to black, the depth of color depending upon the woman's complex-

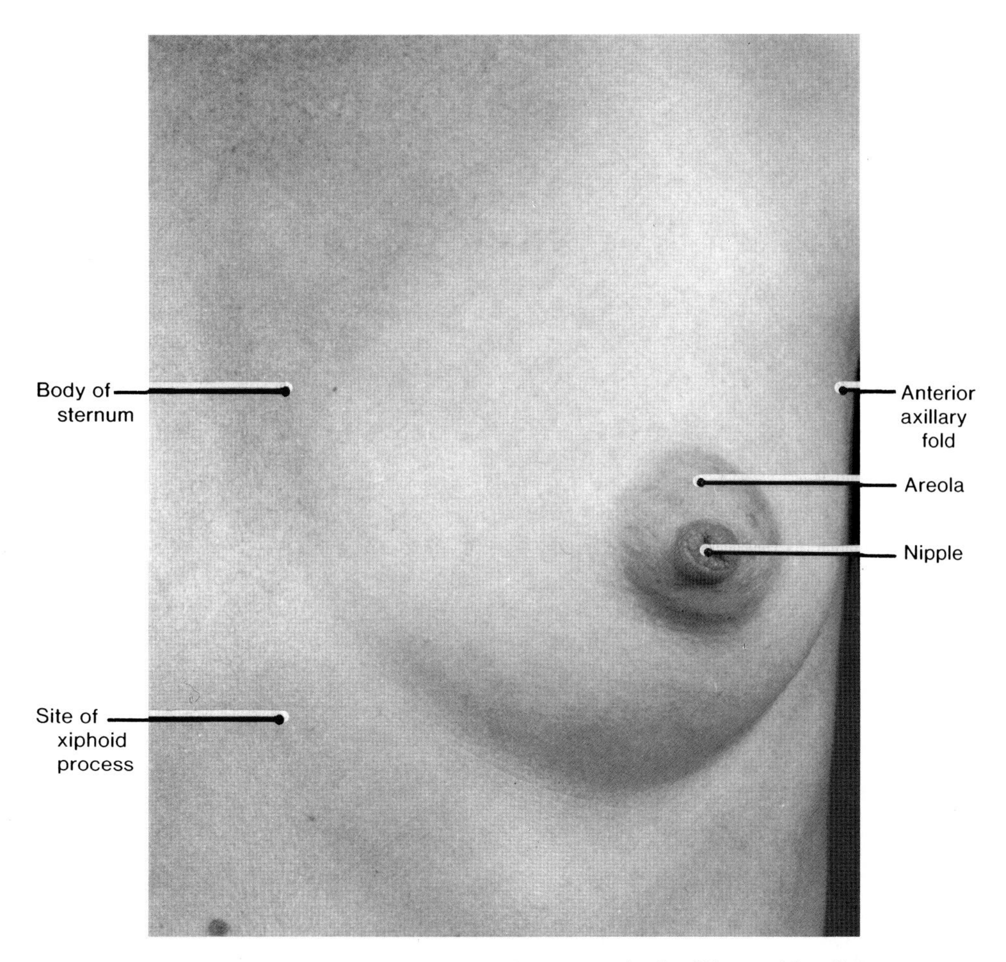

Figure 6-6. Photograph of the smooth conical breast (L. mamma) of a 27-yer-old nulliparous woman. Note that the skin of the nipple is thrown into numerous wrinkles and, on the areola, exhibits many minute, rounded projections or tubercles owing to the underlying areolar glands. The cleavage between the breasts (site of the body of the sternum) is known as the sinus mammarum. The extension of the mammary gland into the anterior axillary fold, known as the "axillary tail", is not obvious in this young woman.

ion. The color diminishes after pregnancy, but the areolae never return to their original rosy-pink color.

As the mammary gland is a modified skin gland, it has no special capsule or sheath. The mammary gland is situated within the **superficial fascia**, anterior to the thorax (Figs. 6-1 and 6-7). Although easily separated from the **deep fascia** covering the pectoralis major and serratus anterior muscles, the mammary gland is firmly attached to the skin of the breast by **suspensory ligaments** (of Cooper). These fibrous bands, which *support the breast*, run between the skin and the deep fascia (Fig. 6-7). The superolateral part of the mammary gland frequently projects into the axilla, forming the **axillary tail of the mammary gland.**

The rounded contour and most of the bulk of the breasts are produced by fat lobules (Fig. 6-7), except during pregnancy and lactation. The shape of the breast varies considerably in different persons and races, and in the same person at different ages.

At puberty the lactiferous ducts undergo branching and thereafter progressive enlargement of the breasts occurs, partly as the result of the increased deposition of fat. During pregnancy the breasts enlarge greatly owing to the formation of new glandular tissue. The secretion of milk begins after delivery of the baby. The milk-secreting cells, referred to as *alveoli*, are arranged in grapelike clusters or lobules.

In **multiparous women** (those who have borne several children), the breasts may be very large and pendulous. In most **elderly women** the breasts are small and wrinkled owing to a decrease in adipose and glandular tissue.

Although breasts vary markedly in size, their roughly circular bases are fairly constant and have the

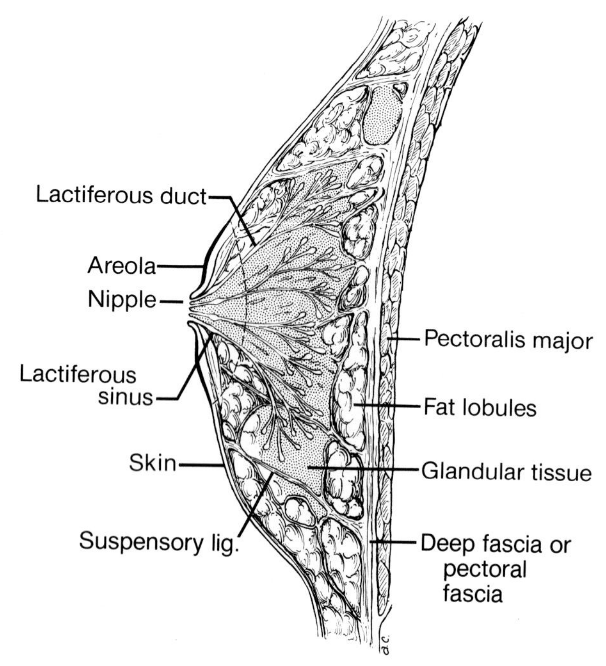

Figure 6-7. Drawing of a sagittal section of a female breast. Observe that it consists of glandular, fibrous, and adipose tissues. Note that the breast is separated from the pectoralis major muscle by deep fascia, often called the pectoral fascia because it covers the pectoralis major muscles. The loose connective tissue between the breast and the deep fascia allows the breast some degree of movement. The female breast overlies the second to sixth ribs. Two-thirds of it rests on the pectoralis major; the other one-third on the serratus anterior (Fig. 6-14). The mammary gland consists of modified sweat gland tissue in the superficial fascia. Note that the mammary gland is firmly attached to the skin at the breast by suspensory ligaments. These fibrous bands, which support the breast, run between the dermis of the skin and the deep fascia covering the pectoralis major muscle.

following limits in a well-developed female: *second to sixth ribs,* and from the *edge of the sternum to the midaxillary line* (Fig. 6-6).

Arterial Supply of the Breast (Figs. 1-16, 6-20, and 6-28). There is an abundant blood supply to the breast. The arteries are derived from: (1) perforating branches of the **internal thoracic artery** (intercostal spaces 2 to 4); (2) lateral mammary branches of the **lateral thoracic artery**, a branch of the axillary; and (3) lateral and anterior cutaneous branches of the **intercostal arteries** (intercostal spaces 3, 4, and 5). The blood supply of the breast is chiefly from superficially lying vessels.

Venous Drainage of the Breast (Figs. 1-16 and 6-40). Veins from the breast drain into the **axillary, internal thoracic, lateral thoracic, and intercostal veins.** There are connections between the intercostal veins and the **vertebral venous plexuses** (Fig. 5-58). *The chief venous drainage is toward the axilla.*

Nerve Supply of the Breast (Figs. 1-15 to 1-17). The breast is supplied by lateral and anterior cutaneous branches of the **second to sixth intercostal nerves.** Although these nerves convey sympathetic fibers to the breast, its secretory activities are chiefly under the control of ovarian and pituitary hormones.

Lymphatic Drainage of the Breast (Figs. 6-8 and 6-43). Most lymph passes along interlobular vessels to a **subareolar plexus.** From it and other parts of the breast, the lymph vessels follow the venous drainage of the breast to the axilla.

Most of the lymphatic drainage of the breast is to the pectoral group of axillary lymph nodes. Some lymph vessels pass to the *apical and subscapular groups of axillary lymph nodes.* In addition, some lymph vessels pass to: (1) the *infraclavicular lymph nodes,* (2) the *opposite breast,* (3) the *parasternal lymph nodes* (Figs. 1-16 and 6-8), and (4) the *abdominal lymph nodes.*

A clear understanding of the lymphatic and venous drainages of the breast must be obtained because of their clinical importance in the spread of ***carcinoma of the breast*** (**breast cancer**), one of the two most common types of female cancer (Fig. 6-9*B*)

Cancer cells are carried by lymph vessels to lymph nodes, chiefly the axillary lymph nodes; they receive more than 75% of the total lymphatic drainage from the breast. The cancer cells lodge in the lymph nodes where they produce nests of tumor cells called **metastases** (G. *meta,* beyond + *stasis,* a placing).

As free communication exists between lymph nodes inferior and superior to the clavicle, and between the axillary and cervical lymph nodes, metastases from the breast may develop in the supraclavicular lymph nodes, the opposite breast, or in the abdomen. In addition, the connections between the intercostal veins and the vertebral venous plexuses allow metastasis to bones and the brain.

The axillary lymph nodes are the most common site of metastases from carcinoma of the breast. Enlargement of axillary lymph nodes in a female therefore suggests the possibility of breast cancer. Cancerous nodes tend to be hard and are not usually tender.

Enlargement of axillary lymph nodes does not necessarily indicate cancerous involvement. Infection of lymphatic vessels (**lymphangitis**),

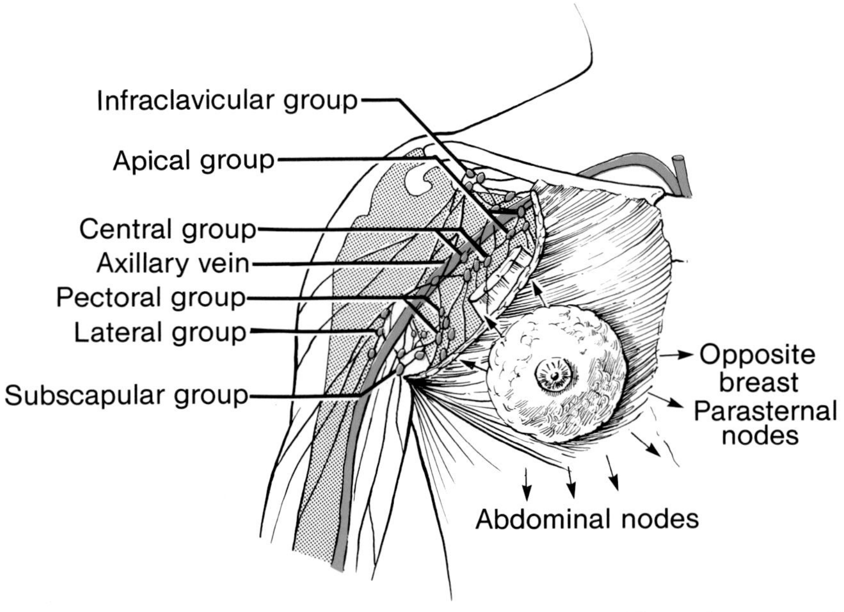

Figure 6-8. Drawing illustrating the lymphatics of the breast and axilla. The main groups of lymph nodes of the upper limb are located in the axilla and are known collectively as the axillary lymph nodes. These nodes may be divided into five groups. The main drainage of the breast is into the axillary lymph nodes, chiefly into the pectoral group (also see Fig. 6-43). About three-quarters of the lymphatic drainage of the breast is to the axillary nodes; most of the remainder enters the parasternal lymph nodes (Fig 1-16).

e.g., resulting from a severe thumb infection, spreads to the axillary lymph nodes causing them to become large and tender (**lymphadenitis**). It is also important to know that the absence of enlarged axillary lymph nodes is no guarantee that metastasis from a breast cancer has not occurred.

Often there is dimpling and a leathery thickening of the skin over the site of a carcinoma of the breast (Fig. 6-9*B*), giving the skin the appearance of an orange peel; this skin change is called *peau d'orange* (Fr., orange skin). Interference with the lymphatic drainage of the breast produces the leathery thickening, whereas dimpling of the skin is mainly caused by infiltration of the cancer cells along the suspensory ligaments of the breast.This invasion shortens the ligaments and causes the skin to invaginate (dimple). **Subareolar cancers** may cause inversion of the nipple by the same mechanism.

Mastectomy (removal of a breast) is not an uncommon operation in females. In *simple mastectomy* the breast (nipple, areola, and glandular, fibrous, and fatty tissues) is removed down to the pectoralis fascia (Fig. 6-6). *Radical mastectomy* is a more extensive operation during which the breast, pectoral muscles, fat, fascia, and all lymph nodes in the axilla and pectoral region are removed.

During mastectomy care must be taken to preserve the long thoracic nerve (Fig. 6-28). Cutting this nerve results in paralysis of the serratus anterior muscle and inability to rotate the scapula superiorly during abduction of the arm. As a result there is difficulty in elevating the arm superior to the head. When the person pushes against a wall, the vertebral border and inferior angle of the scapula protrude posteriorly, producing a "**winged scapula**" (Fig. 6-36). This occurs because the serratus anterior muscle, supplied by the long thoracic nerve, is paralyzed. It normally holds the scapula against the chest wall (Table 6-1).

The foregoing information is intended to emphasize the importance of knowing the structure, blood supply, and lymphatic drainage of the breast. It is not intended to give you sufficient knowledge to determine the significance of a palpable lesion in the breast.

About 1% of breast carcinomas occur in males. The incidence is the same as in women for males with the Klinefelter syndrome. A

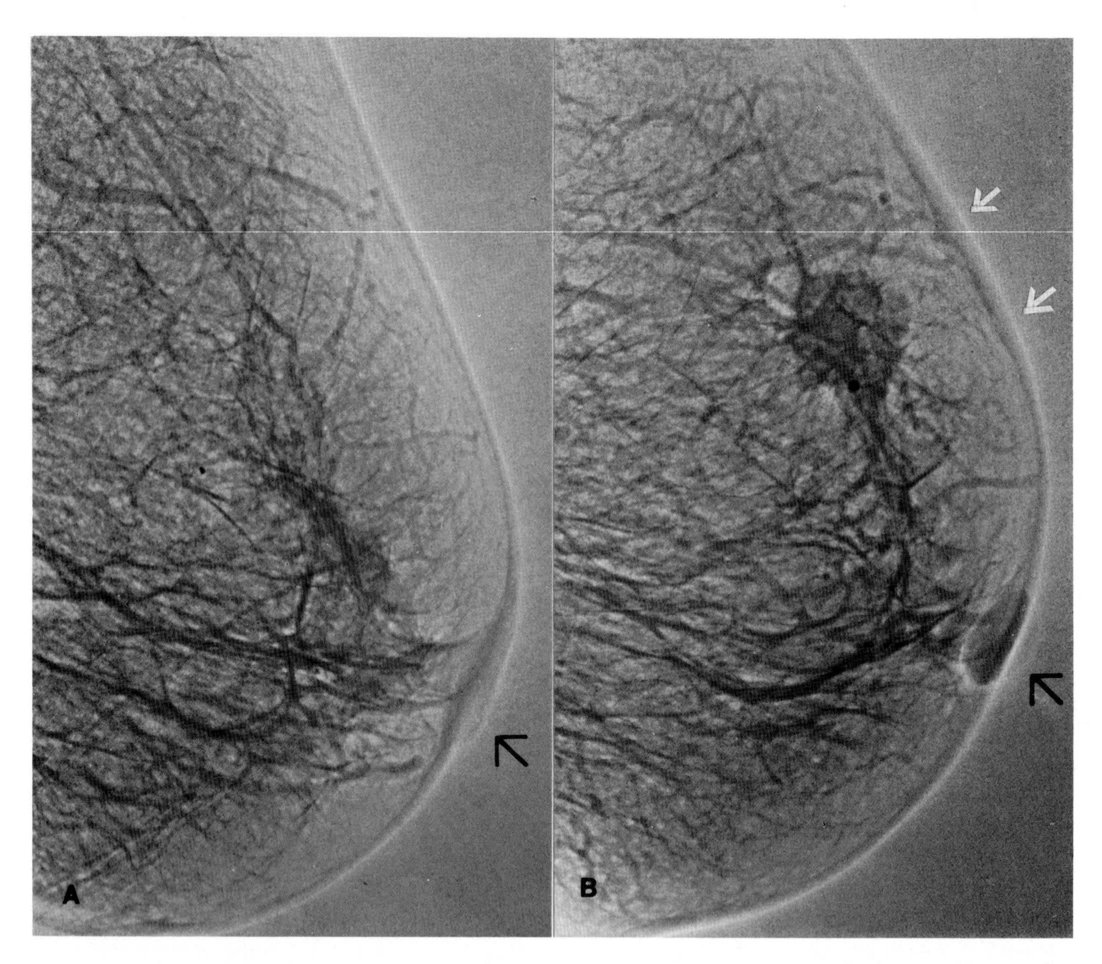

Figure 6-9. Xeromammograms, mediolateral projection. *A*, normal breast showing the suspensory ligaments between the lobules of fat. Veins may also be seen. The skin thickness is uniform except at the nipple (*arrow*). *B*, carcinoma (cancer) appears as a jagged, rounded density. Note the overlying skin thickening (*white arrows*) resulting from impaired lymph drainage. The jagged appearance of the tumor results from the infiltration of cancer cells along the suspensory ligaments and lymphatics. The *black arrow* indicates the nipple. (Courtesy of Dr. T. Connor, Women's College Hospital, Toronto, Ontario, Canada.)

breast tumor in males is often hard and tends to infiltrate the deep fascia, pectoralis major muscle, and axillary lymph nodes.

Accessory breasts (**polymastia**) or nipples (**polythelia**) may occur superior or inferior to the normal breasts. Usually supernumerary "breasts" consist only of a nipple and areola that may be mistaken for a mole or nevus (birthmark). They may appear anywhere along a line extending from the axilla to the groin, the location of the **embryonic mammary ridge.** Accessory mammary tissue may appear elsewhere (*e.g.*, on the neck or genital organs) owing to displacement of parts of the embryonic mammary ridges. See Moore (1982) for illustrations and details.

Mammography is one of the radiographic techniques used to detect breast masses. Mammographs are also used by surgeons to guide them during mastectomies and excision of cysts, abscesses, and tumors (Fig. 6-9).

Thermography, a method of measuring and recording heat radiation emitted by the breast, is sometimes used in conjunction with mammography. Breast tumors emit more heat then normal breast tissue.

Computerized Tomography (CT) is sometimes combined with mammography for cancer detection in the breast. Before the CT scans are taken, an iodide contrast material is given intravenously to the patient. Breast cancer cells have an unusual affinity for iodide and so become recognizable.

THE PECTORAL MUSCLES

The pectoral region contains four muscles, all of which are associated with the upper limb (Table 6-1).

THE PECTORALIS MAJOR MUSCLE (Figs. 6-10, 6-11, 6-13, 6-14, 6-16 to 6-20, and Table 6-1)

This large, thick, fan-shaped muscle covers the superior part of the chest, and its lateral border forms the **anterior axillary fold** and most of the anterior wall of the axilla (Figs. 6-10, 6-11, and 6-13).

The pectoral fascia enclosing the pectoralis major muscle is attached at its origin to the clavicle and sternum (Figs. 6-7, 6-11, and 6-14). The pectoral fascia leaves the lateral border of this muscle to form the **axillary fascia** in the floor of the axilla (Fig. 6-11).

The pectoralis major and deltoid muscles diverge slightly from each other superiorly and, along with the clavicle, form the **deltopectoral triangle** (Figs. 6-14 and 6-17). The *cephalic vein,* one of the two major superficial veins of the upper limb, occupies the furrow between the deltoid and the pectoralis major muscles before it enters the deltopectoral triangle on its way to the *axillary vein* (Figs. 6-14 and 6-40).

Origin (Figs. 6-14 and 6-18). **Anterior surface of medial half of clavicle** (*clavicular head*); anterior surface of **manubrium and body of sternum, costal cartilages of 2nd to 6th ribs, and aponeurosis of external oblique muscle** (*sternocostal head*). The two heads of the pectoralis major muscle meet at the sternoclavicular joint (Fig. 6-14).

Insertion (Figs. 6-10 and 6-18). **Lateral lip of intertubercular groove** or sulcus of the humerus. This groove separates the greater and lesser tubercles of the humerus (Fig. 6-1).

Nerve Supply (Figs. 6-16, 6-19, and 6-20). **Lateral pectoral nerve** from lateral cord of brachial plexus (Fig. 6-24), and **medial pectoral nerve** from medial cord of this plexus.

Actions. Adducts and medially rotates the humerus at the shoulder joint when both parts of the muscles are acting together. Acting alone, the *clavicular head* **helps to flex the humerus** (*i.e.*, draws the arm anteriorly) and from this position the *sternocostal head* **extends the humerus at the shoulder joint** (*i.e.*, carries the arm posteriorly).

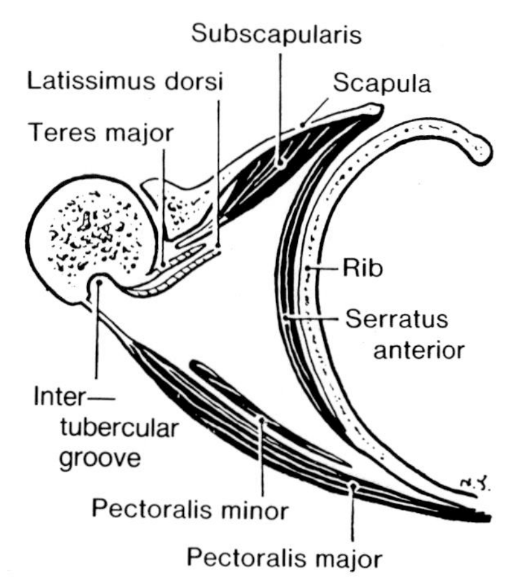

Figure 6-10. Diagram of a cross-section of the axilla showing its walls. Note the pectoral muscles in the *anterior wall*, the scapula and subscapularis muscle in the *posterior wall*, the rib and serratus anterior muscle in the *medial wall*, and the intertubercular groove (sulcus) forming the *lateral wall.*

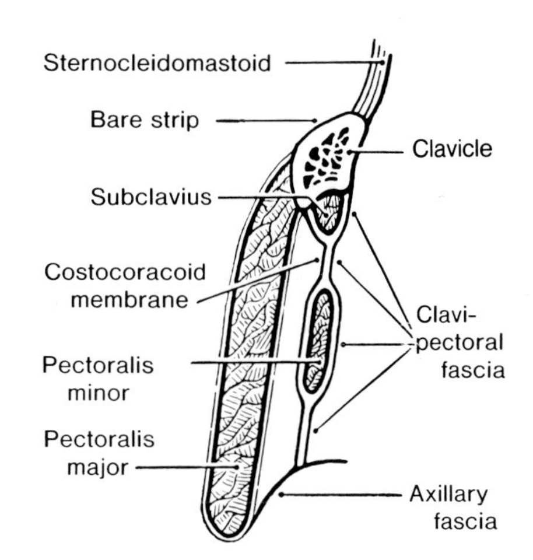

Figure 6-11. Drawing of a sagittal section of the axilla illustrating its anterior wall. The clavicle and three muscles inferior to it (pectoralis major, pectoralis minor, and subclavius) form the anterior wall of the axilla. The clavipectoral fascia is a strong sheet of connective tissue attached to the clavicle and enclosing the subclavius muscle superiorly and the pectoralis minor muscle inferiorly. This fascia then joins the floor of the axilla formed by axillary fascia and skin.

THE PECTORALIS MINOR MUSCLE (Figs. 6-10, 6-11, 6-20, 6-21, and Table 6-1)

This triangular muscle lies in the anterior wall of the axilla, where it is largely covered by the much larger pectoralis major.

The pectoralis minor is the landmark of the axilla as illustrated in Figure 6-23.

Table 6-1
Muscles of the Pectoral Region

Muscle	Origin	Insertion	Nerve Supply	Actions
Pectoralis major (Figs 6-14 and 6-17)	*Clavicular head:* medial half of clavicle (Figs. 6-14 and 6-18) *Sternocostal head:* sternum, costal cartilages of 2nd to 6th ribs, and aponeurosis of external oblique muscle (Fig. 6-18)	Lateral lip of intertubercular groove of humerus (Figs. 6-10 and 6-18)	Lateral and medial pectoral nerves; Clavicular head (C5 and **C6**)[1]; Sternocostal head (**C7**, **C8**, and T1)	Adducts and medially rotates humerus *Acting alone:* Clavicular head flexes humerus and, from this position, sternocostal head extends humerus to side of body Draws scapula anteriorly and inferiorly
Pectoralis minor (Fig. 6-20)	Ribs 3, 4, and 5 (Fig. 6-18)	Medial border of coracoid process of scapula (Fig. 6-18)	Medial and lateral pectoral nerves (C6, **C7**, and C8)	Stabilizes scapula by drawing it inferiorly and anteriorly against thoracic wall
Subclavius (Figs. 6-20 and 6-28)	Junction of rib 1 and its costal cartilage (Fig. 6-20)	Inferior surface of middle third of clavicle (Figs. 6-11 and 6-21)	Nerve to subclavius (**C5** and C6)	May depress lateral end of clavicle
Serratus anterior (Figs. 6-14 and 6-17)	External surfaces of lateral portions of ribs 1-8 (Fig. 6-18).	Anterior surface of medial border of scapula (Fig. 6-49)	Long thoracic nerve (C5, **C6**, and **C7**)	Protracts scapula and holds it against thoracic wall Rotates scapula

[1] In this and subsequent tables, the numbers indicate the spinal cord segmental innervation of the nerves (*e.g.*, C5 and C6 indicate that nerves supplying the clavicular head of the pectoralis major muscle are derived from the 5th and 6th cervical segments of the spinal cord). **Boldface type** indicates the main segmental innervation. Note that the muscles are innervated from more than one segment of the spinal cord. Damage to these segments, or to the motor nerve roots arising from them, results in paralysis of the muscles concerned.

Absence of the pectoralis major, usually its sternocostal part, is rare and when it occurs there is usually no disability. However the anterior axillary fold is absent on the affected side and the nipple is more inferior than usual.

The pectoralis minor is surrounded by the **clavipectoral fascia** (Figs. 6-11 and 6-19), a thin sheet of fibrous tissue that runs from the clavicle to the fascial floor of the axilla.

The connection of the ***clavipectoral fascia*** *with the clavicle supports and suspends the floor of the axilla,* composed of axillary fascia and skin (Fig. 6-11).

Origin (Figs. 6-18 and 6-20). Superior margins and external surfaces of **ribs 3, 4, and 5.** Occasionally part of the muscle also arises from the second rib.

Insertion (Figs. 6-18 and 6-23). **Medial border of coracoid process of scapula.**

Nerve Supply (Figs. 6-16 and 6-20). **Medial and lateral pectoral nerves** (C6, **C7**, and C8).

Actions. Stabilizes scapula by drawing it inferiorly and anteriorly against the thoracic wall.

The pectoralis minor is a useful landmark for many structures in the axilla because, with the coracoid process, it forms an arch deep to which pass the vessels and nerves to the arm (Fig. 6-23).

THE SUBCLAVIUS MUSCLE (Figs. 6-20, 6-21, 6-28, and Table 6-1)

As its name indicates, this small rounded, ***relatively unimportant muscle*** lies inferior to the clavicle. Because of its location it may serve as a protective cushion between a fractured clavicle and the subclavian vessels when this bone is broken (Fig. 6-5).

Origin (Fig. 6-20). **Junction of first rib and its costal cartilage.**

Insertion (Figs. 6-20, 6-21, and 6-29). **Inferior surface of middle third of the clavicle.**

Nerve Supply (Figs. 6-21 and 6-24). **Nerve to subclavius** (**C5** and C6).

Actions. May depress lateral end of clavicle, pull the point of the shoulder anteriorly, and steady the clavicle during shoulder movements.

The actions of the subclavius are not well understood and paralysis of it produces no demonstrable effect.

THE AXILLA

The axilla (armpit) is a roughly pyramidal space at the junction of the arm and thorax (Figs. 6-10 to 6-13). **The axilla provides a passageway for nerves and vessels of the trunk to reach the upper limb** (Fig. 6-20). The axilla has an apex, a base, and four walls.

The apex of the axilla is directed toward the root of the neck and is located at the medial side of the root of the coracoid process of the scapula (Fig. 6-20).

The apex of the axilla is formed by the convergence of the bones in its three major walls; the **clavicle** in its anterior wall, the **scapula** in its posterior wall, and the **first rib** in its medial wall. The interval between these bones is the entrance to the axilla through which all nerves and vessels pass to the upper limb (Fig. 6-20).

The base of the axilla, facing inferiorly, is formed by the fascia and skin of the concave axilla (*i.e.*, the armpit). The skin of the base is normally covered with hair in postpubertal persons. Many women shave the bases of their axillae. The boundaries of the axilla can be visualized best in a cross-section (Fig. 6-10).

The Anterior Wall of the Axilla (Figs. 6-6, 6-11, 6-13, and 6-14). The clavicle and pectoral muscles (pectoralis major, pectoralis minor, and subclavius) form the anterior wall. The lateral border of the pectoralis major forms the **anterior axillary fold** (Fig. 6-13). Posterior to the pectoralis major, the pectoralis minor and subclavius muscles form the deep layer of the anterior wall (Fig. 6-11).

The Posterior Wall of the Axilla (Figs. 6-4, 6-10, 6-12, and 6-13). This wall is formed chiefly by the scapula and the subscapularis muscle. Inferior to the subscapularis is the teres major muscle which combines with the latissimus dorsi to form the **posterior axillary fold** (Fig. 6-13). The tendon of the latissimus dorsi muscle wraps around the lateral part of the teres major (Fig. 6-12), and forms part of the posterior wall of the axilla.

The Medial Wall of the Axilla (Figs. 6-10, 6-14, and 6-15). This wall is formed by the ribs and intercostal muscles which are covered by the serratus anterior muscle.

The Lateral Wall of the Axilla (Figs. 6-1, 6-10, and 6-16). This narrow wall is *formed by the floor of the* ***intertubercular*** **groove** in the humerus (Fig. 6-10), which lodges the tendon of the long head of the biceps muscle (Fig. 6-29).

The axilla contains large important nerves which are branches of the brachial plexus. They pass from the neck to supply the upper limb.

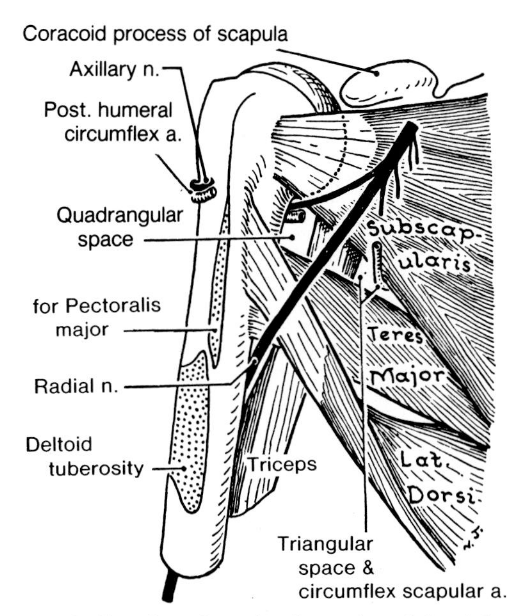

Figure 6-12. Drawing of a dissection of the right axilla showing its posterior wall formed by the scapula and the subscapularis, teres major, and latissimus dorsi muscles. Observe the courses of the radial and axillary nerves, noting in particular the anterior branch of the axillary nerve winding around the surgical neck of the humerus. It may be injured when a fracture occurs in this region (Fig. 6-1).

The axilla also contains the axillary vessels (axillary artery and its branches, axillary vein and its tributaries, and axillary lymph vessels).

The axilla contains several groups of **axillary lymph nodes** (Figs. 6-8 and 6-43) which are of practical importance because of their frequent invasion by cancer cells from the breast.

THE BRACHIAL PLEXUS

This network of nerves extends from the neck into the axilla and supplies motor, sensory, and sympathetic nerve fibers to the upper limb.

The supraclavicular part of the brachial plexus is in the part of the neck known as the posterior triangle (Figs. 6-17 and 6-43), and the infraclavicular part of the brachial plexus is in the axilla (Fig. 6-22).

The brachial plexus is formed by the union of the ventral rami of nerves C5 to C8 and T1 (Fig. 6-22). The ventral rami that form the brachial plexus lie between the scalenus anterior and scalenus medius muscles (Fig. 6-21). These rami are sometimes referred to as the "**roots of the brachial plexus.**" Be careful not to confuse these rami ("roots") with the dorsal and ventral roots which unite to form spinal nerves (Figs. 5-51 and 6-22).

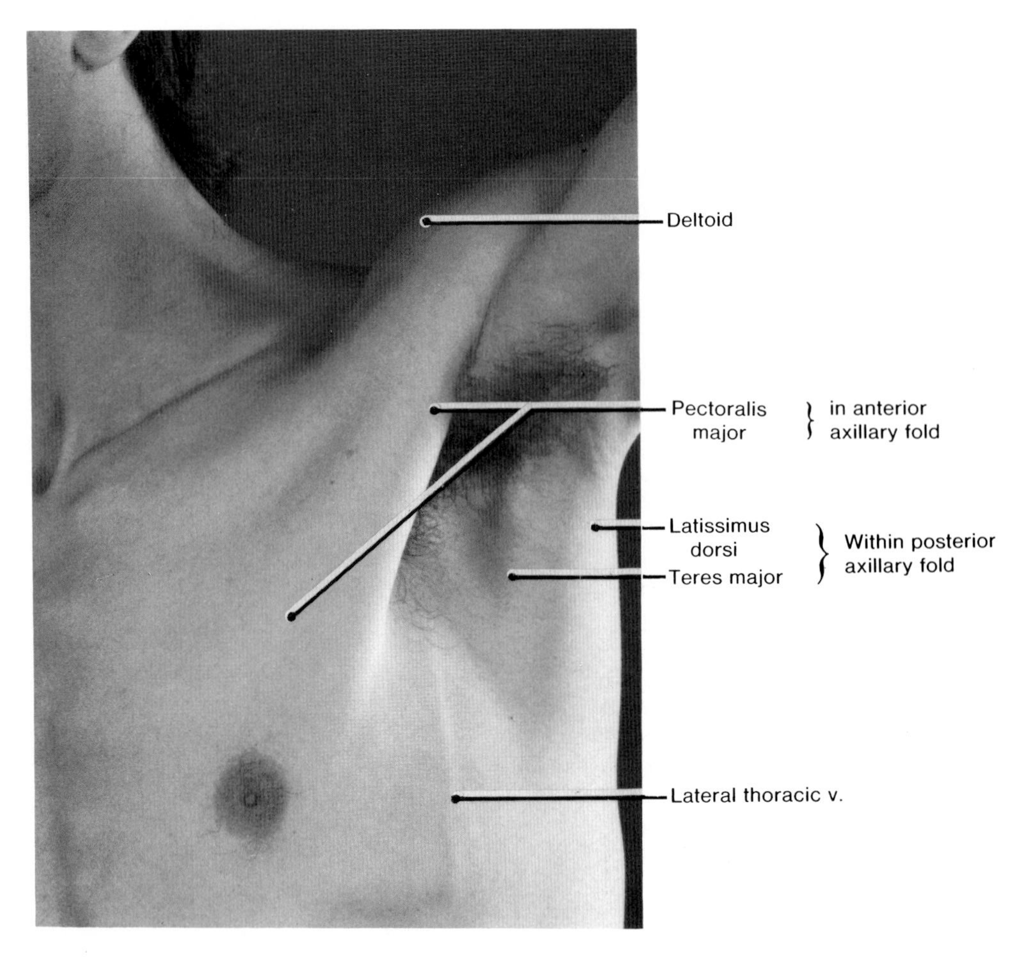

Figure 6-13. Photograph of the left axilla of a 27-year-old man exposed by abducting the upper limb. Observe that the breast is rudimentary, the nipple small, and the areola is surrounded by sparse hairs. The mammary gland (within the breast) is also rudimentary in males. Note that the lateral border of the pectoralis major muscle forms the **anterior axillary fold** and that the teres major and the latissimus dorsi muscles form the **posterior axillary fold**. A vertical line midway between the anterior and posterior axillary folds is called the **midaxillary line** (Fig. 1-14).

The Usual Plan of the Brachial Plexus is as follows (Figs. 6-21 to 6-24):

1. As the ventral primary rami enter the posterior triangle of the neck, those from C5 and C6 unite to form a **superior trunk** (upper trunk). The ventral ramus of C7 continues as a **middle trunk,** and the ventral rami of C8 and T1 unite at the neck of the first rib to form an **inferior trunk** (lower trunk). The inferior trunk lies on the first rib posterior to the subclavian artery (Figs. 6-22 and 6-23).
2. Each of the three trunks then divides into **anterior and posterior divisions** posterior to the clavicle (Fig. 6-22). These divisions are of fundamental significance because the anterior divisions supply anterior (flexor) parts and the posterior divisions supply posterior (extensor) parts of the upper limb.
3. *The three posterior divisions unite* to form the **posterior cord** (Figs. 6-22 to 6-24). The anterior divisions of the superior and middle trunks unite to form the **lateral cord,** and the anterior division of the inferior trunk continues as the **medial cord.**

   In Figure 6-23, observe that the cords of the plexus bear the relationship to the second part of the axillary artery that is indicated by their names (*e.g., the lateral cord is lateral to the axillary artery*).
4. **Each cord of the brachial plexus divides into two terminal branches** (Figs. 6-22 to 6-24).

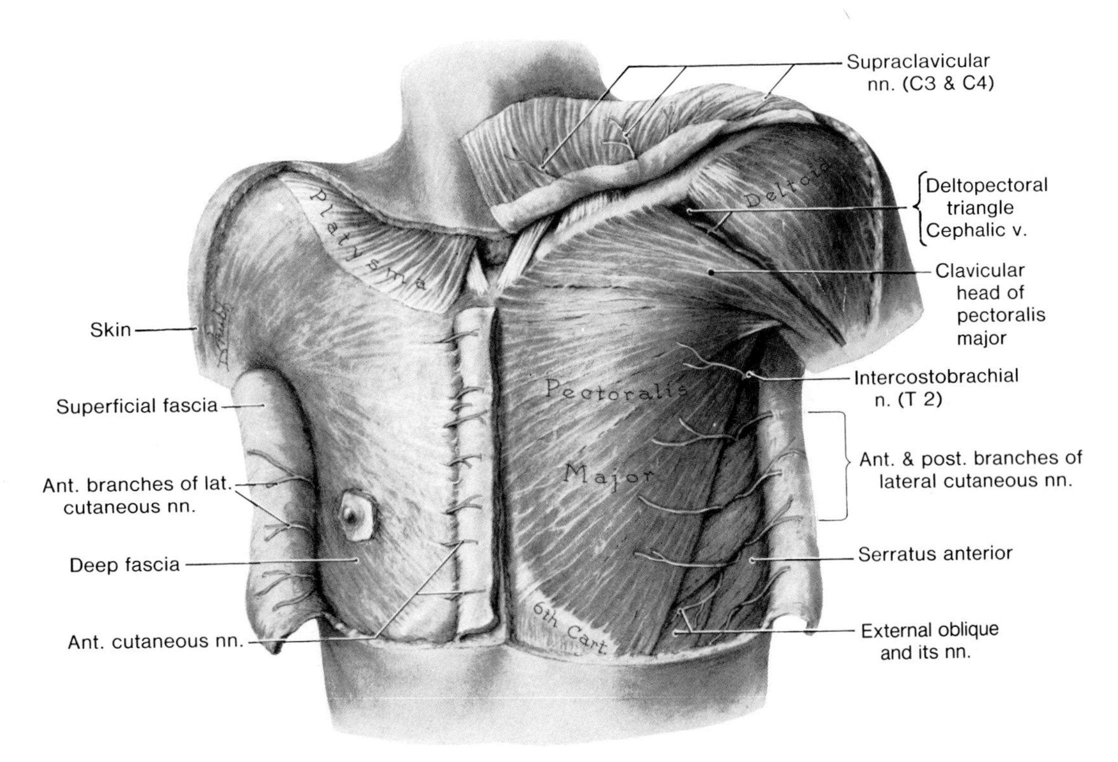

Figure 6-14. Drawing of a superficial dissection of the pectoral region. The platysma, a muscle of facial expression which descends to the second or third rib, is cut short on the left side and, together with the supraclavicular nerves, is turned superiorly to the right side. Note that the two heads of the pectoralis major meet at the sternoclavicular joint. Also observe the cephalic vein traversing the deltopectoral triangle. (The surface anatomy of the pectoral region is shown in Fig. 6-17).

The *lateral cord* divides into the **musculocutaneous nerve and the lateral root of the median nerve.**

The *medial cord* divides into the **ulnar nerve and the medial root of the median nerve.**

The *posterior cord* divides into the **axillary and radial nerves.**

Note that three nerves **(musculocutaneous, median, and ulnar) form the letter M** (Figs. 6-22 to 6-24). This serves as **the key to the brachial plexus** and aids in remembering the parts of the brachial plexus.

Variations of the Brachial Plexus. In addition to the five ventral rami that unite to form the brachial plexus, small contributions may come from the ventral rami of C4 or T2. In some persons trunk divisions or cord formation may be absent in one or other parts of the plexus. However the make-up of the terminal branches is unchanged. In addition the lateral or medial cords may receive fibers from ventral rami inferior or superior to the usual levels, respectively. In some people the median nerve has two medial roots instead of one.

The Branches of the Brachial Plexus. *These may be divided into supraclavicular branches and infraclavicular branches.* Only the infraclavicular branches are approachable through the axilla (Fig. 6-20).

The supraclavicular branches of the rami and trunks of the brachial plexus are as follows (Fig. 6-24 and Table 6-2).

1. **The dorsal scapular nerve** arises from the ventral ramus of **C5** and pierces the scalenus medius muscle (Fig. 6-21) to supply the rhomboid muscles and levator scapulae muscles (Fig. 6-44).
2. **The long thoracic nerve** arises from the ventral rami of C5 to C7 (Fig. 6-24), and passes through the apex of the axilla posterior to the other components of the brachial plexus to supply the serratus anterior muscle (Figs. 6-28 and 6-29). Sometimes the long thoracic nerve receives no contribution from the ventral ramus of C7.
3. **The nerve to the subclavius** arises from the ventral rami of C5 and C6 (Fig. 6-24) and descends anterior to the brachial plexus to supply the subclavius muscle (Fig. 6-21).
4. **The suprascapular nerve** arises from the trunk of the brachial plexus (Fig. 6-24), receiving

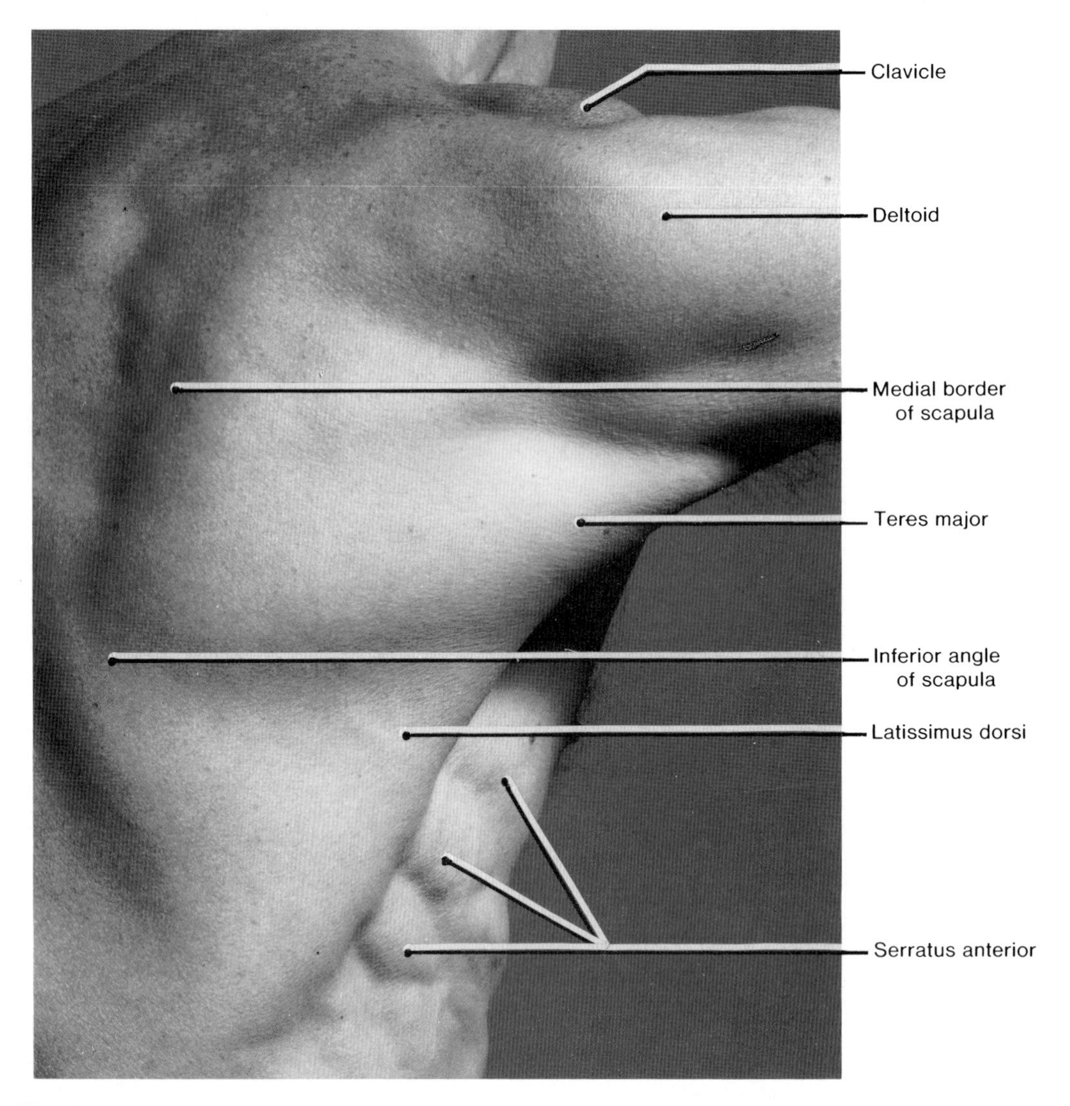

Figure 6-15. Photograph of the posterior aspect of the shoulder and axilla of a 46-year-old man. The teres major and the latissimus dorsi muscles form the posterior axillary fold (also see Fig. 6-12).

fibers from C5 and C6. It passes laterally across the posterior triangle of the neck, superior to the brachial plexus, and **passes through the suprascapular notch** in the scapula (Figs. 6-28 and 6-29). The suprascapular nerve supplies the supraspinatus and infraspinatus muscles and the shoulder joint (Figs. 6-41, 6-52, and 6-54).

The infraclavicular branches of the cords of the brachial plexus are as follows (Table 6-3).

The lateral cord of the brachial plexus has three branches (Fig. 6-24): the lateral pectoral nerve, the musculocutaneous nerve, and the lateral root of the median nerve.

1. **The lateral pectoral nerve** (C5 to C7) pierces the clavipectoral fascia to supply the **pectoralis major** muscle (Figs. 6-16 and 6-20). It sends a branch to the medial pectoral nerve which supplies the pectoralis minor muscle.

   This nerve is called the lateral pectoral nerve because it arises from the lateral cord of the brachial plexus. Remember this, otherwise you will be confused when you observe it running deep to the pectoralis major muscle, more medially than the medial pectoral nerve.
2. **The musculocutaneous nerve** (C5 to C7), supplying the muscles of the anterior aspect of the arm, is *one of the two terminal branches of the lateral cord of the brachial plexus* (Figs. 6-20, 6-22 to 6-25, and 6-29). The musculocutaneous nerve *enters the deep surface of the* ***coracobrachialis*** *muscle* and supplies it, and then continues in the arm to supply the **biceps brachii** and

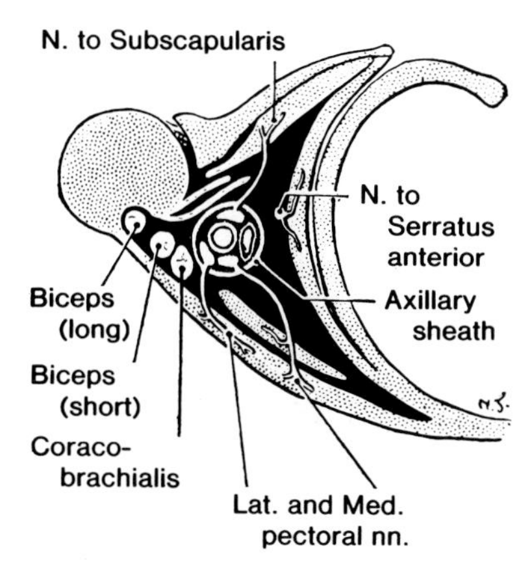

Figure 6-16. Drawing of a cross-section of the axilla showing some of its contents, *e.g.*, the principal vessels and nerves to the upper limb. These nerves arise from the brachial plexus which is located partly in the neck and partly in the axilla (Figs. 6-20 and 6-22). The lateral and medial pectoral nerves supply the pectoralis major and pectoralis minor muscles. Although not shown here, the lateral pectoral nerve sends a filament to the medial pectoral nerve and through it supplies the pectoralis minor muscle (Fig. 6-20).

brachialis muscles (Fig. 6-29). Just proximal to the elbow joint, the musculocutaneous nerve pierces the deep fascia and becomes superficial. From here it is called the **lateral antebrachial cutaneous nerve** (lateral cutaneous nerve of the forearm) and supplies skin on the lateral aspect of the forearm (Figs. 6-67, 6-92, and 6-93).

3. **The lateral root of the median nerve** (Figs. 6-20 and 6-22 to 6-25) is the continuation of the lateral cord; *i.e.*, it is the other terminal branch of the lateral cord of the brachial plexus. It is joined by the medial root of the median nerve, lateral to the axillary artery, to form the **median nerve** (Figs. 6-20 and 6-22 to 6-25). The median nerve passes inferiorly to supply primarily flexor muscles in the forearm, five muscles in the hand, and some skin on the hand (Figs. 6-25 and 6-26).

The medial cord of the brachial plexus has five branches (Fig. 6-24 and Table 6-3).

1. **The medial pectoral nerve** (C8 and T1) enters the deep surface of the **pectoralis minor** muscle (Figs. 6-16 and 6-20), supplying it and part of the **pectoralis major.**

   This nerve is called the medial pectoral nerve because it arises from the medial cord of the brachial plexus. In Figure 6-20, note that it lies lateral to the lateral pectoral nerve.

2. **The medial brachial cutaneous nerve** (medial cutaneous nerve of the arm), is a slender nerve that supplies skin over the medial surface of the arm and the proximal part of the forearm (Figs. 6-92 and 6-93). The medial brachial cutaneous nerve *usually communicates with the* ***intercostobrachial nerve***, which supplies the skin of the floor of the axilla and adjacent regions of the arm.
3. **The medial antebrachial cutaneous nerve** (medial cutaneous nerve of the forearm) runs between the axillary artery and vein to supply skin over the medial surface of the forearm (Fig. 6-92).
4. **The ulnar nerve** (C8 and T1) is a terminal branch of the medial cord of the brachial plexus (Figs. 6-20 and 6-22 to 6-25). The ulnar nerve passes through the arm into the forearm and hand, where it supplies one and one-half muscles in the forearm, most small muscles in the hand, and some skin (Figs. 6-25 and 6-26).
5. **The medial root of the median nerve** is the other terminal branch of the medial cord of the brachial plexus (Figs. 6-22 to 6-25). It joins with the lateral root to form the median nerve which supplies the flexor muscles in the forearm, except the flexor carpi ulnaris (Fig. 6-25), and the skin on the hand (Fig. 6-26).

The posterior cord of the brachial plexus has five branches (Figs. 6-22 to 6-27 and Table 6-3). In general, they supply muscles that extend the joints of the upper limb. These branches also supply cutaneous nerves to the extensor surface of the upper limb.

1. **The upper subscapular nerve** (C5 and C6) is a small nerve that supplies the ***subscapularis muscle*** (Figs. 6-27 and 6-28).
2. **The thoracodorsal nerve** (C6, C7, and C8) arises between the upper and lower subscapular nerves and runs inferolaterally to supply the ***latissimus dorsi muscle*** (Figs. 6-28 and 6-29).
3. **The lower subscapular nerve** (C5 and C6) passes inferolaterally, deep to the subscapular artery and vein, gives a branch to the ***subscapularis muscle***, and ends by supplying the ***teres major muscle*** (Figs. 6-28 and 6-29).
4. **The axillary nerve** (C5 and C6) is a large *terminal branch of the posterior cord of the brachical plexus* (Figs. 6-24 and 6-29). It passes to the posterior aspect of the arm through the *quadrangular space* (quadrilateral space) in company with the posterior circumflex humeral vessels (Figs. 6-28 and 6-29). The axillary nerve supplies articular branches to the shoulder joint.

   On emerging from the ***quadrangular space*** (Fig. 6-52), *the axillary nerve winds around the surgical neck of the humerus* (Fig. 6-12) *to supply the teres minor and deltoid muscles* (Fig. 6-52). The axillary nerve ends as the **upper lateral brachial cutaneous nerve** (Fig. 6-92) and supplies skin over the inferior half of the deltoid and adjacent areas of the arm (Fig. 6-37).

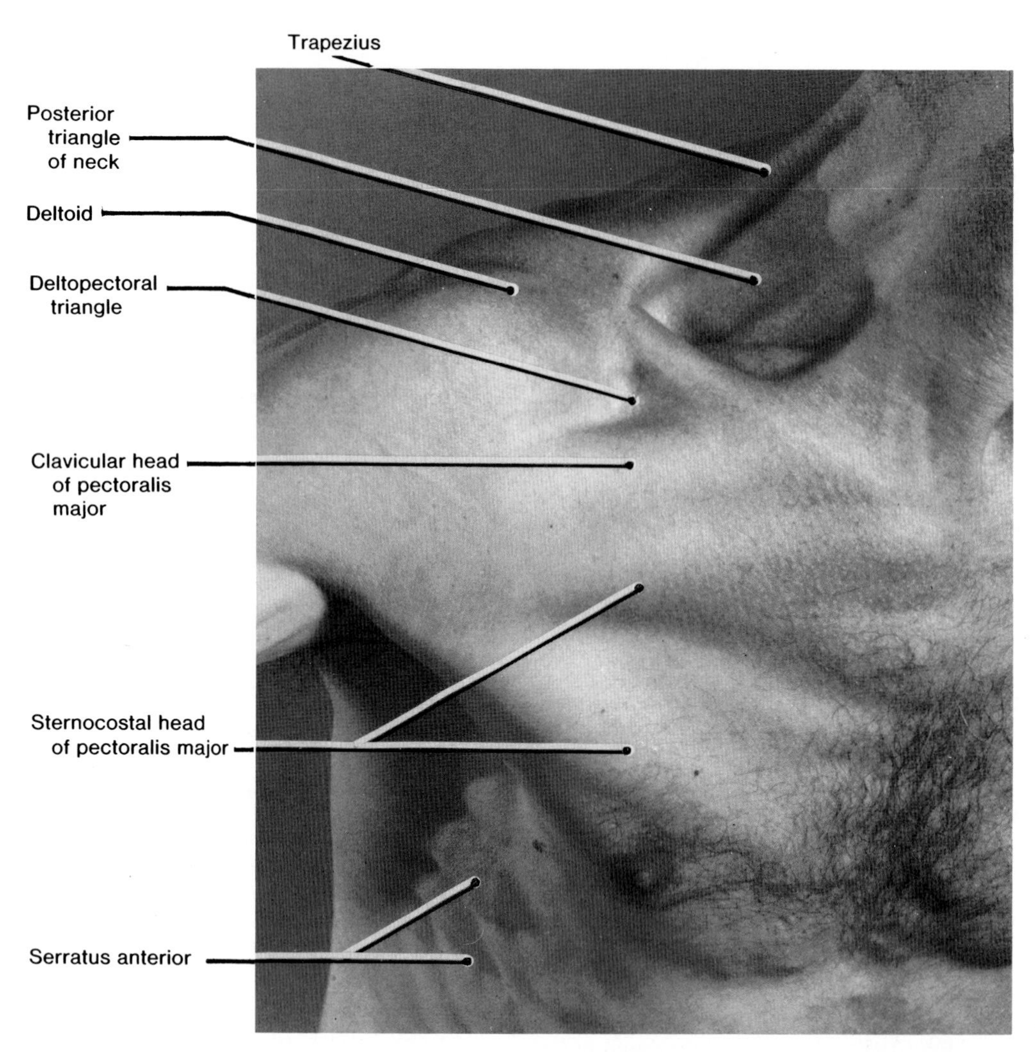

Figure 6-17. Photograph of the cervical, pectoral, and axillary regions of a 46-year-old man showing the principal surface markings. The brachial plexus (Fig. 6-21) is located in the posterior triangle of the neck. The depression in the deltopectoral triangle is called the infraclavicular fossa. See Figure 6-14 for a drawing of a superficial dissection of the pectoral region.

5. **The radial nerve** (C5 to C8 and T1) is the other terminal branch of the posterior cord (Figs. 6-22 to 6-25). *The radial nerve provides the* ***major nerve supply to the extensor muscles*** *of the upper limb* (Fig. 6-27). It also supplies cutaneous sensation to **the skin of the extensor region,** including the hand (Fig. 6-26).

   As it leaves the axilla the radial nerve runs posteriorly, inferiorly, and laterally *between the long and medial heads of the triceps muscle* to enter the **radial groove** of the humerus (Figs. 6-12 and 6-29). The radial nerve gives branches to the triceps, anconeus, brachioradialis muscles, and to the extensor muscles of the forearm (Figs. 6-27 to 6-29).

Injuries to the brachial plexus and its branches are of great importance. The brachial plexus can be injured by disease, stretching, and wounds in the neck or the axilla. The injury may involve: the dorsal and ventral roots of the spinal nerves; the ventral rami of C5 to T1 nerves; the trunks, divisions, cords of the plexus; or the branches of the cords. The signs and symptoms of an injury depend on which part of the brachial plexus is involved.

Injuries to the brachial plexus result in loss of muscular movement (**paralysis**), and often a loss of cutaneous sensation (**anesthesia**). The

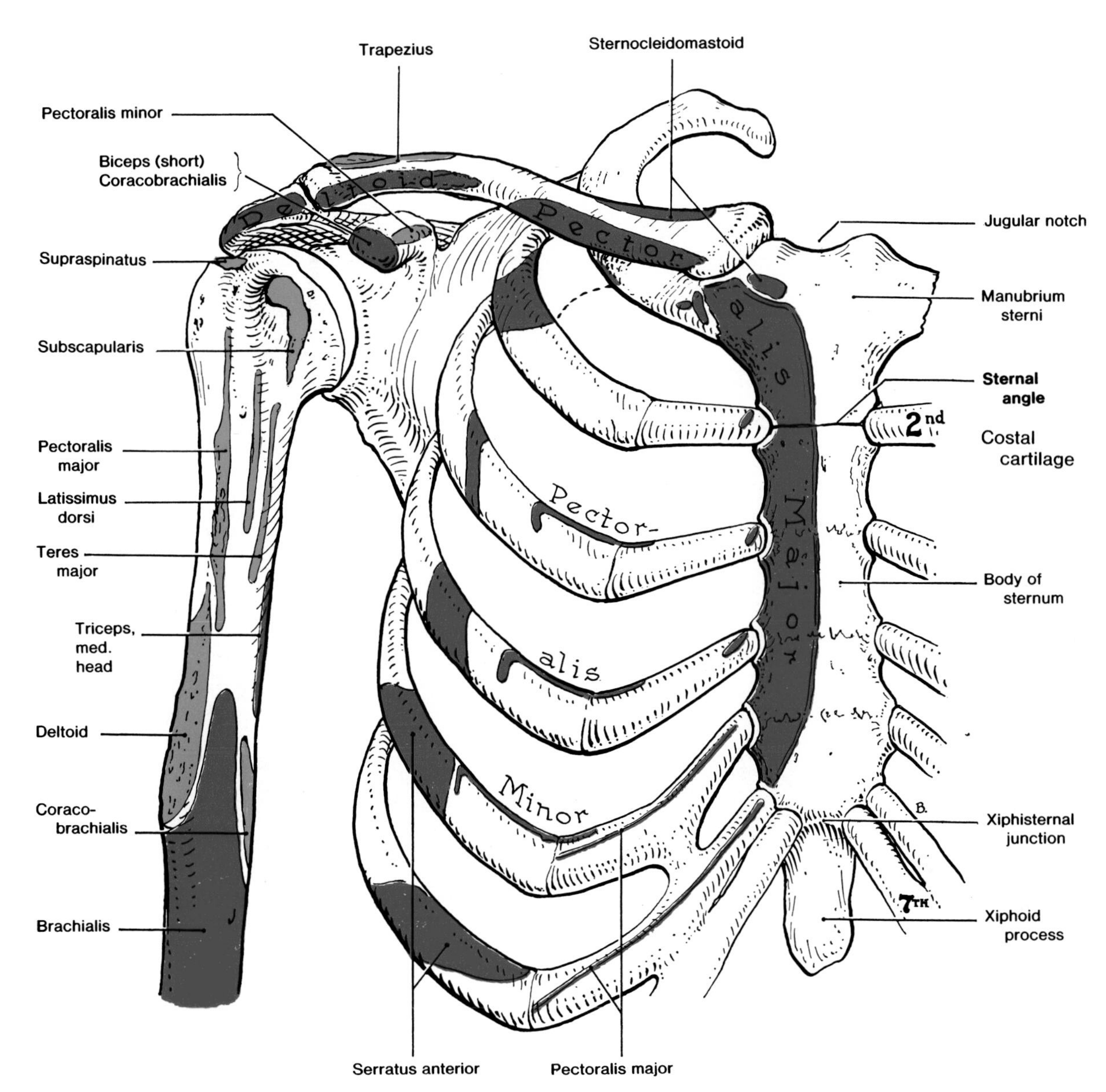

Figure 6-18. Drawing of the bones of the pectoral region and axilla showing the attachments of muscles. Observe that the pectoralis major has a crescentic origin from the medial half of the anterior surface of the clavicle, half of the anterior surface of the sternum, and the superior six costal cartilages. Note that the pectoralis minor arises from the third, fourth, and fifth ribs. *Observe the clinically important sternal angle.* It may be felt as a horizontal ridge at the junction of the manubrium and body of the sternum. The sternal angle is an important bony landmark for the identification of ribs (see p. 59). Origins of muscles are shown in *red*, insertions in *blue.*

degree of paralysis may be assessed by testing the patient's ability to perform movements. In **complete paralysis** no movement can be detected, whereas in **incomplete paralysis** movement can be performed but it is weak compared with that on the normal side. The explanation for this is that in incomplete paralysis not all muscles concerned with the movement are paralyzed. The degree of anesthesia (G. absence of feeling) may be tested by determining the ability of the person to feel pain (*e.g.*, a pin-prick).

Recovery from *minor nerve injuries* (*e.g.*,

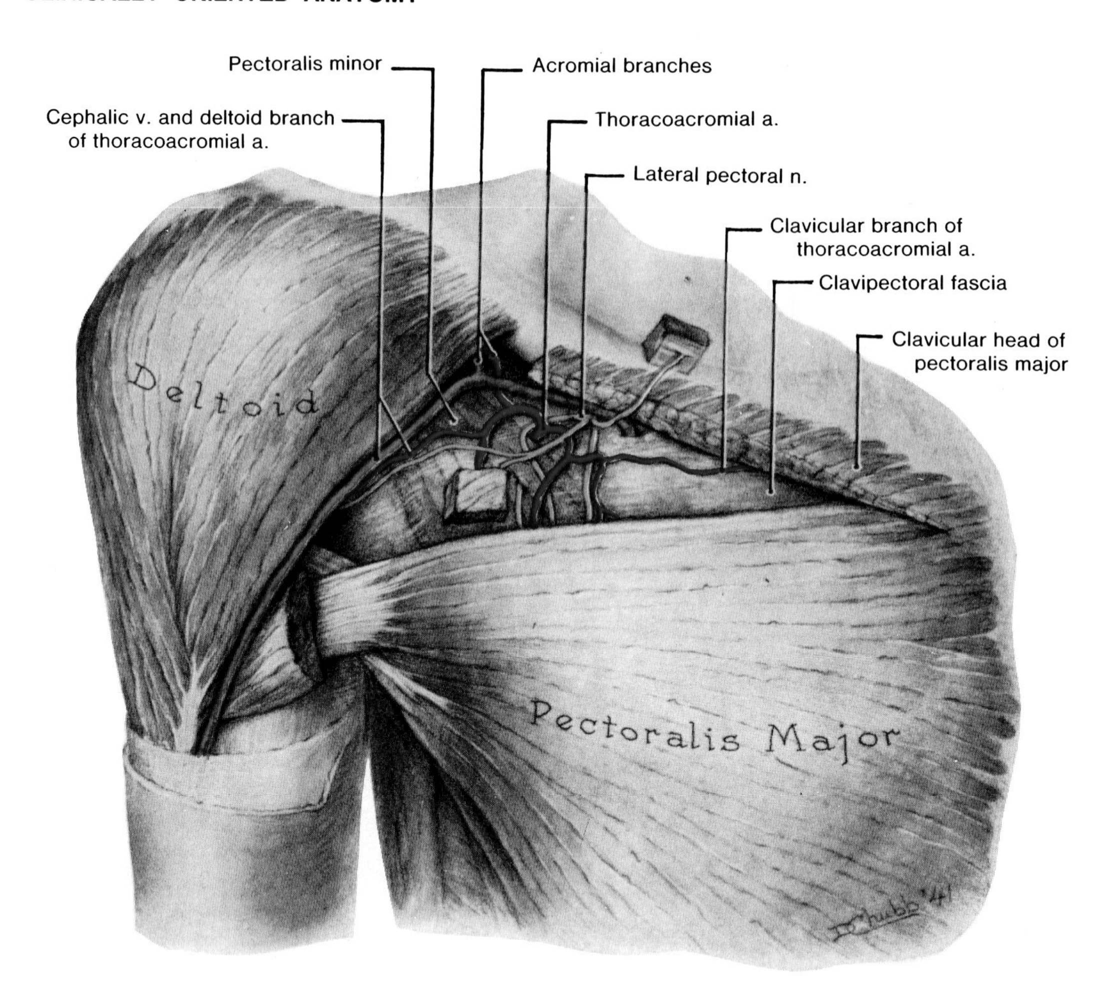

Figure 6-19. Drawing of a dissection of the pectoral region. Most of the clavicular head of the pectoralis major is excised, but cubes of it remain to identify its nerves. Observe the flattened tendon of the pectoralis major on the way to its insertion into the lateral lip of the intertubercular groove (sulcus) of the humerus (Figs. 6-1 and 6-18). Note the course of the cephalic vein as it enters the deltopectoral triangle (Fig. 6-14).

caused by pressure on a nerve) usually occurs in a few weeks, but when a nerve is cut or crushed, **anterograde degeneration** (Wallerian degeneration) occurs distal to the lesion involving the axons, their terminals and myelin sheaths. If the cut ends of the nerve are brought together and sutured soon after injury, partial function may be restored to the part within a few months.

UPPER BRACHIAL PLEXUS INJURIES (Figs. 6-30 to 6-32)

Injuries to superior parts of the brachial plexus usually result from excessive separation of the neck and shoulder. This may occur in football when one tackler is pulling a person's arm as another one hits or pulls the person's head (*e.g.*, pulling on the face mask).

Upper brachial plexus injuries can also occur when a person is thrown from a horse or motorcycle and lands on his/her shoulder in a way that widely separates the neck and shoulder (Fig. 6-30).

Most upper brachial plexus injuries result from motorcycle accidents. When thrown, the patient's shoulder often hits something (*e.g.*, a tree) and stops, but the head and trunk continue to move (Fig. 6-30). Similar damage to the brachial plexus can result from violent stretching an infant's neck during delivery (Fig. 6-31).

In upper brachial plexus injuries, the dorsal and ventral roots of the spinal nerves from C5 and C6 may be pulled out of the spinal cord (Fig. 6-22). In such cases there is paralysis of the

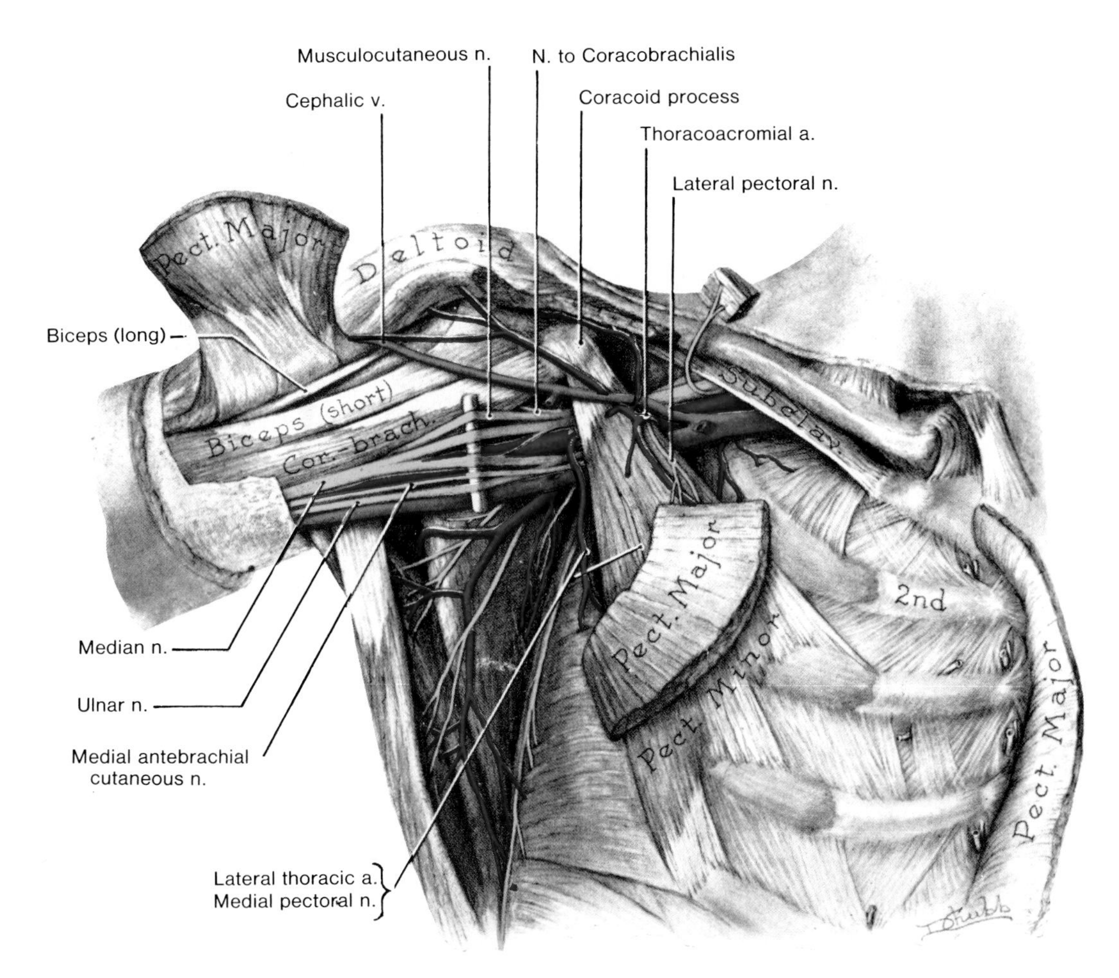

Figure 6-20. Drawing of a dissection of the pectoral region and axilla illustrating the anterior structures of the axilla. The pectoralis major is reflected and the clavipectoral fascia is removed. Observe the subclavius and pectoralis minor, the two deep muscles of the anterior wall (Fig. 6-11). Note the axillary artery lying posterior to the pectoralis minor, a fingerbreadth from the tip of the coracoid process of the scapula, with the lateral cord lateral to it and the medial cord medial to it. Observe that the axillary vein lies medial to the axillary artery. Note the median nerve, followed proximally, leading by its lateral root to the lateral cord and musculocutaneous nerve, and by its medial root to the medial cord and ulnar nerve. These nerves are raised on a stick for display.

scapular muscles (Table 6-3) and loss of sensation over the region of the back supplied by the dorsal primary rami, in addition to **paralysis of muscles and loss of sensation in the upper limb.**

If the lesion is confined to C5, usually no sensory changes can be detected because this nerve is not responsible for the exclusive supply of any area of skin. However when both C5 and C6 are involved, there is usually some loss of sensation on the lateral aspect of the upper limb.

In stab and bullet wounds of the neck, the superior trunk of the brachial plexus may be torn or severed where it emerges between the scalenus anterior and scalenus medius muscles (Fig. 6-21). These injuries result in loss of flexion, abduction, and lateral rotation of the shoulder joint, as well as loss of flexion of the elbow joint.

Injury to the superior trunk of the brachial plexus may be recognized by the characteristic position of the limb (Fig. 6-32). It hangs by the side in medial rotation, a position referred to as the **"waiter's tip" position** because it is the way modest waiters indicate their desire for a tip.

The following muscles that receive nerve fibers from C5 and C6 are most severely affected

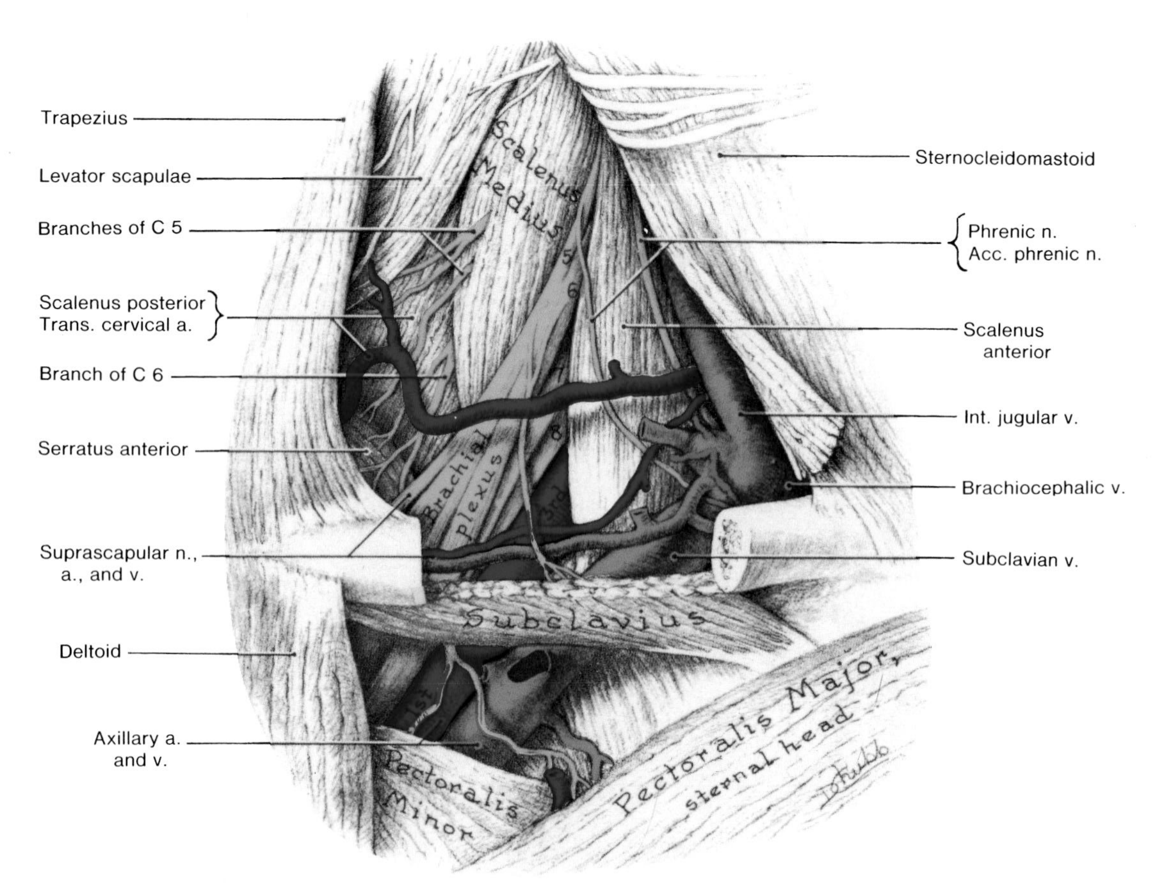

Figure 6-21. Drawing of the posterior triangle of the neck, pectoral region, and axilla, *right side.* Observe the third part of the subclavian artery and the first part of the axillary artery. Note the brachial plexus and subclavian artery appearing between the scalenus medius and scalenus anterior muscles. The most inferior ventral ramus of the plexus (T1) is concealed by the third part of the artery. The suprascapular nerve may be found by following the lateral border of the brachial plexus inferiorly.

when there is an upper brachial plexus injury: deltoid, biceps brachii, brachialis, brachioradialis, supraspinatus, infraspinatus, and teres minor (Fig. 6-25 and Table 6-3).

Poorly fitting crutches (*e.g.*, ones that are too long) **may cause injury to the posterior cord of the brachial plexus.** Often only the radial nerve is affected; as a result the triceps, anconeus, and extensors of the wrist are paralyzed (Fig. 6-27). The person is unable to extend the elbow, the wrist, or the digits. This type of paralysis, produces a **wrist-drop** and an inability to extend the wrist joint and digits (Fig. 6-156*A*).

LOWER BRACHIAL PLEXUS INJURIES (Figs. 6-33 and 6-34)

These injuries are not so common as upper brachial plexus injuries, but they may occur when the upper limb is suddenly pulled superiorly, *e.g.*, a forceful pull of the shoulder during birth (Fig. 6-33). It may also occur when people grasp something to break a fall (Fig. 6-34).

These accidents injure the inferior trunk of the brachial plexus (C8 and T1), often pulling the dorsal and ventral roots of the spinal nerves out of the spinal cord. The paralysis and anesthesia usually affect the muscles and skin supplied by the **ulnar nerve** (C8 and T1). Typically, the chief disabilities are in wrist and finger movements, *e.g.*, impairment of wrist flexion and movements of the intrinsic muscles of the hand (Fig. 6-25 and Tables 6-3 and 6-18 to 6-21). There is likely to be reduced sensation along the ulnar side of the arm, forearm, and hand (Figs. 6-26 and 6-92).

Cervical Rib Syndrome. A rib associated with the seventh cervical vertebra, called a **cer-**

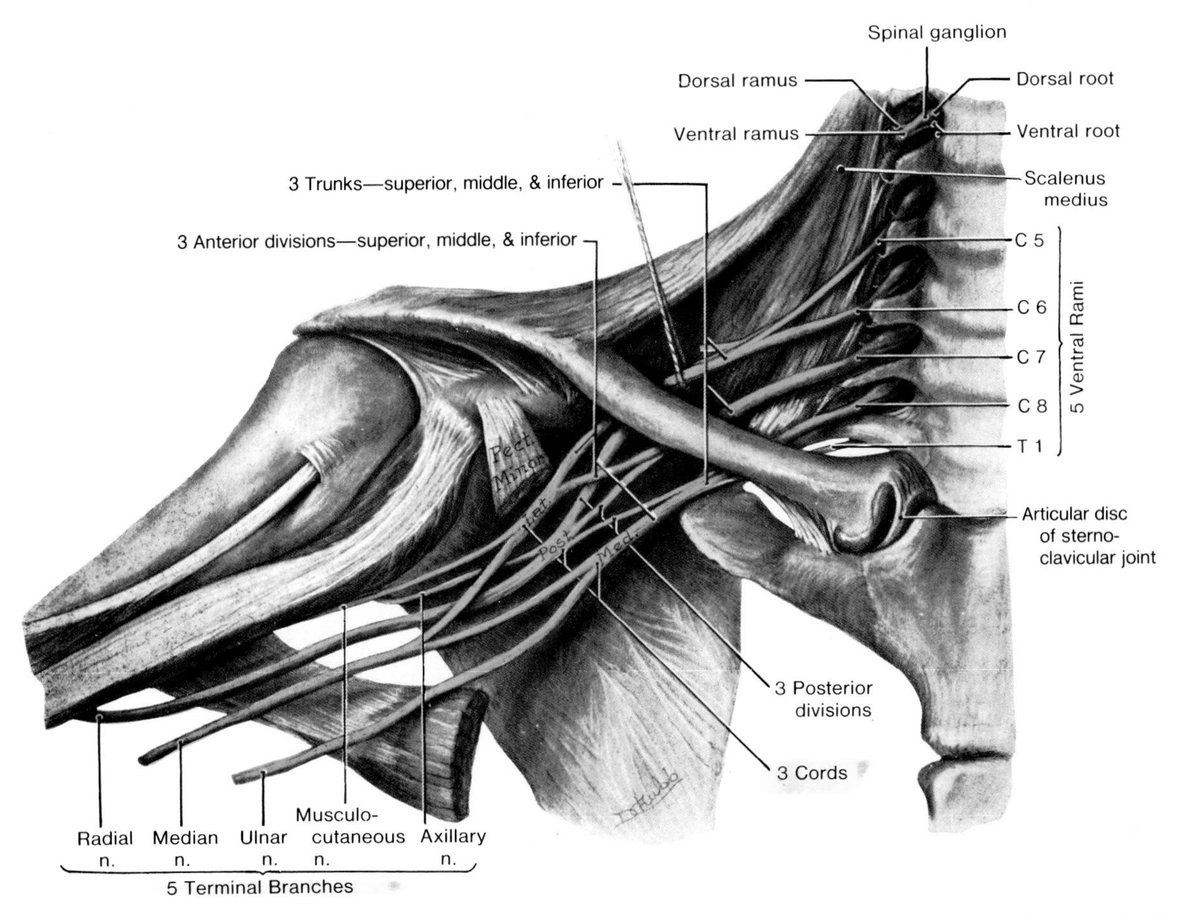

Figure 6-22. Drawing of the brachial plexus, showing that it extends from the neck to the axilla. Observe the five ventral rami of the 5th to 8th cervical nerves and most of the 1st thoracic nerve forming the brachial plexus. Often a small part of the ventral ramus of the 4th cervical nerve also joins the plexus. As there are only seven cervical vertebra, the T1 ramus arises inferior to the pedicle of the 1st thoracic vertebra. Usually five ventral rami unite to form the three trunks of the plexus, and each trunk divides into two divisions, anterior and posterior. From the divisions, observe that three cords lie posterior to the pectoralis minor muscle (also see Fig. 6-23). Understand that the large brachial plexus allows the mingling of nerve fibers from several segments of the spinal cord.

vical rib (Fig. 6-35), can exert pressure on the inferior trunk of the brachial plexus, producing symptoms of nerve compression. The accessory rib exerts pressure on the inferior trunk of the brachial plexus, particularly when the upper limb is pulled inferiorly (*e.g.*, when carrying a heavy suitcase).

A cervical rib probably affects the nerves by compressing the blood vessels supplying them and causing **anoxia of the axons** sufficient to interfere with their function. Sensory fibers are more readily affected by pressure than motor fibers (producing tingling and numbness). When the pressure is released (*e.g.*, when the suitcase is dropped), recovery of sensation and motor function often occurs in a few minutes.

The cervical rib syndrome has an embryological basis. Early in development each vertebra has two costal elements which form ribs in the thoracic region, but usually become parts of the vertebrae in other regions (Fig. 1-5). In up to 1% of persons, the costal elements of the seventh cervical vertebra form **cervical ribs** (Fig. 6-35). The cervical rib may be free anteriorly or attached to the first rib and/or the sternum.

Usually cervical ribs produce no symptoms, *but in some cases the subclavian artery and the inferior trunk of the brachial plexus are kinked* where they pass over the cervical rib. This pressure may produce the **neurovascular compression syndrome**, characterized by tin-

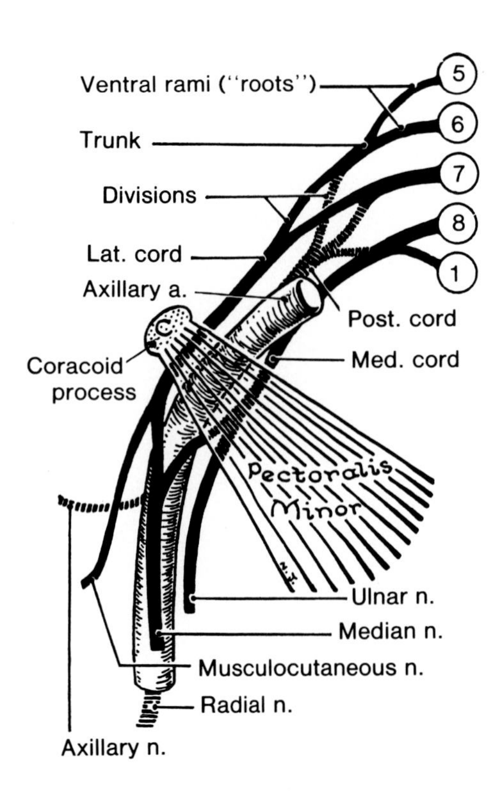

Figure 6-23. Diagram illustrating the usual plan of the brachial plexus. Note that the axillary artery is surrounded by the three cords of the plexus and that they bear the relation indicated by their names to the second part of the axillary artery, *i.e.*, the part posterior to the pectoralis minor. For descriptive purposes it is usual and useful to designate the parts of the brachial plexus as ventral rami, trunks, divisions, and cords (also see Fig. 6-22).

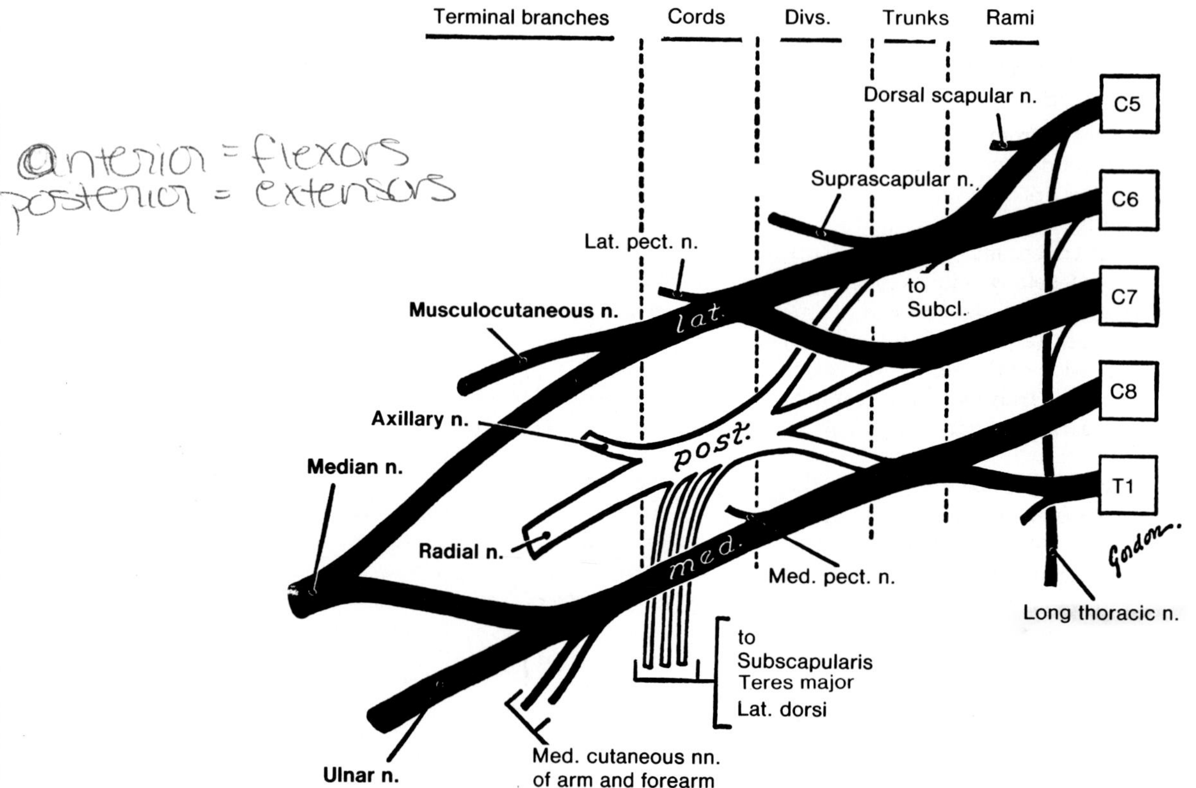

Figure 6-24. Diagram of an anterior view of the right brachial plexus of nerves. Note its parts: five ventral primary rami (C5 to C8 and T1); three trunks (superior, middle, and inferior); six divisions (three anterior and three posterior); three cords (medial, lateral, and posterior); and five terminal nerves (radial, axillary, ulnar, median, and musculocutaneous).

Table 6-2
Supraclavicular Branches of the Brachial Plexus to the Upper Limb

Origin	Nerves	Muscles	Segmental Innervation[1]
From the rami of the plexus	Dorsal scapular (Fig. 6-24)	Rhomboids Levator scapulae	**C4** and **C5** C5
	Long thoracic (Figs. 6-24, 6-28, and 6-29)	Serratus anterior	C5, **C6**, and **C7**
From the trunks of the plexus	Nerve to subclavius (Figs. 6-21 and 6-24)	Subclavius	**C5** and C6
	Suprascapular (Figs. 6-24, 6-28, and 6-29)	Supraspinatus Infraspinatus	**C4, C5**, and C6 **C5** and C6

[1] **Boldface** indicates the main segment innervation.

gling, numbness, and impaired circulation to the upper limb. The symptoms of a cervical rib may not develop until middle life and probably result from the fact that posture becomes more stooped as the tone of muscles of the pectoral girdle relaxes. This allows the shoulder to fall anteriorly and inferiorly, making it easier for the cervical rib to compress the subclavian artery and the inferior trunk of the brachial plexus.

INJURIES TO THE BRANCHES OF THE BRACHIAL PLEXUS

When a nerve arising from the brachial plexus is injured, the effects depend upon the level at which the injury occurs. For example, if the radial nerve is severed superior to the origin of the branches to the triceps muscle (Figs. 6-27 to 6-29), extension of the elbow joint is impossible (Table 6-16). If the radial is cut distal to the origin of these nerves (*e.g.*, at the midshaft of the humerus), extension of the elbow joint is not affected, but there is paralysis of the extensor muscles of the forearm and the digits (Fig. 6-27).

Injuries to the Long Thoracic Nerve (C5 to C7). This nerve lies on the medial wall of the axilla (Figs. 6-28 and 6-29) and may be injured by a stab wound, or during thoracic surgery (*e.g.*, for removal of a lung), or during the removal of cancerous axillary lymph nodes during a mastectomy. Carrying a heavy object on the shoulder (*e.g.*, a steel beam) may also compress this nerve between the clavicle and the lateral part of the first rib (Fig. 6-28).

Crushing or cutting the long thoracic nerve results in paralysis of the serratus anterior muscle and **winging of the scapula** (Fig. 6-36). The medial border and inferior angle of the scapula become unusually prominent. This "winging" of the scapula is accentuated when the person pushes against a wall with both hands. Instead of keeping the scapula applied to the chest wall, as is normal, the paralyzed serratus anterior muscle allows the scapula to move out like a wing. Difficulty is also experienced in flexing or abducting the arm above 45° from the side of the body because the serratus anterior muscle normally protracts and rotates the scapula (Table 6-1), so the glenoid cavity of the scapula faces superiorly when carrying out such a movement.

Injury to the Axillary Nerve (C5 and C6). This nerve passes through the quadrangular space (Figs. 6-28 and 6-29) and its anterior branch winds around the surgical neck of the humerus. It may be injured during fracture of this part of the humerus or during dislocation of the shoulder joint. Following *severance of the axillary nerve*, the deltoid muscle is paralyzed and undergoes **atrophy** (wasting). A loss of sensation (**anesthesia**) may also occur over the lateral side of the proximal part of the arm (Fig. 6-37), and the actions of the deltoid and teres minor muscles would be affected (Tables 6-5 and 6-15).

Injuries of other nerves arising from the brachial plexus are discussed with these nerves when they are described in the forearm and hand.

THE AXILLARY ARTERY (Figs. 6-20, 6-21, and 6-23)

This large vessel begins at the lateral border of the first rib ***as the continuation of the subclavian artery****. The axillary artery ends at the inferior border*

Table 6-3
Infraclavicular Branches of the Brachial Plexus to the Upper Limb

Origin	Nerves	Muscles Supplied	Segmental Innervation[1]
From the lateral cord of the plexus	Lateral pectoral (Figs. 6-16, 6-19, 6-20, and 6-24)	Pectoralis major Pectoralis minor	C5, **C6**, and C7 C5, **C7**, and C8
	Musculocutaneous (Figs. 6-22 to 6-25)	Coracobrachialis Biceps brachii Brachialis	C5, **C6**, and C7 C5 and **C6** C5 and **C6**
	Lateral root of median (Figs. 6-20 and 6-22 to 6-25)	Flexor muscles in forearm (except flexor carpi ulnaris) and 5 muscles in hand	(C5), C6, and C7
From the medial cord of the plexus	Medial pectoral (Figs. 6-16, 6-20, and 6-24)	Pectoralis major Pectoralis minor	**C8** and T1 C6, **C7**, and C8
	Medial brachial cutaneous (Fig. 6-92)		C8 and T1
	Ulnar (Figs. 6-20 and 6-22 to 6-25)	1½ muscles of forearm, most small muscles of hand (Fig. 6-25)	C8 and T1
	Medial root of median (Figs. 6-22 to 6-25)	Flexor muscles in forearm (except flexor carpi ulnaris) and 5 muscles in hand (Fig. 6-25)	C8 and T1
From the posterior cord of the plexus	Upper subscapular (Fig. 6-24 and 6-28)	Subscapularis (Figs. 6-27 and 6-28)	C5 and **C6**
	Thoracordorsal (Figs. 6-24 and 6-27 to 6-29)	Latissimus dorsi (Figs. 6-28 and 6-29)	**C6, C7**, and C8
	Lower subscapular (Figs. 6-24 and 6-27 to 6-29)	Subscapularis (Figs. 6-27 and 6-28) Teres major (Figs. 6-27 and 6-29)	C5 and **C6** **C6** and C7
	Axillary (Figs. 6-24, 6-28, and 6-29)	Teres minor (Figs. 6-27, 6-52, and 6-57) Deltoid (Figs. 6-19, 6-20, and 6-44)	**C5** and C6 **C5** and C6
	Radial (Figs. 6-22 to 6-24 and 6-28)	Triceps and anconeus Brachioradialis and extensor muscles of forearm (Figs. 6-27, 6-52, and 6-62)	C5, **C6, C7**, C8 and T1 C5, **C6**, and C7

[1] **Boldface** indicates the main segmental innervation.

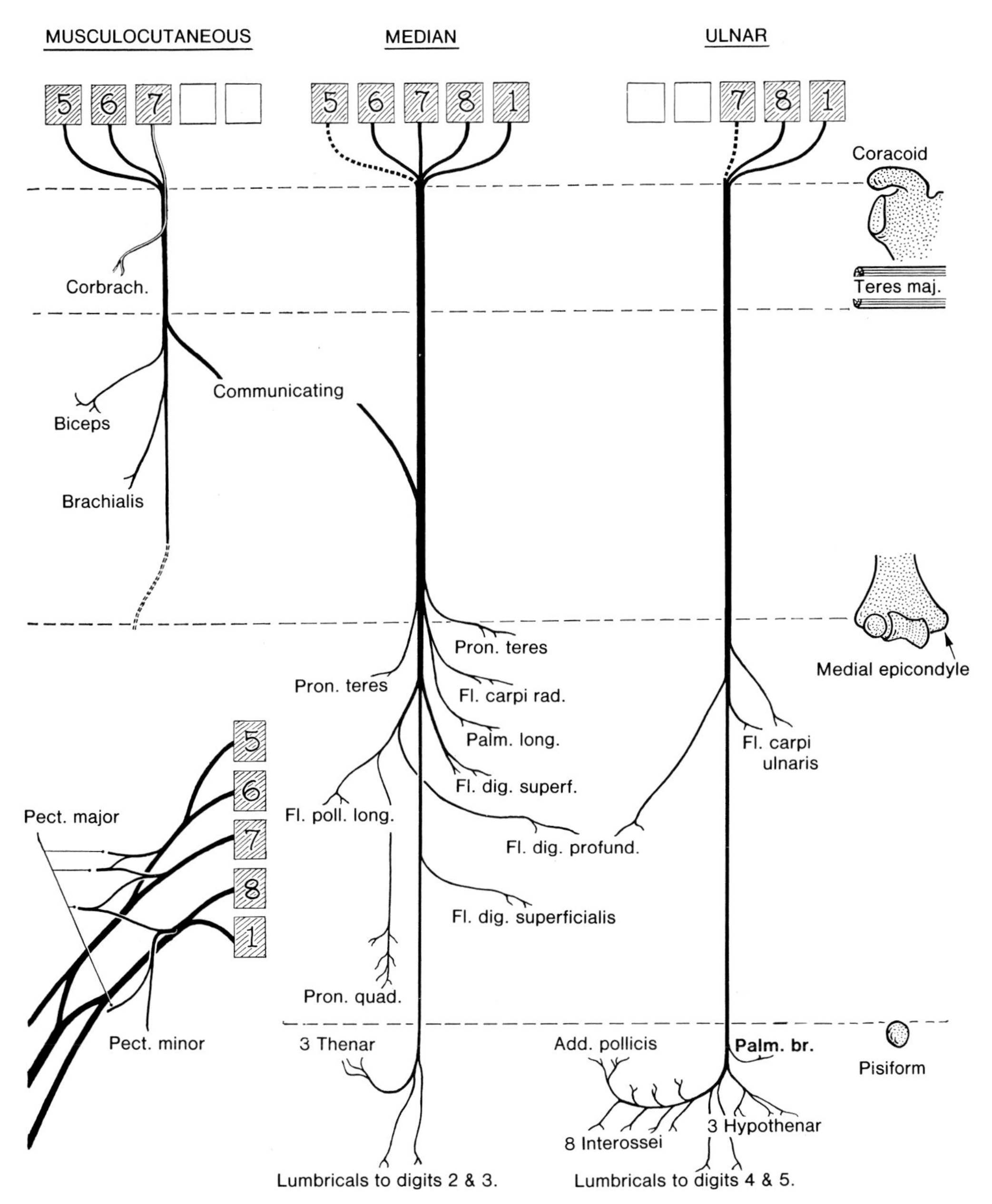

Figure 6-25. Diagrams illustrating the segmental origin and motor distribution of the ventral nerves of the upper limb. The average levels at which the motor branches leave the stems of the main nerves are shown with reference to the inferior border of the axilla (teres major), the elbow joint (medial epicondyle), and the wrist (pisiform bone).

of the teres major muscle. Here it passes into the arm where it **becomes the brachial artery**. During its course through the axilla, the axillary artery passes posterior to the pectoralis minor muscle (Figs. 6-23 and 6-38). For purposes of description, the axillary artery is divided into three parts by this muscle.

The First Part of the Axillary Artery (Figs. 6-23 and 6-38). This part is located between the lateral border of the first rib and the superior border of the pectoralis minor.

The first part of the axillary artery is enclosed in the axillary sheath, along with the axillary vein and the brachial plexus (Fig. 6-41). The first part of the axillary artery has only one branch, the **superior thoracic artery** or highest thoracic artery (Fig. 6-38). This artery helps to supply the first and second inter-

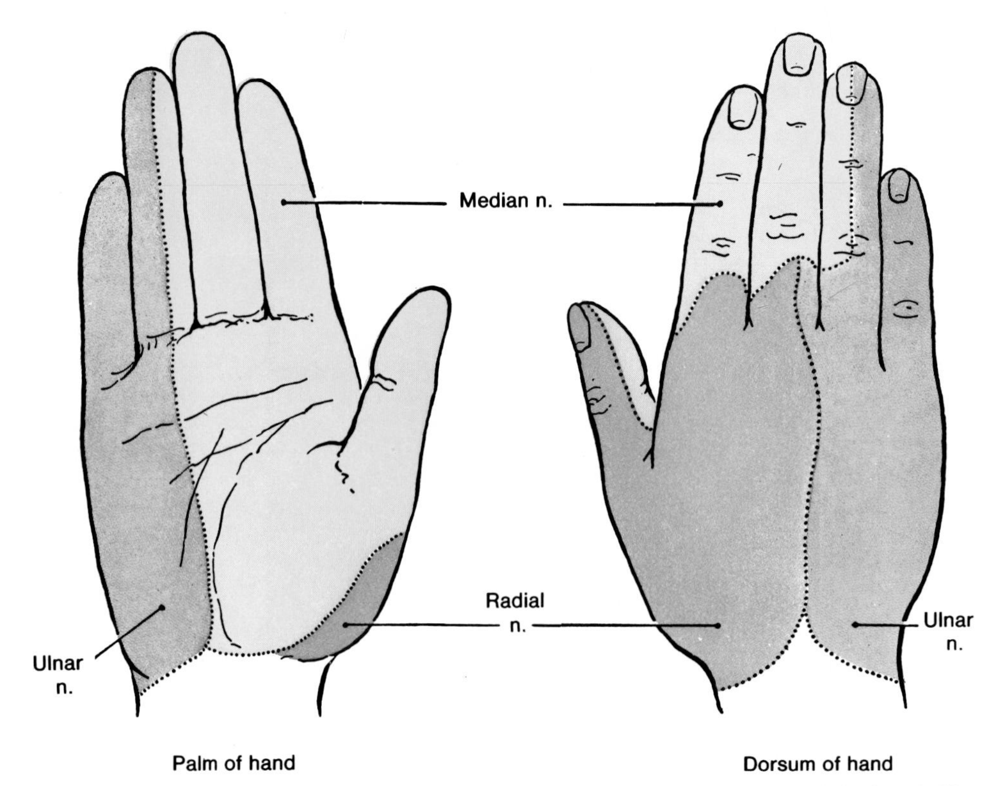

Figure 6-26. Diagrams illustrating the distribution of the cutaneous nerves of the right hand. Note that three major nerves, derived from the brachial plexus, are involved (also see Figs. 6-92 and 6-93).

costal spaces and the superior part of the serratus anterior muscle (Figs. 6-28 and 6-38).

The Second Part of the Axillary Artery (Figs. 6-23 and 6-38). This part lies deep to the pectoralis minor (Figs. 6-23 and 6-38). The lateral cord of the brachial plexus is lateral to the artery, the medial cord is medial to it, and the posterior cord is posterior to it (Fig. 6-23). *The second part of the axillary artery has two branches, the thoracoacromial and the lateral thoracic arteries* (Figs. 6-28 and 6-38).

The thoracoacromial artery, a short wide trunk, arises from the axillary artery, deep to the pectoralis minor and *pierces the costocoracoid membrane,* part of the clavipectoral fascia (Fig. 6-11). It then divides into four branches (acromial, deltoid, pectoral, and clavicular), deep to the clavicular head of the pectoralis major (Fig. 6-19).

The lateral thoracic artery descends along the axillary border of the pectoralis minor muscle (Figs. 6-20 and 6-38). It supplies the pectoral muscles and axillary lymph nodes. In the female, *the lateral thoracic artery is large and* is *an important source of blood to the lateral part of the mammary gland.*

The Third Part of the Axillary Artery (Figs. 6-23 and 6-38). This part extends from the inferior border of the pectoralis minor muscle to the inferior border of the teres major muscle. It has three branches (subscapular, anterior circumflex humeral, and posterior circumflex humeral).

The subscapular artery (Figs. 6-28 and 6-38), *the largest branch of the axillary artery,* descends along the lateral border of the subscapularis muscle and ends as the circumflex scapular and thoracodorsal arteries.

The circumflex scapular artery passes around the lateral border of the scapula to supply muscles on the dorsum of the scapula (Fig. 6-39).

The thoracodorsal artery continues the general course of the subscapular artery to supply adjacent muscles, principally the latissimus dorsi muscle.

The anterior and posterior circumflex humeral arteries pass around the surgical neck of the humerus and anastomose with each other (Fig. 6-38). The larger posterior circumflex humeral artery passes through the posterior wall of the axilla via the quadrangular space with the axillary nerve (Figs. 6-28 and 6-29) to supply the surrounding muscles (*e.g.,* the deltoid and triceps brachii).

The axillary artery can be palpated in the lateral wall of the inferior part of the axilla.

Compression of the axillary artery may be necessary (*e.g.,* in injuries of the axilla). This

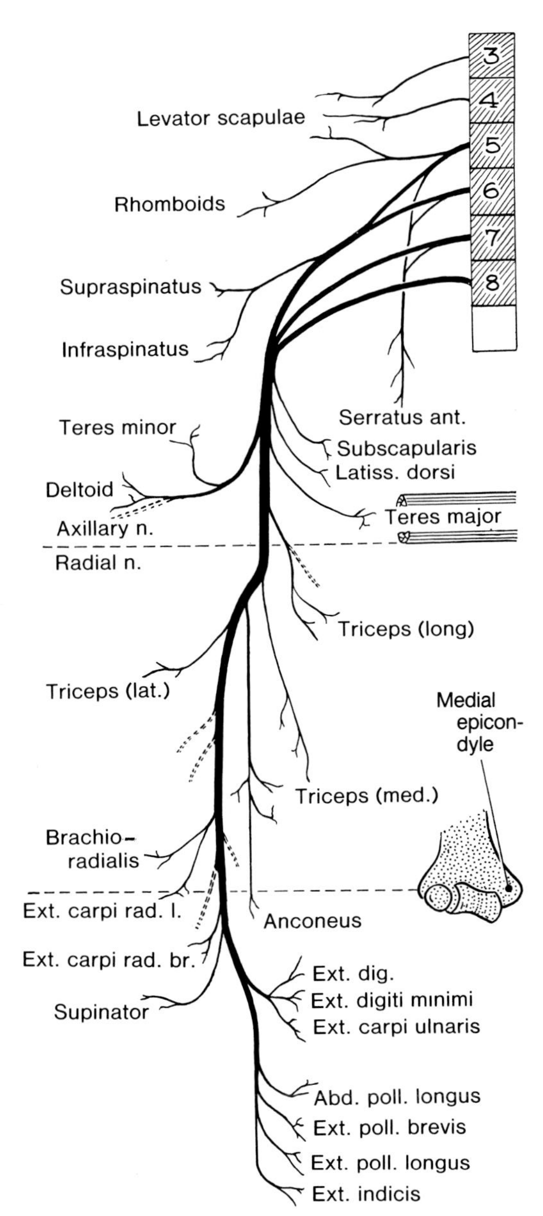

Figure 6-27. Diagram illustrating the motor distribution of the dorsal nerves of the upper limb. The average levels at which the motor branches leave the stems of the main nerves are shown with reference to the inferior border of the axilla (teres major) and the elbow joint (medial epicondyle). See Table 6-3 and Figure 6-24 for details about the nerves supplying these muscles.

can be done in the inferior part of its course by pressing the artery against the humerus (Fig. 6-38).

There is much variation in the branching pattern of the axillary artery in different people.

For this reason, it is important to realize that *the branches of the axillary artery are named according to their distribution rather than by their point of origin.*

There are extensive arterial anastomoses around the scapula (Fig. 6-39). Several vessels join to form networks on both surfaces of the scapula. The surgical importance of the **collateral circulation** that is possible becomes apparent during ligation of an injured axillary or subclavian artery. The axillary artery may be ligated between the thyrocervical trunk and the subscapular artery (Fig. 6-38). In this case, the direction of blood flow in the subscapular artery is reversed and blood reaches the distal portion of the axillary artery first. Note that the subscapular artery receives its blood via several anastomoses with the suprascapular artery, transverse cervical artery, and some intercostal arteries (Figs. 6-38 and 6-39).

Ligation of the axillary artery distal to the subscapular artery cuts off the blood supply to the arm (Fig. 6-38).

An axillary aneurysm produces a fluctuant and pulsatile swelling in the axilla. If the first part of the artery is dilated, the swelling projects anteriorly, inferior to the lateral part of the clavicle. When the aneurysm involves the third part of the artery, the anterior axillary fold is raised and the hollow of the axilla disappears, owing to the presence of the soft pulsating aneurysm.

Because of the thinness of the **axillary sheath** (Figs. 6-41 and 6-42), an aneurysm of the axillary artery commonly enlarges rapidly and compresses the nerves of the brachial plexus. This causes pain and subsequently anesthesia in the areas of the arm and forearm supplied by the nerves concerned. Weakness of the arm may result from pressure of the aneurysm on motor nerves, and edema of the forearm and hand may result from compression of the axillary vein.

THE AXILLARY VEIN (Figs. 6-40 to 6-42)

The large axillary vein lies on the medial side of the axillary artery. It completely overlaps the artery anteriorly when the arm is abducted.

The axillary vein begins at the inferior border of the teres major muscle as the continuation of the basilic vein (Fig. 6-40).

The axillary vein ends at the lateral border of the first rib, where it becomes the subclavian vein. The axillary vein receives tributaries that cor-

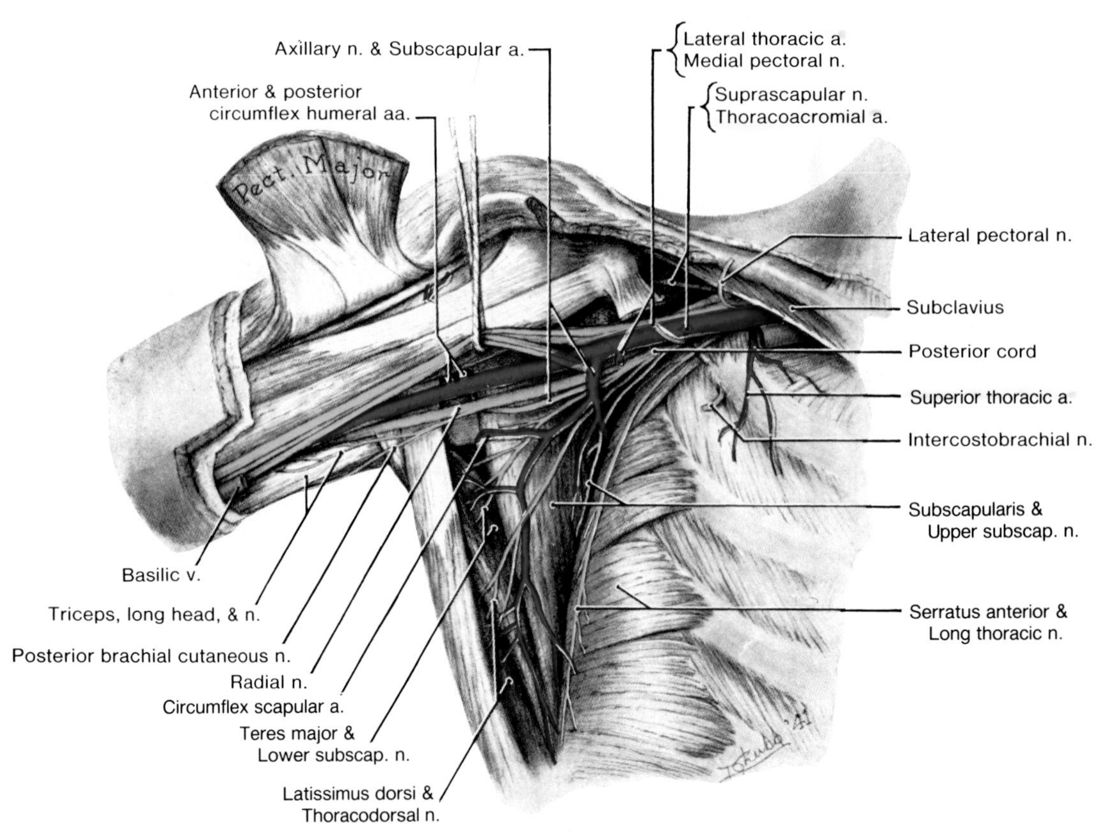

Figure 6-28. Drawing of a dissection of the posterior and medial walls of the axilla. The pectoralis minor is excised, the lateral and medial cords of the brachial plexus are retracted, and the axillary vein is removed. Observe the posterior cord and its two terminal branches (radial and axillary nerves) lying posterior to the axillary artery. Note the nerves to the three posterior muscles: (1) the thoracodorsal nerve (nerve to the latissimus dorsi) enters the deep surface of the muscle; (2) the upper subscapular nerve to the subscapularis lies parallel to but superior to it; and (3) the lower subscapular nerve to the subscapularis and teres major muscles lies parallel to the thoracodorsal nerve, but inferior to it. The nerve to the serratus anterior (long thoracic nerve) is clinging to its muscle throughout. Observe the suprascapular nerve passing toward the root of the coracoid process and the subscapular artery, the largest branch of the axillary artery. Observe the posterior circumflex humeral artery accompanying the axillary nerve through the quadrangular space.

respond to the branches of the axillary artery and at the inferior margin of the subscapularis muscle, the axillary vein **receives the venae comitantes of the brachial artery** (Fig. 6-40). These *accompanying veins* are paired vessels that are united by short branches to form a network around the artery. Superior to the pectoralis minor, *the axillary vein is joined by the cephalic vein* (Figs. 6-20 and 6-40).

Wounds in the axilla often involve the axillary vein owing to its large size and exposed position (Fig. 6-40). A wound in the superior part of the axillary vein, where it is largest, is particularly dangerous not only because of profuse hemorrhage, but also owing to the risk of air entering the vessel.

The walls of the axillary vein tend to be held apart by fibrous expansions from the clavipectoral fascia (Fig. 6-11). For this reason, the axillary vein is isolated and cleared in axillary operations to avoid injuring it during subsequent dissection (*e.g.*, during removal of axillary lymph nodes in a mastectomy related to breast cancer).

THE AXILLARY SHEATH (Figs. 6-41 and 6-42)

The axillary artery and vein and the cords of the brachial plexus are enveloped in a thin fascial sheath. Anterior to the subclavian artery the *prevertebral layer*

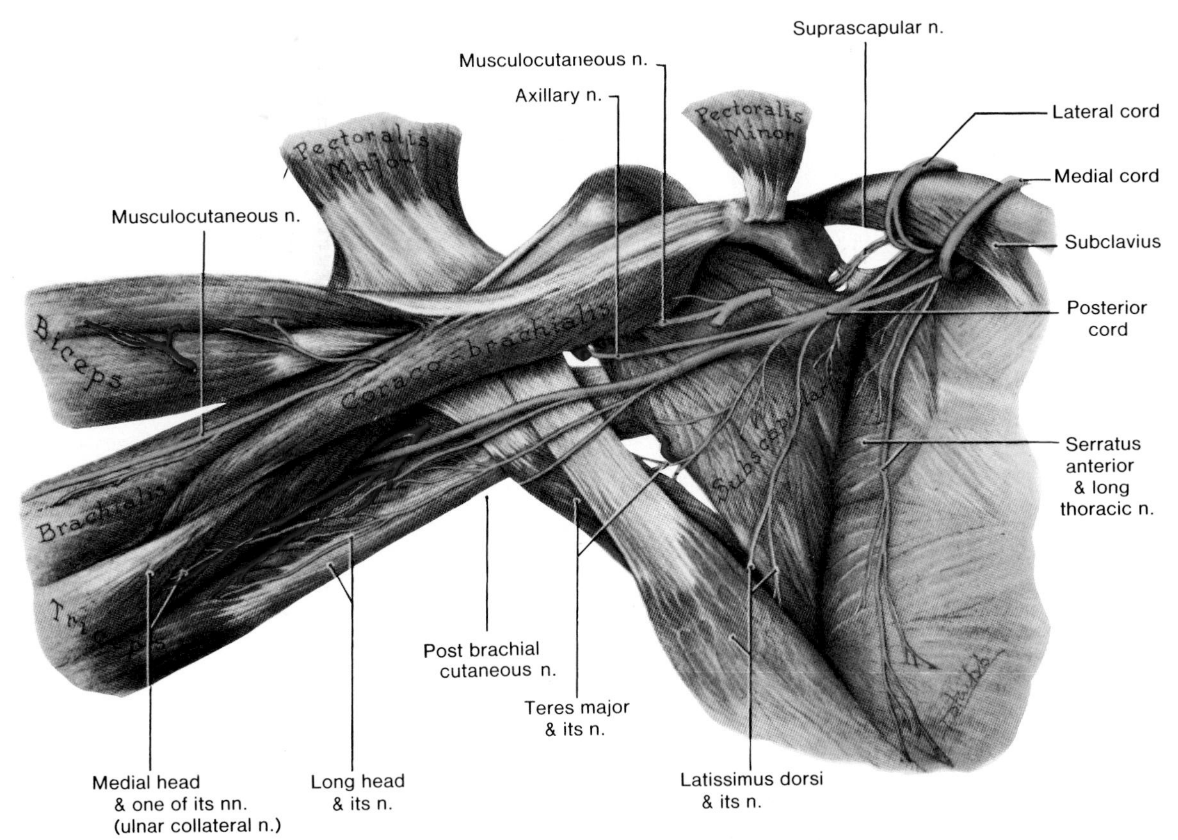

Figure 6-29. Drawing of a dissection of the posterior wall of the axilla demonstrating the posterior cord of the brachial plexus and its branches. The pectoralis major and pectoralis minor muscles are turned laterally; the lateral and medial cords of the brachial plexus are turned superiorly; and the arteries, veins, and median and ulnar nerves are removed. Observe the coracobrachialis muscle arising with the short head of the biceps from the tip of the coracoid process and inserting half way down the humerus. Note the musculocutaneous nerve piercing the coracobrachialis and supplying it, the biceps and brachialis before becoming cutaneous. Observe the posterior cord of the brachial plexus, formed by the union of the three posterior divisions, supplying the three muscles of the posterior wall of the axilla and ending as the radial and axillary nerves. Note that the radial nerve in the axilla, supplies the nerve to the long head of the triceps and also a cutaneous branch. In this specimen, note there is an ulnar collateral branch to the medial head of the triceps. It then enters the radial groove of the humerus with the profunda brachii artery. Note the axillary nerve passing through the quadrangular space with the posterior circumflex humeral artery.

Figure 6-30. Drawing illustrating how the neck and shoulder can be violently separated during a fall on the point of the shoulder formed by the acromion of the scapula. Excessive separation of the neck and shoulder produces an upper brachial plexus injury.

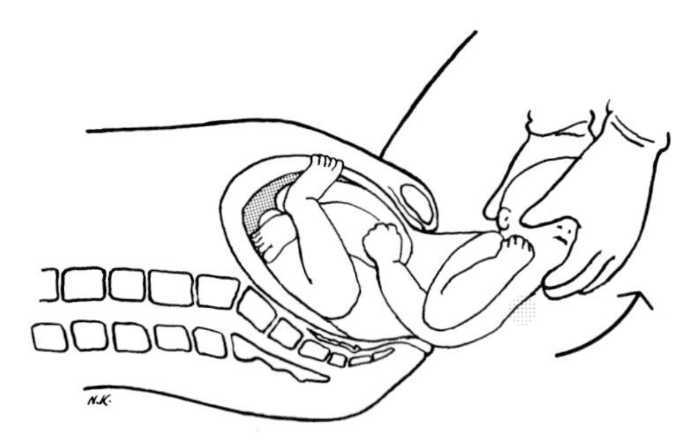

Figure 6-31. Drawing illustrating how the superior part of the brachial plexus may be injured by violent stretching of the neck of a baby during delivery. Excessive stretching of the neck can cause an upper brachial plexus injury.

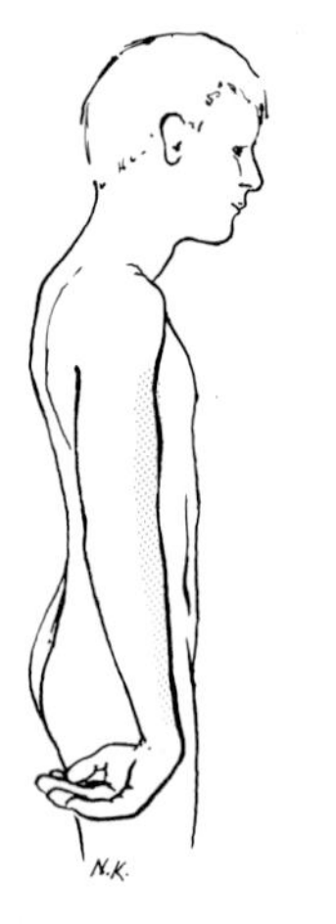

Figure 6-32. An illustration of the position of the upper limb of a young man with an upper brachial plexus injury. Note that his limb is hanging by his side in medial rotation, the characteristic waiter's tip position. The area of skin that is usually anesthetic is shown in *red*. Lesions of the upper brachial plexus (C5 and C6) are sometimes called Erb's or Duchenne-Erb palsy or paralysis.

of cervical fascia (fascia of the neck, Fig. 8-35) is prolonged laterally, where it forms the axillary sheath.

THE AXILLARY LYMPH NODES (Figs. 6-8 and 6-43)

There are 20 to 30 lymph nodes in the fibrofatty connective tissue of the axilla; *they are the main lymph nodes of the upper limb.*

The axillary lymph nodes are arranged in *five principal groups*, four of which lie inferior to the pectoralis minor tendon and one, the apical group, superior to it.

The pectoral group of axillary lymph nodes (Fig. 6-8) consists of three to five lymph nodes that lie along the medial wall of the axilla, around the lateral thoracic artery and the inferior border of the pectoralis major (Figs. 6-42 and 6-43).

The pectoral group of axillary lymph nodes receive lymph mainly from the anterior thoracic wall including the breast. The efferent lymph vessels from these nodes pass to the central and apical groups of axillary lymph nodes (Figs. 6-8 and 6-43).

The lateral group of axillary lymph nodes (Fig. 6-8) consists of four to six lymph nodes that lie along the lateral wall of the axilla, medial and posterior to the axillary vein.

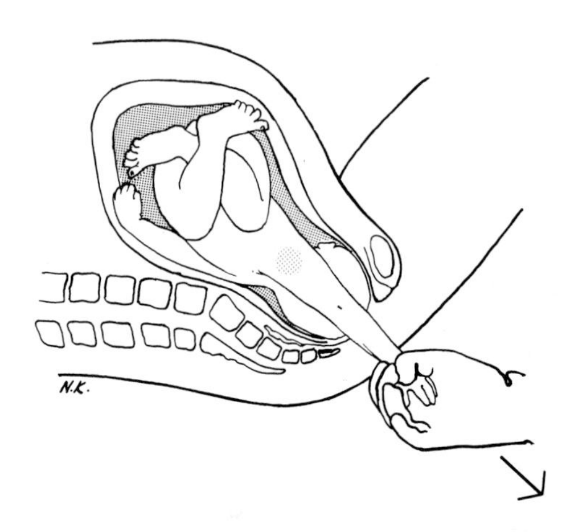

Figure 6-33. Drawing illustrating the way lower type brachial plexus injuries may occur as the result of a forceful pull of the upper limb during birth.

Figure 6-34. Drawing illustrating how a person may stretch and tear the inferior part of his brachial plexus by grasping for a limb during a fall. Lesions of the lower brachial plexus (C8 and T1) are sometimes called Klumpke's or Klumpke-Déjérine palsy or paralysis.

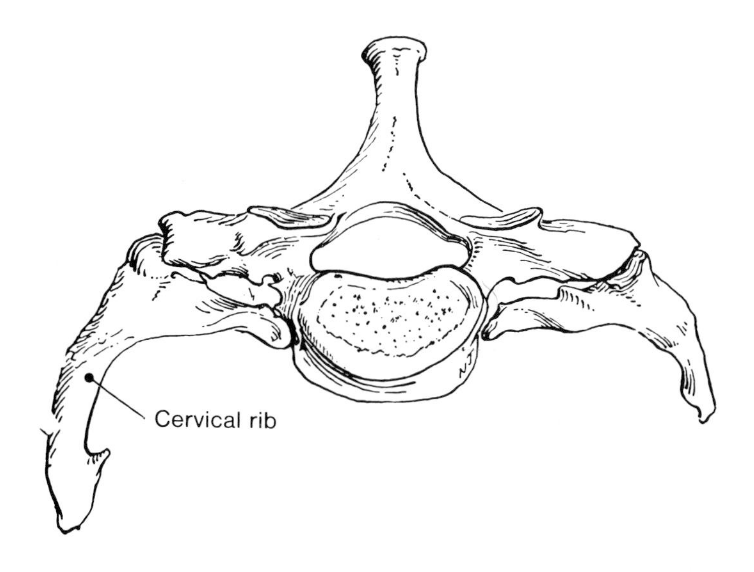

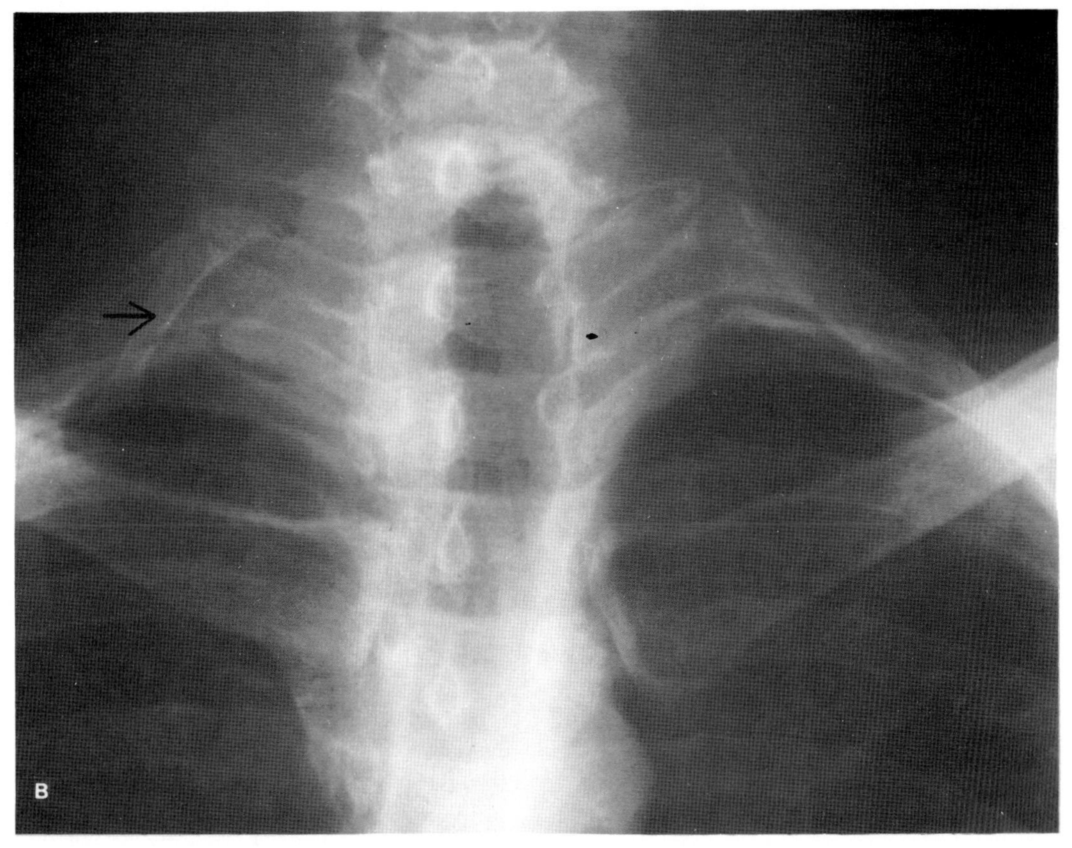

Figure 6-35. *A*, drawing of a seventh cervical vertebra associated with rudimentary cervical ribs. Instead of becoming parts of the vertebra, the costal elements developed as rudimentary ribs. The subclavian artery and inferior trunk of the brachial plexus pass over the rib. *B*. anteroposterior radiograph of the thorax and the cervical region of the vertebral column. Observe the long cervical rib (*arrow*) passing over the first rib. Usually a cervical rib produces no symptoms, but it may exert pressure on the inferior trunk of the brachial plexus.

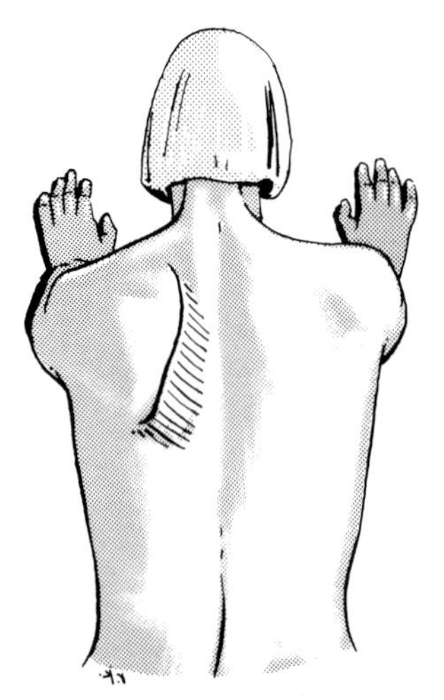

Figure 6-36. Drawing of the back of a woman showing winging of her left scapula resulting from a lesion of the long thoracic nerve. When the patient pushes against a wall with both hands, the medial border of her left scapula stands out like a small wing owing to paralysis of her left serratus anterior muscle. The long thoracic nerve supplying this muscle was damaged during the removal of cancerous lymph nodes from the medial wall of her axilla.

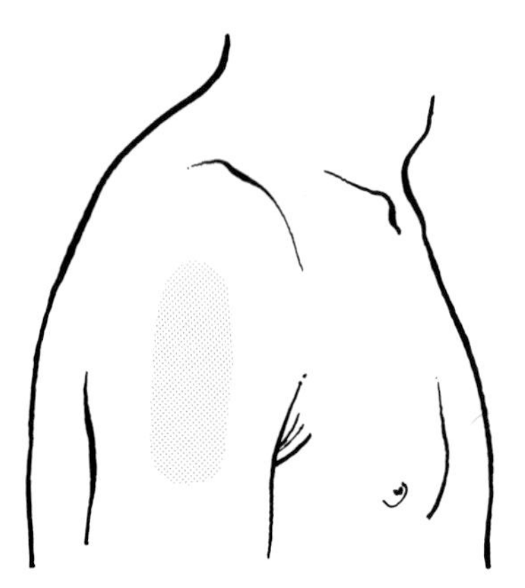

Figure 6-37. Sketch showing the approximate area of skin (*red*) that is usually anesthetic (loses sensation) in lesions of the axillary nerve. The axillary nerve is vulnerable to injury during dislocation of the shoulder and fractures of the surgical neck of the humerus, because it winds around the surgical neck of the humerous (Fig. 6-12).

The lateral group of axillary lymph nodes receive lymph from most of the upper limb; a few vessels accompany the cephalic vein and drain into the infraclavicular lymph nodes.

The lateral group of axillary lymph nodes is the first one to be involved in **lymphangitis** (*e.g.*, inflammation of the lymphatic vessels resulting from a hand infection). Infection can spread along lymphatic vessels to axillary nodes. Lymphangitis is characterized by the development of red, warm tender streaks in the skin. The axillary nodes often become enlarged and tender (Fig. 6-43).

The subscapular group of axillary lymph nodes (Fig. 6-8) consists of six or seven lymph nodes situated along the posterior axillary fold and the **subscapular blood vessels** (Fig. 6-28).

The subscapular group of axillary lymph nodes receive lymph from the posterior aspect of the thoracic wall and scapular region. Efferent vessels pass from them to the central group of axillary lymph nodes (Fig. 6-8)

The central group of axillary lymph nodes (Fig. 6-8) consists of three or four *large lymph nodes* situated near the base of the axilla, in association with the axillary artery (*i.e.*, in the fat deep to the pectoralis minor). As its name indicates, *the central group receives lymph from other groups of axillary lymph nodes* (pectoral, lateral, and subscapular). Efferent vessels from the central group pass to the apical group of axillary lymph nodes (Fig. 6-8).

The apical group of axillary lymph nodes consists of lymph nodes *situated in the apex of the axilla*, along the medial side of the axillary vein and the first part of the *axillary artery* (Figs. 6-8 and 6-43).

The apical group of axillary lymph nodes receive lymph from all other axillary lymph nodes. The efferent vessels from the apical group of axillary lymph nodes unite to form the **subclavian lymphatic trunk** which joins the jugular and bronchomediastinal trunks to form the **right lymphatic duct** (Fig. 41, p. 43). On the left side, the subclavian lymphatic trunk joins the **thoracic duct** (Fig. 1-42).

The axillary lymph nodes frequently become enlarged in **infections of the upper limb.** Infections of adjacent parts (*e.g.*, the pectoral region and the breast) and the superior part of the abdomen can also produce enlargement of the axillary nodes. The axillary lymph nodes are frequently enlarged in later stages of cancers of the skin of the upper limb.

In breast cancer the axillary lymph nodes may be involved in any stage of the disease. In carcinoma of the apical group of axillary lymph nodes, the lymph nodes often adhere to the axillary vein, thereby necessitating

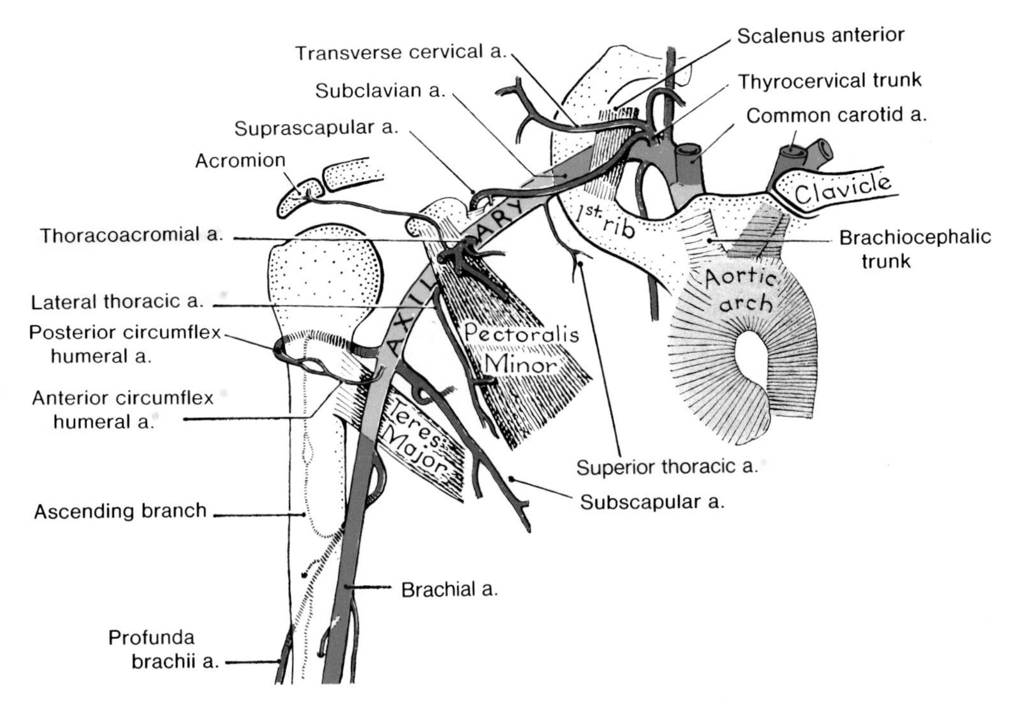

Figure 6-38. Diagram showing the named arteries of the axilla and arm. Note that the subclavian artery becomes the axillary artery at the lateral border of the first rib, and that the axillary artery becomes the brachial artery at the inferior border of the teres major muscle.

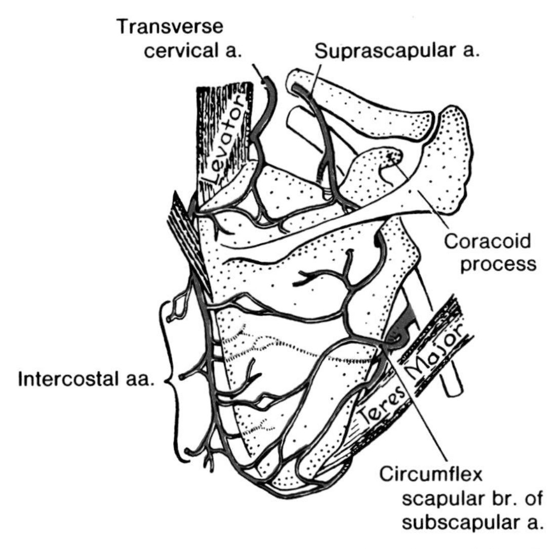

Figure 6-39. Drawing of a posterior view of the right scapular region showing the extensive arterial anastomoses around the scapula. Note that the various branches of the axillary artery take part in many anastomotic junctions with other branches of the axillary artery and with the branches of other arteries.

excision of part of this vessel. Enlargement of the apical group of lymph nodes sometimes obstructs the cephalic vein superior to the pectoralis minor muscle (Figs. 6-11 and 6-20).

THE BACK AND SHOULDER REGION

The back is described in Chapter 5, but the superficial and intermediate groups of muscles (extrinsic back muscles) that attach the upper limb to the axial skeleton are described more extensively in this chapter (Figs. 6-44, 6-45, and 6-47).

The muscles of the shoulder may be divided into three groups: (1) **superficial extrinsic muscles** (trapezius and latissimus dorsi); (2) **deep extrinsic muscles** (levator scapulae, rhomboids, and serratus anterior); and (3) **intrinsic muscles** (deltoid, supraspinatus, infraspinatus, teres minor, teres major, and subscapularis). The muscles in the intrinsic group arise and insert on the skeleton of the upper limb, running from the pectoral girdle to the humerus.

MUSCLES CONNECTING THE UPPER LIMB TO THE VERTEBRAL COLUMN

These muscles (trapezius, latissimus dorsi, levator scapulae, and rhomboids) are **extrinsic muscles of the back** (Table 6-4). *They are supplied by the ventral rami of cervical nerves*, not by the dorsal rami as one would expect. The explanation for this situation is that the superficial back muscles develop as a ventrolateral sheet which migrates posteriorly to gain attachment to the vertebral column.

THE TRAPEZIUS MUSCLE (Figs. 6-44, 6-45, 6-47, and Table 6-4)

This large, flat, triangular muscle covers the posterior aspect of the neck and superior half of the trunk.

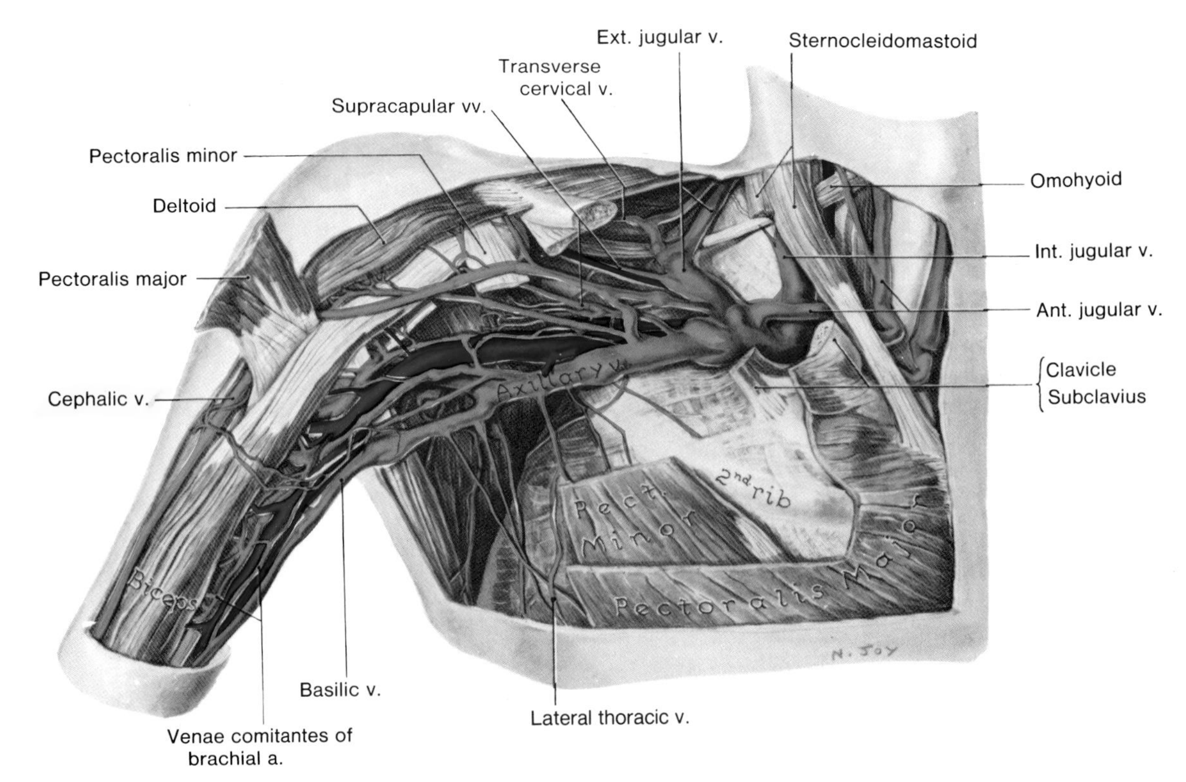

Figure 6-40. Drawing of a dissection of the axilla primarily to show its veins. Observe the basilic vein becoming the axillary vein at the inferior border of the teres major muscle; the axillary vein becoming the subclavian vein at the first rib; and the subclavian vein joining the internal jugular vein to become the brachiocephalic vein posterior to the sternal end of the clavicle. Many venous valves are visible (one in the basilic, three in the axillary, and one in the subclavian). Note the venae comitantes (L. accompanying veins) of the brachial artery uniting and joining the axillary vein. Observe the cephalic vein here bifurcating to end both in the axillary vein and the external jugular vein; usually it drains into the axillary vein.

It was given its name because the muscles of the two sides form a *trapezion* (G. an irregular four-sided figure). The trapezius muscle attaches the pectoral girdle to the skull and the vertebral column and assists in suspending it.

Origin (Figs. 6-44 and 6-47). **Medial third of superior nuchal line, external occipital protuberance, ligamentum nuchae, and spinous processes of C7 to T12 vertebrae.**

Insertion (Fig. 6-46). **Lateral third of clavicle** (superior fibers); **acromion and spine of scapula** (middle fibers); **base of scapular spine** (inferior fibers).

Nerve Supply (Figs. 6-44 and 7-64). **Spinal root of accessory nerve** (CN XI) **and ventral rami of third and fourth cervical nerves.**

Actions (Fig. 6-45). **Elevates, retracts, and rotates scapula** so that the glenoid cavity faces superiorly and anteriorly.

The superior fibers of the trapezius elevate the scapula (*e.g.*, when squaring the shoulders), *the middle fibers retract the scapula* (*i.e.*, pull it posteriorly toward the midline), and *the inferior fibers depress the scapula and lower the shoulder.* The superior and inferior fibers act together in the superior (upward) rotation of the scapula (Table 6-6).

As the trapezius muscles of the two sides brace the shoulders by pulling the scapulae posteriorly, *weakness of the trapezius muscles results in drooping shoulders.*

THE LATISSIMUS DORSI MUSCLE (Figs. 6-12, 6-15, 6-44, 6-45, 6-47, 6-48, and Table 6-4)

The Latin name of this muscle meaning "widest of the back," is a good one because this muscle covers the inferior half of the back (T6 vertebra to iliac crest). This wide, fan-shaped muscle, passing between the trunk and the humerus, acts on the shoulder joint and indirectly on the pectoral girdle.

Origin (Figs. 6-44 and 6-47). **Spinous processes of inferior six thoracic vertebrae, inferior thoracolumbar fascia, iliac crest, and inferior three or four ribs.** The superior part of the muscle usually receives fibers from the inferior angle of the scapula.

Insertion (Figs. 6-1 and 6-49). Floor of **intertubercular groove** (sulcus) of humerus. In Figures 6-12 and 6-29, observe its ribbon-like tendon and that

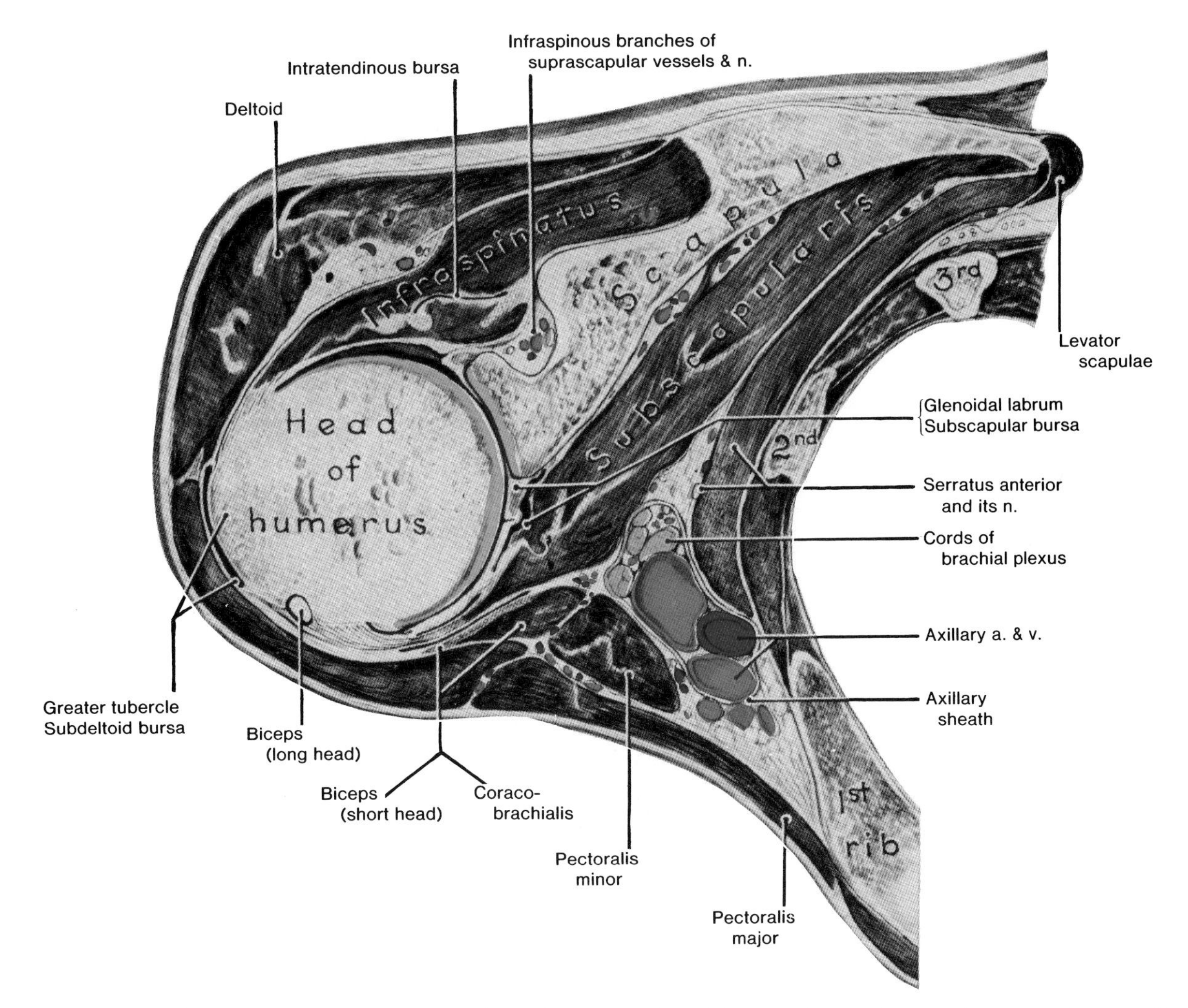

Figure 6-41. Drawing of a cross-section through the shoulder joint and the axilla near its apex. Observe the delicate axillary sheath, enclosing the axillary artery, vein, and cords of the brachial plexus to form a neurovascular bundle. Because of the thinness of the axillary sheath, an aneurysm (circumscribed dilation) of the axillary artery can easily exert pressure on the cords of the brachial plexus. This causes pain, anesthesia, and weakness of the arm which varies according to the nerves that are involved.

it runs in the posterior wall of the axilla before inserting into the humerus.

Nerve Supply (Fig. 6-28). **Thoracodorsal nerve (C6, C7,** and C8) from the posterior cord of the brachial plexus.

Actions. Extends, adducts, and medially rotates the humerus at the shoulder joint. These movements are used when chopping wood, climbing, paddling a canoe, and swimming (particularly during the crawl stroke).

In Figure 6-44, observe that the superior border of the latissimus dorsi and a part of the rhomboid major are overlapped by the trapezius. The triangle formed by the borders of these three muscles is called the **triangle of auscultation** (L. to listen). If the scapula is drawn anteriorly by folding the arms across the chest and the trunk is flexed, the ausculatory triangle enlarges and its lateral side is formed by the medial border of the scapula. Parts of the sixth and seventh ribs and the sixth intercostal space become subcutaneous; consequently, *respiratory sounds may be heard clearly with a stethoscope in the ausculatory triangle* (Fig. 6-45).

The latissmus dorsi and the inferior portion of the pectoralis major form an anteroposterior

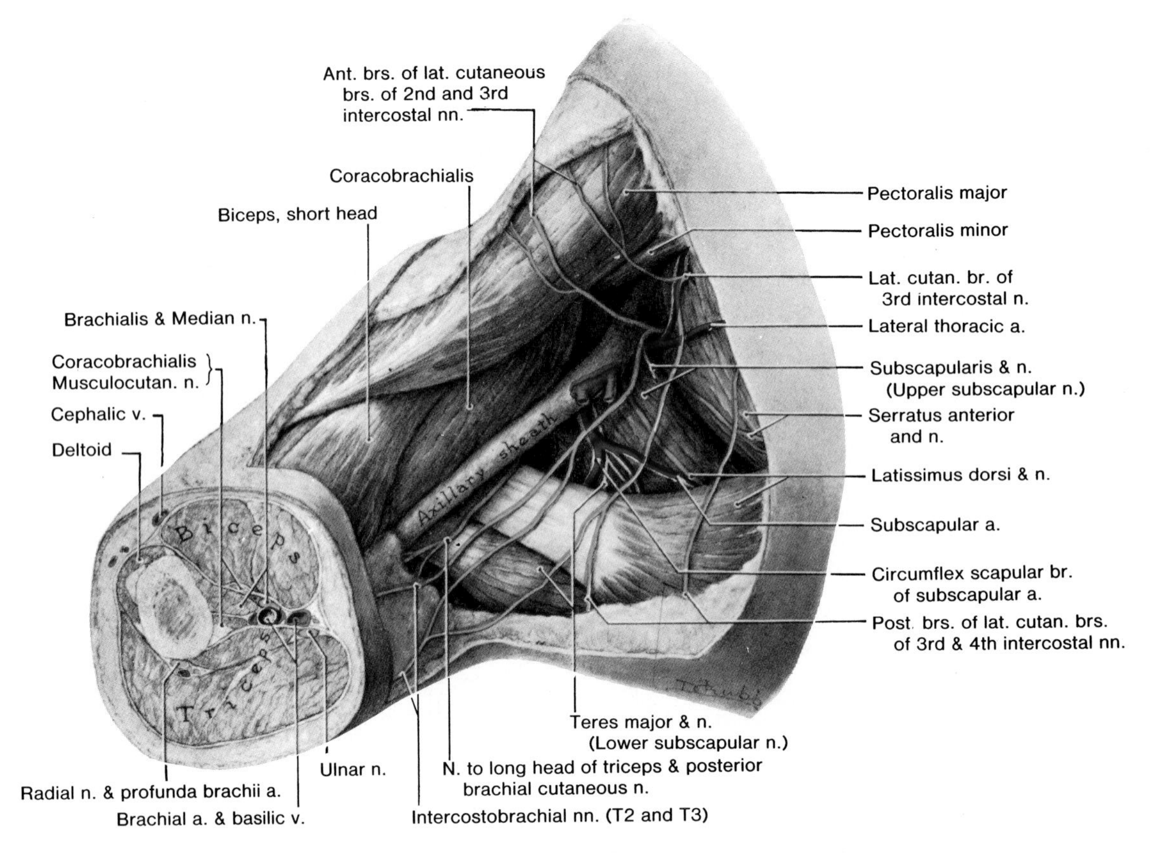

Figure 6-42. Drawing of a dissection of the axilla and a cross-section of the arm. Observe the axillary sheath and the cutaneous nerves crossing the latissimus dorsi muscle. Note that the axillary sheath forms a fascial tube for the axillary artery and vein. It also encloses the three cords of the brachial plexus (Fig. 6-41).

sling from the trunk to arm, but the latissimus dorsi forms the more powerful part of the sling. When the latissimus dorsi is paralyzed, the patient is unable to raise the trunk as occurs during "chinning" one's self. Furthermore, one cannot use crutches when the latissimus dorsi is paralyzed because the shoulder is pushed superiorly by the crutch.

THE LEVATOR SCAPULAE MUSCLE (Figs. 6-44, 6-47, and Table 6-4)

The superior third of this strap-like muscle lies deep to the sternocleidomastoid muscle; the inferior third is deep to the trapezius.

Origin (Fig. 6-47). **Transverse processes of first three or four cervical vertebrae.**

Insertion (Figs. 6-44 and 6-46). **Superior part of medial border of scapula.**

Nerve Supply (Figs. 6-24 and 6-27). **Dorsal scapular nerve** (C5) and **third and fourth cervical nerves.**

Actions. Elevates scapula and helps to tilt its glenoid cavity inferiorly by rotating the scapula. *It also helps to retract the scapula and fix it against the trunk.*

THE RHOMBOID MUSCLES (Figs. 6-44 to 6-47 and Table 6-4)

These two muscles lie deep to the trapezius and are not always distinct from each other. The **rhomboid major** is about two times wider than the **rhomboid minor.** These muscles appear as parallel bands that pass inferolaterally from the vertebrae to the scapula (Fig. 6-44). They have a rhomboid appearance, *i.e.*, they form an oblique parallelogram.

Origin (Figs. 6-44 and 6-47). *Minor:* **ligamentum nuchae and spinous processes of C7 and T1 vertebrae.** *Major:* **spinous processes of T2 to T5 vertebrae.**

Insertion (Fig. 6-46). **Medial border of scapula.** *Minor:* root of scapular spine. *Major:* inferior to scapular spine.

Nerve Supply (Fig. 6-24). **Dorsal scapular nerve** (C4 and **C5**).

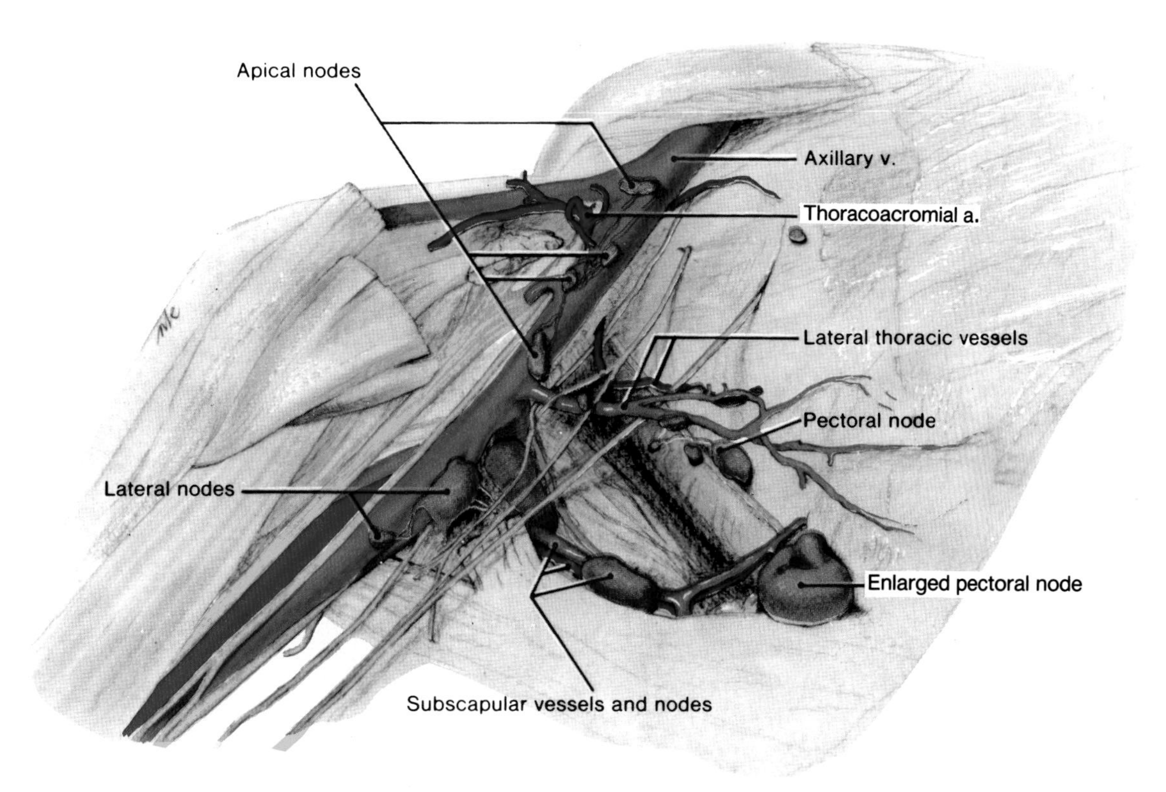

Figure 6-43. Drawing of a dissection of the axillary lymph nodes which receive over 75% of the lymph from the breast. While lymph drainage of the breast is usually into the pectoral axillary lymph nodes, direct drainage from the breast to any group of axillary nodes is possible. (Dissected by Dr. Ross Mackenzie, Associate Professor of Anatomy, University of Toronto).

Actions. Retract the scapula and rotate it to depress the glenoid cavity. They also help the serratus anterior to hold the scapula against the thoracic wall.

The rhomboid muscles are used when forcibly lowering the raised upper limbs, *e.g.,* when using a sledge hammer to drive a stake.

MUSCLES CONNECTING THE UPPER LIMB AND THE THORACIC WALL

This group includes the serratus anterior, pectoralis minor, pectoralis major, and subclavius. All these muscles, except the serratus anterior, have been previously described with the pectoral muscles (see Table 6-1).

THE SERRATUS ANTERIOR MUSCLE (Figs. 6-10, 6-15, 6-47, 6-48, and Table 6-1)

This large, foliate muscle overlies the lateral portion of the thorax and the intercostal muscles. It was given its name (L. *serratus,* a saw) because of the saw-toothed appearance of the fleshy digitations at its origin.

Origin (Figs. 6-47 and 6-48). **External surfaces of first eight ribs,** about midway between their angles and costal cartilages. The muscle runs posteriorly, closely applied to the thorax. Its inferior three digitations interdigitate with the origin of the external oblique muscle of the abdomen (Fig. 2-6).

Insertion (Fig. 6-49). **Entire anterior surface of medial border of scapula.**

Nerve Supply (Fig. 6-29). **Long thoracic nerve** (C5, **C6** and **C7**).

Actions (Tables 6-1 and 6-6). **Protracts the scapula and holds it against the thoracic wall.** Because it acts when punching, it is often called "*the boxer's muscle.*" By fixing the scapula to the thorax, it acts as an anchor for this bone and permits other muscles to use it as a fixed bone for producing movements of the humerus.

The inferior fibers of serratus anterior help to raise the glenoid cavity of scapula (*e.g.,* when the arm is raised above the head).

When the serratus anterior is paralyzed owing to **injury of the long thoracic nerve,** the medial border of the scapula stands out, espe-

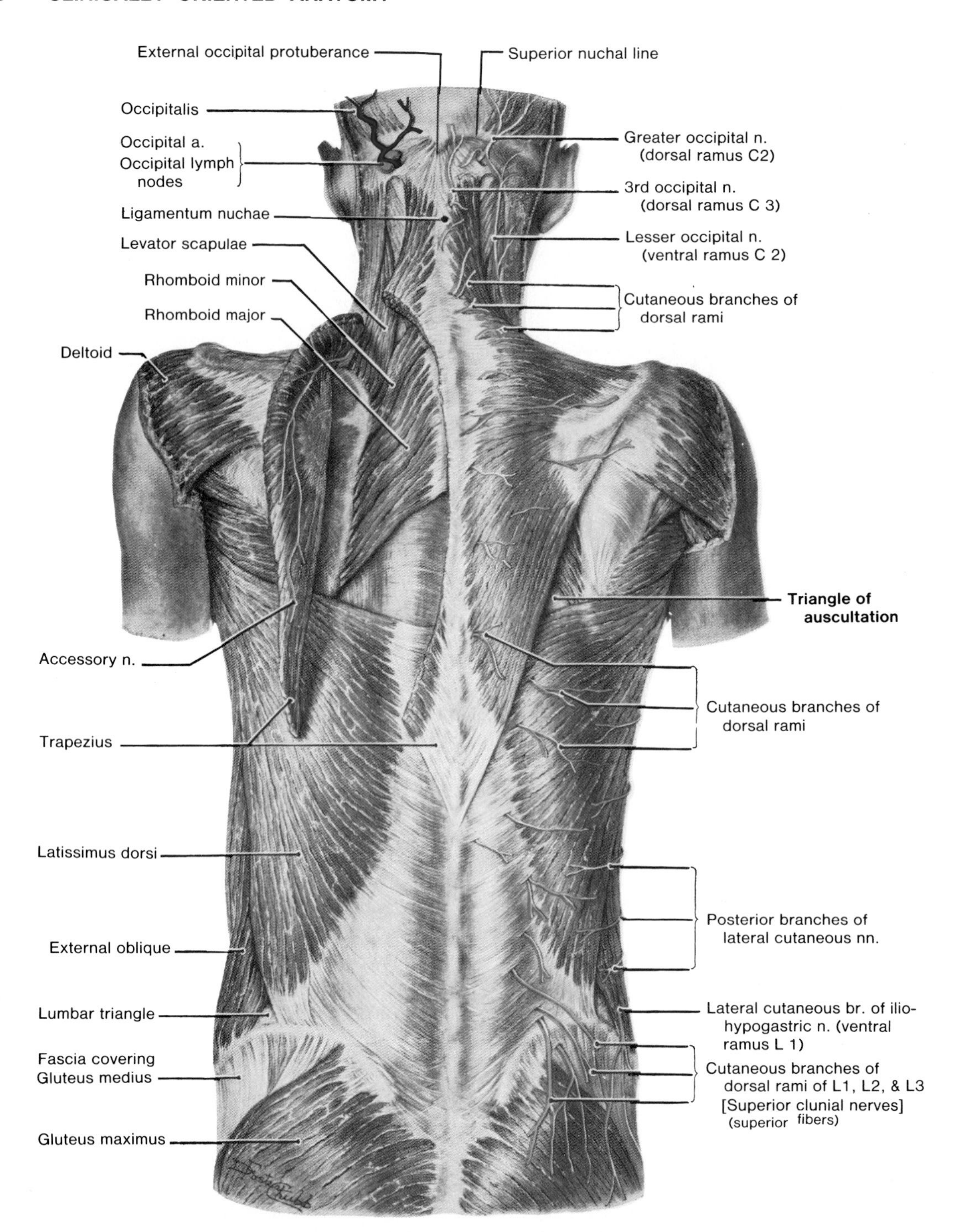

Figure 6-44. Drawing of a dissection of the back and shoulder regions, showing the first two layers of muscle and the cutaneous nerves. The trapezius muscle is severed and reflected on the left side. Observe the cutaneous branches of the dorsal rami of spinal nerves. Note the trapezius and latissimus dorsi of the first layer and the levator scapulae and rhomboids of the second layer of muscles. Observe the *triangle of auscultation* and the lumbar triangle. The ligamentum nuchae is a sheet of fibrous and elastic tissue interposed between the neck muscles of the two sides. It extends from the external occipital protuberance to the spinous processes of C1 to C7 vertebrae.

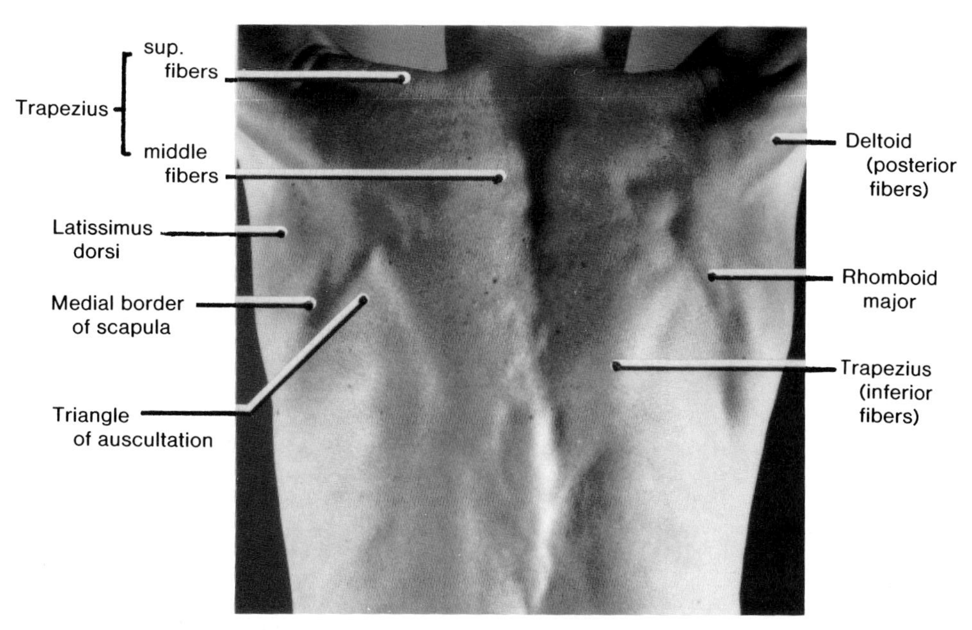

Figure 6-45. Photograph of the superior back region of a 46-year-old man with his arms elevated above his head. Observe the location of the medial border of the scapula. Because the superior fibers of the trapezius are attached to the lateral end of the spine of the scapula and the inferior fibers to its medial end, the scapula rotates when both fiber groups contract. As a result, the glenoid cavity faces superiorly and anteriorly. This rotary action of the scapula is essential for elevating the arms above the head. See Table 6-6 for a summary of the muscles producing movements of the scapula.

cially the inferior angle, giving the appearance of a wing when the person presses anteriorly, *e.g.*, against a wall (Fig. 6-36). This condition is called a "**winged scapula.**" When the arm is raised, the scapula is pulled away from the thoracic wall. In addition the arm cannot be abducted farther than the horizontal position because the serratus anterior is unable to rotate the scapula and raise the glenoid cavity. Consequently, a patient with a paralyzed serratus anterior muscle is unable to raise the arm fully or to push with the upper limb.

THE SCAPULAR MUSCLES

Six short muscles (deltoid, supraspinatus, infraspinatus, subscapularis, teres major and teres minor) pass from the scapula to the humerus and act on the shoulder joint (Tables 6-5 and 6-15).

THE DELTOID MUSCLE (Figs. 6-4, 6-14, 6-15, 6-17, 6-19, 6-44, 6-50, 6-51, and Tables 6-5 and 6-15)

This thick, powerful, triangular muscle **covers the shoulder joint** and forms the rounded contour of the shoulder (Figs. 6-4 and 6-50). As the name "*deltoid*" indicates, this muscle is triangular in outline, *i.e.*, shaped like the Greek letter *delta* (Δ) inverted.

Origin (Figs. 6-18, 6-19, and 6-46). Anteriorly from the **lateral third of the clavicle;** posteriorly from the **spine of the scapula** and, between these origins, from the **acromion** of the scapula.

Insertion (Figs. 6-18, 6-49, and 6-59). **Deltoid tuberosity** on anterolateral surface of humerus.

Nerve Supply (Figs. 6-22, 6-27, 6-29, and 6-52). **Axillary nerve (C5** and C6) from posterior cord of brachial plexus.

Actions. The deltoid may be divided into three parts: anterior, middle, and posterior (Fig. 6-50). *The muscle is capable of acting in part or as a whole.*

The actions of its parts are as follows: (1) the *anterior part* is a strong **flexor and medial rotator of humerus**; (2) the *middle part* is the **chief abductor of humerus**; and (3) the *posterior part* is a strong **extensor and lateral rotator of humerus.** In performing the movements, the deltoid works with other muscles, *e.g.*, the anterior part acts with the pectoralis major and coracobrachialis muscles in flexing the arm, whereas the middle part acts with the supraspinatus in abducting the arm (Table 6-15).

The deltoids are used every day when swinging the upper limbs during walking. The anterior parts flex the arms; the posterior parts extend them.

Table 6-4
Muscles Connecting the Upper Limb to the Vertebral Column

Muscle	Origin	Insertion	Nerve Supply	Actions
Trapezius (Figs. 6-44 and 6-45)	Medial third of superior nuchal line, external occipital protuberance, ligamentum nuchae, and spinous processes of C7 to T12 vertebrae (Figs. 6-44 and 6-47)	Lateral third of clavicle, acromion and spine of scapula (Fig. 6-46)	Spinal root of accessory n. and third and fourth cervical nn. (C3 and C4)	Elevates, retracts and rotates scapula; superior fibers elevate, middle fibers retract, and inferior fibers depress the scapula; superior and inferior fibers act together in superior rotation of the scapula
Latissimus dorsi (Figs. 6-44 and 6-51)	Spinous processes of inferior 6 thoracic vertebrae, thoracolumbar fascia, iliac crest, and inferior 3 or 4 ribs (Figs. 6-44 and 6-47)	Floor of intertubercular groove of humerus (Fig. 6-49)	Thoracodorsal n. (**C6**, **C7**, and C8)	Extends, adducts, and medially rotates humerus
Levator scapulae (Figs. 6-44 and 6-47)	Transverse processes of first three or four cervical vertebrae (Fig. 6-47)	Superior part of medial border of scapula (Fig. 6-46)	Dorsal scapular n. (C5) and third and fourth cervical nn.	Elevates scapula and helps to tilt its glenoid cavity inferiorly by rotating scapula
Rhomboid minor and major (Fig. 6-44)	*Minor:* Ligamentum nuchae and spinous processes of C7 and T1 vertebrae *Major:* Spinous processes of T2 to T5 vertebrae (Fig. 6-47)	Medial border of scapula from level of spine to inferior angle (Fig. 6-46)	Dorsal scapular n. (C4 and **C5**)	Retract scapula and rotate it to depress glenoid cavity Fixes scapula to thoracic wall

The deltoid muscle atrophies when the axillary nerve is damaged (*e.g.*, during a fracture of the surgical neck of the humerus). In these cases the deltoid atrophies and the rounded shape of the shoulder is lost, giving it a flattened appearance similar to that present with dislocation of the shoulder.

To test the strength of the deltoid muscle clinically, the patient's arm is abducted and then he/she is asked to hold it there against resistance. Inability to do this indicates injury to the axillary nerve supplying the deltoid.

THE TERES MAJOR MUSCLE (Figs. 6-4, 6-42, 6-47, 6-51, 6-52, and Table 6-5)

The teres (L. round) major forms a raised oval area on the dorsum of the scapula, beginning at the inferior angle. The inferior border of this rounded muscle forms the inferior border of the posterior wall of the axilla (Figs. 6-10 and 6-12). The teres major and the tendon of the latissimus dorsi form the posterior axillary fold (Figs. 6-13 and 6-51).

Origin (Fig. 6-46). Oval area on **the dorsal surface of the inferior angle of scapula.**

Insertion (Fig. 6-49). **Medial lip of intertubercular groove of humerus.**

Nerve Supply (Fig. 6-28). **Lower subscapular nerve (C6** and C7) from the posterior cord of the brachial plexus.

Actions (Tables 6-5 and 6-15). **Adducts and medially rotates humerus.** It can also help to extend it from the flexed position and is an important stabilizer of the proximal end of the humerus during abduction.

Four muscles joining the scapula to the humerus: *supraspinatus, infraspinatus, teres minor,* and *subscapularis* are referred to as the rotator cuff muscles of the shoulder joint (Figs. 6-52 and 6-54). *They form a musculotendinous cuff for the shoulder joint* which covers all of the shoulder joint, except its inferior aspect. The tendons of these four muscles blend with the articular capsule of the shoulder joint. All of them, except the supraspinatous, are rotators of the humerus; this explains why they are called **rotator cuff muscles.**

The rotator cuff *protects the shoulder joint and gives it stability by holding the head of the humerus in the glenoid cavity of the scapula.*

The bursae around the shoulder joint are of various sizes (Figs. 6-53 and 6-55). They are located between the tendons of the rotator cuff muscles and the fibrous capsule of the shoulder joint. These bursae reduce the friction on tendons passing over bones or other areas of resistance.

THE SUPRASPINATUS MUSCLE (Figs. 6-54, 6-55, 6-57, and Table 6-5)

This rounded muscle lies in the supraspinous fossa of the scapula (Fig. 6-4), deep to the trapezius muscle and the **coracoacromial arch** (Fig. 6-54). Its tendon is covered by the deltoid muscle.

Origin (Fig. 6-46). Medial two-thirds of **supraspinous fossa of scapula.**

Insertion (Figs. 6-18, 6-49, and 6-54). Superior facet on **greater tubercle of humerus.**

Nerve Supply (Fig. 6-28). **Suprascapular nerve** (C4, **C5,** and C6) from the superior trunk of the brachial plexus (Fig. 6-24).

Actions (Table 6-15). **Abducts the humerus and holds the humeral head in the glenoid cavity of the scapula.**

The supraspinatus usually acts with the deltoid during abduction of the arm. The supraspinatous acts strongly when a heavy weight is carried with the upper limb adducted (*e.g.*, carrying a heavy suitcase).

If the deltoid muscle is paralyzed, the supraspinatous can partially abduct the humerus by itself.

The tendon of the supraspinatous is separated from the coracoacromial ligament (Fig. 6-54), the acromion, and the deltoid by the *subacromial bursa* (Fig. 6-55). When this bursa is inflammed (**subacromial bursitis**), abduction of the arm is painful.

The rotator cuff muscles hold the head of the humerus in the glenoid cavity of the scapula. Hence, a torn rotator cuff often results in dislocation of the shoulder joint. The supraspinatous tendon is the most commonly torn part of the rotator cuff.

THE SUBSCAPULARIS MUSCLE (Figs. 6-41, 6-54, 6-56, and Table 6-5)

This thick triangular muscle lies on the costal surface of the scapula and forms part of the posterior wall of the axilla (Figs. 6-10 and 6-12). It crosses the anterior aspect of the shoulder joint (Fig. 6-56) on its way to the humerus.

Origin (Fig. 6-49). **Subscapular fossa** on the costal surface of the scapula.

Insertion (Fig. 6-49). **Lesser tubercle of humerus.** Its tendon is attached to the fibrous capsule of the shoulder joint and is separated from the neck of the scapula by the large **subscapular bursa** (Fig. 6-41).

Nerve Supply (Fig. 6-28). **Upper and lower subscapular** nerves (C-5, **C6**, and C7) from the posterior cord of the brachial plexus (Fig. 6-24).

Actions. Medially rotates the humerus and holds the humeral head in the glenoid cavity (*i.e.*, stabilizes the shoulder joint).

THE TERES MINOR MUSCLE (Figs. 6-52, 6-57, 6-69, and Table 6-5)

This elongated muscle is often inseparable from the infraspinatus muscle which lies along its superior border (Fig. 6-57).

Origin (Figs. 6-46 and 6-57). Superior part of dorsal surface of **lateral border of scapula.**

Insertion (Figs. 6-46 and 6-57). Inferior facet on **greater tubercle of humerus.** The long head of the triceps muscle separates its tendon from the teres major muscle (Fig. 6-52), thereby producing a triangular space and a larger quadrangular space.

Nerve Supply (Figs. 6-28 and 6-29). **Axillary nerve (C5** and C6) from the posterior cord of the brachial plexus (Fig. 6-24).

Actions. Laterally rotates the humerus and holds the humeral head in the glenoid cavity (*i.e.*, stabilizes the shoulder joint).

THE INFRASPINATUS MUSCLE (Figs. 6-52, 6-54, 6-57, and Table 6-5)

This triangular muscle occupies most of the infraspinous fossa (Fig. 6-3).

Origin (Figs. 6-46 and 6-57). **Infraspinous fossa of scapula.**

Insertion (Figs. 6-46 and 6-57). Middle facet on **greater tubercle of humerus.** Its tendon is adherent to the fibrous capsule of the shoulder joint.

Nerve Supply (Fig. 6-28). **Suprascapular nerve (C5** and C6) from the superior trunk of the brachial plexus (Fig. 6-24).

Actions. Laterally rotates the humerus and holds the humeral head in the glenoid cavity (*i.e.*, stabilizes the shoulder joint).

The rotator cuff (formed by the tendons of the supraspinatus, infraspinatus, teres minor, and subscapularis muscles) **holds the head of**

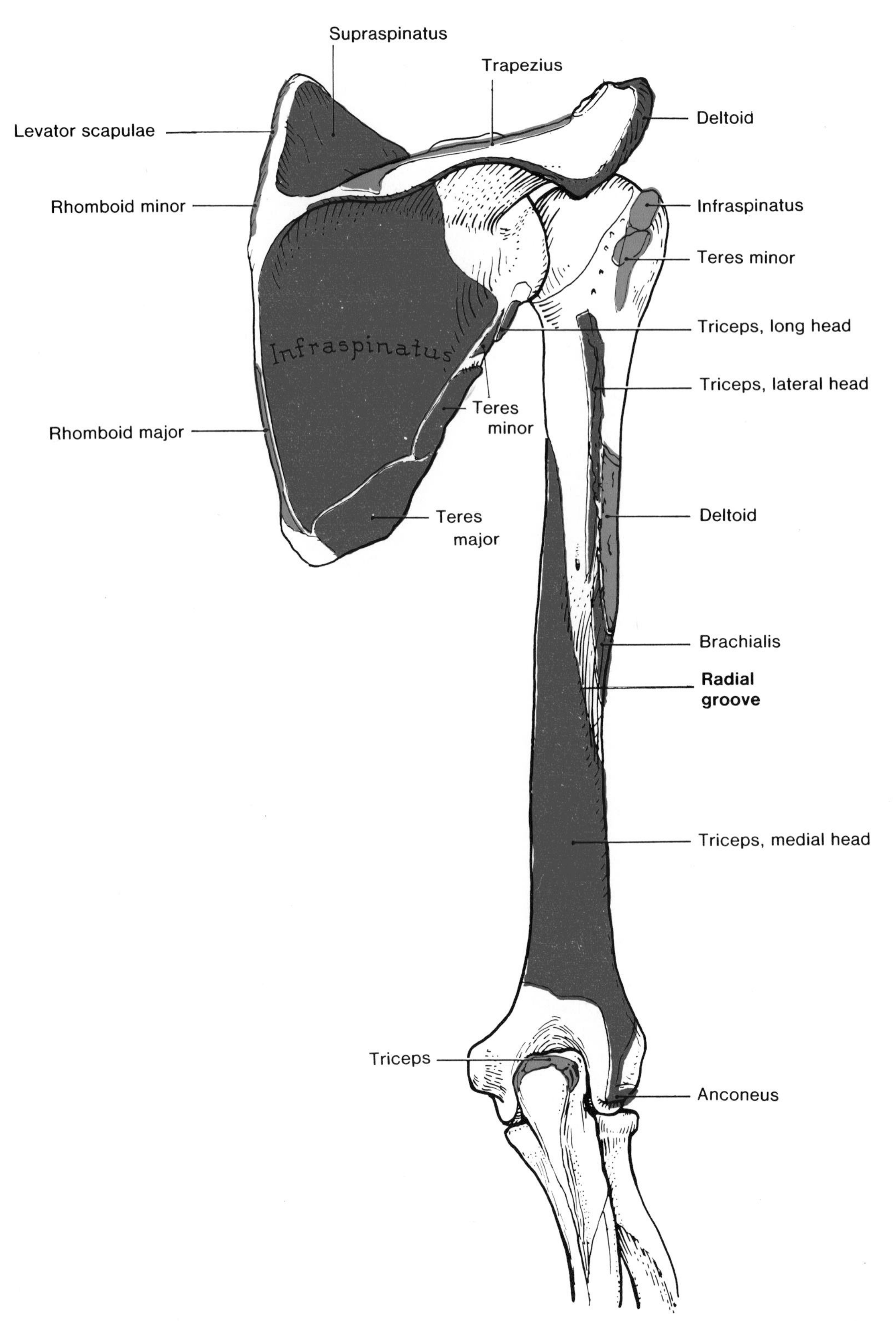

Figure 6-46. Drawing of a posterior view of the scapula and the bones of the upper limb, *right side*, showing the attachment of muscles to them. Origins are shown in *red* and insertions in *blue*. Observe that the lateral head of the triceps arises superior to the radial groove for the radial nerve, and that the medial head arises inferior and medial to the radial groove.

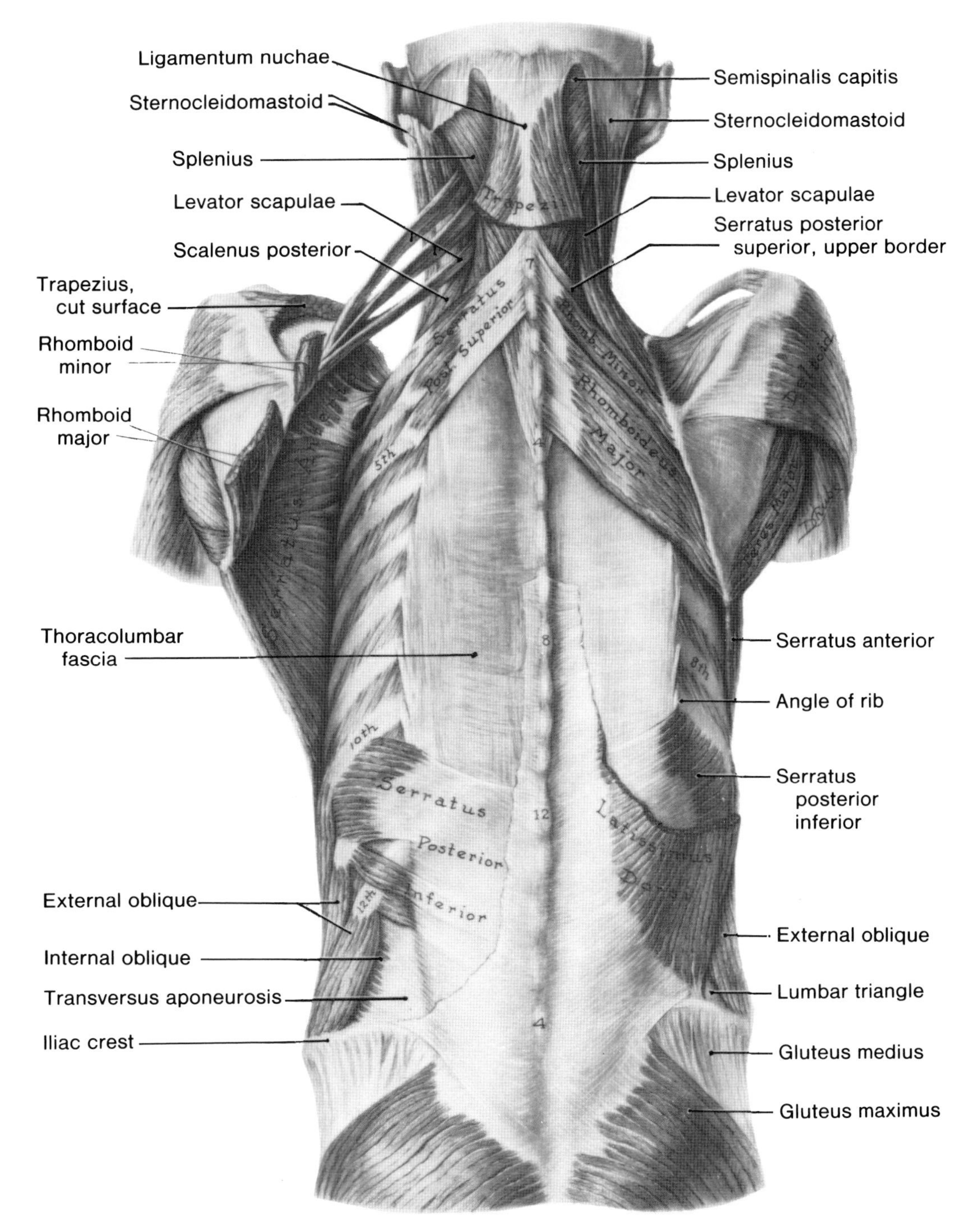

Figure 6-47. Drawing of a dissection of the back showing the intermediate muscles. The trapezius and latissimus dorsi are largely cut away on both sides. On the *right side*, observe the levator scapulae and rhomboids. On the *left side*, the rhomboids are severed allowing the vertebral border of the scapula to separate from the thoracic wall. Note the three digitations (slips) of the levator scapulae. Observe the thoracolumbar fascia, extending laterally to the angles of the ribs, becoming thin superiorly, passing deep to the serratus superior, and reinforced inferiorly by the latissimus dorsi and serratus inferior muscles.

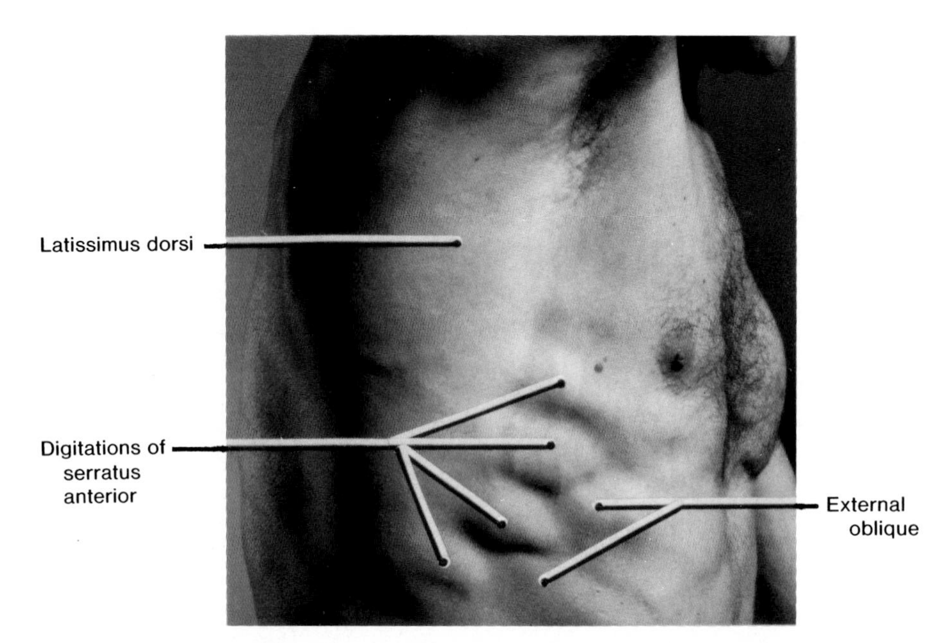

Figure 6-48. Photograph of a lateral view of the thorax of a 46-year-old man. Observe that the inferior three digitations of the serratus anterior muscle interdigitate with the external oblique muscle of the abdomen. As it is inserted into the entire length of the medial margin of the scapula (Fig. 6-49), the serratus anterior is a *powerful protractor of the scapula.* It also holds the scapula against the thoracic wall. Note that the serratus anterior (boxer's muscle) clothes the thoracic wall, posterior to its origin from the true ribs (1–7) at about the midclavicular level.

the humerus in the glenoid cavity of the scapula. This cuff may be damaged by injury or disease, resulting in instability of the shoulder joint. Trauma may tear or rupture one or more of the tendons of the rotator cuff muscles.

Degenerative tendonitis of the rotator cuff is a common disease, especially in older people. Calcium deposits may be demonstrated on radiographs in the supraspinatus tendon and/or in other tendons of the cuff. Tendonitis often leads to adherence of the supraspinatus tendon to the subacromial bursa (Fig. 6-55), and predisposes the tendon to rupture. The **calcium deposits in the supraspinatus tendon** (Fig. 6-135*B*) often rupture into the bursa, producing a painful **chemical bursitis.**

The supraspinatus tendon does not rupture very often in young people because their tendons are usually so strong that they will tear away (avulse) the tip of the greater tubercle of the humerus rather than rupture the tendon.

Tendonitis and inflammation of the subacromial bursa (***subacromial bursitis***) *result in shoulder pain* which is intensified by attempts to abduct the arm (Fig. 6-135*B*).

The main stability of the shoulder joint is provided by the musculotendinous rotator cuff formed by the tendons of the rotator cuff muscles. This cuff fuses with the fibrous capsule of the joint and inserts into the tubercles of the humerus. This musculotendinous rotator cuff strengthens the shoulder joint everywhere, except inferiorly. Consequently, if a person falls when the humerus is abducted, the head of this bone may be levered out of the glenoid cavity of the scapula, producing a **dislocation of the shoulder joint** (Fig. 6-136).

THE ARM

The arm (L. brachium) extends from the shoulder to the elbow. The rounded prominence on the anterior surface of the arm is the **biceps brachii muscle** (Fig. 6-58), which is commonly referred to as the *biceps* (Figs. 6-67 and 6-68).

The body (shaft) of the ***humerus*** *(arm bone) is easy to palpate,* as are its medial and lateral epicondyles (Figs. 6-1 and 6-59).

The triceps brachii muscle, usually called the *triceps,* occupies the posterior part of the arm (Figs. 6-59 and 6-61 to 6-64).

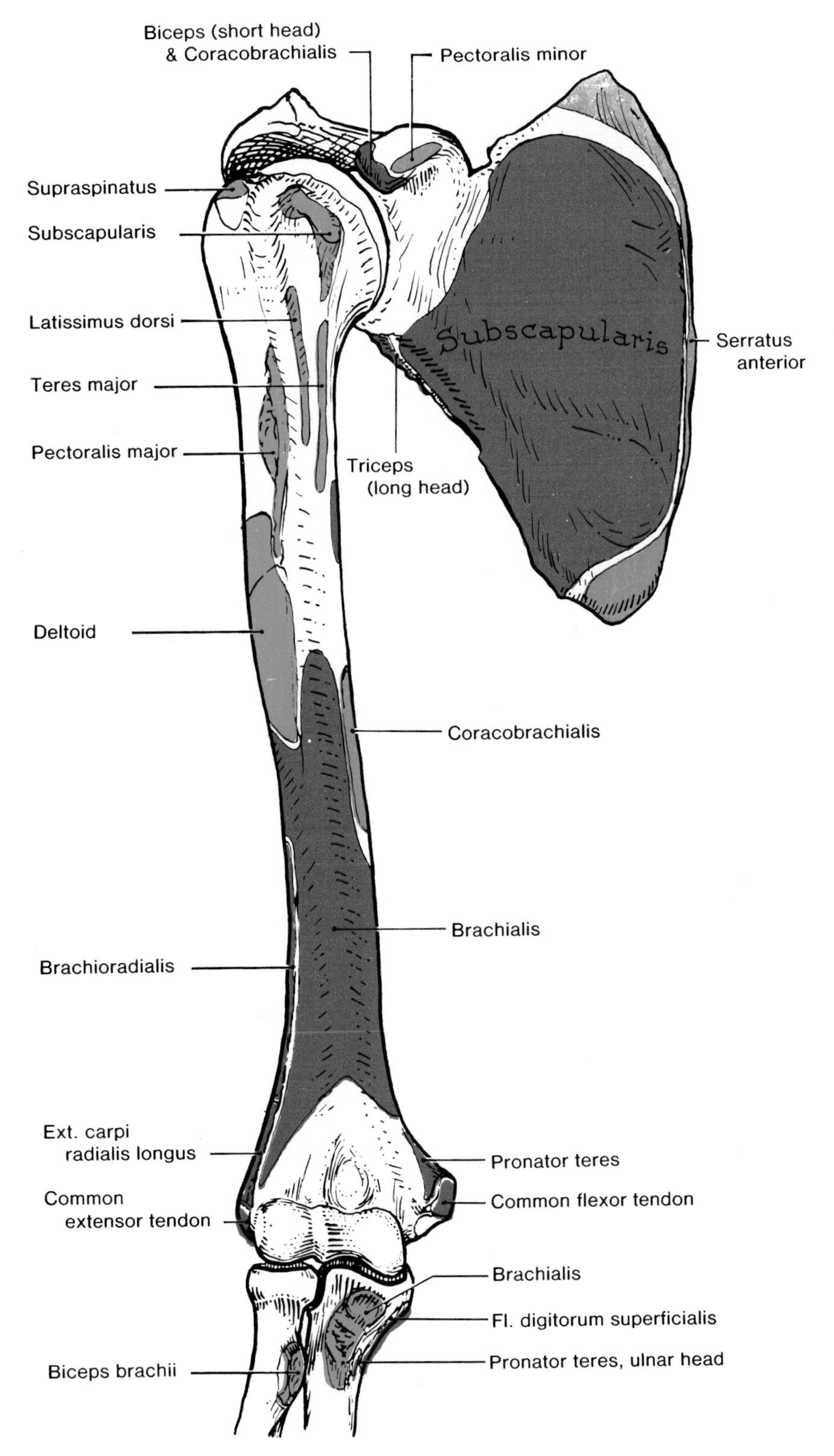

Figure 6-49. Drawing of an anterior view of the scapula and the bones of the upper limb, *right side*, showing the attachment of muscles to them. Origins of muscles are shown in *red*, insertions in *blue*.

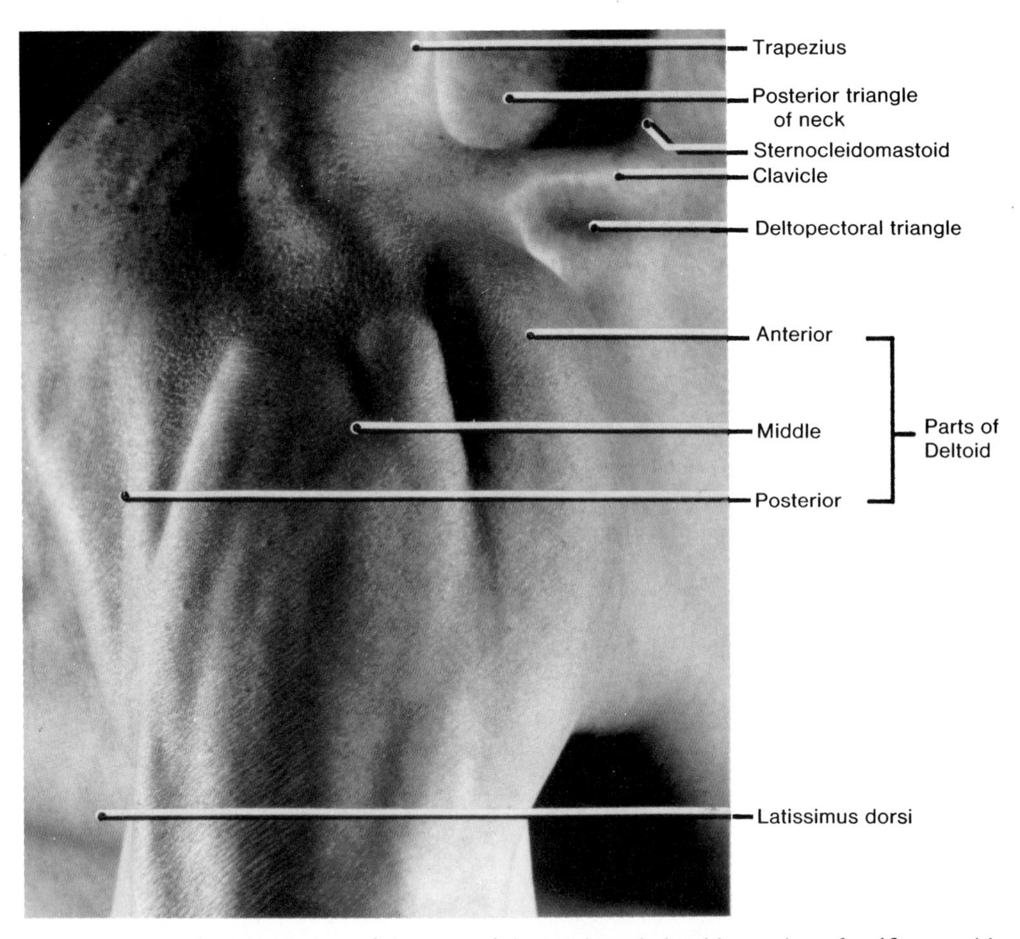

Figure 6-50. Photograph of a lateral view of the root of the neck and shoulder region of a 46-year-old man. To make the parts of his deltoid muscle stand out, he abducted his arm against resistance. The appearance of the deltoid and the way its parts attach to the pectoral girdle give clues to the muscle's actions. The anterior fibers flex the arm and medially rotate it; the posterior fibers reverse these movements; and the middle fibers abduct the arm. See Figure 6-62 for a dissection of this muscle.

THE HUMERUS

The humerus is the largest bone in the upper limb (Figs. 6-1 and 6-59). Its smooth, ball-like head articulates with the glenoid cavity of the scapula. Consequently it is *really the bone of the shoulder and the arm.* Close to the head are the **greater and lesser tubercles** for the insertion of the muscles that surround and move the shoulder joint. The lesser tubercle is separated from the greater tubercle by the **intertubercular groove** or sulcus in which lies the tendon of the long head of the biceps muscle (Fig. 6-129).

The **anatomical neck** separates the head and the tubercles. Distal to the anatomical neck is the **surgical neck** (Fig. 6-59). It is located where the bone narrows to become the body. *This region is called the surgical neck because it is the site of most frequent fracture of the proximal end of the humerus.*

The superior half of the body (shaft) is cylindrical. Anterolaterally there is a roughness known as the **deltoid tuberosity** *for the insertion of the deltoid muscle* (Figs. 6-49 and 6-62). Observe the shallow, oblique **radial groove** (*spiral groove for radial nerve*) that extends inferolaterally on the posterior aspect of the body (Fig. 6-59*B*).

The distal end of the humerus is expanded from side to side. The **trochlea** (L. pulley) fits into the ***trochlear notch*** *of the ulna* (Fig. 6-138*A*), which swings on this pulley (L. trochlea) when the elbow is flexed. Just proximal to the trochlea are the **coronoid fossa** and the **olecranon fossa** for accommodating corresponding parts of the ulna (Figs. 6-59 and 6-139). Adjoining the lateral part of the trochlea is a rounded ball of bone called the **capitulum** (L. little head).

A prominent process, the **medial epicondyle**, projects from the trochlea and the **lateral epicondyle** projects from the capitulum. The epicondyles, being subcutaneous, are easily felt. The medial epicondyle is more prominent (Figs. 6-58*B*, 6-59, and 6-68).

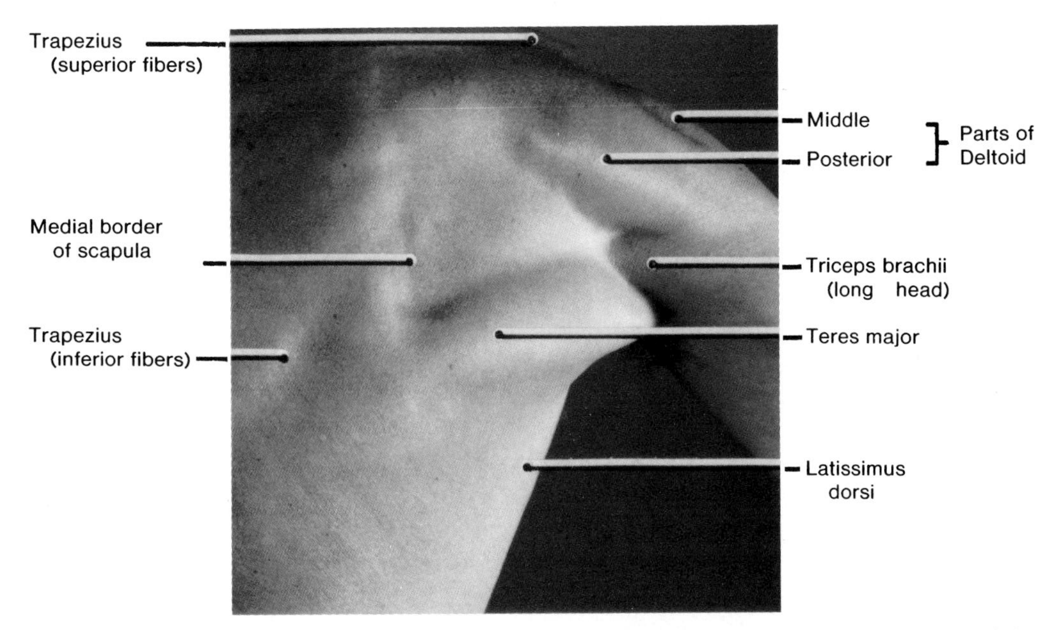

Figure 6-51. Photograph of the back and scapular region of a 46-year-old man showing some of the scapular and arm muscles. The teres major muscle was made to stand out by asking him to adduct his arm against resistance.

From each epicondyle a ridge runs proximally, known as the **medial and lateral supracondylar ridges**, respectively (Fig. 6-59*A*).

Fractures of the superior end of the humerus may be through its anatomical or surgical necks.

Fractures of the surgical neck are common in elderly persons and usually result from falls on the elbow when the arm is abducted. The fracture line occurs superior to the insertion of the pectoralis major, teres major, and latissimus dorsi muscles (Fig. 6-49).

Because they are in contact with the humerus (surgical neck-axillary nerve; radial groove-radial nerve; and medial epicondyle-ulnar nerve), *the axillary, radial, and ulnar nerves may be injured in fractures of the humerus.*

Traumatic separation of the proximal epiphysis of the humerus (Fig. 6-60) can occur in persons under age 20 because this *epiphysis does not fuse with the body* of the humerus until about the 18th year in females and the 20th year in males. *Fracture-separation of the proximal humeral epiphysis* occurs in a child because the articular capsule of the shoulder joint is stronger than the epiphyseal cartilage plate.

The displacement of the head resulting from separation of the epiphysis in young persons is similar to that which occurs in fractures through the surgical neck of the humerus in elderly persons.

THE BRACHIAL FASCIA AND INTERMUSCULAR SEPTA

The arm is enclosed in a sheath of deep fascia known as the **brachial fascia** (Figs. 6-61 and 6-66). It is continuous superiorly with the pectoral and axillary fasciae and with the fascia covering the deltoid and latissimus dorsi muscles (Fig. 6-44).

The brachial fascia is attached inferiorly to the epicondyles of the humerus and the olecranon of the ulna, and it is continuous with the deep fascia of the forearm (Fig. 6-63). Two **fascial intermuscular septa** extend from the sheath of brachial fascia, and are attached to the medial and lateral supracondylar ridges of the humerus (Figs. 6-59 and 6-63).

The medial and lateral intermuscular septa divide the arm into anterior and posterior fascial compartments, each containing muscles, nerves, and blood vessels (Figs. 6-61 and 6-64).

The anterior fascial compartment (flexor compartment) contains three muscles (biceps, brachialis, and coracobrachialis), their nerves and vessels.

Table 6-5
The Scapular Muscles

Muscle	Origin	Insertion	Nerve Supply	Actions
Deltoid (Figs. 6-4, 6-14, 6-15, and 6-50)	Lateral third of clavicle, acromion, and spine of scapula (Figs. 6-18 and 6-46)	Deltoid tuberosity of humerus (Fig. 6-49)	Axillary n. (**C5** and C6)	*Anterior part:* flexes and medially rotates humerus *Middle part:* abducts humerus *Posterior part:* extends and laterally rotates humerus
Supraspinatus[1] (Figs. 6-54 and 6-57)	Supraspinous fossa of scapula (Fig. 6-46)	Superior facet on greater tubercle of humerus (Fig. 6-49)	Suprascapular n. (C4, **C5**, and C6)	Abducts humerus
Infraspinatous[1] (Fig. 6-58)	Infraspinous fossa of scapula (Fig. 6-48)	Middle facet on greater tubercle of humerus (Fig. 6-46)	Suprascapular n. (**C5** and C6)	Laterally rotates humerus
Teres minor[1] (Figs. 6-52 and 6-57)	Superior part of lateral border of scapula (Fig. 6-46)	Inferior facet on greater tubercle of humerus (Fig. 6-46)	Axillary n. (**C5** and C6)	
Teres major (Figs. 6-42 and 6-51)	Dorsal surface of inferior angle of scapula (Fig. 6-46)	Medial lip of intertubercular groove of humerus (Fig. 6-49)	Lower subscapular n. (**C6** and C7)	Adducts and medially rotates humerus
Subscapularis[1] (Figs. 6-54 and 6-56)	Subscapular fossa (Fig. 6-49)	Lesser tubercle of humerus (Fig. 6-49)	Upper and lower subscapular nn. (C5, **C6**, and C7)	Medially rotates humerus

[1] The supraspinatus, infraspinatous, teres minor, and subscapularis muscles are referred to as the **rotator cuff muscles**. They join the scapula to the humerus and have as their prime function the *holding of the head of the humerus in the glenoid cavity* of the scapula. **The rotator cuff muscles stabilize the shoulder joint.**

Table 6-6
Muscles Producing Movements of the Scapula[1]

Elevation	Depression	Protraction	Retraction	Superior Rotation[2]	Inferior Rotation
Trapezius—superior fibers	**Pectoralis minor**	**Serratus anterior**	**Trapezius**	**Trapezius**—superior and inferior fibers	**Levator scapulae**
Levator scapulae	**Latissimus dorsi**	**Pectoralis minor**	**Rhomboids**	**Serratus anterior**—inferior fibers	**Rhomboids**
Serratus anterior-superior fibers	Pectoralis major	Levator scapulae	Latissimus dorsi		**Pectoralis minor**
					Pectoralis major
					Latissimus dorsi

[1] The principal muscles producing these movements appear in **boldface**.
[2] Superior rotation results in the glenoid cavity being elevated so it faces superiorly; inferior rotation of the scapula is the reverse movement.

The posterior fascial compartment (extensor compartment) contains one muscle (triceps), its nerve, and vessels (Table 6-7).

MUSCLES OF THE ARM

There are four muscles in the arm: three flexors in the anterior fascial compartment (biceps brachii, coracobrachialis, and brachialis), supplied by the musculocutaneous nerve, and one extensor in the posterior fascial compartment (triceps) supplied by the radial nerve (Table 6-7).

The anconeus muscle is located chiefly in the forearm, but it will be described with the **brachial muscles** because it is morphologically and functionally related to the triceps muscle.

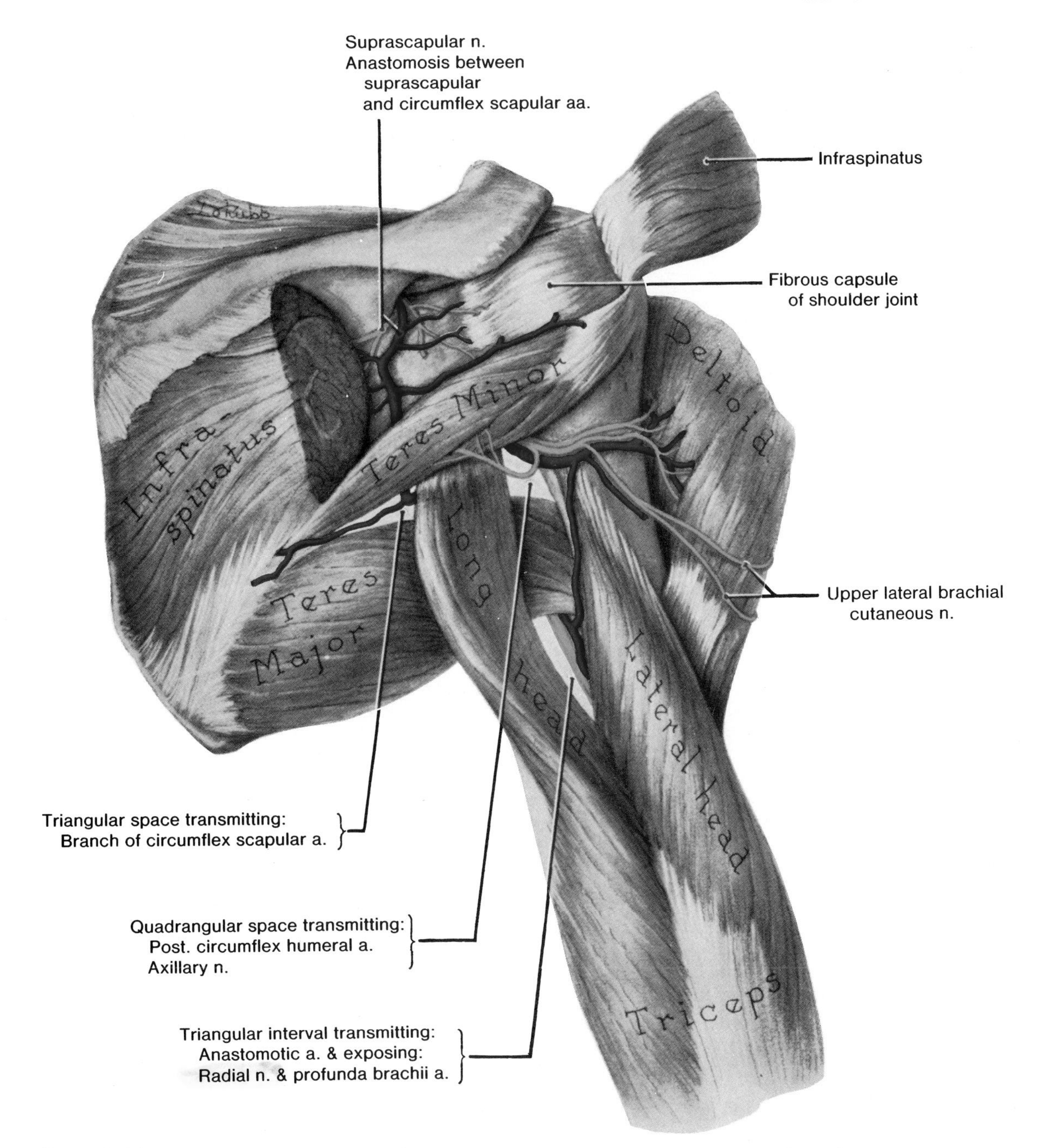

Figure 6-52. Drawing of a dissection of the posterior scapular and subdeltoid regions. Observe the thickness of the infraspinatus muscle which, aided by the teres minor and posterior fibers of the deltoid, rotates the humerus laterally. Note the long head of the triceps muscle passing between the teres minor, a lateral rotator, and the teres major, a medial rotator. Observe the long head of the triceps separating the quadrangular space from the triangular space, and the teres major separating the quadrangular space from another triangular space. Note that the axillary nerve passes through the quadrangular space and around the surgical neck of the humerus (also see Fig. 6-12). Consequently this nerve may be injured in a fracture of the humerus in this region. Note the arterial anastomoses on and around the scapula (also see Fig. 6-39), and the distribution of the suprascapular and axillary nerves.

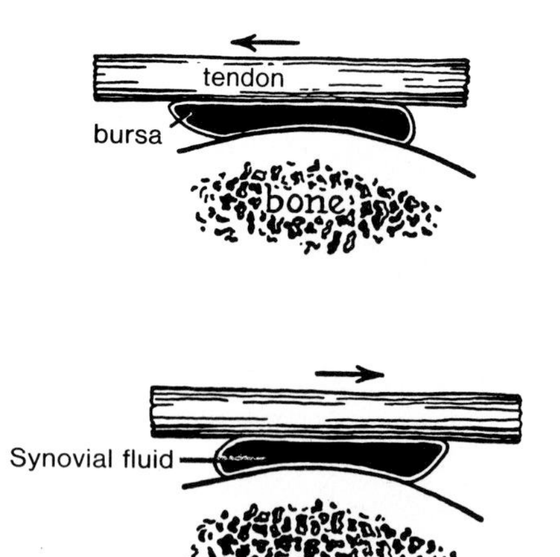

Figure 6-53. Drawings illustrating that a bursa (L. purse) is a device for eliminating friction wherever a muscle or tendon is liable to rub on another muscle, tendon, or bone. Understand that a bursa is a flattened sac and that its walls are separated by only a *capillary film of synovial fluid* which acts as a lubricant, enabling its walls to slide freely over each other.

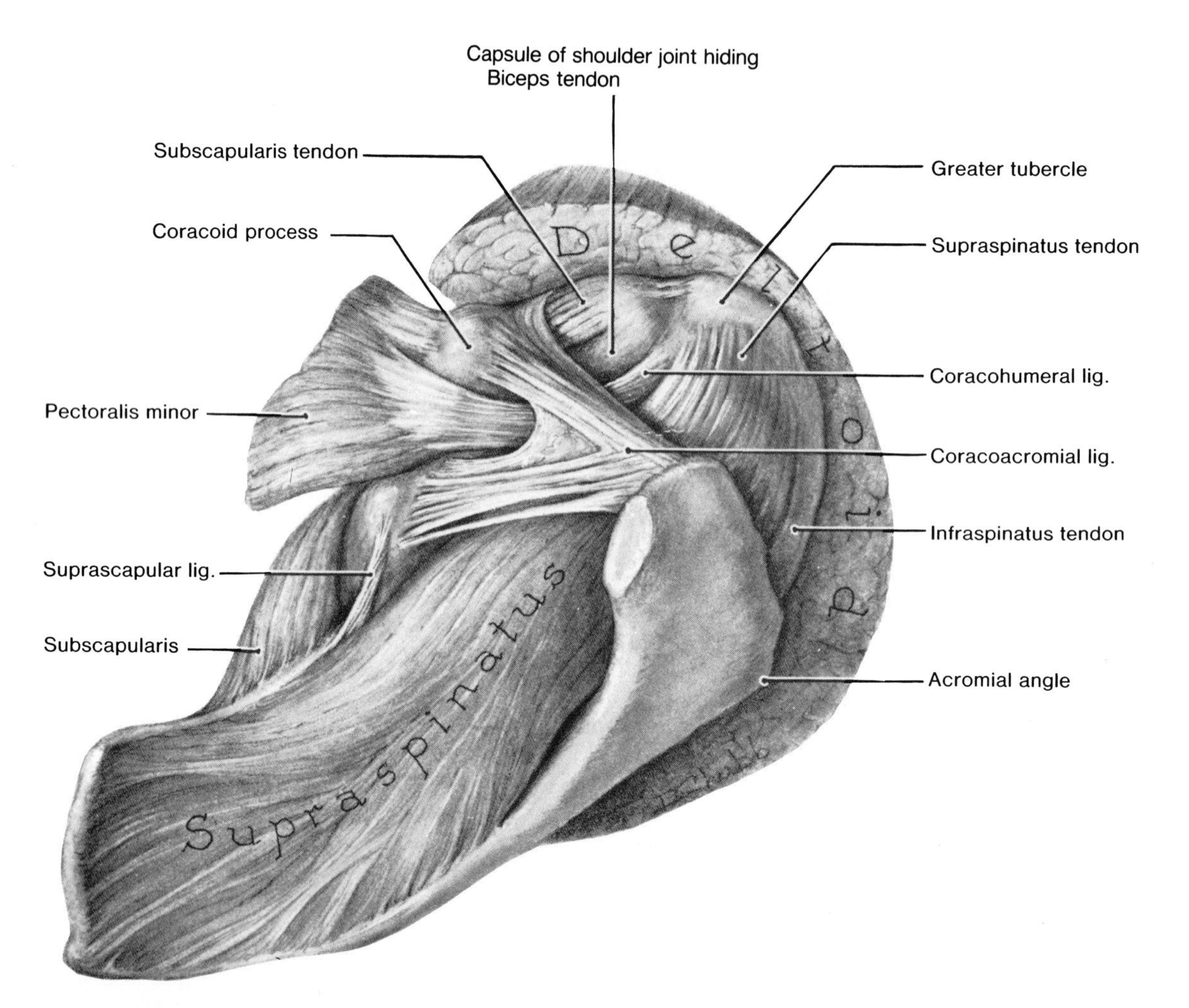

Figure 6-54. Drawing of a dissection of the supraspinous and subdeltoid regions. Observe that the supraspinatous muscle passes inferior to the *coracoacromial arch* formed by the coracoid process, coracoacromial ligament, and acromion. Understand that this muscle lies between the deltoid muscle superiorly and the articular capsule of the shoulder joint inferiorly. The supraspinatus and the middle fibers of the deltoid are abductors of the arm at this joint (Tables 6-5 and 6-15).

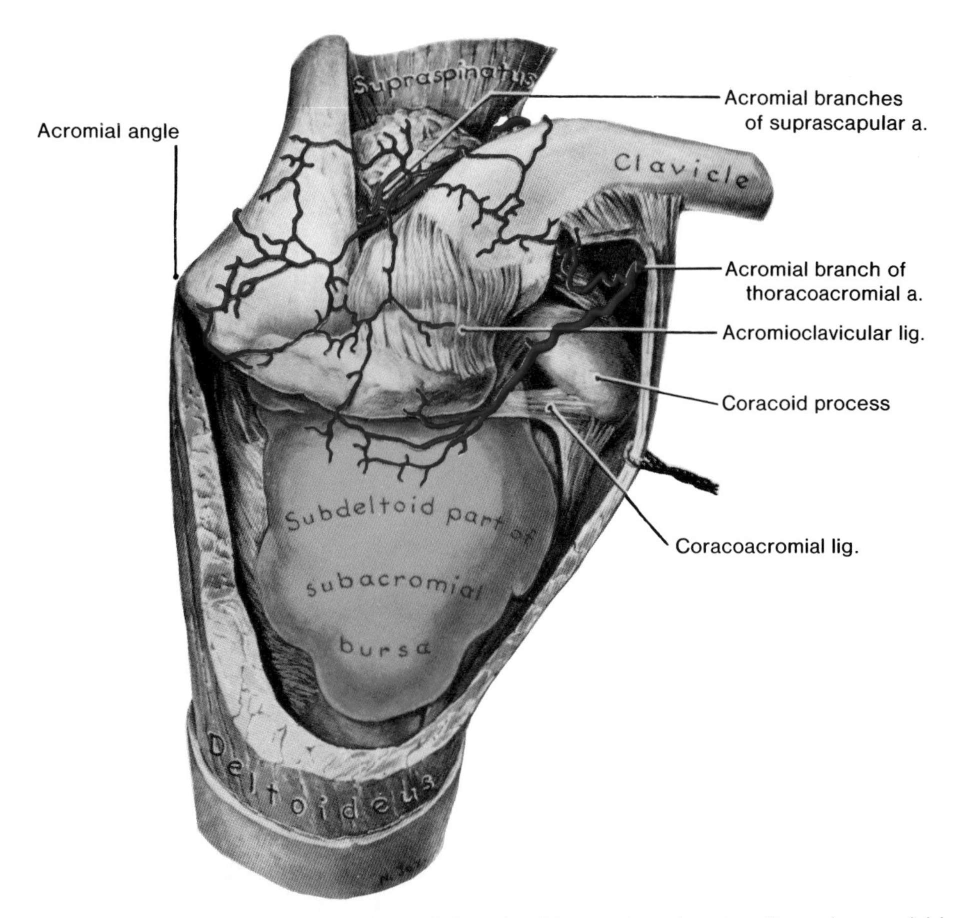

Figure 6-55. Drawing of a superolateral view of the shoulder region showing the subacromial bursa that has been injected with latex. The term "subacromial bursa" is usually understood to include the subdeltoid bursa because the two bursae are generally combined.

THE BICEPS BRACHII MUSCLE (Figs. 6-58, 6-61, 6-62, 6-64, and Table 6-7)

As its name "biceps" indicates, this long fusiform muscle has *two heads of origin (bi,* two + *cipital* or *ceps* from L. *caput,* head). The two bellies of the muscle unite just distal to the middle of the arm. The biceps is located in the anterior aspect of the arm in the anterior fascial compartment (Fig. 6-61).

Origin (Figs. 6-49 and 6-64). *Short head,* **tip of coracoid process of scapula.** *Long head,* **supraglenoid tubercle of scapula.**

The tendon of the long head crosses the head of the humerus within the capsule of the shoulder joint (Fig. 6-41), and descends in the intertubercular groove of the humerus (Figs. 6-59*A* and 6-129).

Insertion (Figs. 6-49 and 6-59). **Tuberosity of radius.** The *bicipitoradial bursa* separates its tendon from the anterior part of the tuberosity (Fig. 6-91).

The biceps brachii also inserts via the bicipital aponeurosis (Figs. 6-58, 6-64, and 6-66), a triangular, membranous band which runs from the biceps tendon across the **cubital fossa** into the deep fascia over the flexor muscles in the medial side of the forearm.

The proximal part of the bicipital aponeurosis can be easily felt where it passes obliquely over the brachial artery and medial nerve (Figs. 6-58, 6-64, and 6-66). The bicipital aponeurosis affords protection for these and other structures in the cubital fossa. It also helps to lessen the pressure of the biceps tendon on the radial tuberosity during pronation and supination (Fig. 6-70*B*).

Nerve Supply (Fig. 6-64). **Musculocutaneous nerve** (C5 and **C6**).

Actions (Tables 6-16 and 6-17). **Flexes and supinates forearm.**

The biceps is a powerful supinator when the forearm is flexed and when more power is needed against resistance (*e.g.,* turning a screw into hard wood). You can easily feel your biceps contract as you supinate your forearm against resistance with the elbow joint flexed at 90°. It may help you to recall the actions of

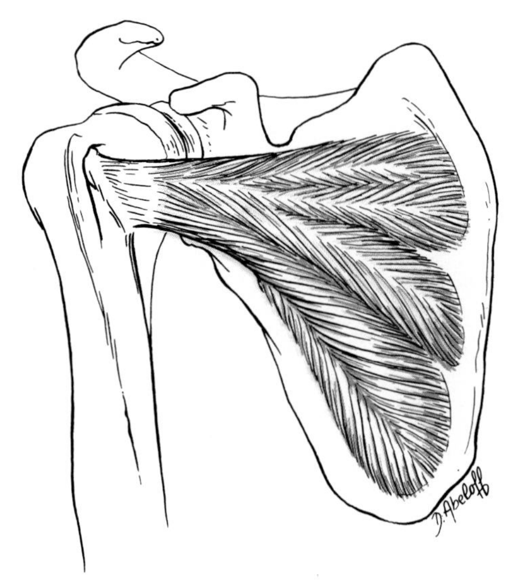

Figure 6-56. Drawing of the anterior aspect of the right scapula and humerus showing how the subscapularis muscle, *one of the rotator cuff muscles*, guards the anterior aspect of the shoulder joint; helps to hold the humeral head in the glenoid cavity of the scapula, and to stabilize the shoulder joint.

the biceps if you remember that it is an important muscle used when inserting a corkscrew and pulling out the cork of a wine bottle.

Sometimes the tendon of the long head of the biceps is dislocated from the intertubercular groove in the humerus. Its normal position in this groove is illustrated in Figure 6-129. When this occurs, the arm is fixed in the abducted position and the head of the humerus can be felt in its normal position. The displaced long tendon of the biceps can be placed in its normal position by flexing the forearm at the elbow joint and rotating the upper limb medially and then laterally.

THE BRACHIALIS MUSCLE (Figs. 6-61, 6-62, 6-64, 6-66, and Table 6-7)

This strong muscle lies posterior to the biceps brachii. *The brachialis is the main flexor of the forearm.*

Origin (Fig. 6-49). **Distal half of anterior surface of humerus** and intermuscular septum.

Insertion (Fig. 6-49). **Coronoid process and tuberosity of ulna.**

Nerve Supply (Fig. 6-64). **Musculocutaneous nerve** (C5 and **C6**). Its lateral part also receives a branch from the radial nerve (C7).

Actions (Table 6-16). **Flexes the forearm at the elbow joint.** Flex your forearm against resistance and feel the brachialis alongside the tendon of the biceps and on each side of the belly of this muscle. The brachialis is the main flexor of the forearm. It always contracts for flexion and is *primarily responsible for maintaining flexion.*

THE CORACOBRACHIALIS MUSCLE (Figs. 6-20, 6-29, 6-42, 6-64, and Table 6-7)

This short, rounded muscle in the superomedial part of the arm (Fig. 6-61) is important mainly as a landmark (*e.g.*, the musculocutaneous nerve pierces it).

Origin (Fig. 6-49). **Tip of coracoid process of scapula,** in common with the short head of the biceps.

Insertion (Fig. 6-49). **Middle third of medial surface of the body of the humerus.**

Nerve Supply (Fig. 6-64). **Musculocutaneous nerve** (C5, **C6**, and C7).

Actions (Table 6-15). **Flexes arm** at shoulder joint and helps to stabilize this joint.

THE TRICEPS BRACHII MUSCLE (Figs. 6-52, 6-58, 6-61 to 6-64, and Table 6-7)

This large muscle is in the posterior compartment of the arm and is the main muscle in it. It is associated with the small anconeus muscle at the elbow (Fig. 6-63). As its name "triceps" indicates it has *three heads of origin*: long, lateral, and medial.

Origin (Figs. 6-46 and 6-49). *Long head*, **infraglenoid tubercle of scapula.** *Lateral head*, **posterior surface of humerus, superior to the radial**

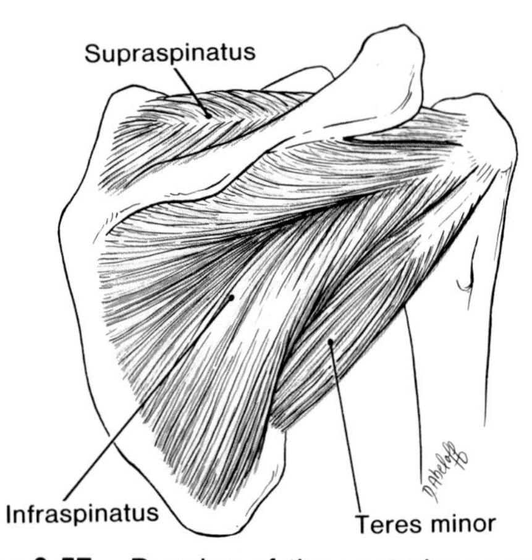

Figure 6-57. Drawing of the posterior aspect of the right scapula and humerus showing how the supraspinatus muscle guards the shoulder joint superiorly and the infraspinatous and teres minor muscles guard it posteriorly. The fourth rotator cuff muscle, the subscapularis, passes anterior to this joint (Fig. 6-56). The four *rotator cuff muscles* have a steadying effect on the head of the humerus, maintaining it in correct apposition to the glenoid cavity of the scapula.

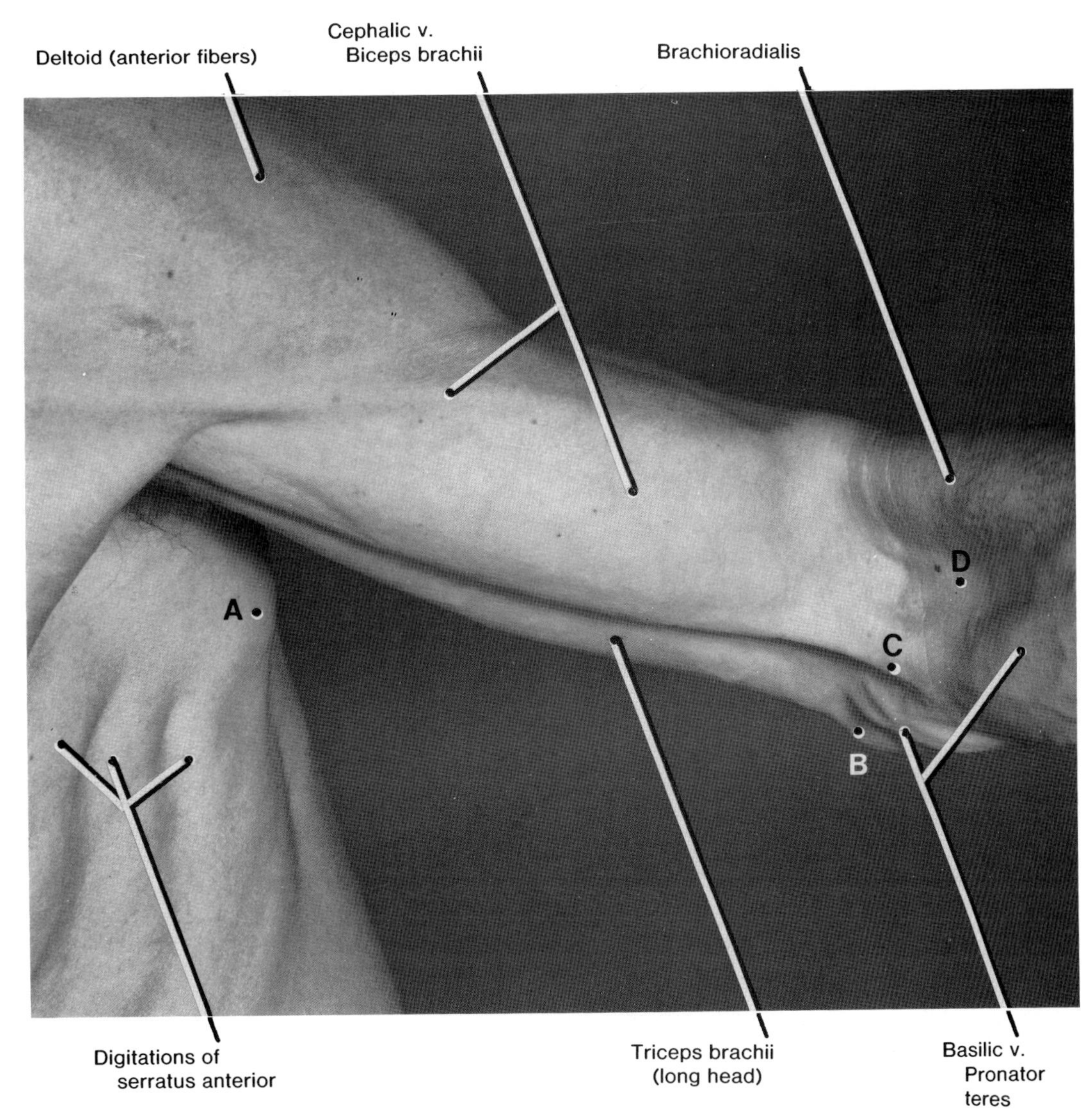

Figure 6-58. Photograph of the anterior surface of the left shoulder, arm, elbow, and lateral chest regions of a 46-year-old man. *A*, inferior angle of scapula. *B*, medial epicondyle of humerus. *C*, bicipital aponeurosis. *D*, cubital fossa.

groove. *Medial head,* **posterior surface of humerus, inferior to radial groove.**

Insertion (Figs. 6-46 and 6-63). **Proximal end of the olecranon** of the ulna and the deep fascia of forearm. Just proximal to its insertion, there is a subtendinous *olecranon bursa* between the triceps tendon and the olecranon (Fig. 6-138*B*).

Nerve Supply (Figs. 6-27, 6-29, 6-52, and 6-69). **Radial nerve** (C6, **C7**, and **C8**).

Actions (Table 6-16). **Extends the forearm at the elbow joint.** The triceps is the chief extensor of the forearm. Because the long head of the triceps crosses the shoulder joint, it also aids in extension and adduction of the arm.

THE ANCONEUS MUSCLE (Figs. 6-63, 6-140, 6-143, and Table 6-10)

This small, triangular muscle is on the lateral part of the posterior aspect of the elbow. It is usually partially blended with the triceps and should be considered as part of this muscle.

Origin (Figs. 6-46 and 6-143). **Posterior aspect of lateral epicondyle of humerus.**

Insertion (Figs. 6-46, 6-140, and 6-143). **Lateral surface of olecranon and superior part of posterior surface of ulna.**

Nerve Supply (Figs. 6-27 and 6-69). **Radial nerve** (C7, C8, and T1).

Actions (Fig. 6-70*B* and Table 6-10). **Abducts ulna during pronation** of forearm, and contracts

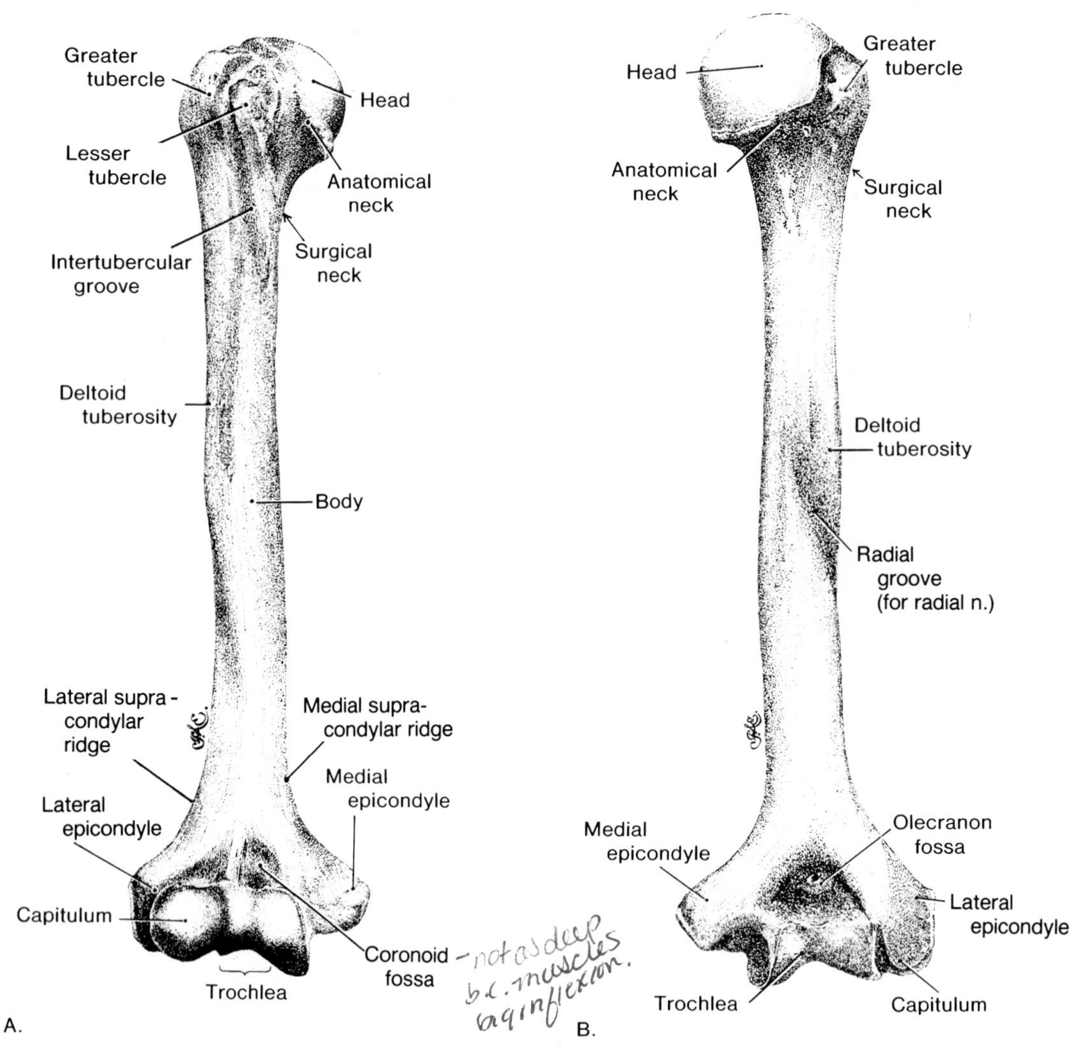

Figure 6-59. Drawings of the right humerus. *A*, anterior view. *B*, posterior view. Note the roughened area on the anterolateral surface of the body (shaft), called the deltoid tuberosity, into which the tendon of the deltoid muscle inserts (Figs. 6-49 and 6-62). Observe the radial groove which marks the course of the radial nerve and the profunda brachii artery (Fig. 6-69).

whenever this joint needs to be stabilized against flexion. **It also assists the triceps during extension of the forearm.**

THE CUBITAL FOSSA

This triangular space or depression is on the anterior surface of the elbow (Figs. 6-58D and 6-71). The cubital fossa is **bounded superiorly** by an imaginary *line connecting the epicondyles* of the humerus, **medially** by the *pronator teres* muscle, and **laterally** by the *brachioradialis* muscle.

The floor of the cubital fossa is formed by the brachialis and supinator muscles of the arm and forearm, respectively (Fig. 6-74).

The roof of the cubital fossa is formed by deep fascia that is strengthened by the bicipital aponeurosis (Figs. 6-64 and 6-66), which is covered by superficial fascia and skin (Fig. 6-68).

CONTENTS OF THE CUBITAL FOSSA (Figs. 6-64 to 6-68 and 6-74)

The cubital fossa is an important area because it contains the ***biceps tendon, brachial artery*** and its terminal branches (***radial and ulnar arteries***), and parts of the ***median and radial nerves.***

ARTERIES OF THE ARM

THE BRACHIAL ARTERY (Figs. 6-38, 6-42, 6-61, 6-64 to 6-66, and 6-74)

This artery provides the main arterial supply to the arm. **The brachial artery begins at the inferior**

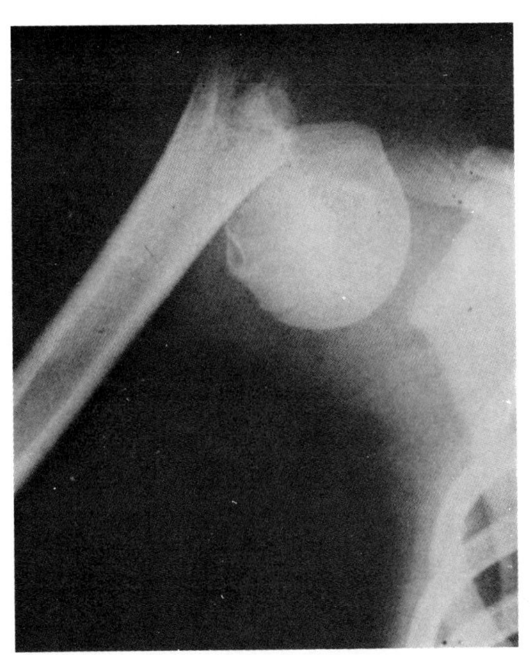

Figure 6-60. Radiograph of the right shoulder of a 14-year-old boy showing separation of the proximal epiphysis of the humerus. Although the humeral head has rotated, it has otherwise retained a fairly normal relationship with the glenoid cavity of the scapula.

border of the teres major as the continuation of the axillary artery (Figs. 6-38 and 6-65). It runs inferiorly and slightly laterally *on the medial side of the biceps* to the cubital fossa, where it **ends opposite the neck of the radius.** Under cover of the bicipital aponeurosis, *the brachial artery divides into the **radial and ulnar arteries*** (Figs. 6-65 and 6-74). Its course through the arm is represented by a line connecting the midpoint of the clavicle with the midpoint of the cubital fossa (Fig. 6-65).

The brachial artery, *superficial and palpable throughout its course,* at first lies medial to the humerus and then anterior to it (Figs. 6-64 to 6-66). It lies anterior to the triceps and brachialis muscles and is overlapped by the coracobrachialis and biceps muscles (Fig. 6-64).

As the brachial artery passes inferiorly and slightly laterally, it **accompanies the median nerve,** which crosses anterior to the artery in the middle of the arm (Fig. 6-64). In the cubital fossa ***the bicipital aponeurosis covers and protects the median nerve and brachial artery*** (Fig. 6-66), and separates them from the medial cubital vein (Fig. 6-67). During its course through the arm, the brachial artery gives rise to many unnamed muscular branches, mainly from its lateral side.

The named branches of the brachial artery are: the ***profunda brachii artery*** (deep brachial artery), the ***nutrient humeral artery***, and the ***superior and inferior ulnar collateral arteries*** (Fig. 6-65).

The Profunda Brachii Artery (Figs. 6-38, 6-42, 6-61, 6-65, and 6-69). **The profunda brachii is the largest branch of the brachial artery** and it has the most superior origin (Fig. 6-65). The profunda brachii artery *accompanies the radial nerve in its posterior course in the **radial groove*** (Figs. 6-59*B*, 6-65, and 6-69). Posterior to the humerus the profunda brachii artery divides into anterior and posterior descending branches, which help to form the arterial *anastomoses of the elbow region* (Fig. 6-65).

The nutrient humeral artery arises from the brachial artery around the middle of the arm and enters the nutrient canal on the anteromedial surface of the humerus. This nutrient canal, like those in the forearm bones, runs toward the elbow.

The superior ulnar collateral artery (Figs. 6-64 and 6-65) arises from the brachial artery near the middle of the arm, and accompanies the ulnar nerve posterior to the medial epicondyle of the humerus (Fig. 6-65). Here it anastomoses with the posterior ulnar recurrent branch of the ulnar artery and the inferior ulnar collateral artery of the brachial (Fig. 6-64).

The inferior ulnar collateral artery (Figs. 6-64 and 6-65) arises from the brachial about 5 cm proximal to the elbow crease, passes inferomedially and anterior to the medial epicondyle of the humerus, where it joins the anastomoses of the elbow region (Fig. 6-65).

The brachial artery is occasionally double during all or part of its course. In these cases one of the arteries lies superficial to the median nerve and is called the ***superficial brachial artery.***

*The arterial anastomoses of the elbow region provide a functionally and surgically important **collateral circulation.*** The brachial artery may be clamped or even ligated distal to the inferior ulnar collateral artery without producing tissue damage. The anatomical basis for this fact is that the ulnar and radial arteries still receive sufficient blood via the anastomoses around the elbow region (Fig. 6-65).

Arterial blood pressure is routinely taken using a **sphygmometer** (sphygmomanometer), consisting of an inflatable (pneumatic) cuff and a mercury manometer. The cuff is placed around the arm and inflated with air until it compresses the brachial artery against the humerus and eventually occludes it (Figs. 6-61 and 6-65). A **stethoscope** is placed over the artery in the cubital fossa, *just medial to the biceps tendon* (Fig. 6-66). As the pressure in the cuff is reduced, blood begins to spurt through the artery. The first audible spurt indicates the *systolic blood pressure.*

Compression of the brachial artery may be produced in almost its entire course (*e.g.,* to control hemorrhage owing to injuries of the forearm). *The best place to compress the brachial*

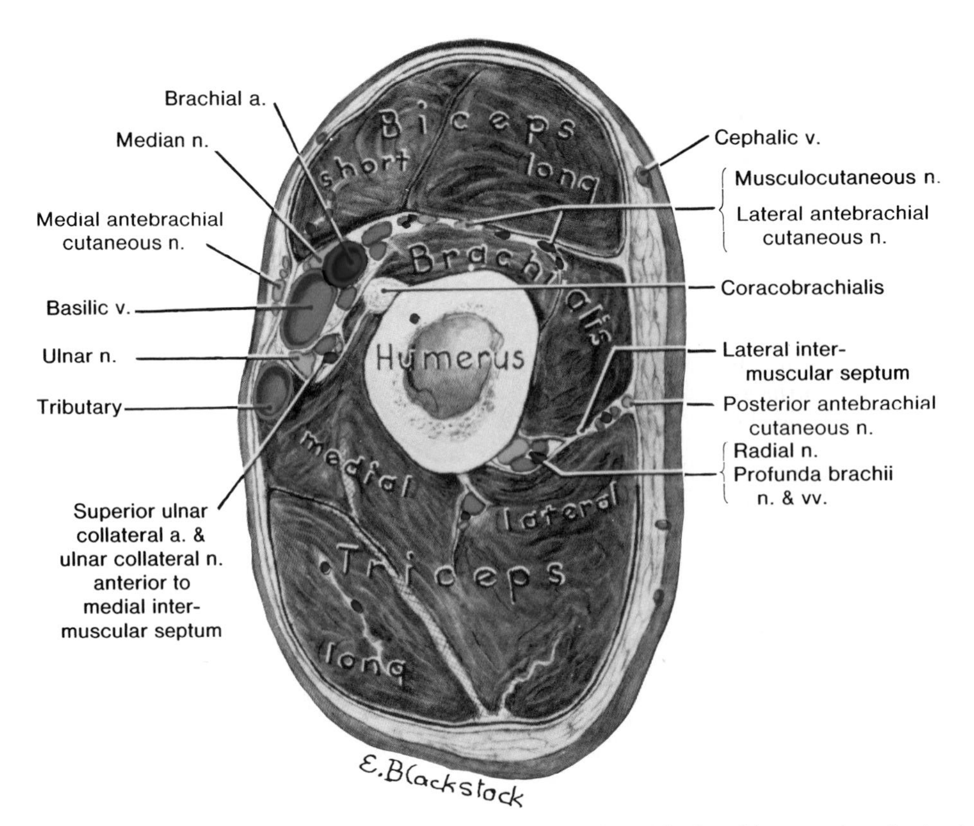

Figure 6-61. Drawing of a cross-section through the arm distal to its midpoint. Observe that the body (shaft) of the humerus is nearly circular; its cortex is thickest here. Note the three heads of the triceps in the posterior fascial compartment of the arm, *i.e.*, posterior to the medial and lateral intermuscular septa. Observe that the radial nerve and its companion vessels are in contact with the humerus. Note the two heads of the biceps, the brachialis, and the insertion of the coracobrachialis in the anterior fascial compartment of the arm, *i.e.*, anterior to the medial and lateral intermuscular septa. Observe the musculocutaneous nerve and its companion vessels in the septum between the biceps and brachialis muscles. Note the median nerve crossing to the medial side of the brachial artery and its venae comitantes. Observe the ulnar nerve moving posteriorly on to the side of the triceps, and that the basilic vein (here as two vessels) has pierced the deep fascia. Note that the skin and subcutaneous tissues are thicker posterolaterally, where they are more exposed to injury, than anteromedially, where they are protected.

artery is near the middle of the arm, where it lies on the tendon of the coracobrachialis muscle medial to the humerus (Figs. 6-61, 6-64, and 6-65). In old people the brachial artery is often tortuous and subcutaneous here and its pulsations may be visible in very thin people. To compress the brachial artery in the inferior part of the arm, pressure has to be directed posteriorly because the artery lies anterior to the humerus in this region (Fig. 6-65).

Occlusion or injury to the brachial artery represents a *surgical emergency* because within a few hours the paralysis resulting from the *associated ischemia of the deep flexors of the forearm* (flexor pollicis longus and flexor digitorum profundus) may be irreversible (Table 6-9).

Often the brachial artery is torn by fragments resulting from a supracondylar fracture of the humerus. Untreated lacerations of the brachial artery causes ischemia of the muscles and nerves resulting in **necrosis** (G. deadness) and paralysis.

Necrotic muscle is replaced by fibrous scar tissue which causes the involved muscles to become permanently shortened. This produces a *flexion deformity* of the wrist and fingers called **Volkmann's ischemic contracture.** There is contraction of the fingers and sometimes of the wrist, with loss of power. *Extended and improper use of a tourniquet can produce this injury,* but it results more often from brachial artery injury (*e.g.*, tearing during a fracture of the elbow). *Impending Volkmann's ischemic contracture, if recognized and treated very early, can be prevented* (Salter, 1983).

Table 6-7
Muscles of the Arm

Muscle	Origin	Insertion	Nerve Supply	Action(s)
Biceps brachii (Figs. 6-58, 6-61, 6-62, and 6-64)	*Short head:* tip of coracoid process of scapula *Long Head:* supraglenoid tubercle of scapula (Figs. 6-49 and 6-64)	Tuberosity of radius and fascia of forearm via bicipital aponeurosis (Figs. 6-49 and 6-66)	Musculocutaneous n. (C5 and **C6**)	Flexes forearm at elbow joint Supinates forearm
Brachialis (Figs. 6-61, 6-62, 6-64, and 6-66)	Distal half of anterior surface of humerus (Fig. 6-49)	Coronoid process and tuberosity of ulna (Fig. 6-49)	Musculocutaneous n. (C5 and **C6**)	Flexes forearm at elbow joint[2]
Coracobrachialis[1] (Fig. 6-64)	Tip of coracoid process of scapula (Fig. 6-49)	Middle third of medial surface of body of humerus (Fig. 6-49)	Musculocutaneous n. (C5, **C6**, and C7)	Flexes arm and helps to stabilize it
Triceps brachii[3] (Figs. 6-52, 6-58, and 6-61 to 6-64)	*Long head:* infraglenoid tubercle of scapula *Lateral head:* posterior surface of humerus, superior to radial groove *Medial head:* posterior surface of humerus, inferior to radial groove	Proximal end of olecranon of ulna and fascia of forearm (Figs. 6-46 and 6-63)	Radial n. (C6, **C7**, and **C8**)	Extends forearm at elbow joint; is the only important extensor muscle of the elbow joint

[1] This muscle is important chiefly as a landmark.
[2] The brachialis is the main flexor of the forearm (see Table 6-16).
[3] The triceps is the main muscle in the posterior aspect of the arm (Fig. 6-61).

VEINS OF THE ARM

The two deep brachial veins accompany the brachial artery (Figs. 6-40, 6-42, 6-61, and 6-66). They begin at the elbow by union of the **venae comitantes** (companion veins) of the ulnar and radial arteries, and end in the axillary vein (Fig. 6-40). The brachial veins contain valves and are connected at intervals by short transverse branches. Not uncommonly, the deep veins join to form one brachial vein for part of their course.

The two main superficial veins of the arm are the ***cephalic and*** **basilic veins** (Figs. 6-40, 6-42, 6-61, and 6-66 to 6-68).

The cephalic vein is located in the superficial fascia (Fig. 6-61) along the anterolateral surface of the biceps and is often visible through the skin (Fig. 6-68). *Superiorly the cephalic vein passes between the deltoid and pectoralis major muscles and through the deltopectoral triangle* (Figs. 6-14 and 6-20) *to empty into the axillary vein* (Fig. 6-40).

The basilic vein, also in the superficial fascia on the medial side of the inferior part of the arm (Figs. 6-61, 6-66, and 6-67). Often the basilic vein is visible through the skin (Fig. 6-68). Near the junction of the middle and inferior thirds of the arm, the basilic vein passes deep to the brachial fascia (Fig. 6-42), runs superiorly into the axilla, where it becomes the axillary vein (Figs. 6-40 and 6-43).

The median cubital vein (Figs. 6-67 and 6-68) is the communication between the basilic and cephalic veins in the cubital fossa; it lies anterior to the bicipital aponeurosis. Because of the prominence and accessibility of the veins in the cubital fossa, they are commonly used for **venipuncture** (taking blood). Considerable variation occurs in the connection of the basilic and cephalic veins in the cubital fossa.

NERVES OF THE ARM

The *four nerves of the arm (median, ulnar, musculocutaneous, and radial)* are terminal branches of the brachial plexus (Figs. 6-20 to 6-24), as is the *axillary nerve* which supplies the skin of the arm over the inferior half of the deltoid and adjacent regions of the arm (Figs. 6-37 and 6-92).

Two brachial nerves (median and ulnar) supply no brachial muscles, but supply the elbow joint and muscles in the anterior aspect of the forearm (Fig. 6-25).

THE MEDIAN NERVE (Figs. 6-20, 6-22 to 6-26, 6-42, 6-61, 6-64, 6-66, and Table 6-3)

This major nerve is formed in the axilla by the union of a lateral root from the lateral cord and a medial root from the medial cord of the brachial plexus. The median

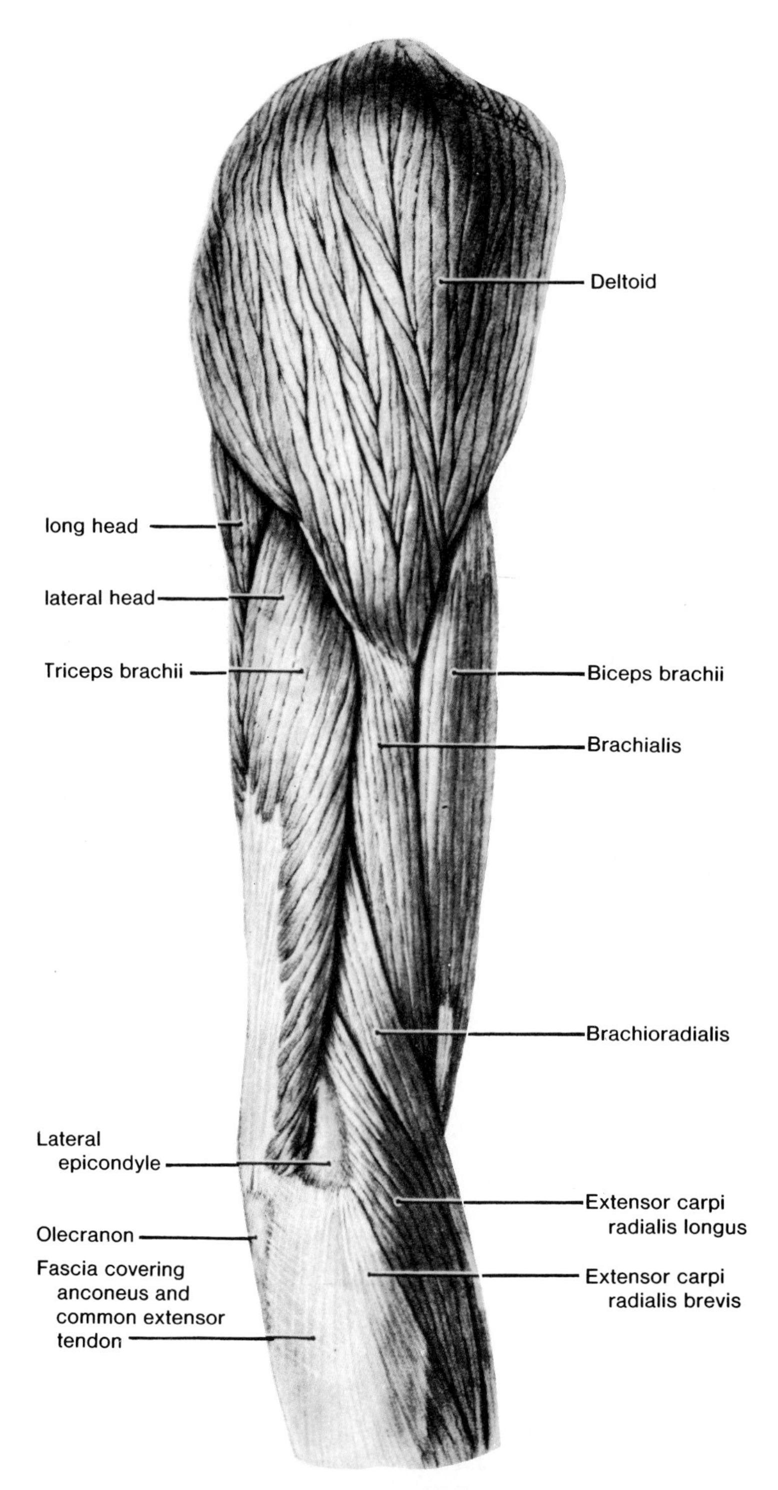

Figure 6-62. Drawing of a lateral view of a dissection of the right arm and proximal part of the forearm. Note particularly the thick triangular deltoid muscle which was given its name because its form is like the Greek letter delta (Δ) inverted. The tendon of this muscle inserts into the deltoid tuberosity of the humerus (Figs. 6-49 and 6-59). The tendon of the triceps brachii inserts on the posterior part of the olecranon and into the deep fascia of the forearm on each side of it.

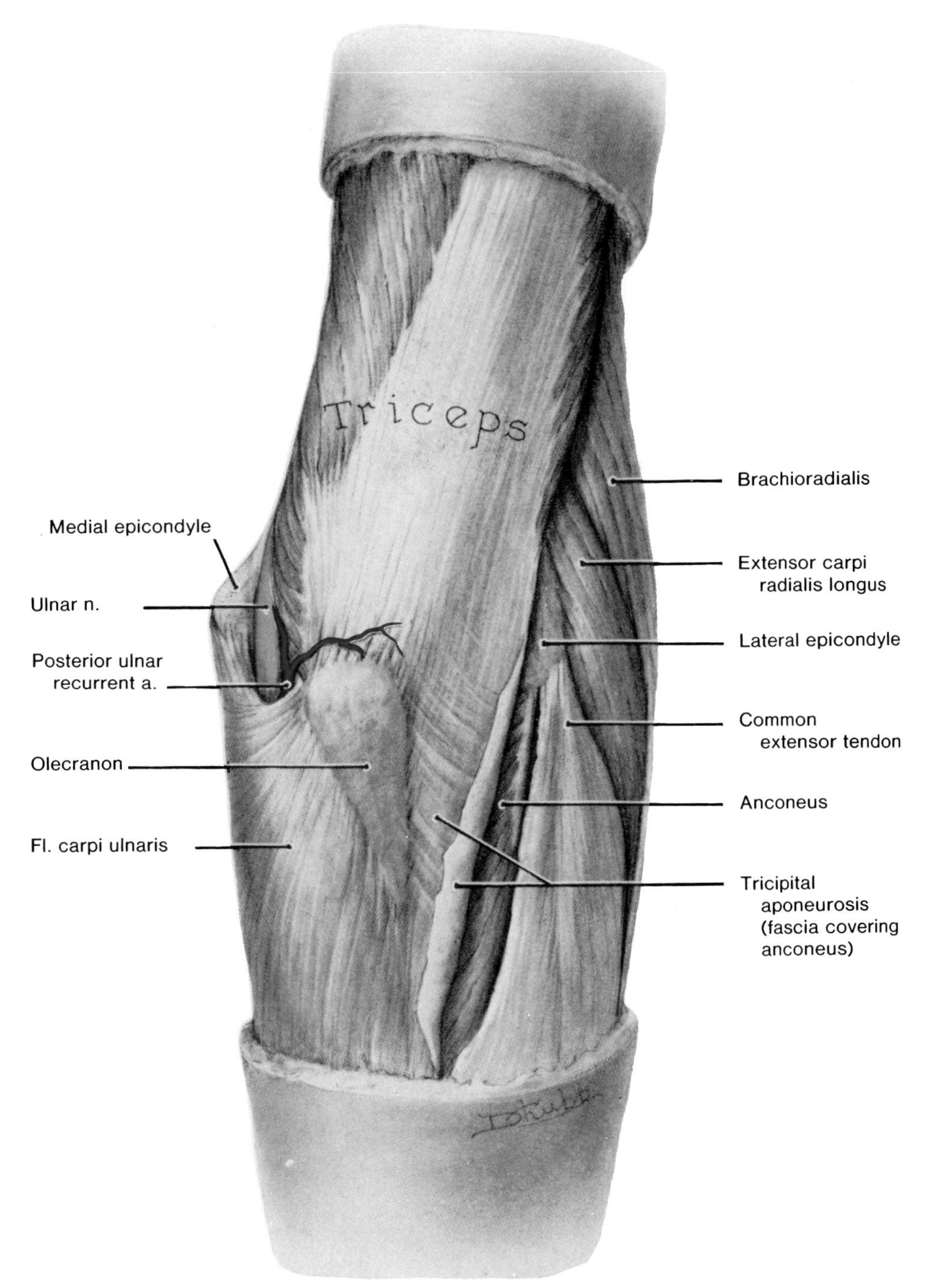

Figure 6-63. Drawing of a posterior view of a dissection of the right elbow region. The fascia covering the anconeus muscle is cut and lifted up. Observe that the triceps is inserted not only into the proximal end of the olecranon, but also via the tricipital aponeurosis into the lateral border of the olecranon. Observe the subcutaneous and palpable posterior surfaces of the medial epicondyle, lateral epicondyle, and olecranon. Note the ulnar nerve, easily palpable, running subfascially posterior to the medial epicondyle. Distal to this point it disappears deep to the two heads of origin of the flexor carpi ulnaris muscle. Note the continuous linear origin from the humerus of the superficial extensor muscles: brachioradialis, extensor carpi radialis longus, common extensor tendon, and anconeus.

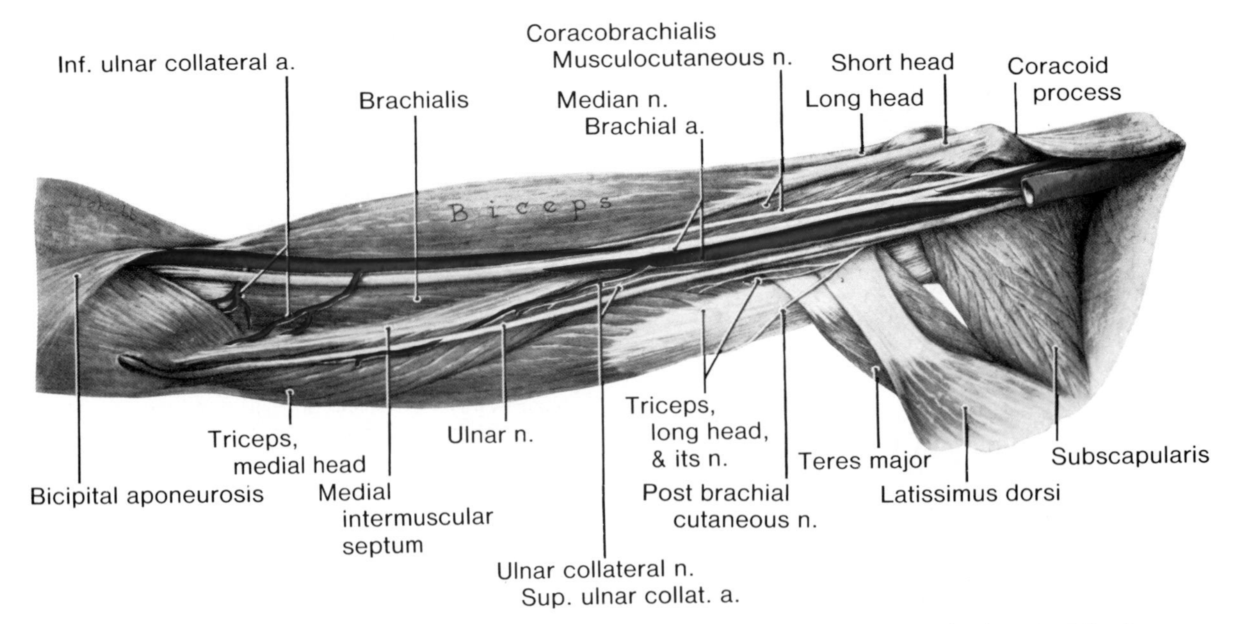

Figure 6-64. Drawing of a dissection of a medial view of the right arm and the proximal part of the forearm. Observe the biceps brachii, coracobrachialis, and brachialis muscles occupying the anterior aspect of the arm, and the triceps brachii occupying the posterior aspect. Note the medial intermuscular septum separating these two muscle groups in the distal two-thirds of the arm. Observe the brachial artery lying near the tip of the coracoid process and applied to the medial side of the coracobrachialis superiorly, and to the anterior surface of the brachialis inferiorly.

nerve runs distally in the arm on the lateral side of the brachial artery (Figs. 6-61, 6-64, and 6-66) until it reaches the middle of the arm, where it crosses to its medial side and contacts the brachialis muscle (Fig. 6-64).

The median nerve descends into the cubital fossa where it lies deep to the bicipital aponeurosis (Fig. 6-66) and the median cubital vein (Fig. 6-67).

The median nerve has no branches in the axilla or in the arm (Fig. 6-25). It passes deeply into the forearm to supply all but one and one-half of the muscles in the anterior part of the forearm (Fig. 6-25 and Table 6-3). It also supplies articular branches to the elbow joint.

Injury to the median nerve proximal to the elbow results in a loss of sensation on the lateral portion of the palm, the palmar surface of the thumb, and the lateral two and one-half fingers (Fig. 6-26). As the median nerve supplies no muscles in the arm, they are not affected. However pronation of the forearm, flexion of the wrist and fingers, and important movements of the thumb are lost or are severely affected because the muscles producing these movements are supplied by the median nerve after it enters the forearm (Fig. 6-25 and Table 6-3).

THE ULNAR NERVE (Figs. 6-20, 6-22 to 6-26, 6-42, 6-61, 6-63, 6-64, and Table 6-3)

This is the larger of the two terminal branches of the medial cord of the brachial plexus (Fig. 6-23). It passes distally anterior to the triceps muscle on the medial side of the brachial artery (Figs. 6-61 and 6-64). Around the middle of the arm, it pierces the medial intermuscular septum and descends between it and the medial head of the triceps muscle (Fig. 6-64).

The ulnar nerve enters the forearm by passing between the medial epicondyle of the humerus and the olecranon (Figs. 6-63 and 6-69). **Posterior to the medial epicondyle of the humerus, the ulnar nerve is superficial and easily palpable.** The ulnar nerve has *no branches in the arm,* but it supplies one and one-half muscles in the forearm (Fig. 6-25 and Table 6-3). It also supplies articular branches to the elbow joint.

Injury to the ulnar nerve in the arm results in *impaired flexion and adduction of the wrist and impaired movement of the thumb, ring, and little fingers* (*e.g.,* poor grasp).

The characteristic clinical sign of ulnar nerve damage is inability to adduct or abduct the medial four digits owing to loss of power of the interosseous muscles (Fig. 6-159). Ulnar nerve inju-

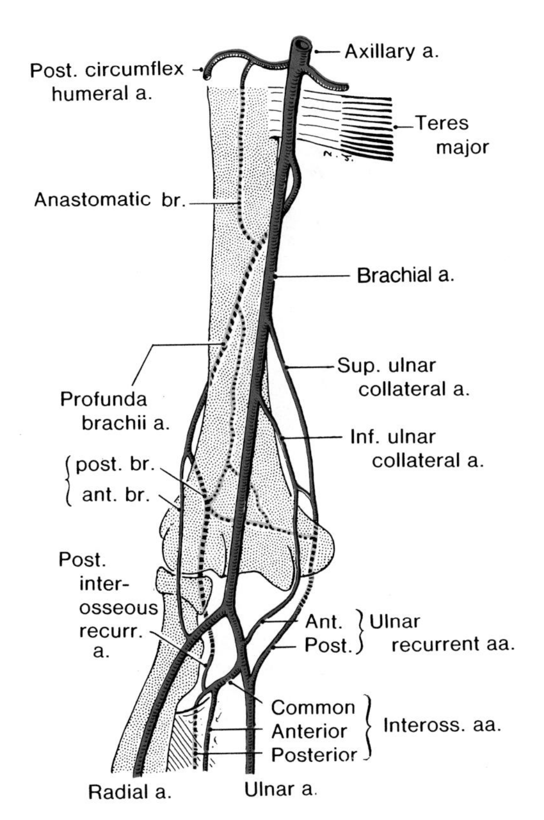

Figure 6-65. Drawing of an anterior view of the arterial supply of the arm and proximal part of the forearm. Observe the clinically important *anastomoses around the elbow region* and that the brachial artery divides opposite the neck of the radius into the ulnar and radial arteries. This bifurcation occurs about 2.5 cm distal to the elbow crease. The profounda brachii artery accompanies the radial nerve through the radial groove of the humerus (Figs. 6-59*B* and 6-69).

ries are discussed in more detail with the forearm and hand.

THE MUSCULOCUTANEOUS NERVE (Figs. 6-20, 6-22 to 6-25, 6-29, 6-42, 6-61, 6-64, and Table 6-3)

This nerve is one of the terminal branches of the lateral cord of the brachial plexus, opposite the inferior border of the pectoralis minor muscle (Figs. 6-22 and 3-23).

The musculocutaneous nerve pierces the coracobrachialis muscle and then continues distally between the biceps and brachialis muscles (Fig. 6-64). It supplies all three of these muscles (Fig. 6-25 and Table 6-3). At the lateral border of the tendon of the biceps, **the musculocutaneous nerve becomes the lateral antebrachial cutaneous nerve** (lateral cutaneous nerve of the forearm) which supplies the skin of the forearm (Figs. 6-66, 6-67, and 6-92).

Injury to the musculocutaneous nerve in the axilla (*e.g.*, from a laceration) before it innervates any muscles (Figs. 6-20 and 6-25) results in paralysis of the coracobrachialis, biceps, and brachialis muscles (Tables 6-3 and 6-6). As a result, **flexion of the elbow joint and supination of the forearm are greatly weakened** (Tables 6-16 and 6-17). There may also be loss of sensation on the lateral surface of the forearm supplied by the lateral antebrachial cutaneous nerve (Figs. 6-67 and 6-92).

THE RADIAL NERVE (Figs. 6-22 to 6-24, 6-26 to 6-29, 6-61, 6-69, and Table 6-3)

This nerve is the direct continuation of the posterior cord of the brachial plexus. The largest branch of the brachial plexus (Fig. 6-24), the radial nerve enters the arm posterior to the brachial artery, medial to the humerus, and anterior to the long head of the triceps (Fig. 6-69).

The radial nerve passes inferolaterally with the profunda brachii artery around the body of the humerus in the radial groove (Figs. 6-59*B* and 6-69). When it reaches the lateral border of this bone, the radial nerve pierces the lateral intermuscular septum. It then continues inferiorly between the brachialis and brachioradialis muscles to the level of the lateral epicondyle of the humerus, where it divides into deep and superficial branches (Fig. 6-74).

The deep branch of the radial nerve is entirely muscular and articular in its distribution (Fig. 6-27 and Table 6-3). The superficial branch supplies sensory fibers to the dorsum of the hand and fingers (Figs. 6-26, 6-92, and 6-93).

Injury to the radial nerve proximal to the origin of the triceps (Fig. 6-27) results in *paralysis of the triceps, brachioradialis, supinator, and extensors of the wrist, thumb, and fingers* (Table 6-3). There would also be loss of sensation in the areas of skin supplied by this nerve (Figs. 6-26 and 6-156*A*).

When the radial nerve is injured in the radial groove, the triceps is not completely paralyzed, but paralysis of other muscles supplied by it occurs (Fig. 6-27).

The characteristic clinical sign of radial nerve injury is wrist-drop, i.e., inability to extend or straighten the wrist (Fig. 6-156*B*).

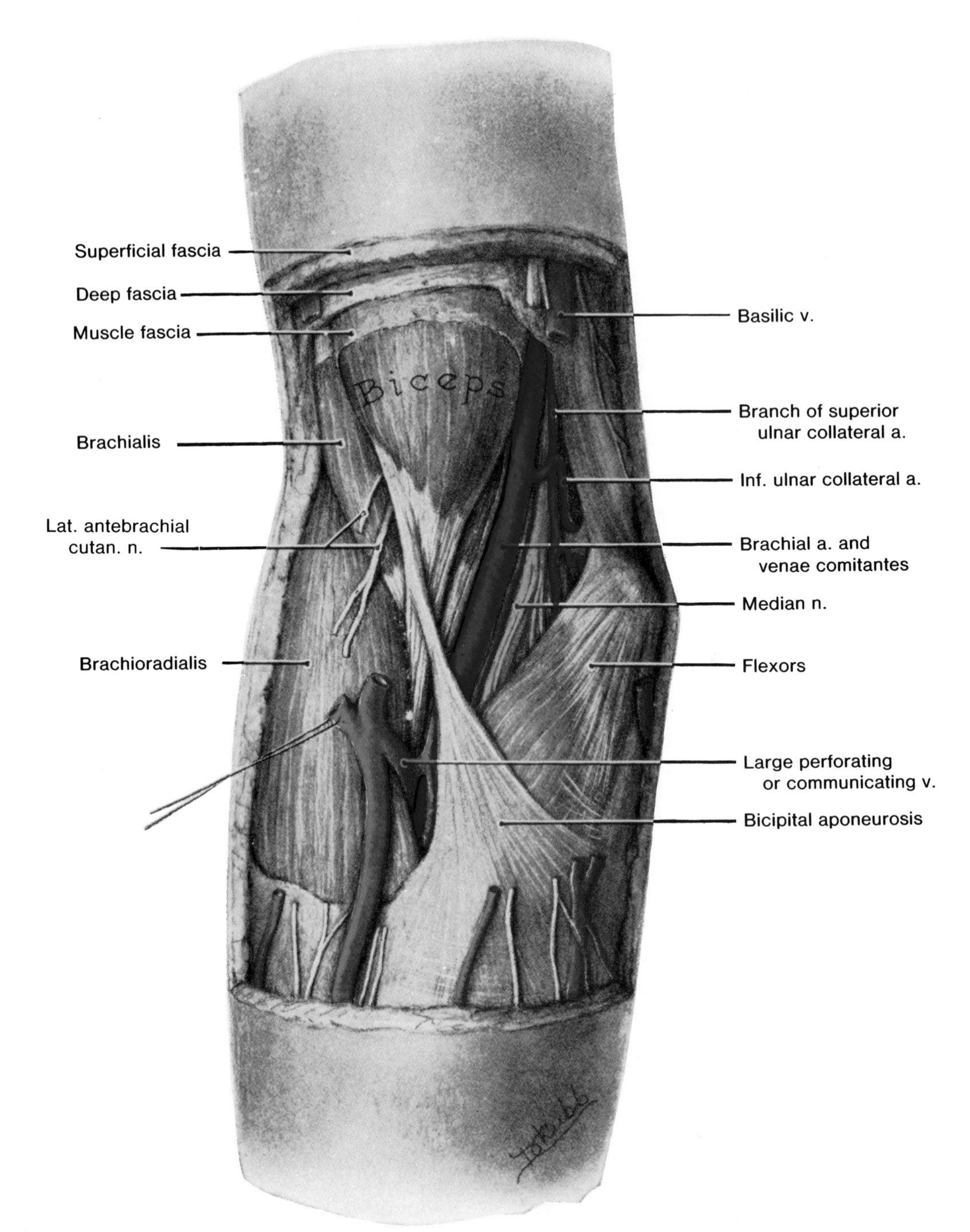

Figure 6-66. Drawing of a dissection of the right *cubital fossa*, the triangular space distal to the elbow crease. It is bounded laterally by the extensor muscles (represented by the brachioradialis) and medially by the flexor muscles (represented by the pronator teres, Fig. 6-80). The apex of the cubital fossa is where these two muscles meet distally. Observe the chief contents of the cubital fossa: biceps tendon, brachial artery, and median nerve. Observe that the brachial artery lies medial to the biceps muscle and its tendon; this is where a stethoscope is placed for taking blood pressure by listening to pulsations of the brachial artery.

THE FOREARM

The forearm (antebrachium) extends from the elbow to the wrist and contains two bones, the **radius** and **ulna**, which are parallel in the anatomical position (Figs. 6-1 and 6-70*A*). In pronation the radius lies across the ulna (Fig. 6-70*B*).

An interosseous membrane joins the radius and ulna (Figs. 6-79 and 6-147). Although thin, *the interosseous membrane is a very strong fibrous sheet.* In addition to tying the forearm bones together, the interosseous membrane provides attachment for some deep forearm muscles (Fig. 6-79 and Table 6-9).

BONES OF THE FOREARM

The forearm bones are called the **radius and the ulna.** The ulna is more firmly connected to the arm bone or humerus, whereas the radius is broadened distally to be more fully in contact with the wrist bones. Also note that *the head of the ulna is at its distal end* (Figs. 6-71 and 6-72*A*), whereas *the head of the radius is at its proximal end* (Figs. 6-72 and 6-73).

THE RADIUS (Figs. 6-1, 6-70, 6-72, and 6-138).

The radius is the shorter and laterally located bone of the two forearm bones. It was given its name because of its resemblance to the spoke of a wheel, which is what the translated *Latin word radius* means.

The proximal end of the radius has a disc-shaped **head,** a smooth cylindrical **neck,** and an oval prominence, the **tuberosity**, distal to the neck (Figs. 6-1 and 6-72*A*).

The body (shaft) of the radius increases in size from the proximal to the distal end; it has a slight lateral convexity or bowing (Fig. 6-72). The body of the radius is concave anteriorly in its proximal three-fourths and flattened in its distal one-fourth. The **anterior oblique line** of the radius runs obliquely across the body from the region of the radial tuberosity to the area of greatest bowing (Figs. 6-1 and 6-72*A*). The medial aspect of the body has a *sharp* ***interosseous border*** for attachment of the interosseous membrane (Figs. 6-79, 6-141, and 6-147). Its lateral border is rounded.

The distal end of the radius has a median **ulnar notch** into which the head of the ulna fits, forming the distal radioulnar joint (Fig. 6-72). Laterally the distal end of the radius tapers abruptly into a prominent pyramidal **styloid process.** The inferior surface of the distal end of the radius is smooth and concave where it articulates with the wrist or carpal bones (Figs. 6-152 and 6-153). Posteriorly there is a prominent **dorsal tubercle** on the distal end of the radius (Fig. 6-72*B*).

A fall on the outstretched hand may result in a fracture of the distal end of the radius (Fig. 6-161*A*). Sometimes there is also a fracture of the styloid process of the ulna.

The common fracture is a *Colles' fracture* in which the distal fragment of the radius is displaced posteriorly and may become impacted (Fig. 6-161*C*). This results in the radial and ulnar styloid processes being at approximately the same horizontal level.

THE ULNA (Figs. 6-1, 6-70, 6-72, and 6-138)

The ulna (L. elbow) is the longer and medially located bone of the forearm. This prismatic bone looks somewhat like a pipe wrench, with the ***olecranon*** resembling the upper jaw, the ***coronoid process*** the lower jaw, and the ***trochlear notch*** the mouth. The olecranon and coronoid processes of the ulna clasp the trochlea of the humerus (Figs. 6-59 and 6-138), somewhat like a pipe wrench clasps a pipe.

Observe that the proximal "wrench-like" end of the ulna is larger than the small, rounded distal end, called the **head** (Figs. 6-71 to 6-73). The lateral side of the *coronoid process* has a small, shallow **radial notch** for the head of the radius (Fig. 6-72*A*). Inferior to the radial notch is a triangular **supinator fossa** (Fig. 6-138*A*), which gives origin to the supinator muscle (Figs. 6-74 and 6-75). This fossa is bounded posteriorly by a distinct **supinator crest** (Fig. 6-72*B*). The irregular anterior surface of the coronoid process is rough and ends distally in the **tuberosity of the ulna** (Fig. 6-72*B*) onto which the brachialis, the chief flexor muscle of the forearm, inserts (Figs. 6-75 and 6-138*A*).

The body (shaft) of the ulna is thick proximally. Its prominent lateral edge is the **interosseous border** (Fig. 6-72*A*), where the *interosseous membrane* is attached (Figs. 6-79 and 6-147).

The small slender distal end of the ulna is composed of a **rounded head** and a conical **styloid process** (Figs. 6-1 and 6-72). The styloid process projects distally, about 1 cm proximal to the styloid process of the radius. *This relationship is clinically important* (see the clinical comments following the description of the radius).

The distal end of the ulna has a convex articular surface on its lateral side for articulation with the ulnar notch of the radius (Fig. 6-72*A*).

SURFACE ANATOMY OF THE FOREARM

The head of the radius can be palpated and felt to rotate in the depression on the posterolateral aspect of the extended elbow joint, just distal to the lateral epicondyle of the humerus (Fig. 6-73).

The body of the radius is partly palpable and *the radial styloid process can be easily palpated* on the lateral side of the wrist, particularly when the tendons that cover it are relaxed (Fig. 6-71).

The radial styloid process is located about 1 cm more distal than the ulnar styloid (Fig. 6-72). *The relationship of the styloid processes of the radius and ulna is important* in the diagnosis *of certain injuries in the wrist* region (*e.g.*, **Colles' fracture**, Fig. 6-161).

Proximal to the radial styloid process, the anterior, lateral, and posterior surfaces of the radius are pal-

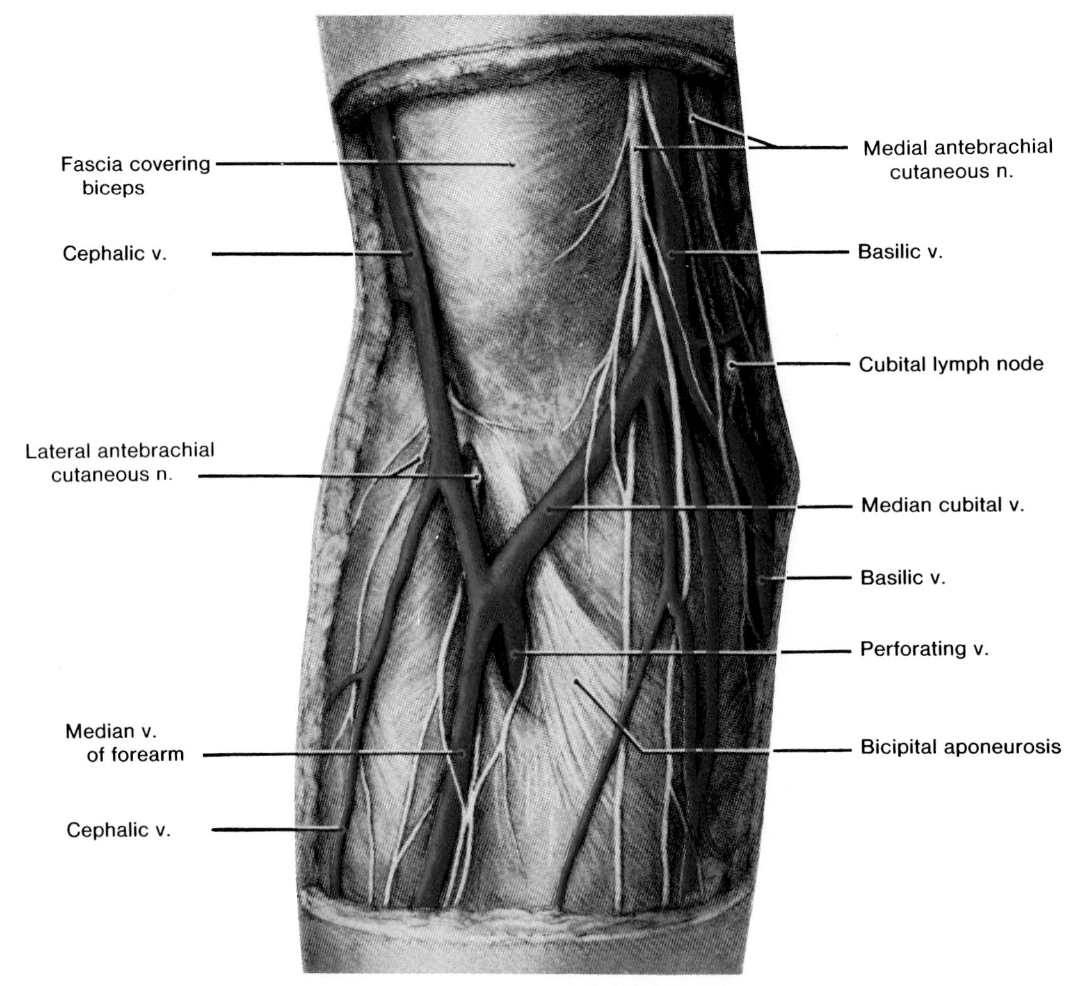

Figure 6-67. Drawing of a dissection of the superficial structures in the anterior aspect of the right elbow region. Observe the superficial veins—cephalic, median, basilic, and their connecting channels—making a variable M-shaped pattern. The median cubital vein is separated from the brachial artery only by the bicipital aponeurosis (see Fig. 6-66). Blood is commonly taken from the cubital veins, usually the median cubital vein. Note that the cephalic and basilic veins in the arm lie on each side of the biceps muscle.

pable for a few centimeters. Note that the anterior and posterior surfaces are deep to the tendons (Figs. 6-80 to 6-84 and 6-96 to 6-101).

The lateral surface of the distal half of the radius is easy to palpate.

The olecranon of the ulna can be easily palpated. The loose skin covering it is often rough (Fig. 6-73) because the elbow frequently rests on it. The subcutaneous posterior border of the ulna (Fig. 6-79) can be palpated along its entire length.

The ulnar nerve can be palpated as a rounded cord where it lies posterior to the medial epicondyle of the humerus (Figs. 6-63 and 6-73). Roll it against the bone and tap it with your finger. You may feel a tingling along the ulnar side of your hand, the area of skin supplied by the ulnar nerve (Fig. 6-26). The sensation resulting from hitting the ulnar nerve here resulted in the term "funny bone" for the medial epicondyle.

The head of the ulna forms a rounded subcutaneous prominence that can be easily seen and felt on the medial part of the dorsal aspect of the wrist (Fig. 6-73), especially when the hand is pronated (Fig. 6-71). The **ulnar styloid process**, also subcutaneous, may be felt slightly distal to the head when the hand is supinated (Fig. 6-70*A*).

MUSCLES OF THE CUBITAL REGION

To facilitate a better understanding of the cubital fossa (Figs. 6-58 and 6-71), the muscles in the cubital region will be described here. They will also be discussed later with the muscles of the forearm.

In the cubital fossa the pronator teres muscle lies medially and the brachioradialis muscle laterally (Fig. 6-74). The floor of the cubital fossa is formed by the brachialis and supinator muscles.

The brachialis muscle (Fig. 6-66 and Table 6-7) was described with the muscles of the arm (p. 680).

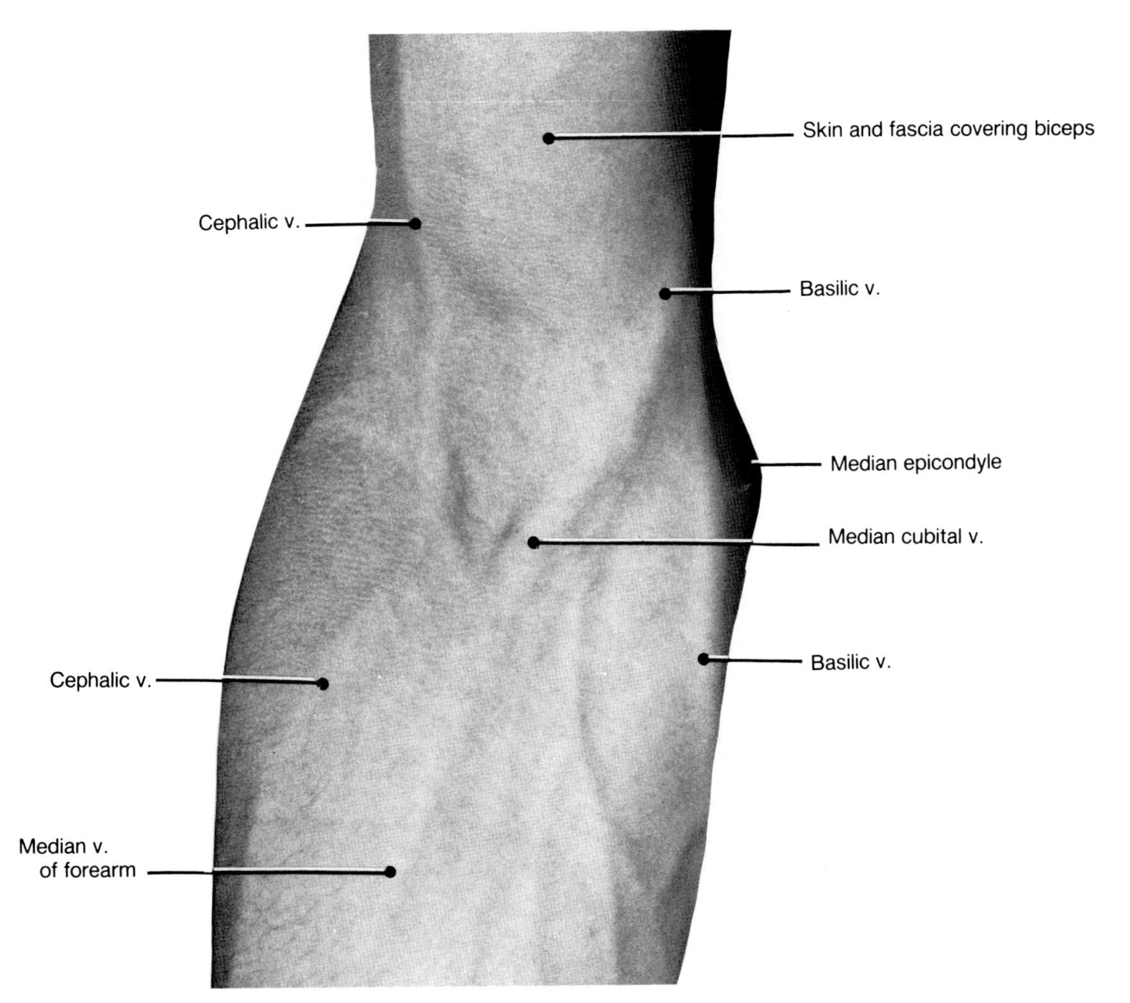

Figure 6-68. Photograph of the superficial veins at the front of the right elbow region of a 27-year-old man. The pattern is similar to that illustrated in Figure 7-67. Note that the basilic and cephalic veins are small and that a median vein of the forearm is present which ends in the basilic vein. The superficial veins of the forearm are extremely variable. The cubital veins are the common site for blood sampling, transfusion, and intravenous injections. The largest vein, usually the median cubital, is commonly selected. Always keep in mind that the brachial artery lies deep to the bicipital aponeurosis.

THE SUPINATOR MUSCLE (Figs. 6-74, 6-85, 6-87, 6-97, and Table 6-11)

This muscle lies deep in the cubital fossa and, along with the brachialis, forms the floor of this triangular fossa.

The humeral and ulnar heads of origin of the supinator muscle envelop the neck and proximal part of the body of the radius (Figs. 6-74 and 6-97).

Origin (Figs. 6-140 and 6-141). **Lateral epicondyle** of the humerus, **radial collateral ligament** of the elbow joint, **anular ligament** of the superior radioulnar joint, **the supinator fossa**, and the **crest of ulna.**

Insertion (Figs. 6-75 and 6-76). Lateral, posterior, and anterior surfaces of **the proximal third of the neck and the body of the radius.**

Nerve Supply (Fig. 6-74). **Deep branch of radial nerve** (C5 and **C6**). This nerve separates the fibers of the supinator into superficial and deep layers.

Action (Fig. 6-70*A* and Table 6-17). **Supinates forearm** (*i.e.*, rotates radius and palm of hand anteriorly). *The biceps brachii assists the supinator in forceful supination*, particularly when the forearm is flexed (*e.g.*, when driving a screw). The supinator acts alone during unopposed supination.

THE BRACHIORADIALIS MUSCLE (Figs. 6-58, 6-62, 6-63, 6-66, 6-74, 6-79, 6-80, and Table 6-10)

This important muscle forms the lateral boundary of the cubital fossa (Fig. 6-66), and is the most superficial muscle on the radial side of the forearm.

Origin (Figs. 6-59*A* and 6-75). **Proximal two-thirds of lateral supracondylar ridge of humerus** and lateral intermuscular septum.

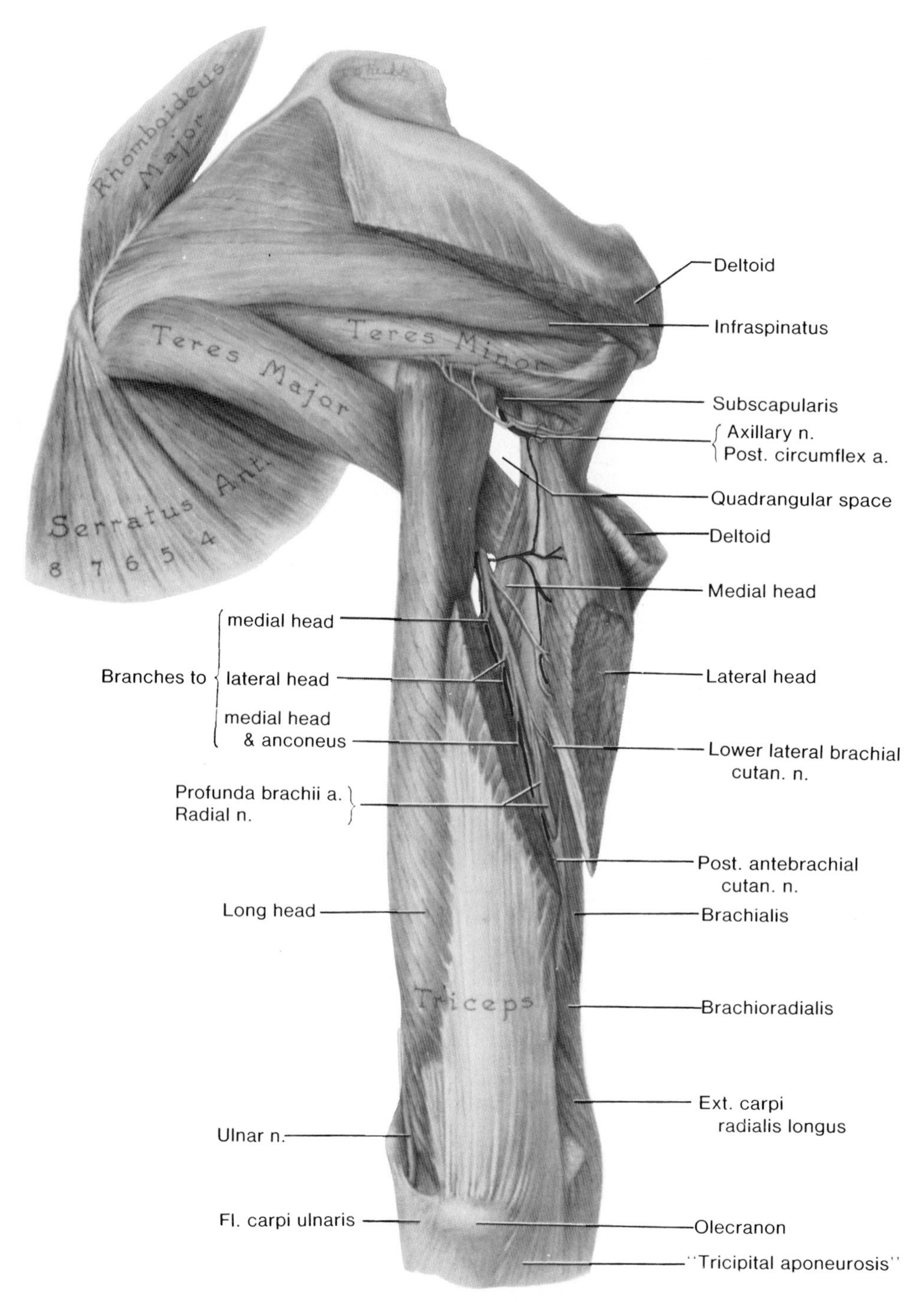

Figure 6-69. Drawing of a posterior view of a dissection of the triceps and its three related nerves (axillary, radial, and ulnar). Observe the radial nerve supplying the lateral and medial heads of the triceps and the anconeus. Note that the triceps is inserted into the proximal end of the olecranon and also into the deep fascia of the forearm. Observe that the teres major, rhomboideus major, and serratus anterior muscles are mainly attached to the inferior angle of the scapula. Rhomboideus major is the Latin name for the rhomboid major muscle (also see Fig. 6-44).

Insertion (Figs. 6-75 and 6-76). **Lateral surface of distal end of radius,** just proximal to the styloid process.

Nerve Supply (Figs. 6-27 and 6-74). **Radial nerve** (C5, **C6,** and C7). *Although a flexor of the elbow joint, the brachioradialis is supplied by the nerve of the extensor muscles (i.e.,* the radial).

Action (Tables 6-10 and 6-16). **Flexes the forearm at the elbow joint.** It is used to give power and speed and acts to best advantage when the forearm is in the midprone position. It is therefore capable of initiating both pronation and supination.

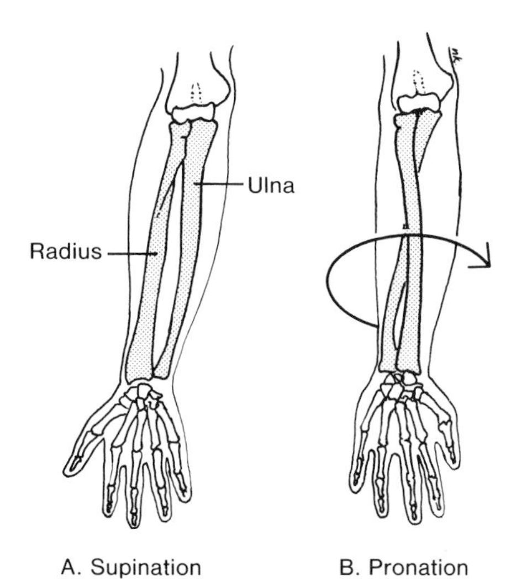

Figure 6-70. Drawing of the bones of the forearm and hand showing how the positions of the radius and ulna change during supination (*A*) and pronation (*B*). The strong supinators are the biceps brachii and supinator muscles (Table 6-17). The chief pronators are the pronator teres and pronator quadratus muscles (Fig. 6-89). Both supination and pronation are most powerful when the elbow is flexed to a right angle.

THE PRONATOR TERES MUSCLE (Figs. 6-58, 6-74, 6-79, 6-80, 6-84, 6-89, 6-96, and Table 6-8)

This fusiform muscle forms the medial boundary of the cubital fossa (Fig. 6-58). *The pronator teres has two heads of origin.*

Origin (Figs. 6-49, 6-75, and 6-89). *Humeral head,* **medial epicondyle of humerus** by common flexor tendon. *Ulnar head,* **coronoid process of ulna.** The pronator teres passes obliquely across the forearm (Figs. 6-80 and 6-84).

Insertion (Figs. 6-75 and 6-76). **Lateral surface of the radius** near its middle.

Nerve Supply (Fig. 6-85). **Median nerve** (C6 and **C7).**

Actions (Fig. 6-70*B* and Tables 6-16 and 6-17). **Pronates the forearm and flexes the elbow joint.** It assists the pronator quadratus in pronation and is a weak flexor of the forearm at the elbow joint.

BONES OF THE WRIST AND HAND

To understand the insertions of the forearm muscles, it is necessary to describe the bones of the hand here, rather than subsequently with the hand.

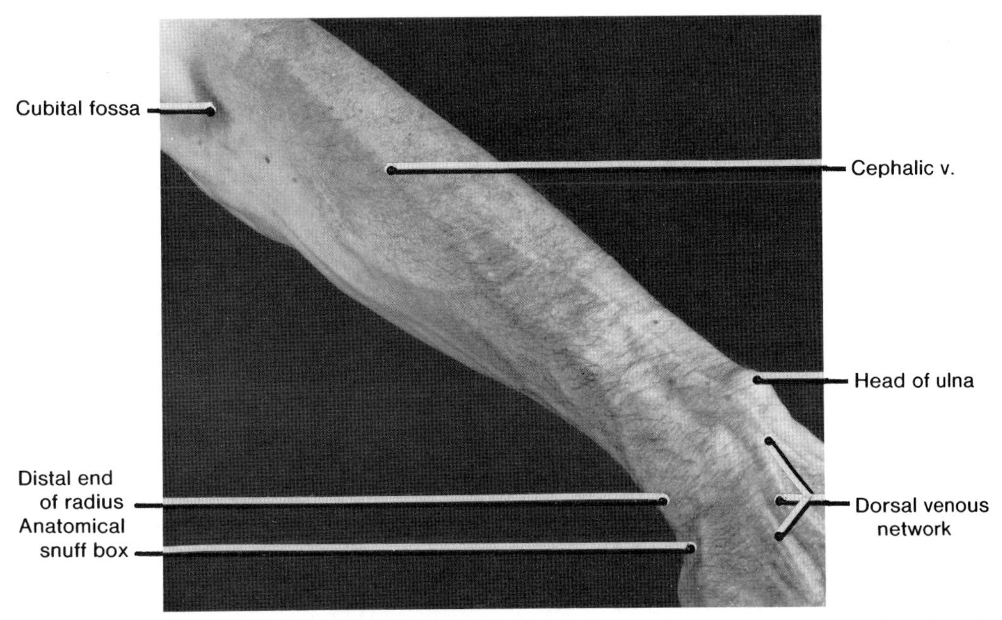

Figure 6-71. Photograph of the pronated left forearm and hand of a 46-year-old man showing the principal surface landmarks. The head of the ulna is a rounded subcutaneous prominence that is easily seen and felt on the medial part of the dorsal aspect of the wrist region when the hand is pronated. The styloid process of the ulna is also subcutaneous and can be felt slightly distal and ventral to the head of the ulna. The styloid process of the radius is not subcutaneous; it is overlaid by the tendons of the abductor pollicis longus and extensor pollicis brevis in the proximal part of the anatomical snuff box. For a close-up of this man's snuff box, see Figure 6-100.

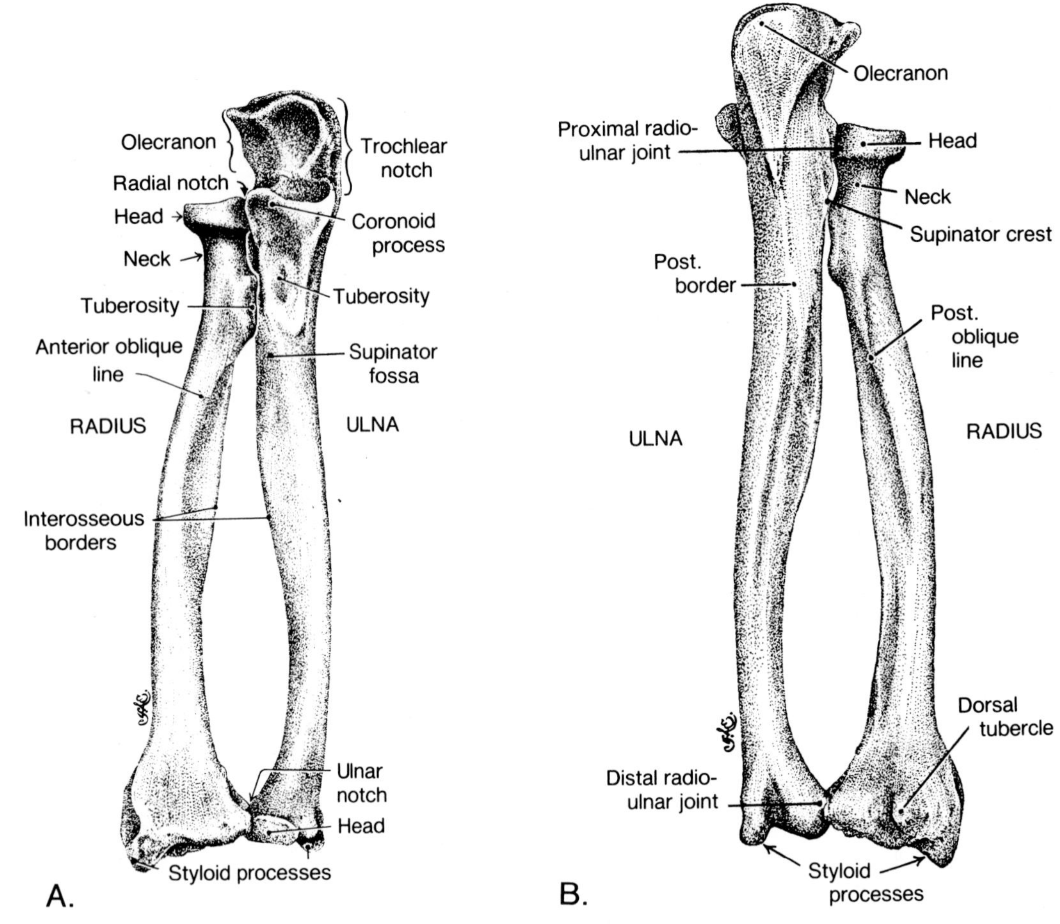

Figure 6-72. Drawings of the radius and ulna. *A*, anterior view. *B*, posterior view. Note the disc-shaped head of the radius. It can be felt through the skin in the depression that is visible on the lateral side of the posterior surface of the extended elbow (Fig. 6-73). In *A*, observe that the proximal end of the ulna resembles a spanner or pipe wrench. The trochlear notch represents the "mouth" of the wrench. Note that the head of the radius is at the elbow, whereas the head of the ulna is at the wrist. In *B*, observe the groove between the head and styloid process of the ulna for the tendon of the extensor carpi ulnaris muscle (Fig. 6-101).

THE CARPUS (Figs. 6-1 and 6-75 to 6-78)

The eight small bones of the wrist called the **carpal bones,** are referred to collectively as the *carpus* (L. wrist). They are arranged in proximal and distal rows, each containing four bones.

The proximal row of carpal bones (lateral to medial) consists of the **scaphoid** (navicular), **lunate, triquetral** (triquetrum), and **pisiform** (Fig. 6-77).

The scaphoid is the largest bone of the proximal row and was given its name because of its resemblance to a boat (G. *scaphe*, rowboat). The pea-shaped pisiform (L. *pisum*, pea + *forma*, appearance) is included in the proximal row, even though it is a *sesamoid bone* in the tendon of the flexor carpi ulnaris muscle (Figs. 6-82 and 6-83).

The pisiform bone is a clinically important landmark that is easily palpable (Fig. 6-82).

The distal row of carpal bones (lateral to medial) consists of the **trapezium, trapezoid, capitate, and hamate.** The hamate can be identified by its prominent process, the **hook of the hamate,** which projects anteriorly (Fig. 6-77*A*). You should also be able to identify the tubercles of the scaphoid and trapezium.

The carpal bones articulate with each other at synovial joints and are bound together with ligaments to form a compact mass. The carpus has an anterior concavity known as the **carpal groove** (sulcus). The carpal groove is converted into an osseofibrous **carpal tunnel** (canal) by the flexor retinaculum, which is attached to the scaphoid and trapezium bones laterally and to the pisiform and the hook of the hamate bone medially (Fig. 6-78). The carpal tunnel is completely filled with tendons and the median nerve (Figs. 6-81 and 6-87). Compression of the median nerve in the carpal tunnel produces the **carpal tunnel syndrome** (see p. 711).

The scaphoid and trapezium lie in the floor of the **anatomical snuff box** (Figs. 6-71 and 6-96 to 6-100). *The scaphoid is the most frequently* *fractured carpal bone* (Fig. 6-160). Injury to this bone results in localized tenderness in the anatomical snuff box.

Initial radiographs may not reveal a fracture of the scaphoid. Repeat radiographs taken two to three weeks after an injury to the wrist may reveal a fracture owing to resorption of bone at the fracture site.

Although fracture of the lunate is rare, anterior dislocation of it is not uncommon. A displaced lunate may compress the median nerve against the flexor retinaculum (Fig. 6-87). The effects of median nerve injury are discussed on page 711.

THE METACARPUS (Figs. 6-1 and 6-77)

The five bones of the hand (**metacarpal bones**) are miniature long bones. They extend from the carpus to the digits and are numbered from the lateral side. In Figure 6-77, note that the first metacarpal is shorter than the others. Although covered with tendons, the metacarpals can be easily palpated throughout their whole length on the dorsum of the hand.

The heads of the metacarpals are at their distal ends (Fig. 6-77*A*), where they articulate with the phalanges (bones of the digits). *The heads of the metacarpals form the knuckles of the hand,* which become visible when the fist is clenched. On the dorsal surface of each head there is a small **tubercle** on each side for attachment of collateral ligaments and joint capsules (Figs. 6-77*A*).

The bodies (shafts) of the metacarpals are slightly concave on their medial and lateral sides, where the dorsal interosseous muscles attach (Fig. 6-117).

The bases of the metacarpals are arranged in a fan-shaped manner from the distal row of carpal bones (Fig. 6-77*B*).

THE PHALANGES (Figs. 6-1 and 6-77)

Each phalanx is a minature long bone consisting of a **body** (shaft), a larger proximal end or **base**, and a smaller distal end or **head.** The thumb (digit 1) has two phalanges (proximal and distal), and each finger (digits 2 to 5) has three phalanges (proximal, middle, and distal). The phalanges of the thumb are shorter and broader than those of the fingers. The proximal phalanges of the digits are the longest and the distal ones the shortest.

MUSCLES OF THE FOREARM

The muscles of the forearm act on the elbow joint, the wrist joint, and the joints of the digits. In the proximal part of the forearm, the muscles form fleshy masses inferior to the medial and lateral epicondyles of the humerus (Figs. 6-79 and 6-80). The tendons of these muscles pass to the distal part of the forearm and continue into the hand (Figs. 6-80, 6-82, 6-83, and 6-87). The muscles of the forearm can easily be divided into flexor and extensor groups.

The flexor-pronator group arises by a common flexor tendon from the medial epicondyle (Fig. 6-75). This is referred to as the **common flexor origin.**

The extensor-supinator group arises by a common extensor tendon from the lateral epicondyle (Figs. 6-75 and 6-76). This is referred to as the **common extensor origin.**

Distal to the elbow joint, the deeper flexor and extensor muscles originate from the anterior and posterior aspects of the bodies (shafts) of the ulna and radius, respectively (Figs. 6-75 and 6-76). The dividing line on the posterior aspect of the forearm between the extensor and flexor groups is the posterior border of the ulna (Fig. 6-79) which is palpable from the olecranon to the wrist (Fig. 6-73).

All the flexor tendons are located on the anterior surface of the wrist and most of them are held in place by the **flexor retinaculum**, a thickening of the deep fascia of the forearm (Figs. 6-78, 6-87, and 6-88).

The eight muscles in the anterior aspect of the forearm are **flexor muscles** (Tables 6-8 and 6-9), and they can be organized into *three functional groups* (Figs. 6-79 and 6-80): (1) **muscles that rotate the radius on the ulna,** *e.g.*, pronate the forearm and hand (pronator teres and pronator quadratus, (2) **muscles that flex the hand** (flexor carpi radialis, flexor carpi ulnaris, and palmaris longus); and (3) **muscles that flex the digits** (flexor digitorum superficialis, flexor digitorum profundus, and flexor pollicis longus).

The anterior forearm (antebrachial) muscles can be divided into three layers (Tables 6-8 and 6-9) as follows: (1) **a superficial layer** (pronator teres, flexor carpi radialis, palmaris longus, and flexor carpi ulnaris); (2) **an intermediate layer** (flexor digitorum superficialis); and (3) **a deep layer** (flexor digitorum profundus, flexor pollicis longus, and pronator quadratus).

A septum separates the deep layer of flexor muscles from the superficial and intermediate layers of flexor muscles (Fig. 6-79). *Within the septum are located the ulnar artery and ulnar nerve.*

THE PRONATOR TERES MUSCLE (Figs. 6-74, 6-79, 6-80, 6-89, and Table 6-8)

This muscle, a pronator of the forearm and hand, and weak flexor of the elbow joint was described with the muscles of the cubital region because the lateral border of the pronator teres *forms the medial boundary of the cubital fossa.* Table 6-8 gives the important facts about this muscle. (Also see p. 695).

THE FLEXOR CARPI RADIALIS MUSCLE (Figs. 6-79 to 6-85 and Table 6-8)

This long narrow muscle lies medial to the pronator teres. In the middle of the forearm its fleshy belly is replaced by a long, flattened tendon that becomes cord-like when it approaches the wrist (Figs. 6-80 and 6-83), where it is readily palpable (Fig. 6-82).

Origin (Fig. 6-75). **Medial epicondyle of the humerus** by the *common flexor tendon.*

Insertion (Fig. 6-75). **Base of second metacarpal bone.** It also sends a slip to the third metacarpal bone. To reach its insertion, the long tendon of the flexor

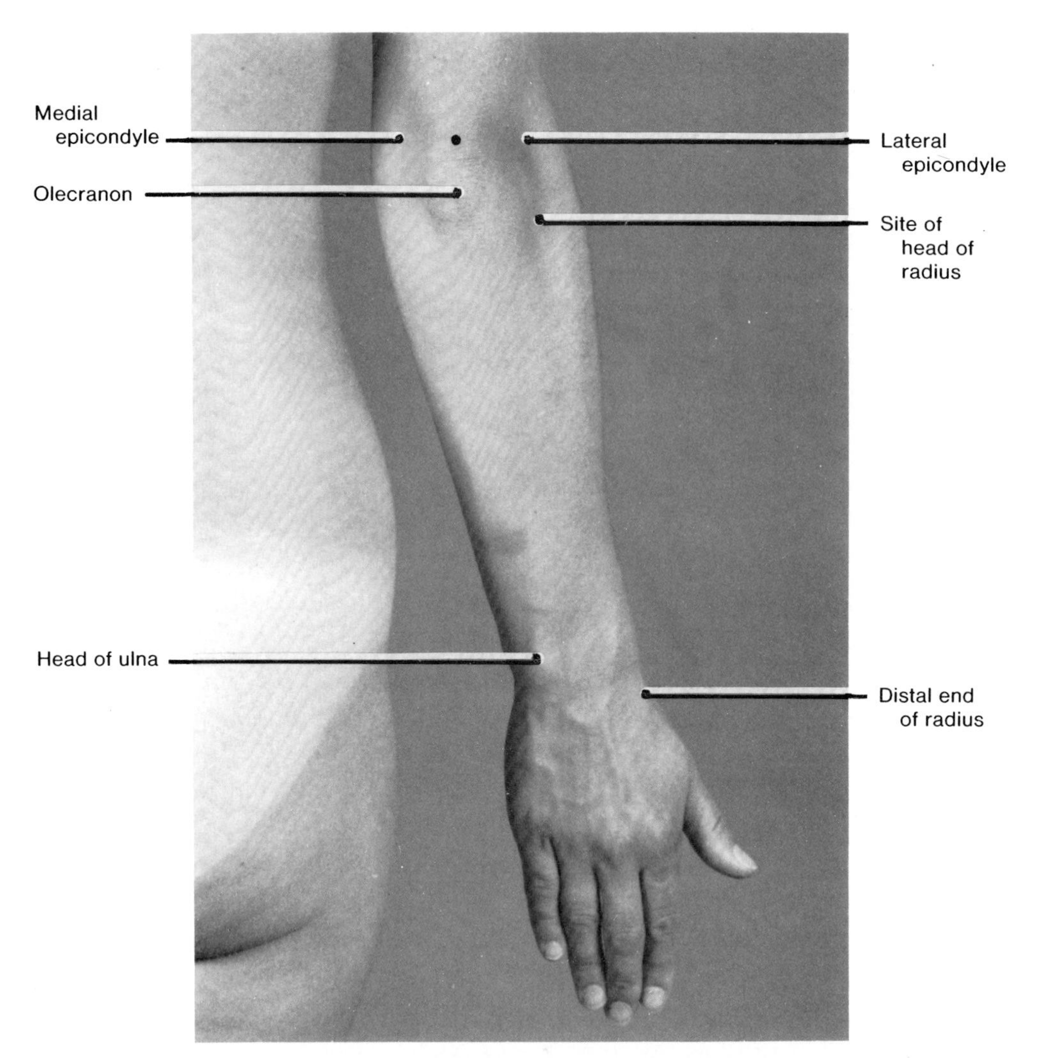

Figure 6-73. Photograph of the posterior aspect of the forearm and hand of a 27-year-old woman showing the principal surface landmarks of the bones (Figs. 6-72 and 6-76). In full extension of the elbow joint, as here, the top of the olecranon (*black dot*) and the two humeral epicondyles are in a straight line. This relationship is important in the diagnosis of certain injuries of the elbow region, *e.g.*, dislocation of the elbow. The head of the radius may be palpated in the depression on the lateral side of the elbow, and the styloid process of the radius can be felt at the distal end of the radius (Fig. 6-72*B*).

carpi radialis passes through a canal in the lateral part of the flexor retinaculum (Fig. 6-81), and through the vertical groove in the trapezium (Fig. 6-78). **The tendon of the flexor carpi radialis may be used as a guide to the radial artery** which lies just lateral to it (Figs. 6-80, 6-82, and 6-83).

Nerve Supply (Fig. 6-80). **Median nerve** (C6 and **C7**).

Actions (Table 6-18). **Flexes wrist and abducts hand.**

The flexor carpi radialis is tested with the posterior aspect of the forearm flat on a table. The patient is asked to flex the wrist against resistance while the examiner feels the tendon of the muscle.

THE PALMARIS LONGUS MUSCLE (Figs. 6-79 to 6-85 and Table 6-8)

Although this small fusiform muscle is absent on one or both sides in about 21% of people, its actions are not missed. Its absence seems to be determined by hereditary factors. When present, the palmaris longus tendon is readily palpable (Figs. 6-82 and 6-84).

Origin (Fig. 6-75). **Medial epicondyle of the humerus** by the *common flexor tendon.*

Insertion (Fig. 6-80). **Palmar aponeurosis.** Its long thin *tendon passes superficial to the flexor retinaculum* (Figs. 6-80 and 6-81) **The palmaris longus tendon may be used as a guide to the median**

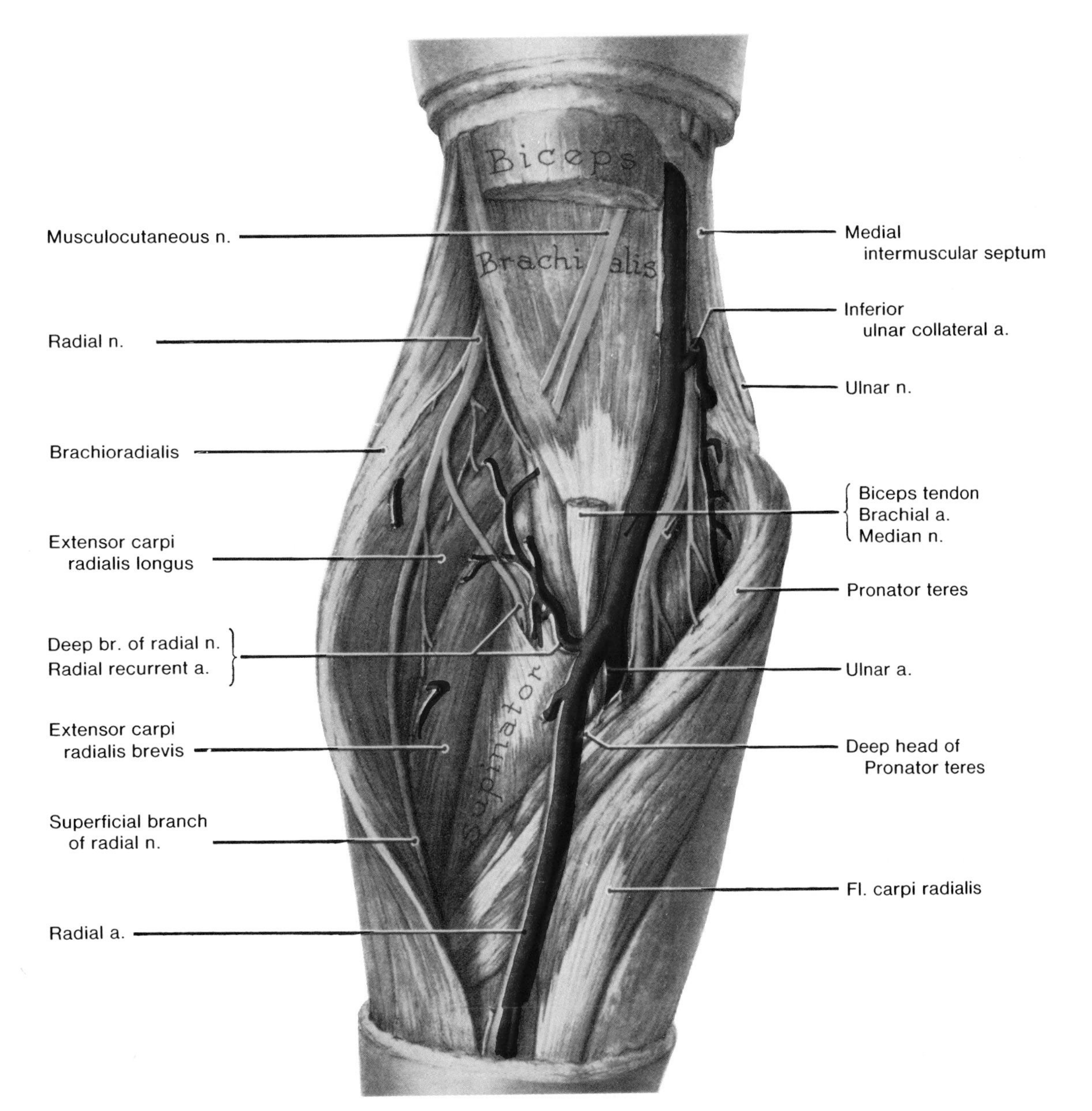

Figure 6-74. Drawing of a dissection of the deep structures in the anterior aspect of the right elbow. Part of the biceps is excised and the cubital fossa (Fig. 6-71) is opened widely. Note that the brachialis and supinator muscles form the floor of the cubital fossa. Observe the brachial artery lying between the biceps tendon and the median nerve and dividing into two nearly equal branches, the ulnar and radial arteries. Note the median nerve supplying flexor muscles and that its motor branches arise from its medial side, except for the twig to the deep head of the pronator teres. Note also *the radial nerve supplying extensor muscles* and that its motor branches arise from its lateral side, except for the twig to the brachialis. The radial nerve has been displaced laterally so its lateral branches appear to run medially in the drawing. Observe the deep branch of the radial nerve piercing the supinator muscle. It appears in the posterior fascial compartment of the forearm as the posterior interosseous nerve (Fig. 6-97).

nerve which passes lateral to it at the wrist (Figs. 6-80 and 6-83).

Nerve Supply (Fig. 6-80). **Median nerve** (C7 and C8).

Action (Table 6-18). **Flexes the wrist** and may act as a tensor of the palmar fascia.

In lacerations of the wrist and on practical examinations, the tendon of the palmaris longus can be mistaken for the median nerve. In Figures 6-80 to 6-83, note that the median nerve is larger

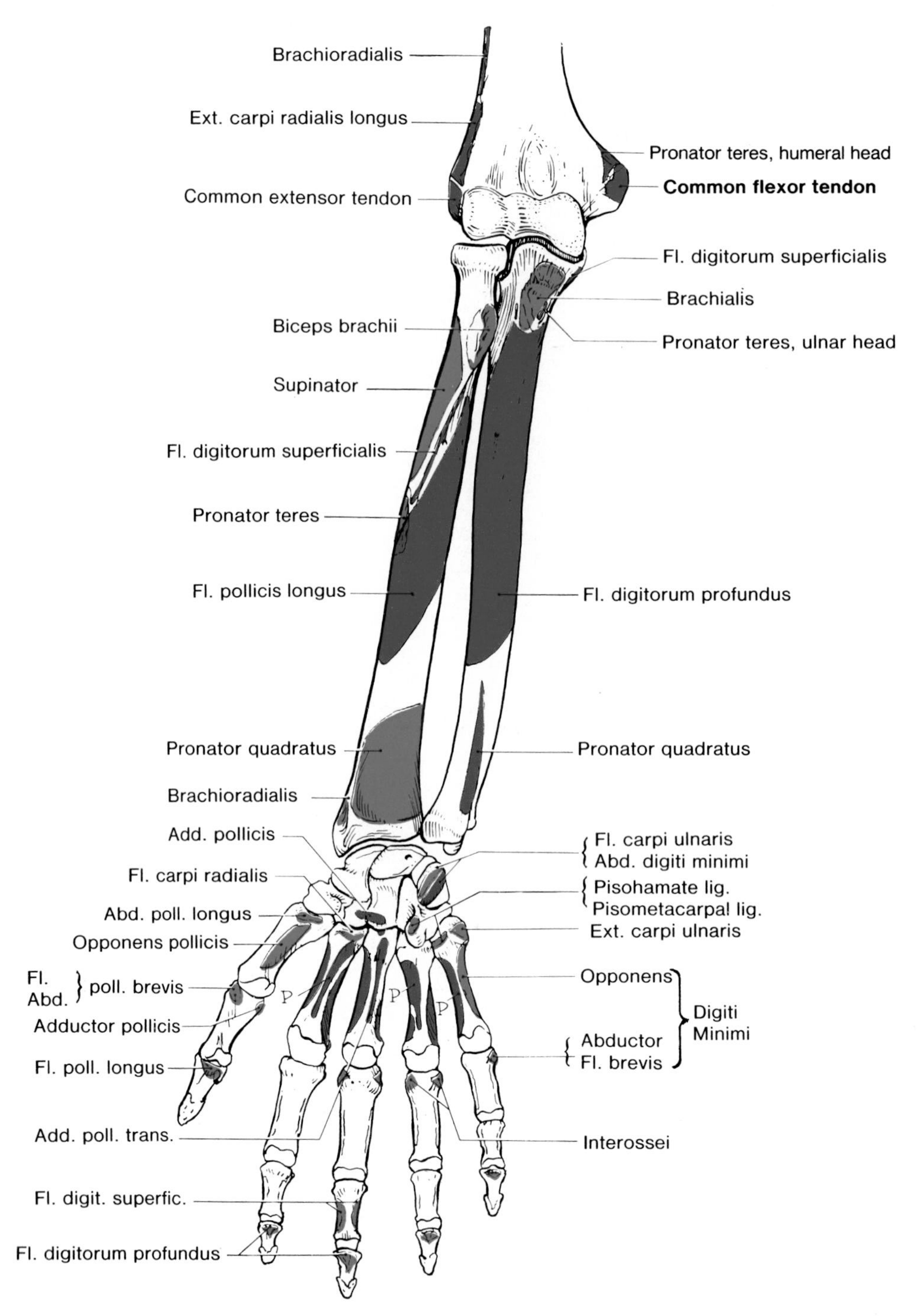

Figure 6-75. Drawing of an anterior view of the bones of the arm, forearm, and hand showing the attachment of muscles. Origins are shown in *red*; insertions in *blue.* Note the origin of the *common flexor tendon* from the medial epicondyle, from which the superficial flexor muscles arise and diverge like a narrow fan (Figs. 6-80 and 6-84).

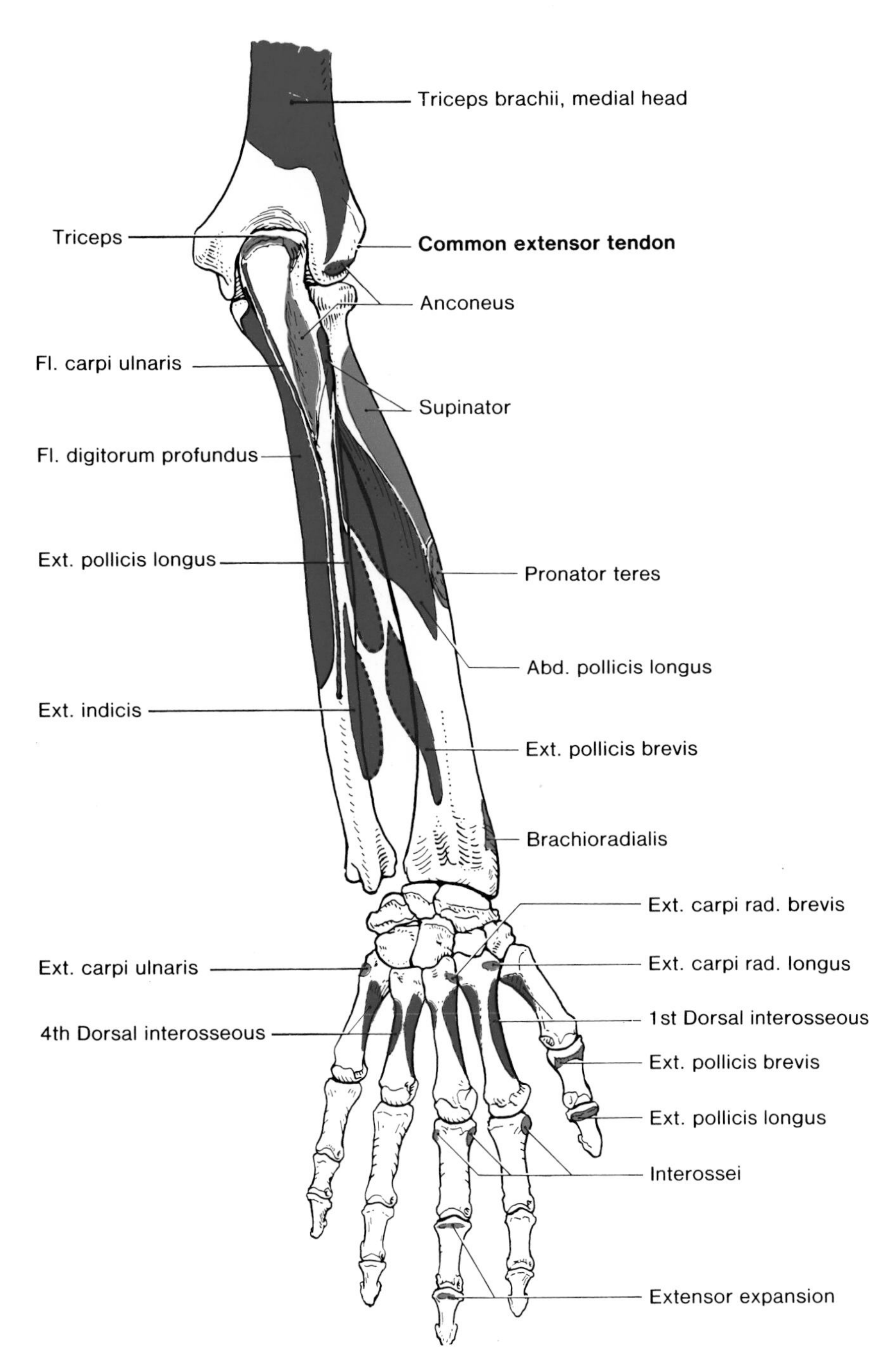

Figure 6-76. Drawing of a posterior view of the bones of the right arm, forearm, and hand showing the attachments of muscles. Note particularly the origin of the *common extensor tendon* from the lateral epicondyle of the humerus. Four superficial extensor muscles arise by this common tendon (Table 6-10).

and lies deep to the palmaris longus tendon (Fig. 6-81).

THE FLEXOR CARPI ULNARIS MUSCLE (Figs. 6-79 to 6-85, 6-87, and Table 6-8)

This is the most medial of the superficial flexor muscles. It has *two heads of origin* between which the ulnar nerve passes distally in the forearm (Fig. 6-85). Its tendon is readily palpable (Fig. 6-82).

Origin (Figs. 6-75 and 6-76). *Humeral head,* **medial epicondyle of the humerus** by the common flexor tendon. *Ulnar head,* **medial border of the olecranon and the proximal two thirds of the posterior border of the ulna.**

Insertion (Figs. 6-75 and 6-83). **Pisiform bone** and through two strong ligaments (pisohamate and pisometacarpal) into the **hook of the hamate and the base of the fifth metacarpal bone,** respectively.

The tendon of the flexor carpi ulnaris is a good guide to the ulnar nerve and artery, which are on its lateral side at the wrist (Fig. 6-83).

Nerve Supply (Fig. 6-85). **Ulnar nerve** (C7 and **C8**).

Actions (Table 6-18). **Flexes the wrist and adducts the hand.** The flexor carpi ulnaris is tested with the posterior aspect of the forearm on a flat table. The patient is asked to flex the wrist against resistance while the examiner feels the tendon of the muscle.

THE FLEXOR DIGITORUM SUPERFICIALIS MUSCLE (Figs. 6-79 to 6-83, 6-85, and Table 6-8)

This is the largest superficial muscle in the forearm; **it has two heads of origin.** The flexor digitorum superficialis is *more deeply located than the other superficial muscles.* It forms an intermediate layer between the superficial and deep groups of muscle.

Origin (Figs. 6-75 and 6-85). *Humeroulnar head,* **medial epicondyle of the humerus** by the common flexor tendon, **ulnar collateral ligament** of the elbow joint, and **coronoid process of the ulna.** *Radial head,* superior half of the **anterior border of the radius.** The radial head, arising from the oblique line on the radius, is broader but thinner than the humeroulnar head. The median nerve and ulnar artery pass more deeply between the two heads of this muscle (Fig. 6-85).

Insertion (Fig. 6-75). **Palmar aspect of bodies of the middle phalanges of the medial four digits** (fingers 2 to 5). As the wrist is approached, the flexor digitorum superficialis gives rise to four tendons which pass deep to the flexor retinaculum (Figs. 6-81 and 6-85), where they are surrounded by a **common flexor synovial sheath** (Fig. 6-88). The superficial pair of tendons passes to the middle and ring fingers within ***synovial sheaths*** in osseofibrous tunnels in the fingers (Figs. 6-86 and 6-88).

Nerve Supply (Fig. 6-85). **Median nerve** (C7, **C8**, and T1).

Actions (Tables 6-19 and 6-20). **Flexes the middle phalanges of the medial four digits** (*i.e.,* flexes proximal interphalangeal joints). *In continued action, it flexes the metacarpophalangeal and wrist joints.*

The Deep Group of Flexor Muscles in the Forearm (Table 6-9) is composed of the flexor digitorum profundus, flexor pollicis longus, and pronator quadratus. None of these muscles arises from the humerus; they all arise from the radius or ulna (Fig. 6-75).

THE FLEXOR DIGITORUM PROFUNDUS MUSCLE (Figs. 6-79 to 6-81, 6-85, 6-87, 6-91, and Table 6-9)

This long, thick, *deep* (L. profundus) muscle is the only one which can flex the distal interphalangeal joints of the fingers, *i.e., it flexes all joints of the fingers,* (Tables 6-19 and 6-20). It has an extensive origin from the ulna and the interosseous membrane.

Origin (Fig. 6-57). **Proximal three-fourths of the medial and anterior surfaces of the ulna and the medial half of the interosseous membrane** (Fig. 6-79).

The flexor digitorum profundus divides into four parts which end in four tendons that pass posterior to the tendons of the flexor digitorum superficialis and the flexor retinaculum (Figs. 6-80, 6-81, and 6-87). Each tendon enters the fibrous sheath of its digit posterior to the tendon of the flexor digitorum superficialis (Figs. 6-88 and 6-117).

Insertion (Fig. 6-75). **Palmar surfaces of the bases of the distal phalanges of the medial four digits** (fingers 2 to 5). The portion of the muscle going to the index finger usually separates from the rest of the muscle for some distance in the distal part of the forearm (Fig. 6-87).

Nerve Supply (Figs. 6-25 and 6-87). The innervation of the flexor digitorum profundus is double. The **medial part**, associated with the little and ring fingers, is supplied by the **ulnar nerve (C8** and T1), and the **lateral part**, associated with the index and middle fingers, is supplied by the **median nerve (C8** and T1) via its anterior interosseous branch.

Actions (Tables 6-19 and 6-20). **Flexes the distal phalanges of the fingers** (digits 2 to 5), after the flexor digitorum superficialis muscle has flexed the middle phalanges. Each tendon flexes two interphalangeal joints, a metacarpophalangeal joint and the wrist joint.

The flexor digitorum profundus is the only muscle that can flex the distal interphalangeal joints of the medial four digits (Tables 6-19 and 6-20). **This is clinically important.** Its action is to flex the distal interphalangeal joints of the fingers, but it helps to flex all joints crossed by its tendons (*i.e.,* wrist joint,

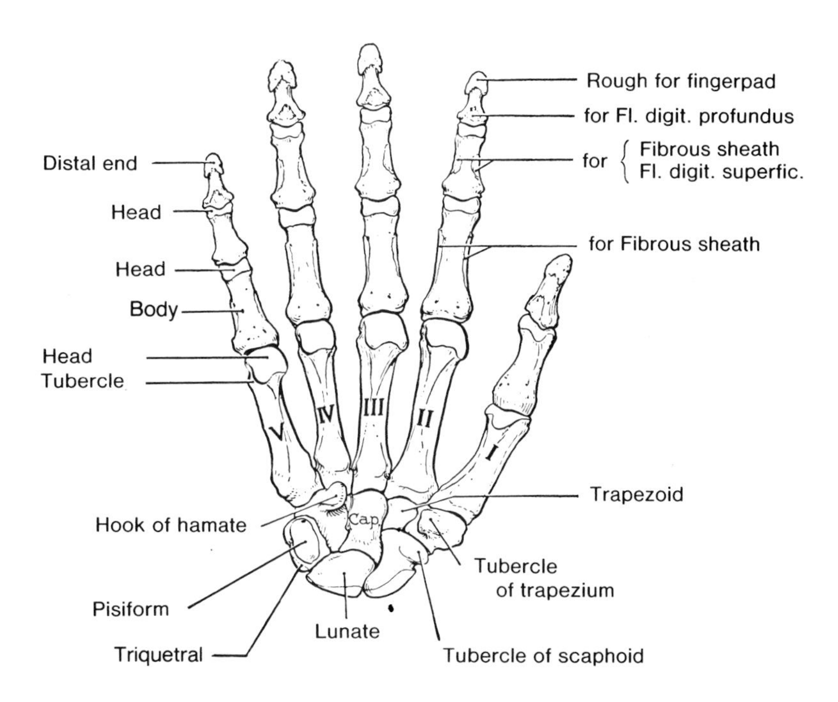

A. PALMAR ASPECT

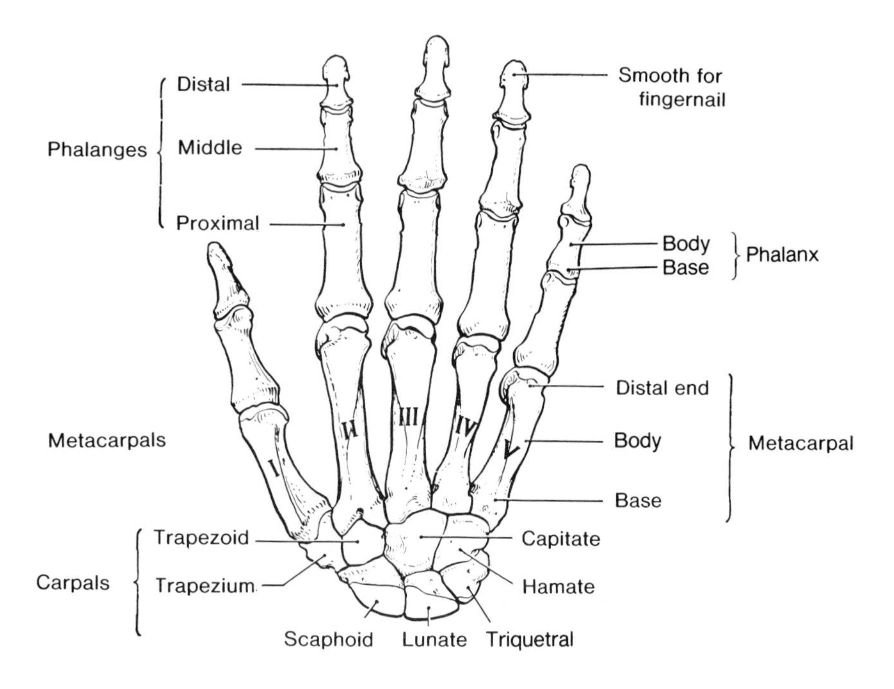

B. DORSAL ASPECT

Figure 6-77. Drawings of the bones of the hand. The skeleton consists of three segments: (1) the carpal bones of the wrist, (2) the metacarpal bones of the hand, and (3) the phalanges of the digits (fingers and thumb). The heads of the metacarpals form the distinctive knuckles of the hand. The phalanges may be felt easily; their heads form the knuckles of the fingers.

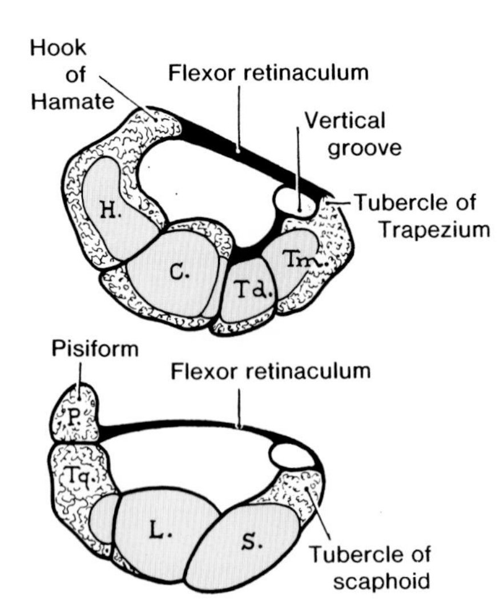

Figure 6-78. Drawings of the distal and proximal rows of carpal bones (Fig. 6-77*A*) showing how the flexor retinaculum stretches between the ends of the concavity of the carpal bones and forms an osseofibrous carpal tunnel through which pass several tendons and the median nerve to the hand (Fig. 6-81). The tendon of the flexor carpi radialis runs through a small tunnel called the vertical groove.

metacarpophalangeal joints, and two interphalangeal joints).

THE FLEXOR POLLICIS LONGUS MUSCLE (Figs. 6-79 to 6-83, 6-87, 6-88, and Table 6-9)

This long flexor of thumb (L. *pollex*) lies lateral to the flexor digitorum profoundus.

Origin (Figs. 6-72*A* and 6-75). **Anterior surface of the radius,** between the anterior oblique line and the pronator quadratus. It also arises from the **interosseous membrane** (Fig. 6-79). Its flat tendon passes deep to the flexor retinaculum (Fig. 6-87), enveloped in its own synovial sheath on the lateral side of the common flexor synovial sheath (Fig. 6-88).

Insertion (Fig. 6-75). **Palmar surface of the base of the distal phalanx of the thumb.**

Nerve Supply (Figs. 6-25 and 6-91). Anterior interosseous branch of the **median nerve (C8** and T1). The anterior interosseous nerve arises in the distal half of the cubital fossa.

Actions (Table 6-21). **Flexes the phalanges of the thumb.** *The flexor pollicis longus is the only muscle that flexes the interphalangeal joint of the thumb.* It also flexes the metacarpophalangeal and carpometa-

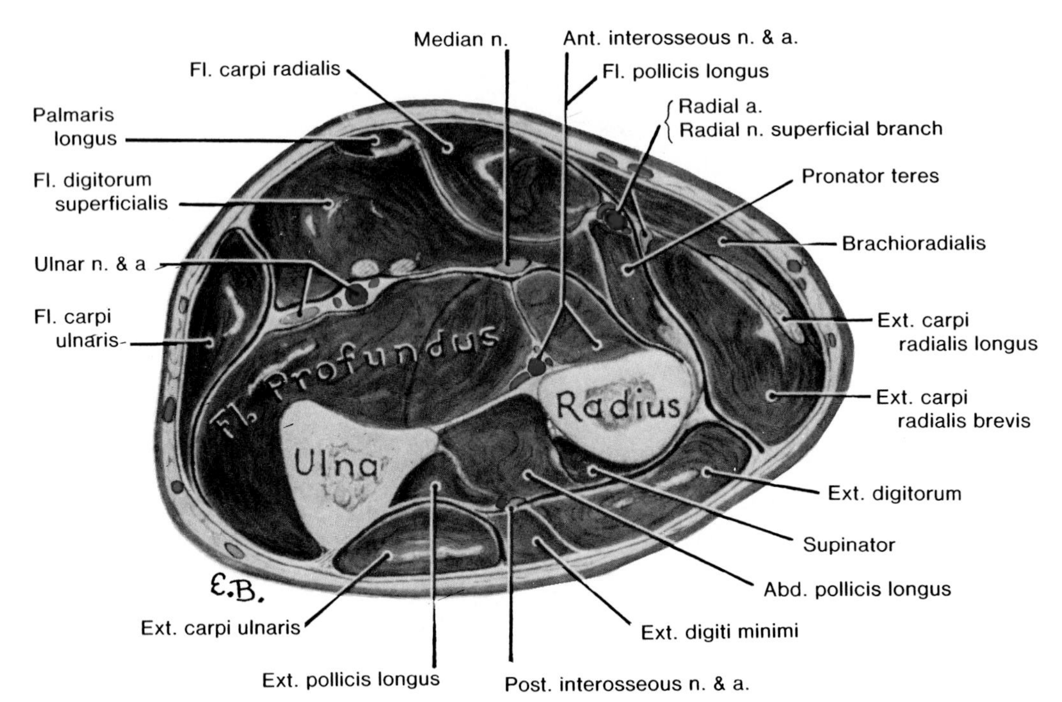

Figure 6-79. Drawing of a cross-section through the middle of the forearm at the level of insertion of the pronator teres (Fig. 6-76). Observe the interosseous membrane stretching from the interosseous border of the ulna to the interosseous border of the radius (also see Fig. 6-147). Note the ulnar nerve and artery and the median nerve lying in the septum between the superficial and deep digital flexor muscles. Observe the flexor digitorum profundus and flexor pollicis longus around the medial and anterior surfaces of the ulna and the anterior surface of the radius. Note the pronator teres inserted into the lateral surface of the radius (see Fig. 6-76).

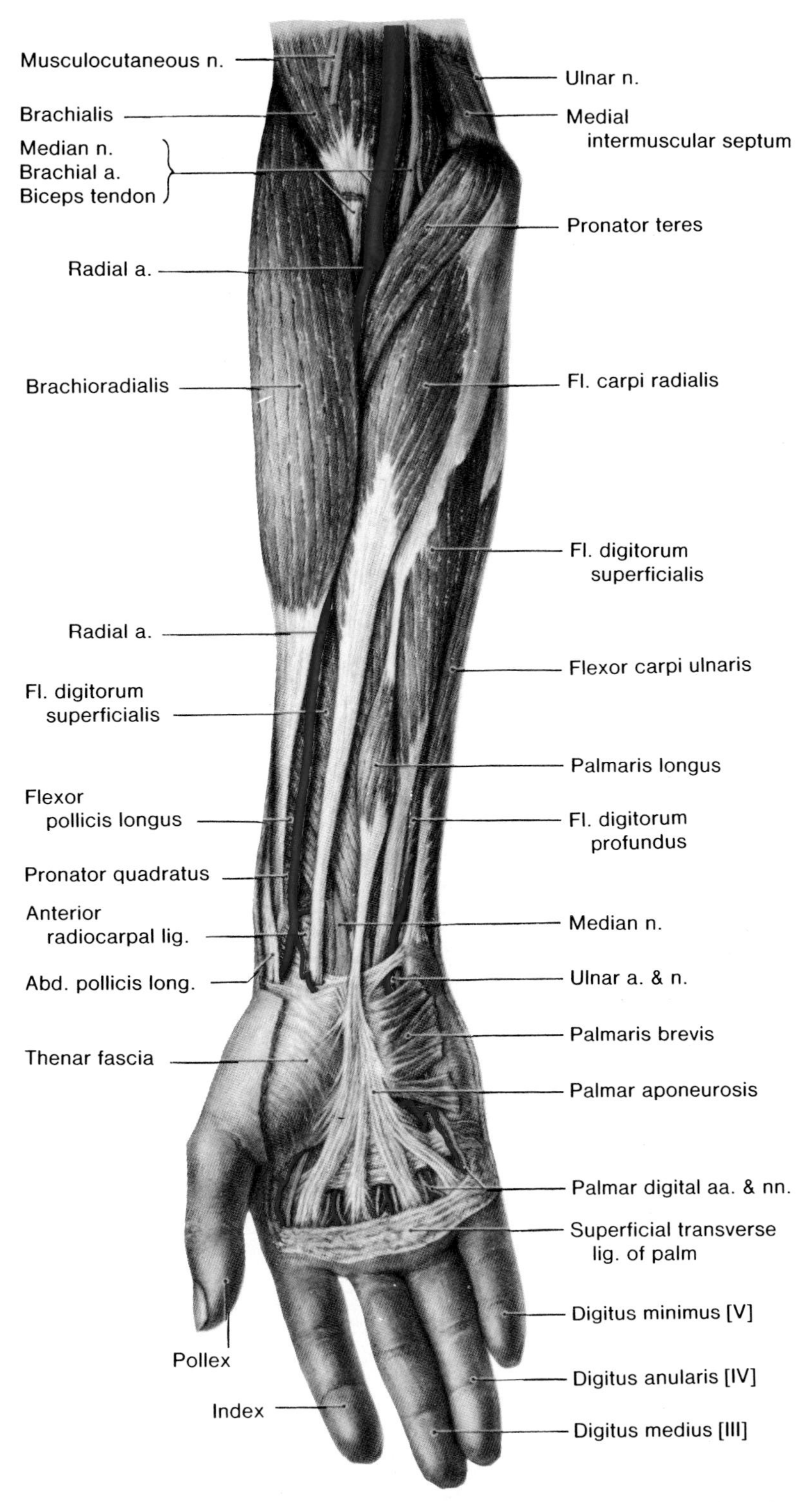

Figure 6-80. Drawing of a dissection of the superficial muscles in the anterior aspect of the forearm. *At the elbow* observe the brachial artery lying between the biceps tendon and the median nerve and bifurcating into the radial and ulnar arteries. *In the forearm* observe the radial artery lying between two muscle groups. Note that the palmaris longus is continued into the palm as the palmar aponeurosis which receives an accession of fibers from the flexor retinaculum and divides into four longitudinal bands, one for each finger. *At the wrist* note the radial artery lateral to the flexor carpi radialis tendon and the ulnar artery lateral to the flexor carpi ulnaris tendon.

carpal joints of the thumb, and it may assist in flexion of the wrist joint (Table 6-18).

THE PRONATOR QUADRATUS MUSCLE (Figs. 6-80, 6-81, 6-85, 6-87, 6-89, and Table 6-9)

As its name indicates, this small muscle is *quadrangular* (*i.e.,* four angles and four sides). It cannot be palpated or observed, except in dissections, because it is the deepest muscle in the anterior aspect of the forearm and comprises the fourth layer.

The pronator quadratus is the only muscle that arises only from the ulna and inserts only into the radius (Fig. 6-89).

Origin (Figs. 6-75 and 6-89). **Distal fourth of the anterior surface of the ulna.** In Figure 6-89 note that the pronator quadratus passes almost transversely, but has a slight distal slant.

Insertion (Figs. 6-75, 6-89, and 6-91). **Distal fourth of the anterior surface of the radius.**

Nerve Supply (Fig. 6-87). Anterior interosseous branch of the **median nerve** (**C8** and T1).

Actions (Figs. 6-70*B*, 6-89, and Table 6-17). **Pronates the forearm** at the superior and inferior radio-ulnar joints. The pronator quadratus initiates pronation and receives assistance from the pronator teres when more speed and power are needed.

The pronator quadratus also helps the interosseous membrane to hold the radius and ulna together (Fig. 6-147), particularly when upward thrusts are transmitted through the wrist *e.g.*, during a fall on the outstretched hand.

SUMMARY OF THE MUSCLES IN THE ANTERIOR PART OF THE FOREARM (Tables 6-8 and 6-9)

All muscles in the anterior part of the forearm are supplied by the median and ulnar nerves, except the brachioradialis (Figs. 6-25 and 6-27). The **brachioradialis muscle** forms the lateral boundary of the cubital fossa (Fig. 6-74). Although belonging to the extensor group (Table 6-10), it lies in the lateral part of the anterior aspect of the forearm.

Although functionally a flexor of the forearm, the brachioradialis is supplied by the radial nerve (Table 6-10). Hence this muscle is the one major exception to the rule that the radial nerve supplies only extensor muscles (Fig. 6-27).

The long flexors of the fingers (flexor digitorum superficialis and flexor digitorum profundus) also flex the metacarpophalangeal and wrist joints (Tables 6-18 and 6-19). The flexor digitorum profundus flexes the fingers in slow action, but this activity is reinforced by the flexor digitorum superficialis when speed and flexion against resistance are required.

When the wrist, metacarpophalangeal, and interphalangeal joints are flexed, the flexor muscles are shortened and their action is consequently weakened. In addition some weakening results from the ligamentous action of the extensor muscles. Verify this by flexing your wrist and gripping a pencil, and then extending your wrist and gripping it again. Note that your grip is firmer in the first position.

THE ANTERIOR NERVES OF THE FOREARM

The nerves of the forearm or antebrachium are the **median, ulnar, and radial.** The median nerve is the principal nerve of the anterior fascial compartment (Fig. 6-79). Although the radial nerve appears in the cubital region (Fig. 6-74), it soon enters the posterior fascial compartment. Aside from cutaneous branches, the nerves of the anterior aspect of the forearm are only two in number: the median and ulnar.

THE MEDIAN NERVE (Figs. 6-74, 6-79 to 6-83, 6-85, 6-87, 6-92, and 6-120)

The median nerve enters the forearm with the brachial artery; it lies on the surface of the brachialis muscle. *It passes between the two heads of the pronator teres* muscle (Fig. 6-74), giving branches to them. The median nerve descends deep to the flexor digitorum superficialis (Fig. 6-80), to which it is closely attached by the muscle's fascial sheath (Fig. 6-79). It continues distally between this muscle and the flexor digitorum profundus (Fig. 6-85).

Near the wrist the median nerve becomes superficial by passing between the tendons of the flexor digitorum superficialis and the flexor carpi radialis, deep to the tendon of the palmaris longus (Figs. 6-80, 6-81, 6-83, and 6-90).

Branches of the Median Nerve. The median nerve has no branches in the arm (Fig. 6-25). They arise in the forearm and hand as follows:

1. **Articular branches** pass to the elbow joint as the median nerve passes it (Fig. 6-87).
2. **Muscular branches** supply the pronator teres, pronator quadratus, and all the flexor muscles (Fig. 6-25), *except* the flexor carpi ulnaris and the medial half of the flexor digitorum profundus, which are supplied by the ulnar nerve (Table 6-8).
3. **The anterior interosseous nerve** (Fig. 6-87) arises from the median nerve in the distal part of the cubital fossa. *It passes inferiorly on the interosseous membrane with the anterior interosseous branch of the ulnar artery* (Fig. 6-91).

The anterior interosseous nerve runs between the flexor digitorum profundus and flexor pollicis longus muscles to reach the pronator quadratus. It supplies all three of these muscles, although the ulnar nerve supplies half of the flexor digitorum profundus (Fig. 6-25). The anterior inter-

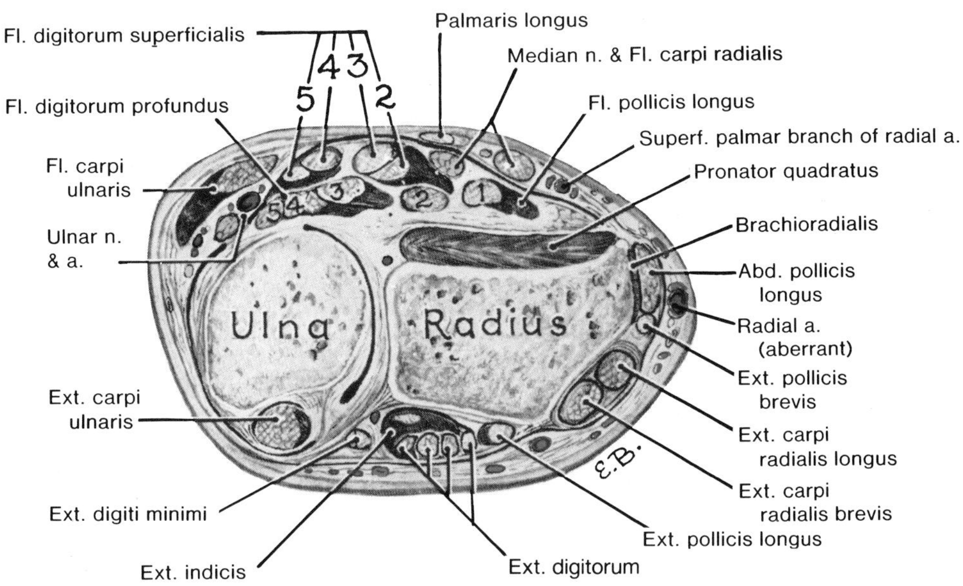

Figure 6-81. Drawing of a cross-section through the forearm superior to the wrist. Observe the flexor carpi radialis, palmaris longus, and flexor carpi ulnaris tendons constituting a surface layer of flexors. Deep to these, the long flexors of the digits: (1) the four tendons of the flexor digitorum superficialis, those to the middle and ring fingers being anterior to those to the index and little fingers; and (2) the five tendons of the deep digital flexors, lying side by side, those to the thumb and index finger being free. Note that the ulnar nerve and artery are deep to the flexor carpi ulnaris; hence the pulse of the artery cannot be felt here. Note the median nerve at the midpoint on the anterior aspect of the wrist, *deep to the palmaris longus* tendon, and at the lateral border of the flexor digitorum superficialis. Note the three large extensor tendons on the dorsum of the wrist; they are inserted into the metacarpal bones and work as synergists with the powerful flexors of the digits, whereas the remaining tendons, being extensors of the digits, are slender.

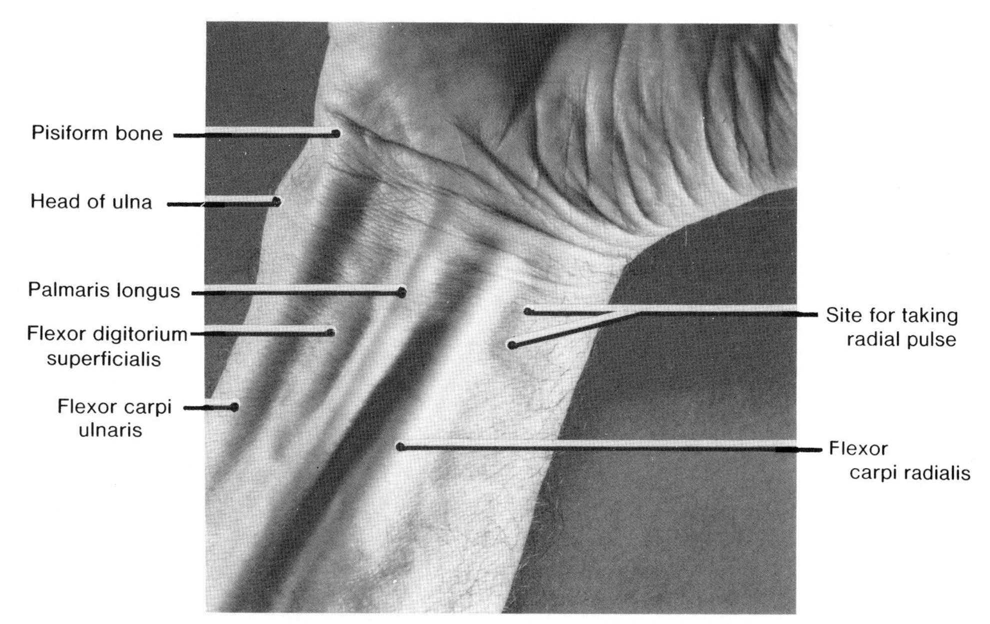

Figure 6-82. Photograph of the anterior aspect of the wrist of a 46-year-old man showing the principal surface markings and the favorite site for taking the pulse of the radial artery. The cord-like tendon of the flexor carpi radialis serves as a guide to the radial artery which lies lateral to it. The long slender tendon of the palmaris longus (absent on one or both sides in about 21% of people) serves as a guide to the median nerve which is deep and often lateral to it. The flexor carpi ulnaris muscle overlies the ulnar nerve throughout its length, but at the wrist its tendon may be used as a guide to the ulnar nerve and artery which are just lateral to it.

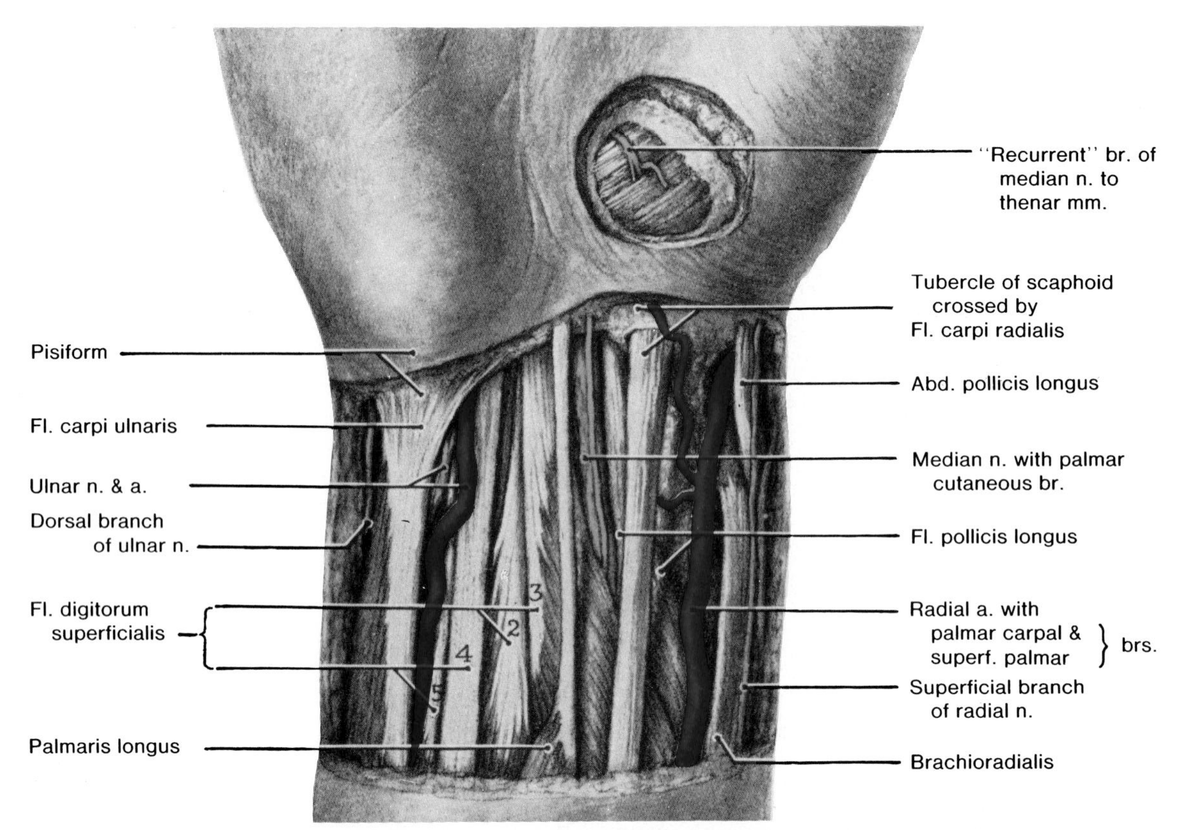

Figure 6-83. Drawing of a dissection of the structures in the anterior aspect of the wrist. The distal transverse skin incision was made along the distal wrist crease (Fig. 6-84). This crease crosses the pisiform bone to which the flexor carpi ulnaris is attached. Observe the palmaris longus tendon bisecting the distal skin crease at the middle of the wrist. *Note that the median nerve is deep to the lateral margin of the palmaris longus tendon.* Observe also that the ulnar nerve and artery are sheltered by the flexor carpi ulnaris tendon and by the expansion this tendon gives to the flexor retinaculum. Note the flexor digitorum superficialis tendons to digits 3 and 4 are somewhat anterior to those to digits 2 and 5. Observe the "recurrent" branch of the median nerve to the thenar muscles lying within a circle, the center of which is 2 to 4 cm distal to the tubercle of the scaphoid (Fig. 6-77*A*). This nerve may be severed by lacerations in this region resulting in impairment of movements of the thumb (Table 6-21).

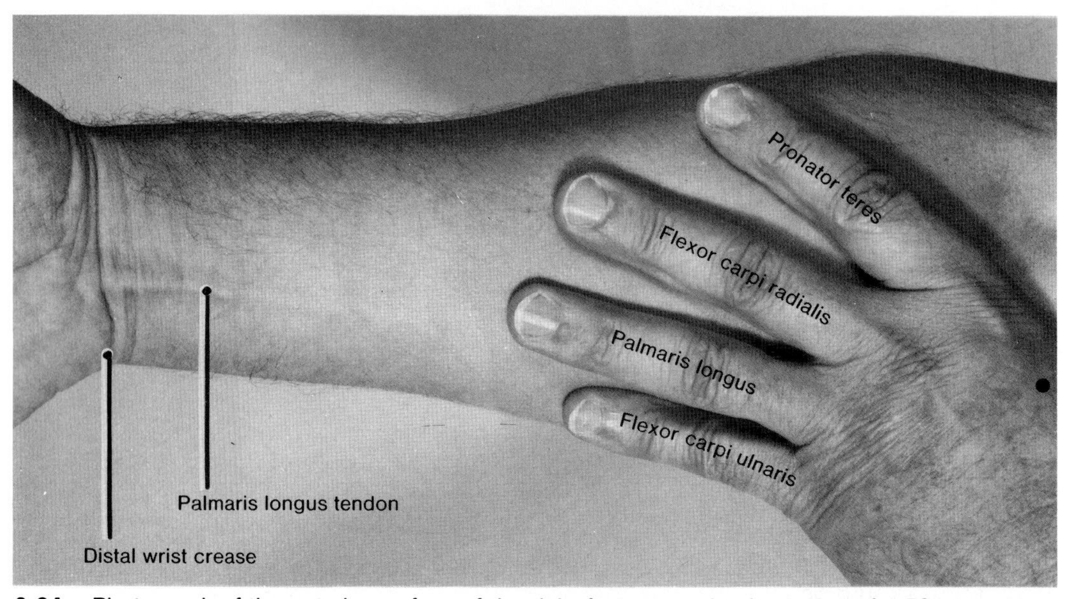

Figure 6-84. Photograph of the anterior surface of the right forearm and wrist region of a 53-year-old man who is showing how to locate the position of the four superficial flexor muscles. The thumb of the person's left hand is placed posterior to the elbow around the medial epicondyle of the humerus (deep to *black dot*), from which arises the common flexor tendon of these superficial muscles (Fig. 6-75).

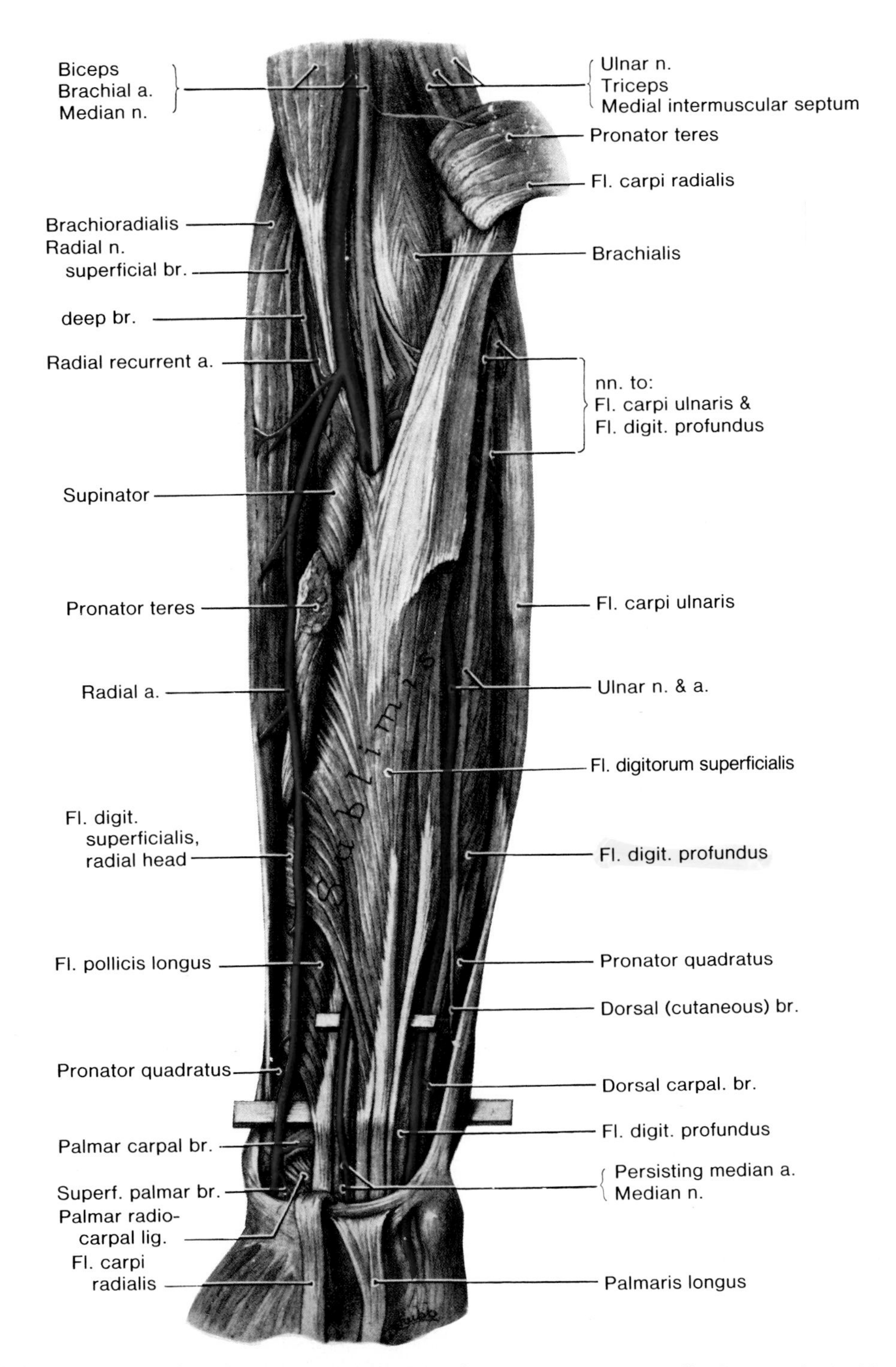

Figure 6-85. Drawing of a dissection showing the flexor digitorum superficialis muscle (sublimis) and related structures. Observe the oblique origin of the superficialis from (1) the medial epicondyle of the humerus, (2) the ulnar collateral ligament of the elbow joint, (3) the tubercle of the coronoid process, and (4) the superior half of the anterior border of the radius (Fig. 6-75). The superficialis, like the three muscles anterior to it and the two and one-half muscles posterior to it, is supplied by the median nerve. Observe the ulnar artery descending obliquely posterior to the flexor digitorum superficialis to join the ulnar nerve. Note the median nerve descending vertically posterior to the superficialis, clinging to it, and appearing at its lateral border.

Table 6-8
Superficial Muscles of the Anterior Aspect of the Forearm[1]

Muscle	Origin	Insertion	Nerve Supply	Action(s)
Pronator teres (Figs. 6-74, 6-79, and 6-80)	Medial epicondyle of humerus and coronoid process of ulna (Fig. 6-75)	Middle of lateral surface of radius (Fig. 6-76)	Median n. (C6 and **C7**)	Pronates forearm and flexes elbow joint
Flexor carpi radialis (Figs. 6-79 to 6-84 and 6-118)	Medial epicondyle of humerus (Fig. 6.75)	Base of 2nd metacarpal bone (Fig. 6-75)	Median n. (C6 and **C7**)	Flexes wrist and abducts hand
Palmaris longus (Figs. 6-79 to 6-84)	Medial epicondyle of humerus (Fig. 6-75)	Palmar aponeurosis (Fig. 6-80)	Median n. (C7 and C8)	Flexes wrist
Flexor carpi ulnaris[2] (Figs. 6-79 to 6-85)	*Humeral head:* medial epicondyle of humerus *Ulnar head:* olecranon and posterior border of ulna (Figs. 6-75 and 6-76)	Pisiform bone, hook of hamate, and 5th metacarpal bone (Figs. 6-75 and 6-83)	Ulnar n. (C7 and **C8**)	Flexes wrist and adducts hand
Flexor digitorum superficialis (Figs. 6-79 to 6-83)	*Humeroulnar head:* medial epicondyle of humerus, ulnar collateral ligament, and coronoid process of ulna *Radial head:* superior half of anterior border of radius (Figs. 6-75 and 6-85)	Bodies of middle phalanges of medial four digits (Fig. 6-75)	Median n. (C7, **C8**, and T1)	Flexes middle phalanges of medial four digits

[1] These superficial muscles arise from the **common flexor origin** (the anterior surface of the medial epicondyle) by a **common flexor tendon** (Fig. 6-75).
[2] In contrast to the other superficial flexors, *the flexor carpi ulnaris is supplied by the ulnar nerve.*

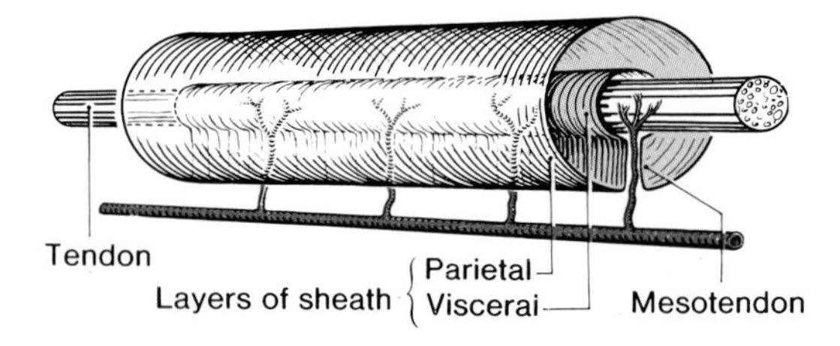

Figure 6-86. Illustration of the structure of a digital synovial sheath of a long flexor tendon in a digit (Fig. 6-88). This tubular sheath (bursa) is a lubricating device that envelops the long digital tendons where they pass through the osseofibrous tunnels in the fingers (Fig. 6-118). The layers of the synovial sheath are separated by a capillary layer or film of synovial fluid. Note that the mesotendon conveys small blood vessels to the tendons. In the fingers these mesotendons are represented by fibrous cords called vinculae (Fig. 6-117).

osseous nerve passes deep to the pronator quadratus muscle and ends by sending articular branches to the wrist joint.

4. **The palmar cutaneous branch** (Fig. 6-83) arises from the median nerve just proximal the flexor retinaculum and becomes cutaneous between the tendons of the palmaris longus and flexor carpi radialis muscles. It passes superficial to the flexor retinaculum to supply the skin of the lateral part of the palm (Figs. 6-26 and 6-92).

Occasionally there are communications between the median and ulnar nerves in the forearm. The communicating branches may be represented by several slender nerves, but these

Table 6-9
Deep Muscles of the Anterior Aspect of the Forearm

Muscle	Origin	Insertion	Nerve Supply	Action(s)
Flexor digitorum profundus (Figs. 6-79 to 6-81, and 6-91)	Proximal three-fourths of medial and anterior surfaces of ulna and interosseous membrane (Figs. 6-75 and 6-79)	Bases of distal phalanges of medial four digits (Fig. 6-75)	*Medial part*, Ulnar n. (**C8** and T1) *Lateral part*, Median n. (**C8** and T1)	Flexes distal phalanges of fingers
Flexor pollicis longus (Figs. 6-79 to 6-83, 6-83, 6-87, and 6-88	Anterior surface of radius and adjacent interosseous membrane (Figs. 6-75 and 6-79)	Base of distal phalanx of thumb (Fig. 6-75)	Anterior interosseous branch of median n. (**C8** and T1)	Flexes phalanges of thumb
Pronator quadratus (Figs. 6-80, 6-89, 6-91, and 6-118)	Distal fourth of anterior surface of ulna (Fig. 6-75)	Distal fourth of anterior surface of radius (Fig. 6-75)		Pronates forearm (Fig. 6-70*B*)

communications are important clinically in that **even with a complete lesion of the median nerve some muscles may not be paralyzed.** This may lead to an erroneus conclusion that the median nerve has not been damaged.

The median nerve may be injured by wounds to the forearm. When it is severed in the elbow region, there is loss of flexion of the proximal interphalangeal joints of all the digits (thumb and fingers). There is also loss of flexion of the distal interphalangeal joints of the index and middle fingers (Table 6-20).

Flexion of the distal interphalangeal joints of the ring and little fingers is not affected because the medial part of the flexor digitorum profundus producing these movements is supplied by the ulnar nerve (Fig. 6-25 and Table 6-20). The ability to flex the metacarpophalangeal joints of the index and middle fingers will be affected because the digital branches of the median nerve supply the first and second lumbrical muscles (Figs. 6-25, 6-110, and Table 6-19).

Most commonly the median nerve is injured just proximal to the flexor retinaculum (Figs. 6-87 and 6-90), owing to the frequency of wrist slashing in suicide attempts. Although severance of the palmaris longus tendon is common in these cases owing to its location superficial to the median nerve (Fig. 6-81), the loss of function of this muscle is not missed. In a **lacerated wrist** the palmaris longus tendon may be mistaken for the median nerve if care is not practiced.

THE CARPAL TUNNEL SYNDROME

The median nerve enters the palm through the osseofibrous carpal tunnel (Figs. 6-78 and 6-81), close to the deep surface of the flexor retinaculum (Fig. 6-87). The tendons of the long flexor muscles of the digits also pass through this rather restricted passage.

Any lesion that significantly reduces the size of the carpal tunnel (*e.g.*, inflammation of the flexor retinaculum, anterior dislocation of the lunate bone, arthritic changes, or tenosynovitis of the tendons and tendon sheaths) **may cause compression of the median nerve.** As this nerve has two terminal branches (lateral and medial) that supply skin of the hand (Fig. 6-26), there is often tingling (**paresthesia**), absence of tactile sensation (**anesthesia**), or diminished sensation (**hypoesthesia**) in the digits. Because the median nerve sends a palmar cutaneous branch superficial to the flexor retinaculum to supply most of the palm, there is often no sensory impairment of this area (Figs. 6-83 and 6-92).

Often a progressive loss of coordination and strength in the thumb occurs if the cause of the median nerve compression is not alleviated. This

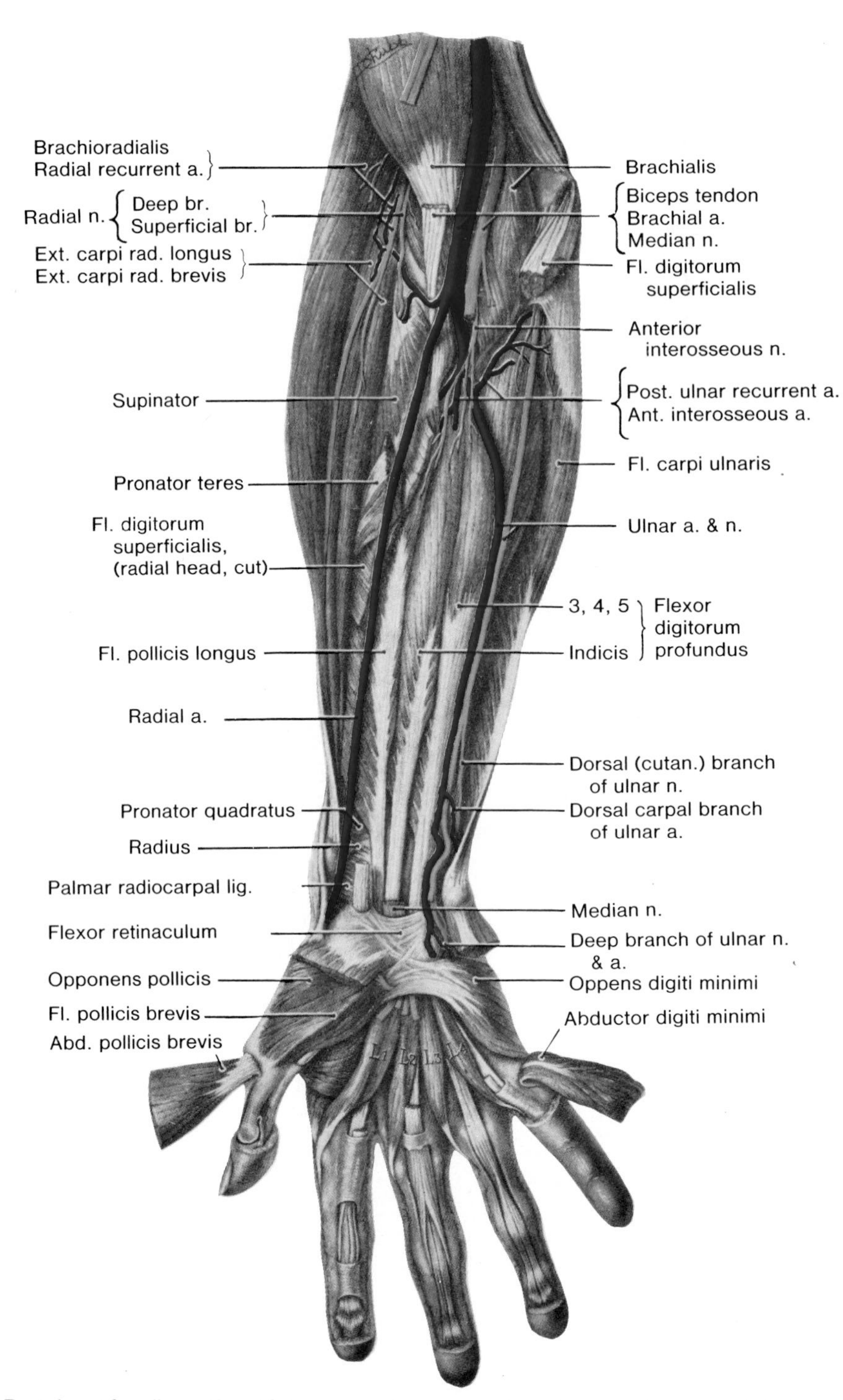

Figure 6-87. Drawing of a dissection of the deep flexor muscles of the digits and related structures. Observe that two deep digital flexor muscles, flexor pollicis longus and flexor digitorum profundus, form a sheet of muscle that arises from the flexor aspects of the radius, the interosseous membrane, and the ulna (Fig. 6-75). Note that the portion of the profundus for the index finger (*i.e.*, indicis) is free, superior to the wrist, and that the portions for digits 3, 4, and 5 are fused.

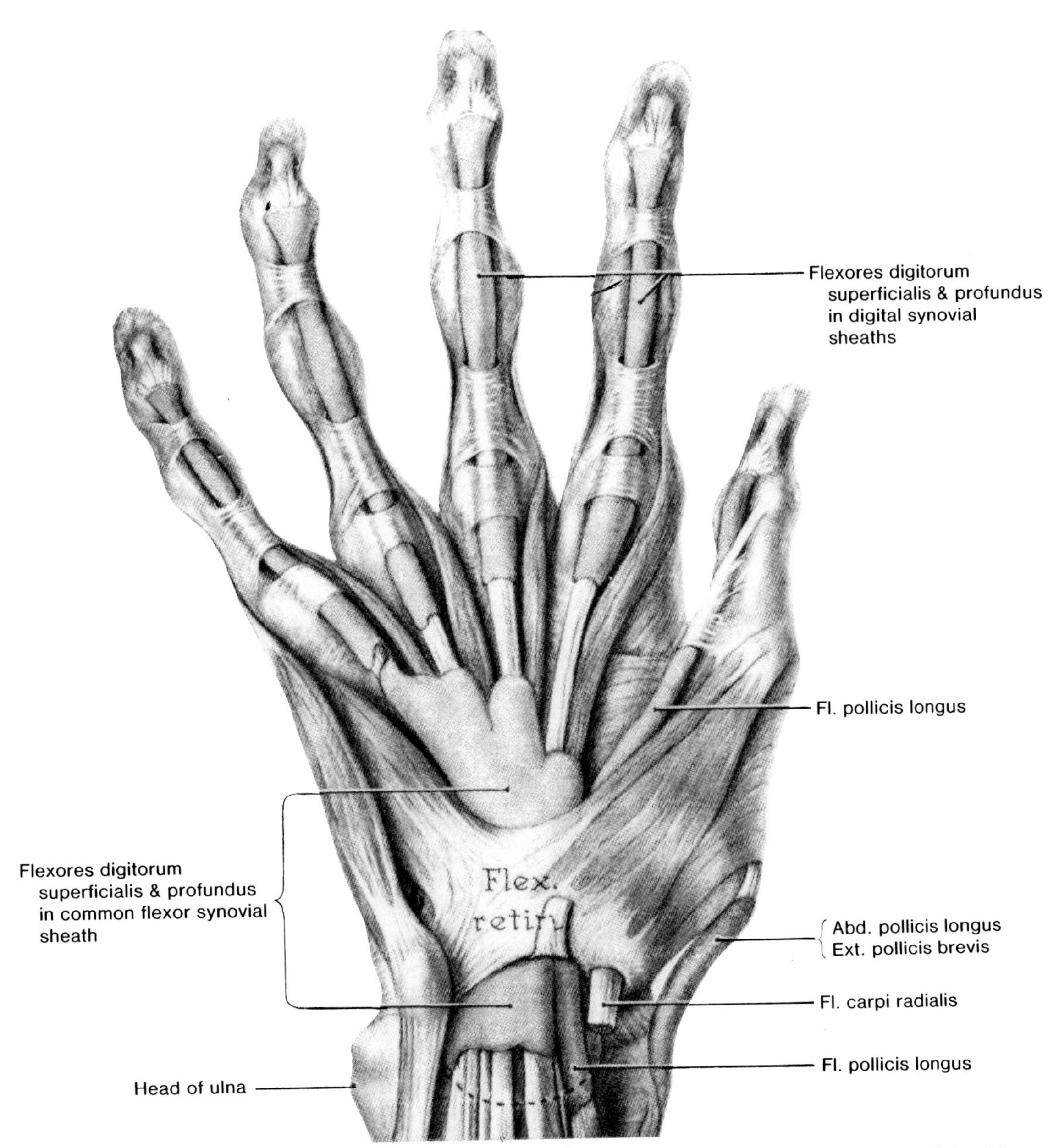

Figure 6-88. Drawing of a dissection of the synovial sheaths of the long flexor tendons of the digits. There are two sets: (1) proximal or carpal posterior to the flexor retinaculum, and (2) distal or digital posterior to the fibrous sheaths of the digital flexors. The carpal synovial sheaths of the flexor tendons of the fingers unite with one another to form a *common flexor synovial sheath.* The carpal sheath of the thumb tendon usually communicates with the common flexor sheath, which extends 1 to 2.5 cm proximal and distal to the flexor retinaculum. The marginal metacarpals being the shortest, the common flexor sheath extends to and is continuous with the digital synovial sheaths of the thumb and little finger.

results in difficulty in performing fine movements (Fig. 6-115 and Table 6-21). As the thenar (G. palm of hand) muscles and the lateral two lumbrical muscles of the fingers are also supplied by the median nerve (Fig. 6-25), the usefulness of the thumb, index, and middle fingers may be diminished.

In cases of severe compression of the median nerve, there may be wasting (atrophy) of the thenar muscles (see discussion of Case 6-6

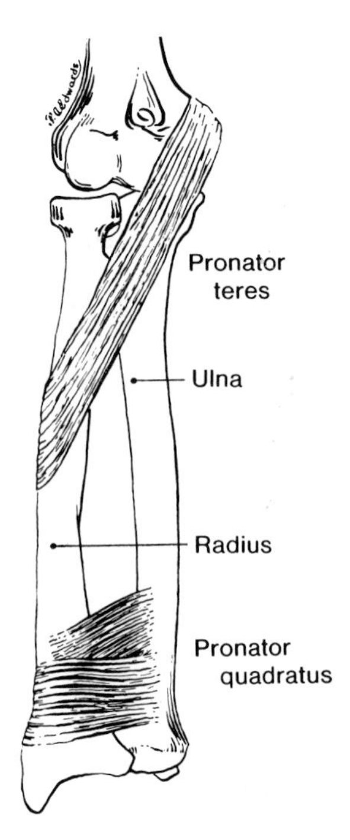

Figure 6-89. Drawing of the right pronator teres and pronator quadratus muscles showing their attachments. *The pronator quadratus is the principal pronator of the forearm*; it is reinforced by the pronator teres during rapid and/or forceful pronation. Observe that the pronator teres descends obliquely from its origin (Fig. 6-75), over the forearm bones, and inserts into the main curve of the radius. This arrangement enables the pronator teres to rotate the radius.

on p. 791). To relieve symptoms of the carpal tunnel syndrome, partial or complete **division of the flexor retinaculum** may be necessary. This operation is called a *carpal tunnel release.*

THE ULNAR NERVE (Figs. 6-63, 6-64, 6-79 to 6-81, 6-83, 6-85, 6-87, and 6-90 to 6-93)

After passing posterior to the medial epicondyle of the humerus (Fig. 6-63), *the ulnar nerve enters the forearm by passing between the two heads of the flexor carpi ulnaris muscle* (Fig. 6-91). It then descends deep to this muscle on the flexor digitorum profundus, where *it accompanies the ulnar artery near the middle of the forearm* (Fig. 6-87). It then passes on the medial side of this artery and the lateral side of the tendon of the flexor carpi ulnaris (Figs. 6-83 and 6-91).

In the distal part of the forearm the ulnar nerve becomes relatively superficial, covered only by fascia and skin (Fig. 6-83). Near the pisiform bone it pierces the deep fascia and **passes superficial to the flexor retinaculum** (Figs. 6-81, 6-87 and 6-91), where it ends by dividing into superficial and deep branches (Figs. 6-92 and 6-93).

Branches of the Ulnar Nerve. The ulnar nerve has no branches in the arm (Fig. 6-25). They arise in the forearm and hand as follows:

1. **Articular branches** pass to the elbow joint while the nerve is in the groove between the olecranon and medial epicondyle (Figs. 6-63, 6-64, 6-74, and 6-91).
2. **Muscular branches** supply the flexor carpi ulnaris and the medial half of the flexor digitorum profundus muscles (Figs. 6-25, 6-85, 6-87, and Tables 6-8 and 6-9).
3. **The palmar cutaneous branch** arises from the ulnar nerve near the middle of the forearm and pierces the deep fascia in its distal third to supply skin on the medial part of the palm (Figs. 6-26 and 6-92).
4. **The dorsal cutaneous branch** (Fig. 6-87) arises from the ulnar nerve in the distal half of the forearm and passes posteroinferiorly between the ulna and the flexor carpi ulnaris. It supplies the posterior surface of the medial part of the hand (Figs. 6-26 and 6-93).

Ulnar nerve injury may occur in wounds of the forearm, the most common site being where the nerve passes posterior to the medial epicondyle of the humerus (Figs. 6-63 and 6-91). Often this occurs when the elbow hits a hard surface and the medial epicondyle is fractured.

Ulnar nerve injury may result in extensive motor and sensory loss to the hand. There is impaired power of adduction and when an attempt is made to flex the wrist joint, the hand is drawn to the radial side by the flexor

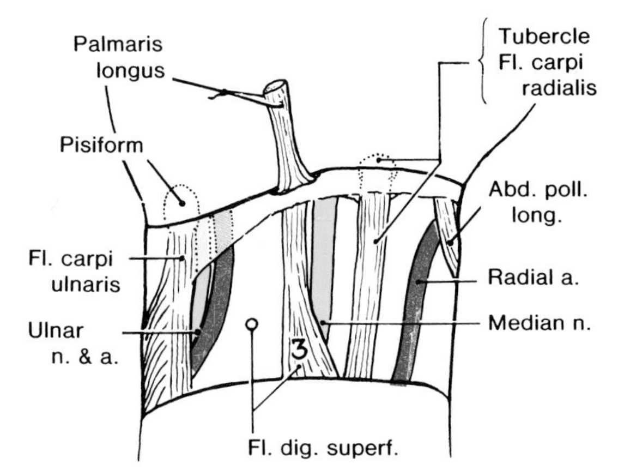

Figure 6-90. Drawing of the anterior surface of the wrist. As the wrist is a common site for suicide attempts, these structures are often lacerated.

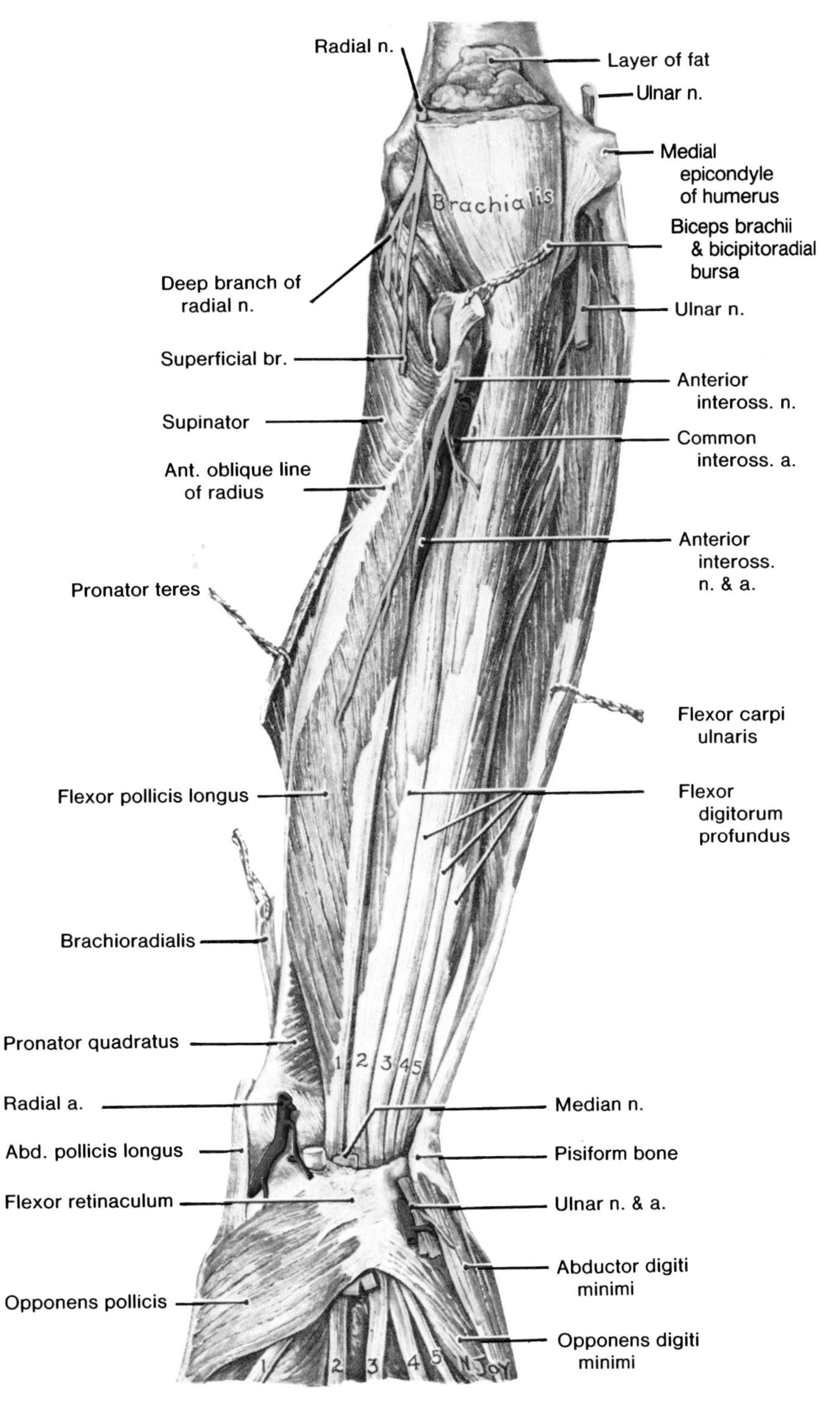

Figure 6-91. Drawing of a dissection of the muscles attached to the anterior aspect of the radius and ulna (Fig. 6-75). Observe that the anterior aspect of the ulna is covered by the brachialis which inserts into the coronoid process of the ulna (Fig. 6-72*A*). Distal to this the ulna is covered by the flexor digitorum profoundus. Note that the profundus also arises from the proximal three-fourths of the medial aspect of the ulna. Observe that the anterior aspect of the radius is covered with the supinator, which is superior to the anterior oblique line, and distal to this, with the flexor pollicis longus. Observe the five tendons of the deep digital flexors converging on the *carpal tunnel* and, after having traversed it, diverging to pass to the five terminal phalanges. Note that the biceps is inserted into the medial aspect of the radius, hence it can rotate it laterally (*i.e.,* supinate it, Fig. 6-70*A*), whereas the pronator teres is attached to the lateral surface and can rotate it medially (*i.e.,* pronate, it, Fig. 6-70*B*).

carpi radialis (Fig. 6-25 and Tables 6-18 to 6-21).

Following ulnar nerve injury, patients are likely to have difficulty in making a fist because they cannot flex their fourth and fifth digits at the distal interphalangeal joints. This characteristic appearance of the hand is known as *main en griffe* or **clawhand** (Fig. 6-94).

THE RADIAL NERVE (Figs. 6-74, 6-83, 6-85, 6-87, 6-91, 6-92, and 6-93)

The radial nerve descends between the brachialis and brachioradialis muscles and crosses the anterior aspect of the lateral epicondyle of the humerus (Fig. 6-87). Soon after it enters the forearm, the radial nerve divides into its terminal branches, superficial and deep (Fig. 6-74).

The superficial branch of the radial nerve (Figs. 6-74, 6-85, and 6-87), the smaller of the two terminal branches, is *the direct continuation of the radial nerve.* It passes distally, anterior to the pronator teres muscle and under cover of the brachioradialis.

In the distal one-third of the forearm, the superficial branch of the radial nerve passes posteriorly, deep to the tendon of the brachioradialis, and enters the posterior fascial compartment of the forearm.

The superficial branch of the radial nerve pierces the deep fascia 3 to 4 cm proximal to the wrist and ***supplies skin on the dorsum of the wrist, hand, thumb, and lateral one (or two) and one-half fingers*** (Figs. 6-26, 6-93, and 6-98).

In Figure 6-85 observe that *the superficial branch of the radial nerve can be exposed from the elbow to near the wrist without cutting a muscle or a tendon.* It can also be exposed on the dorsal aspect of the hand by cutting only the tendon of the brachioradialis muscle (Figs. 6-26 and 6-98).

The deep branch of the radial nerve, the larger of the two terminal branches, is *entirely muscular and articular in its distribution.* As it passes posteroinferiorly, it gives branches to the extensor carpi radialis brevis and supinator muscles (Figs. 6-27 and 6-74). It then **pierces the supinator muscle,** giving additional branches to it and curving around the lateral side of the radius to enter the posterior fascial compartment of the forearm (Fig. 6-79).

On reaching the posterior aspect of the forearm, the deep branch of the radial nerve *gives many branches to the extensor muscles* (Figs. 6-27, 6-79, 6-97, and Table 6-11). One of these branches, the **posterior interosseous nerve**, accompanies the posterior interosseous artery and supplies the deep extensor muscles (Fig. 6-97 and Table 6-11).

Radial nerve injury may occur in deep wounds of the forearm. Severance of the deep branch of the radial nerve produces *inability to extend the thumb and the metacarpophalangeal joints of the fingers* (Tables 6-20 and 6-21). There is no loss of sensation because the deep branch of the radial is entirely muscular and articular in distribution. See Figure 6-27 to determine the muscles that would be paralyzed when there is radial nerve injury.

When the radial nerve is severed, or its superficial branch is cut, *sensation should be lost on the posterior surface of the forearm, hand, and proximal phalanges of the lateral three and one-half digits* (Figs. 6-26 and 6-93). However, you must be aware that the **absence of cutaneous anesthesia is not necessarily indicative of an intact radial nerve** because there is so much overlap between the cutaneous nerves of the hand.

ARTERIES OF THE FOREARM

The two main arteries in the forearm are the **radial and ulnar arteries.** Each of these has several branches.

The brachial artery ends opposite the neck of the radius in the inferior part of the cubital fossa by dividing into its two terminal branches, the radial and ulnar arteries.

THE RADIAL ARTERY (Figs. 6-65, 6-74, 6-80, 6-83, 6-85, 6-87, and 6-90)

The radial artery begins in the cubital fossa, *just medial to the biceps tendon at the level of the neck of the radius* (Figs. 6-65 and 6-80).

The radial artery is the smaller of the two terminal branches of the brachial and continues the direct line of this vessel (Figs. 6-74 and 6-87). The course of the radial artery in the forearm can be represented by a line connecting the midpoint of the cubital fossa to a point just medial to the tip of the styloid process of the radius (Figs. 6-80 and 6-87).

The proximal part of the radial artery is overlapped by the fleshy belly of the brachioradialis muscle (Fig. 6-80), which can be pulled laterally to reveal the entire length of the artery in the forearm (Fig. 6-85).

The radial artery lies on muscle until it comes into contact with the distal end of the radius, where it is covered only by superficial and deep fasciae and skin (Figs. 6-82 and 6-83). **This is the common site for taking the pulse.**

The radial artery leaves the forearm by winding around the lateral aspect of the radius, and passing posteriorly between the lateral collateral ligament of the wrist joint and the tendons of the abductor pollicis longus and extensor pollicis brevis (Fig. 6-97).

The radial artery crosses the floor of the anatomical snuff box, formed by the scaphoid and trapezium bones (Figs. 6-96 to 6-99).

The radial artery ends by completing the deep palmar arch in conjunction with the ulnar artery (Figs. 6-121 and 6-122).

Branches of the Radial Artery in the Forearm.

1. **The radial recurrent artery** (Fig. 6-85) arises from the lateral side of the radial artery, just distal to its origin, and ascends between the brachioradialis and brachialis muscles. It supplies these muscles and the elbow joint and *anastomoses with the radial collateral artery*, a branch of the profunda brachii, thereby participating in the *anastomoses around the elbow region* (Fig. 6-65).
2. **The muscular branches** supply muscles on the lateral side of the forearm (Fig. 6-85).
3. **The palmar carpal branch**, a small artery, arises near the distal border of the pronator quadratus (Fig. 6-85), and runs across the wrist deep to the flexor tendons. Here it anastomoses with the carpal branch of the ulnar artery to form the *palmar carpal arch* (Figs. 6-95 and 6-121).
4. **The superficial palmar branch** (Figs. 6-85 and 6-95) arises just proximal to the wrist and passes through, sometimes over, the muscles of the thenar eminence which it supplies. It usually anastomoses with the terminal part of the ulnar artery to form a **superficial palmar arch** (Figs. 6-113, 6-121, and 6-122).
5. **The dorsal carpal branch** (Fig. 6-96) runs medially across the dorsal surface of the wrist, deep to the extensor tendons, where it anastomoses with the dorsal carpal branch of the ulnar artery to form the **dorsal carpal arch** (Figs. 6-96 and 6-105).

The common place for taking the pulse is where the radial artery lies on the anterior surface of the distal end of the radius, lateral to the tendon of the flexor carpi radialis muscle (Figs. 6-82, 6-83, and 6-95). Here it is covered only by deep and superficial fasciae and skin. About 4 cm of this artery can be compressed against the distal end of the radius.

When taking the **radial pulse,** do not use the pulp of your thumb because it has its own pulse which could be interpreted as the patient's pulse. If the pulse cannot be felt, try the other wrist because an *aberrant radial artery* on one side may make the pulse difficult to palpate.

THE ULNAR ARTERY (Figs. 6-65, 6-74, 6-79 to 6-81, 6-83, 6-85, 6-87, and 6-90)

The ulnar artery, the larger of the two terminal branches of the brachial, makes a gentle curve as it **passes from the cubital fossa to the medial side of the forearm** (Fig. 6-87).

Beginning near the neck of the radius, just medial to the biceps tendon (Figs. 6-74 and 6-85), it passes inferomedially deep to the pronator teres muscle (Fig. 6-80).

In company with the median nerve, *the ulnar artery passes between the ulnar and radial heads of the flexor digitorum superficialis* (Figs. 6-79 and 6-85).

About midway between the elbow and wrist, the ulnar artery crosses posterior to the median nerve to reach the medial side of the forearm, where it lies on the flexor digitorum profundus (Figs. 6-85 and 6-87).

In the distal two-thirds of the forearm, the ulnar artery lies lateral to the ulnar nerve (Figs. 6-87 and 6-91). It leaves the forearm by passing superficial to the flexor retinaculum (Fig. 6-81) on the lateral side of the pisiform bone (Fig. 6-83).

At the wrist the ulnar artery and nerve lie lateral to the tendon on the flexor carpi ulnaris, where they are covered only by fascia and skin (Figs. 6-83 and 6-90).

The pulsations of the ulnar artery can sometimes be felt where it passes anterior to the head of the ulna (Figs. 6-82 and 6-83).

Branches of the Ulnar Artery in the Forearm.

1. **The anterior ulnar recurrent artery** (Fig. 6-65) arises from the ulnar just inferior to the elbow joint, and runs superiorly between the brachialis and pronator teres muscles. It supplies these muscles and anastomoses with the inferior ulnar collateral artery, a branch of the brachial, thereby *participating in the anastomoses around the elbow region* (Fig. 6-65).
2. **The posterior ulnar recurrent artery** (Figs. 6-65 and 6-87) arises distal to the anterior ulnar recurrent artery. It passes superiorly, posterior to the medial epicondyle, where it lies deep to the tendon of the flexor carpi ulnaris (Fig. 6-87). It supplies adjacent muscles and then *takes part in the anastomoses around the elbow region* (Fig. 6-65).
3. **The common interosseous artery** arises in the distal part of the cubital fossa (Figs. 6-65, 6-87, and 6-147), and divides into anterior and posterior interosseous arteries.

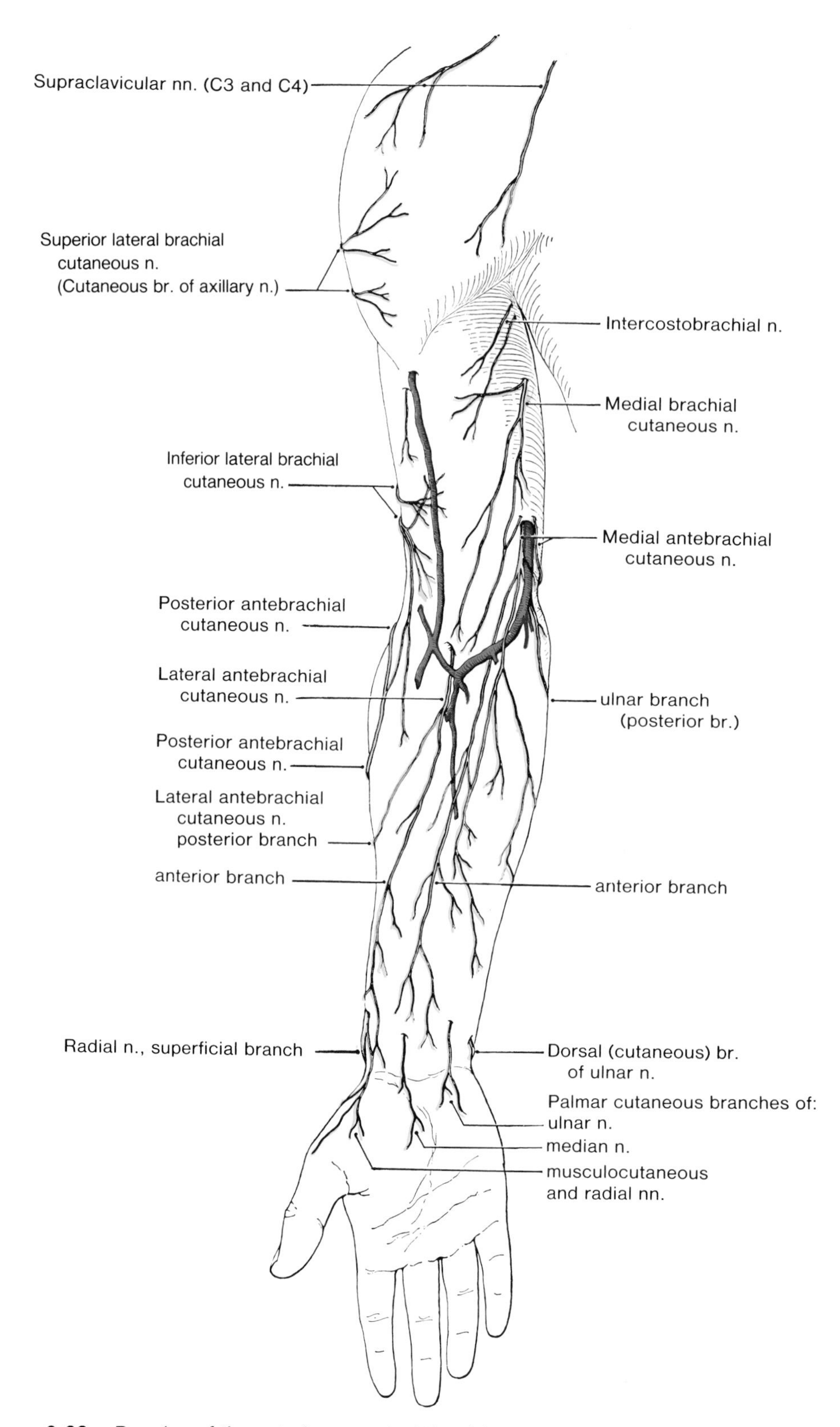

Figure 6-92. Drawing of the anterior aspect of the right upper limb showing its cutaneous nerves.

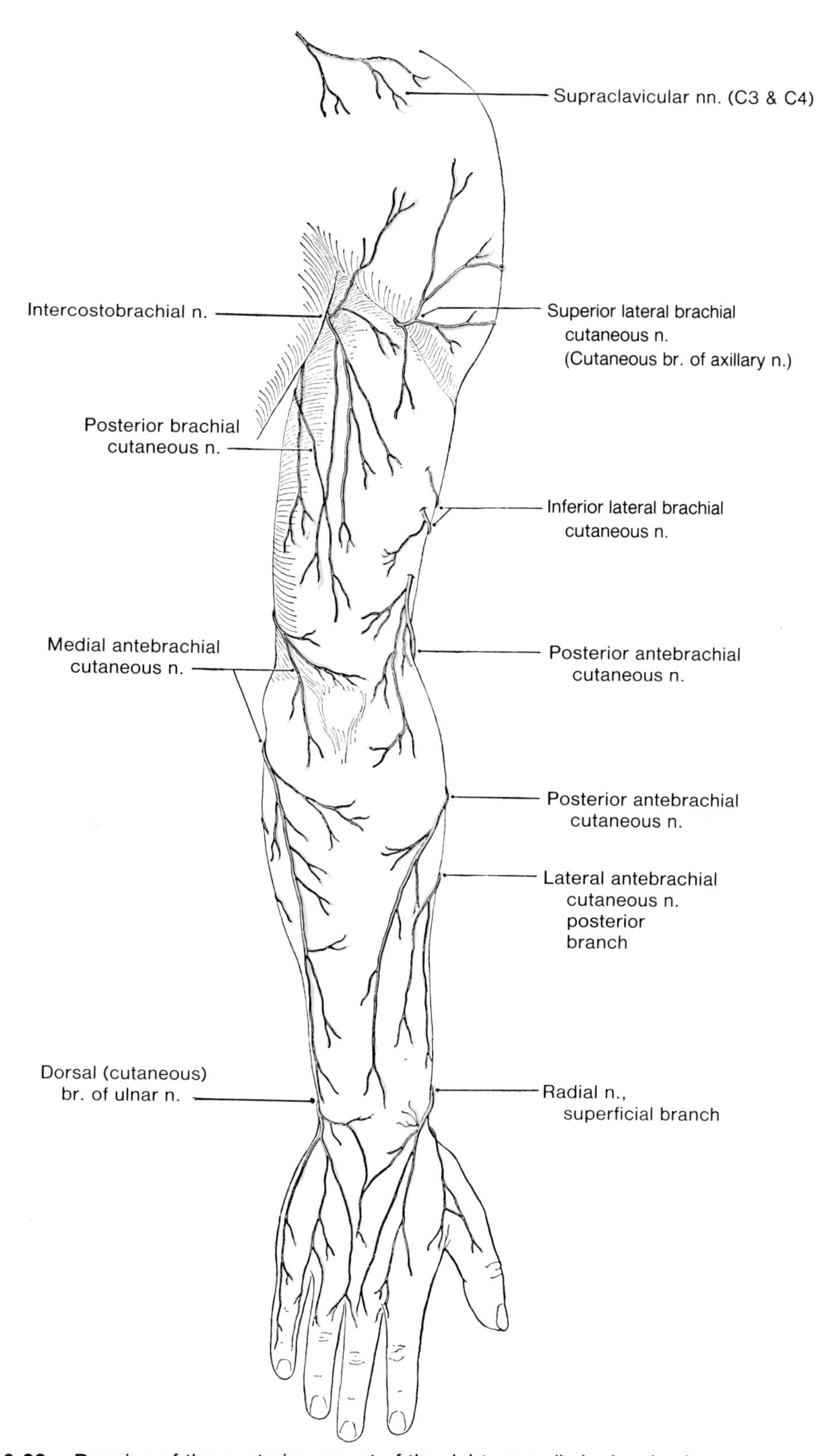

Figure 6-93. Drawing of the posterior aspect of the right upper limb showing its cutaneous nerves.

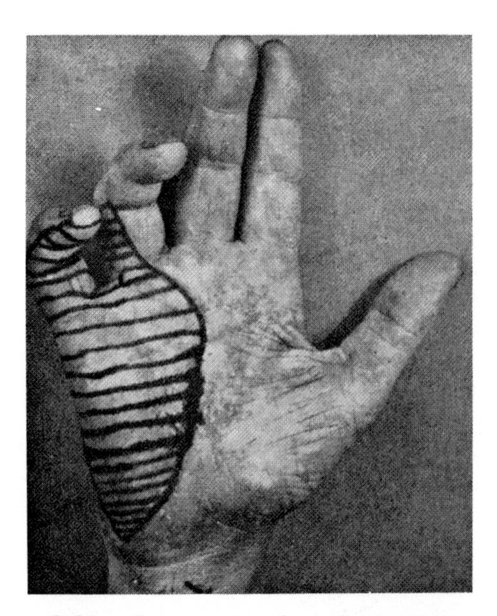

Figure 6-94. Photograph of a clawhand (main en griffe) resulting from severance of the ulnar nerve during a laceration at the wrist. Note that the ring and little fingers are hyperextended at the metacarpophalangeal joints because the medial two lumbrical muscles (Fig. 6-113) are paralyzed and the extensor digitorum is unopposed. (Reprinted with permission from Haymaker W, Woodhall B: *Peripheral Nerve Injuries*, ed 2. Philadelphia, WB Saunders Co, 1953).

The **anterior interosseous artery** (Figs. 6-87 and 6-147) passes distally on the interosseous membrane to the proximal border of the pronator quadratus. Here it pierces the membrane and continues distally to join the dorsal carpal arch (Figs. 6-96 and 6-105).

The **posterior interosseous artery** (Fig. 6-97) passes posteriorly between the bones of the forearm, just proximal to the interosseous membrane (Fig. 6-147). It supplies adjacent muscles and then gives off the **posterior interosseous recurrent artery** (Fig. 6-97), which passes superiorly, posterior to the lateral epicondyle and *participates in the anastomoses around the elbow region* (Fig. 6-65).

4. **The muscular branches** supply muscles on the medial side of the forearm.
5. **The palmar carpal branch** of the ulnar artery is a small branch that runs across the anterior aspect of the wrist, deep to the tendons of the flexor digitorum profundus. The palmar carpal branch of the ulnar artery anastomoses with the palmar carpal branch of the radial artery to form the *palmar carpal arch* (Figs. 6-95, 6-122, and 6-123).
6. **The dorsal carpal branch** of the ulnar artery (Fig. 6-96) arises just proximal to the pisiform bone. It passes across the dorsal surface of the wrist, deep to the extensor tendons, where it anastomoses with the deep carpal branch of the radial artery to form the **dorsal carpal arch** (Figs. 6-96, 6-97, and 6-105).

Sometimes the brachial artery divides at a more proximal level than usual. In other cases the ulnar artery passes superficial to the flexor muscles within the superficial fascia. These variations must be kept in mind when performing **venesections** (incisions into a vein) at the elbow, *e.g.*, for inserting a metal cannula or polyethylene catheter into a vein for intravenous injection of fluids, blood, or medication.

If an aberrant artery is mistaken for a vein and certain drugs are injected into it, the result may be disastrous, *e.g.*, gangrene, resulting in partial or total loss of the hand.

EXTENSOR MUSCLES OF THE FOREARM

The eleven muscles in the posterior part of the forearm are classified as extensors (Tables 6-10 and 6-11). They can be organized into functional groups as follows: (1) *muscles that extend the hand at the wrist* (extensor carpi radialis longus, extensor carpi radialis brevis, and extensor carpi ulnaris); (2) *muscles that extend the fingers* (extensor digitorum, extensor indicis, and extensor digiti minimi); and (3) *muscles that extend the thumb* (abductor pollicis longus, extensor pollicis brevis, and extensor pollicis longus). For purposes of description, the extensor muscles of the forearm are usually divided into superficial and deep groups (Tables 6-10 and 6-11).

Four of the superficial extensors (extensor carpi radialis brevis, extensor digitorum, extensor digiti minimi, and extensor carpi ulnaris) arise from a flattened **common extensor tendon** (Figs. 6-63 and 6-76), which is attached to the **lateral epicondyle**, adjacent fascia, and lateral supracondylar ridge of the humerus (Fig. 6-59*A*). This is known as **the common extensor origin** of the common extensor tendon (Figs. 6-75 and 6-76). The brachioradialis and extensor carpi radialis longus muscles arise from the superior and inferior parts of the lateral supracondylar ridge, respectively (Fig. 6-75).

The brachioradialis, a flexor of the elbow joint, is included with the extensor muscles because it is supplied by the radial nerve. It is also described with the muscles of the cubital region because it forms the lateral boundary of the cubital fossa (Figs. 6-71 and 6-74).

The extensor tendons occupy the radial side as well

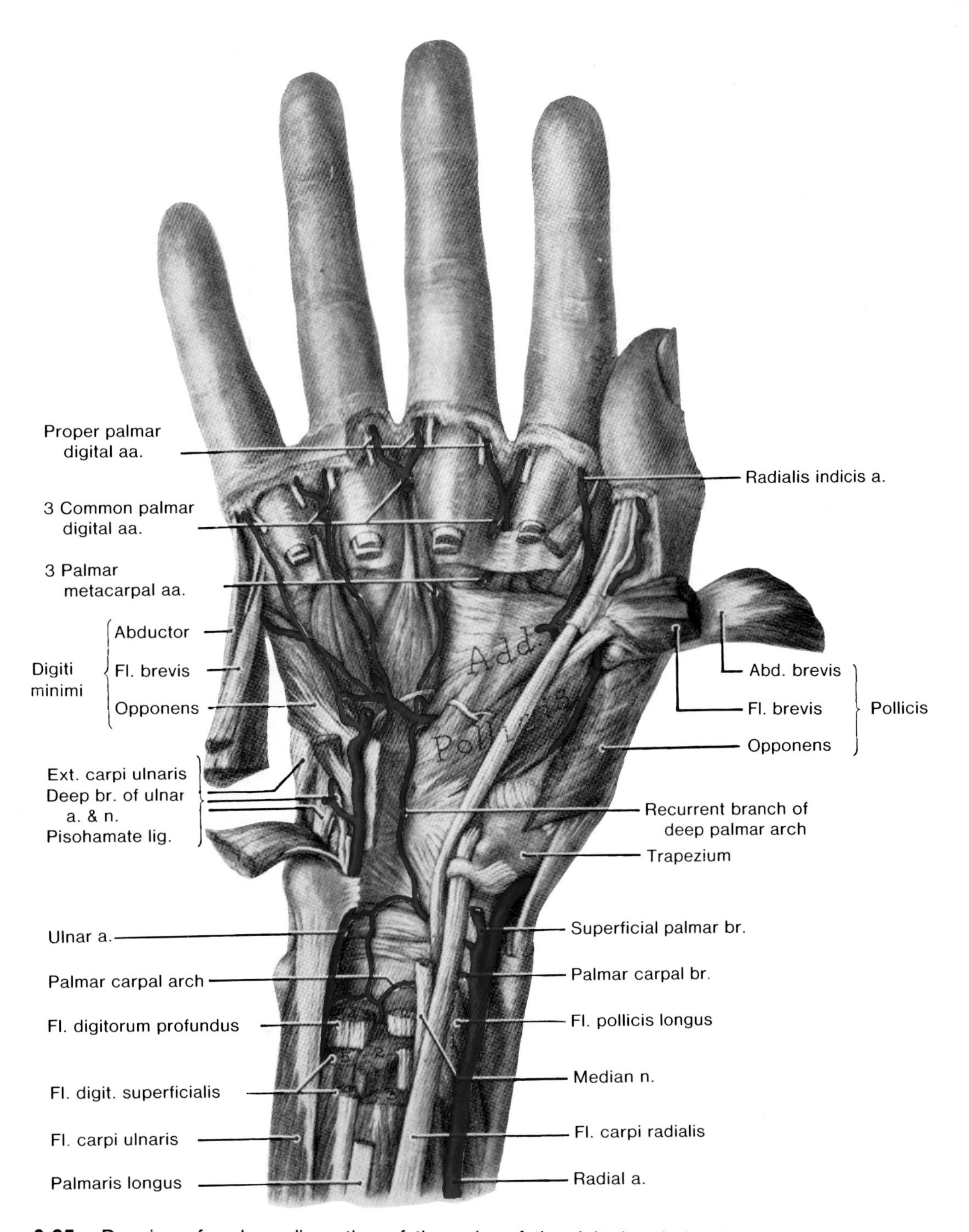

Figure 6-95. Drawing of a deep dissection of the palm of the right hand showing its muscles and arteries. Observe the palmar carpal arch and its connections, and the deep branch of the ulnar artery joining the radial artery to form the deep palmar arch.

as the dorsum of the wrist, where they **lie within synovial sheaths in bony grooves** (Fig. 6-101). The extensor tendons are held in place by a strong fibrous band, the **extensor retinaculum** (Fig. 6-96), which is attached laterally to the distal end of the radius and medially to the styloid process of the ulna and the triquetral and pisiform bones (Fig. 6-77).

The extensor retinaculum prevents bowstringing of

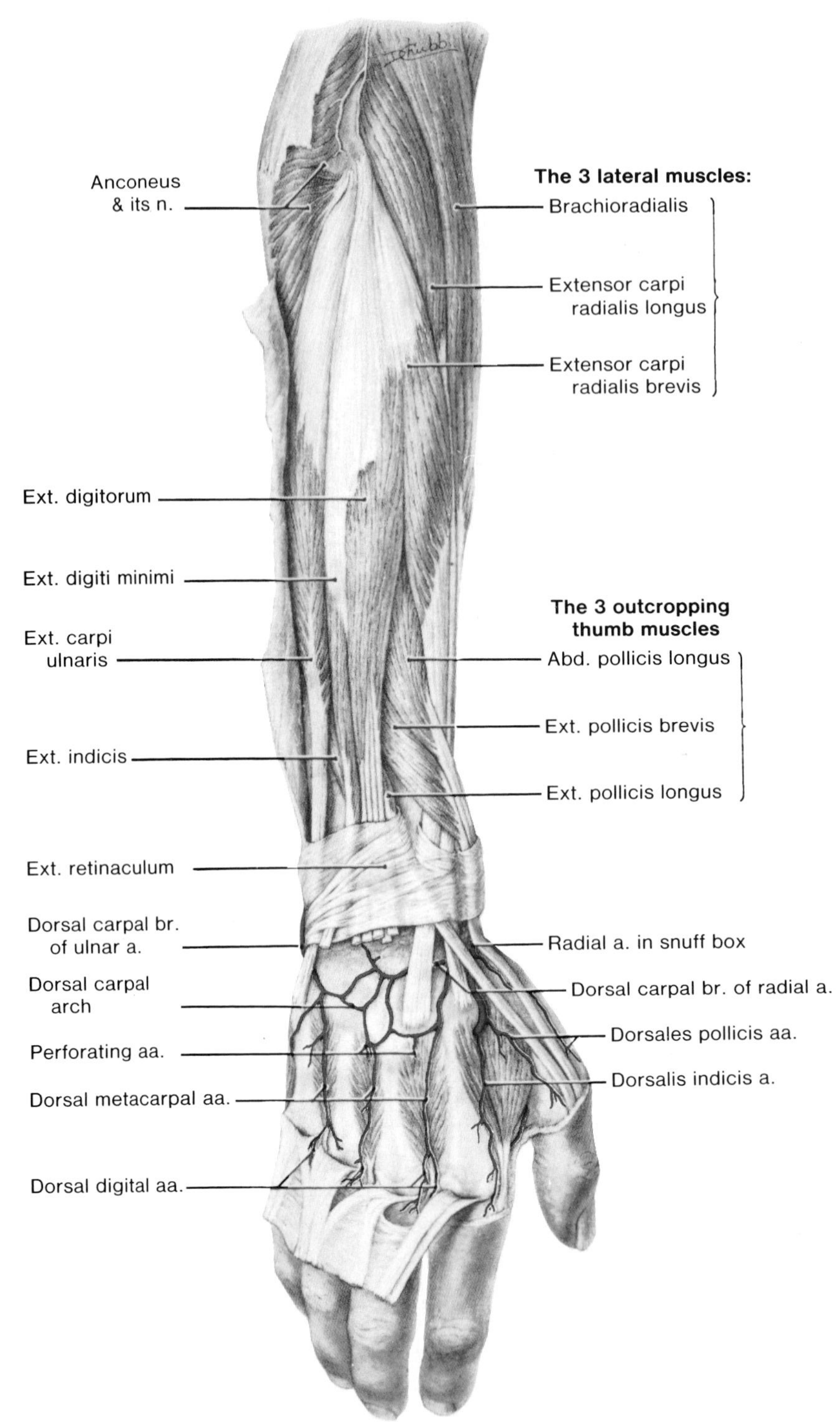

Figure 6-96. Drawing of a dissection of the extensor muscles (dorsal and lateral aspects) of the forearm, and the arteries on the dorsum of the right hand. The extensor tendons of the fingers are reflected. Observe the three muscles of the thumb outcropping between the extensor carpi radialis brevis and extensor digitorum muscles. Note the radial artery, the dorsal carpal branch of the ulnar, and their branches. All these are undisturbed by the removal of muscles because they lie on the skeletal plane (*i.e.*, on bone, ligament, or fascia). Observe the radial artery disappearing between the two heads of the first dorsal interosseous muscle, where it is in series with the perforating arteries. Note that no muscle is attached to the dorsal aspect of any carpal bones.

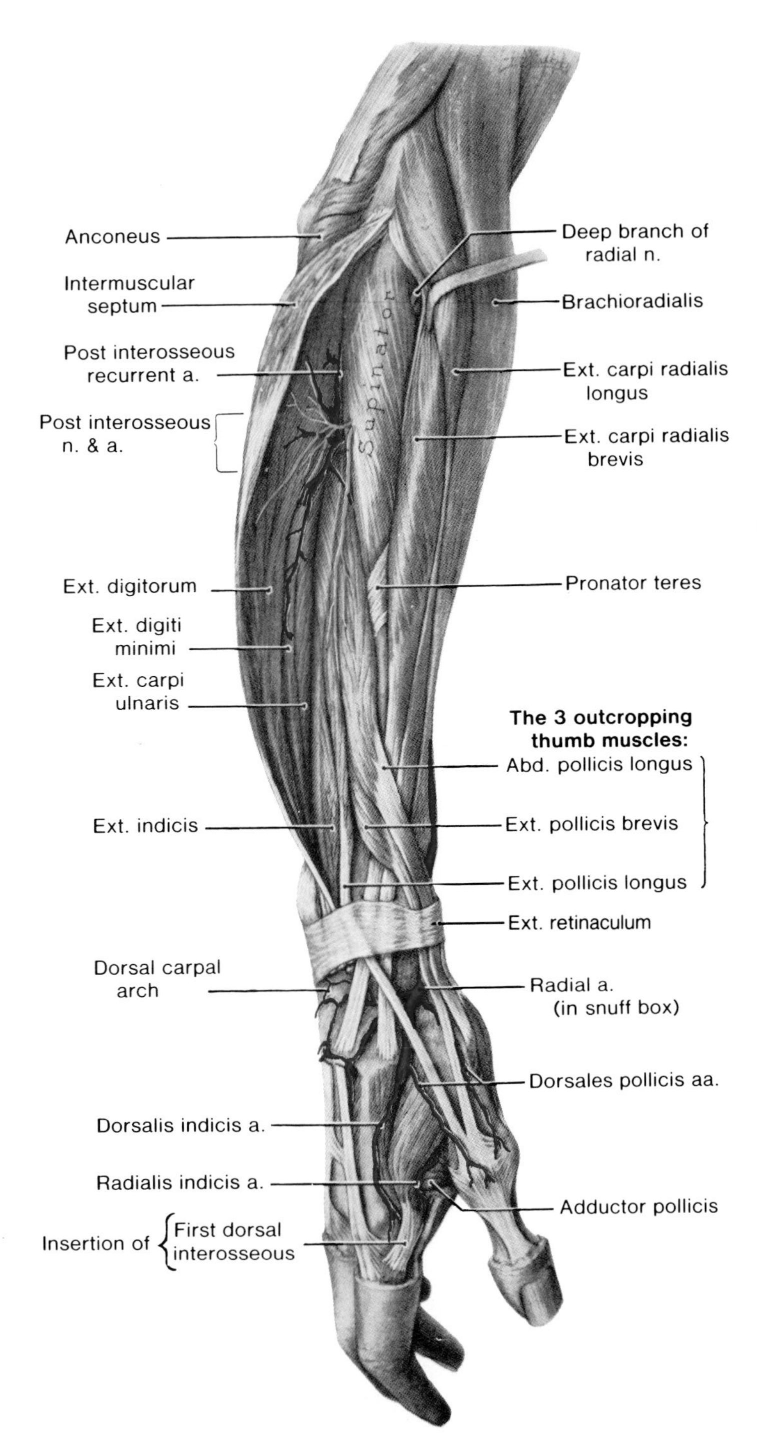

Figure 6-97. Drawing of a posterolateral view of a dissection of the deep structures in the posterior part of the right forearm and hand. Observe the three outcropping thumb muscles. The furrow from which the three muscles outcrop has been opened widely. It crosses the supinator and is a "*line of safety*" because the three laterally retracted muscles are supplied by the posterior interosseous nerve at its proximal end, whereas the others are supplied about 6 cm distal to the head of the radius. Observe that after the deep branch of the radial nerve emerges from the supinator, it is called the posterior interosseous nerve.

the long extensor tendons when the hand is hyperextended at the wrist joint.

Sometimes a cystic round, usually nontender, swelling appears on the dorsum of the wrist or hand. Usually the swelling is the size of a grape, but they can be as large as a plum. Flexion of the wrist makes the swelling get larger and extension of the wrist tends to make it smaller. Clinically this type of swelling is called a ganglion (G. knot); *anatomically this is a misnomer* because a ganglion is a collection of nerve cells outside the central nervous system (*e.g.*, a spinal ganglion). These cystic swellings ("ganglia") which appear on the dorsum of the wrist are near to, and often communicate with, the synovial sheaths on the dorsum of the wrist (Fig. 6-101).

The insertion of the extensor capri radialis brevis tendon into the base of the third metacarpal bone (Figs. 6-76 and 6-101) is a common site for a cystic swelling ("ganglion") of the tendon sheath.

THE EXTENSOR CARPI RADIALIS LONGUS MUSCLE (Figs. 6-74, 6-79, 6-87, 6-96 to 6-100, and Table 6-10)

This muscle is partly overlapped by the brachioradialis with which it is often blended (Fig. 6-96).

Origin (Fig. 6-75). **Distal third of the lateral supracondylar ridge of the humerus** and the lateral intermuscular septum.

Insertion (Fig. 6-76). **Dorsum of the base of the second metacarpal bone.** Its flat tendon runs deep to the extensor retinaculum and passes through the anatomical snuff box (Fig. 6-99).

Nerve Supply (Fig. 6-27). **Radial nerve** (C6 and C7).

Actions (Tables 6-10 and 6-18). **Extends and abducts the hand** at the wrist joint.

THE EXTENSOR CARPI RADIALIS BREVIS MUSCLE (Figs. 6-74, 6-79, 6-87, 6-96 to 6-99, and Table 6-10)

As its name indicates, this muscle is shorter than the extensor carpi radialis longus which covers it.

Origin (Figs. 6-75, 6-76, 6-96, and 6-141). **Lateral epicondyle of the humerus** by the *common extensor tendon* and the radial collateral ligment of the elbow joint.

Insertion (Fig. 6-76). **Dorsal surface of the base of the third metacarpal bone**. Its tendon passes deep to the extensor retinaculum and through the anatomical snuff box (Figs. 6-96 to 6-99).

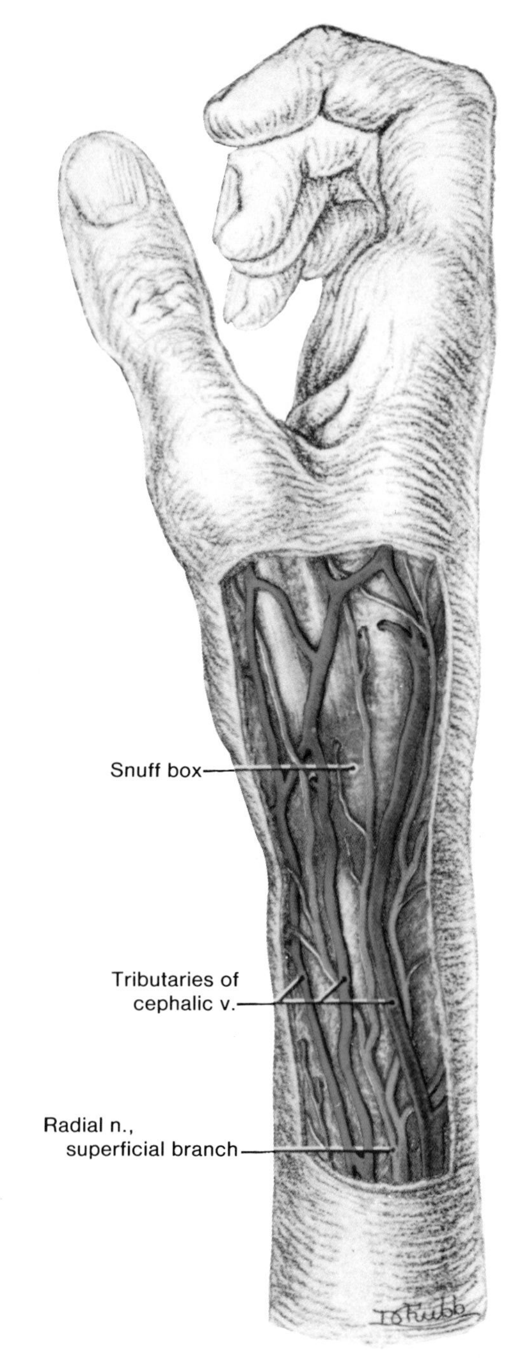

Figure 6-98. Drawing of a dissection of the lateral aspect of the right forearm and hand. Observe the superficial veins and nerves crossing the anatomical snuff box. Superficial branches of the radial nerve are sometimes injured during puncture of the tributaries of the cephalic vein for intravenous transfusion.

Nerve Supply (Fig. 6-87). **Deep branch of the radial nerve** (C7 and C8).

Actions (Tables 6-10 and 6-18). **Extends and abducts the hand** at the wrist joint. This muscle and

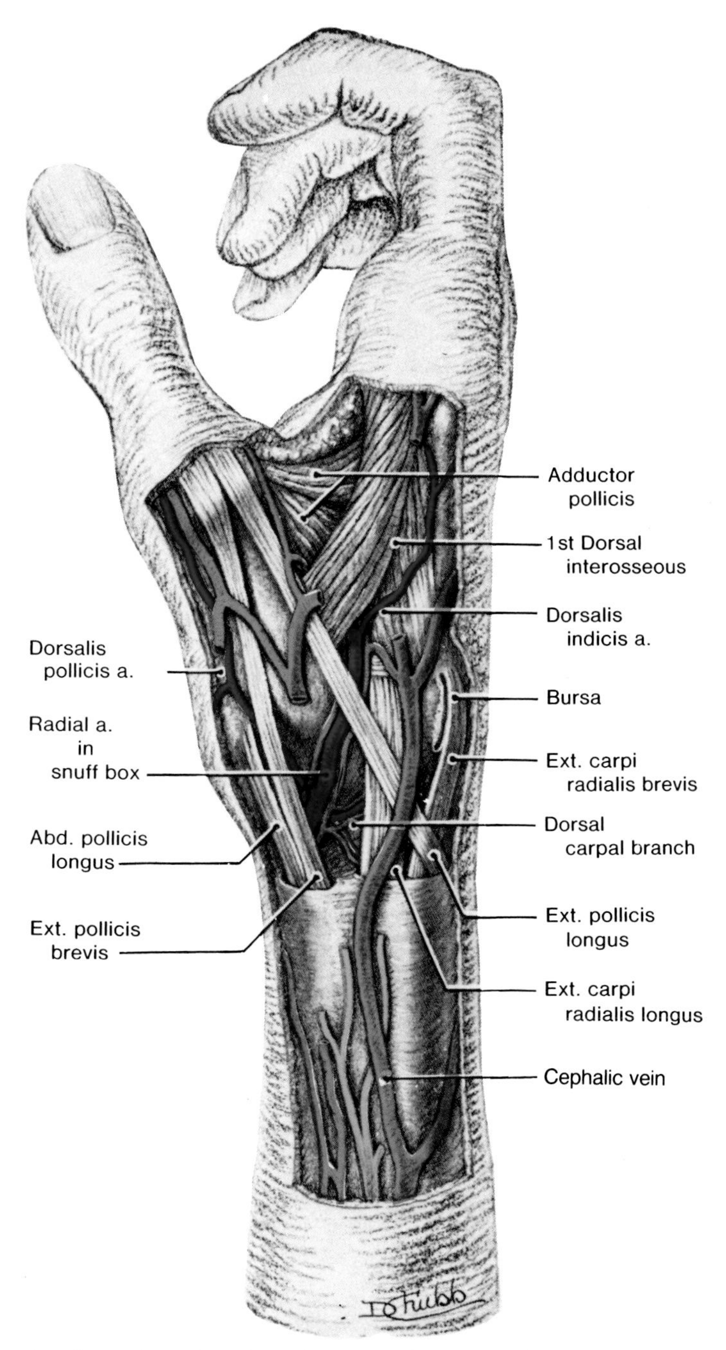

Figure 6-99. Drawing of a dissection of the lateral aspect of the right forearm and hand. Observe the three long tendons of the thumb forming the sides of the triangular hollow known as the *anatomical snuff box.* The abductor pollicis longus and extensor pollicis brevis form the anterior boundary of the snuff box and the extensor pollicis longus forms its posterior boundary. Observe that the radial artery passes on the dorsal aspect of the carpus between the tendons of the abductor pollicis longus and extensor pollicis brevis, and crosses the scaphoid and trapezium in the floor of the snuff box.

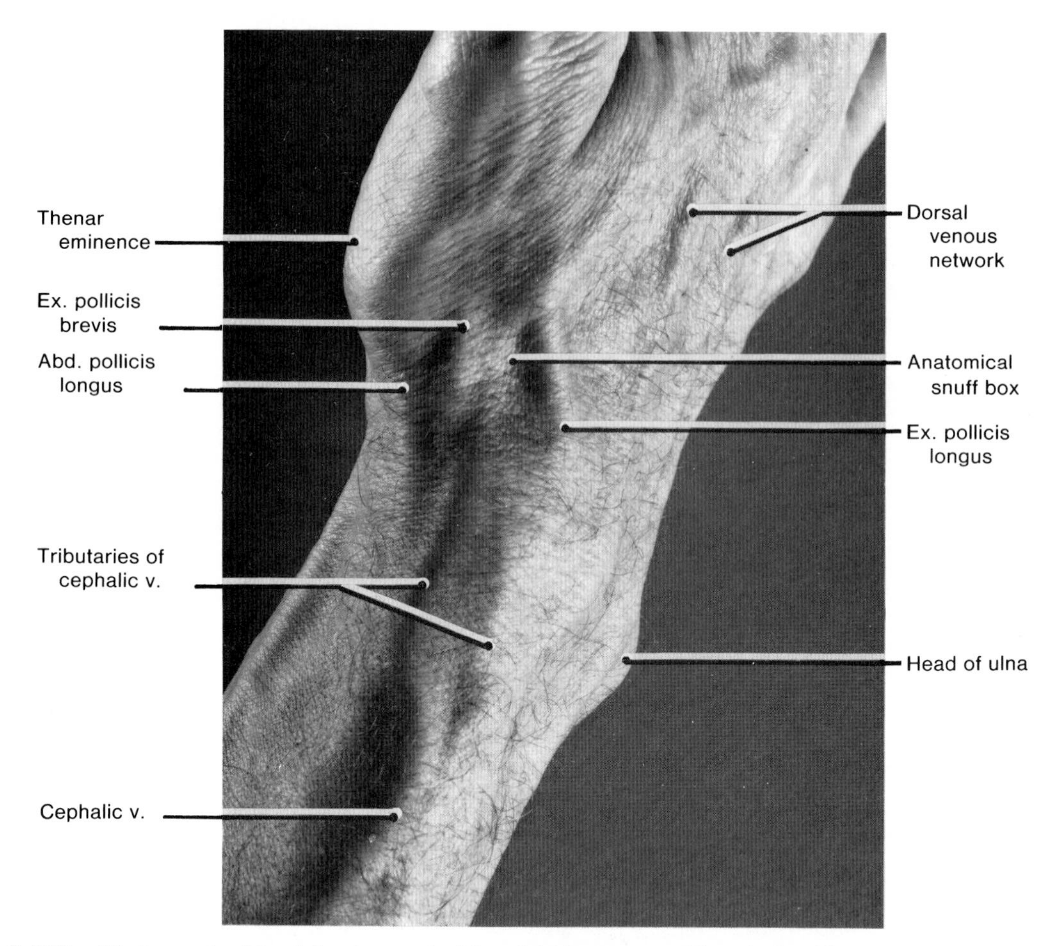

Figure 6-100. Photograph of the lateral aspect of the right forearm and hand of a 46-year-old man. Observe the hollow called the *anatomical snuff box* because of the use to which it was once put. The abductor pollicis longus and extensor pollicis brevis bound the snuff box anteriorly, and the extensor pollicis longus bounds it posteriorly. Compare this photograph with the dissections of the radial aspect of the wrist (Figs. 6-98 and 6-99). Note that the tendons of the abductor pollicis longus and extensor pollicis brevis diverge in this man as they proceed distally. If you have difficulty distinguishing one from the other on your wrist, move your thumb anteriorly and posteriorly and insert your fingernail between them when you see them move. The cephalic vein usually lies in the superficial fascia just posterior to the styloid process of the radius (Fig. 6-99).

the extensor carpi radialis longus act together to steady the wrist during flexion of the fingers.

THE EXTENSOR DIGITORUM MUSCLE (Figs. 6-79, 6-96, 6-97, 6-101, and Table 6-10)

This ***principal extensor of the fingers*** occupies much of the posterior surface of the forearm (Figs. 6-79 and 6-96). It *divides into four tendons* proximal to the wrist which pass through a **common synovial sheath**, deep to the extensor retinaculum with the extensor indicis (Figs. 6-101 and 6-103).

Origin (Fig. 6-76). **Lateral epicondyle of the humerus** by the *common extensor tendon* and the intermuscular septa.

Insertion (Figs. 6-101 to 6-103). **Extensor expansions of the fingers.**

Nerve Supply (Fig. 6-97). **Posterior interosseous nerve (C7** and C8), a branch of the radial.

Actions (Tables 6-19 and 6-20). **Extends the fingers** at the metacarpophalangeal and interphalangeal joints. It can also extend the hand at the wrist joint.

THE EXTENSOR DIGITI MINIMI MUSCLE (Figs. 6-96, 6-97, 6-101, and Table 6-10)

This slip of muscle is a partially detached part of the extensor digitorum. Its tendon runs through a separate compartment in the extensor retinaculum (Fig. 6-101). The tendon divides into two slips; the lateral one is joined to the tendon of the extensor digitorum (Fig. 6-101).

Origin (Figs. 6-75 and 6-76). **Lateral epicondyle of the humerus** by the *common extensor tendon.*

Insertion (Figs. 6-101 and 6-103). **Extensor expansion of the little finger.**

Nerve Supply (Fig. 6-97). **Posterior interosseous nerve (C7** and C8), a branch of the radial.

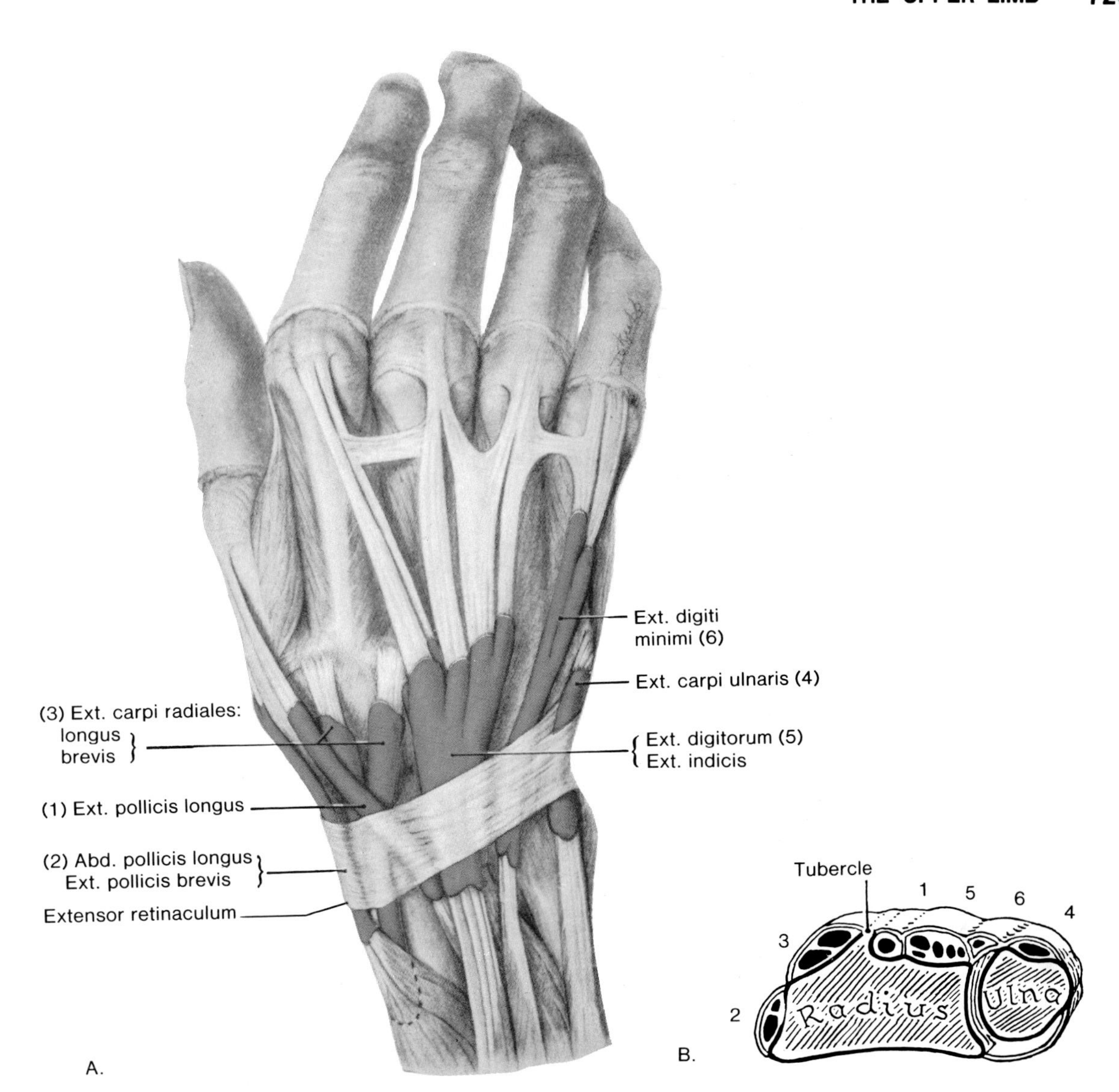

Figure 6-101. *A*, drawing of the synovial sheaths on the dorsum of the right wrist. Observe that the six sheaths occupy six osseofibrous tunnels deep to the extensor retinaculum. They contain nine tendons: three for the thumb in sheaths (one and two); three for the extensors of the wrist in two sheaths (three and four); and three for the extensors of the fingers in two sheaths (five and six). The tendons of the extensors of the wrist are the strongest because they work synergically with the flexors of the digits. Note the bands proximal to the knuckles that connect the tendons of the digital extensors and thereby restrict independent action of the fingers. *B*, diagram of a transverse section of the wrist showing the tendons and their synovial sheaths on the dorsum of the distal ends of the radius and ulna.

Actions (Tables 6-19 and 6-20). **Extends the little finger at the metacarpophalangeal and interphalangeal joints,** allowing separate extension of the little finger.

THE EXTENSOR CARPI ULNARIS MUSCLE (Figs. 6-79, 6-81, 6-96, 6-97, 6-101, and Table 6-10)

This long thin muscle is located on the ulnar border of the forearm. Its tendon runs in a groove between the head and styloid process of the ulna (Fig. 6-72*B*), within a special compartment of the extensor retinaculum (Fig. 6-101).

Origin (Figs. 6-75, 6-76, and 6-96). **Lateral epicondyle of the humerus** by the *common extensor tendon,* **and the posterior border of the ulna.**

Insertion (Figs. 6-76 and 6-101*A*). **Base of the fifth metacarpal bone.**

Nerve Supply (Fig. 6-97). **Posterior interosseous nerve** (C7 and **C8**), a branch of the radial.

Actions (Table 6-18). **Extends and adducts the hand** at the wrist joint. Acting with the extensor carpi

Table 6-10
Superficial Muscles of the Posterior Aspect of the Forearm

Muscle	Origin	Insertion	Nerve Supply	Action(s)
Brachioradialis[1] (Figs. 6-58, 6-74, and 6-96)	Proximal two-thirds of supracondylar ridge of humerus (Fig. 6-75)	Lateral surface of distal end of radius (Fig. 6-75)	Radial n. (C5, **C6**, and C7)	Flexes forearm at elbow joint
Extensor carpi radialis longus (Figs. 6-74, 6-96, and 6-98)	Lateral supracondylar ridge of humerus (Fig. 6-75)	Base of 2nd metacarpal bone (Fig. 6-76)	Radial n. (C6 and C7)	Extend and abduct hand at wrist joint
Extensor carpi radialis brevis (Figs. 6-74, 6-87, and 6-96 to 6-98)	Lateral epicondyle of humerus (Figs. 6-75, 6-76, and 6-97)	Base of 3rd metacarpal bone (Fig. 6-76)	Deep branch of radial n. (**C7** and C8)	
Extensor digitorum (Figs. 6-96, 6-97, and 6-100)	Lateral epicondyle of humerus (Fig. 6-76)	Extensor expansions of fingers (Figs. 6-100 to 6-102)		Extends fingers at metacarpophalangeal and interphalangeal joints; extends hand at wrist joint
Extensor digiti minimi (Figs. 6-96 and 6-97)	Lateral epicondyle of humerus (Figs. 6-75 and 6-76)	Extensor expansion of little finger (Figs. 6-101 and 6-102)	Posterior interosseous n. (**C7** and C8), a branch of the radial n.	Extends little finger at metacarpophalangeal and interphalangeal joints
Extensor carpi ulnaris (Figs. 6-79, 6-81, 6-96, 6-97, and 6-101)	Lateral epicondyle of humerus and posterior border of ulna (Figs. 6-75, 6-76, and 6-96)	Base of 5th metacarpal bone (Figs. 6-76 and 6-101*A*)		Extends and adducts hand at wrist joint
Anconeus[2] (Figs. 6-6, 6-140, and 6-143)	Lateral epicondyle of humerus (Figs. 6-46 and 6-143)	Lateral surface of olecranon and superior part of posterior surface of ulna (Figs. 6-46, 6-140, and 6-143)	Radial n. (C7, C8, and T1)	Assists triceps in extending elbow joint; stabilizes elbow joint; abducts ulna during pronation (Fig. 6-70*B*)

[1] The brachioradialis muscle is described with the muscles of the cubital fossa (p. 693) because it forms the lateral boundary of this space in the anterior aspect of the elbow (Figs. 6-71 and 6-74).

[2] This small triangular muscle should be considered as part of the triceps (Fig. 6-63). It does not belong to the posterior fascial compartment of the forearm, but for convenience and as is the custom, it is listed with them.

radialis, it extends the hand; acting with the flexor carpi ulnaris it adducts the hand.

The Deep Extensor Muscles (Figs. 6-96 to 6-101). The deep extensors of the forearm consist of **three muscles that act on the thumb** (*abductor pollicis longus, extensor pollicis brevis, and extensor pollicis longus*) **and the extensor indicis, which helps to extend the index finger.** These muscles originate from the radius, ulna, and interosseous membrane (Fig. 6-76).

The abductor pollicis longus, extensor pollicis brevis, and extensor pollicis longus arise deep to the superficial extensors, and appear from concealment (*i.e., they crop out) along a furrow that divides the*

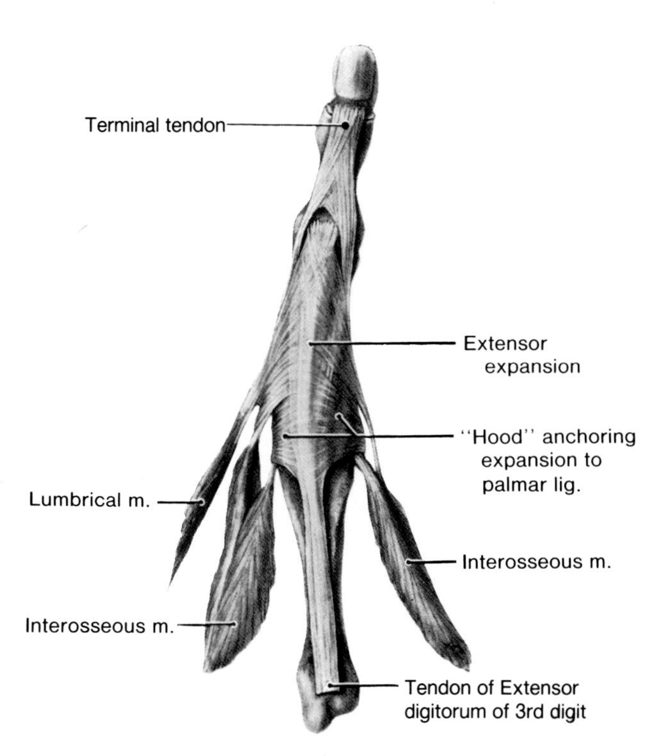

Figure 6-102. Drawing of a dorsal view of a dissection of the middle finger showing its extensor expansion. Observe that the expansion extends to the bases of the middle and distal phalanges. It also gives a strong band to the base of the proximal phalanx (not illustrated here).

extensor muscles into lateral and medial groups. Because of this characteristic, they are referred to as the **three outcropping thumb muscles** (Figs. 6-96 and 6-97). The tendons of these muscles appear from under the lateral border of the extensor digitorum and pass superficial to the tendons of the extensor carpi radialis longus and extensor carpi radialis brevis.

BOUNDARIES OF THE ANATOMICAL SNUFF BOX (Figs. 6-96 to 6-101)

The tendons of the abductor pollicis longus and extensor pollicis brevis bound the anatomical snuff box anteriorly, and the tendon of the extensor pollicis longus bounds it posteriorly. Extend your thumb and examine your anatomical snuff box (Fig. 6-100). Put the tip of your finger into this depression and feel the pulsations of the radial artery as it crosses the floor of the snuff box (Figs. 6-96, 6-97, and 6-99).

THE ABDUCTOR POLLICIS LONGUS MUSCLE (Figs. 6-96 to 6-101, 6-103, and Table 6-11)

This ***long abductor of the thumb*** (L. pollex) lies just distal to the supinator (Fig. 6-97), and is closely related to the extensor pollicis brevis muscle.

Origin (Fig. 6-76). **Posterior surfaces of the ulna and radius and the interosseous membrane.** Its origin is inferior to the supinator muscle. Although deeply situated, the abductor pollicis longus emerges at the wrist as one of the outcropping muscles (Fig. 6-96). Its tendon passes deep to the extensor retinaculum in a common synovial sheath with the tendon of the extensor pollicis brevis (Fig. 6-101). Distal to the extensor retinaculum these tendons form the anterior boundary of the anatomical snuff box (Figs. 6-96 to 6-100).

Insertion (Fig. 6-75). **Base of the first metacarpal bone.**

Nerve Supply (Fig. 6-97). **Posterior interosseous nerve** (C7 and **C8**), a branch of the radial.

Actions (Figs. 6-115, 6-119, and Table 6-21). **Abducts and extends the thumb** at the carpometacarpal joint. The abductor pollicis longus acts with the abductor pollicis brevis during abduction of the thumb, and with the extensor pollicis in extending the thumb.

THE EXTENSOR POLLICIS BREVIS MUSCLE (Figs. 6-96 to 6-101, 6-103, and Table 6-11)

This ***short extensor of the thumb*** lies distal to the long abductor of the thumb (abductor pollicis longus) and is partly covered by it.

Origin (Fig. 6-76). **Posterior surface of the radius and interosseous membrane,** inferior to the abductor pollicis longus. Its tendon lies in contact

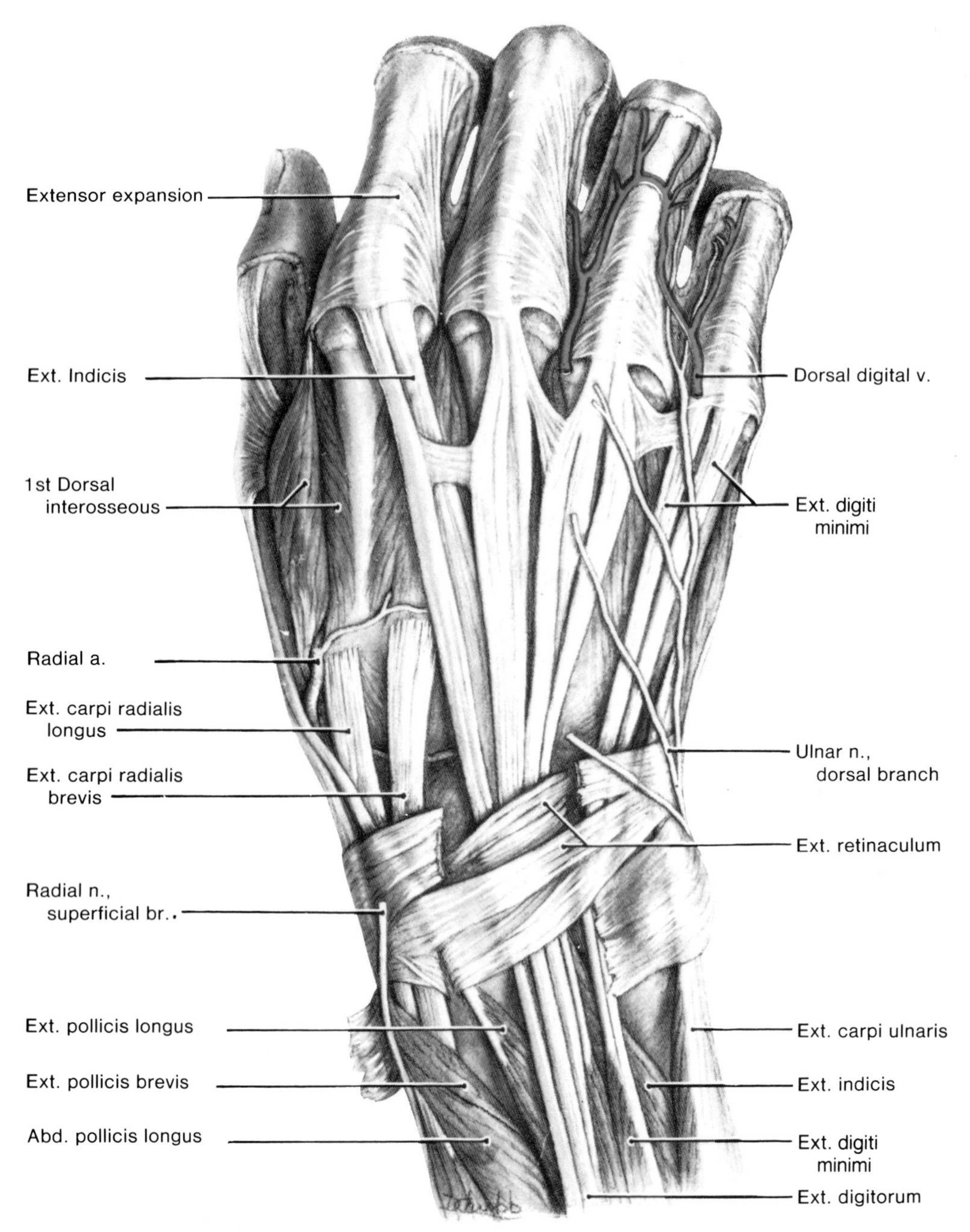

Figure 6-103. Drawing of a dissection of the extensor tendons passing deep to the extensor retinaculum on the dorsum of the right hand. Observe that this strong band stretches obliquely from one ridge on the radius to another. Medially it passes distal to the ulna to be attached to the pisiform and triquetral bones (Fig. 6-77). Note the bands proximal to the knuckles that connect the tendons of the digital extensor and thereby restrict the independent action of the fingers.

with the abductor pollicis longus tendon as they pass deep to the extensor retinaculum (Fig. 6-96).

Insertion (Fig. 6-76). **Base of the proximal phalanx of the thumb.**

Nerve Supply (Fig. 6-97). **Posterior interosseous nerve** (C7 and **C8**), a branch of the radial.

Action (Fig. 6-115 and Table 6-21). **Extends the thumb at the carpometacarpal and metacarpophalangeal joints.** The extensor pollicis brevis extends the proximal phalanx of the thumb and, in continued action, helps to extend the metacarpal bone of the thumb.

Table 6-11
Deep Muscles of the Posterior Aspect of the Forearm

Muscle	Origin	Insertion	Nerve Supply	Action(s)
Supinator[1] (Figs. 6-74, 6-87, and 6-97)	Lateral epicondyle of humerus, radial collateral ligament of elbow joint, anular ligament of superior radioulnar joint, supinator fossa, and crest of ulna (Figs. 6-76 and 6-141)	Lateral, posterior, and anterior surfaces of the proximal third of the radius (Figs. 6-75 and 6-76)	Deep branch of radial n. (C5 and **C6**)	Supinates forearm *i.e.*, rotates radius so as to turn the palm anteriorly (Fig. 6-70*A*)
Abductor pollicis longus (Figs. 6-96 to 6-100 and 6-103)	Posterior surfaces of ulna and radius and interosseous membrane (Fig. 6-76)	Base of 1st metacarpal bone (Fig. 6-75)	Posterior interosseous n. (C7 and **C8**)[2]	Abducts and extends thumb at carpometacarpal joint (Fig. 6-115)
Extensor pollicis brevis (Figs. 6-96 to 6-100 and 6-103)	Posterior surface of radius and interosseous membrane (Fig. 6-76)	Base of proximal phalanx of thumb (Fig. 6-76)		Extends thumb at carpometacarpal and metacarpophalangeal joints (Fig. 6-115)
Extensor pollicis longus (Figs. 6-96 to 6-100 and 6-103)	Posterior surface of middle third of ulna and interosseous membrane (Fig. 6-76)	Base of distal phalanx of thumb (Fig. 6-75)		Extends metacarpophalangeal and interphalangeal joints of thumb (Fig. 6-115)
Extensor indicis (Figs. 6-96, 6-97, 6-100, and 6-103)	Posterior surface of ulna and interosseous membrane (Fig. 6-76)	Extensor expansion of index finger (Fig. 6-103)		Helps to extend index finger

[1] The supinator muscle is described with the muscles of the cubital fossa because it forms the floor of this space in the anterior aspect of the elbow (Figs. 6-71 and 6-74).
[2] The terminal branch of the deep branch of the radial nerve.

THE EXTENSOR POLLICIS LONGUS MUSCLE (Figs. 6-96 to 6-101, 6-103, and Table 6-11)

This ***long extensor of the thumb*** is larger and its tendon is longer than that of the extensor pollicis brevis.

Origin (Fig. 6-76). **Posterior surface of the middle third of the ulna and interosseous membrane.** Its tendon passes through a special compartment of the extensor retinaculum (Fig. 6-101), where it turns laterally around the **dorsal tubercle of the radius** (Fig. 6-72*B*) to form the posterior boundary of the anatomical snuff box (Figs. 6-96 to 6-101).

Insertion (Fig. 6-75). **Base of the distal phalanx of the thumb.**

Nerve Supply (Fig. 6-97). **Posterior interosseous nerve** (C7 and **C8**), a branch of the radial.

Action (Fig. 6-115 and Table 6-21). **Extends the metacarpophalangeal and the interphalangeal joints of the thumb.** In continued action, the extensor pollicis longus adducts the extended thumb and rotates it laterally.

THE EXTENSOR INDICIS MUSCLE (Figs. 6-96, 6-97, 6-101, 6-103, 6-107, and Table 6-11)

This narrow, elongated muscle lies medial to, and alongside, the extensor pollicis longus.

Origin (Fig. 6-76). **Posterior surface of the ulna and interosseous membrane,** inferior to the extensor pollicis longus. Its tendon passes deep to the extensor retinaculum within the same synovial sheath as the tendons of the extensor digitorum (Fig. 6-101).

Insertion (Fig. 6-103). **Extensor expansion of the index finger.** Opposite the second metacarpal bone, its tendon joins the tendon of the extensor digitorum to the other fingers.

Nerve Supply (Fig. 6-97). **Posterior interosseous nerve** (C7 and **C8**), a branch of the radial.

Action (Tables 6-19 and 6-20). **Helps to extend**

the index finger and can assist in extending the wrist.

Tennis elbow (lateral epicondylitis), characterized by pain and point tenderness at or just distal to the lateral epicondyle of the humerus, (Fig. 6-104), appears to result from premature *degeneration of the common extensor origin of the superficial extensor muscles of the forearm* (*i.e.*, the origin of the common extensor tendon, Figs. 6-75 and 6-76).

Tennis elbow is a painful musculoskeletal condition that may follow repetitive forceful pronation-supination of the forearm (Fig. 6-70). *The elbow joint and olecranon are not involved.* The pain is aggravated by activities that put tension on the common extensor tendon (e.g., active extension of the wrist while grasping something such as a tennis racquet. Pain also occurs during passive flexion of the wrist against resistance.

Tennis elbow is common in persons who play tennis because of the repeated strenuous contraction of the extensor muscles, especially during the backhand stroke. This repeated movement *strains the common extensor tendon of these muscles* and produces inflammation of the lateral epicondyle.

Tennis elbow is not confined to those who play tennis. It may develop following an injury to the elbow or result from any continuous activity that involves extensive use of the superficial extensor muscles of the forearm (*e.g.*, lifting and throwing snow with a shovel).

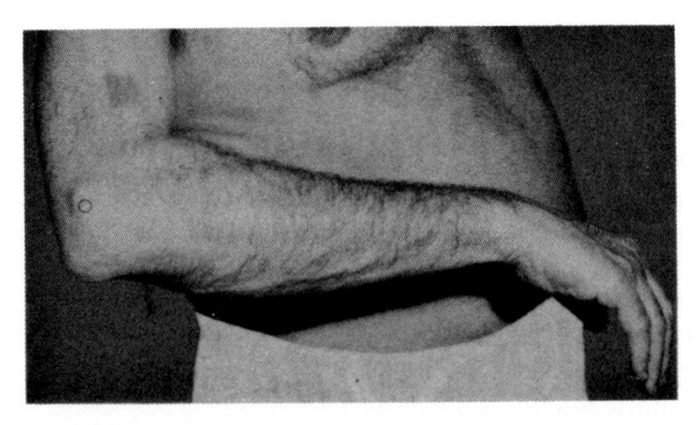

Figure 6-104. Photograph of the right upper limb of a man with lateral epicondylitis (tennis elbow). The *circle* marks the point of local tenderness over the lateral epicondyle of his humerus. Point tenderness is found at or just distal to the lateral epicondyle, where the extensor muscles of the forearm originate (Fig. 6-76).

THE POSTERIOR NERVES OF THE FOREARM

The radial nerve and injuries of it were described with the anterior nerves of the forearm (p. 716), because the radial nerve enters the cubital fossa (Figs. 6-74 and 6-87). At, or inferior to, the lateral epicondyle, the radial nerve divides into deep and superficial branches (Fig. 6-74).

The deep branch of the radial nerve supplies the extensor carpi radialis brevis and supinator muscles before entering the latter muscle (Fig. 6-87). When the deep branch of the radial emerges from the supinator muscle and enters the posterior compartment of the forearm, it is referred to as the **posterior interosseous nerve** (Fig. 6-97). It passes deep to the extensor pollicis longus muscle and lies on the interosseous membrane, where it is *accompanied by the* ***posterior interosseous artery***. The posterior interosseous nerve terminates on the dorsum of the wrist, where it sends articular branches to the distal radioulnar joint and the joints of the hand.

The deep branch of the radial nerve supplies the extensor carpi radialis brevis and supinator muscles (Figs. 6-27 and 6-87). The posterior interosseous nerve, the terminal branch of the deep radial, supplies the extensor digitorum, the extensor digiti minimi, the extensor carpi ulnaris, the extensor indicis, and the three outcropping thumb muscles (Fig. 6-27 and Tables 6-10 and 6-11).

The superficial branch is the continuation of the radial nerve (Fig. 6-85). It passes distally, deep to the brachioradialis muscle. At the wrist it divides into four or five digital nerves (Figs. 6-93 and 6-120).

The distribution of the superficial branch of the radial nerve is cutaneous and articular. It supplies the lateral two-thirds of the posterior surface of the hand and the posterior surface of the lateral two and one-half digits over the proximal phalanx (Figs. 6-26, 6-92, and 6-93).

Radial nerve injuries were discussed previously (see p. 716).

THE POSTERIOR ARTERIES OF THE FOREARM

The posterior aspect of the forearm and hand is supplied by the **posterior interosseous artery** (Fig. 6-97), which *arises from the common interosseous branch of the ulnar artery* (Fig. 6-147). The posterior interosseous artery passes posteriorly between the radius and ulna, just proximal to the interosseous membrane, and appears in the posterior part of the forearm between the supinator and abductor pollicis longus muscles (Fig. 6-97). It then descends between the superficial and deep muscles, supplying muscles in the posterior part of the forearm.

As it reaches the posterior aspect of the wrist, the

posterior interosseous artery becomes very small and ends by anastomosing with the termination of the anterior interosseous artery and the **dorsal carpal arch** (Figs. 6-96, 6-97, and 6-123).

THE WRIST

The wrist or carpal region is between the forearm and hand. A "wrist" watch is usually not worn around the wrist; commonly the band encircles the distal end of the forearm, just proximal to the head of the ulna (Fig. 6-71). Movements of the hand occur primarily at the wrist joint (Table 6-18). The wrist or **carpal bones** were described previously (p. 696) and are illustrated in Figures 6-1, 6-77, and 6-109.

The ***antebrachial fascia*** (deep fascia of forearm) is thickened posteriorly at the wrist to form a transverse band known as the **extensor retinaculum** (Figs. 6-96 and 6-97). This fibrous band retains the extensor tendons in position, thereby increasing their efficiency. The deep fascia is also thickened anteriorly at the wrist to form the **flexor retinaculum** (Fig. 6-87), a fibrous band that converts the anterior concavity of the carpus into a *carpal tunnel* (Fig. 6-78) through which the flexor tendons pass (Figs. 6-81, 6-87, and 6-88).

Observe the **distal wrist crease** (Fig. 6-84) which indicates the proximal border of the flexor retinaculum (Fig. 6-91), and the level of the wrist joint (Fig. 6-109).

When a person falls on the outstretched hand with the forearm pronated, the main force of the fall is transmitted via the carpus to the distal ends of the forearm bones, particularly the radius, and then proximally to the humerus, scapula, and clavicle. During such falls fractures may occur in the wrist, forearm, or clavicle.

In older people the radius tends to break about 2.5 cm proximal to the wrist joint (**Colles' fracture**). In this injury the distal fragment of the radius is often comminuted (broken into pieces) and the fragments are usually displaced posteriorly and superiorly, producing shortening of the radius (Fig. 6-161*C*). When there is a single fragment it may be impacted (*i.e.*, the jagged ends of bone are driven into each other). Displacement of the distal part of the radius often breaks off the ulnar styloid process, owing to the violent pull of the articular disc connnecting the radius and ulna (Figs. 6-150 and 6-151).

Fracture of the scaphoid bone is a common carpal injury (Fig. 6-160), especially when the person falls on the palm with the hand abducted. Commonly the bone fractures at its narrow part or "waist," producing two fragments. As the scaphoid lies in the floor of the anatomical snuff box, **the clinical sign of fracture of the scaphoid is tenderness in the snuff box** or on the anterior aspect of the wrist over the tubercle of the scaphoid (Fig. 6-151).

THE HAND

We pray with our hands. We use them to eat, to work, and to make love. We employ them as marvellously sophisticated instruments of flexibility and strength, and when they are damaged we anguish.

The hand (L. *manus*) forms the distal part of the upper limb; it includes the metacarpus and digits (thumb and fingers). Because of the importance of manual dexterity in occupational and recreational activities, a good understanding of the structure and function of the hand is essential for all who are involved in maintaining or restoring its activities (free motion, power grip, precision handling, and pinching).

The skeleton of the hand is illustrated in Figures 6-1, 6-77, and 6-109.

Fractures and dislocations of the hand are common and disability can result if normal relationships are not restored.

Street fighters commonly fracture the distal end of the body of their fifth metacarpal bone; this region is referred to clinically as the *neck of the bone* (Fig. 6-106). In this type of fracture the head of the metacarpal (Fig. 6-77*A*) is bent toward the palm. Although commonly called a **boxer's fracture**, this injury is better referred to as *street fighter's fracture* because it results from an unskillful punch with the clenched fist. Because a trained boxer punches so that the second and third metacarpals take the strain, he rarely fractures the more mobile fifth metacarpal bone.

SURFACE ANATOMY OF THE HAND

DORSUM OF THE HAND (Figs. 6-100, 6-107, and 6-124)

The skin covering this region is thin and loose when the hand is relaxed. Hair is present on the dorsum of the hand and on the proximal parts of the digits, especially in males.

If the dorsum of the hand is examined with the wrist extended and the fingers abducted, the extensor tendons of the fingers stand out clearly, particularly in thin persons (Fig. 6-107). These tendons are not visible far beyond the knuckles because they flatten here to form the extensor expansions (Fig. 6-103). The knuckles that become visible when a fist is made are produced by the heads of the metacarpal bones (Fig. 6-77*A*).

Under the loose subcutaneous tissue and extensor tendons on the dorsum of the hand, you can palpate the metacarpal bones. A prominent feature of the dorsum of the hand is the **dorsal venous network** (Figs. 6-71 and 6-124).

PALM OF THE HAND (Figs. 6-108 and 6-112)

The skin on the palm is thick because it is required to withstand the wear and tear of work and play. The palmar skin is **richly supplied with sweat glands,** but it contains no hair or sebaceous glands.

The skin of the palm presents several more or less **constant longitudinal and transverse flexion creases,** where the skin is firmly bound to the deep fascia (Fig. 6-112). Usually four major palmar creases form a variable M-shaped pattern (Fig. 6-108). The creases indicate where folding of the skin occurs during flexion of the hand. Observe that the longitudinal creases deepen when the thumb is opposed (Fig. 6-115), and that the transverse creases deepen when you flex your metacarpophalangeal joints.

The Palmar Flexion Creases (Fig. 6-108). Some creases are useful surface landmarks. The **radial longitudinal crease** partially encircles the *thenar eminence* (ball of thumb) formed by the short muscles of the first digit. The **midpalmar crease** indicates the *hypothenar eminence* (ball of little finger) formed by the short muscles of the fifth digit.

The proximal transverse crease commences on

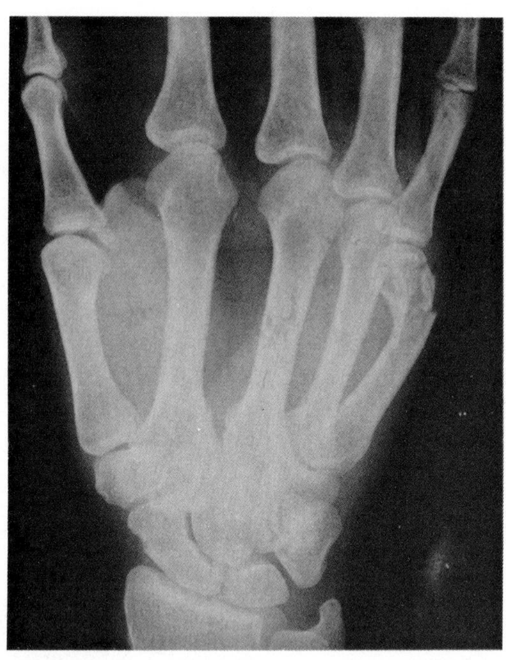

Figure 6-106. Radiograph of a slightly oblique view of the right hand showing an angulated fracture of the distal end of the body ("neck") of the fifth metacarpal bone. This injury occurred during a "fist fight."

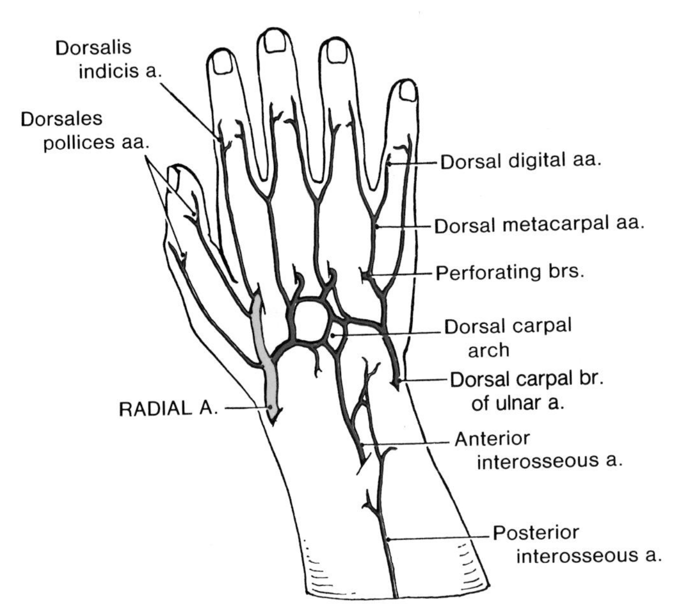

Figure 6-105. Diagram of the arteries of the dorsum of the right forearm and hand. Note the radial artery emerging from the anatomical snuff box (also see Fig. 6-98). Observe that the deep carpal branch of the radial artery anastomoses with the deep carpal branch of the ulnar to form the dorsal carpal arch.

the lateral border of the palm, in common with the radial longitudinal crease (Fig. 6-108), and superficial to the head of the second metacarpal bone (Fig. 6-109). It extends medially and slightly proximally across the palm, superficial to the bodies of the third to fifth metacarpal bones.

The distal transverse crease begins at or near the cleft between the index and middle fingers (Fig. 6-108) and crosses the palm with a slight convexity, superficial to the heads of the second to fourth metacarpal bones (Fig. 6-109).

The Digital Flexion Creases (Fig. 6-108). Each

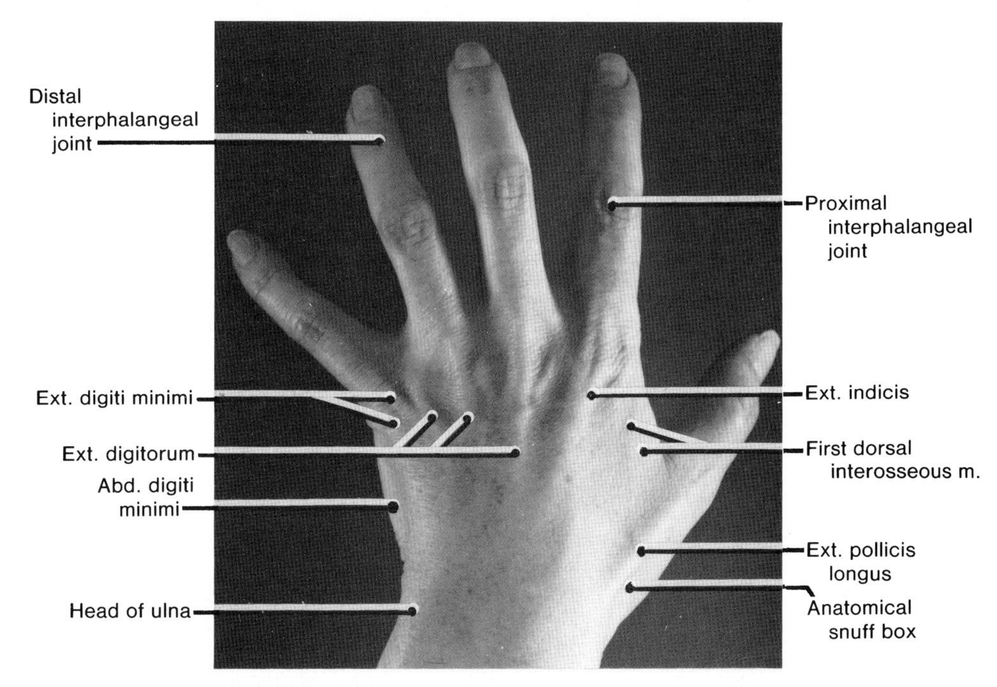

Figure 6-107. Photograph of the dorsum of the hand of a 36-year-old woman with her fingers abducted and her thumb extended. Observe the extensor tendons and the location of the anatomical snuff box. Compare with the dissection of the tendons on the dorsum of the hand shown in Figure 6-103.

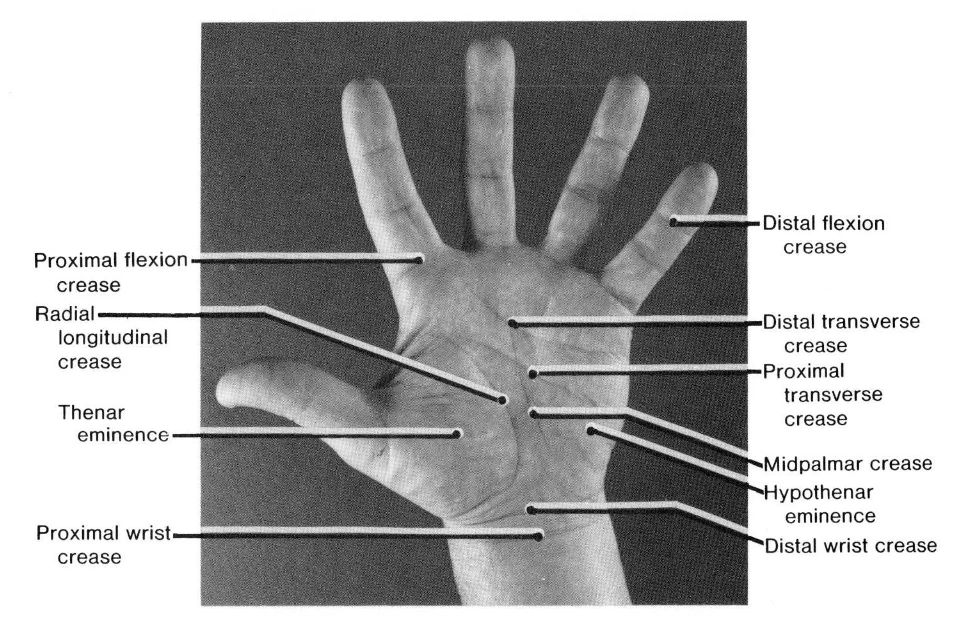

Figure 6-108. Photograph of the palm of the left hand of a 53-year-old man showing its surface landmarks. The distal skin crease on the wrist crosses (medial to the lateral) the pisiform, the tubercle of the scaphoid, and the styloid process of the radius. Compare with the radiograph shown in Figure 6-109.

of the medial four digits usually has three transverse flexion creases. The **proximal flexion crease** is located at the root of the finger, about 2 cm distal to the metacarpophalangeal joint. There are two **middle flexion creases**: the proximal one lies over the proximal interphalangeal joint, and the distal flexion crease lies proximal to the distal interphalangeal joint.

The thumb, having two phalanges, has only two flexion creases. Like other digital creases, they deepen when the thumb is flexed. The proximal flexion crease crosses the thumb obliquely, proximal to the first metacarpophalangeal joint (Fig. 6-109). The distal flexion crease on the thumb, comparable in position to the middle digital creases in the fingers, lies proximal to the interphalangeal joint.

The ridges on the ventral ends of the fingers, known as **fingerprints**, are used for identification (*e.g.*, in criminal investigations) because of their unique patterns. Their anatomical function is to reduce slippage when grasping objects.

Because the skin is firmly bound to the subcutaneous tissues at the skin creases (Fig. 6-110), these creases are avoided when making surgical incisions. In addition, scar tissue may form where an incision crosses a crease and subsequently reduce the range of movement at the nearby joint.

The science of studying the configurations of the dermal ridge patterns of the palm is known as **dermatoglyphics**. The most important landmarks are the patterns on the distal phalanges. Dermatoglyphics can be a valuable extension of the conventional physical examination of patients with certain congenital malformations and genetic diseases. For example, persons with 21-trisomy syndrome (**Down syndrome**) often have only one transverse palmar crease, usually referred to as a **simian crease**. However it is important to know that about 1% of the general population has this crease with no other clinical features of the syndrome.

In patients with the **trisomy 18 syndrome**, the little finger frequently has only one middle flexion crease. Examination of the palmar and digital creases in patients suspected of having chromosomal abnormalities is often helpful in deciding whether or not chromosomal investigations are indicated. For more information about dermatoglyphics and its use in studying chromosomes and their disorders, see Behrman and Vaughan (1983).

FASCIA OF THE PALM (Figs. 6-110 and 6-112)

The deep fascia of the palm is continuous proximally with the antebrachial fascia (fascia of the forearm), and at the borders of the palm with the fascia on the dorsum of the hand.

The fascia is thin over the thenar and hypothenar eminences (***thenar* and *hypothenar fasciae***), but is thick in the palm where it forms the **palmar aponeurosis** (Figs. 6-80 and 6-112), and in the fingers where it forms the *fibrous digital sheaths* (Fig. 6-111).

The Palmar Aponeurosis (Figs. 6-80, 6-110, and 6-112). This strong, well defined *triangular part of the deep fascia* covers the soft tissues of the hand and overlies the long flexor tendons of the palm. The proximal end of the palmar aponeurosis is continuous with the **flexor retinaculum** (Figs. 6-87 and 6-118). The distal end of the palmar aponeurosis divides at the roots of the fingers into four longitudinal bands which pass to the fingers. Each band is attached to the base of the proximal phalanx and is fused with the fibrous digital sheaths (Figs. 6-80, 6-110, and 6-111).

Dupuytren's contracture is a progressive fibrosis (increase in fibrous tissue) of the palmar aponeurosis, resulting in shortening and thickening of the fibrous bands that extend from the aponeurosis to the bases of the phalanges (Fig. 6-80). These fibrotic bands pull the fingers into such marked flexion at the metacarpophalangeal joints that they cannot be straightened (Fig. 6-114). Usually the first sign of Dupuytren's contracture is a thickened plaque overlying the flexor tendon to the ring finger.

Usually the ring and little fingers are affected intially. In more advanced cases the proximal interphalangeal joints of these fingers are also flexed so that the distal phalanges are pulled close to the palm. The etiology of this disease is unknown, but there appears to be a hereditary predisposition (multiple genetic factors appear to be involved).

FASCIAL COMPARTMENTS OF THE PALM (Fig. 6-112)

A fibrous ***medial septum*** extends deeply from the medial border of the palmar aponeurosis to the fifth metacarpal bone. Medial to this septum is the medial or **hypothenar compartment containing the three** *hypothenar muscles*, concerned with movements of the little finger (Figs. 6-112 and 6-116).

Similarly a fibrous ***lateral septum*** extends deeply

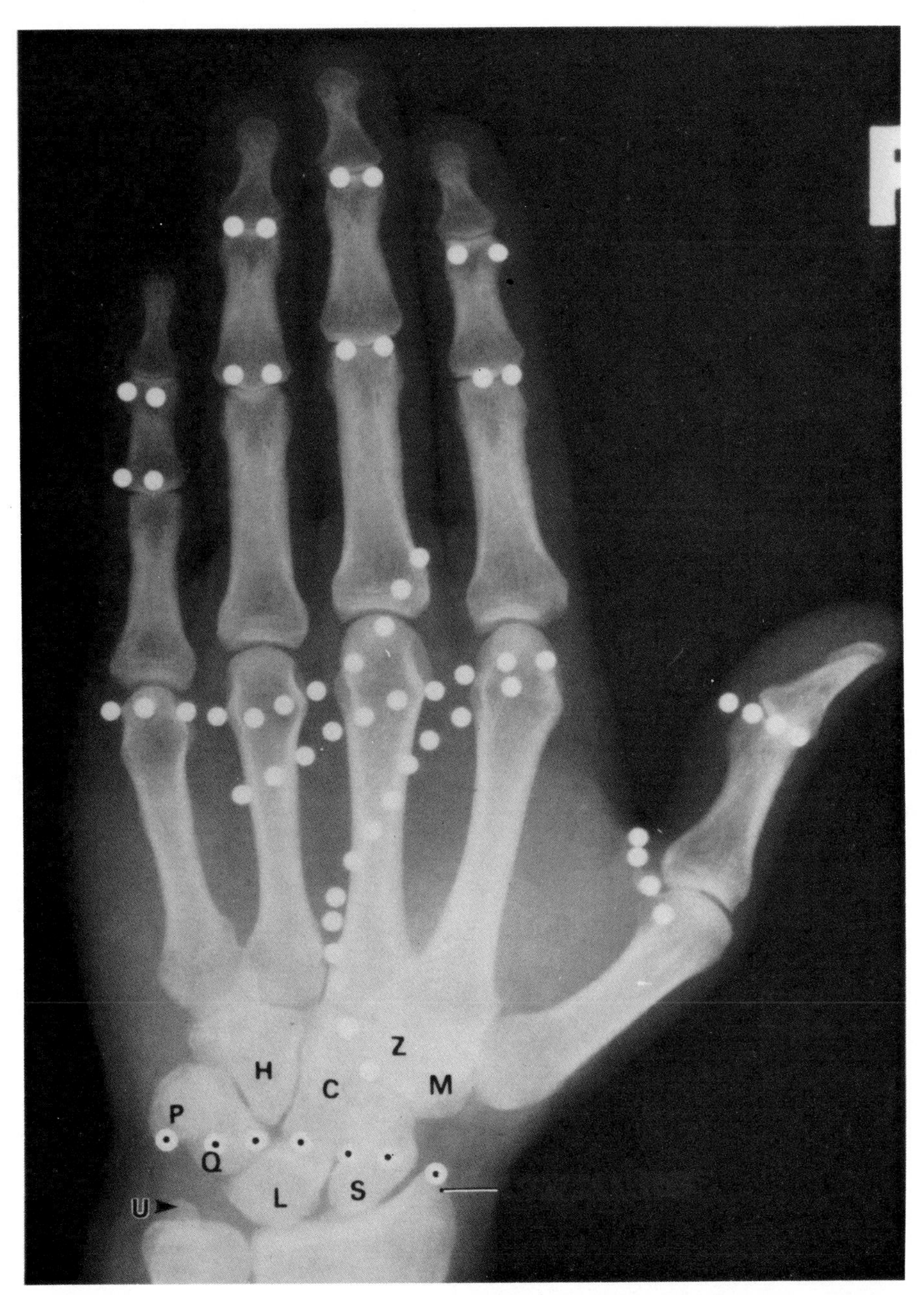

Figure 6-109. Radiograph of the left hand (posteroanterior projection) showing the posterior surfaces of the fingers and the *lateral surface of the thumb*. Compare with Figure 6-107. Place your hand palm down on a table and you will see why this occurs. Observe the two rows of carpal bones. In the distal row, note the hamate (*H*), capitate (*C*), trapezoid (*Z*), and trapezium (*M*) which forms a saddle-shaped joint with the metacrapal (*I*). In the proximal row note the scaphoid (*S*), lunate (*L*), and pisiform (*P*) superimposed on the triquetral (*Q*). Observe the ulnar styloid process (*U* with *arrow*). Lead shot has been placed along the palmar and digital flexion creases to show their relationship to the joints. Note that the distal wrist crease (*black dots inside white circles*) crosses the pisiform, the joint between the capitate and the lunate, the tubercle of the scaphoid, and the radial styloid process.

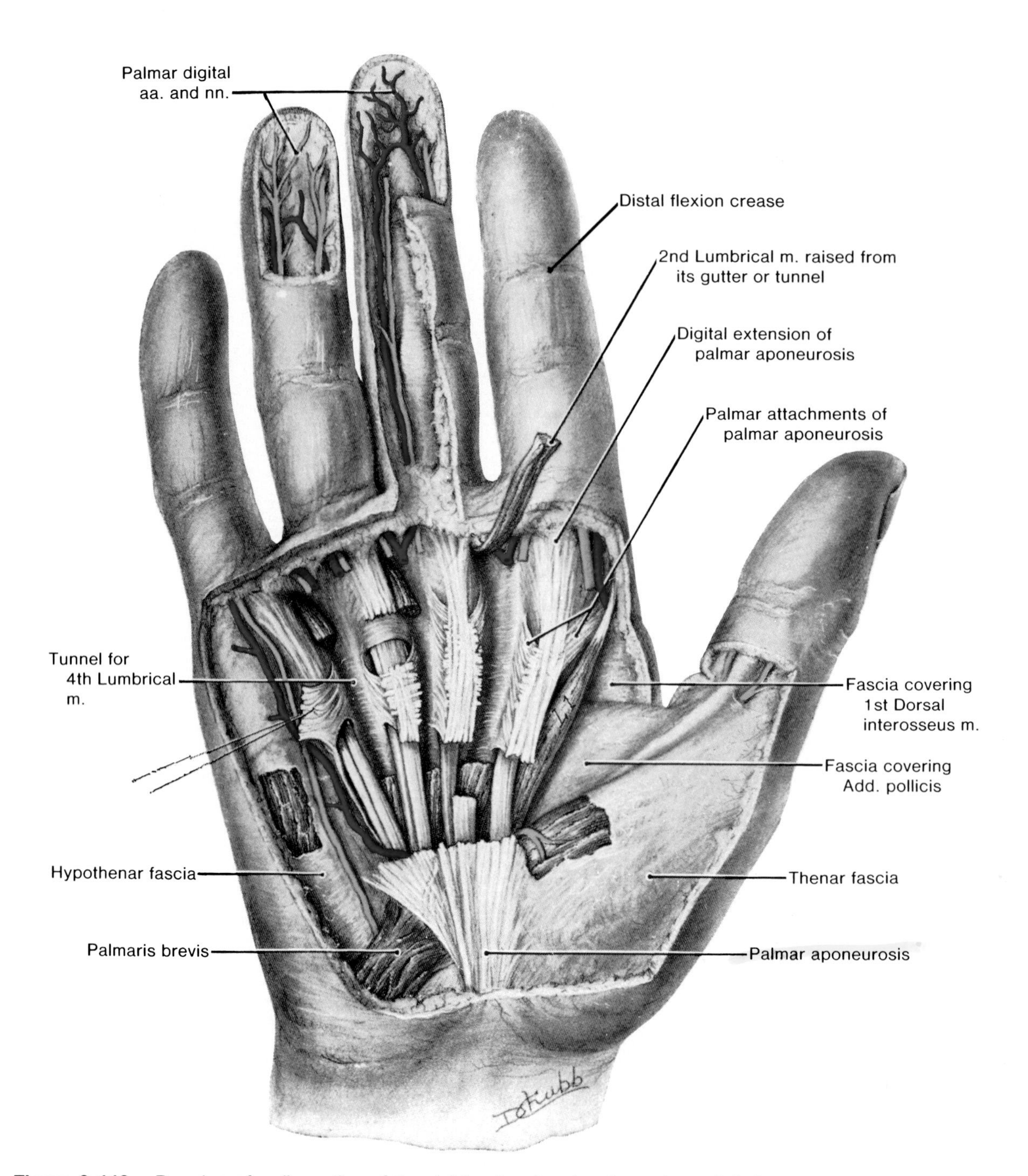

Figure 6-110. Drawing of a dissection of the right palm showing the palmar digital vessels and nerves and the attachments of the palmar aponeurosis. Observe the two sets of tunnels in the distal half of the palm: those for the long flexor tendons and those for the lumbrical muscles and digital vessels and nerves. The former are continued into the fingers; the latter open on the dorsum of the hand posterior to the web.

from the lateral border of the palmar aponeurosis to the first metacarpal bone. Lateral to this septum is the lateral or **thenar compartment** containing the thenar muscles concerned with movements of the thumb (Figs. 6-112 and 6-116).

Between the thenar and hypothenar compartments is the intermediate or **central compartment** containing the flexor tendons and their sheaths, the superficial palmar arch (Fig. 6-122), and branches of the median and ulnar nerves (Figs. 6-112, 6-113, and 6-116).

From the lateral border of the palmar aponeurosis, another fibrous septum passes obliquely posteriorly to the third metacarpal bone. This creates potential medial and lateral *midpalmar spaces* (Fig. 6-112).

The **adductor compartment** is the deepest muscular plane of the palm of the hand. It contains the adductor pollicis muscle (Figs. 6-112 and 6-113).

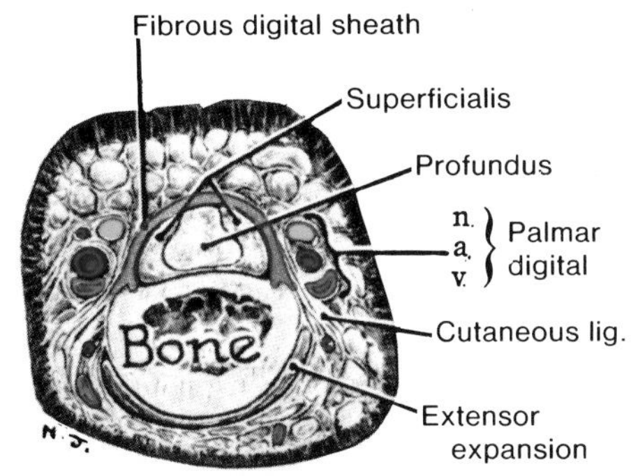

Figure 6-111. Drawing of a transverse section through the proximal phalanx of a finger. Note the osseofibrous tunnel containing the tendons of the flexor muscles of the digit. Observe that the skin is thickest on the palmar surface and that the palmar digital nerves and vessels are applied to the fibrous digital sheath, not to the bone. Note the ligaments that attach the skin to the bone.

The potential fascial spaces of the palm are clinically important because they may become infected (*e.g.*, following a puncture wound of the hand). If the infection occurs between the interosseous muscles and metacarpal bones, the spread of pus is restrained by the fibrous septum connected to the third metacarpal bone (Fig. 6-112).

Depending on the site of infection, pus will accumulate in the thenar, hypothenar, or adductor compartments. Owing to the widespread use

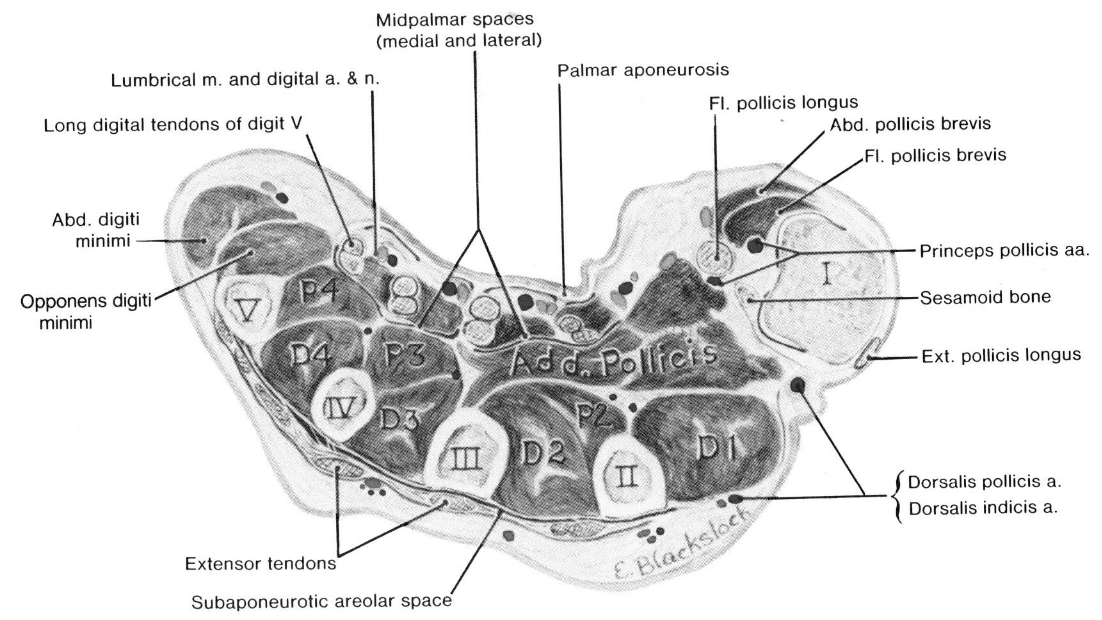

Figure 6-112. Drawing of a transverse section through the middle of the right palm, passing through the head of the first metacarpal bone and therefore distal to the opponens pollicis muscle (Fig. 6-116). **Observe the fascial compartments of the palm**: (1) a lateral *thenar compartment* containing vessels, nerves, and the thenar muscles; (2) an intermediate *central compartment* containing vessels, nerves, and the flexor tendons and their sheaths; (3) a medial *hypothenar compartment* containing vessels, nerves, and the hypothenar muscles; and (4) an *adductor compartment* containing the adductor pollicis muscle. Observe the dorsal interosseous muscles (D1 to D4) between the metacarpal bones and the palmar interossei (P2 to P4). The first palmar interosseous muscle (formerly called the deep head of flexor pollicis brevis) is not visible here. Also observe the long flexor tendons (superficial and deep) of the four fingers, the four lumbricals, and the palmar digital nerves and arteries.

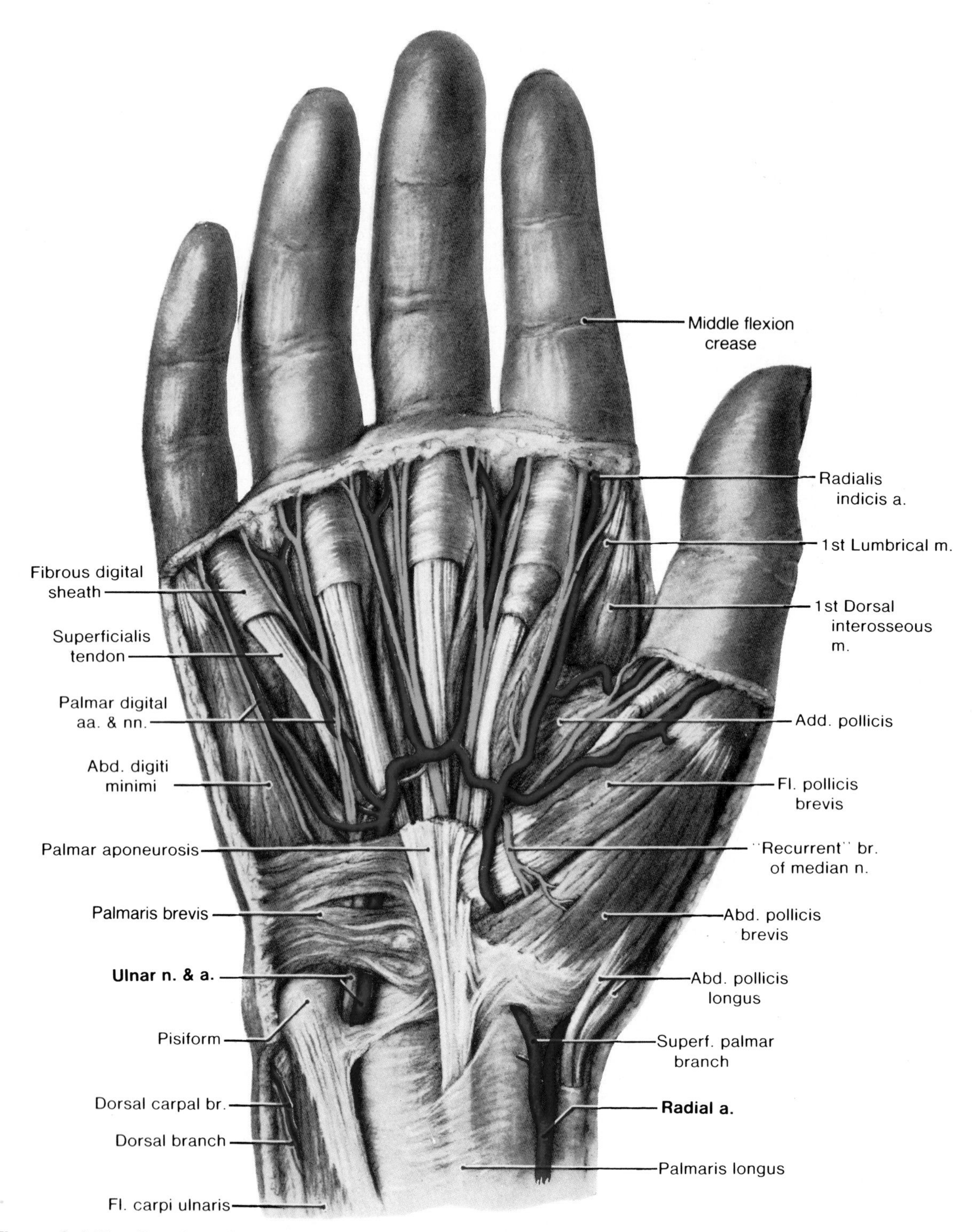

Figure 6-113. Drawing of a superficial dissection of the right palm. The skin and superficial fascia have been removed, as have the palmar aponeurosis and the thenar and hypothenar fasciae (Fig. 6-110). Observe the **superficial palmar arch** formed by the ulnar artery (also see Fig. 6-121). *Note that the pisiform bone protects the ulnar nerve and artery* as they pass into the palm.

of antibiotics, infection rarely spreads from one of these fascial compartments, but an untreated infection can spread proximally from them into the forearm, anterior to the pronator quadratus muscle and its fascia (Figs. 6-80 and 6-118).

MUSCLES OF THE HAND

The intrinsic muscles of the hand are on the palmar aspect and are innervated by branches of the ulnar or median nerves (Fig. 6-113). They can be divided into three groups (Figs. 6-112, 6-113, and 6-116 to 6-118): (1) **the thumb or thenar muscles** in the thenar compartment; (2) **the little finger or hypothenar muscles** in the hypothenar compartment; and (3) **the lumbrical muscles** in the central compartment and the **interosseous muscles** between the metacarpal bones.

The long flexor tendons of the extrinsic muscles of the hand arise in the forearm and pass to the digits. They are located in the central compartment of the palm with the lumbrical muscles (Fig. 6-112). The slender lumbrical muscles arise from the sides of the tendons of the flexor digitorum profundus as they traverse the palm (Figs. 6-116 and 6-117).

THE SHORT MUSCLES OF THE THUMB (Figs. 6-113, 6-116, and Table 6-12)

The three **thenar muscles** (abductor pollicis brevis, flexor pollicis brevis, and opponens pollicis) are **chiefly responsible for the movement termed opposition of the thumb** (Fig. 6-115). The pressure which the opposed thumb can exert on the fingertips is increased by the reinforcing action of the adductor pollicis and flexor pollicis longus muscles.

The three short thenar muscles of the thumb lie in the thenar compartment of the palm (Fig. 6-112). *They are all supplied by the recurrent branch of the median nerve* (Figs. 6-83 and 6-113).

THE ABDUCTOR POLLICIS BREVIS MUSCLE (Figs. 6-112, 6-113, 6-116, and Table 6-12)

This thin, relatively broad muscle forms the anterolateral part of the thenar eminence.

Origin (Figs. 6-87 and 6-116). **Flexor retinaculum and the tubercles of the scaphoid and trapezium bones.**

Insertion (Figs. 6-75 and 6-116). Lateral side of **the base of the proximal phalanx of the thumb.**

Nerve Supply (Fig. 6-113). **Recurrent branch of median nerve (C8** and T1).

Actions (Figs. 6-115, 6-119, and Table 6-21). **Abducts the thumb** and assists the opponens pollicis muscle during the early stages of opposition of the thumb.

THE FLEXOR POLLICIS BREVIS MUSCLE (Figs. 6-113, 6-116, and Table 6-12)

This muscle is located medial to the abductor pollicis brevis.

Origin (Figs. 6-87, 6-113, and 6-116). **Flexor retinaculum and the tubercle of the trapezium.**

Insertion (Figs. 6-75, 6-113, and 6-116). **Lateral side of the base of the proximal phalanx of the thumb**, medial to abductor pollicis brevis.

Nerve Supply (Fig. 6-113). Recurrent branch of **median nerve (C8** and T1).

Actions (Fig. 6-115 and Table 6-21). **Flexes thumb** at carpometacarpal and metacarpophalangeal joints and aids in opposition of the thumb.

THE OPPONENS POLLICIS MUSCLE (Figs. 6-87, 6-116, and Table 6-12)

This muscle lies deep to the abductor pollicis brevis and lateral to the flexor pollicis brevis.

Origin (Figs. 6-87 and 6-116). **Flexor retinaculum and the tubercle of the trapezium.**

Insertion (Figs. 6-75 and 6-87). Whole length of **the lateral side of the palmar surface of the first metacarpal bone.**

Nerve Supply (Fig. 6-113). Recurrent branch of **median nerve (C8** and T1)

Actions (Fig. 6-115 and Table 6-21). **Opposes the thumb.**

Opposition is the most important movement of the thumb. In Figure 6-115, note that the tip of the thumb is brought into contact with the palmar surface of the little finger. The thumb can also be opposed to the other fingers. Opposition involves extension initially, then abduction, flexion, and medial rotation and usually adduction. Several muscles are involved in this complex movement (Table 6-21).

Because of the complexity of opposition of the thumb, it may be affected by most nerve injuries in the upper limb. Obviously injuries to the nerves supplying the intrinsic muscles of the hand, especially the median nerve (Fig. 6-25), have the most severe effects on this movement.

If the median is severed in the forearm or at the wrist, the thumb cannot be opposed. However, the intact abductor pollicis longus and adductor pollicis muscles, supplied by the posterior interosseous and ulnar nerves, respectively, may imitate the action of opposition.

The recurrent branch of the median nerve supplying the thenar muscles lies superficially (Figs. 6-83 and 6-113), and may be severed by relatively minor lacerations of the palm involving the thenar eminence. If this nerve is severed,

Table 6-12
The Short Muscles of the Thumb or Thenar Muscles[1]

Muscle	Origin	Insertion	Nerve Supply	Action(s)
Abductor pollicis brevis (Figs. 6-112, 6-113, and 6-116)	Flexor retinaculum and tubercles of scaphoid and trapezium bones (Figs. 6-87 and 6-116)	Lateral side of base of proximal phalanx of thumb (Fig. 6-75)	Recurrent branch of median n. (**C8** and T1)	Abducts thumb (Fig. 6-115)
Flexor pollicis brevis (Figs. 6-113 and 6-116)	Flexor retinaculum and tubercle of trapezium (Figs. 6-87, 6-113, and 6-116)	Lateral side of base of proximal phalanx of thumb (Fig. 6-75)	Recurrent branch of median n. (**C8** and T1)	Flexes thumb (Fig. 6-115)
Opponens pollicis (Figs. 6-87 and 6-116)	Flexor retinaculum and tubercle of trapezium (Figs. 6-87, 6-113, and 6-116)	Lateral side of 1st metacarpal bone	Recurrent branch of median n. (**C8** and T1)	Draws thumb toward center of palm and rotates it medially, *i.e.*, opposes it (Fig. 6-115)
Adductor pollicis (Figs. 6-95, 6-97, 6-98, 6-112, and 6-113)	*Oblique head:* bases of 2nd and 3rd metacarpals, capitate, and adjacent carpal bones *Transverse head:* anterior surface of body of 3rd metacarpal (Fig. 6-75)	Medial side of base of proximal phalanx of thumb (Fig. 6-75)	Deep branch of ulnar n. (C8 and **T1**)	Adducts thumb (Fig. 6-115)

[1] The three thenar muscles (abductor pollicis brevis, flexor pollicis brevis, and opponens pollicis) form the thenar eminence or ball of the thumb (Fig. 6-108). The actions of the three muscles are indicated by their names to some extent, but they are all involved in opposition (Fig. 6-115), the pincer-like grip between the thumb and index finger that is an indispensable movement to most people.

the thenar muscles are paralyzed and the thumb loses much of its usefulness.

Injury to the median nerve at the wrist (e.g., during wrist slashing) results in paralysis of the thenar muscles, the first two lumbrical muscles (Fig. 6-118 and Tables 6-12 and 6-14). There would probably be sensory impairment in the digits supplied by this nerve (Fig. 6-26). As stated previously, the chief effect of lesions here is inability to oppose the thumb. In a few weeks wasting or **atrophy of the thenar muscles** may occur, producing a characteristic flattening of the thenar eminence, producing what is often referred to as an "**ape hand.**"

THE ADDUCTOR POLLICIS MUSCLE (Figs. 6-95, 6-97, 6-99, 6-112, 6-113, and Table 6-12)

This fan-shaped muscle *lies in the interosseous adductor compartment of the hand* (Fig. 6-112). It has ***two heads of origin*** that are separated by a gap through which the radial artery passes (Fig. 6-99).

Origin (Fig. 6-75). *Oblique head,* **bases of the second and third metacarpal bones, the capitate and adjacent carpal bones**. *Transverse head,* anterior surface of **the body of the third metacarpal bone.**

Insertion (Fig. 6-75). **Medial side of the base of the proximal phalanx of the thumb.** The two heads converge and insert by a tendon that contains a sesamoid bone.

Nerve Supply (Figs. 6-25 and 6-87). Deep branch of **ulnar nerve** (C8 and **T1**).

Actions (Fig. 6-115 and Table 6-21). **Adducts the thumb** (draws the thumb toward the palm). It gives power to the grasp.

THE SHORT MUSCLES OF THE LITTLE FINGER (Figs. 6-113, 6-116, and Table 6-13)

The three ***hypothenar muscles*** are concerned with movements of the little finger. This digit is not nearly so mobile as the thumb.

The three short muscles of the little finger lie in the hypothenar compartment of the palm with the fifth metacarpal bone (Fig. 6-112). These muscles ***produce the hypothenar eminence*** or ball of the little finger (Fig. 6-108). *They are all supplied by the deep branch of the ulnar nerve.*

THE ABDUCTOR DIGITI MINIMI MUSCLE (Figs. 6-107, 6-113, 6-116, and Table 6-13)

This is the most superficial of the three muscles forming the hypothenar eminence.

Origin (Fig. 6-75). **Pisiform bone.** It is frequently continuous with the tendon of the flexor carpi ulnaris.

Insertion (Figs. 6-75, 6-87, and 6-116). **Medial side of the base of the proximal phalanx of the little finger** and the extensor expansion.

Nerve Supply (Figs. 6-25 and 6-87). Deep branch of **ulnar nerve** (C8 and **T1**).

Actions (Table 6-19). **Abducts the little finger** and helps to flex its metacarpophalangeal joint.

THE FLEXOR DIGITI MINIMI BREVIS MUSCLE (Fig. 6-116 and Table 6-13)

This muscle is variable in size and lies lateral to the abductor digiti minimi.

Origin (Figs. 6-77*A* and 6-116). **Hook of the hamate and the flexor retinaculum.**

Insertion (Figs. 6-75 and 6-116). **Medial side of the base of the proximal phalanx of the little finger** with the abductor digiti minimi.

Nerve Supply (Figs. 6-25 and 6-87). Deep branch of **ulnar nerve** (C8 and **T1**).

Action (Table 6-19). **Flexes the little finger** at the metacarpophalangeal joint.

THE OPPONENS DIGITI MINIMI MUSCLE (Figs. 6-87, 6-116, and Table 6-13)

This muscle lies deep to the abductor and flexor muscles of the little finger.

Origin (Figs. 6-87 and 6-118). **Hook of the hamate and the flexor retinaculum.**

Insertion (Figs. 6-75 and 6-87). Whole length of the **medial border of the fifth metacarpal bone.**

Nerve Supply (Figs. 6-25 and 6-87). Deep branch of **ulnar nerve** (C8 and **T1**).

Table 6-13
The Short Muscles of the Little Finger or Hypothenar Muscles[1]

Muscle	Origin	Insertion	Nerve Supply	Action(s)
Abductor digiti minimi (Figs. 6-107, 6-113, and 6-116)	Pisiform bone (Fig. 6-75)			Abducts little finger (Fig. 6-119)
Flexor digiti minimi brevis (Fig. 6-116)	Hook of hamate bone and flexor retinaculum (Fig. 6-116)	Medial side of base of proximal phalanx of little finger	Deep branch of ulnar n. (C8 and **T1**)	Flexes little finger
Opponens digiti minimi (Figs. 6-87 and 6-116)		Medial border of 5th metacarpal bone		Draws 5th metacarpal anteriorly and rotates it, bringing the little finger into opposition with the thumb (Fig. 6-115)

[1] These three hypothenar muscles form the hypothenar eminence or ball of the little finger (Fig. 6-108).

Actions (Fig. 6-115). **Draws the fifth metacarpal bone anteriorly and rotates it laterally**, thereby deepening the hollow of the palm and bringing the little finger into opposition with the thumb.

THE PALMARIS BREVIS MUSCLE (Figs. 6-80, 6-110, and 6-113)

This small, thin, quadrilateral muscle lies in the fascia over the hypothenar eminence formed by the hypothenar muscles (Table 6-13). It is *a relatively unimportant muscle, except that it covers and protects the ulnar nerve and artery.*

Origin (Figs. 6-80 and 6-113). **Flexor retinaculum and the palmar aponeurosis.**

Insertion (Figs. 6-110 and 6-113). **Skin on the medial side of the palm.**

Nerve Supply (Fig. 6-25). Superficial branch of **ulnar nerve** (C8 and **T1**).

Actions. Wrinkles the skin on the medial side of the palm and deepens the hollow of the palm, thereby aiding the grip.

There are 12 short muscles in the hand (four lumbrical and eight interosseous muscles). The lumbricals act only on the fingers, but the interossei act on all five digits (*i.e.*, the fingers and thumb).

THE LUMBRICAL MUSCLES (Figs. 6-110, 6-113, 6-116, 6-117, and Table 6-14)

The four slender lumbricals (L. *lumbricus*, earthworm), one for each finger, were named because of their elongated, worm-like form.

Origin (Figs. 6-116 and 6-117). **Tendons of the flexor digitorum profundus muscle.** The lateral two lumbricals usually arise by one head and the medial two by two heads.

Insertion (Fig. 6-117). **Lateral sides of the extensor expansions** of digits 2 to 5, distal to metacarpophalangeal joints.

Nerve Supply (Figs. 6-25, 6-116, and 6-118). *Lateral two* digits (2nd and 3rd) by **median nerve** (C8 and **T1**). *Medial* two digits (4th and 5th) by deep branch of **ulnar nerve** (C8 and **T1**).

Actions (Tables 6–19 and 6–20). **Flex the digits at the metacarpophalangeal joints and extend the interphalangeal joints.** It is helpful in remembering the action of these muscles to note that they place the fingers in the writing or billiard cue position.

THE INTEROSSEOUS MUSCLES (Figs. 6-110, 6-112, 6-113, 6-117, 6-118, and Table 6-14)

***Eight interosseous muscles* or interossei** are located between the metacarpal bones. For some time the first palmar interosseous muscle was called the deep head of the flexor pollicis brevis; hence only seven interossei were described.

*The interossei are **arranged in two layers**,* four palmar and four dorsal muscles. As their name indicates, they are located *between bones* (*i.e.*, the metacarpals).

Origin (Fig. 6-118). *Dorsal interossei,* **adjacent sides of two metacarpal bones.** *Palmar interossei,* **palmar surfaces of first, second, fourth, and fifth metacarpal bones.**

Insertion (Figs. 6-75, 6-76, and 6-117). **Extensor expansions of digits and bases of proximal phalanges.**

Table 6-14
The Short Muscles of the Hand

Muscle	Origin	Insertion	Nerve Supply	Action(s)
Lumbricals 1 and 2	Lateral two tendons of flexor digitorum profundus	Lateral sides of extensor expansions of digits 2 to 5	*1st and 2nd,* median n. (C8 and **T1**)	Flex digits at metacarpophalangeal joints and extend interphalangeal joints
Lumbricals 3 and 4 (Figs. 6-116 and 6-117)	Medial three tendons of flexor digitorum profundus		*3rd and 4th,* deep branch of ulnar (C8 and **T1**)	
Dorsal interossei 1 to 4 (Figs. 6-102 and 6-103)	Adjacent sides of two metacarpal bones	Extensor expansions and bases of proximal phalanges of digits 2 to 4	Deep branch of ulnar n. (C8 and **T1**)	Abduct fingers (Fig. 6-119)
Palmar interossei 1 to 4 (Fig. 6-118)	Palmar surfaces of 1st, 2nd, 4th, and 5th metacarpal bones	Extensor expansions of digits and bases of proximal phalanges of digits 1, 2, 4, and 5		Adduct fingers (Fig. 6-119)

Nerve Supply (Figs. 6-116 and 6-118). Deep branch of the **ulnar nerve** (C8 and **T1**).

Actions (Fig. 6-119). **Dorsal interossei abduct fingers** (***DAB*** is the key, *i.e.*, ***D***orsal ***AB***duct) and **palmar interossei adduct fingers** (***PAD*** is the key, *i.e.*, ***P***almar ***AD***uct). They also assist the lumbrical muscles in bringing about flexion of the metacarpophalangeal joints and extension of the interphalangeal joints. These are important movements in typing, writing, and playing the piano.

THE LONG FLEXOR TENDONS OF THE EXTRINSIC MUSCLES OF THE HAND (Figs. 6-88, 6-116, and 6-118)

The tendons of the flexor digitorum superficialis and flexor digitorum profundus muscles pass in a common sheath deep to the flexor retinaculum (Fig. 6-88). They then pass deep to the palmar aponeurosis (Figs. 6-80 and 6-110) and enter the osseofibrous digital tunnels (Figs. 6-111 to 6-113).

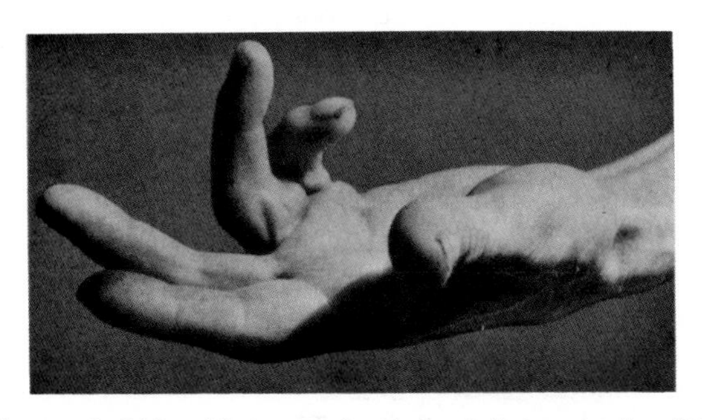

Figure 6-114. Photograph of the left hand of a 56-year-old man with *Dupuytren's contracture*. Note the marked flexion of the ring and little fingers at the metacarpophalangeal and proximal interphalangeal joints, resulting from fibrosis and shortening of the digital slips of the palmar aponeurosis that pass to these fingers (Figs. 6-80 and 6-110).

Observe that there are two tendons in each osseofibrous tunnel (Fig. 6-118). In order that these tendons can slide freely over each other during movements of the fingers, each tendon is covered with synovial membrane (Figs. 6-86 and 6-88). Near the base of the proximal phalanx, the tendon of the flexor digitorum superficialis splits and surrounds the tendon of the flexor digitorum profundus (Figs. 6-117 and 6-118). The halves of the tendon of the flexor digitorum superficialis insert into the margins of the middle phalanx (Fig. 6-75). The tendon of the flexor digitorum profundus, after passing through the split in the tendon of the flexor digitorum superficialis, passes distally to insert into the base of the distal phalanx (Figs. 6-75, 6-117, and 6-118).

The long flexor tendons are supplied with blood by small blood vessels that pass from the periosteum of the phalanges within special folds of connective tissue (Fig. 6-86). These folds are called **vincula tendinum.** There are two kinds of vincula (L. fetter or chain): *vincula brevia and vincula longa* (Fig. 6-117). These folds connect the tendons of the flexor digitorum superficialis and flexor digitorum profundus muscles to the fibrous digital sheaths in the fingers (Figs. 6-111 and 6-117).

The tendon of the flexor pollicis longus passes to the thumb, deep to the flexor retinaculum, within its own synovial sheath (Fig. 6-88), and enters the osseofibrous digital tunnel in the thumb. At the head of the metacarpal, the tendon runs between two sesamoid bones, one in the combined tendon of the flexor pollicis brevis and abductor pollicis brevis, and the other in the tendon of the adductor pollicis (Fig. 6-113).

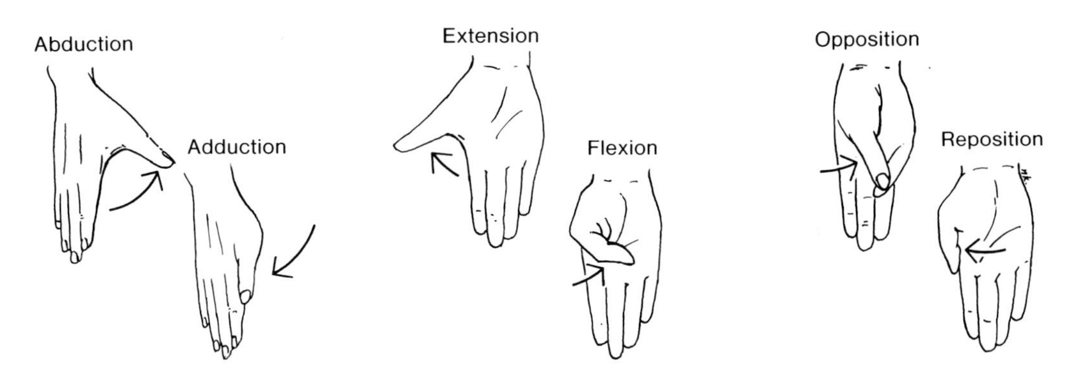

Figure 6-115. Drawings illustrating movements of the thumb. Circumduction is also possible and medial rotation occurs during the complex movement termed opposition. Because the metacarpal of the thumb is set at a right angle to those of the other digits, all movements of the thumb take place at right angles to the corresponding movements of the fingers. *Movements of the thumb are very important; you are unable to do much with your hand without them.* Thumb movements account for about 50% of hand movements.

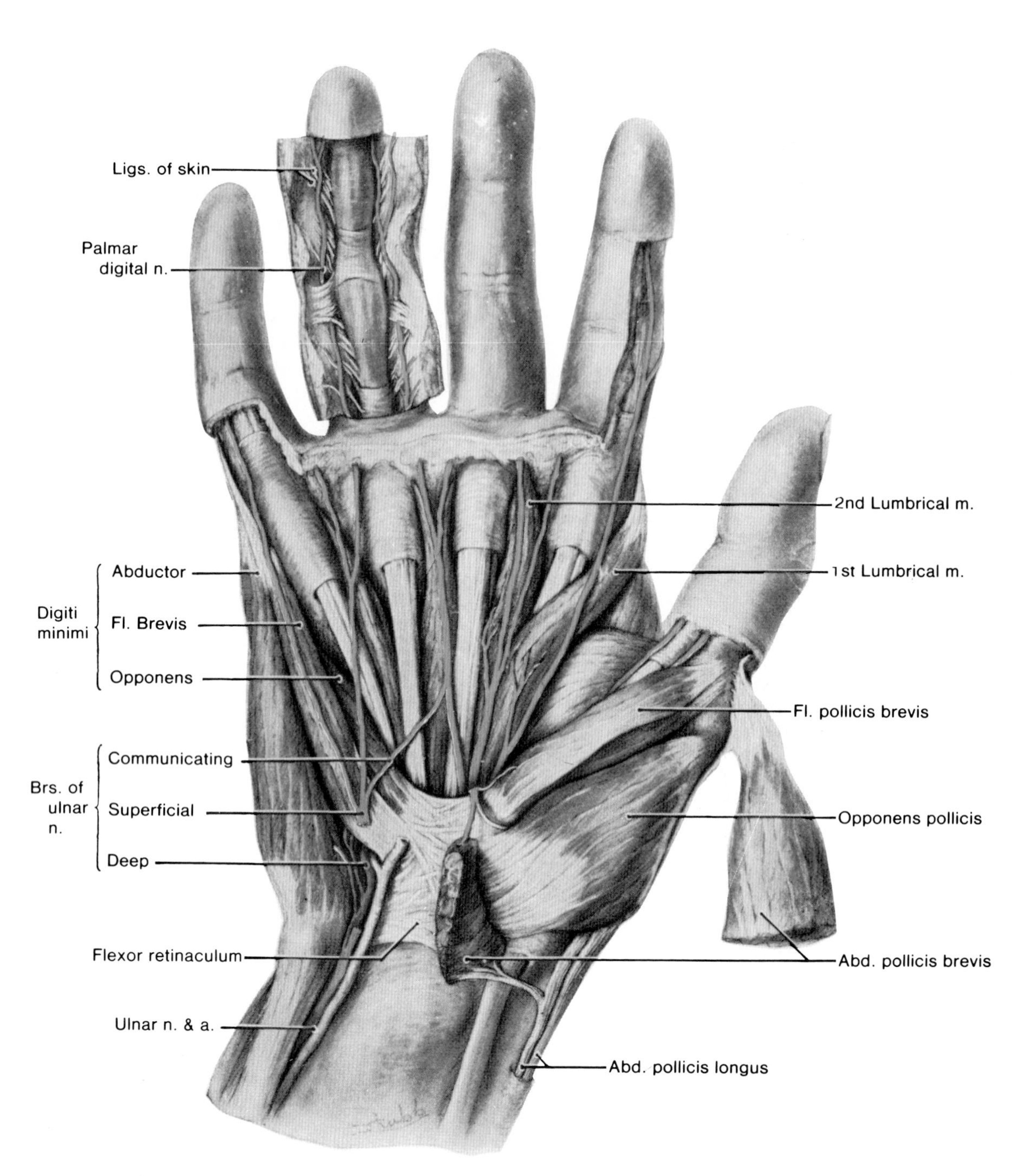

Figure 6-116. Drawing of a superficial dissection of the right palm showing the three thenar and three hypothenar muscles arising from the flexor retinaculum, and the four marginal carpal bones united by it (Fig. 4-78). Note the four lumbrical muscles arising from the lateral sides of the four profundus tendons and inserting into the lateral sides of the extensor expansions of the corresponding digits (Fig. 6-117). Observe that the median nerve supplies five muscles (three thenar and two lumbricals), and provides cutaneous branches to three and one-half digits including parts of their dorsal aspects. Note that the ulnar nerve supplies all other short muscles in the hand and provides cutaneous branches to one and one-half digits (Fig. 6-26).

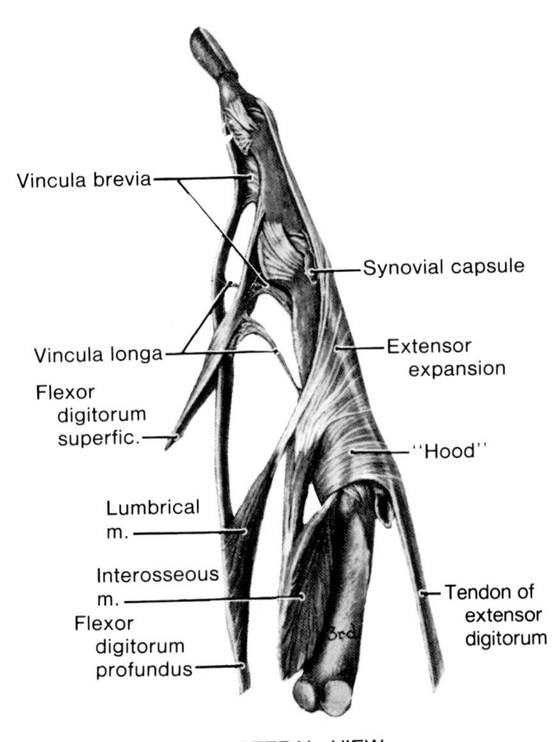

Figure 6-117. Drawing of a dissection of the right middle finger showing its extensor expansion, a unique structure on the dorsal aspect of each digit through which the extensor tendons, interosseous muscles, and lumbrical muscles gain insertion. Observe the interossei inserted in part into the base of the proximal phalanx and in part into the extensor expansion. Note that the lumbrical muscle is inserted into the lateral side of the extensor expansion. Observe the "hood" covering the head of the metacarpal which is moored to the palmar ligament; thus medial, lateral, and posterior bowstringing of the extensor tendon and extensor expansion is prevented. Note that the extensor expansion extends to the bases of the middle and distal phalanges.

The synovial sheaths of the flexor tendons of the hand may become infected (*e.g.*, by entry of a foreign object into a finger, such as a rusty nail). When **tenosynovitis** (inflammation of the tendon and the digital synovial sheath) occurs, the finger swells and movement of it becomes painful. As the tendons of the index, middle, and ring fingers nearly always have separate digital synovial sheaths (Fig. 6-88), the infection is usually confined to the finger concerned. In neglected infections, however, the proximal ends of these sheaths may rupture and infection may spread to the midpalmar fascial spaces (Fig. 6-112).

As the synovial sheaths of the thumb and little finger are often continuous with the **common flexor synovial sheath** (Fig. 6-88), tenosynovitis in these digits may spread to the common sheath. Because there are variations in the connections between the common flexor synovial sheath and the digital sheaths, the degree of spreading of infections from the fingers depends upon whether or not there are connections between them.

Infections of the second, third, and fourth digits are likely to remain localized because their digital synovial sheaths are connected with the common flexor synovial sheath in only about 10% of cases.

NERVES OF THE HAND

The median, ulnar, and radial nerves supply the hand (Figs. 6-25, 6-26, 6-113, 6-116, and 6-118).

THE MEDIAN NERVE (Figs. 6-25, 6-26, 6-81, 6-83, 6-90, 6-92, and 6-120)

The median nerve enters the hand via the carpal tunnel, deep to the flexor retinaculum, between the tendons of the flexor digitorum superficialis and the tendon of the flexor carpi radialis (Figs. 6-81 and 6-83).

The median nerve supplies motor fibers to the three thenar muscles and the first and second lumbrical muscles (Fig. 6-25). It sends cutaneous sensory fibers to the entire palmar surfaces and sides of the thumb; index finger; middle and lateral half of the ring finger; and to the dorsum of the distal halves of these fingers (Figs. 6-26, 6-113, 6-116, and 6-120).

Median nerve injury frequently occurs just proximal to the flexor retinaculum, the common site for suicide attempts (Fig. 6-90). The median nerve can be palpated here, just before it passes deep to the flexor retinaculum (Fig. 6-80), and enters the *carpal tunnel* (Figs. 6-78, 6-81, and 6-83).

When the median nerve is severed at the wrist, the patient will be unable to oppose his/her thumb owing to paralysis of the thenar muscles (Tables 6-12 and 6-21). The first and second lumbrical muscles will also be paralyzed resulting in weakening of flexion of the metacarpophalangeal joints and extension of the interphalangeal joints of the second and third

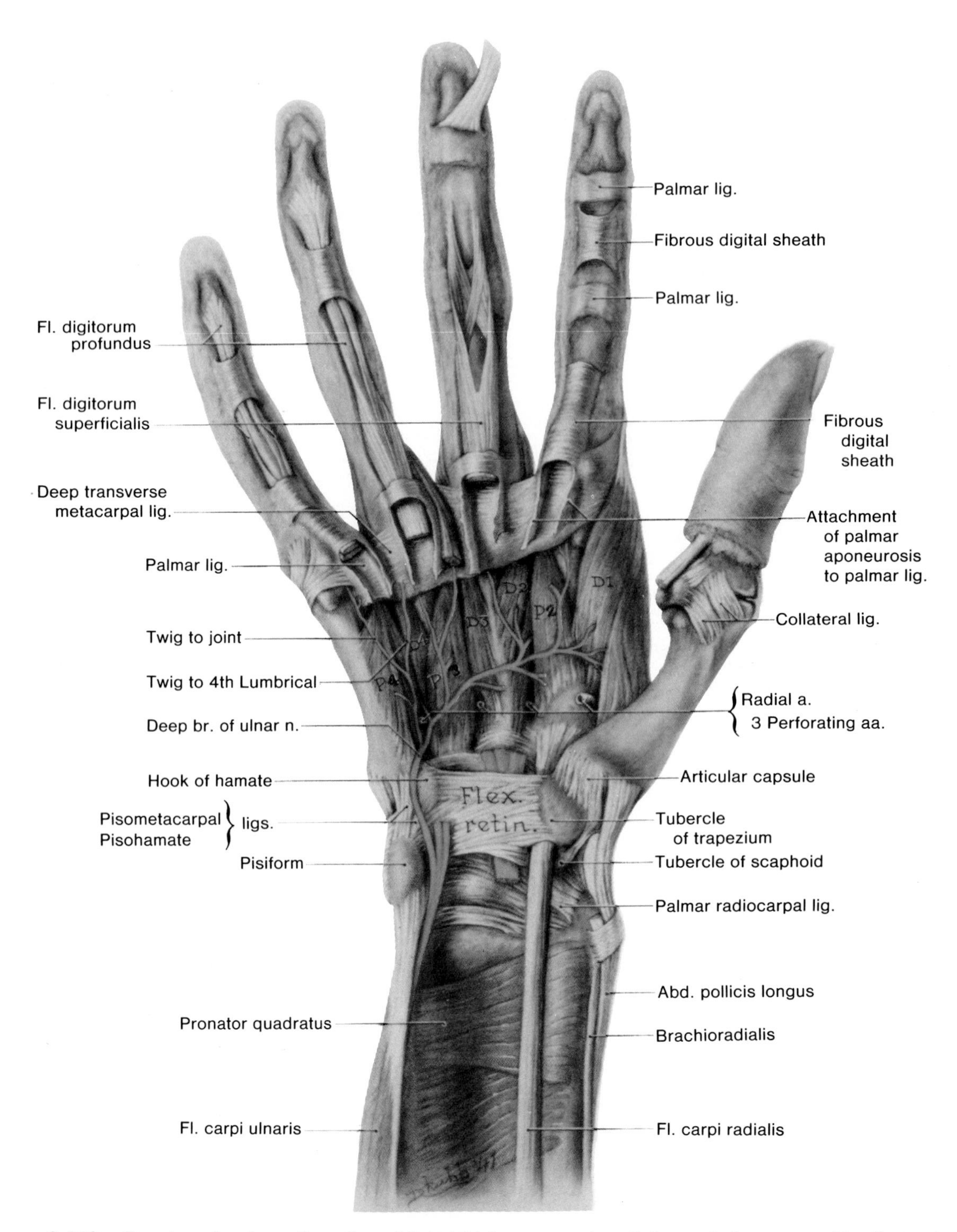

Figure 6-118. Drawing of a deep dissection of the right forearm, palm, digits, and ulnar nerve. The first palmar interosseous muscle, together with the small muscles of the thumb, have been removed. Observe the interosseous muscles located between the metacarpals, and the deep branch of the ulnar nerve passing medial to the hook of the hamate to be distributed to the hypothenar muscles, all the interossei, two lumbricals, the adductor pollicis, and several joints (also see Fig. 6-25).

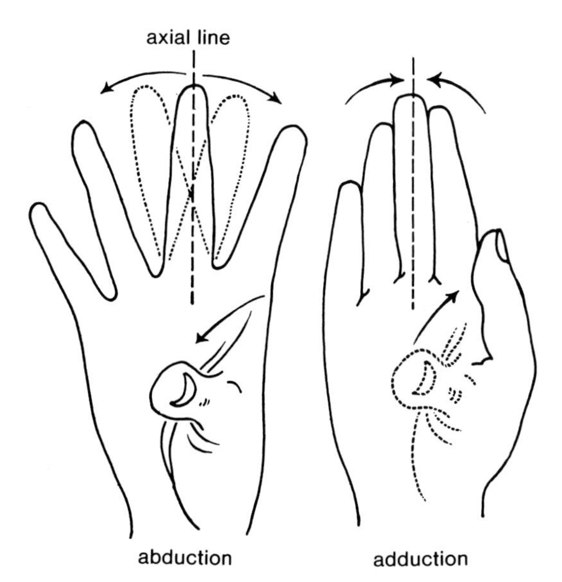

Figure 6-119. Drawings illustrating abduction and adduction of the digits. Observe that the median plane or axial line of the hand passes through the middle digit. Note that abduction of the fingers is movement away from the axial line and that adduction of them is movement toward it. Note also that the middle digit can be abducted to either side of the axial line. Adduction restores it to the axial line. Observe that in abduction of the thumb it stands anteriorly, away from the palm at a right angle and that adduction closes the thumb on the index finger.

digits (Tables 6-19 and 6-20). The fine control of movements of these digits will also be lost.

There will also be a loss of cutaneous sensation in the lateral portion of the palm, the palmar surface of the thumb, and the lateral two and a half fingers including the nail beds of these digits (Fig. 6-26). The areas of anesthesia may not be sharply defined owing to variations in the distribution of the nerves in the hand.

If paralysis of the median nerve has existed for some time, there may be *wasting of the muscles involved, especially the thenar muscles.* In addition the thumb will be held close to the index finger and the two lateral fingers will be slightly hyperextended at the metacarpophalangeal joints and slightly flexed at the interphalangeal joints. These deformities constitute the so-called "**ape hand.**"

THE ULNAR NERVE (Figs. 6-25, 6-26, 6-83, 6-87, 6-92, 6-113, 6-116, 6-118, and 6-120)

Just proximal to the wrist the ulnar nerve gives off a *palmar cutaneous branch* (Fig. 6-92) which passes superficial to the flexor retinaculum and palmar aponeurosis to supply the skin of the medial side of the palm (Fig. 6-26). It also gives off a *dorsal cutaneous branch* (Figs. 6-93 and 6-120) which supplies the medial half of the dorsum of the hand, the little finger, and the medial half of the ring finger (Fig. 6-26).

The ulnar nerve ends by dividing into a superficial and a deep branch (Fig. 6-116).

The superficial branch of the ulnar nerve supplies cutaneous fibers to the anterior surfaces of the medial one and a half digits (Figs. 6-26 and 6-116).

The deep branch of the ulnar nerve supplies motor fibers to the hypothenar muscles, the medial two lumbrical muscles, the adductor pollicis muscle, and all the interossei (Figs. 6-25, 6-116, and 6-118). The deep branch of the ulnar nerve also supplies several joints (wrist, intercarpal, carpometacarpal, and intermetacarpal).

The ulnar nerve is sometimes called the nerve of fine movements because it innervates muscles that are mainly concerned with fine movements of the hand (Tables 6-13 and 6-14).

Ulnar nerve injury is common because this nerve lies superficially in the distal part of the forearm and at the wrist (Figs. 6-87 and 6-90). It may be injured by lacerations at these sites, producing *sensory alteration in the medial part of the hand,* the little finger, and the medial half of the ring finger (Fig. 6-26).

There is also impaired power of adduction and abduction of the fingers owing to paralysis of the interossei (Fig. 6-118 and Table 6-19). Some adduction of the fingers may be possible on flexion of the fingers by the long flexor muscles.

Adduction of the thumb is lost owing to paralysis of the adductor pollicis (Fig. 6-25 and Table 6-21), but other movements of the thumb are normal.

After an ulnar nerve injury, the ring and little fingers are hyperextended at the metacarpophalangeal joints, and somewhat flexed at the interphalangeal joints because the medial two lumbrical muscles are paralyzed (Table 6-19). This deformity, called **clawhand** (Fig. 6-94), does not become obvious until a considerable time after the injury to the ulnar nerve.

THE RADIAL NERVE (Figs. 6-26, 6-27, 6-92, 6-93, 6-98, and 6-120)

The radial nerve supplies no muscles in the hand. Its terminal branches, superficial and deep, arise in the cubital fossa (Fig. 6-87).

The deep branch of the radial is muscular and articular in its distribution (*e.g.,* the elbow joint).

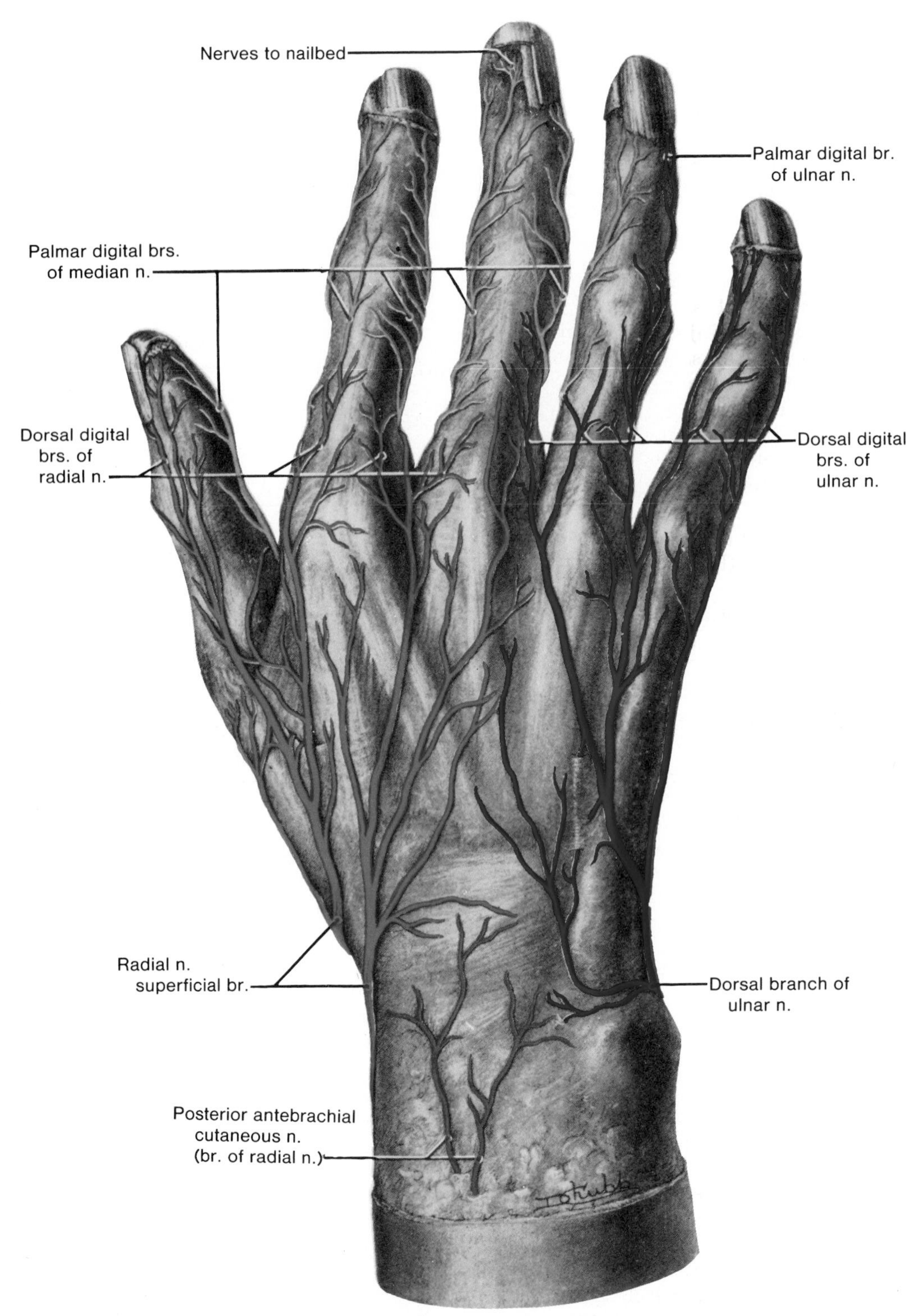

Figure 6-120. Drawing of a superficial dissection of the dorsum of the right wrist and hand showing the cutaneous nerves. Observe that the radial nerve and the dorsal branch of the ulnar nerve are distributed nearly equally and symmetrically on the dorsum of the hand and digits. The radial nerve supplies the lateral half of the dorsum and two and one-half digits. The dorsal branch of the ulnar nerve is distributed similarly on the ulnar half. Note that the palmar digital branches of the median and ulnar nerves alone supply the distal halves of the three middle digits, including the nailbeds.

The superficial branch of the radial is the direct continuation of the radial nerve along the anterolateral side of the forearm and is entirely sensory. It pierces the deep fascia near the dorsum of the wrist to supply skin and fascia over the lateral two-thirds of the dorsum of the hand, the dorsum of the thumb, and proximal parts of the lateral one and one-half fingers (Figs. 6-26 and 6-120).

The radial nerve supplies no muscles in the hand, but **radial nerve injury** in the arm or forearm produces serious disability of the hand. **The characteristic handicap in all injuries to the radial nerve is inability to extend the wrist** owing to paralysis of the extensor muscles of the forearm (Tables 6-10 and 6-11).

The wrist cannot be extended against gravity when the radial nerve is injured. The hand is flexed at the wrist and lies flaccid, a condition known as **wrist-drop** (Fig. 6-156*A*). The fingers are also flexed at the metacarpophalangeal joints. The interphalangeal joints can be extended weakly through the action of the intact lumbrical and interosseous muscles which are supplied by the median and ulnar nerves.

The radial nerve has only a small area of exclusive cutaneous supply on the hand. The extent of anesthesia is minimal even in serious radial nerve injuries, and is usually confined to a small area on the lateral part of the dorsum of the hand (Fig. 6-156*A*).

ARTERIES OF THE HAND

The radial and ulnar arteries provide all the blood to the hand.

THE RADIAL ARTERY (Figs. 6-65, 6-74, 6-91, 6-96, 6-99, 6-110, 6-113, and 6-121 to 6-123)

The radial artery is the smaller of the two branches of the brachial artery and it continues the direct line of this vessel (Fig. 6-80).

Just before passing from the anterior to the posterior surface of the wrist, the radial artery gives off a **superficial palmar branch** (Fig. 6-113). This artery passes through the thenar muscles and runs superficial to the long flexor tendons, where it joins the **superficial palmar arterial arch**, the continuation of the ulnar artery (Figs. 6-113, 6-121, and 6-122).

As the radial artery curves dorsally over the wrist, it passes deep to the tendons of the abductor pollicis longus and extensor pollicis brevis muscles (Figs. 6-97 and 6-99). It then crosses the floor of the anatomical snuff box and enters the palm of the hand by passing between the heads of the first dorsal interosseous muscle (Figs. 6-96 and 6-99).

The radial artery gives off the **princeps pollicis artery** and the **radialis indicis artery** (Figs. 6-121 and 6-122), the **digital arteries** to the thumb and the lateral side of the index finger, respectively.

The radial artery then passes between the two heads of the adductor pollicis muscle and joins the deep branch of the ulnar artery to form the **deep palmar arterial arch** (Figs. 6-121 and 6-122). This *arterial arch* is located between the long flexor tendons and the metacarpal bones. Three **palmar metacarpal arteries** arise from the deep palmar arch and run distally, where they join the common palmar digital arteries arising from the superficial palmar arch (Fig. 6-121).

THE ULNAR ARTERY (Figs. 6-85, 6-87, 6-91, 6-95, 6-105, 6-113, 6-121, and 6-122)

The ulnar artery, the larger of the two terminal branches of the brachial artery (Figs. 6-65 and 6-74), enters the palm on the lateral side of the ulnar nerve (Figs. 6-90 and 6-113), superficial to the flexor retinaculum of the wrist (Fig. 6-87).

The ulnar artery passes lateral to the pisiform bone (Fig. 6-91) and then gives off a deep palmar branch before continuing across the palm as the superficial palmar arterial arch (Fig. 6-122).

The deep palmar branch of the ulnar artery (Figs. 6-121 and 6-122) passes through the hypothenar muscles and anastomoses with the radial artery, thereby completing the **deep palmar arterial arch** or deep palmar arch (Figs. 6-113, 6-121, and 6-122).

The superficial palmar arterial arch (superficial palmar arch) *is formed mainly by the ulnar artery* (Figs. 6-121 and 6-122). The arch is usually completed by the superficial palmar branch of the radial artery. A clear understanding of the palmar arterial arches is important because of the frequency of lacerations of the hand.

The Superficial Palmar Arterial Arch (Figs. 6-113, 6-121, and 6-122). *The superficial palmar arch,* ***formed mainly by the ulnar artery****, is located distal to the deep palmar arch.* It is convex toward the fingers and the middle of its convexity lies deep to the center of the proximal transverse crease of the palm (Fig. 6-108).

The superficial palmar arterial arch gives rise to three common palmar digital arteries (Figs. 6-110, 6-113, 6-121, and 6-122) which anastomose with the palmar metacarpal arteries from the deep palmar arch. Each common palmar digital artery divides into a pair of **proper palmar digital arteries**, which run along the sides of the index, middle, ring, and little fingers (Fig. 6-121).

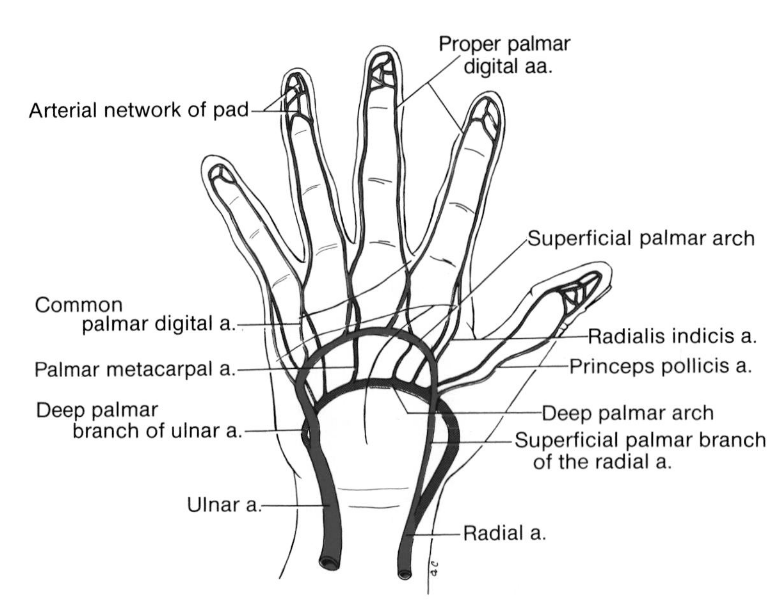

Figure 6-121. Drawing of the palmar arterial arches and their branches. The superficial palmar arterial arch, the larger of the two palmar arches, is formed by the distal curved portion of the ulnar artery; it is usually completed by the superficial palmar branch of the radial artery. Note that the deep palmar arterial arch is formed by the terminal part of the radial artery in conjunction with the deep branch of the ulnar artery. Note that the deep palmar arch is located proximal to the superficial palmar arch and that it is at a deeper plane.

The Deep Palmar Arterial Arch (Figs. 6-121 and 6-122). *The deep palmar arch, **formed mainly by the radial artery***, lies across the metacarpal bones, just distal to their bases. It is about a fingerbreadth closer to the wrist than the superficial palmar arch.

The deep palmar arch gives rise to three palmar metacarpal arteries, which run distally and join the common palmar digital arteries from the superficial palmar arch (Fig. 6-121).

The palmar arches shown in Figures 6-121 and 6-122 are typical, but several variations occur (see Anson and McVey, 1971). For example, *sometimes the superficial palmar arch is formed by the ulnar artery alone.*

Because of the number of arteries in the hand, bleeding is usually profuse when it is lacerated. Often both ends of the bleeding artery must be tied to stop the hemorrhage, because there are four transversely placed arterial arches that communicate with each other (Fig. 6-123).

In lacerations of the palmar arterial arches, it may be useless to ligate only one of the forearm arteries because they usually have numerous communications in the forearm and hand. Even simultaneous clamping of the ulnar and radial arteries proximal to the wrist sometimes fails to stop all bleeding.

To obtain a bloodless operating field in the hand for treating complicated injuries, *it is necessary to compress the brachial artery* and its branches proximal to the elbow (*e.g.*, using a pneumatic tourniquet). This prevents blood from reaching the arteries of the forearm and hand through the anastomoses around the elbow (Fig. 6-65).

Clench your fist tightly and then firmly compress your radial artery at the wrist (Fig. 6-82). While still occluding the radial, unclench your fist and relax it. Note that at first your palm is pale, but gradually turns pink owing to blood entering the superficial palmar arch from the ulnar artery. Persistence of pallor in the palm when one artery is compressed indicates occlusion of the other. This test is used clinically to test for patency of the radial and ulnar arteries.

VEINS OF THE HAND

The superficial and deep palmar arterial arches are accompanied by **venae comitantes**, known as the superficial and deep venous arches, respectively.

The ***dorsal digital veins*** (Figs. 6-98, 6-99, and 6-103) drain into three ***dorsal metacarpal veins*** (Fig.

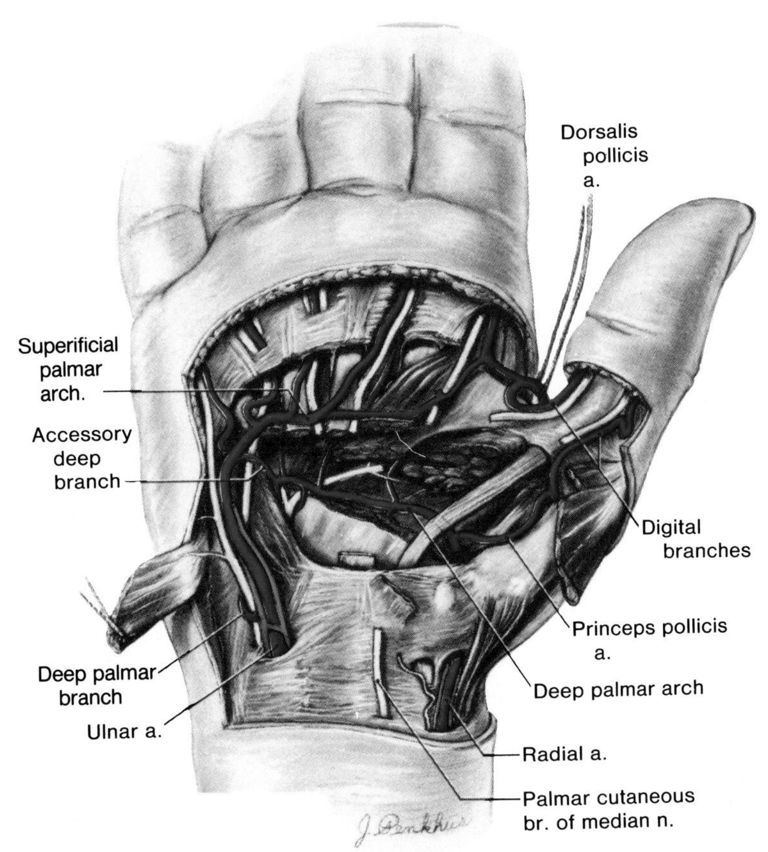

Figure 6-122. Drawing of a deep dissection of the palm of the hand showing the deep branch of the ulnar artery joining the radial artery to form the deep palmar arch. Note that the superficial palmar arch is formed by the direct continuation of the ulnar artery and lies across the center of the palm.

6-124) which unite to form a **dorsal venous network.** It is located superficial to the metacarpus (Figs. 6-100 and 6-124). This network is prolonged proximally as the **cephalic vein**, which winds superiorly around the lateral border of the forearm to its anterior surface (Figs. 6-67, 6-68, 6-98, 6-99, and 6-124).

JOINTS OF THE UPPER LIMB

The ***pectoral girdle*** or shoulder girdle (clavicle and scapula) connects the upper limb to the trunk (Fig. 6-125); therefore its articulations are included with those of the upper limb. The clavicle and scapula articulate at the *acromioclavicular joint.*

THE STERNOCLAVICULAR JOINT

This is the **saddle type of synovial joint** and is the only bony articulation between the upper limb and the axial skeleton (Fig. 6-125). The main function of the clavicle is to hold the upper limb away from the trunk, *i.e.*, it acts as a strut for keeping the shoulder away from the chest in order to give the upper limb the maximum freedom of motion.

The sternoclavicular joint can be readily palpated because the medial end of the clavicle lies superior to the manubrium sterni (Figs. 6-2, 6-22, and 6-125). Put your thumb in the **jugular notch of the sternum** (Fig. 6-2) and your index and middle fingers over the medial end of the clavicle. Convince yourself that the clavicular articular surface is larger than the sternochondral articular surface (Fig. 6-125).

The clavicle is incomplete or absent in some people. This uncommon congenital abnormality is often associated with delayed ossification of the skull. The combined condition, known as **cleidocranial dysostosis**, is characterized by drooping and excessive mobility of the shoulders (Fig. 6-126). Sometimes only the middle of the clavicle is absent and its two ends are joined by fibrous tissue. In other cases only the lateral part of the clavicle is absent.

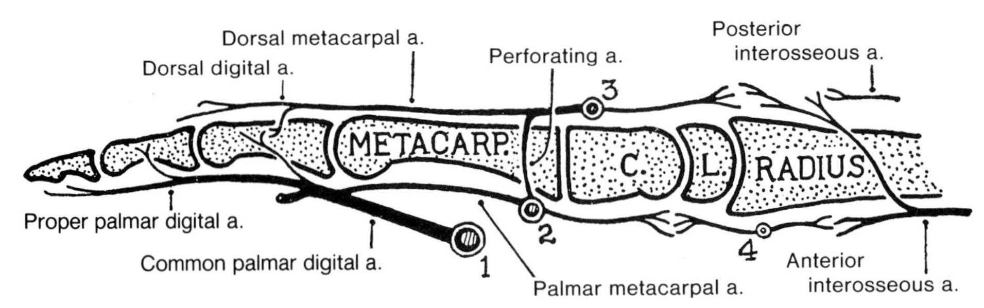

Figure 6-123. Drawing of the arteries of the hand showing the four transversely placed arterial arches numbered in order of size. *1*, the superficial palmar arch lying deep to the palmar aponeurosis; *2*, the deep palmar arch; *3*, the dorsal carpal arch (Fig. 6-105); and *4*, the palmar carpal arch. In common usage, the proper palmar digital arteries are referred to as the digital arteries.

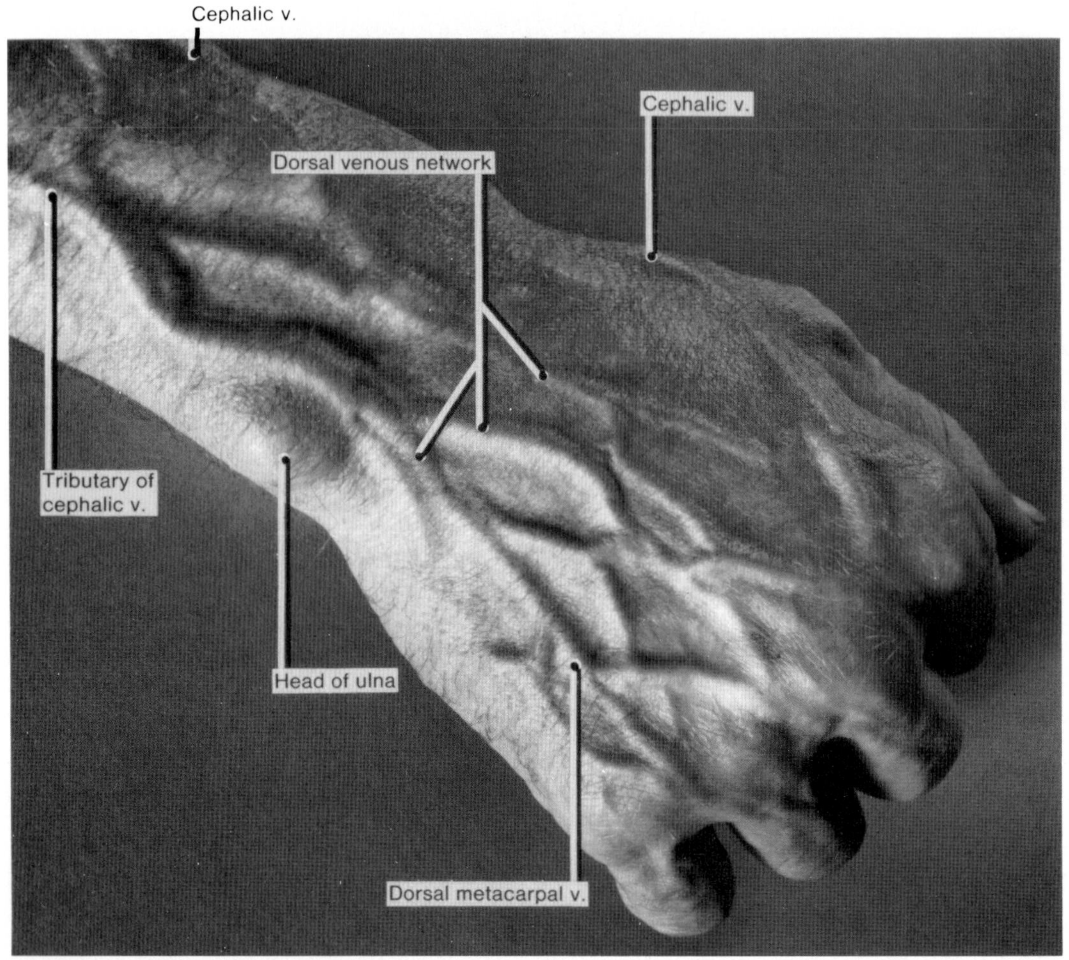

Figure 6-124. Photograph of the dorsum of the distal end of the forearm and hand of a 46-year-old man, showing the superficial veins and the head of the ulna. Observe that the cephalic vein winds from the dorsal venous network around the superiorly lateral border of the forearm, receiving tributaries from both surfaces. *Understand that the superficial venous network and its drainage are variable*. Also see Figure 6-71.

THE ARTICULAR SURFACES OF THE STERNOCLAVICULAR JOINT (Figs. 1-7, 1-9, 6-22, 6-125, and 6-127)

The enlarged medial end of the clavicle articulates in a shallow socket formed by the superolateral part of the manubrium sterni and the medial part of the first costal cartilage (Fig. 1-7). Unlike most articular surfaces, the articular cartilage is fibrocartilage.

MOVEMENTS OF THE STERNOCLAVICULAR JOINT

Despite the saddle-like form of its articular surfaces, the sternoclavicular joint moves in many directions

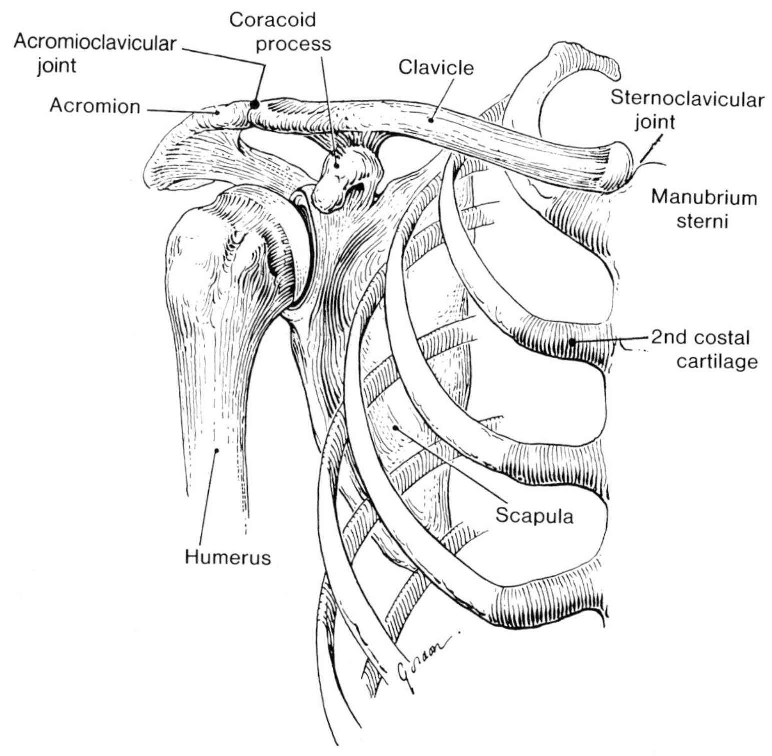

Figure 6-125. Drawing of the bones of the right shoulder region. Note that the clavicle is the only bone uniting the upper limb to the axial skeleton. Observe that the superior part of the medial end of the clavicle lies superior to the manubrium sterni. The clavicle and scapula form the pectoral girdle (shoulder girdle).

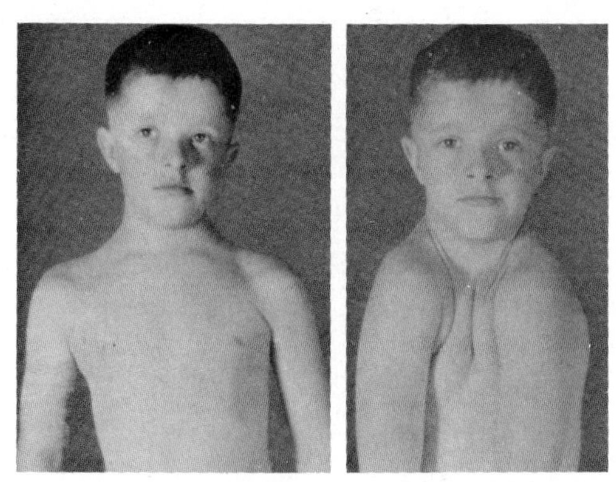

Figure 6-126. Photographs of an 8-year-old boy with a congenital abnormality known as *cleidocranial dysostosis*. It is characterized by delayed closure of the fontanelles, abnormal shape of the skull, late eruption of the teeth, and absence or rudimentary development of the clavicles. This boy has excessive mobility of the shoulders owing to the congenital absence of clavicles and can almost bring them together. He has no significant disability and no treatment was necessary when the congenital abnormality was diagnosed.

like a ball and socket joint. Verify this by elevating and depressing your shoulder and protracting and retracting it. Elevate your upper limb as far as possible and verify by palpation that your clavicle is raised to about a 60° angle from its usual position.

THE ARTICULAR DISC OF THE STERNOCLAVICULAR JOINT (Figs. 6-22 and 6-127)

This strong, thick, fibrocartilaginous disc is located inside the joint and divides it into two synovial cavities. The disc is attached superiorly to the medial end of the clavicle and inferiorly to the junction of the sternum and the first costal cartilage. The articular disc is continuous with the anterior and posterior ***sternoclavicular ligaments*** (Fig. 6-127).

The articular disc prevents medial displacement of the clavicle; it is also an important shock absorber of forces transmitted along the clavicle.

THE ARTICULAR CAPSULE OF THE STERNOCLAVICULAR JOINT (Fig. 6-127)

The fibrous capsule surrounds the entire joint, including the epiphysis at the medial end of the clavicle. Although thin inferiorly, other parts of the fibrous capsule are strong because they are ***reinforced by the anterior and posterior sternoclavicular ligaments and superiorly by the* interclavicular ligament**. The sternoclavicular and interclavicular ligaments are thickenings of the fibrous capsule of the sternoclavicular joint. The interclavicular ligament extends across the jugular notch of the sternum.

The synovial capsule lines the fibrous capsule. Be-

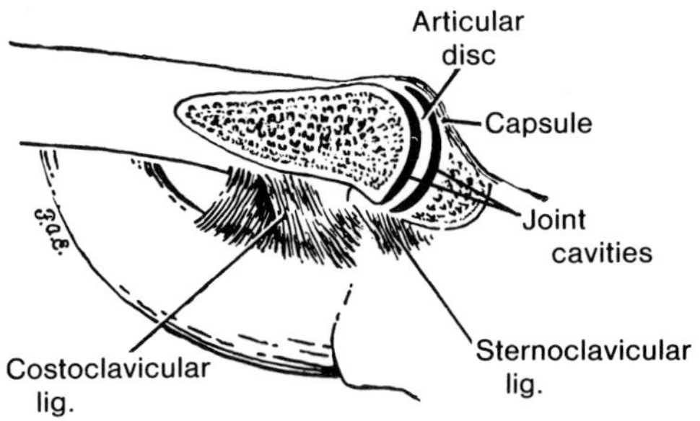

Figure 6-127. Drawing of a coronal section of the right sternoclavicular joint. Note that the fibrocartilaginous articular disc completely divides the joint. In addition to cushioning the articular surfaces from shocks transmitted along the clavicle, this disc prevents such forces from driving the medial end of the clavicle out of its socket and superior to the manubrium of the sternum.

cause the articular disc divides the joint into two cavities (Fig. 6-127), there are two synovial membranes. The lateral one reflects from the articular margin of the medial end of the clavicle to the margins of the articular disc. The medial one lines the capsule between its sternal attachments and the disc.

The Costoclavicular Ligament (Fig. 6-127). This strong ligament ascends from the first rib and its costal cartilage to the inferior margin of the medial end of the clavicle. The costoclavicular ligament reinforces the sternoclavicular joint laterally and limits elevation of the medial end of the clavicle.

STABILITY OF THE STERNOCLAVICULAR JOINT

Because the bony surfaces involved are rather incongruent and the surrounding muscles offer little support, the sternoclavicular joint depends on its ligaments and articular disc for stability.

The strong articular disc is largely responsible for preventing medial displacement of the clavicle (*e.g.*, when carrying a heavy suitcase). The disc prevents the medial end of the clavicle from being pushed out of its socket and superior to the manubrium (Figs. 6-22 and 6-127). The sternoclavicular ligaments also help to prevent displacement of the medial end of the clavicle.

Excessive protraction and elevation of the clavicle are also restrained by the sternoclavicular ligaments, and probably by the subclavius muscle (Figs. 6-20 and 6-21).

BLOOD SUPPLY OF THE STERNOCLAVICULAR JOINT (Figs. 1-16, 1-17, 6-38, and 6-39)

The articular arteries are branches of the *internal thoracic and suprascapular arteries.*

NERVE SUPPLY OF THE STERNOCLAVICULAR JOINT (Figs. 1-16, 1-17, 6-38, and 6-39)

The articular nerves are branches of the *medial supraclavicular nerve* and the nerve to the subclavius.

The rarity of dislocation of the sternoclavicular joint attests to its strength, which results from its ligaments and articular disc. Even when a hole develops in the central part of the disc, which often occurs with old age, the joint rarely dislocates.

When a blow is received to the acromion of the scapula, or when a force is transmitted from the upper limb during a fall on the outstretched hand, the force of the blow is usually transmitted along the long axis of the clavicle. The clavicle may break near the junction of its middle and lateral thirds (Fig. 6-5), but the sternoclavicular joint rarely dislocates.

THE ACROMIOCLAVICULAR JOINT

This is the **plane type of synovial joint** and is located between the lateral end of the clavicle and the acromion of the scapula (Figs. 6-125, 6-128, 6-129, 6-131, and 6-134). It is located 2 to 3 cm medial to the acromion which projects anteriorly from the lateral end of the spine of the scapula (Figs. 6-1, 6-3, 6-125, 6-131, and 6-134). The acromion forms a palpable and sometimes visible prominence (Fig. 6-2). Lay people refer to it as the "point of the shoulder."

THE ARTICULAR SURFACES OF THE ACROMIOCLAVICULAR JOINT (Figs. 6-1, 6-125, 6-128, 6-129, and 6-134)

The small oval articular facet on the lateral end of the clavicle articulates with a similar facet on the anterior part of the medial surface of the medial end of the acromion. Both articular surfaces are covered with fibrocartilage. Both articular surfaces slope inferomedially so that *the clavicle tends to override the acromion* and usually projects over it (Fig. 6-131).

MOVEMENTS OF THE ACROMIOCLAVICULAR JOINT

This articulation allows the acromion to rotate on the clavicle and to move anteriorly and posteriorly. These movements are associated with movement of the scapula and with those at the sternoclavicular joint.

THE ARTICULAR DISC OF THE ACROMIOCLAVICULAR JOINT (Fig. 6-128)

A wedge-shaped, incomplete, fibrocartilaginous disc projects into the joint from the superior part of the articular capsule. The disc, when present, is located in the superior part of the joint and partially divides the joint cavity into two parts.

THE ARTICULAR CAPSULE OF THE ACROMIOCLAVICULAR JOINT (Figs. 6-128, 6-129, and 6-131)

The fibrous capsule enclosing the joint is attached to the margins of its articular surfaces. The fibrous capsule is weak, but is strengthened superiorly by the **acromioclavicular ligament** (Figs. 6-55 and 6-129) and by fibers from the trapezius muscle (Fig. 6-21). This ligament extends from the superior part of the lateral end of the clavicle to the superior surface of the acromion.

The synovial capsule lines the fibrous capsule.

The Coracoclavicular Ligament (Figs. 6-125 and 6-129). This ligament anchors the lateral part of the clavicle to the coracoid process of the scapula. It is the strongest of the ligaments that bind the clavicle to the scapula. It consists of two parts, the *conoid and trapezoid ligaments*, which are directed in such a way that they enable the clavicle to hold the scapula and upper limb laterally.

STABILITY OF THE ACROMIOCLAVICULAR JOINT (Fig. 6-129)

The coracoclavicular ligament, an extrinsic ligament of the acromioclavicular joint, is principally responsible for providing stability to the articulation. It prevents the clavicle from losing contact with the acromion of the scapula (Fig. 6-125).

BLOOD SUPPLY OF THE ACROMIOCLAVICULAR JOINT (Figs. 6-38, 6-39, 6-42, and 6-55)

The articular arteries are branches of the suprascapular and thoracoacromial arteries.

NERVE SUPPLY OF THE ACROMIOCLAVICULAR JOINT (Figs. 6-19, 6-20, and 6-28)

The articular nerves are branches of the supraclavicular, lateral pectoral, and axillary nerves.

In contact sports such as football and hockey, it is not uncommon for dislocation of the acromioclavicular joint to result from a hard fall on the shoulder, or to occur when a hockey player is "driven viciously into the boards." This injury, often inaccurately called a "**shoulder separation**," is serious when both the acromioclavicular and coracoclavicular ligaments are torn. When the coracoclavicular ligament ruptures, the shoulder falls away from the clavicle owing to the weight of the upper limb. The fibrous capsule of the acromioclavicular joint also ruptures, resulting in the acromion passing inferior to the lateral end of the clavicle.

Dislocation of the acromioclavicular joint ("separation of the shoulder") makes the acromion more obvious.

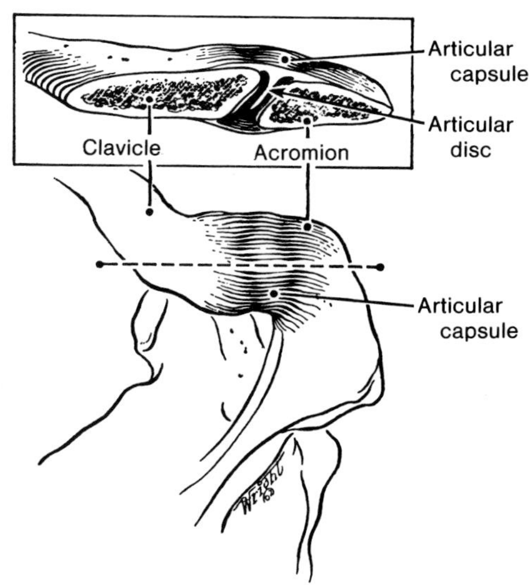

Figure 6-128. Drawing of a superior view of the right acromioclavicular joint. *Inset above* is a coronal section of the joint showing its wedge-shaped articular disc.

THE SHOULDER JOINT

This is a multiaxial ball and socket type of synovial joint which permits a wide range of movement. However, *mobility is gained at the expense of stability and dislocation of the shoulder joint is common, especially in body contact sports and car accidents.*

THE ARTICULAR SURFACES OF THE SHOULDER JOINT (Figs. 6-41, 6-125, and 6-129 to 6-134)

The spheroidal **head of the humerus** (the ball) articulates with the shallow **glenoid cavity of the scapula** (the socket). Both articular surfaces are covered with hyaline cartilage (Fig. 6-41).

The shallow glenoid cavity accepts little more than a third of the large humeral head (Figs. 6-130 to 6-132

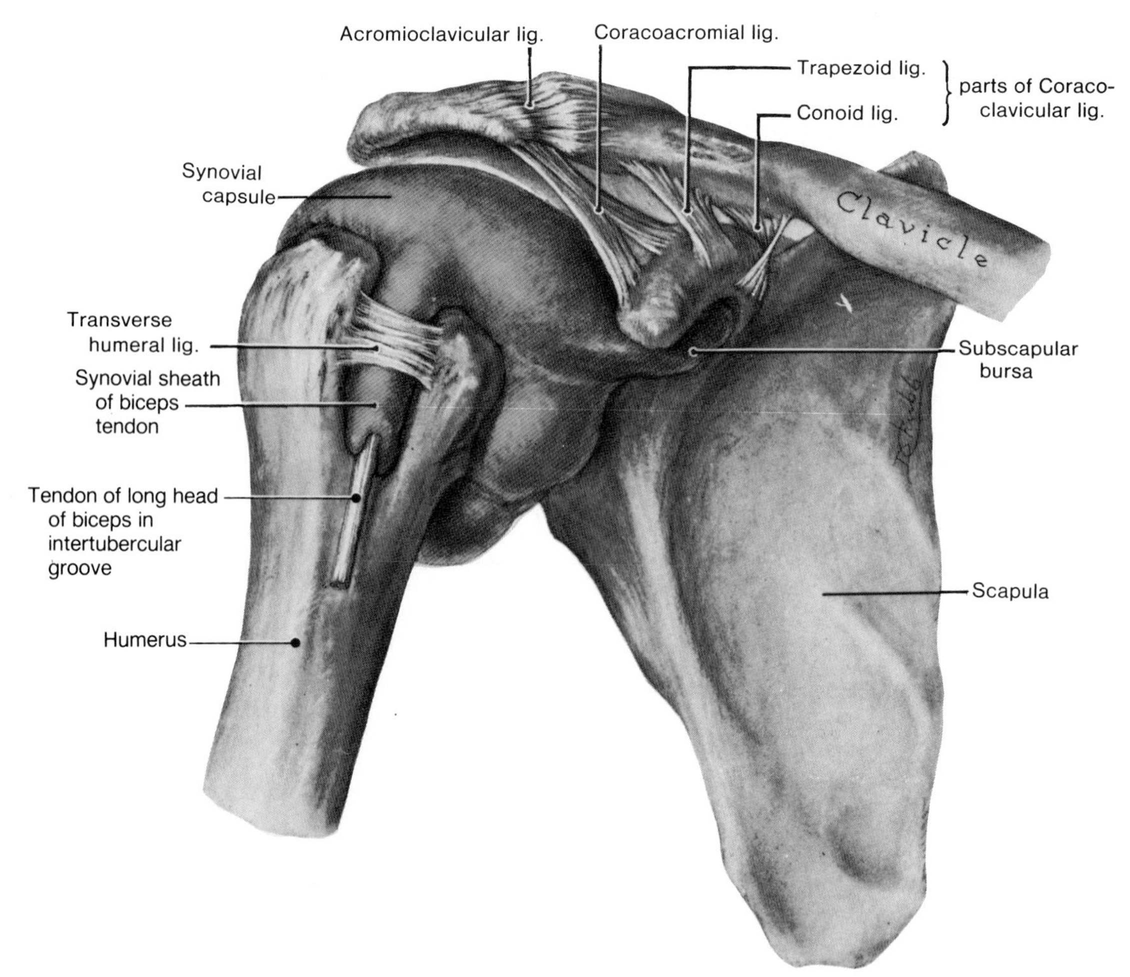

Figure 6-129. Drawing of an anterior view of the synovial capsule of the right shoulder joint and the ligaments at the lateral end of the clavicle. The capsule cannot extend on to the lesser and greater tubercles of the humerus because four short rotator cuff muscles (subscapularis, supraspinatus, infraspinatus, and teres minor) are inserted there (Fig. 6-49), but it can and does extend inferiorly on to the surgical neck of the humerus (Fig. 6-1). Note that the synovial capsule has two prolongations: (1) where it forms a synovial sheath for the tendon of the long head of the biceps, and (2) inferior to the coracoid process where it forms the subscapular bursa between the subscapularis tendon and the margin of the glenoid cavity. Observe that the conoid and trapezoid ligaments, parts of the coracoclavicular ligament, are directed so that the clavicle holds the scapula and humerus laterally.

and 6-134), but the glenoid cavity is deepened slightly and enlarged by a *fibrocartilaginous rim* called the **glenoidal labrum** (L. lip). The superior portion of the glenoidal labrum blends with the tendon of the long head of the biceps brachii muscle (Fig. 6-131).

MOVEMENTS OF THE SHOULDER JOINT (Table 6-15)

The shoulder joint has more freedom of movement than any other joint in the body. This freedom of movement results from the laxity of the joint's articular capsule and the large size of the humeral head compared with the small size of the glenoid cavity.

The shoulder articulation is a **multiaxial ball-and-socket joint** that allows movements around three axes and *permits flexion-extension, abduction-adduction, circumduction, and rotation.*

Most movements of the shoulder joint are illustrated in Figures 11 (p. 12) and 6-135*A*. In *circumduction*, the distal end of the humerus describes the base of a cone, the apex of which is at the head of the humerus.

THE ARTICULAR CAPSULE OF THE SHOULDER JOINT (Figs. 6-54, 6-129, 6-130, 6-132, and 6-133)

The fibrous capsule enclosing the shoulder joint is thin and loose; thus it allows a wide range of movement. The capsule is attached medially to the glenoid

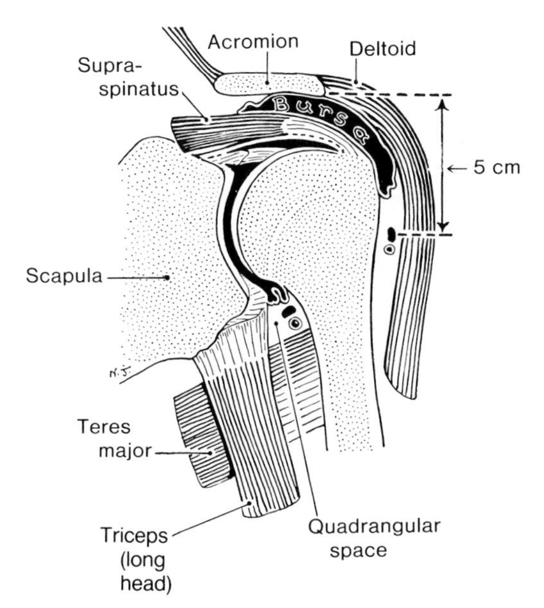

Figure 6-130. Drawing of a coronal section of the left shoulder region showing that the shoulder articulation is a ball-and-socket type of synovial joint. Note that the large head of the humerus articulates with the glenoid cavity of the scapula which is only a little more than one-third its size.

cavity, beyond the glenoidal labrum (Fig. 6-130). Superiorly it encroaches on the root of the coracoid process (Fig. 6-54) so that *the fibrous capsule encloses the origin of the long head of the biceps muscle within the joint* (Figs. 6-41, 6-54, 6-129, and 6-132). Laterally the fibrous capsule is attached to the anatomical neck of the humerus (Figs. 6-59, 6-129, and 6-130).

The inferior part of the capsule is its weakest area (Fig. 6-130). The capsule is lax and lies in folds when the arm is adducted, but it becomes taut when the arm is abducted (Figs. 6-133 and 6-135*A*).

There are two apertures in the articular capsule of the shoulder joint. The opening between the tubercles of the humerus is for passage of the tendon of the long head of the biceps brachii muscle (Fig. 6-129). The other opening in the capsule is situated anteriorly, inferior to the coracoid process. It allows communication between the **subscapular bursa** and the synovial cavity of the joint (Figs. 6-41, 6-129, and 6-132).

The synovial capsule lines the fibrous capsule and is reflected from it on to the glenoidal labrum and the neck of the humerus, as far as the articular margin of the head (Figs. 6-129 and 6-130).

The synovial capsule forms a tubular sheath for the tendon of the long head of the biceps brachii muscle (Figs. 6-41 and 6-129), where it passes into the joint cavity and lies in the intertubercular groove, extending as far as the surgical neck of the humerus (Figs. 6-59 and 6-129).

Intrinsic Ligaments of the Capsule of the Shoulder Joint (Figs. 6-129 and 6-131 to 6-133). *These ligaments are thickenings of the fibrous capsule which strengthen the shoulder joint.*

The glenohumeral ligaments (Fig. 6-133) are thickenings of the anterior part of the fibrous capsule. The superior, middle, and inferior glenohumeral ligaments run from the supraglenoid tubercle of the scapula to the lesser tubercle and the anatomical neck of the humerus (Fig. 6-59*A*). The glenohumeral ligaments are frequently indistinct or absent.

The transverse humeral ligament (Fig. 6-129) is a broad band of transverse fibers passing from the greater to the lesser tubercles of the humerus. *It forms a bridge over the superior end of the intertubercular groove.* This converts the intertubercular groove into a canal which holds the tendon of the long head of the biceps as it emerges from the capsule of the shoulder joint (Fig. 6-129).

The coracohumeral ligament (Figs. 6-54 and 6-131) *is a strong broad band that strengthens the superior part of the capsule of the shoulder joint.* It passes from the lateral side of the base of the coracoid process of the scapula to the anatomical neck of the humerus, adjacent to the greater tubercle (Fig. 6-54).

The Coracoacromial Arch (Figs. 6-54 and 6-131). The coracoid process, coracocromial ligament, and acromion form a protective arch for the shoulder joint. When force is transmitted superiorly along the humerus (*e.g.*, when standing at a desk and partly supporting the body with the outstretched limbs), the head of the humerus is pressed against this arch. *The coracoacromial arch prevents displacement of the humeral head superiorly from the glenoid cavity.*

The **supraspinatous muscle** passes under the coracoacromial arch and lies between the deltoid muscle and the capsule of the shoulder joint (Figs. 6-54 and 6-130). The supraspinatous tendon, passing to the greater tubercle of the humerus (Fig. 6-54), is separated from the coracoacromial arch by the **subacromial bursa** (Figs. 6-55 and 6-130).

The Coracoacromial Ligament (Figs. 6-54, 6-55, and 6-131) is a strong triangular ligament, the base of which is attached to the lateral border of the coracoid process; its apex is inserted into the edge of the acromion. Superiorly the coracoacromial ligament is covered by the deltoid muscle (Fig. 6-54).

STABILITY OF THE SHOULDER JOINT

The free movement of the shoulder joint leads to instability. *The shallowness of the glenoid cavity and the laxity of the fibrous capsule also result in a considerable loss of stability* (Fig. 6-130).

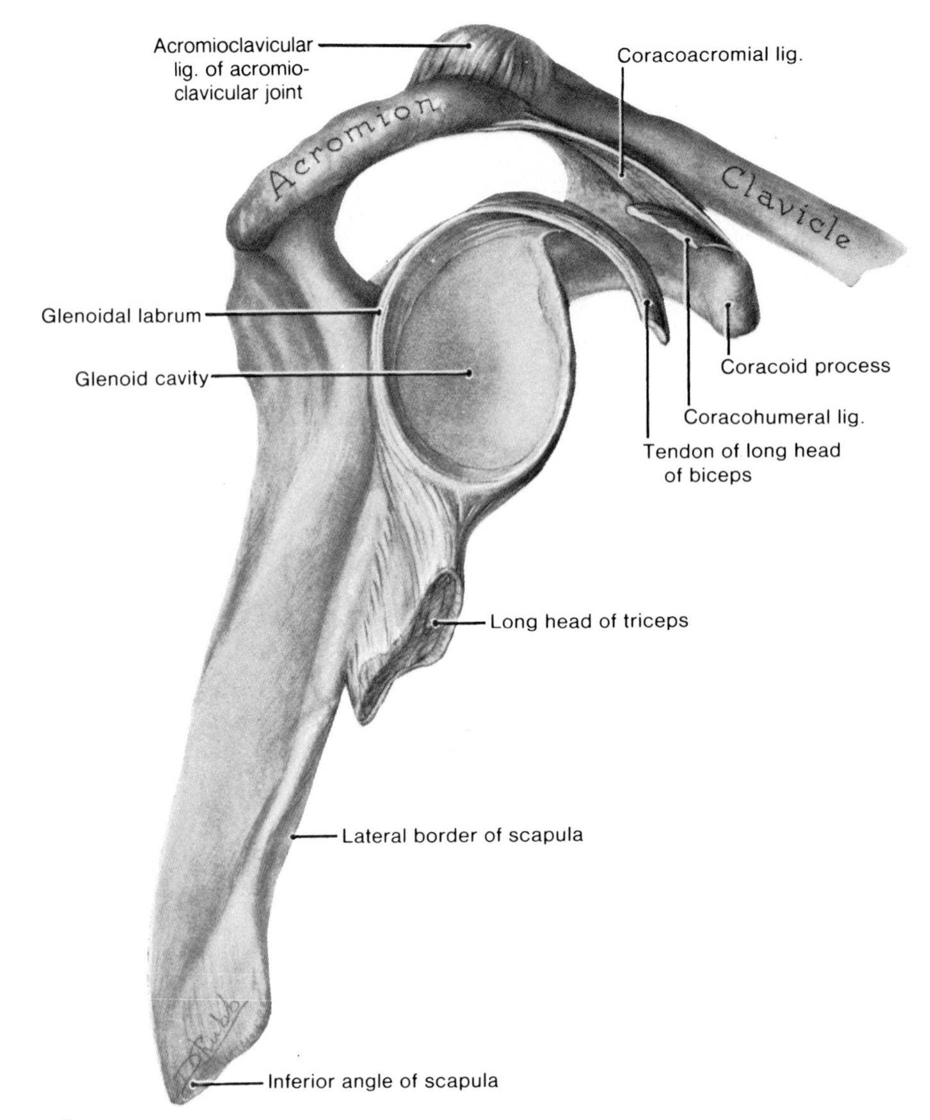

Figure 6-131. Drawing of a lateral view of the right scapula. Observe that the glenoid cavity is deepened by the *glenoidal labrum*, a dense fibrocartilaginous lip (L. labrum) that is attached to the rim of the glenoid cavity. Note the *coracoacromial arch* formed by the coracoid process, coracoacromial ligament, and acromion. This arch prevents superior displacement of the head of the humerus. The acromioclavicular ligament strengthens the articular capsule of the acromioclavicular joint.

The strength of the shoulder joint results mainly from the muscles which surround it (Figs. 6-52 and 6-54 to 6-57), particularly the **rotator cuff muscles** (supraspinatus, infraspinatus, teres minor, and subscapularis). These four scapular muscles, joining the scapula to the humerus (Table 6-5), are attached near the articular areas of the articulation and are closely related to the fibrous capsule of the joint (Fig. 6-52).

Although they have separate functions (Table 6-5), **the rotator cuff muscles work as a group in holding the head of the humerus in the glenoid cavity.** They give stability to the shoulder joint in several positions, especially when the arm is abducted.

The supraspinatus muscle and the coracoacromial arch guard the shoulder joint superiorly; the infraspinatus and teres minor muscles stabilize the shoulder joint posteriorly (Fig. 6-52), and the subscapularis muscle protects it anteriorly (Figs. 6-29 and 6-56). In Figure 6-132, observe that no tendons support the shoulder joint inferiorly; consequently this is where it usually dislocates (Fig. 6-136).

BURSAE AROUND THE SHOULDER JOINT (Figs. 6-41, 6-55, 6-130, and 6-132)

There are several bursae, or flattened sacs containing capillary films of synovial fluid (Fig. 6-53), in the

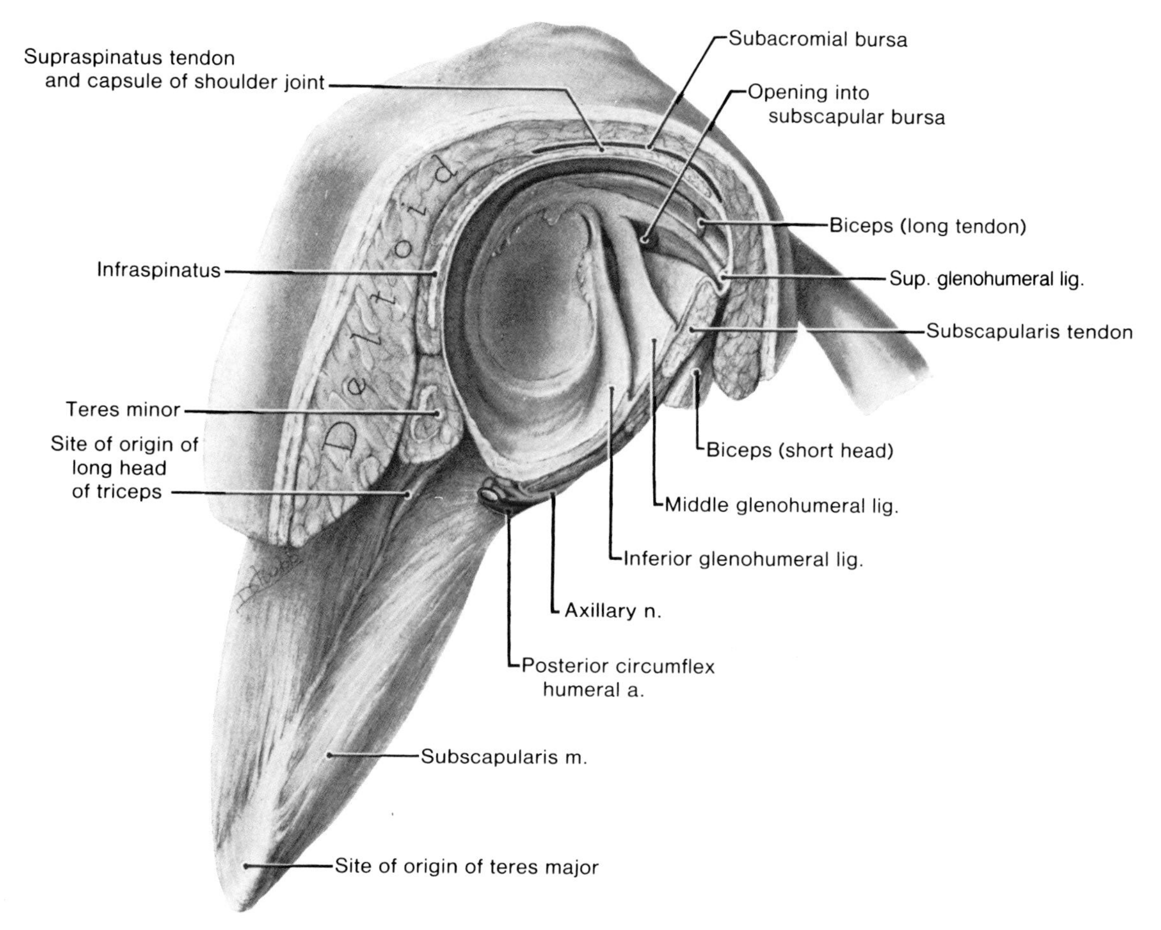

Figure 6-132. Drawing of a dissection of the right glenoid cavity as viewed from the anterolateral aspect. Observe that the articular capsule of the joint is thickened anteriorly by the three glenohumeral ligaments which converge from the humerus to be attached with the long tendon of the biceps muscle to the supraglenoid tubercle of the scapula (Fig. 6-133). *Note the four short rotator cuff muscles* (teres minor, infraspinatus, supraspinatus, and subscapularis) crossing the joint and blending with the capsule. Their prime function is to hold the head of the humerus in the glenoid cavity of the scapula (Table 6-5). Observe the subacromial bursa between the acromion and deltoid muscle superiorly and the tendon of the supraspinatus muscle inferiorly.

vicinity of the shoulder joint. Bursae are located where tendons rub against bone, ligaments, or other tendons, and where skin moves over a bony prominence.

The bursae around the shoulder joint are clinically important. Some of them communicate with the shoulder joint cavity (*e.g.*, the subscapular bursa); hence to open a bursa may mean entering the joint cavity. This is potentially dangerous because of the possibility of infection.

The Subscapular Bursa (Figs. 6-41, 6-129, and 6-132). This bursa is located between the tendon of the subscapularis muscle and the neck of the scapula (p. 669). The bursa protects this tendon where it passes inferior to the root of the coracoid process and over the neck of the scapula. It usually communicates with the cavity of the shoulder joint through an opening in its fibrous capsule (Fig. 6-132); thus it is really an extension of the cavity of the shoulder joint.

The Subacromial Bursa (Figs. 6-55, 6-130, and 6-132). This large bursa lies between the deltoid muscle, the supraspinatus tendon, and the fibrous capsule of the shoulder joint. Its size varies. It does not normally communicate with the cavity of the shoulder joint. The subacromial bursa is located inferior to the acromion and the coracoacromial ligament, between them and the supraspinatus muscle (Fig. 6-132).

The subacromial bursa facilitates movement of the deltoid muscle over the fibrous capsule of the shoulder joint and the supraspinatus tendon.

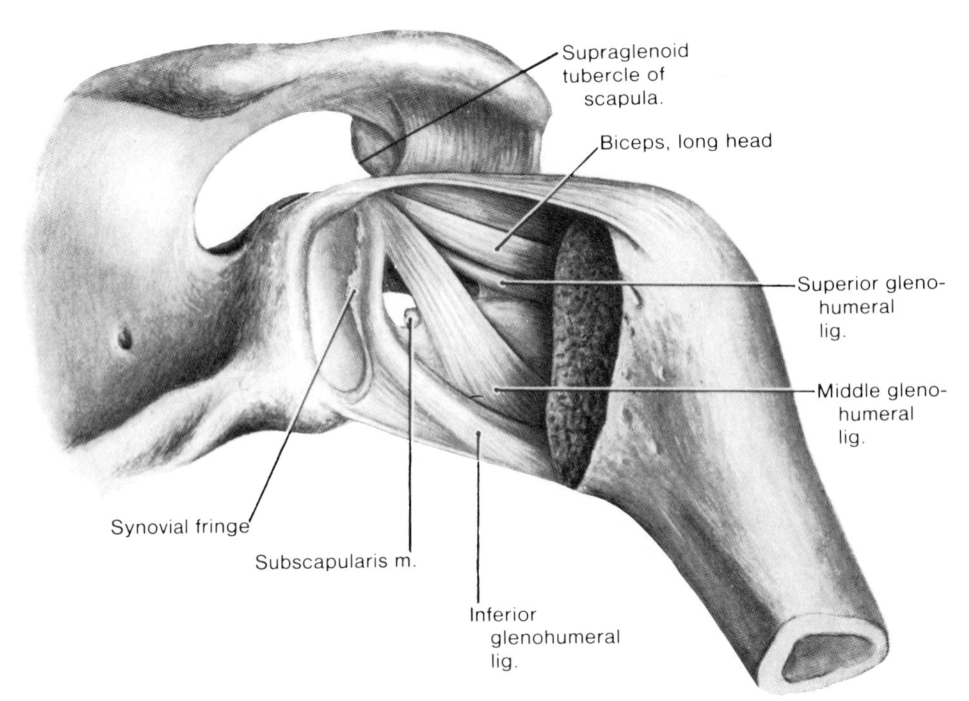

Figure 6-133. Drawing of a dissection of the interior of the right shoulder joint, exposed posteriorly by cutting away the posterior part of the articular capsule and sawing off the head of the humerus. Observe the three thickenings of the anterior part of the fibrous capsule, called the superior, middle, and inferior *glenohumeral ligaments*. They are visible within the joint when the synovial capsule is removed. Note how these three ligaments and the long tendon of the biceps brachii muscle converge on the supraglenoid tubercle of the scapula.

BLOOD SUPPLY OF THE SHOULDER JOINT (Figs. 6-28, 6-38, 6-39, 6-52, 6-55, and 6-132)

The articular arteries to the shoulder joint are branches of the *anterior and posterior circumflex humeral arteries* from the axillary and the *suprascapular artery* from the subclavian (Fig. 6-38).

NERVE SUPPLY OF THE SHOULDER JOINT (Figs. 6-28, 6-29, 6-52, 6-69, and 6-132)

The articular nerves are branches of the suprascapular, axillary, and lateral pectoral nerves.

CALCIFIC SUPRASPINATUS TENDINITIS (Fig. 6-135)

In calcific tendinitis (tendonitis), there is inflammation and calcification of the subacromial or subdeltoid bursa. This results in pain, tenderness, and limitation of movement of the shoulder joint. The condition is also called *calcific bursitis* and *scapulohumeral bursitis.*

Deposition of calcium in the supraspinatus portion of the musculotendinous rotator cuff of the shoulder joint is common. This condition causes increased local pressure and may cause pain during abduction of the arm (Fig. 6-135*A*). The calcium deposit may irritate the overlying subacromial bursa (Figs. 6-55, 6-130 and 6-132), producing an inflammatory reaction known as **subacromial bursitis**; this causes increasing pain. So long as the shoulder joint is adducted, there is usually no pain because in this position the painful lesion is away from the acromion. In most patients pain occurs during 50 to 130° of abduction (Fig. 6-135*A*) because during this arc the supraspinatus tendon is in intimate contact with the inferior surface of the acromion.

RUPTURE OF THE ROTATOR CUFF OF THE SHOULDER JOINT

When an older person strains to lift something (*e.g.*, a window that is stuck), a previously degenerated musculotendinous rotator cuff may rupture. Often the strain also tears the capsule of the shoulder joint. As a result the joint cavity communicates with the subacromial bursa.

Tears of the rotator cuff also occur in athletes owing to strain on the shoulder joint (*e.g.*, base-

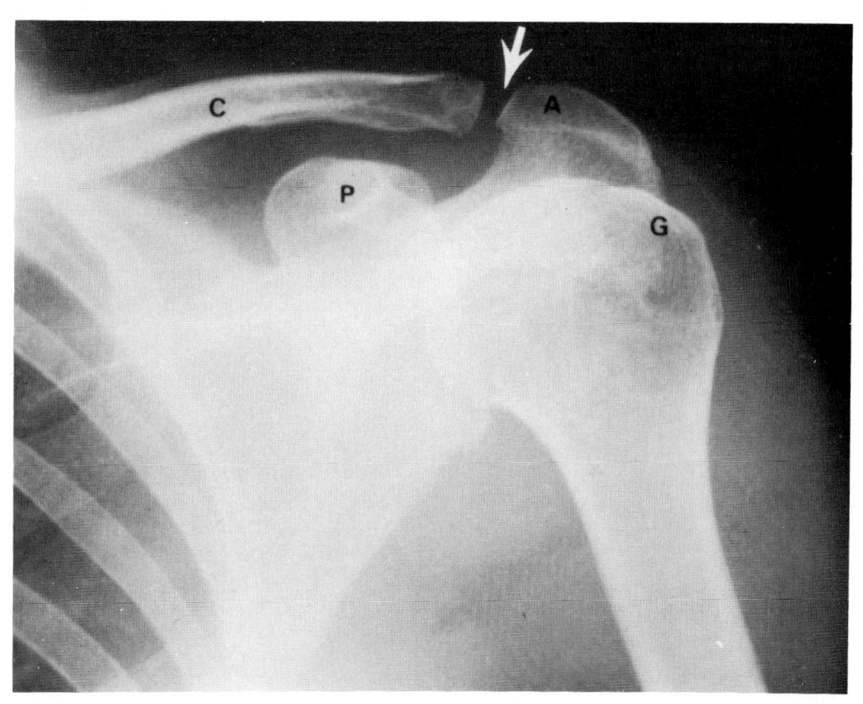

Figure 6-134. Radiograph of the left shoulder with the arm at the side and the humerus rotated laterally (anteroposterior projection; *i.e.*, an AP). The *white arrow* indicates the acromioclavicular joint. The apparent space represents the cartilages on the ends of the clavicle and acromion and the articular disc (Fig. 6-128) which are radiolucent. Observe the relatively small part of the head of the humerus that is within the glenoid cavity of the scapula (also see Fig. 6-130). This partly accounts for the instability of the shoulder joint. *C*, clavicle. *A*, acromion. *P*, coracoid process. *G*, greater tubercle of humerus. Figure 6-125 shows a drawing of the bones of the right shoulder similar to those visible in this radiograph.

ball pitchers). *This causes pain in the shoulder region when the arm is moved* (*e.g.*, to throw a ball).

DISLOCATION OF THE SHOULDER JOINT (Fig. 6-136)

Because of its freedom of movement and instability, the shoulder joint is dislocated more often than any other joint in adults. The dislocation may result from direct or indirect injury.

Anterior dislocation of the shoulder joint occurs most often in young adults, particularly athletes. It is usually caused by excessive extension and lateral rotation of the humerus. The head of the humerus is driven anteriorly and usually the fibrous capsule and glenoidal labrum are stripped from the anterior aspect of the glenoid cavity.

A hard blow to the humerus when the shoulder joint is fully abducted (*e.g.*, when a quarterback is about to release a football), tilts the head of the humerus inferiorly onto the inferior weak part of the articular capsule. This may tear the capsule and dislocate the shoulder so that the humeral head comes to lie inferior to the glenoid cavity (Fig. 6-136*B*). The strong flexor and abductor muscles of the shoulder joint (Table 6-15) usually pull the humeral head anterosuperiorly into a subcoracoid position.

Unable to use the arm, the patient commonly supports it with the other hand. The diagnosis is confirmed by radiographic examination (Fig. 6-136*B*).

The axillary nerve is often injured when the shoulder is dislocated because it is in close relation to the inferior part of the articular capsule of this joint (Fig. 6-132).

The subglenoid displacement of the head of the humerus into the quadrangular space damages the axillary nerve (Figs. 6-52 and 6-69). This nerve injury is indicated by **paralysis of the deltoid muscle** and loss of skin sensation in the shoulder region (Fig. 6-37).

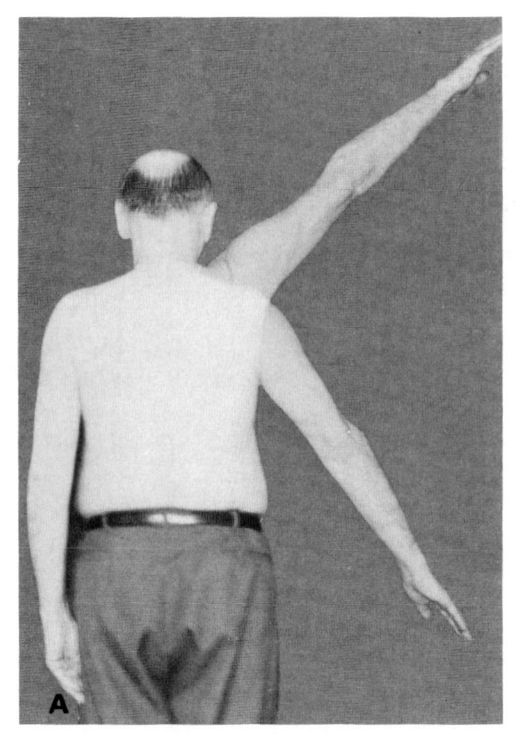

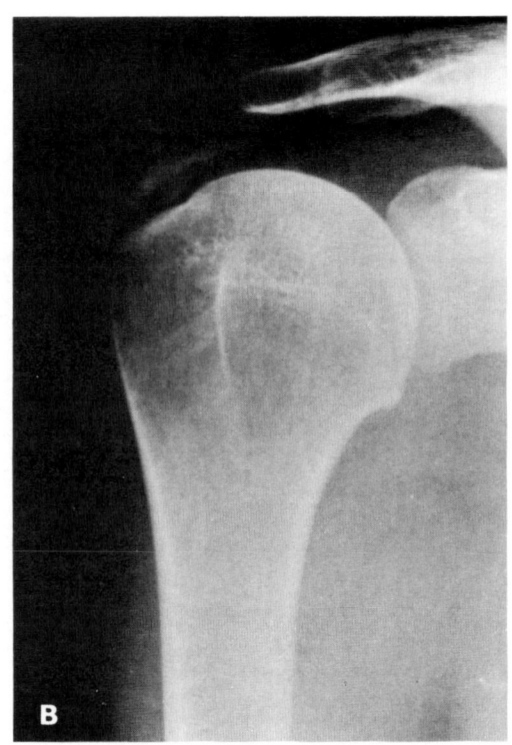

Figure 6-135. *A*, double exposure photograph of a middle-aged man demonstrating *the painful arc syndrome* associated with calcific supraspinatus tendinitis in the right shoulder. Abduction of the shoulder joint from about 50 to 130° causes severe pain owing to the tendinitis and the associated subacromial bursitis. *B*, radiograph of the right shoulder joint showing calcium deposits in the supraspinatous tendon of the musculotendinous rotator cuff, close to its insertion into the humerus.

THE ELBOW JOINT

This is a **hinge type of synovial joint** formed by the distal end of the humerus with the proximal ends of the radius and ulna (Fig. 6-137). The elbow is a *uniaxial joint* and its movements consist of flexion and extension (Table 6-16).

THE ARTICULAR SURFACES OF THE ELBOW JOINT (Figs. 6-137 to 6-142)

The ***trochlea and capitulum*** *of the humerus articulate with the* ***trochlear notch of the ulna*** *and the* ***head of the radius***, respectively. The articular surfaces, covered with hyaline cartilage, are most fully in contact when the forearm is in a position midway between pronation and supination, and is flexed to a right angle.

Extend your forearm completely and palpate the medial and lateral epicondyles of your humerus and the olecranon of your ulna. In this position the epicondyles and the tip of olecranon are in a straight line (Figs. 6-63 and 6-73). Flex your forearm and note that these bony points form the points of an equilateral triangle.

Using an articulated skeleton of the upper limb, verify that *the elbow joint consists of three different articulations* (Figs. 6-1, 6-137, and 6-138):

1. **The humeroulnar articulation** is between the trochlea of the humerus and the trochlear notch of the ulna. They form a **uniaxial hinge joint** permitting movement in one axis: flexion and extension.
2. **The humeroradial articulation** is between the capitulum of the humerus and the head of the radius. The capitulum fits into the slightly cupped surface of the head (Fig. 6-137*A*).
3. **The proximal radioulnar joint** (superior radioulnar joint) is between the head of the radius and the radial notch of the ulna. This is a **pivot joint** permitting rotation of the radius about the ulna (Fig. 6-137*A*).

The articular capsule of the elbow joint invests the three articulations just described. You may hear this compound articulation referred to as the cubital joint.

MOVEMENTS OF THE ELBOW JOINT (Tables 6-15 and 6-16)

The elbow can be flexed or extended. The main flexor is the **brachialis** and the chief extensor is the **triceps** .

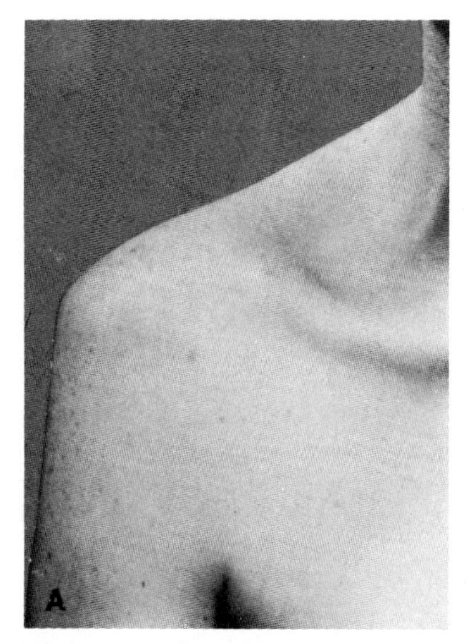

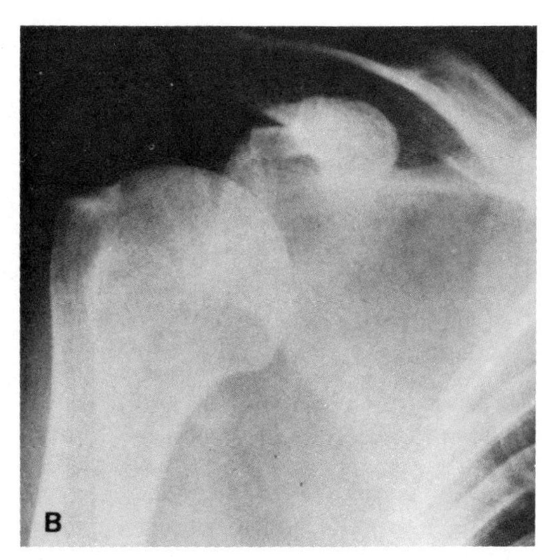

Figure 6-136. *A*, photograph of a young man with an anterior dislocation of the right shoulder. Observe that the shoulder appears square and that its normal round contour is absent because the greater tubercle of the humerus is no longer bulging laterally beneath the deltoid muscle (Fig. 6-134). *B*, radiograph of the shoulder anteroposterior (AP projection) showing that the humeral head is not in articulation with the glenoid cavity, but is in a subcoracoid position (see Fig. 6-125 for orientation).

Flexion is limited by apposition of the anterior surfaces of the forearm and arm, by tension of the posterior muscles, and by the radial and ulnar collateral ligaments (Figs. 6-141 and 6-142).

Several muscles may produce flexion of the elbow joint (Table 6-16), but *the brachialis is the main flexor of the forearm.* The flexion of the forearm, or twitch of the biceps that occurs following tapping of the bicipital aponeurosis without movement, is known as the **biceps jerk.** The reflex center is in C5 and C6 segments of the spinal cord. Although the biceps is an important flexor of the elbow joint, the brachialis muscle is the chief flexor (Tables 6-7 and 6-16). The biceps becomes active against resistance.

Extension *is limited by impingement of the olecranon of the ulna on the olecranon fossa of the humerus* (Figs. 6-137 and 6-138*B*), and by tension of the anterior muscles and collateral ligaments (Figs. 6-140 to 6-142). The main muscle producing extension of the forearm at the elbow joint is the **triceps brachii** (Table 6-16). The anconeus muscle stabilizes the elbow joint and may assist in its extension (Fig. 6-63).

The extension of the forearm, or twitch of the triceps without movement, that occurs following tapping of the triceps tendon (Fig. 6-69) is known as the **triceps jerk**. The reflex center is in C6, C7, and C8 segments of the spinal cord.

When your forearm is fully extended and supinated in the anatomical position, your arm and forearm are not in the same line. Normally the forearm is directed laterally forming a "**carrying angle**" of about 163°. The carrying angle does not differ significantly in males and females. The carrying angle of the elbow permits the extended forearm to clear the side of the hip in swinging movements of the upper limb and when carrying heavy loads.

The "carrying angle" is diminished when the forearm is pronated (Fig. 6-70*B*). This is the usual position of the upper limb during pushing and pulling movements.

Knowledge of the normal carrying angle of the forearm is essential for aligning the bones of the arm and forearm during reduction of a fracture and/or dislocation of the elbow joint (Figs. 6-145 and 6-146).

An increase in the normal carrying angle at the elbow is known as cubitus valgus. This is one

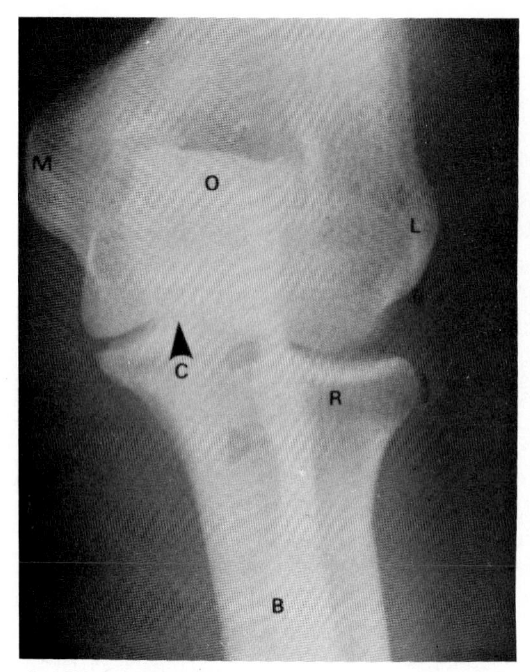

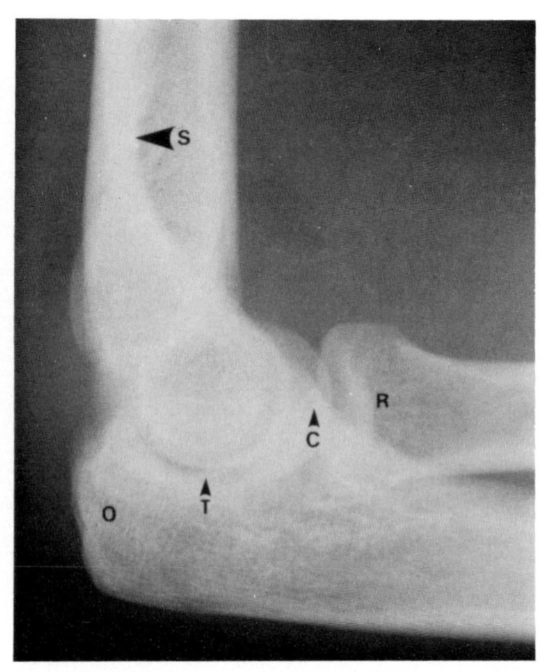

Figure 6-137. Radiographs of the elbow region (AP projection) on the *left* with the elbow extended, and a lateral projection on the *right* with the elbow flexed. *On the humerus* observe the medial (*M*) and lateral (*L*) epicondyles and supracondylar ridge (*S*). *On the ulna* observe the olecranon (*O*), coronoid process (*C*), and trochlear notch (*T*). Observe the head (*R*) and tuberosity (*B*) of the radius. Compare this lateral radiograph of the elbow with Figure 6-141. During flexion and extension of the forearm, the trochlea and capitulum of the humerus (Fig. 6-142*A*) articulate with the trochlear notch of the ulna and the head of the radius. The elbow joint consists of three different articulations: (1) the humeroulnar articulation; (2) the humeroradial articulation; and (3) the proximal radioulnar joint.

of the clinical manifestations of patients with **Turner syndrome** (45, XO chromosome constitution). Other characteristics are *webbing of the neck, broad chest, and short stature.*

THE ARTICULAR CAPSULE OF THE ELBOW JOINT (Figs. 6-140 to 6-143)

The fibrous capsule completely encloses the joint. Its anterior and posterior parts are thin and weak, but its sides are strengthened by collateral ligaments.

The fibrous capsule of the elbow joint is attached to the proximal margins of the ***coronoid and radial fossae*** anteriorly (Fig. 6-138*A*), but not quite to the superior limit of the ***olecranon fossa*** posteriorly (Fig. 6-138*B*). Distally the fibrous capsule is attached to the margins of the ***trochlear notch***, the anterior border of the ***coronoid process***, and the ***anular ligament*** (Figs. 6-138 and 6-139).

The Collateral Ligaments (Figs. 6-141 and 6-142). These strong triangular bands are medial and lateral *thickenings of the fibrous capsule*; hence they are intrinsic ligaments.

The radial collateral ligament is a strong triangular band (Fig. 6-141). Its **apex** is attached proximally to the ***lateral epicondyle*** of the humerus and its **base** blends with the **anular ligament** (annular ligament) of the radius.

The ulnar collateral ligament is also triangular in shape (Fig. 6-142). It is composed of anterior and posterior parts which are connected by a thinner, relatively weak oblique band. Its ***apex*** is attached to the *medial epicondyle* of the humerus. The strong, cord-like ***anterior part*** is attached to the tubercle on the *coronoid process* of the ulna, and the weaker, fan-like ***posterior part*** is attached to the medial edge of the *olecranon* (Fig. 6-142).

The ulnar nerve passes posterior to the medial epicondyle and is closely applied to the ulnar collateral ligament (Figs. 6-63 and 6-140). It enters the forearm between the heads of the flexor carpi ulnaris muscle.

The ulnar nerve is commonly injured at the elbow. The injuries are usually associated with fractures of the medial epicondyle. For the effects of this nerve injury, see page 714.

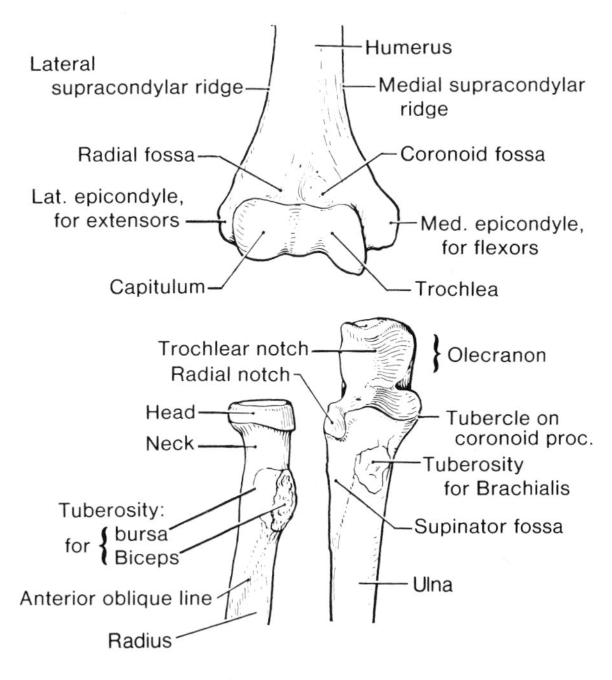

A. ANTERIOR VIEW

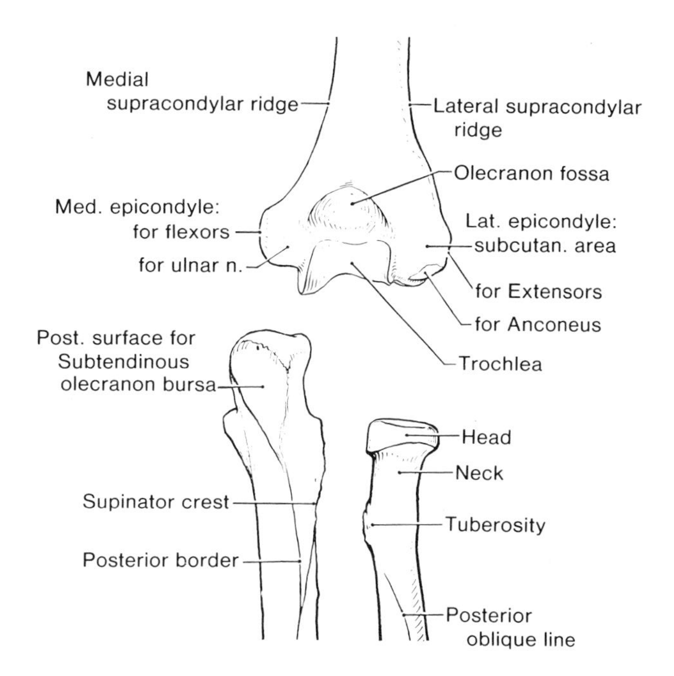

B. POSTERIOR VIEW

Figure 6-138. Drawings illustrating the bones of the right elbow joint. The *capitulum of the humerus* articulates with the superior aspect of the *head of the radius*, forming the humeroradial joint, and the *trochlea of the humerus* articulates with the *trochlear notch of the ulna*, forming the humeroulnar joint.

Table 6-15
Muscles Producing Movements of the Shoulder Joint[1]

Movements	Muscles	Reference Tables	Chief Nerve Supply
Flexion	**Pectoralis major**, clavicular head	6–1	Lateral and medial pectoral (C5 and **C6**)
	Deltoid, anterior fibers	6–5	Axillary (C5 and **C6**)
	Coracobrachialis	6–7	Musculocutaneous (C5, **C6**, and C7)
	Biceps brachii (short head)	6–7	Musculocutaneous (C5 and **C6**)
Extension	**Pectoralis major**, sternocostal head	6–1	Lateral and medial pectoral (**C7, C8**, and T1)
	Deltoid, posterior fibers	6–5	Axillary (**C5** and C6)
	Latissimus dorsi	6–4	Thoracodorsal (**C6, C7**, and C8)
	Teres major	6–5	Lower subscapular (**C6** and C7)
	Triceps brachii, long head	6–7	Radial (C5, **C6**, and C7)
Abduction	**Deltoid**	6–5	Axillary (**C5** and C6)
	Supraspinatus	6–5	Suprascapular (C4, **C5**, and C6)
Adduction	**Pectoralis major**, both heads	6–1	Lateral and medial pectoral (**C7, C8**, and T1)
	Latissimus dorsi	6–4	Thoracodorsal (**C6, C7**, and C8)
	Teres major	6–5	Lower subscapular (**C6** and C7)
	Coracobrachialis	6–7	Musculocutaneous (C5, **C6**, and C7)
	Triceps brachii, long head	6–7	Radial (C5, **C6, C7**)
Medial Rotation	**Pectoralis major**, both heads	6–1	Lateral and medial pectoral (**C7, C8**, and T1)
	Latissimus dorsi	6–4	Thoracodorsal (**C6, C7**, and C8)
	Deltoid, anterior fibers	6–5	Axillary (**C5** and C6)
	Subscapularis	6–5	Upper and lower subscapular (C5, **C6**, and C7)
	Teres major	6–5	Lower subscapular (**C6** and C7)
Lateral Rotation	**Deltoid**, posterior fibers	6–5	Axillary (**C5** and C6)
	Teres minor	6–5	
	Infraspinatus	6–5	Suprascapular (**C5** and C6)

[1] The principal muscles producing these movements are printed in **boldface**. The main spinal cord segmental innervation of the nerves also appears in **boldface**.

The synovial capsule of the elbow joint (Figs. 6-139 and 6-140) lines the fibrous capsule and is reflected onto the humerus, lining the coronoid and radial fossae anteriorly, and the olecranon fossa posteriorly (Figs. 6-138 and 6-139).

The synovial capsule of the elbow joint is continued into the proximal radioulnar joint. A redundant fold of the synovial capsule emerges distal to the anular ligament, called the **sacciform recess** (Fig. 6-139), which facilitates rotation of the head of the radius, *e.g.*, during pronation and supination of the hand (Fig. 6-70).

The escape of synovial fluid from the joint cavity (**effusion**) generally occurs posteriorly, as do *dislocations* of the *elbow joint* (Fig. 6-146). The elbow joint is also most easily approached surgically from its posterior aspect (Fig. 6-140).

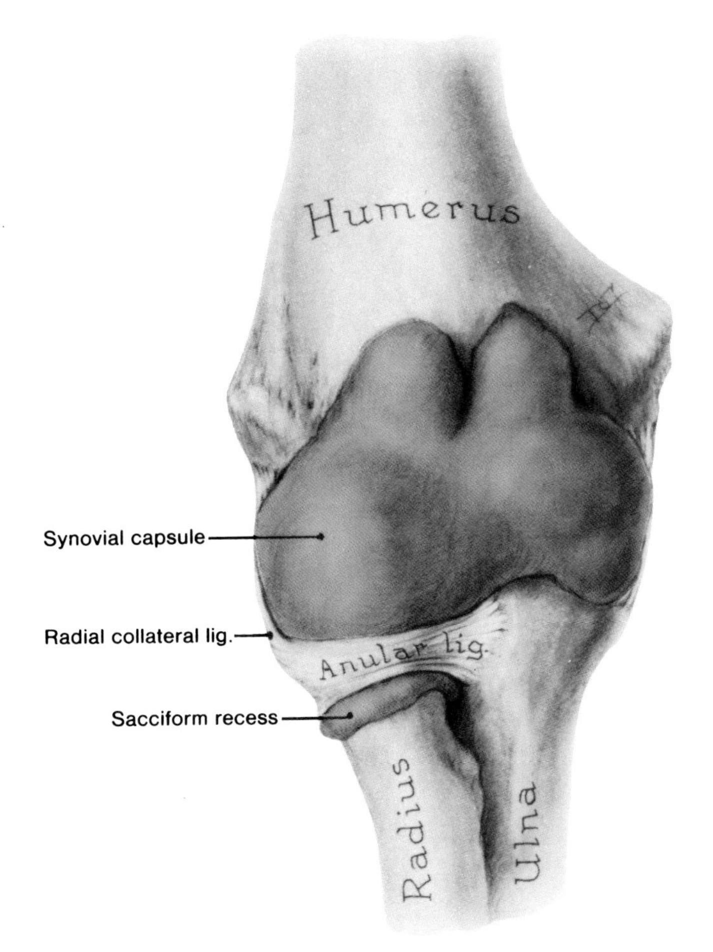

Figure 6-139. Drawing of an anterior view of a dissection of the synovial capsule lining the fibrous capsule of the elbow joint. To display the capsule, the joint cavity was distended with wax and then the fibrous capsule was removed. Note the saccular recess of the joint cavity between the head of the radius and the anular ligament. The joint cavity of the elbow is continuous with the joint cavity of the proximal radioulnar joint. The anular ligament keeps the head of the radius close to the radial notch of the ulna and with it, forms a cup-shaped socket (Fig. 6-148).

STABILITY OF THE ELBOW JOINT

The elbow joint is quite stable because of the hinge-like arrangement formed by the jaw-like trochlear notch of the ulna into which the spool-shaped trochlea of the humerus fits. In addition, the joint is strengthened by very strong ulnar and radial collateral ligaments (Figs. 6-141 and 6-142).

The elbow joint of children is not so stable owing to the late fusion of the epiphyses of the ends of the bones involved in the articulation (humerus, radius, and ulna). For example, the top of the olecranon fuses with the body of the ulna during the 16th to 19th years, and the head of the radius fuses with its body at 15 to 17 years. As a consequence, separation of the epiphyses can occur during a fall on the elbow because the epiphyseal cartilage plate is weaker than the surrounding bone (Fig. 6-145).

BURSAE AROUND THE ELBOW JOINT (Fig. 6-143)

Only a few of the many bursa around the elbow are clinically important. There are two olecranon bursae.

The subcutaneous olecranon bursa (Fig. 6-143) is located in the subcutaneous connective tissue over the olecranon, whereas **the *subtendinous olecranon bursa*** is located between the tendon of the triceps muscle and the olecranon, just proximal to its insertion into the olecranon (Figs. 6-69, 6-76, and 6-138*B*).

The subcutaneous olecranon bursa is exposed to injury during falls on the elbow and to infection from abrasions of the skin covering the olecranon (Fig. 6-73). Repeated excessive fric-

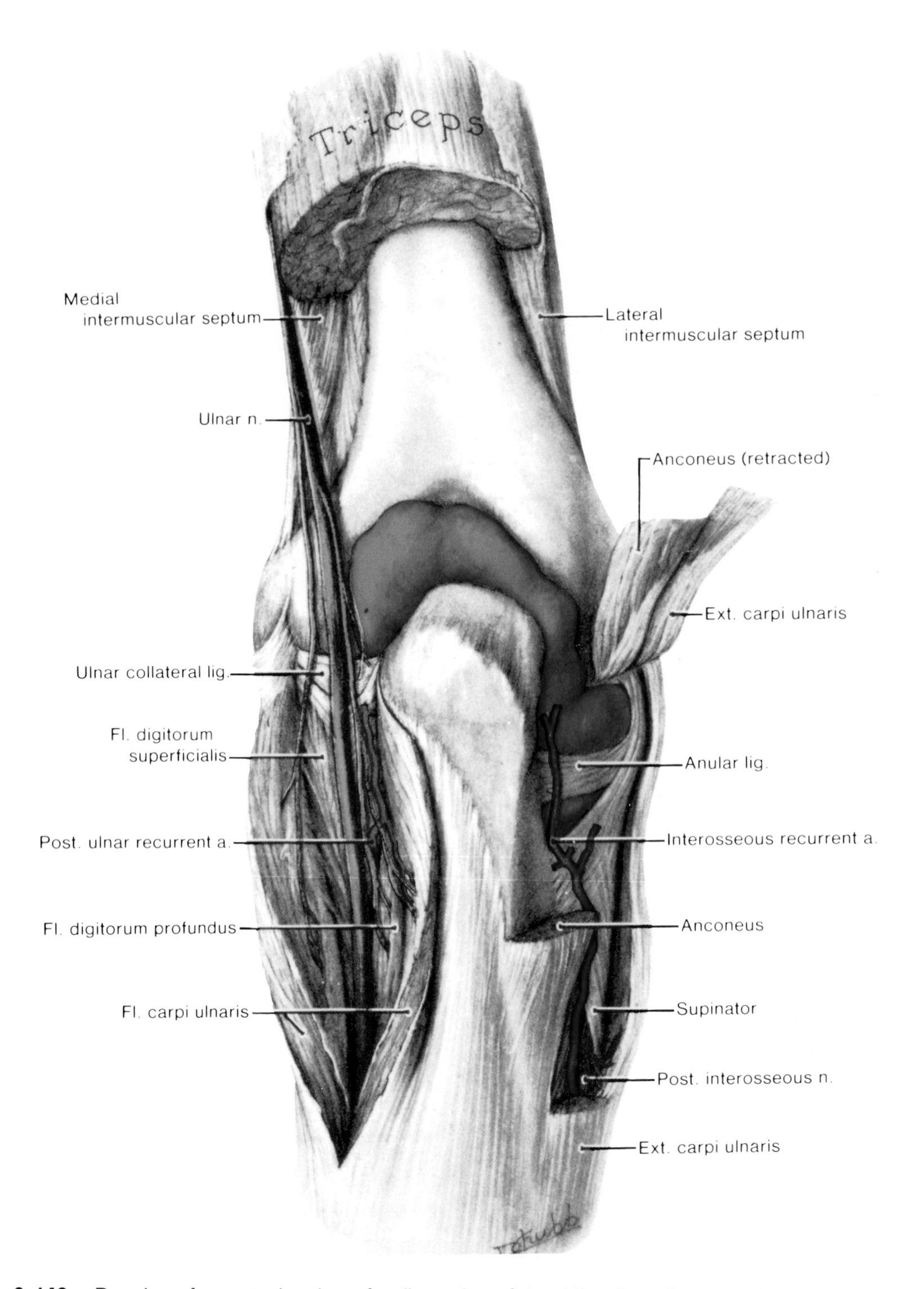

Figure 6-140. Drawing of a *posterior view* of a dissection of the right elbow from which the distal portion of the triceps muscle has been removed. Observe the ulnar nerve descending (1) subfascially within the posterior compartment of the arm, applied to the medial head of the triceps, and posterior to the medial epicondyle; (2) applied to the ulnar collateral ligament of the elbow joint; and (3) between the flexor carpi ulnaris and flexor digitorum profundus muscles. Observe the synovial capsule (*blue*) protruding between the head of the radius and the anular ligament (also see Fig. 6-139).

Table 6-16
Muscles Producing Movements of the Elbow Joint[1]

Movements	Muscles	Reference Tables	Chief Nerve Supply
Flexion	**Brachialis**[2] **Biceps brachii** }	6–7	Musculocutaneous (C5 and **C6**)
	Brachioradialis	6–10	Radial (C5, **C6**, and C7)
	Extensor carpi radialis longus[3]	6–10	Radial (C6 and C7)
	Pronator teres[3] Flexor carpi radialis[3] }	6–8	Median (C6 and **C7**)
Extension	**Triceps brachii** (particularly medial head)	6–7	Radial (C5, **C6**, and **C7**)
	Anconeus	6–10	Radial (C7 and C8) and (T1)
	Brachioradialis[4]	6–10	Radial (C5, **C6**, and C7)

[1] The principal muscles producing these movements are printed in **boldface**.
[2] The main flexor of the forearm at the elbow joint.
[3] Although not normally used to flex the forearm, these muscles can assist with the movement if necessary, *e.g.*, the pronator teres acts as a flexor when flexion is resisted.
[4] During active extension the brachioradialis can assist with this extension of the forearm.

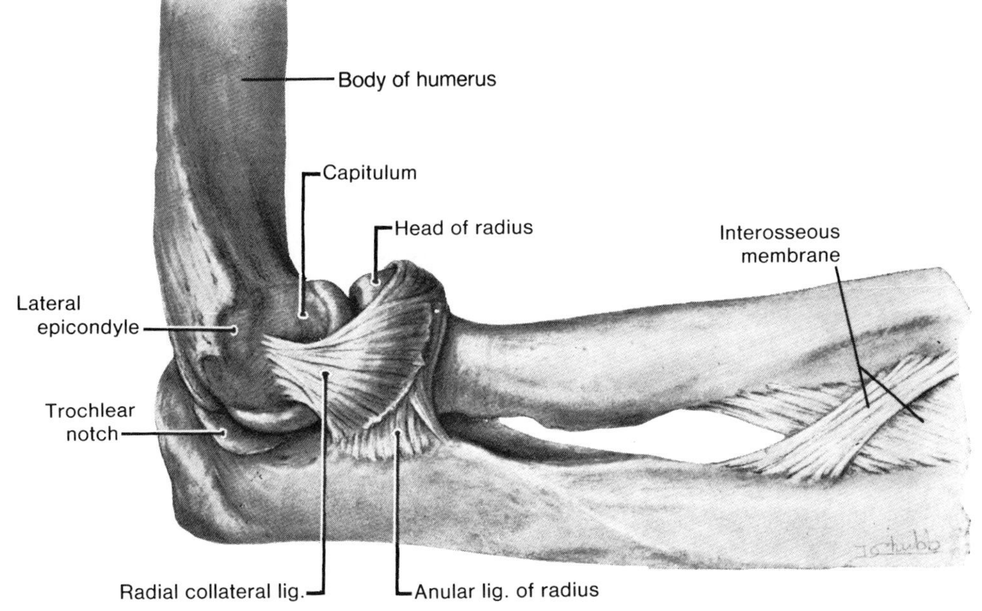

Figure 6-141. Drawing of a lateral view of the bones of the right arm (humerus) and forearm (ulna and radius) in articulation at the elbow joint. Observe that the fan-shaped radial collateral ligament is attached to the anular ligament of the radius. Note that the anular ligament encircles the head of the radius and attaches to the margins of the radial notch of the ulna. For better view of this ligament, see Figures 6-142, 6-147, and 6-148.

tion may cause this bursa to become inflamed, producing a **friction bursitis**, *e.g.*, "student's elbow" (Fig. 6-144).

Inflammation of the subtendinous olecranon bursa is much less common. This type of bursitis may result from excessive friction between the triceps tendon and the olecranon, *e.g.*, resulting from repeated flexion-extension of the forearm

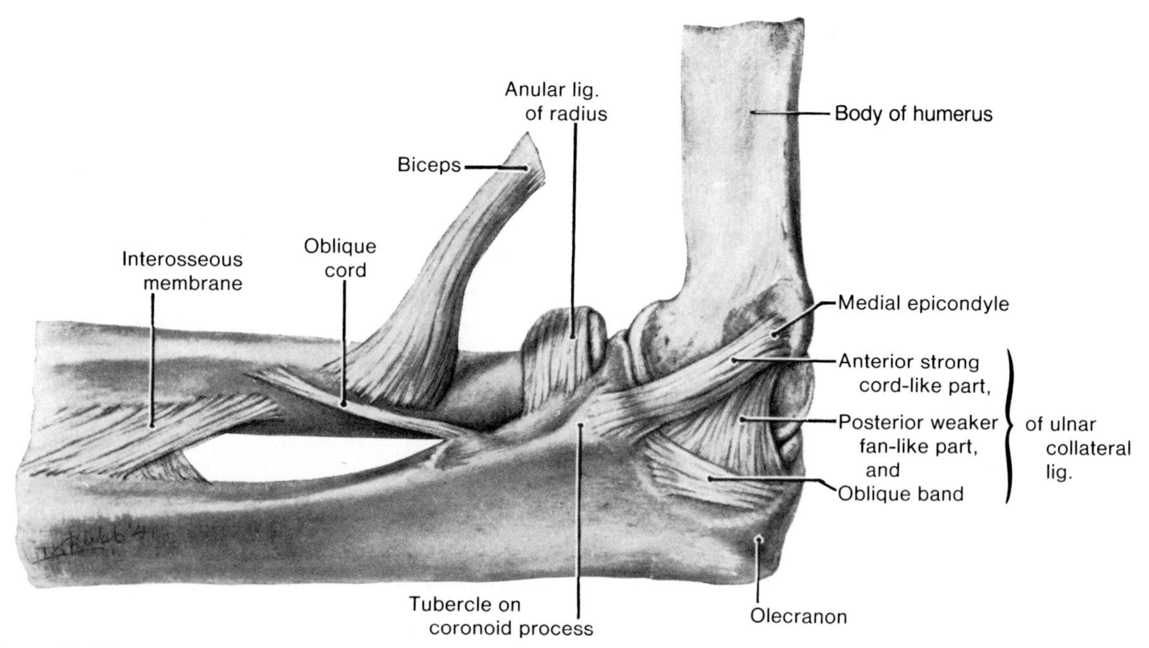

Figure 6-142. Drawing of a medial view of the bones of the right arm and forearm in articulation at the elbow joint. Observe the ulnar collateral ligament and verify that its strong anterior cord-like part becomes taut in extension and that its posterior fan-like part is taut in flexion (as here). The oblique band of fibers deepen the socket for the tróchlea of the humerus. The elbow joint is strong because of its hinge-like ararangement and very strong ulnar and radial collateral ligaments.

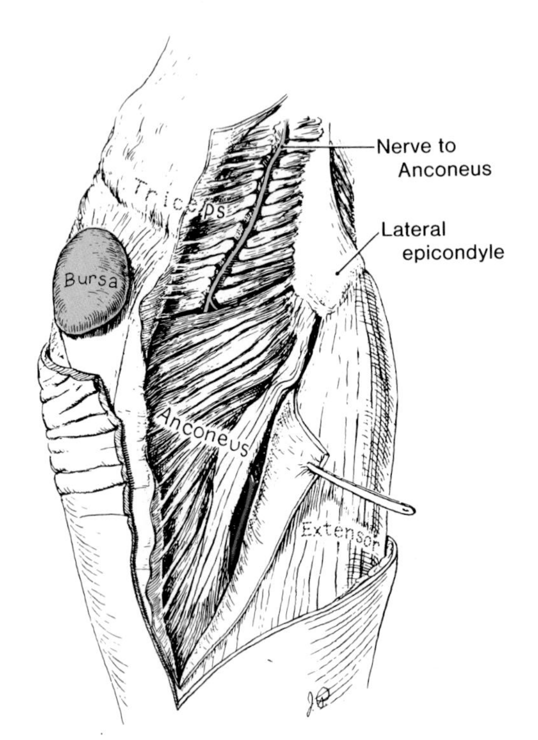

Figure 6-143. Drawing of a dissection of the posterolateral aspect of the right elbow. Observe the subcutaneous olecranon bursa lying on the tendinous expansion of the triceps muscle.

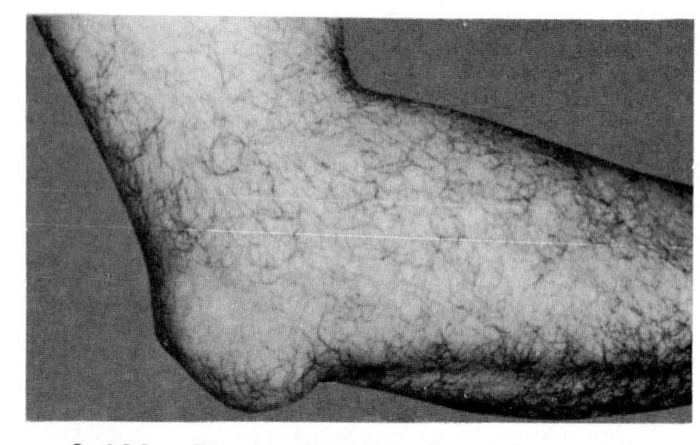

Figure 6-144. Photograph of a lateral view of a student's right elbow exhibiting enlargement of the subcutaneous olecranon bursa illustrated in Figure 6-143. This condition, known as *olecranon bursitis* ("student's elbow) results from repeated friction (*e.g.*, against a desk top), or from a single severe trauma to the olecranon region.

as occurs during certain assembly line jobs. The pain would be most severe during flexion of the forearm because of pressure exerted on the inflamed subtendinous olecranon bursa by the triceps tendon.

The radioulnar bursa lies between the extensor digitorum, the radiohumeral joint, and the supinator muscle. The *bursa* lies posterior to the supinator muscle, lateral to the tendon of the biceps muscle, and medial to the ulna.

Radioulnar bursitis may result from repeated or violent extension of the wrist with the forearm pronated as occurs during the backhand stroke in tennis. It **may be associated with "tennis elbow"** (p. 732). Pain occurs on elbow extension with the forearm pronated.

The bicipitoradial bursa (biceps bursa) lies between the biceps tendon and the anterior part of the tuberosity of the radius (Fig. 6-91).

Bicipitoradial bursitis pain occurs when the elbow joint is flexed and the forearm is supinated. Recall that the biceps is a flexor and supinator of the forearm (Tables 6-7 and 6-17).

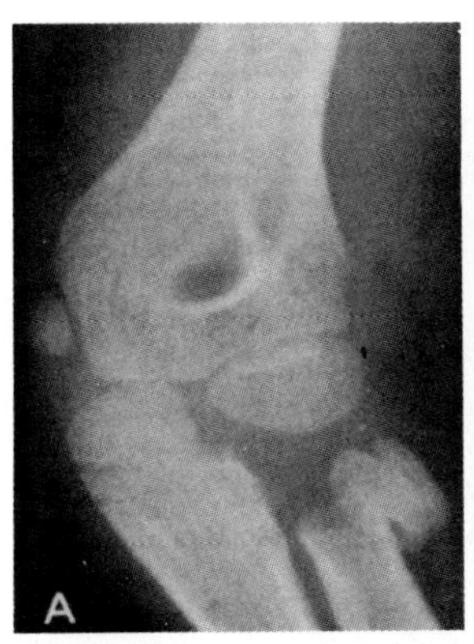

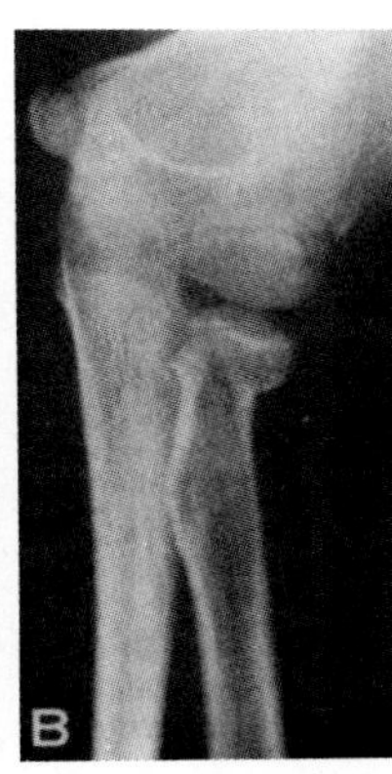

Figure 6-145. *A*, radiograph of the elbow region of a child showing a fracture-separation of the proximal radial epiphysis. Note that the head of the radius is displaced and is not in contact with the capitulum of the humerus. *B*, radiograph taken after reduction of the fracture-dislocation showing the radial head in its normal position. In both radiographs, observe the epiphyses for the distal end of the humerus and the proximal ends of the radius and ulna. Fusion of the head of the radius with the body is complete radiographically at 13 years in females and 16 in males.

BLOOD SUPPLY OF THE ELBOW JOINT
(Figs. 6-63, 6-65, 6-74, and 6-87)

The articular arteries are derived from the **anastomoses around the elbow** which are formed by collateral branches of the brachial and recurrent branches of the ulnar and radial arteries.

NERVE SUPPLY OF THE ELBOW JOINT
(Figs. 6-66, 6-74, 6-87, and 6-91)

The articular nerves are derived mainly from the musculocutaneous and radial nerves, but the ulnar, median, and anterior interosseous nerves may also supply articular branches.

Fracture-separation of the proximal radial epiphysis (Fig. 6-145*A*) can result when a young person falls and exerts compression and abduction forces on the elbow joint. The anatomical basis of the injury is the late fusion (15 to 17 years) of the proximal epiphysis of the head with the body of the radius.

Avulsion of the medial epicondyle of the humerus can also result in children from a fall that causes abduction of the extended elbow joint, *an abnormal movement of this articulation.* The resulting traction on the ulnar collateral ligament (Fig. 6-142) pulls the medial epicondyle distally.

The anatomical basis of avulsion of the medial epicondyle is that the epiphysis for the medial epicondyle may not fuse with the distal end of the humerus until up to the 20th year. Usually fusion is complete radiographically at 14 years in females and 16 in males. A wrong diagnosis of a fracture of the medial epicondyle could be made by a person unaware of this late fusion.

Traction injury of the ulnar nerve is a frequent complication of the abduction type of avulsion of the medial epicondyle. The anatomical basis for this stretching of the ulnar nerve is that it passes posterior to the medial epicondyle before entering the forearm (Figs. 6-63 and 6-91).

Posterior dislocation of the elbow joint may occur when children fall on their hands with their elbows flexed. The distal end of the humerus is driven through the weak anterior portion of the fibrous capsule of the elbow joint as the radius and ulna dislocate posteriorly (Fig. 6-146).

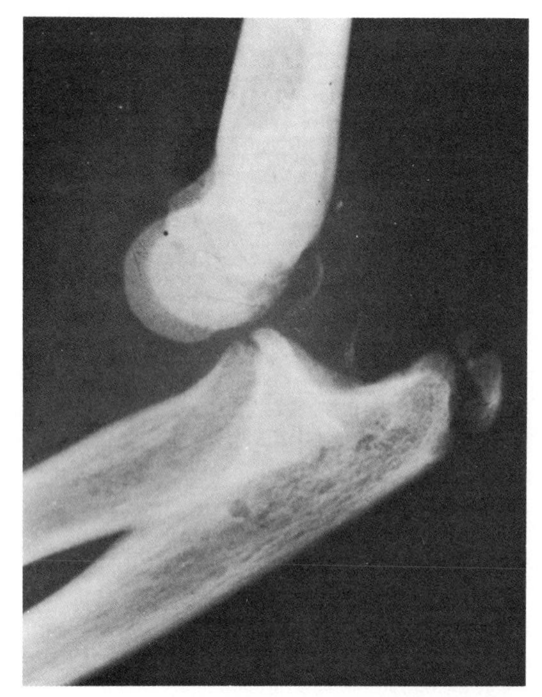

Figure 6-146. Radiograph of the elbow region of a child showing a posterior dislocation of the elbow joint. Note that the distal end of the humerus has been driven anteriorly and is not in articulation with the radius and ulna. Note the *apparently* separated fragment of bone at the proximal end of the olecranon. This is not a fracture; the epiphysis of the proximal end of the ulna has not fused with the olecranon. This usually occurs at 13 years in females and 15 in males.

THE RADIOULNAR JOINTS

The radius and ulna articulate with each other at their proximal and distal ends at synovial joints, called the proximal and distal radioulnar joints (also called the superior and inferior radioulnar joints). These articulations are the **pivot type of synovial joint** which produce pronation and supination of the hand (Fig. 6-70).

The interosseous borders of the radius and ulna are connected by an interosseous membrane (Figs. 6-79 and 6-147). They are also joined by an oblique cord (Fig. 6-142).

The interosseous membrane of the forearm (Figs. 6-79, 6-141, 6-142, and 6-147) is a strong, broad fibrous sheet which stretches between the interosseous borders of the radius and the ulna, commencing 2 to 3 cm distal to the tuberosity of the radius (Figs. 6-141 and 6-147). In addition to providing a flexible and strong attachment between the forearm bones, it provides origins for the deep muscles of the forearm (Table 6-11). A thin fibrous layer, called the ***quadrate ligament***, extends between the radial notch of the ulna (Fig. 6-138*A*) and the medial surface of the neck of the radius. The quadrate ligament covers the synovial membrane and probably supports it.

The oblique cord (Fig. 6-142) is a fibrous band that extends inferolaterally from the lateral border of the tuberosity of the ulna to the radius, just distal to its tuberosity (Figs. 6-138*A* and 6-142). The oblique cord is not always present and is not known to be of much functional significance.

THE PROXIMAL RADIOULNAR JOINT (Figs. 6-139 to 6-141 and 6-147)

This is the ***pivot type of synovial joint*** which allows movement of the radius on the ulna.

The Articular Surfaces of the Proximal Radioulnar Joint (Figs. 6-72, 6-138*A*, 6-139, 6-140, 6-147, and 6-148). The **radial head articulates with the radial notch of the ulna.** The head of the radius is held in position by the strong **anular ligament** (Figs. 6-139 to 6-142), *a U-shaped fibrous collar* which is attached to the anterior and posterior margins of the radial notch (Fig. 6-148).

The Articular Capsule of the Proximal Radioulnar Joint (Figs. 6-139 to 6-142). This articulation is enclosed within the articular capsule of the elbow joint. The **fibrous capsule** enclosing the joint is continuous with the fibrous capsule of the elbow joint.

The **synovial capsule,** lining the fibrous capsule, is an inferior prolongation of the synovial capsule of the elbow joint (Figs. 6-139 and 6-140). The deep surface of the anular ligament is lined with synovial membrane which continues distally as a sac-like structure on the neck of the radius (Figs. 6-139 and 6-140). This arrangement allows the radius to rotate within the anular ligament without tearing the synovial capsule.

The synovial cavities of the elbow and proximal radioulnar joints are in free communication with each other.

Preschool children, especially 1 to 3 year-olds, are particularly vulnerable to an injury usually known as "**pulled elbow**" (Fig. 6-149). Synonyms for this minor injury, known clinically as **subluxation of the head of the radius** (*i.e.*, incomplete dislocation) are: "slipped elbow," and "nursemaid's elbow." The last term is a particularly inappropriate one because it implies that it is the nursemaid that is injured.

The history of these cases is typical (Fig. 6-149). The child is suddenly lifted by the upper limb when the forearm is pronated (*e.g.*, when lifting children into a bus or pulling them away

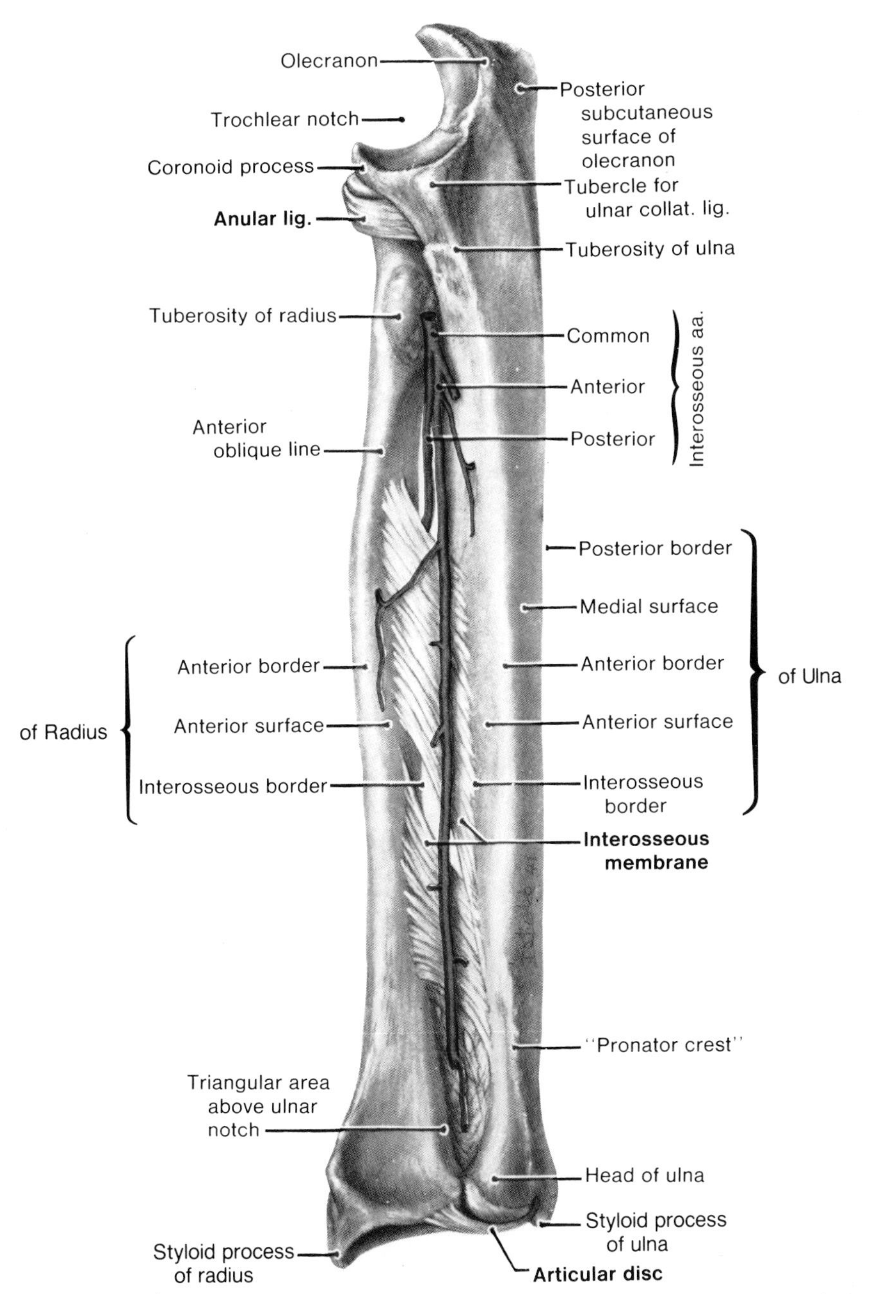

Figure 6-147. Drawing of anterior and lateral views of a dissection of the right radius and ulna, respectively, showing the radioulnar articulations and interosseous arteries. The ligament of the proximal radioulnar joint is the anular ligament; the bond of union at the distal joint is the articular disc. Observe the interosseous membrane and the direction of its fibers which are attached to the interosseous borders of the radius and ulna (Fig. 6-79). Note that the posterior interosseous artery passes posteriorly, superior to the superior border of the interosseous membrane and then runs inferiorly to supply muscles on the posterior surface of the forearm and hand. Observe that the anterior interosseous artery descends on the anterior surface of the interosseous membrane to supply muscles on the anterior surface of the forearm. Note that this artery pierces the interosseous membrane about 5 cm superior to the distal end of the radius.

from danger). The child cries out and refuses to use the limb, which he/she protects by holding it with the elbow flexed and the forearm pronated.

The sudden pulling of the upper limb tears the distal attachment of the anular ligament where it is loosely attached to the neck of the radius. The radial head is pulled distally, partially out of the torn anular ligament. The proximal part of the ligament may become trapped between the head of the radius and the capitulum of the humerus.

THE DISTAL RADIOULNAR JOINT (Figs. 6-72, 6-147, and 6-150 to 6-152)

This is also a **pivot type of synovial joint.** The radius moves around the relatively fixed inferior end of the ulna.

The Articular Surfaces of the Distal Radioulnar Joint (Figs. 6-72, 6-147, 6-150, and 6-151). The rounded side of the **head of the ulna** articulates with the ulnar notch in the **distal end of the radius** (Fig. 6-72*A*).

A fibrocartilaginous **articular disc** binds the ends of the ulna and radius together and is the main uniting structure of the joint (Figs. 6-150 to 6-152). The **base of the articular disc** is attached to the medial edge of the ulnar notch of the radius, and the **apex of the articular disc** is attached to the lateral side of the base of the styloid process of the ulna. The proximal surface of this triangular disc articulates with the distal aspect of the head of the ulna. Hence the joint cavity is L-shaped in a coronal section (Fig. 6-150). The articular disc separates the cavity of the distal radioulnar joint from the cavity of the wrist joint.

The Articular Capsule of the Distal Radioulnar Joint (Figs. 6-150 to 6-152). The **fibrous cap-**

Figure 6-149. Illustration showing how transient subluxation of the head of the radius ("pulled elbow") may occur in a child. This is a common injury in preschool children. The sudden pull on the pronated forearm results in transient subluxation (incomplete dislocation) of the head of the radius. Part of the anular ligament becomes trapped between the radial head and the capitulum of the humerus (Fig. 6-138*A*).

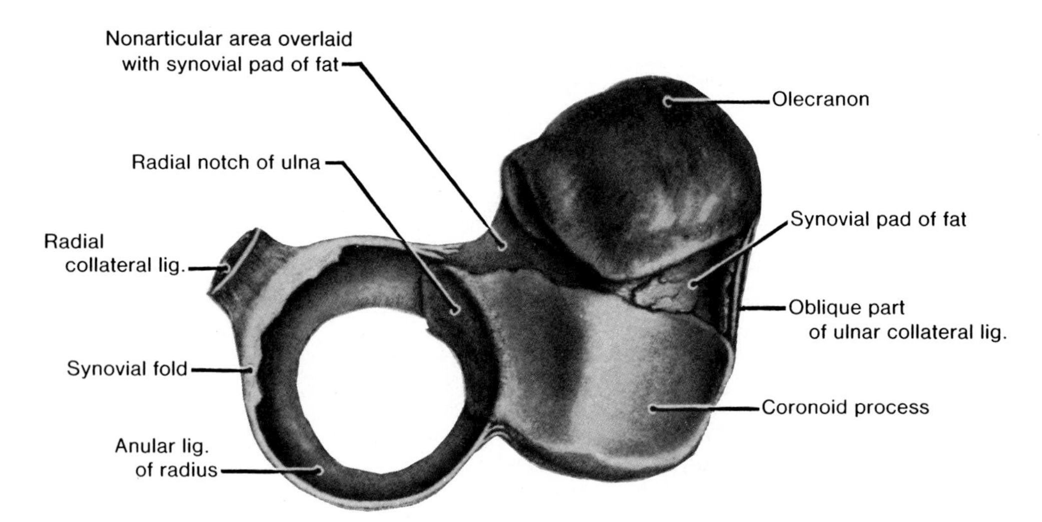

Figure 6-148. Drawing of a superior view of a dissection of the socket for the head of the radius and the trochlea of the humerus. The anular ligament keeps the head of the radius applied to the radial notch of the ulna. Note that the anular ligament and the radial notch of the ulna form a cup-shaped osseofibrous ring that usually prevents the radial head from being pulled distally. Note that the anular (L. ring) ligament is bound to the humerus by the radial collateral ligament of the elbow (also see Figs. 6-139 to 6-142).

sule encloses the joint. It is formed by relatively weak transverse bands which extend from the radius to the ulna across the anterior and posterior surfaces of the joint.

The **synovial capsule** (Figs. 6-150 and 6-151) lines the fibrous capsule and the proximal surface of the articular disc. The synovial capsule extends proximally a short distance between the radius and ulna as the **sacciform recess** (Figs. 6-150 and 6-151). This redundancy of the synovial capsule accommodates the twisting of the capsule that occurs when the distal end of the radius travels around the relatively fixed distal end of the ulna during pronation of the forearm (Fig. 6-70*B*).

MOVEMENTS OF THE RADIOULNAR JOINTS (Fig. 6-70 and Table 6-17)

Movements at these joints make pronation and supination of the hand possible. **Pronation** is rotation so as to turn the palm posteriorly, or inferiorly when the forearm is flexed. **Supination** carries the palm anteriorly, or superiorly when the forearm is flexed.

The axis for these movements passes proximally through the center of the head of the radius and distally through the site of attachment of the apex of

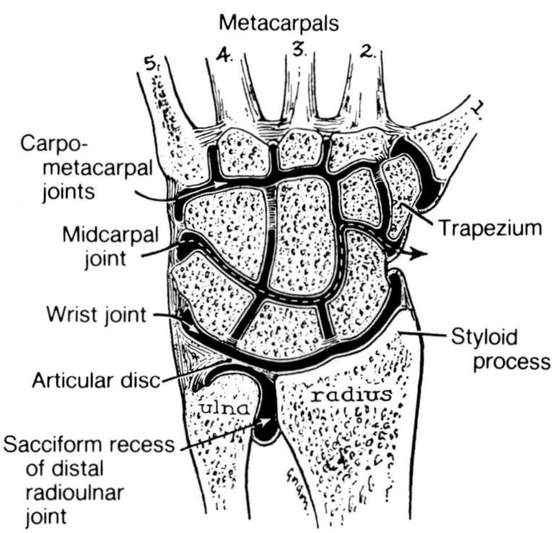

Figure 6-150. Drawing of a coronal section of the distal end of the right forearm and hand, showing the distal radioulnar, wrist, intercarpal, carpometacarpal, and intermetacarpal joints. Note that the cavities of the distal radioulnar and wrist joints are separated by the articular disc of the distal radioulnar joint. Observe the important carpometacarpal joint of the thumb, where the first metacarpal bone articulates with the trapezium at a saddle type of synovial joint. This articulation allows the thumb a considerable range of movement.

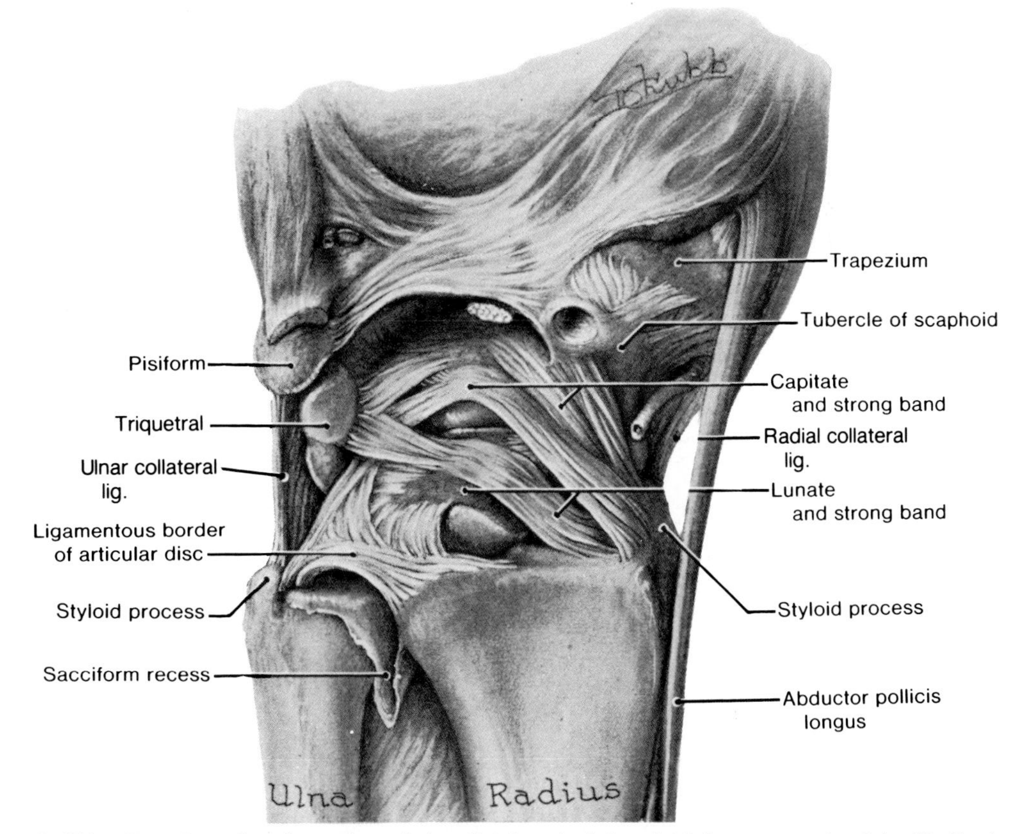

Figure 6-151. Drawing of a dissection of the distal end of the right forearm and wrist with the hand forcibly extended. This dissection gives an anterior view of the ligaments of the distal radioulnar, radiocarpal, and intercarpal joints. Observe the sacciform recess of the synovial capsule of the distal radioulnar joint and the ligamentous anterior border of the triangular articular disc.

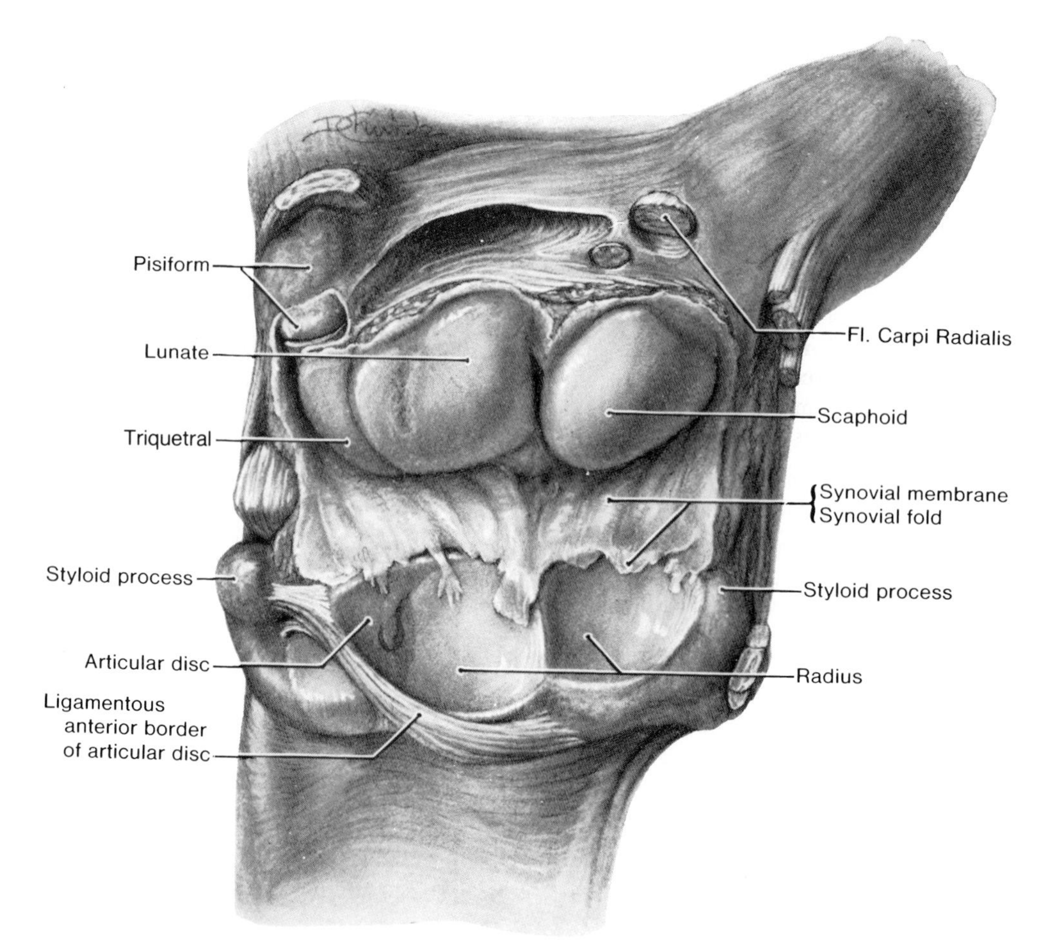

Figure 6-152. Drawing of a dissection of the right wrist joint opened anteriorly. Observe the nearly equal proximal articular surfaces of the scaphoid and lunate bones. Note that the lunate articulates with the radius and the articular disc. Only during adduction of the wrist does the triquetral bone come into articulation with the disc. Observe that the pisotriquetral joint communicates with the wrist joint.

Table 6-17
Muscles Producing Movements of the Radioulnar Joints[1]

Movements	Muscles	Reference Tables	Chief Nerve Supply
Supination	**Supinator**[2]	6–11	Radial (C5 and **C6**)
	Biceps Brachii	6–7	Musculocutaneous (C5 and **C6**)
Pronation	**Pronator quadratus**[3]	6–9	Ant. interosseous (**C8** and T1)
	Pronator teres, Flexor carpi radialis	6–8	Median (C6 and **C7**)
	Anconeus	6–10	Radial (C7, C8, and T1)

[1] The principal muscles producing these movements are printed in **boldface**.
[2] The supinator is the chief supinator of the forearm and hand (Fig. 6-70*A*). It is aided by the biceps during fast or resisted movement.
[3] The pronator quadratus is the chief pronator of the forearm and hand (Fig. 6-70*B*). It is assisted by the pronator teres during fast or resisted movement.

the articular disc to the head of the ulna (Fig. 6-147). *During pronation and supination it is mainly the radius that rotates.* Its head rotates within the cup-shaped ring formed by the anular ligament and the radial notch on the ulna (Figs. 6-147 and 6-148). Distally the end of the radius rotates around the head of the ulna (Figs. 6-70 and 6-147).

Supination is the movement used by right-handed persons to drive a screw. It is more powerful than pronation, the action used to remove a screw.

Pronation and supination are special movements occurring at the proximal and distal radioulnar joints. These movements can occur in different ways, but the usual way is by rotating the distal end of the radius over the distal end of the ulna, while at the same time abducting the ulna with the anconeus muscle (Fig. 6-63 and Table 6-10) so that the ulna takes the place of the distal end of the radius.

The principal supinator muscles of the forearm are the *biceps brachii and supinator* (Figs. 6-66, 6-74, and Tables 6-7 and 6-11). The biceps is the more powerful muscle and is particularly important when force is required, especially when the forearm is flexed (*e.g.*, driving a large screw into hard wood).

The principal pronator muscles of the forearm are the *pronator teres and pronator quadratus* (Figs. 6-74, 6-80, 6-89, 6-91, 6-118, and Tables 6-8 and 6-9). The action of the pronator quadratus is reinforced by the pronator teres during rapid and forceful pronation.

BLOOD SUPPLY OF THE RADIOULNAR JOINTS (Figs. 6-65, 6-91, 6-97, and 6-147)

The articular arteries supplying the **proximal radioulnar joint** are derived from the *anastomoses around the elbow region* (Fig. 6-65), whereas those supplying the **distal radioulnar joint** are derived from the *anterior and posterior interosseous arteries.*

NERVE SUPPLY OF THE RADIOULNAR JOINTS (Figs. 6-74, 6-85, 6-87, and 6-91)

The articular nerves to the *proximal radioulnar joint* are derived mainly from the **musculocutaneous, median, and radial nerves.** The articular nerves to the *distal radioulnar joint* are derived from the **anterior and posterior interosseous nerves.**

THE WRIST JOINT

The wrist or radiocarpal joint is between the distal end of the radius and the carpus. It is a **condyloid or ellipsoid type of synovial joint.**

THE ARTICULAR SURFACES OF THE WRIST JOINT (Figs. 6-109 and 6-150 to 6-153)

The distal end of the radius and the articular disc of the distal radioulnar joint articulate with the proximal row of carpal bones (scaphoid, lunate, and triquetral, Fig. 6-77). The convex surfaces formed by the carpal bones fit into the concave surfaces of the distal end of the radius and the articular disc (Fig. 6-152).

MOVEMENTS OF THE WRIST JOINT (Table 6-18)

The following movements are possible: **adduction, abduction, flexion, extension, and circumduction.**

Rotation of the wrist joint is impossible because the articular surfaces are ellipsoid in shape; however pronation and supination of the hand compensate for the absence of this movement (Fig. 6-70).

THE ARTICULAR CAPSULE OF THE WRIST JOINT (Figs. 6-150 and 6-151)

The fibrous capsule encloses the joint and is attached proximally to the distal ends of the radius and ulna, and distally to the proximal row of carpal bones. It is strengthened by **dorsal and palmar radiocarpal ligaments** (Fig. 6-151); these run obliquely distally and medially from the radius. The fibrous capsule is also strengthened by the *radial and ulnar collateral ligaments* (Fig. 6-151).

The synovial capsule lines the fibrous capsule and is attached to the margins of the articular surfaces of the wrist joint. It presents numerous folds, especially dorsally.

BLOOD SUPPLY OF THE WRIST JOINT (Figs. 6-96, 6-97, and 6-121 to 6-123)

The articular arteries are derived from the dorsal and palmar carpal arches.

NERVE SUPPLY OF THE WRIST JOINT (Figs. 6-87, 6-91, 6-97, and 6-113)

The articular nerves are derived from the anterior interosseous branch of the **median nerve**, the posterior interosseous branch of the **radial nerve,** and the dorsal and deep branches of the **ulnar nerve.** This is a good example of **Hilton's law**, *i.e.*, nerves that supply muscles acting on a joint usually send sensory fibers to it (Table 6-18).

Normally the cavity of the wrist joint does not communicate with either the distal radioulnar or the midcarpal joint (Fig. 6-150); however perforation of the articular disc of the distal radioulnar joint often occurs with age resulting in communication between the two joints (Fig. 6-152).

Table 6-18
Muscles Producing Movements of the Wrist and Midcarpal Joints[1]

Movements	Muscles	Reference Tables	Chief Nerve Supply
Flexion	**Flexor carpi radialis**	6–8	Median (C6 and **C7**)
	Flexor carpi ulnaris	6–8	Ulnar (C7 and **C8**)
	Palmaris longus	6–8	Median (C7 and C8)
	Abductor pollicis longus	6–11	Posterior and interosseous (C7 and **C8**)
	Flexor digitorum profundus	6–9	Medial part, ulnar Lateral part, median (**C8** and T1)
	Flexor digitorum superficialis	6–8	Median (C7, **C8**, and T1)
	Flexor pollicis longus	6–9	Anterior interosseous (**C8** and T1)
Extension	**Extensor carpi radialis longus**	6–10	Radial (C6 and C7)
	Extensor carpi radialis brevis	6–10	Radial (**C6** and C7)
	Extensor carpi ulnaris		
	Extensor digitorum	6–10	Posterior interosseous (C7 and C8)
	Extensor pollicis longus	6–11	
	Extensor indicis		
	Extensor digiti minimi	6–10	Posterior interosseous (**C7** and C8)
Abduction	**Extensor carpi radialis longus**	6–10	Radial (C6 and C7)
	Extensor carpi radialis brevis	6–10	Radial (**C7** and C8)
	Flexor carpi radialis	6–8	Median (C6 and **C7**)
	Abductor pollicis longus		
	Extensor pollicis brevis	6–11	Posterior interosseous (C7 and C8)
	Extensor pollicis longus		
Adduction	**Flexor carpi ulnaris**	6–8	Ulnar (C7 and **C8**)
	Extensor carpi ulnaris	6–10	Posterior interosseous (C7 and **C8**)

[1] The principal muscles producing these movements are printed in **boldface**. The wrist joint is mainly concerned with extension and adduction of the hand, whereas the midcarpal joint is largely concerned with flexion and abduction of the hand. However movements at the wrist and midcarpal joints are best considered together because these joints form parts of the same mechanism and are acted on by the same muscle groups.

***Wrist fractures** (e.g., Colles' fracture) involving the distal end of the radius, are the most common type of fracture in persons over 50 years* (Fig. 6-161*A*). This fracture commonly results when the person slips or trips and, in attempting to break the fall, lands on the outstretched hand with the forearm pronated.

There is a complete transverse fracture of the

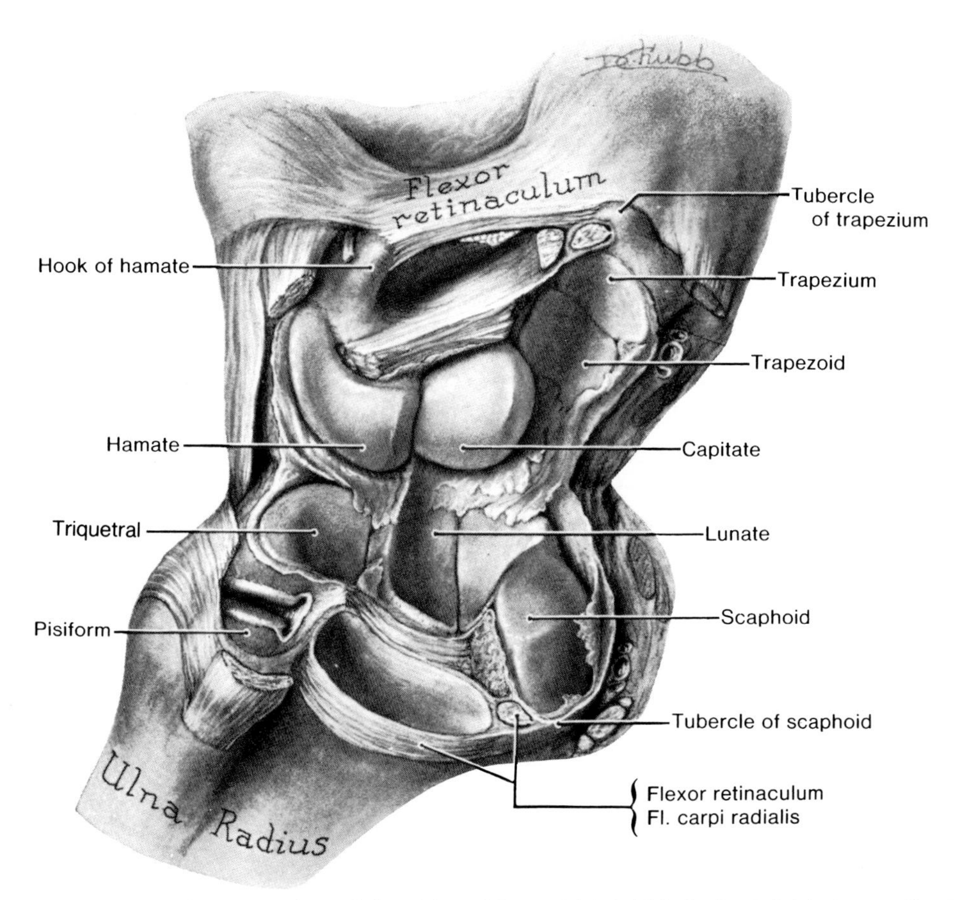

Figure 6-153. Drawing of a dissection of the right midcarpal joint. This is the joint between the two rows of carpal bones (Figs. 6-77 and 6-150). The flexor retinaculum has been divided and the midcarpal joint has been exposed by extending the wrist. Observe the surfaces of the opposed bones: the trapezium and trapezoid together presenting a concave, oval surface to the scaphoid; the capitate and hamate together presenting a convex surface to the scaphoid, lunate, and triquetral. The midcarpal joint is surrounded by a fibrous capsule made up, anteriorly and posteriorly, of irregular bands which run between the two rows of carpal bones. These bands constitute the palmar and dorsal intercarpal ligaments (Fig. 6-151). The principal movements between the carpal bones are those that occur at the midcarpal joint (Table 6-18), shown here and in Figure 6-150.

distal 2-3 cm of the radius, and the fragment is displaced proximally causing shortening of the radius (Fig. 6-161*C*). The fragment is usually tilted posteriorly, producing a characteristic hump described as the silver fork or **"dinner fork" deformity** (Fig. 6-161*A*).

THE INTERCARPAL JOINTS

These are the *plane type of synovial joint which permit gliding and sliding movements.*

THE ARTICULAR SURFACES OF THE INTERCARPAL JOINTS (Figs. 6-150 and 6-153)

These joints are between the carpal bones. **The midcarpal joint** is between the proximal and distal rows of carpal bones (Fig. 6-77).

The pisiform bone rests on the palmar surface of the triquetral bone, forming a separate, small synovial joint with it, called the *pisotriquetral joint.*

Osteoarthritis of the pisotriquetral joint is particularly disabling to typists and pianists.

MOVEMENTS OF THE INTERCARPAL JOINTS (Table 6-18)

Movements of the intercarpal joints increase the range of movements at the wrist joint. Movement of the head of the capitate in its socket and the gliding movement of the bones on each side of it result in considerable flexion of the hand. The midcarpal joint is largely concerned with flexion and abduction of the hand. It also increases the range of abduction of the hand.

Extension of the wrist joint and flexion of the intercarpal joints improve the grasp of the hand.

THE ARTICULAR CAPSULE OF THE INTERCARPAL JOINTS (Figs. 6-150, 6-151, and 6-153)

The fibrous capsule encloses these joints and helps to unite the articulating carpal bones. The bones of each row are connected to each other by dorsal, palmar, and interosseous ligaments. The two rows are connected by dorsal, palmar, ulnar, and radial ligaments (Fig. 6-151).

The synovial capsule lines the fibrous capsule and is attached to the margins of the articular surfaces of the carpal bones. In Figure 6-150 observe that the cavity of the midcarpal joint is part of the general joint cavity, extending between the bones of each row. The common joint space includes the carpometacarpal and intermetacarpal joints.

BLOOD VESSELS OF THE INTERCARPAL JOINTS (Figs. 6-87, 6-95, and 6-121 to 6-123)

The articular arteries are derived from the palmar and dorsal carpal arches.

NERVE SUPPLY OF THE INTERCARPAL JOINTS (Figs. 6-91, 6-97, and 6-116)

The articular nerves are derived from the *anterior interosseous nerve* of the **median,** the *posterior interosseous nerve* of the **radial,** and the dorsal and deep branches of the **ulnar nerve.**

THE CARPOMETACARPAL AND INTERMETACARPAL JOINTS

These are plane synovial joints that permit a small amount of gliding movement. They share a common joint cavity with the intercarpal joints (Fig. 6-150). The articulating bones are united by dorsal, palmar, and interosseous ligaments.

THE CARPOMETACARPAL JOINT OF THE THUMB (Fig. 6-150 and Table 6-21)

This articulation is a separate *saddle type of synovial joint.*

Articular Surfaces of the Carpometacarpal Joint of the Thumb (Figs. 6-77*B* and 6-150). The trapezium articulates with the saddle-shaped base of the first metacarpal bone.

Movements of the Carpometacarpal Joint of the Thumb (Figs. 6-115, 6-119, and Table 6-21). This joint permits angular movements in any plane and a restricted amount of axial rotation. Only ball and socket joints are more mobile.

The following thumb movements are possible: **flexion, extension, abduction, adduction, and opposition.** The functional importance of the thumb lies in its ability to be opposed to the fingers (Fig. 6-115).

The Articular Capsule of the Carpometacarpal Joint of the Thumb (Fig. 6-150). The *fibrous capsule encloses the joint and is attached to the margins of the articular surfaces.* The looseness of its capsule facilitates its important movement of opposition.

The **synovial capsule** lines the fibrous capsule and forms a separate joint cavity from the rest of the carpus (Fig. 6-150).

Blood Supply of the Carpometacarpal Joint of the Thumb (Figs. 6-96 and 6-122). The articular arteries are derived from the dorsal and palmar metacarpal arteries and from the dorsal carpal and deep palmar arches. These vessels are branches of the ulnar and radial arteries.

Nerve Supply of the Carpometacarpal Joint of the Thumb (Figs. 6-91, 6-97, and 6-116). The articular nerves are derived from the *anterior interosseous nerve* of the **median,** the *posterior interosseous nerve* of the **radial,** and the dorsal and deep branches of the **ulnar nerve.**

THE METACARPOPHALANGEAL JOINTS

These articulations are the *condyloid (knuckle-like) type of synovial joint* that allow movement in two directions.

ARTICULAR SURFACES OF THE METACARPOPHALANGEAL JOINTS (Figs. 6-150 and 6-154)

The heads of the metacarpal bones articulate with the bases of the proximal phalanges. The unique feature of their bony surfaces is that they both have oval articular surfaces.

MOVEMENTS OF THE METACARPOPHALANGEAL JOINTS (Fig. 6-154 and Tables 6-19 and 6-21)

The following movements occur at these articulations: flexion, extension, abduction, adduction, and circumduction.

THE ARTICULAR CAPSULE OF THE METACARPOPHALANGEAL JOINTS (Figs. 6-150 and 6-154)

A fibrous capsule encloses each joint. They are strengthened on each side by a triangular **collateral ligament.** It extends from the sides of the head of the proximal bone to the sides of the base of the distal bone.

The **palmar ligaments** are strong thick plates that are firmly attached to the phalanx and loosely at-

Table 6-19
Muscles Producing Movements of the Fingers at the Metacarpophalangeal joints[1]

Movements	Muscles	Reference Tables	Chief Nerve Supply
Flexion	**Flexor digitorum profundus**	6–9	Medial part, ulnar Lateral part, median (**C8** and T1)
	Flexor digitorum superficialis	6–8	Median (C7, **C8**, and T1)
	Flexor digiti minimi Abductor digiti minimi	6–13	Ulnar (C8 and **T1**)
	Interossei	6–14	Ulnar (C8 and T1)
	Lumbricals	6–14	Median (1st and 2nd) Ulnar (3rd and 4th) (C8 and T1)
Extension	**Extensor digitorium** (middle and ring fingers) **Extensor digiti minimi**	6–10	Posterior interosseous (**C7** and C8)
	Extensor indicis	6–11	Posterior interosseous (C7 and **C8**)
Abduction	**Dorsal interossei**[2]	6–14	Ulnar (C8 and T1)
	Extensor digitorum **Extensor digiti minimi**	6–10	Posterior interosseous (**C7** and C8)
	Extensor indicis	6–11	Posterior interosseous (C7 and **C8**)
	Abductor digiti minimi	6–13	Ulnar (C8 and T1)
Adduction	**Palmar interossei**[2]	6–14	Ulnar (C8 and T1)
	Flexor digitorum superficialis	6–8	Median (C7, **C8**, and T1)
	Flexor digitorum profundus	6–9	Medial part, ulnar Lateral part, median (**C8** and T1)

[1] The principal muscles producing the movements are printed in **boldface**. The main spinal cord segmental innervation of the nerves also appears in **boldface**.
[2] In the extended fingers.

tached to the metacarpal (Fig. 6-154). The palmar ligaments of the second to fifth joints are united by **deep transverse metacarpal ligaments** which hold the heads of the metacarpals together (Fig. 6-118).

The synovial capsule lines the fibrous capsule of each joint and is attached to the margins of the articular surfaces.

BLOOD SUPPLY OF THE METACARPOPHALANGEAL JOINTS (Figs. 6-110 and 6-121 to 6-123)

The articular arteries are branches of the digital arteries arising from the superficial palmar arch.

NERVE SUPPLY OF THE METACARPOPHALANGEAL JOINTS (Figs. 6-110 and 6-116)

The articular nerves are derived from the digital nerves which arise from the *ulnar and median nerves.*

THE INTERPHALANGEAL JOINTS

These articulations are uniaxial **hinge joints** which permit only flexion and extension (Fig. 6-154 and Table 6-20). These joints join the head of one phalanx with the base of the more distal one. They are structurally similar to the metacarpophalangeal joints and are reinforced dorsally by the **extensor expansions** of the fingers (Fig. 6-117).

The articular arteries and nerves are derived from the adjacent digital arteries and nerves (Figs. 6-69, 6-97, 6-110, 6-113, 6-116, and 6-121 to 6-123).

Sudden tension on a long extensor tendon which is inserted into a phalanx may avulse part of its insertion (Fig. 6-117). The most common

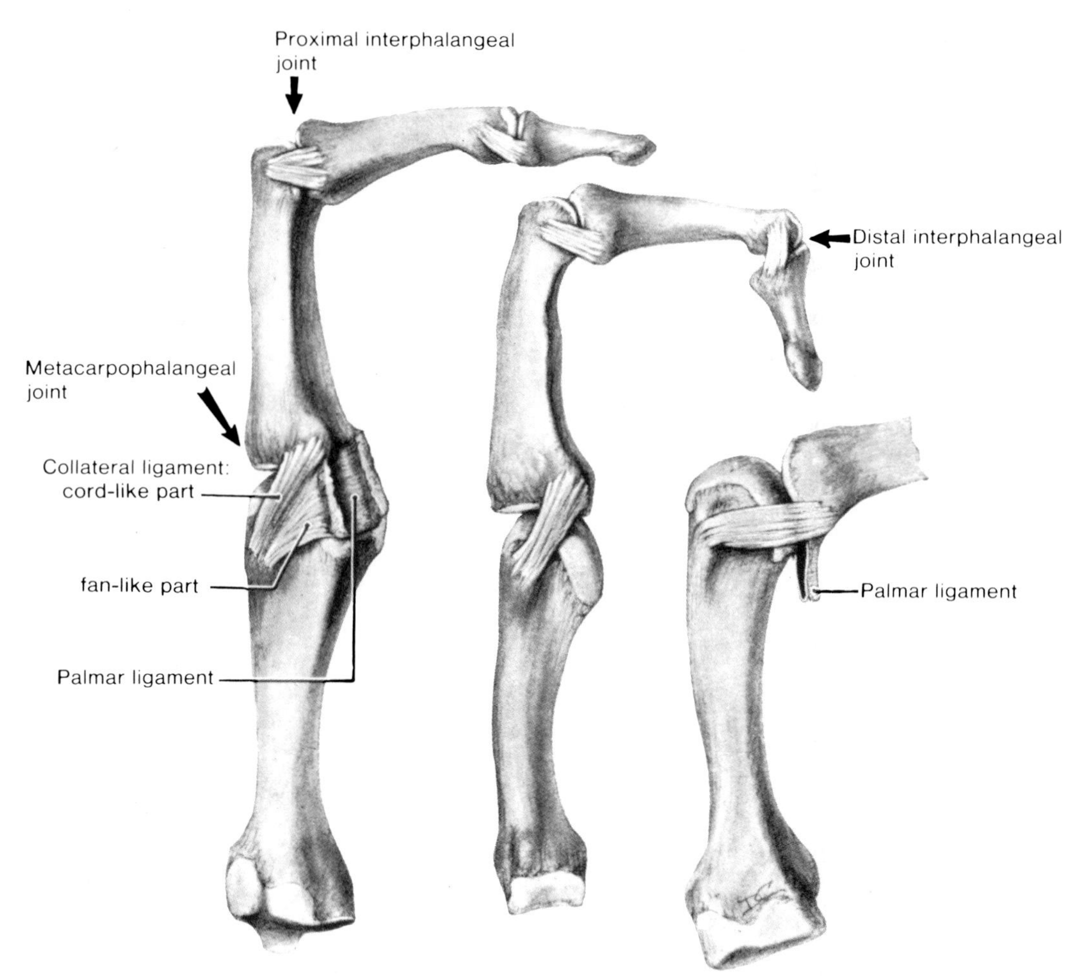

Figure 6-154. Drawings of dissections of the metacarpophalangeal and interphalangeal joints. Observe the palmar ligament, the fibrocartilaginous plate hanging from the base of the proximal phalanx. It is fixed to the head of the metacarpal by the weaker, fan-like part of the collateral ligament. The palmar ligament moves like a visor across the head of the metacarpal. The extremely strong, cord-like parts of the collateral ligaments of this joint are attached eccentrically to the metacarpal heads. These ligaments are slack during extension and taut during flexion; hence the fingers cannot be abducted unless the hand is open. Observe that the interphalangeal joints have corresponding ligaments and that the distal ends of the first and second phalanges are flattened anteroposteriorly.

Table 6-20
Muscles Producing Movements at the Interphalangeal Joints of the Fingers

Movements	Muscles	Reference Tables	Chief Nerve Supply
Flexion	**Flexor digitorum profoundus** (proximal and distal joints)	6–9	Medial part, ulnar (C7 and **C8**) Lateral part, median (**C8** and T1)
	Flexor digitorum superficialis (proximal joints only)	6–8	Median (C7, **C8**, and T1)
Extension	**Extensor digitorum** } **Extensor digiti minimi** }	6–10	Posterior interosseous (**C7** and C8)
	Extensor indicis	6–11	Posterior interosseous (C7 and **C8**)
	Lumbricals	6–14	Median (1st and 2nd) Ulnar (3rd and 4th)
	Interossei	6–14	Ulnar (C8 and T1)

Table 6-21
Muscles Producing Movements of the Thumb[1]

Movements	Muscles	Reference Tables	Chief Nerve Supply
1. *At the Carpometacarpal Joint*			
Flexion	**Flexor pollicis brevis**	6–12	Median (C8 and **T1**)
	Flexor pollicis longus	6–9	Anterior interosseous (**C8** and T1)
	Opponens pollicis	6–12	Median (**C8** and T1)
Extension	**Extensor pollicis longus** **Extensor pollicis brevis** **Abductor pollicis longus**	6–11	Posterior interosseous (C7 and **C8**)
Abduction	**Abductor pollicis longus**	6–11	Posterior interosseous (C7 and **C8**)
	Abductor pollicis brevis	6–12	Median (C8 and **T1**)
Adduction	**Adductor pollicis**	6–12	Ulnar (C8 and **T1**)
	1st dorsal interosseous	6–14	Ulnar (C8 and T1)
	Extensor pollicis longus	6–11	Posterior interosseous (C7 and **C8**)
	Flexor pollicis longus	6–9	Anterior interosseous (**C8** and T1)
Opposition	**Opponens pollicis** **Abductor pollicis brevis** **Flexor pollicis brevis**	6–12	Median (C8 and **T1**)
	Flexor pollicis longus	6–9	Anterior interosseous (**C8** and T1)
	Adductor pollicis	6–12	Ulnar (C8 and **T1**)
2. *At the Metacarpophalangeal Joint*			
Flexion	**Flexor pollicis longus**	6–9	Anterior interosseous (**C8** and T1)
	Flexor pollicis brevis	6–12	Median (**C8** and T1)
	1st palmar interosseous	6–14	Ulnar (C8 and T1)
	Abductor pollicis brevis	6–12	Median (C8 and **T1**)
Extension	**Extensor pollicis longus** **Extensor pollicis brevis**	6–11	Posterior interosseous (C7 and **C8**)
Abduction	**Abductor pollicis brevis**	6–12	Median (C8 and **T1**)
Adduction	**Adductor pollicis**	6–12	Ulnar (C8 and **T1**)
	1st palmar interosseous	6–14	Ulnar (C8 and T1)
3. *At the Interphalangeal Joint*			
Flexion	**Flexor pollicis longus**	6–9	Anterior interosseous (**C8** and T1)
Extension	**Extensor pollicis longus**	6–11	Posterior interosseous (C7 and **C8**)
	Abductor pollicis brevis	6–12	Median (C8 and T1)
	Adductor pollicis	6–12	Ulnar (C8 and **T1**)

[1] The principal muscles producing the movements are printed in **boldface**. The main spinal cord segmental innervation of the nerves also appears in **boldface**.

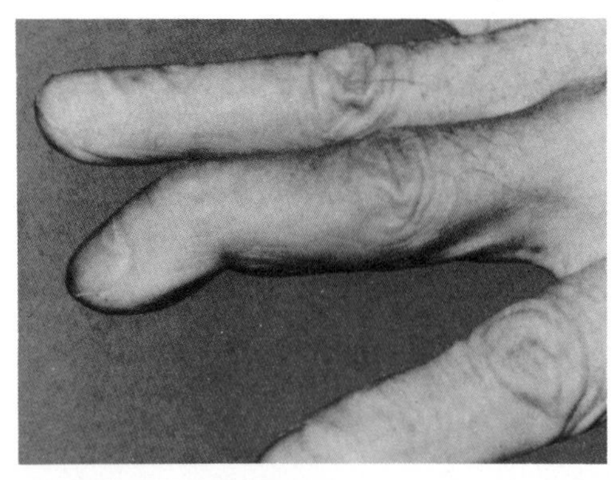

Figure 6-155. Photograph of a mallet or ball finger resulting from avulsion of the attachment of the long tendon of the extensor digitorum from the base of the distal phalanx of the middle finger (Fig. 6-102).

injury is called **mallet finger** (also known as baseball finger or cricket finger), resulting from the distal interphalangeal joints suddenly being forced into extreme flexion (*i.e.*, **hyperflexion**). This avulses the insertion of the terminal tendon into the base of the distal phalanx (Fig. 6-102). As a result the patient is unable to extend the distal interphalangeal joint (Fig. 6-155).

PRESENTATION OF PATIENT ORIENTED PROBLEMS

Case 6-1

A 20-year-old man complained that he was unable to raise his right upper limb. He held it limp at his side with the hand pronated like a waiter hinting for a tip (Fig. 6-32).

During questioning he stated that he had been *thrown from his motorcycle about 2 weeks previously and that he had hit his shoulder against a tree.* He also recalled that his neck felt sore shortly after the accident.

On examination it was found that **the patient was *unable to flex, abduct, or laterally rotate his arm.*** In addition, there was *loss of flexion of the elbow* joint. A lack of sensation was detected on the lateral surface of his arm and forearm.

Problems. Using your anatomical knowledge of the nerve supply to the upper limb, discuss the probable cause of this patient's loss of motor and sensory functions. What muscles are paralyzed? Is he likely to recover use of his paralyzed limb? *These problems are discussed on page 788.*

Case 6-2

One of your classmates injured his shoulder during a hockey game when he was driven heavily into the boards, hitting his shoulder. As you assisted him to the dressing room, you noted that his injury was very painful.

When his sweater and shoulder pads were removed, you observed that *the lateral end of his clavicle produced an abnormal prominence.* At first you thought he may have what sportswriters call a **"shoulder pointer."**

Later the team doctor informed you that your classmate had a **dislocation of the acromioclavicular joint** (or what laymen call a *"shoulder separation"*) and would be out of the lineup for several weeks.

Problems. Explain what sportswriters mean by the terms *"shoulder pointer"* and *"shoulder separation."* How would you explain the structure of the shoulder joint to a nonmedical student? What ligaments would be torn? What makes the patient's shoulder fall? *These problems are discussed on page 788.*

Case 6-3

A 12-year-old boy fell off his skateboard, hitting his elbow on the sidewalk. Because he was suffering considerabe elbow pain and some *numbness in his hand,* his mother took him to a doctor.

The boy told the doctor. "I fell on my funny bone and right away my little finger began to tingle."

The doctor noted that the boy showed *no response to pin-prick over the little finger and the medial border of the palm.* He was unable to grip a piece of paper placed between his fingers.

Suspecting a fracture of the elbow and peripheral nerve damage, the doctor arranged to have the boy's elbow radiographed. The radiographs showed a *separation of the epiphysis of the medial epicondyle of the humerus.*

Problems. Explain the numbness of the boy's little finger and his inability to hold a piece of paper between his fingers. Drawing on your knowledge of degeneration and regeneration of peripheral nerves, make an attempt to forecast the probable degree of recovery of the boy's motor and sensory functions that may occur. *These problems are discussed on page 789.*

Case 6-4

A young man was kicked very hard in the midhumeral region of his left arm. He presented with signs of tenderness, swelling, deformity, and abnormal movements of his left upper limb.

The physical examination revealed a **wrist drop** (Fig. 6-156*A*), an *inability to extend the fingers at the metacarpophalangeal joints*, and loss of sensation on a small area of skin on the dorsum of the hand proximal to the thumb and index finger (Fig. 6-156*A*). There was also *weakness of extension of the interphalangeal joints.* Measurement of the limb indicated that there was some shortening.

Radiographs showed the presence of a **fracture of the humerus** just distal to the attachment of the deltoid muscle, and that the proximal fragment of bone was abducted and the distal fragment displaced proximally.

Problems. Using your anatomical knowledge, determine *what peripheral nerve has been severed,* and what artery may have been torn. Would elbow flexion be weakened? Explain the observed effects of this peripheral nerve injury. Why are the fragments of humerus displaced in the manner described? *These problems are discussed on page 789.*

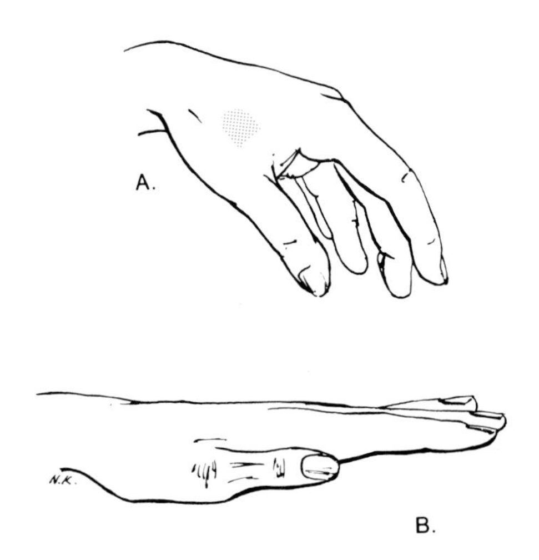

Figure 6-156. *A*, sketch of the patient's left hand showing a wrist drop. Note the small area of sensory loss (*red*). *B*, the patient is unable to extend his wrist and fingers as shown here.

Case 6-5

While you were playing touch football, you *fell on your open hand with your wrist hyperextended and laterally deviated.* You told your friends that you had just sprained your wrist and did not pay much attention to the injury for about 2 weeks. You sought medical advice then because the **wrist pain** was still present.

When the doctor deeply palpated your anatomical snuff box, there was *localized tenderness.* You experienced most pain on the lateral side of your wrist, particularly when he asked you to extend it as far as you could. Suspecting a fracture, he took several radiographs of your wrist which revealed a small *hairline fracture of one of the carpal bones.*

Problems. Which carpal bones lie in the floor of the anatomical snuff box? The distal end of which forearm bone is in the floor of this depression? Which carpal bone was most likely fractured? *These problems are discussed on page 790.*

Case 6-6

A 15-year-old girl was rushed to the emergency room of a hospital while you were visiting the chief resident. The girl had *slashed her wrists with a razor blade.* The moderate bleeding from the left wrist was soon stopped with slight pressure. The small spurts of blood coming from the lateral side of the right wrist were more difficult to stop.

Examination of her left hand and wrist revealed that her hand movements were normal and that there was no loss of sensation.

The following observations were made on her right wrist and hand: two superficial tendons and a large midline nerve were cut; *she could adduct her thumb but was unable to oppose* it; she had lost some fine control of the movements of her second and third digits; and there was anesthesia over the lateral half of her palm and digits.

Problems. Which tendon was almost certainly severed? What nerve was cut? Which tendon may have been severed? What superficial artery appears to have been lacerated? Would flexion of her wrist be affected? *These problems are discussed on page 791.*

Case 6-7

An elderly lady slipped on a patch of ice and *fell on the palm of her outstretched hand.* She told you that she heard her wrist crack and that it was very sore. As you helped her to her feet, you noticed that *the dorsal aspect of her wrist was unduly prominent and resembled a dinner fork.*

In view of these signs and symptoms, you decided to take her to a nearby doctor. When you informed the doctor that you were a first year medical student, he explained that the appearance of her wrist was typical of this kind of fracture and described it as the "*dinner fork deformity.*"

Problems. What bone in the forearm is commonly fractured in persons over the age of 50, particularly women? What do you call this kind of fracture? Explain the cause of the dinner fork appearance of the patient's wrist. *These problems are discussed on page 791.*

Case 6-8

A patient reported to her doctor that she had detected a *lump in her breast* several months ago, but fearing that it was cancer had delayed doing anything about it.

Inspection revealed localized retraction of skin in the superolateral quadrant of the left breast, and the skin in this area was dimpled like an orange peel.

Palpation revealed a hard mass in the superolateral quadrant of the breast that was attached to the skin and the tissues deep to it. The skin over the mass had a leathery feel and a group of *enlarged lymph nodes* was felt in the axilla.

Xeromammograms revealed a jagged, rounded density in the breast and that the skin overlying the mass was thickened (Fig. 6-9*B*). A diagnosis of **breast cancer** was made.

Problems. Name the deep structure to which the tumor would be fixed. What probably caused the leathery thickening of the skin over the tumor? Explain the anatomical basis of the skin retraction.

Based on your anatomical knowledge, where do you think the most common sites for metastases from a carcinoma of the breast would be? Which group of lymph nodes would be primarily involved in the present case? What other nodes might be involved? *These problems are discussed on page 791.*

DISCUSSION OF PATIENT ORIENTED PROBLEMS

Case 6-1

When the patient was thrown from his motorcycle and hit a tree, his right shoulder was pulled violently away from his head (Fig. 6-30). This pulled on the superior trunk of the brachial plexus (Figs. 6-21 and 6-22), stretching or tearing the ventral primary rami of cervical nerves 5 and 6. As a result, the nerves arising from these rami and the superior trunk are affected and the muscles supplied by them are paralyzed.

The muscles involved are the deltoid, biceps, brachialis, brachioradialis, supraspinatus, infraspinatus, teres minor, and supinator (Figs. 6-25 and 6-27).

The patient's arm was medially rotated because the infraspinatus and teres minor muscles (lateral rotators of the shoulder) were paralyzed.

His forearm was pronated because the supinators were paralyzed, notably the biceps muscles.

Flexion of his elbow was weak because of paralysis of the brachialis and biceps muscles.

The inability of the patient to flex his humerus resulted from paralysis of the deltoid and coracobrachialis muscles, and probably the clavicular head of the pectoralis major muscle. Loss of abduction of the humerus resulted from paralysis of the supraspinatus and deltoid muscles.

The paralysis would be permanent if the rootlets making up the rami (C5 and C6) have been pulled out of the spinal cord. As these rootlets cannot be sutured back into the cord, the axons of *the nerves will not regenerate and the muscles supplied by them will soon undergo* ***atrophy.***

Movements of the shoulder and elbow will be greatly affected, *e.g.*, the patient will always have difficulty lifting a glass to his mouth with his right arm.

The *loss of sensation in his arm* resulted from damage to sensory fibers of C5 and C6 that are conveyed in the upper lateral brachial cutaneous nerve (from the axillary), the lower lateral brachial cutaneous nerve (from the radial), and the lateral antebrachial cutaneous nerve (from the musculocutaneous nerve).

Case 6-2

A "*shoulder pointer*" is a sportswriter's term for a contusion over the "point of the shoulder," *i.e.*, the acromion.

To explain a "*shoulder separation*," first you should make a simple diagram of the scapula and clavicle, showing the ligaments attaching these bones together (Fig. 6-129).

Emphasize that it is the **coracoclavicular ligament** which provides most stability to the acromioclavicular joint. You should explain that the scapula and clavicle are parts of the upper limb and make up what is called the **pectoral girdle,** and that the clavicle articulates laterally with the acromion to form the acromioclavicular joint.

Explain that the scapula and clavicle are held together by the acromioclavicular and coracoclavicular ligaments (Fig. 6-129).

You should explain that when he hit the boards with the "point of his shoulder" (acromion), the acromioclavicular and coracoclavicular ligaments were torn (Fig. 6-157). As a result, the shoulder fell under

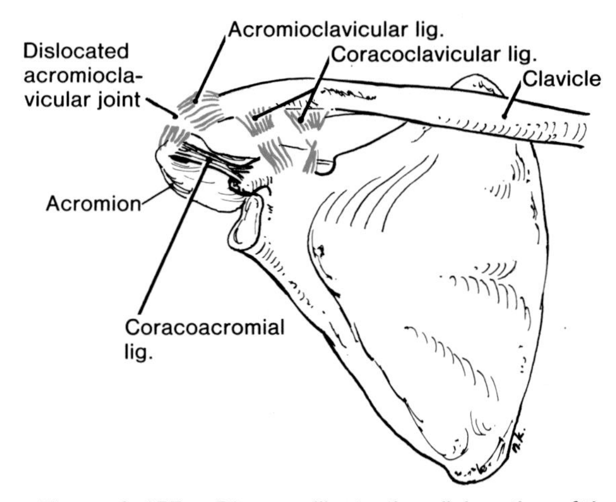

Figure 6-157. Diagram illustrating dislocation of the acromioclavicular joint ("separation of the shoulder"), following tearing of the acromioclavicular and coracoclavicular ligaments. Compare with Figure 6-129.

the weight of the upper limb and the acromion was pulled inferiorly relative to the clavicle. Also the lateral end of the clavicle was displaced superiorly, relative to the acromion and so produced an obvious prominence.

The expression "separation of the shoulder" is a misnomer. Explain that it is the acromioclavicular joint that is separated, not the shoulder joint. Rupture of the acromioclavicular ligament alone is not a serious injury, but when combined with rupture of the coracoclavicular ligament, the dislocation is complicated because the scapula and clavicle are separated and the scapula and upper limb are displaced inferiorly.

Case 6-3

The medial epicondyle of the humerus does not completely fuse with the side of the diaphysis until the 16th year in males (Fig. 6-158).

Although an epiphyseal separation is sometimes called "an epiphyseal fracture" or a fracture-dislocation, it is best to refer to this injury as a **separation of the epiphysis for the medial epicondyle.** Had this accident occurred in a person over 16, a fracture of the medial epicondyle might have occurred.

Because the epiphyseal cartilage plate is weaker than the surrounding bone in children and adolescents, a direct blow that causes a fracture in adults is likely to cause an **epiphyseal cartilage plate injury** in children.

The ulnar nerve passes posterior to the medial epicondyle (Fig. 6-63), between it and the olecranon. Consequently it is vulnerable to injuries at the elbow. In the present case, it is likely that the ulnar nerve was crushed and that the axons were damaged at the site of the injury. This kind of injury causes paralysis of muscles and some loss of sensation in the area of skin supplied by the ulnar nerve (Fig. 6-26).

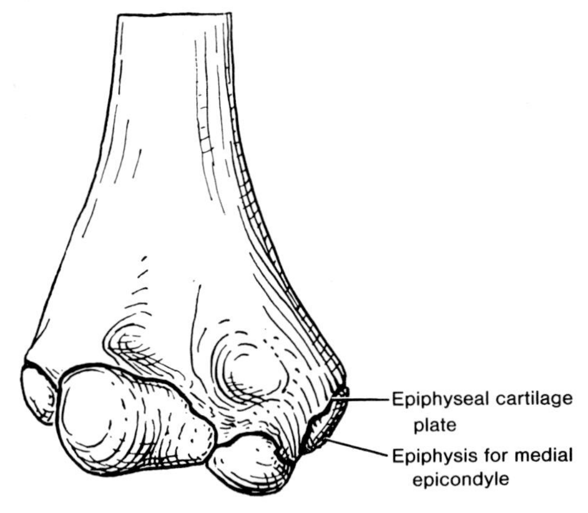

Figure 6-158. Anterior view of the distal end of the right humerus of a child, showing the epiphysis for the medial epicondyle. It appears during 6th year and fuses with the diaphysis of the humerus during the 14th year in females and the 16th year in males.

Appreciation of light touch is usually lost over the medial one and one-half fingers and response to pinprick is lost over the little finger and the medial border of the palm. Undoubtedly it was these sensory changes that first suggested ulnar nerve injury to the doctor.

Knowing that the interosseus muscles are supplied by the ulnar nerve, he probably decided to test them for weakness by placing a piece of paper between the boy's fully extended fingers and asking him to grip it as tightly as possible while he pulled on it (Fig. 6-159).

Inability to adduct the fingers is a classic sign of paralysis of the palmar interosseous muscles and ulnar nerve injury.

Had the doctor tested other movements he would likely have detected inability to abduct the fingers (*paralysis of the dorsal interossei*), loss of adduction of the thumb (paralysis of the adductor pollicis), weakness of flexion of the ring and little fingers at the metacarpophalangeal joints (*paralysis of the medial two lumbricals*), impaired flexion and adduction of the wrist (paralysis of the flexor carpi ulnaris), a poor grasp in the ring and little fingers, and inability to flex the distal interphalangeal joints of the fourth and fifth digits (paralysis of the lumbricals, interossei, and part of the flexor digitorum profundus).

Because all but five of the intrinsic muscles of the hand are supplied by the ulnar nerve, lesions of it at the elbow have their primary effect in the hand. As the nerve was only crushed, the nerve does not require suturing because new axons can grow down into the part of the nerve distal to the injury within the original endoneurial tubes and neurolemmal sheaths and reinnervate the paralyzed muscles. Hence after a crush injury, as in this case, restoration of function should occur in a few months' time with proper physiotherapy.

Case 6-4

The inability of the patient to extend his hand at the wrist indicates **injury to the radial nerve.** As

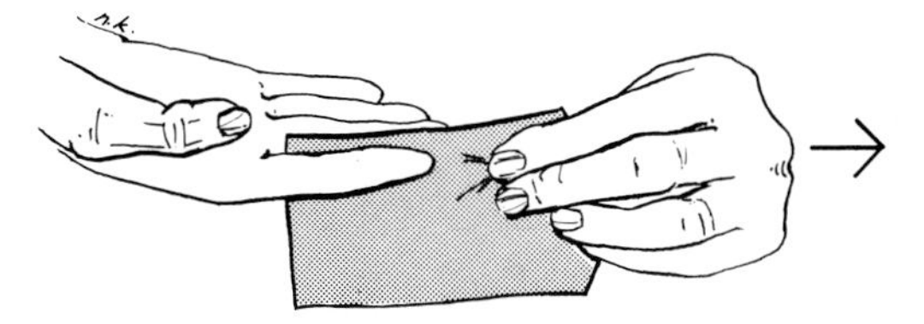

Figure 6-159. Demonstration of the method for testing the interosseous muscles. The examiner is attempting to pull the paper away from the patient in the direction of the *arrow.* Inability to adduct the fingers is a sign of paralysis of the palmar interosseous muscles and ulnar nerve injury.

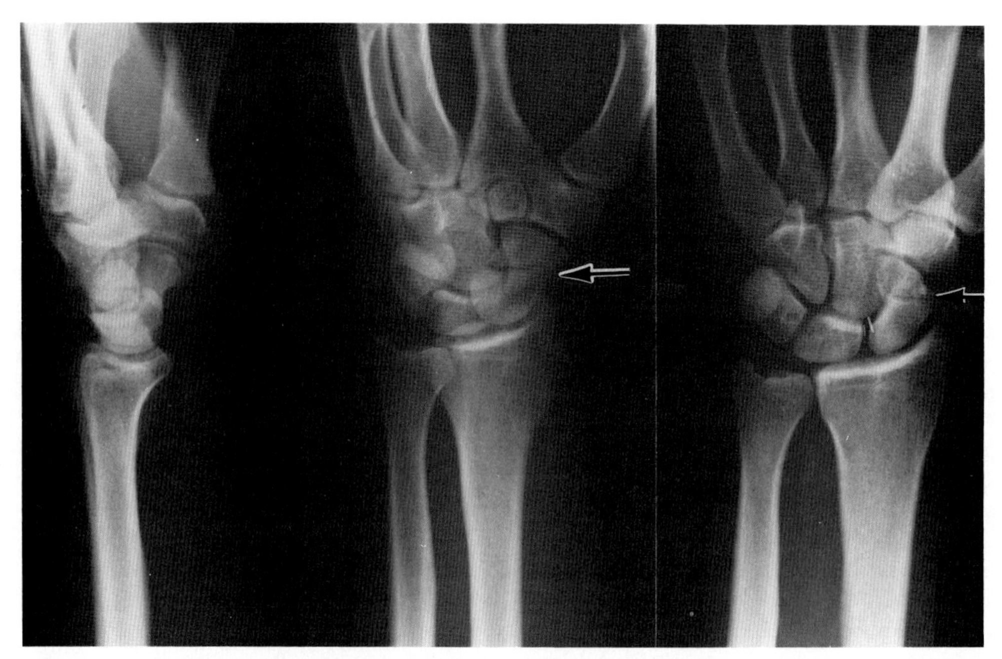

Figure 6-160. *Left to right*, lateral, oblique, and frontal radiographs of the left wrist showing a fracture of the scaphoid bone (*arrows*). In the frontal projection observe that there is some displacement of the distal fragment.

the fracture is in the middle of the humerus, it is likely the radial nerve was damaged where it passes diagonally across the humerus in the radial groove (Figs. 6-12 and 6-29). The nerve is particularly susceptible to injury in this location because of its close relationship to the humerus.

Severing the radial nerve totally paralyzes the extensor muscles of the forearm and hand (Fig. 6-27). As a result, extension of the wrist is impossible and the hand assumes the flexed position referred to clinically as **wrist-drop** (Fig. 6-156*A*).

The radial nerve supplies no muscles in the hand, but it supplies muscles whose tendons pass into the hand; hence the patient is unable to extend his metacarpophalangeal joints.

Because the lumbrical muscles (supplied by the median and ulnar nerves) and the interossei (supplied by the ulnar nerve) are intact, the patient is able to flex his metacarpophalangeal joints and extend his interphalangeal joints. However he would not have normal power of extension of his fingers.

Elbow flexion would be very painful and would be weakened when the forearm is in the position midway between pronation and supination. Recall that the radial nerve innervates the brachioradialis muscle, a strong flexor of the elbow in this position.

The area of sensory loss is often minimal following radial nerve injury (Fig. 6-156*A*) because its area of exclusive supply is very small. The degree of sensory loss varies from patient to patient, depending on the extent to which the territory is overlapped by adjacent nerves. Sometimes there may be no detectable loss of sensation.

The radial nerve arises from the posterior cord of the brachial plexus (C5 to C8 and T1) and is the largest branch of this plexus (Figs. 6-24 and 6-27).

The shortening of the patient's arm occurred because the broken fragments of bone are pulled apart. Contraction of the deltoid muscle abducts the proximal part of the humerus; the proximal contraction of the triceps, biceps, and coracobrachialis muscles pulls the distal fragment superiorly.

Although the profunda brachii artery accompanies the radial nerve through the radial groove (Fig. 6-69), and may be severed by bone fragments, the muscles and structures supplied by this artery (*e.g.,* the humerus) are not likely to show **ischemia** because the radial recurrent artery anastomoses with the profunda brachii artery (Fig. 6-65). This communication should provide sufficient blood for the structures supplied by the damaged artery.

Case 6-5

The lateral marginal bones of the carpus, *scaphoid and trapezium, lie in the floor of the anatomical snuff box* (Fig. 6-109). This depression at the base of the thumb is limited proximally by the styloid process of the radius and distally by the base of the first metacarpal bone.

In Figure 6-160, observe the **fracture of the sca-**

phoid near the middle of the bone, producing two fragments.

Fracture of the scaphoid is the most common type of carpal injury and usually results from a fall on the hand. Because of the position of the scaphoid and its small size, it is a difficult bone to immobilize. Continued movement of the wrist often results in nonunion of the fragments of bone. Usually there is displacement and tearing of ligaments which may interfere with the blood supply to one of the fragments.

Ischemic necrosis of part of the scaphoid bone may result. Usually the bone is supplied by two nutrient arteries, one to the proximal and one to the distal half. Occasionally both vessels supply the distal half; the separated proximal half receives no blood. The resulting **ischemia** may result in delay or lack of union of the fragments.

The characteristic clinical sign of fracture of the scaphoid is acute tenderness in the anatomical snuff box, particularly when the wrist is extended.

Case 6-6

Obviously the patient had not cut her wrist deeply on the left side; the slight bleeding was probably from severed superficial veins. On the right side, she would have certainly *cut the tendon of her palmaris longus* (Figs. 6-83 and 6-84). She probably also *cut the tendon of her flexor carpi radialis muscle.*

In view of the clinical findings, **it is obvious that her median nerve was severed**. At the wrist this nerve lies deep to and lateral to the tendon of the palmaris longus muscle (Figs. 6-80 to 6-83).

The slight spurting of blood in her right wrist suggests that **she probably cut the superficial palmar branch of her radial artery.** This artery arises from the radial just proximal to the wrist (Fig. 6-83). Had she severed her radial artery, the bleeding would have been severe.

Cutting the median nerve at her wrist resulted in paralysis of her thenar muscles and first two lumbricals (Tables 6-13 and 6-14).

Paralysis of the thenar muscles explains her inability to oppose her thumb. As the posterior interosseus nerve (*branch of the radial*) was not affected, she could abduct her thumb with her abductor pollicis longus, but there would be some impairment of this movement owing to *paralysis of the abductor pollicis brevis*, supplied by the recurrent branch of the median nerve. The patient could extend her thumb normally using her extensor pollicis longus and brevis muscles. As the nerve supply to her adductor pollicis muscle (deep branch of ulnar) is intact, she can also adduct her thumb. Owing to **paralysis of her first two lumbrical muscles** and the loss of sensation over the thumb and adjacent two and one-half fingers and the radial two-thirds of her palm (Fig. 6-26), *fine control of movements of her second and third digits is lacking.* Thus, **cutting the median nerve produces a serious disability of the hand.**

In a few weeks there will be atrophy of the thenar muscles. Because of this and the action of her intact adductor pollicis muscle, her thumb will be held close to the base of the lateral surface of her index finger. Her index and middle fingers will probably be hyperextended at the metacarpophalangeal joints and slightly flexed at the interphalangeal joints owing to the paralysis of her first two lumbrical muscles.

Atrophy of the thenar muscles gives a characteristic flattening of the thenar region producing a deformity often referred to as "ape hand", because the hand is flattened and ape-like.

The cutting of the tendons of the palmaris longus and flexor carpi radialis would weaken flexion of her wrist. In addition, if she attempted to flex her wrist, her hand would be pulled to the ulnar side by the flexor carpi ulnaris, which is unaffected because it is supplied by the ulnar nerve.

Obviously the cut ends of the tendons and the median nerve would have to be brought together and fastened by sutures, or held together by some other means. After a considerable time, function should be partially restored.

Case 6-7

The common injury of the wrist in persons over 50, particularly women, is fracture of the distal end of the radius, known as a **Colles' fracture** (Fig. 6-161).

The doctor radiographed her wrist and showed you how the distal fragment of the radius had tilted posteriorly, producing the "**dinner fork deformity**" of her wrist (Fig. 6-161*A*). Note that the styloid processes of the ulna and radius are at the same level (Fig. 6-161*C*), instead of the radial styloid being more distal than the ulnar styloid as is normal (Fig. 6-72). There is also subluxation or *partial dislocation of the distal radioulnar joint.*

Case 6-8

Many carcinomas of the breast occur in the superolateral quadrant of the breast. In advanced cases the tumor is fixed to the **pectoralis fascia,** *the deep fascia covering the pectoralis major muscle* (Fig. 6-7). Interference with the lymphatic drainage of the breast produced the leathery thickening and orange-peel appearance of the skin over the malignant tumor (**neoplasm**). The retraction of the skin is mainly caused by infiltration of cancer cells along the suspensory ligaments of the breast. This shortens the ligaments and causes them to retract the skin.

The most common sites for metastases from a carcinoma of the breast are the axillary lymph nodes (Fig. 6-8). Lymph vessels from the superolateral quadrant

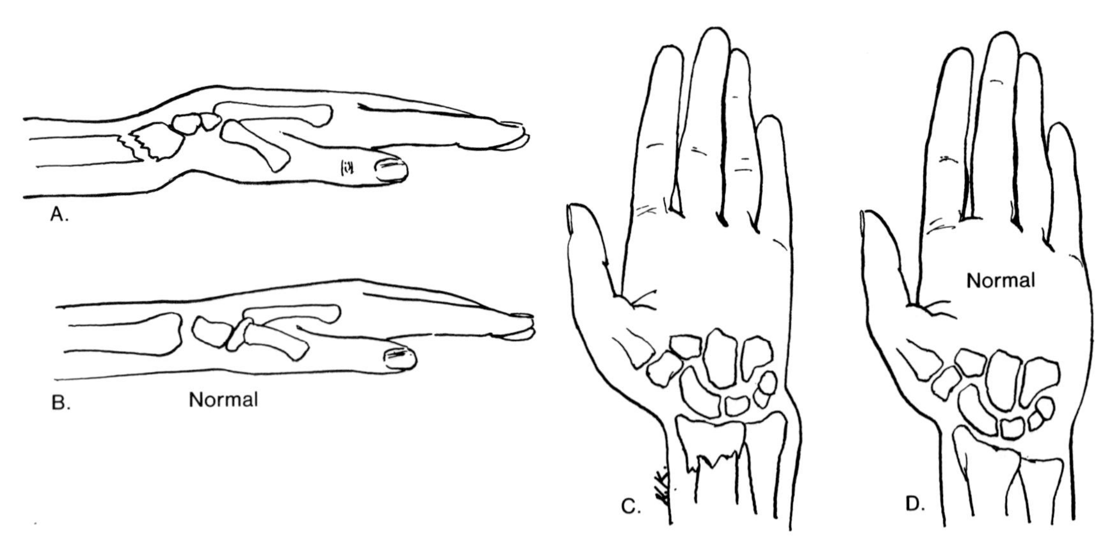

Figure 6-161. *A*, drawing of the patient's wrist showing an obvious bend in the wrist, producing the typical "*dinner fork deformity.*" This abnormality results from fracture of the distal end of the radius and posterior tilting of the distal fragment. Often the hand tends to be radially deviated owing to shortening of the radius. *B*, schematic drawing showing the normal position of the radius and carpal bones. *C*, anterior view of the wrist before reduction of the fracture. Note that the distal ends of the radius and ulna are at the same level. *D*, during reduction of the radial fracture, the shortening was corrected by placing the displaced fragment in its correct position. Observe that the radial styloid is now distal to the ulnar styloid, which is its normal position (Fig. 6-72).

of the breast, the site of the tumor in the present case, mainly drain to the **pectoral group of axillary lymph nodes** (Figs. 6-8 and 6-43). The lymph transports cancer cells to the nodes where *new cancerous growths develop*. It is very probable that the subscapular group of lymph nodes would also be involved.

Although the spread of cancer cells would most likely be as described, it is possible for any group of axillary lymph nodes to be involved (Fig. 6-8).

Cancer cells can also spread via the blood; blood borne metastases may occur in the ovaries, adrenal glands, pituitary, lungs, and vertebral column.

Suggestions for Additional Reading

1. Anson BJ, McVay CB: *Surgical Anatomy*, ed 5. Philadelphia, WB Saunders Co, 1971.

   Anatomy is used in this book as the groundwork for surgical techniques. The rationale for operations, rather than the detailed steps of surgery, is stressed. Descriptions are given of common variations in the attachment of muscles, in the origin and branching of arteries, in the pattern of anastomosis and termination of veins, and in the course and relationship of nerves.

2. Behrman RE, Vaughan III VC: *Nelson's Textbook of Pediatrics*, ed 12. Philadelphia, WB Saunders Co, 1983.

   A classic textbook of pediatrics with a very good section on chromosomes and their abnormalities.

3. Ellis H: *Clinical Anatomy*. A Revision and Applied Anatomy for Clinical Students, ed 7. Oxford, Blackwell Scientific Publications, 1983.

   This classic book integrates the anatomy that is learned in the dissecting laboratory with that used on the wards and in the operating theaters. It highlights features of anatomy which are of clinical importance in medicine and surgery.

4. Haymaker W, Woodhall B: *Peripheral Nerve Injuries. Principles and Diagnosis*, ed 2. Philadelphia, WB Saunders Co, 1953.

   This monograph deals with the fundamentals of diagnosis of peripheral nerve injuries. It contains good descriptions and excellent illustrations of the methods used in the assessment of muscle function in the upper limb and in other regions. It contains many photographs of patients who sustained nerve injuries. Aids to the investigation of peripheral nerve injuries and characteristic clinical features of nerve injuries in the upper limb are clearly presented.

5. Healey JE: *A Synopsis of Clinical Anatomy*. Philadelphia, WB Saunders Co, 1969.

   This book presents many clinical applications of gross anatomy. The essential anatomy of each region is presented before clinical considerations are discussed, using very good illustrations. For example, the various surgical approaches for the open reduction of fractures are well described and illustrated, and the anatomical bases of the techniques are explained.

6. Moore KL: *The Developing Human. Clinically Oriented Embryology*, ed 3. Philadelphia, WB Saunders Co, 1982.

Chapter 17 in this comprehensive text describes normal and abnormal development of the limbs.

7. Salter RB: *Textbook of Disorders and Injuries of the Musculoskeletal System,* ed 2. Baltimore, Williams & Wilkins, 1983.

This textbook was written especially for undergraduate medical students as an introduction to orthopaedics, rheumatology, metabolic bone disease, rehabilitation, and fractures.

CHAPTER 7

The Head

*Few complaints are more common than **headache** and **head pain**.* The term headache is used to describe all painful sensations in the forehead, vertex (crown of the head), temples, and the back of the head, whereas localized head pains are given specific names, *e.g.*, facial pain, **earache** (otalgia), sinus pain, and **toothache** (odontalgia).

Headache often accompanies fever, tension, and/or fatigue, but sometimes it indicates a serious intracranial problem (*e.g.*, **a brain tumor**, subarachnoid hemorrhage or meningitis).

Consequently, all health care professionals must have a sound knowledge of the anatomy of the head to understand the anatomical basis of headache and head pain.

The head contains a number of important structures, the diseases of which form the basis of several medical and surgical specialties: **neurology** (study of the nervous system and its disorders); **neuroradiology** (study of the skull and nervous system using imaging techniques; **neuropsychiatry** (study of organic and functional diseases of the brain); **neurosurgery** (surgery of the nervous system); **ophthalmology** (study of the eye and its disorders); **otology** (study of diseases of the ear and related structures); **rhinology** (study of the nose and its diseases); **maxillofacial surgery** (surgery of the face and jaws), **oral surgery**, (surgery of the mouth), and **dentistry** (study and treatment of the oral-facial complex, especially the teeth).

THE SKULL

The skull is the most complex bony structure in the body because it (1) **encloses the brain** which is irregular in shape; (2) **houses the organs of special senses** (for seeing, hearing, tasting, and smelling); and (3) **encloses the openings into the digestive and respiratory tracts.**

Five views of the exterior of the skull are used in anatomical descriptions; each is spoken of as a **norma.** In each case *the view refers to the skull in the anatomical position*, and each view is from a position that is at right angles to one of the three planes of the body (frontal, median, and horizontal).

In the anatomical position, the skull is oriented so that the inferior margins of the orbits (eye sockets) and the superior margins of the external acoustic meatus (auditory canals) are horizontal. This is called the *orbitomeatal plane* (Frankfurt plane).

ANTERIOR ASPECT OF THE SKULL (NORMA FRONTALIS)

The anterior aspect of the skull comprises the anterior part of the calvaria (brain case) superiorly and the skeleton of the face inferiorly (Figs. 7-1 and 7-2). Notable features of the anterior aspect are: the **forehead**, formed by the frontal bone; the orbits for the eyeballs; the **prominences of the cheek**, formed by the zygomatic bones; the **anterior nasal apertures** which the external nose surrounds; the paired **maxillae** (upper jaw) containing the maxillary (upper) teeth; and the **mandible** (lower jaw) containing the mandibular (lower) teeth.

The anterior aspect of the skull may be divided into five areas: frontal, maxillary, nasal, orbital, and mandibular.

POSTERIOR ASPECT OF THE SKULL (NORMA OCCIPITALIS)

The posterior aspect of the skull (Figs. 7-3 and 7-4), noticeably convex, *is mainly formed by the parietal and occipital bones* which meet the mastoid parts of the temporal bones laterally. The most prominent feature of the posterior aspect of the skull is the rounded posterior pole, called the **occiput.**

The external occipital protuberance (Fig. 7-3) is a median projection that is easily palpable. It is identifiable in living persons at the superior end of the median furrow in the posterior aspect of the neck. The center of the external occipital protuberance and its most prominent projection is called the **inion** (Fig. 7–10).

Curved **superior nuchal lines** run laterally from

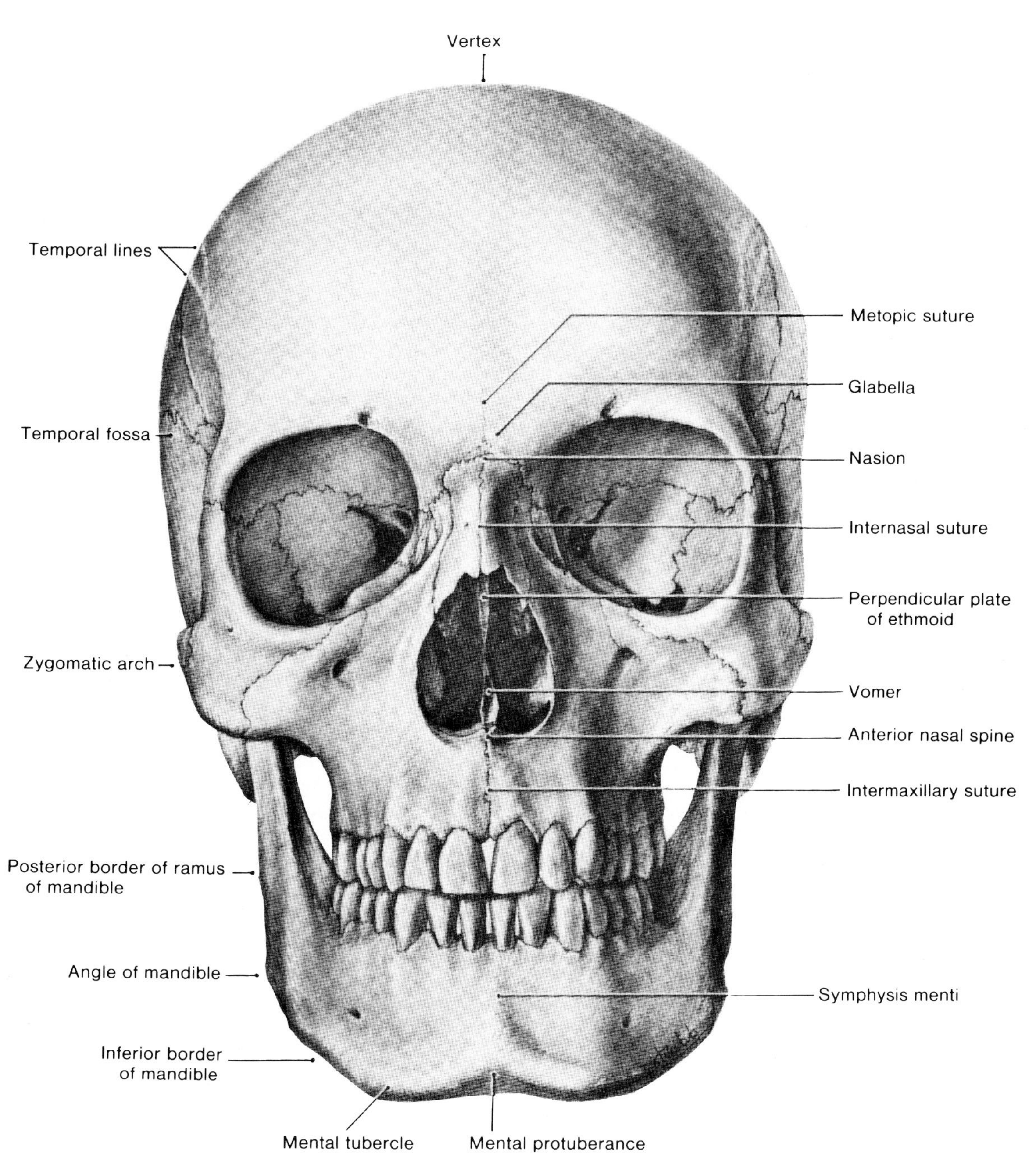

Figure 7-1. Drawing of the anterior aspect (*norma frontalis*) of an adult skull showing marginal features on the right side and midline features on the left side. In this specimen the remains of the metopic or frontal suture is visible; this remnant is present in about 8% of people. The frontal bone, of membranous origin, ossifies from two centers, one on each side. The halves of the frontal bone are separated at birth by the metopic or frontal suture (Fig. 7-5*A*). *Obvious features* of the anterior aspect of the skull are: the *orbits*, so named because the eyes rotate in them; the bony *anterior nasal aperture*, and the upper and lower jaws (*maxillae and mandible*, respectively).

the external occipital protuberance to the **mastoid processes** (Figs. 7-3, 7-10, and 7-11). The superior nuchal lines represent the superior limit of the neck. The Latin word for the posterior aspect of the neck is *nucha*. The posterior part of the *sagittal suture* and the entire *lambdoid suture* are visible posteriorly.

The lambda is where the lambdoidal and sagittal sutures intersect (Figs. 7-3 and 7-10). It is an important anthropological landmark.

SUPERIOR ASPECT OF THE SKULL (NORMA VERTICALIS)

The superior aspect or "*top of the skull*" is rounded or ovoid in many people and is broadened posteriorly by **parietal eminences** or tubers (Figs. 7-5*C* and 7-11). Four bones are visible from this aspect of the adult skull: the two parietal bones are separated by the sagittal suture (Fig. 7-6); the frontal and parietal bones are separated by the coronal suture; and the parietal bones are separated from the occipital bone by the **lambdoid suture** (Fig. 7-3).

The intersection of the sagittal and coronal sutures is called the **bregma.** It is an important anthropological landmark. At the bregma and lambda the skull is membranous at birth, constituting the anterior and posterior **fontanelles or fonticuli** (Fig. 7-5).

The vertex (Fig. 7-1), the most superior part of the skull, is located near the center of the sagittal suture. A **parietal foramen** is located in the parietal bone on each side of the sagittal suture (Fig. 7-3). These foramina transmit *emissary veins* which connect the intracranial venous sinuses with the veins covering the skull (Figs. 7-39 and 7-50).

INFERIOR ASPECT OF THE SKULL (NORMA BASALIS)

The external surface of the *base of the skull*, with the mandible removed (Fig. 7-8), shows the inferior surface of the *maxillae* (upper jaw) anteriorly, with the bony *palate* and *maxillary teeth* (Fig. 7-9), and the zygomatic arches curving posteriorly on each side. The **zygomatic arch** is formed by the *temporal process of the zygomatic bone* and the *zygomatic process of the temporal bone* (zygoma). Centrally the inferior surface of the skull is irregular owing to the many foramina, processes, and articulations. Laterally the base of the skull exhibits the **temporal bones** with their prominent *mastoid and styloid processes.*

The foramen magnum, one of the most conspicuous features of the base of the skull (Fig. 7-8), is bordered anterolaterally by the **occipital condyles,** which articulate with the atlas or first cervical vertebra (Fig. 7-54). The medulla of the brain stem passes through the foramen magnum to become the spinal cord.

Because of its many foramina and thin areas of bone, *the base of the skull is fragile and fractures of it resulting from severe head injuries are common.*

LATERAL ASPECT OF THE SKULL (NORMA LATERALIS)

The lateral aspect of the skull (Figs. 7-10 to 7-12) **includes parts of the temporal bone and the temporal and infratemporal fossae** (Figs. 7-1 and 7-11). The lateral aspect of the skull indicates clearly the division of the skull into the large ovoid *cranial vault* (brain case) and the smaller uneven *facial skeleton.*

The cranial vault is formed by the frontal bone anteriorly, the parietal bones laterally, and the occipital bones posteriorly. Inferolaterally the walls of the cranial vault are formed by the temporal bones and the greater wings of the sphenoid bone (Figs. 7-11 and 7-16). The **calvaria** (often incorrectly called the calvarium) is the superior dome-like portion of the skull which forms the roof of the cranial vault. The skullcap, removed at autopsy and during dissection of the head, usually represents most of the calvaria.

A very important clinical landmark of the lateral aspect of the skull is the pterion, where four bones articulate. Usually the group of sutures joining the bones form an H-shape (Fig. 7-10) and the bones somewhat resemble wings (G. *pteron*, wing).

The area known as the pterion is located about 4 cm superior to the zygomatic arch and 3 cm posterior to the zygomatic process of the frontal bone (Figs. 7-10, 7-11, and 7-16).

The area of the **pterion** (temple) is *very important clinically* because the frontal (anterior) branch of the **middle meningeal artery** lies in a groove on the internal aspect of the lateral wall of the cranial vault (Figs. 7-16, 7-50, and 7-55). Here it is vulnerable to tearing in fractures of the bones forming the pterion. The groove for the artery is usually very deep in this area (Fig. 7-55) and is sometimes roofed over with bone. In these cases, it more liable to tear in cases of skull fractures.

The pterion is also a useful landmark for locating some parts of the brain. An oblique line drawn from the frontozygomatic suture to the pterion (Fig. 7-10) is level with the inferior surface of the frontal lobe (Fig. 7-51).

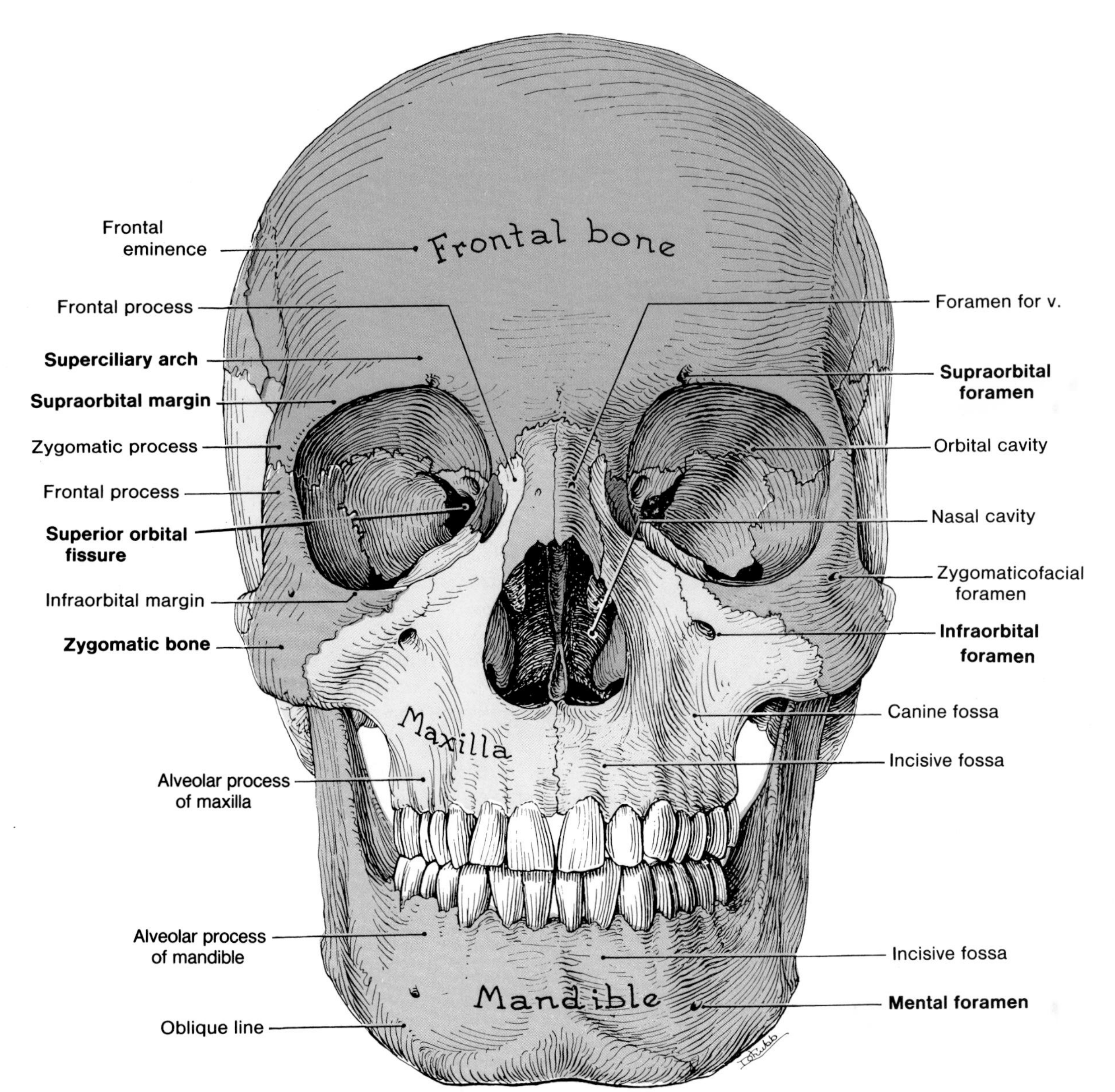

Figure 7-2. Drawing of the anterior aspect of an adult skull showing the surface features on the right side and the foramina, fossae, and cavities on the left side. Note that the skull exhibits a more or less oval outline, wider superiorly than inferiorly. Verify that you can draw an imaginary vertical line through the supraorbital, infraorbital, and mental foramina. In a living person, this line also passes through the pupil of the eye when one looks straight ahead. The term *skull* denotes the skeleton of the head, with or without the mandible. The term *cranium* means the skull without the mandible, and *calvaria* refers to the superior aspect of the skull or skullcap (Fig. 7-6). In skulls used for laboratory study, most of the calvaria has been removed with a circumferential saw cut. Notable features of the anterior aspect of the skull appear in **boldface** or are printed on the drawing.

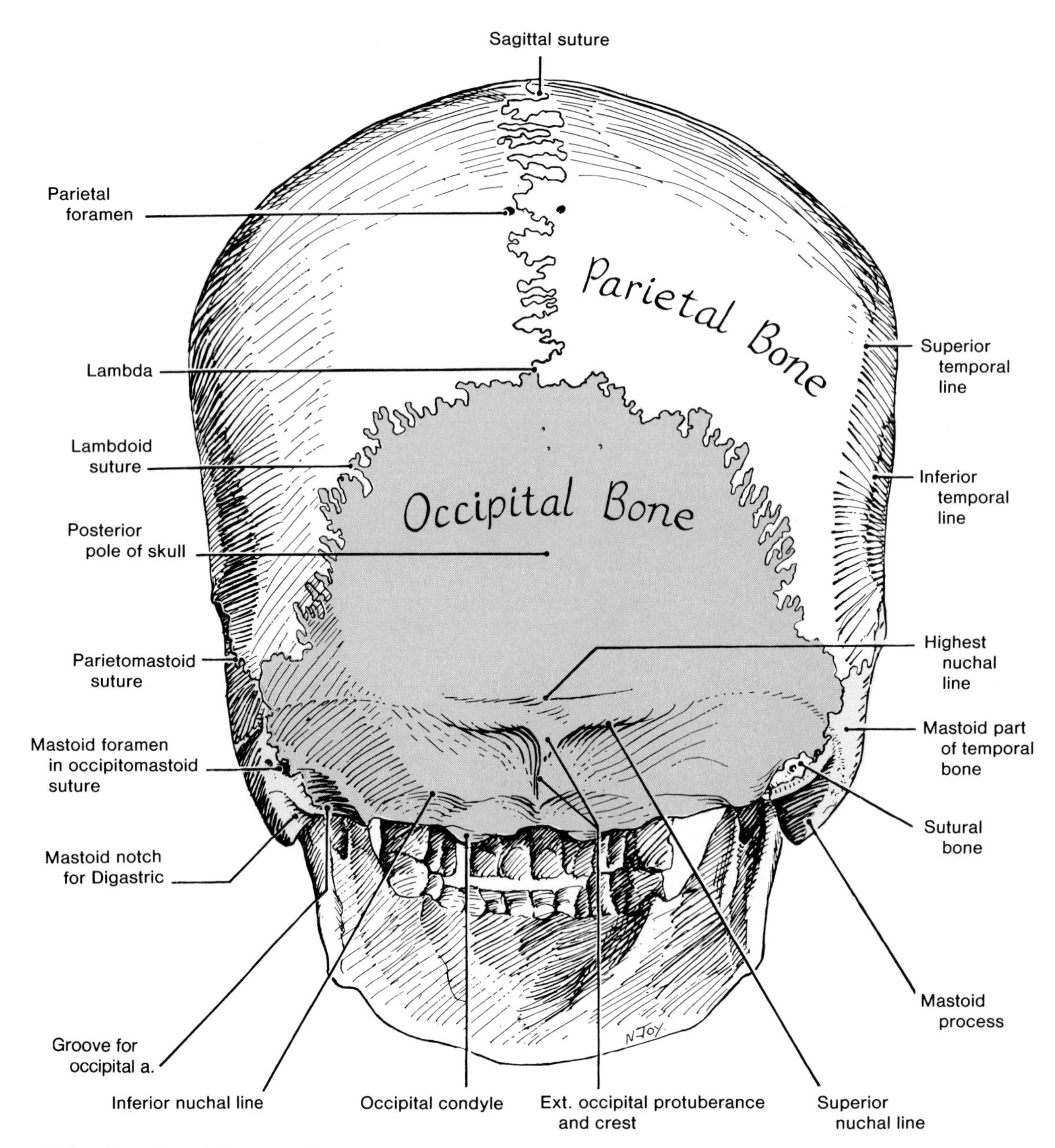

Figure 7-3. Drawing of the posterior aspect (*norma occipitalis*) of an adult skull. Note that the outline of the skull is horseshoe-shaped from the tip of one mastoid process over the vertex (Fig. 7-1) to the other process. The occipital bone (*blue*) forms much of the posterior aspect of the skull. The external occipital protuberance is usually easy to palpate, particularly in males.

The posterior end of this line is close to the anterior end of the lateral sulcus of the brain. Moreover, the **motor speech area** (Broca's area) of the brain lies about one fingerbreadth superior to this oblique line on the left side of the head in all, or in nearly all, right-handed persons.

Other obvious features of the lateral aspect of the skull are: the **external acoustic meatus** (external auditory meatus); **zygomatic bone**; **zygomatic arch**; **mastoid process** of the temporal bone; and **mandible.** In Figure 7-11 observe the body, ramus, coronoid process, head, and neck of the mandible.

The mastoid process projects anteroinferiorly, medial to the lobule of the auricle (Fig. 7-79), where it is readily palpable. *The size of the mastoid process*

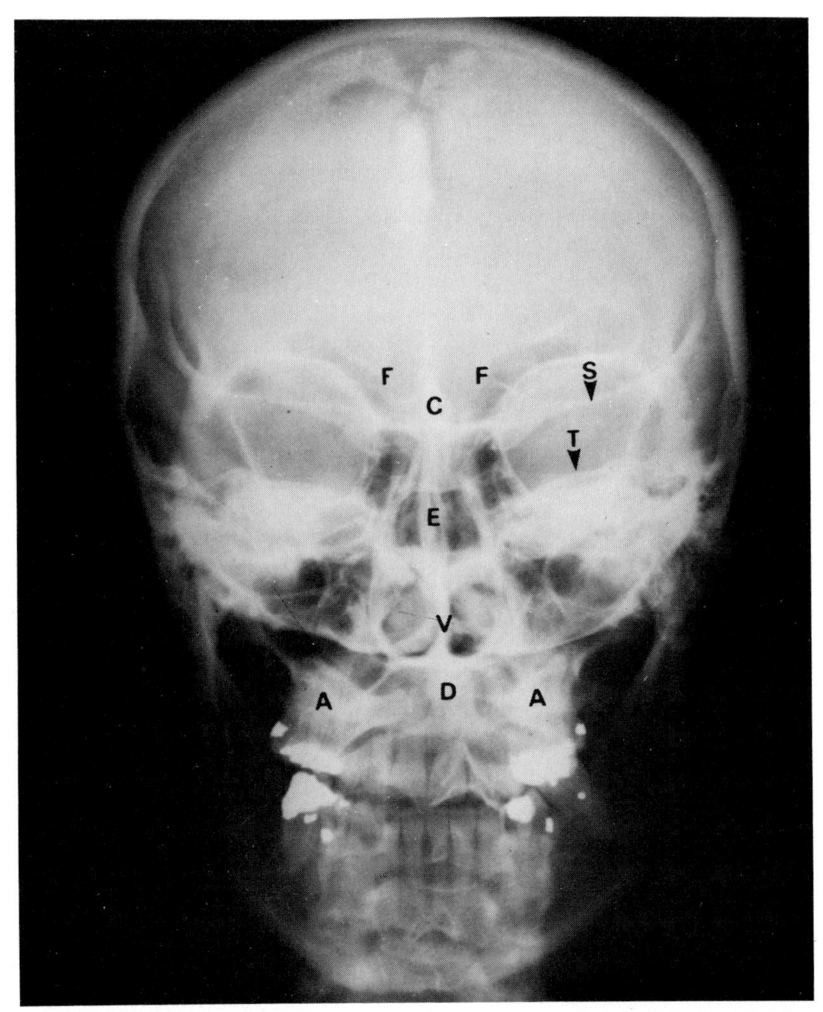

Figure 7-4. Posteroanterior or **PA radiograph of the skull** in which the rays traverse the patient's head from the posterior surface (**P**) to the anterior surface (**A**). Observe that (1) the orbital outline is divided into three horizontal parts by the lesser wings of the sphenoid bone (*S*) and the superior surface of the petrous part of the temporal bone (*T*); (2) the bony nasal septum is formed by the perpendicular plate of the ethmoid (*E*) and the vomer (*V*); (3) the crista galli (*C*) and the frontal sinuses (*F*) are recognizable; and (4) the dens (*D*) of the axis (Fig. 5-16) is superimposed on the facial skeleton, as are the lateral masses of the atlas (*A*).

varies with the age and muscularity of the person. It is not present at birth (Fig. 7-5*C*) and it is small during childhood. After puberty the mastoid process enlarges and is composed of **mastoid cells** (Figs. 7-12 and 7-79). The mastoid process is part of the insertion of the sternocleidomastoid muscle (Figs. 8-6 and 8-10).

The mental protuberance is an obvious feature in lateral views of the skull in most people (Figs. 7-10 to 7-12). It consists of a triangular prominence of bone with its apex directly superiorly toward the **incisive fossa** of the mandible (Figs. 7-2 and 7-11).

INTERNAL ASPECT OF THE SKULL

To expose the interior of the skull, the calvaria or cranial vault has to be sawn circumferentially and removed. The bones that can be seen in the internal aspect of the base of the skull are the frontal, ethmoid, sphenoid, temporal, and occipital bones (Figs. 7-13 to 7-15).

The internal surface of the calvaria is fairly smooth and very concave (Fig. 7-7), particularly from side to side. The striking features of this surface are the grooves in the parietal bones made by the frontal branch of the middle meningeal artery and its vein (Figs. 7-7, 7-16, and 7-55).

The sutures of the skull are more distinct on the external surface (Fig. 7-6) than on the internal surface (Fig. 7-7) because fusion begins on the inside between the ages of 20 and 30 years (10 years earlier than on the external surface).

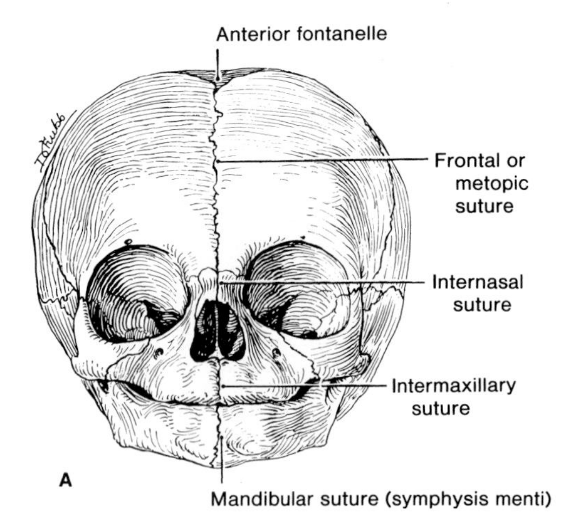

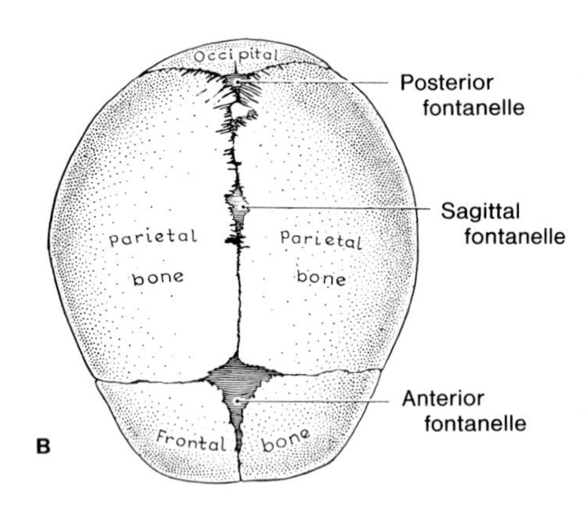

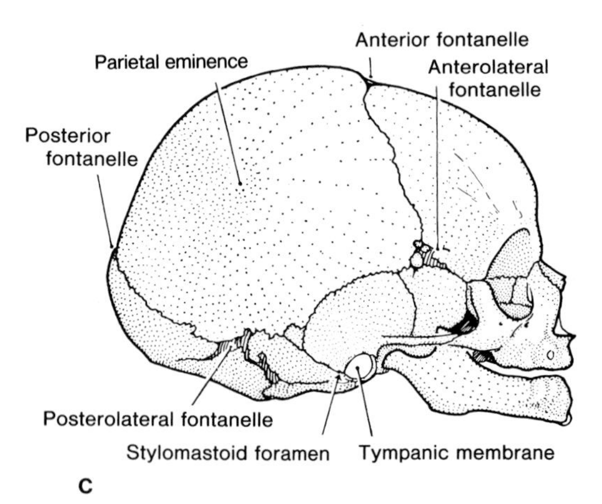

Figure 7-5. Drawings of the skull at birth. Note that the calvaria is large and the face is small. *A*, anterior aspect. *B*, superior aspect. *C*, lateral aspect.

Arachnoid granulations (Fig. 7-67) project into the venous sinuses, particularly into the lacunae at the side of the superior sagittal sinus (Fig. 7-50). They protrude sufficiently to indent the inner table of bone of the calvaria (Fig. 7-7), forming small pits along the groove formed by the superior sagittal sinus. These pits increase in size and number with age.

The internal aspect of the base of the skull, called the floor of the cranial cavity, presents three distinct tiered areas: *the anterior, middle, and posterior cranial fossae* (Fig. 7-15). These are described subsequently, beginning on page 845.

WALLS OF THE CRANIAL CAVITY

The walls of the cranial cavity vary in thickness in different regions of the cranium and in different persons. The skull is usually thinner in females than in males and is thinner in children and the aged. The bone tends to be thinnest in areas that are well covered with muscles, *e.g.*, the squamous part of the temporal bone (Fig. 7-11) and the posteroinferior part of the skull, posterior to the foramen magnum (Fig. 7-13). You can observe these thin areas of bone if you remove the skullcap and hold the remainder of the skull up to a light.

Most bones of the calvaria consist of inner and outer tables (cortices) of compact bone, separated by spongy **diploë** (Figs. 7-13 and 7-46). This is cancellous bone containing **red bone marrow** during life, through which run the channels formed by the **diploic veins** (Fig. 7-49). Examine the diploë in the skullcap from a laboratory specimen. It is not red in the dried skull because the protein was removed during preparation of the specimen. Also observe that the inner table of bone is thinner than the outer table and that in some areas of the skull, there is a thin plate of compact bone with no diploë (Fig. 7-13).

The convexity of the calvaria serves to distribute and thereby minimize the effects of a blow to it, however hard contact with a flat surface (*e.g.*, a concrete side-walk) can produce splits in the scalp that resemble lacerations.

Hard blows to the head in areas where the calvaria is thin (*e.g.*, in the temporal fossa, particularly in the region of the **pterion**, Figs. 7-1 and 7-10) are likely to produce fractures (Fig. 7-26). Fractures involving the bony grooves in the inner table formed by blood vessels (Fig. 7-16) are liable to tear the vessels, resulting in **intracranial bleeding**, a serious complication. In depressed fractures of the skull, the inner table of the calvaria is often more extensively fractured than the outer table.

BONES OF THE SKULL

The skull is composed of many bones which are closely fitted together. The number is unimportant,

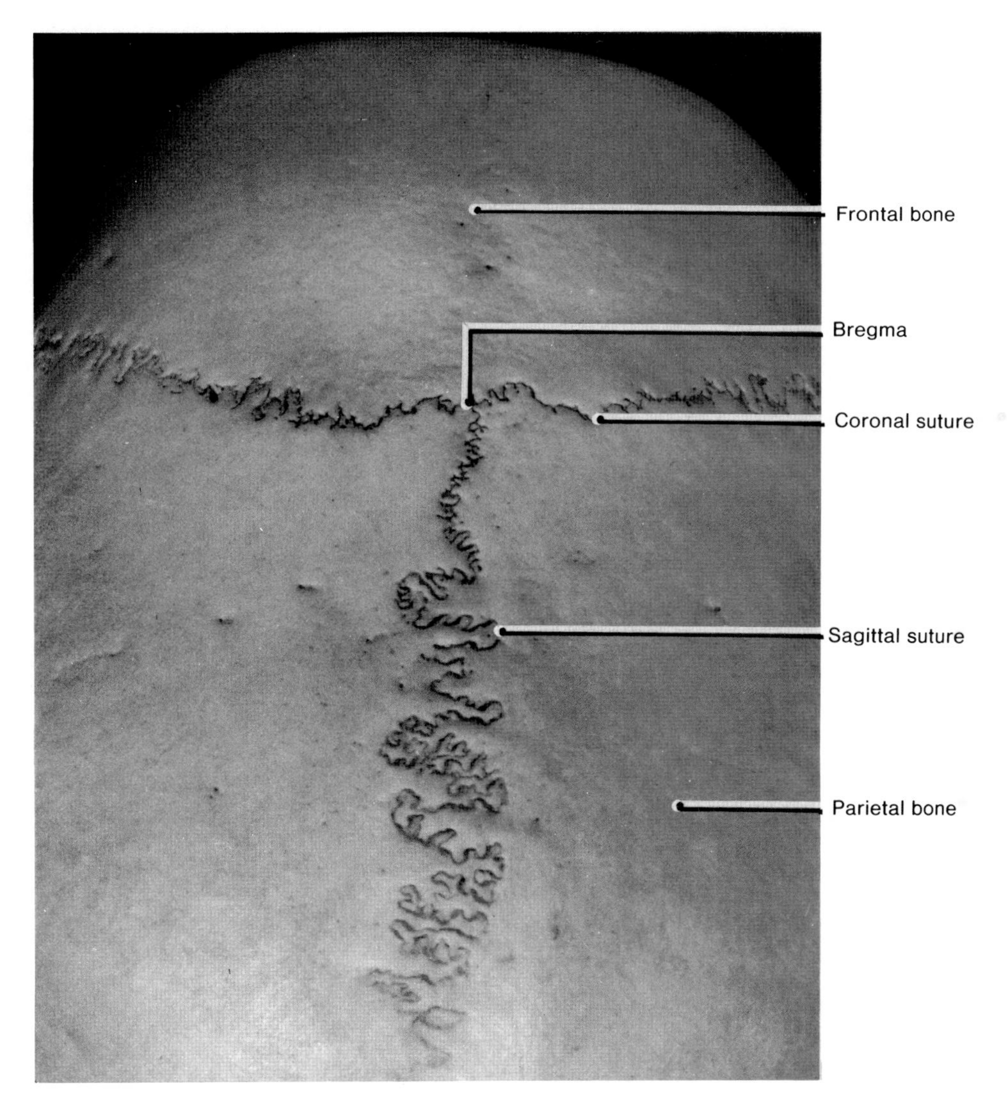

Figure 7-6. Photograph of the anterior two-thirds of the superior aspect (*norma verticalis*) of an adult skull, showing a close-up view of the sagittal and coronal sutures of the calvaria. The bones of the superior aspect of the skull are the frontal, parietal, and occipital (Fig. 7-5*B*). These bones form the calvaria. The meeting place of the coronal and sagittal sutures is called the bregma. In the fetal skull (Fig. 7-5), it is the site of the anterior fontanelle or fonticulus, a membrane-filled gap. Note the variation in the form and degree of interlocking of the cranial bones. Observe that the sutures are serrated like the teeth of a saw and that some projections from the edge of one bone are surrounded by bone from the edge of the opposing bone, not unlike the pieces of a jigsaw puzzle. Although there is no bony union, the bones are locked together in such a way that they cannot be spread apart.

but it is important for you to understand how the skull is constructed (Figs. 7-2, 7-3, 7-8, 7-11, and 7-14).

Except for the mandible and the ossicles of the middle ear, the bones of the adult skull are joined by rigid sutures. In effect, the cranium of a mature adult is essentially a single complex bone.

Although the adult skull is rigid, the bones of the cranium of infants and children grow as individual bones and undergo remodelling. Furthermore, relationships among the various bones are constantly changing during infancy and childhood.

Brief descriptions of the important bones of the skull follow. Only information that is useful clinically is given.

THE FRONTAL BONE (Figs. 7-1, 7-2, 7-5, 7-6, and 7-10 to 7-17)

The forehead (L. *frons*) is formed by the smooth, broad, convex plate of bone called the **frontal squama**. In fetal and infantile skulls, the halves of the frontal squama are divided by a **metopic suture** (G. *metōpon*, forehead) or **frontal suture** (Fig. 7-5*A*).

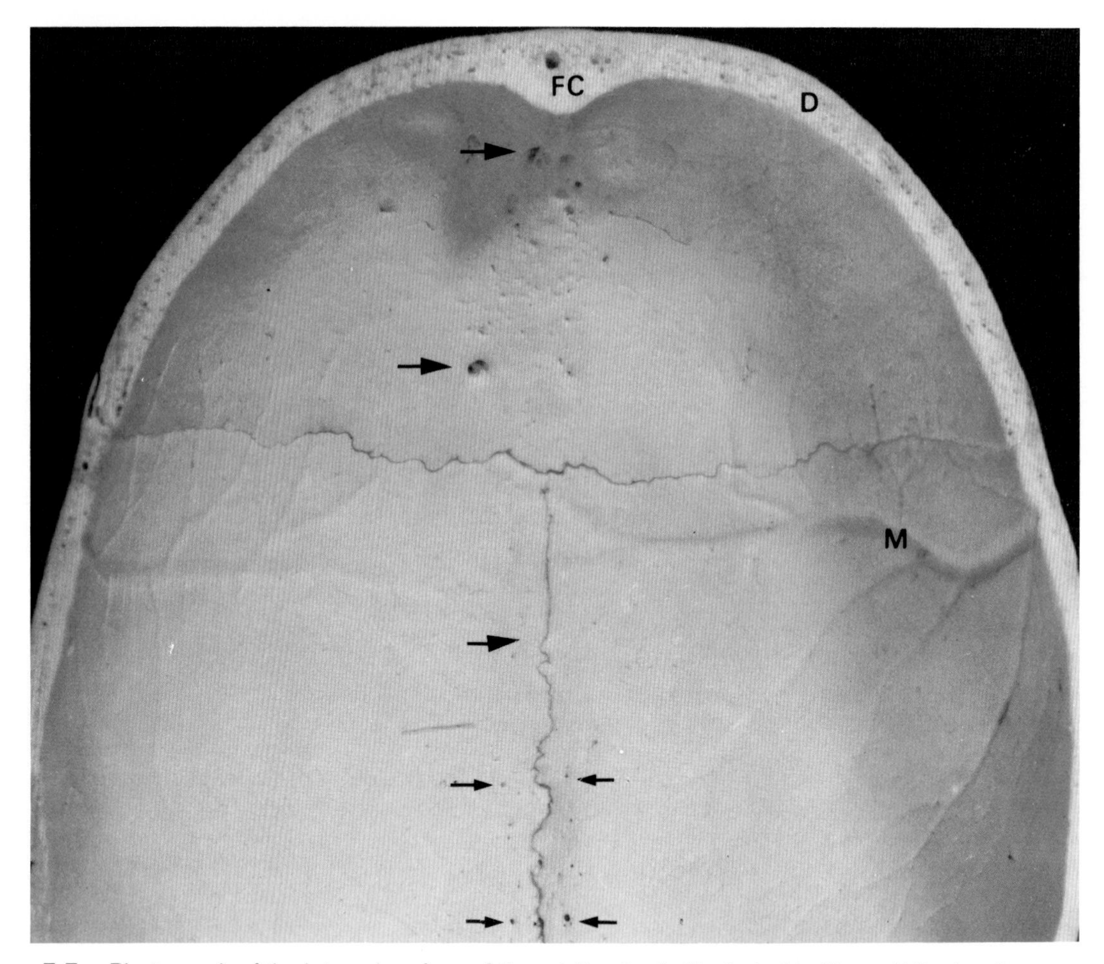

Figure 7-7. Photograph of the internal surface of the adult calvaria illustrated in Figure 7-6, showing parts of the frontal and parietal bones. Note the coronal suture between the frontal and parietal bones and the sagittal suture between the parietal bones. Observe the pits in the frontal bone (*large arrows*) produced by arachnoid granulations, which are hypertrophied aggregations of arachnoid villi (Fig. 7-50). On each side of the sagittal suture, note the parietal foramina (*small arrows*), through which emissary veins passed to connect the superior sagittal sinus with veins in the diploë and scalp (Fig. 7-49). The spongy diploë (*D*), or cancellous bone, that contained red marrow in life is visible in the frontal region of the sectioned calvaria (Fig. 7-46*A*). Note also the sinuous groove (*M*) formed by the frontal branch of the middle meningeal artery (Fig. 7-16). Note the frontal crest (*FC*) to which the falx cerebri is attached (Figs. 7-53 and 7-54).

In most people the halves of the frontal bone begin to fuse during infancy and the suture between them is usually not visible after the 6th year.

In about 8% of adult skulls (Fig. 7-1), a remnant of the inferior part of the metopic or frontal suture is visible and may be mistaken in radiographs for a fracture line by inexperienced observers (Fig. 19, p. 20).

The frontal bone forms the thin roof of the orbits (Figs. 7-88 and 7-89). Just superior to and parallel with each supraorbital margin is a ridge, the **superciliary arch** (Fig. 7-2), which overlies the frontal sinus (Figs. 7-4, 7-12, and 7-17). The superciliary arch is more pronounced in males. Between these arches there is a gently, rounded, median elevation called the **glabella** (Fig. 7-1); this term derives from the Latin word *glabellus* meaning smooth and hairless. In most people the skin over the glabella is hairless.

The slight prominences of the forehead on each side, superior to the superciliary arches, are called the **frontal eminences** or tubers (Figs. 7-2 and 7-11). The **supraorbital foramen** (occasionally a notch), which transmits the supraorbital vessels and nerve (Fig. 7-28), is located in the medial part of the supraorbital margin (Fig. 7-2).

The frontal bone articulates with the two parietal

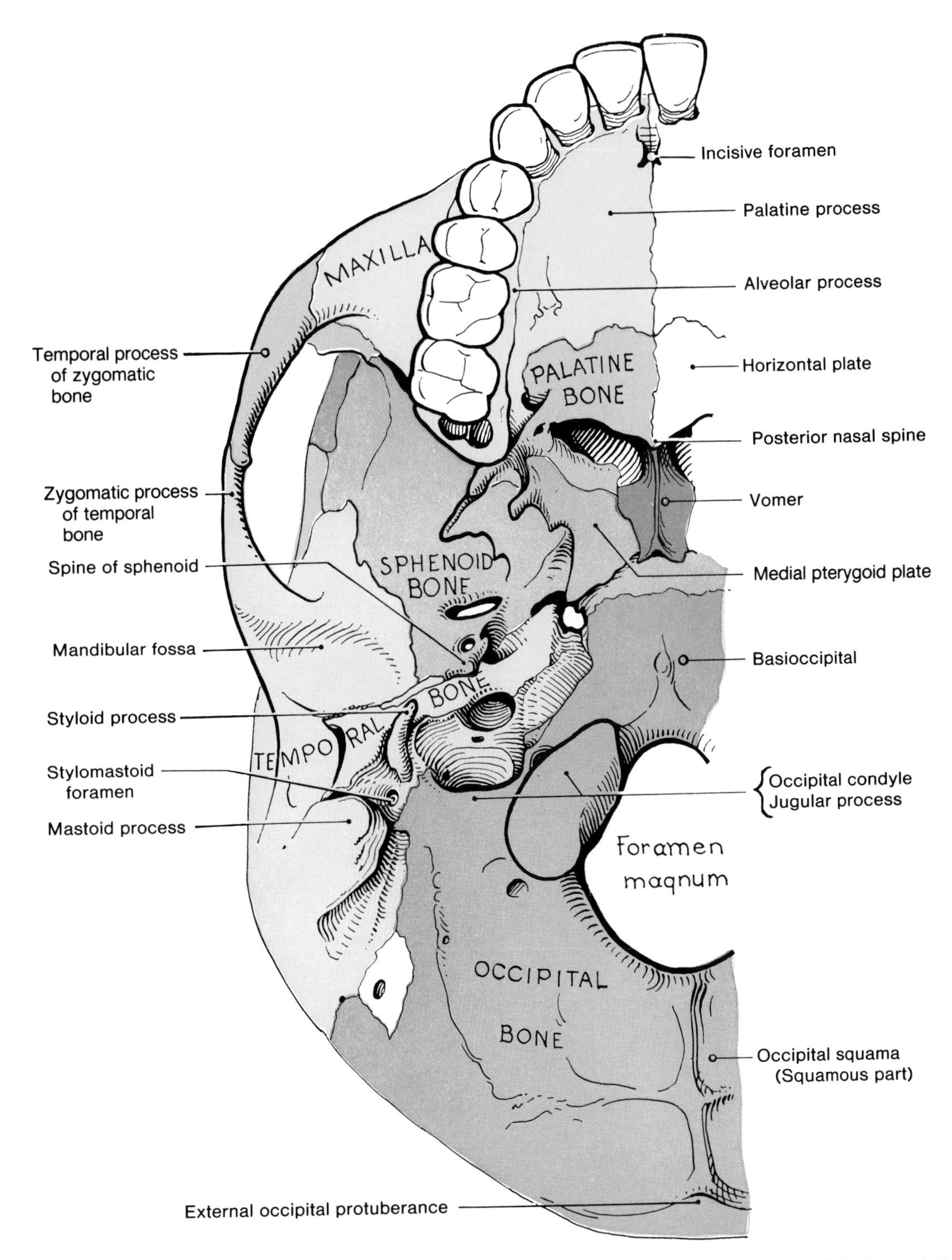

Figure 7-8. Drawing of slightly more than half of the external surface of the base (*norma basalis*) of an adult male skull. The various bones taking part in its formation are colored differently. Note that the temporal process of the zygomatic bone unites with the zygomatic process of the temporal bone to form the zygomatic arch (Fig. 7-11). Note the pharyngeal tubercle, a small midline elevation, located a short distance anterior to the foramen magnum on the basioccipital bone (basal portion of occipital bone). It marks the superior attachment of the median raphe of the pharynx (see Fig. 8-44).

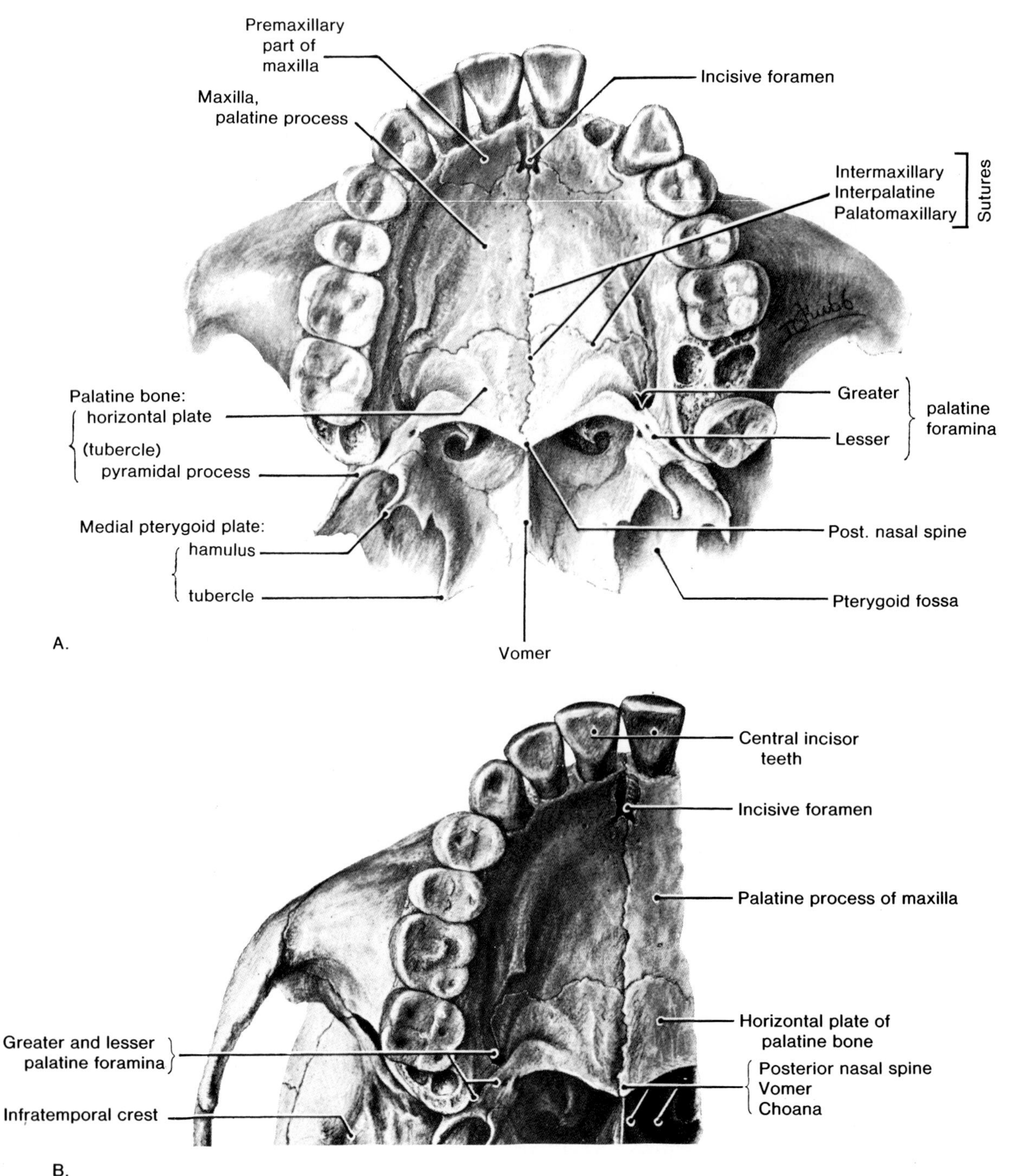

Figure 7-9. Drawings of inferior views of the hard palate and maxillary teeth. Observe that the hard palate is formed by the palatine processes of the maxillae and the horizontal plates of the palatine bones. Note the sutures between these processes and plates and between the maxillae and palatine bones. The suture between the premaxillary part of the maxilla and the fused palatine processes of the maxillae is visible in *A*, a skull from a young person. It is not visible in the hard palates of most dried skulls from older adults (*B*).

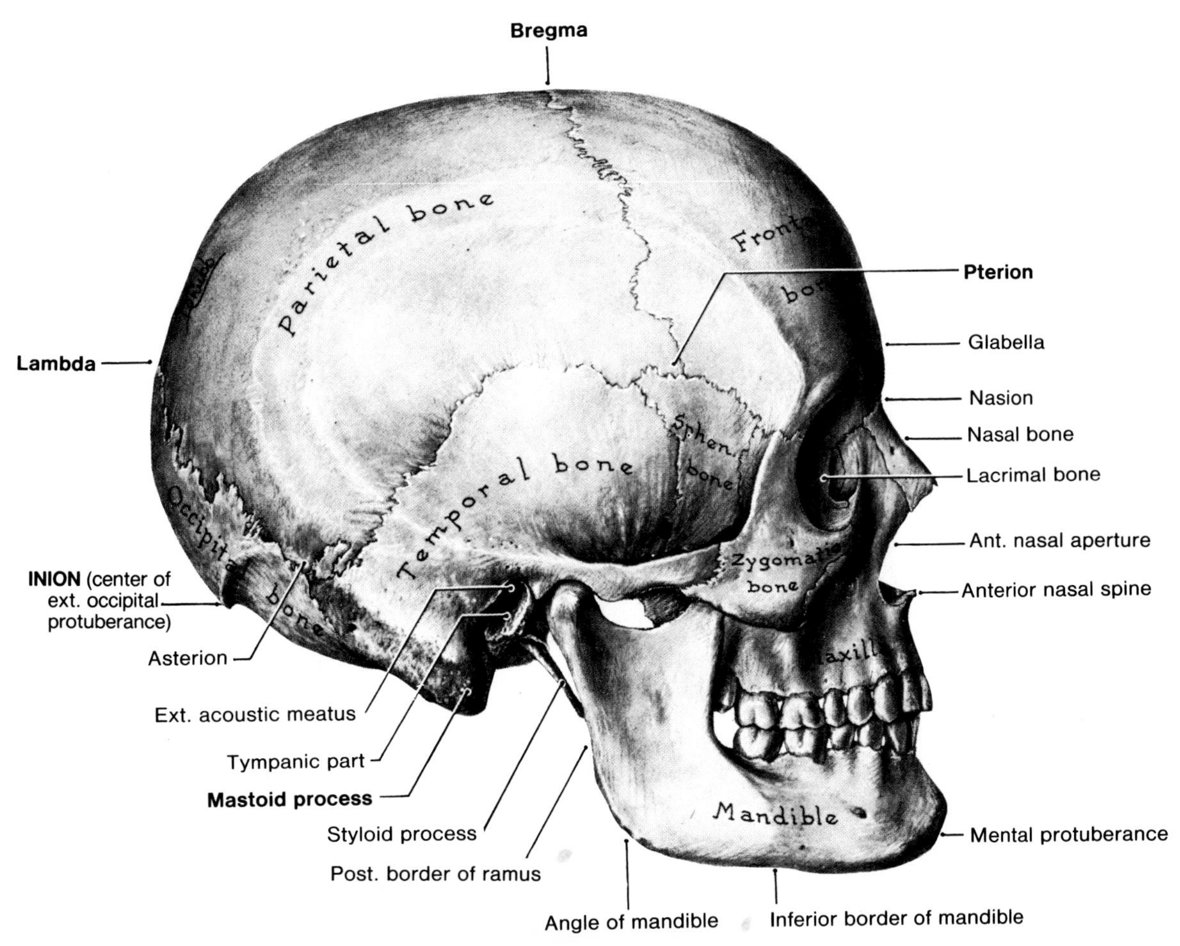

Figure 7-10. Drawing of a lateral view (*norma lateralis*) of an adult skull, naming the bones and various outline features. Observe that the center of the pterion is about two fingerbreadths superior to the zygomatic arch (also see Fig. 7-16). Some of the important anthropological landmarks (pterion, bregma, lambda, and inion) used for *craniometry* (measurement of the skull) are labeled. The external size of the skull and cranial capacity vary considerably in different persons. The shape and size of the cranium is used as a means of comparing skulls. The skulls of women are, on the whole, smaller than those of men and conform with the smaller size of their skeletons.

bones at the ***coronal suture*** (Figs. 7-6 and 7-11). It may help you to remember the plane of this suture if you observe that the crown-like ornament, known as a tiara, fits over the coronal suture (L. *corona*, a crown). As the coronal suture is in a frontal plane, the terms frontal plane and coronal plane are often used interchangeably.

The frontal bone also articulates with the ***nasal bones*** **at the** ***frontonasal suture*** (Figs. 7-1, 7-2, and 7-10). At the point where this suture crosses the **internasal suture** in the median plane, there is an anthropological landmark called the **nasion** (L. *nasus*, nose). This depression is located at the *root of the nose*, where it joins the cranium.

The frontal bone also articulates with the zygomatic, lacrimal, ethmoid, and sphenoid bones.

As the superciliary arches are relatively sharp ridges of bone, a blow to them often lacerates the skin and causes bleeding. This is the reason why "butting" is an infraction of the rules in boxing.

Bruising of the skin over a superciliary arch causes tissue fluid and blood to accumulate in the surrounding connective tissue, and gravitate into the upper eyelid. This results in swelling and a "**black eye.**"

Compression of the **supraorbital nerve** as it emerges from this foramen (Fig. 7-28) causes considerable pain, a fact that is used by anesthetists to determine the depth of anesthesia

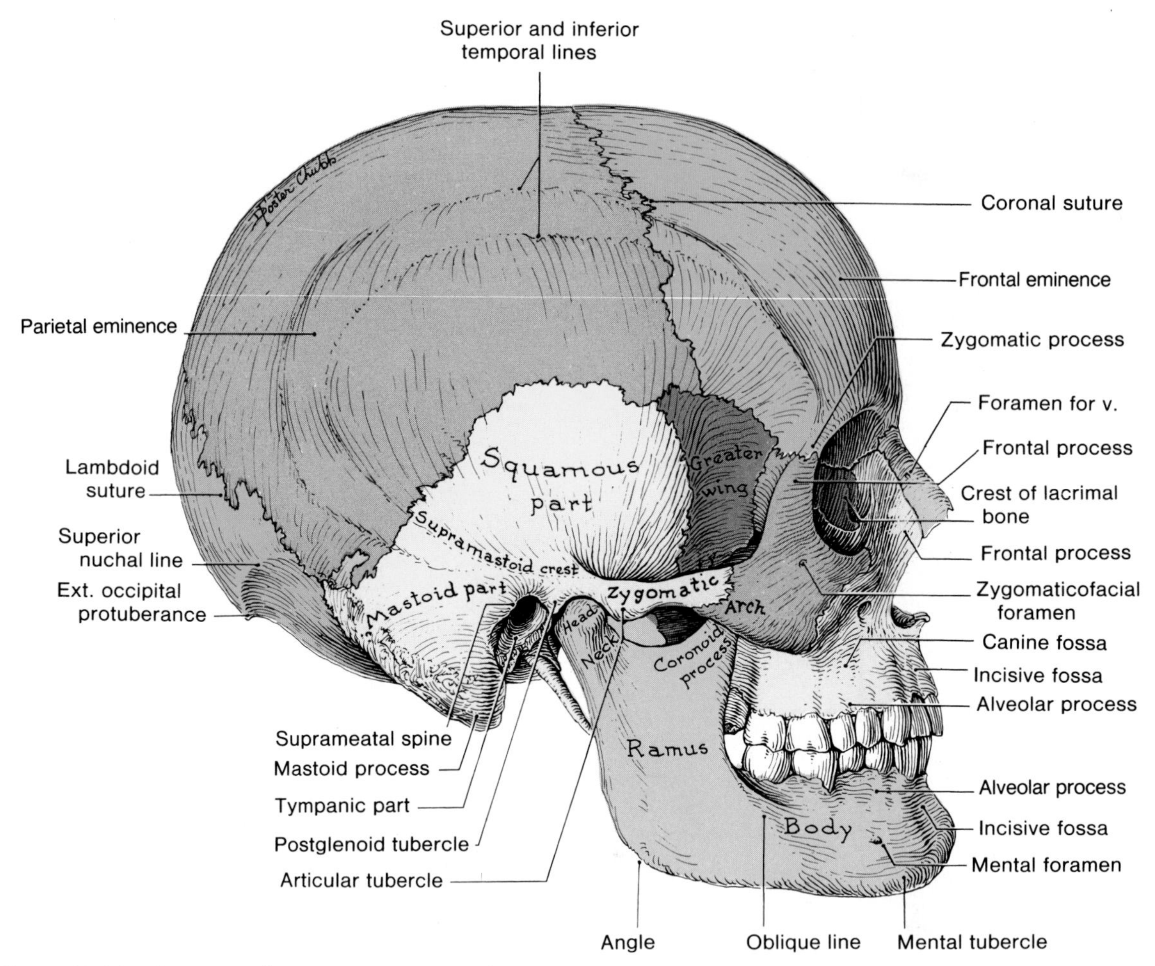

Figure 7-11. Drawing of the lateral aspect of an adult skull. Note that the temporal process of the zygomatic bone (*pink*) and the zygomatic process of the temporal bone (*yellow*) are united at an oblique suture to form the zygomatic arch. Zygomatic is an appropriate adjective because it is derived from the Greek word *zygōtos*, meaning yoked. Observe that this arch forms a bridge that extends horizontally from the zygomatic bone (Fig. 7-10) to the superior attachment of the auricle to the head. Verify by palpation that the zygomatic arch is subcutaneous throughout. In emaciated persons this arch appears unusually prominent. The temporal fossa is bounded by the temporal lines superiorly and the zygomatic arch inferiorly. The infratemporal fossa lies medial to the ramus of the mandible (Fig. 7-19).

and by doctors attempting to arouse a moribund (L. dying) patient.

THE PARIETAL BONES (Figs. 7-3 to 7-6 and 7-10 to 7-14)

The two parietal bones (L. *paries*, wall) form parts of the walls of the calvaria. On the outside of these smooth convex bones, there is a slight elevation near the center called the **parietal eminence** or tuber (Fig. 7-11).

The middle of the lateral surfaces of the parietal bones is crossed by two curved lines, the superior and inferior **temporal lines** (Figs. 7-1, 7-3, and 7-11). The superior temporal line indicates the attachment of the **temporal fascia** (Fig. 7-31); the inferior temporal line marks the superior limit of the **temporalis muscle** (Fig. 7-108).

The parietal bones articulate with each other in the midline at the ***sagittal suture*** (Figs. 7-3, 7-5, and 7-6). The sagittal suture (L. *sagitta*, arrow) was probably given this name because of its resemblance to an arrow in the skull of a newborn (Fig. 7-5*B*). The diamond-shaped membranous interval at the junction of the coronal, sagittal, and frontal sutures, known as the **anterior fontanelle** (F. fountain) or **anterior fon-**

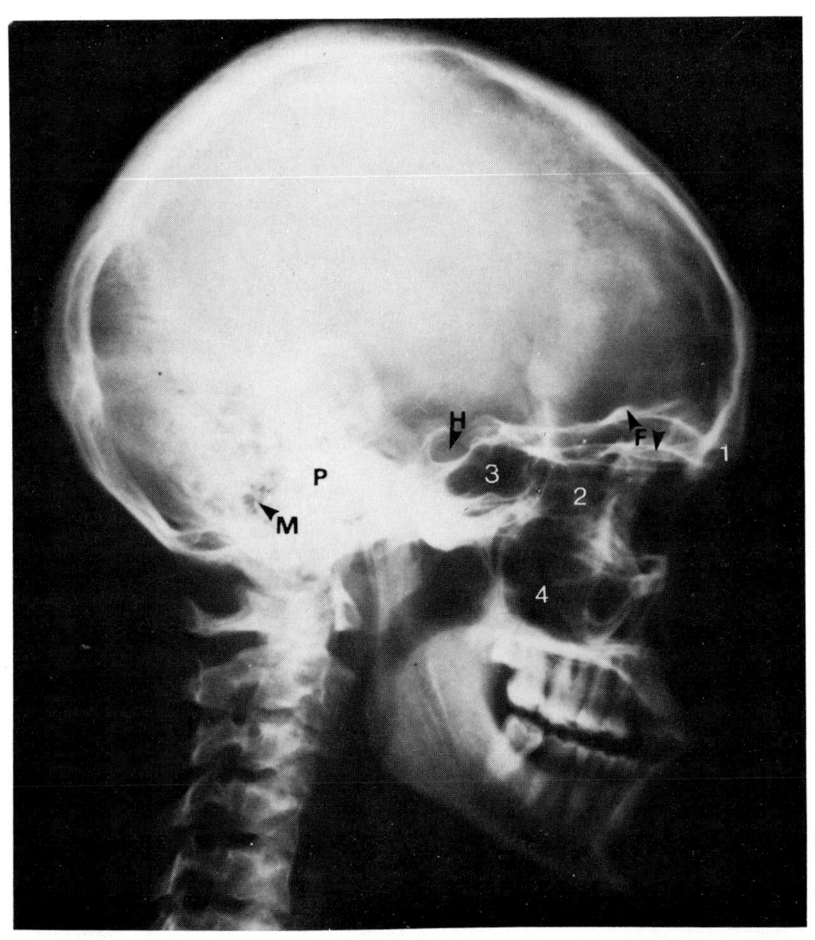

Figure 7-12. Lateral radiograph of the skull. Observe the paranasal sinuses: (1) frontal, (2) ethmoidal, (3) sphenoidal, and (4) maxillary. Also note the hypophysial fossa (*H*), the great density of the petrous parts of the temporal bones (*P*), and the mastoid cells (*M*). The right and left orbital plates of the frontal bone (*F*) are not superimposed, thus the floor of the anterior cranial fossa appears as two lines. Also see Figures 7-13 and 7-15.

ticulus (L. dim. of *fons*, fountain or spring), resembles an arrowhead. The sagittal suture represents the shaft of the arrow.

The **median plane** passes through the sagittal suture; hence the terms midsagittal plane and median plane are used interchangeably. The junction of the sagittal and coronal sutures is called the **bregma** (G. the forepart of the head). See Figure 7-6.

The inverted V-shaped suture between the parietal bones and the occipital bone is called the **lamboid suture** (Figs. 7-3 and 7-11) because of its resemblance to the letter lambda Λ in the Greek alphabet. The point where the parietal and occipital bones join is a useful reference point called the **lambda** (Fig. 7-3). It can be felt as a depression in some people.

In addition to articulating with each other and the frontal and occipital bones, the parietal bones articulate with the temporal bones (Figs. 7-10 and 7-11), and the greater wings of sphenoid bone (Figs. 7-11 and 7-13 to 7-15).

In fetal and infant skulls, the bones of the calvaria are separated by dense connective tissue membranes at fibrous joints called **sutures** (Fig. 7-5). The large fibrous areas where several sutures meet are called **fontanelles or fonticuli.** The softness of the bones and the looseness of their connections at these sutures enable the calvaria to undergo changes of shape during birth called **molding.** The frontal bones become flat; the occipital bone becomes drawn out; and the parietal bones slightly overlap each other. Within a day or so after birth, the shape of the calvaria returns to normal.

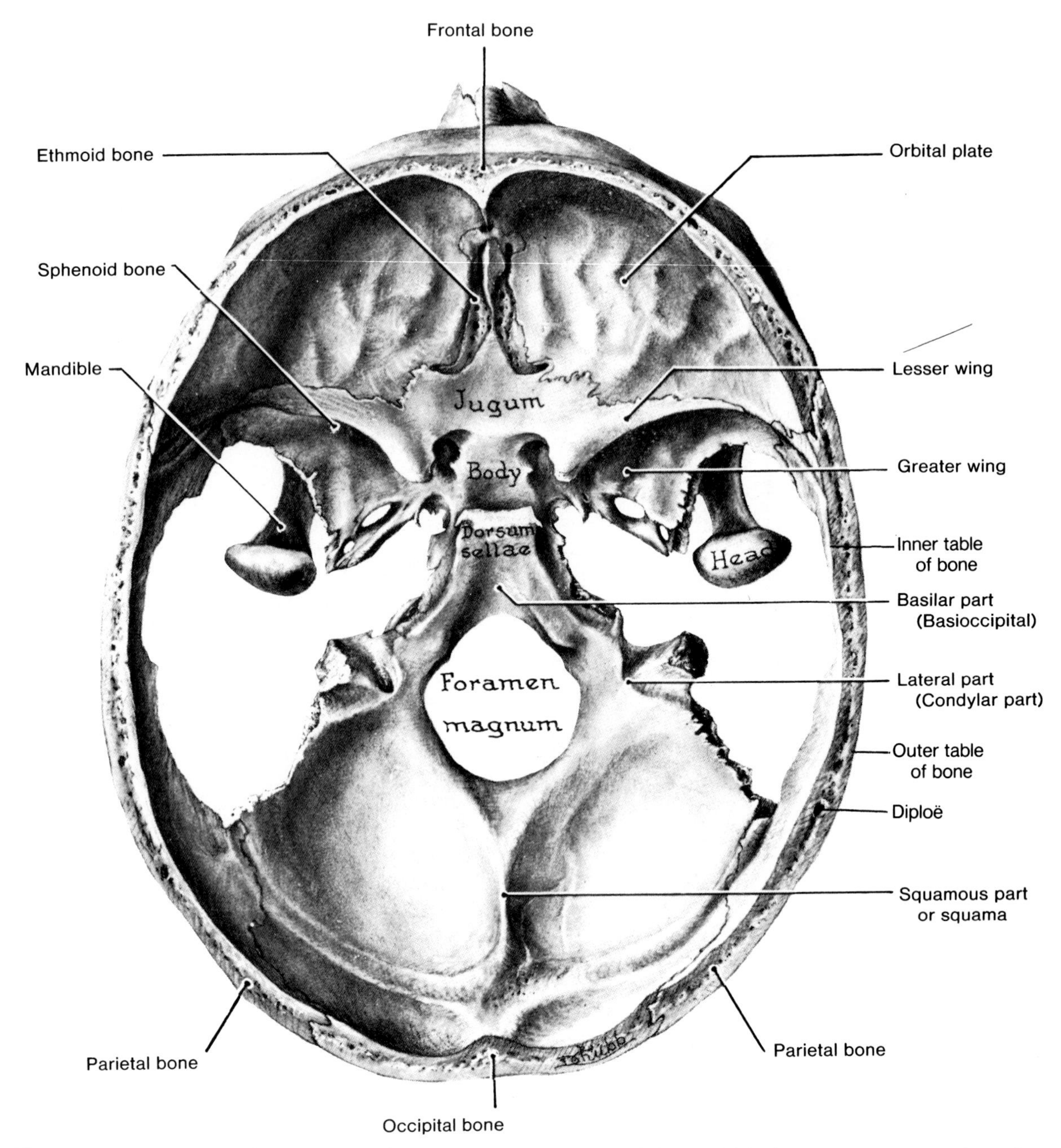

Figure 7-13. Drawing of the interior of the base of an adult skull from which the temporal bones have been removed. Observe that the splenoid bone resembles a bat with its wings outstretched. However, it was so named because of its wedge shape (G. *sphen*, wedge). Observe the large foramen magnum through which the spinal cord becomes continuous with the medulla of the brain stem (Fig. 7-54). Note that the orbital plates of the frontal bone show shallow, sinuous convolutional depressions or brain markings which were formed by the convolutions of the frontal lobe gyri (Fig. 7-52).

The loose construction of the newborn calvaria also allows the skull to enlarge and undergo remodeling during infancy and childhood. Relationships between the various bones are constantly changing during the active growth period.

The increase in the size of the cranium is greatest during the first 2 years of life, the period

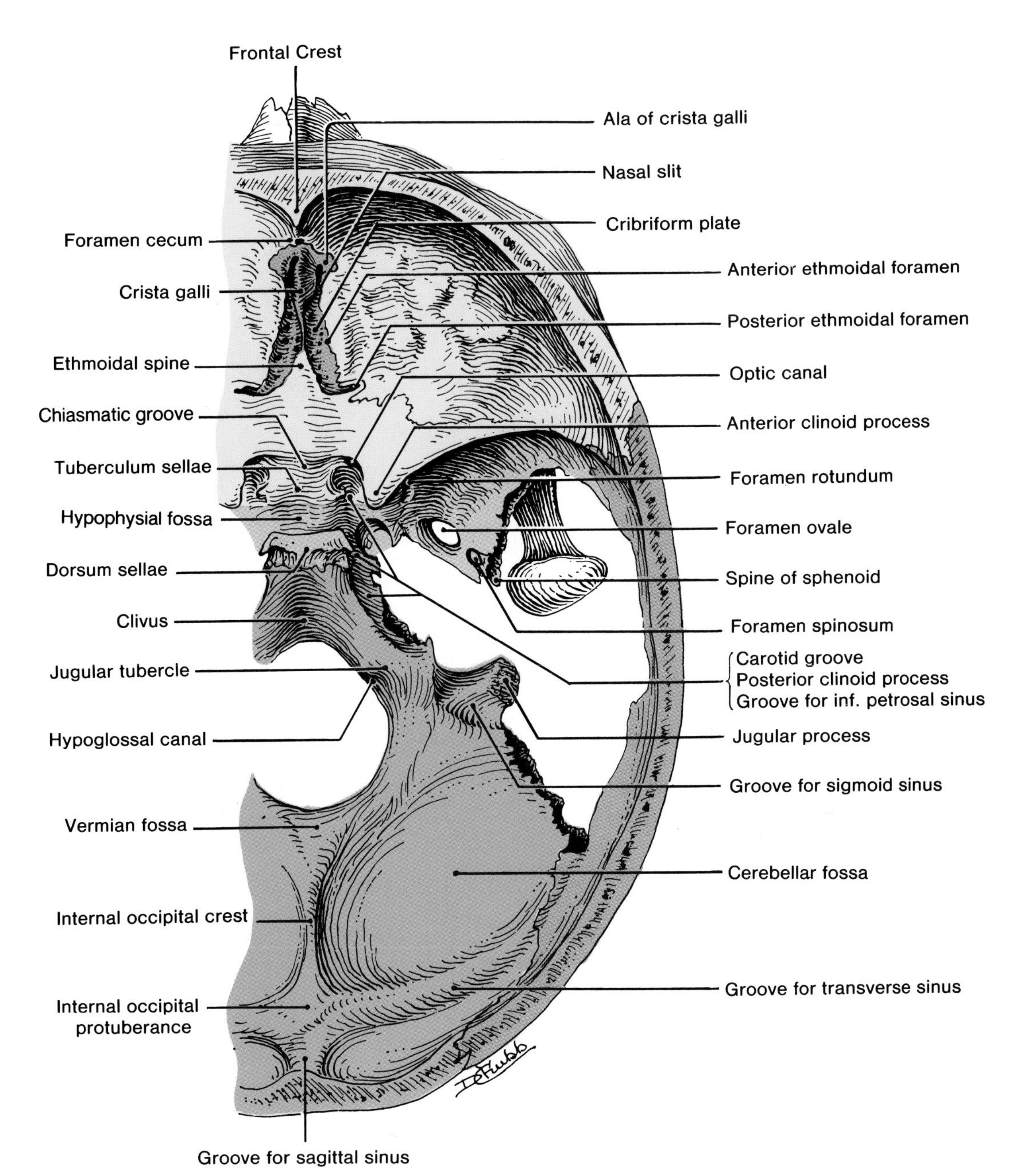

Figure 7-14. Drawing of slightly more than the right half of the interior of the base of the adult skull. Observe the many features in the median plane, noting particularly the crista galli and cribriform plate (*red*) in the anterior cranial fossa (*yellow*). The anterior cranial fossa is limited anteriorly and on each side by the frontal bone. Its floor is formed by the orbital plates of the frontal bone (Fig. 7-13), the cribriform plate of the ethmoid bone, and the lesser wings and anterior part of the body of the sphenoid bone. Most of the middle cranial fossa (*blue*) is missing because the temporal bones have been removed. Note the large cerebellar fossa in the posterior cranial fossa (*green*) in which the cerebellar hemisphere lies (Fig. 7-51). Note the clivus (L. slope), a sloping surface from the dorsum sellae to the foramen magnum. Observe that it is composed of part of the body of the sphenoid bone and part of the basilar part of the occipital bone.

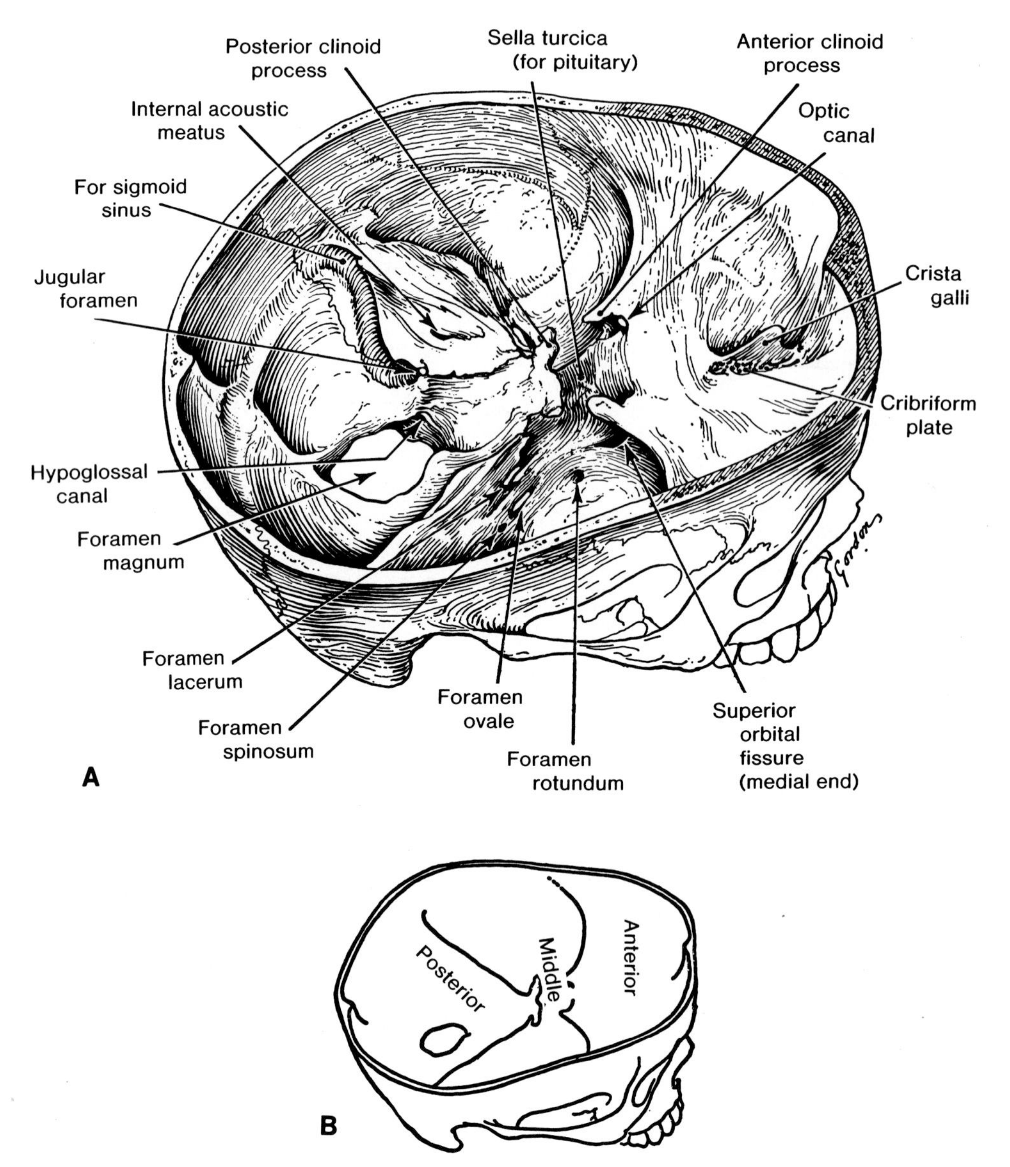

Figure 7-15. Drawing of the internal surface of the base of the cranium showing the three tiered areas; the anterior, middle, and posterior cranial fossae. (For the relationship of the brain to these fossae, see Fig. 7-51.)

of most rapid postnatal growth of the brain. The cranium normally increases in capacity until about the 15th or 16th year; after that the cranium usually increases slightly in size because its bones thicken for 3 to 4 years.

The anterior fontanelle or fonticulus (Fig. 7-5*A*), called the "soft spot" by laymen, is usually obliterated by the end of the 2nd year and its former site is indicated in the adult cranium by the **bregma** (Figs. 7-6 and 7-10).

The posterior fontanelle or fonticulus (Fig. 7-5*B*), the other "soft spot" in an infant's calvaria, is usually obliterated by the end of the 9th month and its former site is indicated by the **lambda** of the adult cranium (Figs. 7-3 and 7-10).

The anterior fontanelle, the larger of the two fibrous areas, is frequently used clinically. During parturition it is palpated to determine the position of the fetal head in a vertex presentation. During infancy it is used to estimate intracranial pressure, *e.g.*, tenseness suggests increased intracranial pressure, as occurs with **meningitis** (inflammation of the membranes covering the brain).

The state of closure of the anterior fontanelle

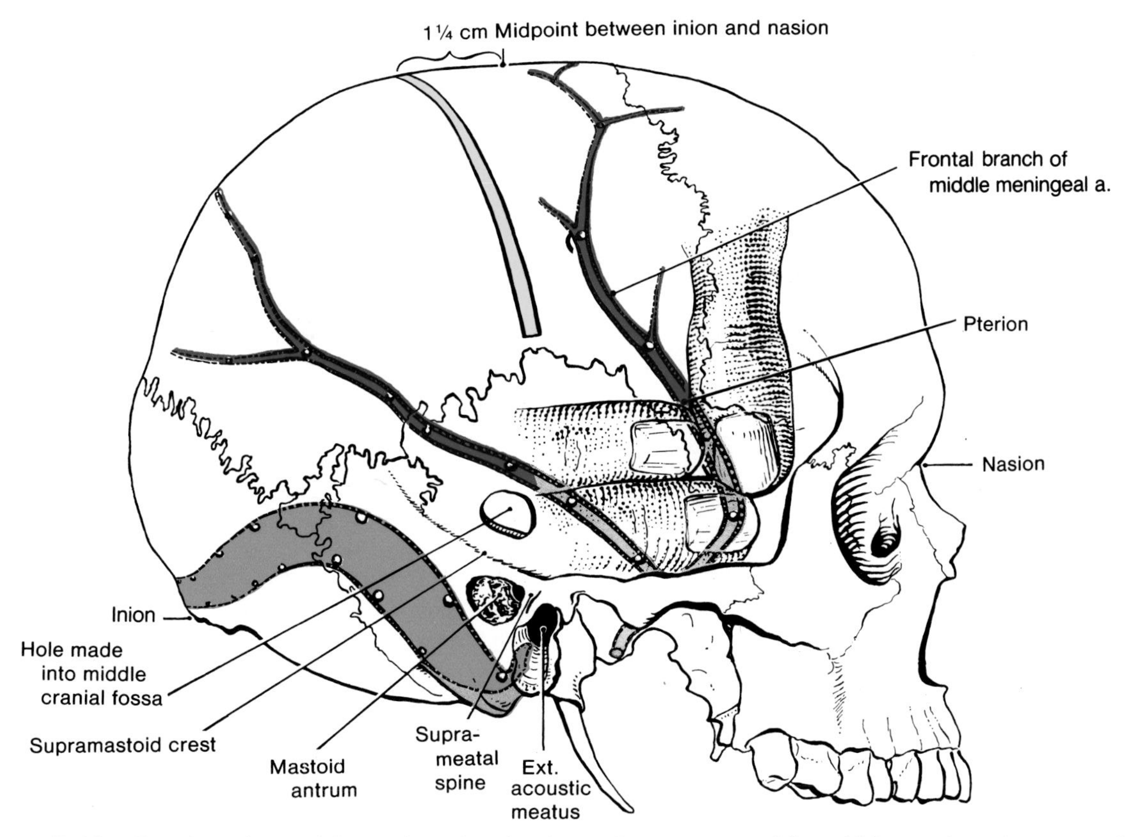

Figure 7-16. Drawing of an adult cranium showing its surface anatomy. The middle meningeal artery and the dural venous sinuses were related to the exterior of the cranium by drilling holes along the grooves produced by these structures on the inside of the cranium. The angle formed by the thumb and fingers indicates the site of the frontal branch of the middle meningeal artery (*red*). *Observe that it crosses the pterion*. The location of the central sulcus of the brain (Fig. 7-51) is shown in *yellow* and the transverse and sigmoid sinuses are shown in *blue.* Because it is liable to rupture from blows on the side of the head, resulting in bleeding between the skull and the dura mater (Fig. 7-175), the surface markings of the middle meningeal artery are of surgical importance.

may also be used to estimate the degree of brain development. Blood may also be withdrawn from the underlying **superior sagittal sinus** (Fig. 7-50) by inserting a needle through the membranous anterior fontanelle.

Several rare **skull deformities** may result from premature closure of the cranial sutures. The type of skull deformity formed depends upon which of the various sutures closes prematurely. If the sagittal suture closes early, the skull becomes long and narrow with a ridge along the closed sagittal suture, a condition called **scaphocephaly** (Fig. 7-18). Premature closure of the lambdoid and coronal sutures results in a tower-like skull, called **turricephaly**, or a sharp pointed head, known as **acrocephaly**. If the lambdoid or coronal suture closes prematurely on one side, the skull is asymmetrical, a condition known as **plagiocephaly**. For illustrations of these skull deformities, see Moore (1982).

THE TEMPORAL BONES (Figs. 7-1 to 7-5 and 7-10 to 7-12)

The sides and base of the skull are formed partly by the temporal bones. Each temporal bone consists of four morphologically distinct parts that fuse during development (squamous, petromastoid, and tympanic parts, and the styloid process).

The flat **squamous part** or squama temporalis (Fig. 7-11) is external to the lateral surface of the temporal lobe of the brain (Fig. 7-51). The **petromastoid part** encloses the internal ear and mastoid cells (Fig. 7-12), and forms part of the base of the skull (Figs. 7-8 and 7-11). The **tympanic part** of the temporal bone contains the bony passage from the auricle, called the **external acoustic meatus** (Figs. 7-10 and 7-162).

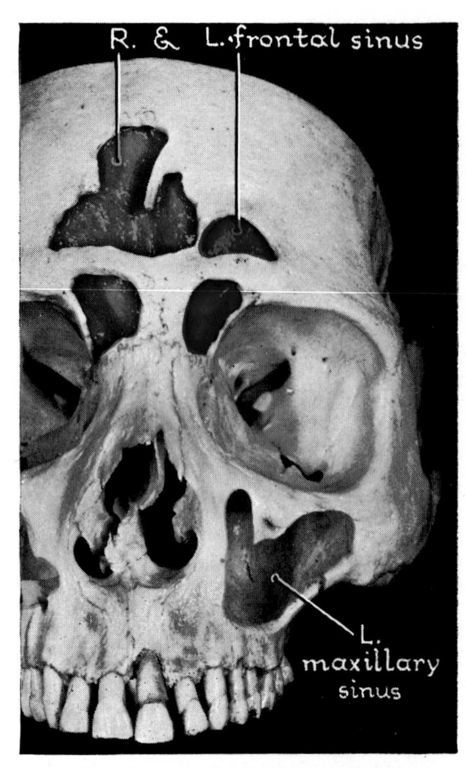

Figure 7-17. Photograph of an anterior view of an adult skull that has been specially prepared to demonstrate the paranasal sinuses. Observe the crooked bony nasal septum in this specimen. This deformity may have resulted from an injury.

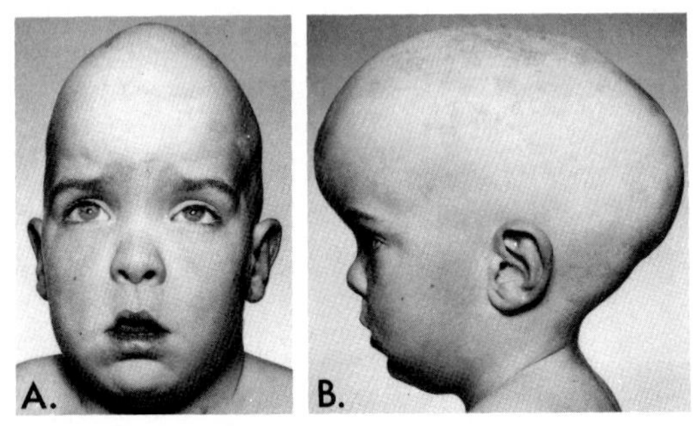

Figure 7-18. Photographs of a boy with scaphocephaly resulting from premature closure of the sagittal suture. *A*, anterior view. *B*, lateral view. (From Laurence KM, Weeks R: Abnormalities of the central nervous system. In Norman AP (ed): *Congenital Abnormalities in Infancy*, ed 2. Oxford, Blackwell Scientific Publications, 1971.

The petromastoid part also forms a portion of the bony wall of the tympanic cavity. The meatus and tympanic cavity are concerned with the transmission of sound waves (Fig. 7-162). The slender, pointed **styloid process** of the temporal bone (Figs. 7-8 and 7-10) gives attachment to certain ligaments and muscles (*e.g.*, the stylohyoid muscle which elevates the hyoid bone).

The temporal bones articulate at sutures with the parietal, occipital, sphenoid, and zygomatic bones. The **zygomatic process of the temporal bone** unites with the *temporal process of the zygomatic bone* to form the **zygomatic arch** (Figs. 7-8 and 7-11). The zygomatic arches form the widest parts of the face.

The head of the mandible (Figs. 7-11 and 7-19) articulates with the **mandibular fossa** (Fig. 7-8) on the inferior surface of the zygomatic process of the temporal bone. Anterior to the mandibular fossa is the **articular tubercle** (Fig. 7-11).

Because the zygomatic arches are the widest parts of the face and are such prominent facial features, they are commonly fractured and depressed. A fracture of the temporal process of the zygomatic bone would involve the lateral wall of the orbit and could injure the eye.

THE SPHENOID BONE (Figs. 7-8 to 7-15)

This wedge-shaped bone (G. *sphēn*, wedge) is located in the base of the skull, anterior to the temporal bones (Fig. 7-8). Its form somewhat resembles a bat with its wings outstretched (Figs. 7-13 and 7-14).

The sphenoid bone is a key bone in the cranial skeleton because it articulates with eight bones (frontal, parietal, temporal, occipital, vomer, zygomatic, palatine, and ethmoid). Its main parts are the central **body** and the **greater and lesser wings** which spread laterally from the body (Figs. 7-4, 7-13, and 7-14).

The superior surface of the body of the sphenoid is shaped like a Turkish saddle (L. *sella*, a saddle); hence its name **sella turcica** (Fig. 7-15*A*). The "seat of the saddle" or sella turcica forms the **hypophysial fossa** (Fig. 7-12) which contains the *hypophysis cerebri* or *pituitary gland* (Fig. 7-54). Inside the body of the sphenoid bone, there are right and left **sphenoidal sinuses** (Figs. 7-12 and 7-154). The floor of the sella turcica forms the roof of these sinuses when they are well developed.

Studies of the sella turcica and hypophysial fossa in radiographs (Fig. 7-12) are important because they may reflect pathological changes such as a **pituitary tumor** or a localized dilation or an **aneurysm of the internal carotid artery**. Decalcification of the dorsum sellae is one of the signs of a generalized increase in intracranial pressure.

THE OCCIPITAL BONE (Figs. 7-3, 7-5, 7-8, and 7-10 to 7-16)

This bone forms much of the base and posterior aspect of the skull. It has a large oval opening in it called the **foramen magnum**, through which the cranial cavity communicates with the vertebral canal. It is also the opening where the spinal cord becomes continuous with the medulla of the brain stem (Fig. 7-54).

The occipital bone is saucer-shaped and can be divided into four parts: a squamous part or *squama*, a basilar part, and two lateral parts (Fig. 7-13). These parts develop separately around the foramen magnum and at about the age of 6 years unite to form one bone.

On the inferior surfaces of the lateral parts of the occipital bone are **occipital condyles**, where the skull articulates with the *atlas*, the first cervical vertebra, at the atlantooccipital joints (Fig. 5-35).

The internal aspect of the squamous part of the occipital bone is divided into four fossae: the superior two for the occipital poles of the cerebral hemispheres, and the inferior two for the cerebellar hemispheres (Figs. 7-14 and 7-51).

OTHER BONES OF THE CRANIUM (Figs. 7-1 to 7-15)

The previously described cranial bones are the principal ones. Other bones in the *Nomina Anatomica* classified as part of the cranium are the unpaired **vomer and ethmoid bones**, and the paired inferior **nasal conchae, lacrimal, and nasal bones**. These bones are often described with the facial skeleton.

The nasal bones may be easily felt because they form the bridge of the nose. The right and left bones articulate with each other at the **internasal suture** (Fig. 7-1). The nasal bones also articulate with the frontal bones, the maxillae, and the ethmoid bones (Figs. 7-2, 7-10, and 7-11).

The other bones forming parts of the skull are discussed subsequently with the face, orbit, and nasal cavities.

The mobility of the anteroinferior portion of the nose, supported only by cartilages, serves as a partial protection against injury (*e.g.*, a punch in the nose). However a hard blow to the bony portion of the nose (*e.g.*, with a hockey stick or a baseball), may fracture the nasal bones producing a **broken nose**. Often the bones are displaced sideways and/or posteriorly.

The outer table of bone forming the cranium of a living person is somewhat resilient, unlike that in a dried skull from which the proteins have been removed. This elasticity, especially in infants and children, tends to prevent many blows to the head from producing **skull fractures**. Furthermore, in places where the bone is very thin (e.g., the squamous part of the temporal bone, often called the **temporal squama** (Fig. 7-11), the overlying muscles afford some assistance in cushioning blows (Fig. 7-108).

Fractures of the skull are caused by external violence. **Linear fractures** are the most frequent type. The fracture usually occurs at the point of impact, but fracture lines may radiate away from it in two or more directions (*e.g.*, the way a window breaks when hit by a stone). The directions of the radiating fracture lines are determined by the thick and thin areas of the skull. Sometimes there is no fracture at the point of impact, but one occurs at the opposite side of the skull. This is called a **contrecoup fracture.** *Contrecoup* is a French word meaning counterblow.

Depressed fractures of the calvaria are invariably caused by direct violence. Usually the inner table of bone (Fig. 7-13) is more extensively fractured than the outer table. In infants and children, depressed fractures are similar to the indentation that occurs in a ping-pong ball when it is pressed with the thumb. Using appropriate techniques, depressed fragments can be elevated, but in many cases one of the fragments has to be removed from the dura mater (Fig. 7-50) so it will not injure the brain.

THE FACIAL SKELETON

Most of the facial skeleton is formed by seven bones (Figs. 7-2 and 7-9 to 7-11): three paired (**zygomatic, maxilla, and palatine**) and one unpaired (**mandible**).

The calvaria of the newborn infant is large compared with the face (Fig. 7-5). The relatively small facial skeleton results from the small size of the jaws and the almost complete absence of the maxillary and other **paranasal sinuses** in the newborn skull. The sinuses form large spaces in the adult facial skeleton (Figs. 7-12 and 7-17). As the teeth develop and the sinuses develop during infancy and childhood, the facial bones enlarge (Fig. 7-24). The growth of the maxillae between the ages of 6 and 12 years accounts for the vertical elongation of the face.

THE MAXILLAE (Figs. 7-1, 7-2, 7-5, 7-8 to 7-12, and 7-17 to 7-20)

The skeleton of the face between the mouth and eyes is formed by the two maxillae which are united in the median plane at the **intermaxillary suture**

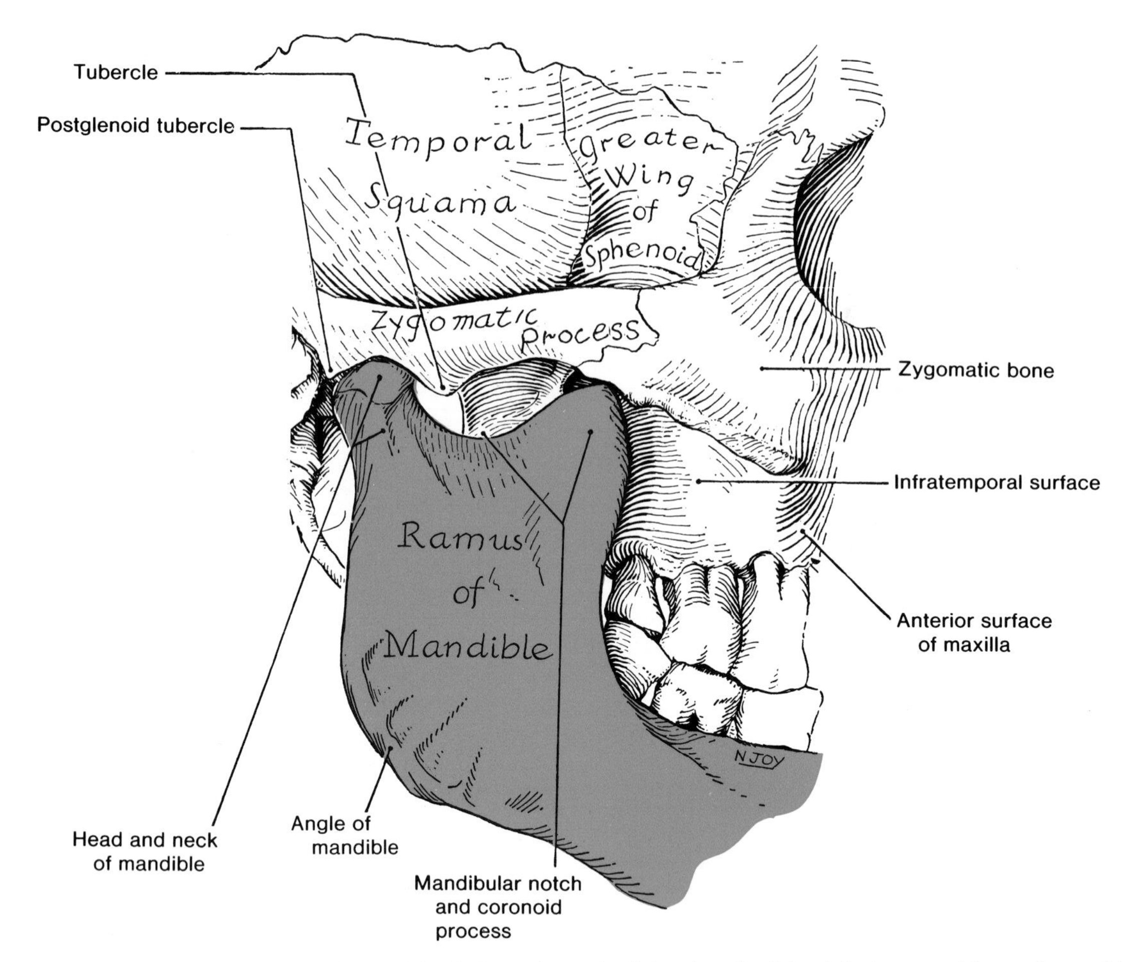

Figure 7-19. Drawing of part of an adult skull to show the lateral wall of the infratemporal fossa formed by the ramus of the mandible. The infratemporal fossa is discussed on page 910.

(Fig. 7-1) to form the entire upper jaw. This suture is also visible in the hard palate (Fig. 7-9*A*), where the palatine processes of the maxillae unite.

Each adult maxilla consists of: a hollow **body** containing a large ***maxillary sinus*** (Figs. 7-12 and 7-17); a **zygomatic process** that articulates with the zygomatic bone; a **frontal process** that articulates with the frontal and nasal bones; a **palatine process** that articulates with its mate on the other side to form most of the hard palate (Fig. 7-9); and an **alveolar process** that bears the upper or maxillary teeth (Figs. 7-2 and 7-8 to 7-10). The maxillae also articulate with the vomer, lacrimal, sphenoid, palatine bones, and the inferior nasal conchae.

The body of the maxilla has a **nasal surface** that contributes to the lateral wall of the nasal cavity (Fig. 7-2); an **orbital surface** that forms most of the floor of the orbit (Figs. 7-2 and 7-12); an **infratemporal surface** (Fig. 7-19) that forms the anterior wall of the infratemporal fossa (Fig. 7-11); and an **anterior surface**, facing partly anteriorly and partly anterolaterally, which is covered with facial muscles (Fig. 7-28).

The relatively large **infraorbital foramen**, which faces inferomedially, is located about 1 cm inferior to the infraorbital margin (Fig. 7-2); it transmits the infraorbital vessels and nerve (Figs. 7-20 and 7-32).

In old age, or earlier if the teeth are removed, the bone of the alveolar processes of the maxillae is absorbed. As a result, the maxillae become smaller (*i.e.*, decreased in height) and the shape of the face changes. Owing to absorption of the alveolar processes of the jaws, there is a marked reduction in the height of the lower face which produces deep creases in the facial skin that pass posteriorly from the corners of the mouth.

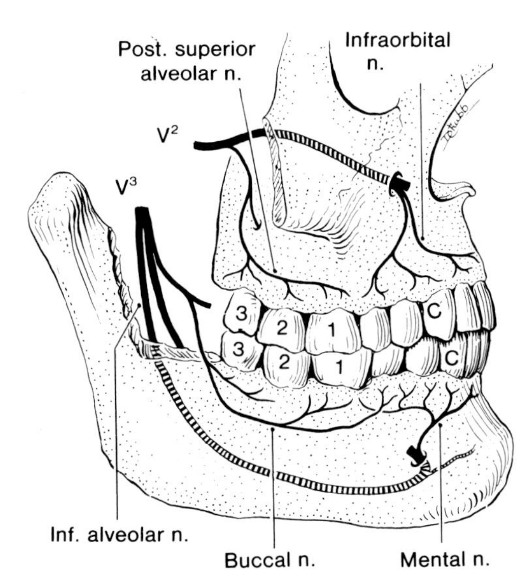

Figure 7-20. Drawing of the jaws illustrating the distribution of the maxillary and mandibular divisions (CN V^2 and CN V^3) of the trigeminal nerve (also see Fig. 7-33). Awareness of the distribution of these nerves is essential for planning local anesthesia for pain control. As we all know, pain control in these areas is very important in dentistry. *C*, canine teeth; molar teeth: 1, 2, 3.

THE MANDIBLE (Figs. 7-1 to 7-5, 7-10 to 7-12, and 7-19 to 7-25)

The mandible forms the lower jaw and face. It is the *largest and strongest bone of the face*. The lower or mandibular teeth project superiorly from their sockets in the alveolar process of the mandible.

The mandible (L. *mandere*, to masticate) consists of two parts (Fig. 7-11): a horizontal part called the **body**, and two vertical oblong parts, the **rami**. The right and left parts of the body are fused anteriorly at the *symphysis menti* (Fig. 7-1) to form a U-shaped bone (Fig. 7-25). The halves of the newborn mandible articulate at the symphysis menti (Fig. 7-5*A*). Fusion of this joint occurs in the 2nd year.

Each ramus ascends almost vertically from the posterior aspect of the body (Figs. 7-11 and 7-19). The superior part of the ramus has two processes: a posterior **condylar process** with a head and neck, and a sharp anterior **coronoid process** (Figs. 7-11, 7-21, and 7-22). The condylar process is separated from the coronoid process by the **mandibular notch** (Fig. 7-21), which forms the concave superior border of the mandible. Viewed from the superior aspect, the mandible is horseshoe-shaped (Fig. 7-127), whereas each half is L-shaped when viewed from the side (Fig. 7-21). The rami and the body meet posteriorly at the **angle of the mandible** (Figs. 7-11, 7-19, and 7-21).

Inferior to the second premolar tooth on each side, there is a **mental foramen** (Figs. 7-2, 7-11, and 7-21) for transmission of the mental (L. *mentum*, chin) vessels and nerve (Figs. 7-20 and 7-28).

In the anatomical position, the rami of the mandible are almost vertical, except during infancy (Fig. 7-5*C*) and in edentulous (toothless) persons (Fig. 7-23).

On the internal aspect of the ramus of the mandible, there is a large **mandibular foramen** (Fig. 7-22). It is the oblong entrance to the **mandibular canal** which transmits the inferior alveolar vessels and nerve to the roots of the mandibular teeth (Fig. 7-20). Branches of these vessels and the **mental nerve** emerge from the mandibular canal at the **mental foramen** (Figs. 7-20 and 7-28).

Sometimes it is necessary to anesthetize the mental nerve and the incisive branch of the inferior alveolar nerve (*e.g.*, to suture a lacerated lower lip and chin). **A mental and incisive nerve block** is done by injecting the anesthetic fluid into the mouth of the mental foramen. As this foramen opens superiorly and slightly posteriorly, access to it is somewhat difficult. During injection the mental artery and vein must be avoided so that the anesthetic agent will not be injected into them.

Anterior to the mandibular foramen there is a thin, tongue-like projection (spur) of bone, called the **lingula of the mandible** (L. *lingua*, tongue), that somewhat overlaps and guards the superoanterior border of the foramen like a tongue or shield (Fig. 7-22). The *sphenomandibular ligament* attaches to the lingula (Fig. 7-115).

The lingula may interfere with injections of anesthetic fluid into the mandibular foramen for performing an **inferior alveolar nerve block**. Of all the routine nerve blocks in the mouth, the inferior alveolar block is probably the most difficult to perform satisfactorily. In addition to avoiding the inferior alveolar vessels that run with the nerve (Fig. 7-20), one must be careful not to push the needle too far posteriorly into the parotid gland and block the facial nerve.

Running inferiorly and slightly anteriorly on the internal surface of the mandible from the mandibular foramen is a small **mylohyoid groove** (sulcus), indicating the course taken by the mylohyoid nerve and vessels (Figs. 7-22 and 7-114). These structures arise from the inferior alveolar nerve and vessels just before they enter the mandibular foramen.

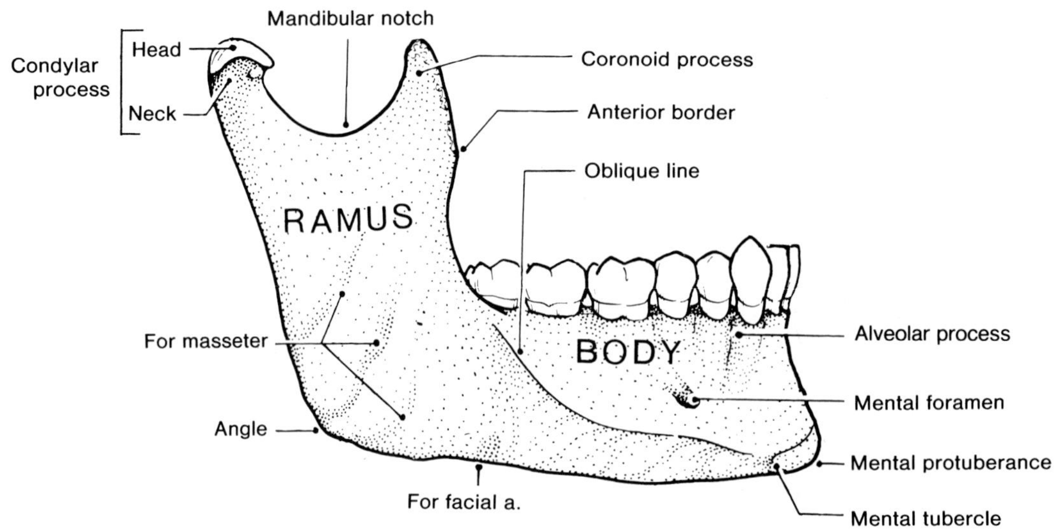

Figure 7-21. Drawing of the lateral aspect of the right half of an adult mandible. Because the ramus of the mandible is covered by the masseter muscle (Fig. 7-29), only its posterior border can be easily palpated. The neck of the mandible lies just anterior to the lobule of the auricle (Fig. 7-160).

The internal surface of the mandible is divided into two areas by the ***mylohyoid line*** (Fig. 7-22), which commences posterior to the third molar tooth. Just superior to the anterior end of the mylohyoid line is a small irregular elevation, called the **mental spine** (genial tubercle). In some cases there are two or four spines. The mental spine serves as an attachment for muscles (Figs. 8-14 and 8-16).

The size and shape of the mandible and the number of teeth it bears change with age. **In the newborn**, the mandible consists of two halves united in the median plane (***symphysis menti***) by fibrous tissue at the mandibular suture (Fig. 7-5*A*). The body of the newborn's mandible is a mere shell, enclosing in each half five primary teeth in their sockets like peas in a pod.

The teeth usually begin to erupt in infants of about 6 months (Table 7-7). The body of the mandible elongates, particularly posterior to the mental foramen, to accommodate the eight secondary teeth (Fig. 7-24), which begin to erupt during the 6th year. Eruption of the teeth is not complete until early adulthood, but the third molars ("wisdom teeth") may be malposed and impacted so they are unable to erupt properly (Fig. 7-118).

In old age or earlier, with loss of teeth, the bone of the alveolar ridges of the mandible absorb and the mental and mandibular foramina come to lie near the superior border of the body of the bone (Fig. 7-23). In extreme cases the mental foramen and part of the mandibular canal may disappear, *exposing the mental and inferior alveolar nerves to injury.* **Pressure of a dental prosthesis**, (*e.g.*, a denture on an exposed nerve in an edentulous jaw) may produce pain during eating.

The bones of the face are often fractured by direct violence (*e.g.*, during fights and when the face crashes into a car's dashboard). The common sites of **fractures of the mandible** are explained and illustrated in Figure 7-25. Usually there are two fractures and they are frequently on opposite sides; thus if one fracture is observed, a search should be made for another. For example, a hard blow to the jaw often fractures the neck of the mandible and its body in the region of the opposite canine tooth. Displacement of bone fragments at the fracture lines results in **malocclusion** (deviation from the normal contact of teeth).

THE ZYGOMATIC BONES (Figs. 7-2, 7-10 to 7-12, and 7-19)

The prominences of the cheeks (L. *mala*), the anterolateral rims of the orbits, and much of the infraorbital margins of the orbits, are formed by the zygomatic bones (malar bones). These prominences formed by the zygomatic bones are what laymen call the *cheek bones*.

The zygomatic bone on each side articulates with the frontal, maxilla, sphenoid, and temporal bones. The frontal process of the zygomatic bone passes

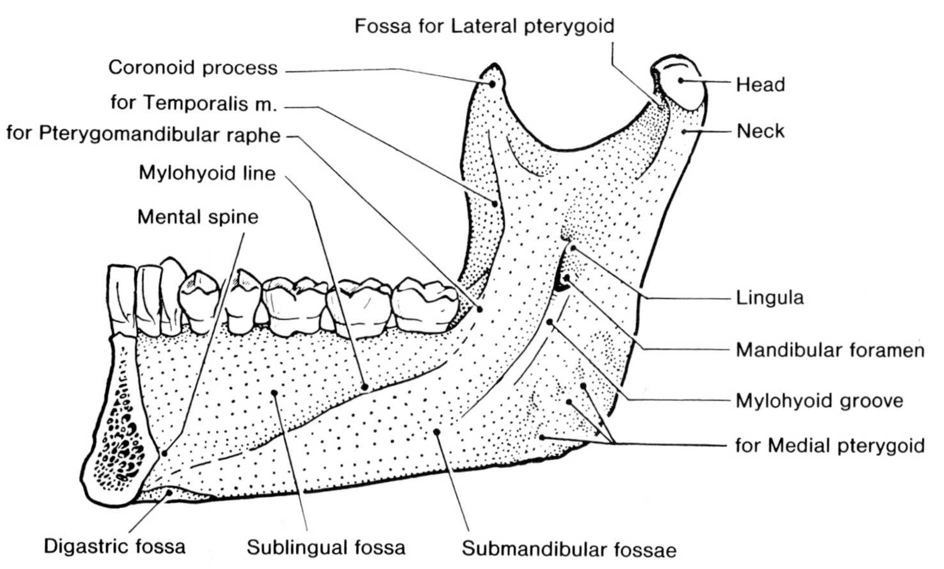

Figure 7-22. Drawing of the medial (internal) aspect of the right half of an adult mandible. The inferior alveolar nerve enters the mandibular canal via the mandibular foramen (Fig. 7-20). The tongue-like lingula, a clinically important bony landmark, may interfere with anesthetic injections at the mouth of the mandibular foramen for anesthetizing the inferior alveolar nerve (*e.g.*, prior to dental procedures).

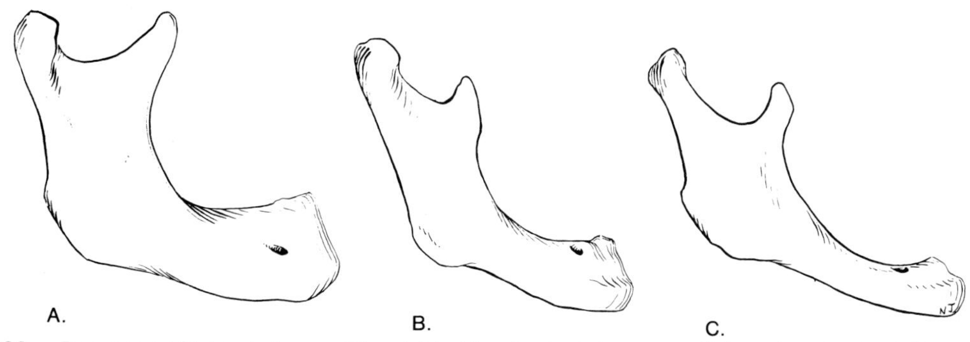

Figure 7-23. Drawings of lateral views of the right side of edentulous mandibles, showing how the position of the mental foramen varies with the extent of absorption of the alveolar process (also see Fig. 7-21). *A*, several months after removal of all the permanent mandibular teeth. Observe that the angle between the body and the ramus is about normal, *i.e.*, approaches a right angle (Fig. 7-21). Note that the mental foramen is about midway between the superior and inferior borders of the bone. *B*, several years after loss of the teeth. Observe that much of the alveolar bone has been absorbed and that the angle of the mandible has increased. Note that the mental foramen is now near the superior border of the mandible. *C*, many years after loss of the mandibular teeth showing that the alveolar process is entirely absorbed (an example of disuse atrophy) and that the body of the mandible consists entirely of the inferior part. The mental foramen now lies at the superior border of the mandible. Note that the angle between the body and the ramus has further increased and that the neck of the condylar process is bent posteriorly. Because of the exposed position of the mental nerve emerging from the mental foramen, probably a dental prosthesis would exert pressure on it producing pain in the skin of the chin and lower lip, including the labial gingiva (gum) in this region.

superiorly, forming the lateral border of the orbit and articulating with the frontal bone at the lateral edge of the supraorbital margin (Fig. 7-2). The maxilla lies mainly inferomedial to the zygomatic bone.

The zygomatic bones articulate medially with the greater wings of the ***sphenoid bone***. The site of their articulation may be observed on the lateral wall of the orbit (Figs. 7-2, 7-11, and 7-19). On the anterolateral aspect of the zygomatic bone, near the infraorbital margin, there is a small **zygomaticofacial foramen** (often double) for the nerve and vessels of the same name (Figs. 7-2 and 7-11). The posterior surface of the zygomatic bone, near the base of its frontal process, is pierced by a small **zygomaticotemporal foramen** for the nerve of the same name. The zygomaticofacial and zygomaticotemporal nerves, leaving the

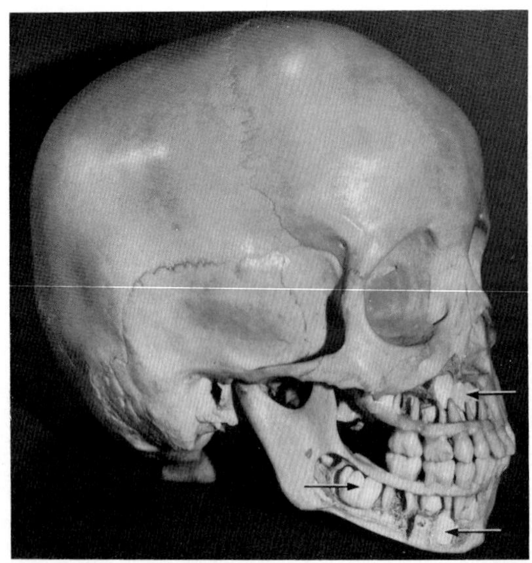

Figure 7-24. Photograph of the skull of a 4-year-old child. Alveolar bone has been ground away and the jaws have been dissected to show the relations of the developing permanent teeth (*arrows*) to the deciduous teeth. Between the 6th and 12th years, the 20 primary teeth are shed as the permanent teeth erupt. By the age of 12, 28 permanent teeth are present. The last four (third molars or "wisdom teeth") erupt during adolescence, early adulthood, or never. Table 7-7 shows the order and time of eruption of teeth and when the decidiuous teeth are shed. (Reprinted with permission from Moore KL: *The Developing Human: Clinically Oriented Embryology*, ed 3. Philadelphia, WB Saunders, 1982).

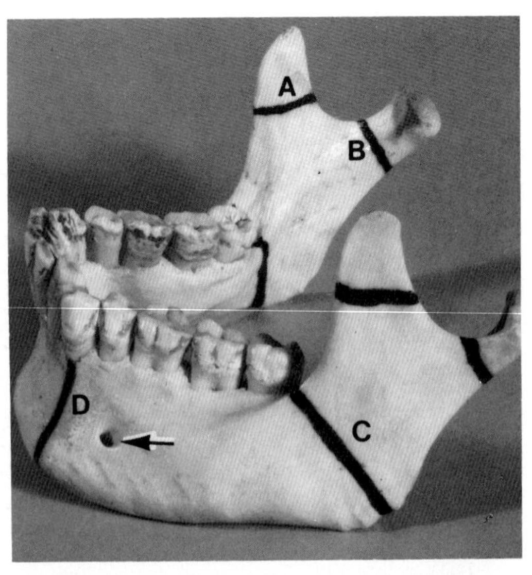

Figure 7-25. Photograph of the lateral aspect of an adult mandible from the left side, indicating the common sites of fracture. *Usually there are two fractures*, frequently on opposite sides. *A*, fractures of the coronoid process are usually single. *B*, fractures of the neck are often transverse and may be associated with dislocation of the temporomandibular joint on the same side. *C*, fractures of the angle of the mandible are usually oblique and may involve the socket of the third molar tooth. Fractures through the sockets of the teeth are, or potentially are, compound fractures. *D*, fractures of the body of the mandible frequently pass through the socket of the canine tooth. The *arrow* indicates the mental foramen.

orbit through the previously mentioned foramina, enter the zygomatic bone through small **zygomaticoorbital foramina** that pierce its orbital surface.

*The **temporal process** of the zygomatic bone unites with the **zygomatic process** of the temporal bone to form the **zygomatic arch*** (Figs. 7-1, 7-8, and 7-10 to 7-12). This arch can be easily palpated on the side of the head, posterior to the zygomatic prominence (malar eminence), and inferior to the temporal fossa (temple). The zygomatic arches, especially prominent in emaciated persons, form one of the useful landmarks for determining the location of the pterion (Figs. 7-10 and 7-16). A horizontal plane passing medially from the zygomatic arch separates the temporal fossa superiorly from the infratemporal fossa inferiorly (Figs. 7-1, 7-10, and 7-19).

You may hear the term **malar flush**; this is redness of the skin of the zygomatic prominences (malar eminences) that often occurs with tuberculosis, mitral stenosis (narrowing of the left atrioventricular opening), and sometimes in rheumatic and other fevers.

The common variants of fractures of the maxilla and zygomatic bones were classified by Le Fort, a Paris surgeon. The three types of fracture are remarkably constant; they are described and illustrated in Figure 7-26.

THE FACE

The face is the part of the head that is visible in a frontal view (Fig. 7-28), *i.e.*, all that is anterior to the external ears and between the hairline and tip of the chin.

The term **facies** (L. face) refers to the appearance of the face. Some diseases produce a typical facial appearance or facies *e.g.*, the mask-like, expressionless facies of patients with **Parkinson's disease** (Parkinson's syndrome), a disturbance of motor function that is an objective and verifiable sign of the disease.

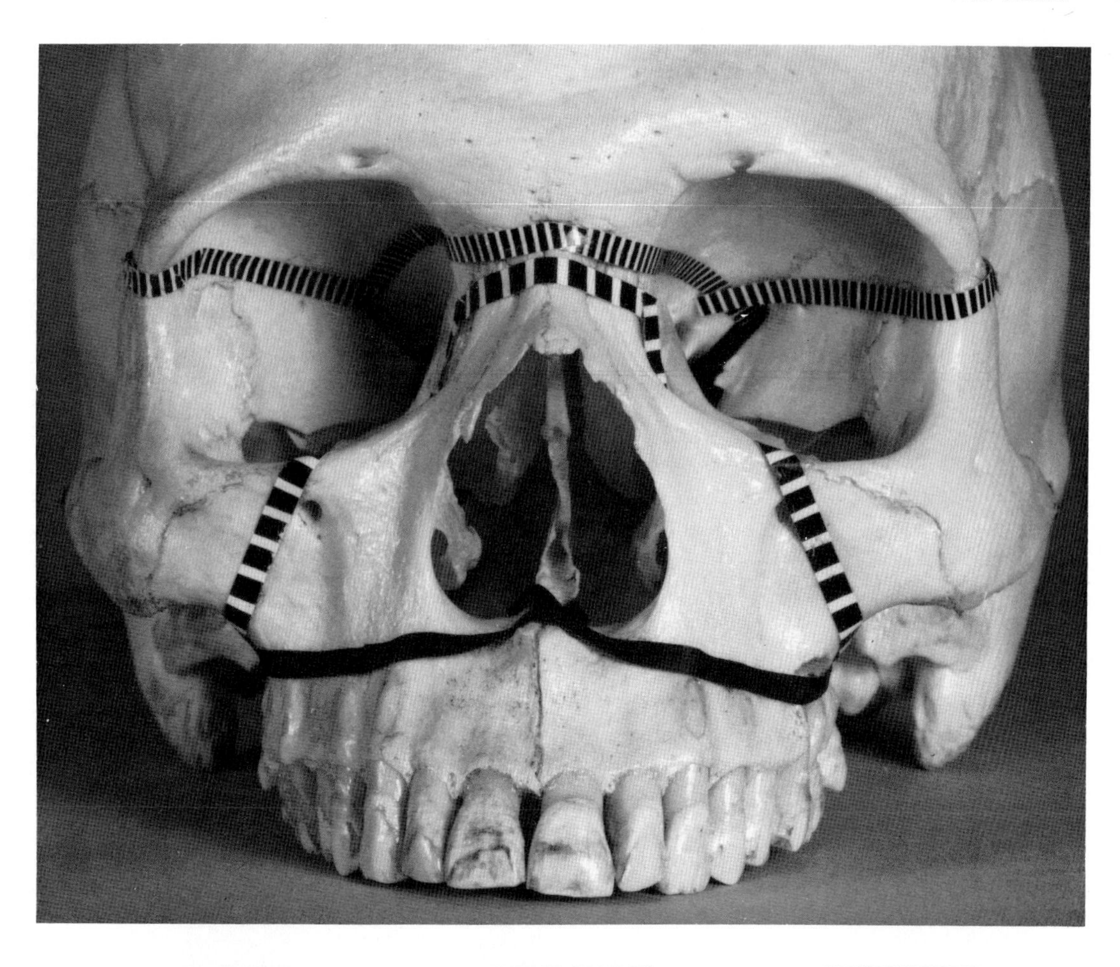

Figure 7-26. Photograph of a skull showing the common variants of fractures of the maxillae and concurrent fracturing of other bones. *A*, **Le Fort I**. A horizontal fracture of the maxillae, located just superior to the alveolar process, crossing the bony nasal septum and the pterygoid plates of the sphenoid bone (also see Fig. 7-113). The upper jaw (maxillae) becomes movable. *B*, **Le Fort II**. A fracture that passes from the posterolateral parts of the maxillary sinuses, superomedially through the infraorbital notch, the lacrimals, or the ethmoid to the bridge of the nose. As a result, the entire central part of the face, including the hard palate and the alveolar processes, is separated from the rest of the skull. *CSF rhinorrhea* (leakage of cerebrospinal fluid from the nose) is common because of fracturing of the ethmoid bone and tearing of the meninges (membranes) covering the brain. *C*, **Le Fort III**. A horizontal fracture that passes through the superior orbital fissures, the ethmoid and nasal bones, and extends laterally through the greater wings of the sphenoid bone and the frontozygomatic sutures. As there is concurrent fracturing of the zygomatic arches, the maxillae and zygomatic bones are separated from the rest of the skull.

At birth the face is small compared with the rest of the head (Fig. 7-5, *A* and *C*), as discussed previously, because the jaws are not developed fully and most of the sinuses in the facial bones have not started to form. As these structures develop during infancy and childhood, the face becomes longer, the zygomatic prominences (cheek bones) more prominent, and the cheeks contain less fat. The cheeks of an infant appear full because of the "sucking" or *buccal pad of fat* (Fig. 7-28) between the buccinator muscle and the superficial facial muscles. This fat-pad is present in adults, but is relatively much smaller.

MUSCLES OF THE FACE

The muscles of facial expression are attached to the skin of the face and lie in the subcutaneous tissue (Figs. 7-27 to 7-29 and 7-31). They enable us to move our skin and change our **facial expression**. Most facial muscles arise from bone or fascia.

All muscles of facial expression are supplied

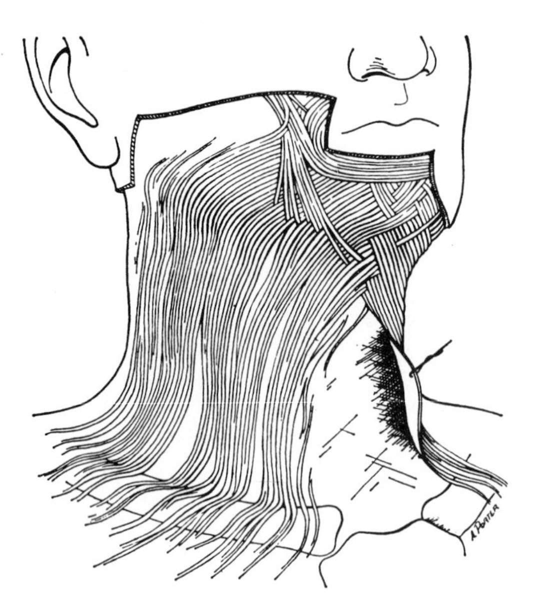

Figure 7-27. Drawing of a dissection of the platysma muscle and some of the inferior facial muscles. Observe that the platysma is a thin extensive sheet of vertical muscle fibers that lies in the subcutaneous tissue of the neck and lower face. Note that it extends from the pectoral region (also see Fig. 6-14) and inserts into the margin of the mandible. Observe that its fibers cross the mandible and blend with the other facial muscles (also see Fig. 7-29). This muscle is used chiefly to tense the skin of the neck (*e.g.*, while shaving).

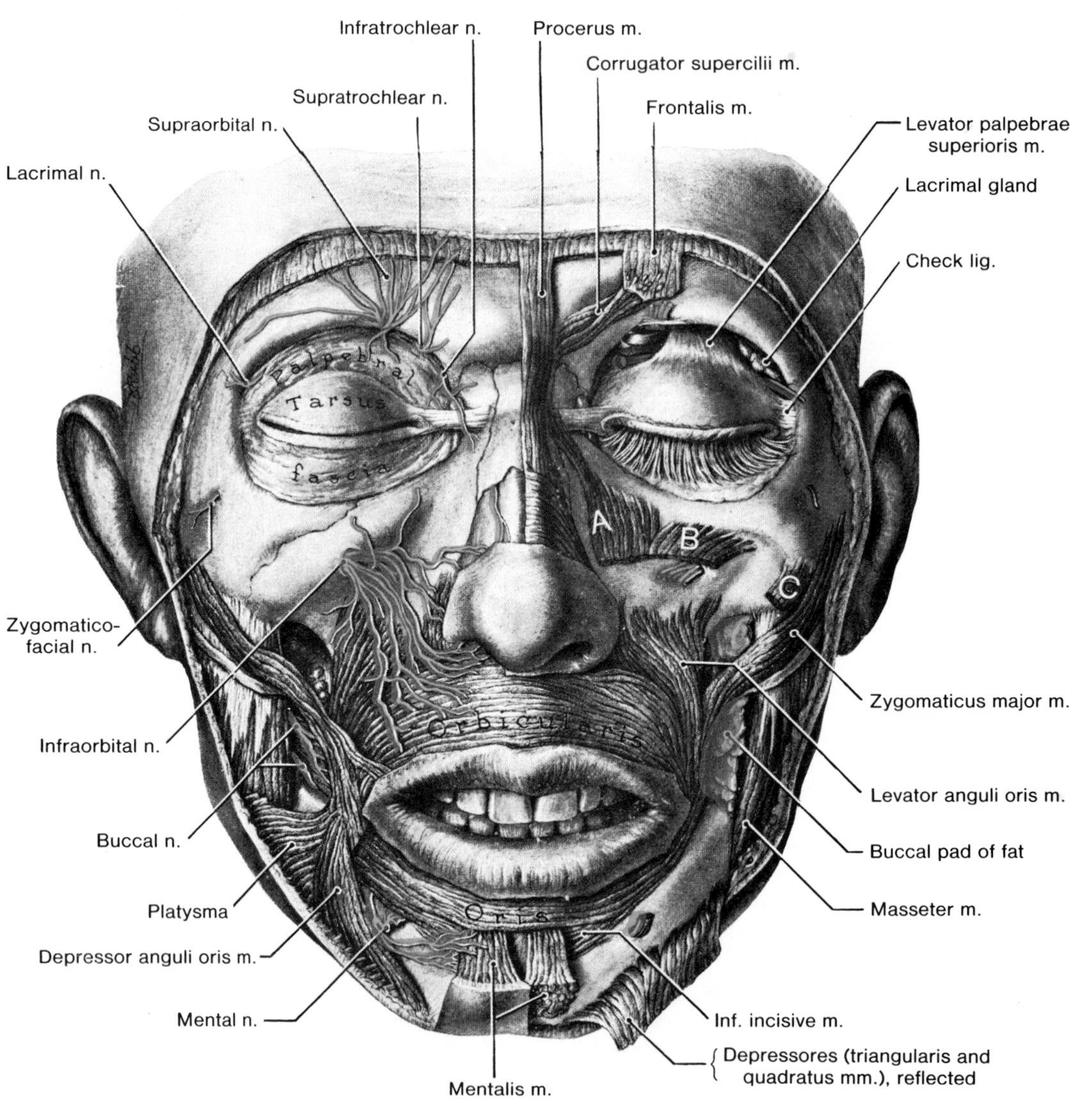

Figure 7-28. Drawing of a dissection of the face showing the cutaneous branches of the trigeminal nerve (*yellow*) and several facial muscles. A *greenish-colored pin* has been inserted posterior to the aponeurosis of the levator palpebrae superioris muscle to demonstrate its fan-shaped attachment to the superior tarsus (thin plate of condensed fibrous tissue in the eyelid). Some fibers of its tendon insert into the skin of the upper eyelid. *A*, levator labii superioris alaeque nasi. *B*, levator labii superioris. *C*, zygomaticus major.

by the facial nerve (**CN VII**). The facial muscles surround the facial orifices: the mouth, eyes, nose, and ears and act as **sphincters** (G. a band) and **dilators,** *i.e.*, they close and open these orifices.

The facial muscles develop from the second branchial arch (hyoid arch) as part of a subcutaneous muscle sheet in the head and neck known as the **platysma** (Figs. 6-14 and 7-27). It spreads like a sheet over the face during embryonic development, bringing branches of the facial nerve with it. Because of their common origin, the sheet-like platysma and the facial muscles are often fused and their fibers of insertion into the skin are frequently intermingled (Figs. 7-27 and 7-28).

Because the face has no definite deep fascia and the superficial fascia between the cutaneous attachments of the facial muscles is loose, **facial lacerations tend to gape** (part widely). Consequently the skin must be sutured (L. *sutura*, a sewing) with great care to prevent scarring. The looseness of the superficial fascia also enables much fluid and blood to accumulate in the loose connective tissue following bruising of the face. Similarly, facial inflammation causes considerable swelling (*e.g.*, the swelling resulting from a bee sting on the bridge of the nose can close both eyes).

MUSCLE OF THE FOREHEAD

The **frontalis muscle** (Fig. 7-31) is part of the scalp muscle called the occipitofrontalis (Fig. 7-46*B*). The frontalis muscle elevates the eyebrows (*e.g.*, with a surprised look) and produces the transverse wrinkles in the forehead when one frowns.

MUSCLES OF THE MOUTH

There are several muscles that alter the shape of the mouth and lips (*e.g.*, during whistling, speaking, and mimicry). Some of them will be described because they produce oral movements which are affected when there is a lesion of the facial nerve.

The depressor anguli oris muscle (Figs. 7-28, 7-29, and 7-31), as its name indicates, depresses the corner of the mouth. Posterior fibers of the platysma assist with this movement.

The zygomaticus major muscle (Figs. 7-28, 7-29, and 7-31), extending from the zygomatic bone to the angle of the mouth, draws the angle of the mouth superolaterally as in laughing.

The levator labii superioris muscle (Fig. 7-31), descending from the infraorbital margin to the upper lip (L. *labium*, lip), raises the upper lip. A few people are able to use this muscle to evert their upper lip as chimpanzees often do.

The orbicularis oris muscle (Fig. 7-28) is the sphincter muscle of the mouth. Its fibers lying within the lips, encircle the mouth and blend with other facial muscles, particularly the buccinator (Fig. 7-29). The orbicularis oris **closes the mouth, purses the lips** (as in whistling and sucking), **and plays an important role in articulation and mastication** (chewing). In association with the buccinator muscle, *it helps to hold food between the teeth during mastication.*

MUSCLES OF THE CHEEK

The buccinator (Fig. 7-29) is a thin, flat muscle that *aids mastication by pressing the cheeks against the teeth* during chewing. This pushes the food against the occlusal surfaces of the teeth. The buccinator is also used when sucking by forcing the cheeks against the molar teeth. The buccinator muscle was given its name because it also compresses the cheeks (L. *buccae*) during blowing, *e.g.*, when a musician plays a wind instrument such as a trumpet (L. *buccinator*, trumpeter).

Some trumpeters have stretched their buccinator and other buccal muscles so much that their cheeks balloon out when they blow forcibly on their instrument.

MUSCLES OF THE EYELIDS

The function of the eyelids is to protect the eye and to keep the cornea moist (Fig. 7-92).

The orbicularis oculi muscle (Figs. 7-29 and 7-31) is the sphincter muscle of the eye. Its fibers sweep in concentric circles around the orbital margin and in the eyelids. Contraction of its fibers narrows the orbital opening and encourages the flow of tears by helping to empty the **lacrimal sac** or tear sac (Fig. 7-96).

The orbicularis oculi muscle *consists of three parts*: (1) a thick *orbital part* for closing the eyes to protect against the glare of light and dust in the air; (2) a thin *palpebral part* for closing the eyelids lightly (L. *palpebra*, eyelid) to keep the cornea from drying; and (3) a *lacrimal part* for drawing the eyelids and ***lacrimal puncta*** medially (Figs. 7-92 and 7-96).The third part lies deep to the palpebral part and is often considered to be part of it. The lacrimal part of the muscle encloses the ***lacrimal canaliculi*** and passes posterior to the lacrimal sac (Fig. 7-96).

When all three parts of the orbicularis oris contract,

the eyes are firmly closed and the adjacent skin becomes wrinkled. Similar wrinkling occurs when a person scrutinizes something. By 30 to 35 years these wrinkles or creases around the eyes become permanent. Many people call them "crow's feet." Their presence in children and young adults may indicate a visual defect.

The levator palpebrae superioris muscle (Fig. 7-28), as its name indicates (L. *levare*, to raise), **raises the upper eyelid**. It is supplied by the oculomotor nerve (CN III).

Injury to the facial nerve (CN VII) or to some of its branches (Figs. 7-29 and 7-100) produces paresis (weakness) or paralysis (loss of voluntary movement) in all or some of the facial muscles on the affected side (Fig. 7-100).

The facial nerve is most often affected as it passes through the facial canal in the petromastoid part of the temporal bone (Figs. 7-55 and 7-168). In some cases the pressure on the nerve is caused by swelling (edema) of the nerve resulting from a virus infection.

Paralysis of the facial nerve for no obvious reason, known as **Bell's palsy**, often occurs after exposure to a cold draft. Patients with this condition are unable to close their lips and eye on the affected side. Thus they are unable to whistle, to blow a wind instrument, or to chew effectively. Because their buccinator muscle is also paralyzed, food dribbles out that side of the mouth or collects in its vestibule (Fig. 7-124).

Incomplete or complete paralysis of the facial muscles is particularly noticeable around the mouth because its corner sags on the affected side. Complete paralysis of the entire side of the face indicates that the nerve has been injured between its origin in the brain stem and its point of branching in the parotid gland (Figs. 7-29 and 7-35).

Paresis (incomplete paralysis) of some or all of the facial muscles suggests injury to one or more branches of the facial nerve within or beyond the parotid gland (Fig. 7-29), because there is some overlapping of the nerve supply to the muscles of the forehead.

In persons with **paralysis agitans** (Parkinson's disease) *resulting from neuronal degeneration in the* ***substantia nigra*** (a large motor nucleus in the midbrain), emotional changes of facial expression are typically lost. This gives patients a mask-like, expressionless facies.

MUSCLES OF THE NOSE

All muscles of the nose are supplied by the facial nerve (CN VII). The **procerus** (Fig. 7-28) is a small slip of muscle that is continuous with the occipitofrontalis muscle (Fig. 7-46*B*). The procerus muscle passes from the forehead over the bridge of the nose, where it is inserted into the skin.

The procerus draws the medial part of the eyebrow inferiorly, producing transverse wrinkles over the bridge of the nose. This probably reduces the glare of bright sunlight.

The **nasalis muscle** (Fig. 7-31) consists of transverse (compressor naris) and alar (dilator naris) parts. The *compressor naris* passes from the maxilla, superior to the incisor teeth, to the dorsum of the nose.

The compressor naris muscle compresses the anterior nasal aperture (nostril). The dilator naris arises from the maxilla superior to the transverse part and inserts into the alar cartilages of the nose (Fig. 7-150). **The dilator naris muscle widens the anterior nasal aperture**.

Marked action of the dilator naris in sick infants strongly suggests respiratory distress, e.g., as occurs with pneumonia.

The **depressor septi muscle**, often regarded as part of the dilator naris, arises from the maxilla superior to the central incisor tooth, and inserts into the mobile part of the nasal septum. *The depressor septi assists the dilator naris muscle in widening the nasal aperture during deep inspiration.*

SENSORY NERVES OF THE FACE

The innervation of the skin of the face is largely through the three branches of the trigeminal nerve (Figs. 7-28 and 7-32). Some skin over the angle of the mandible and anterior and posterior to the auricle is supplied by the ***great auricular nerve*** from the cervical plexus (Fig. 7-29). Some cutaneous fibers of the facial nerve probably also supply skin on both sides of the auricle (Fig. 7-35).

THE TRIGEMINAL NERVE (Figs. 7-32, 7-33, and Tables 7-1 and 7-2)

The trigeminal or *fifth cranial nerve* (CN V) is the largest of the 12 cranial nerves. *It is the principal general sensory nerve to the head*, particularly the face, and **it is the motor nerve to the muscles of mastication** (masseter, temporalis, and several others).

The cell bodies of most primary sensory neurons of the trigeminal nerve are located in the **trigeminal ganglion** (Fig. 7-33). The other cell bodies are in the **mesencephalic nucleus** (the midbrain nucleus of this nerve). The peripheral processes of the unipolar cells in the trigeminal ganglion form the ophthalmic

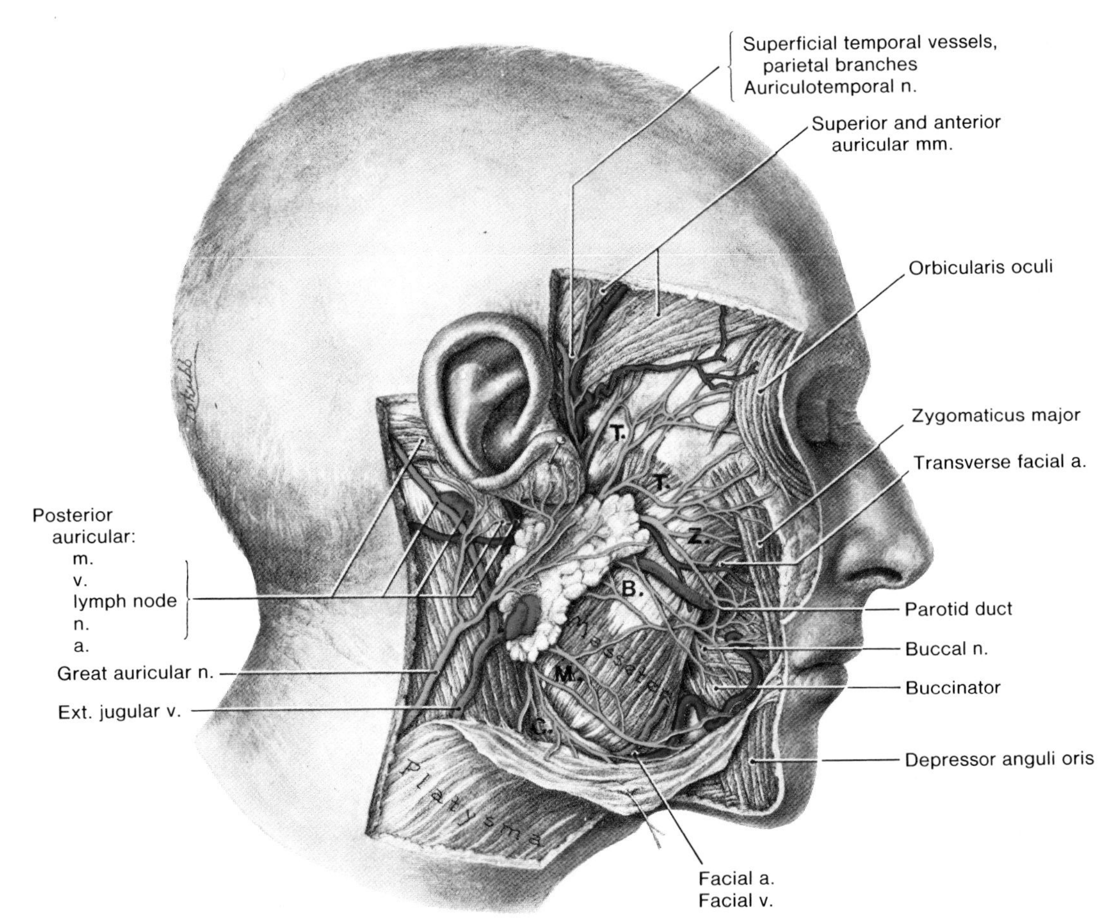

Figure 7-29. Drawing of a lateral view of a dissection of the right side of the head showing the great auricular nerve (C2 and C3), and terminal branches of the facial nerve (CN VII): *T*, temporal, *Z*, zygomatic, *B*, buccal, *M*, mandibular and *C*, cervical. Obviously the facial nerve, motor to the muscles of facial expression, is in jeopardy during surgery of the parotid gland. The masseter is a muscle of mastication that is derived from mesenchyme of the second branchial arch and is supplied by the nerve of that arch (CN V^3). The parotid lymph nodes are shown in *green* (also see Fig. 7-40). Note that the parotid duct turns medially at the anterior border of the masseter muscle to pierce the buccal pad of fat (Fig. 7-28) and the buccinator muscle. It enters the oral cavity opposite the crown of the second maxillary molar tooth (Fig. 7-41).

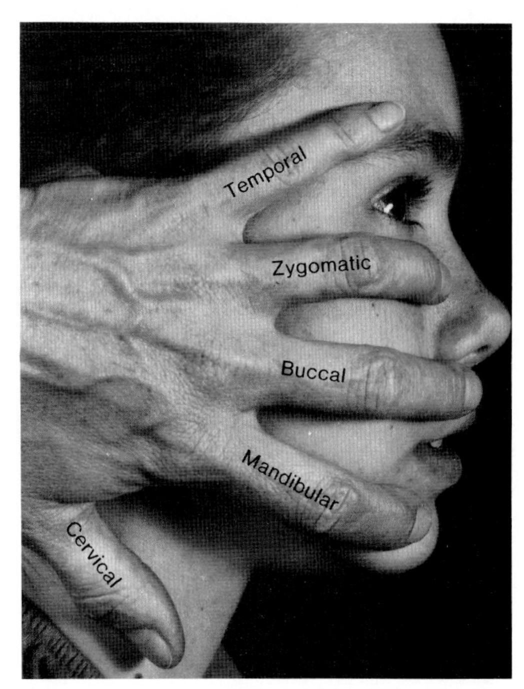

Figure 7-30. Photograph of a lateral view of the face of a 12-year-old girl illustrating a simple method for remembering the general course of the five main branches of the facial nerve (CN VII) that are motor to the muscles of facial expression. (also see Fig. 7-29).

Table 7-1
Summary of the Cranial Nerves[1]

Nerve	Name	Special Sense	Sensory	Motor	Parasympathetic
CN I	Olfactory	*			
CN II	Optic	*			
CN III	Oculomotor			*	*
CN IV	Trochlear			*	
CN V	Trigeminal		*	*	
CN VI	Abducens			*	
CN VII	Facial	*	*	*	*
CN VIII	Vestibulocochlear	*			
CN IX	Glossopharyngeal	*	*	*	*
CN X	Vagus	*	*	*	*
CN XI	Accessory			*	
CN XII	Hypoglossal			*	

[1] Note that four sensory modalities may be carried by the cranial nerves and that three nerves carry special sense only (CN I, CN II, CN VIII) and have no motor component. Note also that four nerves (CN III, CN VII, CN IX, and CN X) carry parasympathetic fibers to smooth muscles and glands.

(CN V^1), the maxillary (CN V^2), and the sensory part of the mandibular nerve (CN V^3).

The Ophthalmic Nerve (Figs. 7-32 and 7-33). *This is the superior division of the trigeminal* and is the smallest of the three branches of this nerve. **It is wholly sensory** and supplies the area of skin derived from the embryonic frontonasal prominence (process).

The ophthalmic nerve divides into three branches: *nasociliary, frontal, and lacrimal,* just before entering the orbit through the superior orbital fissure. Five branches from these nerves take part in the sensory supply to the skin of the forehead, upper eyelid, and nose (Fig. 7-33).

The nasociliary nerve supplies the tip of the nose via the external nasal branch of the anterior ethmoidal nerve, and the root of the nose via the infratrochlear nerve (Fig. 7-32).

The frontal nerve, the direct continuation of the ophthalmic nerve, *divides into two branches*: supratrochlear and supraorbital (Fig. 7-32). The **supratrochlear nerve** supplies the middle part of the forehead and the **supraorbital nerve** supplies the lateral part of the forehead and the front of the scalp.

The lacrimal nerve (Fig. 7-32) emerges over the superolateral orbital margin to supply the lateral part of the upper eyelid.

The Maxillary Nerve (Figs. 7-32 and 7-33). *This is the intermediate division of the trigeminal.* It has three cutaneous branches that supply the area of skin derived from the embryonic maxillary prominence (process).

The infraorbital nerve (Figs. 7-28, 7-32, and 7-33), the large terminal branch of the maxillary nerve, passes through the infraorbital foramen and breaks up into *branches which convey sensation from skin on the lateral aspect of the nose, the upper lip, and the lower eyelid.*

The zygomaticofacial nerve (Fig. 7-32*C*), a small branch of the maxillary, emerges from the zygomatic bone through a small foramen with the same name. It supplies the skin of the face over the zygomatic bone (*i.e.*, the zygomatic prominence).

The zygomaticotemporal nerve (Fig. 7-32*C*) emerges from the zygomatic bone through a foramen of the same name and supplies the skin over the temporal region.

For local anesthesia of the face, the infraorbital nerve is often infiltrated with an anesthetic agent at the **infraorbital foramen** (Fig. 7-28), or inside the canal (*e.g.*, for treatment of wounds of the upper lip or for repairing the upper incisor teeth). You can easily determine the site of emergence of this nerve by exerting pressure on your zygomatic bone in the region of the infraorbital foramen and nerve. Pressure on the nerve causes considerable pain.

Care must be exercised when performing an **infraorbital nerve block** because companion infraorbital vessels (artery and vein) leave the infraorbital foramen with the nerve. Careful aspiration of the syringe during injection prevents inadvertant injection of the anesthetic fluid into a blood vessel. In Figure 7-2 note that the orbit is located just superior to the injection site. A careless injection could result in the passage of anesthetic fluid into the orbit, causing temporary paralysis of the extraocular muscles.

The Mandibular Nerve (Figs. 7-28, 7-32, and 7-33). *This is the inferior division of the trigeminal.* It

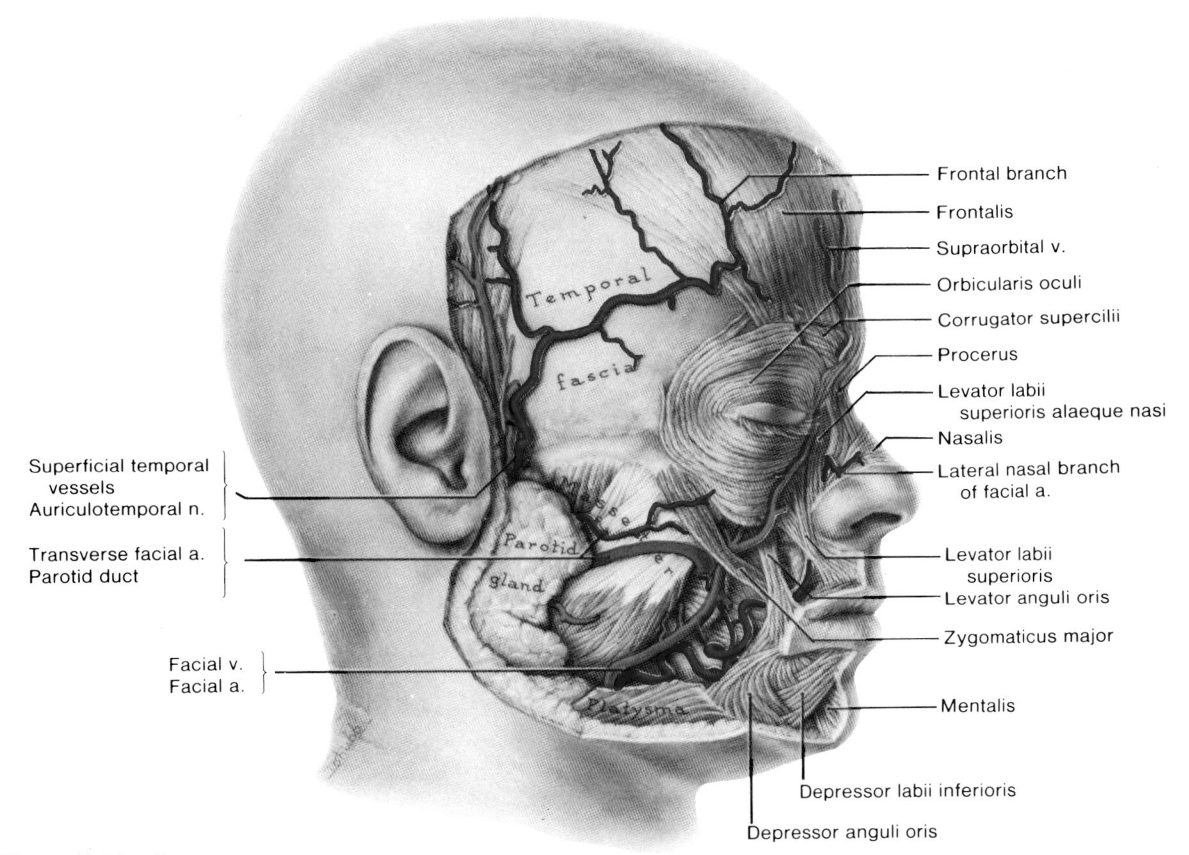

Figure 7-31. Drawing of a lateral view of a dissection of the face of a young man exposing the muscles of facial expression (all supplied by CN VII) and the arteries of the face. The masseter muscle, one of the muscles of mastication, a powerful closer of the jaw, is also shown; it is supplied by CN V^3 (the mandibular division of the trigeminal nerve).

has three sensory branches that supply the area of skin derived from the embryonic mandibular prominence (process) of the first branchial arch. It also **supplies motor fibers to the muscles of mastication.** Of the three divisions of the trigeminal, *CN V^3 is the only division that carries motor fibers.*

The mental nerve (Figs. 7-20, 7-28, and 7-32), *a branch of the inferior alveolar nerve*, emerges from the mental foramen and divides into three branches that supply the skin of the chin and the skin and mucous membrane of the lower lip and gingiva (gum).

The buccal nerve (Figs. 7-20, 7-28, 7-29, 7-32, and 7-33) is a small branch of the mandibular nerve that emerges from deep to the ramus of the mandible to supply the skin of the cheek over the buccinator muscle. It also supplies the mucous membrane lining the cheek, and the posterior part of the buccal surface of the gingiva.

The auriculotemporal nerve (Figs. 7-29, 7-31, and 7-32), the third sensory branch of the mandibular nerve, passes medial to the neck of the mandible and then turns superiorly, posterior to its head and anterior to the auricle.

The auriculotemporal nerve crosses over the root of the zygomatic process of the temporal bone, deep to the superficial temporal artery (Fig. 7-31). As its name suggests, it supplies parts of the auricle, external acoustic meatus, tympanic membrane (eardrum), and skin in the temporal area.

Dentists frequently anesthetize the inferior alveolar nerve before repairing or removing the premolar or molar teeth of the mandible (Fig. 7-20). As the mental nerve is one of its two terminal branches, it is understandable why one's chin and lower lip on the affected side also lose sensation (Fig. 7-32).

The mental nerve can be blocked by injecting the anesthetic fluid around the nerve as it emerges from the mental foramen (Fig. 7-28). The mental artery and vein must be avoided.

Trigeminal neuralgia (tic douloureaux) is a condition characterized by sudden attacks of excruciating pain, brought on by a mere touch in the area of distribution of one of the divisions

Table 7-2
An Overview of the Cranial Nerves

Nerve	Efferent		Afferent		Special Senses
	Striated muscles	Smooth and cardiac muscles and glands	Skin	Mucous membranes and organs	
CN I					Smell
CN II					Sight
CN III	Supplies all muscles of eyeball except sup. oblique and lat. rectus	Muscles of lens and iris of eye		Proprioceptive fibers from eye muscles	
CN IV	Sup. oblique muscle of eyeball			Proprioceptive fibers from eye muscles	
CN V	Muscles of mastication and tensors of tympanum and palate	Carries parasympathetic ganglia for preganglionic nerve fibers of CN III, CN VII, and CN IX	Face and anterior part of scalp	Teeth and mucous membranes of tongue, mouth, nose, and eye	Taste (fibers from chorda tympani) to ant. two-thirds of tongue
CN VI	Lat. rectus muscle of eyeball			Proprioceptive fibers from lat. rectus muscle	
CN VII	Muscles of facial expression	Nervus intermedius; glands of mouth, nose, and palate; lacrimal gland; submandibular and sublingual glands (see CN V)	External ear		Nervus intermedius, taste, ant. two-thirds of tongue
CN VIII					Hearing and equilibrium
CN IX	Stylopharyngeus muscle	Parotid gland (see CN V)		Tympanic membrane, middle ear, pharynx, and tongue (post. one-third)	Taste, post. one-third of tongue
CN X	Muscles of pharynx	Organs in neck, thorax, and abdomen	Ext. acoustic meatus and tympanic membrane	Organs in neck, thorax, and abdomen	Taste, epiglottis
CN XI	Soft palate, pharynx, and larynx; sternocleidomastoid and trapezius muscles				
CN XII	Extrinsic and intrinsic muscles of tongue				

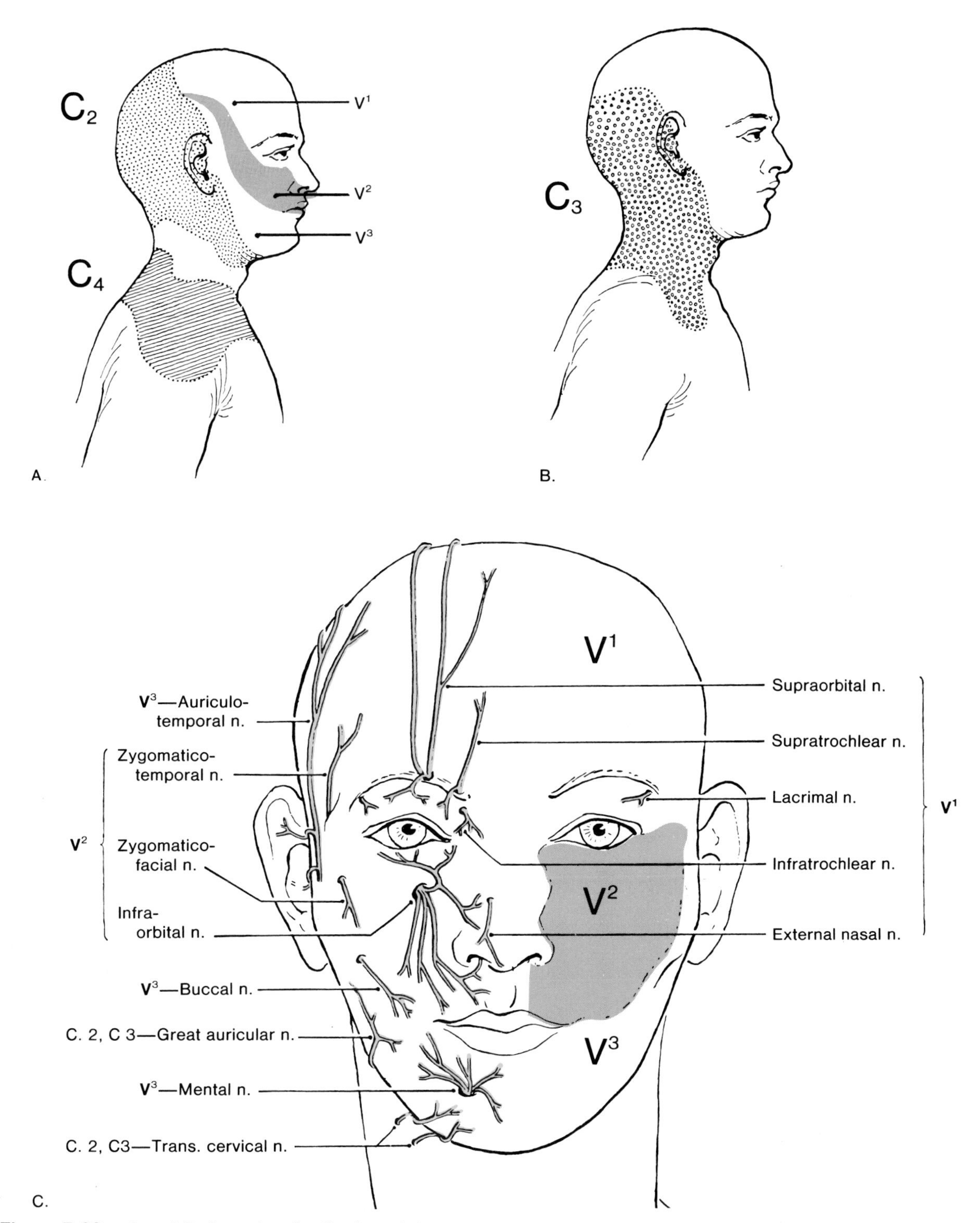

Figure 7-32. *A* and *B* show the distribution of the cutaneous areas (dermatomes) supplied by spinal nerves. In *A* observe that the trigeminal nerve (CN V) is responsible for general sensation from the skin of the face, forehead, and scalp as far posterior as the vertex. In *C*, the distribution of the three divisions of the trigeminal nerve (CN V) correspond *roughly* to the three embryological regions of the face.

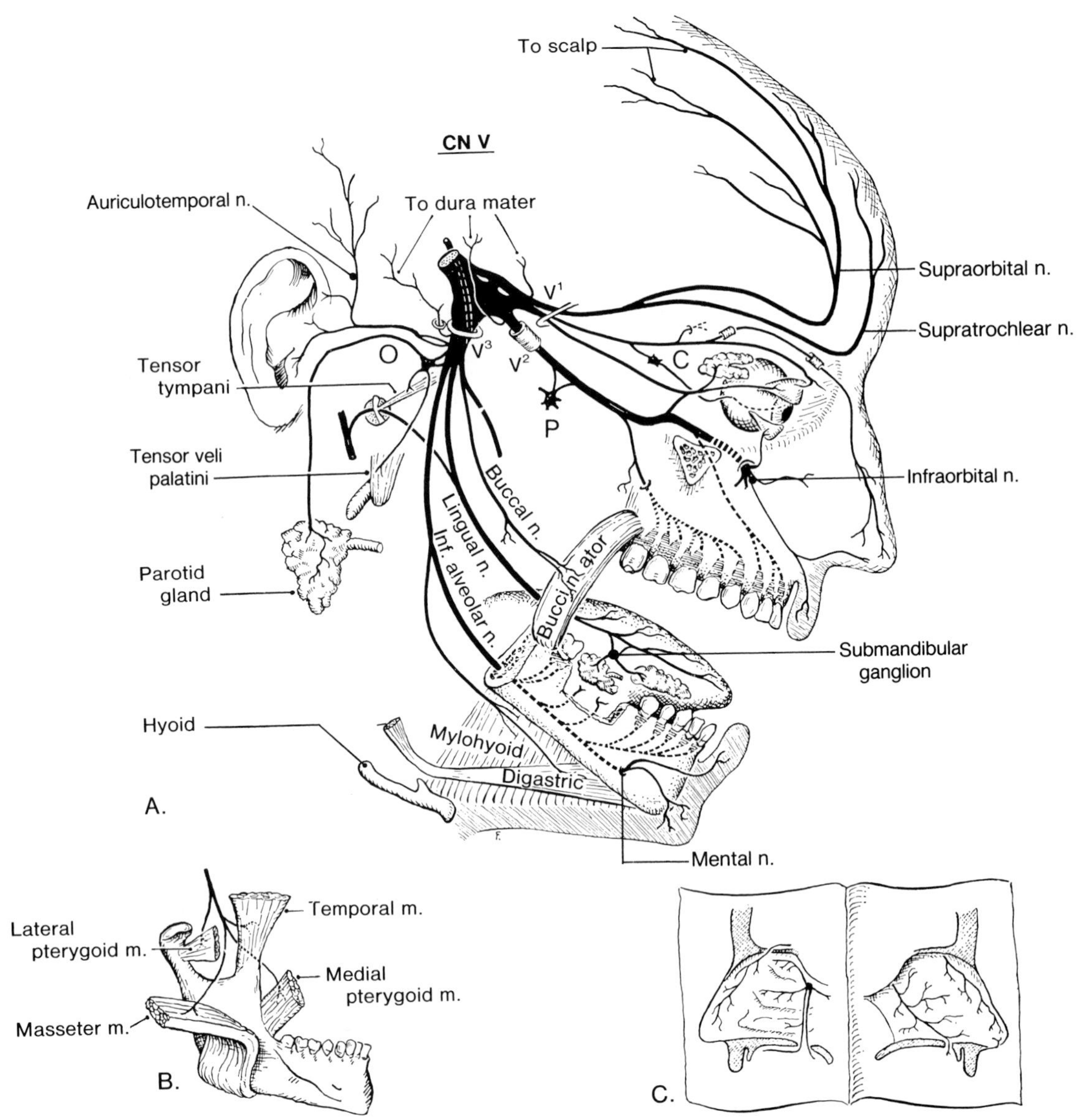

Figure 7-33. Drawing illustrating the distribution of the *trigeminal nerve* (CN V). In *A* observe its three divisions (V^1, V^2, and V^3) arising from the large trigeminal ganglion. Note that the maxillary nerve (CN V^2) gives off two pterygopalatine nerves which suspend the small pterygopalatine ganglion (*P*). Note also the ciliary (*C*) and otic (*O*) ganglia. Each of the three divisions of the trigeminal nerve is connected with a parasympathetic ganglion: V^1 with the ciliary, V^2 with the pterygopalatine, and V^3 with the submandibular and otic. These ganglia are excitor cell stations whose preganglionic fibers travel with nerves CN III, CN VII, and CN IX. The postganglionic fibers are distributed with the branches of CN V to smooth muscles of the eyeball and glands. Note that the ophthalmic nerve (CN V^1) is sensory to: (1) the eyeball and the cornea via the ciliary nerves; hence if paralyzed, the bulbar conjunctiva is insensitive to touch; (2) the frontal, ethmoidal, and sphenoidal sinuses via the supraorbital and ethmoidal nerves; and (3) the skin and conjunctival surfaces of the upper eyelid and to the skin and mucous surfaces of the external part of the nose. *B* illustrates the motor fibers passing to the muscles of mastication. *C* shows the sensory and secretory fibers to the nasal mucosa and palate.

of the trigeminal nerve, usually CN V^2. The cause of the neuralgia (G. nerve pain) is unknown.

The sensory and motor nuclei of the trigeminal nerve may be involved in degenerative or other lesions in the brain stem, or intracranial portions of the nerve may be affected by trauma or a tumor. If the motor fibers are affected, the muscles of mastication will be weakened or paralyzed, causing deviation of the mandible to the affected side when the mouth is opened.

A patient with his/her eyes closed who is unable to feel wisps of cotton touching the forehead, cheeks, and chin has anesthesia to light touch (Fig. 7-34). Other modalities of sensitivity may be tested using pinpricks and warm and cold objects.

Lesions of the peripheral branches of the trigeminal nerve are not common, but they may result from: (1) traumatic injury to the face, (2) tumors, and (3) fractures of the bones of the skull (Figs. 7-25, 7-26, and Table 7-3).

The **infraorbital nerve** (Figs. 7-20 and 7-32*C*) is commonly injured in fractures of the maxilla (Fig. 7-26), and *the mandibular nerve may be damaged by a fracture of the ramus of the mandible* (Fig. 7-25).

A lesion of the entire trigeminal nerve causes *anesthesia of*: (1) the corresponding anterior *half of the scalp* (Fig. 7-32); (2) *the face,* except for the area around the angle of the mandible; (3) *the cornea and conjunctiva*; and (4) *the mucous membranes of the nose, mouth, and tongue* (anterior two-thirds). Paralysis and atrophy of the muscles of mastication also occur so that when the mouth is opened the mandible moves to the paralyzed side.

Herpes zoster, a virus infection, also affects the trigeminal ganglion (Fig. 7-33). Inflammation of the ganglion may result in necrosis of some ganglion cells, which usually produces typical herpetic eruptions in one or more of the three divisions of the nerve.

Herpes zoster of the face usually involves the region supplied by the ophthalmic nerve (herpes zoster ophthalmicus); hence the cornea is often involved (Fig. 7-33). No doubt it is the inflammatory changes affecting the ganglion cells and their fibers that cause the skin lesions. In some cases there is partial paralysis or paresis of the ocular muscles indicating that the infection has involved CN III, CN IV, and/or CN VI, the nerves supplying the muscles that move the eye (Tables 7-2 to 7-4).

Aneurysm of the internal carotid artery may involve CN V, particularly lesions inferior to the anterior clinoid process (Fig. 7-62). When the aneurysm is located near the foramen lacerum (Fig. 7-15), it often affects all three divisions of the trigeminal nerve (Fig. 7-33).

MOTOR NERVES OF THE FACE

CN VII, *the facial nerve,* supplies the muscles of facial expression (Fig. 7-35) and CN V^3, *the mandibular nerve,* supplies the muscles of mastication (Fig. 7-33).

THE FACIAL NERVE (Figs. 7-29 to 7-31, 7-35, 7-42, and Tables 7-1 and 7-2)

CN VII supplies the superficial muscles of the neck, *i.e.*, the platysma (Figs. 7-27 to 7-29), muscles of facial expression, auricular muscles (Fig. 7-29), scalp muscles (Fig. 7-31), and certain other muscles derived from mesoderm of the embryonic second branchial arch.

The facial nerve is the sole motor supply to the muscles of facial expression and is sensory to the taste buds in the anterior two-thirds of the tongue (Fig. 7-35). It also conveys general sensation from a small area around the external acoustic meatus and is secretomotor to the submandibular, sublingual, and intralingual salivary glands (Fig. 7-35).

The facial nerve emerges from the skull through the stylomastoid foramen (Fig. 7-8), between the mastoid and styloid processes of the temporal bone, and almost immediately **enters the parotid gland** (Fig. 7-29). The facial nerve runs superficially within this gland before giving rise to **five terminal branches** (*temporal, zygomatic, buccal, mandibular, and cervical*), which emerge from the superior, anterior, and inferior margins of the gland. These branches spread out like the abducted fingers of the hand (Figs. 7-29 and 7-30) to **supply the muscles of facial expression** or mimetic (G. imitative) muscles.

The motor component of the facial nerve, supplying muscles of facial expression and certain other muscles, is the most important part of the nerve clinically. The motor nucleus of the facial nerve is located in the pons of the brain stem (Fig. 7-35).

The main clinical tests for the facial nerve concern the facial muscles; hence, you should have a general idea where the five major branches of the facial nerve go (Figs. 7-29 and 7-30). *The names of the nerves indicate the regions they supply.*

Do not attempt to memorize the individual muscles supplied by the branches of the facial nerve. It is not worthwhile because many facial muscles are supplied by more than one branch.

The temporal branches of the facial nerve (Figs. 7-29 and 7-30) cross the zygomatic arch to supply all the superficial facial muscles superior to it, including the orbital and forehead muscles (Figs. 7-29 and 7-31).

The zygomatic branches of the facial nerve (Figs. 7-29 and 7-30) pass transversely over the zygomatic bone to supply muscles in the zygomatic, orbital, and infraorbital regions.

The buccal branches of the facial nerve (Figs. 7-29 and 7-30) pass horizontally, external to the masseter muscle, to supply the buccinator and the muscles of the upper lip.

The mandibular branch of the facial nerve

Table 7-3
Summary of Lesions Involving Cranial Nerves

Nerve	Frequency	Type and/or Site of Lesion	Abnormal Findings
CN I	Uncommon	Fracture of cribriform plate or in ethmoid area	Anosmia (loss of smell); CNS rhinorrhea
CN II	Common	Direct trauma to orbit or eyeball; fracture involving optic canal	Loss of pupillary constriction
	Common	Pressure on optic pathway; laceration or intracerebral clot in temporal, parietal, or occipital lobes	Absence of blink reflex indicating visual field defect
CN III	Common	Pressure of herniating uncus on nerve just before it enters cavernous sinus; fracture involving cavernous sinus; aneurysms	Dilated pupil, ptosis, eye turns down and out; direct pupil reflex absent
CN IV	Uncommon	Course of nerve around brain stem or fracture of orbit	Inability to look down and in during convergence
CN V	Uncommon	Injury to terminal branches, particularly CN V^2 in roof of maxillary sinus (e.g., fracture of bones of face; pathological processes affecting trigeminal ganglion	Loss of pain and touch sensations; paresthesia; masseter and temporalis muscles do not contract; deviation of mandible to side of lesion when mouth is opened.
CN VI	Common	Base of brain or fracture involving cavernous sinus or orbit	Eye fails to move laterally; diplopia on lateral gaze
CN VII	Common	Laceration or contusion of parotid region	Paralysis of facial muscles; eye remains open; angle of mouth droops; forehead does not wrinkle
	Common	Fracture of temporal bone	As above, plus associated involvement of cochlear n. and chorda tympani; dry cornea and loss of taste on ant. two-thirds of tongue
	Common	Intracranial hematoma ("stroke")	Forehead wrinkles because of bilateral innervation of frontalis muscle; otherwise paralysis of contralateral facial muscles
CN VIII	Common	Eighth nerve tumor (acoustic neuroma)	Progressive unilateral hearing loss; tinnitus (noises in ear)
CN IX	Rare	Brain stem or deep laceration of neck	Loss of taste on post. one-third of tongue; loss of sensation on affected side of soft palate
CN X	Rare	Brain stem or deep laceration of neck	Sagging of soft palate; deviation of uvula to normal side; hoarseness owing to paralysis of vocal fold
CN XI	Rare	Laceration of neck	Sternocleidomastoid and superior fibers of trapezius muscles fail to contract; drooping of shoulder
CN XII	Rare	Neck laceration often with major vessel damage; basal skull fractures	Protruded tongue deviates toward affected side; moderate dysarthria

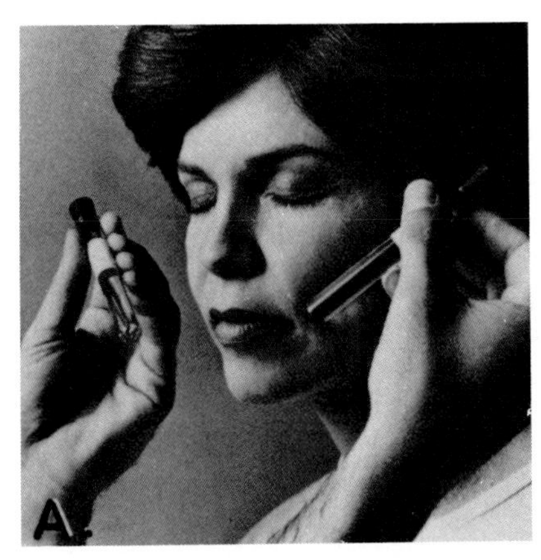

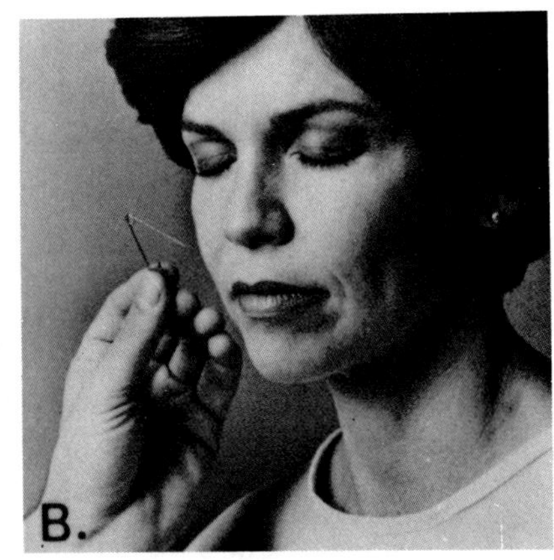

Figure 7-34. Photographs of a 33-year-old woman on whom the examiner is illustrating two ways the trigeminal nerve is tested during a neurological examination. Note that CN V tests are conducted with the patient's eyes closed. *A*, test tubes filled with warm and cold fluid are pressed alternately against her cheeks. Differences in response on opposite sides of the face indicate increased or decreased sensitivity to temperature. *B*, differences in response to pinpricks on opposite sides of the face indicate increased or decreased sensitivity to pain. (Used with permission from Smith Kline Corporation: *Essentials of Neurological Examination*, 1978).

(Figs. 7-29 and 7-30) supplies the muscles of the lower lip and chin.

The cervical branch of the facial nerve (Figs. 7-29 and 7-30) supplies the platysma, the superficial muscle of the neck.

If the motor nucleus of the facial nerve is involved in a disease process or a lesion in the brain stem, *there will be paralysis of the ipsilateral (L. same side) muscles of facial expression.* In time the facial muscles atrophy (*i.e.*, waste away).

Peripheral facial paralysis may be caused by chilling of the face, middle ear infections, tumors, fractures, and other disorders. About 75% of all facial nerve lesions are of this type. The types of signs and symptoms depend upon the location of the lesion (Table 7-3).

As the facial nerve runs superficially within the parotid gland, it may be infiltrated by malignant cells (*e.g.*, from a **carcinoma of the parotid gland**). This commonly results in incomplete paralysis (**paresis of the muscles of facial expression**). Benign tumors usually do not infiltrate and cause facial nerve paralysis, but care must be taken to preserve the facial nerve and its branches during excision of these tumors. On rare occasions, parotid inflammation (*e.g.*, mumps) may temporarily affect the facial nerve.

As the mastoid process of the temporal bone is not present at birth, the facial nerve may be easily injured by forceps during delivery of a baby. Similarly, because the mastoid process is not well developed in children, the facial nerve may be more easily injured at its site of emergence from the stylomastoid foramen than it is in adults.

In view of the fan-like distribution of the branches of the facial nerve (Fig. 7-29), it is unwise from an anatomical standpoint to make vertical incisions in the parotid gland (*e.g.*, for drainage of pus). Short incisions, parallel to the courses of the branches are less likely to produce a noticeable effect on the functioning of the facial muscles. Weakness rather than complete paralysis of a muscle usually results from injury to the branches of the facial nerve because of the overlapping distribution of its branches.

Following a stroke, or **cerebrovascular accident (CVA)**, one side of the body is paralyzed; however there is usually some preservation of movement of muscles of the forehead on the paralyzed side. The explanation for this is that the superior part of the motor nucleus of the facial nerve is bilaterally controlled from the cerebral cortex; hence *forehead muscles are relatively spared in a unilateral brain lesion superior to the level of origin of the motor component of the facial nerve.*

All functions of the facial nerve are lost if the nerve is damaged proximal to the geniculate ganglion (Fig. 7-35). Hence, in addition to facial paralysis, there is loss of sensation in the anterior two-thirds of the tongue and the palate on the affected side. There is also loss of secretion of the submandibular, sublingual, and lacrimal glands, including the mucous membrane on the side of the lesion.

To test the facial nerve, the patient is asked to look at the ceiling, wrinkle the forehead, frown, smile, and raise the eyebrows. All the muscles producing these movements are controlled by the facial nerve. Consequently, inability to perform a movement is of clinical significance.

To test the strength of the orbicularis oculi muscles (Figs. 7-29 and 7-31), the sphincter muscles of the eyelids, the patient is asked to try to keep his/her eyes closed while the examiner attempts to open them (Fig. 7-36).

When CN VII is damaged, the patient cannot close the eyelid tightly on the affected side. Sounds are also very loud in the affected ear owing to **paralysis of the stapedius muscle,** supplied by the facial nerve.

ARTERIES OF THE FACE

The face is richly supplied with blood from several arteries, the terminal branches of which anastomose freely (Figs. 7-29 and 7-31). *The arteries of the face are either derived from the internal or external carotid arteries* (Fig. 7-37).

THE FACIAL ARTERY

The facial artery, the chief artery of the face, arises from the external carotid (Fig. 7-37), and appears at the inferior border of the mandible, just

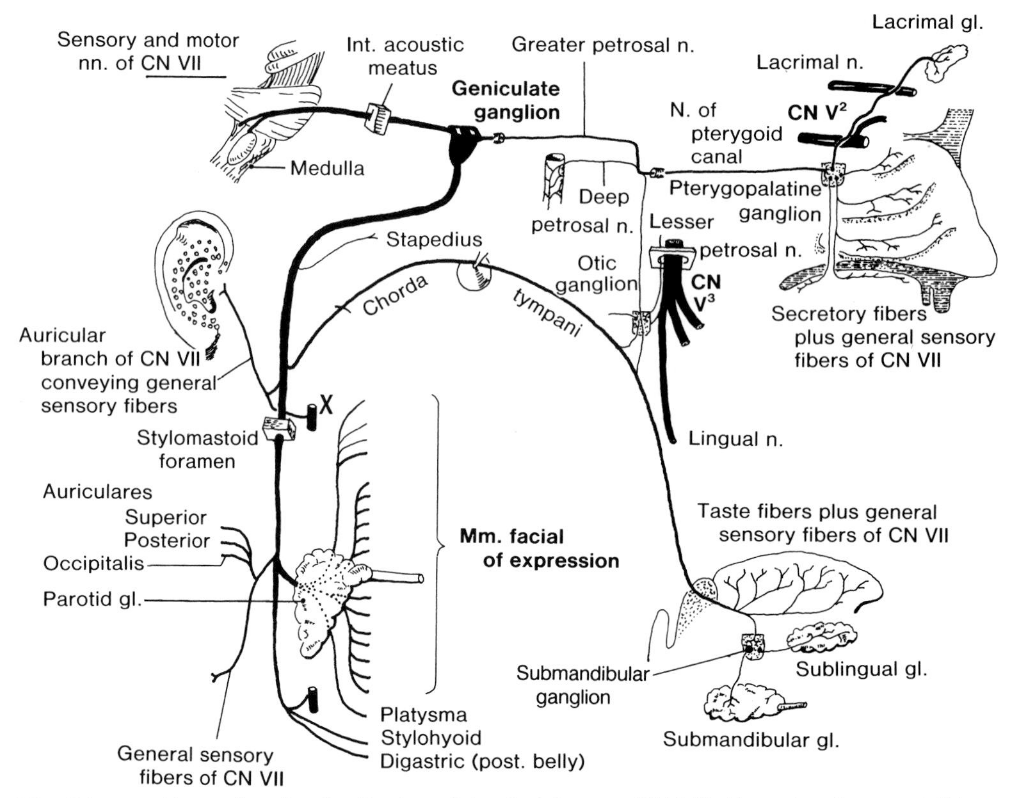

Figure 7-35. Diagram illustrating the distribution of the *facial nerve* (CN VII). Note that four modalities are carried by the facial nerve: (1) *motor* to the muscles of facial expression; it also supplies the stylohyoid muscle and the posterior belly of the digastric, as well as the stapedius muscle; and (2) *special sense* (taste) fibers with cell stations in the geniculate ganglion pass from the palate through the pterygopalatine ganglion, the nerve of the pterygoid canal, and the greater petrosal nerve to the geniculate ganglion. Fibers from the anterior two-thirds of the tongue pass via two routes: (1) the chorda tympani to the facial nerve and then to the geniculate ganglion, and (2) by a branch of the chorda tympani that traverses the otic ganglion to join the greater petrosal nerve and then to the geniculate ganglion. Evidence of this double route is the fact that the chorda tympani may be cut without loss of taste, whereas cutting the greater petrosal nerve may result in loss of taste. The submandibular ganglion, illustrated in this diagram, is clearly shown in Figure 8-14.

anterior to the masseter muscle (Fig. 7-29). As it hooks around the inferior border of the mandible, it usually grooves the bone (Figs. 7-21 and 7-37). Because the artery lies superficially here, immediately beneath the platysma, its pulsations can be easily felt in most people.

In its course over the face to the **inner canthus** (medial angle) of the eye, the facial artery crosses the mandible, the buccinator, and the maxilla (Fig. 7-29), but lies deep to the zygomaticus major and levator labii superioris muscles (Fig. 7-31).

Near the termination of its sinuous course, the facial artery passes about a fingerbreadth lateral to the angle of the mouth and ends by sending branches to the lip and the side of the nose (Figs. 7-31 and 7-37). The part of the artery that runs along the nose to the inner angle of the eye is often called the *angular artery.*

There are numerous anastomoses between the branches of the facial artery and other arteries of the face; hence, compressing the facial artery against the mandible on one side does not stop all bleeding from a lacerated facial artery or one of its branches.

In lacerations of the lip, pressure must be applied on both sides of the cut to stop the bleeding (*e.g.*, by compressing both parts of the cut lip between the index finger and thumb). In general, *facial wounds bleed freely and heal quickly.*

Because branches of the external carotid artery anastomose so freely with each other, and with those of the external carotid artery on the other side, one external carotid can be clamped in order to minimize bleeding and facilitate extensive surgery on one side of the head. Owing to the numerous anastomoses, there would be little or no impairment of functional circulation or wound healing.

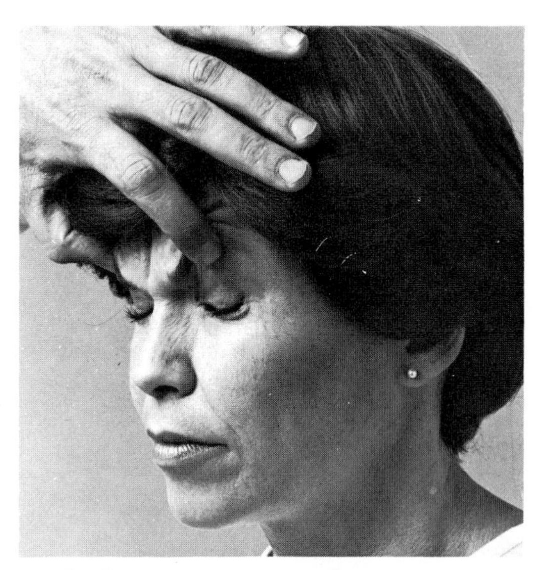

Figure 7-36. Photograph of a 33-year-old woman attempting to keep her eyes closed while the examiner attempts to open them. The neurologist is testing for paresis (weakness) of the orbicularis oculi muscles supplied by CN VII. When a facial nerve is paralyzed the patient is unable to close the eyelids on the affected side owing to paralysis of the orbicularis oculi, the sphincter muscle of the eyelids. (Used with permission from Smith Kline Corporation: *Essentials of Neurological Examination*, 1978).

THE SUPERFICIAL TEMPORAL ARTERY (Figs. 7-29, 7-31, and 7-37)

The superficial temporal artery is the smaller of the two terminal branches of the external carotid artery; the other terminal branch is the maxillary artery (Fig. 7-37). The superficial temporal artery begins deep to the parotid gland, posterior to the neck of the mandible, and ascends superficial to the posterior part of the zygomatic process of the temporal bone to enter the **temporal fossa** (Figs. 7-1 and 7-37).

The superficial temporal artery ends in the scalp by dividing into frontal and parietal branches (Fig. 7-31). Pulsations of this artery can be felt when it is compressed against the root of the zygomatic process of the temporal bone (Fig. 7-37) and they are often visible, particularly in thin elderly persons.

THE TRANSVERSE FACIAL ARTERY (Figs. 7-29 and 7-37)

This small artery of the face arises from the superficial temporal artery before it emerges from the parotid gland. It crosses the face, superficial to the masseter muscle, about a fingerbreadth inferior to the zygomatic arch in company with one or two branches of the facial nerve. It divides into numerous branches which supply the parotid gland and duct, the masseter muscle, and the skin of the face. It anastomoses with branches of the facial artery.

The pulse of the superficial temporal and facial arteries are often taken when it is not convenient to take the pulse of other arteries. Anesthetists, sitting at the head of the operating table, often take the **temporal pulse** just anterior to the ear, where the superficial temporal artery crosses the root of the zygomatic process of the temporal bone (Fig. 7-37). They also take the **facial pulse** where the facial artery winds around the inferior border of the mandible.

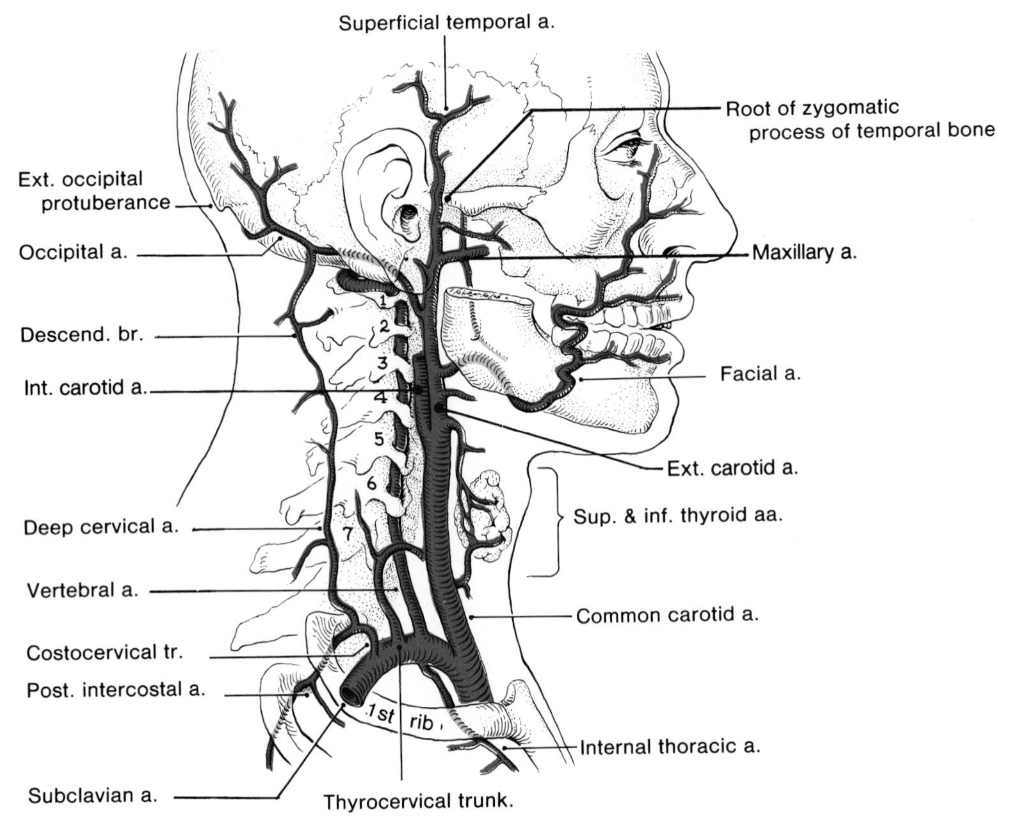

Figure 7-37. Diagram showing the arteries of the head. Note particularly the tortuous course of the facial artery. The temporal pulse can be taken anterior to the auricle where the superficial temporal artery crosses the root of the zygomatic process of the temporal bone. The facial pulse may be felt where the facial artery curves around the inferior border of the mandible. Observe the important vertebral artery arising from the first part of the subclavian artery and ascending through the foramina transversaria of the upper six cervical vertebrae. Note that on reaching the base of the skull, it winds around the lateral mass of the atlas (C1). Also see Figure 7-77. It may be difficult to distinguish between the external and internal carotid arteries at operation. It will help if you remember that the external carotid is anterior and is the only carotid vessel which gives off branches in the neck.

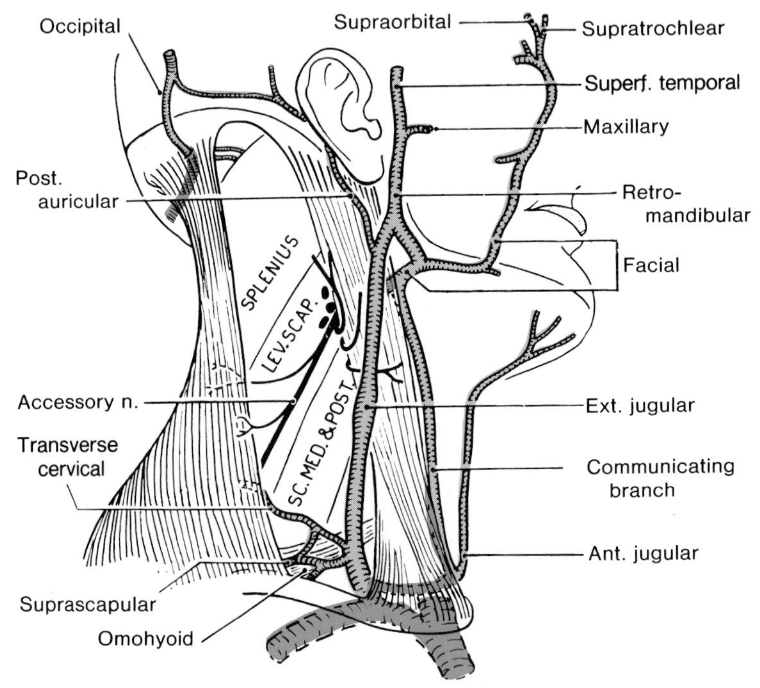

Figure 7-38. Diagram of the superficial veins of the right side of the head and neck. Observe that the facial vein is formed by the union of the supraorbital and supratrochlear veins at the medial angle of the eye. The first part of the facial vein is often referred to clinically as the angular vein.

VEINS OF THE FACE

The facial veins anastomose freely and are drained by veins accompanying the arteries of the face.

THE FACIAL VEIN (Figs. 7-29, 7-31, 7-38, 7-39, and 7-42)

The facial vein provides the *major venous drainage of the face.* It begins at the medial angle of the eye by the union of the supraorbital and supratrochlear veins. The superior part of the facial vein is often called the **angular vein.**

The facial vein runs inferoposteriorly through the face, posterior to the facial artery, but it takes a straighter and more superficial course than the artery. Inferior to the margin of the mandible, the facial vein is joined by the anterior branch of the **retromandibular vein** (Fig. 7-38). The facial vein ends by draining into the internal jugular vein deep to the sternocleidomastoid muscle (Fig. 7-65).

The facial vein makes clinically important connections with the cavernous sinus (Fig. 7-39) through the **superior ophthalmic vein** and the **pterygoid plexus** via the deep facial vein.

Blood from the inner canthus of the eye, the nose, and the lips usually drains inferiorly through the facial vein, especially when the patient is erect. However, as *the facial vein has no valves,* blood may run through it in the opposite direction when pressure conditions are different. Blood may then enter the cavernous sinus, one of the venous sinuses of the dura mater (Fig. 7-39).

In patients with **thrombophlebitis of the facial vein,** pieces of an infected clot may extend into the intracranial venous system. Here it may produce **thrombophlebitis of the cortical veins or of the cavernous sinuses.** Infection of the facial veins spreading to the dural venous sinuses may be initiated by squeezing pustules on the side of the nose and upper lip. Consequently this area is often called **the danger triangle of the face** (Fig. 7-43).

THE SUPERFICIAL TEMPORAL VEIN (Figs. 7-29, 7-31, and 7-38)

The superficial temporal vein drains the forehead and scalp and receives tributaries from the veins of the temple and face. In the region of the temporomandibular joint, it enters the parotid gland and unites with the maxillary vein to form the retromandibular vein (Fig. 7-38).

THE RETROMANDIBULAR VEIN (Figs. 7-38 and 7-42)

The retromandibular vein is formed by the union of the superficial temporal and maxillary veins, posterior to the neck of the mandible (Fig. 7-42). It descends within the parotid gland, superficial to the external carotid artery, but deep to the facial nerve. It divides into an anterior branch that unites with the facial vein and a posterior branch that joins the posterior auricular vein to form the external jugular vein.

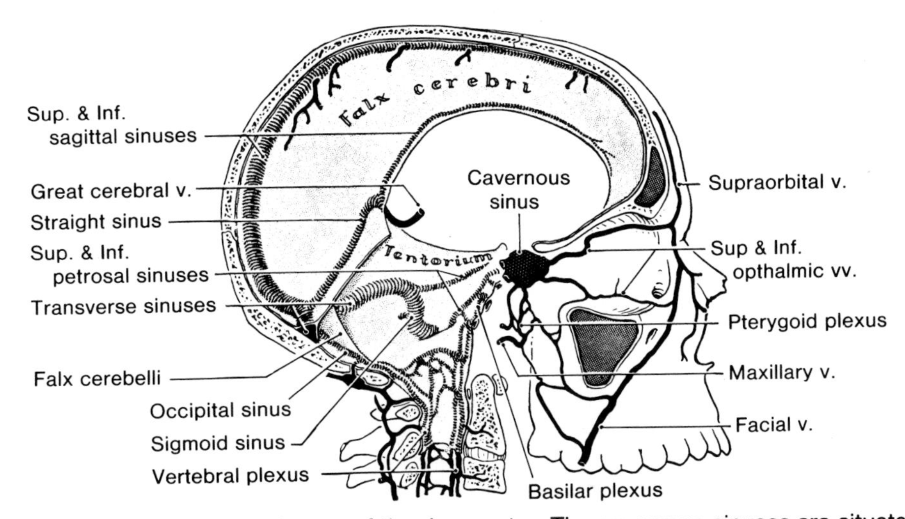

Figure 7-39. Diagram of the venous sinuses of the dura mater. The cavernous sinuses are situated one on each side of the sphenoid bone. The cavernous sinus drains into the transverse sinus through the superior petrosal sinus. Note particularly that the cavernous sinus communicates with the veins of the face via the opthalmic veins and the pterygoid plexus. Also see Figures 7-79 and 7-80.

LYMPHATIC DRAINAGE OF THE FACE

The lymphatic vessels in the forehead and anterior part of the face accompany the other facial vessels and drain into the **submandibular lymph nodes**, located along the inferior border of the mandible (Fig. 7-40).

The lymph vessels from the lateral part of the face, including the eyelids, drain inferiorly toward the **superficial parotid lymph nodes** (Figs. 7-29 and 7-40). The superficial parotid nodes drain into deep parotid nodes which in turn drain into the **deep cervical lymph nodes** (Figs. 7-40 and 8-35).

Lymphatics in the upper lip and in lateral parts of the lower lip drain into the **submandibular lymph nodes**, whereas lymphatics in the central part of the lower lip and in the chin drain into the **submental lymph nodes**, from which lymph may drain directly into the **juguloomohyoid lymph nodes**.

Carcinomas of the lip most commonly involve the lower lip (Fig. 7-44). Overexposure to sunshine over many years, as occurs in outdoor workers, is a common feature of the history in these cases. Chronic irritation from pipe smoking also appears to be a factor in lip cancer, and may be related to long-term contact with tobacco tar.

Cancer cells from the central part of the lip (Fig. 7-44), the floor of the mouth, and the tip of the tongue spread to the **submental lymph nodes**, whereas *cancer cells from lateral parts of the lip* drain to the **submandibular lymph nodes** (Fig. 7-40).

The lymph vessels draining an infected area of the face may become inflamed, resulting in acute or chronic **lymphangitis** (inflammation of lymph vessels). **Lymphadenitis** (inflammation of lymph nodes) also occurs. Blockage of lymphatic channels may also result from the spread of malignant cells, a process known as **lymphogenous metastasis**. Knowledge of the lymphatic drainage of the face enables one to predict to which area(s) cancer cells may spread.

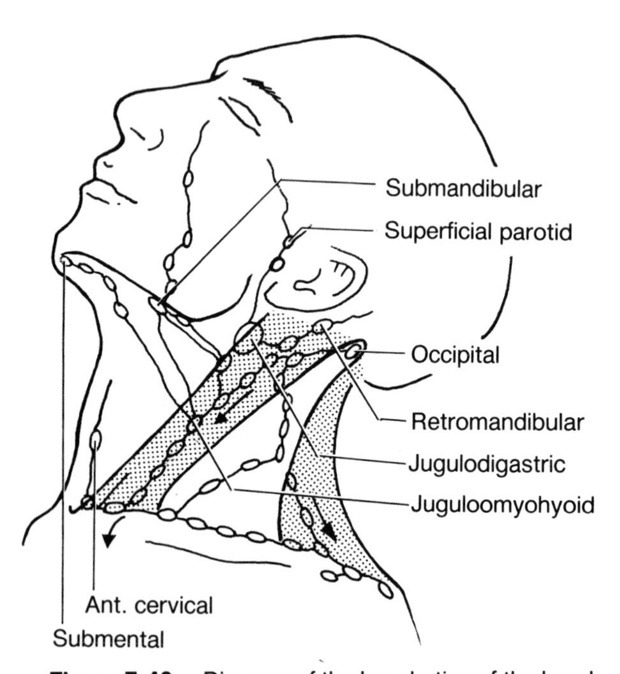

Figure 7-40. Diagram of the lymphatics of the head and neck illustrating the lymphatic drainage of the face. Note that there is a ring of superficial lymph nodes around the junction of the head and the neck composed as follows: parotid, occipital, retroauricular (mastoid), submandibular, and submental lymph nodes. Lymph from the anterior part of the face drains via the submandibular and submental lymph nodes into the juguloomohyoid node.

THE PAROTID GLAND

The parotid is a salivary gland. It is located near the auricle or external ear (G. *para*, near + *otis*, the ear), where it is wedged between the ramus of the mandible and the mastoid process (Figs. 7-29, 7-31, 7-33, and 7-35). The parotid gland is wrapped with a fibrous capsule (**parotid fascia**) that is continuous with the deep investing fascia of the neck. It occupies the side of the face anterior and inferior to the auricle. You can determine the general location of the parotid gland by making a triangle, the angles of which are at the mastoid process, the angle of the mandible, and the midpoint of the zygomatic arch.

In living persons the parotid gland is an irregular, lobulated, yellowish mass which is closely related to the ramus of the mandible (Fig. 7-31). It is irregular in shape because during development it grew into the cervical fascia between the mandible and the mastoid process, enclosing structures in the area it invaded (*e.g.*, the facial nerve). There is a waist-like constriction of the parotid gland, sometimes called the *isthmus*, between the ramus of the mandible and the masseter muscle anteriorly, and the posterior belly of the digastric muscle posteriorly.

The parotid gland is the largest of the paired salivary glands and, like the other two, develops as an outgrowth from the primitive mouth. As viewed superficially, the parotid gland is somewhat triangular in shape (Fig. 7-29), with its apex posterior to the angle of the mandible and its base along the zygomatic arch (Figs. 7-29 and 7-31). Note that it overlaps the posterior part of the masseter muscle (Fig. 7-29).

The parotid duct (Stensen's duct), about 5 cm long (Figs. 7-29, 7-31, and 7-41) and 5 mm in diameter, passes horizontally from the anterior edge of the gland. At the anterior border of the masseter muscle, the

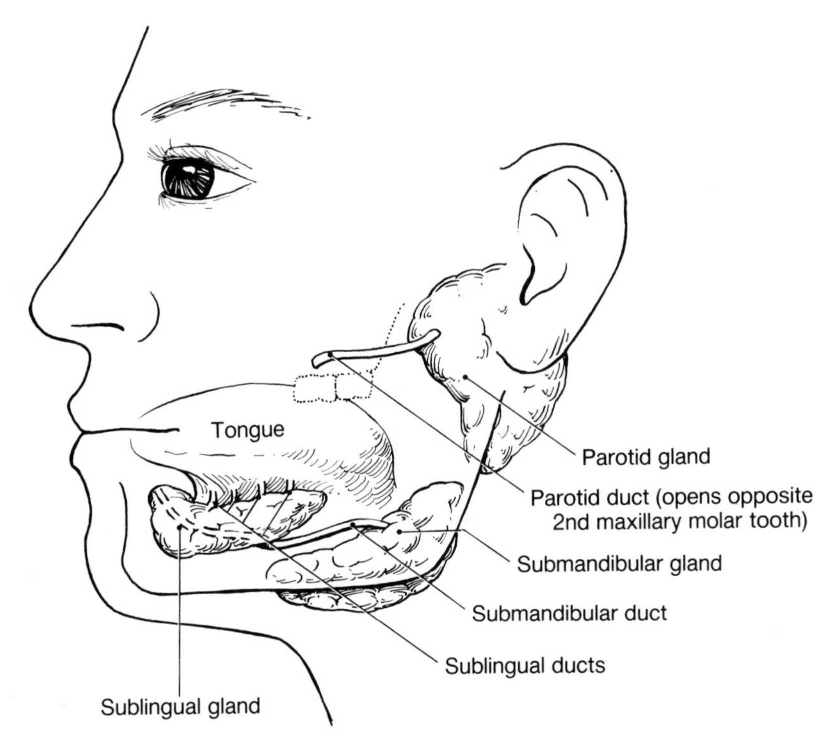

Figure 7-41. Diagram illustrating the location of the three salivary glands. Note that the parotid gland, the largest of them, is wrapped around the neck of the mandible. The parotid duct is a thick-walled tube that drains the saliva from the gland and carries it to the vestibule of the mouth (also see Figs. 7-29 and 7-31).

parotid duct turns medially and **pierces the buccinator muscle** (Figs. 7-29 and 7-112).

The parotid duct enters the oral cavity opposite the crown of the second maxillary molar tooth (Fig. 7-41). To locate the approximate course of the parotid duct, place your index finger along the inferior border of the **zygomatic arch** and point it toward the upper lip (Fig. 7-45). The parotid duct courses along the inferior border of your finger and then curves medially, approximately at your fingertip (also see Fig. 7-29). If you tense your masseter muscle by clenching your teeth, you may be able to palpate the thick-walled parotid duct in your cheek, about a fingerbreadth inferior to your zygomatic arch, as you roll it with your finger over the anterior border of the masseter muscle.

Look into someone's mouth with a flashlight and observe the opening of the parotid duct opposite the second maxillary molar tooth. If the person is asked to suck a lemon slice, you may be able to see saliva flowing out of the duct. Where the duct opens into the mouth, there is a small papilla.

BLOOD VESSELS OF THE PAROTID GLAND (Figs. 7-29, 7-31, 7-37, and 7-38)

The parotid gland is supplied by branches from the external carotid and superficial temporal arteries.

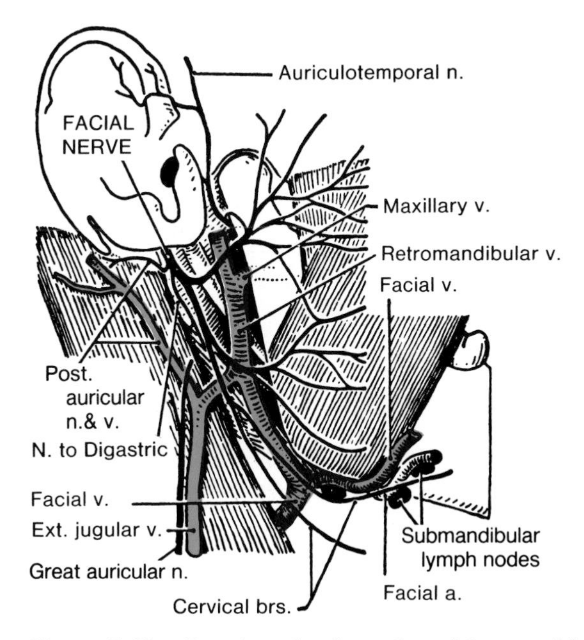

Figure 7-42. Drawing of a dissection of the parotid region showing the location of the facial nerve and veins. The facial nerve and its branches are foremost in importance and are in jeopardy during surgery of the parotid gland. Compare with the superficial dissection illustrated in Figure 7-29.

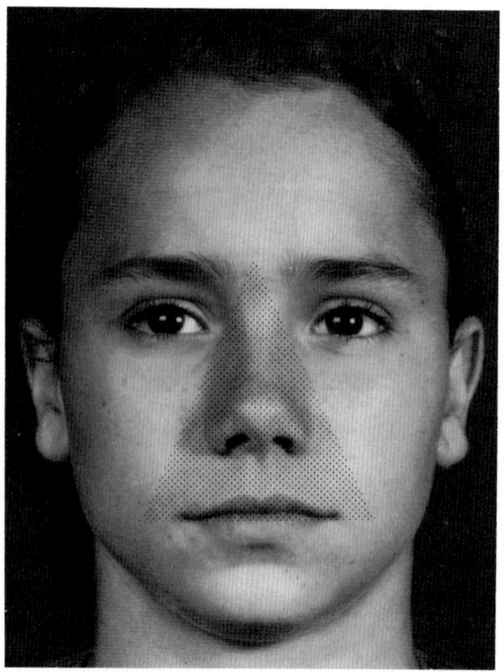

Figure 7-43. Photograph of a 12-year-old girl illustrating the danger triangle of the face. Veins in this area communicate with the superior and inferior ophthalmic veins which drain into the cavernous sinus posterior to the orbit. The squeezing of pustules (pimples) in the danger area may result in spreading of infection to this venous sinus of the dura mater (Fig. 7-39).

Figure 7-44. Drawing of a patient showing a cancerous lesion on his lower lip. Cancer cells from the lateral portion of the lip spread to the submandibular lymph nodes (Fig. 7-40).

The veins from the parotid gland drain into the external jugular vein (Fig. 7-38).

LYMPH VESSELS OF THE PAROTID GLAND (Figs. 7-29 and 7-40)

The lymph vessels of the parotid end in the superficial and deep *cervical lymph nodes* (Fig. 8-35). There are two or three lymph nodes on the surface of the parotid gland and in the substance of the gland.

NERVES OF THE PAROTID GLAND (Figs. 7-29, 7-31 to 7-33, and 7-35)

The nerves of the parotid gland are derived from the *auriculotemporal nerve* and from sympathetic and parasympathetic sources.

The parasympathetic fibers are derived from the glossopharyngeal nerve, i.e., CN IX (Fig. 7-111). Stimulation of the parasympathetic fibers cause a thin watery saliva to flow from the parotid duct.

The sympathetic fibers are derived from the cervical ganglia via the external carotid plexus (plexus of sympathetic nerve fibers on the external carotid artery). Stimulation of the sympathetic fibers produces a thick, mucous saliva.

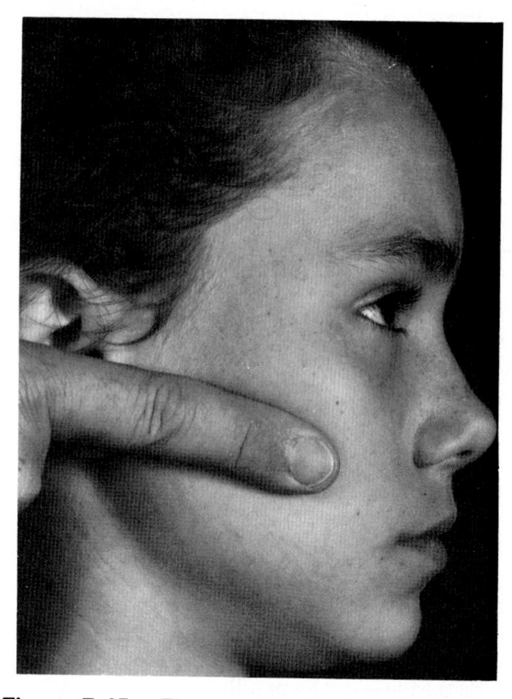

Figure 7-45. Photograph of a lateral view of the right side of the face of a 12-year-old girl illustrating how to determine the course of the parotid duct. The examiner's finger is placed along the inferior border of her zygomatic arch. The parotid duct runs along the inferior border of the finger and ends internally opposite the second maxillary molar tooth, indicated by the tip of the finger (see Figs. 7-31 and 7-41). With the jaws clenched you can palpate the duct by rolling it with the fingertip against the anterior border of the underlying masseter muscle (also see Figs. 7-29 and 7-31).

Although the facial nerve passes through the parotid gland (Fig. 7-35), and carries parasympathetic fibers (Fig. 7-111), it does not supply the parotid gland. Occasionally some parasympathetic fibers are derived from CN VII.

As the facial nerve passes into the substance of the parotid gland and divides into numerous branches (Figs. 7-29 and 7-42), it is in jeopardy during surgery in the parotid region (*e.g.*, for removal of a **carcinoma of the parotid gland**). Injury to the facial nerve results in partial or complete paralysis of the ipsilateral muscles of facial expression (Fig. 7-100).

Removal of the parotid gland would be a relatively simple procedure were it not for the many branches of the facial nerve. The surgeon must carefully remove the gland; otherwise facial paralysis will result.

During illness, particularly debilitating ones, *a bacterial infection may spread from the mouth along the parotid duct to the gland.* Extremely poor oral hygiene can also result in passage of an oral infection to the parotid gland. Bacterial infection, a noncontagious condition, often results in a **parotid abscess**.

The parotid gland may also become infected via the blood stream, *e.g.*, as occurs in **mumps** (*epidemic parotiditis* or *parotitis*), an acute communicable **viral disease** affecting the salivary glands, particularly the parotid.

Infection of the parotid gland causes inflammation of it (**parotiditis**). Severe pain soon occurs because the gland's capsule, derived from the **cervical fascia**, limits swelling. The pain results from stretching of the parotid capsule. Often the pain is worse during chewing because the gland is wrapped around the posterior border of the ramus of the mandible (Fig. 7-31), and is compressed against the mastoid process when the mouth is opened.

The mumps virus may also cause inflammation of the parotid duct, producing redness of its papilla. As mumps, an old English word meaning lumps, may be confused with a **toothache**, redness of the papilla of the parotid duct is often an early sign indicating disease involving the parotid gland and not the teeth. Sucking a lemon is also used as a **test for parotiditis** because lemon juice, being acid, stimulates the secretion of saliva (*i.e.*, it is a **salivant**). This increases the swelling of the gland and stimulates the sensory nerve fibers in its capsule.

Parotid gland disease often causes pain in the auricle, external acoustic meatus, temple, and temporomandibular joint because the pain may be referred to these structures from the parotid gland by the **auriculotemporal nerve** (Figs. 7-29 and 7-31), a superficial branch of the mandibular division of the trigeminal nerve (CN V^3). It supplies sensory fibers to the skin of the temple and carries parasympathetic motor fibers to the gland (Fig. 7-33).

Physicians and dentists frequently must determine whether swellings of the cheek result from infection of the parotid gland (*e.g.*, bacterial or viral infection), or from an **abscessed tooth**.

A *radiopaque material can be injected into the duct system of the parotid gland* via a cannula inserted into the parotid duct. This technique, followed by radiography, is called **sialography**. ***Parotid sialograms***, (G. *sialon*, saliva + *graphō* to write) will demonstrate parts of the duct system that have been displaced or dilated by disease.

The parotid duct may become blocked by a calcified deposit called a sialolith or a **calculus** (L. a pebble). The resulting pain in the parotid gland is made worse by eating. Again, sucking a lemon slice is painful because of the build-up of saliva in the proximal part of the duct, owing to the blockage.

THE SCALP

The scalp consists of five layers of soft tissue covering the calvaria (Fig. 7-46). It extends from the *superior nuchal line* at the posterior aspect of the head (Figs. 7-3 and 7-46*B*) to the *supraorbital margins* (Fig. 7-2), *i.e.*, over the forehead to the eyebrows. Laterally, the scalp extends into the temporal fossae to the level of the zygomatic arches (Fig. 7-1).

The scalp is of clinical importance mainly because it covers the calvaria which encloses the brain (Figs. 7-51 and 7-54).

Scalp lacerations are the most common type of head injury requiring surgical care. Scalp wounds bleed profusely because arteries enter all around the periphery of the scalp (Fig. 7-48), and they do not retract because the scalp is tough. Hence, *unconscious patients may bleed to death from scalp lacerations* if the bleeding is not controlled.

If scalp wounds are not treated appropriately, a scalp infection may develop and spread into the underlying bones of the calvaria, causing **osteomyelitis**. The infection can also spread to the cranial cavity, producing an **extradural abscess** (collection of pus outside the dura ma-

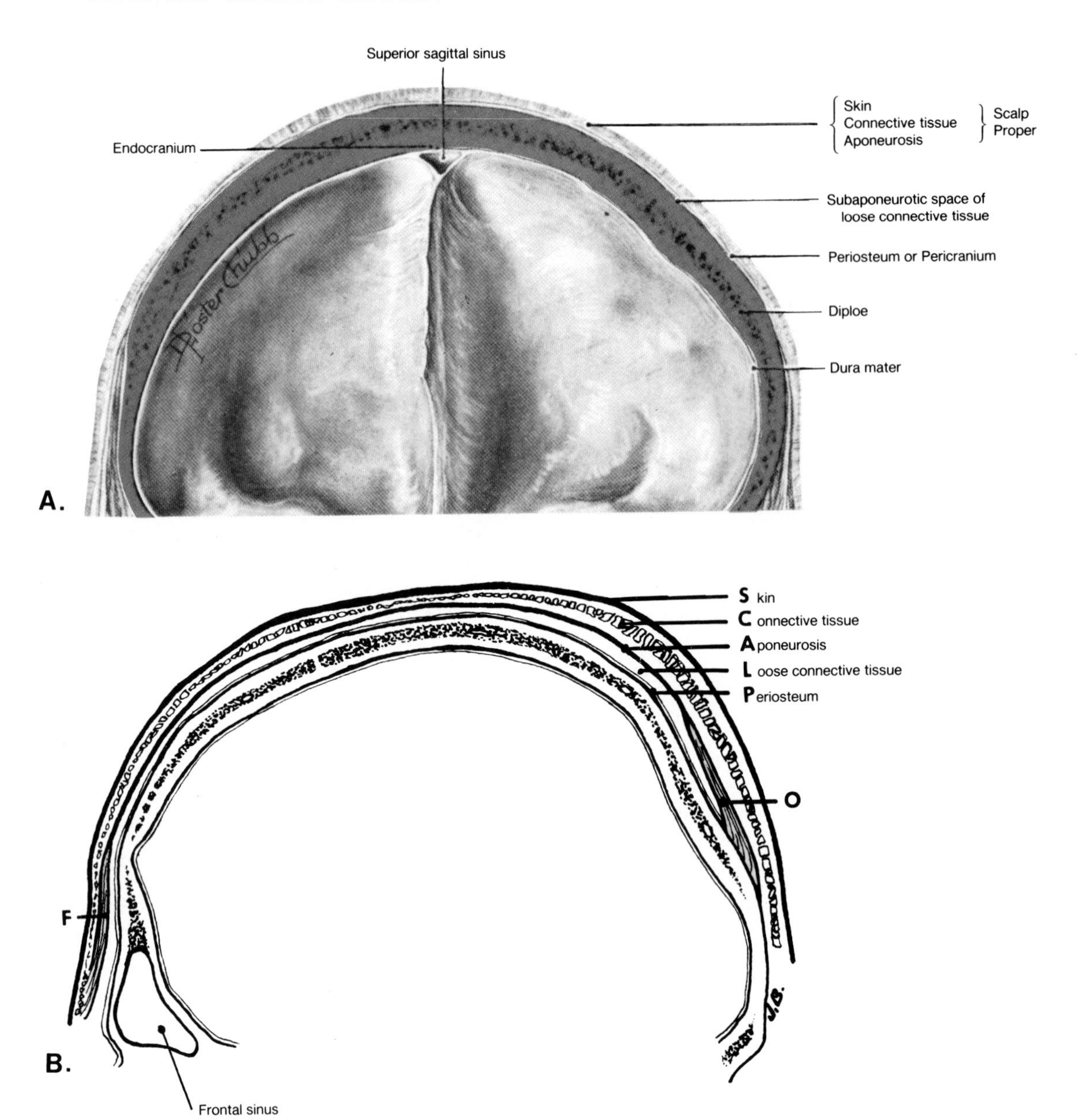

Figure 7-46. Sections through the scalp and calvaria. *A*, a coronal section showing the layers of the scalp and the dura mater. *B*, a sagittal section showing the key to remembering the five layers of the scalp as demonstrated. *S*, skin; *C*, connective tissue (dense); *A*, aponeurosis epicranialis or galea aponeurotica, including the frontalis and occipitalis muscles at its anterior and posterior ends, respectively; *L*, loose connective tissue; *P*, periosteum or pericranium. The skin (*S*) and connective tissue (*C*) are firmly bound to the epicranial aponeurosis (*A*), which is attached to the skull laterally and to two fleshy muscles, frontalis (*F*) and occipitalis (*O*). Blood from a torn vessel may spread widely over the calvaria, deep to the aponeurosis epicranialis. It can only escape anteriorly and appear as bruising in the area of the eyelids (*i.e.*, forming "black eyes.")

ter) and/or **meningitis** (inflammation of the membranes covering the brain).

LAYERS OF THE SCALP

The scalp consists of five layers (Fig. 7-46), but clinically the first three layers, called *the scalp proper*, are usually regarded as a single layer because they remain together when a scalp flap is made during a **craniotomy** (surgical opening of the cranium), or is torn off in accidents. It is difficult to separate the skin and the subcutaneous tissues of the scalp from the dense connective tissue of the **epicranial aponeurosis** or galea aponeurotica (L. *galea*, helmet).

The scalp proper is composed of three fused

layers (Fig. 7-46*A*) It is separated from the pericranium (periosteum of the cranium) by *loose connective tissue that forms a cleavage plane.* Because of this potential areolar space, the scalp is fairly mobile. Verify this by massaging your scalp or by moving your scalp on your skull by contracting your frontalis muscle (Figs. 7-31 and 7-46). As your scalp moves, note in the mirror that your eyebrows and the skin over the root of your nose rise and that your forehead wrinkles. As this scalp muscle is used to express surprise, it is a muscle of facial expression. The looseness of the scalp proper explains why a person can be scalped so easily (*e.g.*, during automobile and industrial accidents).

Each letter of the word S C A L P serves as a memory key for one of the five layers of the scalp (Fig. 7-46*B*): (1) **S***kin*; (2) **C***onnective tissue*; (3) **A***poneurosis epicranalis*; (4) **L***oose connective tissue*; and (5) **P***eriosteum* or pericranium.

1. SKIN (Figs. 7-46 to 7-48)

Most of the scalp skin is thin; it is thick in the occipital region. The scalp contains many sweat glands, sebaceous glands, and hair follicles. Hair covers the scalp in most people. The skin of the scalp has an abundant arterial supply and a good venous and lymphatic drainage.

The ducts of the ***sebaceous glands*** *associated with hair follicles may become obstructed*, resulting in the retention of secretions and the formation of **sebaceous cysts** (*wens*). As they are in the skin and do not invade the subcutaneous tissue, *sebaceous cysts move with the scalp.*

Hair follicles in the scalp go through alternate growing and resting phases, *i.e.*, the hairs grow and eventually drop out of their follicles, *e.g.*, during combing. After a while new hairs begin to grow in the same follicles.

Hair can be lost from the scalp without being replaced owing to various factors, *e.g.*, disease. **Alopecia** (baldness or loss of hair) can be congenital or develop after puberty. Some men have an inherent tendency to become bald (*alopecia heriditaria*), *i.e.*, they have a genetic predisposition to baldness that requires the presence of male sex hormone for its expression. There is also a type of *marginal alopecia* (hair loss at the margins of the frontal, temporal, and parietal areas of the scalp) that may be caused by the continuous contraction of a "pony tail" or by tight hair braiding. The hair loss is typically symmetrical and triangular in shape, but hair loss can occur anywhere depending on the type of hairdo. The anatomical basis for this type of hair loss is interference with the blood supply to the scalp which enters around the periphery of the scalp (Fig. 7-48). This process of hair loss is usually reversible, but if the stress is continued for a long time the hair follicles may atrophy and the hair loss may be permanent.

Sometimes synthetic fiber hairs are sewn into the scalp or a hairpiece (toupee) is attached to it with sutures. Generally these procedures are unacceptable owing to the high incidence of scalp infection that follows.

Hair transplanting performed by dermatologists (physicians who specialize in the diagnosis and treatment of skin lesions) rarely become infected. A small punch is used to remove hair-bearing skin from the base of the hairline; these *hair plugs* are then implanted into the bald area, where they remain viable because they are self-produced, *i.e.*, they are **autogenous grafts.**

2. CONNECTIVE TISSUE (Figs. 7-46 to 7-48)

This is a thick, subcutaneous layer of connective tissue which is **richly vascularized and well supplied with nerves.** Its collagenous and elastic fibers crisscross in all directions (Fig. 7-46*B*), attaching the skin to the aponeurosis epicranialis. ***Fat*** *is enclosed in lobules between the connective fibers*, much as it is in the palms of the hands and the soles of the feet. The amount of subcutaneous fat in the scalp is relatively constant, varying little in **emaciation** or **obesity**, but decreases with advancing age. As a result, the scalp proper is thinner in old people.

Properly treated scalp lacerations normally heal quickly. Superficial infections of the scalp tend to remain superficial owing to the density of the connective tissue layer.

3. APONEUROSIS EPICRANIALIS (Figs. 7-46 and 7-47)

The aponeurosis epicranialis or **epicranial aponeurosis** (galea aponeurotica) is *a strong membranous sheet that covers the central portion of the calvaria.* The term **galea** (L. helmet) is sometimes used to indicate the helmet-like nature of the epicranial aponeurosis.

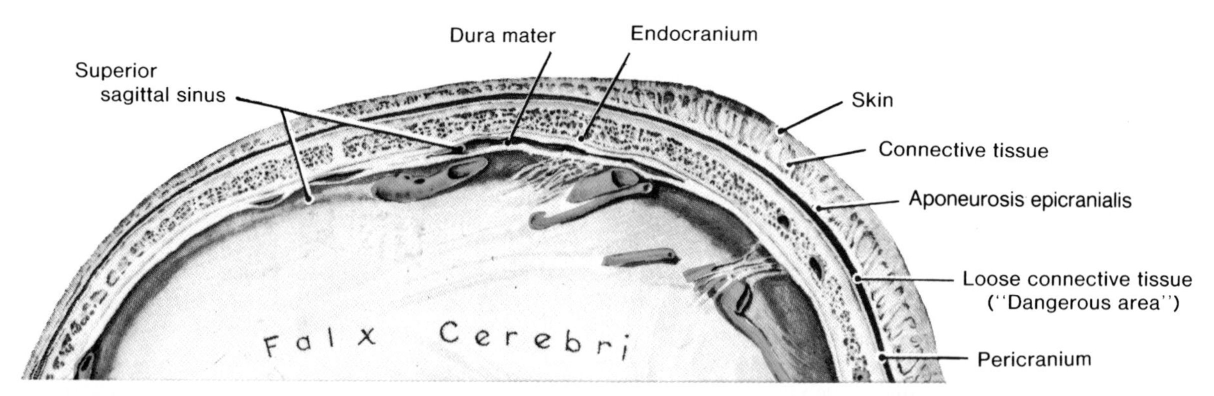

Figure 7-47. Median section of the superior part of the head showing the scalp proper (first three layers) separated from the pericranium (5th layer) by loose connective tissue (4th layer). Emissary veins pass through apertures in the cranium (Fig. 7-7) and establish communication between the dural venous sinuses (*e.g.*, the superior sagittal sinus) inside the cranium and the veins external to it. If lacerated, blood from these veins travels freely within the loose connective tissue layer and is limited only by its attachments (highest nuchal line posteriorly and zygomatic arches laterally). Anteriorly this blood may enter the eyelids because the frontalis muscle attaches to skin, not bone (Fig. 7-46*B*).

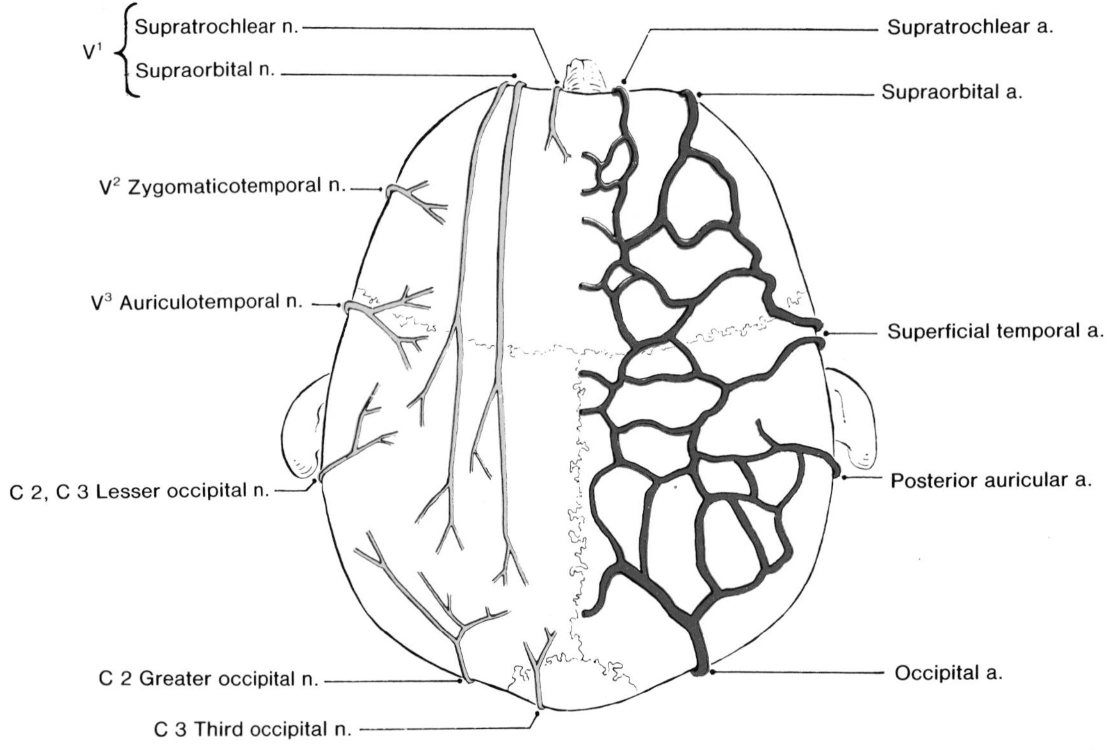

Figure 7-48. Diagram showing the arteries and nerves of the scalp. Note that the arteries anastomose freely. The supraorbital and supratrochlear arteries are derived from the internal carotid artery via the ophthalmic artery. The other three arteries are branches of the external carotid artery (Fig. 7-37). Because of the rich anastomosis of arteries, bleeding from scalp wounds is usually profuse. The nerves appearing in sequence are: CN V^1, CN V^2, and CN V^3 (branches of the trigeminal nerve); ventral rami of C2 and C3; and dorsal rami of C2 and C3. (For anterior and lateral views of the sensory nerves of the face and scalp, see Fig. 7-32.)

The epicranial aponeurosis is the membranous tendon for the fleshy bellies of the epicranius muscle, consisting of frontal and occipital parts, located in the forehead and occipital regions, respectively, (Fig. 7-46*B*). During embryonic development, the **epicranius** was part of a broad sheet of muscle that was derived from the second branchial arch. This muscular sheet included the muscles of facial expression and the platysma (Figs. 7-27 to 7-29). Hence the epicranius or occipitofrontalis muscle is also supplied by the **facial nerve** (CN VII), the nerve of the second branchial arch in the embryo.

The epicranius muscle consists of the **occipitofrontalis**, a broad musculoaponeurotic layer that covers the calvaria from the highest nuchal line posteriorly (Fig. 7-3) to the supraorbital margins anteriorly (Fig. 7-2). The epicranius muscle consists of four parts: two occipital bellies (**occipitalis**) and two frontal bellies (**frontalis**) connected by the epicranial aponeurosis (galea aponeurotica). The four muscle bellies, connected by the aponeurosis, move the scalp and wrinkle the forehead. The frontalis portion pulls the scalp anteriorly and wrinkles the forehead transversely, whereas the occipital portion pulls the scalp posteriorly.

Each occipital belly arises from the lateral two-thirds of the highest nuchal line and the mastoid part of the temporal bone (Fig. 7-3) and ends in the epicranial aponeurosis (Fig. 7-43).

Each frontal belly arises from the epicranial aponeurosis and inserts into the skin and dense subcutaneous connective tissue at about the level of the eyebrows.

The frontalis muscle has no bony attachments. All four parts of the occipitofrontalis are **supplied by the facial nerve (CN VII)**, the occipitalis by posterior auricular branches and the frontalis by temporal branches (Fig. 7-29).

The epicranial aponeurosis is continuous laterally with the fascia over the temporalis muscle (**temporal fascia**), which is attached to the zygomatic arch (Fig. 7-31).

The epicranial aponeurosis or galea aponeurotica is a clinically important layer of the scalp. Owing to its strength, a superficial cut in the skin does not gape because its margins are held together by this aponeurosis. When suturing a superficial laceration, deep sutures are not necessary because the epicranial aponeurosis does not allow separation. Careful suturing of the epicranial aponeurosis following **craniotomy** or deep lacerations prevents disruption of the skin incision.

Scalp wounds gape widely when the epicranial aponeurosis is split or cut because of the pull of the frontal and occipital parts of the epicranius muscle in different directions (anteriorly and posteriorly, respectively). Obviously coronal lacerations gape most widely. Bleeding from scalp wounds is severe because the arteries cannot retract owing to the dense connective tissue.

4. LOOSE CONNECTIVE TISSUE (Figs. 7-46 and 7-47)

This subaponeurotic layer is somewhat like a collapsed sponge because it contains innumerable potential spaces that are capable of being distended with fluid. This loose connective tissue layer allows free movement of the scalp proper (the first three layers) as one sheet.

The loose connective tissue layer is sometimes called the dangerous area because pus or blood in it can spread easily. Infection in it can also be transmitted to the cranial cavity via **emissary veins** that pass from this layer through apertures in the cranial bones (*e.g.*, the **parietal foramina**, Fig. 7-7). The emissary veins connect with the intracranial venous sinuses (*e.g.*, the superior sagittal sinus, Fig. 7-47).

Infections in the loose connective tissue layer may produce inflammatory processes in the emissary veins, leading to **thrombophlebitis of the intracranial venous sinuses** and the cortical veins (Fig. 7-39).

Cortical infarction (an area of necrosis in the cerebral cortex) could follow. Cortical infarcts may produce permanent neurological defects.

Awareness of the limits of the loose connective tissue layer (subaponeurotic space) is important so that the possible spread of an infection can be anticipated. **Posteriorly**, an infection is unable to spread into the neck because the occipitalis muscle is attached to the highest nuchal line of the occipital bone and the mastoid part of the temporal bone (Fig. 7-3). **Laterally**, an infection is unable to spread beyond the zygomatic arches because the epicranial aponeurosis is continuous with the temporal fascia, which is attached to the zygomatic arches (Fig. 7-31). **Anteriorly**, an infection or fluid can enter the eyelids and the root of the nose because the frontalis is inserted into the skin and dense subcutaneous tissue, not into the frontal bone.

Because of the free movement permitted by the loose connective tissue layer, it is through it that the scalp separates in accidents (*e.g.*, when the hair is caught in machinery).

5. PERIOSTEUM OR PERICRANIUM (Figs. 7-46 and 7-47)

The periosteum of the cranium or pericranium is fairly firmly attached to the outer surface of the calvaria by connective tissue fibers known as **Sharpey's fibers**. They penetrate the bones and *firmly anchor the periosteum to the calvaria.* The pericranium can be stripped fairly easily from the cranial bones of living persons, except where it is continuous with the fibrous tissue of the cranial sutures. Here the pericranium passes inwardly to be continuous with the *endocranium* on the inside of the calvaria (Fig. 7-46*A*).

The pericranium is a dense layer of specialized connective tissue that has relatively poor osteogenic properties in adults.

During birth, bleeding sometimes occurs between the pericranium and the calvaria, usually over one parietal bone. The bleeding results from rupture of multiple, minute periosteal arteries that enter and nourish the bones of the calvaria. The resulting swelling, which develops several hours after birth, is called a **cephalohematoma**. This *"blood cyst" of the scalp* in infants does not spread beyond the margins of the bone concerned because the pericranium is closely bound to the fibrous tissue connecting the edges of the bones at the cranial sutures.

Cephalohematoma should not be confused with **caput succedaneum**, which is an edematous swelling of the soft tissues of the scalp that may develop during birth. This condition occurs in the area that presents in the birth canal (usually the scalp in the region of the lambda), as this region is the only part of the infant that is not being compressed by the walls of the birth canal (cervix of uterus and vagina). This swelling, being in the second layer of the scalp, the connective tissue layer of the subcutaneous tissue, is not localized to any particular bone as occurs with a cephalohematoma.

Because the adult pericranium has poor osteogenic properties, there is very little regeneration if bone loss occurs, e.g., when pieces of bone have to be removed following a fracture. Surgically produced **bone flaps** are usually put back into place and wired to the other parts of the calvaria. Traumatic defects in the adult calvaria (*e.g.,* owing to severe trauma) usually do not fill in, necessitating the insertion of a metal or plastic plate to protect the brain.

NERVES OF THE SCALP

The sensory innervation of the scalp anterior to the auricles (ears) is via nerves that are branches of all three divisions of the trigeminal nerve (Figs. 7-32 and 7-48). Posterior to the auricles, the nerve supply consists of the spinal cutaneous nerves of the neck from the **cervical plexus** (C2 and C3). The area of distribution of the trigeminal and cervical nerves is usually about equal (Fig. 7-32).

ARTERIES OF THE SCALP

The blood supply of the scalp is from the **external carotid arteries** via the occipital, posterior auricular, and superficial temporal arteries (Figs. 7-37 and 7-48), and from the **internal carotid arteries** via the supratrochlear and the supraorbital arteries. *All these arteries anastomose freely with each other* (Fig. 7-48).

The arteries of the scalp are in the dense subcutaneous connective tissue layer (second layer). Very few branches of these arteries cross the subaponeurotic loose connective tissue layer (fourth layer) of the scalp to supply the calvaria. Its bones are supplied mainly by the middle meningeal arteries (Figs. 7-16 and 7-50), but they receive some blood from pericranial vessels that enter the cranium through **Volkmann's canals** (vascular canals in bones).

VEINS OF THE SCALP

The **vena comitantes** accompany the arteries and have the same names. The **supraorbital and supratrochlear veins** (visible on the foreheads of some people) unite at the medial angle of the eye to form the **facial vein** (Figs. 7-31 and 7-38). At this point it communicates with the **superior ophthalmic vein** (Fig. 7-39), thereby making a link which may allow facial infections to reach the **cavernous sinus** (Fig. 7-39).

The superficial temporal vein (Figs. 7-31 and 7-38) joins the **maxillary vein** posterior to the neck of the mandible to form the **retromandibular vein** (Figs. 7-38 and 7-42).

The posterior auricular vein drains the scalp posterior to the auricle and often receives a mastoid emissary vein from the **sigmoid sinus** (Fig. 7-39), an intracranial venous sinus.

LYMPHATIC DRAINAGE OF THE SCALP

Most lymph from the scalp and forehead drains to **the superficial "collar chain" or ring of lymph nodes**, located at the junction of the head and neck (Fig. 7-40); however some vessels drain directly into the **deep cervical lymph nodes** (Figs. 8-35 and 8-41).

These lymph nodes form a number of groups which are named according to their position: *submental, submandibular, parotid,* (preauricular), *retromandibular* (mastoid), and *occipital* (suboccipital). These lymph nodes drain the superficial tissues of the head and drain to the cervical lymph nodes.

The deep cervical lymph nodes (Figs. 8-35 and 8-41) are the most important ones. They are located along the internal jugular vein and lymph from them passes via the jugular trunk into the *thoracic duct* or the right lymphatic duct (Fig. 1-42).

The superficial cervical lymph nodes lie along the external jugular vein. Efferent vessels from the parotid gland and the inferior part of the auricle drain into these lymph nodes (Fig. 7-40).

Lymph vessels from the frontal region of the head, superior to the root of the nose, drain into the **submandibular nodes**, one of the deep group of lymph nodes. The occipital region drains into the **occipital nodes**; the temporoparietal region into the **retroauricular nodes**; and the frontoparietal region into the superficial **parotid lymph nodes**. *Lymph from the entire scalp eventually drains into the deep cervical group of lymph nodes* (Figs. 8-35 and 8-41).

As the nerves and vessels of the scalp enter inferiorly (Fig. 7-48) and ascend through the connective tissue (second layer of the scalp) into the skin, *surgical pedicle flaps of the scalp are made so that they remain attached inferiorly in order to preserve the nerves and vessels and to promote good healing.* The term **pedicle flap** is used to describe a detached mass of tissues cut away from underlying parts, but attached at the edge or *pedicle* which contains its blood vessels and nerves.

The anastomoses of blood vessels are so numerous in the scalp that partially detached portions of the scalp can usually be replaced without necrosis occurring.

Accidental scalping can occur during automobile accidents (*e.g.,* when the head goes through a windshield), or during industrial accidents (*e.g.,* when a person's long hair becomes entangled in machinery). Scalping does not result in necrosis of the cranial bones because their blood is supplied by the pericranial and **middle meningeal arteries** (Fig. 7-50).

The abundant blood supply of the scalp often leads to ***extensive hemorrhage*** *when it is lacerated because the vessels are unable to retract or contract,* owing to the denseness of the subcutaneous connective tissue surrounding them. However, if the skin adjacent to a scalp laceration is compressed between the fingers and the cranium, bleeding can usually be stopped.

In ***extensive scalp wounds****, hemorrhaging can be temporarily stopped by applying a tourniquet around the base of the scalp.* This procedure applies the anatomical knowledge that the vessels enter the scalp inferiorly (Fig. 7-48).

Reflection (L. *reflexio,* a bending back) of the scalp and removal of the skullcap exposes the **meninges** (G. membranes) enveloping the brain (Fig. 7-50). This is done routinely in autopsies by pulling the anterior half of the scalp over the face, and the posterior half over the back of the neck, and cutting the skullcap horizontally, about 5 cm superior to the **external acoustic meatus** (Figs. 7-10 and 7-15). Following sectioning of the cranial nerves and all attachments of the meninges to the cranium, the brain can be removed intact by severing the spinal cord just inferior to the foramen magnum (Fig. 7-52).

THE CRANIAL FOSSAE

When the skullcap and brain are removed, the internal aspect of the base of the skull is exposed (Fig. 7-15). This bowl-shaped area that supports the brain has three levels called the **anterior, middle, and posterior cranial fossae** (L. depressions).

On a dried skull note that each fossa is at a slightly more inferior level than the one rostral to it. The posterior cranial fossa is the largest and deepest of the three fossae. The floors of the cranial fossae are irregular owing to projections of some bones in the base of the skull.

THE ANTERIOR CRANIAL FOSSA

The inferior part and anterior extremities of the frontal lobes of the cerebral hemispheres, known as the **frontal poles** (Fig. 7-51), occupy the anterior cranial fossa, the shallowest of the three fossae (Figs. 7-15 and 7-68).

The anterior cranial fossa is largely formed by the **frontal bone**. Most of its floor is composed of the convex ***orbital plates*** of this bone (Fig. 7-13), which constitute the bony roofs of the **orbits** (F. circles) or eye sockets. The orbital plates show sinuous, shallow depressions called convolutional impressions or *brain markings*, which are formed by the **orbital gyri** (convolutions) of the frontal lobes of the brain (Figs. 7-13 and 7-52). Between these convolutional impressions,

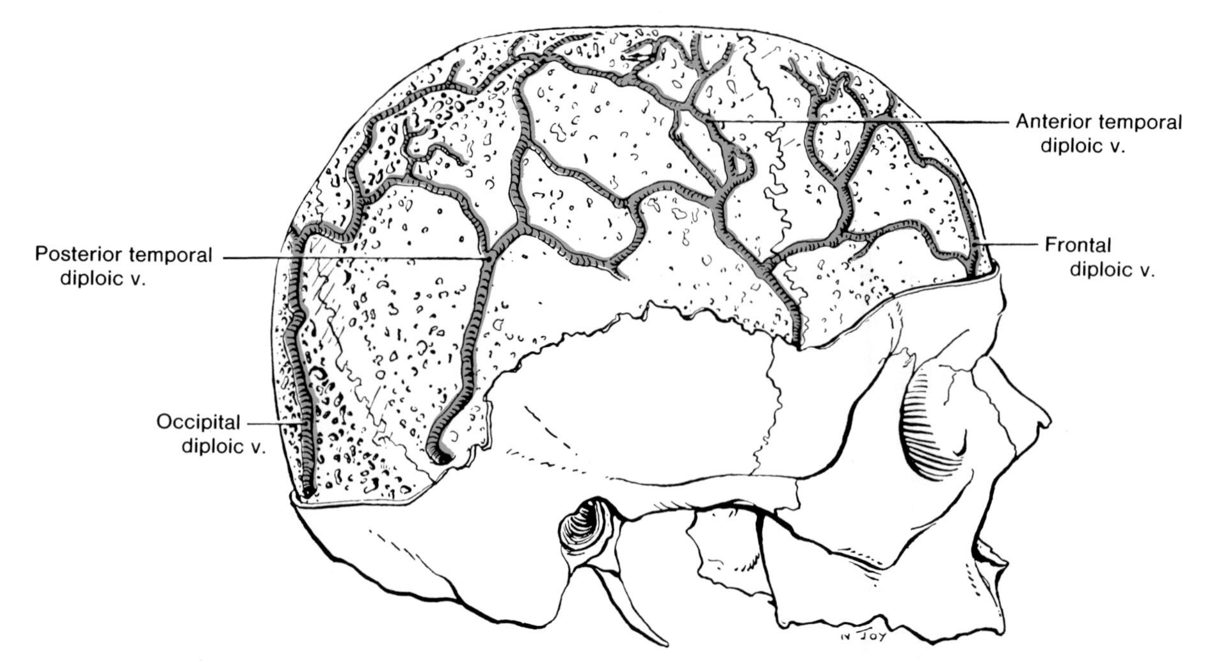

Figure 7-49. Drawing of the diploic veins which have been displayed by filing away the outer table of compact bone, thereby opening the channels that contained them. Of the four paired diploic veins, the frontal diploic vein, opens into the supraorbital vein at the supraorbital notch; the anterior temporal diploic vein opens into the sphenoparietal sinus; the posterior temporal and the occipital diploic veins both open into the transverse sinus (Fig. 7-57), but they may open into veins of the scalp. The connections of the diploic veins with intracranial and extracranial venous channels allow an infection to spread from the scalp, through the skull to the meninges and the brain (Fig. 7-50). There are no accompanying diploic arteries. The blood supply for the bones of the skull is derived from the meningeal and pericranial arteries.

low sinuous ridges are produced by the sulci (L. furrows) between the gyri on the inferior surface of the frontal lobe (Fig. 7-52).

The **crista galli** (L. *crista*, crest + *gallus*, a cock) is a median process or crest resembling a cock's comb that extends superiorly from the **ethmoid bone** (Figs. 7-14 and 7-15). On each side of the crista galli there is a perforated **cribriform plate** (L. *cribrum*, sieve), through which the *olfactory nerves* pass to enter the **olfactory bulbs** (Figs. 7-52 and 7-152). The crista galli, together with the **frontal crest**, gives attachment to a midline septum (fold) of dura mater, called the **falx cerebri** (Figs. 7-47, 7-53, and 7-54). The falx cerebri lies between the cerebral hemispheres of the brain.

The lesser wings of the sphenoid bone (Fig. 7-13), which articulate with the orbital plates of the frontal bone and the jugum of the sphenoid (which joins the wings), form the posterior part of the floor of the anterior cranial fossa. The lesser wings present sharp posterior margins, called **sphenoidal ridges**, which overhang the anterior part of the middle cranial fossa and project into the anterior part of the lateral sulci of the cerebral hemispheres (Fig. 7-51). Each lesser wing ends medially in an **anterior clinoid process** (Figs. 7-14 and 7-15), which gives attachment to a dural septum called the **tentorium cerebelli** (Figs. 7-53 and 7-54). The clinoid processes (G. *klinē*, bed + *eidos*, resemblance) reminded Greek anatomists of bedposts.

THE MIDDLE CRANIAL FOSSA

The rounded anterior extremities of the temporal lobes of the cerebral hemispheres, known as the **temporal poles** (Fig. 7-51), and about half of the inferior surface of the temporal lobe (Fig. 7-51), fit into the middle cranial fossa (Fig. 7-15).

The middle cranial fossa, which is posterior and inferior to the anterior cranial fossa, is marked off from the posterior cranial fossa by a median rectangular bony projection, the **dorsum sellae** (L. seat or saddle), with its knob-like projections called the **posterior clinoid processes** (Figs. 7-14 and 7-15). The dorsum sellae is a square portion of bone on the body of the **sphenoid bone**, posterior to the *sella turcica* (Fig. 7-15).

More laterally, the middle cranial fossa is also separated from the posterior cranial fossa by crests or prominences formed by the superior borders of the *petrous parts of the* ***temporal bones*** (Fig. 7-55). Ex-

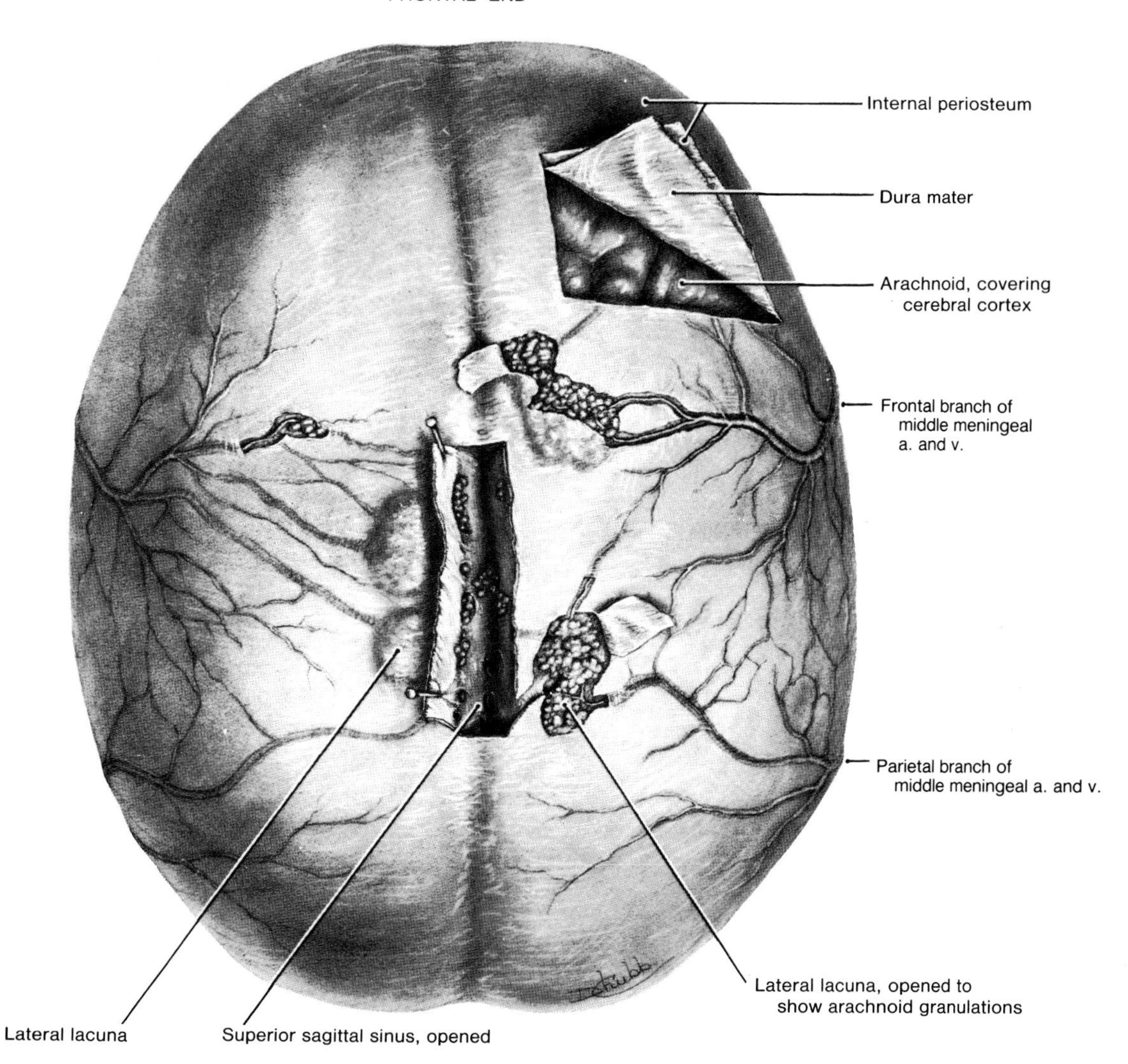

Figure 7-50. Drawing of the external surface of the dura mater indicating the location of the arachnoid covering the cerebral cortex. The roof of the superior sagittal sinus (a dural venous sinus) has been opened and pinned back. Observe the arachnoid granulations (small granular bodies) which lie mainly along the sides of the sinus. They are protrusions of arachnoid through apertures in the dura (Fig. 7-67). These granulations are normal enlargements of the arachnoid villi present in younger persons. In old age they may become very large and even erode the inner table of calvarial bone (Fig. 7-7). When the brain and its meninges were removed from the skull, the internal periosteum (endocranium) was also stripped from the calvaria. Near the frontal end of this specimen, an angular flap of the internal periosteum and the dura mater has been turned anterolaterally. Some authors consider the internal periosteum to be the outer layer of the dura. The middle meningeal artery is of considerable surgical importance because it may be torn in fractures of the temporal region, resulting in bleeding between the skull and the dura mater (Fig. 7-175). This results in symptoms of compression of the brain.

amine the floor of the middle cranial fossa and note that the part formed by the sphenoid (G. wedge) bone resembles a butterfly (Fig. 7-13). The body of the "butterfly" is formed by the body of the sphenoid bone and the wings are formed by the greater wings. The remainder of the floor of the middle cranial fossa is formed by the **petrous and squamous parts of the temporal bone** (Fig. 7-55).

The midline saddle-like part of the sphenoid bone, between the anterior and posterior clinoid processes, is known as the **sella turcica** (L. Turkish saddle). It is composed of three parts (Figs. 7-13 and 7-15): (1) anteriorly an olive-shaped swelling, the **tuberculum sellae** (the knob-like protuberance at the anterior end of the sella turcica); (2) a seat-like depression, called the **hypophysial fossa**, for the **hypophysis cerebri** (Figs. 7-13 and 7-54); and (3) the posterior part of the "saddle" or sella turcica, known as the **dorsum sellae** (Figs. 7-13, 7-14, and 7-60).

THE POSTERIOR CRANIAL FOSSA

The posterior cranial fossa is the largest and deepest of the three cranial fossae. It lodges the cerebellum, pons, and medulla. *It is formed largely by the inferior and anterior parts of the* ***occipital bone*** (Figs. 7-13 to 7-15, and 7-51), but the body of the sphenoid and the petrous and mastoid parts of the temporal bones also contribute to its formation.

The occipital lobes of the cerebral hemispheres lie on the tentorium cerebelli (Figs. 7-53, 7-54, and 7-68), superior to the posterior cranial fossa. In Figure 7-14, observe the broad grooves formed by the **transverse sinuses** which lie between diverging folds of the peripheral attachments of the tentorium (Figs. 7-39 and 7-57). The groove for the right sinus is usually larger because the superior sagittal sinus commonly enters the transverse sinus on that side.

The **tentorium** (L. tent), a tent-shaped dural septum, roofs over most of the posterior cranial fossa, intervening between the occipital lobes of the cerebral hemispheres and the cerebellum (Figs. 7-51 and 7-54). Between the anteromedial parts of the right and left leaves of the tentorium is an oval opening, called the ***tentorial incisure*** (notch), for the brain stem as it passes from the middle to the posterior cranial fossa (Fig. 7-58).

In the midline inferior, posterior to the edge of the foramen magnum, there is a prominent bony ridge, the **internal occipital crest** (Fig. 7-14), which partly divides the posterior cranial fossa into two **cerebellar fossae** for the cerebellar hemispheres (Figs. 7-51 and 7-56). This crest ends superiorly and posteriorly in an irregular elevation called the **internal occipital protuberance** (Fig. 7-14), which more or less matches the **external occipital protuberance** (Figs. 7-3 and 7-10).

The **vermis** (L. worm) or midline portion of the cerebellum (Fig. 7-59) lies in the **vermian fossa** (Fig. 7-14). Directly anterior to this fossa is the largest opening in the base of the skull, the **foramen magnum** (Fig. 7-13), through which the spinal cord passes to join the medulla (Fig. 7-54). Rostral to the foramen magnum, the basilar part of the occipital bone rises to meet the body of the sphenoid bone (Fig. 7-13) or **basisphenoid**. This inclined bony surface, called the *clivus* (L. slope), is located anterior to the pons and medulla of the brain stem (Figs. 7-14 and 7-60).

THE CRANIAL FORAMINA

Many foramina or apertures (L. openings) perforate the base of the skull. Most of them are for transmission of the 12 cranial nerves, but some are for the passage of blood vessels. The presence of so many foramina and thin areas of bone in the floor of the cranium makes the base of the skull fragile and vulnerable to fracture.

FORAMINA IN THE ANTERIOR CRANIAL FOSSA (Figs. 7-14 and 7-15)

The sieve-like appearance of the cribriform plates of the ethmoid on each side of the crista galli is caused by the numerous small foramina in them. ***Axons of olfactory cells*** in the olfactory epithelium in the roof of each nasal cavity converge to form about 20 small olfactory nerves (Fig. 7-149). These bundles of axons pass through the tiny foramina in the cribriform plates of the ethmoid bone to enter the olfactory bulbs of the brain (Fig. 7-152).

The foramina for the **anterior ethmoidal nerve** (Fig. 7-149) and artery to the nasal cavity are located lateral to the anterior end of a slit-like fissure in the anterior end of the cribriform plate (Fig. 7-14). These structures enter the cranial cavity after arising from the nasociliary nerve and ophthalmic artery in the orbit. There are other foramina for the posterior ethmoidal nerves and arteries which are small and relatively unimportant clinically.

Between the frontal crest and the crista galli there is often a small **foramen cecum** (Fig. 7-14), through which an emissary vein passes from the superior sagittal sinus to the veins of the frontal sinus and nose. It is present in children and in some adults, but is relatively unimportant clinically.

FORAMINA IN THE MIDDLE CRANIAL FOSSA (Figs. 7-14, 7-15, 7-55, and 7-61 to 7-63)

In the anteromedial part of the middle cranial fossa, the **optic canals** establish communication with the orbits. The optic canals are traversed by the **optic nerves** (CN II) and the **ophthalmic arteries.**

On each side of the base of the body of the sphenoid bone (**basisphenoid**) in the greater wing of the sphenoid bone, there is an important **crescent of foramina**

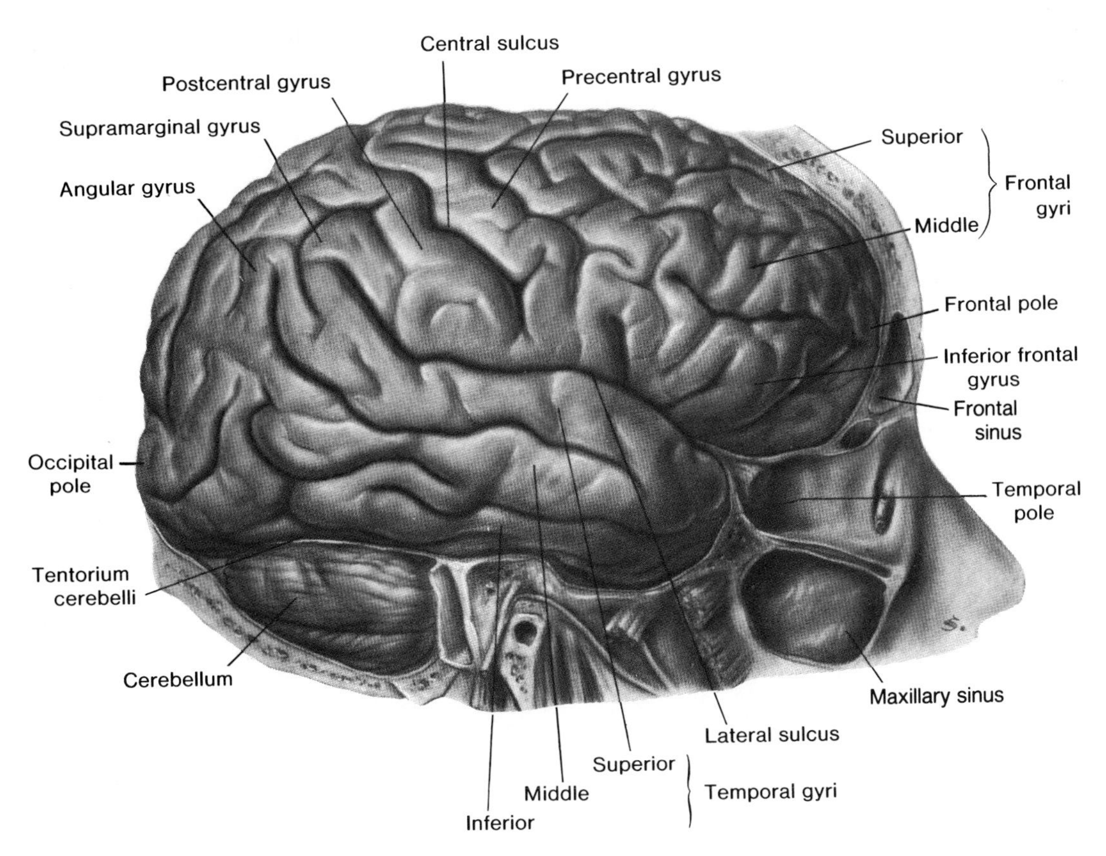

Figure 7-51. Lateral view of the brain exposed in the skull to show its relationship to the cranial cavity. The meninges (Fig. 7-50) have been removed. Observe that (1) the frontal poles occupy the anterior cranial fossa; (2) the temporal poles are lodged in the middle cranial fossa; and (3) the cerebellum, pons, and medulla are located in the posterior cranial fossa. Consequently, in frontal impacts the frontal and temporal poles of the hemispheres may be bruised, whereas in occipital impacts the occipital poles and cerebellum may be damaged. The cranial fossae are illustrated in Figure 7-15.

(Fig. 7-61). Only the *four constant foramina* and the nerves and vessels traversing them warrant special note.

The superior orbital fissure is an elongated slit between the greater and lesser wings of the sphenoid bone (Figs. 7-2, 7-15, 7-61, and 7-62). *It permits communication between the middle cranial fossa and the orbit.*

The superior orbital fissure is traversed by four cranial nerves (CN III, CN IV, the nasociliary, frontal, and lacrimal branches of CN V, and CN VI) and the ophthalmic vein(s).

The foramen rotundum is located in the greater wing of the sphenoid bone (Figs. 7-14 and 7-61), immediately inferior and a little posterior to the medial end of the superior orbital fissure. It ***transmits the maxillary nerve*** (Fig. 7-62), the second division of the trigeminal (CN V^2). Although the name of this foramen (L. *rotundus*, round) indicates that it is circular, it is frequently oval rather than round and is often a short canal leading anteriorly rather than a foramen (L. an aperture).

The foramen ovale is the next largest aperture along the crescent of foramina (Figs. 7-14, 7-15*A*, and 7-61 to 7-63). As its name indicates, it is oval in form. It passes through the greater wing of the sphenoid bone, posterior and slightly lateral to the foramen rotundum, before opening into the **infratemporal fossa** (Fig. 7-113). The foramen ovale **transmits the mandibular nerve** (Fig. 7-62), the third division of the trigeminal (CN V^3).

The foramen spinosum is the smallest of the important foramina in the crescent of foramina (Figs. 7-14, 7-15, 7-61, and 7-63). It was named the foramen spinosum because it enters the skull close to the spine of the sphenoid bone. The foramen spinosum, located posterolateral to the foramen ovale, ***transmits the middle meningeal artery*** (Figs. 7-16, 7-50, 7-55, and 7-62), the largest of the meningeal arteries.

The foramen lacerum, a ragged foramen (really a

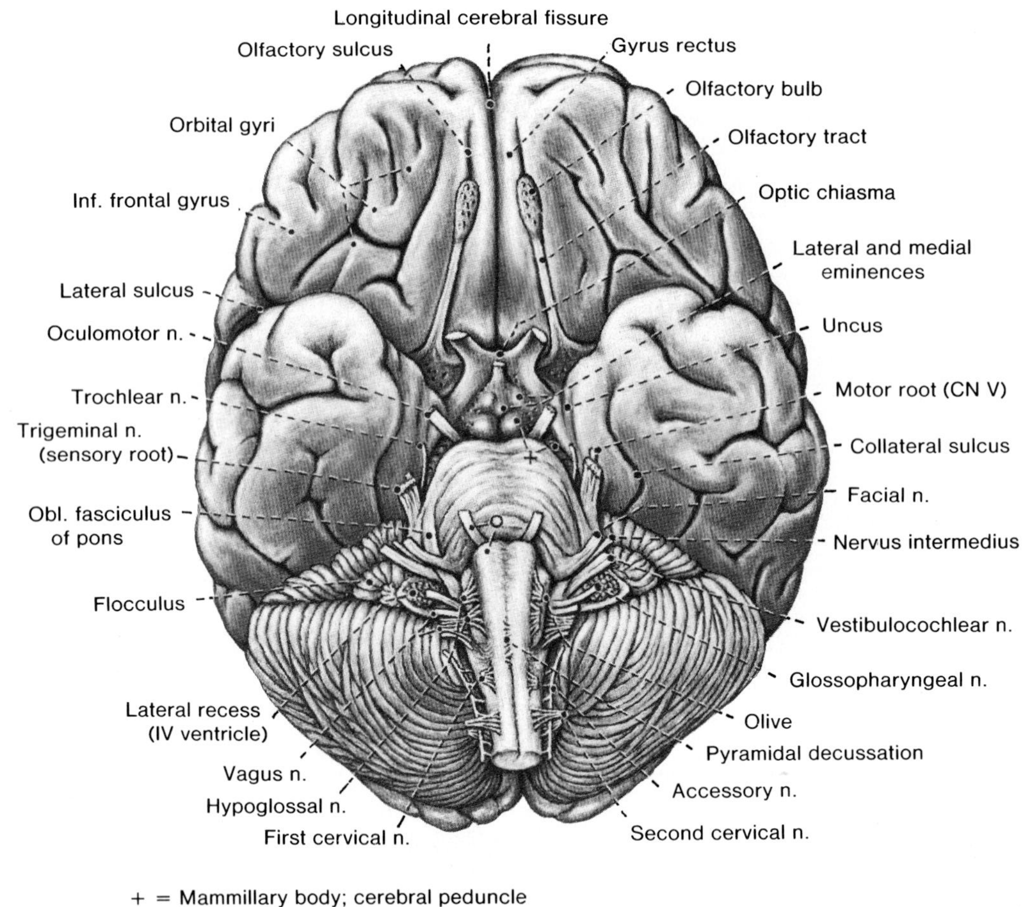

Figure 7-52. Drawing of the inferior surface of the brain showing the cranial nerves. Also see Tables 7-1 and 7-2.

canal), is not part of the crescent of foramina (Figs. 7-15*A* and 7-61). It is located between the basisphenoid and the apex of the petrous part of the temporal bone (Figs. 7-55 and 7-63), posteromedial to the foramen ovale. The carotid and pterygoid canals open into the foramen lacerum which is filled with cartilage, except in dried skulls. The superior end of the foramen lacerum ***contains the internal carotid artery*** and its accompanying sympathetic and venous plexuses.

Hiatuses for the Petrosal Nerves (Figs. 7-14 and 7-63). Extending posteriorly and laterally from the foramen lacerum, there is a narrow groove (sulcus) for the **greater petrosal nerve** on the anterior surface on the petrous part of the temporal bone. At the lateral end of this sulcus, there is a small hiatus (slit), called the **greater petrosal hiatus** (foramen). It ***transmits the greater petrosal nerve*** as it leaves the *geniculate ganglion of the facial nerve* (Fig. 7-35). Sometimes there is also a groove for the **lesser petrosal nerve**, but its hiatus is usually so small that it cannot be identified with certainty.

FORAMINA IN THE POSTERIOR CRANIAL FOSSA (Figs. 7-13, 7-15, 7-60, and 7-63)

There are four clinically important foramina and canals in the posterior cranial fossa.

The foramen magnum (Figs. 7-8, 7-15*A*, 7-55, and 7-63) is *unique for at least three reasons*: (1) it is the largest foramen in the skull; (2) the junction of the medulla and spinal cord occurs within it (Fig. 7-54); and (3) it is unpaired. The foramen magnum, usually oval, lies at the most inferior part of the posterior cranial fossa, midway between the mastoid processes (Fig. 7-8). The posterior cranial fossa communicates with the vertebral canal via the foramen magnum.

Structures passing through the foramen magnum are: (1) the junction of the medulla and spinal cord; (2) the spinal roots of the accessory nerves, CN

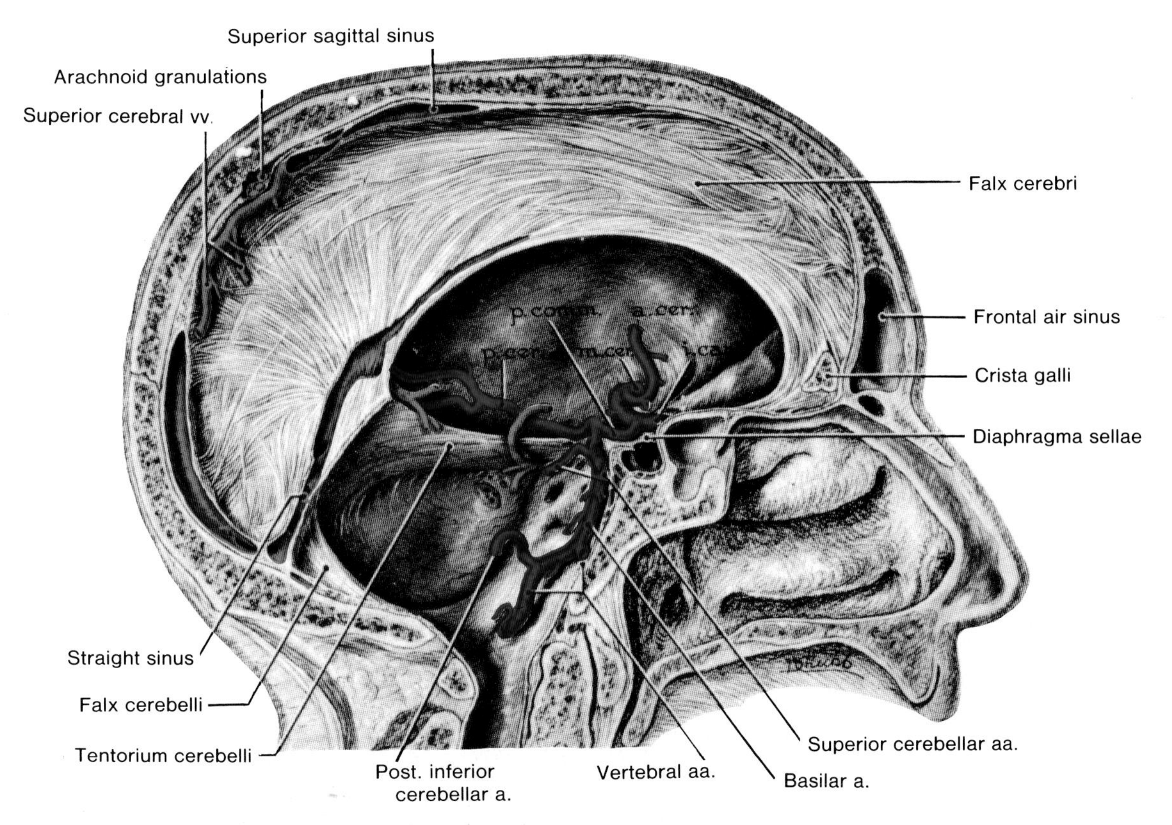

Figure 7-53. Drawing of a median section of the head showing the four folds (septa) of the dura mater: two sickle-shaped folds, the falx cerebri and the falx cerebelli, which lie vertically in the median plane; and two roof-like folds, the tentorium cerebelli and the diaphragma sellae. The tentorium is a wide, sloping tent-like fold of dura mater that covers the cerebellar hemispheres. The tentorium is traversed by the midbrain and the diaphragma sellae, and by the stalk of the hypophysis cerebri, called the infundibulum (Fig. 7-72). Note the two paired arteries that supply the brain, the internal carotid (*i. car.*) and the vertebral arteries. The right vertebral artery has been cut off (also see Fig. 7-77).

XI (Fig. 7-64); (3) the meningeal branches of the upper cervical nerves (C1 to C3); (4) the meninges (coverings of the brain and spinal cord); (5) the **vertebral arteries** ascending to supply parts of the brain (Fig. 7-37); and (6) the anterior and posterior **spinal arteries** descending to supply the superior part of the spinal cord (Fig. 5-56).

The jugular foramen (Figs. 7-60 and 7-63) is an interosseous foramen, located between the occipital bone and the petrous part of the temporal bone. It lies at the posterior end of the **petrooccipital suture.** The posterior part of the jugular foramen contains the superior bulb of the internal jugular vein (Fig. 7-65) into which the **sigmoid sinus** enters (Figs. 7-39 and 7-57).

The tympanic body or jugular glomus is a small ovoid body, consisting of *chemoreceptor tissue*, which is enclosed in the adventitia of the jugular bulb. Anteromedial to the internal jugular vein, several structures descend through the jugular foramen: (1) the glossopharyngeal (**CN IX**); (2) the vagus (**CN X**); (3) the accessory (**CN XI**); and (4) the inferior petrosal sinus on its way to the superior end of the internal jugular vein (Figs. 7-39 and 7-57*B*).

A tumor of the tympanic body (jugular glomus) may cause symptoms owing to involvement of the neighbouring cranial nerves (CN IX, CN X and CN XI) and the middle ear.

The hypoglossal canal (Figs. 7-14, 7-60, and 7-63) *transmits the hypoglossal nerve* (Fig. 7-66). Its internal opening lies between the jugular foramen and the occipital condyle. Part of the hypoglossal canal is frequently divided into two canals by a bony septum which separates the two roots of the hypoglossal nerve (Fig. 7-66). Pass a coarse thread through the hypoglossal canal and you will understand how CN XII reaches the tongue to supply its intrinsic muscles.

The condylar canal (Fig. 7-63). This canal, located at the inferior end of the groove for the sigmoid sinus, is inconstant. It transmits an emissary vein from the sigmoid sinus to the vertebral veins in the neck (Fig. 7-39). The condylar canal passes inferiorly

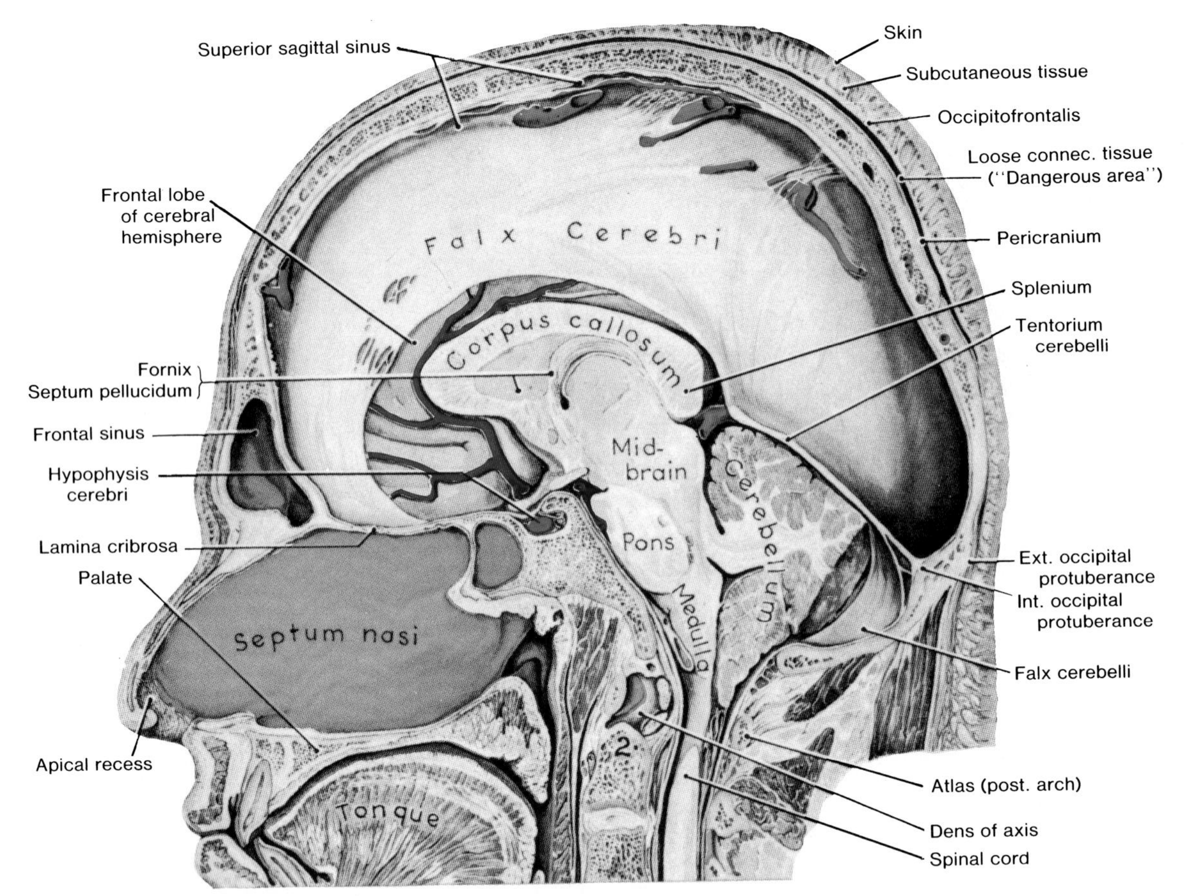

Figure 7-54. Drawing of a median section of the head. Note the attachments of the dural septa (reflections) and their relationship to the parts of the brain. There are two major septa, one vertical (falx cerebri) and one sloping like the roof of a tent (tentorium cerebelli). Note the superior sagittal sinus (a dural venous sinus) in the base of the falx cerebri. Observe that part of the cerebellum has herniated into the foramen magnum in this cadaveric specimen. Note that the sella turcica containing the hypophysis cerebri is roofed over by a fold of dura called the diaphragma sellae (see Fig. 7-53). Note the anterior cerebral artery, the smaller of the two terminal branches of the internal carotid artery (Fig. 7-77*A*). Observe that it curves around the genu of the corpus callosum and runs posteriorly along the superior surface of this large commissure.

and posteriorly to emerge just posterior to the occipital condyle.

The internal acoustic meatus (Figs. 7-60 and 7-63) *lies superior to the anterior part of the jugular foramen* in the petrous part of the temporal bone. Running to it, in a transverse direction from the brain stem, are the **facial (CN VII) and vestibulocochlear (CN VIII) nerves, the nervous intermedius**, and the labyrinthine vessels. The internal acoustic meatus is closed laterally by a perforated plate of bone which separates it from the internal ear.

Fractures of the anterior part of the skull (*e.g.*, "**a facial smash**" during a head-on collision) *may involve the cribriform plates of the ethmoid bone* and may tear the **meninges** (Fig. 7-50), allowing cerebrospinal fluid (CSF) to escape. The CSF drains from the nose or back into the pharynx, depending on the patient's posture. A **CSF leak** indicates a compound fracture and the possibility that **meningitis** (inflammation of the meninges) may develop.

Fractures of the middle part of the skull may extend inferiorly into the floor of the middle cranial fossa and involve important structures: *e.g.*, fracture of the petrous part of the temporal bone with resultant *otorrhagia* (bleeding from the ear); and/or *CSF otorrhea* (discharge of CSF from the ear).

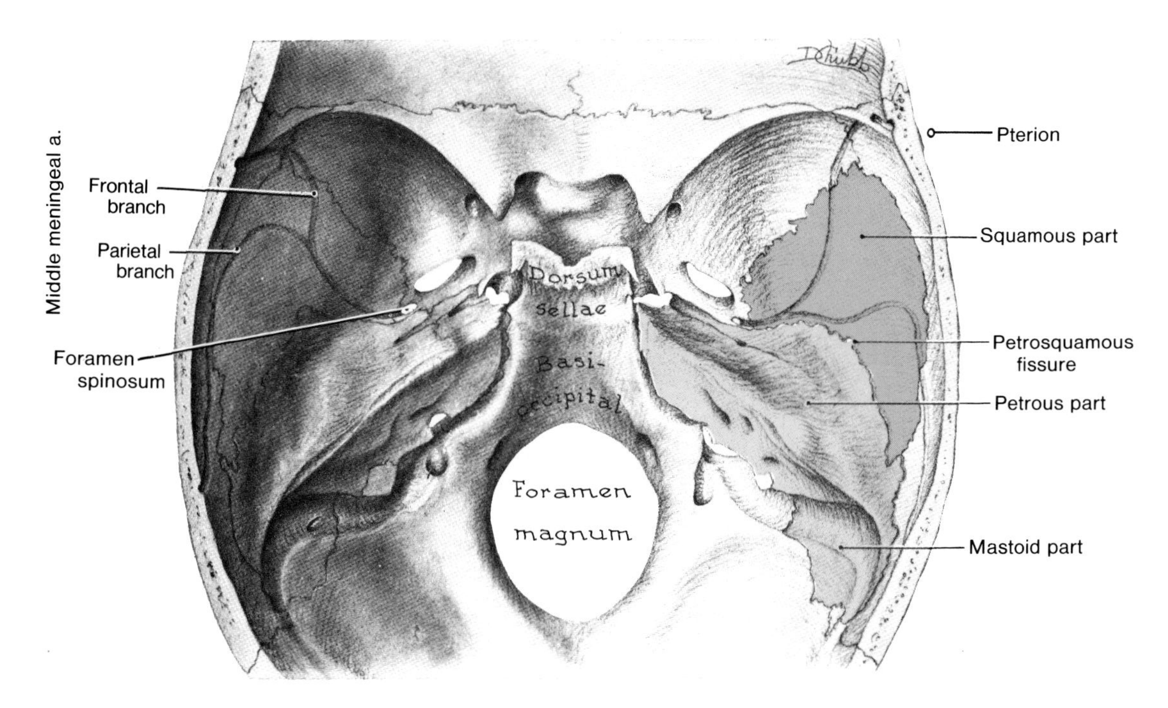

Figure 7-55. Drawing of the interior of the base of the skull showing the middle cranial fossa and most of the posterior cranial fossa. The parts of the temporal bone are colored. Observe the bone markings formed by the branches of the middle meningeal artery; usually a corresponding vein runs in the groove with the artery. These vessels run between the dura mater and the skull and may be torn when there is a blow to the temple in the region of the pterion, particularly if the bones are fractured. This bleeding is known as an extradural hemorrhage (Fig. 7-75).

The most important intracranial structure to consider in skull fractures is the brain. The various types of skull fracture have been discussed previously (Fig. 7-26), as have the nerves traversing the many foramina in the cranial fossae. During your clinical studies you will learn the symptoms and signs resulting from fractures in each of the three cranial fossae (Table 7-3). In view of the possible consequences, it is *no wonder safety rules for many workers require that hard hats be worn!*

CRANIAL MENINGES AND CSF

A clear understanding of the membranes (L. meninges) covering the brain and their relationship to the cerebrospinal fluid or CSF is an essential basis for the understanding of intracranial disease and head injuries.

The soft fresh brain tends to flatten when removed from the skull. In the living body, this vital organ is *enveloped by three membranes* (Fig. 7-50): (1) an external, thick, **tough dura mater** (L. *dura*, hard + *mater*, mother); (2) an intermediate, thin, **cobweb-like arachnoid mater** (G. *arachne*, spider + *eidos*, resemblance); and (3) an internal, delicate, **vascular pia mater** (L. *pius*, tender). These three membranes, known collectively as the **cranial meninges** (Fig. 7-50), are *continuous with the* ***spinal meninges*** covering the spinal cord (Fig. 5-53). In common usage the arachnoid mater is referred to as the arachnoid.

The main function of the cranial meninges and the CSF is to provide support and protection for the brain, in addition to that afforded by the calvaria.

The dura mater is occasionally called the *pachymenix* (G. *pachys*, thick + *menix*, membrane); hence you may hear the term **pachymeningitis**, meaning inflammation of the dura mater. Since the advent of antibiotics, this condition is uncommon.

The pia mater and the arachnoid mater develop from a single layer of mesoderm that surrounds the embryonic brain. During development, fluid-filled spaces develop within this layer which divide it into two layers. The spaces coalesce and give rise to the subarachnoid space containing CSF (Fig. 7-67). The embryonic ori-

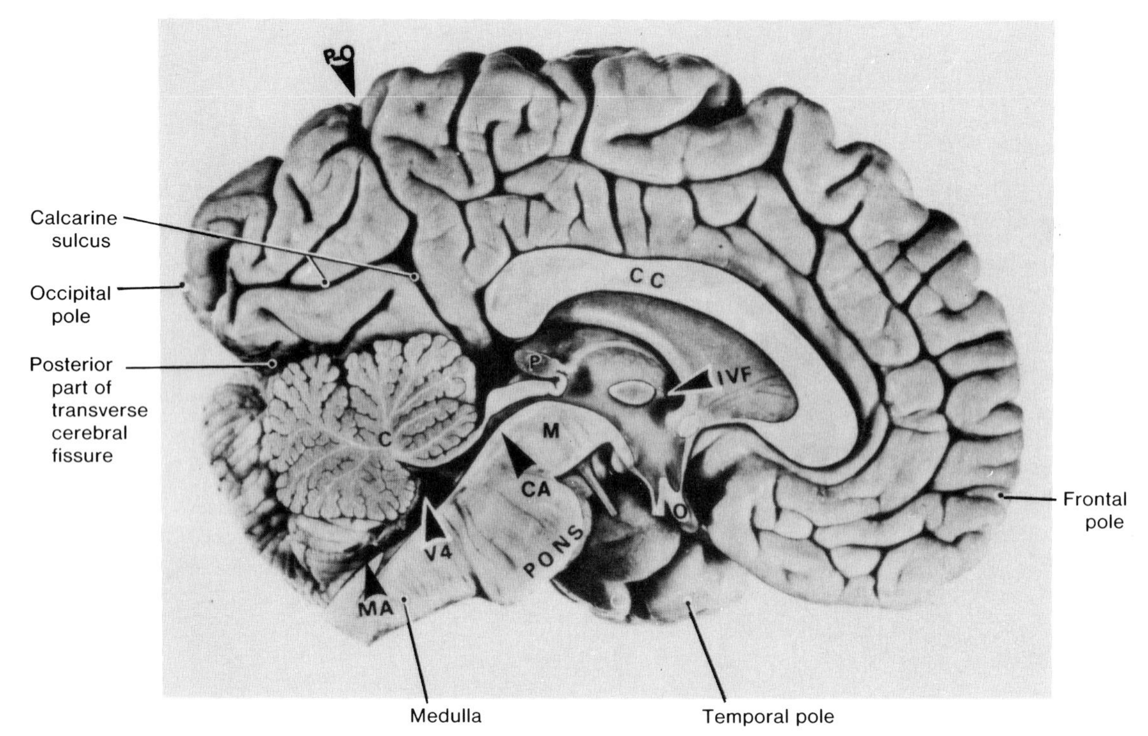

Figure 7-56. Photograph of a median section of a brain. Note (1) the parietooccipital sulcus (*PO*) separating the parietal and occipital lobes; (2) the cerebellum (*C*) inferior to the occipital lobe; (3) the corpus callosum (*CC*), a broad curved band of fibers, connecting the two cerebral hemispheres; (4) the optic chiasma (*O*); (5) the brain stem; (6) the interventricular foramen (*IVF*) connecting the lateral and third ventricles; (7) the median aperture (*MA*) of the fourth ventricle (*V4*), which connects the ventricular system with the subarachnoid space; (8) the cerebral aqueduct (*CA*) which runs through the midbrain (*M*) connecting the third and fourth ventricles; and (9) the pineal body (*P*) that is located inferior to the splenium of the corpus callosum (*CC*). When calcified the pineal body (pineal gland) is easily identified on skull radiographs. In these cases it may give the important radiographic sign of lateral displacement that indicates a space-occupying lesion of the cerebral hemisphere.

gin of the pia mater and arachnoid mater from one layer is reflected by the numerous connective tissue fibers called **trabeculae** (L. beams), passing between them.

The pia and arachnoid together are referred to as the **leptomeninges** (G. *leptos*, slender + *meninges*, membranes). Later, when you hear of **leptomeningeal arteries**, you will know that they supply the piarachnoid. Also, when your clinical instructors refer to piarachnitis, **leptomeningitis**, or simply meningitis, you should realize that all these terms mean the same thing (*i.e.*, inflammation of the piarachnoid or leptomeninges).

THE DURA MATER

This thick, tough membrane consists of collagenous connective tissue. You will find, when you remove the calvaria during dissection to expose the brain, that the skullcap can be pried loose from the membranes without difficulty. When you do this, you also strip the internal periosteum (endocranium) from the inner surface of the bones of the calvaria (Figs. 7-47 and 7-50).

Although the cranial dura consists of only one layer (like the spinal dura), it is sometimes described as consisting of two layers because the dura adheres so closely to the internal periosteum, except where there are dural venous sinuses and venous lacunae (Figs. 7-50 and 7-67). The dura also extends inwardly to form dural folds or septa, *e.g.*, the falx cerebri (Figs. 7-53 and 7-54).

It is important that you understand that the so-called "outer layer of dura" is really the internal periosteum of the calvaria. It is attached to the bones of the calvaria by **Sharpey's fibers** that penetrate the cranial bones. The "outer layer of the dura" is also attached to the outer pericranium, where the Sharpey's fibers traverse the cranial sutures (*e.g.*, the sagittal suture). This periosteal layer ("outer layer of dura") is also continuous with the outer periosteum of

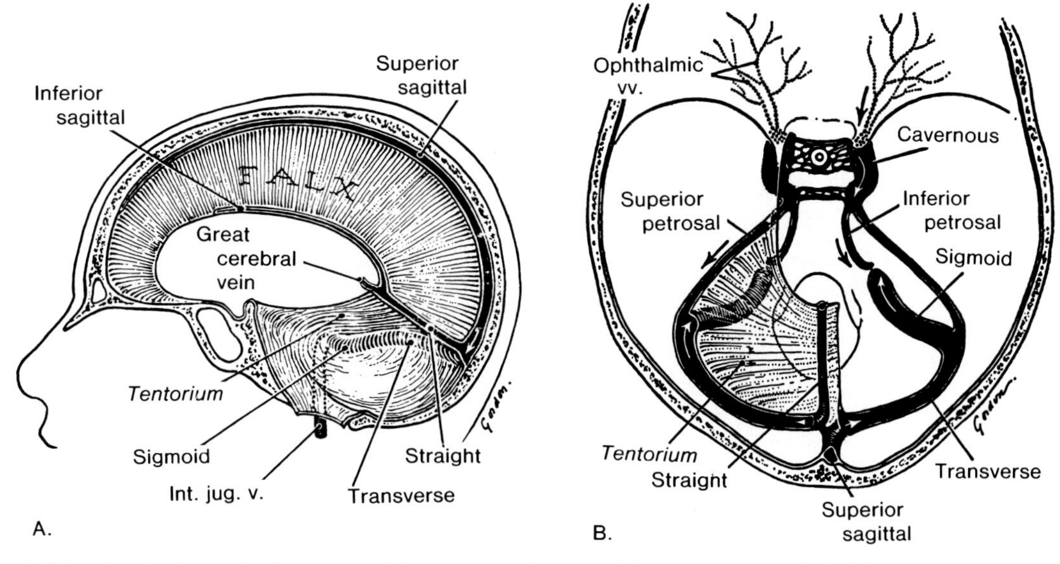

Figure 7-57. Semischematic diagrams of the venous sinuses of the dura mater. *A*, in falx cerebri and tentorium cerebelli. *B*, opened from above. *Arrows* show the direction of the blood flow. See Figures 7-39 and 7-78 for other views of these sinuses.

the cranial bones at the margins of the foramen magnum and at the smaller foramina for nerves and vessels.

The cranial dura, referred to as the meningeal layer of the dura by those who describe it as consisting of two layers, is continuous with the spinal dura at the foramen magnum. *The cranial dura also provides* ***tubular sheaths for the cranial nerves*** *as they pass through the foramina* in the floors of the **cranial fossae** (Fig. 7-68). Once outside the cranium, the dural sheaths fuse with the epineurium of the cranial nerves.

The dural sheaths of the cranial nerves extend approximately to the ganglia; *e.g.*, the trigeminal ganglion of the fifth cranial nerve (CN V) is surrounded by an extension of the meninges. The dura-enclosed space at the **trigeminal impression** in the petrous part of the temporal bone (Fig. 7-63) is occupied by the trigeminal ganglion. This is called the **trigeminal cave** (Fig. 7-69).

The dural sheath of the optic nerve is continuous with the cranial dura and the sclera of the eye (Fig. 7-93). *This relationship is clinically important.*

Part of the sensory root of the trigeminal nerve (CN V) is sometimes sectioned (**trigeminal rhizotomy**) to alleviate the pain of *trigeminal neuralgia* (tic douloureux). This syndrome is characterized by severe pains in the area of distribution of one or more branches of the trigeminal nerve (Figs. 7-32 and 7-33). During trigeminal rhizotomy involving the mandibular nerve, the electrode passes through the cheek and the foramen ovale into the trigeminal cave occupied by the trigeminal ganglion (Figs. 7-68 and 7-69).

The attachment of the dura mater to the bones in the floors of the cranial fossae (Fig. 7-68) is firmer than it is to the calvaria. Thus, a blow on the head can detach the dura from the calvaria without fracturing the bones, whereas a basal fracture usually tears the dura, resulting in leakage of CSF into the soft tissues of the neck, nose, ear, or nasopharynx.

Dural Septa or Dural Reflections (Figs. 7-51, 7-53, 7-54, and 7-67). During development of the brain, the dura is duplicated or reflected to form four inwardly projecting folds or septa.

The dural septa divide the cranial cavity into three intercommunicating compartments, one subtentorial and two supratentorial. These septa provide support for parts of the brain, particularly the cerebral hemispheres.

The falx cerebri (L. *falx*, a sickle) is a large sickle-shaped, vertical partition in the longitudinal fissure between the two cerebral hemispheres (Figs. 7-53, 7-54, 7-67, and 7-68). This large dural fold is attached in the median plane to the inner surface of the calvaria from the frontal crest of the frontal bone and the **crista galli** of the ethmoid bone anteriorly to the **internal occipital protuberance** posteriorly (Fig. 7-

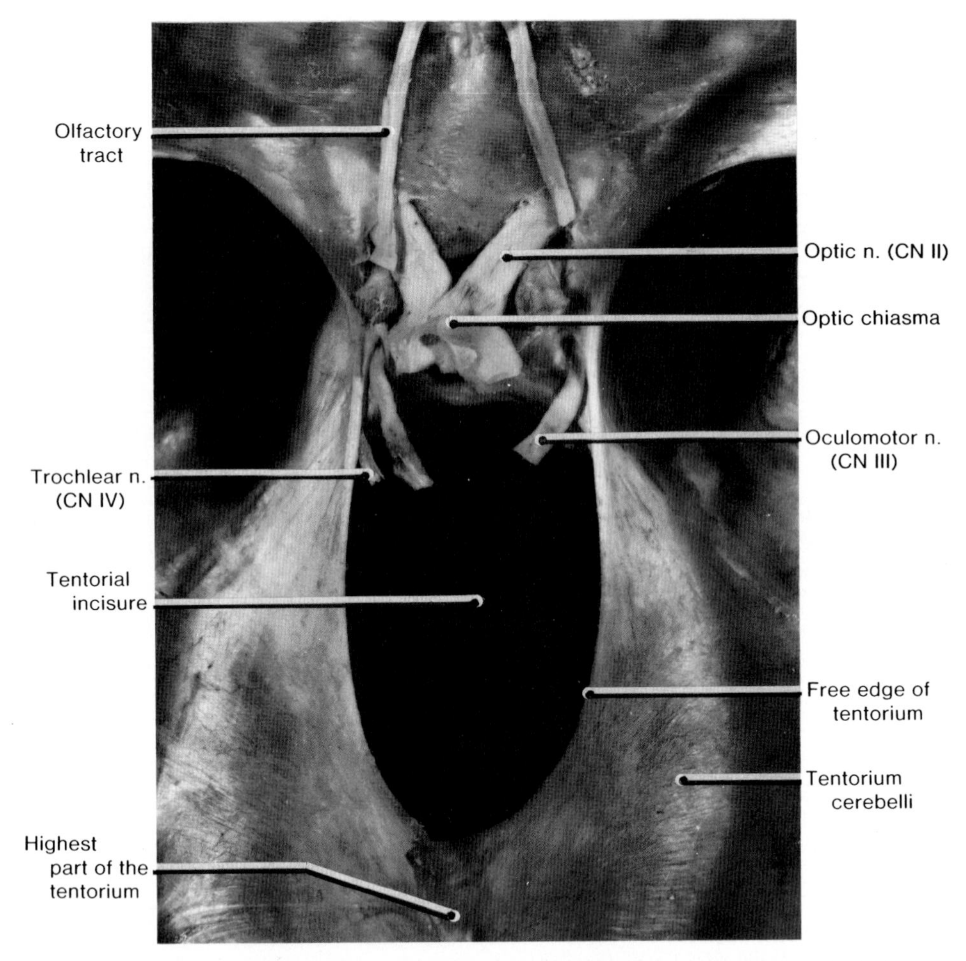

Figure 7-58. Photograph of a dissection of the tentorium cerebelli showing the typical shape of the tentorial incisure (*notch*), the opening in the tentorium cerebelli through which the brain stem extends from the middle to the posterior cranial fossa. (For orientation, see Figs. 7-68 and 7-69).

14). The falx cerebri is also attached to the midline of the **tentorium cerebelli**, another dural fold that lies between the occipital lobes of the cerebral hemispheres and the cerebellum (Figs. 7-51, 7-53, 7-54, and 7-68). At the superior convex border of the falx cerebri, its two layers separate to enclose the **superior sagittal sinus** (Figs. 7-46*A*, 7-50, and 7-67).

The falx cerebri increases in depth anteroposteriorly so that its inferior edge comes close to the **splenium**, the thickened posterior end of the *corpus callosum* (Fig. 7-54).

The **inferior sagittal sinus** (Figs. 7-39 and 7-57*A*) lies enclosed within the free inferior edge of the falx cerebri.

Throughout life the falx cerebri forms a rigid partition between the cerebral hemispheres which reduces side-to-side movement of the cerebral hemispheres.

The tentorium cerebelli (L. *tentorium*, tent) is a wide crescentic, arched fold of dura mater that separates the occipital lobes of the cerebral hemispheres from the cerebellum (Figs. 7-51, 7-53, 7-54, and 7-68). The attachment of the falx cerebri to the midline portion of the tentorium cerebelli holds the latter fold up like the ridgepole of a tent.

The tentorium cerebelli is attached anterolaterally to the superior edges of the petromastoid parts of the temporal bones, and to the **anterior and posterior clinoid processes** (Fig. 7-15*A*). Posteriorly, the tentorium is attached to the occipital bone along the grooves for the transverse sinuses, which it encloses (Figs. 7-14 and 7-68). Its concave anteromedial border is free, and between it and the dorsum sellae of the sphenoid bone there is an opening, called the **tentorial incisure** or notch (Figs. 7-58 and 7-68). This oval opening surrounds the midbrain as it passes from the middle to the posterior cranial fossa. The tentorial incisure is also closely related to the anterior part of the superior surface of the **vermis** of the cerebellum

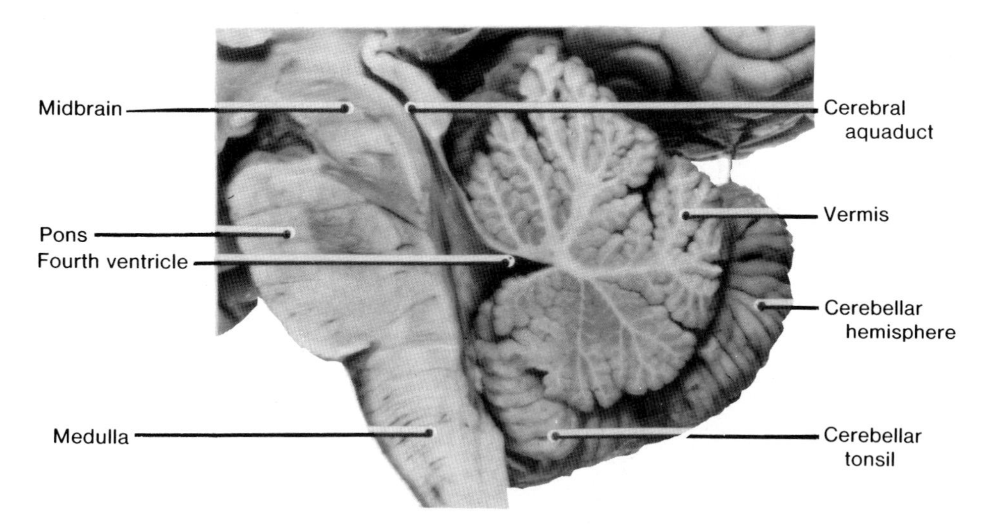

Figure 7-59. Photograph of a median section through the brain stem, fourth ventricle, and cerebellum. The cerebellar hemispheres, being lateral to the vermis that lies in the midline, are not cut in this median section.

(Fig. 7-59), and to the *uncus* of each temporal lobe (Fig. 7-52).

CSF from the lateral ventricles of the brain passes *inferiorly through the tentorial incisure via the* ***cerebral aqueduct*** (Fig. 7-70*A*) into the fourth ventricle. After leaving the ventricular system, CSF passes *superiorly through the tentorial incisure* in the subarachnoid space surrounding the midbrain. Hence, there are two chances of obstructing the flow of ventricular fluid and/or CSF at the tentorial incisure, and each produces a different type of **hydrocephalus** (excessive accumulation of ventricular fluid and/or CSF resulting in enlargement of the head).

The tentorium cerebelli prevents inferior displacement of the occipital lobes of the brain (Fig. 7-51) when a person is erect.

The tentorial incisure is slightly larger than is necessary to accommodate the midbrain; hence, when intracranial pressure superior to the tentorium is considerably higher than that inferior to it (*e.g.*, when a **supratentorial tumor or a hematoma** is present, part of the adjacent ipsilateral temporal lobe may herniate through the tentorial incisure. Often it is the **uncus** (L. a hook), the hooked extremity of the rostral end of the parahippocampal gyrus (Fig. 7-52), that herniates through the tentorial incisure. During **tentorial herniation** the temporal lobe may be lacerated by the tough taut tentorium, and CN III, **the oculomotor nerve** (Figs. 7-68 and 7-72), may be stretched and/or compressed.

Compression of CN III leads to third nerve palsy, causing absence of pupillary constriction and a fixed, dilated pupil (Fig. 7-87 and Table 7-3).

Increased intracranial pressure may also force the cerebellar tonsils, the rounded lobules on the inferior surface of the cerebellar hemispheres (Fig. 7-59), inferomedially through the foramen magnum (Fig. 7-54). **Herniation of the cerebellar tonsils** compresses the medulla containing the vital respiratory and cardiovascular centers, and produces a life-threatening situation.

The falx cerebelli (Figs. 7-53 and 7-54) is a small, sickle-shaped midline dural fold in the posterior part of the posterior cranial fossa, extending almost vertically, inferior to the inferior surface of the tentorium cerebelli. Its free edge projects slightly between the cerebellar hemispheres. The ***occipital venous sinus*** is located in the base of the falx cerebelli (Fig. 7-39).

The diaphragma sellae is a small circular, horizontal dural fold that roofs over the **hypophysial fossa** in the sella turcica (Figs. 7-15 and 7-53). This fold is formed by the dura surrounding the hypophysis cerebri and encircling the stalk of this gland. The diaphragma sellae covers the **hypophysis cerebri** or pituitary gland, which lies in the hypophysial fossa (Fig. 7-54). The diaphragma sellae has a central ap-

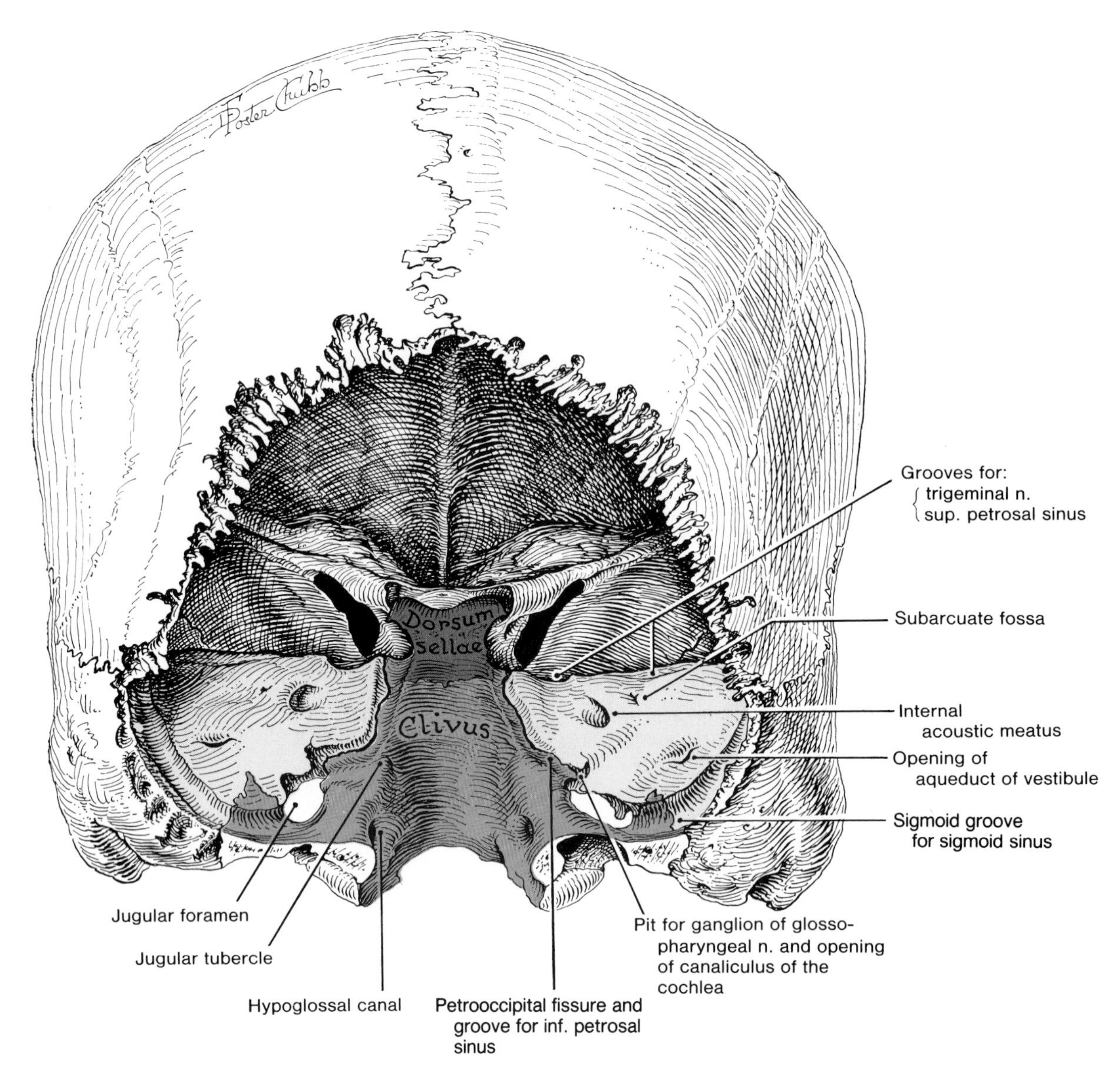

Figure 7-60. Drawing of the anterior wall of the posterior cranial fossa (*colored*) after removal of the squamous part of the occipital bone. Observe the flat plate of bone known as the dorsum sellae on the body of the sphenoid bone. It forms the posterior wall of the sella turcica or hypophysial fossa. The clivus is a sloping piece of bone that is formed by the basilar part of the occipital bone (basiocciput) and the body of the sphenoid bone (basisphenoid). Note the large superior oribital fissures (*black*) in the middle cranial fossa (also see Figs. 7-2 and 7-88).

erture for passage of the hypophysial veins and the hypophysial stalk or **infundibulum** (Fig. 7-72), connecting the hypothalamus and the hypophysis.

Tumors of the hypophysis cerebri (**pituitary tumors**) may extend superiorly through the aperture in the **diaphragma sellae** and/or cause bulging of it. Pituitary tumors often expand the diaphragma sellae and may produce endocrine symptoms early or late (*i.e.*, before or after enlargement of the diaphragma sellae). Superior extension of a tumor causes visual symptoms owing to pressure on the **optic chiasma** (Figs. 7-72, 7-68, and 7-73), the point of crossing of the optic nerve fibers.

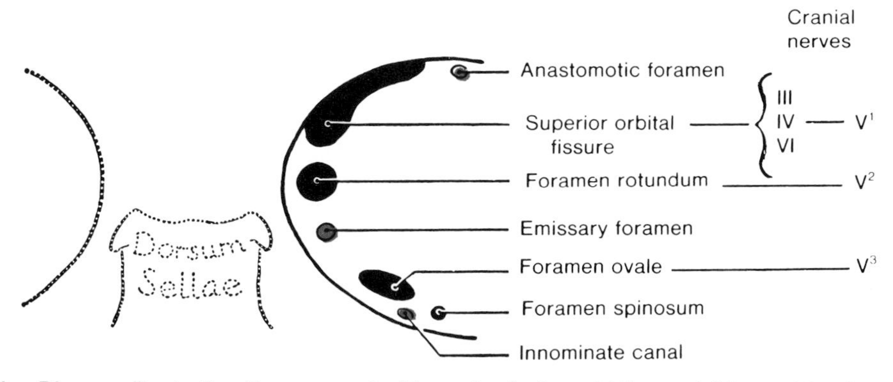

Figure 7-61. Diagram illustrating the crescent of foramina in the middle cranial fossa. The four foramina which are constant are shown in *black*, whereas those that are inconstant are *red*. These foramina all open from the sphenoid bone into the middle cranial fossa and are also shown in the drawing of this fossa (Fig. 7-63).

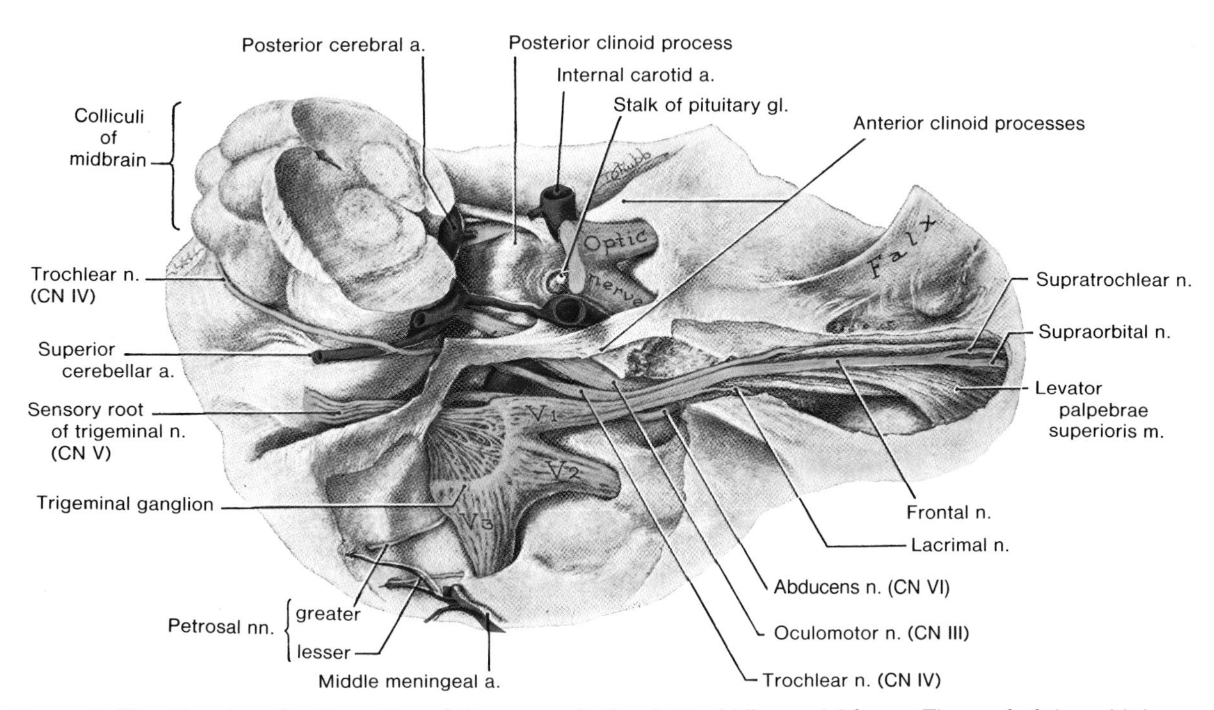

Figure 7-62. Drawing of a dissection of the nerves in the right middle cranial fossa. The roof of the orbit has been partly removed and the tentorium has been removed on the right side to show the trochlear nerve (CN IV) and the sensory root of the trigeminal (CN V). It may help you to orientate this drawing if you place the left hand side of the illustration toward you. The large trigeminal ganglion forms an impression on the anterior surface of the petrous part of the temporal bone (Fig. 7-63). This ganglion occupies a recess called the trigeminal cave (Fig. 7-69) in the dura mater covering the trigeminal impression. Observe the mandibular nerve (CN V^3) passing through the foramen ovale to enter the infratemporal fossa (Fig. 7-115). The cerebral arterial circle is illustrated (also see Fig. 7-77).

THE ARACHNOID MATER (Figs. 7-50 and 7-67)

This is a delicate, transparent membrane composed of *cobweb-like tissue*. The arachnoid forms the intermediate covering of the brain and is separated from the dura mater by a film of fluid in the potential **subdural space**. The arachnoid does not form a close investment of the brain, but passes over the sulci (L. furrows or ditches) and fissures without dipping into them. The term arachnoid is derived from Greek words meaning "resembling a spider's web."

The arachnoid is partly separated from the pia mater by the ***subarachnoid space*** *containing CSF.* Numerous trabeculae, delicate strands of connective tissue, pass from the arachnoid to the pia (Fig. 7-67), giving it a cobweb-like structure.

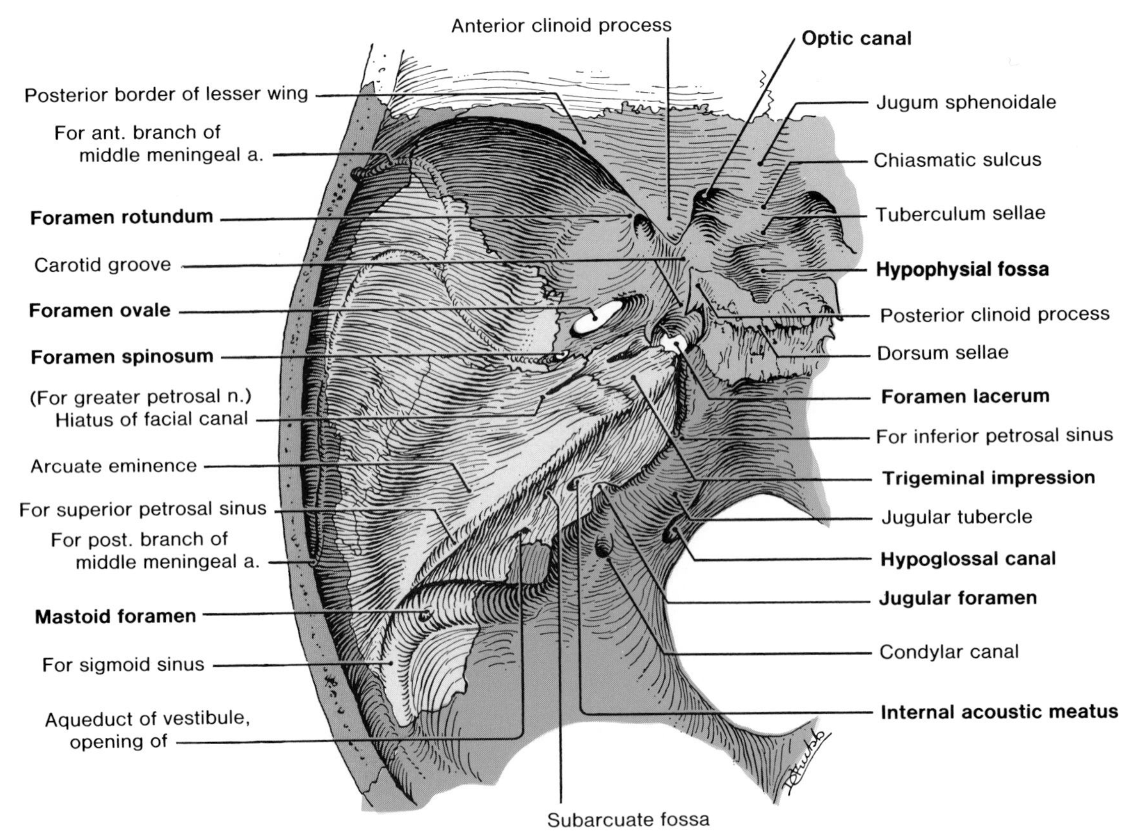

Figure 7-63. Drawing of a superior view of the left middle and posterior cranial fossae, showing the location of several foramina. Note the distinct impression posterolateral to the foramen lacerum formed by the trigeminal ganglion (the ganglion of CN V). This trigeminal impression is on the anterior surface of the petrous part of the temporal bone.

THE PIA MATER (Fig. 7-67)

This membrane is very thin (L. *pius*, tender) but it is thicker than the arachnoid. The pia, the innermost of the three layers of meninges, is a **highly vascularized, loose connective tissue membrane** that adheres closely to the surface of the brain. It dips into all sulci and fissures, carrying small blood vessels with it.

The cerebral veins run on the pia mater within the subarachnoid space (Figs. 7-47, 7-53, and 7-67). When branches of cerebral vessels penetrate the brain, the pia follows them for a short distance, forming a sleeve of pia mater. Hence the **perivascular spaces** (Virchow-Robin spaces) are continuous with the subarachnoid space. They extend in an increasingly attenuated form as far as the arterioles and venules in the brain.

Meningiomas (neoplastic growths of meningeal tissues) make up about 25% of primary intracranial tumors. As they enlarge, they compress adjacent neural tissues. One common site for meningiomas is along the **sphenoidal ridge**, formed by the sharp posterior margin of the lesser wing of the sphenoid bone (Fig. 7-13).

The central artery and vein of the retina cross the subarachnoid space to become enclosed in the space around the posterior part of the optic nerve (Figs. 7-73 and 7-93). Consequently an increase in CSF pressure around the optic nerve slows the return of venous blood, resulting in **edema of the retina**. This is most apparent as a swelling of the optic papilla (**papilledema**). Thus, inspection of the **ocular fundus** is an important part of neurological examinations (Fig. 7-98).

In cases of ruptured aneurysm of the circulus arteriosus cerebri (Figs. 7-76 and 7-77), there is usually hemorrhage into the subarachnoid space. Blood in the CSF results in an increase in intracranial pressure.

An increase in intracranial pressure is transmitted to the subarachnoid spaces around the optic nerves, compressing the retinal vein where

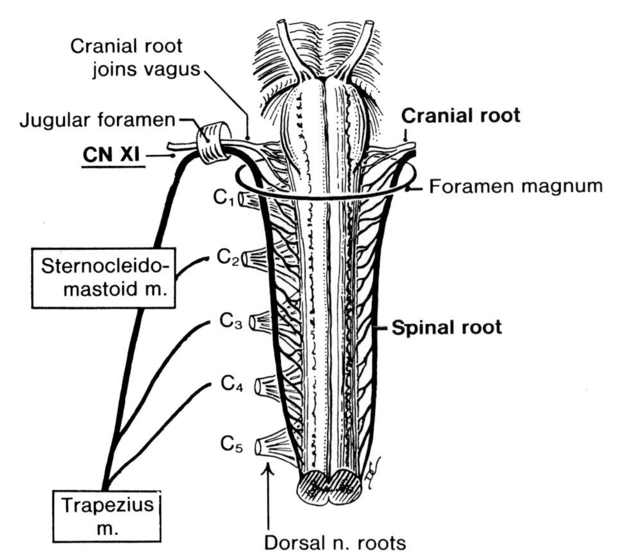

Figure 7-64. Scheme of the distribution of the spinal root of the *accessory nerve* (CN XI). Note that this root is joined by fibers from the ventral ramus of C2 and supplies the sternocleidomastoid muscle and is joined by fibers from the ventral rami of C3 and C4 and supplies the trapezius muscle. There is clinical evidence (both surgical and medical) that these contributions from C2 to C4 convey motor as well as sensory fibers.

it runs in the subarachnoid space around the optic nerve (Figs. 7-73 and 7-93). This results in increased pressure in the retinal capillaries and **subhyaloid hemorrhages between the retina and vitreous body.**

ARTERIES OF THE DURA MATER (Figs. 7-16, 7-50, 7-55, and 7-67)

There are many ***meningeal arteries*** in the periosteum which are inappropriately named because they supply more blood to the bones of the calvaria than to the dura of the meninges. Only very fine branches of these arteries are distributed to the dura. The arteries to the brain are branches of the paired internal carotid and vertebral arteries; they lie in the pia mater within the subarachnoid space (Fig. 7-67).

The middle meningeal artery (Figs. 7-16, 7-50, and 7-62) is the largest of the meningeal arteries. It is

---

The spinal root of the accessory nerve usually passes through the spinal ganglion of C1 and may receive sensory fibers from it.

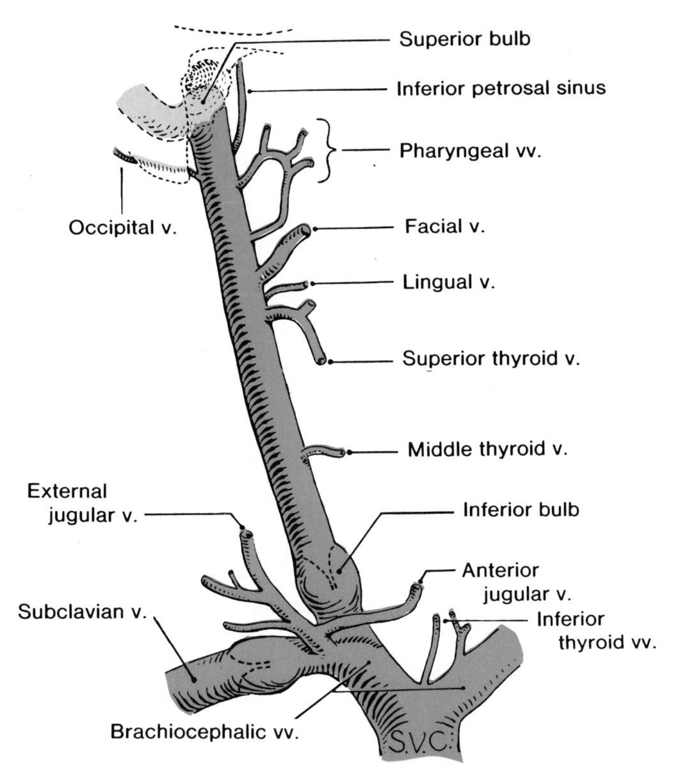

Figure 7-65. Drawing of an anterior view of the right internal jugular vein and its tributaries. It begins at the superior margin of the jugular foramen and dilates within it to form the superior bulb (jugular bulb). Note that the inferior petrosal sinus joins the internal jugular vein just inferior to the skull. Recall that it drains the posterior end of the cavernous sinus and passes through the jugular foramen to reach the internal jugular vein (Figs. 7-57 and 7-65).

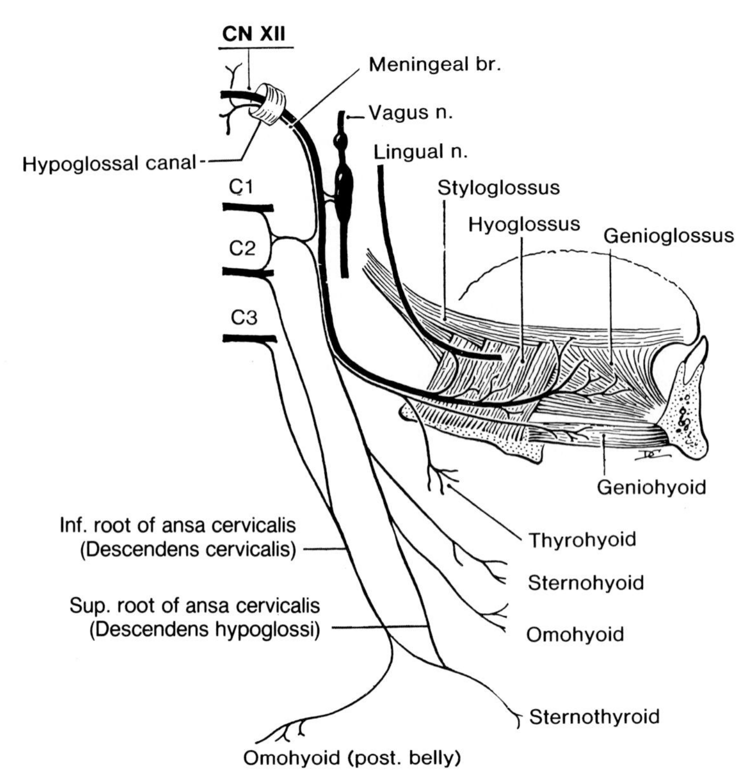

Figure 7-66. Scheme of the distribution of the *hypoglossal nerve* (CN XII). Note that this efferent nerve supplies all the intrinsic (longitudinal, transverse, and vertical) and extrinsic (styloglossus, hyoglossus, and genioglossus) muscles of the tongue, except the palatoglossus. Observe that it receives a mixed (motor and sensory) branch from the loop between the ventral rami of C1 and C2. The sensory or afferent fibers in part take a recurrent course and end in the dura mater of the posterior cranial fossa. Note that the motor or efferent branch supplies the geniohyoid and thyrohyoid muscles and, via the superior root of ansa cervicalis which unites with the inferior root of ansa cervicalis to form the ansa cervicalis, the remaining depressor muscles of the larynx. Observe that most of the muscular branches are true hypoglossal fibers, but those to the geniohyoid muscle arise from the first cervical nerve (C1). The *ansa cervicalis* is a loop (L. *ansa*) in the cervical plexus consisting of fibers from the first three cervical nerves. Note that some of them accompany the hypoglossal nerve for a short distance.

a branch of the maxillary artery and is clinically important largely because it is often torn when the overlying skull is fractured. These vessels, **located within the periosteum**, often *form distinct grooves on the internal surface of the calvaria* (Fig. 7-7). Note the middle meningeal artery enters the cranial cavity through the **foramen spinosum** in the floor of the middle cranial fossa (Figs. 7-61 and 7-62), and soon divides into frontal and parietal branches. These branches ramify on the internal surface of the calvaria.

A fracture in the temporal region of the skull (*e.g.*, at the pterion) may tear one or more branches of this artery (Fig. 7-16), producing an **extradural hematoma** (epidural hematoma).

These fractures may be followed by profuse hemorrhage between the calvaria and the dura mater (Fig. 7-175). This produces symptoms of compression of the brain which may require trephination (removal of a circular piece of the calvaria) for removal of the hematoma (mass of clotted or partly clotted blood).

VEINS OF THE DURA MATER (Figs. 7-47, 7-50, 7-52, and 7-67)

The meningeal veins accompany the meningeal arteries; hence they may also be torn in fractures of the calvaria and add to the extradural bleeding.

NERVE SUPPLY TO THE DURA MATER (Figs. 7-32, 7-66, and 7-74)

The rich sensory supply to the dura is largely through the three divisions of the trigeminal nerve (**CN V**), but

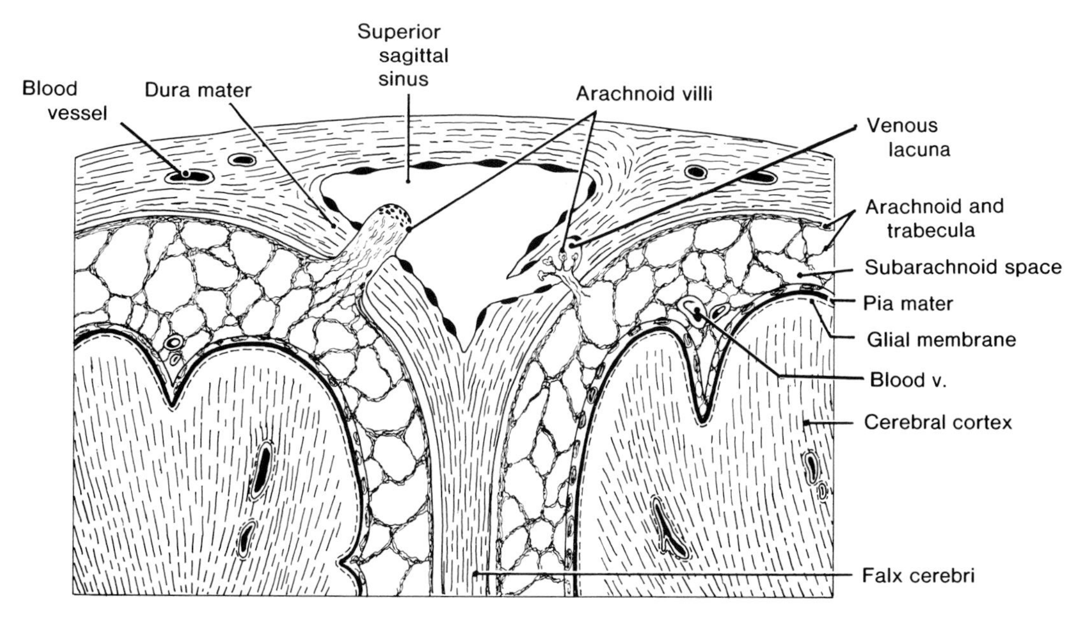

Figure 7-67. Diagram of the meningeal-cortical relationships. Note that the arachnoid does not form a close investment of the cerebral cortex and that it is attached to the pia by delicate strands (trabeculae). The subarachnoid space (containing CSF) is shown of greater width than is normal to illustrate the trabeculae crossing it and connecting the pia mater and arachnoid. Note that the pia is firmly anchored to the cerebral cortex. The main site of passage of CSF into venous blood is through the arachnoid villi projecting into the dural venous sinuses, especially the superior sagittal sinus and adjacent venous lacunae (also see Fig. 7-50). Arachnoid villi that have become hypertrophied with age (*e.g.*, the one on the left) are called arachnoid granulations. These granulations produce pits in the bones of the calvaria (Fig. 7-7).

sensory branches are also received from the vagus nerve (**CN X**), and the superior three **cervical nerves** via the hypoglossal nerve (CN XII). The sensory endings are more numerous in the dura mater along each side of the superior sagittal sinus and in the tentorium cerebelli (Fig. 7-53) than they are in the floor of the cranium. *Pain fibers are also numerous where arteries and veins pierce the dura mater.*

The brain is insensitive to pain, but the cerebral dura mater is sensitive to pain, especially around the venous sinuses of the dura (Fig. 7-39). Consequently, pulling on arteries at the base of the skull, or on veins in the vertex or base where they pierce the dura mater, causes pain. Although there are many causes of headache, distention of the scalp and/or meningeal vessels is believed to be one cause.

Many headaches appear to be dural in origin; for example the headache that occurs during and after a **lumbar puncture** (Fig. 5-60) for the removal of CSF is thought to result from stimulation of sensory nerve endings in the dura. When CSF is removed the brain sags slightly, pulling on the dura mater. This is thought to cause pain and headache. It is for this reason that, following lumbar punctures, patients are asked to keep their heads down to minimize or prevent headaches, *i.e.*, use of a pillow is not advisable.

THE MENINGEAL SPACES

The extradural space is superficial to the dura mater, *i.e.*, between the bone and the internal periosteum (Fig. 7-50). Because the dura is attached intimately to the periosteum of the calvaria, *the extradural space is only a potential space* which becomes real when blood accumulates in it from torn meningeal vessels (**extradural hemorrhage**), *i.e.*, resulting from a fractured calvaria (Fig. 7-175).

The subdural space is deep to the dura mater, *i.e.*, between the dura and arachnoid (Figs. 7-50 and 7-67). It is also a potential space with only a thin film of subdural fluid in it. It is not obliterated except at places where it is pierced by arteries, veins, and nerves.

Air or oil injected into the **spinal subdural space** in the lumbar region (Fig. 5-60) for radi-

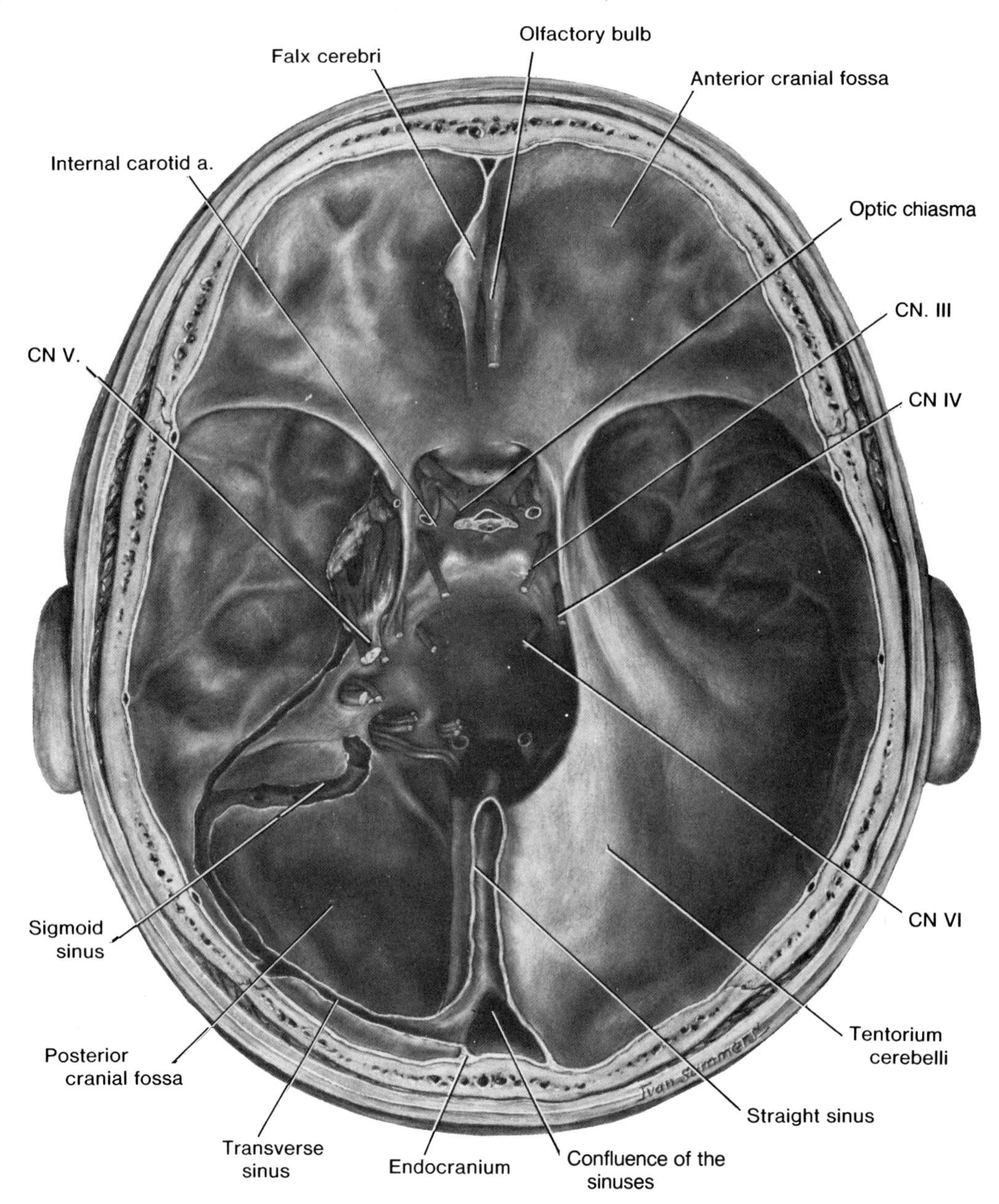

Figure 7-68. Drawing of the interior of the base of the skull showing the dura mater. Observe that the dura and the endocranium are separated by the venous sinuses (*e.g.*, at the confluence of the sinuses). The falx cerebri has been cut close to its anterior attachment and most of it has been removed. The tentorium cerebelli has been cut away on the left to expose the posterior cranial fossa. Also see Figure 7-53.

ological studies may enter the **cranial subdural space**, but this is inadvertent.

Head injuries may be associated with various types of ***intracranial bleeding*** (hemorrhage).

EXTRADURAL HEMORRHAGE

Bleeding between the dura and the internal periosteum of the calvaria (*i.e.*, superficial to the dura) may follow a blow to the head. Typically there is a ***brief concussion*** (loss of consciousness), followed by a ***lucid interval*** of some hours. This is succeeded by *drowsiness and* ***coma*** (profound unconsciousness).

When there is extradural hemorrhage, there is tearing of meningeal vessels.

The artery most commonly injured is the mid-

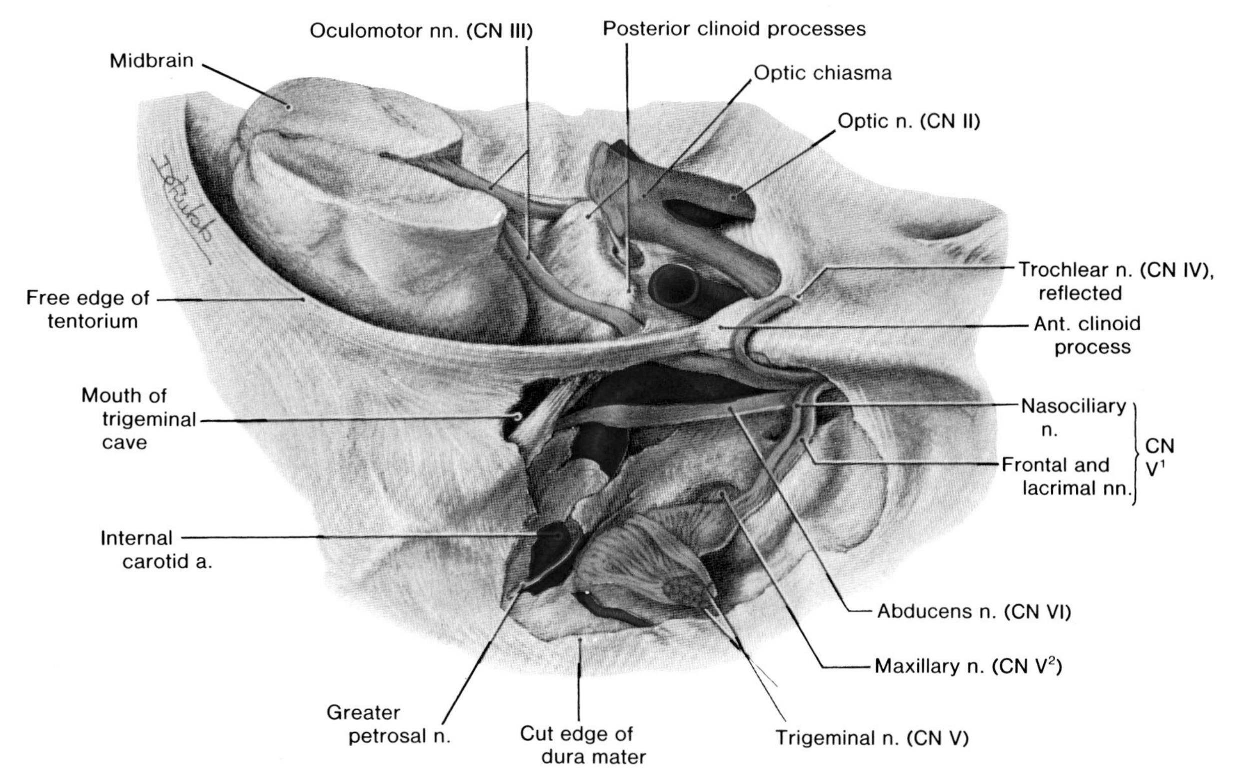

Figure 7-69. Drawing showing the nerves in the middle cranial fossa and the sinuous course of the internal carotid artery. The trigeminal nerve is divided and turned anterolaterally from the mouth of the trigeminal cave. Observe the midbrain and the oculomotor nerves in the tentorial notch. Also see Figures 7-58 and 7-68.

dle meningeal artery or one of its branches, often the frontal one (Fig. 7-50). Similarly, the superior sagittal sinus (Figs. 7-50 and 7-54) may be torn in fractures of the superior part of the calvaria.

Although veins and venous sinuses may be torn, most bleeding is usually from torn arteries. Because the internal periosteum is firmly attached to calvaria, there is a slow, localized accumulation of blood; this forms an **extradural hematoma** (Fig. 7-175). As this blood mass increases in size, compression of the brain occurs. This necessitates evacuation of the fluid and occlusion of the bleeding vessels (*e.g.*, trephination).

The formation of an extradural hematoma superior to the **tentorium cerebelli** (Fig. 7-51), causes a rise in supratentorial pressure which, if high enough, produces herniation of the uncus (Fig. 7-52) of the temporal lobe through the tentorial incisure adjacent to CN III (Figs. 7-58 and 7-68). This causes **third nerve palsy** and dilation of the pupil on the affected side (Fig. 7-87). If the pressure rises very high, the cerebellar tonsils may be forced through the foramen magnum (Fig. 7-54), compressing the medulla. This may cause fatal effects owing to interference with the respiratory and cardiovascular centers.

SUBDURAL HEMORRHAGE

Bleeding between the dura and arachnoid may follow a **blow to the head** which jerks the brain inside the skull. Such a displacement of the brain is greatest in elderly people in whom some shrinkage of the brain has occurred. Subdural hemorrhage commonly results from tearing of **cerebral veins** where they enter the superior sagittal sinus (Fig. 7-52).

Subdural hemorrhage is usually classified as acute or chronic. At first the subdural blood is spread over the cerebral hemispheres, but after a few days or weeks it localizes, forming a subdural hematoma (Fig. 7-176), most often deep to the parietal eminence (Fig. 7-11). Being hypertonic, the collection of blood slowly expands; this explains why symptoms may not occur for weeks after a trivial injury.

SUBARACHNOID HEMORRHAGE

Bleeding into the subarachnoid space usually follows the **rupture of a berry aneurysm** of an intracranial artery (Figs. 7-75 and 7-76). These thin-walled outpouchings or evaginations occur chiefly at bifurcations of the arteries at the base of the brain (Fig. 7-77), where two pulse waves meet (*e.g.*, where the wave from the internal carotid artery meets the wave from the posterior cerebral artery). Subarachnoid hemorrhages are also associated with skull fractures and cerebral lacerations.

Subarachnoid hemorrhage results in meningeal irritation, which produces a severe headache, stiff neck, and often loss of consciousness.

INTRACEREBRAL HEMORRHAGE

Bleeding into the brain, often from one of the **branches of the middle cerebral artery** (Figs. 7-77 and 7-85) going to the corpus striatum (Fig. 7-70*B*), is frequent in persons with hypertension. One of these arteries was named the **artery of cerebral hemorrhage** by Charcot because he observed that it hemorrhages into the internal capsule, producing a **paralytic stroke**. Paralysis occurs because of the interruption of motor pathways from the motor cortex to the brain stem and spinal cord.

THE VENOUS SINUSES OF THE DURA MATER (Figs. 7-39, 7-53, 7-54, 7-57, and 7-78 to 7-80)

These sinuses and lacunae are ***venous channels*** *located between the dura mater and the internal periosteum lining the cranium*, usually along the lines of attachment of the dural septa. The venous sinuses are lined with endothelium that is continuous with that of the cerebral veins entering them (Fig. 7-53). These sinuses have no valves and no muscle in their walls.

The venous sinuses drain all the blood from the brain. Several of the sinuses are triangular in cross-section (Fig. 7-67) because their bases are on bone and their side walls are formed by the origins of the dural folds.

The Superior Sagittal Sinus (Figs. 7-39, 7-46*A*, 7-50, 7-53, 7-54, 7-57*A*, and 7-78). This *venous sinus of the dura mater* lies in the midline, along the attached border of the falx cerebri (Fig. 7-54). It *begins at the* ***crista galli*** (Fig. 7-14) and runs the entire length of the superior attached portion of the falx cerebri (Fig. 7-52), and **ends at the internal occipital protuberance** (Fig. 7-54).

In about 60% of cases the superior sagittal sinus ends by becoming the right transverse sinus (Figs. 7-57 and 7-78). In other cases, it turns to the left and ends in the left transverse sinus or ends in both.

At the termination of the superior sagittal sinus there is a dilation, known as the **confluence of the sinuses** (Figs. 7-68 and 7-78). Occasionally you will hear this dilation referred to as the sinus confluens, or the *torcular Herophili* (L. winepress of Herophilus, an early anatomist and surgeon). Five sinuses often communicate at the confluence; however considerable variation in venous pathways occurs.

The superior sagittal sinus is triangular in cross-section (Figs. 7-46*A* and 7-67), the superior wall being formed by the internal periosteum or endocranium lining the calvaria and the lateral walls by the dura mater (Figs. 7-46*A*, 7-50 and 7-68). The superior sagittal sinus becomes larger as it passes posteriorly and receives more and more **superior cerebral veins** (Fig. 7-53). These veins enter the sinus in an anterior direction, resulting from their fixation to the superior sagittal sinus in early fetal life. During later development their more distal parts are carried posteriorly by the growth of the brain.

The superior sagittal sinus communicates on each side via slit-like openings with venous spaces in the dura mater, called **lateral lacunae**, into which some of the arachnoid villi project (Figs. 7-50 and 7-67). Usually there are three venous lacunae on each side of the superior sagittal sinus. In very old people, these lacunae often coalesce to form one long sinus on each side.

The main site of passage of CSF into venous blood is through the arachnoid villi, especially those projecting into the superior sagittal sinus and the adjacent venous lacunae. *Hypertrophied aggregations of arachnoid villi*, associated with advancing age, are called **arachnoid granulations** (Pacchionian bodies). These granulations often produce erosion or pitting of the internal surface of the superior portions of the frontal and parietal bones (Fig. 7-7).

The superior sagittal sinus also communicates with veins in the scalp via ***emissary veins*** *which pass through the parietal foramina* (Fig. 7-7).

The Inferior Sagittal Sinus (Figs. 7-39 and 7-57*A*). This *venous sinus of the dura mater* is much smaller than the superior sagittal sinus. It occupies the posterior two-thirds of the free inferior edge of the **falx cerebri** (Fig. 7-53). The inferior sagittal sinus ends by joining the **great cerebral vein** (of Galen) to form the straight sinus (Figs. 7-39, 7-53, and 7-57*A*). The inferior sagittal receives cortical veins from the medial aspects of the cerebral hemispheres.

The Straight Sinus (Figs. 7-39, 7-53, 7-57, and 7-68). This *venous sinus of the dura mater* is formed by the union of the inferior sagittal sinus with the great cerebral vein (Fig. 7-39). It runs inferoposteriorly along the line of attachment of the **falx cerebri** to the tentorium cerebelli, where it becomes continuous with one of the transverse sinuses, usually the left.

The Transverse Sinuses (Figs. 7-39, 7-57*B*, and 7-78). These *venous sinuses of the dura mater* pass laterally from the confluence of the sinuses in the attached border of the tentorium cerebelli, grooving the occipital bones and the posteroinferior angles of the parietal bones (Fig. 7-14). They then leave the tentorium and change their names to the sigmoid sinuses, because they curve inferiorly and then medially in S-shaped curves (Fig. 7-63).

The Sigmoid Sinuses (G. *sigma*, the letter S). These *venous sinuses of the dura mater* follow S-shaped courses in the posterior cranial fossa (Figs. 7-39, 7-57, 7-68, and 7-79), forming deep grooves in the inner surface of the posterior part of the mastoid parts of the temporal bones, and in the lateral surfaces of the jugular tubercles of the occipital bone (Fig. 7-60). The sigmoid sinuses then turn anteriorly and enter venous enlargements called the **superior bulbs of the internal jugular veins** (Fig. 7-65), which occupy the **jugular foramina** (Fig. 7-60). These large bulbs of the internal jugular veins receive the inferior petrosal sinuses and continue as the **internal jugular veins** (Figs. 7-39, 7-65, and 7-79).

All the venous sinuses of the dura mater eventually deliver most of their blood to the sigmoid sinuses and then to the internal jugular veins (Fig. 7-39), except for the inferior petrosal sinuses which enter these veins directly (Fig. 7-65).

The Occipital Sinus (Fig. 7-39). This is the smallest of the dural venous sinuses; it begins near the posterior margin of the foramen magnum as two or more venous channels around the edges of this large aperture. These channels usually join to form an unpaired sinus that lies in the attached border of the falx cerebelli. The occipital sinus communicates inferiorly with the **internal vertebral plexus** (Figs. 5-58 and 7-39), and ends superiorly in the confluence of the sinuses (Figs. 7-68 and 7-78).

The Cavernous Sinuses (L. *caverna*, cave). These *large sinuses of the dura mater* are about 2 cm long and 1 cm wide (Figs. 7-39, 7-57*B*, 7-79, and 7-80). They are located on each side of the sella turcica and the body of the sphenoid bone (Figs. 7-13 and 7-15). They were named cavernous sinuses because of their cave-like appearance resulting from the many blood channels formed by numerous trabeculae.

Each cavernous sinus extends from the superior orbital fissure anteriorly to the apex of the petrous part of the temporal bone posteriorly (Fig. 7-84). Here it joins the posterior ***intercavernous sinus*** and the superior and inferior petrosal sinuses (Figs. 7-39 and 7-57*B*).

Each cavernous sinus receives blood from the superior and inferior **ophthalmic veins**, the superficial **middle cerebral vein** in the lateral fissure of the cerebral hemisphere, and the **sphenoparietal sinus** (Fig. 7-57*B*). The latter sinus lies inferior to the edge of the lesser wing of the sphenoid bone.

The cavernous sinuses communicate with each other through ***intercavernous sinuses*** that pass anterior and posterior to the hypophysial stalk or infundibulum (Figs. 7-57*B* and 7-62). The cavernous sinuses drain posteriorly and inferiorly via the superior and inferior petrosal sinuses and the pterygoid plexuses (Figs. 7-39, 7-57*B*, and 7-79).

Located in the lateral wall of each cavernous sinus, from superior to inferior, are the following structures (Fig. 7-80): (1) the oculomotor nerve (**CN III**); the trochlear nerve (**CN IV**); the ophthalmic and maxillary divisions of the trigeminal nerve (**CN V^1** and **CN V^2**).

Inside each cavernous sinus are the **internal carotid artery** (Figs. 7-80 and 7-84), with its sympathetic plexus and the abducens nerve (**CN VI**).

The Superior Petrosal Sinuses (Figs. 7-39, 7-57*B*, and 7-79). *These venous sinuses of the dura mater are small channels that drain the cavernous sinuses. They run from the posterior ends of the cavernous sinuses to the* ***transverse sinuses***, at the points where they curve inferiorly to form the **sigmoid sinuses** (Fig. 7-57*B*). Each superior petrosal sinus lies in the attached margin of the **tentorium cerebelli**, running in a small groove on the superior margin of the petrous part of the temporal bone (Figs. 7-57*B*, 7-60, and 7-63).

The Inferior Petrosal Sinuses (Figs. 7-39, 7-57*B*, 7-65, and 7-79). *These are also small venous sinuses of the dura mater that drain the cavernous sinuses directly into the internal jugular veins*, just inferior to the skull (Figs. 7-57 and 7-65). Commencing at the posterior end of the cavernous sinus, each inferior petrosal sinus runs posteriorly, laterally, and inferiorly (Fig. 7-79) in a groove between the petrous part of the temporal bone and the basilar part of the occipital bone (Figs. 7-60 and 7-63).

The inferior petrosal sinus enters the jugular foramen and joins the superior bulb of the internal jugular vein, but it usually enters this vein inferior to the skull (Figs. 7-75*A* and 7-65). It receives cerebellar and labyrinthine (internal auditory) veins.

The Basilar Sinus. This sinus consists of several interconnecting venous channels on the **clivus** (Fig. 7-60), the posterior surfaces of the basiocciput, and the basisphenoid. It connects the two inferior petrosal sinuses and communicates inferiorly with the *internal vertebral venous plexus* (Figs. 5-58 and 7-39).

Emissary veins connect the intracranial venous sinuses with veins outside the cranium. Although they are valveless and blood may flow in both directions, *the flow in the emissary veins is usually* *away from the brain.* The size and number of emissary veins vary. Only those which are clinically important are mentioned.

In children (and in some adults) a ***frontal emissary vein*** passes through the foramen cecum of the skull (Fig. 7-14), connecting the superior sagittal sinus with the veins of the frontal sinus and/or with those of the nasal cavities.

The parietal emissary veins, one on each side, pass through the parietal foramina in the calvaria (Fig. 7-7), and connect the **superior sagittal sinus** with the veins external to it, particularly those in the scalp (Fig. 7-54). They may also communicate with the diploic veins (Fig. 7-49).

A **mastoid emissary vein** connects each sigmoid sinus through the mastoid foramen with the occipital or the posterior auricular vein. A *posterior condylar emissary vein* may also be present and pass through the condylar canal (Figs. 7-60 and 7-63), connecting the sigmoid sinus with the suboccipital plexus of veins.

Although most blood leaves the skull and brain via the internal jugular veins (Fig. 6-65), there are several other routes (*e.g.*, the emissary veins) that are important clinically.

The basilar sinus and the occipital sinus, via the foramen magnum, communicate with the internal vertebral plexuses (Figs. 5-58 and 7-39). Consequently blood may pass to or from the **vertebral venous system.** As these venous channels are valveless, compression of the thorax, abdomen, or pelvis, as occurs during heavy coughing and straining, may force venous blood from these regions into the vertebral venous system and from it into the venous sinuses of the dura mater (Fig. 7-39). As a result, tumor cells and abscesses of the head or inside the thorax, abdomen, or pelvis may spread to the vertebrae and the brain.

These connections of the vertebral plexuses with the dural venous sinuses also explain how a blood clot or thrombus (*e.g.*, from the pelvic veins after childbirth) can reach the dural venous sinuses without passing through the heart and lungs. Such a thrombus may block a dural venous sinus and could result in a venous infarction of the cerebral cortex.

Infections outside the skull may pass via emissary veins into the venous sinuses of the dura mater. An infection in the scalp can be transmitted to the superior sagittal sinus via the **parietal emissary veins.** The bacteria may produce inflammatory processes in the walls of the emissary veins that could cause **thrombophlebitis of the dural venous sinuses** and later of the cortical veins. *Cortical infarction* (death of nervous tissue) could follow.

The facial veins connect with the cavernous sinuses through communications with the ophthalmic veins or with their tributaries, the supraorbital veins (Figs. 7-39 and 7-57*B*). Consequently, **thrombophlebitis of the facial vein may extend into the dural venous sinuses.** Thrombophlebitis in one cavernous sinus commonly spreads to the other one because they are connected by **intercavernous sinuses** (Fig. 7-57*B*). Furthermore, thrombosis could spread from the cavernous sinus into the superior and inferior petrosal sinuses, the transverse sinus, the sigmoid sinus, and the internal jugular vein. In cases of **mastoiditis** (uncommon since the advent of antibiotics), thrombosis of the sigmoid sinus may occur owing to its proximity to the mastoid cells (Fig. 7-79).

Excision of an internal jugular vein is sometimes necessary, *e.g.*, in a radical neck dissection, to remove the **deep cervical lymph nodes** (Fig. 7-40). These lymph nodes are commonly involved with malignant tumors of the head and neck. In most cases circulatory complications are minimal because the connections between the dural venous sinuses and veins outside the skull (*e.g.*, the vertebral venous plexus and the facial veins) enlarge enough to carry all the blood from intracranial structures.

Although emissary veins cannot enlarge quickly because their size is limited by the foramina in the cranial bones, they can enlarge and carry considerable blood in certain conditions (*e.g., with a chronic increase in intracranial pressure*).

When a dural venous sinus is torn in association with a skull fracture, bleeding may persist instead of diminishing as in most other vessels. The reason for this is that the patient is usually lying down, which increases the pressure in the dural venous sinuses. In addition the dural venous sinuses are noncontractile. If there is bleeding into the subdural or extradural spaces, it will continue. If it is an open fracture (*i.e.*, compound fracture of the calvaria), blood may drain from a dural venous sinus through the wound.

The fact that the cavernous sinus envelops the internal carotid artery and several cranial nerves is of clinical significance (Figs. 7-80 and 7-84). In fractures of the base of the skull, the internal carotid artery may tear within the cavernous sinus, producing an **arteriovenous fistula**. In such cases arterial blood rushes into the cavernous sinus, enlarging it and forcing blood out into the connecting veins (Fig. 7-39), especially the ophthalmic veins, which normally drain the orbital cavity (Fig. 7-57). As a result, the eye protrudes (**exophthalmos**) and the conjunctiva becomes engorged (**chemosis**) on the side of the torn artery. In these circumstances, the bulging eye pulsates in synchrony with the radial pulse (Fig. 6-82). Owing to this phenomenon, the condition is called **pulsating exophthalmos**.

Because cranial nerves III, IV, V^1, V^2, and VI lie in or close to the lateral wall of the cavernous sinus (Fig. 7-80), they may also be affected when injuries or infections of this sinus occur (Table 7-3).

THE BRAIN

The gross features of the brain are described in this gross anatomy book to help you understand the relationship of the brain to the meninges and the calvaria. A clear understanding of these relationships is needed so that you will know when an intracranial lesion has altered the appearance of structures within the head (*e.g.*, **displacement of the pineal body** (Fig. 7-71) or distortion of the cerebral arteries or ventricles in radiographs. In addition, accurate analysis of computerized tomograms (**CT scans**) of the head depends mainly on a knowledge of transverse sections of this region (Fig. 7-81).

Normally the brain is within the cranial cavity, but space-occupying lesions may result in displacement of parts of it through the foramen magnum (Fig. 7-54). When the calvaria and dura mater are removed (Fig. 7-51), you can observe the convolutions (**gyri**) and grooves (**sulci**) of the cerebral cortex through the delicate piarachnoid (Fig. 7-50).

The complicated folding of the surface of the cerebral hemispheres greatly increases the surface area of the brain (Figs. 7-51 and 7-52).

MAJOR FISSURES AND SULCI OF THE BRAIN

There are six main fissures and sulci: (1) the **longitudinal cerebral fissure**; (2) the **transverse cerebral fissure**; (3) the **lateral sulcus** (fissure); (4) the **central sulcus**; (5) the **parietooccipital sulcus**; and (6) the **calcarine sulcus**.

THE LONGITUDINAL CEREBRAL FISSURE

The longitudinal cerebral fissure (sagittal fissure) partially separates the cerebral hemispheres (Figs. 7-53 and 7-72). *In situ* this contains the **falx cerebri** (Figs. 7-53 and 7-54). The longitudinal cerebral fissure separates the cerebral hemispheres in the frontal and occipital regions, but between these parts of the brain, the fissure extends only as far as the **corpus callosum** (Figs. 7-54 and 7-56).

THE TRANSVERSE CEREBRAL FISSURE

The transverse cerebral fissure separates the cerebral hemispheres superiorly from the cerebellum, midbrain, and diencephalon inferiorly. The ***tentorium cerebelli*** lies in the posterior part of this fissure (Fig. 7-51).

THE LATERAL SULCUS

The lateral sulcus begins inferiorly on the inferior surface of the cerebral hemisphere as a deep furrow (Fig. 7-52) and extends posteriorly, separating the frontal and temporal lobes (Fig. 7-51). Posteriorly, the lateral sulcus separates parts of the parietal and temporal lobes.

THE CENTRAL SULCUS (Fig. 7-51)

The central sulcus is a prominent groove running inferoanteriorly from about the middle of the superior margin of the cerebral hemisphere, stopping just short of the lateral sulcus.

The central sulcus is an important landmark of the cerebral cortex because the **motor cortex** (precentral gyrus) lies anterior to it and the **general sensory cortex** (postcentral gyrus) lies posterior to it. The superior end of the central sulcus is located about 1¼ cm posterior to the midpoint of a line joining the **inion** and the **nasion** (Fig. 7-16), and its inferior end is about 5 cm superior to the external acoustic meatus.

THE PARIETOOCCIPITAL SULCUS

The parietooccipital sulcus, as its name indicates, separates the parietal and occipital lobes of the brain on the medial aspect of the brain (Fig. 7-56). It extends from the calcarine sulcus to the superior border and continues for a short distance on the superolateral surface (Fig. 7-51).

THE CALCARINE SULCUS

The calcarine sulcus is on the medial surface of the brain (Fig. 7-56). It commences near the occipital pole and runs anteriorly, taking a curved course and joining the parietooccipital sulcus at an acute angle.

MAIN LOBES OF THE CEREBRAL HEMISPHERES

There are four main lobes of the cerebral hemispheres.

THE FRONTAL LOBES (Figs. 7-51, 7-52, 7-56, and 7-72)

The frontal lobes are the largest of all the lobes of the brain. They form the anterior parts of the cerebral hemispheres. The frontal lobes lie anterior to the central sulci and superior to the lateral sulci. Their lateral and superior surfaces extend posterior to the coronal suture. The basal surfaces of the frontal lobes rest on the orbital part of the frontal bone in the **anterior cranial fossa** (Figs. 7-13 and 7-51). The **orbital gyri** (Fig. 7-52) form impressions in the orbital plates of the frontal bone. The **olfactory bulbs** rest on the **cribriform plates** (Fig. 7-13).

THE PARIETAL LOBES (Fig. 7-51)

The parietal lobes are related to the internal aspects of the posterior and superior parts of the parietal bones. Each parietal lobe is bounded anteriorly by the central sulcus (Fig. 7-51) and posteriorly by the superior part of the line joining the parietoocipital sulcus (Fig. 7-56) and the **preoccipital notch**. The inferior boundary of each lobe is indicated by an imaginary line extending from the posterior ramus of the lateral sulcus to the inferior end of the posterior boundary.

THE TEMPORAL LOBES (Figs. 7-51 and 7-52)

The temporal lobes lie inferior to the lateral sulci. Their convex anterior ends, called **temporal poles,** fit into the anterior and lateral parts of the middle cranial fossa (Figs. 7-15 and 7-51). Their posterior parts lie against the middle one-third of the inferior part of the parietal bone.

THE OCCIPITAL LOBES (Figs. 7-51 and 7-56)

The occipital lobes are relatively small and are located posterior to the parietooccipital sulci (Fig. 7-56). They *rest on the* ***tentorium cerebelli,*** superior to the posterior cranial fossa (Figs. 7-51 and 7-68). Although small, ***the occipital lobes are important because they contain the visual cortex*** (Fig. 7-70).

MAIN PARTS OF THE BRAIN

To understand the origin of cranial nerves and the structure of the ventricular system (Fig. 7-71), the major divisions of the brain must be understood. As the names of some structures are derived from embry-

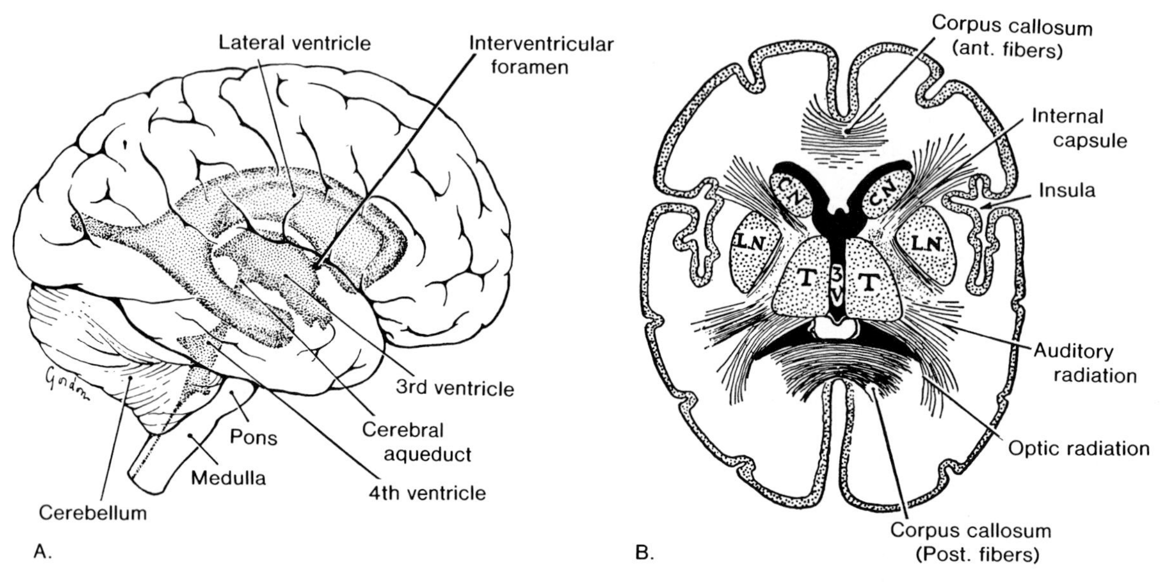

Figure 7-70. *A*, schematic drawing of a lateral view of the right side of the brain illustrating the ventricular system. *B*, diagramatic horizontal section of the brain. Between the right and left thalami (*T*) is the third ventricle (*3V*). The internal capsule separates the thalamus (*T*) and the caudate nucleus (*CN*) from the lentiform nucleus (*LN*) which lies deep to the insula in the depth of the lateral sulcus (Fig. 7-51). The corpus callosum consists of nerve fibers connecting the cerebral hemispheres. The body of the corpus callosum (Figs. 7-71 and 7-86), its largest part, would be visible in horizontal sections superior to this one. The basal nuclei of the telencephalon, which include the corpus striatum (caudate and lentiform nuclei), the claustrum, and the amygdaloid nuclei, are often referred to as the basal "ganglia." This is a misnomer because ganglia are collections of nerve cells outside the central nervous system.

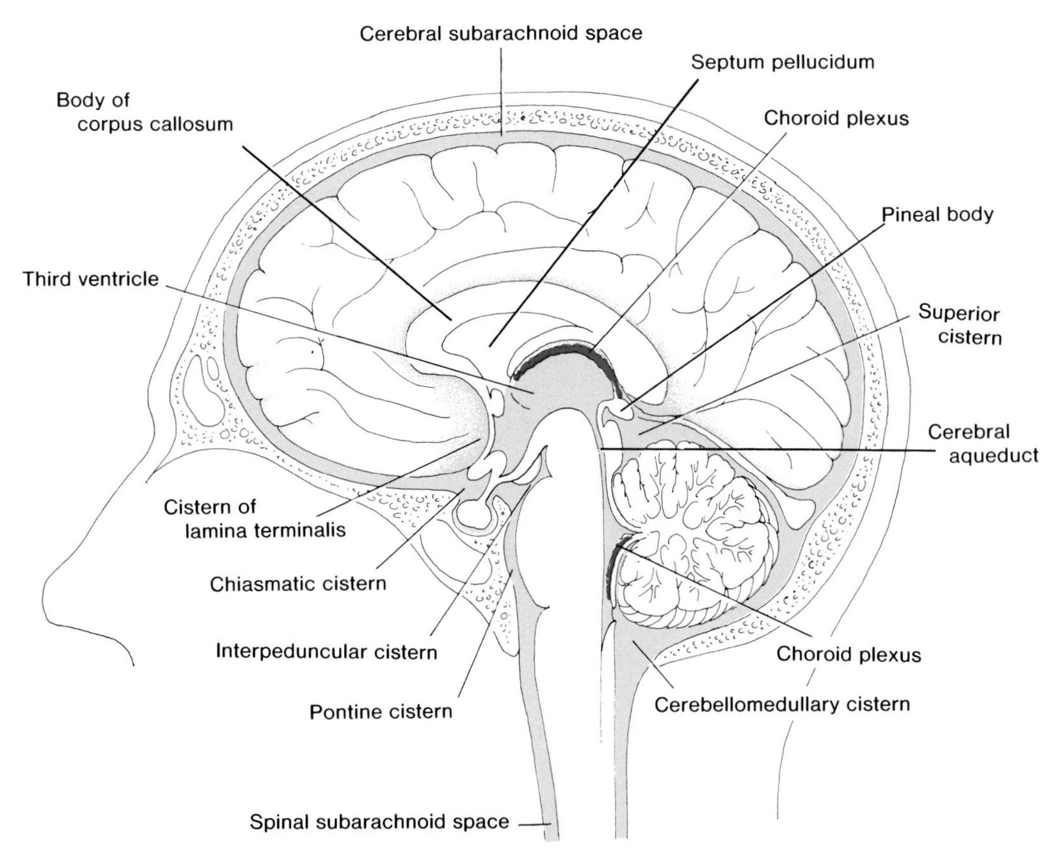

Figure 7-71. Diagram of the subarachnoid spaces and cisterns (*blue*) as seen in a median section of the brain. The superior cistern (located dorsal to the midbrain) together with the subarachnoid space at the sides of the midbrain are referred to clinically as the cisterna ambiens. The superior cistern is important because it contains internal cerebral veins which join caudally to form the great cerebral vein (Fig. 7-57). It also contains the posterior cerebral and superior cerebellar arteries (Fig. 7-77). The choroid plexuses in the roof of the third and fourth ventricles are shown in *red*.

onic terms, a basic knowledge of brain development is also required.

The brain develops from the walls of embryonic brain vesicles that form from the cranial end of the ***neural tube***. The names of some walls of the brain vesicles are used to describe parts of the mature brain (*e.g.*, the **telencephalon and diencephalon**, derivatives of the embryonic forebrain, and the **mesencephalon**, the name of the embryonic midbrain).

Encephalon is derived from the Greek word *enkephalos*, meaning brain. It forms the basis of many medical terms (*e.g.*, **encephalitis**, inflammation of the brain and **encephalocele**, herniation of the brain). The Latin word for the brain is **cerebrum.** It is used to refer to derivatives of the **telencephalon** (*e.g.*, the cerebral hemispheres).

THE CEREBRAL HEMISPHERES (Figs. 7-50 to 7-52, 7-56, 7-70, and 7-71)

The cerebral hemispheres occupy the anterior and middle cranial fossae and extend posteriorly over the tentorium cerebelli and the cerebellum to the internal occipital protuberance (Figs. 7-14, 7-15, and 7-51).

The cerebral hemispheres comprise the cerebral cortex, the basal nuclei (basal "ganglia"), their fiber connections, and the lateral ventricles (Fig. 7-70).

The cerebral hemispheres form the largest part of the cerebrum and of the entire brain. The cavity in each cerebral hemisphere, known as a **lateral ventricle** (Figs. 7-70 and 7-81), is part of the ventricular system of the brain (Fig. 7-71).

THE DIENCEPHALON (Figs. 7-52, 7-56, 7-70, and 7-72)

The diencephalon, composed of the thalamus, hypothalamus, epithalamus, and subthalamus, surrounds the third ventricle of the brain. It forms the central core of the brain and is surrounded by the cerebral hemispheres. Only the basal or inferior surface of the diencephalon is exposed to view in the diamond-shaped area containing the **infundibulum** and **mammillary bodies** (Figs. 7-52 and 7-72).

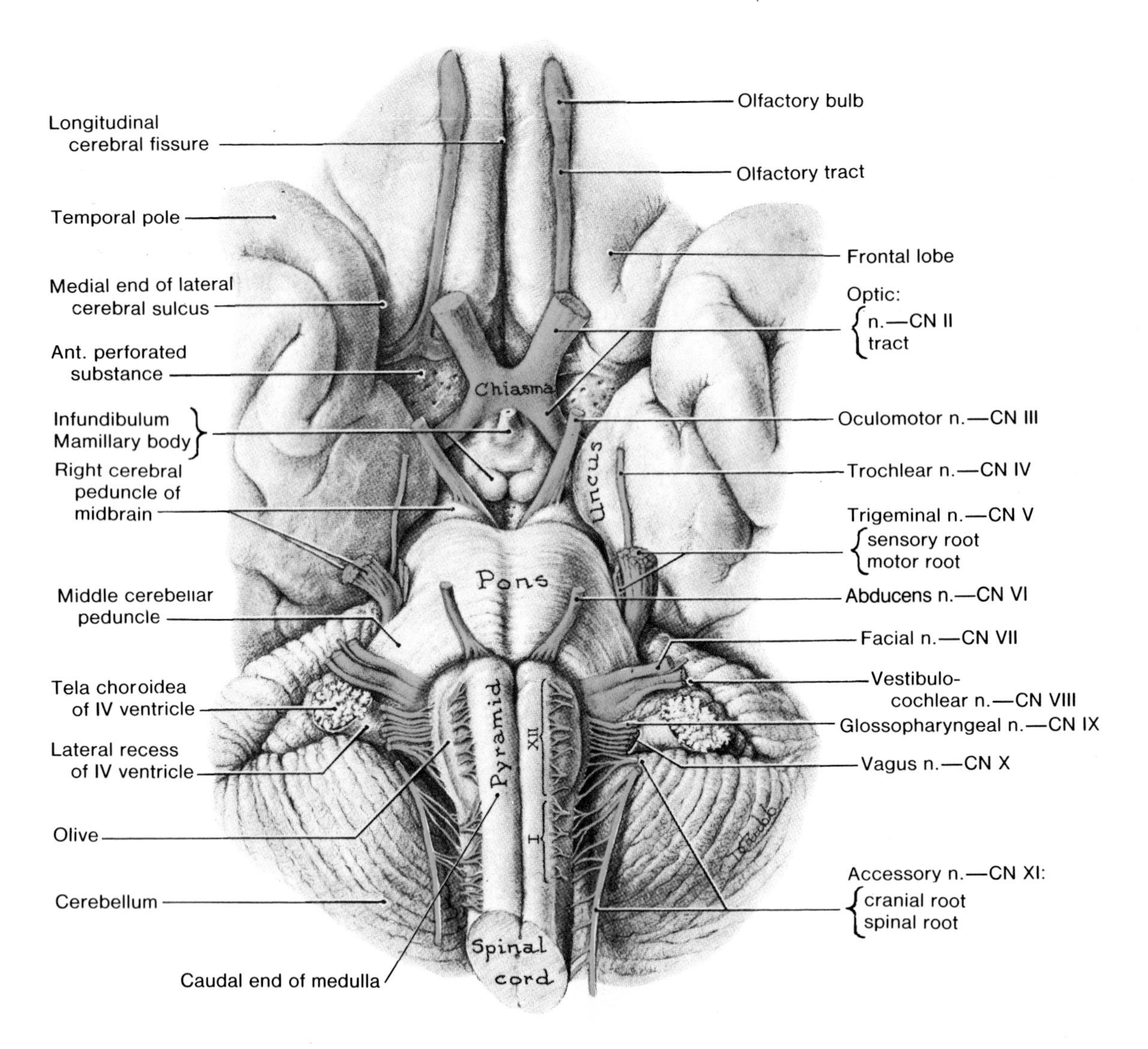

Figure 7-72. Drawing of the inferior surface of the brain showing the superficial origin of the 12 cranial nerves. The olfactory bulbs, into which the olfactory nerves (CN I) end, are related to the cribriform plates (Figs. 7-14 and 7-52). The infundibulum and mammillary bodies are in the middle cranial fossa, as are the temporal lobes lateral to them. The pons, medulla, and cerebellum are in the posterior cranial fossa (Fig. 7-51). I, indicates rootlets of the first cervical segment of the spinal cord. The brain stem is the stem-like portion of the brain connecting the cerebral hemispheres of the spinal cord. It comprises the midbrain, pons, and medulla.

The two thalami make up four-fifths of the diencephalon (Fig. 7-70*B*). Although the **hypothalamus** forms only a small part of the diencephalon, it is very important to life. The cavity of the diencephalon is the narrow **third ventricle** lying between the right and left thalami (Fig. 7-70*B*). CSF enters the third ventricle from the lateral ventricles through the **interventricular foramina** (Figs. 7-56 and 7-70*A*).

THE MIDBRAIN (Figs. 7-59 and 7-69 to 7-71)

This is the smallest part of the brain. Very little of it is visible in an unsectioned brain. The midbrain lies at the junction of the middle and posterior cranial fossae, lying partly in each and in the tentorial incisure (Fig. 7-58).

The cavity of the midbrain is represented by a narrow canal, called the **cerebral aqueduct** (Figs. 7-70*A* and 7-71). It conducts CSF from the lateral and third ventricles to the fourth ventricle.

THE PONS (Figs. 7-54, 7-56, 7-59, 7-62, 7-69, 7-70, and 7-72)

The pons is a Latin word meaning a **bridge**, an appropriate name for this part of the brain because all of it that is visible from the inferior and lateral surfaces is a wide, bridge-like transverse band of nerve fibers.

Although it looks somewhat like a bridge between the cerebellar hemispheres, it is not. *The pontine fibers connect one cerebral hemisphere with the opposite cerebellar hemisphere.* The cavity in the pons forms the superior part of the **fourth ventricle.** The inferior part of this ventricle is formed by the cavity of the medulla.

The fourth ventricle is largely roofed by the cerebellum (Figs. 7-56, 7-59, 7-70, and 7-71). It receives **CSF** from the lateral and third ventricles via the cerebral aqueduct. The pons lies in the anterior part of the posterior cranial fossa, posterior to the superior part of the **clivus** and the posterior surface of the **dorsum sellae** (Figs. 7-15 and 7-60).

THE MEDULLA OBLONGATA (Figs. 7-52, 7-54, 7-56, 7-59, 7-64, 7-70, and 7-72)

This is the most caudal part of the brain stem, composed of the midbrain, pons, and medulla oblongata. In common usage, the medulla oblongata is referred to as the medulla. It is located in the posterior cranial fossa with its ventral aspect facing the clivus (Fig. 7-60).

The medulla is continuous with the spinal cord at the foramen magnum (Fig. 7-54), but the transition is gradual. The distinctive characteristics of its ventral surface are the elongated **pyramids**, containing the corticospinal tracts from the cerebral cortex.

The medulla contains the cardiovascular and respiratory centers for the automatic control of heart beat and respiration, respectively. The cavity of the medulla forms the inferior part of the **fourth ventricle** (Figs. 7-56, 7-59, 7-70*A*, and 7-71).

CSF enters the fourth ventricle from the cerebral aqueduct, and leaves it through three openings in its roof: one median and two lateral (Figs. 7-56 and 7-70*A*). The **median aperture** (foramen of Magendie) is just dorsal to the inferior part of the medulla; the **lateral apertures** (foramina of Luschka) are located at the ends of the lateral recesses of the fourth ventricle.

All three apertures of the fourth ventricle open into the subarachnoid space and are the only communications between the ventricular system and the subarachnoid space. If these foramina are occluded, all the ventricles become enlarged (**internal hydrocephalus**, Fig. 7-82) because CSF production continues but is unable to leave the ventricular system. Hydrocephalus is discussed on page 877.

THE CEREBELLUM (Figs. 7-51, 7-52, 7-54, 7-56, 7-59, and 7-70 to 7-72)

This part of the brain overlies the posterior aspect of the pons and medulla and extends laterally beneath the tentorium cerebelli. **It occupies most of the posterior cranial fossa** (Figs. 7-14 and 7-51).

The name cerebellum is the Latin diminutive of the *cerebrum* (L. brain); hence the cerebellum is the "little brain." The cerebellum consists of a midline portion, the **vermis** (L. worm), and two lateral lobes or **cerebellar hemispheres.** These hemispheres lie posterior to the petrous parts of the temporal bones and rest in the concavities of the occipital bones (Figs. 7-14 and 7-68).

The cerebellum is mainly concerned with motor functions that regulate: posture, muscle tone, and muscular coordination.

Cerebellar lesions result in disturbance of one or more of the motor functions of the cerebellum producing *unsteady gait, hypotonia* (G. *hypo,* under + *tonos,* tone), tremor, nystagmus (rhythmical oscillation of the eyes), and *dysarthria* (disturbance of articulation).

Brain injuries are commonly associated with severe head injuries. The main purpose of the calvaria and CSF is to protect the brain. Helmets are used by motorcycle riders, construction workers, soldiers, some athletes, and others for additional protection.

Various terms are used to describe injuries to the brain resulting from trauma.

Cerebral Concussion. A concussion is an *abrupt but transient loss of consciousness* immediately following a blow to the head. A concussion can also follow the sudden stopping of the moving head, as occurs during rear end collisions (Fig. 5-62). These accidents result in the brain hitting the stationary skull (Fig. 5-67). When a person is hit (*e.g.*, by a club or baseball), they may be stunned and fall to the ground. The concussion appears to result from a mechanical disturbance of cerebral cells because **there is no visible bruising of the brain** on gross or microscopic examination. However, repeated cerebral concussions, as may occur in professional fighters, lead to **cerebral atrophy**; thus some submicroscopic damage appears to occur with each concussion.

The punch-drunk syndrome, caused by repeated cerebral concussion, is characterized by weakness in the lower limbs, unsteady gait, slowness of muscular movements, tremors of the hands, hesitancy of speech, and slow cerebration (the process of using one's brain).

Cerebral Contusion (L. *contusio,* a bruising). *There is visible bruising of the brain* owing to trauma and blood leaking from small vessels. The pia mater is stripped from the brain in the injured area and may be torn, allowing blood to enter the subarachnoid space. Contusions are frequently found at the frontal, temporal, and occipital poles (Figs. 7-51 and 7-56). The bruising results either from the sudden impact of the moving brain against the stationary skull, or the skull against the brain when there is a blow to it (Fig. 5-67). The living calvaria is somewhat flexible and may sometimes bend without breaking. As a consequence, brusing of the brain may occur when there is no fracture.

A contusion usually results in an extended loss of consciousness (several minutes to many hours).

Blows to the forehead tend to produce contusions of the frontal poles of the brain (Fig. 7-51) and perhaps the temporal poles, whereas *blows to the occiput (back of the head) generally produce contusions of the occipital poles.* Contusion is soon followed by **edema** (swelling) of the brain, which contributes to a rise in intracranial pressure, a serious complication that may result in a life-threatening situation, as discussed subsequently under *cerebral compression.*

Cerebral Laceration (L. *lacero,* to tear). *Tearing of the brain* is often associated with a depressed skull fracture or a gunshot wound. Cerebral lacerations result in rupture of large blood vessels with bleeding into the brain and subarachnoid space. This leads to the formation of a **cerebral hematoma**, edema, and an increase in local and general intracranial pressure.

Cerebral Compression. Pressure on the brain may be produced by: (1) intracranial collections of blood (**hematomas**) or other fluids; (2) obstruction of the circulation or absorption of CSF; (3) intracranial tumors or abscesses; and (4) edema of the brain (*e.g.*, owing to contusions).

Cerebral compression is associated with increased intracranial pressure. A marked increase in intracranial pressure results in herniation of parts of the brain. Often the **uncus** of the temporal lobe (Figs. 7-52 and 7-72) herniates through the **tentorial incisure,** causing pressure on the midbrain and kinking of the oculomotor nerves on the edge of the tentorium (Figs. 7-58 and 7-68). This results in **third nerve palsy,** usually commencing with progressive enlargement of the pupil and fixation of it to light (Fig. 7-87). In some cases the **cerebellar tonsils** herniate through the foramen magnum (Fig. 7-54), compressing the medulla and disturbing the vital cardiovascular and respiratory centers.

Tonsillar herniation carries the cerebellar tonsils (Fig. 7-59) and the median aperture of the fourth ventricle into the superior part of the vertebral canal (Fig. 7-54). This compromises the flow of fluid out of the fourth ventricle and into the subarachnoid space.

Internal hydrocephalus results. Immediate surgical intervention is usually necessary to relieve the high intracranial pressure because *death can result from disruption of the cardiovascular and respiratory centers.*

THE VENTRICULAR SYSTEM AND CEREBROSPINAL FLUID (CSF)

FORMATION OF CSF (Figs. 7-70 and 7-71)

The main source of CSF is the choroid plexuses of the lateral, third, and fourth ventricles.

Choroid plexuses are located in the roofs of the third and fourth ventricles and on the floors of the bodies and inferior horns of the lateral ventricles.

The choroid plexuses in the lateral ventricles are the largest and most important. They are continuous with the choroid plexus in the roof of the third ventricle via the interventricular foramina (Fig. 7-56).

The **lateral apertures of the fourth ventricle** are also partially occupied by parts of the choroid plexuses which protrude through them from the fourth ventricle, and secrete CSF into the subarachnoid space.

Each choroid plexus (G. *choroid*, delicate membrane + *eidos*, resemblance) is composed of highly vascular pia mater called **tela choroidea** (L. *a web* + G. *chorioeidēs*, like a membrane), covered by a simple cuboid or low columnar epithelium.

The flow of CSF is from the lateral ventricles into the third ventricle through the **interventricular foramina**, and then into the fourth ventricle via the **cerebral aqueduct** of the midbrain (Fig. 7-71). CSF passes from the fourth ventricle through the **median aperture** (Fig. 7-56) and **lateral apertures** into the subarachnoid space around the brain, spinal cord, and cauda equina (Figs. 5-53 and 7-71).

Ventricular fluid and CSF do not have the same composition. The fluid in the ventricles is more dilute and has a different protein content than the CSF in the subarachnoid space.

THE SUBARACHNOID CISTERNS (Fig. 7-71)

At certain places, mainly at the base of the brain, the arachnoid is widely separated from the pia mater forming large ***pools of CSF*** **called subarachnoid cisterns** (L. cisterns or reservoirs). These cisterns communicate freely with each other, with the subarachnoid space of the spinal cord and covering all surfaces of the cerebral hemispheres.

The cerebellomedullary cistern, or cisterna

magna, is located in the space between the cerebellum and the inferior part of the medulla (Fig. 7-71). It receives CSF from the **median aperture** of the fourth ventricle (Fig. 7-56) and is continuous with the large *subarachnoid space* around the brain and spinal cord (Fig. 7-71).

The pontine cistern is the extensive space along the ventral and lateral surfaces of the pons, containing the basilar artery and some of the cranial nerves (Figs. 7-71 and 7-77). It is continuous inferiorly with the cerebellomedullary cistern and superiorly with the interpeduncular cistern.

The interpeduncular cistern, between the cerebral peduncles (Figs. 7-71 and 7-72), contains the posterior part of the **arterial circle** (circle of Willis). This circle is formed by the major arteries supplying the cerebrum (Fig. 7-77). The interpeduncular cistern is continuous anterosuperiorly with the **chiasmatic cistern**, which continues as the **cistern of the lamina terminalis** (Fig. 7-71). This cistern continues into the shallow **cistern of the corpus callosum** which is located, anterosuperior to this large commissure.

The cistern of the lateral sulcus, containing the *middle cerebral artery* (Fig. 7-77), is located anterior to each temporal lobe where the arachnoid covers the lateral sulcus (Fig. 7-51).

The superior cistern or cistern of the great cerebral vein (Fig. 7-71), lies between the splenium of the corpus callosum and the superior surface of the cerebellum. It contains the great cerebral vein (Figs. 7-39 and 7-57*A*) and the ***pineal body***. Because this cistern lies dorsal to the **colliculi of the midbrain** (corpora quadrigemina), it is often referred to clinically as the *quadrigeminal cistern* (Figs. 7-62 and 7-71).

The pineal body (pineal gland) or epiphysis (Figs. 7-56 and 7-71) is an endocrine gland that has the shape of a pine cone. It produces **melatonin.**

The pineal body also serves as a useful neuroradiological and neurosurgical landmark. It becomes calcified in adolescence and is visible on ordinary skull radiographs of approximately 75% of North American adults. Charts showing its usual position on lateral skull radiographs are used to diagnose displacement of the pineal gland in patients with suspected expanding intracranial lesions. Side to side displacement of the pineal body can be measured with a ruler.

Samples of CSF for laboratory studies are usually obtained by lumbar puncture (Fig. 5-60), but CSF may be obtained from the cerebellomedullary cistern (Fig. 7-71), a procedure known as **cisternal puncture**. In a person with a spina bifida cystica in the lumbar region (Fig. 5-61), a lumbar puncture is not practical; hence a **cisternal puncture** would likely be performed.

Samples of ventricular fluid are obtained via an **anterior fontanelle puncture** in infants (Fig. 7-5*B*), or through burr holes drilled in the calvaria of adults.The subarachnoid space or the ventricular system are also entered for measuring or monitoring CSF pressure, injecting antibiotics, or administering contrast media for radiography (**ventriculography**).

CIRCULATION OF CSF (Figs. 7-70 and 7-71)

CSF produced in the lateral and third ventricles passes via the cerebral aqueduct into the fourth ventricle, where more CSF is produced by the choroid plexuses in its roof (Fig. 7-71).

CSF leaves the fourth ventricle through its median and lateral apertures and passes into the subarachnoid space, where it collects in the cerebellomedullary and pontine cisterns. From these cisterns some CSF passes inferiorly into the spinal subarachnoid space around the spinal cord, and posterosuperiorly over the cerebellum; however, *most CSF flows superiorly through the tentorial incisure into the subarachnoid space around the midbrain* (**interpeduncular cistern** and **superior cistern**).

From the various cisterns, the CSF spreads superiorly through the sulci and fissures on the medial and superolateral surfaces of the cerebral hemispheres (Fig. 7-71). This flow of CSF is probably aided by pulsations of the cerebral arteries and cerebral hemispheres.

CSF also passes into the extensions of the subarachnoid space around the cranial nerves, the most important of which are those surrounding the optic nerves (Figs. 7-73 and 7-93).

ABSORPTION OF CSF (Figs. 7-50 and 7-67)

The main site of absorption or passage of CSF into venous blood is through the arachnoid villi. These villi are protrusions of the arachnoid into the dural venous sinuses, especially the superior sagittal sinus and its adjacent lateral lacunae.

The rate of CSF absorption is pressure dependent and the arachnoid villi appear to act as one-way "valves." When CSF pressure is greater than venous pressure, the valves open and CSF passes into the blood in the venous sinuses of the dura mater (Fig. 7-57). However when venous pressure is higher than CSF pressure, the valves close, preventing blood from entering the CSF.

Some CSF appears to be absorbed by the ependymal

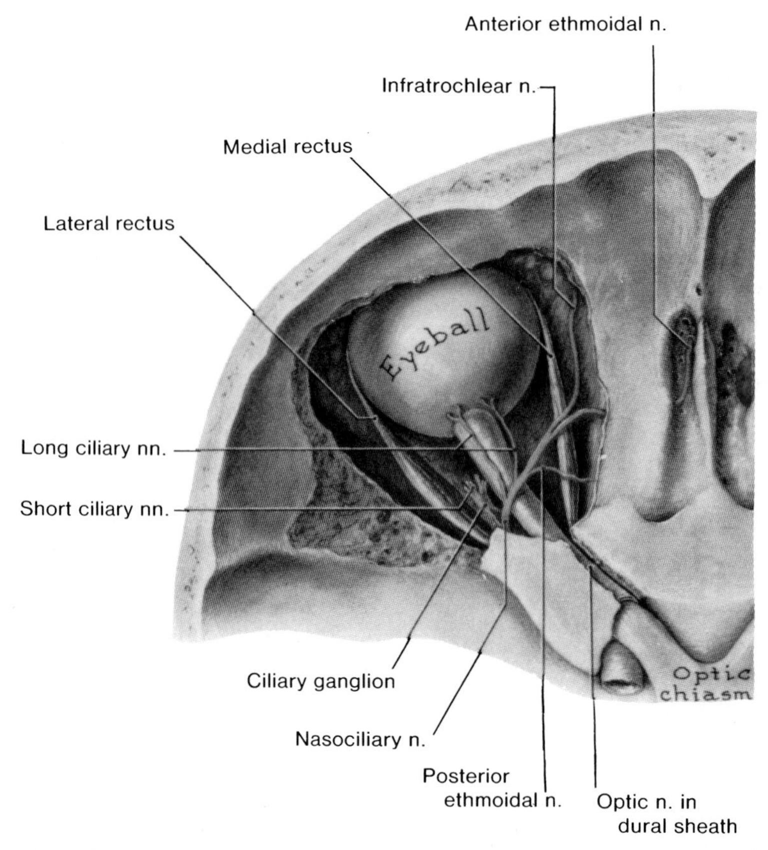

Figure 7-73. Drawing of a dissection of the left orbital cavity, superior view, showing the nasociliary nerve and its branches to the ocular muscles. Observe the optic nerve in its dural sheath which is continuous with the dura mater covering the brain (also see Fig. 7-93).

lining of the ventricles (**ependyma**), in the spinal subarachnoid space, and through the walls of capillaries in the pia mater. In addition, some CSF is probably absorbed into the lymphatics adjacent to the subarachnoid space around cerebrospinal nerves (*e.g.*, the optic nerves, Fig. 7-93).

By replacing some CSF by contrast media of various types, certain *parts of the brain can be outlined by radiography.* Air may be used to replace CSF to demonstrate the ventricular system and the subarachnoid spaces, and to detect or exclude displacement or deformities of their walls.

Ventriculography (Fig. 7-83). Before taking radiographs, a hole is made in the posterior part of the frontal bone, or one of the parietal bones with a **trephine** (cylindrical or crown saw) or a burr. Then a special **ventricular needle** or cannula is passed through the hole and a relatively safe part of the cerebrum into one of the lateral ventricles (Fig. 7-70). *Air is then exchanged for ventricular fluid,* the needle or cannula removed or replaced by a fine caliber soft catheter, the head bandaged, and a series of radiographs made. By suitably positioning the patient's head, air can be moved into various parts of the ventricular system if there is no obstruction.

Sometimes ***radiopaque oil*** *is injected* and, with or without fluoroscopic control, the oil is maneuvered into the third ventricle, the cerebral aqueduct, or the fourth ventricle, depending on the site of the lesion under investigation.

Encephalography (Fig. 7-83). First a lumbar puncture is performed with the patient sitting up (Fig. 5-60). Then the CSF pressure is measured. If it is not abnormally high, air is injected and CSF is removed alternately in small amounts. This is called **pneumoencephalography** (*i.e.*, radiography of the brain that is outlined with air). The radiograph that is taken is called a **pneumoencephalogram.**

The air bubbles pass superiorly in the spinal subarachnoid space and enter the cerebellomedullary cistern and fourth ventricle because the patient's head is strongly flexed. Frontal and lateral films are usually made at this time to show structures in the posterior cranial fossa. Later more air is injected with the head less flexed. The air then passes superiorly in the ventricular system, partly draining it, and also superiorly around the brain stem and through the tentorial incisure. The air finally passes onto the medial and dorsolateral surfaces of the cerebral hemispheres.

The lumbar puncture needle is removed and the patient lies on the x-ray table while a series of radiographs are taken. By suitably positioning the patient's head, air can be made to outline the ventricular system of the brain, *e.g.*, in **brow-up films** (Fig. 7-83), air outlines anterior parts of the ventricular system and subarachnoid spaces, whereas in **brow-down films**, air outlines posterior parts of the ventricular system.

Radiographic study of the brain by ventriculography or encephalography enables radiologists to locate atrophic or space-occupying lesions (e.g., a brain tumor). These radiological methods may also be used to determine the site of an obstruction in the circulation of CSF.

As the above techniques have certain disadvantages, they have been almost completely replaced by **computerized tomography** (see p. 17). CT scans (Fig. 7-81*B*) are so sensitive that much more information can be obtained than by ventriculography or encephalography. CT scans will show the presence of hemorrhage, infarction, tumors, cysts and other lesions.

Hydrocephalus (Fig. 7-82). Overproduction of CSF, obstruction of its flow, or interference with its absorption results in an excess of CSF in the head, a condition known as **hydrocephalus** (G. *hydór*, water + *kephalē*, head). The condition is marked by an excessive accumulation of fluid that dilates the ventricles of the brain and causes a separation of the bones of the calvaria in infants.

Internal hydrocephalus *is a hydrocephalus in which the accumulation of fluid is confined to the ventricles.* A blockage of the median aperture of the fourth ventricle results in the thin wall of the fourth ventricle herniating through the foramen magnum into the superior part of the vertebral canal.

External hydrocephalus *is a hydrocephalus in which the accumulation of fluid is in the subarachnoid spaces of the brain.* There may also be an accumulation of fluid in the subdural space owing to a communication between the subarachnoid and subdural spaces. The **subdural space** is a potential space between the dura mater and the external surface of the arachnoid (Fig. 7-67).

Meningitis (inflammation of the meninges) may produce pus in the subarachnoid space of the brain, including the cisterns (Fig. 7-71); this could cause obstruction of CSF circulation.

A tumor may block the cerebral aqueduct, stopping the flow of CSF to the fourth ventricle. Hydrocephaly during infancy often results from **congenital aqueductal stenosis** in which the cerebral aqueduct is narrow or consists of several minute channels.

Blockage of CSF circulation results in dilation of the ventricles superior the point of obstruction, and in pressure on the cerebral hemispheres. This squeezes the brain between the ventricular fluid and the bones of the calvaria. In infants the internal pressure results in expansion of the brain and calvaria (Fig. 7-82) because the sutures and fontanelles are still open (Fig. 7-5). It is possible to form an artificial drainage system (**ventriculoatrial shunt**) to allow the CSF to escape, thereby lessening damage to the brain.

Hydrocephalus usually refers to internal hydrocephalus in which all or most of the ventricular system is enlarged. All ventricles are enlarged if the apertures of the fourth ventricle or the subarachnoid spaces are blocked, whereas the lateral and third ventricles are dilated when the cerebral aqueduct is obstructed. Although rare, obstruction of one interventicular foramen produces dilation of one lateral ventricle.

FUNCTIONS OF CSF

Along with the meninges and calvaria, *CSF protects the brain by providing a cushion against blows to the head.* As the brain is slightly heavier than CSF, the gyri on the basal surface of the brain are in contact with the cranial fossae in the floor of the cranial cavity (Figs. 7-15 and 7-68) when a person is erect. In many places at the base of the brain, only the cranial meninges intervene between the brain and the cranial bones. In this position the CSF is in the subarachnoid cisterns and in the sulci on the superior and lateral parts of the brain. Hence *CSF normally separates the superior part of the brain from the calvaria or skullcap.*

The brain is kept from sagging by the cerebral veins

as they enter the dural venous sinuses (Figs. 7-53 and 7-54). Of course the relations of CSF to the calvaria and the brain are reversed when a person stoops or does a headstand.

There are small rapidly recurring changes in intracranial pressure owing to the heartbeat, as well as slow recurring changes resulting from unknown causes. In addition, *there are momentarily large changes in intracranial pressure during coughing and straining.* Any change in the volume of the intracranial contents (*e.g.*, a brain tumor, ventricular fluid resulting blockage of the cerebral aqueduct, and blood owing to hemorrhage) will be reflected by a change in intracranial pressure. This is called the **Monro-Kellie doctrine** *which states that the cranial cavity is a closed rigid box and that therefore a change in the quantity of intracranial blood can occur only through the displacement or replacement of CSF.*

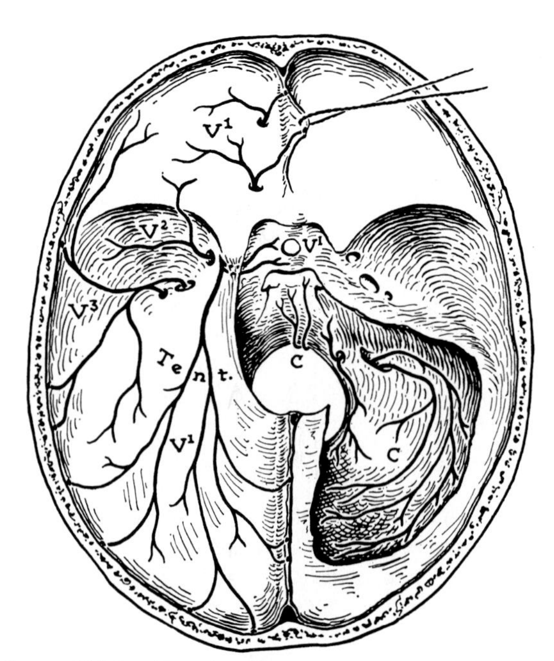

Figure 7-74. Drawing illustrating the nerves of the cranial dura mater. The nerves to the dura, the dural vessels, and the dural sinuses are derived mainly from the three divisions of the trigeminal nerve (CN V^1, CN V^2, and CN V^3), but branches are also received from the vagus (CN X) and the superior cervical nerves (*C*). Also see Figure 7-33.

Subarachnoid hemorrhage (bleeding into the subarachnoid space) often impedes CSF circulation by partly obstructing the cisterns, sulci, and arachnoid villi. This results in an increase in intracranial pressure.

Contusions of the brain resulting from sudden acceleration or deceleration of the head, are frequent at the base of the brain and at the frontal, temporal, and occipital poles, partly because of the paucity of CSF in these areas when a person is in the erect position. This situation allows the brain to strike the cranial bones with only the meninges intervening (Fig. 5-67).

Fractures of the floor of the middle cranial fossa *may result in leakage of CSF from the ear* (**CSF otorrhea**), if the meninges superior the middle ear and mastoid antrum (Fig. 7-164) are torn and the tympanic membrane (eardrum) is also ruptured.

Fractures of the floor of the anterior cranial fossa *may involve the cribriform plate of the ethmoid bone (Fig. 7-14), resulting in leakage of CSF through the nose* (**CSF rhinorrhea**). In these cases the subarachnoid space is in communication with the outside via the nose as the result of a tear in the meninges.

CSF otorrhea and CSF rhinorrhea present a risk of **meningitis** because an infection may spread from the ear or the nose, especially if the nose is blown hard.

BLOOD SUPPLY OF THE BRAIN

The brain is supplied through an extensive system of branches from two pairs of vessels, the internal carotid arteries and the vertebral arteries (Figs. 7-37, 7-53, 7-62, 7-68, 7-69, 7-77, and 7-85).

THE INTERNAL CAROTID ARTERIES (Figs. 7-37, 7-68, 7-69, 7-77, 7-84, and 7-85)

Each internal carotid arises in the neck from the common carotid artery opposite the superior border of the thyroid cartilage.

The cervical part of the internal carotid artery ascends almost vertically to the base of the skull (Fig. 7-37), where it turns and enters *the carotid canal* in the petrous part of the temporal bone (Fig. 7-62).

The petrous part of the internal carotid artery enters the middle cranial fossa through the superior portion of the **foramen lacerum** (Fig. 7-63), and then runs anteriorly *in the cavernous sinus* (Figs. 7-80 and 7-84).

The cavernous part of the internal carotid artery is covered by the endothelium of this sinus. *At the anterior end of the cavernous sinus, the internal carotid artery makes a hairpin turn* and leaves the sinus to enter the subarachnoid space (Figs. 7-80, 7-84, and 7-85).

The cerebral part of the internal carotid artery (supracavernous or supraclinoid part) immediately gives off the important **ophthalmic artery**, which supplies the eye (Figs. 7-77*A*, 7-80, and 7-99).

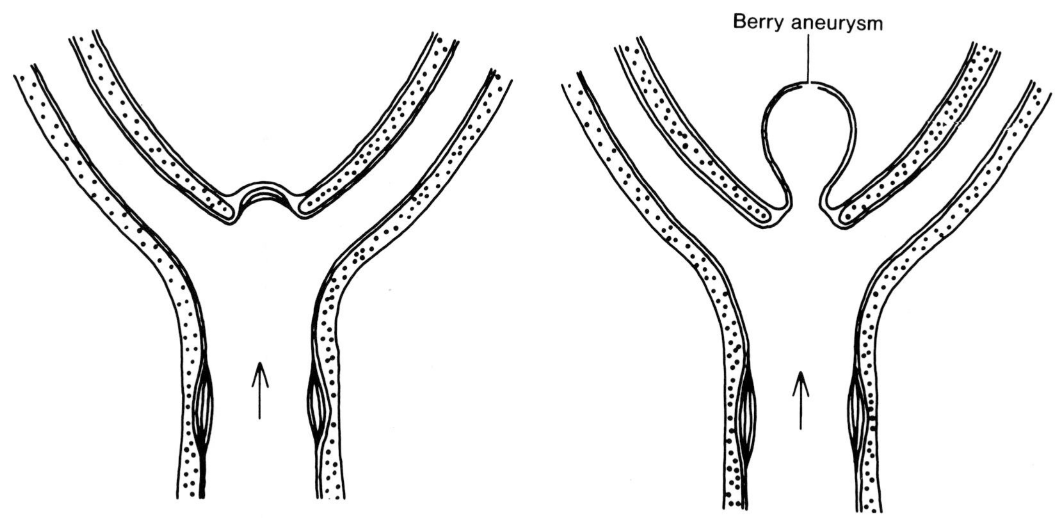

Figure 7-75. Diagrams illustrating the mechanism of development of a *berry aneurysm* at the site of a weakness in an arterial wall, usually at the bifurcation of an artery. Evagination or outpouching of the wall gradually occurs producing a saccular aneurysm or localized dilation.

The internal carotid artery then passes inferior to the optic nerve. Finally it turns obliquely superiorly, lateral to the **optic chiasma** (Figs. 7-62 and 7-69) for a variable distance, before branching into the **anterior and middle cerebral arteries** at the medial end of the lateral sulcus (Figs. 7-77 and 7-85).

The sinuous course taken by the cavernous and cerebral parts of the internal carotid artery forms a U-shaped bend (Figs. 7-84 and 7-85), often called the "**carotid siphon**" (G. a bent tube). Within the cranial cavity, the internal carotid artery and its branches supply the **hypophysis cerebri** (pituitary), the **orbit**, and much of the supratentorial **part of the brain** (Figs. 7-53, 7-54, and 7-77).

THE VERTEBRAL ARTERIES (Figs. 5-36, 7-37, 7-53, 7-77, and 7-78)

Each vertebral artery begins in the root of the neck as a branch of the first part of the subclavian artery (Fig. 7-37). It ascends vertically through the foramina transversaria of C6 to C3 vertebrae, and then *inclines laterally in the foramen transversarium of C2* (Fig. 7-78). Superior to this foramen, the vertebral artery ascends vertically into the foramen transversarium of C1 (Fig. 7-78). It then bends posteriorly at right angles and *winds around the superior part of the lateral mass of the atlas* (Figs. 5-36 and 7-37).

The vertebral artery pierces the posterior atlantooccipital membrane (Figs. 5-34 and 5-36), the dura mater, and the arachnoid. It *enters the **subarachnoid space** of the cerebellomedullary cistern* at the level of the foramen magnum (Fig. 7-71). The vertebral artery runs anteriorly on the anterolateral surface of the medulla and unites with its fellow of the opposite side at the caudal border of the pons to form the basilar artery.

THE BASILAR ARTERY (Figs. 7-77 and 7-86)

This artery is **formed by the union of the two vertebral arteries.** *It runs within the pontine cistern* (Figs. 6-71 and 6-86) to the superior border of the pons, where it **ends by dividing into the two posterior cerebral arteries** (Fig. 7-77).

To investigate the vertebral and basilar arteries using **angiography** (Fig. 8-21), injections of contrast material are made into them by various routes, *e.g.*, via a femoral catheter advanced into the inferior part of the vertebral artery, or by retrograde injections into the brachial artery in the cubital fossa (Fig. 6-74).

Occlusion or stenosis of one vertebral artery results in the brain stem being dependent mainly on the other vertebral artery for its blood supply. If that supply is reduced by injury, brain stem symptoms may occur such as dizziness, fainting, spots before the eyes, and transient **diplopia** (double vision).

THE CEREBRAL ARTERIAL CIRCLE (Figs. 7-62 and 7-77)

The cerebral arterial circle or circulus arteriosus cerebri (circle of Willis) is **an important anastomosis between the four arteries that supply the brain** (the two vertebral and the two internal carotid

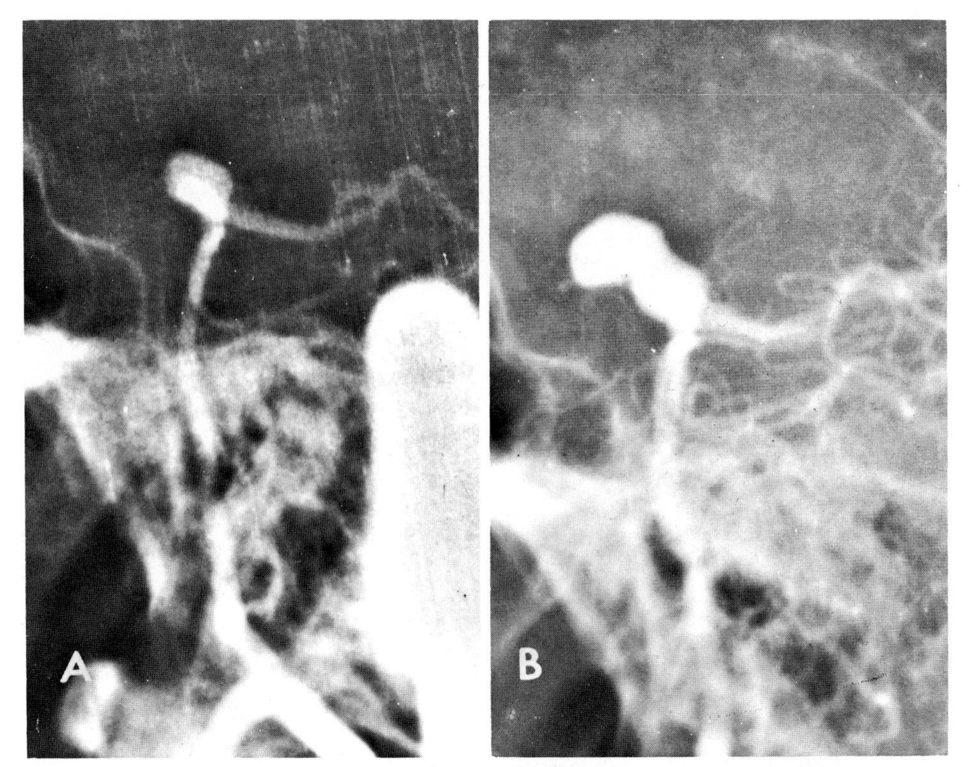

Figure 7-76. Cerebral angiograms demonstrating a berry aneurysm at the bifurcation of the basilar artery into the two posterior cerebral arteries. *A*, initial examination. *B*, 3 months later. Note its enlargement which probably indicates deterioration of the wall of the aneurysm and danger of rupture.

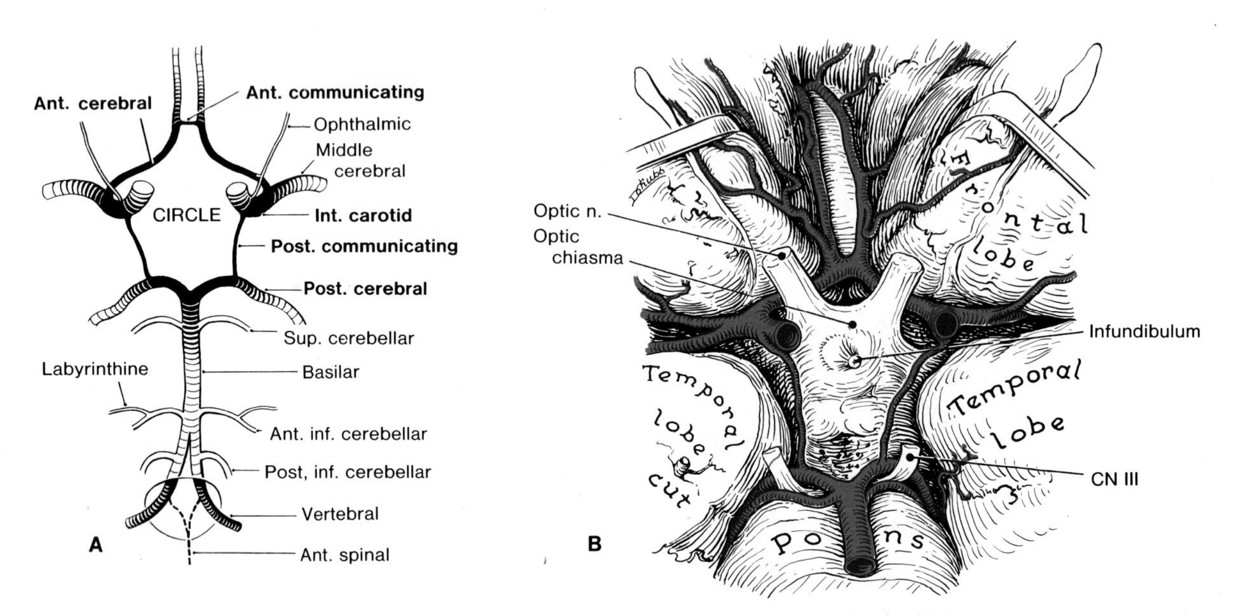

Figure 7-77. *A*, diagram showing the two internal carotid arteries and the two vertebral arteries that supply the brain. The internal carotids enter the cranium through the carotid canals in the temporal bones; the vertebral arteries enter through the foramen magnum. The names of *the arteries contributing to the circulus arteriosus cerebri appear in* **bold face** . Note that the middle cerebral artery is not a part of the arterial circle and that it is the continuation of the internal carotid artery. *B*, drawing of the anastomosis of major arteries supplying the brain. They join together at the base of the brain to form the cerebral arterial circle. This diagram shows the classical arterial circle, similar to that drawn by Sir Christopher Wren in 1664, but there are many variations of this typical configuration. The cerebral arterial circle was first described by Dr. Thomas Willis, an English physician. Many clinicians still refer to it as the "the circle of Willis," but this terminology is not recommended by Nomina Anatomica.

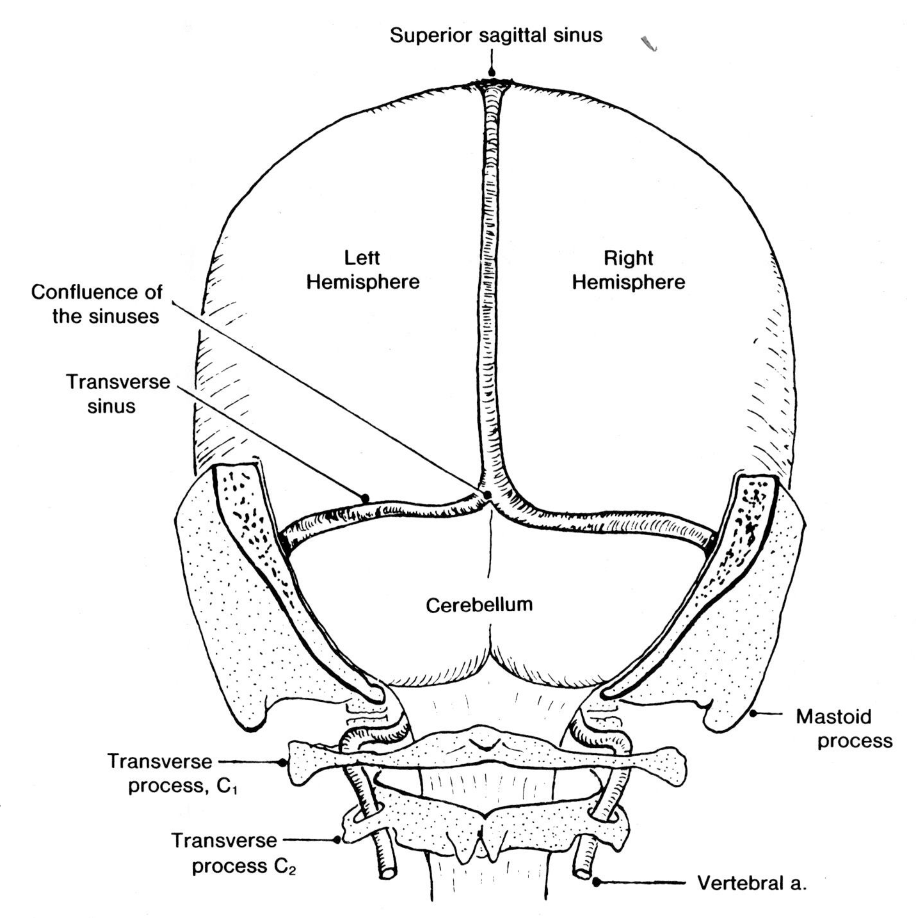

Figure 7-78. Schematic drawing of a posterior view of some of the venous sinuses of the dura mater. The calvaria and a large wedge of the occipital bone have been removed. The straight and occipital sinuses (not shown) also enter the confluence of sinuses (confluens sinuum). Observe the vertebral arteries passing through the foramina transversaria in the transverse processes of the axis (C2) and the atlas (C1). Note that once they pass through these foramina they turn posteriorly and then medially. They enter the skull through the foramen magnum and join to form the basilar artery (Fig. 7-77).

arteries). It is formed by the posterior cerebral, posterior communicating, internal carotid, anterior cerebral, and anterior communicating arteries. (Fig. 7-77).

The cerebral arterial circle is located at the base of the brain, **principally in the interpeduncular fossa** (Figs. 7-71, 7-72, and 7-77). The arterial circle extends from the superior border of the pons to the longitudinal fissure between the cerebral hemispheres (Fig. 7-72).

In Figures 7-62 and 7-77, observe that the cerebral arterial circle *encircles the* ***optic chiasma****, the* ***infundibulum****, and the* ***mammillary bodies*** (Fig. 7-72). Two types of branches, central and cortical, arise from the cerebral arterial circle and the main cerebral arteries.

Central arteries penetrate the substance of the brain and supply deep structures (*e.g., the* ***basal nuclei).***

The cortical branches pass in the pia mater and supply the more superficial parts of the brain. In general each of the cerebral arteries, (anterior, middle, and posterior) supplies a surface and a pole of the brain as follows: (1) the **anterior cerebral artery** supplies most of the *medial and superior surface and the frontal pole*; (2) the **middle cerebral artery** supplies the *lateral surface and the temporal pole*; and (3) the **posterior cerebral artery** supplies the *inferior surface and the occipital pole.*

The anterior, middle, and posterior cerebral arteries communicate with each other via small terminal branches. These important anastomoses occur on the surface of the brain (Fig. 7-85). If occlusion of one of the three major cerebral

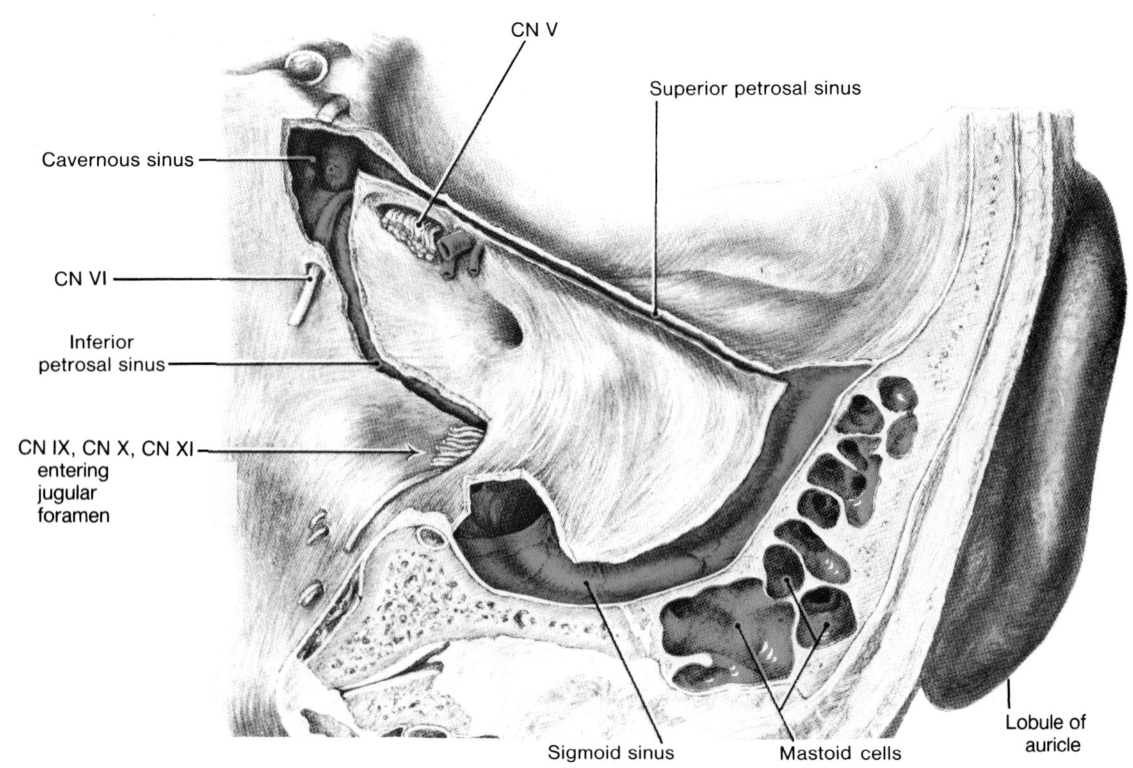

Figure 7-79. Drawing of a dissection of the anterior wall of the right posterior cranial fossa, as viewed posteriorly. Some of the venous sinuses of the dura mater are illustrated. Note that the posterior surface of the petrous part of the temporal bone is encircled by three sinuses: sigmoid, superior petrosal, and inferior petrosal. These sinuses have no valves. The two petrosal sinuses drain the cavernous sinus. Observe the mastoid cells (air-filled spaces) lined with mucus membrane (*pink*) in the diploë of the mastoid process of the temporal bone. Observe the sigmoid sinus, cranial nerves IX, X, and XI, and the inferior petrosal sinus entering the jugular foramen.

arteries occurs, the remaining two patent ones carry some blood to the area that would otherwise be bloodless. Unfortunately, not enough blood is supplied by these collateral routes in most cases to prevent **infarction** of at least part of the area supplied by the occluded artery. The effect of the infarction depends on the area involved, *e.g.*, a lesion involving the motor area results in weakness of the opposite side of the body.

In the typical cerebral arterial circle (Fig. 7-77*B*), there is usually little exchange of blood between the cerebral arteries via the anterior and posterior communicating arteries. However, **the cerebral arterial circle forms an important means of collateral circulation in the event of obstruction of one of the major arteries entering the circle.** In elderly persons, the anastomoses forming the circle are often inadequate when there is a **sudden occlusion of a large artery**, such as the internal carotid. Consequently, a stroke occurs.

Vascular insufficiency of the brain usually results in irreversible neurological damage in about 5 minutes. If the whole brain is involved, (*e.g.*, resulting from cardiac arrest) unconsciousness occurs in about 10 seconds.

For radiographic diagnostic purposes, radiopaque material may be injected into the carotid or vertebral arteries in the neck by direct **arterial puncture**, or by catheterization of them from a distant site such as the femoral artery. Thus, the condition of the extracranial and intracranial carotid and/or vertebral artery systems can be studied (Fig. 7-85).

Variations in the size of the vessels forming the arterial circle and in the configuration of the circle are common, e.g., **the posterior cerebral artery is a branch of the internal carotid in about 20% of persons.** Sometimes one anterior cerebral artery is very small and the anterior communicating artery is larger than usual to compensate for this.

A stroke or cerebrovascular accident

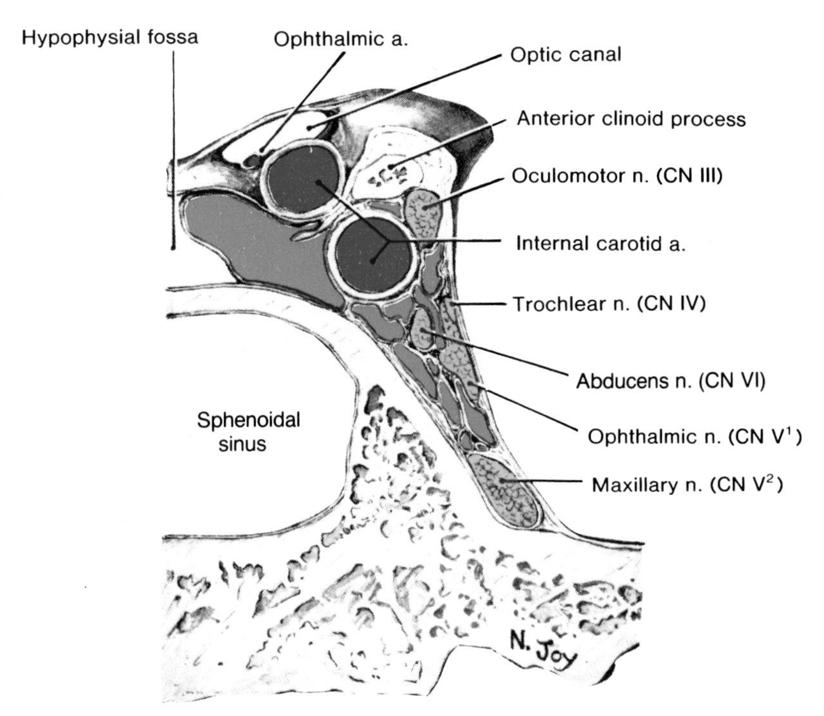

Figure 7-80. Drawing of a coronal section of the cavernous sinus. Observe that this dural venous sinus is situated at the side of the sphenoidal sinus and of the hypophysial fossa. Note that cranial nerves III, IV, V^1, and V^2 are in a sheath in the lateral wall of the cavernous sinus. Observe that the internal carotid artery and the abducens nerve pass through the cavernous sinus and so are vulnerable when there is thrombosis of the cavernous sinus. Note that the internal carotid artery, having made a hairpin bend, is cut twice, once in the cavernous sinus and again superior to the cavernous sinus in the subarachnoid space. Refer to Figure 7-39 for help in interpreting this drawing.

(CVA) results from either *sudden hemorrhage into the brain, or sudden stoppage of the blood supply to a part of the brain.* The term "**stroke**" was coined to indicate the abruptness with which symptoms appear following the arrest or sudden insufficiency of circulation in an artery supplying a part. Death of nervous tissue, called **infarction** occurs in the area of brain supplied by the blocked artery.

Hemorrhagic stroke follows from rupture of an arteriosclerotic artery or an **aneurysm**, resulting in bleeding into the brain substance (Figs. 7-75 and 7-76).

Thrombotic stroke results from ***thrombosis*** (G. a clotting) which is the formation of a **thrombus or clot** in an artery supplying part of the brain. Thrombotic stroke may also result from an **embolus** (G. a plug) which is a mass of material that is carried from a distant site by the blood stream to a small artery.

Emboli may be (1) ***blood clots*** (usually from the heart), (2) ***masses of atheromatous material*** (from ulcerated atheromata in large or medium sized arteries), (3) ***aggregations of platelets*** (swept off the walls of medium sized arteries), or (4) ***gas bubbles***, either nitrogen (in divers or tunnel workers) or air (when large veins of the head and neck are opened at operation or during an accident).

Temporary strokes or transient ischemic attacks (TIAs) often result from soft emboli which gradually disintegrate and/or slowly pass through small arteries producing ***temporary neurological deficits***. Such attacks warrant immediate investigation to determine if a remedial cause is present that can be removed before a second, possibly permanent, episode occurs.

Localized abnormal dilation of arteries supplying the brain are the most commonly encountered aneurysms in the body. Usually the lesions are located where division points occur in the arteries at the base of the brain (Figs. 7-75 and 7-76). Pathologists classify aneurysms according to their form as saccular, fusiform, or tubular. ***Most aneurysms develop as the result of***

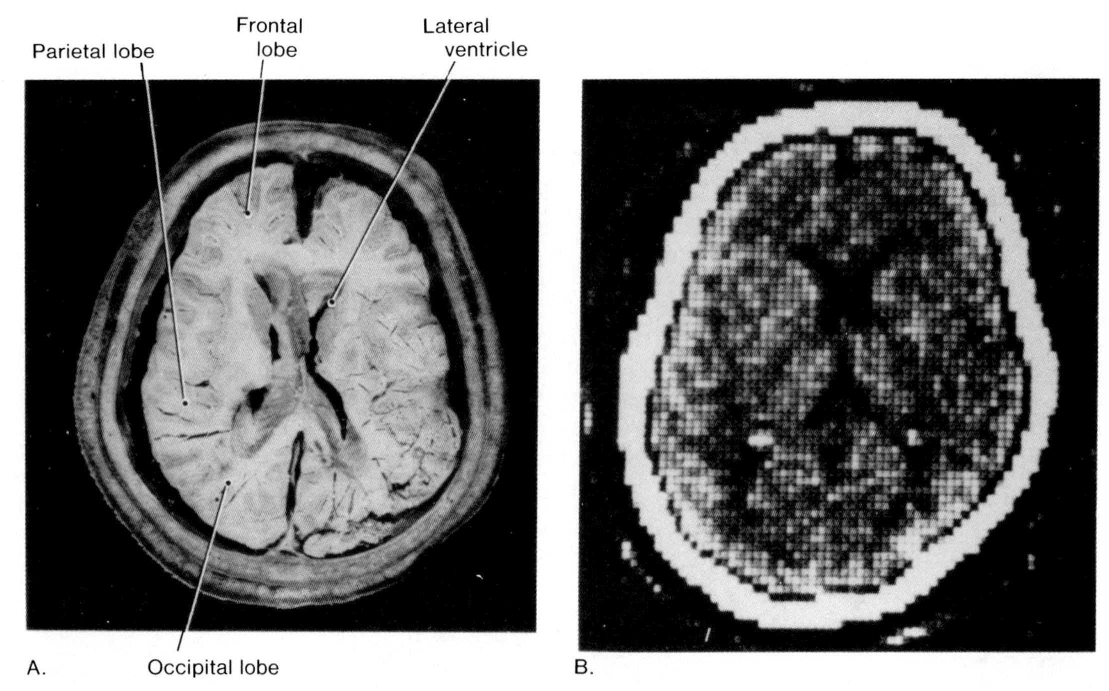

Figure 7-81. *A*, horizontal section through the head of a cadaver. *B*, computerized tomograph (CT scan) of a living person's head scanned in a horizontal plane at a more superior level. Note (1) the reduced density (*dark*) in the area of the cerebral ventricles; (2) the reduced density in the subarachnoid spaces outside the brain; (3) the dense (*white*) skull outline; and (4) the intermediate density of the brain.

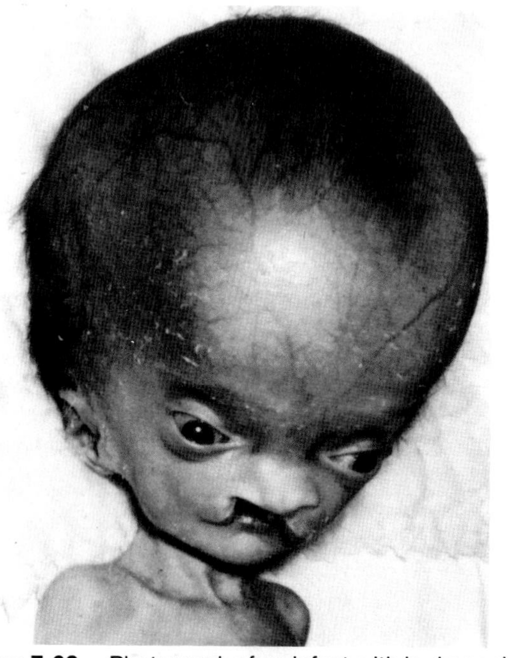

Figure 7-82. Photograph of an infant with hydrocephalus (excess of CSF) and bilateral cleft lip. (From Moore KL: *The Developing Human: Clinically Oriented Embryology*, ed 3. WB Saunders Co, 1982.)

congenital weakness of the arterial wall, permitting a localized berry-like evagination to occur.

The most common type of aneurysm is the saccular **berry aneurysm**, arising from the cerebral arterial circle and the medium-sized arteries at the base of the brain (Figs. 7-75 to 7-77). In time, especially in persons with high blood pressure (**hypertension**), the weak part of the wall of the artery expands and may rupture.

Rupture of a berry aneurysm results in blood escaping into: (1) the subarachnoid space (***spontaneous subarachnoid hemorrhage***); (2) the brain substance (***intracerebral hemorrhage***) or; (3) into the subdural space (***subdural hemorrhage***). The sites where the blood collects and forms a hematoma depend on the relationship of the aneurysm to the cranial meninges.

Unruptured cerebral aneurysms are usually asymptomatic, but intermittent enlargements of them may cause symptoms (*e.g.*, severe throbbing headaches) and sometimes neurological signs (*e.g.*, **third nerve palsy** associated with an aneurysm of the posterior communicating artery).

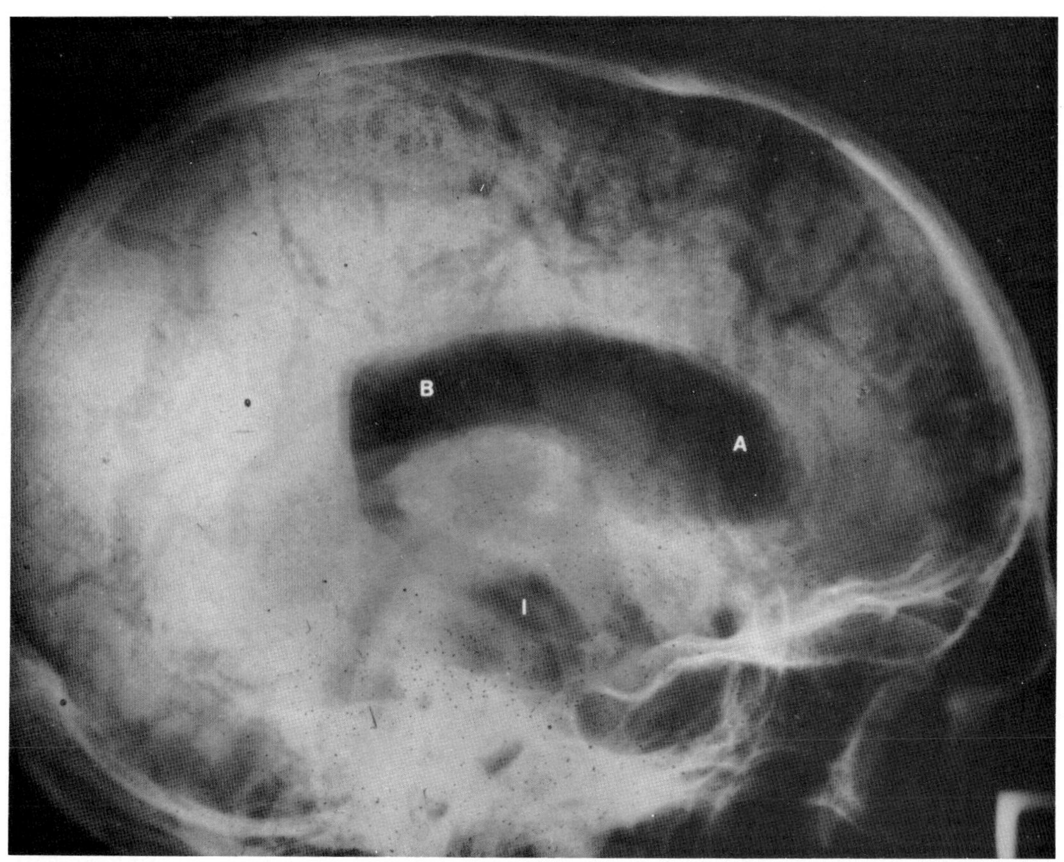

Figure 7-83. Brow-up lateral radiograph of the head after lumbar injection of air (*pneumoencephalogram*). Compare with Figure 7-70*A*. Note that the anterior horns (*A*) and bodies (*B*) of both lateral ventricles are filled with air and that the right and left images overlap. One inferior horn (*I*) is also filled with air. The posterior horns are filled with fluid because the patient is in the *supine position* (lying on the back); if the patient is put in the *prone position* (lying face down), air will fill these horns. Air is also present in the third ventricle, the cerebral aqueduct, and the fourth ventricle, but is difficult to see because these parts of the ventricular system are narrow. Note that some air is visible in the basal cisterns and in some frontal sulci.

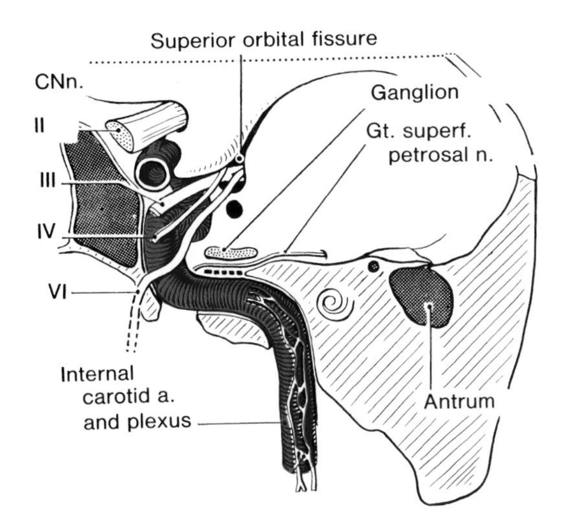

Figure 7-84. Drawing showing the course of the internal carotid artery in the temporal bone. Note that it takes an inverted L-shaped course from the inferior surface of the petrous part of the temporal bone to its apex. At the superior end of the foramen lacerum, its internal carotial artery enters the cavernous sinus, turns anteriorly, makes a hairpin turn, and enters the subarachnoid space.

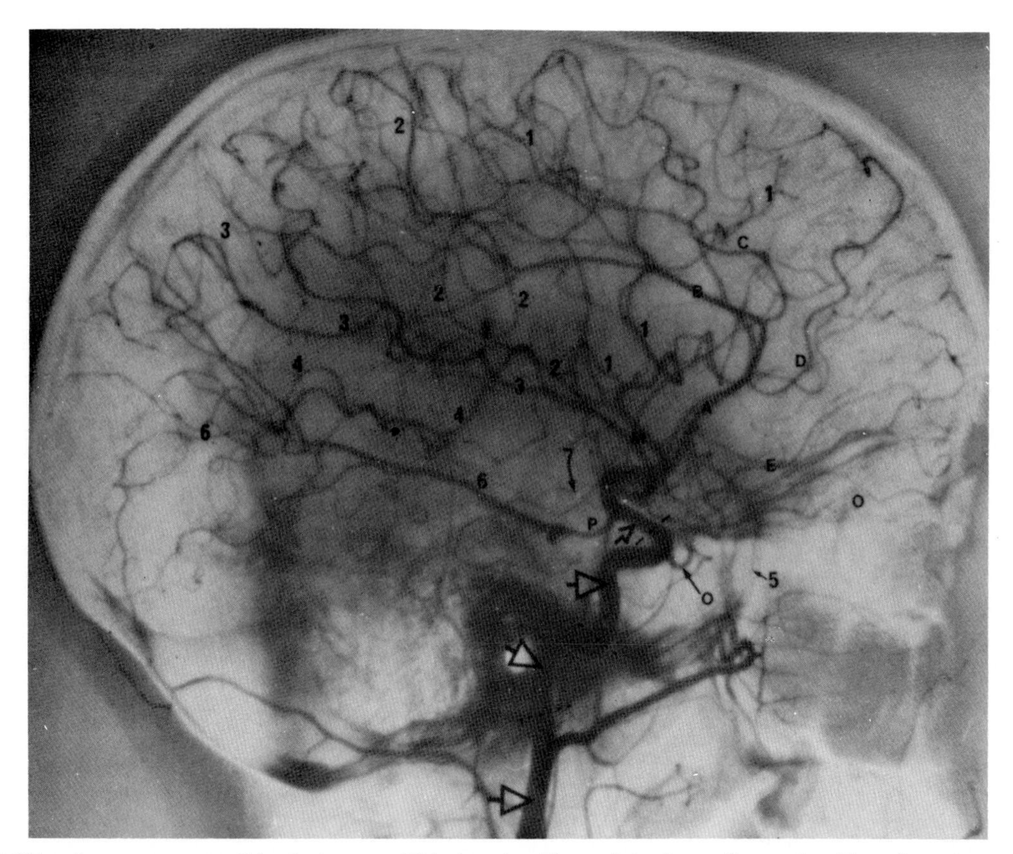

Figure 7-85. A common carotid arteriogram. This is a positive print of a radiograph with radiopaque material in the internal carotid artery. (Some external carotid branches are also visible.) *Four arrows* indicate parts of the internal carotid: *cervical*, before entering the skull; *petrous*, within the petrous part of the temporal bone; *cavernous*, within this venous sinus (note the hairpin turn inferior to the broken line); and *supracavernous* or cerebral, within the cranial subarachnoid space. *A*, anterior cerebral artery; *M*, middle cerebral artery; *P*, other posterior communicating artery connecting the internal carotid to the posterior cerebral artery. The other letters and numbers refer to small branches of arteries that are of concern to neurosurgeons and neuroradiologists.

Third nerve palsy results in ptosis, dilated pupil, and a divergent squint (Fig. 7-87). The throbbing headaches probably result from stretching of the meninges, but in some cases they may result from slight bleeding of the aneurysm into the subarachnoid space.

Sudden rupture of a cerebral aneurysm usually produces a very sudden severe, almost unbearable headache owing to gross bleeding into the subarachnoid space. The subarachnoid hemorrhage produces a sudden increase in intracranial pressure and **meningeal irritation**. The pain may be so severe that the patient is maniacal (wild, furious).

*Rupture of a berry aneurysm is one of the causes of a "**stroke.**"*

THE ORBIT

The orbits of the skull appear as two **bony recesses or sockets**, when viewed from the anterior aspect (Figs. 7-1, 7-2, and 7-88). The orbits almost surround the eyes (eyeballs, globes), protecting them and their associated muscles, nerves, and vessels, together with most of the lacrimal (tear) apparatus (Fig. 7-90). The orbital aperture is quadrilateral.

The orbit is shaped somewhat like a four-sided pyramid lying on its side, with its apex pointing posteriorly and its base anteriorly. It has a roof, a floor, a medial wall, and a lateral wall. Each of these areas is roughly triangular.

The bones forming the orbit are lined with periosteum, called **the periorbita** (orbital periosteum). At the optic canal and superior orbital fissure, the periorbita is continuous with the periosteum lining the in-

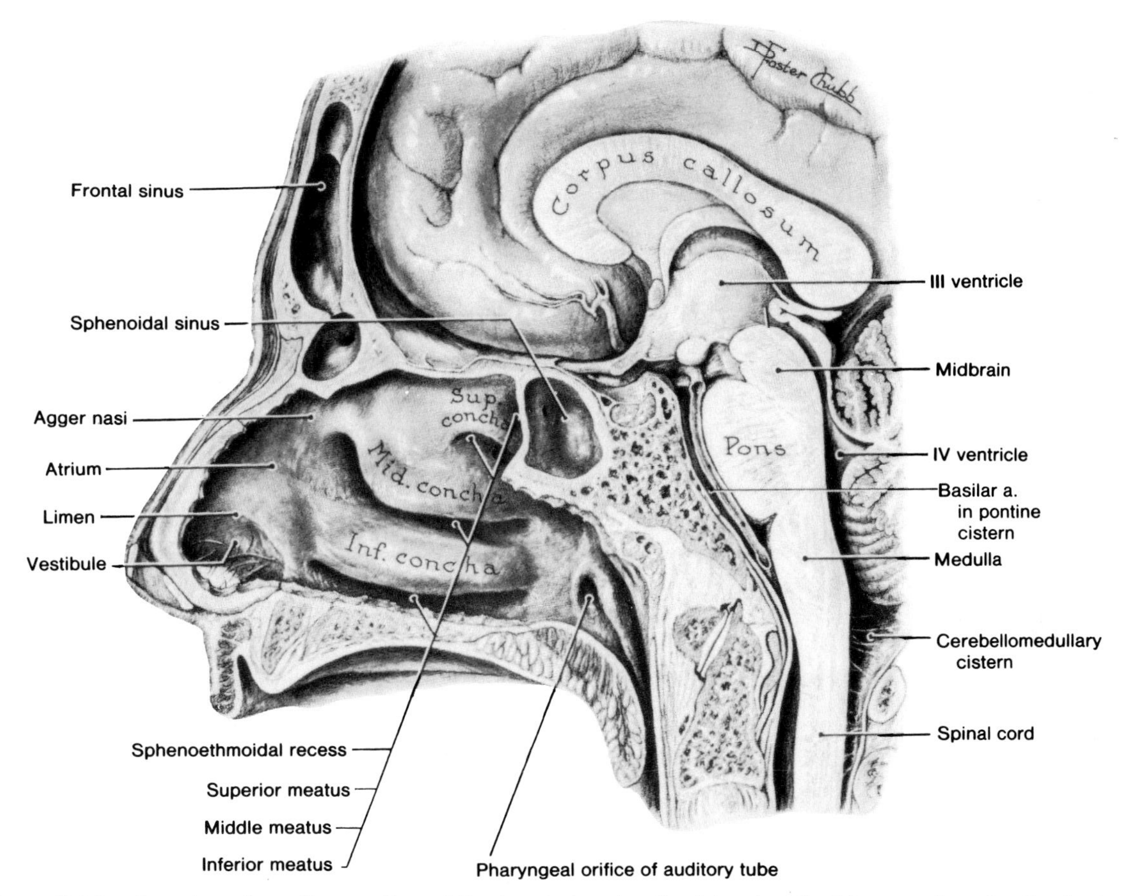

Figure 7-86. Drawing of median section of the head showing the lateral wall of the nasal cavity, the brain, and the spinal cord. The midbrain, pons, and medulla collectively are referred to as the brain stem. Note that the basilar artery in the pontine cistern has been sectioned longitudinally. The sphenoethmoidal recess is a small cleft-like pocket superior to the superior concha in which are found the openings of the sphenoidal air cells making up the sphenoidal sinus (also see Fig. 7-154).

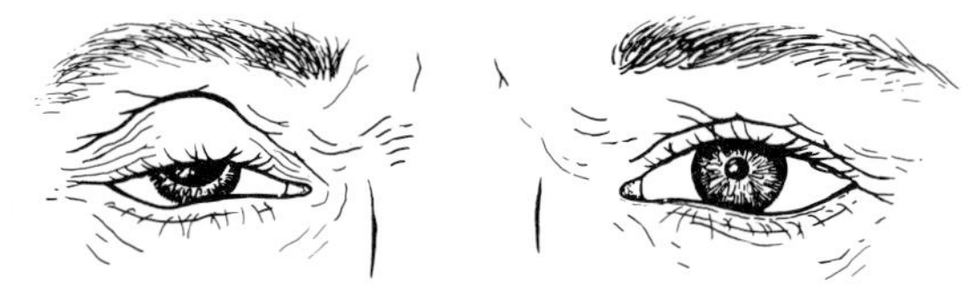

Figure 7-87. Drawing showing *right third nerve palsy*. Observe that her right upper eyelid droops (**ptosis**) owing to paralysis of the levator palpebrae superioris. Also observe that her *right pupil is dilated* owing to paralysis of the pupilloconstrictor fibers of CN III.

terior of the skull (**endocranium**). The periorbita is continuous over the orbital margins and through the inferior orbital fissure with the periosteum covering the external surface of the skull (**pericranium**).

The periorbita forms a funnel-shaped sheath which encloses the orbital contents. It is tough and may be easily detached, especially from the roof and the medial wall of the orbit. This is important from a surgical viewpoint.

THE ORBITAL MARGIN

*The **frontal, maxilla, and zygomatic bones** contribute about equally to the formation of the orbital margin* (Figs. 7-2 and 7-88). The bone of the orbital margin is strong.

The supraorbital margin is formed entirely by the frontal bone. At the junction of its medial and middle thirds is the *supraorbital foramen* (or notch, as the case may be), which *transmits the supraorbital nerve and vessels* (Fig. 7-28).

The lateral orbital margin is formed almost entirely by the frontal process of the zygomatic bone (Figs. 7-2 and 7-88). The zygomatic bone laterally and the maxilla medially share in the formation of the *infraorbital margin*. Both these margins are sharp and can easily be palpated.

The *medial orbital margin* is formed superiorly by the frontal bone and inferiorly by the *lacrimal crest* of the frontal process of the maxilla (Figs. 7-2 and 7-88). The medial orbital margin is sharp and distinct in its inferior half only.

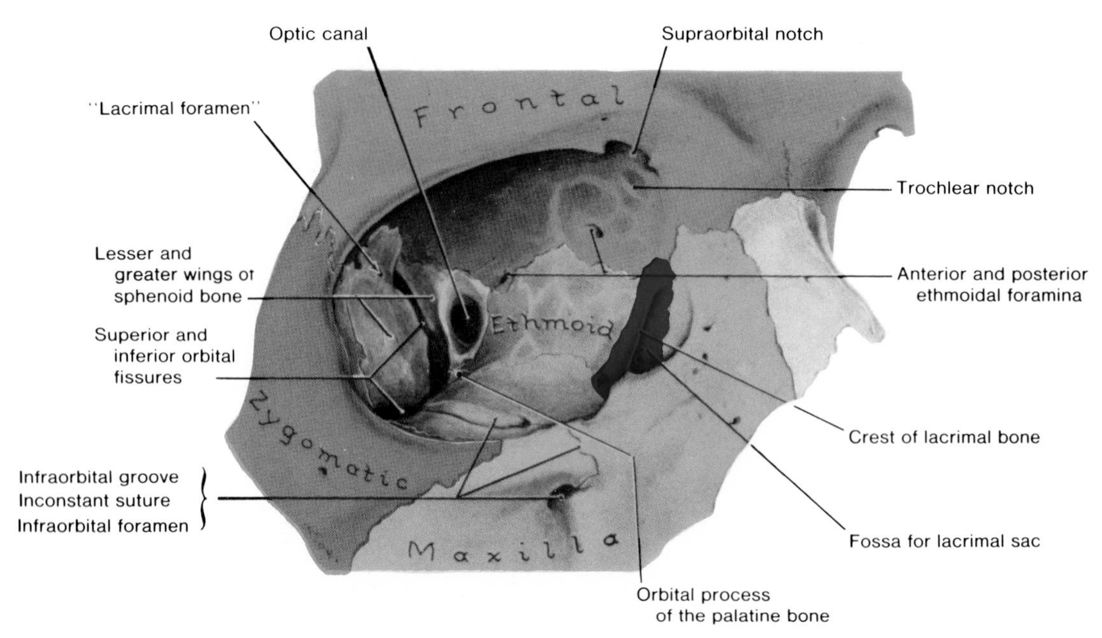

Figure 7-88. Drawing of the bony walls of the orbit (orbital cavity). The orbital margin (brim) surrounds its entrance. Note that the supraorbital margin is formed entirely by the frontal bone and that the lateral margin is formed almost entirely by the frontal process of the zygomatic bone, but is completed superiorly by the zygomatic process of the frontal bone. The zygomatic bone and the maxilla share in the formation of the infraorbital margin, whereas the medial margin is formed by the frontal bone superiorly and the lacrimal crest of the frontal process of the maxilla inferiorly. The printed word *ethmoid* on the drawing covers the thin orbital lamina (plate) that separates the orbit from the ethmoidal sinuses, which are visible through it.

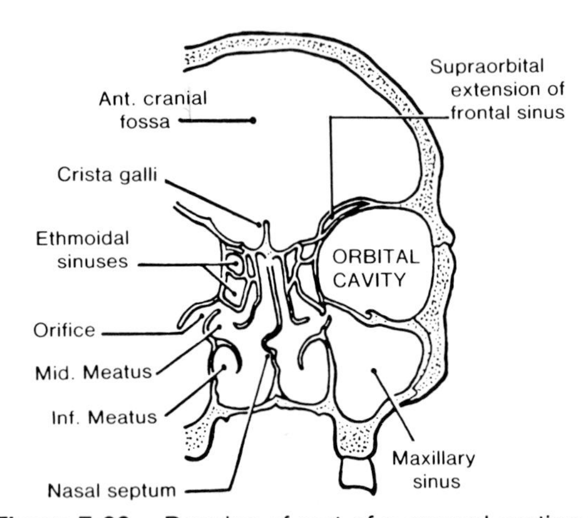

Figure 7-89. Drawing of part of a coronal section of the cranium showing the relations of the orbital cavity. Note that there is a supraorbital extension of the frontal sinus in this specimen. Note also that the aperture of the maxillary sinus is high on its medial wall. Secretions from this sinus drain to the nasal cavity. Because the nasal mucosa is continuous with the paranasal sinuses, nasal infection may spread to the sinuses. The opening of the maxillary sinus can be cannulated *in vivo* through the nostril. For more details, see the subsequent sections on the nasal cavities and the paranasal sinuses.

WALLS OF THE ORBIT

Each orbit has four walls: superior (roof), medial, inferior (floor), and lateral (Figs. 7-2, 7-88, and 7-89). The medial walls are almost parallel to each other, with the superior parts of the nasal cavities separating them, whereas the lateral walls are at approximately right angles to each other.

THE SUPERIOR WALL OF THE ORBIT (Figs. 7-2, 7-12 to 7-14, 7-88, and 7-89)

The superior wall or roof of the orbit is formed almost completely by the *orbital plate of the frontal bone*. Posteriorly the superior wall is formed by the lesser wing of the *sphenoid bone*.

The roof of the orbit is a thin, translucent, gently arched plate of bone which **intervenes between the orbital cavity and the anterior cranial fossa** (Figs. 7-51 and 7-89). Often there is an extension of the frontal sinus into the orbital part of the frontal bone (Figs. 7-4, 7-17, 7-89, and 7-90).

The optic canal (Fig. 7-88), located in the posterior part of the roof, transmits the **optic nerve** (CN II) and its meninges, along with the **ophthalmic artery** (Figs. 7-62, 7-69, 7-73, 7-90*A*, 7-91, and 7-99).

THE MEDIAL WALL OF THE ORBIT (Figs. 7-2, 7-88, and 7-89)

The paper-thin medial wall of the orbit is formed by the **orbital lamina** (lamina papyracea) of the ethmoid bone, along with contributions from the frontal, lacrimal, and sphenoid bones. The Latin word *papyraceus* means "made of papyrus" (parchment paper).

There is a vertical **lacrimal groove** in the medial wall of the orbit, which is formed anteriorly by the maxilla and posteriorly by the lacrimal bone (Fig. 7-88). It forms a fossa for the **lacrimal sac** and the adjacent part of the ***nasolacrimal duct*** (Fig. 7-96).

Along the suture between the ethmoid and frontal bones, are located two small foramina, the anterior and posterior **ethmoidal foramina** (Figs. 7-14 and 7-88) which transmit vessels and nerves of the same name.

THE INFERIOR WALL OF THE ORBIT (Figs. 7-2, 7-88, and 7-89)

The thin inferior wall or floor of the orbit is formed mainly by the orbital surface of the maxilla, and partly by the zygomatic bone and the orbital process of the palatine bone.

The inferior wall or floor of the orbit is partly separated from the lateral wall of the orbit by the **inferior orbital fissure** (Fig. 7-88) which transmits various structures, the most important of which is the **maxillary nerve** (Fig. 7-91). *The thin floor of the orbit forms the roof of the* ***maxillary sinus*** (Fig. 7-89).

THE LATERAL WALL OF THE ORBIT (Figs. 7-2, 7-88, and 7-89)

The lateral wall of the orbit is thick, especially the posterior part separating the orbit from the **middle cranial fossa** (Fig. 7-15). The lateral wall of the orbit is formed by the frontal process of the **zygomatic bone** (Fig. 7-2) and the greater wing of the **sphenoid bone** (Figs. 7-11 and 7-88).

Anteriorly the lateral wall lies between the orbit and the temporal fossa (Fig. 7-1). The lateral wall and the roof of the orbit are partially separated posteriorly by the **superior orbital fissure** (Figs. 7-2 and 7-88). This cleft communicates with the middle cranial fossa and transmits the **oculomotor** (CN III), **trochlear** (CN IV), and **abducens** (CN VI) nerves, and the terminal branches of the **ophthalmic nerve** (CN V^1), as well as the superior ophthalmic vein (Figs. 7-97 and 7-103).

The apex of the orbit is at the medial ends of the superior and inferior orbital fissures (Fig. 7-88). These fissures form a V-shaped area which encloses the orbital part of the greater wing of the sphenoid bone.

Owing to the thinness of the medial and inferior walls of the orbital cavity, a blow to the eye may produce a **blow-out fracture of the orbit** as the result of the sudden increase in intraorbital pressure. Because the medial wall of the orbit is so delicate, surgery of the ethmoidal sinuses (Fig. 7-89) must be performed with extreme care.

Although the superior wall (roof) of the orbit is stronger than the medial and inferior walls, it is thin enough to be translucent and may be readily penetrated. Thus a sharp object, even a pencil, may pass through it into the frontal lobe of the brain (Fig. 7-51).

One type of prefrontal **lobotomy or leukotomy** (G. *leukos*, white + *tome*, a cutting), used in the past in attempting to modify the behavior of severely psychotic patients, utilized this anatomical fact by driving a sterile instrument (**leukotome**) through the roof of the orbit and sectioning nerve fiber connections in the anterior part of the frontal lobe.

An object accidentally pushed through the anteromedial part of the roof of the orbit may penetrate the **frontal sinus** (Figs. 7-17 and 7-89), whereas an object pushed through its floor would enter the **maxillary sinus**. Similarly, a foreign object (*e.g.*, a piece of wire or a bullet) penetrating the eye in an anteroposterior direction could traverse the superior orbital fissure (Fig. 7-88), and enter the middle cranial fossa and the frontal lobe of the brain (Fig. 7-51). Because the petrous part of the temporal bone is so hard, it might stop the bullet so that it would remain in the middle cranial fossa.

Owing to the closeness of the optic nerve to the sphenoidal and posterior ethmoidal sinuses (Figs. 7-89 and 7-155), a **malignant tumor** in these sinuses may erode the thin bony walls and compress the optic nerve and orbital contents. Tumors in the orbit produce bulging of the eyeball (**proptosis or exophthalmos**). The easiest entrance to the orbital cavity for a tumor in the middle cranial fossa is through the **superior orbital fissure**, whereas tumors in the temporal or infratemporal fossa gain access to the orbital cavity through the **inferior orbital fissure**.

Although the lateral wall of the orbit is nearly as long as the medial wall, it does not reach so far anteriorly; thus, nearly 2.5 cm of the eyeball is exposed when the pupil is turned medially as far as possible. This anatomical fact is why the lateral side affords the best approach for operations on the eyeball.

A person attempting suicide by shooting through the anterior part of their temple could destroy the optic nerve and/or the eyeball, but miss the brain (Figs. 7-51 and 7-90 to 7-92).

THE ORBITAL CONTENTS

The most important contents of the orbit are the eyeball and the optic nerve, CN II (Figs. 7-73 and 7-90). The orbit also contains the muscles of the eyeball, their nerves and vessels, the lacrimal gland, and other nerves (Fig. 7-91).

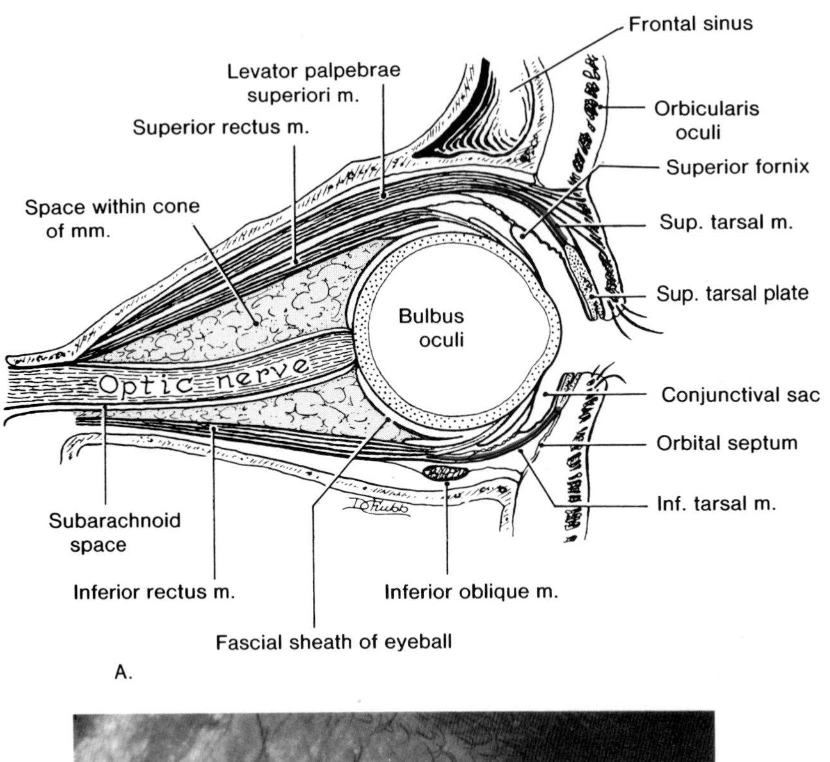

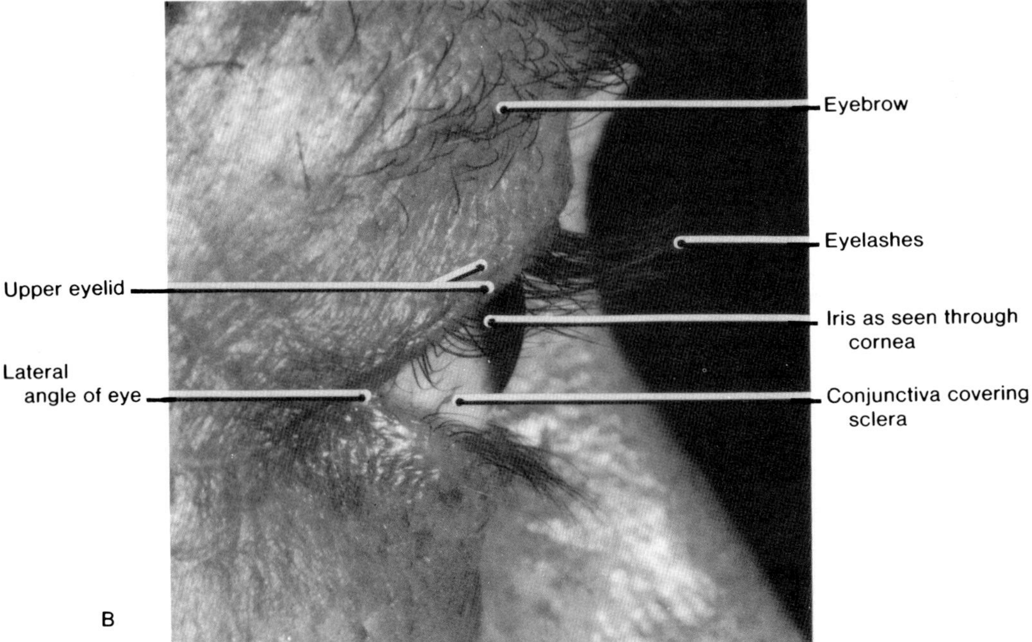

Figure 7-90. *A*, drawing of a sagittal section of the orbit and its contents. Note that the eyeball is sunken or retracted in this cadaveric specimen owing to embalming and loss of fluid. *B*, photograph of a lateral view of the eye of a 46-year-old man. The most prominent anterior bulge is produced by the transparent cornea, which becomes continuous at its margins with the white sclera.

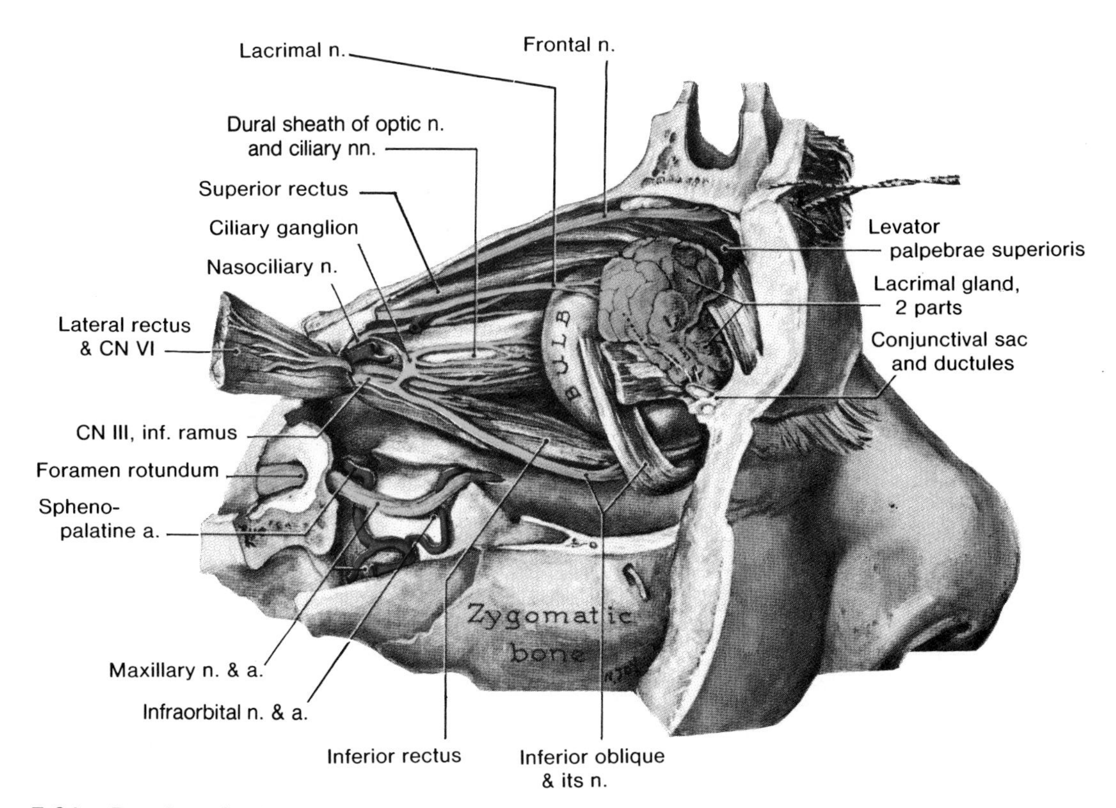

Figure 7-91. Drawing of a lateral view of a dissection of the orbit. Note in particular the 8 to 10 short ciliary nerves going to the eyeball, the orbital muscles, and the lacrimal gland. Compare with Figure 6-90.

TERMINOLOGY

The terms eye, eyeball, oculus, bulbus oculi, bulb, and globe all refer to the same structure, *i.e.*, the **visual organ**.

The eye is called **oculus** in Latin and **ophthalmos** in Greek; hence the terms *oculist* and *ophthalmologist* are interchangeable and indicate physicians who specialize in **ophthalmology**, the branch of medical science that deals with the eye, its diseases, and refractive errors.

You will encounter many other terms derived from these Latin and Greek words, *e.g.*, **oculomotor** (related to movements of the eye) and **ophthalmoscope** (an instrument for examining the interior of the eye).

THE LIVING EYE

The "white of the eye" (Figs. 7-90 and 7-92) or anterior aspect of the **sclera** (G. *sclēros*, hard) is continuous with the dural sheath (outer sheath) of the optic nerve and the **dura mater** covering the brain (Fig. 7-93). This tough opaque part of the external tunic of the eye appears slightly blue in infants and children and has a yellow hue in many older people.

The anterior transparent part of the eye is the **cornea**; it is continuous at its margins with the sclera (Figs. 7-92 and 7-93). The dark circular aperture that you see through the cornea is the **pupil**. This opening is surrounded by a circular, pigmented diaphragm known as the **iris**. The white sclera that you see in a mirror, or when looking at someone else's eye, is covered by a thin moist, mucous membrane called the **bulbar conjunctiva** (Figs. 7-92 and 7-102). As it is transparent, you can see the white sclera deep to it.

The bulbar conjunctiva is reflected off the sclera on to the deep surface of the eyelids (L. **palpebrae**), lining them to their margins. This **palpebral conjunctiva** becomes continuous with the skin of the eyelid (Fig. 7-92). As the palpebral conjunctiva is normally red and very vascular, it is commonly examined in cases of suspected **anemia**, a blood condition that is commonly manifested by pallor of the mucous membranes.

As the bulbar conjunctiva is continuous with the anterior epithelium of the cornea and with the palpebral conjunctiva, it forms a **conjunctival sac** (Figs. 7-90*A* and 7-91). The opening between the eyelids, called the **palpebral fissure**, is the mouth of the conjunctival sac; hence when the eyelids are closed, the bulbar and palpebral conjunctivae form a closed sac. Sensory innervation of the conjunctiva is from the trigeminal nerve (Figs. 7-28 and 7-32*C*) through the infratrochlear, maxillary, and lacrimal nerves.

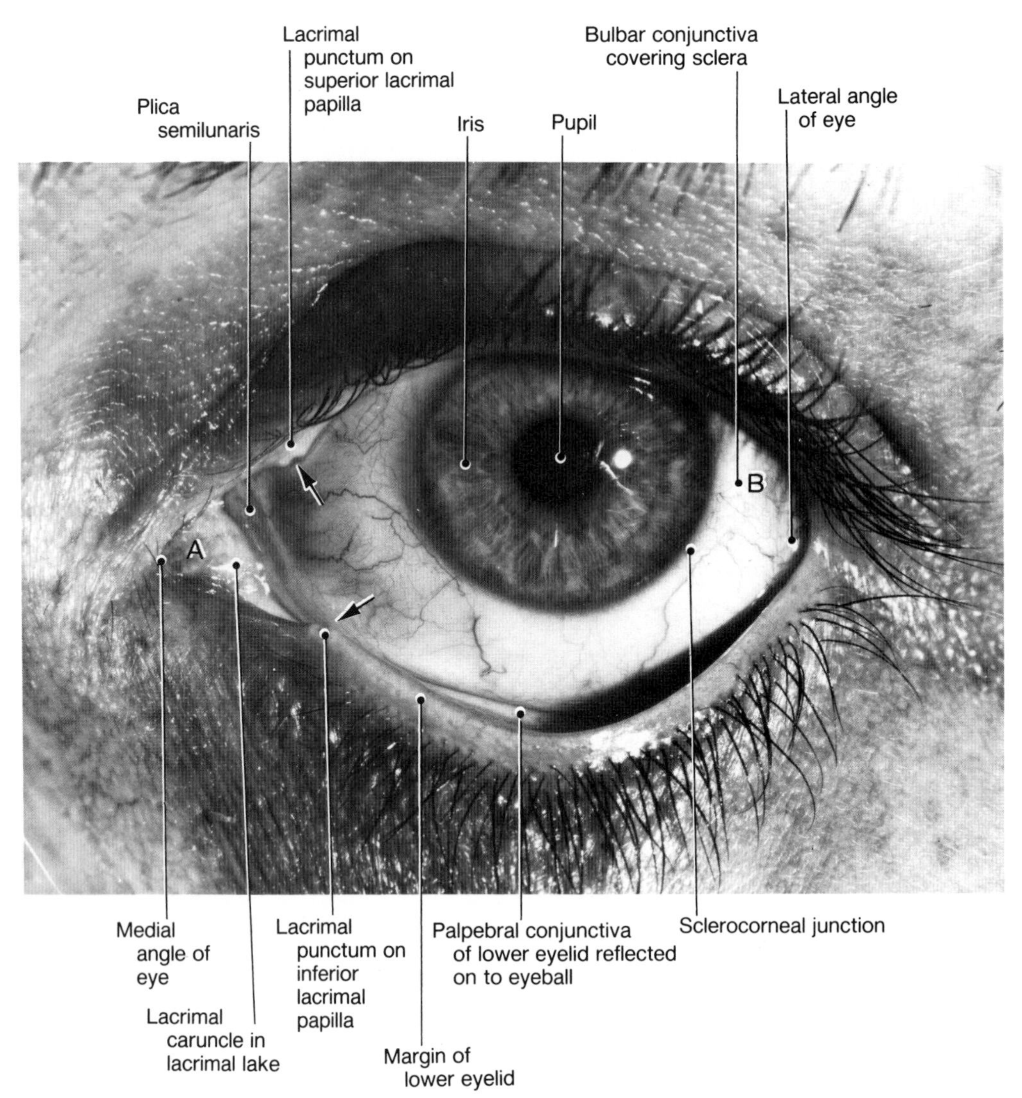

Figure 7-92. Photograph of the left eye of a 36-year-old woman with her eyelids slightly everted to show the structures at the medial angle or canthus. Note that the pupil, an aperture in the iris, appears black because you are looking through the pupil toward the pigmented posterior aspect of the eye. Understand that you are looking through the transparent cornea to see the iris and pupil. Observe the fine vascular network of the bulbar conjunctiva covering the sclera. The line of reflection of the palpebral conjunctiva from the eyelids on to the eyeball is called the conjunctival fornix; hence there are superior and inferior fornices (Fig. 7-90*A*). The white spot on the iris is a highlight from the photographer's lamp. Note that her upper eyelid partly covers her iris; this is normal. The *arrows* indicate the lacrimal puncta on the upper and lower lacrimal papillae. *A*, lacrimal caruncle in the lacrimal lake. *B*, ocular conjunctiva covering the sclera.

The bulbar conjunctiva is colorless, except when its vessels are dilated and congested (*e.g., bloodshot eyes*). This **hyperemia of the bulbar conjunctiva** is caused by local irritations (*e.g.,* dust, chlorine, and smoke) and infections (*e.g.,* pinkeye or *conjunctivitis*).

Subconjunctival hemorrhages are common and are manifested by bright or dark red patches deep to and in the bulbar conjunctiva. Subconjunctival hemorrhages may result from injury or inflammation. The conjunctiva is frequently inflamed owing to infection, a condition called conjunctivitis.

Acute conjunctivitis is common in children, especially in newborn infants who have prophylactic silver nitrate instilled into their eyes. This

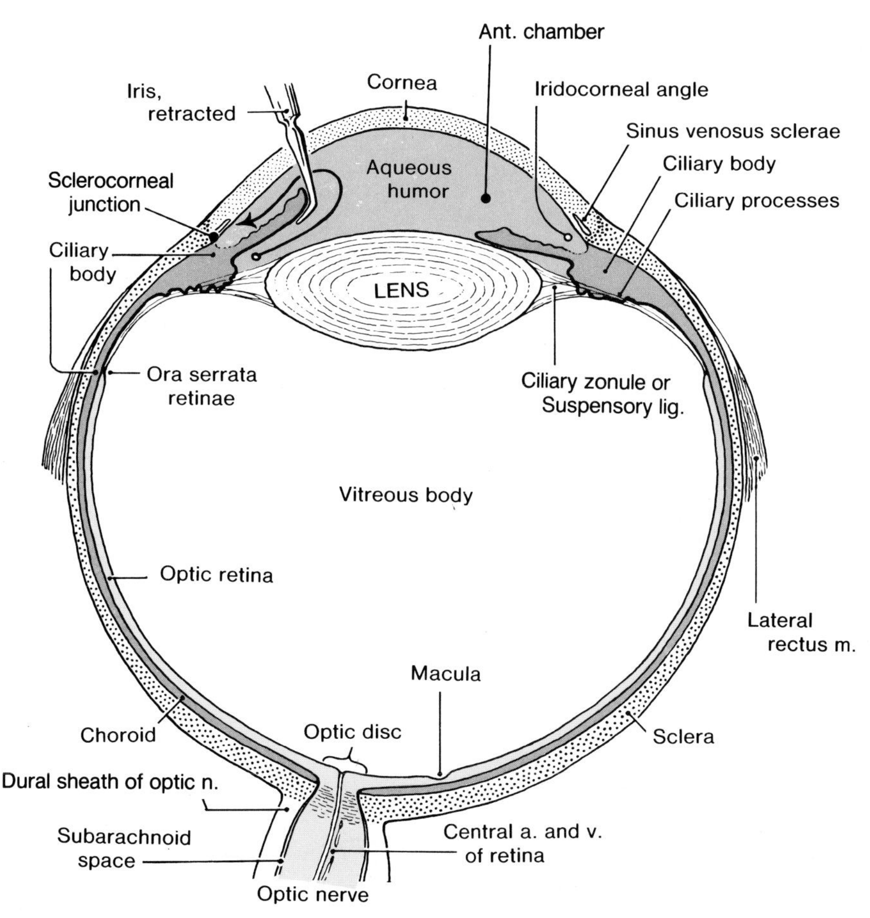

Figure 7-93. Drawing of a horizontal section of the eyeball. Observe its *three coats*: (1) external or fibrous coat (sclera and cornea); (2) middle or vascular coat (choroid, ciliary body, and iris); and (3) internal or retinal layer. The *four refractive media* are: (1) the cornea; (2) the aqueous humor; (3) the lens; and (4) the vitreous body. The *arrow on the left* indicates the flow of aqueous humor from the posterior chamber to the anterior chamber. The aqueous humor is a thin watery medium that is formed in the posterior chamber by the ciliary processes.

is a preventative measure in case they were exposed to venereal disease during birth.

Examine your eyes in a mirror. Note that from the anterior aspect (Fig. 7-92), most of the eyeball appears to be in the orbit, but from the side much of it protrudes between the eyelids, *i.e.*, through the palpebral fissure (Fig. 7-90*B*).

THE EYELIDS

The eyelids *protect the eyes from injury and excessive light and they keep the cornea moist.* Look into a mirror and slowly move your finger toward your eye. Note that your eyelids close when your finger gets close to your eye. Verify that the upper eyelid is larger and more movable than the lower one, and that the *upper eyelid partly covers the iris* (Figs. 7-90 and 7-92), whereas the entire inferior half of the eye is normally uncovered.

The eyelids are essentially movable folds that are covered externally by thin skin and internally by the *highly vascular palpebral conjunctiva* (Figs. 7-92 and 7-94). The palpebral conjunctiva is reflected onto the eyeball where it is continuous with the bulbar conjunctiva. The palpebral conjunctiva forms deep recesses known as the *superior and inferior fornices* (Fig. 7-90A).

Each eyelid is strengthened by a dense connective tissue band, about 2.5 cm wide, called the **tarsal plate or tarsus** (Figs. 7-28, 7-90*A*, and 7-94). In the connective tissue between this plate and the skin of the eyelid are fibers of the **orbicularis oculi muscle** (Figs. 7-29, 7-90*A*, and 7-94).

Embedded in the tarsal plates of the upper and lower eyelids are a number of **tarsal glands**, the fatty

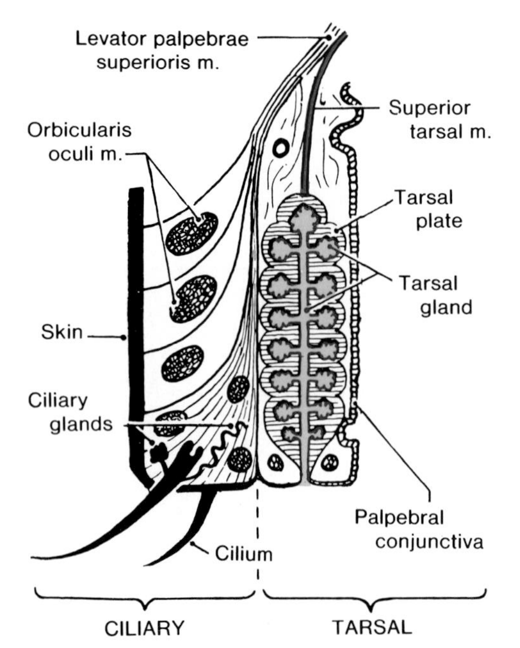

Figure 7-94. Drawing of a sagittal section of an upper eyelid. The eyelid is essentially a movable fold of skin which protects the eye from injury and excessive light.

secretion of which lubricates the edges of the eyelids, preventing them from sticking together and sealing them when the eyelids are closed.

The eyelashes (cilia) are in the margins of the

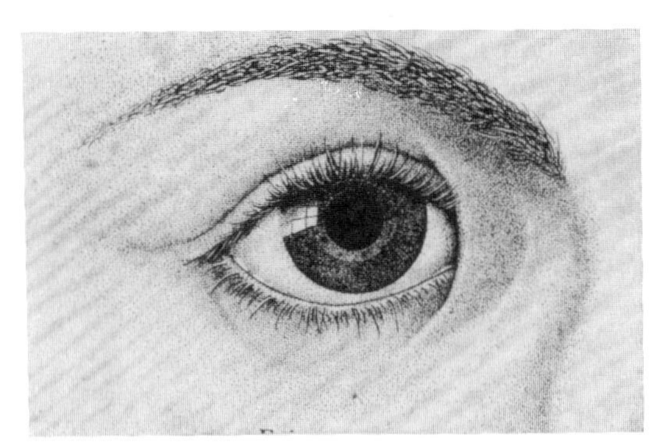

Figure 7-95. Drawing of the right eye of a person with a large inner epicanthal fold. Note that it obscures the medial canthus of the eye. The epicanthal fold gives an oblique appearance to the palpebral fissure. This fold is present in certain races, notably Mongolians. It is sometimes well formed in persons with the Down syndrome (trisomy 21).

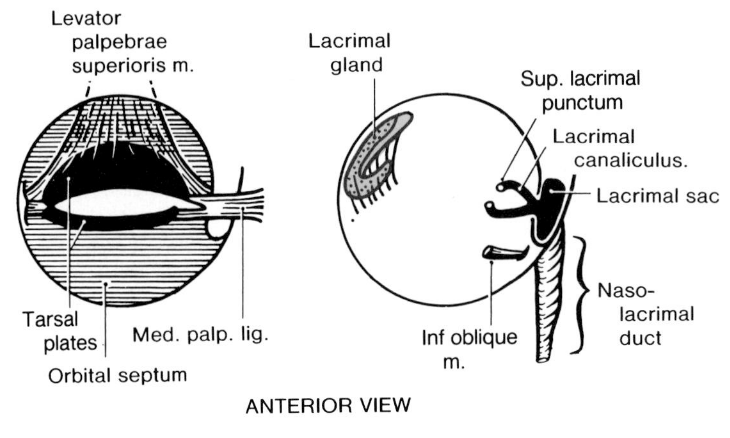

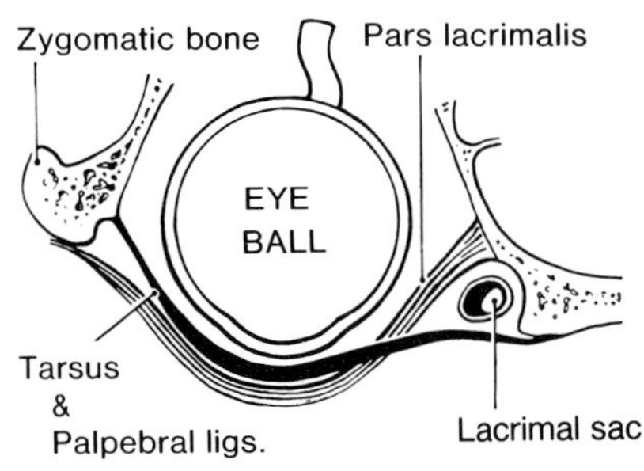

Figure 7-96. Drawings illustrating the parts of the lacrimal apparatus. Tears are secreted from the lacrimal gland into the superolateral angle of the conjunctival sac. After passing over the cornea, the tears drain into the lacrimal puncta in the upper and lower eyelids (also see Fig. 7-92). The puncta open into the lacrimal canaliculi (little canals) through which the lacrimal fluid is transported to the lacrimal sac. Medially the superior and inferior canaliculi usually meet to form a common canaliculus which carries the tears to the lacrimal sac (*upper right drawing*). This sac drains into the nasolacrimal duct, which empties into the inferior meatus of the nose (Fig. 7-86).

eyelids (Figs. 7-90 to 7-92 and 7-94) and are arranged in two or three irregular rows. The large sebaceous glands associated with the eyelashes are known as **ciliary glands** (Fig. 7-94). Between the hair follicles there are modified sweat glands.

The place where the two eyelids meet is called the **angle** or **canthus** (G. *kanthos*, corner of eye). Thus each eye has medial and lateral angles or canthi (Figs. 7-90*A* and 7-92). When the eyelids are closed the **palpebral fissure** is nearly in the horizontal plane, except in certain races (*e.g.*, Mongolian) where there is a slight superior slant of the palpebral fissure toward the nose because the medial ends of the upper eyelids project superomedially (Fig. 7-95). Furthermore, the medial canthi of their eyes are covered by an extra skin fold called the **epicanthal fold**, which varies in size.

Slanted palpebral fissures and inner epicanthal folds are also present in persons with the **Down syndrome** or the trisomy 21 chromosomal abnormality and with several other syndromes (*e.g.*, the **cri du chat syndrome**), resulting from a terminal deletion of chromosome number 5. (For more information on these chromosomal aberrations, see Moore, 1982 and Thompson and Thompson, 1980).

In the medial angle of your eye, observe the reddish area known as the **lacrimal lake** or lacus lacrimalis (Fig. 7-92), within which there is a small hillock, called the **lacrimal caruncle** (L. a small fleshy mass). Lateral to the caruncle is a vertical curved fold of conjunctiva, called the **semilunar fold** or plica semilunaris. This fold slightly overlaps the eyeball and is a *remnant of the nictitating membrane present in some animals* (*e.g.*, amphibians).

Evert the edge of your lower eyelid and at its medial end locate a small black pit, the **lacrimal punctum** (L. a point), on the summit of a small elevation called the **lacrimal papilla** (Figs. 7-92 and 7-96). There is a similar punctum and papilla on the upper eyelid (Fig. 7-92). The lacrimal punctum is the opening of a slender canal called the **lacrimal canaliculus** (L., canal), which carries the tears to the **lacrimal sac** (Fig. 7-96). From here, they are conducted to the nose via the **nasolacrimal duct**.

The lacrimal sac has some fibers of the orbicularis oris muscle posterior to it which insert into the crest of the lacrimal bone, called the **lacrimal crest** (Fig. 7-88). When these muscle fibers contract, the lacrimal sac is squeezed, forcing the tears into the nasolacrimal duct which opens into the nasal cavity (Fig. 7-96).

Observe that the lacrimal puncta face posteriorly and so are able to suck up the tears. Press your fingertip between your nose and the medial canthus of your eye. You should feel a horizontal cord, the **medial palpebral ligament** (Fig. 7-96). It connects the eyelids, including their muscles, to the medial margin of the orbit. if you pull your eyelid laterally, this ligament may raise a small skin fold. A similar **lateral palpebral ligament** attaches the eyelids to the lateral margin of the orbit. The medial and lateral palpebral ligaments are connected by **tarsal plates** (Figs. 7-94 and 7-96). Because of these dense connective tissue plates, the eyelids can be everted.

It is easy to evert the lower eyelid, but difficult to evert the upper eyelid because its tarsal plate is rigid. Hence, when the upper eyelid is everted (*e.g.*, over a wooden match stick), it tends to stay that way until it is turned inferiorly.

The palpebral fascia is a thin fibrous membrane which connects the tarsal plates to the margins of the orbit and with them forms an orbital septum. This septum passes posterior to the lacrimal sac and is pierced by the levator palpebrae superioris muscle (Fig. 7-96).

On the deep surface of your eyelids you may be able to see the **tarsal glands** (Fig. 7-94) because they appear as yellowish streaks through the palpebral conjunctiva. The ducts of these glands open on the flat free margin of the eyelid near its posterior edge (Fig. 7-94).

Tears (lacrimal fluid) are produced by a small, almond-shaped lacrimal gland, located in the superolateral part of the orbit (Figs. 7-91 and 7-96). Its three to nine excretory ducts open into the **superior fornix of the conjunctival sac** (Figs. 7-90*A* and 7-91). In addition to the main gland, there are numerous accessory lacrimal glands. As the lacrimal fluid drains into the lacrimal sac, it moistens the corneal surface. When the cornea becomes dry, the eye blinks and the eyelids carry a film of lacrimal fluid over the cornea, somewhat like windshield wipers on a car. In this way, foreign material (*e.g.*, dust) is carried to the medial canthus of the eye where you can remove it, *e.g.*, with a tissue. When you cry, the excess tears cause overflowing of the lacrimal lakes and the tears roll down your cheeks.

Awareness of the normal position of the eyelids is clinically important because their pattern changes when the nerves that control their movement or tone are damaged.

In **third nerve palsy** (Fig. 7-87), the upper eyelid droops (**ptosis**) and cannot be voluntarily raised. This results from damage to the superior division of the **oculomotor nerve** (CN III), which supplies the **levator palpebrae superioris** muscle (Figs. 7-28, 7-32, 7-91, and 7-94). You can tell by its name that this muscle normally elevates the upper lid.

When the facial nerve (CN VII) is damaged (Table 7-3), the eyelids cannot be closed owing to paralysis of the **orbicularis oculi muscle**, which closes the eyelids. In this case the levator palpebrae superioris (acting alone) keeps the upper eyelid elevated, even during sleep.

When the facial nerve is paralyzed the eyelids cannot be closed and protective blinking of the eye is lost. As a result, tears cannot be washed across the cornea. Irritation of the unprotected eyeball results in excessive lacrimation (tear formation). Excessive tearing also occurs when the lacrimal drainage apparatus is obstructed, or when the lower eyelid is lax and everted, thereby preventing the tears from reaching the inferior lacrimal punctum (Figs. 7-92 and 7-96).

Any of the glands in the eyelid (Fig. 7-94) may become inflamed and swollen owing to infection or obstruction of their ducts. If the ducts of the ciliary glands become obstructed or inflamed, a painful red swelling, known as a **sty**, develops on the eyelid. They are common in children and nearly always are caused by staphylococcal infections.

Cysts of the sebaceous glands of the eyelid, called **chalazia**, may also form. An obstruction of a tarsal gland produces an inflammation, called a **tarsal chalazion**, that protrudes toward the eyeball and rubs against it as the eyelids blink. Usually chalazia are more painful than sties.

THE EYEBALL

The eyeball (about 2.5 cm long) is like a miniature camera suspended in the anterior half of the orbital cavity in such a way that the six ocular muscles can move it in all directions (Figs. 7-91 and 7-97). In Figure 7-90*A* observe that there is a large space posterior to the eyeball that is occupied by muscles, nerves, and fat.

Do not be misled by the position of the eyeballs in cadavers; they are sunken owing to dehydration and postmortem atrophy of the fat and muscles in the orbit. Sunken eyes are also characteristic of emaciated and/or dehydrated living people, owing to the scantiness of fat in their orbital cavities. **The eyeball has three concentric coats.**

1. EXTERNAL OR FIBROUS COAT (Fig. 7-93)

This external or supporting coat of the eye consists of a white, opaque posterior five-sixths, **the sclera**, and a transparent anterior one-sixth, **the cornea**. The sclera is a firm, smooth fibrous layer. The cornea is discussed with the "*Refractive Media*" on page 897.

2. MIDDLE OR VASCULAR COAT (Fig. 7-93)

This heavily pigmented and vascular layer consists, from posterior to anterior, of **the choroid, ciliary body, and iris.**

The choroid is a dark brown membrane located between the sclera and retina. It forms the largest part of the middle coat of the eye and lines most of the sclera. It terminates anteriorly in the ciliary body. The choroid is firmly attached to the retina, but it can easily be stripped from the sclera.

The choroid contains many venous plexuses and layers of capillaries that are responsible for nutrition of the adjacent layers of the retina.

The ciliary body connects the choroid with the circumference of the iris; it is continuous posteriorly with the choroid. The ciliary body has protrusions or folds on its internal surface, called **ciliary processes** (Fig. 7-93), which *secrete aqueous humor*, a watery fluid that fills the **anterior and posterior chambers of the eye**. These chambers are the fluid-filled spaces anterior and posterior to the iris. The direction of flow of the aqueous humor is shown by the arrow hooking around the retracted iris in Figure 7-93.

Externally the ciliary body contains **ciliary muscle**, which on contraction *permits the lens to bulge by relaxing its suspensory ligament.*

The iris is a contractile diaphragm situated anterior to the lens. It has a central, circular aperture for transmitting light, called the **pupil** (Fig. 7-92). The iris is between the cornea and the lens (Fig. 7-93). When awake, the size of the pupil is continually varying in order to regulate the amount of light entering the eye through the lens.

Eye color *depends on the amount and the distribution of pigment-containing cells (chromatophores) in the iris.* The pupil appears black because one looks through the pupil toward the posterior aspect of the eye at the **optic fundus** (Fig. 7-98), which is heavily pigmented. In persons with blue eyes, the pigment is limited to the posterior surface of the iris, whereas in persons with dark brown eyes the pigment is scattered throughout the loose connective tissue stroma of the iris.

3. INTERNAL OR RETINAL COAT (Fig. 7-93)

The retina of the eyeball is a very thin, delicate membrane. It is covered externally by the choroid and internally by the vitreous body.

The retina is composed of two layers: an **outer pigment cell layer** and an **inner neural layer**. The light-sensitive neural layer of the retina ends at the posterior edge of the ciliary body in a wavy border, called the **ora serrata retinae** (Fig. 7-93). A thin, insensitive layer continues anteriorly onto the ciliary body and the iris.

In the posterior portion of the interior of the eye, called the **optic fundus**, there is a circular depressed, white to pink area in the retina, known as the optic papilla or **optic disc** (Figs. 7-93 and 7-98). The optic disc (disk) is where the optic nerve enters the eyeball and its fibers spread out in the neural layer of the retina. Because the optic disc contains nerve fibers and no photoreceptor cells, it is *insensitive to light.* For this reason, it is sometimes referred to as the **blind spot**.

Just lateral to the optic disc is a small oval, yellowish area called the **macula lutea** (L. yellow spot). The central depressed part of the macula, known as the **fovea centralis,** *is the area of most acute vision.* The retina is more adherent to the choroid at the optic disc and the ora serrata than elsewhere.

The retina is supplied by the central artery of the retina, a branch of the ophthalmic artery, which enters the eyeball with the optic nerve (Figs. 7-93 and 7-99). Pulsation is usually visible in the retinal arteries through an **ophthalmoscope** (Fig. 7-98).

A corresponding system of retinal veins unites to form the **central vein of the retina**. The retinal arteries and veins usually accompany each other and sometimes cross one another (Fig. 7-98).

One sign of hypertension (high blood pressure) is *nicking of the retinal veins* where the retinal arteries cross them; this is visible through an ophthalmoscope.

The retina and the optic nerve develop from an outgrowth of the embryonic forebrain, known as the **optic vesicle**, which takes its covering of meninges with it. Hence the meningeal layers and subarachnoid space extend around the optic nerve to its attachment to the eyeball (Fig. 7-93).

The central artery and vein of the retina, which are branches of the ophthalmic artery and vein, run within the anterior part of the optic nerve and cross the extension of subarachnoid space around it. Consequently an increase in CSF pressure slows venous return from the retina, causing **edema** as the result of fluid accumulation. This edema is obvious on **ophthalmoscopy** as a swelling of the optic disc or optic papilla (*i.e.,* **papilledema**). For this reason, inspection of the optic fundus (**funduscopy**) is an essential part of a neurological examination.

The **pigment epithelium** of the retina develops from the **outer layer of the optic cup,** a derivative of the embryonic optic vesicle, and the **neural layer** develops from the **inner layer of the optic cup.** *As they are first separate in the embryo and then fuse during the early fetal period, these layers of the retina are separated by a potential intraretinal space.* Although the pigment layer becomes firmly fixed to the choroid, its attachment to the neural layer is not so firm. **Detachment of the retina,** which may follow a blow to the eye, is separation of the pigment layer from the neural layer, as in the embryo.

REFRACTIVE MEDIA

On their way to the retina, light waves must pass through a number of structures with different densities: the cornea, aqueous humor, lens, and vitreous body (Fig. 7-93). These structures constitute the refracting media of the eye.

THE CORNEA (Figs. 7-90, 7-92, and 7-93)

The cornea forms the *clear, circular area* of the anterior part of the outer fibrous coat of the eyeball. *The cornea is transparent and avascular, i.e., it has no blood vessels.* It is largely responsible for refraction of the light that enters the eye.

The cornea is continuous peripherally with the sclera at the *sclerocorneal junction* (Figs. 7-92 and 7-93), and consists chiefly of a special kind of dense connective tissue. The degree of curvature of the cornea varies in different people and is greater in young than in old persons.

The most frequent injury of the eye is the deposit of a foreign body on the cornea. Because the cornea is exposed to the exterior, it is also subject to cuts, abrasions, and other kinds of trauma. A stick poked into the eye may tear the cornea or produce a **corneal abrasion**. These injuries are painful and may take several months to heal.

In **facial nerve paralysis** (Fig. 7-100), the orbicularis oculi muscle is paralyzed and the eyelids remain open. As a result, there is no blinking to keep the cornea moist and the tears simply flow out of the eye.

A *dry cornea may become ulcerated and if the condition is not treated the eyesight may be lost.*

Ointment applied to the eye will slow the drying process, but sometimes the eyelids have to be sutured together to keep the cornea moist until the facial nerve regenerates.

Homologous corneal transplants can be done for patients with scarred or opaque corneas with considerable success. The surface epithelium is regenerated by the host and covers the transplant in a few days. **Corneal implants** of nonreactive plastic material can also be used.

As the central part of the cornea receives its oxygen from the air, soft **plastic lenses** worn for long periods must be gas permeable.

THE AQUEOUS HUMOR (Fig. 7-93)

This is a clear watery fluid in the anterior and posterior chambers of the eye. *Aqueous humor is continuously produced by the ciliary processes.* After passing through the pupil from the posterior chamber into the anterior chamber, the aqueous humor is drained off through spaces at the **iridocorneal angle** (filtration angle). These spaces, visible through an instrument known as a **slitlamp**, open into a circular venous canal, called the **sinus venosus sclerae** (canal of Schlemm). This sinus drains via aqueous veins into the scleral venous plexuses. The aqueous humor provides nutrients for the avascular cornea and lens.

If the rates of production and absorption of aqueous humor become unequal, or there is **a blockage of the sinus venosus sclerae**, the pressure of the aqueous humor rises. This produces a condition known as **glaucoma**.

The presence of the **rubella virus** in the embryo during the critical stages of eye development can result in atresia or abnormal formation of the sinus venous sclerae. This causes a condition known as **congenital glaucoma** (see Moore, 1982 for details).

THE LENS (Fig. 7-93)

The lens is a transparent, flexible, biconvex structure enclosed in a transparent capsule. It is encircled by the ciliary processes and is located posterior to the iris and anterior to the vitreous humor. Like the cornea, the lens is both transparent and avascular.

About 1 cm in diameter, the highly elastic lens is held in position by a series of radially arranged fibers known collectively as the ciliary zonule or **suspensory ligament of the lens** (Fig. 7-93). The fibers are attached to the capsule of the lens and laterally to the ciliary processes. The curvatures of the surfaces of the lens, particularly the anterior surface, are constantly varying in order to focus near or distant objects on the retina.

As one gets older, the lens becomes harder and more flattened. In old age the lens gradually acquires a yellow tint. These changes gradually reduce the person's focussing power, a condition known as **presbyopia.** In some old people, there is also a loss of transparency of the lens; this opaqueness of the lens is known as a **cataract.** Many lens opacities are inherited, but some are caused by noxious agents affecting early development of the lens. **Congenital cataracts** may develop when the mother contracts **German measles** (rubella) during the early part of the first trimester.

THE VITREOUS BODY (Fig. 7-93)

The vitreous body fills the eyeball posterior to the lens. It consists of a jelly-like substance in which there is a meshwork of fine collagenic fibrils. This colorless, transparent gel occupies the **vitreous chamber**, the space between the lens and the retina. It consists of about 99% water and forms about four-fifths of the eyeball. In addition to transmitting light, the vitreous body holds the retina in place and provides support for the lens.

In contrast to the aqueous humor, *the vitreous body is not continuously replaced.* It forms during the embryonic period and is not exchanged. At the periphery of the vitreous body, there is a **vitreous membrane** which is formed by a condensation of the gel.

A narrow passage, called the **hyaloid canal** (Cloquet's canal), runs from the optic disc to the posterior surface of the lens. In the fetus this canal was traversed by the hyaloid artery. This vessel normally disappears before birth. A remnant of this artery may be visible during ophthalmoscopic examinations as a very small corkscrew-like structure hanging from the posterior aspect of the lens.

THE FASCIAL SHEATH OF THE EYEBALL

The fascia forms a thin cup-like sheath around the eyeball, except for its corneal part, and separates it from the fat and other contents in the orbit (Fig. 7-90*A*). The fascial sheath (bulbar fascia, vagina bulbi, Tenon's capsule) is attached posteriorly to the sclera close to the optic nerve and anteriorly, just posterior to the cornea (***sclerocorneal junction***). The tendons of the muscles that rotate the eyeball pierce the fascial sheath on their way to their places of insertion. The fascial sheath also blends with the fascial sheaths of the ocular muscles. There is a potential space between the eyeball and the fascial sheath which allows the eyeball to move inside this cup-shaped fascia.

There are triangular expansions from the sheaths of the medial and lateral rectus muscles (Figs. 7-93 and 7-97), which are attached to the lacrimal and zygomatic bones, respectively. As they check (prevent excessive movement) the actions of these muscles, they are called the medial and lateral **check ligaments**. There is also a check ligament associated with the levator palpebrae superioris muscle (Fig. 7-91).

The fascial sheath is perforated posteriorly by the ciliary nerves (Fig. 7-91) and vessels and fuses with the dural sheath of the optic nerve (Fig. 7-93). A thickening of the inferior part of the fascial sheath of the eyeball, called the **suspensory ligament of the eye**, is attached to the anterior parts of the medial and lateral walls of the orbit. As it is a sling-like hammock inferior to the eyeball, it supports this organ.

The cup-like fascial sheath of the eyeball helps to form a socket for an artificial eye when the eyeball is removed (**enucleation**). After this operation, the eye muscles cannot retract far because their fascial sheaths are attached to the fascial sheath of the eyeball.

Because the suspensory ligament of the eye supports the eyeball, it is preserved when surgical removal of the floor of the orbit is carried out (*e.g.*, during the removal of a tumor).

MUSCLES OF THE ORBIT

There are seven voluntary muscles of the orbit, all of which move the eyeball, except the levator palpebrae superioris which elevates the upper eyelid (Fig. 7-28).

The six extraocular or extrinsic ocular muscles are capable of rotating the eyeball (Table 7-4). There are four straight muscles (L. *rectus*) and two oblique muscles.

LEVATOR PALPEBRAE SUPERIORIS MUSCLE (Figs. 7-28, 7-90*A*, 7-94, 7-96, and 7-97)

This thin, triangular-shaped muscle elevates the upper eyelid. The Latin word levator means "**a lifter.**"

The levator palpebrae superioris muscle does not insert into the eyeball and so cannot move it. **It arises from the inferior surface of the lesser wing of the sphenoid bone**, superior and anterior to the optic canal.

This thick muscle fans out into a wide aponeurosis that inserts into the ***skin of the upper eyelid***. The inferior part of the aponeurosis contains some smooth muscle fibers that form the **superior tarsal muscle** (Figs. 7-90*A* and 7-94). These involuntary muscle fibers insert into the tarsal plate. The superior tarsal muscle presumably assists the levator palpebrae superioris muscle in elevating the eyelid. It is responsible for the wide-eyed stare of a frightened person.

The levator palpebrae superioris is innervated by the superior branch of the oculomotor nerve (Fig. 7-97), *whereas the superior tarsal muscle is innervated by sympathetic fibers from the superior cervical ganglion* (Fig. 8-28).

Injury to sympathetic fibers supplying the superior tarsal muscle (*e.g.*, inadvertently during thyroidectomy) and/or surgical excision of the superior cervical ganglion (*e.g.*, during removal of a malignant tumor in the neck) results in a slight lowering (**ptosis**) of the upper eyelid owing to loss of tone and/or *paralysis of the* ***superior tarsal muscle***. In these cases, more of the iris is covered by the upper eyelid than is usual (Fig. 7-92). This type of ptosis (G. a falling) of the eyelid is a characteristic of the ***Horner syndrome***, which also includes pupillary constriction.

Patients with hemisection of the cervical region of the spinal cord also have the Horner syndrome on the side of the lesion owing to *interruption of descending autonomic fibers in the lateral white column of the spinal cord.*

In **third nerve palsy** (Fig. 7-87), the levator palpebrae superioris muscle is involved and the upper eyelid is in complete ptosis, *i.e.*, *the eye is partially closed and the upper lid cannot be raised voluntarily.*

In **seventh nerve paralysis** (Fig. 7-100), the levator palpebrae superioris muscle, acting unopposed, (because the orbicularis oculi muscle used for gentle closing of the eye is paralyzed) elevates the upper eyelid, keeping the eye open, even during sleeping. *Protective blinking is lost and irritation of the unprotected and dry cornea results,* which may lead to ulceration of the cornea.

THE RECTUS MUSCLES: SUPERIOR, INFERIOR, MEDIAL, AND LATERAL

The four rectus (L. straight) muscles *arise by a tough tendinous cuff,* called the **common tendinous ring**

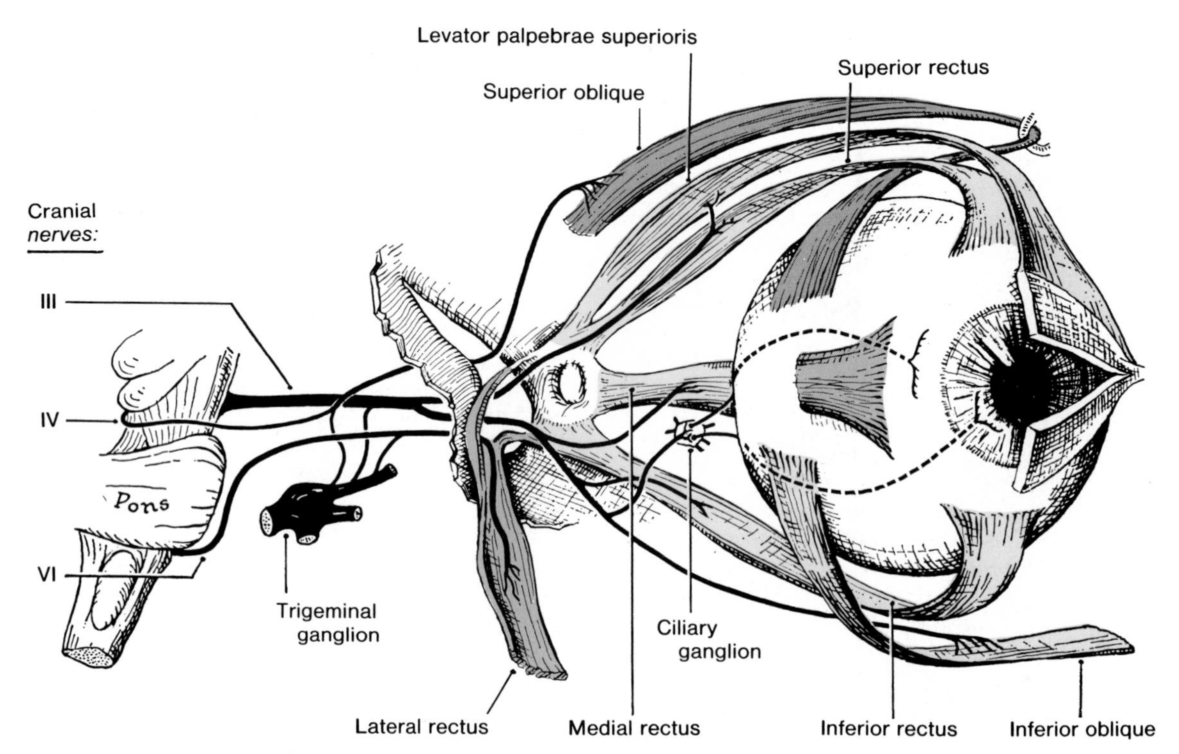

Figure 7-97. Diagram illustrating the distribution of the oculomotor (CN III), trochlear (CN IV), and abducens (CN VI) nerves to the muscles of the eyeball, which enter the orbit through the superior orbital fissure (shown here and in Fig. 7-88). Note that CN IV supplies the superior oblique, CN VI supplies the lateral rectus, and CN III supplies the remaining five muscles. The ciliary ganglion is a parasympathetic ganglion. Its postsynaptic fibers pass in the short ciliary nerves (- - - -) to the eyeball (also see Fig. 7-33).

(common ring tendon), which surrounds the optic canal and the junction of the superior and inferior orbital fissures (Figs. 7-91, 7-97, 7-101, and 7-102).

From their common origin, the four rectus muscles run anteriorly, close to the walls of the orbit and *insert into the eyeball, just posterior to the sclerocorneal junction* (Figs. 7-92 and 7-97). Each muscle passes anteriorly in the position implied by its name.

All the rectus muscles are supplied by the oculomotor nerve (CN III), except the lateral rectus (Table 7-4), which is supplied by the abducens nerve (CN VI).

Note that the lateral and medial rectus muscles lie in the same horizontal plane, whereas the superior and inferior rectus muscles lie in the same vertical plane (Figs. 7-97 and 7-101).

THE OBLIQUE MUSCLES: SUPERIOR AND INFERIOR (Figs. 7-91, 7-97, and 7-101 to 7-103)

The superior oblique muscle is fusiform and arises from the body of the sphenoid bone, superomedial to the common tendinous ring, and passes anteriorly, superior and medial to the superior and medial rectus muscles. It ends in a round tendon which runs through a pulley-like loop, called the **trochlea** (Figs. 7-102 and 7-103), which is attached to the superomedial angle of the orbital wall. After passing through the trochlea (L. pulley), the tendon of *the superior oblique turns posterolaterally and inserts into the sclera* at the posterosuperior aspect of the lateral side of the orbit.

The inferior oblique muscle is a thin, narrow muscle that arises from the maxilla in the floor of the orbit (Figs. 7-101 and 7-102). It passes laterally and posteriorly, inferior to the inferior rectus muscle, and *inserts into the sclera at the posteroinferior aspect of the lateral side of the orbit* (Fig. 7-97).

SUMMARY OF THE NERVE SUPPLY OF THE MUSCLES OF THE ORBIT (Table 7-4)

All three cranial nerves supplying the muscles of the eyeball (oculomotor, **CN III**; trochlear, **CN IV**; and abducens, **CN VI**) *enter the orbit through the* ***superior orbital fissure*** (Fig. 7-97).

Cranial nerves IV and VI each supply one muscle, whereas CN III supplies the remaining five muscles.

CN IV supplies the superior oblique (SO); **CN VI** supplies the lateral rectus (LR); and **CN III** supplies the levator palpebrae superioris (muscle of eyelid, not of eyeball), superior rectus (SR), medial rectus (MR), inferior rectus (IR), and inferior oblique (IO). Hence, all three nerves carry fibers which are motor to extraocular muscles (muscles of the eyeball).

In summary, all orbital muscles are supplied by CN

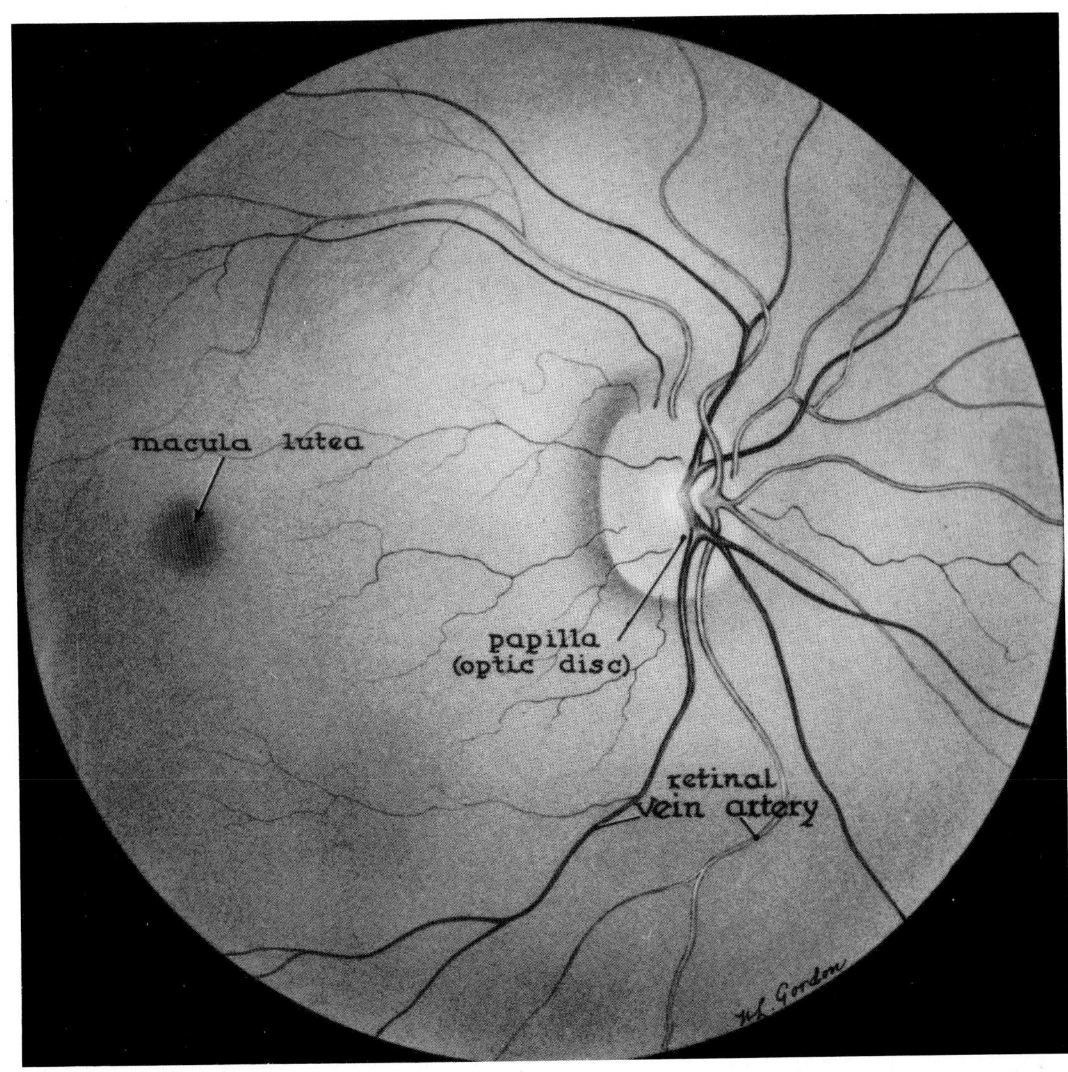

Figure 7-98. Drawing of the fundus of the right eye as seen through an ophthalmoscope. (See Fig. 7-93 for orientation). The papilla, usually called the optic disc, is a blind spot because it contains only optic nerve fibers. The macula lutea or central area in line with the visual axis is specialized for visual acuity. (Reprinted with permission from Ham AW, Cormack DH: Histology, ed 8. JB Lippincott Co, 1979.)

III except the superior oblique and the lateral rectus, which are supplied by CN IV and VI, respectively, (*i.e.*, SO IV, LR VI, all others III). The following "formula" is worth committing to memory: $SO_4(LR_6)_3$.

ACTIONS OF THE SIX OCULAR MUSCLES (Table 7-4)

The six extraocular muscles rotate the eyeball in the orbit about three axes (sagittal, horizontal, and vertical). Verify that the four rectus muscles are arranged around the orbital axis, not around the anteroposterior, sagittal, or optic axis; thus the medial, superior, and inferior recti, respectively, are adductors.

Knowing the actions of the ocular muscles is important not only for understanding their effects on the field of vision, but also for clinical testing of the cranial nerves supplying them.

The paralysis of one or more ocular muscles, owing to injury of the nerves supplying them, results in **diplopia** (*double vision*). Paralysis of a muscle of the eyeball is noted by the limitation of movement of the eye in the field of action of the paralyzed muscle, and by the production of two images when an attempt is made to use the paralyzed muscle.

Table 7-4
Actions and Nerve Supply of the Ocular Muscles

Muscle	Action(s) on Eyeball	Nerve Supply
Medial rectus[1]	**Adducts**	CN III
Lateral rectus[1]	**Abducts**	CN VI[2]
Superior rectus	**Elevates, adducts,** and rotates medially	CN III
Inferior rectus	**Depresses, adducts,** and rotates laterally	CN III
Superior oblique[3-4]	**Depresses medially rotated eye,** abducts, and rotates medially	CN IV[2]
Inferior oblique[3-4]	**Elevates medially rotated eye,** abducts, and rotates laterally	CN III

1. The medial rectus and the lateral rectus move the eyeball in one axis only, whereas each of the other four muscles moves it in all three axes (see Fig. 7-97).
2. Cranial nerves IV and VI each supply one muscle, whereas CN III supplies the other four muscles.
3. The two oblique muscles protrude the eyeball, whereas the four rectus muscles retract it.
4. The superior and inferior oblique muscles are used with the medial rectus muscle in turning both eyes medially for near vision. This movement, accompanied by pupillary constriction, is known as **accommodation** and is used when examining a near object and when reading.

When the abducens nerve is paralyzed, the patient is unable to abduct the eye on the affected side (Table 7-3 and Fig. 7-104). Usually there is also double vision (diplopia).

For clinical testing, each muscle is examined in its position of greatest efficiency, i.e., when its action is at a right angle to the axis around which it is moving the eyeball. The patient is asked to look: superolaterally (upward and outward) to test the superior rectus; inferolaterally (downward and outward) to test the inferior rectus; superomedially (upward and inward) to test the inferior oblique; inferomedially (downward and inward) to test the superior oblique; medially (inward) to test the medial rectus; and laterally (outward) to test the lateral rectus.

The eyes of patients with Graves' disease (a form of hyperthyroidism) commonly protrude, a condition known as **exophthalmos** or ***proptosis*** (G. a falling forward). The association of hyperthyroidism and exophthalmos was first described by Dr. Graves, but the cause of the protrusion is not precisely known. A considerable increase in the size of the orbital muscles can be observed in CT scans, but no increase in orbital fat is obvious.

THE ORBITAL VESSELS

The orbital contents are supplied chiefly by the ophthalmic artery (Figs. 7-99 and 7-105). The infraorbital artery, the continuation of the maxillary, also contributes to the supply of this region.

Venous drainage is through the two ophthalmic veins, superior and inferior, which pass through the superior orbital fissure to enter the **cavernous sinus** (Figs. 7-39, 7-57, and 7-105).

THE OPHTHALMIC ARTERY (Fig. 7-99)

The ophthalmic artery arises from the internal carotid artery as it emerges from the cavernous sinus (Fig. 7-80). **The ophthalmic artery passes through the optic foramen within the dural sheath of the optic nerve** (Figs. 7-105 and 7-106). It runs anteriorly close to the superomedial wall of the orbit, giving off branches to structures in the orbit and to the ethmoid bone (Fig. 7-99).

The central artery of the retina (Figs. 7-93 and 7-99), one of the smallest but most important branches of the ophthalmic artery, arises inferior to the optic nerve (Fig. 7-105). It runs within the dural sheath of the optic nerve until it approaches the eyeball; then it pierces the optic nerve and runs within it to emerge through the optic disc (Fig. 7-93).

The central artery of the retina spreads over the internal surface of the retina and supplies it (Fig. 7-98). Twigs of this artery anastomose with the ciliary arteries (Fig. 7-99), but *the terminal branches of the central artery of the retina are essentially end arteries.*

Because the retinal artery is essentially an end artery, obstruction of it by an **embolus or thrombosis** leads to instant and ***total blindness*** in the eye concerned.

THE CILIARY ARTERIES (Fig. 7-99)

The ciliary arteries, branches of the ophthalmic, supply the sclera, choroid, ciliary body, and iris. Two

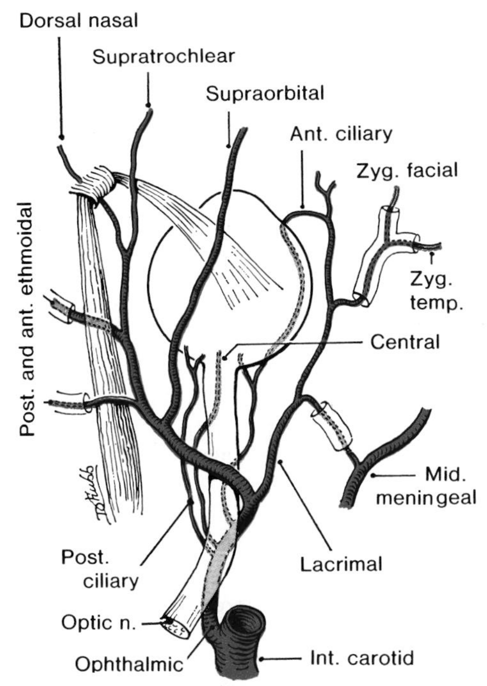

Figure 7-99. Diagram of the ophthalmic artery and its branches supplying the contents of the orbit. Note its relationship to the optic nerve with which it enters the orbit via the optic canal. Also see Figure 7-105.

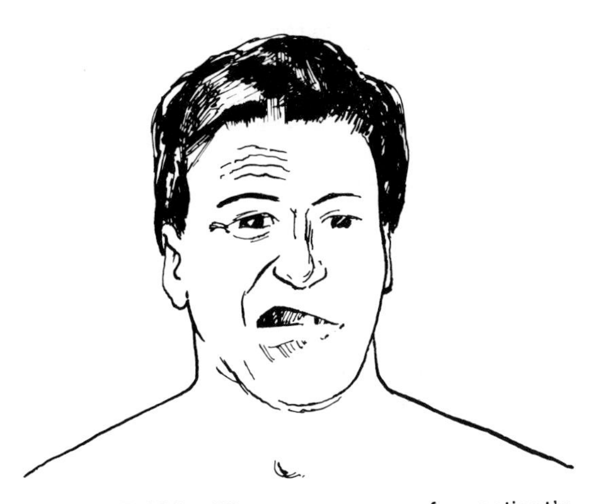

Figure 7-100. The appearance of a patient's face with seventh nerve (CN VII) paralysis when he tries to examine his teeth. Paralysis of the left side of his face is obvious as shown by the lack of wrinkling on that side, and the failure to expose the left teeth. Note that the skin of his left forehead does not wrinkle.

long posterior ciliary arteries pierce the sclera and supply the ciliary body and iris (Figs. 7-92 and 7-99). Several short posterior ciliary arteries pierce the sclera and supply the choroid.

THE LACRIMAL ARTERY (Fig. 7-99)

The lacrimal artery supplies the lacrimal gland, conjunctiva, and eyelids. A ***recurrent meningeal branch*** anastomoses with the middle meningeal artery; hence there is *an anastomosis between branches of the internal and external carotid arteries.*

The muscular branches of the ophthalmic artery to the eye muscles frequently arise from a common trunk and accompany the branches of the oculomotor nerve. Muscular branches also give rise to **anterior ciliary arteries** (Fig. 7-99), which give branches to the conjunctiva and then pierce the sclera to supply the iris (Fig. 7-92).

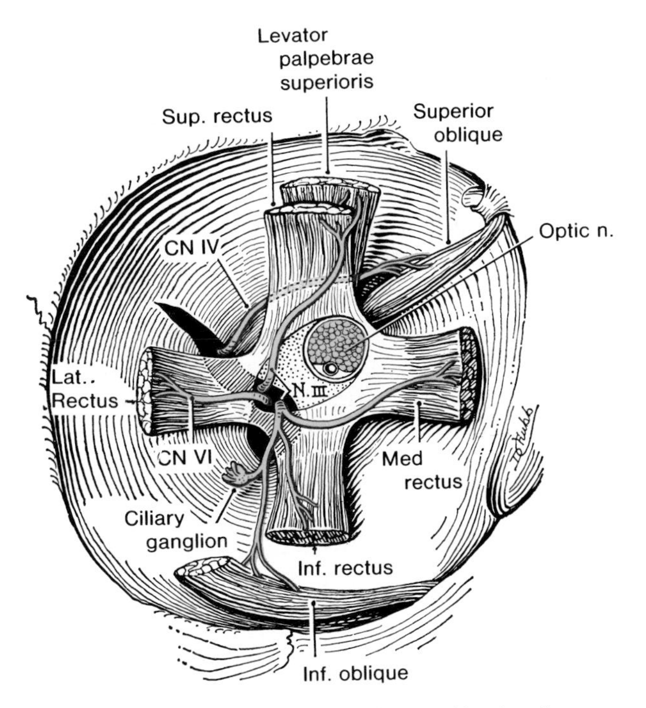

Figure 7-101. Dissection of the orbit showing the common tendinous ring and the motor nerves of the orbit. Observe the large optic nerve surrounded by its meningeal layers. Note the four rectus muscles arising from the fibrous cuff, known as the common tendinous ring. Observe that this ring encircles the dural sheath of the optic nerve (CN II), the abducens nerve (CN VI), and the superior and inferior divisions of the oculomotor nerve (CN III). The nasociliary nerve (not shown) also passes through this cuff, but the trochlear nerve (CN IV) clings to the bony roof of the cavity and is outside the common tendinous ring. CN IV and CN VI supply one muscle each and CN III supplies the remaining five orbital muscles: two via its superior division and three via its inferior division. The oculomotor nerve (CN III) via the ciliary ganglion (Fig. 7-33) supplies parasympathetic fibers to the ciliary muscle and the sphincter pupillae muscle.

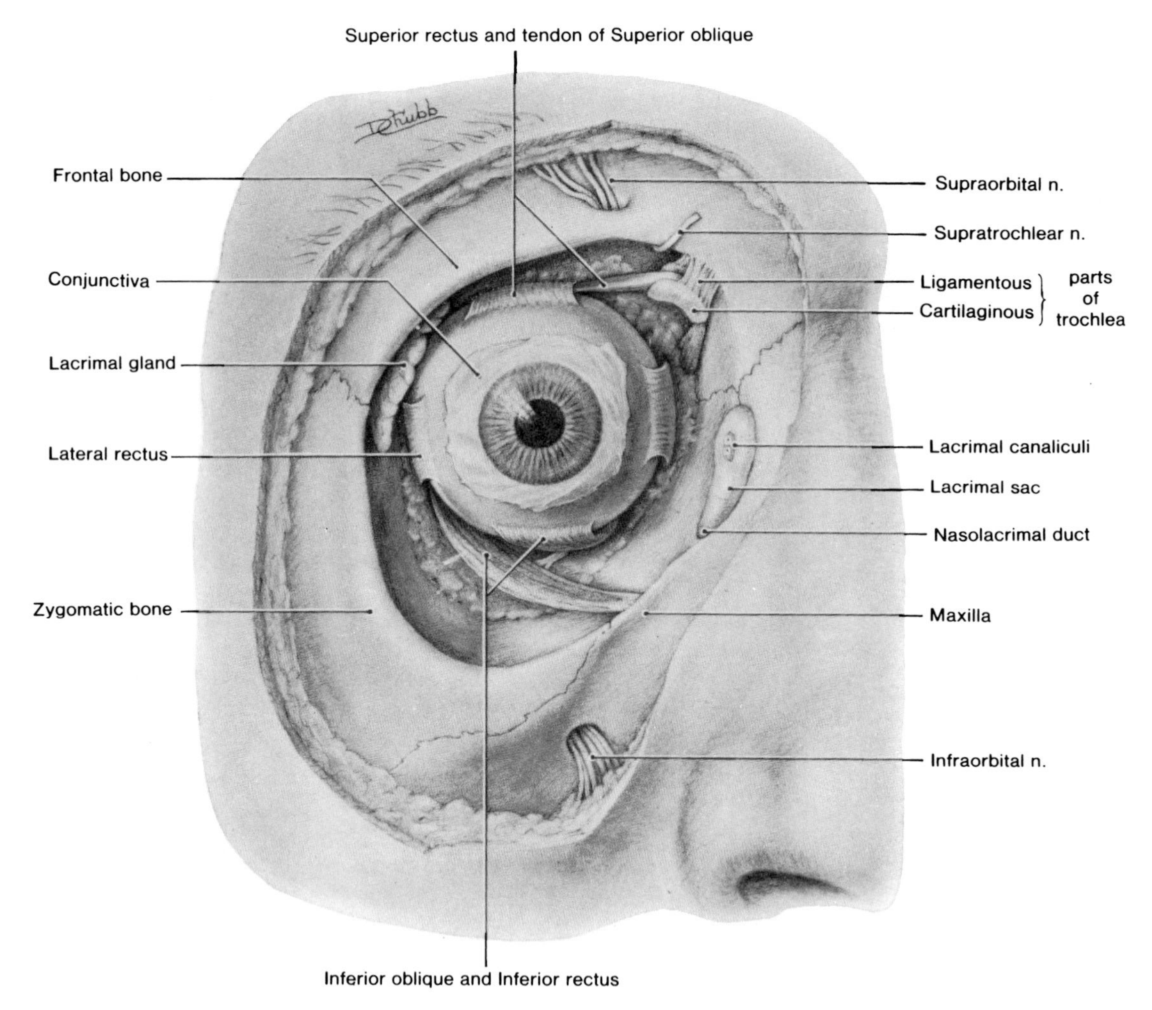

Figure 7-102. Drawing of a dissection of an anterior view of the orbital cavity. The eyelids, orbital septum, levator palpebrae superioris muscle, and some fat have been removed. Observe the aponeurotic insertion of the four rectus muscles, inserted 6 to 8 mm posterior to the sclerocorneal junction. Note the superior and inferior oblique muscles crossing inferior to the corresponding superior and inferior rectus muscles, and the tendon of the superior oblique running through a cartilaginous pulley or trochlea, which is fixed by ligamentous fibers just posterior to the superomedial angle of the orbital margin.

Five other branches of the ophthalmic artery leave the orbit with correspondingly named nerves and anastomose with branches of the external carotid artery. They are the *supraorbital, supratrochlear and dorsal nasal arteries* (Figs. 7-48 and 7-99), which end on the forehead or face. The anterior and posterior *ethmoidal arteries* (Fig. 7-99) enter the skull, but end in the nasal mucosa (Fig. 7-153).

THE OPHTHALMIC VEINS (Figs. 7-39 and 7-105)

The superior ophthalmic vein anastomoses with the facial vein; *as it has no valves, blood can flow in either direction.* It crosses superior to the optic nerve, passes through the superior orbital fissure, and ends in the cavernous sinus (Fig. 7-39).

The inferior ophthalmic vein begins as a plexus on the floor of the orbit. It communicates through the inferior orbital fissure with the **pterygoid plexus** (Fig. 7-39), crosses inferior to the optic nerve, and ends in either the superior ophthalmic vein or in the cavernous sinus.

The central vein of the retina (Fig. 7-93) usually enters the cavernous sinus directly, but it may join one of the ophthalmic veins, usually the superior ophthalmic vein (Fig. 7-39).

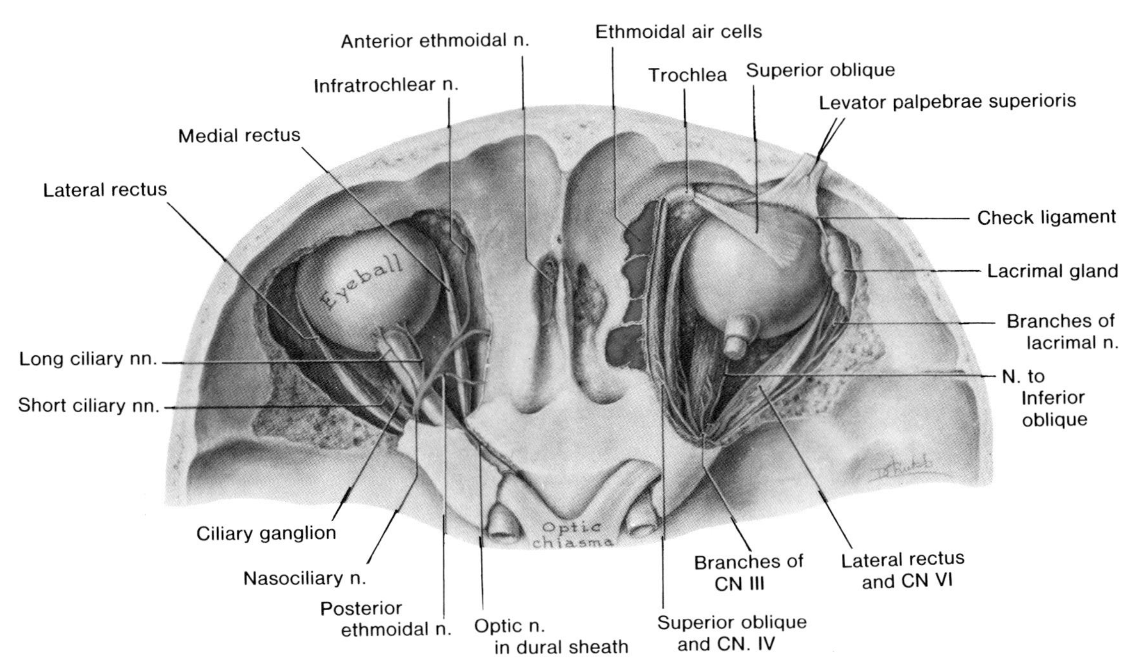

Figure 7-103. Dissection of a superior view of the orbital cavities. On the *right side*, observe the nerves to the six ocular muscles (four rectus and two oblique muscles), the trochlear to obliquus superior, the abducens to rectus lateralis, and the oculomotor to the remaining four eye muscles and to the levator palpebrae superioris muscle.

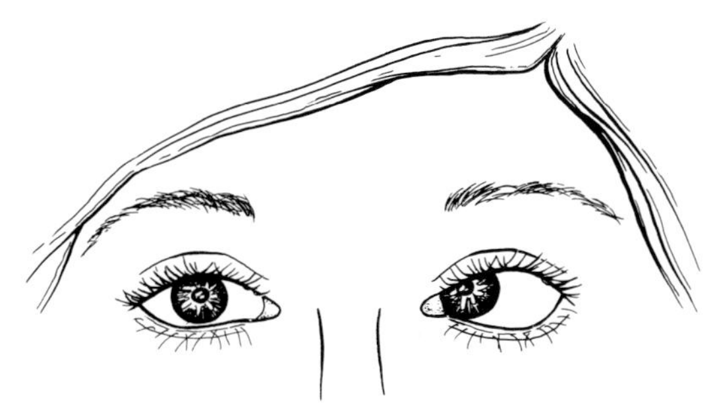

Figure 7-104. Drawing of a woman's eyes, illustrating *paralysis of the right lateral rectus* muscle owing to injury to the abducens nerve. Observe that her right eye does not move laterally when she looks to her right.

As the cavernous sinuses of the dura mater communicate with the veins of the face via the valveless superior and inferior ophthalmic veins, **thrombophlebitis of a facial vein**, resulting from infections of the face in the area drained by these veins (Figs. 7-38 and 7-43), may spread to the cavernous sinus (Fig. 7-39).

As the central vein of the retina enters either the cavernous sinus or the superior ophthalmic vein, thrombosis may sometimes extend along the central vein of the retina and produce thromboses in the small retinal veins. For this reason *pustules or furuncles (boils) on the face should never be squeezed.*

THE OPTIC NERVE

This is the second cranial nerve and the nerve of sight (Figs. 7-69, 7-72, 7-84, 7-90*A*, 7-93, 7-99, 7-101, 7-103, 7-105, and 7-106). It is about 5 cm in length and extends from the **optic chiasma** (G. a crossing) to the eyeball. The optic nerve is slightly longer than the distance it travels; thus it permits free movement of the eyeball.

Most fibers of the optic nerve are afferent and arise from ganglion cells in the retina. The fibers converge on the optic disc (Figs. 7-93 and 7-98). The optic nerve passes posteromedially within a cone formed by the extraocular muscles (Figs. 7-90*A* and 7-101), and *leaves the orbit through the optic canal* to enter the optic chiasma slightly superior and anterior to the tuberculum sellae (Figs. 7-14 and 7-62).

During eye development the optic nerve formed within sheaths of dura mater, arachnoid, and pia mater as it extended anterolaterally. These layers enclose extensions of the subdural and subarachnoid spaces (Figs. 7-90*A*, 7-93, 7-105, and 7-106) as far as the posterior aspect of the eyeball.

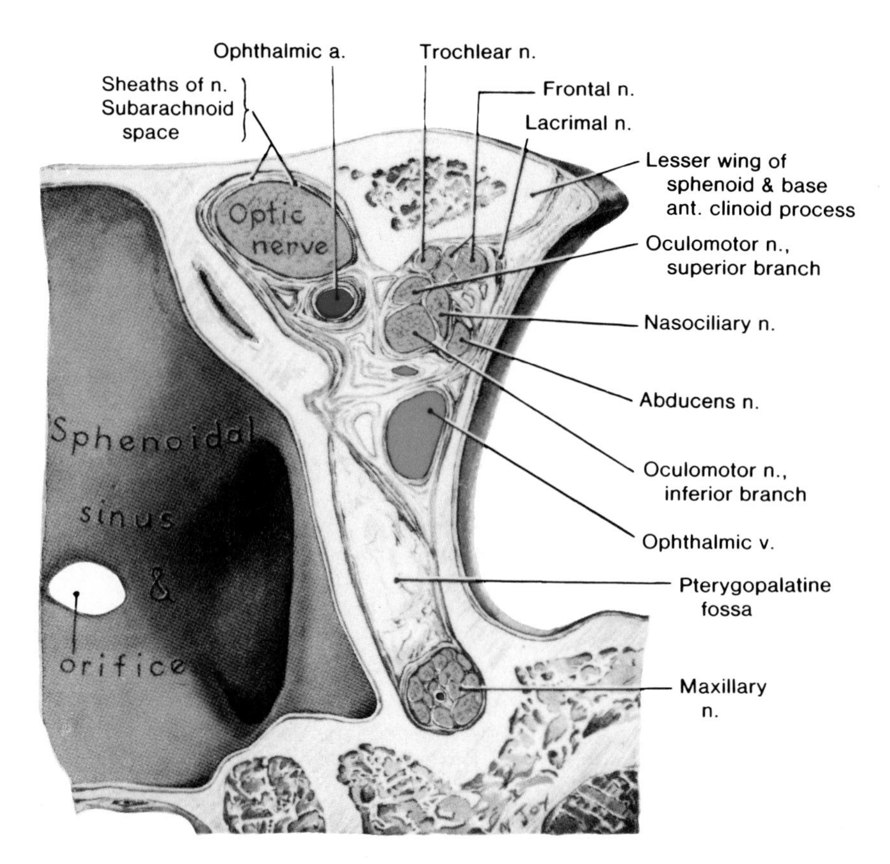

Figure 7-105. Drawing of a coronal section of the apex of the orbital cavity. Observe the optic nerve surrounded by its sheaths (extensions of the cranial meninges), the subarachnoid space, and the ophthalmic artery, a branch of the internal carotid artery, as they enter the optic canal. Note that the nerves to the orbit are crowded together as they pass through the medial end of the superior orbital fissure. Note also the ophthalmic vein, about to open into the cavernous sinus. (See Fig. 7-39 for orientation.)

The optic nerve is a special sensory nerve. In reality it *is a fiber tract of the brain because developmentally the retina is a part of the brain.* Posterior to the optic chiasma, the optic nerves are continued as the **optic tracts** which pass to the lateral geniculate bodies and the **superior colliculi** (Fig. 7-62).

Within the optic chiasma there is a partial decussation of the optic nerve fibers. Fibers from the nasal half of each retina cross to the opposite side, whereas those from the temporal half of each retina are uncrossed. Thus, *fibers from the right half of the retina of both eyes form the right optic tract and those from the left halves form the left tract.* This crossing of nerve fibers results in the right optic tract conveying impulses from the left visual field and vice versa. For details about this, consult a neuroanatomy text.

Injury to any part of the optic pathway results in visual defects, the nature of which depends on the location and extent of the injury (Table 7-3). **Severe degenerative disease** or a complete lesion of the optic nerve (*e.g.,* sectioning) causes **total blindness** in the corresponding eye.

A lesion at the lateral border of the optic chiasma results in nasal hemianopia or nasal hemianopsia (G. *hemi,* half + *an,* no + *opsis,* vision), *i.e.,* loss of the nasal half of the visual field of the eye on the same side as the lesion. A localized dilation or **aneurysm of the internal carotid artery**, superior to the cavernous sinus, could exert this kind of pressure on the optic chiasma and cause nasal hemianopsia.

The optic nerve is peculiarly liable to neuritis (inflammation of a nerve). **Optic neuritis** results in atrophy. *Retrobulbar neuritis* involving the optic nerve or tract is commonly caused by *multiple sclerosis* (MS). **Optic atrophy** results in diminished visual acuity or blindness. Primary (simple) optic atrophy is caused by processes that involve the optic nerve, such as adjacent tumors.

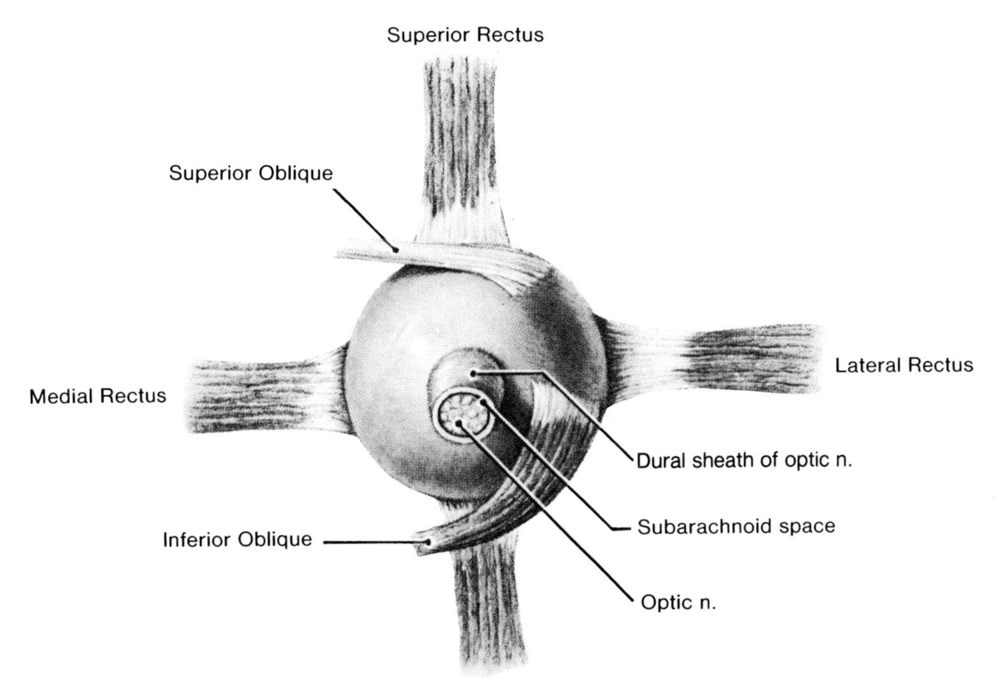

Figure 7-106. Drawing of a posterior view of the eyeball showing the insertions of the superior and inferior oblique muscles. Observe that the optic nerve is surrounded by a dural sheath and a subarachnoid space; these are extensions from the meninges and subarachnoid space covering the brain (Figs. 7-50 and 7-71).

THE PAROTID REGION

This region is mainly the space between the mastoid process of the temporal bone and the neck and ramus of the mandible. It contains the parotid gland as well as the structures related to it.

THE PAROTID GLAND AND PAROTID BED

The parotid gland and parotid duct (Figs. 7-29, 7-31, and 7-41) were described with the face (p. 836).

The parotid bed, occupied by the parotid gland, is a small space between the mastoid process posteriorly and the ramus of the mandible anteriorly (Figs. 7-107 to 7-110).

Superiorly, the parotid bed is bounded by the floor of the *external acoustic meatus* and the *zygomatic process of the temporal bone.* **Medially** (deeply), there is the *styloid process* of the temporal bone and its associated muscles. **Laterally** (superficially), the parotid bed is bounded by the superficial layer of the parotid fascia and skin (Fig. 7-31).

The posterior wall of the parotid bed extends between the mastoid and styloid processes; thus the muscles attached to them (sternocleidomastoid, posterior belly of digastric, and stylohyoid) are closely related to the parotid gland (Figs. 7-107 and 7-108).

The anterior wall of the parotid bed is formed by the ramus of the mandible and the two muscles (masseter and medial pterygoid) attached to it (Fig. 7-107).

The parotid gland, enclosed within fascia called the ***parotid sheath****, occupies all the space available to it in the area around the posterior margin of the mandible.*

STRUCTURES WITHIN THE PAROTID GLAND (Figs. 7-29, 7-42, 7-107, and 7-109 to 7-110)

From superficial to deep, structures traversing the parotid gland are the **facial nerve**, the ***retromandibular vein***, and the **external carotid artery**. There are also parotid lymph nodes on the parotid fascia and within the gland (Fig. 7-29).

The facial nerve (CN VII) is unique in traversing the parotid gland (Figs. 7-29 and 7-107), an occurrence of considerable clinical significance. During its early development, the gland lies between the two major branches of the facial nerve. As the gland enlarges, it envelops these branches and then superficial and deep parts of the gland fuse with each other between and around the branches of the facial nerve (Fig. 7-29).

The facial nerve emerges from the stylomastoid foramen and can be exposed in the notch between the mastoid process and the external acoustic meatus (Fig. 7-110). Beyond this, the nerve enters the posterior part of the parotid gland and

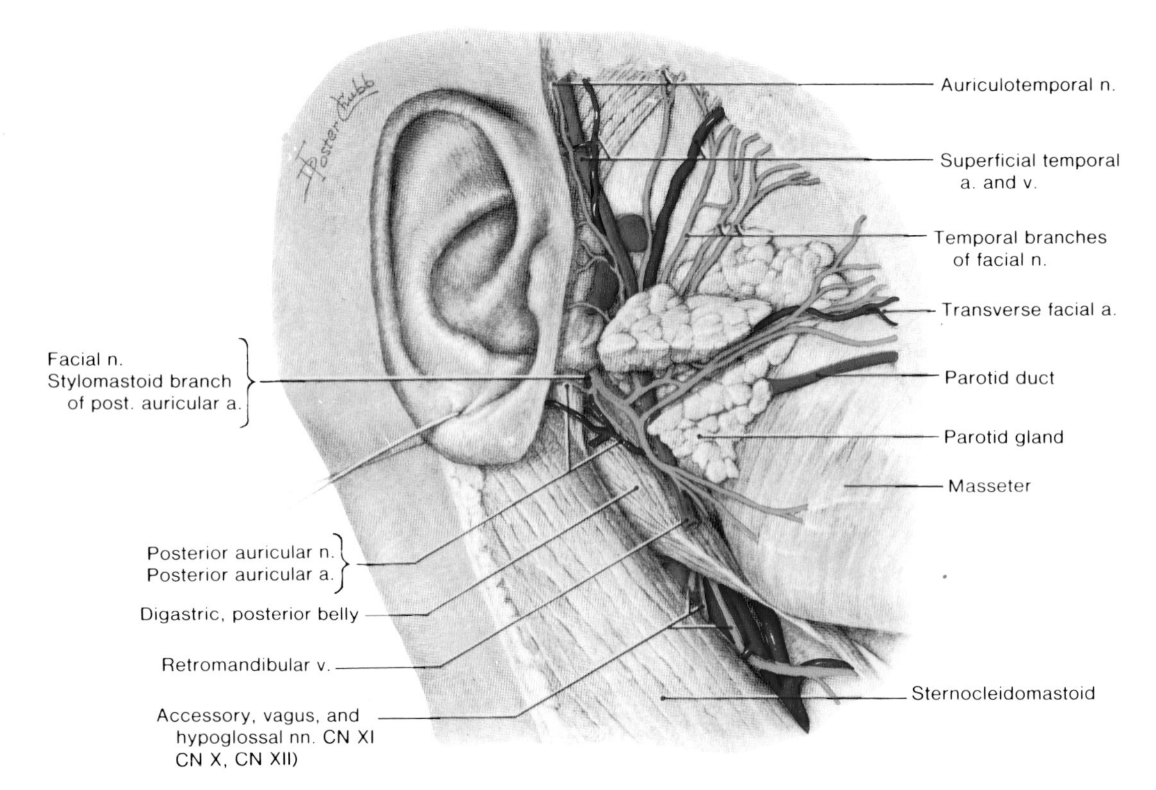

Figure 7-107. Drawing of a dissection of the parotid region. Part of the parotid gland has been removed to expose the facial nerve (CN VII). For a more superficial dissection, see Figure 7-29. Observe the facial nerve descending from the stylomastoid foramen and then curving anteriorly to penetrate the deep part of the parotid gland. Note the superficial position of the posterior belly of the digastric muscle and the two preauricular lymph nodes (*green*).

divides almost at once into superior and inferior divisions. These divisions give rise to temporal, zygomatic, and buccal branches and to mandibular and cervical branches, respectively (Fig. 7-29). The branches of the facial nerve emerge on the anterior aspect of the periphery of the parotid gland and lie on the lateral surface of the **masseter muscle** (Fig. 7-109). From here, they pass to the muscles of facial expression which they supply (Figs. 7-28 and 7-35).

The retromandibular vein (Figs. 7-38 and 7-42), formed by the union of the temporal and maxillary veins, descends in the parotid gland, superficial to the external carotid artery, but deep to the facial nerve (Fig. 7-109). Blood drains from the parotid gland into the retromandibular vein, which *joins the posterior auricular vein to form the external jugular vein* via its posterior branch, *and the facial vein* via its anterior branch (Fig. 7-38).

The external carotid artery (Figs. 7-37 and 7-108 to 7-110) enters the deep surface of the parotid gland, where it lies deep to the facial nerve and retromandibular vein. *At the neck of the mandible, the external carotid artery* ***divides into the superficial temporal and maxillary arteries*** (Figs. 7-37 and 7-107). The external carotid and its branches supply blood to the parotid gland.

NERVES NEAR THE PAROTID GLAND (Figs. 7-29, 7-33, 7-42, 7-109, and 7-110)

The Auriculotemporal Nerve. A branch of the mandibular division of the trigeminal nerve (CN V^3), it is in the parotid space, but passes superior to the superior part of the parotid gland. It communicates with the facial nerve, usually via two branches (Fig. 7-107).

The Great Auricular Nerve. This is a superficial ascending branch of the **cervical plexus** (Figs. 7-29 and 8-10). It passes external to the parotid gland where it divides into anterior and posterior branches, but it usually does not enter the parotid gland.

VESSELS OF THE PAROTID GLAND (Figs. 7-37, 7-42, 7-109, and 7-110)

The parotid gland is supplied by the **external carotid artery** and its terminal branches (*superficial temporal and maxillary arteries*). They arise within the gland.

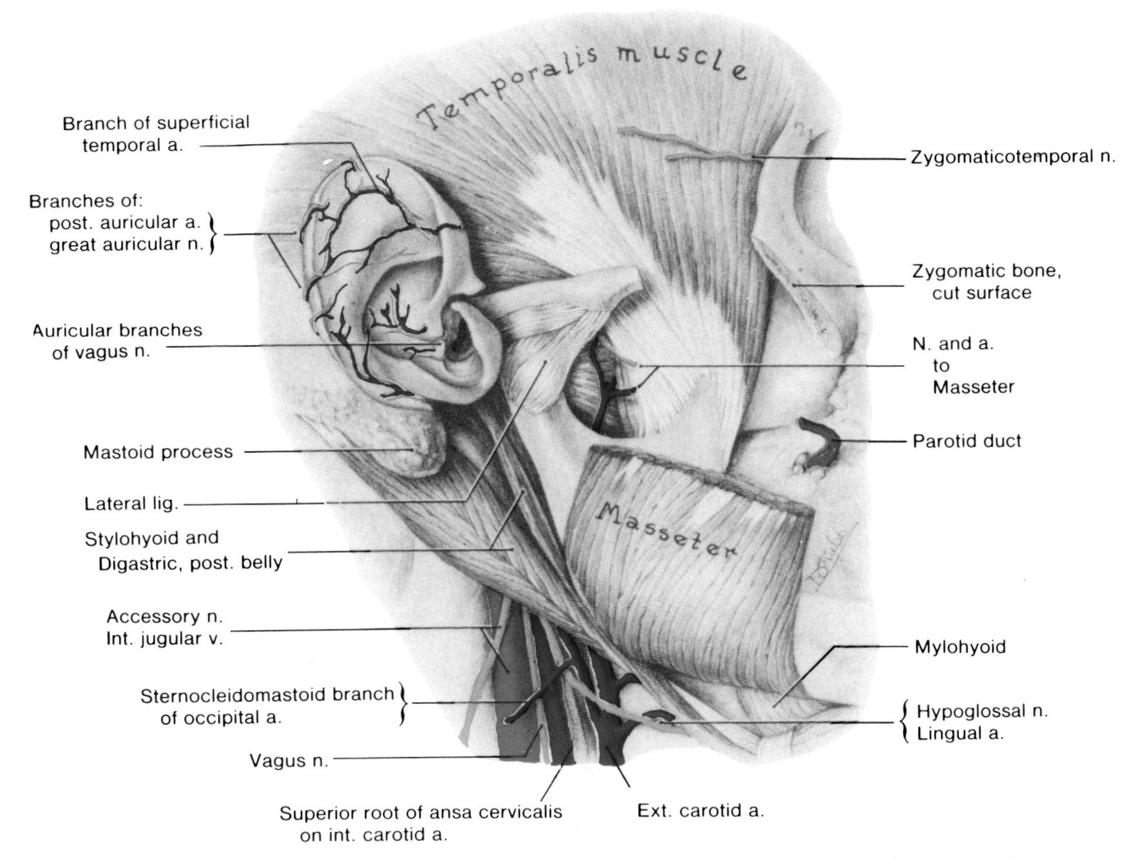

Figure 7-108. Drawing of a dissection of the parotid bed, the temporalis muscle, and the auricular vessels and nerves. The sternocleidomastoid, splenius capitus, and longissimus capitus muscles have been removed from the mastoid process. Note the posterior belly of the digastric muscle arising deep to the mastoid process and passing deep to the angle of the mandible. This muscle is closely related to the parotid gland as shown in Figure 7-110.

The veins from the parotid gland drain into the *retromandibular vein* and then into the *external jugular vein* (Figs. 7-29 and 7-38).

NERVE SUPPLY TO THE PAROTID GLAND (Figs. 7-33, 7-35, and 7-111)

The *parasympathetic component of the glossopharyngeal nerve* (**CN IX**) supplies secretory fibers to the parotid gland via the **otic ganglion**. Stimulation of these fibers produces a thin, watery saliva. Sensory nerve fibers pass to the gland via the great auricular and the auriculotemporal nerves (Fig. 7-29). Sympathetic fibers also pass to the gland via a nerve plexus associated with the external carotid artery.

LYMPH DRAINAGE OF THE PAROTID GLAND (Figs. 7-29, 7-40, and 7-120)

The parotid lymph nodes are located on or deep to the parotid fascia and within the gland. They receive lymph from the forehead, lateral parts of the eyelids, temporal region, lateral surface of the auricle, and the anterior wall of the external acoustic meatus. The lymph nodes within the gland also receive lymph from the middle ear. Lymph from the parotid nodes drains into the superficial and deep **cervical lymph nodes** (Figs. 7-40 and 8-35).

THE TEMPORAL REGION

The oval temporal fossa is bounded superiorly and posteriorly by the temporal lines and anteriorly by the frontal and zygomatic bones (Figs. 7-1 and 7-11). The **temporal fascia** (Fig. 7-31) stretches over the temporal fossa and the **temporalis muscle** (Fig. 7-108). Inferiorly the temporalis fascia splits into two layers, superficial and deep. The superficial layer is attached to the superior margin of the zygomatic arch, and the deep layer passes medial to the arch to become continuous with the fascia deep to the masseter muscle.

The floor of the temporal fossa (Figs. 7-10 and 7-11), which gives origin to the temporalis muscle, is usually formed by portions of four bones: parietal, frontal, greater wing of the sphenoid, and squamous part of the temporal bone. The area where these bones meet is called the **pterion** (Figs. 7-10 and 7-16); see p. 796 for a discussion of this clinically important area.

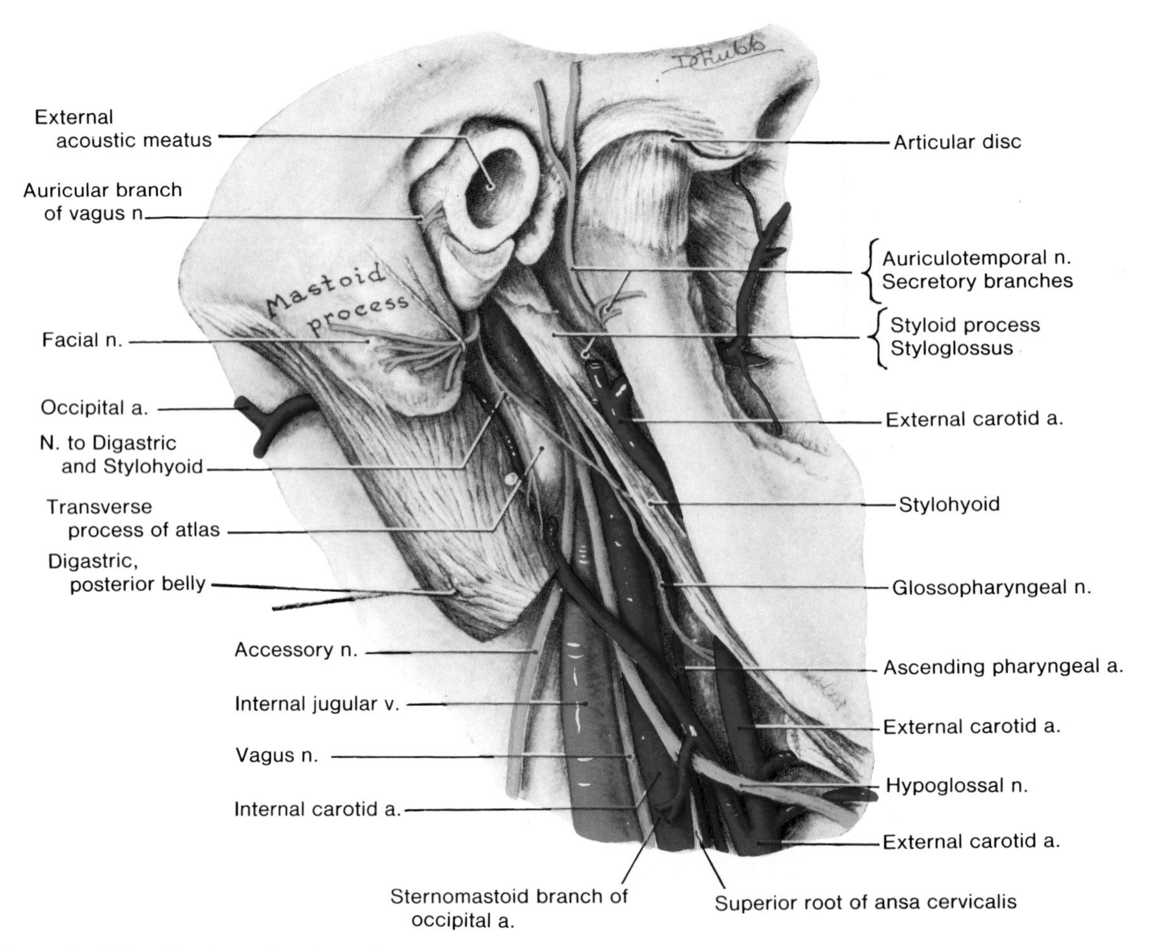

Figure 7-109. Drawing of a dissection of the structures deep to the parotid bed. The facial nerve, the posterior belly of the digastric, and the nerve to it are retracted. Observe the internal jugular vein, the internal carotid artery, and the last four cranial nerves crossing anterior to the transverse process of the atlas and deep to the styloid process.

The temporal fossa contains the fan-shaped temporalis muscle, the "handle" of which passes deep to the zygomatic arch (Figs. 7-108 and 7-112). The temporal fossa is deepest where the **temporalis muscle** is thickest (*i.e.*, anteroinferiorly, Figs. 7-11 and 7-108), where it inserts into the coronoid process of the mandible and the anterior surface of the ramus (Figs. 7-22 and 7-108).

The masseter muscle (Figs. 7-28, 7-29, 7-31, 7-107, and 7-108) is quadrilateral and consists of three superimposed layers that blend anteriorly. It covers the lateral surface of the ramus of the mandible, but not its neck region (Figs. 7-21 and 7-29).

THE INFRATEMPORAL REGION

The infratemporal fossa is an irregularly shaped space posterior to the maxilla (Figs. 7-107 and 7-113 to 7-114). This fossa communicates with the temporal fossa via the interval between the zygomatic arch and the skull, which is traversed by the temporalis muscle and the deep temporal nerves and vessels.

BONES AND WALLS OF INFRATEMPORAL FOSSA

The lateral wall of this fossa is the *ramus of the mandible* (Fig. 7-11). Its **medial wall** is formed by the *lateral pterygoid plate* (Fig. 7-113).

The anterior wall of the infratemporal fossa is formed by the *infratemporal surface of the maxilla* (Fig. 7-113). This wall is limited superiorly by the inferior orbital fissure and medially by the pterygomaxillary fissure. Hold a dried skull up to a light and observe the eggshell thickness of anterior wall of the infratemporal fossa, which is formed by the posterior wall of the maxillary sinus (Figs. 7-89 and 7-146).

The posterior wall of the infratemporal fossa is formed by the *anterior surface of the condylar process*

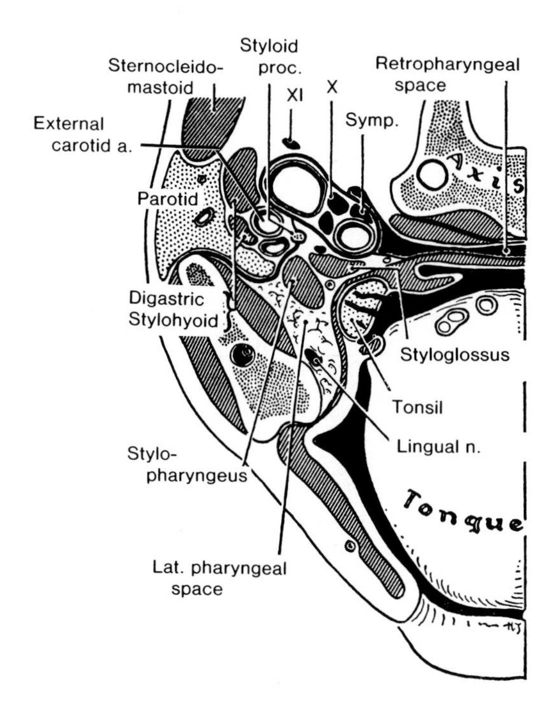

Figure 7-110. Drawing of a cross-section at the level of the parotid gland. Observe that this gland fills the wedge-shaped parotid bed, and that the digastric and stylohyoid muscles intervene between it and the great vessels and nerves of the neck. Note also the masseter muscle inserted into the lateral surface of the ramus of the mandible, and the medial pterygoid muscle inserted into its medial surface.

of the mandible (Figs. 7-19 and 7-22) and the styloid process of the temporal bone (Fig. 7-10).

The roof of the infratemporal fossa is flat and is formed mainly by the inferior surface of the *greater wing of the sphenoid bone* (Figs. 7-11 and 7-19). The roof is separated from the temporal fossa by a ragged edge called the *infratemporal crest* (Fig. 7-113).

Observe the foramen ovale in the roof of the infratemporal fossa (Figs. 7-15*A* and 7-113). *The lateral pterygoid plate is the bony guide to this foramen.* The foramen ovale transmits the mandibular division of the trigeminal nerve (Figs. 7-33 and 7-69).

The inferior boundary of the infratemporal fossa is the point where the medial pterygoid muscle inserts into the medial aspect of the mandible near the angle (Fig. 7-114).

CONTENTS OF THE INFRATEMPORAL FOSSA

This fossa contains the inferior part of the ***temporalis muscle***, the medial and lateral ***pterygoid muscles***, the ***maxillary artery***, the ***pterygoid venous plexus***, the ***mandibular*** **and** ***chorda tympani nerves***, the ***otic ganglion***, the ***inferior alveolar, lingual***, **and** ***buccal nerves***.

The temporalis muscle (Fig. 7-112) is described subsequently with the temporomandibular joint.

The Maxillary Artery (Figs. 7-114 to 7-116). This vessel is the larger of the two terminal branches of the **external carotid artery.** It arises posterior to the neck of the mandible. The maxillary artery passes anteriorly, deep to the neck of the mandibular condyle and *traverses the infratemporal fossa.* It passes superficial to the lateral pterygoid muscle (Fig. 7-114) and then disappears in the infratemporal fossa.

The maxillary artery is divided into three parts by the lateral pterygoid muscle (Fig. 7-116).

The branches of the first part of the maxillary artery are: (1) the **deep auricular artery** to the external acoustic meatus; (2) the **anterior tympanic artery** to the tympanic membrane (eardrum); (3) the **middle meningeal artery** (4) **accessory meningeal arteries** to the cranial cavity via the foramen spinosum and the foramen ovale, respectively; and (5) the **inferior alveolar artery** to the mandible, gingivae (gums), and teeth.

The middle meningeal artery is of considerable sur-

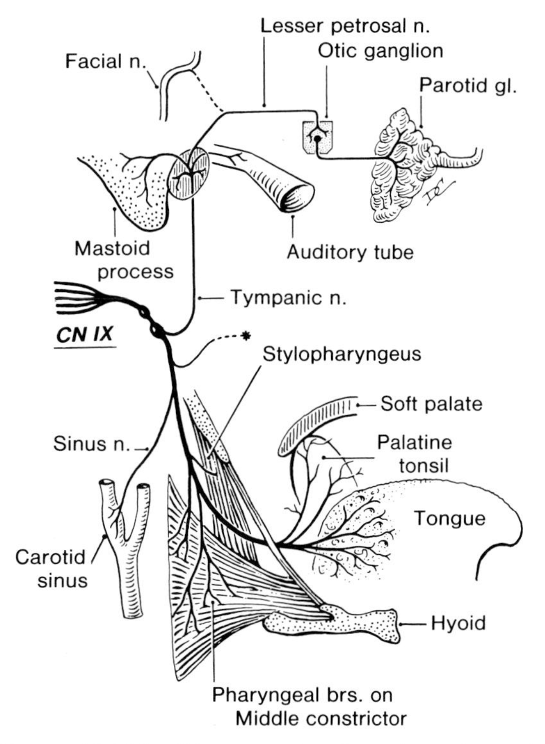

Figure 7-111. Diagram showing the glossopharyngeal nerve (CN IX) sending secretomotor fibers to the parotid gland via its tympanic branch, the lesser petrosal nerve, the otic ganglion, and the auriculotemporal nerve.

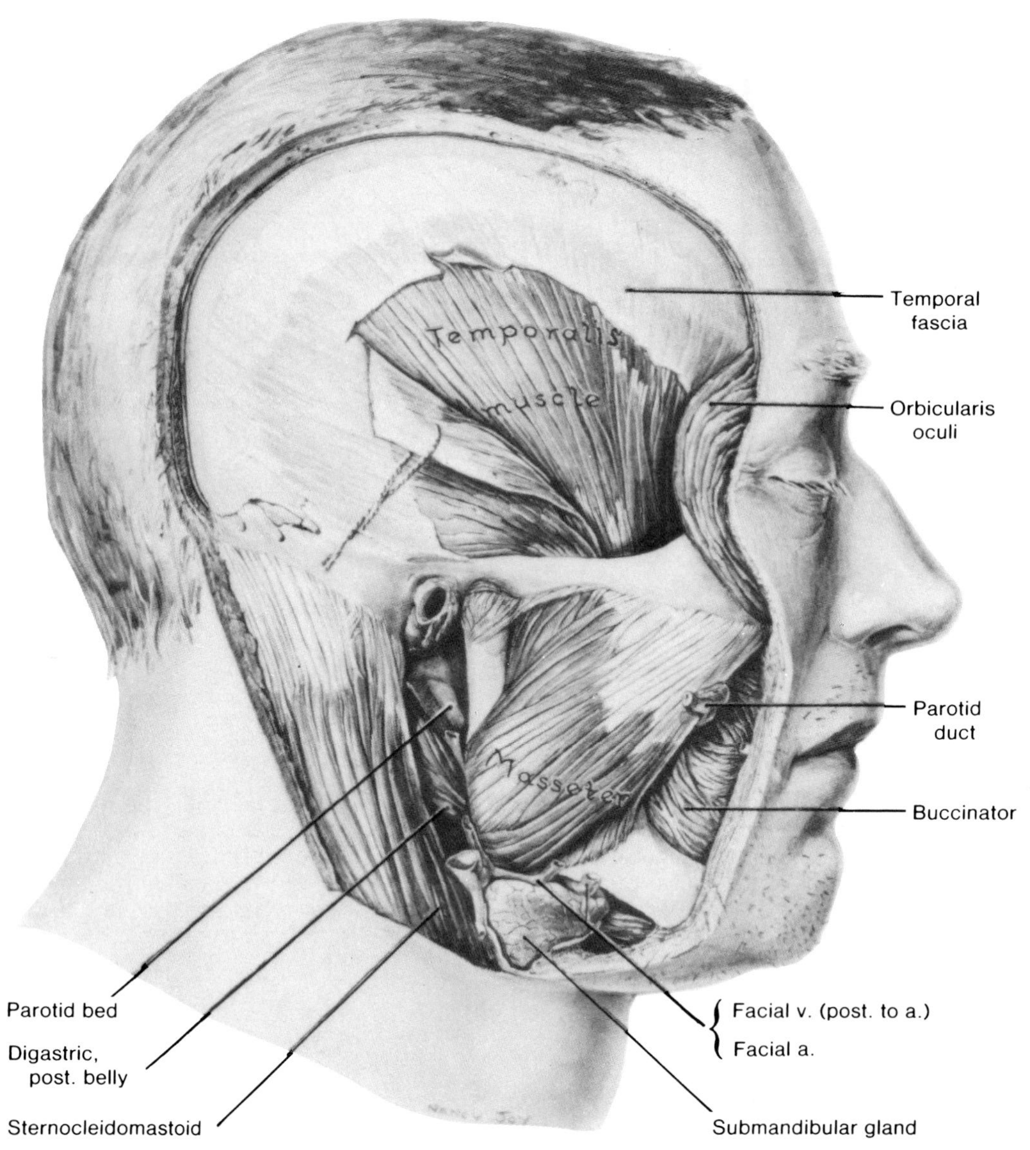

Figure 7-112. Drawing of the large muscles on the side of the head, especially the temporalis and masseter muscles. Both are supplied by the trigeminal nerve and both close the jaw. The temporalis arises in part from the overlying temporal fascia. Most of the parotid duct has been removed (see it in Figs. 7-29 and 7-31).

gical importance (Fig. 7-50). The largest of the meningeal arteries, **the middle meningeal is the principal artery to the calvaria.** It ascends between the two roots of the auricotemporal nerve (Figs. 7-115 and 7-116) and enters the skull via the foramen spinosum (Figs. 7-62 and 7-63) to *supply the dura mater and the interior of the calvaria.*

Owing to the relatively thin bone in the floor of the temporal fossa, fractures in this area are not uncommon. In these cases, *lacerations of the middle meningeal artery may occur.* This results in an **extradural hemorrhage** (see p. 864).

The branches of the second part of the maxillary artery supply muscles through masseteric, deep temporal, pterygoid, and buccal branches (Fig. 7-116).

The branches of the third or pterygopalatine part of the maxillary artery arise just before and after it enters the pterygopalatine fossa (Fig. 7-116): (1) *posterior superior alveolar,* (2) *middle superior alveolar,* (3) *infraorbital,* (4) *descending palatine,* (5)

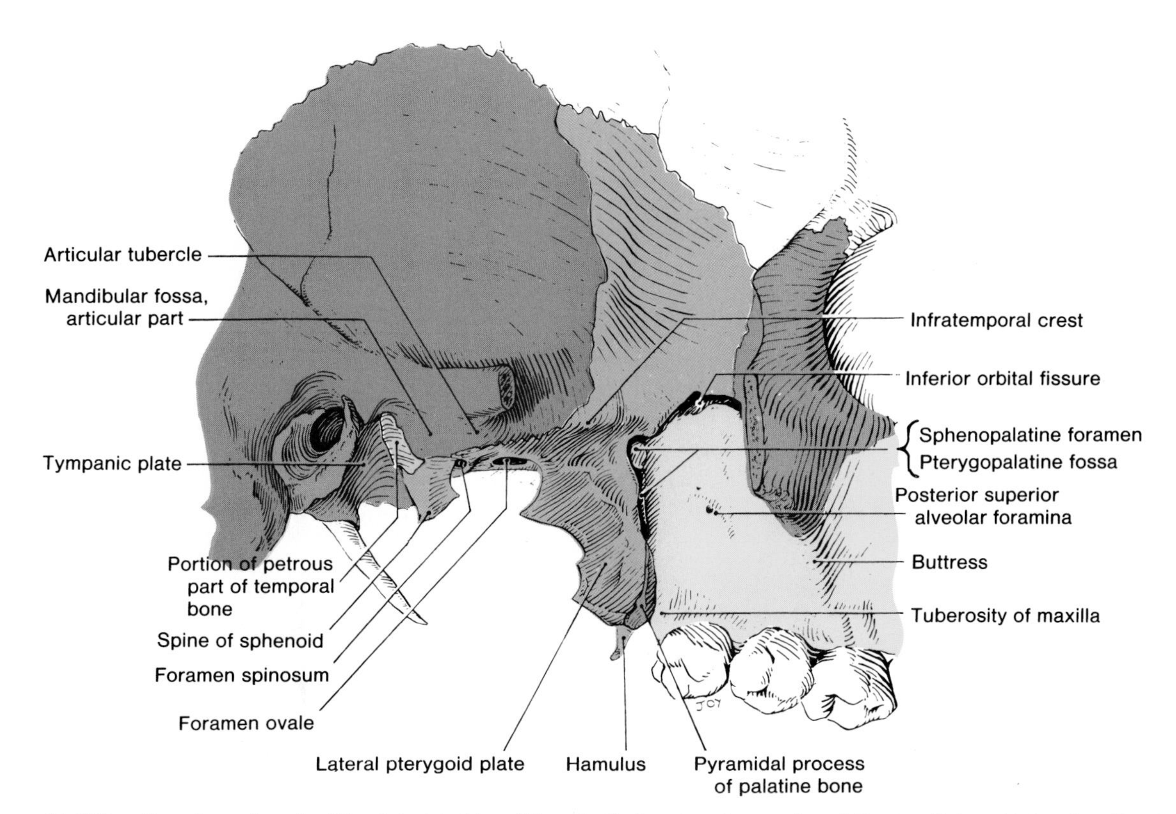

Figure 7-113. Drawing of part of the lateral side of the skull showing the roof and the medial and lateral walls of the infratemporal fossa. The ramus of the mandible which forms the lateral wall of the fossa is not illustrated (see Fig. 7-19). Note that the posterior free border of the lateral pterygoid plate, when followed superiorly, leads to the foramen ovale in the roof of the fossa. Superiorly the anterior border of the lateral pterygoid plate forms the posterior limit of the pterygomaxillary fissure, which is the entrance to the pterygopalatine fossa. The pterygopalatine part of the maxillary artery passes through the pterygomaxillary fissure and enters the pterygopalatine fossa (Figs. 7-114 and 7-115).

artery of the pterygoid canal, (6) *pharyngeal,* and (7) *sphenopalatine.*

The Pterygoid Plexus (Fig. 7-39). This important venous plexus is located partly between the temporalis and lateral pterygoid muscles, and partly between the two pterygoid muscles. It has connections with the facial vein via the cavernous sinus.

The Mandibular Nerve (Figs. 7-33, 7-62, 7-115, and Table 7-5). All nerves of the infratemporal region except the chorda tympani, a branch of the facial (CN VII), are derived from this inferior division of the trigeminal nerve. Descending through the foramen ovale into the infratemporal fossa from the medial part of the middle cranial fossa (Figs. 7-15*A* and 7-62), the mandibular nerve (CN V^3) divides at once into sensory and motor fibers (Fig. 7-33).

Branches of the mandibular nerve supply the four muscles of mastication (temporalis, masseter, and medial and lateral pterygoids), but not the buccinator, which is supplied by the facial nerve (CN VII).

Table 7-5
Branches of the Mandibular Nerve (CN V^3)

Muscular Branches	Sensory Branches	Other Branches
Temporalis and Masseter	Auriculotemporal	Taste
Medial and Lateral pterygoids	Inferior alveolar	Secretory
Tensor veli palatini and Tensor tympani	Lingual	Articular
Mylohyoid and Digastric (anterior belly)	Buccal	

The area of facial skin derived from the mandibular prominence (process) of the first embryonic branchial arch, which CN V^3 supplies, is illustrated in Figure 7-32.

The three sensory branches of the mandibular nerve

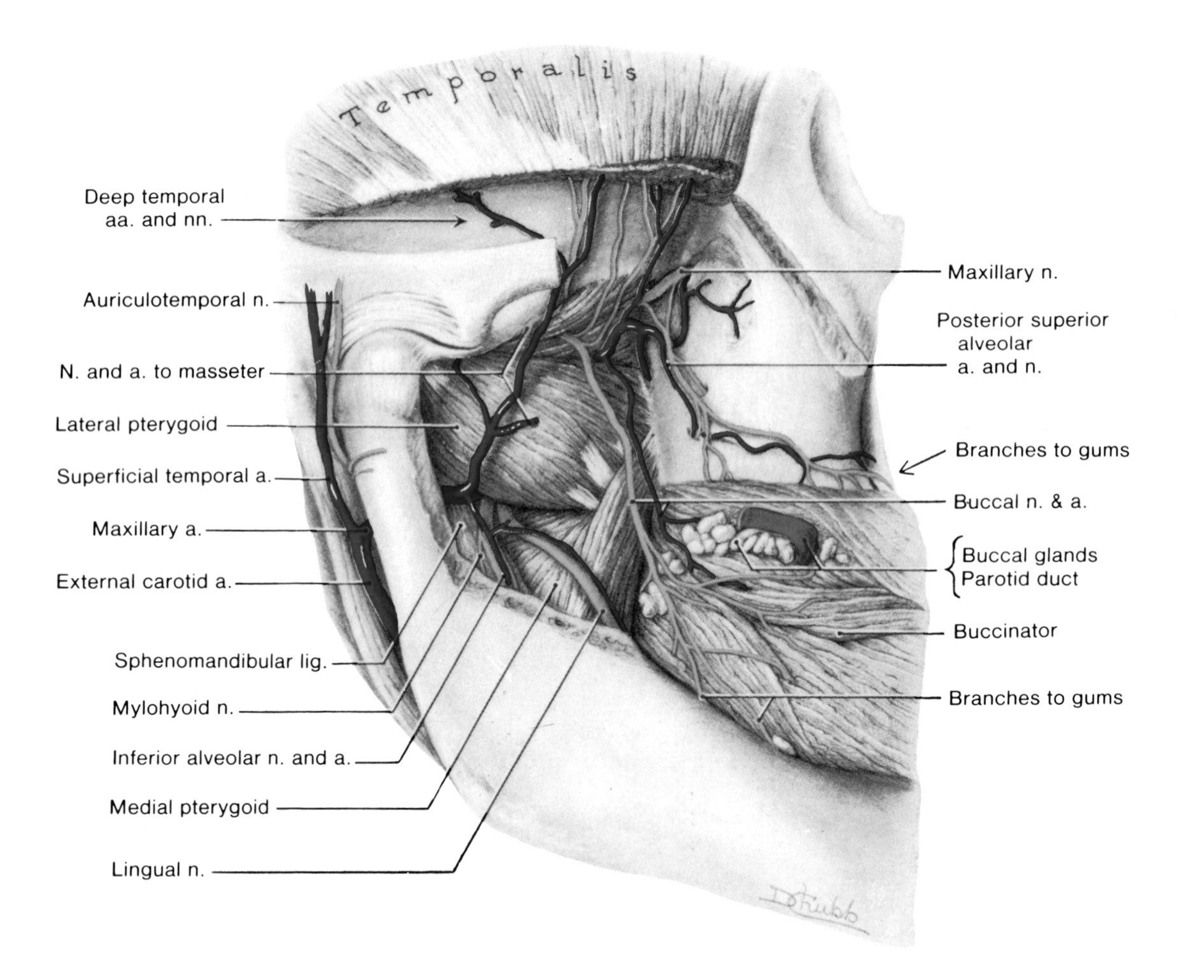

Figure 7-114. Drawing of a superficial dissection of the infratemporal region. Observe the maxillary artery, the larger of the two end branches of the external carotid, running anteriorly, deep to the neck of the mandible. Note that the maxillary artery disappears deep to the lateral pterygoid muscle and reappears between its two heads before plunging into the pterygopalatine fossa.

to the face are the **auriculotemporal, inferior alveolar, and buccal nerves** (Figs. 7-32 and 7-33). The mandibular nerve also supplies sensory fibers to the gingivae of the mandible via the lingual (medial side) and buccal (lateral side) nerves. The mandibular teeth are supplied by the inferior alveolar nerve (Fig. 7-33).

The Auriculotemporal Nerve (Figs. 7-29, 7-31 to 7-33, 7-107, 7-109, and 7-114) encircles the middle meningeal artery and breaks up into numerous branches, the largest of which passes posteriorly, medial to the neck of the mandible and *supplies sensory fibers to the auricle and the temporal region.* The auriculotemporal nerve also *sends articular fibers to the temporomandibular joint, and secretomotor fibers (parasympathetic) to the parotid gland* (Fig. 7-33).

The Buccal Nerve (Figs. 7-20, 7-29, 7-33, 7-114, and 7-115), a long sensory nerve, usually runs between the two heads of the lateral pterygoid muscle. It descends through the deep fibers of the temporalis muscle, where its branches spread out over the lateral surface of the buccinator muscle (Figs. 7-114 and 7-115).

The buccal nerve supplies sensory fibers to the skin and mucous membrane of the cheek, as well as to the mandibular buccal gingiva in the molar region (Fig. 7-33).

The Inferior Alveolar Nerve (Figs. 7-20, 7-33, 7-114, 7-115, and 7-117) *is commonly anesthetized in dental practice.* It enters the mandibular foramen, passes through the mandibular canal and ***appears on the face as the mental nerve.*** While in the mandibular canal, it sends nerves to all teeth in the mandible on its side.

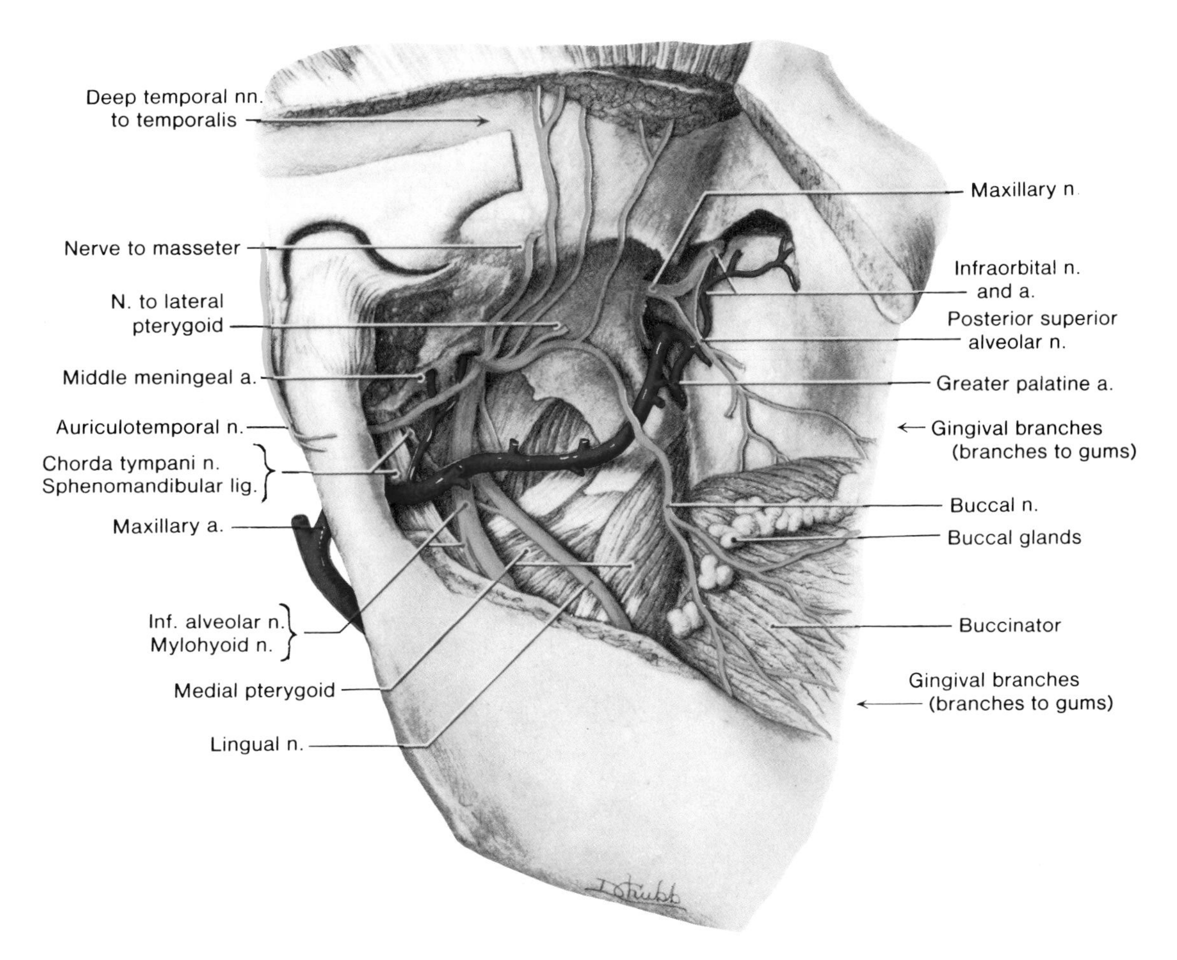

Figure 7-115. Drawing of a dissection of the infratemporal region, deeper than the one illustrated in Figure 7-114. The lateral pterygoid muscle and most branches of the maxillary artery have been removed. Observe the maxillary artery and auriculotemporal nerve passing between the sphenomandibular ligament and the neck of the mandible. Note that the mandibular nerve (CN V^3) enters the infratemporal fossa through its roof via the foramen ovale. Observe the inferior alveolar and lingual nerves descending on the medial pterygoid muscle, the former giving off the mylohyoid nerve (to the mylohyoid muscle and the anterior belly of digastric), the latter receiving the chorda tympani nerve which carries secretory fibers and fibers of taste (Fig. 7-139). Note that the buccal branch of the mandibular nerve (CN V^3) is sensory. The buccal branch of the facial nerve (CN VII) is the motor supply to the buccinator muscle.

The Mental Nerve, a branch of the inferior alveolar nerve (Figs. 7-28 and 7-33), passes to the face through the mental foramen, which is generally in the region of the second premolar tooth (Fig. 7-20). The mental nerve supplies the skin and mucous membrane of the lower lip, the skin of the chin, and the vestibular gingiva of the mandibular incisors (Fig. 7-33).

The Lingual Nerve (Figs. 7-33, 7-114, 7-115, and 7-117), a long nerve, lies anterior to the inferior alveolar nerve. It is sensory to the tongue (L. *lingual*, tongue), the floor of the mouth, and the gingivae. It enters the mouth between the medial pterygoid muscle and the ramus of the mandible and passes anteriorly under cover of the oral mucosa, just inferior to the third molar tooth.

Because the lingual nerve passes submucously along the lingual aspect of the aveolar area of the third molar tooth (Figs. 7-115 and 7-117), care must be taken not to damage it during operative procedures in the mouth (*e.g.*, removal of a third molar or "wisdom" tooth).

The **inferior alveolar nerve**, as it passes through the mandibular canal, is close to the root of the third molar tooth (Fig. 7-118). Hence, it may also be injured during removal of a malposed and impacted third molar tooth.

Local anesthesia may be applied to the mandibular nerve, a procedure called a **mandibular**

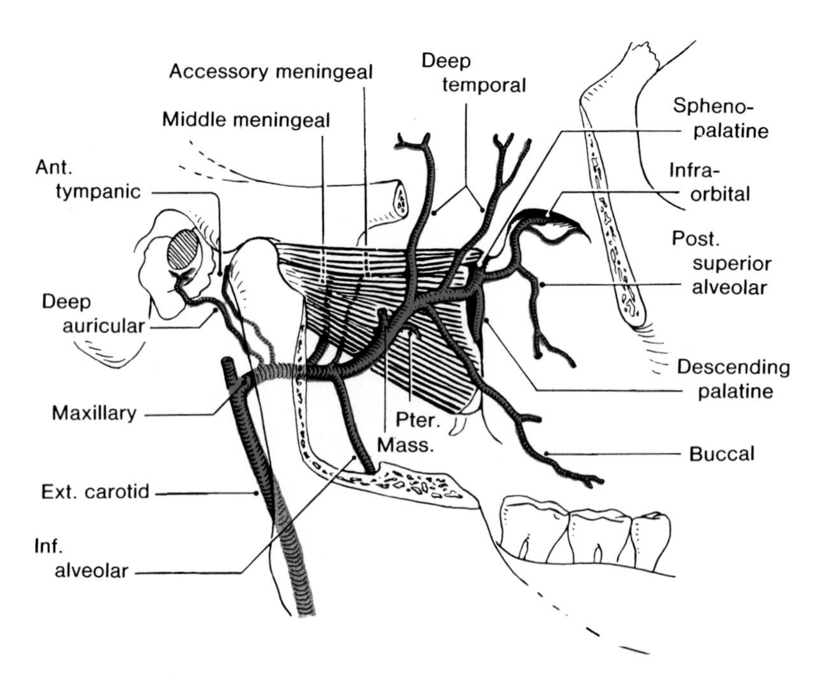

Figure 7-116. Diagram of the maxillary artery, the larger of the two terminal branches of the external carotid artery. Observe that it arises at the neck of the mandible and is divided into three parts by the lateral pterygoid muscle.

nerve block, as it emerges from the foramen ovale and enters the infratemporal fossa. The injection needle is passed (about 5 cm) through the mandibular notch (Figs. 7-21 and 7-108) into the infratemporal fossa (**extraoral approach**), where the local anesthetic agent is injected. The following nerves are usually anesthetized: the auriculotemporal, inferior alveolar, lingual, and buccal. Thus, all areas of skin innervated by the mandibular nerve (CN V^2) and its branches are anesthetized (Figs. 7-32 and 7-33).

More commonly, dentists and oral surgeons plan only to block some branches of the mandibular nerve (*e.g.*, the inferior alveolar and lingual nerves). The anesthetic fluid blocks the transmission of pain stimuli to the brain. As a result, the nerve's area of supply lacks sensation (*i.e.*, it's numb)

Inferior Alveolar Nerve Block. This procedure is performed regularly and routinely in dental offices; it is a relatively safe procedure because the patient remains awake. The inferior alveolar nerve is anesthetized by injecting the anesthetic fluid around the mouth of the mandibular canal (*i.e.*, the mandibular foramen).

When an inferior alveolar nerve block is done properly, all the mandibular teeth are anesthetized to the midline. The skin and mucous membrane of the lower lip, the labial alveolar mucosa and gingivae, and the skin of the chin are also anesthetized because they are supplied by the mental branch of the inferior alveolar nerve (Fig. 7-33).

The Chorda Tympani Nerve (Figs. 7-35, 7-115, and 7-165). The lingual nerve is joined by the chorda tympani branch of the facial nerve (CN VII). It leaves this nerve in the facial canal and crosses the medial aspect of the tympanic membrane to join the lingual nerve posteriorly, near the inferior border of the lateral pterygoid muscle. The chorda tympani nerve comes through the petrotympanic fissure in the mandibular fossa of the temporal bone.

The Otic Ganglion (Figs. 7-33 and 7-111). This *parasympathetic ganglion* is located in the infratemporal fossa, just inferior to the foramen ovale, medial to the mandibular nerve, and posterior to the medial pterygoid muscle.

Preganglionic parasympathetic fibers, derived mainly from the ***glossopharyngeal nerve*** (Fig. 7-111), synapse in the otic ganglion. Postganglionic parasympathetic fibers, which are secretory to the parotid gland, pass from this ganglion to the auriculotemporal nerve (Fig. 7-33).

THE TEMPOROMANDIBULAR JOINT

The temporomandibular articulation is a **synovial joint**, by which the mandible articulates with the cranium (Figs. 7-119 to 7-121). The temporomandibular joint is between (1) **the head of the mandible** inferiorly and (2) **the articular tubercle** and the **mandibular fossa** of the temporal bone superiorly. Between the two bones involved (mandible and temporal), there is an oval **fibrocartilaginous articular disc** which divides the joint cavity into superior and inferior compartments, The articular disc of the joint is fused with the articular capsule surrounding the joint, and through this attachment it is bound superiorly to the limits of the temporal articular surface and inferiorly to the neck of the mandible.

The articular disc of the temporomandibular joint is more firmly attached to the mandible than it is to the temporal bone. Thus, when the head of the mandible slides anteriorly on the articular tubercle as the mouth is opened (Fig. 7-121*B*), the articular disc slides

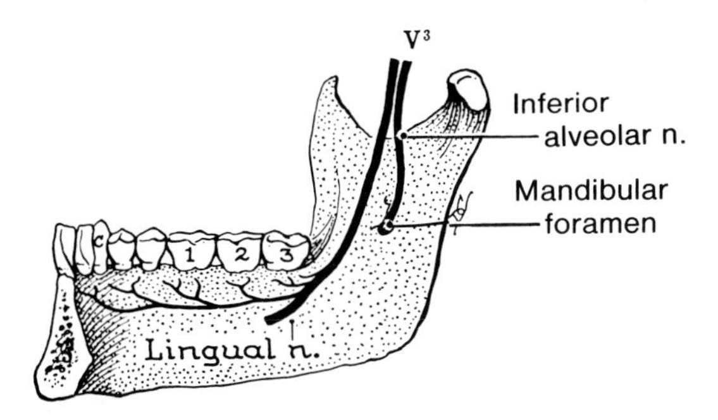

Figure 7-117. Drawing of the medial aspect of the mandible to show the inferior alveolar and lingual nerves, branches of the mandibular division of the trigeminal nerve (CN V^3). The inferior alveolar nerve enters the mandibular foramen and passes through the mandibular canal in the mandible, whereas the lingual nerve passes deep to the mucosa.

anteriorly against the posterior surface of the articular tubercle.

The articular capsule of the temporomandibular joint is loose. The **fibrous capsule** is attached to the margins of the articular area on the temporal bone and around the neck of the mandible. The fibrous capsule is thickened laterally to form the **lateral ligament** or temporomandibular ligament (Fig. 7-108). The base of this triangular ligament is attached to the zygomatic process of the temporal bone and the articular tubercle; its apex is fixed to the lateral side of the neck of the mandible.

Two other ligaments connect the mandible to the cranium, but neither of them provides much strength to the joint.

The **stylomandibular ligament**, a thickened band of deep cervical fascia, runs from the styloid process of the temporal bone to the angle of the mandible separating the parotid and submandibular salivary glands.

The **sphenomandibular ligament** (Figs. 7-115 and 7-120), *a remnant of the first branchial arch cartilage* (Meckel's cartilage), is a long membranous band that lies medial to the joint. The sphenomandibular ligament runs from the spine of the sphenoid bone to the lingula on the medial aspect of the mandible (Fig. 7-22).

The **synovial membrane** lines the fibrous capsule and is reflected superiorly on to the neck of the mandible to the margin of the articular cartilage.

MOVEMENTS OF THE TEMPOROMANDIBULAR JOINT (Fig. 7-121 and Table 7-6)

Two movements occur at this joint, an anterior gliding and a hinge-like rotation. When the mandible

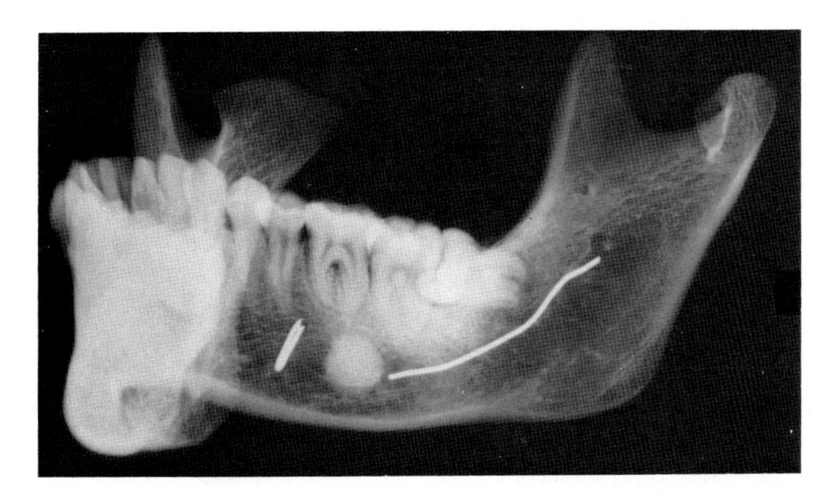

Figure 7-118. Radiograph of the mandible. A copper wire was inserted through the mandibular foramen into the mandibular canal before the radiograph shown was taken. Note how close the inferior alveolar nerve (represented by the wire) passes to the root of the third molar tooth (also see Fig. 7-117). Observe the malposed and impacted third molar tooth and the direction of the emerging wire from the mental foramen (the verticle white line). For another view of this impacted tooth, see Figure 7-127.

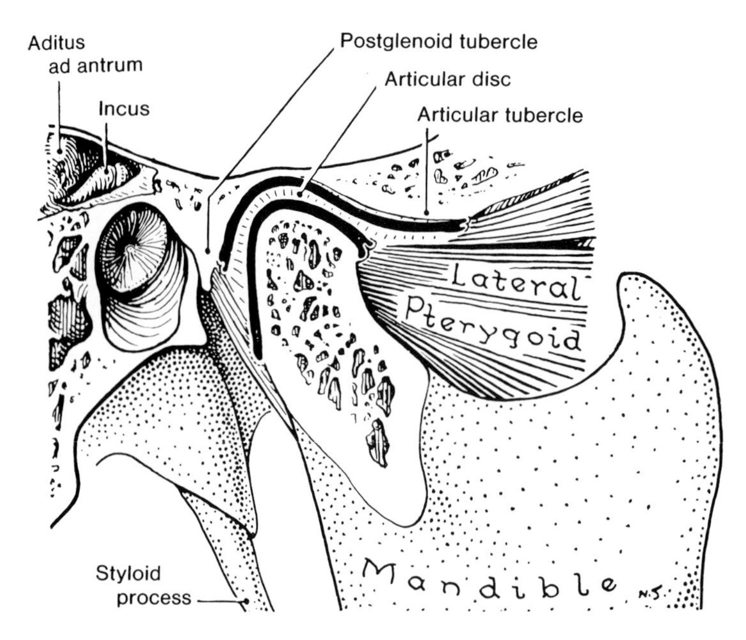

Figure 7-119. Drawing of a sagittal section of the temporomandibular joint. When the mouth opens, the head of the mandible passes anteriorly to a point directly inferior to the articular tubercle in normal persons (Fig. 7-121*B*).

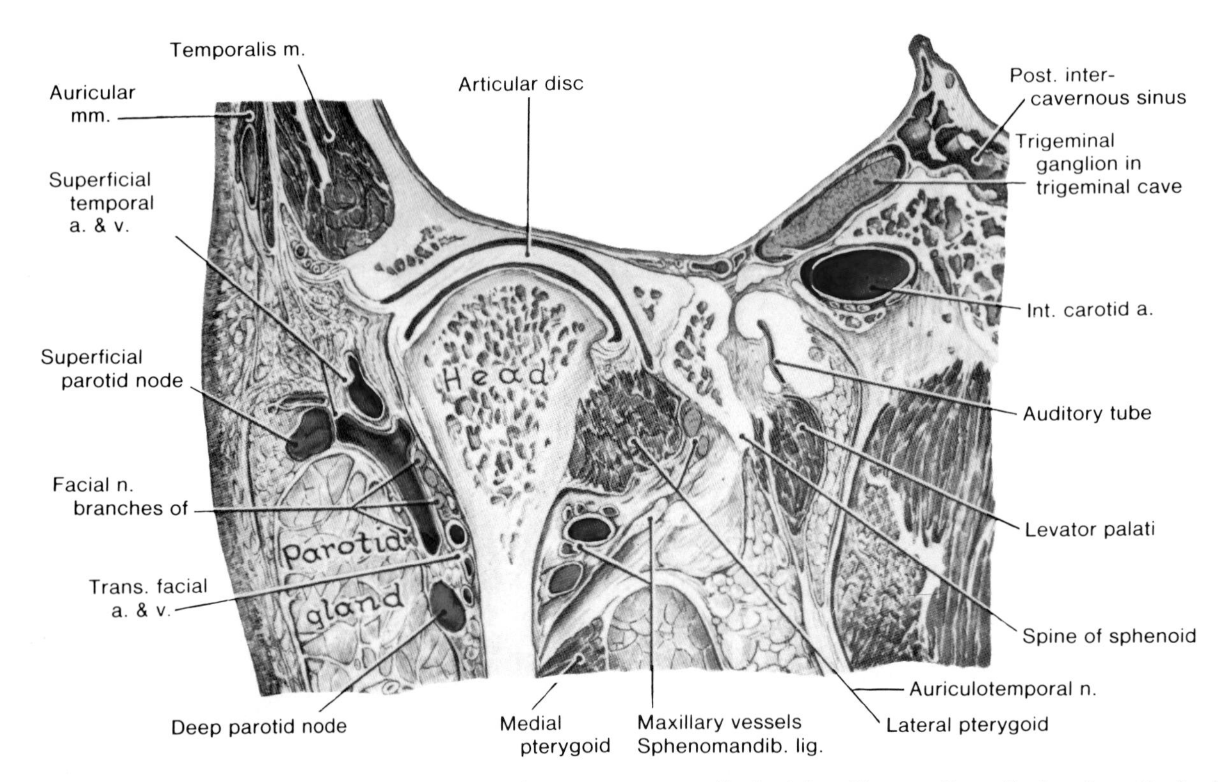

Figure 7-120. Drawing of a coronal section of the temporomandibular joint. Observe the articular disc attached to the neck of the mandible medially and laterally, partly in conjunction with the lateral pterygoid muscle. Note the articular disc dividing the articular cavity into superior and inferior compartments, and the lateral pterygoid muscle inserted in part into the anterior part of the disc. Note also that the roof of the mandibular fossa, separating the head of the mandible and the articular disc from the middle cranial fossa, is thin centrally but thick elsewhere. Also observe the maxillary vessels crossing the neck of the mandible on its medial side and the superficial and deep parotid lymph nodes.

Table 7-6
Movements of the Mandible and the Chief Muscles Involved

Depress (Open Mouth)	Elevate (Close Mouth)	Protract (Protrude Chin)	Side-to-Side (Grinding, Chewing)
Lateral pterygoid Suprahyoid Infrahyoid	Temporalis Masseter Medial pterygoid	Masseter Lateral pterygoid Medial pterygoid	Temporalis of same side Pterygoids of opposite side Masseter

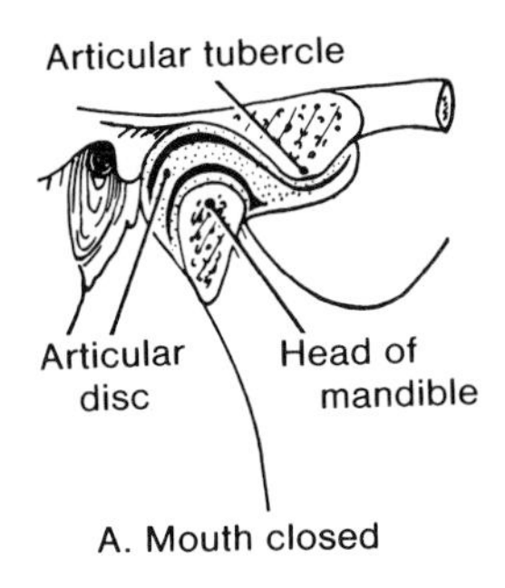

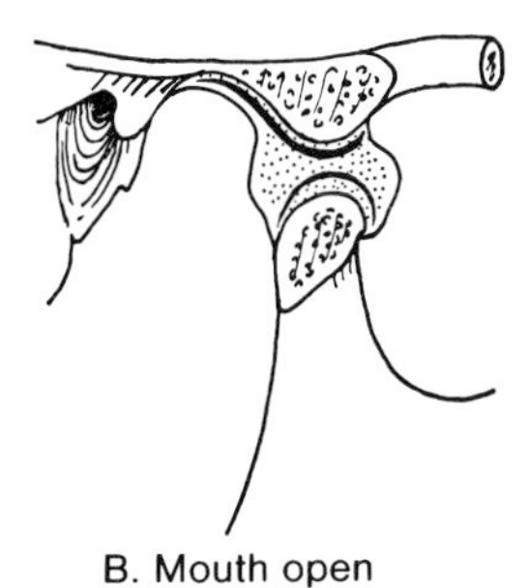

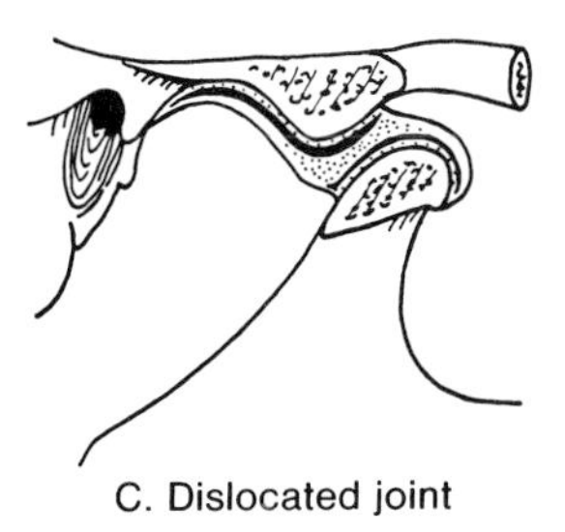

Figure 7-121. Diagrams illustrating the changing position of the head of the mandible and the temporal bone when *A*, the mouth is closed, *B*, the mouth is open, and *C*, the temporomandibular joint is dislocated. Note that the head of the mandible and the articular disc slide anteriorly over the articular tubercle as the head rotates on the disc.

is depressed during opening of the mouth, the head of the mandible and articular disc move anteriorly on the articular surface until the head lies inferior to the articular tubercle (Fig. 7-121*B*).

As this anterior gliding occurs, the head of the mandible rotates on the inferior surface of the articular disc (Fig. 7-121*B*). This permits simple chewing or grinding movements over a small range. During this kind of chewing, both types of movement of the joint occur, *i.e.*, anterior gliding and hinge-like rotation of the head of the mandible.

The axes of these two movements are different. In simple opening of the mouth, the axis of the anterior gliding movement passes approximately through the mandibular foramen, whereas in hinge-like rotation the axis is through the head of the mandible. Because of these axes of movement, the vessels and nerves entering the mandible are not excessively stretched when the mouth is wide open.

In **protraction** and **retraction** of the mandible, the head and articular disc slide anteriorly and posteriorly on the articular surface of the temporal bone, both sides moving together. The **grinding movement** occurs when protraction and retraction of the mandible alternate on the two sides.

MUSCLES ACTING ON THE TEMPOROMANDIBULAR JOINT

The mandible may be depressed or elevated, protracted or retracted (Table 7-6); considerable rotation also occurs. These movements are controlled mainly by the muscles acting on the joint. The various movements result from the cooperative activity of several muscles bilaterally or unilaterally.

Movements of the temporomandibular joint result chiefly from the action of the **muscles of mastication** (temporalis, masseter, and medial and lateral pterygoids). The temporalis, masseter, and medial pterygoid muscles produce the biting movement (*i.e.*, elevate the mandible and close the mouth). The mandible is protracted by the lateral pterygoid muscles with help from the medial pterygoids, and retracted largely by the posterior fibers of the temporalis muscle. Gravity is sufficient to depress the mandible, but if there is resistance (*e.g.*, a person wearing a chin strap), the muscles listed in Table 7-6 and the mylohoid and anterior digastric muscles are activated.

The Temporalis Muscle (Figs. 7-108, 7-112, 7-114, and 7-120). This extensive fan-shaped muscle, covering the temporal region, is a powerful masticatory (biting) muscle that can easily be seen and felt during closure of the mandible.

Origin. Floor of **the temporal fossa and the temporal fascia.**

Insertion (Fig. 7-22). **Coronoid process** and anterior border of the **ramus of the mandible.**

Actions (Fig. 7-121 and Table 7-6). **Elevates mandible** (closes mouth) **and retracts mandible** (posterior fibers) after protraction.

Nerve Supply (Figs. 7-114 and 7-115). Deep temporal branches of the **mandibular nerve** (CN V^3).

The Masseter Muscle (Figs. 7-29, 7-31, 7-107, 7-108, and 7-112). This quadrangular muscle covers the lateral aspect of the ramus and coronoid process of the mandible. The Greek word *masētēr* means masticator or chewer. Place your fingers on your cheek and tense your masseter by clenching your teeth. You should have no difficulty feeling this muscle because of its proximity to the skin.

Origin (Fig. 7-112). Inferior margin and deep surface of the **zygomatic arch.**

Insertion (Fig.7-21). Lateral surface of the **ramus** and **coronoid process** of the mandible.

Actions (Table 7-6). **Elevates the mandible, clenches the teeth, and helps to protract the mandible.**

Nerve Supply (Figs. 7-108 and 7-114). **Mandibular nerve** (CN V^3) via a branch that enters its deep surface.

The Lateral Pterygoid Muscle (Figs. 7-114, 7-116, 7-119, and 7-120). This short thick muscle has *two heads of origin.*

Origin (Fig. 7-113). *Superior head,* **infratemporal surface of the greater wing of the sphenoid bone;** *inferior head,* lateral surface of **the lateral pterygoid plate.**

Insertion (Fig. 7-22). **Neck of the mandible** and **the articular disc** of temporomandibular joint.

Actions (Table 7-6). Acting together, the lateral pterygoid muscles **protrude the mandible and depress the chin.** Acting alone and alternately, they produce **side-to-side movement** of the mandible.

Nerve Supply (Figs. 7-33, 7-108, and 7-114). **Mandibular nerve** (CN V^3) via a branch from anterior trunk that enters its deep surface.

The Medial Pterygoid Muscle (Figs. 7-114, 7-115, and 7-120). This thick, quadrilateral muscle also has *two heads of origin* which embrace the inferior head of the lateral pterygoid muscle. The medial pterygoid muscle is located deep to the ramus of the mandible.

Origin (Fig. 7-113). *Deep head* (most of muscle), medial surface of the **lateral pterygoid plate.** *Superficial head,* **tuberosity of the maxilla.**

Insertion (Fig. 7-22). **Medial surface of the mandible,** near its angle.

Actions (Fig. 7-6). **Assists in elevating and protruding the mandible.** Acting alone, it pulls the chin to the opposite side; when the muscles act alternately, they produce a grinding motion.

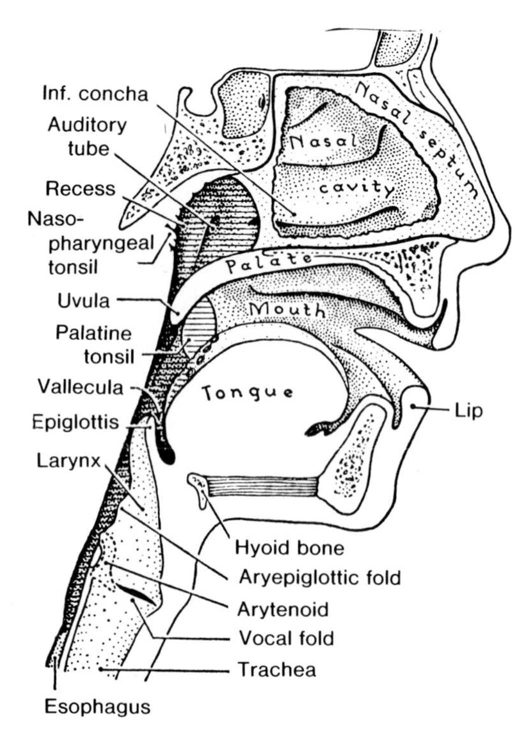

Figure 7-122. Drawing of a sagittal section of the anterior part of the head and neck showing the nose, mouth, pharynx, and larynx.

Nerve Supply (Figs. 7-33, 7-108, and 7-114). Branch from **mandibular nerve** (CN V^3).

The temporomandibular joint usually dislocates anteriorly (Fig. 7-121*C*). During normal opening of the mouth, the head of the mandible and the articular disc move anteriorly to the articular tubercle of the zygomatic process of the temporal bone (Fig. 7-121*B*). During yawning or taking a large bite, contraction of the lateral pterygoid muscles may cause the heads of the mandible to dislocate, *i.e.,* pass anterior to the articular tubercle (Fig. 7-121*C*). In this position, the mandible remains wide open and the person is unable to close it.

Dislocation of the temporomandibular joint can also occur during extraction of teeth if the mandible is depressed excessively. Most commonly, the joint is dislocated by a blow to the chin when the mouth is open (*e.g.,* when a person is laughing, gaping, or yawning). The displacement is usually bilateral and the mandible projects with the mouth fixed in an open position.

Fractures of the mandible (Fig. 7-25) may be accompanied by dislocation of the temporo-

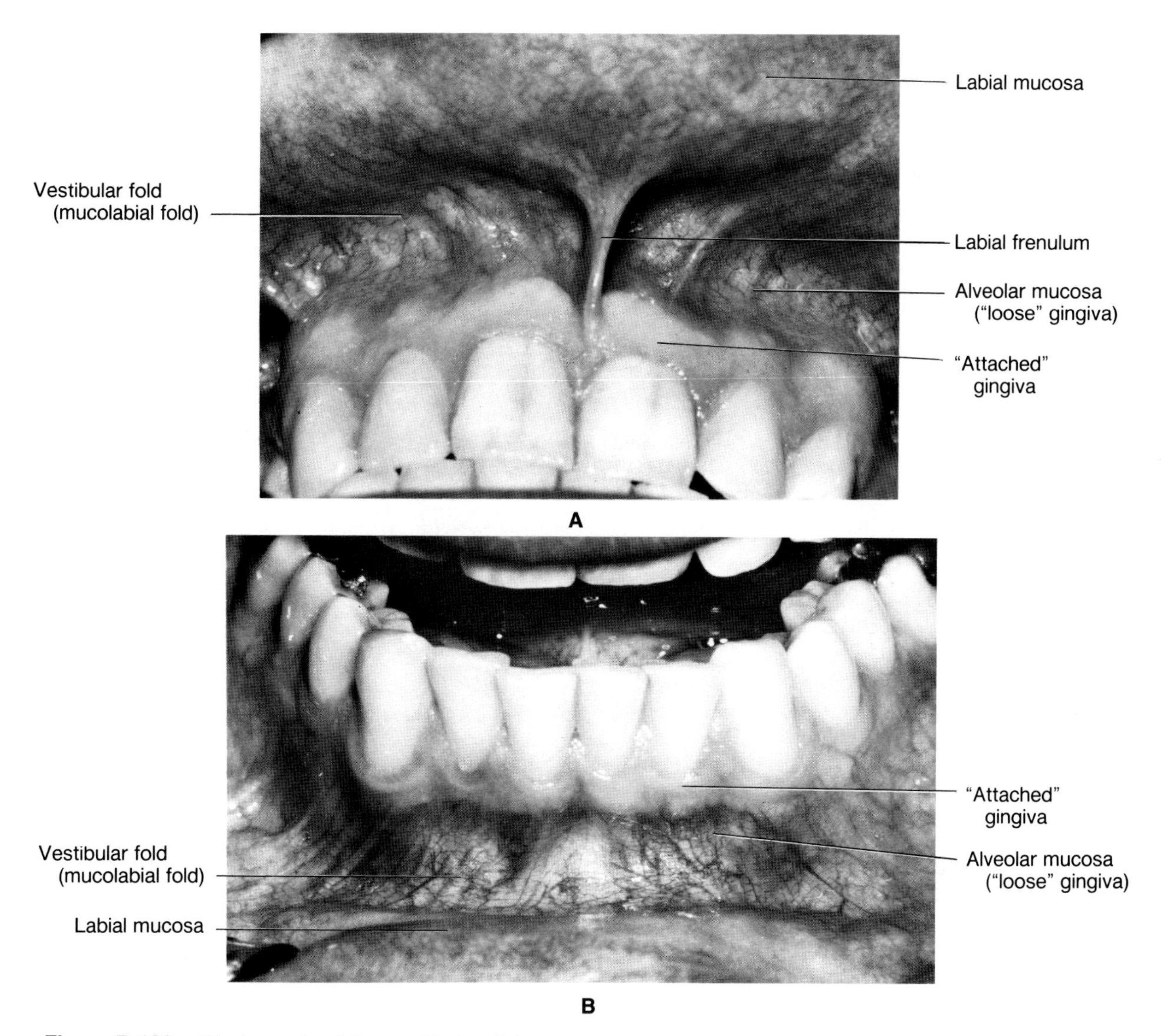

Figure 7-123. Photograph of the vestibule of the mouth and the gingivae. *A*, of the maxilla. *B*, of the mandible (Reprinted with permission from Liebgott B: *The Anatomical Basis of Dentistry*. WB Saunders Company, Philadelphia, 1982.)

mandibular joint(s), a possibility that is not to be overlooked in treating these fractures.

Because of the close relationship of the facial (CN VII) and auriculotemporal (branch of CN V^3) nerves to the temporomandibular joint (Figs. 7-107 and 7-120), care must be taken during operations on the joint to preserve the branches of the facial nerve overlying it, and the articular branches of the auriculotemporal nerve that enter the posterior part of the joint. Injury to the facial nerve causes **facial paralysis** (Fig. 7-100).

Injury to the articular branches of the auriculotemporal nerve supplying the temporomandibular joint, associated with traumatic dislocation and rupture of the articular capsule and/or the lateral ligament, leads to *joint laxity and instability of the joint.*

THE MOUTH

The mouth is the first part of the digestive tube (Fig. 7-122). It is also used in breathing. The **oral cavity** (mouth cavity) consists of a small external part, the **vestibule,** and larger internal part, **the oral cavity proper** (Figs. 7-123 and 7-124). The oral cavity is bounded externally by the cheeks and lips. The aper-

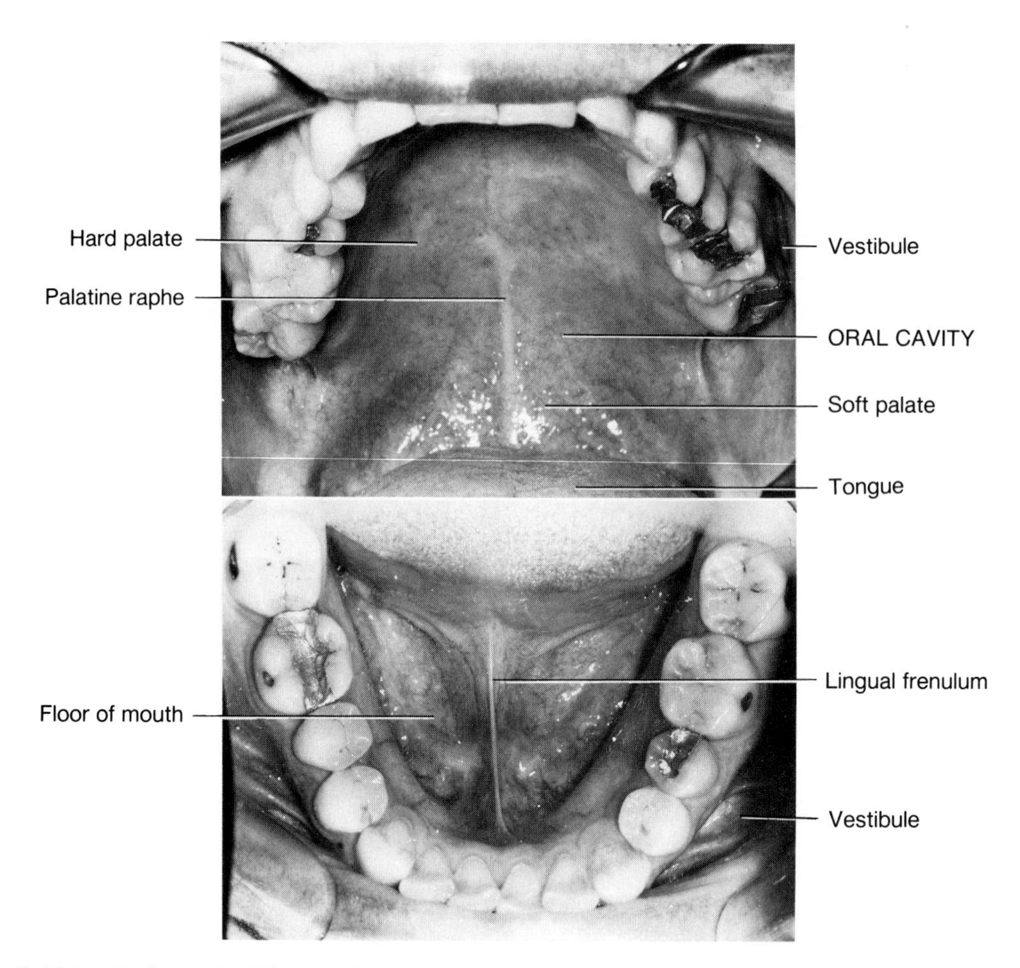

Figure 7-124. Photograph of the mouth or oral cavity. The oral cavity proper is the space between the superior and inferior dental arches. The roof of the oral cavity is formed by the hard palate. The anterior or oral part of the tongue is in contact with the hard palate when the mouth is closed. The floor of the mouth is visible when the tongue is raised. (From Liebgott B: *The Anatomical Basis of Dentistry*, WB Saunders Co, 1982.)

ture between the lips is called the **oral orifice** or opening of the mouth.

The horseshoe-shaped space between the teeth and the inner mucosal lining of the cheeks and lips is called the **oral vestibule.**

The roof of the oral cavity is formed by the ***hard and soft palates*** *and the median, conical process called the* ***uvula*** (L. a grape), in which the soft palate ends (Figs. 7-122, 7-130, and 8-55).

Posteriorly, the oral cavity communicates with the **oropharynx**, the oral part of the pharynx, which is bounded by the soft palate superiorly, the **epiglottis** inferiorly (Fig. 7-130), and the palatoglossal arches laterally (Fig. 7-131).

THE LIPS (Figs. 7-28, 7-122, 7-123, and 7-130)

These muscular folds surrounding the mouth or oral cavity are covered externally by skin and internally by mucous membrane. In between these layers are the muscles of the lips (especially the **orbicularis oris muscle**) and the superior and inferior labial branches of the facial arteries (Figs. 7-28 and 7-31). Recall that the labial arteries anastomose with each other to form an arterial ring (Figs 7-31, 7-37, and 7-153). The pulsations of these arteries can be palpated by grasping the lip lightly between the index finger and the thumb.

Labial salivary glands are located around the orifice of the mouth, between the mucous membrane and the orbicularis oris muscle. These small glands resemble mucous salivary glands in structure. Their ducts open into the vestibule.

The upper and lower lips are attached to the gingivae in the median plane by raised folds of mucous membrane, called the **labial frenula** (Mod. L. dim of L. *frenum*, a little bridle). You can see your labial frenulum when you pull your lower lip inferiorly or your upper lip superiorly (Fig. 7-123*A*).

The junction of the upper lip and the cheek is clearly demarcated by the **nasolabial sulcus**, running laterally from the margin of the nose to the angle of the mouth. This sulcus is particularly obvious during smiling and in older persons. The **mentolabial sulcus** indicates the junction of the lower lip and the chin.

The upper lip has a median, shallow, vertical groove called the **philtrum** (G. *philtron*, a love-charm). Examine your lip, identifying the following: the cutaneous zone, the vermilion border (the red zone), and the mucosal zone.

The vermilion border of the lip (to which lipstick is sometimes applied) *is a distinctive characteristic of humans.* It appears red because of the presence of capillary loops close to the surface which is composed of thin skin.

The sensory nerves of the upper and lower lips are the infraorbital and mental nerves, branches of the maxillary (CN V^2) and mandibular (CN V^3) nerves, respectively (Figs. 7-32 and 7-33).

Lymph vessels from both lips drain into the submandibular lymph nodes (Figs. 7-40 and 7-42). In addition, lymph from the central part of the lower lip drains into the **submental lymph nodes.** The submental nodes drain to either the submandibular lymph nodes or to the juguloomohyoid node.

The submandibular lymph nodes drain to the deep cervical chain of lymph nodes (Figs. 8-35 and 8-41).

Pustules of the upper lip are potentially dangerous because the infection may extend intracranially via the superior labial vein, angular vein, and supraorbital vein (Fig. 7-31), and enter the **cavernous sinus** (Fig. 7-39). See Figure 7-43 for a note on the *danger triangle of the face* and the condition known as **thrombophlebitis of the cavernous sinus** (p. 835). Structures lying in and around it may be involved (Figs. 7-79 and 7-80).

Persons with facial palsy (paralysis of facial nerve) are unable to whistle because the air blows out through their paralyzed lips on that side (Fig. 7-100). When persons with unilateral facial paralysis are asked to show their teeth, the nasolabial fold does not form on the injured side, and the angle of the mouth on that side does not rise (Fig. 7-100). This results from paralysis of the facial muscles, including the orbicularis oris, supplied by the facial nerve and the levator anguli oris, (Figs. 7-28, 7-29, and 7-31).

Cleft lip ("Hare Lip") is a congenital malformation of the upper lip that occurs about once in 1000 births (Fig. 7-125). The clefts vary from a small notch in the vermilion border to ones that extend through the lip into the nose. In severe cases the cleft extends deeper and is continuous with a cleft in the palate.

Cleft lip may be unilateral or bilateral (Fig. 7-125). Unilateral cleft lip is the more common deformity and results from failure of the maxillary prominence on the affected side to fuse with the merged medial nasal prominences during the embryonic period of facial development (For details and illustrations, see Moore, 1982).

Cancer or carcinoma of the lip accounts for about 15% of all head and neck malignancies. It is more common in the lower lip and in males (Fig. 7-44). A major etiologic (causative) factor appears to be extensive exposure to intense sunlight. Using your knowledge of the lymphatic drainage of the lips, you can predict that these tumors will metastasize to the submandibular and/or submental lymph nodes depending on

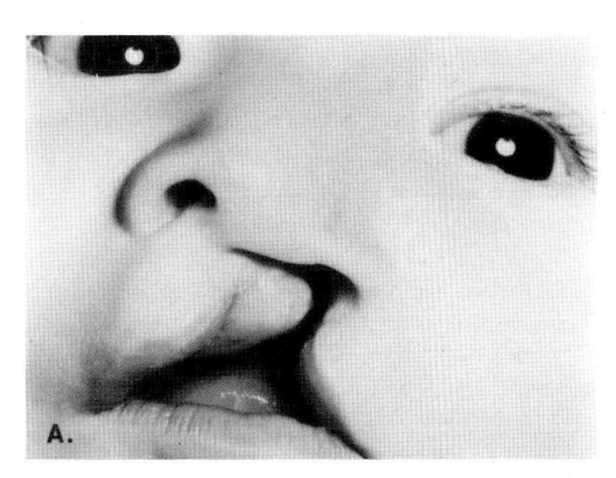

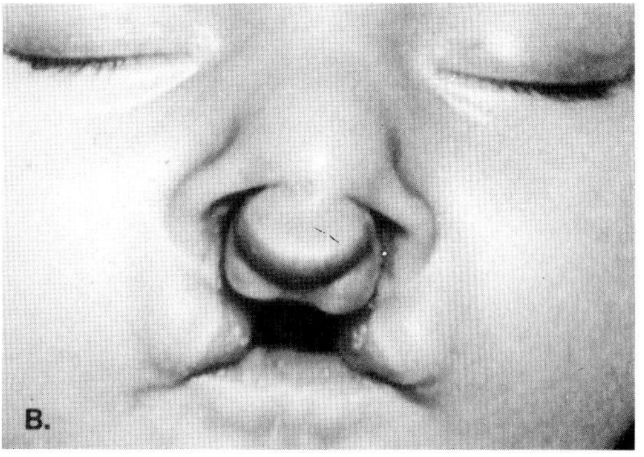

Figure 7-125. Photographs of infants with clefts of the upper lip. *A*, unilateral. *B*, bilateral. In both cases there is loss of continuity of the orbicularis oris muscle; see Figure 7-28. (From Moore KL: *The Developing Human: Clinically Oriented Embryology*, ed 3. WB Saunders Co, Philadelphia, 1982.)

the site of the lesion (Fig. 7-40). *In advanced cases the deep cervical lymph nodes will also be involved* (Figs. 8-35 and 8-41).

THE CHEEKS

The lateral walls of the oral cavity, formed by the cheeks (L. buccae), have essentially the same structure as the lips, with which they are continuous. In common usage, the cheek includes not only the movable parts, but also the prominence of the cheek over the zygomatic bone and arch (Fig. 7-11).

The muscle of the cheek is the **buccinator** (Figs. 7-29, 7-112, and 7-115). Superficial to the fascia covering the buccinator muscle is the **buccal fat pad** that gives the cheeks their rounded contour (Fig. 7-28), particularly in infants. This relatively large fat deposit in infants is thought to reinforce the cheeks and keep them from collapsing during sucking (*e.g.*, on the nipple). For this reason, you will hear this fat deposit referred to as the *buccal sucking fat pad.*

The **buccal glands** (Figs. 7-114 and 7-115) are small mucous glands that are situated between the mucous membrane and the buccinator muscle. There are groups of these glands around the terminal part of the parotid duct, which runs anteriorly from the gland, *external to the masseter muscle* (Figs. 7-29 and 7-31). The parotid duct opens on a small papilla on the oral surface of the cheek, *opposite the crown of the second maxillary molar tooth* (Figs. 7-29, 7-31, and 7-41).

The sensory nerves of the cheeks are branches of the maxillary and mandibular nerves (CN V^2 and CN V^3). They supply the skin of the cheek and the mucous membrane lining the cheek (Figs. 7-32 and 7-33).

The lips and cheeks function as a unit (*e.g.*, during sucking, blowing, eating, and kissing). They act as an oral sphincter in pushing food from the vestibule to the oral cavity proper. The tongue and the buccinator muscles keep the food between the molar teeth during chewing.

In paralysis of the facial nerve, which supplies the muscles of the cheeks and lips, food tends to accumulate in the vestibule (Fig. 7-124) on the affected side. In addition, these patients are unable to puff out the cheek on the paralyzed side. As a consequence, saliva and food dribble out of the corners of their mouths.

THE GINGIVAE (Fig. 7-123)

The gingivae or gums are composed of fibrous tissue which is covered with a mucous membrane. They are attached to the margins of the alveolar processes of the jaws and to the necks of the teeth (Figs. 7-2 and 7-128).

The **alveolar mucosa** ("unattached or loose gingiva") is normally shiny red and nonkeratinizing (Fig. 7-123).

The **gingiva proper** ("attached gingiva") is normally pink, stippled, and keratinizing. It is firmly anchored to the underlying alveolar bone and to the neck of the tooth (Figs. 7-123 and 7-128).

In addition to their nerve supply from the nerves supplying the teeth, the gingivae receive nerve fibers from adjacent sensory nerves (buccal, infraorbital, greater palatine, and mental, Fig. 7-33).

Improper oral hygiene results in food deposits in tooth and gingival crevices. This can cause inflammation of the gingivae or **gingivitis.** As a result, the gingivae swell and become red. If untreated, the disease spreads to other supporting structures, including the alveolar bone. This produces **periodontitis** (inflammation and destruction of the alveolar bone and the *periodontal ligament*, which holds the teeth in their sockets). *Periodontitis, if untreated, causes increasing mobility and eventual loss of the teeth* (see Liebgott, 1982 for details).

THE TEETH

Ten deciduous teeth (primary or "milk" teeth) usually develop in each jaw of children (Fig. 7-126 and Table 7-7). The first tooth usually erupts at 6 to 8 months and the last by 20 to 24 months. As everyone knows, this eruption process is called "teething or cutting teeth." The deciduous teeth are usually shed from the 6th to the 12th year, as they are replaced by the permanent teeth (Figs. 7-24 and 7-126).

Eruption of the permanent teeth (normally 16 in each jaw) is usually complete by the 18th year, except for the third molars ("wisdom" teeth). If they are malposed and/or impacted (Figs. 7-118 and 7-127), they may not erupt. Table 7-7 gives the average times for the eruption and shedding of deciduous teeth and the eruption of permanent teeth.

The 16 teeth in each adult jaw (Fig. 7-127) consist of four **incisors** ("cutters"), two **canines** ("piercers"), four **premolars**, and six **molars** ("grinders"). Each tooth has a **crown**, visible in the oral cavity (Figs. 7-123 and 7-124), a **neck** embedded in the gingiva (Figs. 7-123 and 7-128), and a **root** fixed in an alveolus (socket) in the alveolar process of the jaw by a fibrous, **periodontal membrane** (Fig. 7-129).

Most of the tooth is composed of **dentine** which is covered by **enamel** over the crown and **cementum**

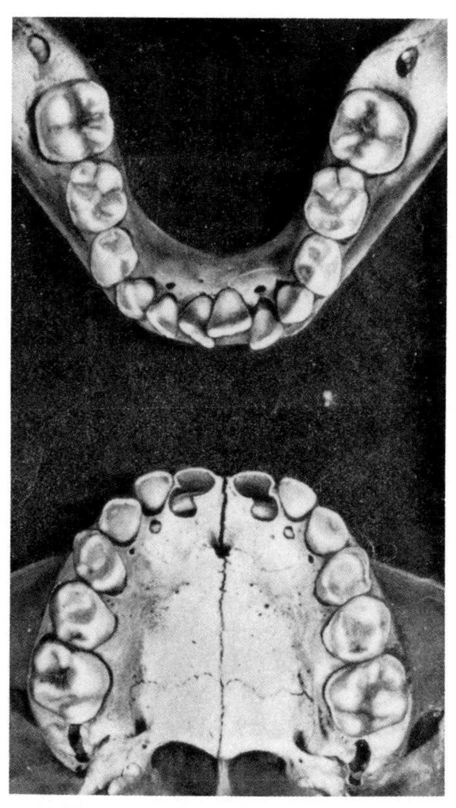

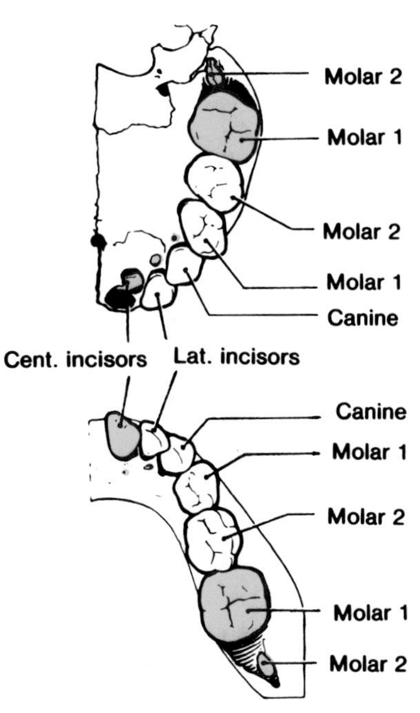

Figure 7-126. *Left*, photographs of the jaws of a child 6 to 7 years old. Note that the maxillary deciduous central incisors are absent (*i.e.*, they have been shed). *Right*, drawings of halves of these jaws giving the names of the teeth. Permanent teeth are colored *yellow*. Observe that the permanent first molars (6-year molars) are fully erupted and that the deciduous central incisors have been shed. Note that the mandibular central incisors have nearly fully erupted, whereas the maxillary central incisors have not erupted, but the buds of the permanent incisors are beginning to move inferiorly into the empty sockets. See Table 7-7 for the order and time of eruption of teeth.

Table 7-7
Order and Time of Eruption of Teeth and Time of Shedding of Deciduous Teeth

	DECIDUOUS TEETH				
	Medial Incisor	Lateral Incisor	Canine	First Molar	Second Molar
Eruption (months)	6 to 8	8 to 10	16 to 20	12 to 16	20 to 24
Shedding (years)	6 to 7	7 to 8	10 to 12	9 to 11	10 to 12

	PERMANENT TEETH							
	Medial Incisor	Lateral Incisor	Canine	First Premolar	Second Premolar	First Molar	Second Molar	Third Molar
Eruption (years)	7 to 8	8 to 9	10 to 12	10 to 11	11 to 12	6 to 7	12	13 to 25

over the root (Fig. 7-128). The **pulp cavity** contains connective tissue, blood vessels, and nerves. It is continuous with the periodontal tissue through the **root canal** and the apical foramen.

For purposes of description, incisor and canine teeth have lingual (L. *lingua*, tongue) and labial (L. *labium*, lip) surfaces and incisal (L. *incido*, to cut into) edges. The premolar and molar teeth have buccal (L. *bucca*, cheek), lingual, and occlusal (L. *occlusio*, to close up) surfaces.

Teeth vary considerably in shape (Figs. 7-9 and 7-129). The **incisors** (cutters) have chisel-like edges (Fig. 7-127) and usually the maxillary ones overlap the mandibular ones (Fig. 7-2). The **canines** (L. *canis*, a

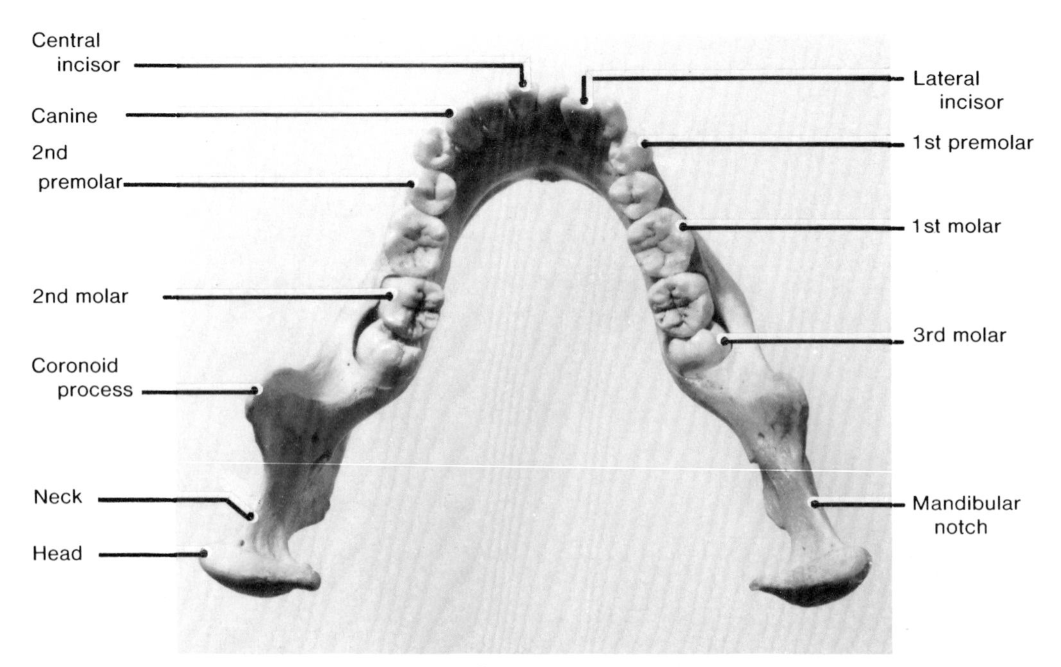

Figure 7-127. Photograph of a superior view of a mandible showing the permanent mandibular teeth of an adult male. Note that the third molar is malposed (*i.e.*, pointing anteriorly); it probably would not have erupted because it is also impacted.

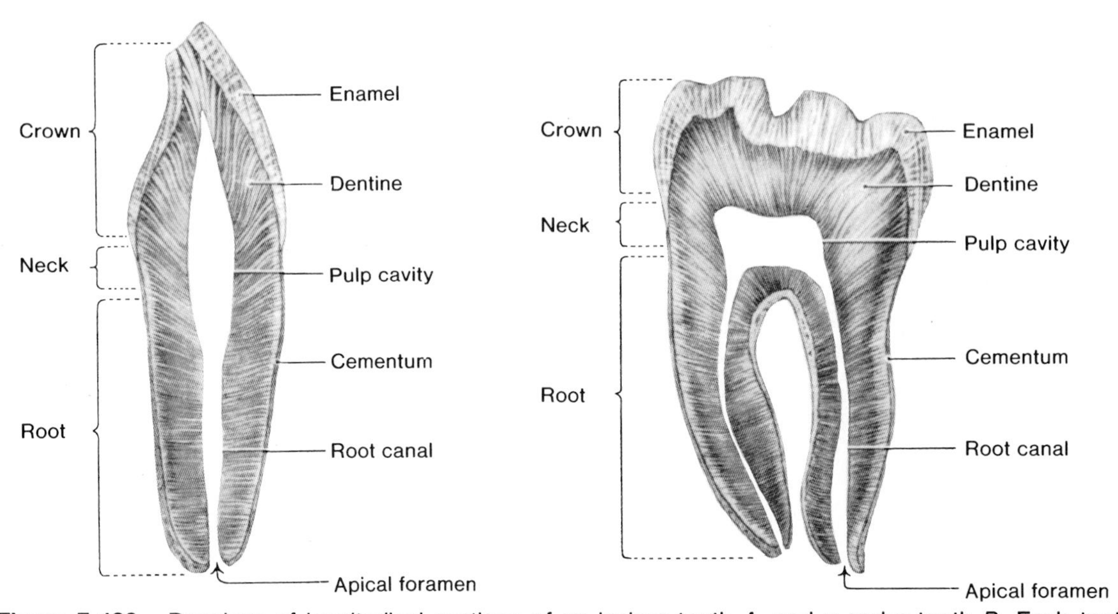

Figure 7-128. Drawings of longitudinal sections of an incisor tooth *A*, and a molar tooth *B*. Each tooth is composed of connective tissue, the *pulp*, covered by three calcified tissues: *dentine*, *enamel*, *and cementum*. The pulp occupies the pulp cavity. The cementum is attached to the alveolar bone by *periodontium* to form a fibrous joint between the tooth and its socket or *alveolus*. The tooth is held firmly in its socket by the *periodontal ligament*.

dog) are conical, but they are poorly developed in man compared to carnivorous (flesh-eating) animals. As premolars have two cusps (L. points) on the crown, they are often referred to as **bicuspids**, whereas maxillary molars have four cusps and mandibular molars five.

The roots of the teeth vary (Figs. 7-12, 7-128, and 7-129). The root of a tooth is separated from the cortical plate of its socket by the **periodontal membrane.** This membrane, composed of densely packed collagen fibers, attaches the tooth to its bony alveolus or socket.

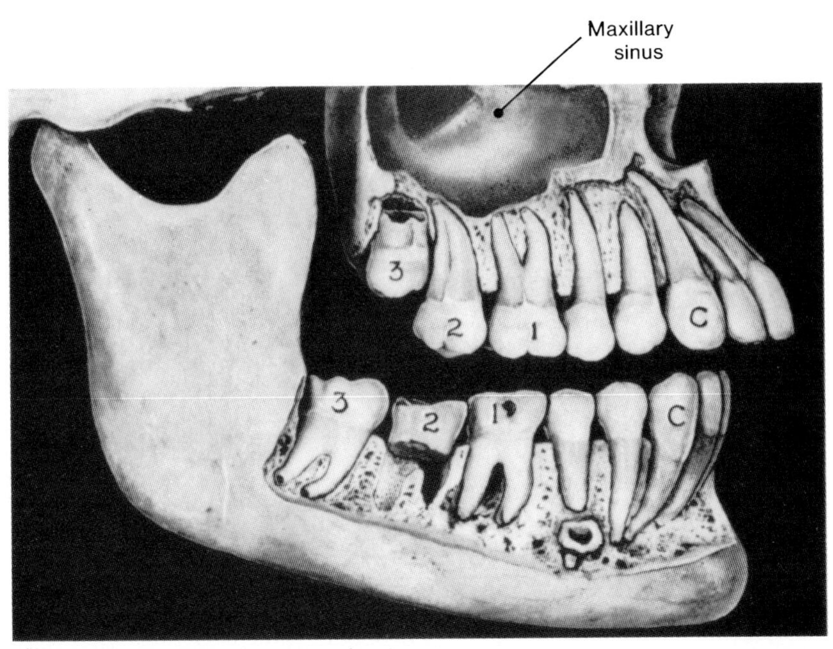

Figure 7-129. Drawing of the permanent teeth on the right side of an adult. The alveolar bone has been ground away to expose the roots of the teeth. Note that the maxillary canine tooth has the longest root and that the roots of the three maxillary molars are very close to the maxillary sinus. The roots of the second mandibular molar have been removed to show the cribriform (sieve-like) nature of the wall of the socket owing to collagen fibers (Sharpey's fibers) of the periodontal membrane entering the bone.

As doctors have numerous opportunities to detect dental problems, and may have to deal with them, a basic knowledge of the teeth and how diseases affect them is required. Disease that occurs while the teeth are developing can produce distinctive abnormalities that may be useful diagnostic signs.

Congenital syphilis affects differentiation of the permanent teeth, resulting in **barrel-shaped incisor teeth** which often have central notches in their incisal edges.

A *marked delay in the eruption of teeth commonly indicates a nutritional disturbance,* or a severe childhood disease.

Transverse ridges on the teeth indicate temporary arrests of tooth development. These localized disturbances of calcification can often be correlated with periods of illness, malnutrition, or trauma (*e.g.,* **neonatal lines** on the teeth resulting from a traumatic birth).

Discoloration of teeth may follow the administration of tetracycline antibiotics, before and after birth. Tetracyclines become incorporated into the teeth and may produce brownish-yellow discoloration. As the enamel is not completely formed on the first and second molar teeth until the 8th year, tetracycline therapy can affect the teeth of unborn infants and of children up to 8 years of age.

Infections of the pulp cavity of a tooth lead to infection in the periodontal ligament, destroying it and the compact layer of bone lining the alveolus. The abscess that forms causes swelling of the adjacent soft tissues ("**gum boil**"). Pus may escape from the abscess and pass between the periosteum and soft tissues, or between the periosteum and the jaw. For example, an abscess associated with an impacted third mandibular molar tooth (Fig. 7-118) may penetrate the tissues at the angle of the mandible and form a large abscess in the submandibular region.

Pus from abscesses of the maxillary molar teeth may extend into the nasal cavity or the maxillary sinus (Fig. 7-129). Note that the roots of the maxillary molar teeth are closely related to the floor of this sinus. As a consequence, infection in the pulp cavity of a tooth may cause **sinusitis,** or a sinusitis may stimulate nerves entering the teeth and simulate toothache.

Pus from abscesses of the maxillary canine teeth often penetrate the facial region, just in-

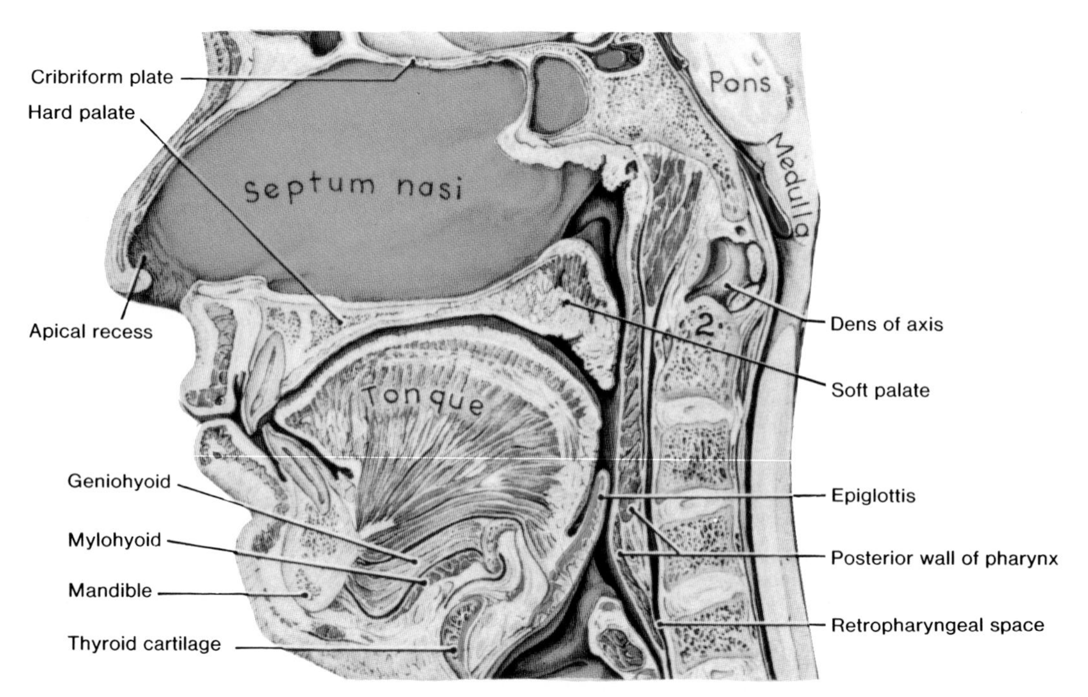

Figure 7-130. Drawing of a median section of part of the head showing the nasal septum, palate, tongue, pharynx, jaws, lips, and various other structures.

ferior to the medial canthus of the eye (Fig. 7-92). Such swelling may obstruct drainage from the **angular vein** (Fig. 7-38) and allow infected material to pass via the superior ophthalmic vein to the cavernous sinus (Fig. 7-39).

Trigeminal neuralgia (tic douloureaux), a syndrome of the trigeminal nerve (CN V), is characterized by extremely severe, unilateral, stabbing pain of the face usually in the lips, gingivae, cheek, or chin. Consequently, it may be confused with diseases of the jaws, teeth, or sinuses.

As people get older their teeth *appear* to get longer owing to the slow recession of their gingivae; hence the expression "*long in the tooth.*" **Gingival recession** exposes the sensitive cementum of the teeth (Fig. 7-128). This process occurs faster in persons who do not have the **tartar** (yellowish-brown deposit) removed from their teeth. It forms at or inferior to the gingival margin of the teeth (Fig. 7-123), and is removed by a procedure called **scaling**. Often a *dental hygienist* removes the tartar from the teeth before a dentist examines them.

Periodontitis (inflammation of the periodontium) results in inflammation of the gingivae and may result in absorption of alveolar bone and recession of the gingivae. As a consequence of gingival recession, exposure of the periodontal membrane occurs allowing microorganisms to invade and destroy it. **Granulation tissue** (vascular connective tissue forming projections on the gingival surface) and foreign matter collect in the space formerly occupied by the periodontal membrane. The gingivae become inflamed and may bleed; sometimes pus exudes, a condition often called **pyorrhea alveolaris** (G. *pyon*, pus + *rhoia*, a flow). Periodontitis and pyorrhea are a common cause of **halitosis** (bad breath).

Decay of the hard tissues of the teeth results in the formation of **dental carries** (cavities). Restoration of a tooth consists of removal of the decayed tissue, followed by restoration of the tooth with a suitable dental material. This is called restorative dentistry.

Toothache or pain in a tooth results from involvement of the pulp cavity or periodontal membrane, as the result of caries (cavities), infection, or trauma. Sometimes it is not practical to restore a tooth owing to extreme tooth destruction. The usual alternative in such cases is to remove the tooth.

THE PALATE

The palate forms the roof of the mouth and the floor of the nasal cavities (Figs. 7-9, 7-122, and 7-124).

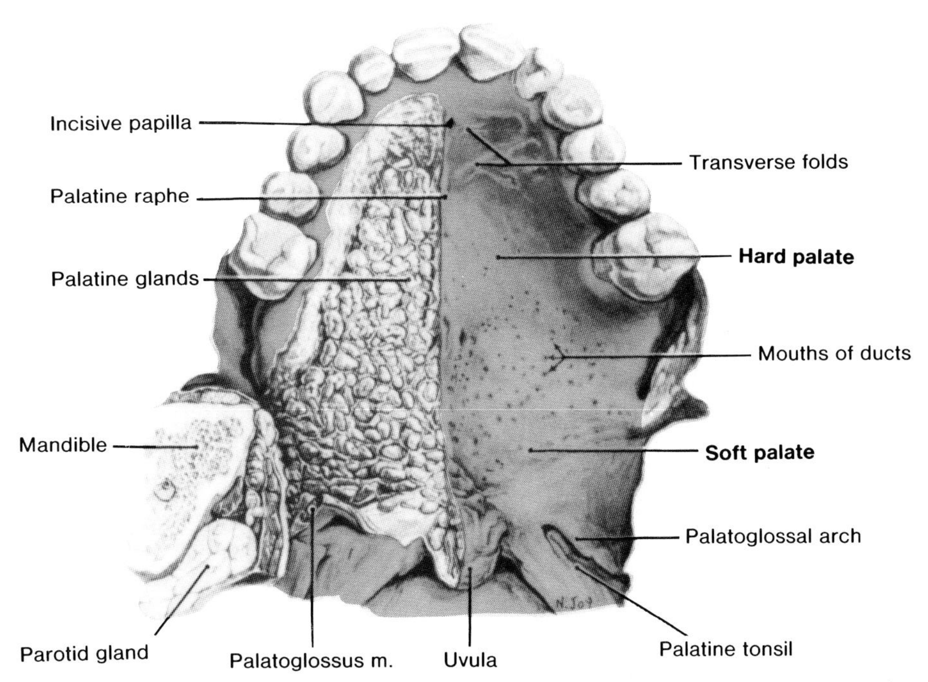

Figure 7-131. Drawing of the palate of an adult. Note that the second and third maxillary molars are missing and that the epithelium has been removed on the right side of the palate. Observe the transverse palatine folds (rugae) and the orifices of the ducts of the palatine glands, which give the mucous membrane an orange-peel appearance. Note that the palatine glands form a very thick layer in the soft palate (Fig. 7-134), but not in the hard palate. Observe that the palate ends posteriorly in the uvula and on each side in the palatopharyngeal arch. Observe the palatoglossus muscle and the palatoglossal arch extending from the inferior surface of the soft palate to the pharynx. See Figure 8-55 for a color photograph of the palate.

Hence it separates the oral cavity from the nasal cavities and the nasal part of the pharynx or nasopharynx (Fig. 7-130).

The palate consists of two regions, the anterior two-thirds or bony part, called the **hard palate** (Figs. 7-9 and 7-124), and the mobile posterior one-third or fibromuscular part, known as the **soft palate.**

The palate is arched anteroposteriorly and transversely. The arch of the palate is more pronounced anteriorly in the region of the hard palate. The depth and breadth of the palatine vault is subject to considerable variation.

THE HARD PALATE

The hard palate is formed by the palatine processes of the maxillae and the horizontal plates of the palatine bones (Figs. 7-8, 7-9, and 7-130). Anteriorly and laterally, the hard palate is bounded by the alveolar processes (Figs. 7-2, 7-8, and 7-9) and the gingivae (Fig. 7-124). Posteriorly the hard palate is continuous with the soft palate.

An **incisive foramen** is located posterior to the maxillary, central incisor teeth (Figs. 7-8 and 7-9). Within this foramen, there are two to four foramina. The two constant ones are the orifices of the incisive canals which transmit the **nasopalatine nerves and arteries** (Fig. 7-132). The arteries arise from the sphenopalatine arteries and anastomose with the **greater palatine arteries.**

In hard palates from young people (Figs. 7-9 and 7-126), a suture line is usually visible between the premaxillary part of the maxilla and the palatine processes of the maxillae. This suture represents the site of fusion of the median and lateral palatine processes during the 12th week of prenatal development. In clefts of the anterior palate, the defect passes along the site of this suture line (Fig. 7-136).

Medial to the third molar tooth, observe the **greater palatine foramen** (Fig. 7-9), piercing the lateral border of the bony palate. It is the inferior orifice of the **greater palatine canal.** The greater palatine vessels and nerve emerge from this foramen (Figs. 7-132 and 7-149) and run anteriorly in two grooves on the palate. The **lesser palatine foramina** (Fig. 7-9) transmit the lesser palatine nerves and vessels to the soft palate and adjacent structures (Fig. 7-149).

The hard palate is covered by mucous membrane (oral mucosa) which is intimately connected to the periosteum (Figs. 7-124, 7-131, and 7-149). Deep to the mucosa are mucus secreting **palatine glands.** The orifices of the ducts of these glands give the mucous membrane of the palate an orange-peel appearance (Figs. 7-124 and 7-131).

The mucous membrane of the anterior part of the hard palate, has three or four **transverse palatine folds** (**rugae**) of the mucous membrane (Figs. 7-124 and 7-131). The **palatine raphe** runs posteriorly from the **incisive papilla**. *The palatine raphe indicates the site of fusion of the palatine processes during the 12th week of development.*

Dentists anesthetize the nasopalatine nerves (Figs. 7-132 and 7-149) by injecting anesthetic fluid into the mouth of the **incisive canal**, *i.e.*, the incisive foramen (Figs. 7-8, 7-9, and 7-131). The needle is inserted posterior to the **incisive papilla**, a slight elevation of the mucosa that covers the incisive foramen at the anterior end of the palatine raphe (Fig. 7-131). Branches of the nasopalatine nerve leave this foramen and pass laterally and posteriorly (Figs. 7-132 and 7-149).

A **nasopalative nerve block** is done before performing certain operative procedures in the oral cavity (*e.g.*, preparing an incisor tooth for capping or extracting one of them. As both right and left nerves emerge through the incisive foramen, both nerves are anesthetized with the same injection. The tissues anesthetized are: the palatal mucosa; the lingual gingivae of the six anterior maxillary teeth; the lingual plate of the alveolar bone; and the hard palate associated with the six maxillary teeth just mentioned.

Injections into the palate are painful because the mucosa is tightly bound to the hard palate. Because of this, injections should be given very slowly.

THE SOFT PALATE

The soft palate, or velum palatinum (L. *velum*, sail or veil), is the curtain-like posterior part of the palate (Figs. 7-122, 7-124, 7-130, 7-131, and 8-55).

The soft palate has no bony framework, but it contains a membranous aponeurosis or "aponeurotic palate." The soft palate is a movable fold that is attached to the posterior edge of the hard palate. It extends posteroinferiorly to a curved free margin from which hangs a conical process, the **uvula** (Fig. 7-122). At rest the soft palate hangs from the posterior aspect of the palate into the cavity of the pharynx (Figs. 7-122 and 7-130). It separates the nasopharynx superiorly from the oropharynx inferiorly.

During swallowing the soft palate moves posteriorly against the wall of the pharynx, thereby preventing regurgitation of food into the nasal cavities.

Laterally the soft palate is continuous with the wall of the pharynx and is joined to the tongue and the pharynx by the **palatoglossal and palatopharyngeal arches**, respectively, (Figs. 7-131 and 8-55). The

Table 7-8
Muscles of the Soft Palate

Muscle	Origin	Insertion	Nerve Supply	Action
Levator veli palatini (Fig. 7-133)	Cartilage of auditory tube and petrous part of temporal bone	Palatine aponeurosis (Fig. 7-132)	Pharyngeal plexus by fibers from the cranial root of CN XI (Fig. 7-64)	Elevates soft palate
Tensor veli palatini (Figs. 7-132 to 7-134)	Scaphoid fossa of medial pterygoid and spine of sphenoid bone	Palatine aponeurosis (Fig. 7-132)	Mandibular n. (Fig. 7-33)	Tenses soft palate
Palatoglossus (Fig. 7-131)	Palatine aponeurosis (Fig. 7-132)	Side of tongue	Pharyngeal plexus by fibers from the cranial root of CN XI (Fig. 7-64)	Elevates posterior part of tongue and draws soft palate on to tongue (*i.e.*, closes off oral cavity)
Palatopharyngeus (Fig. 8-51).	Hard palate and palatine aponeurosis (Fig. 7-132)	Lateral wall of pharynx	Pharyngeal plexus by fibers from the cranial root of CN XI (Fig. 7-64)	Tenses soft palate and pulls walls of pharynx superiorly, anteriorly, and medially during swallowing
Musculus uvulae (Figs. 7-134 and 8-51)	Posterior nasal spine and palatine aponeurosis	Mucosa of the uvula	Pharyngeal plexus by fibers from the cranial root of CN XI (Fig. 7-64)	Shortens uvula and pulls it superiorly

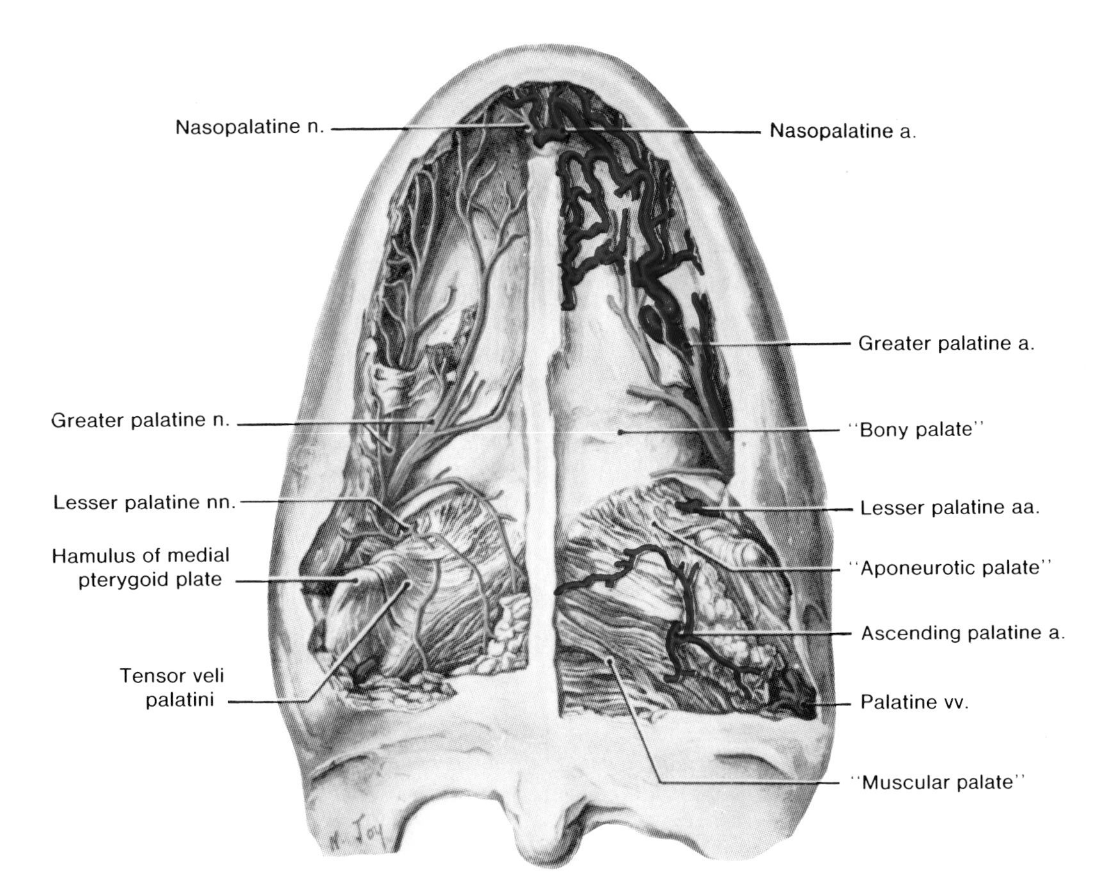

Figure 7-132. Drawing of a dissection of the inferior surface of the palate. Observe that the palate has bony, aponeurotic, and muscular parts. Note the tensor veli palatini muscle hooking around the hamulus of the medial pterygoid plate of the sphenoid bone to join the palatine aponeurosis ("aponeurotic palate"). Also observe the palatine vessels and nerves.

palatopharyngeal arches used to be called the "pillars of the throat" or **fauces** (L. pl. of *faux*, "a gorge or narrow passage").

The **palatine tonsil** is located in the triangular interval between the palatoglossal and palatopharyngeal arches (Figs. 7-131 and 8-55). Deep to the palatal mucosa there are mucous glands (Fig. 7-131).

The soft palate is strengthend by the **palatine aponeurosis**, formed by the *expanded tendon of the tensor veli palatini muscle* (Figs. 7-132 to 7-134). This aponeurosis, attached to the posterior margin of the hard palate, is thick anteriorly and very thin posteriorly. *All other muscles of the soft palate are attached to the palatine aponeurosis.*

The anterior part of the soft palate consists mainly of the aponeurosis (Figs. 7-132 and 7-134), whereas the posterior part is muscular.

The Muscles of the Soft Palate (Figs. 7-132, 7-133, and Table 7-8). The *five muscles of the soft palate*, arise from the base of the skull and descend to the palate. They produce various movements of the muscular flap called the soft palate. The posterior part may be raised so that it becomes a horizontal continuation of the plane of the hard palate and is in contact with the posterior wall of the pharynx (Fig. 7-122). The soft palate can also be drawn inferiorly so that its inferior surface is in contact with the posterior part of the superior surface of the tongue (Fig. 7-130). During quiet breathing the soft palate is curved posteroinferiorly. Movements of the palate during speech and swallowing are discussed in Chapter 8.

The Levator Veli Palatini (levator palati) is a cylindrical muscle (Figs. 7-133 and 7-134) that *arises from the cartilage of the auditory tube and the petrous part of the temporal bone.* It runs inferoanteriorly, spreading out in the soft palate, where it inserts into the **palatine aponeurosis.** As its name indicates, the levator veli palatini *elevates the soft palate* in order to seal the nasopharynx from the oropharynx (Table 7-8).

The Tensor Veli Palatini (tensor palati), a thin triangular muscle (Figs. 7-132 to 7-134), arises from the **scaphoid fossa** at the base of the medial pterygoid process, the **spine of the sphenoid bone**, and the

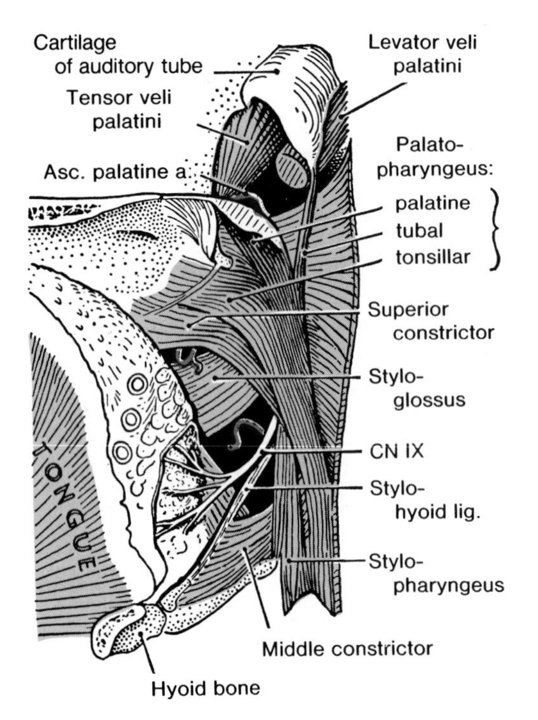

Figure 7-133. Drawing of a dissection of the lateral wall of the pharynx.

lateral aspect of the **auditory tube**. As it passes inferiorly, the tendon of this muscle hooks around the **hamulus** of the medial pterygoid plate (Figs. 7-113 and 7-132), before inserting into the **palatine aponeurosis**.

The tensor veli palatini, as its Latin name indicates, **tenses the soft palate** by using the hamulus as a pulley. It also pulls open the membranous portion of the auditory tube (Fig. 7-133). Acting together, the levator veli palatini muscles raise the soft palate as the two tensor veli palatini muscles tense it (Table 7-8). This forces the soft palate against the posterior wall of the pharynx, a movement which occurs during swallowing.

The Palatoglossus Muscle arises from the palatine aponeurosis, passes laterally and inferiorly in the palatoglossal arch, and inserts into the side of the tongue (Fig. 7-131). The palatoglossal muscle, covered with mucous membrane, forms the **palatoglossal arch** (Figs. 7-131 and 8-55).

The palatoglossus muscle elevates the posterior part of the tongue and draws the soft palate inferiorly on to the tongue. These movements close off the oral cavity from the oral part of the pharynx (*i.e.*, they close the oropharyngeal isthmus).

The Palatopharyngeus Muscle (Fig. 8-51) arises

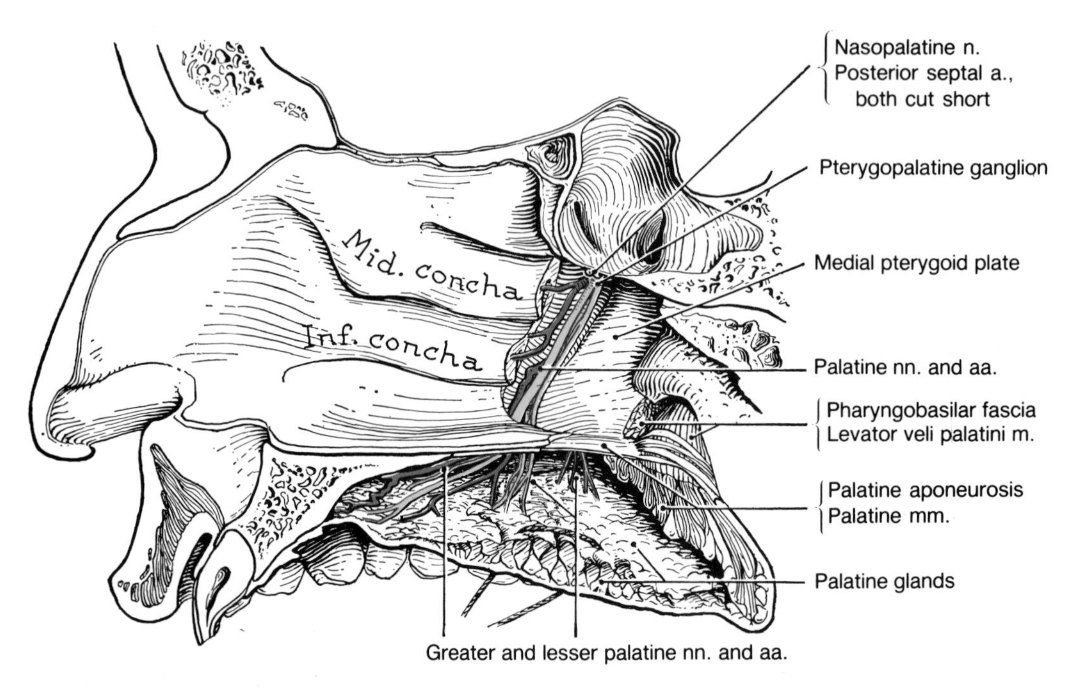

Figure 7-134. The mucous membrane, containing a layer of mucous glands, has been separated by blunt dissection. The layer of glands is thin on the bony palate; it is thickest on the aponeurotic part; and is less thick on the muscular part. The posterior ends of the middle and inferior conchae are cut through; these and the mucoperiosteum are peeled off the side wall of the nose as far as the posterior border of the medial pterygoid plate. The "papery" perpendicular plate of the palatine bone was broken through to expose the palatine nerves and arteries.

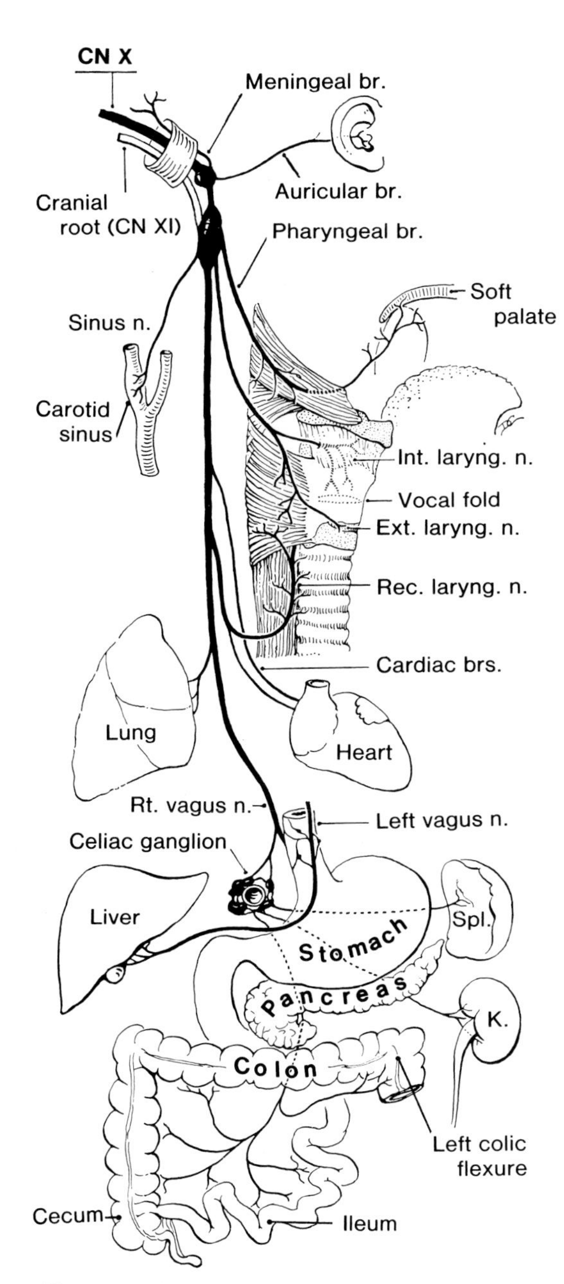

Figure 7-135. Scheme of distribution of the *vagus nerve* (CN X). The vagus (L. the wanderer) is a mixed nerve. The fibers received from the cranial root of CN XI supply the muscles of the pharynx (except for the stylopharyngeus which is supplied by CN IX), the palate (tensor veli palatini excepted, but including the palatoglossus), and the larynx. The foregoing are all skeletal muscles.

from the posterior border of the hard palate and the palatine aponeurosis and passes posteroinferiorly in the **palatopharyngeal arch** (Fig. 8-53). It inserts into the lateral wall of the pharynx. The palatopharyngeal muscle, covered by mucous membrane, forms the **palatopharyngeal arch**. The palatopharyngeus muscle tenses the soft palate and pulls the walls of the pharynx superiorly and medially during swallowing (Table 7-8).

The Musculus Uvulae (Figs. 7-134 and 8-51), consisting of two small slips of muscle, arises from the posterior nasal spine and the palatine aponeurosis. It passes posteriorly on each side of the midline and **inserts into the mucosa of the uvula**. When the uvular muscle contracts it shortens the uvula and pulls it superiorly. *This movement assists in closing the nasopharynx during swallowing.*

The nerve supply of the palatal muscles is through the pharyngeal plexus, formed by fibers derived from the cranial root of the accessory nerve (**CN XI**), which joins the vagus nerve (Fig. 7-64), except for the tensor veli palatini, which is supplied by the mandibular nerve (Fig. 7-33).

NERVES OF THE PALATE (Figs 7-132 and 7-134)

The sensory nerves of the palate, branches of the *pterygopalatine ganglion* (Figs. 7-33 and 7-134), are the greater and lesser palatine nerves. They accompany the arteries through the greater and lesser palatine foramina, respectively.

The **greater palatine nerve** supplies the gingivae, the mucous membrane, and the glands of the hard palate. The **lesser palatine nerve** supplies the soft palate. The **nasopalatine nerve**, another branch of the pterygopalatine ganglion, emerges from the incisive foramen (Figs. 7-9 and 7-33) and supplies the mucous membrane of the anterior part of the hard palate (Figs. 7-132 and 7-134).

VESSELS OF THE PALATE (Figs. 7-132 to 7-134)

The palate has a rich blood supply from branches of the maxillary artery (Figs. 7-37, 7-114, and 7-116), chiefly the **greater palatine artery**, a branch of the descending palatine artery. This artery passes through the **greater palatine foramen** and runs anteriorly and medially. The **lesser palatine artery** enters via the lesser palatine foramen and anastomoses with the **ascending palatine artery** (Fig. 7-132), *a branch of the facial artery* (Fig. 7-37).

About once in 2500 births, infants are born with a **cleft palate** (Fig. 7-136); it may or may not be associated with a **cleft lip.** The cleft in the palate may involve only the uvula or it may extend through the entire palate. In severe cases associated with cleft lip, the cleft extends through the alveolar process of the maxilla and the whole palate (Fig. 7-136). In these infants

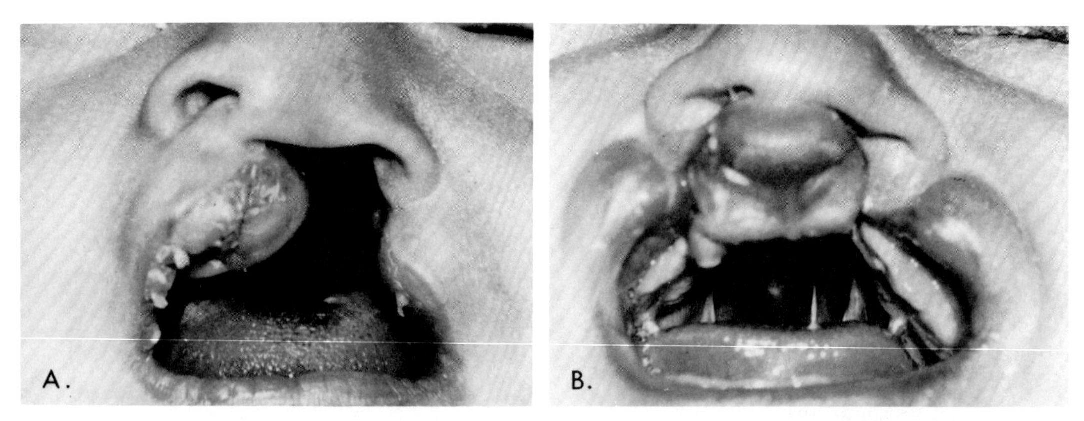

Figure 7-136. Photographs of infants with cleft lips and palates. *A*, complete unilateral cleft of the upper lip and alveolar process of the maxilla (left side). *B*, complete bilateral cleft of the upper lip and alveolar process of the maxilla, associated with a bilateral cleft of the anterior palate. (From Moore KL: *The Developing Human: Clinically Oriented Embryology*, ed 3. WB Saunders Co, Philadelphia, 1982.)

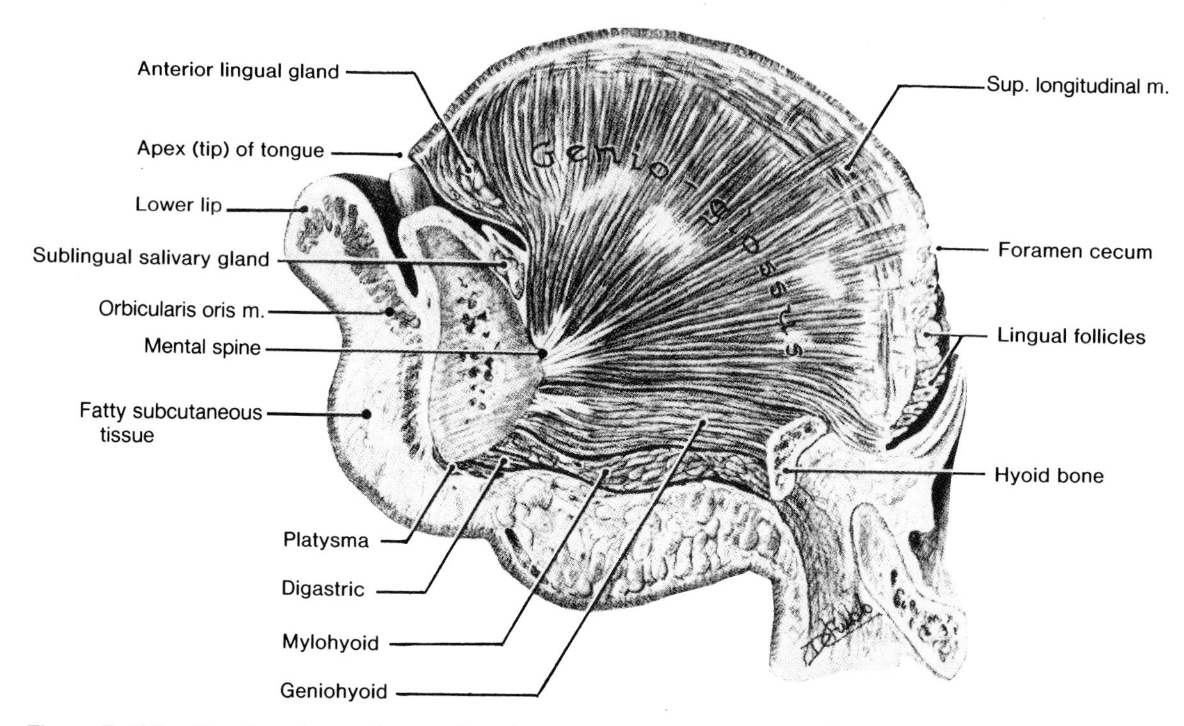

Figure 7-137. Drawing of a median section of the tongue and the floor of the mouth. Observe that the tongue is composed mainly of muscle. The extrinsic ones alter the position of the tongue and the intrinsic ones alter its shape. In this illustration, extrinsic muscles are represented by the genioglossus and intrinsic ones by the superior longitudinal muscle.

the nasal and oral cavities communicate. An **artificial palate** (prosthesis) is inserted until the palate can be repaired surgically.

The embryological basis of unilateral cleft palate is failure of the mesenchymal masses of the lateral palatine processes to fuse with each other, with the nasal septum, and/or with the posterior margin of the primary palate. (For details and explanatory illustrations, see Moore, 1982).

Paralysis of the soft palate sometimes results from diphtheria, a rare and preventable disease. In these patients a **neurotoxin** is produced that destroys nerve cells; this can produce temporary or permanent paralysis of muscles. When the palatal muscles are affected, the soft palate will not rise during swallowing. As a result, food and fluids may be forced into the nasopharynx and run out the nose.

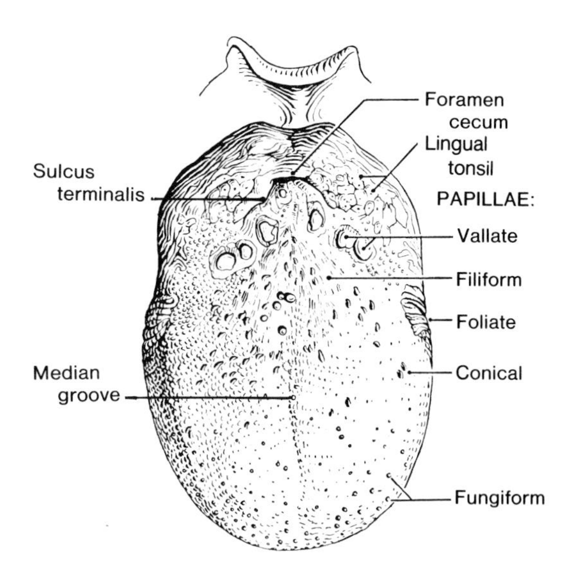

Figure 7-138. Drawing of the dorsum of the tongue. Observe the foramen cecum, the patent superior end of the embryonic thyroglossal duct, and the V-shaped sulcus terminalis. This sulcus demarcates the developmentally different pharyngeal or posterior one-third of the tongue from the oral or anterior two-thirds. These parts differ in the structure of their mucous membrane and in their nerve supply. These anatomical differences are related to their different developmental origins. The median groove (sulcus) indicates the plane of fusion of the distal tongue buds in the embryo.

THE TONGUE

The tongue (L. *lingua*; G. *glossa*) is located in the floor of the mouth. *The tongue is a highly mobile muscular organ* that one can vary greatly in shape. At rest it fills most of the oral cavity proper (Figs. 7-130 and 7-142). The tongue is concerned with mastication, taste, deglutition (swallowing), articulation (speech), and oral cleansing, but **its main functions are squeezing food into the pharynx when swallowing, and forming words during speaking**.

The tongue is situated partly in the mouth and partly in the oropharynx (Figs. 7-122 and 7-130). It is mainly composed of muscles and is covered by a mucous membrane on its dorsum, tip, and sides (Figs. 7-137 and 7-142).

GROSS FEATURES OF THE TONGUE

The dorsum of the tongue is divided by the **sulcus terminalis** (Fig. 7-138) into oral (palatine) and pharyngeal parts. The **oral part** (constituting the body) forms about two-thirds, and the pharyngeal part (making up the root) about one-third of the dorsum of the tongue.

At the apex of the terminal sulcus is a small median pit, the **foramen cecum** (Figs. 7-137 and 7-138); this is the remnant of the opening of the embryonic *thyroglossal duct*. The thyroid gland in the embryo was attached to the tongue by this duct which normally disappears leaving only this small pit in the tongue.

The oral part of the dorsum of the tongue is freely movable, but it is loosely attached to the floor of the mouth by the **lingual frenulum** (Fig. 7-124). On each side of the frenulum there is a **deep lingual vein** (Fig. 7-65). It begins at the tip of the tongue and runs posteriorly near the median plane. The lingual veins lie close to the mucous membrane on the inferior surface of the tongue. All the veins of one side of the tongue unite at the posterior border of the hyoglossus muscle to form the **lingual vein**, which joins either the facial vein (Fig. 7-42), or the internal jugular vein (Fig. 7-65). The inferior surface and sides of the tongue are covered with smooth, thin mucous membrane.

On the dorsum of the oral part of the tongue, there is a **median groove** (Fig. 7-138). The median groove (L. *sulcus*, a furrow or ditch) is inconspicuous in some people. The **dorsal lingual veins** drain the dorsum and sides of the tongue and join the lingual veins, which accompany the lingual artery.

Lingual Papillae and Taste Buds (Fig. 7-138). The mucous membrane on the oral part of the tongue is rough, owing to the presence of numerous papillae.

The filiform papillae (L. *filum*, thread) are numerous and thread-like. They are arranged in rows parallel to the sulcus terminalis. Feel these papillae on your tongue with your index finger, verifying that they are rough. Their roughness facilitates the licking of semisolid foods (*e.g.*, ice cream). *The filiform papillae contain afferent nerve endings that are sensitive to touch.*

The fungiform papillae are small and mushroom-

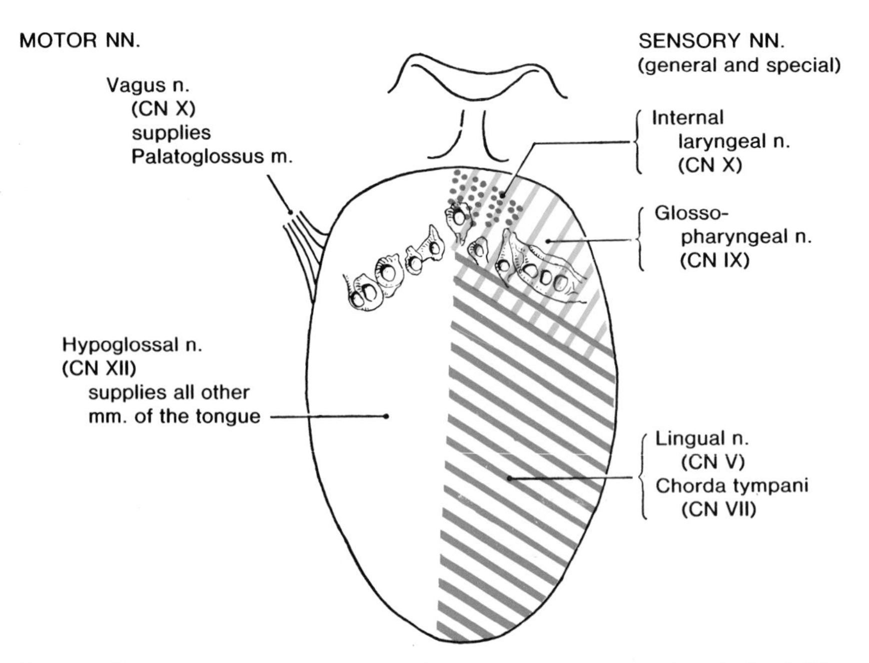

Figure 7-139. Diagram illustrating the nerve supply of the tongue. (Also see Figs. 7-33, 7-35, and 7-114).

shaped. Examine your tongue in a mirror and observe these on the tip and margins of your tongue. They usually appear as pink or red spots.

The vallate papillae (circumvallate papillae) are relatively large (1 to 2 mm in diameter); they are the largest taste buds. They lie just anterior to the sulcus terminalis. They appear like flat-topped, short cylinders sunken into the mucosa. They are surrounded by a deep trench or trough, the walls of which are studded with taste buds.

The foliate papillae are small lateral folds of the lingual mucosa; they are poorly developed in humans.

The vallate, foliate, and most of the fungiform papillae contain *taste receptors*; they are in the *taste buds.* The underlying serous glands secrete a watery fluid which keeps the trough around the vallate papillae full and flushed out. This fluid dissolves the substances to be tasted and clears away debris and lingering tastes.

The pharyngeal part of the dorsum of the tongue lies posterior to the terminal sulcus and the palatoglossal arches (Fig. 7-131). Its mucous membrane has no papillae. However the underlying **nodules of lymphoid tissue** give part of the tongue a *cobblestone appearance.* These lymphoid nodules (lingual follicles) are collectively known as the **lingual tonsil** (Fig. 7-138). In Figures 7-130 and 7-137, note that the pharygeal part of the dorsum of the tongue faces posteriorly.

When quick absorption of a drug is desired (*e.g.*, nitroglycerine used as a vasodilator in **angina pectoris**, p. 119), it is placed under the tongue where it dissolves and enters the lingual veins in less than a minute. This sublingual surface is covered with thin, transparent mucosa through which one can see the underlying vessels.

Occasionally the lingual frenulum of infants extends almost to the tip of the tongue and interferes with its protrusion. This condition is known as *"tongue-tie"* or **ankyloglossia.** Usually the frenulum grows sufficiently during the 1st year of life so that surgical correction is unnecessary.

Fissured tongue (scrotal tongue), often present in people with the **Down syndrome,** is marked by multiple deep fissures owing to *chronic glossitis* (inflammation of the tongue).

Thyroglossal duct cysts may form anywhere along the course followed by the embryonic thyroglossal duct through the tongue during descent of the thyroid gland (Fig. 8-40). Usually the thyroglossal duct atrophies and disappears, but remnants of it may persist and give rise to cysts in the tongue. In very rare instances, the developing thyroid gland fails to descend, resulting in a **lingual thyroid.** (For illustrations and descriptions of the developmental bases for these congenital abnormalities, see Moore, 1982.)

The embryology of the tongue is clinically important because the anterior two-thirds (oral part) and the posterior one-third (pharyngeal part) not only differ developmentally, but structurally and in their nerve supply and lymphatic drainage as well.

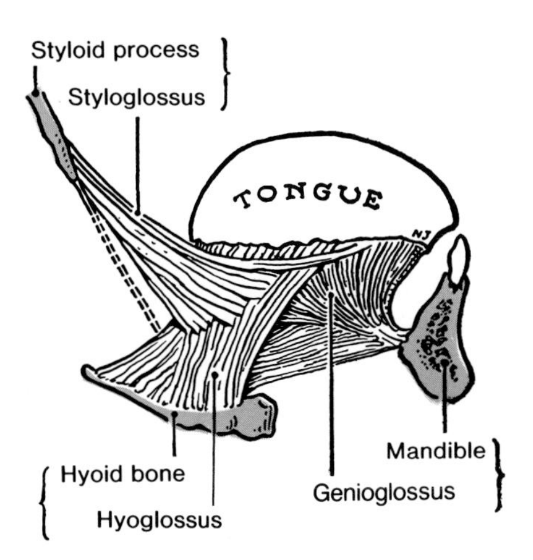

Figure 7-140. Drawing of three extrinsic muscles of the tongue showing their bony origins, shapes, and directions. The styloid process is part of the temporal bone (Figs. 7-8 and 7-10).

MUSCLES OF THE TONGUE

The tongue is divided into halves by a median fibrous **lingual septum** that lies deep to the median groove (Figs. 7-138 and 7-142). In each half of the tongue, there are four extrinsic and four intrinsic muscles.

The lingual muscles are supplied by CN XII, the hypoglossal nerve (Fig. 7-66), except for the palatoglossus, which is supplied from the **pharyngeal plexus** by fibers from the cranial root of CN XI (Fig. 7-64).

The Extrinsic Muscles of the Tongue (Figs.7-137, 7-140, and Table 7-9). This group of four muscles (genioglossus, hyoglossus, styloglossus, and palatoglossus) originates outside the tongue and insert into it. The extrinsic muscles mainly move the tongue, but they can alter its shape as well. The **hyoid bone** (Figs. 7-137 and 7-140) is attached to the posterior part of the tongue and moves with it.

The Genioglossus Muscle (Figs. 7-137, 7-140, and 7-142). This is a bulky, fan-shaped muscle that arises by a short tendon ("the handle") from the **superior mental spine** of the mandible (Figs. 7-22, 7-137), and 7-140). It fans out as it enters the tongue inferiorly and its fibers insert into the entire dorsum of the tongue. Its most inferior fibers insert into the body of the hyoid bone (Figs. 7-137 and 7-141).

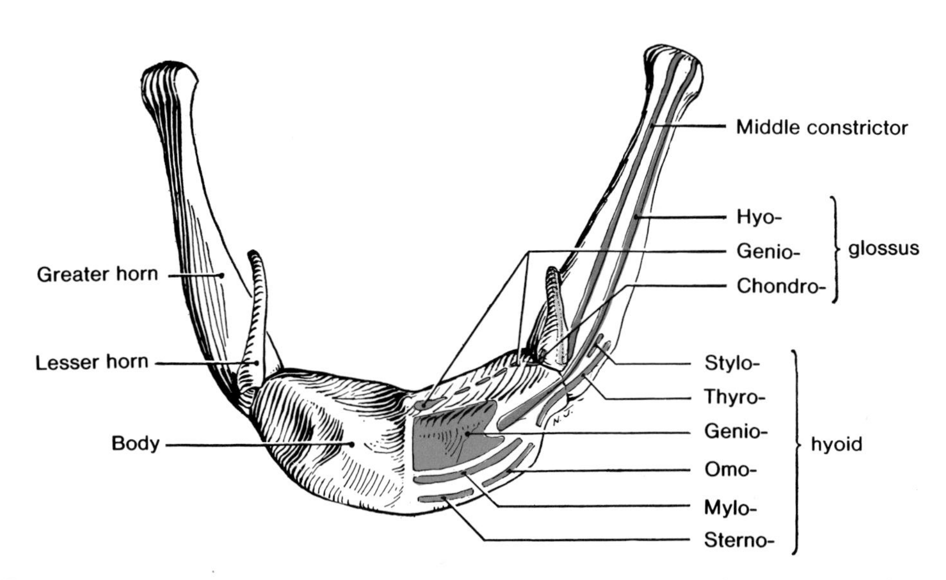

Figure 7-141. Drawing of the hyoid bone showing its parts and the sites of attachments of muscles. Observe that the hyoid is a horseshoe-shaped bone with a central body and greater and lesser horns. The origin of muscles is shown in *red* and the insertion of muscles is indicated in *blue*.

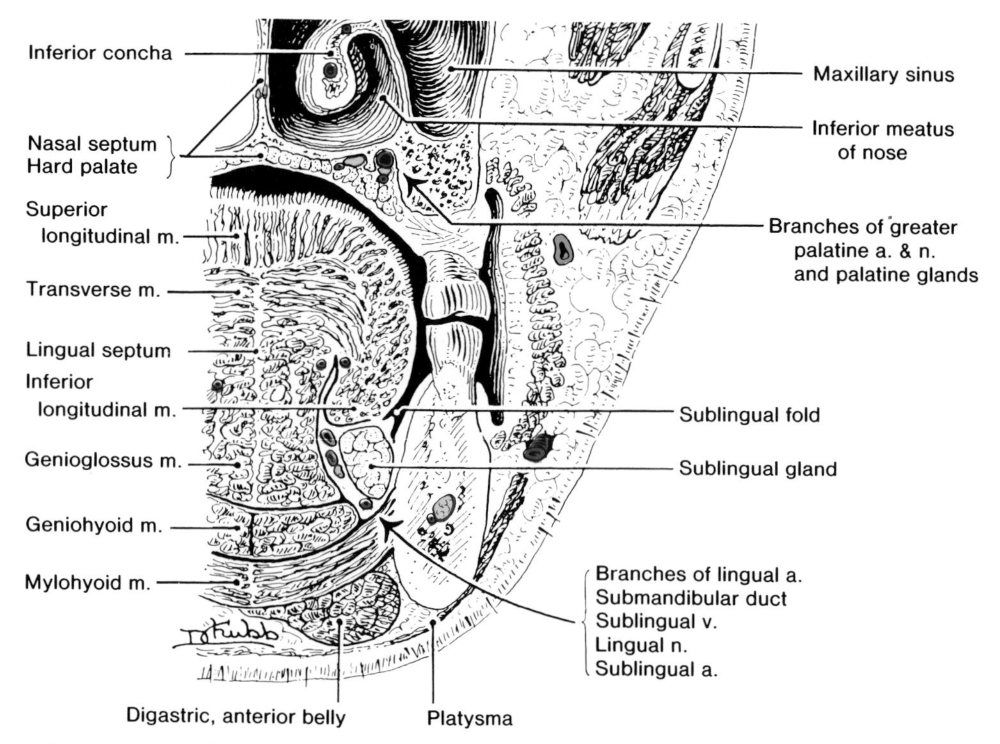

Figure 7-142. Drawing of a coronal section through part of the head, particularly to show the intrinsic muscles of the tongue and the muscles in the floor of the mouth.

The genioglossus muscle depresses the tongue and its posterior part protrudes the tongue.

When the genioglossus muscle is paralyzed, *the tongue has a tendency to fall posteriorly and obstruct the vital airway of the oropharynx*, presenting the risk of suffocation.

Total relaxation of the genioglossus muscles occurs during general anesthesia; therefore the tongue of the anesthetized patient must be prevented from relapsing by exerting pressure on the mandible and inserting an airway which extends from the lips to the laryngopharynx. This device does not enter the larynx or trachea (Fig. 7-122).

The Hyoglossus Muscle (Fig. 7-140). This muscle is thin and quadrilateral in shape. *It arises from the body and greater horn of the hyoid bone* (Fig. 7-141), and passes superoanteriorly to insert into the side and inferior aspect of the tongue.

The hyoglossus muscle depresses the tongue, pulling its sides inferiorly. The hyoglossus muscle also aids in retraction of the tongue (Table 7-9).

The Styloglossus Muscle (Fig. 7-140). This is a small, short muscle that *arises from the anterior border of the styloid process near its tip and from the stylohyoid ligament.* It passes inferoanteriorly to insert into the side and inferior aspect of the tongue. Its fibers interdigitate with those of the hyoglossus muscle.

The styloglossus retracts the tongue and draws up its sides to create a trough for swallowing (Table 7-9).

The Palatoglossus Muscle (Fig. 7-131). This muscle originates in the palatine aponeurosis of the soft palate (Figs. 7-132 and 7-134), and enters the lateral part of the tongue with the styloglossus. It passes almost transversely through the tongue with the transverse intrinsic muscle fibers, and inserts into the side of the tongue. *The palatoglossus elevates the posterior part of the tongue* (Table 7-9).

The Intrinsic Muscles of the Tongue (Figs. 7-137 and 7-142). There are four pairs of intrinsic muscles: superior and inferior longitudinal, transverse, and vertical.

The intrinsic muscles are mainly concerned with altering the shape of the tongue, e.g., making it broad or narrow. The intrinsic muscles run in three directions.

The superior longitudinal muscle of the tongue (Fig. 7-142) forms a thin layer deep to the mucous membrane on the dorsum of the tongue, running from its tip to its root. It arises from the submucous fibrous

Table 7-9
Extrinsic Muscles of the Tongue

Muscle	Origin	Insertion	Nerve Supply	Action(s)
Genioglossus (Fig. 7-137)	Superior mental spine of mandible (Figs. 7-22 and 7-137)	Dorsum of tongue and body of hyoid bone (Fig. 7-141)	Hypoglossal n., CN XII (Fig. 7-66)	Depresses tongue; posterior part protrudes tongue
Hyoglossus (Fig. 7-140)	Body and greater horn of hyoid bone (Fig. 7-141)	Side and inferior aspect of tongue	Hypoglossal n., CN XII (Fig. 7-66)	Depresses and retracts tongue
Styloglossus (Fig. 7-140)	Styloid process and stylohyoid ligament (Fig. 7-140)	Side and inferior aspect of tongue	Hypoglossal n., CN XII (Fig. 7-66)	Retracts tongue and draws up sides to create trough for swallowing
Palatoglossus (Fig. 7-131)	Palatine aponeurosis of soft palate (Fig. 7-132)	Side of tongue	Pharyngeal plexus by fibers from cranial root of CN XI (Fig. 7-64)	Elevates posterior part of tongue

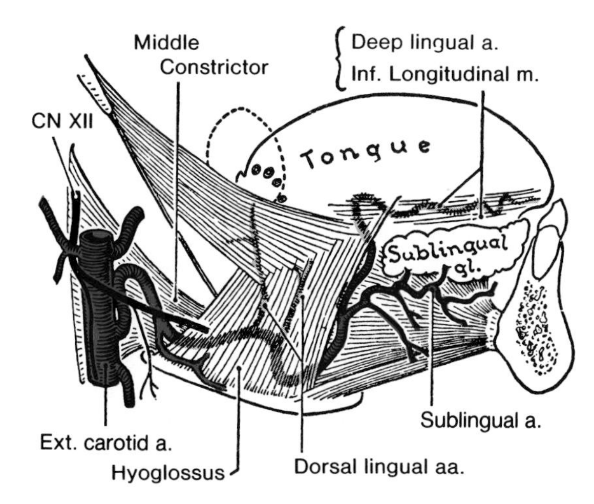

Figure 7-143. Drawing of the lingual artery and its branches. Note the dorsal lingual branches to the posterior part and the deep lingual artery to the anterior part of the tongue. The lingual artery, the main artery of the tongue, is a branch of the external carotid artery.

layer and the **lingual septum** and inserts mainly into the mucous membrane.

The superior longitudinal muscle curls the tip and sides of the tongue superiorly, making the dorsum of the tongue concave.

The inferior longitudinal muscle of the tongue (Fig. 7-142) consists of a narrow band close to the inferior surface of the tongue. It extends from the tip to the root of the tongue. Some of its fibers attach to the hyoid bone.

The inferior longitudinal muscle curls the tip of the tongue inferiorly, making the dorsum of the tongue convex.

The transverse muscle of the tongue (Fig. 7-142), lying inferior to the superior longitudinal muscle, arises from the fibrous **lingual septum** and runs lateral to its right and left margins. Its fibers are inserted into the submucous fibrous tissue. *The transverse muscle narrows and increases the height of the tongue.*

The vertical muscle of the tongue runs inferolaterally from the dorsum of the tongue. **It *flattens and broadens the tongue*.** Acting with the transverse muscle, the vertical muscle ***increases the length of the tongue***.

NERVES OF THE TONGUE

The muscles and mucous membrane of the tongue have separate nerve supplies. The reason for this difference is embryological. The mucous membrane of the tongue is derived from the floor of the primitive pharynx, whereas most of its muscles originate from the **occipital myotomes**. Myoblasts (primitive muscle cells) from myotomes in the occipital region migrate into the tongue and differentiate into its muscles. The hypoglossal nerve (CN XII) accompanies the myoblasts during their migration.

Innervation of the Muscles of the Tongue (Figs. 7-66 and 7-139). All the muscles of the tongue, except the palatoglossus, are supplied by CN XII, the **hypoglossal nerve**. The palatoglossus, although an extrinsic muscle of the tongue, is more closely associated with the soft palate. Like most muscles of the soft palate, *the palatoglossus* is supplied by nerve fibers from the **pharyngeal plexus** via the **vagus nerve** (Fig. 7-36).

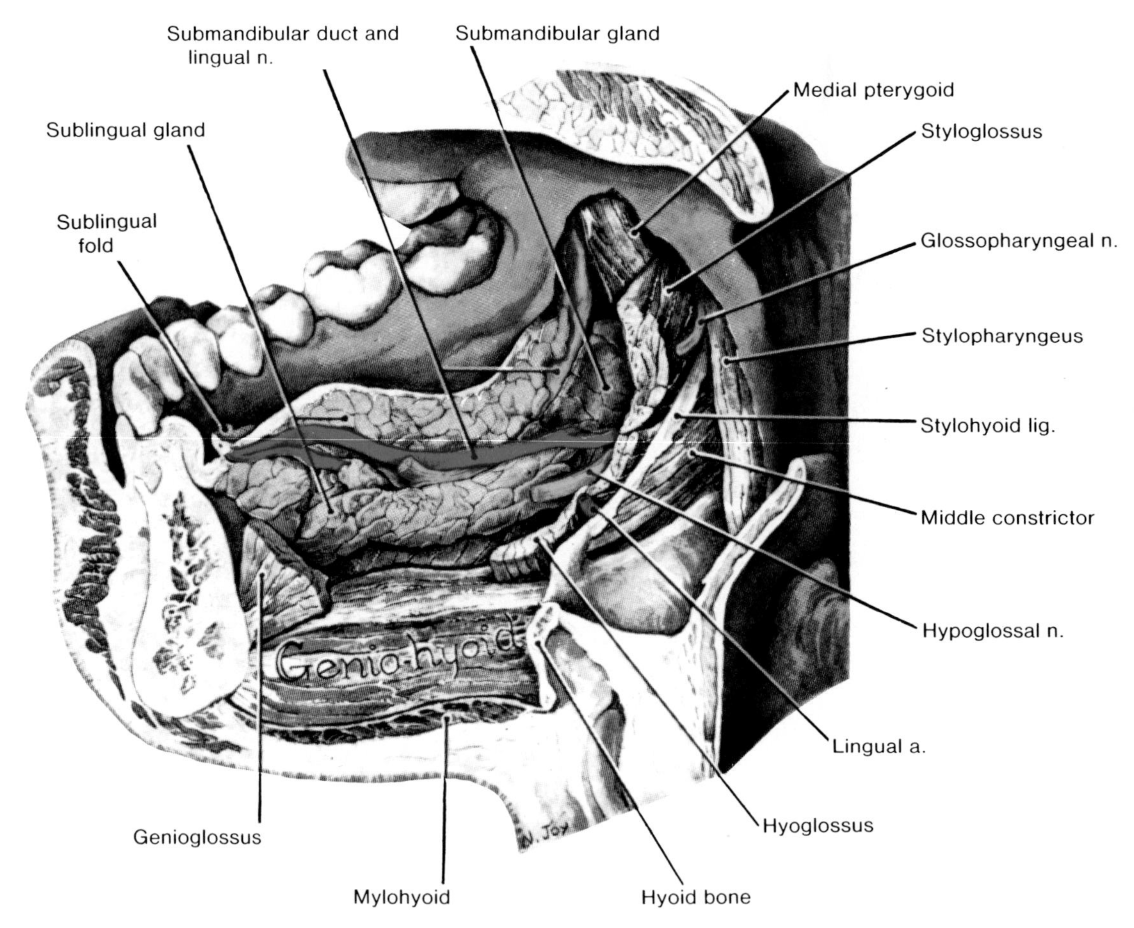

Figure 7-144. Drawing of a dissection of the floor of the mouth from which the tongue has been excised. Note the deep or oral part of the submandibular gland in the angle between the lingual nerve and the submandibular duct, which separates it from the sublingual gland. The orifice of the submandibular duct is at the anterior end of the sublingual fold. Also observe that the submandibular duct adheres to the medial side of the sublingual gland and here receives (as it sometimes does) a large accessory duct from the inferior part of the sublingual gland. This communication is of importance in sialography (radiographic demonstration of the salivary ducts) because the sublingual duct that sometimes opens into the submandibular duct may be injected, resulting in visualization of some of the sublingual ducts as well as the submandibular duct.

The hypoglossal nerve (CN XII) is the *motor nerve of the tongue* (Figs. 7-66 and 7-144). It descends from the medulla of the brain (Fig. 7-72), exits through the hypoglossal canal (Figs. 7-14 and 7-15*A*), and passes laterally between the internal jugular vein and internal and external carotid arteries. The hypoglossal nerve then curves anteriorly to enter the tongue. It runs anteriorly on the inferior part of the hyoglossus muscle and passes on to the lateral aspect of the genioglossus muscle (Fig. 7-144). It continues in the substance of the tongue as far as its tip, distributing branches to the tongue muscles.

Section or injury of a hypoglossal nerve results in paralysis and eventual atrophy of one side of the tongue. The tongue deviates to the paralyzed side during protrusion because of the action of the unaffected genioglossus muscle on the other side. Hence, *asking a person to stick out the tongue is a good test of the function of the hypoglossal nerve* (Fig. 7-145).

Trauma, such as a fractured jaw (Fig. 7-25), may injure the hypoglossal nerve and cause paralysis of the tongue musculature. In these cases the nonaffected muscles pull the tongue to the same side.

Neck lacerations, often with major vessel damage, and **basal fractures of the skull** (Table 7-3) also cause hypoglossal nerve injuries. *Moderate dysarthria* (G. *dys*, difficult + *arthroun*, to articulate) may result.

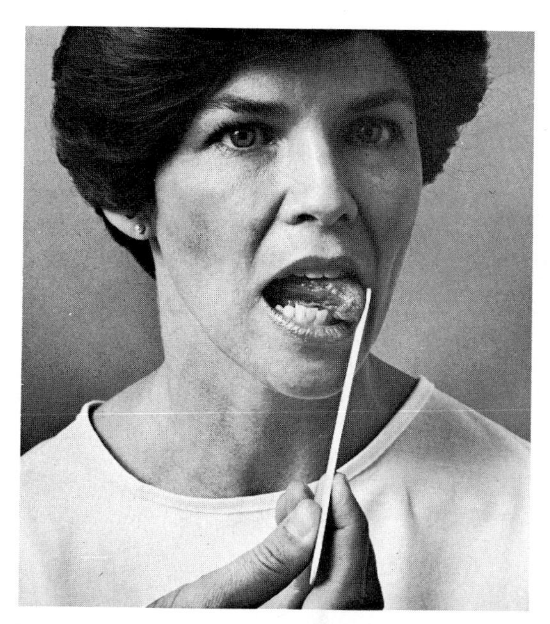

Figure 7-145. Photograph of a 33-year-old woman whose hypoglossal nerve is being tested. The strength of her tongue was determined by asking her to protrude her tongue and move it from side to side against the tongue depressor. If her left hypoglossal nerve was paralyzed, it would deviate to the left as shown here and she would be unable to move her tongue to the right. (From Smith Kline Corporation: *Essentials in Neurological Examination*, 1978).

Sensory Nerves of the Tongue (Fig. 7-139). The nerve supply to the mucous membrane of the tongue is related to its origin from the floor of the embryonic pharynx. For details and illustrations, see Moore (1982).

For general sensation, the mucosa of the *anterior two-thirds* of the tongue is supplied by the *lingual nerve*, a branch of the **mandibular division of the trigeminal** (CN V^3), the nerve of the first branchial arch (Fig. 7-33).

For special sensation (taste), the anterior two-thirds of the tongue, except for the vallate papillae, is supplied via the ***chorda tympani*** from the **facial nerve** (CN VII), which is the nerve of the second branchial arch (Fig. 7-35). The chorda tympani joins the lingual nerve and runs anteriorly in its sheath.

The mucosa of the posterior one-third of the tongue is supplied by the lingual branch of the glossopharyngeal nerve (CN IX), the nerve of the third branchial arch, for general and special (taste) sensation, including the vallate papillae. Twigs of the internal laryngeal nerve, a branch of the **vagus nerve** (CN X), the nerve of the posterior branchial arches, supplies a small area of the mucosa of the tongue, just anterior to the epiglottis (Fig. 7-139).

The sensory nerves to the tongue also carry ***parasympathetic secretomotor fibers*** *to the serous glands in the tongue.* Parasympathetic fibers from the chorda tympani *travel with the lingual nerve* to the ***submandibular*** **and** ***sublingual salivary glands*** (Figs. 7-142 to 7-144). These nerve fibers synapse in the **submandibular ganglion** (Figs. 7-33 and 7-35) that hangs from the lingual nerve and rests on the hyoglossus muscle.

ARTERIES OF THE TONGUE

The arterial supply to the tongue is chiefly through the lingual artery (Figs. 7-108, 7-143, and 7-144). This artery arises from the external carotid opposite the tip of the greater horn of the hyoid bone (Fig. 8-5).

On entering the tongue, the lingual artery passes deep to the hyoglossus muscle and sends **dorsal lingual branches** to the tongue muscles and to the mucosa of the posterior one-third of the tongue.

At the tip of the tongue, *the terminal part of the lingual artery*, called the **deep lingual artery** (profunda linguae artery) forms an anastomotic loop by joining the artery on the other side. The **sublingual artery** arises from the lingual artery at the anterior border of the hyoglossus muscle (Figs. 7-142 and 7-143). It runs anterosuperiorly to supply the sublingual gland and the adjacent muscles.

VEINS OF THE TONGUE

Two **lingual veins** accompany the lingual artery (venae comitantes). The **deep lingual vein**, the principle one of the tongue, begins at its tip and runs posteriorly in the median plane. Usually this vein can be seen at the side of the lingual frenulum (Fig. 7-124). The lingual veins receive the dorsal lingual vein and then terminate in the internal jugular vein (Fig. 7-65).

The vena comitans nervi hypoglossi, often larger than the lingual veins, begins near the tip of the tongue and passes posteriorly on the hyoglossus muscle. *This vein accompanies the hypoglossal nerve* and ends by entering the lingual vein, the facial vein, or the internal jugular vein (Fig. 7-65).

All veins of one side of the tongue unite to form the **lingual vein**, which joins either the facial vein or the internal jugular vein (Fig. 7-65).

LYMPHATIC DRAINAGE OF THE TONGUE
(Figs. 7-40 and 8-60)

There is a submucous network or plexus of lymph vessels in the tongue. Lymph from the mucosa of the tongue takes four routes: (1) *the tip of the tongue* drains to the **submental lymph nodes**; (2) lymph from the *lateral aspects of the remainder of the anterior two-thirds of the tongue* drains to the **submandibular lymph nodes**, and from them into the deep cervical lymph nodes; (3) lymph from the *medial part of the anterior two-thirds of the tongue* drains directly to the

inferior deep cervical lymph nodes; and (4) lymph from the *posterior one-third of the tongue* drains into the **superior deep cervical lymph nodes on both sides** (Figs. 8-35 and 8-40).

There are many anastomoses between the lymph vessels in the posterior one-third of the tongue. However, there are few anastomoses between the lymph vessels in the anterior two-thirds of the tongue. Consequently, lymph from the anterior two-thirds of the tongue usually drains unilaterally, whereas lymph from the posterior one-third drains bilaterally.

A clear understanding of the lymphatic drainage of the tongue is important in predicting the early spread of carcinoma of the tongue. **Lingual tumors of the posterior one-third of the tongue** readily metastisize to the superior deep cervical lymph nodes on both sides (Figs. 7-40 and 8-60), whereas tumors of the anterior two-thirds more than 10–12 mm from the median plane of the tongue usually do not metastisize to the inferior deep cervical lymph nodes until late in the disease.

Carcinoma of the tongue is uncommon on the dorsum where it can be readily noticed, except in males using excessive amounts of tobacco and alcohol. *Lingual tumors usually occur on its free borders* at the junction of the heavily papillated mucosa of the dorsum and the smooth mucosa of the inferior surface. Cancer of the tongue occurs more often in its posterior one-third and on its inferior surface. Any ulcers or nodules that persist for two or three weeks are considered with suspicion.

THE SALIVARY GLANDS

The salivary glands are three large paired masses: the parotid, submandibular, and sublingual glands.

The Parotid Glands (Figs. 7-29, 7-31, and 7-41). These are the largest of the three salivary glands. Each gland is wedged between the mandible and the sternocleidomastoid muscle, and partially covers them. These glands have been described previously (p. 836).

The Submandibular Glands (Fig. 7-144). Each salivary gland, shaped like a U, lies along the body of the mandible. It is partly superior and partly inferior to the posterior half of the base of the mandible, and partly superficial and partly deep to the mylohyoid muscle. *The submandibular gland is palpable* as a soft mass over the posterior portion of the mylohyoid muscle when it is made tense by forcing the tip of the tongue against the maxillary incisor teeth.

The submandibular duct (Figs. 7-41 and 7-144) arises from the portion of the gland that lies between the mylohyoid and hyoglossus muscles. The duct passes deep and then superficial to the lingual nerve. It then opens by one to three orifices on a small sublingual papilla beside the lingual frenulum (Fig. 7-124). Its orifice is readily visible and usually saliva can be seen trickling from it.

The submandibular lymph nodes (Fig. 7-40) lie partly embedded in the submandibular gland, and partly between it and the mandible (Fig. 8-18).

The arterial supply of the submandibular gland is from the submental branch of the facial artery. The veins accompany the arteries. The submandibular gland is supplied by parasympathetic, secretomotor fibers from the **submandibular ganglion** (Fig. 7-35).

Excision of the submandibular gland owing to a calculus in its duct or a tumor in the gland is not uncommon. The skin crease incision must be placed at least 2.5 cm inferior to the angle of the mandible in order to avoid the mandibular branch of the facial nerve (Fig. 7-29).

Persons who have not learned to palpate the submandibular gland are apt to mistake it for an enlarged submandibular lymph node (Fig. 8-18).

Swelling of the submandibular gland sometimes occurs with mumps, a contagious, generalized viral disease. The swelling is usually an ovoid enlargement that extends anteroinferiorly from the angle of the mandible as would be expected from its anatomical position (Fig. 7-41).

The Sublingual Glands (Figs. 7-41, 7-142, 7-143, and 7-144). These are the smallest of the three paired salivary glands and are the most deeply situated. Each almond-shaped gland lies in the floor of the mouth between the mandible and the genioglossus muscle. The glands on each side unite to form a horseshoe-shaped glandular mass around the lingual frenulum (Fig. 7-124). Numerous small ducts (10 to 12) open into the floor of the mouth (Fig. 7-41). Sometimes one of the ducts opens into the submandibular duct.

The arteries supplying the sublingual glands are the sublingual and submental arteries (Fig. 7-143), branches of the lingual and facial arteries.

The nerves accompany those of the submandibular gland to the sublingual gland (Fig. 7-35). They are derived from the lingual and chorda tympani nerves, and from the sympathetic. The parasympathetic secretomotor fibers are from the parasympathetic ganglion (Fig. 7-33).

In addition to the main salivary glands just described, there are small *accessory salivary glands* scattered over the palate, lips, cheeks, tonsils, and tongue.

Calculi (L. pebbles) may also form in the ducts of the sublingual glands and produce pain, particularly during eating, because the saliva cannot escape from the glands. *The calculi are usually composed of calcium salts of inorganic or organic acids or other material.* Often they can be detected on plain radiographs.

The parotid and submandibular salivary glands may be examined radiographically following the injection of a contrast medium into their ducts. This special type of radiograph, called a **sialogram**, demonstrates the salivary ducts and some of the secretory units. Owing to the small size of the ducts of the sublingual glands, injections of contrast medium cannot be made into them.

THE PTERYGOPALATINE FOSSA

The pterygopalatine fossa is a small, elongated, pyramidal space inferior to the apex of the orbit (Figs. 7-113 and 7-146). Its superior and large end opens into the **superior orbital fissure** (Fig. 7-113), and its inferior end is closed except for the palatine foramina (Fig. 7-9). Laterally, the pterygopalatine fossa opens into the infratemporal fossa and thus can be partly inspected without separating the bones (Fig. 7-113).

The pterygopalatine fossa was given its name because it is located between the **pterygoid process** of the sphenoid bone posteriorly, and the **palatine** bone medially. The maxilla lies anteriorly and the fragile vertical plate of the palatine bone forms its medial wall.

The roof of the pterygopalatine fossa, which is incomplete, is formed by the *greater wing of the sphenoid bone* (Figs. 7-11, 7-19, and 7-113).

The pterygopalatine fossa has several communications (Fig. 7-146): (1) ***laterally*** *with the infratemporal fossa* through the **pterygomaxillary fissure**; (2) ***medially*** *with the nasal cavity* through the **sphenopalatine foramen; (3)** ***anteriorly*** *with the orbit* through the **inferior orbital fissure**; and (4) ***posterosuperiorly*** *with the middle cranial fossa* through the **foramen rotundum** (Figs. 7-14 and 7-15*A*).

CONTENTS OF THE PTERYGOPALATINE FOSSA (Figs. 7-115 and 7-147)

This fossa contains the terminal branches of the **maxillary artery**, the **maxillary nerve** (Fig. 7-115) and the **pterygopalatine ganglion** (Figs. 7-35 and 7-147). The most important structure in the pterygopalatine fossa is the maxillary nerve.

The maxillary nerve (CN V^2) enters the pterygopalatine fossa through the foramen rotundum (Figs. 7-15 and 7-62), and runs anterolaterally in the posterior part of the pterygopalatine fossa (Fig. 7-147). Verify this course of the maxillary nerve by passing a toothpick through the foramen rotundum of a dried skull into the pterygopalatine fossa.

Within the pterygopalatine fossa, the maxillary nerve gives off the **zygomatic nerve**, which divides into *zygomaticofacial and zygomaticotemporal* nerves (Fig. 7-147). The nerves emerge from the zygomatic bone through foramina of the same name (Figs. 7-2 and 7-11), and supply the lateral region of the cheek and the temple (Fig. 7-32).

While in the pterygopalatine fossa, the maxillary nerve also gives off two pterygopalatine nerves which suspend the **pterygopalatine ganglion** (Figs. 7-33 and 7-147), a *parasympathetic ganglion.* It is situated in the superior part of the pterygopalatine fossa. Fibers of the maxillary nerve pass through this ganglion without synapsing. The parasympathetic fibers to the pterygopalatine ganglion come from the facial nerve (CN VII) via the greater petrosal nerve (Fig. 7-35). The sensory fibers, passing through the pterygopalatine ganglion without synapsing, supply the nose, palate, tonsil, and gingivae.

The maxillary nerve leaves the pterygopalatine fossa through the inferior orbital fissure, after which it is known **as the infraorbital nerve** (Figs. 7-28, 7-32C, and 7-147). Within the orbit the infraorbital nerve passes through the infraorbital groove and canal in the floor of the orbit (Figs. 7-88 and 7-147). The infraorbital nerve appears on the face by passing through the **infraorbital foramen** (Figs. 7-2, 7-28, 7-33, and 7-147) and supplies the skin of the lower eyelid, the side of the nose, and the anterior portion of the cheek (Fig. 7-32). Just before it enters the orbit, and during its passage through it, the maxillary nerve gives off the **superior alveolar nerves.** They supply the maxillary teeth (Figs. 7-33, 7-114, and 7-115).

The third or pterygopalatine part of the maxillary artery (Figs. 7-116 and 7-148) passes through the *pterygomaxillary fissure* into the pterygopalatine fossa (Figs. 7-113 to 7-116), where it lies anterior to the pterygopalatine ganglion (Figs. 7-147 and 7-148). The artery breaks up into branches which accompany all the nerves in the fossa with the same names.

The posterior superior alveolar artery arises from the maxillary artery as it enters the pterygopalatine fossa (Fig. 7-114). It descends on the infratemporal surface of the maxilla and supplies the molar and premolar teeth, and the lining of the maxillary sinus (Figs. 7-129 and 7-146).

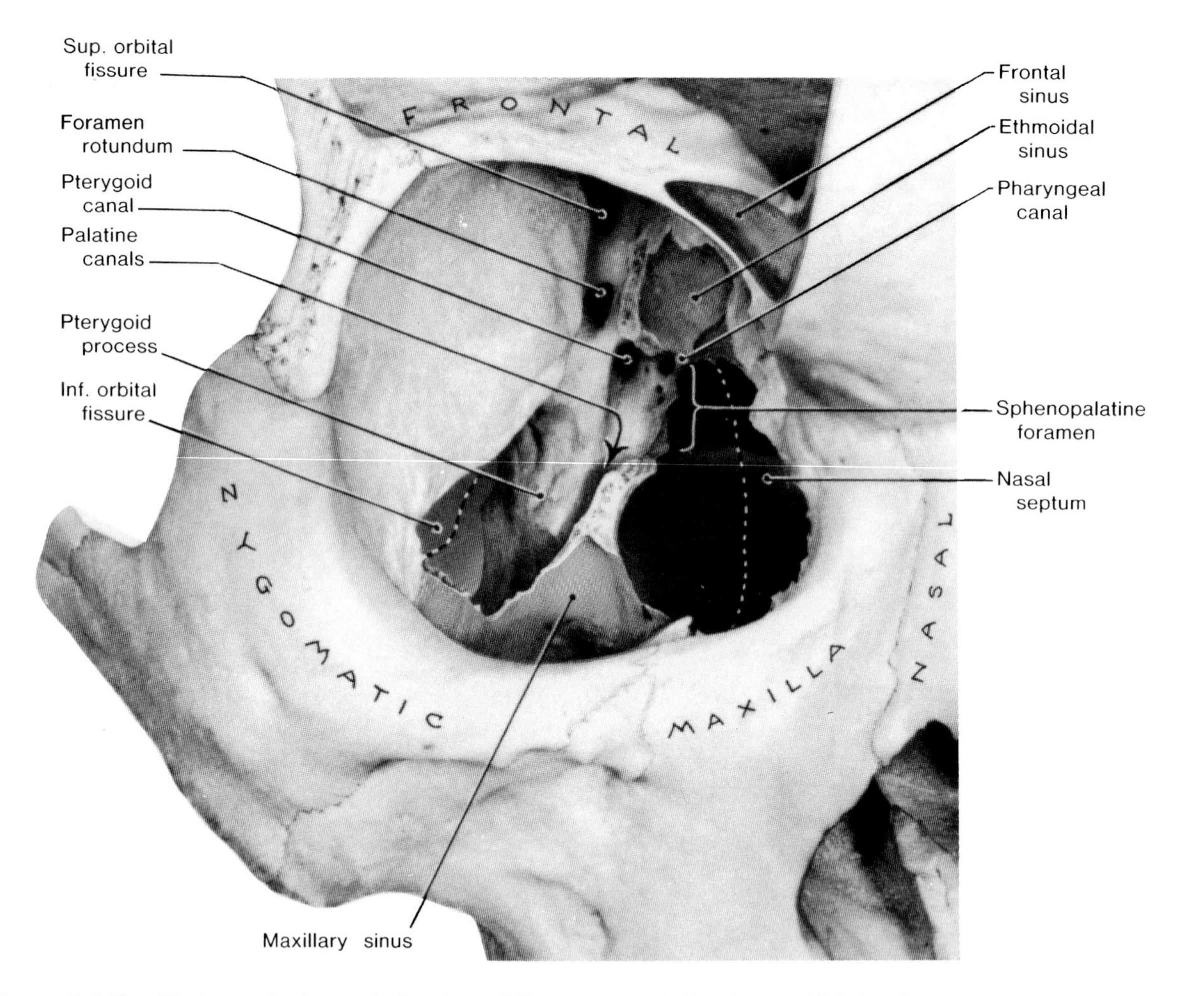

Figure 7-146. Photograph of an anterior view of the pterygopalatine fossa which has been exposed through the floor of the orbit and the maxillary sinus. For a lateral view of the pterygopalatine fossa, see Figure 7-113.

The infraorbital artery (Figs. 7-114 and 7-115), often arising in conjunction with the posterior superior alveolar artery, enters the orbital cavity (Fig. 7-91). The infraorbital artery passes anteriorly in the infraorbital groove and canal (Figs. 7-88 and 7-148). Within the canal it gives off the anterior superior alveolar artery. The infraorbital artery exits onto the face through the infraorbital foramen (Figs. 7-2 and 7-148) and sends branches to the lower eyelid, the lateral aspect of the external nose, and the upper lip.

THE NOSE

The nose is the superior part of the respiratory tract (Fig. 1-11) and contains the peripheral organ of smell (Fig. 7-149). The nose is divided into right and left **nares** by the nasal septum. *Each naris is divisible into an olfactory area and a respiratory area.*

The functions of the nose and nasal cavities are: (1) respiration, (2) olfaction, (3) filtration of dust, (4) humidification of inspired air, and (5) the reception of secretions from the paranasal sinuses (Figs. 7-86 and 7-89) and the nasolacrimal duct (Figs. 7-96 and 7-102).

THE EXTERNAL NOSE

Noses vary considerably in size and shape, mainly as a result of differences in the nasal cartilages and the depth of the glabella (Figs. 7-1 and 7-151). The nose projects anteroinferiorly from the face to which its root is joined just inferior to the forehead (Figs. 7-10 and 7-29).

The **dorsum of the nose** extends from the **root** to the **apex** (tip). The inferior surface of the nose is pierced by two apertures, called the **anterior nares** (L. nostrils), which are separated from each other by the **nasal septum** or septum nasi (Figs. 7-130, 7-150, 7-151, and 7-155). Each naris is bounded laterally by an **ala** (L. wing), *i.e.*, the side of the nose. The posterior nasal apertures or **choanae** (Fig. 7-9*B*) open into the nasopharynx (Fig. 7-122).

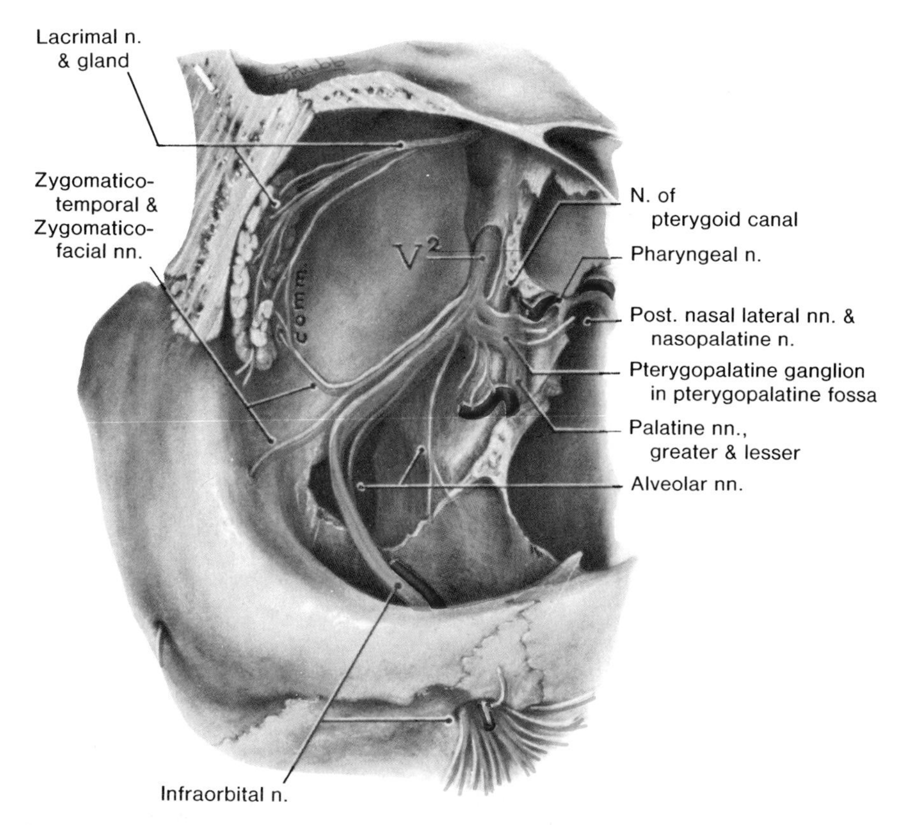

Figure 7-147. Drawing of a dissection of the maxillary nerve (CN V^2). Also see Figure 7-33. This afferent nerve supplies the territory extending from the skin laterally (*pink area* in Fig. 7-32C) to the nasal septum medially. The lateral and medial branches of the nerve are separated by the maxillary sinus (Fig. 7-146). The greater petrosal nerve, via the nerve of the pterygoid canal, brings parasympathetic fibers to the pterygopalatine ganglion in the pterygopalatine fossa. They are relayed and distributed with branches of CN V^2 as secretomotor fibers (Figs. 7-33 and 7-35).

THE SKELETON OF THE NOSE (Figs. 7-10, 7-150, and 7-151)

The immovable ***bridge of the nose,*** the superior bony part of the nose, consists of the **nasal bones,** the frontal processes of the **maxillae,** and the nasal part of the **frontal bone.** The movable cartilaginous part of the nose consists of five main cartilages and a few smaller ones (Figs. 7-150, 7-151, and 7-155).

The nasal cartilages are composed of hyaline cartilage and are connected with one another and the nasal bones by the continuity of the perichondrium and periosteum, respectively. The U-shaped alar nasal cartilages are free and movable (Figs. 7-150 and 7-151). The alar cartilages can dilate or constrict the external nares by contraction of the muscles acting on the external nose (Fig. 7-31 and p. 822).

Structure of the Nasal Septum (Figs. 7-130, 7-150, and 7-151). The partly bony and partly cartilaginous nasal septum divides the nasal cavity or chamber of the nose into two narrow **nasal cavities.** The bony part of the septum is usually located in the median plane until the 7th year; thereafter it often bulges to one or other side (Fig. 7-150), more frequently to the right.

The nasal septum has three main components: (1) ***the perpendicular plate*** *of the ethmoid bone*; (2) ***the vomer***; and (3) ***the septal cartilage.*** The perpendicular plate of the ethmoid (Fig. 7-151), forming the superior part of the septum, is very thin and descends from the cribriform plate of the ethmoid bone (Figs. 7-14 and 7-152).

The **vomer** is a thin flat bone (Fig. 7-151); it forms the posteroinferior part of the nasal septum. The vomer articulates with the perpendicular plate of the ethmoid and with the septal cartilage (Figs. 7-150 and 7-151).

Fractures of the nose are quite common because its bony parts, the perpendicular plate of the ethmoid bone and the vomer are thin bones. Usually the fractures are transverse (Fig. 7-26). If the injury results from a direct blow,

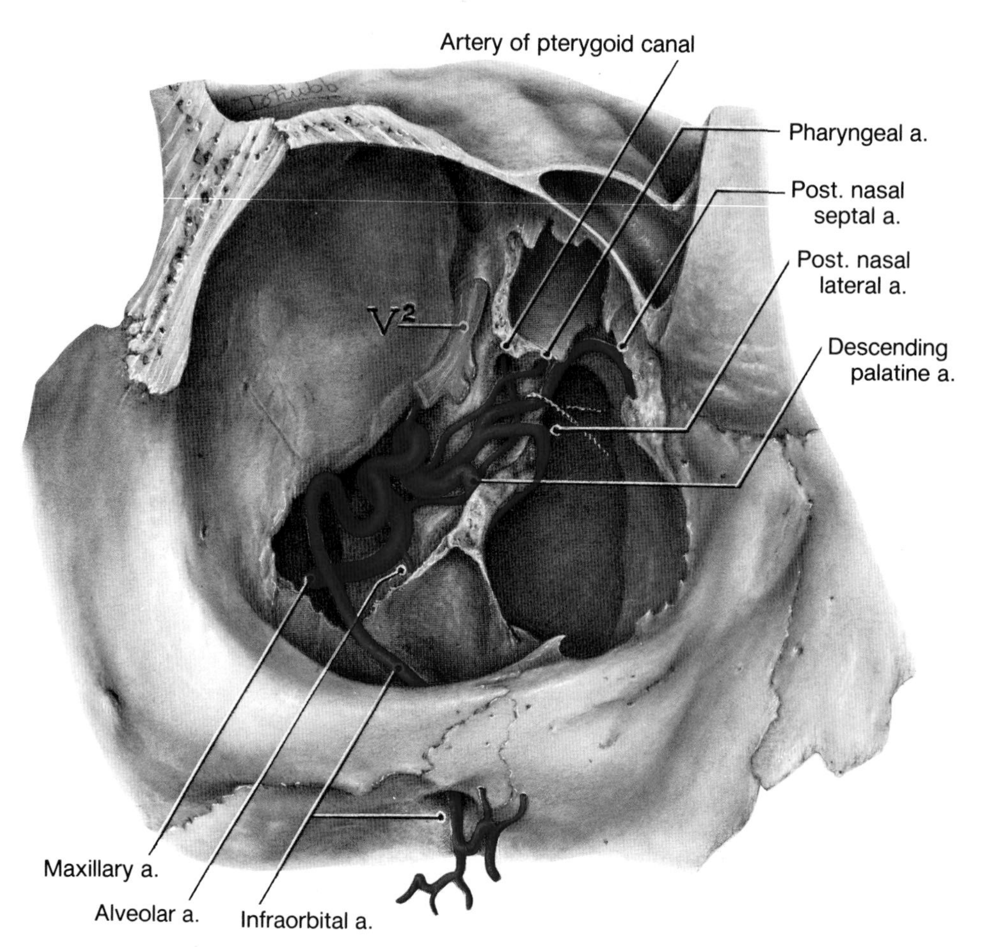

Figure 7-148. Drawing illustrating the third or pterygopalatine part of the maxillary artery. Observe that the stem of this artery, arising at the neck of the mandible, is divided into three parts by the lateral pterygoid muscle (Fig. 7-116). The branches of the third part arise just before and after it enters the pterygopalatine fossa: (a) infraorbital, (b) posterior superior alveolar, (c) descending palatine, (d) artery of pterygoid canal, (e) pharyngeal, and (f) sphenopalatine arteries. The descending palatine artery divides into the greater and lesser palatine arteries (Fig. 7-132). The third part of the maxillary artery, often very tortuous, lies anterior to the maxillary nerve and its branches.

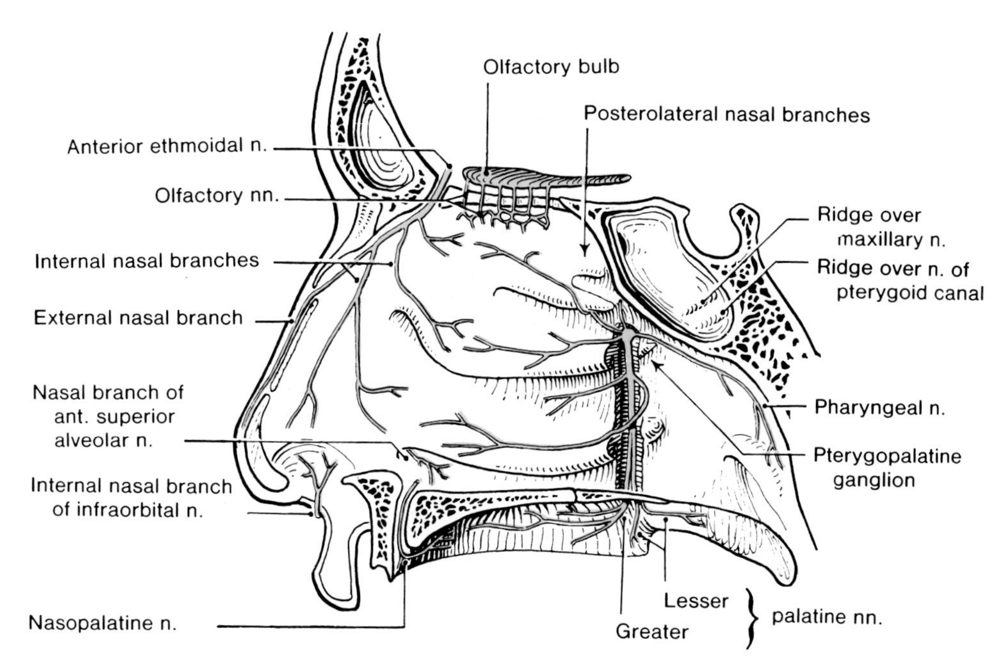

Figure 7-149. Drawing showing the nerve supply to the lateral wall of the nasal cavity. Observe the olfactory nerves (CN 1) passing through the cribriform plate (Figs. 7-14 and 7-152) and entering the olfactory bulb. The olfactory neurons, usually called olfactory receptor cells (Fig. 7-152), are located in the olfactory area situated in the superior one-third of the nasal mucosa (Fig. 1-11). The axons of these cells constitute the olfactory nerves. They are collected into about 20 bundles which pass through the foramina in the cribriform plate (Fig. 7-152). These nerves pierce the dura mater and arachnoid of the brain and enter the olfactory bulbs in the anterior cranial fossa (Figs. 7-51 and 7-52). Observe that the greater palatine nerve passes inferiorly in the pterygopalatine fossa and exits via the greater palatine foramen to supply the hard palate.

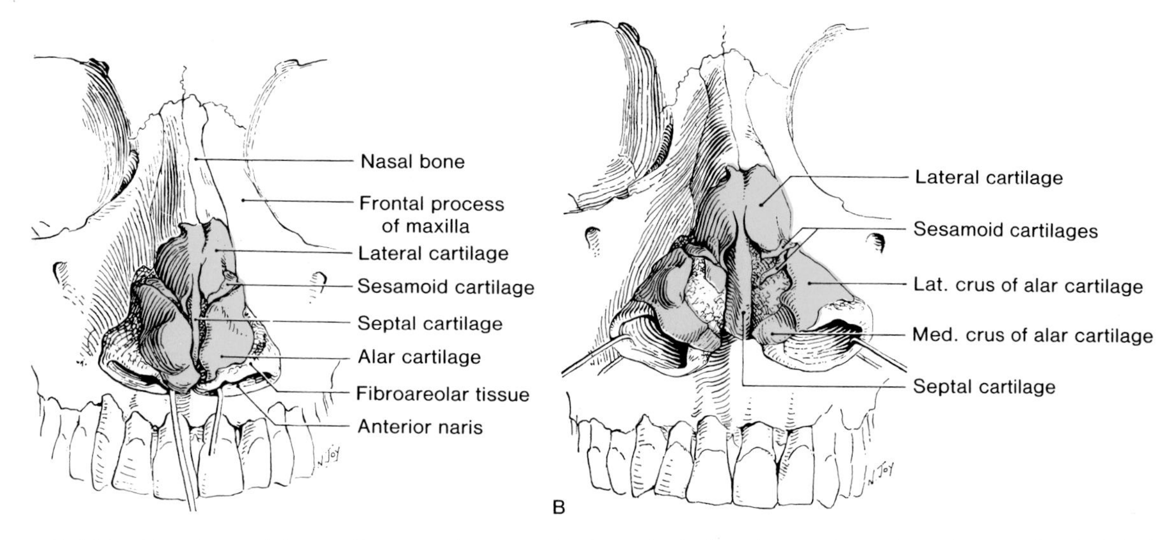

Figure 7-150. Drawings of the cartilages of the nose (*yellow*). In *A*, the alar cartilages have been pulled inferiorly to expose the sesamoid cartilages. In *B*, the alar cartilages are separated by dissection and retracted laterally.

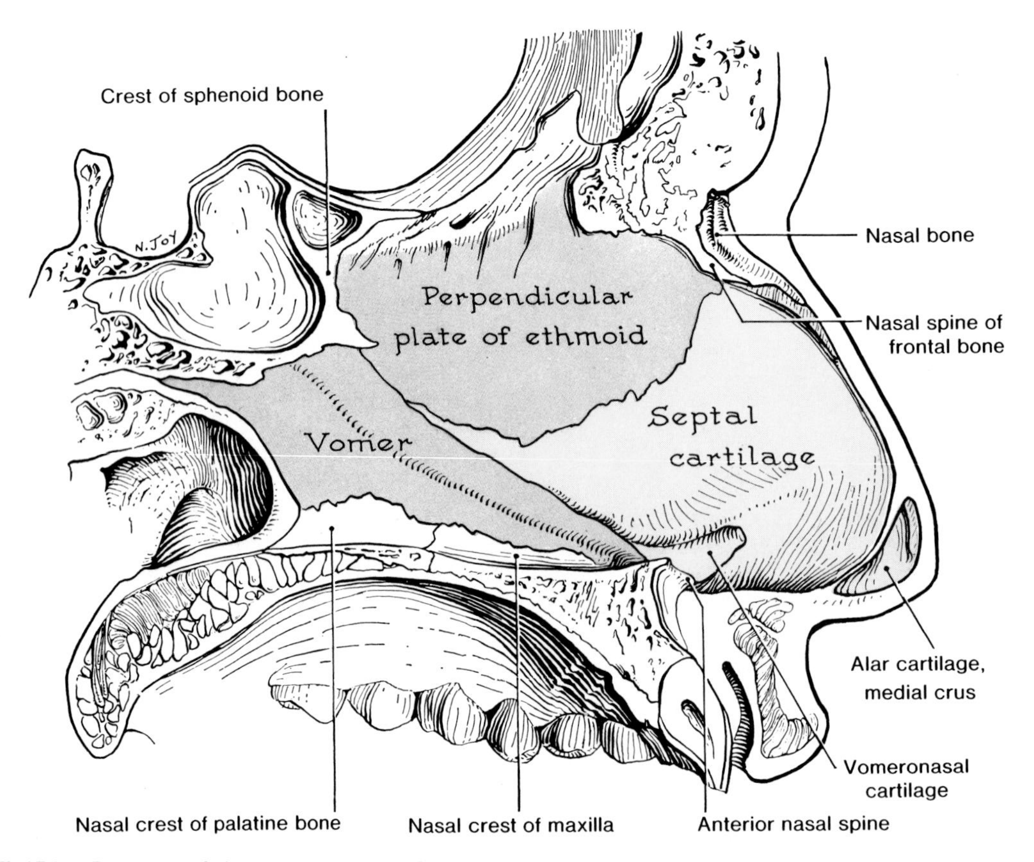

Figure 7-151. Drawing of the anterior part of a bisected skull showing the nasal septum. Note that, like the palate, the nasal septum has a hard part (here partly bony and partly cartilaginous) and a soft or mobile part (also see Fig. 7-155). The skeleton of the hard septum consists of three parts: perpendicular plate of ethmoid, septal cartilage, and vomer. Around the circumference of these, the adjacent bones (frontal, nasal, maxillary, palatine, and sphenoid) make minor contributions. The soft or mobile part of the septum is composed of the alar cartilages, covered by skin and soft tissues between the tip of the nose and the anterior nasal spine.

the cribriform plate of the ethmoid bone is often fractured.

Congenital malformations of the nose are not common, except for those associated with cleft lip and cleft palate (Figs. 7-125 and 7-136). Complete absence of the external nose may occur with only one aperture present. Sometimes there is normal development of one-half of the nose with absence or malformation of the other side.

The nasal septum may be displaced or deviate from the median plane as the result of a birth injury or a congenital malformation. More often the deviation is caused by postnatal trauma (*e.g.*, during a fist fight).

Sometimes the deviation is so severe that the nasal septum comes into contact with the lateral wall of the nasal cavity. As this obstructs breathing, surgical repair is usually necessary.

THE NASAL CAVITY AND MUCOUS MEMBRANE

The nasal cavities are entered through the nostrils or anterior nares. They open into the **nasopharynx** through the **choanae** (Figs. 7-9*B* and 7-122).

Mucosa lines the entire nasal cavity, except the vestibule (Figs. 7-86 and 7-153). *The entrance to the nose or vestibule is lined with skin* on both the septum side and the lateral wall. The skin has hairs called **vibrissae**, a name derived from the French verb *vibro,* to quiver (Fig. 7-153*B*). The nasal mucosa is firmly bound to the periosteum and perichondrium of the supporting structures of the nose (Figs. 7-130 and 7-154). It is continuous with the lining of all the chambers with which the nasal cavities communicate: the nasopharynx posteriorly (Fig. 7-130), the paranasal sinuses superiorly and laterally (Figs. 7-86 and 7-89), and the lacrimal sac and conjunctiva of the orbit (Figs. 7-96 and 7-102).

The nasal septum is covered with mucous membrane, except for the part in the vestibule. The inferior two-thirds of the nasal mucosa is called the **respiratory area** and the superior one-third is the **olfactory area** (Fig. 7-149). Air passing over the respiratory area is warmed and moistened before it passes through the rest of the upper respiratory tract to the lungs.

The Olfactory Area of the Nasal Mucosa (Fig. 7-149). Yellowish in living persons, this area contains the *peripheral organ of smell* (Fig. 7-152). Sniffing draws air to the olfactory area. The functional cells concerned with olfaction (L. *olfacere*, to smell) are called **olfactory receptor cells** (Fig. 7-152). They are located in the mucosa of the olfactory area of the nose (Fig. 7-149).

The olfactory cell is a primitive type of sensory receptor. Its dendrite extends to the surface of the olfactory epithelium and ends as an exposed bulbous enlargement which has cilia up to 100 μ in length. Hence an odoriferous substance has direct access to the neuron without the intervention of non-nervous tissue. It has been estimated that there are about 25 million olfactory cells in each half of the olfactory mucosa of young adults.

The axons of the olfactory cells, constituting the **olfactory nerves**, are collected into 18 to 20 bundles which pass through the foramina in the cribriform plates of the ethmoid bone in the anterior cranial fossa (Figs. 7-14, 7-149, and 7-152). The olfactory nerves are unmyelinated and, after traversing the cribriform plates, they pierce the dura mater and arachnoid covering the brain. Surrounded by a thin tube of pia mater, the *olfactory nerves cross the subarachnoid space containing cerebrospinal fluid* (*CSF*), and enter the inferior aspects of the olfactory bulbs of the brain (Figs. 7-72, 7-149, and 7-152).

Although nasal discharges are commonly associated with upper respiratory tract infections, **nasal discharge associated with a head injury may actually be CSF** (an occurrence known as CSF rhinorrhea) owing to fracture of the cribriform plate, tearing of the meninges and leakage of CSF. This condition may be diagnosed by injecting a radioactive tracer into the CSF and later detecting it in pieces of cotton that have been inserted into the nostrils.

Unlike many mammals, we do not rely heavily on the primitive sense of smell for information about our environment. However we are able to detect certain noxious substances and our sense of smell adds much to our enjoyment of food, wine, and each other (thanks to lotions and perfumes). Even newborn infants react to certain obnoxious smells by grimacing.

Anosmia (loss of sense of smell) is not however a serious handicap and occurs gradually with age. Olfactory receptor cells degenerate at the rate of about 1% per year in most old people.

Disorders of olfaction may result from conditions affecting: (1) the olfactory receptor cells in the nasal mucous membrane; (2) the secondary olfactory neurons in the olfactory bulb and tract; or (3) their intracranial connections, e.g., the lateral olfactory stria (Fig. 7-152).

Olfactory disorders are usually caused by: (1) lesions of the nasal cavity, *e.g.*, owing to inflammation; (2) fractures involving the anterior cranial fossa (Fig. 7-26 and Table 7-3); (3) tumors of adjacent parts of the brain; (4) **meningitis** (p. 853); (5) arteriosclerosis; (6) certain drug intoxications; and (7) cerebrovascular accidents (*e.g.*, effusion of blood into the base of the frontal lobe).

Unilateral anosmia may be of diagnostic significance in localizing brain lesions (*e.g.*,

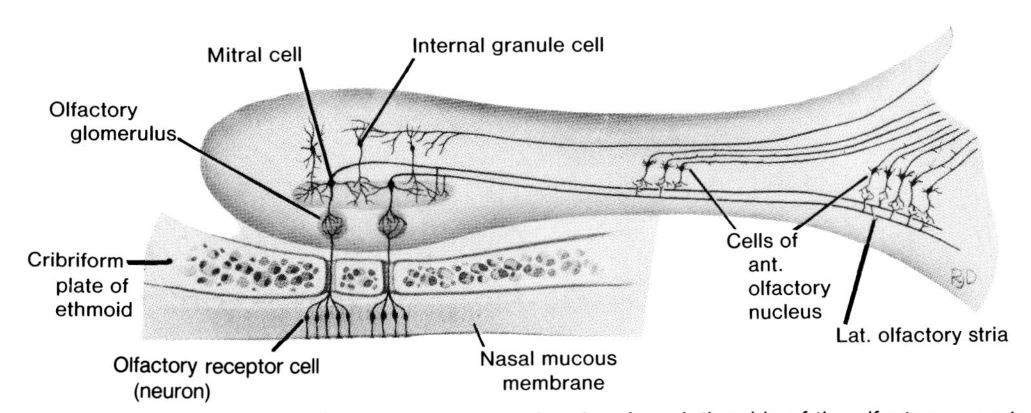

Figure 7-152. Diagram of the olfactory bulb and tract showing the relationship of the olfactory receptors cells or neurons in the nasal mucosa with cells in the olfactory bulb of the brain (Fig. 7-72). The processes of olfactory receptor cells or neurons are located in the nasal mucous membrane and are in contact with the odor-producing chemicals in the inhaled air.

a tumor at the base of the frontal lobe). Before **testing the olfactory nerves**, it must be determined that there is no obstruction of the nasal passages. With the eyes closed, the patient is asked to identify familiar odors (*e.g.*, coffee). *Comparisons between the two sides are important*; thus each nostril is tested separately by occluding the other one.

Unilateral brain lesions do not usually result in a loss of the sense of smell, unless both olfactory tracts are injured. Lesions of one olfactory tract produce unilateral anosmia.

Tumors of the inferior surface of the frontal lobe and trauma to the external nose and the forehead are the common causes of neurogenic anosmia.

Nerves of the Nasal Mucosa (Figs. 7-33, 7-35, and 7-149). The lining of the roof and adjacent surfaces of the nasal septum, constituting the **olfactory area**, is served by twigs of the *olfactory nerves* (CN 1).

The **respiratory area**, the inferior two-thirds of the nasal mucosa, is supplied chiefly by the trigeminal nerve. The mucous membrane of the nasal septum is supplied chiefly by the **nasopalatine nerve**, a branch of the maxillary division of the trigeminal nerve (*i.e.*, CN V^2). Its anterior portion is supplied by the **anterior ethmoidal nerve** (a branch of the *nasociliary nerve*, (Fig. 7-69), derived from the ophthalmic division of the trigeminal nerve (*i.e.*, CN V^1).

The lateral wall of the nasal cavity is supplied by nasal branches from the maxillary nerve (Fig. 7-33*C*), by branches of the greater palatine nerve, and by the anterior ethmoidal nerve (Fig. 7-149).

Arteries of the Nasal Mucosa (Fig. 7-153). The blood supply of the mucosa of the nasal septum is derived from the **maxillary artery** (Fig. 7-116). There is a rich blood supply to the mucosa of the nasal cavity.

The sphenopalatine artery, *a branch of the maxillary artery* (Fig. 7-116), **supplies most of the blood to the nasal mucosa**, (Fig. 7-153). It enters via the sphenopalatine foramen (Figs. 7-113 and 7-116) and sends branches to posterior regions of the lateral wall and to the nasal septum.

The greater palatine artery (Fig. 7-115), another branch of the maxillary artery, passes through the incisive foramen (Figs. 7-132 and 7-134) to supply the nasal septum. The sphenopalatine and greater palatine arteries anastomose in the anteroinferior part of the nasal septum (Fig. 7-153).

The anterior and posterior ethmoidal arteries (Fig. 7-99), *branches of the ophthalmic artery*, supply the anterosuperior part of the mucosa of the lateral wall of the nasal cavity and the nasal septum (Fig. 7-153). Three branches of the **facial artery** (superior labial, ascending palatine, and laternal nasal) also supply anterior parts of the nasal mucosa (Figs. 7-29, 7-37, and 7-153).

The nasal mucosa becomes swollen and inflamed (**rhinitis**) during upper respiratory infections and with some allergies (*e.g., hayfever*). The Greek word *rhis* (rhin-) means nose; hence *rhinitis* indicates inflammation of the nasal mucous membrane. Swelling of this membrane occurs readily because of its vascularity. When swelling is associated with increased mucus secretion, the common "stopped-up nose" or "runny nose" occurs. The spread of infection from the nose and paranasal sinuses to the meninges, although rare, is dangerous because meningitis can be fatal.

The skin of the nose contains many sebaceous glands which may become infected and blocked. As discussed previously, *infections about the nose may spread to the cavernous sinus via connections between the facial and ophthalmic veins* (Fig. 7-39); hence the nose is part of the danger triangle of the face (Fig. 7-43).

Because of its relations, infections of the nasal cavities may spread to: (1) the *anterior cranial fossa* via the cribriform plate of the ethmoid bone (Figs. 7-15 and 7-152) (2) the *nasopharynx and retropharyngeal soft tissues* (Figs. 7-122 and 7-130); (3) the *middle ear* via the auditory tube (Figs. 7-86 and 7-122); (4) the *paranasal sinuses* (Figs. 7-17 and 7-89); and (5) the *lacrimal apparatus and conjunctiva* (Figs. 7-96 and 7-102).

Veins of the Nasal Mucosa. The veins of the nasal cavity form a rich venous network or plexus in the connective tissue of the nasal mucosa, especially over the inferior part of the septum. Some of the veins open into the **sphenopalatine vein** and drain to the *pterygoid plexus* (Fig. 7-39). Others join the facial vein and some empty into the ophthalmic veins and drain into the cavernous sinus (Fig. 7-57*B*).

Epistaxis (nasal hemorrhage; nosebleed) is relatively common owing to the richness of the blood supply to the nasal mucosa (Fig. 7-153). In most cases the cause of epistaxis (G. fr. *epistazō*, to bleed at the nose) is trauma and the

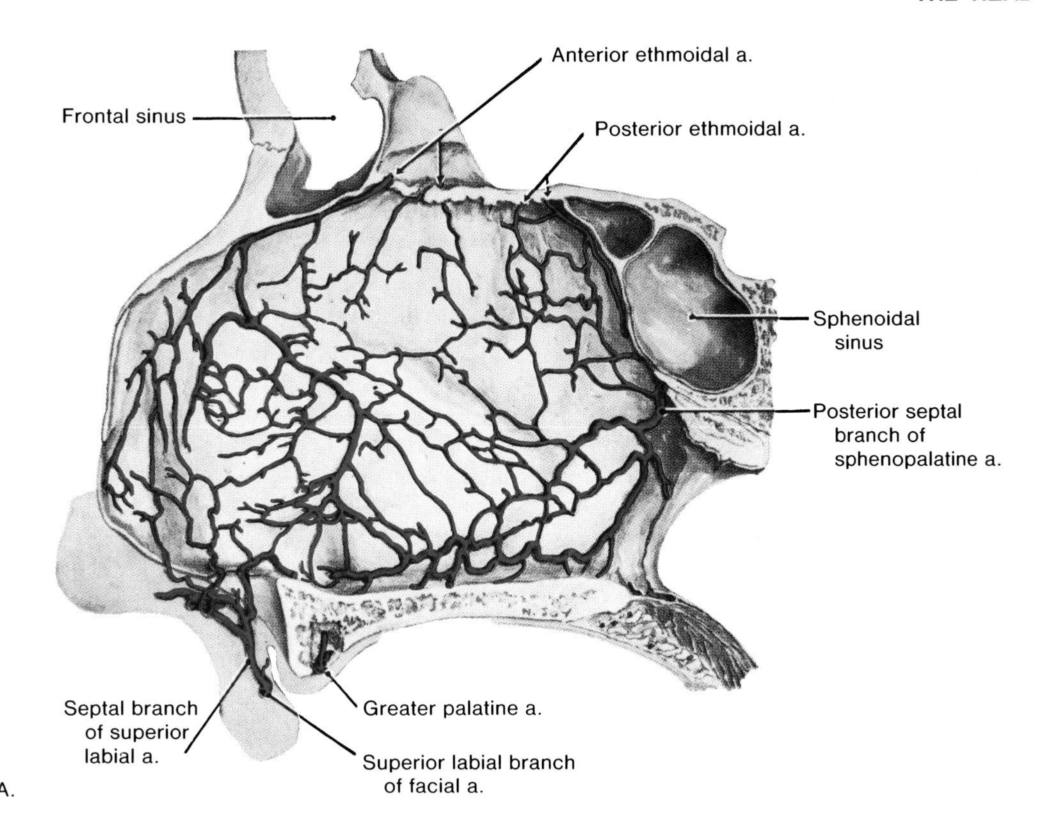

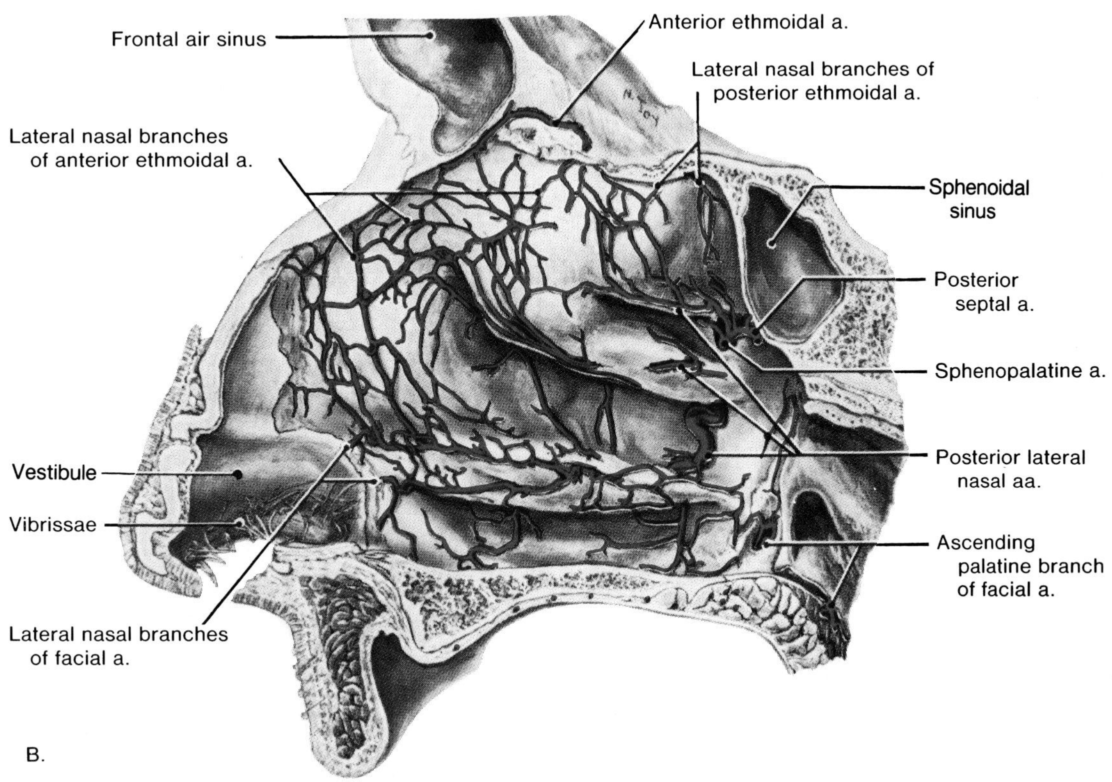

Figure 7-153. Drawings of the medial wall of the nasal septum (*A*) and of the lateral wall of the nasal cavity (*B*), showing the arterial supply of the mucosa. The sphenopalatine artery is the main supply. Entering through the sphenopalatine foramen (Fig. 7-132), it sends lateral nasal branches anteriorly on both surfaces of the conchae (Fig. 7-154), partly in bony canals.

bleeding is located in the anterior third of the nose.

Mild epistaxis often results from nose picking which tears the veins in the vestibule of the nose around the anterior nares (Fig. 7-153). However epistaxis is also associated with infections (*e.g.*, typhoid fever) and *hypertension.*

Spurting of blood from the nose results from rupture of arteries, particularly at the site of anastomosis of the sphenopalatine and greater palatine arteries (Fig. 7-153). If nasal bleeding is so profuse that it cannot be stopped by usual treatments, the external carotid artery is sometimes clamped and/or ligated in the neck; this is the source of the blood passing to the nose via the maxillary and sphenopalatine arteries (Figs. 7-37, 7-115, and 7-116).

The roof of the nasal cavity is curved and narrow, except at the posterior end. It is divided into three parts, frontonasal, ethmoidal, and sphenoidal, which indicate the bones forming them.

The floor of the nasal cavity is wider than the roof. It is formed by the *palatine process of the* ***maxilla*** and the *horizontal plates of the palatine bone* (Figs. 7-8 and 7-9).

Walls of the Nasal Cavity (Figs. 7-86, 7-122, 7-149, 7-151, and 7-153 to 155).

The medial wall of the nasal cavity is formed by the nasal septum; it is usually smooth.

The lateral wall of the nasal cavity is uneven owing to three longitudinal, scroll-shaped elevations, called **conchae** (L. shells) or *turbinates* (L. shaped like a top). These elevations are called superior, middle, and inferior **nasal conchae** (Figs. 7-86 and 7-122), according to their position on the lateral wall of the nasal cavity.

The superior and middle conchae are parts of the ethmoid bone, whereas the inferior concha is a separate bone. The inferior and middle conchae project medially and inferiorly, producing air passageways called the **inferior and middle meatus** (L., a going or passage). The plural of meatus is the same as the singular, although it is not uncommon to see "meatuses." The short superior concha conceals the **superior meatus.** The space posterosuperior to the superior concha into which the *sphenoidal sinus* opens is called the ***sphenoethmoidal recess*** (Figs. 7-86 and 7-154).

The superior meatus (Fig. 7-86) is a narrow passageway between the superior and middle nasal conchae into which the **posterior ethmoidal sinuses** open by one or more orifices (Figs. 7-89, 7-154, and 7-155).

The middle meatus (Fig. 7-86) is longer and wider than the superior meatus. The anterosuperior part of this meatus leads into a funnel-shaped opening, called the **infundibulum** (Fig. 7-157), through which it communicates with the frontal sinus. The passage that leads inferiorly from the frontal sinus to the infundibulum of the middle meatus is called the **frontonasal duct** (Fig. 7-154). There is one duct for each frontal sinus and, as there may be several frontal sinuses on each side, there may be several frontonasal ducts.

When the middle concha is raised or removed, a rounded elevation, called the **ethmoidal bulla** (L. a bubble), is visible (Fig. 7-154). This bulge is formed by the **middle ethmoidal cells** (Fig. 7-155) which constitute the ethmoidal sinuses (Fig. 7-89). The middle ethmoidal cells open on the surface of the ethmoidal bulla. Inferior to the ethmoidal bulla is a semicircular groove called the **hiatus semilunaris** (Fig. 7-154). Anterosuperiorly, the frontal sinus opens into this hiatus. Near this are the openings of the anterior ethmoidal sinuses (Fig. 7-155). The maxillary sinus opens into the middle meatus (Figs. 7-89 and 7-155).

The inferior meatus (Fig. 7-86) is a horizontal passage, inferolateral to the inferior nasal concha. The ***nasolacrimal duct*** opens into the anterior part of this meatus (Figs. 7-96, and 7-102). Usually the orifice of this duct is wide and circular (Fig. 7-154).

THE PARANASAL SINUSES

The paranasal sinuses (paranasal air sinuses) are air-filled extensions of the nasal cavities **(pneumatic areas)** in the following cranial bones: frontal, ethmoid, sphenoid, and maxilla. (Figs. 7-4, 7-12, 7-17, 7-86, 7-155, and 7-156). Owing to the presence of these air-filled cavities, the bones mentioned are sometimes referred to as **pneumatic bones**. The sinuses are named according to the bones in which they are located. Hence there are **frontal, ethmoidal, sphenoidal, and maxillary sinuses.** They vary considerably in size and form in different people and in different races (*e.g.*, the frontal sinuses are generally small in oriental people).

The paranasal sinuses develop as outgrowths from the nasal cavities mainly after birth. The original openings of the outgrowths persist as orifices in the nasal cavities. Hence all the sinuses drain directly or indirectly into the nasal cavity. Secretions from the mucosa of the sinuses eventually drain through these openings into the nasal cavity (Figs. 7-154 and 7-155). The mucosal lining of the sinuses is also continuous with the nasal mucosa; this results from their origin as outgrowths from the nasal cavities.

Rhinitis, as previously stated, is a common symptom of an *upper respiratory infection* or a

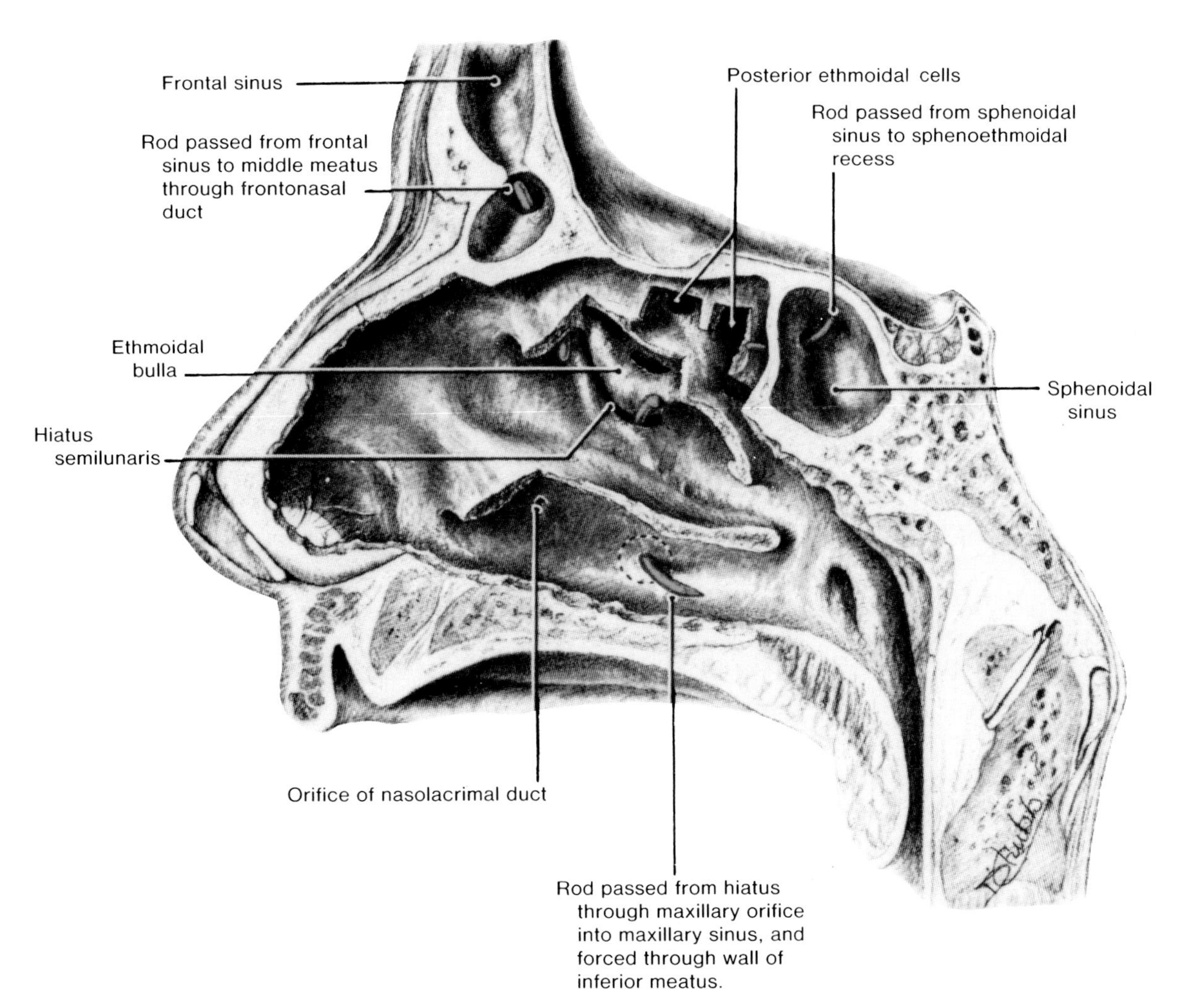

Figure 7-154. Drawing of a dissection of the lateral wall of the nasal cavity. Parts of the superior, middle, and inferior conchae are cut away. Observe the sphenoidal sinus in the body of the sphenoid bone and its orifice, superior to the middle of its anterior wall, that opens into the sphenoethmoidal recess (also see Fig. 7-86). Note the orifices of the posterior ethmoidal cells opening into the superior meatus. Also observe the orifice of the nasolacrimal duct, a short distance inferolateral to the inferior concha (also see Fig. 7-102). The hiatus semilunaris is a curved depression in the middle meatus. Its superior and inferior ends are open and lead to the paranasal sinuses. The superior opening leads to the infundibulum (Fig. 7-157) into which the frontonasal duct opens. The ethmoidal bulla is a swelling on the superior border of the hiatus semilunaris caused by the underlying anterior ethmoidal air cells (Fig. 7-155). The sphenoethmoidal recess is a small cleftlike pocket superior to the superior concha (see Fig. 7-86).

"head cold." The infection may spread to one or more of the sinuses, producing **sinusitis.**

Most of the sinuses are rudimentary or absent in newborn infants. There are no frontal or sphenoidal sinuses present at birth, but there are usually a few ethmoidal cells and tiny maxillary sinuses. These sinuses enlarge during childhood. The frontal and sphenoidal sinuses develop during childhood and adolescence.

Growth of the paranasal sinuses is important in altering the size and shape of the face during infancy and childhood, and in adding resonance to the voice during adolescence.

As has been stated, *the paranasal sinuses are lined with mucous membrane which is continuous with that of the nasal cavities.* However, the mucosa of the sinuses is thinner, less vascular, and not so adherent to the bony walls of the nasal cavities as is the nasal

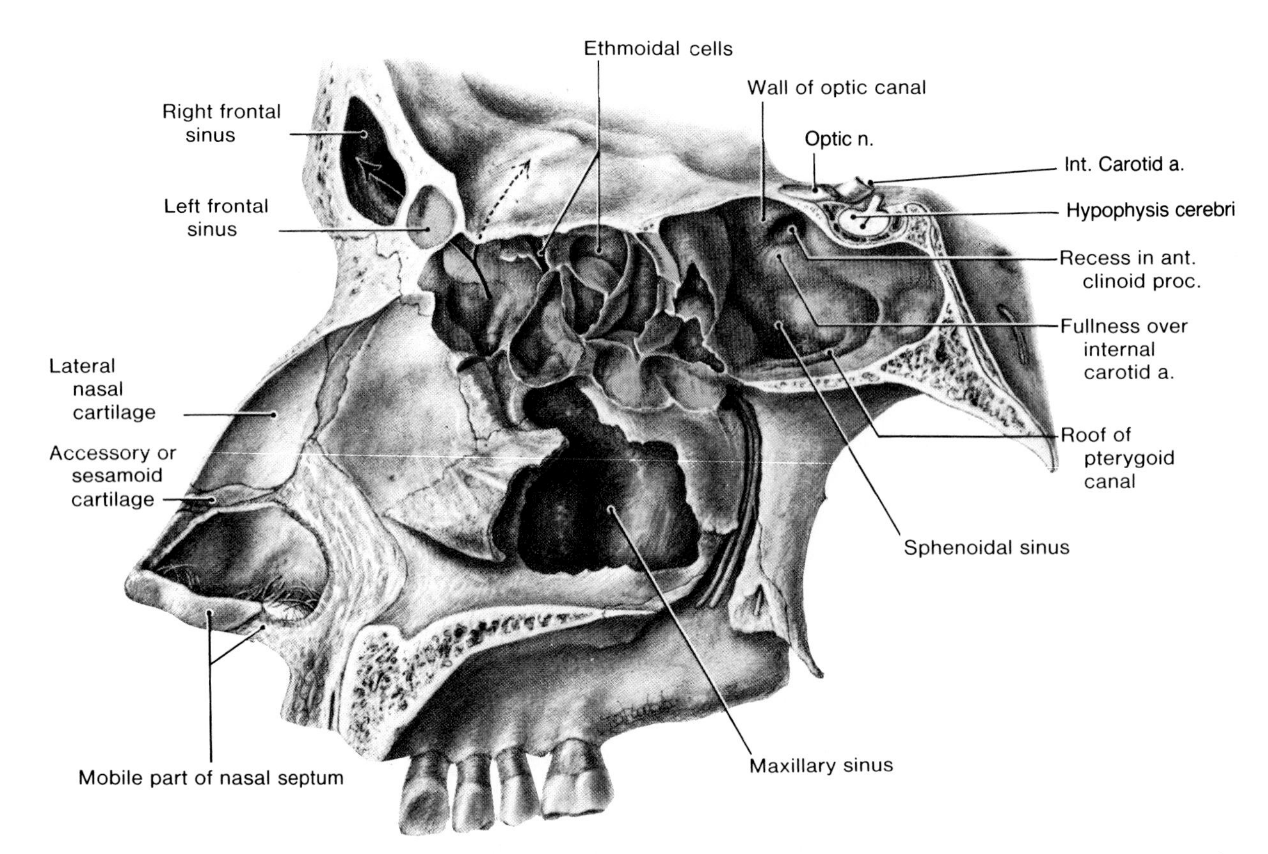

Figure 7-155. Drawing of a sagittal section of the nasal cavity and palate. Bone has been removed to show the paranasal sinuses (opened). The ethmoidal cells (*pink*), collectively called the ethmoidal sinus, appear like a honeycomb. An anterior ethmoidal cell (*blue*) is invading the diploë of the frontal bone to become a frontal sinus. It is ethmoidal in origin, but frontal in location. An offshoot (*broken arrow*) invades the orbital plate of the frontal bone (Fig. 7-12). The sphenoidal sinus (*blue*) in this specimen is very extensive, extending (1) posteriorly, inferior to the hypophysis cerebri, to the clivus (Fig. 7-60), (2) laterally, inferior to the optic nerve, into the anterior clinoid process, and (3) inferior to the pterygoid process, but leaving the pterygoid canal and rising as a ridge on the floor of the sinus. The maxillary sinus (*yellow*) is pyramidal in shape. Its base contributes to the lateral wall of the nasal cavity, its apex is in the zygomatic process, and its orifice is at its highest point. In "blow-out fractures" of the orbit, resulting from a blow to the eyeball, the thin medial wall (Fig. 7-88) and the floor (maxillary sinus) of the orbit are the walls that most often fracture. As a consequence, orbital tissues may enter the ethmoid and maxillary sinuses.

mucosa. Mucus secreted by the glands in the mucous membrane of the paranasal sinuses passes into the nasal cavities through the openings in the lateral walls of the nasal cavities (Figs. 7-86, 7-89, and 7-155).

THE FRONTAL SINUSES

These air chambers are located between the outer and inner tables of the ***frontal bone,*** posterior to the ***superciliary arches*** (Fig. 7-2) and the root of the nose (Figs. 7-4, 7-12, 7-17, 7-86, and 7-153 to 7-157).

The size of the superciliary arches varies in degree of development (Fig. 7-2); however, *the prominence of the superciliary arches is no indication of the size of the subjacent frontal sinuses.* Only a radiographic examination or transillumination (discussed subsequently on p. 958) can reveal their actual size in any given person. Usually the frontal sinuses are detectable in radiographs of children by the 7th year.

Understand also that *the right and the left frontal sinuses are rarely of equal size* in the same person, and that the *septum between the right and left sinuses is rarely situated entirely in the median plane* (Figs. 7-17 and 7-156). Often a frontal sinus has two parts: (1) a **vertical part** *in the squamous part* of the frontal bone, and (2) a **horizontal part** *in the orbital part* of the frontal bone. One or both parts may be large or small. When the supraorbital part is large, its roof forms the floor of the anterior cranial fossa and its floor forms the roof of the orbit (Figs. 7-4, 7-12, 7-15, 7-146, and 7-156).

The frontal sinuses vary in size from about 5 mm

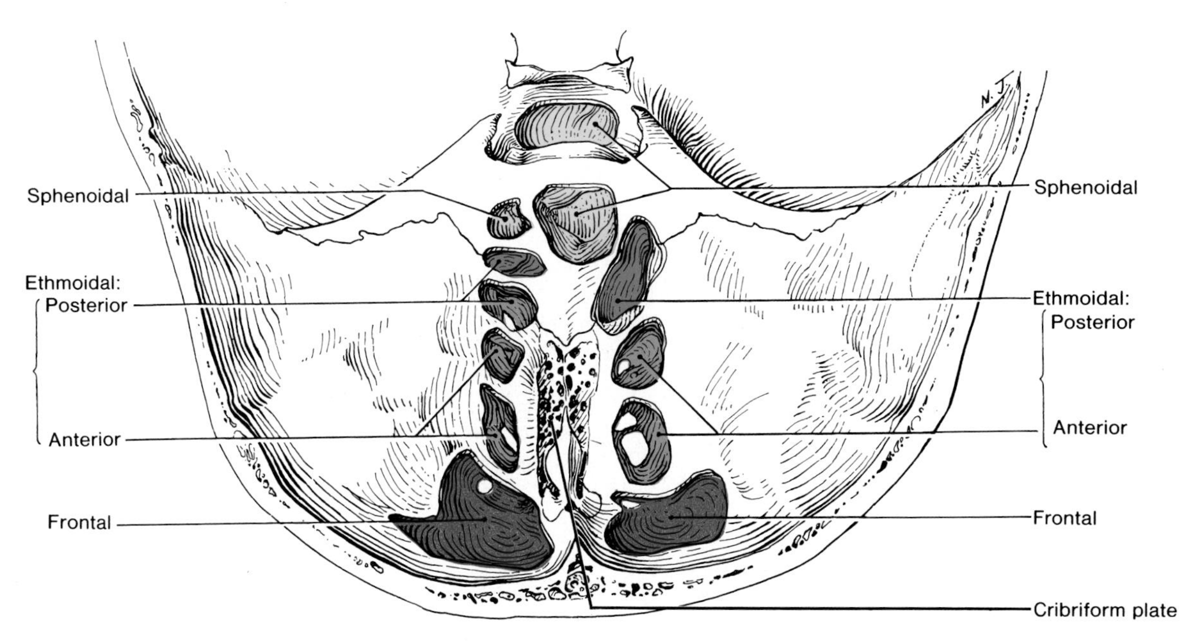

Figure 7-156. Drawing of the floor of the anterior cranial fossa showing the paranasal sinuses surrounding the cribriform plates (also see Fig. 7-14). The exposed parts of the frontal sinuses are the supraorbital portions. (For other views of these sinuses, see Figs. 7-4, 7-12, 7-86, 7-89, 7-142, 7-146, 7-154, and 7-155). All paranasal sinuses drain to the nasal cavity through apertures in the lateral nasal wall. These openings are obscured by the nasal conchae (Fig. 7-86). They can only be seen in the cadaver when the conchae are snipped away (Fig. 7-154).

(pea size) to large spaces extending laterally into the greater wings of the sphenoid bone. The frontal sinuses may be multiple on each side and each of these may have a separate **frontonasal duct** (Fig. 7-154). This duct drains inferiorly into the funnel-shaped *infundibulum* (Fig. 7-157), which eventually opens into the *hiatus semilunaris* of the middle meatus (Figs. 7-154 and 7-157). Usually the frontal sinus drains via one frontonasal duct on each side (Fig. 7-154).

The frontal sinuses are innervated by branches of the supraorbital nerves (Figs. 7-28, 7-32, and 7-62), which are branches of the ophthalmic divisions of the trigeminal nerves (CN V^1).

THE ETHMOIDAL SINUSES

The ethmoidal sinuses comprise several small cavities, called ethmoidal cells, within the ethmoidal labyrinth (G. *labyrinthos*, a maze) of the lateral mass of the **ethmoid bone** (Figs. 7-12, 7-89, and 7-154 to 7-158).

The ethmoidal cells form the labyrinth of the ethmoid bone, *located between the nasal cavity and the orbit* (Figs. 7-89, 7-146, 7-157, and 7-158). The number of cells varies from 3 to 18; they are larger when the number is small. Extremely thin septa of bone, covered with mucosa, form a variable number of interconnecting ethmoidal compartments (cells), which ultimately drain into the lateral wall of the nasal cavity (Figs. 7-89, 7-157, and 7-158). Hold a skull up to a light and observe the thin-walled ethmoidal cells through the delicate orbital lamina of the ethmoid bone (Fig. 7-88), which forms the medial wall of the bony orbit (Figs. 7-89 and 7-158).

Usually the ethmoidal sinuses are not visible in radiographs before the age of 2 years, and they do not grow rapidly until 6 to 8 years. For purposes of description, **the ethmoidal sinuses are divided into anterior, middle, and posterior groups of cells.** The middle ethmoidal cells open directly into the middle meatus. There are several openings on the ethmoidal bulla and on the lateral surface of the middle meatus (Fig. 7-154). The anterior ethmoidal cells may drain indirectly into the middle meatus via the infundibulum (Fig. 7-157). The posterior ethmoidal cells open directly into the superior meatus (Fig. 7-154). The middle ethmoidal cells are sometimes called "bullar cells" because they form the ethmoidal bulla (L. bubble), a swelling on the superior border of the hiatus semilunaris (Fig. 7-154).

The anterior and posterior ethmoidal branches of the **nasociliary nerves** (Figs. 7-69, 7-73, 7-103, and 7-105), branches of the **ophthalmic nerves (CNV^1)** divisions of the trigeminal nerves, *supply the ethmoidal sinuses.*

The anterior and posterior **ethmoidal branches of the ophthalmic artery** supply blood to the ethmoidal sinuses (Figs. 7-99 and 7-153).

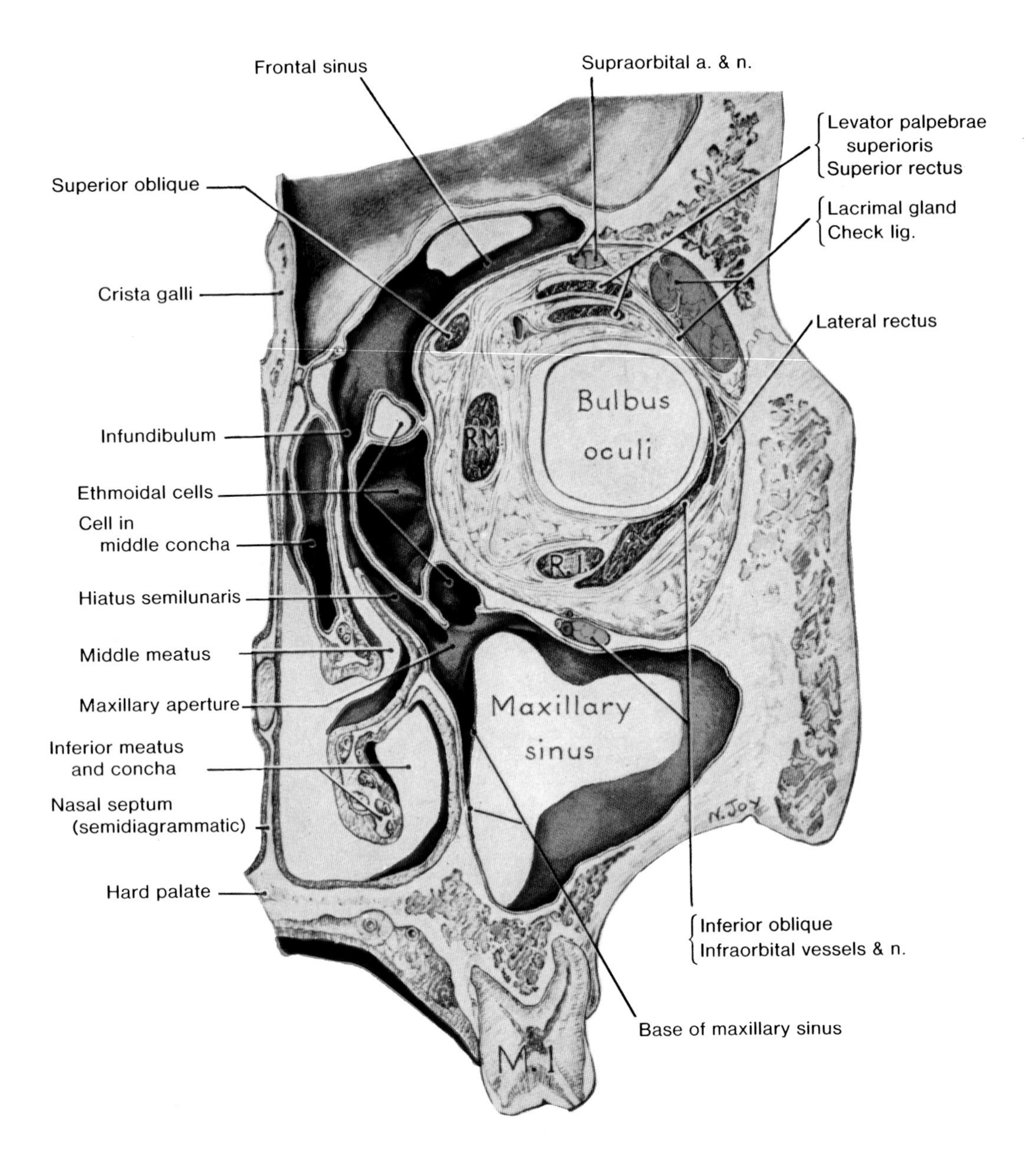

Figure 7-157. Drawing of a coronal section of the right side of the head (posterior view), showing the orbital contents, the nasal cavities, and the paranasal sinuses. Observe the eyeball within the somewhat circular orbital cavity. The orbit has a stout, thick, lateral bony wall, a roof, a medial wall, and a floor, which are surrounded by paranasal sinuses (frontal, ethmoidal, and maxillary). The middle concha in this specimen contains a cell (cavity) and the horizontal part of the frontal sinus has produced two plates of bone between the frontal lobe of the brain and the orbit. Note the entrance to the frontal sinus through the infundibulum, which is at the most inferior point of the sinus. Observe the entrance to the maxillary sinus through the hiatus semilunaris, which is at the level of the roof of the sinus. The most inferior point of the sinus is inferior to the level of the floor of the nasal cavity. Note that the nasal wall of the sinus is very thin in the inferior meatus, and well superior to the floor of the nose. Also note the relationship of the maxillary sinus to the first maxillary molar tooth (*M.1*).

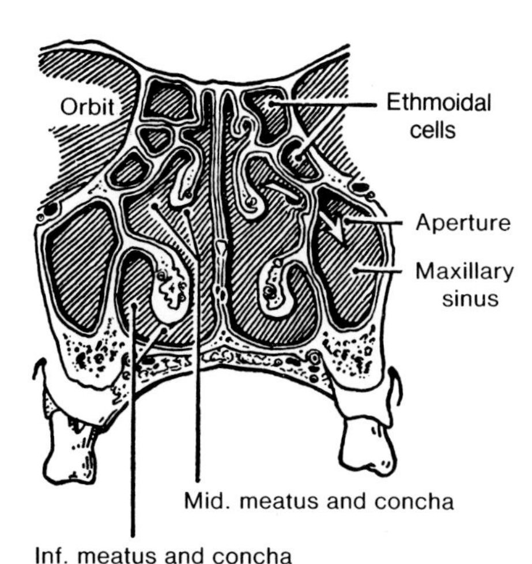

Figure 7-158. Drawing of a coronal section of the nasal cavities showing their relationship to the paranasal sinuses. Note the superior position of the aperture of the maxillary sinus. Note that the ethmoidal sinus consists of two cells (small cavities); the number of cells in the ethmoidal sinuses varies from 3 to 18.

If nasal drainage is blocked, infections of the ethmoidal cells may break through the fragile medial wall of the orbit (Figs. 7-88 and 7-158) into the orbit, producing a **cellulitis** (inflammation of cellular tissue or connective tissue). Severe infections resulting from this source may cause blindness because some posterior ethmoidal cells lie close to the optic canal (Figs. 7-103 and 7-155). Spread of infection from them could affect dural nerve sheath of the optic nerve and cause **optic neuritis** (p. 906).

THE SPHENOIDAL SINUSES

These cavities are in the body of the sphenoid bone (Fig. 7-156). They are located posterior to the superior part of the nasal cavity (Figs. 7-12 and 7-154 to 7-156). *The two sphenoidal sinuses are separated by a bony septum* which is usually not in the median plane (Fig. 7-156). Owing to the presence of the sphenoidal sinuses, the body of the sphenoid bone is a fragile hollow structure (Fig. 7-12). Only thin plates of bone separate the sinuses from important structures.

The sphenoidal sinuses are intimately related to the optic nerves and optic chiasma, the hypophysis cerebri, the internal carotid arteries, the cavernous sinuses, and intercavernous sinuses (Figs. 7-12, 7-39, 7-57, 7-80, 7-108, and 7-155).

Although it is sometimes stated that the sphenoidal sinuses are present at birth (although minute), this is not generally accepted because *the sphenoidal sinuses are not observable in skull films of newborn infants.* The current view is that sphenoidal sinuses are derived from a posterior ethmoidal cell (Fig. 7-155), which begins to invade the sphenoid bone at about 2 years of age. In some people, several posterior ethmoidal cells invade the sphenoid bone, giving rise to multiple sphenoidal sinuses that open separately into the **sphenoethmoidal recess** (Figs. 7-86 and 7-154).

The sphenoidal sinuses occupy a variable amount of the sphenoid bone; they are very large in the specimen shown in Figure 7-155, but they may extend into the wings of the sphenoid and even into the basiocciput (Figs. 7-13 and 7-60).

The posterior ethmoidal nerve (Fig. 7-103) *and posterior ethmoidal artery* (Fig. 7-153*A*) *supply the sphenoidal sinuses.*

THE MAXILLARY SINUSES

This pair of sinuses in the maxillae is the largest of the paranasal sinuses (Figs. 7-12, 7-89, 7-142, 7-146, 7-155, 7-157, and 7-158). They are pyramidal-shaped cavities occupying the entire bodies of the maxillae. The apex of the maxillary sinus extends toward and often into the zygomatic bone (Figs. 7-17 and 7-155). The base of the maxillary sinus forms the inferior part of the lateral wall of the nasal cavity (Figs. 7-157 and 7-158). The roof of the maxillary sinus is formed by the floor of the orbit (Fig. 7-89), and its narrow floor is formed by the alveolar process of the maxilla (Figs. 7-2 and 7-17). The roots of the maxillary teeth, particularly the first two molars, often produce conical elevations in the floor of the maxillary sinus (Fig. 7-129).

The maxillary sinuses are very small at birth and grow slowly until puberty. They are not fully developed until all the permanent teeth have erupted (up to 25th year; see Table 7-7).

The maxillary sinus drains into the middle meatus of the nasal cavity via the hiatus semilunaris (Figs. 7-154 and 7-157) through an aperture in the superior part of its base (Figs. 7-157 and 7-158). Because of the location of this opening, it is impossible for fluid in the sinus to drain when the head is erect until the sinus is nearly full.

The innervation of the maxillary sinus is from the ***anterior, middle, and posterior superior alveolar nerves***, branches of the maxillary nerve (Figs. 7-33*C* and 7-147).

The blood supply of the maxillary sinus is from the **superior alveolar branches of the maxillary artery** (Fig. 7-116). Branches of the *greater palatine artery* (Figs. 7-132 and 7-153*A*) contribute to the blood supply of the floor of the maxillary sinus.

The proximity of the maxillary molar teeth to the floor of the maxillary sinus (Fig. 7-129) poses potentially serious problems for the dentist. During removal of a maxillary molar tooth, a fracture of one of the roots may occur. If proper retrieval methods are not used, the broken piece of the root may be driven superiorly into the maxillary sinus. This may introduce oral bacteria into the sinus, creating **sinusitis** in the maxillary sinus. In addition, a communication may be created between the oral cavity and the maxillary sinus which interferes with the nasal-oral seal that is necessary for sucking and blowing. Infection can also spread to the maxillary sinus from an **abscessed maxillary molar tooth** (Figs. 7-129 and 7-157).

As each paranasal sinus is continuous with the nasal cavity through an aperture that opens into a meatus of the nasal cavity (Figs. 7-86, 7-157, and 7-158), *infection may spread from the nasal cavities producing inflammation and swelling of the mucosa of the sinuses* **(sinusitis)** and local pain. Sometimes several sinuses are inflamed (*pansinusitis*) and the swelling of the mucosa may result in blockage of one or more openings of the sinuses into the nasal cavities.

Acute infections of the frontal or maxillary sinuses often result in localized tenderness to pressure in the area over the infected sinus.

The frontal sinus may be palpated by pressing the finger superiorly at the medial end of the superior orbital margin. Although such pressure is uncomfortable, it is normally not painful unless the sinus is infected.

The ethmoidal sinuses may also be palpated with the thumb in one inner canthus and the index finger in the other, and pushing posteriorly, posterior to the lacrimal bone, and squeezing. *If you try this on a colleague, do so carefully*, keeping in mind the fragile medial wall of the orbit overlying these cells (Figs. 7-88 and 7-158). As the ethmoidal air cells are separated from the orbital cavity by only the thin orbital plate of the ethmoid bone, infection may spread from these sinuses into the orbit, producing **orbital cellulitis**. Although rare, this condition is apt to be dangerous.

Usually the paranasal sinuses are radiolucent. See Figures 18 (p. 19), 19 (p. 20), 7-4, and 7-12. Diseased sinuses show varying degrees of opacity. Radiographs may also reveal thickening of the mucous membranes and excess fluid in the sinuses.

The maxillary and frontal sinuses can also be examined by ***transillumination*** *in a dark room.* To examine the maxillary sinuses, a very bright light is placed in the patient's mouth. A normal maxillary sinus is revealed as a red glow on the anterior aspect of the cheek, the area occupied by the sinus (Fig. 7-17). A diseased sinus does not transilluminate.

To examine the frontal sinus, the light is placed against the superomedial angle of the orbit. If a normal frontal sinus is present, a red glow appears on the forehead over the vertical part of the sinus (Fig. 7-17).

Pus formation (suppuration) in the paranasal sinuses is not uncommon. Pus oozing from the frontal or anterior ethmoidal sinuses may be directed into the opening of the maxillary sinus by the hiatus semilunaris (Figs. 7-154 and 7-157). One can often determine whether the pus is coming from the frontal or the maxillary sinus by placing the patient's head between his/her legs. Pus from the maxillary sinus, but not usually from the frontal sinus, will drain in this position.

Because the superior alveolar nerves (Fig. 7-33), branches of the maxillary nerve (CN V^2), supply both the maxillary teeth and the mucous membrane of the maxillary sinus, inflammation of the mucosa of the sinus is frequently accompanied by the sensation of toothache, especially when bone is very thin in the inferior part of the wall of this sinus (Fig. 7-129).

The maxillary sinus is the one most commonly involved in infection, probably because its aperture is located superior to the floor of the sinus (Figs. 7-157 and 7-158), and is ***so poorly located for natural drainage of the sinus***. In addition, when the mucous membrane of the sinus is congested, the maxillary aperture may be obstructed. Gravity drainage from the maxillary sinus is best when one is lying on the side opposite the affected sinus. Surgical drainage may occasionally be necessary and is done by opening the lateral wall of the inferior meatus of the nasal cavity (Figs. 7-157 and 7-158).

Infants and children commonly put peanuts, candies, small toys, etc. into their nasal cavities. Because of the shelf-like conchae and the deep meatus on their lateral walls (Figs. 7-86 and 7-158), it is easy for these foreign bodies to become impacted. To locate such objects, the nasal cavities are examined with a **nasal speculum** inserted through the external nares (a procedure known as **anterior rhinoscopy).**

The posterior nasal apertures or **choanae** (Fig. 7-9) may also be examined by a special mirror placed in the nasopharynx (a procedure known as **posterior rhinoscopy**).

Patients with fractures of the frontal, ethmoid,

maxillary, or nasal bones* should be warned against blowing their noses because of the possibility of expelling air from their paranasal sinuses and/or nasal cavities into the subcutaneous tissues, the cranium, or the orbit. In addition, **CSF rhinorrhea may occur** (p. 878).

THE EAR

The ear consists of three anatomical parts: external, middle, and internal. The external and middle parts are concerned mainly with the transference of sound to the internal ear, which contains the **vestibulocochlear organ** that is concerned with equilibration and hearing.

The **tympanic membrane** (eardrum) separates the external ear from the middle ear. The **auditory tube** (pharyngotympanic tube, Eustachian tube) joins the middle ear cavity to the nasopharynx (Figs. 7-86, 7-122, and 7-164).

THE EXTERNAL EAR

The external ear is composed of the oval **auricle** or pinna (L. a feather, wing, or fin), and the **external acoustic meatus** (external auditory meatus, ear canal).

THE AURICLE (Figs. 7-159 and 7-160)

The auricle (L. *auris,* ear) is the shell-like part of the external ear. It consists of a single elastic cartilage covered with thin skin contains hairs, sweat glands and sebaceous glands. The cartilage is irregularly ridged and hollowed, giving the auricle its shell-like form. The cartilage is prolonged medially, where it is continuous with the cartilage of the external acoustic meatus (Fig. 7-159). The auricle is connected with the fascia on the side of the skull by the auricular muscles (Fig. 7-29), which are supplied by the facial nerve (CN VII).

Many terms are used to describe the parts of the auricle, but only a few of them are commonly used. They are named in Figure 7-160.

The lobule (ear lobe) consists of fibrous tissue, fat, and blood vessels, covered with skin. The auricular vessels are derived from the superficial temporal artery and vein (Figs. 7-29 and 7-31).

The lymph vessels of the lateral surface of the superior half of the auricle drain to the *superficial parotid lymph nodes* (Fig. 7-40), which lie just anterior to the tragus (Fig. 7-160). Lymph from the cranial surface of the superior half of the auricle drains to the retroauricular and deep cervical lymph nodes (Figs. 8-40 and 8-55). Lymph from the remainder of the auricle, including the lobule, drains into the superficial cervical lymph nodes (Fig. 7-40).

The auricle *projects from the side of the head and* **acts as a collective trumpet for sound waves** (although not quite so effectively as in animals), directing them into the relatively narrow external acoustic meatus (Fig. 7-159). The shape of the auricle varies considerably in different people; usually this is of no clinical significance.

Most minor abnormalities of the auricle are of no clinical significance. In some abnormally developed persons (*e.g.,* with the trisomy 18 syndrome), the auricles are abnormally formed and low set.

A pinhead-sized blind depression, called a **preauricular pit**, is sometimes found in a triangular area anterior to the tragus. A deep pit is referred to as a **preauricular sinus;** usually they also have pinpoint openings. An obstructed preauricular sinus may eventually give rise to a **preauricular cyst**, which may have to be removed sugically.

Most preauricular pits and sinuses result from imperfect fusion of the separate embryonic swellings, called **auricular hillocks**, that develop around the margins of the first branchial groove and fuse to form the auricle. The derivative of the first branchial groove is the *external acoustic meatus.*

Auricular appendages or tags are also most often observed anterior to the tragus. They result from the development of more than the usual six auricular hillocks. Minor malformations of the auricle are common, but serious ones are fortunately rare. For illustrations and more details about abnormalities of the auricle, see Moore (1982).

THE EXTERNAL ACOUSTIC MEATUS (Figs. 7-10, 7-159, and 7-160)

The external acoustic meatus is a canal that extends from the **concha** (L. a shell) of the auricle to the tympanic membrane (L. *tympanum,* a tambourine), a distance of about 2.5 cm in an adult. The lateral third of this S-shaped passage is cartilaginous, whereas the medial two-thirds is bony. *In infants, the external acoustic meatus is almost entirely cartilaginous.*

The lateral third of the meatus is lined with the skin of the auricle that contains hair follicles, sebaceous glands, and **ceruminous glands,** that produce a waxy exudate called **cerumen** (L, *cera,* wax).

The medial two-thirds of the meatus is lined with a thin layer of stratified squamous epithelium. This layer is continuous with the skin lining the external third of the meatus; it also covers the tympanic membrane, forming its external layer.

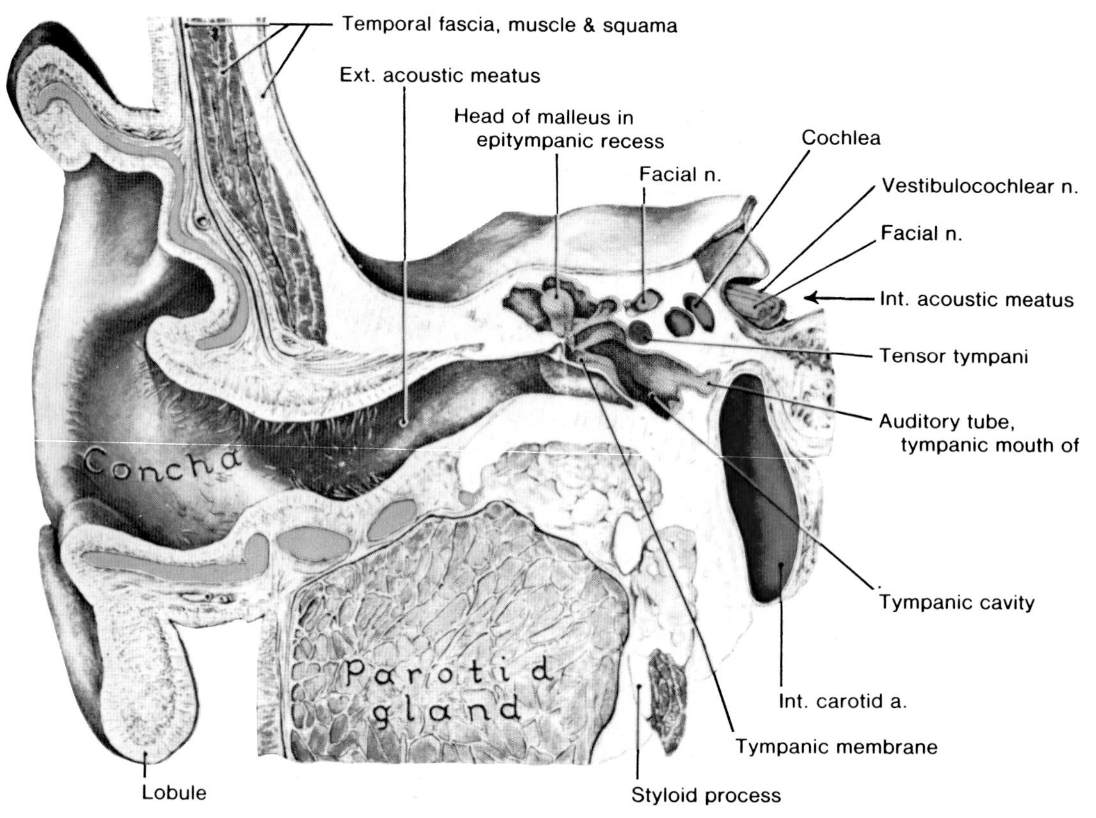

Figure 7-159. Drawing of an anterior view of a coronal section of the right ear. The internal ear is tinted *blue*; the mucous membrane of the middle ear or tympanic cavity is *pink*. The external acoustic meatus is one-third cartilaginous (yellow) and two-thirds bony. It is narrowest near the tympanic membrane owing to the rise on the floor; hence there is a "well" where fluid can collect at the medial end of the meatus during swimming. The cartilaginous part of the external acoustic meatus is lined with thick skin, and has hairs and many ceruminous glands that secrete wax (cerumen). The bony internal part of the external acoustic meatus is lined with a thin epithelium that also forms the external layer of the tympanic membrane. Note the obliquity of the tympanic membrane which meets the roof of the meatus at an obtuse angle and the floor at an acute one. The handle of the malleus (an ear bone) is attached to the internal layer of the tympanic membrane (see Figs. 7-161 and 7-162).

The lateral end of the external acoustic meatus, its widest part, is about the width of a pencil. The external acoustic meatus becomes narrow at the medial end of its cartilaginous portion, and in the bony part about 4 mm from its medial end. This constriction is called the **isthmus.** The inferior wall of the meatus is about 5 mm longer than its superior wall, owing to the obliquity of the tympanic membrane (Fig. 7-159).

The arteries supplying the external acoustic meatus are: the posterior auricular branch of the **external carotid artery** (Fig. 7-29); the deep auricular branch of the **maxillary artery** (7-116); and the auricular branches of the **superficial temporal artery** (Figs. 7-31 and 7-37).

The nerves supplying the external acoustic meatus are derived from two cranial nerves (Figs. 7-32 and 7-33): (1) the auricular branch of the **auriculotemporal nerve**, derived from the *mandibular nerve* (*CN* V^3); and (2) the auricular branch of the *vagus nerve* (Fig. 7-135).

The veins from the external acoustic meatus drain into the *external jugular and maxillary veins* (Figs. 7-38 and 7-42), and into the *pterygoid plexus* (Fig. 7-39).

The lymphatic drainage is the same as described for the auricle.

The external acoustic meatus is directed somewhat anteriorly as well as medially. As the stethoscope tips are angulated to conform to this shape, you must put the **stethoscope** to your auricles with the tips pointing anteriorly.

The anatomy of the external acoustic meatus also has to be considered when using an **auriscope** or ear speculum to examine the meatus and the tympanic membrane. The instrument can be inserted more readily when the auricle is pulled posterosuperiorly.

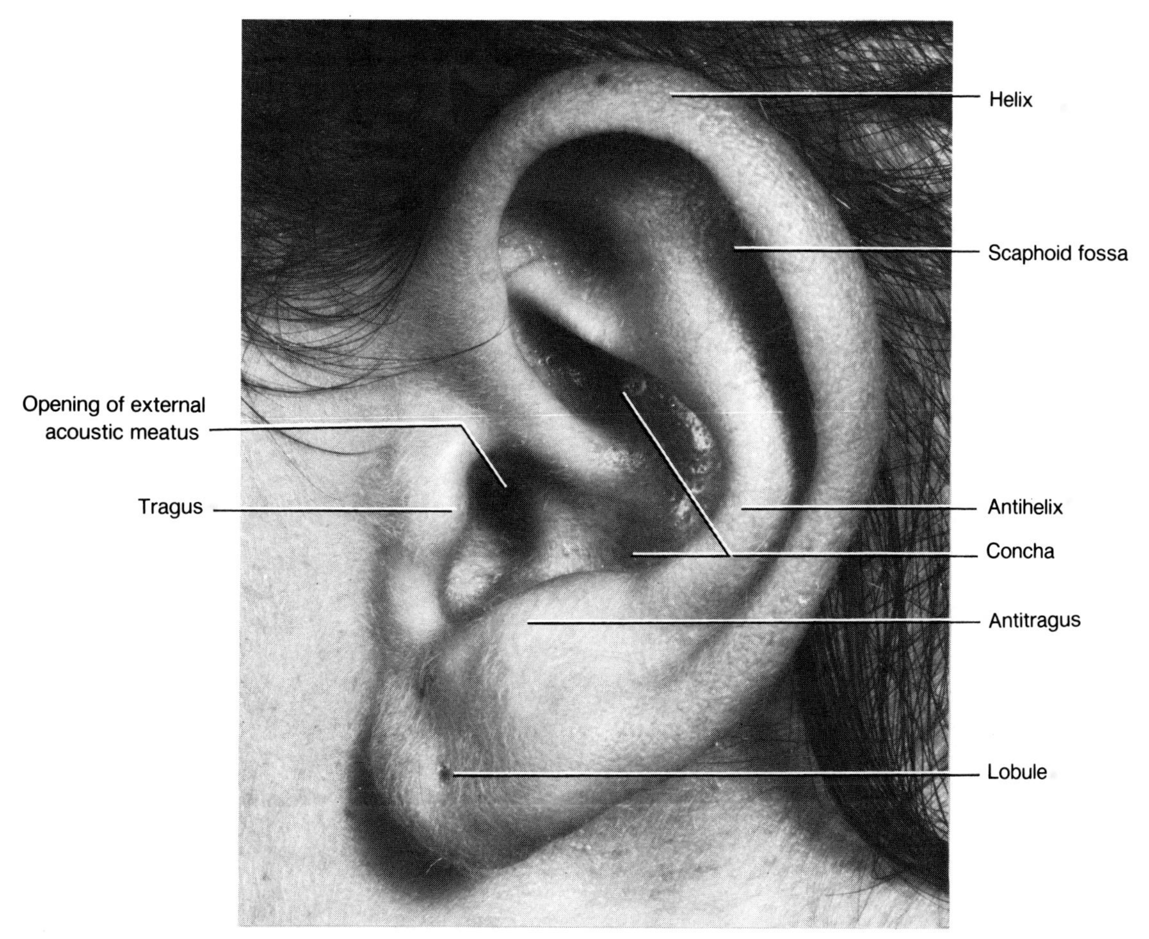

Figure 7-160. Photograph of the left auricle of a 12-year-old girl. Note that her lobule has been pierced (indicated by end of pointer) for an earring. The names given here are the ones commonly used in clinical descriptions of the external ear. The external ear, which conducts sound and protects deeper parts of the ear, consists of the auricle and the external acoustic meatus (also see Fig. 7-159). The central depression leading to the opening of the external acoustic meatus is called the concha because of its resemblance to a conch's shell (L. *concha*, a shell). The tragus (G. a goat) was given its name because hairs project from it (Fig. 7-159) which thicken in older men, giving the appearance of a goatee. The lobule is the only part of the auricle that is not supported by cartilage.

The external acoustic meatus is relatively short in infants, therefore extra care must be exercised so that the tympanic membrane will not be damaged.

Foreign bodies (*e.g.*, beans, candies, *etc.*) inserted in the meatus by children usually become lodged in the constriction where its cartilaginous and bony parts meet (Fig. 7-159).

As the condylar process of the mandible is close to the external acoustic meatus (Figs. 7-21 and 7-119), a hard blow on the chin may drive the head of the mandible into the meatus and injure it or even fracture it in adults.

The cerumen prevents maceration or softening of the lining of the external acoustic meatus by trapped water (*e.g.*, after swimming). It is also thought to discourage insects from entering the external acoustic meatus.

Overproduction or prolonged retention of cerumen may block the meatus, or when adjacent to the tympanic membrane it may cause vibratory responses. An accumulation of hard cerumen can cause discomfort, pain, and even temporary deafness. Any acute inflammatory process of the meatus may cause pain.

Atresia of the external acoustic meatus (*i.e.*, congenital blockage) is often associated with severe abnormalities of the auricle.

THE TYMPANIC MEMBRANE

This thin, semitransparent, oval membrane (8–9 mm in diameter) is at the medial end of the external acoustic meatus, separating it from the middle ear (Figs. 7-159 and 7-161 to 7-165). It is a thin, fibrous membrane covered with very thin skin externally and mucous membrane internally. In the adult it is oblique, sloping inferomedially (Fig. 7-159).

In living persons, the *tympanic membrane normally appears pearly gray and shiny.* Its parts and their relationship to the auditory ossicles of the middle ear are illustrated in Figures 7-159, 7-161, and 7-162. The tympanic membrane shows a concavity towards the meatus with a central depression, the **umbo**, (Fig. 7-161*A*) formed by the end of the handle of the malleus, one of three middle ear bones, called **auditory ossicles** (Fig. 7-166). From the umbo a bright area, re-

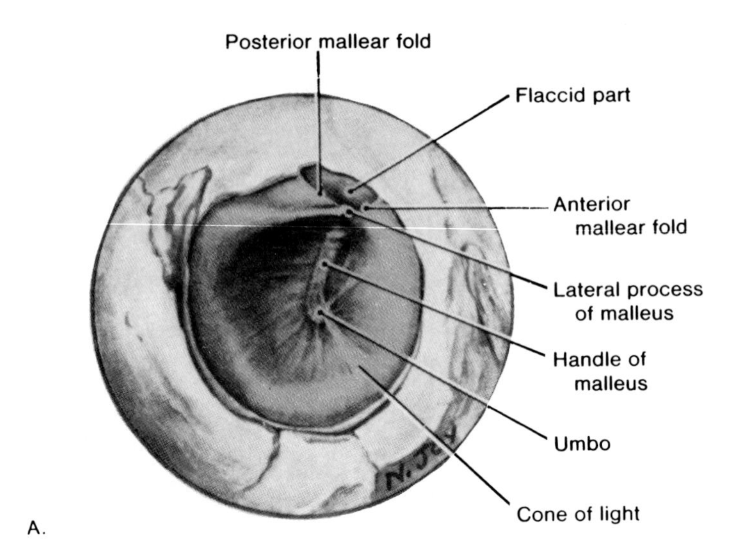

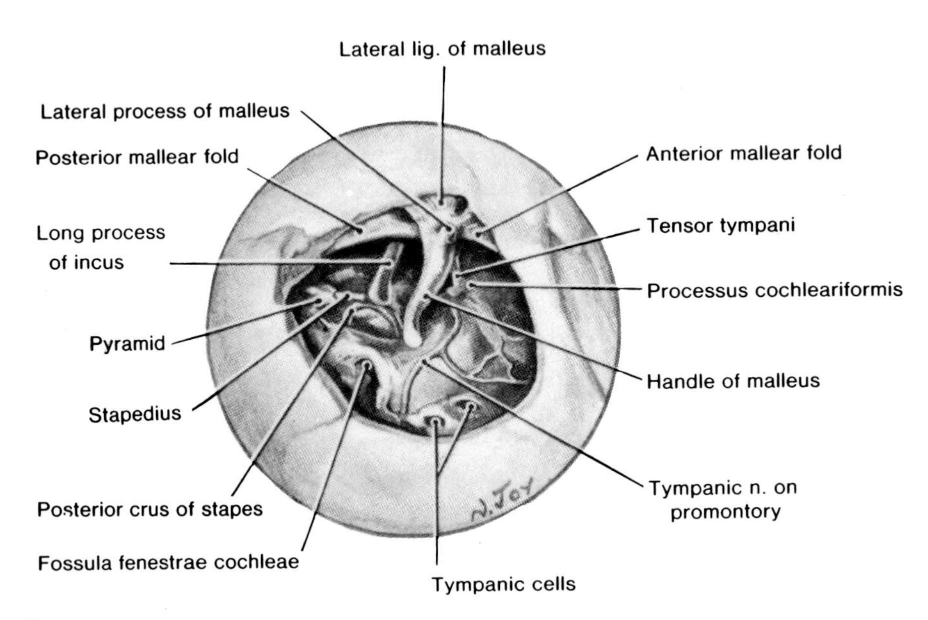

Figure 7-161. *A*, drawing of a lateral view of the right tympanic membrane. *B*, drawing of inferolateral view of the tympanic cavity after removal of the tympanic membrane. Observe that the tympanic membrane is oval rather than round. Note that it is shaped like a funnel with a rolled rim and a depressed part, called the umbo, at the tip of the handle of the malleus. Superior to the lateral process of the malleus the membrane is thin and is called the flaccid part (pars flaccida). The flaccid part lacks the radial and circular fibers present in the remainder of the membrane (tense part). The cone of light is a reflection of light from the tympanic membrane. Observe the tendon of the stapedius in *B*, passing the neck of the stapes on which it pulls.

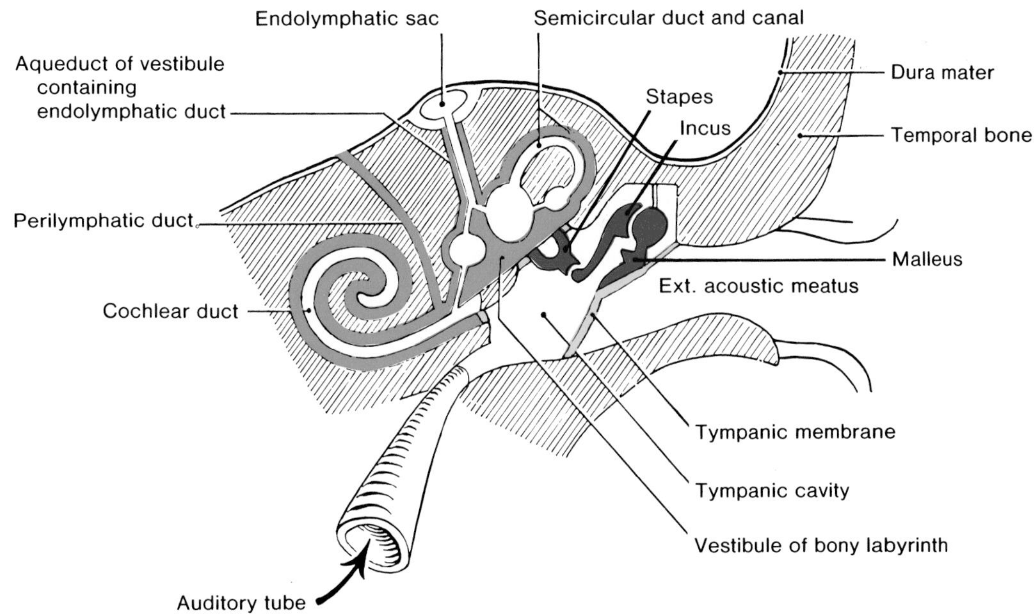

Figure 7-162. Diagram showing the three parts of the ear. The middle ear or tympanic cavity lies between the tympanic membrane and the internal ear. Three ossicles (*red*), malleus, incus, and stapes, stretch from the lateral wall to the medial wall of the tympanic cavity. Of these, the malleus is attached to the tympanic membrane; the stapes is attached by an anular ligament to an oval opening in the wall of the bony vestibule of the inner ear, called the fenestra vestibuli (Fig. 7-169); and the incus connects these two ossicles. The auditory tube opens into the anterior wall of the tympanic cavity; the aditus ad antrum opens from the epitympanic recess posteriorly to the mastoid antrum. The internal ear is contained in the petrous part of the temporal bone (Fig. 7-55). The internal ear is usually described in two parts, a bony labyrinth and a membranous labyrinth. The bony labyrinth is a series of interconnected spaces and canals containing the organs for hearing and balancing. The organ of hearing is housed in a spiral tube, the cochlea, the base of which connects with the vestibule, the globular middle part of the bony labyrinth. The organ of equilibrium or balance is in three semicircular canals and structures in the vestibule (Fig. 7-172). The bony labyrinth is filled with a fluid called perilymph which is continuous with the CSF through a small duct. Inside the bony labyrinth and largely surrounded by perilymph (*blue*), is a complicated system of tubes and spaces containing endolymph which bathes the special end organs for hearing and balancing.

ferred to as **the cone of light**, radiates anteroinferiorly.

The tympanic membrane moves in response to air vibrations that pass to it via the external acoustic meatus. The vibrations are transmitted from this membrane by the auditory ossicles through the middle ear or tympanic cavity to the internal ear (Fig. 7-162).

Blood Vessels of the Tympanic Membrane (Figs. 7-29, 7-37, and 7-116). The deep auricular branch of the **maxillary artery** supplies the external surface of the tympanic membrane. This surface is also supplied by the stylomastoid branch of the **posterior auricular artery** (Fig. 7-29), and by the *tympanic branch of the maxillary artery* (Fig. 7-116).

The veins of the external surface of the tympanic membrane open into the external jugular vein (Fig. 7-38). The veins of the internal surface drain into the transverse sinus and veins of the dura mater (Fig. 7-39).

Nerve Supply of the Tympanic Membrane (Figs. 7-29, 7-33, and 7-135). The external surface of the tympanic membrane is supplied by the **auriculotemporal nerve**, a branch of the mandibular division of the trigeminal nerve (CN V^3). Some innervation is supplied by a small **auricular branch of the vagus nerve** (CN X). This nerve may also contain some glossopharyngeal and facial nerve fibers.

Rupture of the tympanic membrane is one of the several causes of middle ear deafness. Perforation of the tympanic membrane may result from foreign bodies, excessive pressure (*e.g.*, during scuba diving), or from infection.

Severe bleeding and/or escape of CSF through a ruptured tympanic membrane and the external acoustic meatus **(CSF otorrhea)** may occur following a severe blow on the head. Either of these conditions is indicative of a **skull fracture** and results from the close relation of the tympanic cavity, mastoid antrum, mastoid cells, and the bony external acoustic meatus to the

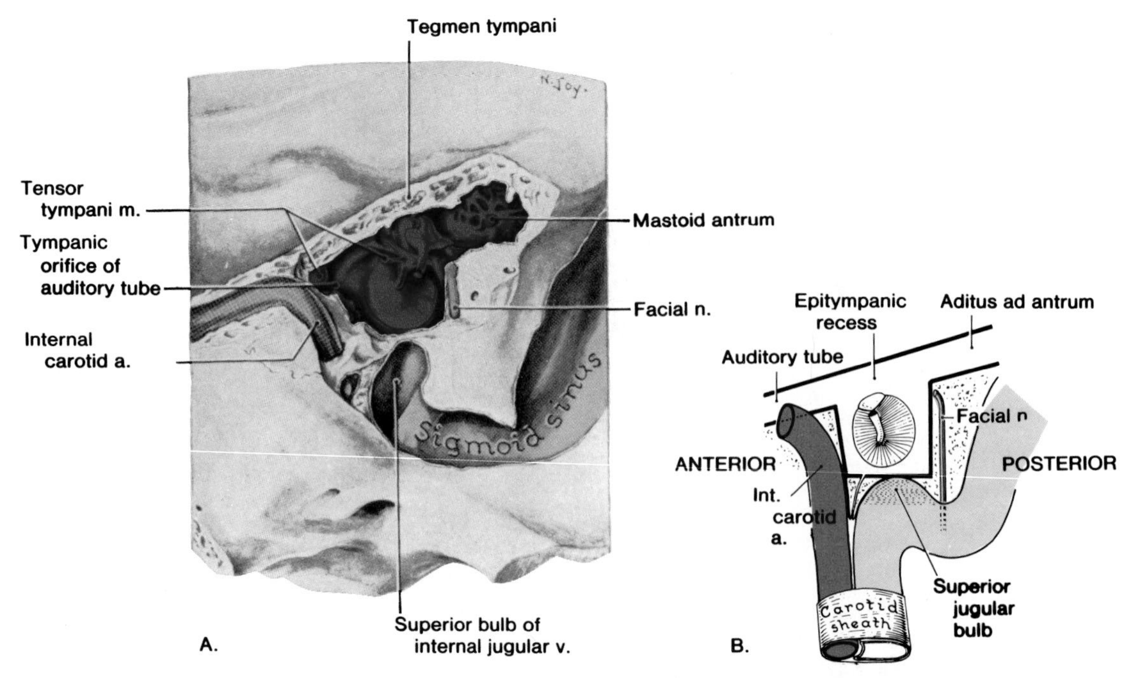

Figure 7-163. *A*, drawing of a dissection of the middle ear to show the walls of the tympanic cavity. Note that the tegmen tympani forms the roof of the mastoid antrum and part of the middle ear. Also observe the internal carotid artery (the main feature of the anterior wall), the internal jugular vein (the main feature of the floor), and the facial nerve (the main feature of the posterior wall). *B*, inset drawing that simplifies the dissection shown in *A* and emphasizes the relations of the tympanic cavity.

meninges of the brain (Fig. 7-163). Sometimes a skull fracture passes through the bony part of the external acoustic meatus and causes bleeding and/or loss of CSF via the external acoustic meatus, even with an intact tympanic membrane.

On rare occasions it is necessary to incise the tympanic membrane to allow pus to escape from the middle ear. Because the superior half of the tympanic membrane is much more vascular than the inferior half, incisions are made posteroinferiorly through the membrane (Fig. 7-161). This site also avoids the **chorda tympani nerve** (Fig. 7-35) and the auditory ossicles (Fig. 7-161).

THE MIDDLE EAR

The middle ear is a rather large cavity in the temporal bone (Fig. 7-162). This middle ear or **tympanic cavity** contains air, three **auditory ossicles**, a nerve, and two small muscles (Figs. 7-161 to 7-163). The middle ear is separated from the external acoustic meatus by the tympanic membrane. Vibrations of this membrane are transmitted through the tympanic cavity by the chain of three auditory ossicles (Fig. 7-161).

The tympanic cavity is connected anteriorly with the nasopharynx by the **auditory tube** (Figs. 7-162 to 7-165). Posterosuperiorly the tympanic cavity connects with the **mastoid cells** through the mastoid antrum (Figs. 7-164 and 7-167).

The tympanic cavity is lined with a mucous membrane which is continuous with that lining the auditory tube, mastoid cells, and mastoid antrum (Fig. 7-163).

The tympanic cavity consists of two parts (Figs. 7-159, 7-162, and 7-163): the **tympanic cavity proper**, opposite the tympanic membrane; and the **epitympanic recess**, superior to the level of this membrane.

CONTENTS OF THE MIDDLE EAR OR TYMPANIC CAVITY (Figs. 7-159 and 7-161 to 7-165)

The middle ear contains the ***auditory ossicles*** (malleus, incus, and stapes); the ***stapedius and tensor tympani muscles***; the ***chorda tympani nerve*** (a branch of the facial, CN VII), and the ***tympanic plexus*** of nerves.

WALLS OF THE MIDDLE EAR OR TYMPANIC CAVITY (Fig. 7-163)

The tympanic cavity is shaped like a narrow, six-sided box which has convex medial and lateral walls. *It has the shape of a biconcave lens or red blood cell in cross-section.* It is 15 mm in vertical diameter, 2 mm across the center, 4 mm at the floor, and 6 mm at the

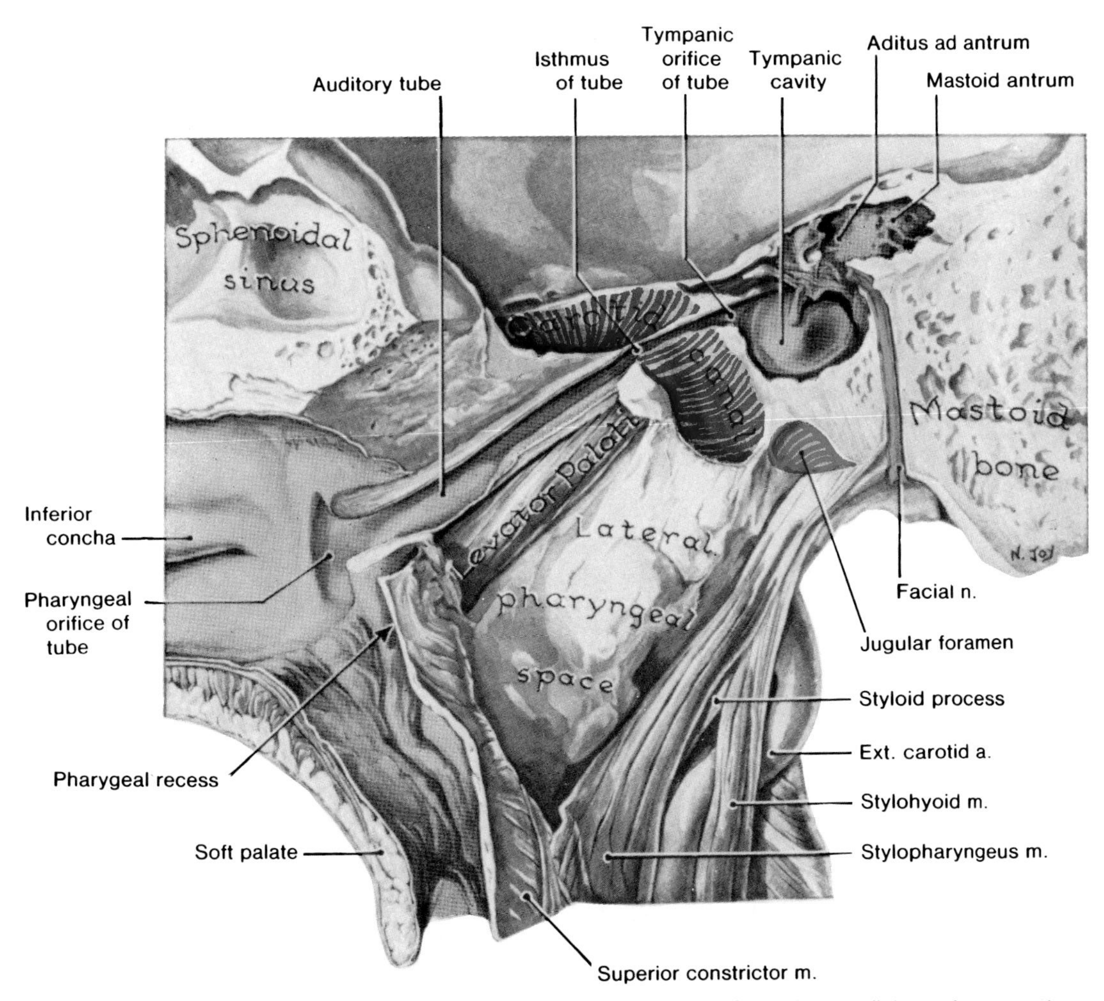

Figure 7-164. Drawing of a dissection that exposes the auditory tube from the medial or pharyngeal aspect. Observe that this tube passes superiorly, posteriorly, and laterally from the nasopharynx to the tympanic cavity. Note the funnel-shaped pharyngeal orifice of the auditory tube, situated posterior to the inferior concha of the nose (Fig. 7-86). Observe the bony part of the tube (passing lateral to the carotid canal) narrows at the isthmus, where it joins the cartilaginous part. Note the course of the facial nerve and its relationship to the mastoid antrum and the mastoid cells (also see Fig. 7-79).

roof. The general shape of the tympanic cavity is shown diagrammatically in Figure 7-163*B*.

The roof or tegmental wall of the tympanic cavity (Fig. 7-163) is formed by a thin plate of bone, called the **tegmen tympani** (L. *tegmen*, a roof). It separates the tympanic cavity from the dura mater on the floor of the middle cranial fossa. The tegmen tympani also covers the mastoid antrum.

The floor or jugular wall of the tympanic cavity (Fig. 7-163) is thicker than the roof, but it is formed by a layer of bone which may be thick or thin. It separates the tympanic cavity from the superior bulb of the internal jugular vein (Figs. 7-65 and 7-163*B*). In Figure 7-163*B* note that as the internal jugular vein and internal carotid artery pass superiorly within the carotid sheath, they diverge at the floor of the tympanic cavity. The **tympanic nerve** (Fig. 7-111), a branch of the glossopharyngeal (CN IX), passes through an aperture in the floor of the tympanic cavity and branches to form the tympanic plexus.

The lateral wall or membranous wall of the tympanic cavity is formed almost entirely by the tympanic membrane (Fig. 7-163). Superiorly it is formed by the lateral bony wall of the **epitympanic recess.** The handle of the malleus is incorporated in the tympanic membrane and its head extends into the epitympanic recess (Fig. 7-161*A*).

The medial wall or labyrinthine wall of the tympanic cavity separates the tympanic cavity from the inner ear, consisting of a membranous labyrinth (semicircular ducts and cochlear duct) encased in a bony labyrinth (Figs. 7-162, 7-168, and 7-169).

The medial wall of the tympanic cavity exhibits several important features. Centrally, opposite the tym-

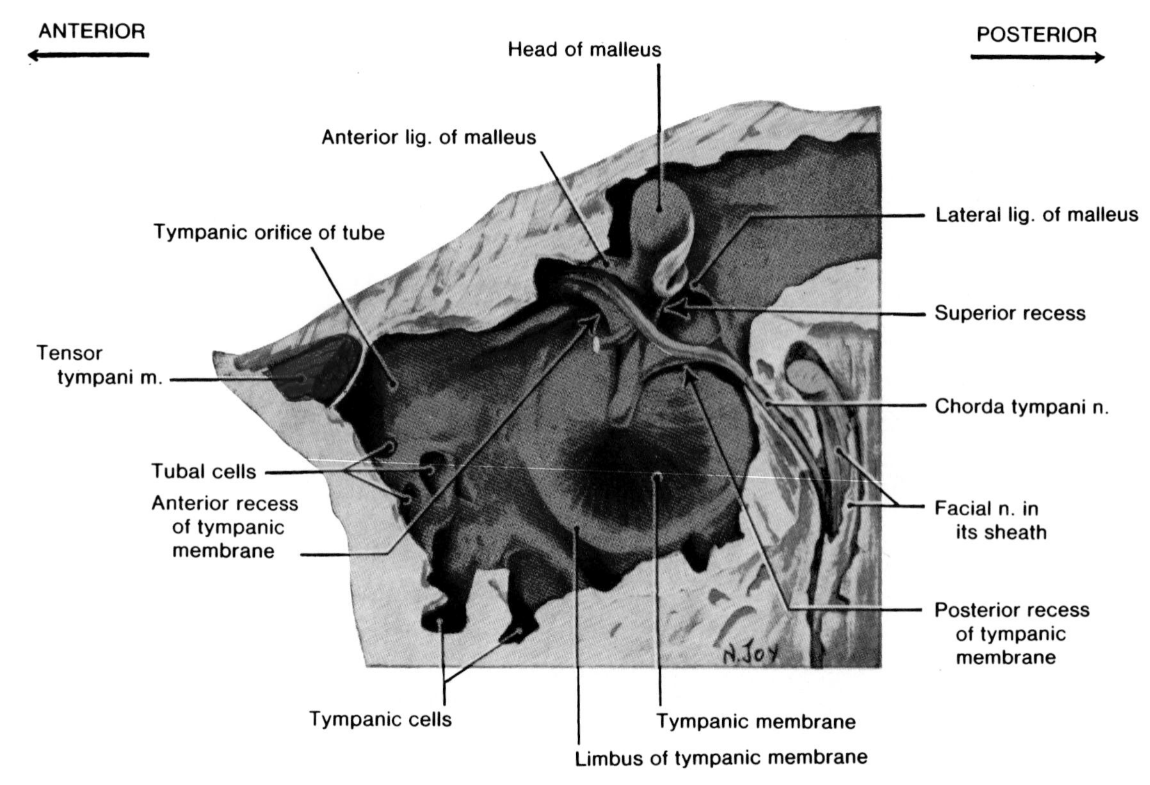

Figure 7-165. Medial view of a dissection of the right middle ear to show the lateral wall of the tympanic cavity. Observe that the lateral wall is formed almost entirely by the tympanic membrane. Note that this membrane has a greater vertical than horizontal diameter and that the handle of the malleus is incorporated in the membrane, its end being at the umbo (Fig. 7-161*A*). Note the facial nerve within its periosteal tube and the chorda tympani leaving the facial nerve (also see Fig. 7-35). Note the head of the malleus in the epitympanic recess. Here its posterior surface articulates within the body of the incus. Also note the chorda tympani nerve passing anteriorly across the medial aspect of the superior part of the tympanic membrane. *It is incorporated in the tympanic membrane.*

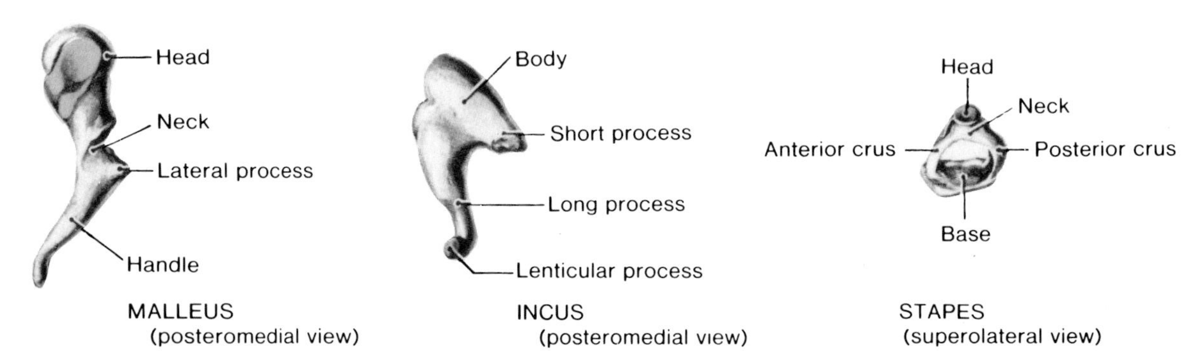

Figure 7-166. Drawings of the auditory ossicles (×5). The three ossicles (ear bones) form a chain between the tympanic membrane and the fenestra vestibuli, as shown in Figure 7-162 (also see (Fig. 7-169).

panic membrane, there is a rounded **promontory** (**L.** an eminence) formed by the large first turn of the cochlea (Fig. 7-168).

The tympanic plexus of nerves, lying on the promontory, is formed by fibers of the facial and glossopharyngeal nerves (CN VII and CN IX) that pass through an aperture in the floor of the tympanic cavity within the tympanic nerve (Fig. 7-111).

The medial wall of the tympanic cavity has two small apertures or windows (Fig. 7-169). The **fenestra vestibuli** is closed by the base (footplate) of the stapes (Fig. 7-166), which is bound to its margins by an

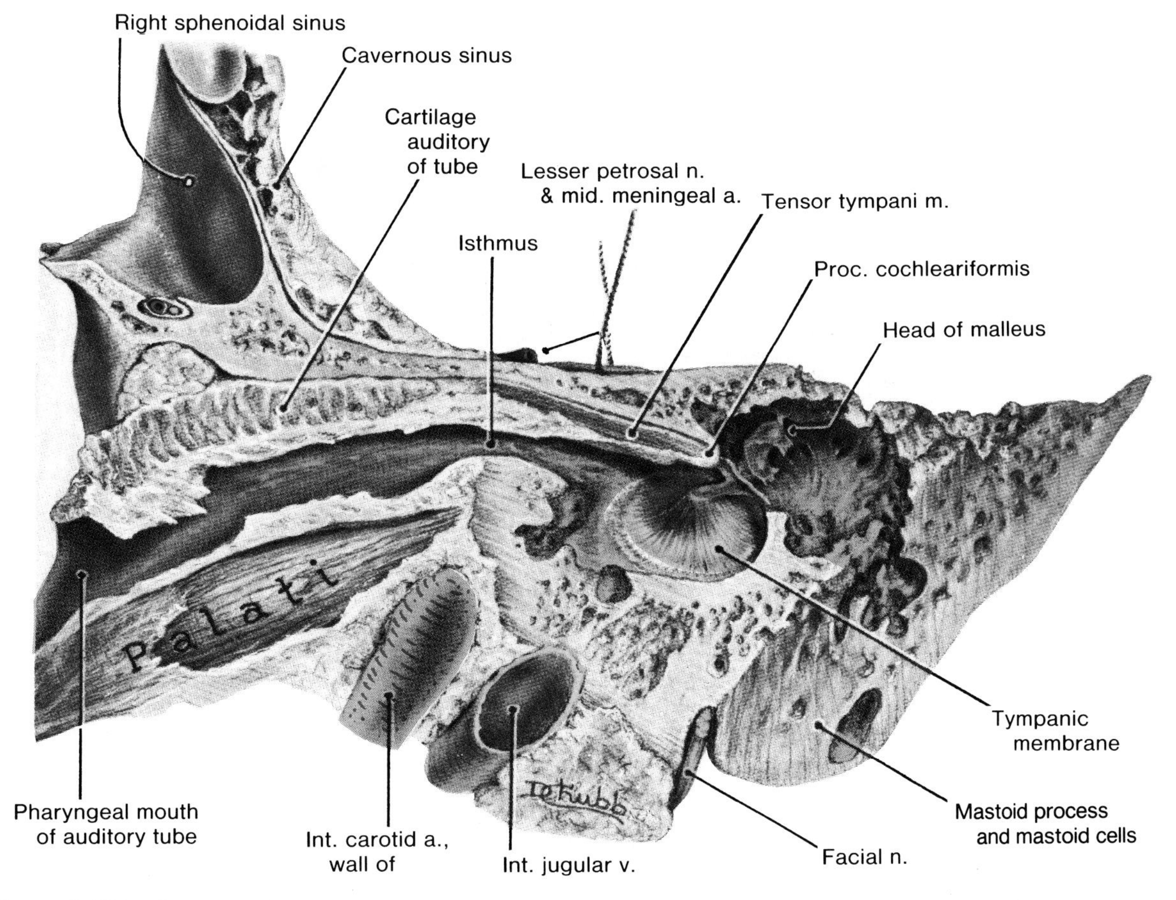

Figure 7-167. Drawing of a dissection of the auditory tube and tympanic cavity. The medial part of a longitudinally split specimen is shown. Note the tensor tympani muscle arising from the auditory tube and adjacent bones. Its tendon is shown turning around the processus cochleariformis before inserting into the handle of the malleus.

anular ligament. Through this window, vibrations of the stapes are transmitted to the perilymph within the bony labyrinth of the inner ear (Fig. 7-162). The **fenestra cochleae**, inferior to the fenestra vestibuli, is closed by a **secondary tympanic membrane.** This membrane allows the perilymph to move slightly in response to impulses from the base of the stapes. Thus, both apertures are related to the middle ear or tympanic cavity. See the legend of Figure 7-162 for a general description of their roles in the over-all scheme of the ear.

In **otosclerosis** (G. *ous* (*ōt*), ear + *sklērosis*, hardening), there is a new formation of spongy bone around the stapes and fenestra vestibuli, resulting in progressively increasing deafness. This bony overgrowth may stop movement of the base of the stapes and/or the membrane of the fenestra cochlea. If either or both of these are immobilized, they must be freed to restore the hearing.

The posterior wall or mastoid wall of the tympanic cavity has several openings in it (Figs. 7-163 and 7-164). In its superior part, there is the **aditus to the mastoid antrum or aditus ad antrum**, which leads posteriorly from the **epitympanic recess** into the mastoid antrum, and beyond into the **mastoid cells** (Fig. 7-167). Inferiorly there is a pinpoint aperture on the apex of a tiny, hollow projection of bone, called the **pyramidal eminence** or pyramid. The pyramid contains the stapedius muscle (Fig. 7-161*B*); its aperture transmits the tendon of the stapedius which enters the tympanic cavity and inserts into the stapes. Lateral to the pyramid, there is an aperture through which the **chorda tympani nerve**, a branch of the facial (CN VII), enters the tympanic cavity.

The anterior wall or carotid wall of the tym-

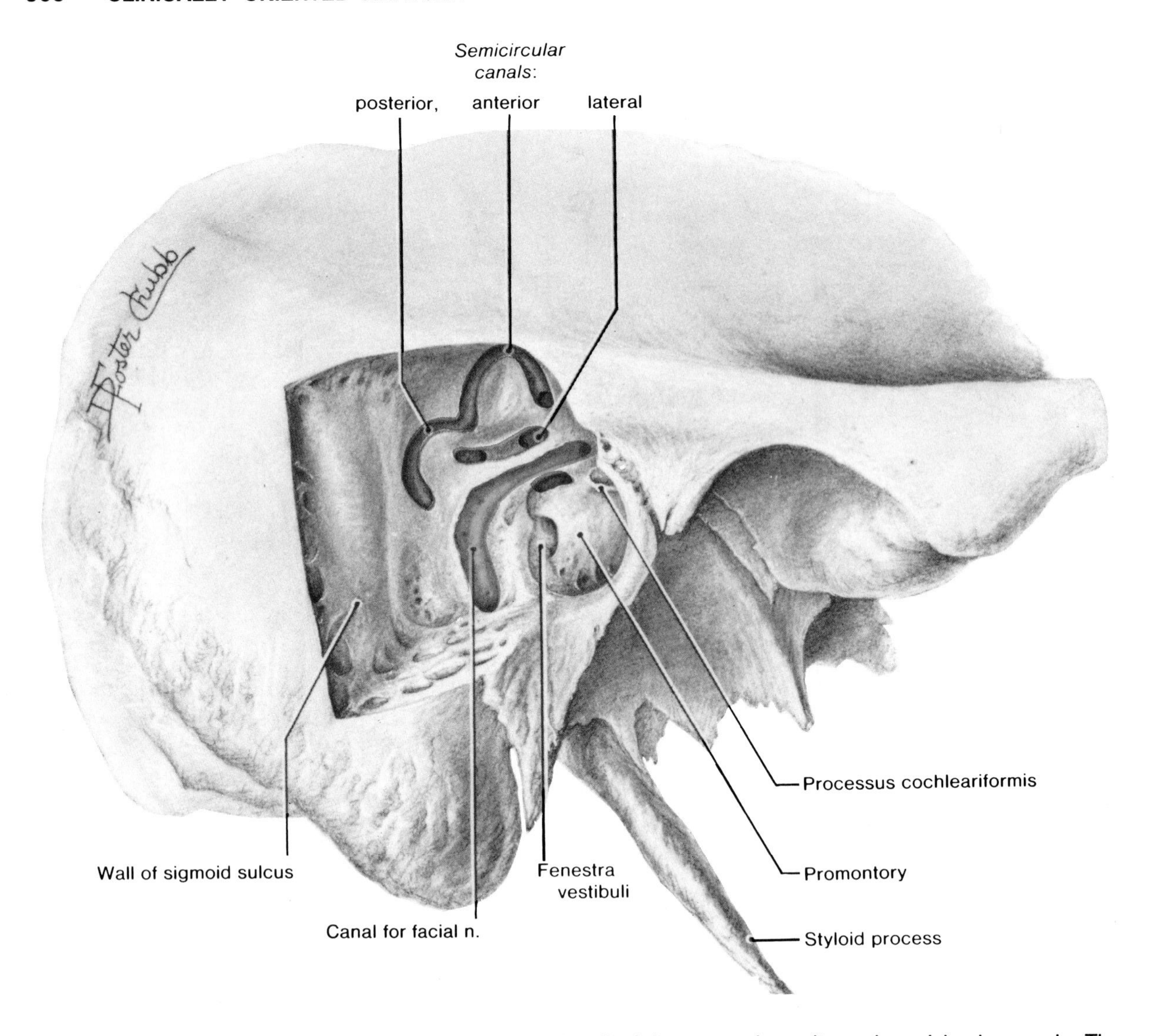

Figure 7-168. Drawing of a lateral view of the medial wall of the tympanic cavity and semicircular canals. The posterior wall of the external acoustic meatus and the mastoid antrum have been removed between the fenestra vestibuli (oval window) and the lateral semicircular canal. Note the following features of the medial wall of the tympanic cavity: (1) the promontory lying 2 mm deep to the umbo (Fig. 7-161) and overlying the basal turn of the cochlea; (2) the processus cochleariformis at the end of the canal for the tensor tympani that acts as a pulley for the tensor; (3) the fenestra vestibuli posterior to the pulley and medial to it; and (4) the facial canal (opened) running horizontally and posteriorly between the vestibular window and the lateral semicircular canal to the junction of the medial and posterior walls, then descending in the posterior wall to its orifice, the stylomastoid foramen (Fig. 7-8).

panic cavity is narrow because the medial and lateral walls converge anteriorly. There are two openings in the anterior wall. The superior opening communicates with a canal occupied by the **tensor tympani muscle** (Fig. 7-167). Its tendon inserts into the handle of the malleus and keeps the tympanic membrane tense. Inferiorly the tympanic cavity communicates with the nasopharynx via the **auditory tube** (Figs. 7-164 and 7-167). It runs inferomedially and anteriorly to open into the nasopharynx posterior to the inferior meatus of the nasal cavity (Fig. 7-86).

Inflammatory conditions in the tympanic cavity (*i.e.*, **otitis media**) may sometimes spread through the thin tegmen tympani (Fig. 7-163*A*). This causes inflammation of the meninges **(meningitis)** and brain **(cerebritis)**, and **a brain abscess**. In infants and children the unossified *petrosquamous fissure* (Fig. 7-55) may allow direct spread of infection from the tympanic cavity to the meninges of the brain. In the

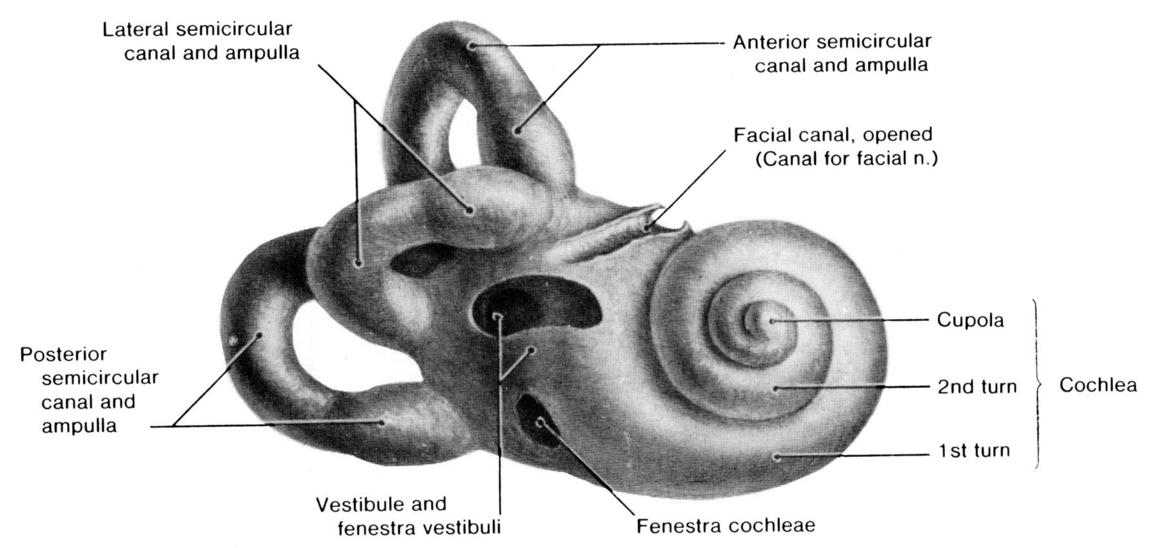

Figure 7-169. Drawing of a lateral view of the right side of the bony labyrinth, as it would appear if it was possible to dissect it from the petrous part of the temporal bone. Observe the three parts of the bony labyrinth: cochlea anteriorly, vestibule in the middle, and the semicircular canals posteriorly. Note that the cochlea makes 2½ coils. In life the fenestra vestibuli (oval window) and fenestra cochleae (round window) are closed by the base of the stapes (Fig. 7-166) and the secondary tympanic membrane, respectively. Observe the three semicircular canals: anterior, posterior, and lateral. Note that the anterior and posterior canals are set at a right angle to each other and that the lateral canal is set horizontally and at a right angle to the other two. Each canal forms about two-thirds of a circle and has an ampulla at one end. The lateral canal is shortest and the posterior one the longest.

adult, veins pass through the *petrosquamous suture* to the **superior petrosal sinus** (Figs. 7-57*B* and 7-79). Thus, infection can spread via these veins to the dural venous sinuses (Fig. 7-39).

Earache is a common symptom which has multiple causes, some in the ear and others at a distance. **Otitis externa** (inflammation of the external acoustic meatus) is one cause. ***Movement of the tragus*** (Fig. 7-160) ***results in increased pain*** because the cartilage in it is continuous with that in the external acoustic meatus (Fig. 7-159).

Otitis media (middle ear infection) is of more concern. Movement of the tragus does not increase the pain in adults, but may do so in infants and children. In young patients, the external acoustic meatus is almost entirely cartilaginous so that movement of the tragus is transmitted to the tympanic membrane.

Earache may be referred pain from distant lesions, e.g., the mouth **(dental abscess or cancer of the tongue)** via the mandibular nerve (CN V^3), or the pharynx and larynx via the vagus nerve (Fig. 7-135).

Some patients refer to **temporomandibular joint disease** as ear pain or an earache, because of the proximity of this articulation to the tragus of the ear (Figs. 7-119 to 7-121).

THE AUDITORY TUBE

This funnel-shaped tube connects the nasopharynx to the tympanic cavity (Fig. 7-164), and beyond this to the mastoid cells through the mastoid antrum. Its wide end is toward the nasopharynx, where it opens posterior to the inferior meatus of the nasal cavity (Figs. 7-86 and 8-53). The auditory tube is 3.5 to 4 cm long. Its posterior one-third is bony and the other two-thirds is cartilaginous. The bony part lies in a groove on the inferior aspect of the base of the skull, between the petrous part of the temporal bone and the greater wing of the sphenoid bone (Fig. 7-55).

The auditory tube is lined by mucous membrane that is continuous posteriorly with that of the tympanic cavity, and anteriorly with that of the nasopharynx (Fig. 7-164). *A collection of lymphoid tissue,* called the **tubal tonsil,** is located near the pharyngeal orifice of the auditory tube (also see Chap. 8).

The function of the auditory tube is to equalize pressure in the middle ear with the atmospheric pressure, thereby allowing free movement of the tympanic membrane. By allowing air to enter and leave the cavity, it balances the pressure on both sides of the membrane.

The arteries of the auditory tube are derived from: (1) the *ascending pharyngeal artery*, a branch of the **external carotid artery** (Fig. 7-37); and (2) the *middle meningeal artery* and the *artery of the pterygoid canal*, branches of the **maxillary artery** (Fig. 7-116).

The veins of the auditory tube drain into the *pterygoid plexus* (Fig. 7-39).

The nerves of the auditory tube arise from the *tympanic plexus*, which is formed by fibers of the facial (*CN VII*) and glossopharyngeal (*CN IX*) nerves (Fig. 7-111). The auditory tube also receives fibers from the **pterygopalatine ganglion** (Fig. 7-147).

The cartilaginous part of the auditory tube remains closed except during swallowing or yawning. The tube is opened by the simultaneous contraction of the tensor veli palatini and salpingopharyngeus muscles (Figs. 7-133 and 8-51), which are attached to opposite sides of the tube.

When the cartilaginous part of the auditory tube opens (*i.e.*, during swallowing), it prevents excessive pressure in the tympanic cavity. We purposefully swallow during the descent of an aircraft to relieve the pressure in our middle ears.

The auditory tube forms a route through which infections may pass from the nasopharynx to the tympanic cavity. The auditory tube is easily blocked by swelling of its mucous membrane, even by mild infections, because the walls of its cartilaginous part are normally in apposition. When the auditory tube is blocked, residual air in the tympanic cavity is usually absorbed into the mucosal blood vessels. This results in lowering of pressure in the tympanic cavity, retraction of the tympanic membrane, and interference with its free movement. As a result hearing is affected. Under these circumstances fluid may exude into the tympanic cavity **(serous otitis)**.

THE AUDITORY OSSICLES

These little ear bones **(malleus, incus, and stapes)** *form a chain across the tympanic cavity from the tympanic membrane to the fenestra vestibuli* (Figs. 7-159, 7-161, and 7-162). The malleus is attached to the tympanic membrane and the stapes occupies the fenestra vestibuli (Figs. 7-168 and 7-169). The incus is located between these two bones and articulates with them. The ossicles are covered with the mucous membrane that lines the tympanic cavity.

The Malleus (L. a hammer). Its rounded superior part, the **head** (Fig. 7-166), lies in the epitympanic recess (Fig. 7-163*B*). The **neck** of the malleus lies against the flaccid part of the tympanic membrane (Fig. 7-161). The **handle** of the malleus is embedded in the tympanic membrane and so moves with it (Fig. 7-161*A*). **The head of the malleus articulates with the incus and the tendon of the tensor tympani muscle inserts into its handle** (Fig. 7-167). *The chorda tympani nerve crosses the medial surface of the neck of the malleus* (Fig. 7-165).

The Incus (L. an anvil). Its large body lies in the epitympanic recess where it articulates with the head of the malleus (Figs. 7-162 and 7-163). **Its long process articulates with the stapes** and its short one is connected by a ligament to the posterior wall of the tympanic cavity.

The Stapes (L. a stirrup). The ***base*** (footplate) of this little bone, the smallest ossicle, **fits into the fenestra vestibuli** on the medial wall of the tympanic cavity (Figs. 7-162, 7-168, and 7-169). Its ***head***, directed laterally, articulates with the lenticular process of the incus (Figs. 7-162 and 7-166).

The malleus functions as a lever with the longer of its two arms attached to the tympanic membrane. The base of the stapes is considerably smaller than the tympanic membrane. As a result of these anatomical facts, *the vibratory force of the stapes is about 10 times that of the tympanic membrane.* Thus, **the auditory ossicles increase the force but decrease the amplitude of the vibrations transmitted from the tympanic membrane.**

Abnormal development of the malleus and incus is often associated with abnormal transformation of the first branchial arch into adult structures. The first two pairs of embryonic branchial arches give rise to the jaws, the auditory ossicles, and the auricles. The first pair of pharyngeal pouches becomes the auditory tubes. This explains why infants with the **first arch syndrome** have multiple malformations (deformed auricle, abnormal development of the cheek and mandible, and hearing defects). See Moore (1982) for details and illustrations.

Muscles Moving the Auditory Ossicles and the Tympanic Membrane. Two muscles produce the movements of the auditory ossicles and the tympanic membrane.

The Tensor Tympani Muscle (Figs. 7-163 and 7-167) is only about 2 cm long.

Origin (Fig. 7-167). Superior surface of the cartilaginous part of the **auditory tube,** the greater wing of the **sphenoid bone**, and the petrous part of **temporal bone.**

Insertion (Figs. 7-167 and 7-168). **Handle of the**

malleus. Its tendon turns around the processus cochleariformis before inserting into the handle of malleus.

Nerve Supply (Fig. 7-33). **Mandibular nerve** (CN V^3) via fibers that pass through the otic ganglion.

Actions. Pulls the handle of the malleus medially; thus it **tenses the tympanic membrane and reduces the amplitude of its oscillations.** This tends to prevent damage to the inner ear when one is exposed to loud sounds.

The Stapedius Muscle (Fig. 7-161*B*) is a tiny muscle in the pyramidal eminence or pyramid.

Origin. Pyramidal eminence on the posterior wall of the tympanic cavity. Its tendon enters the tympanic cavity by traversing a pinpoint foramen in the apex of the pyramid (Fig. 7-161*B*).

Insertion (Fig. 7-161). **Neck of the stapes.**

Nerve Supply (Fig. 7-35). **Facial** nerve (CN VII).

Actions. Pulls the stapes posteriorly and tilts its base in the fenestra vestibuli, thereby tightening the anular ligament and reducing the oscillatory range. **The stapedius also prevents excessive movement of the stapes.**

The tympanic muscles (muscles of the auditory ossicles) have a protective action in that they dampen large vibrations of the tympanic membrane resulting from loud noises. Thus, paralysis of the stapedius muscle (*e.g.,* resulting from a lesion of the facial nerve) is associated with excessive acuteness of hearing (a condition known as **hyperacusia**). This condition (G. *hyper,* over, above, + *akousis,* a hearing) results from uninhibited movements of the stapes.

Congenital fixation of the stapes results in severe conduction deafness at birth. This deformity appears to result from failure of normal differentiation of the anular ligament which attaches the base of the stapes to the fenestra vestibuli. As a result, the stapes becomes fixed to the bony labyrinth and is immovable.

THE MASTOID ANTRUM AND MASTOID CELLS (Figs. 7-79, 7-163, 7-164, and 7-167)

The mastoid antrum is a spherical sinus, slightly smaller than the tympanic cavity. It is located in the petromastoid part of the temporal bone (Figs. 7-55 and 7-79), posterior to the **epitympanic recess** (Fig. 7-163*B*). The mastoid antrum is connected to the tympanic cavity by the **aditus to the mastoid antrum,** and is separated from the middle cranial fossa by a thin roof, the **tegmen tympani** (Fig. 7-163*A*). Its floor has a number of apertures through which the mastoid antrum communicates with the **mastoid cells** (Figs. 7-79 and 7-164).

Anteroinferiorly the mastoid antrum is related to the canal for the facial nerve (Fig. 7-164). The lateral wall of the mastoid antrum is only 1 mm thick at birth, but it increases about 1 mm a year until it is about 15 mm thick. Unlike the paranasal sinuses, **the mastoid antrum is well developed at birth and is almost adult size.**

No mastoid cells or mastoid processes are present at birth. (Compare Figs. 7-5*C* and 7-10.) As the mastoid processes form, mastoid cells invade from the mastoid antrum. By 2 years of age these cells have bulged the bone laterally and inferiorly, forming a small mastoid process, the internal structure of which resembles a honeycomb (Fig. 7-79).

Infections of the mastoid antrum and mastoid cells always begin in the middle ear (*otitis media*). Infections may spread superiorly toward the middle cranial fossa via the petrosquamous fissure in young children (Fig. 7-55), or because of **osteomyelitis** (bone infection) of the tegmen tympani (Fig. 7-163).

Extradural pus is produced and occasionally a **temporal lobe abscess** forms. In other patients the infection may spread posteriorly into the posterior cranial fossa, producing osteomyelitis of the bone forming the groove for the sigmoid sinus, thrombophlebitis of the sigmoid sinus, and occasionally a **cerebellar abscess.**

Since the advent of antibiotics, mastoiditis as a complication of middle ear infection is rare. **During operations for mastoiditis, surgeons have to be conscious of the course of the facial nerve** (Figs. 7-163 to 7-165), so that it will not be injured. One access to the tympanic cavity is through the mastoid antrum. In a child, only a thin plate of bone needs to be removed from the lateral wall of the mastoid antrum in the suprameatal region to expose the tympanic cavity. In adults, however, bone must be penetrated for 15 mm or more to reach the mastoid antrum. At present, most mastoidectomies are **endaural** (within the auricle) *i.e.,* through the posterior wall of the external acoustic meatus.

Pneumatization (formation of mastoid cells) may be arrested during childhood; thus, in about 20% of persons, very few mastoid cells are visible in radiographs (Fig. 7-12). In rare cases no mastoid cells are visible.

THE INTERNAL EAR

The internal ear (the vestibulocochlear organ) is concerned with the reception of sound and the maintenance of balance. It is buried in the petrous part of the temporal bone (Figs. 7-55, 7-162, and 7-168). The internal ear consists of the sacs and the ducts of the **membranous labyrinth.** This membranous system contains endolymph and the end organs for hearing and balancing.

The membranous labyrinth, surrounded by perilymph, is suspended within the **bony labyrinth** (Fig. 7-162). The bony labyrinth is visible on radiographs (Fig. 7-12).

THE BONY LABYRINTH (Fig. 7-169)

The bony labyrinth, composed of three parts (**cochlea, vestibule, and semicircular canals)**, occupies much of the lateral portion of the petrous part of the temporal bone (Figs. 7-55 and 7-60).

The Cochlea (Fig. 7-169). This shell-like part of the bony labyrinth contains the *part of the internal ear that is concerned with hearing i.e.*, the membranous cochlea or cochlear duct (Fig. 7-172). **The cochlea** (L. *snail shell*) **has the shape of a snail's shell.** It makes two-and-one-half turns around a bony core, called the **modiolus** (L. nave of a wheel), in which there are canals for blood vessels and nerves. It is the large basal turn of the cochlea that produces the **promontory** on the medial wall of the tympanic cavity (Fig. 7-168).

The axis of the modiolus is across the long axis of the petrous part of the temporal bone; thus the apex of the cochlea, called the **cupola**, points anterolaterally (Fig. 7-169). A small shelf of bone, the **osseous spiral laminal**, protrudes from the modiolus like the thread on a screw. This starts at the vestibule and continues to the apex. The **basilar membrane** is attached to the osseous spiral lamina (Fig. 7-170). The osseous spiral lamina and the basilar membrane divide the cochlear canal into two halves or **scalae** (scala vestibuli and scala tympani), which communicate at the apex of the cochlea.

The modiolus is pierced by several longitudinal channels which turn outward toward the spiral lamina and enter the **spiral canal** of the modiolus. This canal runs in the base of the spiral lamina and contains the sensory **cochlear ganglion** (Fig. 7-170). Cells in this ganglion send their peripheral processes to the **spiral organ** (organ of Corti), which is concerned with hearing.

The scala vestibuli was so-named because it opens into the vestibule of the bony labyrinth (Fig. 7-162). Through its opening, perilymph can be exchanged freely between the vestibule and the scala vestibuli. The scala tympani, which also contains perilymph, is related to the tympanic cavity at the fenestra cochleae (Fig. 7-169), which is closed by the **secondary tympanic membrane.** The scala vestibuli communicates with the scala tympani through a small aperture, the **helicotrema,** at the apex of the cochlea.

The perilymph in the scalae is similar in composition to CSF. This similarity is understandable because there is a narrow connection between the perilym-

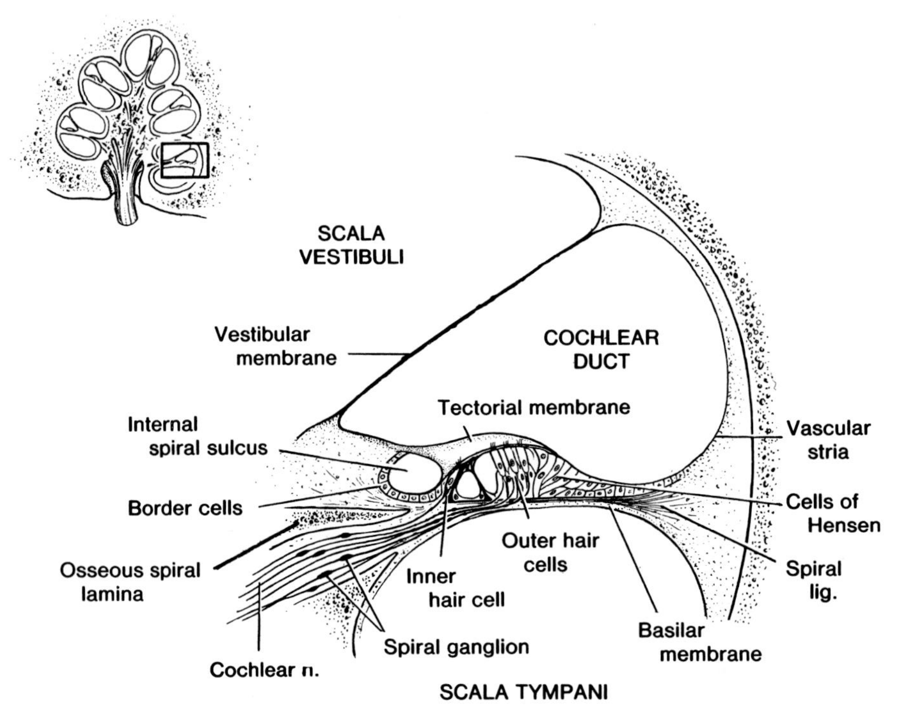

Figure 7-170. Drawing of a radial section through the cochlea showing the cochlear duct, basilar membrane, spiral organ, and tectorial membrane. In the *upper left* is a small drawing of an axial section of the cochlea. The large drawing shows details of the area enclosed in the rectangle.

phatic spaces of the bony labyrinth and the subarachnoid space, called the **perilymphatic duct** (Figs. 7-162 and 7-172). This connection consists of a canal in the petrous part of the temporal bone, running from the scala tympani in the basal turn of the cochlea to an extension of the subarachnoid space around the glossopharyngeal, vagus, and accessory nerves (CN IX, CN X, and CN XI, respectively).

The Vestibule (Figs. 7-162 and 7-169). This oval bony chamber (about 5 mm in length), contains the **utricle and saccule**, parts of the balancing apparatus (Fig. 7-172). The vestibule is continuous anteriorly with the bony cochlea, posteriorly with the semicircular canals, and with the posterior cranial fossa by the **aqueduct of the vestibule** (Fig. 7-162). This canal extends to the posterior surface of the petrous part of the temporal bone, where it opens posterolateral to the internal acoustic meatus (Figs. 7-60, 7-63, and 7-159). It contains the **endolymphatic duct** and two small blood vessels (Figs. 7-162 and 7-172). The endolymphatic duct emerges through the bone of the posterior cranial fossa and expands into a blind pouch, the **endolymphatic sac**. It is located under cover of the dura mater on the posterior surface of the petrous part of the temporal bone (Fig. 7-162).

The endolymphatic sac is a storage reservoir for excess endolymph formed by the blood capillaries within the membranous labyrinth.

The Semicircular Canals. These canals communicate with the vestibule of the bony labyrinth (Figs. 7-162 and 7-169). The three semicircular canals (anterior, posterior, and lateral) lie posterosuperior to the vestibule into which they open, and are set at right angles to each other. They occupy three planes in space. Each semicircular canal forms about two-thirds of a circle and is about 1.5 mm in diameter, except at one end where there is a swelling called the **ampulla** (Fig. 7-169). The three semicircular canals have only five openings into the vestibule because the anterior and posterior canals have one stem common to both.

The anterior semicircular canal (superior semicircular canal) lies at a right angle to the posterior surface of the petrous part of the temporal bone, and is closely related to the floor of the middle cranial fossa (Figs. 7-15, 7-168, 7-169, and 7-171). It forms a transversely rounded elevation, called the **arcuate eminence** (Figs. 7-63 and 7-171), on the superior surface of the petrous process, just medial to the tegmen tympani (Fig. 7-163*A*).

The posterior semicircular canal lies in the long axis of the petrous part of the temporal bone, immediately deep to the posterior or cerebellar surface (Figs. 7-168 and 7-171), close to the sigmoid sinus.

The lateral semicircular canal is horizontal and its arch is directed horizontally, posteriorly, and laterally (Fig. 7-168). It lies deep to the medial wall of the **aditus to the mastoid antrum** or aditus ad antrum (Figs. 7-163*B* and 7-164), and runs superior to the *canal for the facial nerve* (Figs. 7-164 and 7-165). The lateral semicircular canal of one ear is in the same plane as that in the other.

In one type of operation performed to restore hearing, an additional opening is made into the internal ear (this procedure is called **fenestration**). The opening is made in the lateral semicircular canal at its ampullary end (Fig. 7-169). This operation produces another fenestra (L. window) between the tympanic cavity and the internal ear. Fenestration is performed when the stapes in the fenestra vestibuli has become immovable owing to **otosclerosis** (bone forms around the stapes and the fenetra vestibuli, resulting in *progressively increasing deafness*).

THE MEMBRANOUS LABYRINTH (Fig. 7-172)

The membranous labyrinth is a series of communicating membranous sacs and ducts that are contained in the cavities of the bony labyrinth (Figs. 7-162 and 7-169). The membranous labyrinth generally follows the form of the bony labyrinth, but it is much smaller. The membranous labyrinth contains a watery fluid, called **endolymph**, which differs in composition from the **perilymph** around it in the bony labyrinth (Fig. 7-162). In these terms "lymph" does not indicate a relationship to the lymph circulating in lymph vessels.

The membranous labyrinth consists of: (1) two small communicating sacs, **the *utricle* and *saccule*** in the vestibule; (2) **the *three semicircular ducts*** in the semicircular canals; and (3) **the *cochlear duct*** in the cochlea.

The vestibular parts of the membranous labyrinth are suspended in the bony vestibule of the bony labyrinth by trabeculae of connective tissue; however the cochlear duct is firmly attached along two sides of the wall of the cochlear canal. Except at the ampulae, *the semicircular ducts are much smaller than the semicircular canals*, but the convexity of each duct is adherent to the bony canal in which it is located.

The various parts of the membranous labyrinth form a closed system of sacs and ducts which communicate with one another (Figs. 7-162 and 7-172). The semicircular ducts open into the utricle through five openings, and the utricle communicates with the saccule through the **utriculosaccular duct**, which also joins the *endolymphatic duct*. The saccule is continuous with the cochlear duct through a narrow communication known as the **ductus reuniens** (Fig. 7-172).

The Utricle and Saccule (Figs. 7-162 and 7-172). Each of these dilations has a specialized area of sensory epithelium called a macula. The **macula utriculi** is in the floor of the utricle, parallel with the base of

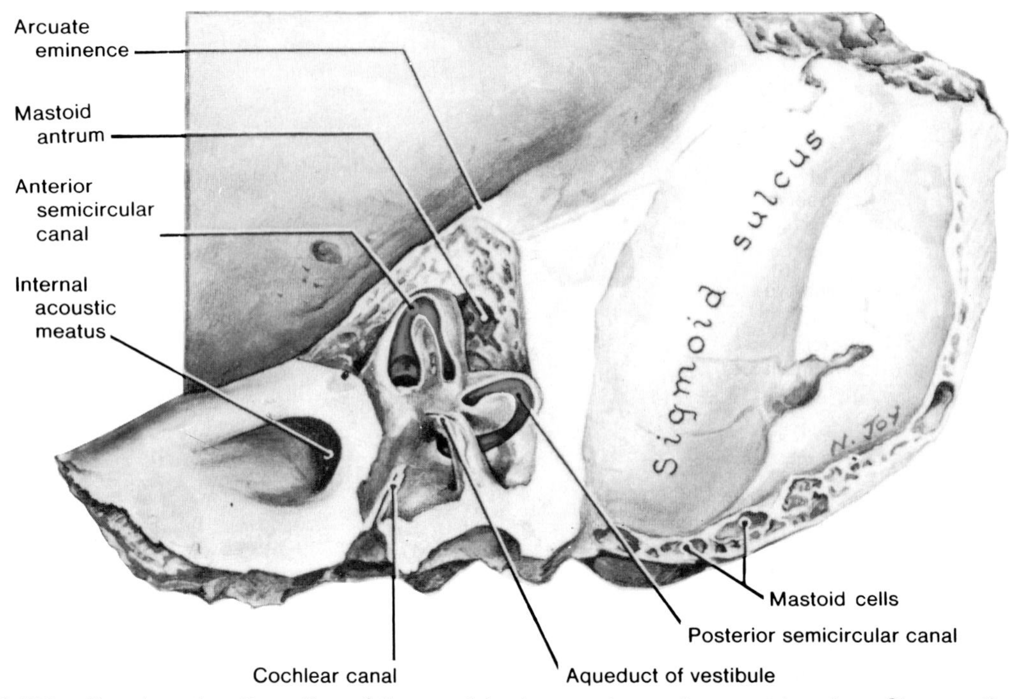

Figure 7-171. Drawing of a dissection of the semicircular canals, posterosuperior view. Observe the anterior semicircular canal set vertically and inferior to the arcuate eminence, and making a right angle with the posterior surface of the petrous bone. Note that the posterior semicircular canal is nearly parallel to the posterior surface of the bone, about 5 mm from the sigmoid sulcus or groove for the sigmoid sinus (also see (Fig. 7-60). Observe the aqueduct of the vestibule containing the endolymphatic duct (Fig. 7-172). Note the cochlear duct containing the perilymphatic duct (Fig. 7-162) which opens into the subarachnoid space at the apex of the depression for the ganglion of CN IX.

the skull, whereas the **macula sacculi** is vertically placed on the medial wall of the saccule.

*The **hair cells** in the maculae are innervated by fibers of the vestibular division of the **vestibulocochlear nerve*** (Fig. 7-173). The primary sensory neurons are in the **vestibular ganglion,** which is in the **internal acoustic meatus** (Figs. 7-60, 7-63, and 7-171).

The maculae are primarily static organs for signaling the position of the head in space, but they also respond to quick tilting movements and to linear acceleration and deceleration. **Motion sickness** results mainly from prolonged, fluctuating stimulation of the maculae.

The Semicircular Ducts (Figs. 7-172 and 7-173). Each duct has an **ampulla** or expansion at one end, containing a sensory area called a **crista ampullaris.**

The cristae are sensors of movement, recording movements of the endolymph in the ampulla that result from rotation of the head in the plane of the duct. The **hair cells of the cristae,** like those of the maculae, are supplied by primary sensory neurons whose cell bodies are in the **vestibular ganglion** in the internal acoustic meatus (Figs. 7-60 and 7-171).

The Cochlear Duct (Figs. 7-170 and 7-172). The duct of the cochlea is a spiral, blind tube which is firmly fixed to the internal and external walls of the cochlear canal within the bony labyrinth (Figs. 7-162 and 7-171). It is triangular in transverse section and lies between the **osseus spiral lamina** and the external wall of the cochlear canal.

*The roof of the spiral cochlear duct is formed by the **vestibular membrane**, and its floor is formed by the **basilar membrane** and the external part of the osseus spiral lamina.*

The receptor of auditory stimuli is the spiral organ (organ of Corti), situated on the basilar membrane. This complex sensory organ developed from the epithelium of the cochlear duct. Sound waves in the air pass into the external acoustic meatus and cause movements of the tympanic membrane and auditory ossicles (Fig. 7-162). Stronger waves are transmitted to the perilymph at the fenestra vestibuli by the base of the stapes. Vibrations of the perilymph are transmitted to the **basilar membrane,** displacement of which in response to acoustic stimuli, causes bending of the hair-like projections of the sensory hair cells of the spiral organ (Fig. 7-170). These cells are in contact with the **tectorial membrane** (L. *tectum*, roof) which forms a roof over the hair cells. These cells are innervated by peripheral fibers or bipolar primary sensory neurons in the **spiral ganglion** (Fig.

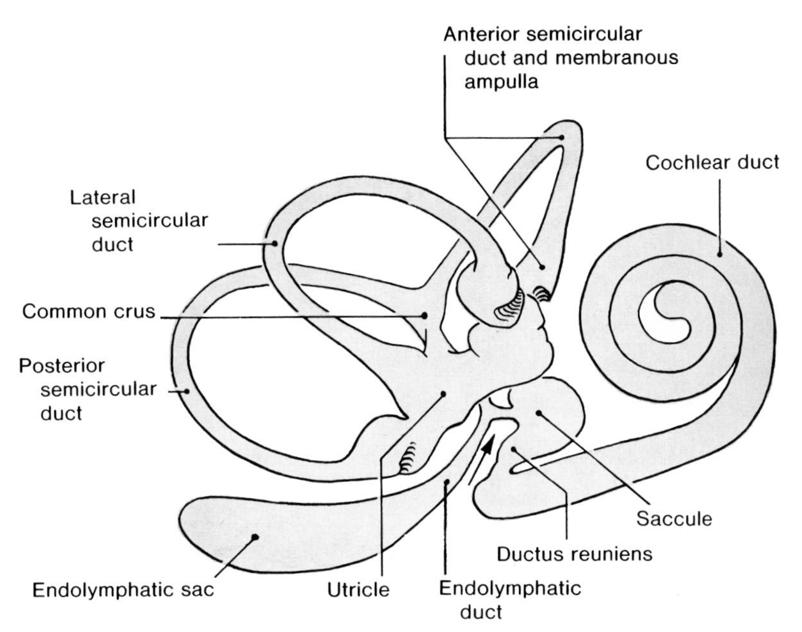

Figure 7-172. Drawing of a lateral view of the membranous labyrinth, right side, *In vivo* it is contained within the bony labyrinth (Fig. 7-169). The membranous labyrinth is a closed system of ducts and chambers filled with endolymph and bathed in perilymph (Fig. 7-162). Observe its three parts—the cochlear duct within the cochlea; the saccule and the utricle within the vestibule; and the three semicircular ducts within the three semicircular canals. One end of the cochlear duct is closed; the other end communicates with the saccule through the ductus reuniens. Note that the utricle communicates with the saccule via the utriculosaccular duct (indicated by *arrow*). The lateral semicircular duct lies in the horizontal plane and is more horizontal than it appears in this drawing.

7-170), situated around the modiolus of the cochlea. The central processes of these nerve cells form the **cochlear nerve**, part of the vestibulocochlear nerve (Fig. 7-173).

Persistent exposure to excessively loud sounds causes degenerative changes in the spiral organ at the base of the cochlea, resulting in **high tone deafness**. This type of hearing loss commonly occurs in workers who are exposed to loud noises and do not wear protective ear muffs (*e.g.*, persons working for long periods around jet engines or farm tractors).

Acoustic trauma disease is sometimes called "boiler-maker's disease" because it used to be detected in workers in boiler factories, owing to injury to the cochlear nerve incident to riveting the inside of boilers.

Injury to the ear by an imbalance in pressure between ambient (surrounding) air and the air in the middle ear is called **otic barotrauma** (*baro* is a combining form relating to pressure). This type of injury occurs in flyers, divers, caisson workers, and battered infants. Some parents who beat their children "box their ears," thereby injuring them.

THE INTERNAL ACOUSTIC MEATUS (Figs. 7-60, 7-171, and 7-174)

The internal acoustic meatus is a *narrow canal running laterally for about 1 cm within the petrous part of the temporal bone.* The opening of the meatus is in the posteromedial part of this bone, **opposite the external acoustic meatus**, at a depth of about 5 cm.

The internal acoustic meatus is closed laterally by a thin, perforated plate of bone which separates it from the internal ear. Through this plate pass the facial nerve (CN VII), branches of the vestibulocochlear nerve (CN VIII), and blood vessels.

The **vestibulocochlear nerve** divides near the lateral end of the internal acoustic meatus into an anterior or **cochlear portion** and a posterior or **vestibular portion**. As has been described, *the cochlear nerve is the nerve of hearing and the vestibular nerve is the nerve of balance.*

An abnormal increase in the amount of endolymph is called *hydrops of the internal ear* or Ménière's disease (syndrome). This excess fluid produces **recurrent vertigo** (dizziness) that is accompanied in later stages by **tinnitus** (L. a jingling) or noises in the ear and deafness. The pathological changes resulting from dilation of

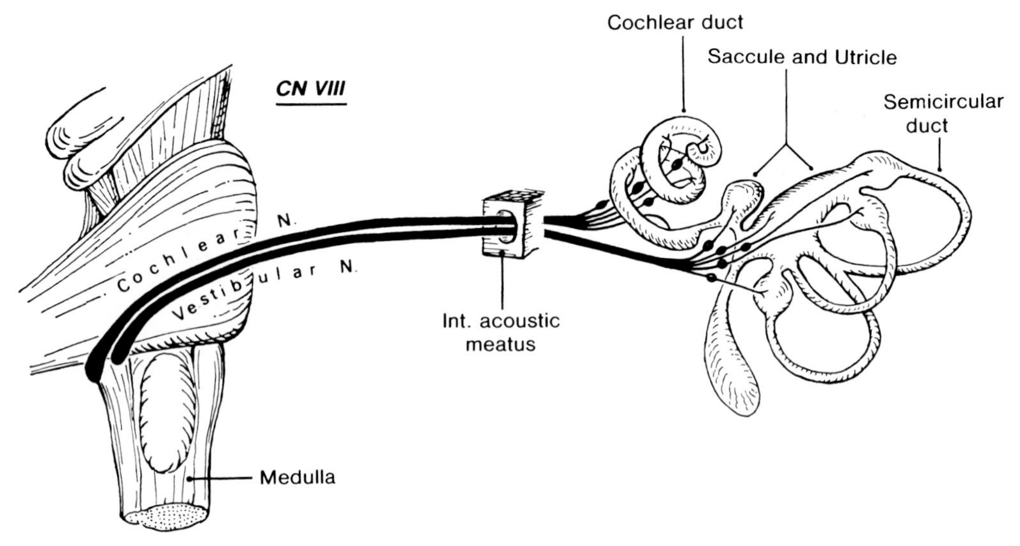

Figure 7-173. Scheme of the distribution of the *vestibulocochlear nerve* (CN VIII). This nerve has two parts: (1) the *cochlear nerve*, the nerve of hearing, whose fibers transmit impulses from the spiral organ in the cochlear duct, and (2) the *vestibular nerve*, the nerve of balance, whose fibers transmit impulses from the maculae of the saccule and utricle, and the ampullae of the semicircular ducts (also see Fig. 7-170).

the endolymphatic system are **degeneration of hair cells** in the maculae of the vestibule and in the spiral organ. The relation of these changes to paroxysms (sudden onsets) of vertigo (dizziness) is unknown.

Other causes of vertigo are: **labyrinthitis** (inflammation of the membranous labyrinth), **trauma resulting in hemorrhage into the internal ear**, certain **brain stem infarcts** (areas of necrotic nerve tissue), and **tumors of the vestibulocochlear nerve.**

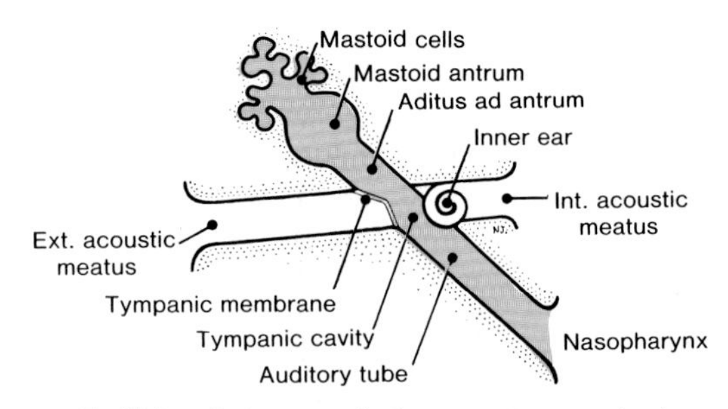

Figure 7-174. Scheme of the meatus and airway. Observe that the line of the external and internal meatus (meatuses) intersect at the tympanic cavity with the line of the airway from the mastoid cells to the nasopharynx.

PRESENTATION OF PATIENT ORIENTED PROBLEMS

Case 7-1

The wife of a 45-year-old commercial traveler was awakened by the unusual nature of his snoring, and was puzzled when she noticed that he was sleeping with his left eye open. In the morning she observed that the left side of his face was drooping. When he tried to look at his teeth, he found that his lips were also paralyzed on that side. He was also unable to whistle or to puff out his cheek because the air blew out through his paralyzed lips on the left side. He also found that he was unable to raise his eyebrow or to frown on that side.

During breakfast he had trouble chewing his food, as it dribbled out of the left side of his mouth. Fearing that he may have poliomyelitis or have had a mild stroke during the night, he made an appointment to see his doctor.

During his examination the doctor made the following observations: at rest the left side of his face appeared flattened and expressionless; there were no lines on the left side of his forehead; there was sagging in the left lower half of his face, and saliva drooled from the left corner of his mouth. In addition there was a loss of taste sensation on the anterior two-thirds of the left side of his tongue, and an absence of voluntary control of left facial and platysma muscles. When the patient smiled, the lower portion of his face was pulled to the normal side and the right corner of his mouth was raised, but the left corner was not.

During questioning the patient related how he had driven home late the night before and, because of drowsiness, had rolled the window down part way. He also recalled that he had had a severe head cold and an ear infection a few days previously, and that the doctor who treated him had described his illness as a viral infection.

Problems. Paralysis of what nerve would produce the signs exhibited by this patient? Why did his left eye remain open, even when he was sleeping? Why was there loss of taste sensation on the anterior two-thirds of the left side of his tongue? Where would the lesion of the nerve probably be located? *These problems are discussed on page 979.*

Case 7-2

A 62-year-old man complained to his dentist about sudden short bouts of excruciating pain on the left side of his face. They were of about 2 months' duration and had been increasing in severity. Following examination, the dentist informed him that there was no dental cause for the pain. He stated that it was probably a neurological disorder and that he should see a physician.

The man told the doctor that the **stabbing pains**, lasting 15 to 20 seconds, occurred several times a day and were so severe that he had once contemplated suicide. He said that the onset of pain seemed to be **triggered by chewing** or by a cold wind blowing on his upper lip.

When the doctor asked him to point out the area where the pains occurred, he carefully pointed to his left upper lip, left cheek, and inferior to his left eye. He said the pain also radiated to his lower eyelid, the lateral side of his nose, and the inside of his mouth.

The doctor applied firm steady pressure over the patient's left cheek and over his infraorbital area, but detected no tenderness indicative of inflammation of the maxillary sinus, or defects in the orbital margin caused by a lesion of the maxilla. *Radiographs of the patient's skull showed nothing abnormal.*

On evaluation the doctor detected an acuteness of sensitivity to touch (**hyperesthesia**) on the left upper lip, and to pin pricking over the entire left maxillary region, but he found no abnormality of sensation in the forehead or mandibular regions.

Problems. Which branch of what major nerve supplies the area of skin and mucous membrane where the **paroxysms** (sudden recurring attacks) of stabbing pain were felt? Where does this nerve leave the skull? What are its branches and how are they distributed? *These problems are discussed on page 979.*

Case 7-3

A 55-year-old farmer complained to his doctor about a sore that had been on his lower lip for 6 months. He stated that he first thought it was a cold sore and then he became worried because this one looked different. Furthermore, it did not respond to his usual treatment with salve.

On examination the doctor observed the patient's ulcerated, indurated (L. *indurare*, to harden) lesion was present on the central portion of his lower lip. The man's face was darkly tanned. Systematic palpation of all the lymph nodes of the patient's neck revealed **enlarged, hard submental lymph nodes.** None of the submandibular or deep cervical lymph nodes was enlarged. Examination of a small biopsy from the edge of the lesion revealed a **squamous cell carcinoma** (malignant tumor of epithelial origin).

Problems. Where are the *submental lymph nodes* located? Between the bellies of which muscle do they lie? What structures, in addition to the central portion of the lip, do afferent lymph vessels of these nodes drain? To which lymph nodes do lymph vessels from lateral portions of the lip pass? If the cancer had spread from the submental lymph nodes, where would you expect to find **metastases** (new tumors). *These problems are discussed on page 980.*

Case 7-4

A 22-year-old medical student was struck by a puck on the left "temple" during an interfaculty hockey game. *He fell to the ice unconscious, but regained consciousness in about 1 minute.* There was some bleeding from a laceration located about two finger-breadths superior to his left zygomatic arch. The gash extended from the top of his auricle almost to his eyebrow.

As you helped him to the bench, he said that he felt rather weak and unsteady. Realizing that he may have sustained a skull fracture, you asked a classmate to call a doctor while you took him to the dressing room. The deep tendon reflexes in his upper and lower limbs were equal. His pupils were equal in size and both contracted to light. As you waited, you observed that the injury site started to swell, but your friend otherwise seemed well. In about half an hour he said that he was sleepy and wanted to lie down.

His left pupil was now moderately dilated and reacted sluggishly to light. By the time the doctor arrived, he was unconscious. The pupil on the left was widely dilated and did not respond to light, whereas the pupil on the right was slightly dilated but showed a normal reaction to light. The doctor said, "*We must get him to the hospital right away*!"

In hospital, several skull radiographs were made and a **CT scan** (computerized tomographic scan) was done. As the doctor was almost certain that there was an intracranial hemorrhage, he called a neurosurgeon.

When this specialist arrived, the radiologist reported that there was a **fracture of the temporal squama**, posterior to the pterion, and that the CT scan showed an extradural hematoma (Fig. 7-175).

Problems. Where is the temple? Define the area known as the **pterion**. In what part of the ***temporal fossa*** is it located? Why is the pterion clinically

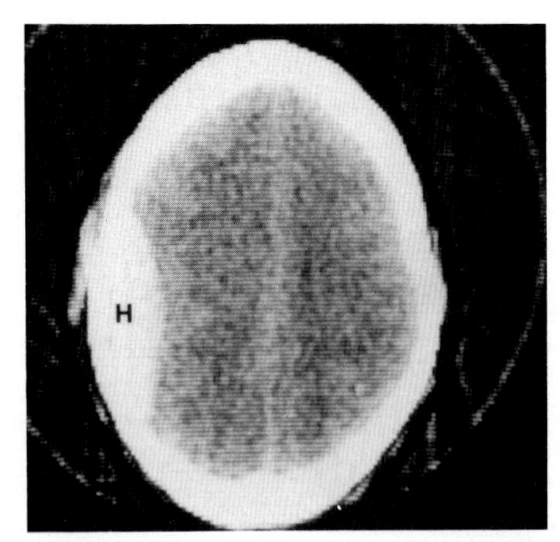

Figure 7-175. Computerized tomographic (CT) scan showing an extradural hematoma (*H*) in the left middle cranial fossa. (Reprinted with permission from Norman D, Korobkin M, Newton TH: *Computed Tomography*. St. Louis, CV Mosby Co, 1977.)

important? What artery was most likely torn? What other vessel(s) may have been torn? Where would the blood collect? Differentiate between an extradural and a subdural hemorrhage. *These problems are discussed on page 980.*

Case 7-5

While cleaning the bathtub, a 49-year-old woman developed a **throbbing headache.** It lasted for about 30 minutes and then slowly faded away. Similar headaches occurred occasionally for the next week.

One day as she was lifting a heavy chair, she experienced a **sudden, severe headache** which was accompanied by nausea, vomiting, and a general feeling of weakness. Her friend decided that she should see her doctor immediately.

The physical examination revealed **nuchal rigidity** (a stiff neck) and an elevation in blood pressure. Visualization of her optic fundus through the ophthalmoscope showed **subhyaloid hemorrhages** (bleeding between the retina and vitreous body). Her deep tendon reflexes were symmetrical and all modalities of sensation were normal.

On the basis of these distinct signs and symptoms, **the doctor made a diagnosis of subarachnoid hemorrhage**. He suggested that the bleeding was probably caused by the rupture of an **aneurysm of the cerebral arterial circle** (Figs. 7-76 and 7-77).

Arteriograms showed a saccular aneurysm of the anterior communicating artery. A **lumbar puncture** or "spinal tap" was performed (Fig. 5-60). Examination demonstrated bloody CSF. After centrifugation, the supernatant fluid was **xanthochromatic** (yellow-colored).

Problems. Where would blood from the ***ruptured aneurysm*** most likely go? How do you explain anatomically the formation of **subhyaloid hemorrhages**? Why was the supernatant part of the CSF xanthochromatic? *These problems are discussed on page 981.*

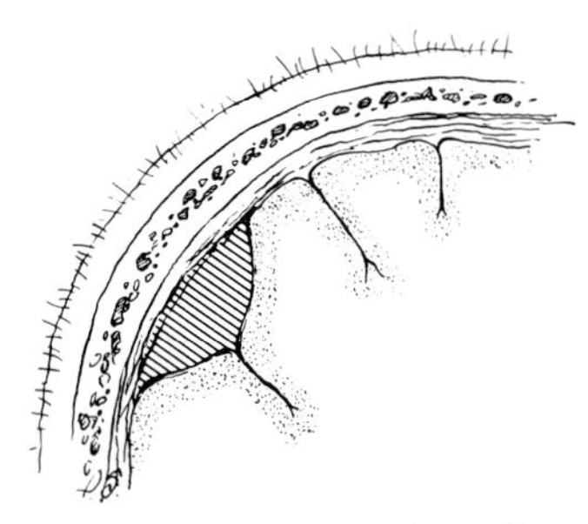

Figure 7-176. Illustrative drawing of a subdural hematoma. Observe that the collection of blood is on the surface of the brain, *i.e.*, deep to the dura mater (L. *sub*, under). Acute subdural hematomas are associated with a lacerated brain (see p. 865).

Case 7-6

A 23-year-old man went to a dentist to have a badly **decayed mandibular third molar** ("wisdom") tooth extracted. The dentist explained that there was likely to be considerable pain associated with removal of the tooth, and informed the patient that he was going to inject a "local" (anesthetic agent) to desensitize the tooth and associated soft tissues. When agreeing to the extraction, the patient requested that plenty of anesthetic be given because he was extremely sensitive to pain.

The dentist inserted the needle through the mucous membrane on the inside of the patient's mouth, where the needle came to rest near the **lingula**, a bony projection on the medial surface of the ramus of the mandible (Fig. 7-22). In a few minutes the patient stated that his gum (gingiva), lip, chin, and tongue on the affected side were numb (anesthetized). During the extraction procedure the patient said he felt pain; hence the dentist injected more anesthetic. The tooth was removed without further incident.

As the patient was preparing to leave, he happened to look in the mirror. **The patient found that he was unable to close his eye and lips on the affected side** and that his mouth sagged on this side, particularly when he attempted to expose his teeth. He also found that his ear lobule was numb. When he reported these unusual symptoms, the dentist drew a sketch of the nerves of the face and explained that, because of the large amount of anesthetic injected, other nerves in addition to those supplying the teeth had been anesthetized. He assured the patient that all these effects would disappear in 3 to 4 hours.

Problems. Name the nerve supplying the mandibular molar and premolar teeth. Why was the patient's chin, lower lip, and tongue on the injected side also anesthetized? When anesthetizing this nerve, what others might be affected? What probably caused the patient's **facial paralysis** and loss of sensation in the lobule his auricle? *These problems are discussed on page 981.*

DISCUSSION OF PATIENT ORIENTED PROBLEMS

Case 7-1

Sudden facial paralysis often follows exposure to the cold; thus **Bell's palsy** is the most probable diagnosis in this case. The characteristic facial appearance results from **a lesion of the facial nerve (CN VII)**. In the present patient, the motor supply to the muscles of the left face, forehead, and eyelids were most severely affected (Fig. 7-100).

Paralysis of the muscles of facial expression on the left side explains the expressionless look on that side of his face, and his inability to whistle, puff his cheek, or close his left eye.

When the facial nerve is paralyzed, the levator palpebrae superioris (acting unopposed) causes the eye to remain open, even during sleep. The drooling and difficulty in chewing result from paralysis of the orbicularis oris and buccinator muscles (Figs. 7-28 and 7-29). Loss of taste sensation on the anterior two-thirds of the left side of his tongue is understandable anatomically because this region of the tongue receives taste fibers via the **chorda tympani** branch of CN VII (Figs. 7-35 and 7-139). This symptom also indicates that the nerve lesion is proximal to the origin of this nerve in the facial canal (Figs. 7-164 and 7-165).

Because of paralysis of the orbicularis oculi muscle, the lacrimal puncta are no longer in contact with the cornea. As a result, tears tend to flow over the left lower eyelid onto the cheek. In addition, the cornea may dry out during sleep (if an ointment is not used) because the eyelids on the affected side remain open.

Drying of the cornea can also occur during the day owing to the inability to blink; this dryness could result in **corneal ulceration**.

The site of the lesion is most likely in the facial canal in the petrous part of the temporal bone. The **paresis or paralysis of the facial muscles** are thought to be caused by inflammation of the facial nerve superior to the stylomastoid foramen (Fig. 7-8). The etiology is generally thought to be a **viral infection** which causes edema of the facial nerve and compression of its fibers in the **facial canal** or at the stylomastoid foramen. If the lesion is complete, all the facial muscles on that side are affected equally. Voluntary, emotional, and associated movements are all affected.

In most cases the nerve fibers are not permanently damaged and nerve degeneration is incomplete. As a result, recovery is very slow but generally good. **Some facial asymmetry may persist** (*e.g.*, slight sagging of the left corner of the mouth).

Case 7-2

The area of skin and mucosa in which the stabbing pain was felt is supplied by the **maxillary nerve** (Figs. 7-32 and 7-33), the second division of the trigeminal nerve (**CN V^2**). This wholly sensory nerve leaves the skull through the **foramen rotundum** (Figs. 7-61 and 7-62). At its termination as the infraorbital nerve, it gives rise to branches that supply the ala (side) of the nose, the lower eyelid, and the skin and mucous membrane of the cheek and upper lip (Figs. 7-28 and 7-32). Branches of the maxillary nerve also innervate the teeth in the maxilla and the mucous membranes of the nasal cavities, palate, mouth, and tongue (Fig. 7-33).

The symptoms described by this patient are characteristic of **trigeminal neuralgia** (tic douloureaux). It occurs most often in middle-aged and elderly persons. The pain may be so intense that the patient winces; hence the common term "tic" (twitch).

In some cases the pain may be so severe that mental changes occur; there may be depression and even suicide attempts. The maxillary nerve distribution, as in the present case, is most frequently involved, then the mandibular, and least frequently the ophthalmic. The **paroxysms of sudden stabbing pain**, as in the present case, are of sudden onset and are often set off by touching the face, brushing the teeth, drinking, or chewing. Often there is an especially sensitive "*trigger zone*," *e.g.*, the left upper lip in the present case.

The complete cause of trigeminal neuralgia is unknown. Some persons believe the condition is caused by a pathological process affecting neurons in the trigeminal ganglion (Fig. 7-62), whereas others believe that neurons in the nucleus of the spinal tract may be involved. Medical and/or surgical treatment is used to alleviate the pain. Only the anatomical aspects of these treatments will be discussed here.

Attempts were made to block the nerve at the infraorbital foramen by using alcohol; this usually gives temporary relief of pain. The simplest surgical procedure is **avulsion** or *cutting of the branches of the nerve at the infraorbital foramen* (Fig. 7-28).

Radiofrequency selective coagulation of the trigeminal ganglion via a needle electrode passing through the cheek and the foramen ovale is also used. *To prevent regeneration of nerve fibers, the sensory root of the trigeminal nerve* may be partially cut between the ganglion and the brain stem (**rhizotomy**). Although the axons may regenerate, they do not do so within the brain stem. Attempts are made to differentiate and cut only the sensory fibers to the division of the trigeminal nerve involved.

The same result may be achieved by sectioning the spinal tract of CN V (**tractotomy**). After this operation the sensation of pain, temperature, and simple (light) touch are lost over the area of skin and mucous membrane supplied by the maxillary nerve (Fig. 7-32). This may be annoying to the patient who does not recognize the presence of food on the lip and cheek, or feel it within the mouth on the side of the nerve section, but these disabilities are preferable to the excruciating pain.

If sensation is lost in the eye, traumatic lesions (*e.g.*, scarring) of the cornea, followed by inflammation (**keratitis**), may occur which sometimes result in loss of vision in that eye.

"Decompression" of the trigeminal ganglion may be done by removing the dura mater over it in the **trigeminal cave** (Figs. 7-69 and 7-120), and then massaging it. This often relieves the pain for several years without affecting other sensations.

Case 7-3

Carcinomas of the lip most commonly involve the lower lip. *Overexposure to sunshine over many years,* as occurs in outdoor workers such as farmers, is a common feature of the history in these cases. Chronic irritation from pipe smoking appears to be a factor also, and may be related to long-term contact with tobacco tar (Fig. 7-44).

The submental lymph nodes lie on the fascia covering the mylohyoid muscle between the anterior bellies of the right and left digastric muscles (Figs. 7-40 and 8-60). The central part of the lip, the floor of the mouth, and the tip of the tongue drain to the submental lymph nodes, whereas lateral parts of the lip drain to the **submandibular lymph nodes** (Figs. 7-40 and 8-18).

If cancer cells had spread further, metastases would have developed in the submandibular lymph nodes because efferents from the submental lymph nodes pass to them. In addition, lymph vessels from the submental lymph nodes pass directly to the **juguloomohyoid node** (Fig. 7-40). As the submandibular nodes are situated beneath the deep cervical fascia in the submandibular triangle, the patient's chin may have to be lowered to slacken this fascia before these enlarged nodes can be palpated.

Because all parts of the head and neck drain into the **deep cervical nodes** (Figs. 7-40 and 8-35), they might also be sites of metastases. As the juguloomohyoid node drains the submental and submandibular lymph nodes, it could be involved in the spread of tumor cells from a carcinoma of the lip. It is located where the omohyoid muscle crosses the internal jugular vein (Figs. 7-40 and 8-18).

Case 7-4

The temple is the area between the temporal line and the zygomatic arch (Fig. 7-10), where the skull is thin and is covered by the temporalis muscle and the temporal fascia. The blood vessels of the temple are very numerous. The **pterion** is a somewhat variable H-shaped area (Fig. 7-9) that lies deep to the temporalis muscle. Here four bones approach each other or meet (frontal, parietal, temporal, and sphenoid). The pterion is an important bony landmark because it indicates the location of the **frontal branch of the middle meningeal artery** (Fig. 7-16). The center of the pterion is 4.0 cm superior to the zygomatic arch and 3.5 cm posterior to the frontozygomatic suture. It lies in the anterior part of the temporal fossa.

The thin squamous part of the temporal bone is grooved by the **middle meningeal artery** and its branches (Fig. 7-7). The temporal squama is easy to fracture and the broken pieces may tear the artery and its branches as they pass superiorly on the external surface of the dura mater. This results in a slow accumulation of blood in the extradural space (Fig. 7-175), forming an **extradural hematoma** (epidural hematoma). The hematoma forms relatively slowly because the dura is firmly attached to the bone by **Sharpey fibers**. These fibers resist stripping of the dura from the bone to a certain extent.

A **subdural hematoma** is a localized mass of extravasated blood that is *located on the surface of the brain,* deep to the dura mater (Fig. 7-176).

The middle meningeal artery, a branch of the first part of the maxillary artery (Fig. 7-116), enters the skull through the **foramen spinosum** (Figs. 7-61 and 7-62). It divides within the first 4 or 5 cm of its intracranial course. The frontal branch passes superiorly from the **pterion**, more or less parallel to the coronal suture of the skull (Figs. 7-16 and 7-50). The parietal branch passes posterosuperiorly, its exact site depending on its point of origin. In the present case, the frontal branch of the middle meningeal artery was almost certainly torn. This artery is usually accompanied by a meningeal vein which may have also been torn (Fig. 7-50).

The lucid interval which followed the patient's recovery from the brief loss of consciousness resulting from cerebral concussion occurs because of the slow formation (up to 12 hours) of the extradural hematoma (Fig. 7-175). In addition, this kind of a space-occupying intracranial lesion can be tolerated for a short time because some blood and CSF are squeezed out of the calvaria through the veins and subarachnoid space. However, as the cranium is nonexpansile, the intracranial pressure soon rises, producing drowsiness and then **coma** (G. *koma*, deep sleep).

The increased intracranial pressure forces the

supratentorial part of the brain, usually the uncus, through the **tentorial incisure** (Figs. 7-58 and 7-68), squeezing the oculomotor nerve (CN III) between the brain and the sharp, free edge of the tentorium. **Compression of CN III causes third nerve palsy** (Fig. 7-87) which results in a dilated, nonreacting pupil on the side of the lesion. An extradural hemorrhage in the characteristic position, illustrated by the present case, primarily causes **compression of the temporal lobe** underlying the pterion (Figs. 7-16 and 7-51). Immediate surgical intervention is necessary to relieve the intracranial pressure so that further compression of the brain will not occur, which could cause death by interfering with the **cardiac and respiratory centers** in the medulla.

Case 7-5

*Unruptured **saccular aneurysms** are usually asymptomatic* (Fig. 7-76*A*). In the present case the initial headaches were probably caused by **intermittent enlargement of the aneurysm**, or by slight bleeding from it into the subarachnoid space (the so-called "**warning leak**").

Her subsequent severe, almost unbearable headache, was the result of gross bleeding from the aneurysm into the subarachnoid space (Fig. 7-67).

Blood in the CSF causes meningeal irritation which produces a headache. As the anterior communicating artery is in the longitudinal fissure, rostral to the optic chiasma (Fig. 7-77*B*), blood escaping from the ruptured aneurysm would enter the **chiasmatic cistern** (Fig. 7-71) and other subarachnoid spaces around the brain and spinal cord. This explains why there was blood in the CSF obtained by lumbar puncture.

*Some authorities would recommend against a **lumbar puncture** in a case of subarachnoid hemorrhage* that is so obvious as the present case, because of the possibility of causing herniation of the brain. The lowering of CSF pressure in the spinal subarachnoid space by removing CSF might cause inferior movement of the brain resulting in herniation (*e.g.*, of part of the *cerebellar tonsils,* Figs. 7-54 and 7-59).

Rupture of an aneurysm of the anterior communicating artery into the adjacent part of one frontal lobe may cause symptoms of a mass lesion in one hemisphere. In some cases, the **intracranial hematoma** may break into the ventricular system, causing an acute expansion of the ventricle and probably death.

Blockage of subarachnoid spaces by large amounts of blood in the CSF could impair circulation of this fluid, resulting in a further increase in intracranial pressure. This could force the medial part of the temporal lobe, usually the uncus, through the **tentorial notch** (Fig. 7-68), and the cerebellar tonsils through the foramen magnum.

Herniation of the uncus leads to third nerve palsy, which is indicated by drooping of the upper eyelid (Fig. 7-87) and paralysis of pupillary constriction. This results in a fixed dilated pupil. (*Fixed* means not reacting to light or accommodation).

Herniation of the cerebellar tonsils compresses the medulla containing the vital respiratory and cardiovascular centers and produces a life-threatening situation. Surgical intervention would be required to lower the intracranial pressure (e.g., ventricular drainage via a burr hole in the skull will sometimes reduce the intracranial pressure).

The **subhyaloid hemorrhages** observed during *funduscopy* resulted from the abrupt rise in intracranial pressure transmitted to the subarachnoid space around the optic nerve (Fig. 7-93). This compressed and obstructed the central retinal vein where it crosses this space. **This results in increased pressure in the retinal capillaries and hemorrhages between the retina and vitreous body**. After centrifugation of the CSF, the supernatant fluid was yellow because it contained serum bilirubin and products of hemolized red blood cells.

Case 7-6

The inferior alveolar nerve supplies the mandibular molar and premolar teeth (Fig. 7-33), and then branches to supply the canine and incisor teeth. The **mental nerve** supplies the skin of the chin and the lower lip on that side. Hence the **inferior alveolar nerve** supplies all teeth in one half of the mandible. Anesthetization of this nerve also anesthetizes the chin and lower lip because the mental nerve supplying these structures is a terminal branch of the inferior alveolar nerve.

As the **lingual nerve** descends just anterior to the inferior alveolar nerve near the mandibular foramen (Fig. 7-117), it was also anesthetized. This is advantageous because in addition to supplying the tongue, the lingual nerve also supplies sensory fibers to the mandibular gingiva.

Because of the relatively large amount of anesthetic solution that was injected, it must have spread into the parotid gland. **Paralysis of the muscles of facial expression resulted from anesthetization of branches of the facial nerve** (CN VII). As the parotid gland and these nerves occupy the space around the posterior margin of the ramus of the mandible (Fig. 7-29), they could easily be affected as the anesthetic agent infiltrated the area. Probably the injection was made posteriorly so that the anesthetic solution passed through the **stylomandibular ligament**, a sheet of fascia condensed between the parotid and submandibular glands which is continuous with the fascia covering the parotid gland. Like the anesthesia of the teeth and gingivae, these effects on the muscles of facial expression and of mastication would disappear in a few hours.

His ear lobule was numb because the intermediate branches of the **great auricular nerve** were also anesthetized. The anterior branches of this nerve supply skin on the posteroinferior part of the face and its intermediate branches supply the inferior part of the auricle on both surfaces.

Suggestions for Additional Reading

1. Barr ML, Kiernan JA: *The Human Nervous System: An Anatomical Viewpoint*, ed 4. Hagerstown, Harper & Row Publishers, Inc, 1983.
   This textbook is recommended for obtaining the sound basis of neurological anatomy that is necessary for the interpretation of signs and symptoms of lesions in the head, particularly in the brain.
2. Bertram EG, Moore KL: *An Atlas of the Human Brain and Spinal Cord.* Baltimore, Williams & Wilkins, 1982.
   A clinically oriented atlas of the brain and spinal cord with many photographs illustrating dissections of important parts of the brain and spinal cord.
3. Carpenter MB: *Core Text of Neuroanatomy*, ed 3. Baltimore, Williams & Wilkins, 1985.
   Another highly recommended textbook for 1st year students that presents a synthesis of basic concepts of neuroanatomy. Good attempts are made to correlate the structure and function of the nervous system and to demonstrate clinical applications of the structures concerned.
4. Hanaway JL, Scott WR, Strother CM: *Atlas of the Human Brain and the Orbit for Computed Tomography.* St. Louis, Warren H. Green, Inc, 1977.
   CT scans are commonly used for visualizing the cross-sectional anatomy of the brain, the ventricular system, and the subarachnoid cisterns. This atlas shows cross-sections of the head with CT scans. It is recommended for persons learning computed tomography of the head for the first time. The quality of the illustrations is very good.
5. Liebgott B: *The Anatomical Basis of Dentistry.* Philadelphia, WB Saunders Co, 1982.
   This text was written to meet the needs of dental students. For example, there are good accounts of the anatomical aspects of needle insertions for regional anesthesia of structures in the oral cavity, especially the teeth and gingivae.
6. Moore KL: *The Developing Human: Clinically Oriented Embryology*, ed 3. Philadelphia, WB Saunders Co, 1982.
   This text has a style similar to that used in the present book. Students wishing details of normal and abnormal development of the head, briefly referred to in the present text, should consult this clinically oriented embryology book.

CHAPTER 8

The Neck

The neck contains vessels, nerves, and other *structures connecting the head and the trunk* (*e.g.*, the carotid arteries, jugular veins, vagus nerves, lymphatics, vertebrae, muscles, esophagus, and trachea). *The neck also contains very important endocrine glands (e.g.*, the thyroid and parathyroid glands).

The common word "*collar*" is derived from the Latin word *collum*, meaning the neck. Many medical words are also derived from it; *e.g.*, **torticollis** (L. *tortus*, twisted + *collum*, neck) is the medical term for wryneck (Fig. 8-9). *Cervix* is another Latin word for the neck; hence the **cervical plexus** is a network of nerves in the neck (Fig. 8-10) and the anterior and posterior **cervical triangles** are triangular areas of the neck (Fig. 8-6) used for descriptive purposes.

As most lymphatic vessels in the head drain into **cervical lymph nodes** (Figs. 7-40 and 8-35), their enlargement may indicate a tumor in the head, but the cancer may be in the thorax or abdomen because the neck connects the trunk and the head. There are several causes of **a pain in the neck** (*e.g.*, inflamed lymph nodes, muscle strain, and *protrusion of a cervical intervertebral disc*).

This brief introduction indicates the clinical importance of the neck and the need for you to acquire a clear understanding of its structure and functions. **Many vital structures are crowded together in the neck**, but there are good landmarks for helping you to locate them. Learn them well because missing one could result in a misdiagnosis or make a surgical procedure difficult and/or dangerous.

SURFACE ANATOMY

The surface anatomy of the posterior aspect or back of the neck was described with the back (p. 570). Review this material, noting that *the spinous process of the* ***axis*** *(C2) is the first bony point that can be felt in the midline inferior to the external occipital protuberance.* The spinous process of the **vertebra prominens** (C7) is easily palpable and is usually clearly visible when the neck is flexed (Fig. 5-10).

The laryngeal prominence ("Adam's apple")[1] is an important surface feature in the anterior part of the neck (Fig. 8-1). It is formed by the **thyroid cartilage**, the largest one in the laryngeal skeleton (Figs. 8-2 and 8-3). The superior part of the thyroid cartilage is the most prominent and is more noticeable in men than in women and children, and in some people more than in others.

The sex difference in the angle formed by the laminae of the thyroid cartilage explains why the thyroid cartilage is more distinct in males. In Figure 8-2, *C* and *D*, note that the angle formed by the convergence of the laminae in the median plane is greater in females than in males. Also observe that each lamina in the male has a greater anteroposterior breadth than that in the female.

Because of these anatomical differences, the superior border of the thyroid cartilage in most males projects anteriorly (Figs. 8-14 and 8-25), producing a distinct **laryngeal prominence** (Fig. 8-1). Sexual differences in the thyroid cartilages develop during puberty (13 to 16 years).

The thyroid cartilage (Fig. 8-2) lies at the level of the fourth and fifth cervical vertebrae. It consists of two quadrilateral plates called laminae which can easily be felt. Grasp the laminae between your index finger and thumb and move your thyroid cartilage from side to side. Also note that it rises when you swallow.

The vocal folds (true vocal cords) lie about level with the midpoint of the anterior border of the thyroid cartilage (Figs. 8-2*B* and 8-25).

The hyoid bone (Figs. 8-3 and 8-4) lies just superior to the thyroid cartilage. The U-shaped hyoid bone is located at the level of the body of the third cervical vertebra. You can feel the body of the hyoid bone in the angle between the floor of the mouth and the

[1] Supposedly this is where the forbidden fruit stuck in Adam's throat. It is a fanciful way of accounting for the prominence of the male thyroid cartilage. The anatomical basis for this sex difference is illustrated in Figure 8-2.

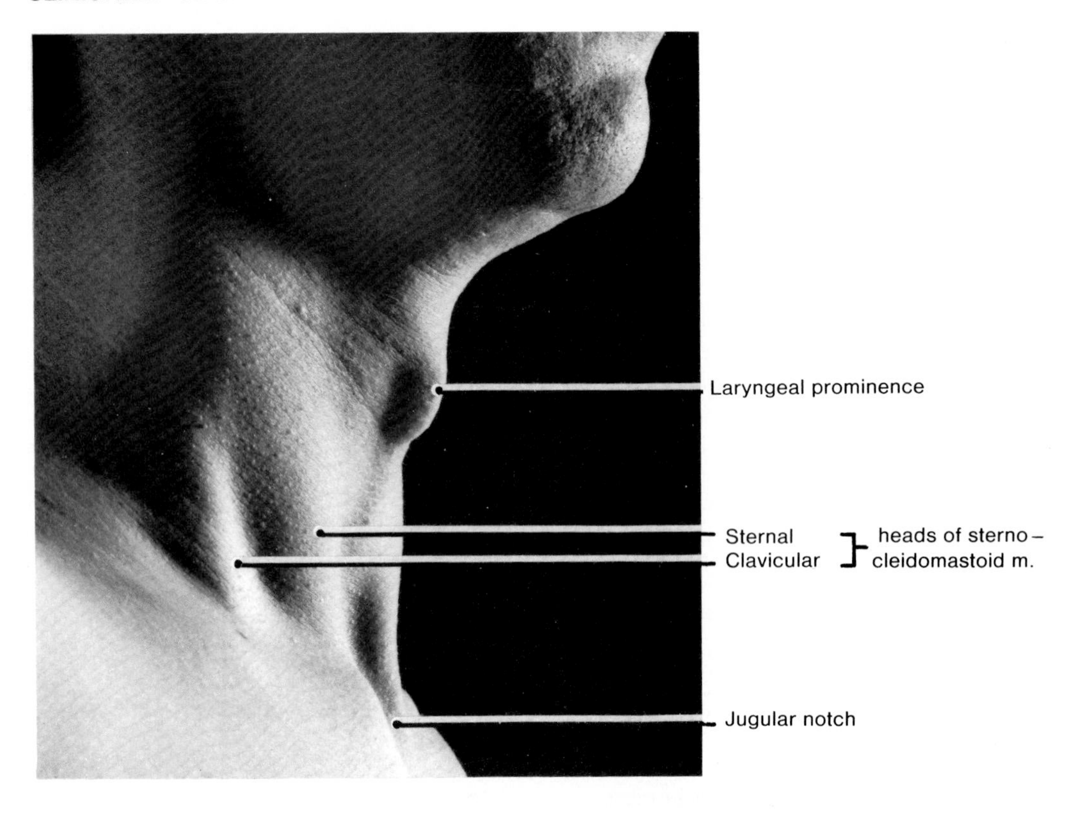

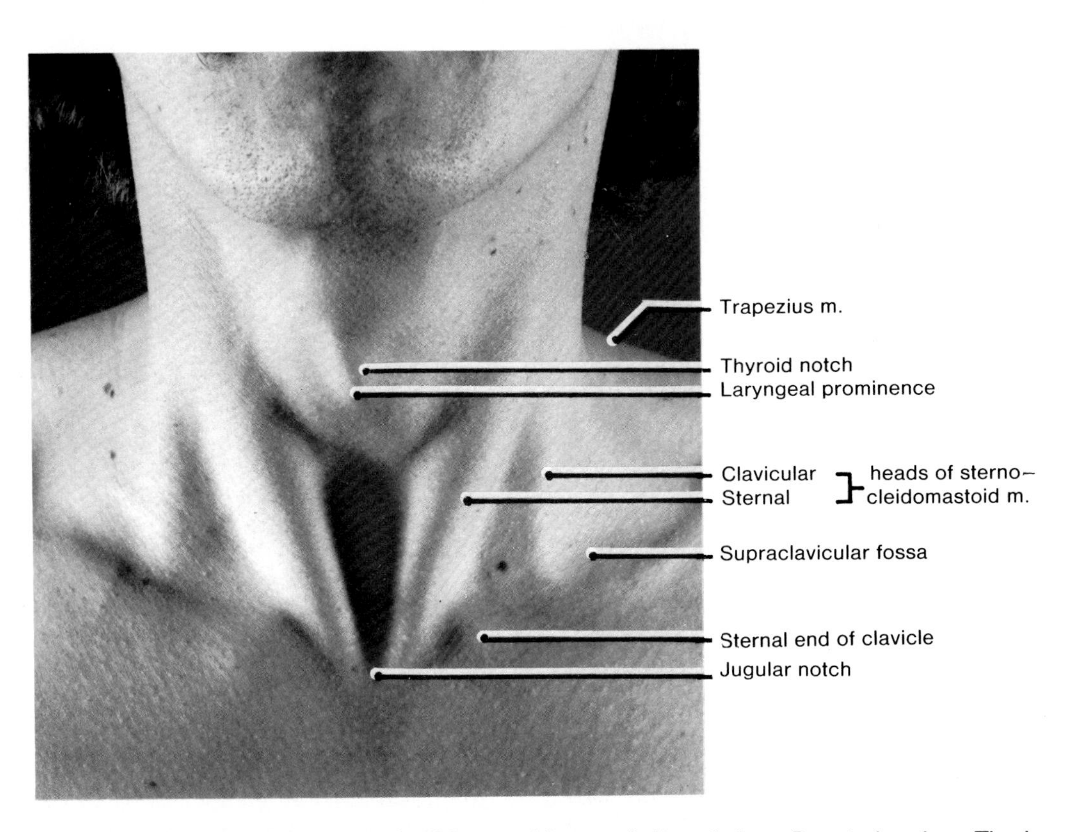

Figure 8-1. Surface features of the neck of a 28-year-old man. *A*, lateral view. *B*, anterior view. The laryngeal prominence, produced by the thyroid cartilage (Fig. 8-25), is clearly visible in this person. The thyroid cartilage of the larynx is not prominent in most females (Fig. 8-33).

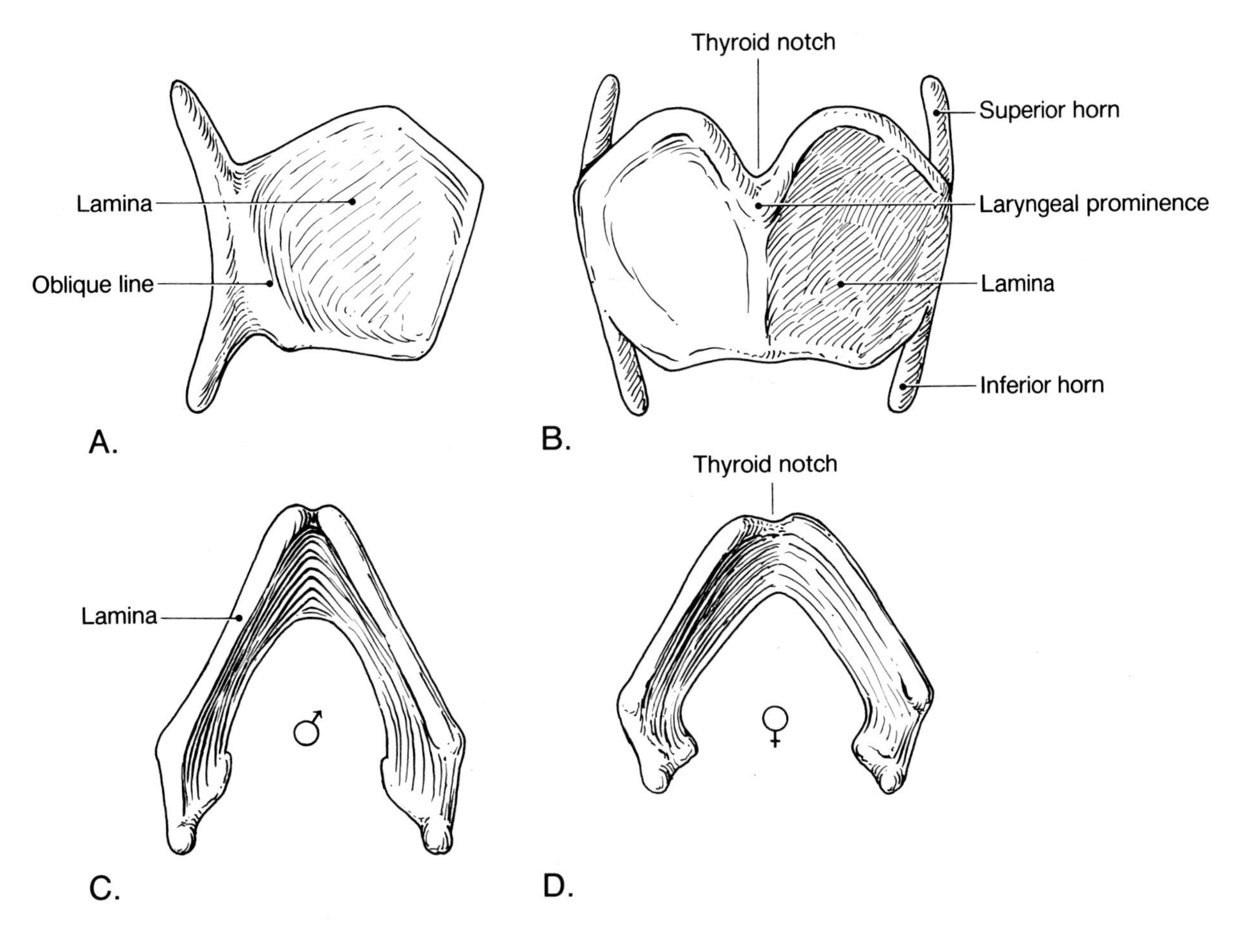

Figure 8-2. Drawings of the thyroid cartilage. *A*, right lateral view. *B*, anterior view. The two quadrilateral laminae are fused together in the median plane to form the laryngeal prominence, which is more pronounced in males (Fig. 8-1). Immediately superior to the prominence, the thyroid laminae diverge to form a V-shaped thyroid notch. *C* and *D*, superior views of male and female thyroid cartilages illustrating the sex difference in the angle at which the laminae meet.

anterior aspect of the neck. *It is the first resistant structure felt in the midline inferior to the chin* (Fig. 8-3). You can feel it more easily during swallowing. Grasp your hyoid bone between your index finger and thumb and move it from side to side.

When your neck is relaxed, you can palpate the tip of the **greater horn** (L. *cornu*) on one side of the hyoid bone if you steady the opposite side, as illustrated in Figure 8-4. Verify by palpation that the tip of the greater horn is fairly close to the anterior border of the sternocleidomastoid muscles (Fig. 8-6*B*).

The tip of the greater horn of the hyoid bone lies midway between the laryngeal prominence and the mastoid process of the temporal bone (Figs. 7-10 and 8-3); thus, **the tip of the greater horn is an important surgical landmark for locating the lingual artery** (Figs. 7-143, 8-5, and 8-20), which arises from the external carotid artery, posteroinferior to the tip of the greater horn of the hyoid. Ligation of this artery is sometimes necessary, *e.g.*, during radical resection (L. cutting off) of the tongue because of the presence of a carcinoma (cancer) in it.

The tips of the transverse processes of the atlas (C1 vertebra) can also be felt by deep palpation. Press superiorly with your index finger between the angle of your mandible and a point 1 cm anteroinferior to the tip of the mastoid process (Fig. 8-3). As you do this, rotate your head slowly from side to side. During this

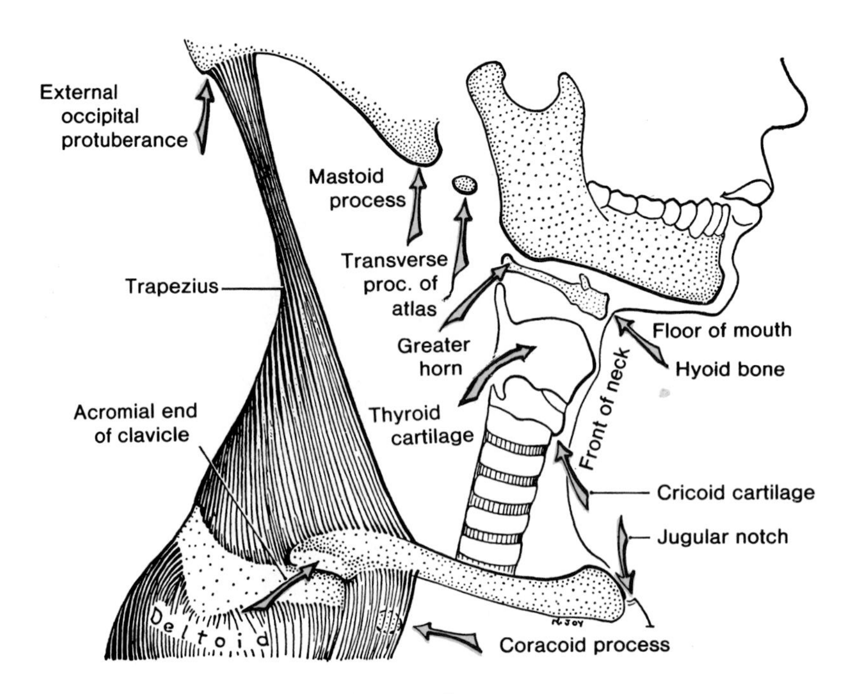

Figure 8-3. Drawing illustrating the bony landmarks of the neck. See the jugular notch in Figures 8-1 and 8-27.

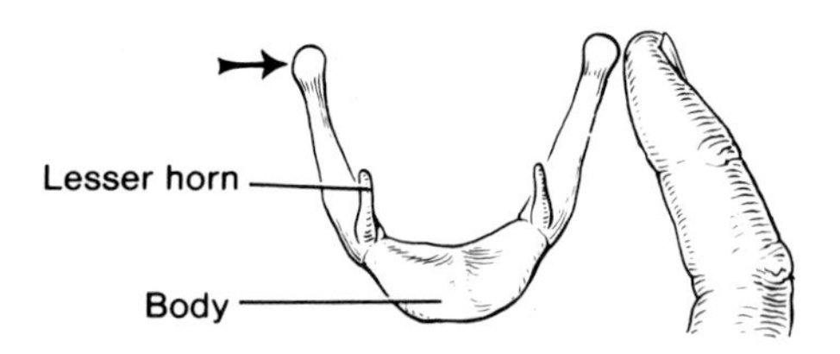

Figure 8-4. Diagram illustrating how to palpate the greater horn (cornu) of the hyoid bone (*arrow*). Note the small, conical lesser horns.

movement, the skull and the atlas rotate as a unit on the axis (Figs. 5-34 and 5-35).

The cricoid cartilage, part of the laryngeal skeleton, lies inferior to the laminae of the thyroid cartilage (Fig. 8-3). It can easily be felt inferior to the laryngeal prominence (Figs. 8-1, 8-3, 8-25, and 8-26). Extend your neck and run your fingertip inferiorly from your chin over your hyoid bone and your thyroid and cricoid cartilages. Note that after you pass the cricoid, your fingertip sinks in because the arch of the cricoid projects anteriorly beyond the rings of the trachea (Figs. 8-3 and 8-26).

The cricoid cartilage is *shaped like a signet ring* with its wide part (lamina) posterior and its narrow part (arch) anterior (Fig. 8-69). Ensure that you can palpate the cricoid cartilage because it is a clinically important landmark, *e.g.*, during a **tracheotomy** (Fig. 8-38).

The cricoid cartilage lies at the level of the sixth cervical vertebra (Fig. 8-31), where the pharynx joins the esophagus and the larynx and trachea join each other (Fig. 8-46).

Some of the tracheal rings may be palpable in the inferior part of the neck (Fig. 8-3). The tracheal rings are usually not palpable just inferior to the cricoid cartilage, because the **isthmus of the thyroid gland** lies anterior to them (Fig. 8-37). Grasp the trachea between your index finger and thumb, just superior to the **jugular notch** (Figs. 8-1 and 8-8). Verify that it moves superiorly during swallowing.

The lobes of the thyroid gland may be palpable, particularly in females in whom it enlarges slightly during menstruation and pregnancy. The isthmus of the thyroid gland, where the lobes are connected across the midline (Fig. 8-37), may be felt as a soft cushion-like mass about a fingerbreadth inferior to the cricoid cartilage.

The isthmus of the thyroid usually lies over the second and third tracheal rings (Figs. 8-31, 8-36, and 8-37). Feel it slip superiorly as you swallow and oscillate as you speak.

The jugular notch (suprasternal notch) is easily palpable between the medial ends of the clavicle, and is clearly visible (Figs. 8-1 and 8-8). It is a rounded depression in the superior border of the **manubrium sterni** (Fig. 8-27). Put your index finger in your jugular notch and press posteriorly until you feel your trachea. Turn your head to the right and move your finger to the left; it will encounter the narrow tendi-

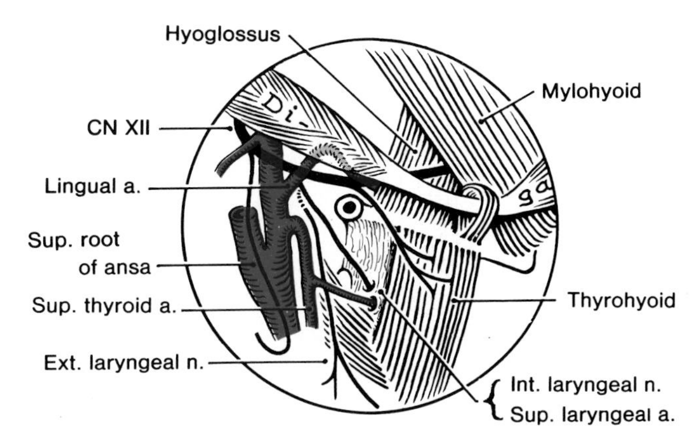

Figure 8-5. Drawing of the hyoid region of the anterior triangle of the neck (see Figs. 8-6*B* and 8-19 for a more extensive view of the area). The tip of the greater horn of the hyoid bone (*black dot in circle*) is an important reference point for many muscles, nerves, and arteries in the neck (*e.g.*, note that the lingual artery arises from the external carotid artery, just posterior to the tip of the greater horn). Also see Figure 8-24.

nous sternal head of the **sternocleidomastoid muscle** (Fig. 8-1). Let your finger pass over this tendon into a slight depression between the two heads of origin of the sternocleidomastoid. Now run your finger over the clavicular head of this muscle and you will feel a triangular depression, called the **supraclavicular fossa** (Fig. 8-1). *This fossa is clinically important because it contains the* ***pressure point for the subclavian artery*** (Fig. 8-12), which lies deep to it in the *supraclavicular triangle or subclavian triangle* (Fig. 8-6*B*).

In an emergency, *the subclavian artery can be compressed in the supraclavicular triangle against the first rib* (Figs. 1-16 and 1-17), preventing hemorrhage in the upper limb.

The medial or sternal ends of the clavicles are clearly visible at the root of the neck (Figs. 6-2 and 8-1), particularly in thin persons. Palpate one of your clavicles, verifying that it is practically subcutaneous throughout its length. Observe its curves and general form on a skeleton as you palpate your clavicle.

The bony landmarks forming the superior limit of the neck are: the *inferior margin of the mandible*, the *mastoid process* of the temporal bone, and the *external occipital protuberance* (Fig. 8-3).

Put your hand on your shoulder and then shrug it as you feel the rounded edge of the large **trapezius muscle** (Figs. 8-1, 8-3, 8-6, and 8-10 to 8-12). Verify by palpation that it extends from the external occipital protuberance (Fig. 8-3) to the bones of the shoulder or **pectoral girdle**, composed of the clavicle and scapula. The trapezius, a muscle of the upper limb, extends over the posterior aspect of the neck and attaches the pectoral girdle to the skull and vertebral column (Figs. 6-44 and 6-45). The anterior border of the trapezius marks the posterior limit of the side of the neck and the midline of the neck demarcates the anterior limit of the neck (Figs. 8-1 and 8-6).

The sternocleidomastoid muscle divides the lateral side of the neck into anterior and posterior triangles (Fig. 8-6). Although you can define the boundaries of these triangles by palpation, they can be also seen during dissection after the skin and the platysma muscle have been reflected (Fig. 8-10).

The sternocleidomastoid muscle forms an important landmark in the neck. When contracted, the sternocleidomastoid muscle forms a prominent ridge (Figs. 8-1 and 8-33).

SUPERFICIAL STRUCTURES OF THE NECK

The skin of the neck is thin and pliable. The subcutaneous connective tissue contains cutaneous nerves and superficial veins (Fig. 8-10). The platysma muscle is superficial to the main parts of these veins and nerves (Figs. 6-14 and 8-7).

MUSCLES OF THE NECK

THE PLATYSMA MUSCLE (Figs. 8-7 and 8-8)

This wide, thin, subcutaneous sheet of striated muscle is located in the superficial fascia of the neck. The platysma (G. a flat plate) covers the superior part of the anterior triangle and the anteroinferior part of the posterior triangle of the neck (Figs. 8-6 to 8-8 and 8-10). Its fibers blend superiorly with the facial muscles (Figs. 7-28 and 7-29).

Origin (Figs. 8-7 and 8-8). **Fascia and skin** over the pectoralis major and deltoid muscles.

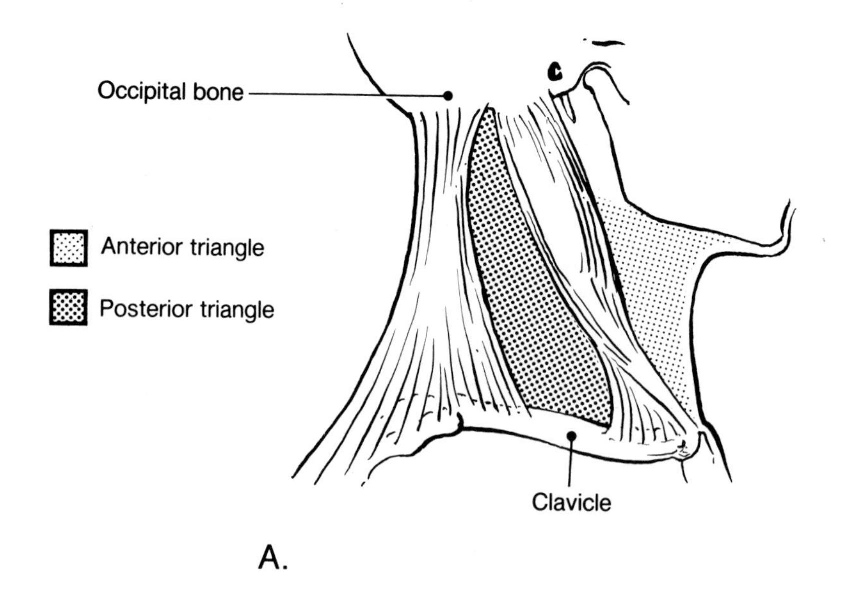

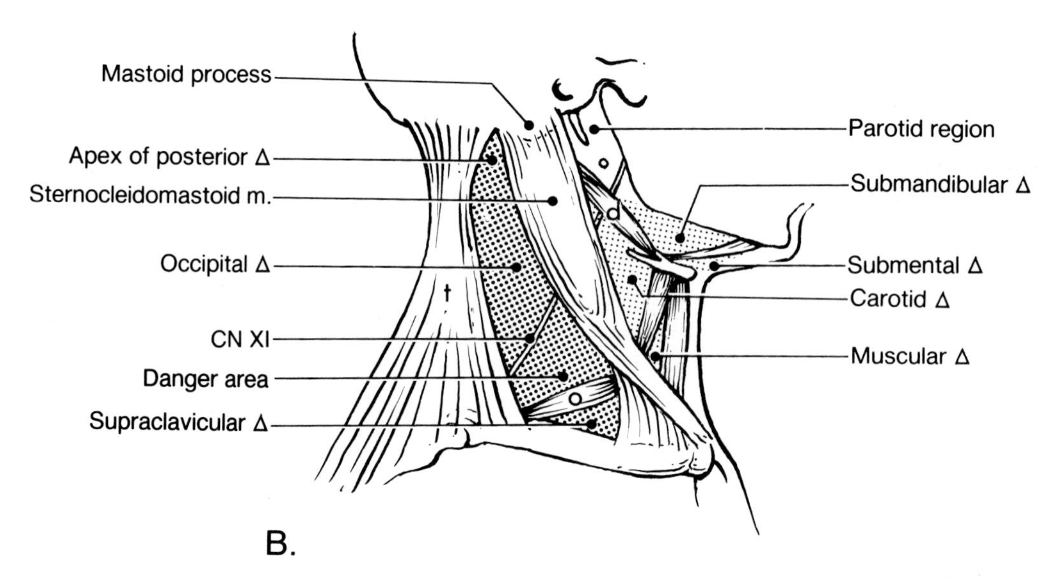

Figure 8-6. Drawings of the lateral aspect of the neck (right side) showing the triangles of the neck used for purposes of description. *A*, the boundaries of the posterior triangle are: the clavicles, the posterior border of the sternocleidomastoid muscle, and the anterior border of the trapezius muscle. The anterior triangle is bounded by the anterior border of the sternocleidomastoid muscle, the inferior margin of the mandible, and the anterior midline of the neck. In *B*, note that the anterior and posterior triangles are subdivided into smaller triangles. *O*, omohyoid. *d*, digastric; *t*, trapezius. Note the important accessory nerve (CN XI) dividing the posterior triangle into superior and inferior parts. The inferior part contains many important structures (Figs. 8-10 to 8-13); thus, careful dissection is required during surgery so that these structures will be preserved. Note that the course of the accessory nerve is marked by two points: one slightly superior to the middle of the sternocleidomastoid muscle, and one about 5 cm superior to the clavicle at the anterior border of the trapezius. Observe that the inferior belly of the omohyoid muscle divides the posterior triangle into a large occipital triangle (superiorly) and a smaller supraclavicular triangle (inferiorly). The subclavian artery (Fig. 8-12) may be compressed against the first rib by deep pressure in the supraclavicular triangle, which lies deep to the supraclavicular fossa (Fig. 8-1).

Insertion (Figs. 7-29 and 8-7). **Inferior border of the mandible and skin of the lower face.**

Nerve Supply (Fig. 8-10). **Cervical branch of the facial nerve** (CN VII).

Actions (Fig. 8-8). **Tenses the skin of the neck** (*e.g.*, during shaving), **draws the corners of the mouth inferiorly**, and assists in depressing the mandible.

When the entire muscle contracts, it wrinkles the skin of the neck in an oblique direction and widens the aperture of the mouth (Fig. 8–8).

The platysma is one of the muscles of facial expression we use to express sadness, horror, or fright. Men use this muscle when attempting to ease the pressure of a tight collar and when shaving. It also acts during violent deep inspiration (*e.g.*, as occurs after a 200 meter race).

Although the platysma is thin or absent in some people, it tenses the skin of the neck in most people. The platysma is well developed in animals and it is used to shake flies off their necks. When the platysma is very thin in man, the external jugular veins lie subcutaneously and are clearly visible, particularly when the intrathoracic pressure is raised (Fig. 8-33).

If a well developed platysma is paralyzed owing to injury of the cervical branch of the facial nerve (*e.g.*, owing to an invasion by a malignant lesion from the parotid or submandibular area, or by a laceration of the neck inferior to the angle of the mandible), the skin tends to fall away from the neck in slack folds. Hence, during surgical procedures performed in this region, *e.g.*, for removal of a branchial cyst (Fig. 8-77), care is taken to preserve the mandibular and cervical branches of the facial nerve (Figs. 7-29, 8-10, and 8-18).

Injury to the mandibular branch produces a noticeable facial deformity, whereas *damage to the cervical branch of the facial nerve produces unsightly postoperative defects (e.g., the skin of the neck droops).* Similarly, surgeons carefully suture the edges of the playtsma after neck surgery, *e.g.*, a thyroidectomy (removal of the thyroid gland), to prevent gaping of the skin incision.

THE STERNOCLEIDOMASTOID MUSCLE

The official name of this muscle is an informative one because it indicates its complete origin and insertion. The "*cleido*" part of its name is derived from the Greek word *kleis* meaning clavicle. Its old name "sternomastoid muscle" does not give a clue to the origin of its clavicular head.

The sternocleidomastoid muscle is the key muscular landmark in the neck. Running superolaterally from the sternum and clavicle to the lateral surface of the mastoid process, *the sternocleidomastoid muscle divides the side of the neck into anterior and posterior triangles*, which are useful for descriptive purposes (Fig. 8-6). Hence, swellings and other lesions of the neck can be described with reference to these **cervical triangles.** When it contracts, the sternocleidomastoid muscle stands out as a well defined prominence between the anterior and posterior triangles (Figs. 8-6 and 8-33).

The sternocleidomastoid (sternomastoid) muscle is crossed by the platysma and the external jugular vein (Fig. 8-10). **The sternocleidomastoid muscle covers the great vessels of the neck and the cervical plexus of nerves** (Figs. 8-10 and 8-11).

Origin (Figs. 8-1, 8-6, and 8-10). *Sternal head:* **anterior surface of the manubrium sterni.** *Clavicular head:* **superior surface of the medial third of the clavicle.**

Insertion (Figs. 7-3, 8-3, 8-6, and 8-12). Lateral surface of **the mastoid process of the temporal bone and the lateral half of the superior nuchal line of the occipital bone.**

Nerve Supply (Figs. 8-10 and 8-38). **Accessory nerve (CN XI)** and the ventral ramus of the **second cervical nerve (C2)**; sometimes the ventral ramus of C3 is also involved.

Actions. *Acting alone,* **the sternocleidomastoid tilts the head to its own side** (*i.e.*, laterally bends the neck) **and rotates it** so the face is turned superiorly toward the opposite side. *Acting together,* **the sternocleidomastoids flex the neck** (*e.g.*, when raising the head from a pillow against gravity). In performing this action, they assist the longus colli muscles (see p. 1014 and Fig. 8-28).

Occasionally the sternocleidomastoid muscle is injured at birth, resulting in a condition known as **congenital torticollis** or wryneck (Fig. 8-9). There is fixed rotation and tilting of the head owing to **contracture of the sternocleidomastoid muscle**. *The stiff neck results from fibrosis and shortening of the muscle on one side.* Because torticollis (L. *tortus*, twisted + *collum*, neck) is a readily correctable condition, it is rarely seen in a more advanced form than that illustrated in Figure 8-9.

Most cases of congenital torticollis result from

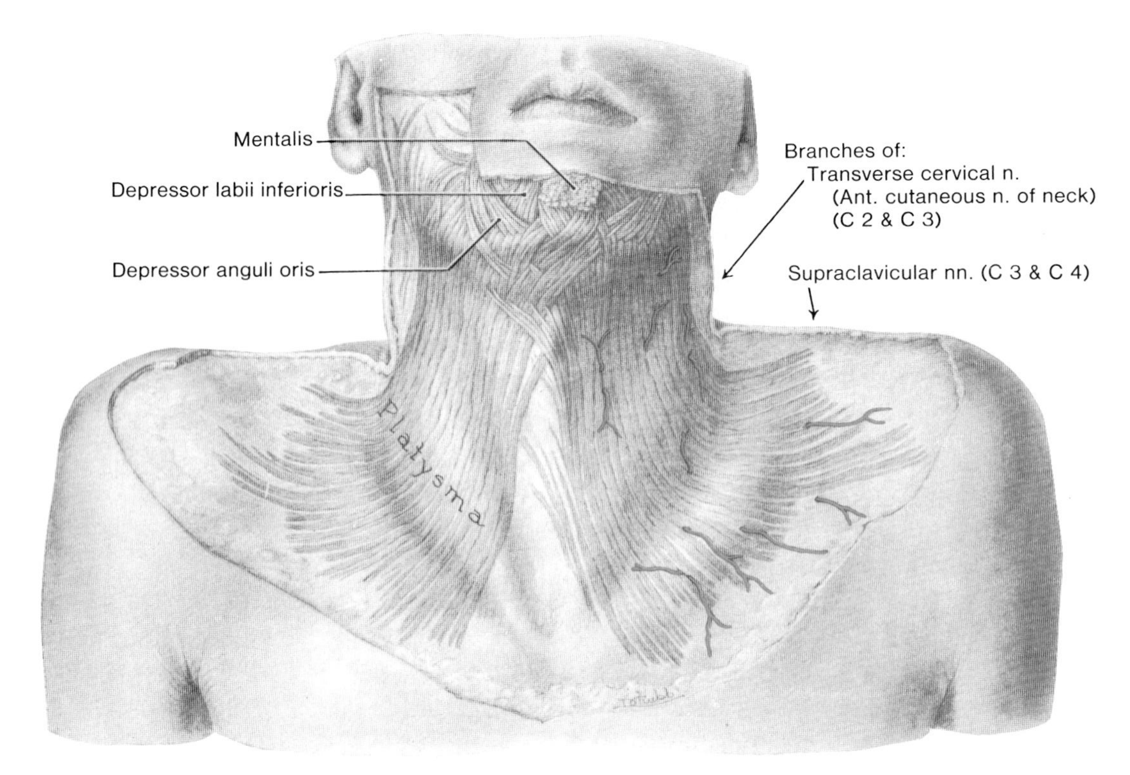

Figure 8-7. Drawing of a dissection of the platysma muscle. Observe that it spreads subcutaneously like a sheet and is pierced by the cutaneous nerves (also see Fig. 6-14). Note that it crosses the whole length of the inferior border of the mandible superiorly, and the entire length of the clavicle inferiorly, extending to the level of the first or second rib. Observe that the platysma does not cover the median part of the neck, but overlies the superior part of the anterior triangle (Fig. 8-6), and most of the posterior triangle.

tearing of fibers of the sternocleidomastoid when pulling the head excessively during a difficult birth, particularly in a breech presentation. Bleeding into the muscle may occur diffusely and/or in localized areas, the latter forming a small swelling (**hematoma**). Later a firm mass develops owing to necrosis of muscle fibers and the formation of fibrous tissue (**fibrosis**), which is part of the reparative or reactive process. Although this mass may disappear, it usually results in shortening of the muscle and torticollis by 3 to 4 years of age as the neck elongates.

Contracture of the sternocleidomastoid muscle causes the typical lateral bending of the head to the affected side and the slight turning away of the chin from the side of the short muscle (Fig. 8-9). Usually daily massaging, turning, and tilting of the head for several months (**physiotherapy**) stretches the affected muscle and corrects the condition, but in some cases surgery in addition to physiotherapy may be required to reduce the pull on the head. Failure to correct this abnormal condition results in asymmetry of the skull, distortion of the face on the affected side, and an inability to turn the head normally.

Spasmotic torticollis may develop in adults. It commences with tonic or clonic **spasm of one sternocleidomastoid muscle,** and is followed by a spasm of the trapezius muscle on the same side.

POSTERIOR TRIANGLE OF THE NECK

BOUNDARIES OF THE POSTERIOR TRIANGLE (Fig. 8-6)

The posterior cervical triangle is bounded: ***anteriorly*** *by the posterior border of the* ***sternocleidomastoid muscle*** (Figs. 8-6, 8-10, and 8-11); ***posteriorly*** *by the anterior border of the* ***trapezius muscle***; *and* ***inferiorly*** *by the middle third of the* ***clavicle*** (Figs. 8-6, 8-10, and 8-11).

The clavicle forms the base of the posterior triangle; *the apex of the posterior triangle is formed*

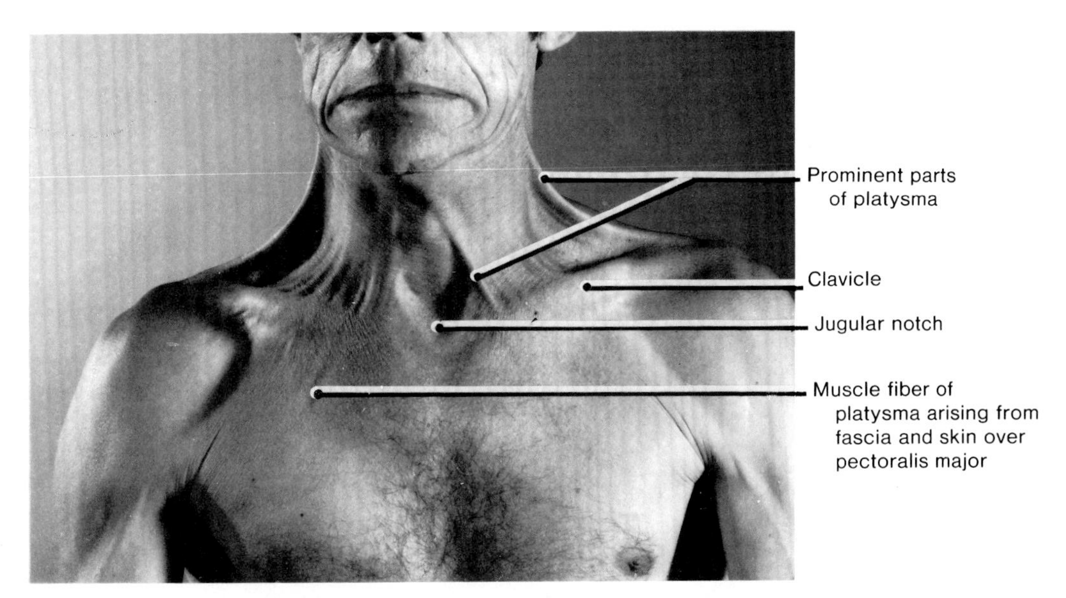

Figure 8-8. Photograph of a 46-year-old man showing wrinkling of the skin in an oblique direction when the entire platysma is in action, as occurs during violent clenching of the jaws and drawing down of the corners of the mouth. Observe the fibers arising from the fascia and skin over the superior pectoral and anterior deltoid regions, particularly on the right. Note that the sheet of muscle passes over the clavicle and the anterolateral aspects of the neck, and that the clavicle stands out during contraction of the platysma. This muscle of facial expression may be used to indicate sadness and fright. Men use it to tense the skin of the neck during shaving.

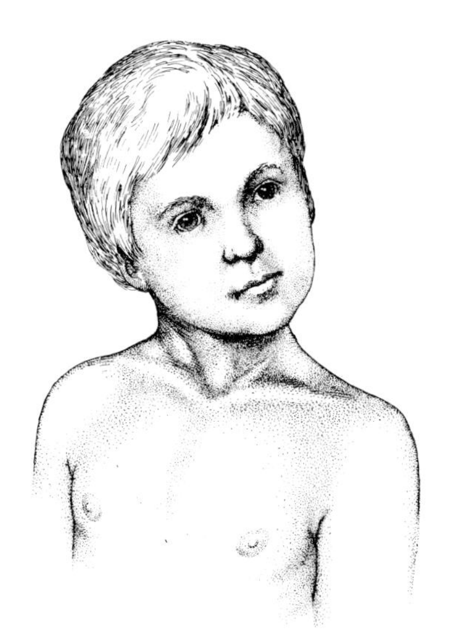

Figure 8-9. Drawing of an 8-year-old boy with congenital torticollis (wryneck). Contraction of the right sternocleidomastoid muscle, resulting from fibrosis, has drawn the head to the right and rotated it so that his chin points to the left.

where the borders of the sternocleidomastoid and trapezius muscles meet on the superior nuchal line (Figs. 7-3, 8-6*B*, and 8-10). Note that **the occipital artery passes through the apex of the posterior triangle**, before it ascends over the posterior aspect of the head (Figs. 7-37 and 8-10).

ROOF OF THE POSTERIOR TRIANGLE OF THE NECK (Figs. 8-7, 8-8, and 8-10)

The posterior cervical triangle is covered by the deep fascia which covers the space between the trapezius and sternocleidomastoid muscles. Superficial to the **deep fascial roof** are the superficial fascia, platysma, superficial veins, cutaneus nerves, and skin (Figs. 8-7 and 8-8).

FLOOR OF THE POSTERIOR TRIANGLE OF THE NECK (Figs. 7-38 and 8-10 to 8-13)

The muscular floor of the posterior cervical triangle is formed by the **splenius capitis, levator scapulae, scalenus medius, and scalenus posterior muscles.** The four muscles forming the floor are "carpeted" or covered by deep **cervical fascia.** This "fascial carpet" of the posterior triangle (Fig. 8-10) is a lateral prolongation of the *prevertebral fascia* (Figs. 8-30 and 8-35).

The splenius capitis muscle, the larger superior part of the splenius muscle (Fig. 5-40), **arises from the *ligamentum nuchae*** (Fig. 6-47) *and the spinous processes of the superior* ***thoracic vertebrae.*** *The splenius capitis* runs superolaterally and **inserts into the *mastoid process*** of the temporal bone *and the lateral third of the* ***superior nuchal line*** (Figs. 6-44 and 7-3).

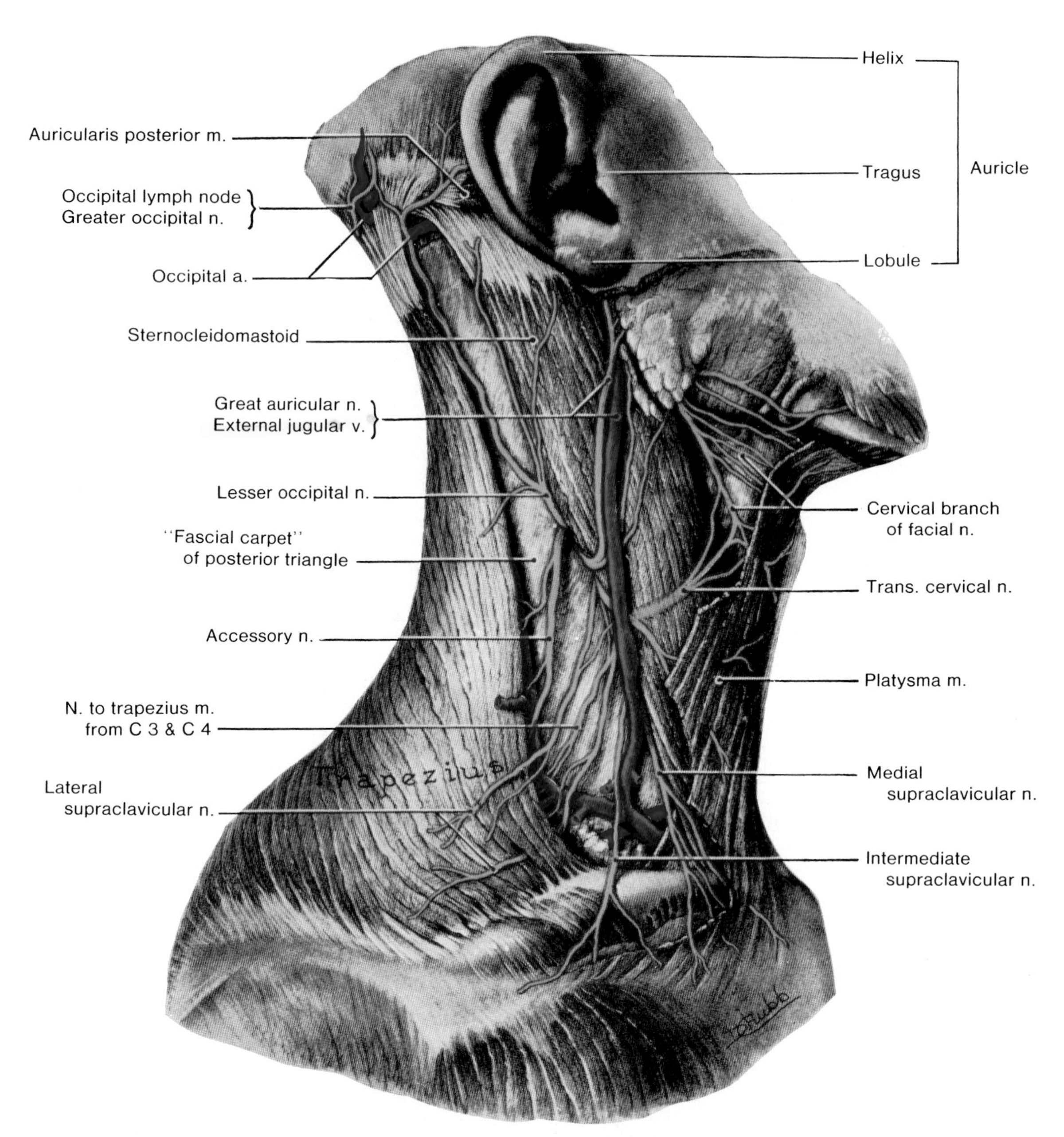

Figure 8-10. Drawing of a superficial dissection of the right side of the neck, primarily to illustrate the posterior triangle (Fig. 8-6). Observe its boundaries: the anterior border of trapezius, the posterior border of sternocleidomastoid, and the middle third of clavicle (its base). The apex of the posterior triangle is where the aponeuroses of the two muscles blend just inferior to the superior nuchal line (Figs. 6-43 and 7-3). The superficial fascia and the superficial layer of the deep fascia have been removed. The deep layer of cervical fascia forms the floor of the posterior triangle. This deep "fascial carpet" covers the muscular floor of the posterior triangle (Fig. 7-38). Note the accessory nerve, the only motor nerve superficial to the "fascial carpet", descending within the deep fascia and disappearing two fingerbreadths or more superior to the clavicle. Observe that the part of the platysma muscle which has been cut off covers the inferior part of the posterior triangle (also see Fig. 8-7).

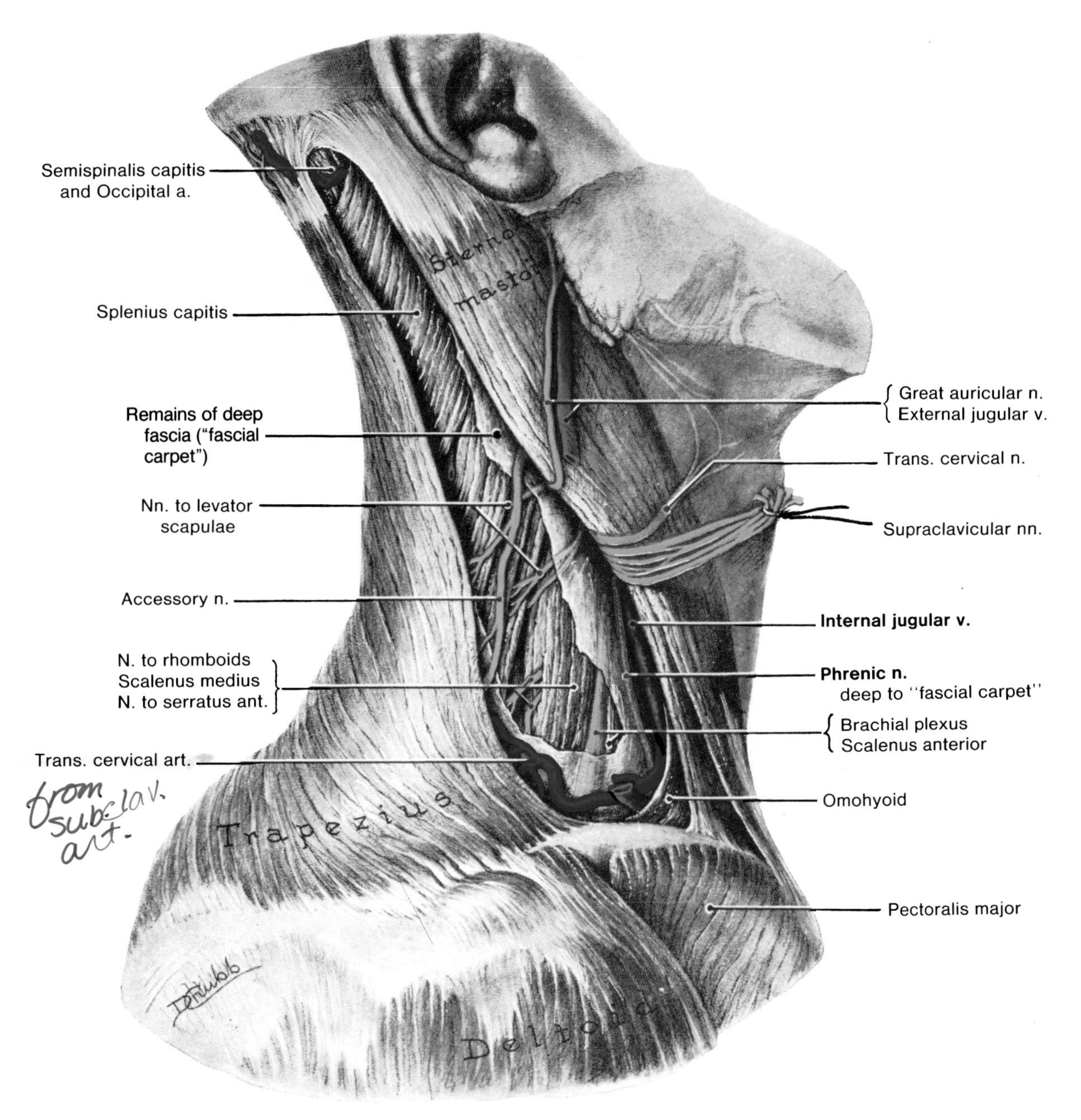

Figure 8-11. Drawing of a dissection of the right side of the neck showing the nerves deep to the "fascial carpet" on the floor of the posterior triangle, formed by the deep fascia. Observe the muscles forming the floor of the triangle (splenius capitis, levator scapulae, and scalenes). Also see Figure 7-38. Note the accessory nerve to the sternocleidomastoid and trapezius muscles, lying along the levator scapulae muscle, but separated from it by the "fascial carpet." Note the three motor nerves to upper limb muscles: to the levator scapulae (C3 and C4), to the rhomboids (C5), and to the serratus anterior (C5 and C6). *Observe the two structures of surgical importance* situated just beyond the geometrical confines of the posterior triangle: (1) the **phrenic nerve** to the diaphragm (C3, **C4**, and C5); and (2) the **internal jugular vein.**

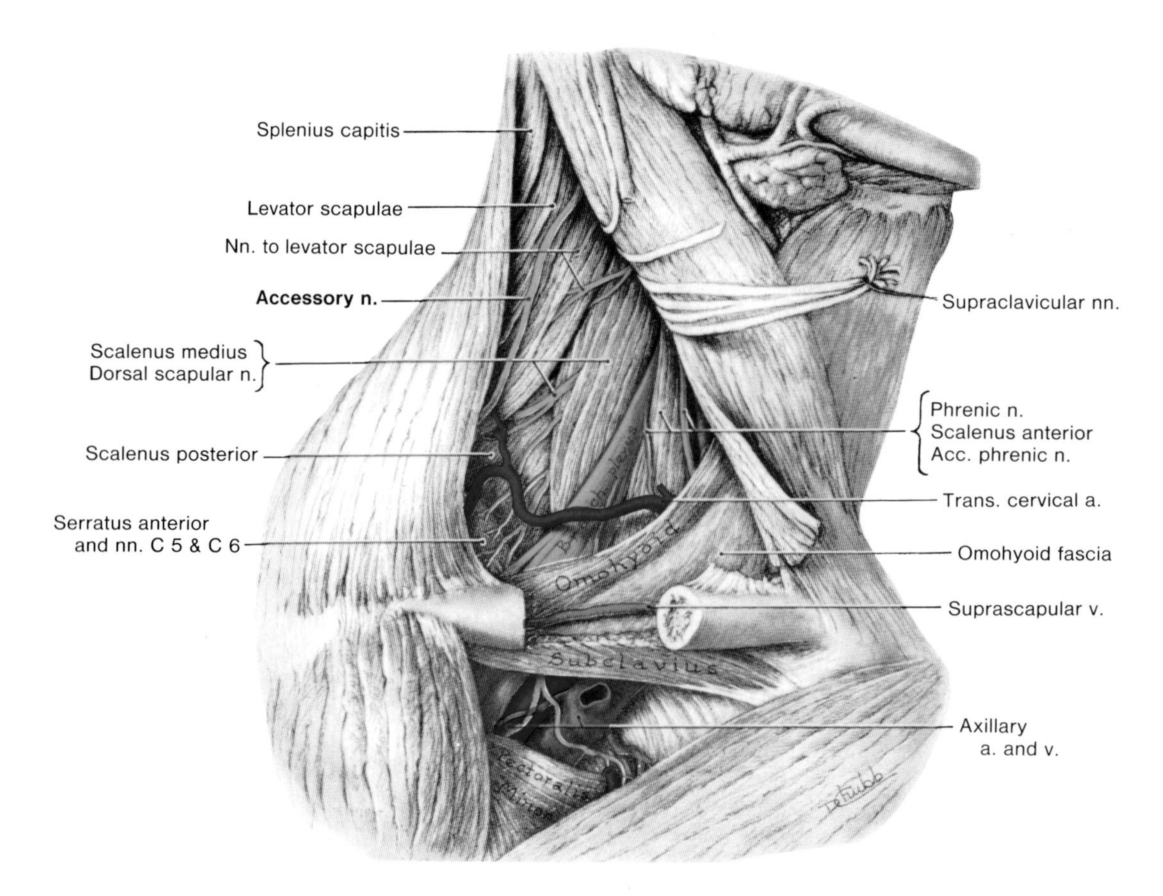

Figure 8-12. Drawing of a dissection of the right side of the neck and superolateral part of the thorax, showing the muscles forming the floor of the posterior triangle (also see Fig. 7-38). The deep fascia covering the floor (Fig. 8-10) and part of the clavicle have been removed. Note the phrenic nerve that supplies the diaphragm is closely related to the anterior surface of the scalenus anterior muscle. See Chapter 6 for more illustrations and a full description of the brachial plexus.

Acting together, *the splenius capitis muscles extend the neck and the head on it.* They also slightly rotate the head.

The levator scapulae muscle arises from the *transverse processes of the **first four cervical vertebrae***, passes deep to the trapezius muscle, and **inserts into the medial border of the scapula** (Fig. 6-46). As its name levator (L. lifter) indicates, *the levator scapulae muscle elevates the scapula.*

The scalenus posterior muscle arises from the *posterior tubercles of the transverse processes of the **fourth, fifth, and sixth cervical vertebrae***, and inserts into the external border of the **second rib.** *The scalenus posterior muscle elevates the second rib and flexes the cervical region of the vertebral column.*

The scalenus medius muscle arises from the *posterior tubercles of the transverse processes of **all cervical vertebrae***, and inserts into the superior surface of the posterior part of the **first rib.** *The scalenus medius muscle helps to elevate the first rib and flexes and rotates the cervical region of the vertebral column to the opposite side.* In Figure 8-12, note that ***the scalenus medius lies posterior to the roots of the brachial plexus*** and the third part of the subclavian artery. The brachial plexus is discussed in Chapter 6 (p. 639).

CONTENTS OF THE POSTERIOR TRIANGLE OF THE NECK

The posterior cervical triangle contains mostly vessels and nerves connecting the neck and upper limb.

Veins in the Posterior Triangle (Figs. 7-38, 8-10, and 8-11). The external jugular vein begins near the angle of the mandible, just inferior to the lobule of the auricle (Fig. 8-10), by the union of the posterior division of the retromandibular vein with the posterior auricular vein (Fig. 7-38).

The external jugular vein crosses the sternocleidomastoid muscle in the superficial fascia, and then pierces the fascial roof of the posterior tri-

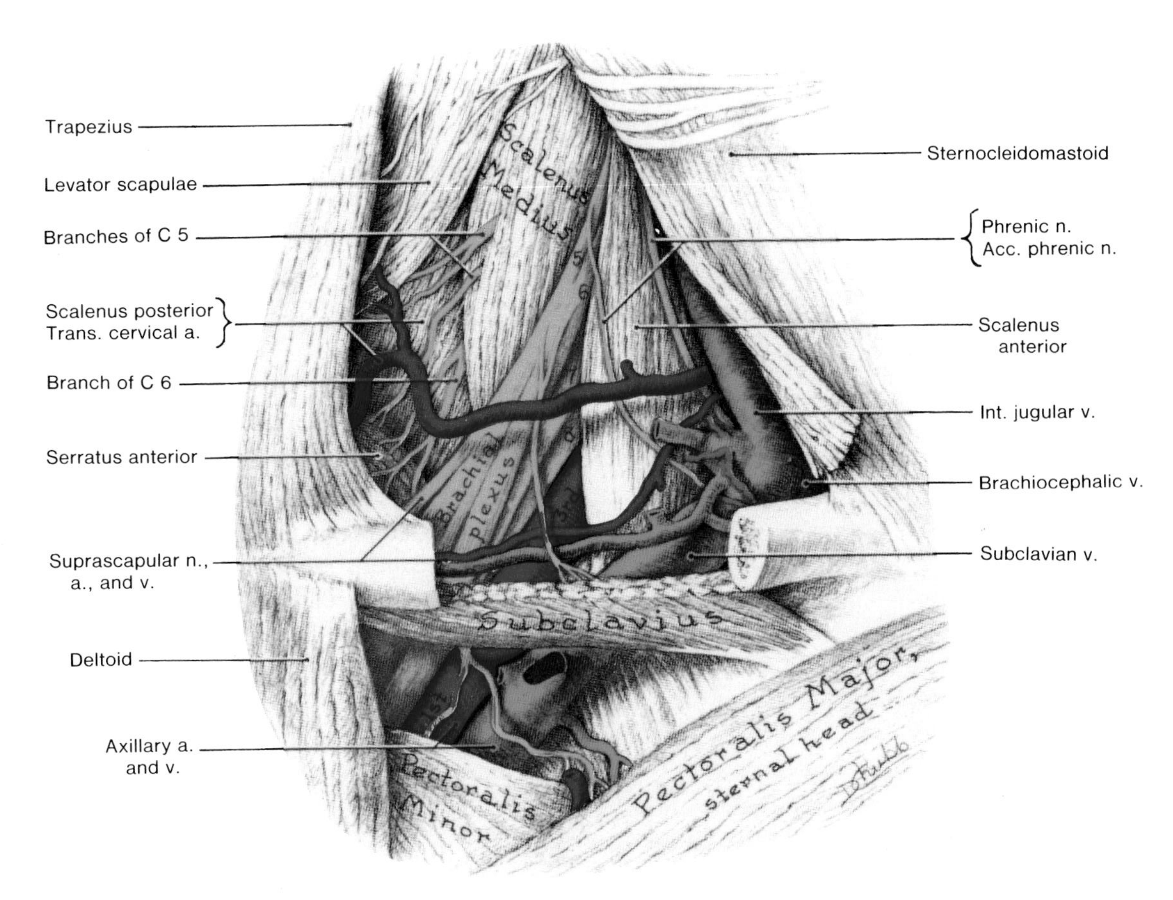

Figure 8-13. Drawing of a dissection of the right side of the neck and superolateral part of the thorax, showing the brachial plexus and parts of the subclavian vessels in the posterior triangle of the neck. Compare with Figure 8-12, noting that the omohyoid muscle and its fascia have been removed. Observe the third part of the subclavian artery, the first part of the axillary artery, and the muscles forming the floor of the inferior part of the triangle (scalene muscles and the first digitation of the serratus anterior). Note the brachial plexus and subclavian artery appearing between the scalenus medius and scalenus anterior muscles, and that the most inferior root of the plexus (T1) is concealed by the third part of the artery. Observe that the subclavian vein hardly rises superior to the level of the clavicle and is separated from the second part of the subclavian artery by the scalenus anterior. Note that the internal jugular vein (usually the largest in the neck) is not in the posterior triangle, but is very close to it.

angle at the posterior border of this muscle, about 5 cm superior to the clavicle. The external jugular vein passes obliquely through the inferior part of the posterior triangle, and usually ends by emptying into the **subclavian vein** about 2 cm superior to the clavicle (Figs. 7-38 and 8-10). The subclavian vein is the major venous channel draining the upper limb (Fig. 6-40). It lies posterior to the clavicle and so is not really in the posterior triangle. *The external jugular vein drains most of the scalp and face on the same side.*

When venous pressure is within the normal range, the external jugular vein is either invisible or observable for only a short distance superior to the clavicle (Fig. 8-33). However, *when venous pressure is raised* (*e.g.*, owing to **heart failure**), the external jugular vein becomes prominent throughout its course along the side of the neck (Fig. 8-10). Consequently, routine observation of this vessel during physical examinations may give diagnostic signs of heart failure, *obstruction of the superior vena cava* (*e.g.*, by a tumor), enlarged supraclavicular lymph nodes, or increased intrathoracic pressure.

Opera singers commonly exhibit bilateral enlargement of their external jugulars owing to the prolonged periods of high intrathoracic pressure required in their strenuous type of singing. Although uncommon, one or both external jugular veins may be absent. If present, they will become obvious when you compress their inferior ends just superior to the midpoints of the clavicles (Fig. 8-10), and when the person takes a deep breath and holds it (Fig. 8-33).

Increased blood volume resulting from the administration of too much intravenus fluid will cause engorgement of the external and anterior jugular veins (Fig. 7-38). For this reason, the neck of a patient receiving intravenous therapy is left uncovered so this sign of increased blood volume will be readily visible to all who care for the patient.

Should an external jugular vein be lacerated where it pierces the roof of the posterior triangle, along the posterior border of the sternocleidomastoid about 5 cm superior to the clavicle (Fig. 8-10), *air may be sucked into the vein during inspiration.* This may occur because the vein does not retract (collapse) at this site owing to the attachment of its walls to the deep fascia.

A **venous air embolism**, produced in this way, fills the right heart with froth and practically stops blood flow through it. This results in **dyspnea** (G. bad breathing), **cyanosis** (G. dark blue color), and sometimes death.

ARTERIES IN THE POSTERIOR TRIANGLE (Figs. 8-11 to 8-13, 8-26, 8-28, and 8-35)

The third part of the subclavian artery, the large vessel supplying blood to the upper limb, begins about a fingerbreadth superior to the clavicle, opposite the posterior border of the scalenus anterior muscle. As previously stated, it lies hidden in the anteroinferior part of the posterior triangle, barely qualifying as one of its contents.

The transverse cervical artery *arises from the thyrocervical trunk* (Figs. 8-11, 8-12, 8-28, and 8-35), a branch of the subclavian artery (Fig. 8-26). The transverse cervical artery runs superficially and laterally across the posterior triangle, 2 to 3 cm superior to the clavicle (Fig. 8-35), deep to the omohyoid muscle, to supply muscles in the scapular region (Fig. 6-39).

The suprascapular artery (Figs. 6-38, 8-12, and 8-35), *another branch of the thyrocervical trunk*, passes inferolaterally across the inferior part of the posterior triangle, just superior to the clavicle. It thens runs posterior to the clavicle to supply muscles around the scapula (Fig. 6-39). Not infrequently the suprascapular artery arises from the third part of the subclavian artery, lateral to the scalenus anterior muscle, and passes anterior to the inferior and middle trunks of the brachial plexus and posterior to the superior trunk.

The occipital artery (Figs. 7-37, 8-10, 8-11, and 8-26), *a branch of the external carotid artery*, enters the apex of the posterior triangle before ascending over the posterior aspect of the head to supply the posterior half of the scalp (Fig. 7-48).

NERVES IN THE POSTERIOR TRIANGLE (Figs. 8-6, 8-10 to 8-13, 8-28, 8-34, and 8-35)

The accessory nerve (CN XI) divides the posterior triangle into nearly equal superior and inferior parts. The superior part contains only the *lesser occipital nerve*, which supplies the scalp (Fig. 7-48), whereas the inferior part contains numerous important nerves, *e.g., the ventral primary rami of the brachial plexus.*

The accessory nerve (Figs. 7-64, 8-6*B*, 8-10, 8-11, and 8-13) enters the posterior triangle at or inferior to the junction of the superior and middle thirds of the posterior border of the sternocleidomastoid muscle (Figs. 8-10 and 8-11).

The accessory nerve passes posteroinferiorly through the posterior triangle, and then disappears deep to the anterior border of the trapezius muscle at the junction of its superior two-thirds with its inferior one-third (Fig. 8-6B).

The accessory nerve* (CN XI) is a motor nerve consisting of *spinal and cranial roots (Fig. 7-64).

The spinal root of CN XI is composed of fibers which arise from the cervical segments of the spinal cord (C1 to C5), and pass superiorly in the subarachnoid space to enter the posterior cranial fossa through the foramen magnum. Here it joins the *cranial root of the accessory nerve* (Fig. 7-72), the fibers of which originate in the medulla of the brain stem. Both roots leave the skull through the jugular foramen (Figs. 7-60 and 8-29).

The spinal root of the accessory nerve separates immediately from the cranial root and passes posteroinferiorly. ***The spinal root of CN XI supplies the sternocleidomastoid muscle*** (Figs. 8-10 and 8-11). It then crosses the posterior triangle, superficial to the deep fascia covering its floor ***and supplies the trapezius muscle***. It passes deep to this muscle about 5 cm superior to the clavicle.

Lesions of the accessory nerve are rare (Table 7-3). CN XI may be damaged by: (1) traumatic injury; (2) tumors at the base of the skull; (3) fractures involving the jugular foramen; and (4) neck lacerations. Although contraction of one sternocleidomastoid muscle turns the head to one side, a unilateral lesion of CN XI usually does not produce an abnormality in the position of the head. However, weakness in turning the head to the opposite side against resistance can be detected in patients with a lesion of CN XI.

Unilateral paralysis of the trapezius muscle is evident by the patient's inability to elevate and retract the shoulder, and by difficulty is elevating the arm superior to the horizontal level. The normal nuchal ridge formed by the trapezius muscle is also depressed and **drooping of the shoulder is an obvious sign of injury to the spinal root of CN XI** (Table 7-3).

The functions of CN XI may also be interfered with by inflamed lymph nodes in the neck (Fig. 8-60). This can cause acute **torticollis** (wryneck), or drawing of the head to one side. During extensive dissections in the posterior triangle of the neck (Figs. 8-10 to 8-12), *e.g.*, for the removal of malignant lymph nodes, the accessory nerve is isolated in order to preserve it.

The Cervical Plexus (Figs. 8-10, 8-11, and 8-28). *This is a network of nerves formed by the communications between the ventral rami of the superior four cervical nerves.* The cervical plexus lies deep to the internal jugular vein and the sternocleidomastoid muscle (Fig. 8-18), immediately ***deep to the accessory nerve,*** **CN XI** (Figs. 8-11 to 8-13).

Cutaneous branches from the cervical plexus emerge around the middle of the posterior border of the sternocleidomastoid to supply the skin of the neck and scalp, between the auricle and external occipital protuberance (Fig. 7-32, A and B).

The following nerves are derived from the ventral rami of C2 to C4 via the cervical plexus.

The lesser occipital nerve (C2) ascends a short distance along the posterior border of the sternocleidomastoid muscle (Fig. 8-10), before dividing into several branches that supply the skin of the neck and scalp posterior to the auricle (Fig. 7-32*A*). *The greater occipital nerve is not in the posterior triangle* (Fig. 8-10). It is from the dorsal ramus of C2 and supplies the skin on the posterior aspect of the scalp (Fig. 7-48).

The great auricular nerve (C2 and C3) curves over the posterior border of the sternocleidomastoid muscle, and ascends vertically toward the **parotid gland** (Figs. 7-29, 8-10, and 8-11). The great auricular nerve supplies branches to the skin of the neck, and then divides into anterior and posterior branches that supply the skin on the posterior aspect of the auricle and an area extending from the mandible to the mastoid process (Figs. 7-29, 7-32, and 8-10).

The transverse cervical nerve (transverse nerve of the neck), from **C2 and C3**, passes transversely across the middle of the sternocleidomastoid muscle to supply the skin over the anterior triangle of the neck (Figs. 7-32, 8-6*A*, 8-10, and 8-11).

The supraclavicular nerves (C3 and C4) arise as a single trunk (Figs. 8-10 and 8-11) which divides into medial, intermediate, and lateral branches. They send small branches to the skin of the neck and then pierce the deep fascia, just superior to the clavicle, to supply the skin over the anterior aspect of the chest and shoulder (Figs. 6-14 and 8-10). The medial and lateral supraclavicular nerves also supply the sternoclavicular and acromioclavicular joints, respectively (Figs. 6-128 and 6-129).

The Phrenic Nerve, *the sole motor nerve supply to the diaphragm* (Fig. 1-87), arises from the ventral primary rami of the *third,* ***fourth,*** *and fifth cervical nerves. The phrenic nerve, an important muscular branch of the cervical plexus,* curves around the lateral border of the scalenus anterior muscle (Figs. 8-11 and 8-12). It then descends obliquely across its anterior surface, deep to the transverse cervical and suprascapular arteries (Figs. 8-12 and 8-35). The phrenic nerve enters the thorax by crossing the origin of the internal thoracic artery, between the subclavian artery and vein (Figs. 1-46, 8-12, and 8-32).

The contribution to the phrenic nerve from the fifth cervical nerve may be derived from an **accessory phrenic nerve**. It lies lateral to the main phrenic nerve (Fig. 8-12) and descends posterior to, or sometimes inferior to, the subclavian vein. The accessory phrenic nerve, really a part of the phrenic, joins it in the root of the neck or in the thorax.

Severance of the phrenic nerve in the neck results in complete paralysis and atrophy of all the muscle of the corresponding half of the diaphragm. If an *accessory phrenic nerve* is present (Fig. 8-12), severance or crushing of the main nerve only will not result in complete paralysis of the corresponding half of the diaphragm. Injury to both nerves causes **dyspnea**. Other lesions in the neck irritate the phrenic nerves causing **hiccups** (singultus).

Close to their origin the nerves of the cervical plexus receive rami communicantes, most of which descend from the **superior cervical ganglion** which is located in the superior part of the neck (Fig. 8-28). The rami may pass through the muscles and be difficult to identify. Injury to the sympathetic trunk or the superior cervical trunk results in the **Horner syndrome** (p. 1022).

The supraclavicular part of the brachial plexus is located in the posterior triangle, lying immediately anterior to the scalenus medius muscle and the first digitation of the serratus anterior muscle (Figs. 8-11, 8-12, and 8-28). The infraclavicular part of this important plexus of nerves to the upper limb is located in the axilla (Figs. 6-20 and 6-22).

The brachial plexus is formed by the ventral primary rami of the inferior four cervical and the first thoracic nerves. There is often a small contribution from the fourth cervical and second thoracic nerves. The plan of the brachial plexus and the nerves derived from it are discussed in Chapter 6 (p. 640).

Branches of the ventral primary rami of cervical nerves supply the rhomboid, serratus anterior, and nearby prevertebral muscles. Along the lateral border of the brachial plexus, the **suprascapular nerve** (Fig. 8-12) runs across the posterior triangle to supply the supraspinatus and infraspinatus muscles (Fig. 6-57), which join the scapula to the humerus (Table 6-4).

The **nerve point of the neck** is in the region around the midpoint of the posterior border of the sternocleidomastoid muscle. Several nerves lie superficially here, deep to the platysma (Figs. 8-10 and 8-11).

Slash wounds of the neck may sever these relatively superficial nerves, resulting in loss of cutaneous sensation in the neck and posterior part of the scalp (Fig. 7-32, *A* and *B*). When extensive surgical dissections are done in the posterior triangle of the neck, the accessory nerve is located, if present, and isolated in order to preserve it.

For regional anesthesia prior to surgery, nerve blocks (temporary arrest of nerve impulses) are performed by injecting anesthetic solutions around the nerves of the cervical and brachial plexuses (Figs. 8-10 and 8-12).

Cervical Plexus Block (Figs. 8-10 to 8-13). The anesthetic solution is injected at several points along the posterior border of the sternocleidomastoid muscle. The main site of injection is at the junction of the superior and middle thirds of this muscle, *i.e.*, around the nerve point of the neck, where the accessory nerve begins its passage across the posterior triangle (Figs. 8-6*B* and 8-10). Because the phrenic nerve (supplying half the diaphragm) is usually paralyzed by a cervical nerve block (Fig. 8-12), they are not performed on patients with pulmonary and/or cardiac disease.

Brachial Plexus Block (Figs. 8-10 to 8-13). For anesthesia of the upper limb, the anesthetic solution is injected around the brachial plexus of nerves. The main site of injection is superior to the midpoint of the clavicle (Figs. 6-22 and 8-12); the needle is directed medially and inferiorly toward the first rib. The subclavian artery (Fig. 8-12) is located by palpation prior to making the injections in order to avoid entering it.

Muscles in the Posterior Triangle (Figs. 7-38, 8-11 to 8-13, and 8-17 to 8-20). The muscles forming the floor of the posterior triangle have been discussed (p. 991). The slender inferior belly of the strap-like **omohyoid muscle** (G. *omos*, shoulder), to be described with the infrahyoid muscles (Fig. 8-17 and Table 8-2), passes within the fascial roof of the anterior and posterior triangles (Figs. 8-6 and 8-35). The inferior belly of the omohyoid runs one to two fingerbreadths superior to the clavicle to which it is attached by a fascial sling. *The omohyoid muscle is an important landmark in the neck; it can often be seen contracting when thin people speak.*

SUBDIVISIONS OF THE POSTERIOR TRIANGLE OF THE NECK

The inferior belly of the omohyoid muscle divides the posterior triangle into a large ***occipital triangle*** superior to it (Fig. 8-6*B*), and a much smaller ***supraclavicular triangle*** inferior to it.

The occipital triangle was given this name because its apex contains a portion of the occipital bone (Fig. 8-6), and possibly because the occipital artery appears in the superior part of this triangle. *The most important nerve crossing the occipital triangle is the accessory nerve (CN XI).*

The supraclavicular triangle (Fig. 8-6*B*), the much smaller division of the posterior triangle, is indicated on the surface of the neck by the supraclavicular fossa (Fig. 8-1). **The external jugular vein crosses the supraclavicular triangle superficially** (Fig. 8-10), and **the subclavian artery lies deep in it** (Fig. 8-12). These vessels are covered by the omohyoid fascia (Fig. 8-35). Because of the presence of the subclavian artery, you may hear the supraclavicular triangle called the *subclavian triangle.*

Pressure in the supraclavicular fossa will occlude blood flow in the subclavian artery in the supraclavicular triangle, where it passes over the superior surface of the first rib (Fig. 8-28). In the event of **hemorrhage in the upper limb**, pressure in this fossa will control the bleeding because the subclavian artery supplies blood to vessels of the upper limb.

The accessory nerve may be used to divide the

posterior triangle into a carefree area superiorly and a danger area inferiorly (Fig. 8-6*B*). Care is essential during surgical dissection inferior to the accessory nerve because of the presence of many important vessels and nerves in the area (Figs. 8-10 to 8-13).

ANTERIOR TRIANGLE OF THE NECK

BOUNDARIES OF THE ANTERIOR TRIANGLE (Fig. 8-6)

The anterior cervical triangle is bounded by the *anteromedian line of the neck, the inferior border of the mandible, and the anterior border of the sternocleidomastoid muscle.*

The apex of the anterior triangle *is at the jugular notch,* and its **base** is formed by the *inferior border of the mandible* and a line drawn from the angle of the mandible to the mastoid process (Figs. 8-1, 8-3, and 8-6).

THE HYOID MUSCLES

The hyoid bone is held in place by several muscles that are attached to the mandible, the skull, the thyroid cartilage, the manubrium sterni, and the medial end of the scapula (Figs. 8-15, 8-17, and Tables 8-1 and 8-2).

The hyoid muscles are primarily concerned with steadying or moving the hyoid bone and the larynx. For purposes of description, they are divided into suprahyoid and infrahyoid muscles.

The Suprahyoid Muscles (Figs. 7-108, 8-14 to 8-17, and Table 8-1). *As their group name indicates, these muscles lie superior to the hyoid bone.* They include the mylohyoid, geniohyoid, stylohyoid, and digastric muscles. The suprahyoid muscles connect the hyoid bone to the skull.

The Mylohyoid Muscles (Figs. 7-108 and 8-14 to 8-17) are thin, flat muscles that form a sling inferior to the tongue which supports the floor of the mouth. The Greek word *mylē* means "a mill" and denotes the relationship of this muscle to the mouth where the food is ground.

Origin (Figs. 7-22 and 8-17). **Mylohyoid line on the medial aspect of the mandible.**

Insertion (Figs. 7-141 and 8-19). **Raphe and body of the hyoid bone.**

Nerve Supply (Figs. 7-33*A* and 8-19). **Mylohyoid nerve,** a branch of the inferior alveolar nerve.

Actions. Elevates the hyoid, floor of mouth, and tongue in swallowing and speaking. Palpate these muscles in the floor of your mouth as you press the tip of your tongue against your maxillary incisor teeth.

The Geniohyoid Muscles (Figs. 7-66, 8-14, and 8-16) are short narrow muscles that contact each other in the midline. This pair of muscles lies superior to the mylohyoid muscles and reinforces the floor of the mouth.

Origin (Figs. 7-22 and 8-14). **Inferior mental spine of the mandible.**

Insertion (Fig. 7-141). **Body of the hyoid bone.**

Nerve Supply (Fig. 7-66). Ventral ramus of first cervical nerve, *i.e.*, **C1 via hypoglossal nerve** (CN XII).

Actions. Pulls the hyoid anterosuperiorly, thereby shortening the floor of the mouth and widening the pharynx for receiving food during swallowing.

The Stylohyoid Muscles (Figs. 7-108, 8-15, and 8-16), form a small slip on each side, which is nearly parallel to the posterior belly of the digastric muscle.

Origin (Figs. 7-10, 7-108, and 8-15). **Styloid process of the temporal bone.**

Insertion (Fig. 7-141). **Body of the hyoid bone.**

Nerve Supply (Figs. 7-35 and 8-18). **Facial nerve (CN VII).**

Actions. Elevates and retracts the hyoid bone, thereby elongating the floor of the mouth in swallowing.

The Digastric Muscle (Figs. 8-5, 8-6, 8-14 to 8-16, and 8-17) has **two bellies** (G. *gastēr*, belly or stomach). The bellies are joined by an intermediate tendon.

Origin and Insertion (Figs. 7-22, 8-15, and 8-19). *The anterior belly* of the digastric muscle arises from the **digastric fossa** on the internal surface of the inferior border of the mandible, close to the symphysis menti (Fig. 7-1).

The posterior belly of the digastric muscle arises from the mastoid notch on the medial side of the mastoid process of the temporal bone. The two bellies descend toward the hyoid bone and are joined by an **intermediate tendon.** This tendon is connected to the body and greater horn of the hyoid bone by a strong loop or sling of fibrous connective tissue (Figs. 8-5 and 8-17). This fibrous pulley allows the tendon to slide anteriorly and posteriorly.

Nerve Supply (Figs. 7-33 and 7-35). **The two bellies of the digastric muscle are supplied by different nerves:** the *anterior belly is supplied by the mylohyoid nerve,* a branch of the ***inferior alveolar nerve,*** and the *posterior belly is supplied by the* ***facial nerve*** *(CN VII).*

This difference in nerve supply of the two bellies of the digastric muscle results from the embryological origin of the anterior and posterior bellies from the first and second branchial arches, respectively. CN V is the nerve to the first branchial arch and CN VII supplies the second branchial arch.

Actions. Acting together both bellies of the digastric muscle raise the hyoid bone and steady it during swallowing and speaking.

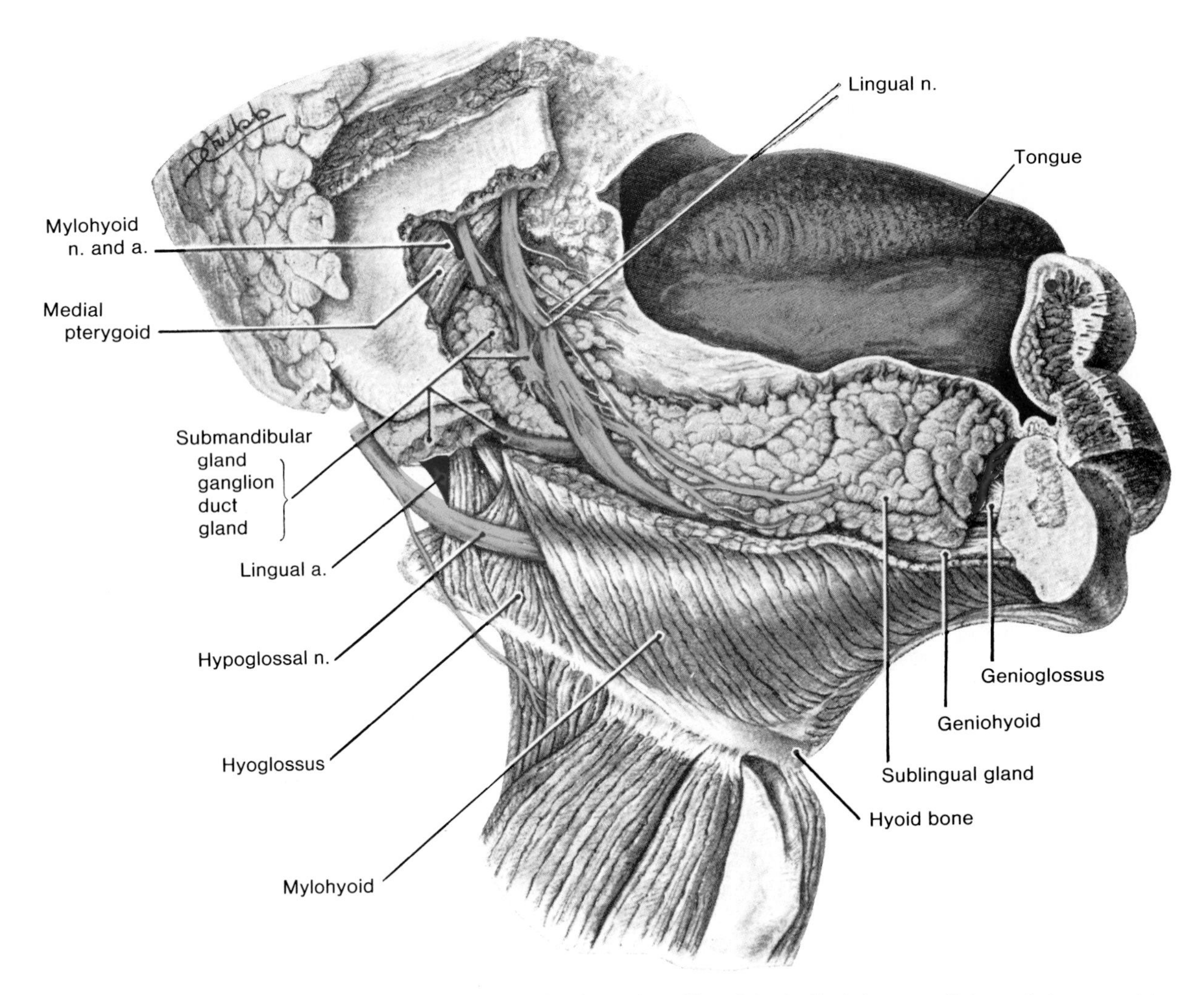

Figure 8-14. Drawing of a dissection of the suprahyoid region. The right half of the mandible and the superior part of the mylohyoid muscle have been removed. Observe that the cut surface of the mylohyoid muscle becomes progressively thinner as it is traced anteriorly. Note the mylohyoid nerve and artery (cut short) and the lingual nerve (clamped) between the medial pterygoid muscle and the ramus of the mandible.

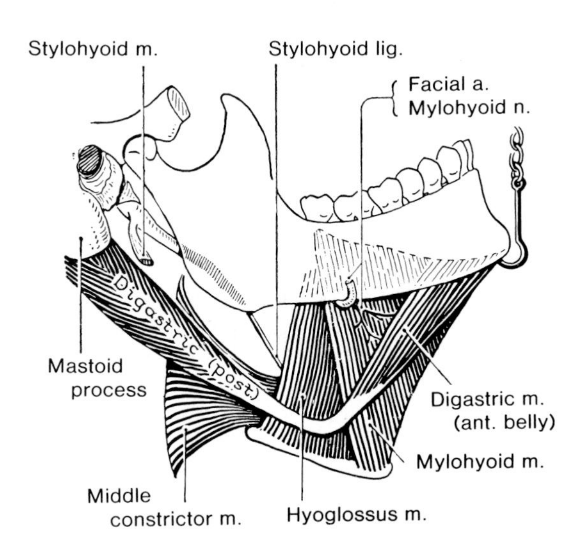

Figure 8-15. Drawing of the suprahyoid muscles showing that they are arranged in layers (also see Table 8-1). Observe the facial artery hooking around the inferior border of the mandible to enter the face (also see Figs. 7-31 and 7-37).

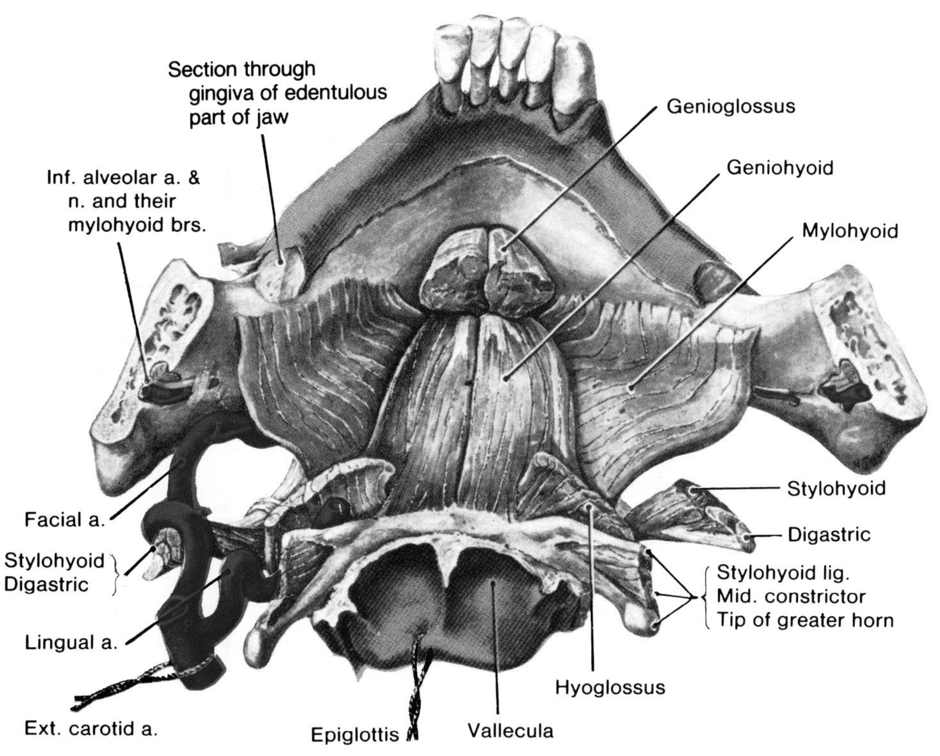

Figure 8-16. Drawing of a dissection of the muscles of the floor of the mouth. Observe the paired, triangular geniohyoid muscles occupying a horizontal plane, with their apex at the mental spine of the mandible (Fig. 7-22), and their base at the body of the hyoid bone. The two mylohyoid muscles form a muscular floor for the oral cavity which supports the tongue, sublingual glands, and other structures shown in Figure 8-14. Note the mylohyoid muscle arising from the whole length of the mylohyoid line of the mandible (also see Fig. 7-22).

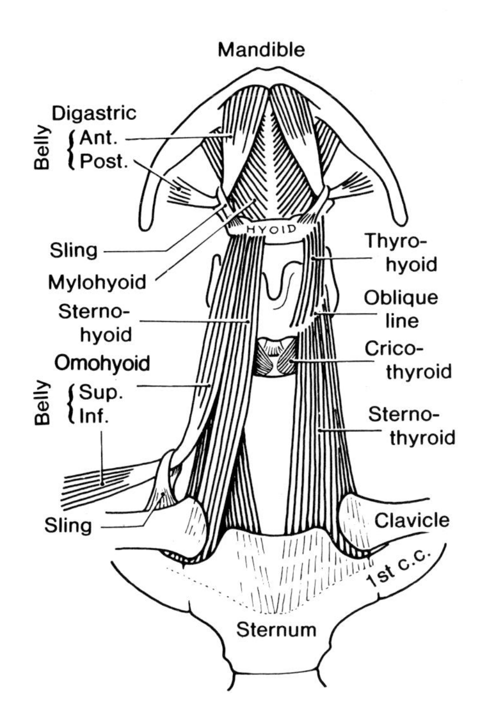

Figure 8-17. Drawing of the hyoid muscle, mainly to illustrate the infrahyoid muscles (often called the strap muscles). Note that there are three flattened strap muscles in the anterior triangle (also see Fig. 8-6*B* and Table 8-2).

Table 8-1
The Suprahyoid Muscles[1]

Muscle	Origin	Insertion	Nerve Supply	Actions
Mylohyoid (Figs. 8-14 to 8-16)	Mylohyoid line of mandible (Figs. 7-22 and 8-17)	Raphe and body of hyoid bone (Fig. 7-141)	Mylohyoid n., a branch of inf. alveolar n. (Figs. 7-33 and 8-19)	Elevates hyoid, floor of mouth, and tongue
Geniohyoid (Figs. 8-14 and 8-16)	Inf. mental spine (genial tubercle) of mandible (Fig. 7-22)	Body of hyoid bone (Fig. 7-141)	C1 via hygoglossal n. (CN XII)	Pulls hyoid anterosuperiorly and shortens floor of mouth
Stylohyoid (Figs. 7-108, 8-15, and 8-16)	Styloid process of temporal bone (Figs. 7-10 and 8-15)	Body of hyoid bone (Fig. 7-141)	Facial n. (CN VII), cervical branch (Fig. 8-18)	Elevates and retracts hyoid
Digastric (Figs. 7-108, 8-15, and 8-17 to 8-20)	*Ant. belly*: digastric fossa of mandible (Fig. 7-22) *Post. belly*: mastoid notch of temporal bone (Figs. 8-15 and 8-19)	Intermediate tendon to body and greater horn of hyoid bone (Figs. 8-15 and 8-17)	*Ant. belly*: mylohyoid n., branch of inf. alveolar n. (Figs. 7-33 and 8-19) *Post. belly*: facial n. (Fig. 7-35)	Depresses mandible Raises hyoid and steadies it

[1] The suprahyoid muscles connect the hyoid bone to the skull (Fig. 8-17).

Acting posteriorly, they open the mouth and depress the mandible. To feel the hyoid being elevated by the digastric muscles, grasp your hyoid between your thumb and index finger as you swallow.

The Infrahyoid Muscles (Figs. 8-17 to 8-20, 8-22, and Table 8-2). Because of their ribbon-like appearance, **the infrahyoid muscles are often called the strap muscles.** The infrahyoid muscles anchor the hyoid bone.

As their group name indicates, they lie inferior to the hyoid bone. All four of them act to depress the hyoid bone and larynx during swallowing and speaking.

The Sternohyoid Muscle (Figs. 8-17, 8-18, and 8-25) is a thin narrow strap muscle. It is superficial, except inferiorly, where it is covered by the sternocleidomastoid muscle.

Origin (Fig. 8-17). **Posterior surface of the manubrium sterni and the medial end of the clavicle.**

Insertion (Fig. 7-141). **Inferior border of the body of the hyoid bone.**

Nerve Supply (Figs. 7-66 and 8-19). Ventral rami of first three cervical nerves, *i.e.*, **C1 to C3 via the ansa cervicalis** (L. *ansa*, a loop), a slender nerve root in the cervical plexus.

Action. Depresses the hyoid bone and larynx after it has been elevated during swallowing.

The Sternothyroid Muscle (Figs. 8-17 and 8-19) is a thin muscle located deep to the sternohyoid muscle. It is shorter and wider than the sternohyoid muscle.

Origin (Fig. 8-17). **Posterior surface of the manubrium sterni** and the first costal cartilage.

Insertion (Figs. 8-2*A* and 8-17). **Oblique line of the thyroid cartilage.**

Nerve Supply (Figs. 7-66 and 8-19). Branches from the **ansa cervicalis (C1 to C3).**

Action. Depresses hyoid bone and larynx after it has been elevated during swallowing and vocal movements. The sternothyroid pulls the thyroid cartilage away from the hyoid bone, thereby opening the laryngeal orifice.

The Thyrohyoid Muscle (Figs. 8-17 and 8-19) appears as the superior continuation of the sternothyroid muscle.

Origin (Figs. 8-2*A* and 8-17). **Oblique line of the thyroid cartilage.**

Insertion (Figs. 7-141 and 8-17). **Inferior border of the body and greater horn of the hyoid bone.**

Nerve Supply (Fig. 7-66). Ventral ramus of **first cervical nerve via the hypoglossal nerve** (CN XII).

Actions. Depresses the hyoid bone and elevates the thyroid cartilage. The thyrohyoid muscle is mainly responsible for closing the laryngeal orifice and in preventing food from entering the larynx during swallowing.

The Omohyoid Muscle (Figs. 8-6, 8-12, 8-17 to 8-20, and 8-35) has **two bellies** which are *united by an intermediate tendon.* It is connected to the clavicle by a **fascial sling** (Fig. 8-17). The prefix *omo* in this

Table 8-2
The Infrahyoid Muscles[1]

Muscle	Origin	Insertion	Nerve Supply	Actions
Sternohyoid (Figs. 8-17, 8-18, and 8-25)	Manubrium sterni and medial end of clavicle	Body of hyoid bone (Figs. 7-141 and 8-17)	C1, C2, and C3 from ansa cervicalis (Fig. 8-19)	Depresses hyoid and larynx
Omohyoid (Figs. 8-6, 8-13, 8-17 to 8-20, and 8-35)	Superior border of scapula near suprascapular notch	Inferior border of hyoid bone (Figs. 7-141 and 8-17)	C2 and C3 from ansa cervicalis (Fig. 8-19)	Depresses, retracts, and steadies hyoid
Sternothyroid (Figs. 8-17 and 8-19)	Posterior surface of manubrium sterni	Oblique line of thyroid cartilage (Figs. 8-2 and 8-17)	C1 to C3 from ansa cervicalis (Fig. 8-19)	Depresses hyoid and larynx
Thyrohyoid (Figs. 8-17 and 8-19)	Oblique line of thyroid cartilage (Figs. 8-2 and 8-17)	Inferior border of body and greater horn of hyoid bone (Figs. 7-141 and 8-17)	C1 via hypoglossal n. (CN XII)	Depresses hyoid and elevates thyroid cartilage

[1] The infrahyoid muscles are four strap-like muscles that anchor the hyoid bone (*i.e.*, they fix and steady it). They are concerned with the suprahyoid muscles (Fig. 8-17 and Table 8-1) in movements of the tongue, hyoid bone, and larynx in both swallowing and speaking.

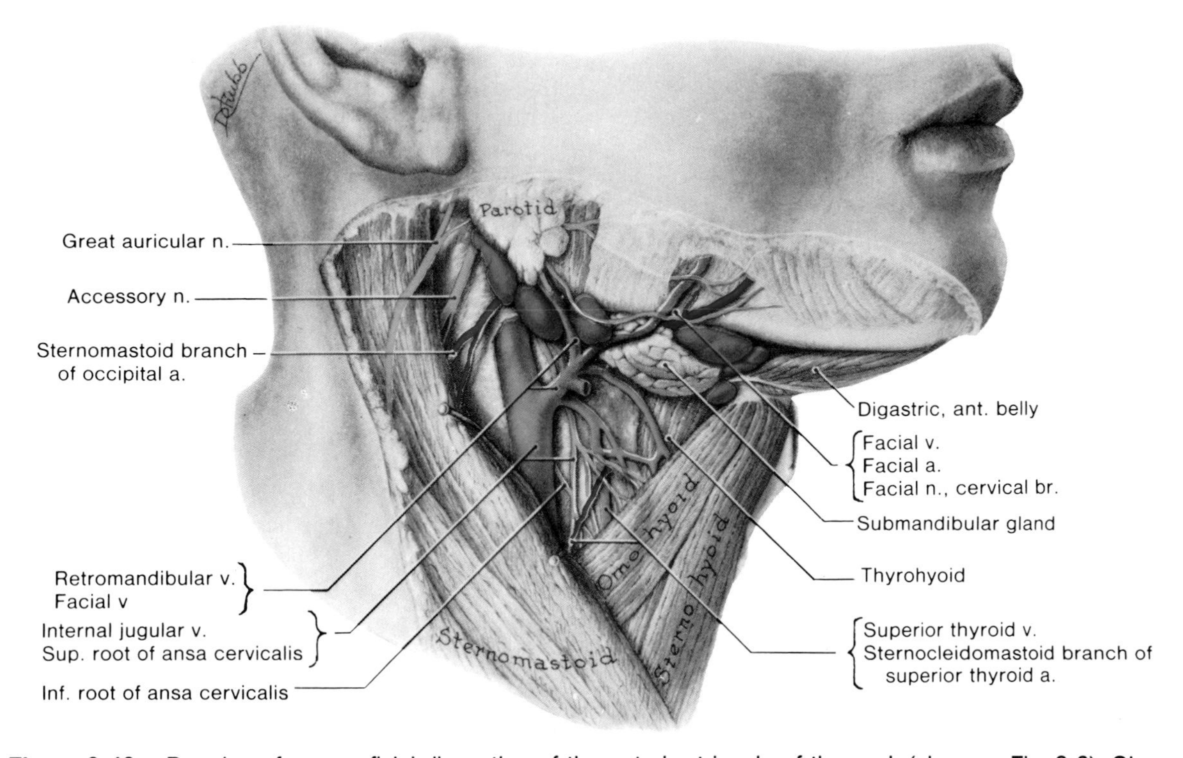

Figure 8-18. Drawing of a superficial dissection of the anterior triangle of the neck (also see Fig. 8-6). Observe the submandibular gland, lymph nodes (*green*), and the cervical branch of the facial nerve in the submandibular triangle. Note the retromandibular and facial veins running superficial to the submandibular gland (also see Fig. 7-37).

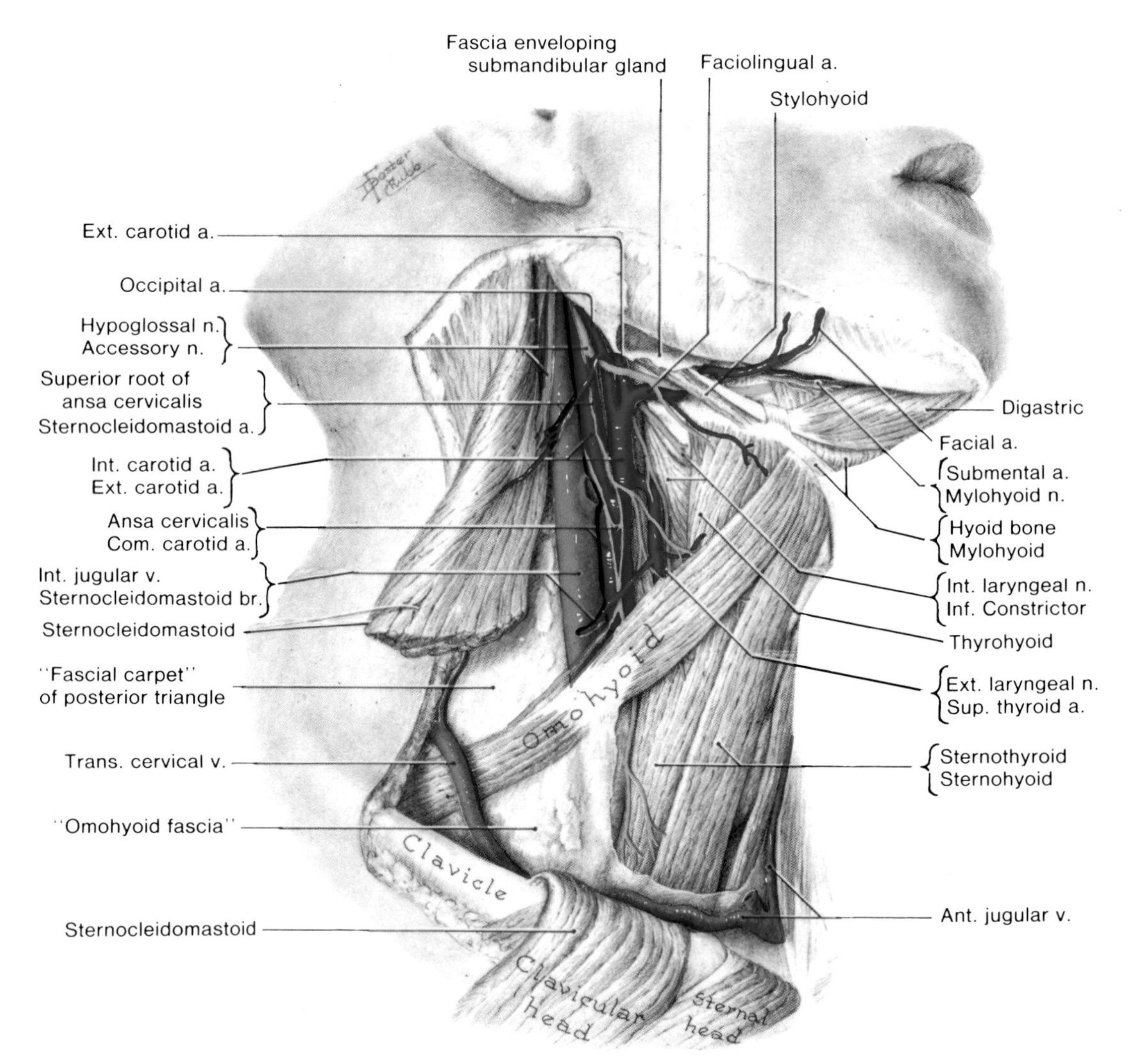

Figure 8-19. Drawing of a deep dissection of the right side of the neck showing the anterior triangle. Note that the intermediate tendon of the digastric muscle is attached to the hyoid bone by a fascial sling (also see Fig. 8-5), and the intermediate tendon of omohyoid muscle is connected to the clavicle by a sling (also see Fig. 8-17). Observe the facial and lingual arteries, here arising by a common stem, passing deep to the stylohyoid and digastric muscles to enter the submandibular triangle. Examine the common carotid artery in the carotid triangle and its branches in the superior end of it. Also note the hypoglossal nerve curving in and out of the carotid triangle and passing deep to the digastric muscle as it passes through the digastric triangle on its way to the tongue muscles. (For a description and drawing of the ansa cervicalis, see Fig. 7-66).

muscle's name is derived from the Greek word meaning shoulder. The clavicle and scapula form the pectoral or shoulder girdle (Figs. 6-1 and 6-3).

The omohyoid muscle is an important landmark in the neck (Fig. 8-6), and it divides the posterior triangle into occipital and supraclavicular triangles.

Origin and Insertion (Figs. 7-141, 8-6, 8-18, and 8-19). The *inferior belly arises from* **the superior border of the scapula** near the suprascapular notch. It ends in the **intermediate tendon**, from which the *superior belly* arises; this belly inserts into the **inferior border of hyoid bone.**

Nerve Supply (Figs. 7-66, 8-18, and 8-19). Both bellies are supplied by branches of the **superior root of the ansa cervicalis** (C1) and by the ansa cervicalis itself (C2 and C3).

Actions. Depresses, retracts, and steadies the hyoid bone in swallowing and speaking. In thin per-

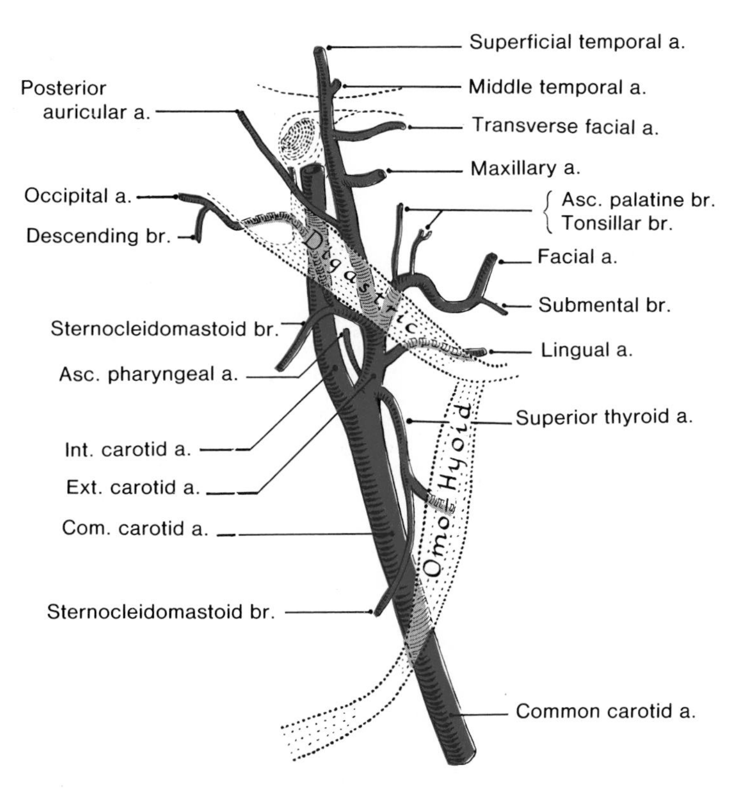

Figure 8-20. Diagram of the carotid arteries and their branches in the right side of the neck (for better orientation, see Fig. 8-19). Note that the common carotid divides into two terminal branches, the internal and the external carotid arteries, and that neither the common carotid nor the internal carotid arteries gives off branches in the neck. The carotid triangle (Fig. 8-6*B*) is of considerable surgical significance because it contains these parts of the carotid system of arteries.

sons, the inferior belly of the omohyoid muscle can often be seen contracting when they are speaking.

SUBDIVISIONS OF THE ANTERIOR TRIANGLE OF THE NECK (Fig. 8-6*B*)

In addition to the *unpaired submental triangle*, each anterior triangle is divisible into three smaller triangles: *submandibular, carotid, and muscular.*

The Submandibular Triangle (Figs. 8-6*B*, 8-18, and 8-19). This ***glandular area*** on each side lies between the inferior border of the mandible and the anterior and posterior bellies of the digastric muscle. For this reason, you may hear it referred to as the digastric triangle.

The submandibular gland nearly fills the submandibular triangle (Figs. 8-14 and 8-18). It wraps itself around the free posterior border of the mylohyoid muscle, not unlike a capital letter U on its side (⊂). This thin sheet of muscle separates the superficial and deep parts of the submandibular gland.

The submandibular gland is about half the size of the parotid gland (Fig. 7-41), and is usually palpable as a soft mass between the body of the mandible and the mylohyoid muscle (Fig. 8-18). It is easily felt when you tense your mylohyoid muscle by forcing the tip of your tongue against your maxillary incisor teeth.

The submandibular gland may become inflamed along with the parotid gland (*e.g.*, owing to mumps). You can see the swellings formed by the enlarged glands. Because of the location of these glands (Figs. 7-41 and 8-18), it becomes painful to open the mouth and to eat.

In Figure 8-18, note the position of the submandibular lymph nodes. They lie on the submandibular gland and along the inferior border of the mandible (Fig. 7-40). Be certain you know what a normal submandibular gland feels like so you will later be able to differentiate it from enlarged submandibular lymph nodes, *e.g.*, owing to **metastases from a lip cancer** (Fig. 7-44). Tense your platysma as in Figure 8-8 and verify that the submandibular gland is deep to this superficial muscle.

The submandibular duct, about 5 cm in length, passes from the deep process of the gland (Figs. 7-41 and 8-14), parallel to the tongue, to open by one to three orifices into the oral cavity. The orifices are on the **sublingual papilla** at the side of the lingual frenulum (Figs. 7-124, 7-142, and 7-144).

The hypoglossal nerve (CN XII) which is motor to the intrinsic and extrinsic muscles of the tongue, passes into the submandibular triangle, (Figs. 8-14 and 8-19). The **mylohyoid nerve**, a branch of the inferior alveolar nerve (Fig. 7-33), and parts of the facial artery and vein also pass through the submandibular triangle (Fig. 8-18).

The Carotid Triangle (Figs. 8-6 and 8-18 to 8-20). *This vascular area is bounded by the superior belly of* **the omohyoid**, the posterior belly of **the digastric**, *and the anterior border of* **the sternocleidomastoid muscle.**

The carotid triangle is an important area because the ***common carotid artery*** *ascends into it.* Its pulse can be auscultated (L. to listen to) with a stethoscope or it can palpated by placing the index finger in the triangle and compressing the artery lightly against the transverse processes of the cervical vertebrae. At the level of the superior border of the thyroid cartilage, ***the common carotid artery divides within the carotid triangle*** **into the internal and external carotid arteries** (Figs. 8-5 and 8-26).

The Carotid Sinus (Figs. 8-21, 8-26, and 8-31) is a slight dilation of the proximal part of the internal carotid artery; it may involve the common carotid. The carotid sinus, ***a blood pressure regulating area***, is innervated principally by the glossopharyngeal nerve (CN IX) through a branch, the **carotid sinus nerve** (Fig. 7-111). The carotid sinus is also supplied by the vagus nerve (Fig. 7-135) and the sympathetic division of the autonomic nervous system.

The carotid sinus reacts to changes in arterial blood pressure and effects appropriate modifications reflexly.

The *carotid pulse* (neck pulse) is routinely checked in **cardiopulmonary resuscitation** (CPR), because it is easily felt in the side of the neck. The carotid artery lies in a groove created by the trachea and the strap muscles (Fig. 8-19 and Table 8-2).

The rescuer puts the tips of the fingers gently on the trachea, then slides the fingers to the side nearest him/her, gently pressing toward the trachea. If there is a pulse, it can easily be felt. **The absence of a carotid pulse indicates cardiac arrest and a lift-threatening situation.**

Although the common carotid artery can be occluded by compressing it against the carotid tubercle of the sixth cervical vertebra (Fig. 8-31), *e.g.*, for control of hemorrhage in the neck, it is recommended that you not practice this on your colleagues because **syncope** may result (*i.e.*, the person may faint owing to deficiency of blood to the brain, **cerebral anemia**).

The carotid sinus responds to an increase in arterial pressure (Fig. 8-21), slowing the heart owing to parasympathetic outflow from the brain via the vagus nerve (Fig. 7-135). Pressure on the carotid sinus may cause syncope, and if the person happens to have a **supersensitive carotid sinus**, it may cause cessation of the heart beat (temporary or permanent).

In elderly people with asymptomatic occlusion of one internal carotid artery owing to ***atherosclerosis*** (a common type of **arteriosclerosis or hardening of the arteries**), occlusion of the common carotid on the other side will deprive the brain of a major source of blood and result in unconsciousness in 10 to 12 sec. As soon as the pressure is released, the patient should regain consciousness, but a permanent neurological deficit is likely to be produced if the artery is occluded for several minutes.

Carotid Endarterectomy. Atherosclerotic thickening of the intima of arteries supplying the brain will cause partial or complete obstruction of the blood flow. The resulting symptoms depend on the vessel obstructed, the degree of obstruction, and the amount of collateral blood flow from other vessels, *e.g.*, from those of ***the cerebral arterial circle*** (Fig. 7-77).

The obstruction can be relieved partly or completely by opening the artery (**arteriotomy**), and stripping off the offending plaque with the adjacent intima. A common site for a carotid endarterectomy is the **internal carotid artery**, just superior to its origin (Fig. 8-20). After the operation, drugs are used to inhibit clot formation at the operated area until the endothelium has regrown.

During a carotid endarterectomy, two nerves in the carotid triangle are in danger of injury: **the vagus and recurrent laryngeal nerves** (Figs. 8-19, 8-24, 8-27, 8-41, and 8-75). Damage to these nerves may produce an alteration in the voice. Temporary paralysis of these nerves may occur as the result of postoperative edema.

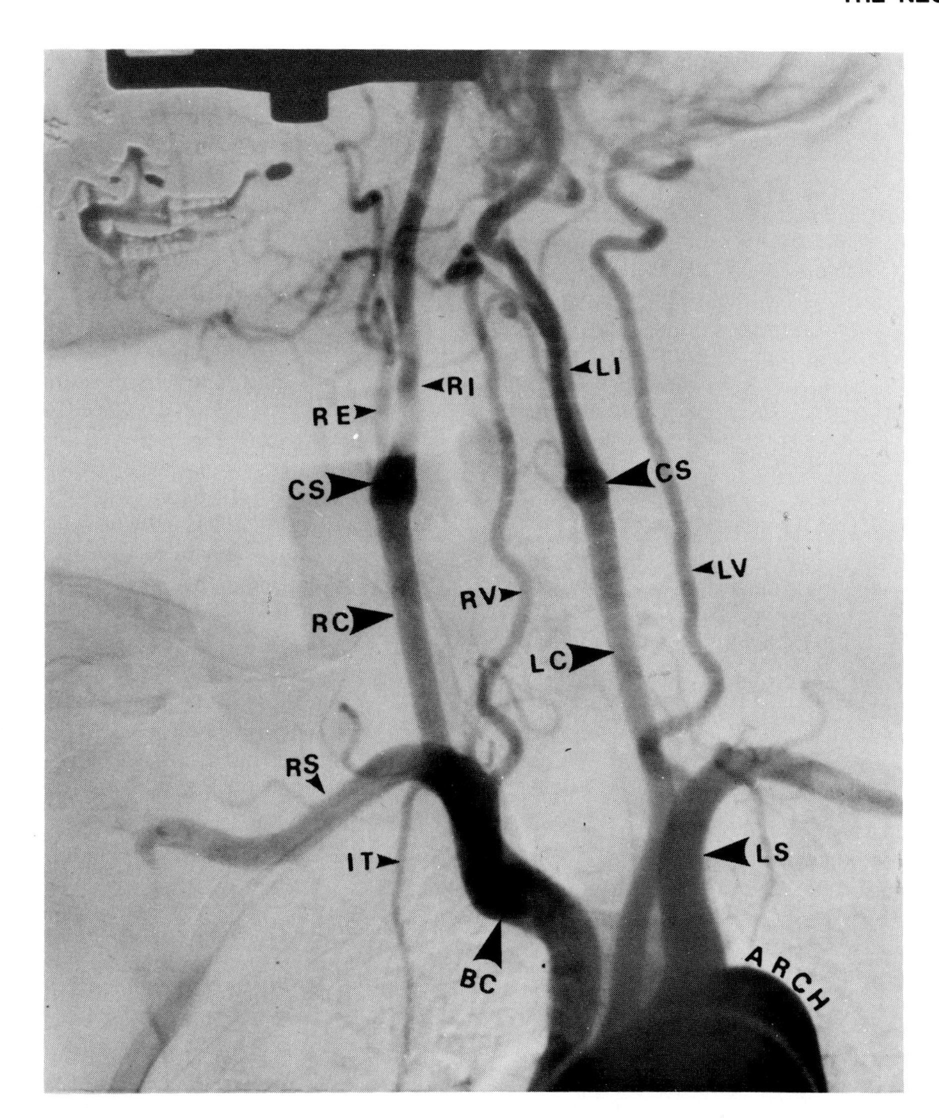

Figure 8-21. Arteriogram of the neck. The arteries were injected with a radiopaque material before the film was taken. Note that the brachiocephalic artery (*BC*) divides into right subclavian (*RS*) and right common carotid (*RC*) arteries, and that the common carotid artery divides into right external (*RE*) and right internal (*RI*) carotid arteries. Observe the internal thoracic artery (*IT*) arising from the subclavian arteries as is the right vertebral artery (*RV*). *On the left* observe the common carotid (*LC*), internal carotid (*LI*), and the tortuous course of the left vertebral artery (*LV*). On both sides observe the carotid sinus (*CS*). Distention of the wall of this pressure receptor stimulates nerve endings in it which results in a reflex slowing of the heart and a fall in blood pressure. When this radiograph was taken, the patient was lying supine in a right posterior oblique position (*i.e.*, right scapula on table and left shoulder elevated) with the face to the right. *ARCH* indicates the arch of the aorta (Fig. 1-73). Compare this arteriogram with Figure 7-37.

The Carotid Body (Fig. 8-22). This is a small, reddish-brown, ovoid mass of tissue located at the bifurcation of the common carotid artery, in close relation to the carotid sinus (Fig. 7-111).

The carotid body is a chemoreceptor that responds to changes in the chemical composition of the blood. It is supplied mainly by the **carotid sinus nerve**, a branch of CN IX (Fig. 7-111), but is also supplied by CN X (Fig. 7-135) and sympathetic fibers.

The carotid body responds to either increased carbon dioxide tension or to decreased oxygen tension in the blood. A fall in the oxygen content or an increase in the carbon dioxide content of the blood circulating through the carotid body initiates reflexes through the glossopharyngeal and vagus nerves that stimulate respiration.

The Carotid Sheath (Figs. 8-23, 8-36, and 8-41) is a tubular, fascial condensation that extends from the

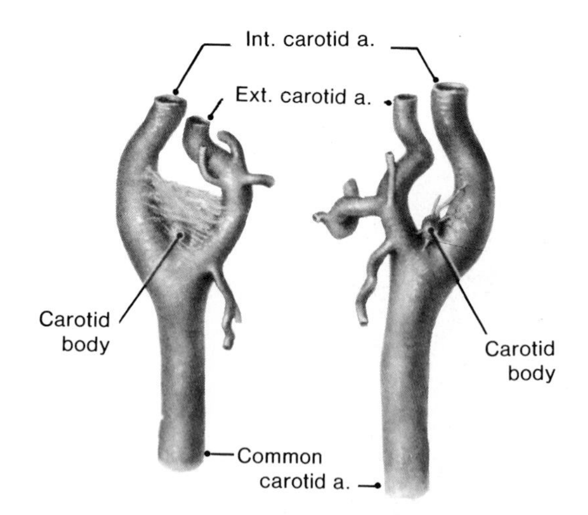

Figure 8-22. Drawing of two stages of a posterior view of a dissection of the carotid body. This particular body, black from engorged superficial veins, was easily recognized. Normally it is reddish-brown *in vivo.* The carotid body acts as a chemoreceptor, responding to changes in the chemical composition of the blood.

base of the skull to the root of the neck. It is formed by fascial extensions of the cervical fascia which fuse with the prevertebral fascia (Fig. 8-23).

The carotid sheath encloses several clinically important structures: (1) *the common and internal carotid arteries* medially; (2) *the internal jugular vein* laterally; and (3) the *vagus nerve (CN X)* posteriorly. The superior root of the **ansa cervicalis** (Figs. 8-18 and 8-19) descends between the common carotid artery and the internal jugular vein; it is sometimes embedded in the carotid sheath.

Many **deep cervical lymph nodes** (Figs. 8-35 and 8-41) *lie along the carotid sheath and internal jugular vein,* and between this vein and the common carotid artery.

The cervical part of the ganglionated ***sympathetic trunk*** *runs posterior to the carotid sheath* (Figs. 8-24 and 8-30).

The Muscular Triangle (Figs. 8-6*B*, 8-17, 8-19, and 8-24). *This triangle is bounded by the superior belly of the omohyoid muscle* (separating it from the carotid triangle), *the anterior border of the sternocleidomastoid muscle, and the midline of the neck. The muscular triangle contains the four infrahyoid muscles and the neck viscera.*

The Submental Triangle (Figs. 8-6*B*, 8-17 to 8-19, and 8-22). *This is an unpaired area* that is bounded inferiorly by the body of the hyoid bone and laterally by the right and left anterior bellies of the digastric muscles.

The floor of the submental triangle is formed by the two ***mylohyoid muscles***, which meet in a median fibrous raphe (Figs. 8-14 and 8-25).

The apex of the submental triangle is at the inferior end of the symphysis menti, and its base is formed by the hyoid bone (Figs. 8-17 and 8-25).

The submental triangle contains the submental lymph nodes (Figs. 8-25 and 8-60). They receive lymph from the tip of the tongue, the floor of the mouth, the

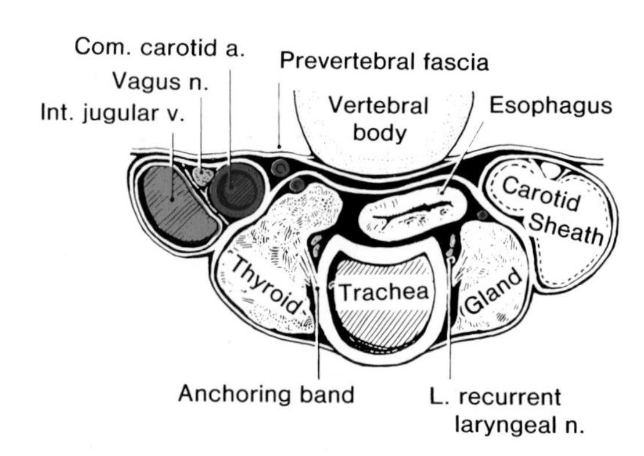

Figure 8-23. Drawing of a cross-section of the neck showing structures in its anterior part. Observe in particular the carotid sheath and its contents. This tubular condensation of cervical fascia is thicker around the artery than the vein. Observe that it both separates and encloses the common carotid artery, the internal jugular vein, and the vagus nerve (CN X). (Also see Fig. 8-41).

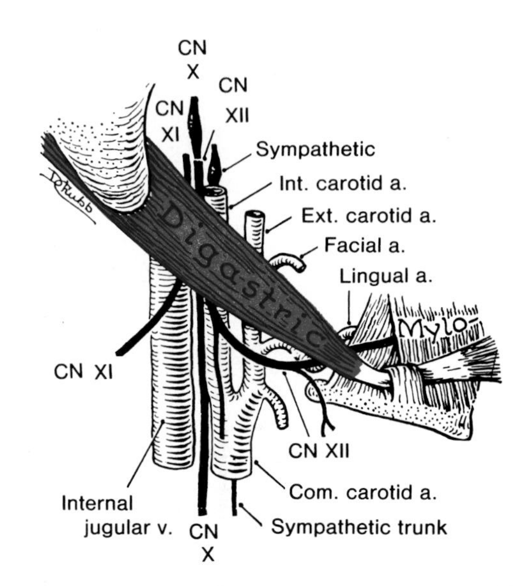

Figure 8-24. Drawing of structures in the carotid triangle (Fig. 8-6*B*), related to the posterior belly of the digastric muscle. The carotid sheath has been removed. Note the key position of the posterior belly of the digastric muscle, running from the mastoid process to the hyoid bone, deep to the angle of the mandible. Observe that all the vessels and nerves cross deep to the posterior belly of this muscle. (Also see Fig. 8-19).

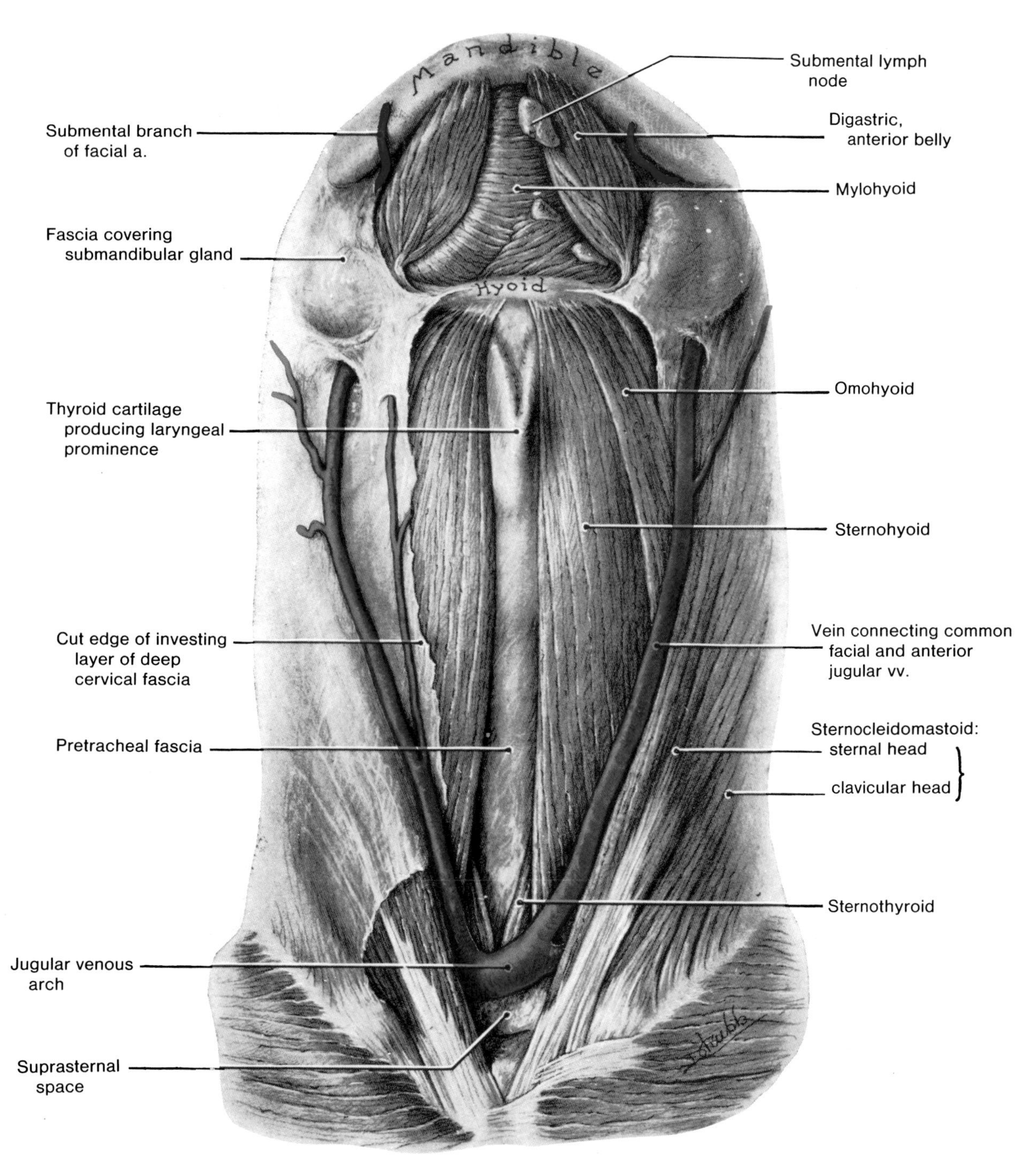

Figure 8-25. Drawing of a dissection of the anterior aspect of the neck. Note the submental triangle (Fig. 8-6*B*) bounded inferiorly by the body of the hyoid bone and laterally by the right and left anterior bellies of the digastric muscles. Observe that the floor of the submental triangle is formed by the two mylohyoid muscles and that it contains some submental lymph nodes. Actually the submental triangle is part of the floor of the mouth. Note the pronounced laryngeal prominence in this male specimen. The thyroid gland (not visible here; see Fig. 8-42) is enclosed in a fascial compartment formed by the sheath of pretracheal fascia, which fixes it firmly to the trachea and larynx (Fig. 8-41) via its attachment to the oblique line of the thyroid cartilage (Fig. 8-62).

mandibular incisor teeth and associated gingivae, the central part of the lower lip, and the skin of the chin. **Lymph from the submental lymph nodes drains into the submandibular and deep cervical lymph nodes** (Figs. 7-40, 8-18, 8-35, and 8-60). The submental triangle also contains small veins that unite to form the anterior jugular vein (Fig. 7-38).

Most cancers of the lip occur on the lower lip (Fig. 7-44), and tend to spread via the lymphatics. Depending on the site of the lesion, metastases spread to the submental nodes from the central part of the lower lip and to the submandibular nodes from other parts of the lip. In advanced cancers of the central part of the lip, the submandibular and deep cervical lymph nodes would also be involved because they receive lymph from the submental nodes (Figs. 7-40, 8-25, and 8-60).

A discharging sinus on the point of the chin often results from an *abscess of a mandibular incisor tooth.* The pus from the infected tooth passes to the apex of the **submental triangle,** located at the inferior end of the symphysis menti (Figs. 7-1 and 8-25), where it forms a sinus from which pus escapes.

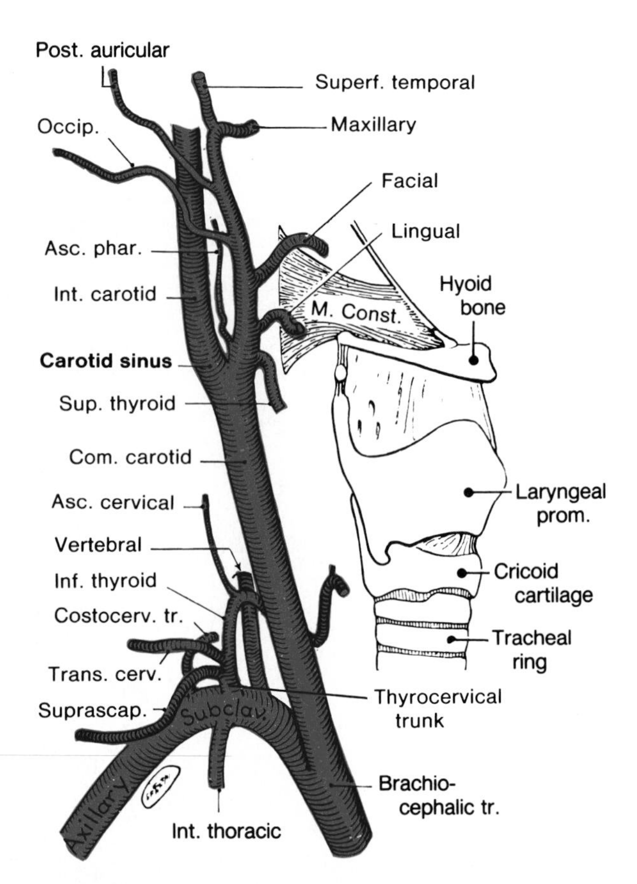

Figure 8-26. Drawing of the subclavian and carotid arteries and their branches. Note that the terminal part of the right common carotid and the proximal part of the internal carotid artery are dilated for about 1 cm to form the carotid sinus. The walls of this region, which are important in blood pressure regulation, are especially elastic and contain many sensory nerve endings (Fig. 7-111) that respond to changes in blood pressure and bring about appropriate modifications reflexly.

Most arteries in the anterior triangle of the neck arise from the common carotid artery or one of its branches.

THE COMMON CAROTID ARTERIES (Figs. 8-19 to 8-24, 8-26, 8-27, and 8-30 to 8-32)

The right common carotid artery begins at the bifurcation of the brachiocephalic trunk, posterior to the right sternoclavicular joint.

The left common carotid artery arises from the arch of the aorta (Fig. 8-21) and ascends into the neck, posterior to the left sternoclavicular joint.

Each common carotid artery ascends within the **carotid sheath** (Figs. 8-23, 8-36, and 8-41) to the level of the superior border of the thyroid cartilage (Fig. 8-42), where it terminates by dividing into the internal and external carotid arteries.

THE INTERNAL CAROTID ARTERY (Figs. 8-19 to 8-23, 8-24, 8-26, and 8-42)

This vessel is the direct continuation of the common carotid artery; *it has no branches in the neck.* As its name indicates, it *supplies structures within the skull.* The internal carotid arteries are two of the four major arteries supplying blood to the brain (Figs. 7-77 and 7-85).

The internal carotid artery arises from the common carotid artery at the level of the superior border of the thyroid cartilage (Figs. 8-26 and 8-42) and passes superiorly, almost in the vertical plane, to enter the **carotid canal** in the petrous part of the temporal bone (Fig. 7-62). A plexus of sympathetic fibers accompanies it (Fig. 7-84). During its course through the neck, the internal carotid artery lies on the longus capitis muscle and the sympathetic trunk (Figs. 8-24 and 8-28). The vagus nerve (CN X) lies posterolateral to it (Fig. 8-47).

The internal carotid artery enters the middle cranial fossa beside the dorsum sellae of the sphenoid bone (Figs. 7-13 and 7-62). Within the cranial cavity, the internal carotid artery and its branches supply the hypophysis cerebri, the orbit, and most of the supratentorial part of the brain (Figs. 7-77 and 7-85).

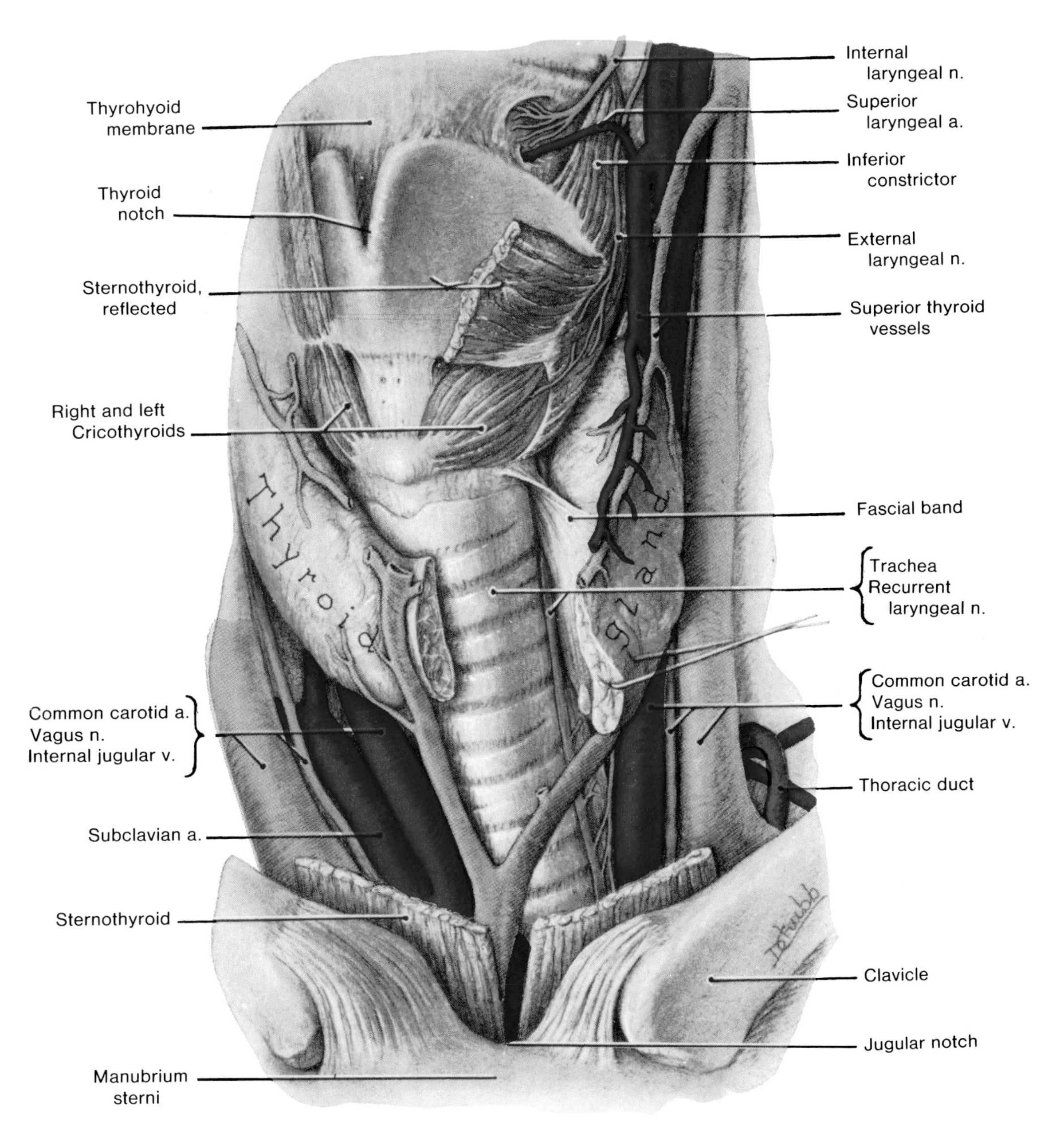

Figure 8-27. Drawing of a dissection of the anterior aspect of the neck. The isthmus of the thyroid gland is divided and its left lobe is retracted. Observe the left recurrent laryngeal nerve on the side of the trachea, just anterior to the groove between the trachea and esophagus and posterior to the retaining band. Note also the internal laryngeal nerve running along the superior border of the inferior constrictor muscle, piercing the thyrohyoid membrane, and dividing into several branches. Observe the external laryngeal nerve applied to the inferior constrictor, running along the anterior border of the superior thyroid artery, passing deep to the insertion of the sternothyroid, giving twigs to the inferior constrictor, and piercing it before ending in the cricothyroid muscle.

THE EXTERNAL CAROTID ARTERY (Figs. 7-37, 8-20 to 8-22, 8-24, and 8-26)

This vessel begins at the bifurcation of the common carotid, at the level of the superior border of the thyroid cartilage (Fig. 8-42). As its name indicates, it *supplies structures external to the skull.* The external carotid artery runs posterosuperiorly to the region between the neck of the mandible and the lobule of the auricle (Figs. 7-37 and 8-20). **The external carotid artery terminates by dividing into two branches, the maxillary and superficial temporal arteries** (Fig. 7-37). The external carotid artery has several other branches, some within the carotid

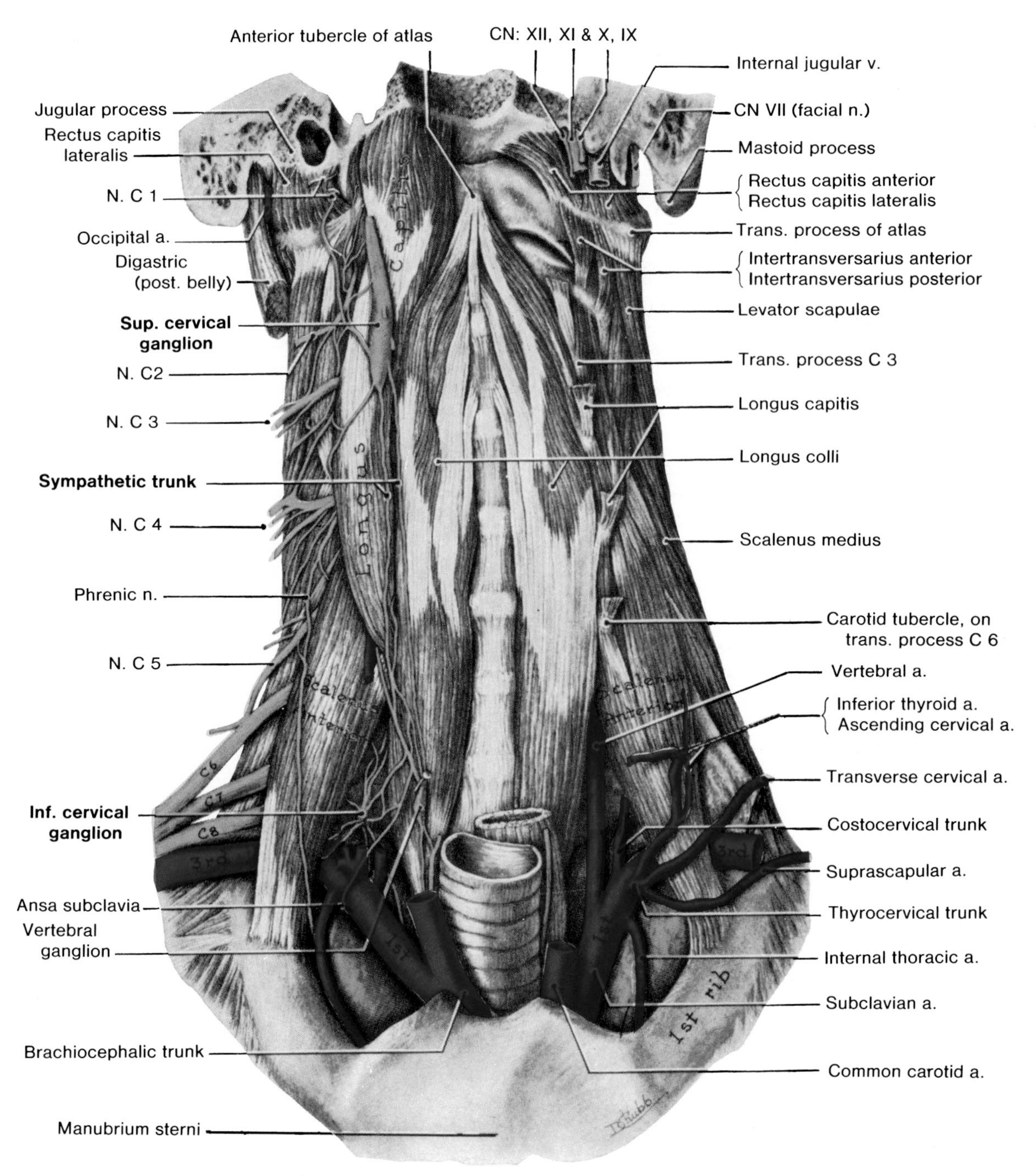

Figure 8-28. Drawing of a dissection of the prevertebral region and the root of the neck. The prevertebral fascia has been removed and the longus capitis muscle has been excised on the left side. Note that the internal jugular vein crosses these structures. Observe the cervical plexus arising from ventral rami C1, C2, C3, and C4, and the brachial plexus from C5, C6, C7, C8, and T1. Note the sympathetic trunk, sympathetic ganglia, and the gray rami communicantes. Examine the subclavian artery and its branches. Although the inferior part of the ansa subclavia is colored *red* here, it is a nerve filament or cord that forms a loop (L. *ansa*) around the subclavian artery. The ansa subclavia is shown better in Figure 8-34.

triangle (Fig. 8-24), some outside it (Figs. 8-19 to 8-21 and Table 8-3).

The stems of most of the ***six branches of the external carotid*** *artery are in the carotid triangle* (Fig. 8-24). The three important branches of this artery are described first (Table 8-3).

The Superior Thyroid Artery (Figs. 8-5, 8-20, 8-26, and 8-27). This is the most inferior of the three

Table 8-3
Branches of the External Carotid Artery

Surface	Branches	Figure References
Anterior[1]	**Superior thyroid**	8-5, 8-20, and 8-27
	Lingual	7-143, 8-24, and 8-26
	Facial	7-31, 8-19, and 8-20
Posterior	Occipital	7-37, 8-19, and 8-26
	Posterior auricular	7-29, 8-20, and 8-26
Medial	Ascending pharyngeal	8-20 and 8-26

[1] The anterior branches, printed in **boldface type**, are of major importance.

anterior branches of the external carotid; it arises close to the origin of the vessel, just inferior to the greater horn of the hyoid (Figs. 8-5 and 8-27).

The superior thyroid artery runs anteroinferiorly, deep to the infrahyoid muscles, to reach the superior pole of the **thyroid gland** (Figs. 8-27 and 8-36). In addition to supplying the thyroid gland, the superior thyroid artery gives muscular branches to the sternocleidomastoid and the infrahyoid muscles (Table 8-2), and **gives off the superior laryngeal artery** (Figs. 8-27 and 8-74). The superior laryngeal artery pierces the thyrohyoid membrane in company with the internal laryngeal nerve, and supplies the larynx (Figs. 8-27 and 8-74).

The Lingual Artery (Figs. 7-143, 8-24, and 8-26). This vessel arises from the external carotid artery as it lies on the middle constrictor muscle of the pharynx (Figs. 7-143 and 8-26). It arches superoanteriorly, about 5 mm superior to the tip of the greater horn of the hyoid bone (Fig. 8-5), and then passes deep to the hypoglossal nerve (CN XII), the stylohyoid muscle, and the posterior belly of the digastric muscle. It disappears deep to the hyoglossus muscle (Fig. 7-143); at the anterior border of this muscle, it turns superiorly and ends by becoming the **deep lingual artery**.

The Facial Artery (Figs. 7-29, 7-37, 8-18 to 8-20, 8-24, and 8-26). This vessel arises from the external carotid, either in common with the lingual artery or immediately superior to it. In the neck the facial artery gives off its important **tonsillar branch** and branches to the palate and the submandibular gland.

The facial artery then passes superiorly under cover of the digastric and stylohyoid muscles and the angle of the mandible (Figs. 8-19, 8-20, and 8-24). The facial artery loops anteriorly, and enters a deep groove in the submandibular gland (Fig. 8-14).

The facial artery hooks around the inferior border of the mandible to enter the face (Fig. 7-37); here pulsations of the facial artery can easily be felt.

The Ascending Pharyngeal Artery (Figs. 8-20 and 8-26). This artery is the first or second branch of the external carotid artery. This small vessel ascends on the pharynx, deep to the internal carotid artery, and gives branches to the pharynx, prevertebral muscles, middle ear, and meninges.

The Occipital Artery (Figs. 7-37, 8-11, 8-19, 8-20, and 8-26). It arises from the posterior surface of the external carotid at the level of the facial artery. It passes posteriorly along the inferior border of the posterior belly of the digastric muscle to end in the posterior part of the scalp (Figs. 7-48 and 8-11).

The occipital artery is often palpable as it crosses the superior nuchal line between the trapezius and sternocleidomastoid muscles (Figs. 6-44 and 8-11). During this course it passes superficial to the internal carotid artery (Fig. 8-19) and three cranial nerves (CN IX, CN X, and CN XI).

The Posterior Auricular Artery (Figs. 7-29, 8-20, and 8-26). This small posterior branch of the external carotid artery, arises from it at the superior border of the posterior belly of the digastric muscle (Fig. 8-20). It ascends posterior to the external acoustic meatus and supplies adjacent muscles, the parotid gland, the facial nerve, structures in the temporal bone, the auricle, and the scalp (Figs. 7-29 and 7-48).

THE INTERNAL JUGULAR VEIN (Figs. 7-65, 8-11, 8-12, 8-18, 8-19, 8-23, and 8-24)

Usually the *largest vein in the neck*, the internal jugular drains blood from the brain and superficial parts of the face and neck. Its course corresponds to a line drawn from a point immediately inferior to the external acoustic meatus to the medial end of the clavicle.

The internal jugular vein commences at the jugular foramen in the posterior cranial fossa (Figs. 7-60, 7-63, and 7-65), *as the direct continuation of the sigmoid sinus* (Fig. 7-57*A*). From the dilation at its origin, called the **superior bulb of the internal jugular vein** (Fig. 7-65), this large vein runs inferiorly through the neck in the carotid sheath (Figs. 8-23, 8-36, and 8-41). It shares the **carotid sheath**, a tubular condensation of fascia, with the internal carotid artery (later with the common carotid) and the vagus nerve (CN X). The artery is medial, the vein lateral, and the nerve posterior in the angle between these vessels (Fig. 8-41).

The internal jugular vein leaves the anterior triangle of the neck by passing deep to the sternocleidomastoid muscle (Fig. 8-18). Posterior to the sternal end of the clavicle, **the internal jugular vein unites with the subclavian to form the brachiocephalic vein** (Figs. 7-65 and 8-12). Near its termination is the **inferior bulb of the internal jugular vein** (Fig. 7-65), which contains a bicuspid valve like that in the subclavian vein.

The internal jugular vein is usually larger on the right side than on the left side because of the greater volume of blood entering it from the superior sagittal sinus via the sigmoid sinus (Figs. 7-39 and 7-57*A*).

Deep cervical lymph nodes (Figs. 8-35, 8-41, and 8-60) lie along the course of the internal jugular vein, many on its superficial surface. These nodes may be small and scattered; thus they are often difficult to identify in dissections.

Tributaries of the Internal Jugular Vein (Fig. 7-65). This large vein is joined at its origin by the **inferior petrosal sinus** (Fig. 7-39), the facial, lingual, pharyngeal, superior and middle thyroid veins, and often by the occipital vein (Fig. 7-65).

Superiorly the internal jugular vein lies posterolateral to the internal carotid artery, with cranial nerves CN IX to CN XII between them (Figs. 8-19, 8-27, and 8-28). When **thrombophlebitis** occurs in the superior bulb of the internal jugular vein (Fig. 7-65), for example, associated with middle ear infection (**otis media**), these cranial nerves may cease to conduct impulses owing to pressure from this large congested vein.

Pulsations of the internal jugular vein resulting from contraction of the right ventricle of the heart may be palpable and visible at the root of the neck (Fig. 8-35). Because there are no valves in the brachiocephalic vein or the superior vena cava, a wave of contraction passes up these vessels to the internal jugular vein.

The systolic venous pulse, palpable in the internal jugular vein, is considerably increased in certain conditions (*e.g.*, disease of the mitral valve leading to increased pressure in the pulmonary circulation, the right side of the heart, and the great veins).

All lymphatic vessels from the head and neck drain into the deep cervical lymph nodes (Figs. 7-40 and 8-60), many of which lie in the carotid sheath (Figs. 8-36 and 8-41). They lie along the internal jugular vein, between it and the common carotid artery.

The deep cervical lymph nodes are very important when there are metastases from tumors (*e.g.*, carcinoma of the mouth, larynx, or other structures in the head and neck). **Removal of cancerous deep cervical lymph nodes** is more difficult when they adhere closely to the internal jugular vein and the carotid sheath.

DEEP STRUCTURES OF THE NECK

SKELETON OF THE NECK (Figs. 5-1, 5-18, and 5-66)

The cervical vertebrae and joints of the neck are described in Chapter 5 with the back. Revise your knowledge of these bones, particularly the atlas and axis, and review the joints associated with them (Figs. 5-34 and 5-35). Recall that the joints between the atlas and axis permit rotation, whereas those between the skull and atlas are structured to allow nodding movements of the head on the cervical region of the vertebral column.

MUSCLES OF THE NECK (Figs. 5-39 to 5-41, 6-44, 6-47, 8-6, 8-7, and 8-10 to 8-13)

The superficial and the lateral cervical muscles (**platysma, trapezius, and sternocleidomastoid**) were described previously, as were the hyoid muscles (Fig. 8-17 and Tables 8-1 and 8-2) and the scalene muscles (Fig. 8-13).

The Anterior Vertebral Muscles (Figs. 8-28, 8-29, and 8-41). These deep prevertebral muscles are covered anteriorly by prevertebral fascia (Figs. 8-23 and 8-35). **They all flex the neck and the head on the neck,** and they are all supplied by ventral primary rami of the cervical nerves.

The longus colli muscle (longus cervicis), the longest and most medial of the prevertebral muscles, *extends from the anterior tubercle of the atlas to the body of the third thoracic vertebra* (Fig. 8-28). It is also attached to the bodies of the vertebrae between C1 and C3, and to the transverse processes of the third to sixth cervical vertebrae.

The longus capitus muscle (Fig. 8-28), broad and thick superiorly, *arises from the anterior tubercles of the third to sixth cervical transverse processes and inserts into the base of the skull.*

The rectus capitus anterior (Fig. 8-28), a short wide muscle, *arises from the anterior surface of the lateral mass of the atlas and inserts into the base of the skull,* just anterior to the occipital condyle.

The rectus capitis lateralis (Figs. 8-28 and 8-29), a short flat muscle, *arises from the transverse process of the* ***atlas*** *and inserts into the* ***jugular process*** *of the occipital bone.* This muscle and the rectus capitus anterior, in addition to flexing the head on the neck, help to stabilize the skull on the cervical region of the vertebral column.

THE ROOT OF THE NECK

The thoracocervical region or the root of the neck is the junctional area between the thorax and the neck. It includes **the superior thoracic aperture** (p. 67) through which pass all structures going from the head to the thorax and vice versa.

BOUNDARIES OF THE ROOT OF THE NECK (Figs. 8-17, 8-27, 8-28, and 8-32)

The thoracocervical region is bounded *laterally* by the **first pair of ribs** and their costal cartilages (Figs. 8-17, 8-28, and 8-32), *anteriorly* by the **manubrium sterni** (Figs. 8-17, 8-27, and 8-28), and *posteriorly* by the body of the **first thoracic vertebra**.

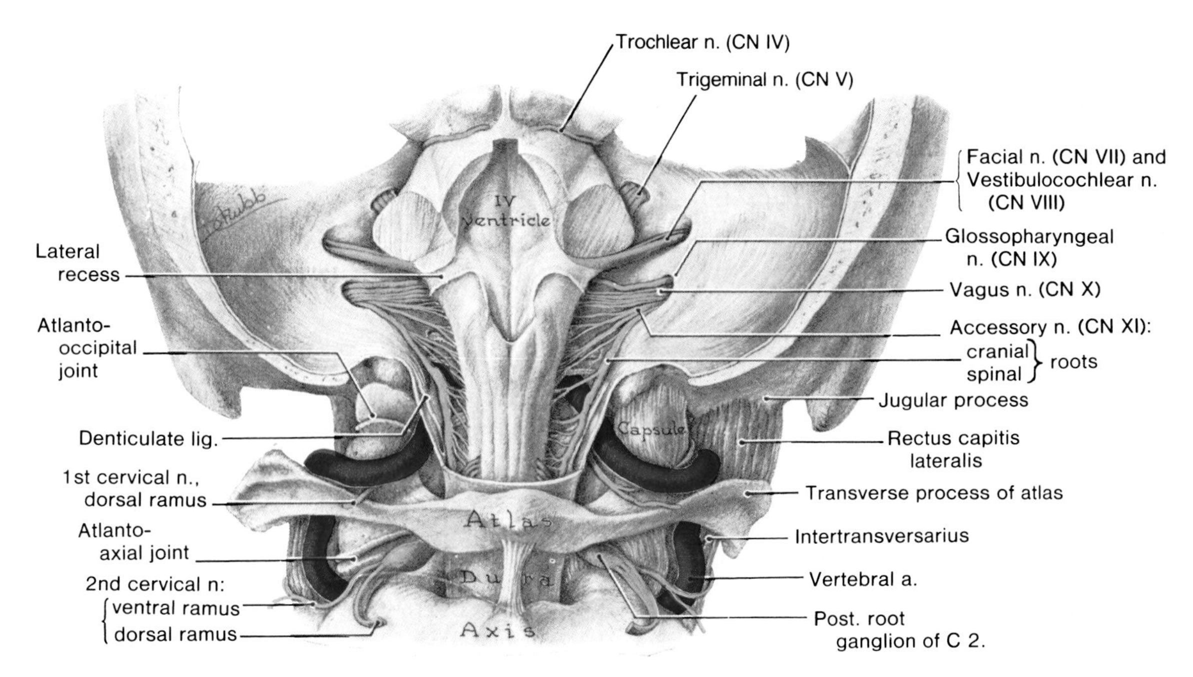

Figure 8-29. Dissection showing a posterior view of the cranial nerves. Observe that the transverse process of the atlas is joined to the jugular process of the occipital bone by the rectus capitis lateralis muscle; morphologically it is an intertransverse muscle. Note that the vertebral arteries are raised from their "beds" on the posterior arch of the atlas. The vertebral arteries join to form the basilar artery (see Fig. 7-77).

ARTERIES AT THE ROOT OF THE NECK (Figs. 8-27, 8-28, and 8-32)

The arteries in this junctional area originate from the arch of the aorta. They are the brachiocephalic trunk on the right side and the common carotid and subclavian arteries on the left side.

The Brachiocephalic Trunk (Figs. 1-73, 8-26, 8-28, 8-30, and 8-42). The brachiocephalic trunk (brachiocephalic artery) is *the largest branch of the arch of the aorta.* It is 4 to 5 cm in length and arises posterior to the center of the manubrium sterni. It passes superiorly and to the right, posterior to the **right sternoclavicular joint**, where it *divides into the right common carotid and right subclavian arteries.* The brachiocephalic trunk is covered anteriorly by the sternohyoid and sternothyroid muscles (Fig. 8-42). At first, the brachiocephalic trunk lies on the trachea and then to its right side.

Usually the brachiocephalic trunk has no branches, other than its terminal ones (right subclavian and common carotid arteries), but sometimes a small artery, the **thyroid ima** (L. lowest), arises from it and ascends anterior to the trachea to the isthmus of the thyroid gland (Fig. 8-37).

The Subclavian Arteries (Figs. 7-37, 8-28, and 8-30 to 8-32). *These are the arteries of the upper limbs,* but they also supply branches to the neck and brain.

The **right subclavian artery**, one of the terminal branches of the brachiocephalic trunk, arises posterior to the right sternoclavicular joint.

The **left subclavian artery** arises from the arch of the aorta (Fig. 1-73), and enters the root of the neck by passing superiorly, posterior to the left sternoclavicular joint (Fig. 8-28).

Each subclavian artery arches superiorly, posteriorly, and laterally, grooving the pleura and lung and then passes inferiorly posterior to the midpoint of the clavicle (Figs. 8-28, 8-32, and 8-34). As these arteries rise 2 to 4 cm into the root of the neck, they are crossed anteriorly by the scalenus anterior muscles (Figs. 8-28, 8-30, and 8-31).

For purposes of description, *the scalenus anterior muscle divides the subclavian artery into three parts:* the first part medial to the muscle, the second part posterior to it, and the third part lateral to it (Figs. 8-28 and 8-30).

Branches of the subclavian arteries (Figs. 8-26, 8-31, and 8-32) are the vertebral, thyrocervical trunk, and internal thoracic *from the first part*; the costocervical trunk *from the second part*; and occasionally the suprascapular and/or the dorsal scapular artery *from the third part.*

The Vertebral Artery (Figs. 7-37, 8-28, 8-29, 8-31, 8-32, and 8-34). The vertebral artery arises from the first part of the subclavian and ascends through the foramina transversaria of the cervical vertebrae, except for C7 vertebra (Fig. 7-37). After winding around the lateral mass of the atlas, *the vertebral artery enters the skull through the foramen magnum* (Fig. 8-29).

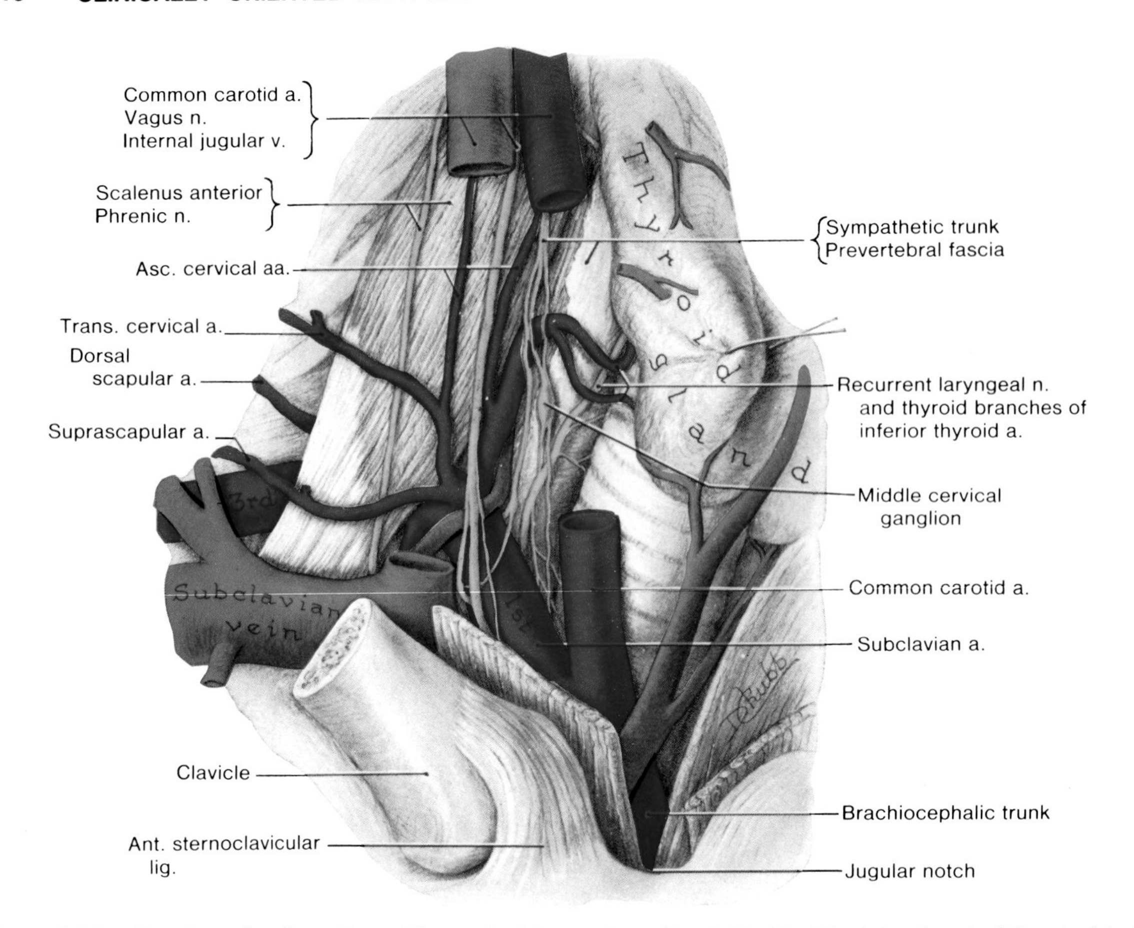

Figure 8-30. Drawing of a dissection of the root of the neck on the right side. The lateral part of the clavicle has been removed and sections have been taken from the common carotid artery and the internal jugular vein. The right lobe of the thyroid gland is retracted to expose the vagus nerve crossing the first part of the subclavian artery and giving off the recurrent laryngeal nerve. Note that this nerve hooks inferior to the subclavian artery and crosses posterior to the common carotid artery on its way to the side of the trachea. Observe that it gives twigs to the trachea and esophagus and receives twigs from the sympathetic trunk. Note the close relationship of the recurrent laryngeal nerve to the branches of the inferior thyroid artery; hence this nerve is vulnerable to injury during ligation of this vessel during thyroidectomy.

At the inferior border of the pons, **the vertebral arteries join to form the basilar artery** (Fig. 7-77), a very important artery supplying the brain.

The Thyrocervical Trunk (Figs. 7-37, 8-26, 8-28, and 8-32). The thyrocervical trunk arises from the first part of the subclavian artery, just medial to the scalenus anterior muscle. It gives rise to several branches, the largest and most important of which is the **inferior thyroid artery**, which passes to the inferior pole of the thyroid gland.

Other branches of the thyrocervical trunk are (Fig. 8-26): the *suprascapular artery* (except when it arises from the third part of the subclavian artery) supplying muscles around the scapula, and the *transverse cervical artery*, sending branches to the muscles in the posterior triangle of the neck (Figs. 8-12 and 8-13).

The Internal Thoracic Artery (Figs. 1-16, 1-46, 7-37, 8-26, and 8-28). The internal thoracic artery arises from the inferior aspect of the subclavian artery and passes inferomedially into the thorax. It runs parallel to the sternum (about 2.5 cm lateral to it) and gives off anterior intercostal branches to the first six intercostal spaces (Fig. 1-17).

The Costocervical Trunk (Figs. 7-37, 8-26, and 8-28). The costocervical trunk arises from the posterior aspect of the subclavian artery and passes superoposteriorly over the **cervical pleura**. It divides into the superior intercostal and deep cervical arteries, which supply the first two intercostal spaces and muscles in the neck.

VEINS AT THE ROOT OF THE NECK

The external jugular vein, receiving blood mostly from the scalp and the face, is described on page 994 and is illustrated in Figures 7-38 and 8-10.

The Anterior Jugular Vein (Figs. 7-38, 8-19, and

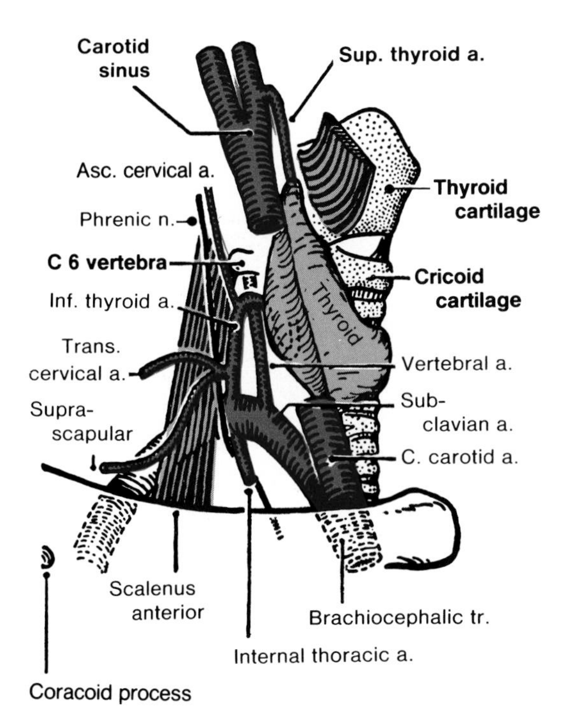

Figure 8-31. Drawing of the arteries in the root of the neck. Note that the course of the subclavian artery forms a curved line, extending from the sternoclavicular joint, 2 to 4 cm superior to the clavicle and then crossing posterior to this bone near its middle. Also observe the scalenus anterior muscle which is used to divide the subclavian artery into three parts for descriptive purposes (Fig. 8-28).

8-25). This vein is usually the *smallest of the jugular veins and is very variable.* The anterior jugular vein arises from the **submental venous plexus** (Fig. 7-38) and descends in the superficial fascia between the anteromedian line and the anterior border of the sternocleidomastoid muscle.

At the root of the neck the anterior jugular vein turns laterally, posterior to the sternocleidomastoid muscle (Fig. 8-19), and opens into the termination of the external jugular, or directly into the subclavian vein (Fig. 7-38). Just superior to the sternum, the right and left anterior jugular veins are united by a large transverse trunk, the **jugular venous arch** (Fig. 8-25).

The anterior jugular veins have no valves and they may be replaced by a single trunk descending in the midline of the neck. The anterior jugular veins, like the external jugular veins, can often be made visible by "blowing" with the mouth closed.

The Subclavian Vein (Figs. 6-40, 7-34, 7-65, 8-12, 8-13, and 8-30). *This large vein is the continuation of the axillary vein.* It begins at the lateral border of the first rib and ends at the medial border of the scalenus anterior muscle, where it unites with the **internal jugular** (Figs. 7-65 and 8-35), posterior to the medial end of the clavicle to form the **brachiocephalic vein.**

The subclavian vein, lying in the concavity of the subclavian artery superior to the clavicle (Fig. 8-30), has a bicuspid valve near its termination (Fig. 7-65). It usually has only one named tributary, the **external jugular vein** (Fig. 8-10), which is often visible in the neck (Fig. 8-33). Observe that the subclavian vein passes over the first rib parallel to the subclavian artery, but is separated from it by the scalenus anterior muscle (Fig. 8-30). The subclavian vein crosses the first rib anterior to the scalene tubercle (Fig. 1-4). This tubercle separates the groove in the first rib for the subclavian vein from the groove for the subclavian artery.

The Internal Jugular Vein (Figs. 7-57*A*, 7-65, 8-30, 8-32, 8-35, 8-36, and 8-41). This vein, described on page 1013, ends posterior to the medial end of the clavicle by uniting with the subclavian to form the **brachiocephalic vein**. *Throughout its course, the internal jugular vein is enclosed within the* ***carotid sheath*** (Figs. 8-23 and 8-41).

NERVES AT THE ROOT OF THE NECK

Several important nerves are located at the root of the neck (Figs. 8-27, 8-28, 8-30, 8-32, 8-34, 8-36, and 8-41).

The Vagus Nerve (CN X). The vagus (L. wandering), so-named because of its wide distribution (Fig. 7-135), leaves the skull through the **jugular foramen** with the internal jugular vein and cranial nerves IX and XI (Figs. 7-60 and 7-79). The vagus, **the main parasympathetic nerve to the organs of the thorax and abdomen,** passes inferiorly in the posterior part of the **carotid sheath** (Figs. 8-23, 8-24, and 8-41) in the angle between and posterior to the internal jugular vein and carotid artery (first internal carotid and then the common carotid artery).

On the right side, the vagus nerve crosses the origin of the subclavian artery, posterior to the brachiocephalic vein and the sternoclavicular joint, to enter the thorax (Figs. 8-27 and 8-30).

The recurrent laryngeal nerve (Figs. 8-30, 8-34, 8-36, and 8-75), a branch of the vagus, *hooks around the subclavian artery on the right* side and *around the arch of the aorta on the left side* (Figs. 8-45 and 8-75). Review the embryological basis for this clinically important difference in the course of the recurrent laryngeal nerves (Moore, 1982).

After looping, both recurrent laryngeal nerves then pass superiorly to reach the posteromedial aspect of the inferior pole of the thyroid gland (Fig. 8-36), where they ascend in the *tracheoesophageal groove* to supply all the intrinsic muscles of the larynx except the cricothyroid (Fig. 8-71 and Table 8-5).

The cardiac nerves (also branches of the vagus,

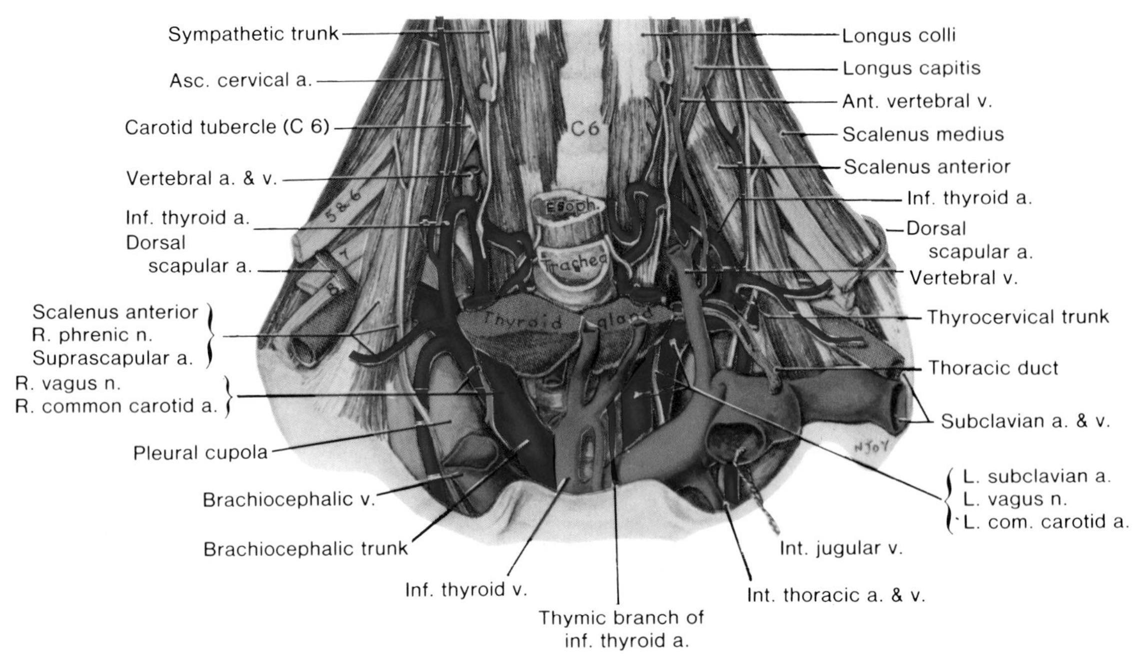

Figure 8-32. Drawing of a superior view of a dissection of the root of the neck, viewed obliquely from above. Observe the "triangle of the vertebral artery", bounded laterally by the scalenus anterior muscle and medially by the longus colli muscle. Note that the apex of the triangle is where these two muscles meet at the carotid tubercle (anterior tubercle of transverse process of C6), and the base of the triangle is formed by the first part of the subclavian artery. Observe the vertebral artery ascending from the base to the apex and dividing the triangle into two nearly equal parts.

CN X) originate in the neck and thorax (Fig. 7-135), and run along the arteries to the arch of the aorta and the ***cardiac plexuses*** (Figs. 1-63, 1-70, and p. 118). When stimulated they slow the heartbeat, reducing its force, and in unusual cases may produce cardiac arrest. Cardiac branches also arise from the vagus where it lies on the subclavian artery; they descend beside the trachea to the deep cardiac plexus.

The Phrenic Nerve (Figs. 1-46, 8-28, 8-30, 8-35, and 8-41). This nerve, usually about 30 cm long, is the *sole motor nerve to the diaphragm* (Fig. 2-120).

The phrenic nerve arises chiefly from the fourth cervical nerve (with contributions from the third and fifth cervical nerves). The phrenic nerve is formed at the superior part of the lateral border of the scalenus anterior muscle, at the level of the superior border of the thyroid cartilage (Fig. 8-31), and superolateral to the internal jugular vein (Fig. 8-13). The phrenic nerve descends obliquely with this vein across the scalenus anterior muscle, deep to prevertebral fascia and the transverse cervical and suprascapular arteries (Fig. 8-35).

On the left, the phrenic nerve crosses the first part of the subclavian artery, but *on the right* it lies on the scalenus anterior muscle which covers the second part of this artery (Fig. 8-30). The phrenic nerve crosses posterior to the subclavian vein on both sides, and anterior to the internal thoracic artery to enter the thorax (Figs. 8-30 and 8-32).

The unexpected innervation of the diaphragm by *cervical nerve roots* has an embryological basis and clinical significance. During the 5th week of development, ventral rami from C3, C4, and C5 grow into the **septum transversum,** the primordium of the central tendon of the diaphragm, when it is in the cervical region. As the developing diaphragm migrates caudally, it carries with it the phrenic nerves, formed by the nerve roots.

Severance of a phrenic nerve in the root of the neck (or elsewhere) results in paralysis of the corresponding half of the diaphragm.

Injuries to the inferior cervical region of the spinal cord (*e.g.,* C7), severe enough to cause paralysis of the upper limbs, have little effect on breathing because the phrenic nerves arise from more cranial segments of the spinal cord (C3, **C4**, and C5). Breathing would not be normal however, because the intercostal muscles would be paralyzed.

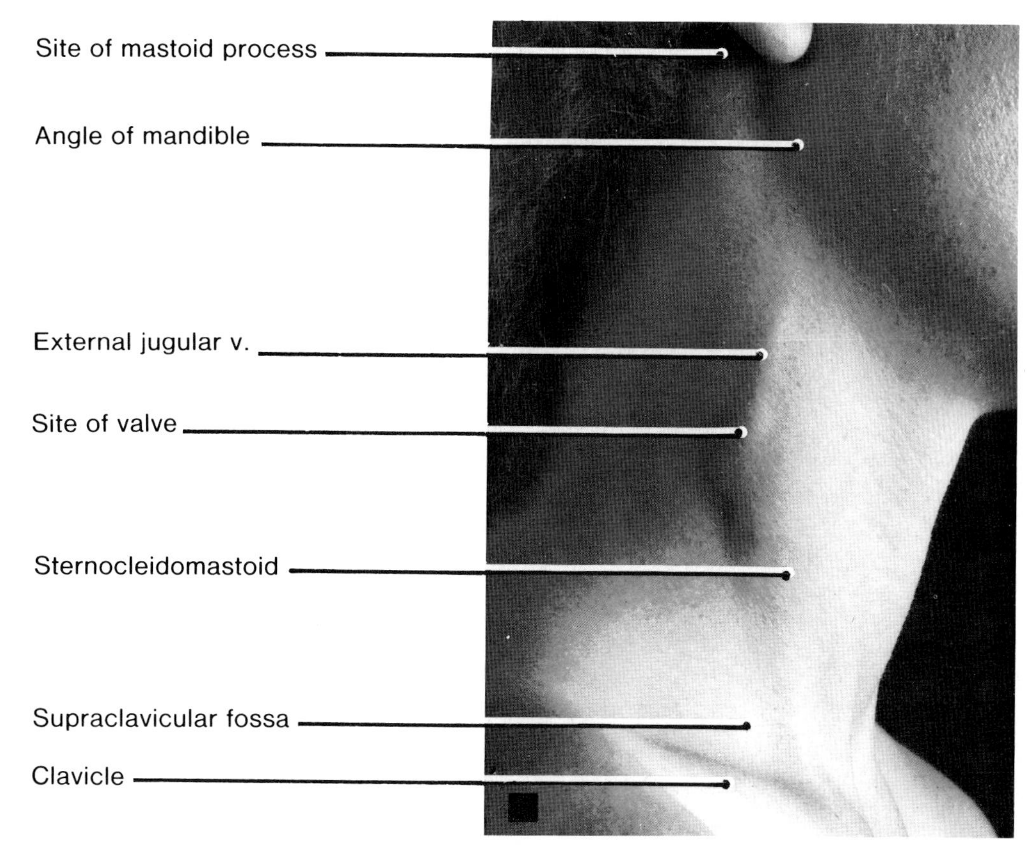

Figure 8-33. Photograph of the right side of the neck of a 27-year-old woman. To make her external jugular vein stand out, she was asked to take a deep breath and to hold it. The slight swelling in her vein about 4 cm superior to the clavicle indicates the site of one of the valves in this vein. You are unable to see the inferior part of her vein because it pierces the roof of the posterior triangle and passes deep to it in this triangle.

To produce temporary therapeutic **paralysis of one half the diaphragm** (*e.g.*, to interrupt a severe case of hiccoughs or spasmodic sharp contractions of the diaphragm), a **phrenic nerve block** is sometimes done. The anesthetic solution is injected around the phrenic nerve where it lies on the anterior surface of the middle third of the scalenus anterior muscle, about 3 cm superior to the clavicle (Figs. 8-12, 8-30, and 8-35).

To produce a longer period of paralysis of half of the diaphragm, *e.g.*, for several months after the surgical repair of a diaphragmatic hernia, a **phrenic crush** may be performed. In this case the phrenic nerve is crushed with a hemostat for up to 1 cm of its length. In other cases a **phrenicotomy** is performed, during which the phrenic nerve is sectioned. The phrenic nerve is exposed by making a 2 to 3 cm incision in a skin crease superior to the clavicle. Posterior to the sternocleidomastoid muscle (Fig. 8-11), the phrenic nerve is isolated, grasped, and cut where it passes across the scalenus anterior muscle posterior to the prevertebral fascia, and between the transverse cervical and suprascapular arteries (Fig. 8-35).

To produce permanent paralysis of the hemidiaphragm [*e.g.,* **after a pneumonectomy** (removal of a lung) to allow the abdominal viscera to push the hemidiaphragm superiorly and to help obliterate the pleural space], a **phrenicectomy** is done *by excising a portion of the phrenic nerve.* Usually the nerve is avulsed, *i.e.*, the distal stump is pulled until it tears, but some surgeons pull up 3 to 4 cm of the nerve and cut it.

An accessory phrenic nerve (Figs. 8-12 and 8-13) occurs in 20 to 30% of persons. It is frequently derived from the fifth cervical nerve as a *branch of the nerve to the subclavius muscle.* It lies lateral to the main phrenic nerve and usually joins it in the root of the neck, or in the superior part of the thorax. Therefore, should an accessory phrenic nerve be present, section-

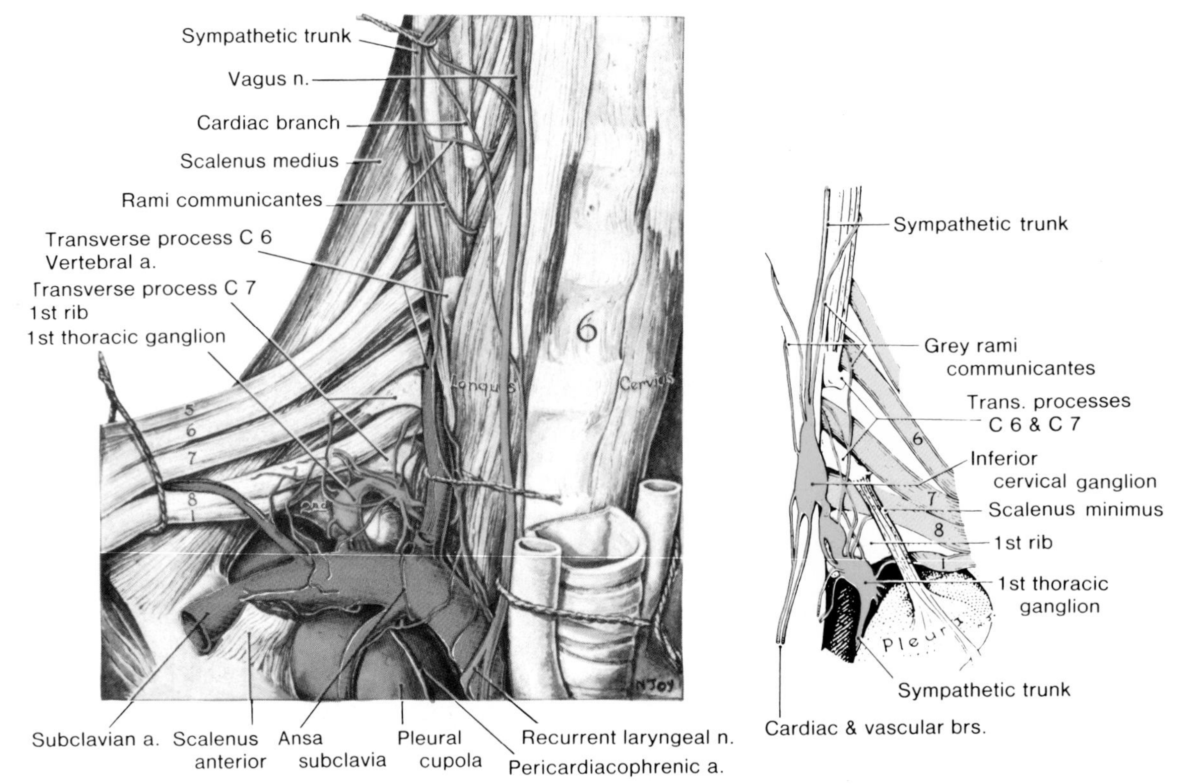

Figure 8-34. Drawing of a dissection of the neck on the right side. Observe that the inferior trunk of the brachial plexus (C8 and T1) has been raised from the groove it occupies on the first rib, posterior to the subclavian artery. Note that the sympathetic trunk (retracted laterally) is sending a communicating branch to the vagus, and that the gray rami communicantes (postganglionic fibers) pass to the roots of the cervical nerves. The cardiac branches are also shown. The vertebral artery has been retracted medially to uncover the stellate or cervicothoracic ganglion (combined inferior cervical and first thoracic ganglia) which rests posterior to it on the first and second ribs. The illustration on the right, a tracing of a photograph of the left side of the same specimen, reveals a very different pattern. Thus, the inferior cervical ganglion occupies its more usual position between the transverse process of C7 and the first rib. The ganglion (T1) lies on and inferior to the first rib. The ansa subclavia consists of nerve filaments connecting the middle and cervicothoracic ganglion (stellate ganglion). It forms a loop (L. ansa) around the subclavian artery.

ing or crushing the phrenic nerve alone in the neck will not produce complete paralysis of the corresponding half of the diaphragm. However, ***avulsion of the cervical nerve roots of the phrenic nerve ruptures the accessory phrenic nerve also***, and results in complete paralysis of the hemidiaphragm.

THE SYMPATHETIC TRUNKS (Figs. 8-24, 8-28, 8-30, 8-34, and 8-36)

These longitudinal strands of nerve fibers and sympathetic ganglia *lie anterolateral to the vertebral column* from the level of the first cervical vertebra.

The sympathetic trunks in the neck receive no white rami communicantes (Figs. 8-28 and 8-34), but they contain **three cervical sympathetic ganglia** (superior, middle, and inferior). These ganglia receive their preganglionic fibers from the superior thoracic spinal nerves via white rami communicantes whose fibers leave the spinal cord in the ventral roots of thoracic spinal nerves (Figs. 45 and 46, pp. 47 and 48).

From the sympathetic trunk in the neck, fibers pass to cervical structures as postganglionic fibers in cervical spinal nerves, or leave as direct visceral branches (*e.g.*, to the thyroid gland). Branches to the head run with the arteries, especially the internal and external carotid arteries (Fig. 8-24).

The inferior cervical ganglion (Figs. 8-28, 8-34, and 8-41) lies at the level of the superior border of the neck of the first rib, where it is wrapped around the posterior aspect of the vertebral artery. It is usually fused with the first thoracic ganglion (and sometimes the second) to form a large ganglion, known as **the cervicothoracic ganglion**. It is often called the *stellate ganglion* even though it is not star-shaped in

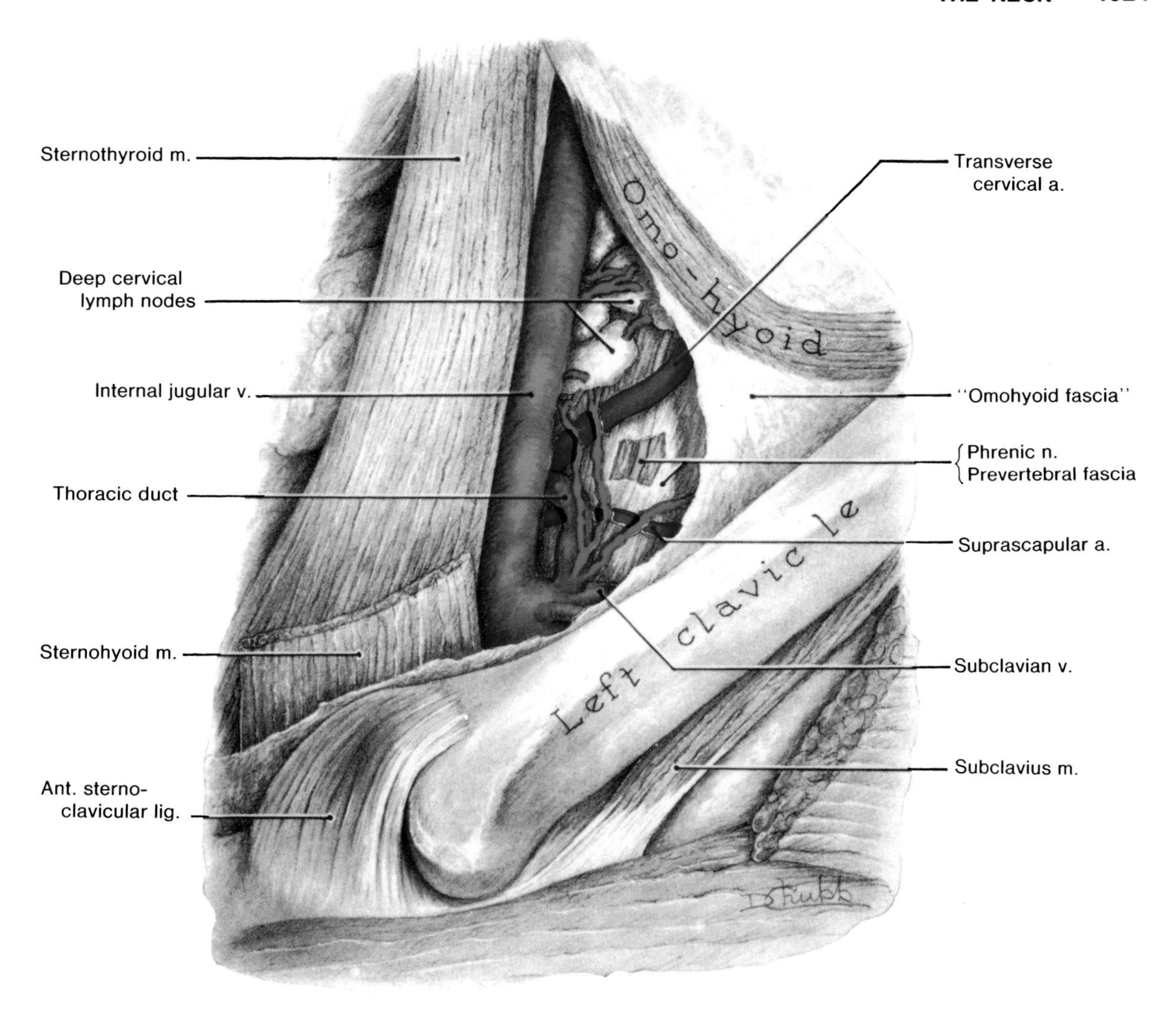

Figure 8-35. Drawing of a dissection of the root of the neck on the left side showing the deep cervical lymph nodes and the termination of the thoracic duct. Note that the thoracic duct receives a tributary from the nodes of the neck and the jugular lymph trunk and ends in the angle between the internal jugular and subclavian veins. Observe that the deep cervical lymph nodes are arranged along the blood vessels, for the most part lateral and posterior to the internal jugular vein (also see Fig. 8-41). The inferior deep cervical nodes, inferior to the omohyoid muscle, that are shown here, are often referred to as the supraclavicular or scalene nodes because they lie superior to the clavicle and on the scalenus anterior muscle. To produce temporary paralysis of the diaphragm for medical reasons, the phrenic nerve is crushed where it passes inferiorly across the scalenus anterior muscle.

humans. The inferior cervical ganglion lies anterior to the transverse process of the vertebra prominens (C7), just superior to the neck of the first rib on each side, posterior to the origin of the vertebral artery (Fig. 8-34). Some postganglionic fibers from the inferior cervical ganglion pass into the seventh and eighth cervical nerves and to the heart; other fibers contribute to the **vertebral plexus** around the vertebral artery (Fig. 8-34, left side).

The middle cervical ganglion (Fig. 8-30) is small and lies on the anterior aspect of the inferior thyroid artery, at about the level of the cricoid cartilage and the transverse process of C6 vertebra (Fig. 8-31), just anterior to the vertebral artery. Postganglionic branches pass from it to the fifth and sixth cervical nerves and to the heart and thyroid gland.

The superior cervical ganglion (Fig. 8-28) is large (*2 to 3 cm long*). It is located at the level of the atlas and axis and because of its size, *it forms a good landmark for locating the sympathetic trunk in the neck.* Postganglionic branches from it pass along the internal carotid artery and enter the cranial cavity. It

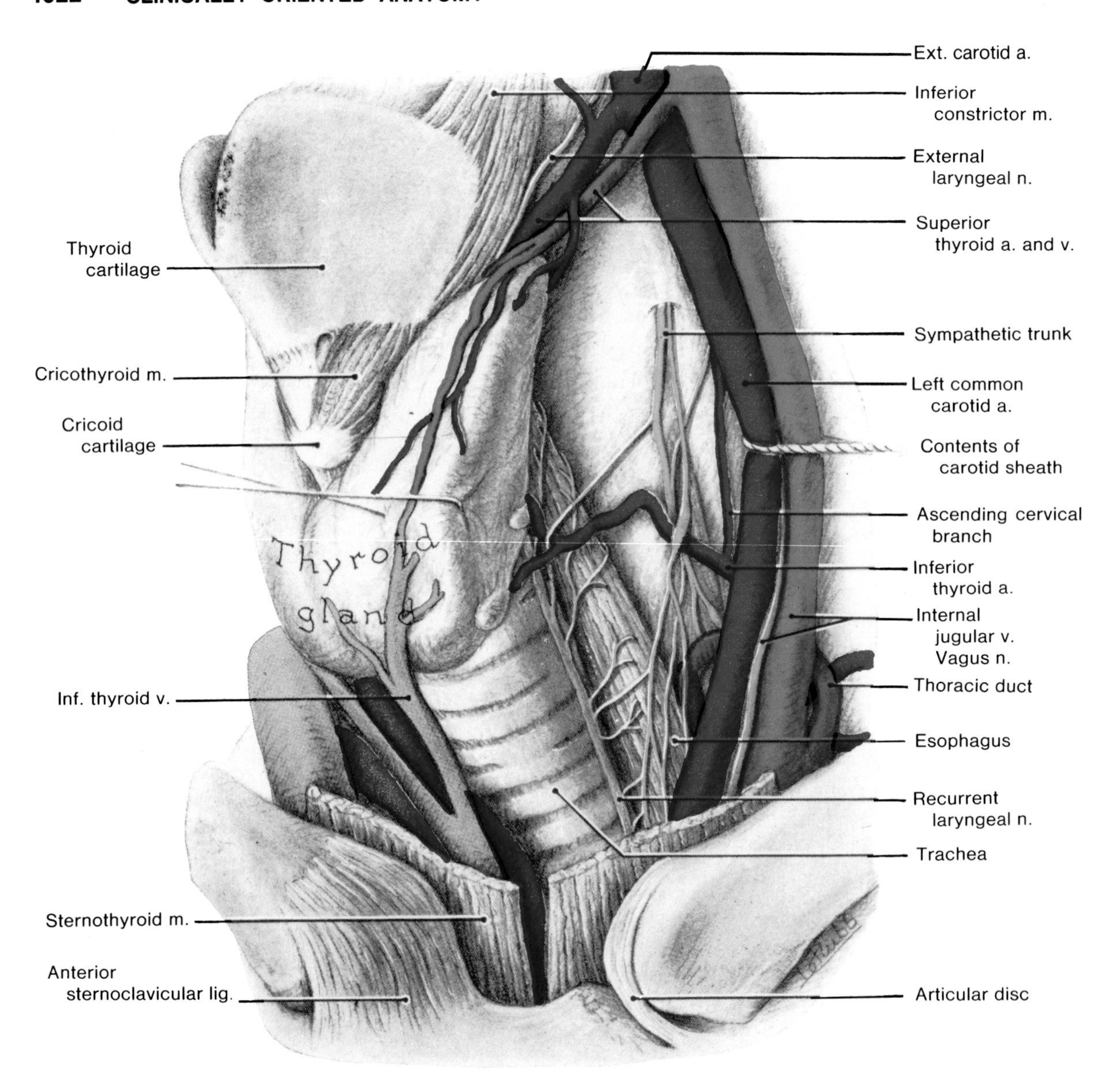

Figure 8-36. Drawing of a dissection of the left side of the root of the neck. Observe the esophagus, a thick muscular tube, which begins at the inferior border of the cricoid cartilage and passes to the left of the trachea. Note the recurrent laryngeal nerve ascending on the side of the trachea just anterior to the angle between the trachea and esophagus. Note that it gives twigs to the esophagus and trachea (not in view) and receives twigs from the sympathetic trunk. Observe that the thoracic duct passes from the side of the esophagus to its termination and arches immediately posterior to the three structures contained in the carotid sheath (internal jugular vein, common carotid artery, and vagus nerve), which are retracted with a cord.

also sends branches to the external carotid artery and into the superior four cervical nerves. Other postganglionic fibers pass to the **cardiac plexus** (Fig. 1-63).

If a sympathetic trunk is severed in the neck, interruption of the sympathetic nerve supply to the head on that side occurs. Patients have a sympathetic disturbance known as the **Horner syndrome** consisting of: (1) **pupillary constriction** owing to paralysis of the dilator pupillae muscle (Fig. 7-97); (2) **ptosis** (slight lowering of the upper eyelid) owing to paralysis of the smooth muscle in the levator palpebrae superioris (Figs. 7-28, 7-94, and 7-96); (3) slight sinking in of the eye, possibly resulting from paralysis of the orbitalis muscle, a scanty sheet of smooth muscle in the orbit; and (4) **vasodilation and absence of sweating** on the face and neck owing to lack of a sympathetic nerve supply to the blood vessels and sweat glands.

Hemisection of the spinal cord in the cervical region also produces the Horner syndrome. In these cases, the disturbance results from interruption of descending autonomic fibers in the spinal cord.

Anesthetic fluid injected around the cervicothoracic ganglion will block the transmission of stimuli through the cervical and superior thoracic ganglia. A cervicothoracic ganglion or **stellate ganglion block** is performed to relieve vascular spasms involving the brain and the upper limb.

LYMPHATICS AT THE ROOT OF THE NECK

Several large lymph nodes are arranged within the carotid sheath along the blood vessels of the neck, particularly the internal jugular vein (Figs. 7-40, 8-35, 8-41, and 8-60). Another group of lymph nodes is found along the transverse cervical artery (Figs. 8-30 and 8-35). All these are called **deep cervical lymph nodes** because they are deep to the deep cervical fascia (*e.g.*, the "omohyoid fascia" shown in Fig. 8-35).

For descriptive purposes, the deep cervical lymph nodes are often divided into superior and inferior groups according to their relationship to the point of crossing of the omohyoid muscle over the internal jugular vein (Figs. 8-19 and 8-60). The deep cervical lymph nodes receive lymph from the **superficial cervical lymph nodes** and from the entire head and neck (Fig. 7-40). From the inferior end of the deep group of cervical lymph nodes, a **jugular lymph trunk** emerges and joins the venous system near the junction of the internal jugular and the subclavian veins (Figs. 1-42 and 8-35). On the left side the jugular lymph trunk may empty into the **thoracic duct** (Figs. 1-42, 8-32, and 8-36).

The deep cervical lymph nodes, particularly those located along the **transverse cervical artery** (Fig. 8-35), may become involved in the spread of cancer from the abdomen and/or the thorax. As enlargement of them may give the first clue to cancer in these regions, they are often referred to as the **cervical sentinel lymph nodes.**

THE THORACIC DUCT (Figs. 1-42, 1-73, 8-27, 8-32, 8-35, and 8-36)

This large lymphatic channel, draining lymph into the venous system, passes superiorly from the thorax through the superior thoracic aperture (p. 67) at the left border of the esophagus (Fig. 8-36). It then arches laterally in the root of the neck, posterior to the **carotid sheath** (Figs. 8-23 and 8-41), and anterior to the sympathetic trunk and the vertebral and subclavian arteries (Fig. 8-32). The thoracic duct enters the left brachiocephalic vein at the junction of the subclavian and internal jugular veins (Figs. 1-42, 8-27, and 8-35).

The thoracic duct drains lymph from the entire body, except the right side of the head and neck, the right upper limb, and the right side of the thorax. These areas drain via the **right lymphatic duct**, a 1 to 2 cm long vessel, which empties into the venous system at or near the junction of the right internal jugular and right subclavian veins (Fig. 1-42).

***Blockage of the thoracic duct* owing to the permeation of tumor cells** (*e.g.*, from an abdominal carcinoma) usually produces no symptoms. The lymph apparently enters the venous system via other lymphatic channels.

Malignant cells pass from an abdominal cancer via the thoracic duct into the root of the neck. Some tumor cells enter the venous system and **other tumor cells extend by retrograde permeation into the inferior deep cervical lymph nodes** or supraclavicular nodes. The cancer cells proliferate here, forming **metastases** (new malignant tumors in parts of the body remote from the site of the primary tumor).

The **supraclavicular nodes**, particularly on the left side, may be enlarged with carcinoma of the bronchus (p. 95), stomach, or any other abdominal organ.

THE CERVICAL VISCERA

THE ESOPHAGUS (Figs. 1-42, 1-72, 1-73, 8-23, 8-28, 8-36, and 8-41)

This thick, distensible, ***muscular tube*** extends from the pharynx to the stomach (about 25 cm). It begins in the midline at the inferior border of the cricoid cartilage and ends at the cardiac orifice of the stomach (Fig. 2-46). Some lay people use the term "*gullet*" to refer to the pharynx and esophagus. Gullet is derived from the Latin word "*gulla*" meaning *throat.*

In the neck the esophagus lies between the trachea and the anterior longitudinal ligament on the anterior surfaces of the vertebral bodies (Fig. 8-32). *On the right side,* the esophagus is in contact with the **cervical pleura** at the root of the neck, whereas *on the* *left side*, posterior to the subclavian artery, the thoracic duct lies between the pleura and the esophagus. Also see the discussion of the esophagus in Chapter 2 (pp. 192 and 201).

THE TRACHEA

The walls of this wide tube ("windpipe") are supported by incomplete **cartilaginous tracheal rings** which are deficient posteriorly where the trachea is related to the esophagus (Figs. 8-27, 8-28, and 8-41). Hence, the posterior wall of the trachea is flat. The cartilaginous rings keep the trachea patent (L. open).

The trachea extends from the larynx to the roots of the lungs, a distance of about 12 cm (Figs. 1-30 and 1-33). The isthmus of the **thyroid gland** usually lies over the second and third tracheal rings (Figs. 8-37, 8-42, and 8-46). Inferior to the isthmus of the thyroid gland are: the jugular venous arch (Fig. 8-25), the inferior thyroid veins (Figs. 8-36 and 8-42), and occasionally a **thyroid ima artery** (see p. 1026).

The brachiocephalic trunk is related to the right side of the trachea at the root of the neck (Figs. 8-26, 8-28, 8-30, 8-31, and 8-42). Lateral to the trachea are the common carotid arteries and the lobes of the thyroid gland (Figs. 8-23, 8-37, and 8-41).

A surgical incision through the neck and the anterior wall of the trachea (**tracheotomy,** Fig. 8-38*A*), is often performed in patients with laryngeal obstruction in order to establish an adequate airway, either as an emergency lifesaving measure or as an elective procedure. A **tracheotomy tube** is inserted into the trachea to keep it patent (Fig. 8-38*C*).

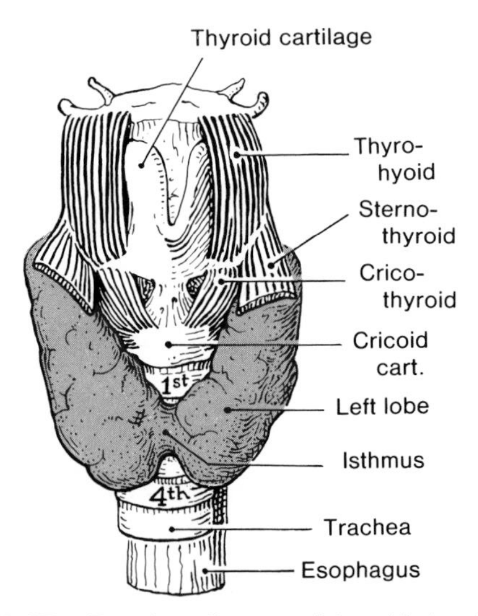

Figure 8-37. Drawing of a normal thyroid gland showing its relationship to the trachea, esophagus, and cricoid cartilage. The sternothyroid muscles have been cut to expose the thyroid gland. Note that the isthmus of the thyroid gland lies anterior to the second and third tracheal rings. The horseshoe-shaped gland illustrated here is typically found in men and thin, elderly women. In pregnant women the contour is more rounded.

Sometimes a round or square opening is made in the neck (Fig. 8-38*B*) rather than a slit, and the tracheal mucosa is brought into continuity with the skin. This operation is referred to as a **tracheostomy** (L. *ostium*, mouth). If the anticipated duration of *endotracheal intubation* is short, a tracheotomy is usually performed, but when supportive measures to maintain respiratory functions are expected to be longer than 72 hours (*e.g.*, a **comatose patient**), a tracheostomy is usually done (Fig. 8-38*C*). When a permanent opening in the trachea is necessary [*e.g.*, after removal of the larynx (**laryngectomy**) owing to cancer], the tracheotomy tube is removed when the skin and epithelium of the trachea have united. The opening is covered with a gauze square for cosmetic purposes and to keep foreign bodies from entering the trachea.

Obstruction of the larynx can result from inhaled foreign bodies (*e.g.*, a piece of steak), *laryngotracheobronchitis* (inflammation of the larynx, trachea, and bronchi), *allergic reactions*, tumors of the larynx, *neurological disorders*, ***epiglotitis,*** and laryngeal diphtheria (now uncommon).

Tracheotomy is also performed for the evacuation of excessive secretions (*e.g.*, resulting from a postoperative chest infection in a patient who is too weak to cough adequately), and for prolonged artificial ventilation in patients with respiratory problems (*e.g.*, related to neurological disorders or drug overdosage).

Although ***emergency tracheotomies*** *are occasionally performed without anesthesia and surgical instruments, a tracheotomy is not a simple operation.* The common approach to the trachea through the skin, subcutaneous tissue, and deep cervical fascia is made by a transverse incision in the neck (Fig. 8-38*A*), usually midway between the laryngeal prominence and the jugular notch (Figs. 8-1 and 8-25).

Surgical opinions vary concerning the best site for making the tracheal incision, but it is commonly made through the second and third tracheal rings. The first ring is not cut because of the danger of narrowing of the trachea during healing. As the isthmus of the thyroid gland covers the second and third tracheal rings (Fig. 8-37), it is retracted inferiorly or superiorly or divided between clamps.

Because of the shortness of the neck in in-

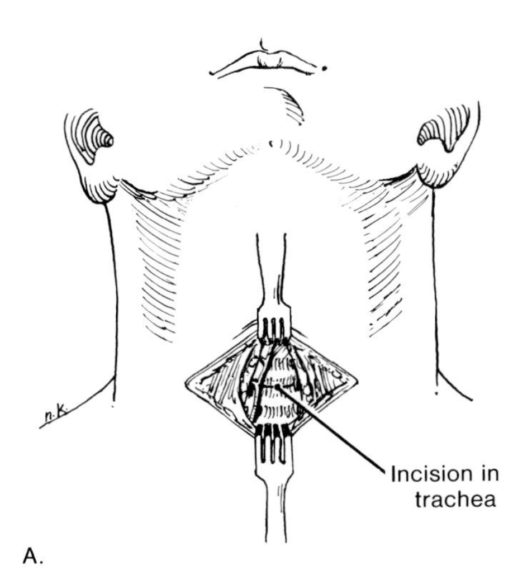

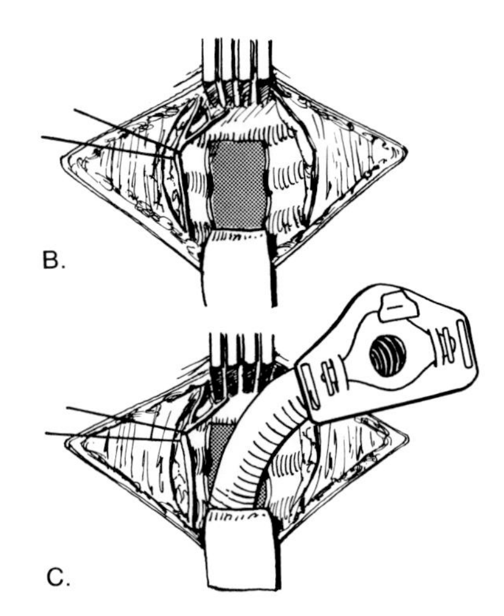

Figure 8-38. *A*, drawing of a transverse incision in the neck and trachea (tracheotomy). *B*, the second and third tracheal rings have been cut, creating a larger opening and a tracheal flap that is sutured to the skin. This facilitates removal and reinsertion of the tracheotomy tube. *C*, a tracheotomy tube is being inserted into the trachea.

fants, the incision in the trachea may be made through the **median cricothyroid ligament** in an emergency (Fig. 8-62). The wound is repaired after the life-threatening situation is over and a formal tracheotomy has been done.

During a tracheotomy inferior to the thyroid gland, *the following anatomical facts must be kept in mind* in order to avoid possible damage to important structures: (1) the ***inferior thyroid veins*** form a plexus anterior to the trachea (Fig. 8-42); (2) a small ***thyroid ima artery*** (lowest thyroid artery) is present in 10% of people which ascends to the inferior border of the isthmus; (3) the ***left brachiocephalic vein*** (Fig. 8-32), jugular venous arch (Fig. 8-25), and the pleurae (Fig. 1-25), may be encountered, particularly in infants and children; (4) the **thymus gland** covers the inferior part of the trachea in infants and children; and (5) the *trachea is small, mobile, and soft in infants,* making it easy to cut through its posterior wall and damage the esophagus (Fig. 8-23).

THE THYROID GLAND

This is an important endocrine gland. Brownish-red during life (Fig. 8-37), it consists of right and left lobes united by a narrow **isthmus of glandular tissue** that extends across the trachea (Figs. 8-32 and 8-37). Rarely there is no isthmus; in these cases the thyroid gland is in two separate parts.

The size of the thyroid gland varies greatly, but it usually weighs about 25 g. It is relatively larger and heavier in women, in whom it becomes slightly larger during menstruation and pregnancy.

A pyramidal lobe is present in about 40% of people. It ascends from the isthmus of the thyroid gland toward the hyoid bone (Fig. 8-39). This lobe may be attached to the hyoid bone by fibrous or muscular tissue (levator glandulae thyroideae). In some cases the pyramidal lobe contains accessory thyroid tissue (Fig. 8-39). The pyramidal lobe represents a persistent portion of the inferior end of the **thyroglossal duct** (Fig. 8-40). In the embryo this duct opened into the foramen cecum of the tongue. This foramen persists in the adult as a minute pit at the apex of the sulcus terminalis (Figs. 7-137 and 7-138).

The thyroid gland lies deep to the sternothyroid and sternohyoid muscles (Figs. 8-25 and 8-41). Its isthmus usually covers the second and third tracheal rings (Fig. 8-37), but its size varies. Each lobe of the thyroid gland extends inferiorly on each side of the trachea, often to the level of the sixth tracheal ring posteriorly, and extends along the sides of the esophagus (Fig. 8-41).

The thyroid gland is surrounded by a fibrous capsule. External to this is a sheath of **pretracheal fascia** (Fig. 8-25) which is attached to the arch of the cricoid cartilage and to the oblique line of the thyroid cartilage (Fig. 8-62). Hence, the thyroid gland moves with the larynx during swallowing and oscillates during speaking. Between the capsule and pretracheal fascia (Fig. 8-41) lie the vessels supplying the gland.

Arterial Supply of the Thyroid Gland (Figs. 8-27, 8-30, 8-31, 8-36, 8-41, and 8-42). This ***highly vascular gland*** usually receives its blood from two rather large arteries: *the superior and inferior thyroid arteries.*

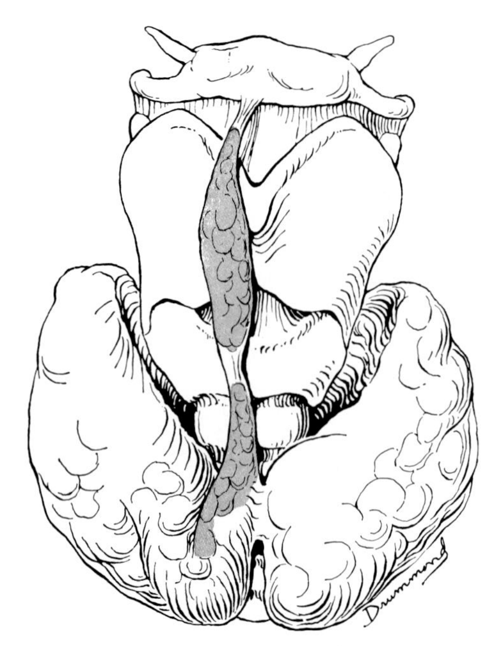

Figure 8-39. Drawing of a thyroid gland with a slender elongated process known as the pyramidal lobe, extending from the isthmus of the thyroid gland to the hyoid bone. In this case it contains accessory thyroid tissue (*blue*). In other cases the pyramidal lobe is composed of fibrous and/or muscular tissue.

The superior thyroid artery, the first branch of the *external carotid artery* (Fig. 8-26), descends to the superior pole of the gland, pierces the pretracheal fascia, and then divides into two or three branches (Fig. 8-27).

The inferior thyroid artery, a branch of the *thyrocervical trunk* (Fig. 8-26), runs superomedially posterior to the carotid sheath to reach the posterior aspect of the gland. It divides into several branches which pierce the pretracheal fascia to supply the inferior pole of the thyroid gland (Fig. 8-36).

Occasionally a third vessel, the unpaired **thyroid ima artery** (arteria thyroidea ima) supplies the thyroid gland. This small artery may arise from the arch of the aorta, the brachiocephalic trunk (most commonly), or the left common carotid artery. It ascends anterior to the trachea and *supplies the isthmus of the thyroid gland.* The thyroid ima artery is normally present in the embryo and persists in about 10% of adults.

The possible presence of a thyroid ima artery must be remembered when incising the trachea inferior to the isthmus. As it runs anterior to the trachea, it is a potential source of serious bleeding.

As well as the named thyroid arteries, numerous small vessels pass to the thyroid gland from those supplying the pharynx and trachea. Hence, the thyroid gland still oozes blood during a subtotal thyroidectomy (discussed on p. 1029), even when the two main arteries supplying it are ligated.

All the thyroid arteries anastomose with each other on and in the substance of the thyroid gland (Fig. 8-42), but there is little anastomosis across the median plane except for the branches of the superior thyroid artery.

Venous Drainage of the Thyroid Gland (Figs. 8-27, 8-30, 8-32, 8-36, and 8-42). Usually three pairs of veins drain the venous plexus on the surface of the thyroid gland.

The superior thyroid veins drain the superior poles of the thyroid gland, and the **middle thyroid veins** drain its lateral parts. The superior and middle thyroid veins empty into the *internal jugular veins* (Fig. 8-36).

The inferior thyroid veins drain the inferior poles of the thyroid gland, and empty into the *brachiocephalic veins* (Fig. 8-32). Often they unite to form a single vein (Fig. 8-42) that opens into one of the brachiocephalic veins.

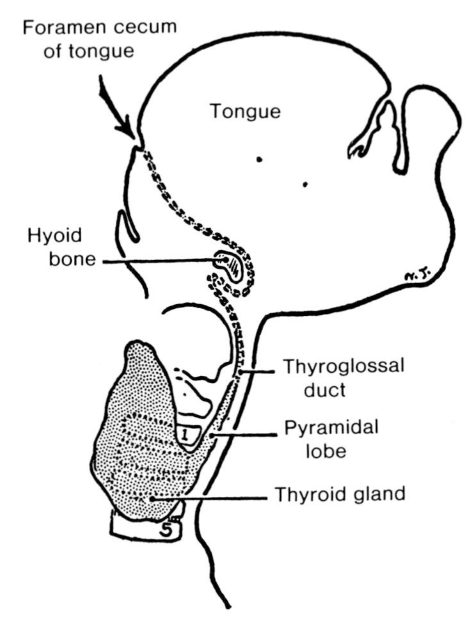

Figure 8-40. Drawing showing the course of the thyroid gland as it developed and descended through the neck. The thyroglossal duct usually degenerates, but sometimes it persists and forms a pyramidal lobe (Fig. 8-39). Remnants of it may also persist and give rise to cysts or to a fistula (Fig. 8-43).

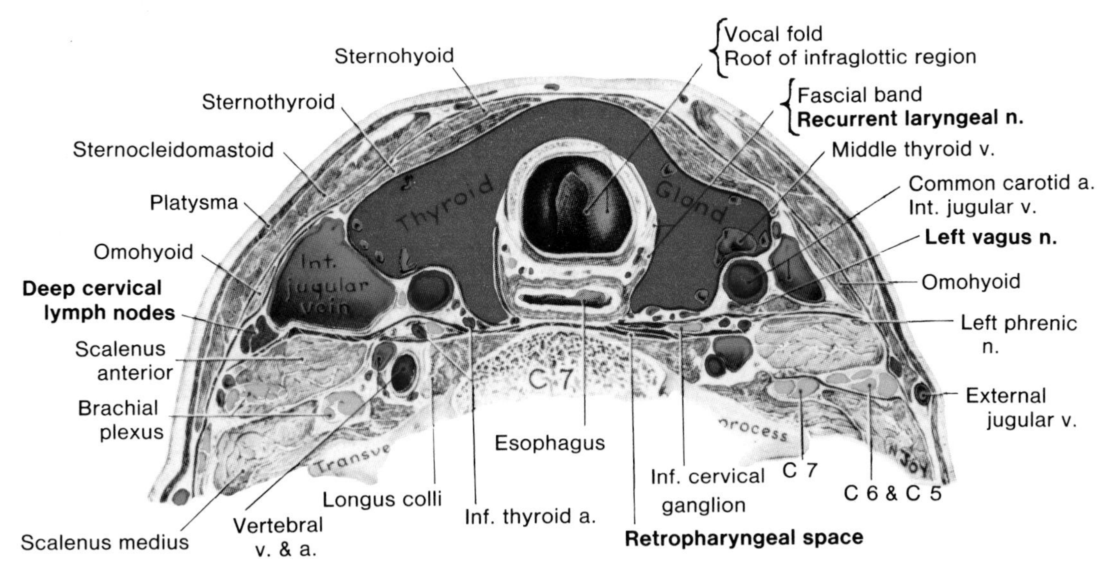

Figure 8-41. Drawing of a cross-section of the neck at the level of C7 vertebra. Observe the thyroid gland, viewed inferiorly as in CT scans. Note that the fascia surrounding the esophagus is separated from the prevertebral fascia by a retropharyngeal space (also see Fig. 8-46). Observe the thyroid gland within its sheath, asymmetrically enlarged and covering the carotid sheath and its contents (common carotid artery, internal jugular vein, and vagus nerve) on the *right side.* Note the fascial band (sheath) that attaches the thyroid gland to the cricoid and thyroid cartilages. Observe the recurrent laryngeal nerve which is at risk during a thyroidectomy.

As they cover the anterior surface of the trachea, inferior to the isthmus of the thyroid gland (Fig. 8-36), the inferior thyroid veins are potential sources of bleeding during a tracheotomy.

Lymph Drainage of the Thyroid Gland (Figs. 8-32, 8-35, 8-41, and 8-44). The lymph vessels accompany the arteries and pass to the **inferior deep cervical lymph nodes** and the paratracheal lymph nodes (Fig. 8-44). Some vessels anterior to the trachea drain to the *pretracheal lymph nodes,* which are connected with the *parasternal lymph nodes* (Fig. 1-16). Other lymph vessels may empty directly into the thoracic duct (Fig. 8-35).

Nerve Supply to the Thyroid Gland (Figs. 8-28, 8-30, 8-34, and 8-36). The nerves are derived from the **superior, middle, and inferior cervical sympathetic ganglia** (Fig. 8-28). They reach the thyroid gland via the cardiac and laryngeal branches of the vagus nerve (Fig. 7-135), which run along the arteries supplying the gland (Fig. 8-30). These postganglionic fibers are vasomotor and affect the gland indirectly through their action on the blood vessels.

Thyroglossal duct cysts may develop from remnants of the thyroglossal duct (Fig. 8-40), anywhere along the course taken by the duct during embryonic descent of the thyroid gland (Fig. 8-43). The cysts may be in the tongue or in the midline of the neck, usually just inferior to the hyoid bone. In some cases an opening to the skin develops as a result of perforation following infection of a cyst. **Thyroglossal duct sinuses** usually open in the *midline of the neck,* anterior to the thyroid cartilage (Fig. 8-43*A*).

Rarely the thyroid gland fails to descend during development resulting in the development of a **lingual thyroid gland,** or a superior cervical thyroid gland in the region of the hyoid bone. **Accessory thyroid gland** tissue may also develop from remnants of the thyroglossal duct (Fig. 8-42).

Abnormal enlargement of the thyroid gland is called a goiter. The enlarged gland causes a swelling in the anterior part of the neck. It may exert pressure on the trachea or the recurrent laryngeal nerves (Fig. 8-41). There are various types of goiter, *e.g.,* **exophthalmic goiter** is a disorder caused by an excessive production of the thyroid hormone. One sign of this disease, bulging eyeballs, is called *exophthalmos* (Fig. 8-78).

In some cases of **hyperthyroidism** the enlarged thyroid gland may cause **stridor** (a harsh, high-pitched respiratory sound), **dyspnea,** and **dysphagia** (difficulty in swallowing) owing to compression of the trachea and esophagus (Figs. 8-23 and 8-41).

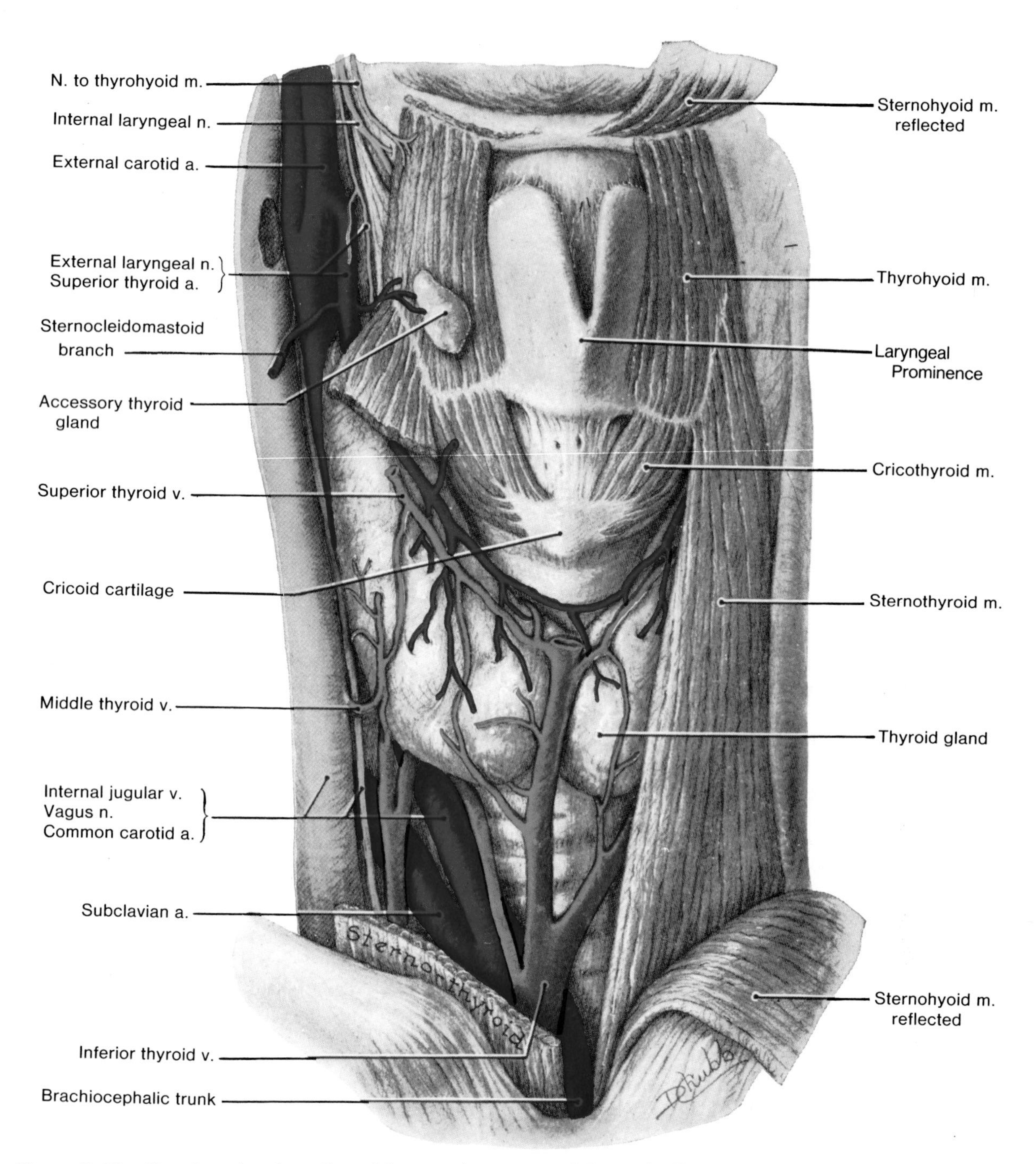

Figure 8-42. Drawing of a dissection of the anterior aspect of the neck. Observe the lobes of the thyroid gland which are united across the median plane by an isthmus like that shown in Figure 8-43. Note the network of veins on the gland which is drained by the superior, middle, and inferior thyroid veins. Observe that the right lobe of the gland overlies the common carotid artery. In this specimen there is an accessory thyroid gland on the right, lying on the thyrohyoid muscle, lateral to the thyroid cartilage. The inferior thyroid veins have united to form a single vessel that ends in the right brachiocephalic vein (see Fig. 8-32).

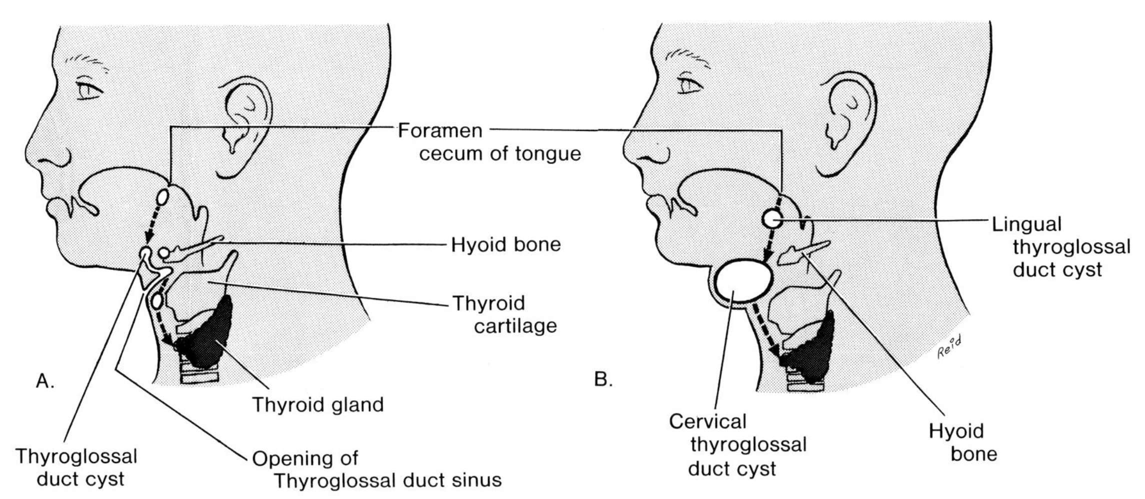

Figure 8-43. *A*, diagrammatic sketch of the head showing the possible locations of thyroglossal duct cysts. A thyroglossal duct sinus is also illustrated. The broken line (– – –) indicates the course taken by the thyroglossal duct during descent of the thyroid gland from the foramen cecum to its final position anterior to the trachea (also see Fig. 8-40). *B*, similar sketch illustrating lingual and cervical thyroglossal duct cysts. Most thyroglossal duct cysts develop near the hyoid bone. (From Moore KL: *The Developing Human: Clinically Oriented Embryology*, ed 3. Philadelphia, WB Saunders Co, 1982.)

Narrowing of the trachea is commonly found with carcinoma of the thyroid gland, and *retrosternal goiter* (extension of an enlarged thyroid gland posterior to the sternum).

It is sometimes necessary to remove the thyroid gland (***total thyroidectomy***), *e.g.*, during excision of a carcinoma of the thyroid gland.

In the surgical treatment of hyperthyroidism, usually the posterior portion of each lobe of the enlarged thyroid gland is left (**subtotal thyroidectomy**). A sound knowledge of the anatomy of the thyroid gland and its relations is essential during a thyroidectomy, otherwise serious complications may arise during and/or after the operation.

An enlarged thyroid gland may stretch the strap muscles (sternohyoid and sternothyroid) making them very thin. To obtain a good exposure of the thyroid gland, it may be necessary to divide these muscles. To prevent injury to their nerves, the muscles are divided at the level of the gland because the nerves mainly enter superior and inferior parts of these muscles (Fig. 8-19).

Injury to the recurrent laryngeal nerves is not common; however, as the risk of injuring them during surgery is ever present, a good knowledge of their relationship to the trachea, carotid arteries, thyroid gland, and inferior thyroid arteries is essential (Figs. 8-27, 8-30, 8-36, and 8-40).

Near the inferior pole of the thyroid gland, **the right recurrent laryngeal nerve is intimately related to the inferior thyroid artery** (Fig. 8-30). This nerve may cross anterior or posterior to the artery, or it may pass between its branches. Because of this close relationship, *the inferior thyroid artery is ligated some distance lateral to the thyroid gland, where it is not so close to the right recurrent laryngeal nerve.* Although the danger of injuring the left recurrent laryngeal nerve during surgery is not so great, it must be remembered that the artery and nerve are also closely associated near the inferior pole of the thyroid gland (Fig. 8-36). Damage to a recurrent laryngeal nerve may be partial or complete.

Temporary disturbance of phonation (voice production) and laryngeal spasm not uncommonly follow partial thyroidectomy, as the result of handling the nerves during the operation, or owing to the pressure of accumulated blood and/or serous exudate after the operation.

Injury to the external laryngeal nerve is uncommon during thyroidectomy; however, as *it is closely related to the superior thyroid artery during part of its course to the superior pole of the thyroid gland* (Fig. 8-36), care must be taken not to damage it when the superior thyroid artery is ligated and sectioned.

Injury to external laryngeal nerve results in the voice becoming monotonous in character be-

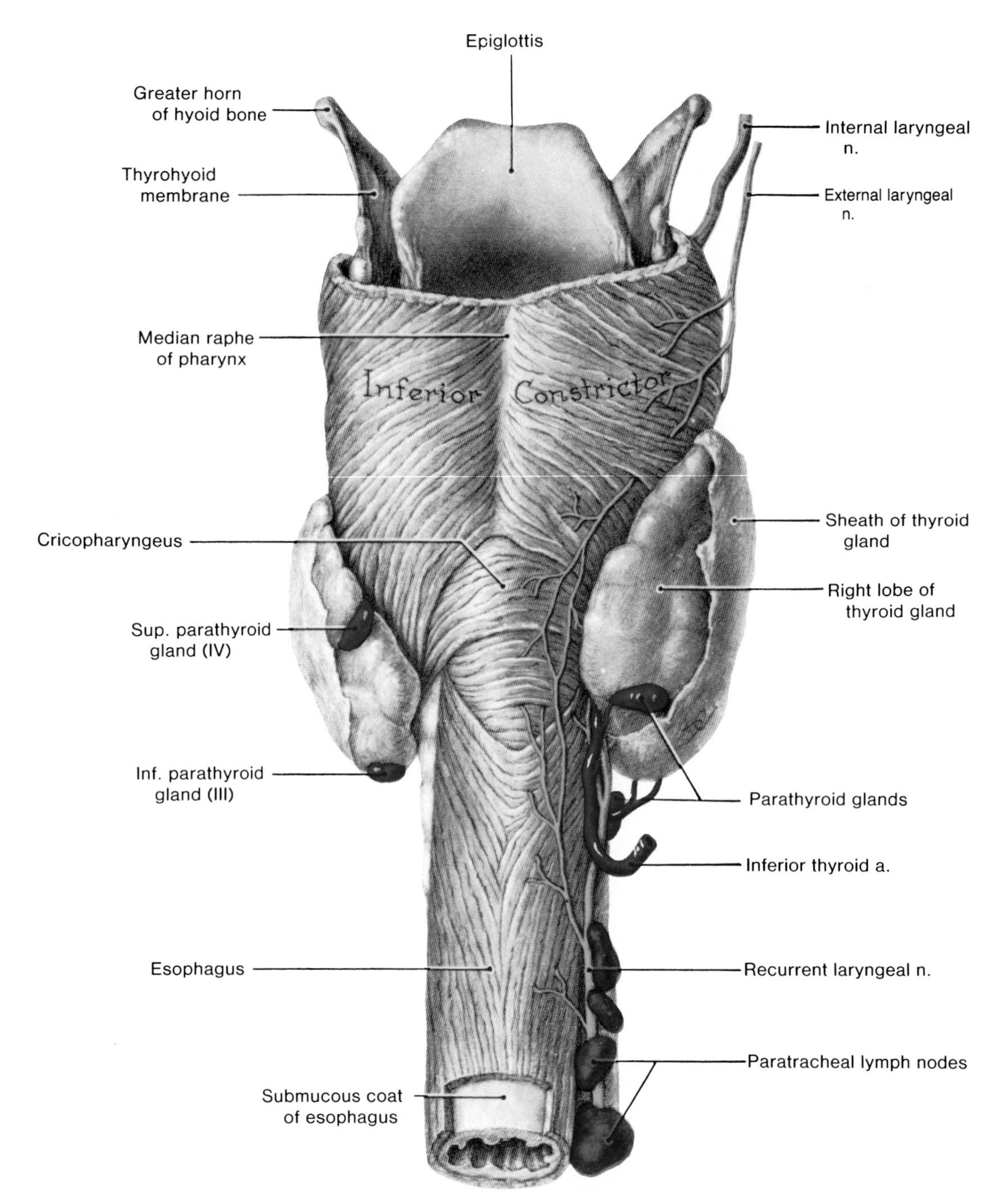

Figure 8-44. Drawing of a dissection of the posterior surface of the thyroid and parathyroid glands. Observe that the right and left lobes of the thyroid gland are unequal in size in this specimen, and that these lobes are applied to the inferior constrictor muscle of the pharynx, trachea, and esophagus. *On the right side*, both parathyroids are rather low, the inferior gland being inferior to the thyroid gland. The Roman numerals after the glands indicate that they developed from the IVth and IIIrd pharyngeal pouches, respectively. The inferior parathyroid glands (III) were pulled caudal to the superior parathyroid glands (IV) by the descending thymus gland, another derivative of the third pair of pharyngeal pouches. After giving off branches (motor and sensory) to the trachea, esophagus, and pharynx, the continuation of the recurrent laryngeal nerve is known as the inferior laryngeal nerve.

cause the paralyzed cricothyroid muscle is unable to vary the length and tension of the **vocal fold** (Fig. 8-72).

To avoid injury to the external laryngeal nerve, the superior thyroid artery is ligated and sectioned near the superior pole of the thyroid gland, where it is not so closely related to the nerve as it is at its origin (Fig. 8-36).

Because an enlarged thyroid gland may itself be the cause of impaired innervation of the larynx as the result of compression, it is common practice to examine the vocal folds prior to an operation. In this way, damage to the larynx or its nerves resulting from a surgical mishap may be distinguished from a pre-existing injury.

THE PARATHYROID GLANDS

There are *usually four* small (about 6 × 3 × 2 mm) parathyroid glands. Yellowish-brown during life, these ovoid **endocrine glands** lie along the posterior border of the thyroid gland, between its capsule and sheath (Figs. 8-44 and 8-47). Usually there are two glands associated with each lobe of the thyroid gland, but the total number usually varies between two and six.

The parathyroid glands are named according to their positions as the **superior and inferior parathyroid glands** (Fig. 8-44). The superior ones are more constant in position than the inferior ones. They are usually located near the middle of the posterior surface of the lobes of the thyroid gland at the level of the inferior border of the cricoid cartilage (Figs. 8-41, 8-44, and 8-45).

The inferior parathyroid glands are variable in position. They are usually located near the inferior surface of the thyroid gland, but they may lie some distance inferior to it (Fig. 8-44), even in the superior mediastinum (Fig. 1-43). The best guide to their location is to follow their small arteries.

Blood Supply of the Parathyroid Glands (Figs. 8-44, 8-45, and 8-47). These glands are usually supplied by the **inferior thyroid arteries**, but they may be supplied by the superior thyroid arteries, or from the longitudinal anastomosis between the superior and inferior thyroid arteries.

Venous Drainage of the Parathyroid Glands (Fig. 8-42). The veins from the parathyroid glands drain into the **thyroid plexus of veins** on the anterior surface of the thyroid gland and trachea.

Lymph Drainage of the Parathyroid Glands (Figs. 8-35, 8-44, and 8-60). The lymph vessels drain with those of the thyroid gland into the *inferior* ***deep cervical lymph nodes*** *and into the paratracheal lymph nodes*. These lymph vessels end in the thoracic duct and the right lymphatic duct (Fig. 8-35).

Nerve Supply of the Parathyroid Glands (Figs. 8-30 and 8-36). The nerves to these glands are derived

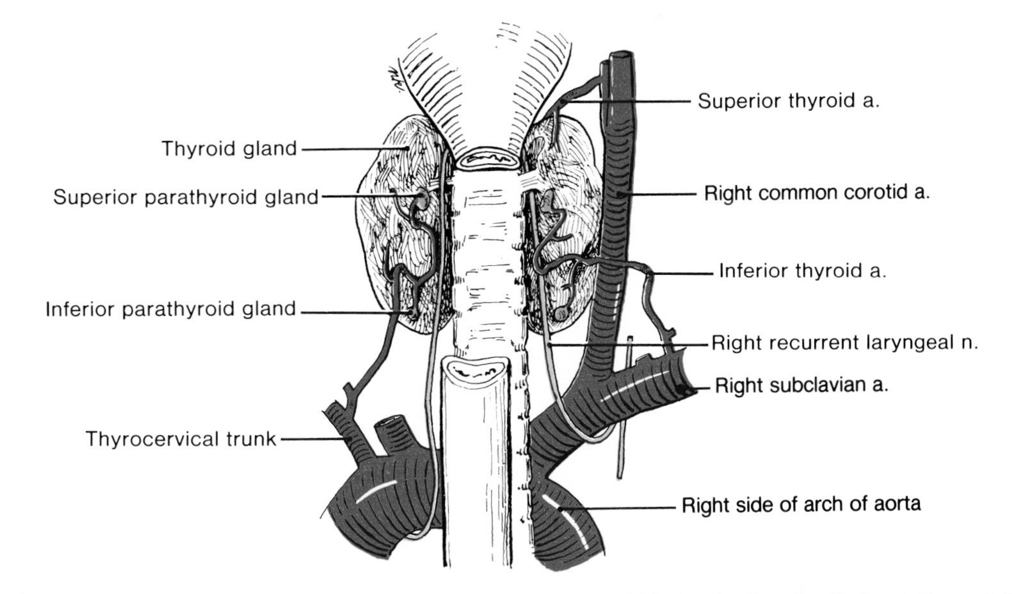

Figure 8-45. Drawing of a posterior view of the thyroid and parathyroid glands showing their relations. A thorough knowledge of the appearance, location, and blood supply of the parathyroid glands is essential for their preservation during operations on the thyroid gland. The superior thyroid artery is a branch of the external carotid (Fig. 8-26), and the inferior thyroid artery is a branch of the thyrocervical trunk (Fig. 8-26). The superior parathyroid glands are the more constant in position and are therefore usually easy to find. The inferior parathyroid glands are often close to the inferior surface of the lobes of the thyroid gland (as here), but they may lie inferior to them (see Figs. 8-44 and 8-47).

from the **sympathetic trunks**, either directly from the superior or middle cervical sympathetic ganglia (Figs. 8-28 and 8-30), or from the plexus surrounding the superior and inferior thyroid arteries.

An awareness of the close relationship between the parathyroid glands, the thyroid gland, and the recurrent laryngeal nerves is essential knowledge. The possibility of injuring the recurrent laryngeal nerves during thyroidectomy has been discussed. The small parathyroid glands are also in danger of being damaged or removed during this operation, but they are usually safe during subtotal thyroidectomy because the posterior part of the thyroid gland is preserved.

The variability in the position of the parathyroid glands, particularly the inferior ones, creates a problem in thyroid and parathyroid surgery. The parathyroids may be as far superior as the thyroid cartilage, and the inferior ones may be as inferior as the **superior mediastinum** (Fig. 1-43). Sometimes an inferior parathyroid gland is embedded within the inferior end of the thyroid gland. These possible aberrant sites are of concern when searching for an abnormal parathyroid gland (*e.g.*, a parathyroid adenoma) in hyperparathyroidism.

The variability in position of the inferior parathyroid glands has an embryological basis. They develop in association with the thymus gland and are carried caudally with it during its embryonic descent through the neck **from the third pair of pharyngeal pouches.** If the thymus gland fails to descend to its usual position, the inferior parathyroids usually lie at the level of the thyroid cartilage.

If the inferior parathyroids do not disassociate from the thymus and become associated with the thyroid gland, they may be carried with it inferior to the thyroid and even into the thorax (*i.e.*, into the superior mediastinum, Fig. 1-43). Consequently it may be very difficult to distinguish between superior and inferior parathyroid glands or to find all four of them; *rarely are there more than four parathyroid glands.*

If the parathyroid glands atrophy or are inadvertently removed during surgery, the patient suffers from a severe *convulsive disorder* known as tetany.[2] The generalized convulsive spasms result from a lowered serum calcium level.

In tetany, there is nervousness, twitching, and spasms in the facial and limb musculature. If the respiratory and laryngeal muscles are also affected, death may occur if medical treatment is not given (*e.g.*, by injecting a calcium and/or a parathyroid extract).

To safeguard the parathyroid glands during thyroidectomy, the posterior portion of each lobe of the thyroid gland is usually not removed. In rare instances when it is necessary to do a **total thyroidectomy** (*e.g.*, in malignant disease), the parathyroid glands are isolated with their blood vessels intact before the thyroid gland is removed.

FASCIAL PLANES OF THE NECK

Although the layers of cervical fascia have been mentioned previously, their surgical importance warrants that they be discussed in more detail.

The deep fascia of the neck is described in three layers: investing, pretracheal, and prevertebral (Figs. 8-23, 8-25, and 8-41).

THE INVESTING FASCIA

The superficial layer of cervical fascia encircles the neck and surrounds the structures in the neck (Figs. 8-25 and 8-41). It is attached superiorly to the **superior nuchal line** (Figs. 6-44 and 7-3), the **mastoid process**, the **zygomatic arch**, the inferior border of the **mandible, hyoid bone**, and the **spinous processes of the cervical vertebrae.**

Inferiorly, the investing fascia is attached to the **manubrium sterni** (Fig. 8-28), the **clavicle** (Fig. 8-35), and the acromion and spine of the **scapula.**

Immediately superior to the sternum, the investing fascia divides into two layers; they are attached to the anterior and posterior surfaces of the manubrium sterni, respectively. The interval between these two layers of fascia is called the **suprasternal space** (Figs. 8-25 and 8-46). It encloses the sternocleidomastoid and trapezius muscles, the jugular venous arch, and an occasional lymph node.

Together with the skin, the investing fascia also forms the roof of the anterior and posterior triangles of the neck (Figs. 8-6 and 8-18).

The superficial investing layer of cervical fascia tends to prevent the extension of abscesses toward the surface. Pus beneath it

[2] Note that *tetany* and *tetanus* are not the same conditions. Tetanus is an infectious disease in which tonic muscle spasm results in "lockjaw" and similar spasm of other muscle groups.

usually extends laterally in the neck, but if the abscess is in the anterior triangle, the pus may pass inferiorly, producing a swelling in the suprasternal space (Figs. 8-25 and 8-46). It could also pass into the anterior mediastinum (Fig. 1-43).

THE PRETRACHEAL FASCIA

This thin layer of cervical fascia is limited to the anterior aspect of the neck. It extends from the thyroid cartilage and the arch of the cricoid cartilage inferiorly into the thorax (Figs. 8-23, 8-25, and 8-41). Note that this fascia is more extensive than its name implies. *The pretracheal fascia splits to enclose* the *thyroid gland, trachea,* and *esophagus and blends laterally with the carotid sheath* (Figs. 8-23 and 8-41).

Infections in the head or cervical region of the vertebral column can spread inferiorly, posterior to the esophagus, into the posterior mediastinum (Fig. 1-77). They can also spread inferiorly, anterior to the trachea, entering the anterior mediastinum (Fig. 1-43).

Air from a **ruptured trachea**, bronchus, esophagus, or so-called "spontaneous" **mediastinal emphysema**, can pass superiorly into the neck. An unusual source of air in the face and neck is from a dentist's drill via a tooth socket. Similarly a small slit in the buccal mucosa, followed by hard blowing with the mouth closed, may result in subcutaneous **cervicofacial emphysema.** This unusual kind of swollen neck has been reported in glass blowers, players of wind instruments, and malingerers.

THE PREVERTEBRAL FASCIA

This layer of cervical fascia covers the prevertebral muscles and is continuous with the deep fascia covering the muscular floor of the posterior triangle of the neck (Figs. 8-23, 8-35, and 8-41). It is part of the strong "fascial sleeve" that envelops the deep muscles of the neck.

The prevertebral fascia extends from the base of the skull to the third thoracic vertebra; **it fuses with the anterior longitudinal ligament of the vertebral column** (Figs. 5-28 and 5-34).

The prevertebral fascia extends inferiorly and laterally as the **axillary sheath** (Fig. 6-42), which surrounds the axillary vessels and brachial plexus (Figs. 8-12 and 8-35). The adjective "prevertebral" is misleading because *the prevertebral fascia surrounds the vertebral column and its associated musculature* (Figs. 8-23 and 8-41).

Pus from an abscess located posterior to the prevertebral fascia may perforate anteriorly in the **retropharyngeal space** (Figs. 8-41 and 8-46). However it is more likely to extend into the lateral parts of the neck deep to the prevertebral fascia, on the floor of the posterior triangle (Fig. 8-35), and form a swelling that reaches the surface, posterior to the sternocleidomastoid muscle.

The fascial planes of the neck are important to surgeons because they form natural **lines of cleavage** through which the tissues may be separated, and because they limit the spread of pus resulting from infections in the neck.

THE RETROPHARYNGEAL SPACE

This potential space consists of loose connective tissue between the prevertebral fascia and the buccopharyngeal fascia (Figs. 8-41, 8-46, and 8-57). The buccopharyngeal fascia surrounds the pharynx superficially (Fig. 8-46).

The retropharyngeal space is a potential space, permitting movement of the pharynx, larynx, trachea, and esophagus during swallowing. This "space" is closed superiorly by the base of the skull and on each side by the carotid sheath (Figs. 8-41 and 8-57). The retropharyngeal space opens inferiorly into the superior mediastinum (Figs. 1-43 and 8-46), which contains the thymus, the great vessels of the heart, the trachea, and the esophagus (Fig. 1-71).

The retropharyngeal space is of considerable surgical interest because of the structures related to it. Pus located posterior to the prevertebral fascia (Fig. 8-23) may perforate this layer and enter the retropharyngeal space (Figs. 8-41, 8-46, and 8-57). This produces a bulge in the pharynx known as a **retropharyngeal abscess,** which causes difficulty in swallowing and speaking (*i.e.,* dysphagia and dysarthria). Infections in this retropharyngeal space may extend inferiorly into the superior mediastinum, (Fig. 1-43), producing **mediastinitis** (inflammation of the tissue in the mediastinum).

THE PHARYNX

The pharynx is the *continuation of the digestive cavity from the oral cavity.* **It is a funnel-shaped fibromuscular tube that is the common route for air and food.** In its superior part, the pharynx receives the posterior openings of the nasal cavities, called **choanae** (Figs. 7-9*B* and 8-52). The pharynx is

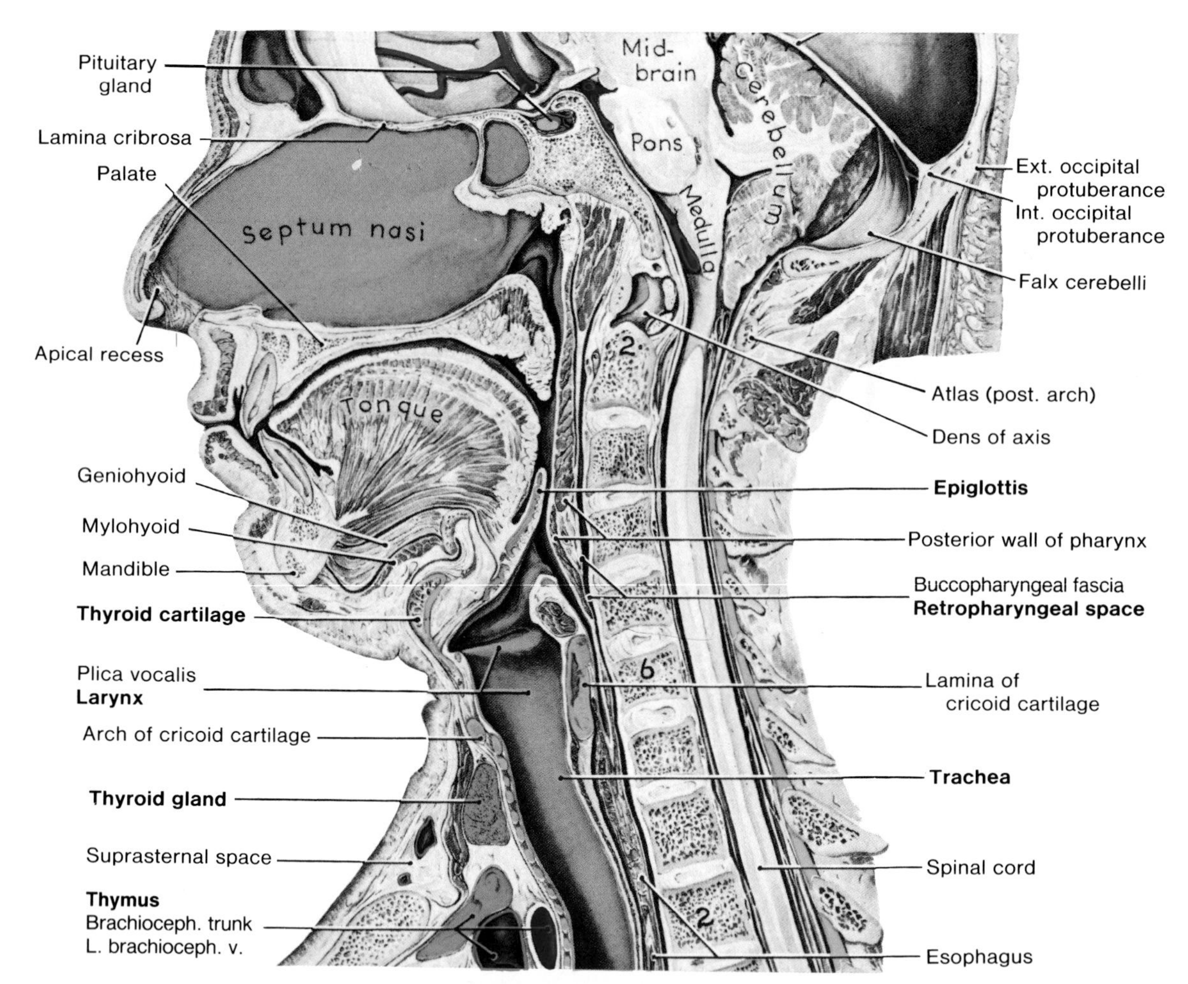

Figure 8-46. Drawing of a median section of the head and neck. Observe the pharynx extending from the base of the skull to the level of the body of the sixth cervical vertebra, where it is continuous with the esophagus. Observe that the pharynx lies posterior to the nose, mouth, and larynx. Note the lamina of the cricoid cartilage at the level of the body of the sixth cervical vertebra and that at its inferior border, the larynx becomes the trachea and the pharynx becomes the esophagus. Observe the *retropharyngeal space* extending from the level of the atlas inferiorly into the superior mediastinum. Observe the laryngopharynx (laryngeal portion of the pharynx) lying posterior to the aperture and the posterior wall of the larynx.

located posterior to the nasal and oral cavities and the larynx (Figs. 8-46 and 8-52).

The pharynx is the superior end of the respiratory and digestive tubes and is continuous inferiorly with the esophagus. The pharynx conducts food to the esophagus and air to the larynx and lungs.

For convenience of description, the pharynx is divided into three parts (Figs. 8-46 and 8-52): (1) **the nasopharynx**, posterior to the nose and superior the soft palate; (2) **the oropharynx**, posterior to the mouth; and (3) **the laryngopharynx**, posterior to the larynx.

*The pharynx, about 15 cm long, extends from the **base of the skull** to the inferior border of the **cricoid cartilage** anteriorly, and to the inferior border of the **sixth cervical vertebra** posteriorly.* It is widest (about 5 cm) opposite the hyoid bone and narrowest (about 1.5 cm) at its inferior end (Fig. 8-46), where it is continuous with the esophagus. The posterior wall of the pharynx lies against the **prevertebral fascia** (Fig. 8-23), with the potential *retropharyngeal space* between them (Figs. 8-46 and 8-57).

In Figure 8-46 observe that the pathways for food and air cross each other in the pharynx. Hence food sometimes enters the respiratory tract and causes choking, and air may enter the digestive tract and produce gas in the stomach. Because this is uncomfortable it results in **eructation** (belching).

The pharyngeal wall is composed of five layers. From internal to external, they are: (1) **a mucous membrane** that lines the pharynx and is continuous with all chambers with which it communicates (Figs. 8-46, 8-51, and 8-52); (2) **a submucosa**; (3) **a fibrous layer** forming the *pharyngobasilar fascia* (Figs. 8-47

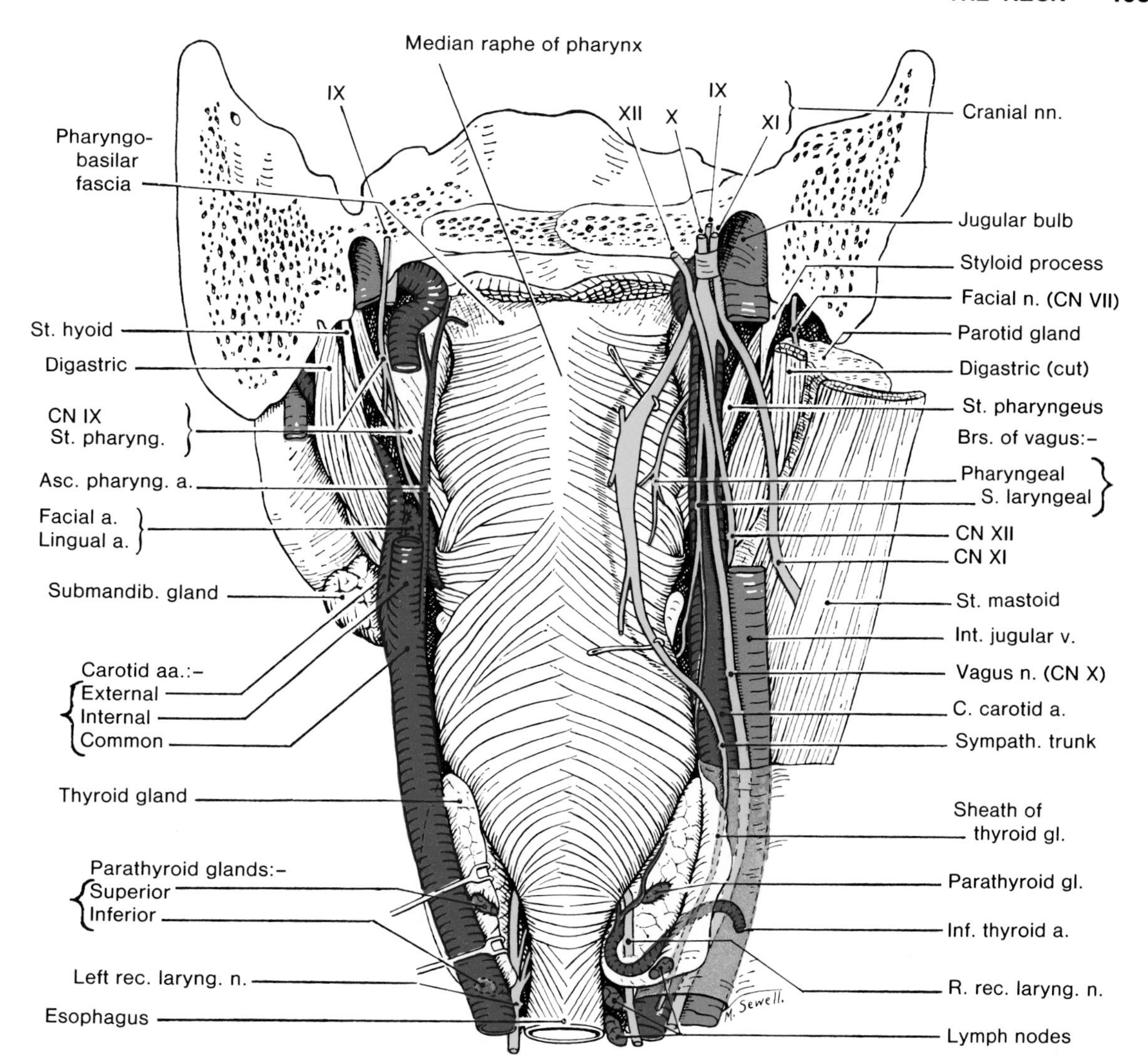

Figure 8-47. Drawing of a dissection of the posterior surface of the pharynx. Note that the pharynx begins at the skull and ends inferiorly in the esophagus. Observe the thyroid and parathyroid glands and that the parathyroid glands are closely applied to the posterior aspect of the thyroid gland. Although not visible, they are embedded in the posterior surface of the fibrous capsule of this gland. The parathyroid glands, particularly the inferior ones, may be located on the lateral or anterior surface of the thyroid or be imbedded in it. Note that the left inferior parathyroid in this specimen is located posterior to the common carotid artery. *The inferior parathyroid glands are in an abnormal position in about 10% of persons.* Obviously a thorough knowledge of the parathyroid glands is necessary for their preservation during operations on the thyroid, and in searching for an abnormal parathyroid gland.

and 8-48) which is attached to the skull; (4) **a muscular layer** composed of *inner longitudinal and outer circular parts* (Fig. 8-61); and (5) **a loose connective tissue layer forming the buccopharyngeal fascia** (Fig. 8-46). This fascia is continuous with the epimysium covering of the buccinator and pharyngeal muscles. The areolar layer permits movements of the pharynx and contains the pharyngeal plexus of nerves and veins (Figs. 8-47 and 8-59).

EXTERNAL MUSCLES OF THE PHARYNX (Figs. 8-47 to 8-50 and Table 8-4)

The external circular part of the muscular layer of the wall of the pharynx is formed by the paired superior, middle, and inferior constrictor muscles, which overlap one another. These constrictor muscles are arranged so that the superior one is innermost and the inferior one is outermost. As their names indicate, *they all constrict the pharynx in swallowing.* They contract involuntarily in a way that results in contraction taking place sequentially from the superior to the inferior end of the pharynx. This action propels the food into the esophagus.

All three constrictors of the pharynx are supplied by the ***pharyngeal plexus of nerves*** (Figs. 8-47 and 8-50), which lies on the lateral wall of the pharynx, mainly on the middle constrictor muscle. This plexus is formed by the *pharyngeal branches of the glossopha-*

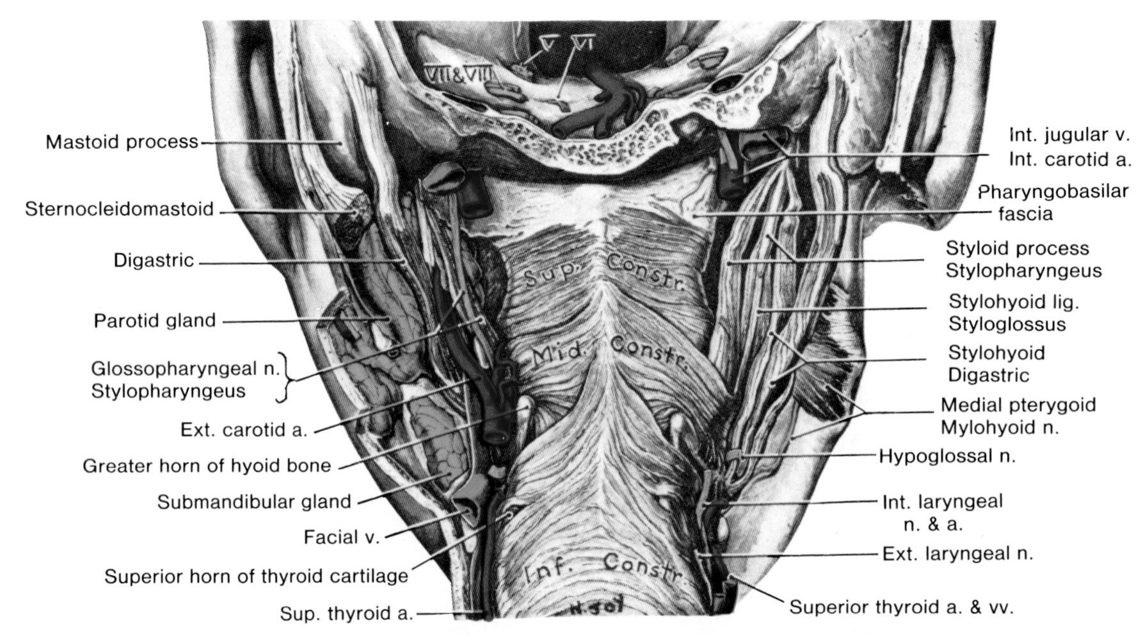

Figure 8-48. Drawing of a dissection of the posterior aspect of the pharynx and parotid gland. Observe the pharyngobasilar fascia which suspends the pharynx from the basioccipital bone, and note the three constrictor muscles nestled within each other like stacked roof tiles or boats. Observe that the posterior aspect of the pharynx is flat or slightly concave from side to side where it is applied to the prevertebral region. Note the stylopharyngeus muscle and glossopharyngeal nerve (CN IX) passing from the medial side of the styloid process anteromedially through the gap between the superior and middle constrictor muscles.

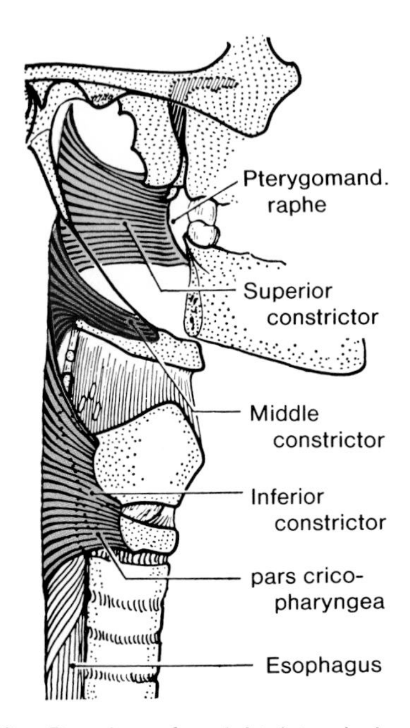

Figure 8-49. Drawing of a right lateral view of the constrictor muscles of the pharynx, showing their attachments and borders. Note that each constrictor muscle is fan-shaped and that the narrow ends of the fans are fixed anteriorly. Posteriorly the constrictor muscles meet in the median raphe of the pharynx (Figs. 8-44 and 8-47). Observe the continuous origin of the inferior constrictor muscle from the thyroid and cricoid cartilages.

ryngeal (CN IX) and vagus (CN X) nerves (Figs. 7-111 and 7-135).

The Superior Constrictor Muscle (Figs. 8-47 to 8-51 and Table 8-4). This muscle is broad and arises from the **pterygoid hamulus** (Fig. 7-9*A*), the **pterygomandibular raphe** (Fig. 8-49), and **the mylohyoid line of the mandible** (Fig. 7-22), posterior to the third molar tooth. It also arises by a few fibers from the side of the tongue.

The pterygomandibular raphe is the fibrous line of junction between the buccinator and superior constrictor muscles (Fig. 8-49). Fibers of the superior constrictor muscle curve posteriorly around the pharynx to insert into the **median raphe of the pharynx** (Fig. 8-47). The most superior fibers reach the pharyngeal tubercle of the skull.

The Middle Constrictor Muscle (Figs. 8-47 to 8-51 and Table 8-4). This muscle arises from the inferior end of the **stylohyoid ligament** and the **greater and lesser horns of the hyoid bone** (Figs. 7-141 and 8-49). Its fibers curve posteriorly around the pharynx to insert into its *median raphe* (Fig. 8-47).

The Inferior Constrictor Muscle (Figs. 8-47 to 8-51 and Table 8-4). This muscle has a continuous origin from the **oblique line of the thyroid cartilage**, the fascia over the cricothyroid muscle, and the side of the **cricoid cartilage**. Its fibers pass superiorly, horizontally, and inferiorly, overlapping those of the middle constrictor, to insert into the **median raphe of the pharynx** (Fig. 8-47). The fibers arising

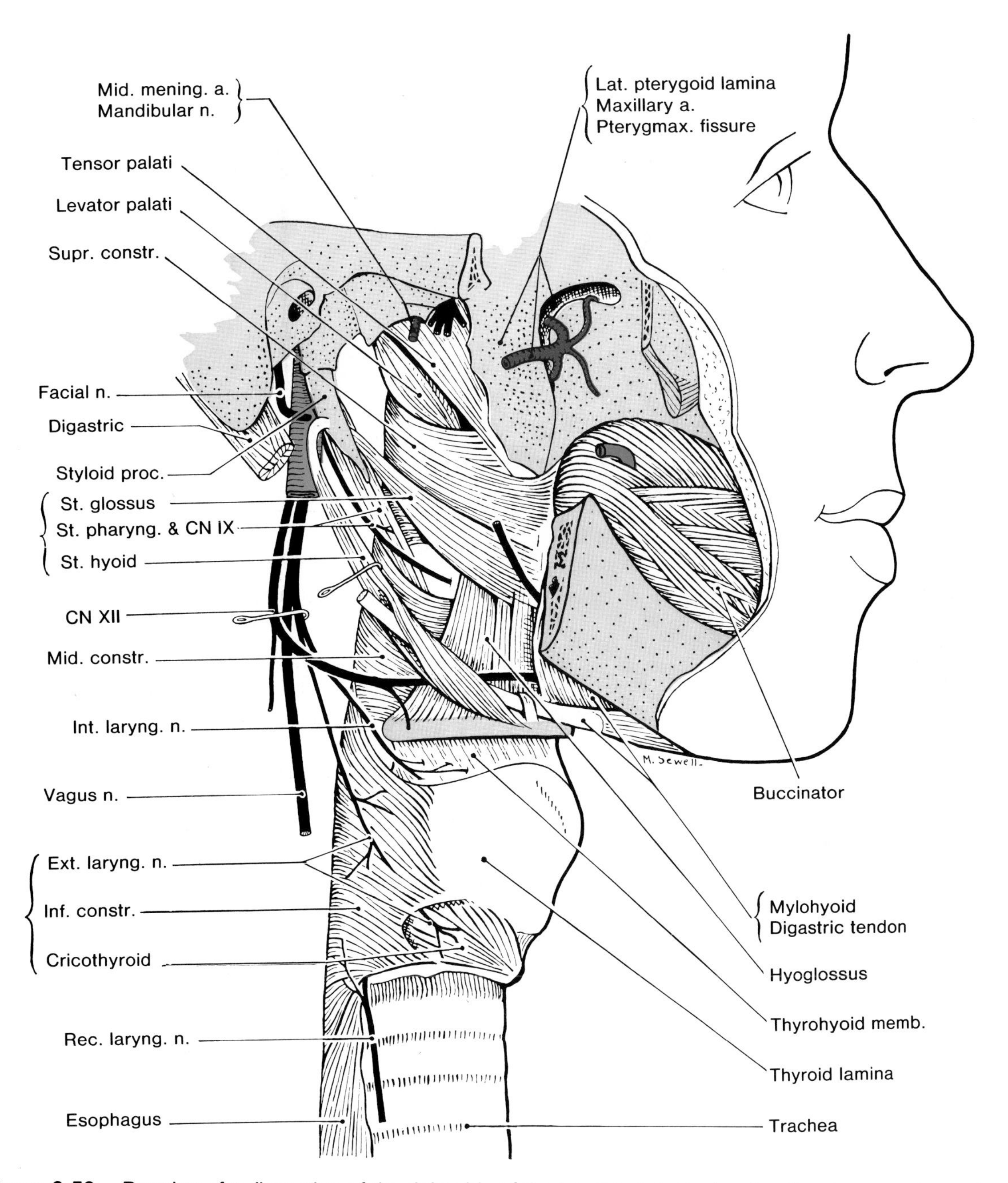

Figure 8-50. Drawing of a dissection of the right side of the head and neck showing the pharyngeal muscles and the buccinator muscle. Observe that the superior constrictor and buccinator muscles arise from opposite sides of the pterygomandibular raphe, and that the middle constrictor muscle is overlapped by the hyoglossus muscle, which is in turn overlapped by the myohyoid muscle. Observe the gaps between the constrictor muscles through which vessels and nerves pass. Note that the recurrent laryngeal nerve enters the pharyngeal wall inferior to the free inferior border of inferior constrictor muscle.

from the cricoid cartilage (cricopharyngeus fibers) are believed to act as a sphincter, preventing air from entering the esophagus; they relax during swallowing.

The arrangement of the constrictor muscles leaves four deficiencies or gaps in the pharyngeal musculature for structures to enter the pharynx (Fig. 8-50).

1. **Superior to the superior constrictor muscle,** the levator veli palatini muscle, the auditory tube, and the ascending palatine artery pass through the gap between the superior constrictor muscle and the skull (Figs. 8-49 and 8-50). Superior to the superior border of the superior constrictor, the pharyngobasilar fascia

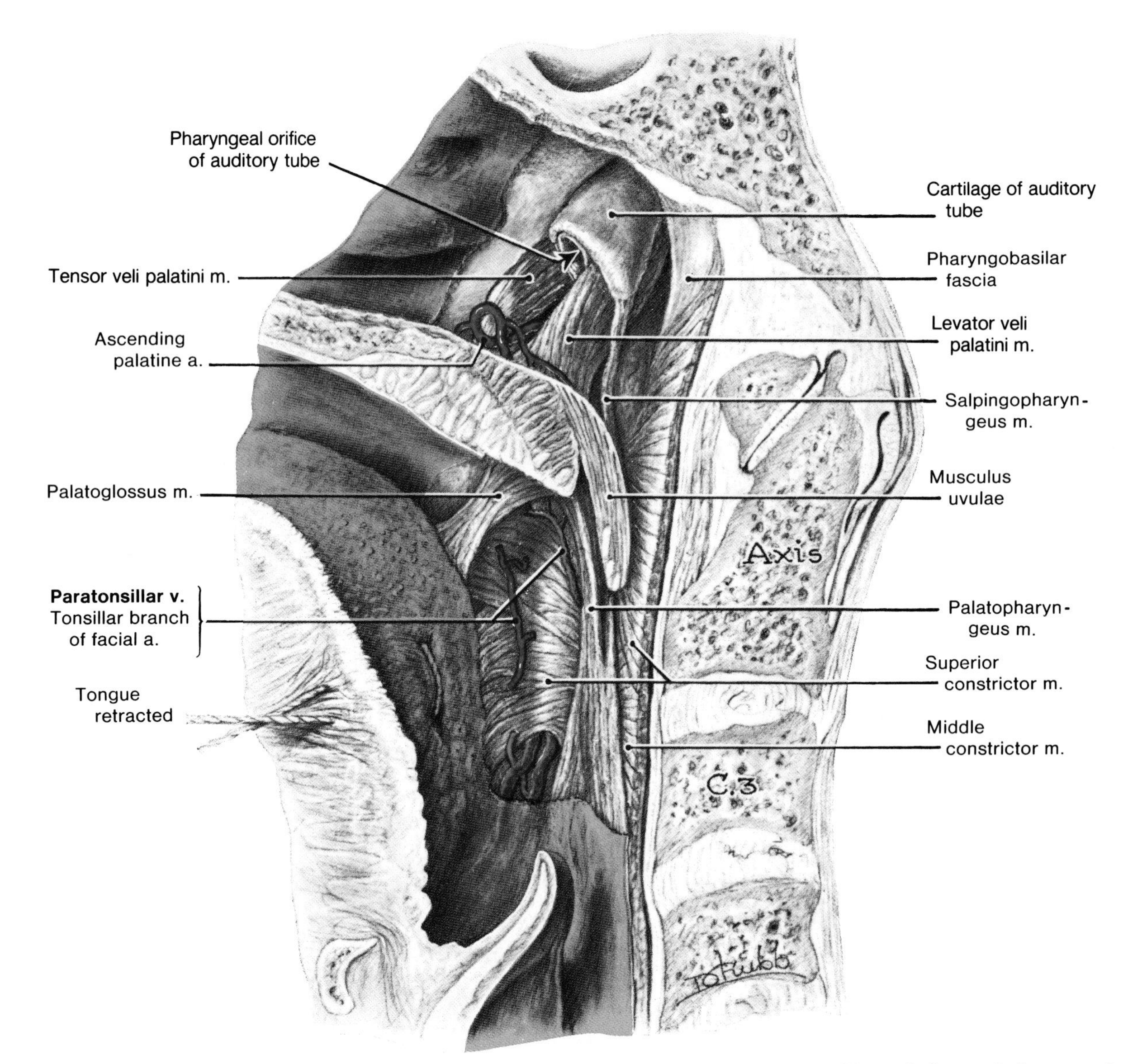

Figure 8-51. Drawing of a lateral view of a dissection of the interior of the pharynx. The palatine and pharyngeal tonsils and the mucous membrane have been removed. Observe the remaining part of the submucous pharyngobasilar fascia which attaches the pharynx to the basilar part of the occipital bone. Examine the curved cartilage of the auditory tube, its free superior and posterior lips, and its pharyngeal orifice. Note the salpingopharyngeus muscle descending from the posterior lip of the auditory tube to join the palatopharyngeus muscle. Observe the tonsillar bed from which a thin sheet of pharyngobasilar fascia has been removed to expose the palatopharyngeus and superior constrictor muscles. Note that the bed of the palatine tonsil extends far into the soft palate. The tonsillar branch of the facial artery is long and large here. The *paratonsillar vein*, descending from the soft palate to join the pharyngeal plexus of veins, is a close lateral relation of the tonsil and is the chief source of hemorrhage in tonsil operations (*e.g.*, tonsillectomies).

blends with the buccopharyngeal fascia to form, with the mucous membrane, the thin wall of the ***pharyngeal recess*** (Figs. 8-52 and 8-54).

2. **Between the superior and middle constrictor muscles** (Figs. 8-48 and 8-49) is the *gateway to the mouth*, through which pass the stylopharyngeus muscle, the glossopharyngeal nerve (CN IX), the styloglossus muscle, the lingual nerve, the lingual artery, and the hypoglossal nerve (CN XII).

3. **Between the middle and inferior constrictor muscles** (Figs. 8-48 to 8-50), the internal laryngeal nerve and the superior laryngeal artery and vein pass to the larynx.

4. **Inferior to the inferior constrictor muscle,**

Table 8-4
Muscles of the Pharynx

Muscle	Origin	Insertion	Nerve Supply	Action
Superior constrictor (Figs. 8-47 to 8-51)	Pterygoid hamulus, pterygomandibular raphe, posterior end of mylohyoid line of mandible, and side of tongue (Figs. 7-9*A*, 7-22, and 8-49)	Median raphe of pharynx (Fig. 8-47) and pharyngeal tubercle (Fig. 7-8)		Constrict wall of pharynx in swallowing
Middle constrictor (Figs. 8-47 to 8-51)	Stylohyoid ligament and greater and lesser horns of hyoid (Figs. 7-141 and 8-49)	Median raphe of pharynx (Fig. 8-47)	Pharyngeal and superior laryngeal branches of the vagus (CN X) through the pharyngeal plexus	
Inferior constrictor (Figs. 8-47 to 8-51)	Oblique line of thyroid cartilage and side of cricoid cartilage			
Palatopharyngeus (Figs. 8-51, 8-57, and 8-59)	Hard palate and palatine aponeurosis (Figs. 7-132 and 8-59)	Posterior border of lamina of thyroid cartilage (Fig. 8-59) and side of pharynx and esophagus		
Salpingopharyngeus (Figs. 8-51 and 8-57)	Cartilaginous part of auditory tube (Fig. 8-51)	Blends with palatopharyngeus muscle (Fig. 8-51)		Elevate pharynx and larynx[1]
Stylopharyngeus (Figs. 8-48, 8-50, and 8-51)	Styloid process (Figs. 7-8, 8-47, and 8-48)	Posterior and superior borders of thyroid cartilage with the palatopharyngeus muscle (Fig. 8-48)	Glossopharyngeal n. (Fig. 7-111)	

[1] The salpingopharyngeus muscle also opens the auditory tube.

the recurrent laryngeal nerve and the inferior laryngeal artery pass superiorly into the larynx (Figs. 8-50 and 8-61).

INTERNAL MUSCLES OF THE PHARYNX (Figs. 8-48, 8-50, 8-51, and Table 8-4)

The internal, chiefly longitudinal muscular layer, consists of three muscles: the stylopharyngeus, palatopharyngeus, and salpingopharyngeus. They all elevate the larynx and pharynx in swallowing and speaking.

The Stylopharyngeus Muscle (Figs. 8-47, 8-48, 8-50, 8-59, and Table 8-4). This long thin, conical muscle descends inferiorly between the external and internal carotid arteries. It enters the wall of the pharynx between the superior and middle constrictors muscles.

Origin (Fig. 8-48). Medial surface of the **styloid process** of the temporal bone.

Insertion (Fig. 8-48). Posterior and superior borders of the **thyroid cartilage** with the palatopharyngeus muscle with which it is continuous.

Nerve Supply (Fig. 7-111). **Glossopharyngeal nerve (CN IX).** *The stylopharyngeus is the only muscle of the pharynx that is not supplied by the pharyngeal plexus* of nerves.

Actions. Elevates the pharynx and larynx and expands the sides of the pharynx, thereby aiding in pulling the pharyngeal wall over a bolus of food during swallowing.

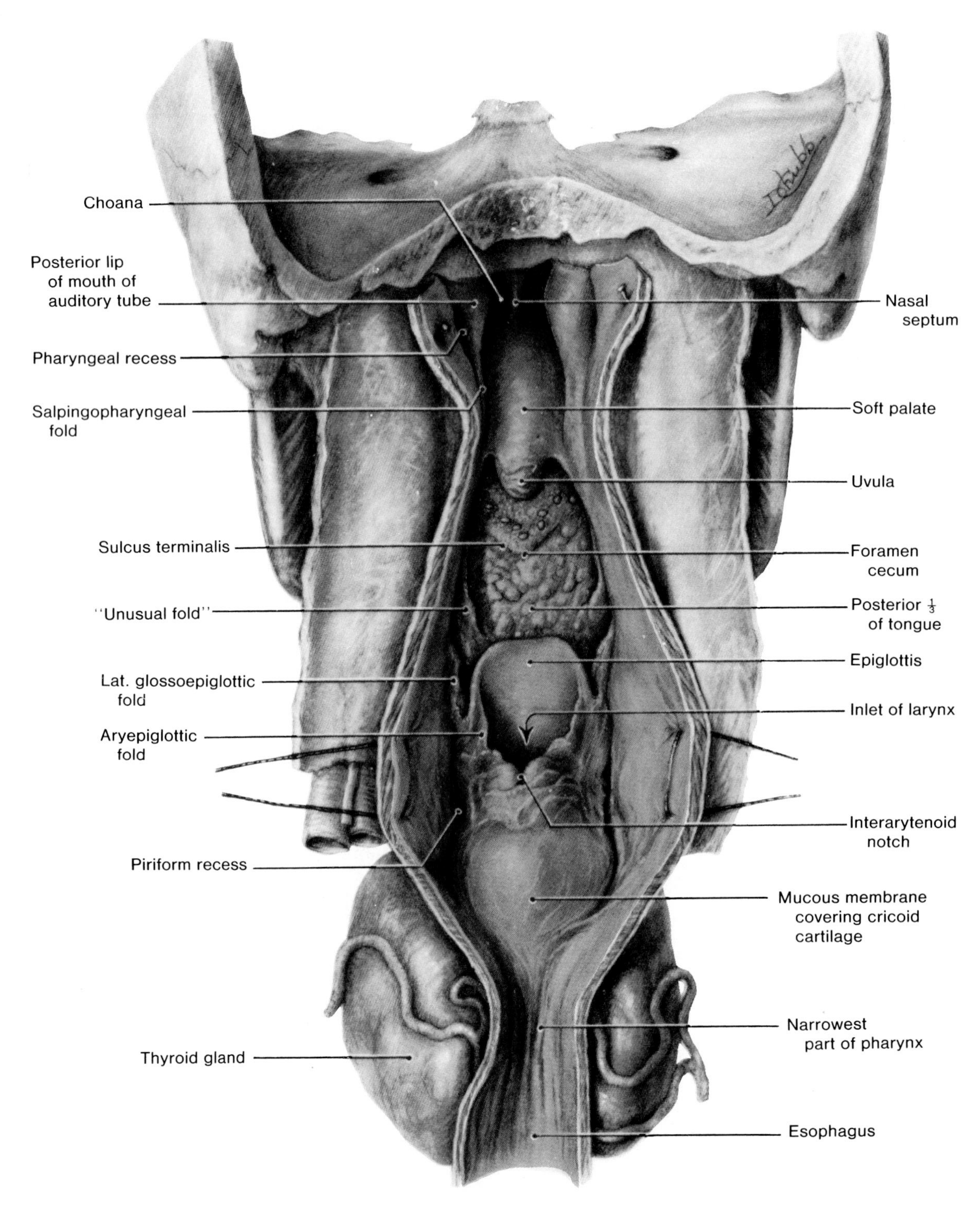

Figure 8-52. Drawing of a posterior view of a dissection of the interior of the pharynx. Observe the pharynx extending from the base of the skull to the inferior border of the cricoid cartilage, where it narrows to become the esophagus. Note the three parts of the pharynx; nasal, oral, and laryngeal. The nasal part of the pharynx or *nasopharynx* lies superior to the level of the soft palate and is continuous anteriorly through the choanae with the nasal cavities. The oral part of the pharynx or *oropharynx* lies between the levels of the soft palate and larynx and communicates anteriorly with the oral cavity. It has the posterior one-third of the tongue as its anterior wall. Note that this part of the tongue is studded with lymph follicles, collectively called the lingual tonsil, and is demarcated from the anterior two-thirds by the foramen cecum and the V-shaped sulcus terminalis (also see Fig. 7-138). The lingual follicles constitute the lingual tonsil. The laryngeal part of the pharynx or *laryngopharynx* lies posterior to the larynx and communicates with the cavity of the larynx through the inlet of the larynx. On each side of the inlet, and separated from it by the aryepiglottic fold, observe a piriform recess.

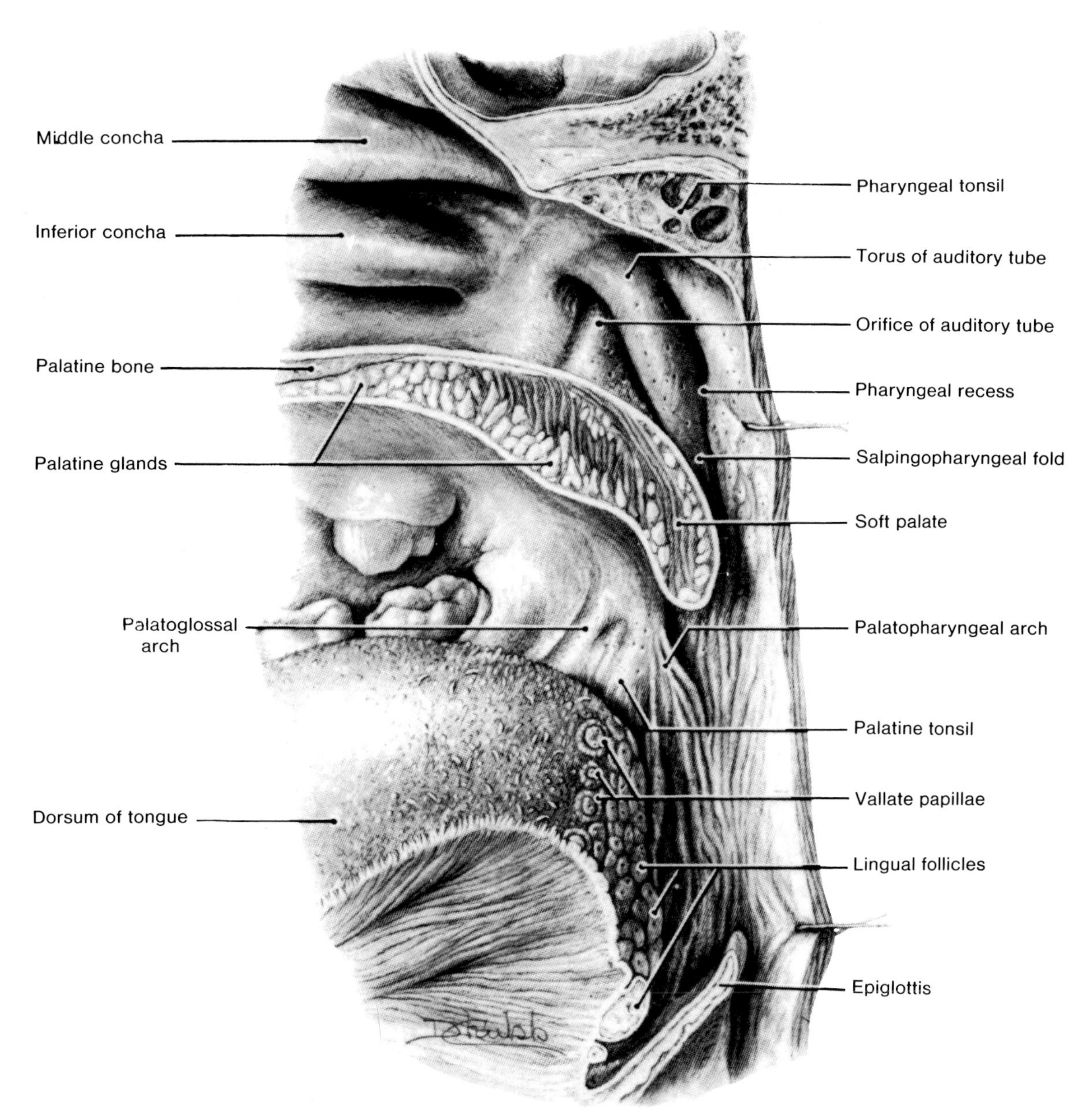

Figure 8-53. Drawing of a lateral view of a dissection of the pharynx. Observe the prominent torus (L. swelling) of the auditory tube and the salpingopharyngeal fold which descends from the torus. Note the location of the orifice of the auditory tube (about 1.5 cm posterior to the inferior concha) and the deep pharyngeal recess posterior to the torus of the tube. Note the palatine tonsils which consist of masses of lymphoid tissue in the lateral walls of the oropharynx (also see Fig. 8-55). The tonsils reach their maximum size during early childhood and usually begin to diminish in size after puberty. Observe the lingual follicles which are known collectively as the lingual tonsil (also see Figs. 7-138 and 8-52).

The Palatopharyngeus Muscle (Figs. 8-51, 8-57, 8-59, and Table 8-4). *This muscle and the overlying mucosa form the* ***palatopharyngeal arch*** (Figs. 8-53, 8-55, 8-57, and 8-58).

Origin (Figs. 7-132 and 8-59). Posterior border of **the hard palate and the palatine aponeurosis.**

Insertion (Fig. 8-59). **Posterior border of the lamina of the thyroid cartilage** and the side of the pharynx and esophagus.

Actions. Elevates the pharynx and larynx, thereby shortening the pharynx in swallowing. It also produces constriction of the palatopharyngeal arch.

The Salpingopharyngeus Muscle (Figs. 8-51, 8-57, and Table 8-4). This slender muscle descends in the lateral wall of the pharynx and is covered by the **salpingopharyngeal fold** of mucous membrane (Fig. 8-53).

Origin (Fig. 8-51). **Cartilaginous portion of the auditory tube** at its pharyngeal end.

Insertion (Fig. 8-51). Descends inferiorly inside the

constrictor muscles and **blends with the palatopharyngeus muscle.**

Actions. Elevates the pharynx and larynx and opens the pharyngeal orifice of the auditory tube during swallowing.

THE INTERIOR OF THE PHARYNX (Figs. 8-51 to 8-57)

The interior of the pharynx communicates with three cavities; the nose, the mouth, and the larynx.

The nasal part of the pharynx or nasopharynx has a respiratory function (Figs. 7-49, 7-152, and 8-46). It lies superior to the soft palate and is a posterior extension of the nasal cavities. The nose opens into the nasopharynx via two large posterior apertures called the internal nares or **choanae** (Figs. 7-9*B*, 7-151, and 8-52). They are separated by the bony nasal septum.

The roof and posterior wall of the nasopharynx form a continuous surface that lies inferior to the body of the sphenoid bone and the basilar part of the occipital bone. In the mucous membrane of the roof and posterior wall of the nasopharynx, there is a *collection of lymphoid tissue,* known as the **pharyngeal tonsil** (Figs. 8-53 and 8-54). When enlarged, the pharyngeal tonsil is commonly called the **"adenoids."**

The pharyngeal orifice of the auditory tube is on the lateral wall of the nasopharynx, 1 to 1.5 cm posterior to the inferior concha (Figs. 8-51 and 8-53), at the level of the superior border of the palate. The auditory tube is about 3.5 cm long (one-third osseous, two-thirds cartilaginous). Its pharyngeal orifice is directed inferiorly.

There is a hood-like *tubal elevation* over the pharyngeal orifice of the auditory tube, called the **torus of the auditory tube** (Fig. 8-53) or the **torus tubarius** (L. *torus,* swelling). The torus is produced by the projection of the base of the cartilaginous part of the auditory tube. Extending inferiorly from the torus of the tube is a vertical fold of mucous membrane, known

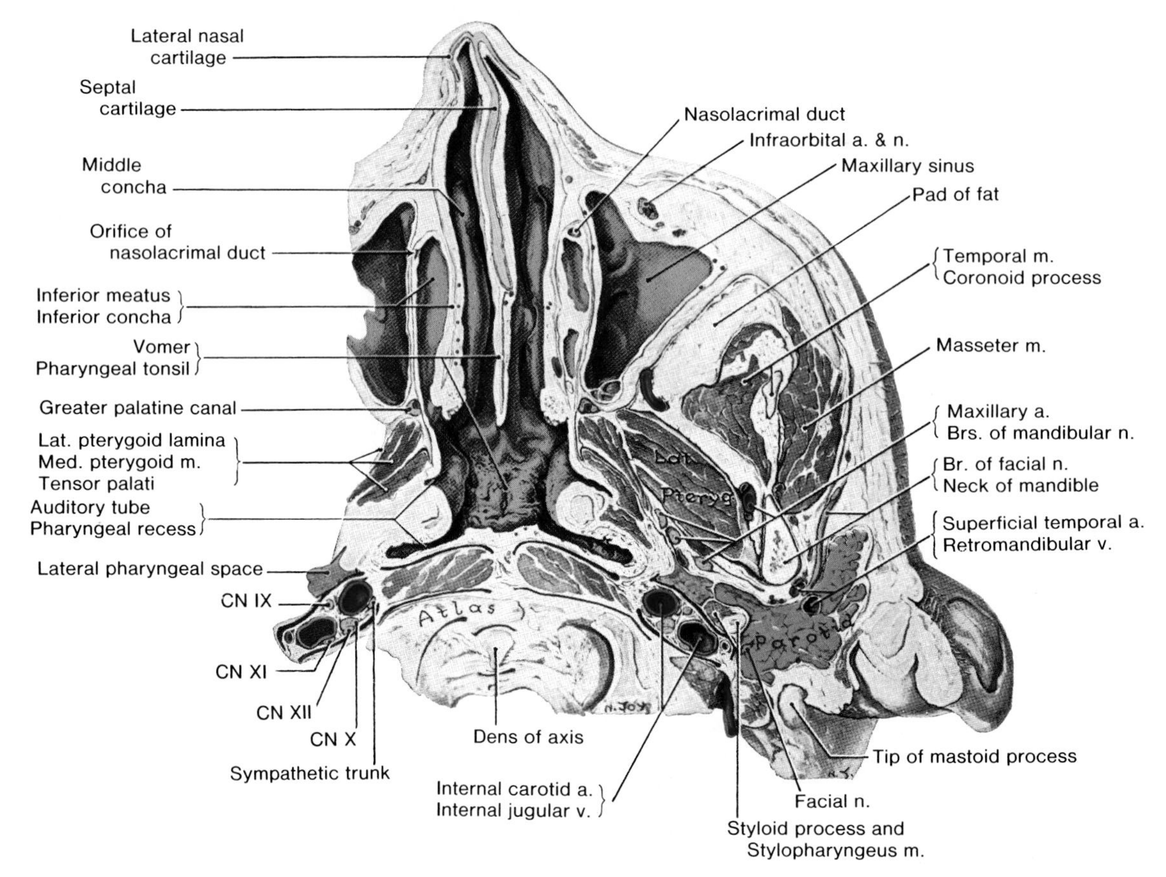

Figure 8-54. Drawing of a cross-section, inferior view, that passes through the nasal cavities and nasopharynx. Observe the slightly enlarged pharyngeal tonsil ("adenoids") in the roof and posterior wall of the nasopharynx. Note the close relationship of the pharyngeal tonsil to the choanae and to the orifices of the auditory tubes (also see (Fig. 8-52). Consequently, hypertrophy of the pharyngeal tonsil often interferes with the passage of air through the nose and obstructs the auditory tubes. This commonly results in mouth breathing and rhinitis (inflammation of the nasal mucous membrane). When these symptoms persist, the adenoids are sometimes removed (adenoidectomy). For a description of the lateral pharyngeal space, see Figure 8-57.

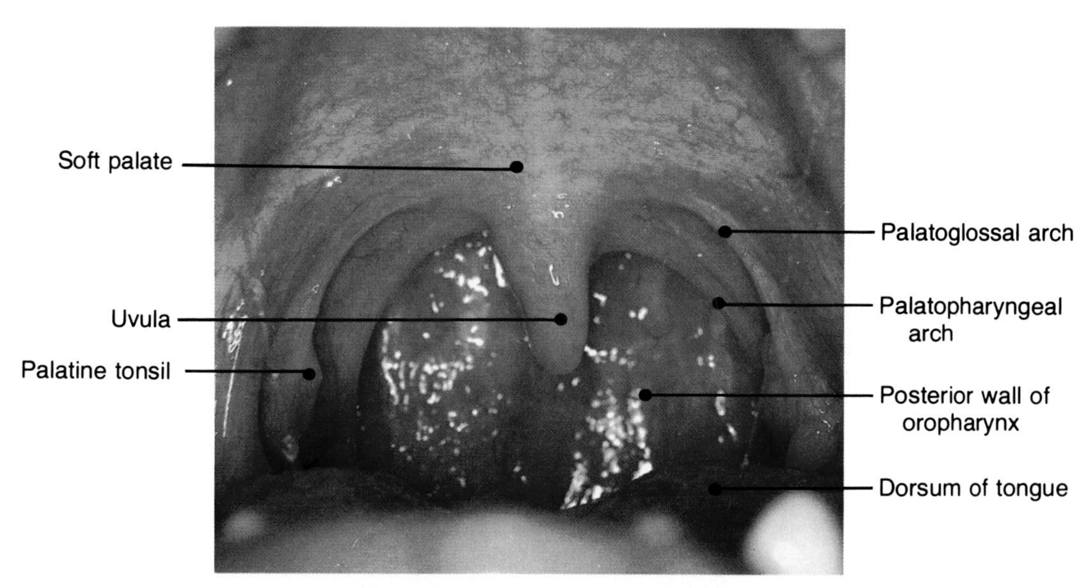

Figure 8-55. Photograph of the oral cavity of a young adult woman, taken with the mouth wide open and the tongue protruding as far as possible. The oral cavity (L. *oris*, mouth) is lined by mucous membrane. Its roof is formed by the palate and its floor is largely occupied by the tongue (also see Fig. 8-46). The fauces (L. throat) is the opening of the mouth into the pharynx. Note that the lateral walls are formed by two arches ("pillars of the fauces") between which the palatine tonsils lie. The palatine tonsils are rounded masses of lymphoid tissue (also see Fig. 8-56). The palatine tonsils are large during childhood and begin to atrophy during puberty. The tonsils are normal in this woman. Inflammation of the tonsils (tonsillitis) causes them to enlarge (hypertrophy). Extremely hypertrophic palatine tonsils may meet medially and cause dysphagia and dyspnea. (From Liebgott B: *The Anatomical Basis of Dentistry*. Philadelphia, WB Saunders Co, 1982).

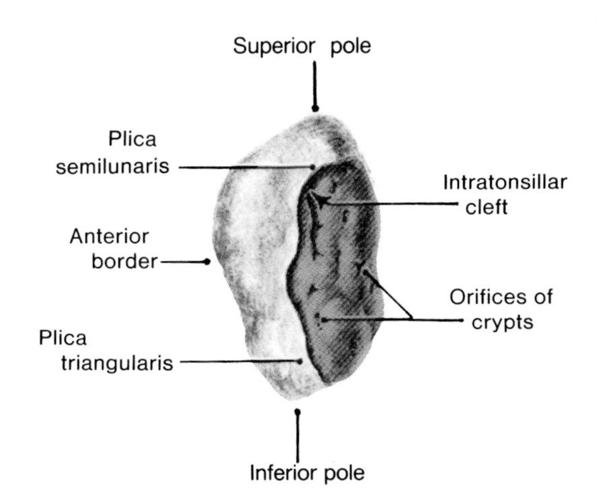

Figure 8-56. Drawing of a medial view of a palatine tonsil removed from its bed (Figs. 8-58 and 8-59). Observe that its thin fibrous capsule extends around the anterior border and slightly over the medial surface as a thin, free fold (L. *plica*) which is covered with mucous membrane on both surfaces. The superior part of this fold is called the plica semilunaris, and its inferior part is called the plica triangularis. On the surface of the tonsil, note the orifices of the tonsillar crypts. Observe the intratonsillar cleft that extends toward the superior pole. The intratonsillar cleft is the remnant of the second pharyngeal pouch in the embryo (see Moore, 1982 for details).

as the **salpingopharyngeal fold** (Fig. 8-53). It covers the salpingopharyngeus muscle which opens the pharyngeal orifice of the auditory tube during swallowing (Figs. 8-51 to 8-53). The collection of lymphoid tissue in the submucosa posterior to the orifice of the auditory tube is known as the **tubal tonsil.**

Posterior to the torus of the auditory tube and the salpingopharyngeal fold, there is a slit-like lateral projection of the pharynx that is called the **pharyngeal recess** (Figs. 8-53, 8-54, and 8-57), which extends laterally and posteriorly. The auditory tube is also discussed with the ear. For a description of its functions and the spread of infection through it to the middle ear, see page 969.

Because of the close relationship of the pharyngeal tonsil to the choanae and the orifices of the auditory tubes (Figs. 8-53 and 8-54), inflammation of this lymphoid tissue (**adenoiditis**) may lead to obstruction of the orifice of the auditory tube and to infection in the middle ear (Figs. 7-162 and 7-164).

An enlarged pharyngeal tonsil (**adenoids**) can obstruct the passage of air from the nasal cavities through the choanae into the nasopharynx,

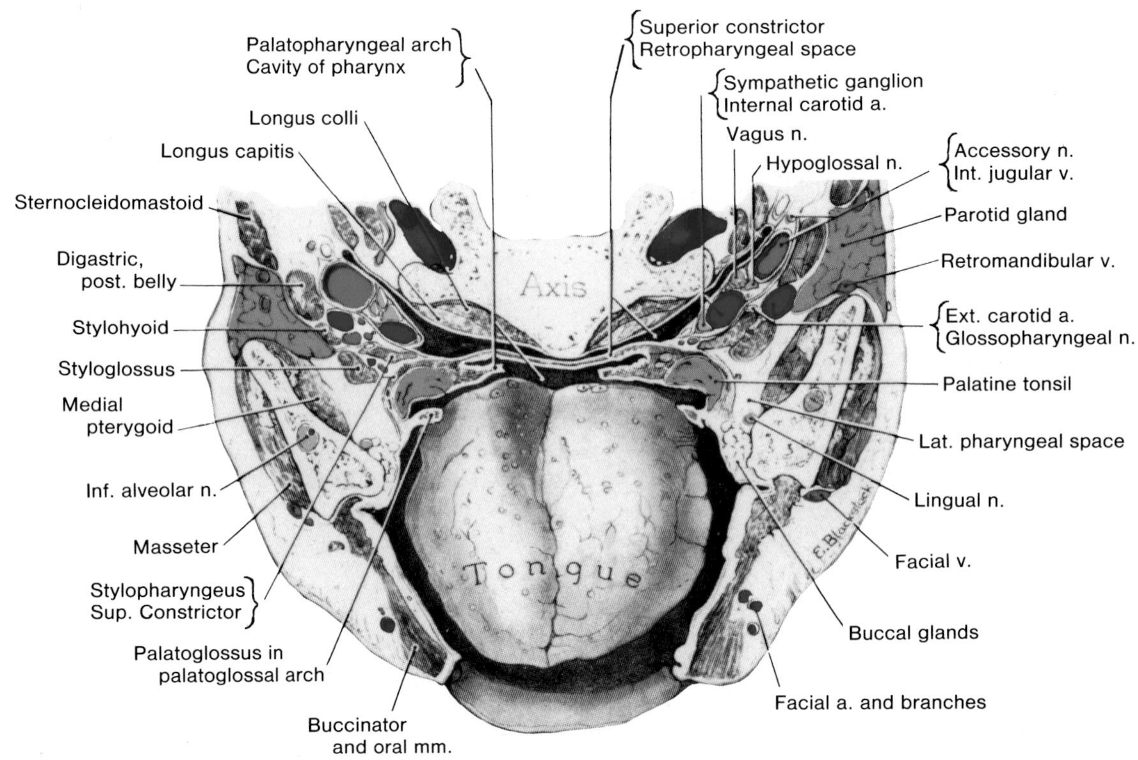

Figure 8-57. Drawing of a transverse section of the head that passes through the oral cavity. Anterior to the ribbon-like palatoglossus muscle and its arch is the oral cavity and posterior to it is the pharynx. Note that the tonsillar bed is formed by the superior constrictor and palatopharyngeus muscles with a potential space intervening, and that the bed is limited anteriorly and posteriorly by the palatine arches. Observe the carotid arteries posterior to the tonsillar bed. Examine the potential retropharyngeal space, here opened up, which allows the pharynx to contract and relax during swallowing. This space is closed laterally at the carotid sheath and is limited posteriorly by the prevertebral fascia (also see Figs. 8-23, 8-46, and 8-51). The lateral pharyngeal space is lined with connective tissue, filled with fat (also see Fig. 8-54), and contains branches of the maxillary nerve and vessels. Infection may spread to this space from the palatine tonsil. From this space, the infection may spread via the veins to the internal jugular vein and the inferior petrosal sinus (Figs. 7-39 and 7-65).

making mouth breathing necessary. In chronic cases the patient develops a characteristic facial expression called the **adenoid facies.** The open mouth and protruding tongue give the person a dull expression.

Infection from the adenoids may spread to the tubal tonsil (**tubal tonsillitis**), causing swelling and closure of the auditory tube. Infection spreading from the nasopharynx to the tympanic cavity causes **otitis media** (middle ear infection), which may result in a temporary or permanent hearing loss. (For more information, see p. 968).

The oral part of the pharynx or *oropharynx* *has a digestive function.* It is continuous with the oral cavity via the **oropharyngeal isthmus**.

The oropharynx is bounded by the **soft palate** superiorly, the **base of the tongue** inferiorly, and the **palatoglossal and palatopharyngeal arches** laterally. (Fig. 8-55). The oropharynx extends from the soft palate to the superior border of the epiglottis (Figs. 8-46, 8-52, 8-53, and 8-55).

The Palatine Tonsils (Figs. 8-53 and 8-55 to 8-60), usually referred to as "*the tonsils*", lie on each side of the oropharynx in the triangular interval between the palatine arches, once called the pillars of the fauces (L. throat). Each palatine tonsil is a collection of lymphoid tissue deep to the mucous membrane of the oropharynx. These tonsils are called *palatine tonsils* because their superior one-third extends into the soft palate and because of their relationship to the palatine arches (Fig. 8-55).

The palatine tonsils vary in size from person to person. In children the palatine tonsils tend to be large, whereas in older persons they are usually small and often inconspicuous.

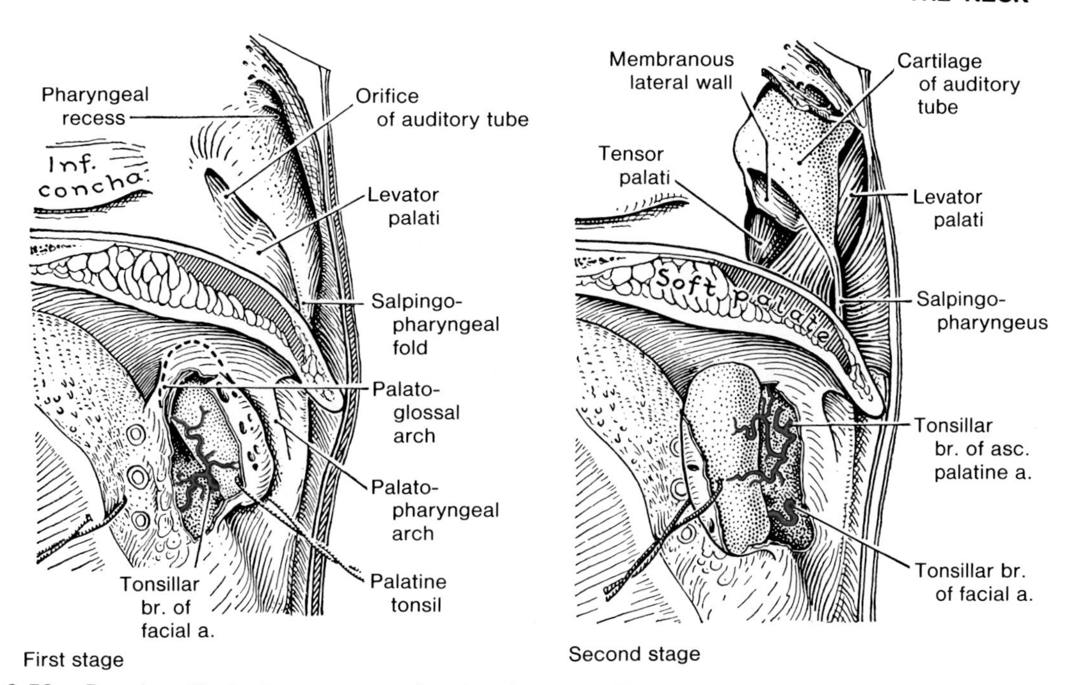

Figure 8-58. Drawings illustrating *one way* of performing a tonsillectomy (removing the palatine tonsils). First the mucous membrane is incised along the palatoglossal arch and then the capsule of the tonsil is incised. With the point and rounded handle of a scalpel the anterior border of the tonsil is freed and shelled out. The mucous membrane along the palatopharyngeal arch is then cut through. Bleeding from the tonsillar vessels, particularly the veins may occur at tonsillectomy (Figs. 8-51 and 8-59).

The visible part of the tonsil is no guide to its actual size because much of it may be hidden by the tongue and buried in the soft palate. Commonly the palatine tonsil is about 2 cm in its greatest dimension (Fig. 8-56) and usually does not fill the space between the palatine arches (Fig. 8-55). Part of the remaining space, called the **intratonsillar cleft** (Fig. 8-56), penetrates the tonsil and may pass into the soft palate (Fig. 8-55). This cleft is the remains of the second pharyngeal pouch in the embryo. The exposed free surface of the palatine tonsil is characterized by the slit-like orifices of the mouths of the ***tonsillar crypts*** (Fig. 8-56).

The tonsillar bed, in which the palatine tonsil lies, is between the palatoglossal and palatopharyngeal arches (Figs. 8-53 and 8-55). The thin, fibrous sheet covering the tonsillar bed is part of the **pharyngobasilar fascia** (Figs. 8-47 and 8-48).

The tonsillar bed is composed of two muscles, the palatopharyngeus and the superior constrictor (Figs. 8-51, 8-57, and Table 8-4), which form part of the muscular coat of the pharynx.

The tonsillar artery, a branch of the facial artery (Fig. 7-31), passes through the superior constrictor muscle and enters the inferior pole of the tonsil (Figs. 8-58 and 8-59). The tonsillar bed also receives small arterial twigs from the ascending palatine, lingual, descending palatine, and ascending pharyngeal arteries (Fig. 8-26).

The large external palatine vein (Fig. 8-59) descends from the soft palate and passes close to the lateral surface of the tonsil before entering the **pharyngeal plexus of veins**. One or more veins leave the inferior part of the deep aspect of the tonsil and open into this pharyngeal plexus and the facial vein (Figs. 7-38 and 7-42).

The nerves of the tonsil are derived from the *tonsillar plexus* formed by branches of the glossopharyngeal and vagus nerves (Figs. 7-111 and 7-135). Other branches are from the **pharyngeal plexus**, a network of fine nerve fibers in the fascia covering the middle constrictor muscle (Figs. 8-47 and 8-50).

The lymph vessels from the tonsil pass laterally and inferiorly to the lymph nodes near the angle of the mandible and to the **jugulodigastric node** (Figs. 7-40, 8-18, and 8-60). *Because of the frequent enlargement of jugulodigastric node in tonsillitis*, it is often referred to as the *tonsillar lymph node* (tonsillar node).

The palatine, lingual, and pharyngeal tonsils form a circular band of lymphoid tissue at the oropharyngeal isthmus called the tonsillar ring (Figs. 8-54 and 8-57). The anteroinferior part of the ring is formed by the **lingual tonsil**, a collection of lymphoid tissue in the posterior part of the tongue (Figs. 7-138, 8-52, and 8-53). Lateral parts of the ring are formed by the palatine and tubal tonsils, and the posterior and superior parts of the ring are formed by the pharyngeal tonsil.

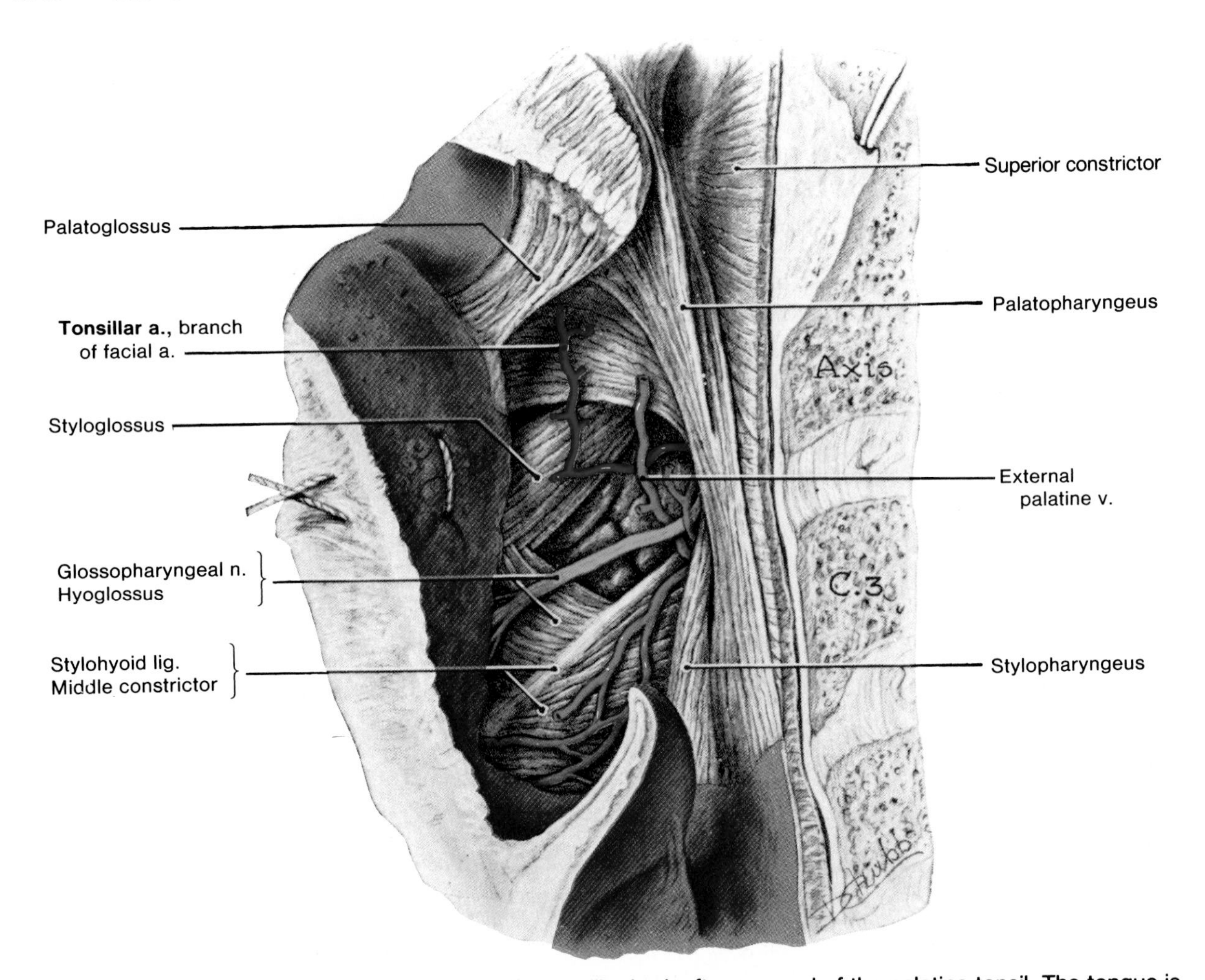

Figure 8-59. Drawing of a deep dissection of the tonsillar bed, after removal of the palatine tonsil. The tongue is pulled anteriorly and the inferior or lingual origin of the superior constrictor muscle is cut away. Observe the tonsillar branch of the facial artery, here sending a large branch (cut short) to accompany the glossopharyngeal nerve to the tongue (Fig. 7-111). Lateral to this artery and the external palatine vein (paratonsillar vein) is the submandibular gland (Figs. 8-18 and 8-48). These veins may be a source of bleeding during tonsillectomy.

The tonsillar ring of lymphoid tissue does not form a strong defense system against the spread of infection from the oral and nasal cavities to the lower respiratory organs (Fig. 1-11).

*Infection of the tonsils (***tonsillitis***)* is often associated with a *sore throat* and **pyrexia** (G. feverishness). Usually the jugulodigastric lymph node or "*tonsillar lymph node*" (Figs. 7-40 and 8-60) in the deep cervical chain is enlarged and tender.

A **peritonsillar abscess (quinsy)** may develop in the loose connective tissue external the capsule of the tonsil, owing to proliferation of pyogenic organisms within the **intratonsillar cleft** (Fig. 8-56). Airway obstruction may occur. Following frequent attacks of tonsillitis, especially when associated with abscess formation, the tonsils may be removed.

Tonsillectomy is carried out by dissection (Fig. 8-58), or by a **guillotine operation.** In each case the tonsil and the fascial sheet covering the tonsillar bed are removed. Some people believe this operation to be a simple one, but **considerable bleeding and other complications can follow removal of the tonsils.** Owing to its *abundant blood supply*, bleeding may arise from the tonsillar artery or other arterial twigs (Figs. 8-58 and 8-59), but more commonly bleeding comes from the **external palatine vein** (Fig. 8-59). This vein, which descends from

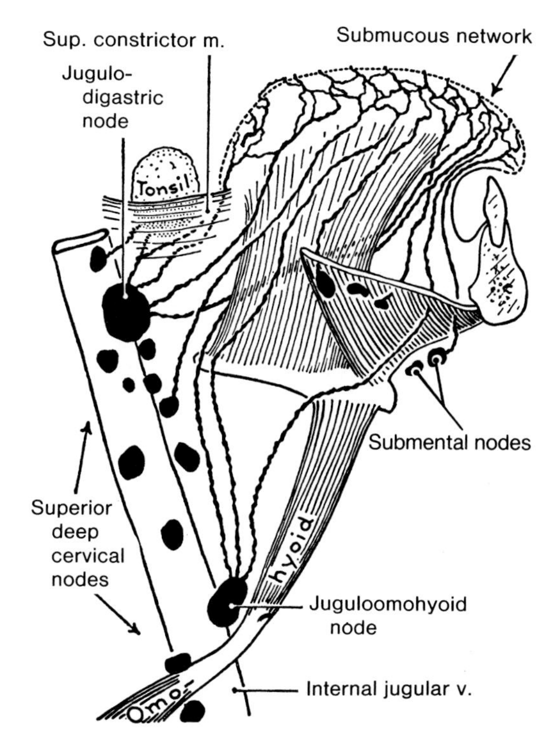

Figure 8-60. Diagram showing the lymphatics of the tongue and the palatine tonsil. Note the jugulodigastric node, one of the most superior lymph nodes of the deep cervical chain (Fig. 8-35). It is located at about the level of the hyoid bone and the posterior belly of the digastric muscle. It mainly drains the palatine tonsil and is often enlarged in cases of tonsillitis (also see Fig. 7-40). Lymph nodes inferior to the omohyoid muscle are called the inferior deep cervical lymph nodes.

the soft palate is *immediately related to the lateral surface of the tonsil.*

Bleeding from the external palatine vein is often the chief source of hemorrhage during and/or after operations on the tonsils.

The glossopharyngeal nerve (Fig. 7-111) to the tongue accompanies the tonsillar artery on the lateral wall of the pharynx (Fig. 8-59). As the wall is thin, CN IX is vulnerable to injury. Also, edema around this nerve following tonsillectomy may result in temporary loss of taste (Fig. 7-139).

Careless removal of a tonsil could injure the lingual nerve. It does not supply the tonsil, but it passes lateral to the pharyngeal wall near the anterior part of the tonsil (Figs. 8-14 and 8-57).

The internal carotid artery should be safe during tonsillectomy (Fig. 8-57), but it can be injured if adjacent tissues are damaged in attempting to ensure that all tonsillar tissue is removed. The internal carotid is especially vulnerable when it is tortuous and lies directly lateral to the tonsil (Fig. 8-48).

Bleeding from the internal carotid artery results in severe hemorrhage which can be controlled by compressing the common carotid artery against the anterior tubercle and the anterior surface of the sixth cervical transverse process (Figs. 8-31 and 8-32).

Congenital sinuses and fistulae of the oropharynx generally result from failure of the second pharyngeal pouch and/or branchial groove to obliterate.

External branchial sinuses frequently open laterally along the anterior border of the sternocleidomastoid muscle, usually in the inferior third of the neck.

Internal branchial sinuses opening into the oropharynx are rare, but usually they represent failure of the second pharyngeal pouch to obliterate normally. These sinuses open into the **intratonsillar cleft** (Fig. 8-56) and pass inferiorly through the neck for a variable distance between the external and internal carotid arteries.

A branchial fistula opens into the intratonsillar cleft and on the side of the neck. Usually they result from persistence of remnants of the second pharyngeal pouch and second branchial groove. The fistula ascends from its cervical opening through the subcutaneous tissue, the platysma, and the deep fascia of the neck to enter the **carotid sheath** (Figs. 8-23 and 8-41). It then pases between the internal and external carotid arteries on its way to its opening in the intratonsillar cleft.

Branchial cysts, often called *lateral cervical cysts* (Fig. 8-77), develop from remnants of the second branchial groove or the *cervical sinus.* These embryonic structures normally obliterate as the neck forms in the fetus.

For explanatory drawings, photographs, and more information about the embryological bases of branchial cysts, sinuses, and fistulae, see Moore, 1982.

The laryngeal part of the pharynx or laryngopharynx lies posterior to the larynx (Figs. 8-46 and 8-52). It extends from the superior border of the **epiglottis** to the inferior border of the cricoid cartilage, where it narrows to become continuous with the esophagus.

Posteriorly the laryngopharynx is related to the bodies of the fourth to sixth cervical vertebrae (Fig. 8-46). Its posterior and lateral walls are formed by the middle and inferior constrictor muscles (Figs. 8-47 and 8-48), with the palatopharyngeus and stylopharyngeus internally (Fig. 8-51).

The laryngopharynx communicates with the larynx through the ***inlet of the larynx*** *(laryngeal aditus).*

The piriform recess (Fig. 8-52) is a small depression of the laryngopharyngeal cavity on each side of the inlet of the larynx. This rather deep, mucosa-lined fossa (L. a trench or ditch) is separated from the inlet

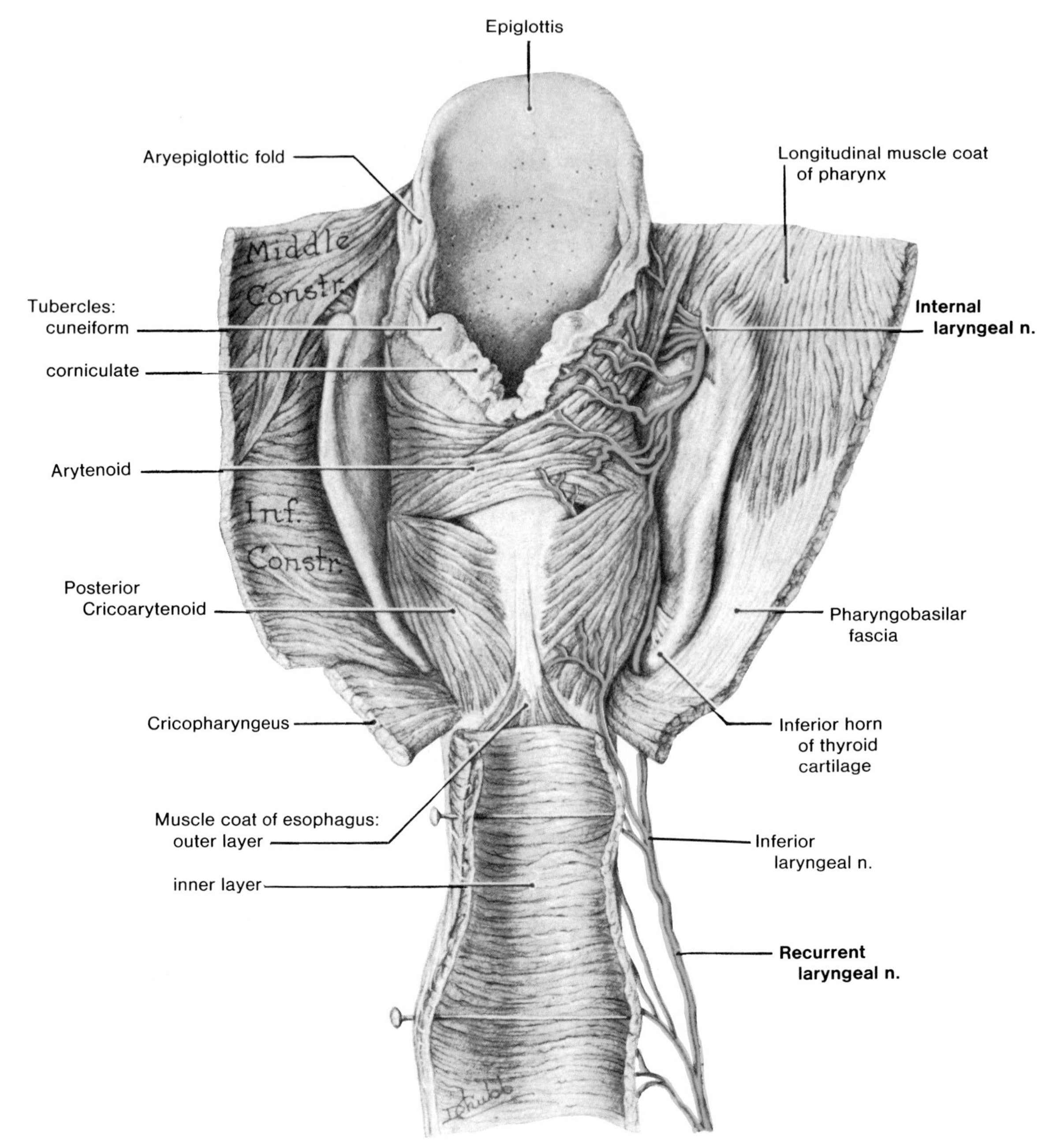

Figure 8-61. Drawing of a dissection showing a posterior view of the muscles of the laryngopharynx, larynx, and esophagus. The mucous membrane of this part of the pharynx and esophagus has been removed; the left palatopharyngeus muscle has also been removed. Deep to the mucosa of the piriform recess, observe the continuation of the recurrent laryngeal nerve, called the inferior laryngeal nerve. Note that it ascends deep to the inferior border of the inferior constrictor muscle of the pharynx, and divides into anterior and posterior branches. Within the larynx it communicates with the internal laryngeal nerve, a branch of the superior laryngeal nerve. The recurrent laryngeal nerve, a branch of CN X, supplies all the intrinsic muscles of the larynx except the cricothyroid. It also supplies the mucous membrane of the larynx inferior to the vocal folds.

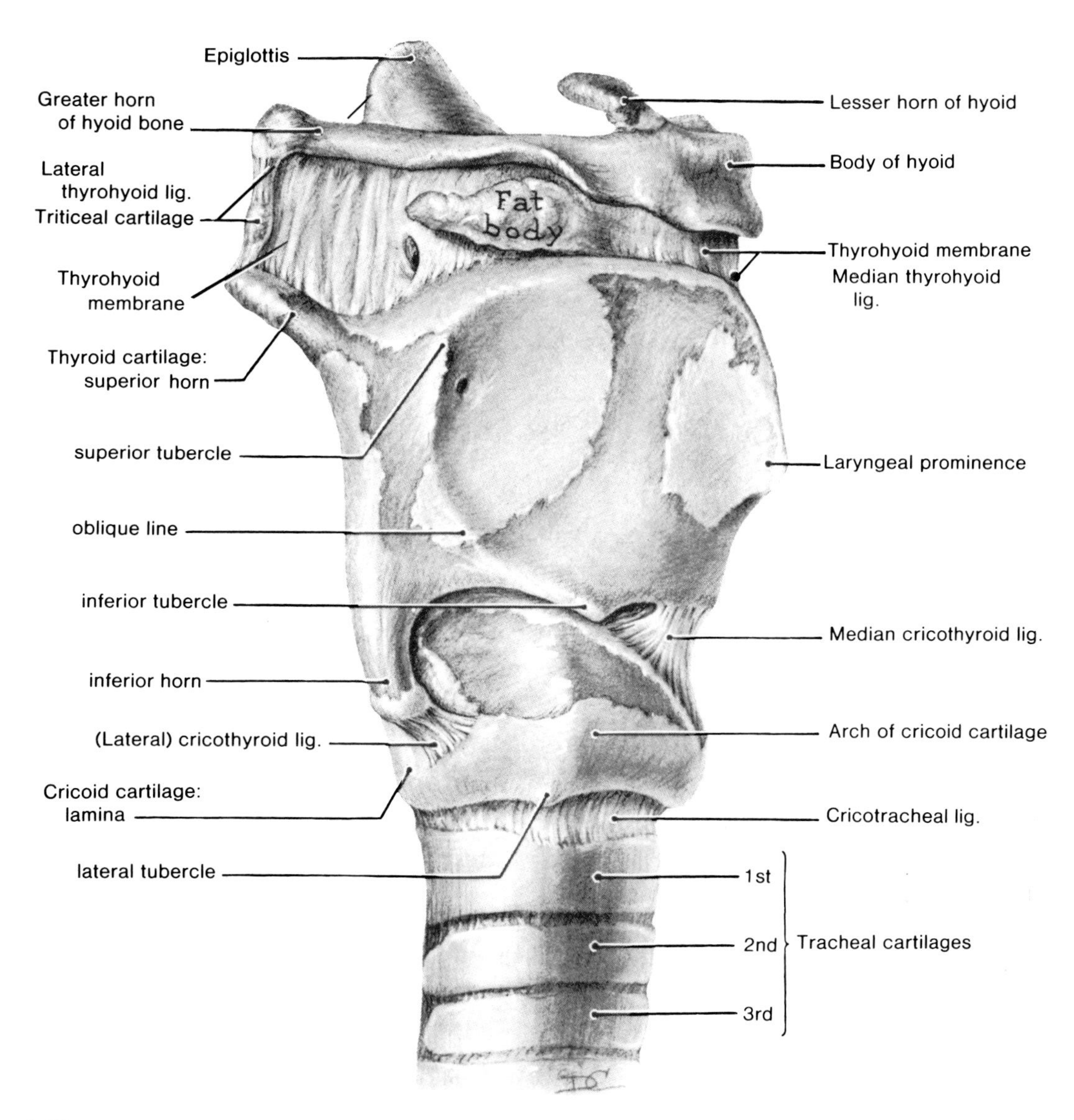

Figure 8-62. Drawing of a lateral view of the skeleton of the larynx. Note that the larynx extends vertically from the tip of the epiglottis to the inferior border of the cricoid cartilage. *The hyoid bone is not a part of the larynx.* Observe the right lamina of the thyroid cartilage projecting anteriorly, superior to the point of union with its fellow to form the laryngeal prominence (also see Figs. 8-1 and 8-25). Note that its posterior border is prolonged into a superior and an inferior horn. Note that the inferior horn articulates with the cricoid cartilage. Observe that the thyrohyoid membrane is: (1) attached to the whole length of the superior border of the thyroid lamina and to the superior and inner border of the body and greater horn of the hyoid bone; (2) thickened anteriorly to form the median thyrohyoid ligament; and (3) thickened posteriorly to form the lateral thyrohyoid ligament which contains a nodule of cartilage. Examine the median cricothyroid ligament uniting the median parts of the adjacent borders of the cricoid and thyroid cartilages.

of the larynx by the **aryepiglottic fold** (Fig. 8-52). Laterally the piriform recess is bounded by the medial surfaces of the thyroid cartilage and the thyrohyoid membrane (Fig. 8-62). In Figure 8-61, observe that branches of *the internal laryngeal and recurrent laryngeal nerves lie deep to the mucous membrane of the piriform recess.*

INNERVATION OF THE PHARYNX (Figs. 8-28, 8-44, 8-47, 8-50, and 8-61)

The motor and most of the sensory supply to the pharynx is derived from the **pharyngeal plexus of nerves** on the surface of the pharynx. This plexus is formed by pharyngeal branches of the vagus nerves (Fig. 7-135), glossopharyngeal nerves (Fig. 7-111), and sympathetic branches from the **superior cervical ganglion** (Figs. 8-28 and 8-47).

The motor fibers in the pharyngeal plexus are derived from the cranial root of the accessory nerve (Fig. 7-64), and are carried by the vagus to all muscles of the pharynx and soft palate (Fig. 7-135), except the stylopharyngeus (supplied by CN IX) and the tensor veli palatini (supplied by CN V^2).

The sensory fibers in the pharyngeal plexus are derived from the glossopharyngeal nerve (CN IX). They supply most of the mucosa of all three parts of the pharynx (Fig. 7-111). The sensory nerve supply of the mucous membrane of the nasopharynx is mainly from the maxillary nerve (CN V^2), a purely sensory nerve (Fig. 7-33).

The piriform recess can be viewed radiographically after the patient has been given a barium swallow of "meal" (p. 287 and Fig. 2-33).

Foreign bodies (*e.g.*, chicken bones and "safety pins") entering the pharynx may become lodged in the piriform (L. pear-shaped) recess. If sharp, they may pierce the mucous membrane and injure the **internal laryngeal nerve** (Fig. 8-61). This may result in anesthesia of the laryngeal mucous membrane as far inferiorly as the vocal folds (Fig. 8-67). Similarly, the nerve may be injured if the instrument used to remove the foreign body pierces the mucous membrane, because the internal laryngeal nerve is just external to the mucous membrane (Fig. 8-48).

DEGLUTITION (SWALLOWING)

Although we swallow without thinking, deglutition is a complex process whereby food is transferred from the mouth through the pharynx and esophagus into the stomach. The term **bolus** (L. fr. G. bōlos, lump or clot) is used to describe the mass of food or quantity of liquid that is swallowed at one time. Solid food is masticated (chewed) and mixed with saliva to form a soft bolus during chewing. Deglutition is described in three stages: (1) in the mouth, (2) in the pharynx, and (3) in the esophagus.

The first stage of swallowing is voluntary, during which the bolus is pushed from the mouth into the oropharynx, mainly by movements of the tongue. The tongue is raised and pressed against the hard palate by the intrinsic muscles of the tongue (Figs. 7-137 and 7-142).

The second stage of swallowing is involuntary and is usually rapid. It involves contraction of the walls of the pharynx. Breathing and chewing stop and successive contractions of the three constrictor muscles (Table 8-4) move the food through the oral and laryngeal parts of the pharynx.

The bolus of food is prevented from entering the nasopharynx by elevation of the soft palate. The *tensor veli palatini and levator veli palatini muscles* (Figs. 7-132, 7-133, and Table 7-8) tense and elevate the soft palate against the posterior wall of the pharynx. These actions close the **pharyngeal isthmus**, thereby preventing food from entering the nasopharynx. Should a person happen to laugh during this stage, the muscles of the soft palate relax, and may allow some food to enter the nasopharynx. In these cases the food, especially if it is liquid, is expelled through the nose.

As the bolus of food passes through the oropharynx, the walls of the pharynx are raised. The contraction of the pharyngeal muscles (Fig. 8-51) elevate the pharynx and larynx (Table 8-4).

Watch someone swallow, particularly a man, and observe that the **laryngeal prominence** (Fig. 8-1) rises. The palatopharyngeus and the stylopharyngeus muscles elevate the larynx and pharynx in swallowing (Fig. 8-51 and Table 8-4). Palpate your hyoid bone with your thumb and index finger as you swallow and verify that it also rises. The hyoid bone is raised and fixed in swallowing by contraction of the geniohyoid, mylohyoid, digastric, and stylohyoid muscles (Fig. 8-14 and Table 8-1). Verify that elevation and anterior movement of the hyoid bone precedes elevation of the larynx.

During deglutition the vestibule of the larynx (Fig. 8-70) is closed, the epiglottis is bent posteriorly over the inlet of the larynx (Figs. 8-46 and 8-52), and the **aryepiglottic folds** are approximated (Fig. 8-52). These folds provide lateral food channels that guide the bolus of food from the sides of the epiglottis. The food now passes over the oral surface of the epiglottis and the closed inlet of the larynx into laryngopharynx. All these actions are designed to prevent food from entering the larynx.

The third and final stage of swallowing squeezes or "milks" the bolus from the laryngopharynx into the esophagus. This is effected by the inferior constrictor muscle of the pharynx (Figs. 8-47 to 8-50).

If the recurrent laryngeal nerves are injured (*e.g.*, during a **thyroidectomy**), paralysis of the muscles in the aryepiglottic folds occur. As a result the inlet of the larynx does not close completely during swallowing, the aryepiglottic folds fail to come together, and food may enter the larynx.

***Choking on food* is a common cause of laryngeal obstruction**, particularly in persons who have consumed excessive amounts of alcohol, or who have **bulbar palsy** (degeneration of motor neurons in brain stem nuclei of CN IX and CN X which supply the muscles of deglutition).

Difficulty in swallowing is called dysphagia. Diagnostic studies of swallowing are done by **cinefluoroscopy** (x-ray motion pictures of the passage of contrast material through the pharynx and esophagus). Often fluoroscopic examinations of the act of swallowing are not helpful because of the speed of the process. Experienced observers can however detect abnormalities in deglutition.

Children swallow a variety of objects, most of which reach the stomach and pass through the gastrointestinal tract without difficulty. In some cases the foreign body stops at the inferior end of the laryngopharynx, its narrowest part (Figs. 8-46 and 8-52), or in the esophagus just inferior to the cricopharyngeus muscle, part of the inferior constrictor (Fig. 8-61).

Foreign bodies in the pharynx are removed under direct vision through an **pharyngoscope** (an instrument for examining the interior of the pharynx).

Radiographic examinations will also reveal the presence of a *radiopaque foreign body* in the pharynx (*e.g.*, a pin in the piriform recess).

THE LARYNX

The larynx (G. "upper end of the trachea") communicates with the mouth and nose through the laryngeal and oral parts of the pharynx (Fig. 8-46). Although it is part of the air passages, the larynx acts as a valve for preventing swallowed food and foreign bodies from entering the lower respiratory passages. The larynx is specifically designed for voice production (**phonation**).

The larynx is the phonating mechanism (G. *phōnē*, voice). **Phonation is defined as the utterance of sounds with the aid of the vocal folds.** Through movements of its cartilages, the larynx varies the opening between the vocal folds (Fig. 8-72), thereby varying the pitch of sounds produced by the passage of air through them. These sounds are translated into intelligible speech by articulatory and resonating structures (*e.g.*, the tongue and mouth).

The larynx is located in the anterior portion of the neck. In adult males it is about 5 cm in length (Fig. 8-2) and is related posteriorly to the bodies of the third to sixth cervical vertebrae (Fig. 8-46). The larynx is shorter in women and children and is situated slightly more superiorly in the neck. This sex difference in the larynx normally develops at puberty in males (Fig. 8-2), at which time all its cartilages enlarge.

THE SKELETON OR FRAMEWORK OF THE LARYNX (Fig. 8-62)

The laryngeal skeleton is formed by nine cartilages that are joined by various ligaments and membranes. Three of the cartilages are single (thyroid, cricoid, and epiglottis), and three are paired (arytenoid, corniculate, and cuneiform).

The Thyroid Cartilage (Figs. 8-2 and 8-62 to 8-64). The thyroid cartilage (G. *thyroideus*, resembling a shield) is the largest of the laryngeal cartilages. It is composed of two quadrilateral **laminae** (L. thin plates). The inferior two-thirds of these laminae are fused anteriorly in the midline to form a subcutaneous projection, called the **laryngeal prominence** (Figs. 8-1, 8-2, 8-25, 8-62, and 8-63).

The larynx is more prominent in postpubertal males (Fig. 8-1) because the angle at which the laminae meet is smaller in males (Fig. 8-2*C*), and the anteroposterior diameter of the laminae is greater. These sex differences in the thyroid cartilage are usually evident by the 16th year.

Immediately superior to the laryngeal prominence, the two thyroid laminae diverge to form a V-shaped **thyroid notch** (Figs. 8-2 and 8-63). The posterior border of each lamina projects superiorly as the superior horn and inferiorly as the inferior horn (Figs. 8-62 and 8-63). The superior border of the thyroid cartilage is attached to the hyoid bone by the **thyrohyoid membrane** (Fig. 7-62).

The inferior horns of the thyroid cartilage articulate with the cricoid cartilage at special facets that allow the thyroid cartilage to tilt or glide anteriorly or posteriorly in a visor-like manner (Figs. 8-62 to 8-64).

The lateral surface of each lamina of the thyroid cartilage is marked by an **oblique line** (Figs. 8-2*A*, 8-62, and 8-63). It provides attachment for the inferior constrictor muscle of the pharynx and the sternothyroid and thyrohyoid muscles (Figs. 8-17, 8-49, and 8-61).

The Cricoid Cartilage (Figs. 8-62 to 8-65). The cricoid (G. ring) is *shaped like a signet ring* with its band facing anteriorly (Fig. 8-26). The posterior (signet) part of the cricoid is called the **lamina** and the anterior (band) part is termed the **arch**. Although much smaller than the thyroid cartilage, the cricoid is thicker and stronger. Being the most inferior of the cartilages of the larynx, the cricoid forms the inferior parts of the anterior and lateral walls and most of the posterior wall of the larynx. The cricoid cartilage is attached to the inferior margin of the thyroid cartilage by the **cricothyroid ligaments** (Figs. 8-62 and 8-63) and to the first tracheal ring by the **cricotracheal ligament** (Fig. 8-64).

The Arytenoid Cartilages (Figs. 8-64 and 8-65). These paired cartilages, shaped like three-sided pyramids, articulate with the lateral parts of the superior border of the lamina of the cricoid cartilage. Each

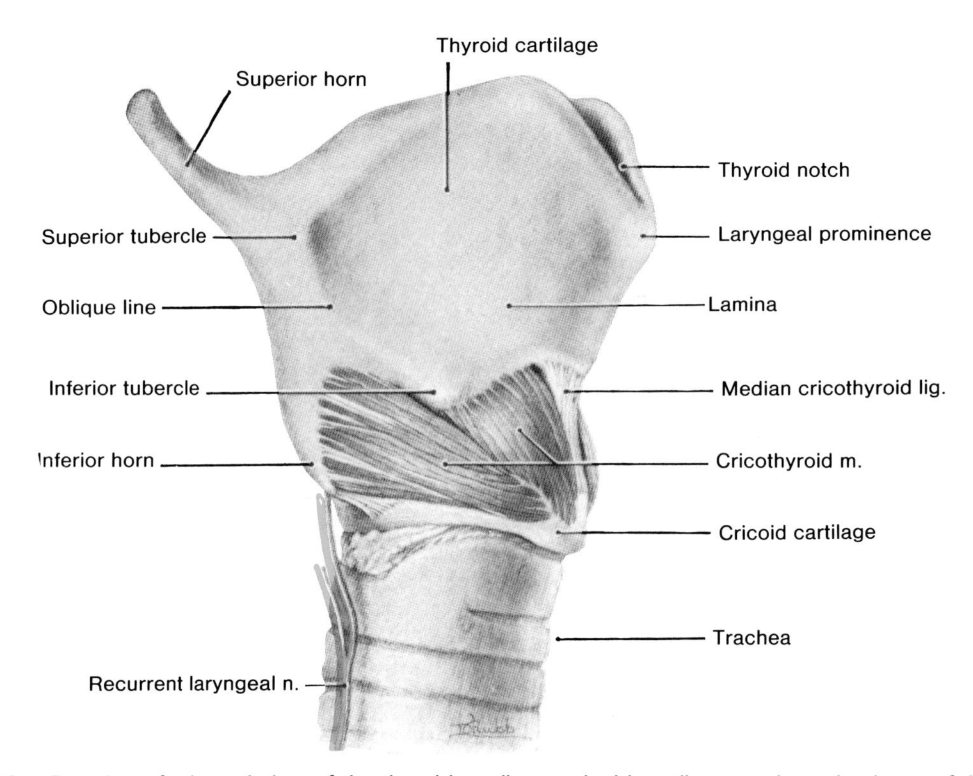

Figure 8-63. Drawing of a lateral view of the thyroid cartilage, cricoid cartilage, and proximal part of the trachea. Observe that the cricothyroid muscle arises from the external surface of the arch of the cricoid cartilage, and that it has a straight part inserted into the inferior border of the lamina of the thyroid cartilage, and an oblique part inserted into the anterior border of the inferior horn. *Observe the course of the recurrent laryngeal nerve* which supplies all intrinsic muscles of the larynx, except the one shown here (the cricothyroid muscle). Also see Figure 8-61.

arytenoid cartilage has an **apex** superiorly, a **vocal process** anteriorly, and a **muscular process** laterally (Fig. 8-65). The apex is attached to the aryepiglottic fold (Fig. 8-52), the vocal process to the **vocal ligament** (Figs. 8-65 and 8-67), and the muscular process to the posterior and lateral ***cricoarytenoid muscles*** (Figs. 8-71 and 8-72).

The Corniculate and Cuneiform Cartilages (Figs. 8-64 and 8-65). These small cartilaginous nodules are in the posterior part of the aryepiglottic folds (Fig. 8-52). The corniculate cartilages are attached to the apices of the arytenoid cartilages and serve to prolong them. The cuneiform (L. wedge-shaped) cartilages lie in the aryepiglottic folds and are approximated to the tubercle of the epiglottis when the inlet of the larynx is closed during swallowing.

The Epiglottic Cartilage (Figs. 8-61 to 8-65). This thin cartilage, shaped like a leaf or bicycle saddle, gives flexibility to the epiglottis. Situated posterior to the root of the tongue and hyoid bone, and anterior to the inlet of the larynx (Fig. 8-46), the epiglottic cartilage forms the superior part of the anterior wall and the superior margin of the inlet of the larynx (Fig. 8-52). Its broad superior end is free and its tapered inferior end is attached to the **thyroepiglottic ligament** (Fig. 8-65), located in the angle formed by the thyroid laminae. The anterior surface of the epiglottic cartilage is attached to the hyoid bone by the **hyoepiglottic ligament** (Fig. 8-66).

The mucous membrane covering the epiglottis is united to the posterior part of the tongue by a median and two lateral **glossoepiglottic folds** (Fig. 8-52). Between the median and lateral folds are depressions called **epiglottic valleculae** (Fig. 8-16). *Vallecula* is a Latin word meaning a little ditch, an appropriate term because saliva collects in the vallecula from the surface of the tongue.

The inferior part of the posterior surface of the epiglottic cartilage that projects posteriorly is called the **epiglottic tubercle** (Fig. 8-68). The mucous membrane covering this tubercle bulges into the larynx and comes into contact with the cuneiform cartilages during swallowing.

The thyroid, cricoid, and most parts of the arytenoid cartilages often calcify with age.

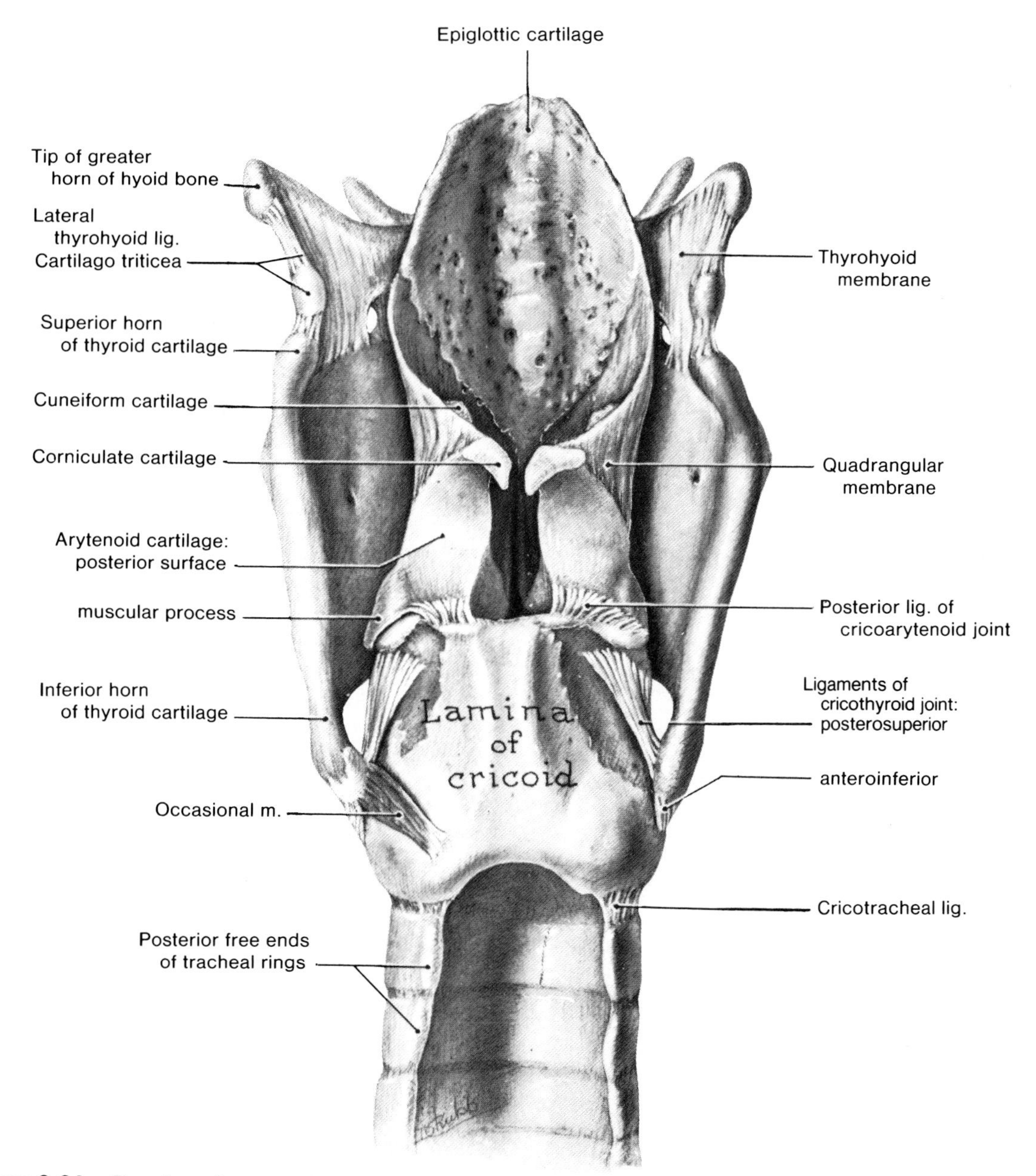

Figure 8-64. Drawing of a posterior view of the skeleton of the larynx. Observe that the thyroid cartilage shields the smaller cartilages of the larynx (epiglottic, arytenoid, corniculate, and cuneiform). The hyoid bone, although not a part of the larynx, also shields the superior part of the epiglottic cartilage. Note that the rounded posterior border of the thyroid cartilage is prolonged into superior and inferior horns. The inferior horns articulate with the cricoid cartilage at synovial joints (cricothyroid joints).

Fractures of the laryngeal skeleton may result from blows received during boxing, karate, or from compression by a shoulder strap during an automobile accident. These fractures produce submucous hemorrhage and edema, respiratory obstruction, hoarseness, and sometimes an inability to speak.

JOINTS OF THE LARYNX

Some of the laryngeal cartilages articulate freely, thereby allowing them to move during voice production. There are two pairs of synovial joints in the larynx.

The Cricothyroid Joints (Figs. 8-62 and 8-64.) These articulations are between the facets on the lateral surfaces of the cricoid cartilage and the inferior horns of the thyroid cartilage. Each joint has a fibrous

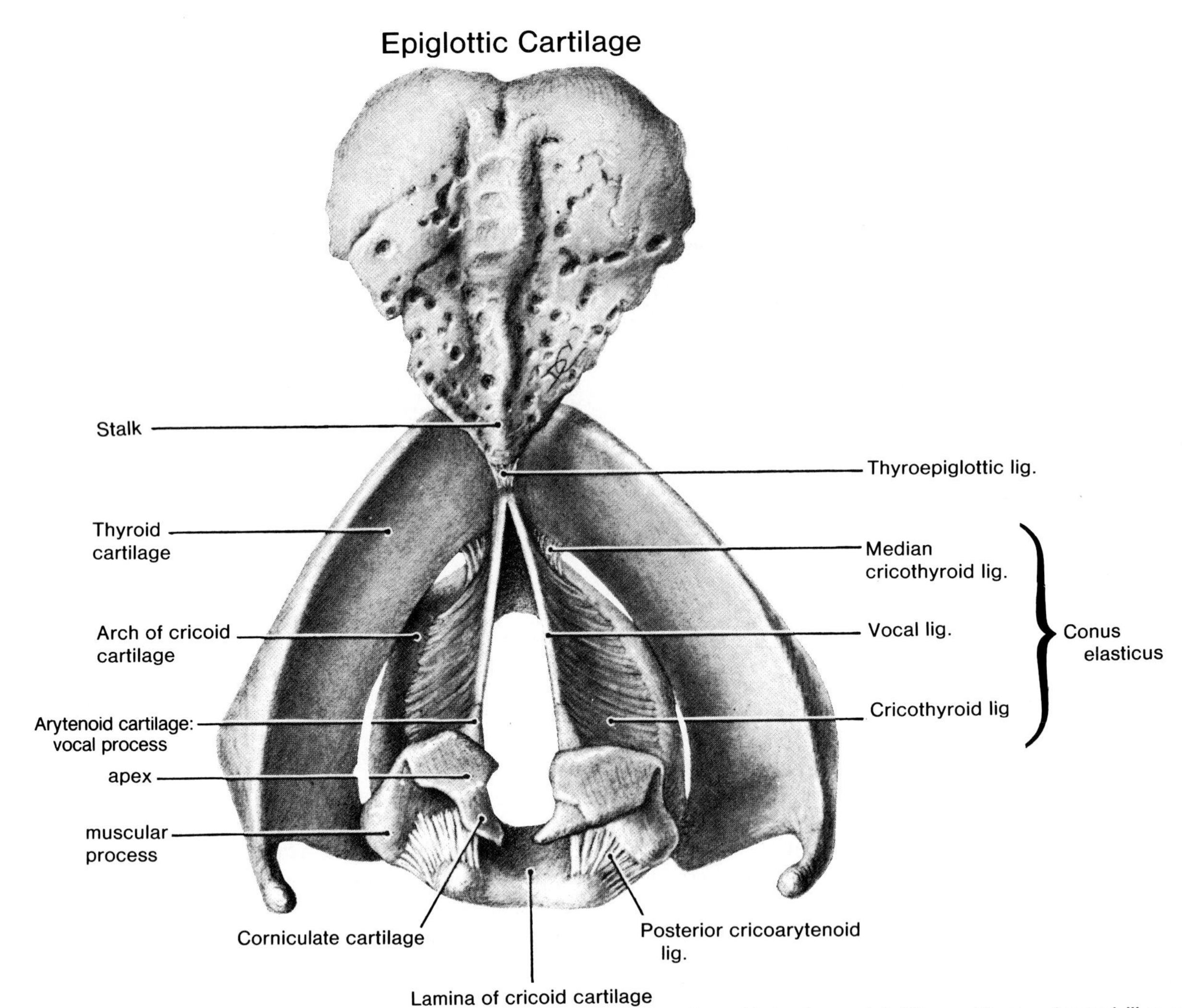

Figure 8-65. Drawing of the skeleton of the larynx, superior view. Note the epiglottic cartilage, shaped like a bicycle seat and showing pits for the mucous glands. Observe that it is attached at its apex by ligamentous fibers to the angle of the thyroid cartilage superior to the vocal ligaments. Observe the paired arytenoid cartilages which have a blunt apex prolonged as the corniculate cartilage; a rounded, lateral, basal angle called the muscular process; and a sharp, anterior basal angle called the vocal process, for the attachment of the vocal ligament. Note the strong posterior cricoarytenoid ligament which prevents the arytenoid cartilage from falling into the larynx. Observe the vocal ligament, which forms the skeleton of the vocal fold, extending from the vocal process to the "angle" of the thyroid cartilage, and there joining its fellow inferior to the thyroepiglottic ligament. Note that the cricothyroid ligament blends anteriorly with the median cricothyroid ligament and sweeps superiorly from the superior border of the arch of the cricoid cartilage to the vocal ligament. Hence, when the vocal ligaments are in apposition, the membranes of opposite sides form a roof for the infraglottic section of the larynx inferior to them.

capsule which is lined by a synovial membrane. The main movements at these joints are rotation and gliding of the thyroid cartilage at the cricothyroid joints. These movements result in changes in the length of the vocal folds. They also slacken or tighten the **vocal ligaments** which pass between the arytenoid cartilages and the thyroid cartilage (Fig. 8-65).

The Cricoarytenoid Joints (Fig. 8-64). These articulations are between the bases of the arytenoid cartilages and the superior sloping surfaces of the laminae of the cricoid cartilage. The cricoarytenoid joints permit the following movements of the arytenoid cartilages: (1) sliding toward or away from one another; (2) tilting anteriorly and posteriorly; and (3) rotary motion. These movements are important in approximating, tensing, and relaxing the vocal folds.

LIGAMENTS AND MEMBRANES OF THE LARYNX

The various laryngeal cartilages are united by several ligaments and membranes.

The Thyrohyoid Membrane (Figs. 8-62 and 8-

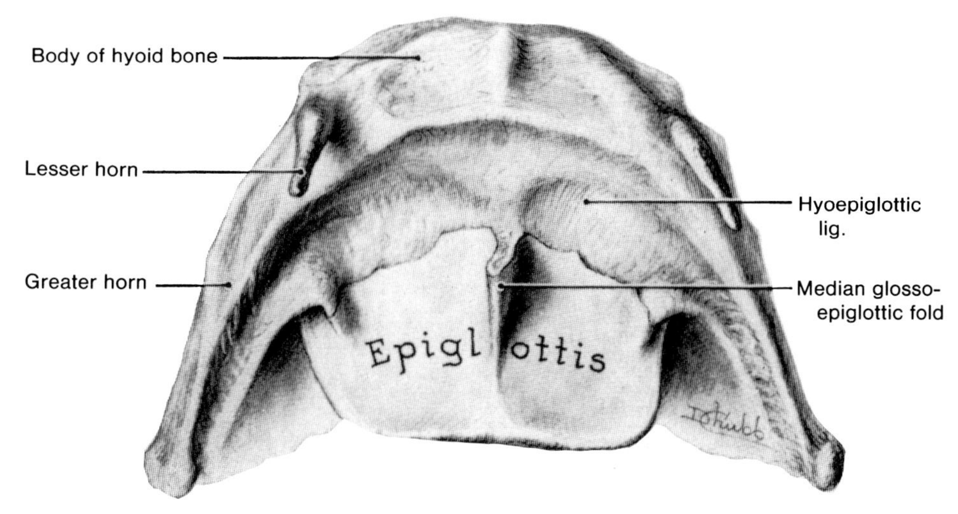

Figure 8-66. Drawing of the hyoepiglottic ligament, superior view. Note that it unites the epiglottic cartilage to the hyoid bone. On the *left side* observe the three parts of the hyoid bone. Also note the asymmetry of the greater and lesser horns of opposite sides.

64). This is an extrinsic ligament that connects the thyroid cartilage and the hyoid bone, thereby suspending the larynx. It is separated from the posterior surface of the body of the hyoid by a bursa. Its thicker median part is called the **median thyrohyoid ligament**, and its thickened lateral parts are called the lateral thyrohyoid ligaments. **The lateral thyrohyoid ligaments** connect the tips of the superior horns of the thyroid cartilage to the tips of the greater horns of the hyoid bone. They each contain a *triticeal* (G. kernel-like) *cartilage* which helps to close the inlet of the larynx during swallowing.

The Cricothyroid and Cricotracheal Ligaments (Figs. 8-62 and 8-65) connect the cricoid cartilage with the arch of the thyroid cartilage and the first tracheal ring, respectively.

If food (*e.g.*, a piece of steak) or some other foreign object enters the larynx, the laryngeal muscles go into spasm; this tenses the vocal folds. As a result the **rima glottidis** (Fig. 8-68) closes and no air can enter the trachea, bronchi, and lungs. Obviously the person is in danger of **asphyxiation** ("choking to death").

If the foreign object cannot be dislodged, emergency therapy must be given to open the airway. The procedure used depends on the condition of the patient, the facilities available, and the experience of the person giving first aid.

Often a large bore needle is inserted through the median cricothyroid ligament (Fig. 8-62) to permit fast entry of air. Later a **cricothyrotomy** (***inferior laryngotomy***) is performed, during which an incision is made through the skin and cricothyroid ligament for more adequate relief of the respiratory obstruction. This procedure may be followed by a **tracheotomy** (Fig. 8-38) and insertion of a short curved metal tube (**tracheotomy tube**) into the trachea.

The Vocal Ligament, The Vocal Fold, and The Conus Elasticus (Figs. 8-65, 8-67, and 8-69). The elastic vocal ligament on each side extends from the junction of the laminae of the thyroid cartilage anteriorly to the vocal process of the arytenoid cartilage posteriorly. **The vocal ligament** is the fibrous core of the **vocal fold** and is the free edge of the **conus elasticus**. This elastic membrane extends superiorly from the cricoid cartilage to the vocal ligament.

The Quadrangular Membrane and the Vestibular Ligament (Figs. 8-64 and 8-67). The quadrangular membrane is a thin submucosal sheet of connective tissue that extends from the arytenoid cartilage to the cartilage of the epiglottis. The free inferior margin of this membrane constitutes the **vestibular ligament**. It is covered loosely by a **vestibular fold** of mucous membrane (Fig. 8-67). This fold lies superior to the vocal fold and extends from the thyroid cartilage to the arytenoid cartilage. The cricothyroid ligament and the quadrangular membrane (Fig. 8-67), although separated by the interval between the vocal and vestibular ligaments, are referred to as the *fibroelastic membrane of the larynx*.

The Ligaments of the Epiglottis (Figs. 8-65, 8-66, and 8-69). The epiglottis has several attachments that were described with the epiglottic cartilage. Briefly, the epiglottis is attached to the hyoid bone by the ***hyoepiglottic ligament***, to the posterior part of

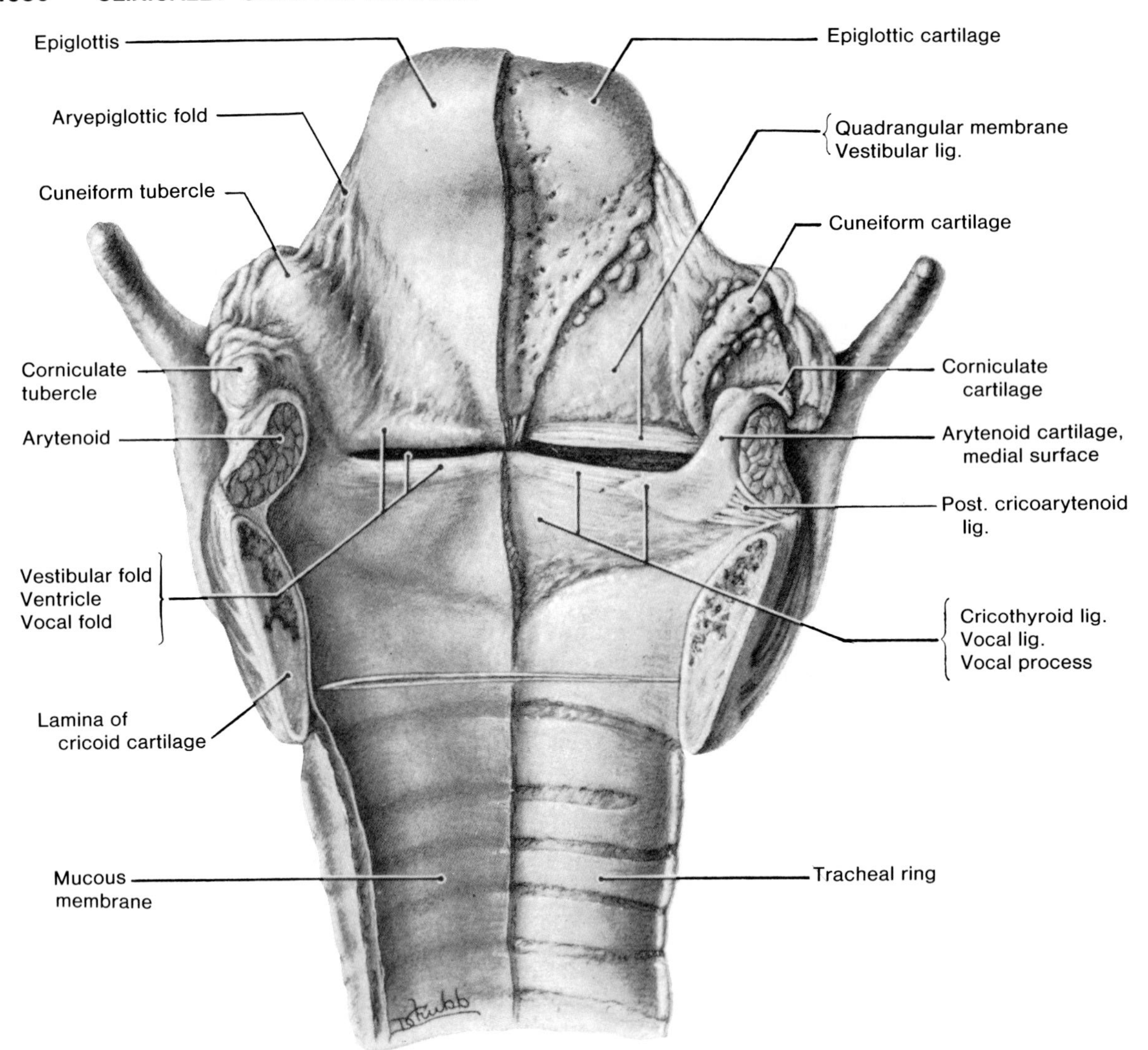

Figure 8-67. Drawing of a posterior view of a dissection of the internal surface of the larynx. The posterior wall of the larynx is split in the median plane and the two sides are held apart by a glass rod. *On the left side* the mucous membrane is intact; *on the right side* the mucous and submucous coats are peeled off and the next coat, consisting of cartilages, ligaments, and the fibroelastic membrane, is laid bare. Note the *three compartments of the larynx*: (1) the superior compartment or *vestibule*, superior to the level of the vestibular folds; (2) the middle compartment between the levels of the vestibular and vocal folds which has right and left depressions, the *ventricles*; and (3) the inferior or *infraglottic cavity* inferior to the level of the vocal folds (also see Fig. 8-70). Note that the superior part, the quadrangular membrane, is thickened inferiorly to form the vestibular ligament and that the inferior part, the cricothyroid ligament, ends superiorly as the vocal ligament. Between the vocal and vestibular ligaments, the membrane lined with mucous membrane is evaginated to form the wall of the laryngeal ventricle. The *two horizontal lines* crossing the cricoid cartilage represent the glass rod used to separate the lamina of this ring-like cartilage.

the tongue by the ***median glossoepiglottic fold***, to the sides of the pharynx by the ***lateral glossoepiglottic fold***, and to the thyroid cartilage by the ***thyroepiglottic ligament***.

THE INTERIOR OF THE LARYNX (Figs. 4-6, 8-65, and 8-67 to 8-70)

The cavity of the larynx extends from the **inlet of the larynx**, through which it communicates with the laryngopharynx (Fig. 8-46), to the level of the inferior border of the cricoid cartilage, where it is continuous with the cavity of the trachea. The inlet of the larynx lies in an almost vertical plane (Fig. 8-52).

The larynx is divided into three parts by superior and inferior projecting folds of mucous membrane on each side, called **vestibular folds**. Superior to vestibular folds, the cavity of the larynx is called the **vestibule** (Fig. 8-70). Between the vestibular folds, su-

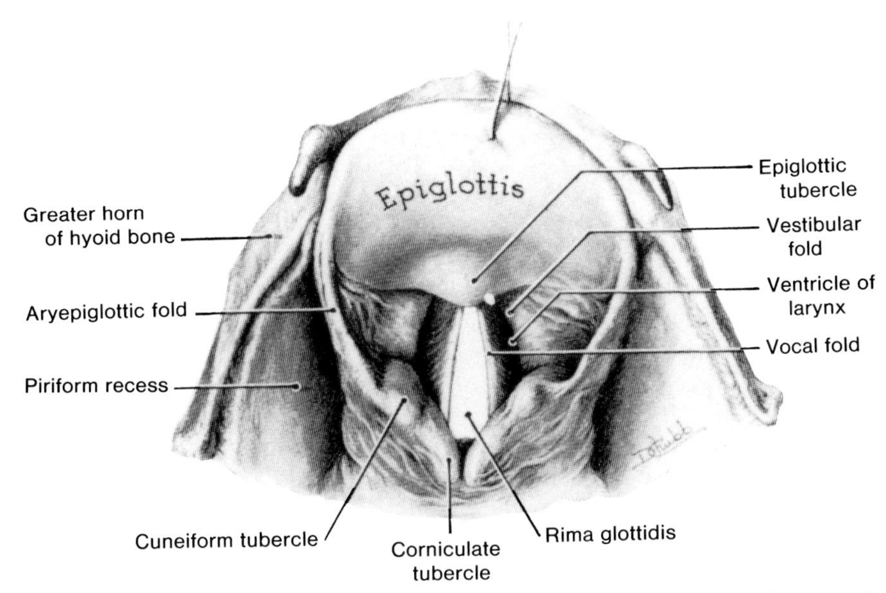

Figure 8-68. Drawing of the larynx, superior view. See Fig. 8-52 for orientation. Observe the inlet of the larynx (laryngeal aditus) bounded (1) anteriorly by the free curved edge of the epiglottis; (2) posteriorly by the arytenoid cartilages, the corniculate cartilages which cap them, and the interarytenoid fold which unites them; and (3) on each side, by the aryepiglottic fold which contains the superior end of the cuneiform cartilage. Note that the vocal folds are closer together than the vestibular folds and, therefore, are visible inferior to them. Observe the sharpness of the vocal folds and the fullness of the vestibular folds.

perior to the vocal folds, is the **ventricle of the larynx** (Figs. 8-68 to 8-70). This is the smallest of the three cavities of the larynx. It extends laterally between the two folds as the **sinus of the larynx**. From each sinus, a ***saccule of the larynx*** passes superiorly between the vestibular fold and the thyroid lamina (Figs. 8-70 and 8-71). The inferior cavity of the larynx, called the **infraglottic cavity**, extends from the vocal folds to the inferior border of the cricoid cartilage, where it is continuous with the cavity of trachea.

The Vocal Folds (Figs. 8-67 to 8-70, 8-72, and 8-73). These folds are concerned with the production of sound (phonation). The apex of each wedge-shaped vocal fold (vocal cord) projects medially into the laryngeal cavity, and its base lies against the lamina of the thyroid cartilage. Each vocal fold consists of the vocal ligament, the conus elasticus, muscle fibers, and a covering of mucous membrane.

The rima glottidis (Fig. 8-68) is the aperture between the vocal folds. The term **glottis** refers to the vocal folds, the rima glottidis, and the narrow part of the larynx at the level of the vocal folds (Fig. 8-70). The glottis is the part of the larynx most directly concerned with voice production (*i.e.*, **the glottis is the vocal apparatus**). The term glottis is sometimes inaccurately used synonymously with the inlet of the larynx (G. *glottis*).

The shape of the rima glottidis varies according to the position of the vocal folds (Fig. 8-72). During ordinary breathing, the rima glottidis is narrow and wedge-shaped; it is wide during forced respiration. The vocal folds are closely approximated during speaking so that the rima glottidis appears as a linear slit.

Variation in the tension and length of the vocal folds, in the width of the rima glottidis, and in the intensity of the expiratory effort produces changes in the **pitch of the voice**. The lower range of pitch in the male voice results from the greater length of his vocal folds.

The Vestibular Folds (Figs. 8-67 to 8-70). These folds, extending between the thyroid and the arytenoid cartilages, play little or no part in voice production. They consist of two thick *folds of mucous membrane* enclosing the **vestibular ligaments**. As they may be confused with the vocal folds (vocal cords), they used to be called the *false vocal cords*. The space between the vestibular ligaments is called the **rima vestibuli**. The vestibular folds are part of the protective mechanism by which the larynx is closed during swallowing to prevent the entry of food and foreign particles into it.

THE MUSCLES OF THE LARYNX (Figs. 8-17, 8-19, 8-42, 8-63, 8-71 to 8-73, and Table 8-5)

The muscles of the larynx are divided into extrinsic and intrinsic groups for descriptive purposes.

The extrinsic muscles move the larynx as a whole. The omohyoid, sternohyoid, and sternothyroid muscles (the infrahyoid muscles, Fig. 8-17) are *depressors of the hyoid bone and larynx* (Table 8-2), whereas the stylohyoid, digastric, mylohyoid, geniohyoid (the suprahyoid muscles) (Fig. 8-17), and the stylo-

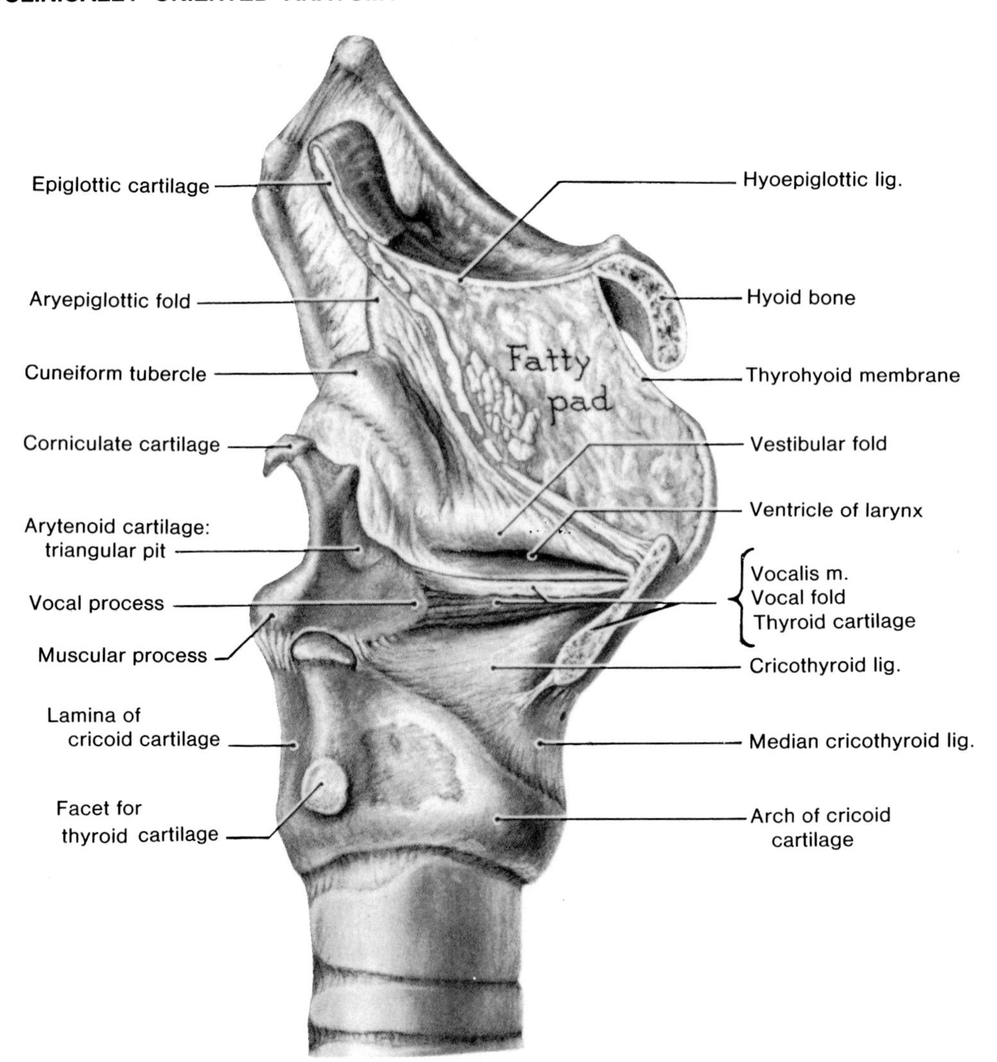

Figure 8-69. Drawing of a lateral view of the larynx. Superior to the vocal folds, the larynx is sectioned near the median plane and the interior of its left side is seen. Inferior to this level, the right side of the larynx is dissected. Observe the hyoepiglottic ligament and the thyrohyoid membrane, both attached to the superior part of the body of the hyoid bone. Note the lateral aspect of the cricoid cartilage and the raised circular facet for the inferior horn of the thyroid cartilage, separating the lamina from the arch. Superior to this, observe the sloping facet for the arytenoid cartilage. Examine the triangular membrane, called the cricothyroid ligament, which has the vocal ligament for its superior border and blends with the median cricothyroid ligament anteroinferiorly.

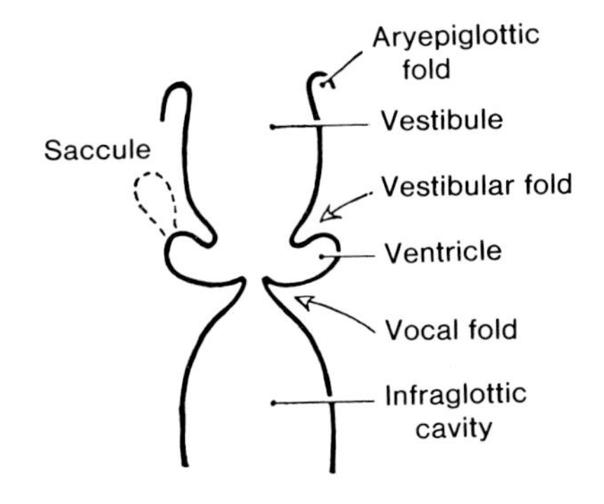

Figure 8-70. Sketch of a coronal section of the larynx showing its compartments: a vestibule, a middle compartment having right and left ventricles, and an infraglottic cavity.

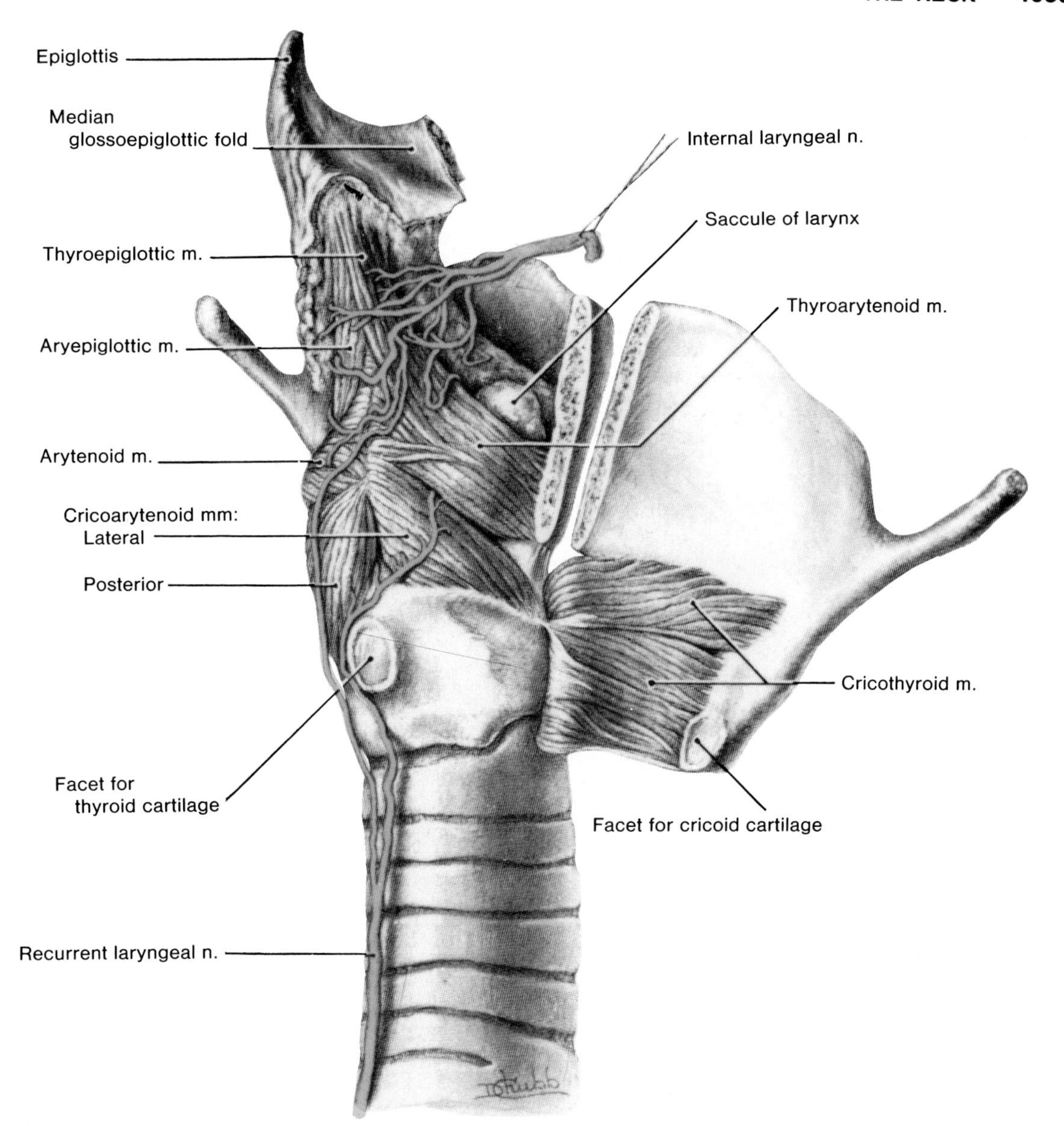

Figure 8-71. Drawing of a dissection of the muscles and nerves of the larynx. The thyroid cartilage is sawn through on the right of the median plane. The cricothyroid joint is laid open and the right lamina of the thyroid cartilage is turned anteriorly, stripping the cricothyroid muscles off the arch of the cricoid cartilage. Observe the lateral cricoarytenoid muscle arising from the superior border of the arch of the cricoid cartilage, and inserting with the posterior cricoarytenoid into the muscular process of the arytenoid cartilage. Note that the thyroarytenoid muscle is inserted with the arytenoid muscle into the lateral border of the arytenoid cartilage, and that its superior fibers continue to the epiglottis as the thyroepiglottic muscle.

pharyngeus are *elevators of the hyoid bone and larynx* (Tables 8-1 and 8-4). The thyrohyoid muscle (Table 8-2) draws the hyoid bone and the thyroid cartilage together *i.e.*, it depresses the hyoid bone and elevates the thyroid cartilage.

The intrinsic muscles are concerned with movements of the laryngeal parts, making alterations in the length and tension of the vocal folds, and in the size and shape of the rima glottidis in voice production (Fig. 8-72).

All intrinsic muscles of the larynx are supplied by the recurrent laryngeal nerve, a branch of CN X (Figs. 7-135, 8-71, and Table 8-5), except the cricothyroid muscle which is supplied by the **external laryngeal nerve** (Fig. 8-75). The muscles of the larynx can best be understood if they are considered as functional groups.

Muscles of the Inlet of the Larynx (Fig. 8-61). These muscles have a sphincteric action and *close the inlet of the larynx* as a protective mechanism during

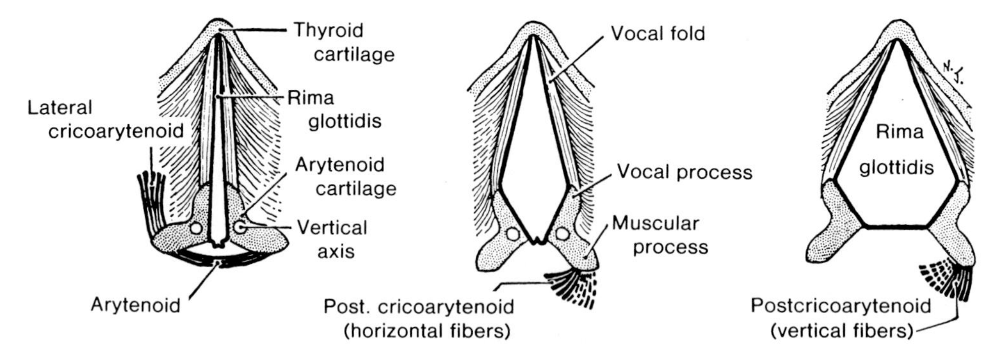

Figure 8-72. Drawings illustrating the scheme of the glottis (superior view) and the actions of the lateral and posterior cricoarytenoid muscles.

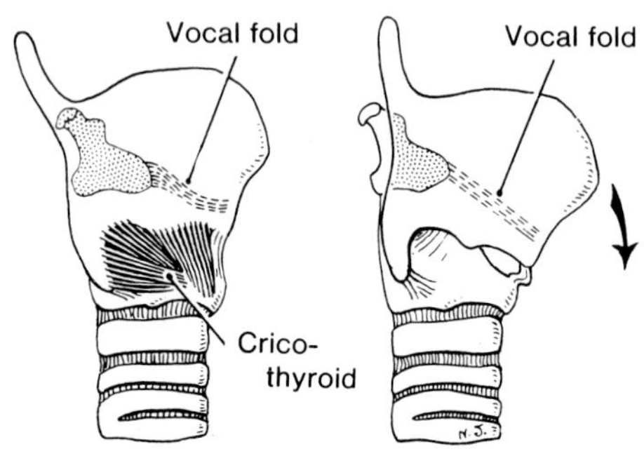

Figure 8-73. Diagrams illustrating the action of cricothyroid muscles. They act on the cricothyroid joints, tightening the vocal folds and raising the pitch of the voice.

swallowing. Contraction of the **transverse and oblique arytenoid muscles and the aryepiglottic muscles** (prolongations of the oblique arytenoids) brings the aryepiglottic folds together and pulls the arytenoid cartilages toward the epiglottis. These movements help close the inlet of the larynx (Fig. 8-52).

The Transverse Arytenoid Muscle (Fig. 8-61 and Table 8-5), the only unpaired muscle of the larynx, covers the arytenoid cartilages posteriorly. It extends from the posterior aspect of one arytenoid cartilage to the same region of the opposite arytenoid.

The Oblique Arytenoid Muscle (Fig. 8-61 and Table 8-5), superficial to the transverse arytenoid muscle, consists of two fasciculi which cross each other in an X-like fashion. Some oblique fibers continue as the **aryepiglottic muscle** (Fig. 8-71). The continuity of these muscles ensures that the arytenoid cartilages are brought together at the same time as the epiglottis is pulled inferiorly toward these cartilages. The inlet of the larynx is closed in two ways during swallowing, thereby preventing food from entering the larynx.

The Thyroepiglottic Muscles (Fig. 8-71 and Table 8-5) arise from the anteromedial surface of the laminae of the thyroid cartilage, and insert on the lateral margin of the epiglottic cartilage. *The thyroepiglottic muscles widen the inlet of the larynx.*

Muscles of the Vocal Folds (Figs. 8-61, 8-65, 8-69, 8-71, and 8-72). ***These muscles open and close the rima glottidis.***

1. *Adductors of the Vocal Folds* (Table 8-5). The **lateral cricoarytenoid muscles** arise from the lateral portions of the cricoid cartilage and insert into the muscular processes of the arytenoid cartilages.

Actions. Pull the muscular processes anteriorly, rotating the arytenoids so that their vocal processes swing medially. *These movements adduct the vocal folds and close the rima glottidis.* This action is reinforced by the transverse arytenoid muscle, which pulls the arytenoid cartilages together.

2. *Abductors of the Vocal Folds* (Table 8-5). The principal abductors of the vocal folds are the **posterior cricoarytenoid muscles**. These muscles arise on each side from the posterior surface of the lamina of the cricoid cartilage and pass laterally and superiorly to insert into the muscular processes of the arytenoid cartilages.

Action. Rotate the arytenoid cartilages, thereby deviating them laterally and widening the rima glottidis.

3. *Tensors of the Vocal Folds* (Table 8-5). The main tensors of these folds are the **cricothyroid muscles**. They are located on the external surface of the larynx between the cricoid and thyroid cartilages. The muscle on each side arises from the anterolateral part of the **cricoid cartilage** and inserts into the inferior margin and anterior aspect of the inferior horn of the thyroid cartilage.

Actions. Tilt or pull anteriorly the thyroid cartilage on the cricoid cartilage, increasing the distance between the thyroid and arytenoid cartilages. As a result, the vocal ligaments are elongated and tightened and the pitch of the voice is raised.

4. *Relaxors of the Vocal Folds* (Table 8-5). The principal relaxors of these folds are the **thyroarytenoid muscles**, which arise from the posterior surface of the thyroid cartilage near the midline, and insert into the anterolateral surfaces of the arytenoid cartilages. One band of fibers of each muscle, called the **vocalis muscle** (Fig. 8-69), arises from the vocal

Table 8-5
Muscles of the Larynx

Muscle	Origin	Insertion	Nerve Supply	Action(s)
Cricothyroid (Fig. 8-63)	Anterolateral part of cricoid cartilage	Inferior margin and inferior horn of thyroid cartilage	External laryngeal n. (Figs. 8-19, 8-27, and 8-42)	Lengthens, tenses, and adducts vocal fold
Posterior cricoarytenoid (Figs. 8-61, 8-65, 8-69, and 8-72)	Posterior surface of laminae of cartilage	Muscular process of arytenoid cartilage		Abducts vocal fold
Lateral cricoarytenoid (Figs. 8-61, 8-65, 8-69, and 8-71)	Arch of cricoid cartilage			
Transverse and oblique arytenoid (Fig. 8-61)	One arytenoid cartilage	Opposite arytenoid cartilage	Recurrent laryngeal n. (Figs. 8-63, 8-71, and 8-75)	Adduct vocal fold
Thyroarytenoid[1] (Fig. 8-71)	Posterior surface of thyroid cartilage	Muscular process of arytenoid process		Relaxes vocal fold
Vocalis (Fig. 8-69)	Angle between laminae of thyroid cartilage	Vocal process of arytenoid cartilage		Alters vocal fold during phonation

[1] The superior fibers of the thyroarytenoid muscle pass into aryepiglottic fold (Fig. 8-52) and some of them reach the epiglottic cartilage. These fibers constitute the thyroepiglottic muscle which widens the inlet of the larynx.

ligament and passes to the vocal process of the arytenoid cartilage.

Actions. Pull the arytenoid cartilages anteriorly, thereby slackening the vocal ligaments. *The vocalis muscle produces minute adjustments of the vocal ligaments* (*e.g.*, as required in whispering).

BLOOD SUPPLY OF THE LARYNX (Figs. 8-26, 8-27, 8-31, and 8-74)

The superior and inferior laryngeal arteries supply the larynx. They are branches of the superior and inferior thyroid arteries, respectively.

The superior laryngeal artery runs with the internal branch of the superior laryngeal nerve through the thyrohyoid membrane, and then branches to supply the internal surface of the larynx (Fig. 8-27).

The inferior laryngeal artery runs with the inferior laryngeal nerve and supplies the mucous membrane and muscles of the inferior aspect of the larynx.

NERVES OF THE LARYNX (Figs. 8-5, 8-47, 8-61, and 8-75)

The nerves are derived from the vagus (CN X) via the internal and external branches of the **superior laryngeal nerve** (Fig. 8-75), and from the **recurrent laryngeal nerve** (Fig. 8-61). Fibers from the sympathetic nerves supply the blood vessels and glands (Fig. 8-47).

The Superior Laryngeal Nerve (Figs. 7-135 and 8-75). This branch of the vagus (CN X) arises from the middle of the inferior ganglion of the vagus at the superior end of the carotid triangle (Figs. 8-5 and 8-6*B*). It divides within the carotid sheath into two terminal branches, the internal laryngeal nerve (sensory and autonomic) and the external laryngeal nerve (motor).

The internal laryngeal nerve (Figs. 8-5, 8-27, 8-42, 8-44, and 8-75) is the larger of the two terminal branches of the superior laryngeal nerve. It pierces the thyrohyoid membrane with the superior laryngeal artery. *The internal laryngeal nerve supplies sensory fibers to the laryngeal mucous membrane superior to the vocal folds*, including the superior surface of these folds.

The external laryngeal nerve (Figs. 8-5, 8-27, 8-36, 8-44, 8-48, 8-50, and 8-75) is the smaller of the two terminal branches of the superior laryngeal nerve. It descends posterior to the sternothyroid muscle in company with the superior thyroid artery. At first it lies

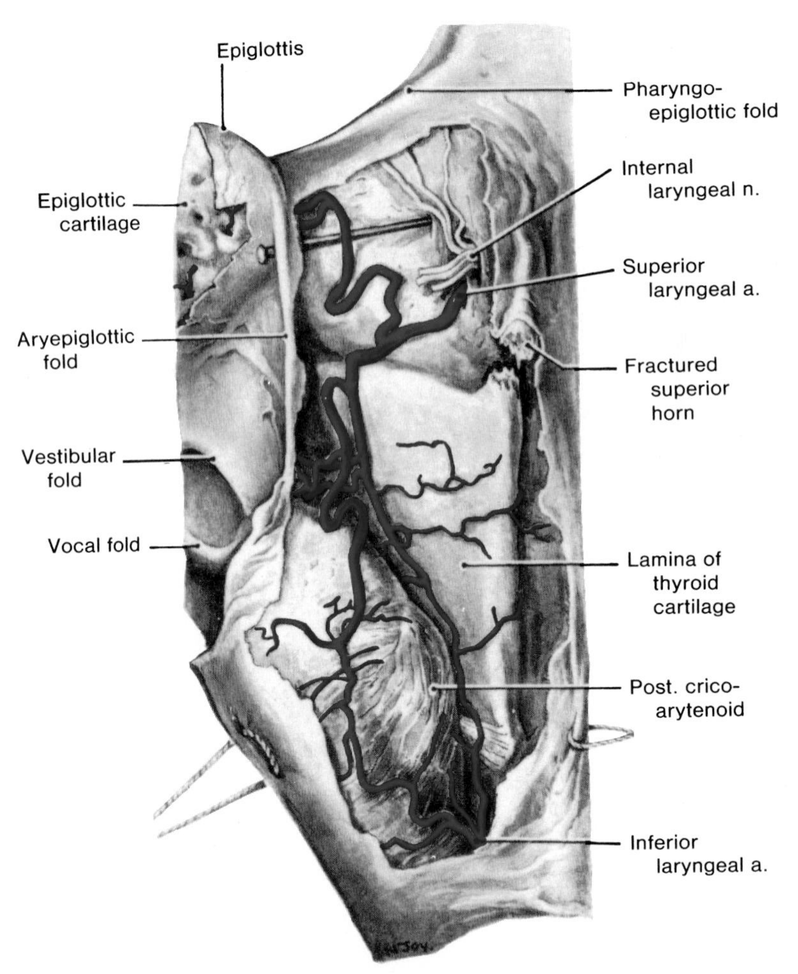

Figure 8-74. Drawing illustrating the blood supply of the larynx. Observe the anastomoses between the superior and inferior laryngeal arteries (branches of the superior and inferior thyroid arteries, respectively). Arterial twigs pierce the epiglottic cartilage at the sites of the pits for the glands (Fig. 8-65).

on the inferior constrictor muscle of the pharynx, and then pierces it to supply this muscle and the cricothyroid. *The cricothyroid is the only intrinsic laryngeal muscle that is not supplied by the recurrent laryngeal nerve* (Table 8-5).

The Recurrent Laryngeal Nerve (Figs. 7-135, 8-44, 8-63, 8-71, and 8-75). This *very important nerve* ascends in the groove between the trachea and esophagus, where it is intimately related to the medial surface of the thyroid gland (Fig. 8-36). It gives branches to the pharynx, esophagus, and trachea.

The recurrent laryngeal nerve is vulnerable to injury during thyroidectomy, carotid endarterectomy, and other operations in the anterior triangle of the neck. It enters the larynx by passing posterior to the inferior horn of the thyroid cartilage, and gives *branches to all muscles of the larynx, except the cricothyroid muscle* (Table 8-5).

The recurrent laryngeal nerve also *supplies sensory fibers to the mucous membrane of the larynx inferior to the vocal folds,* including the inferior surface of these folds. The terminal part of the recurrent laryngeal is known as the **inferior laryngeal nerve** (Fig. 8-61). It enters the larynx by passing deep to the inferior border of the inferior constrictor muscle of the pharynx (Figs. 8-50 and 8-61) and divides into anterior and posterior branches. In Figure 8-44, observe that these branches accompany the inferior laryngeal artery into the larynx.

LYMPHATICS OF THE LARYNX (Figs. 7-40, 8-35, 8-44, and 8-60)

The lymph vessels superior to the vocal folds accompany the superior laryngeal artery through the thyrohyoid membrane and drain into the **superior deep cervical lymph nodes** (Fig. 8-35).

The lymph vessels *inferior to the vocal folds* drain into the **inferior deep cervical lymph nodes** (supraclavicular nodes) via the prelaryngeal, pretracheal, and paratracheal lymph nodes (Figs. 8-44 and 8-60).

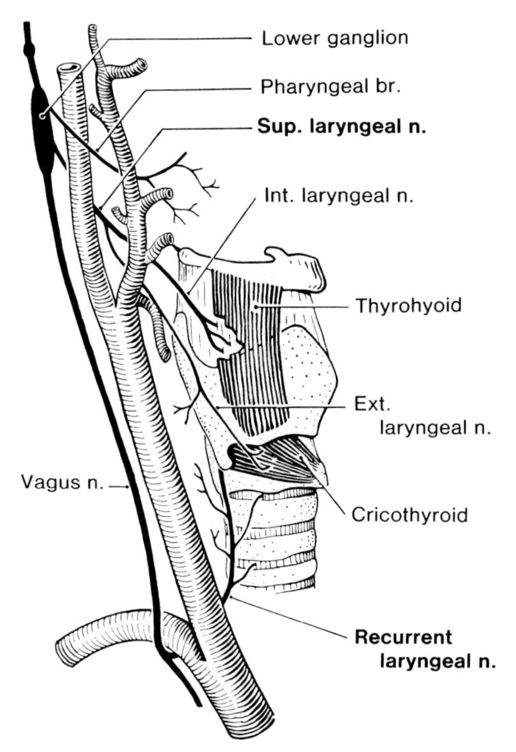

Figure 8-75. Drawing illustrating the laryngeal branches of the right vagus nerve (also see Fig. 7-135). Note that the nerves of the larynx are the internal and external branches of the superior laryngeal nerve and the terminal branches of the recurrent laryngeal nerve. Laryngeal nerves are also derived from the sympathetic trunks (Figs. 8-36 and 8-47).

The lymphatics from the vocal folds do not communicate across the midline, but vessels in the posterior wall of the larynx anastomose submucously.

The larnyx may be examined by indirect laryngoscopy (using a laryngoscopic mirror), or it can be viewed by **direct laryngoscopy** (using a tubular instrument called a **laryngoscope**). The vestibular folds normally appear pink, whereas the vocal folds are pearly white in color. The size of the larynx varies somewhat from person to person and is not dependent on stature. This largely explains the difference in pitch of the voice in different persons.

The larynx is larger in adult males than in adult females and the vocal ligaments are longer. The larynx is not so prominent in females as in males (Fig. 8-1). In most men the vocal folds are longer than in women and children; as a result the voice of most men is deeper than most women.

Hoarseness is the most common symptom of disorders of the larynx (*e.g.*, cancer of the vocal folds). In cases requiring *laryngectomy* (removal of the larynx), **esophageal speech** (regurgitation of ingested air) and other rehabilitative speech techniques can be learned.

Inhalation of foreign bodies into the larynx rapidly produces symptoms (*e.g.*, choking). If the inhaled object is sharp (*e.g.*, a chicken bone), there is usually sharp pain and progressive obstruction to breathing owing to **inflammation of the larynx**. This may cause edema of the tissues superior to the glottis; however, because of the close attachment of the mucous membrane at that level, the edema does not extend inferior to the vocal folds.

The mucosa of the larynx superior to the vocal folds is extremely sensitive and contact with a foreign body immediately induces explosive coughing. This is a protective mechanism that is designed to keep foreign bodies out of the larynx.

AGE CHANGES IN LARYNX

The larynx grows until about the 3rd year, after which little growth occurs until about the 12th year. Prior to this, there are no major laryngeal sex differences.

At puberty, particularly in males (13 to 16 years), the walls of the larynx become strengthened, the laryngeal cavity enlarges, the vocal folds lengthen and thicken, and the laryngeal prominence becomes conspicuous in most males (Fig. 8-1).

The length of the vocal folds increases gradually in both sexes up to puberty. *During puberty the increase in the length of the vocal folds in the male is abrupt.* The pitch of the voice lowers by an octave in boys. The change in the length of the vocal folds is largely responsible for the **voice changes** occurring in boys that are familiar to everyone.

The pitch of the voice of **eunuchs**, persons in whom testes have not developed (*agonadal males*), or whose testes have been removed (*castrated males*) during childhood, does not become lower unless male hormones are administered. Similarly, laryngeal changes do not occur in males with **seminiferous tubule degeneration** (47, XXY males with the **Klinefelter syndrome**) who have an inadequate production of androgens.

PRESENTATION OF PATIENT ORIENTED PROBLEMS

Case 8-1

A 22-year-old woman consulted her physician about **a swelling in the anterior midline of her neck**

(Fig. 8-76). Although painless, she was concerned because it seemed to be slowly getting larger.

Physical examination revealed that the swelling was located just inferior to the hyoid bone and that it was cystic and freely movable. The doctor grasped the swelling between his index finger and thumb and requested the patient to open her mouth and stick out her tongue. Feeling some movement of the mass, the doctor requested the patient to stick her tongue out as far as possible and then retract it. The doctor noted a definite superior tug on the mass as the patient's tongue protruded. *The swelling moved superiorly during swallowing.* Fluid was aspirated from the swelling for laboratory investigation. Subsequently a diagnosis of **thyroglossal duct cyst** was made.

Problems. Explain the embryological basis of this cyst. Where are these cysts likely to be found? What is the anatomical basis for movement of the cyst superiorly when the patient protrudes her tongue and swallows? What would this condition be called if there had also been a midline cervical opening into the cyst? *These problems are discussed on page* 1066.

Case 8-2

A 27-year-old 2nd year medical student consulted her clinical instructor about a painless, **plum-shaped swelling in her neck,** *inferior to the angle of her right mandible* (Fig. 8-77). As her mandibular third molar teeth ("wisdom teeth") had not erupted, she thought the swelling might be caused by a dental abscess in the submandibular triangle. She also feared that the firm swelling might be caused by a tumor of the submandibular gland, or of the juguloomohyoid lymph node (Fig. 8-60).

Radiographs of her mandible showed the crown of her left third mandibular molar tooth was in contact with the posterior surface of her second molar tooth. The instructor recommended that her impacted tooth be surgically removed after consulting her doctor about the swelling in her submandibular region. He stated that her impacted molar tooth was not the cause of the swelling in her neck.

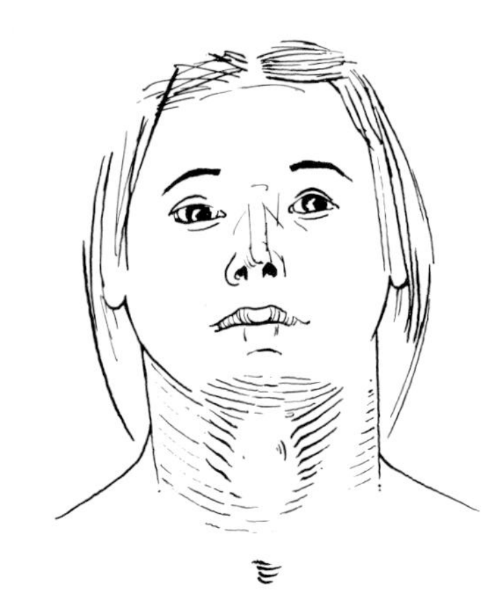

Figure 8-76. Drawing showing the infrahyoid swelling in the midline of the woman's neck.

On examination the doctor found that **the swelling was caused by a painless fluctuant cyst**, located anterior to the superior one-third of her sternocleidomastoid muscle (Fig. 8-77). Further investigation resulted in a diagnosis of a **branchial cyst** (branchiogenic cyst). During excision of the cyst, it was discovered that a sinus tract passed superiorly from it.

Problems. Explain the embryological basis of the branchial cyst. Where does the sinus tract probably terminate? What nerve might be damaged during excision of this cyst? What signs would be present if this nerve were damaged? If the sinus tract had passed inferiorly, where would it probably open? *These problems are discussed on page* 1066.

Case 8-3

After completing your first anatomy exam, your father decided to celebrate and take you out for a steak dinner. Even though he had obviously had a few drinks before meeting you, he still had three drinks before dinner. You noted that his speech was slurred and that he was eating his steak very rapidly.

While telling you an off-color story and laughing loudly, you noticed your father's face change suddenly. He had a terrified look and then collapsed on the floor. At first you suspected that he had passed out from too much drinking, but as you examined him more closely you thought perhaps that he was having a stroke, a heart attack, or some other seizure.

His pulse was strong and then his face began to turn blue (**cyanosis**). You then realized that your father was suffering from **asphyxia**.

You opened his mouth widely and observed that a large piece of steak was caught in the posterior part of his throat. First you reached into his mouth with your finger and tried to pull it out. On being unsuccessful, you rolled him into the prone position and, with your hands interlocked against his epigastrium, you gave him a forceful bear-hug, exerting pressure on his abdomen inferior to his thorax. This increased his intraabdominal pressure and moved his diaphragm superiorly, forcing the air out of his lungs and expelling the piece of steak. He soon recovered and when he resumed eating, you cautioned him to cut his steak into small pieces and to stop drinking.

Problems. Where was the piece of steak most likely lodged? If the "*Heimlich* maneuver" had not been successful and a doctor at another table had come to help you, what life-saving measures do you think he might have taken? Discuss so-called "restaurant deaths." *These problems are discussed on page* 1067.

Case 8-4

A 30-year-old woman frequently complained of ***a midline lump in her neck, nervousness, and loss***

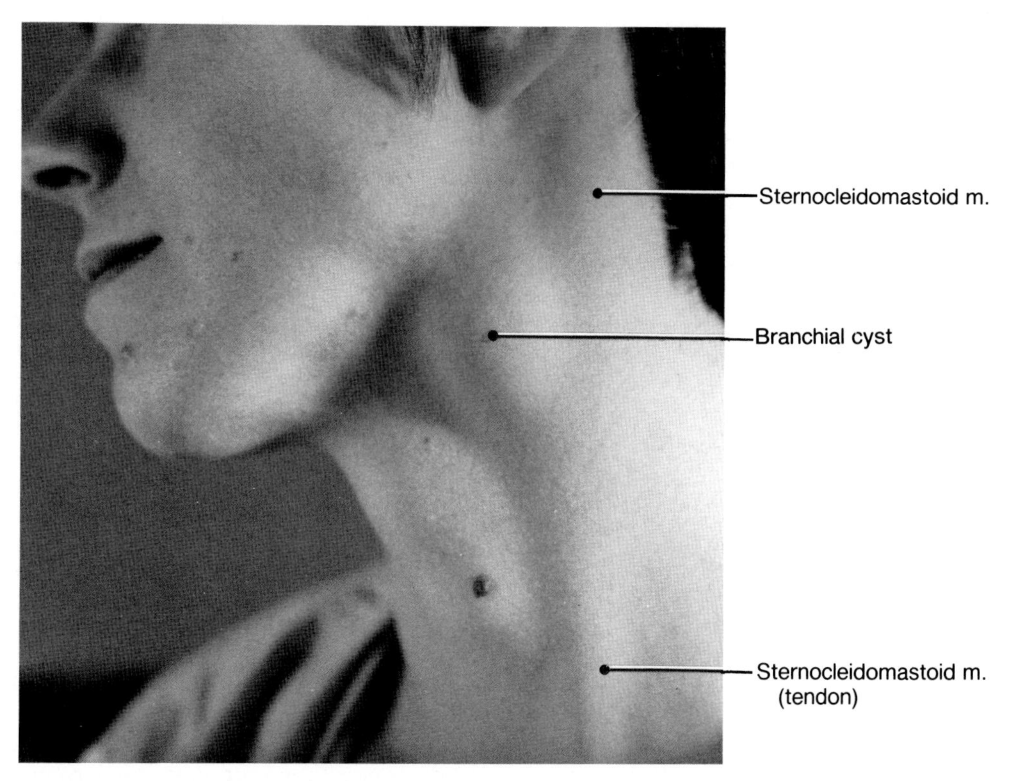

Fligure 8-77. Drawing of a 27-year-old woman showing a swelling inferior to the angle of her left mandible. Note that it is anterior to her sternocleidomastoid muscle.

of weight. She stated that her family complains that she is irritable, excitable, and cries easily.

On examination a smooth swelling was apparent on each side of the midline of her neck, inferior to the larynx (Fig. 8-78). During palpation of the patient's neck from a posterior position, the doctor felt an ***enlarged thyroid gland*** and noted that it moved up and down during deglutition. The following signs were also detected: protrusion of the eyes (Fig. 8-78), rapid pulse, tremor of the fingers, moist palms, and loss of weight.

A diagnosis of hyperthyroidism (exophthalmic goiter, Graves' disease) was made. When the patient did not respond to medical treatment, *a subtotal thyroidectomy was performed*. After the operation the patient complained of hoarseness.

Problems. What is the anatomical basis for the swelling moving up and down during deglutition? As the patient's thyroid gland was enlarged, what nerves might have been compressed or displaced? *If a total thyroidectomy had been done, what other endocrine glands might inadvertently have been removed along with the thyroid*? What would result from this error? What was the probable cause of the patient's hoarseness? *These problems are discussed on page* 1067.

Figure 8-78. Drawing of a 30-year-old woman exhibiting the characteristic clinical features of hyperthyroidism. Note the slight protrusion of her eyes (exophthalmos) and that a rim of white (the sclera) shows superior and inferior to her iris, giving her an alarmed, staring expression.

Case 8-5

A 10-year-old boy was admitted to hospital with a ***sore throat and earache***. He had a high fever (temperature 40.5°C or 105°F), and rapid pulse and respirations. Examination of his throat revealed diffuse redness and swelling of the pharynx, especially of the palatine tonsils. His left tympanic membrane was bulging.

The history revealed that the boy had had chronic symptoms of inflammation of the nasal mucous membrane (**rhinitis**), including the pharyngeal tonsils (**tonsillitis**), resulting in persistent mouth breathing. On one occasion he had had a *peritonsillar abscess* or **quinsy**.

Following antibiotic treatment, the boy's infection cleared up. In view of his history, it was decided to readmit him 3 or 4 months later for a **T&A** (tonsillectomy and adenoidectomy).

Problems. What is meant by the term tonsils? Explain the anatomical basis of the boy's earache. What lymph node in particular might be swollen and tender in this case? What is the probable source of hemorrhage in tonsillectomy? **Compression of what vessel would control severe arterial bleeding in the tonsillar bed**? *These problems are discussed on page* 1068.

DISCUSSION OF PATIENT ORIENTED PROBLEMS

Case 8-1

A thyroglossal duct cyst is derived from the embryonic thyroglossal duct that connects the thyroid gland with the base of the tongue (Fig. 8-43).

Normally the thyroglossal duct atrophies and degenerates as the thyroid gland reaches its final site in the neck. Remnants of this duct may persist anywhere along the midline of the neck between the foramen cecum of the tongue and the thyroid gland (Fig. 8-40).

These remnants may give rise to cysts in the tongue or in the midline of the neck, *usually just inferior to the hyoid bone.* Often the cyst is in intimate contact with the anterior part of this bone. It may be connected superiorly by a duct with the foramen cecum of the tongue and/or inferiorly with the pyramidal lobe (Figs. 8-39 and 8-40), or the isthmus of the thyroid gland (Fig. 8-37). These connections explain why thyroglossal duct cysts move up and down during **deglutition** and when the tongue is protruded.

Sometimes a thyroglossal duct cyst develops an opening onto the surface of the neck (**thyroglossal fistula**). This results from erosion of cervical tissues following infection and rupture of the cyst (Fig. 8-43).

On physical examination it may be difficult to differentiate a thyroglossal duct cyst from an abnormally positioned thyroid gland (Fig. 8-79). The location and external appearance of the mass may be similar. The aspiration of fluid from the cyst in the present case indicated that thyroid tissue was probably not present.

As the thyroid gland develops from a thickening in the floor of the primitive pharynx, where the tongue forms and normally descends into the neck, an aberrant thyroid or thyroglossal duct may be found anywhere along its usual path of descent (Fig. 8-79).

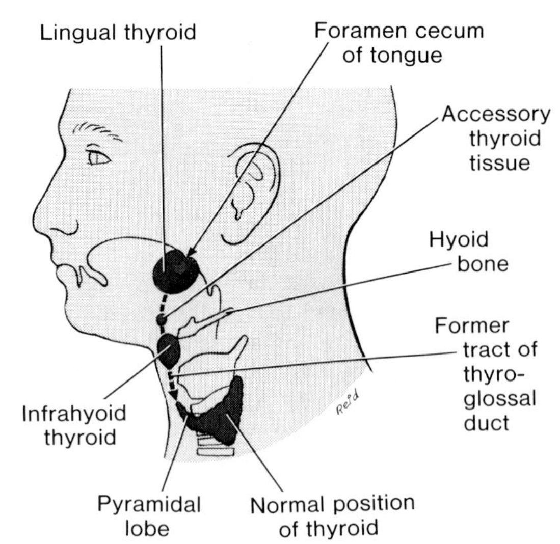

Figure 8-79. Diagrammatic sketch of the head showing the usual sites of ectopic thyroid tissue. ---- indicates the path followed by the thyroid gland during its descent and the former tract of the thyroglossal duct. Normally it reaches its usual site by the 7th week of embryonic development. (From Moore KL: *The Developing Human: Clinically Oriented Embryology*, ed 3. Philadelphia, WB Saunders Co, 1982.)

Case 8-2

All the conditions that came to the student's mind could have caused the swelling in the side of her neck. Branchial cysts may be derived from remnants of parts of the **cervical sinus**, the second branchial groove, or the second pharyngeal pouch. Although they may be associated with branchial sinuses, as in the present case, and drain through them, these cysts often lie free in the neck just inferior to the angle of the mandible. *Branchial cysts may develop at any level in the neck. Invariably they develop along the anterior border of the sternocleidomastoid muscle.* The cyst usually extends deep to this muscle and involves other structures.

In the present case **the cyst was probably derived from a remnant of the second pharyngeal pouch**. The sinus tract running superiorly from it probably passed between the internal and external carotid arteries, just superior to the hypoglossal nerve and the bifurcation of the common carotid artery (Fig. 8-24). Probably it terminated in the **intratonsillar cleft** (Fig. 8-56), the adult derivative of the cavity of the second pharyngeal pouch.

During excision of these cysts, the hypoglossal nerve may be injured, causing unilateral **lingual paralysis**. This would be indicated by hemiatrophy of the tongue and deviation of the tongue to the paralyzed side when it was protruded. This results from the unopposed action of the tongue muscles on the other side (Fig. 7-145).

If the sinus tract had passed inferiorly, it probably would have opened in the inferior third of the neck, along the anterior border of the sternocleidomastoid muscle. Branchial sinuses that open externally are sometimes called **branchial cleft sinuses** because they are derived from remnants of the second branchial groove (branchial cleft).

Case 8-3

Probably the piece of steak was lodged in the inlet of the larynx. Choking on food is a common cause of laryngeal obstruction, particularly in children, in persons who have consumed too much alcohol, and in persons with neurological impairment.

Many "restaurant deaths", thought to be caused by heart attacks, have been shown to result from choking. Persons with dentures and/or who are drunk are less able to chew their food properly and to detect a bite that is too large.

The mucous membrane of the superior part of the larynx is very sensitive and contact by a foreign body (*e.g.*, a piece of steak) causes immediate explosive coughing to expel it. However, if there is neurological impairment or the person is drunk, this response may be reduced or absent.

In rare cases, the foreign body passes through the larynx and becomes lodged in the trachea or a main bronchus. Usually, as in the present case, the piece of steak is only partly in the larynx, but entry of air into the trachea and lungs is largely prevented. The patient would likely have died within minutes, almost certainly before there was time to get him to hospital, if the piece of steak had not been dislodged using the *Heimlich maneuver*, enabling adequate respiration to be re-established.

Had the emergency procedure not been successful, the doctor would likely have first tried to get the piece of steak out of the patient's larynx with his finger, a long spoon, or a fork. If these procedures had failed, he would likely have done a life-saving, emergency **inferior laryngotomy**. If he happened to have a large bore needle with him, he would have inserted it through the **median cricothyroid ligament** (Fig. 8-62). If not, he probably would have used a penknife or a steak knife to make an incision through the midline of the neck into the cricothyroid ligament (**cricothyrotomy**).

In emergency situations, it is safer to incise the cricothyroid ligament than the trachea because the isthmus of the thyroid gland and many blood vessels cover the anterior surface of the trachea (Figs. 8-32, 8-37, and 8-42). Probably the doctor would have inserted a large plastic straw, or a tube of some sort (*e.g.*, an empty ballpoint pen), to enable the patient to breathe while he was being taken to the hospital for removal of the piece of steak from his larynx and repair of the cervical wound.

Case 8-4

The tongue, hyoid, and larynx rise and fall during swallowing. As the thyroid gland is attached to the larynx by pretracheal fascia, it also moves up and down during swallowing.

Physiological enlargement of the thyroid gland is commonly seen at puberty and during pregnancy; otherwise, any enlargement of the thyroid is called a **goiter** (L. *guttur*, throat). In the present case the patient's goiter resulted from hyperthyroidism. The association of hyperthyroidism with protrusion of the eyes (**exophthalmos**) was first described by an Irish physician, Dr. R.J. Graves.

The cause of exophthalmos is not precisely known; however, a considerable increase in the size of the orbital muscles is certainly a factor.

In the surgical treatment of hyperthyroidism (Graves' disease), part of each lobe of the thyroid is removed (**subtotal thyroidectomy**), thereby leaving less glandular tissue to secrete hormones.

As the four small **parathyroid glands** typically lie on the posterior surface of the thyroid gland (Fig. 8-45), posterior parts of the lobes are left so that these glands will not be inadvertently removed. At least one of them is essential for secretion of parathyroid hormones which maintain the normal level of calcium in the blood and body fluids.

If the parathyroid glands are removed during surgery, the patient soon develops a convulsive disorder known as ***tetany***. The signs are nervousness, twitching, and spasms in the facial and limb muscles.

When the thyroid gland is being removed, there is danger that the important laryngeal nerves may be injured. Near the inferior pole of the thyroid gland, **the recurrent laryngeal nerves are intimately related to the inferior thyroid arteries** (Fig. 8-45). The nerves may cross anterior or posterior to this artery or between its branches before ascending in or near the groove between the trachea and esophagus.

Because of the close realtionship between the recurrent laryngeal nerves and the inferior thyroid arteries, the risk of injuring them during surgery is ever present. These nerves supply all muscles of the larynx except the cricothyroids (Table 8-5). If one of the nerves is damaged or cut, there is likely to be a serious effect on speech (*e.g.*, hoarseness as in the present case), or a change in the quality of the voice (*e.g.*, brassy sound). Some patients also have difficulty clearing their throats.

Temporary paralysis of the recurrent laryngeal nerves may also result from postoperative edema affecting them. It must be remembered also that a common cause of temporary hoarseness after surgery is **trauma to the mucous membrane of the larynx** by the endotracheal tube inserted by the anesthetist as an airway.

The recurrent laryngeal nerve is also vulnerable to injury during operations in the carotid triangle of the neck (Figs. 8-5 and 8-6*B*), *e.g.*, **carotid endarterectomy**. If both nerves are completely destroyed, breathing will be severely impaired and speech will be difficult because the vocal folds remain partly abducted (the position of complete paralysis of the intrinsic muscles). Thus the **rima glottidis** is not fully open. If the nerves are compressed as a result of inflammation or the accumulation of fluid, the breathing and speech defects will normally disappear following healing and drainage of the operative site.

Case 8-5

The term tonsil usually refers to the palatine tonsil (Fig. 8-55). The other tonsils are the lingual, pharyngeal, and tubal tonsils. All these tonsils form the **tonsillar ring around the faucial isthmus** leading from the oral cavity into the nasopharynx. Although it is often stated that the tonsillar ring acts as a barrier to infection, its function is not clearly understood. However, it is certain that *this lymphatic tissue is important in the immune reaction to infection.*

The infection in this case had spread along the auditory tube into the middle ear (Figs. 7-162 and 7-164), producing **otitis media** and bulging of the tympanic membrane. *This would be the chief cause of the boy's earache.*

The tonsils are supplied by twigs from the glossopharyngeal nerve (CN IX) and, as the tympanic branch of this nerve supplies the mucous membrane of the tympanic cavity, some of the pain related to the tonsillitis may have also been referred to the ear. When the opening of the auditory tube is closed, as it probably was in the present case, pressure changes in the middle ear can also cause earache.

The numerous lymphatic vessels of the tonsil penetrate the pharyngeal wall and terminate principally in the **jugulodigastric node** of the deep cervical chain of lymph nodes (Fig. 8-60). Because its enlargement is commonly associated with tonsillitis, it is often called the **tonsillar node.**

The external palatine vein is usually the chief source of hemorrhage following tonsillectomy. This important and sometimes large vein descends from the soft palate and is immediately related to the lateral surface of the tonsil, before it pierces the superior constrictor muscle of the pharynx (Figs. 8-51 and 8-59).

In cases of severe and uncontrolled bleeding (*e.g.*, from the tonsillar branch of the facial artery), hemorrhage may be controlled by compressing or clamping the external carotid artery at its origin, because this vessel supplies blood to the tonsillar arteries (Fig. 8-26).

Suggestions for Additional Reading

1. Basmajian, JB: *Grant's Method of Anatomy*, ed 10. Baltimore, Williams & Wilkins, 1980.
   A concise account of the neck that is well illustrated with simple, easily reproducible line drawings. Direct applications of anatomy to problems in medicine and surgery are described.
2. Fletcher G, Jing B: *The Head and Neck. An Atlas of Tumor Radiology*. Chicago, Year Book Medical Publishers Inc, 1968.
   Perusal of this book will indicate how important the anatomy of the neck is to the radiologist. The discussions of diagnostic radiographic studies of the pharynx, larynx, trachea, and esophagus are fascinating, particularly those used for demonstrating tumors of the vocal folds.
3. Hung W: The growth and development of the thyroid. In Davis JA, Dobbing J (eds): *Scientific Foundations of Paediatrics.* Philadelphia, WB Saunders Co, 1974.
   This section includes a discussion of the growth and development of the thyroid gland in the fetus, newborn child, and adolescent.
4. Moore KL: *The Developing Human: Clinically Oriented Embryology*, ed 3. Philadelphia, WB Saunders Co, 1982.
   Most congenital malformations of the neck develop during transformation of the branchial apparatus into adult derivatives (*e.g.*, cervical cysts, ectopic thyroid, and branchial sinuses). These developmental abnormalities of the neck are fully described.

INDEX

Main entries (*e.g.*, **Abdomen**) and page references are printed in **bold type**. In most cases items are listed under nouns rather than under descriptive adjectives; *e.g.*, the deltoid muscle will not be found under deltoid but under the general heading **Muscles**. Similarly all nerves, veins, and arteries are listed under Nerves, Veins, and Arteries, respectively.

Widely used *eponyms* are also listed so that students can determine the meaning of terms they may hear clinicians use, *e.g.*, the pouch of Douglas for the rectouterine pouch. Commonly used old terms are also given so that clinicians can learn the new terminology, *e.g.*, the internal mammary artery is now called the internal thoracic artery.

A

C

E

G

I

J

K

L

N

O

S

U

V